## POWER OF TEN PREFIXES

| Prefix | Abbreviation | Value |
|---|---|---|
| Exa | E | $10^{18}$ |
| Peta | P | $10^{15}$ |
| Tera | T | $10^{12}$ |
| Giga | G | $10^{9}$ |
| Mega | M | $10^{6}$ |
| Kilo | k | $10^{3}$ |
| Hecto | h | $10^{2}$ |
| Deka | da | $10^{1}$ |
| Deci | d | $10^{-1}$ |
| Centi | c | $10^{-2}$ |
| Milli | m | $10^{-3}$ |
| Micro | $\mu$ | $10^{-6}$ |
| Nano | n | $10^{-9}$ |
| Pico | p | $10^{-12}$ |
| Femto | f | $10^{-15}$ |
| Atto | a | $10^{-18}$ |

## SOME USEFUL FACTS

Area of circle (radius $r$) $\pi r^2$

Area of sphere (radius $r$) $4\pi r^2$

Volume of sphere $\frac{4}{3}\pi r^3$

Trig definitions:

$\sin\theta = $ (opposite side)/(hypotenuse)

$\cos\theta = $ (adjacent side)/(hypotenuse)

$\tan\theta = $ (opposite side)/(adjacent side)

Quadratic equation:

$$0 = ax^2 + bx + c,$$

$$\text{where } x = \frac{-b \pm \sqrt{b^2 - 4ac}}{2a}$$

# COLLEGE PHYSICS

# COLLEGE PHYSICS

**EUGENIA ETKINA**
*Rutgers University*

**MICHAEL GENTILE**
*Rutgers University*

**ALAN VAN HEUVELEN**
*Rutgers University*

PEARSON

Boston Columbus Indianapolis New York San Francisco Upper Saddle River
Amsterdam Cape Town Dubai London Madrid Milan Munich Paris Montréal Toronto
Delhi Mexico City São Paulo Sydney Hong Kong Seoul Singapore Taipei Tokyo

Publisher: Jim Smith
Executive Editor: Becky Ruden
Project Managers: Katie Conley and Beth Collins
Managing Development Editor: Cathy Murphy
Associate Content Producer: Kelly Reed
Assistant Editor: Kyle Doctor
Team Lead, Program Management, Physical Sciences: Corinne Benson
Full-Service Production and Composition: PreMediaGlobal
Copy Editor: Joanna Dinsmore
Illustrator: Rolin Graphics
Photo Researcher: Eric Shrader
Image Lead: Maya Melenchuk
Manufacturing Buyer: Jeff Sargent
Marketing Manager: Will Moore
Text Designer: tani hasegawa
Cover Designer: Tandem Creative, Inc.
Cover Photo Credit: © Markus Altmann/Corbis

Credits and acknowledgments borrowed from other sources and reproduced, with permission, in this textbook appear on p. C-1.



**Library of Congress Cataloging-in-Publication data**

Etkina, Eugenia.
College physics / Eugenia Etkina, Michael Gentile, Alan Van Heuvelen.
pages cm
ISBN-13: 978-0-321-71535-7
ISBN-10: 0-321-71535-7
1. Physics—Textbooks. I. Gentile, Michael J. II. Van Heuvelen, Alan. III. Title.
QC21.3.E85 2012
530—dc23

2012035388

4 16

PEARSON

www.pearsonhighered.com

**ISBN 10:** 0-321-71535-7; **ISBN 13:** 978-0-321-71535-7 (Student edition)
**ISBN 10:** 0-321-90181-9; **ISBN 13:** 978-0-321-90181-1 (Instructor's resource copy)
**ISBN 10:** 0-321-87970-8; **ISBN 13:** 978-0-321-87970-7 (Books a la carte edition)

# Brief Contents

# About the Authors

**Eugenia Etkina** holds a PhD in physics education from Moscow State Pedagogical University and has more than 30 years experience teaching physics. She currently teaches at Rutgers University, where she received the highest teaching award in 2010 and the New Jersey Distinguished Faculty award in 2012. Professor Etkina designed and now coordinates one of the largest programs in physics teacher preparation in the United States, conducts professional development for high school and university physics instructors, and participates in reforms to the undergraduate physics courses. In 1993 she developed a system in which students learn physics using processes that mirror scientific practice. That system serves as the basis for this textbook. Since 2000, Professors Etkina and Van Heuvelen have conducted over 60 workshops for physics instructors and co-authored *The Physics Active Learning Guide* (a companion edition to *College Physics* is now available). Professor Etkina is a dedicated teacher and an active researcher who has published over 40 peer-refereed articles.

**Michael Gentile** is an Instructor of Physics at Rutgers University. He has a masters degree in physics from Rutgers University, where he studied under Eugenia Etkina and Alan Van Heuvelen, and has also completed postgraduate work in education, high energy physics, and cosmology. He has been inspiring undergraduates to learn and enjoy physics for more than 15 years. Since 2006 Professor Gentile has taught and coordinated a large-enrollment introductory physics course at Rutgers where the approach used in this book is fully implemented. He also assists in the mentoring of future physics teachers by using his course as a nurturing environment for their first teaching experiences. Since 2007 his physics course for the New Jersey Governor's School of Engineering and Technology has been highly popular and has brought the wonders of modern physics to more than 100 gifted high school students each summer.

**Alan Van Heuvelen** holds a PhD in physics from the University of Colorado. He has been a pioneer in physics education research for several decades. He taught physics for 28 years at New Mexico State University where he developed active learning materials including the *Active Learning Problem Sheets* (the *ALPS Kits)* and the *ActivPhysics* multimedia product. Materials such as these have improved student achievement on standardized qualitative and problem-solving tests. In 1993 he joined Ohio State University to help develop a physics education research group. He moved to Rutgers University in 2000 and retired in 2008. For his contributions to national physics education reform, he won the 1999 AAPT Millikan Medal and was selected a fellow of the American Physical Society. Over the span of his career he has led over 100 workshops on physics education reform. In the last ten years, he has worked with Professor Etkina in the development of the Investigative Science Learning Environment (*ISLE*), which integrates the results of physics education research into a learning system that places considerable emphasis on helping students develop science process abilities while learning physics.

# About the Authors

# SET THE WHEELS IN MOTION

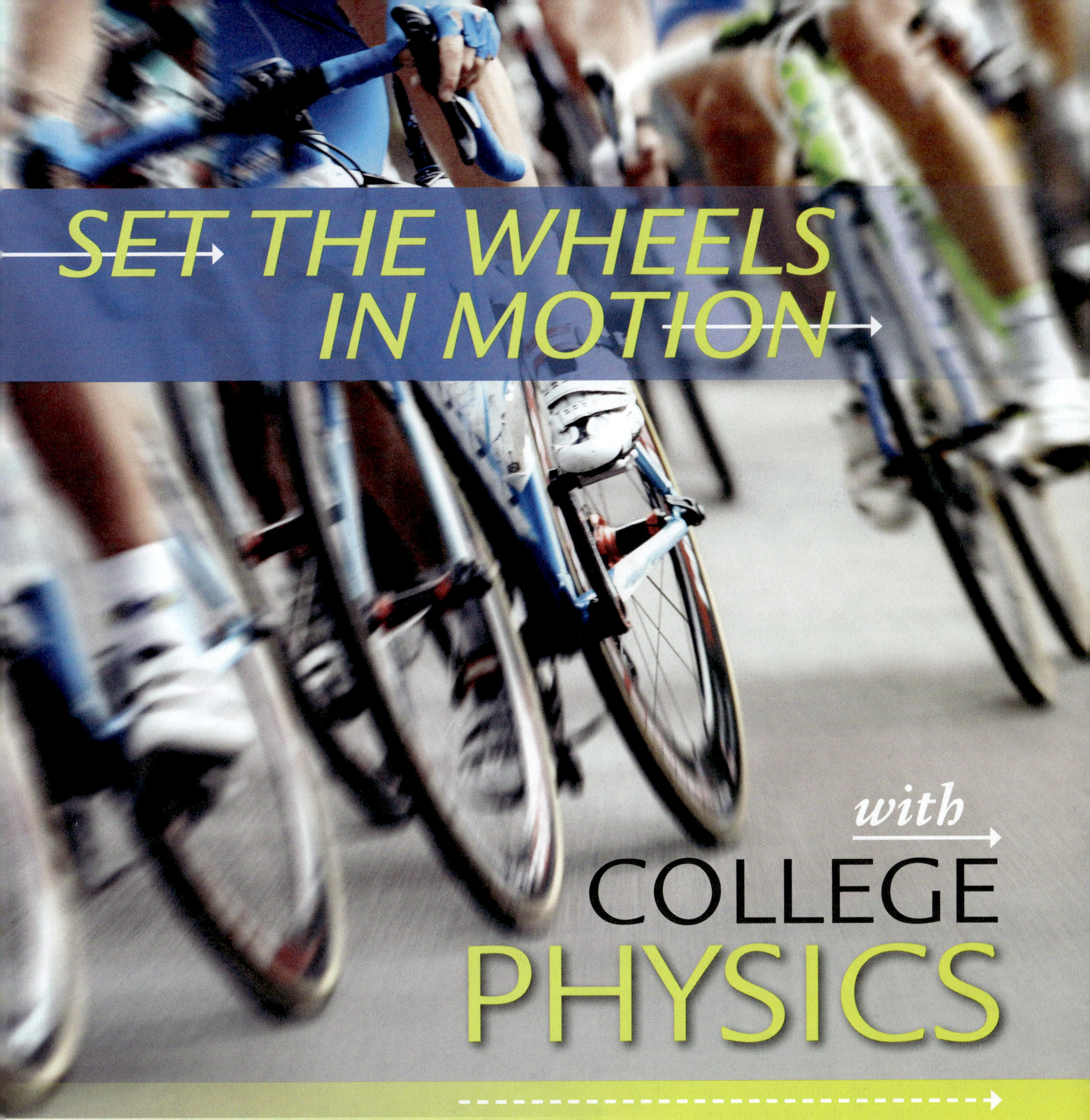

## *with* COLLEGE PHYSICS

"This is an excellent way to teach physics. The approach is so logical that students will feel they are a) discovering physics themselves, and b) reaching the best conclusions... The style is approachable, consistent, systematic, engaging. I think [this textbook] teaches more than physics—it also gets at the core of the scientific process and that will be just as valuable for the students as any of the physics content."

—Andy Richter, *Valparaiso University*

# BUILD A DEEP UNDERSTANDING OF PHYSICS AND THE SCIENTIFIC PROCESS

An active learning approach encourages students to construct an understanding of physics concepts and laws in the same ways that scientists acquire knowledge. Students learn physics by doing physics.

## OBSERVATIONAL EXPERIMENT TABLES

Observational Experiment Tables engage students through active discovery. Students make observations, analyze data, and identify patterns.

Scan this QR code with your smartphone to view the video that accompanies this table.

**OBSERVATIONAL EXPERIMENT TABLE**

**2.3 Two observers watch the same coffee mug.**

VIDEO 2.3

| Observational experiment | Analysis done by each observer |
|---|---|
| **Experiment 1.** Observer 1 is slouched down in the passenger seat of a car and cannot see outside the car. Suddenly he observes a coffee mug sliding toward him from the dashboard. | Observer 1 creates a motion diagram and a force diagram for the mug as he observes it. On the motion diagram, increasingly longer $\vec{v}$ arrows indicate that the mug's speed changes from zero to nonzero as seen by observer 1 even though no external object is exerting a force on it in that direction. 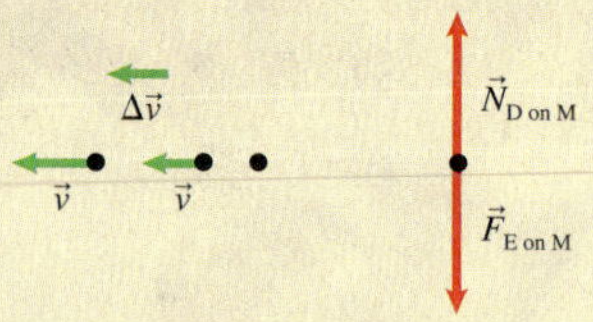  |
| **Experiment 2.** Observer 2 stands on the ground beside the car. She observes that the car starts moving forward at increasing speed and that the mug remains stationary with respect to her. | Observer 2 creates a motion diagram and force diagram for the mug as she observes it. There are no $\vec{v}$ or $\Delta\vec{v}$ arrows on the diagram and the mug is at rest relative to her. 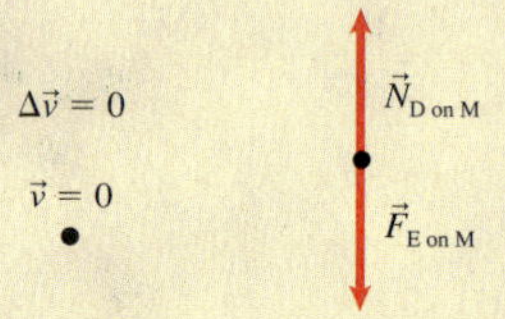  |

**Pattern**

**Observer 1:** The forces exerted on the mug by Earth and by the dashboard surface add to zero. But the velocity of the mug increases as it slides off the dashboard. This is inconsistent with the rule relating the sum of the forces and the change in velocity.
**Observer 2:** The forces exerted on the mug by Earth and by the dashboard surface add to zero. Thus the velocity of the mug should not change, and it does not. This is consistent with the rule relating the sum of the forces and the change in velocity.

## VIDEOS

Physics demonstration videos, accessed by QR codes in the text or through the MasteringPhysics® Study Area, accompany most of the Observational and Testing Experiment Tables. Students can observe the exact experiment described in the table.

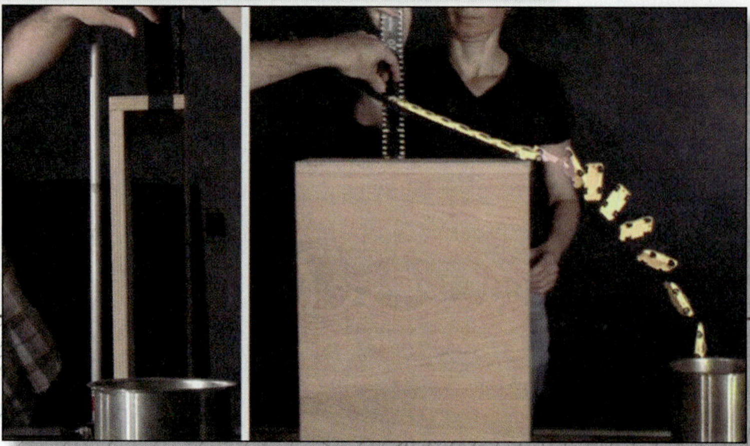

## TESTING EXPERIMENT TABLES

Each testing experiment evaluates a hypothesis arising from the observational experiment, and includes the experimental setup, one or more predictions, and the outcome of the experiment. A conclusion summarizes the result of the experimental process.

Scan this QR code with your smartphone to view the video shown below.

**TESTING EXPERIMENT TABLE**

**5.2 Testing the idea that $\Sigma m\vec{v}$ in an isolated system remains constant (all velocities are with respect to the track).**

VIDEO 5.2

| Testing experiment | Prediction | Outcome |
|---|---|---|
| Cart A (0.40 kg) has a piece of modeling clay attached to its front and is moving right at 1.0 m/s. Cart B (0.20 kg) is moving left at 1.0 m/s. The carts collide and stick together. Predict the velocity of the carts after the collision. | 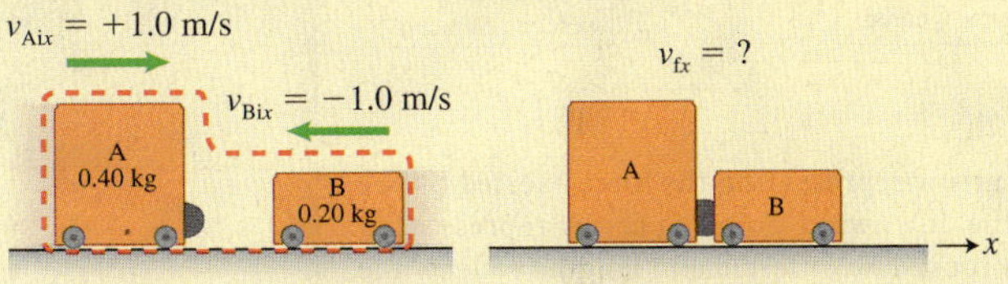<br>The system consists of the two carts. The direction of velocity is noted with a plus or minus sign of the velocity component:<br>$(0.40\text{ kg})(+1.0\text{ m/s}) + (0.20\text{ kg})(-1.0\text{ m/s}) = (0.40\text{ kg} + 0.20\text{ kg})v_{fx}$<br>or<br>$v_{fx} = (+0.20\text{ kg}\cdot\text{m/s})/(0.60\text{ kg}) = +0.33\text{ m/s}$<br>After the collision, the two carts should move right at a speed of about 0.33 m/s. | After the collision, the carts move together toward the right at close to the predicted speed. |

**Conclusion**

Our prediction matched the outcome. This result gives us increased confidence that this new quantity $m\vec{v}$ might be the quantity whose sum is constant in an isolated system.

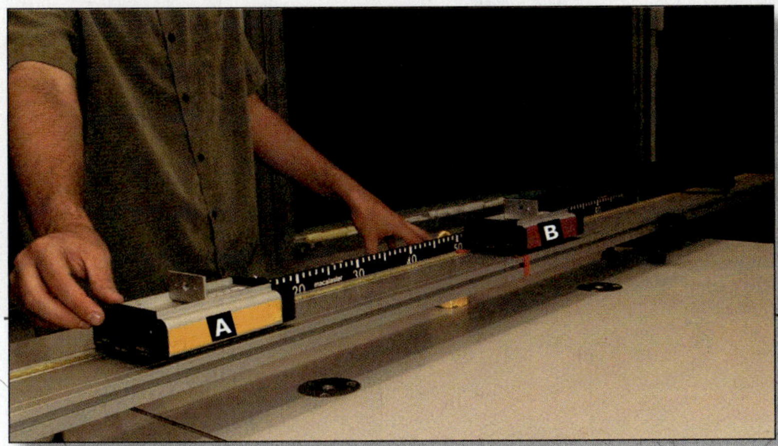

# DEVELOP ADVANCED PROBLEM-SOLVING SKILLS

Students learn to represent physical phenomena in multiple ways using words, figures, and equations, including qualitative diagrams and innovative bar charts that create a foundation for quantitative reasoning and problem solving.

**REASONING SKILL** Constructing a force diagram

1. Sketch the situation (a rock sinking into sand).

2. Circle the system (the rock).

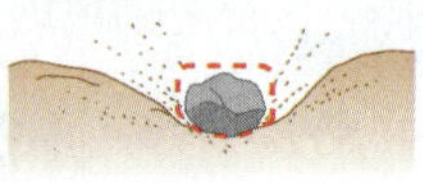

3. Identify external interactions:
- The sand pushes up on the rock.
- Earth pulls down on the rock.
- We assume that the force that the air exerts on the rock is small in comparison and can be ignored.

4. Place a dot at the side of the sketch, representing the system object.

5. Draw force arrows to represent the external interactions.

6. Label the forces with a subscript with two elements.

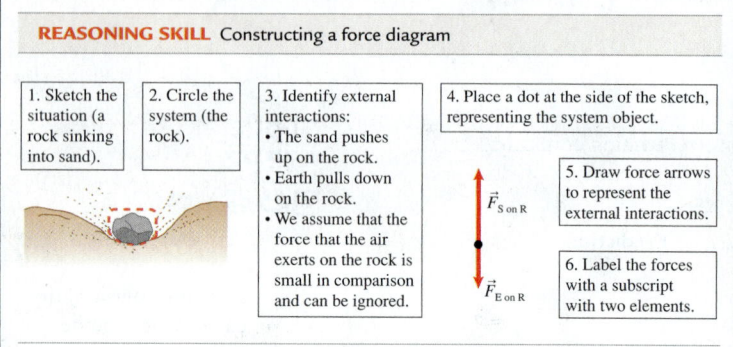

Notice that the upward-pointing arrow representing the force exerted by the sand on the rock is longer than the downward-pointing arrow representing the force exerted by Earth on the rock. The difference in lengths reflects the difference in the magnitudes of the forces. Later in the chapter we will learn why they have different lengths. For now, we just need to include arrows for all external forces exerted on the system object (the rock).

## REASONING SKILL BOXES

These boxes reinforce a particular skill, such as drawing a motion diagram, force diagram, or work-energy bar chart.

## BAR CHARTS

Innovative bar charts help to create a foundation for quantitative reasoning and problem solving.

**Figure 5.7** A bar chart analysis of the collision of car 2 with car 1.

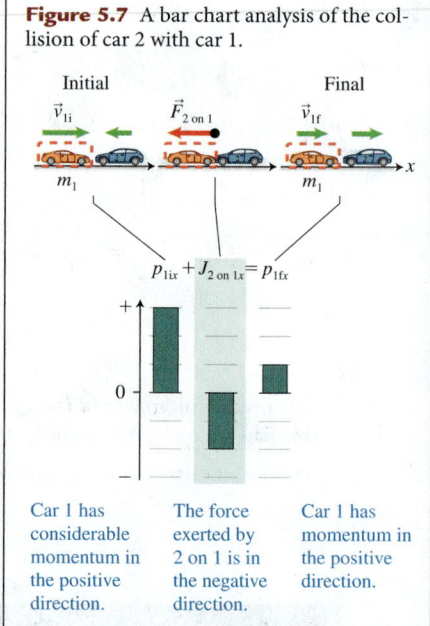

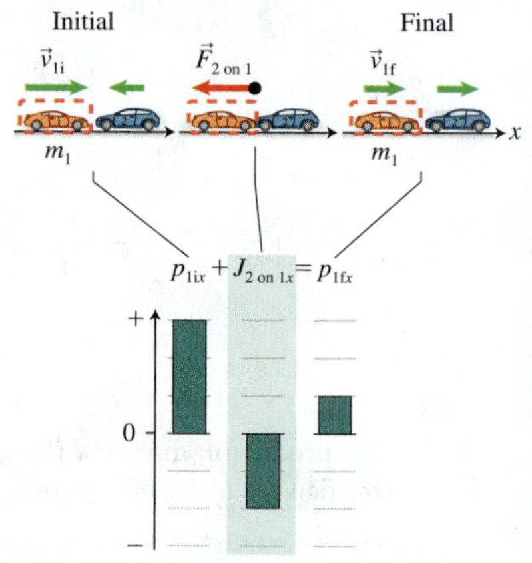


## PROBLEM-SOLVING STRATEGY

The Problem-Solving Strategy boxes walk students step-by-step through the process of solving a worked example, applying concepts covered in the text.

**PROBLEM-SOLVING STRATEGY** Applying Static Equilibrium Conditions

**EXAMPLE 7.6 Use the biceps muscle to lift**

Imagine that you hold a 6.0-kg lead ball in your hand with your arm bent. The ball is 0.35 m from the elbow joint. The biceps muscle attaches to the forearm 0.050 m from the elbow joint and exerts a force on the forearm that allows it to support the ball. The center of mass of the 12-N forearm is 0.16 m from the elbow joint. Estimate the magnitude of (a) the force that the biceps muscle exerts on the forearm and (b) the force that the upper arm exerts on the forearm at the elbow.

**Sketch and translate**

- Construct a labeled sketch of the situation. Include coordinate axes and choose an axis of rotation.
- Choose a system for analysis.

We choose the axis of rotation to be where the upper arm bone (the humerus) presses on the forearm at the elbow joint. This will eliminate from the torque equilibrium equation the unknown force that the upper arm exerts on the forearm.

We choose the system of interest to be the forearm and hand.

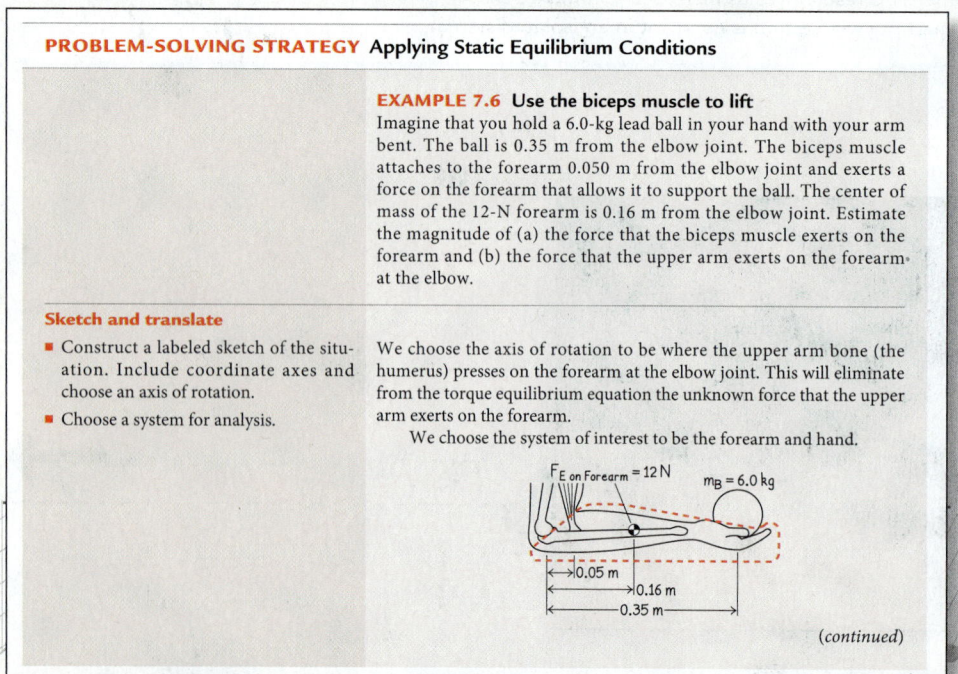

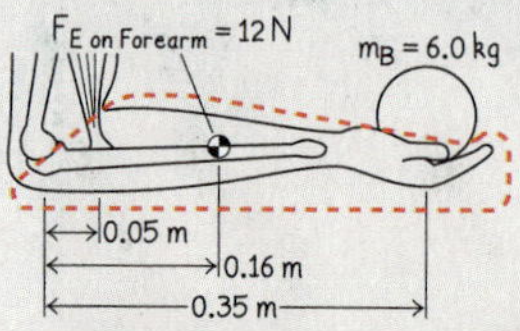


*(continued)*

# INSPIRE HIGHER-LEVEL REASONING

Innovative, widely praised examples, exercises, and problems engage students, assess learning, and promote higher-level reasoning.

* **Equation Jeopardy 1** The equation below describes a rotational dynamics situation. Draw a sketch of a situation that is consistent with the equation and construct a word problem for which the equation might be a solution. There are many possibilities.

$$-(2.2\,\text{N})(0.12\,\text{m}) = [(1.0\,\text{kg})(0.12\,\text{m})^2]\alpha$$

* **Equation Jeopardy 2** The equation below describes a rotational dynamics situation. Draw a sketch of a situation that is consistent with the equation and construct a word problem for which the equation might be a solution. There are many possibilities.

$$-(2.0\,\text{N})(0.12\,\text{m}) + (6.0\,\text{N})(0.06\,\text{m}) = [(1.0\,\text{kg})(0.12\,\text{m})^2]\alpha$$

## JEOPARDY-STYLE END-OF-CHAPTER PROBLEMS

Unique, Jeopardy-style end-of-chapter problems ask students to work backwards from an equation to craft a problem statement. Chapters also include "what if" problems, estimating problems, and qualitative/quantitative multi-part problems.

## REVIEW QUESTIONS

Questions at the end of each section of the chapter encourage critical thinking and synthesis rather than recall.

**Review Question 1.5** Why is the following statement true? "Displacement is equal to the area between a velocity-versus-time graph line and the time axis with a positive or negative sign."

## ESTIMATION PROBLEMS

Estimation problems ask students to make reasonable assumptions and estimates in problem solving as a scientist would do.

**EXAMPLE 5.6 Bone fracture estimation**[1]

A bicyclist is watching for traffic from the left while turning toward the right. A street sign hit by an earlier car accident is bent over the side of the road. The cyclist's head hits the pole holding the sign. Is there a significant chance that his skull will fracture?

**Sketch and translate** The process is sketched at the right. The initial state is at the instant that the head initially contacts the pole; the final state is when the head and body have stopped. The person is the system. We have been given little information, so we'll have to make some reasonable estimates of various quantities in order to make a decision about a possible skull fracture.

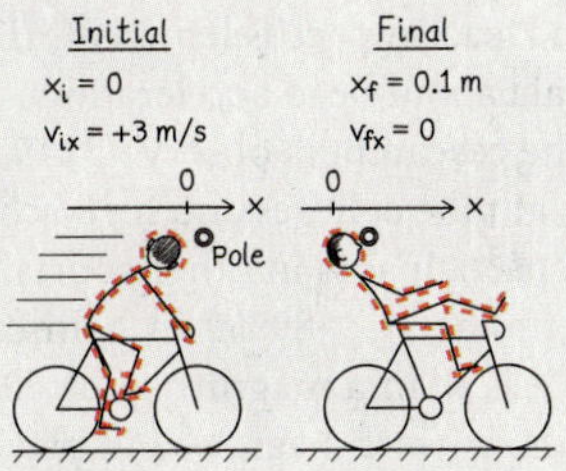


**Simplify and diagram** The bar chart illustrates the momentum change of the system and the impulse exerted by the pole that caused the change. The person was initially moving in the horizontal *x*-direction with respect to Earth, and not moving after the collision. The pole exerted an impulse in the negative *x*-direction on the cyclist. We'll need to estimate the following quantities: the mass and speed of the cyclist in this situation, the stopping time interval, and the area of contact. Let's assume that this is a 70-kg cyclist moving at about 3 m/s. The person's body keeps moving forward for a short distance after the bone makes contact with the pole. The skin indents some during the collision. Because of these two factors, we assume

[1]This is a true story—it happened to one of the book's authors, Alan Van Heuvelen.

# MOTIVATE WITH REAL-WORLD APPLICATIONS

Real-world applications relate physics concepts and laws to everyday experiences and apply them to problems in diverse fields such as biology, medicine, and astronomy.

## 8.7 Rotational motion: Putting it all together

We can use our knowledge of rotational motion to analyze a variety of phenomena that are part of our world. In this section, we consider two examples—the effect of the tides on the period of Earth's rotation (the time interval for 1 day) and the motion of bowling (also called pitching) in the sport of cricket.

### Tides and Earth's day

The level of the ocean rises and falls by an average of 1 m twice each day, a phenomenon known as the tides. Many scientists, including Galileo, tried to explain this phenomenon and suspected that the Moon was a part of the answer. Isaac Newton was the first to explain how the motion of the Moon actually creates tides. He noted that at any moment, different parts of Earth's surface are at different distances from the Moon and that the distance from a given location on Earth to the Moon varied as Earth rotated. As illustrated in **Figure 8.18**, point A is closer to the Moon than the center of Earth or point B are, and therefore the gravitational force exerted by the Moon on point A is greater than the gravitational force exerted on point B. Due to the difference

**Figure 8.18** The ocean bulges on both sides of Earth along a line toward the Moon.

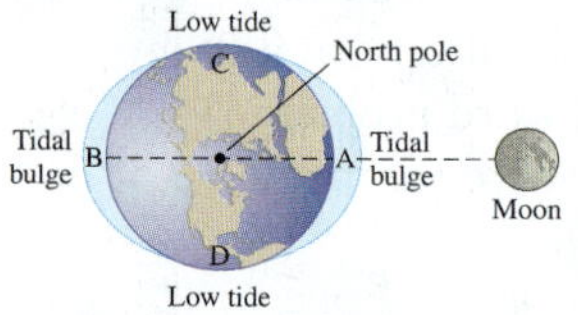


## PUTTING IT ALL TOGETHER

These sections help students synthesize chapter content within real-world applications such as avoiding "the bends" in scuba diving (Chapter 10), making automobiles more efficient (Chapter 13), and building liquid crystal displays (Chapter 24).

## Reading Passage Problems

**BIO Muscles work in pairs** Skeletal muscles produce movements by pulling on tendons, which in turn pull on bones. Usually, a muscle is attached to two bones via a tendon on each end of the muscle. When the muscle contracts, it moves one bone toward the other. The other bone remains in nearly the original position. The point where a muscle tendon is attached to the stationary bone is called the *origin*. The point where the other muscle tendon is attached to the movable bone is called the *insertion*. The origin is like the part of a door spring that is attached to the doorframe. The insertion is similar to the part of the spring that is attached to the movable door.

During movement, bones act as levers and joints act as axes of rotation for these levers. Most movements require several skeletal muscles working in groups, because a muscle can only exert a pull and not a push. In addition, most skeletal muscles are arranged in opposing pairs at joints. Muscles that bring two limbs together are called flexor muscles (such as the biceps muscle in the upper arm in **Figure 7.26**). Those that cause the limb to extend outward are called extensor muscles (such as the triceps muscle in the upper arm). The flexor muscle is used when you hold a heavy object in your hand; the extensor muscle can be used, for example, to extend your arm when you throw a ball.

**Figure 7.26** Muscles often come in flexor-extensor pairs.

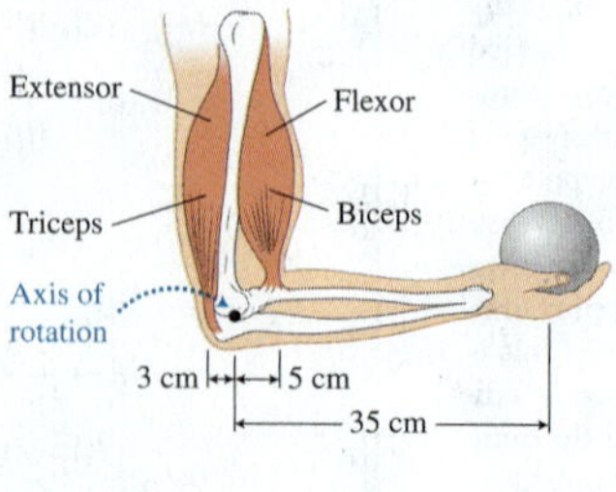


## MCAT-STYLE READING PASSAGE PROBLEMS

Help students prepare for the MCAT exam. Because so many students who take this course are planning to study medicine, each chapter includes MCAT-style reading passages and related multiple-choice questions to help prepare students for this important test.

## BIOLOGICAL AND MEDICAL EXAMPLES

Examples throughout the text provide relevance for life science majors and include topics such as understanding the effect of radon on the lungs (Chapter 5), controlling body temperature (Chapter 12), and measuring the speed of blood flow (Chapter 20).

## Reading Passage Problems

**BIO Head injuries in sports** A research group at Dartmouth College has developed a Head Impact Telemetry (HIT) System that can be used to collect data about head accelerations during impacts on the playing field. The researchers observed 249,613 impacts from 423 football players at nine colleges and high schools and collected collision data from participants in other sports. The accelerations during most head impacts ($>89\%$) in helmeted sports caused head accelerations less than a magnitude of $400\ m/s^2$. However, a total of 11 concussions were diagnosed in players whose impacts caused accelerations between 600 and $1800\ m/s^2$, with most of the 11 over $1000\ m/s^2$.

# REINFORCE SCIENTIFIC THINKING

***Active Learning Guide for College Physics***

by Eugenia Etkina, Michael Gentile, and Alan Van Heuvelen

© 2014 • Paper • 400 pages

978-0-321-86445-1 • 0-321-86445-X

Discovery-based activities supplement the knowledge-building approach of the textbook. This workbook is organized in parallel with the textbook's chapters.

Blue labels, located in the text's margins, link the discovery-based activities in the *Active Learning Guide* to concepts covered in *College Physics*.

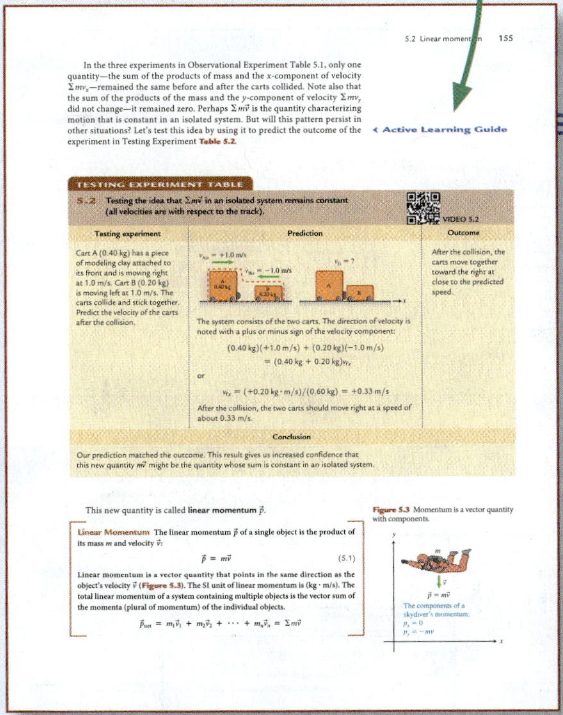

5.2 Linear momentum 155

In the three experiments in Observational Experiment Table 5.1, only one quantity—the sum of the products of mass and the x-component of velocity $\Sigma mv_x$—remained the same before and after the carts collided. Note also that the sum of the products of the mass and the y-component of velocity $\Sigma mv_y$ did not change—it remained zero. Perhaps $\Sigma m\vec{v}$ is the quantity characterizing motion that is constant in an isolated system. But will this pattern persist in other situations? Let's test this idea by using it to predict the outcome of the experiment in Testing Experiment Table 5.2.

< Active Learning Guide

TESTING EXPERIMENT TABLE

5.2 Testing the idea that $\Sigma m\vec{v}$ in an isolated system remains constant (all velocities are with respect to the track).

VIDEO 5.2

| Testing experiment | Prediction | Outcome |
|---|---|---|
| Cart A (0.40 kg) has a piece of modeling clay attached to its front and is moving right at 1.0 m/s. Cart B (0.20 kg) is moving left at 1.0 m/s. The carts collide and stick together. Predict the velocity of the carts after the collision. | The system consists of the two carts. The direction of velocity is noted with a plus or minus sign of the velocity component: $(0.40\ kg)(+1.0\ m/s) + (0.20\ kg)(-1.0\ m/s) = (0.40\ kg + 0.20\ kg)v_{fx}$ or $v_{fx} = (+0.20\ kg \cdot m/s)/(0.60\ kg) = +0.33\ m/s$. After the collision, the two carts should move right at a speed of about 0.33 m/s. | After the collision, the carts move together toward the right at close to the predicted speed. |

Conclusion

Our prediction matched the outcome. This result gives us increased confidence that this new quantity $m\vec{v}$ might be the quantity whose sum is constant in an isolated system.

This new quantity is called **linear momentum** $\vec{p}$.

Linear Momentum The linear momentum $\vec{p}$ of a single object is the product of its mass $m$ and velocity $\vec{v}$:

$$\vec{p} = m\vec{v} \quad (5.1)$$

Linear momentum is a vector quantity that points in the same direction as the object's velocity $\vec{v}$ (Figure 5.3). The SI unit of linear momentum is (kg · m/s). The total linear momentum of a system containing multiple objects is the vector sum of the momenta (plural of momentum) of the individual objects.

$$\vec{p}_{net} = m_1\vec{v}_1 + m_2\vec{v}_2 + \cdots + m_n\vec{v}_n = \Sigma m\vec{v}$$

Figure 5.3 Momentum is a vector quantity with components.

The components of a skydiver's momentum: $p_x = 0$, $p_y = -mv$

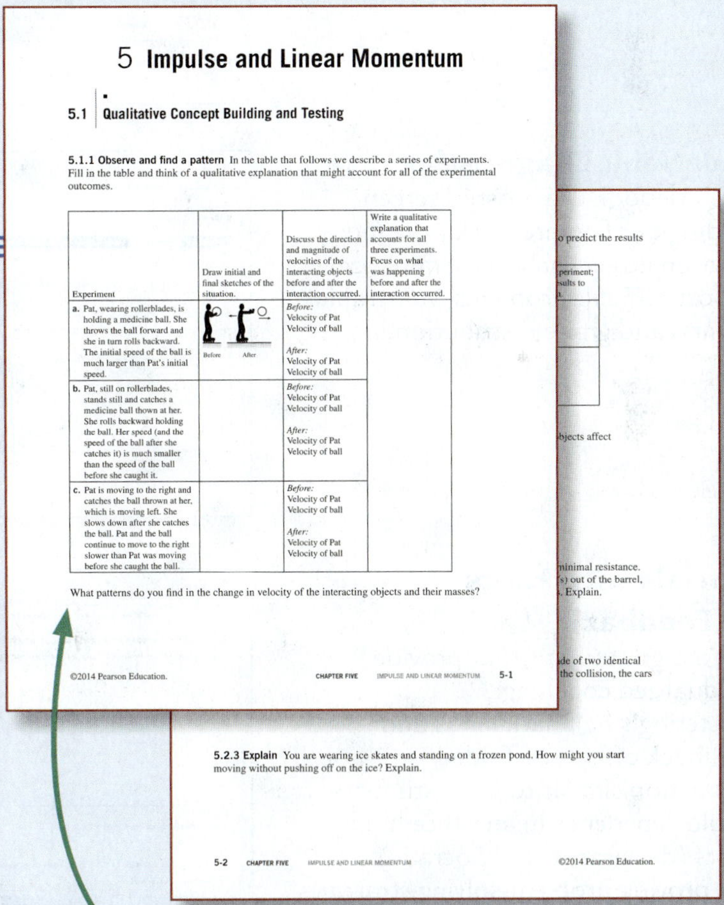

5 **Impulse and Linear Momentum**

5.1 **Qualitative Concept Building and Testing**

**5.1.1 Observe and find a pattern** In the table that follows we describe a series of experiments. Fill in the table and think of a qualitative explanation that might account for all of the experimental outcomes.

| Experiment | Draw initial and final sketches of the situation. | Discuss the direction and magnitude of velocities of the interacting objects before and after the interaction occurred. | Write a qualitative explanation that accounts for all three experiments. Focus on what was happening before and after the interaction occurred. |
|---|---|---|---|
| **a.** Pat, wearing rollerblades, is holding a medicine ball. She throws the ball forward and she in turn rolls backward. The initial speed of the ball is much larger than Pat's initial speed. | Before After | *Before:* Velocity of Pat Velocity of ball *After:* Velocity of Pat Velocity of ball | |
| **b.** Pat, still on rollerblades, stands still and catches a medicine ball thown at her. She rolls backward holding the ball. Her speed (and the speed of the ball after she catches it) is much smaller than the speed of the ball before she caught it. | | *Before:* Velocity of Pat Velocity of ball *After:* Velocity of Pat Velocity of ball | |
| **c.** Pat is moving to the right and catches the ball thrown at her, which is moving left. She slows down after she catches the ball. Pat and the ball continue to move to the right slower than Pat was moving before she caught the ball. | | *Before:* Velocity of Pat Velocity of ball *After:* Velocity of Pat Velocity of ball | |

What patterns do you find in the change in velocity of the interacting objects and their masses?

©2014 Pearson Education. CHAPTER FIVE IMPULSE AND LINEAR MOMENTUM 5-1

**5.2.3 Explain** You are wearing ice skates and standing on a frozen pond. How might you start moving without pushing off on the ice? Explain.

5-2 CHAPTER FIVE IMPULSE AND LINEAR MOMENTUM ©2014 Pearson Education.

These activities provide an opportunity for further observation, testing, sketching, and analysis.

# MAKE A DIFFERENCE WITH MasteringPhysics®

The Mastering platform is the most effective and widely used online homework, tutorial, and assessment system for physics. It delivers self-paced tutorials that focus on instructors' course objectives, provides individualized coaching, and responds to each student's progress. The Mastering system helps instructors maximize class time with easy-to-assign, customizable, and automatically graded assessments that motivate students to learn outside of class and arrive prepared for lecture and lab.

**www.masteringphysics.com**

## Prelecture Concept Questions

Assignable Prelecture Concept Questions encourage students to read the textbook prior to lecture so they're more engaged in class.

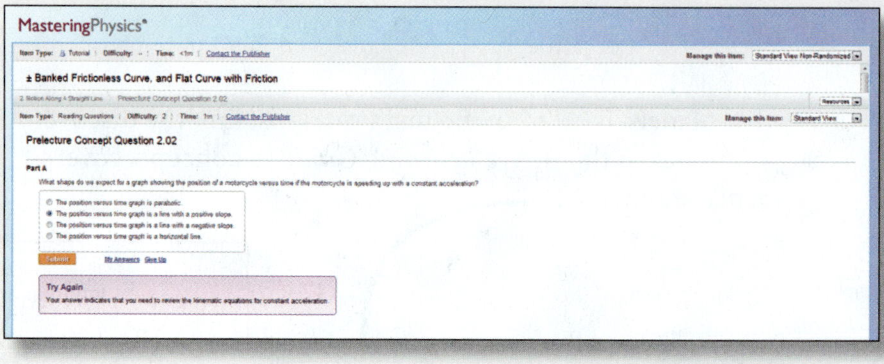


## Gradebook Diagnostics

The Gradebook Diagnostics screen provides your favorite weekly diagnostics. With a single click, charts summarize the most difficult problems, vulnerable students, and grade distribution.

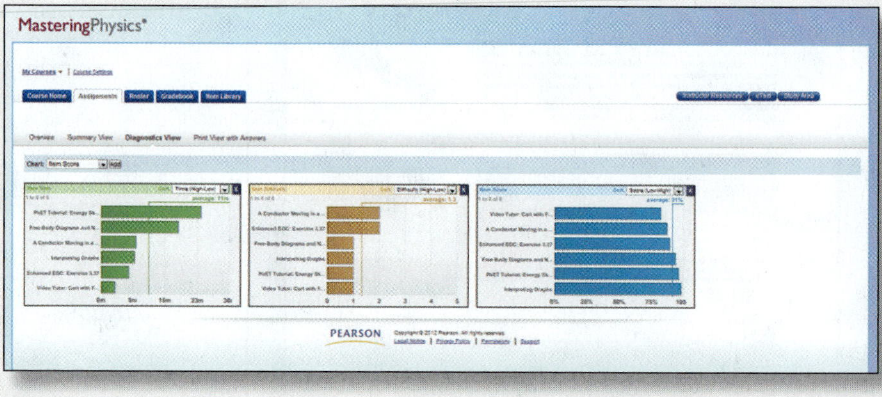


## Tutorials with Hints and Feedback

Easily assign tutorials that provide individualized coaching.

- Mastering's hallmark Hints and Feedback offer "scaffolded" instruction similar to what students would experience in an office hour.
- Hints (declarative and Socratic) can provide problem-solving strategies or break the main problem into simpler exercises.
- Wrong-answer-specific feedback gives students exactly the help they need by addressing their particular mistake without giving away the answer.

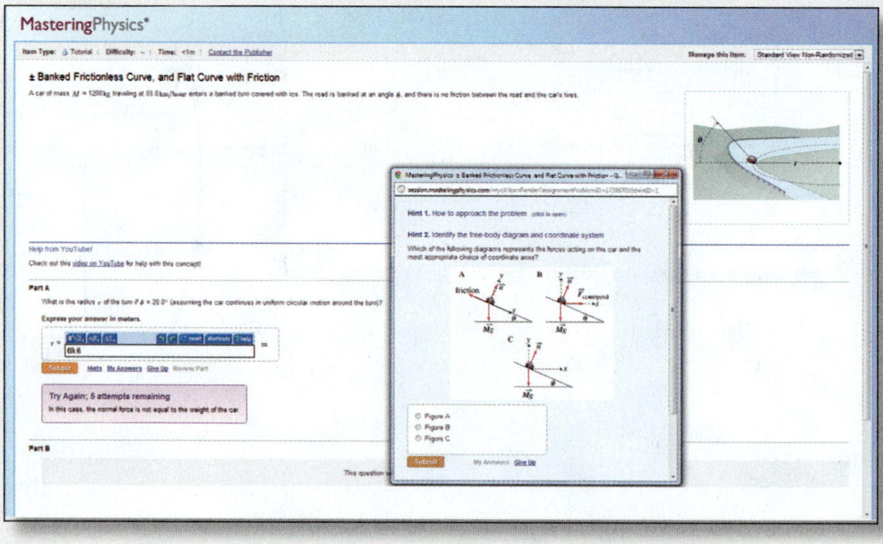

# Preface

## To the student

*College Physics* is more than just a book. It's a learning *companion.* As a companion, the book won't just tell you about physics; it will act as a guide to help you build physics ideas using methods similar to those that practicing scientists use to construct knowledge. The ideas that you build will be *yours*, not just a copy of someone else's ideas. As a result, the ideas of physics will be much easier for you to use when you need them: to succeed in your physics course, to obtain a good score on exams such as the MCAT, and to apply to everyday life.

Although few, if any, textbooks can honestly claim to be a pleasure to read, *College Physics* is designed to make the process interesting and engaging. The physics you learn in this book will help you understand many real-world phenomena, from why giant cruise ships are able to float to how telescopes work.

A great deal of research has been done over the past few decades on how students learn. We, as teachers and researchers, have been active participants in investigating the challenges students face in learning physics. We've developed unique strategies that have proven effective in helping students think like physicists. These strategies are grounded in *active learning*, deliberate, purposeful action on your part to learn something new. It's not passively memorizing so that you can repeat it later. When you learn actively you engage with the material. You relate it to what you already know. You think about it in as many different ways as you can. You ask yourself questions such as "Why does this make sense?" and "Under what circumstances does this not apply?"

This book (your learning companion) includes many tools to support the active learning process: each problem-solving strategies tool, worked example, observational experiment, testing experiment, review question, and end-of-chapter question and problem is designed to help you build your understanding of physics. To get the most out of these tools and the course, stay actively engaged in the process of developing ideas and applying them. When things get challenging, don't give up.

At this point you should turn to the chapter Introducing Physics and begin reading. That's where you'll learn the details of the approach that the book uses, what physics is, and how to be successful in the physics course you are taking.

## To the instructor

In writing *College Physics,* our main goal was to produce an effective learning companion for students that incorporates results from the last few decades of physics education research. This research has shown that there is a dramatic difference between how physicists construct new ideas and how students traditionally learn physics. Students often leave their physics course thinking of physics as a disconnected set of facts that has little to do with the real world, rather than as a framework for understanding it.

To address this problem we have based this book on a framework known as ISLE (Investigative Science Learning Environment) developed by authors Etkina and Van Heuvelen. In ISLE, the construction of new ideas begins with observational experiments. Students are explicitly presented with simple experiments from which they discern patterns using available tools (diagrams, graphs, bar charts, etc.). To explain the patterns, students devise explanations (hypotheses) for their observations. They then use these explanations in testing experiments to make predictions about the outcomes of these new experiments. If the prediction does not match the outcome of the experiment, the explanation needs to be reevaluated. Explanations that survive this testing process are the physics ideas in which we then have more confidence.

The goal of this approach is to help students understand physics as a process by which knowledge of the natural world is constructed, rather than as a body of given laws and facts. This approach also helps students reason using the tools that physicists and physics educators have developed for the analysis of phenomena—for example, motion and force diagrams, kinematics and thermodynamics graphs, energy and momentum bar charts, and many other visual representations. Using these tools helps students bridge the gap between words and mathematical equations. Along the way, they develop independent and critical thinking skills that will allow them to build their own understanding of physics principles.

All aspects of *College Physics* are grounded in ISLE and physics education research. As a result, all of the features of the text have been designed to encourage students to investigate, test ideas, and apply scientific reasoning.

## Key learning principles

To achieve these goals we adhere to five key learning principles:

1. **Concept first, name second:** The names we use for physics concepts have everyday-life meanings that may differ from the meanings they have when used in physics. For example, in physics *flux* refers to the amount to which a directed quantity (such as the magnetic field) points through a surface, but in everyday-life *flux* refers to continuous change. Confusion over the meaning of terms can get in the way of learning. We address this difficulty by developing the concept first and only then assigning a name to it.

2. **Careful language**: The vernacular physicists use is rooted in history and tradition. While physicists have an internal "dictionary" that lets them understand the meaning of specific terms, students do not. We are extremely careful to use language that promotes understanding. For example: physicists would say that "heat flows from a hot object to a cool object." Heat isn't a substance that objects possess; heat is the *flow of energy*. In this book we only use the word *heat* to refer to the process of energy transfer.
3. **Bridging words and mathematics**: Words and mathematics are very abstract representations of physical phenomena. We help students translate between these abstractions by using concrete representations such as force diagrams and energy bar charts as intermediate steps.
4. **Making sense of mathematics**: We explicitly teach students how to evaluate the results of their quantitative reasoning so they can have confidence in that reasoning. We do this by building qualitative understanding first and then explicitly teaching students how to use that understanding to check for quantitative consistency. We also guide students to use limiting cases to evaluate their results.
5. **Moving away from plug-and-chug problem solving approaches**: In this book you will find many non-traditional examples and end-of-chapter problems that require students to use higher-level reasoning skills and not just plug numbers into equations that have little meaning for them. Jeopardy problems (where a solution is given and students must invent a problem that leads to it), "tell-all" problems (where students must determine everything possible), and estimation problems (where students do not have quantities given to them) are all designed to encourage higher reasoning and problem solving skills.

These key principles are described in greater detail in the Introduction to the *Instructor's Guide* that accompanies *College Physics*—please read that introduction. It elaborates on the implementation of the methodology that we use in this book and provides guidance on how to integrate the approach into your course.

While our philosophy informs *College Physics*, you need not fully subscribe to it to use this textbook. We've organized the book to fit the structure of most algebra-based physics courses: We begin with kinematics and Newton's laws, then move on to conserved quantities, statics, gases, fluids, thermodynamics, electricity and magnetism, vibrations and waves, optics, and finally modern physics. The structure of each chapter will work with any method of instruction. You can assign all of the innovative experimental tables and end-of-chapter problems, or only a few. The text provides thorough treatment of fundamental principles, supplementing this coverage with experimental evidence, new representations, an effective approach to problem solving, and interesting and motivating examples.

## Real-world applications

To effectively teach physics, especially to the non-physics student, a textbook must actively engage the student's interest. To that effect, *College Physics* includes a wealth of real-world applications. Each chapter begins with a brief vignette designed to intrigue the reader. For example, Chapter 11 opens with a description of plaque build-up in arteries that can lead to stroke. Chapters also open with a set of motivating questions that are answered as students read subsequent sections. In each chapter, worked examples and exercises cover such topics as what keeps a car on the road when spinning around a circular track (Chapter 4), why air bags are so effective (Chapter 5), and why your ears pop when you change altitude (Chapter 10). A Putting it all together section applies concepts from the chapter to complex phenomena such as collisions (Chapter 6), lightning (Chapter 15), and the Doppler Effect (Chapter 20). Many applications are grounded in biology or medicine. Approximately eight percent of end-of-chapter problems are on biomedical topics. A complete list of applications appears on pages xxiv–xxvi.

## Chapter features

*Chapter-opening features engage the student in the chapter topic.*

Each chapter opens with a bridge from the concepts and skills that students will have learned in previous chapters. This bridge takes the form of **"Be sure you know how to"** statements with cross-references to the relevant material in previous chapters.

Each chapter also includes a set of **Motivating questions** to capture student interest. These questions are answered within the chapter content. Two examples are, "Why do people snore?" and "How does a refrigerator stay cold inside?"

A brief **vignette** opens each chapter with a real-world story related to one of the motivating questions.

In-chapter features encourage the active construction of knowledge about physics and support students as they read and review the material.

**Experimental tables** help students explore science as a process of inquiry (e.g., making observations, analyzing data, identifying patterns, testing hypotheses, etc.) and develop reasoning skills they can use to solve physics problems.

- **Observational experiment tables** engage students in an active discovery process as they learn about key physics ideas. By analyzing and finding patterns for the experiments, students learn the process of science.

- **Testing experiment tables** allow students to test hypotheses by predicting an outcome, conducting the experiment, and forming a conclusion that compares the prediction to the outcome and summarizes the results.

Many of these tables are accompanied by videos of the experiments. Students can view them through a QR code on the table using their smartphone, or online in the MasteringPhysics study area.

- **Section review questions** encourage critical thinking and synthesis rather than recall. Answers to review questions are given at the end of each chapter.

*Three types of worked examples guide students through the problem-solving process.*

**Examples** are complete problems that utilize the four-step problem-solving strategy.

- **Sketch and Translate:** This step teaches students to translate the problem statement into the language of physics. Students read the problem, sketch the situation and include known values, and identify the unknown(s).
- **Simplify and Diagram:** Students simplify the physics problem with an appropriate physical representation, a force diagram or other representation that reflects the situation in the problem and helps them construct a mathematical equation to solve it.
- **Represent Mathematically:** In this step, students apply the relevant mathematical equations. For example, they use the force diagram to apply Newton's second law in component form.
- **Solve and Evaluate:** In the last step, students rearrange equations and insert known values to solve for the unknown(s).

**Conceptual exercises** focus on developing students' conceptual understanding. These exercises utilize two of the problem-solving steps: *Sketch and Translate,* and *Simplify and Diagram.*

**Quantitative exercises** develop students' ability to solve unknowns quantitatively. These exercises utilize the problem-solving steps *Represent Mathematically* and *Solve and Evaluate.*

Each worked example ends with a **Try It Yourself** question, an additional exercise that builds on the worked example and asks students to solve a similar problem without the scaffolding.

*The text also includes additional support for problem solving.*

**Problem-solving strategy** boxes work through an example, explaining as well as applying the four-step strategy.

**Reasoning skill** boxes summarize the use of a particular skill, such as drawing a motion diagram, a force diagram, or a work-energy bar chart.

**Tips** within the text encourage the use of particular strategies or caution the reader about common misconceptions.

**Key equations** and **definitions boxes** highlight important laws or principles that govern the physics concepts developed in the chapters.

*The Putting it all together section (found in most chapters) focuses on two to four real-world applications of the physics learned in the chapter.*

**Putting it all together** contains conceptual explanations and capstone worked examples that will often draw on more than one principle. The goal of this section is to help students synthesize what they have learned and broaden their understanding of the phenomena they are exploring.

## End-of-chapter features

The **chapter summary** reviews key concepts presented in the chapter. The summary utilizes the multiple representation approach, displaying the concepts in words, figures, and equations.

Each chapter includes the authors' widely-lauded and highly creative problem sets, including their famous Jeopardy-style problems, "tell-all" problems, and estimating problems. The authors have written every end-of-chapter item themselves. Each question and problem is thoroughly grounded in their deep understanding of how students learn physics.

**Questions:** The question section includes both multiple choice and conceptual short answer questions, to help build students' fluency in the words, symbols, pictures, and graphs used in physics. Most of the questions are qualitative.

**Problems:** Chapters include an average of 60 section-specific problems. Approximately eight percent of the problems are drawn from biology or medicine; other problems relate to astronomy, geology, and everyday life.

**General Problems:** These challenging problems often involve multiple parts and require students to apply conceptual knowledge learned in previous chapters. As much as possible, the problems have a real-world context to enhance the connection between physics and students' daily lives.

**MCAT reading passage and related multiple choice questions and problems:** Because so many students who take this course are planning to study medicine, each chapter includes MCAT-style reading passages with related multiple-choice questions to help prepare students for their MCAT exam.

## Instructor supplements

The ***Instructor's Guide***, written by Eugenia Etkina, Alan Van Heuvelen, and highly respected physics education researcher David Brookes, walks you through the innovative approaches they take to teaching physics. Each chapter of the *Instructor's Guide* contains a roadmap to assigning chapter content, *Active Learning Guide* assignments, homework, and videos of the demonstration experiments. In addition, the authors call out common pitfalls to mastering physics concepts and describe techniques that will help your students identify and overcome their misconceptions. Tips include how to manage the complex vocabulary of physics, when to use classroom-response tools, and how to organize lab, lecture, and small group learning time. Drawing from their extensive experience as teachers and researchers, the authors give you the support you need to make *College Physics* work for you.

The cross-platform **Instructor Resource DVD** (ISBN 0-321-88897-9) provides invaluable and easy-to-use resources for your class, organized by textbook chapter. The contents include a comprehensive library of more than 220 applets from **ActivPhysics OnLine**™, as well as all figures, photos, tables, and summaries from the textbook in JPEG and PowerPoint formats. A set of editable **Lecture Outlines** and **Classroom Response System "Clicker" Questions** on PowerPoint will engage your students in class.

**MasteringPhysics**® (www.masteringphysics.com) is a powerful, yet simple, online homework, tutorial, and assessment system designed to improve student learning and results. Students benefit from wrong-answer specific feedback, hints, and a huge variety of educationally effective content while unrivalled gradebook diagnostics allow an instructor to pinpoint the weaknesses and misconceptions of their class.

NSF-sponsored published research (and subsequent studies) show that MasteringPhysics has dramatic educational results. MasteringPhysics allows instructors to build wide-ranging homework assignments of just the right difficulty and length and provides them with efficient tools to analyze in unprecedented detail both class trends and the work of any student.

In addition to the textbook's end-of-chapter problems, MasteringPhysics for *College Physics* also includes tutorials, prelecture concept questions, and Test Bank questions for each chapter. MasteringPhysics also now has the following learning functionalities:

- **Prebuilt Assignments:** These offer instructors a mix of end-of-chapter problems and tutorials for each chapter.
- **Learning Outcomes:** In addition to being able to create their own learning outcomes to associate with questions in an assignment, professors can now select content that is tagged to a large number of publisher-provided learning outcomes. They can also print or export student results based on learning outcomes for their own use or to incorporate into reports for their administration.
- **Quizzing and Testing Enhancements:** These include options to hide item titles, add password protection, limit access to completed assignments, and randomize question order in an assignment.
- **Math Remediation:** Found within selected tutorials, special links provide just-in-time math help and allow students to brush up on the most important mathematical concepts needed to successfully complete assignments. This new feature links students directly to math review and practice, helping students make the connection between math and physics.

The **Test Bank** contains more than 2,000 high-quality problems, with a range of multiple-choice, true/false, short-answer, and regular homework-type questions. Test files are provided in both TestGen® (an easy-to-use, fully networkable program for creating and editing quizzes and exams) and Word format, and can be downloaded from www.pearsonhighered.com/educator.

## Student supplements

The ***Active Learning Guide*** workbook by Eugenia Etkina, Michael Gentile, and Alan Van Heuvelen consists of carefully-crafted activities that provide an opportunity for further observation, sketching, analysis, and testing. Marginal "Active Learning Guide" icons throughout *College Physics* indicate content for which a workbook activity is available. Whether the activities are assigned or not, students can always use this workbook to reinforce the concepts they have read about in the text, to practice applying the concepts to real-world scenarios, or to work with sketches, diagrams, and graphs that help them visualize the physics.

**Physics demonstration videos**, accessed with a smartphone through QR codes in the text or online in the MasteringPhysics study area, accompany most of the Observational and Testing Experiment Tables. Students can observe the exact experiment described in the table.

**MasteringPhysics**® (www.masteringphysics.com) is a powerful, yet simple, online homework, tutorial, and assessment system designed to improve student learning and results. Students benefit from wrong-answer specific feedback, hints, and a huge variety of educationally effective content while unrivalled gradebook diagnostics allow an instructor to pinpoint the weaknesses and misconceptions of their class. The individualized, 24/7 Socratic tutoring is recommended by 9 out of 10 students to their peers as the most effective and time-efficient way to study.

**Pearson eText** is available through MasteringPhysics, either automatically when MasteringPhysics is packaged with new books, or available as a purchased upgrade online. Allowing students access to the text wherever they have access to the Internet, Pearson eText comprises the full text, including figures that can be enlarged for better viewing. Within eText, students are also able to pop up definitions and terms to help with vocabulary and the reading of the material. Students can also take notes in eText using the annotation feature at the top of each page.

**Pearson Tutor Services** (www.pearsontutorservices.com) Each student's subscription to MasteringPhysics also contains complimentary access to Pearson Tutor services, powered by Smarthinking, Inc. By logging in with their MasteringPhysics ID and password, they will be connected to highly qualified e-instructors™ who provide additional, interactive online tutoring on the major concepts of physics. Some restrictions apply; offer subject to change.

## Acknowledgments

We wish to thank the many people who helped us create this textbook and its supporting materials. First and foremost, we want to thank our team at Pearson Higher Education, especially Cathy Murphy, who provided constructive, objective, and untiring feedback on every aspect of the book; Jim Smith, who has been a perpetual cheerleader for the project; Katie Conley and Beth Collins, who shepherded the book through production, and Becky Ruden and Kelly Reed, who oversaw the media component of the program. Kyle Doctor attended to many important details, including obtaining reviews of the text and production of the *Active Learning Guide*. Special thanks to Margot Otway, Brad Patterson, Gabriele Rennie, and Michael Gillespie, who helped shape the book's structure and features in its early stages. We also want to thank Adam Black for believing in the future of the project. We are indebted to Frank Chmely and Brett Kraabel, who checked and rechecked every fact and calculation in the text. Sen-Ben Liao and Brett Kraabel prepared detailed solutions for every end-of-chapter problem for the *Instructor's Solutions Manual*. We also want to thank all of the reviewers who put their time and energy to providing thoughtful, constructive, and supportive feedback on every chapter.

Our infinite thanks go to Suzanne Brahmia who came up with the Investigative Science Learning Environment acronym "ISLE," and was and is an effective user of the ISLE learning strategy. Her ideas about relating physics and mathematics are reflected in many sections of the book. We thank Kruti Singh for her help with the class testing of the book and all of Eugenia's students for providing feedback and ideas. We also want to thank David Brookes, co-author of the *Instructor's Guide*, whose research affected the treatment of language in the textbook; Gorazd Planinšič, who provided feedback on many chapters; Paul Bunson, for using first drafts of the book's chapters in his courses; and Dedra Demaree, who has supported ISLE for many years. Dedra also prepared the extensive set of multiple-choice Clicker questions and the PowerPoint lecture outline available on our Instructor Resource DVD.

We have been very lucky to belong to the physics teaching community. Ideas of many people in the field contributed to our understanding of how people learn physics and what approaches work best. These people include Arnold Arons, Fred Reif, Jill Larkin, Lillian McDermott, David Hestenes, Joe Redish, Jim Minstrell, David Maloney, Fred Goldberg, David Hammer, Andy Elby, Tom Okuma, Curt Hieggelke, and Paul D'Alessandris. We thank all of them and many others.

### Personal notes from the authors

We wish to thank Valentin Etkin (Eugenia's father), an experimental physicist whose ideas gave rise to the ISLE philosophy many years ago, Inna Vishnyatskaya (Eugenia's mother), who never lost faith in the success of our book, and Dima and Sasha Gershenson (Eugenia's sons), who provided encouragement to Eugenia and Alan over the years. While teaching Alan how to play violin, Alan's uncle Harold Van Heuvelen provided an instructional system very different from that of traditional physics teaching. In Harold's system, many individual abilities (skills) were developed with instant feedback and combined over time to address the process of playing a complex piece of music. We tried to integrate this system into our ISLE physics learning system.

—Eugenia Etkina and Alan Van Heuvelen

First, thanks to my co-authors Alan and Eugenia for bringing me onto this project and giving me the opportunity to fulfill a life goal of writing a book, and for being cherished friends and colleagues these many years. Thanks also to my students and teaching assistants in Physics for Sciences at Rutgers University these last three years for using the book as their primary text and giving invaluable feedback. Thanks eternally to my parents for unquestioning support in all my endeavors, and gifting me with the attitude that all things can be achieved.

For my beloved partner Christine, as with this project and all other challenges in life: Team Effort. Best Kind.

—Mike Gentile

## Reviewers and classroom testers

Ricardo Alarcon
*Arizona State University*
Eric Anderson
*University of Maryland, Baltimore County*
James Andrews
*Youngstown State University*
David Balogh
*Fresno City College*
Linda Barton
*Rochester Institute of Technology*
Ian Beatty
*University of North Carolina at Greensboro*
Robert Beichner
*North Carolina State University, Raleigh*
Aniket Bhattacharya
*University of Central Florida*
Luca Bombelli
*University of Mississippi*
Scott Bonham
*Western Kentucky University*
Gerald Brezina
*San Antonio College*
Paul Bunson
*Lane Community College*
Hauke Busch
*Georgia College & State University*
Rebecca Butler
*Pittsburg State University*
Paul Camp
*Spelman College*
Amy Campbell
*Georgia Gwinnett College*
Juan Catala
*Miami Dade College North*
Colston Chandler
*University of New Mexico*
Soumitra Chattopadhyay
*Georgia Highlands College*
Betsy Chesnutt
*Itawamba Community College*
Chris Coffin
*Oregon State University*
Lawrence Coleman
*University of California, Davis*
Michael Crivello
*San Diego Mesa College*
Elain Dahl
*Vincennes University*
Charles De Leone
*California State University, San Marcos*
Carlos Delgado
*Community College of Southern Nevada*
Christos Deligkaris
*Drury University*
Dedra Demaree
*Oregon State University*
Karim Diff
*Santa Fe Community College*
Kathy Dimiduk
*Cornell University*
Diana Driscoll
*Case Western Reserve University*
Raymond Duplessis
*Delgado Community College*
Taner Edis
*Truman State University*
Bruce Emerson
*Central Oregon Community College*
Xiaojuan Fan
*Marshall University*
Nail Fazleev
*University of Texas at Arlington*
Gerald Feldman
*George Washington University*
Jane Flood
*Muhlenberg College*
Tom Foster
*Southern Illinois University, Edwardsville*
Richard Gelderman
*Western Kentucky University*
Anne Gillis
*Butler Community College*
Martin Goldman
*University of Colorado, Boulder*
Greg Gowens
*University of West Georgia*
Michael Graf
*Boston College*
Alan Grafe
*University of Michigan, Flint*
Recine Gregg
*Fordham University*
Elena Gregg
*Oral Roberts University*
John Gruber
*San Jose State University*
Arnold Guerra
*Orange Coast College*
Edwin Hach III
*Rochester Institute of Technology*
Steve Hagen
*University of Florida, Gainesville*
Thomas Hemmick
*State University of New York, Stony Brook*
Scott Hildreth
*Chabot College*
Zvonimir Hlousek
*California State University, Long Beach*
Mark Hollabaugh
*Normandale Community College*
Klaus Honscheid
*Ohio State University*
Kevin Hope
*University of Montevallo*
Joey Huston
*Michigan State University*
Richard Ignace
*East Tennessee State University*
Doug Ingram
*Texas Christian University*
George Irwin
*Lamar University*
Darrin Johnson
*University of Minnesota*
Adam Johnston
*Weber State University*
Mikhail Kagan
*Pennsylvania State University*
David Kaplan
*Southern Illinois University, Edwardsville*
James Kawamoto
*Mission College*
Julia Kennefick
*University of Arkansas*
Casey King
*Horry-Georgetown Technical College*
Patrick Koehn
*Eastern Michigan University*
Victor Kriss
*Lewis-Clark State College*
Peter Lanagan
*College of Southern Nevada, Henderson*
Albert Lee
*California State University, Los Angeles*
Todd Leif
*Cloud County Community College*
Eugene Levin
*Iowa State University*
Jenni Light
*Lewis-Clark State College*
Curtis Link
*Montana Tech of The University of Montana*
Donald Lofland
*West Valley College*
Susannah Lomant
*Georgia Perimeter College*
Rafael Lopez-Mobilia
*University of Texas at San Antonio*
Kingshuk Majumdar
*Berea College*
Gary Malek
*Johnson County Community College*
Eric Mandell
*Bowling Green State University*

Lyle Marschand
*Northern Illinois University*
Donald Mathewson
*Kwantlen Polytechnic University*
Mark Matlin
*Bryn Mawr College*
Timothy McKay
*University of Michigan*
David Meltzer
*Arizona State University*
William Miles
*East Central Community College*
Rabindra Mohapatra
*University of Maryland, College Park*
Enrique Moreno
*Northeastern University*
Joe Musser
*Stephen F. Austin State University*
Charles Nickles
*University of Massachusetts, Dartmouth*
Gregor Novak
*United States Air Force Academy*
John Ostendorf
*Vincennes University*
Philip Patterson
*Southern Polytechnic State University*
Jeff Phillips
*Loyola Marymount University*
Francis Pichanick
*University of Massachusetts, Amherst*
Dmitri Popov
*State University of New York, Brockport*
Matthew Powell
*West Virginia University*
Roberto Ramos
*Indiana Wesleyan University*
Greg Recine
*Fordham University*
Edward Redish
*University of Maryland, College Park*
Lawrence Rees
*Brigham Young University*
Lou Reinisch
*Jacksonville State University*
Andrew Richter
*Valparaiso University*
Melodi Rodrigue
*University of Nevada, Reno*
Charles Rogers
*Southwestern Oklahoma State University*
David Rosengrant
*Kennesaw State University*
Alvin Rosenthal
*Western Michigan University*
Lawrence Rowan
*University of North Carolina at Chapel Hill*
Roy Rubins
*University of Texas at Arlington*
Otto Sankey
*Arizona State University*
Rolf Schimmrigk
*Indiana University, South Bend*
Brian Schuft
*North Carolina Agricultural and Technical State University*
Sara Schultz
*Montana State University, Moorhead*
Bruce Schumm
*University of California, Santa Cruz*
David Schuster
*Western Michigan State University*
Bart Sheinberg
*Houston Community College, Northwest College*
Carmen Shepard
*Southwestern Illinois College*
Douglas Sherman
*San Jose State University*
Chandralekha Singh
*University of Pittsburgh*
David Snoke
*University of Pittsburgh*
David Sokoloff
*University of Oregon*
Mark Stone
*Northwest Vista College*
Bernhard Stumpf
*University of Idaho*
Tatsu Takeuchi
*Virginia Polytechnic Institute and State University*
Julie Talbot
*University of West Georgia*
Colin Terry
*Ventura College*
Beth Ann Thacker
*Texas Tech University*
John Thompson
*University of Maine, Orono*
Som Tyagi
*Drexel University*
David Ulrich
*Portland Community College*
Eswara P. Venugopal
*University of Detroit, Mercy*
James Vesenka
*University of New England*
William Waggoner
*San Antonio College*
Jing Wang
*Eastern Kentucky University*
Tiffany Watkins
*Boise State University*
Laura Weinkauf
*Jacksonville State University*
William Weisberger
*State University of New York, Stony Brook*
John Wernegreen
*Eastern Kentucky University*
Daniel Whitmire
*University of Louisiana at Lafayette*
Brian Woodahl
*Indiana University-Purdue University Indianapolis*
Gary Wysin
*Kansas State University, Manhattan*

# Real-World Applications

## Chapter 12



## Chapter 13



## Chapter 14



## Chapter 15



## Chapter 16



## Chapter 17



## Chapter 18



## Chapter 19



## Chapter 20



## Chapter 21



## Chapter 22

# Contents

15
25
15

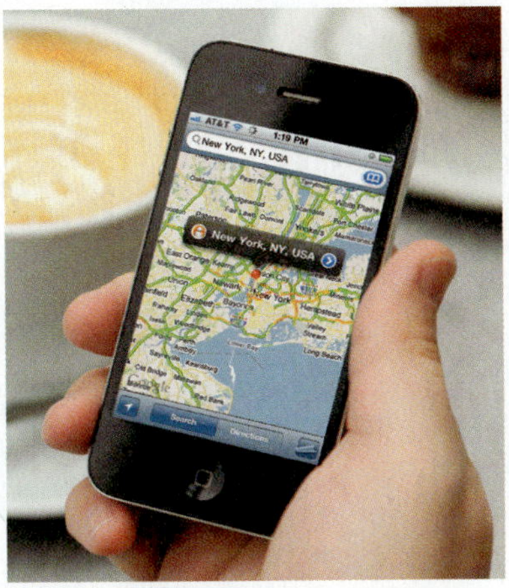

# Introducing Physics

**Why do we need to use models to explain the world around us?**

**How is the word "law" used differently in physics than in the legal system?**

**How do we solve physics problems?**

## I.1 What is physics?

Physics is a fundamental experimental science encompassing subjects such as mechanical motion, waves, light, electricity, magnetism, atoms, and nuclei. Knowing physics allows you to understand many aspects of the world, from why bending over to lift a heavy load can injure your back to why Earth's climate is changing. Physics explains the very small—atoms and subatomic particles—and the very large—planets, galaxies, and celestial bodies such as white dwarfs, pulsars, and black holes.

In each chapter, we will apply our knowledge of physics to other fields of science and technology such as biology, medicine, geology, astronomy, architecture, engineering, agriculture, and anthropology. For instance, in this book you will learn about techniques used by archeologists to determine the age of

**Physics in the field.** Archaeologists applied principles from physics to determine that this skeleton of *Australopithecus afarensis*, nicknamed "Lucy," lived about 3.2 million years ago.

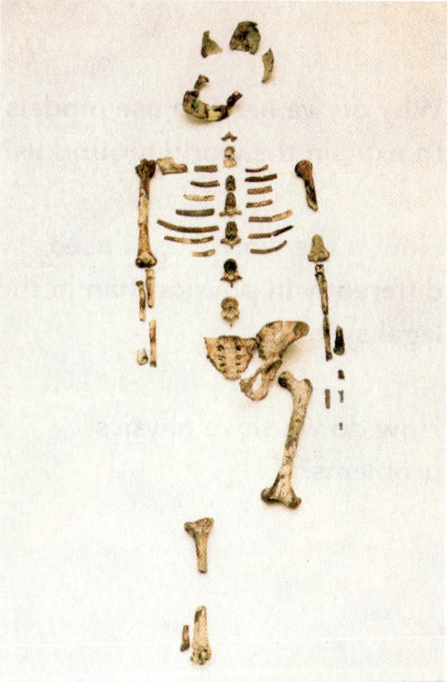

bones, about electron microscopes and airport metal detectors, about ways in which thermal energy is gained and lost in homes, about the development of stresses and tensions in body muscles, and why high blood pressure indicates problems with the circulatory system.

In this book we will concentrate not only on developing an understanding of the important basic laws of physics but also on the processes that physicists employ to discover and use these laws. The processes (among many) include:

- Collecting and analyzing experimental data.
- Making explanations and experimentally testing them.
- Creating different representations (pictures, graphs, bar charts, etc.) of physical processes.
- Finding mathematical relations between different variables.
- Testing those relations in new experiments.

## The search for rules

Physicists search for general rules or **laws** that bring understanding to the chaotic behavior of our surroundings. In physics the word *law* means a mathematical relation between variables inferred from the data or through some reasoning process. The laws, once discovered, often seem obvious, yet their discovery usually requires years of experimentation and theorizing. Despite being called "laws," these laws are temporary in the sense that new information often leads to their modification, revision, and, in some cases, abandonment.

For example, in 200 B.C. Apollonius of Perga watched the Sun and the stars moving in arcs across the sky and adopted the concept that Earth occupied the center of a revolving universe. Three hundred years later, Ptolemy provided a theory to explain the complicated motion of the planets in that Earth-centered universe. Ptolemy's theory, which predicted with surprising accuracy the changing positions of the planets, was accepted for the next 1400 years. However, as the quality of observations improved, discrepancies between the predictions of Ptolemy's theory and the real positions of the planets became bigger and bigger. A new theory was needed. Copernicus, who studied astronomy at the time that Columbus sailed to America, developed a theory of motion for the heavenly bodies in which the Sun resided at the center of the universe while Earth and the other planets moved in orbits around it. More than 100 years later the theory was revised by Johannes Kepler and later supported by careful experiments by Galileo Galilei. Finally, 50 years after Galileo's death, Isaac Newton formulated three simple laws of motion and the universal law of gravitation, which together provided a successful explanation for the orbital motion of Earth and the other planets. These laws also allowed us to predict the positions of new planets, which at the time were not yet known. For nearly 300 years Newton's ideas went unaltered until Albert Einstein made several profound improvements to our understanding of motion and gravitation at the beginning of the 20th century.

Newton's inspiration provided not only the basic resolution of the 1800-year-old problem of the motion of the planets but also a general framework for analyzing the mechanical properties of nature. Newton's simple laws give us the understanding needed to guide rockets to the moon, to build skyscrapers, and to lift heavy objects safely without injury.

It is difficult to appreciate the great struggles our predecessors endured as they developed an understanding that now seems routine. Today, similar struggles occur in most branches of science, though the questions being investigated have changed. How does the brain work? What causes Earth's magnetism? What is the nature of the pulsating sources of X-ray radiation in our galaxy? Is the recently discovered accelerated expansion of the universe really caused by a mysterious "dark energy," or is our interpretation of the observations of distant supernovae that revealed the acceleration incomplete?

**Newton's laws.** Thanks to Newton, we can explain the motion of the Moon. We can also build skyscrapers.

## Does this understanding make the world a better place?

The pursuit of basic understanding often seems greatly removed from the activities of daily living. If J. J. Thomson's peers had asked him in 1897 if there was any practical application for his discovery of the electron, he probably could not have provided a satisfactory answer. Yet a little over a century later, the electron plays an integral part in our everyday technology. Moving electrons in electric circuits produce light for reading and warmth for cooking. Knowledge of the electron has made it possible for us to transmit the information that we see as images on our smart phones and hear as sound from our MP3 players. Could the discovery of dark energy mentioned above lead to a similar technological revolution sometime in the future? It is certainly possible, but the details of that revolution would be very difficult to envision today, just as Thomson could not envision the impact his discovery of the electron would have throughout the 20th and 21st centuries.

## The processes for devising and using new rules

Physics is an experimental science. To answer questions, physicists do not just think and dream in their offices but constantly engage in experimental investigations. Physicists use special measuring devices to observe phenomena (natural and planned), describe their observations (carefully record them using words, numbers, graphs, etc.), find repeating features called patterns (for example, the distance traveled by a falling object is directly proportional to the square of the time in flight), and then try to explain these patterns. By doing this, physicists answer the questions of "why" or "how" the phenomenon happened and then deduce the rules that explain the phenomenon.

However, a deduced rule is not automatically accepted as true. Every rule needs to undergo careful testing. When physicists test a rule, they use the rule to predict the outcomes of new experiments. As long as there is no experiment whose outcome is inconsistent with predictions made using the rule, the rule is not disproved. The rule is consistent with all experimental evidence gathered so far. However, a new experiment could be devised tomorrow whose outcome is not consistent with the prediction made using the rule. The point is that there is no way to "prove" a rule once and for all. At best, the rule just hasn't been disproven yet.

A simple example will help you understand some processes that physicists follow when they study the world. Imagine that you walk into the house of your acquaintance Bob and see 10 tennis rackets of different quality and sizes. This is an **observational experiment**. During an observational experiment a scientist collects data that seem important. Sometimes it is an accidental or unplanned experiment. The scientist has no prior expectation of the outcome. In this case the number of tennis rackets and their quality and sizes represent the data. Having so many tennis rackets seems unusual to you, so you try to explain the data you collected (or, in other words, to explain why Bob has so many rackets) by devising several hypotheses. A **hypothesis** is an explanation of some sort that usually is based on some mechanism that is behind what is going on. One hypothesis is that Bob has lots of children and they all play tennis. A second hypothesis is that he makes his living by fixing tennis rackets. A third hypothesis is that he is a thief and he steals tennis rackets.

How do you decide which hypothesis is correct? You reason: if Bob has many children, and I walk around the house checking the sizes of clothes that I find, then I will find clothes of different sizes. Checking the clothing sizes is a new experiment, called a **testing experiment**. A testing experiment is different from an observational experiment. In a testing experiment, a specific hypothesis is being "put on trial." This hypothesis is used to construct a clear expectation of the outcome of the experiment. This clear expectation (based on the hypothesis being tested) is called a **prediction**. So, you conduct the testing experiment and walk around the house checking the closets. You do find clothes of different sizes. This is the outcome of your testing experiment. Does it mean for absolute certain that Bob has the rackets because all of his children play tennis? He could still be a racket repairman or a thief. Therefore, if the outcome of the testing experiment matches the prediction based on your hypothesis, you cannot say that you proved the hypothesis. All you can say is that you failed to disprove it. However, if you walk around the house and do not find any children's clothes, you can say with more confidence that the number of rackets in the house is not due to Bob having lots of children who play tennis. Still, this conclusion would only be valid if you made an **assumption:** Bob's children live in the house and wear clothes of different sizes. Generally, in order to reject a hypothesis you need to check the additional assumptions you made and determine if they are reasonable.

Imagine you have rejected the first hypothesis (you didn't find any children's clothes). Next you wish to test the hypothesis that Bob is a thief. This is your reasoning: If Bob is a thief (the hypothesis), and I walk around the house checking every drawer (the testing experiment), I should not find any receipts for the tennis rackets (the prediction). You perform the experiment and you find no receipts. Does it mean that Bob is a thief? He might just be a disorganized father of many children or a busy repairman. However, if you find all of the receipts, you can say with more confidence that he is not a thief (but he could still be a repairman). Thus it is possible to disprove (rule out) a hypothesis, but it is not possible to prove it once and for all. The process that you went through to create and test your hypotheses is depicted in **Figure I.1**. At the end of your investigation you might be left with a hypothesis that you failed to disprove. As a physicist you would now have some confidence in this hypothesis and start using it to solve other problems.

> **TIP** Notice the difference between a hypothesis and a prediction. A *hypothesis* is an idea that explains why or how something that you observe happens. A *prediction* is a statement of what should happen in a particular experiment if the hypothesis being tested were true. The prediction is based on the hypothesis and cannot be made without a specific experiment in mind.

Using this book you will learn physics by following a process similar to that described above. Throughout the book are many observational experiments and descriptions of the patterns that emerge from the data. After you read about the observational experiments and patterns, think about possible

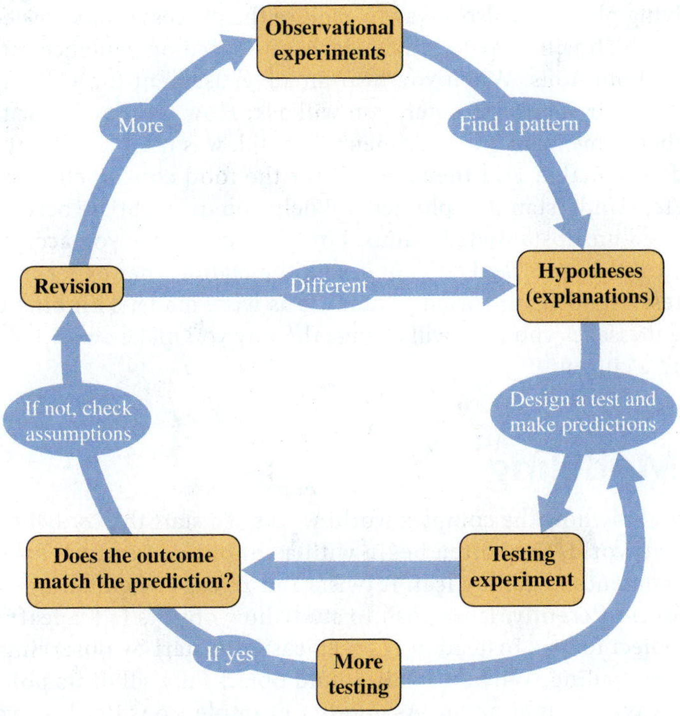


**Figure I.1** Science is a cyclical process for creating and testing knowledge.

explanations for these patterns. The book will then describe possible experiments to test the proposed explanations and also the predicted outcomes based on the hypotheses being tested. Then it will describe the outcomes of the actual experiments. Sometimes the outcomes of the actual experiments will match the predicted outcomes, and sometimes they will not. Based on the experimental results and the analysis of the assumptions that were made, the book will help you make a judgment about the hypothesis being tested.

## What language do physicists use?

Physicists use words and the language of mathematics to express ideas about the world. But they also represent these ideas and the world itself in other ways—sketches, diagrams, and even cut-out paper models (James Watson made a paper model of DNA when trying to determine its structure). In physics, however, the ultimate goal is to understand the mechanisms behind physical phenomena and to devise mathematical rules that allow for quantitative predictions of new phenomena. Thus, a big part of physics is identifying measurable properties of the phenomena (**physical quantities**, such as mass, speed, force), collecting quantitative data, and finding the patterns in that data.

## How will learning physics change your interactions with the world?

Even if you do not plan on becoming a professional physicist, learning physics can change the way you think about the world. For example, why do you feel cold when you wear wet clothes? Why is it safe to sit in a car during a lightning storm? Why, when people age, do they have trouble reading small-sized fonts? Why are parts of the world experiencing more extreme climate events? Knowing physics will also help you understand what underlies many important technologies. How does an MRI work? How can a GPS know your present position and guide you to a distant location? How do power plants generate electric energy?

Studying physics is also a way to acquire the processes of knowledge construction, which will help you make decisions based on evidence rather than on personal opinions. When you hear an advertisement for a shampoo that makes your hair 97.5% stronger, you will ask: How do they know this? Did this number come from an experiment? If it did, was it tested? What assumptions did they make? Did they control for the food consumed, exercise, air quality, etc.? Understanding physics will help you differentiate between actual evidence and unsubstantiated claims. For instance, before you accept a claim, you might ask about the data supporting the claim, what experiments were used to test the idea, and what assumptions were made. Thinking critically about the messages you hear will change the way you make decisions as a consumer and a citizen.

**Modeling.** Physicists often model complex structures as point-like objects.

(a)

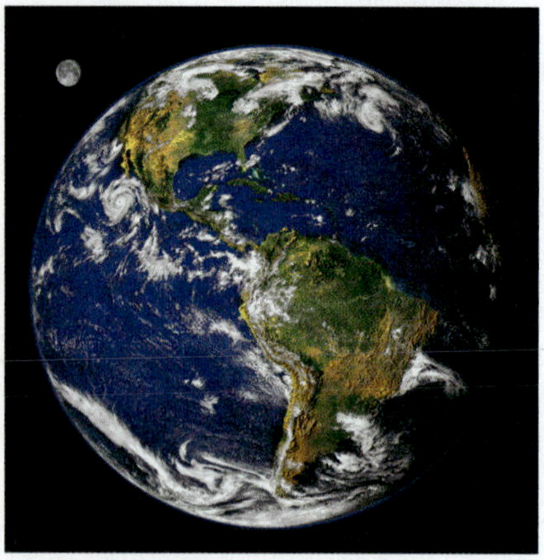

(b)

(c)

## I.2 Modeling

Physicists study how the complex world works. To start the study of some aspect of that world, they often begin with a simplified version. Take a common phenomenon: a falling leaf. It twists and zigzags when falling. Different leaves move differently. If we wish to study how objects fall, a leaf is a complicated object to use. Instead, it is much easier to start by observing a small, round object falling. When a small, round object falls, all of its points move the same way—straight down. As another example, consider how you move your body when you walk. Your back foot on the pavement lifts and swings forward, only to stop for a short time when it again lands on the pavement, now ahead of you. Your arms swing back and forth. The trunk of your body moves forward steadily. Your head also moves forward but bobs up and down slightly, especially if you run. It would be very difficult to start our study of motion by analyzing all these complicated parts and movements. Thus, physicists create in their minds simplified representations (called **models**) of physical phenomena and then think of the phenomena in terms of those models. Physicists begin with very simple models and then add complexity as needed to investigate more detailed aspects of the phenomena.

### A simplified object

To simplify real objects, physicists often neglect both the dimensions of objects (their sizes) and their structures (the different parts) and instead regard them as single **point-like objects**.

Is modeling a real object as a point-like object a good idea? Imagine a 100-meter race. The winner is the first person to get a body part across the finish line. It might be a runner's toe or it might be the head. The judge needs to observe the movement of all body parts (or a photo of the parts) across a very small distance near the finish line to decide who wins. Here, that very small distance near the finish line is small compared to the size of the human body. This is a situation where modeling the runners as point-like objects is not reasonable. However, if you are interested in how long it takes a person to run 100 meters, then the movement of different body parts is not as important, since 100 meters is much larger than the size of a runner. In this case, the runners can be modeled as point-like objects. Even though we are talking about the same situation (a 100-meter race), the aspect of the situation that interests us determines how we choose to model the runners.

Consider an airplane landing on a runway (**Figure I.2a**). We want to determine how long it takes for it to stop. Since all of its parts move together, the part we study does not matter. In that case it is reasonable to model the airplane as a point-like object. However, if we want to build a series of gates for planes to unload passengers (Figure I.2b), then we need to consider the

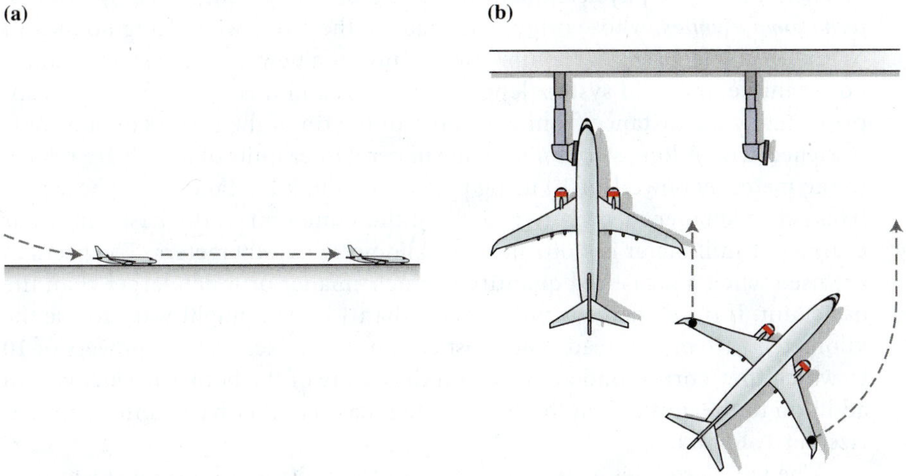

**Figure I.2** An airplane can be considered a point-like object (a) when landing, (b) but not when parking.

motion of the different parts of the airplane. For example, there must be enough room for an airplane to turn while maneuvering into and out of the gate. In this case the airplane cannot be modeled as a point-like object.

> **Point-like object** A point-like object is a simplified representation of a real object. As a rule of thumb, you can model a real object as a point-like object when one of the following two conditions are met: (a) when all of its parts move in the same way, or (b) when the object is much smaller than the other relevant lengths in the situation. The same object can be modeled as a point-like object in some situations but not in others.

## Modeling

The process that we followed to decide when a real object could be considered a point-like object is an example of what is called **modeling**. The modeling of objects is a first step that physicists use when they study natural phenomena. In addition to simplifying the objects that they study, scientists simplify the interactions between objects and also the processes that are occurring in the real world. Then they add complexity as their understanding grows. Galileo Galilei is believed to be the first scientist to consciously model a phenomenon. In his studies of falling objects in the early 17th century, he chose to simplify the real phenomenon by ignoring the interactions of the falling objects with the air.

> **Modeling** A model is a simplified representation of an object, a system (a group of objects), an interaction, or a process. A scientist creating the model decides which features to include and which to neglect.

# I.3 Physical quantities

To describe physical phenomena quantitatively, physicists construct **physical quantities:** features or characteristics of phenomena that can be measured experimentally. Measurement means comparing the characteristic to an assigned **unit** (a chosen standard).

## Units of measure

Physicists describe physical quantities using the **SI system**, or ***Le Système international d'unités***, whose origin goes back to the 1790s when King Louis XVI of France created a special commission to invent a new metric system of units. For example, in the SI system length is measured in meters. One meter is approximately the distance from your nose to the tip of the fingers of your outstretched arm. A long step is about one meter. Other units of length are related to the meter by powers of 10 using prefixes (milli, kilo, nano, ...). These prefixes relate smaller or bigger versions of the same unit to the basic unit. For example, 1 millimeter is 0.001 meter; 1 kilometer is 1000 meters. The prefixes are used when a measured quantity is much smaller or much larger than the basic unit. If the distance is much larger than 1 m, you might want to use the kilometer $(10^3\ \mathrm{m})$ instead. The most common prefixes and the powers of 10 to which they correspond are given on the inside of the book's back cover. In addition to the unit of length, the SI system has six other basic units, summarized in **Table I.1**.

The table provides a "feel" for some of the units but does not say exactly how each unit is defined. More careful definitions are important in order that measurements made by scientists in different parts of the world are consistent. However, to understand the precise definitions of these units, one needs to know more physics. We will learn how each unit is precisely defined when we investigate the concepts on which the definition is based.

## Measuring instruments

Physicists use a measuring instrument to compare the quantity of interest with a standardized unit. Each measuring instrument is calibrated so that it reads in multiples of that unit. Some examples of measuring instruments are a thermometer to measure temperature (calibrated in degrees Celsius or degrees Fahrenheit), a watch to measure time intervals (calibrated in seconds), and a

**Table I.1 Basic SI physical quantities and their units.**

| Physical quantity | Unit name and symbol | Physical description |
|---|---|---|
| Time | Second, s | One second is the time it takes for the heart to beat once. |
| Length | Meter, m | One meter is the length of one stride. |
| Mass | Kilogram, kg | One kilogram is the mass of 1 liter of water. |
| Electric current | Ampere, A | One ampere is the electric current through a 100-watt lightbulb in an American household |
| Temperature | Kelvin, K | One Kelvin degree is the same as 1 degree on the Celsius scale or about 2 degrees on the Fahrenheit scale. |
| Amount of matter | Mole, mol | One mole of oxygen is about 32 g. |
| Intensity of light | Candela, cd | One candela is the intensity of light produced by a relatively large candle at a distance of 1 m. |

meter stick to measure the height of an object (calibrated in millimeters). We can now summarize these ideas about physical quantities and their units.

**Physical quantity** **A physical quantity is a feature or characteristic of a physical phenomenon that can be measured in some unit. A measuring instrument is used to make a quantitative comparison of this characteristic with a unit of measure. Examples of physical quantities are your height, your body temperature, the speed of your car, and the temperature of air or water.**

## Significant digits

When we measure a physical quantity, the instrument we use and the circumstances under which we measure it determine how precisely we know the value of that quantity. Imagine that you wear a pedometer (a device that measures the number of steps that you take) and wish to determine the number of steps on average that you take per minute. You walk for 26 min (as indicated by your analog wristwatch) and see that the pedometer shows 2254 steps. You divide 2254 by 26 using your calculator, and it says 86.692307692307692. If you accept this number, it means that you know the number of steps per minute within plus or minus 0.000000000000001 steps/min. If you accept the number 86.69, it means that you know the number of steps to within 0.01 steps/min. If you accept the number 90, it means that you know the number of steps within 10 steps/min. Which answer should you use?

To answer this question, let's first focus on the measurements. Although your watch indicated that you walked for 26 min, you could have walked for as few as 25 min or for as many as 27 min. The number 26 does not give us enough information to know the time more precisely than that. The time measurement 26 min has two **significant digits**, or two numbers that carry meaning contributing to the precision of the result. The pedometer measurement 2254 has four significant digits. Should the result of dividing the number of steps by the amount of time you walked have two or four significant digits? If we accept four, it means that the number of steps per minute is known more precisely than the time measurement in minutes. This does not make sense. The number of significant digits in the final answer should be the same as the number of significant digits of the quantity used in the calculation that has the *smallest number of significant digits*. Thus, in our example, the average number of steps per minute should be 86, plus or minus 1 step/min: $86 \pm 1$.

Let's summarize the rule for determining significant digits. The precision of the value of a physical quantity is determined by one of two cases. If the quantity is measured by a single instrument, its precision depends on the instrument used to measure it. If the quantity is calculated from other measured quantities, then its precision depends on the least precise instrument out of all the instruments used to measure a quantity used in the calculation.

Another issue with significant digits arises when a quantity is reported with no decimal points. For example, how many significant digits does 6500 have—two or four? This is where scientific notation helps. **Scientific notation** means writing numbers in terms of their power of 10. For example, we can write 6500 as $6.5 \times 10^3$. This means that the 6500 actually has two significant digits: 6 and 5. If we write 6500 as $6.50 \times 10^3$, it means 6500 has three significant digits: 6, 5, and 0. The number 6.50 is more precise than the number 6.5, because it means that you are confident in the number to the hundredths place. Scientific notation provides a compact way of writing large and small numbers and also allows us to indicate unambiguously the number of significant digits a quantity has.

**Measuring and estimating.** In everyday life, rough estimates are often sufficient.

## 1.4 Making rough estimates

Sometimes we are interested in making a rough estimate of a physical quantity. The ability to make rough estimates is useful in a variety of situations, such as the following. (1) You need to decide whether a goal is worth pursuing—for example, can you make a living as a piano tuner in a town that already has a certain number of tuners? (2) You need to know roughly the amount of material needed for some activity—for instance, the food needed for a party or the number of bags of fertilizer needed for your lawn. (3) You want to estimate a number before it is measured—for example, how rapid a time-measuring device should be to detect laser light reflected from a distant object. (4) You wish to check whether a measurement you have made is reasonable—for instance, the measurement of the time for light to travel to a mountain and back or the mass of oxygen consumed by a hummingbird. (5) You wish to determine an unknown quantity—for example, an estimation of the number of cats in the United States or the compression force on the disks in your back when lifting a box of books in different ways.

The procedure for making rough estimates usually means selecting some basic physical quantities whose values are known or can be estimated and then combining the numbers using a mathematical procedure that leads to the desired answer. For example, suppose we want to estimate the number of pounds of food that an average person eats during a lifetime. First, assume that the average person consumes about 2000 calories/day to maintain a healthy metabolism. This food consists of carbohydrates, proteins, and fat. Using the labels of food packaging we find that 1 gram of carbohydrate or 1 gram of protein gives us about 4 calories, and 1 g of fat gives us about 9 calories. Thus, we assume that each gram of food consumed gives us on average 5 calories. Thus, each day, according to our assumptions, the person consumes about

$$(2000 \text{ calories/day})(1 \text{ g/5 calories}) = 400 \text{ g/day}$$

There are 365 days in a year, and we assume that the average life expectancy of a person is 70 years. Thus, the total food consumed during a lifetime is

$$(2000 \text{ calories/day})(1 \text{ g/5 calories})(365 \text{ days/year})(70 \text{ years/lifetime})$$
$$= 10{,}080{,}000 \text{ g/lifetime}$$

From the conversion table on the inside front cover, we see that 2.2 lb is 1000 g. Thus, our estimate of the number of pounds of food consumed in a lifetime is

$$(10{,}080{,}000 \text{ g/lifetime})(2.2 \text{ lb/1000 g}) = 22{,}176 \text{ lb/lifetime}$$

Our estimated result has five significant digits. Is this appropriate? To answer this, we need to look at the number of significant digits in each quantity used in the estimate. The calories/day quantity is probably uncertain by about 500 calories. That means the 2000 calories/day has just one significant digit. The ratio $(1 \text{ g/5 calories}) = 0.20$ could probably be $(1 \text{ g/6 calories}) = 0.17$, or about 0.03 different from our estimate. So that quantity has one or two significant digits. The life expectancy (70 years) could be off by about 10 years. Again, that's just one significant digit. Since the least certain quantity used in the calculation (the calories/day) has one significant digit, the final result should also be reported with just one significant digit. That would be 20,000 lb/lifetime.

## 1.5 Vector and scalar physical quantities

There are two general types of physical quantities—those that contain information about magnitude as well as direction and those that contain magnitude information only. Physical quantities that do not contain information about direction are called **scalar quantities** and are written using *italic* symbols ($m$, $T$, etc). Mass is a scalar quantity, as is temperature. To manipulate scalar quantities, you use standard arithmetic and algebra rules—addition, subtraction, multiplication, division, etc. You add, subtract, multiply, and divide scalars as though they were ordinary numbers.

Physical quantities that contain information about magnitude and direction are called **vector quantities** and are represented by italic symbols with an arrow on top ($\vec{F}$, $\vec{v}$, etc.). The little arrow on top of the symbol always points to the right. The actual direction of the vector quantity is shown in a diagram. For example, force is a physical quantity with both magnitude and direction (direction is very important if you are trying to hammer a nail into the wall). When you push a door, your push can be represented with a force arrow on a diagram; the stronger you push, the longer that arrow must be. The direction of the push is represented by the direction of that arrow (**Figure I.3**). The arrow's direction indicates the direction of the vector, and the arrow's relative length indicates the vector's magnitude. The methods for manipulating vector quantities (adding and subtracting them as well as multiplying a vector quantity by a scalar quantity and multiplying two vector quantities) are introduced as needed in the following chapters. Such manipulations are also summarized in the appendix Working with Vectors.

**Figure I.3** The force that your hand exerts on a door is a vector quantity represented by an arrow.

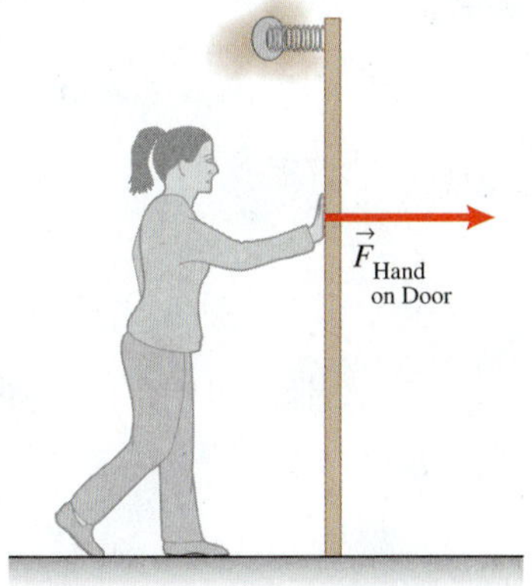


**Scalar quantities.** Temperature is a scalar quantity; it has magnitude, but not direction.

## I.6 How to use this book to learn physics

A textbook is only one part of a learning system, but knowing how to use it most effectively will make it easier to learn and to succeed in the course. This textbook will help you construct understanding of some of the most important ideas in physics, learn to use physics knowledge to analyze physical phenomena, and develop the general process skills that scientists use in the practice of science.

### Learning new material

Read the book as soon as possible after new material is discussed in class while the material is still fresh in your mind. First, scan the relevant sections and, if necessary, the whole chapter. Does it appear that the material involves completely new ideas, or is it just the application of what you have already learned? If the material does involve new ideas, how do the new ideas fit into what you have already learned? Then, read the relevant new section(s) slowly. Keep relating what you read to your current understanding. Pay attention to the Tips—they will help prevent confusion and future difficulties.

The most important strategy that will help you learn better is called **interrogation.** Interrogation means continually asking yourself the same question when reading the text. This question is so important that we put it in the box below:

**Why is this true?**

Make sure that you ask yourself this question as often as possible so that eventually it becomes a habit. Out of all the strategies that are recommended for reading comprehension, this is the one that is directly connected to better learning outcomes. For example, consider the first sentence of the next paragraph: "Solving physics problems is much more than plugging numbers into an equation." Ask yourself, "Why is this true?" Possibly, because one needs to understand what physics concepts are relevant, or what simplifying assumptions are important. There can be other reasons. By just stopping and interrogating yourself as often as possible about what is written in the book you will be able to understand and remember this information better.

## Problem solving

Solving physics problems is much more than plugging numbers into an equation. To use the book for problem-solving practice, focus on the problem-solving steps used in the worked examples.

**Step 1:** *Sketch and Translate* First, read the text of the problem several times slowly to make sure you understand what it says. Next, try to visualize the situation or process described in the text of the problem. Try to imagine what is happening. Draw a sketch of the process and label it with any information you have about the situation. This often involves an initial situation and a final situation. Often, the information in the problem statement is provided in words and you will need to *translate* it into physical quantities. Having the problem information in a visual sketch also frees some of your mind so that you can use its resources for other parts of the problem solving.

**Step 2:** *Simplify and Diagram* Decide how you can simplify the process. How will you model the object of interest (the object you are investigating)? What interactions can you neglect? To diagram means to represent the problem process using some sort of diagram, bar chart, graph, or picture that includes physics quantities. Diagrams bridge the gap between the verbal and sketch representations of the process and the mathematical representation of it.

**Step 3:** *Represent Mathematically* Construct a mathematical description of the process. You will use the sketch from Step 1 and the diagram(s) from Step 2 to help construct this mathematical description and evaluate it to see if it is reasonable. By representing the situation in these multiple ways and learning to translate from one way to the other, you will start giving meaning to the abstract symbols used in the mathematical description of the process.

**Step 4:** *Solve and Evaluate* Finally, solve the mathematical equations and evaluate the results. Do the numbers and signs make sense? Are the units correct? Another method involves evaluating whether the answer holds in extreme cases—you will learn more about this technique as you progress through the book.

Try to solve the example problems that are provided in the chapters by using this four-step strategy without looking at the solution. After finishing, compare your solution to the one described in the book. Then do the *Try It Yourself* part of the example problem and compare your answer to the book's answer. If you are still having trouble, try to use the same strategy to actively solve other example problems in the text or from the Active Learning Guide (if you are using that companion book). Uncover the solutions to these worked examples only after you have tried to complete the problem on your own. Then try the same process on the homework problems assigned by your instructor.

Notice that quantitative exercises and conceptual exercises in the book have fewer steps: *Represent Mathematically* and *Solve and Evaluate* for the quantitative exercises, and *Sketch and Translate* and *Simplify and Diagram* for the conceptual exercises. Sketching the process and representing it in different ways is an important step in solving any problem.

## Summary

We are confident that this book will act as a useful companion in your study of physics and that you will take from the course not just the knowledge of physics but also an understanding of the process of science that will help you in all your scientific endeavors. Learning physics through the approach used in this book builds a deeper understanding of physics concepts and an improved ability to solve difficult problems compared to traditional learning methods. In addition, you will learn to reason scientifically and be able to transfer those reasoning skills to many other aspects of your life.

# 1 Kinematics: Motion in One Dimension

What is a safe following distance between you and the car in front of you?

Can you be moving and not moving at the same time?

Why do physicists say that an upward thrown object is falling?

**Be sure you know how to:**

- Define what a point-like object is (Introducing Physics).
- Use significant digits in calculations (Introducing Physics).

When you drive, you are supposed to follow the 3-second tailgating rule. When the car in front of you passes some fixed sign at the side of the road, your car should be far enough behind so that it takes you 3 seconds to reach the same sign. You then have a good chance of avoiding a collision if the car in front stops abruptly. If you are 3 seconds behind the car in front of you when you see its brake lights, you should be able to step on the brake and avoid a collision. If you are closer than 3 seconds away, a collision is likely. In this chapter we will learn the physics behind the 3-second rule.

**Scientists often ask** questions about things that most people accept as being "just the way it is." For example, in the northern hemisphere, we have more hours of daylight in June than in December. In the southern hemisphere, it's just the opposite. Most people simply accept this fact. However, scientists want explanations for such simple phenomena. In this chapter, we learn to describe a phenomenon that we encounter every day but rarely question—motion.

## 1.1 What is motion?

When describing motion, we need to focus on two important aspects: the object whose motion we are describing (**the object of interest**) and the person who is doing the describing (**the observer**). Consider Observational Experiment **Table 1.1**, which analyzes how the description of an object's motion depends on the observer.

**OBSERVATIONAL EXPERIMENT TABLE**

**1.1 Different observers describe an object's motion.**

| Observational experiment | Analysis |
| --- | --- |
| **Experiment 1.** Jan observes a ball in her hands as she walks across the room. Tim, sitting at a desk, also observes the ball.<br>Jan reaches the other side of the room without taking her eyes from the ball; her head did not turn. Tim's head has turned in order to follow the ball.<br><br> | The two observers (Jan and Tim) see the same object of interest (the ball) differently. With respect to Jan, the ball's position does not change. With respect to Tim, its position does change. |

*(continued)*

| Observational experiment | Analysis |
|---|---|
| **Experiment 2.** Ted and Sue are passengers on the same train. Ted does not have to turn his head to keep his eyes on Sue. Joan, standing on the station platform, turns her head to follow Sue. | Ted and Joan see the same object of interest (Sue) differently. With respect to Ted, Sue's position does not change. With respect to Joan, Sue's position changes. |

| Pattern |
|---|
| Different observers can describe the same process differently, including whether or not motion is even occurring. |

In Table 1.1, we saw that different observers can describe the same process differently. One person sees the object of interest moving while another does not. They are both correct from their own perspectives. In order to describe the motion of something, we need to identify the observer.

**Motion is a change in an object's position relative to a given observer during a certain change in time. Without identifying the observer, it is impossible to say whether the object of interest moved. Physicists say *motion is relative,* meaning that the motion of any object of interest depends on the point of view of the observer.**

Are you moving as you read this book? Your friend walking past you first sees you in front of her, then she sees you next to her, and finally she sees you behind her. Though you are sitting in a chair, you definitely are moving with respect to your friend. You are also moving with respect to the Sun or with respect to a bird flying outside.

What makes the idea of relative motion confusing at first is that people intuitively use Earth as the object of reference—the object with respect to which they describe motion. If an object does not move with respect to Earth, many people would say that the object is not moving. That is why it took scientists thousands of years to understand the reason for days and nights on Earth. An observer on Earth uses Earth as the object of reference and sees the Sun moving in an arc across the sky (**Figure 1.1a**). An observer on a distant spaceship sees Earth rotating on its axis so that different parts of its surface face the Sun at different times (Figure 1.1b).

**Figure 1.1** Motion is relative. Two observers explain the motion of the Sun relative to Earth differently.

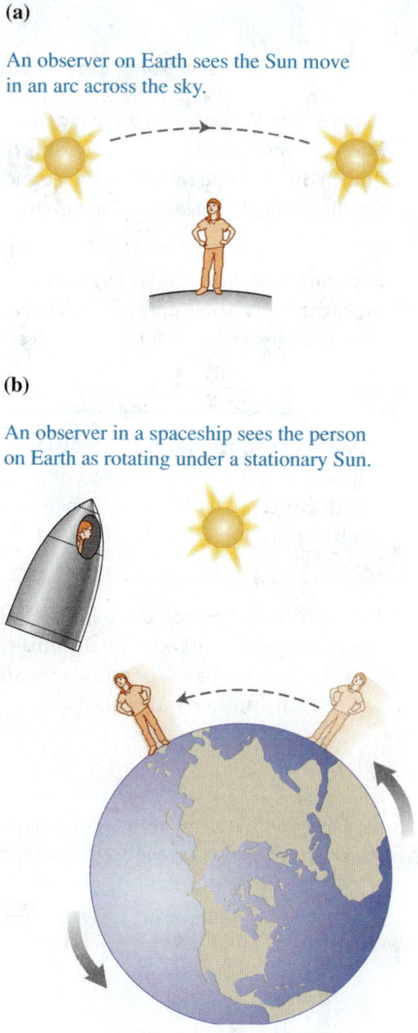

(a)
An observer on Earth sees the Sun move in an arc across the sky.
(b)
An observer in a spaceship sees the person on Earth as rotating under a stationary Sun.


## Reference frames

Specifying the observer before describing the motion of an object of interest is an extremely important part of constructing what physicists call a **reference frame.** A reference frame includes an object of reference, a coordinate system with a scale for measuring distances, and a clock to measure time. If the object of reference is large and cannot be considered a point-like object, it is important to specify where on the object of reference the origin of the coordinate system is placed. For example, if you want to describe the motion of a bicyclist and choose your object of reference to be Earth, you place the origin of the coordinate system at the surface, not at Earth's center.

**Reference frame** A reference frame includes three essential components:
(a) An *object of reference* with a specific *point of reference* on it.
(b) A *coordinate system*, which includes one or more coordinate axes, such as, $x$, $y$, $z$, and an origin located at the point of reference. The coordinate system also includes a unit of measurement (a scale) for specifying distances along the axes.
(c) A *clock*, which includes an origin in time called $t = 0$ and a unit of measurement for specifying times and time intervals.

## Modeling motion

When we model objects, we make simplified assumptions in order to analyze complicated situations. Just as we simplified an object to model it as a point-like object, we can also simplify a process involving motion. What is the simplest way an object can move?

Imagine that you haven't ridden a bike in a while. You would probably start by riding in a straight line before you attempt a turn. This kind of motion is called **linear motion** or **one-dimensional motion.**

**Linear motion** is a model of motion that assumes that an object, considered as a point-like object, moves along a straight line.

For example, we want to model a car's motion along a straight stretch of highway. We can assume the car is a point-like object (it is small compared to the length of the highway) and the motion is linear motion (the highway is long and straight).

**Review Question 1.1** Physicists say, "Motion is relative." Why is this true?

# 1.2 A conceptual description of motion

To describe linear motion more precisely, we start by devising a visual representation. Consider Observational Experiment **Table 1.2**, in which a bowling ball rolls on a smooth floor.

**OBSERVATIONAL EXPERIMENT TABLE**

**1.2 Using dots to represent motion.**

VIDEO 1.2

| Observational experiment | Analysis |
|---|---|
| **Experiment 1.** You push a bowling ball (the object of interest) once and let it roll on a smooth linoleum floor. You place beanbags each second beside the bowling ball. The beanbags are evenly spaced. | We can represent the locations of the bags each second for the slow-moving bowling ball as dots on a diagram. 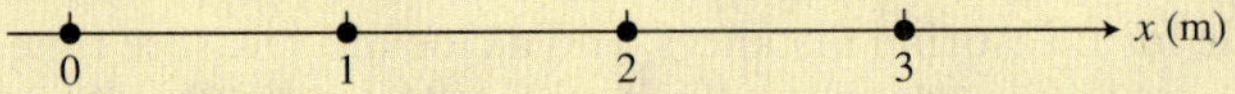  |
| **Experiment 2.** You repeat Experiment 1, but you push the ball harder before you let it roll. The beanbags are farther apart but are still evenly spaced. | The dots in this diagram represent the evenly spaced bags, which are separated by a greater distance than the bags in Experiment 1. 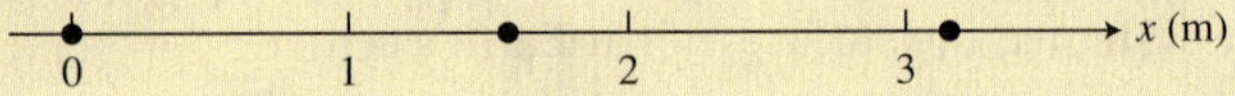  |
| **Experiment 3.** You push the bowling ball and let it roll on a carpeted floor instead of a linoleum floor. The distance between the beanbags decreases as the ball rolls. | The dots in this diagram represent the decreasing distance between the bags as the ball rolls on the carpet. 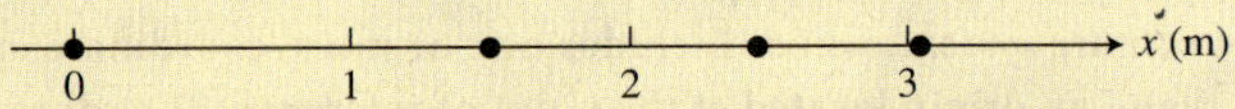  |
| **Experiment 4.** You roll the ball on the linoleum floor and gently and continually push on it with a board. The beanbag separation spreads farther apart as the pushed ball rolls. | The dots in this diagram represent the increasing distance between bags as the ball is continually pushed across the linoleum floor. 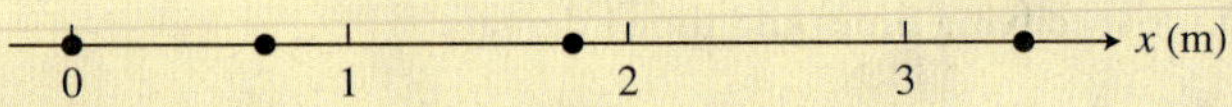  |

**Pattern**

- The spacing of the dots allows us to visualize motion.
- When the object travels without speeding up or slowing down, the dots are evenly spaced.
- When the object slows down, the dots get closer together.
- When the object moves faster and faster, the dots get farther apart.

## Motion diagrams

In the experiments in Table 1.2, the beanbags were an approximate record of where the ball was located as time passed and help us visualize the motion of the ball. We can represent motion in even more detail by adding **velocity arrows** to each dot that indicate which way the object is moving and how fast it is moving as it passes a particular position (see **Figure 1.2**). These new diagrams are called **motion diagrams.** The longer the arrow, the faster the motion. The small arrow above the letter $v$ indicates that this characteristic of motion has a direction as well as a magnitude—called a **vector quantity**. In Figure 1.2a, the dots are evenly spaced, and the velocity arrows all have the same length and point in the same direction. This means that the ball was moving equally fast in the same direction at each point. Similar diagrams with velocity arrows for the other three experiments in Table 1.2 are shown in Figures 1.2b–d.

## Velocity change arrows

In Experiment 4 the bowling ball was moving increasingly fast while being pushed. The velocity arrows in the motion diagram thus got increasingly longer. We can represent this change with a **velocity change arrow** $\Delta\vec{v}$. The $\Delta$ (delta) means a change in whatever quantity follows the $\Delta$, a change in $\vec{v}$ in this case. The $\Delta\vec{v}$ doesn't tell us the exact increase or decrease in the velocity; it only indicates a qualitative difference between the velocities at two adjacent points in the diagram.

**REASONING SKILL** Constructing a motion diagram.

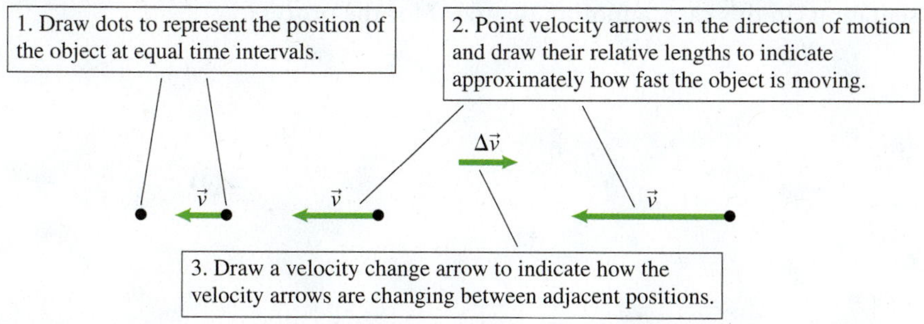
1. Draw dots to represent the position of the object at equal time intervals.
2. Point velocity arrows in the direction of motion and draw their relative lengths to indicate approximately how fast the object is moving.
$\Delta\vec{v}$
$\vec{v}$
$\vec{v}$
$\vec{v}$
3. Draw a velocity change arrow to indicate how the velocity arrows are changing between adjacent positions.

Note that we have redrawn the diagram shown in Figure 1.2d in **Figure 1.3a**. For illustration purposes only, we number the $\vec{v}$ arrows consecutively for each position: $\vec{v}_1$, $\vec{v}_2$, $\vec{v}_3$, etc. To draw the velocity change arrow as the ball moves from position 2 to position 3 in Figure 1.3a, we place the second arrow $\vec{v}_3$ directly above the first arrow $\vec{v}_2$, as shown. The $\vec{v}_3$ arrow is longer than the $\vec{v}_2$ arrow. This tells us that the object was moving faster at position 3 than at position 2. To visualize the change in velocity, we need to think about how arrow $\vec{v}_2$ can be turned into $\vec{v}_3$. We can do it by placing the tail of a velocity change arrow $\Delta\vec{v}_{23}$ at the head of $\vec{v}_2$ so that the head of $\Delta\vec{v}_{23}$ makes the combination $\vec{v}_2 + \Delta\vec{v}_{23}$ the same length as $\vec{v}_3$ (Figure 1.3b). Since they are the same length and in the same direction, the two vectors $\vec{v}_2 + \Delta\vec{v}_{23}$ and $\vec{v}_3$ are equal:

$$\vec{v}_2 + \Delta\vec{v}_{23} = \vec{v}_3$$

Note that if we move $\vec{v}_2$ to the other side of the equation, then

$$\Delta\vec{v}_{23} = \vec{v}_3 - \vec{v}_2$$

Thus, $\Delta\vec{v}_{23}$ is the difference of the third velocity arrow and the second velocity arrow—the change in velocity between position 2 and position 3. (To learn more about vector addition, read the appendix Graphical Addition and Subtraction of Vectors.)

## Making a complete motion diagram

We now place the $\Delta\vec{v}$ arrows above and between the dots in our diagrams where the velocity change occurred (see **Figure 1.4a**). The dots in these more detailed motion diagrams indicate the object's position at equal time intervals; velocity arrows and velocity change arrows are also included. A $\Delta\vec{v}$ arrow points in the same direction as the $\vec{v}$ arrows when the object is speeding up; the $\Delta\vec{v}$ arrow points in the opposite direction of the $\vec{v}$ arrows when the object is slowing down. When velocity changes by the same amount during each consecutive time interval, the $\Delta\vec{v}$ arrows for each interval are the same length. In such cases we need only one $\Delta\vec{v}$ arrow for the entire motion diagram (see Figure 1.4b).

The Reasoning Skill box summarizes the procedure for constructing a motion diagram. Notice that in the experiment represented in this diagram, the object is moving from right to left and slowing down.

**Figure 1.4** Two complete motion diagrams, including position dots, $\vec{v}$ arrows, and $\Delta\vec{v}$ arrows.

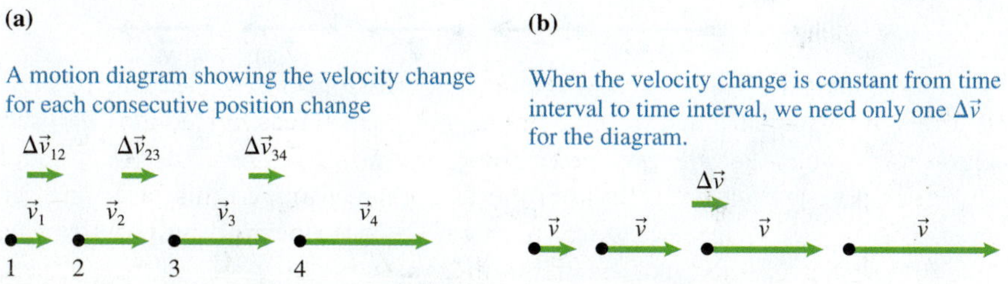
(a)
A motion diagram showing the velocity change for each consecutive position change
$\Delta\vec{v}_{12}$ $\Delta\vec{v}_{23}$ $\Delta\vec{v}_{34}$
$\vec{v}_1$ $\vec{v}_2$ $\vec{v}_3$ $\vec{v}_4$
1 2 3 4
(b)
When the velocity change is constant from time interval to time interval, we need only one $\Delta\vec{v}$ for the diagram.
$\Delta\vec{v}$
$\vec{v}$ $\vec{v}$ $\vec{v}$ $\vec{v}$

**Figure 1.2** Motion diagrams represent the types of motion shown in Table 1.2.

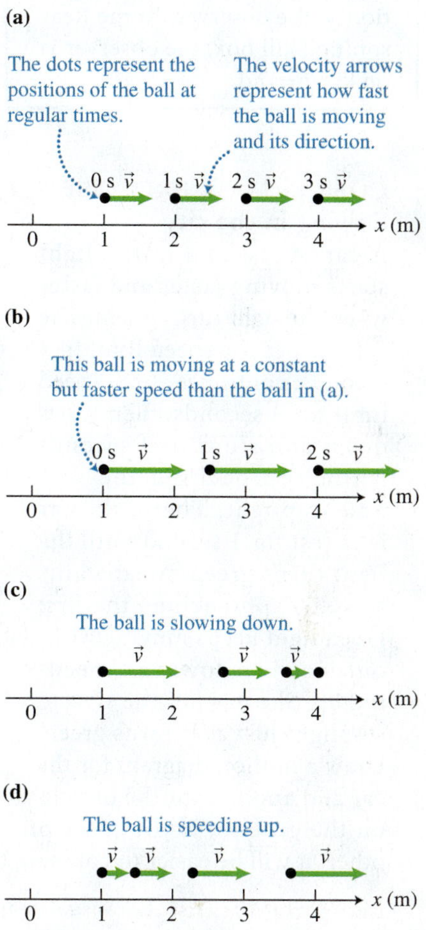
(a)
The dots represent the positions of the ball at regular times.
The velocity arrows represent how fast the ball is moving and its direction.
0 s $\vec{v}$ 1 s $\vec{v}$ 2 s $\vec{v}$ 3 s $\vec{v}$
0 1 2 3 4 x (m)
(b)
This ball is moving at a constant but faster speed than the ball in (a).
0 s $\vec{v}$ 1 s $\vec{v}$ 2 s $\vec{v}$
0 1 2 3 4 x (m)
(c)
The ball is slowing down.
$\vec{v}$ $\vec{v}$ $\vec{v}$
0 1 2 3 4 x (m)
(d)
The ball is speeding up.
$\vec{v}$ $\vec{v}$ $\vec{v}$ $\vec{v}$
0 1 2 3 4 x (m)

**Figure 1.3** Determining the magnitude and the direction of the velocity change arrow in a motion diagram.

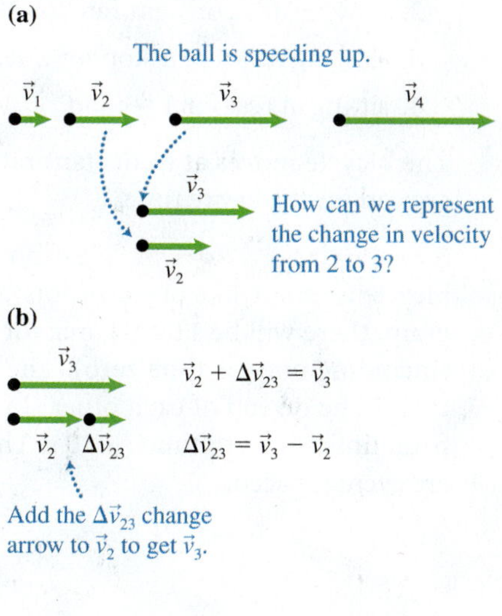
(a)
The ball is speeding up.
$\vec{v}_1$ $\vec{v}_2$ $\vec{v}_3$ $\vec{v}_4$
$\vec{v}_3$
$\vec{v}_2$
How can we represent the change in velocity from 2 to 3?
(b)
$\vec{v}_3$
$\vec{v}_2 + \Delta\vec{v}_{23} = \vec{v}_3$
$\vec{v}_2$ $\Delta\vec{v}_{23}$
$\Delta\vec{v}_{23} = \vec{v}_3 - \vec{v}_2$
Add the $\Delta\vec{v}_{23}$ change arrow to $\vec{v}_2$ to get $\vec{v}_3$.

**TIP** When drawing a motion diagram, always specify the position of the observer. In the Reasoning Skill box, the observer is on the ground.

Read Conceptual Exercise 1.1 several times and visualize the situation. If possible, draw a sketch of what is happening. Then construct a physics representation (in this case, a motion diagram) for the process.

## CONCEPTUAL EXERCISE 1.1

### Driving in the city

A car at rest at a traffic light starts moving faster and faster when the light turns green. The car reaches the speed limit in 4 seconds, continues at the speed limit for 3 seconds, then slows down and stops in 2 seconds while approaching the second stoplight. There, the car is at rest for 1 second until the light turns green. Meanwhile, a cyclist approaching the first green light keeps moving without slowing down or speeding up. She reaches the second stoplight just as it turns green. Draw a motion diagram for the car and another for the bicycle as seen by an observer on the ground. If you place one diagram below the other, it will be easier to compare them.

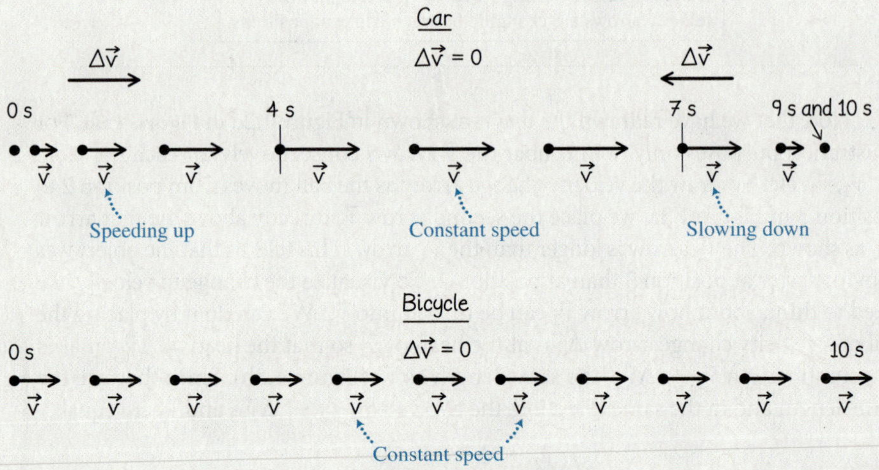


**Sketch and translate** Visualize the motion for the car and for the bicycle as seen by the observer on the ground. The car and the bicycle will be our objects of interest.

The motion of the car has four distinct parts:

1. starting at rest and moving faster and faster for 4 seconds;
2. moving at a constant rate for 3 seconds;
3. slowing down to a stop for 2 seconds; and
4. sitting at rest for 1 second.

The bicycle moves at a constant rate with respect to the ground for the entire time.

**Simplify and diagram** We can model the car and the bicycle as point-like objects (dots). In each motion diagram, there will be 11 dots, one for each second of time (including one for time zero). The last two dots for the car will be on top of each other since the car was at rest from time = 9 s to time = 10 s. The dots for the bicycle are evenly spaced.

**Try it yourself:** Two bowling balls are rolling along a linoleum floor. One of them is moving twice as fast as the other. At time zero, they are next to each other on the floor. Construct motion diagrams for each ball's motion during a time of 4 seconds, as seen by an observer on the ground. Indicate on the diagrams the locations at which the balls were next to each other at the same time. Indicate possible mistakes that a student can make answering the question above.

*Answer:* See the figure below. The balls are side by side only at time zero—the first dot for each ball. It looks like they are side by-side when at the 2-m position, but the slow ball is at the 2-m position at 2 s and the faster ball is there at 1 s. Similar reasoning applies for the 4-m positions—the balls reach that point at different times.

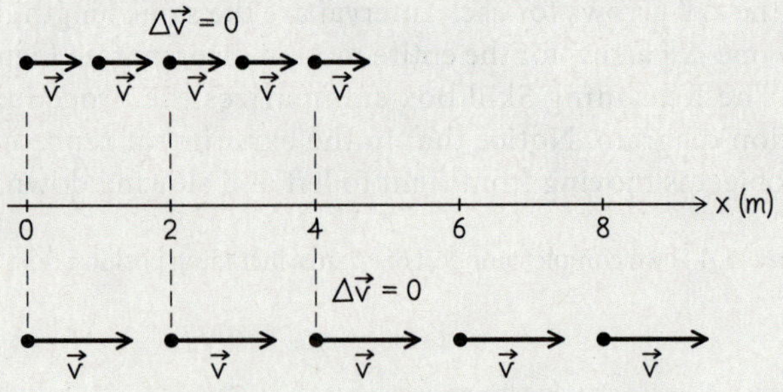

**Review Question 1.2** What information about a moving object can we extract from a motion diagram?

## 1.3 Quantities for describing motion

A motion diagram helps represent motion qualitatively. To analyze situations more precisely, for example, to determine how far a car will travel after the brakes are applied, we need to describe motion quantitatively. In this section, we devise some of the quantities we need to describe linear motion.

### Time and time interval

People use the word "time" to talk about the reading on a clock and how long a process takes. Physicists distinguish between these two meanings with different terms: time (a clock reading) and time interval (a difference in clock readings).

**Time and time interval** ***Time*** **(clock reading)** $t$ **is the reading on a clock or some other time-measuring instrument.** ***Time interval*** **(**$t_2 - t_1$**) or** $\Delta t$ **is the difference of two times. In the SI system (metric units), the unit of time and of time interval is the second. Other units are minutes, hours, days, and years. Time and time interval are both scalar quantities.**

### Position, displacement, distance, and path length

Along with a precise definition for time and time interval, we need to precisely define four quantities that describe the location and motion of an object: position, displacement, distance, and path length.

**Position, displacement, distance, and path length** **The** ***position*** **of an object is its location with respect to a particular coordinate system (usually indicated by** $x$ **or** $y$**). The** ***displacement*** **of an object, usually indicated by** $\vec{d}$**, is a vector that starts from an object's initial position and ends at its final position. The magnitude (length) of the displacement vector is called** ***distance d.*** **The** ***path length l*** **is how far the object moved as it traveled from its initial position to its final position. Imagine laying a string along the path the object took. The length of string is the path length.**

**Figure 1.5a** shows a car's initial position $x_i$ at initial time $t_i$. The car first backs up (moving in the negative direction) toward the origin of the coordinate system at $x = 0$. The car stops and then moves in the positive $x$-direction to its final position $x_f$. Notice that the *initial position* and the *origin* of a coordinate system are not necessarily the same points! The displacement $\vec{d}$ for the whole trip is a vector that points from the starting position at $x_i$ to the final position at $x_f$ (Figure 1.5b). The distance for the trip is the magnitude of the displacement (always a positive value). The path length $l$ is the distance from $x_i$ to 0 plus the distance from 0 to $x_f$ (Figure 1.5c). Note that the path length does not equal the distance.

**Figure 1.5** Position, displacement, distance, and path length for a short car trip.

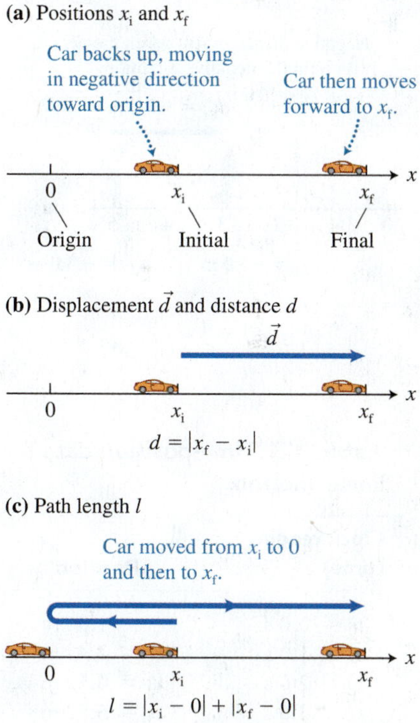


**TIP** Sometimes we use the subscripts 1, 2, and 3 for times and the corresponding positions to communicate a sequence of different and distinguishable stages in any process, and sometimes we use i (initial) and f (final) to communicate the sequence.

### Scalar component of displacement for motion along one axis

To describe linear motion quantitatively we first specify a reference frame. For simplicity we can point one coordinate axis either parallel or antiparallel (opposite in direction) to the object's direction of motion. For linear motion,

**Figure 1.6** Indicating an object's position at a particular time $t$, for example, $x(t)$.

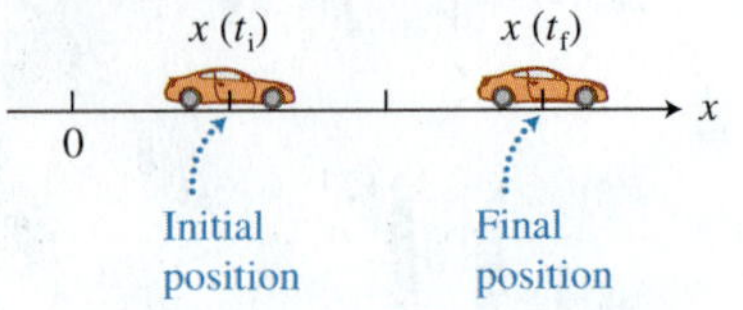


**Figure 1.7** The $x$-component of displacement is (a) positive; (b) negative.

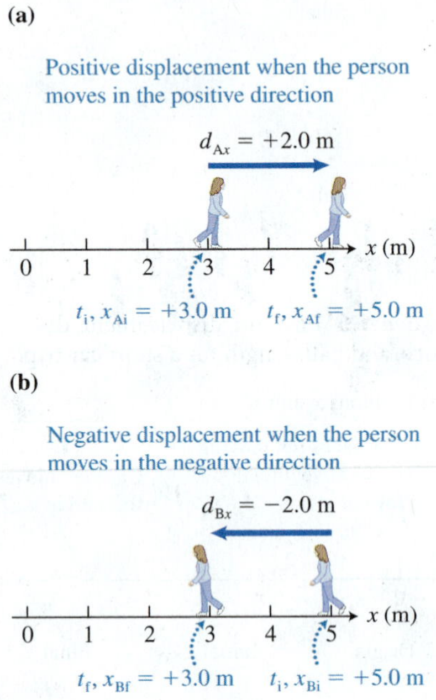


we need only one coordinate axis to describe the object's changing position. In the example of the car trip, at the initial time $t_i$ the car is at position $x(t_i)$, and at the final time $t_f$ the car is at $x(t_f)$ (**Figure 1.6**). The notation $x(t)$ means the position $x$ is a function of clock reading $t$ (spoken "$x$ of $t$"), not $x$ multiplied by $t$. When we need to note a specific value of position $x$ at a specific clock reading $t_1$, instead of writing, $x(t_1)$ we will write $x_1$. The same applies to $x_i$, $x_f$, etc. The vector that points from the initial position $x_i$ to the final position $x_f$ is the **displacement vector.**

The quantity that we determine through the operation $x_f - x_i$ is called the ***x*-scalar component of the displacement** vector and is abbreviated $d_x$ (usually, we drop the term "scalar" and just call this the ***x*-component of the displacement**). **Figure 1.7a** shows that the initial position of person A is $x_{Ai} = +3.0$ m and the final position is $x_{Af} = +5.0$ m; thus the $x$-component of the person's displacement is

$$d_{Ax} = x_{Af} - x_{Ai} = (+5.0 \text{ m}) - (+3.0 \text{ m}) = +2.0 \text{ m}$$

The displacement is positive since the person moved in the positive $x$-direction. In Figure 1.7b, person B moved in the negative direction from initial position of +5.0 m to the final position of +3.0 m; thus the $x$-component of displacement of the person is negative:

$$d_{Bx} = x_{Bf} - x_{Bi} = (+3.0 \text{ m}) - (+5.0 \text{ m}) = -2.0 \text{ m}$$

Distance is always positive, as it equals the absolute value of the displacement $|x_f - x_i|$. In the example above, the displacements for A and B are different, but the distances are both +2.0 m (always positive).

## Significant digits

Note that in Figure 1.7 the positions were written as +3.0 m, +5.0 m, etc. Could we have written them instead as +3 m and +5 m, or as +3.00 m and +5.00 m? The thickness of a human body from the back to the front is about 0.2 m (20 cm). Thus, we should be able to measure the person's location at one instant of time to within about 0.1 m but not to 0.01 m (1 cm). Thus, the locations can reasonably be given as +3.0 m, which implies an accuracy of ±0.1 m. (For more on significant digits, see Chapter I, Introducing Physics.)

**Review Question 1.3** Sammy went hiking between two camps that were separated by about 10 kilometers (km). He hiked approximately 16 km to get from one camp to the other. Translate 10 km and 16 km into the language of physical quantities.

**Table 1.3 Time-position data for linear motion.**

| Clock reading (time) | Position |
|---|---|
| $t_0 = 0.0$ s | $x_0 = 1.00$ m |
| $t_1 = 1.0$ s | $x_1 = 2.42$ m |
| $t_2 = 2.0$ s | $x_2 = 4.13$ m |
| $t_3 = 3.0$ s | $x_3 = 5.52$ m |
| $t_4 = 4.0$ s | $x_4 = 7.26$ m |
| $t_5 = 5.0$ s | $x_5 = 8.41$ m |
| $t_6 = 6.0$ s | $x_6 = 10.00$ m |

**Active Learning Guide>**

# 1.4 Representing motion with data tables and graphs

So far, we have learned how to represent linear motion with motion diagrams. In this section, we learn to represent linear motion with data tables and graphs.

Imagine your friend (the object of interest) walking across the front of your classroom. To record her position, you drop a beanbag on the floor at her position each second (**Figure 1.8a**). The floor is the object of reference. The origin of the coordinate system is 1.00 m from the first beanbag, and the position axis points in the direction of motion. **Table 1.3** shows each bag's positions and the corresponding clock reading. Do you see a pattern in the table's data? One way to determine if there is a pattern is to plot the data on a graph (Figure 1.8b). This graph is called a **kinematics position-versus-time**

**graph.** In physics, the word **kinematics** means description of motion. Kinematics graphs contain more precise information about an object's motion than motion diagrams can.

Time $t$ is usually considered to be the independent variable, as time progresses even if there is no motion, so the horizontal axis will be the $t$ axis. Position $x$ is the dependent variable (position changes with time), so the vertical axis will be the $x$-axis.

A row in Table 1.3 turns into two points, one on each axis. Each point on the horizontal axis represents a time (clock reading). Each point on the vertical axis represents the position of a beanbag. When we draw lines through these points and perpendicular to the axes, they intersect at a single location—a dot on the graph that simultaneously represents a time and the corresponding position of the object. This dot *is not* a location in real space but rather a representation of the position of the beanbag at a specific time.

Is there a trend in the locations of the dots on the graph? We see that the position increases as the time increases. This makes sense. We can draw a smooth best-fit curve that passes as close as possible to the data points—a **trendline** (Figure 1.8c). It looks like a straight line in this particular case—the position is linearly dependent on time.

**Figure 1.8** Constructing a kinematics position-versus-time graph.

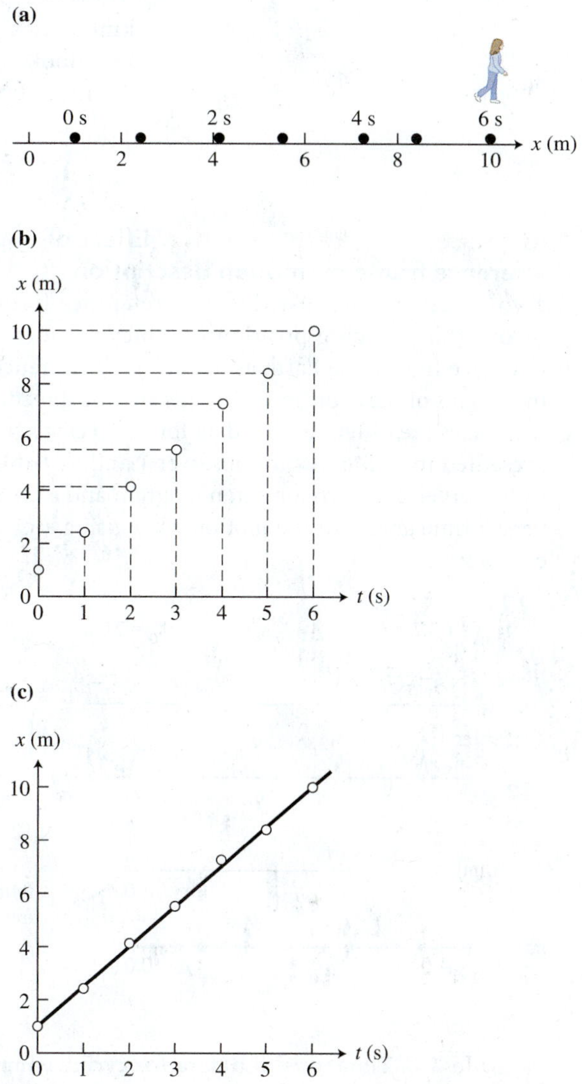


## Correspondence between a motion diagram and position-versus-time graph

To understand how graphs relate to motion diagrams, consider the motion represented by the data in Table 1.3 and in Figure 1.8c. **Figure 1.9** shows a modified motion diagram for the data in Table 1.3 (the dot times are shown and the $\Delta\vec{v}$ arrows have been removed for simplicity) and the corresponding position-versus-time graph. The position of each dot on the motion diagram corresponds to a point on the position axis. The graph line combines the information about the position of an object and the clock reading when this position occurred. Note, for example, that the $t = 4.0$ s dot on the motion diagram at position $x = 7.26$ m is at 7.26 m on the position axis. The corresponding dot on the graph is at the intersection of the vertical line passing through 4.0 s and the horizontal line passing though 7.26 m.

**Figure 1.9** Correspondence between a motion diagram and the position-versus-time graph.

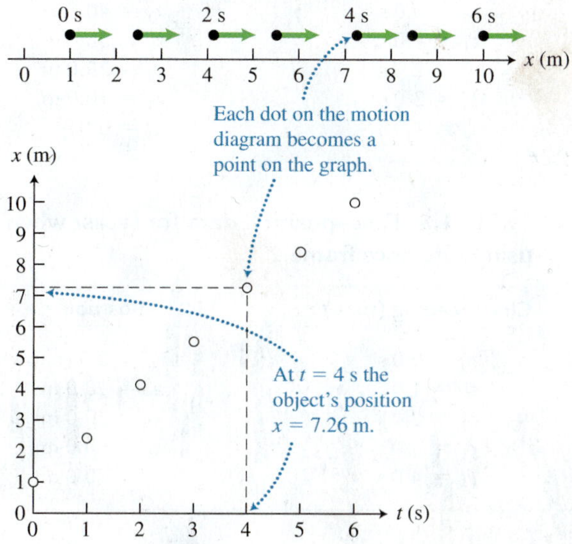


> **TIP** The quantity that appears on the vertical axis of the graph can represent the position of an object whose actual position is changing along a horizontal axis (or along a vertical axis or along an inclined axis). The position on the vertical axis does not mean the object is moving in the vertical direction.

## The role of a reference frame

Always keep in mind that representations of motion (motion diagrams, tables, kinematics graphs, equations, etc.) depend on the reference frame chosen. Let's look at the representations of the motion of a cyclist using two different reference frames.

### CONCEPTUAL EXERCISE 1.2 Effect of reference frame on motion description

Two observers each use different reference frames to record the changing position of a bicycle rider. Both reference frames use Earth as the object of reference, but the origins of the coordinate systems and the directions of the $x$-axes are different. The data for the cyclist's trip are presented in **Table 1.4** for observer 1 and in **Table 1.5** for observer 2. Sketch a motion diagram and a position-versus-time graph for the motion when using each reference frame.

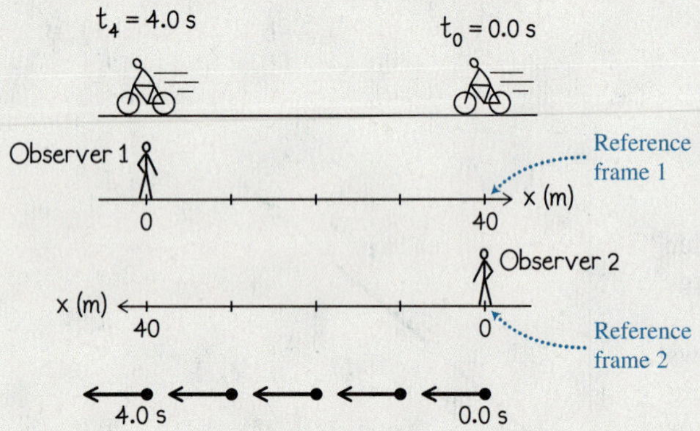


**Table 1.4** Time–position data for cyclist when using reference frame 1.

| Clock reading (time) | Position |
|---|---|
| $t_0 = 0.0$ s | $x_0 = 40.0$ m |
| $t_1 = 1.0$ s | $x_1 = 30.0$ m |
| $t_2 = 2.0$ s | $x_2 = 20.0$ m |
| $t_3 = 3.0$ s | $x_3 = 10.0$ m |
| $t_4 = 4.0$ s | $x_4 = 0.0$ m |

**Table 1.5** Time–position data for cyclist when using reference frame 2.

| Clock reading (time) | Position |
|---|---|
| $t_0 = 0.0$ s | $x_0 = 0.0$ m |
| $t_1 = 1.0$ s | $x_1 = 10.0$ m |
| $t_2 = 2.0$ s | $x_2 = 20.0$ m |
| $t_3 = 3.0$ s | $x_3 = 30.0$ m |
| $t_4 = 4.0$ s | $x_4 = 40.0$ m |

**Sketch and translate** According to Table 1.4, the observer in reference frame 1 sees the cyclist (the object of interest) at time $t_0 = 0.0$ s at position $x_0 = 40.0$ m and at $t_4 = 4.0$ s at position $x_4 = 0.0$ m. Thus, the cyclist is moving in the negative direction relative to the coordinate axis in reference frame 1. Meanwhile, according to Table 1.5, the observer in reference frame 2 sees the cyclist at time $t_0 = 0.0$ s at position $x_0 = 0.0$ m and at time $t_4 = 4.0$ s at position $x_4 = 40.0$ m. Thus, the cyclist is moving in the positive direction relative to reference frame 2.

**Simplify and diagram** Since the size of the cyclist is small compared to the distance he is traveling, we can represent him as a point-like object. The motion diagram for the cyclist is the same for both observers, as they are using the same object of reference. Using the data in the tables, we plot kinematics position-versus-time graphs for each observer, below. Although the graph for observer 1 looks very different from the group for observer 2, they represent the same motion. The graphs look different because the reference frames are different.

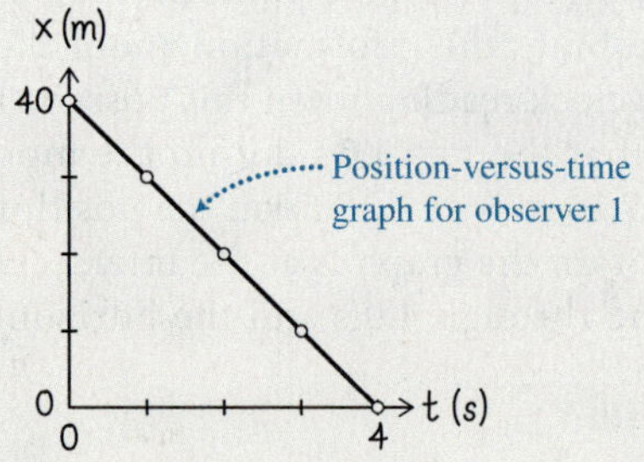


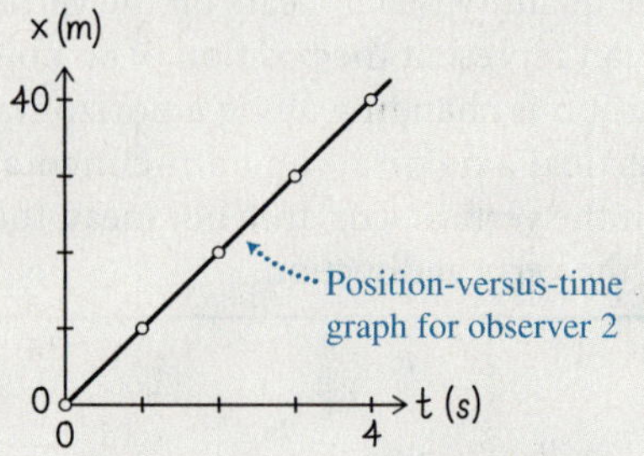

**Try it yourself:** A third observer recorded the values (in **Table 1.6**) for the time and position of the same cyclist. Describe the reference frame of this observer.

*Answer:* The point of reference could be another cyclist moving in the opposite positive direction from the direction in which the first cyclist is traveling, with each covering the same distance relative to the ground during the same time interval.

**Table 1.6 Data collected by the third observer.**

| Clock reading (time) | Position |
|---|---|
| $t_0 = 0.0$ s | $x_0 = 0.0$ m |
| $t_1 = 1.0$ s | $x_1 = -20.0$ m |
| $t_2 = 2.0$ s | $x_2 = -40.0$ m |
| $t_3 = 3.0$ s | $x_3 = -60.0$ m |
| $t_4 = 4.0$ s | $x_4 = -80.0$ m |

**Review Question 1.4** A position-versus-time graph representing a moving object is shown in **Figure 1.10**. What are the positions of the object at clock readings 2.0 s and 5.0 s?

**Figure 1.10** A position-versus-time graph representing a moving object.

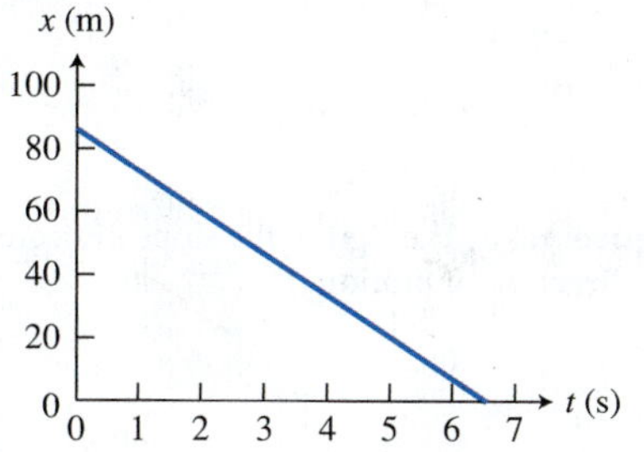


## 1.5 Constant velocity linear motion

In the last section we devised a graphical representation of motion. Here we will connect graphs to mathematical representations using two motorized toy cars racing toward a finish line. At time 0 they are next to each other, but car B is moving faster than car A and reaches the finish line first (**Figure 1.11**). The data that we collect are shown in Observational Experiment **Table 1.7**. Earth is the object of reference. The origin of the coordinate system (the point of reference) is 1.0 m to the left of the position of the cars at $t_0 = 0$. The positive $x$-direction points right in the direction of the cars' motions. Now let's use the data to find a pattern.

**OBSERVATIONAL EXPERIMENT TABLE**

**1.7 Graphing the motion of cars.**

**Observational experiment**

| Data for car A | | Data for car B | |
|---|---|---|---|
| $t_0 = 0.0$ s | $x_0 = 1.0$ m | $t_0 = 0.0$ s | $x_0 = 1.0$ m |
| $t_1 = 1.0$ s | $x_1 = 1.4$ m | $t_1 = 1.0$ s | $x_1 = 1.9$ m |
| $t_2 = 2.0$ s | $x_2 = 1.9$ m | $t_2 = 2.0$ s | $x_2 = 3.0$ m |
| $t_3 = 3.0$ s | $x_3 = 2.5$ m | $t_3 = 3.0$ s | $x_3 = 3.9$ m |
| $t_4 = 4.0$ s | $x_4 = 2.9$ m | $t_4 = 4.0$ s | $x_4 = 5.0$ m |
| $t_5 = 5.0$ s | $x_5 = 3.5$ m | $t_5 = 5.0$ s | $x_5 = 6.0$ m |

**Analysis**

We graph the data with the goal of finding a pattern. The trendlines for both cars are straight lines. The line for car B has a bigger angle with the time axis than the line for car A.

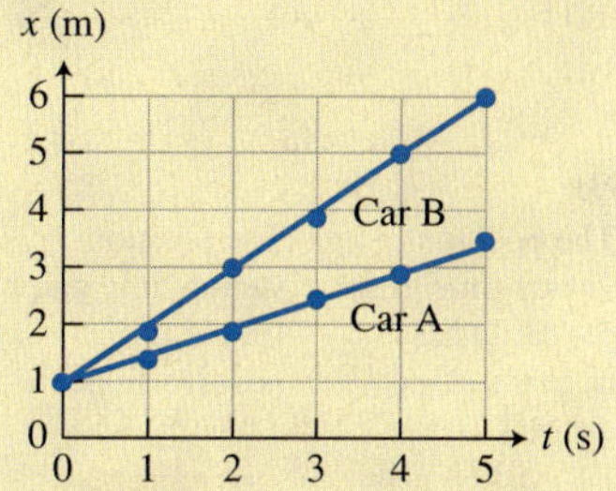


**Pattern**

It looks like a straight line is the simplest reasonable choice for the best-fit curve in both cases (the data points do not have to be exactly on the line).

**Figure 1.11** Positions of cars A and B at 0 s and 5.0 s.

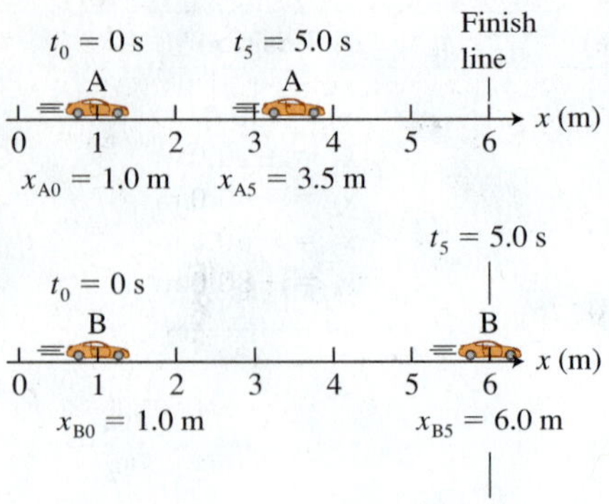


In Table 1.7, the slope of the line representing the motion of car B is greater than the slope of the line representing the motion of car A. What is the physical meaning of this slope? In mathematics the value of a dependent variable is usually written as $y$ and depends on the value of an independent variable, usually written as $x$. A function $y(x) = f(x)$ is an operation that one needs to do to $x$ as an input to have $y$ as the output. For a straight line, the function $y(x)$ is $y(x) = kx + b$, where $k$ is the slope and $b$ is the $y$ intercept—the value of the $y$ when $x = 0$.

In the case of the cars, the independent variable is time $t$ and the dependent variable is position $x$. The equation of a straight line becomes $x(t) = kt + b$, where $b$ is the $x$-intercept of the line, and $k$ is the slope of the line. The $x$-intercept is the $x$-position when $t = 0$, also called the initial position of the car $x_0$. Both cars started at the same location: $x_{A0} = x_{B0} = 1.0$ m.

To find the **slope** $k$ of a straight line, we can choose *any* two points on the line and divide the change in the vertical quantity ($\Delta x$ in this case) by the change in the horizontal quantity ($\Delta t$ in this case):

$$k = \frac{x_2 - x_1}{t_2 - t_1} = \frac{\Delta x}{\Delta t}.$$

For example, for car A the slope of the line is

$$k_A = \frac{3.5\text{ m} - 1.0\text{ m}}{5.0\text{ s} - 0.0\text{ s}} = +0.5\text{ m/s}.$$

The slope of the line for car B is

$$k_B = \frac{6.0\text{ m} - 1.0\text{ m}}{5.0\text{ s} - 0.0\text{ s}} = +1.0\text{ m/s}.$$

**Figure 1.12** The sign of the slope indicates the direction of motion.

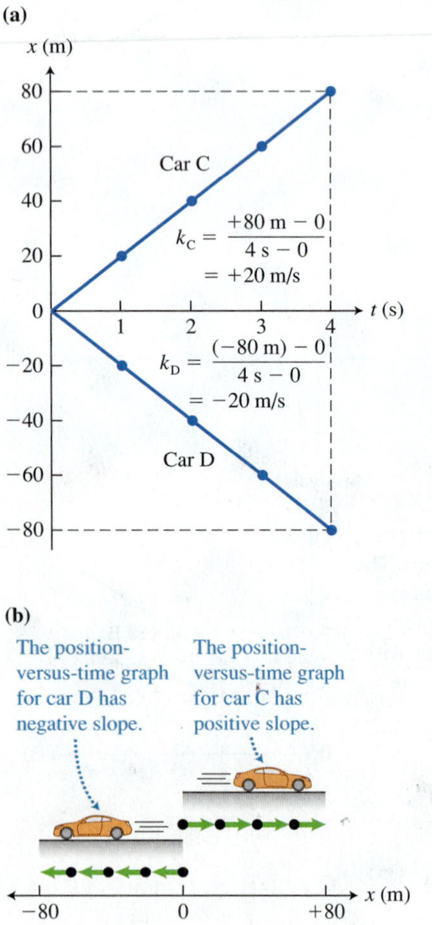


Now we have all the information we need to write mathematical equations that describe the motion of each of the two cars:

$$\text{Car A:} \quad x_A = (+0.5\text{ m/s})t + (1.0\text{ m})$$

$$\text{Car B:} \quad x_B = (+1.0\text{ m/s})t + (1.0\text{ m})$$

Notice that the units of the slope are meters per second. The slope indicates how the object's position changes with respect to time. The slope of the line contains more information than just how fast the car is going. It also tells us the direction of motion relative to the coordinate axis.

Consider the motions represented graphically in **Figure 1.12a**. The slope of the position-versus-time graph for car C is +20 m/s, but the slope of the position-versus-time graph for car D is −20 m/s. What is the significance of the minus sign? Car C is moving in the positive direction, but car D is moving in the negative direction (check the motion diagrams in Figure 1.12b). The magnitudes of the slopes of their position-versus-time graphs are the same, but the signs are different. Thus, in addition to the information about how fast the car is traveling (its **speed**), the slope tells in what direction it is traveling. Together, speed and direction are called **velocity**, and this is what the slope of a position-versus-time graph represents. You are already familiar with the term "velocity arrow" used on motion diagrams. Now you have a formal definition for velocity as a physical quantity.

**Velocity and speed for constant velocity linear motion** **For constant velocity linear motion, the component of velocity $v_x$ along the axis of motion can be found as the slope of the position-versus-time graph or the ratio of the component of the displacement of an object $x_2 - x_1$ during *any* time interval $t_2 - t_1$:**

$$v_x = \frac{x_2 - x_1}{t_2 - t_1} = \frac{\Delta x}{\Delta t} \tag{1.1}$$

**Examples of units of velocity are m/s, km/h, and mi/h (which is often written as mph). *Speed* is the magnitude of the velocity and is always a positive number.**

Note that velocity is a vector quantity. In vector form, motion at constant velocity is $\vec{v} = \vec{d}/\Delta t$. Here we divide a vector by a scalar. As scalars are just numbers that do not have directions, when we multiply or divide a vector by a scalar, all we need to do is to change the magnitude accordingly without changing direction (unless the scalar is negative, in which case it changes the direction of the vector by 180 degrees). Therefore, in our case the velocity vector has the same direction as the displacement vector. This means that the direction for the velocity vector shows the direction of motion (same as the direction of the displacement vector) and the magnitude shows the speed. But since it is difficult to operate mathematically with vectors, we will work with components.

**TIP** Eq. (1.1) allows you to use *any* change in position divided by the time interval during which that change occurred to obtain the same number—as long as the position-versus-time graph is a straight line (the object is moving at constant velocity). Later in the chapter, you will learn how to modify this equation for cases in which the velocity is not constant.

## Equation of motion for constant velocity linear motion

We can rearrange Eq. (1.1) into a form that allows us to determine the position of an object at time $t_2$ knowing only its position at time $t_1$ and the $x$-component of its velocity: $x_2 = x_1 + v_x(t_2 - t_1)$. If we apply this equation for time zero ($t_0 = 0$) when the initial position is $x_0$, then the position $x$ at any later position time $t$ can be written as follows.

**Position equation for constant velocity linear motion**

$$x = x_0 + v_x t \quad (1.2)$$

**where $x$ is the function $x(t)$, position $x_0$ is the position of the object at time $t_0 = 0$ with respect to a particular reference frame, and the (constant) $x$-component of the velocity of the object $v_x$ is the slope of the position-versus-time graph.**

Below you see a new type of task—a quantitative exercise. Quantitative exercises include two steps of the problem-solving process: Represent mathematically and Solve and evaluate. Their purpose is to help you practice using new equations right away.

**QUANTITATIVE EXERCISE 1.3 A cyclist**

In Conceptual Exercise 1.2, you constructed graphs for the motion of a cyclist using two different reference frames. Now construct mathematical representations (equations) for the cyclist's motion for each of the two graphs. Do the equations indicate the same position for the cyclist at time $t = 6.0$ s?

**Represent mathematically** The cyclist moves at constant velocity; thus the general mathematical description of his motion is $x = x_0 + v_x t$, where

$$v_x = \frac{x_2 - x_1}{t_2 - t_1}.$$

**Solve and evaluate** Using the graph for reference frame 1 in Conceptual Exercise 1.2, we see that the cyclist's initial position is $x_0 = +40$ m. The velocity along the $x$-axis (the slope of the graph line) is

$$v_x = \frac{0\text{ m} - 40\text{ m}}{4\text{ s} - 0\text{ s}} = -10\text{ m/s}$$

The minus sign indicates that the velocity points in the negative $x$-direction (toward the left) relative to that axis. The motion of the bike with respect to reference frame 1 is described by the equation

$$x = x_0 + v_x t = 40\text{ m} + (-10\text{ m/s})t$$

Using the graph for reference frame 2, we see that the cyclist's initial position is $x_0 = 0$ m. The $x$-component of the velocity along the axis of motion is

$$v_x = \frac{40\text{ m} - 0\text{ m}}{4\text{ s} - 0\text{ s}} = +10\text{ m/s}$$

*(continued)*

The positive sign indicates that the velocity points in the positive $x$-direction (toward the left). The motion of the bike relative to reference frame 2 is described by the equation

$$x = x_0 + v_x t = 0\,\text{m} + (10\,\text{m/s})t$$

The position of the cyclist at time $t_1 = 6$ s with respect to reference frame 1 is

$$x = 40\,\text{m} + (-10\,\text{m/s})(6\,\text{s}) = -20\,\text{m}.$$

With respect to reference frame 2:

$$x = 0\,\text{m} + (+10\,\text{m/s})(6\,\text{s}) = +60\,\text{m}.$$

How can the position of the cyclist be both −20 m and +60 m? Remember that the description of motion of an object depends on the reference frame. If you put a dot on coordinate axis 1 at the −20 m position and a dot on coordinate axis 2 at the +60 m position, you find that the dots are in fact at the same location, even though that location corresponds to a different position in each reference frame. Both descriptions of the motion are correct and consistent; but each one is meaningful only with respect to the corresponding reference frame.

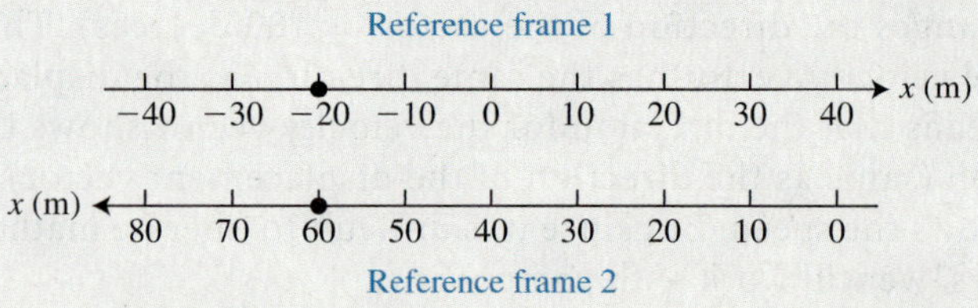


**Try it yourself:** Use the data for the motion of the cyclist as seen by the third observer in the Try It Yourself part of Conceptual Exercise 1.2 to write the equation of motion. Why is the magnitude of the cyclist's velocity different than the 10 m/s in the example above?

*Answer:* $x = (0\,\text{m}) + (-20\,\text{m/s})t$. The observer is moving with respect to Earth at the same speed in the direction opposite to the cyclist.

Below you see another new type of task—a worked example. The worked examples include all four steps of the problem-solving strategy we use in this book. (See Introducing Physics for descriptions of these steps.)

**EXAMPLE 1.4 You chase your sister**

Your young sister is running at 2.0 m/s toward a mud puddle that is 6.0 m in front of her. You are 10.0 m behind her running at 5.0 m/s to catch her before she enters the mud. Will she need a bath?

**Sketch and translate** We start by drawing a sketch of what is happening. Your sister and you are two objects of interest. Next, we choose a reference frame with Earth as the object of reference. The origin of the coordinate system will be your initial position and the positive direction will be toward the right, in the direction that you both run, as shown below.

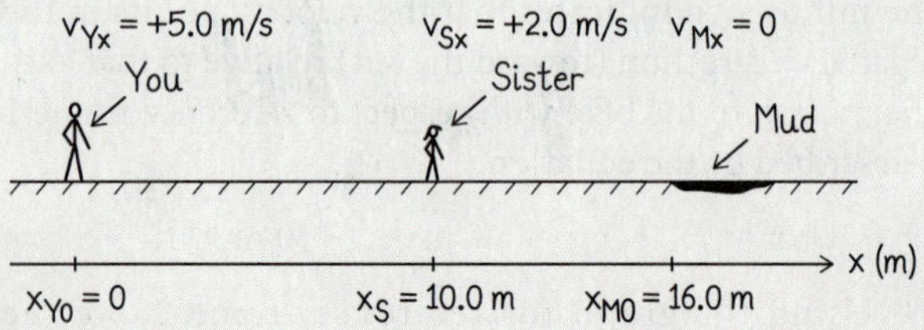


We can now mathematically describe the positions and velocities of you and your sister at the beginning of the process. The initial clock reading is zero at the moment that you are at the origin. Note that you were both running before time zero; this just happened to be the time when we started analyzing the process. We want to know the time when you and your sister are at the same position. This will be the position where you catch up to her.

**Simplify and diagram** We assume that you and your sister are point-like objects. To sketch graphs of the motions, find the sister's position at 1 second by multiplying her speed by 1 second and adding it to her initial position. Do it for 2 seconds and for 3 seconds as well. Plot these values on a graph for the corresponding clock readings (1 s, 2 s, 3 s, etc.) and draw a line that extends through these points. Repeat this for yourself.

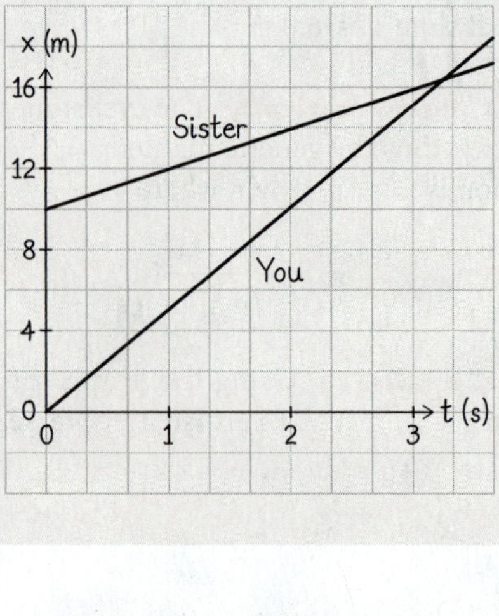

**Represent mathematically** Use Eq. (1.2) to construct mathematical representations of motion. The form of the equation is the same for both ($x = x_0 + v_x t$); however, the values for the initial positions and the components of the velocities along the axis are different.

$$\text{Sister:} \quad x_S = (10.0\text{ m}) + (2.0\text{ m/s})t$$
$$\text{You:} \quad x_Y = (0.0\text{ m}) + (5.0\text{ m/s})t$$

From the graphs, we see that the distance between you and your sister is shrinking with time. Do the equations tell the same story? For example, at time $t = 2.0$ s, your sister is at position

$$x_S(2\text{ s}) = (10.0\text{ m}) + (2.0\text{ m/s})(2.0\text{ s}) = 14.0\text{ m}$$

and you are at

$$x_Y(2\text{ s}) = (0.0\text{ m}) + (5.0\text{ m/s})(2.0\text{ s}) = 10.0\text{ m}.$$

You are catching up to your sister.

**Solve and evaluate** The time $t$ at which the two of you are at the same position can be found by setting $x_S(t) = x_Y(t)$:

$$(10.0\text{ m}) + (2.0\text{ m/s})t = (0.0\text{ m}) + (5.0\text{ m/s})t$$

Rearrange the above to determine the time $t$ when you are both at the same position:

$$(2.0\text{ m/s})t - (5.0\text{ m/s})t = (0.0\text{ m}) - (10.0\text{ m})$$
$$(-3.0\text{ m/s})t = -(10.0\text{ m})$$
$$t = 3.33333333\text{ s}$$

The 3.33333333 s number produced by our calculator has many more significant digits than the givens. Should we round it to have the same number of significant digits as the given quantities? The rule of thumb is that if it is the final result, you need to round this number to 3.3, as the answer cannot be more precise than the given information. However, we do not round the result of an intermediate calculation. We use the result as is to calculate the next quantity needed to get the final answer and then round the final result.

$$\text{Sister:} \quad x_S(t) = (10.0\text{ m}) + (2.0\text{ m/s})(3.33333333\text{ s}) = 16.7\text{ m}$$
$$\text{You:} \quad x_Y(t) = (0.0\text{ m}) + (5.0\text{m/s})(3.33333333\text{ s}) = 16.7\text{ m}$$

Note that if you used the rounded number 3.3 s, you would get 16.5 m for your sister and 16.6 m for you. These would be slightly less than the result calculated above. However, for our purposes it does not matter, as the goal of this example was to decide if you could catch your sister before she reaches the puddle. Since you caught her at a position of about 16.7 m, with the uncertainty of about 0.1 m, this position is slightly greater than the 16.0 m distance to the puddle. Therefore, your sister reaches the puddle before you. This answer seems consistent with the graphical representation of the motion shown above.

**Try it yourself:** Describe the problem situation using a reference frame with the sister (not Earth) as the object and point of reference and the positive direction pointing toward the puddle.

*Answer:* With respect to this reference frame, the sister is at position 0 and at rest; you are initially at −10.0 m and moving toward your sister with velocity +3.0 m/s; and the mud puddle is initially at +6.0 m and moving toward your sister with velocity −2.0 m/s.

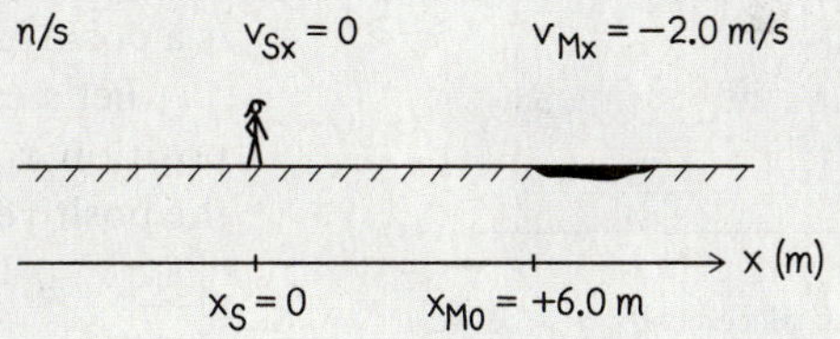


**TIP** In the reference frame we chose in Example 1.4, the positions of you and your sister are always positive, as are the components of your velocities. Also, your initial position will be zero. Thus the calculations are the easiest. Often the description of the motion of object(s) will be simplest in one particular reference frame.

## Graphing velocity

So far, we have learned to make position-versus-time graphs. We could also construct a graph of an object's velocity as a function of time. Consider Example 1.4, in which you chase your sister. Again we will use Earth as the object of reference. You are moving at a constant velocity whose $x$-component is $v_x = +5$ m/s. For your sister, the $x$-component of velocity is +2.0 m/s. We

**Figure 1.13** Velocity-versus-time graphs.

**(a)** $v$-versus-$t$ graph lines with Earth as object of reference

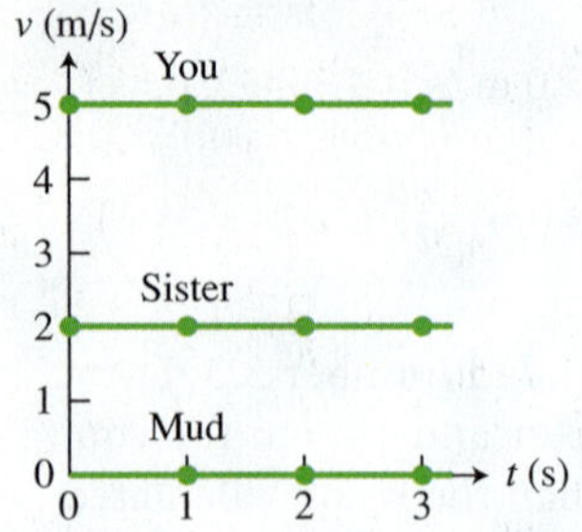


**(b)** $v$-versus-$t$ graph lines with Sister as object of reference

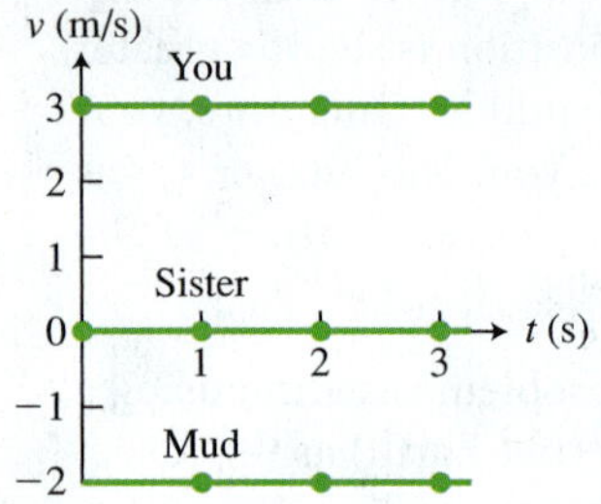


place clock readings on the horizontal axis and the $x$-component of your and your sister's velocities on the vertical axis; then we plot points for these velocities at each time (see **Figure 1.13a**). The best-fit line for each person is a horizontal straight line, which makes sense since neither of their velocities are changing. For you, the equation of the best-fit line is $v_{Yx}(t) = +5.0\ \text{m/s}$, and for your sister, it is $v_{Sx}(t) = +2.0\ \text{m/s}$, where $v_x(t)$ represents the $x$-component of velocity as a function of time.

If instead we choose your sister as the object of reference, her velocity with respect to herself is zero, so the best-fit curve is again a horizontal line but at a value of $0.0\ \text{m/s}$ instead of $+2.0\ \text{m/s}$; your velocity is $+3.0\ \text{m/s}$ (see Figure 1.13b); and the mud's velocity is $-2.0\ \text{m/s}$. The minus sign indicates that from your sister's point of view, the mud is moving in the negative direction toward her at speed $+2.0\ \text{m/s}$.

**TIP** Notice that a horizontal line on a position-versus-time graph means that the object is at rest (the position is constant with time). The same horizontal line on the velocity-versus-time graph means that the object is moving at constant velocity (its velocity does not change with time).

## Finding displacement from a velocity graph

We have just learned to construct a velocity-versus-time graph. Can we get anything more out of such graphs besides being able to represent velocity graphically? As you know, for constant velocity linear motion, the position of an object changes with time according to $x = x_0 + v_x t$. Rearranging this equation a bit, we get $x - x_0 = v_x t$. The left side is the displacement of the object from time zero to time $t$. Now look at the right side: $v_x$ is the vertical height of the velocity-versus-time graph line and $t$ is the horizontal width from time zero to time $t$ (see **Figure 1.14**). We can interpret the right side as the shaded area between the velocity-versus-time graph line and the time axis. In equation $x - x_0 = v_x t$, this area (the right side) equals the displacement of the object from time zero to time $t$ on the left side of the equation. Here the displacement is a positive number because we chose an example with positive velocity.

Let's extend this reasoning to more general cases: an object initially at position $x_1$ at time $t_1$ and later at position $x_2$ at time $t_2$ and moving in either the positive or negative direction.

**Figure 1.14** Using the $v_x$-versus-$t$ graph to determine displacement $x - x_0$.

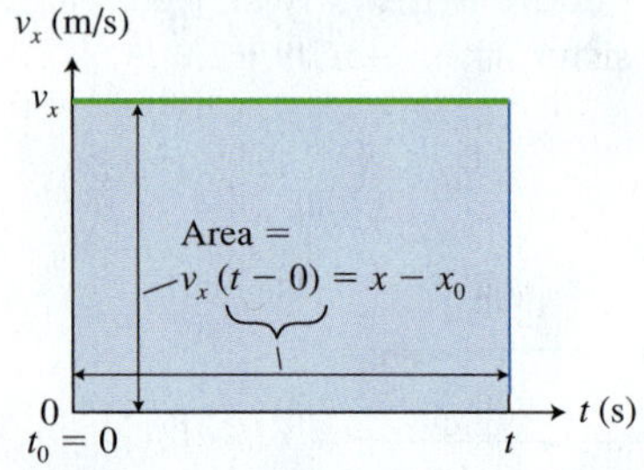


An object's displacement $x - x_0$ between $t_0 = 0$ and time $t$ is the area between the $v_x$-versus-$t$ curve and the $t$ axis.

**Displacement is the area between a velocity-versus-time graph line and the time axis** For motion with constant velocity, the magnitude of the displacement $x_2 - x_1$ (the distance traveled) of an object during a time interval from $t_1$ to $t_2$ is the area between a velocity-versus-time graph line and the time axis between those two clock readings. The displacement is the area with a plus sign when the velocity is positive and the area with a negative sign negative when velocity is negative.

**QUANTITATIVE EXERCISE 1.5 Displacement of you and your sister**

Use the velocity-versus-time graphs shown in Figure 1.13a for you and your sister (see Example 1.4) to find your displacements with respect to Earth for the time interval from 0 to 3.0 s. The velocity components relative to Earth are $+2.0$ m/s for your sister and $+5.0$ m/s for you.

**Represent mathematically** For constant velocity motion, the object's displacement $x - x_0$ is the area of a rectangle whose vertical side equals the object's velocity $v_x$ and the horizontal side equals the time interval $t - t_0 = t - 0$ during which the motion occurred.

**Solve and evaluate** The displacement of your sister heading toward the mud puddle during the 3.0-s time interval is the product of her velocity and the time interval:

$$d_S = (x - x_0)_S = (+2.0\,\text{m/s})(3.0\,\text{s}) = +6.0\,\text{m}$$

She was originally at position 10.0 m, so she is now at position $(10.0 + 6.0)\,\text{m} = +16.0\,\text{m}$. Your displacement during that same 3.0-s time interval is the product of your velocity and the time interval:

$$d_Y = (x - x_0)_Y = (+5.0\,\text{m/s})(3.0\,\text{s}) = +15.0\,\text{m}$$

You were originally at 0.0 m, so you are now at +15.0 m—one meter behind your sister.

**Try it yourself:** Determine the magnitudes of the displacements of you and your sister from time zero to time 2.0 s and your positions at that time. Your initial position is zero and your sister's is 10 m.

*Answer:* The sister's values are $d_S = 4.0\,\text{m}$ and $x_S = 14.0\,\text{m}$, and your values are $d_Y = 10.0\,\text{m}$ and $x_Y = 10.0\,\text{m}$. She is 4.0 m ahead of you at that time.

**Review Question 1.5** Why is the following statement true? "Displacement is equal to the area between a velocity-versus-time graph line and the time axis with a positive or negative sign."

# 1.6 Motion at constant acceleration

In the last section, the function $v_x(t)$ was a horizontal line on the velocity-versus-time graph because the velocity was constant. How would the graph look if the velocity were changing? One example of such a graph is shown in **Figure 1.15**. A point on the curve indicates the velocity of the object shown on the vertical axis at a particular time shown on the horizontal axis. In this case, the velocity is continually changing and is positive.

**Figure 1.15** Velocity-versus-time graph for motion with changing velocity.

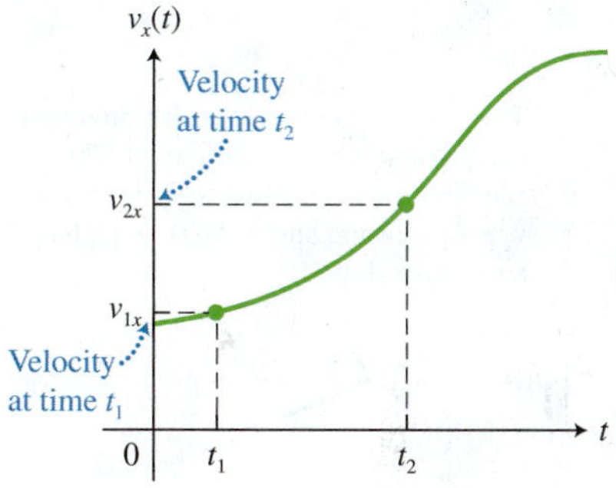


## Instantaneous velocity and average velocity

The velocity of an object at a particular time is called the **instantaneous velocity**. Figure 1.15 shows a velocity-versus-time graph for motion with continually changing instantaneous velocity. When an object's velocity is changing, we cannot use Eq. (1.1) to determine its instantaneous velocity, because the ratio

$$v_x = \frac{x_2 - x_1}{t_2 - t_1} = \frac{\Delta x}{\Delta t}$$

is not the same for different time intervals the way it was when the object was moving at constant velocity. However, we can still use the this equation to determine the **average velocity**, which is the ratio of the change in position and the time interval during which this change occurred. For motion at constant velocity, the instantaneous and average velocity are equal; for motion with changing velocity, they are not.

When an object moves with changing velocity, its velocity can change quickly or slowly. To characterize the rate at which the velocity of an object is changing, we need a new physical quantity.

## Acceleration

To analyze motion with changing velocity, we start by looking for the simplest type of linear motion with changing velocity. This occurs when the velocity of the object increases or decreases by the same amount during the same time interval (a *constant rate* of change). Imagine that a cyclist is speeding up so that his velocity is increasing at a constant rate with respect to an observer on the ground.

**Figure 1.16** Velocity-versus-time graph when the velocity is changing at a constant rate.

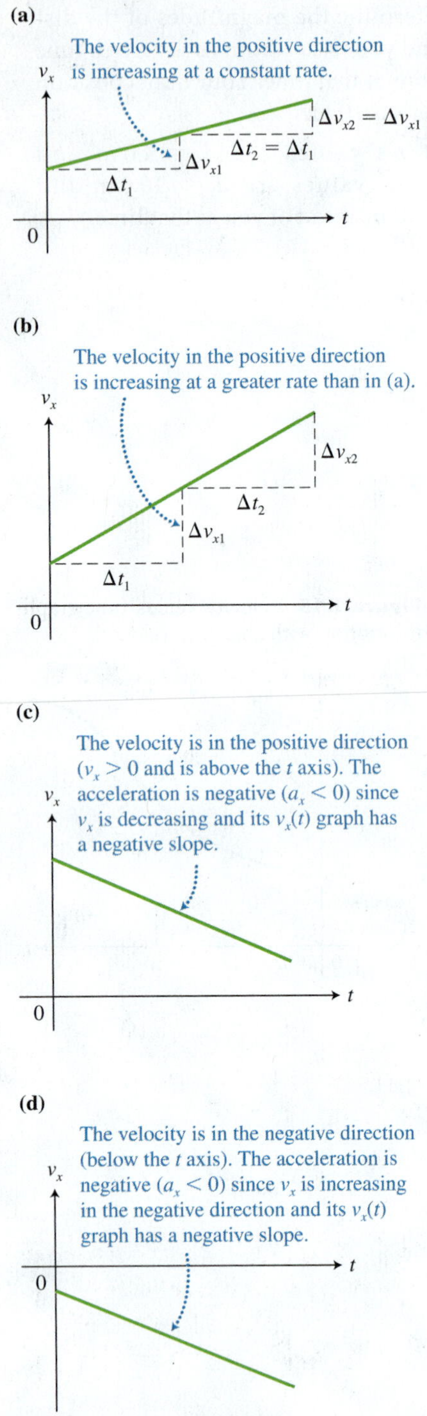
(a)
The velocity in the positive direction is increasing at a constant rate.
$v_x$
$\Delta v_{x2} = \Delta v_{x1}$
$\Delta t_2 = \Delta t_1$
$\Delta v_{x1}$
$\Delta t_1$
0
$t$
(b)
The velocity in the positive direction is increasing at a greater rate than in (a).
$v_x$
$\Delta v_{x2}$
$\Delta t_2$
$\Delta v_{x1}$
$\Delta t_1$
0
$t$
(c)
The velocity is in the positive direction ($v_x > 0$ and is above the $t$ axis). The acceleration is negative ($a_x < 0$) since $v_x$ is decreasing and its $v_x(t)$ graph has a negative slope.
$v_x$
0
$t$
(d)
The velocity is in the negative direction (below the $t$ axis). The acceleration is negative ($a_x < 0$) since $v_x$ is increasing in the negative direction and its $v_x(t)$ graph has a negative slope.
$v_x$
0
$t$

Graphically, this process is represented as the velocity-versus-time graph shown in **Figure 1.16a**. The graph is a straight line that is not horizontal. Now imagine that a car next to the cyclist is also speeding up, but at a greater rate than the biker. Its velocity-versus-time graph is shown in Figure 1.16b. The larger slope indicates that the car's velocity increases at a faster rate. The physical quantity that characterizes the change in velocity during a particular time interval is called **acceleration** $\vec{a}$. When the object is moving along a straight line and the slope of the velocity-versus-time graph is constant, the acceleration of an object is equal to the slope:

$$a_x = \frac{v_{2x} - v_{1x}}{t_2 - t_1} = \frac{\Delta v_x}{\Delta t}$$

The acceleration can be either positive or negative. In the previous examples, if the cyclist or the car had been slowing down, their velocity-versus-time graph would instead have a negative slope, which corresponds to a decreasing speed and a negative acceleration, as $v_{2x}$ is smaller than $v_{1x}$ (Figure 1.16c). However, an object can have a negative acceleration and speed up! This happens when the object is moving in the negative direction and has a negative component of velocity, but its speed in the negative direction is increasing in magnitude (see Figure 1.16d).

Because velocity is a vector quantity and the acceleration shows how quickly the velocity changes as time progresses, acceleration is also a vector quantity. We can define acceleration in a more general way. The **average acceleration** of an object during a time interval is the following:

$$\vec{a} = \frac{\vec{v}_2 - \vec{v}_1}{t_2 - t_1} = \frac{\Delta \vec{v}}{\Delta t}$$

To determine the acceleration, we need to determine the velocity change vector $\Delta\vec{v} = \vec{v}_2 - \vec{v}_1$. This equation involves the subtraction of vectors. You can think about the same equation as addition by rearranging it to be $\vec{v}_1 + \Delta\vec{v} = \vec{v}_2$. Note that $\Delta\vec{v}$ is the vector that we add to $\vec{v}_1$ to get $\vec{v}_2$ (**Figure 1.17**). We did this when making motion diagrams in Section 1.3, only then we were not concerned with the exact lengths of the vectors. The acceleration vector $\vec{a} = \Delta\vec{v}/\Delta t$ is in the same direction as the velocity change vector $\Delta\vec{v}$, as the time interval $\Delta t$ is a scalar quantity.

**Acceleration** **An object's average acceleration during a time interval $\Delta t$ is the change in its velocity $\Delta\vec{v}$ divided by that time interval:**

$$\vec{a} = \frac{\vec{v}_2 - \vec{v}_1}{t_2 - t_1} = \frac{\Delta \vec{v}}{\Delta t} \qquad (1.3)$$

**If $\Delta t$ is very small, then the acceleration given by this equation is the instantaneous acceleration of the object. For one-dimensional motion, the component of the average acceleration along a particular axis (for example, for the $x$-axis) is**

$$a_x = \frac{v_{2x} - v_{1x}}{t_2 - t_1} = \frac{\Delta v_x}{\Delta t} \qquad (1.4)$$

**The unit of acceleration is $(\mathrm{m/s})/\mathrm{s} = \mathrm{m/s^2}$.**

**TIP** If it is difficult for you to think about velocity and acceleration in abstract terms, try calculating the acceleration for simple integer velocities.

Note that if an object has an acceleration of $+6\ \mathrm{m/s^2}$, it means that its velocity changes by $+6$ m/s in 1 s, or by $+12$ m/s in 2 s $[(+12\ \mathrm{m/s})/(2\ \mathrm{s}) = +6\ \mathrm{m/s^2}]$.

It is possible for an object to have a zero velocity and nonzero acceleration—for example, at the moment when an object starts moving from rest. An object can also have a nonzero velocity and zero acceleration—for example, an object moving at constant velocity. Note that the acceleration of an object depends on the observer. For example, a car is accelerating for an observer on the ground but is not accelerating for the driver of that car.

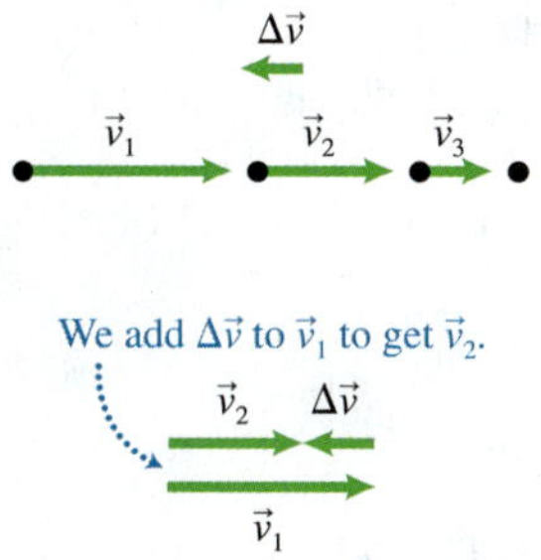


**Figure 1.17** How to determine the change in velocity $\Delta\vec{v}$

## Determining the velocity change from the acceleration

If, at $t_0 = 0$ for linear motion, the $x$-component of the velocity of some object is $v_{0x}$ and its acceleration $a_x$ is constant, then its velocity $v_x$ at a later time $t$ can be determined by substituting these quantities into Eq. (1.4):

$$a_x = \frac{v_x - v_{0x}}{t - 0}$$

Rearranging, we get an expression for the changing velocity of the object as a function of time:

$$v_x = v_{0x} + a_x t \tag{1.5}$$

For one-dimensional motion, the directions of the vector components $a_x$, $v_x$, and $v_{0x}$ are indicated by their signs relative to the axis of motion—positive if in the positive $x$- direction and negative if in the negative $x$-direction.

**EXAMPLE 1.6 Bicycle ride**

Suppose that you are sitting on a bench watching a cyclist riding a bicycle on a flat, straight road. In a 2.0-s time interval, the velocity of the bicycle changes from −4.0 m/s to −7.0 m/s. Describe the motion of the bicycle as fully as possible.

**Sketch and translate** We can sketch the process as shown below. The bicycle (the object of interest) is moving in the negative direction with respect to the chosen reference frame. The components of the bicycle's velocity along the axis of motion are negative: $v_{0x} = -4.0\text{ m/s}$ at time $t_0 = 0.0\text{ s}$ and $v_x = -7.0\text{ m/s}$ at $t = 2.0\text{ s}$. The speed of the bicycle (the magnitude of its velocity) increases. It is moving faster in the negative direction. We can determine the acceleration of the bicycle and describe the changes in its velocity.

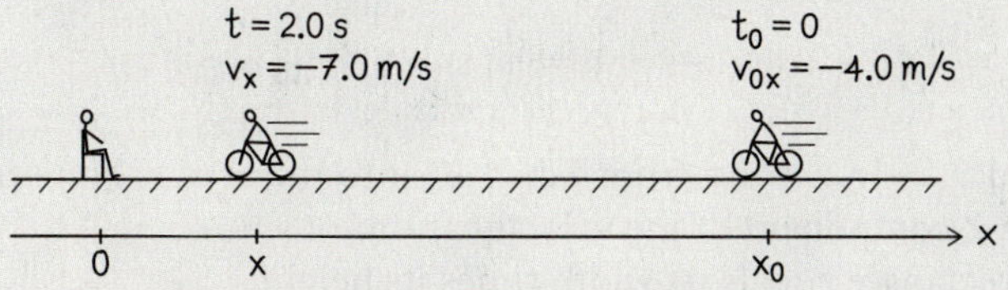


**Simplify and diagram** Our sketch of the motion diagram for the bicycle is shown below. Note that $\Delta\vec{v}$ points in the negative $x$-direction.

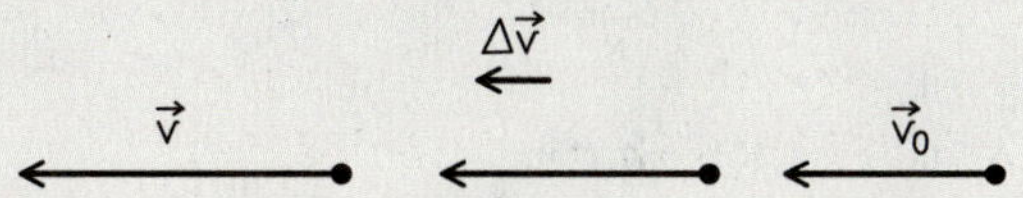


**Represent mathematically** We apply Eq. (1.4) to determine the acceleration:

$$a_x = \frac{v_x - v_{0x}}{t - t_0}$$

**Solve and evaluate** Substituting the given velocities and times, we get

$$a_x = \frac{(-7.0\text{ m/s}) - (-4.0\text{ m/s})}{2.0\text{ s} - 0.0\text{ s}} = -1.5\text{ m/s}^2$$

The bicycle's $x$-component of velocity at time zero was −4.0 m/s. Its velocity was changing by −1.5 m/s each second. So 1 s later, its velocity was

$$\begin{aligned} v_{1\text{s}\,x} &= v_{0x} + \Delta v_x \\ &= (-4.0\text{ m/s}) + (-1.5\text{ m/s}) \\ &= -5.5\text{ m/s}. \end{aligned}$$

During the second 1-s time interval, the velocity changed by another −1.5 m/s and was now (−5.5 m/s) + (−1.5 m/s) = (−7.0 m/s). In this example, the bicycle was speeding up by 1.5 m/s each second in the negative direction.

**Try it yourself:** A car's acceleration is $-3.0\text{ m/s}^2$. At time 0 its velocity is +14 m/s. What happens to the velocity of the car? What is its velocity after 3 s?

*Answer:* The car's velocity in the positive $x$-direction is decreasing. After 3 s $v_x = 5.0\text{ m/s}$.

**TIP** It is possible for an object to have a positive acceleration and slow down and to have a negative acceleration and speed up. When an object is speeding up, the acceleration vector is in the same direction as the velocity vector, and the velocity and acceleration components along the same axis have the same sign. When an object slows down, the acceleration is in the opposite direction relative to the velocity; their components have opposite signs.

## Displacement of an object moving at constant acceleration

**Active Learning Guide>**

For motion at constant velocity, we know that the area between the velocity-versus-time graph line and the time axis between two clock readings equal the magnitude of the object's displacement. Is this still the case when the velocity is changing?

Consider the displacement of an object during the short shaded time interval $\Delta t$ shown in **Figure 1.18a**. The velocity is almost constant during that time interval. Note that $v_x = \Delta x/\Delta t$ or $v_x \cdot \Delta t = \Delta x$. Thus, the small displacement $\Delta x$ during that time interval $\Delta t$ is the small shaded area $v_x \cdot \Delta t$ between that curve and the time axis (the height times the width of the narrow rectangle). We can repeat the same procedure for many successive short time intervals (Figure 1.18b), building up the area between the curve and the time axis as a sum of areas of small rectangles of different heights. The total area, shown in Figure 1.18c, is the total displacement of the object during the time interval between the initial time $t_0$ and the final time $t$. A negative area (the rectangle is below the time axis) corresponds to a negative displacement.

**Figure 1.18** The magnitude of the object's displacement is the area under a velocity-versus-time graph.

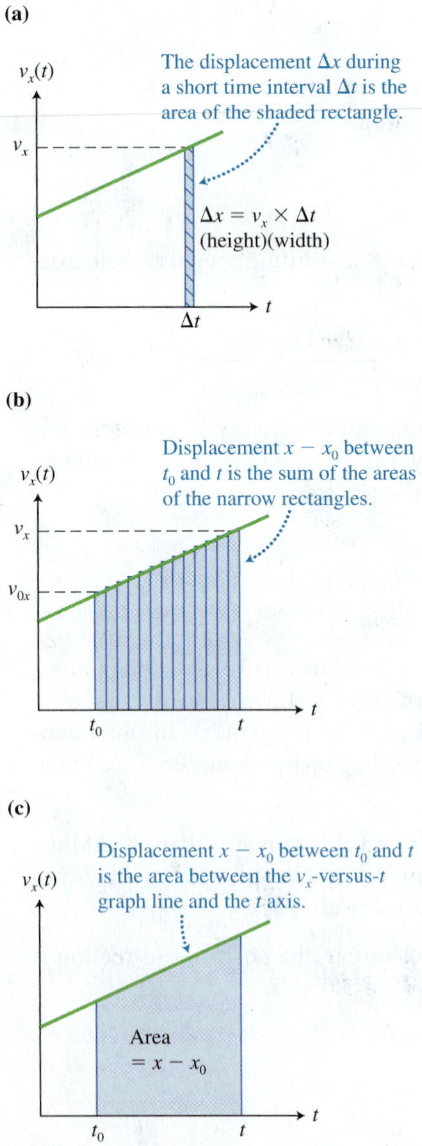
(a)
$v_x(t)$
The displacement $\Delta x$ during a short time interval $\Delta t$ is the area of the shaded rectangle.
$v_x$
$\Delta x = v_x \times \Delta t$
(height)(width)
$\Delta t$
$t$
(b)
$v_x(t)$
Displacement $x - x_0$ between $t_0$ and $t$ is the sum of the areas of the narrow rectangles.
$v_x$
$v_{0x}$
$t_0$
$t$
(c)
$v_x(t)$
Displacement $x - x_0$ between $t_0$ and $t$ is the area between the $v_x$-versus-$t$ graph line and the $t$ axis.
Area
$= x - x_0$
$t_0$
$t$

**Displacement from a *v*-versus-*t* graph** **The magnitude of the displacement $x - x_0$ (distance) of an object during a time interval $t - t_0$ is the area between the velocity-versus-time curve and the time axis between those time readings. The displacement is negative for areas below the time axis and positive for areas above.**

## Equation of motion—position as a function of time

We can use the preceding idea to find an equation for the position $x$ of an object at different times $t$. Consider **Figure 1.19a**. We can find the area between this curve and the time axis by breaking the trapezoidal area into a triangle on top and a rectangular below (Figure 1.19b). The rectangle represents the displacement for motion at constant velocity. The triangle represents the additional displacement caused by acceleration. The area of a triangle is $\frac{1}{2} \times$ base $\times$ height. The base of the triangle is $t - 0$ and the height is $v_x - v_{0x}$. So the area of the triangle is

$$A_{\text{triangle}} = \frac{1}{2}(t - 0)(v_x - v_{0x}) = \frac{1}{2}(t)(a_x t) = \frac{1}{2}a_x t^2,$$

where we substituted $v_x - v_{0x} = a_x t$ from Eq. (1.5) into the above. Note that $v_x$ is the value of the $x$-component of the velocity at time $t$.

The area of the rectangle equals its width times its height:

$$A_{\text{rectangle}} = v_{x0}(t - 0)$$

The total area between the curve and the time axis (the displacement $x - x_0$ of the object) is

$$x - x_0 = A_{\text{rectangle}} + A_{\text{triangle}} = v_{0x}t + \frac{1}{2}a_x t^2$$

$$\Rightarrow x = x_0 + v_{0x}t + \frac{1}{2}a_x t^2$$

(The symbol $\Rightarrow$ means that this equation follows from the previous equation.)

Does the above result make sense? Consider a limiting case, for example, when the object is traveling at a constant velocity (when $a_x = 0$). In this case the equation should reduce to the result from our investigation of linear motion with constant velocity ($x = x_0 + v_{x0}t$). It does. We can also check the units of each term in this equation for consistency (when terms in an equation are added or subtracted, each of those terms must have the same units). Each term has units of meters, so the units also check.

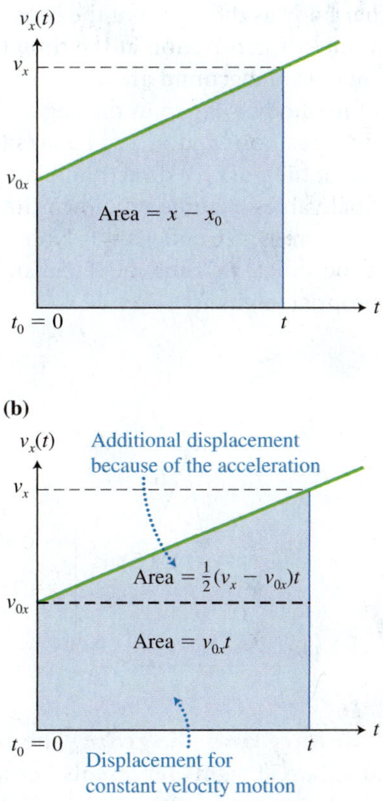


**Figure 1.19** The magnitude of the total displacement $x - x_0$ equals the sum of the two areas.

**Position of an object during linear motion with constant acceleration** For any initial position $x_0$ at clock reading $t_0 = 0$, we can determine the position $x$ of an object at any later time $t$, provided we also know the initial velocity $v_{0x}$ of the object and its constant acceleration $a_x$:

$$x = x_0 + v_{0x}t + \frac{1}{2}a_x t^2 \tag{1.6}$$

Since $t^2$ appears in Eq. (1.6), the position-versus-time graph for this motion will not be a straight line but will be a parabola (a parabola is a graph line for a quadratic function) (**Figure 1.20a**). Unlike the position-versus-time graph for constant velocity motion where $\Delta x/\Delta t$ is the same for any time interval, the position-versus-time graph line for accelerated motion is not a straight line; it does not have a constant slope. At different times, the change in position $\Delta x$ has a different value for the same time interval $\Delta t$ (Figure 1.20b). The line tangent to the position-versus-time graph line at a particular time has a slope $\Delta x/\Delta t$ that equals the velocity $v_x$ of the object at that time. The slopes of the tangent lines at different times for the graph line in Figure 1.20b differ—they are greater when the position changes more during the same time interval.

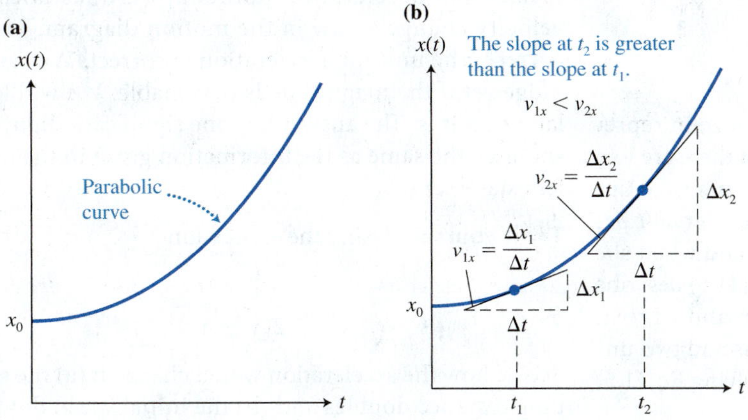


**Figure 1.20** (a) A position-versus-time graph for constant acceleration motion has a parabolic shape. (b) The velocity $v_x$ is the slope of the position-versus-time graph.

The next example involves vertical motion. Thus, we will use a vertical $y$-coordinate axis and apply the equations of motion.

**EXAMPLE 1.7 Acceleration estimate for stopping a falling person**

A woman jumps off a large boulder. She is moving at a speed of about 5 m/s when she reaches the ground. *Estimate* her acceleration during landing. Indicate any other quantities you use in the estimate.

**Sketch and translate** To solve the problem, we assume that the woman's knees bend while landing so that her body travels 0.4 m closer to the ground compared to the standing upright position. That is, 0.4 m is the stopping distance for the main part of her body. Below we have sketched the initial and final situations during the landing.

*(continued)*

We chose the central part of her body as the object of interest. We use a vertical reference with a $y$-axis pointing down and the final location of the central part of her body as the origin of the coordinate axis. The initial values of her motion at the time $t_0 = 0$ when she first touches the ground are $y_0 = -0.4$ m (the central part of her body is 0.4 m in the negative direction above her final position) and $v_{0y} = +5$ m/s (she is moving downward, the positive direction relative to the $y$-axis). The final values at some unknown time $t$ at the instant she stops are $y = 0$ and $v_y = 0$. Note: It is important to have a coordinate axis and the initial and final values with appropriate signs.

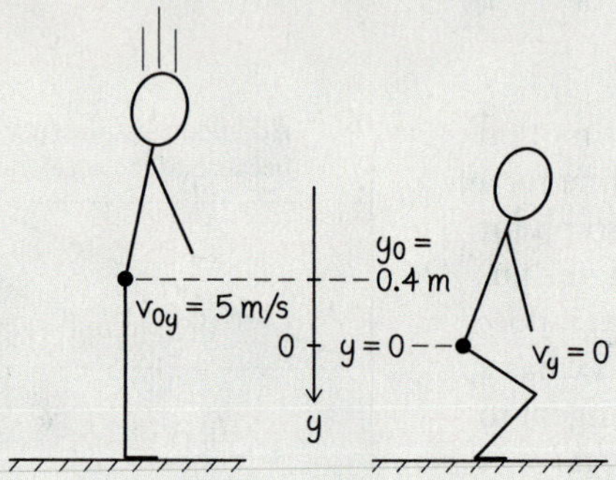


**Simplify and diagram** The motion diagram at right\ represents her motion while stopping, assuming constant acceleration. We cannot model the woman as a point-like object in this situation, so we will focus on the motion of her midsection.

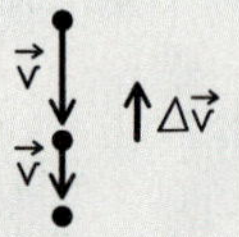


**Represent mathematically** The challenge in representing this situation mathematically is that there are two unknowns: the magnitude of her acceleration $a_y$ (the unknown we wish to determine) and the time interval $t$ between when she first contacts the ground and when she comes to rest. However, both Eqs. (1.5) and (1.6) describe linear motion with constant acceleration and have $a_y$ and $t$ in them. Since we have two equations and two unknowns, we can handle this challenge by solving Eq. (1.5) for the time $t$

$$t = \frac{v_y - v_{0y}}{a_y}$$

and substitute it into rearranged Eq. (1.6):

$$y = y_0 + v_{0y}t + \frac{1}{2}a_y t^2$$

The result is:

$$y - y_0 = v_{0y}\left(\frac{v_y - v_{0y}}{a_y}\right) + \frac{a_y(v_y - v_{0y})^2}{2a_y{}^2}$$

Using algebra, we can simplify the above equation:

$$\Rightarrow 2a_y(y - y_0) = 2v_{0y}(v_y - v_{0y}) + (v_y - v_{0y})^2$$
$$\Rightarrow 2a_y(y - y_0) = (2v_{0y}v_y - 2v_{0y}^2) + (v_y^2 - 2v_y v_{0y} + v_{0y}^2)$$
$$\Rightarrow 2a_y(y - y_0) = v_y^2 - v_{0y}^2$$
$$\Rightarrow a_y = \frac{v_y^2 - v_{0y}^2}{2(y - y_0)}$$

**Solve and evaluate** Now we can use the above equation to find her acceleration:

$$a_y = \frac{v_y^2 - v_{0y}^2}{2(y - y_0)} = \frac{(0)^2 - (5\text{ m/s})^2}{2[0 - (-0.4\text{ m})]} = -31\text{ m/s}^2 \approx -30\text{ m/s}^2$$

Is the answer reasonable? The sign is negative. This means that acceleration points upward, as does the velocity change arrow in the motion diagram. This is correct. The unit for acceleration is correct. We cannot judge yet if the magnitude is reasonable. We will learn later that it is. The answer has one significant digit, as it should—the same as the information given in the problem statement.

**Try it yourself:** Using the expression

$$a_y = \frac{v_y^2 - v_{0y}^2}{2(y - y_0)},$$

decide how the acceleration would change if (a) the stopping distance doubles and (b) the initial speed doubles. Note that the final velocity is $v_y = 0$.

*Answer:* (a) $a_y$ would be half the magnitude; (b) $a_y$ would be four times the magnitude.

In the Represent Mathematically step above, we developed a new equation for the acceleration by combining Eqs. (1.5) and (1.6). This useful equation is rewritten and described briefly below.

**Alternate equation for linear motion with constant acceleration:**

$$2a_x(x - x_0) = v_x^2 - v_{0x}^2 \qquad (1.7)$$

**This equation is useful for situations in which you do not know the time interval during which the changes in position and velocity occurred.**

**Review Question 1.6** (a) Give an example in which an object with negative acceleration is speeding up. (b) Give an example in which an object with positive acceleration is slowing down.

# 1.7 Skills for analyzing situations involving motion

To help analyze physical processes involving motion, we will represent processes in multiple ways: the words in the problem statement, a sketch, one or more diagrams, possibly a graph, and a mathematical description. Different representations have to agree with each other; in other words, they need to be consistent.

## Motion at constant velocity

### EXAMPLE 1.8 Two walking friends

You stand on a sidewalk and observe two friends walking at constant velocity. At time zero Jim is 4.0 m east of you and walking away from you at speed 2.0 m/s. At time zero, Sarah is 10.0 m east of you and walking toward you at speed 1.5 m/s. Represent their motions with an initial sketch, with motion diagrams, and mathematically.

**Sketch and translate** We choose Earth as the object of reference with your position as the reference point. The positive direction will point to the east. We have two objects of interest here: Jim and Sarah. Jim's initial position is $x_0 = +4.0$ m and his constant velocity is $v_x = +2.0$ m/s. Sarah's initial position is $X_0 = +10.0$ m and her constant velocity is $V_x = -1.5$ m/s (the velocity is negative since she is moving westward). In our sketch we are using capital letters to represent Sarah and lowercase letters to represent Jim.

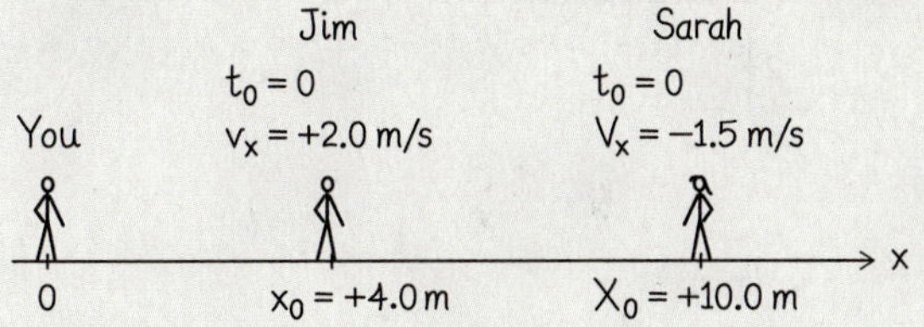


**Simplify and diagram** We can model both friends as point-like objects since the distances they move are somewhat greater than their own sizes. The motion diagrams below represent their motions.

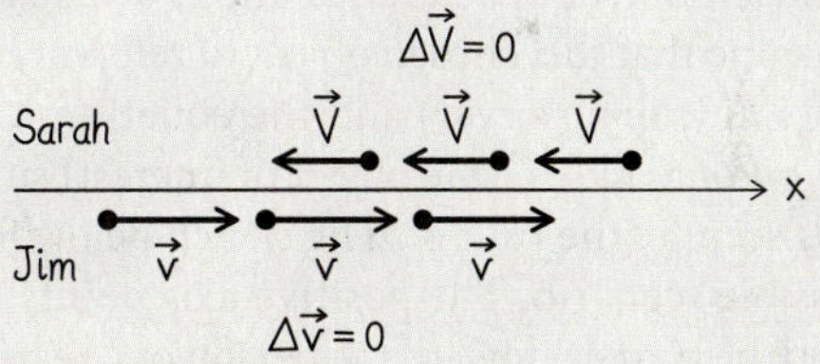


**Represent mathematically** Now construct equations to represent Jim's and Sarah's motion:

Jim: $x = +4.0\text{ m} + (2.0\text{ m/s})t$

Sarah: $X = +10.0\text{ m} + (-1.5\text{ m/s})t$

**Solve and evaluate** We were not asked to solve for any quantity. We will do it in the Try it yourself exercise.

**Try it yourself:** Determine the time when Jim and Sarah are at the same position, and where that position is.

*Answer:* They are at the same position when $t = 1.7$ s and when $x = X = 7.4$ m.

# Equation Jeopardy problems

Learning to read the mathematical language of physics with understanding is an important skill. To help develop this skill, this text includes Jeopardy-style problems. In this type of problem, you have to work backwards: you are given one or more equations and are asked to use them to construct a consistent sketch of a process. You then convert the sketch into a diagram of a process that is consistent with the equations and sketch. Finally, you invent a word problem that the equations could be used to solve. Note that there are often many possible word problems for a particular mathematical description.

**CONCEPTUAL EXERCISE 1.9 Equation Jeopardy**

The following equation describes an object's motion:

$$x = (5.0\text{ m}) + (-3.0\text{ m/s})t$$

Construct a sketch, a motion diagram, kinematics graphs, and a verbal description of a situation that is consistent with this equation. There are many possible situations that the equation describes equally well.

**Sketch and translate** This equation looks like a specific example of our general equation for the linear motion of an object with constant velocity: $x = x_0 + v_x t$. The minus sign in front of the 3.0 m/s indicates that the object is moving in the negative $x$-direction. At time zero, the object is located at position $x_0 = +5.0$ m with respect to some chosen object of reference and *is already moving.* Let's imagine that this chosen object of reference is a running person (the observer) and the equation represents the motion of a person (the object of interest) sitting on a bench as seen by the runner. The sketch below illustrates this possible scenario. The positive axis points from the observer (the runner) toward the person on the bench, and at time zero the person sitting on it is 5.0 m in front of the runner and coming closer to the runner as time elapses.

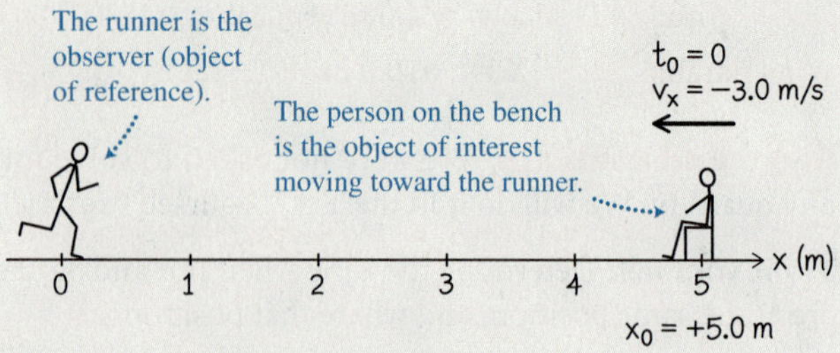


**Simplify and diagram** Model the object of interest as a point-like object. A motion diagram for the situation is shown below. The equal spacing of the dots and the equal lengths of velocity arrows both indicate that the object of interest is moving at constant velocity with respect to the observer.

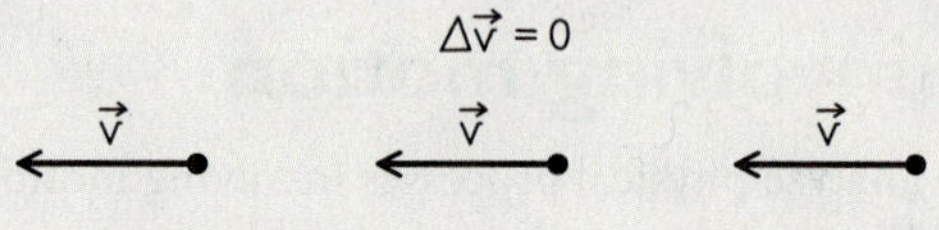

Motion diagram for bench relative to runner

Position-versus-time and velocity-versus-time kinematics graphs of the process are shown in below. The position-versus-time graph has a constant $-3.0$ m/s slope and a $+5.0$ m intercept with the vertical (position) axis. The velocity-versus-time graph has a constant value ($-3.0$ m/s) and a zero slope (the velocity is not changing). The following verbal description describes this particular process: A jogger sees a person on a bench in the park 5.0 m in front of him. The bench is approaching at a speed of 3.0 m/s as seen in the jogger's reference frame. The direction pointing from the jogger to the bench is positive.

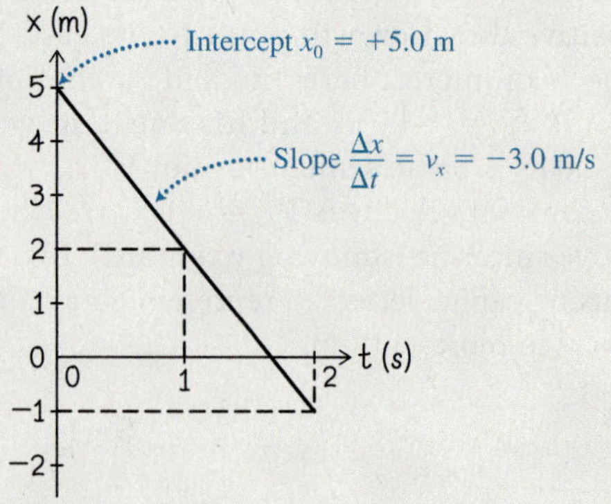


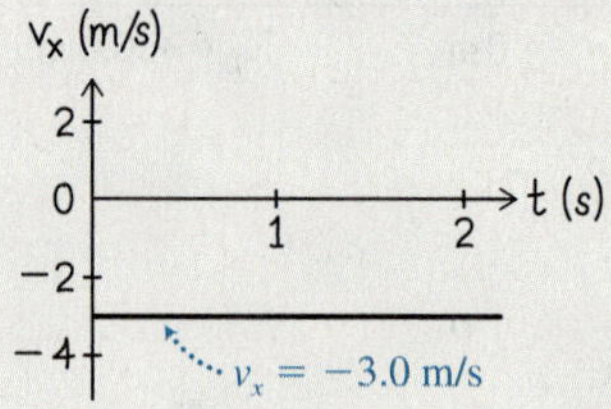

**Try it yourself:** Suppose we switch the roles of observer and object of reference in the last example. Now the person on the bench is the object of reference and observes the runner. We choose to describe the process by the same equation as in the example:

$$x = (5.0\,\text{m}) + (-3.0\,\text{m/s})\,t$$

Construct an initial sketch and a motion diagram that are consistent with the equation and with the new observer and new object of reference.

*Answer:* An initial sketch for this process and a consistent motion diagram are shown to the right.

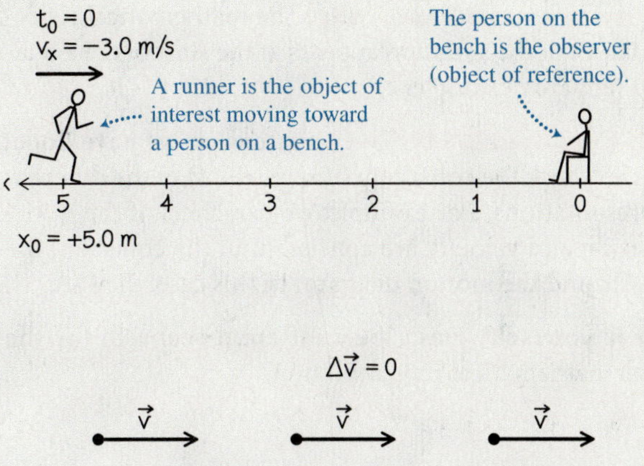

Motion diagram for runner relative to person on bench

## Motion at constant nonzero acceleration

Now let's apply some representation techniques to linear motion with constant (nonzero) acceleration.

**EXAMPLE 1.10 Equation Jeopardy**

A process is represented mathematically by the following equation:

$$x = (-60\,\text{m}) + (10\,\text{m/s})\,t + (1.0\,\text{m/s}^2)\,t^2$$

Use the equation to construct an initial sketch, a motion diagram, and words to describe a process that is consistent with this equation.

**Sketch and translate** The above equation appears to be an application of Eq. (1.6), which we constructed to describe linear motion with constant acceleration, if we assume that the 1.0 m/s² in front of $t^2$ is the result of dividing 2.0 m/s² by 2:

$$x = x_0 + v_{0x}t + \frac{1}{2}a_x t^2$$

$$x = (-60\,\text{m}) + (10\,\text{m/s})\,t + \frac{1}{2}(2.0\,\text{m/s}^2)\,t^2$$

It looks like the initial position of the object of interest is $x_0 = -60$ m, its initial velocity is $v_{0x} = +10$ m/s, and its acceleration is $a_x = +2.0\,\text{m/s}^2$. Let's imagine that this equation describes the motion of a car passing a van in which you, the observer, are riding on a straight highway. The car is 60 m behind you and moving 10 m/s faster than your van. The car speeds up at a rate of 2.0 m/s² with respect to the van. The object of reference is you in the van; the positive direction is the direction in which the car and van are moving. A sketch of the situation appears below.

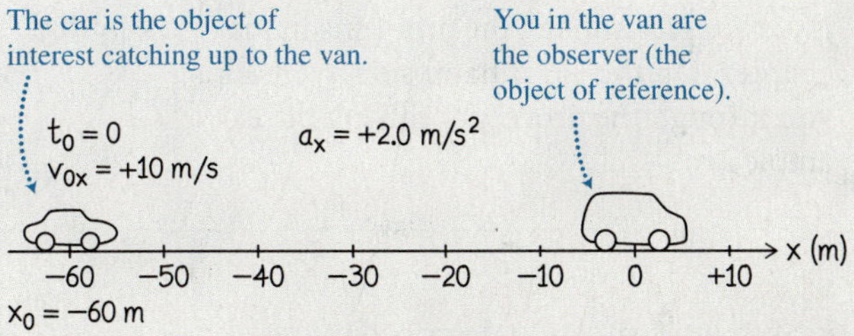


**Simplify and diagram** The car can be considered a point-like object—much smaller than the dimensions of the path it travels. The car's velocity and acceleration are both positive. Thus, the car's velocity in the positive $x$-direction is increasing as it moves toward the van (toward the origin). Below is a motion diagram for the car's motion as seen from the van. The successive dots in the diagram are spaced increasingly farther apart as the velocity increases; the velocity arrows are drawn increasingly longer. The velocity arrow (and the acceleration) point in the positive $x$-direction, that is, in same direction as the velocity arrows.

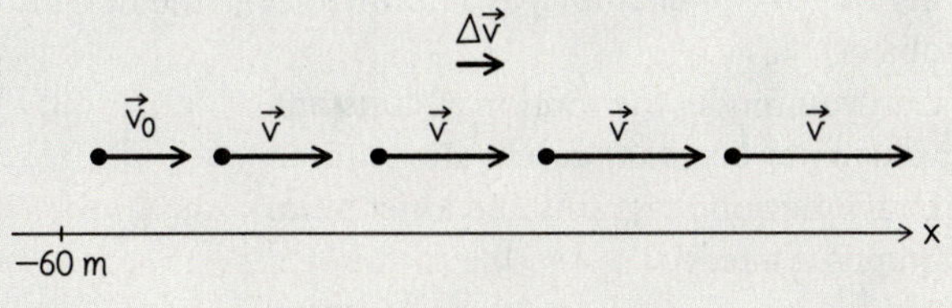


*(continued)*

**Represent mathematically** The mathematical representation of the situation appears at the start of the Equation Jeopardy example.

**Solve and evaluate** To evaluate what we have done, we can check the consistency (agreement) of the different representations. For example, we can check if the initial position and velocity are consistent in the equation, the sketch, and the motion diagram. In this case, they are.

**Try it yourself:** Describe a different scenario for the same mathematical representation.

*Answer:* This mathematical representation could describe the motion of a cyclist moving on a straight path as seen by a person standing on a sidewalk 60 m in front of the cyclist. The positive direction is in the direction the cyclist is traveling. When the person starts observing the cyclist, she is moving at an initial velocity of $v_{0x} = +10\ \text{m/s}$ and speeding up with acceleration $a_x = +2.0\ \text{m/s}^2$.

## PROBLEM-SOLVING STRATEGY Kinematics

Our four-step problem-solving procedure uses a multiple representation strategy that has proven successful in solving physics problems. In this chapter and many others we will walk you through the strategy with an example problem. In the left-hand column, you will find general guidelines for solving the problems in that chapter. On the right-hand side, we walk you through the process of solving the example problem.

### EXAMPLE 1.11 Car arriving at a red light

The velocity-versus-time graph shown here represents a car's motion. What time interval is needed for the car to stop, and how far does it travel while stopping?

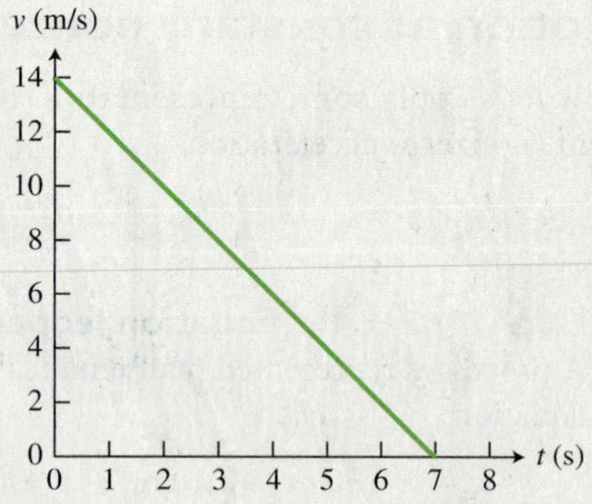


**Sketch and translate**

- Sketch the situation described in the problem. Choose the object of interest.
- Include an object of reference and a coordinate system. Indicate the origin and the positive direction.
- Label the sketch with relevant information.

From the graph, we see that the car's velocity at time zero is $v_{0x} = +14\ \text{m/s}$. The car is the object of interest. The object of reference is the ground. The car's initial position is unknown—we'll choose to place it at location $x_0 = 0$ at $t_0 = 0$. The plus sign means the car is moving in the positive $x$-direction. From the graph, we see that the car's velocity in the positive $x$-direction decreases by 2.0 m/s for each second; thus the slope of the graph $\Delta v_x/\Delta t$ is $(-2.0\ \text{m/s})/\text{s}$. We create an initial sketch.

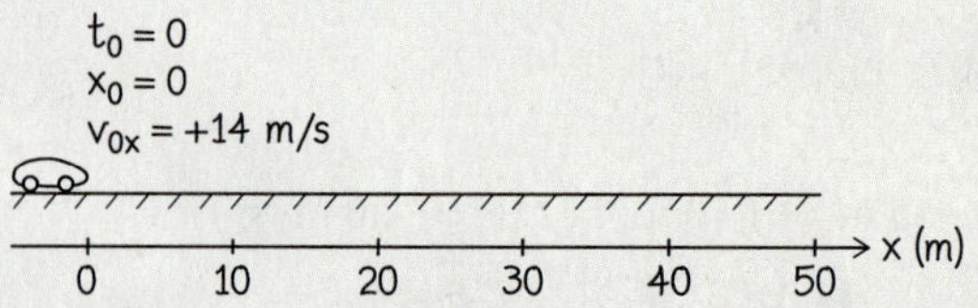


**Simplify and diagram**

- Decide how you will model the moving object (for example, as a point-like object).
- Can you model the motion as constant velocity or constant acceleration?
- Draw motion diagrams and kinematics graphs if needed.

We model the car as a point-like object moving along a straight line at constant acceleration. The velocity arrows get increasingly smaller since the magnitude of the velocity is decreasing. We draw a motion diagram.

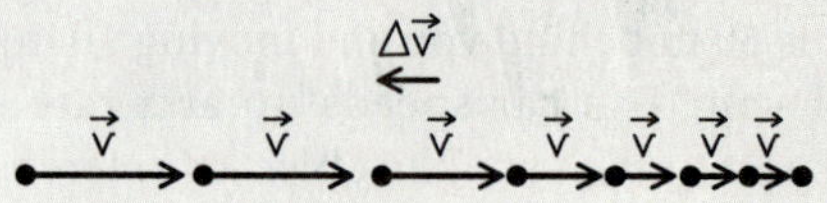

**Represent mathematically**

- Use the sketch(es), motion diagram(s), and kinematics graph(s) to construct a mathematical representation (equations) of the process. Be sure to consider the sign of each quantity.

Rearrange Eq. (1.5) to determine the time at which the velocity decreases to zero:

$$t = \frac{v_x - v_{0x}}{a_x}$$

We can then determine the position of the car at that time using the position equation of motion:

$$x = 0 + (14\ \text{m/s})t + \frac{1}{2}(-2.0\ \text{m/s}^2)t^2.$$

**Solve and evaluate**

- Solve the equations to find the answer to the question you are investigating.
- Evaluate the results to see if they are reasonable. Check the units and decide if the calculated quantities have reasonable values (sign, magnitude). Check limiting cases: Examine whether the final equation leads to a reasonable result if one of the quantities is zero or infinity. This strategy applies when you derive a new equation while solving a problem.

Substituting the known information in the first equation above:

$$t = \frac{(0\ \text{m/s}) - (14\ \text{m/s})}{-2.0\ \text{m/s}^2} = 7.0\ \text{s}$$

The car stops after a time interval $(t - t_0) = 7.0$ s.

The car's position when it stops is

$$x = 0 + (14\ \text{m/s})(7.0\ \text{s}) + \frac{1}{2}(-2.0\ \text{m/s}^2)(7.0\ \text{s})^2 = 49\ \text{m}$$

The units are correct and the magnitudes are reasonable.

In the limiting case of zero acceleration, the car should never stop. Our equation for the time it takes to stop gives

$$t = \frac{v_x - v_{0x}}{a_x} = \frac{v_x - v_{0x}}{0}$$

The result of dividing of a nonzero quantity by zero is infinity. Thus our equation predicts that it takes an infinite time for the car to stop. The limiting case checks out.

**Try it yourself:** A cyclist is moving in the negative $x$-direction at a speed of 6.0 m/s. He sees a red light and stops in 3.0 s. What is his acceleration?

*Answer:* $a_x = +2.0\ \text{m/s}^2$. The acceleration is positive even though the cyclist's speed decreased.

**Review Question 1.7** A car's motion with respect to the ground is described by the following function:

$$x = (-48\ \text{m}) + (12\ \text{m/s})t + (-2.0\ \text{m/s}^2)t^2$$

Mike says that its original position is $(-48\ \text{m})$ and its acceleration is $(-2.0\ \text{m/s}^2)$. Do you agree? If yes, explain why; if not, explain how to correct his answer.

## 1.8 Free fall

In this chapter we have learned about two simple models of motion—linear motion with constant velocity and linear motion with constant acceleration. Now we will look at a special case of linear motion—the motion of falling objects.

Let's start with the following observational experiment. Tear out a sheet of paper from your notebook and hold the paper in one hand. Hold a textbook parallel to the floor in the other hand and then drop both side by side.

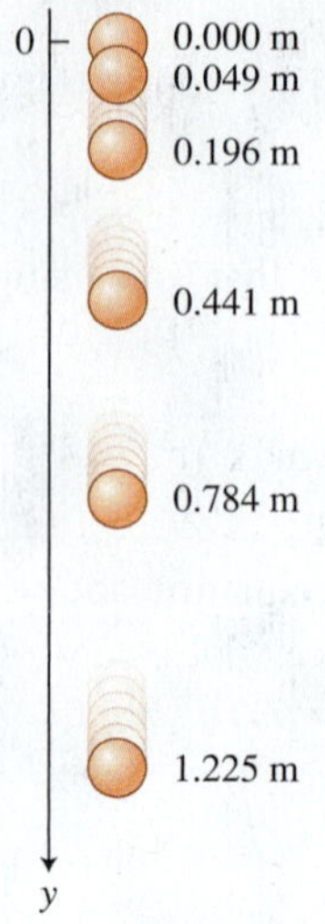


**Figure 1.21** The position of a falling ball every 0.100 s.

The book lands first. Next, crumple the paper into a tight ball. Now drop the book and the crumpled paper side by side. This time they land at about the same time. Does the motion depend on how heavy the objects are or on their shapes? Galileo Galilei (1564–1642) was the first to realize that it was easier to answer this question if he considered first the motion of falling objects in the absence of air. This hypothetical motion was *a model of the real process* and became known as **free fall.** Galileo hypothesized, based on a series of experiments, that free fall occurs exactly the same way for all objects regardless of their mass and shape.

Based on observations of falling objects, Galileo thought that the speed of freely falling objects was increasing as they moved closer to the surface of Earth. He hypothesized that the speed increases in the simplest way—linearly with time of flight or, in other words, the acceleration of free-falling objects was constant. Galileo did not have a video camera or a watch with a second hand to test his hypothesis. But we do! Imagine that we videotape a small metal ball that is dropped beside a ruler (**Figure 1.21**). Using the small ball allows us to create a situation very close to the model as the air has very little effect on the ball's motion. If the hypothesis is correct, the speed of the ball should increase linearly with time. After recording the fall, we step through the video frame by frame and record the ball's position every 0.100 s (**Table 1.8**). Earth is our object of reference; the origin of the coordinate axis is at the initial location of the ball. The positive direction points down.

**Active Learning Guide>**

To determine the average velocity during each time interval, we calculate the displacement of the ball between consecutive times and then divide by the time interval. For example, the average speed between 0.100 s and 0.200 s is $(0.196\text{ m} - 0.049\text{ m})/(0.200\text{ s} - 0.100\text{ s}) = 1.47\text{ m/s}$. These calculated velocities in the last row are associated with the clock readings $t^*$ at the middle of each time interval (the third row). Finally, we use these velocities and $t^*$ times to make a velocity-versus-time graph (**Figure 1.22**). The best-fit curve for this data is a straight line. Therefore, we model the motion of the metal ball as motion with constant acceleration. The slope of the line equals $9.8\text{ m/s}^2$.

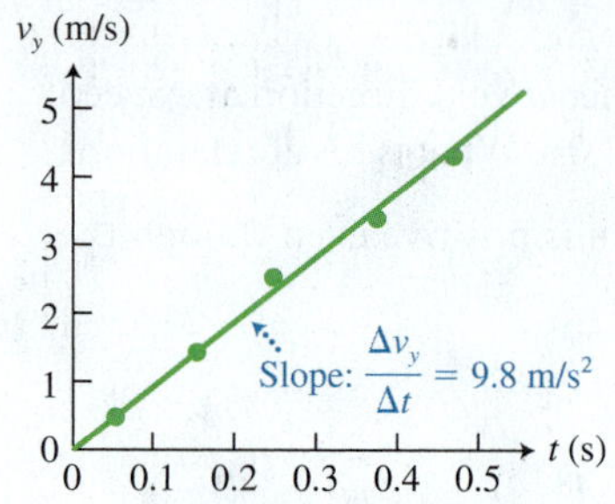


**Figure 1.22** A velocity-versus-time graph for a falling ball.

We can represent the motion of a falling ball mathematically using the equations of motion for constant acceleration [Eqs. (1.5) and (1.6)] with $a_y = 9.8\text{ m/s}^2$:

$$v_y = v_{0y} + a_y t = v_{0y} + (9.8\text{ m/s}^2)\,t \tag{1.8}$$

$$y = y_0 + v_{0y}t + \frac{1}{2}(9.8\text{ m/s}^2)\,t^2 \tag{1.9}$$

where $y_0$ and $v_{0y}$ are the position and instantaneous velocity, respectively, of the object at the clock reading $t_0 = 0$. These equations apply if the positive $y$-direction is down. When using an upward pointing $y$-axis, we place a minus sign in front of the $9.8\text{ m/s}^2$. The magnitude of the object's acceleration while falling without air resistance is given a special symbol, $g$, where $g = 9.8\text{ m/s}^2$.

**Table 1.8** **Position and time data for a small falling ball.**

| | | | | | | |
|---|---|---|---|---|---|---|
| **$t$ (s)** | 0.000 | 0.100 | 0.200 | 0.300 | 0.400 | 0.500 |
| **$y$ (m)** | 0.000 | 0.049 | 0.196 | 0.441 | 0.784 | 1.225 |
| **$t^*$ (s)** | 0.050 | 0.150 | 0.250 | 0.350 | 0.450 | |
| **$v_{av}$ (m/s)** | 0.49 | 1.47 | 2.45 | 3.43 | 4.41 | |

If we videotape a small object thrown upward and then use the data to construct a velocity-versus-time graph, we find that its acceleration is still the same at all clock readings. For an upward-pointing axis, the object's acceleration is $-9.8\ \text{m/s}^2$ on the way up, $-9.8\ \text{m/s}^2$ on the way down, and even $-9.8\ \text{m/s}^2$ at the instant when the object is momentarily at rest at the highest point of its motion. At all times during the object's flight, its velocity is changing at a rate of $-9.8\ \text{m/s}$ each second. A motion diagram and the graphs representing the position-versus-time, velocity-versus-time, and acceleration-versus-time are shown in **Figure 1.23**. The positive direction is up. Notice that when the position-versus-time graph is at its maximum (object is at maximum height), the velocity is instantaneously zero (the slope of the position-versus-time graph is zero). The acceleration is never zero, even at the moment when the velocity of the object is zero.

It might be tempting to think that at the instant an object is not moving, its acceleration must be zero. This is only true for an object that is at rest and remains at rest. In the case of an object thrown upward, if its acceleration at the top of the flight is zero, it would never descend (it would remain at rest at its highest point).

**Figure 1.23** Free-fall motion for an upward thrown ball: (a) a motion diagram, (b) position-versus-time graph, (c) velocity-versus-time graph, and (d) acceleration-versus-time graph.

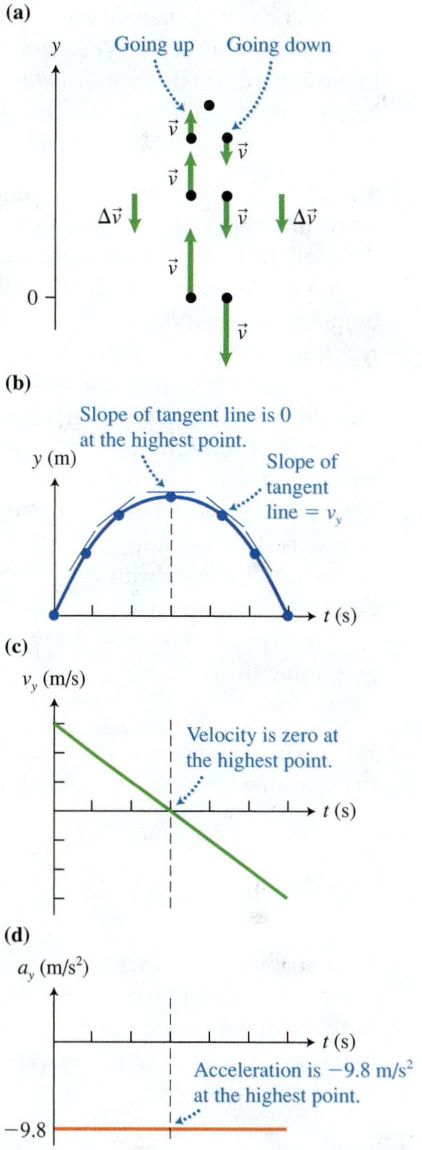


**TIP** Physicists say that an object is in a state of free fall even when it is thrown upward, because its acceleration is the same on the way up as on the way down.

**Review Question 1.8** Free-fall acceleration can be both positive or negative. Why is this true?

## 1.9 Tailgating: Putting it all together

Drivers count on their ability to apply the brakes in time if the car in front of them suddenly slows. However, if you are following too closely behind another car, you may not be able to stop in time. Let's look at the motion of two vehicles in what appears to be a safe driving situation.

**EXAMPLE 1.12 An accident involving tailgating**

A car follows about two car lengths (10.0 m) behind a van. At first, both vehicles are traveling at a conservative speed of 25 m/s (56 mi/h). The driver of the van suddenly slams on the brakes to avoid an accident, slowing down at $9.0\ \text{m/s}^2$. The car driver's reaction time is 0.80 s and the car's maximum acceleration while slowing down is also $9.0\ \text{m/s}^2$. Will the car be able to stop before hitting the van?

**Sketch and translate** At right we represent this situation for each vehicle (we have two objects of interest). We'll use capital letters to indicate quantities referring to the van and lowercase letters for quantities referring to the car. We use the coordinate system shown with the

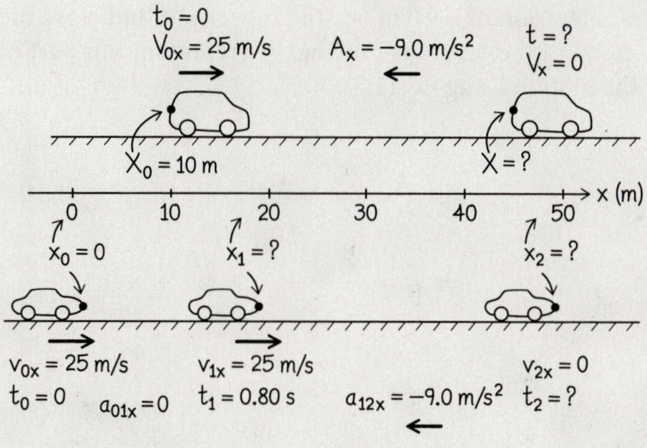


*(continued)*

origin of the coordinates at the initial position of the car's front bumper. The positive direction is in the direction of motion.

The process starts when the van starts braking. It moves at constant negative acceleration throughout the entire problem process. We separate the motion of the car into two parts: (1) the motion before the driver applies the brakes (constant positive velocity) and (2) its motion after the driver starts braking (constant negative acceleration).

**Simplify and diagram** We model each vehicle as a point-like object, but since we are trying to determine if they collide, we need to be more specific about their positions. The position of the car will be the position of its front bumper. The position of the van will be the position of its rear bumper. We look at the motion of each vehicle separately. If the car's final position is greater than the van's final position, then a collision has occurred at some point during their motion. Assume that the vehicles have constant acceleration so that we can apply our model of motion with constant acceleration. A velocity-versus-time graph line for each vehicle is shown at right.

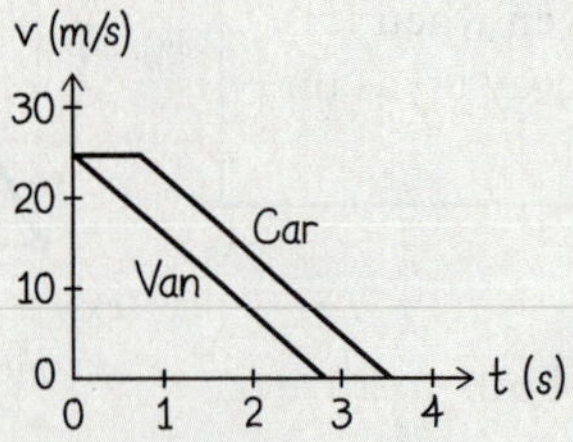


**Represent mathematically** Equation (1.7) can be used to determine the distance the van travels while stopping:

$$2A_x(X - X_0) = V_x^2 - V_{0x}^2$$

$$\Rightarrow X = \frac{V_x^2 - V_{0x}^2}{2A_x} + X_0$$

**The car part 1** Since the car is initially traveling at constant velocity, we use Eq. (1.2). The subscript 0 indicates the moment the driver sees the van start slowing down. The subscript 1 indicates the moment the car driver starts braking.

$$x_1 = x_0 + v_{0x}t_1$$

**The car part 2** After applying the brakes, the car has an acceleration of $-9.0\ \text{m/s}^2$. The subscript 2 indicates the moment the car stops moving. We represent this part of the motion using Eq. (1.7):

$$2a_x(x_2 - x_1) = v_{2x}^2 - v_{1x}^2$$

$$\Rightarrow x_2 = \frac{v_{2x}^2 - v_{1x}^2}{2a_x} + x_1$$

$$\Rightarrow x_2 = \frac{v_{2x}^2 - v_{1x}^2}{2a_x} + (x_0 + v_{0x}t_1)$$

The last step came from inserting the result from Part 1 for $x_1$.

**Solve and evaluate** The van's initial velocity is $V_0 = +25\ \text{m/s}$, its final velocity is $V_x = 0$, and its acceleration is $A = -9.0\ \text{m/s}^2$. Its initial position is two car lengths in front of the front of the car, so $X_0 = 2 \times 5.0\ \text{m} = 10\ \text{m}$. The final position of the van is

$$X = \frac{V_x^2 - V_{0x}^2}{2A_x} + X_0$$

$$= \frac{0^2 - (25\ \text{m/s})^2}{2(-9.0\ \text{m/s}^2)} + 10\ \text{m}$$

$$= 45\ \text{m}$$

The car's initial position is $x_0 = 0$, its initial velocity is $v_{0x} = 25\ \text{m/s}$, its final velocity is $v_{2x} = 0$, and its acceleration when braking is $a_x = -9.0\ \text{m/s}^2$. The car's final position is

$$x_2 = \frac{v_{2x}^2 - v_{1x}^2}{2a_x} + (x_0 + v_{0x}t_1)$$

$$= \frac{0^2 - (25\ \text{m/s})^2}{2(-9.0\ \text{m/s}^2)} + [0\ \text{m} + (25\ \text{m/s})(0.8\ \text{s})]$$

$$= 55\ \text{m}$$

The car would stop about 10 m beyond where the van would stop. There will be a collision between the two vehicles.

This analysis illustrates why tailgating is such a big problem. The car traveled at a 25-m/s constant velocity during the relatively short 0.80-s reaction time. During the same 0.80 s, the van's velocity decreased by $(0.80\ \text{s})(-9.0\ \text{m/s}^2) = 7.2\ \text{m/s}$ from 25 m/s to about 18 m/s. So the van was moving somewhat slower than the car when the car finally started to brake. Since they were both slowing down at about the same rate, the tailgating vehicle's velocity was always greater than that of the vehicle in front until they hit.

**Try it yourself:** Two cars, one behind the other, are traveling at 30 m/s (13 mi/h). The front car hits the brakes and slows down at the rate of $10\ \text{m/s}^2$. The driver of the second car has a 1.0-s reaction time. The front car's speed has decreased to 20 m/s during that 1.0 s. The rear car traveling at 30 m/s starts braking, slowing down at the same rate of $10\ \text{m/s}^2$. How far behind the front car should the rear car be so it does not hit the front car?

*Answer:* The rear car should be at least 30 m behind the front car.

**Review Question 1.9** Explain, using physics terms, why tailgating accidents occur.

## Summary

| Words | Pictorial and physical representations | Mathematical representation |
|---|---|---|

A **reference frame** consists of an object of reference, a point of reference on that object, a coordinate system whose origin is at the point of reference, and a clock. (Section 1.1)

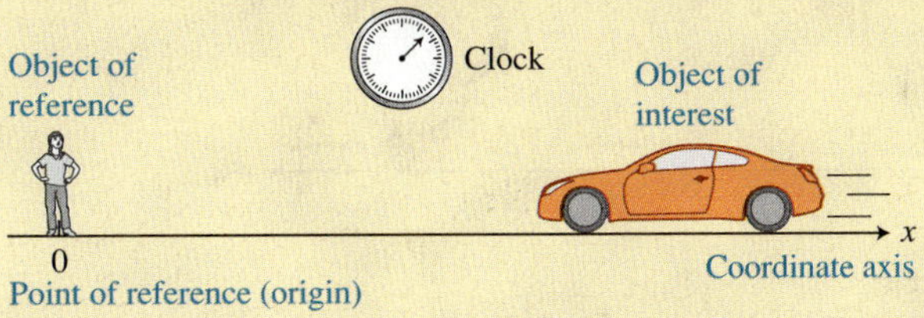


---

- **Time or clock reading** $t$ (a scalar quantity) is the reading on a clock or another time measuring instrument. (Section 1.1)

Time: $t$

- **Time interval** $\Delta t$ (a scalar quantity) is the difference of two times. (Section 1.3)

Time interval:

$$\Delta t = t_2 - t_1$$

---

- **Position** $x$ (a scalar quantity) is the location of an object relative to the chosen origin. (Section 1.3)
- **Displacement** $\vec{d}$ is a vector drawn from the initial position of an object to its final position. The $x$-component of the displacement $d_x$ is the change in position of the object along the $x$-axis. (Section 1.3)
- **Distance** $d$ (a scalar quantity) is the magnitude of the displacement and is always positive. (Section 1.3)
- **Path length** $l$ is the length of a string laid along the path the object took. (Section 1.3)

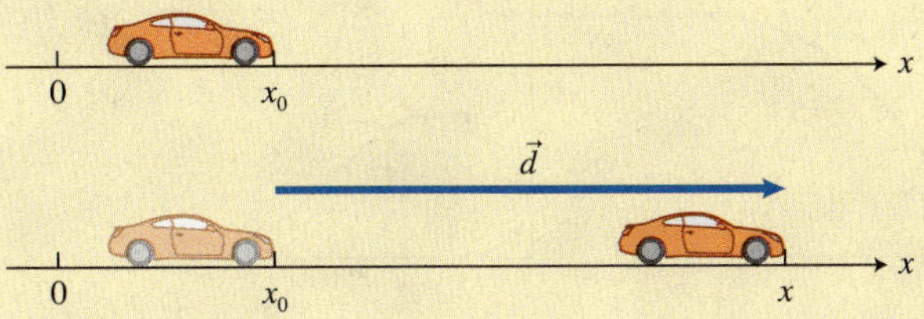


$d_x = x - x_0 > 0$ if $\vec{d}$ points in the positive direction of $x$-axis.

$d_x = x - x_0 < 0$ if $\vec{d}$ points in the negative direction.

$d = |x - x_0|$

---

- **Velocity** $\vec{v}$ (a vector quantity) is the displacement of an object during a time interval divided by that time interval. The velocity is *instantaneous* if the time interval is very small and *average* if the time interval is longer. (Section 1.5)
- **Speed** $v$ (a scalar quantity) is the magnitude of the velocity. (Section 1.5)

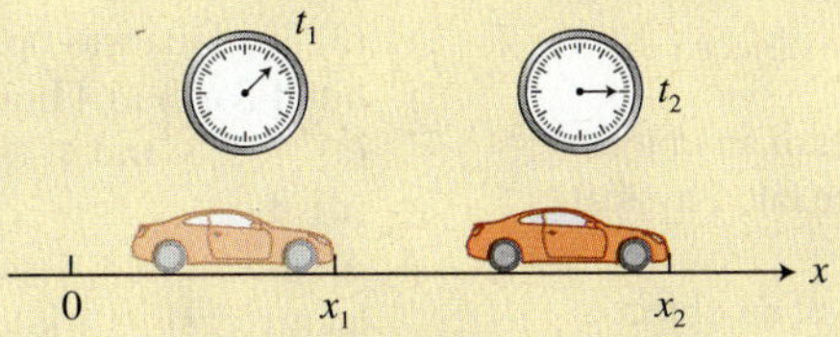


For linear constant velocity, motion $\vec{v} = \vec{d}/\Delta t$

$$v_x = \frac{\Delta x}{\Delta t} = \frac{x_2 - x_1}{t_2 - t_1} \quad \text{Eq. (1.1)}$$

$$v = \left|\frac{\Delta x}{\Delta t}\right|$$

(*continued*)

| Words | Pictorial and physical representations | Mathematical representation |
|---|---|---|
| ■ **Acceleration** $\vec{a}$ (a vector quantity) is the change in an object's velocity $\Delta\vec{v}$ during a time interval $\Delta t$ divided by the time interval. The acceleration is *instantaneous* if the time interval is very small and *average* if the time interval is longer. (Section 1.6) | 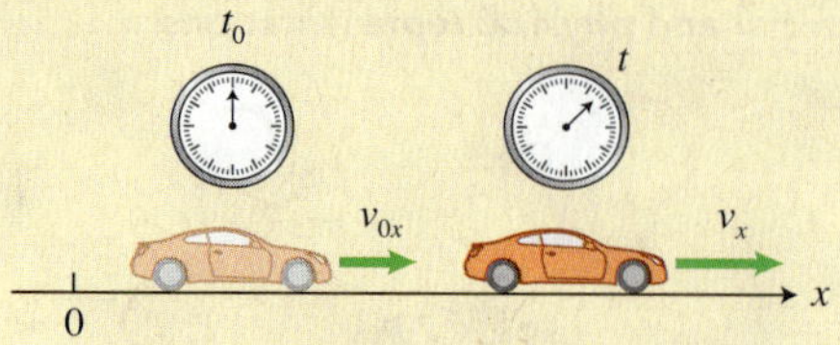 | $a_x = \dfrac{v_{2x} - v_{1x}}{t_2 - t_1} = \dfrac{\Delta v_x}{\Delta t}$ Eq. (1.4)<br>(rearranged)<br>$a_x = \dfrac{\Delta v_x}{\Delta t} = \dfrac{v_x - v_{0x}}{t - t_0}$ Eq. (1.4) |
| Motion with constant velocity or constant acceleration can be represented with a sketch, a motion diagram, kinematics graphs, and mathematically. (Sections 1.5–1.6) | 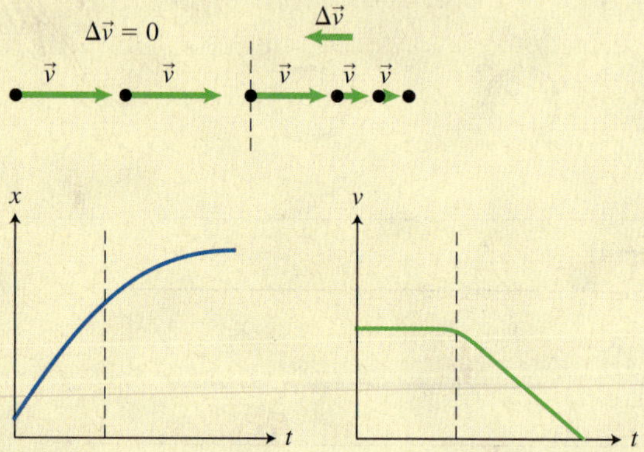 | $v_x = v_{0x} + a_x t$ Eq. (1.5)<br>$x = x_0 + v_{0x}t + \frac{1}{2}a_x t^2$ Eq. (1.6)<br>$2a_x(x - x_0) = v_x^2 - v_{0x}^2$ Eq. (1.7)<br>(rearranged)<br>$x - x_0 = \dfrac{v_x^2 - v_{0x}^2}{2a_x}$ Eq. (1.7) |

MP® For instructor-assigned homework, go to **MasteringPhysics.**

# Questions

## Multiple Choice Questions

1. Match the general elements of physics knowledge (left) with the appropriate examples (right).

| | |
|---|---|
| Model of a process | Free fall |
| Model of an object | Acceleration |
| Physical quantity | Rolling ball |
| Physical phenomenon | Point-like object |

   (a) Model of a process—Acceleration; Model of an object—Point-like object; Physical quantity—Free fall; Physical phenomenon—Rolling ball.
   (b) Model of a process—Rolling ball; Model of an object—Point-like object; Physical quantity—Acceleration; Physical phenomenon—Free fall.
   (c) Model of a process—Free fall; Model of an object—Point-like object; Physical quantity—Acceleration; Physical phenomenon—Rolling ball.

2. Which group of quantities below consists only of scalar quantities?
   (a) Average speed, displacement, time interval
   (b) Average speed, path length, clock reading
   (c) Temperature, acceleration, position
3. Which of the following are examples of time interval? (1) I woke up at 7 am. (2) The lesson lasted 45 minutes. (3) Svetlana was born on November 26. (4) An astronaut orbited Earth in 4 hours.
   (a) 1, 2, 3, and 4 (b) 2 and 4 (c) 2
   (d) 4 (e) 3
4. A student said, "The displacement between my dorm and the lecture hall is 1 kilometer." Is he using the correct physical quantity for the information provided? What should he have called the 1 kilometer?
   (a) Distance (b) Path length (c) Position
   (d) Both a and b are correct.

5. An object moves so that its position depends on time as $x = +12 - 4t + t^2$. Which statement below is not true?
   (a) The object is accelerating.
   (b) The speed of the object is always decreasing.
   (c) The object first moves in the negative direction and then in the positive direction.
   (d) The acceleration of the object is $+2\ \text{m/s}^2$.
   (e) The object stops for an instant at 2.0 s.
6. Choose a correct approximate velocity-versus-time graph for the following hypothetical motion: a car moves at constant velocity, and then slows to a stop and without a pause moves in the opposite direction with the same acceleration (**Figure Q1.6**).

**Figure Q1.6**

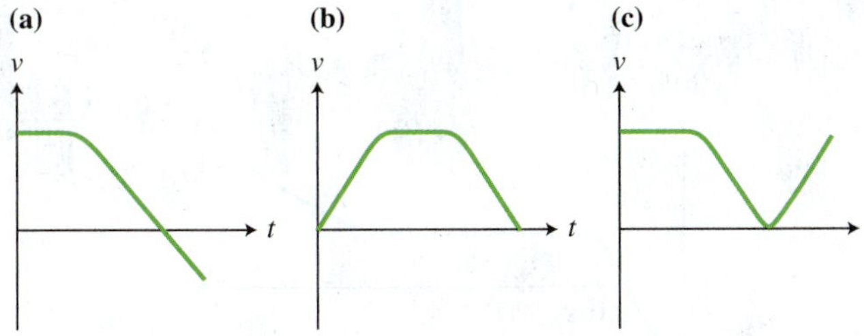


7. In which case are average and instantaneous velocities the same? Explain.
   (a) When the object moves at constant velocity
   (b) When the object moves at constant acceleration
   (c) When the object does not move
   (d) a and c
   (e) a, b, and c
8. You drop a small ball, and then a second small ball. When you drop the second ball, the distance between them is 3 cm. What statement below is correct? Explain.
   (a) The distance between the balls stays the same.
   (b) The distance between the balls decreases.
   (c) The distance between the balls increases.
9. Your car is traveling west at 12 m/s. A stoplight (the origin of the coordinate axis) to the west of you turns yellow when you are 20 m from the edge of the intersection (see **Figure Q1.9**). You apply the brakes and your car's speed decreases. Your car stops before it reaches the stoplight. What are the signs for the components of kinematics quantities?

**Figure Q1.9**

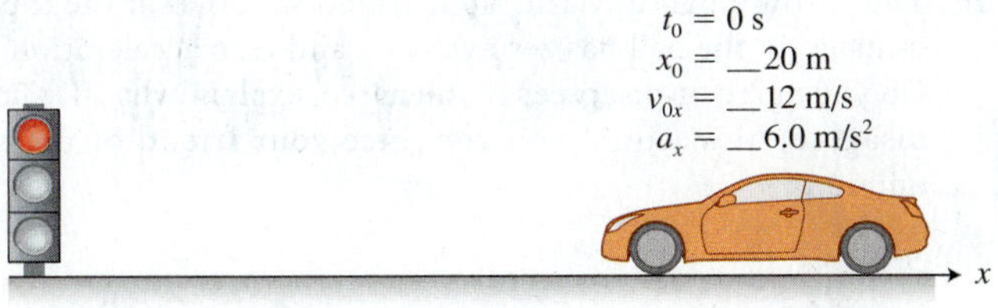


| | $x_0$ | $v_{0x}$ | $a_x$ |
|---|---|---|---|
| (a) | + | − | − |
| (b) | + | − | + |
| (c) | + | + | − |
| (d) | + | + | + |
| (e) | − | − | + |

10. Which velocity-versus-time graph in **Figure Q1.10** best describes the motion of the car in the previous problem (see Figure Q1.9) as it approaches the stoplight?

**Figure Q1.10**

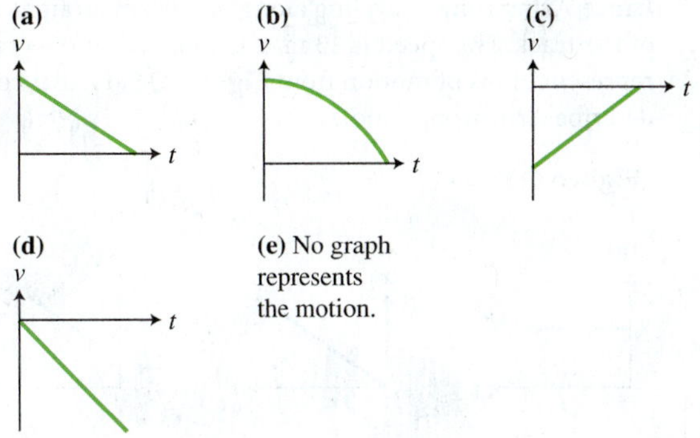


11. Suppose that (c) in Figure Q1.10 represents the velocity-versus-time graph for a moving object. Which of the following gives the correct signs for the velocity and acceleration components [positive (+) or negative (−)] of this motion?

| | $v_x$ | $a_x$ |
|---|---|---|
| (a) | + | + |
| (b) | + | − |
| (c) | − | + |
| (d) | − | − |

12. A sandbag hangs from a rope attached to a rising hot air balloon. The rope connecting the bag to the balloon is cut. How will two observers see the motion of the sandbag? Observer 1 is in the hot air balloon and observer 2 is on the ground.
   (a) Both 1 and 2 will see it go down.
   (b) 1 will see it go down and 2 will see it go up.
   (c) 1 will see it go down and 2 will see it go up and then down.
13. An apple falls from a tree. It hits the ground at a speed of about 5.0 m/s. What is the approximate height of the tree?
   (a) 2.5 m (b) 1.2 m (c) 10.0 m (d) 2.4 m
14. You have two small metal balls. You drop the first ball and throw the other one in the downward direction. Choose the statements that are not correct.
   (a) The second ball will spend less time in flight.
   (b) The first ball will have a slower final speed when it reaches the ground.
   (c) The second ball will have larger acceleration.
   (d) Both balls will have the same acceleration.
15. You throw a small ball upward. Then you throw it again, this time at twice the initial speed. Choose the correct statement.
   (a) The second time, the ball travels twice as far up as the first time.
   (b) The second time, the ball has twice the magnitude of acceleration while in flight that it did the first time.
   (c) The second time, the ball spends twice as much time in flight.
   (d) All of the choices are correct.
16. You throw a small ball upward and notice the time it takes to come back. If you then throw the same ball so that it takes twice as much time to come back, what is true about the motion of the ball the second time?
   (a) Its initial speed was twice the speed in the first experiment.
   (b) It traveled an upward distance that is twice the distance of the original toss.
   (c) It had twice as much acceleration on the way up as it did the first time.
   (d) The ball stopped at the highest point and had zero acceleration at that point.

## Conceptual Questions

17. Lance Armstrong is cycling along an 800-m straight stretch of the track. His speed is 13 m/s. Choose all of the graphical representations of motion from **Figure Q1.17** that correctly describe Armstrong's motion.

**Figure Q1.17**

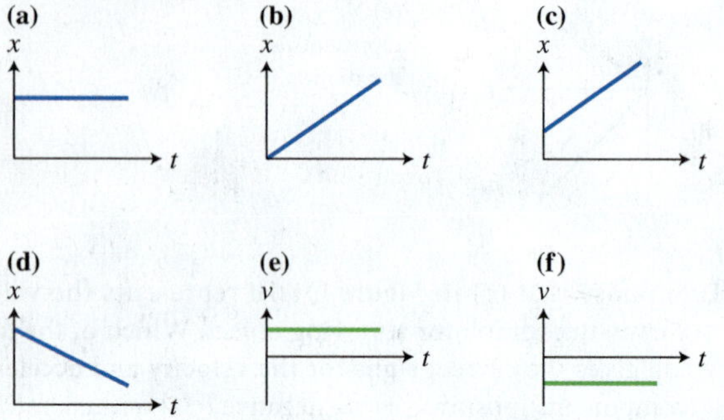


18. In what reasonable ways can you represent or describe the motion of a car traveling from one stoplight to the next? Construct each representation for the moving car.
19. What is the difference between speed and velocity? Between path length and distance? Between distance and displacement? Give an example of each.
20. What physical quantities do we use to describe motion? What does each quantity characterize? What are their SI units?
21. Devise stories describing each of the motions shown in each of the graphs in **Figure Q1.21**. Specify the object of reference.

**Figure Q1.21**

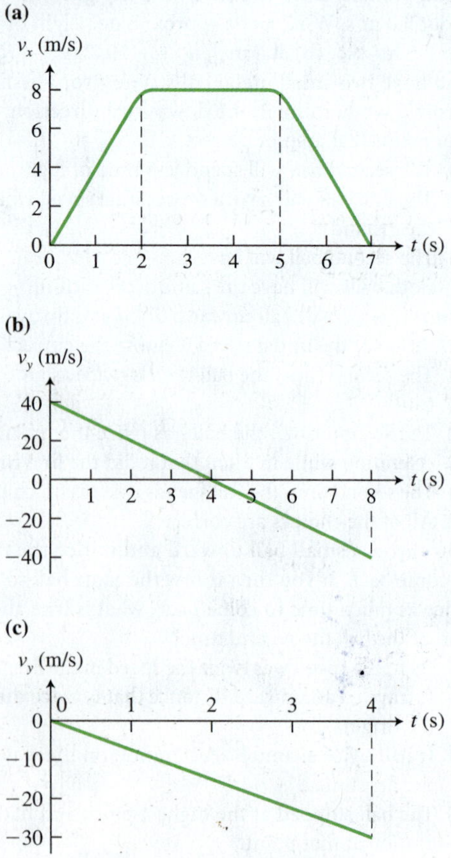


22. For each of the position-versus-time graphs in **Figure Q1.22**, draw velocity-versus-time graphs and acceleration-versus-time graphs.

**Figure Q1.22**

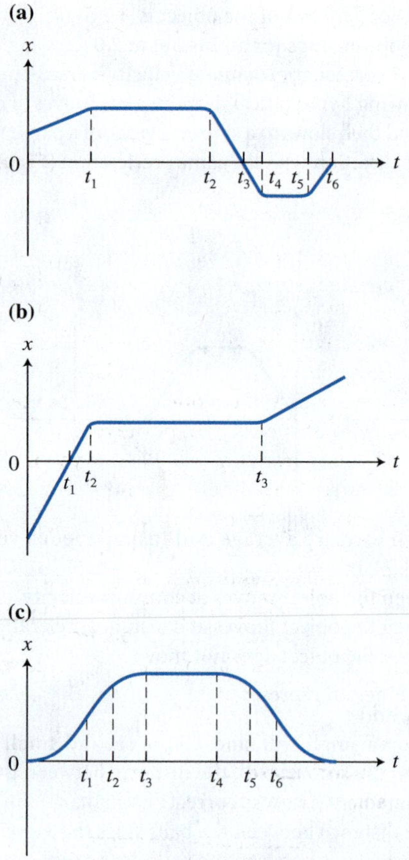


23. Can an object have a nonzero velocity and zero acceleration? If so, give an example.
24. Can an object at one instant of time have zero velocity and nonzero acceleration? If so, give an example.
25. Your little sister has a battery-powered toy truck. When the truck is moving, how can you determine whether it has constant velocity, constant speed, constant acceleration, or changing acceleration? Explain in detail.
26. You throw a ball upward. Your friend says that at the top of its flight the ball has zero velocity and zero acceleration. Do you agree or disagree? If you agree, explain why. If you disagree, how would you convince your friend of your opinion?

# Problems

Below, **BIO** indicates a problem with a biological or medical focus. Problems labeled **EST** ask you to estimate the answer to a quantitative problem rather than derive a specific answer. Asterisks indicate the level of difficulty of the problem. Problems marked with ✎ require you to make a drawing or graph as part of your solution. Problems with no * are considered to be the least difficult. A single * marks moderately difficult problems. Two ** indicate more difficult problems.

### 1.2 A conceptual description of motion

1. A car starts at rest from a stoplight and speeds up. It then moves at constant speed for a while. Then it slows down until reaching the next stoplight. Represent the motion with a motion diagram as seen by the observer on the ground.
2. * ✎ You are an observer on the ground. (a) Draw two motion diagrams representing the motions of two runners moving at the same constant speeds in opposite directions toward you. Runner 1, coming from the east, reaches you in 5 s, and runner 2 reaches you in 3 s. (b) Draw a motion of diagram for the second runner as seen by the first runner.
3. * A car is moving at constant speed on a highway. A second car catches up and passes the first car 5 s after it starts to speed up. Represent the situation with a motion diagram. Specify the observer with respect to whom you drew the diagram.
4. * ✎ A hat falls off a man's head and lands in the snow. Draw a motion diagram representing the motion of the hat as seen by the man.

### 1.3 and 1.4 Quantities for describing motion and Representing motion with data tables and graphs

5. * You drive 100 km east, do some sightseeing, and then turn around and drive 50 km west, where you stop for lunch. (a) Represent your trip with a displacement vector. Choose an object of reference and coordinate axis so that the scalar component of this vector is (b) positive; (c) negative; (d) zero.
6. * Choose an object of reference and a set of coordinate axes associated with it. Show how two people can start and end their trips at different locations but still have the same displacement vectors in this reference frame.
7. The scalar $x$-component of a displacement vector for a trip is $-70$ km. Represent the trip using a coordinate axis and an object of reference. Then change the axis so that the displacement component becomes $+70$ km.
8. * You recorded your position with respect to the front door of your house as you walked to the mailbox. Examine the data presented in **Table 1.9** and answer the following questions: (a) What instruments did you use to collect data? (b) What are the uncertainties in your data? (c) Represent your motion using a position-versus-time graph. (d) Tell the story of your motion in words. (e) Show on the graph the displacement, distance, and path length.

**Table 1.9**

| $t$ (s) | 1 | 2 | 3 | 4 | 5 | 6 | 7 | 8 | 9 |
|---|---|---|---|---|---|---|---|---|---|
| $x$ (steps) | 2 | 4 | 9 | 13 | 18 | 20 | 16 | 11 | 9 |

### 1.5 Constant velocity linear motion

9. * You need to determine the time interval (in seconds) needed for light to pass an atomic nucleus. What information do you need? How will you use it? What simplifying assumptions about the objects and processes do you need to make? What approximately is that time interval?
10. A speedometer reads 65 mi/h. (a) Use as many different units as possible to represent the speed of the car. (b) If the speedometer reads 100 km/h, what is the car's speed in mi/h?
11. Convert the following record speeds so that they are in mi/h, km/h, and m/s. (a) Australian dragonfly—36 mi/h; (b) the diving peregrine falcon—349 km/h; and (c) the Lockheed SR-71 jet aircraft—980 m/s (about three times the speed of sound).
12. **EST Hair growth speed** Estimate the rate that your hair grows in meters per second. Indicate any assumptions you made.
13. * **EST** ✎ A kidnapped banker looking through a slit in a van window counts her heartbeats and observes that two highway exits pass in 80 heartbeats. She knows that the distance between the exits is 1.6 km (1 mile). (a) Estimate the van's speed. (b) Choose and describe a reference frame and draw a position-versus-time graph for the van.
14. **EST** Make a simplified map of the path from where you live to your physics classroom. (a) Label your path and your displacement. (b) Estimate the time interval that you need to reach the classroom from where you live and your average speed.
15. * **Equation Jeopardy** Two observers observe two different moving objects. However, they describe their motions mathematically with the same equation: $x(t) = 10\text{ km} - (4\text{ km/h})t$. (a) Write two short stories about these two motions. Specify where each observer is and what she is doing. What is happening to the moving object at $t = 0$? (b) Use significant digits to determine the interval within which the initial position is known.
16. * Your friend's pedometer shows that he took 17,000 steps in 2.50 h during a hike. Determine everything you can about the hike. What assumptions did you make? How certain are you in your answer? How would the answer change if the time were given as 2.5 h instead of 2.50 h?
17. During a hike, two friends were caught in a thunderstorm. Four seconds after seeing lightning from a distant cloud, they heard thunder. How far away was the cloud (in kilometers)? Write your answer as an interval using significant digits as your guide. Sound travels in air at about 340 m/s.
18. Light travels at a speed of $3.0 \times 10^8$ m/s in a vacuum. The approximate distance between Earth and the Sun is $150 \times 10^6$ km. How long does it take light to travel from the Sun to Earth? What are the margins within which you know the answer?
19. Proxima Centauri is $4.22 \pm 0.01$ light-years from Earth. Determine the length of 1 light-year and convert the distance to the star into meters. What is the uncertainty in the answer?
20. * Spaceships traveling to other planets in the solar system move at an average speed of $1.1 \times 10^4$ m/s. It took Voyager about 12 years to reach the orbit of Uranus. What can you learn about the solar system using these data? What assumption did you make? How did this assumption affect the results?

21. ** **Figure P1.21** shows a velocity-versus-time graph for the bicycle trips of two friends with respect to the parking lot where they started. (a) Determine their displacements in 20 s. (b) If Xena's position at time zero is 0 and Gabriele's position is 60 m, what time interval is needed for Xena to catch Gabriele? (c) Use the information from (b) to write function $x(t)$ for Gabriele with respect to Xena.

**Figure P1.21**

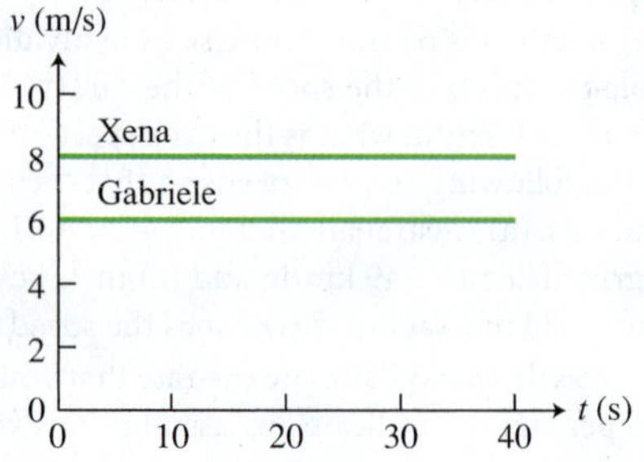


22. * **Table 1.10** shows position and time data for your walk along a straight path. (a) Tell everything you can about the walk. Specify the object of reference. (b) Draw a motion diagram, draw a graph $x(t)$, and write a function $x(t)$ that is consistent with the data and the chosen reference frame.

**Table 1.10**

| Time (s) | Position (m) |
|---|---|
| 0 | 80 |
| 10 | 40 |
| 20 | 0 |
| 30 | −40 |
| 40 | −80 |
| 50 | −120 |

23. * **Table 1.11** shows position and time data for your friend's bicycle ride along a straight bike path. (a) Tell everything you can about his ride. Specify the observer. (b) Draw a motion diagram, draw a graph $x(t)$, and write a function $x(t)$ that is consistent with the ride.

**Table 1.11**

| Time (s) | Position (m) |
|---|---|
| 0 | −200 |
| 10 | −120 |
| 20 | −40 |
| 30 | 40 |
| 40 | 120 |
| 50 | 200 |

24. * You are walking to your physics class at speed 1.0 m/s with respect to the ground. Your friend leaves 2.0 min after you and is walking at speed 1.3 m/s in the same direction. How fast is she walking with respect to you? How far does your friend travel before she catches up with you? Indicate the uncertainty in your answers. Describe any assumptions that you made.

25. * Gabriele enters an east–west straight bike path at the 3.0-km mark and rides west at a constant speed of 8.0 m/s. At the same time, Xena rides east from the 1.0-km mark at a constant speed of 6.0 m/s. (a) Write functions $x(t)$ that describe their positions as a function of time with respect to Earth. (b) Where do they meet each other? In how many different ways can you solve this problem? (c) Write a function $x(t)$ that describes Xena's motion with respect to Gabriele.

26. * Jim is driving his car at 32 m/s (72 mi/h) along a highway where the speed limit is 25 m/s (55 mi/h). A highway patrol car observes him pass and quickly reaches a speed of 36 m/s. At that point, Jim is 300 m ahead of the patrol car. How far does the patrol car travel before catching Jim?

27. * You hike two thirds of the way to the top of a hill at a speed of 3.0 mi/h and run the final third at a speed of 6.0 mi/h. What was your average speed?

28. * Olympic champion swimmer Michael Phelps swam at an average speed of 2.01 m/s during the first half of the time needed to complete a race. What was his average swimming speed during the second half of the race if he tied the record, which was at an average speed of 2.05 m/s?

29. * A car makes a 100-km trip. It travels the first 50 km at an average speed of 50 km/h. How fast must it travel the second 50 km so that its average speed is 100 km/h?

30. * Jane and Bob see each other when 100 m apart. They are moving toward each other, Jane at speed 4.0 m/s and Bob at speed 3.0 m/s with respect to the ground. What can you determine about this situation using these data?

31. * The graph in **Figure P1.31** represents four different motions. (a) Write a function $x(t)$ for each motion. (b) Use the information in the graph to determine as many quantities related to the motion of these objects as possible. (c) Act out these motions with two friends. (Hint: think of what each object was doing at $t = 0$.)

**Figure P1.31**

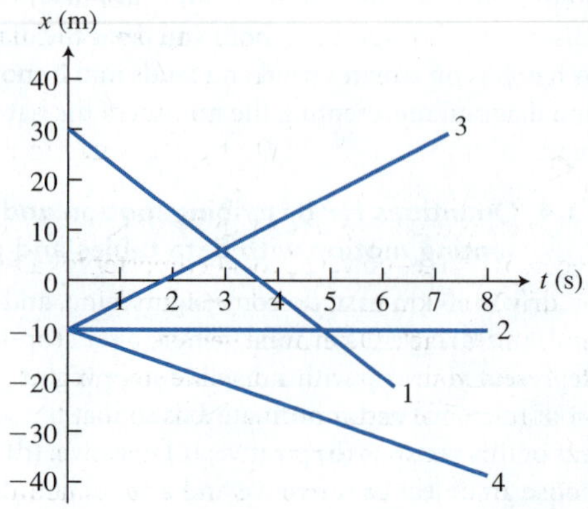


## 1.6 Motion at constant acceleration

32. A car starts from rest and reaches the speed of 10 m/s in 30 s. What can you determine about the motion of the car using this information?

33. A truck is traveling east at +16 m/s. (a) The driver sees that the road is empty and accelerates at $+1.0\ m/s^2$ for 5.0 s. What can you determine about the truck's motion using these data? (b) The driver then sees a red light ahead and decelerates at $-2.0\ m/s^2$ for 3.0 s. What can you determine about the truck's motion using these data? (c) Determine the values of the quantities you listed in (a) and (b).

34. **Bumper car collision** On a bumper car ride, friends smash their cars into each other (head-on) and each has a speed *change* of 3.2 m/s. If the magnitudes of acceleration of each car during the collision averaged $28\ m/s^2$, determine the time interval needed to stop and the stopping distance for each car while colliding. Specify your reference frame.

35. A bus leaves an intersection accelerating at $+2.0\ \text{m/s}^2$. Where is the bus after 5.0 s? What assumption did you make? If this assumption is not valid, would the bus be closer or farther away from the intersection compared to your original answer? Explain.
36. A jogger is running at $+4.0$ m/s when a bus passes her. The bus is accelerating from $+16.0$ m/s to $+20.0$ m/s in 8.0 s. The jogger speeds up with the same acceleration. What can you determine about the jogger's motion using these data?
37. * The motion of a person as seen by another person is described by the equation $v = -3.0\ \text{m/s} + (0.5\ \text{m/s}^2)t$. (a) Represent this motion with a motion diagram and position-, velocity-, and acceleration-versus-time graphs. (b) Say everything you can about this motion and describe what happens to the person when his speed becomes zero.
38. **Tour de France** While cycling at speed of 10 m/s, Lance Armstrong starts going downhill with an acceleration of magnitude $1.2\ \text{m/s}^2$. The descent takes 10.0 s. What can you determine about Lance's motion using these data? What assumptions did you make?
39. An automobile engineer found that the impact of a truck colliding at 16 km/h with a concrete pillar caused the bumper to indent only 6.4 cm. The truck stopped. Determine the acceleration of the truck during the collision.
40. BIO **Squid propulsion** *Lolliguncula brevis* squid use a form of jet propulsion to swim—they eject water out of jets that can point in different directions, allowing them to change direction quickly. When swimming at a speed of 0.15 m/s or greater, they can accelerate at $1.2\ \text{m/s}^2$. (a) Determine the time interval needed for a squid to increase its speed from 0.15 m/s to 0.45 m/s. (b) What other questions can you answer using the data?
41. **Dragster record on the desert** In 1977, Kitty O'Neil drove a hydrogen peroxide–powered rocket dragster for a record time interval (3.22 s) and final speed (663 km/h) on a 402-m-long Mojave Desert track. Determine her average acceleration during the race and the acceleration while stopping (it took about 20 s to stop). What assumptions did you make?
42. * Imagine that a sprinter accelerates from rest to a maximum speed of 10.8 m/s in 1.8 s. In what time interval will he finish the 100-m race if he keeps his speed constant at 10.8 m/s for the last part of the race? What assumptions did you make?
43. ** Two runners are running next to each other when one decides to accelerate at a constant rate of $a$. The second runner notices the acceleration after a short time interval $\Delta t$ when the distance between the runners is $\Delta x$. The second runner accelerates at the same acceleration. Represent their motions with a motion diagram and position-versus-time graph (both graph lines on the same set of axes). Use any of the representations to predict what will happen to the distance between the runners—will it stay $\Delta x$, increase, or decrease? Assume that the runners continue to have the same acceleration for the duration of the problem.
44. **Meteorite hits car** In 1992, a 14-kg meteorite struck a car in Peekskill, NY, leaving a 20-cm-deep dent in the trunk. (a) If the meteorite was moving at 500 m/s before striking the car, what was the magnitude of its acceleration while stopping? Indicate any assumptions you made. (b) What other questions can you answer using the data in the problem?
45. BIO **Froghopper jump** A spittlebug called the froghopper (*Philaenus spumarius*) is believed to be the best jumper in the animal world. It pushes off with muscular rear legs for 0.0010 s, reaching a speed of 4.0 m/s. Determine its acceleration during this launch and the distance that the froghopper moves while its legs are pushing.
46. **Tennis serve** The fastest server in women's tennis is Venus Williams, who recorded a serve of 130 mi/h (209 km/h) in 2007. If her racket pushed on the ball for a distance of 0.10 m, what was the average acceleration of the ball during her serve? What was the time interval for the racket-ball contact?
47. * **Shot from a cannon** In 1998, David "Cannonball" Smith set the distance record for being shot from a cannon (56.64 m). During a launch in the cannon's barrel, his speed increased from zero to 80 km/h in 0.40 s. While he was being stopped by the catching net, his speed decreased from 80 km/h to zero with an average acceleration of $180\ \text{m/s}^2$. What can you determine about Smith's flight using this information?
48. **Col. John Stapp's final sled run** Col. John Stapp led the U.S. Air Force Aero Medical Laboratory's research into the effects of higher accelerations. On Stapp's final sled run, the sled reached a speed of 284.4 m/s (632 mi/h) and then stopped with the aid of water brakes in 1.4 s. Stapp was barely conscious and lost his vision for several days but recovered. Determine his acceleration while stopping and the distance he traveled while stopping.
49. * Sprinter Usain Bolt reached a maximum speed of 11.2 m/s in 2.0 s while running the 100-m dash. (a) What was his acceleration? (b) What distance did he travel during this first 2.0 s of the race? (c) What assumptions did you make? (d) What time interval was needed to complete the race, assuming that he ran the last part of the race at his maximum speed? (e) What is the total time for the race? How certain are you of the number you calculated?
50. * Imagine that Usain Bolt can reach his maximum speed in 1.7 s. What should be his maximum speed in order to tie the 19.5-s record for the 200-m dash?
51. A bus is moving at a speed of 36 km/h. How far from a bus stop should the bus start to slow down so that the passengers feel comfortable (a comfortable acceleration is $1.2\ \text{m/s}^2$)?
52. * EST You want to estimate how fast your car accelerates. What information can you collect to answer this question? What assumptions do you need to make to do the calculation using the information?
53. * In your car, you covered 2.0 m during the first 1.0 s, 4.0 m during the second 1.0 s, 6.0 m during the third 1.0 s, and so forth. Was this motion at constant acceleration? Explain.
54. (a) Determine the acceleration of a car in which the velocity changes from $-10$ m/s to $-20$ m/s in 4.0 s. (b) Determine the car's acceleration if its velocity changes from $-20$ m/s to $-18$ m/s in 2.0 s. (c) Explain why the sign of the acceleration is different in (a) and (b).

## 1.7 Skills for analyzing situations involving motion

55. * Use the velocity-versus-time graph lines in **Figure P1.55** to determine the change in the position of each car from 0 s to 60 s. Represent the motion of each car mathematically as a function $x(t)$. Their initial positions are A (200 m) and B (−200 m).

**Figure P1.55**

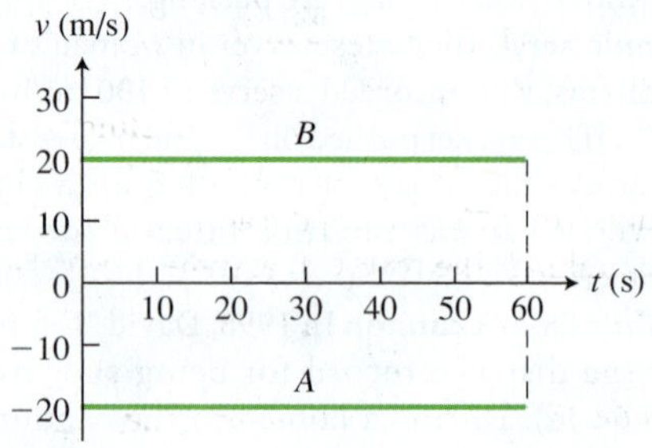


56. * An object moves so that its position changes in the following way: $x = 10\text{ m} - (4\text{ m/s})t$. (a) Describe all of the known quantities for this motion. (b) Invent a story for the motion. (c) Draw a position-versus-time graph, and use the graph to determine when the object reaches the origin of the reference frame. (e) Act out the motion.

57. ** An object moves so that its position changes in the following way: $x(t) = -100\text{ m} + (30\text{ m/s})t + (3.0\text{ m/s}^2)t^2$. (a) What kind of motion is this (constant velocity, constant acceleration, or changing acceleration)? (b) Describe all of the known quantities for this motion. (c) Invent a story for the motion. (d) Draw a velocity-versus-time graph, and use it to determine when the object stops. (e) Use equations to determine when and where it stops. Did you get the same answer using graphs and equations?

58. ** The position of an object changes according to the functions listed below. For each case, determine the known quantities concerning the motion, devise a story describing the motion consistent with the functions, and draw position-versus-time, velocity-versus-time, and acceleration-versus-time graphs: (a) $x(t) = 15.0\text{ m} - (-3.0\text{m/s}^2)t^2$; (b) $x(t) = 30.0\text{ m} - (1.0\text{ m/s})t$; and (c) $x = -10\text{ m}$.

59. * The positions of objects A and B with respect to Earth depend on time as follows: $x(t)_A = 10.0\text{ m} - (4.0\text{ m/s})t$; $x(t)_B = -12\text{ m} + (6\text{ m/s})t$. Represent their motions on a motion diagram and graphically (position-versus-time and velocity-versus-time graphs). Use the graphical representations to find where and when they will meet. Confirm the result with mathematics.

60. * Two cars on a straight road at time zero are beside each other. The first car, traveling at speed 30 m/s, is passing the second car, which is traveling at 24 m/s. Seeing a cow on the road ahead, the driver of each car starts to slow down at 6.0 m/s². Represent the motions of the cars mathematically and on a velocity-versus-time graph from the point of view of a pedestrian. Where is each car when it stops?

61. * The changing velocity of a car is represented in the velocity-versus-time graph shown in **Figure P1.61**. (a) Describe everything you can about the motion of the car using the graph. (b) What is the displacement of the car between times 10 s and 20 s? (c) What was the average speed of the car?

**Figure P1.61**

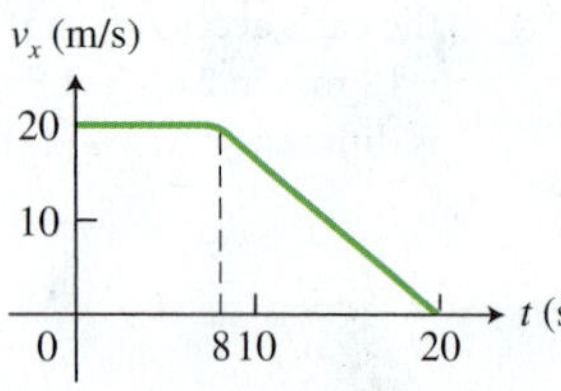


62. * The changing velocity of a car is represented in the velocity-versus-time graph shown in **Figure P1.62**. (a) Describe everything you can about the motion of the car using the graph. (b) What is the displacement of the car between times 0 s and 45 s? What is the path traveled? (c) What is the average speed of the car during all 70 s? What is the average velocity?

**Figure P1.62**

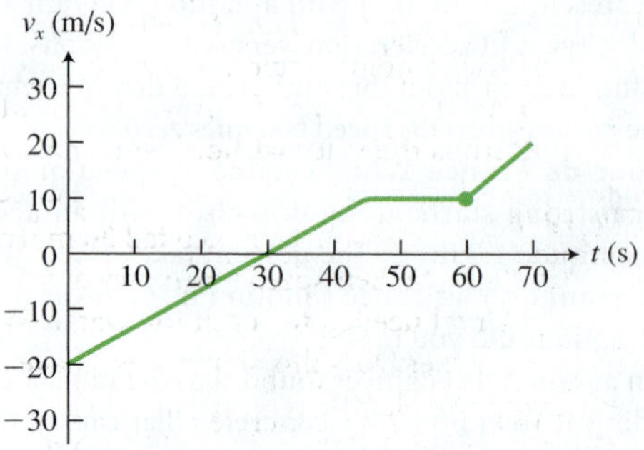


63. A diagram representing the motion of two cars is shown in **Figure P1.63**. The number near each dot indicates the clock reading in seconds when the car passes that location. (a) Indicate times when the cars have the same speed. (b) Indicate times when they have the same position.

**Figure P1.63**

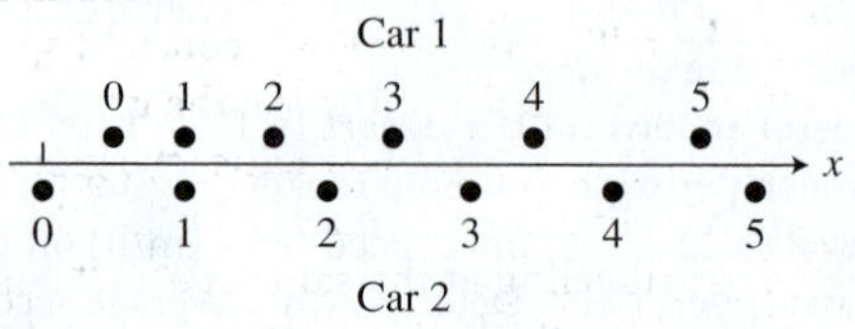


64. ** Solve the equations below for the unknown quantities and then describe a possible process that is consistent with the equations. There are many possibilities. The object is moving on an inclined surface. This is a two-part process.

Part I: $x_1 = 0 + (0)t_1 + (2.5\text{ m/s}^2)t_1^2$

Part II: $x_2 = x_1 + (20\text{ m/s})(0.40\text{ s}) + (1/2)a_{x2}(0.40\text{ s})^2$

## 1.8 and 1.9 Free fall and Tailgating: Putting it all together

65. You accidentally drop an eraser out the window of an apartment 15 m above the ground. (a) How long will it take for the eraser to reach the ground? (b) What speed will it have just before it reaches the ground? (c) If you multiply the time interval answer from (a) and the speed answer from (b), why is the result much more than 15 m?

66. What is the average speed of the eraser in the previous problem?

67. You throw a tennis ball straight upward. The initial speed is about 12 m/s. Say everything you can about the motion of the ball. Is 12 m/s a realistic speed for an object that you can throw with your hands?

68. While skydiving, your parachute opens and you slow from 50.0 m/s to 8.0 m/s in 0.80 s. Determine the distance you fall while the parachute is opening. Some people faint if they experience acceleration greater than 5 $g$ (5 times 9.8 m/s²). Will you feel faint? Explain and discuss simplifying assumptions inherent in your explanation.

69. After landing from your skydiving experience, you are so excited that you throw your helmet upward. The helmet rises 5.0 m above your hands. What was the initial speed of the helmet when it left your hands? How long was it moving from the time it left your hands until it returned?

70. You are standing on the rim of a canyon. You drop a rock and in 7.0 s hear the sound of it hitting the bottom. How deep is the canyon? What assumptions did you make? Examine how each assumption affects the answer. Does it lead to a larger or smaller depth than the calculated depth? (The speed of sound in air is about 340 m/s.)

71. You are doing an experiment to determine your reaction time. Your friend holds a ruler. You place your fingers near the sides of the lower part of the ruler without touching it. The friend drops the ruler without warning you. You catch the ruler after it falls 12.0 cm. What was your reaction time?

72. EST **Cliff divers** Divers in Acapulco fall 36 m from a cliff into the water. Estimate their speed when they enter the water and the time interval needed to reach the water. What assumption did you make? Does this assumption make the calculated speed larger or smaller than actual speed?

73. * Galileo dropped a light rock and a heavy rock from the Leaning Tower of Pisa, which is about 55 m high. Suppose that Galileo dropped one rock 0.50 s before the second rock. With what initial velocity should he drop the second rock so that it reaches the ground at the same time as the first rock?

74. A person holding a lunch bag is moving upward in a hot air balloon at a constant speed of 7.0 m/s. When the balloon is 24 m above the ground, she accidentally releases the bag. What is the speed of the bag just before it reaches the ground?

75. A parachutist falling vertically at a constant speed of 10 m/s drops a penknife when 20 m above the ground. What is the speed of the knife just before it reaches the ground?

76. * You are traveling in your car at 20 m/s a distance of 20 m behind a car traveling at the same speed. The driver of the other car slams on the brakes to stop for a pedestrian who is crossing the street. Will you hit the car? Your reaction time is 0.60 s. The maximum acceleration of each car is 9.0 $m/s^2$.

77. * You are driving a car behind another car. Both cars are moving at speed 80 km/h. What minimum distance behind the car in front should you drive so that you do not crash into the car's rear end if the driver of that car slams on the brakes? Indicate any assumptions you made.

78. A driver with a 0.80-s reaction time applies the brakes, causing the car to have 7.0-$m/s^2$ acceleration opposite the direction of motion. If the car is initially traveling at 21 m/s, how far does the car travel during the reaction time? How far does the car travel after the brakes are applied and while skidding to a stop?

79. ** Some people in a hotel are dropping water balloons from their open window onto the ground below. The balloons take 0.15 s to pass your 1.6-m-tall window. Where should security look for the raucous hotel guests? Indicate any assumptions that you made in your solution.

80. ** BIO EST **Avoiding injury from hockey puck** Hockey players wear protective helmets with facemasks. Why? Because the bone in the upper part of the cheek (the zygomatic bone) can fracture if the acceleration of a hockey puck due to its interaction with the bone exceeds 900 *g* for a time lasting 6.0 ms or longer. Suppose a player was not wearing a facemask. Is it likely that the acceleration of a hockey puck when hitting the bone would exceed these numbers? Use some reasonable numbers of your choice and estimate the puck's acceleration if hitting an unprotected zygomatic bone.

81. ** EST A bottle rocket burns for 1.6 s. After it stops burning, it continues moving up to a maximum height of 80 m above the place where it stopped burning. Estimate the acceleration of the rocket during launch. Indicate any assumptions made during your solution. Examine their effect.

82. * **Data from state driver's manual** The state driver's manual lists the reaction distances, braking distances, and total stopping distances for automobiles traveling at different initial speeds (**Table 1.12**). Use the data determine the driver's reaction time interval and the acceleration of the automobile while braking. The numbers assume dry surfaces for passenger vehicles.

**Table 1.12** Data from driver's manual.

| Speed (mi/h) | Reaction distance (m) | Braking distance (m) | Total stopping distance (m) |
|---|---|---|---|
| 20 | 7 | 7 | 14 |
| 40 | 13 | 32 | 45 |
| 60 | 20 | 91 | 111 |

83. ** EST Estimate the time interval needed to pass a semi-trailer truck on a highway. If you are on a two-lane highway, how far away from you must an approaching car be in order for you to safely pass the truck without colliding with the oncoming traffic? Indicate any assumptions used in your estimate.

84. * Car A is heading east at 30 m/s and Car B is heading west at 20 m/s. Suddenly, as they approach each other, they see a one-way bridge ahead. They are 100 m apart when they each apply the brakes. Car A's speed decreases at 7.0 m/s each second and Car B decreases at 9.0 m/s each second. Do the cars collide?

## Reading Passage Problems

BIO **Head injuries in sports** A research group at Dartmouth College has developed a Head Impact Telemetry (HIT) System that can be used to collect data about head accelerations during impacts on the playing field. The researchers observed 249,613 impacts from 423 football players at nine colleges and high schools and collected collision data from participants in other sports. The accelerations during most head impacts ($>89\%$) in helmeted sports caused head accelerations less than a magnitude of 400 $m/s^2$. However, a total of 11 concussions were diagnosed in players whose impacts caused accelerations between 600 and 1800 $m/s^2$, with most of the 11 over 1000 $m/s^2$.

85. Suppose that the magnitude of the head velocity change was 10 m/s. Which time interval below for the collision would be closest to producing a possible concussion with an acceleration of 1000 $m/s^2$?
(a) 1 s (b) 0.1 s (c) $10^{-2}$ s
(d) $10^{-3}$ s (e) $10^{-4}$ s

86. Using numbers from the previous problem, which answer below is closest to the average speed of the head while stopping?
(a) 50 m/s (b) 10 m/s (c) 5 m/s
(d) 0.5 m/s (e) 0.1 m/s

87. Suppose the average speed while stopping was 4 m/s (not necessarily the correct value) and the collision lasted 0.01 s. Which answer below is closest to the head's stopping distance (the distance it moves while stopping)?
(a) 0.04 m (b) 0.4 m (c) 4 m
(d) 0.02 m (e) 0.004 m

88. Use Eq. (1.7) and the numbers from Problem 85 to determine which stopping distance below is closest to that which would lead to a 1000 $m/s^2$ head acceleration.
(a) 0.005 m (b) 0.5 m (c) 0.1 m
(d) 0.01 m (e) 0.05 m

89. Choose from the list below the changes in the head impacts that would *reduce* the acceleration during the impact.
1. A shorter impact time interval
2. A longer impact time interval
3. A shorter stopping distance
4. A longer stopping distance
5. A smaller initial speed
6. A larger initial speed
(a) 1, 4, 6 (b) 1, 3, 5 (c) 1, 4, 5
(d) 2, 4, 5 (e) 2, 4, 6

**Sending rockets to observe X-ray sources** Before 1962, few astronomers believed that the universe contained celestial bodies that were hot enough to emit X-rays—about 10,000 times hotter than the surface of the Sun.

Because the atmosphere absorbs the X-rays produced by such sources, they can only be detected beyond Earth's atmosphere, 200 km or more above Earth's surface. Before satellites were available in the 1970s, scientists searched for X-ray sources by launching rockets (the first in 1962 from White Sands Missile Range in New Mexico) that contained detectors that could sample the skies for the short time interval that the rocket remained above the atmosphere—less than 10 min. Such a Terrier-Sandhawk rocket was flown on May 11, 1970 from the Kauai Test Range in Hawaii. Modern satellites can collect data continuously. Satellite observations and analysis have now identified several types of celestial bodies that emit X-rays, including X-ray pulsars in the constellations of Cygnus and Hercules, supernovae remnants, and quasars.

90. Detectors on rockets moving above Earth's atmosphere can detect X-ray sources, but similar detectors on Earth cannot because
(a) light from the Sun overwhelms the X-ray signals in the detectors.
(b) air in the atmosphere absorbs the X-rays before they reach Earth-based detectors.
(c) the rocket can see the X-ray sources more easily because it is nearer them.
(d) Earth is much heaver than a rocket, and hence the X-rays affect it less.

91. During fuel burn, the vertically launched Terrier-Sandhawk rocket had an acceleration of 300 $m/s^2$ (30 times free-fall acceleration—called 30 $g$). The fuel burned for 8 s. About how fast was the rocket moving at the end of the burn?
(a) 2400 m/s (b) 40 m/s (c) 240 $g$ (d) 4 $g$

92. Which answer below is closest to the height of the Terrier-Sandhawk rocket at the end of fuel burn?
(a) 20,000 m (b) 10,000 m (c) 1000 m (d) 300 m

93. Which number below is closest to the time interval after blast-off that the Terrier-Sandhawk rocket reached its maximum height?
(a) 19,000 s (b) 2400 s (c) 250 s (d) 10 s

94. Which number below is closest to the maximum height reached by the Terrier-Sandhawk rocket?
(a) 300,000 m (b) 200,000 m (c) 12,000 m (d) 9600 m

# Newtonian Mechanics

# 2

**Why do seat belts and air bags save lives?**

**If you stand on a bathroom scale in a moving elevator, does its reading change?**

**Can a parachutist survive a fall if the parachute does not open?**

Seat belts and air bags save about 250,000 lives worldwide every year by preventing a seated driver or passenger from flying forward into the hard-surfaced steering wheel or dashboard after a vehicle stops abruptly. Air bags combined with seat belts significantly reduce the risk of injury (belts alone by about 40% and belts with air bags by about 54%). How do seat belts and air bags provide this protection?

**Be sure you know how to:**

- Draw a motion diagram for a moving object (Section 1.2).
- Determine the direction of acceleration using a motion diagram (Section 1.6).
- Add vectors graphically and by components for one-dimensional motion (Section 1.2 and Mathematics Review appendix).

**In the last chapter,** we learned to *describe* motion—for example, to determine a car's acceleration when stopped abruptly during a collision. However, we did not discuss the causes of the acceleration. In this chapter, we will learn *why* an object has a particular acceleration. This knowledge will help us *explain* the motion of many objects: cars, car passengers, elevators, skydivers, and even rockets.

## 2.1 Describing and representing interactions

What causes objects to accelerate or maintain a constant velocity? Consider a simple experiment—standing on Rollerblades® on a horizontal floor. No matter how hard you swing your arms or legs you cannot start moving by yourself; you need to either push off the floor or have someone push or pull you. Physicists say that the floor or the other person *interacts* with you, thus changing your motion. Objects can interact directly, when they touch each other, or at a distance—a magnet attracts or repels another magnet without touching it.

### Choosing a system in a sketch of a process

We learned (in Chapter 1) that the first step in analyzing any process is sketching it. **Figure 2.1** shows a sketch of a car skidding to avoid a collision with a van. In this and later chapters we choose in the sketch one particular object for detailed analysis. We call this object the **system**. All other objects that are not part of the system can interact with it (touch it, pull it, and push it) and are in the system's **environment**. Interactions between the system object and objects in the environment are called **external interactions**.

**Figure 2.1** A sketch of a car skidding to avoid a collision with a van.

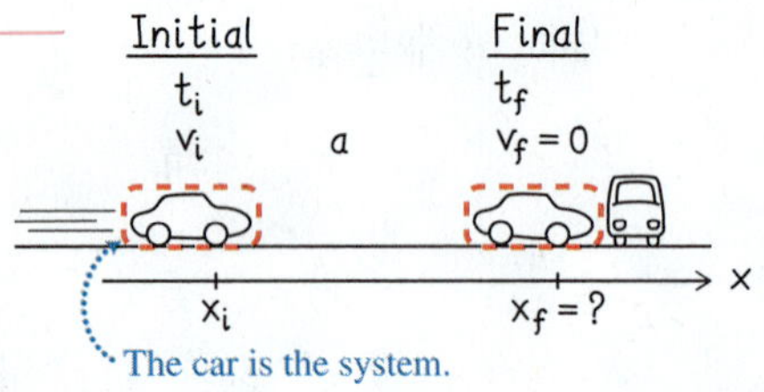


> **System** **A system is the object that we choose to analyze. Everything outside that system is called its environment and consists of objects that might interact with the system (touch, push, or pull it) and affect its motion through external interactions.**

On the sketch we make a light boundary (a closed dashed line) around the system object to emphasize the system choice (the skidding car in Figure 2.1). In this chapter systems consist of only one object. In later chapters systems can have more than one object. Sometimes, a single system object has parts—like the wheels on the car and its axles. The parts interact with each other. Since both parts are in the system, these are called **internal interactions**. In this chapter we will model an object like a car as a point-like object and ignore such internal interactions.

### Representing interactions

External interactions affect the motion or lack of motion of a system object. Consider holding a bowling ball in one hand and volleyball in the other. Each ball is considered as a system object (**Figure 2.2a**). What objects interact with each ball? Your hand pushes up hard to keep the bowling ball steady and much less to keep the volleyball steady (Figure 2.2b). We use an arrow to represent the upward push exerted by each hand on one of the balls. Notice that the arrow is longer for the interaction of the hand with the bowling ball than for the hand with the volleyball.

The arrow represents a "force" that is exerted by the hand on the ball. A **force** is a physical quantity that characterizes how hard and in what direction an external object pushes or pulls on the system object. The symbol for force is $\vec{F}$ with a subscript that identifies the external object that exerts the force and the system object on which the force is exerted. For example, the hand pushing on the bowling ball is represented as $\vec{F}_{\text{H on B}}$. The force that the hand exerts on the volleyball is $\vec{F}_{\text{H on V}}$. The arrow above the symbol indicates that force is a vector quantity with a magnitude *and* direction. The SI unit of a force is called a newton (N). When you hold a 100-g ball, you exert an upward force that is a little less than 1.0 N. We will devise a formal definition for force later in this chapter.

Do any other objects exert forces on the balls? Intuitively, we know that something must be pulling down to balance the upward force your hands exert on the balls. The word **gravity** represents the interaction of planet Earth with the ball. Earth pulls downward on an object toward Earth's center. Because of this interaction we need to include a second arrow representing the force that Earth exerts on the ball $\vec{F}_{\text{E on B}}$ (see Figure 2.2c). Intuitively, we know that the two arrows for each ball should be of the same length, since the ball is not moving anywhere.

Do other objects besides the hand and Earth interact with the ball? Air surrounds everything close to Earth. Does it push down or up on the balls? Let's hypothesize that the air does push down. If air pushes down, then our hands have to push up harder to balance the combined effect of the downward push of the air and the downward pull of Earth on the balls. Let's test the hypothesis that air pushes down on the balls.

**Figure 2.2** Representing external interactions (forces exerted on a system).

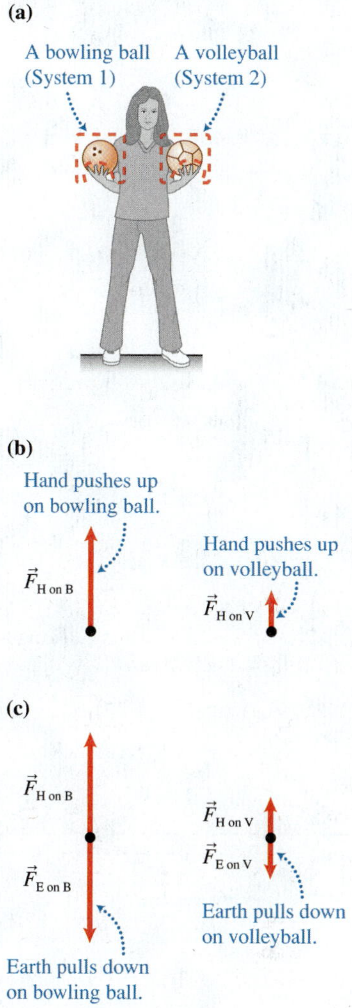


## Testing a hypothesis

To test a hypothesis in science means to first accept it as a true statement (even if we disagree with it); then design an experiment whose outcome we can predict using this hypothesis (a testing experiment); then compare the outcome of the experiment and the prediction; and, finally, make a preliminary judgment about the hypothesis. If the outcome matches the prediction, we can say that the hypothesis has not been disproved by the experiment. When this happens, our confidence in the hypothesis increases. If the outcome and prediction are inconsistent, we need to reconsider the hypothesis and possibly reject it.

To test the hypothesis that the air exerts a downward force on objects, we attach a ball to a spring and let it hang; the spring stretches (**Figure 2.3a**). Next we place the ball and spring inside a large jar that is connected to a vacuum pump and pump the air out of the inside of the jar. We predict that *if* the air

**Figure 2.3** A testing experiment to determine the effect of air on the ball.

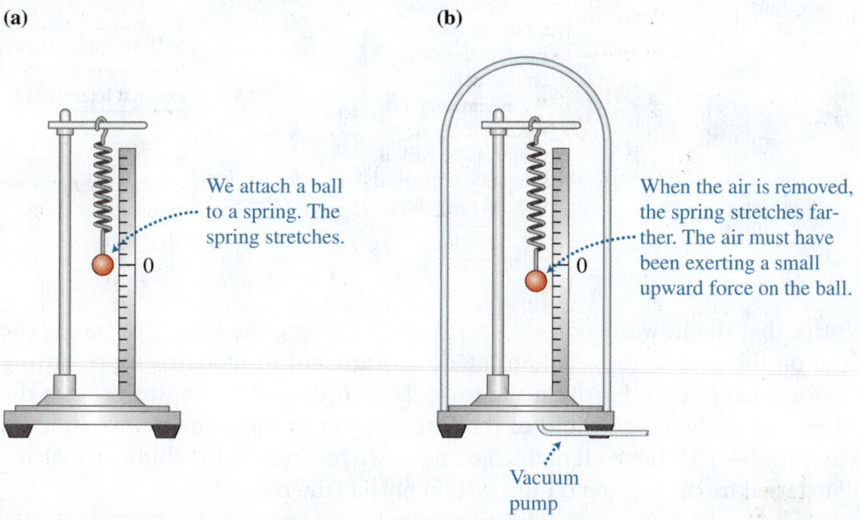

inside the jar pushes down on the ball (the hypothesis), *then* when we pump the air out of the jar, it should be easier to support the ball—the spring should stretch less (the prediction that follows from the hypothesis).

When we do the experiment, the outcome does not match the prediction—the spring actually stretches *slightly more* when the air is pumped out of the jar (Figure 2.3b). Evidently the air does not push down on the ball; instead, it helps support the ball by exerting an upward force on the ball. This outcome is surprising. When you study fluids, you will learn the mechanism by which air pushes up on objects.

## Reflect

Let's reflect on what we have done here. We formulated an initial hypothesis—air *pushes down* on objects. Then we designed an experiment whose outcome we could predict using the hypothesis—the ball on a spring in a vacuum jar. We used the hypothesis to make a prediction of the outcome of the testing experiment—the spring should stretch less in a vacuum. We then performed the experiment and found that something completely different happened. We revised our hypothesis—air *pushes up slightly* on objects. Note that air's upward push on the ball is very small. For many situations, the effect of air on objects can be ignored.

## Drawing force diagrams

A force diagram (sometimes called a free-body diagram) represents the forces that objects in a system's environment exert on it (see Figure 2.2c). We represent the system object by a dot to show that we model it as a point-like object. Arrows represent the forces. Unlike a motion diagram, a force diagram does not show us how a process changes with time; it shows us only the forces at a single instant. For processes in which no motion occurs, this makes no difference. But when motion does occur, we need to know if the force diagram is changing as the object moves.

Consider a rock dropped from above and sinking into sand, making a small crater. We construct a force diagram for shortly after the rock touches the sand but before it completely stops moving.

**REASONING SKILL** Constructing a force diagram

1. Sketch the situation (a rock sinking into sand).

2. Circle the system (the rock).

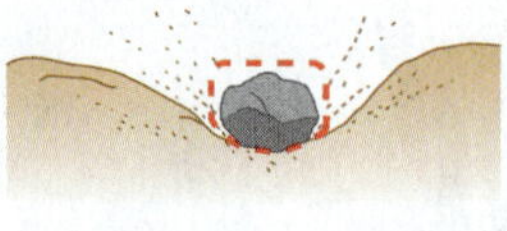

3. Identify external interactions:
- The sand pushes up on the rock.
- Earth pulls down on the rock.
- We assume that the force that the air exerts on the rock is small in comparison and can be ignored.

4. Place a dot at the side of the sketch, representing the system object.

5. Draw force arrows to represent the external interactions.

6. Label the forces with a subscript with two elements.

Notice that the upward-pointing arrow representing the force exerted by the sand on the rock is longer than the downward-pointing arrow representing the force exerted by Earth on the rock. The difference in lengths reflects the difference in the magnitudes of the forces. Later in the chapter we will learn why they have different lengths. For now, we just need to include arrows for all external forces exerted on the system object (the rock).

**TIP** Remember that on the force diagram, you only draw forces exerted *on* the system object. Do not draw forces that the system object exerts on other objects! For example, the rock exerts a force on the sand, but we do not include this force in the force diagram since the sand is not part of the system.

**CONCEPTUAL EXERCISE 2.1 Force diagram for a book**

Book A sits on a table with book B on top of it. Construct a force diagram for book A.

**Sketch and translate** We sketch the situation below. We choose book A as the system object. Notice that the dashed line around book A passes between the table and book A, and between book B and book A. It's important to be precise in the way you draw this line so that the separation between the system and the environment is clear. In this example, Earth, the table, and book B are external environmental objects that exert forces on book A.

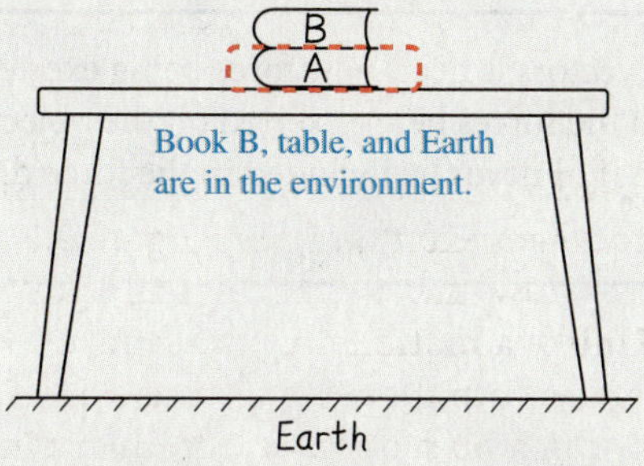

B
A
Book B, table, and Earth are in the environment.
Earth

**Simplify and diagram** Draw a force diagram for book A, which is represented by a dot. Two objects in the environment touch book A. The table pushes up on the bottom surface of the book, exerting a force $\vec{F}_{\text{T on A}}$, and book B pushes down on the top surface of book A, exerting force $\vec{F}_{\text{B on A}}$. In addition, Earth exerts a downward force on book A $\vec{F}_{\text{E on A}}$.

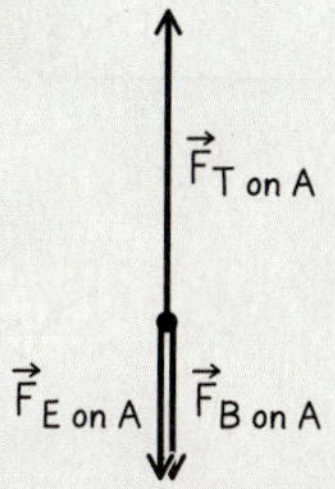


**Try it yourself:** Construct a force diagram for book A assuming that another book C is placed on top of book B.

*Answer:* The same three objects interact with book A. Earth exerts the same downward force on book A ($\vec{F}_{\text{E on A}}$). C does not directly touch A and exerts no force on A. However, C does push down on B, so B exerts a greater force on A ($\vec{F}_{\text{B on A}}$). Because the downward force of B on A is greater, the table exerts a greater upward force on book A ($\vec{F}_{\text{T on A}}$).

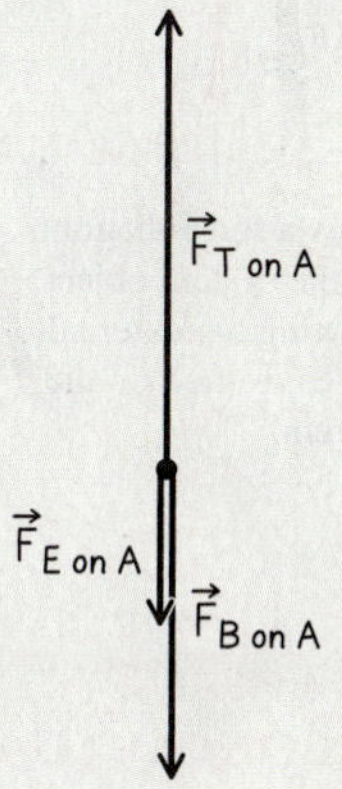


## Normal force

In the previous example, the force that the table exerts on book A and the force that book B exerts on book A are both perpendicular to their touching surfaces with A. Such perpendicular touching forces are called **normal forces.** Normal does not mean vertical, although in this example they were vertical forces. In the future, these forces will be labeled using the letter $N$ instead of $F$. Normal forces are *contact forces* (due to touching objects), as opposed to the force that Earth exerts on the book.

**Review Question 2.1** You slide toward the right at decreasing speed on a horizontal wooden floor. Choose yourself as the system and list the external objects that interact with and exert forces on you.

# 2.2 Adding and measuring forces

Most often, more than one environmental object exerts a force on a system object. How can we add them to find the total or net force exerted on the system object? In this chapter we restrict our attention to forces that are exerted and point along one axis. Consider the process of lifting a suitcase.

**Figure 2.4** The sum of the forces (the net force) exerted on the suitcase.

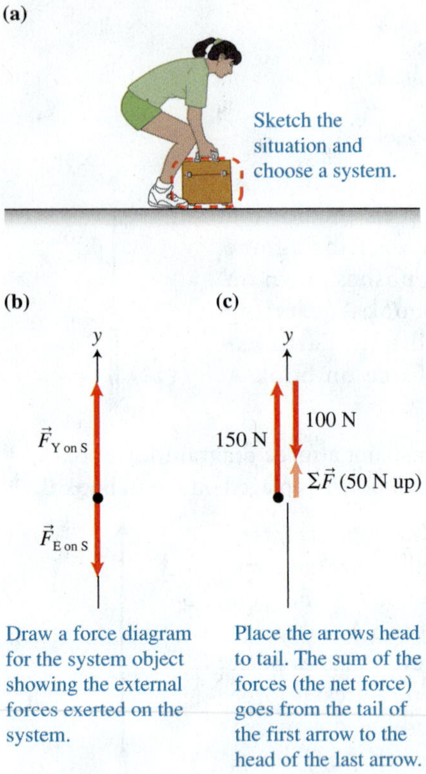


## Adding forces graphically

You lift a suitcase straight up (**Figure 2.4a**). Earth pulls down on the suitcase, exerting a force of magnitude $F_{\text{E on S}} = 100\text{ N}$, and you exert an upward force of magnitude $F_{\text{Y on S}} = 150\text{ N}$ (Figure 2.4b). What is the magnitude and direction of the total force exerted on the suitcase? The net effect of the two forces exerted along a vertical axis is the same as a 50-N force pointed straight up. Why? Remember that force is a vector. To add two vectors to find the sum $\Sigma\vec{F} = \vec{F}_{\text{E on S}} + \vec{F}_{\text{Y on S}}$, we place them head to tail (see Figure 2.4c) and draw the vector that goes from the tail of the first vector to the head of the second vector. This new vector is the sum vector, often called the resultant vector. In the case of forces it is called the **net force**. In our example, the net force points upward and its length is 50 N. If we assign the upward $y$-direction to be positive, then the sum of the forces has a $y$-component $\Sigma F_y = +50\text{ N}$. If the positive direction is down, the sum of the forces has a $y$-component $\Sigma F_y = -50\text{ N}$.

If several external objects in the environment exert forces on the system object, we still use vector addition to find the *sum* $\Sigma$ of the forces exerted on the object:

$$\Sigma\vec{F}_{\text{on O}} = \vec{F}_{1\text{ on O}} + \vec{F}_{2\text{ on O}} + \cdots + \vec{F}_{n\text{ on O}} \qquad (2.1)$$

**TIP** The sum of the force vectors is not a new force being exerted. Rather, it is the combined effect of all the forces being exerted on the object. Because of this, the resultant vector should never be included in the force diagram for that object.

**CONCEPTUAL EXERCISE 2.2 Measuring forces**

A very light plastic bag hangs from a light spring. The spring is not stretched. We place one golf ball into the bag and observe that the spring stretches to a new length. We add a second ball and observe that the spring stretches twice as far. We add a third ball and observe that the spring stretches three times as far. How can we use this experiment to develop a method to measure the magnitude of a force?

**Sketch and translate** First, draw sketches to represent the four situations as shown below. On each sketch, carefully show the change in the length of the spring. Choose the bag with the golf balls as the system and analyze the forces exerted in each case.

**Simplify and diagram** Assume that Earth exerts the same force on each ball $\vec{F}_{\text{E on 1B}}$ independent of the presence of other balls. Thus, the total force exerted by Earth on the three-ball system is three times greater than the force exerted on the one-ball system. Draw a force diagram for each case. Assume that in each case the spring exerts a force on the system $\vec{F}_{\text{S on \#B}}$ that is equal in magnitude and opposite in direction to the force that Earth exerts on the system $\vec{F}_{\text{E on \#B}}$ so that the sum of the forces exerted on the system with a number # of golf balls is zero.

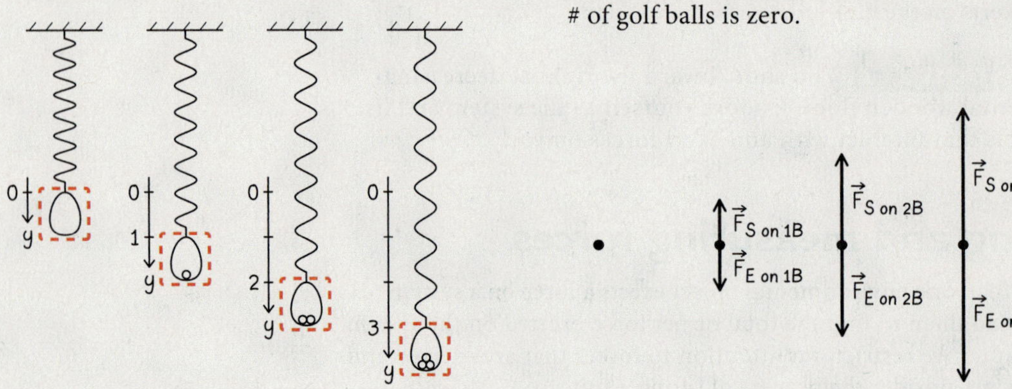

**Try it yourself:** Represent the relation between the force that the spring exerts on the bag with a number # of golf balls ($F_{S\text{ on }\#B}$) and spring stretch ($y$) with an $F$-versus-$y$ graph. Draw a trendline.

*Answer:* Based on the graph's trendline, we see that the spring elongates until the force it exerts on the system object balances the force that Earth exerts on it.

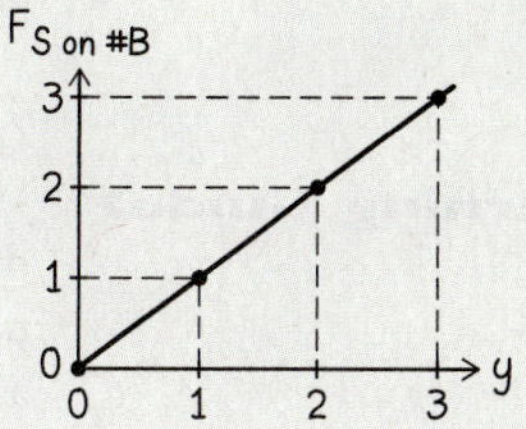


## Measuring force magnitudes

Force is a vector physical quantity with a magnitude and a direction. In the next conceptual exercise, we develop a method for determining the magnitude of a force.

Conceptual Exercise 2.2 provides us with one method to measure an unknown force that an object exerts on a system. We calibrate a spring in terms of some standard force, such as Earth's pull on one or more golf balls. Then if some unknown force is exerted on a system object, we can use the spring to exert a balancing force on that object. The unknown force is equal in magnitude to the force exerted by the spring and opposite in direction. In this case, we would be measuring force in units equal to Earth's pull on a golf ball. We could use any spring to balance a known standard force (1 N or approximately the force that Earth exerts on a 100-g object) and then calibrate this spring in newtons by placing marks at equal stretch distances as we pull on its end with increasing force. We thus build a spring scale—a simple instrument to measure forces (**Figure 2.5**).

**Figure 2.5** A spring scale.

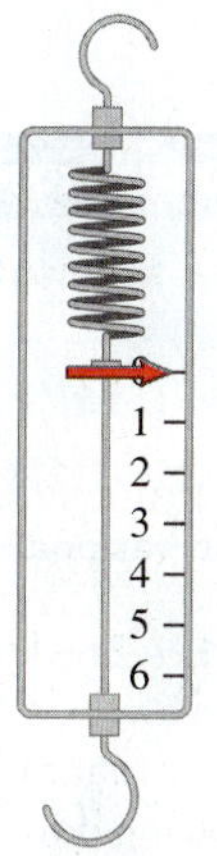


**TIP** In physics, force is a physical quantity that characterizes an interaction between two objects—its direction and magnitude. For a force to exist there must be two objects that interact, just like a hug requires the interaction of two people. Force does not reside in an object. However, in everyday language the idea that force resides in an object remains very strong; people say, "The truck's force caused a lot of damage to the telephone pole." We will be careful in this book to always identify the two interacting objects when speaking about any force. Remember, if you are thinking about a force that is exerted on a moving object and cannot find another object that interacts with it, then you are thinking of something else, not force.

**Review Question 2.2** A book bag hanging from a spring scale is partially supported by a platform scale. The platform scale reads about 36 units of force and the spring scale reads about 28 units of force. What is the magnitude of the force that Earth exerts on the bag? Explain.

## 2.3 Conceptual relationship between force and motion

**Active Learning Guide›**

When we drew a force diagram for a ball held by a person, we intuitively drew the forces exerted on the ball as being equal in magnitude. What if the person throws the ball upward? Or slowly lifts it upward? Or if she catches a ball falling from above? Would she still need to exert a force on the ball of magnitude equal to that Earth exerts on the ball? In other words, is there a relationship between the forces that are exerted on an object and the way the object moves?

Consider the three simple experiments in Observational Experiment **Table 2.1** involving a bowling ball (the system) rolling on a smooth surface. Is there is a pattern between its motion diagram and the force diagram?

**OBSERVATIONAL EXPERIMENT TABLE**

**2.1 How are motion and forces related?**

VIDEO 2.1

| Observational experiment | Analysis: Motion diagram | Analysis: Force diagrams for first and third positions |
|---|---|---|
| **Experiment 1.** A bowling ball *B* rolls on a very hard, smooth surface *S* without slowing down. 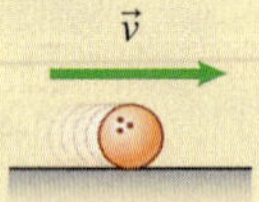  | 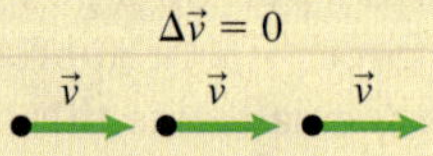  | 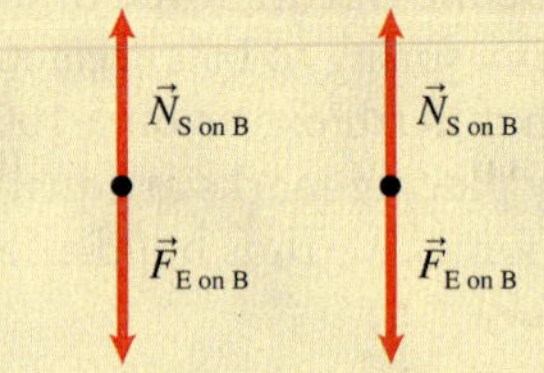  |
| **Experiment 2.** A ruler *R* lightly pushes the rolling bowling ball opposite the ball's direction of motion. The ball continues to move in the same direction, but slows down. 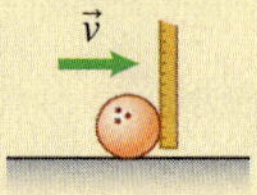  | 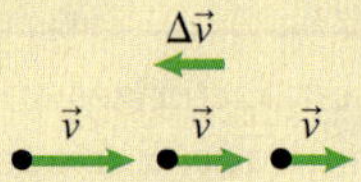  | 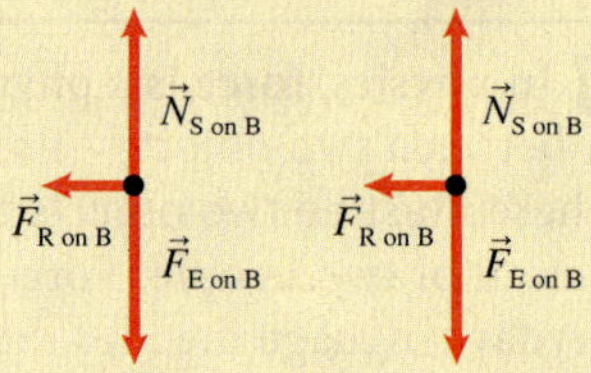  |
| **Experiment 3.** A ruler *R* lightly pushes the rolling bowling ball in the direction of its motion.   | 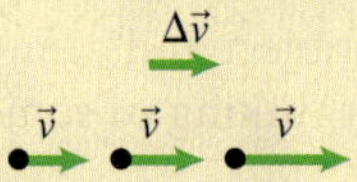  | 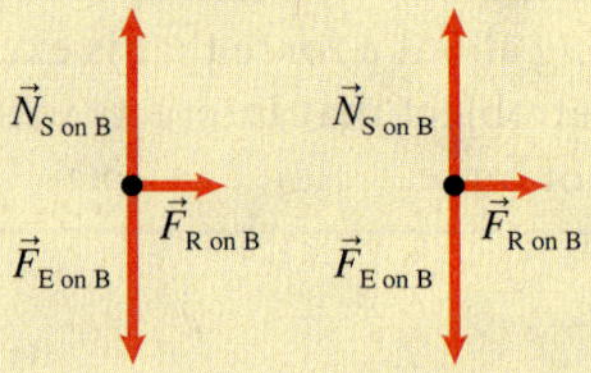  |

**Pattern**

- In all the experiments, the vertical forces add to zero and cancel each other. We consider only forces exerted on the ball in the horizontal direction.
- In the first experiment, the sum of the forces exerted on the ball is zero; the ball's velocity remains constant.
- In the second and third experiments, when the ruler pushes the ball, the velocity change arrow ($\Delta\vec{v}$ arrow) points in the same direction as the sum of the forces.

**Summary:** The $\Delta\vec{v}$ arrow in all experiments is in the same direction as the sum of the forces. Notice that there is no pattern relating the *direction* of the velocity $\vec{v}$ to the direction of the sum of the forces. In Experiment 2, the velocity and the sum of the forces are in opposite directions, but in Experiment 3, they are in the same direction.

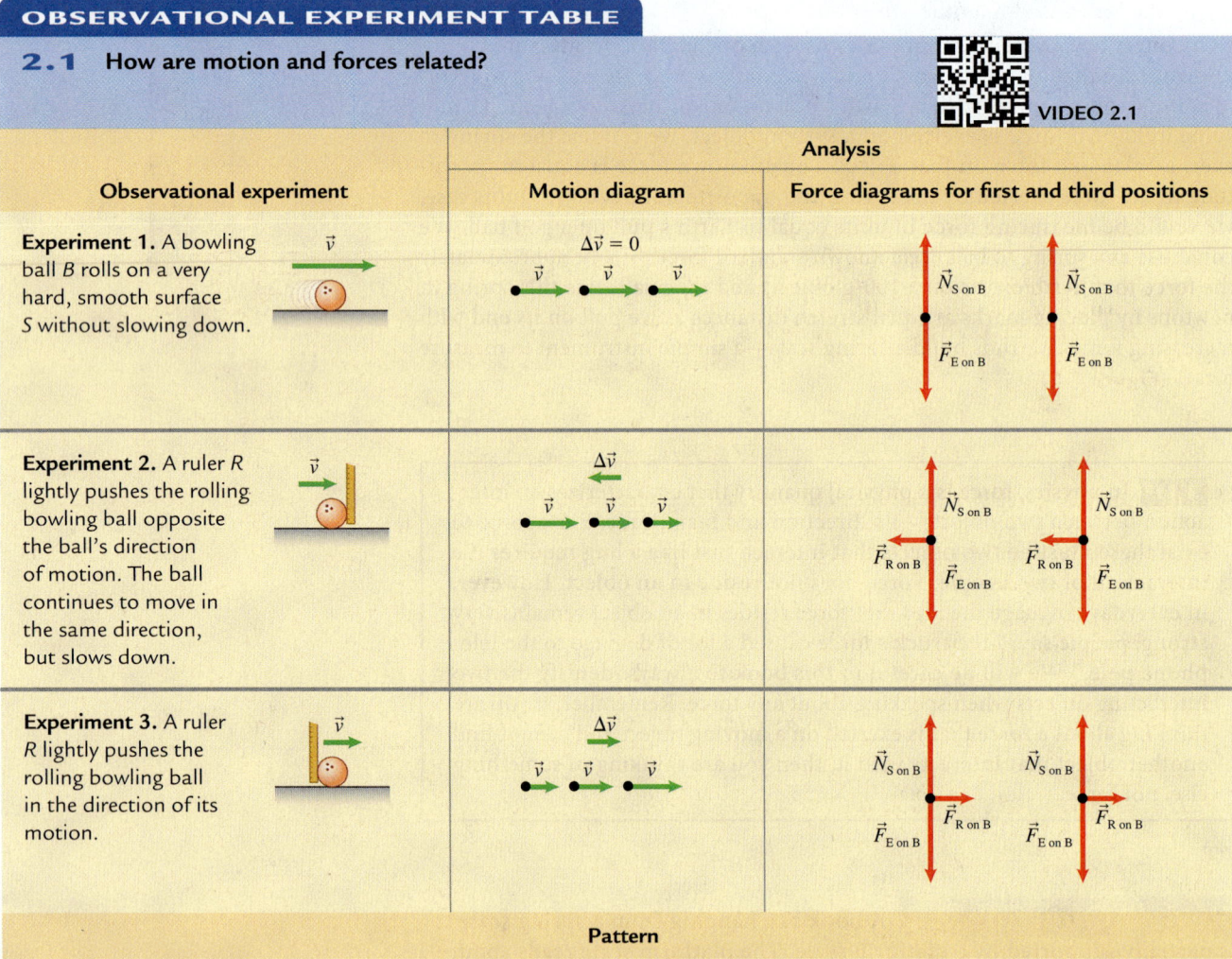

In each of the experiments in Table 2.1, the $\Delta\vec{v}$ arrow for a system object and the sum of the forces $\Sigma\vec{F}$ that external objects exert on that object are in the same direction. That is one idea. A second idea is that we often observe that the $\vec{v}$ arrow for a system object (the direction the object is moving) is in the same direction as the sum of the forces exerted on it. For example, a grocery cart moves in the direction the shopper pushes it and a soccer ball moves in the direction the player kicks it. We should test both ideas.

## Testing possible relationships between force and motion

We have two possible ideas that relate motion and force:

1. An object's velocity $\vec{v}$ always points in the direction of the sum of the forces $\Sigma\vec{F}$ that other objects exert on it.
2. An object's velocity *change* $\Delta\vec{v}$ always points in the direction of the sum of the forces $\Sigma\vec{F}$ that other objects exert on it.

To test these two relationships, we use each to predict the outcome of the experiments in Testing Experiment **Table 2.2**. Then we perform the experiments and compare the outcomes with the predictions. From this comparison, we decide if we can reject one or both of the relationships.

**TESTING EXPERIMENT TABLE**

**2.2 Testing ideas of how velocity and the sum of the forces $\Sigma\vec{F}$ are related.**

VIDEO 2.2

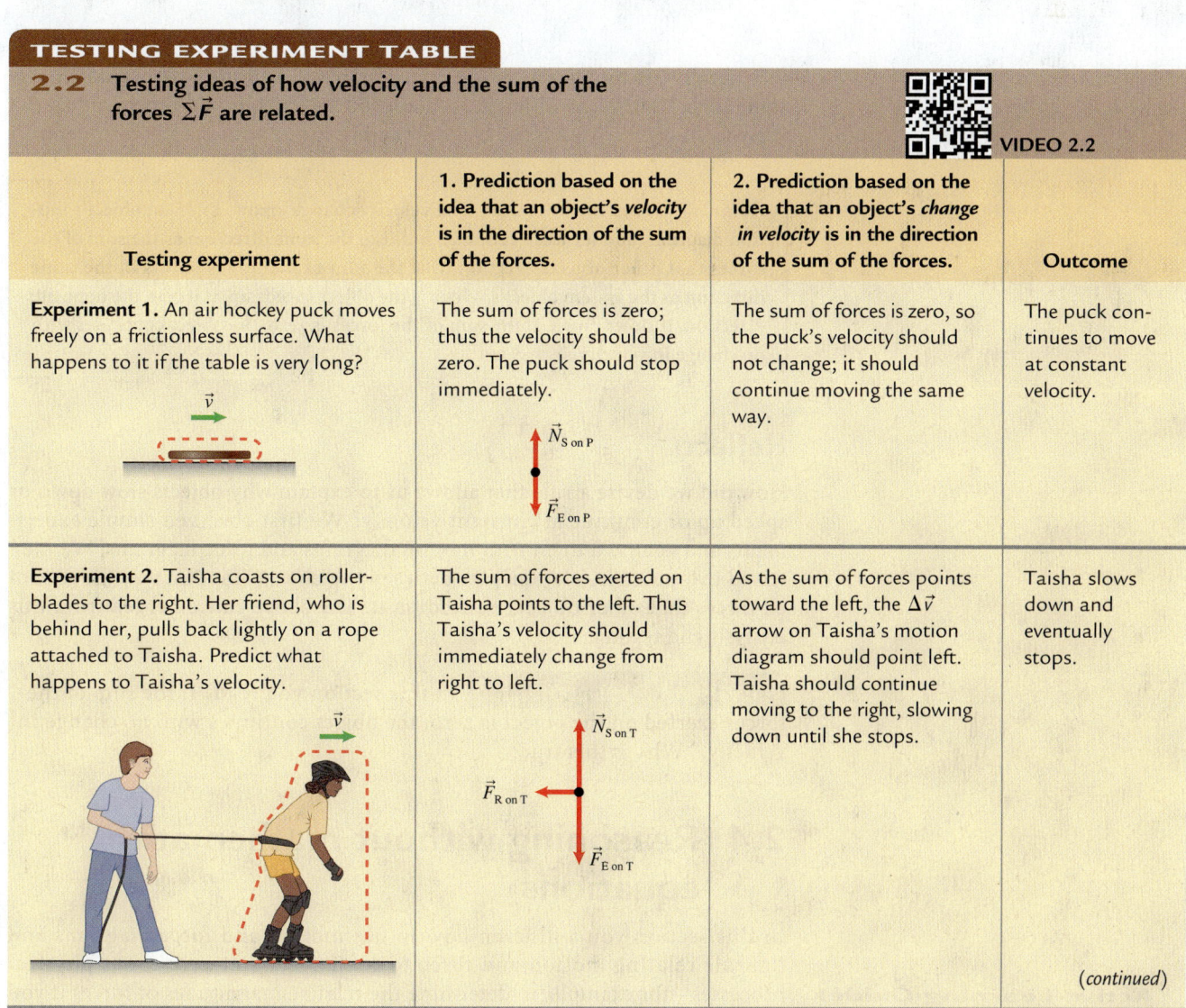

| Testing experiment | 1. Prediction based on the idea that an object's *velocity* is in the direction of the sum of the forces. | 2. Prediction based on the idea that an object's *change in velocity* is in the direction of the sum of the forces. | Outcome |
|---|---|---|---|
| **Experiment 1.** An air hockey puck moves freely on a frictionless surface. What happens to it if the table is very long? 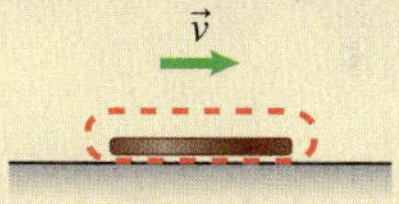  | The sum of forces is zero; thus the velocity should be zero. The puck should stop immediately.   | The sum of forces is zero, so the puck's velocity should not change; it should continue moving the same way. | The puck continues to move at constant velocity. |
| **Experiment 2.** Taisha coasts on rollerblades to the right. Her friend, who is behind her, pulls back lightly on a rope attached to Taisha. Predict what happens to Taisha's velocity. 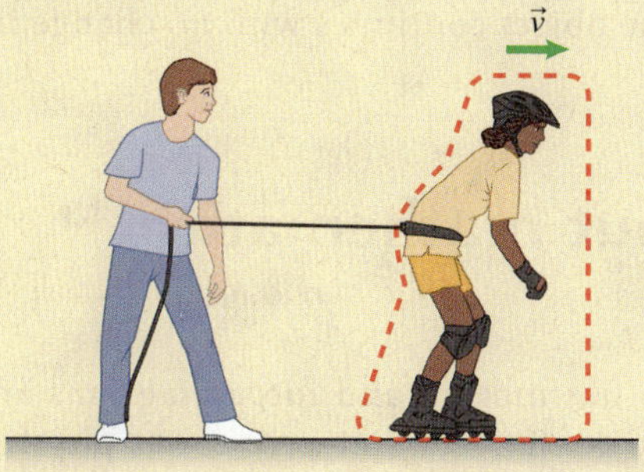  | The sum of forces exerted on Taisha points to the left. Thus Taisha's velocity should immediately change from right to left. 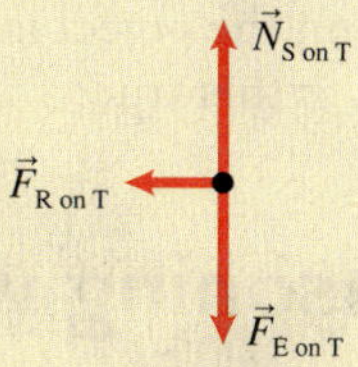  | As the sum of forces points toward the left, the $\Delta\vec{v}$ arrow on Taisha's motion diagram should point left. Taisha should continue moving to the right, slowing down until she stops. | Taisha slows down and eventually stops. |

(*continued*)

| Testing experiment | Prediction 1 | Prediction 2 | Outcome |
|---|---|---|---|
| **Experiment 3.** You throw a ball upward. What happens to the ball after it leaves your hand? 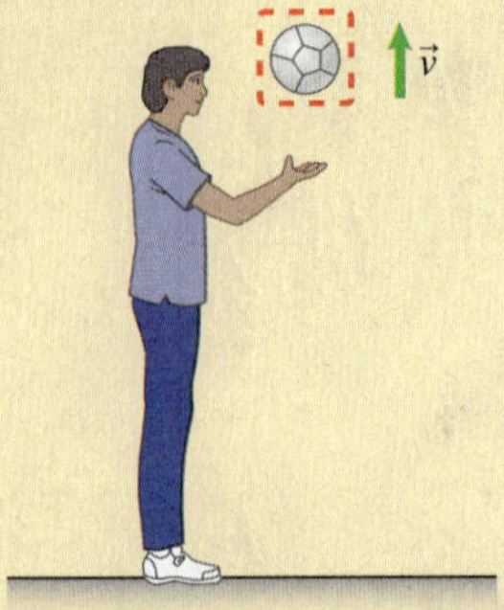  | The only force exerted on the ball after it leaves your hand points down. Thus the ball should immediately begin moving downward after you release it. 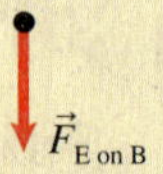  | The only force exerted on the ball after it leaves your hand points down. The $\Delta\vec{v}$ arrow on the motion diagram should point down, and the ball should slow down until it stops and then start moving back down at increasing speed. | The ball moves up at decreasing speed and then reverses direction and starts moving downward. |

**Conclusion**

- All outcomes contradict the predictions based on idea 1—we can reject it.
- All outcomes are consistent with the predictions based on idea 2. This does not necessarily mean it is true, but it does mean our confidence in the idea increases.

Recall that the direction of the $\Delta\vec{v}$ arrow in a motion diagram is in the same direction as the object's acceleration $\vec{a}$. Thus, based on this idea and these testing experiments, we can now accept idea 2 with greater confidence.

**Relating forces and motion** **The velocity change arrow $\Delta\vec{v}$ in an object's motion diagram (and its acceleration $\vec{a}$) point in the same direction as the sum of the forces that other objects exert on it. If the sum of the forces points in the same direction as the system object's velocity, the object speeds up; if it is in the opposite direction, it slows down. If the sum of the forces is zero, the object continues with no change in velocity.**

## Reflect

How did we devise a rule that allows us to explain why objects slow down or speed up or continue at constant velocity? We first observed simple experiments and analyzed them with motion diagrams and force diagrams. We then tested two possible relationships between the objects' motion and the sum of all forces that other objects exerted on it. The above rule emerged from this analysis and testing.

**Review Question 2.3** In this section you read "If the sum of the forces exerted on the object is zero, the object continues with no change in velocity." Why is this true?

## 2.4 Reasoning without mathematical equations

In this section you will learn how to use motion and force diagrams and the rule relating motion and force to reason qualitatively about physical processes—for example, to determine the relative magnitudes of forces if you

Active Learning Guide›

have information about motion or to estimate velocity changes if you have information about forces. The key here is to make sure that the two representations (the force diagrams and the motion diagrams) are consistent.

In the next conceptual exercise, we use information about the forces that external objects exert on a woman to answer a question about her motion. In the Try it Yourself question we reverse the process—we use known information about the motion to answer a question about an unknown force. Try to answer the questions yourself before looking at the solutions.

### CONCEPTUAL EXERCISE 2.3

**Diagram Jeopardy**

The force diagram shown here describes the forces that external objects exert on a woman (in this scenario, the force diagram does not change with time). Describe three different types of motion that are consistent with the force diagram.

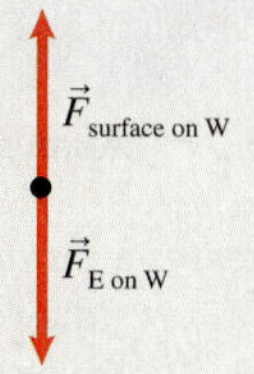


**Sketch and translate** Two equal-magnitude oppositely directed forces are being exerted on the woman ($\Sigma\vec{F} = 0$). Thus, a motion diagram for the woman must have a zero velocity change ($\Delta\vec{v} = 0$).

**Simplify and diagram** Three possible motions consistent with this idea are shown at the right.

1. She stands at rest on a horizontal surface.

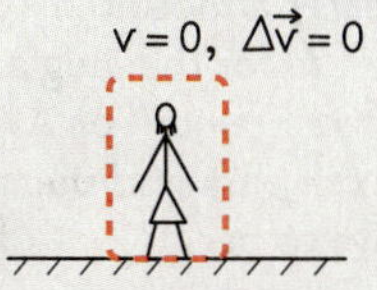


2. She glides at constant velocity on rollerblades on a smooth horizontal surface.

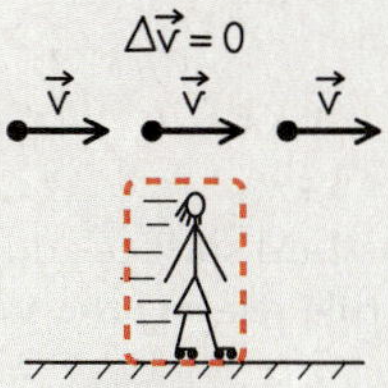


3. She stands on the floor of an elevator that moves up or down at constant velocity.

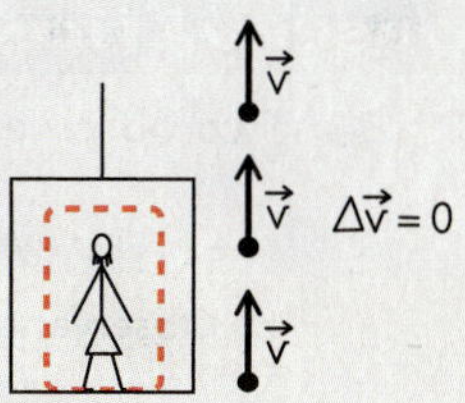


Note that in all three of the above, the velocity change arrow is zero. This is consistent with the sum of the forces being zero.

**Try it yourself:** Suppose that the elevator described above was moving up at decreasing speed instead of at constant speed. How then would the force diagram be different?

*Answer:* A velocity change $\Delta\vec{v}$ arrow for her motion would now point down opposite the direction of her velocity. Thus, the sum of the forces $\Sigma\vec{F}$ that other objects exert on her must also point down. This means that the magnitude of the upward force $\vec{F}_{\text{S on W}}$ that the elevator floor (surface) exerts on her must now be less than the magnitude of the downward force $\vec{F}_{\text{E on W}}$ that Earth exerts on her ($F_{\text{S on W}} < F_{\text{E on W}}$).

**Review Question 2.4** An elevator in a tall office building moves downward at constant speed. How does the magnitude of the upward force exerted by the cable on the elevator $\vec{F}_{\text{C on El}}$ compare to the magnitude of the downward force exerted by Earth on the elevator $\vec{F}_{\text{E on El}}$? Explain your reasoning.

‹Active Learning Guide

## 2.5 Inertial reference frames and Newton's first law

Our description of the motion of an object depends on the observer's reference frame. However, in this chapter we have tacitly assumed that all observers were standing on Earth's surface. For example, in Section 2.3, we analyzed several experiments and concluded that if the forces exerted on one object by other objects add to zero, then the chosen object moves at constant velocity. Are there any observers who will see a chosen object moving with changing velocity even though the sum of the forces exerted on the object appears to be zero?

## Inertial reference frames

In Observational Experiment **Table 2.3**, we consider two different observers analyzing the same situation.

**OBSERVATIONAL EXPERIMENT TABLE**

**2.3 Two observers watch the same coffee mug.**

VIDEO 2.3

| Observational experiment | Analysis done by each observer |
|---|---|
| **Experiment 1.** Observer 1 is slouched down in the passenger seat of a car and cannot see outside the car. Suddenly he observes a coffee mug sliding toward him from the dashboard. | Observer 1 creates a motion diagram and a force diagram for the mug as he observes it. On the motion diagram, increasingly longer $\vec{v}$ arrows indicate that the mug's speed changes from zero to nonzero as seen by observer 1 even though no external object is exerting a force on it in that direction. 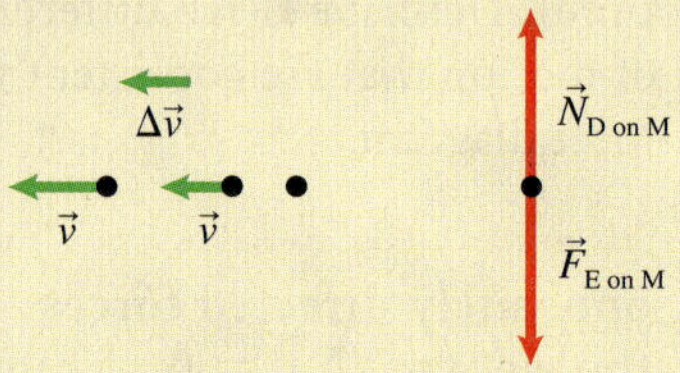 |
| **Experiment 2.** Observer 2 stands on the ground beside the car. She observes that the car starts moving forward at increasing speed and that the mug remains stationary with respect to her. | Observer 2 creates a motion diagram and force diagram for the mug as she observes it. There are no $\vec{v}$ or $\Delta\vec{v}$ arrows on the diagram and the mug is at rest relative to her. 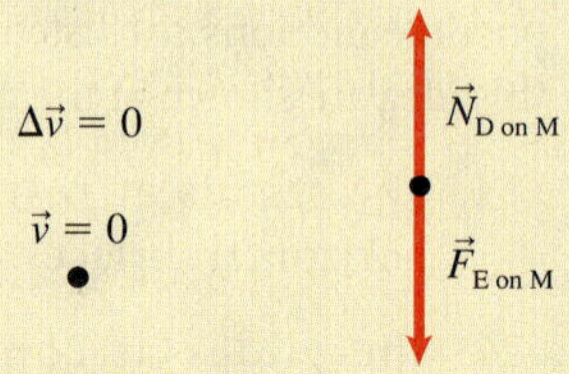 |

**Pattern**

**Observer 1:** The forces exerted on the mug by Earth and by the dashboard surface add to zero. But the velocity of the mug increases as it slides off the dashboard. This is inconsistent with the rule relating the sum of the forces and the change in velocity.
**Observer 2:** The forces exerted on the mug by Earth and by the dashboard surface add to zero. Thus the velocity of the mug should not change, and it does not. This is consistent with the rule relating the sum of the forces and the change in velocity.

Observer 2 in Table 2.3 can account for what is happening using the rule relating the sum of the forces and changing velocity, but observer 1 cannot. For observer 1, the mug's velocity changes for no apparent reason.

Similarly, a passenger on a train (observer 1) might suddenly see her laptop computer start to slide forward off her lap. A person on the platform (observer 2) can explain this event using the rule we developed. The train's velocity started decreasing as it approached the station, but the computer continued forward at constant velocity.

It appears that the applicability of the rule depends on the reference frame of the observer. Observers (like observer 2) who *can* explain the behavior of the mug and the computer by using the rule relating the sum of the forces and changing velocity are said to be observers in **inertial reference**

**frames**. Those (like observer 1) who *cannot* explain the behavior of the mug and the computer using the rule are said to be observers in **noninertial reference frames.**

**Inertial reference frame** **An inertial reference frame is one in which an observer sees that the velocity of the system object does not change if no other objects exert forces on it or if the sum of all forces exerted on the system object is zero. For observers in noninertial reference frames, the velocity of the system object can change even though the sum of forces exerted on it is zero.**

A passenger in a car or train that is speeding up or slowing down with respect to Earth is an observer in a noninertial reference frame. When you are in a car that abruptly stops, your body jerks forward—yet nothing is pushing you forward. When you are in an airplane taking off, you feel pushed back into your seat, even though nothing is pushing you in that direction. In these examples, you are an observer in a noninertial reference frame. Observers in inertial reference frames can explain the changes in velocity of objects by considering the forces exerted on them by other objects. Observers in noninertial reference frames cannot. From now on, we will always analyze phenomena from the point of view of observers in inertial reference frames. This idea is summarized by Isaac Newton's first law.

**Isaac Newton.** Isaac Newton (1643–1727) invented differential and integral calculus, formulated the law of universal gravitation, developed a new theory of light, and put together the ideas for his three laws of motion. His work on mechanics was presented in the book entitled *Philosophiae Naturalis Principia Mathematica (Mathematical Principles of Natural Philosophy).*

**Newton's first law of motion** **For an observer in an inertial reference frame when no other objects exert forces on a system object or when the forces exerted on the system object add to zero, the object continues moving at constant velocity (including remaining at rest). Inertia is the phenomenon in which a system object continues to move at constant velocity when the sum of the forces exerted on it by other objects is zero.**

Physicists have analyzed the motion of thousands of objects from the point of view of observers in inertial reference frames and found no contradictions to the rule. Newton's first law of motion limits the reference frames with respect to which the other laws that you will learn in this chapter are valid—these other laws work only for the observers in inertial reference frames.

**Review Question 2.5** What is the main difference between inertial and noninertial reference frames? Give an example.

## 2.6 Newton's second law

Our conceptual analyses in Sections 2.3 and 2.4 indicated that an object does not change velocity and does not accelerate when the sum of all forces exerted on it is zero. We also learned that an object's velocity change and acceleration are in the same direction as the sum of the forces that other objects exert on it. In this section we will learn how to predict the magnitude of an object's acceleration if we know the forces exerted on it. The experiments in Observational Experiment **Table 2.4** will help us construct this quantitative relationship.

## OBSERVATIONAL EXPERIMENT TABLE

### 2.4 Forces and resulting acceleration.

VIDEO 2.4

| Observational experiment | Analysis |
|---|---|
| **Experiment 1.** A cart starts at rest on a low-friction horizontal track. A force probe continuously exerts one unit of force in the positive direction. The same experiment is repeated five times. Each time, the force probe exerts one more unit of force on the cart (up to five units, shown in the last diagram). A computer records the value of the force, and a motion detector on the track records the cart's speed and acceleration. | Using this information, we create velocity-versus-time and acceleration-versus-time graphs for two of the five different magnitudes of force. Note that the greater the force, the greater the acceleration. |

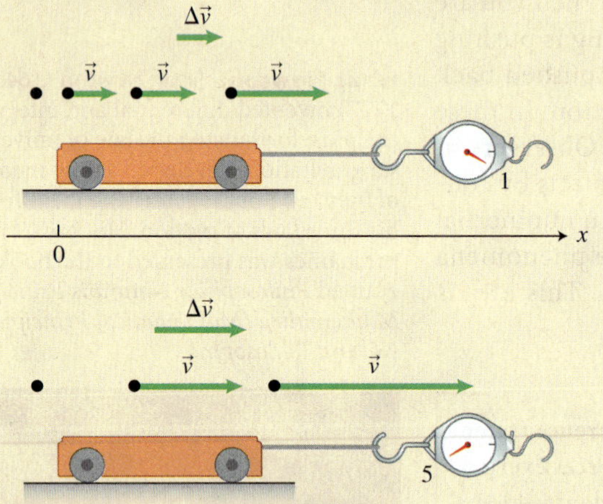


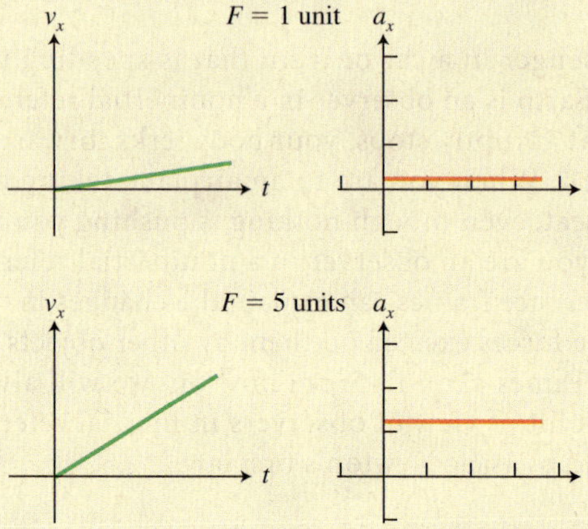


| Observational experiment | Analysis |
|---|---|
| **Experiment 2.** We repeat the same five experiments, only this time the cart is moving in the positive direction, and the probe pulls back on the cart in the negative direction so that the cart slows down. | We create velocity-versus-time and acceleration-versus-time graphs for the cart when forces of two different magnitudes oppose the cart's motion. |

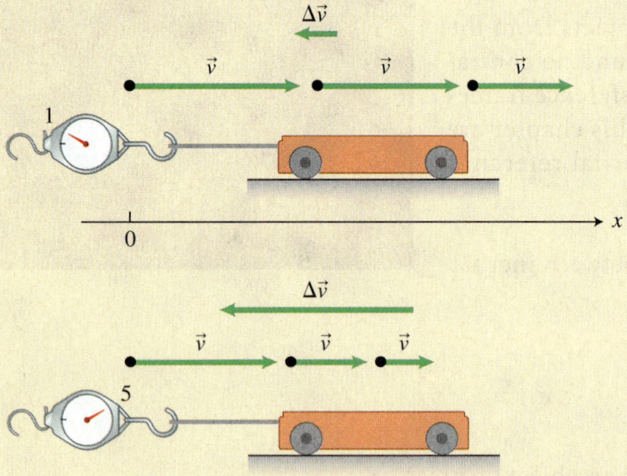


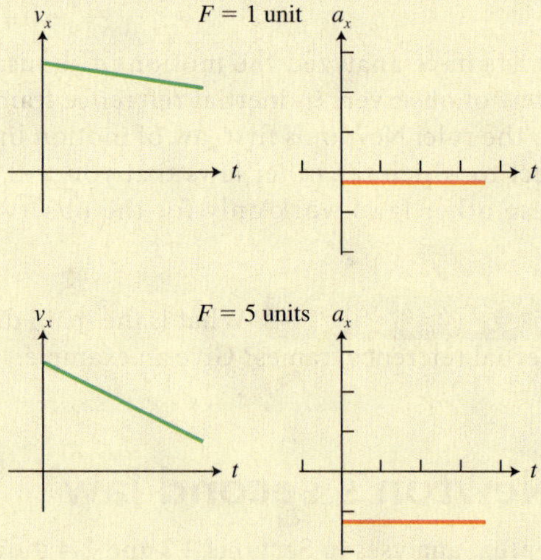


**Pattern**

- When the sum of the forces exerted on the cart is constant, its acceleration is constant—the cart's speed increases at a constant rate.
- Plotting the acceleration-versus-force using the five positive and five negative values of the force, we obtain the graph at the right.
- The acceleration is directly proportional to the force exerted by the force probe (in this case it is the sum of all forces) and points in the direction of the force.

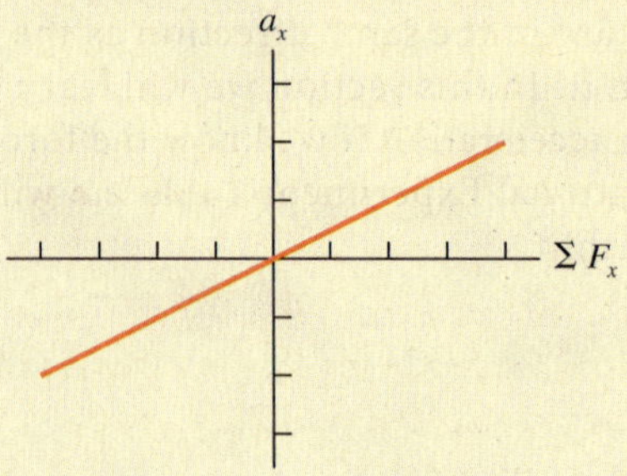

The outcome of these experiments expressed mathematically is as follows:

$$\vec{a} \propto \Sigma \vec{F} \qquad (2.1)$$

where $\Sigma \vec{F}$ is the sum of all the forces that other objects exert on the object (not an additional force), and $\vec{a}$ is the object's acceleration. The symbol $\propto$ means "is proportional to." In other words, if the sum of the forces doubles, then the acceleration doubles. When the sum of the forces is zero, the acceleration is zero. When the sum of the forces exerted on an object is constant, the object's resulting acceleration (not velocity) is constant.

**Figure 2.6** A strong man pulls a bus.

## Mass, another physical quantity

Do other physical quantities affect acceleration? Note that the strong man shown in **Figure 2.6** can only cause a very small acceleration while pulling the bus but could easily cause a large acceleration when pulling a small wagon. The amount of matter being pulled must affect the acceleration.

Let's perform another experiment to find the quantitative effect of the amount of *matter* being pulled. We use the force probe to pull one cart, then two carts stacked on top of each other, and then three and four carts on top of each other. In each case, the force probe exerts the same force on the carts, regardless of how many carts are being pulled. The experiment is summarized in Observation Experiment **Table 2.5**.

‹Active Learning Guide

**OBSERVATIONAL EXPERIMENT TABLE**

### 2.5 Amount of matter and acceleration.

**Observational experiment**

We pull the indicated number of stacked carts using an identical pulling force and measure the acceleration with a motion detector.

| Number *n* of carts | Acceleration $(\text{m/s}^2)$ |
|---|---|
| 1 | 1.00 |
| 2 | 0.49 |
| 3 | 0.34 |
| 4 | 0.25 |

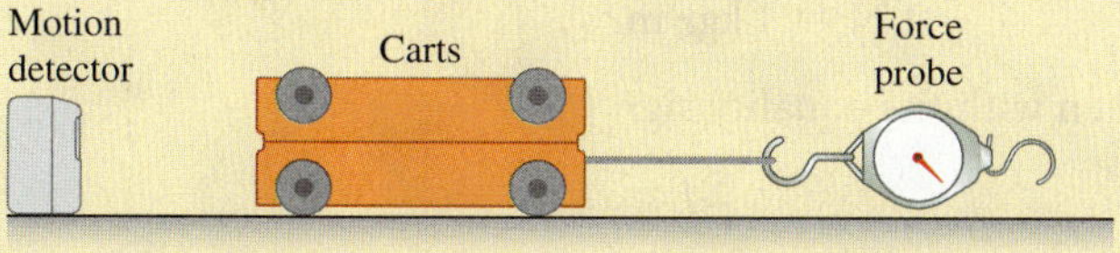


**Analysis**

We can graph the acceleration-versus-number of carts for constant pulling force.

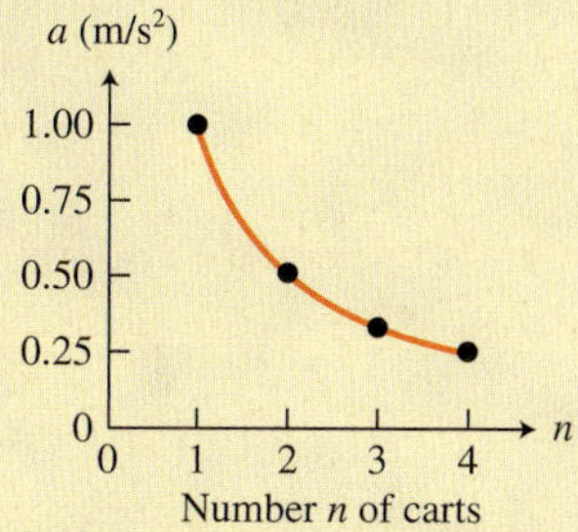


From the graph we see that increasing the number of carts decreases the acceleration. To check whether this relationship is inversely proportional, plot $a$ versus $\frac{1}{n}$.

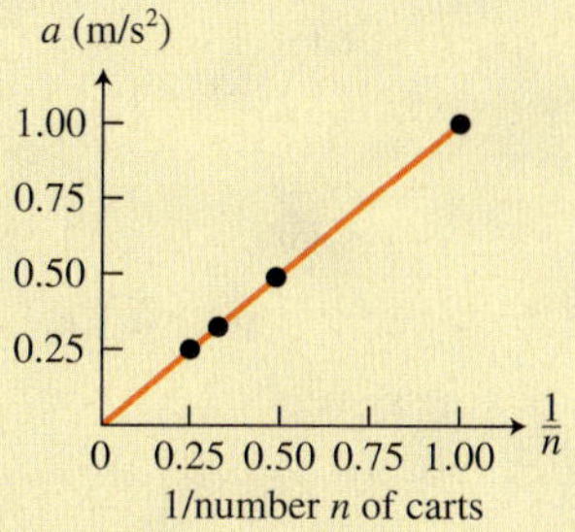


**Pattern**

Since the graph $a$ versus $\dfrac{1}{n}$ is a straight line, we conclude that $a \propto \dfrac{1}{n}$.

From the pattern observed in Table 2.5, we conclude that the greater the amount of matter being pulled, the smaller the object's acceleration when the same force is exerted on it. This property of an object, which affects its acceleration, is called **mass**.

To measure the mass of an object quantitatively, you first define a standard unit of mass. The choice for the unit of mass is arbitrary, but after the unit has been chosen, the masses of all other objects can be determined from this unit. The SI standard of mass, the kilogram (kg), is a cylinder made of a platinum-iridium alloy stored in a museum of measurements near Paris. Copies of this cylinder are available in most countries. A quart of milk has a mass of about 1 kg. Suppose, for example, that you exert a constant pulling force on a 1.0-kg object (and that all other forces exerted on this object are balanced), and you measure its acceleration. You then exert the same pulling force on another object of unknown mass. Your measurement indicates that it has half the acceleration of the standard 1.0-kg object. Thus, its mass is twice the standard mass (2.0 kg). This method is not practical for everyday use. Later we will learn another method to measure the mass of an object, a method that is simple enough to use in everyday life.

Our experiments indicate that when the same force is exerted on two objects, the one with the greater mass will have a smaller acceleration. Mathematically:

$$a \propto \frac{1}{m} \tag{2.2}$$

**Mass** ***m*** **characterizes the amount of matter in an object. When the same unbalanced force is exerted on two objects, the object with greater mass has a smaller acceleration. The unit of mass is called the kilogram (kg). Mass is a scalar quantity, and masses add as scalars.**

## Newton's second law

We have found that the acceleration $\vec{a}$ of a system is proportional to the vector sum of the forces $\Sigma\vec{F}$ exerted on it by other objects [Eq. (2.1)] and inversely proportional to the mass $m$ of the system [Eq. (2.2)]. We can combine these two proportionalities into a single equation.

$$\vec{a}_{\text{System}} \propto \frac{\Sigma\vec{F}_{\text{on System}}}{m_{\text{System}}} \tag{2.3}$$

Rearrange the above to get $m_{\text{System}}\vec{a}_{\text{System}} \propto \Sigma\vec{F}_{\text{on System}}$. We can turn this into an equation if we choose the unit of force to be $\text{kg}\cdot\text{m/s}^2$. Because force is such a ubiquitous quantity, physicists have given the force unit a special name called a newton (N). A force of 1 newton (1 N) causes an object with a mass of 1 kg to accelerate at $1\ \text{m/s}^2$.

$$1\ \text{N} = 1\ \text{kg}\cdot\text{m/s}^2. \tag{2.4}$$

Eq. (2.3), rewritten with the equality sign

$$\vec{a}_{\text{System}} = \frac{\Sigma\vec{F}_{\text{on System}}}{m_{\text{System}}}$$

is called Newton's second law. As noted earlier in the chapter, the symbol $\Sigma$ (the Greek letter sigma) means that you must add what follows the $\Sigma$.

**Newton's second law** **The acceleration $\vec{a}_{\text{System}}$ of a system object is proportional to the vector sum of all forces being exerted on the object and inversely proportional to the mass *m* of the object:**

$$\vec{a}_{\text{System}} = \frac{\Sigma\vec{F}_{\text{on System}}}{m_{\text{System}}} = \frac{\vec{F}_{1\text{ on System}} + \vec{F}_{2\text{ on System}} + \cdots}{m_{\text{System}}} \tag{2.5}$$

**The vector sum of all the forces being exerted on the system by other objects is**

$$\Sigma\vec{F}_{\text{on System}} = \vec{F}_{1\text{ on System}} + \vec{F}_{2\text{ on System}} + \cdots$$

**The acceleration of the system object points in the same direction as the vector sum of the forces.**

**TIP** Notice that the "vector sum of the forces" mentioned in the definition above does not mean the sum of their magnitudes. Vectors are not added as numbers; their directions affect the magnitude of the vector sum.

Does this new equation make sense? For example, does it work in extreme cases? First, imagine an object with an infinitely large mass. According to the law, it will have zero acceleration for any process in which the sum of the forces exerted on it is finite:

$$\vec{a}_{\text{System}} = \frac{\Sigma \vec{F}_{\text{on System}}}{\infty} = 0$$

This seems reasonable, as an infinitely massive object would not change motion due to finite forces exerted on it. On the other hand, an object with a zero mass will have an infinitely large acceleration when a finite magnitude force is exerted on it:

$$\vec{a}_{\text{System}} = \frac{\Sigma \vec{F}_{\text{on System}}}{0} = \infty$$

Both extreme cases make sense. Newton's second law is a so-called *cause-effect relationship.* The right side of the equation (the sum of the forces being exerted) is the cause of the effect (the acceleration) on the left side.

$$\underbrace{\vec{a}_{\text{System}}}_{\text{Effect}} = \underbrace{\frac{\Sigma \vec{F}_{\text{on System}}}{m_{\text{System}}}}_{\text{Cause}}$$

On the other hand, $\vec{a} = \Delta\vec{v}/\Delta t$ is called an *operational definition* of acceleration. It tells us how to determine the quantity acceleration but does not tell us *why* it has a particular value. For example, suppose that an elevator's speed changes from 2 m/s to 5 m/s in 3 s as it moves vertically along a straight line in the positive $x$-direction. The elevator's acceleration (using the definition of acceleration) is

$$a_x = \frac{5\text{ m/s} - 2\text{ m/s}}{3\text{ s}} = +1\text{ m/s}^2$$

This operational definition does not tell you the reason for the acceleration. If you know that the mass of the elevator is 500 kg and that Earth exerts a 5000-N downward force on the accelerating elevator and the cable exerts a 5500-N upward force on it, then using the cause-effect relationship of Newton's second law:

$$\frac{5500\text{ N} + (-5000\text{ N})}{500\text{ kg}} = +1\text{ m/s}^2$$

Thus, you obtain the same number using two different methods—one from kinematics (the part of physics that *describes* motion) and the other from dynamics (the part of physics that *explains* motion).

## Force components used for forces along one axis

When the vector sum of the forces exerted on a system object points along one direction (for example, the $x$-direction), you can use the component form of the Newton's second law equation for the $x$-direction instead of the vector equation [Eq. (2.5)]:

$$a_{\text{System }x} = \frac{\Sigma F_{\text{on System }x}}{m_{\text{System}}} \quad (2.6)$$

A similar equation applies for the $y$-direction, if the forces are all along the $y$-axis. To use Eq. (2.6) (or the $y$-version of the equation), you first need to identify the positive direction of the axis. Then find the components of all the forces being exerted on the system. Forces that point in the positive direction have a positive component, and forces that point in the negative direction have a negative component.

In this chapter, we will analyze situations in which (a) the forces that external objects exert on the system are all along the $y$-axis or (b) the forces pointing along the $y$-axis balance and don't contribute to the acceleration of the system along the $x$-axis, allowing us to analyze the situation along the $x$-axis only. Consider first a situation where all forces are along the $y$-axis.

**EXAMPLE 2.4 Lifting a suitcase**

Earth exerts a downward 100-N force on a 10-kg suitcase. Suppose you exert an upward 120-N force on the suitcase. If the suitcase starts at rest, how fast is it traveling after lifting for 0.50 s?

**Sketch and translate** First, we make a sketch of the initial and final states of the process, choosing the suitcase as the system object. The sketch helps us visualize the process and also brings together all the known information, letting our brains focus on other aspects of solving the problem. One common aspect of problems like this is the use of a two-step strategy. Here, we use Newton's second law to determine the acceleration of the suitcase and then use kinematics to determine the suitcase's speed after lifting 0.50 s.

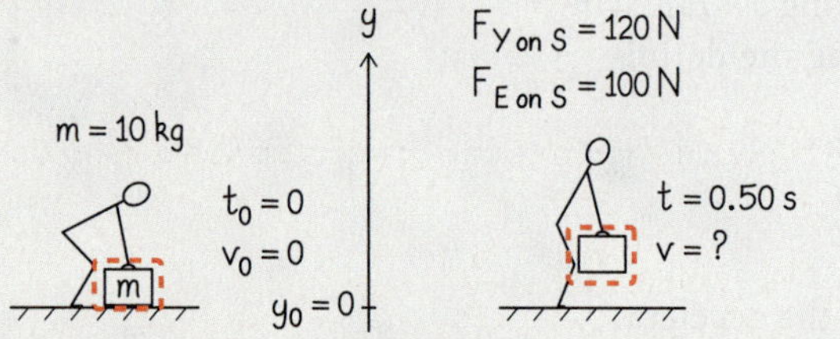


**Simplify and diagram** Next, we construct a force diagram for the suitcase while being lifted. The $y$-components of the forces exerted on the suitcase are your upward pull on the suitcase $F_{\text{Y on S}\,y} = +F_{\text{Y on S}} = +120\text{ N}$ and Earth's downward pull on the suitcase $F_{\text{E on S}\,y} = -F_{\text{E on S}} = -100\text{ N}$. Because the upward force is larger, the suitcase will have an upward acceleration $\vec{a}$.

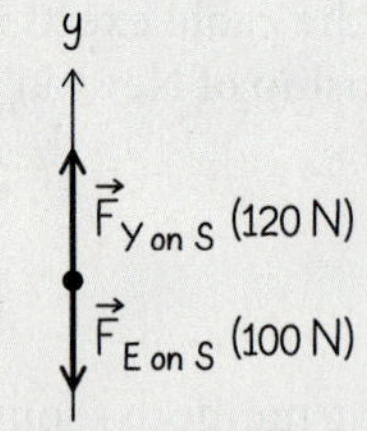


**Represent mathematically** Since all the forces are along the $y$-axis, we apply the $y$-component form of Newton's second law to determine the suitcase's acceleration (notice how the subscripts in the equation below change from step to step):

$$a_{\text{S}\,y} = \frac{\Sigma F_{\text{on S}\,y}}{m_\text{S}} = \frac{F_{\text{Y on S}\,y} + F_{\text{E on S}\,y}}{m_\text{S}}$$

$$= \frac{(+F_{\text{Y on S}}) + (-F_{\text{E on S}})}{m_\text{S}}$$

$$= \frac{F_{\text{Y on S}} - F_{\text{E on S}}}{m_\text{S}}$$

After using Newton's second law to determine the acceleration of the suitcase, we then use kinematics to determine the suitcase's speed after traveling upward for 0.50 s:

$$v_y = v_{0y} + a_y t$$

The initial velocity is $v_{0y} = 0$.

**Solve and evaluate** Now substitute the known information in the Newton's second law $y$-component equation above to find the acceleration of the suitcase:

$$a_{\text{S}\,y} = \frac{F_{\text{Y on S}} - F_{\text{E on S}}}{m_\text{S}} = \frac{120\text{ N} - 100\text{ N}}{10\text{ kg}} = +2.0\text{ m/s}^2$$

Insert this and other known information into the kinematics equation to find the vertical velocity of the suitcase after lifting for 0.50 s:

$$v_y = v_{0y} + a_y t = 0 + (+2.0\text{ m/s}^2)(0.50\text{ s}) = +1.0\text{ m/s}$$

The unit for time is correct and the magnitude is reasonable.

**Try it yourself:** How far up did you pull the suitcase during this 0.50 s?

*Answer:* The average speed while lifting it was $(0 + 1.0\text{ m/s})/2 = 0.50\text{ m/s}$. Thus you lifted the suitcase $y - y_0 = (0.50\text{ m/s})(0.50\text{ s}) = 0.25\text{ m}$.

Now let's consider a case in which an object moves on a horizontal surface, Earth exerts a downward gravitational force on it, and the surface exerts an upward normal force of the same magnitude.

**EXAMPLE 2.5 Pulling a lawn mower**

You pull horizontally on the handle of a lawn mower that is moving across a horizontal grassy surface. The lawn mower's mass is 32 kg. You exert a force of magnitude 96 N on the mower. The grassy surface exerts an 83-N resistive friction-like force on the mower. Earth exerts a downward force on the mower of magnitude 314 N, and the grassy surface exerts an upward normal force of magnitude 314 N. What is the acceleration of the mower?

**Sketch and translate** We choose the mower as the system and sketch the situation, as shown below.

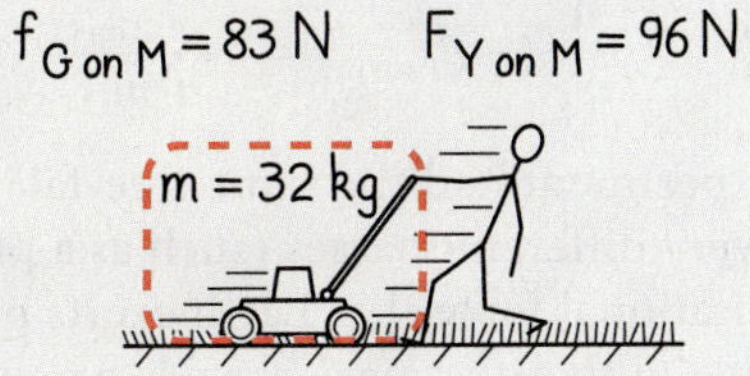


**Simplify and diagram** We model the mower as a point-like object and draw a force diagram for it. As the forces do not change during motion, the diagram can represent the motion at any instant. Three external objects interact with the system—the grassy surface, Earth, and you. You exert a horizontal force on the mower $\vec{F}_{\text{Y on M}}$; Earth exerts a downward gravitational force on the mower $\vec{F}_{\text{E on M}}$. The grassy surface exerts a force on the mower that we will represent with two force arrows—a normal force $\vec{N}_{\text{G on M}}$ perpendicular to the surface (in this case it points upward) and a horizontal friction force $\vec{f}_{\text{G on M}}$ opposite the direction of motion.

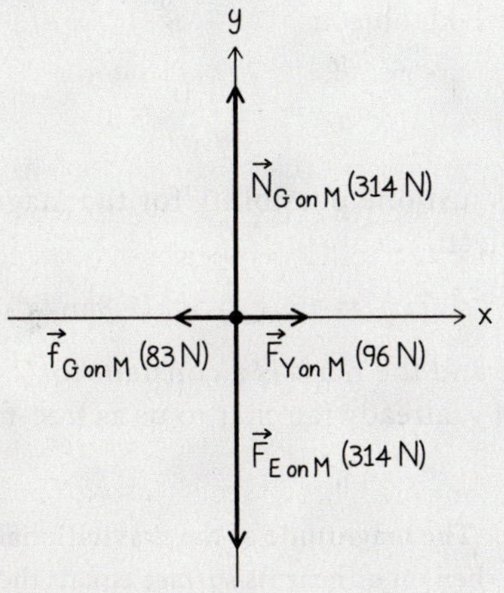


The downward force $\vec{F}_{\text{E on M}}$ and the upward normal force $\vec{N}_{\text{G on M}}$ have the same magnitudes and point in opposite directions. This means that these two forces cancel and do not contribute to the horizontal acceleration of the mower. We can ignore them.

**Represent mathematically** Since the acceleration of the system is along the $x$-axis and the forces perpendicular to the grassy surface cancel, we can use the $x$-component form of Newton's second law to determine the acceleration of the mower:

$$a_{\text{M}\,x} = \frac{\Sigma F_x}{m_{\text{M}}} = \frac{F_{\text{Y on M}\,x} + f_{\text{G on M}\,x}}{m_{\text{M}}}$$

The $x$-component of the force you exert on the mower is positive since that force points in the positive $x$-direction ($F_{\text{Y on M}\,x} = +F_{\text{Y on M}}$). The $x$-component of the friction force exerted by the grass on the mower is negative since that force points in the negative $x$-direction ($f_{\text{G on M}\,x} = -f_{\text{G on M}}$). Therefore, the acceleration of the mower is

$$a_{\text{M}\,x} = \frac{F_{\text{Y on M}} + (-f_{\text{G on M}})}{m_{\text{M}}} = \frac{F_{\text{Y on M}} - f_{\text{G on M}}}{m_{\text{M}}}$$

**Solve and evaluate** We can now determine the mower's acceleration by substituting the known information into the preceding equation:

$$a_{\text{M}\,x} = \frac{F_{\text{Y on M}} - f_{\text{G on M}}}{m_{\text{M}}} = \frac{96\text{ N} - 83\text{ N}}{32\text{ kg}}$$

$$= +0.40625\text{ N/kg} = +0.41\text{m/s}^2$$

The known information had two significant digits, so we rounded the answer to two significant digits. The units for acceleration are correct, and the magnitude is reasonable.

**Try it yourself:** Imagine that you are pulling the mower as described above. After the mower has accelerated to a velocity with which you are comfortable, what is the magnitude of the force you should exert on the mower so that it now moves at constant velocity?

*Answer:* 83 N.

**Review Question 2.6** Jim says that $m\vec{a}$ is a special force exerted on an object and it should be represented on the force diagram. Do you agree or disagree with Jim? Explain your answer.

## 2.7 Gravitational force law

In the last example, we were given the mass of the lawn mower (32 kg) and the magnitude of the force that Earth exerts on the mower (314 N). Is it possible to determine the magnitude of this force by just knowing the mass of the mower? In fact, it is.

Imagine that we evacuate all the air from a 3.0-m-long Plexiglas tube, place a motion sensor at the top, and drop objects of various sizes, shapes, and compositions through the tube. The measurements taken by the motion sensor reveal that all objects fall straight down with the same acceleration, $9.8\ \text{m/s}^2$. Earth exerts the only force on the falling object $\vec{F}_{\text{E on O}}$ during the entire flight. If we choose the positive $y$-axis pointing down and apply the $y$-component form of Newton's second law, we get

$$a_{\text{O}\,y} = \frac{1}{m_\text{O}} F_{\text{E on O}\,y} = \frac{1}{m_\text{O}}(+F_{\text{E on O}}) = +\frac{F_{\text{E on O}}}{m_\text{O}}$$

Every object dropped in our experiment had the same free-fall acceleration, $g = 9.8\ \text{m/s}^2$, even those with very different masses (such as a ping-pong ball and a lead ball). Thus, the gravitational force that Earth exerts on each object must be proportional to its mass so that the mass cancels when we calculate the acceleration. Earth must exert a force on a 10-kg object that is 10 times greater than that on a 1-kg object:

$$a_{\text{O}\,y} = \frac{F_{\text{E on O}}}{m_\text{O}} = g$$

This reasoning leaves just one possibility for the magnitude of the force that Earth exerts on an object:

$$F_{\text{E on O}} = m_\text{O} g = m_\text{O}(9.8\ \text{m/s}^2)$$

The ratio of the force and the mass is a constant for all objects—the so-called gravitational constant $g$, already familiar to us as free-fall acceleration.

**Gravitational force** **The magnitude of the gravitational force that Earth exerts on any object $F_{\text{E on O}}$ when on or near its surface equals the product of the object's mass $m$ and the constant $g$:**

$$F_{\text{E on O}} = m_\text{O} g \qquad (2.7)$$

**where $g = 9.8\ \text{m/s}^2 = 9.8\ \text{N/kg}$ on or near Earth's surface. This force points toward the center of Earth.**

The value of the free-fall acceleration $g$ in the above gravitational force law [Eq. (2.7)] does not mean that the object is actually falling. The $g$ is just used to determine the magnitude of the gravitational force exerted on the object by Earth whether the object is falling or sitting at rest on a table or moving down a water slide. To avoid confusion, we will use $g = 9.8\ \text{N/kg}$ rather than $9.8\ \text{m/s}^2$ when calculating the gravitational force.

We learn in the chapter on circular motion (Chapter 4) that the gravitational constant $g$ at a particular point depends on the mass of Earth and on how far this point is from the center of Earth. On Mars or the Moon, the gravitational constant depends on the mass of Mars or the Moon, respectively. The gravitational constant is 1.6 N/kg on the Moon and 3.7 N/kg on Mars. You could throw a ball upward higher on the Moon since $g$ is smaller there, resulting in a smaller force exerted downward on the ball.

Active Learning Guide›

**Review Question 2.7** Newton's second law says that the acceleration of an object is inversely proportional to its mass. However, the acceleration with which all objects fall in the absence of air is the same. How can this be?

**The language of physics**

The *weight* of an object on a planet is the force that the planet exerts on the object. We will not use the term "weight of an object" because it implies that weight is a property of the object rather than an interaction between two objects.

## 2.8 Skills for applying Newton's second law for one-dimensional processes

In this section we develop a strategy that can be used whenever a process involves force and motion. We will introduce the strategy by applying it to the 2007 sky dive of diving champion Michael Holmes. After more than 1000 successful jumps, Holmes jumped from an airplane 3700 m above Lake Taupo in New Zealand. His main parachute failed to open, and his backup chute became tangled in its cords. The partially opened backup parachute slowed his descent to about 36 m/s (80 mi/h) as he reached a 2-m-high thicket of wild shrubbery. Holmes plunged through the shrubbery, which significantly decreased his speed before he reached the ground. Holmes survived with a collapsed right lung and a broken left ankle. The general steps of a problem-solving strategy for force-motion problems are described on the left side of Example 2.6 and applied to Holmes's landing in the shrubbery on the right side.

**‹Active Learning Guide**

**PROBLEM-SOLVING STRATEGY** Applying Newton's Laws For One-Dimensional Processes

**EXAMPLE 2.6** Holmes's sky dive

Michael Holmes (70 kg) was moving downward at 36 m/s (80 mi/h) and was stopped by 2.0-m-high shrubbery and the ground. Estimate the average force exerted by the shrubbery and ground on his body while stopping his fall.

**Sketch and translate**

- Make a sketch of the process.
- Choose the system object.
- Choose a coordinate system.
- Label the sketch with everything you know about the situation.

We sketch the process, choosing Holmes as the system object H. We want to know the average force that the shrubbery and ground S-G exert on him from when he first touches the shrubbery to the instant when he stops. We choose the *y*-axis pointing up and the origin at the ground where Holmes comes to rest. We use kinematics to find his acceleration while stopping and Newton's second law to find the average force that the shrubbery and ground exerted on him while stopping him.

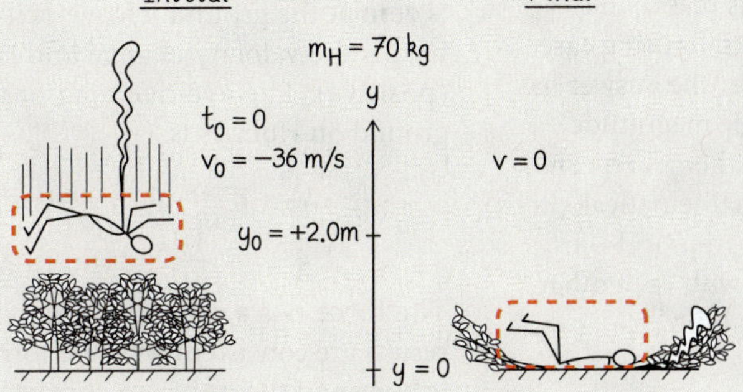


(*continued*)

**Simplify and diagram**

- Make appropriate simplifying assumptions about the process. For example, can you neglect the size of the system object or neglect frictional forces? Can you assume that forces or acceleration are constant?
- Then represent the process with a motion diagram and/or a force diagram(s). Make sure the diagrams are consistent with each other.

We model Holmes as a point-like object and assume that the forces being exerted on him are constant so that they lead to a constant acceleration. A motion diagram for his motion while stopping is shown along with the corresponding force diagram. To draw the force diagram we first identify the objects interacting with Holmes as he slows down: the shrubbery and the ground (combined as one interaction) and Earth. The shrubbery and ground exert an upward normal force $\vec{N}_{\text{S-G on H}}$ on Holmes. Earth exerts a downward gravitational force $\vec{F}_{\text{E on H}}$. The force diagram is the same for all points of the motion diagram because the acceleration is constant. On the force diagram the arrow for $\vec{N}_{\text{S-G on H}}$ must be longer to match the motion diagram, which shows the velocity change arrow pointing up.

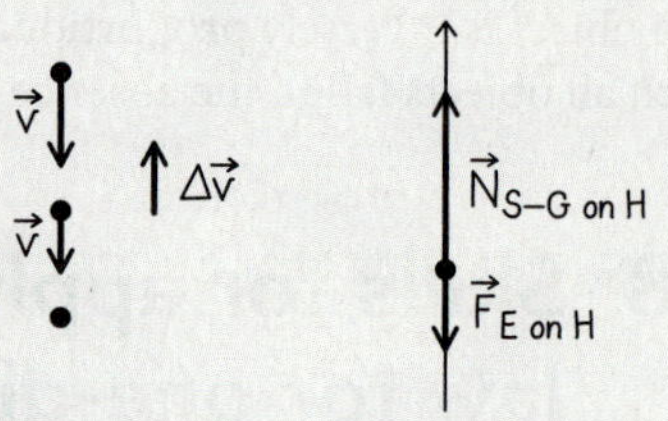


**Represent mathematically**

- Convert these qualitative representations into quantitative mathematical descriptions of the situation using kinematics equations and Newton's second law for motion along the axis. After you make the decision about the positive and negative directions, you can determine the signs for the force components in the equations. Add the force components (with either positive or negative signs) to find the sum of the forces.

The $y$-component of Holmes's average acceleration is

$$a_y = \frac{v_y^2 - v_{0y}^2}{2(y - y_0)}$$

The $y$-component of Newton's second law with the positive $y$-direction up is

$$a_y = \frac{\Sigma F_{\text{on H}\,y}}{m_{\text{H}}}$$

The $y$-component of the force exerted by the shrubbery-ground on Holmes is $N_{\text{S-G on H}\,y} = +N_{\text{S-G on H}}$ and the $y$-component of the force exerted by Earth is $F_{\text{E on H}\,y} = -F_{\text{E on H}} = -m_{\text{H}}g$. Therefore,

$$a_y = \frac{N_{\text{S-G on H}\,y} + F_{\text{E on H}\,y}}{m_{\text{H}}} = \frac{(+N_{\text{S-G on H}}) + (-F_{\text{E on H}})}{m_{\text{H}}} = \frac{+N_{\text{S-G on H}} - m_{\text{H}}g}{m_{\text{H}}}$$

$$\Rightarrow N_{\text{S-G on H}} = m_{\text{H}}a_y + m_{\text{H}}g$$

**Solve and evaluate**

- Substitute the known values into the mathematical expressions and solve for the unknowns.
- Finally, evaluate your work to see if it is reasonable (check units, limiting cases, and whether the answer has a reasonable magnitude). Check whether all representations—mathematical, pictorial, and graphical—are consistent with each other.

Holmes's average acceleration was

$$a_y = \frac{0^2 - (-36\text{ m/s})^2}{2(0 - 2.0\text{ m})} = +324\text{ m/s}^2$$

Holmes's initial velocity is negative, since he is moving in the negative direction. His initial position is +2.0 m at the top of the shrubbery, and his final position is zero at the ground. His velocity in the negative direction is decreasing, which means the velocity change and the acceleration point in the opposite direction (positive). The average magnitude of the force exerted by the shrubbery and ground on Holmes is

$$N_{\text{S-G on H}} = m_{\text{H}}a_y + m_{\text{H}}g = (70\text{ kg})(324\text{ m/s}^2) + (70\text{ kg})(9.8\text{ N/kg})$$

$$= 22{,}680\text{ kg}\cdot\text{m/s}^2 + 686\text{ N} = 23{,}366\text{ N} = 23{,}000\text{ N}$$

The force has a magnitude greater than the force exerted by Earth—thus the results are consistent with the force diagram and motion diagram. The magnitude is huge and the units are correct. A limiting case for zero acceleration gives us a correct prediction—the force exerted on Holmes by the shrubbery and ground equals the force exerted by Earth.

**Try it yourself:** Use the strategy discussed above to estimate the average force that the ground would have exerted on Holmes if he had stopped in a conservative 0.20 m with no help from the shrubbery.

*Answer:* 230,000 N (over 50,000 lb).

The force when Holmes landed in the shrubbery had a magnitude of 23,000 N, greater than 5000 lb! The force was exerted over a significant area of his body, thanks to the shrubbery. Most of all, the shrubbery increased the stopping distance. If Holmes had landed directly on dirt or something harder that would compress less than the shrubbery, his acceleration would have been much greater, as would the force exerted on him.

> **TIP** In the last equation we used an $N$ in italics to indicate the magnitude of the normal force exerted by the shrubbery and ground on Holmes. On the right side, we used an N in roman type to indicate the unit of force. Be careful not to confuse these two similar looking notations.

## An elevator ride standing on a bathroom scale

In Example 2.7, we consider a much less dangerous process, one you could try the next time you ride an elevator. When you stand on a bathroom scale, the scale reading indicates how hard you are pushing on the scale. The normal force that it exerts on you balances the downward force that Earth exerts on you (called your *weight*, in everyday language), resulting in your zero acceleration. What will the scale read if you stand on it in a moving elevator?

### EXAMPLE 2.7 Elevator ride

You stand on a bathroom scale in an elevator as it makes a trip from the first floor to the tenth floor of a hotel. Your mass is 50 kg. When you stand on the scale in the stationary elevator, it reads 490 N (110 lb). What will the scale read (a) early in the trip while the elevator's upward acceleration is $1.0\ \text{m/s}^2$, (b) while the elevator moves up at a constant speed of 4.0 m/s, and (c) when the elevator slows to a stop with a downward acceleration of $1.0\ \text{m/s}^2$ magnitude?

**Sketch and translate** We sketch the situation as shown at right, choosing you as the system object. The coordinate axis points upward with its origin at the first floor of the elevator shaft. Your mass is $m_Y = 50$ kg, the magnitude of the force that Earth exerts on you is $F_{\text{E on Y}} = m_Y g = 490$ N, and your acceleration is (a) $a_y = +1.0\ \text{m/s}^2$ (the upward velocity is increasing); (b) $a_y = 0$ ($v$ is a constant 4.0 m/s upward); and (c) $a_y = -1.0\ \text{m/s}^2$ (the upward velocity is decreasing, so the acceleration points in the opposite, negative direction).

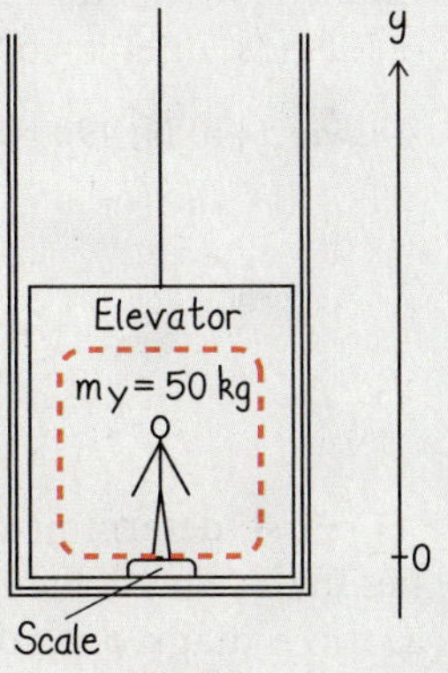

y
Elevator
$m_Y = 50$ kg
0
Scale

**Simplify and diagram** We model you as a point-like object and represent you as a dot in both the motion and force diagrams, shown for each part of the trip in Figures a, b, and c. On the diagrams, E represents Earth, Y

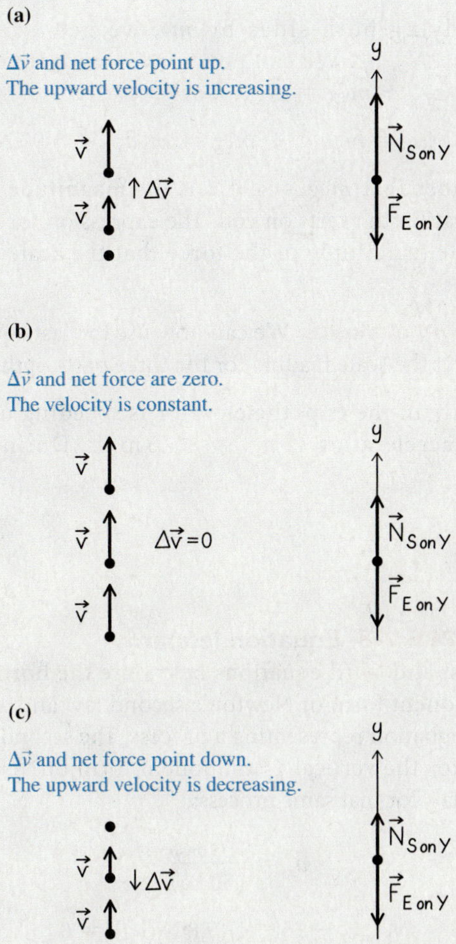

(a)
$\Delta\vec{v}$ and net force point up.
The upward velocity is increasing.
$\vec{v}$
$\Delta\vec{v}$
y
$\vec{N}_{\text{S on Y}}$
$\vec{F}_{\text{E on Y}}$
(b)
$\Delta\vec{v}$ and net force are zero.
The velocity is constant.
$\Delta\vec{v} = 0$
(c)
$\Delta\vec{v}$ and net force point down.
The upward velocity is decreasing.

*(continued)*

is you, and S is the scale. The magnitude of the downward force that Earth exerts does not change (it equals $m_Y\vec{g}$, and neither $m_Y$ nor $\vec{g}$ change). Notice that the force diagrams and motion diagrams are consistent with each other for each part of the trip. The length of the normal force arrows representing the force that the scale exerts on you changes from one case to the next so that the sum of the forces point in the same direction as your velocity change arrow.

**Represent mathematically** The motion and the forces are entirely along the vertical $y$-axis. Thus, we use the vertical $y$-component form of Newton's second law [Eq. (2.6)] to analyze the process. There are two forces exerted on you (the system object) so there will be two vertical $y$-component forces on the right side of the equation: the $y$-component of the force that Earth exerts on you, $F_{\text{E on Y}\,y} = -m_Y g$, and the $y$-component of the normal force that the scale exerts on you, $N_{\text{S on Y}\,y} = +N_{\text{S on Y}}$:

$$a_{Y\,y} = \frac{\Sigma F_y}{m_Y} = \frac{F_{\text{E on Y}\,y} + N_{\text{S on Y}\,y}}{m_Y} = \frac{-m_Y g + N_{\text{S on Y}}}{m_Y}.$$

Multiplying both sides by $m_Y$, we get $a_{Y\,y} m_Y = -m_Y g + N_{\text{S on Y}}$. We can now move $-m_Y g$ to the left side: $m_Y a_{Y\,y} + m_Y g = N_{\text{S on Y}}$, or

$$N_{\text{S on Y}} = m_Y a_{Y\,y} + m_Y g = m_Y a_{Y\,y} + 490\ \text{N}$$

Remember that $m_Y g = 490\ \text{N}$ is the magnitude of the force that Earth exerts on you. The expression for $N_{\text{S on Y}}$ gives the magnitude of the force that the scale exerts on you.

**Solve and evaluate** We can now use the last equation to predict the scale reading for the three parts of the trip.

(a) Early in the trip, the elevator is speeding up and its acceleration is $a_{Y\,y} = +1.0\ \text{m/s}^2$. During that time interval, the force exerted by the scale on you should be

$$\begin{aligned} N_{\text{S on Y}} &= m_Y\, a_{Y\,y} + 490\ \text{N} \\ &= (50\ \text{kg})(+1.0\ \text{m/s}^2) + 490\ \text{N} = 540\ \text{N} \end{aligned}$$

(b) In the middle of the trip, when the elevator moves at constant velocity, your acceleration is zero and the scale should read:

$$\begin{aligned} N_{\text{S on Y}} &= m_Y\, a_{Y\,y} + 490\ \text{N} \\ &= (50\ \text{kg})(0\ \text{m/s}^2) + 490\ \text{N} = 490\ \text{N} \end{aligned}$$

(c) When the elevator is slowing down near the end of the trip, its acceleration points downward and is $a_y = -1.0\ \text{m/s}^2$. Then the force exerted by the scale on you should be

$$\begin{aligned} N_{\text{S on Y}} &= m_Y\, a_{Y\,y} + 490\ \text{N} \\ &= (50\ \text{kg})(-1.0\ \text{m/s}^2) + 490\ \text{N} = 440\ \text{N} \end{aligned}$$

When the elevator is at rest or moving at constant speed, the scale reading equals the magnitude of the force that Earth exerts on you. When the elevator accelerates upward, the scale reads more. When it accelerates downward, even if you are moving upward, the scale reads less. What is also important is that the motion and force diagrams in Figures a, b, and c are consistent with each other and the force diagrams are consistent with the predicted scale readings—an important consistency check of the motion diagrams, force diagrams, and math.

**Try it yourself:** What will the scale read when the elevator starts from rest on the tenth floor and moves downward with increasing speed and a downward acceleration of $-1.0\ \text{m/s}^2$, then moves down at constant velocity, and finally slows its downward trip with an acceleration of $+1.0\ \text{m/s}^2$ until it stops at the first floor?

*Answer:* 440 N, 490 N, and 540 N.

## EXAMPLE 2.8 Equation Jeopardy

The first and third equations below are the horizontal $x$-component form of Newton's second law and a kinematics equation representing a process. The second equation is for the vertical $y$-component form of Newton's second law for that same process:

$$a_x = \frac{-f_{\text{S on O}}}{(50\ \text{kg})}$$

$$N_{\text{S on O}} - (50\ \text{kg})(9.8\ \text{N/kg}) = 0$$

$$0 - (20\ \text{m/s})^2 = 2a_x(+25\ \text{m})$$

First, determine the values of all unknown quantities in the equations. Then work backward and construct a force diagram and a motion diagram for a system object and invent a process that is consistent with the equations (there are many possibilities). Reverse the problem-solving strategy you used in Example 2.6.

**Solve** The second equation can be solved for the normal force:

$$N_{\text{S on O}} = (50\ \text{kg})(9.8\ \text{N/kg}) = 490\ \text{N}$$

We solve the third equation for the acceleration $a_x$:

$$a_x = \frac{-(20\text{ m/s})^2}{2(25\text{ m})} = -8.0\text{ m/s}^2$$

Now substitute this result in the first equation to find the magnitude of the force $f_{\text{S on O}}$:

$$-f_{\text{S on O}} = (50\text{ kg})(-8.0\text{ m/s}^2) = -400\text{ N}$$

**Represent mathematically** The third equation in the problem statement looks like the application of the following kinematics equation to the problem process:

$$v_x^2 - v_{0x}^2 = 2a_x(x - x_0)$$

By comparing the above to the third provided equation, we see that the final velocity was $v_x = 0$, the initial velocity was $v_{0x} = +20\text{ m/s}$, and the displacement of the object was $(x - x_0) = +25\text{ m}$. The first equation indicates that there is only one force exerted on the system object in the horizontal $x$-direction. It has a magnitude of 400 N and points in the negative direction (the same direction as the object's acceleration), which is opposite the direction of the initial velocity. The second equation indicates that the system object has a 50-kg mass. It could be some kind of a friction force that causes a 50-kg object to slow down and stop while it is traveling 25 m horizontally.

**Simplify and diagram** We can now construct a motion diagram and a force diagram for the object (see **a** and **b**).

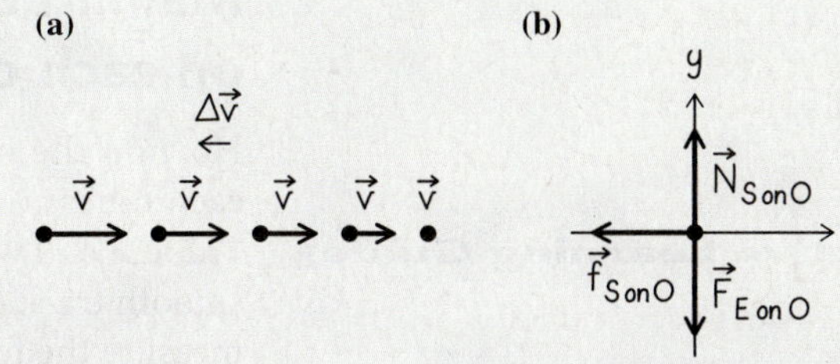


**Sketch and translate** This could have been a 50-kg snowboarder moving at 20 m/s after traveling down a steep hill and then skidding to a stop in 25 m on a horizontal surface (see below). You might have been asked to determine the average friction force that the snow exerted on the snowboarder while he was slowing down.

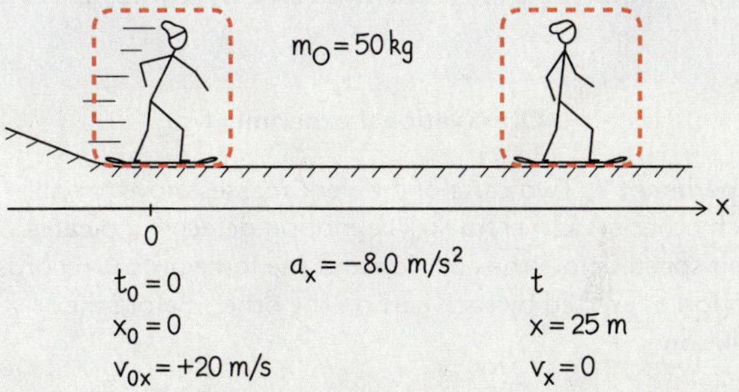


**Try it yourself:** Suppose the friction force remained the same but the snowboarder's initial speed was 10 m/s instead of 20 m/s. In what other ways would the trip be affected?

*Answer:* The snowboarder would stop in $(25\text{ m})/4 = 6\text{ m}$.

**Review Question 2.8** Three friends argue about the type of information a bathroom scale reports: Eugenia says that it reads the weight of a person, Alan says that it reads the sum of the forces exerted on the person by Earth and the scale, and Mike says that the scale reads the force that the person exerts on the scale. Who do you think is correct? Why?

## 2.9 Forces come in pairs: Newton's third law

So far, we have analyzed a system's acceleration due to the forces exerted on it by external objects. What effect does the system have on these other external objects? To help answer this question, we observe the interaction of two objects and analyze what happens to each of them. Suppose you wear rollerblades and push abruptly on a wheeled cart that is loaded with a heavy box. If you and the cart are on a hard smooth floor, the cart starts moving away (it accelerates), and you also start to move and accelerate in the opposite direction. Evidently, you exerted a force on the cart and the cart exerted a force on you. Since the accelerations were in opposite directions, the forces must point in opposite directions. Let's consider more quantitatively the effect of such mutual interactions between two objects.

## Magnitudes of the forces that two objects exert on each other

**Active Learning Guide>**

How do the magnitudes of the forces that two interacting objects exert on each other compare? Consider the experiments in Observational Experiment **Table 2.6**. Two dynamics carts with very low-friction wheels roll freely on a smooth track before colliding. We mount force probes on each cart in order to measure the forces that each cart exerts on the other while colliding. A motion sensor on each end of the track records the initial velocity of each cart before the collision.

**OBSERVATIONAL EXPERIMENT TABLE**

**2.6 Analyze the forces that two dynamics carts exert on each other.**

VIDEO 2.6

| Observational experiment | Analysis |
|---|---|
| **Experiment 1.** Two carts of different masses move toward each other on a level track. The motion detector indicates their speed before the collision and the force probes record the forces exerted by each cart on the other. Before the collision:<br>$m_1 = 2\text{ kg}, v_{1x} = +2\text{ m/s}$<br>$m_2 = 1\text{ kg}, v_{2x} = -2\text{ m/s}$ | As both carts changed velocities due to the collision, they must have exerted forces on each other. The computer recordings from the force probes show that the forces that the carts exert on each other vary with time and at each time have the same magnitude and point in the opposite direction. Cart 1 exerts a force on cart 2 toward the right, and cart 2 exerts a force on cart 1 toward the left.<br>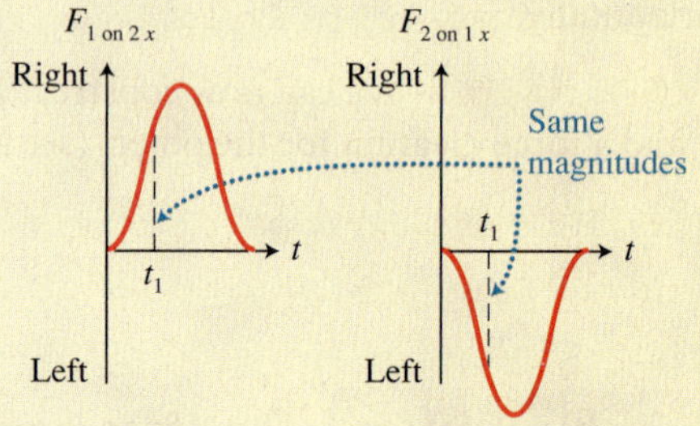<br> |
| **Experiment 2.** Cart masses and velocities before collision:<br>$m_1 = 2\text{ kg}, v_{1x} = 0\text{ m/s}$ (at rest)<br>$m_2 = 1\text{ kg}, v_{2x} = -1\text{ m/s}$ | Although the forces that the carts exert on each other are smaller than in the first experiment, the magnitudes of the forces at each time are still the same.<br>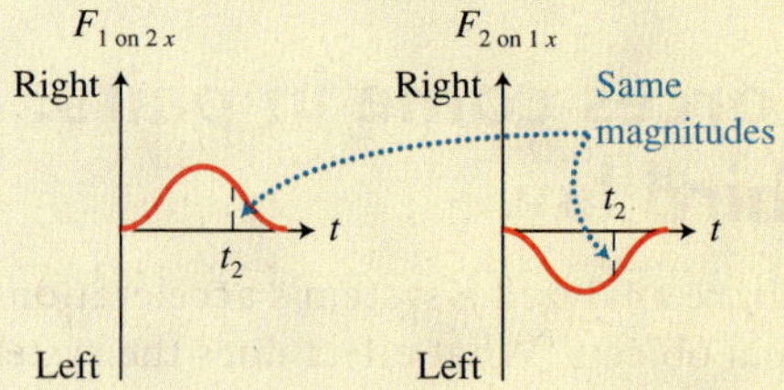<br> |
| **Experiment 3.** Cart masses and velocities before collision:<br>$m_1 = 2\text{ kg}, v_{1x} = +2\text{ m/s}$<br>$m_2 = 1\text{ kg}, v_{2x} = -1\text{ m/s}$ | The same analysis applies.<br>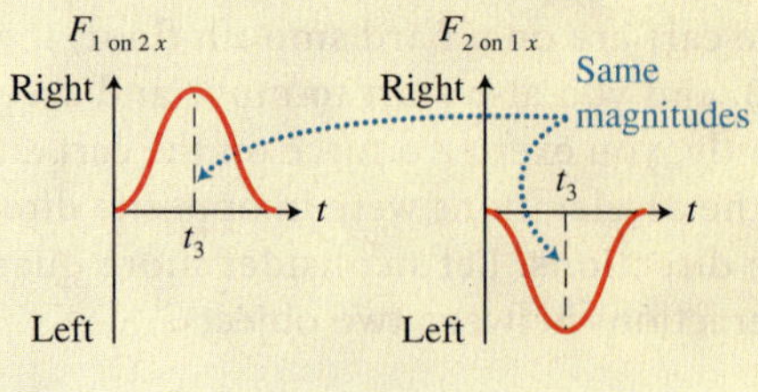<br> |

**Pattern**

In each experiment, independent of the masses and velocities of the carts before the collisions, at every instant during the collision the force that cart 1 exerted on cart 2 $\vec{F}_{1\text{ on }2}$ had the same magnitude as but pointed in the opposite direction from the force that cart 2 exerted on cart 1 $\vec{F}_{2\text{ on }1}$.

The cart collisions in Table 2.6 indicate that the force that one cart exerts on the other is equal in magnitude and opposite in direction to the force that the other cart exerts on the first.

$$\vec{F}_{\text{object 1 on object 2}} = -\vec{F}_{\text{object 2 on object 1}}$$

Will the pattern that we found allow us to correctly predict the results of some new experiment?

## Testing the idea

Imagine that you have two spring scales. Attach one scale to a hook on the wall and pull on its other end with a second scale (**Figure 2.7**). If the preceding equation is correct, then the scale you pull should have the same reading as the scale fixed to the wall. When you do the experiment, you find that the scales do indeed have the same readings. If you reverse the scales and repeat the experiment, you find that they *always* have the same readings.

**Figure 2.7** Spring scales exert equal-magnitude forces on each other.

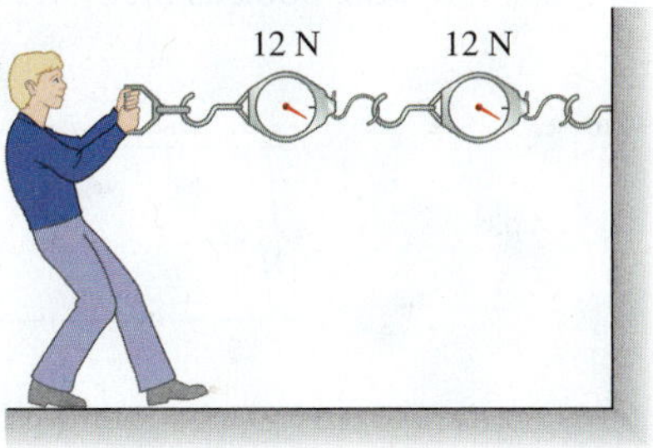


The outcome of the experiment matched the prediction we made based on the preceding equation. The objects exert equal-magnitude, oppositely directed forces on each other. To be convinced of the validity of this outcome, we need many more testing experiments. So far, physicists have found no experiments involving the dynamics of everyday processes that violate this equation. This rule is called Newton's third law.

**Newton's third law** **When two objects interact, object 1 exerts a force on object 2. Object 2 in turn exerts an equal-magnitude, oppositely directed force on object 1:**

$$\vec{F}_{\text{object 1 on object 2}} = -\vec{F}_{\text{object 2 on object 1}} \tag{2.8}$$

**Note that these forces are exerted on different objects and *cannot* be added to find the sum of the forces exerted on one object.**

It seems counterintuitive that two interacting objects always exert forces of the same magnitude on each other. Imagine playing ping-pong. A paddle hits the ball and the ball flies rapidly toward the other side of the table. However, the paddle seems to move forward with little change in motion. How is it possible that the light ball exerted a force of the same magnitude on the paddle as the paddle exerted on the ball?

To resolve this apparent contradiction, think about the masses of the interacting objects and their corresponding accelerations. If the forces are the same, the object with larger mass has smaller magnitude acceleration than the object with smaller mass:

$$a_{\text{paddle}} = \frac{F_{\text{ball on paddle}}}{m_{\text{paddle}}} \quad \text{and} \quad a_{\text{ball}} = \frac{F_{\text{paddle on ball}}}{m_{\text{ball}}}$$

Because the mass of the ball is so small, the same force leads to a large change in velocity. The paddle's mass, on the other hand, is much larger. Thus, the same magnitude force leads to an almost zero velocity change. We observe the velocity change and incorrectly associate that alone with the force exerted on the object.

**TIP** Remember that the forces in Newton's third law are exerted on two different objects. This means that the two forces will never show up on the same force diagram, and they should not be added together to find the sum of the forces. You have to choose the system object and consider only the forces exerted on *it*!

### CONCEPTUAL EXERCISE 2.9 A book on the table

A book sits on a tabletop. Identify the forces exerted on the book by other objects. Then, for each of these forces, identify the force that the book exerts on another object. Explain why the book is not accelerating.

**Sketch and translate** Draw a sketch of the situation and choose the book as the system.

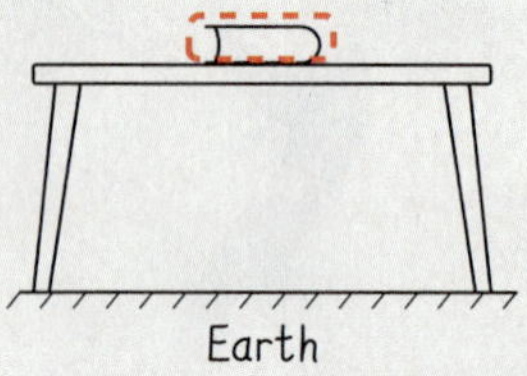


**Simplify and diagram** Assume that the tabletop is perfectly horizontal and model the book as a point-like object. A force diagram for the book is shown at right. Earth exerts a downward gravitational force on the book $\vec{F}_{\text{E on B}}$, and the table exerts an upward normal (contact) force on the book $\vec{N}_{\text{T on B}}$. Newton's second law explains why the book is not accelerating; the forces exerted on it by other objects are balanced and add to zero.

The subscripts on each force identify the two objects involved in the interaction. The Newton's third law pair force will have its subscripts reversed. For example, Earth exerts a downward gravitational force on the book ($\vec{F}_{\text{E on B}}$). According to Newton's third law, the book must exert an equal-magnitude upward gravitational force on Earth ($\vec{F}_{\text{B on E}} = -\vec{F}_{\text{E on B}}$), as shown at right. The table exerts an upward contact force on the book ($\vec{N}_{\text{T on B}}$), so the book must exert an equal-magnitude downward contact force on the table ($\vec{N}_{\text{B on T}} = -\vec{N}_{\text{T on B}}$).

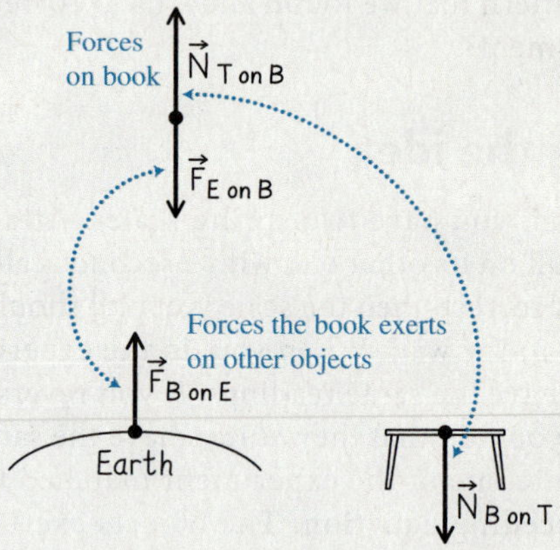


**Try it yourself:** A horse pulls on a sled that is stuck in snow and not moving. Your friend Chris says this happens because the horse exerts on the sled the same magnitude force that the sled exerts on the horse. Since the sum of the forces is zero, there is no acceleration. What is wrong with Chris' reasoning?

*Answer:* Chris added the forces exerted on two different objects and did not consider all forces exerted on the sled. If you choose the sled as the system object, then the horse pulls forward on the sled, and the snow exerts a backward, resistive force. If these two horizontal forces happen to be of the same magnitude, they add to zero, and the sled does not accelerate horizontally. If, on the other hand, we choose the horse as the system, the ground exerts a forward force on the horse's hooves (since the horse is exerting a force backward on the ground), and the sled pulls back on the horse. If those forces have the same magnitude, the net horizontal force is again zero, and the horse does not accelerate.

## EXAMPLE 2.10 Froghopper jump

The froghopper (*Philaenus spumarius*), an insect about 6 mm in length, is considered by some scientists to be the best jumper in the animal world. The froghopper can jump 0.7 m vertically, over 100 times higher than its length.

The froghopper's speed can change from zero to 4 m/s in about 1 ms $= 1 \times 10^{-3}$ s—an acceleration of 4000 m/s$^2$. The froghopper achieves this huge acceleration using its leg muscles, which occupy 11% of its 12-mg body mass. What is the average force that the froghopper exerts on the surface during the short time interval while pushing off during its jump?

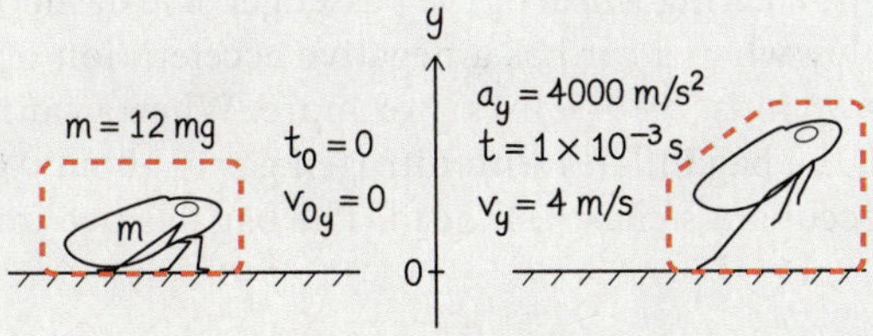


**Sketch and translate** We construct an initial-final sketch of the process, choosing the froghopper as the system object. We use a vertical $y$-axis with the positive direction pointing up and the origin of the coordinate system at the surface. We know the froghopper's acceleration while pushing off.

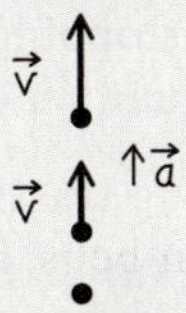


**Simplify and diagram** We can make a motion diagram for the pushing off process and a force diagram. The surface $S$ exerts an upward normal force $\vec{N}_{\text{S on F}}$ on the froghopper, and Earth exerts a downward gravitational force $\vec{F}_{\text{E on F}}$. Note that because the acceleration is upward, the sum of the $y$-components of the forces must also be upward; thus, the upward normal force exerted by the surface is greater in magnitude than the downward gravitational force exerted by Earth.

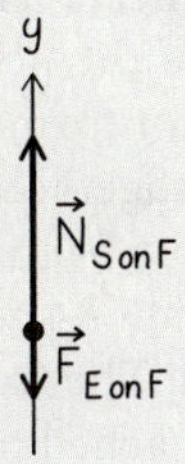


The force diagram shows the normal force exerted by the surface on the froghopper $\vec{N}_{\text{S on F}}$. We are asked to determine the force that the froghopper exerts on the surface $\vec{N}_{\text{F on S}}$. According to Newton's third law, the magnitude of $\vec{N}_{\text{S on F}}$ will be the same as the magnitude of $\vec{N}_{\text{F on S}}$.

**Represent mathematically** The process occurs along the vertical direction, so we apply the vertical $y$-component form of Newton's second law:

$$a_y = \frac{\Sigma F_y}{m_{\text{F}}} = \frac{F_{\text{E on F}\,y} + N_{\text{S on F}\,y}}{m_{\text{F}}} = \frac{-m_{\text{F}}g + N_{\text{S on F}}}{m_{\text{F}}}$$

Rearranging the above, we get an expression for the force of the surface on the froghopper while pushing off:

$$N_{\text{S on F}} = m_{\text{F}}a_y + m_{\text{F}}g$$

**Solve and evaluate** We can now substitute the known information in the above, converting the froghopper mass into kg:

$$12\text{ mg} = (12\text{ mg})\left(\frac{1\text{ g}}{10^3\text{ mg}}\right)\left(\frac{1\text{ kg}}{10^3\text{ g}}\right) = 12 \times 10^{-6}\text{ kg}$$

$$\begin{aligned} N_{\text{S on F}} &= (12 \times 10^{-6}\text{ kg})(4000\text{ m/s}^2) \\ &\quad + (12 \times 10^{-6}\text{ kg})(9.8\text{ m/s}^2) \\ &= 0.048\text{ N} + 0.00012\text{ N} = 0.048\text{ N} \approx 0.05\text{ N} \end{aligned}$$

According to Newton's third law, the magnitude of the force exerted on the surface by the froghopper is also 0.05 N.

The force that the surface exerts on the froghopper is 400 times greater than the force that Earth exerts on the froghopper. Also, notice that the force magnitudes and both the motion and force diagrams are consistent with each other—a nice check on our work.

**Try it yourself:** Suppose an 80-kg basketball player could push off a gym floor exerting a force (like the froghopper's) that is 400 times greater than the force Earth exerts on him and that the push off lasted 0.10 s (unrealistically short). How fast would he be moving when he left contact with the floor?

***Answer:*** About 400 m/s (900 mi/h). He would be moving faster if he took longer to push off while exerting the same force.

**Review Question 2.9** Identify force pairs for the following interactions and compare the force magnitudes: A rollerblader and the floor; you pushing a refrigerator across the kitchen floor; and a tow truck pulling a car.

**Figure 2.8** An air bag stops a crash test dummy during a collision.

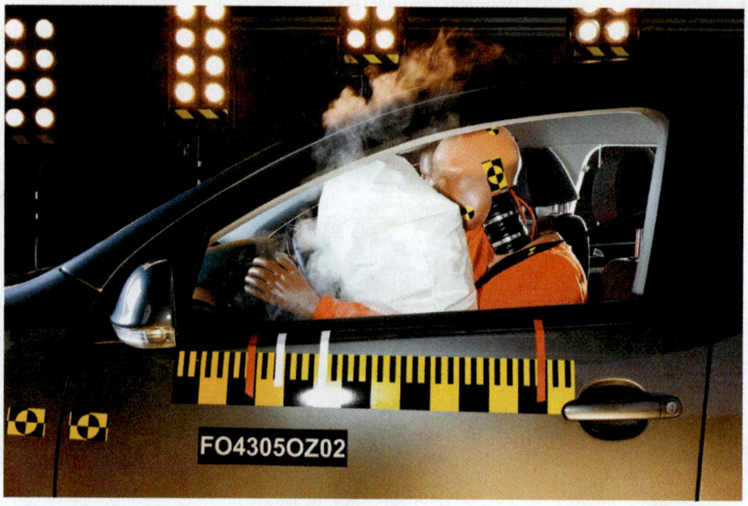


## 2.10 Seat belts and air bags: Putting it all together

At the beginning of the chapter we posed a question about airbags. How do they save lives? We now have all the physics needed to investigate this question. Consider **Figure 2.8**. An air bag is like a balloon with heavy-walled material that is packed in a small box in the steering wheel or the passenger side dashboard. Air bags are designed to deploy when a car has a negative acceleration of magnitude $10g[10(9.8\text{ m/s}^2) = 98\text{ m/s}^2 \approx 100\text{ m/s}^2]$ or more. When a car has such a rapid decrease in speed, the bag inflates with nitrogen gas in about 0.04 s and forms a cushion for the occupant's chest and head. The bag has two important effects:

1. It spreads the force that stops the person over a larger area of the body.
2. It increases the stopping distance, and consequently the stopping time interval, thus reducing the average force stopping the occupant.

Why is spreading the stopping force over the air bag an advantage? If a person uses only seat belts, his head is not belted to the seat and tends to continue moving forward during a collision, even though his chest and waist are restrained. To stop the head without an air bag, the neck must exert considerable force on the head. This can cause a dangerous stretching of the spinal cord and muscles of the neck, a phenomenon known as "whiplash." The air bag exerts a more uniform force across the upper body and head and helps make all parts stop together.

How does the air bag increase the stopping distance? Suppose a test car is traveling at a constant speed of 13.4 m/s (30 mi/h) until it collides head-on into a concrete wall. The front of the car crumples about 0.65 m. A crash test dummy is rigidly attached to the car's seat and is further protected by the rapidly inflating airbag. The dummy also travels about 0.65 m before coming to rest. Without an air bag or a seat belt, the dummy would continue to move forward at the initial velocity of the car. The dummy would then crash into the steering wheel or windshield of the stopped car and stop in a distance much less than 0.65 m—like flying into a rigid wall. The smaller the stopping distance, the greater the acceleration, and therefore the greater the force that is exerted on the dummy. Let's estimate the average force exerted by the air bag on the body during a collision.

**Active Learning Guide›**

**EXAMPLE 2.11 Force exerted by air bag on driver during collision**

A 60-kg crash test dummy moving at 13.4 m/s (30 mi/h) stops during a collision in a distance of 0.65 m. Estimate the average force that the air bag and seat belt exert on the dummy.

**Sketch and translate** We sketch and label the situation as shown below. choosing the crash test dummy as the system object. The positive $x$-direction will be in the direction of motion, and the origin will be at the position of the dummy at the start of the collision.

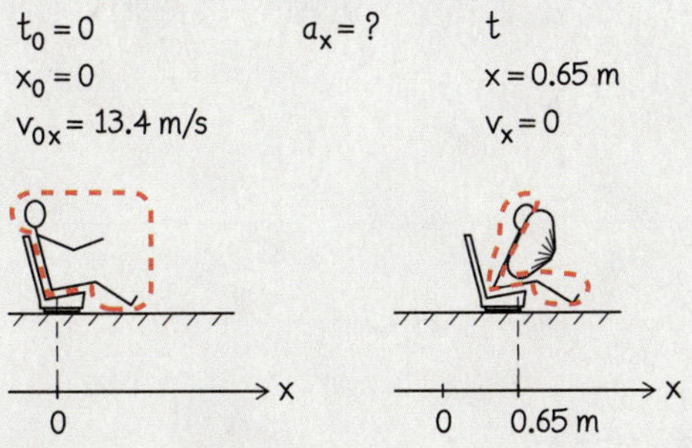


**Simplify and diagram** We model the dummy D as a point-like object and assume that the primary force exerted on the dummy while stopping is due to the air bag and seat belt's $\vec{F}_{\text{A on D}}$, shown in the force diagram. We can ignore the downward gravitational force that Earth exerts on the dummy $\vec{F}_{\text{E on D}}$ and the upward normal force that the car seat exerts on the dummy $\vec{N}_{\text{S on D}}$ since they balance and do not contribute to the acceleration.

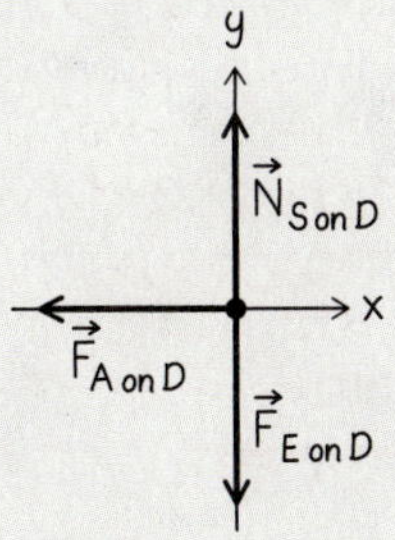


**Represent mathematically** To determine the dummy's acceleration, we use kinematics:

$$a_x = \frac{v_x^2 - v_{0x}^2}{2(x - x_0)}$$

Once we have the dummy's acceleration, we apply the $x$-component form of Newton's second law to find the force exerted by the air bag and seat belts on the dummy:

$$a_x = \frac{F_{\text{A on D}\,x}}{m_{\text{D}}} = \frac{-F_{\text{A on D}}}{m_{\text{D}}} = -\frac{F_{\text{A on D}}}{m_{\text{D}}}$$

$$\Rightarrow F_{\text{A on D}} = -m_{\text{D}} a_x$$

**Solve and evaluate** We know that $v_{0x} = +13.4\text{ m/s}$ and $v_x = 0$ (the dummy has stopped). The initial position of the dummy is $x_0 = 0$ and the final position is $x = 0.65\text{ m}$. The acceleration of the dummy while in contact with the air bag and seat belt is:

$$a_x = \frac{0^2 - (13.4\text{ m/s})^2}{2(0.65\text{ m} - 0\text{ m})} = -138\text{ m/s}^2$$

Thus, the *magnitude* of the average force exerted by the air bag and seat belt on the dummy is

$$F_{\text{A on D}} = -(60\text{ kg})(-138\text{ m/s}^2) = 8300\text{ N}$$

This force $[8300\text{ N}(1\text{ lb}/4.45\text{ N}) = 1900\text{ lb}]$ is almost 1 ton. Is this estimate reasonable? The magnitude is large, but experiments with crash test dummies in the real world are consistent with a force this large in magnitude, a very survivable collision.

**Try it yourself:** Find the acceleration of the dummy and the magnitude of the average force needed to stop the dummy if it is not belted, has no air bag, and stops in 0.1 m when hitting a hard surface.

*Answer:* $-900\text{ m/s}^2$ and 54,000 N.

**Review Question 2.10** Explain how an air bag and seat belt reduce the force exerted on the driver of a car during a collision.

## Summary

**Words** | **Pictorial and physical representations** | **Mathematical representation**

**System and environment** A system object is circled in a sketch of a process. Environmental objects that are not part of the system are external and might interact with the system and affect its motion. (Section 2.1)

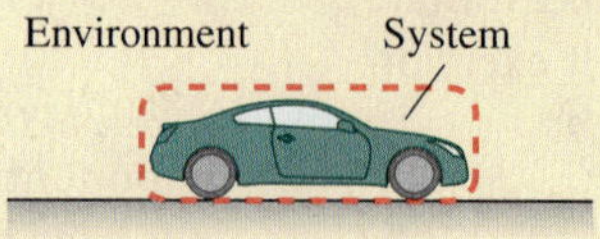


**The force** that one object exerts on another characterizes an interaction between the two objects (a pull or a push) denoted by a symbol $\vec{F}$ with a subscript with two elements indicating the two interacting objects. (Section 2.1)

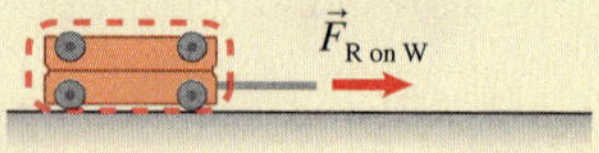


$\vec{F}_{\text{E on O}}$

Interacting objects

**A force diagram** represents the forces that external objects exert on the system object. The arrows in the diagram point in the directions of the forces, and their lengths indicate the relative magnitudes of the forces. The unit of force is the newton (N); $1\text{ N} = (1\text{ kg})(1\text{ kg}\cdot\text{m/s}^2)$. (Sections 2.1 and 2.6)

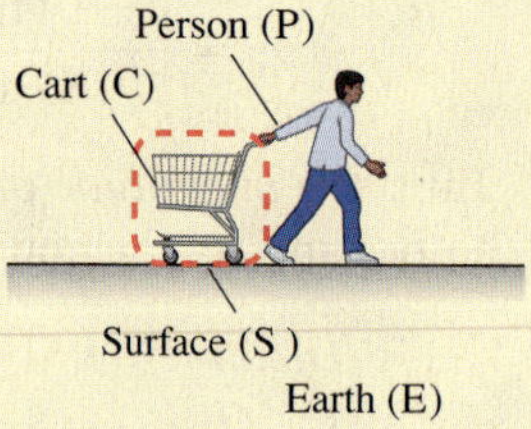


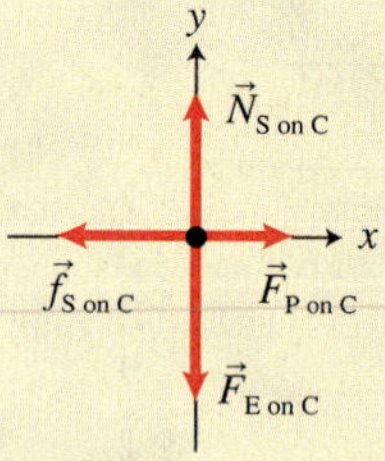


**Relating forces and motion** The $\Delta\vec{v}$ arrow in an object's motion diagram is in the same direction as the sum of the forces that other objects exert on it. (Section 2.3)

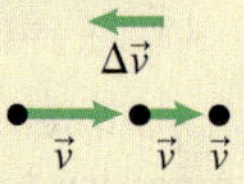


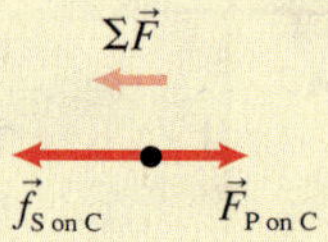


In an **inertial reference frame,** the velocity of a system object does not change if no other objects exert forces on it or if the sum of all the forces exerted on the system object is zero. (Section 2.5)

**Newton's first law of motion** If no other objects exert forces on a system object or if the forces on the system object add to zero, then the object continues moving at constant velocity (as seen by observers in inertial reference frames). (Section 2.5)

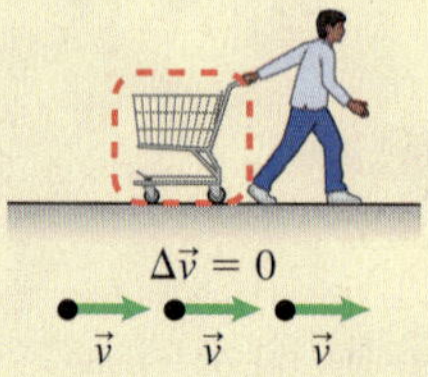


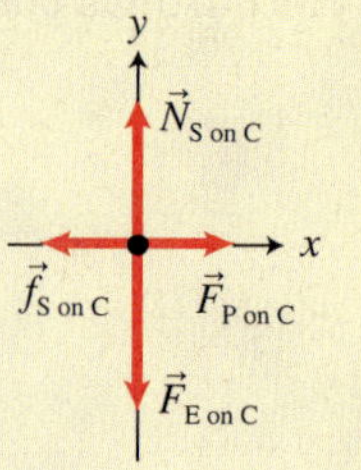


If $\sum F_x = 0$, then $v_x =$ constant (same for $y$-direction)

**Mass** $m$ characterizes the amount of "material" in an object and its resistance to a change in motion. (Section 2.6)

**Newton's second law** The acceleration $a_O$ of a system object is proportional to the sum of the forces that other objects exert on the system object and inversely proportional to its mass $m$. (Section 2.6)

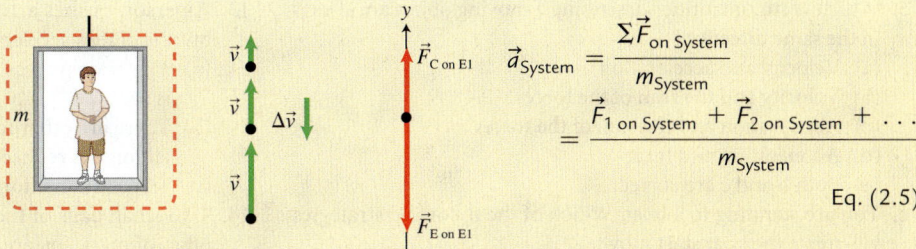

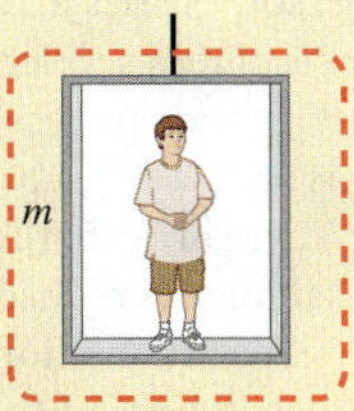


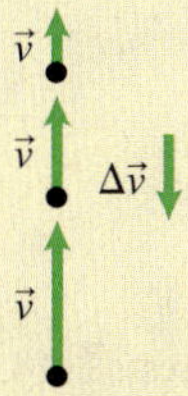


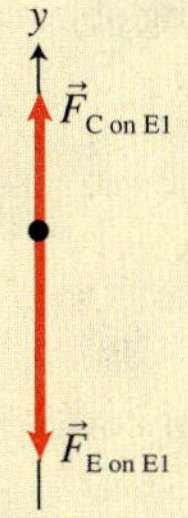


$$\vec{a}_{\text{System}} = \frac{\Sigma\vec{F}_{\text{on System}}}{m_{\text{System}}} = \frac{\vec{F}_{1\text{ on System}} + \vec{F}_{2\text{ on System}} + \ldots}{m_{\text{System}}} \qquad \text{Eq. (2.5)}$$

In the **component form of Newton's second law,** the force components along each axis are included with + or − signs times their magnitudes. (Section 2.6)

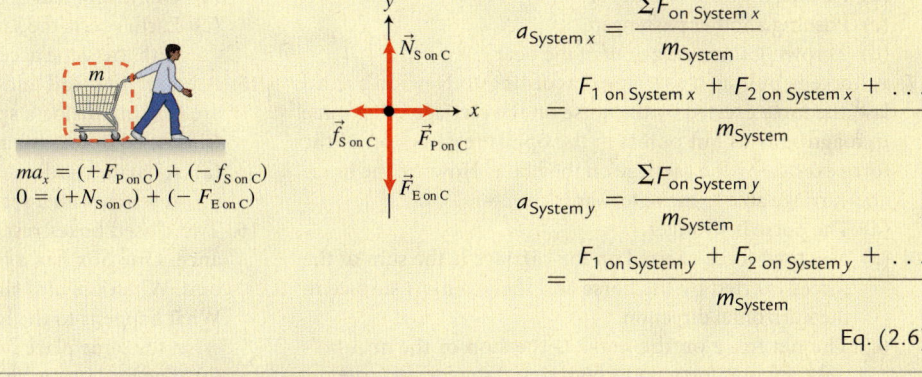

$ma_x = (+F_{\text{P on C}}) + (-f_{\text{S on C}})$
$0 = (+N_{\text{S on C}}) + (-F_{\text{E on C}})$

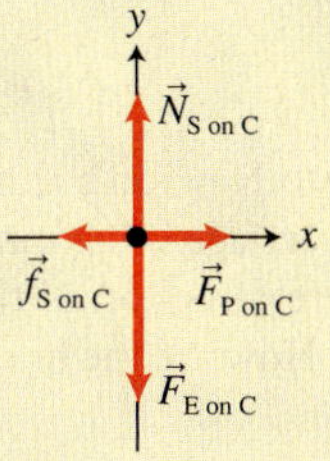


$$a_{\text{System}\,x} = \frac{\Sigma F_{\text{on System}\,x}}{m_{\text{System}}} = \frac{F_{1\text{ on System}\,x} + F_{2\text{ on System}\,x} + \ldots}{m_{\text{System}}}$$

$$a_{\text{System}\,y} = \frac{\Sigma F_{\text{on System}\,y}}{m_{\text{System}}} = \frac{F_{1\text{ on System}\,y} + F_{2\text{ on System}\,y} + \ldots}{m_{\text{System}}} \qquad \text{Eq. (2.6)}$$

**Newton's third law** Two objects exert equal-magnitude and opposite direction forces of the same type on each other. (Section 2.9)

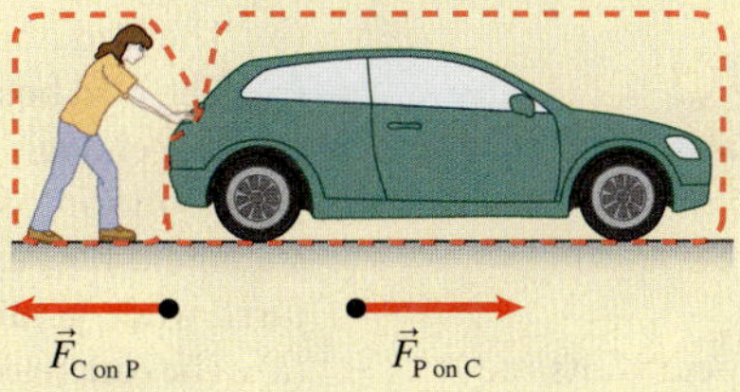


$$\vec{F}_{1\text{ on }2} = -\vec{F}_{2\text{ on }1} \qquad \text{Eq. (2.8)}$$

**The gravitational force** $\vec{F}_{\text{E on O}}$ that Earth exerts on an object of mass $m$ when on or near Earth's surface depends on the gravitational constant $g$ of Earth. If on or near another planet or the Moon, the gravitational constant near those objects is different. (Section 2.7)

Magnitude

$$F_{\text{E on O}} = m_O g \qquad \text{Eq. (2.7)}$$

where

$$g = 9.8\ \text{m/s}^2 = 9.8\ \text{N/kg}$$

when the object is on or near Earth's surface.

MP® For instructor-assigned homework, go to MasteringPhysics.

# Questions

## Multiple Choice Questions

1. An upward-moving elevator slows to a stop as it approaches the top floor. Which answer below best describes the relative magnitudes of the upward force that the cable exerts on the elevator $\vec{F}_{\text{C on El}}$ and the downward gravitational force that Earth exerts on the elevator $\vec{F}_{\text{E on El}}$?
   (a) $F_{\text{C on El}} > F_{\text{E on El}}$ (b) $F_{\text{C on El}} = F_{\text{E on El}}$
   (c) $F_{\text{C on El}} < F_{\text{E on El}}$ (d) None of these
2. You apply the brakes of your car abruptly and your book starts sliding off the front seat. Three observers explain this differently. Observer A says that the book continued moving and the car accelerated from underneath it. Observer B says that the car pushed forward on the book. Observer C says that she must be in a noninertial reference frame because the book started moving without any extra objects interacting with it. Which of the observers is correct?
   (a) A (b) B (c) C
   (d) A and C (e) All of the observers
3. Which of the statements below explains why a child lurches forward in a stroller when you abruptly stop the stroller?
   (a) The child does not lurch forward but instead continues her motion.
   (b) Your pull on the stroller causes the child to move in the opposite direction.
   (c) Newton's third law
4. Which observers can explain the phenomenon of whiplash, which occurs when a car stops abruptly using Newton's laws?
   (a) The driver of the car (b) A passenger in the car
   (c) An observer on the sidewalk beside the car and road

5. Which vector quantities describing a moving object are always in the same direction?
   (a) Velocity and acceleration
   (b) Velocity and the sum of the forces
   (c) Acceleration and the sum of the forces
   (d) Acceleration and force
   (e) Both b and c are correct.
6. You are standing in a boat. Which of the following strategies will make the boat start moving?
   (a) Pushing its mast
   (b) Pushing the front of the boat
   (c) Pushing another passenger
   (d) Throwing some cargo out of the boat
7. A horse is pulling a carriage. According to Newton's third law, the force exerted by the horse on the carriage is the same in magnitude as but points in the opposite direction of the force exerted by the carriage on the horse. How are the horse and carriage able to move forward?
   (a) The horse is stronger.
   (b) The total force exerted on the carriage is the sum of the forces exerted by the horse and the ground's surface in the horizontal direction.
   (c) The net force on the horse is the sum of the ground's static friction force on its hooves and the force exerted by the carriage.
   (d) Both b and c are correct.
8. A book sits on a tabletop. Which of Newton's laws explains its equilibrium?
   (a) First (b) Second (c) Third
   (d) Both the first and the second
9. A spaceship moves in outer space. What happens to its motion if there are no external forces exerted on it? If there is a constant force exerted on it in the direction of its motion? If something exerts a force opposite its motion?
   (a) It keeps moving; it speeds up with constant acceleration; it slows down with constant acceleration.
   (b) It slows down; it moves with constant velocity; it slows down.
   (c) It slows down; it moves with constant velocity; it stops instantly.
10. A 0.10-kg apple falls on Earth, whose mass is about $6 \times 10^{24}$ kg. Which is true of the gravitational force that Earth exerts on the apple?
    (a) It is bigger than the force that the apple exerts on Earth by almost 25 orders of magnitude.
    (b) It is the same magnitude.
    (c) We do not know the magnitude of the force the apple exerts on Earth.
11. A man stands on a scale and holds a heavy object in his hands. What happens to the scale reading if the man quickly lifts the object upward and then stops lifting it?
    (a) The reading increases, returns briefly to the reading when standing stationary, then decreases.
    (b) The reading decreases, returns briefly to the reading when standing stationary, then increases.
    (c) Nothing, since the mass of the person with the object remains the same. Thus the reading does not change.
12. You stand on a bathroom scale in a moving elevator. What happens to the scale reading if the cable holding the elevator suddenly breaks?
    (a) The reading will increase.
    (b) The reading will not change.
    (c) The reading will decrease a little.
    (d) The reading will drop to 0 instantly.
13. A person pushes a 10-kg crate exerting a 200-N force on it, but the crate's acceleration is only 5 m/s$^2$. Explain.
    (a) The crate pushes back on the person, thus the total force is reduced.
    (b) There are other forces exerted on the crate so that the total force is reduced.
    (c) Not enough information is given to answer the question.
14. Two small balls of the same material, one of mass $m$ and the other of mass $2m$, are dropped simultaneously from the Leaning Tower of Pisa. On which ball does Earth exert a bigger force?
    (a) On the $2m$ ball (b) On the $m$ ball
    (c) Earth exerts the same force on both balls because they fall with the same acceleration.
15. A box full of lead and a box of the same size full of feathers are floating inside a spaceship that has left the solar system. Choose equipment that you can use to compare their masses.
    (a) A balance scale (b) A digital scale
    (c) A watch with a second hand and a meter stick
16. Two closed boxes rest on the platforms of an equal arm balance. One box has a ball in it and the other box contains a bird. When the bird sits still in the box, the scale is balanced. What happens to the balance if the bird flies in the box (hovers at the same place inside the box)?
    (a) The bird box is lighter.
    (b The bird box is heavier.
    (c) The bird box is the same.
    (d) Not enough information is provided.
17. A person jumps from a wall and lands stiff-legged. Which statement best explains why the person is less likely to be injured when landing on soft sand than on concrete?
    (a) The concrete can exert a greater force than the sand.
    (b) The person sinks into the sand, increasing the stopping distance.
    (c) The upward acceleration of the person in the sand is less than on concrete, thus the force that the sand exerts on the person is less.
    (d) b and c (e) a, b, and c
18. What do objects that are already in motion on a smooth surface need in order to maintain the same motion?
    (a) A constant push exerted by another object
    (b) A steadily increasing push (c) Nothing

## Conceptual Questions

19. **Figure Q2.19** is a velocity-versus-time graph for the vertical motion of an object. Choose a correct combination (a, b, or c) of

**Figure Q2.19**

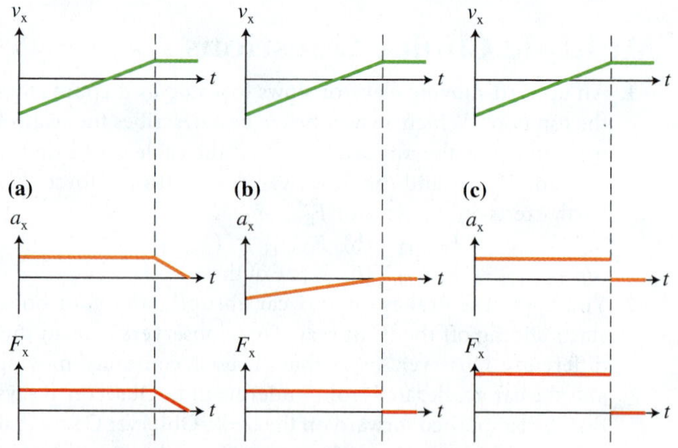

an acceleration-versus-time graph and a force-versus-time graph for the object.

20. Explain the purpose of crumple zones, that is, the front of a car that collapses during a collision.
21. Explain why when landing on a firm surface after a fall you should not land with stiff legs.
22. A small car bumps into a large truck. Compare the forces that the truck exerts on the car and the car exerts on the truck if before the collision (a) the truck was stationary and the car was moving; (b) the car and the truck were moving in opposite directions; (c) the car and the truck were moving in the same direction.
23. You are pulling a sled. Compare the forces that you exert on the sled and the sled exerts on you if you (a) move at constant velocity; (b) speed up; (c) slow down.
24. ✎ You stand on a bathroom scale in a moving elevator. The elevator is moving up at increasing speed. The acceleration is constant. Draw three consecutive force diagrams for you.

# Problems

Below, BIO indicates a problem with a biological or medical focus. Problems labeled EST ask you to estimate the answer to a quantitative problem rather than derive a specific answer. Problems marked with ✎ require you to make a drawing or graph as part of your solution. Asterisks indicate the level of difficulty of the problem. Problems with no * are considered to be the least difficult. A single * marks moderately difficult problems. Two ** indicate more difficult problems.

## 2.1 and 2.2 Describing and representing interactions and Adding and measuring forces

1. In **Figure P2.1** you see unlabeled force diagrams for balls in different situations. Match the diagrams with the following descriptions. (1) A ball is moving upward after it leaves your hand. (2) You hold a ball in your hand. (3) A ball is falling down. (4) You are throwing a ball (still in your hand) straight up. (5) You are lifting a ball at a constant pace. Explain your choices. Label the forces on the diagrams.

**Figure P2.1**

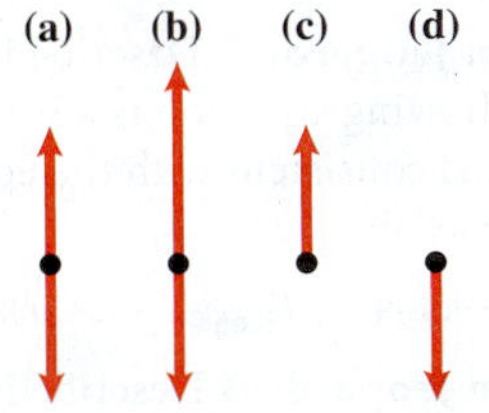


2. ✎ Draw a force diagram (a) for a bag hanging at rest from a spring; (b) for the same bag sitting on a table; and (c) for the same bag that you start to lift so it moves up faster and faster.
3. ✎ For each of the following situations, draw the forces exerted on the moving object and identify the other object causing each force. (a) You pull a wagon along a level floor using a rope oriented horizontally. (b) A bus moving on a horizontal road slows down in order to stop. (c) You lift your overnight bag into the overhead compartment on an airplane.
4. ✎ You hang a book bag on a spring scale and place the bag on a platform scale so that the platform scale reads 25.7 N and the spring scale reads 17.6 N. (a) Draw a force diagram to represent the situation. (b) What is the magnitude of the force that Earth exerts on the bag?

## 2.3 and 2.4 Conceptual relationship between force and motion and Reasoning without mathematical equations

5. ✎ A block of dry ice slides at constant velocity along a smooth, horizontal surface (no friction). (a) Construct a motion diagram. (b) Draw position- and velocity-versus-time graphs. (c) Construct a force diagram for the block for three instances represented by dots on the motion diagram. Are the diagrams consistent with each other?
6. * ✎ You throw a ball upward. (a) Draw a motion diagram and two force diagrams for the ball on its way up and another motion diagram and two force diagrams for the ball on its way down. (b) Represent the motion of the ball with a position-versus-time graph and velocity-versus-time graph.
7. ✎ A string pulls horizontally on a cart so that it moves at increasing speed along a smooth, frictionless, horizontal surface. When the cart is moving medium-fast, the pulling is stopped abruptly. (a) Describe in words what happens to the cart's motion when the pulling stops. (b) Illustrate your description with motion diagrams, force diagrams, and position-versus-time and velocity-versus-time graphs. Indicate on the graphs when the pulling stopped. What assumptions did you make?
8. * Solving the previous problem, your friend says that after the string stops pulling, the cart starts slowing down. (a) Give a reason for his opinion. (b) Do you agree with him? Explain your opinion. (c) Explain how you can design an experiment to test his idea.
9. * ✎ A string pulls horizontally on a cart so that it moves at increasing speed along a smooth, frictionless, horizontal surface. When the cart is moving medium-fast, the magnitude of the pulling force is reduced to half its former magnitude. (a) Describe what happens to the cart's motion after the reduction in the string pulling. (b) Illustrate your description with motion diagrams, force diagrams, and position-versus-time and velocity-versus-time graphs.
10. * Solving the previous problem, your friend says that if the string pulls half as hard, the cart should move half as fast compared to the speed it moved when the string was pulling twice as hard. (a) Explain why your friend would think this way. (b) Do you agree with his opinion? (c) Explain how you would convince him that he is incorrect.
11. ✎ Three motion diagrams for a moving elevator are shown in **Figure P2.11**. Construct two force diagrams for the elevator

for *each* motion diagram. Be sure that the lengths of the force arrows are the appropriate relative lengths and that there is consistency between the force diagrams and the motion diagrams. What assumptions did you make?

**Figure P2.11**

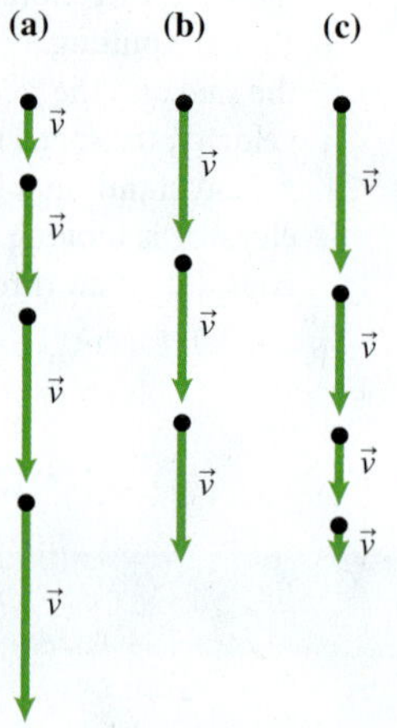


12. * An elevator is pulled upward so it moves with increasing upward speed—the force exerted by the cable on the elevator is constant and greater than the downward gravitational force exerted by Earth. When the elevator is moving up medium-fast, the force exerted by the cable on the elevator changes abruptly to just balance the downward gravitational force of Earth—the sum of the forces that the cable and Earth exert on the elevator is now zero. Now what happens to the elevator? Explain. Represent your answer with position-versus-time and velocity-versus-time graphs. What assumptions did you make?
13. * Solving the previous problem, your friend says that the elevator will stop if the two forces have equal magnitudes. (a) Why would he think this way? (b) Do you agree with his opinion? (c) If you disagree, how would you convince him that he is incorrect?
14. A block of dry ice slides at a constant velocity on a smooth, horizontal, frictionless surface. A second block of dry ice slides twice as fast on the same surface (at a higher constant velocity). Compare the resultant forces exerted on each block. Explain your reasoning.
15. An elevator moves downward at constant velocity. Construct a motion diagram and three consecutive force diagrams for the elevator (ignore the resistive force exerted by the air on the elevator) as it is moving. Make the relative lengths of the force arrows consistent with the motion diagram.
16. * Figures P2.11a, b, and c show three motion diagrams for an elevator moving downward. (a) For each diagram, say everything you can about the elevator's motion. (b) Draw a force diagram for each motion diagram. (c) Could you draw a different motion diagram for each force diagram? Explain how it is possible.

## 2.5 Inertial reference frames and Newton's first law

17. * Your friend has a pie on the roof of his van. You are standing on the ground and observe the van stopping abruptly for a red light. The pie does not slip off the roof. (a) Construct a motion diagram and a force diagram for the pie as the van approaches the red light, from your frame of reference and from the driver's frame of reference. (b) Repeat part (a) for the case when the light turns green. Be sure to specify the observer and identify the other object causing each force. (c) Are the motion diagrams consistent with the force diagrams for each case?
18. * A train traveling from New York to Philadelphia is passing a station. A ball is sitting on the floor of the train not moving with respect to the train. (a) Draw a force diagram and a motion diagram for the ball as seen by the observers on the train and on the platform. (b) The ball now starts accelerating forward relative to the floor. Draw force and motion diagrams for the ball as seen by the observers on the train and on the platform. Which of the observers can use Newton's first law to explain the ball's acceleration? Explain.
19. * Explain the phenomenon of whiplash from two points of view: that of an observer on the ground and an observer in the car.

## 2.6 Newton's second law

20. An astronaut exerts a 100-N force pushing a beam into place on the International Space Station. The beam accelerates at $0.10\ \text{m/s}^2$. Determine the mass of the beam. What is the percent uncertainty in your answer?
21. Four people participate in a rope competition. Two of them pull the rope right, exerting forces of magnitude 330 N and 380 N. The other two pull left, exerting forces of magnitude 300 N and 400 N. What is the sum of the forces exerted on a short section in the middle of the rope?
22. * **Shot put throw** During a practice shot put throw, the 7.0-kg shot left world champion C. J. Hunter's hand at speed 13 m/s. While making the throw, his hand pushed the shot a distance of 1.7 m. Describe all the physical quantities you can determine using this information. Describe the assumptions you need to make to determine them.
23. * You know the sum of the forces $\Sigma\vec{F}$ exerted on an object of mass $m$ during $\Delta t$ seconds. The object is at rest at the beginning of the time interval. List three physical quantities that you can determine about that object's motion using this information. Then explain how you will determine them.
24. * You record the displacement of an object as a constant force is exerted on it. (a) If the time interval during which the force is exerted doubles, how does the object's displacement change? Indicate all the assumptions that you made. (b) Explain how your answer changes if one of the assumptions is not valid.
25. * **Equation Jeopardy 1** Invent a problem for which the following equation can be a solution:

$$200\ \text{N} - 40\ \text{N} = (40\ \text{kg})a_x$$

## 2.7 and 2.8 Gravitational force law and Skills for applying Newton's second law for one-dimensional processes

26. * **Equation Jeopardy 2** Describe in words a problem for which the following equation is a solution and draw a force diagram that is consistent with the equation (specify the direction of the axis):

$$+29.4\ \text{N} - F_{\text{R on O}} = (3.0\ \text{kg})(3.0\ \text{m/s}^2)$$

27. * **Equation Jeopardy 3** Describe in words a problem for which the following equation is a solution and draw a force diagram that is consistent with the equation (specify the direction of the axis):

$$100\ \text{N} - f_{\text{S on O}} = (30\ \text{kg})(-1.0\ \text{m/s}^2)$$

28. * **Equation Jeopardy 4** Describe in words a problem for which the following equation is a solution and draw a force diagram that is consistent with the equation (specify the direction of the axis):

$$-196\ \text{N} + F_{\text{P on O}} = (20\ \text{kg})(-2.0\ \text{m/s}^2)$$

29. * **Spider-Man** Spider-Man holds the bottom of an elevator with one hand. With his other hand, he holds a spider cord attached to a 50-kg box of explosives at the bottom of the cord. Determine the force that the cord exerts on the box if (a) the elevator is at rest; (b) the elevator accelerates up at $2.0\ \text{m/s}^2$; (c) the upward-moving elevator's speed decreases at a rate of $2.0\ \text{m/s}^2$; and (d) the elevator falls freely.

30. * A farmer pushes his 500-kg wagon along a horizontal level icy road, exerting a 125-N horizontal force on the wagon. (a) Determine the acceleration of the wagon. How certain are you about your answer? What assumptions did you make? Would the number be higher or lower if you did not make those assumptions? (b) If the wagon started at rest, how fast was it moving after being pushed for 5.0 s?

31. * **Stuntwoman** The downward acceleration of a 60-kg stuntwoman near the end of a fall from a very high building is $7.0 \text{ m/s}^2$. What resistive force does the air exert on her body at that point?

32. EST Estimate the average force that a baseball pitcher's hand exerts on a 0.145-kg baseball as he throws a 40 m/s (90 mi/h) pitch. Indicate all of the assumptions you made.

33. * **Super Hornet jet takeoff** A $2.1 \times 10^4$-kg F-18 Super Hornet jet airplane (see **Figure P2.33**) goes from zero to 265 km/h in 90 m during takeoff from the flight deck of the USS Nimitz aircraft carrier. What physical quantities can you determine using this information? Make a list and determine the values of three of them.

**Figure P2.33**

34. * **Lunar Lander** The Lunar Lander of mass $2.0 \times 10^4$ kg made the last 150 m of its trip to the Moon's surface in 120 s, descending at approximately constant speed. The *Handbook of Lunar Pilots* indicates that the gravitational constant on the Moon is 1.633 N/kg. Using these quantities, what can you learn about the Lunar Lander's motion?

35. * A Navy Seal of mass 80 kg parachuted into an enemy harbor. At one point while he was falling, the resistive force of air exerted on him was 520 N. What can you determine about his motion?

36. * **Astronaut** Karen Nyberg, a 60-kg astronaut, sits on a bathroom scale in a rocket that is taking off vertically with an acceleration of 3 *g*. What does the scale read?

37. * A 0.10-kg apple falls off a tree branch that is 2.0 m above the grass. The apple sinks 0.060 m into the grass while stopping. Determine the force that the grass exerts on the apple while stopping it. Indicate any assumptions you made.

38. ** An 80-kg fireman slides 5.0 m down a fire pole. He holds the pole, which exerts a 500-N steady resistive force on the fireman. At the bottom he slows to a stop in 0.40 m by bending his knees. What can you determine using this information? Determine it.

### 2.9 Forces come in pairs: Newton's third law

39. Earth exerts a 1.0-N gravitational force on an apple as it falls toward the ground. (a) What force does the apple exert on Earth? (b) Compare the accelerations of the apple and Earth due to these forces. The mass of the apple is about 100 g and the mass of Earth is about $6 \times 10^{24}$ kg.

40. * You push a bowling ball down the lane toward the pins. Draw force diagrams for the ball (a) just before you let it go; (b) when the ball is rolling (for two clock readings); (c) as the ball is hitting a bowling pin. (d) For each force exerted on the ball in parts (a)–(c), draw the Newton's third law force beside the force diagram, and indicate the object on which these third law forces are exerted.

41. * EST (a) A 50-kg skater initially at rest throws a 4-kg medicine ball horizontally. Describe what happens to the skater and to the ball. (b) Estimate the acceleration of the ball during the throw and of the skater using a reasonable value for the force that a skater can exert on the medicine ball. (c) The skater moving to the right catches the ball moving to the left. After the catch, both objects move to the right. Draw force diagrams for the skater and for the ball while the ball is being caught.

42. ** EST Basketball player LeBron James can jump vertically over 0.9 m. Estimate the force that he exerts on the surface of the basketball court as he jumps. (a) Compare this force with the force that the surface exerts on James. Describe all assumptions used in your estimate and state how each assumption affects the result. (b) Repeat the problem looking at the time interval when he is landing back on the floor.

43. * EST The Scottish Tug of War Association contests involve eight-person teams pulling on a rope in opposite directions. Estimate the force that the rope exerts on each team. Indicate any assumptions you made and include a force diagram for a short section of the rope.

44. * A bowling ball hits a pin. (a) Draw a force diagram and a motion diagram for the ball during the collision and separate diagrams for the pin. (b) Explain why your friend who has not taken physics would insist that the bowling ball hits the pin harder than the pin hits the ball.

### 2.10 Seat belts and air bags: Putting it all together

45. * **Car safety** The National Transportation Safety Bureau indicates that a person in a car crash has a reasonable chance of survival if his or her acceleration is less than $300 \text{ m/s}^2$. (a) What magnitude force would cause this acceleration in such a collision? (b) What stopping distance is needed if the initial speed before the collision is 20 m/s (72 km/h or 45 mi/h)? (c) Indicate any assumptions you made.

46. * A 70-kg person in a moving car stops during a car collision in a distance of 0.60 m. The stopping force that the air bag exerts on the person is 8000 N. Name at least three physical quantities describing the person's motion that you can determine using this information, and then determine them.

## General Problems

47. * BIO EST **Left ventricle pumping** The lower left chamber of the heart (the left ventricle) pumps blood into the aorta. According to biophysical studies, a left ventricular contraction lasts about 0.20 s and pumps 88 g of blood. This blood starts at rest and after 0.20 s is moving through the aorta at about 2 m/s. (a) Estimate the force exerted on the blood by the left ventricle. (b) What is the percent uncertainty in your answer? (c) What assumptions did you make? Did the assumptions increase or decrease the calculated value of the force compared to the actual value?

48. ** EST **Acorn hits deck** You are sitting on a deck of your house surrounded by oak trees. You hear the sound of an acorn hitting the deck. You wonder if an acorn will do much damage if instead of the deck it hits your head. Make appropriate estimations and assumptions and provide a reasonable answer.

49. ** EST **Olympic dive** During a practice dive, Olympic diver Fu Mingxia reached a maximum height of 5.0 m above the water. She came to rest 0.40 s after hitting the water. Estimate the average force that the water exerted on her while stopping her.

50. ** EST The brakes on a bus fail as it approaches a turn. The bus is traveling at the speed limit before it moves about 23 m across grass before hitting a wall. A bicycle on the bike rack on the front of the bus is crushed, but there is little damage to the bus. (a) Draw force diagrams for the bus and the wall during the collision. (b) Estimate the average force that the bicycle and bus exert on the wall while stopping. Indicate any assumptions made in your estimate.

51. ** EST You are doing squats on a bathroom scale. You decide to push off the scale and jump up. Estimate the reading as you push off and as you land. Indicate any assumptions you made.

52. ** EST Estimate the horizontal speed of the runner shown in **Figure P2.52** at the instant she leaves contact with the starting blocks. Indicate any assumptions you made.

**Figure P2.52**

53. ** EST Estimate the maximum acceleration of Earth if all people got together and jumped up simultaneously.

54. ** EST Estimate how much Earth would move during the jump described in Problem 53.

## Reading Passage Problems

**Col. John Stapp crash tests** From 1946 through 1958, Col. John Stapp headed the U.S. Air Force Aero Medical Laboratory's studies of the human body's ability to tolerate high accelerations during plane crashes. Conventional wisdom at the time indicated that a plane's negative acceleration should not exceed 180 $m/s^2$ (18 times gravitational acceleration, or 18 $g$). Stapp and his colleagues built a 700-kg "Gee Whiz" rocket sled, track, and stopping pistons to measure human tolerance to high acceleration. Starting in June 1949, Stapp and other live subjects rode the sled. In one of Stapp's rides, the sled started at rest and 360 m later was traveling at speed 67 m/s when its braking system was applied, stopping the sled in 6.0 m. He had demonstrated that 18 $g$ was not a limit for human deceleration.

55. In an early practice run while the rocket sled was stopping, a passenger dummy broke its restraining device and the window of the rocket sled and stopped after skidding down the track. What physics principle best explains this outcome?
(a) Newton's first law (b) Newton's second law
(c) Newton's third law (d) The first and second laws
(e) All three laws

56. Which answer below is closest to Stapp's 67 m/s speed in miles per hour?
(a) 30 mi/h (b) 40 mi/h (c) 100 mi/h
(d) 120 mi/h (e) 150 mi/h

57. Which answer below is closest to the magnitude of the acceleration of Stapp and his sled as their speed increased from zero to 67 m/s?
(a) 5 $m/s^2$ (b) 6 $m/s^2$ (c) 10 $m/s^2$
(d) 12 $m/s^2$ (e) 14 $m/s^2$

58. Which answer below is closest to the magnitude of the acceleration of Stapp and his sled as their speed decreased from 67 m/s to zero?
(a) 12 $g$ (b) 19 $g$ (c) 26 $g$
(d) 38 $g$ (e) 48 $g$

59. Which answer below is closest to the average force exerted by the restraining system on 80-kg Stapp while his speed decreased from 67 m/s to zero in a distance of 6.0 m?
(a) 10,000 N (b) 20,000 N (c) 30,000 N
(d) 40,000 N (e) 50,000 N

60. Which answer below is closest to the time interval for Stapp and his sled to stop as their speed decreased from 67 m/s to zero?
(a) 0.09 s (b) 0.18 s (c) 0.34 s
(d) 5.4 s (e) 10.8 s

**Using proportions** A proportion is defined as an equality between two ratios; for instance, $a/b = c/d$. Proportions can be used to determine the expected change in one quantity when another quantity changes. Suppose, for example, that the speed of a car doubles. By what factor does the stopping distance of the car change? Proportions can also be used to answer everyday questions, such as whether a large container or a small container of a product is a better buy on a cost-per-unit-mass basis.

Suppose that a small pizza costs a certain amount. How much should a larger pizza of the same thickness cost? If the cost depends on the amount of ingredients used, then the cost should increase in proportion to the pizza's area and not in proportion to its diameter:

$$Cost = k(Area) = k(\pi r^2) \qquad (2.9)$$

where $r$ is the radius of the pizza and $k$ is a constant that depends on the price of the ingredients per unit area. If the area of the pizza doubles, the cost should double, but $k$ remains unchanged.

Let us rearrange Eq. (2.9) so the two variable quantities (cost and radius) are on the right side of the equation and the constants are on the left:

$$k\pi = \frac{Cost}{r^2}$$

This equation should apply to any size pizza. If $r$ increases, the cost should increase so that the ratio $Cost/r^2$ remains constant. Thus, we can write a proportion for pizzas of different sizes:

$$k\pi = \frac{Cost}{r^2} = \frac{Cost'}{r'^2}$$

For example, if a 3.5-in.-radius pizza costs \$4.00, then a 5.0-in. radius pizza should cost

$$Cost' = \frac{r'^2}{r^2}Cost = \frac{(5.0\text{ in})^2}{(3.5\text{ in})^2}(\$4.00) = \$8.20$$

This process can be used for most equations relating two quantities that change while all other quantities remain constant.

61. The downward distance $d$ that an object falls in a time interval $t$ if starting at rest is $d = \frac{1}{2}at^2$. On the Moon, a rock falls 10.0 m in 3.50 s. How far will the object fall in 5.00 s, assuming the same acceleration?
(a) 14.3 m (b) 20.4 m (c) 4.90 m
(d) 7.00 m (e) 10.0 m

62. The downward distance $d$ that an object falls in a time interval $t$ if starting at rest is $d = \frac{1}{2}at^2$. On the Moon, a rock falls 10.0 m in 3.50 s. What time interval $t$ is needed for it to fall 15.0 m, assuming the same acceleration?
(a) 2.33 s (b) 2.86 s (c) 3.50 s
(d) 4.29 s (e) 5.25 s

63. A car's braking distance $d$ (the distance it travels if rolling to a stop after the brakes are applied) depends on its initial speed $v_0$, the maximum friction force $\vec{f}_{\text{R on C}}$ exerted by the road on the car, and the car's mass $m$ according to the equation

$$\frac{2f_{s\,\max}}{m}d = v_0^2.$$

Suppose the braking distance for a particular car and road surface is 26 m when the initial speed is 18 m/s. What is the braking distance when traveling at 27 m/s?
(a) 59 m (b) 39 m (c) 26 m
(d) 17 m (e) 12 m

64. You decide to open a pizza parlor. The ingredients require that you charge \$4.50 for a 7.0-in.-diameter pizza. How large should you make a pizza whose price is \$10.00, assuming the cost is based entirely on the cost of ingredients?
(a) 1.4 in. (b) 3.1 in. (c) 7.0 in.
(d) 10 in. (e) 16 in.

65. A circular wool quilt of 1.2 m diameter costs \$200. What should the price of a 1.6-m-diameter quilt be if it is to have the same cost per unit area?
(a) \$110 (b) \$150 (c) \$270 (d) \$360

# 3 Applying Newton's Laws

**How does knowing physics help human cannonballs safely perform their tricks?**

**Would an adult or a small child win a race down a water slide?**

**How does friction help us walk?**

**Be sure you know how to:**

- Use trigonometric functions (Mathematics appendix).
- Use motion diagrams and mathematical equations to describe motion (Sections 1.2, 1.5, and 1.6).
- Identify a system, construct a force diagram for it, and use the force diagram to apply Newton's second law (Sections 2.1, 2.6, and 2.8).

On March 10, 2011, David Smith, Jr., set the record for the distance traveled by a human cannonball—59 meters, improving the record set by his father by more than 2 meters. A human cannonball is a performance trick in which a person is ejected from a cannon by a compressed spring or compressed air and lands either on a horizontal net or on an inflated bag. To make a human cannonball shot successful, a designer must apply Newton's laws and kinematics equations to the complex process. Most importantly, the designer must be able to predict where the cannonball will land to make sure there is a supporting cushion there. We'll learn how to do it in this chapter.

**When we first** studied force and motion dynamics processes (in Chapter 2), we learned that a system object's acceleration depends on the sum of the forces that other objects exert on it and on the mass of the object. We considered processes in which the forces exerted on an object by other objects were mainly along one axis—the axis along which motion occurred. These processes are common, but most everyday life processes involve forces that are *not* all directed only along the axis of motion. In this chapter we will learn to apply Newton's laws to those more complex processes.

## 3.1 Force components

To apply Newton's second law in situations in which the force vectors do not all point along the coordinate axes, we need to learn how to break forces into their components. We start by using a vector quantity with which we are already very familiar—displacement $\vec{d}$, a vector that starts from an object's initial position and ends at its final position.

### Graphical vector addition and components of a vector

Suppose you want to take a trip and can reach your final destination in two different ways (see **Figure 3.1**). Route 1 is a direct path that is represented by a displacement vector $\vec{C}$. Its tail represents your initial location and its head represents your final destination. Route 2 goes along two roads represented by displacement vectors $\vec{A}$ and $\vec{B}$ that are perpendicular to each other. You first travel along $\vec{A}$, then along $\vec{B}$, and end at the same final destination.

Route 2 is an example of how to add displacement vectors graphically. You place the tail of $\vec{A}$ at your initial position; then place the tail of $\vec{B}$ at the head of $\vec{A}$. In this example, the head of $\vec{B}$ will be at your final position. Because the initial and final positions are the same for either route, we say that $\vec{C} = \vec{A} + \vec{B}$. Any displacement vector (for example, $\vec{C}$) can be replaced by two perpendicular displacement vectors (for example, $\vec{A}$ and $\vec{B}$) if these two vectors graphically add to equal $\vec{C}$, as illustrated in Figure 3.1.

Because forces are vectors, we can add them graphically as well. As we did with the travel routes above, we can replace a force $\vec{F}$ by two perpendicular forces $\vec{F}_x$ and $\vec{F}_y$ that graphically add to equal the original force. Suppose, for example, we place a small box on a very smooth surface. The box is stationary. Then we attach strings and spring scales to the box and pull in opposite directions, exerting a 5-N force on each string at angles of 37° relative to the plus or minus $x$-axis (**Figure 3.2a**). These two forces balance, and the box remains stationary.

Now, let's replace the string and spring scale pulling at 37° relative to the positive $x$-axis with two strings and two spring scales, one pulling along the $x$-axis and the other along the $y$-axis (Figure 3.2b). We find that if the $x$-axis scale pulls exerting a 4-N force and the $y$-axis scale pulls exerting a 3-N force, the box again remains stationary. Thus, these two forces have the same effect on the box as the single 5-N force exerted on the box at an angle of 37° above the positive $x$-direction.

**Figure 3.1** Two routes to get to the same destination.

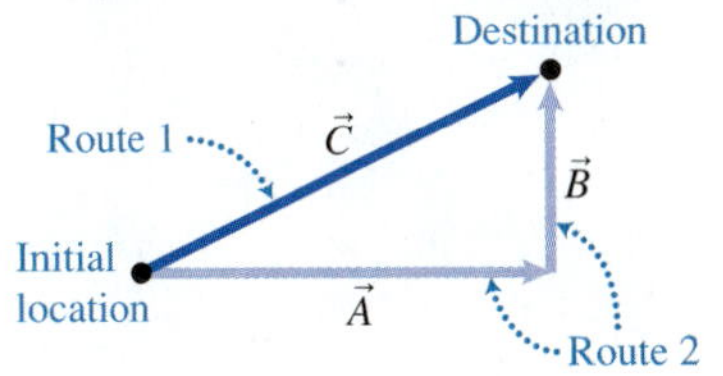

Destination
Route 1
$\vec{C}$
$\vec{B}$
Initial location
$\vec{A}$
Route 2

**Figure 3.2** Graphically adding force components. We can always replace any force $\vec{F}$ with two perpendicular forces $\vec{F}_x$ and $\vec{F}_y$, as long as the perpendicular forces graphically add to $\vec{F}$.

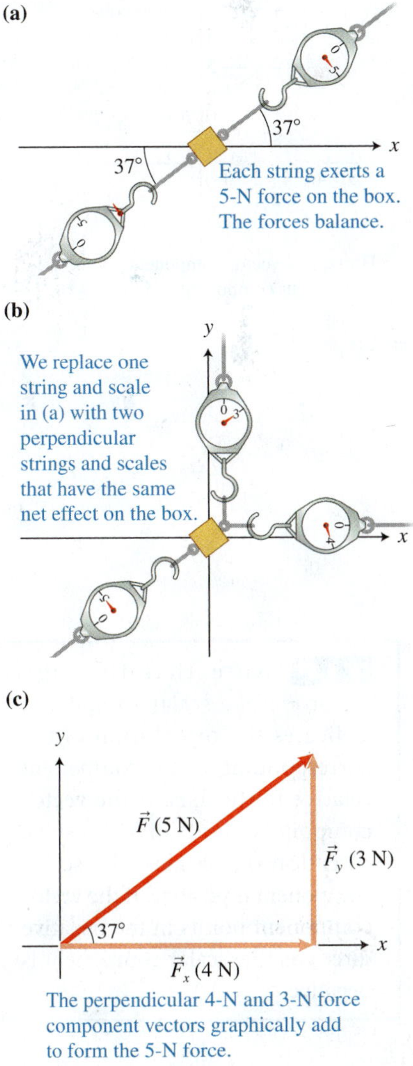

(a)
37°
x
37°
Each string exerts a 5-N force on the box. The forces balance.
(b)
y
We replace one string and scale in (a) with two perpendicular strings and scales that have the same net effect on the box.
x
(c)
y
$\vec{F}$ (5 N)
$\vec{F}_y$ (3 N)
37°
x
$\vec{F}_x$ (4 N)
The perpendicular 4-N and 3-N force component vectors graphically add to form the 5-N force.

This experimental result is consistent with graphical vector addition, where we find that a 4-N force in the positive $x$-direction plus a 3-N force in the positive $y$-direction add to produce a 5-N force at 37° relative to the positive $x$-axis (Figure 3.2c). The vectors form a "3-4-5" right triangle.

In summary, we can always replace any force $\vec{F}$ with two perpendicular forces $\vec{F}_x$ and $\vec{F}_y$, as long as the perpendicular forces graphically add to $\vec{F}$ (Figure 3.2c). In this case, the perpendicular forces are along the perpendicular $x$- and $y$-axes and are called the $x$- and $y$-**vector components** of the original force $\vec{F}$. Since the vector components are also vectors, they are written with a vector symbol above them. Note that $\vec{F}_x + \vec{F}_y = \vec{F}$, just as displacement $\vec{C}$ was identical to the displacement $\vec{A} + \vec{B}$ in Figure 3.1.

## Scalar components of a vector

When working with $x$- and $y$-axes that are perpendicular to each other, we do not need to use the vector components of $\vec{F}$ but can instead provide the same information about the vector by specifying what are called the **scalar components** $F_x$ and $F_y$ of the force $\vec{F}$. The advantage of the scalar components is that they are numbers with signs, which can be added and subtracted more easily than vector quantities. **Figure 3.3** shows how to find the scalar components. The force vector $\vec{F}$ is 5 N and points 37° above the negative $x$-axis. The $x$-vector component of force $\vec{F}$ is 4 N long and points in the negative $x$-direction. Thus, the $x$-scalar component of $\vec{F}$ is $F_x = -4$ N. Similarly, the $y$-vector component of $\vec{F}$ is 3 N in the positive $y$-direction—it pulls upward in the positive $y$-direction. Thus, we can specify the $y$-scalar component of $\vec{F}$ as $F_y = +3$ N. Thus, the $x$- and $y$-components $F_x = -4$ N and $F_y = +3$ N tell us everything we need to know about the force $\vec{F}$. Note that scalars are *not* written with a vector symbol above them.

**Figure 3.3** The vector and scalar components of a vector.

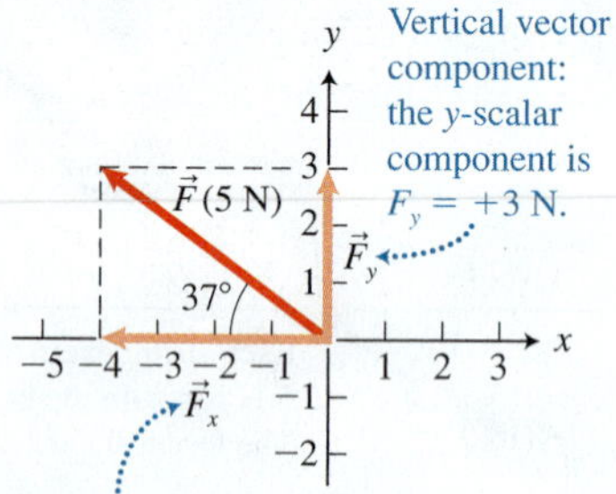
Vertical vector component: the $y$-scalar component is $F_y = +3$ N.
$\vec{F}$ (5 N)
37°
$\vec{F}_y$
$\vec{F}_x$
$y$
$x$
Horizontal vector component: the $x$-scalar component is $F_x = -4$ N.

## Finding the scalar components of a force from its magnitude and direction

If we know the magnitude of a force $F$ and the angle $\theta$ that the force makes above or below the positive or negative $x$-axis, we can determine its scalar components. The Skill box summarizes how to calculate the scalar components of the force.

**REASONING SKILL** Determining the scalar components of a vector.

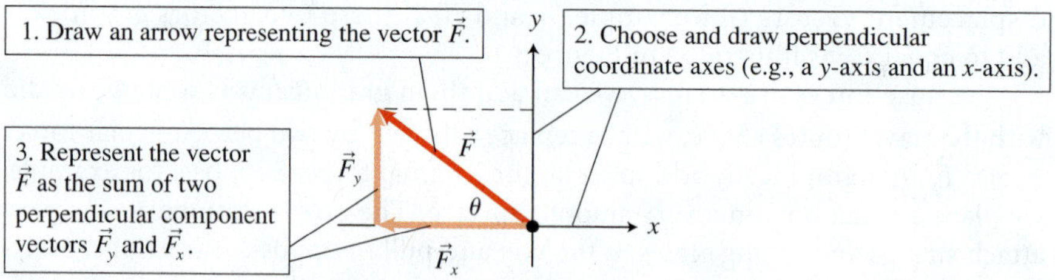
1. Draw an arrow representing the vector $\vec{F}$.
2. Choose and draw perpendicular coordinate axes (e.g., a $y$-axis and an $x$-axis).
3. Represent the vector $\vec{F}$ as the sum of two perpendicular component vectors $\vec{F}_y$ and $\vec{F}_x$.
$\vec{F}$
$\vec{F}_y$
$\vec{F}_x$
$\theta$
$y$
$x$

4. Finally, use the right triangle and trigonometry to calculate the values of the scalar components:

$$F_x = \pm F \cos\theta \qquad (3.1x)$$
$$F_y = \pm F \sin\theta \qquad (3.1y)$$

where $F$ is the magnitude of the force vector and $\theta$ is the angle (90° or less) that $\vec{F}$ makes with respect to the $\pm x$-axis. $F_x$ is positive if in the positive $x$-direction and negative if in the negative $x$-direction. $F_y$ is positive if in the positive $y$-direction and negative if in the negative $y$-direction.

**TIP** Note that the sign (+ or −) of a scalar component indicates the orientation of the corresponding vector component relative to the axis. If the vector component points in the positive direction of the axis, the scalar component is positive. If the vector component points in the negative direction, the scalar component is negative.

**QUANTITATIVE EXERCISE 3.1 Components of forces exerted on a knot**

Three ropes pull on the knot shown below. We place the force diagram on a grid so that you can see the components of the three forces that the ropes exert on the knot. Use the diagram to visually determine the $x$- and $y$-scalar components of the force that rope 2 exerts on the knot. Then use the mathematical method described in the Skill box to calculate the components of this force. The results of using the two methods should be consistent with each other.

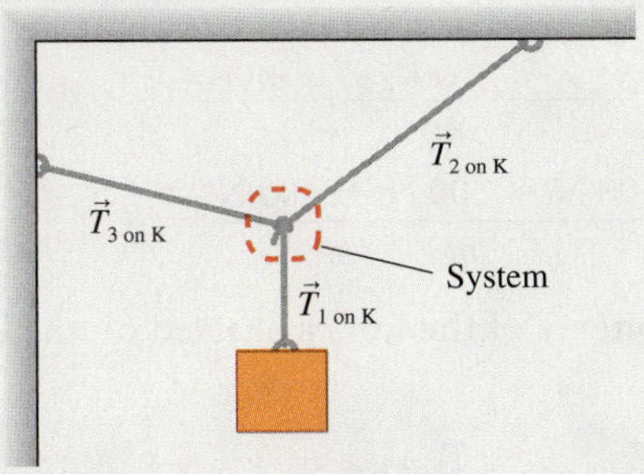


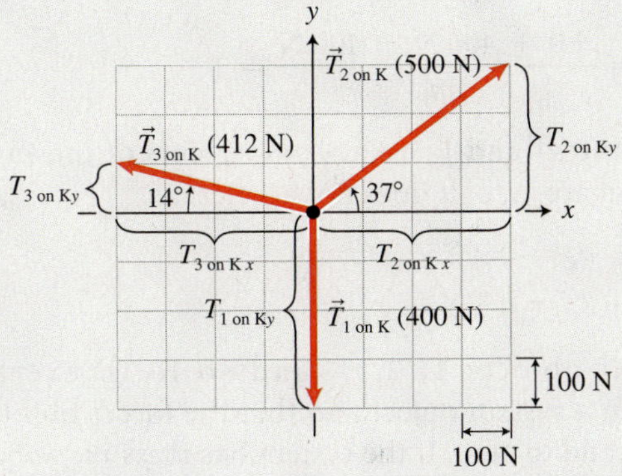


**Represent mathematically** We label forces that ropes exert on the knot using a letter $T$ instead of $F$. This is a common way of noting the forces that ropes or strings exert on an object—tension forces. From the force diagram, we see that rope 2 pulls on the knot at a 37° angle above the positive $x$-axis, exerting a force of magnitude $T_{2\text{ on K}} = 500\text{ N}$. The $x$-scalar component of this force is $+400\text{ N}$ (4 grid units in the positive $x$-direction) and the $y$-scalar component of this force is $+300\text{ N}$ (3 grid units in the positive $y$-direction). In symbols, $T_{2\text{ on K}x} = +400\text{ N}$ and $T_{2\text{ on K}y} = +300\text{ N}$.

**Solve and evaluate** Now use the mathematical method to calculate the components of the force that rope 2 exerts on the knot:

$$T_{2\text{ on K}x} = +(500\text{ N})\cos 37^\circ = +400\text{ N}$$
$$T_{2\text{ on K}y} = +(500\text{ N})\sin 37^\circ = +300\text{ N}$$

The diagrammatical and the mathematical methods are consistent.

**Try it yourself:** Use the same two methods to determine the components of the forces exerted by ropes 1 and 3 on the knot.

*Answer:* $T_{3\text{ on K}x} = -400\text{ N}$ and $T_{3\text{ on K}y} = +100\text{ N}$, and $T_{1\text{ on K}x} = 0$ and $T_{1\text{ on K}y} = -400\text{ N}$.

**TIP** Be careful when applying Eqs. (3.1$x$) and (3.1$y$). The angles that appear in those equations must be measured with respect to the positive or negative $x$-axis.

**Review Question 3.1** When does a vector have a positive scalar component? When does a vector have a negative scalar component?

## 3.2 Newton's second law in component form

Now that we have learned how to work with vectors and their components, we can start to apply Newton's second law to more complex situations in which one or more of the forces exerted on the system do not point along one of the coordinate axes. The figure in Quantitative Exercise 3.1 is an example of such a situation. We know that the knot is not accelerating. Thus, the sum of the forces exerted on the knot is zero:

$$\Sigma \vec{F}_{\text{on K}} = \vec{T}_{1\text{ on K}} + \vec{T}_{2\text{ on K}} + \vec{T}_{3\text{ on K}} = 0$$

It is difficult to use this vector equation to analyze the situation further—for example, to determine one of the forces that is unknown if you know the other two forces. However, we can do such tasks if this situation is represented in scalar component form.

Notice in the force diagram in Quantitative Exercise 3.1 that the $y$-scalar components of the forces that the three ropes exert on the knot add to zero. Ropes 2 and 3 exert forces that have positive $y$-scalar components $+300\text{ N} + 100\text{ N} = +400\text{ N}$, and rope 1 exerts a force with a negative $y$-scalar component $-400$ N. Consequently:

$$a_{\text{K}y} = \frac{T_{1\text{ on K}y} + T_{2\text{ on K}y} + T_{3\text{ on K}y}}{m_{\text{K}}}$$

$$= \frac{-400\text{ N} + 300\text{ N} + 100\text{ N}}{m_{\text{K}}} = 0$$

Similarly, the $x$-scalar components of the forces exerted on the knot also add to zero:

$$a_{\text{K}x} = \frac{T_{1\text{ on K}x} + T_{2\text{ on K}x} + T_{3\text{ on K}x}}{m_{\text{K}}}$$

$$= \frac{+0 + 400\text{ N} - 400\text{ N}}{m_{\text{K}}} = 0$$

Thus, we infer that when the $x$- and $y$-scalar components of the sum of the forces exerted on the system are zero, it does not accelerate:

$$a_x = 0 \text{ if } \Sigma F_{\text{on K}x} = 0$$

$$a_y = 0 \text{ if } \Sigma F_{\text{on K}y} = 0$$

Suppose that in general objects 1, 2, 3, and so forth exert forces $\vec{F}_{1\text{ on S}}, \vec{F}_{2\text{ on S}}, \vec{F}_{3\text{ on S}}, \ldots$ on a system object and that the forces point in arbitrary directions and do not add to zero. If the system has mass $m_{\text{S}}$, it has an acceleration $\vec{a}_{\text{S}}$ as a consequence of these forces (Newton's second law):

$$\vec{a}_{\text{S}} = \frac{\vec{F}_{1\text{ on S}} + \vec{F}_{2\text{ on S}} + \vec{F}_{3\text{ on S}} + \cdots}{m_{\text{S}}}$$

Because this equation involves vectors, we can't work with it directly. However, we can split it into its $x$- and $y$-scalar component forms to use Newton's second law to determine the system's acceleration.

**Newton's second law rewritten in scalar component form becomes:**

$$a_{\text{S}x} = \frac{F_{1\text{ on S}x} + F_{2\text{ on S}x} + F_{3\text{ on S}x} + \cdots}{m_{\text{S}}} \tag{3.2 x}$$

$$a_{\text{S}y} = \frac{F_{1\text{ on S}y} + F_{2\text{ on S}y} + F_{3\text{ on S}y} + \cdots}{m_{\text{S}}} \tag{3.2 y}$$

In practice, we usually choose the axes so that the object accelerates along only one of these axes and has zero acceleration along the other axis. The process of analyzing a situation using Newton's second law in component form is illustrated in the Reasoning Skill box on the next page. (Note that for convenience we can drop the word *scalar* and use the simpler term *component form*).

**REASONING SKILL** Using Newton's second law in component form.

1. Draw a force diagram (the force arrows represent the forces that other objects exert on the system).

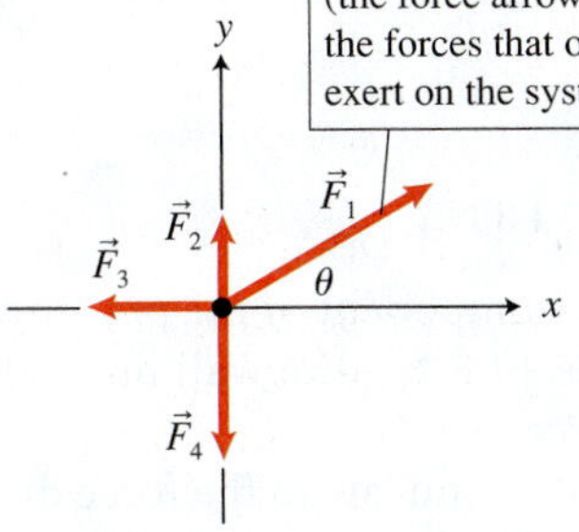


2. Visualize the $x$-components of the forces and apply the $x$-component form of Newton's second law to the force diagram:

$$a_x = \frac{\Sigma F_x}{m} = \frac{F_{1x} + F_{2x} + F_{3x} + F_{4x}}{m} = \frac{+F_1 \cos\theta + 0 + (-F_3) + 0}{m}$$

Notice that:
$F_{1x} = +F_1 \cos\theta$
$F_{2x} = +F_2 \cos 90° = 0$
$F_{3x} = -F_3 \cos 0° = -F_3$
$F_{4x} = +F_4 \cos 90° = 0$

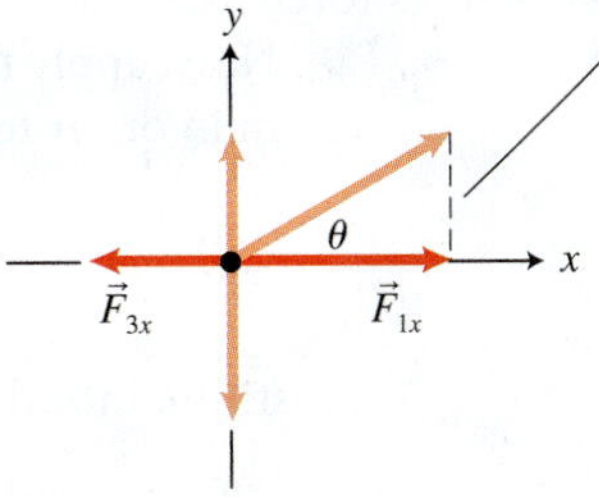


3. Visualize the $y$-components of the forces and apply the $y$-component form of Newton's second law to the force diagram:

$$a_y = \frac{\Sigma F_y}{m} = \frac{F_{1y} + F_{2y} + F_{3y} + F_{4y}}{m} = \frac{+F_1 \sin\theta + F_2 + 0 + (-F_4)}{m}$$

Notice that:
$F_{1y} = +F_1 \sin\theta$
$F_{2y} = +F_2 \sin 90° = +F_2$
$F_{3y} = -F_3 \sin 0° = 0$
$F_{4y} = -F_4 \sin 90° = -F_4$

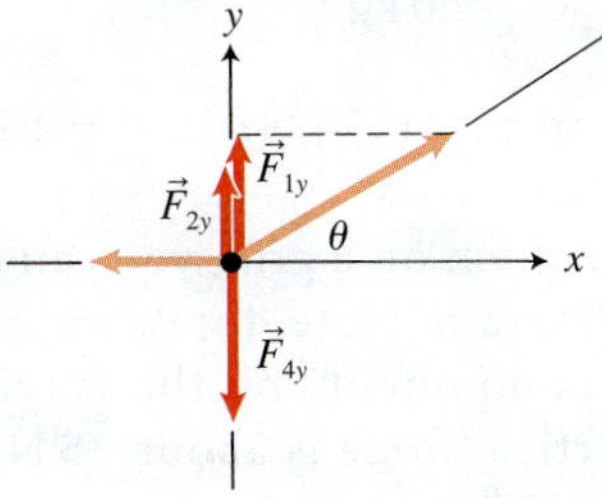


## Using components to solve a problem

Now, let's use this force component method to analyze the following process. Alice pushes on a 2.0-kg book, exerting a 21.0-N force as it slides down a slippery vertical wall (**Figure 3.4a**). She pushes 45° above the horizontal. We use the above reasoning skill to determine the magnitude of the force that the wall exerts on the book and the acceleration of the book down the wall's surface.

Choose the book as the system, model it as a point-like object, and draw a force diagram for it (Figure 3.4b). The force diagram includes three forces: the 21.0-N force that Alice exerts on the book $\vec{F}_{\text{A on B}}$, the downward gravitational force that Earth exerts on the book $\vec{F}_{\text{E on B}}$ of magnitude $mg = (2.0\text{ kg})(9.8\text{ N/kg}) = 19.6\text{ N}$, and the normal force $\vec{N}_{\text{W on B}}$ that the wall exerts on the book (its magnitude is not yet known—we did not include in the force diagram the end of the arrow representing that force). Notice that the normal force that the wall exerts on the book is perpendicular to the wall's surface and points horizontally away from the wall (the normal force is not vertical).

The book does not leave contact with the wall; therefore, the $x$-component of its acceleration is zero, as is the $x$-component of the sum of the forces exerted on the book. We apply the $x$-component form of Newton's second law to the force diagram to determine the magnitude of the normal force so that the book stays on the surface:

$$a_{\text{B}\,x} = \frac{1}{m_{\text{B}}}\Sigma F_x = \frac{F_{\text{A on B}\,x} + N_{\text{W on B}\,x} + F_{\text{E on B}\,x}}{m_{\text{B}}} = 0$$

**Figure 3.4** Calculating the normal force. (a) Alice lowers a book down the wall; (b) a force diagram for the situation.

(a)

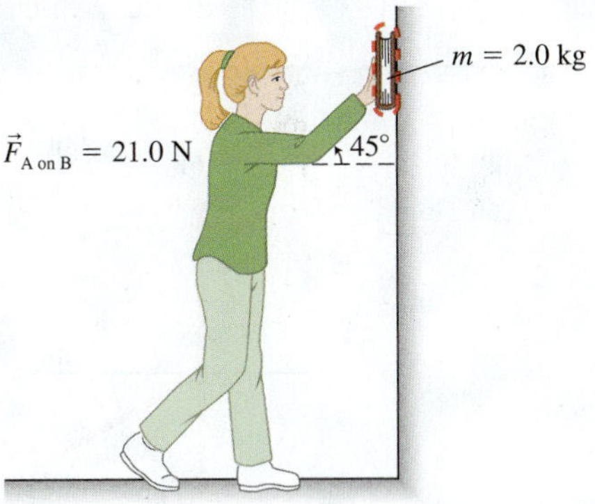


(b)

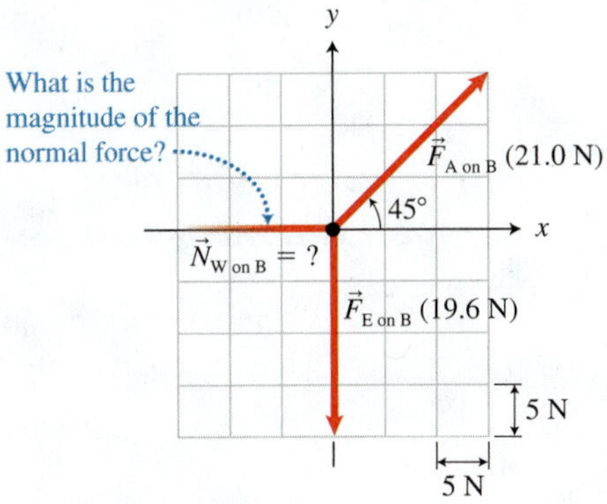

Now, insert expressions for the $x$-scalar components into the above equation:

$$+(21.0\,\text{N})\cos 45^\circ + (-N_{\text{W on B}}\cos 0^\circ) + (19.6\,\text{N})\cos 90^\circ = 0$$

Since $\cos 90^\circ = 0$ and $\cos 0^\circ = 1.0$, we find that

$$(21.0\,\text{N})\cos 45^\circ - N_{\text{W on B}} + 0 = 0$$

or

$$N_{\text{W on B}} = (21.0\,\text{N})\cos 45^\circ = 14.8\,\text{N}$$

Notice in the force diagram with the grid that the $x$-component of the pushing force that Alice exerts on the book has magnitude 14.8 N—the wall normal force balances the $x$-component of that pushing force.

Next, apply the $y$-component form of Newton's second law to the force diagram in order to determine the $y$-component of the acceleration of the book:

$$a_{\text{B}\,y} = \frac{\Sigma F_y}{m_\text{B}} = \frac{F_{\text{A on B}\,y} + N_{\text{W on B}\,y} + F_{\text{E on B}\,y}}{m_\text{B}}$$

Substitute into this equation expressions for the $y$-components of the forces:

$$a_{\text{B}\,y} = \frac{+(21.0\,\text{N})\sin 45^\circ + N_{\text{W on B}}\sin 0^\circ + [-(19.6\,\text{N})\sin 90^\circ]}{2.0\,\text{kg}}$$

Noting that $N_{\text{W on B}} = 14.8\,\text{N}$, $\sin 0^\circ = 0$, and $\sin 90^\circ = 1.0$, we get

$$a_{\text{B}\,y} = \frac{+(21.0\,\text{N})\sin 45^\circ + (14.8\,\text{N})(0) + (-19.6\,\text{N})(1)}{2.0\,\text{kg}}$$

$$= \frac{14.8\,\text{N} + 0 - 19.6\,\text{N}}{2.0\,\text{kg}} = -2.4\,\text{N/kg} = -2.4\,\text{m/s}^2$$

For this problem with the force diagram on a grid, you can check your work visually by looking at the force diagram. Note, for example, that the pushing force has a +14.8-N vertical component and the gravitational force is $-19.6 \approx -20\,\text{N}$. Thus, the net vertical force is about $-5\,\text{N}$ and the vertical acceleration should be about $-5\,\text{N}/(2.0\,\text{kg}) \approx 2.5\,\text{m/s}^2$, as calculated.

**TIP** Perhaps the part of this procedure that causes the most difficulty is the construction of the force diagram. Be sure to include in the diagram only forces exerted *on* the system by external objects (outside the system). Do *not* include forces that the system exerts on external objects (objects that are not included in the system).

**Figure 3.5**

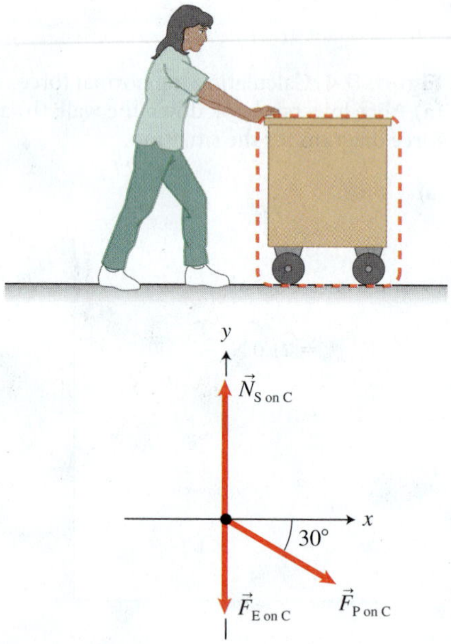


**Review Question 3.2** Apply Newton's second law in component form for the force diagram and process sketched in **Figure 3.5** (both $x$- and $y$-axes).

## 3.3 Problem-solving strategies for analyzing dynamics processes

**Active Learning Guide›**

Our analysis of processes in this and the previous chapter often involves the application of both Newton's second law and kinematics equations, thus relating the external forces exerted on an object and its changing motion. We call these **dynamics processes**. Below we describe a general method for analyzing dynamics processes and illustrate its use for a simple example of pulling a sled.

## PROBLEM-SOLVING STRATEGY Analyzing dynamics processes

Active Learning Guide›

### EXAMPLE 3.2 Pulling a sled

You pull a sled across a hard snowy surface. The sled and the two children sitting on it have a total mass of 60.0 kg. The rope you pull exerts a 100-N force on the sled and is oriented 37° above the horizontal. If the sled starts at rest, how fast is it moving after being pulled 10.0 m?

#### Sketch and translate

- Make a sketch of the process.
- Choose a system.
- Choose coordinate axes with one axis in the direction of acceleration and the other axis perpendicular to that direction.
- Indicate in the sketch everything you know about the process relative to these axes. Identify the unknown quantity of interest.

We first sketch the process.

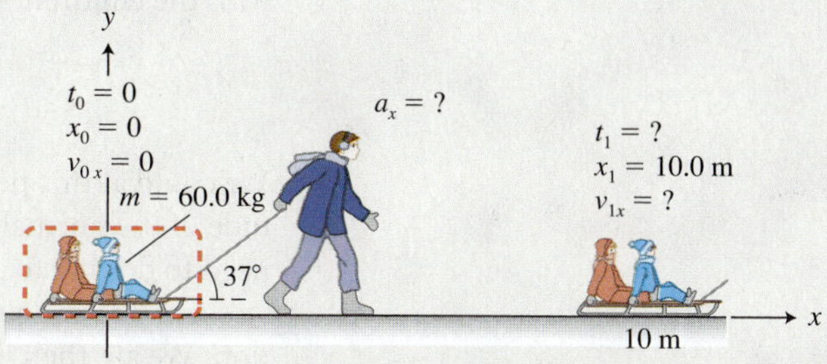


Choose the sled and children as the system.

The process starts with the sled at rest ($v_{0x} = 0$) and ends when it has traveled $x_1 - x_0 = 10.0$ m and has an unknown final speed $v_{1x}$.

#### Simplify and diagram

- Simplify the process. For example, can you model the system as a point-like object? Can you ignore friction?
- Represent the process diagrammatically with a motion diagram and/or a force diagram.
- Check for consistency of the diagrams—for example, is the sum of the forces in the direction of the acceleration?

Consider the system as a point-like object. Since we have no information about friction, assume its effects on the sled are minor.

A motion diagram for the sled is shown below. The sled moves at increasing speed toward the right.

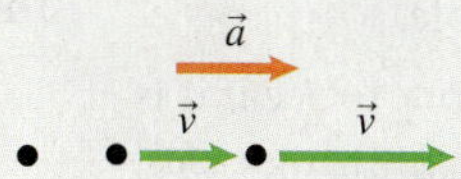


Draw a force diagram for the system. Earth exerts a downward gravitational force on the system $\vec{F}_{\text{E on SL}}$, the rope pulls on the sled $\vec{T}_{\text{R on SL}}$ at a 37° angle above the horizontal, and the snow exerts a normal force on the sled perpendicular to the surface (in this case upward) $\vec{N}_{\text{S on SL}}$.

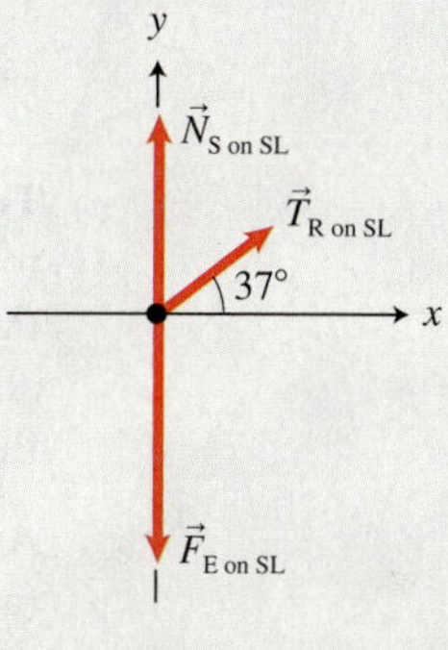


Are the diagrams consistent? The sum of the forces is toward the right in the direction of the acceleration.

*(continued)*

## Represent mathematically

- Convert these qualitative representations into quantitative mathematical descriptions of the process using Newton's second law and kinematics equations.

The horizontal $x$-component form of Newton's second law is

$$ma_x = \Sigma F_x$$

$$m_{SL}a_x = T_{R \text{ on } SL\, x} + N_{S \text{ on } SL\, x} + F_{E \text{ on } SL\, x}$$

Substitute expressions for the $x$-components of these three forces:

$$m_{SL}a_x = T_{R \text{ on } SL} \cos 37° + N_{S \text{ on } SL} \cos 90° + F_{E \text{ on } SL} \cos 90°$$

Noting that $\cos 90° = 0$ and dividing both sides by the mass of the sled with the children, we find that

$$a_x = \frac{T_{R \text{ on } SL} \cos 37° + 0 + 0}{m_{SL}}$$

We could at this point use the $y$-component equation to find the magnitude of the normal force that the surface exerts on the sled, but we don't need to do that to answer the question of interest.

The above equation can be used to determine the sled's acceleration. We are then left with a kinematics problem to determine the speed $v_{1\,x}$ of the sled after pulling it for $x_1 - x_0 = 10.0$ m. We can use kinematics Eq. (1.7):

$$v_{1\,x}^2 = v_{0\,x}^2 + 2a_x(x_1 - x_0)$$

## Solve and evaluate

- Substitute the given values into the mathematical expressions and solve for the unknowns.
- Decide whether the assumptions that you made were reasonable.
- Finally, evaluate your work to see if it is reasonable (check units, limiting cases, and whether the answer has a reasonable magnitude).
- Make sure the answer is consistent with other representations.

The acceleration is

$$a_x = \frac{+T_{R \text{ on } SL} \cos 37°}{m_{SL}} = \frac{+(100\text{ N})(\cos 37°)}{60.0\text{ kg}} = +1.33\text{ m/s}^2$$

The speed of the sled after being pulled for 10.0 m will be

$$v_{1\,x}^2 = v_{0\,x}^2 + 2a_x(x_1 - x_0) = 0^2 + 2(1.33\text{ m/s}^2)(10.0\text{ m} - 0)$$

or

$$v_{1\,x} = 5.2\text{ m/s}$$

This is fast but not unreasonable. In real life, friction between the sled and the snow would probably cause the speed to be slower. The units are correct. If we examine a limiting case, in which the rope pulls vertically, the horizontal acceleration is zero as it should be.

**Try it yourself:** (a) Determine the $y$-component of the gravitational force that Earth exerts on the sled-children system. (b) Then determine the $y$-component of the force that the rope exerts on the system. (c) Finally, use the $y$-component form of Newton's second law to determine the $y$-component of the normal force that the snow exerts on the sled.

*Answers:*

(a) $F_{E \text{ on } SL\, y} = -m_{SL}g \sin 90° = -590$ N;

(b) $T_{R \text{ on } SL\, y} = +T_{R \text{ on } SL} \sin 37° = +60$ N;

(c) $N_{S \text{ on } SL\, y} = +530$ N.

Looking at the force diagram, we see that these numbers make sense. Why is the magnitude of the normal force less than $m_{SL}g$?

**EXAMPLE 3.3 Acceleration of a train**

You carry a pendulum and a protractor with you onto a train. As the train starts moving, the pendulum string swings back to an angle of 8.0° with respect to vertical and remains at that angle. Determine the acceleration of the train at that moment.

**Sketch and translate** We sketch the situation as shown below. Choose the pendulum bob attached to the bottom of the string as the system. The train station is our object of reference. The pendulum bob is accelerating horizontally; thus, the sum of the forces exerted on it should point horizontally. We choose the $x$-axis to point in the horizontal direction and the $y$-axis to point perpendicular in the vertical direction.

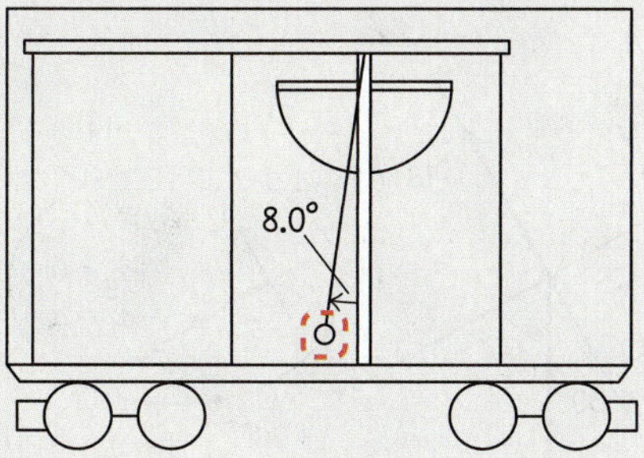


**Simplify and diagram** Assume that the pendulum bob is a point-like object and that its acceleration equals the acceleration of the train. Next, construct a force diagram for the pendulum bob. The string exerts a force $\vec{T}_{\text{S on B}}$ that is oriented at an 8.0° angle relative to the vertical or 82.0° relative to the horizontal. Earth exerts a downward gravitational force on the bob $\vec{F}_{\text{E on B}}$.

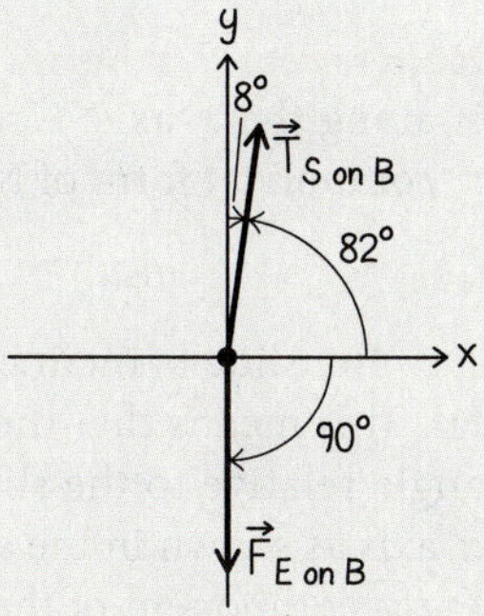


**Represent mathematically** Use the force diagram to help apply Newton's second law in the $x$- and $y$-component forms:

*x-component form of Newton's second law:*

$$m_B a_{Bx} = T_{\text{S on B}\,x} + F_{\text{E on B}\,x}$$
$$m_B a_{Bx} = +T_{\text{S on B}}\cos 82^\circ + F_{\text{E on B}}\cos 90^\circ$$

*y-component form of Newton's second law:*

$$m_B a_{By} = T_{\text{S on B}\,y} + F_{\text{E on B}\,y}$$
$$m_B a_{By} = +T_{\text{S on B}}\sin 82^\circ + (-F_{\text{E on B}}\sin 90^\circ)$$

**Solve and evaluate** Note that the bob's velocity is not changing in the vertical direction; thus, $a_{By} = 0$. The bob's velocity is changing in the horizontal direction, so the $x$-component of acceleration is not zero. Also, recall that the gravitational force that Earth exerts on the bob has magnitude $F_{\text{E on B}} = m_B g$. Finally, note that $\cos 90^\circ = 0$ and $\sin 90^\circ = 1.0$. With these substitutions and a bit of algebra, the above $x$- and $y$-component equations become

$$m_B a_{Bx} = T_{\text{S on B}}\cos 82^\circ$$
$$0 = T_{\text{S on B}}\sin 82^\circ - m_B g$$

We want to determine $a_{Bx}$ but do not know $T_{\text{S on B}}$ and $m_B$. In this particular case, a bit of mathematical creativity helps. Move the $m_B g$ in the second equation to the left side, and divide the left side of the first equation by the left side of the second equation; then divide the right side of the first equation by the right side of the second equation (we can do it as we are dividing each side of the first equation by terms that are equal and nonzero). We get

$$\frac{m_B a_{Bx}}{m_B g} = \frac{T_{\text{S on B}}\cos 82^\circ}{T_{\text{S on B}}\sin 82^\circ}$$

The masses of the bob and the force exerted by the string cancel, thus:

$$a_{Bx} = \frac{\cos 82^\circ}{\sin 82^\circ} g = \cot 82^\circ (9.8\ \text{m/s}^2) = 1.4\ \text{m/s}^2$$

The units are correct, and the magnitude is reasonable. As a limiting case analysis, we find from the above equation that if the acceleration of the train is zero, then the string hangs straight down at a 90° angle relative to the horizontal

$$\frac{\cos 90^\circ}{\sin 90^\circ} = 0 \text{ as } \cos 90^\circ = 0 \quad \text{and} \quad \sin 90^\circ = 1.$$

This is what we expect if the train is parked at the train station.

**Try it yourself:** Suppose the train is moving forward at a constant velocity of magnitude 18 m/s. At what angle will the pendulum bob string hang?

*Answer:* The bob is not accelerating ($a_{Bx} = 0$) and the left side of the final equation in the solution is zero. The bob should hang straight down, as in a stationary train.

## Processes involving inclines

Processes involving inclines are common in everyday life. For example, a skier skis down a slope or a wagon is pulled uphill. How do we use the component form of Newton's second law and kinematics to help analyze such processes?

**EXAMPLE 3.4 Who wins a water slide race?**

A child of mass $m$ and an adult of mass $4m$ simultaneously start to slide from the top of a water slide that has a downward slope of 30° relative to the horizontal. Which person reaches the bottom first?

**Sketch and translate** We can choose either the child or the adult as the system. Our sketch shows the person at one moment during the downhill slide and includes the known information. The person who has the greater acceleration along the slide will win the race, so we will construct an expression for the person's acceleration and see how that acceleration depends on the person's mass. The object of reference is Earth. We choose the axes later when constructing the force diagram.

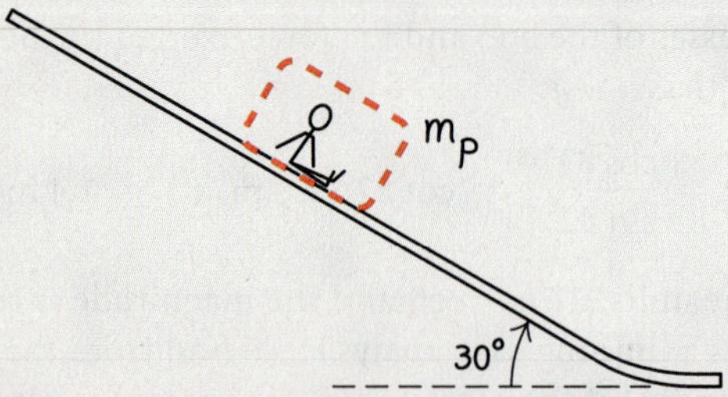


**Simplify and diagram** The person is modeled as a point-like object. Since a water slide is very smooth and slippery, we assume that we can disregard the friction force exerted by the slide on the person.

Next, we build a motion diagram for the person, as shown below. The velocity arrows are parallel to the slide, and the person's velocity is increasing so the velocity change arrow $\Delta\vec{v}$ (and the acceleration $\vec{a}$) points down the slide.

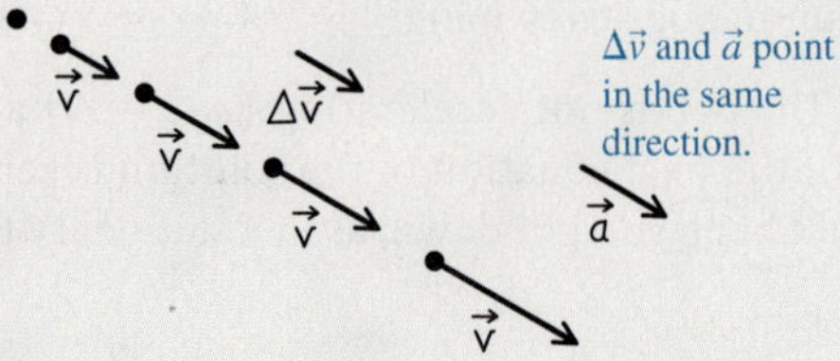


A force diagram shows that Earth exerts a downward gravitational force $\vec{F}_{\text{E on P}}$ on the person and the slide exerts a normal force $\vec{N}_{\text{S on P}}$ on the person perpendicular to the surface. We choose the $x$-axis parallel to the inclined surface and the $y$-axis perpendicular to the inclined surface. With this choice, the acceleration points in the positive $x$-direction. If we had chosen the standard horizontal and vertical directions for our coordinate axes, then the acceleration would have nonzero $x$- and $y$-components, making the situation more difficult to analyze.

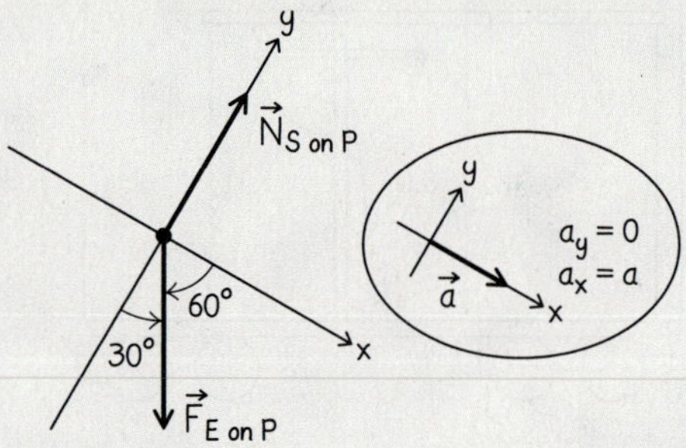


**Represent mathematically** The motion of the person is along the $x$-axis. Use the force diagram to apply the $x$-component form of Newton's second law:

$$m_P a_x = F_{\text{E on P}\,x} + N_{\text{S on P}\,x}$$

The slide is inclined at 30° relative to the horizontal. This means that the gravitational force makes a 60° angle relative to the slide (and therefore relative to the $x$-axis as shown in the above diagram). We can calculate the $x$-component of the gravitational force in two ways: $F_{\text{E on P}\,x} = m_P g \cos 60°$ or $F_{\text{E on P}\,x} = m_P g \sin 30°$. Both methods give the same result. The above component equation becomes

$$m_P a_x = +(m_P g)\cos 60° + N_{\text{S on P}} \cos 90°$$

**Solve and evaluate** Substituting $\cos 90° = 0$, we get

$$m_P a_x = +(m_P g)\cos 60° + 0$$

Dividing by the $m_P$ on each side of the equation, we get $a_x = g\cos 60°$. The person's acceleration is independent of her/his mass! Thus, the acceleration of the child and that of the adult are the same. If they start at the same time at the same position, both initially at rest, and have the same acceleration, they arrive at the bottom at the same time. The race is a tie (see the photo).

**Try it yourself:** Suppose the slide above is 20 m long. How fast will the person be moving when they reach the bottom of the slide?

*Answer:* 14 m/s.

The result of Example 3.4 may surprise you, but it is reasonable. The acceleration depends on both the gravitational force and the mass. The gravitational force that Earth exerts on the adult (only the $x$-component of this force determines the acceleration in this case) is four times that exerted by Earth on the child, but because the adult's mass is four times greater than the child's, the accelerations of the adult and the child are equal. This result is similar to what we learned in Chapters 1 and 2—objects of all different masses have the same free-fall acceleration.

**TIP** Notice in the last example that the inclined surface and the $x$-axis were at a 30° angle relative to the horizontal. The gravitational force exerted by Earth on the system then makes a 60° angle relative to the $x$-axis (60° is the complement of 30°). In general, the angle $\theta$ of the inclined surface relative to the horizontal is the complement $90° - \theta$ of the angle that the gravitational force makes with respect to the inclined surface (see **Figure 3.6**).

**Figure 3.6** Complementary angles.

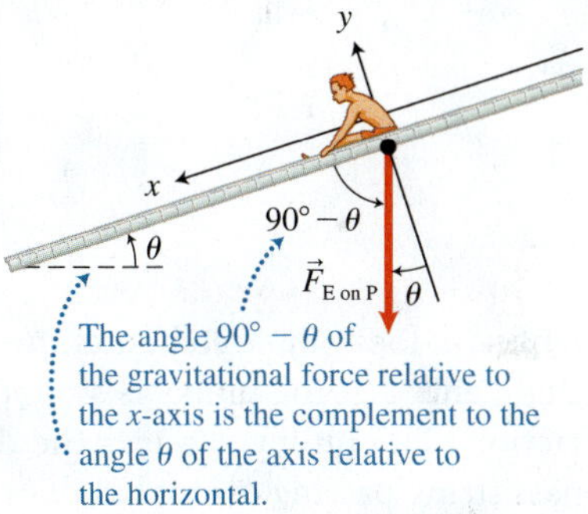
y
x
90° − θ
θ
$\vec{F}_{\text{E on P}}$
θ
The angle 90° − θ of the gravitational force relative to the x-axis is the complement to the angle θ of the axis relative to the horizontal.

## Two objects linked together

We can apply Newton's second law to a process in which two objects are connected together by a cable or rope, such as a van pulling a rope connected to a wagon that is pulling a rope connected to a second wagon. Perhaps the first "scientific" quantitative application of this type involved two blocks of different mass at the ends of a string that passed over a pulley (**Figure 3.7**). George Atwood invented this apparatus (now called the Atwood machine) in 1784. At that time, there were no motion sensors or precision stopwatches—nothing that would allow the accurate measurement of the motion of a rapidly accelerating object (9.8 $m/s^2$ would have been considered a rapid acceleration in those days). Atwood's machine allowed the determination of $g$ despite these difficulties. We'll consider next a modified and simplified version of the Atwood machine, a machine that can serve the same purpose.

**Figure 3.7** An Atwood machine.

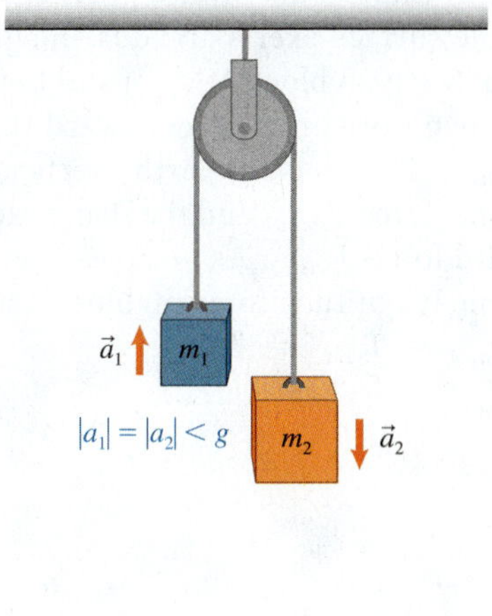
$\vec{a}_1$
$m_1$
$|a_1| = |a_2| < g$
$m_2$
$\vec{a}_2$

**EXAMPLE 3.5 Slowing the fall**

In a modified Atwood machine, one block rests on a horizontal surface and the other hangs off the edge of the surface. A string attached to each block passes over a light, smooth pulley. Determine the force that the string exerts on the hanging block of mass $m_2$ and the magnitude of the acceleration of each block.

**Sketch and translate** We first sketch the modified Atwood machine shown below. For analysis we will use two systems—first block 1 and then block 2. The two blocks have masses $m_1$ and $m_2$. Block 2 will fall and pull the string attached to block 1. Both blocks will move with increasing speed—block 1 on the horizontal surface and block 2 downward. If the string does not stretch, then at every moment they move with the same speed. The sliding block 1 has the same magnitude acceleration as block 2, but pointing toward the right. Our goal is to find the force exerted by the string on block 2 and the magnitudes of the acceleration of the blocks.

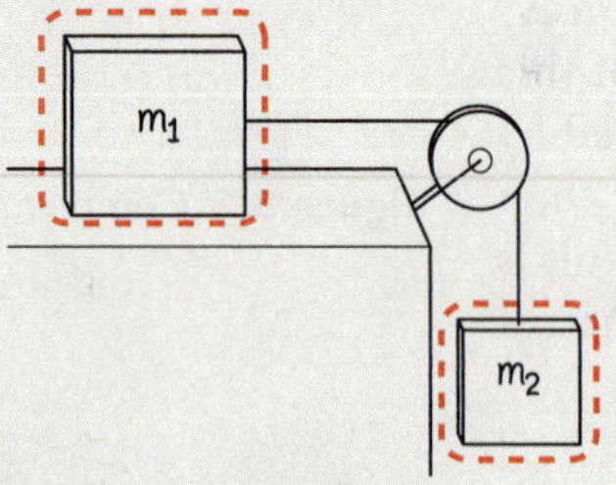


**Simplify and diagram** Assume that the string is very light and does not stretch, that the pulley's mass is very small, and that the pulley is smooth and rotates free of friction. The pulley changes the direction of the low-mass string passing over it but does not change the magnitude of the force that the string exerts on the objects attached at either end. Assume also that there is no friction between the horizontal surface and block 1.

At right we show our force diagrams for each block and our choice of coordinate systems. For block 1, Earth exerts a downward gravitational force $\vec{F}_{\text{E on 1}}$; the table surface exerts an equal-magnitude upward normal force on block 1 $\vec{N}_{\text{T on 1}}$; and the string exerts an unknown horizontal force toward the right on the block $\vec{T}_{\text{S on 1}}$. For block 2, Earth exerts a downward gravitational force $\vec{F}_{\text{E on 2}}$ and the string exerts an unknown upward force $\vec{T}_{\text{S on 2}}$. As noted earlier, the string exerts the same magnitude force on block 1 as it does on block 2: $T_{\text{S on 1}} = T_{\text{S on 2}} = T$.

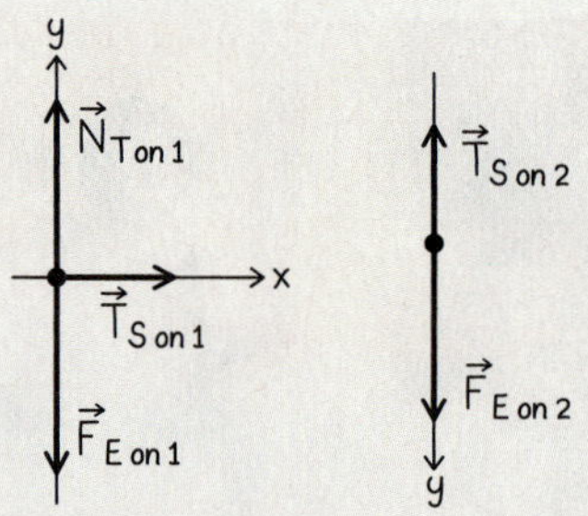


**Represent mathematically** Use the force diagrams to help apply Newton's second law in component form for each block. For block 1, the forces in the vertical direction balance, since the vertical component of its acceleration is zero. We see by inspection of the force diagram that the gravitational force and the normal force have zero $x$-components and so we do not include them in the $x$-component equation. This leaves us with only one force in the horizontal direction—the force exerted by the string. The $x$-component form of Newton's second law becomes

$$m_1 a_{1x} = T_{\text{S on 1}x} = +T \quad (3.3)$$

For block 2, we only need to look at the vertical direction, as no forces are exerted in the horizontal direction. Notice that we've chosen the vertical $y$-axis pointing downward so that the $y$-component of block 2's acceleration will equal the $x$-component of block 1's acceleration. The $y$-component form of Newton's second law for block 2 is

$$m_2 a_{2y} = T_{\text{S on 2}y} + F_{\text{E on 2}y}$$

Noting that $T_{\text{S on 2}y} = -T$ and $F_{\text{E on 2}y} = m_2 g$, we get

$$m_2 a_{2y} = -T + m_2 g \quad (3.4)$$

Because the two blocks are connected together by the string, they move with the same speed and have accelerations of the same magnitude, 1 to the right and 2 down.

**Solve and evaluate** After substituting $a$ for the two accelerations in Eqs. (3.3) and (3.4), the two equations now have the same two unknown quantities $T$ and $a$.

$$m_1 a = T$$

$$m_2 a = -T + m_2 g$$

We substitute the expression for $T$ from the first equation into the second to get

$$m_2 a = -m_1 a + m_2 g$$

After moving the terms containing the acceleration to the left side, factoring $a$ out, and dividing both sides by the sum of the masses, we have

$$a = \frac{m_2 g}{m_1 + m_2}$$

Note that the acceleration is less than $g$! Also note that if the mass $m_1 \gg m_2$ ($m_1$ much greater than $m_2$), then the acceleration is almost zero and can be measured easily. We can now determine the force exerted by the string on the hanging block by inserting the above expression for the acceleration into $m_1 a = T$:

$$T = m_1 a = m_1 \frac{m_2 g}{m_1 + m_2} = \frac{m_1}{m_1 + m_2} m_2 g$$

Are the results reasonable? Consider the equation for acceleration. The only force pulling the hanging block down is the gravitational force $m_2 g$ that Earth exerts on the hanging block. But because the two blocks are connected, this force has to cause the sum of their masses to accelerate $[m_2 g = (m_1 + m_2)a]$, making the acceleration less than $g$.

**Try it yourself:** Determine an expression for acceleration for a regular Atwood machine for which two objects of mass $m_1$ and $m_2$ move in the vertical direction ($m_1 > m_2$).

*Answer:* $a = \dfrac{(m_1 - m_2)g}{m_1 + m_2}$.

We can see now how to use a modified or a regular Atwood machine to determine the acceleration $g$. If you measure the distance that one of the objects travels ($\Delta y$) during a particular time interval ($\Delta t$) after being released from rest, you can use the expression

$$\Delta y = a \frac{\Delta t^2}{2}$$

to determine the acceleration of the objects. Using this value for acceleration and the values for their masses, you can use the expression

$$a = \frac{m_2 g}{m_1 + m_2}$$

for a modified machine or

$$a = \frac{(m_1 - m_2)g}{m_1 + m_2}$$

for a regular one to determine the value of $g$.

**Review Question 3.3** For problems involving objects moving upward or downward along inclined surfaces, we choose the $x$-axis parallel to the surface and the $y$-axis perpendicular to the surface. Why not use horizontal and vertical axes?

## 3.4 Friction

Up to this point, we have assumed that objects move across absolutely smooth surfaces with no friction. In reality, the vast majority of situations involve some degree of friction. In this section we examine the phenomenon of friction conceptually and construct mathematical models that allow us to take friction into account quantitatively.

### Static friction

Consider a simple experiment as shown in Observational Experiment **Table 3.1**. A spring scale exerts an increasing force on the block. Observe carefully what happens to the block.

**‹Active Learning Guide**

## OBSERVATIONAL EXPERIMENT TABLE

### 3.1 Pulling a block with a spring scale.

VIDEO 3.1

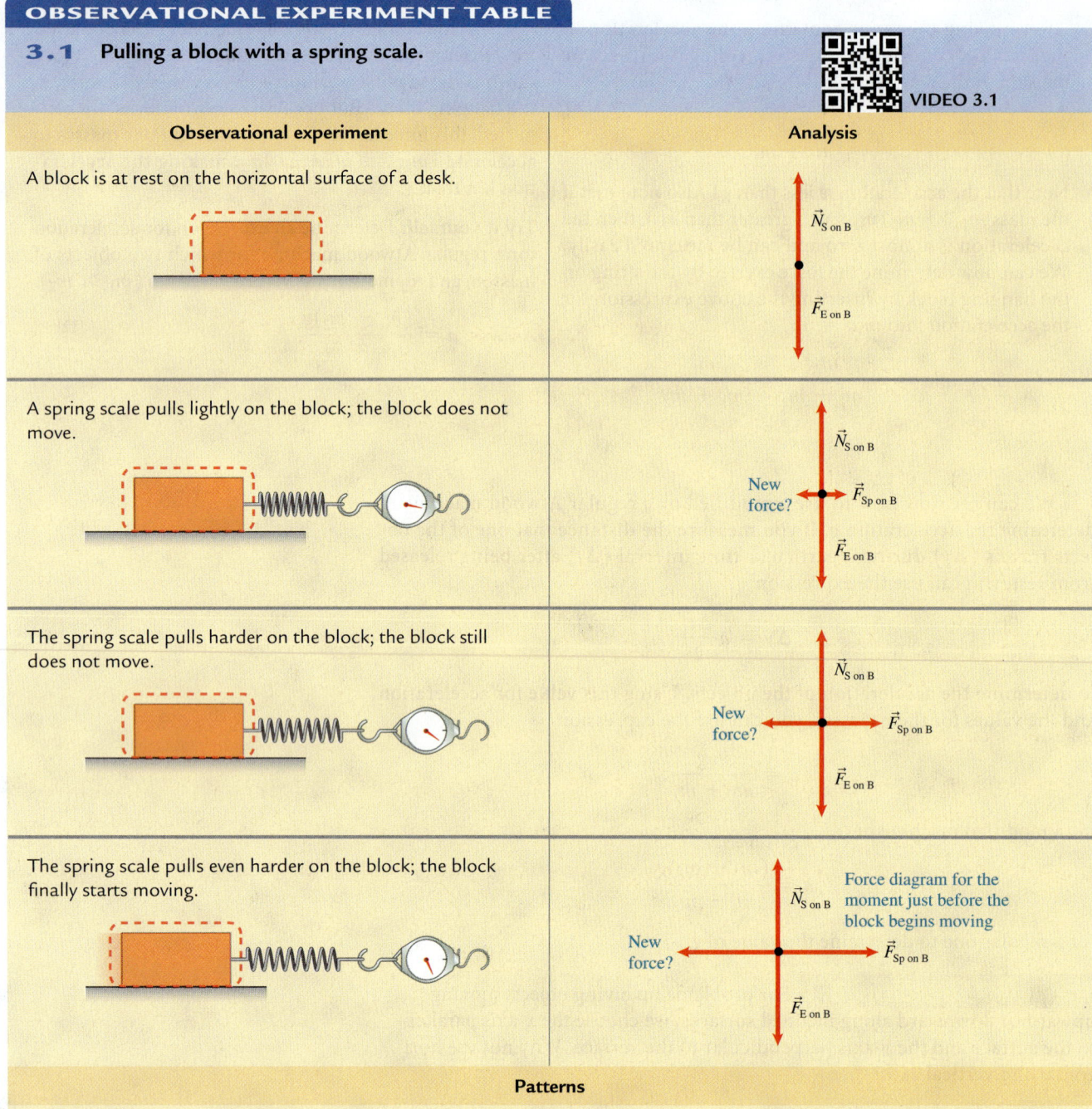

| Observational experiment | Analysis |
|---|---|
| A block is at rest on the horizontal surface of a desk.  |   |
| A spring scale pulls lightly on the block; the block does not move. 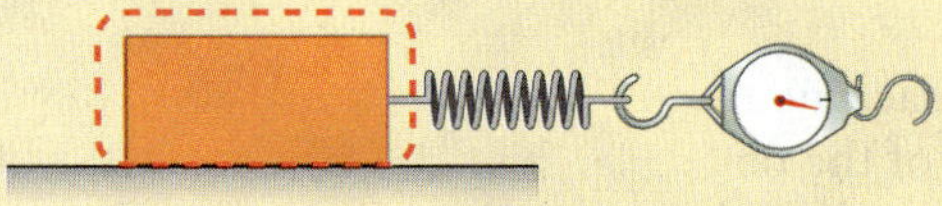 | 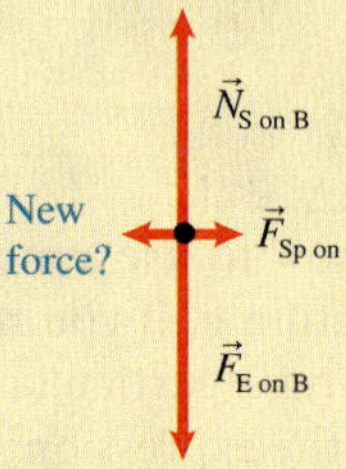  |
| The spring scale pulls harder on the block; the block still does not move. 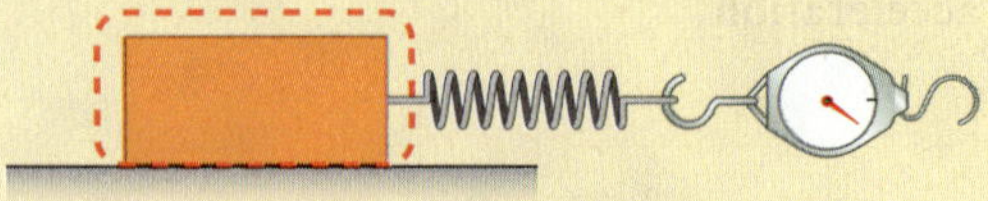 | 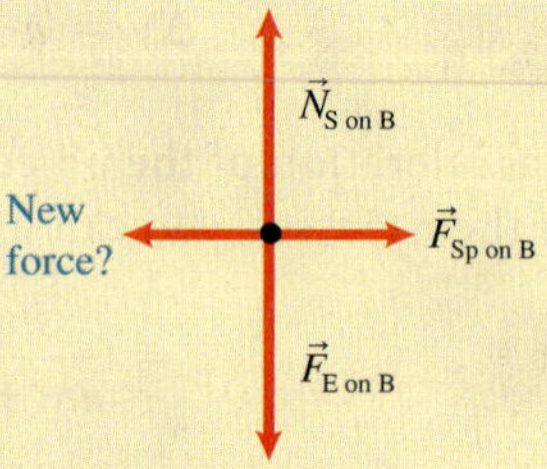  |
| The spring scale pulls even harder on the block; the block finally starts moving. 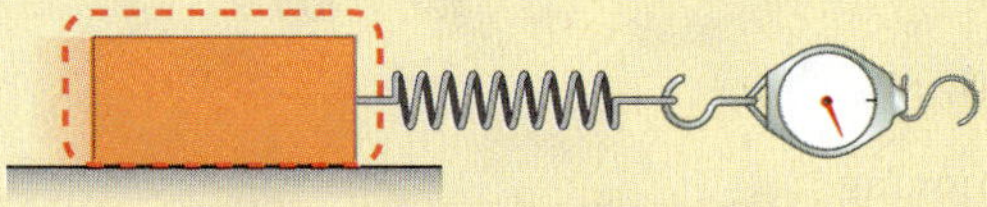 | 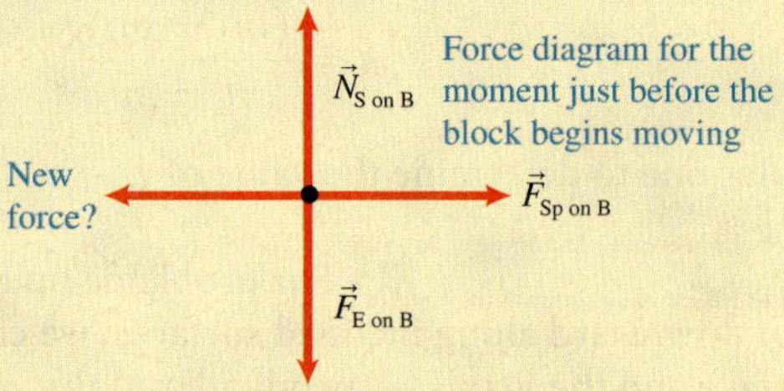  |

**Patterns**

- In each of these experiments, the surface exerted a normal force on the block that balanced the downward gravitational force exerted by Earth on the block.
- As the spring scale exerted an increasing force on the block to the right, the block remained stationary (zero acceleration). The surface must have exerted an additional force—an increasing force on the block toward the left.
- Eventually, the spring scale exerted a strong enough force on the block that the block started sliding. Thus, the resistive force must have a maximum value.

The patterns inferred from the experiments in Table 3.1 demonstrated **static friction force**. This force is parallel to the surfaces of two objects that are not moving in relation to each other and opposes the tendency of one object to move across the other. The static friction force changes magnitude to prevent motion—up to a maximum value. When the external force exceeds this static friction force, the block starts moving. This maximum resistive force that the surface can exert on the block is called the **maximum static friction force**.

**Figure 3.8** The surface exerts a static friction force that helps you walk or run.

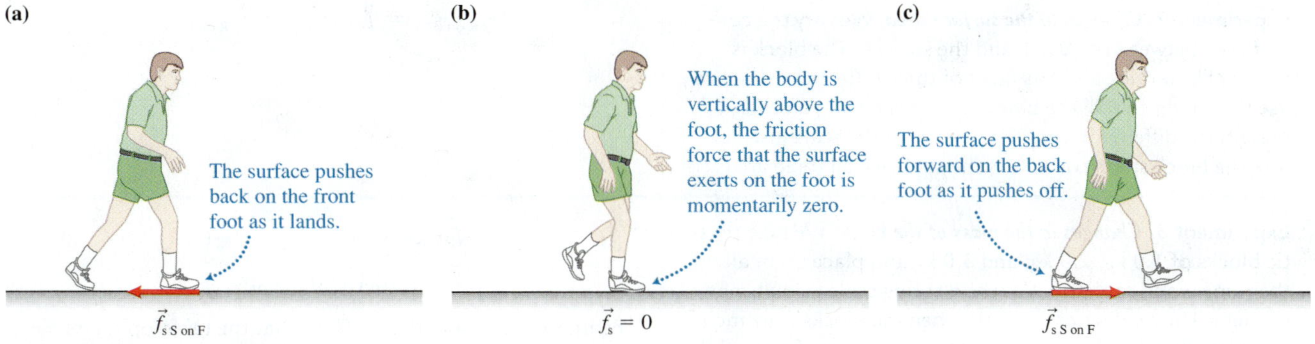
(a)
The surface pushes back on the front foot as it lands.
$\vec{f}_{s\,S\,on\,F}$
(b)
When the body is vertically above the foot, the friction force that the surface exerts on the foot is momentarily zero.
$\vec{f}_s = 0$
(c)
The surface pushes forward on the back foot as it pushes off.
$\vec{f}_{s\,S\,on\,F}$

We often talk about the importance of reducing friction in car engines, in bicycle chains, etc. However, sometimes friction is a necessary phenomenon. For instance, walking on a flat horizontal sidewalk would not be possible if there were no static friction (**Figure 3.8a**). When one foot swings forward and contacts the sidewalk, static friction prevents it from continuing forward and slipping (the way it might on ice). When the front shoe lands, the friction force pushes back on it and prevents it from slipping forward.

When your body is vertically above the foot, the friction force that the sidewalk exerts on your foot is momentarily zero (Figure 3.8b). As your body gets ahead of the back foot, the foot has a tendency to slip backward. However, the static friction force that the sidewalk exerts on your foot now points opposite that slipping direction, that is, forward, in the direction of motion of the body (Figure 3.8c). As long as this foot does not slip, the surface continues exerting a forward static friction force on you, helping your body move forward.

> **TIP** A surface really exerts only one force on an object pressing against it. It is convenient to break this force into two vector components: a component perpendicular to the surface, the normal force $\vec{N}_{S\,on\,O}$, and a component parallel to the surface, the static friction force $\vec{f}_{s\,S\,on\,O}$ (**Figure 3.9**). We can apply Newton's second law in component form to treat these components as separate forces.

**Figure 3.9** The force that the surface exerts on an object.

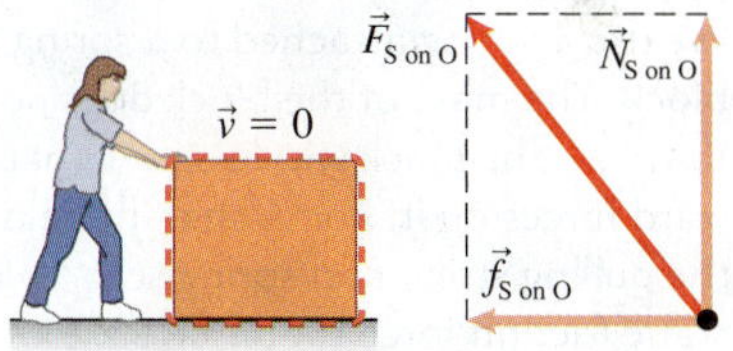
$\vec{F}_{S\,on\,O}$
$\vec{N}_{S\,on\,O}$
$\vec{v} = 0$
$\vec{f}_{S\,on\,O}$

The force of the surface on the block is considered as two forces: the normal force $\vec{N}$ and the friction force $\vec{f}$.

What determines the magnitude of the maximum static friction force? To investigate this question, we use an experimental setup (see Observational Experiment **Table 3.2**) similar to that used in Observational Experiment Table 3.1, except that we will vary the characteristics of the block and the surfaces.

## OBSERVATIONAL EXPERIMENT TABLE

### 3.2 What affects the maximum friction force?

VIDEO 3.2

| Observational experiment | Analysis |
|---|---|
| **Experiment 1.** ***Changes in the smoothness of the surfaces*** We'll use the spring to pull a smooth, plastic block that is resting on three different surfaces: (1) a glass tabletop, (2) a wood tabletop, and (3) a rubber exercise mat. The reading of the spring scale just before the blocks start moving is largest for the rubber mat, next largest for the wood tabletop, and smallest for the glass tabletop. | $\vec{f}_{s\,R\,on\,B} > \vec{f}_{s\,W\,on\,B} > \vec{f}_{s\,G\,on\,B}$ |

(*continued*)

| Observational experiment | Analysis |
|---|---|
| **Experiment 2.** ***Changes in the surface area*** We vary the contact area between the block and the surface. The block is shaped like a brick and has faces of three different areas. We use the spring to pull the block while it is resting on each of these three different faces. The reading of the scale just before the block starts to move is the same for all three areas. | $\vec{f}_{\text{s R on BA1}} = \vec{f}_{\text{s R on BA2}} = \vec{f}_{\text{s R on BA3}}$ |
| **Experiment 3.** ***Changes in the mass of the block*** We take plastic blocks of 1.0 kg, 2.0 kg, and 3.0 kg and place them all on the same wood tabletop. We use a spring scale to pull each of them. The reading of the scale when the blocks start moving is smallest for the 1.0-kg block, twice as large for the 2.0-kg block, and three times as large for the 3.0-kg block. | $\vec{f}_{\text{s W on 3 kg B}} = 3\vec{f}_{\text{s W on 1 kg B}}$<br>$\vec{f}_{\text{s W on 2 kg B}} = 2\vec{f}_{\text{s W on 1 kg B}}$<br>The maximum static friction force that the tabletop exerts on the block is directly proportional to the mass of the block. |

**Patterns**

The maximum static friction force that the surface can exert on the block depends on the roughness of the contacting surfaces and the mass of the block, but does not depend on the area of contact between the surfaces.

The patterns in Table 3.2 make testable hypotheses. To test a hypothesis, we need to design an experiment in which we can vary one of the properties and make a prediction of the outcome based on the hypothesis being tested. We will test one hypothesis in Testing Experiment **Table 3.3**, that the maximum static friction force is directly proportional to the mass of the object.

**TESTING EXPERIMENT TABLE**

**3.3 Does the maximum static friction force depend on mass?**

VIDEO 3.3

| Testing experiment | Prediction | Outcome |
|---|---|---|
| We use a spring attached to a spring scale to pull a 1-kg block. The mass of the block does not change, but we push down on the block with a spring that exerts a series of downward forces on it. For each of these downward forces, we use the pulling string and spring scale to determine the maximum static friction force the surface exerts on the block. | If the friction force is proportional to the mass of the block, the friction force should remain constant during the experiment. | The friction force changes—the harder we press on the block, the higher the maximum static friction force. |

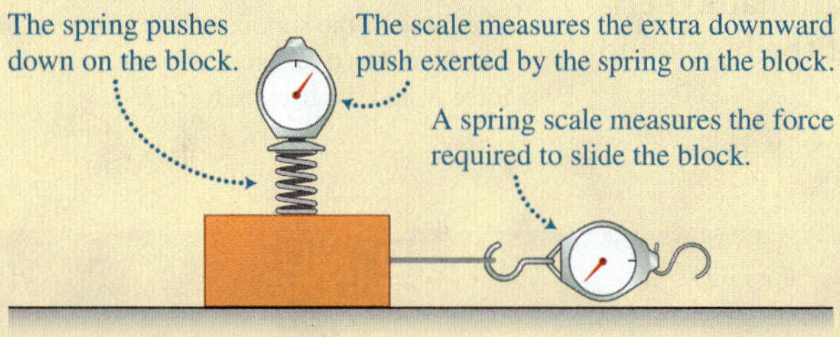
The spring pushes down on the block.
The scale measures the extra downward push exerted by the spring on the block.
A spring scale measures the force required to slide the block.

**Conclusion**

The outcome of the experiment did not match the prediction; the hypothesis requires a revision.

The unexpected outcome of the testing experiment requires further investigation. In **Table 3.4** we present the detailed data using the apparatus in Table 3.3. Remember that we are keeping the roughness of the surfaces and surface areas the same during all of the experiments.

**Table 3.4** **Maximum static friction force when a block presses harder against a surface.**

| Mass of the block | Extra downward force exerted on the 1-kg block | Normal force exerted by the surface on the block | Maximum static friction force | Ratio of maximum static friction force to normal force |
|---|---|---|---|---|
| 1.0 kg | 0.0 N | 9.8 N | 3.0 N | 0.31 |
| 1.0 kg | 5.0 N | 14.8 N | 4.5 N | 0.30 |
| 1.0 kg | 10.0 N | 19.8 N | 6.1 N | 0.31 |
| 1.0 kg | 20.0 N | 29.8 N | 9.1 N | 0.31 |

**Table 3.5** **The coefficients of kinetic and static friction for two different surfaces.**

| Contacting surfaces | Coefficient of static friction | Coefficient of kinetic friction |
|---|---|---|
| Rubber on concrete (dry) | 1 | 0.6–0.85 |
| Steel on steel | 0.74–0.78 | 0.42–0.57 |
| Aluminum on steel | 0.61 | 0.47 |
| Glass on glass | 0.9–1 | 0.4 |
| Wood on wood | 0.25–0.5 | 0.20 |
| Waxed skis on wet snow | 0.14 | 0.1 |
| Teflon on Teflon | 0.04 | 0.04 |
| Greased metals | 0.1 | 0.06 |
| Surfaces in a healthy human joint | 0.01 | 0.003 |

The data in Table 3.4 disprove the hypothesis that the maximum static friction force depends on the mass of the block. We have also found a new pattern. The ratio of the maximum static friction force to the normal force

$$\frac{f_{\text{s S on O max}}}{N_{\text{S on O}}} \approx 0.31$$

and is about the same (within experimental uncertainty) for all the measurements (the last column in the table). It appears that the maximum static friction force is directly proportional to the magnitude of the normal force: $f_{\text{s S on O max}} \propto N_{\text{S on O}}$. This finding makes sense if you look back at Figure 3.9: the normal force and the friction force are two perpendicular components of the same force—the force that a surface exerts on an object! If the normal force exerted by the surface on an object increases, the maximum static friction force the surface exerts on the object increases proportionally. It also makes sense if you think of pulling a sled over snow. Pulling is easier than pushing because when you pull, you lift the sled a little off the surface, reducing the force that it exerts on the snow and thus reducing the normal force the snow exerts on the sled. As a result, the friction force exerted on the sled decreases and it is easier to pull.

If we repeat the Table 3.4 experiments using a different type of block on a different surface, we get similar results. The ratio of the maximum friction force to the normal force is the same for all of the different values of the normal forces. However, the proportionality constant is different for different surfaces; the proportionality depends on the types of contacting surfaces. The proportionality constant is greater for two rough surfaces contacting each other and less for smoother surfaces.

This ratio

$$\mu_s = \frac{f_{\text{s max}}}{N}$$

is called the **coefficient of static friction** $\mu_s$ for a particular pair of surfaces. The coefficient of static friction is a measure of the relative difficulty of sliding two surfaces across each other. The easier it is to slide one surface on the other, the smaller the value of $\mu_s$. You experience different values of $\mu_s$ when you try to walk on ice and on rough snow. Where are you more likely to slide?

The coefficient of static friction $\mu_s$ has no units because it is the ratio of two forces. Although $\mu_s$ usually has values between 0 and 1, the value can be greater than 1. Its value is about 0.8 for rubber car tires on a dry highway surface and is very small for bones in healthy body joints separated by cartilage and synovial fluid. Some values for different surfaces are listed in **Table 3.5**.

**Static friction force** **When two objects are in contact and we try to pull one across the other, they exert a static friction force on each other. This force is parallel to the contacting surfaces of the two objects and opposes the tendency of one object to move across the other. The static friction force changes magnitude to prevent motion—up to a maximum value. This maximum static friction force depends on the roughness of the two surfaces (on the coefficient of static friction $\mu_s$ between the surfaces) and on the magnitude of the normal force $N$ exerted by one surface on the other. The magnitude of the static friction force is always less than or equal to the product of these two quantities:**

$$0 \leq f_s \leq \mu_s N \quad (3.5)$$

Keep in mind the assumptions used when constructing this model of friction. We used relatively light objects resting on relatively firm surfaces. The objects never caused the surfaces to deform significantly (for example, it was not a car tire sinking into mud). Equation (3.5) is only reasonable in situations in which these conditions hold.

## Kinetic friction

If we repeat the previous friction experiments with a block that is already in motion, we find a similar relationship between the resistive friction force exerted by the surface on the block and the normal force exerted by the surface on the block. There are, however, two differences: (1) under the same conditions, the magnitude of the kinetic friction force is always lower than the magnitude of the maximum static friction force; (2) the resistive force exerted by the surface on the moving object does not vary but has a constant value. As with the static friction force $\vec{f}_s$, the magnitude of this **kinetic friction force** $f_k$ depends on the roughness of the contacting surfaces (indicated by a **coefficient of kinetic friction** $\mu_k$) and on the magnitude $N$ of the normal force exerted by one of the surfaces on the other, but not on the surface area of contact. The word *kinetic* indicates that the surfaces in contact are moving relative to each other.

**Kinetic friction force** **When an object slides along a surface, the surfaces exert kinetic friction forces on each other. These forces are exerted parallel to the contacting surfaces and oppose the motion of one surface relative to the other surface. The kinetic friction force depends on the surfaces themselves (on the coefficient of kinetic friction $\mu_k$) and on the magnitude of the normal force $N$ exerted by one surface on the other:**

$$f_k = \mu_k N \quad (3.6)$$

As with any mathematical model, this expression for kinetic friction has its limitations. First, it is applicable only for sliding objects, not rolling objects. Second, it has the same assumption about the rigidity of the surfaces as the model for static friction. Third, predictions based on Eq. (3.6) fail for objects moving at high speed. Although this equation does not have general applicability, it is simple and useful for rigid surfaces and everyday speeds.

## What causes friction?

Let's use the example of Velcro fastening material to help understand the phenomenon of friction. The hooks and loops on the two surfaces of Velcro connect with each other, making it almost impossible to slide one surface across the other. All objects are, in a less dramatic way, like Velcro. Even the slickest surfaces have tiny bumps that can hook onto the tiny bumps on another surface (**Figure 3.10**). Logically, smoother surfaces should have reduced friction. For example, a book with a glossy cover slides farther across a table than a book with a rough, unpolished cover. The friction force exerted by the table on the glossy book is less that on the other book.

However, if the surfaces are too smooth (for example, two polished metal blocks), the friction increases again. Why? All substances are made of particles that are attracted to each other. This attraction between particles keeps a solid object in whatever shape it has. But this attraction between particles works only at very short distances. If the two surfaces are very smooth so that the microscopic particles are close enough to attract each other strongly (almost close enough to form chemical bonds with each other), it becomes more difficult to slide one surface relative to the other (for example, two pieces of plastic wrap or two polished metal blocks).

**Figure 3.10** A microscopic view of contacting surfaces.

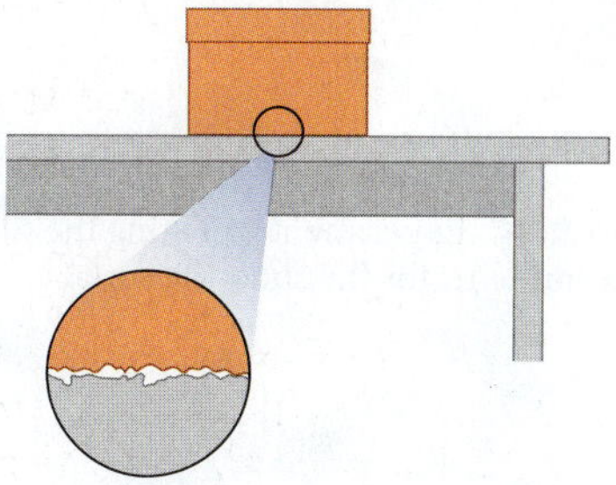

Rough edges on the contacting surfaces cause friction.

‹Active Learning Guide

## Determining friction experimentally

Let's try to determine the coefficient of static friction between a running shoe and a kitchen floor tile using two independent methods. If we only use one method, it will be difficult to decide if the result is reasonable.

***Experiment 1***

In the first experiment, we secure a tile to a horizontal tabletop and place a shoe on top of it. The shoe is pulled horizontally with a spring scale, which exerts an increasingly greater force on it until the shoe begins to slide (**Figure 3.11a**). A force diagram for the instant just before it slides is shown in Figure 3.11b. Since the shoe is not accelerating in the vertical direction, the upward normal force exerted on it by the tile equals the downward gravitational force exerted on it by Earth: $N_{\text{T on S}} = m_S g$.

For the horizontal $x$-direction, only the tension force exerted by the spring scale on the shoe and the static friction force exerted by the tile on the shoe are included in the $x$-scalar component of Newton's second law. The other two forces exerted on the shoe do not have $x$-components. Thus, the $x$-scalar component form of Newton's second law is

$$ma_x = T_{\text{Scale on S}\,x} + f_{\text{s T on S}\,x}$$

Just before the shoe starts to slide, its acceleration is zero, and the scale reads the maximum force of static friction that the tile exerts on the shoe:

$$0 = T_{\text{Scale on S max}} - f_{\text{s T on S max}}$$

Since $T_{\text{max Scale on S}} = f_{\text{s max T on S}} = \mu_s N_{\text{T on S}}$, we can determine the coefficient of static friction:

$$\mu_s = \frac{f_{\text{s T on S max}}}{N_{\text{T on S}}} = \frac{T_{\text{Scale on S max}}}{m_S g}$$

All we need to do is measure the mass of the shoe and record the reading on the spring scale at the moment the shoe starts to slide.

The measured shoe mass is 0.37 kg, and the maximum scale reading is 2.6 N just before the shoe starts to slide. Thus, the coefficient of static friction is

$$\mu_s = \frac{f_{\text{s T on S max}}}{N_{\text{T on S}}} = \frac{T_{\text{Scale on S max}}}{m_S g} = \frac{2.6\ \text{N}}{(0.37\ \text{kg})(9.8\ \text{N/kg})} = 0.72$$

**Figure 3.11** Experiment 1. Determining $\mu_s$ for the shoe-tile surface by pulling the shoe across a horizontal tile.

(a)

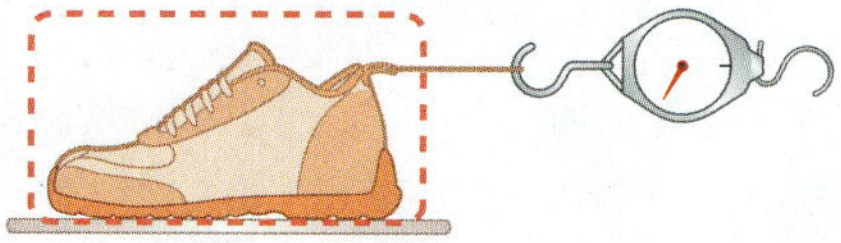

(b)

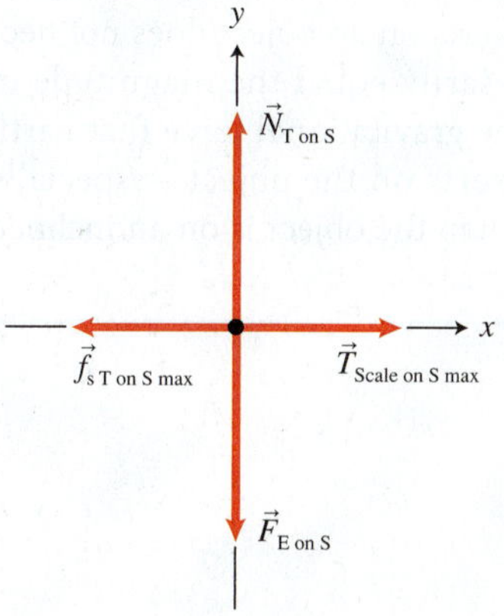

We reported just two significant digits since the measured quantities have just two significant digits. Thus, based on this experiment, the coefficient of static friction is $\mu_s = 0.72 \pm 0.01$.

**Figure 3.12** Experiment 2. Tilting the tile to determine $\mu_s$ for the shoe-tile surface.

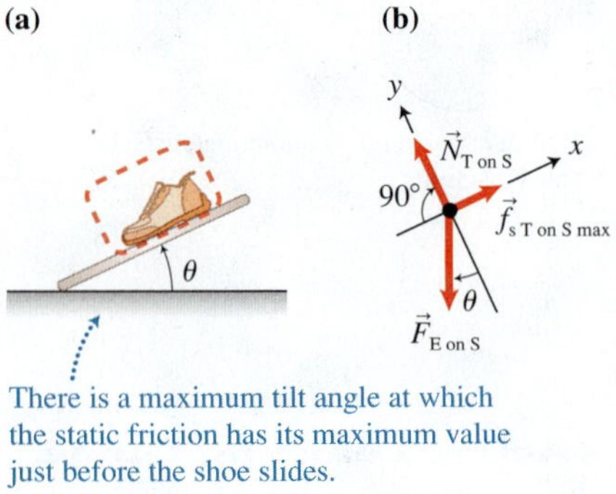


*Experiment 2*

For the second experiment, we place the shoe on the tile and tilt the tile until the shoe starts to slide (**Figure 3.12a**). The static friction force that the tile exerts on the shoe increases as the tilt angle increases. Just before the shoe slides, the static friction has its maximum possible value.

We can now use the force diagram in Figure 3.12b to help apply Newton's second law in component form for the shoe just before it starts sliding. Notice that the $x$-axis of the coordinate system is parallel to the tilted tile and the $y$-axis is perpendicular to the tile. The magnitude of the gravitational force that Earth exerts on the shoe is $F_{E\,on\,S} = m_S g$, and the shoe's acceleration is zero just before it starts to slide.

$$y\text{-component equation:}\quad m_S \cdot 0 = N_{T\,on\,S}\sin 90° - m_S g\cos\theta + f_{s\,T\,on\,S\,max}\sin 0°$$

$$x\text{-component equation:}\quad m_S \cdot 0 = N_{T\,on\,S}\cos 90° - m_S g\sin\theta + f_{s\,T\,on\,S\,max}\cos 0°$$

Computing the values of the known sines and cosines and inserting the expression for the maximum static friction force $f_{s\,max\,T\,on\,S} = \mu_s N_{T\,on\,S}$, we get

$$0 = N_{T\,on\,S} - m_S g\cos\theta$$
$$0 = -m_S g\sin\theta + \mu_s N_{T\,on\,S}$$

We have two equations with two unknowns, $N_{T\,on\,S}$ and $\mu_s$. Since our interest is in the coefficient of static friction, we solve the first equation for the normal force ($N_{T\,on\,S} = m_S g\cos\theta$) and substitute this into the second equation:

$$0 = -m_S g\sin\theta + \mu_s m_S g\cos\theta$$

Cancel the common $m_S g$ from each term and rearrange the above equation to get an expression for $\mu_s$:

$$\mu_s = \frac{\sin\theta}{\cos\theta} = \tan\theta$$

This is an amazing result—the coefficient of static friction between the shoe and the tile can be determined just from the angle of the tile's tilt when the shoe starts sliding.

When we do the experiment, the shoe starts sliding when the tile is at an angle of about $\theta = 36°$. Thus, the coefficient of static friction determined from this experiment is:

$$\mu_s = \frac{\sin\theta}{\cos\theta} = \tan\theta = \tan 36° = 0.73$$

**TIP** Notice that the magnitude of the normal force that a surface exerts on an object does not necessarily equal the magnitude of the gravitational force that Earth exerts on the object—especially when the object is on an inclined surface!

Again, because of the number of significant digits, this is equivalent to $\mu_s = 0.73 \pm 0.01$. This is consistent with the result from the first experiment, since the ranges of possible values overlap. The coefficient of static friction between the shoe and the tile is between 0.72 and 0.73.

Now, let's consider a real-world situation that involves kinetic friction.

## Using skid marks for evidence

When a car stops under normal conditions, it rolls to a stop; the tires do not skid along the road surface. However, if the driver slams on the brakes to stop suddenly, the tires can lock, causing the car to skid. Police officers use the length of skid marks to estimate the speed of the vehicle at the time the driver applied the brakes. Police stations have charts listing the kinetic friction coefficients of various brands of car tires on different types of road surfaces.

**EXAMPLE 3.6 Was the car speeding?**

A car involved in a minor accident left 18.0-m skid marks on a horizontal road. After inspecting the car and the road surface, the police officer decided that the coefficient of kinetic friction was 0.80. The speed limit was 15.6 m/s (35 mi/h) on that street. Was the car speeding?

**Sketch and translate** We first sketch the process (see the figure below). We choose the car as the system. Earth is the object of reference. The coordinate system consists of a horizontal $x$-axis pointing in the direction of the velocity and a vertical $y$-axis pointing upward.

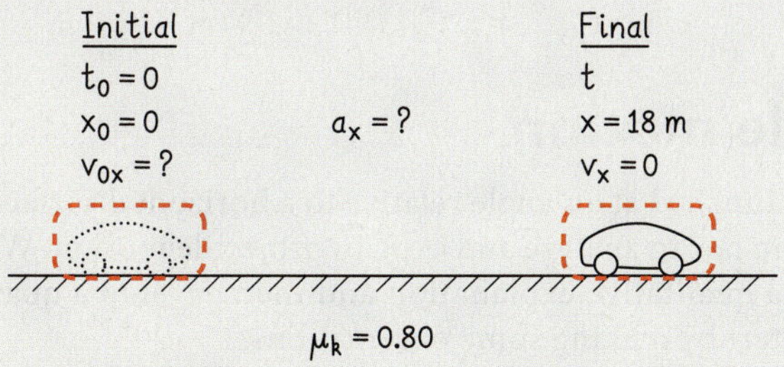


**Simplify and diagram** Assume that the car can be modeled as a point-like object, that its acceleration while stopping was constant, and that the resistive force exerted by the air on the car is small compared with the other forces exerted on it. We can sketch a motion diagram for the car. Our force diagram below shows three forces exerted on the car. Earth exerts a downward gravitational force on the car $\vec{F}_{\text{E on C}}$, the road exerts an upward normal force on the car $\vec{N}_{\text{R on C}}$ (perpendicular to the road surface), and the road also exerts a backward kinetic friction force on the car $\vec{f}_{\text{k R on C}}$ (parallel to the road's surface and opposite the car's velocity). This friction force causes the car's speed to decrease.

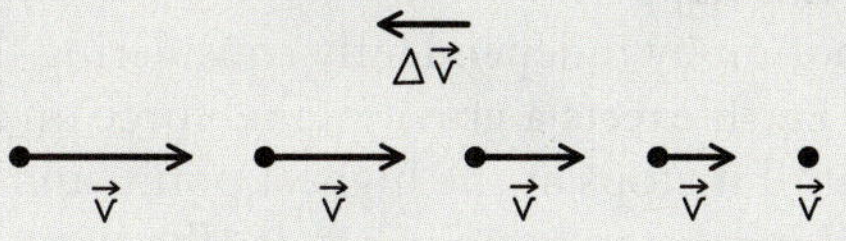


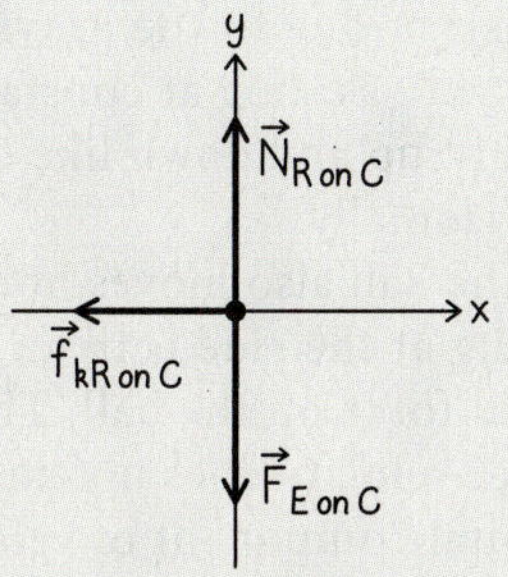


**Represent mathematically** Use the force diagram as shown above to help apply Newton's second law in component form. Use the expression for the kinetic friction force ($f_{\text{k Ron C}} = \mu_k N_{\text{R on C}}$), the expression for the gravitational force ($F_{\text{E on C}} = m_C g$), and one of the kinematics equations. The car remains in contact with the road surface, so the $y$-component of its acceleration $a_y$ is zero.

*y-component equation:*

$$0 = N_{\text{R on C}\,y} + F_{\text{E on C}\,y} + f_{\text{kR on C}\,y}$$

$$0 = N_{\text{R on C}} \sin 90° - m_C g \sin 90° + f_{\text{k R on C}} \sin 0°$$

Note that $\sin 90° = 1.0$ and $\sin 0° = 0$. Thus,

$$N_{\text{R on C}} = m_C g$$

The magnitude of the kinetic friction force is then

$$f_{\text{k R on C}} = \mu_k N_{\text{R on C}} = \mu_k m_C g$$

*x-component equation:*

$$m_C a_x = N_{\text{R on C}\,x} + F_{\text{E on C}\,x} + f_{\text{kR on C}\,x}$$

$$m_C a_x = N_{\text{R on C}} \cos 90° + m_C g \cos 90° - f_{\text{k R on C}} \cos 0°$$

Substitute $\cos 90° = 0$ and $\cos 0° = 1.0$ into the above to get

$$m_C a_x = -f_{\text{kR on C}}$$

Combine the two equations for $f_{\text{k R on C}}$ to get

$$m_C a_x = -\mu_k m_C g$$

or

$$a_x = -\mu_k g$$

Now use kinematics to determine the car's velocity $v_{0x}$ before the skid started:

$$v_x^2 - v_{0x}^2 = 2(x - x_0)a_x$$

**Solve and evaluate** The car's acceleration while stopping was

$$a_x = -\mu_k g = -(0.80)(9.8\ \text{m/s}^2) = -7.84\ \text{m/s}^2$$

We use the kinematics equation $0^2 - v_{0x}^2 = 2(x - x_0)a_x$ to determine the initial speed of the car before the skid started (recall that the final speed $v_x = 0$ and that the stopping distance was 18 m):

$$0^2 - v_{0x}^2 = 2(x - x_0)a_x = 2(18\ \text{m} - 0)(-7.84\ \text{m/s}^2)$$
$$= -282\ \text{m}^2/\text{s}^2 = -(16.8\ \text{m/s})^2$$

$$\Rightarrow v_{0x} = 16.8\ \text{m/s}$$
$$= 16.8\ \text{m/s}(3600\ \text{s/h})(1\ \text{mi}/1609\ \text{m})$$
$$= 38\ \text{mi/h}$$

This is slightly over the 15.6 m/s (35 mi/h) speed limit, but probably not enough for a speeding conviction. Note also that the answer had the correct units.

**Try it yourself:** Your car is moving at a speed 16 m/s on a flat, icy road when you see a stopped vehicle ahead. Determine the distance needed to stop if the effective coefficient of friction between your car tires and the road is 0.40.

*Answer:* 33 m.

## Other types of friction

There are other types of friction besides static and kinetic friction, such as rolling friction. Rolling friction is caused by the surfaces of rolling objects indenting slightly as they turn. This friction is decreased in tires that have been inflated to a higher pressure. In a later chapter (Chapter 11) we learn about another type of friction, the friction that air or water exerts on a solid object moving through the air or water—a so-called drag force.

**Review Question 3.4** What is the force of friction that the floor exerts on your refrigerator? Assume that the mass of the refrigerator is 100 kg, the coefficient of kinetic friction is 0.30, and the coefficient of static friction is 0.35. What assumptions did you make in answering this question?

# 3.5 Projectile motion

**Projectiles** are objects launched at an angle relative to a horizontal surface. We can use Newton's second law to analyze and explain projectile motion. We will begin by constructing a qualitative explanation and then develop a quantitative description. Let's start by making some observations.

## Qualitative analysis of projectile motion

**Figure 3.13** A projectile launched by a moving person.

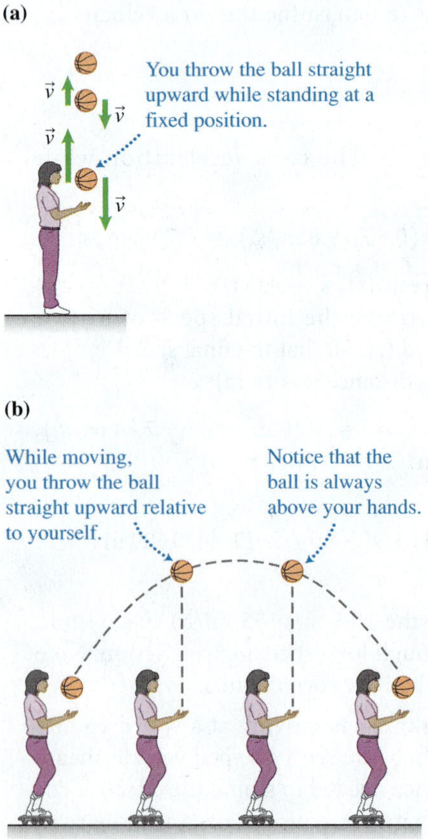
(a)
You throw the ball straight upward while standing at a fixed position.
$\vec{v}$
(b)
While moving, you throw the ball straight upward relative to yourself.
Notice that the ball is always above your hands.

You can easily create your own projectile using a basketball. First, throw the ball straight upward. It moves up and then down with respect to you and with respect to the floor and eventually returns to your hands (**Figure 3.13a**). Next, walk in a straight line at constant speed and throw the ball straight up again. Have a friend videotape you from the side. The ball goes up and returns to your hands as before, but this time you've moved along your walking path. Your friend sees a projectile—a ball traveling in an arc. Finally, put on rollerblades and repeat the experiment several times, each time moving with greater speed. As long as you don't change your speed or direction while the ball is in flight, it lands back in your hands. The video shows that the ball moves in an arc with respect to the ground, and at every frame it is directly above your hands (Figure 3.13b). Why does this happen?

We can analyze projectile motion by independently considering the ball's vertical and horizontal motion. Earth exerts a gravitational force on the ball, so its upward speed decreases until it stops at the highest point, and then its downward speed increases until it returns to your hands (the vertical component of the ball's acceleration is constant due to the gravitational force exerted on it by Earth). With respect to you, skating at constant velocity on rollerblades, the ball simply moves straight up and down like the ball in Figure 3.13a, even when you are moving horizontally.

In addition to this vertical motion, the ball also moves horizontally with respect to the floor. An observer standing at the side is in an inertial reference frame. No object exerts a horizontal force on the ball. Thus, according to Newton's first law, the ball's horizontal velocity does not change once it is released and is the same as your horizontal component of velocity. In every experiment the ball continues moving horizontally as if it were not thrown upward, and it moves up and down as if it does not move horizontally. It seems that the horizontal and vertical motions of the ball are independent of each other. Let's test this explanation in Testing Experiment **Table 3.6**.

**TESTING EXPERIMENT TABLE**

**3.6 Testing the independence of horizontal and vertical motions.**

VIDEO 3.6

| Testing experiment | Prediction | Outcome |
|---|---|---|
| One ball is shot horizontally when a compressed spring is released. Simultaneously, a second ball is dropped. Which ball hits the surface first? 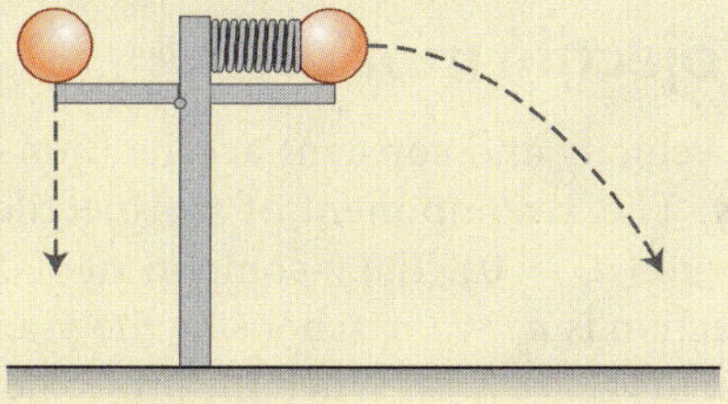 | Both balls start with zero initial vertical speed; thus their vertical motions are identical. Since we think that the vertical motion is independent of the horizontal motion, we predict that they will land at the same time. | When we try the experiment, the balls do land at the same time. |

**Conclusion**

The outcome supports the idea of independent horizontal and vertical motions. We've failed to disprove that idea.

The prediction may seem counterintuitive. However, the result matched the prediction. Since the ball on the right travels a longer path than the one on the left, why doesn't it land later? The vertical motions of both projectiles are identical; thus they land at the same time. The one on the right moves forward at constant velocity while it is falling, but this horizontal motion does not affect the vertical fall.

**CONCEPTUAL EXERCISE 3.7 Throwing a ball**

You throw a tennis ball as a projectile. Draw an arrow or arrows representing its instantaneous velocity and acceleration and the force or forces exerted on the ball by other objects when at the three positions shown below.

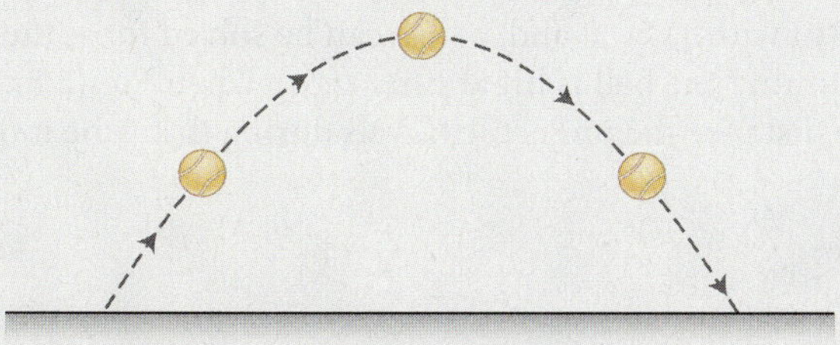

**Sketch and translate** The ball is the system of interest.

**Simplify and diagram** To draw velocity arrows, we consider the velocity at each point as consisting of a constant horizontal vector component and a vertical vector component whose magnitude decreases as the ball moves up and increases as it moves down. The vertical component is zero at the highest point in the trajectory. If we do this very carefully, we find that the velocity arrows are tangent to the ball's path at each position. If we ignore air resistance, only one object exerts a force on the ball—Earth. Thus in our sketch, the force arrows point down at all three positions as shown below. The acceleration arrows point in the same direction as the sum of the forces—downward. Notice that at the top of the path, the velocity of the object is horizontal; however, the acceleration still points downward. The direction of velocity at each point and the direction of the

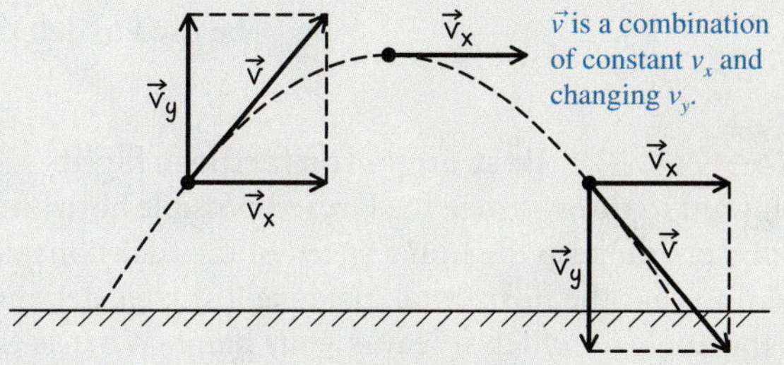


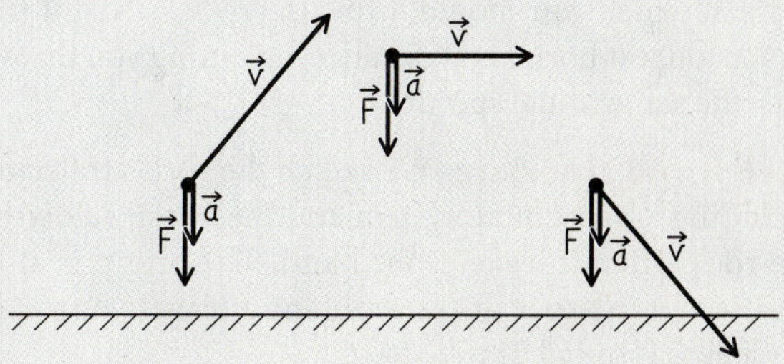


*(continued)*

sum of the forces exerted on the object at each point do not have to point in the same direction.

**Try it yourself:** How does the magnitude of the net force and the acceleration compare at the first and third positions shown in diagram?

*Answer:* The magnitude of the net force is the same at both positions. The magnitude of the acceleration is the same at both positions.

The idea that the motions in two directions are independent of each other will help us to develop a quantitative way to describe projectile motion.

## Quantitative analysis of projectile motion

We can use the equations of motion for velocity and constant acceleration to analyze projectile motion quantitatively. The $x$-component of a projectile's acceleration in the horizontal direction is zero ($a_x = 0$). The $y$-component of the projectile's acceleration in the vertical direction is $a_y = -g$ (choosing the $y$-axis to point up). Consider the projectile shown in **Figure 3.14**. When the projectile is launched at speed $v_0$ at an angle $\theta$ relative to the horizontal, its initial $x$- and $y$-velocity components are $v_{0x} = v_0 \cos\theta$ and $v_{0y} = v_0 \sin\theta$. The $x$-component of velocity remains constant during the flight, and the $y$-component changes in the same way that the velocity of an object thrown straight upward changes. If the motions in the $x$- and $y$-directions are independent of each other, then the equations that describe the motion of a projectile become the following:

**Figure 3.14** The initial velocity component vectors and their magnitudes.

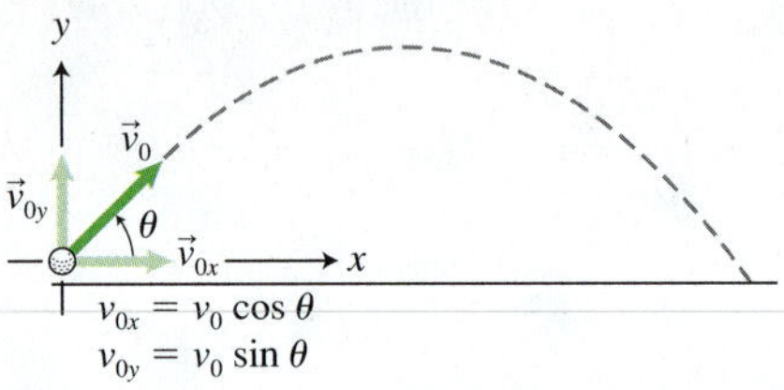


Projectile motion in the $x$-direction

$$(a_x = 0)$$

$$v_x = v_{0x} = v_0 \cos\theta \qquad (3.7x)$$

$$x = x_0 + v_{0x}t = x_0 + (v_0\cos\theta)t \qquad (3.8x)$$

Projectile motion in the $y$-direction

$$(a_y = -g)$$

$$v_y = v_{0y} + a_y t = v_0 \sin\theta + (-g)t \qquad (3.7y)$$

$$y = y_0 + v_{0y}t + \frac{1}{2}a_y t^2 = y_0 + (v_0 \sin\theta)t - \frac{1}{2}gt^2 \qquad (3.8y)$$

Equation (3.8$y$) can be used to determine the time interval for the projectile's flight, and Eq. (3.8$x$) can be used to determine how far the projectile travels in the horizontal direction during that time interval. For example, if the projectile leaves ground level and returns to ground level (for example, when you hit a golf ball), then Eq. (3.8$y$) with $y_0 = 0$ and $y = 0$ can be solved for $t$, the time when the ball lands (assuming the ball is hit at time zero). Then Eq. (3.8$x$) can be used to determine the distance the projectile travels during that time interval.

**EXAMPLE 3.8 Best angle for farthest flight**

You want to throw a rock the farthest possible horizontal distance. You keep the initial speed of the rock constant and find that the horizontal distance it travels depends on the angle at which it leaves your hand. What is the angle at which you should throw the rock so that it travels the longest horizontal distance, assuming you throw it with the same initial speed?

**Sketch and translate** We sketch the rock's trajectory, including a coordinate system and the initial velocity of the rock when it leaves your hand. The origin is at the position of the rock at the moment it leaves your hand. We call that initial time $t_0 = 0$.

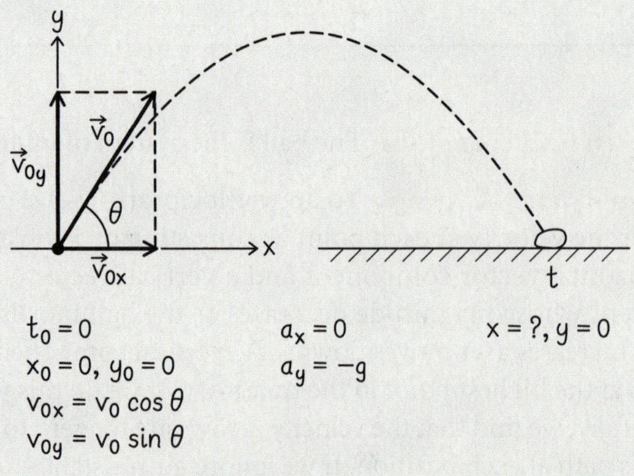

**Simplify and diagram** Assume that the rock leaves at the same elevation as it lands. That's not actually what happens, since the rock leaves your hand several feet above the ground, but it is a reasonable assumption if the rock travels a long distance compared to its initial height above the ground. Assume also that the rock is a point-like object and that the drag force exerted by the air on the rock does not affect its motion.

**Represent mathematically** First, determine the time interval that the rock is in flight using Eq. (3.8$y$). Note that because of the coordinate system we chose, $y_0 = y = 0$ and $t_0 = 0$:

$$0 = 0 + (v_0 \sin\theta)t - \frac{1}{2}gt^2$$

Dividing both sides of the equation by $t$ and then solving for $t$, we obtain

$$t = \frac{2v_0 \sin\theta}{g}$$

This is the time of flight—we see that it depends only on the initial velocity of the rock and the launch angle. To determine the distance that the rock travels in the horizontal direction, insert this expression for $t$ into Eq. (3.8$x$), and note that $x_0 = 0$:

$$x = (v_0 \cos\theta)t = v_0 \cos\theta\left(\frac{2v_0 \sin\theta}{g}\right) = \frac{v_0^2\, 2\sin\theta\cos\theta}{g}$$

**Solve and evaluate** Recall from trigonometry that $2\sin\theta\cos\theta = \sin(2\theta)$. This lets us rewrite the above equation as

$$x = \frac{v_0^2 \sin 2\theta}{g}$$

Our goal is to determine what launch angle will result in the rock traveling the farthest distance. Remember that the sine of an angle has a maximum value of 1 when the angle is 90°. Thus, $\sin 2\theta = 1$ when $2\theta = 90°$ or when $\theta = 45°$. The rock will travel the maximum horizontal distance when you launch it at an angle of 45° above the horizontal.

This answer seems reasonable. If you throw the rock at a smaller angle, the vertical component of the velocity is small and it spends less time in flight. If you launch it at an angle closer to 90°, it spends more time in flight, but its horizontal velocity component is small so that it does not travel very far horizontally. The 45° angle is a nice compromise between long time of flight and large horizontal velocity.

**Try it yourself:** Will the rock travel farther if it is launched at a 60° angle or a 30° angle?

*Answer:* It will travel the same distance in each case (see below.)

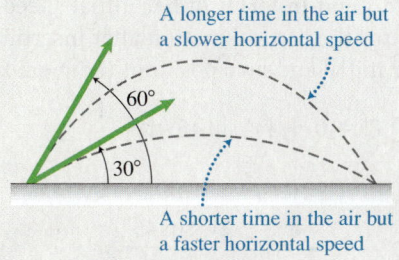

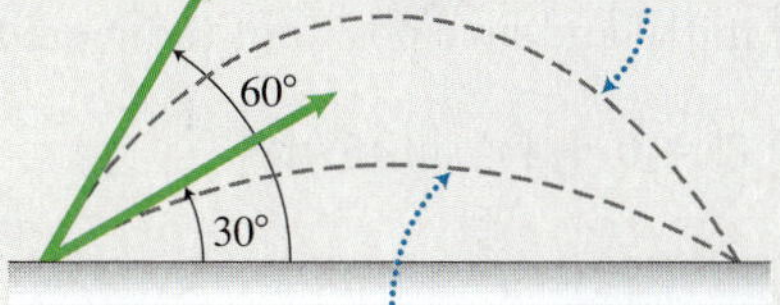
A longer time in the air but
a slower horizontal speed
60°
30°
A shorter time in the air but
a faster horizontal speed

## Human cannonballs

Calculating projectile motion is not always as simple as Eqs. (3.7) and (3.8) would suggest. Human cannonball launches must use complicated formulas to account for the effect of air resistance. Results obtained from Eqs. (3.7) and (3.8) are correct only in the absence of air resistance or if air resistance is small. Even so, we will use the simplified equations in the next example.

**TIP** Notice that the sign in front of $g$ in the equations depends on the chosen direction of the $y$-axis. Try to choose the direction of the axis so that it simplifies the situation as much as possible.

**EXAMPLE 3.9 Shot from a cannon**

Stephanie Smith Havens (sister of David Smith, Jr.) is to be shot from an 8-m-long cannon into a net 40 m from the end of the cannon barrel and at the same elevation (our assumption). The barrel of the cannon is oriented 45 degrees above the horizontal. Estimate the speed with which she needs to leave the cannon to make it to the net.

**Sketch and translate** We sketch the process as shown on the next page. Smith Havens leaves the barrel traveling at an unknown speed $v_0$ at a 45° angle above the horizontal. We choose the origin of the coordinate system to be at the end of the barrel. Time zero will be when she leaves the cannon barrel. Thus, $y_0 = y = 0$, where $y$ is her final elevation. The initial $x$-component of her velocity is $v_{0x} = v_0 \cos 45°$, and the initial $y$-component

*(continued)*

of her velocity is $v_{0y} = v_0 \sin 45°$. The $x$-component of her acceleration is $a_x = 0$ , and the $y$-component of her acceleration is $a_y = -g$.

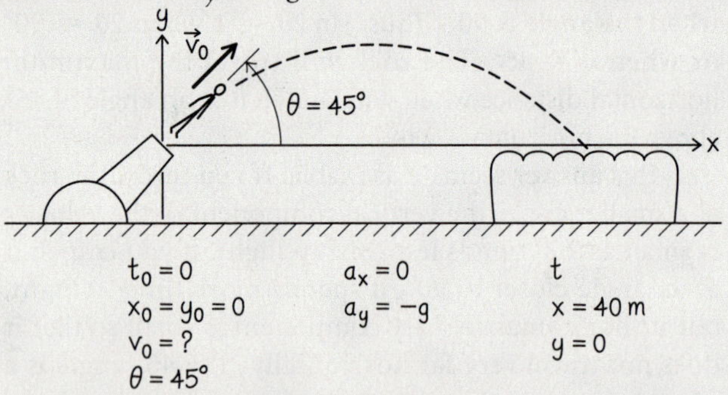


**Simplify and diagram** We model Smith Havens as a point-like object and assume that air resistance does not significantly affect her motion. The only force exerted on her while she is in flight is the gravitational force that Earth exerts on her.

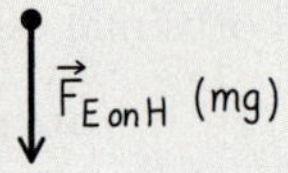


**Represent mathematically** We get an expression for her time of flight in terms of her initial speed by using Eq. (3.8$y$) for the vertical motion after inserting the values for her initial and final positions (both are zero):

$$0 = 0 + (v_0 \sin 45°)t + \frac{1}{2}(-g)t^2$$

or

$$t = \frac{2v_0 \sin 45°}{g}$$

We can substitute this expression into Eq. (3.8$x$) and solve for $v_0$:

$$x = 0 + (v_0 \cos 45°)t$$

$$= 0 + v_0(\cos 45°)\frac{2v_0(\sin 45°)}{g}$$

$$= v_0^2 \frac{2(\cos 45°)(\sin 45°)}{g}$$

**Solve and evaluate** Multiply each side of the previous equation by $g$ and divide both sides by $2(\cos 45°)(\sin 45°)$ to get an expression for the initial speed that Stephanie needs in order to reach the net:

$$v_0^2 = \frac{xg}{2(\cos 45°)(\sin 45°)}$$

$$= \frac{(40\text{ m})(9.8\text{ m/s}^2)}{2(\cos 45°)(\sin 45°)}$$

$$= 392\text{ m}^2/\text{s}^2$$

Taking the positive square root of both sides of this equation, we find that $v_0 \approx 20$ m/s or 44 mi/h. If we had included the effects of air resistance, her initial speed would have had to be greater than our estimate—so the fact that we underestimated the initial speed makes sense. Also notice that the units of our result are correct.

**Try it yourself:** Suppose Smith Havens was shot from a horizontal barrel that was 19.6 m higher than the catching net and that the net was placed 40 m horizontally from the end of the barrel. (a) What time interval is needed for her trip to the net? (b) At what speed does she need to leave the barrel in order to reach the net?

*Answer:* (a) 2.0 s, the time interval needed to fall 19.6 m; (b) 20 m/s, the horizontal speed needed to travel 40 m horizontally in 2.0 s.

**Review Question 3.5** Why do we need to resolve the initial velocity of a projectile into components when we analyze situations involving projectiles?

## 3.6 Using Newton's laws to explain everyday motion: Putting it all together

Each second of our lives is affected by phenomena that can be described and explained using Newton's laws—the cycles of day and night, the four seasonal variations, walking, driving cars, riding bicycles, and throwing and catching balls. We have already explored walking as a function of Newton's laws. In this section, we will analyze another everyday phenomenon, starting and stopping a car.

### Static friction helps a car start and stop

When a car is moving, the wheels turn relative to the road surface. The part of the tire immediately behind the part in contact with the surface is lifting up off the road, and the part of the tire immediately in front of the part in contact is moving down to make new contact with the road. But the part of the tire in contact with the road is at rest with respect to the road—for just a short time interval (**Figure 3.15a**)—just as your foot is stationary relative to the ground as your body moves forward while walking or running.

Increasing or decreasing the car's speed involves static friction between the tire's region of contact and the pavement. How can this be? If you want to move faster, the tire turns faster and pushes back harder on the pavement (Figure 3.15b). The pavement in turn pulls forward more on the tire (Newton's third law) and helps accelerate the car forward. If you want to slow down, the tire turns slower and pulls forward on the pavement (Figure 3.15c). The pavement in turn pushes back on the tire (Newton's third law) and helps the car accelerate backwards. Thus, the static friction force exerted by the road on the car can cause the car to speed up or slow down depending on the direction of that force.

Because the coefficient of static friction between the tire and the road is greater than the coefficient of kinetic friction, stopping is more efficient if your tires do not skid but instead roll to a stop. This is why vehicles are equipped with antilock brakes: to keep the tires rolling on the road instead of locking up and skidding.

Static friction not only helps a car speed up and slow down but also plays an important role in maintaining a car's constant speed. Without this additional force, the car would eventually slow down, because oppositely directed forces, such as air resistance and rolling resistance, point opposite the car's forward motion. A car's engine makes the axle and car tires rotate. The tires push back on the road and the road in turn pushes forward on the car in the opposite direction. The forward static friction force that the road exerts on the car keeps the car moving at constant speed. The presence of air resistance and rolling resistance explain why when you are driving you must continually give the engine gas; otherwise, you would not be able to maintain constant speed and your car would slow down.

**Figure 3.15** Static friction helps a car accelerate.

**(a)** The tire is moving to the right at constant speed.

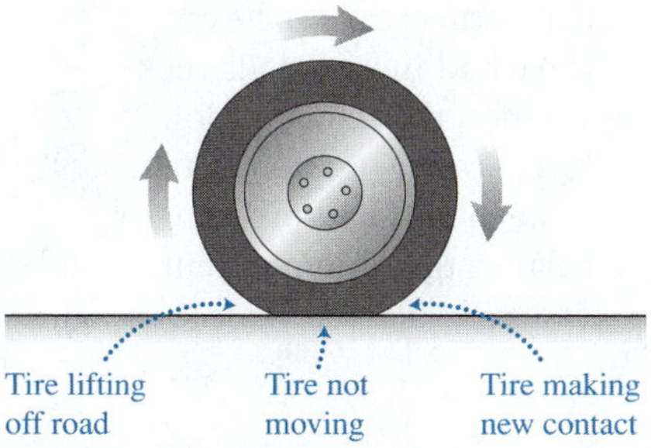


**(b)** The tire is moving to the right and turning faster.

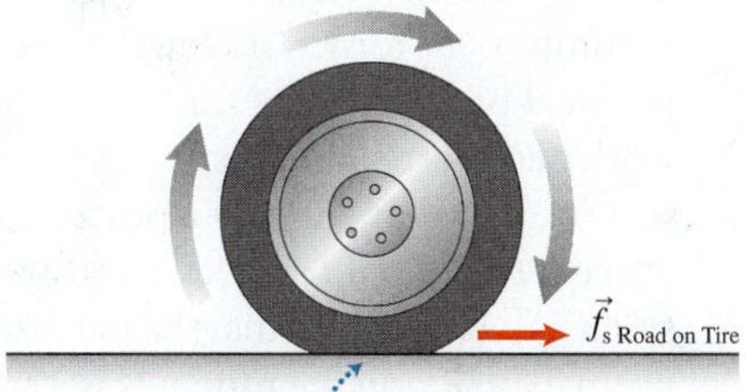


If the tire turns faster, it pushes back harder on the road. The road in turn pushes forward on the tire, causing the car to accelerate to the right.

**(c)** The tire is moving to the right and turning slower.

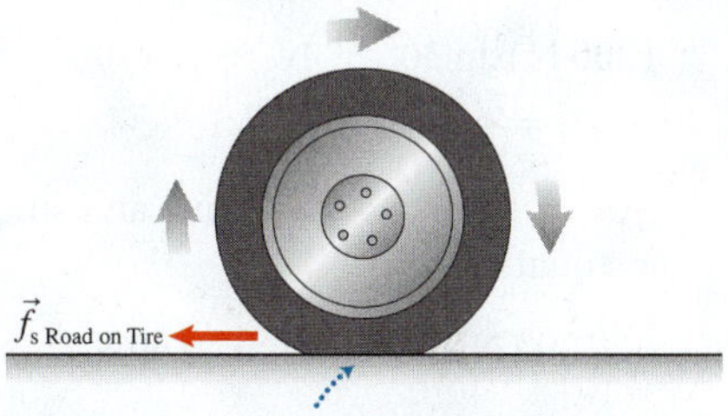


If the tire turns slower, it pulls forward on the road and the road pushes back on the tire, exerting a force that slows the car, causing it to accelerate backward.

**TIP** Some people think that the car's engine exerts a force on the car that starts the car's motion and helps it maintain its constant speed despite the air resistance. However, the forces that the engine exerts on other parts of the car are internal forces. Only external forces exerted by objects in the environment can affect the car's acceleration. The engine does rotate the wheels, and the wheels push forward or back on the ground. However, it is the ground (an external object) that pushes backward or forward on the wheels causing the car to slow down or speed up. The force that is responsible for this backward or forward push is the static friction force that the road exerts on the car tires.

## CONCEPTUAL EXERCISE 3.10 Car going down a hill

Your car moves down a 6.0° incline at a constant velocity of magnitude 16 m/s. Describe everything you can about the friction force that the road exerts on your car.

**Sketch and translate** Sketch the situation as shown. Since the car is moving at constant velocity, the sum of the forces that other objects exert on the car must be zero. We use this information to help construct a force diagram for the car and then determine the direction and relative magnitude of the friction force that the road exerts on the car.

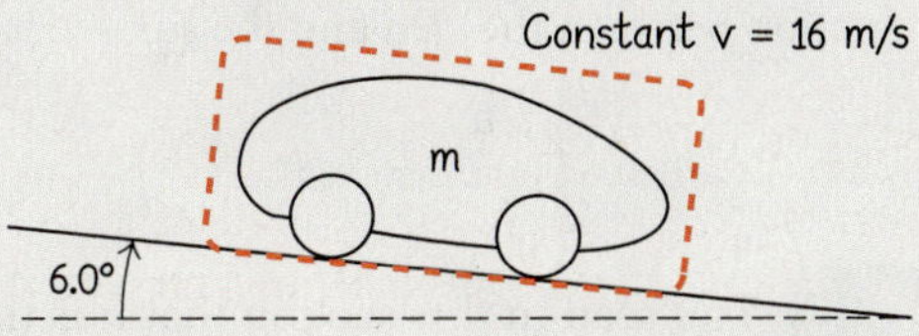


**Simplify and diagram** Model the car as a point-like object and assume that air resistance does not significantly affect the motion of the car. Make a force diagram for the car (as shown on the next page). The $x$-axis points in the direction of motion, and the $y$-axis points perpendicular to the road surface. The road exerts a normal force on the car $\vec{N}_{\text{R on C}}$ perpendicular to the road (it has only a $y$-component) and Earth exerts a downward gravitational force on the car $\vec{F}_{\text{E on C}}$ (it has a negative $y$-component and a small positive $x$-component). For the car to move at constant velocity, the road must exert a static friction force on the car $\vec{f}_{s\text{ R on C}}$ that points in the negative

*(continued)*

$x$-direction to balance the $x$-component of the gravitational force ($mg \cos 84°$) that Earth exerts on the car. If we had included air resistance, the static friction force would not have had to be as large in order to balance the $x$-component of the gravitational force.

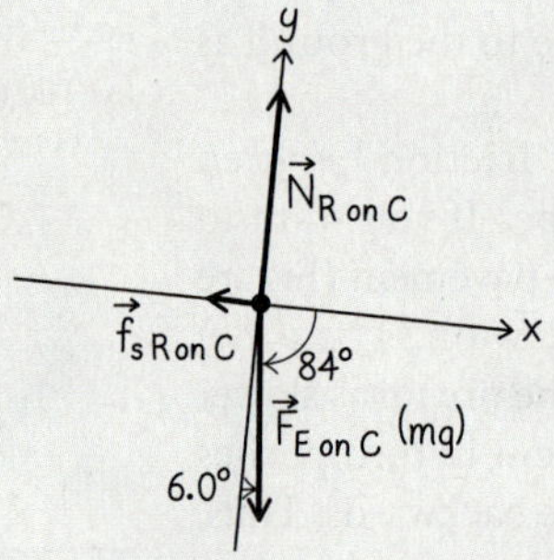


**Try it yourself:** A car moves at a high, constant velocity on a horizontal, level road. Does the road exert a static friction force on the car, and if so, in what direction?

*Answer:* Air resistance becomes more important at high speeds and can exert a several hundred newton force on the car that opposes its motion. Thus, the road now has to exert a static friction force on the car tires that pushes the car in the forward direction—the tires push back on the road, and the road in turn pushes forward on the tires.

### EXAMPLE 3.11 Equation Jeopardy

The equations below are the horizontal $x$- and vertical $y$-component forms of Newton's second law applied to a physical process. Solve for the unknown quantities. Then work backward and construct a force diagram for the system of interest and invent a process and question for which the equations might provide an answer (there are many possibilities). Remember that the italicized $N$ is the symbol for normal force, and the roman N is a symbol for the newton.

$x$-equation:

$$(200\text{ N})\cos 30° + 0 - 0.40N_{\text{S on O}} + 0 = (50\text{ kg})a_x$$

$y$-equation:

$$(200\text{ N})\sin 30° + N_{\text{S on O}} + 0 - (50\text{ kg})(9.8\text{ N/kg}) = (50\text{ kg})0$$

**Solve** Inserting the cosine and sine values, we get $x$-equation:

$$(200\text{ N})0.87 + 0 - 0.40N_{\text{S on O}} + 0 = (50\text{ kg})a_x$$

$y$-equation:

$$(200\text{ N})0.50 + N_{\text{S on O}} + 0 - (50\text{ kg})(9.8\text{ N/kg}) = 0$$

We can solve the $y$-equation for the magnitude of the normal force: $N_{\text{S on O}} = 390\text{ N}$. Inserting this value into the $x$-equation produces the following:

$$(200\text{ N})0.87 + 0 - 0.40(390\text{ N}) + 0 = (50\text{ kg})a_x$$

This can now be solved for $a_x$:

$$a_x = (174\text{ N} - 156\text{ N})/(50\text{ kg}) = +0.36\text{ N/kg} = +0.36\text{ m/s}^2$$

**Simplify and diagram** The equations provide the components for each of the four forces exerted on the system object (since there are four terms on the left side of each equation). Consider the $x$- and $y$-scalar components of each force:

1. A 200-N force oriented 30° above the positive $x$-axis—maybe a rope is exerting a force on an object.
2. A 390-N normal force points along the $y$-axis, perpendicular to a surface.
3. It looks like a $0.40(390\text{ N}) = 156\text{ N}$ friction force points in the negative $x$-direction with a coefficient of friction equal to 0.40.
4. A 490-N gravitational force points in the negative $y$-direction perpendicular to the surface.

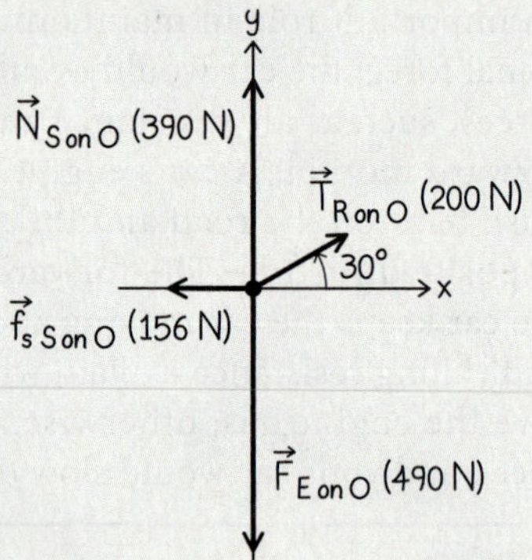


We can now use these forces to construct a force diagram.

**Sketch and translate** The situation could involve a sled or wagon or crate that is being pulled along a horizontal surface as shown. Note that the rope is at an angle with respect to the horizontal, so the force that it exerts has a +100-N $y$-component. It combines with the normal force's +390-N $y$-component to balance the gravitational force's −490-N $y$-component.

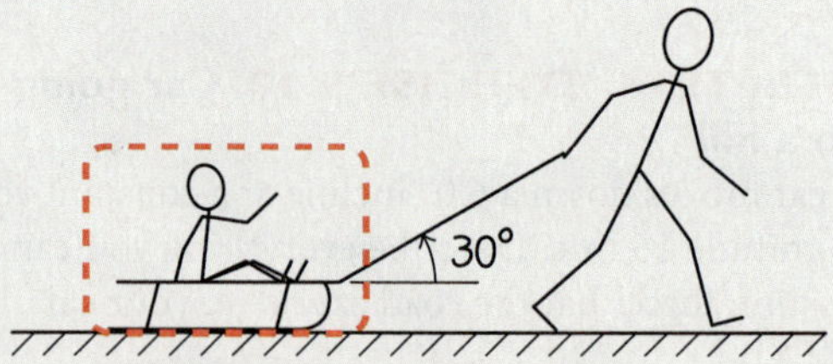


**One possible problem statement** Determine the acceleration of a 50-kg sled being pulled across a horizontal surface by a rope oriented 30° above the horizontal and pulling with a force of 200 N. The coefficient of kinetic friction between the sled and the surface is 0.40.

**Review Question 3.6** You read in this section that it is the road and not the engine that is most directly responsible for a car speeding up. Why is this true?

# Summary

**Words** | **Pictorial and physical representations** | **Mathematical representation**

**Components of a vector quantity** A vector quantity such as force $\vec{F}$ can be broken into its scalar components $F_x$ and $F_y$, which indicate the effect of the force in two perpendicular directions. (Section 3.1)

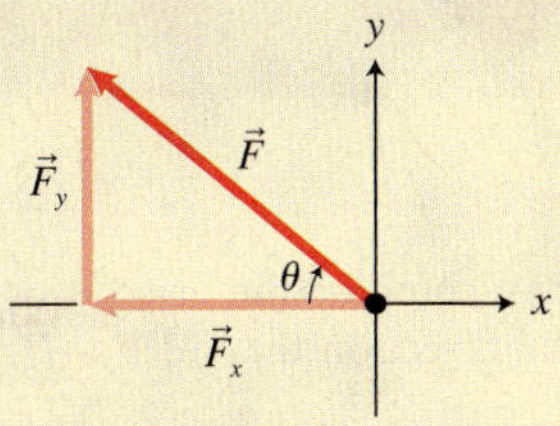


$$F_x = \pm F\cos\theta \quad \text{Eq. (3.1}x\text{)}$$

$$F_y = \pm F\sin\theta \quad \text{Eq. (3.1}y\text{)}$$

$\theta$ is the angle relative to the $\pm x$-axis. A scalar component is positive if it falls along a positive axis and negative if it falls along a negative axis.

**Newton's second law in component form** The acceleration of an object in the $x$-direction is the sum of the $x$-components of the forces exerted on it divided by its mass. The acceleration of the object in the $y$-direction is the sum of the $y$-components of the forces divided by its mass. (Section 3.2)

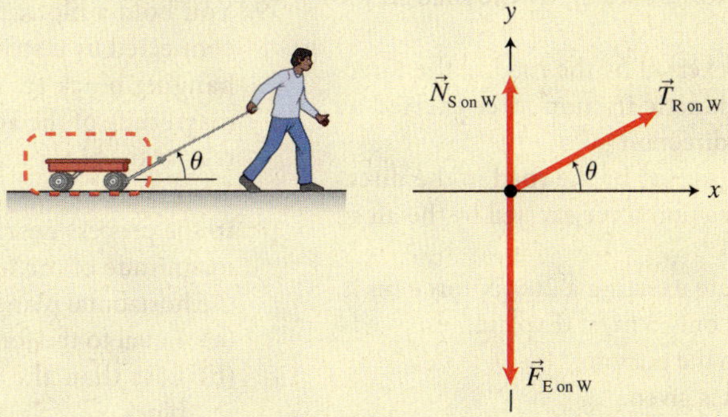


$$a_x = \frac{\Sigma F_x}{m} \quad \text{Eq. (3.2}x\text{)}$$

$$a_y = \frac{\Sigma F_y}{m} \quad \text{Eq. (3.2}y\text{)}$$

**Static friction force** is the force exerted by a surface on another surface (parallel to both surfaces) when they are not moving relative to each other. The force magnitude adjusts up to a maximum static friction force, depending on the force exerted on the object in an effort to start its motion. (Section 3.4)

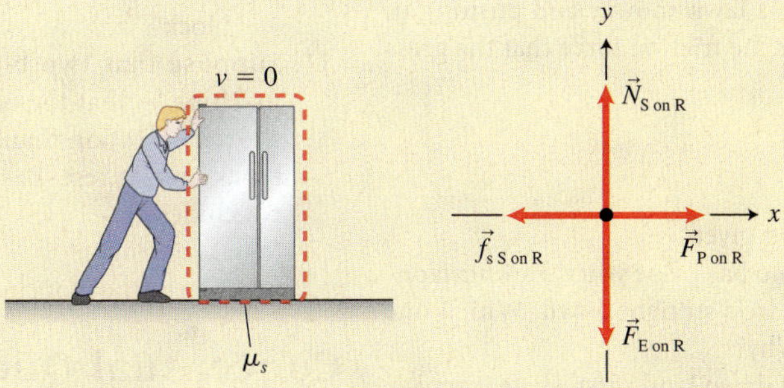


$$f_{s\,\max} \leq \mu_s N \quad \text{Eq. (3.5)}$$

**Kinetic friction force** is the force exerted by a surface on another surface (parallel to both surfaces) when the surfaces are moving relative to each other. (Section 3.4)

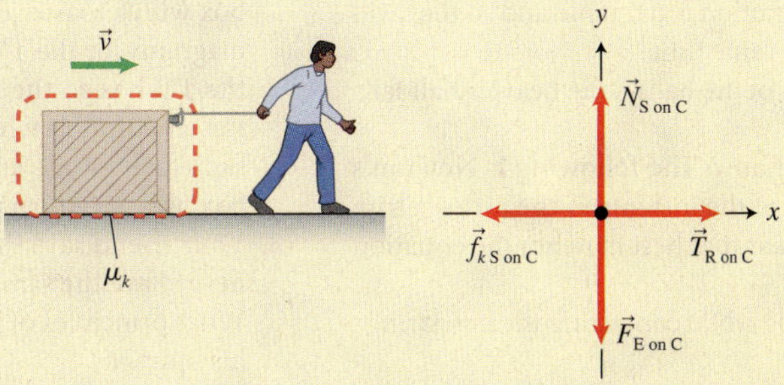


$$f_k = \mu_k N \quad \text{Eq. (3.6)}$$

**Projectile motion** Projectiles are objects launched either horizontally or at an angle with the horizontal. Horizontal and vertical motions of projectiles are independent of each other. If we ignore the resistive force that air exerts on the projectile, then its horizontal acceleration is zero ($a_x = 0$) and its vertical acceleration is the free-fall acceleration ($a_y = -g$). (Section 3.5)

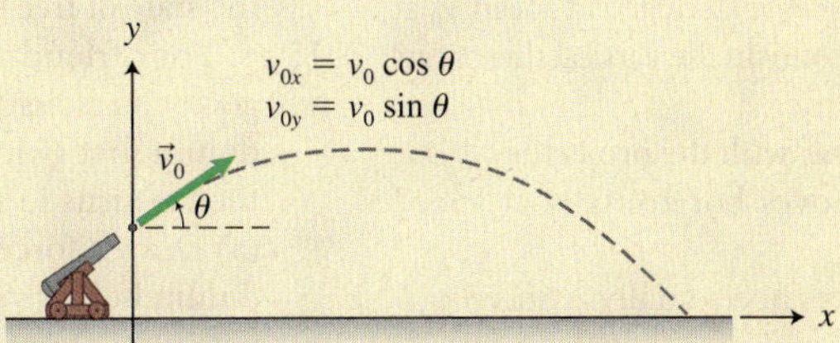


$x$-direction equations:

$$x = x_0 + v_{0x}t$$

$$v_x = v_0\cos\theta = \text{constant}$$

$y$-direction equations: (if positive direction up)

$$y = y_0 + v_{0y}t + \frac{1}{2}(-g)t^2 \quad \text{Eq. (3.8x)}$$

$$v_y = v_0\sin\theta + (-g)t \quad \text{Eq. (3.8y)}$$

MP For instructor-assigned homework, go to MasteringPhysics.

# Questions

## Multiple Choice Questions

1. A car accelerates along a road. Identify the best combination of forces exerted on the car in the horizontal direction that explains the acceleration.
   (a) The force exerted by the engine in the direction of motion and the kinetic friction force exerted by the road in the opposite direction
   (b) The static friction force exerted by the road in the direction of motion and the kinetic friction force exerted by the road in the opposite direction
   (c) The static friction force exerted by the road in the direction of motion and the friction force exerted by the air in the opposite direction
2. A person pushes a 10-kg crate exerting a 200-N force on it, but the crate's acceleration is only 5 $m/s^2$. Explain.
   (a) The crate pushes back on the person.
   (b) Not enough information is given.
   (c) There are other forces exerted on the crate.
3. Compare the ease of pulling a lawn mower and pushing it. In particular, in which case is the friction force that the grass exerts on the mower greater?
   (a) They are the same.
   (b) Pulling is easier.
   (c) Pushing is easier.
   (d) Not enough information is given.
4. You simultaneously release two balls: one you throw horizontally, and the other one you drop straight down. Which one will reach the ground first? Why?
   (a) The ball dropped straight down lands first, since it travels a shorter distance.
   (b) Neither. Their vertical motion is the same and so they will reach the ground at the same time.
   (c) It depends on the mass of the balls—the heavier ball falls faster.
5. You shoot an arrow with a bow. The following is Newton's second law applied for one of the instants in the arrow's trip: $m\vec{a} = m\vec{g}$. Choose the instant that best matches the equation description:
   (a) The arrow is accelerating while contacting the bowstring.
   (b) The arrow is flying up.
   (c) The arrow is slowing down while sinking into the target.
   (d) This equation does not describe any part of the arrow's trip.
6. In what reference frame does a projectile launched at speed $v_0$ at angle $\theta$ above the horizontal move only in the vertical direction?
   (a) There is no such reference frame.
   (b) In a reference frame that moves with the projectile
   (c) In a reference frame that moves horizontally at speed $v_0 \cos \theta$
   (d) In a reference frame that moves horizontally at speed $v_0$
7. You throw a ball vertically upward. From the time it leaves your hand until just before it returns to your hand, where is it located when the magnitude of its acceleration is greatest? Do not neglect air resistance.
   (a) Just after it leaves your hand on the way up
   (b) Just before arriving at your hand on the way down
   (c) At the top
   (d) Acceleration is the same during the entire flight.
8. While running, how should you throw a ball with respect to you so that you can catch it yourself?
   (a) Slightly forward (b) Slightly backward
   (c) Straight up (d) It is impossible.
9. You hold a block on a horizontal, frictionless surface. It is connected by a string that passes over a pulley to a vertically hanging block (a modified Atwood machine). What is the magnitude of the acceleration of the hanging object after you release the block on the horizontal surface?
   (a) Less than $g$ (b) More than $g$ (c) Equal to $g$
10. In the process described in the previous question, what is the magnitude of the force exerted by the string on the block on the horizontal plane after it is released?
    (a) Equal to the force that Earth exerts on the hanging block
    (b) Less than the force exerted by Earth on the hanging block
    (c) More than the force exerted by Earth on the hanging block
11. Suppose that two blocks are positioned on an Atwood machine so that the block on the right of mass $m_1$ hangs at a lower elevation than the block on the left of mass $m_2$. Both blocks are at rest. Based on this observation, what can you conclude?
    (a) $m_1 > m_2$ (b) $m_1 = m_2$ (c) $m_1 < m_2$
    (d) You cannot conclude anything with the given information.

## Conceptual Questions

12. A box with a heavy television set in it is placed against a box with a toaster oven in it. Both sit on the floor. Draw force diagrams for the TV box and the toaster oven box if you push the TV box to the right, exerting a force $\vec{F}_{\text{Y on TV}}$. Repeat for the situation in which you are exerting a force of exactly the same magnitude and direction, but pushing the toaster oven box, which in turn pushes the TV box.
13. Your friend says that two blocks in an Atwood machine can never have the same acceleration, no matter what we assume. What principles of physics is your friend using as the basis for his opinion?
14. How can an Atwood machine be used to determine the acceleration of freely falling objects if none of the objects used is in the state of free fall?
15. Your friend is on rollerblades holding a pendulum. You gently push her forward and let go. You observe that the pendulum first swings in the opposite direction (backward) and then returns to the vertical orientation as she coasts forward. (a) Draw a force diagram to explain the behavior of the pendulum bob as your friend is being pushed. (b) How can you use it to determine the acceleration of your friend while you are pushing her? (c) Why does the pendulum return to its vertical orientation after you stop pushing your friend?
16. Explain why a car starts skidding when a driver abruptly applies the brakes.
17. Explain why old tires need to be replaced.

18. Describe two experiments that policemen can perform to determine the coefficient of kinetic friction between car tires and the road. Why would they need to do two independent experiments for the same set of tires?
19. Explain how friction helps you to walk.
20. Explain why you might fall forward when you stumble.
21. Explain why you might fall backward when you slip.
22. Explain why the tires of your car can "spin out" when you are caught in the mud.
23. You throw two identical balls simultaneously at the same initial speed: one downward and the other horizontally. Describe and compare their motions in as much detail as you can.
24. Your friend says that the vertical force exerted on a projectile when at the top of its flight is zero. Why would he say this? Do you agree or disagree? If you disagree, how would you convince your friend that your opinion is correct?
25. Your friend says that a projectile launched at an angle relative to the horizontal moves forward because it retains the force of the launcher. Why would she say this? Do you agree or disagree? If you disagree, how would you convince your friend that your opinion is correct?
26. An object of mass $m_1$ placed on an inclined plane (angle $\theta$ relative to the horizontal) is connected by a string that passes over a pulley to a hanging object of mass $m_2$ ($m_2 \gg m_1$). Draw a force diagram for each object. Do not ignore friction.
27. An object of mass $m_1$ placed on an inclined plane (angle $\theta$ relative to the horizontal) is connected by a string that passes over a pulley to a hanging object of mass $m_2$ ($m_1 \gg m_2$). Draw a force diagram for each object. Do not assume that friction can be ignored.

# Problems

Below, **BIO** indicates a problem with a biological or medical focus. Problems labeled **EST** ask you to estimate the answer to a quantitative problem rather than derive a specific answer. Problems marked with require you to make a drawing or graph as part of your solution. Asterisks indicate the level of difficulty of the problem. Problems with no * are considered to be the least difficult. A single * marks moderately difficult problems. Two ** indicate more difficult problems.

## 3.1 Force components

1. Determine the $x$- and $y$-components of each force vector shown in **Figure P3.1**.

**Figure P3.1**

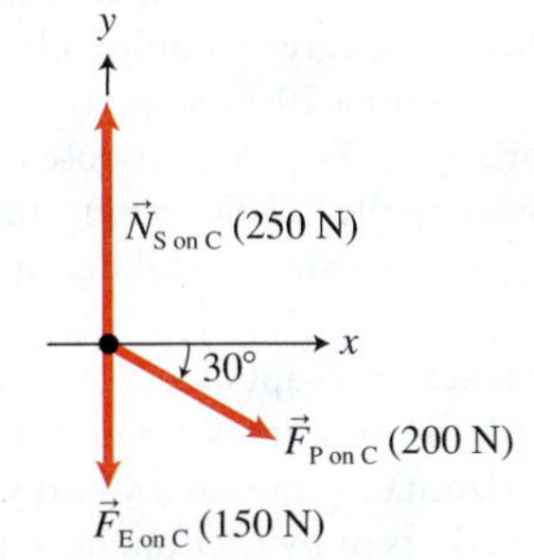


2. Determine the $x$- and $y$-components of each force vector shown in **Figure P3.2**.

**Figure P3.2**

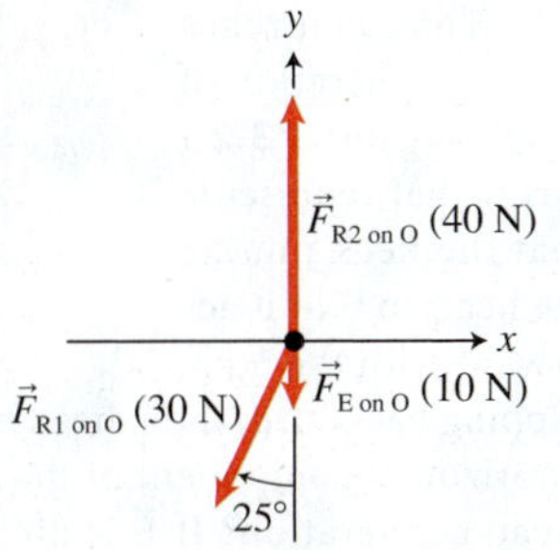


3. Determine the $x$- and $y$-components of each displacement shown in **Figure P3.3**.

**Figure P3.3**

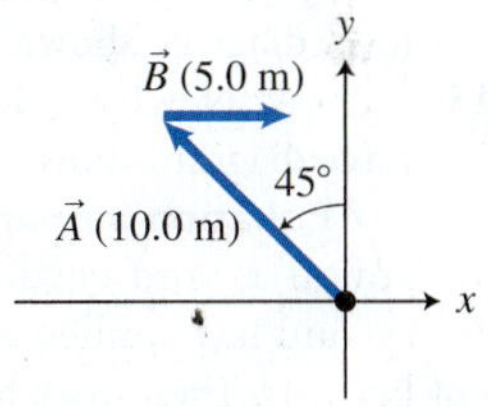


4. * The $x$- and $y$-components of several unknown forces are listed below ($F_x$, $F_y$). For each force, draw on an $x$, $y$ coordinate system the components of the force vectors. Determine the magnitude and direction of each force: (a) (+100 N, −100 N), (b) (−300 N, −400 N), and (c) (−400 N, +300 N).
5. * The $x$- and $y$-scalar components of several unknown forces are listed below ($F_x$, $F_y$). For each force, draw an $x$, $y$ coordinate system and the vector components of the force vectors. Determine the magnitude and direction of each force: (a) (−200 N, +100 N), (b) (+300 N, +400 N), and (c) (+400 N, −300 N).

## 3.2 Newton's second law in component form

6. * Three ropes pull on a knot shown in **Figure P3.6a**. The knot is not accelerating. A partially completed force diagram for the knot is shown in Figure P3.6b. Use qualitative reasoning (no math) to determine the magnitudes of the forces that ropes 2 and 3 exert on the knot. Explain in words how you arrived at your answers.
7. * Solve the previous problem quantitatively using Newton's second law.

**Figure P3.6a**

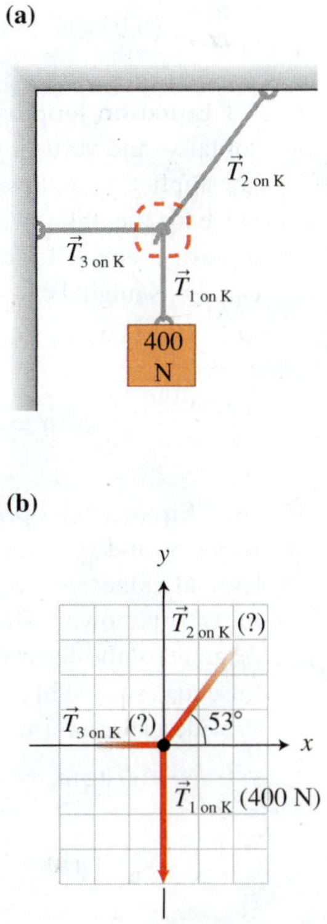

8. For each of the following situations, draw the forces exerted on the moving object and identify the other object causing each force. (a) You pull a wagon along a level floor using a rope oriented 45° above the horizontal. (b) A bus moving on a horizontal road slows in order to stop. (c) You slide down an inclined water slide. (d) You lift your overnight bag into the overhead compartment on an airplane. (e) A rope connects two boxes on a horizontal floor, and you pull horizontally on a second rope attached to the right side of the right box (consider each box separately).

9. * Write Newton's second law in component form for each of the situations described in Problem 8.

10. For the situations described here, construct a force diagram for the block, sled, and skydiver. (a) A cinder block sits on the ground. (b) A rope pulls at an angle of 30° relative to the horizontal on a sled moving on a horizontal surface. The sled moves at increasing speed toward the right (the surface is not smooth.) (c) A rope pulls on a sled parallel to an inclined slope (inclined at an arbitrary angle). The sled moves at increasing speed up the slope. (d) A skydiver falls downward at constant terminal velocity (air resistance is present).

11. * Write Newton's second law in component form for each of the situations described in Problem 10.

12. * Apply Newton's second law in component form for the force diagram shown in Figure P3.1.

13. * Apply Newton's second law in component form for the force diagram shown in Figure P3.2.

14. * **Equation Jeopardy 1** The equations below are the horizontal $x$- and vertical $y$- component forms of Newton's second law applied to a physical process. Solve for the unknowns. Then work backward and construct a force diagram for the object of interest and invent a problem for which the equations might be an answer (there are many possibilities).

$$(5.0\text{ kg})a_x = (50\text{ N})\cos 30° + N\cos 90° + (5.0\text{ kg})(9.8\text{ N/kg})\cos 90°$$

$$(5.0\text{ kg})0 = (-50\text{ N})\sin 30° + N\sin 90° - (5.0\text{ kg})(9.8\text{ N/kg})\sin 90°$$

15. * **Equation Jeopardy 2** The equations below are the horizontal $x$- and vertical $y$-component forms of Newton's second law applied to a physical process for an object on an incline. Solve for the unknowns. Then work backward and construct a force diagram for the object and invent a problem for which the equations might be an answer (there are many possibilities).

$$(5.0\text{ kg})a_x = (30\text{ N})\cos 30° + N\cos 90° - (5.0\text{ kg})(9.8\text{ N/kg})\cos 60°$$

$$(5.0\text{ kg})0 = (30\text{ N})\sin 30° + N\sin 90° - (5.0\text{ kg})(9.8\text{ N/kg})\sin 60°$$

16. ** **Equation Jeopardy 3** The equations below are the horizontal $x$- and vertical $y$-component forms of Newton's second law and a kinematics equation applied to a physical process. Solve for the unknowns. Then work backward and construct a force diagram for the object of interest and invent a problem for which the equations might be an answer (there are many possibilities). Provide all the information you know about your process.

$$(5.0\text{ kg})a_x = (50\text{ N})\cos 30° + N\cos 90° + (5.0\text{ kg})(9.8\text{ N/kg})\cos 90°$$

$$(5.0\text{ kg})0 = (-50\text{ N})\sin 30° + N\sin 90° - (5.0\text{ kg})(9.8\text{ N/kg})\sin 90°$$

$$x - 0 = (2.0\text{ m/s})(4.0\text{ s}) + \frac{1}{2}a_x(4.0\text{ s})^2$$

17. * You exert a force of 100 N on a rope that pulls a sled across a very smooth surface. The rope is oriented 37° above the horizontal. The sled and its occupant have a total mass of 40 kg. The sled starts at rest and moves for 10 m. List all the quantities you can determine using these givens and determine three of the quantities on the list.

18. * You exert a force of a known magnitude $F$ on a grocery cart of total mass $m$. The force you exert on the cart points at an angle $\theta$ below the horizontal. If the cart starts at rest, determine an expression for the speed of the cart after it travels a distance $d$. Ignore friction.

19. * **Olympic 100-m dash start** At the start of his race, 86-kg Olympic 100-m champion Usain Bolt from Jamaica pushes against the starting block, exerting an average force of 1700 N. The force that the block exerts on his foot points 20° above the horizontal. Determine his horizontal speed after the force is exerted for 0.32 s. Indicate any assumptions you made.

20. * **Accelerometer** A string with one 10-g washer on the end is attached to the rearview mirror of a car. When the car leaves an intersection, the string makes an angle of 5° with the vertical. What is the acceleration of the car? [*Hint:* Choose the washer as the system object for your force diagram. Use the vertical component equation of Newton's second law to find the magnitude of the force that the string exerts on the washer. Then continue with the horizontal component equation.]

21. * **Your own accelerometer** A train has an acceleration of magnitude $1.4\text{ m/s}^2$ while stopping. A pendulum with a 0.50-kg bob is attached to a ceiling of one of the cars. Determine everything you can about the pendulum during the deceleration of the train.

## 3.3 Problem-solving strategies for analyzing dynamics problems

22. * **Skier** A 52-kg skier starts at rest and slides 30 m down a hill inclined at 12° relative to the horizontal. List five quantities that describe the motion of the skier, and solve for three of them (at least one should be a kinematics quantity).

23. * **Ski rope tow** You agree to build a backyard rope tow to pull your siblings up a 20-m slope that is tilted at 15° relative to the horizontal. You must choose a motor that can pull your 40-kg sister up the hill. Determine the force that the rope should exert on your sister to pull her up the hill at constant velocity.

24. * **Soapbox racecar** A soapbox derby racecar starts at rest at the top of a 301-m-long track tilted at an average 4.8° relative to the horizontal. If the car's speed were not reduced by any structural effects or by friction, how long would it take to complete the race? What is the speed of the car at the end of the race?

25. * BIO **Whiplash experience** A car sitting at rest is hit from the rear by a semi-trailer truck moving at 13 m/s. The car lurches forward with an acceleration of about $300\text{ m/s}^2$. **Figure P3.25** shows an arrow that represents the force that the neck muscle exerts on the head so that it accelerates forward with the body instead of flipping backward. If the head has a mass of 4.5 kg, what is the horizontal component of the force $\vec{F}$ required to cause this head acceleration? If $\vec{F}$ is directed 37° below the horizontal, what is the magnitude of $\vec{F}$?

**Figure P3.25**

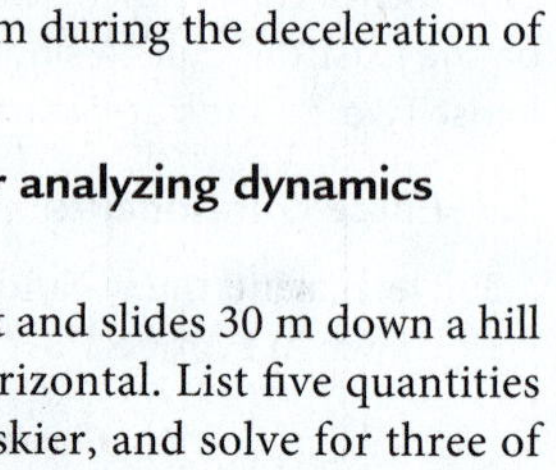

26. * **Iditarod race practice** The dogs of four-time Iditarod Trail Sled Dog Race champion Jeff King pull two 100-kg sleds that are connected by a rope. The sleds move on an icy surface. The dogs exert a 240-N force on the rope attached to the front sled. Find the acceleration of the sleds and the force the rope between the sleds exerts on each sled. The front rope pulls horizontally.

27. * You pull a rope oriented at a 37° angle above the horizontal. The other end of the rope is attached to the front of the first of two wagons that have the same 30-kg mass. The rope exerts a force of magnitude $T_1$ on the first wagon. The wagons are connected by a second horizontal rope that exerts a force of magnitude $T_2$ on the second wagon. Determine the magnitudes of $T_1$ and $T_2$ if the acceleration of the wagons is $2.0\ \text{m/s}^2$.

28. ** Rope 1 pulls horizontally, exerting a force of 45 N on an 18-kg wagon attached by a second horizontal rope to a second 12-kg wagon. Make a list of physical quantities you can determine using this information, and solve for three of them, including one kinematics quantity.

29. * Three sleds of masses $m_1$, $m_2$, $m_3$ are on a smooth horizontal surface (ice) and connected by ropes, so that if you pull the rope connected to sled 1, all the sleds start moving. Imagine that you exert a force of a known magnitude on the rope attached to the first sled. What will happen to all of the sleds? Provide information about their accelerations and all the forces exerted on them. What assumptions did you make?

30. ** Repeat Problem 29, only this time with the sleds on a hill.

31. ** Your daredevil friends attach a rope to a 140-kg sled that rests on a frictionless icy surface. The rope extends horizontally to a smooth dead tree trunk lying at the edge of a cliff. Another person attaches a 30-kg rock at the end of the rope after it passes over the tree trunk and then releases the rock—the rope is initially taut. Determine the acceleration of the sled, the force that the rope exerts on the sled and on the rock, and the time interval during which the person can jump off the sled before it reaches the cliff 10 m ahead. There is no friction between the rope and the tree trunk.

32. * Assume the scenario described in Problem 31, but in this case a hanging rock of unknown mass accelerates downward at $2.7\ \text{m/s}^2$ and pulls the sled with it. Determine the mass of the hanging rock and the force that the rope exerts on the sled. There is no friction between the rope and the tree trunk.

33. ** The 20-kg block shown in **Figure P3.33** accelerates down and to the left, and the 10-kg block accelerates up. Find the magnitude of this acceleration and the force that the cable exerts on a block. There is no friction between the block and the inclined plane, and the pulley is frictionless and light.

**Figure P3.33**

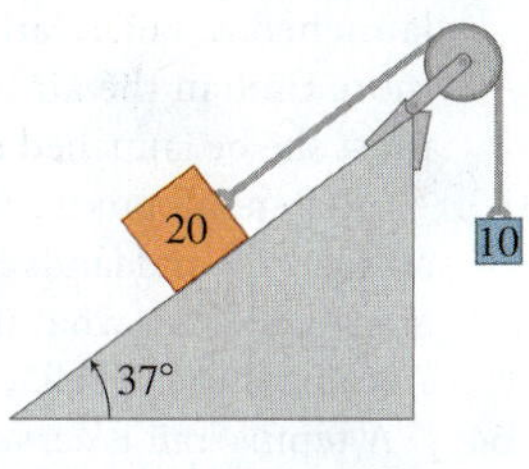


34. ** A person holds a 200-g block that is connected to a 250-g block by a string going over a light pulley with no friction in the bearing (an Atwood machine). After the person releases the 200-g block, it starts moving upward and the heavier block descends. (a) What is the acceleration of each block? (b) What is the force that the string exerts on each block? (c) How long will it take each block to traverse 1.0 m?

35. ** Two blocks of masses $m_1$ and $m_2$ are connected to each other on an Atwood machine. A person holds one of the blocks with her hand. When the system is released, the heavier block moves down with an acceleration of $2.3\ \text{m/s}^2$ and the lighter object moves up with an acceleration of the same magnitude. What is one possible set of masses for the blocks?

## 3.4 Friction

36. * **Equation Jeopardy 4** The equations below are the horizontal $x$- and vertical $y$-component forms of Newton's second law applied to a physical process. Solve for the unknowns. Then work backward and construct a force diagram for the object of interest and invent a problem for which the equations might provide an answer (there are many possibilities).

$$(5.0\ \text{kg})a_x = (50\ \text{N})\cos 30° + N\cos 90° - 0.5\,N\cos 0° + (5.0\ \text{kg})(9.8\ \text{N/kg})\cos 90°$$

$$(5.0\ \text{kg})0 = (-50\ \text{N})\sin 30° + N\sin 90° + 0.5\,N\sin 0° - (5.0\ \text{kg})(9.8\ \text{N/kg})\sin 90°$$

37. * **Equation Jeopardy 5** The equations below are the $x$- and $y$-component forms of Newton's second law applied to a physical process for an object on an incline. Solve for the unknowns. Then work backward and construct a force diagram for the object and invent a problem for which the equations might provide an answer (there are many possibilities).

$$(5.0\ \text{kg})0 = +F\cos 0° + N\cos 90° - 0.50\,N\cos 0° - (5.0\ \text{kg})(9.8\ \text{N/kg})\cos 60°$$

$$(5.0\ \text{kg})0 = +F\sin 0° + N\sin 90° - 0.50\,N\sin 0° - (5.0\ \text{kg})(9.8\ \text{N/kg})\sin 60°$$

38. A 91.0-kg refrigerator sits on the floor. The coefficient of static friction between the refrigerator and the floor is 0.60. What is the minimum force that one needs to exert on the refrigerator to start the refrigerator sliding?

39. A 60-kg student sitting on a hardwood floor does not slide until pulled by a 240-N horizontal force. Determine the coefficient of static friction between the student and floor.

40. * **Racer runs out of gas** James Stewart, 2002 Motocross/Supercross Rookie of the Year, is leading a race when he runs out of gas near the finish line. He is moving at 16 m/s when he enters a section of the course covered with sand where the effective coefficient of friction is 0.90. Will he be able to coast through this 15-m-long section to the finish line at the end? If yes, what is his speed at the finish line?

41. * **Car stopping distance and friction** A certain car traveling at 60 mi/h (97 km/h) can stop in 48 m on a level road. Determine the coefficient of friction between the tires and the road. Is this kinetic or static friction? Explain.

42. * A 50-kg box rests on the floor. The coefficients of static and kinetic friction between the bottom of the box and the floor are 0.70 and 0.50, respectively. (a) What is the minimum force a person needs to exert on the box to start it sliding? (b) After the box starts sliding, the person continues to push it, exerting the same force. What is the acceleration of the box?

43. * Marsha is pushing down and to the right on a 12-kg box at an angle of 30° below horizontal. The box slides at constant velocity across a carpet whose coefficient of kinetic friction with the box is 0.70. Determine three physical quantities using this information, one of which is a kinematics quantity.

44. * A wagon is accelerating to the right. A book is pressed against the back vertical side of the wagon and does not slide down. Explain how this can be.

45. * In Problem 44, the coefficient of static friction between the book and the vertical back of the wagon is $\mu_s$. Determine an expression for the minimum acceleration of the wagon in terms of $\mu_s$ so that the book does not slide down. Does the mass of the book matter? Explain.

46. * A car has a mass of 1520 kg. While traveling at 20 m/s, the driver applies the brakes to stop the car on a wet surface with a 0.40 coefficient of friction. (a) How far does the car travel before stopping? (b) If a different car with a mass 1.5 times greater is on the road traveling at the same speed and the coefficient of friction between the road and the tires is the same, what will its stopping distance be? Explain your results.

47. * A 20-kg wagon accelerates on a horizontal surface at $0.50\ \mathrm{m/s^2}$ when pulled by a rope exerting a 120-N force on the wagon at an angle of 25° above the horizontal. Determine the magnitude of the effective friction force exerted on the wagon and the effective coefficient of friction associated with this force.

48. * You want to use a rope to pull a 10-kg box of books up a plane inclined 30° above the horizontal. The coefficient of kinetic friction is 0.30. What force do you need to exert on the other end of the rope if you want to pull the box (a) at constant speed and (b) with a constant acceleration of $0.50\ \mathrm{m/s^2}$ up the plane? The rope pulls parallel to the incline.

49. * A car with its wheels locked rests on a flatbed of a tow truck. The flatbed's angle with the horizontal is slowly increased. When the angle becomes 40°, the car starts to slide. Determine the coefficient of static friction between the flatbed and the car's tires.

50. * **Olympic skier** Olympic skier Lindsey Vonn skis down a steep slope that descends at an angle of 30° below the horizontal. The coefficient of sliding friction between her skis and the snow is 0.10. Determine Vonn's acceleration, and her speed 6.0 s after starting.

51. * **Another Olympic skier** Bode Miller, 80-kg downhill skier, descends a slope inclined at 20°. Determine his acceleration if the coefficient of friction is 0.10. How would this acceleration compare to that of a 160-kg skier going down the same hill? Justify your answer using sound physics reasoning.

52. * A crate of mass $m$ sitting on a horizontal floor is attached to a rope that pulls at an angle $\theta$ above the horizontal. The coefficient of static friction between the crate and floor is $\mu_s$. (a) Construct a force diagram for the crate when being pulled by the rope but not sliding. (b) Determine an expression for the smallest force that the rope needs to exert on the crate that will cause the crate to start sliding.

53. * **EST** You absentmindedly leave your book bag on the top of your car. (a) Estimate the safe acceleration of the car needed for the bag to stay on the roof. Describe the assumptions that you made. (b) Estimate the safe speed. Describe the assumptions that you made.

54. A book slides off a desk that is tilted 15° degrees relative to the horizontal. What information about the book or the desk does this number provide?

55. * Block 1 is on a horizontal surface with a 0.29 coefficient of kinetic friction between it and the surface. A string attached to the front of block 1 passes over a light frictionless pulley and down to hanging block 2. Determine the mass of block 2 in terms of block 1 so that the blocks move at constant nonzero speed while sliding.

## 3.5 Projectile motion

56. * **Equation Jeopardy 6** The equations below describe a projectile's path. Solve for the unknowns and then invent a process that the equations might describe. There are many possibilities.

$$x = 0 + (20\ \mathrm{m/s})(\cos 0°)t$$

$$0 = 8.0\ \mathrm{m} + (20\ \mathrm{m/s})(\sin 0°) - \frac{1}{2}(9.8\ \mathrm{m/s^2})t^2$$

57. A bowling ball rolls off a table. Draw a force diagram for the ball when on the table and when in the air at two different positions.

58. A ball moves in an arc through the air (see **Figure P3.58**). Construct a force diagram for the ball when at positions (a), (b), and (c). Ignore the resistive force exerted by the air on the ball.

**Figure P3.58**

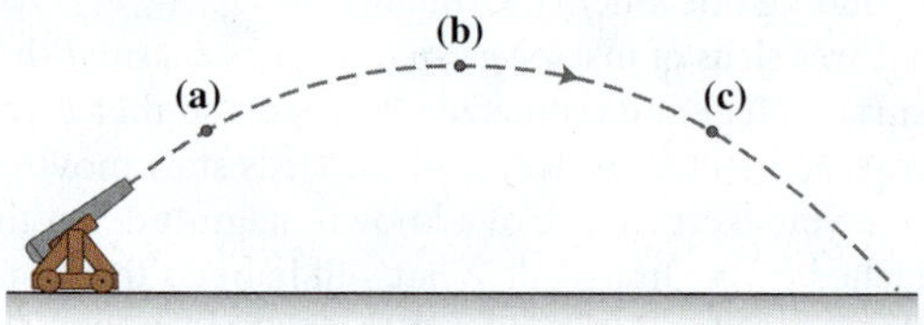


59. A marble is thrown as a projectile at an angle above the horizontal. Draw its path during the flight. Choose six different positions along the path so that one of them is at the highest point. For each position, indicate the direction of the marble's velocity, acceleration, and all of the forces exerted on it by other objects.

60. A baseball leaves a bat and flies upward and toward center field. After it leaves the bat, are any forces exerted on the ball in the horizontal direction? In the direction of motion? In the vertical direction? If so, identify the other object that causes each force. Do not ignore air resistance.

61. * **Robbie Knievel ride** On May 20, 1999, Robbie Knievel easily cleared a narrow part of the Grand Canyon during a world-record–setting long-distance motorcycle jump—69.5 m. He left the jump ramp at a 10° angle above the horizontal. How fast was he traveling when he left the ramp? Indicate any assumptions you made.

62. * Daring Darless wishes to cross the Grand Canyon of the Snake River by being shot from a cannon. She wishes to be launched at 60° relative to the horizontal so she can spend more time in the air waving to the crowd. With what speed must she be launched to cross the 520-m gap?

63. ** A football punter wants to kick the ball so that it is in the air for 4.0 s and lands 50 m from where it was kicked. At what angle and with what initial speed should the ball be kicked? Assume that the ball leaves 1.0 m above the ground.

64. * A tennis ball is served from the back line of the court such that it leaves the racket 2.4 m above the ground in a horizontal direction at a speed of 22.3 m/s (50 mi/h). (a) Will the ball cross a 0.91-m-high net 11.9 m in front of the server? (b) Will the ball land in the service court, which is within 6.4 m of the net on the other side of the net?

65. * **EST** An airplane is delivering food to a small island. It flies 100 m above the ground at a speed of 160 m/s. (a) Where should the parcel be released so it lands on the island? Neglect

air resistance. (b) Estimate whether you should release the parcel earlier or later if there is air resistance. Explain.

66. * If you shoot a cannonball from the same cannon first at 30° and then at 60° relative to the horizontal, which orientation of the cannon will make the ball go farther? How do you know? Under what circumstances is your answer valid? Explain.

67. When you actually perform the experiment described in Problem 66, the ball shot at a 60° angle lands closer to the cannon than the ball shot at a 30° angle. Explain why this happens.

68. ** You can shoot an arrow straight up so that it reaches the top of a 25-m-tall building. (a) How far will the arrow travel if you shoot it horizontally while pulling the bow in the same way? The arrow starts 1.45 m above the ground. (b) Where do you need to put a target that is 1.45 m above the ground in order to hit it if you aim 30° above the horizontal? (c) Determine the maximum distance that you can move the target and hit it with the arrow.

69. * Robin Hood wishes to split an arrow already in the bull's-eye of a target 40 m away. If he aims directly at the arrow, by how much will he miss? The arrow leaves the bow horizontally at 40 m/s.

### 3.6 Using Newton's laws to explain everyday motion: Putting it all together

70. Three force diagrams for a car are shown in **Figure P3.70**. Indicate as many situations as possible for the car in terms of its velocity and acceleration at that instant for each diagram.

**Figure P3.70**

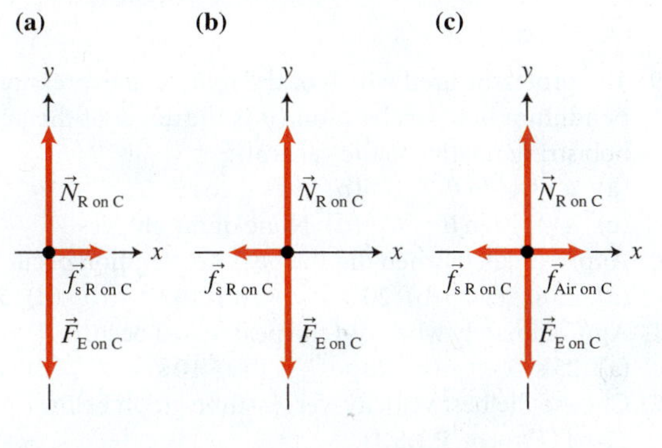


71. * A minivan of mass 1560 kg starts at rest and then accelerates at $2.0\ \text{m/s}^2$. (a) What is the object exerting the force on the minivan that causes it to accelerate? What type of force is it? (b) Air resistance and other opposing resistive forces are 300 N. Determine the magnitude of the force that causes the minivan to accelerate in the forward direction.

72. ** A daredevil motorcycle rider hires you to plan the details for a stunt in which she will fly her motorcycle over six school buses. Provide as much information as you can to help the rider successfully complete the stunt.

## General Problems

73. * **EST** Estimate the range of the horizontal force that a sidewalk exerts on you during every step while you are walking. Indicate clearly how you made the estimate.

74. * Two blocks of masses $m_1$ and $m_2$ hang at the ends of a string that passes over the very light pulley with low friction bearings shown in **Figure P3.74**. Determine an expression in terms of the masses and any other needed quantities for the magnitude of the acceleration of each block and the force that the string exerts on each block. Apply the equation for two cases: (a) the blocks have the same mass, but one is positioned lower than the other and (b) the blocks have different masses, but the heavier block is positioned higher than the light one. What assumptions did you make?

**Figure P3.74**

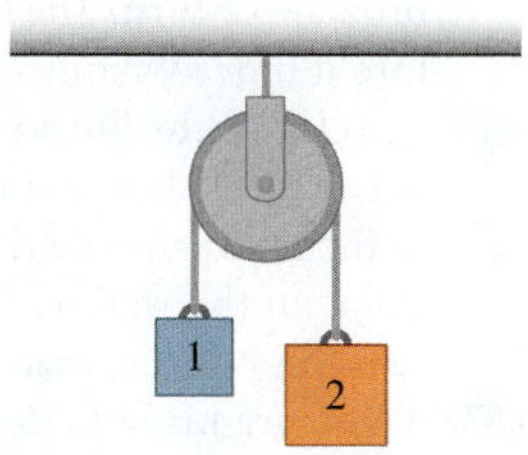


75. ** A 0.20-kg block placed on an inclined plane (angle 30° above the horizontal) is connected by a string going over a pulley to a 0.60-kg hanging block. Determine the acceleration of the system if there is no friction between the block and the surface of the inclined plane.

76. ** A 3.5-kg object placed on an inclined plane (angle 30° above the horizontal) is connected by a string going over a pulley to a 1.0-kg hanging block. (a) Determine the acceleration of the system if there is no friction between the object and the surface of the inclined plane. (b) Determine the magnitude of the force that the string exerts on both objects.

77. ** A 3.5-kg object placed on an inclined plane (angle 30° above the horizontal) is connected by a string going over a pulley to a 1.0-kg object. Determine the acceleration of the system if the coefficient of static friction between object 1 and the surface of the inclined plane is 0.30 and equals the coefficient of kinetic friction.

78. ** An object of mass $m_1$ placed on an inclined plane (angle $\theta$ above the horizontal) is connected by a string going over a pulley to a hanging object of mass $m_2$. Determine the acceleration of the system if there is no friction between object 1 and the surface of the inclined plane. If the problem has multiple answers, explore all of them.

79. ** An object of mass $m_1$ placed on an inclined plane (angle $\theta$ above the horizontal) is connected by a string going over a pulley to a hanging object of mass $m_2$. Determine the acceleration of the system if the coefficient of static friction between object 1 and the surface of the inclined plane is $\mu_s$, and the coefficient of kinetic friction is $\mu_k$. If the problem has multiple answers, explore all of them.

80. ** You are driving at a reasonable constant velocity in a van with a windshield tilted 120° relative to the horizontal (see **Figure P3.80**). As you pass under a utility worker fixing a power line, his wallet falls onto the windshield. Determine the acceleration needed by the van so that the wallet stays in place. When choosing your coordinate axes, remember that you want the wallet's acceleration to be horizontal rather than vertical. What assumptions and approximations did you make?

**Figure P3.80**

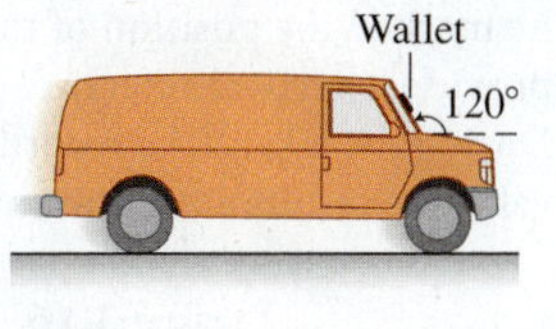


81. ** A ledge on a building is 20 m above the ground. A taut rope attached to a 4.0-kg can of paint sitting on the ledge passes up over a pulley and straight down to a 3.0-kg can of nails on the ground. If the can of paint is accidentally knocked off the ledge, what time interval does a carpenter have to catch the can of paint before it smashes on the floor?

82. ** EST **Bicycle ruined** The brakes on a bus fail as it approaches a turn. The bus was traveling at the speed limit before it moved about 24 m across grass and hit a brick wall. A bicycle attached to a rack on the front of the bus was crushed between the bus and the brick wall. There was little damage to the bus. Estimate the average force that the bicycle and bus exert on the wall while stopping. Indicate any assumptions made in your estimate.
83. * You are hired to devise a method to determine the coefficient of friction between the ground and the soles of a shoe and of its competitors. Explain your experimental technique and provide a physics analysis that could be used by others using this method.
84. * The mass of a spacecraft is about 480 kg. An engine designed to increase the speed of the spacecraft while in outer space provides 0.09-N thrust at maximum power. By how much does the engine cause the craft's speed to change in 1 week of running at maximum power? Describe any assumptions you made.
85. * EST A 60-kg rollerblader rolls 10 m down a 30° incline. When she reaches the level floor at the bottom, she applies the brakes. Use Newton's second law to estimate the distance she will move before stopping. Justify any assumptions you made.
86. Design, perform, and analyze the results of an experiment to determine the coefficient of static friction and the coefficient of kinetic friction between a penny and the cover of this textbook.
87. * **Tell all** A sled starts at the top of the hill shown in **Figure P3.87**. Add any information that you think is reasonable about the process that ensues when the sled goes down the hill and finally stops. Then tell everything you can about this process.

**Figure P3.87**

## Reading Passage Problems

**Professor tests airplane takeoff speed** D. A. Wardle, a professor of physics from the University of Auckland, New Zealand, tested the takeoff speed of a commercial airliner. The pilot had insisted that the takeoff speed had to be 232 km/h.[1] To perform the testing experiment, Wardle used a pendulum attached to stiff cardboard (**Figure 3.16**). Prior to takeoff, when the plane was stationary, he marked the position of the pendulum bob on the cardboard to provide a vertical reference line (the dashed line in Figure 3.16). During the takeoff, he recorded the position of the bob at 5-s intervals. The results are shown in the table.

**Figure 3.16** Wardle's device.

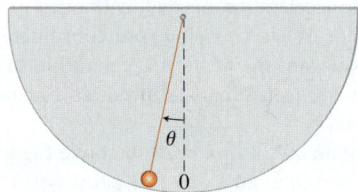

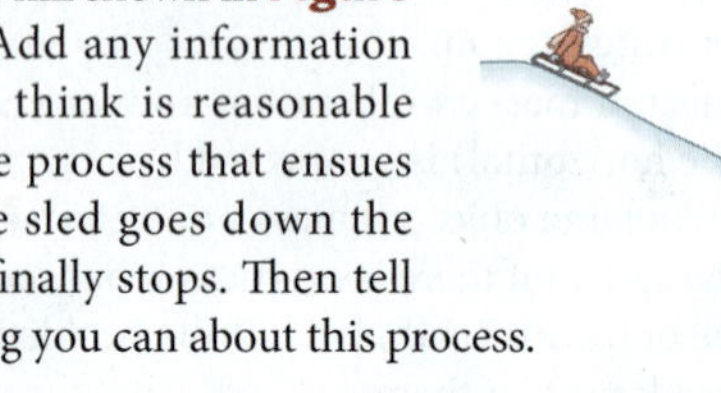


[1]The information is taken from the article by D. A. Wardle "Measurement of aeroplane takeoff speed and cabin pressure" published in *The Physics Teacher*, 37 410–411 (1999).

| $t$ (seconds) | $\theta$ (degrees) |
|---|---|
| 0 | 9.9 |
| 5 | 14.8 |
| 10 | 13.8 |
| 15 | 13.0 |
| 20 | 12.0 |
| 25 | 11.4 |

Using these data, Professor Wardle determined the acceleration at takeoff to be greater than $g/4$. Then he plotted an acceleration-versus-time graph and used it to find the takeoff speed. It turned out to be about 201 km/h. He was very satisfied—the day was windy, and the speed of the breeze was about 15–20 km/h. Thus the takeoff speed predicted by his simple pendulum was 215–220 km/h, very close to what the pilot said.

88. Choose the best force diagram for the pendulum bob as the plane is accelerating down the runway (**Figure P3.88**).

**Figure P3.88**

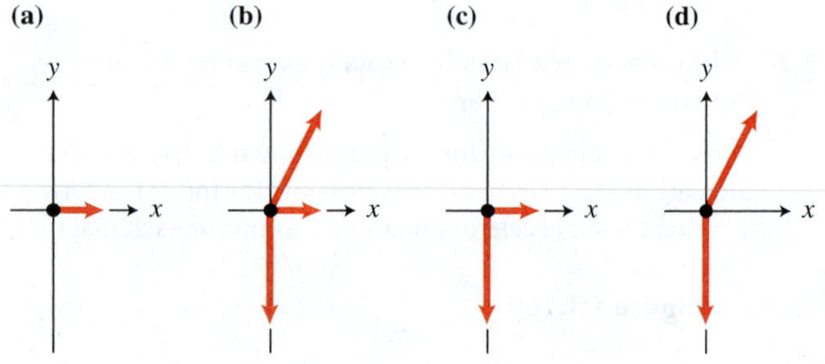


89. The professor used which of the following expression for the pendulum bob acceleration ($\theta$ is the angle of the pendulum bob string relative to the vertical)?
(a) $a = g\sin\theta$ (b) $a = g\cos\theta$
(c) $a = g\tan\theta$ (d) None of the choices
90. Approximately when did the peak acceleration occur?
(a) 25 s (b) 20 s (c) 10 s (d) 5 s
91. Approximately when did the peak speed occur?
(a) 25 s (b) 20 s (c) 10 s (d) 5 s
92. Choose the best velocity-versus-time graph below for the airplane (**Figure P3.92**).

**Figure P3.92**

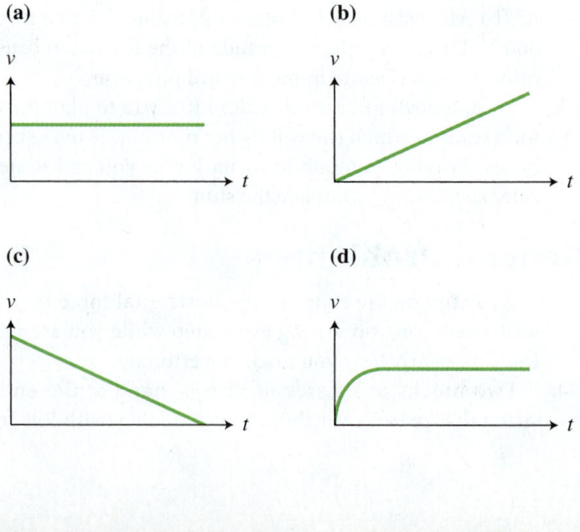

**Ski jumping in Vancouver** The 2010 Olympic ski jumping contest was held at Whistler Mountain near Vancouver (**Figure 3.17**). During a jump, a skier starts near the top of the in-run, the part down which the skier glides at increasing speed before the jump. The Whistler in-run is 116 m long and for the first part is tilted down at about 35° below the horizontal. There is then a curve that transitions into a takeoff ramp, which is tilted 11° below the horizontal. The skier flies off this ramp at high speed, body tilted forward and skis separated (**Figure 3.18**). This position exposes a large surface area to the air, which creates lift, extends the time of the jump, and allows the jumper to travel farther. In addition, the skier pushes off the exit ramp of the in-run to get a vertical component of velocity when leaving the ramp. The skier lands 125 m or more from the end of the in-run. The landing surface has a complex shape and is tilted down at about 35° below the horizontal. The skier moves surprisingly close (2 to 6 m) above the snowy surface for most of the jump. The coefficient of kinetic friction between the skis and the snow on the in-run is about $0.05 \pm 0.02$ and skiers' masses are normally small—about 60 kg. We can make some rough estimates about an idealized ski jump with an average in-run inclination of $(35° + 11°)/2 = 23°$.

**Figure 3.17** Whistler Mountain ski jump in-run and landing area.

**Figure 3.18** A ski jumper exposes a large surface area to the air in order to get more lift.

93. Which answer below is closest to the magnitude of the normal force that the idealized in-run exerts on the 60-kg skier?
(a) 590 N (b) 540 N (c) 250 N (d) 230 N

94. Which numbers below are closest to the magnitudes of the kinetic friction force and the component of the gravitational force parallel to the idealized inclined in-run?
(a) 30 N, 540 N (b) 27 N, 540 N (c) 12 N, 540 N
(d) 30 N, 230 N (e) 27 N, 230 N (f) 12 N, 230 N

95. Which answers below are closest to the magnitude of the skier's acceleration while moving down the idealized in-run and to the skier's speed when leaving its end?
(a) 9.8 m/s$^2$, 48 m/s (b) 4.3 m/s$^2$, 32 m/s
(c) 4.3 m/s$^2$, 28 m/s (d) 3.4 m/s$^2$, 32 m/s
(e) 3.4 m/s$^2$, 28 m/s

96. Assume that the skier left the ramp moving horizontally. Treat the skier as a point-like particle and assume the force exerted by air on him is minimal. If he landed 125 m diagonally from the end of the in-run and the landing region beyond the in-run was inclined $35^0$ below the horizontal for its entire length, which answer below is closest to the time interval that he was in the air?
(a) 1.9 s (b) 2.4 s (c) 3.1 s
(d) 3.8 s (e) 4.3 s

97. Using the same assumptions as stated in Problem 96, which answer below is closest to the jumper's speed when leaving the in-run?
(a) 37 m/s (b) 31 m/s (c) 27 m/s
(d) 24 m/s (e) 21 m/s

98. Which factors below would keep the skier in the air longer and contribute to a longer jump?
   1. The ramp at the end of the in-run is level instead of slightly tilted down.
   2. The skier extends his body forward and positions his skis in a V shape.
   3. The skier has wider and longer than usual skis.
   4. The skier pushes upward off the end of the ramp at the end of the in-run.
   5. The skier crouches in a streamline position when going down the in-run.

(a) 1 (b) 5 (c) 1, 3, 5
(d) 2, 3, 4 (e) 1, 2, 4, 5 (f) 1, 2, 3, 4, 5

# 4 Circular Motion

**Why do pilots sometimes black out while pulling out at the bottom of a power dive?**

**Are astronauts really "weightless" while in orbit?**

**Why do you tend to slide across the car seat when the car makes a sharp turn?**

Kruti Patel, a civilian test pilot, wears a special flight suit and practices special breathing techniques to prevent dizziness, disorientation, and possibly passing out as she comes out of a power dive. These symptoms characterize a blackout, which can occur when the head and brain do not received a sufficient amount of blood. Why would pulling out of a power dive cause a blackout, and why does a special suit prevent it from occurring? Our study of circular motion in this chapter will help us understand blackouts and other interesting phenomena.

**Be sure you know how to:**

- Find the direction of acceleration using a motion diagram (Section 1.6).
- Draw a force diagram (Section 2.1).
- Use a force diagram to help apply Newton's second law in component form (Sections 3.1 and 3.2).

**In previous chapters** we studied motion in situations in which the sum of the forces exerted on a system had a constant magnitude and direction. In fact, in real life we encounter relatively few situations where motion is this simple. More often the forces exerted on an object continually change direction and magnitude as time passes. In this chapter we will focus on circular motion. It is the simplest example of motion in which the sum of the forces exerted on a system object by other objects continually changes.

## 4.1 The qualitative velocity change method for circular motion

Consider the motion of the car shown in **Figure 4.1** as it travels at constant speed around a circular track. The instantaneous velocity of the car is tangent to the circle at every point. Recall from Chapter 1 that an object has acceleration if its velocity changes—in magnitude *or* in direction or in both. Even though the car is moving at constant speed, it is accelerating because the *direction* of its velocity changes from moment to moment.

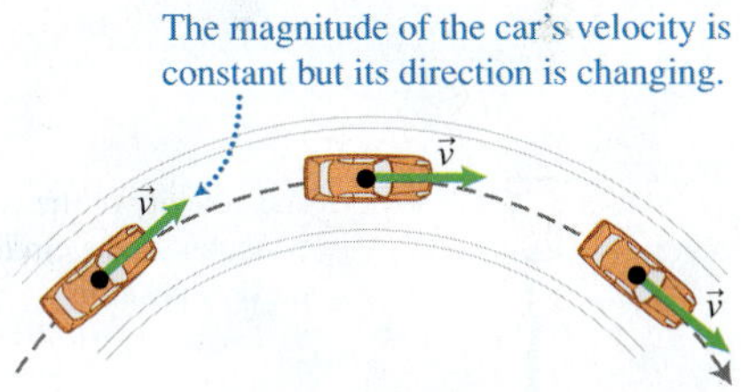

The magnitude of the car's velocity is constant but its direction is changing.
$\vec{v}$
$\vec{v}$
$\vec{v}$

**Figure 4.1** Since the direction of the velocity is changing, the car is accelerating.

**‹ Active Learning Guide**

To estimate the direction of acceleration of any object (including that of the car) while passing a particular point on its path (see **Figure 4.2a**), we use the **velocity change method**. Consider a short time interval $\Delta t = t_f - t_i$ during which the car passes that point. The velocity arrows $\vec{v}_i$ and $\vec{v}_f$ represent the initial and final velocities of the car, a little before and a little after the point where we want to estimate its acceleration (note that the velocity arrows are tangent to the curve in the direction of motion). The velocity change vector is $\Delta\vec{v} = \vec{v}_f - \vec{v}_i$. You can think of $\Delta\vec{v}$ as the vector that you need to add to the initial velocity $\vec{v}_i$ in order to get the final velocity $\vec{v}_f$, that is, $\vec{v}_i + \Delta\vec{v} = \vec{v}_f$. Rearranging this, we get $\Delta\vec{v} = \vec{v}_f - \vec{v}_i$ (the change in velocity).

To estimate $\Delta\vec{v}$, place the $\vec{v}_i$ and $\vec{v}_f$ arrows tail to tail (Figure 4.2b) without changing their magnitudes or directions. $\Delta\vec{v}$ starts at the head of $\vec{v}_i$ and ends at the head of $\vec{v}_f$. The car's acceleration $\vec{a}$ is in the direction of the $\Delta\vec{v}$ arrow (Figure 4.2c) and is the ratio of the velocity change and the time interval needed for that change:

$$\vec{a} = \frac{\Delta\vec{v}}{\Delta t}$$

**Figure 4.2** Estimating the direction of acceleration during two-dimensional motion.

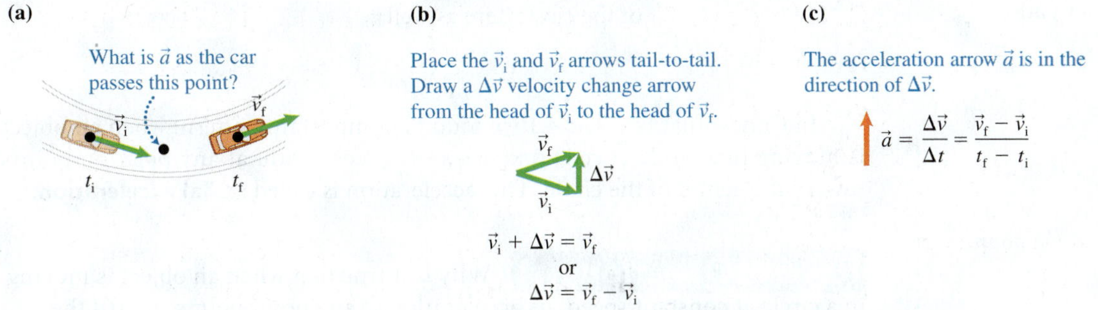

(a)
What is $\vec{a}$ as the car passes this point?
$\vec{v}_i$
$\vec{v}_f$
$t_i$
$t_f$
(b)
Place the $\vec{v}_i$ and $\vec{v}_f$ arrows tail-to-tail. Draw a $\Delta\vec{v}$ velocity change arrow from the head of $\vec{v}_i$ to the head of $\vec{v}_f$.
$\vec{v}_f$
$\Delta\vec{v}$
$\vec{v}_i$
$\vec{v}_i + \Delta\vec{v} = \vec{v}_f$
or
$\Delta\vec{v} = \vec{v}_f - \vec{v}_i$
(c)
The acceleration arrow $\vec{a}$ is in the direction of $\Delta\vec{v}$.
$\vec{a} = \frac{\Delta\vec{v}}{\Delta t} = \frac{\vec{v}_f - \vec{v}_i}{t_f - t_i}$

**TIP** When using this diagrammatic method to estimate the acceleration direction during circular motion, make sure that you choose initial and final points at the same distance before and after the point at which you are estimating the acceleration direction. Draw long velocity arrows so that when you put them tail to tail, you can clearly see the direction of the velocity change arrow. Also, be sure that the velocity change arrow points from the head of the initial velocity to the head of the final velocity.

**CONCEPTUAL EXERCISE 4.1 Direction of racecar's acceleration**

Determine the direction of the racecar's acceleration at points A, B, and C in the figure below as the racecar travels at constant speed on the circular path.

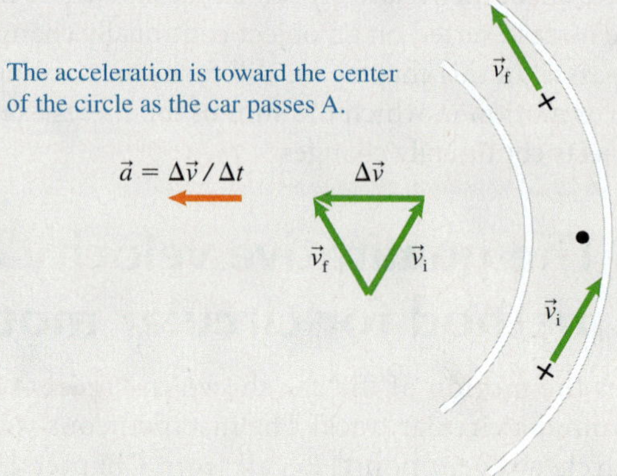


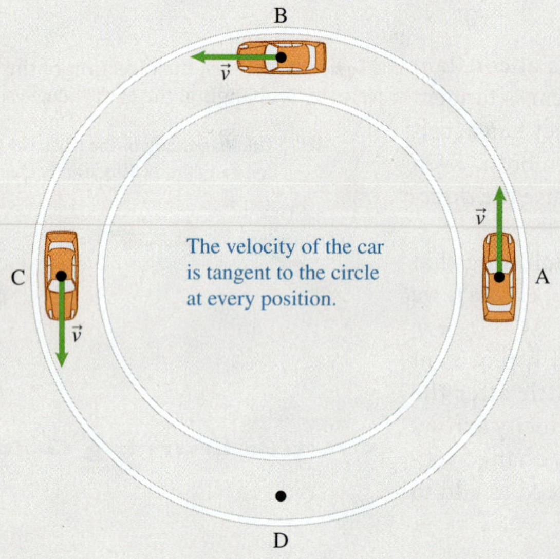


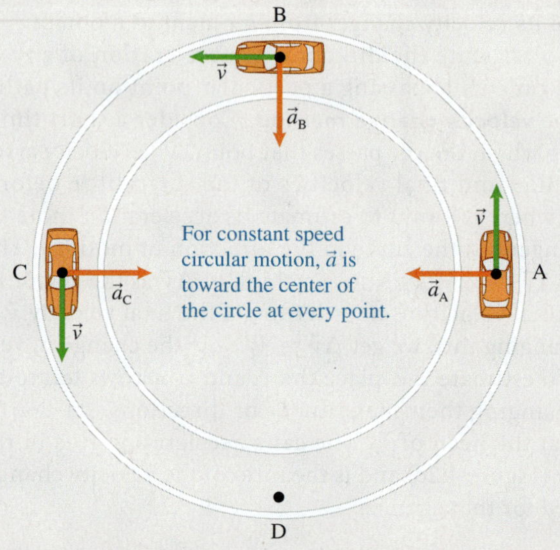


**Sketch and translate** A top view of the car's path is shown above. We are interested in the car's acceleration as it passes points A, B, and C.

**Simplify and diagram** To find the direction of the car's acceleration at each point, we use the velocity change method (shown for point A above, right). When done for all three points, notice that a pattern emerges: the acceleration at different points along the car's path has a different direction, but in every case it points toward the center of the circular path.

**Try it yourself:** Find the direction of the car's acceleration at point D on the track.

*Answer:* The car's acceleration points toward the center of the circle here as well.

In Conceptual Exercise 4.1 we found an important pattern: when an object is moving in a circle at *constant speed,* its acceleration at any position points toward the center of the circle. This acceleration is called **radial acceleration**.

**Review Question 4.1** Why is it true that when an object is moving in a circle at constant speed, its acceleration at any point points toward the center of the circle?

# 4.2 Qualitative dynamics of circular motion

Newton's second law tells us that an object's acceleration during linear motion is caused by the forces exerted on the object and is in the direction of the vector sum of all forces ($\vec{a} = \Sigma\vec{F}/m$). By applying this idea to constant speed circular motion, we can devise the following hypothesis:

> *Hypothesis* The sum of the forces exerted on an object moving at constant speed along a circular path points toward the center of that circle in the same direction as the object's acceleration.

Notice that the hypothesis mentions only the sum of the forces pointing toward the center (and no forces in the direction of motion). Consider the two experiments described in Testing Experiment **Table 4.1**. Are the net forces exerted on these objects consistent with the above hypothesis?

**TESTING EXPERIMENT TABLE**

**4.1 Does the net force exerted on an object moving at constant speed in a circle point toward the center of the circle?**

<table>
<tr><th>Testing experiment</th><th>Predictions based on hypothesis</th><th>Outcome</th></tr>
<tr><td>**Experiment 1.** You swing a pail at the end of a rope in a horizontal circle at a constant speed.<br>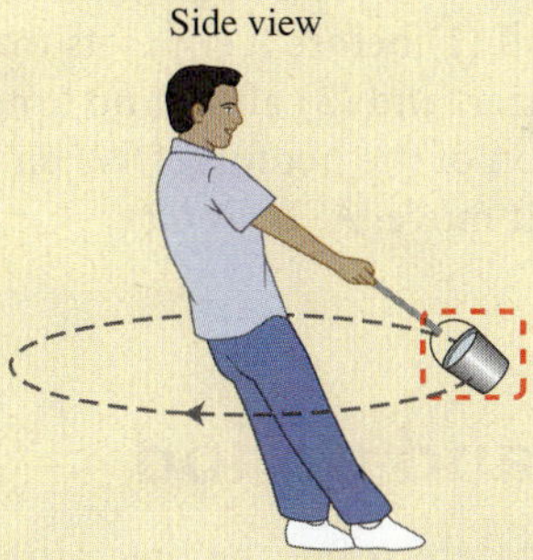<br></td><td>We predict that as the pail passes any point along its path, the sum of the forces exerted on the pail by other objects should point toward the center of the circle—in the direction of the acceleration.</td><td>The vertical force components balance; the sum of the forces points along the radial axis toward the center of the circle in agreement with the prediction. No force pushes the pail in the direction of motion.<br>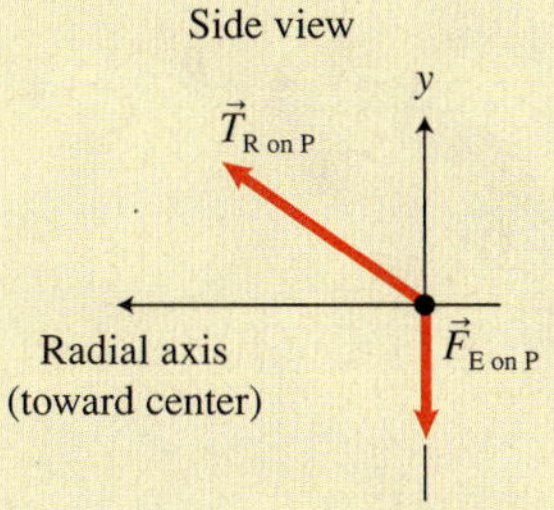<br></td></tr>
<tr><td>**Experiment 2.** A metal ball rolls in a circle on a flat, smooth surface against the inside wall of a metal ring.<br>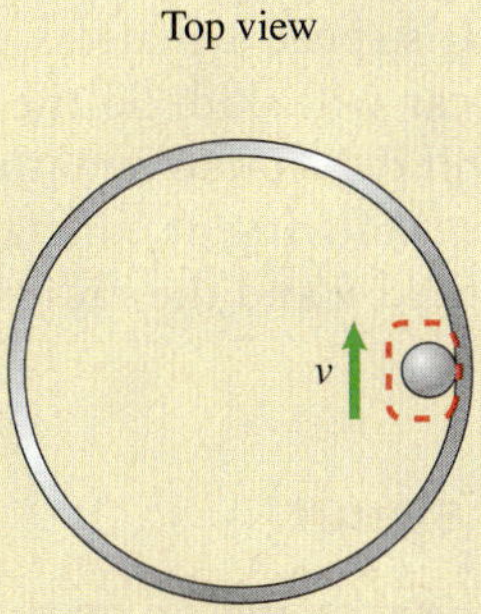<br></td><td>We predict that the sum of the forces that other objects exert on the ball points toward the center of the circle—in the direction of the acceleration.</td><td>The vertical force components balance. The sum of the forces points along the radial axis toward the center of the circle in agreement with the prediction. No force pushes in the direction of motion.<br>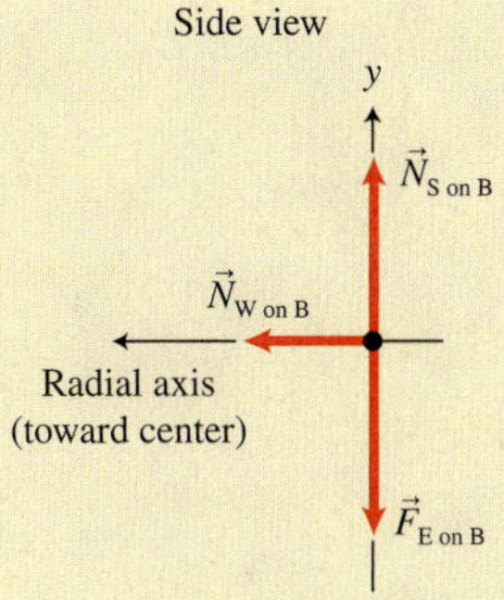<br></td></tr>
</table>

*(continued)*

**Conclusion**

In both cases, the sum of the forces exerted on the system object by other objects points toward the center of the circle in the direction of the acceleration—consistent with the hypothesis.

The results of our experiments were consistent with our hypothesis. If you were to repeat the analysis for other points on the path, you would get the same result. This outcome is consistent with Newton's second law—an object's acceleration equals the sum of the forces (the net force) that other objects exert on it divided by the mass of the object:

$$\vec{a} = \frac{\vec{F}_{1 \text{ on object}} + \vec{F}_{2 \text{ on object}} + \cdots}{m_{\text{object}}}$$

**Figure 4.3** Top view of a ball rolling toward a circular ring.

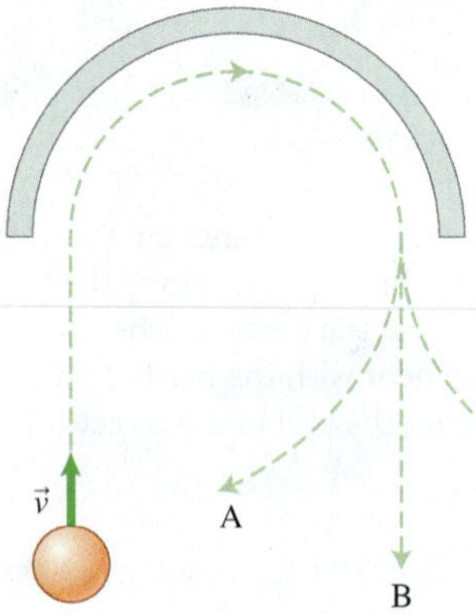


**TIP** Notice that when the object moves at constant speed along the circular path, the net force has no tangential component.

**Review Question 4.2** A ball rolls at a constant speed on a horizontal table toward a semicircular barrier as shown in **Figure 4.3**. Is there a nonzero net force exerted on the ball (1) before it contacts the barrier, (2) while it is in contact with the barrier, and (3) after it no longer contacts the barrier? If so, what is the direction of the net force? When the ball leaves the barrier, in what direction will it move: A, B, or C?

## 4.3 Radial acceleration and period

So far we have learned how to qualitatively determine the direction of the acceleration of an object moving in a circle and how that acceleration relates to the forces exerted on the object. Let's now look at how to determine the magnitude of the object's acceleration. We will begin by thinking about factors that might affect its acceleration.

Imagine a car following the circular curve of a highway. Our experience indicates that the faster a car moves along a highway curve, the greater the risk that the car will skid off the road. So the car's speed $v$ matters. Also, the tighter the turn, the greater the risk that the car will skid. So the radius $r$ of the curve also matters. In this section we will determine a mathematical expression relating the acceleration of an object moving at constant speed in a circular path to these two quantities (its speed $v$ and the radius $r$ of the circular path).

### Dependence of acceleration on speed

Let's begin by investigating the dependence of the acceleration on the object's speed. In Observational Experiment **Table 4.2**, we use the diagrammatic velocity change method to investigate how the acceleration differs for objects moving at speeds $v$, $2v$, and $3v$ while traveling along the same circular path of radius $r$.

## OBSERVATIONAL EXPERIMENT TABLE

### 4.2 How does an object's speed affect its radial acceleration during constant speed circular motion?

**Observational experiment** | **Analysis using velocity change method**

**Experiment 1.** An object moves in a circle at constant speed.

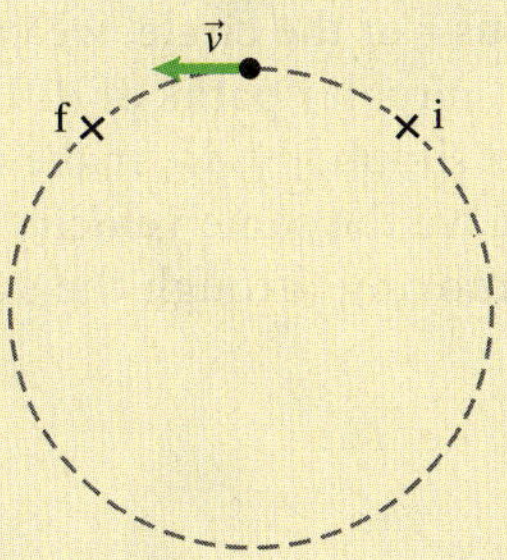


The acceleration is toward the center of the circular path.

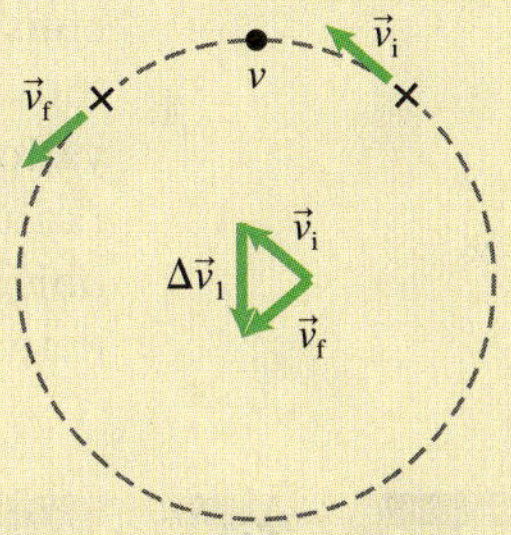


$$a_1 = \frac{\Delta v_1}{\Delta t_1}$$

**Experiment 2.** An object moves in the same circle at a constant speed that is twice as fast as in Experiment 1.

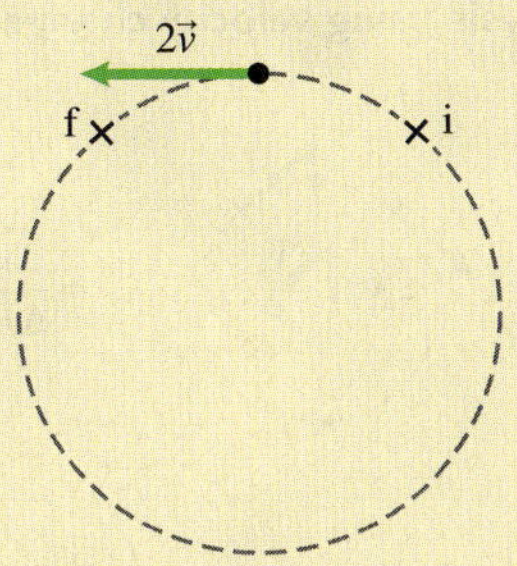


When the object moves twice as fast between the same two points on the circle, the velocity change doubles. In addition, the velocity change occurs in one-half the time interval since it is moving twice as fast. Hence, the acceleration increases by a factor of 4.

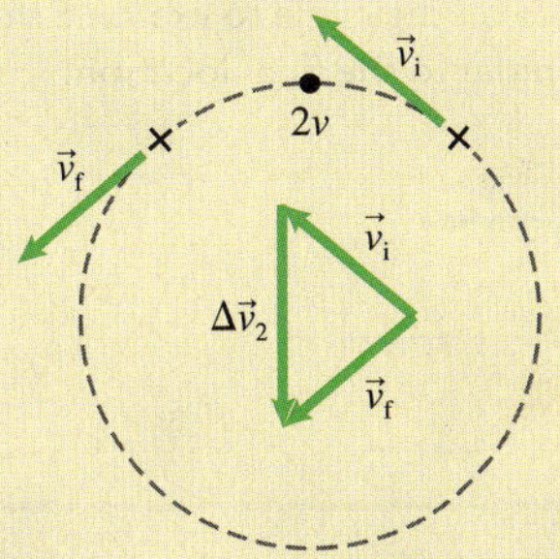


$$a_2 = \frac{\Delta v_2}{\Delta t_2} = \frac{2\Delta v_1}{\Delta t_1/2} = 4\frac{\Delta v_1}{\Delta t_1}$$

**Experiment 3.** An object moves in the same circle at a constant speed that is three times as fast as in Experiment 1.

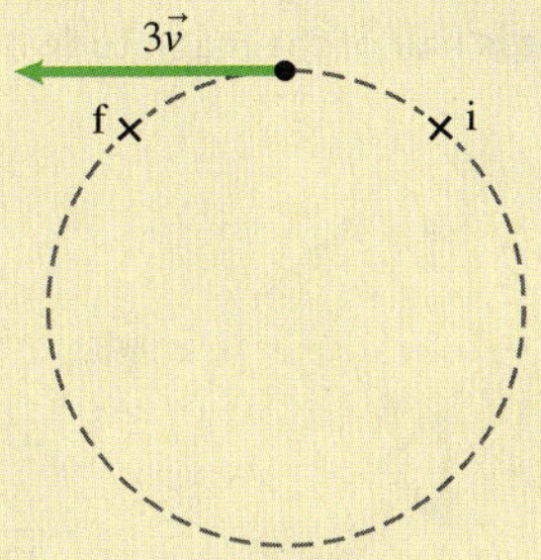


Tripling the speed triples the velocity change and reduces by one-third the time interval needed to travel between the points—the acceleration increases by a factor of 9.

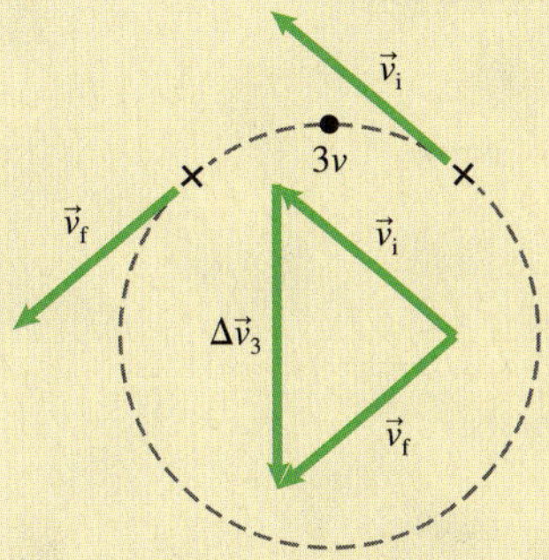


$$a_3 = \frac{\Delta v_3}{\Delta t_3} = \frac{3\Delta v_1}{\Delta t_1/3} = 9\frac{\Delta v_1}{\Delta t_1}$$

**Pattern**

We find that doubling the speed of the object results in a fourfold increase of its radial acceleration; tripling the speed leads to a ninefold increase. Therefore, the radial acceleration of the object is proportional to its speed squared.

$$a_r \propto v^2$$

From the pattern in Table 4.2 we conclude that the magnitude of radial acceleration is proportional to the speed squared. We can express this pattern mathematically:

$$a_r \propto v^2 \tag{4.1}$$

## Dependence of acceleration on radius

To find how the magnitude of acceleration of an object moving in a circle at constant speed depends on the radius $r$ of the circle, we consider two objects moving with the same speed but on circular paths of different radii (Observational Experiment **Table 4.3**). For simplicity, we make one circle twice the radius of the other. We arrange to have the same velocity change for the two objects by considering them while moving through the same angle $\theta$ (rather than through the same distance).

**OBSERVATIONAL EXPERIMENT TABLE**

### 4.3 How does the acceleration depend on the radius of the curved path?

| Observational experiment | Analysis using velocity change method |
| --- | --- |
| **Experiment 1.** An object moves in a circle of radius $r$ at speed $v$. Choose two points on the circle to examine the velocity change from the initial to the final location.<br>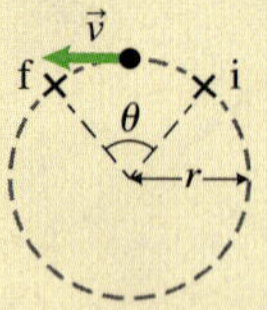 | 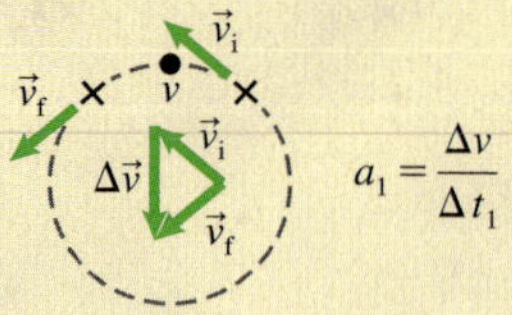 |
| **Experiment 2.** An object moves in a circle of radius $2r$ at speed $v$. Choose the points for the second experiment so that the velocity change is the same as in Experiment 1. This occurs if the radii drawn to the location of the object at the initial position and to the final position make the same angle as they did in Experiment 1.<br>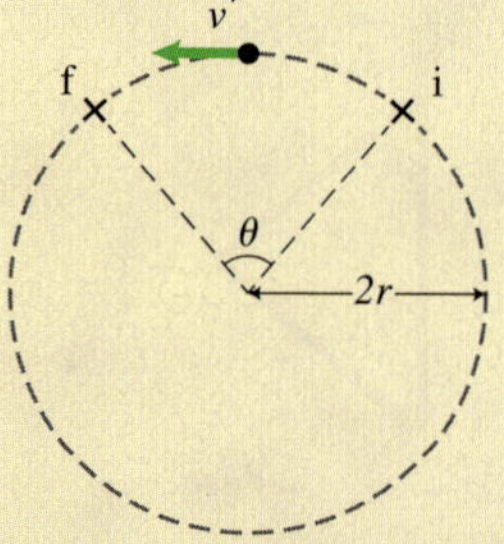 | To have the same velocity change as in Experiment 1, the object has to travel twice the distance, as the radius is twice as long. Since the speed of the object is the same as in the first experiment, it takes the object twice as long to travel that distance. Hence, the magnitude of the acceleration in this experiment is half of the magnitude of the acceleration in Experiment 1.<br>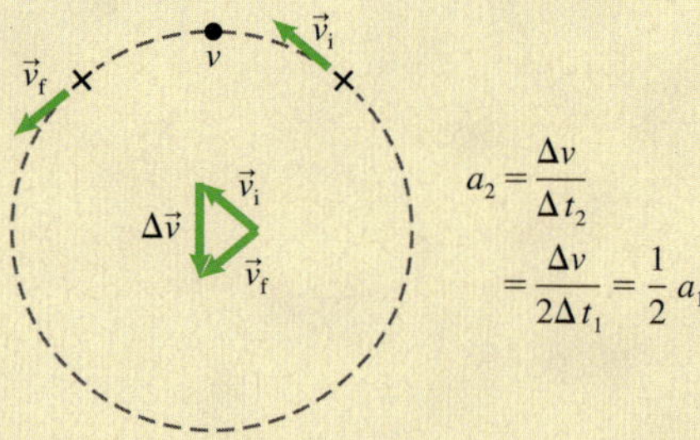 |

**Pattern**

When an object moves in a circle at constant speed, its acceleration decreases by half when the radius of the circular path doubles—the bigger the circle, the smaller the acceleration. It appears that the acceleration of the object is proportional to the inverse of the radius of its circular path:

$$a_r \propto \frac{1}{r}$$

If we carried out the same thought experiment with circular paths of other radii, we would get similar results. The magnitude of the radial component of the acceleration of an object moving in a circular path is inversely proportional to the radius of the circle:

**‹Active Learning Guide**

$$a_r \propto \frac{1}{r} \tag{4.2}$$

We now combine the two proportionalities in Eqs. (4.1) and (4.2):

$$a_r \propto \frac{v^2}{r}$$

If we were to do a detailed mathematical derivation, we would find that the constant of proportionality is 1. Thus, we can turn this into an equation for the magnitude of the radial acceleration.

**Radial acceleration** For an object moving at constant speed $v$ on a circular path of radius $r$, the magnitude of the radial acceleration is

$$a_r = \frac{v^2}{r} \tag{4.3}$$

The acceleration points toward the center of the circle. The units for $v^2/r$ are the correct units for acceleration:

$$\left(\frac{\text{m}^2/\text{s}^2}{\text{m}}\right) = \frac{\text{m}}{\text{s}^2}$$

This expression for radial acceleration agrees with our everyday experience. When a car is going around a highway curve at high speed, a large $v^2$ in the numerator leads to its large acceleration. When it is going around a sharp turn, the small radius $r$ in the denominator also leads to a large acceleration.

Consider a limiting case: when the radius of the circular path is infinite—equivalent to the object moving in a straight line. The velocity of an object moving at constant speed in a straight line is constant, and its acceleration is zero. Notice that the acceleration in Eq. (4.3) is zero if the radius in the denominator is infinite.

**EXAMPLE 4.2 Blackout**

When a fighter pilot pulls out from the bottom of a power dive, his body moves at high speed along a segment of an upward-bending approximately circular path. However, while his body moves up, his blood tends to move straight ahead (tangent to the circle) and begins to fill the easily expandable veins in his legs. This can deprive his brain of blood and cause a blackout if the radial acceleration is 4 $g$ or more and lasts several seconds. Suppose during a dive an airplane moves at a modest speed of $v = 80$ m/s (180 mi/h) through a circular arc of radius $r = 150$ m. Is the pilot likely to black out?

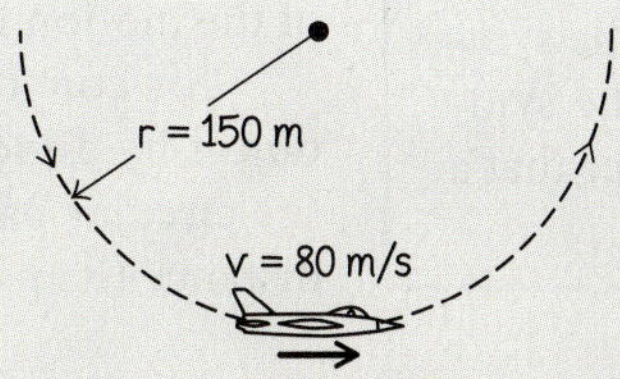


Find $a_r$ and $\Delta t$ to decide if blackout occurs.

**Sketch and translate** We sketch the situation above, right. To determine if blackout occurs, we will estimate the radial acceleration of the pilot as he passes along the lowest point of the circular path.

**Simplify and diagram** Assume that at the bottom of the plane's power dive, the plane is moving in a circle at constant speed and the point of interest is the lowest point on the circle. Assume also that the magnitude of the plane's acceleration is constant for a quarter circle, that is, one-eighth of a circle on each side of the bottom point. The acceleration points up (see the velocity change method shown on the next page).

*(continued)*

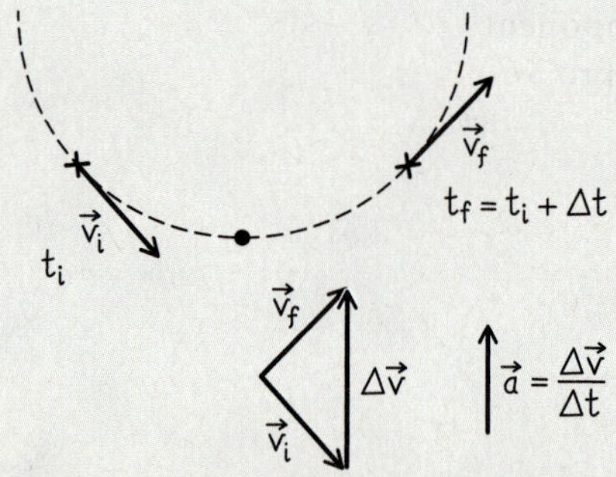


**Represent mathematically** We calculate the magnitude of the radial acceleration using Eq. (4.3):

$$a_r = \frac{v^2}{r}$$

To estimate the time interval that the pilot will experience this acceleration, we calculate the time interval for the plane to move through the part of the power dive represented by a quarter circle (arc length $l = 2\pi r/4$):

$$\Delta t = \frac{l}{v} = \frac{2\pi r/4}{v}$$

**Solve and evaluate** Inserting the given quantities, we find:

$$a_r = \frac{(80\text{ m/s})^2}{150\text{ m}} = 42\text{ m/s}^2$$

The acceleration is more than four times greater than $g$ (more than $4 \times 9.8\text{ m/s}^2$). Thus, the pilot could potentially black out if the acceleration lasts too long. The time interval during which the pilot will experience this acceleration is about

$$\Delta t = \frac{[2\pi(150\text{ m})]/4}{80\text{ m/s}} = \frac{240\text{ m}}{80\text{ m/s}} \approx 3\text{ s}$$

This is long enough that blackout is definitely a concern. Special flight suits are made that exert considerable pressure on the legs during such motion. This pressure prevents blood from accumulating in the veins of the legs.

**Try it yourself:** Imagine that you are a passenger on a roller coaster with a double loop-the-loop—two consecutive loops. You are traveling at speed 24 m/s as you move along the bottom part of a loop of radius 10 m. Determine (a) the magnitude of your radial acceleration while passing the lowest point of that loop and (b) the time interval needed to travel along the bottom quarter of that approximately circular loop. (c) Are you at risk of having a blackout? Explain.

*Answer:* (a) $58\text{ m/s}^2$ or almost six times the $9.8\text{ m/s}^2$ free-fall acceleration; (b) about 0.7 s; (c) the acceleration is more than enough to cause blackout but does not last long enough—you are safe from blacking out and will get a good thrill!

## Period

When an object repeatedly moves in a circle, we can describe its motion with another useful physical quantity, its **period** $T$. The period equals the time interval that it takes an object to travel around the entire circular path one time and has the units of time. For example, suppose that a bicyclist racing on a circular track takes 24 s to complete a circle that is 400 m in circumference. The period $T$ of the motion is 24 s.

**TIP** Do not confuse the symbol $T$ for period with the symbol $T$ for the tension force that a string exerts on an object.

For constant speed circular motion we can determine the speed of an object by dividing the distance traveled in one period (the circumference of the circular path, $2\pi r$) by the time interval $T$ it took the object to travel that distance (its period), or $v = 2\pi r/T$. Thus,

$$T = \frac{2\pi r}{v} \tag{4.4}$$

We can express the radial acceleration of the object in terms of its period by inserting this special expression for speed in terms of period $v = 2\pi r/T$ into Eq. (4.3):

$$a_r = \frac{v^2}{r} = \left(\frac{2\pi r}{T}\right)^2 \frac{1}{r} = \frac{4\pi^2 r^2}{T^2 r} = \frac{4\pi^2 r}{T^2} \tag{4.5}$$

Let's use limiting case analysis to see if Eq. (4.5) is reasonable. For example, if the speed of the object is extremely large, its period would be very short ($T = 2\pi r/v$). Thus, according to Eq. (4.5), its radial acceleration would be very large. That makes sense—high speed and large radial acceleration. Similarly, if the speed of the object is small, its period will be very large and its radial acceleration will be small. That also makes sense.

**QUANTITATIVE EXERCISE 4.3**

**Singapore hotel**

What is your radial acceleration when you sleep in a hotel in Singapore at Earth's equator? Remember that Earth turns on its axis once every 24 hours and everything on its surface actually undergoes constant speed circular motion with a period of 24 hours. A picture of Earth with you as a point at the equator is shown below.

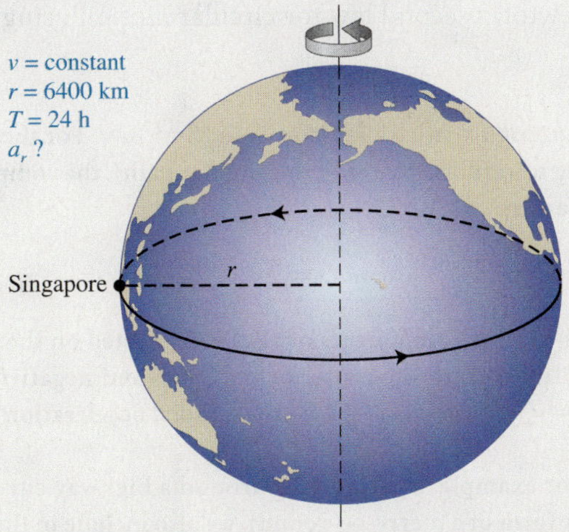


**Represent mathematically** Since you are in constant speed circular motion, Eq. (4.5) can be used to determine your radial acceleration. Your period $T$ is the time interval needed to travel once in this circle (24 h). Thus the magnitude of your radial acceleration is

$$a_r = \frac{v^2}{r} = \frac{4\pi^2 r}{T^2}$$

**Solve and evaluate** At the equator, $r = 6400$ km and $T = 24$ h. So, making the appropriate unit conversions, we get the magnitude of the acceleration:

$$a_r = \frac{4\pi^2(6.4 \times 10^6\,\text{m})}{(24\,\text{h} \times 3600\,\text{s/h})^2} = 0.034\,\text{m/s}^2$$

Is this result reasonable? Compare it to the much greater free-fall acceleration of objects near Earth's surface—$9.8\,\text{m/s}^2$. Your radial acceleration due to Earth's rotation when in Singapore is tiny by comparison. Because it is so small, the radial acceleration due to Earth's rotation around its axis can be ignored under most circumstances.

**Try it yourself:** Use the figure below to estimate what your radial acceleration would be if you were living in Anchorage, Alaska.

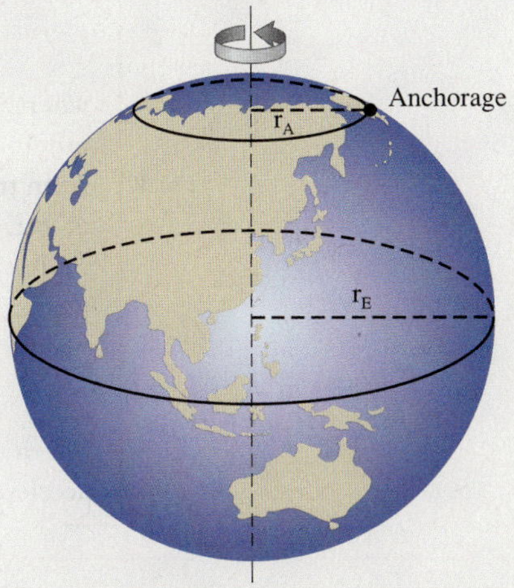


*Answer:* About $0.016\,\text{m/s}^2$. Note that the period is the same, but the distance of Anchorage from the axis of Earth's rotation (the radial distance used in this estimate) is about one-half of Earth's radius.

Remember that Newton's laws were formulated and are valid only for observers in inertial reference frames, that is, for observers who are not accelerating. But we have just discovered that observers on Earth's surface are accelerating due to Earth's rotation. Does this mean that Newton's laws do not apply? As we found in Quantitative Exercise 4.3, the acceleration due to Earth's rotation is much smaller than the accelerations we experience from other types of motion. Thus, in most situations we can assume that Earth is not rotating and therefore *does* count as an inertial reference frame. This means that Newton's laws do apply with a high degree of accuracy when using Earth's surface as a reference frame.

**Review Question 4.3** Use dimensional analysis (inspecting the units) to evaluate whether or not the two expressions for radial acceleration ($v^2/r$ and $4\pi^2 r/T^2$) have the correct units of acceleration.

## 4.4 Skills for analyzing processes involving circular motion

The strategy for analyzing processes involving constant speed circular motion is similar to that used to analyze linear motion processes. However, when we analyze constant speed circular motion processes, we do *not* use traditional $x$- and $y$-axes but instead use a radial $r$-axis and sometimes a vertical $y$-axis. The radial $r$-axis should always point *toward the center of the circle*. The radial acceleration of magnitude $a_r = v^2/r$ is positive along this axis.

The application of Newton's second law for circular motion using a radial axis is summarized below.

**Circular motion component form of Newton's second law** **For the radial direction (the axis pointing toward the center of the circular path), the component form of Newton's second law is**

$$a_r = \frac{\Sigma F_r}{m} \quad \text{or} \quad ma_r = \Sigma F_r \tag{4.6}$$

**where $\Sigma F_r$ is the sum of the radial components of all forces exerted on the object moving in the circle (positive toward the center of the circle and negative away from the center) and $a_r = v^2/r$ is the magnitude of the radial acceleration of the object.**

**For some situations (for example, a car moving around a highway curve or a person standing on the platform of a merry-go-round), we also include in the analysis the force components along a perpendicular vertical $y$-axis:**

$$ma_y = \Sigma F_y = 0 \tag{4.7}$$

**When an object moves with uniform circular motion, both the $y$-component of its acceleration and the $y$-component of the net force exerted on it are zero.**

### PROBLEM-SOLVING STRATEGY Processes involving constant speed circular motion

The following example explains how to solve circular motion problems. We describe the general strategy on the left side of the table and apply it specifically to the problem in Example 4.4 on the right side.

**EXAMPLE 4.4 Driving over a hump in the road**

Josh drives his car at a constant 12 m/s speed over a bridge whose shape is bowed in the vertical arc of a circle. Find the direction and the magnitude of the force exerted by the car seat on Josh as he passes the top of the 30-m-radius arc. His mass is 60 kg.

**Sketch and translate**

- Sketch the situation described in the problem statement. Label it with all relevant information.
- Choose a system object and a specific position to analyze its motion.

The sketch includes all of the relevant information: Josh's speed, his mass, and the radius of the arc along which the car moves. Josh is the system object.

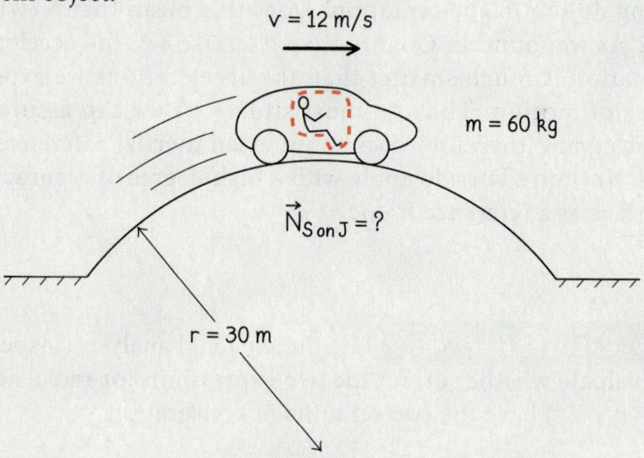

**Simplify and diagram**

- Decide if the system can be modeled as a point-like object.
- Determine if the constant speed circular motion approach is appropriate.
- Indicate with an arrow the direction of the object's acceleration as it passes the chosen position.
- Draw a force diagram for the system object at the instant it passes that position.
- On the force diagram, draw an axis in the radial direction toward the center of the circle.

Consider Josh as a point-like object and analyze him as he passes the highest part of the vertical circle along which he travels.

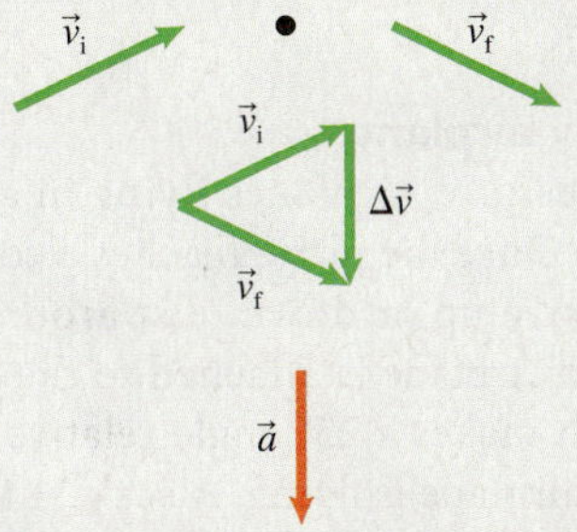


A velocity change diagram as he passes the top of the circular path indicates that he has a downward radial acceleration toward the center of the circle.

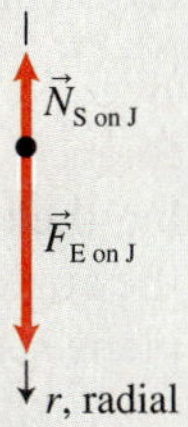


The forces exerted on him are shown in the force diagram. The net force must also point down. Thus, the upward normal force $\vec{N}_{\text{S on J}}$ that the car seat exerts on Josh is smaller in magnitude than the downward gravitational force $\vec{F}_{\text{E on J}}$ that Earth exerts on Josh. A radial $r$-axis points down toward the center of the circle.

**Represent mathematically**

- Convert the force diagram into the radial $r$-component form of Newton's second law.
- For objects moving in a horizontal circle (unlike this example), you may also need to apply a vertical $y$-component form of Newton's second law.

Apply the radial form of Newton's second law $ma_r = \Sigma F_r$.

The radial components of the two forces exerted on Josh are $F_{\text{E on J}\,r} = +F_{\text{E on J}} = +mg$ and $N_{\text{S on J}\,r} = -N_{\text{S on J}}$. The magnitude of the radial acceleration is $a_r = v^2/r$. Therefore, the radial form of Newton's second law is

$$m\frac{v^2}{r} = +mg + (-N_{\text{S on J}})$$

Solving for $N_{\text{S on J}}$, we have

$$N_{\text{S on J}} = mg - m\frac{v^2}{r}$$

**Solve and evaluate**

- Solve the equations formulated in the previous two steps.
- Evaluate the results to see if they are reasonable (the magnitude of the answer, its units, limiting cases).

Substituting the known information into the previous equation, we get

$$N_{\text{S on J}} = (60\text{ kg})(9.8\text{ m/s}^2) - (60\text{ kg})\frac{(12\text{ m/s})^2}{(30\text{ m})} = 588\text{ N} - 288\text{ N} = 300\text{ N}$$

The seat exerts a smaller upward force on Josh than Earth pulls down on him. You have probably noticed this effect when going over a smooth hump in a roller coaster or while in a car or on a bicycle when crossing a hump in the road—it almost feels like you are leaving the seat or starting to float briefly above it. This feeling is caused by the reduced upward normal force that the seat exerts on you.

**Try it yourself:** Imagine that Josh drives on a road that has a dip in it. The speed of the car and the radius of the dip are the same as in Example 4.4. Find the direction and the magnitude of the force exerted by the car seat on Josh as he passes the bottom of the 30-m-radius dip.

*Answer:* The seat exerts an upward force of 880 N. This is almost 50% more than the 590-N force that Earth exerts on Josh—he sinks into the seat.

**Active Learning Guide >**

The next three examples involve processes in which some of the forces have nonzero radial $r$-components *and* some of the forces have nonzero vertical $y$-components. To solve these problems we will use the component form of Newton's second law in both the radial $r$- and vertical $y$-directions.

### EXAMPLE 4.5 Toy airplane

You have probably seen toy airplanes flying in a circle at the end of a string. Once the plane reaches a constant speed, it does not move up or down, just around in a horizontal circle. Our airplane is attached to the end of a 46-cm string, which makes a 25° angle relative to the horizontal while the airplane is flying. A scale at the top of the string measures the force that the string exerts on the airplane. Predict the period of the airplane's motion (the time interval for it to complete one circle).

**Sketch and translate** We sketch the situation, including the known information: the length of the string and the angle of the string relative to the horizontal. The airplane is the system object.

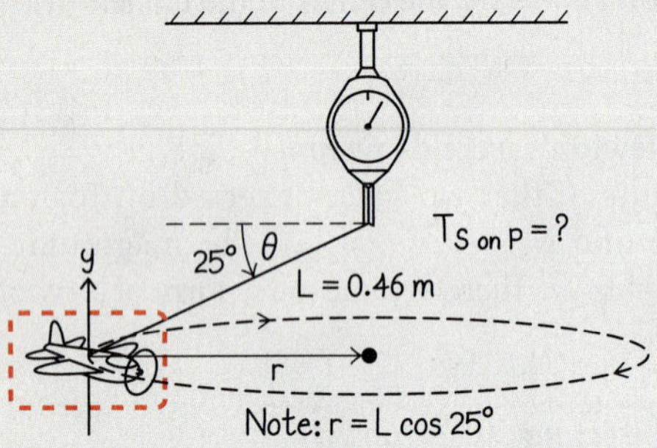


**Simplify and diagram** Neglect the airplane's interaction with the air and assume that it is a point-like particle moving at constant speed in a horizontal circle. To draw a force diagram, identify objects that interact with the airplane—in this case, Earth and the string. Include a vertical $y$-axis and a radial $r$-axis; note that the radial axis points horizontally toward the center of the circle—not along the string. Because the airplane moves at constant speed, its acceleration points toward the center of the circle—it has a radial component but does not have a vertical component. Thus, the vertical component of the tension force that the string exerts on the plane $\vec{T}_{S\text{ on }P\,y}$ must balance the downward force that Earth exerts on the plane $\vec{F}_{E\text{ on }P}$, which has magnitude $m_P g$ (see the force diagram broken into components at right). The radial component of the force that the string exerts on the plane is the only force with a nonzero radial component and causes the plane's radial acceleration.

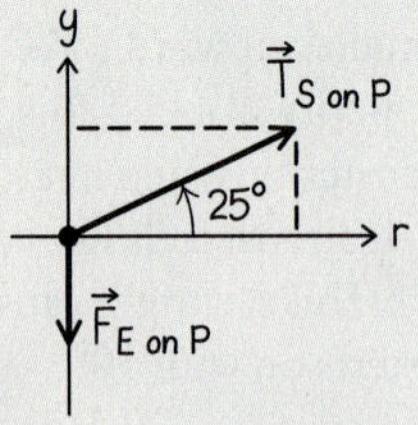


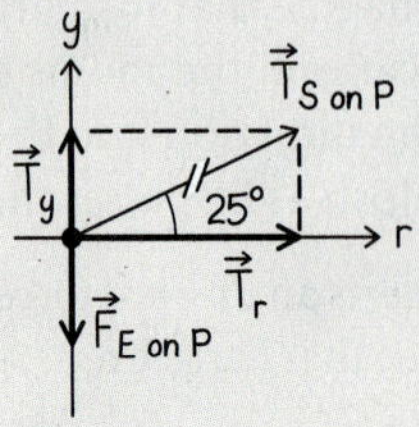


**Represent mathematically** Now, we use the force diagram to help apply Newton's second law in component form. First, apply the vertical $y$-component form of Newton's second law: $m_P a_y = \Sigma F_y$. The $y$-components of the forces exerted on the plane are $T_{S\text{ on }P\,y} = +T_{S\text{ on }P}\sin 25°$ and $F_{E\text{ on }P\,y} = -m_P g$, and the $y$-component of the plane's acceleration is zero. We can now apply the $y$-component form of Newton's second law:

$$0 = +T_{S\text{ on }P}\sin 25° + (-m_P g)$$

Next, apply the radial $r$-component version of Newton's second law: $m_P a_r = \Sigma F_r$. The force exerted by Earth on the plane does not have a radial component. The radial component of the force exerted by the string on the plane is $T_{S\text{ on }P\,r} = +T_{S\text{ on }P}\cos 25°$, and the magnitude of the radial acceleration is $a_r = 4\pi^2 r/T^2$ [Eq. (4.5)]. Thus:

$$m_P\frac{4\pi^2 r}{T^2} = +T_{S\text{ on }P}\cos 25°$$

Note that $T_{S\text{ on }P}$ represents the magnitude of the tension force that the string exerts on the airplane and $T$ represents the period of the plane's motion—the time interval it takes the plane to travel once around its circular path. Note also that the radius of the circular path is not the length of the string but is instead

$$r = L\cos 25° = (0.46\text{ m})\cos 25° = 0.417\text{ m}$$

where $L$ is the length of the string (46 cm).

**Solve and evaluate** The goal is to predict the period $T$ of the plane's circular motion. Unfortunately, the above radial application of Newton's second law has three unknowns: $T_{S\text{ on }P}$, $T$, and the mass of the plane. However, we can rearrange the vertical $y$-equation to get an expression for the magnitude of the tension force exerted by the string on the plane:

$$T_{S\text{ on }P} = \frac{m_P g}{\sin 25°}$$

Insert this expression for $T_{S\text{ on }P}$ into the radial component equation to get an expression that does *not* involve the force that the string exerts on the plane:

$$m_P\frac{4\pi^2 r}{T^2} = \frac{m_P g}{\sin 25°}\cos 25°$$

Divide both sides by $m_P$, multiply both sides by $T^2$, and do some rearranging to get

$$T^2 = \frac{4\pi^2 r\sin 25°}{g\cos 25°}$$

Taking the square root of each side, we find that

$$T = 2\pi\sqrt{\frac{r\sin 25°}{g\cos 25°}}$$

Substituting the known information, we predict the period to be

$$T = 2\pi\sqrt{\frac{(0.417\text{ m})\sin 25°}{(9.8\text{ m/s}^2)\cos 25°}} = 0.88\text{ s}$$

**Try it yourself:** Imagine that another plane of mass 0.12 kg moves with a speed of 6.2 m/s. A spring scale measures the tension force that the string exerts on the plane as 5.0 N. Predict the radius of the plane's orbit.

*Answer:* 0.95 m.

Many carnivals and amusement parks have a "rotor ride" in which people stand up against the wall of a spinning circular room. The room (also called the "drum") spins faster until, at a certain speed, the floor drops out! Amazingly, the people remain up against the wall with their feet dangling. How is this possible?

**EXAMPLE 4.6 Rotor ride**

A 62-kg woman is a passenger in a rotor ride. A drum of radius 2.0 m rotates faster and faster about a vertical axis until it reaches a period of 1.7 s. When the drum reaches this turning rate, the floor drops away but the woman does not slide down the wall of the drum. Imagine that you were one of the engineers who designed this ride. Which characteristics of the ride would ensure that the woman remained stuck to the wall? Justify your answer quantitatively.

**Sketch and translate** Representing the situation in different ways should help us decide which characteristics of the drum are important. We start with a sketch along with all the relevant information. The woman will be the system. Her mass is $m_W = 62$ kg, the radius of her circular path is $r = 2.0$ m, and the period of her circular motion is $T = 1.7$ s.

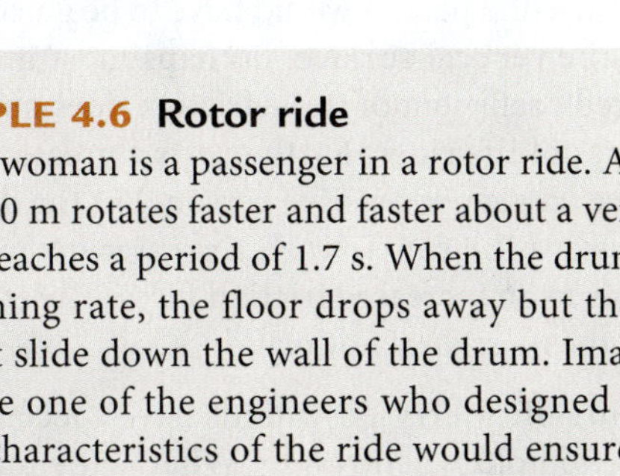

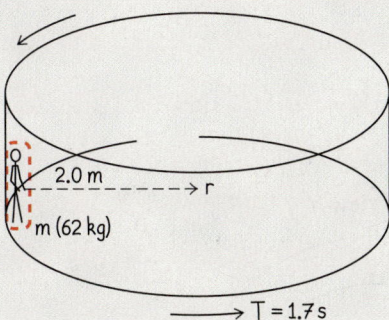


**Simplify and diagram** Model the woman as a point-like object and consider the situation once the drum has reached its maximum (constant) speed. Because she is moving in a circular path at constant speed, her acceleration points toward the center of the circle and we can use our understanding of constant speed circular motion to analyze the situation.

Next, draw a force diagram for the woman as she passes one point along the circular path. She interacts with two objects—Earth and the drum. Earth exerts a downward gravitational force $\vec{F}_{\text{E on W}}$ on her. The drum exerts a force that we can resolve into two vector components: a normal force $\vec{N}_{\text{D on W}}$ perpendicular to the drum's surface and toward the center of the circle and an upward static friction force $\vec{f}_{\text{s D on W}}$ parallel to its surface. Include a radial $r$-axis that points toward the center of the drum and a vertical $y$-axis pointing upward.

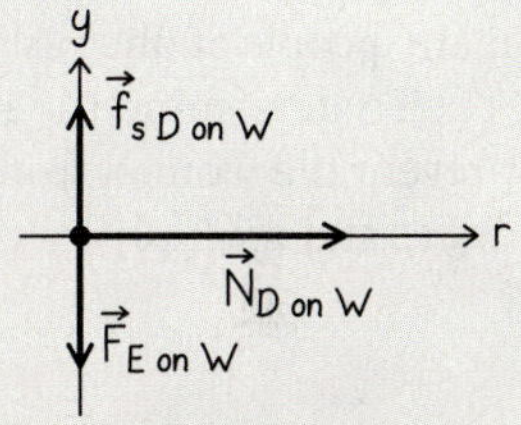


Examining the force diagram, we see that if the maximum upward static friction force is less than the downward gravitational force, the woman will slip. Thus the engineer's rotor ride design problem can be formulated as follows. How large must the coefficient of static friction be so that the static friction force exerted on the woman by the drum balances the force that Earth exerts on her?

**Represent mathematically** Use the force diagram to help apply the component form of Newton's second law for the inward radial $r$-direction and for the vertical $y$-direction. The normal force is the only force that has a nonzero radial component.

*Radial r-equation:*
$$m_W a_r = \Sigma F_r$$
$$\Rightarrow m_W\frac{v^2}{r} = N_{\text{D on W}}$$

Now, consider the $y$-component form of Newton's second law. The static friction force $\vec{f}_{\text{s D on W}}$ points in the positive $y$-direction, and the gravitational force $\vec{F}_{\text{E on W}}$ points in the negative $y$-direction. Since these forces must balance for the woman not to slip, the $y$-component of the woman's acceleration should be zero.

*Vertical y-equation:*
$$m_W a_y = \Sigma F_y$$
$$\Rightarrow 0 = f_{\text{s D on W}} + (-F_{\text{E on W}})$$

Assume that the static friction force is at its maximum possible value, meaning $f_{\text{s D on W}} = \mu_s N_{\text{D on W}}$. We wish

*(continued)*

to find the smallest (minimum) coefficient of static friction $\mu_{s\,min}$ that results in the woman remaining stationary. Substitute this minimum coefficient of static friction and the earlier expression for the normal force exerted by the drum into this expression:

$$f_{s\,max\,D\,on\,W} = \mu_{s\,min}N_{D\,on\,W} = \mu_{s\,min}m_W\frac{v^2}{r}$$

Substituting this expression for $f_{s\,max\,D\,on\,W}$ and $F_{E\,on\,W} = m_W g$ into the vertical $y$-component application of Newton's second law, we get

$$0 = \mu_{s\,min}m_W\frac{v^2}{r} - m_W g$$

Notice that the woman's mass cancels out of this equation. This means that the rotor ride works equally well for any person, independent of the person's mass and for many people at once as well!

To determine the minimum value of $\mu_s$ needed to prevent the woman from sliding, we divide both sides by $m_W v^2$ and multiply by $r$ and rearrange to get

$$\mu_{s\,min} = \frac{gr}{v^2}.$$

We find the woman's speed by using the information we have about the radius and period of her circular motion:

$$v = \frac{2\pi r}{T}.$$

Plug this expression for the speed into the equation above for the minimum coefficient of friction:

$$\mu_{s\,min} = \frac{gr}{v^2} = \frac{grT^2}{4\pi^2 r^2} = \frac{gT^2}{4\pi^2 r}$$

**Solve and evaluate** We can now use the given information to find the minimum coefficient of friction:

$$\mu_{s\,min} = \frac{gT^2}{4\pi^2 r} = \frac{9.8\ \text{m/s}^2 \times (1.7\ \text{s})^2}{4 \times 9.87 \times 2.0\ \text{m}} = 0.36$$

The coefficient of friction is a number with no units (also called *dimensionless*), and our answer also has no units. The magnitude 0.36 is reasonable and easy to obtain with everyday materials (wood on wood ranges from 0.25–0.5). To be on the safe side, we probably want to make the surface rougher, with twice the minimum coefficient of static friction. The limiting case analysis of the final expression also supports the result. If the drum is stationary ($T = \infty$), the coefficient of static friction would have to be infinite—the person would have to be glued permanently to the vertical surface. Perhaps most important, the required coefficient of static friction does not depend on the mass of the rider. Earth exerts a greater gravitational force on a more massive person. But the drum also pushes toward the center with a greater normal force, which leads to an increased friction force. The effects balance, and the mass does not matter.

**Try it yourself:** Merry-go-rounds have special railings for people to hold when riding. What is the purpose of the railings? Answer this question by drawing a force diagram for an adult standing on a rotating merry-go-round.

*Answer:* The railings exert a force on the person in the radial direction. This force helps provide the necessary radial acceleration.

**EXAMPLE 4.7 Texas Motor Speedway**

The Texas Motor Speedway is a 2.4-km (1.5-mile)-long oval track. One of its turns is about 200 m in radius and is banked at 24° above the horizontal. How fast would a car have to move so that no friction is needed to prevent it from sliding sideways off the raceway (into the infield or off the track)?

**Sketch and translate** We start by drawing top view and rear view sketches of the situation. The car is the system. For simplicity, we draw the speedway as a circular track.

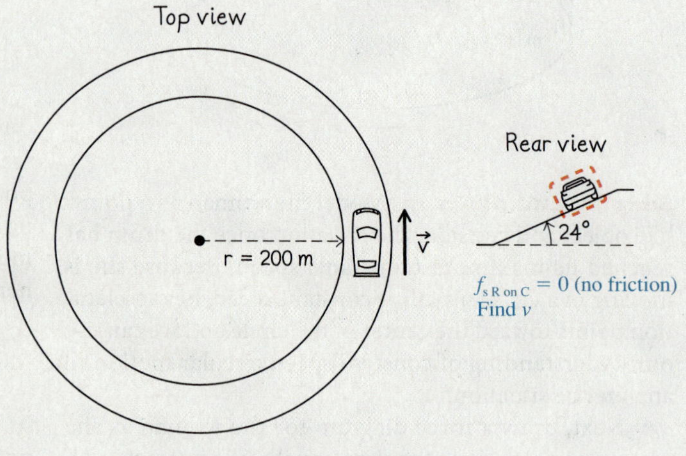

**Simplify and diagram** Model the car as a point-like object that moves along a circular path at constant speed. Earth exerts a downward force $\vec{F}_{\text{E on C}}$. The surface of the road R exerts on the car a normal force $\vec{N}_{\text{R on C}}$ perpendicular to the surface (see the force diagram at right). We assume the sideways friction force exerted by the road on the car is zero (see the problem statement). For constant speed circular motion, the net force exerted on the car should point toward the center of the circle. The horizontal radial component of the normal force $N_{\text{R on C}\,r}$ causes the radial acceleration toward the center. Since the vertical $y$-component of the car's acceleration is zero, the vertical $y$-component of the normal force $N_{\text{R on C}\,y}$ balances the downward gravitational force $\vec{F}_{\text{E on C}}$ that Earth exerts on the car.

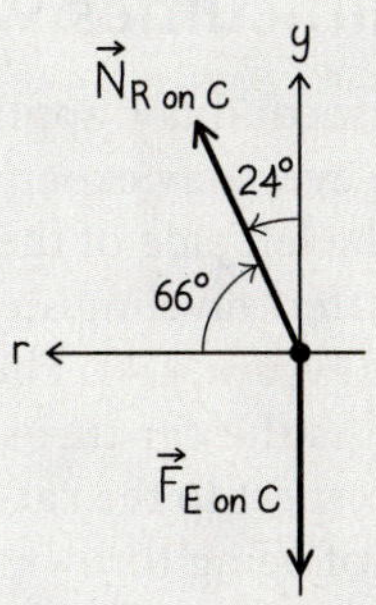


**Represent mathematically** Use the force diagram to help apply the radial $r$- and vertical $y$-component forms of Newton's second law. Let's start with the vertical $y$-equation.

*Vertical y-equation:* $m_C a_y = \Sigma F_y = N_{\text{R on C}\,y} + F_{\text{E on C}\,y}$

Since $a_y = 0$, $N_{\text{R on C}\,y} = N_{\text{R on C}} \sin 66°$, and $F_{\text{E on C}\,y} = -m_C g$, we get $0 = +N_{\text{R on C}} \sin 66° - m_C g$, or

$$N_{\text{R on C}} \sin 66° = m_C g \tag{4.8}$$

Now, apply the radial component equation.

*Radial r-equation:* $m_C a_r = \Sigma F_r = N_{\text{R on C}\,r} + F_{\text{E on C}\,r}$

Since $a_r = v^2/r$, $N_{\text{R on C}\,r} = N_{\text{R on C}} \cos 66°$, and $F_{\text{E on C}\,r} = 0$, we get:

$$m_C \frac{v^2}{r} = N_{\text{R on C}} \cos 66° + 0 \tag{4.9}$$

Now combine Eqs. (4.8) and (4.9) to eliminate $N_{\text{R on C}}$ and determine an expression for the speed of the car. To do this, divide the first equation by the second with the $N_{\text{R on C}}$ sides of each equation on the left:

$$\frac{N_{\text{R on C}} \sin 66°}{N_{\text{R on C}} \cos 66°} = \frac{m_C g}{m_C \frac{v^2}{r}}$$

Canceling $N_{\text{R on C}}$ on the left side and the $m_C$ on the right side and remembering that

$$\frac{\sin 66°}{\cos 66°} = \tan 66°,$$

we get

$$\tan 66° = \frac{gr}{v^2}$$

Rearrange the above to find $v^2$:

$$v^2 = \frac{gr}{\tan 66°}$$

**Solve and evaluate** The road has a radius $r = 200$ m. Thus,

$$v = \sqrt{\frac{gr}{\tan 66°}} = \sqrt{\frac{(9.8\ \text{m/s}^2)(200\ \text{m})}{\tan 66°}}$$

$$= 29.5\ \text{m/s}$$

$$= 30\ \text{m/s} = 66\ \text{mi/h}$$

This is clearly much less than the speed of actual racecars. See the Try It Yourself question.

**Try it yourself:** Construct a force diagram for a racecar that is traveling somewhat faster around the circular track and indicate what keeps the racecar moving in a circle at this higher speed.

*Answer:* If the racecar is going faster than 30 m/s, static friction has to push in on the car parallel to the road surface and toward the center of the track, as shown. The friction force has a negative $y$-component and causes the normal force to be bigger. The combination of the radial components of the increased magnitude normal force and the static friction force provides the much greater net radial force needed to keep the car moving in a circle.

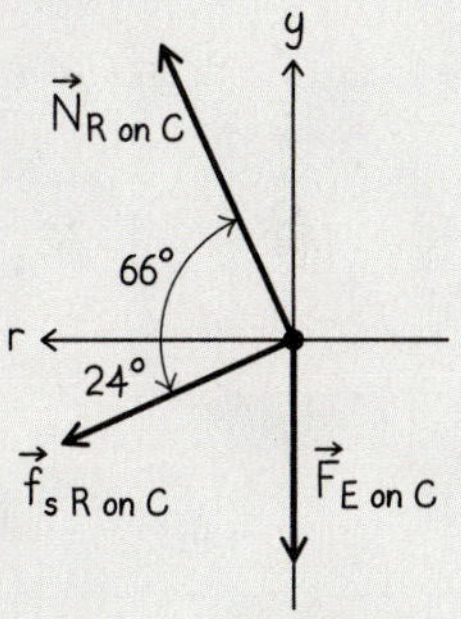


**TIP** Remember that there is no special force that causes the radial acceleration of an object moving at constant speed along a circular path. This acceleration is caused by all of the forces exerted on the system object by other objects (Earth, the surface of a road, a rope...). Add the radial components of these regular forces. This sum is what causes the radial acceleration of the system object.

## Conceptual difficulties with circular motion

Although Newton's second law applies to circular motion just as it applies to linear motion, some everyday experiences seem to contradict that fact. Imagine that you sit on the left side of the back seat of a taxi moving at high speed on a straight road. After traveling straight for a while, the taxi starts to make a high speed *left* turn (**Figure 4.4**). The car seat is slippery and you aren't holding on to anything. As the car turns left, you start sliding across the seat *toward the right* with respect to the car until you hit the door on the other side of the car. This feeling of being thrown outward in a turning car seems inconsistent with the idea that the net force points toward the center of the circle during circular motion. The feeling that there is a force pushing outward on you seems very real, but what object is exerting that force?

**Figure 4.4** An observer's view of a passenger as a taxi makes a high-speed turn.

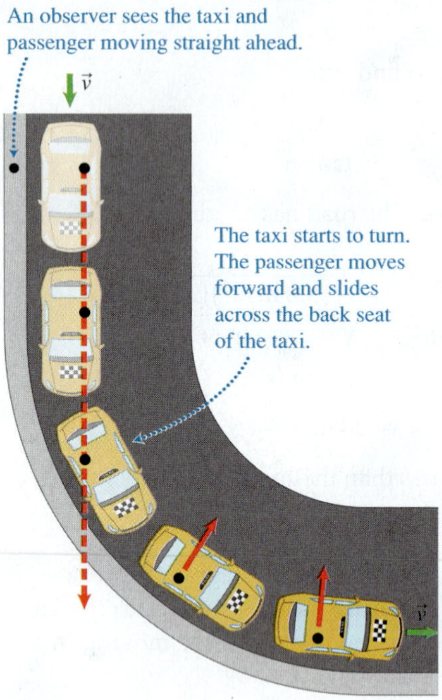
An observer sees the taxi and passenger moving straight ahead.
$\vec{v}$
The taxi starts to turn. The passenger moves forward and slides across the back seat of the taxi.
$\vec{v}$

Remember that Newton's laws explain motion only when made by an observer in an inertial reference frame. Since the car is accelerating as it goes around the curve, it is not an inertial reference frame. To a roadside observer, you and the car were moving at constant velocity before the car started turning. The forces exerted on you by Earth and the car seat balanced. The roadside observer saw the car begin turning to the left and you continuing to travel with constant velocity because the net force that other objects exerted on you was zero (we assume that the seat is slippery and the friction force that the surface of the seat exerted on you was very small). When the door finally intercepted your forward-moving body, it started exerting a force on you toward the center of the circle. At that moment, you started moving with the car at constant speed but changing velocity along the circular path. Your velocity now changes because of the normal force exerted on you by the door.

**Review Question 4.6** Think back to Example 4.6 (the rotor ride). Use your understanding of Newton's laws and constant speed circular motion to explain why the woman in the ride seems to feel a strong force pushing her against the wall of the drum.

# 4.5 The law of universal gravitation

So far, we have investigated examples of circular motion on or near the surface of Earth. Now let's look at circular motion as it occurs for planets moving around the Sun and satellites and our Moon moving around Earth.

## Observations and explanations of planetary motion

By Newton's time, scientists already knew a great deal about the motion of the planets in the solar system. Planets were known to move around the Sun at approximately constant speed in almost circular orbits. (Actually, the orbits have a slightly elliptical shape.) However, no one had a scientific explanation for what was causing the Moon and the planets to travel in their nearly circular orbits.

Newton was among the first to hypothesize that the Moon moved in a circular orbit around Earth because Earth pulled on it, continuously changing the direction of the Moon's velocity. He wondered if the force exerted by Earth on the Moon was the same type of force as the force that Earth exerted on falling objects, such as an apple falling from a tree.

**The Dependence of Gravitational Force on Distance** Newton realized that directly measuring the gravitational force exerted by Earth on another object was impossible. However, his second law ($\vec{a} = \Sigma\vec{F}/m$) provided an indirect way to determine the gravitational force exerted by Earth: measure the acceleration of the object and use the second law to determine the gravitational force. He knew that the Moon was separated from Earth by about $r_M = 3.8 \times 10^8$ m, roughly 60 Earth radii ($60R_E$). With this distance and the Moon's orbital period of 27.3 days, Newton was able to use Eq. (4.5) to determine the Moon's radial acceleration:

$$a_{r\text{ at }R=60R_E} = \frac{4\pi^2 R}{T^2} = \frac{4\pi^2(3.8 \times 10^8\text{ m})}{[(27.3\text{ days})(86{,}400\text{ s/day})]^2}$$

$$= \frac{4\pi^2(3.8 \times 10^8\text{ m})}{(2.36 \times 10^6\text{ s})^2} = 2.69 \times 10^{-3}\text{ m/s}^2$$

He then did a similar analysis for a second situation. He performed a thought experiment in which he imagined what would happen if the Moon were condensed to a small point-like object (while keeping its mass the same) located near Earth's surface. This would place the "Moon particle" at a distance of one Earth radius ($1\ R_E$) from Earth's center. If the Moon only interacted with Earth, it would have the same free-fall acceleration as any object near Earth's surface ($9.8\text{ m/s}^2$), since this acceleration is independent of mass. Now Newton knew the Moon's acceleration at two different distances from Earth's center.

Newton then used this information to determine how the gravitational force that one object exerts on another depended on the separation of the objects, assuming that the force causing the acceleration of the Moon is the gravitational force exerted on it by Earth and that this force changes with the separation of the objects. He took the ratio of these two accelerations using his second law:

$$\frac{a_{r\text{ at }R=60R_E}}{a_{r\text{ at }R=1R_E}} = \frac{F_{\text{E on Moon at }R=60R_E}/m_{\text{Moon}}}{F_{\text{E on Moon at }R=1R_E}/m_{\text{Moon}}} = \frac{F_{\text{E on Moon at }R=60R_E}}{F_{\text{E on Moon at }R=1R_E}}$$

Notice that the Moon's mass cancels when taking this ratio. Substituting the two accelerations he had observed, Newton made the following comparison between the forces:

$$\frac{F_{\text{E on Moon at }R=60R_E}}{F_{\text{E on Moon at }R=1R_E}} = \frac{a_{r\text{ at }R=60R_E}}{a_{r\text{ at }R=1R_E}} = \frac{2.69 \times 10^{-3}\text{ m/s}^2}{9.8\text{ m/s}^2} = \frac{1}{3600} = \frac{1}{60^2}$$

$$\Rightarrow F_{\text{E on Moon at }R=60R_E} = \frac{1}{60^2} F_{\text{E on Moon at }R=1R_E}$$

The force exerted on the Moon when the Moon is at a distance of 60 times Earth's radius is $1/60^2$ times the force exerted on the Moon when near Earth's surface (about 1 times Earth's radius). It appears that as the distance from Earth's center to the Moon *increases* by a factor of 60, the force decreases by a factor of $1/60^2 = 1/3600$. This suggests that the gravitational force that Earth exerts on the Moon depends on the inverse square of its distance from the center of Earth:

$$F_{\text{E on Moon at }r} \propto \frac{1}{r^2} \tag{4.10}$$

**TIP** You might be wondering why if Earth pulls on the Moon, the Moon does not come closer to Earth in the same way that an apple falls from a tree. The difference in these two cases is the speed of the objects. The apple is at rest with respect to Earth before it leaves the tree, and the Moon is moving tangentially. Think of what would happen to the Moon if Earth stopped pulling on it—it would fly away along a straight line.

**The Dependence of Gravitational Force on Mass** Newton next wanted to determine in what way the gravitational force depended on the masses of the interacting objects. Remember that the free-fall acceleration of an object near

Earth's surface does not depend on the object's mass. All objects have the same free-fall acceleration of 9.8 m/s²:

$$a_O = \frac{F_{E\text{ on }O}}{m_O} = g$$

Mathematically, this only works if the magnitude of the gravitational force that Earth exerts on a falling object $F_{E\text{ on }O}$ is proportional to the mass of that object: $F_{E\text{ on }O} = m_O g$. Newton assumed that the gravitational force that Earth exerts on the Moon has this same feature—the gravitational force must be proportional to the Moon's mass:

$$F_{E\text{ on }M} \propto m_{Moon} \quad (4.11)$$

The last question was to decide whether the gravitational force exerted by Earth on the Moon depends only on the Moon's mass or on the mass of Earth as well. Here, Newton's own third law suggested an answer. If Earth exerts a gravitational force on the Moon, then according to Newton's third law the Moon must also exert a gravitational force on Earth that is equal in magnitude (see **Figure 4.5**). Because of this, Newton decided that the gravitational force exerted by Earth on the Moon should also be proportional to the mass of Earth:

**Figure 4.5** Earth and the Moon exert equal magnitude and oppositely directed forces on each other. (Earth and Moon not drawn to scale.)

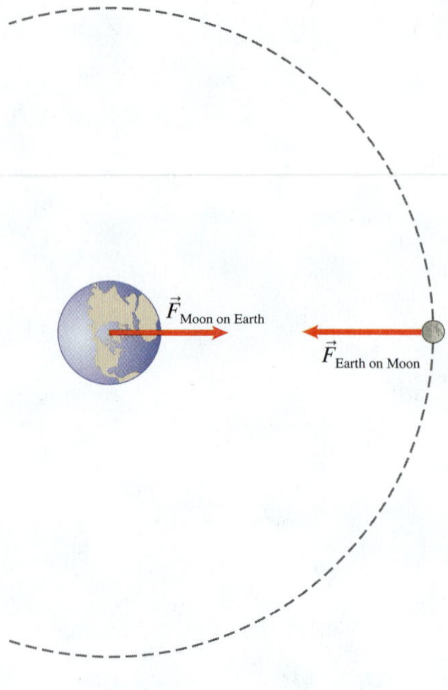


$$F_{E\text{ on }M} = F_{M\text{ on }E} \propto m_{Earth} \quad (4.12)$$

Newton then combined Eqs. (4.10)–(4.12), obtaining a mathematical relationship for the dependence of the gravitational force on all these factors:

$$F_{E\text{ on }M} = F_{M\text{ on }E} \propto \frac{m_{Earth} m_{Moon}}{r^2} \quad (4.13)$$

Newton further proposed that this was not just a mathematical description of the gravitational interaction between Earth and the Moon, but that it was more general and described the gravitational interaction between any two objects; he called this relation the **law of universal gravitation**. Newton's extension of the applicability of Eq. (4.13) to celestial objects was based on his analysis of Kepler's laws, which you will read about in the next section.

When Newton devised Eq. (4.13), he did not know the masses of Earth and the Moon and could not measure the force that one exerted on the other. Thus, the best he could do was to construct the above proportionality relationship. It was much later that scientists determined the value of the constant of proportionality:

$$G = 6.67 \times 10^{-11}\,\text{N}\cdot\text{m}^2/\text{kg}^2$$

$G$, known as **the universal gravitational constant**, allows us to write a complete mathematical equation for the gravitational force that one mass exerts on another:

$$F_{g\,1\text{ on }2} = G\frac{m_1 m_2}{r^2} \quad (4.14)$$

Notice that the value of the universal gravitation constant is quite small. If you take two objects, each with a mass of 1.0 kg and separate them by 1.0 m, the force that they exert on each other equals $6.67 \times 10^{-11}$ N. This means that the gravitational force that objects like us (with masses of about 50–100 kg) exert on each other is extremely weak. However, because the masses of celestial objects are huge (Earth's mass is approximately $6.0 \times 10^{24}$ kg and the Sun's mass is $2.0 \times 10^{30}$ kg), the gravitational forces that they exert on each other are very large, despite the small value of the gravitational constant.

## Newton's third law and the gravitational force

The mass of Earth is approximately $6.0 \times 10^{24}$ kg. The mass of a tennis ball is approximately 50 g, or $5.0 \times 10^{-2}$ kg. How does the gravitational force that

Earth exerts downward on the ball compare to the upward force that the ball exerts on Earth? According to Newton's third law, interacting objects exert forces of equal magnitude and opposite direction on each other. The gravitational force exerted by Earth on the ball has the same magnitude as the force that the ball exerts on Earth. Equation (4.14) is consistent with this idea:

$$F_{\text{E on B}} = G\frac{M_E m_B}{r^2}$$
$$= (6.67 \times 10^{-11}\,\text{N}\cdot\text{m}^2/\text{kg}^2)\frac{(6.0 \times 10^{24}\,\text{kg})(5.0 \times 10^{-2}\,\text{kg})}{(6.4 \times 10^6\,\text{m})^2} = 0.49\,\text{N}$$
$$F_{\text{B on E}} = G\frac{m_B M_E}{r^2}$$
$$= (6.67 \times 10^{-11}\,\text{N}\cdot\text{m}^2/\text{kg}^2)\frac{(5.0 \times 10^{-2}\,\text{kg})(6.0 \times 10^{24}\,\text{kg})}{(6.4 \times 10^6\,\text{m})^2} = 0.49\,\text{N}$$

(In the denominator we use the distance between the center of Earth and the center of the ball.)

It might seem counterintuitive that the ball exerts a gravitational force on Earth. After all, Earth doesn't seem to react every time someone drops something. But according to Newton's second law, if the ball exerts a nonzero net force on Earth, then Earth should accelerate. What is the magnitude of this acceleration?

$$a_E = \frac{F_{\text{B on E}}}{M_E} = \frac{0.49\,\text{N}}{6.0 \times 10^{24}\,\text{kg}} = 8.2 \times 10^{-26}\,\text{m/s}^2$$

Earth's acceleration is so tiny that there is no known way to observe it. On the other hand, the acceleration of the ball is easily noticeable since the ball's mass is so much smaller.

$$a_B = \frac{F_{\text{E on B}}}{m_B} = \frac{0.49\,\text{N}}{(5.0 \times 10^{-2}\,\text{kg})} = 9.8\,\text{m/s}^2$$

Notice that the ball has the familiar free-fall acceleration. In fact, the discussion above allows us to understand why free-fall acceleration on Earth equals $9.8\,\text{m/s}^2$ and not some other number. Use Newton's second law and the law of gravitation for an object falling freely near Earth's surface to determine its acceleration:

$$a_{\text{object}} = \frac{F_{\text{Earth on object}}}{m_{\text{object}}} = \frac{(GM_{\text{Earth}}m_{\text{object}})/R_{\text{Earth}}^2}{m_{\text{object}}} = \frac{GM_{\text{Earth}}}{R_{\text{Earth}}^2}$$

Inserting Earth's mass $M_{\text{Earth}} = 5.97 \times 10^{24}$ kg and its radius $R_{\text{Earth}} = 6.37 \times 10^6$ m, we have

$$a_{\text{object}} = \frac{GM_{\text{Earth}}}{R_{\text{Earth}}^2}$$
$$= \frac{(6.67 \times 10^{-11}\,\text{N}\cdot\text{m}^2/\text{kg}^2)(5.97 \times 10^{24}\,\text{kg})}{(6.37 \times 10^6\,\text{m})^2} = 9.8\,\text{N/kg} = 9.8\,\text{m/s}^2$$

This is exactly the free-fall acceleration of objects near Earth's surface. The fact that the experimentally measured value of $9.8\,\text{m/s}^2$ agrees with the value calculated using the law of gravitation gives us more confidence in the correctness of the law. This consistency check serves the purpose of a testing experiment. The expression $F_{\text{E on O}} = m_O g$ is actually a special case of the law of gravitation used to determine the gravitational force that Earth exerts on objects when near Earth's surface. Using a similar technique we can determine the free-fall acceleration of an object near the surface of the Moon ($m_{\text{Moon}} = 7.35 \times 10^{22}$ kg and $R_{\text{Moon}} = 1.74 \times 10^6$ m) to be $1.6\,\text{N/kg} = 1.6\,\text{m/s}^2$.

## Kepler's laws and the law of universal gravitation

About half a century before Newton's work, Johannes Kepler (1571–1630) studied astronomical data concerning the motion of the known planets and crafted three laws, called Kepler's laws of planetary motion. Newton knew about Kepler's work and applied the force equation developed for Earth-Moon interaction to the Sun and the known planets. By using the law of universal gravitation he could explain Kepler's laws. This fact contributed to scientists' confidence in the law of universal gravitation.

**Figure 4.6** Orbital models used by Kepler to develop his laws of planetary motion.

**(a)** The orbits of all planets are ellipses with the Sun located at one focus.

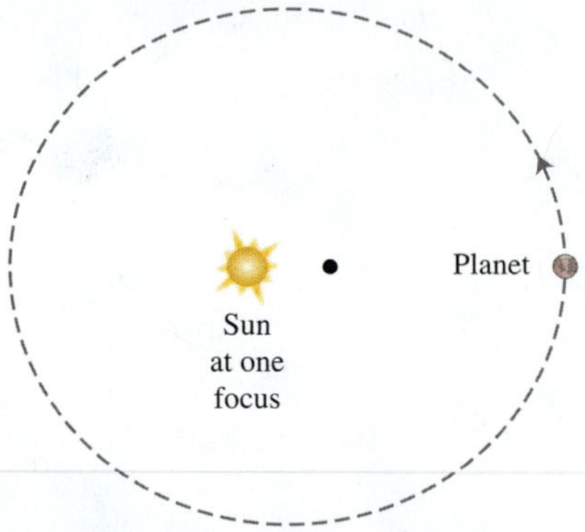


**(b)** An imaginary line connecting the planet and the Sun continually sweeps out the same area during the same time interval.

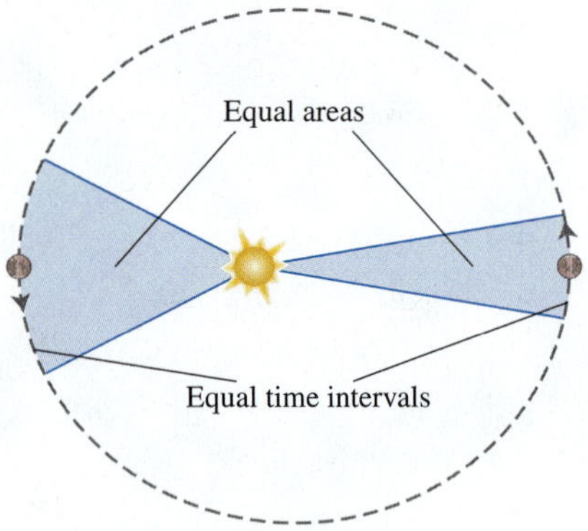


**Kepler's First Law of Planetary Motion** The orbits of all planets are ellipses with the Sun located at one of the ellipse's foci (**Figure 4.6a**).[1]

**Kepler's Second Law of Planetary Motion** When a planet travels in an orbit, an imaginary line connecting the planet and the Sun continually sweeps out the same area during the same time interval, independent of the planet's distance from the Sun (Figure 4.6b).

**Kepler's Third Law of Planetary Motion** The square of the period $T$ of the planet's motion (the time interval to complete one orbit) divided by the cube of the semi-major axis of the orbit (which is half the maximum diameter of an elliptical orbit or the radius $r$ of a circular one) equals the same constant for all the known planets:

$$\frac{T^2}{r^3} = K \tag{4.15}$$

Newton was able use his laws of motion and gravitation to derive all three of Kepler's laws of planetary motion. In addition, astronomers later used Newton's law of universal gravitation to predict the locations of the then-unknown planets Neptune and Pluto. They did it by observing subtle inconsistencies between the observed orbits of known planets and the predictions of the law of universal gravitation. At present, physicists and engineers use the law of universal gravitation to decide how to launch satellites into their desired orbits and how to successfully send astronauts to the Moon and beyond. Newton's law of universal gravitation has been tested countless times and is consistent with observations to a very high degree of accuracy.

**Newton's law of universal gravitation** **The magnitude of the attractive gravitational force that an object with mass $m_1$ exerts on an object with mass $m_2$ separated by a center-to-center distance $r$ is:**

$$F_{g\,1\,\text{on}\,2} = G\frac{m_1 m_2}{r^2} \tag{4.14}$$

**where $G = 6.67 \times 10^{-11}\ \text{N}\cdot\text{m}^2/\text{kg}^2$ is known as the gravitational constant.**

However, just because this relationship is called a "law" does not mean that the law is always valid. Every mathematical expression in physics is applicable only for certain circumstances, and the law of universal gravitation is no exception.

## Limitations of the law of universal gravitation

The law of universal gravitation in the form of Eq. (4.14) only applies to objects with spherical symmetry or whose separation distance is much bigger than their size. An object that is spherically symmetric has the shape of a perfect sphere and a density that only varies with distance from its center. In these cases, we can model the objects as though all of their mass was located

[1] The shape of the planetary orbits is close to circular for most of the planets. In those cases, the foci of the ellipse are very close to the center of the circular orbit that most closely approximates the ellipse.

at a single point at their centers. However, if the objects are not spherically symmetric and are close enough to each other so that they cannot be modeled as point-like particles, then the law of universal gravitation does not apply directly. But using calculus, we can divide each object into a collection of very small point-like objects and use the law of universal gravitation on each pair to add by integration the forces of the small point-like particles on each other.

However, even with calculus, there are details of some motion for which the law cannot account. When astronomers made careful observations of the orbit of Mercury, they noticed that its orbit exhibited some patterns that the law of universal gravitation could not explain. It wasn't until the early 20th century, when Einstein constructed a more advanced theory of gravity, that scientists could predict all of the details of the motion of Mercury.

**Review Question 4.5** Give an example that would help a friend better understand how the magnitude of the universal gravitation constant $G = 6.67 \times 10^{-11}\ \text{N} \cdot \text{m}^2/\text{kg}^2$ affects everyday life.

## 4.6 Satellites and astronauts: Putting it all together

The Moon orbits Earth due to the gravitational force that provides the necessary radial acceleration. It is Earth's natural satellite, a celestial object that orbits a bigger celestial object (usually a planet). Artificial satellites are objects that are placed in orbit by humans. The first artificial satellite was launched in 1957, and since then thousands of satellites have been launched. Earth satellites make worldwide communication possible, allow us to monitor Earth's surface and the weather, help us find our way to unknown destinations with our global positioning systems (GPS), and provide access to hundreds of television stations. The special satellites used in many of these applications must orbit at a specific distance above Earth's surface. How do we determine that distance?

### Satellites

You may have noticed that the satellite TV receiving dishes on residential rooftops never move. They always point at the same location in the sky. This means that the satellite from which they are receiving signals must always remain at the same location in the sky. In order to do so, the satellite must be placed at a very specific altitude that allows the satellite to travel once around Earth in exactly 24 hours while always remaining above the equator. Such satellites are called **geostationary.** An array of such satellites can provide communications to all parts of Earth.

**EXAMPLE 4.8 Geostationary satellite**

You are in charge of launching a geostationary satellite into orbit. At what altitude above the equator must the satellite orbit in order to provide continuous communication to a stationary dish antenna on Earth? The mass of Earth is $5.97 \times 10^{24}$ kg.

**Sketch and translate** A geostationary satellite, such as that shown on the top of the next page, completes one orbit around Earth each 24 hours, making the period of its circular motion the same as that of Earth rotating below it, $T = 24\ \text{h} = 86{,}400\ \text{s}$. We use our understanding of circular motion and the law of universal gravitation to decide the radius $r$ of the satellite's orbit from the center of Earth. Then we can determine the altitude of the satellite's orbit above Earth's surface. The satellite is the system object in this problem.

*(continued)*

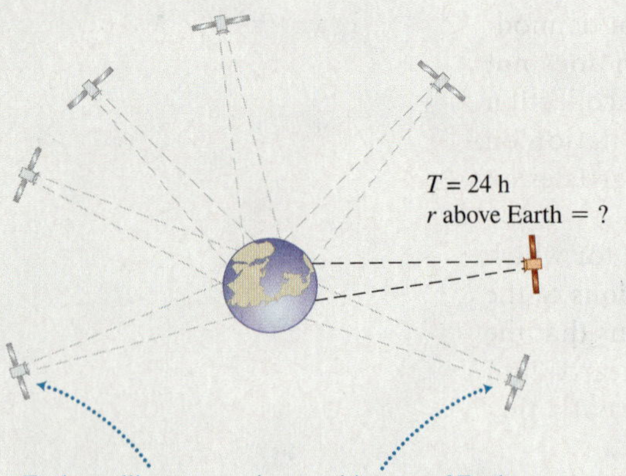


**Simplify and diagram** Model the satellite as a point-like particle and assume that it moves with constant speed circular motion. Make a force diagram for the satellite.

**Represent mathematically** The only force exerted on the satellite is the gravitational force exerted by Earth. Use the radial $r$-component form of Newton's second law and the law of universal gravitation to get an expression for the satellite's radial acceleration:

$$a_{S\,r} = \frac{F_{E\text{ on }S}}{m_S} = \frac{Gm_Em_S/r^2}{m_S} = \frac{Gm_E}{r^2}$$

Since the satellite travels with constant speed circular motion, its acceleration must also be

$$a_{S\,r} = \frac{v^2}{r} = \frac{(2\pi r/T)^2}{r} = \frac{4\pi^2 r}{T^2}$$

Setting the two expressions for acceleration equal to each other, we get

$$\frac{4\pi^2 r}{T^2} = \frac{Gm_E}{r^2}$$

Now, solve this for the radius of the orbit:

$$r^3 = \frac{Gm_E T^2}{4\pi^2} \quad \text{or} \quad r = \left(\frac{Gm_E T^2}{4\pi^2}\right)^{1/3}$$

**Solve and evaluate** Inserting all the relevant information gives an answer:

$$r = \left(\frac{(6.67 \times 10^{-11}\,\text{N}\cdot\text{m}^2/\text{kg}^2)(5.97 \times 10^{24}\,\text{kg})(8.64 \times 10^4\,\text{s})^2}{4\pi^2}\right)^{1/3}$$

$$= (75.3 \times 10^{21}\,\text{m}^3)^{1/3} = 4.2 \times 10^7\,\text{m}$$

This is the distance of the satellite from the center of Earth. Since the radius of Earth is $6.4 \times 10^6\,\text{m} = 0.64 \times 10^7\,\text{m}$, the distance of the satellite above Earth's surface is

$$4.2 \times 10^7\,\text{m} - 0.64 \times 10^7\,\text{m} = 3.6 \times 10^7\,\text{m} = 22{,}000\,\text{mi}$$

**Try it yourself:** Imagine that you want to launch a satellite that moves just above Earth's surface. Determine the speed of such a satellite. (This isn't realistic, since the atmosphere would strongly affect the satellite's orbit, but it's a good practice exercise.)

*Answer:* 7900 m/s.

## Are astronauts weightless?

We are all familiar with videos of astronauts floating in the Space Shuttle or in the International Space Station. News reports commonly say that the astronauts are weightless or experience zero gravity. Is this true?

**QUANTITATIVE EXERCISE 4.9 Are astronauts weightless in the International Space Station?**
The International Space Station orbits about $0.40 \times 10^6$ m (250 miles) above Earth's surface, or $(6.37 + 0.40) \times 10^6\,\text{m} = 6.77 \times 10^6\,\text{m}$ from Earth's center. Compare the force that Earth exerts on an astronaut in the station to the force when he is on Earth's surface.

**Represent mathematically** The gravitational force that Earth exerts on the astronaut (system object) in the space station is

$$F_{E\text{ on A in station}} = G\frac{m_E m_A}{(r_{\text{station-Earth}})^2}$$

where $r_{\text{station-Earth}} = 6.77 \times 10^6$ m is the distance from Earth's center to the astronaut in the space station. The gravitational force exerted by Earth on the astronaut when on Earth's surface is

$$F_{\text{E on A on Earth's surface}} = G\frac{m_E m_A}{(r_{\text{surface-Earth}})^2}$$

where $r_{\text{surface-Earth}} = 6.37 \times 10^6$ m is the distance from the center of Earth to its surface. To compare these two forces, take their ratio:

$$\frac{F_{\text{E on A in station}}}{F_{\text{E on A on Earth's surface}}} = \frac{G\dfrac{m_E m_A}{r^2_{\text{station-Earth}}}}{G\dfrac{m_E m_A}{r^2_{\text{surface-Earth}}}} = \left(\frac{r_{\text{surface-Earth}}}{r_{\text{station-Earth}}}\right)^2$$

**Solve and evaluate** Inserting the appropriate values gives

$$\frac{F_{\text{E on A in station}}}{F_{\text{E on A on Earth's surface}}} = \left(\frac{6.37 \times 10^6\ \text{m}}{6.77 \times 10^6\ \text{m}}\right)^2 = 0.89$$

The gravitational force that Earth exerts on the astronaut when in the space station is just 11% less than the force exerted on him when on Earth's surface. The astronaut is far from "weightless."

**Try it yourself:** At what distance from Earth's center should a person be so that her weight is half her weight when on Earth's surface?

*Answer:* $1.4\ R_E$.

Quantitative Exercise 4.9 shows that astronauts are not actually weightless while in the International Space Station. So why do they float? Remember that Earth exerts a gravitational force on both the astronaut and the space station. This force causes them both to fall toward Earth at the same rate while they fly forward, thus staying on the same circular path. Both the astronaut and the space station are in free fall. As a result, if the astronaut stood on a scale placed on the floor of the orbiting space station, her weight, according to the scale, would be zero. The same thing happens on Earth when you stand on a scale inside an elevator that is in free fall, falling at the acceleration $\vec{g}$. Since both you and the elevator fall with the same acceleration, you do not press on the scale, and the scale does not press on you; it reads zero.

The confusion occurs because in physics weight is a shorthand way of referring to the gravitational force being exerted on an object, not the reading of a scale. The scale measures the normal force it exerts on any object with which it is in contact. So astronauts are not "weightless"; the word is actually being misused.

**Review Question 4.6** A friend says he has heard that the Moon is falling toward Earth. What can you tell your friend to reassure him that all is well?

# Summary

**Words** | **Pictorial and physical representations** | **Mathematical representation**

**Velocity change method to find the direction of acceleration** We can estimate the direction of an object's acceleration as it passes a point along its circular path by drawing an initial velocity vector $v_i$ just before that point and a final velocity vector $v_f$ just after that point. We place the vectors tail to tail and draw a velocity change vector from the tip of the initial to the tip of the final velocity. The acceleration points in the direction of the velocity change vector. (Section 4.1)

Top view

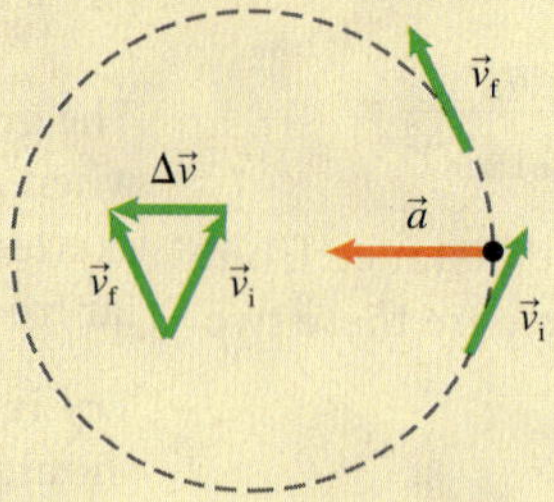


$$\vec{v}_i + \Delta\vec{v} = \vec{v}_f$$

or

$$\Delta\vec{v} = \vec{v}_f - \vec{v}_i$$

$$\vec{a} = \frac{\Delta\vec{v}}{\Delta t} = \frac{\vec{v}_f - \vec{v}_i}{t_f - t_i}$$

**Radial acceleration for constant speed circular motion** An object moving at constant speed along a circular path has acceleration that points toward the center of the circle and has a magnitude $a_r$ that depends on its speed $v$ and the radius $r$ of the circle. (Section 4.3)

Top view

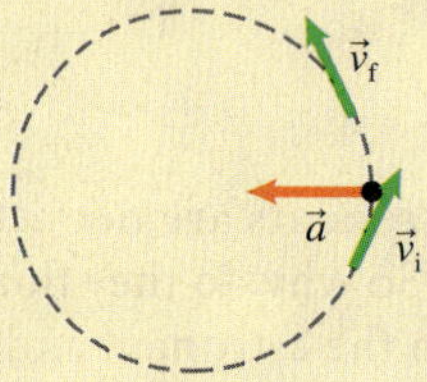


$$a_r = \frac{v^2}{r} \quad \text{Eq. (4.3)}$$

**Period and radial acceleration** The radial acceleration can also be expressed using the period $T$ of circular motion, the time interval needed for an object to complete one trip around the circle. (Section 4.3)

$$v = \frac{2\pi r}{T} \quad \text{Eq. (4.4)}$$

$$a_r = \frac{4\pi^2 r}{T^2} \quad \text{Eq. (4.5)}$$

**Net force for constant speed circular motion** The sum of the forces exerted on an object during constant speed circular motion points in the positive radial direction toward the center of the circle. The object's acceleration is the sum of the radial components of all forces exerted on an object divided by its mass—consistent with Newton's second law. In addition, for horizontal circular motion, you sometimes analyze the vertical $y$-components of forces exerted on an object. (Section 4.4)

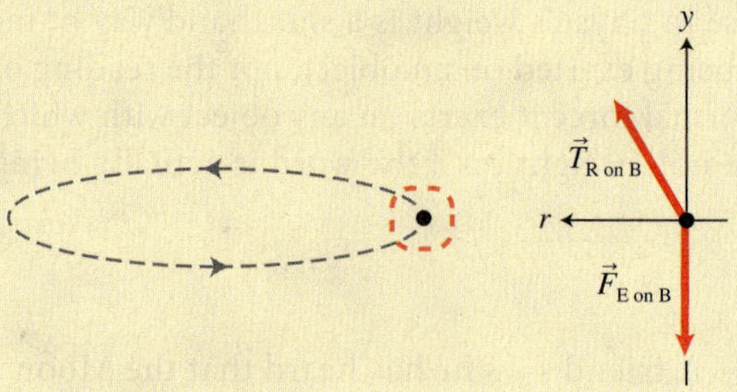


$$a_r = \frac{\Sigma F_{\text{r on Object } r}}{m_{\text{Object}}} \quad \text{Eq. (4.6)}$$

$$a_y = \frac{\Sigma F_{\text{vertical } y\text{-component on Object } y}}{m_{\text{Object}}} \quad \text{Eq. (4.7)}$$

**Law of universal gravitation** This force law is used primarily to determine the magnitude of the force that the Sun exerts on planets or that planets exert on satellites or on moons. The force depends on the masses of the objects and on their center-to-center separation $r$ (Section 4.5).

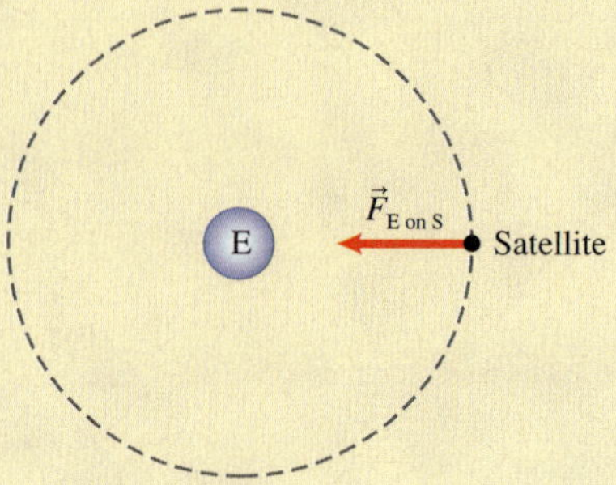


$$F_{g\,1\text{ on }2} = G\frac{m_1 m_2}{r^2} \quad \text{Eq. (4.14)}$$

MP® For instructor-assigned homework, go to MasteringPhysics.

# Questions

## Multiple Choice Questions

1. Which of the objects below is accelerating?
   (a) Object moving at constant speed along a straight line
   (b) Object moving at constant speed in a circle
   (c) Object slowing down while moving in a straight line
   (d) Both b and c describe accelerating objects.

2. The circle in **Figure Q4.2** represents the path followed by an object moving at constant speed. At four different positions on the circle, the object's motion is described using velocity and acceleration arrows. Choose the location where the descriptions are correct.

**Figure Q4.2**

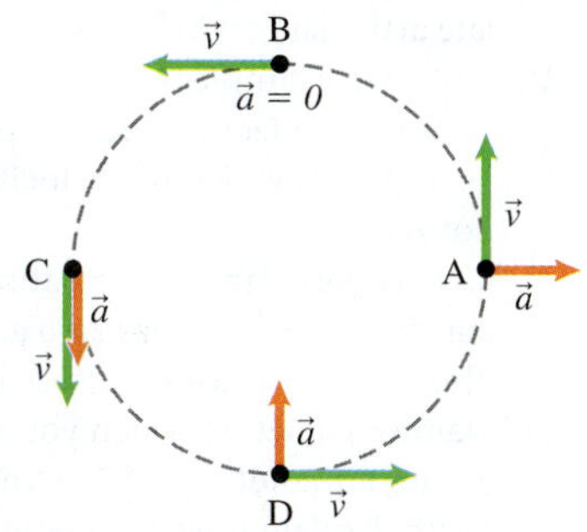


   (a) point A (b) point B (c) point C (d) point D

3. One of your classmates drew a force diagram for a pendulum bob at the bottom of its swing. He put a horizontal force arrow in the direction of the velocity. Evaluate his diagram by choosing from the statements below:
   (a) He is incorrect since the force should point partly forward and tilt partly down.
   (b) He is incorrect since there are no forces in the direction of the object's velocity.
   (c) He is correct since any moving object has a force in the direction of its velocity.
   (d) He is correct but he also needs to add an outward (down) force.

4. Why is it difficult for a high-speed car to negotiate an unbanked turn?
   (a) A huge force is pushing the car outward.
   (b) The magnitude of the friction force might not be enough to provide the necessary radial acceleration.
   (c) The sliding friction force is too large.
   (d) The faster the car moves, the harder it is for the driver to turn the steering wheel.

5. How does a person standing on the ground explain why you, sitting on the left side of a slippery back car seat, slide to the right when the car makes a high-speed left turn?
   (a) You tend to move in a straight line and thus slide with respect to the seat that is moving to the left under you.
   (b) There is a net outward force being exerted on you.
   (c) The force of motion propels you forward.
   (d) The car seat pushes you forward.

6. A pilot performs a vertical loop-the-loop at constant speed. The pilot's head is always pointing toward the center of the circle. Where is the blackout more likely to occur?
   (a) At the top of the loop
   (b) At the bottom of the loop
   (c) At both top and bottom
   (d) None of the answers is correct.

7. Why is the following an inaccurate statement about blackout? "As the *g* forces climb up toward 7 *g*'s, . . . ."
   (a) There is no such thing as a *g* force.
   (b) 7 *g*'s is not a big force.
   (c) *g* forces are constant and do not climb.
   (d) *g* forces are not responsible for blackout but can cause dizziness.

8. Why do you feel that you are being thrown upward out of your seat when going over an upward arced hump on a roller coaster?
   (a) There is an additional force lifting up on you.
   (b) At the top you continue going straight and the seat moves out from under you.
   (c) You press on the seat less than when the coaster is at rest. Thus the seat presses less on you.
   (d) Both b and c are correct.
   (e) a, b, and c are correct.

9. Compare the magnitude of the normal force of a car seat on you with the magnitude of the force that Earth exerts on you when the car moves across the bottom of a dip in the road.
   (a) The normal force is more than Earth's gravitational force.
   (b) The normal force is less than Earth's gravitational force.
   (c) The two forces are equal.
   (d) It depends on whether the car is moving right or left.

10. If you put a penny on the center of a rotating turntable, it does not slip. However, if you place the penny near the edge, it is likely to slip off. Which answer below explains this observation?
    (a) The penny moves faster at the edge and hence needs a greater force to keep it moving.
    (b) The edge is more slippery since the grooves are farther apart.
    (c) The radial acceleration is greater at the edge and the friction force is not enough to keep the penny in place.
    (d) The outward force responsible for slipping is greater at the edge than at the middle.

11. Where on Earth's surface would you expect to experience the greatest radial acceleration as a result of Earth's rotation?
    (a) On the poles
    (b) On the equator
    (c) On the highest mountain
    (d) Since all points of Earth have the same period of rotation, the acceleration is the same everywhere.

12. What observational data might Newton have used to decide that the gravitational force is inversely proportional to the distance squared between the centers of objects?
    (a) The data on the acceleration of falling apples
    (b) The data describing the Moon's orbit (period and radius) and the motion of falling apples
    (c) The data on comets
    (d) The data on moonrise and moonset times

13. What observations combined with his second and third laws helped Newton decide that the gravitational force of one object on another object is directly proportional to the product of the masses of the interacting objects?
    (a) The data on the acceleration of falling apples
    (b) The data on Moon phases
    (c) The data on comets
    (d) The data on moonrise and moonset times

14. What would happen to the force exerted by the Sun on Earth if the Sun shrank and became half its present size while retaining the same mass?
    (a) The force would be half the present force.
    (b) The force would be one-fourth of the present force.
    (c) The force would double.
    (d) The force would stay the same.

## Conceptual Questions

15. Your friend says that an object weighs less on Jupiter than on Earth as Jupiter is far away from the center of Earth. Do you agree or disagree?
16. Your friend says that when an object is moving in a circle, there is a force pushing it out away from the center. Why would he say this? Do you agree or disagree? If you disagree, how would you convince your friend of your opinion?
17. Describe three everyday phenomena that are consistent with your knowledge of the dynamics of circular motion. Specify where the observer is. Then find an observer who will not be able to explain the same phenomena using the knowledge of Newton's laws and circular motion.
18. You place a coin on a rotating turntable. Describe a circular motion experiment to estimate the maximum coefficient of static friction between the coin and the turntable.
19. Astronauts on the space station orbiting Earth are said to be in "zero gravity." Do you agree or disagree? If you disagree, why?
20. In the movies you often see space stations with "artificial gravity." They look like big doughnuts rotating around an axis perpendicular to the plane of the doughnut (see **Figure Q4.21**). People walking on the outer rim inside the turning space station feel the same gravitational effects as if they were on Earth. How does such a station work to simulate artificial gravity?

**Figure Q4.21**

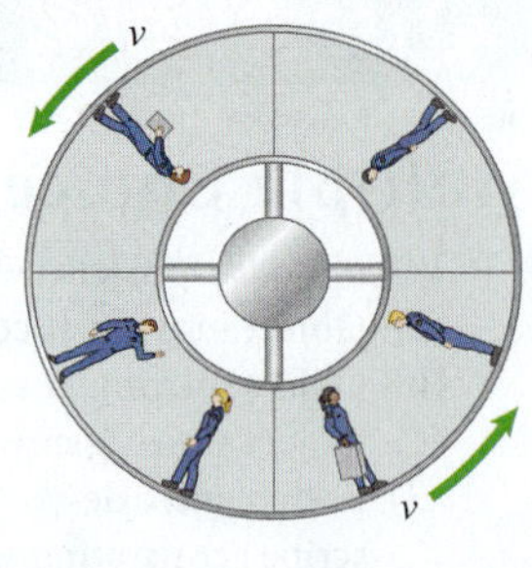


21. Give two examples of situations in which the acceleration of a moving object is zero but the velocity is not zero and two examples in which the velocity is zero but the acceleration is not zero.
22. Give two examples of situations in which an object moves at constant speed and has zero acceleration and two in which the object moves at constant speed and has a nonzero acceleration.
23. Name a planet on which you would weigh less than on Earth. Name a planet on which you would weigh more than on Earth. Explain how you know.

# Problems

Below, **BIO** indicates a problem with a biological or medical focus. Problems labeled **EST** ask you to estimate the answer to a quantitative problem rather than derive a specific answer. Problems marked with ✎ require you to make a drawing or graph as part of your solution. Asterisks indicate the level of difficulty of the problem. Problems with no * are considered to be the least difficult. A single * marks moderately difficult problems. Two ** indicate more difficult problems.

### 4.1–4.3 The qualitative velocity change method for circular motion, Qualitative dynamics of circular motion, and Radial acceleration and period

1. ✎ **Mountain biker** While mountain biking, you first move at constant speed along the bottom of a trail's circular dip and then at constant speed across the top of a circular hump. Assume that you and the bike are a system. Determine the direction of the acceleration at each position and construct a force diagram for each position (consistent with the direction of the acceleration). Compare at each position the magnitude of the force of the surface on the bike with the force Earth exerts on the system.
2. * ✎ You swing a rock tied to a string in a vertical circle. (a) Determine the direction of the acceleration of the rock as it passes the lowest point in its swing. Construct a consistent force diagram for the rock as it passes that point. How does the force that the string exerts on the rock compare to the force that Earth exerts on the rock? Explain. (b) Repeat the above analysis as best you can for the rock as it passes the highest point in the swing. (c) If the string is tied around your finger, when do you feel a stronger pull—when the rock is at the bottom of the swing or at the top? Explain.
3. ✎ **Loop-the-loop** You ride a roller coaster with a loop-the-loop. Compare *as best you can* the normal force that the seat exerts on you to the force that Earth exerts on you when you are passing the bottom of the loop and the top of the loop. Justify your answers by determining the direction of acceleration and constructing a force diagram for each position. Make your answers consistent with Newton's second law.
4. You start an old record player and notice a bug on the surface close to the edge of the record. The record has a diameter of 12 inches and completes 33 revolutions each minute. (a) What are the speed and the acceleration of the bug? (b) What would the bug's speed and acceleration be if it were halfway between the center and the edge of the record?
5. Determine the acceleration of Earth due to its motion around the Sun. What do you need to assume about Earth to make the calculation? How does this acceleration compare to the acceleration of free fall on Earth?
6. The Moon is an average distance of $3.8 \times 10^8$ m from Earth. It circles Earth once each 27.3 days. (a) What is its average speed? (b) What is its acceleration? (c) How does this acceleration compare to the acceleration of free fall on Earth?
7. **Aborted plane landing** You are on an airplane that is landing. The plane in front of your plane blows a tire. The pilot of your plane is advised to abort the landing, so he pulls up, moving in a semicircular upward-bending path. The path has a radius of 500 m with a radial acceleration of 17 $m/s^2$. What is the plane's speed?
8. **BIO Ultracentrifuge** You are working in a biology lab and learning to use a new ultracentrifuge for blood tests. The specifications for the centrifuge say that a red blood cell rotating in the ultracentrifuge moves at 470 m/s and has a radial acceleration of 150,000 *g*'s (that is, 150,000 times 9.8 $m/s^2$). The radius of the centrifuge is 0.15 m. You wonder if this claim is correct. Support your answer with a calculation.

9. Jupiter rotates once about its axis in 9 h 56 min. Its radius is $7.13 \times 10^4$ km. Imagine that you could somehow stand on the surface (although in reality that would not be possible, because Jupiter has no solid surface). Calculate your radial acceleration in meters per second squared and in Earth *g*'s.
10. * Imagine that you are standing on a horizontal rotating platform in an amusement park (like the platform for a merry-go-round). The period of rotation and the radius of the platform are given, and you know your mass. Make a list of the physical quantities you could determine using this information, and describe how you would determine them.
11. * A car moves along a straight line to the right. Two friends standing on the sidewalk are arguing about the motion of a point on the rotating car tire at the instant it reaches the lowest point (touching the road). Jake says that the point is at rest. Morgan says that the point is moving to the left at the car's speed. Justify each friend's opinion. Explain whether it is possible for them to simultaneously be correct.
12. * Three people are standing on a horizontally rotating platform in an amusement park. One person is almost at the edge, the second one is $(3/5)R$ from the center, and the third is $(1/2)R$ from the center. Compare their periods of rotation, their speeds, and their radial accelerations.
13. * Consider the scenario described in Problem 12. If the platform speeds up, who is more likely to have trouble staying on the platform? Support your answer with a force diagram and describe the assumptions that you made.
14. * **Merry-go-round acceleration** Imagine that you are standing on the rotating platform of a merry-go-round in an amusement park. You have a stopwatch and a measuring tape. Describe how you will determine your radial acceleration when standing at the edge of the platform and when halfway from the edge. What do you expect the ratio of these two accelerations to be?
15. **Ferris wheel** You are sitting on a rotating Ferris wheel. Draw a force diagram for yourself when you are at the bottom of the circle and when you are at the top.
16. EST * **Estimate** the radial acceleration of the foot of a college football player in the middle of punting a football.
17. EST * **Estimate** the radial acceleration of the toe at the end of the horizontally extended leg of a ballerina doing a pirouette.

## 4.4 Skills for analyzing processes involving circular motion

18. * Is it safe to drive your 1600-kg car at speed 27 m/s around a level highway curve of radius 150 m if the effective coefficient of static friction between the car and the road is 0.40?
19. * You are fixing a broken rotary lawn mower. The blades on the mower turn 50 times per second. What is the magnitude of the force needed to hold the outer 2 cm of the blade to the inner portion of the blade? The outer part is 21 cm from the center of the blade, and the mass of the outer portion is 7.0 g.
20. * Your car speeds around the 80-m-radius curved exit ramp of a freeway. A 70-kg student holds the armrest of the car door, exerting a 220-N force on it in order to prevent himself from sliding across the vinyl-covered back seat of the car and slamming into his friend. How fast is the car moving in meters per second and miles per hour? What assumptions did you make?
21. How fast do you need to swing a 200-g ball at the end of a string in a horizontal circle of 0.5-m radius so that the string makes a 34° angle relative to the horizontal? What assumptions did you make?
22. ** Christine's bathroom scale in Maine reads 110 lb when she stands on it. Will the scale read more or less in Singapore if her mass stays the same? To answer the question, (a) draw a force diagram for Christine. (b) With the assistance of this diagram, write an expression using Newton's second law that relates the forces exerted on Christine and her acceleration along the radial direction. (c) Decide whether the reading of the scale is different in Singapore. List the assumptions that you made and describe how your answer might change if the assumptions are not valid.
23. ** A child is on a swing that moves in the horizontal circle of radius 2.0 m depicted in **Figure P4.23**. The mass of the child and the seat together is 30 kg and the two cables exert equal-magnitude forces on the chair. Make a list of the physical quantities you can determine using the sketch and the known information. Determine one kinematics and two dynamics quantities from that list.

**Figure P4.23**

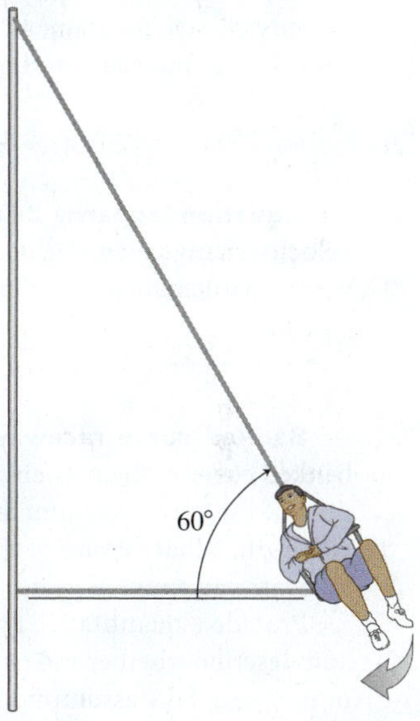


24. * A coin rests on a record 0.15 m from its center. The record turns on a turntable that rotates at variable speed. The coefficient of static friction between the coin and the record is 0.30. What is the maximum coin speed at which it does not slip?
25. **Roller coaster ride** A roller coaster car travels at speed 8.0 m/s over a 12-m-radius vertical circular hump. What is the magnitude of the upward force that the coaster seat exerts on a 48-kg woman passenger?
26. * A person sitting in a chair (combined mass 80 kg) is attached to a 6.0-m-long cable. The person moves in a horizontal circle. The cable angle $\theta$ is 62° below the horizontal. What is the person's speed? Note: The radius of the circle is not 6.0 m.
27. * A car moves around a 50-m-radius highway curve. The road, banked at 10° relative to the horizontal, is wet and icy so that the coefficient of friction is approximately zero. At what speed should the car travel so that it makes the turn without slipping?
28. * A 20.0-g ball is attached to a 120-cm-long string and moves in a horizontal circle (see **Figure P4.28**). The string exerts a force on the ball that is equal to 0.200 N. What is the angle $\theta$?

**Figure P4.28**

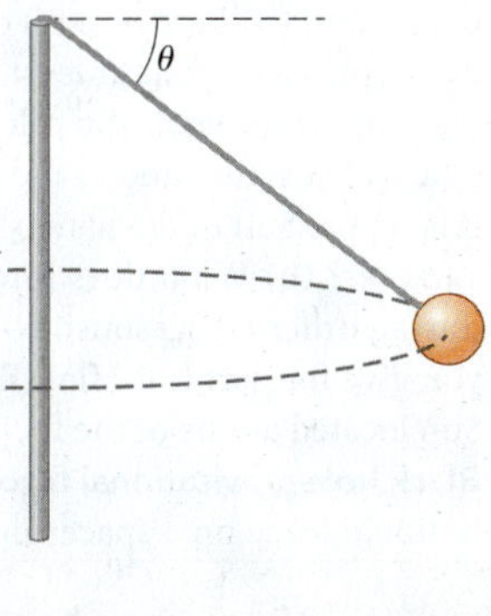

29. A 50-kg ice skater goes around a circle of radius 5.0 m at a constant speed of 3.0 m/s on a level ice rink. What are the magnitude and direction of the horizontal force that the ice exerts on the skates?

30. * A car traveling at 10 m/s passes over a hill on a road that has a circular cross section of radius 30 m. What is the force exerted by the seat of the car on a 60-kg passenger when the car is passing the top of the hill?

31. * A 1000-kg car is moving at 30 m/s around a horizontal level curved road whose radius is 100 m. What is the magnitude of the frictional force required to keep the car from sliding?

32. * **Equation Jeopardy 1** Describe using words, a sketch, a velocity change diagram, and a force diagram two situations whose mathematical description is presented below.

$$700\text{ N} - (30\text{ kg})(9.8\text{ m/s}^2) = \frac{(30\text{ kg})v^2}{12\text{ m}}$$

33. * **Equation Jeopardy 2** Describe using words, a sketch, a velocity change diagram, and a force diagram two situations whose mathematical description is presented below.

$$\frac{(2.0\text{ kg})(4.0\text{ m}^2/\text{s}^2)}{r} = 0.4 \times (2.0\text{ kg}) \times (9.8\text{ N/kg})$$

34. ** **Banked curve raceway design** You need to design a banked curve at the new circular Super 100 Raceway. The radius of the track is 800 m and cars typically travel at speed 160 mi/h. What feature of the design is important so that all racecars can move around the track safely in any weather? (a) Provide a quantitative answer. (b) List your assumptions and describe whether the number you provided will increase or decrease if the assumption is not valid.

35. * A circular track is in a horizontal plane, has a radius of $r$ meters, and is banked at an angle $\theta$ above the horizontal. (a) Develop an expression for the speed a person should rollerblade on this track so that she needs zero friction to prevent her from sliding sideways off the track. (b) Should another person move faster or slower if her mass is 1.3 times the mass of the first person? Justify your answer.

36. ** Design a quantitative test for Newton's second law as applied to constant speed circular motion. Describe the experiment and provide the analysis needed to make a prediction using the law.

37. * **Spin-dry cycle** Explain how the spin-dry cycle in a washing machine removes water from clothes. Be specific.

### 4.5–4.6 The law of universal gravitation and Satellites and astronauts: Putting it all together

38. * Your friend says that the force that the Sun exerts on Earth is much larger than the force that Earth exerts on the Sun. (a) Do you agree or disagree with this opinion? (b) If you disagree, how would you convince him of your opinion?

39. Determine the gravitational force that (a) the Sun exerts on the Moon, (b) Earth exerts on the Moon, and (c) the Moon exerts on Earth. List at least two assumptions for each force that you made when you calculated the answers.

40. * (a) What is the ratio of the gravitational force that Earth exerts on the Sun in the winter and the force that it exerts in the summer? (b) What does it tell you about the speed of Earth during different seasons? (c) How many correct answers can you give for part (a)? Hint: Earth's orbit is an ellipse with the Sun located at one of the foci of the ellipse.

41. **Black hole gravitational force** A black hole exerts a 50-N gravitational force on a spaceship. The black hole is $10^{14}$ m from the ship. What is the magnitude of the force that the black hole exerts on the ship when the ship is one-half that distance from the black hole? [*Hint:* One-half of $10^{14}$ m is not $10^7$ m.]

42. EST * The average radius of Earth's orbit around the Sun is $1.5 \times 10^8$ km. The mass of Earth is $5.97 \times 10^{24}$ kg, and it makes one orbit in approximately 365 days. (a) What is Earth's speed relative to the Sun? (b) Estimate the Sun's mass using Newton's law of universal gravitation and Newton's second law. What assumptions did you need to make?

43. * The Moon travels in a $3.8 \times 10^5$ km radius orbit about Earth. Earth's mass is $5.97 \times 10^{24}$ kg. Determine the period $T$ for one Moon orbit about Earth using Newton's law of universal gravitation and Newton's second law. What assumptions did you make?

44. Determine the ratio of Earth's gravitational force exerted on an 80-kg person when at Earth's surface and when 1000 km above Earth's surface. The radius of Earth is 6370 km.

45. Determine the magnitude of the gravitational force Mars would exert on your body if you were on the surface of Mars.

46. * When you stand on a bathroom scale here on Earth, it reads 540 N. (a) What would your mass be on Mars, Venus, and Saturn? (b) What is the magnitude of the gravitational force each planet would exert on you if you stood on their surface? (c) What assumptions did you make?

47. The free-fall acceleration on the surface of Jupiter, the most massive planet, is $24.79\text{ m/s}^2$. Jupiter's radius is $7.0 \times 10^4$ km. Use Newtonian ideas to determine Jupiter's mass.

48. A satellite moves in a circular orbit a distance of $1.6 \times 10^5$ m above Earth's surface. Determine the speed of the satellite.

49. * Mars has a mass of $6.42 \times 10^{23}$ kg and a radius of $3.40 \times 10^6$ m. Assume a person is standing on a bathroom scale on the surface of Mars. Over what time interval would Mars have to complete one rotation on its axis to make the bathroom scale have a zero reading?

50. * Determine the speed a projectile must reach in order to become an Earth satellite. What assumptions did you make?

51. * Determine the distance above Earth's surface to a satellite that completes two orbits per day. What assumptions did you make?

52. Determine the period of an Earth satellite that moves in a circular orbit just above Earth's surface. What assumptions did you need to make?

53. * A spaceship in outer space has a doughnut shape with 500-m outer radius. The inhabitants stand with their heads toward the center and their feet on an outside rim (see Figure Q4.21). Over what time interval would the spaceship have to complete one rotation on its axis to make a bathroom scale have the same reading for the person in space as when on Earth's surface?

## General Problems

54. * **Loop-the-loop** You have to design a loop-the-loop for a new amusement park so that when each car passes the top of the loop inverted (upside-down), each seat exerts a force against a passenger's bottom that has a magnitude equal to 1.5 times the gravitational force that Earth exerts on the passenger. Choose some reasonable physical quantities so these conditions are met. Show that the loop-the-loop will work equally well for passengers of any mass.

55. ** **A Tarzan swing** Tarzan (mass 80 kg) swings at the end of an 8.0-m-long vine (**Figure P4.55**). When directly under the

**Figure P4.55**

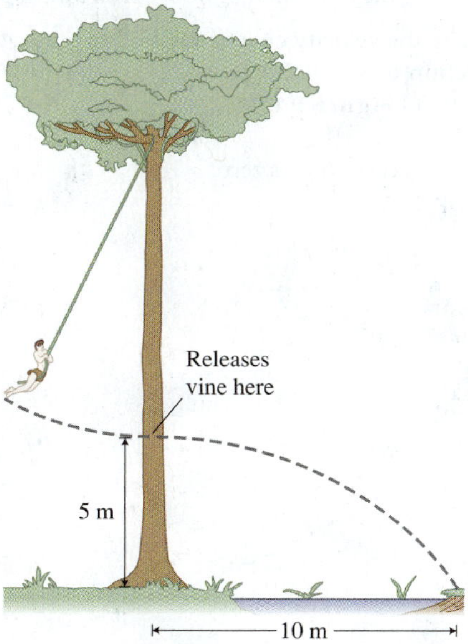


vine's support, he releases the vine and flies across a swamp. When he releases the vine, he is 5.0 m above the swamp and 10.0 m horizontally from the other side. Determine the force the vine exerts on him at the instant before he lets go (the vine is straight down when he lets go).

56. * (a) If the masses of Earth and the Moon were both doubled, by how much would the radius of the Moon's orbit about Earth have to change if its speed did not change? (b) By how much would its speed have to change if its radius did not change? Justify each answer.
57. EST * Estimate the radial acceleration of the tread of a car tire. Indicate any assumptions that you made.
58. EST ** Estimate the force exerted by the tire on a 10-cm-long section of the tread of a tire as the car travels at speed 80 km/h. Justify any numbers used in your estimate.
59. EST ** Estimate the maximum radial force that a football player's leg needs to exert on his foot when swinging the leg to punt the ball. Justify any numbers that you use.
60. **Design 1** Design and solve a circular motion problem for a roller coaster.
61. * **Design 2** Design and solve a circular motion problem for the amusement park ride shown in **Figure P4.61**.

**Figure P4.61**

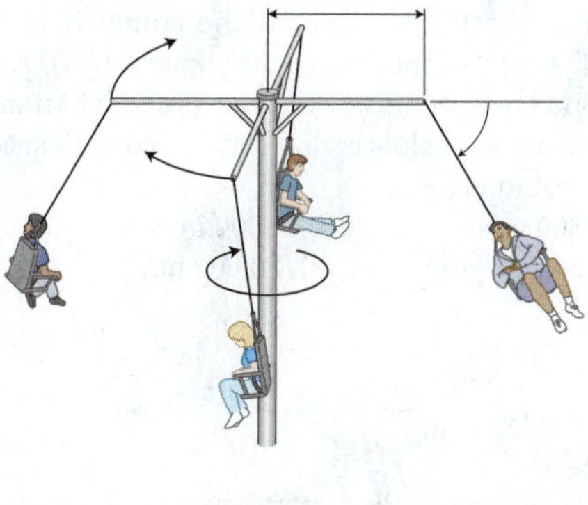

62. ** **Demolition** An old building is being demolished by swinging a heavy metal ball from a crane. Suppose that such a 100-kg ball swings from a 20-m-long wire at speed 16 m/s as the wire passes the vertical orientation. (a) What tension force must the wire be able to withstand in order not to break? (b) Assume the ball stops after sinking 1.5 m into the wall. What was the average force that the ball exerted on the wall? Indicate any assumptions you made for each part of the problem.
63. **Designing a banked roadway** You need to design a banked curve for a highway in which cars make a 90° turn moving at 50 mi/h. Indicate any assumptions you make.
64. * **Evaluation question** You find the following report about blackouts. "The acceleration that causes blackouts in fighter pilots is called the maximum *g*-force. Fighter pilots experience this force when accelerating or decelerating quickly. At high *g*'s, the pilot's blood pressure changes and the flow of oxygen to the brain rapidly decreases. This happens because the pressure outside of the pilot's body is so much greater than the pressure a human is normally accustomed to." Indicate any incorrect physics (including the application of physics to biology) that you find.
65. * Suppose that Earth rotated much faster on its axis—so fast that people were almost weightless when at Earth's surface. How long would the length of a day be on this new Earth?
66. * On Earth, an average person's vertical jump is 0.40 m. What is it on the Moon? Explain.
67. * You read in a science magazine that on the Moon, the speed of a shell leaving the barrel of a modern tank is enough to put the shell in a circular orbit above the surface of the Moon (there is no atmosphere to slow the shell). What should be the speed for this to happen? Is this number reasonable?

## Reading Passage Problems

**Texas Motor Speedway** On Oct. 28, 2000 Gil de Ferran set the single-lap average speed record for the then-named California Speedway—241.428 mi/h. The following year, on April 29, 2001, the Championship Auto Racing Teams (CART) organization canceled the Texas Motor Speedway inaugural Firestone Firehawk 600 race 2 hours before it was to start. During practice, drivers became dizzy when they reached speeds of more than 230 mph on the high-banked track. The Texas Motor Speedway has turns banked at 24° above the horizontal. By comparison, the Indianapolis raceway turns are banked at 9°.

Banking of 24° is unprecedented for Indy-style cars. In qualifying runs for the race, 21 of 25 drivers complained of dizziness and disorientation. They experienced accelerations of 5.5 *g* (5.5 times 9.8 $m/s^2$) that lasted for several seconds. Such accelerations for such long time intervals have caused pilots to black out. Blood drains from their heads to their legs as they move in circles with their heads toward the center of the circle. CART felt that the track was unsafe for their drivers. The Texas Motor Speedway had tested the track with drivers before the race and thought it was safe. The Texas Motor Speedway sued CART.

68. Why did drivers get dizzy and disoriented while driving at the Texas Motor Speedway?
(a) The cars were traveling at over 200 mi/h.
(b) The track was tilted at an unusually steep angle.
(c) On turns the drivers' blood tended to drain from their brains into veins in their lower bodies.
(d) The *g* force pushed blood into their heads.
(e) The combination of a and b caused c.

69. What was the time interval needed for Gil de Ferran's car to complete one lap during his record-setting drive?
(a) 16.8 s (b) 18.4 s (c) 22.4 s
(d) 25.1 s (e) 37.3 s

70. If the racecars had no help from friction, which expression below would describe the normal force of the track on the cars while traversing the 24° banked curves?
(a) $mg\cos 66°$ (b) $mg\sin 66°$ (c) $mg/\cos 66°$
(d) $mg/\sin 66°$ (e) None of these

71. For the racecars to stay on the road while traveling at high speed, how did a friction force need to be exerted?
(a) Parallel to the roadway and outward
(b) Parallel to the roadway and toward the infield
(c) Horizontally toward the center of the track
(d) Opposite the direction of motion
(e) None of the above

72. The average speed reported in the reading passage has six significant digits, implying that the speed is known to within ±0.001 mi/h. If this is correct, which answer below is closest to the uncertainty in the time needed to travel around the 1.5-mi oval track? (Think about the percent uncertainties.)
(a) ±0.0001 s (b) ±0.001 s (c) ±0.01 s
(d) ±0.1 s (e) ±1 s

73. What was the approximate radius of the part of the track where the drivers experienced the 5.5 *g* acceleration?
(a) 40 m (b) 80 m (c) 200 m
(d) 400 m (e) 1000 m

**Halley's Comet** Edmond Halley was the first to realize that the comets observed in 1531, 1607, and 1682 were really one comet (now called Halley's Comet) that moved around the Sun in an elongated elliptical orbit (see **Figure 4.7**). He predicted that the peanut-shaped comet would reappear in 1757. It appeared in March 1759 (attractions to Jupiter and Saturn delayed its trip by 618 days). More recent appearances of Halley's Comet were in 1835, 1910, and 1986. It is expected again in 2061.

**Figure 4.7** The elongated orbit of Halley's Comet.

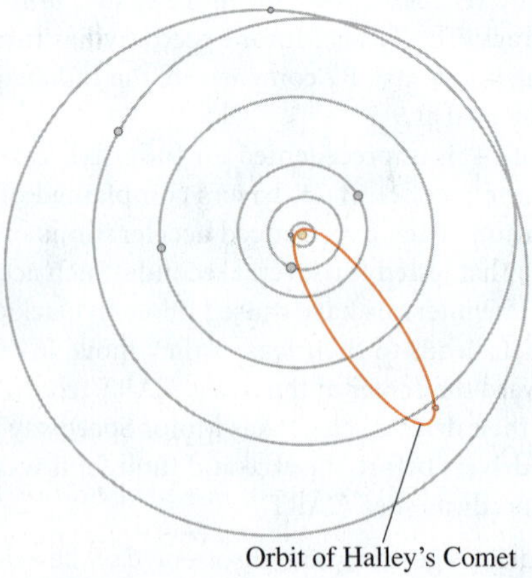


The nucleus of Halley's Comet is relatively small (15 km long, 8 km wide, and 8 km thick). It has a low $2.2 \times 10^{14}$ kg mass with an average density of about 600 kg/m$^3$. (The density of water is 1000 kg/m$^3$.) The nucleus rotates once every 52 h. When Halley's Comet is closest to the Sun, temperatures on the comet can rise to about 77 °C and several tons of gas and dust are emitted each second, producing the long tail that we see each time it passes the Sun.

74. **EST** Use the velocity change method to estimate the comet's direction of acceleration when passing closest to the Sun (position I **Figure P4.74**).
(a) A (b) B (c) C (d) D
(e) The acceleration is zero.

**Figure P4.74**

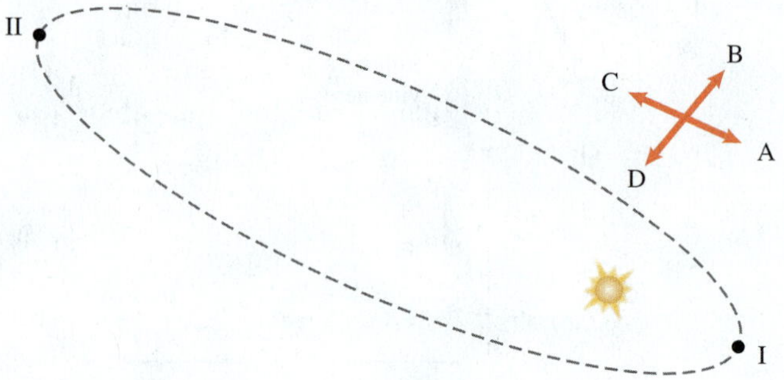


75. What object or objects exert forces on the comet as it passes position I (shown Figure P4.74)?
(a) The Sun's gravitational force toward the Sun
(b) The force of motion tangent to the direction the comet is traveling
(c) An outward force away from the Sun
(d) a and b
(e) a, b, and c

76. Suppose that instead of being peanut shaped, Halley's Comet was spherical with a radius of 5.0 km (about its present volume). Which answer below would be closest to your radial acceleration if you were standing on the "equator" of the rotating comet?
(a) $10^{-5}$ m/s$^2$ (b) $10^{-3}$ m/s$^2$ (c) 0.1 m/s$^2$
(d) 10 m/s$^2$ (e) 1000 m/s$^2$

77. Approximately what gravitational force would the spherical-shaped 5-km radius comet exert on a 100-kg person on the surface of the comet?
(a) 0.06 N (b) 0.6 N (c) 6 N
(d) 60 N (e) 600 N

78. The closest distance that the comet passes relative to the Sun is $8.77 \times 10^{10}$ m (position I in Figure P4.74). Apply Newton's second law and the law of universal gravitation to determine which answer below is closest to the comet's speed when passing position I.
(a) 1000 m/s (b) 8000 m/s (c) 20,000 m/s
(d) 40,000 m/s (e) 800,000 m/s

79. The farthest distance that the comet is from the Sun is $5.25 \times 10^{12}$ m (position II in Figure P4.74). Apply Newton's second law and the law of universal gravitation to determine which answer below is closest to the comet's speed when passing position II.
(a) 800 m/s (b) 5000 m/s (c) 10,000 m/s
(d) 50,000 m/s (e) 80,000 m/s

# Impulse and Linear Momentum

# 5

**How does jet propulsion work?**

**How can you measure the speed of a bullet?**

**Would a meteorite collision significantly change Earth's orbit?**

In previous chapters we discovered that the pushing interaction between car tires and the road allows a car to change its velocity. Likewise, a ship's propellers push water backward; in turn, water pushes the ship forward. But how does a rocket, far above Earth's atmosphere, change velocity with no object to push against?

Less than 100 years ago, rocket flight was considered impossible. When U. S. rocket pioneer Robert Goddard published an article

**Be sure you know how to:**

- Construct a force diagram for an object (Section 2.1).
- Use Newton's second law in component form (Section 3.2).
- Use kinematics to describe an object's motion (Section 1.7).

in 1920 about rocketry and even suggested a rocket flight to the Moon, he was ridiculed by the press. A *New York Times* editorial dismissed his idea, saying, ". . . even a schoolboy knows that rockets cannot fly in space because a vacuum is devoid of anything to push on." We know now that Goddard was correct—but why? What does the rocket push on?

We can use Newton's second law ($\vec{a} = \Sigma\vec{F}/m$) to relate the acceleration of a system object to the forces being exerted on it. However, to use this law effectively we need quantitative information about the forces that objects exert on each other. Unfortunately, if two cars collide, we don't know the force that one car exerts on the other during the collision. When fireworks explode, we don't know the forces that are exerted on the pieces flying apart. In this chapter you will learn a new approach that helps us analyze and predict mechanical phenomena when the forces are not known.

**Figure 5.1** The mass is the same in the closed flask (an isolated system) (a) before burning the steel wool and (b) after burning the steel wool. (c) However, the mass increases when the steel wool is burned in the open flask (a non-isolated system).

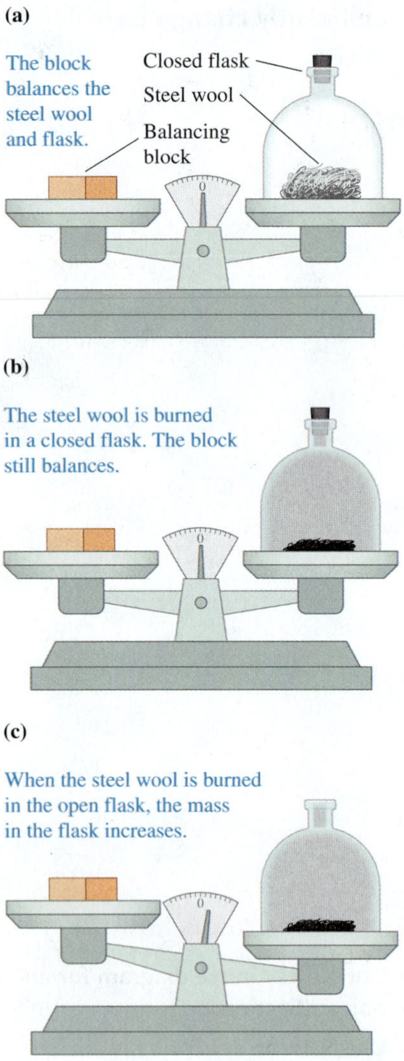
(a)
The block balances the steel wool and flask.
Closed flask
Steel wool
Balancing block
0
(b)
The steel wool is burned in a closed flask. The block still balances.
0
(c)
When the steel wool is burned in the open flask, the mass in the flask increases.
0

## 5.1 Mass accounting

We begin our investigation by analyzing the physical quantity of mass. Earlier (in Chapter 2), we found that the acceleration of an object depended on its mass—the greater its mass, the less it accelerated due to an unbalanced external force. We ignored the possibility that an object's mass might change during some process. Is the mass in a system always a constant value?

You have probably observed countless physical processes in which mass seems to change. For example, the mass of a log in a campfire decreases as the log burns; the mass of a seedling increases as the plant grows. What happens to the "lost" mass from the log? Where does the seedling's increased mass come from?

A system perspective helps us understand what happens to the burning log. If we choose only the log as the system, the mass of the system decreases as it burns. However, air is needed for burning. What happens to the mass if we choose the surrounding air and the log as the system?

Suppose that we place steel wool in a closed flask on one side of a balance scale and a metal block of equal mass on the other side (**Figure 5.1a**). In one experiment, we burn the steel wool in the closed flask (the flask also contains air), forming an oxide of iron. We find that the total mass of the closed flask containing burned steel wool (iron oxide) is the same as the mass of the balancing metal block (Figure 5.1b). Next, we burn the steel wool in an open flask and observe that the mass of that flask increases (Figure 5.1c). The steel wool in the open flask burns more completely and absorbs some external oxygen from the air as it burns.

Eighteenth-century French chemist Antoine Lavoisier actually performed such experiments. He realized that the choice of the system was very important. Lavoisier defined an **isolated system** as a group of objects that interact with each other but not with external objects outside the system. The mass of an isolated system is the sum of the masses of all objects in the system. He then used the concept of an isolated system to summarize his (and our) experiments in the following way:

**Law of constancy of mass** When a system of objects is isolated (a closed container), its mass equals the sum of the masses of its components and does not change—it remains constant in time.

When the system is not isolated (an open container system), the mass might change. However, this change is not random—it is always equal to the amount of mass leaving or entering the system from the environment.

Thus, even when the mass of a system is not constant, we can keep track of the changes if we take into account how much is leaving or entering the system:

$$\begin{pmatrix} \textit{initial} \text{ mass of} \\ \text{system at earlier} \\ \text{clock reading} \end{pmatrix} + \begin{pmatrix} \textit{new} \text{ mass entering or} \\ \text{leaving system between} \\ \text{the two clock readings} \end{pmatrix} = \begin{pmatrix} \textit{final} \text{ mass of} \\ \text{system at later} \\ \text{clock reading} \end{pmatrix}$$

The above equation helps describe the change of mass in any system. The mass is constant if there is no flow of mass in or out of the system, or the mass changes in a predictable way if there is some flow of mass between the system and the environment. Basically, mass cannot appear from nowhere and does not disappear without a trace. Imagine you have a system that has a total mass of $m_i = 3\text{ kg}$ (a bag of oranges). You add some more oranges to the bag ($\Delta m = 1\text{ kg}$). The final mass of the system equals exactly the sum of the initial mass and the added mass: $m_i + \Delta m = m_f$ or $3\text{ kg} + 1\text{ kg} = 4\text{ kg}$ (**Figure 5.2a**). We can represent this process with a bar chart (Figure 5.2b). The bar on the left represents the initial mass of the system, the central bar represents the mass added or taken away, and the bar on the right represents the mass of the system in the final situation. As a result, the height of the left bar plus the height of the central bar equals the height of the right bar. The bar chart allows us to keep track of the changes in mass of a system even if the system is not isolated.

**Figure 5.2** (a) The initial mass of the oranges plus the mass of the oranges that were added (or subtracted) equals the final mass of the oranges. (b) The mass change process is represented by a mass bar chart.

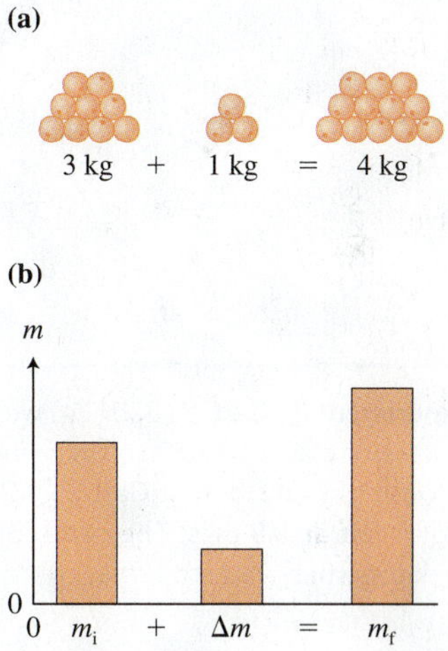


Mass is called a **conserved** quantity. A conserved quantity is constant in an isolated system. When the system is not isolated, we can account for the changes in the conserved quantity by what is added to or subtracted from the system.

Just as with every idea in physics, the law of constancy of mass in an isolated system does not apply in all cases. We will discover later in this book (Chapters 28 and 29) that in situations involving atomic particles, mass is not constant even in an isolated system; instead, what is constant is a new quantity that includes mass as a component.

**Review Question 5.1** When you burn a log in a fire pit, the mass of wood clearly decreases. How can you define the system so as to have the mass of the objects in that system constant?

## 5.2 Linear momentum

‹ Active Learning Guide

We now know that mass is an example of a conserved quantity. Is there a quantity related to motion that is conserved? When you kick a stationary ball, there seems to be a transfer of motion from your foot to the ball. When you knock bowling pins down with a bowling ball, a similar transfer occurs. However, motion is not a physical quantity. What physical quantities describing motion are constant in an isolated system? Can we describe the changes in these quantities using a bar chart?

Let's conduct a few experiments to find out. In Observational Experiment **Table 5.1** we observe two carts of different masses that collide on a smooth track. For these experiments, the system will include both carts. A collision is a process that occurs when two (or more) objects come into direct contact with each other. The system is isolated since the forces that the carts exert on each other are internal, and external forces are either balanced (as the vertical forces are) or negligible (the horizontal friction force).

**OBSERVATIONAL EXPERIMENT TABLE**

## 5.1 Collisions in a system of two carts (all velocities are with respect to the track).

VIDEO 5.1

| Observational experiment | Analysis |
|---|---|

**Experiment 1.** Cart A (0.20 kg) moving right at 1.0 m/s collides with cart B (0.20 kg), which is stationary. Cart A stops and cart B moves right at 1.0 m/s.

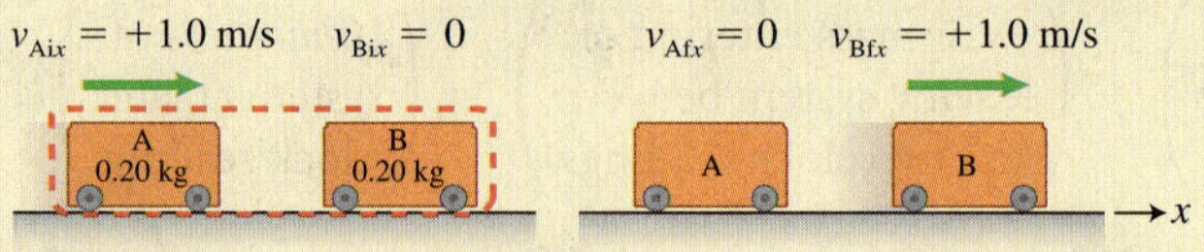


The direction of motion is indicated with a plus and a minus sign.

- *Speed:* The sum of the speeds of the system objects is the same before and after the collision: 1.0 m/s + 0 m/s = 0 m/s + 1.0 m/s.
- *Mass · speed:* The sum of the products of mass and speed is the same before and after the collision: $0.20\text{ kg}(1.0\text{ m/s}) + 0.20\text{ kg}(0\text{ m/s}) = 0.20\text{ kg}(0\text{ m/s}) + 0.20\text{ kg}(1.0\text{ m/s})$.
- *Mass · velocity:* The sum of the products of mass and the *x*-component of velocity is the same before and after the collision: $0.20\text{ kg}(+1.0\text{ m/s}) + 0.20\text{ kg}(0) = 0.20\text{ kg}(0) + 0.20\text{ kg}(+1.0\text{ m/s})$.

**Experiment 2.** Cart A (0.40 kg) moving right at 1.0 m/s collides with cart B (0.20 kg), which is stationary. After the collision, both carts move right, cart B at 1.2 m/s, and cart A at 0.4 m/s.

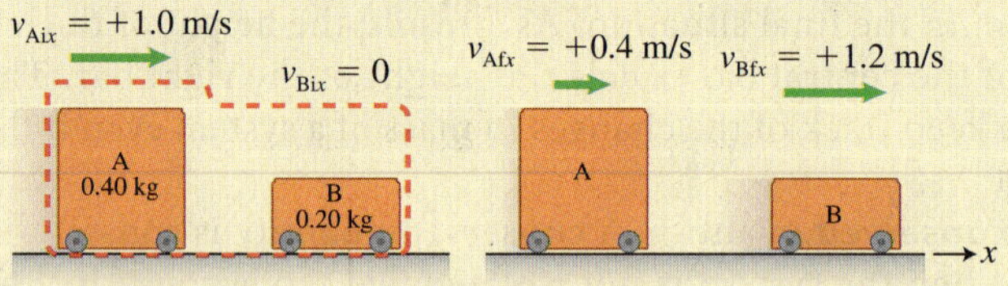


- *Speed:* The sum of the speeds of the system objects is not the same before and after the collision: 1.0 m/s + 0 m/s ≠ 0.4 m/s + 1.2 m/s.
- *Mass · speed:* The sum of the products of mass and speed is the same before and after the collision: $0.40\text{ kg}(1.0\text{ m/s}) + 0.20\text{ kg}(0\text{ m/s}) = 0.40\text{ kg}(0.4\text{ m/s}) + 0.20\text{ kg}(1.2\text{ m/s})$.
- *Mass · velocity:* The sum of the products of mass and the *x*-component of velocity is the same before and after the collision: $0.40\text{ kg}(+1.0\text{ m/s}) + 0.20\text{ kg}(0) = 0.40\text{ kg}(+0.4\text{ m/s}) + 0.20\text{ kg}(+1.2\text{ m/s})$.

**Experiment 3.** Cart A (0.20 kg) with a piece of clay attached to the front moves right at 1.0 m/s. Cart B (0.20 kg) moves left at 1.0 m/s. The carts collide, stick together, and stop.

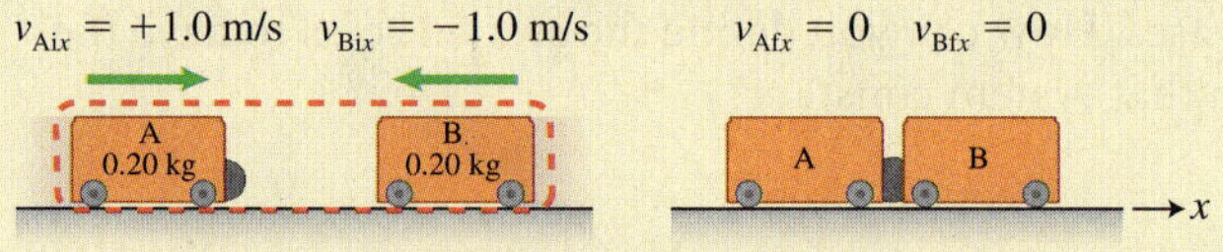


- *Speed:* The sum of the speeds of the system objects is not the same before and after the collision: 1.0 m/s + 1.0 m/s ≠ 0 m/s + 0 m/s.
- *Mass · speed:* The sum of the products of mass and speed is not the same before and after the collision: $0.20\text{ kg}(1.0\text{ m/s}) + 0.20\text{ kg}(1.0\text{ m/s}) \neq 0.20\text{ kg}(0\text{ m/s}) + 0.20\text{ kg}(0\text{ m/s})$.
- *Mass · velocity:* The sum of the products of mass and the *x*-component of velocity is the same before and after the collision: $0.20\text{ kg}(+1.0\text{ m/s}) + 0.20\text{ kg}(-1.0\text{ m/s}) = 0.20\text{ kg}(0\text{ m/s}) + 0.20\text{ kg}(0\text{ m/s})$.

**Patterns**

One quantity remains the same before and after the collision in each experiment—the sum of the products of the mass and *x*-velocity component of the system objects.

In the three experiments in Observational Experiment Table 5.1, only one quantity—the sum of the products of mass and the $x$-component of velocity $\Sigma mv_x$—remained the same before and after the carts collided. Note also that the sum of the products of the mass and the $y$-component of velocity $\Sigma mv_y$ did not change—it remained zero. Perhaps $\Sigma m\vec{v}$ is the quantity characterizing motion that is constant in an isolated system. But will this pattern persist in other situations? Let's test this idea by using it to predict the outcome of the experiment in Testing Experiment **Table 5.2**.

‹ Active Learning Guide

**TESTING EXPERIMENT TABLE**

**5.2 Testing the idea that $\Sigma m\vec{v}$ in an isolated system remains constant (all velocities are with respect to the track).**

VIDEO 5.2

| Testing experiment | Prediction | Outcome |
|---|---|---|
| Cart A (0.40 kg) has a piece of modeling clay attached to its front and is moving right at 1.0 m/s. Cart B (0.20 kg) is moving left at 1.0 m/s. The carts collide and stick together. Predict the velocity of the carts after the collision. | 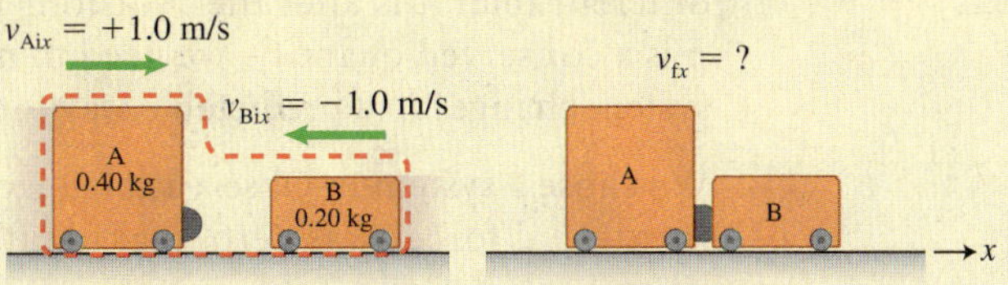 <br>The system consists of the two carts. The direction of velocity is noted with a plus or minus sign of the velocity component:<br>$(0.40\text{ kg})(+1.0\text{ m/s}) + (0.20\text{ kg})(-1.0\text{ m/s}) = (0.40\text{ kg} + 0.20\text{ kg})v_{fx}$<br>or<br>$v_{fx} = (+0.20\text{ kg}\cdot\text{m/s})/(0.60\text{ kg}) = +0.33\text{ m/s}$<br>After the collision, the two carts should move right at a speed of about 0.33 m/s. | After the collision, the carts move together toward the right at close to the predicted speed. |

**Conclusion**

Our prediction matched the outcome. This result gives us increased confidence that this new quantity $m\vec{v}$ might be the quantity whose sum is constant in an isolated system.

This new quantity is called **linear momentum $\vec{p}$.**

**Linear Momentum** The linear momentum $\vec{p}$ of a single object is the product of its mass $m$ and velocity $\vec{v}$:

$$\vec{p} = m\vec{v} \tag{5.1}$$

Linear momentum is a vector quantity that points in the same direction as the object's velocity $\vec{v}$ (**Figure 5.3**). The SI unit of linear momentum is (kg · m/s). The total linear momentum of a system containing multiple objects is the vector sum of the momenta (plural of momentum) of the individual objects.

$$\vec{p}_{\text{net}} = m_1\vec{v}_1 + m_2\vec{v}_2 + \cdots + m_n\vec{v}_n = \Sigma m\vec{v}$$

**Figure 5.3** Momentum is a vector quantity with components.

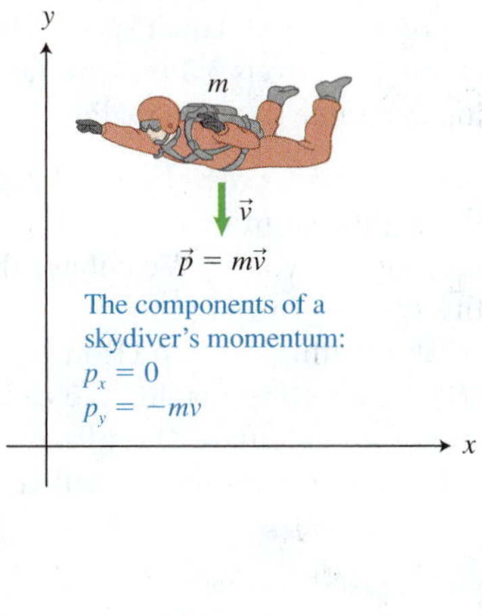

Note the following three important points.

1. Unlike mass, which is a scalar quantity, $\vec{p} = m\vec{v}$ is a vector quantity. Therefore, it is important to consider the direction in which the colliding objects are moving before and after the collision. For example, because cart B in Table 5.2 was moving left along the $x$-axis, the $x$-component of its momentum was negative before the collision.
2. Because momentum depends on the velocity of the object, and the velocity depends on the choice of the reference frame, different observers will measure different momenta for the same object. As a passenger, the momentum of a car with respect to you is zero. However, it is not zero for an observer on the ground watching the car move away from him.
3. We chose an isolated system (the two carts) for our investigation. The sum of the products of mass and velocity $\Sigma m\vec{v}$ of all objects in the isolated system remained constant even though the carts collided with each other. However, if we had chosen the system to be just one of the carts, we would see that the linear momentum $\vec{p} = m\vec{v}$ of the cart before the collision is different than it is after the collision. Thus, to establish that momentum $\vec{p}$ is a conserved quantity, we need to make sure that the momentum of a system changes in a predictable way for systems that are not isolated.

We chose a system in Observational Experiment Table 5.1 so that the sum of the external forces was zero, making it an isolated system. Based on the results of Table 5.1 and Table 5.2, it appears that the total momentum of an isolated system is constant.

**Momentum constancy of an isolated system** **The momentum of an isolated system is constant. For an isolated two-object system:**

$$m_1\vec{v}_{1i} + m_2\vec{v}_{2i} = m_1\vec{v}_{1f} + m_2\vec{v}_{2f} \qquad (5.2)$$

**Because momentum is a vector quantity and Eq. (5.2) is a vector equation, we will work with its $x$- and $y$-component forms:**

$$m_1 v_{1i\,x} + m_2 v_{2i\,x} = m_1 v_{1f\,x} + m_2 v_{2f\,x} \qquad (5.3x)$$

$$m_1 v_{1i\,y} + m_2 v_{2i\,y} = m_1 v_{1f\,y} + m_2 v_{2f\,y} \qquad (5.3y)$$

For a system with more than two objects, we simply include a term on each side of the equation for each object in the system. Let's test the idea that the momentum of an isolated system is constant in another situation.

**EXAMPLE 5.1 Two rollerbladers**

Jen (50 kg) and David (75 kg), both on rollerblades, push off each other abruptly. Each person coasts backward at approximately constant speed. During a certain time interval, Jen travels 3.0 m. How far does David travel during that same time interval?

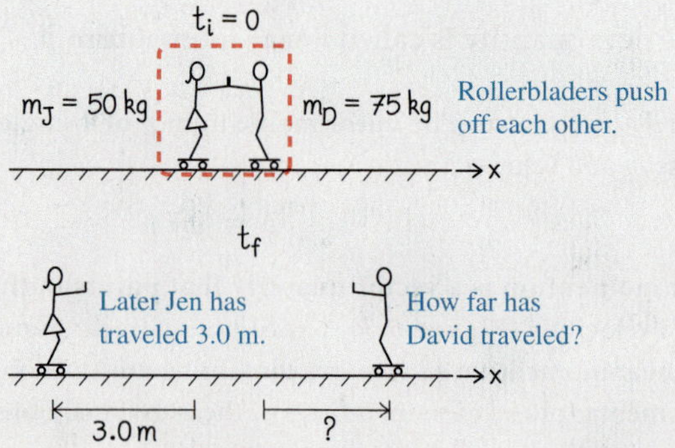


**Sketch and translate** The process is sketched at the right. All motion is with respect to the floor and is along the $x$-axis. We choose the two rollerbladers as the system. Initially, the two rollerbladers are at rest. After pushing off, Jen (J) moves to the left and David (D) moves to the right. We can use momentum constancy to calculate David's velocity component and predict the distance he will travel during that same time interval.

**Simplify and diagram** We model each person as a point-like object and assume that the friction force exerted on the rollerbladers does not affect their motion. Thus there are no horizontal external forces exerted on the system. In addition, the two vertical forces, an upward normal force $\vec{N}_{\text{F on P}}$ that the floor exerts on each person and an equal-magnitude downward gravitational force $\vec{F}_{\text{E on P}}$ that Earth exerts on each person, cancel, as we see in the force diagrams. Since the net external force exerted on the system is zero, the system is isolated. The forces that the rollerbladers exert on each other are internal forces and should not affect the momentum of the system.

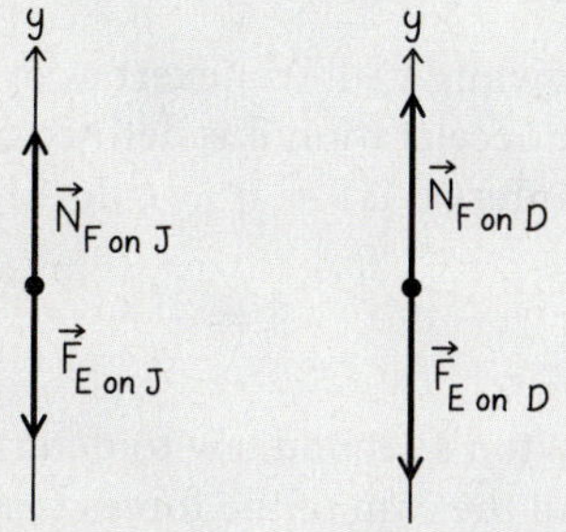


**Represent mathematically** The initial state (i) of the system is before they start pushing on each other, and the final state (f) is when Jen has traveled 3.0 m.

$$m_J v_{Ji\,x} + m_D v_{Di\,x} = m_J v_{Jf\,x} + m_D v_{Df\,x}$$

We choose the positive direction toward the right. Because the initial velocity of each person is zero, the above equation becomes

$$0 + 0 = m_J v_{Jf\,x} + m_D v_{Df\,x}$$

or

$$m_D v_{Df\,x} = -m_J v_{Jf\,x}$$

**Solve and evaluate** The $x$-component of Jen's velocity after the push-off is $v_{Jf\,x} = -(3.0\text{ m})/\Delta t$, where $\Delta t$ is the time interval needed for her to travel 3.0 m. We solve the above equation for David's final $x$-velocity component to determine how far David should travel during that same time interval:

$$v_{Df\,x} = -\frac{m_J v_{Jf\,x}}{m_D} = -\frac{m_J}{m_D} v_{Jf\,x}$$

$$= -\frac{(50\text{ kg})}{(75\text{ kg})}\frac{(-3.0\text{ m})}{\Delta t} = \frac{(2.0\text{ m})}{\Delta t}$$

Since momentum is constant in this isolated system, we predict that David will travel 2.0 m in the positive direction during $\Delta t$. The measured value is very close to the predicted value.

**Try it yourself:** Estimate the magnitude of your momentum when walking and when jogging. Assume your mass is 60 kg.

*Answer:* When walking, you travel at a speed of about 1 to 2 m/s. So the magnitude of your momentum will be $p = mv \approx (60\text{ kg})(1.5\text{ m/s}) \approx 90\text{ kg}\cdot\text{m/s}$. When jogging, your speed is about 2 to 5 m/s or a momentum of magnitude $p = mv \approx (60\text{ kg})(3.5\text{ m/s}) \approx 200\text{ kg}\cdot\text{m/s}$.

Notice that in Example 5.1 we were able to determine David's velocity by using the principle of momentum constancy. We did not need any information about the forces involved. This is a very powerful result, since in all likelihood the forces they exerted on each other were not constant. The kinematics equations we have used up to this point have assumed constant acceleration of the system (and thus constant forces). Using the idea of momentum constancy has allowed us to analyze a situation involving nonconstant forces.

So far, we have investigated situations involving isolated systems. In the next section, we will investigate momentum in nonisolated systems.

**Review Question 5.2** Two identical carts are traveling toward each other at the same speed. One of the carts has a piece of modeling clay on its front. The carts collide, stick together, and stop. The momentum of each cart is now zero. If the system includes both carts, did the momentum of the system disappear? Explain your answer.

## 5.3 Impulse and momentum

‹ Active Learning Guide

So far, we have found that the linear momentum of a system is constant if that system is isolated (the net external force exerted on the system is zero). How do we account for the change in momentum of a system when the net external force exerted on it is not zero? We can use Newton's laws to derive an expression relating forces and momentum change.

## Impulse due to a force exerted on a single object

When you push a bowling ball, you exert a force on it, causing the ball to accelerate. The average acceleration $\vec{a}$ is defined as the change in velocity $\vec{v}_f - \vec{v}_i$ divided by the time interval $\Delta t = t_f - t_i$ during which that change occurs:

$$\vec{a} = \frac{\vec{v}_f - \vec{v}_i}{t_f - t_i}$$

We can also use Newton's second law to determine an object's acceleration if we know its mass and the sum of the forces that other objects exert on it:

$$\vec{a} = \frac{\Sigma \vec{F}}{m}$$

We now have two expressions for an object's acceleration. Setting these two expressions for acceleration equal to each other, we get

$$\frac{\vec{v}_f - \vec{v}_i}{t_f - t_i} = \frac{\Sigma \vec{F}}{m}$$

Now multiply both sides by $m(t_f - t_i)$ and get the following:

$$m\vec{v}_f - m\vec{v}_i = \vec{p}_f - \vec{p}_i = \Sigma \vec{F}(t_f - t_i) \tag{5.4}$$

The left side of the above equation is the change in momentum of the object. This change depends on the product of the net external force and the time interval during which the forces are exerted on the object (the right side of the equation). Note these two important points:

1. Equation (5.4) is just Newton's second law written in a different form—one that involves the physical quantity momentum.
2. Both force *and* time interval affect momentum—the longer the time interval, the greater the momentum change. A small force exerted for a long time interval can change the momentum of an object by the same amount as a large force exerted for a short time interval.

The product of the external force exerted on an object during a time interval and the time interval gives us a new quantity, the **impulse** of the force. When you kick a football or hit a baseball with a bat, your foot or the bat exerts an impulse on the ball. The forces in these situations are not constant but instead vary in time (see the example in **Figure 5.4**). The shaded area under the varying force curve represents the impulse of the force. We can estimate the impulse by drawing a horizontal line that is approximately the average force exerted during the time interval of the impulse. The area under the rectangular average force-impulse curve equals the product of the height of the rectangle (the average force) and the width of the rectangle (the time interval over which the average force is exerted). The product $F_{av}(t_f - t_i)$ equals the magnitude of the impulse.

**Figure 5.4** The impulse of a force is the area under the *F*-versus-*t* graph line.

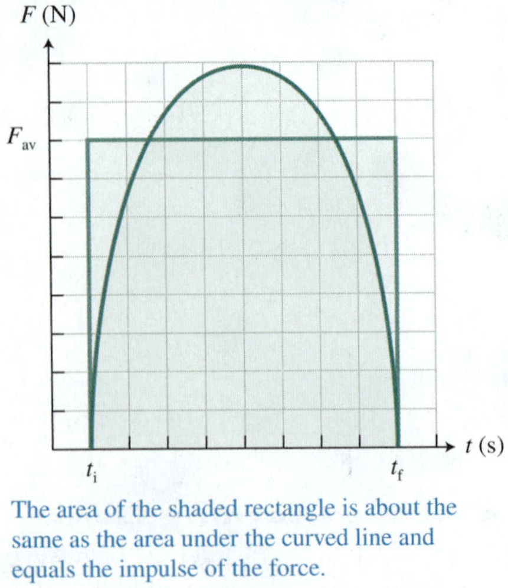
F (N)
$F_{av}$
t (s)
$t_i$
$t_f$
The area of the shaded rectangle is about the same as the area under the curved line and equals the impulse of the force.

**Impulse** **The impulse $\vec{J}$ of a force is the product of the average force $\vec{F}_{av}$ exerted on an object during a time interval $(t_f - t_i)$ and that time interval:**

$$\vec{J} = \vec{F}_{av}(t_f - t_i) \tag{5.5}$$

**Impulse is a vector quantity that points in the direction of the force. The impulse has a plus or minus sign depending on the orientation of the force relative to a coordinate axis. The SI unit for impulse is $N \cdot s = (kg \cdot m/s^2) \cdot s = kg \cdot m/s$, the same unit as momentum.**

It is often difficult to measure directly the impulse of the net average force during a time interval. However, we can determine the net force on the right side of Eq. (5.4) indirectly by measuring or calculating the momentum change on the left side of the equation. For this reason, the combination of impulse and momentum change provides a powerful tool for analyzing interactions between objects. We can now write Eq. (5.4) as the impulse-momentum equation for a single object.

**Impulse-momentum equation for a single object** **If several external objects exert forces on a single-object system during a time interval $(t_f - t_i)$, the sum of their impulses $\Sigma\vec{J}$ causes a change in momentum of the system object:**

$$\vec{p}_f - \vec{p}_i = \Sigma\vec{J} = \Sigma\vec{F}_{\text{on System}}(t_f - t_i) \qquad (5.6)$$

**The $x$- and $y$-scalar component forms of the impulse-momentum equation are**

$$p_{fx} - p_{ix} = \Sigma F_{\text{on System }x}(t_f - t_i) \qquad (5.7x)$$

$$p_{fy} - p_{iy} = \Sigma F_{\text{on System }y}(t_f - t_i) \qquad (5.7y)$$

A few points are worth emphasizing. First, notice that Eq. (5.6) is a vector equation, as both the momentum and the impulse are vector quantities. Vector equations are not easy to manipulate mathematically. Therefore, we will use the scalar component forms of Eq. (5.6)—Eqs. (5.7$x$) and (5.7$y$).

Second, the time interval in the impulse-momentum equation is very important. When object 2 exerts a force on object 1, the momentum of object 1 changes by an amount equal to

$$\Delta\vec{p}_1 = \vec{p}_{1f} - \vec{p}_{1i} = \vec{F}_{2\text{ on }1}(t_f - t_i) = \vec{F}_{2\text{ on }1}\Delta t$$

The longer that object 2 exerts the force on object 1, the greater the momentum change of object 1. This explains why a fast-moving object might have less of an effect on a stationary object during a collision than a slow-moving object interacting with the stationary object over a longer time interval. For example, a fast-moving bullet passing through a partially closed wooden door might not open the door (it will just make a hole in the door), whereas your little finger, moving much slower than the bullet, could open the door. Although the bullet moves at high speed and exerts a large force on the door, the time interval during which it interacts with the door is very small (milliseconds). Hence, it exerts a relatively small impulse on the door—too small to significantly change the door's momentum. A photo of a bullet shot through an apple illustrates the effect of a short impulse time (**Figure 5.5**). The impulse exerted by the bullet on the apple was too small to knock the apple off its support.

**Figure 5.5** The bullet's time of interaction with the apple is very short, causing a small impulse that does not knock the apple over.

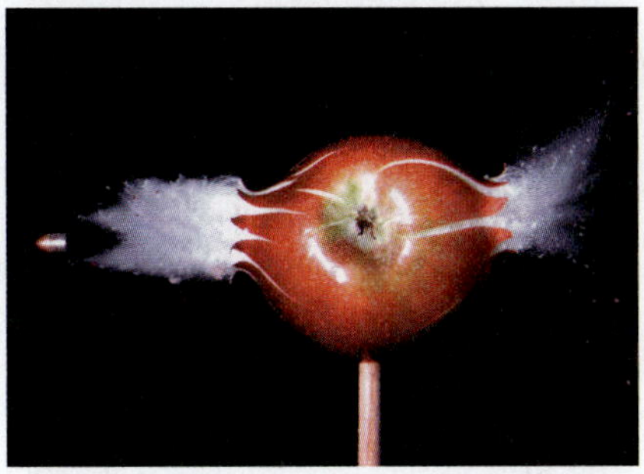

Third, if the magnitude of the force changes during the time interval considered in the process, we use the average force.

Finally, if the same amount of force is exerted for the same time interval on a large-mass object and on a small-mass object, the objects will have an equal change in momentum (the same impulse was exerted on them). However, the small-mass object would experience a greater change in velocity than the large-mass object.

‹ Active Learning Guide

### EXAMPLE 5.2 Abrupt stop in a car

A 60-kg person is traveling in a car that is moving at 16 m/s with respect to the ground when the car hits a barrier. The person is not wearing a seat belt, but is stopped by an air bag in a time interval of 0.20 s. Determine the *average* force that the air bag exerts on the person while stopping him.

**Sketch and translate** First we draw an initial-final sketch of the process. We choose the person as the system since we are investigating a force being exerted on him.

*(continued)*

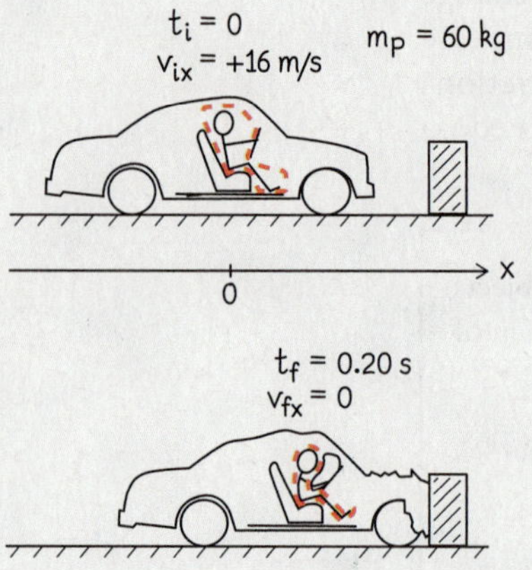


**Simplify and diagram** The force diagram shows the average force $\vec{F}_{\text{B on P}}$ exerted in the negative direction by the bag on the person. The vertical normal force and gravitational forces cancel.

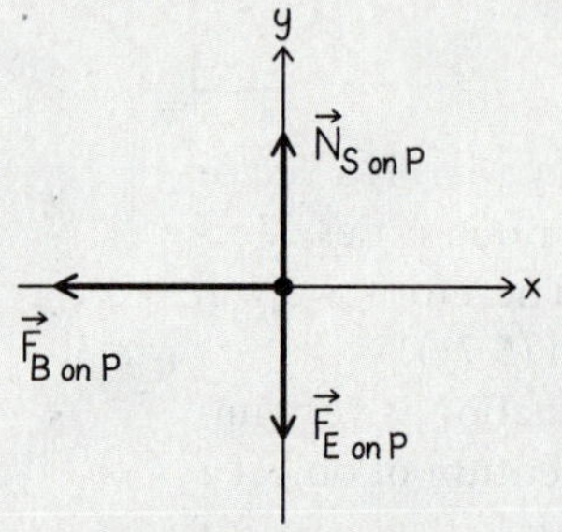


**Represent mathematically** The $x$-component form of the impulse-momentum equation is

$$m_P v_{Pix} + F_{\text{B on P}x}(t_f - t_i) = m_P v_{Pfx}$$

**Solve and evaluate** Solve for the force exerted by the air bag on the person:

$$F_{\text{B on P}x} = \frac{m_P(v_{Pfx} - v_{Pix})}{(t_f - t_i)}$$

The person's initial $x$-component of velocity $v_{Pix} = +16$ m/s decreases to the final $x$-component of velocity $v_{Pfx} = 0$ in a time interval $(t_f - t_i) = 0.20$ s. Thus the average force exerted by the air bag on the person in the $x$-direction is

$$F_{\text{B on P}x} = \frac{(60\text{ kg})(0 - 16\text{ m/s})}{(0.20\text{ s} - 0)} = -4800\text{ N}$$

The negative sign in –4800 N indicates that the average force points in the negative $x$-direction. The magnitude of this force is about 1000 lb!

**Try it yourself:** Suppose a 60-kg crash test dummy is in a car traveling at 16 m/s. The dummy is not wearing a seat belt and the car has no air bags. During a collision, the dummy flies forward and stops when it hits the dashboard. The stopping time interval for the dummy is 0.02 s. What is the average magnitude of the stopping force that the dashboard exerts on the dummy?

*Answer:* The average force that the hard surface exerts on the dummy would be about 50,000 N, extremely unsafe for a human. Note that the momentum change of the person in Example 5.2 was the same. However, since the change for the dummy occurs during a shorter time interval (0.02 s instead of 0.20 s), the force exerted on the dummy is much greater. This is why air bags save lives.

## Using Newton's laws to understand the constancy of momentum in an isolated system of two or more objects

**Figure 5.6** Analyzing the collision of two carts in order to develop the momentum constancy idea.

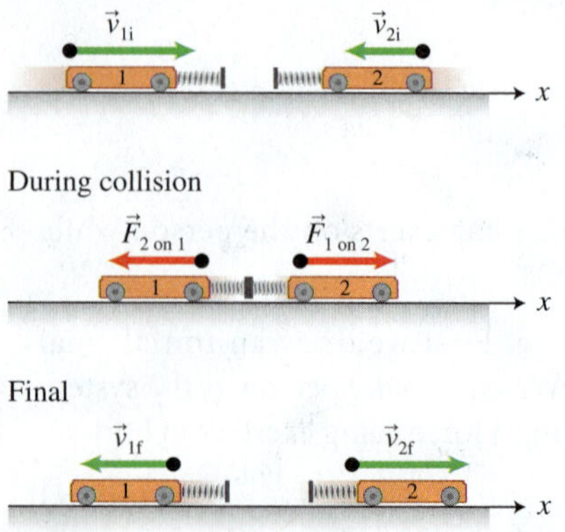


Let's apply the impulse-momentum equation Eq. (5.4) to the scenario we described in Observational Experiment Table 5.1 in order to explore momentum constancy in a two-object isolated system.

Two carts travel toward each other at different speeds, collide, and rebound backward (**Figure 5.6**). We first analyze each cart as a separate system and then analyze them together as a single system. Assume that the vertical forces exerted on the carts are balanced and that the friction force exerted by the surface on the carts does not significantly affect their motion.

*Cart 1:* In the initial state, before the collision, cart 1 with mass $m_1$ travels in the positive direction at velocity $\vec{v}_{1i}$. In the final state, after the collision, cart 1 moves with a different velocity $\vec{v}_{1f}$ in the opposite direction. To determine the effect of the impulse exerted by cart 2 on cart 1, we apply the impulse-momentum equation to cart 1 only:

$$m_1(\vec{v}_{1f} - \vec{v}_{1i}) = \vec{F}_{2\text{ on }1}(t_f - t_i)$$

*Cart 2:* We repeat this analysis with cart 2 as the system. Its velocity and momentum change because of the impulse exerted on it by cart 1:

$$m_2(\vec{v}_{2f} - \vec{v}_{2i}) = \vec{F}_{1 \text{ on } 2}(t_f - t_i)$$

Newton's third law provides a connection between our analyses of the two carts; interacting objects at each instant exert equal-magnitude but oppositely directed forces on each other:

$$\vec{F}_{1 \text{ on } 2} = -\vec{F}_{2 \text{ on } 1}$$

Substituting the expressions for the forces from above and simplifying, we get

$$\frac{m_2(\vec{v}_{2f} - \vec{v}_{2i})}{t_f - t_i} = -\frac{m_1(\vec{v}_{1f} - \vec{v}_{1i})}{t_f - t_i}$$

$$m_2(\vec{v}_{2f} - \vec{v}_{2i}) = -m_1(\vec{v}_{1f} - \vec{v}_{1i})$$

We now move the initial momentum for both objects to the left side and the final momentum for both objects to the right side:

$$\underbrace{m_1\vec{v}_{1i} + m_2\vec{v}_{2i}}_{\text{Initial momentum}} = \underbrace{m_1\vec{v}_{1f} + m_2\vec{v}_{2f}}_{\text{Final momentum}}$$

This is the same equation we arrived at in Section 5.2, where we observed and analyzed collisions to understand the constant momentum of an isolated system. Here we have reached the same conclusions using only our knowledge of Newton's laws, momentum, and impulse.

**Review Question 5.3** An apple is falling from a tree. Why does its momentum change? Specify the external force responsible. Find a system in which the momentum is constant during this process.

## 5.4 The generalized impulse-momentum principle

We can summarize what we have learned about momentum in isolated and non-isolated systems. The change in momentum of a system is equal to the net external impulse exerted on it. If the net impulse is zero, then the momentum of the system is constant. This idea, expressed mathematically as the **generalized impulse-momentum principle,** accounts for situations in which the system includes one or more objects and may or may not be isolated. The generalized impulse-momentum principle means that we can treat momentum as a conserved quantity.

**Generalized impulse-momentum principle** **For a system containing one or more objects, the initial momentum of the system plus the sum of the impulses that external objects exert on the system objects during the time interval $(t_f - t_i)$ equals the final momentum of the system:**

$$\underbrace{(m_1\vec{v}_{1i} + m_2\vec{v}_{2i} + \cdots)}_{\text{Initial momentum of the system}} + \underbrace{\Sigma\vec{F}_{\text{on Sys}}(t_f - t_i)}_{\text{Net impulse exerted on the system}} = \underbrace{(m_1\vec{v}_{1f} + m_2\vec{v}_{2f} + \cdots)}_{\text{Final momentum of the system}} \quad (5.8)$$

**The $x$- and $y$-component forms of the generalized impulse-momentum principle are**

$$(m_1 v_{1ix} + m_2 v_{2ix} + \cdots) + \Sigma F_{\text{on Sys}\,x}(t_f - t_i) = (m_1 v_{1fx} + m_2 v_{2fx} + \cdots) \quad (5.9x)$$

$$(m_1 v_{1iy} + m_2 v_{2iy} + \cdots) + \Sigma F_{\text{on Sys}\,y}(t_f - t_i) = (m_1 v_{1fy} + m_2 v_{2fy} + \cdots) \quad (5.9y)$$

***Note:*** **If the net impulse exerted in a particular direction is zero, then the component of the momentum of the system in that direction is constant.**

Equations (5.8) and (5.9) are useful in two ways. First, any time we choose to analyze a situation using the ideas of impulse and momentum, we can start from a single principle, regardless of the situation. Second, the equations remind us that we need to consider all the interactions between the environment and the system that might cause a change in the momentum of the system.

## Impulse-momentum bar charts

We can describe an impulse-momentum process mathematically using Eqs. (5.9*x* and *y*). These equations help us see that we can represent the changes of a system's momentum using a bar chart similar to the one used to represent the changes of a system's mass. The Reasoning Skill box shows the steps for constructing an **impulse-momentum bar chart** for a simple system of two carts of equal mass traveling toward each other.

**REASONING SKILL** Constructing a qualitative impulse-momentum bar chart.

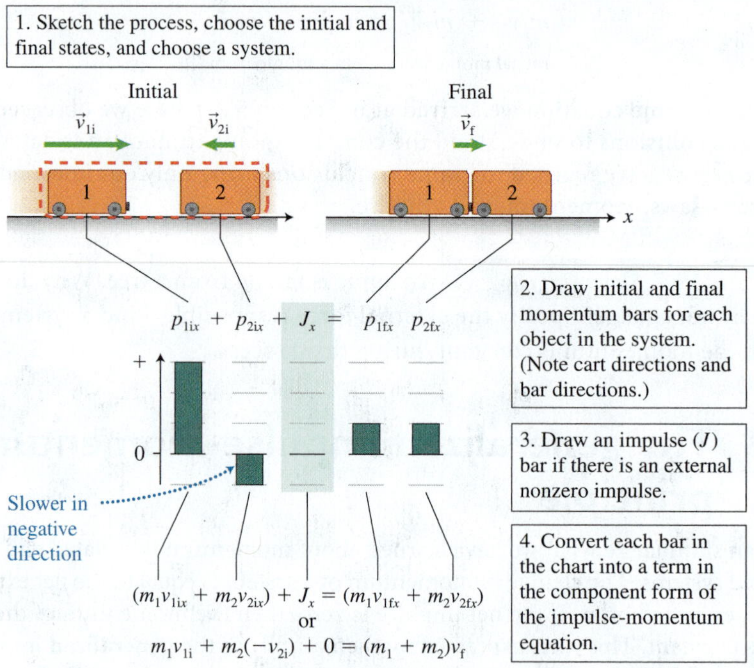
1. Sketch the process, choose the initial and final states, and choose a system.
Initial
Final
$\vec{v}_{1i}$
$\vec{v}_{2i}$
$\vec{v}_f$
1
2
1
2
x
$p_{1ix} + p_{2ix} + J_x = p_{1fx}\ p_{2fx}$
+
0
Slower in negative direction
2. Draw initial and final momentum bars for each object in the system. (Note cart directions and bar directions.)
3. Draw an impulse (J) bar if there is an external nonzero impulse.
4. Convert each bar in the chart into a term in the component form of the impulse-momentum equation.
$(m_1v_{1ix} + m_2v_{2ix}) + J_x = (m_1v_{1fx} + m_2v_{2fx})$
or
$m_1v_{1i} + m_2(-v_{2i}) + 0 = (m_1 + m_2)v_f$

Note that before constructing the bar chart, we represent the process in an initial-final sketch (Step 1 in the Skill box). We then use the sketch to help construct the impulse-momentum bar chart. The lengths of the bars are *qualitative* indicators of the relative magnitudes of the momenta. In the final state in the example shown, the carts are stuck together and are moving in the positive direction. Since they have the same mass and velocity, they each have the same final momentum.

The middle shaded column in the bar chart represents the net external impulse exerted on the system objects during the time interval $(t_f - t_i)$—there is no impulse for the process shown. The shading reminds us that impulse does not reside in the system; it is the influence of the external objects on the momentum of the system. Notice that the sum of the heights of the bars on the left plus the height of the shaded impulse bar should equal the sum of the heights of the bars on the right. This "conservation of bar heights" reflects the conservation of momentum.

We can use the bar chart to apply the generalized impulse-momentum equation (Step 4). Each nonzero bar corresponds to a nonzero term in the equation; the sign of the term depends on the orientation of the bar.

## Using impulse-momentum to investigate forces

Can we use the ideas of impulse and momentum to learn something about the forces that two objects exert on each other during a collision? Consider a collision between two cars (**Figure 5.7**).

To analyze the force that each car exerts on the other, we will define the system to include only one of the cars. Let's choose car 1 and construct a bar chart for it. Car 2 exerts an impulse on car 1 during the collision that changes the momentum of car 1. If the initial momentum of car 1 is in the positive direction, then the impulse exerted by car 2 on car 1 points in the negative direction. Because of this, the impulse bar on the bar chart points downward. Note that the total height of the initial momentum bar on the left side of the chart and the height of the impulse bar add up to the total height of the final momentum bar on the right side. Using the bar chart, we can apply the component form of the impulse-momentum equation:

$$m_1 v_{1ix} + J_x = m_1 v_{1fx}$$

The components of the initial and final momentum are positive. As the force is exerted in the negative direction, the $x$-component of the impulse is negative and equal to $-F_{2\text{ on }1}\Delta t$. Thus,

$$+m_1 v_{1i} + (-F_{2\text{ on }1}\Delta t) = +m_1 v_{1f}$$

If we know the initial and final momentum of the car and the time interval of interaction, we can use this equation to determine the magnitude of the average force that car 2 exerted on car 1 during the collision.

**Figure 5.7** A bar chart analysis of the collision of car 2 with car 1.

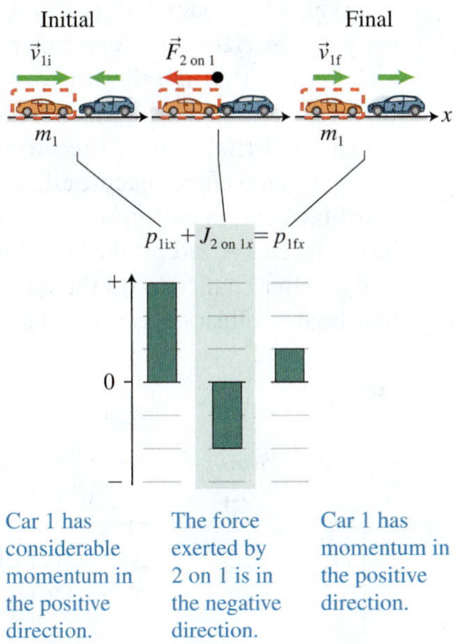


**‹ Active Learning Guide**

> **TIP** When you draw a bar chart, always specify the reference frame (the object of reference and the coordinate system). The direction of the bars on the bar chart (up for positive and down for negative) should match the direction of the momentum or impulse based on the chosen coordinate system.

**EXAMPLE 5.3 Happy and sad balls**

You have two balls of identical mass and size that behave very differently. When you drop the so-called "sad" ball, it thuds on the floor and does not bounce at all. When you drop the so-called "happy" ball from the same height, it bounces back to almost the same height from which it was dropped. The difference in the bouncing ability of the happy ball is due its internal structure; it is made of different material. You hang each ball from a string of identical length and place a wood board on its end directly below the support for each string. You pull each ball back to an equal height and release the balls one at a time. When each ball hits the board, which has the best chance of knocking the board over: the sad ball or the happy ball?

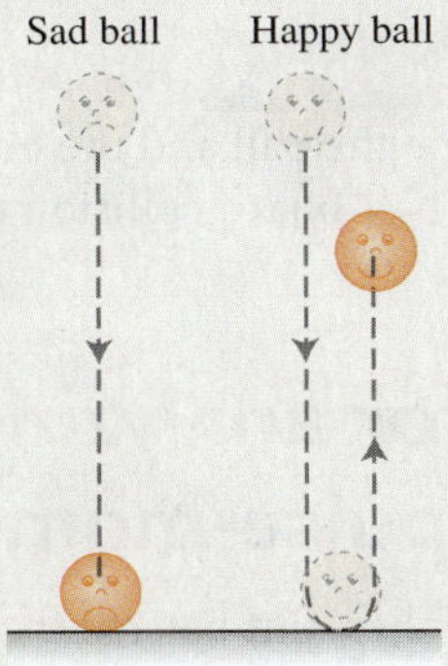


**Sketch and translate** Initial and final sketches of the process are shown at the right. The system is just the ball. In the initial state, the ball is just about to hit the board, moving horizontally toward the left (the balls are moving equally fast). The final state is just after the collision with the board. The happy ball (H) bounces back, whereas the sad ball (S) does not.

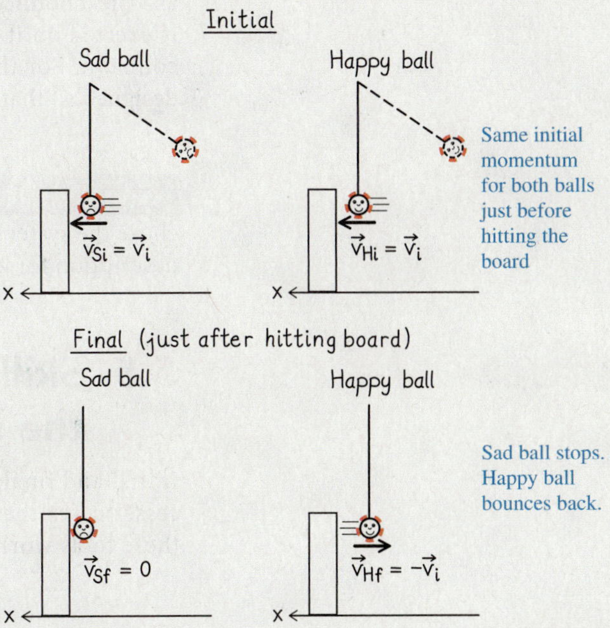


*(continued)*

**Simplify and diagram** Assume that the collision time interval $\Delta t$ for each ball is about the same. We analyze only the horizontal $x$-component of the process, the component that is relevant to whether or not each of the boards is knocked over. Each board exerts an impulse on the ball that causes the momentum of the ball to change. Therefore, each ball, according to Newton's third law, exerts an impulse on the board that it hits. A larger force exerted on the board means a larger impulse and a better chance to tip the board. A bar chart for each ball-board collision is shown below.

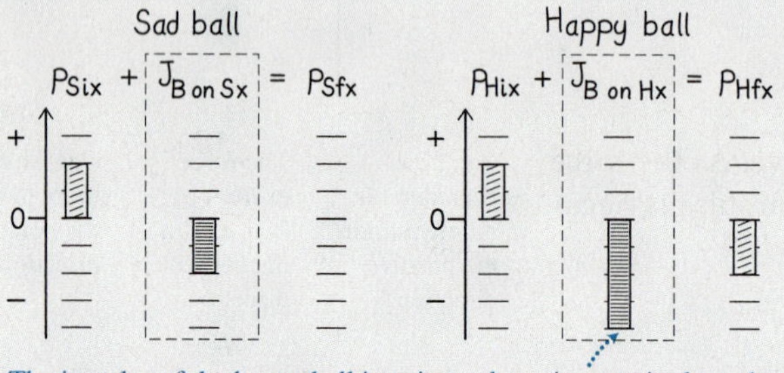


The impulse of the happy ball is twice as large in magnitude as that of the sad ball and causes twice as large a momentum change.

**Represent mathematically** The $x$-component form of the impulse-momentum Eq. (5.5$x$) applied to each ball is as follows:

*Sad ball:* $mv_i + F_{B\,on\,S\,x}\Delta t = m \cdot 0$

*Happy ball:* $mv_i + F_{B\,on\,H\,x}\Delta t = m(-v_i)$

Note that the $x$-component of the final velocity of the sad ball is $v_{Sfx} = 0$ (it does not bounce) and that the $x$-component of the final velocity of the happy ball is $v_{Hfx} = -v_i$ (it bounces).

**Solve and evaluate** We can now get an expression for the force exerted by each board on each ball:

*Sad ball:* $$F_{B\,on\,S\,x} = \frac{m(0 - v_i)}{\Delta t} = -\frac{mv_i}{\Delta t}.$$

*Happy ball:* $$F_{B\,on\,H\,x} = \frac{m[(-v_i) - v_i]}{\Delta t} = -\frac{2mv_i}{\Delta t}.$$

Because we assumed that the time of collision is the same, the board exerts twice the force on the happy ball as on the sad ball, since the board causes the happy ball's momentum to change by an amount twice that of the sad ball. According to Newton's third law, this means that the happy ball will exert twice as large a force on the board as the sad ball. Thus, the happy ball has a greater chance of tipping the board.

**Try it yourself:** Is it less safe for a football player to bounce backward off a goal post or to hit the goal post and stop?

*Answer:* Although any collision is dangerous, it is better to hit the goal post and stop. If the football player bounces back off the goal post, his momentum will have changed by a greater amount (like the happy ball in the last example). This means that the goal post exerts a greater force on him, which means there is a greater chance for injury.

The pattern we found in the example above is true for all collisions—when an object bounces back after a collision, we know that a larger magnitude force is exerted on it than if the object had stopped and did not bounce after the collision. For that reason, bulletproof vests for law enforcement agents are designed so that the bullet embeds in the vest rather than bouncing off it.

**Review Question 5.4** If in solving the problem in Example 5.3 we chose the system to be the ball and the board, how would the mathematical description for each ball-board collision change?

## 5.5 Skills for analyzing problems using the impulse-momentum equation

Initial and final sketches and bar charts are useful tools to help analyze processes using the impulse-momentum principle. Let's investigate further how these tools work together. A general strategy for analyzing such processes is described on the left side of the table in Example 5.4 and illustrated on the right side for a specific process.

## PROBLEM-SOLVING STRATEGY Applying the impulse-momentum equation

**Active Learning Guide>**

### EXAMPLE 5.4 Bullet hits wood block

A 0.020-kg bullet traveling horizontally at 250 m/s embeds in a 1.0-kg block of wood resting on a table. Determine the speed of the bullet and wood block together immediately after the bullet embeds in the block.

**Sketch and translate**

- Sketch the initial and final states and include appropriate coordinate axes. Label the sketches with the known information. Decide on the object of reference.
- Choose a system based on the quantity you are interested in; for example, a multi-object isolated system to determine the velocity of an object, or a single-object nonisolated system to determine an impulse or force.

The left side of the sketch below shows the bullet traveling in the positive $x$-direction with respect to the ground; it then joins the wood. All motion is along the $x$-axis; the object of reference is Earth. The system includes the bullet and wood; it is an isolated system since the vertical forces balance. The initial state is immediately before the collision; the final state is immediately after.

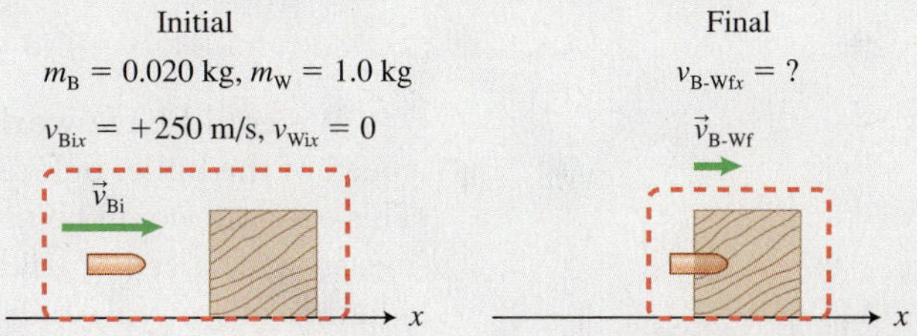


**Simplify and diagram**

- Determine if there are any external impulses exerted on the system. Drawing a force diagram could help determine the external forces and their directions.
- Draw an impulse-momentum bar chart for the system for the chosen direction(s) to help you understand the situation, formulate a mathematical representation of the process, and evaluate your results.

Assume that the friction force exerted by the tabletop on the bottom of the wood does not change the momentum of the system during the very short collision time interval. The bar chart represents the process. The bar for the bullet is shorter than that for the block—their velocities are the same after the collision, but the mass of the bullet is much smaller. We do not draw a force diagram here, as the system is isolated.

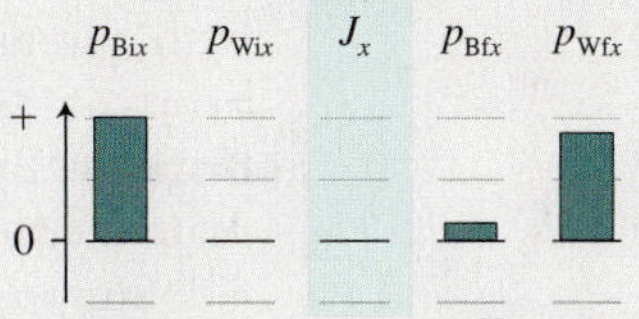


**Represent mathematically**

- Use the bar chart to apply the generalized impulse-momentum equation along the chosen axis. Each nonzero bar becomes a nonzero term in the equation. The orientation of the bar determines the sign in front of the corresponding term in the equation.
- Remember that momentum and impulse are vector quantities, so include the plus or minus signs of the components based on the chosen coordinate system.

$$m_B v_{Bix} + m_W \cdot 0 + (J_x) = m_B v_{B\text{-}Wfx} + m_W v_{B\text{-}Wfx}$$

$$\text{Since } J_x = 0, \qquad v_{B\text{-}Wfx} = \frac{m_B v_{Bix}}{(m_B + m_W)}$$

*(continued)*

**Solve and evaluate**

- Insert the known information to determine the unknown quantity.

$$v_{\text{B-W}fx} = \frac{(0.020\ \text{kg})(250\ \text{m/s})}{(0.020\ \text{kg} + 1.0\ \text{kg})} = +4.9\ \text{m/s}$$

- Check if your answer is reasonable with respect to sign, unit, and magnitude. Also make sure it applies for limiting cases, such as objects of very small or very large mass.

The magnitude of the answer seems reasonable given how fast the bullet was initially traveling. The plus sign indicates the direction, which makes sense, too. The units are also correct (m/s). We can test this using a limiting case: if the mass or speed of the bullet is zero, the block remains stationary after the collision.

**Try it yourself:** A 0.020-kg bullet is fired horizontally into a 2.00-kg block of wood resting on a table. Immediately after the bullet joins the block, the block and bullet move in the positive $x$-direction at 4.0 m/s. What was the initial speed of the bullet?

*Answer:* 400 m/s.

We could have worked Example 5.4 backward to determine the initial speed of the bullet before hitting the block (like the Try It Yourself question). This exercise would be useful since the bullet travels so fast that it is difficult to measure its speed. Variations of this method are used, for example, to decide whether or not golf balls conform to the necessary rules. The balls are hit by the same mechanical launching impulse and the moving balls embed in another object. The balls' speeds are determined by measuring the speed of the object they embed in.

## Determining the stopping time interval from the stopping distance

When a system object collides with another object and stops—a car collides with a tree or a wall, a person jumps and lands on a solid surface, or a meteorite collides with Earth—the system object travels what is called its **stopping distance**. By estimating the stopping distance of the system object, we can estimate the stopping time interval.

Suppose that a car runs into a large tree and its front end crumples about 0.5 m. This 0.5 m, the distance that the center of the car traveled from the beginning of the impact to the end, is the car's stopping distance. Similarly, the depth of the hole left by a meteorite provides a rough estimate of its stopping distance when it collided with Earth. However, to use the impulse-momentum principle, we need the stopping time interval associated with the collision, not the stopping distance. Here's how we can use a known stopping distance to estimate the stopping time interval.

- Assume that the acceleration of the object while stopping is constant. In that case, the average velocity of the object while stopping is just the sum of the initial and final velocities divided by 2: $v_{\text{average}\,x} = (v_{fx} + v_{ix})/2$.
- Thus, the stopping displacement $(x_f - x_i)$ and the stopping time interval $(t_f - t_i)$ are related by the kinematics equation

$$x_f - x_i = v_{\text{average}\,x}(t_f - t_i) = \frac{(v_{fx} + v_{ix})}{2}(t_f - t_i)$$

- Rearrange this equation to determine the stopping time interval:

$$t_f - t_i = \frac{2(x_f - x_i)}{v_{fx} + v_{ix}} \tag{5.10}$$

Equation (5.10) provides a method to convert stopping distance $x_f - x_i$ into stopping time interval $t_f - t_i$. Equation (5.10) can be applied to horizontal or vertical stopping.

**< Active Learning Guide**

**EXAMPLE 5.5 Stopping the fall of a movie stunt diver**

The record for the highest movie stunt fall without a parachute is 71 m (230 ft), held by 80-kg A. J. Bakunas. His fall was stopped by a large air cushion, into which he sank about 4.0 m. His speed was about 36 m/s (80 mi/h) when he reached the top of the air cushion. Estimate the average force that the cushion exerted on his body while stopping him.

**Sketch and translate** We focus only on the part of the fall when Bakunas is sinking into the cushion. The situation is sketched below. We choose Bakunas as the system and the $y$-axis pointing up. The initial state is just as he touches the cushion at position $y_i = +4.0$ m, and the final state is when the cushion has stopped him, at position $y_f = 0$. All motion is with respect to Earth. The other information about the process is given in the figure. Be sure to pay attention to the signs of the quantities (especially the initial velocity).

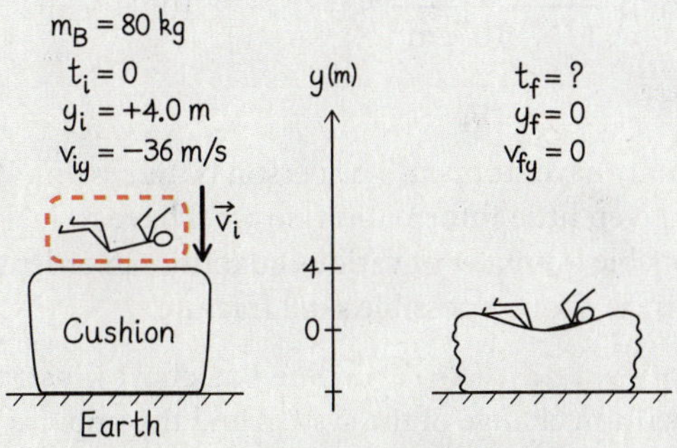


**Simplify and diagram** We draw a force diagram top right, modeling Bakunas as a point-like object. Since Bakunas's downward speed decreases, the cushion must be exerting an upward force on Bakunas of greater magnitude than the downward force that Earth exerts on him. Thus, the net force exerted on him points upward, in the positive $y$-direction. Using this information, we can draw a qualitative impulse-momentum bar chart for the process.

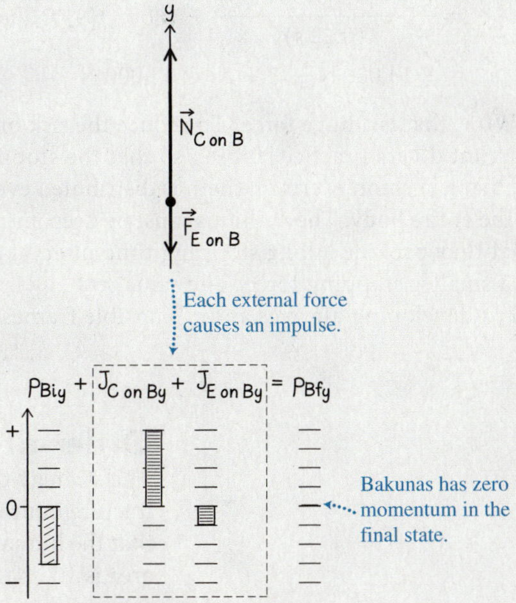


**Represent mathematically** Since all motion and all of the forces are in the vertical direction, we use the bar chart to help construct the vertical $y$-component form of the impulse-momentum equation [Eq. (5.6$y$)] to determine the force that the cushion exerts on Bakunas as he sinks into it:

$$m_B v_{iy} + (N_{C\,on\,B\,y} + F_{E\,on\,B\,y})(t_f - t_i) = m_B v_{fy}$$

Using the force diagram, we see that the $y$-components of the forces are $N_{C\,on\,B\,y} = +N_{C\,on\,B}$ and $F_{E\,on\,B\,y} = -F_{E\,on\,B} = -m_B g$, where $N_{C\,on\,B}$ is the magnitude of the average normal force that the cushion exerts on Bakunas, the force we are trying to estimate. Noting that $v_{fy} = 0$ and substituting the force components into the above equation, we get

$$m_B v_{iy} + [(+N_{C\,on\,B}) + (-m_B g)](t_f - t_i) = m_B \cdot 0$$

$$\Rightarrow m_B v_{iy} + (N_{C\,on\,B} - m_B g)(t_f - t_i) = 0.$$

We can find the time interval that the cushion takes to stop Bakunas using Eq. (5.10) and noting that $v_{fy} = 0$:

$$t_f - t_i = \frac{2(y_f - y_i)}{0 + v_{iy}}$$

*(continued)*

**Solve and evaluate** The stopping time interval while Bakunas sinks 4.0 m into the cushion is

$$t_f - t_i = \frac{2(0 - 4.0\text{ m})}{0 + (-36\text{ m/s})} = 0.22\text{ s}$$

Solving for $N_{\text{C on B}}$, we get

$$N_{\text{C on B}} = \frac{-m_B v_{iy}}{(t_f - t_i)} + m_B g$$
$$= \frac{-(80\text{ kg})(-36\text{ m/s})}{(0.22\text{ s})} + (80\text{ kg})(9.8\text{ N/kg})$$
$$= +13{,}000\text{ N} + 780\text{ N} = 14{,}000\text{ N}$$

Wow, that is a huge force! To reduce the risk of injury, stunt divers practice landing so that the stopping force that a cushion exerts on them is distributed evenly over the entire body. The cushions must be deep enough so that they provide a long stopping time interval and thus a smaller stopping force. The same strategy is applied to developing air bags and collapsible frames for automobiles to make them safer for passengers during collisions.

Notice four important points. First, we've included only two significant digits since that is how many the data had. Second, it is very easy to make sign mistakes. A good way to avoid these is to draw a sketch that includes a coordinate system and labels showing the values of known physical quantities, including their signs. Third, the impulse due to Earth's gravitational force is small in magnitude compared to the impulse exerted by the air cushion. Lastly, the force exerted by the air cushion would be even greater if the stopping distance and consequently the stopping time interval were shorter.

**Try it yourself:** Suppose that the cushion in the last example stopped Bakunas in 1.0 m instead of 4.0 m. What would be the stopping time interval and the magnitude of the average force of the cushion on Bakunas?

*Answer:* The stopping time interval is 0.056 s, and the average stopping force is approximately 50,000 N.

## Order-of-magnitude estimate—will bone break?

The strategy that we used in the previous example can be used to analyze skull fracture injuries that might lead to concussions. Laboratory experiments indicate that the human skull can fracture if the compressive force exerted on it per unit area is $1.7 \times 10^8\text{ N/m}^2$. The surface area of the skull is much smaller than $1\text{ m}^2$, so we will use square centimeters, a more reasonable unit of area for this discussion. Since $1\text{ m}^2 = 1 \times 10^4\text{ cm}^2$, we convert the compressive force per area to

$$(1.7 \times 10^8\text{ N/m}^2)\left(\frac{1\text{ m}^2}{1 \times 10^4\text{ cm}^2}\right) = 1.7 \times 10^4\text{ N/cm}^2.$$

**EXAMPLE 5.6 Bone fracture estimation**[1]

A bicyclist is watching for traffic from the left while turning toward the right. A street sign hit by an earlier car accident is bent over the side of the road. The cyclist's head hits the pole holding the sign. Is there a significant chance that his skull will fracture?

**Sketch and translate** The process is sketched at the right. The initial state is at the instant that the head initially contacts the pole; the final state is when the head and body have stopped. The person is the system. We have been given little information, so we'll have to make some reasonable estimates of various quantities in order to make a decision about a possible skull fracture.

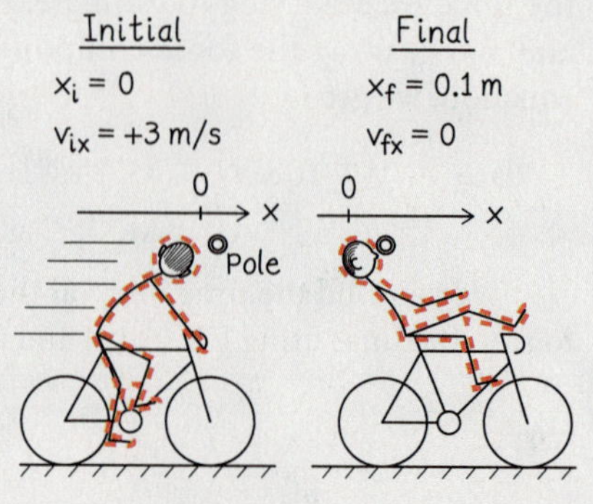


**Simplify and diagram** The bar chart illustrates the momentum change of the system and the impulse exerted by the pole that caused the change. The person was initially moving in the horizontal $x$-direction with respect to Earth, and not moving after the collision. The pole exerted an impulse in the negative $x$-direction on the cyclist. We'll need to estimate the following quantities: the mass and speed of the cyclist in this situation, the stopping time interval, and the area of contact. Let's assume that this is a 70-kg cyclist moving at about 3 m/s. The person's body keeps moving forward for a short distance after the bone makes contact with the pole. The skin indents some during the collision. Because of these two factors, we assume

[1] This is a true story—it happened to one of the book's authors, Alan Van Heuvelen.

a stopping distance of about 10 cm. Finally, we assume an area of contact of about 4 cm$^2$. All of these numbers have large uncertainties and we are not worrying about significant figures, because this is just an estimate.

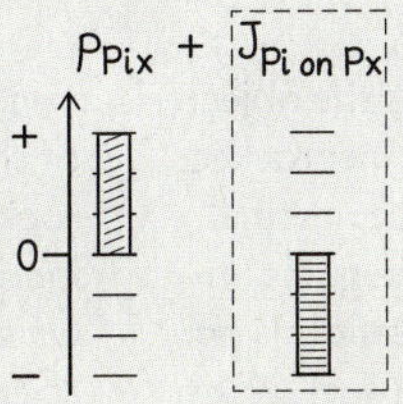
$p_{Pix}$ + $J_{Pi\ on\ Px}$ = $p_{Pfx}$
+
0
–
The pole's negative impulse on the person causes the momentum to decrease.

**Represent mathematically** We now apply the generalized impulse-momentum principle:

$$m_{\text{Person}} v_{\text{Person i}x} + (F_{\text{Pole on Person }x})(t_f - t_i) = m_{\text{Person}} v_{\text{Person f}x} = 0$$

$$\Rightarrow F_{\text{Pole on Person }x} = -\frac{m_{\text{Person}} v_{\text{Person i}x}}{t_f - t_i}$$

We can use the strategy from the last example to estimate the stopping time interval $t_f - t_i$ from the stopping distance $x_f - x_i$:

$$t_f - t_i = \frac{2(x_f - x_i)}{v_{fx} + v_{ix}}$$

where $v_{ix}$ is the initial velocity of the cyclist and $v_{fx} = 0$ is his final velocity.

**Solve and evaluate** Substituting the estimated initial velocity and the stopping distance into the above, we get an estimate for the stopping time interval:

$$t_f - t_i = \frac{2(x_f - x_i)}{v_{fx} + v_{ix}} = \frac{2(0.1\text{ m} - 0)}{0 + 3\text{ m/s}} = 0.067\text{ s}$$

Since this stopping time interval is an intermediate calculated value, we don't need to worry about its number of significant digits. When we complete our estimate, though, we will keep just one significant digit.

We can now insert our estimated values of quantities in the expression for the force exerted by the pole on the person:

$$F_{\text{Pole on Person }x} = -\frac{m_{\text{Person}} v_{\text{Person i}x}}{(t_f - t_i)} = -\frac{(70\text{ kg})(3\text{ m/s})}{(0.067\text{ s})} = -3000\text{ N}$$

Our estimate of the force per area is

$$\frac{\text{Force}}{\text{Area}} = \frac{3000\text{ N}}{4\text{ cm}^2} \approx 800\text{ N/cm}^2$$

Is the person likely to fracture his skull? The force per area needed to break a bone is about $1.7 \times 10^4\text{ N/cm}^2 = 17{,}000\text{ N/cm}^2$. Our estimate could have been off by at least a factor of 10. The force per area is still too little for a fracture.

**Try it yourself:** What would be the magnitude of the force exerted on the cyclist if he bounced back off the pole instead of stopping, assuming the collision time interval remains the same?

*Answer:* 6000 N.

**Review Question 5.5** As the bullet enters the block in Example 5.4, the block exerts a force on the bullet, causing the bullet's speed to decrease to almost zero. Why did we not include the impulse exerted by the block on the bullet in our analysis of this situation?

## 5.6 Jet propulsion

Cars change velocity because of an interaction between the tires and the road. Likewise, a ship's propellers push water backward; in turn, water pushes the ship forward. Once the ship or car is moving, the external force exerted by the water or the road has to balance the opposing friction force or the vehicle's velocity will change.

What does a rocket push against in empty space to change its velocity? Rockets carry fuel that they ignite and then eject at high speed out of the exhaust nozzles (see **Figure 5.8**). Could this burning fuel ejected from the rocket provide the push to change its velocity? Choose the system to be the rocket and fuel together. If the rocket and fuel are at rest before the rocket fires its engines, then its momentum is zero. If there are no external impulses, then even after the rocket fires its engines, the momentum of the rocket-fuel system should still be zero. However, the burning fuel is ejected backward at high velocity from the exhaust nozzle and has a backward momentum. The rocket must now have a nonzero forward velocity. We test this idea quantitatively in Testing Experiment **Table 5.3**.

**Figure 5.8** A rocket as it expels fuel.

Expelled fuel moves left.
Rocket moves right.

**TESTING EXPERIMENT TABLE**

## 5.3 Rocket propulsion.

### Testing experiment

You are traveling through space in a rocket and observe another rocket moving with equal velocity next to you. All of a sudden you notice a burst of burning fuel that is ejected from it. Predict what happens to that rocket's velocity.

### Prediction

Choose the other rocket and its fuel as the system. Your rocket serves as the object of reference and the +*x*-direction is in the direction of its motion. The other rocket has zero velocity in the initial state with respect to the object of reference. Its final state is just after it expels fuel backward at high speed; the rocket in turn gains an equal magnitude of momentum in the forward direction. We can represent this process with an initial-final sketch and a momentum bar chart for the rocket-fuel system.

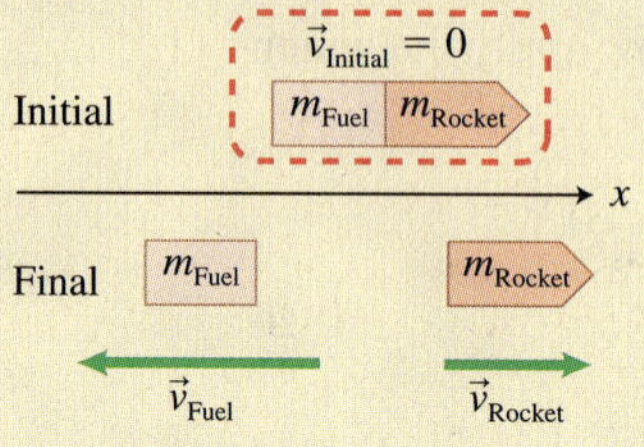


Fuel and rocket system

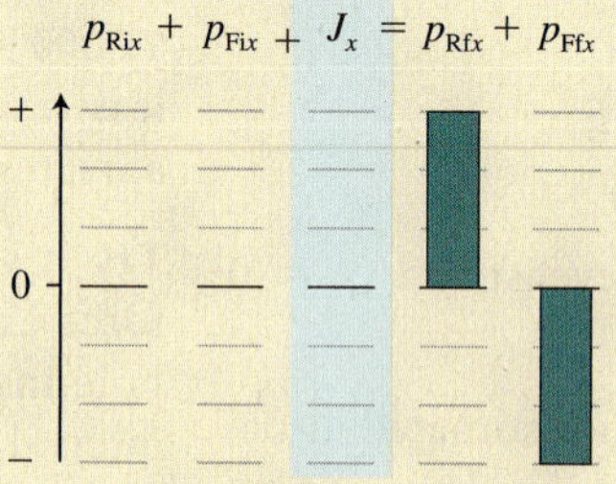


Assuming that fuel is ejected all at once at constant speed, the velocity of the rocket should be

$$0 = m_{Rocket} v_{Rocket\,x} + m_{Fuel} v_{Fuel\,x}$$

$$\Rightarrow v_{Rocket\,x} = -\frac{m_{Fuel} v_{Fuel\,x}}{m_{Rocket}}$$

Here $m_{Rocket}$ is the mass of the rocket without the fuel. We can also choose the rocket alone as the system. The rocket pushes back on the fuel, expelling it backward at high speed ($v_{Fuel\,x} < 0$); the fuel in turn pushes forward against the rocket, exerting an impulse that causes the rocket's momentum and the velocity (assuming the mass of the rocket itself does not change) to increase ($v_{Rocket\,x} > 0$).

Rocket alone as system

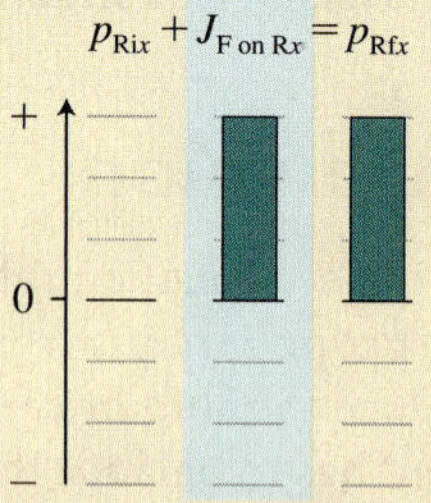


### Outcome

The velocity of the other rocket does increase, and we see it move ahead of our rocket.

### Conclusion

The outcome of the experiment is consistent with the prediction, supporting the generalized impulse-momentum principle. We have learned that, independent of the choice of the system, when a rocket expels fuel in one direction, it gains velocity and therefore momentum in the opposite direction. This mechanism of accelerating a rocket or spaceship is called jet propulsion.

The force exerted by the fuel on a rocket during jet propulsion is called **thrust.** Typical large rocket thrusts measure in mega-newtons ($10^6$ N), and exhaust speeds are more than 10 times the speed of sound. Thrust provides the necessary impulse to change a rocket's momentum. You can observe the principles of jet propulsion using a long, narrow balloon. Blow up the balloon; then open the valve and release it. The balloon will shoot away rapidly in the opposite direction of the air streaming out of the balloon's valve.

In reality, a rocket burns its fuel gradually rather than in one short burst; thus its mass is not a constant number but changes gradually. However, the same methods we used in Testing Experiment Table 5.3, together with some calculus, can be applied to determine the change in the rocket's velocity.

The main idea behind the jet propulsion method is that when an object ejects some of its mass in one direction, it accelerates in the opposite direction. This means that the same method that is used to speed up a rocket can also be used to slow it down. To do this, the fuel needs to be ejected in the same direction that the rocket is traveling.

**TIP** You can become your own jet propulsion machine by standing on rollerblades or a skateboard and throwing a medicine ball or a heavy book forward or backward.

**Review Question 5.6** The following equation is a solution for a problem. State a possible problem.

$$(2.0\ \text{kg})(-8.0\ \text{m/s}) + 0 = (2.0\ \text{kg} + 58\ \text{kg})v_x.$$

# 5.7 Meteorites, radioactive decay, and two-dimensional collisions: Putting it all together

In this section we apply impulse-momentum ideas to analyze meteorites colliding with Earth, radioactive decay of radon in the lungs, and two-dimensional car collisions. We start by analyzing a real meteorite collision with Earth that occurred about 50,000 years ago.

## Canyon Diablo Crater

In this example we use two separate choices of systems to answer different questions about a meteorite collision with Earth.

### EXAMPLE 5.7 Meteorite impact

Arizona's Meteor Crater (also called Canyon Diablo Crater), shown in **Figure 5.9**, was produced 50,000 years ago by the impact of a $3 \times 10^8$-kg meteorite traveling at $1.3 \times 10^4$ m/s (29,000 mi/h). The crater is about 200 m deep. Estimate (a) the change in Earth's velocity as a result of the impact and (b) the average force exerted by the meteorite on Earth during the collision.

**Sketch and translate** A sketch of the process is shown on the next page. To analyze Earth's motion, we choose a coordinate system at rest with respect to Earth. The origin of the coordinate axis is at the point where the meteorite first hits Earth. We keep track of the dot at the bottom of the meteorite. The axis points in the direction of the meteorite's motion.

**Figure 5.9** Canyon Diablo Crater, site of a meteorite impact 50,000 years ago.

To answer the first question, we choose Earth and the meteorite as the system and use momentum constancy to determine Earth's change in velocity due to the

*(continued)*

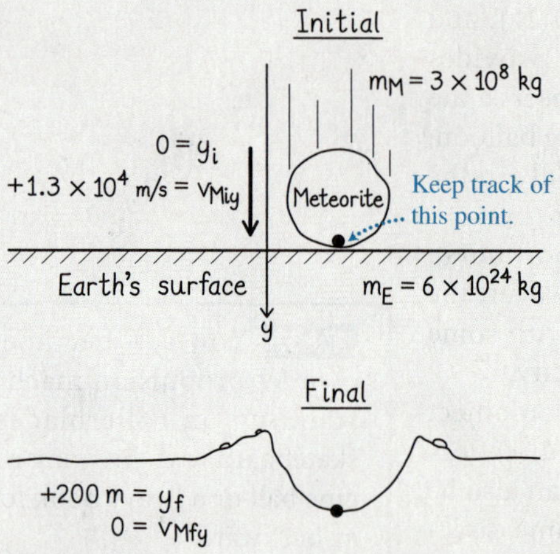


collision. To estimate the average force that the meteorite exerted on Earth during the collision (and that Earth exerted on the meteorite), we choose the meteorite alone as the system and use the impulse-momentum equation to answer that question.

**Simplify and diagram** Assume that the meteorite hits perpendicular to Earth's surface in the positive $y$-direction. The first impulse-momentum bar chart below represents the process for the Earth-meteorite system to answer the first question. The second bar chart represents the meteorite alone as the system during its collision with Earth to answer the second question.

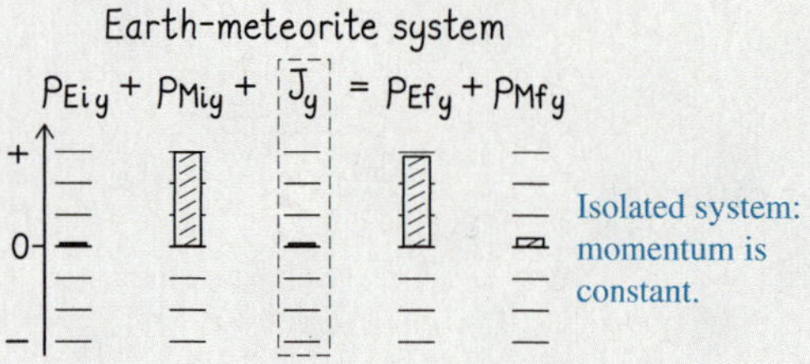


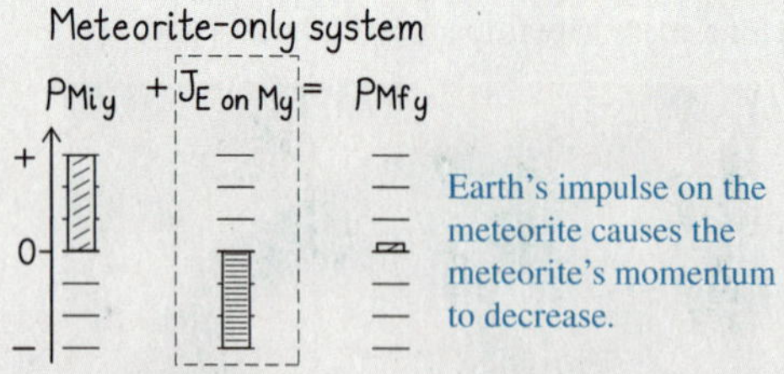


**Represent mathematically** The $y$-component of the meteorite's initial velocity is $v_{\text{Mi}y} = +1.3 \times 10^4\ \text{m/s}$. Earth's initial velocity is zero (with respect to the object of reference). The $y$-component of the meteorite's final velocity equals Earth's since the meteorite embeds in Earth. The meteorite's mass is about $3 \times 10^8$ kg and Earth's mass is $6 \times 10^{24}$ kg. We use momentum constancy to determine the speeds of Earth and the meteorite after they join together:

$$(m_{\text{E}} \cdot 0 + m_{\text{M}} v_{\text{Mi}y}) + [0(t_{\text{f}} - t_{\text{i}})] = (m_{\text{E}} v_{\text{Ef}y} + m_{\text{M}} v_{\text{Mf}y})$$

$$\Rightarrow m_{\text{M}} v_{\text{Mi}y} = m_{\text{E}} v_{\text{Ef}y} + m_{\text{M}} v_{\text{Mf}y} = (m_{\text{E}} + m_{\text{M}}) v_{\text{f}y}$$

To estimate the force that the meteorite exerts on Earth during the collision, we use the $y$-component form of the impulse-momentum equation with the meteorite alone as the system:

$$m_{\text{M}} v_{\text{Mi}y} + F_{\text{E on M}\,y}(t_{\text{f}} - t_{\text{i}}) = m_{\text{M}} v_{\text{Mf}y}$$

The time interval required for the collision [using Eq. (5.9)] is

$$t_{\text{f}} - t_{\text{i}} = \frac{2(y_{\text{f}} - y_{\text{i}})}{v_{\text{Mf}y} + v_{\text{Mi}y}}$$

**Solve and evaluate** To answer the first question, we solve for the final velocity of Earth and meteorite together:

$$v_{\text{f}y} = \frac{m_{\text{M}}}{m_{\text{E}} + m_{\text{M}}} v_{\text{Mi}y}$$

$$= \frac{3 \times 10^8\ \text{kg}}{6 \times 10^{24}\ \text{kg} + 3 \times 10^8\ \text{kg}} (1.3 \times 10^4\ \text{m/s})$$

$$= 7 \times 10^{-13}\ \text{m/s}$$

This is so slow that it would take Earth about 50,000 years to travel just 1 m. Since Earth is so much more massive than the meteorite, the meteorite's impact has extremely little effect on Earth's motion.

For the second question, the time interval for the impact is about

$$t_{\text{f}} - t_{\text{i}} = \frac{2(y_{\text{f}} - y_{\text{i}})}{v_{\text{Mf}y} + v_{\text{Mi}y}}$$

$$= \frac{2(200\ \text{m})}{(1.3 \times 10^4\ \text{m/s} + 7 \times 10^{-13}\ \text{m/s})}$$

$$= 0.031\ \text{s}$$

Like most impulsive collisions, this one was over quickly! Note that we've estimated the displacement of the meteorite to be the depth of the crater. Rearranging the impulse-momentum equation as applied to the collision, we find the average force exerted by Earth on the meteorite:

$$F_{\text{E on M}\,y} = \frac{m_{\text{M}}(v_{\text{Mf}y} - v_{\text{Mi}y})}{(t_{\text{f}} - t_{\text{i}})}$$

$$= \frac{(3 \times 10^8\ \text{kg})(7 \times 10^{-13}\ \text{m/s} - 1.3 \times 10^4\ \text{m/s})}{(0.031\ \text{s})}$$

$$= -1.3 \times 10^{14}\ \text{N} \approx -1 \times 10^{14}\ \text{N}$$

The force exerted by Earth on the meteorite is negative—it points opposite the direction of the meteorite's initial velocity. According to Newton's third law, the force that the meteorite exerts on Earth is positive and has the same magnitude:

$$F_{\text{M on E}\,y} = +1 \times 10^{14}\,\text{N}$$

This sounds like a very large force, but since the mass of Earth is $6 \times 10^{24}$ kg, this force will cause an acceleration of a little over $10^{-11}$ m/s², a very small number.

**Try it yourself:** Estimate the change in Earth's velocity and acceleration if it were hit by a meteorite traveling at the same speed as in the last example, stopping in the same distance, but having mass of $6 \times 10^{19}$ kg instead of $3 \times 10^{8}$ kg.

*Answer:* About 0.1 m/s and 4 m/s².

**TIP** Notice how the choice of system in Example 5.7 was motivated by the question being investigated. Always think about your goal when deciding what your system will be.

## An object breaks into parts (radioactive decay)

We will learn in the chapter on nuclear physics (Chapter 28) that the nuclei of some atoms are unstable and spontaneously break apart. In a process called alpha decay, the nucleus of the atom breaks into a daughter nucleus that is slightly smaller and lighter than the original parent nucleus and an even smaller alpha particle (symbolized by $\alpha$; actually a helium nucleus). For example, radon decays into a polonium nucleus (the daughter) and an alpha particle.

Radon is produced by a series of decay reactions starting with heavy elements in the soil, such as uranium. Radon diffuses out of the soil and can enter a home through cracks in its foundation, where it can be inhaled by people living there. Once in the lungs, the radon undergoes alpha decay, releasing fast-moving alpha particles that may cause mutations that could lead to cancer. In the next example, we will analyze alpha decay by radon by using the idea of momentum constancy.

### EXAMPLE 5.8 Radioactive decay of radon in lungs

An inhaled radioactive radon nucleus resides more or less at rest in a person's lungs, where it decays to a polonium nucleus and an alpha particle. With what speed does the alpha particle move if the polonium nucleus moves away at $4.0 \times 10^{5}$ m/s relative to the lung tissue? The mass of the polonium nucleus is 54 times greater than the mass of the alpha particle.

**Sketch and translate** An initial-final sketch of the situation is shown at the right. We choose the system to be the radon nucleus in the initial state, which converts to the polonium nucleus and the alpha particle in the final state. The coordinate system has a positive $x$-axis pointing in the direction of motion of the alpha particle, with the object of reference being the lung tissue. The initial velocity of the radon nucleus along the $x$-axis is 0 and the final velocity component of the polonium daughter nucleus is $v_{\text{Po f}x} = -4.0 \times 10^{5}$ m/s. The final velocity component of the alpha particle $v_{\alpha fx}$ is unknown. If the mass of the alpha particle is $m$, then the mass of the polonium is $54m$.

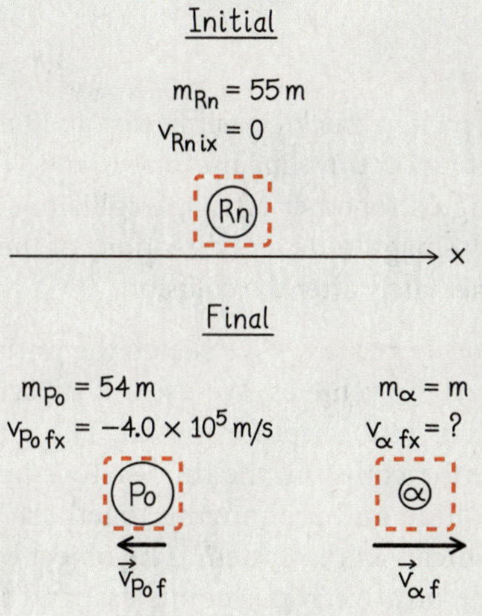


*(continued)*

**Simplify and diagram** Assume that there are no external forces exerted on the system, meaning that the system is isolated and thus its momentum is constant. The impulse-momentum bar chart below represents the process.

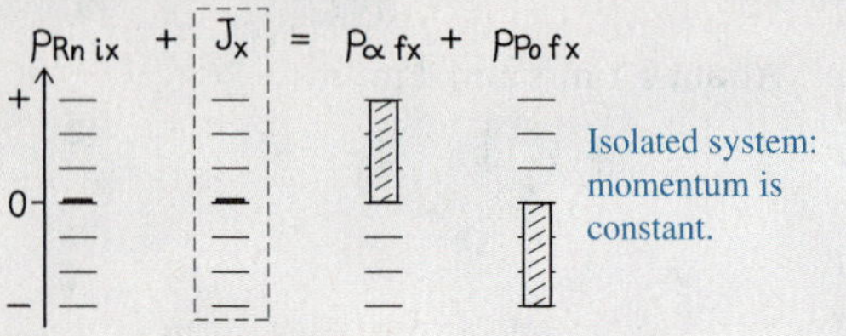


**Represent mathematically** Use the bar chart to help apply the impulse-momentum equation for the process:

$$m_{\text{Rn}}(0) + (0)(t_{\text{f}} - t_{\text{i}}) = m_{\text{Po}}v_{\text{Pof}x} + m_{\alpha}v_{\alpha\text{f}x}$$

$$\Rightarrow 0 = m_{\text{Po}}v_{\text{Pof}x} + m_{\alpha}v_{\alpha\text{f}x}$$

**Solve and evaluate** Rearranging, we get an expression for the final velocity of the alpha particle in the $x$-direction:

$$v_{\alpha\text{f}x} = -\frac{m_{\text{Po}}v_{\text{Pof}x}}{m_{\alpha}}$$

The $x$-component of the velocity of the alpha particle after radon decay is

$$v_{\alpha\text{f}x} = -\frac{m_{\text{Po}}v_{\text{Pof}x}}{m_{\alpha}} = -\frac{(54m_{\alpha})(-4.0 \times 10^5\ \text{m/s})}{m_{\alpha}}$$

$$= +2.2 \times 10^7\ \text{m/s}$$

The sign indicates that the alpha particle is traveling in the positive $x$-direction opposite the direction of the polonium. The magnitude of this velocity is huge—about one-tenth the speed of light! The speeding alpha particle passes through lung tissue and collides with atoms and molecules, dislodging electrons and creating ions. Radon exposure causes approximately 15,000 cases of lung cancer each year.

**Try it yourself:** Francium nuclei undergo radioactive decay by emitting either an alpha particle or a beta particle (an electron). The alpha particle is about 8000 times more massive than a beta particle. If the particles are emitted with the same speed, in which case is the recoil speed of the nucleus that is left after an alpha or a beta particle is emitted greatest?

*Answer:* Since the mass of the alpha particle is much greater than the mass of the beta particle, and they are traveling with the same speed, the momentum of the alpha particle is much greater than the momentum of the beta particle. Therefore, the nucleus that is left would have a greater recoil speed during alpha decay.

## Collisions in two dimensions

So far, the collisions we have investigated have occurred along one axis. Often, a motor vehicle accident involves two vehicles traveling along perpendicular paths. For these two-dimensional collisions, we can still apply the ideas of impulse and momentum, but we will use one impulse-momentum equation for each coordinate axis.

**EXAMPLE 5.9**

A 1600-kg pickup truck traveling east at 20 m/s collides with a 1300-kg car traveling north at 16 m/s. The vehicles remain tangled together after the collision. Determine the velocity (magnitude and direction) of the combined wreck immediately after the collision.

**Sketch and translate** We sketch the initial and final situations of the vehicles. We use a P subscript for the pickup and a C subscript for the car. The initial state is just before the collision; the final state is just after the vehicles collide and are moving together. We choose the two vehicles as the system. The object of reference is Earth; the positive $x$-axis points east and the positive $y$-axis points north.

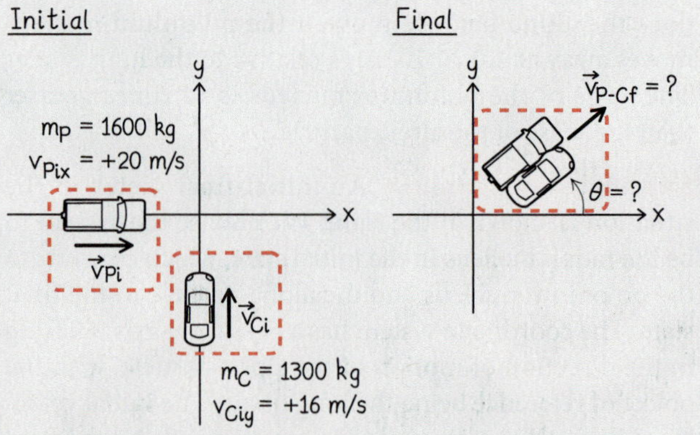


**Simplify and diagram** Force diagrams represent the side view for each vehicle just before the collision. We assume that the friction force exerted by the road is very small compared to the force that each vehicle exerts on the other. Thus, we ignore the impulse due to surface friction during the short collision time interval of about 0.1 s. We then apply momentum constancy in each direction. Impulse-momentum bar charts for the $x$-direction and for the $y$-direction are shown below.

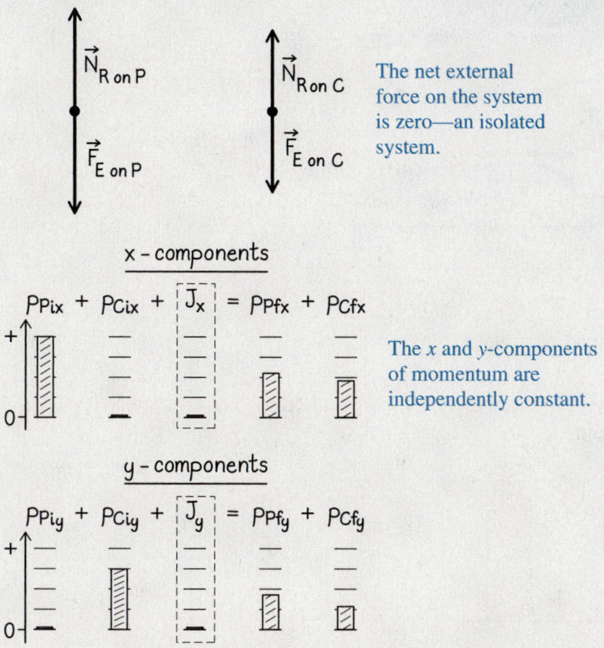


**Represent mathematically** Now, convert each momentum bar in the $x$-component bar chart into a term in the $x$-component form of the impulse-momentum equation [Eq. (5.7$x$)] and each bar in the $y$-component bar chart into a term in the $y$-component form of the impulse-momentum equation [Eq. (5.7$y$)]. Notice that the $x$-component of the final velocity vector is $v_{\text{P-Cf}\,x} = v_{\text{P-Cf}} \cos\theta$ and the $y$-component is $v_{\text{P+Cf}\,x} = v_{\text{P-Cf}} \sin\theta$:

*x-component equation:*

$$m_{\text{P}} v_{\text{Pi}\,x} + m_{\text{C}} v_{\text{Ci}\,x} = (m_{\text{P}} + m_{\text{C}}) v_{\text{P-Cf}} \cos\theta$$

*y-component equation:*

$$m_{\text{P}} v_{\text{Pi}\,y} + m_{\text{C}} v_{\text{Ci}\,y} = (m_{\text{P}} + m_{\text{C}}) v_{\text{P-Cf}} \sin\theta$$

We have two equations and two unknowns ($v_{\text{P-Cf}}$ and $\theta$). We can solve for both unknowns.

**Solve and evaluate**

*x-component equation:*

$$(1600\text{ kg})(20\text{ m/s}) + (1300\text{ kg})(0\text{ m/s}) = (2900\text{ kg}) v_{\text{P-Cf}} \cos\theta$$

*y-component equation:*

$$(1600\text{ kg})(0\text{ m/s}) + (1300\text{ kg})(16\text{ m/s}) = (2900\text{ kg}) v_{\text{P-Cf}} \sin\theta$$

Divide the left side of the second equation by the left side of the first equation and the right side of the second equation by the right of the first, and cancel the 2900 kg and $v_{\text{P+Cf}}$ on the top and bottom of the right side. We get

$$\frac{(1300\text{ kg})(16\text{ m/s})}{(1600\text{ kg})(20\text{ m/s})} = \frac{\sin\theta}{\cos\theta} = \tan\theta = 0.65$$

A 33° angle has a 0.65 tangent. Thus, the vehicles move off at 33° above the $+x$-axis (the north of east direction). We can now use this angle with either the $x$-component equation or the $y$-component equation above to determine the speed of the two vehicles immediately after the collision. Using the $x$-component equation, we get

$$v_{\text{P-Cf}} = \frac{(1600\text{ kg})(20\text{ m/s})}{(2900\text{ kg})\cos 33°} = 13\text{ m/s}$$

From the $y$-component equation, we have

$$v_{\text{P-Cf}} = \frac{(1300\text{ kg})(16\text{ m/s})}{(2900\text{ kg})\sin 33°} = 13\text{ m/s}$$

The two equations give the same result for the final speed, a good consistency check. For collisions in which vehicles lock together like this, police investigators commonly use the lengths of the skid marks along with the direction of the vehicles after the collision to determine their initial speeds. This allows them to decide whether either vehicle was exceeding the speed limit before the collision.

**Try it yourself:** Use a limiting case analysis and the $x$- and $y$-component forms of the impulse-momentum equation to predict what would happen during the collision if the pickup had infinite mass. Is the answer reasonable?

*Answer:* If we place ∞ in

$$\frac{(1300\text{ kg})(16\text{ m/s})}{(1600\text{ kg})(20\text{ m/s})} = \frac{\sin\theta}{\cos\theta} = \tan\theta$$

in place of the 1600-kg mass of the pickup, the left side of the equation becomes zero. Then $\tan\theta = 0$. The pickup would move straight ahead when hitting the car. In other words, the collision with the car would not change the direction of travel of the pickup. The result seems reasonable if the mass of the pickup was large compared to the mass of the car.

**Review Question 5.7** When a meteorite hits Earth, the meteorite's motion apparently disappears completely. How can we claim that momentum is conserved?

# Summary

| Words | Pictorial and physical representations | Mathematical representation |
|---|---|---|
| **Isolated system** An isolated system is one in which the objects interact only with each other and not with the environment, or the sum of external forces exerted on it is zero. (Sections 5.1–5.2) | Isolated system<br><br>Nonisolated system<br> | |
| **Conservation of mass** If the system is isolated, its mass is constant. If the system is not isolated, the change in the system's mass equals the mass delivered to the system or taken away from it. (Section 5.1) | 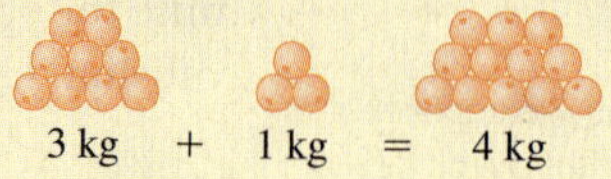<br><br>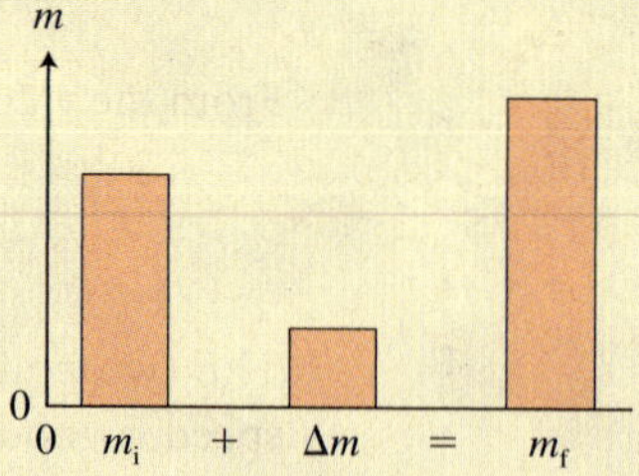<br> | $m_i + \Delta m = m_f$ |
| **Linear momentum** $\vec{p}$ is a vector quantity that is the product of an object's mass $m$ and velocity $\vec{v}$. The total momentum of the system is the sum of the momenta of all objects in the system. (Section 5.2) | 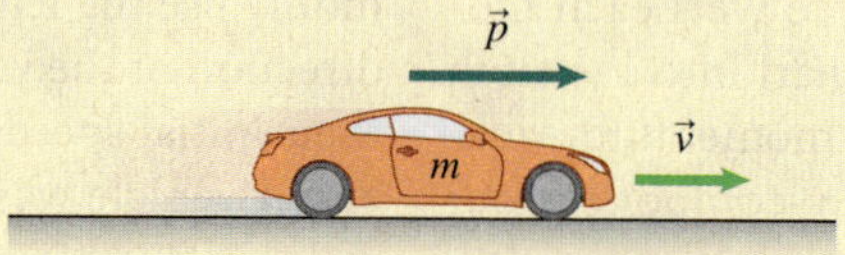<br> | $\vec{p} = m\vec{v}$ Eq. (5.1)<br>$\vec{p}_{system} = \vec{p}_1 + \vec{p}_2 + \cdots$ |
| **Impulse** $\vec{J}$ Impulse is the product of the average external force $\vec{F}_{av}$ exerted on an object during a time interval $\Delta t$ and that time interval. (Section 5.3) |  | $\vec{J} = \vec{F}_{av}(t_f - t_i)$ Eq. (5.5) |

**Generalized impulse-momentum principle** If the system is isolated, its momentum is constant. If the system is not isolated, the change in the system's momentum equals the sum of the impulses exerted on the system during the time interval $\Delta t = (t_f - t_i)$. (Section 5.4)

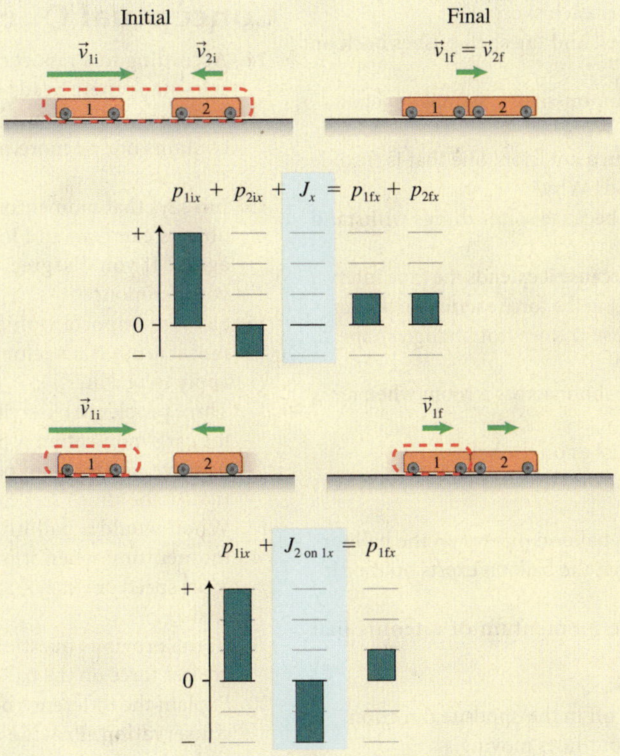

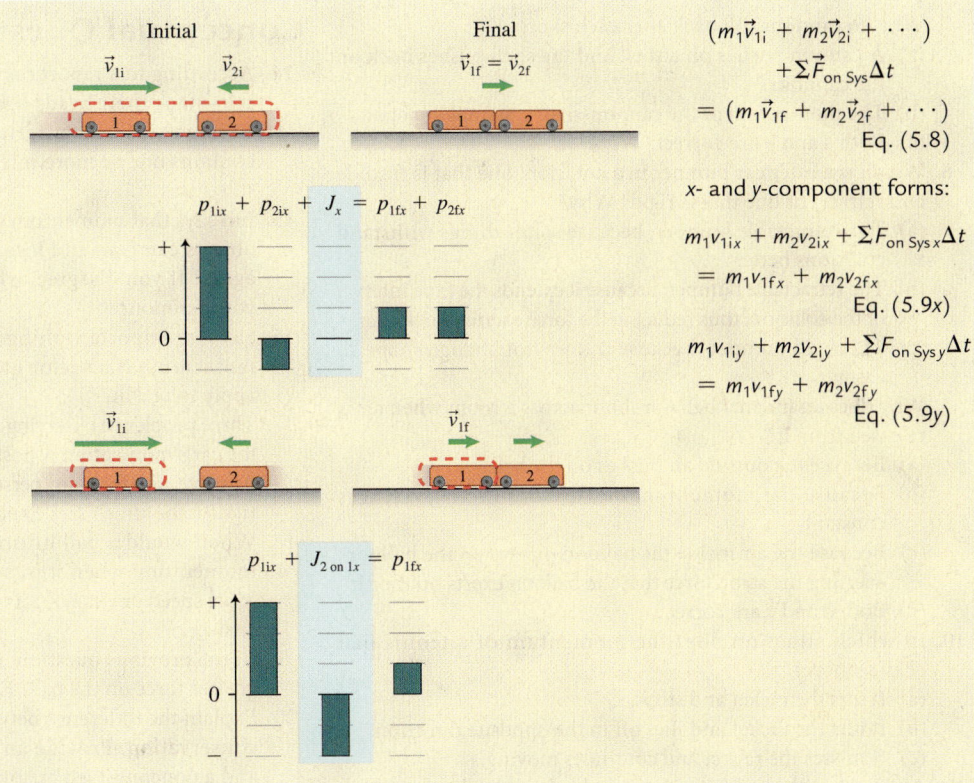

$$(m_1\vec{v}_{1i} + m_2\vec{v}_{2i} + \cdots) + \Sigma\vec{F}_{\text{on Sys}}\Delta t = (m_1\vec{v}_{1f} + m_2\vec{v}_{2f} + \cdots) \quad \text{Eq. (5.8)}$$

$x$- and $y$-component forms:

$$m_1 v_{1ix} + m_2 v_{2ix} + \Sigma F_{\text{on Sys}\,x}\Delta t = m_1 v_{1fx} + m_2 v_{2fx} \quad \text{Eq. (5.9x)}$$

$$m_1 v_{1iy} + m_2 v_{2iy} + \Sigma F_{\text{on Sys}\,y}\Delta t = m_1 v_{1fy} + m_2 v_{2fy} \quad \text{Eq. (5.9y)}$$

MP® **For instructor-assigned homework, go to MasteringPhysics.**

# Questions

## Multiple Choice Questions

1. The gravitational force that Earth exerts on an object causes an impulse of +10 N · s in one experiment and +1 N · s on the same object in another experiment. How can this be?
   (a) The mass of the object changed.
   (b) The time intervals during which the force was exerted are different.
   (c) The magnitudes of the force were different.
2. A bullet fired at a door makes a hole in the door but does not open it. Your finger does not make a hole in the door but does open it. Why?
   (a) The bullet is too small.
   (b) The force exerted by the bullet is not enough to open the door.
   (c) A finger exerts a smaller force but the time interval is much longer.
   (d) The bullet goes through the door and does not exert a force at all.
3. How would you convince somebody that the momentum of an isolated system is constant?
   (a) It is a law; thus, you do not need to convince anybody.
   (b) Use an example from a textbook to show that the sum of the initial and final velocities of the objects involved in a collision are the same.
   (c) Derive it from Newton's second and third laws.
   (d) Use it to make predictions about a new experiment, and then compare the outcome to the prediction.
   (e) Both (c) and (d) will work.
4. A wagon full of medicine balls is rolling along a street. Suddenly one medicine ball (3 kg) falls off the wagon. What happens to the speed of the wagon?
   (a) The wagon slows down.
   (b) The speed of the wagon does not change.
   (c) The wagon speeds up.
   (d) Additional information about the ball's motion is needed to answer.
5. When can you apply the idea that momentum is constant to solve a problem?
   (a) When the system is isolated
   (b) When the system is not isolated but the time interval when the external forces are exerted is very small
   (c) When the external forces are much smaller than the internal forces
6. Choose an example in which the momentum of a system is not constant.
   (a) A bullet shot from a rifle, with the rifle and the bullet as the system
   (b) A freely falling metal ball, with the ball as the system
   (c) A freely falling metal ball, with the ball and Earth as the system
   (d) It is not possible to give an example since the momentum of a system is always constant.

7. Why do cannons roll back after each shot?
   (a) A cannon pushes on a shell and the shell pushes back on the cannon.
   (b) The momentum of the cannon-shell system is constant.
   (c) Both a and b are correct.
8. Which is a safer car bumper in a collision: one that is flexible and retracts or one that is rigid? Why?
   (a) The retractable bumper, because softer things withstand collisions better
   (b) The retractable bumper, because it extends the time interval of the collision, thus reducing the force exerted on the car
   (c) The rigid bumper, because it does not change shape so easily
9. Why does an inflated balloon shoot across a room when air is released from it?
   (a) Because the outside air pushes on the balloon
   (b) Because the momentum of the balloon-air system is constant
   (c) Because the air inside the balloon pushes on the balloon, exerting the same force that the balloon exerts on the air
   (d) Both b and c are correct.
10. In which situation does the momentum of a tennis ball change more?
    (a) It hits the racket and stops.
    (b) It hits the racket and flies off in the opposite direction.
    (c) It misses the racket and continues moving.
11. A toy car with very low friction wheels and axles rests on a level track. In which situation will its speed increase more?
    (a) It is hit from the rear by a wad of clay that sticks to the car.
    (b) It is hit by a rubber ball with the same mass and velocity of the clay that rebounds in the opposite direction after hitting the car.
12. A meteorite strikes Earth and forms a crater, decreasing the meteorite's momentum to zero. Does this phenomenon contradict the conservation of momentum? Choose as many answers as you think are correct.
    (a) No, because the meteorite system is not isolated
    (b) No, because in the meteorite-Earth system, Earth acquires momentum lost by the meteorite
    (c) No, because the meteorite brings momentum from space
    (d) Yes, because the meteorite is not moving relative to a medium before the collision
13. A 1000-kg car traveling east at 24 m/s collides with a 2000-kg car traveling west at 21 m/s. The cars lock together. What is their velocity immediately after the collision?
    (a) 3 m/s east (b) 3 m/s west
    (c) 6 m/s east (d) 6 m/s west
    (e) 15 m/s east

## Conceptual Questions

14. According to a report on traumatic brain injury, woodpeckers smack their heads against trees at a force equivalent to 1200 *g*'s without suffering brain damage. This statement contains one or more mistakes. Identify the mistakes in this statement.
15. Jim says that momentum is not a conserved quantity because objects can gain and lose momentum. Do you agree or disagree? If you disagree, what can you do to convince Jim of your opinion?
16. Say five important things about momentum (for example, momentum is a vector quantity). How does each statement apply to real life?
17. Three people are observing the same car. One person claims that the car's momentum is positive, another person claims that it is negative, and the third person says that it is zero. Can they all be right at the same time? Explain.
18. When would a ball hitting a wall have a greater change in momentum: when it hits the wall and bounces back at the same speed or when it hits and sticks to the wall? Explain your answer.
19. In the previous question, in which case does the wall exert a greater force on the ball? Explain.
20. Explain the difference between the concepts of constancy and conservation. Provide an example of a conserved quantity and a nonconserved quantity.
21. Why do you believe that momentum is a conserved quantity?
22. A heavy bar falls straight down onto the bed of a rolling truck. What happens to the momentum of the truck at the instant the bar lands on it? Explain. How many correct answers do you think are possible? Make sure you think of what "falls straight down" means.
23. Construct impulse-momentum bar charts to represent a falling ball in (a) a system whose momentum is not constant and (b) a system whose momentum is constant. In the initial state, the ball is at rest; in the final state, the ball is moving.
24. A person moving on rollerblades throws a medicine ball in the direction opposite to her motion. Construct an impulse-momentum bar chart for this process. The person is the system.
25. A person moving on rollerblades drops a medicine ball straight down relative to himself. Construct an impulse-momentum bar chart for the system consisting of the ball and Earth for this process. The rollerblader is the object of reference, and the final state is just before the ball hits the ground.

# Problems

Below, **BIO** indicates a problem with a biological or medical focus. Problems labeled **EST** ask you to estimate the answer to a quantitative problem rather than derive a specific answer. Problems marked with require you to make a drawing or graph as part of your solution. Asterisks indicate the level of difficulty of the problem. Problems with no * are considered to be the least difficult. A single * marks intermediate difficult problems. Two ** indicate more difficult problems.

### 5.2 Linear momentum

1. You and a friend are playing tennis. (a) What is the magnitude of the momentum of the 0.057-kg tennis ball when it travels at a speed of 30 m/s? (b) At what speed must your 0.32-kg tennis racket move to have the same magnitude momentum as the ball? (c) If you run toward the ball at a speed of 5.0 m/s, and the ball is flying directly at you at a speed of 30 m/s, what

is the magnitude of the total momentum of the system (you and the ball)? Assume your mass is 60 kg. In every case specify the object of reference.

2. You are hitting a tennis ball against a wall. The 0.057-kg tennis ball traveling at 25 m/s strikes the wall and rebounds at the same speed. (a) Determine the ball's original momentum (magnitude and direction). (b) Determine the ball's change in momentum (magnitude and direction). What is your object of reference?
3. A ball of mass $m$ and speed $v$ travels horizontally, hits a wall, and rebounds. Another ball of the same mass and traveling at the same speed hits the wall and sticks to it. Which ball has a greater change in momentum as a result of the collision? Explain your answer.
4. (a) A 145-g baseball travels at 35 m/s toward a baseball player's bat (the bat is the object of reference) and rebounds in the opposite direction at 40 m/s. Determine the ball's momentum change (magnitude and direction). (b) A golfer hits a 0.046-kg golf ball that launches from the grass at a speed of 50 m/s. Determine the ball's change in momentum.
5. * A 1300-kg car is traveling at a speed of 10 m/s with respect to the ground when the driver accelerates to make a green light. The momentum of the car increases by 12,800 kg · m/s. List all the quantities you can determine using this information and determine three of those quantities.
6. * The rules of tennis specify that the 0.057-kg ball must bounce to a height of between 53 and 58 inches when dropped from a height of 100 inches onto a concrete slab. What is the change in the momentum of the ball during the collision with the concrete? You will have to use some free-fall kinematics to help answer this question.
7. A cart of mass $m$ moving right at speed $v$ with respect to the track collides with a cart of mass $0.7m$ moving left. What is the initial speed of the second cart if after the collision the carts stick together and stop?
8. A cart of mass $m$ moving right collides with an identical cart moving right at half the speed. The carts stick together. What is their speed after the collision?
9. EST Estimate your momentum when you are walking at your normal pace.

## 5.3 Impulse and momentum

10. A 100-g apple is falling from a tree. What is the impulse that Earth exerts on it during the first 0.50 s of its fall? The next 0.50 s?
11. * The same 100-g apple is falling from the tree. What is the impulse that Earth exerts on it during the first 0.50 m of its fall? The next 0.50 m?
12. Why does Earth exert the same impulse during the two time intervals in Problem 10 but different impulses during the same distances traveled in Problem 11?
13. * **Van hits concrete support** In a crash test, a van collides with a concrete support. The stopping time interval for the collision is 0.10 s, and the impulse exerted by the support on the van is $7.5 \times 10^3$ N · s. (a) Determine everything you can about the collision using this information. (b) If the van is constructed to collapse more during the collision so that the time interval during which the impulse is exerted is tripled, what is the average force exerted by the concrete support on the van?
14. BIO **Force exerted by heart on blood** About 80 g of blood is pumped from a person's heart into the aorta during each heartbeat. The blood starts at rest with respect to the body and has a speed of about 1.0 m/s in the aorta. If the pumping takes 0.17 s, what is the magnitude of the average force exerted by the heart on the blood?
15. * The train tracks on which a train travels exert a $2.0 \times 10^5$ N friction force on the train, causing it to stop in 50 s. (a) Determine the average force needed to stop the train in 25 s. (b) Determine the stopping time interval if the tracks exert a $1.0 \times 10^5$-N friction force on the train.
16. ** EST Your friend is catching a falling basketball after it has passed through the basket. Her hands move straight down 0.20 m while catching the ball. Estimate (a) the time interval for the ball to stop as she catches it and (b) the average force that her hands exert on the ball while catching it. Indicate any assumptions or estimates you have to make in order to answer the questions.
17. * BIO **Traumatic brain injury** According to a report on traumatic brain injury, the force that a professional boxer's fist exerts on his opponent's head is equivalent to being hit with a 5.9 kg bowling ball traveling at 8.9 m/s that stops in 0.018 s. Determine the average force that the fist exerts on the head.
18. * A 65-kg astronaut pushes against the inside back wall of a 2000-kg spaceship and moves toward the front. Her speed increases from 0 to 1.6 m/s. (a) If her push lasts 0.30 s, what is the average force that the astronaut exerts on the wall of the spaceship? (b) If the spaceship was initially at rest, with what speed does it recoil? (c) What was the object of reference that you used to answer parts (a) and (b)?
19. * You decide to use your garden hose to wash your garage door. The water shoots out at a rate of 10 kg/s and a speed of 16 m/s with respect to the hose. When the water hits the garage, its speed decreases to zero. Determine the force that the water exerts on the wall. What assumptions did you make?
20. * The air in a windstorm moves at a speed of 30 m/s. When it hits a stop sign, the air stops momentarily. The mass of air hitting the stop sign each second is about 2.0 kg. Make a list of physical quantities you can determine using this information and determine three of them.
21. * An egg rolls off a kitchen counter and breaks as it hits the floor. How large is the impulse that the floor exerts on the egg, and how large is the force exerted on the egg by the floor when stopping it? The counter is 1.0 m high, the mass of the egg is about 50 g, and the time interval during the collision is about 0.010 s.
22. ** **Retractable car bumper** A car bumper exerts an average force on a car as it retracts a certain distance during a collision. Using the impulse-momentum equation, show that the magnitude of the force and the retraction distance are related by the equation $F\Delta x = 0.5mv_0^2$. What assumptions did you make?
23. ** **Proportional reasoning** Use proportional reasoning and the equation from Problem 22 to determine (a) the necessary percent change in the retraction distance so that the average force required to stop a car is reduced by 20% and (b) the percent change in initial to final speed that would produce the same reduction in force.
24. (a) What force is required to stop a 1500-kg car in a distance of 0.20 m if it is initially moving at 2.2 m/s? (b) What if the car is moving at 4.5 m/s?

25. * A boxer delivers a punch to his opponent's head, which has a mass of 7.0 kg. Use the graph in **Figure P5.25** to estimate (a) the impulse of the force exerted by the boxer and (b) the speed of the head after the punch is delivered. What assumptions did you make?

**Figure P5.25**

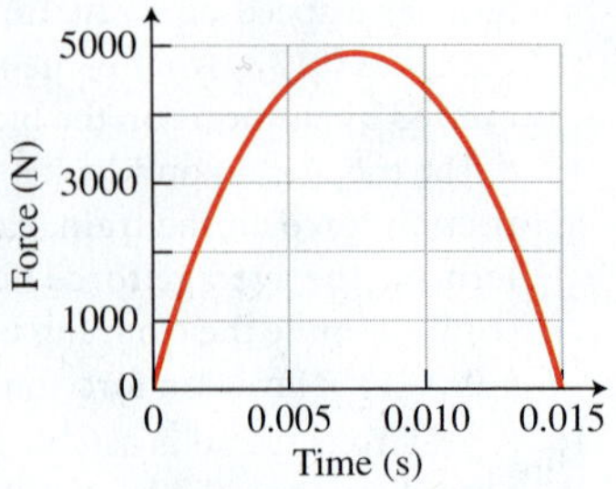


26. * **Air bag force on head** The graph in **Figure P5.26** shows the time variation of the force that an automobile's air bag exerts on a person's head during a collision. The mass of the head is 8.0 kg. Determine (a) the total impulse of the force exerted by the air bag on the person's head and (b) the person's speed just before the collision occurred.

**Figure P5.26**

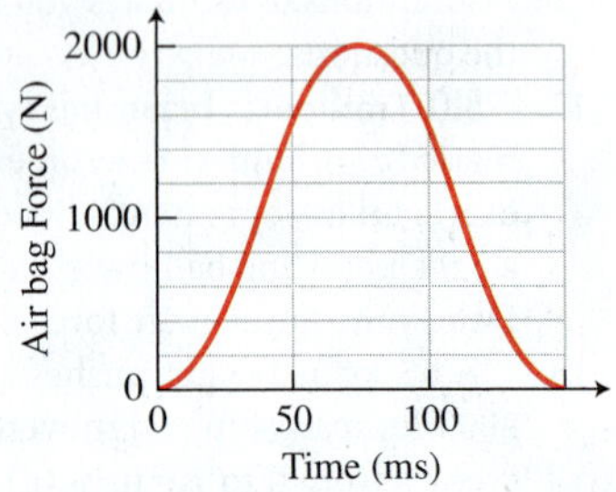


27. * **Equation Jeopardy 1** Invent a problem for which the solution is

$$(27\text{ kg})(-3.0\text{ m/s}) + (30\text{ kg})(+4.0\text{ m/s}) = (27\text{ kg} + 30\text{ kg})v.$$

28. * **Equation Jeopardy 2** Invent a problem for which the solution is

$$(0.020\text{ kg})(300\text{ m/s}) - (10\text{ N})(0.40\text{ s}) = (0.020\text{ kg})(100\text{ m/s}).$$

29. * Write a general impulse-momentum equation that describes the following process: a person skating on rollerblades releases a backpack that falls toward the ground (the process ends before the backpack hits the ground). What is the system, and what are the physical quantities you will use to describe the process?

## 5.4 Generalized impulse-momentum principle

30. * Two carts (100 g and 150 g) on an air track are separated by a compressed spring. The spring is released. Represent the process with a momentum bar chart (a) with one cart as the system and (b) with both carts as the system. (c) Write expressions for all of the physical quantities you can from this information. Identify your object of reference.

31. * A tennis ball of mass $m$ hits a wall at speed $v$ and rebounds at about the same speed. Represent the process with an impulse-momentum bar chart for the ball as the system. Using the bar chart, develop an expression for the change in the ball's momentum. What is the object of reference?

32. * A tennis ball traveling at a speed of $v$ stops after hitting a net. Represent the process with an impulse-momentum bar chart for the ball as the system. Develop an expression for the ball's change in momentum. What is the object of reference?

33. * You drop a happy ball and a sad ball of the same mass from height $h$ (see Example 5.3). One ball hits the ground and rebounds almost to the original height. The other ball does not bounce. Represent each process with a bar chart, starting just before the balls hit the ground to just after the first ball rebounds and when the other ball stops. Choose the ball as the system.

34. * You experiment again with the balls from Problem 33. You drop them from the same height onto a ruler that is placed on the edge of a table (**Figure P5.34**). One ball knocks the ruler off; the other does not. Represent each process with an impulse-momentum bar chart with (a) the ball as a system and (b) the ball and the ruler as the system. The process starts just before the balls hit the ruler and ends immediately after they hit the ruler. Use the bar charts to help explain the difference in the results of the experiment.

**Figure P5.34**

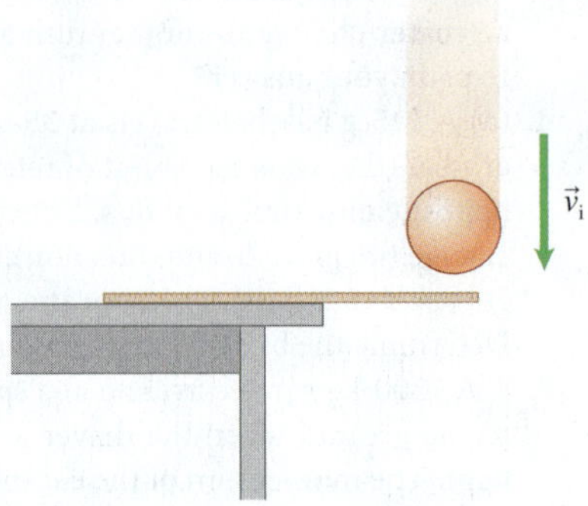


35. ** You demonstrate hitting a board in a karate class. The speed of your hand as it hits the thick board is 14 m/s with respect to the board, and the mass of your hand is about 0.80 kg. How deep does your hand go into the board before stopping if the collision lasts for $2.0 \times 10^{-3}$ s? What assumptions did you make? What other quantities can you determine using this information?

36. * You hold a beach ball with your arms extended above your head and then throw it upward. Represent the motion of the ball with an impulse-momentum bar chart for (a) the ball as the system and (b) the ball and Earth as the system.

37. * A basketball player drops a 0.60-kg basketball vertically so that it is traveling at 6.0 m/s when it reaches the floor. The ball rebounds upward at a speed of 4.2 m/s. (a) Determine the magnitude and direction of the ball's change in momentum. (b) Determine the average net force that the floor exerts on the ball if the collision lasts 0.12 s.

38. * **Bar chart Jeopardy** Invent a problem for each of the bar charts shown in **Figure P5.38**.

**Figure P5.38**

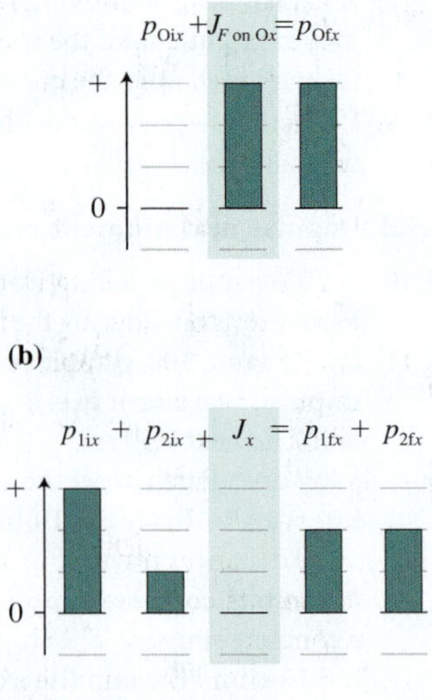


## 5.5 Skills for solving impulse-momentum problems

39. * A baseball bat contacts a 0.145-kg baseball for $1.3 \times 10^{-3}$ s. The average force exerted by the bat on the ball is 8900 N. If the ball has an initial velocity of 36 m/s toward the bat and the force of the bat causes the ball's motion to reverse direction, what is the ball's speed as it leaves the bat?

40. * A tennis ball traveling horizontally at a speed of 40.0 m/s hits a wall and rebounds in the opposite direction. The time interval for the collision is about 0.013 s, and the mass of the ball is 0.059 kg. Make a list of quantities you can determine using this information and determine four of them. Assume that the ball rebounds at the same speed.
41. A cannon mounted on the back of a ship fires a 50-kg cannonball in the horizontal direction at a speed of 150 m/s. If the cannon and ship have a combined mass of 40000 kg and are initially at rest, what is the speed of the ship just after shooting the cannon? What assumptions did you make?
42. * A team in Quebec is playing ice baseball. A 72-kg player who is initially at rest catches a 145-g ball traveling at 18 m/s. If the player's skates are frictionless, how much time is required for him to glide 5.0 m after catching the ball?
43. A 10-kg sled carrying a 30-kg child glides on a horizontal, frictionless surface at a speed of 6.0 m/s toward the east. The child jumps off the back of the sled, propelling it forward at 20 m/s. What was the child's velocity in the horizontal direction relative to the ground at the instant she left the sled?
44. A 10,000-kg coal car on the Great Northern Railroad coasts under a coal storage bin at a speed of 2.0 m/s. As it goes under the bin, 1000 kg of coal is dropped into the car. What is the final speed of the loaded car?
45. * **Avoiding chest injury** A person in a car during a sudden stop can experience potentially serious chest injuries if the combined force exerted by the seat belt and shoulder strap exceeds 16,000 N. Describe what it would take to avoid injury by estimating (a) the minimum stopping time interval and (b) the corresponding stopping distance, assuming an initial speed of 16 m/s. Indicate any other assumptions you made.
46. * **Bruising apples** An apple bruises if a force greater than 8.0 N is exerted on it. Would a 0.10-kg apple be likely to bruise if it falls 2.0 m and stops after sinking 0.060 m into the grass? Explain.
47. * **Fast tennis serve** The fastest server in women's tennis is Venus Williams, who recorded a serve of 204 km/h at the French Open in 2007. Suppose that the mass of her racket was 328 g and the mass of the ball was 57 g. If her racket was moving at 200 km/h when it hit the ball, approximately what was the racket's speed after hitting the ball? Indicate any assumptions you made.
48. * You are in an elevator whose cable has just broken. The elevator is falling at 20 m/s when it starts to hit a shock-absorbing device at the bottom of the elevator shaft. If you are to avoid injury, the upward force that the floor of the elevator exerts on your upright body while stopping should be no more than 8000 N. Determine the minimum stopping distance needed to avoid injury (do not forget to include your mass in the calculations). What assumptions did you make? Do these assumptions make the stopping distance smaller or larger than the real-world value?
49. * You jump from the window of a burning hotel and land in a safety net that stops your fall in 1.0 m. Estimate the average force that the net exerts on you if you enter the net at a speed of 24 m/s. What assumptions did you make? If you did not make these assumptions, would the stopping distance be smaller or larger?
50. * **Skid marks** A car skids to a stop. The length of the skid marks is 50 m. What information do you need in order to decide whether the car was speeding before the driver hit the brakes?
51. * BIO **Leg injuries during car collisions** During a car collision, the knee, thighbone, and hip can sustain a force no greater than 4000 N. Forces that exceed this amount could cause dislocations or fractures. Assume that in a collision a knee stops when it hits the car's dashboard. Also assume that the mass of the body parts stopped by the knee is about 20% of the total body mass. (a) What minimum stopping time interval in needed to avoid injury to the knee if the person is initially traveling at 15 m/s (34 mi/h)? (b) What is the minimum stopping distance?
52. * BIO **Bone fracture** The zygomatic bone in the upper part of the cheek can be fractured by a 900-N force lasting 6.0 ms or longer. A hockey puck can easily exert such a force when hitting an unprotected face. (a) What change in velocity of a 0.17-kg hockey puck is needed to provide that impulsive force? What assumptions did you make? (b) A padded face-mask doubles the stopping time. By how much does it change the force on the face? Explain.
53. * An impulse of 150 N · s stops your head during a car collision. (a) A crash test dummy's head stops in 0.020 s, when the cheekbone hits the steering wheel. What is the average force that the wheel exerts on the dummy's cheekbone? (b) Would this crash fracture a human cheekbone (see Problem 52)? (c) What is the shortest impact time that a person could sustain without breaking the bone?
54. A cart is moving on a horizontal track when a heavy bag falls vertically onto it. What happens to the speed of the cart? Represent the process with an impulse-momentum bar chart.
55. * A cart is moving on a horizontal track. A heavy bag falls off the cart and moves straight down relative to the cart. Describe what happens to the speed of the cart. Represent your answer with the impulse-momentum bar chart. [*Hint:* What reference frame will you use when you draw the bar chart?]

## 5.6 and 5.7 Jet propulsion and Putting it all together

56. Your friend shoots an 80-g arrow through a 100-g apple balanced on William Tell's head. The arrow has a speed of 50 m/s before passing through the apple and 40 m/s after. Determine the final speed of the apple.
57. * BIO **Potassium decay in body tissue** Certain natural forms of potassium have nuclei that are radioactive. Each radioactive potassium nucleus decays to a slightly less massive daughter nucleus and a high-speed electron called a beta particle. If after the decay the daughter nucleus is moving at speed 200 m/s with respect to the decaying material, how fast is the electron (the beta particle) moving? Indicate any assumptions you made. The mass of the daughter is about 70,000 times greater than the mass of the beta particle.
58. ** **Meteorite impact with Earth** About 65 million years ago a 10-km-diameter $2 \times 10^{15}$-kg meteorite traveling at about 10 km/s crashed into what is now the Gulf of Mexico. The impact produced a cloud of debris that darkened Earth and led to the extinction of the dinosaurs. Estimate the speed Earth gained as a result of this impact and the average force that the meteorite exerted on Earth during the collision. Indicate any assumptions made in your calculations.
59. ** Three friends play beach volleyball. The 280 g ball is flying east at speed 8.0 m/s with respect to the ground when one of the players bumps the ball north. The force exerted by the wrist on the ball has an average magnitude of 84 N and lasts for 0.010 s. Determine the ball's velocity (magnitude and direction) following the bump. Does your answer make sense?

60. * **Car collision** A 1180-kg car traveling south at 24 m/s with respect to the ground collides with and attaches to a 2470-kg delivery truck traveling east at 16 m/s. Determine the velocity (magnitude and direction) of the two vehicles when locked together just after the collision.

61. * **Ice skaters collide** While ice skating, you unintentionally crash into a person. Your mass is 60 kg, and you are traveling east at 8.0 m/s with respect to the ice. The mass of the other person is 80 kg, and he is traveling north at 9.0 m/s with respect to the ice. You hang on to each other after the collision. In what direction and at what speed are you traveling just after the collision?

62. **Drifting space mechanic** An astronaut with a mass of 90 kg (including spacesuit and equipment) is drifting away from his spaceship at a speed of 0.20 m/s with respect to the spaceship. The astronaut is equipped only with a 0.50-kg wrench to help him get back to the ship. With what speed and in what direction relative to the spaceship must he throw the wrench for his body to acquire a speed of 0.10 m/s and direct him back toward the spaceship? Explain.

63. * **Astronaut flings oxygen tank** While the astronaut in Problem 62 is trying to get back to the spaceship, his comrade, a 60-kg astronaut, is floating at rest a distance of 10 m from the spaceship when she runs out of oxygen and fuel to power her back to the spaceship. She removes her oxygen tank (3.0 kg) and flings it away from the ship at a speed of 15 m/s relative to the ship. (a) At what speed relative to the ship does she recoil toward the spaceship? (b) How long must she hold her breath before reaching the ship?

64. **Rocket stages** A 5000-kg rocket ejects a 10,000-kg package of fuel. Before ejection, the rocket and the fuel travel together at a speed of 200 m/s with respect to distant stars. If after the ejection, the fuel package travels at 50 m/s opposite the direction of its initial motion, what is the velocity of the rocket?

65. * ✎ A rocket has just ejected fuel. With the fuel and the rocket as the system, construct an impulse-momentum bar chart for (a) the rocket's increase in speed and (b) the process of a rocket slowing down due to fuel ejection. (c) Finally, draw bar charts for both situations using the rocket without the fuel as the system.

66. ** You have two carts, a force probe connected to a computer, a motion detector, and an assortment of objects of different masses. Design three experiments to test whether momentum is a conserved quantity. Describe carefully what data you will collect and how you will analyze the data.

## General Problems

67. ** EST Estimate the recoil speed of Earth if all of the inhabitants of Canada and the United States simultaneously jumped straight upward from Earth's surface (reaching heights from several centimeters to a meter or more). Indicate any assumptions that you made in your estimate.

68. * A cart of mass $m$ traveling in the negative $x$-direction at speed $v$ collides head-on with a cart that has triple the mass and is moving at 60% of the speed of the first cart. The carts stick together after the collision. In which direction and at what speed will they move?

69. ** Two cars of unequal mass moving at the same speed collide head-on. Explain why a passenger in the smaller mass car is more likely to be injured than one in the larger mass car. Justify your reasoning with the help of physics principles.

70. * **Restraining force during collision** A 1340-kg car traveling east at 13.6 m/s (20 mi/h) has a head-on collision with a 1930-kg car traveling west at 20.5 m/s (30 mi/h). If the collision time is 0.10 s, what is the force needed to restrain a 68-kg person in the smaller car? In the larger car?

71. ** EST A carpenter hammers a nail using a 0.40-kg hammerhead. Part of the nail goes into a board. (a) Estimate the speed of the hammerhead before it hits the nail. (b) Estimate the stopping distance of the hammerhead. (c) Estimate the stopping time interval. (d) Estimate the average force that the hammerhead exerts on the nail.

72. ** A 0.020-kg bullet traveling at a speed of 300 m/s embeds in a 1.0-kg wooden block resting on a horizontal surface. The block slides horizontally 4.0 m on a surface before stopping. Determine the coefficient of kinetic friction between the block and surface.

73. ** EST Nolan Ryan may be the fastest baseball pitcher of all time. The *Guinness Book of World Records* clocked his fastball at 100.9 mi/h in a 1974 game against the Chicago White Sox. Use the impulse-momentum equation to estimate the force that Ryan exerted on the ball while throwing that pitch. Include any assumptions you made.

74. ** A record rainstorm produced 304.8 mm (approximately 1 ft) of rain in 42 min. Estimate the average force that the rain exerted on the roof of a house that measures 10 m $\times$ 16 m. Indicate any assumptions you made.

75. ** EST The U.S. Army special units MH-47E helicopter has a mass of 23,000 kg, and its propeller blades sweep out an area of 263 $m^2$. It is able to hover at a fixed elevation above one landing point by pushing air downward (the air pushes up on the helicopter blades). Choose a reasonable air mass displaced downward each second and the speed of that air in order for the helicopter to hover. Indicate any assumptions you made.

76. ** A 2045-kg sports utility vehicle hits the rear end of a 1220-kg car at rest at a stop sign. The car and SUV remain locked together and skid 4.6 m before stopping. If the coefficient of friction between the vehicles and the road is 0.70, what was the SUV's initial velocity?

77. ** A car of mass $m_1$ traveling north at a speed of $v_1$ collides with a car of mass $m_2$ traveling east at a speed of $v_2$. They lock together after the collision. Develop expressions for the direction and the distance the cars will move until they stop if the coefficient of kinetic friction $\mu_k$ between the cars' tires and the road is about the same for both cars.

78. ** **Force exerted by wind on Willis Tower** A 10.0-m/s wind blows against one side of the Willis Tower in Chicago. The building is 443 m tall and approximately 80 m wide. Estimate the average force of the air on the side of the building. The density of air is approximately 1.3 $kg/m^3$. Indicate any assumptions that you made.

79. * **Write your own problem.** Write and solve a problem that requires using the law of conservation of momentum in which it is important to know that momentum is a vector quantity.

## Reading Passage Problems

BIO **Heartbeat detector** A prisoner tries to escape from a Nashville, Tennessee prison by hiding in the laundry truck. The prisoner is surprised when the truck is stopped at the gate. A guard enters the truck and handcuffs him. "How did you know I was here?" the prisoner asks. "The heartbeat detector," says the guard.

A heartbeat detector senses the tiny vibrations caused by blood pumped from the heart. With each heartbeat, blood is pumped upward to the aorta, and the body recoils slightly, conserving the momentum of the blood-body system. The body's vibrations are transferred to the inside of the truck. Vibration sensors on the outside of the truck are linked to a geophone, or signal amplifier, attached to a computer. A wave analyzer program in the computer compares vibration signals from the truck to wavelets produced by heartbeat vibrations. The wave analyzer distinguishes a person's heartbeat from other vibrations in the truck or in the surrounding environment, allowing guards to detect the presence of a human in the truck.

80. What does the heartbeat detector sense?
 (a) Electric signals caused by electric dipole charges produced on the heart
 (b) Body vibrations caused by blood pumped from the heart
 (c) Sound caused by breathing
 (d) Slight uncontrollable reflexive motions of an enclosed person
 (e) All of the above
81. A heartbeat detector relies on a geophone placed against the exterior of a truck or car. What can cause the vibrations of the truck or car?
 (a) Wind
 (b) Ground vibrations due to other moving cars or trucks
 (c) Vibrations of passengers in the car or truck
 (d) All of the above
82. What can be used to analyze the motion of a person's body hidden inside a car or truck (choosing the body and blood as a system)?
 (a) The idea that mass of an isolated system is constant
 (b) The idea that momentum of an isolated system is constant
 (c) The impulse-momentum principle
 (d) a and b
 (e) b and c
83. During each heartbeat, about 0.080 kg of blood passes the aorta in about 0.16 s. This blood's velocity changes from about 0.8 m/s upward toward the head to 0.8 m/s down toward the feet. What is the blood's acceleration?
 (a) zero (b) $5\ \text{m/s}^2$ up (c) $5\ \text{m/s}^2$ down
 (d) $10\ \text{m/s}^2$ up (e) $10\ \text{m/s}^2$ down
84. Suppose 0.080 kg of blood moving upward in the aorta at 0.8 m/s reverses direction in 0.16 s when it reaches the aortic arch. If a prisoner is trying to escape from prison by hiding in a laundry truck, and the mass of his body is 70 kg, which is the closest to the speed his body is moving immediately after the blood changes direction passing through the aortic arch?
 (a) 0.0009 m/s (b) 0.002 m/s (c) 0.8 m/s
 (d) 0.08 m/s (e) 0.01 m/s

**Space Shuttle launch** The mass of the Space Shuttle at launch is about $2.1 \times 10^6$ kg. Much of this mass is the fuel used to move the orbiter, which carries the astronauts and various items in the shuttle's payload. The Space Shuttle generally travels from $3.2 \times 10^5$ m (200 mi) to $6.2 \times 10^5$ m (385 mi) above Earth's surface. The shuttle's two solid fuel boosters (the cylinders on the sides of the shuttle) provide 71.4% of the thrust during liftoff and the first stage of ascent before being released from the shuttle 132 s after launch at 48,000-m above sea level. The boosters continue moving up in free fall to an altitude of approximately 70,000 m and then fall toward the ocean to be recovered 230 km from the launch site. The shuttle's five engines together provide $3.46 \times 10^7$ N of thrust during liftoff.

85. Which number below is closest to the acceleration of the shuttle during liftoff? [Hint: Remember the gravitational force that Earth exerts on the shuttle.]
 (a) $3.3\ \text{m/s}^2$ (b) $6.6\ \text{m/s}^2$ (c) $9.8\ \text{m/s}^2$
 (d) $16\ \text{m/s}^2$ (e) $33\ \text{m/s}^2$
86. Which number below is closest to the average vertical acceleration of the shuttle during the first 132 s of its flight?
 (a) $3.3\ \text{m/s}^2$ (b) $5.5\ \text{m/s}^2$ (c) $9.8\ \text{m/s}^2$
 (d) $14\ \text{m/s}^2$ (e) $360\ \text{m/s}^2$
87. The boosters are released from the shuttle 132 s after launch. How do their vertical components of velocity compare to that of the shuttle at the instant of release?
 (a) The boosters' vertical component of velocity is zero.
 (b) The boosters' vertical component of velocity is about $-9.8$ m/s.
 (c) The vertical component of velocity of the boosters and that of the Shuttle are the same.
 (d) There is too little information to decide.
88. What is the approximate impulse of the jet engine thrust exerted on the shuttle during the first 10 s of flight?
 (a) $980\ \text{N}\cdot\text{s}$ downward (b) $980\ \text{N}\cdot\text{s}$ upward
 (c) $3.4 \times 10^7\ \text{N}\cdot\text{s}$ upward (d) $3.4 \times 10^8\ \text{N}\cdot\text{s}$ upward
 (e) $3.4 \times 10^8\ \text{N}\cdot\text{s}$ downward
89. What is the approximate impulse of Earth's gravitational force exerted on the shuttle during the first 10 s of flight?
 (a) $980\ \text{N}\cdot\text{s}$ downward (b) $980\ \text{N}\cdot\text{s}$ upward
 (c) $2.1 \times 10^7\ \text{N}\cdot\text{s}$ upward (d) $2.1 \times 10^8\ \text{N}\cdot\text{s}$ upward
 (e) $2.1 \times 10^8\ \text{N}\cdot\text{s}$ downward
90. What is the momentum of the Space Shuttle 10 s after liftoff closest to?
 (a) $2.1 \times 10^6\ \text{kg}\cdot\text{m/s}$ down (b) $2.1 \times 10^6\ \text{kg}\cdot\text{m/s}$ up
 (c) $2.1 \times 10^7\ \text{kg}\cdot\text{m/s}$ up (d) $1.3 \times 10^8\ \text{kg}\cdot\text{m/s}$ up
 (e) $1.3 \times 10^8\ \text{kg}\cdot\text{m/s}$ down
91. What answer below is closest to the speed of the shuttle and boosters when they are released? Assume that the free-fall gravitational acceleration at this elevation is about $9.6\ \text{m/s}^2$ down.
 (a) 100 m/s (b) 300 m/s (c) 600 m/s
 (d) 1000 m/s (e) 1400 m/s

# 6 Work and Energy

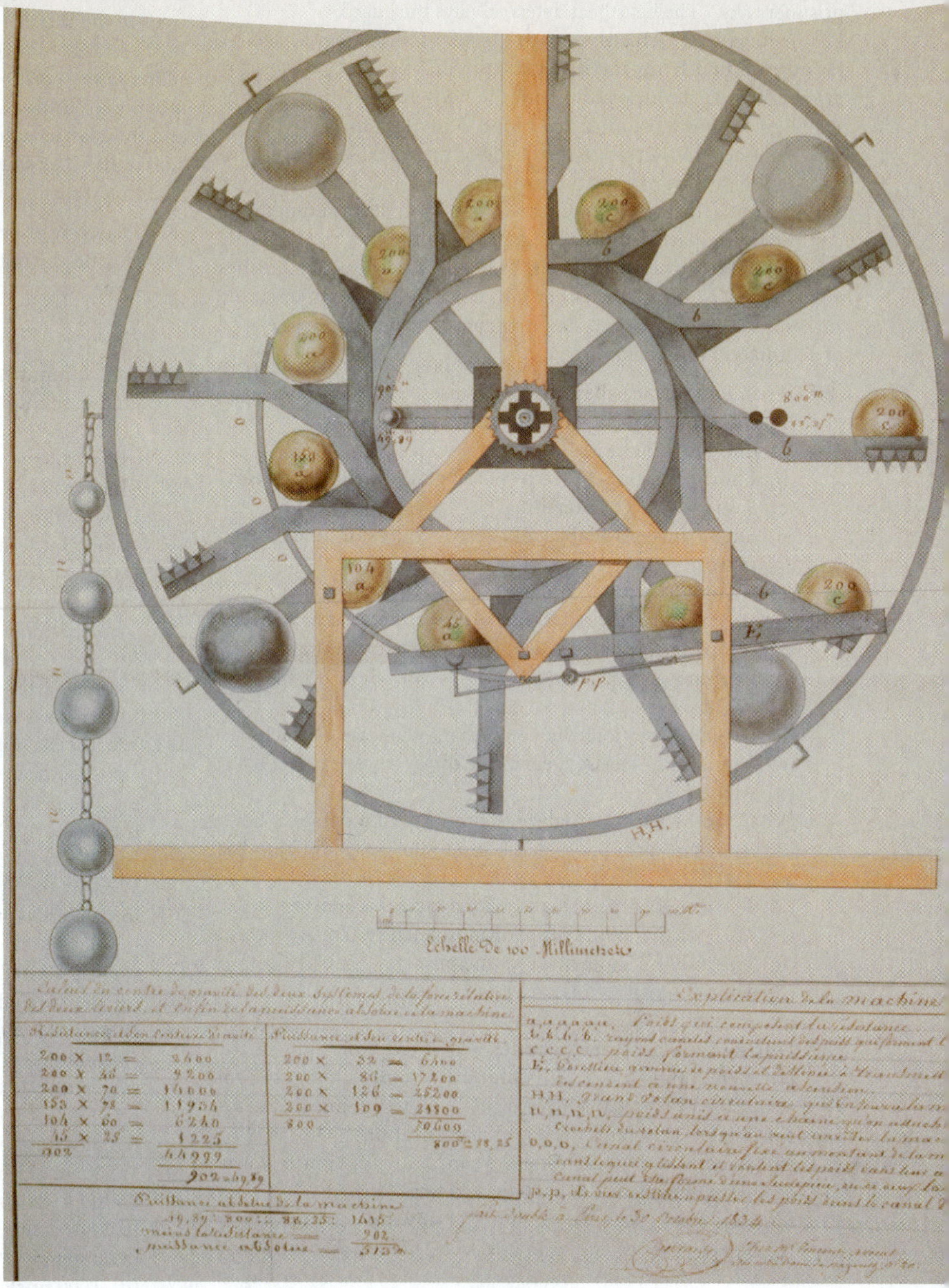

**Why it is impossible to build a perpetual motion machine?**

**Why does blood pressure increase when the aorta walls thicken?**

**If Earth were to become a black hole, how big would it be?**

**Be sure you know how to:**

- Choose a system and the initial and final states of a physical process (Sections 5.2–5.4).
- Use Newton's second law to analyze a physical process (Section 3.3).
- Use kinematics to describe motion (Section 1.7).

For centuries people have dreamed of making a perpetual motion machine. Such a machine would be able to function forever without any sort of power source. Imagine owning a laptop computer that ran continuously year after year with no external power. Although claims of perpetual motion machines go back as far as the 8th century, so far no one has succeeded in inventing one. In 1775 the Royal Academy of Sciences in Paris decreed that the Academy "will no longer accept or deal with proposals concerning perpetual motion." Why has no one been able to build a perpetual motion machine? We will begin investigating this question in this chapter.

**We have had good** success at analyzing a variety of everyday phenomena using vector quantities (acceleration, force, impulse, and momentum). When working with such vector quantities, we need to apply the component forms of principles such as Newton's second law and the impulse-momentum equation. Can we analyze interesting everyday phenomena using a different type of thinking that depends less on vector quantities? That might make our analysis somewhat easier.

## 6.1 Work and energy

We will begin our new approach by conducting several experiments and looking for a pattern to help explain what we observe. In each experiment we choose a system of interest, its initial state, its final state, and an external force that is causing the system to change from its initial state to its final state. This change involves a displacement of one of the system objects from one position to another. In the analysis we draw vectors indicating the external force causing the displacement and the resulting displacement of the system object. In Observational Experiment **Table 6.1**, below, we look at the effect of external forces on system objects.

An important pattern in all four observational experiments in Table 6.1 is that an external force $\vec{F}$ was exerted on an object in the system, causing a displacement $\vec{d}$ of the object in the *same direction* as the external force. Physicists

**OBSERVATIONAL EXPERIMENT TABLE**

**6.1 External forces and system changes.**

VIDEO 6.1

| Observational experiment | | Analysis |
| --- | --- | --- |
| **Experiment 1.** You hold a heavy block just above a piece of chalk (the initial state) and then release the block. The chalk does not break. Now you lift the block about 30 cm above the chalk (the final state). If you release the block from the higher elevation final state position, the block falls and smashes the chalk. | 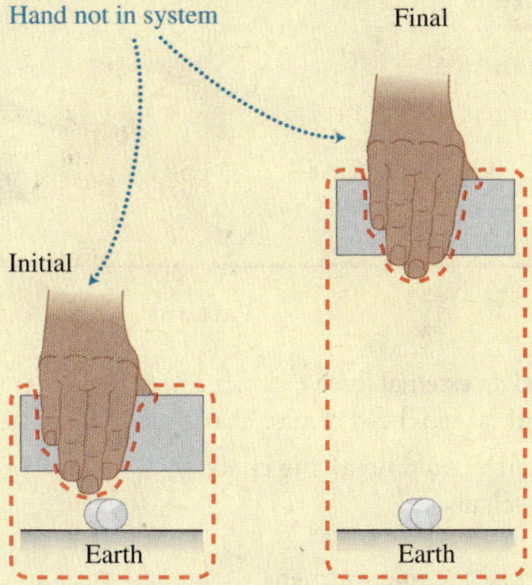 | The force you exerted and the block's displacement while being lifted were in the same direction and caused an increase in the block's elevation and in its ability to break the chalk. 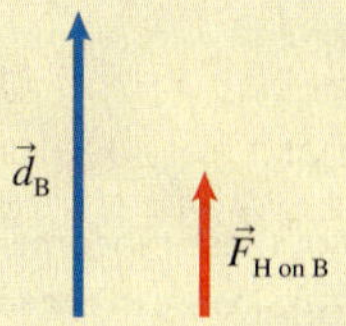 |

*(continued)*

| Observational experiments | Analysis |
| --- | --- |
| **Experiment 2.** You push a cart initially at rest (the initial state) until it is moving fast about two-thirds of the way across a smooth track (the final state is where you stop pushing the cart). A piece of chalk is taped to the end of the track. The fast-moving cart (no longer being pushed) collides with the piece of chalk and breaks the chalk. 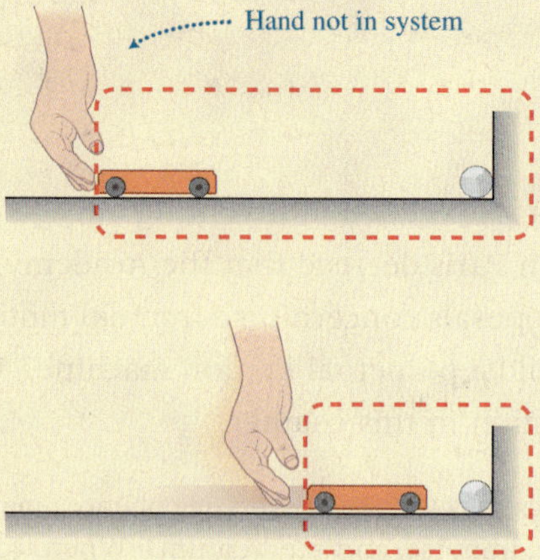 | The force exerted on the cart and the cart's displacement are in the same direction and increased the cart's speed so it could break the chalk. 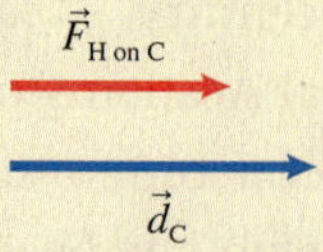 |
| **Experiment 3.** A piece of chalk rests in the hanging sling of a slingshot (the initial state). You then pull it back until the slingshot is fully stretched (the final state). When released from the stretched sling, the chalk flies across the room, hits the wall, and smashes. 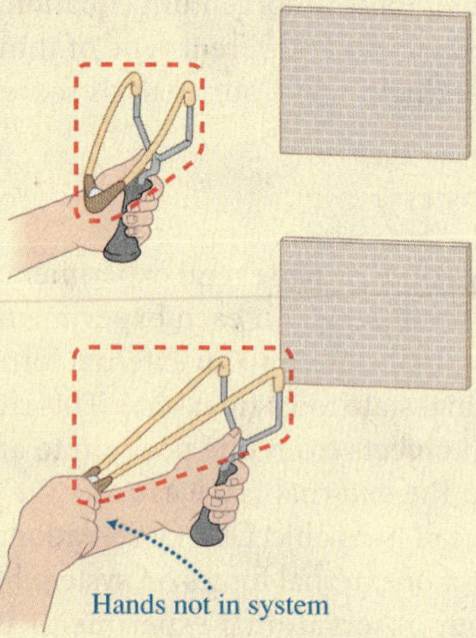 | The force that you exerted on the sling and its displacement are in the same direction and made it possible for the stretched sling to break the chalk. 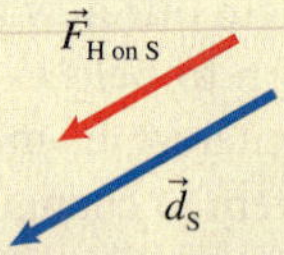 |
| **Experiment 4.** A heavy box sits on a shag carpet (the initial state). You pull the box across the carpet to a position several meters from where it started (the final state). When you reach the final position, you touch the bottom of the box and the carpet—they feel slightly warmer. 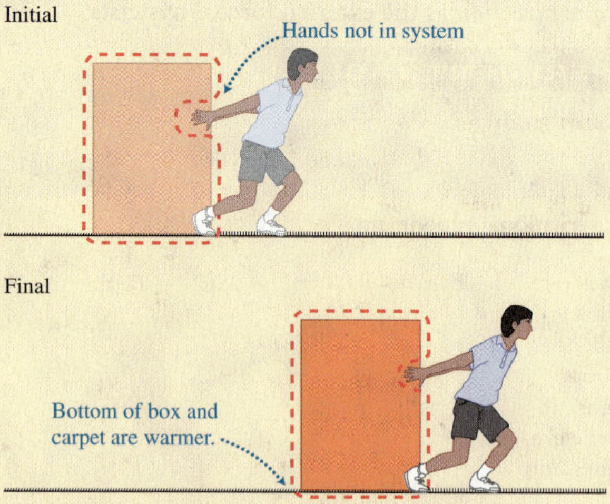 | You exerted a force on the box in the direction of its displacement. After several meters of travel across the carpet, the bottom became warmer. 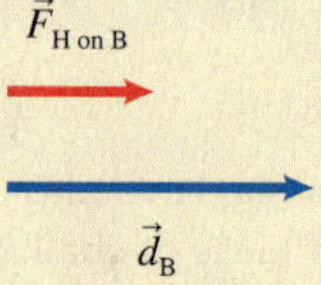 |

**Patterns**

In each of these experiments, you exerted an external force $\vec{F}$ on an object in a system. The force $\vec{F}_{\text{You on Object}}$ and the object's displacement $\vec{d}_{\text{Object}}$ were in the *same direction* and caused the system to change so that the system gained the potential to do something new:

1. The *block at higher elevation* above Earth could break the chalk.
2. The *fast-moving cart* could break the chalk.
3. The *stretched slingshot* could break the chalk.
4. The *box and the carpet it was pulled across* became warmer.

say that an object in the environment (in this case, you) did *positive* **work** on the system. In these experiments, the system changed because of the positive work done on it by the external forces. We call those *changes* in the system's **energy**. Four types of energy changed in the systems in the Table 6.1 experiments.

- *Gravitational potential energy* In Experiment 1, the block in the final state was at a *higher elevation* with respect to Earth than in the initial state. The energy of the object-Earth system associated with the elevation of the object above Earth is called **gravitational potential energy** $U_g$. The higher above Earth, the greater is the gravitational potential energy.
- *Kinetic energy* In Experiment 2, the cart was *moving faster* in the final state than when at rest in the initial state. The energy due to an object's motion is called **kinetic energy** $K$. The faster the object is moving, the greater its kinetic energy.
- *Elastic potential energy* In Experiment 3, the slingshot in the final state was *stretched more* than in the initial state. The energy associated with an elastic object's degree of stretch is called **elastic potential energy** $U_s$. The greater the stretch (or compression), the greater is the object's elastic potential energy.
- *Internal energy* In Experiment 4, the bottom surface of the box *became warmer* as it was pulled across the rough surface. Also, there can be structural changes in the surfaces—part of them can come off. The energy associated with both temperature and structure is called **internal energy** $U_{int}$. You will learn later that internal energy is the energy of motion and interaction of the microscopic particles making up the objects in the system.

## Negative and zero work

Is it possible to devise a process in which an external force causes the energy of a system to decrease or possibly causes no energy change at all? Let's try more experiments to investigate these questions.

Observational Experiment **Table 6.2** shows us that external forces exerted on system objects can have negative or zero effect on a system.

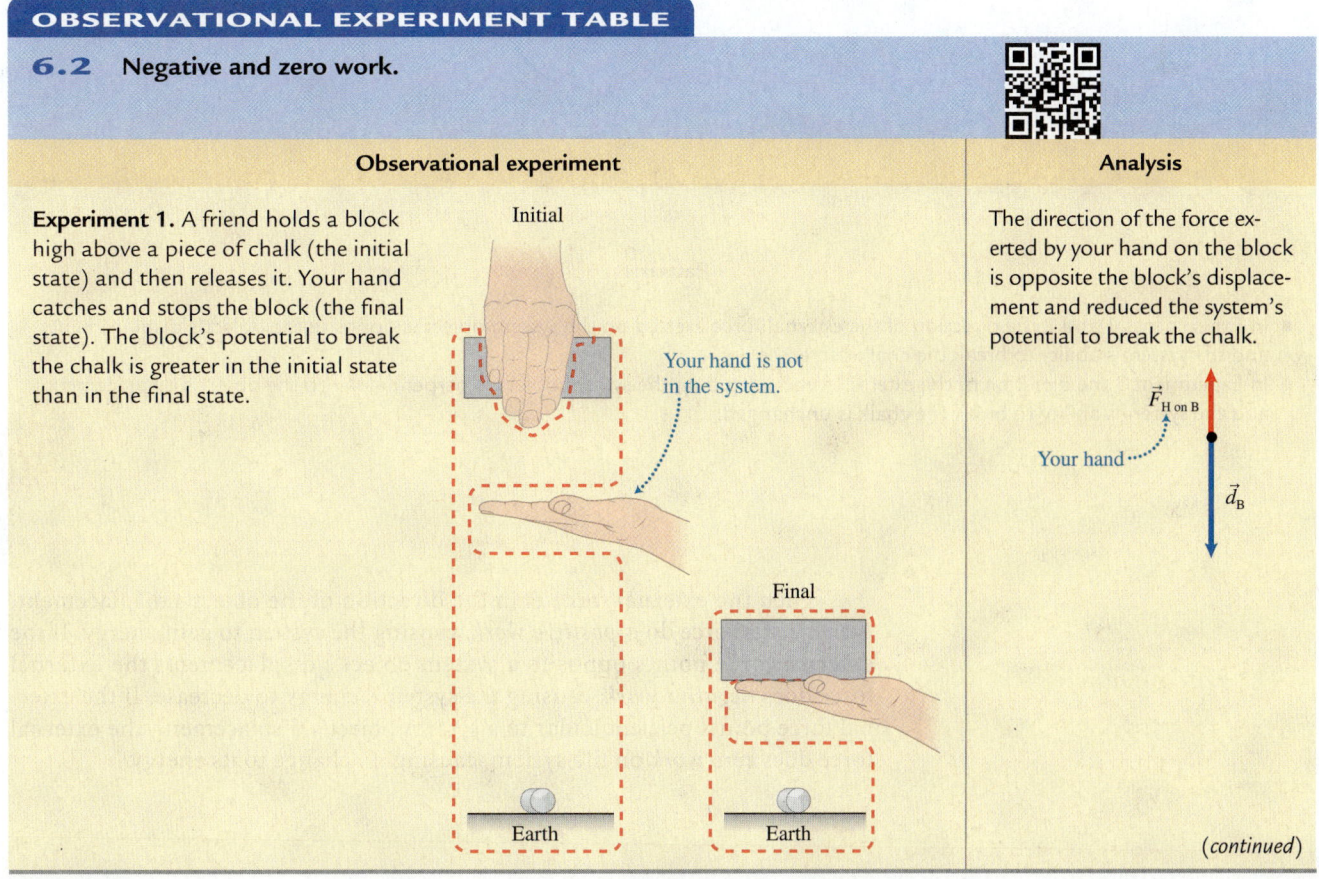

**OBSERVATIONAL EXPERIMENT TABLE**

### 6.2 Negative and zero work.

| Observational experiment | Analysis |
| --- | --- |
| **Experiment 1.** A friend holds a block high above a piece of chalk (the initial state) and then releases it. Your hand catches and stops the block (the final state). The block's potential to break the chalk is greater in the initial state than in the final state. 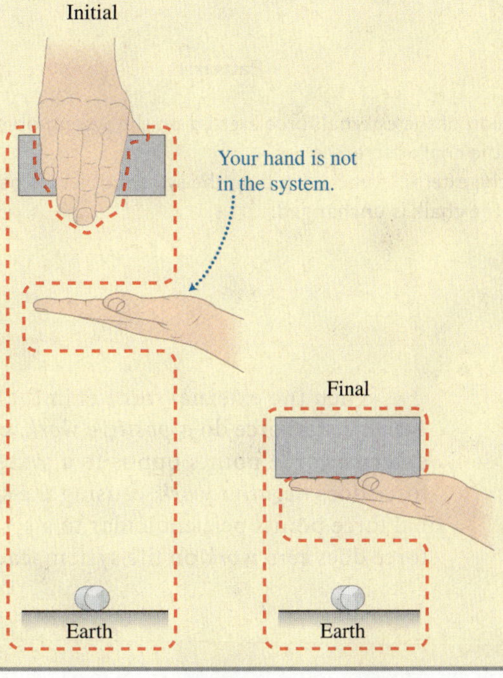  | The direction of the force exerted by your hand on the block is opposite the block's displacement and reduced the system's potential to break the chalk. 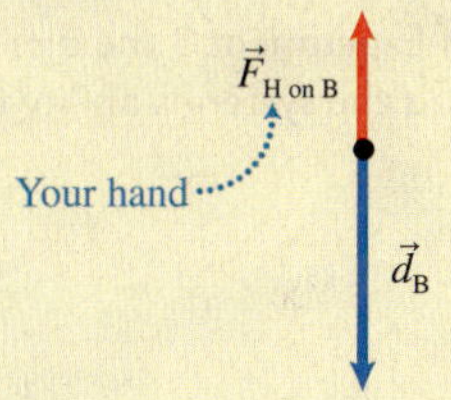  |

*(continued)*

| Observational experiments | Analysis |
|---|---|
| **Experiment 2.** A cart is moving fast (the initial state) toward a piece of chalk taped on the wall. While it is moving, you push lightly on the moving cart opposite the direction of its motion, causing it to slow down and stop (final state). The cart's potential to break the chalk is greater in the initial state than in the final state. 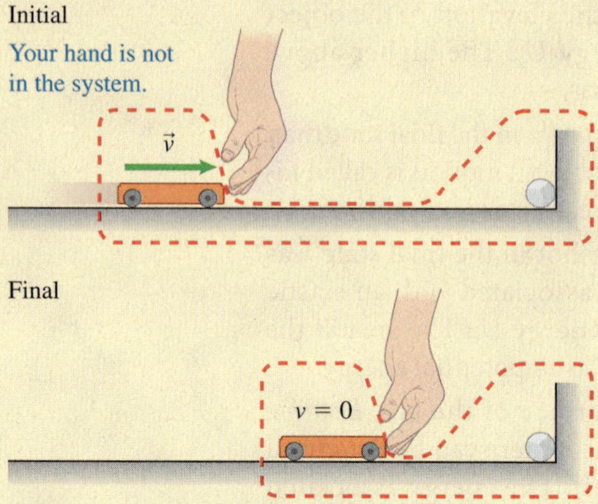 | The direction of the force exerted by your hand on the cart is opposite the cart's displacement and caused the moving cart to slow down and stop, thus reducing its potential to break the chalk. 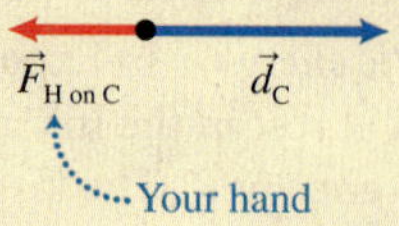 |
| **Experiment 3.** Your hand holds a block less than 1 cm above a piece of chalk (the initial state)—so close that the chalk would not break if the block is released. Your hand slowly moves the block to the right, keeping the block just above the tabletop until the block is less than 1 cm above a second piece of chalk (the final state), which also would not break if the block were released. 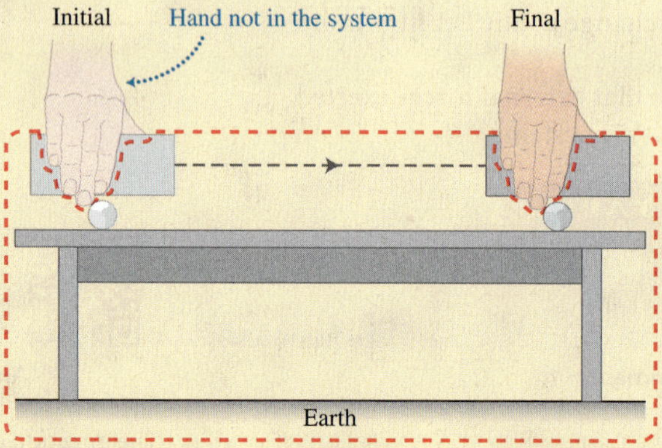 | The direction of the force exerted by your hand on the block is perpendicular to the block's displacement and caused no change in the block's potential to break the chalk. 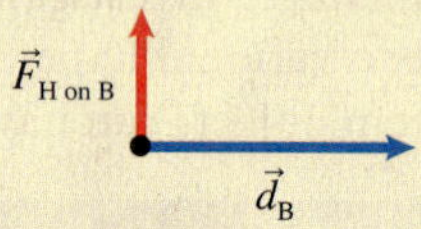 |









**Patterns**

- In Experiments 1 and 2 the direction of the external force exerted on the system object is opposite the object's displacement, and the system's ability to break the chalk decreases.
- In Experiment 3 the direction of the external force exerted on the system object is perpendicular to the object's displacement, and the system's ability to break the chalk is unchanged.

When the external force is in the direction of the object's displacement, the external force does *positive work*, causing the system to gain energy. If the external force points opposite a system object's displacement, the external force does *negative work*, causing the system's energy to decrease. If the external force points perpendicular to a system object's displacement, the external force does *zero work* on the system, causing no change to its energy.

## Defining work as a physical quantity

We can now create an equation to determine how much work a particular external force does on a system. The equation should be consistent with our observational experiments.

**Work** The work done by a constant external force $\vec{F}$ exerted on a system object while that system object undergoes a displacement $\vec{d}$ is

$$W = Fd\cos\theta. \quad (6.1)$$

where $F$ is the magnitude of the force in newtons (always positive), $d$ is the magnitude of the displacement in meters (always positive), and $\theta$ is the angle between the direction of $\vec{F}$ and the direction of $\vec{d}$. The sign of $\cos\theta$ determines the sign of the work. Work is a scalar physical quantity. The unit of work is the joule (J); $1\text{ J} = 1\text{ N}\cdot\text{m}$ (see **Figure 6.1**).

**Figure 6.1** The definition of work.

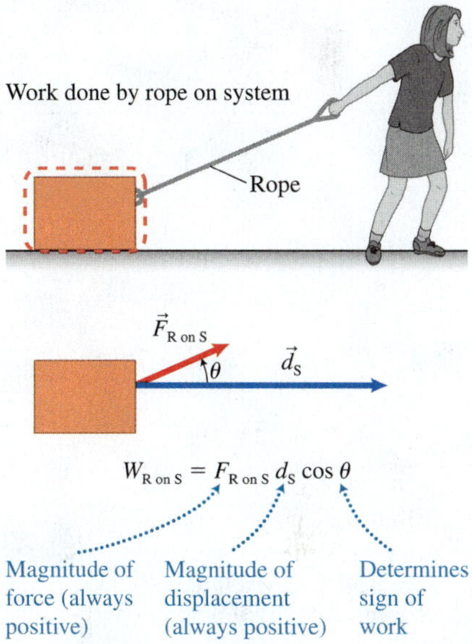


The joule is named in honor of James Joule (1818–1889), one of many physicists who contributed to our understanding of work-energy relationships.

Note that in the four experiments in Table 6.1, the force and displacement were in the same direction: $\theta = 0°$, and $\cos 0° = +1.0$. Positive work was done. In the first two experiments in Table 6.2, the force and displacement were in opposite directions: $\theta = 180°$, and $\cos 180° = -1.0$. Negative work was done. Finally, in Experiment 3 in Table 6.2, the force and displacement were perpendicular to each other: $\theta = 90°$, and $\cos 90° = 0$. Zero work was done.

**TIP** It is tempting to equate the work done on a system with the force that is exerted on it. However, in physics, there must be a displacement of a system object in order for an external force to do work. Force and work are not the same thing.

**‹Active Learning Guide**

**QUANTITATIVE EXERCISE 6.1 Pushing a bicycle uphill**

Two friends are cycling up a hill inclined at 8°—steep for bicycle riding. The stronger cyclist helps his friend up the hill by exerting a 50-N pushing force on his friend's bicycle and parallel to the hill while the friend moves a distance of 100 m up the hill. As you can see in the figure below, the force exerted on the weaker cyclist and the displacement are in the same direction. Determine the work done by the stronger cyclist on the weaker cyclist.

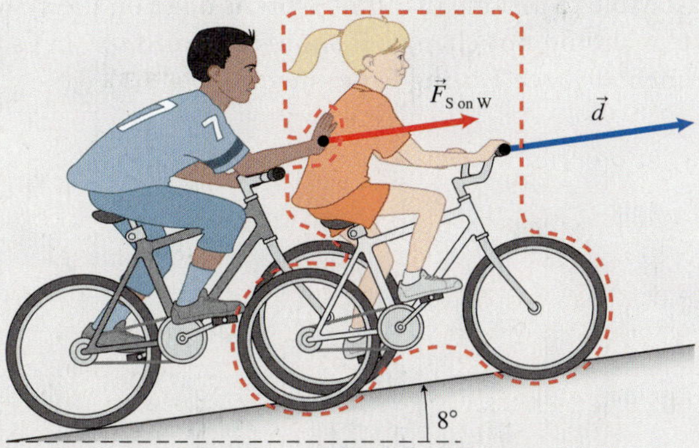


**Represent mathematically** Choose the system to be the weaker cyclist. The external force that the stronger cyclist (S) exerts on the weaker cyclist (W) $\vec{F}_{\text{S on W}}$ is parallel to the hill, as is the 100-m displacement of the weaker cyclist. The work done by the stronger cyclist on the weaker cyclist is $W = Fd\cos\theta$. The hill is inclined at 8° above the horizontal. Before reading on, decide what angle you would insert in this equation.

**Solve and evaluate** Note that the angle between $\vec{F}_{\text{S on W}}$ and $\vec{d}$ is 0° and not 8°. In this case, the angle of the hill is not relevant to solving the problem because the force exerted on the cyclist is parallel to the person's displacement. Thus:

$$\begin{aligned} W_{\text{S on W}} = F_{\text{S on W}}\, d\cos\theta &= (50\text{ N})(100\text{ m})\cos 0° \\ &= +5000\text{ N}\cdot\text{m} \\ &= +5000\text{ J} \end{aligned}$$

**Try it yourself:** You pull a box 20 m up a 10° ramp. The rope is oriented 20° above the surface of the ramp as shown on the next page. The force that the rope exerts on the box is 100 N. What is the work done by the rope on the box?

*(continued)*

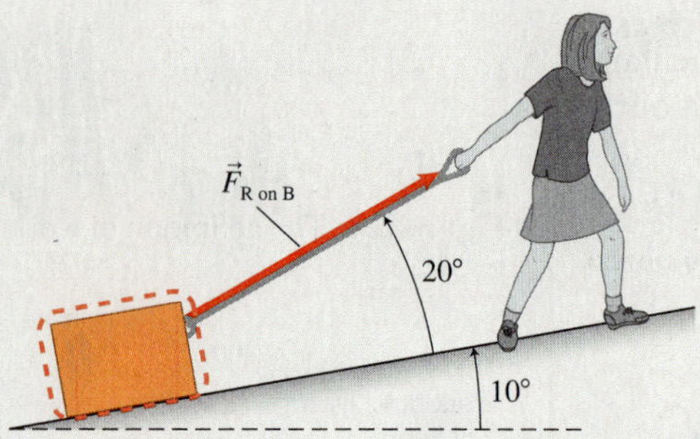


*Answer:* $W_{R\,on\,B} = (100\ N)(20\ m)\cos 20° = 1900\ J$. Note that you use the angle between the displacement (parallel to the ramp) and the force the rope exerts on the box (20° above the ramp). You can disregard the angle of the ramp itself.

**TIP** Remember, the angle that appears in the definition of work is the angle between the external force and the displacement of the system object. It is useful when calculating work to draw tail-to-tail arrows representing the external force doing the work and the system object displacement. Then note the angle between the arrows.

**Review Question 6.1** Describe two processes in which an external force is exerted on a system object and no work is done on the system. Explain why no work is done.

## 6.2 Energy is a conserved quantity

We have found that the work done on a system object by an external force results in a change of one or more types of energy in the system: kinetic energy, gravitational potential energy, elastic potential energy, and internal energy. We can think of the *total energy U* of a system as the sum of all these energies in the system:

$$\text{Total energy} = U = K + U_g + U_s + U_{int} \tag{6.2}$$

The energy of a system can be converted from one form to another. For example, the elastic potential energy of a stretched slingshot is converted into the kinetic energy of the chalk when the sling is released. Similarly, the gravitational potential energy of a separated block-Earth system is converted into kinetic energy of the block when the block falls. What happens to the amount of energy when it is converted from one form to another? So far we have one mechanism through which the energy of the system changes—that mechanism is work. Thus it is reasonable to assume that if no work is done on the system, the energy of the system should not change; it should be *constant*. Let's test this hypothesis experimentally (see Testing Experiment **Table 6.3**).

**TESTING EXPERIMENT TABLE**

## 6.3 Is the energy of an isolated system constant?

VIDEO 6.3

### Testing experiment

You have a toy car and a frictionless track. The bottom of the track is horizontal to the edge of a table. You can tilt the track at different angles to make the track steeper or shallower. When the car reaches the end of the track, it flies horizontally off the table. Where should you release a car on the track so the car always lands the same distance from the table regardless of the slope of the track?

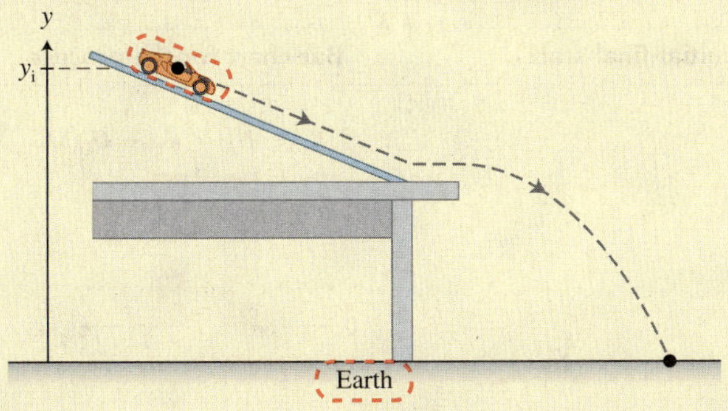


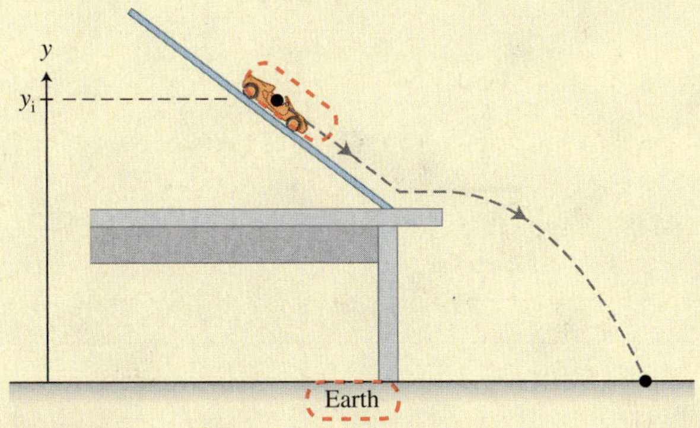


### Prediction

Consider car+Earth as a system. The initial state is just before we release the car and the final state is just as the car leaves the horizontal track. For the car to land on the floor at the same distance from the table's edge, the car needs to have the same horizontal velocity and hence the same kinetic energy when leaving the track. If energy is constant in the isolated system, to get the same kinetic energy at the bottom of the track, the car-Earth system must have the same initial gravitational energy when it starts. Since the gravitational potential energy depends on the separation between the car and Earth, the car should start at the same vertical elevation independent of the tilt angle of the track, $y_i = h$. Our prediction is based on the assumption that the gravitational potential energy of the system is only converted to the kinetic energy of the car and not to internal energy.

### Outcome

When released from the same vertical height with respect to the table, the car lands the same distance from the table.

### Conclusion

The outcome of the experiment matches the prediction and thus supports our hypothesis that the energy of an isolated system is constant.

So far we have experimental support for a qualitative hypothesis that the energy of an isolated system is constant and different processes inside the system convert energy from one form to another. We also reason that work is a mechanism through which the energy of a nonisolated system changes. This sounds a lot like our discussion of linear momentum (Chapter 5). In that chapter we described a *conserved quantity* as constant in an isolated system. We also said that when the system is not isolated, we can account for the changes in a conserved quantity by what is added to or subtracted from the system. Based on this reasoning we can hypothesize that energy is a conserved quantity—it is constant in an isolated system and changes as a result of work done on a nonisolated system.

## Work-energy bar charts

We can represent work-energy processes with bar charts that are similar to the impulse-momentum bar charts we used to describe momentum. A work-energy bar chart indicates with vertical bars the relative amount of a system's different types of energy in the initial state of a process, the work done on the system by external forces during the process, and the relative amount of different types of energy in the system at the end of the process. The area for the work bar is shaded to emphasize that work does not reside in the system. In **Table 6.4** you see three examples of energy changing from one form to another—represented by words, sketches, and bar charts.

**Table 6.4** **Three examples of system energy conversions.**

| Description of process | Sketch of the system and initial-final state | Bar chart for the process |
|---|---|---|
| (1) A girl starts at rest at the top of a smooth water slide and is moving fast at the bottom.<br>$U_g \rightarrow K$<br>The system's gravitational potential energy is converted to the girl's kinetic energy as she moves down the slide. | 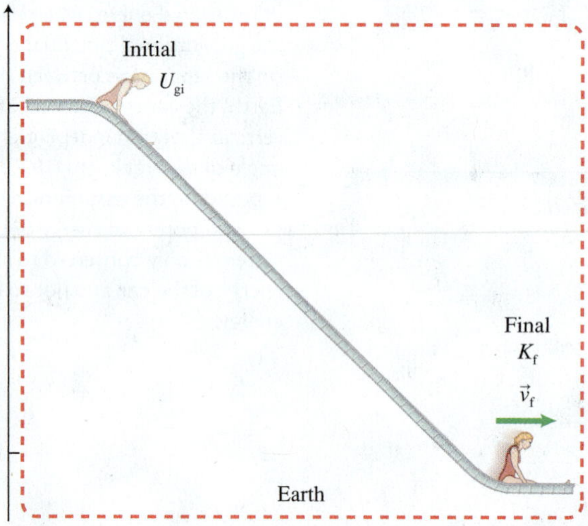  | 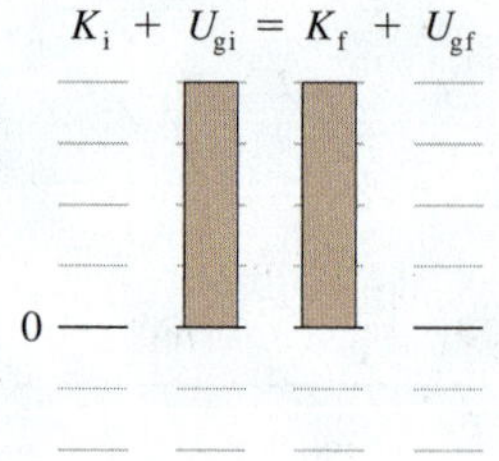  |
| (2) A fast-moving car skids to a stop on a level road.<br>$K \rightarrow U_{int}$<br>The system's kinetic energy is converted to internal thermal energy due to friction. | 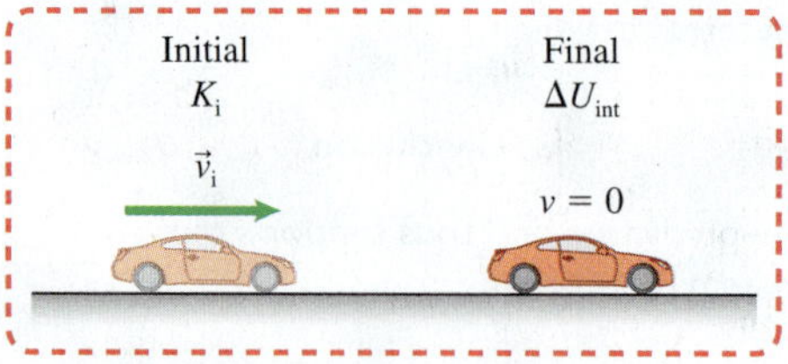  | 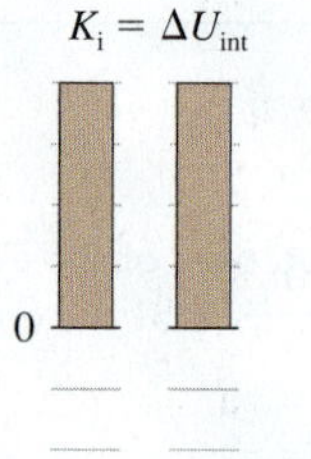  |
| (3) A pop-up toy is compressed and when released pops up to a maximum height of 0.50 m.<br>$U_s \rightarrow U_g$<br>The system's elastic potential energy is converted to gravitational potential energy. | 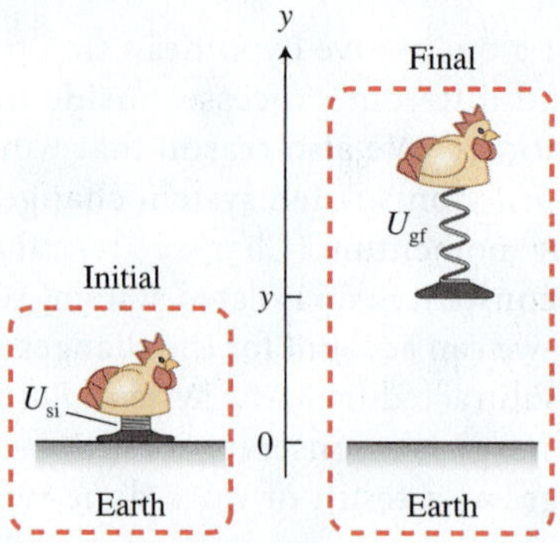  | 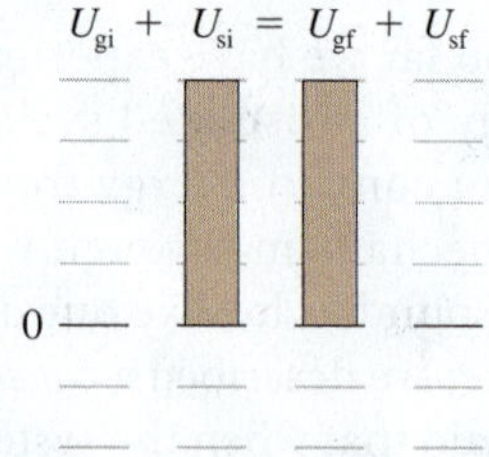  |

**REASONING SKILL** Constructing a qualitative work-energy bar chart.

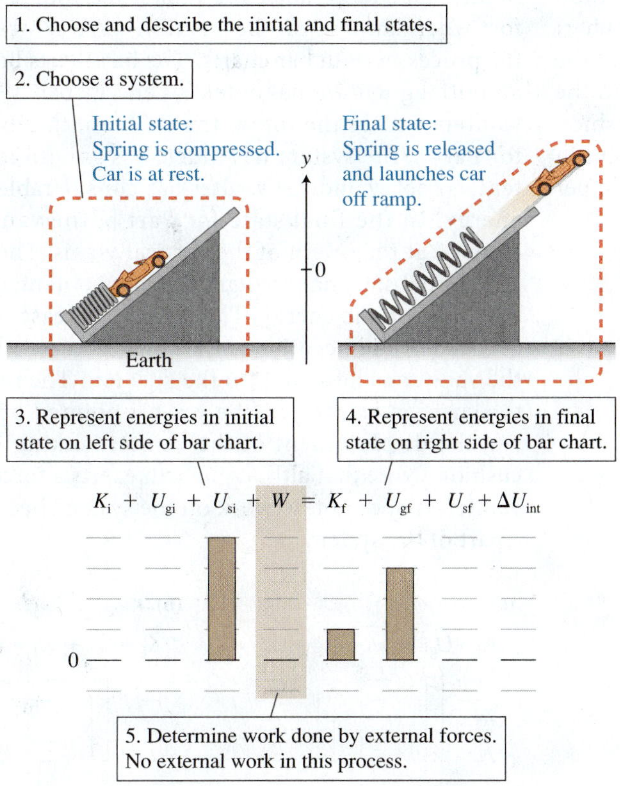
1. Choose and describe the initial and final states.
2. Choose a system.
Initial state:
Spring is compressed.
Car is at rest.
Final state:
Spring is released
and launches car
off ramp.
y
0
Earth
3. Represent energies in initial state on left side of bar chart.
4. Represent energies in final state on right side of bar chart.
$K_i + U_{gi} + U_{si} + W = K_f + U_{gf} + U_{sf} + \Delta U_{int}$
0
5. Determine work done by external forces. No external work in this process.

**TIP** In the Skill box for work-energy bar charts, the system is isolated. No work is done on it. If we do not include Earth in the system, then Earth will do negative work on the cart. However, such a system will not have gravitational potential energy.

## The generalized work-energy principle

We can summarize what we have discovered about work and energy in the generalized work-energy principle.

**Generalized work-energy principle** **The sum of the initial energies of a system plus the work done on the system by external forces equals the sum of the final energies of the system:**

$$U_i + W = U_f \quad (6.3)$$

**or**

$$(K_i + U_{gi} + U_{si}) + W = (K_f + U_{gf} + U_{sf} + \Delta U_{int}) \quad (6.3)$$

**Note that we have moved $U_{int\,i}$ to the right hand side ($\Delta U_{int} = U_{int\,f} - U_{int\,i}$) since values of internal energy are rarely known, while internal energy changes are.**

‹Active Learning Guide

The generalized work-energy principle allows us to define the total energy of a system as the sum of the different types of energy in the system. Total energy is measured in the same units as work and changes when work is done on the system.

The work-energy principle also gives insight into why perpetual motion machines cannot exist. A *perpetual motion machine* is a mechanical device that, once set in motion, continuously and indefinitely does useful things by transferring energy to the environment. This seems impossible because this energy transfer by negative work causes the system's energy to decrease. The machine cannot continue forever.

**CONCEPTUAL EXERCISE 6.2 Pole vaulter**

A pole vaulter crosses the bar high above the cushion below. Construct a sketch and a work-energy bar chart for two processes relative to the vaulter's jump: (a) the initial state is at the highest point in the jump and the final state is just before the vaulter reaches the cushion below, and (b) the initial state is at the highest point in the jump and the final state is at the instant the jumper has stopped after sinking into the cushion.

**Sketch and translate** We sketch the processes with different final states (a) just before he hits the cushion and (b) where he is stopped by the cushion. The system includes the vaulter and Earth, but not the cushion. The zero point of the vertical axis is at the vaulter's position after he has sunk into the cushion and stopped.

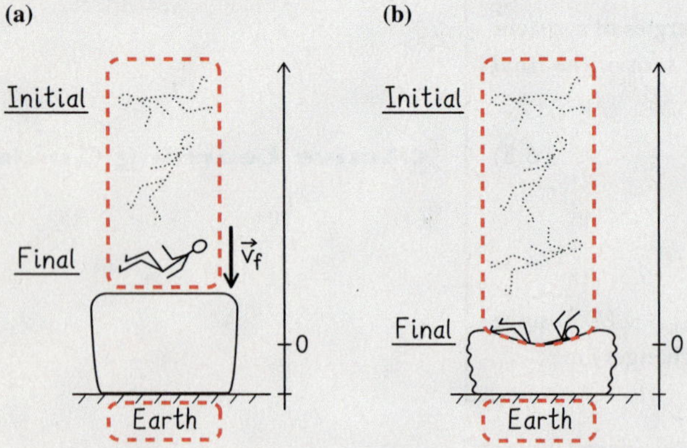


**Simplify and diagram** We assume that we can ignore air resistance and that the vaulter has zero speed (zero kinetic energy) at the top of the flight. We assume also that the internal energy of the vaulter does not change significantly (in Part b there is a small increase in internal energy, since the vaulter feels some discomfort when landing on the cushion). We can represent the processes with bar charts. The bar charts both have an initial gravitational potential energy bar. When the vaulter reaches the top of the cushion, the final state for Part a, the system has much less gravitational potential energy and the vaulter has considerable kinetic energy. In the final state for Part b, the vaulter has stopped at the origin of the vertical $y$-axis. The system in the final state has zero gravitational potential energy and zero kinetic energy. The energy decrease occurred because of the negative work done by the cushion on the vaulter as he sank into the cushion. The force that the cushion exerted on the vaulter pointed up, opposite the displacement of the vaulter sinking into the cushion. Note that although Earth exerts a force on the vaulter, it does not do work on the system, because it is a part of the system.

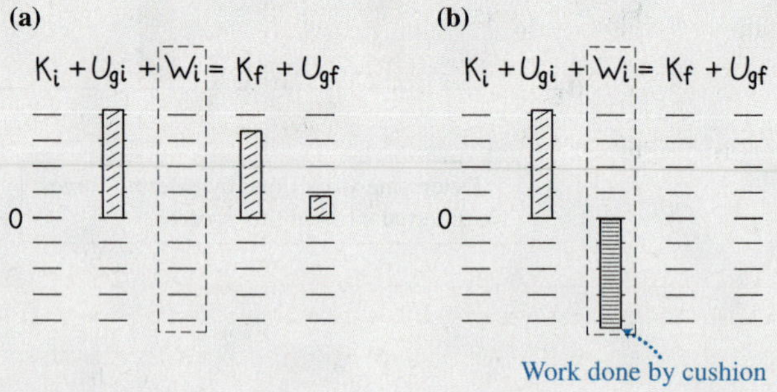


**Try it yourself:** You throw a ball straight upward as shown below. The system is the ball and Earth (but not your hand). Ignore interactions with the air. Draw a work-energy bar chart starting when the ball is at rest in your hand and ending when the ball is at the very top of its flight.

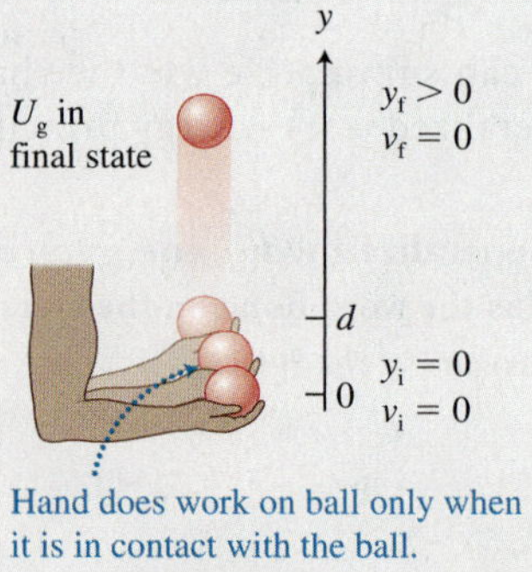


*Answer:* See the diagram at the right. Note that there is no kinetic energy represented in the bar chart because the ball was not moving in either the initial state or the final state. It does not matter that it was moving in the time interval between those two states.

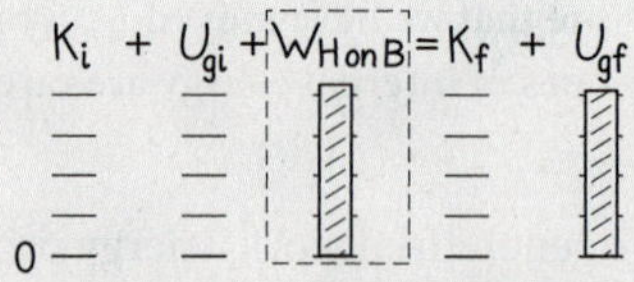

**TIP** Note that the amount of gravitational potential energy in a system depends on where the origin is placed on the vertical $y$-axis. This placement is arbitrary. The important thing is the change in position and the corresponding change in gravitational potential energy.

You might be wondering what objects to include in a system and what objects not to include. Generally, it is preferable to have a larger system so that the changes occurring can be included as energy changes within the system rather than as the work done by external forces. However, often it is best to exclude something like a motor from a system because its energy changes are complex (**Figure 6.2**). For instance, if you are studying a moving elevator, you might choose to exclude from the system the motor that turns the cable while lifting the elevator. However, you can include the motor's effect on the process by determining the magnitude of the force that the cable exerts when pulling up on the elevator.

**Figure 6.2** Energy changes in the motor pulling up on the elevator cable are very complicated. It is best to exclude the motor from the system.

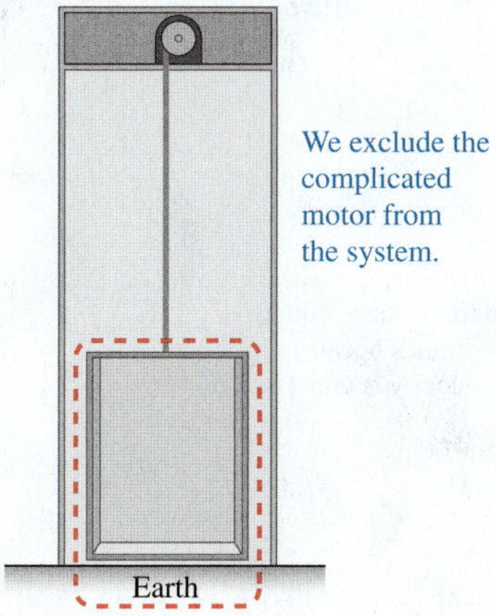
We exclude the complicated motor from the system.
Earth

**Review Question 6.2** A system can possess energy but it cannot possess work. Why?

# 6.3 Quantifying gravitational potential and kinetic energies

In this section we will use what we know about work, forces, and kinematics to devise mathematical expressions for two different types of energy—gravitational potential energy and kinetic energy. We start with gravitational potential energy.

## Gravitational potential energy

**<Active Learning Guide**

Imagine that a rope lifts a heavy box upward at a constant negligible velocity (**Figure 6.3a**). The rope is attached to a motor above, which is not shown in the figure. First, we choose only the box as the system and apply Newton's second law to find the magnitude of the force that the rope exerts on the box. Since the box moves up at constant velocity, the upward tension force $\vec{T}_{\text{R on B}}$ exerted by the rope on the box is equal in magnitude to the downward gravitational force $\vec{F}_{\text{E on B}}$ exerted by Earth on the box (see the force diagram in Figure 6.3b). Since the magnitude of the gravitational force is $m_{\text{B}}g$, we find that the magnitude of the tension force for this process is $T_{\text{R on B}} = m_{\text{B}}g$.

To derive an expression for gravitational potential energy, we must change the boundaries of the system to include the box and Earth (if Earth is not included in the system, the system does not have gravitational potential energy). The origin of a vertical $y$-axis is the ground directly below the box with the positive direction upward. The initial state of the system is the box at position $y_{\text{i}}$ moving upward at a negligible speed $v_{\text{i}} \approx 0$. The final state is the box at position $y_{\text{f}}$ moving upward at the same negligible speed $v_{\text{f}} \approx 0$. According to work-energy Eq. (6.3):

$$U_{\text{i}} + W = U_{\text{f}}$$

The rope does work on the box, lifting the box from vertical position $y_{\text{i}}$ to $y_{\text{f}}$:

$$W_{\text{R on B}} = T_{\text{R on B}}d\cos\theta = T_{\text{R on B}}(y_{\text{f}} - y_{\text{i}})\cos 0° = mg(y_{\text{f}} - y_{\text{i}})$$

**Figure 6.3** Lifting a box at a negligible constant speed. Notice that the system is box-Earth. (a) Initial and final states; (b) force diagram for the box as the system; (c) energy bar chart for the process.

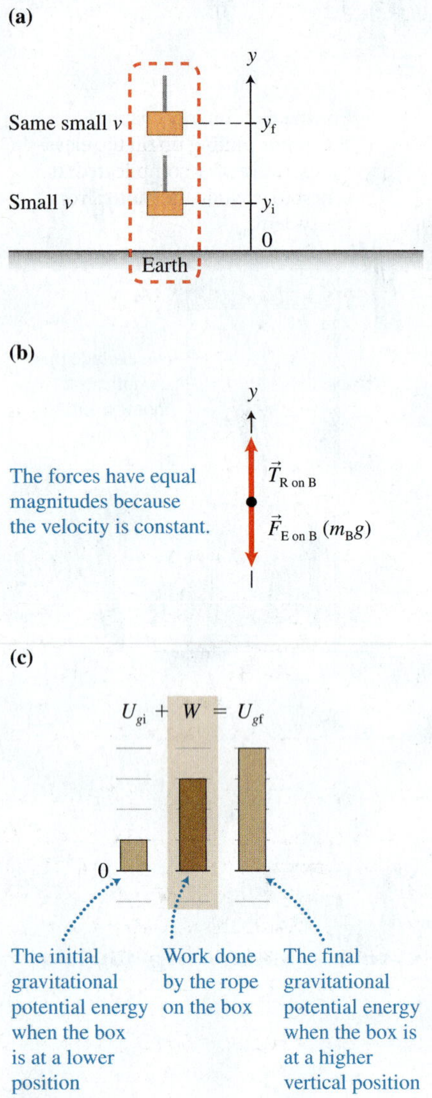


where we substituted $T_{R\text{ on }B} = mg$ and $\cos 0° = 1$. The kinetic energy did not change. Substituting this expression for work into the work-energy equation, we get

$$U_{g\,i} + mg(y_f - y_i) = U_{g\,f}$$

Figure 6.3c represents this information with an energy bar chart. We now have an expression for the change in the gravitational potential energy of the system: $U_{g\,f} - U_{g\,i} = mgy_f - mgy_i$. This suggests the following definition for the gravitational potential energy of a system.

> **Gravitational potential energy** **The gravitational potential energy of an object-Earth system is**
>
> $$U_g = mgy \tag{6.4}$$
>
> **where $m$ is the mass of the object, $g = 9.8\ \text{N/kg}$, and $y$ is the position of the object with respect to the zero of a vertical coordinate system (the origin of the coordinate system is our choice). The units of gravitational potential energy are kg(N/kg)m = N·m = J (joule), the same unit used to measure work and the same unit for every type of energy.**

## Kinetic energy

Next, we analyze a simple thought experiment to determine an expression for the kinetic energy of a system that consists of a single object. Imagine that your hand exerts a force $\vec{F}_{H\text{ on }C}$ on a cart of mass $m$ while pushing it toward the right a displacement $\vec{d}$ on a horizontal frictionless surface (**Figure 6.4a**). A bar chart for this process is shown in Figure 6.4b. There is no change in gravitational potential energy. The kinetic energy changes from the initial state to the final state because of the work done by the external force exerted by your hand on the cart:

$$K_i + W_{H\text{ on }C} = K_f$$

or

$$W_{H\text{ on }C} = K_f - K_i$$

We know that the work in this case equals $F_{H\text{ on }C}d\cos 0° = F_{H\text{ on }C}d$. Thus,

$$F_{H\text{ on }C}d = K_f - K_i$$

This does not look like a promising result—the kinetic energy change on the right equals quantities on the left side that do not depend on the mass or speed of the cart. However, we can use dynamics and kinematics to get a result that does depend on these properties of the cart. The horizontal component form of Newton's second law is

$$m_C a_C = F_{H\text{ on }C}$$

**Figure 6.4** The work done by the hand causes the cart's kinetic energy to increase.

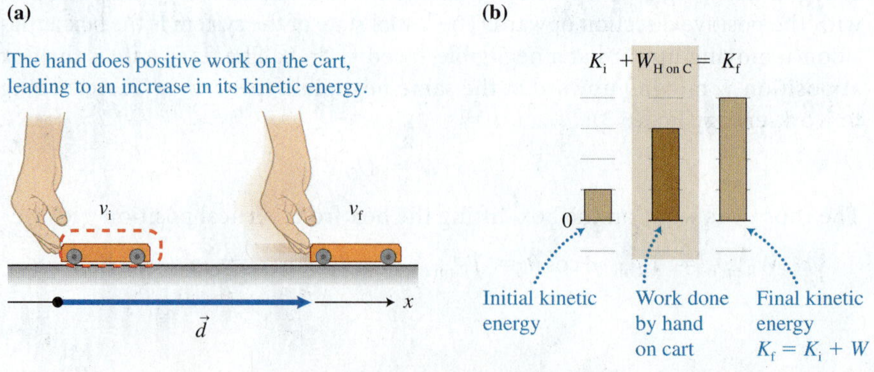

m·g   600·9.81 = 5886N



We can rearrange a kinematics equation ($v_f^2 = v_i^2 + 2ad$) to get an expression for the displacement of the cart in terms of its initial and final speeds and its acceleration:

$$d = \frac{v_f^2 - v_i^2}{2a_C}$$

Now, insert these expressions for force and displacement into the left side of the equation $F_{H\text{ on }C}d = K_f - K_i$:

$$F_{H\text{ on }C}d = (m_C a_C)\left(\frac{v_f^2 - v_i^2}{2a_C}\right) = m_C\left(\frac{v_f^2}{2} - \frac{v_i^2}{2}\right) = \frac{1}{2}m_C v_f^2 - \frac{1}{2}m_C v_i^2$$

We can now insert this result into the equation $F_{H\text{ on }C}d = K_f - K_i$:

$$\frac{1}{2}m_C v_f^2 - \frac{1}{2}m_C v_i^2 = K_f - K_i$$

It appears that $\frac{1}{2}m_C v^2$ is an expression for the kinetic energy of the cart.

**Kinetic energy** **The kinetic energy of an object is**

$$K = \frac{1}{2}mv^2 \qquad (6.5)$$

**where $m$ is the object's mass and $v$ is its speed relative to the chosen coordinate system.**

To check whether the unit of kinetic energy is the joule (J), we use Eq. (6.5) with the units $\frac{\text{kg} \cdot \text{m}^2}{\text{s}^2} = \left(\frac{\text{kg} \cdot \text{m}}{\text{s}^2}\right)\text{m} = \text{N} \cdot \text{m} = \text{J}$.

## EXAMPLE 6.3 An acorn falls

You sit on the deck behind your house. Several 5-g acorns fall from the trees high above, just missing your chair and head. Use the work-energy equation to estimate how fast one of these acorns is moving just before it reaches the level of your head.

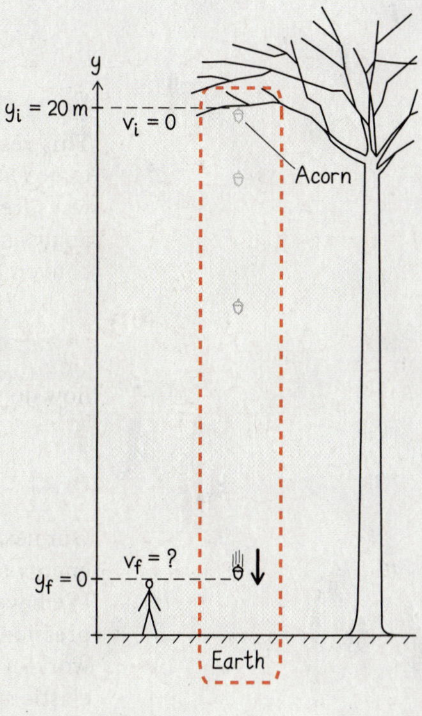


**Sketch and translate** First, we draw a sketch of the process. The system will be the acorn and Earth. The origin of a vertical $y$-axis will be at your head with the positive $y$-axis pointing up. The acorn is about 20 m above your head as it begins to fall. We keep track of kinetic energy and gravitational potential energy to find the acorn's speed when it reaches the level of your head. The initial state will be the instant the acorn leaves the tree. The final state is when it reaches the level of your head.

*(continued)*

**Simplify and diagram** The acorn is small, and we assume that the air does no significant work on the acorn as it falls. We represent the process with a bar chart.

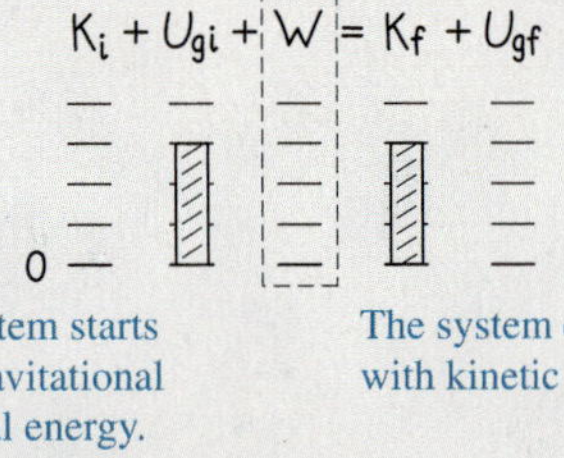
$K_i + U_{gi} + W = K_f + U_{gf}$
0
The system starts with gravitational potential energy.
The system ends with kinetic energy.

**Represent mathematically** Use the bar chart to apply the work-energy equation:

$$0 + U_{gi} + 0 = K_f + 0$$

$$mgy_i + 0 = \frac{1}{2}mv_f^2$$

Cancelling the common $m$ on each side and rearranging, we get

$$v_f = \sqrt{2(gy_i)}$$

**Solve and evaluate** Our estimate of the final speed of the acorn is

$$v_f = \sqrt{2(gy_i)} = \sqrt{2(9.8\ \text{m/s}^2)(20\ \text{m})} = 20\ \text{m/s}$$

That's 45 mi/h—that seems reasonable based on the sound it makes when it hits the deck.

**Try it yourself:** If you throw an acorn upward at a speed $v_i$, what is the maximum height above its launching position that the acorn will reach before it starts descending?

*Answer:* $\frac{v_i^2}{2g}$.

**Active Learning Guide›**

What if in Example 6.3 we had chosen only the acorn as the system of interest? In that case the system would not have any gravitational potential energy. Instead, Earth would be an external object doing positive work on the acorn system. Since the acorn is at rest initially, the system has no initial energy. Earth does work on the system. In the final state, just as the acorn reaches the level of your head, the acorn system has kinetic energy:

$$0 + W = K_f$$

$$Fd\cos\theta = \frac{1}{2}mv_f^2$$

$$(mg)y_i\cos 0° = \frac{1}{2}mv_f^2$$

$$v_f = \sqrt{2(gy_i)}$$

This result is the same as the one we arrived at using the acorn-Earth system. *The choice of system did not affect the result of the analysis.* We are always free to choose the system of interest so that it best suits the goal of our analysis, just as we are free to choose whichever coordinate system is most convenient.

**Review Question 6.3** When we use the work-energy equation, how do we incorporate the force that Earth exerts on an object?

## 6.4 Quantifying elastic potential energy

Our next goal is to construct a mathematical expression for the elastic potential energy stored by an elastic object when it has been stretched or compressed. We have a special problem in deriving this expression. When we derived expressions for gravitational potential and kinetic energies, a constant force did work in changing those energies. However, when you stretch or compress an elastic spring-like object, you have to pull or push harder the more the object is stretched or compressed. The force is not constant. How does the force you exert to stretch a spring-like object change as the object stretches?

**Table 6.5 Result of pulling on springs while exerting an increasing force.**

| Force $F$ exerted by the scale on the spring | Spring 1 stretch distance $x$ | Spring 2 stretch distance $x$ |
|---|---|---|
| 0.00 N | 0.000 m | 0.000 m |
| 1.00 N | 0.050 m | 0.030 m |
| 2.00 N | 0.100 m | 0.060 m |
| 3.00 N | 0.150 m | 0.090 m |
| 4.00 N | 0.200 m | 0.120 m |

**Figure 6.5** Measuring spring stretch $x$ caused by a scale pulling with force $F$ on a spring. Notice the difference in thickness and stiffness of the two springs.

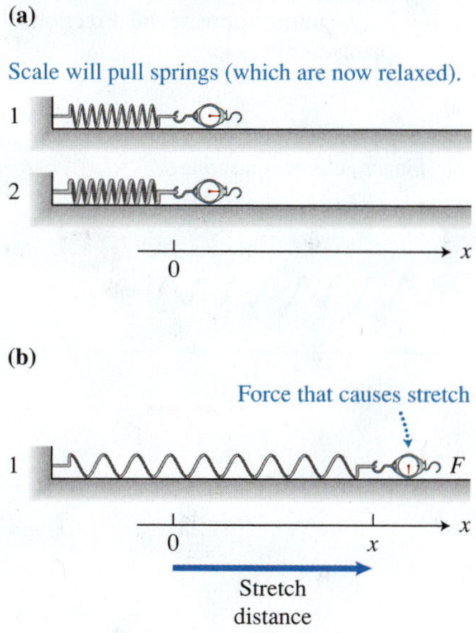


## Hooke's law

To answer this question, we use two springs of the same length: a thinner and less stiff spring 1 and a thicker and stiffer spring 2 (**Figure 6.5a**). The springs are attached at the left end to a rigid object and placed on a smooth surface. We use a scale to pull on the right end of each spring, exerting a force whose magnitude $F$ can be measured by the scale (Figure 6.5b). We record $F$ and the distance $x$ that each spring stretches from its unstretched position (see **Table 6.5**).

**Figure 6.6** shows a graph of the Table 6.5 data. We use stretch distance $x$ as an independent variable and the magnitude of the force $\vec{F}_{\text{Scale on Spring}}$ as a dependent variable. The magnitude of the force exerted by the scale on each spring is proportional to the distance that each spring stretches.

$$F_{\text{Scale on Spring}} = kx$$

The coefficient of proportionality $k$ (the slope of the $F$-versus-$x$ graph) is called the **spring constant.** The slope for the stiffer spring 2 is larger $(33\text{ N/m})$ than the slope for spring 1 $(20\text{ N/m})$. In other words, to stretch spring 1 by 1.0 m we would have to exert only a 20-N force, but we would need a 33-N force for spring 2.

Often we are interested not in the force that something exerts on the spring, but in the force that the spring exerts on something else, $\vec{F}_{\text{Spring on Scale}}$. Using Newton's third law we have $\vec{F}_{\text{Spring on Scale}} = -\vec{F}_{\text{Scale on Spring}}$. The $x$-component of this force is

$$F_{\text{Spring on Scale }x} = -kx$$

Note that if an object stretches the spring to the right in the positive direction, the spring pulls back on the object in the opposite negative direction (**Figure 6.7a**, in this case the object is your finger). If the object compresses the spring to the left in the negative direction, the spring pushes back on the object in the opposite positive direction (Figure 6.7b). These observations are the basis for a rule first developed by Robert Hooke (1635–1703), called **Hooke's law.**

**Figure 6.6** A graph of the stretch of springs 1 and 2 when the same force is exerted on each spring. The bigger slope indicates a smaller stretch distance when the same force is exerted on the spring.

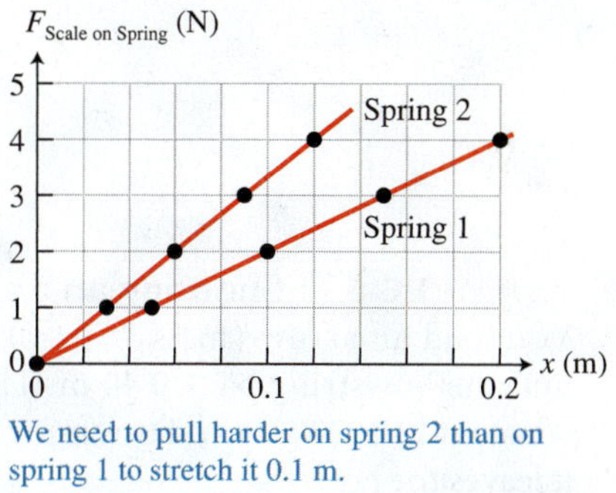


**Elastic force (Hooke's law)** If any object causes a spring to stretch or compress, the spring exerts an elastic force on that object. If the object stretches the spring along the $x$-direction, the $x$-component of the force the spring exerts on the object is

$$F_{\text{S on O }x} = -kx \quad (6.6)$$

The spring constant $k$ is measured in newtons per meter and is a measure of the stiffness of the spring (or any elastic object); $x$ is the distance that the object has been stretched/compressed (not the total length of the object). The elastic force exerted by the spring on the object points in a direction opposite to the direction it was stretched (or compressed)—hence the negative sign in front of $kx$. The object in turn exerts a force on the spring:

$$F_{\text{O on S }x} = +kx$$

**Figure 6.7** Hooke's law. (a) The force that the spring exerts on the finger stretching the spring $F_{S\text{ on }F}$ points opposite the direction the spring stretches. (b) The force that the spring exerts on the finger compressing the spring $F_{S\text{ on }F}$ points opposite the direction of the spring compression.

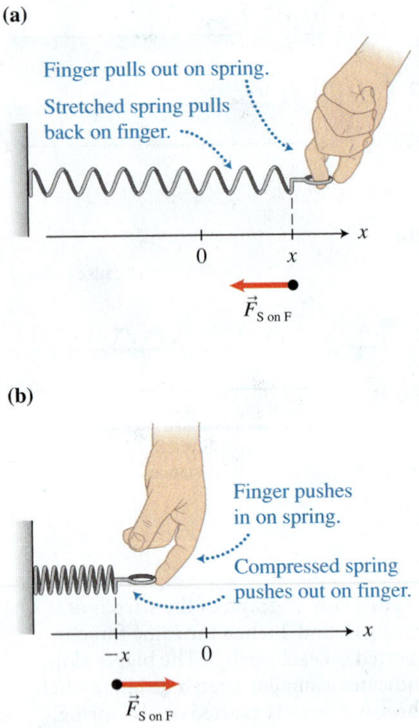


## Elastic potential energy

Our goal now is to develop an expression for the elastic potential energy of an elastic stretched or compressed object (such as a stretched spring). Consider the constant slopes of the lines in the graph shown in Figure 6.6. While stretching the spring with your hand from zero stretch ($x = 0$) to some arbitrary stretch distance $x$, the magnitude of the force your hand exerts on the spring changes in a linear fashion from zero when unstretched to $kx$ when stretched.

To calculate the work done on the spring by such a variable force, we can replace this variable force with the average force $(F_{\text{H on S}})_{\text{average}}$:

$$(F_{\text{H on S}})_{\text{average}} = \frac{0 + kx}{2}$$

The force your hand exerts on the spring is in the same direction as the direction in which the spring stretches. Thus the work done by this force on the spring to stretch it a distance $x$ is

$$W = (F_{\text{H on S}})_{\text{average}}x = \left(\frac{1}{2}kx\right)x = \frac{1}{2}kx^2$$

This work equals the change in the spring's elastic potential energy. Assuming that the elastic potential energy of the unstretched spring is zero, the work we calculated above equals the final elastic potential energy of the stretched spring.

**Elastic potential energy** **The elastic potential energy of a spring-like object with a spring constant $k$ that has been stretched or compressed a distance $x$ from its undisturbed position is**

$$U_s = \frac{1}{2}kx^2 \tag{6.7}$$

**Just like any other type of energy, the unit of elastic potential energy is the joule (J).**

### EXAMPLE 6.4 Shooting an arrow

You load an arrow (mass = 0.090 kg) into a bow and pull the bowstring back 0.40 m. The bow has a spring constant $k = 900\ \text{N/m}$. Determine the arrow's speed as it leaves the bow.

**Sketch and translate** We sketch the process, as shown below. The system is the bow and arrow. In the initial state, the bowstring is pulled back 0.40 m. In the final state, the string has just relaxed and the arrow has left the string.

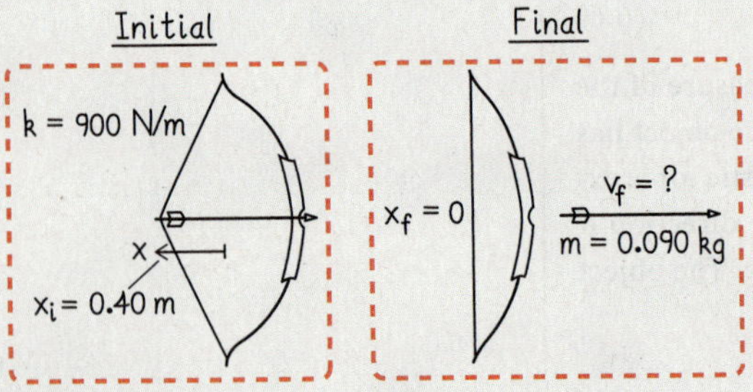


**Simplify and diagram** Since the arrow moves horizontally, we do not need to keep track of gravitational potential energy. The initial elastic potential energy of the bow is converted into the final kinetic energy of the arrow. We represent the process with the bar chart.

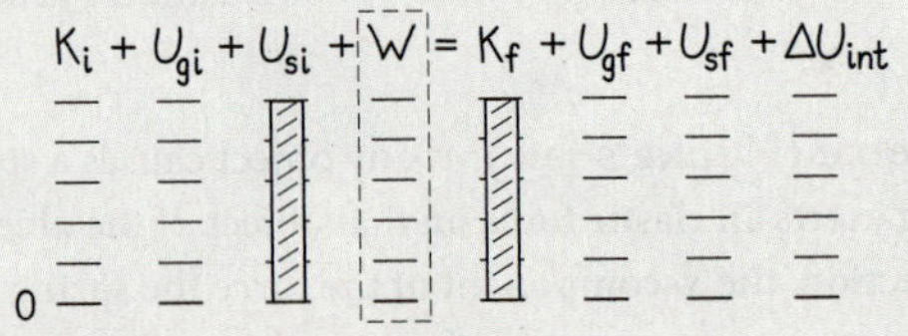


**Represent mathematically** Use the bar chart to help apply the work-energy equation:

$$U_i + W = U_f$$

$$\Rightarrow K_i + U_{si} + 0 = K_f + U_{sf}$$

$$\Rightarrow 0 + \frac{1}{2}kx^2 = \frac{1}{2}mv^2 + 0$$

Multiply both sides of the above by 2, divide by $m$, and take the square root to get

$$v = \sqrt{\frac{k}{m}}x$$

**Solve and evaluate**

$$v = \sqrt{\frac{k}{m}}x = \sqrt{\frac{900\text{ N/m}}{0.090\text{ kg}}}(0.40\text{ m}) = 40\text{ m/s}.$$

This is reasonable for the speed of an arrow fired from a bow. The units of $\sqrt{\frac{k}{m}}x$ are equivalent to $\frac{\text{m}}{\text{s}}$:

$$\sqrt{\frac{\text{N}}{\text{m}\cdot\text{kg}}}(\text{m}) = \sqrt{\frac{\text{kg}\cdot\text{m}}{\text{s}^2\cdot\text{m}\cdot\text{kg}}}(\text{m}) = \frac{\text{m}}{\text{s}}$$

**Try it yourself:** If the same arrow were shot vertically, how high would it go?

*Answer:* 82 m.

**Review Question 6.4** If the magnitude of the force exerted by a spring on an object is $kx$, why is it that the work done to stretch the spring a distance $x$ is not equal to $kx \cdot x = kx^2$?

## 6.5 Friction and energy conversion

In nearly every mechanical process, objects exert friction forces on each other. Sometimes the effect of friction is negligible (for example, in an air hockey game), but most often friction is important (for example, a driver applying the brakes to avoid a collision). Our next goal is to investigate how we can incorporate friction into work and energy concepts. Let's analyze a car skidding to avoid an accident.

### Can friction force do work?

‹Active Learning Guide

Imagine that the car's brakes have locked, and the tires are skidding on the road surface (**Figure 6.8a**). Let's first choose the system of interest to be the car. How much work does the road's friction force do on the car? There are three forces exerted on the car by external objects, and two of them cancel—the gravitational force exerted by Earth and the normal force exerted by the road surface. The only force left is the kinetic friction force $f_k$ exerted by the road surface on the car, which does work on the car and slows it to a stop:

$$W_{\text{friction}} = f_k d \cos 180° = -f_k d$$

The negative work done by the friction force causes the car's kinetic energy to decrease to zero (Figure 6.8b). Mathematically,

$$K_i + (-f_k d) = 0$$

But we've left out a very important feature of friction. If you touched the car's tires just after the car came to rest, they would be warm to the touch. You would also notice black skid marks on the road—some of the rubber had been scraped off the tires. Thus, the internal energy of the system increased ($\Delta U_{\text{int}} > 0$). There should be a term on the right side of the above equation indicating this increase in internal energy of the car. But the two terms on the left side of the above equation add to zero, so we would get $0 = \Delta U_{\text{int}}$, which is not possible.

This same difficulty occurs with another process. Imagine that you pull a rope attached to a box (the system object) so that the box moves at constant very slow velocity on a rough carpet (**Figure 6.9a**). The box is moving so slowly that we will ignore its kinetic energy. A force diagram for the box is shown in Figure 6.9b.

The bar chart (Figure 6.9c) shows that the force exerted by the rope does positive work on the box system; the friction force does negative work. These

**Figure 6.8** Friction. If we use the car alone as the system for this energy analysis involving friction, we cannot account for the gain in the internal thermal and chemical energy (removing tread) of the tire.

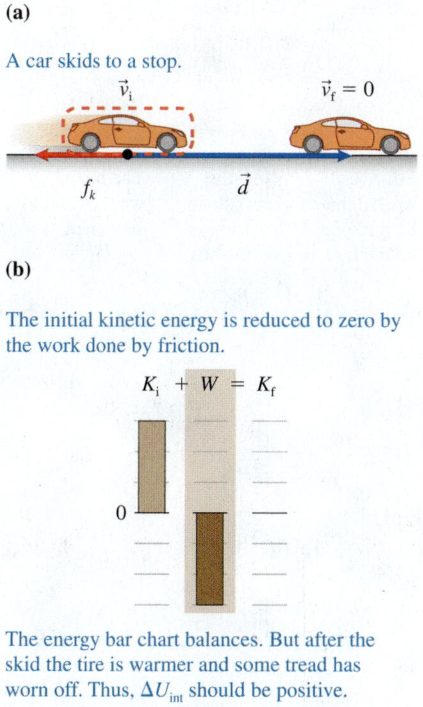

(a)
A car skids to a stop.
$\vec{v}_i$
$\vec{v}_f = 0$
$f_k$
$\vec{d}$
(b)
The initial kinetic energy is reduced to zero by the work done by friction.
$K_i + W = K_f$
0
The energy bar chart balances. But after the skid the tire is warmer and some tread has worn off. Thus, $\Delta U_{\text{int}}$ should be positive.

**Figure 6.9** Pulling a box across a rough carpet. The total work done on the box is zero but its internal energy changes. You cannot account for the internal energy change of the box if it alone is the system.

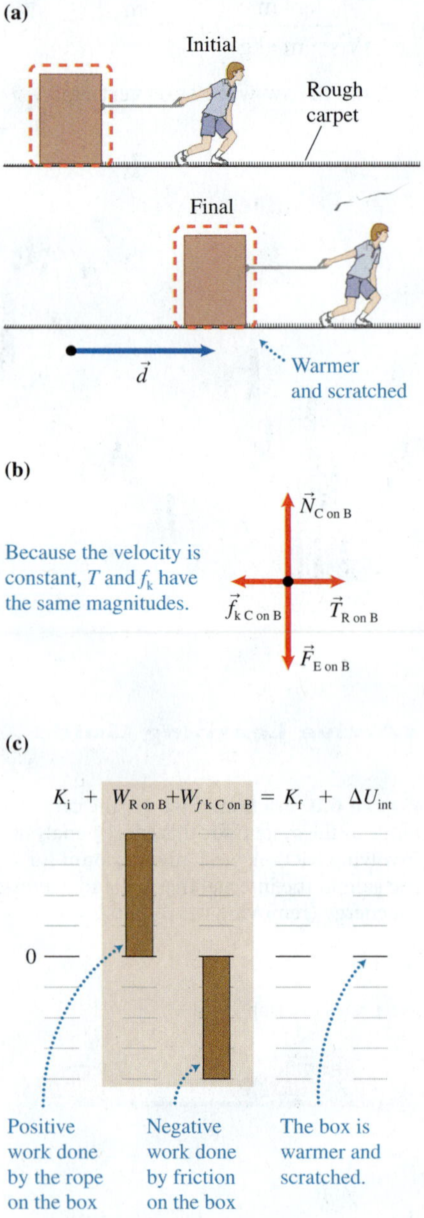

(a)
Initial
Rough carpet
Final
$\vec{d}$
Warmer and scratched
(b)
Because the velocity is constant, $T$ and $f_k$ have the same magnitudes.
$\vec{N}_{C\text{ on }B}$
$\vec{f}_{k\,C\text{ on }B}$
$\vec{T}_{R\text{ on }B}$
$\vec{F}_{E\text{ on }B}$
(c)
$K_i + W_{R\text{ on }B} + W_{f\,k\,C\text{ on }B} = K_f + \Delta U_{int}$
0
Positive work done by the rope on the box
Negative work done by friction on the box
The box is warmer and scratched.

horizontal forces have exactly the same magnitude (the box is moving at constant velocity). Thus, the sum of the work done by those forces on the system is zero and the energy of the system should not change.

$$T_{\text{Rope on Box}}d\cos 0° + f_{\text{k Carpet on Box}}d\cos 180° = T_{\text{Rope on Box}}d - f_{\text{k Carpet on Box}}d = 0$$

However, if you touch the bottom of the box at its final position, you find that it is warmer than before you started pulling and the box has scratches on its bottom. Again, the internal energy of the box increased. We get $0 = \Delta U_{\text{int}}$, where $\Delta U_{\text{int}}$ is greater than zero. Again, the zero work done on the box does *not* equal the positive increase in internal energy. This is a contradiction of the work-energy principle. How can we resolve this?

## The effect of friction as a change in internal energy

The key to resolving this problem is to change the system. The new system will include both surfaces that are in contact, for example, the car and the road, or the box and the carpet. The friction force is then an internal force and therefore does no work on the system. But there is a change in the internal energy of the system caused by friction between the two surfaces.

When the rope is pulling the box across the rough carpet at constant velocity, we know that if the rope pulls horizontally on the box, the magnitude of the force that it exerts on the box $T_{\text{R on B}}$ must equal the magnitude of the friction force that the carpet exerts on the box $f_{\text{k C on B}}$. But now the box and carpet are both in the system—so the force exerted by the rope is the only external force. Thus the work done on the box-carpet system is

$$W = T_{\text{R on B}}d\cos 0°$$

Substituting $T_{\text{R on B}} = f_{\text{k C on B}}$ and $\cos 0° = 1.0$ into the above, we get

$$W = +f_{\text{k C on B}}d$$

The only system energy change is its internal energy $\Delta U_{\text{int}}$. The work-energy equation for pulling the box across the surface is

$$W = \Delta U_{\text{int}}$$

After inserting the expression for the work done on the box and rearranging, we get

$$\Delta U_{\text{int}} = +f_k d$$

We have constructed an expression for the change in internal energy of a system caused by the friction force that the two contacting surfaces in the system exert on each other when one object moves a distance $d$ across the other.

**Increase in the system's internal energy due to friction**

$$\Delta U_{\text{int}} = +f_k d \quad (6.8)$$

**where $f_k$ is the magnitude of the average friction force exerted by the surface on the object moving relative to the surface and $d$ is the distance that the object moves across that surface. The increase in internal energy is shared between the moving object and the surface.**

Including friction in the work-energy equation as an increase in the system's internal energy produces the same result as calculating the work done by friction force. In this new approach, there is an increase in internal energy $\Delta U_{\text{int}} = +f_k d$ in a system that includes the two surfaces rubbing against each other (this expression goes on the right side of the work-energy equation).

In the work done by friction approach, the negative work done by friction $W_{\text{friction}} = -f_k d$ is included if one of the surfaces is not in the system (this term goes on the left side of the work-energy equation). So mathematically, they have the same effect. When we include the two surfaces in the system, we can see why a skidding tire or the moving box gets warmer and why there might be structural changes. If we include only the car or box in the system and consider work done by friction, the change in the internal energy of the rubbing surfaces is a mystery. In this book we prefer to include both surfaces in the system and consider the increase in internal energy caused by friction.

**EXAMPLE 6.5 Skidding to a stop**

You are driving your car when another car crosses the road at an intersection in front of you. To avoid a collision, you apply the brakes, leaving 24-m skid marks on the road while stopping. A police officer observes the near collision and gives you a speeding ticket, claiming that you were exceeding the 35 mi/h speed limit. She estimates your car's mass as 1390 kg and the coefficient of kinetic friction $\mu_k$ between your tires and this particular road as about 0.70. Do you deserve the speeding ticket?

**Sketch and translate** We sketch the process. We choose your car and the road surface as the system. We need to decide if your car was traveling faster than 35 mi/h at the instant you applied the brakes.

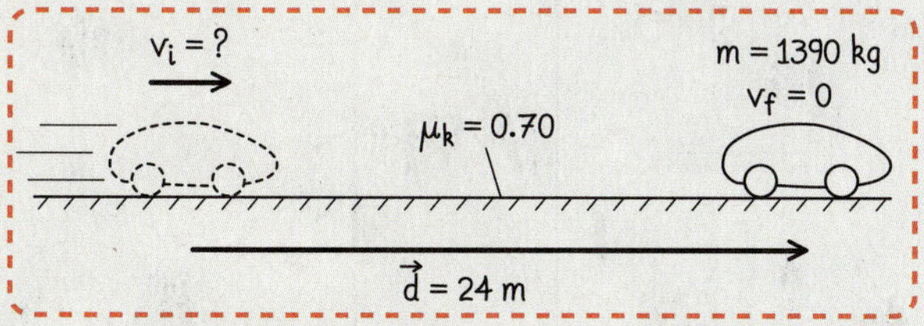


**Simplify and diagram** Assume that the process occurs on a horizontal level road and neglect interactions with the air. The initial state is just before the brakes are applied. The final state is just after your car has come to rest. We draw the energy bar chart to represent the process. In the initial state, the system has kinetic energy. In the final state, the system has no kinetic energy and has increased internal energy due to friction.

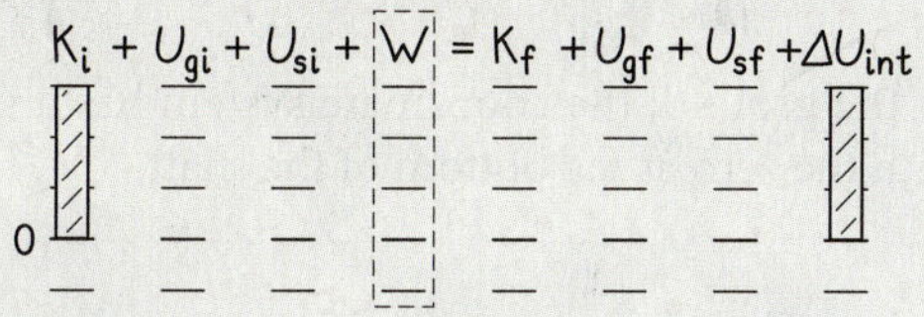


**Represent mathematically** Convert the bar chart into an equation:

$$K_i + 0 = U_{\text{int f}}$$

$$\Rightarrow \frac{1}{2}m_C v_i^2 = f_{k\,\text{R on C}}\,d$$

$$\Rightarrow \frac{1}{2}m_C v_i^2 = (\mu_k N_{\text{R on C}})d$$

The magnitude of the upward normal force $N_{\text{R on C}}$ that the road exerts on the car equals the magnitude of the downward gravitational force $F_{\text{E on C}} = m_C g$ that Earth exerts on the car: $N_{\text{R on C}} = m_C g$. Thus the above becomes

$$\frac{1}{2}m_C v_i^2 = \mu_k m_C g d$$

**Solve and evaluate** Rearranging the above and canceling the car mass, we get

$$v_i = \sqrt{2\mu_k g d} = \sqrt{2(0.70)(9.8\ \text{N/kg})(24\ \text{m})}$$

$$= (18.1\ \text{m/s})\left(\frac{1\ \text{mi/h}}{0.447\ \text{m/s}}\right)$$

$$= 41\ \text{mi/h}$$

It looks like you deserve the ticket. We ignored the resistive drag force that air exerts on the car, which could be significant for a car traveling at 41 mi/h. Thus, air resistance helps the car's speed decrease, which means you actually were traveling faster than 41 mi/h.

**Try it yourself:** Imagine the same situation as above, only you are driving a 2000-kg minivan. Will the answer for the initial speed increase or decrease?

*Answer:* The speed does not change.

**Review Question 6.5** Why, when friction cannot be neglected, is it useful to include both surfaces in the system when analyzing processes using the energy approach?

## 6.6 Skills for analyzing processes using the work-energy principle

**Active Learning Guide›**

In this section, we use a general problem-solving strategy to analyze work-energy processes. The general strategy is described on the left side of the table in Example 6.6 and illustrated on the right side for a specific process.

### PROBLEM-SOLVING STRATEGY Applying the work-energy principle

**EXAMPLE 6.6 An elevator slows to a stop**

A 1000-kg elevator is moving downward. While moving down at 4.0 m/s, its speed decreases steadily until it stops in 6.0 m. Determine the magnitude of the tension force that the cable exerts on the elevator ($T_{\text{C on El}}$) while it is stopping.

**Sketch and translate**

- Sketch the initial and final states of the process, labeling known and unknown information.
- Choose the system of interest.
- Include the object of reference and the coordinate system.

- The elevator and Earth are in the system. We exclude the cable—its effect will be included as the work done by the cable on the elevator.

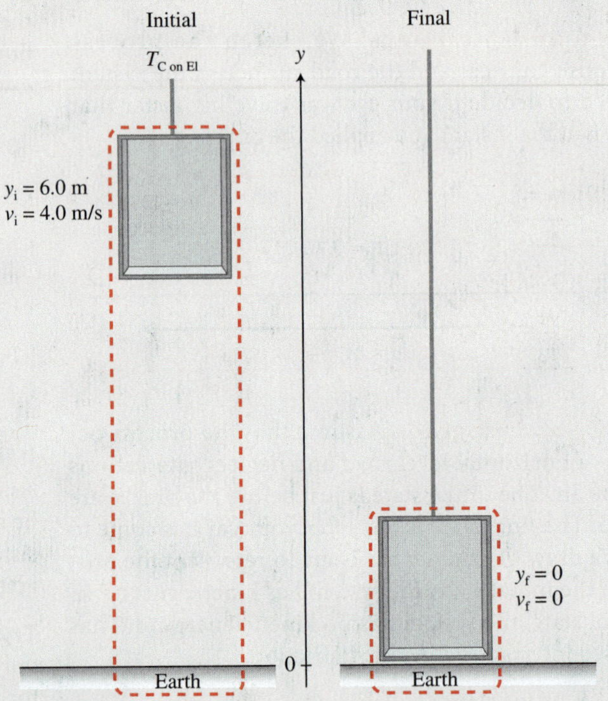


- The observer is on the ground; the coordinate system has a vertical axis pointing up with the zero at the bottom of the shaft.

**Simplify and diagram**

- What simplifications can you make to the objects, interactions, and processes?
- Decide which energy types are changing.
- Are external objects doing work?

- We assume that the cable exerts a constant force on the elevator and that the elevator can be considered a point-like object since all of its points move the same way (no deformation).
- We will keep track of kinetic energy (since the elevator's speed changes) and gravitational potential energy (since the elevator's vertical position changes).

- Use the initial-final sketch to help draw a work-energy bar chart. Include work bars (if needed) and initial and final energy bars for the types of energy that are changing. Specify the zero level of gravitational potential energy.

- The tension force exerted by the cable on the elevator does negative work (the tension force points up, and the displacement of the elevator points down).

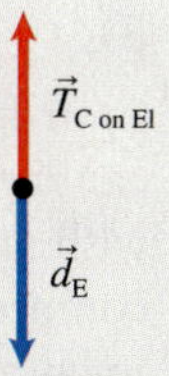


- The zero gravitational potential energy is at the lowest position of the elevator. In its initial state the system has kinetic energy and gravitational potential energy. In its final state the system has no energy.

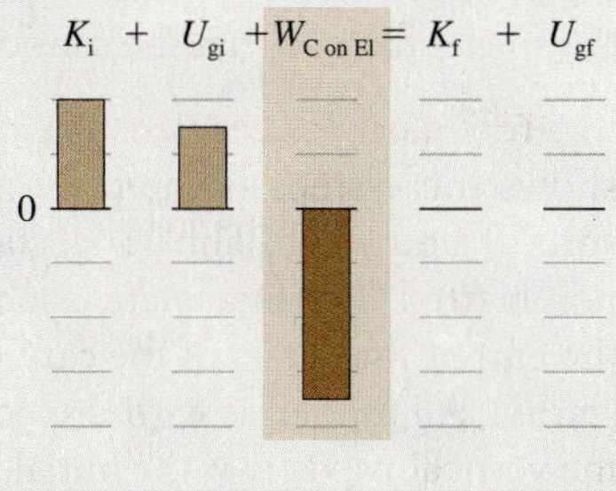


**Represent mathematically**

- Convert the bar chart into a mathematical description of the process. Each bar in the chart will appear as a single term in the equation.

$$U_i + W = U_f$$

$$\left(\frac{1}{2}\right)mv_i^2 + mgy_i + T_{\text{C on El}}(y_i - 0)\cos 180° = 0$$

$$\left(\frac{1}{2}\right)mv_i^2 + mgy_i - T_{\text{C on El}}\, y_i = 0$$

**Solve and evaluate**

- Solve for the unknown and evaluate the result.
- Does it have the correct units? Is its magnitude reasonable? Do the limiting cases make sense?

$$T_{\text{C on El}} = mg + \frac{mv^2}{2y_i}$$

$$= (1000\text{ kg})(9.8\text{ N/kg}) + \frac{(1000\text{ kg})(4.0\text{ m/s})^2}{2(6.0\text{ m})}$$

$$= 11{,}000\text{ N}$$

- The result has the correct units. The force that the cable exerts is more than the 9800-N force that Earth exerts. This is reasonable since the elevator slows down while moving down; thus the sum of the forces exerted on it should point up.
- Limiting case: If the elevator slowed down over a much longer distance ($y_f$ = very large number instead of 6.0 m), then the force would be closer to 9800 N.

**Try it yourself:** Solve the same problem using Newton's second law and kinematics.

*Answer:* 11,000 N.

## Human cannonball

The Try It Yourself exercise demonstrates that we can obtain the same result for Example 6.6 using Newton's second law and kinematics. However, the energy approach is often easier and quicker. Knowing that, let's revisit the human cannonball problem using the ideas of work and energy. (We first analyzed this using Newton's laws and kinematics in Example 3.9.)

**EXAMPLE 6.7 The human cannonball again**

In order to launch a 60-kg human so that he leaves the cannon moving at a speed of 15 m/s, you need a spring with an appropriate spring constant. This spring will be compressed 3.0 m from its natural length when it is ready to launch the person. The cannon is oriented at an angle of 37° above the horizontal. What spring constant should the spring have so that the cannon functions as desired?

**Sketch and translate** We draw the sketch first. It shows the system as the person, the cannon (with the spring), and Earth. The initial state is just before the cannon is fired. The final state is when the person is leaving the end of the barrel of the cannon 3.0 m from where he started. All motion is with respect to Earth. The origin of the vertical $y$-axis is at the initial position of the person.

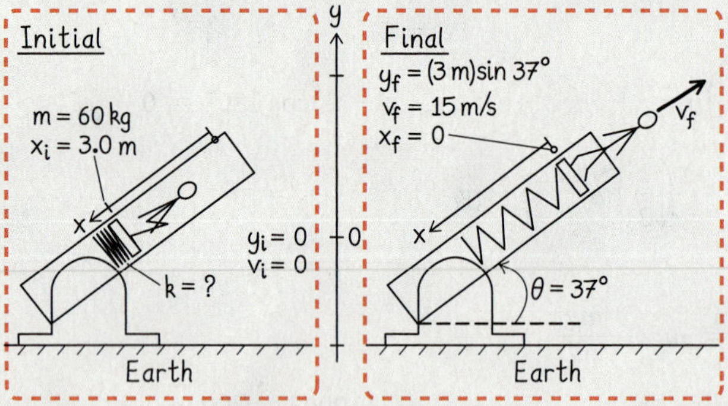


**Simplify and diagram** Assume that the spring obeys Hooke's law and neglect the relatively small amount of friction between the person and the inside barrel walls of the cannon and between the person and the air. We need to keep track of kinetic energy (the person's speed changes), gravitational potential energy (Earth is in the system and the person's vertical position changes relative to the vertical $y$-axis), and elastic potential energy changes (the spring compression changes relative to a special $x$-axis used to keep track of elastic energy). No external forces are being exerted on the system, so the total energy of the system is constant. We draw the bar chart to represent the process.

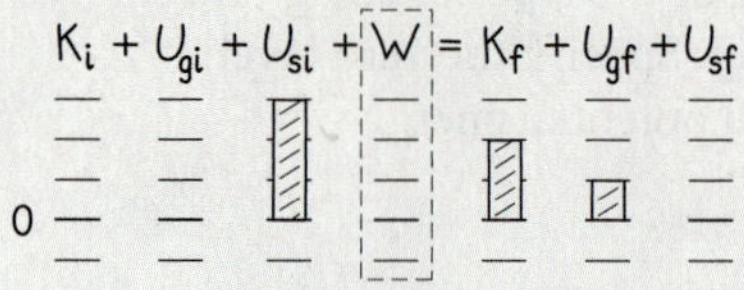


**Represent mathematically** The work-energy equation for this process is

$$U_i + W = U_f$$

$$\Rightarrow U_{si} + 0 = K_f + U_{gf}$$

$$\Rightarrow \frac{1}{2}kx_i^2 = \frac{1}{2}mv_f^2 + mgy_f$$

**Solve and evaluate** Dividing both parts of the equation by $(1/2)x_i^2$ we get

$$k = \frac{2\left(\frac{1}{2}mv_f^2 + mgx_i \sin 37°\right)}{x_i^2}$$

$$= \frac{2\left[\frac{1}{2}(60\text{ kg})(15\text{ m/s})^2 + (60\text{ kg})(9.8\text{ N/kg})(3.0\text{ m})\sin 37°\right]}{(3.0\text{ m})^2}$$

$$= 1736\text{ N/m} \approx 1700\text{ N/m}$$

The units of the answer are correct for a spring constant. The spring constant is always a positive number and we obtained a positive number. The magnitude of $k$ is quite large, which means this is a stiff spring. This makes sense given that it is launching a person.

**Try it yourself:** What should the spring constant of a spring be if there is a 100-N friction force exerted on a human cannonball while he is moving up the barrel?

*Answer:* 1800 N/m.

## High blood pressure

We have used the work-energy approach to examine several real-world phenomena. Let's use it to examine how the body pumps blood.

**BIO CONCEPTUAL EXERCISE 6.8 Stretching the aorta**

Every time your heart beats, the left ventricle pumps about 80 cm$^3$ of blood into the aorta, the largest artery in the human body. This pumping action occurs during a very short time interval, about 0.13 s. The elastic aorta walls stretch to accommodate the extra volume of blood. During the next 0.4 s or so, the walls of the aorta contract, applying pressure on the blood and moving it out of the aorta into the rest of the circulatory system. Represent this process with a qualitative work-energy bar chart.

**Sketch and translate** We sketch the process, as shown below. Choose the system to be the aorta and the 80 $cm^3$ of blood that is being pumped. The left ventricle is not in the system. Choose the initial state to be just before the left ventricle contracts. Choose the final state to be when the blood is in the stretched aorta before moving out into the rest of the circulatory system.

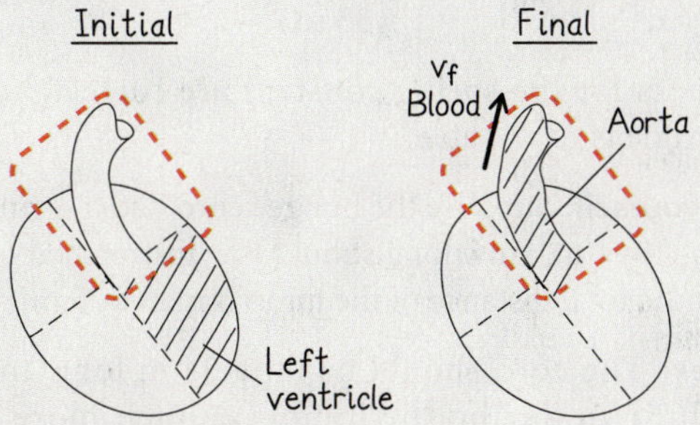


**Simplify and diagram** We keep track of the kinetic energy (the blood speed changes) and the elastic potential energy (the aorta wall stretches). We ignore the slight increase in the vertical elevation of the blood. The left ventricle is not in the system and does positive work pushing the blood upward in the direction of the blood's displacement. The work-energy bar chart below left represents this process for an aorta with flexible walls.

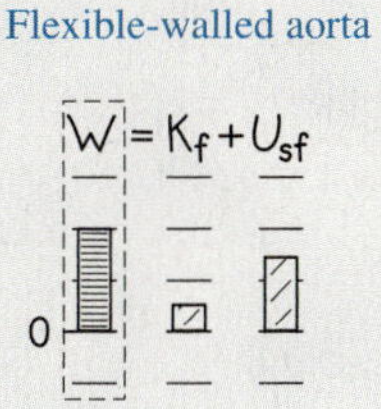


**Try it yourself:** Modify the work-energy bar chart for a person with hardened and thickened artery walls.

*Answer:* If the person has stiff, thick arteries (a condition called atherosclerosis), more energy than normal is required to stretch the aorta walls. This process is represented by the bar chart below. In this case, the blood pressure will be higher than it would be for a healthy cardiovascular system because the heart has to do more work to stretch the walls while pushing blood into the aorta.

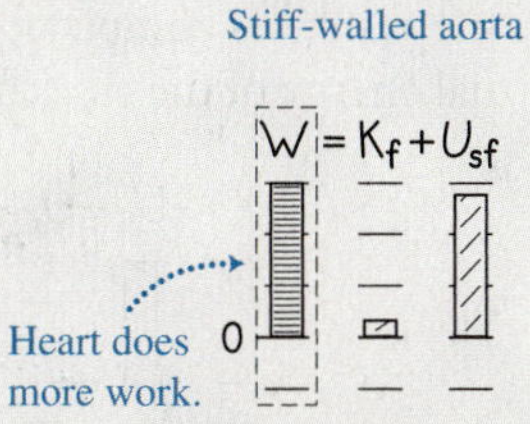


## Bungee jumping

On April 1, 1979, four members of the Oxford University Dangerous Sport Club made the first modern bungee jump. They jumped from the 76-m-high (250 ft) Clifton Suspension Bridge in Bristol, England while tied to the bridge with a rubber bungee cord. Let's analyze their jump using work-energy principles.

**EXAMPLE 6.9 Bungee jumping**

We estimate that the Oxford team used a 40-m-long bungee cord that had stretched another 35 m when the jumper was at the very lowest point in the jump, 1.0 m above the ground. We estimate that the jumper's mass is 70 kg. Imagine that your job is to buy a bungee cord that would provide a safe jump with the above specifications. Specifically, you need to determine the spring constant $k$ of the cord you need to buy.

**Sketch and translate** We sketch the process, as shown at right. The initial state is just before the jumper jumps, and the final state is when the cord is fully stretched and the jumper is momentarily at rest at the lowest position. The motion is with respect to Earth. We choose a coordinate system with the positive $y$-direction pointing up and the origin at the jumper's final position. The system is the jumper, the cord, and Earth.

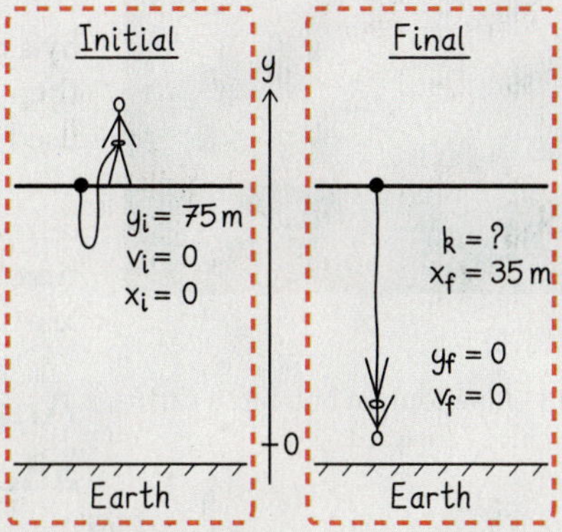


**Simplify and diagram** We are interested in the gravitational potential energy (the jumper's elevation changes) and the elastic potential energy (the cord stretch changes).

*(continued)*

In both the initial and final states the system has no kinetic energy. Assume that the bungee cord obeys Hooke's law and that its mass is negligible compared to the mass of the jumper. There are no external forces doing work on the system; thus, the energy of the system is constant. We represent the process with a bar chart, as shown below. It shows us that the initial gravitational potential energy of the system is completely converted into the elastic potential energy of the stretched bungee cord.

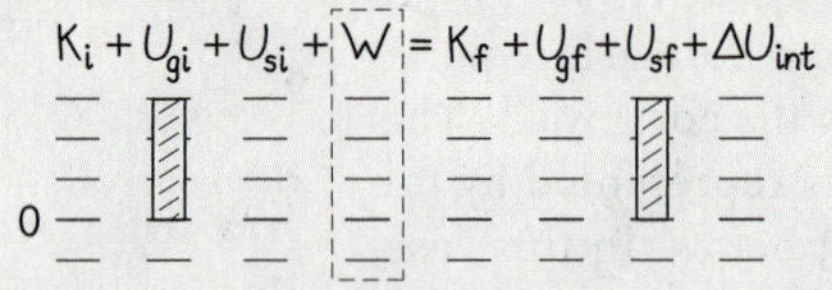


**Represent mathematically** We apply the work-energy equation with one term for each bar in the bar chart:

$$U_{gi} = U_{sf}$$

$$\Rightarrow mgy_i = \frac{1}{2}kx_f^2$$

**Solve and evaluate** Solving the above for the spring constant gives

$$k = \frac{2mgy_i}{x_f^2}$$

The length of the cord is $L = 40$ m and the cord stretches $x_f = 35$ m. Thus, the distance of the person's initial *position* $y_i$ above the final $y_f = 0$ position is

$$y_i = L + x_f = 40\text{ m} + 35\text{ m} = 75\text{ m}$$

This means that the spring constant has the value

$$k = \frac{2mgy_i}{x_f^2} = \frac{2(70\text{ kg})(9.8\text{ N/kg})(75\text{ m})}{(35\text{ m})^2} = 84\text{ N/m}$$

The units for the spring constant are correct, and the magnitude is reasonable.

**Try it yourself:** Suppose the bungee cord had a spring constant of 40 N/m. How long should the unstretched cord be so that the total distance of the jump remains 75 m?

*Answer:* The cord should be only 24 m long and will stretch 51 m during the jump—a much more easily stretched cord.

**Review Question 6.6** What would change in the solution to the problem in Example 6.9 if we did not include Earth in the system? How would the answer be different?

**Figure 6.10** A baseball is compressed while being hit by a bat.

## 6.7 Collisions: Putting it all together

A collision is a process that occurs when two (or more) objects are in direct contact with each other for a short time interval, such as when a baseball is hit by a bat (**Figure 6.10**). The ball compresses during the first half of the collision, then decompresses during the second half of the collision. We have already used impulse and momentum principles to analyze collisions (in Chapter 5). We learned that the forces that the two colliding objects exert on each other during the collision are complicated, nonconstant, and exerted for a very brief time interval—roughly 1 ms in the case of the baseball-bat collision. Can we learn anything new by analyzing collisions using the work-energy principle?

**Figure 6.11** A pendulum bob hits a cart, causing it to move forward.

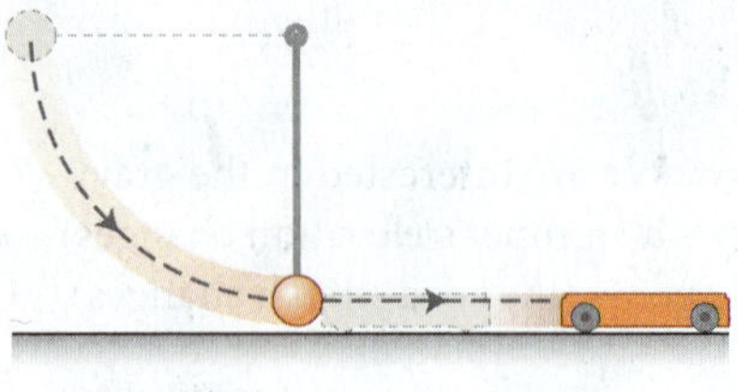

### Analyzing collisions using momentum and energy principles

**Table 6.6** shows three different observational experiments involving collisions. In each case a 1.0-kg object attached to the end of a string (the bob of a pendulum) swings down and hits a 4.0-kg wheeled cart at the lowest point of its swing (**Figure 6.11**). In each experiment, the pendulum bob and cart start at the same initial positions, but the compositions of the pendulum bob and cart are varied. After each collision, the cart moves at a nearly constant speed because of the smoothness of the surface on which it rolls.

## OBSERVATIONAL EXPERIMENT TABLE

### 6.6 Analyzing energy and momentum during collisions.

VIDEO 6.6

| Observational experiment | Analysis |
|---|---|
| **Experiment 1.** A 1.0-kg metal bob swings and hits a 4.0-kg metal cart. Their velocity components just before and just after the collision are shown below:<br>$m_1 = 1.0\ \text{kg};\ v_{1ix} = 10\ \text{m/s}$<br>$v_{1fx} = -6.0\ \text{m/s}$<br>$m_2 = 4.0\ \text{kg};\ v_{2ix} = 0\ \text{m/s}$<br>$v_{2fx} = 4.0\ \text{m/s}$ | **Momentum:**<br>Before collision:<br>$(1.0\ \text{kg})(+10\ \text{m/s}) + (4.0\ \text{kg})\,0 = +10\ \text{kg}\cdot\text{m/s}$<br>After collision:<br>$(1.0\ \text{kg})(-6.0\ \text{m/s}) + (4.0\ \text{kg})(+4.0\ \text{m/s}) = +10\ \text{kg}\cdot\text{m/s}$<br>**Kinetic energy:**<br>Before collision:<br>$(1/2)(1.0\ \text{kg})(+10\ \text{m/s})^2 + (1/2)(4.0\ \text{kg})\,0^2 = 50\ \text{J}$<br>After collision:<br>$(1/2)(1.0\ \text{kg})(-6.0\ \text{m/s})^2 + (1/2)(4.0\ \text{kg})(+4.0\ \text{m/s})^2 = 50\ \text{J}$ |
| **Experiment 2.** A 1.0-kg sand-filled balloon swings and hits a 4.0-kg flimsy cardboard cart. After the collision, the damaged cart moves across the table at constant speed. The side of the balloon that hit the cart is flattened in the collision. Their velocity components just before and just after the collision are shown below:<br>$m_1 = 1.0\ \text{kg};\ v_{1ix} = 10\ \text{m/s}$<br>$v_{1fx} = -4.2\ \text{m/s}$<br>$m_2 = 4.0\ \text{kg};\ v_{2ix} = 0\ \text{m/s}$<br>$v_{2fx} = 3.55\ \text{m/s}$ | **Momentum:**<br>Before collision:<br>$(1.0\ \text{kg})(+10\ \text{m/s}) + (4.0\ \text{kg})\,0 = +10\ \text{kg}\cdot\text{m/s}$<br>After collision:<br>$(1.0\ \text{kg})(-4.2\ \text{m/s}) + (4.0\ \text{kg})(+3.55\ \text{m/s}) = +10\ \text{kg}\cdot\text{m/s}$<br>**Kinetic energy:**<br>Before collision:<br>$(1/2)(1.0\ \text{kg})(+10\ \text{m/s})^2 + (1/2)(4.0\ \text{kg})\,0^2 = 50\ \text{J}$<br>After collision:<br>$(1/2)(1.0\ \text{kg})(-4.2\ \text{m/s})^2 + (1/2)(4.0\ \text{kg})(+3.55\ \text{m/s})^2 = 34\ \text{J}$ |
| **Experiment 3.** A 1.0-kg sand-filled balloon covered with Velcro swings down and sticks to a 4.0-kg Velcro-covered cardboard cart. The string holding the balloon is cut by a razor blade immediately after the balloon contacts the cart. The damaged cart and flattened balloon move off together across the table. Their velocity components just before and just after the collision are shown below:<br>$m_1 = 1.0\ \text{kg};\ v_{1ix} = 10\ \text{m/s}$<br>$v_{1fx} = 2.0\ \text{m/s}$<br>$m_2 = 4.0\ \text{kg};\ v_{2ix} = 0\ \text{m/s}$<br>$v_{2fx} = 2.0\ \text{m/s}$ | **Momentum:**<br>Before collision:<br>$(1.0\ \text{kg})(+10\ \text{m/s}) + (4.0\ \text{kg})\,0 = +10\ \text{kg}\cdot\text{m/s}$<br>After collision:<br>$(1.0\ \text{kg})(+2.0\ \text{m/s}) + (4.0\ \text{kg})(+2.0\ \text{m/s}) = +10\ \text{kg}\cdot\text{m/s}$<br>**Kinetic energy:**<br>Before collision:<br>$(1/2)(1.0\ \text{kg})(+10\ \text{m/s})^2 + (1/2)(4.0\ \text{kg})\,0^2 = 50\ \text{J}$<br>After collision:<br>$(1/2)(1.0\ \text{kg})(+2.0\ \text{m/s})^2 + (1/2)(4.0\ \text{kg})(+2.0\ \text{m/s})^2 = 10\ \text{J}$ |

**Patterns**

Two important patterns emerge from the data collected from these different collisions.

- The momentum of the system is constant in all three experiments.
- The kinetic energy of the system is constant when no damage is done to the system objects during the collision (Experiment 1 but not in Experiments 2 and 3).

We use momentum and energy principles to analyze the results of the experiments. In all cases, the system is the pendulum bob and the cart. The initial state of the system is the moment just before the collision. The final state is the moment just after the collision ends. The momentum of the system is in the positive $x$-direction. We keep track of the kinetic energy of both the pendulum bob and the cart. In some of the experiments, one or more objects in the system are deformed; we will discuss internal energy changes after completing our experiment. We do not keep track of gravitational potential energy since the vertical position of the system objects does not change between the initial and final states.

We can understand the first pattern in Table 6.6 using our knowledge of impulse-momentum. The $x$-component of the net force exerted on the system in all three cases was zero; hence the $x$-component of momentum should be constant.

What about the second pattern? In Experiment 1, the system objects were very rigid, but in Experiments 2 and 3, they were more fragile and as a result were deformed during the collision. Using the data we can determine the amount of kinetic energy that was converted to internal energy. But it is impossible to predict this amount ahead of time. Unfortunately, this means that in collisions where any deformation of the system objects occurs, we cannot make predictions about the amount of kinetic energy converted to internal energy. However, we now know that even in collisions where the system objects become damaged, the momentum of the system still remains constant. So, even though the work-energy equation is less useful in these types of collisions, the impulse-momentum equation is still very useful.

## Types of collisions

The experiments in Observational Experiment Table 6.6 are examples of the three general collision categories: elastic collisions (Experiment 1), inelastic collisions (Experiment 2), and totally inelastic collisions (Experiment 3). **Table 6.7** summarizes these three types of collisions.

## Measuring the speed of a fast-moving projectile

We have already encountered a ballistic pendulum (in Chapter 5), a device that measures the speed of fast projectiles, such as golf balls hit in a ball-testing device or bullets fired from a gun. Let's use a ballistic pendulum to learn more about collisions, momentum, work, and energy.

**Table 6.7** Types of collisions.

| Elastic collisions | Inelastic collisions | Totally inelastic collisions |
|---|---|---|
| Both the momentum and kinetic energy of the system are constant. The internal energy of the system does not change. The colliding objects never stick together. Examples: There are no perfectly elastic collisions in nature, although collisions between very rigid objects (such as billiard balls) come close. Collisions between atoms or subatomic particles are almost exactly elastic. | The momentum of the system is constant but the kinetic energy is not. The colliding objects do not stick together. Internal energy increases during the collisions. Examples: A volleyball bouncing off your arms, or you jumping on a trampoline. | These are inelastic collisions in which the colliding objects stick together. Typically, a large fraction of the kinetic energy of the system is converted into internal energy in this type of collision. Examples: You catching a football, or a car collision where the cars stick together. |

**EXAMPLE 6.10 A ballistic pendulum**

A gun is several centimeters from a 1.0-kg wooden block hanging at the end of strings. The gun fires a 10-g bullet that embeds in the block, which swings upward a height of 0.20 m. Determine the speed of the bullet when leaving the gun.

**Sketch and translate** We sketch the process, below. We use subscripts b for the bullet and B for the block of wood; when they join together, the subscript is bB. The process involves two parts. Part I is the collision of the bullet with the wood block. We know that this is a totally inelastic collision because the bullet combines with the block. Part II is the swinging of the block with the embedded bullet upward to its maximum height.

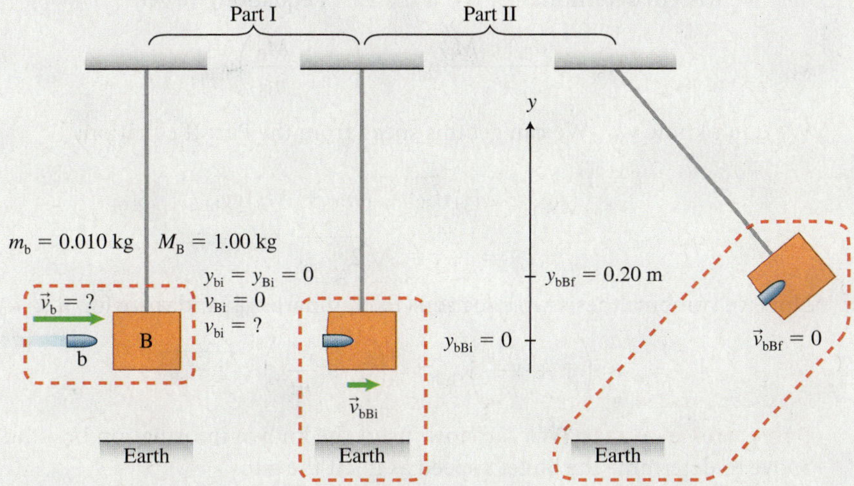


For the Part I collision, sketched on the left side of the figure above, the system is the bullet and the block. The kinetic energy of the system is not constant, but the momentum is. So we use the constant momentum of an isolated system to analyze the collision. The initial state is the instant just before the collision begins, and the final state is the instant just after the bullet joins the block. In the initial state, only the bullet has momentum. In the final state, both the bullet and the wood block have momentum, and their velocities are equal.

The details for the Part II upward swing of the block and bullet to its maximum height are sketched on the right side of the figure above. We choose the bullet, block, and Earth as the system. The initial state is just after the collision is over (the final state of Part I), and the final state is when the block reaches its maximum height.

The two parts are analyzed separately and then combined to determine the bullet's speed before hitting the block.

**Simplify and diagram** Part I: We draw bar charts to represent each process. The momentum bar chart below, left, represents the $x$-components of momentum for Part I.

Part II: The string does no work on the block, because it is perpendicular to the block's velocity at every instant. The energy bar chart below, right, shows that the initial kinetic energy of the bullet and block is converted to the gravitational potential energy of the bullet-block-Earth system.

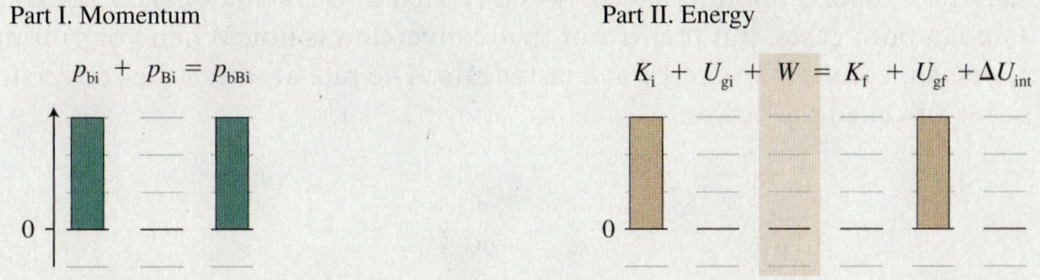


*(continued)*

**Represent mathematically** Use the bar charts to help apply momentum constancy to Part I and energy constancy to Part II.

Part I: $mv_{\text{bi}} = (m_{\text{b}} + M_{\text{B}})v_{\text{bBi}}$

Part II: $\frac{1}{2}(m_{\text{b}} + M_{\text{B}})v_{\text{bBi}}^2 = (m_{\text{b}} + M_{\text{B}})gy_{\text{bBf}}$

where $m_{\text{b}}$ is the mass of the bullet, $M_{\text{B}}$ is the mass of the block, $v_{\text{bi}}$ is the initial speed of the bullet, $v_{\text{bBi}}$ is the speed of the bullet + block immediately after the collision, and $y_{\text{bBf}}$ is the $y$-coordinate of the bullet + block at its highest point.

We wish to determine $v_{\text{bi}}$. From the Part I equation, we get

$$v_{\text{bi}} = \frac{m_{\text{b}} + M_{\text{B}}}{m_{\text{b}}}v_{\text{bBi}} = \left(1 + \frac{M_{\text{B}}}{m_{\text{b}}}\right)v_{\text{bBi}}$$

We don't know $v_{\text{bBi}}$. We can get this speed from the Part II equation:

$$\frac{1}{2}(m_b + M_B)v_{\text{bBi}}^2 = (m_b + M_B)gy_{\text{bBf}}$$

$$\Rightarrow v_{\text{bBi}} = \sqrt{2gy_{\text{bBf}}}$$

Now we combine these two equations to eliminate $v_{\text{bBi}}$ and solve for $v_{\text{bi}}$:

$$v_{\text{bi}} = \left(1 + \frac{M_B}{m_b}\right)v_{\text{bBi}} = \left(1 + \frac{M_B}{m_b}\right)\sqrt{2gy_{\text{bBf}}}$$

**Solve and evaluate** We can now insert the known information into the above to determine the bullet's speed as it left the gun:

$$v_{\text{bi}} = \left(1 + \frac{1.0\text{ kg}}{0.010\text{ kg}}\right)\sqrt{2(9.8\text{ N/kg})(0.20\text{ m})} = 200\text{ m/s}$$

That's close to 450 mi/h—very fast but reasonable for a bullet fired from a gun.

**Try it yourself:** Determine the initial kinetic energy of the bullet in this example, the final gravitational potential energy of the block-bullet system, and the increase in internal energy of the system.

*Answers:* $K_{\text{i}} = 200$ J; $U_{\text{gf}} = 2$ J; and $\Delta U_{\text{int}} = 198$ J.

**Review Question 6.7** Imagine that a collision occurs. You measure the masses of the two objects before the collision and measure the velocities of the objects both before and after the collision. Describe how you could use this data to determine which type of collision had occurred.

## 6.8 Power

Why is it harder for the same person to run up a flight of stairs than to walk if the change in gravitational potential energy of the system person-Earth is the same? The *amount* of internal energy converted into gravitational energy is the same in both cases, but the *rate* of that conversion is not. When you run upstairs you convert the energy at a faster rate. The rate at which the conversion occurs is called the **power**.

**Power** **The power of a process is the amount of some type of energy converted into a different type divided by the time interval $\Delta t$ in which the process occurred:**

$$\text{Power} = P = \left|\frac{\Delta U}{\Delta t}\right| \tag{6.9}$$

**If the process involves external forces doing work, then power can also be defined as the magnitude of the work $W$ done on or by the system divided by the time interval $\Delta t$ needed for that work to be done:**

$$\text{Power} = P = \left|\frac{W}{\Delta t}\right| \tag{6.10}$$

**The SI unit of power is the watt (W). 1 watt is 1 joule/second ($1\ \text{W} = 1\ \text{J/s}$).**

A lightbulb with a power of 60 W converts electrical energy into light and internal energy at a rate of 60 J/s. A cyclist in good shape pedaling at moderate speed will convert about 400–500 J of internal chemical energy each second (400–500 W) into kinetic, gravitational potential, and thermal energies.

Power is sometimes expressed in horsepower (hp): $1\ \text{hp} = 746\ \text{W}$. Horsepower is most often used to describe the power rating of engines or other machines. A 50-hp gasoline engine (typical in cars) converts the internal energy of the fuel into other forms of energy at a rate of $50 \times 746\ \text{W} = 37{,}300\ \text{W}$, or 37,300 J/s.

### EXAMPLE 6.11 Lifting weights

Xueli is doing a dead lift. She lifts a 13.6-kg (30-lb) barbell from the floor to just below her waist (a vertical distance of 0.70 m) in 0.80 s. Determine the power during the lift.

**Sketch and translate** First, we sketch the process, shown below. The system is the barbell and Earth. The initial state is just before Xueli starts lifting. The final state is just after she finishes lifting. A vertical $y$-axis is used to indicate the change in elevation.

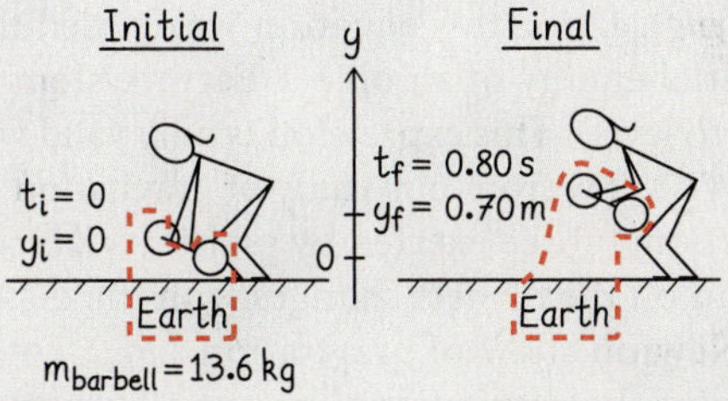


**Simplify and diagram** We assume that Xueli lifts the barbell so slowly that kinetic energy is zero during the process. She does work on the barbell, causing the system's gravitational potential energy to increase, but no change in kinetic energy. The origin is at the initial position of the barbell. Since the barbell is moving at a small constant velocity, the external force exerted by Xueli on the system is very nearly constant and equals the gravitational force exerted by Earth on the barbell. We can represent this process with an energy bar chart, as shown.

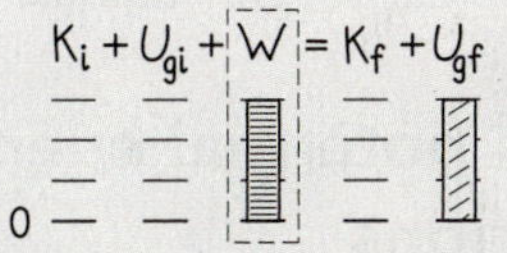


**Represent mathematically** The power of this process is

$$P = \left|\frac{W}{\Delta t}\right| = \left|\frac{Fd\cos\theta}{\Delta t}\right| = \left|\frac{mgd\cos\theta}{\Delta t}\right|$$

**Solve and evaluate**

$$P = \left|\frac{mgd\cos\theta}{\Delta t}\right| = \frac{(13.6\ \text{kg})(9.8\ \text{N/kg})(0.70\ \text{m})\cos 0^\circ}{0.80\ \text{s}}$$

$$= 120\ \text{W}$$

This is a reasonable power for lifting a barbell. If you use an exercise machine that displays the power output, compare what you can achieve to this number.

**Try it yourself:** Xueli performs an overhead press—lifting the same barbell from her shoulders to above her head. Estimate the power of this process. The length of her arm is approximately 49 cm. It takes her 1.0 s to lift the bar.

*Answer:* 65 W.

**EXAMPLE 6.12 Power and driving**

A 1400-kg car is traveling on a level road at a constant speed of 27 m/s (60 mi/h). The drag force exerted by the air on the car and the rolling friction force exerted by the road on the car tires add to a net force of 680 N pointing opposite the direction of motion of the car. Determine the rate of work done by air and the road on the car. Express the results in watts and in horsepower.

**Sketch and translate part** First, we sketch the system, choosing the car alone as the system. The initial state is the moment the car passes position $x$ on the $x$-axis and the final state is a short time interval $\Delta t$ later when the car has had a displacement $\Delta x$ parallel to the road.

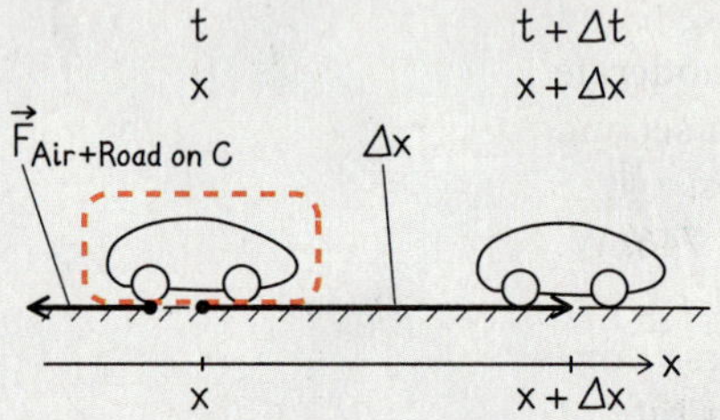


**Simplify and diagram** We need to determine the magnitude of work per unit time by the combined resistive air and rolling friction forces ($\vec{F}_{\text{Air+Road on C}}$).

**Represent mathematically** We determine the work done during displacement $\Delta x$ and divide by the time interval $\Delta t$ needed to complete that displacement. The power is the magnitude of that ratio:

$$P = \left|\frac{W}{\Delta t}\right| = \left|\frac{(F_{\text{Air+Road on C}})\Delta x \cos 180^\circ}{\Delta t}\right|$$
$$= \left|-(680\text{ N})\frac{\Delta x}{\Delta t}\right| = (680\text{ N})v$$

**Solve and Evaluate** Substitute the speed into the above to determine the power:

$$P = (680\text{ N})v = (680\text{ N})(27\text{m/s})$$
$$= 1.8 \times 10^4\text{ W} = 25\text{ hp}$$

That's a relatively small power.

**Try it yourself:** What is the average power needed to cause a 1400-kg car's speed to increase from 20 to 27 m/s in 5 s? Ignore any resistive forces.

*Answer:* $P = 4.6 \times 10^4$ W or 62 hp.

**Review Question 6.8** Jim (mass 80 kg) rollerblades on a smooth linoleum floor a distance of 4.0 m in 5.0 s. Determine the power of this process.

## 6.9 Improving our model of gravitational potential energy

So far, we have assumed that the gravitational force exerted by Earth on an object is constant ($F_{\text{E on O}} = mg$). Using this equation we devised the expression for the gravitational potential energy of an object-Earth system as $U_g = mgy$ (with respect to a chosen zero level). This expression is only valid when an object is close to Earth's surface. We know from our study of gravitation (at the end of Chapter 4) that the gravitational force exerted by planetary objects on moons and satellites and by the Sun on the planets changes with the distance between the objects according to Newton's law of gravitation ($F_{1\text{ on }2} = G\,m_1 m_2/r^2$) How would the expression for the gravitational potential energy change when we take into account that the gravitational force varies with distance?

### Gravitational potential energy for large mass separations

Imagine that a "space elevator" has been built to transport supplies from the surface of Earth to the International Space Station (ISS). The elevator moves at constant velocity, except for the very brief acceleration and deceleration at the beginning and end of the trip. How much work must be done to lift the supplies from the surface to the ISS?

The initial state is the moment just as the supplies leave the surface. The final state is the moment just as they arrive at the ISS (**Figure 6.12a**). We choose Earth and the supplies as the system. The force that the elevator cable exerts on the supplies is an external force that does positive work on the system. We keep track of gravitational potential energy only; it is the only type of energy that changes between the initial and final states. Since the supplies are moving at constant velocity, the force exerted by the elevator cable on the supplies is equal in magnitude to the gravitational force exerted by Earth on the supplies (Figure 6.12b). Earth is the object of reference, and the origin of the coordinate system is at the center of Earth. The process can be described mathematically as follows:

$$U_{gi} + W = U_{gf}$$
$$\Rightarrow W = U_{gf} - U_{gi}$$

At this point it would be tempting to say that the work done by the elevator cable on the system is $W = Fd\cos\theta$, where $F$ is the constant force that the cable exerts on the supplies. However, as the supplies reach higher and higher altitudes, the force exerted by the elevator cable decreases—Earth exerts a weaker and weaker force on the supplies.

Determining the work done by a variable force requires a complex mathematical procedure. The outcome of this procedure is

$$W = \left(-G\frac{m_E m_S}{R_E + h_{ISS}}\right) - \left(-G\frac{m_E m_S}{R_E}\right)$$

where $m_E$ is the mass of Earth, $R_E$ is the radius of Earth, $m_S$ is the mass of the supplies, $h_{ISS}$ is the altitude of the ISS above Earth's surface, and $R_E + h_{ISS}$ is the distance of the ISS from the center of Earth. Before you move on, check whether this complicated equation makes sense (for example, check the units).

The work is written as the difference in two quantities; the latter describes the initial state and the former the final state. If we compare the above result with $W = U_{gf} - U_{gi}$, we see that each term is an expression for the gravitational potential energy of the Earth-object system for a particular object's distance from the center of Earth.

**Figure 6.12** A cable lifts supplies to the International Space Station via a space elevator. (a) Determine the work required to lift the supplies. (b) The cable exerts the force in the direction of motion. In this force diagram, the supplies are the system.

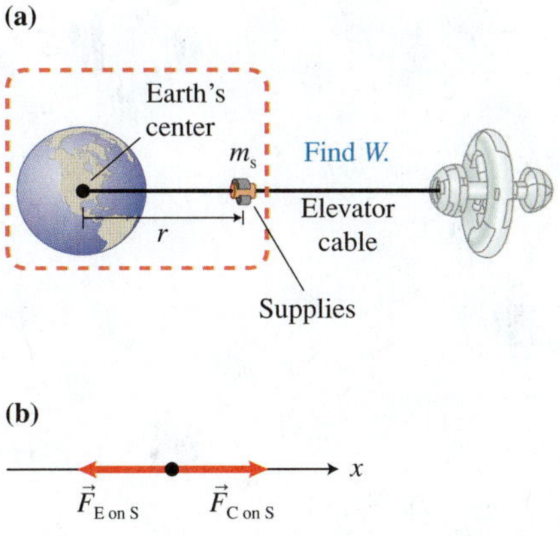


**Gravitational potential energy of a system consisting of Earth and any object**

$$U_g = -G\frac{m_E m_O}{r_{E\text{-}O}} \quad (6.11)$$

**where $m_E$ is the mass of Earth ($5.97 \times 10^{24}$ kg), $m_O$ is the mass of the object, $r_{E\text{-}O}$ is the distance from the center of Earth to the center of the object, and $G = 6.67 \times 10^{-11}\ \text{N}\cdot\text{m}^2/\text{kg}^2$ is Newton's universal gravitational constant.**

We can use Eq. (6.11) to find the gravitational potential energy of the system of any two spherical or point-like objects if we know the masses of the objects and the distance between their centers.

Note that the cable did positive work on the system while pulling the supplies away from Earth. When the object (the supplies in this case) is infinitely far away, the gravitational potential energy is zero. The only way to add positive energy to a system and have it become zero is if it started with negative energy, for example, $-5 + 5 = 0$. Thus, for the case of zero energy at infinity, the gravitational potential energy is a negative number when the object is closer to Earth. We can represent the process of pulling an object from the surface of Earth to infinity using a work-energy bar chart (**Figure 6.13**). The initial state is when the object is near Earth and the final state is when it is infinitely far away.

**Figure 6.13** A bar chart representing the work needed to take an object from near Earth to infinitely far away. The system is the supplies and Earth. Note that the final gravitational potential energy of the system is zero.

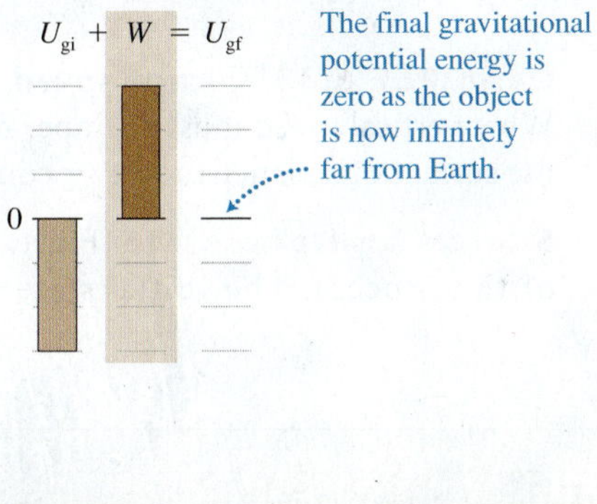

Now we can determine the amount of work needed to raise 1000 kg of supplies to the International Space Station.

$$W = \left(-G\frac{m_E m_S}{R_E + h_{ISS}}\right) - \left(-G\frac{m_E m_S}{R_E}\right)$$

$$= -Gm_E m_S\left(\frac{1}{R_E + h_{ISS}} - \frac{1}{R_E}\right)$$

$$= -(6.67 \times 10^{-11}\,\mathrm{N \cdot m^2/kg^2})(5.97 \times 10^{24}\,\mathrm{kg})(1000\,\mathrm{kg})$$

$$\times\left(\frac{1}{6.37 \times 10^6\,\mathrm{m} + 3.50 \times 10^5\,\mathrm{m}} - \frac{1}{6.37 \times 10^6\,\mathrm{m}}\right)$$

$$= 3.26 \times 10^9\,\mathrm{J}$$

Let's compare this with what would have been calculated had we used our original expression for gravitational potential energy. Choose the zero level at the surface of Earth.

$$W = m_s g y_f - m_s g y_i = m_s g y_f - 0$$

$$= (1000\,\mathrm{kg})(9.8\,\mathrm{N/kg})(3.50 \times 10^5\,\mathrm{m}) = 3.43 \times 10^9\,\mathrm{J}$$

This differs by only about 5% from the more accurate result. Remember that $U_g = mgy$ is reasonable when the distance above the surface of Earth is a small fraction of the radius of Earth. The altitude of the ISS (350 km) is a small fraction of the radius of Earth (6371 km), so $U_g = mgy$ is still reasonably accurate.

## Escape speed

The best Olympic high jumpers can leap over bars that are about 2.5 m (8 ft) above Earth's surface. Let's estimate a jumper's speed when leaving the ground in order to attain that height. We choose the jumper and Earth as the system and the zero level of gravitational potential energy at ground level. The kinetic energy of the jumper as he leaves the ground is converted into the gravitational potential energy of the system.

$$\frac{1}{2}mv^2 = mgy$$

$$\Rightarrow v = \sqrt{2gy} = \sqrt{2(9.8\,\mathrm{N/kg})(2.5\,\mathrm{m})} = 7.0\,\mathrm{m/s}$$

How high could he jump if he were on the Moon? The gravitational constant of objects near the Moon's surface is $g_M = \dfrac{Gm_M}{{R_M}^2} = 1.6\,\mathrm{N/kg}$. Using the equation above,

$$\frac{1}{2}mv^2 = mg_M y$$

$$\Rightarrow y = \frac{v^2}{2g_M} = \frac{(7.0\,\mathrm{m/s})^2}{2(1.6\,\mathrm{N/kg})} = 15.3\,\mathrm{m}$$

That's about 50 feet!

Is it possible to jump entirely off a celestial body—jumping up and never coming down? What is the minimum speed you would need in order to do this? This minimum speed is called **escape speed**.

**EXAMPLE 6.13 Escape speed**

What vertical speed must a jumper have in order to leave the surface of a planet and never come back down?

**Sketch and translate** First, we draw a sketch of the process. The initial state will be the instant after the jumper's feet leave the surface. The final state will be when the jumper has traveled far enough away from the planet to no longer feel the effects of its gravity (at $r = \infty$). Choose the system to be the jumper and the planet.

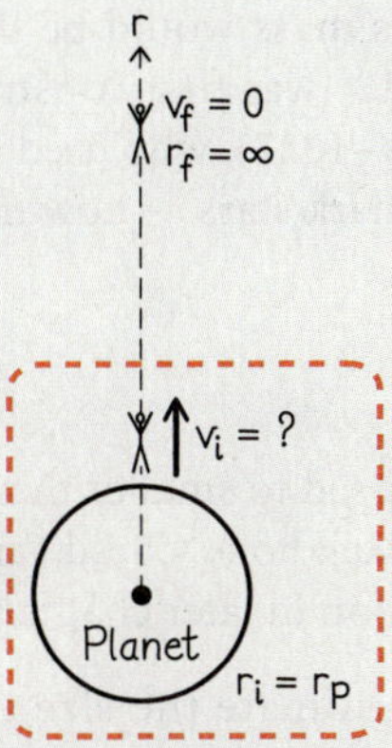


**Simplify and diagram** We represent the process with the bar chart. In the initial state, the system has both kinetic and gravitational potential energy. In the final state, both the kinetic energy and the gravitational potential energy are zero.

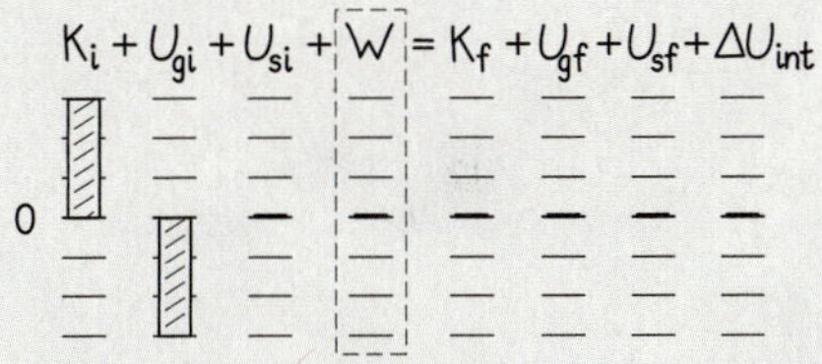


**Represent mathematically** Using the generalized work-energy equation and the bar chart:

$$U_i + W = U_f$$
$$\Rightarrow K_i + U_{gi} + 0 = K_f + U_{gf}$$
$$\Rightarrow \frac{1}{2}m_J v^2 + \left(-G\frac{m_P m_J}{r_P}\right) + 0 = 0 + 0$$

where $m_P$ is the mass of the planet, $m_J$ is the mass of the jumper, $r_P$ is the radius of the planet, and $G = 6.67 \times 10^{-11}\,\text{N}\cdot\text{m}^2/\text{kg}^2$ is the gravitational constant.

**Solve and evaluate** Solving for the escape speed of the jumper,

$$v = \sqrt{\frac{2Gm_P}{r_P}} \tag{6.12}$$

We can use the above equation to determine the escape speed for any celestial body. For example, the escape speed for the Moon is

$$v = \sqrt{\frac{2(6.67 \times 10^{-11}\,\text{N}\cdot\text{m}^2/\text{kg}^2)(7.35 \times 10^{22}\,\text{kg})}{1.74 \times 10^6\,\text{m}}}$$
$$= 2370\,\text{m/s}$$

The escape speed for Earth is

$$v = \sqrt{\frac{2Gm_E}{r_E}}$$
$$= \sqrt{\frac{2(6.67 \times 10^{-11}\,\text{N}\cdot\text{m}^2/\text{kg}^2)(5.97 \times 10^{24}\,\text{kg})}{6.37 \times 10^6\,\text{m}}}$$
$$= 11{,}200\,\text{m/s}$$
$$= 11.2\,\text{km/s}$$

**Try it yourself:** What is the escape speed of a particle near the surface of the Sun? The mass of the Sun is $2.0 \times 10^{30}$ kg and its radius is 700,000 km.

*Answer:* 620 km/s.

## Black holes

Equation (6.12) for the escape speed suggests something amazing. If the mass of a star or planet were large enough and/or its radius small enough, the escape speed could be made arbitrarily large. What if the star's escape speed were greater than light speed ($c = 3.00 \times 10^8\,\text{m/s}$)? What would this star look like? Light leaving the star's surface would not be moving fast enough to escape the star. The star would be completely dark.

Let's imagine what would happen if Earth started shrinking so that its material were compressed into a smaller volume. How small would Earth have to be for its escape speed to be greater than light speed? Use Eq. (6.12) with $v = c$ to answer this question.

$$v = c = \sqrt{\frac{2Gm_P}{r_P}} \tag{6.13}$$

or

$$r_P = \frac{2Gm_P}{c^2} = \frac{2(6.67 \times 10^{-11}\,\text{N}\cdot\text{m}^2/\text{kg}^2)(5.97 \times 10^{24}\,\text{kg})}{(3.00 \times 10^8\,\text{m/s})^2} = 8.85\,\text{mm}$$

**TIP** Notice that the escape speed does not depend on the mass of the escaping object—a tiny speck of dust and a huge boulder would need the same initial speed to leave Earth. Why is that?

Earth would have to be smaller than 9 millimeters! It's difficult to imagine Earth compressed to the size of a marble. Its mass would be the same but it would now be incredibly dense. Equation (6.13) was first constructed by the brilliant astronomer Pierre-Simon Laplace (1749–1827), who used classical mechanics to predict the presence of what he called "dark stars"—now more commonly called black holes.

**QUANTITATIVE EXERCISE 6.14 Sun as a black hole**

How small would our Sun need to be in order for it to become a black hole?

**Represent Mathematically** The mass of the Sun is $1.99 \times 10^{30}$ kg. All we need to do is use Eq. (6.13) to find the radius of this black hole:

$$r_{\text{Sun}} = \frac{2Gm_{\text{Sun}}}{c^2}$$

**Solve and Evaluate**

$$r_{\text{Sun}} = \frac{2(6.67 \times 10^{-11}\,\text{N}\cdot\text{m}^2/\text{kg}^2)(1.99 \times 10^{30}\,\text{kg})}{(3.00 \times 10^8\,\text{m/s})^2}$$
$$= 2.95 \times 10^3\,\text{m} \approx 3\,\text{km}$$

So, if the Sun collapsed to smaller than 3 km in radius, it would become a black hole. Could this happen? We will return to this question in later chapters.

**Try it yourself:** Estimate the size to which a human would need to shrink to become invisible in the same sense that a black hole is invisible.

*Answer:* About $10^{-25}$ m.

We've been talking about objects whose escape speed is larger than light speed. Physicists once thought that since light has zero mass, the gravitational force exerted on it would always be zero. At the beginning of the 20th century, Albert Einstein's theory of general relativity improved greatly on Newton's law of universal gravitation. According to general relativity, light is affected by gravity. Amazingly, the theory predicts the same size for a black hole that is provided by the Newtonian theory.

**Review Question 6.9** In this section you read that the gravitational potential energy of two large bodies (for example, the Sun and Earth) is negative. Why is this true?

# Summary

| Words | Pictorial and physical representations | Mathematical representation |
|---|---|---|
| **Work** ($W$) is a way to change the energy of a system. Work is done on a system when an external object exerts a force of magnitude $F$ on an object in the system as it undergoes a displacement of magnitude $d$. The work depends on the angle $\theta$ between the directions of $\vec{F}$ and $\vec{d}$. It is a scalar quantity. (Section 6.1) | 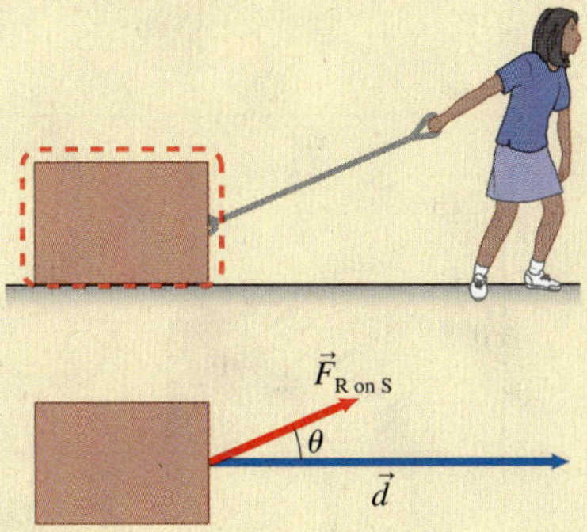 | $W = Fd\cos\theta$ Eq. (6.1) |
| **Gravitational potential energy** ($U_g$) is the energy that a system has due to the relative separation of two objects with mass. It is a scalar quantity. $U_g$ depends on the gravitational interaction of the objects. A single object cannot have gravitational potential energy. | | $U_g = mgy$ Eq. (6.4)<br>(near Earth's surface, zero level is at the surface)<br>$U_g = -G\dfrac{m_A m_B}{r_{AB}}$ Eq. (6.11)<br>(general expression; zero level is at infinity) |
| **Kinetic energy** ($K$) is the energy of an object of mass $m$ moving at speed $v$. It is a scalar quantity. | | $K = \dfrac{1}{2}mv^2$ Eq. (6.5) |
| **Elastic potential energy** ($U_s$) is the energy of a stretched or compressed elastic object (e.g., coils of a spring or a stretched bow string). | | $U_s = \dfrac{1}{2}kx^2$ Eq. (6.7) |
| **Internal energy** ($U_{int}$) is the energy of motion and interaction of the microscopic particles making up the objects in the system. The internal energy of the system changes when the surfaces of the system objects rub against each other. (Section 6.1) | | $\Delta U_{int} = f_k d$ Eq. (6.8)<br>(conversion of mechanical energy to internal due to friction) |

*(continued)*

| Words | Pictorial and physical representations | Mathematical representation |
|---|---|---|
| **Total energy** ($U$) is the sum of all the energies of the system. (Section 6.2) | | $U = K + U_g + U_s + U_{int} + \cdots$ Eq. (6.2) |
| **Work-energy principle** The energy of a system changes when external forces do work on it. Internal forces do not change the energy of the system. When there are no external forces doing work on the system, the system's energy is constant. (Section 6.2) | 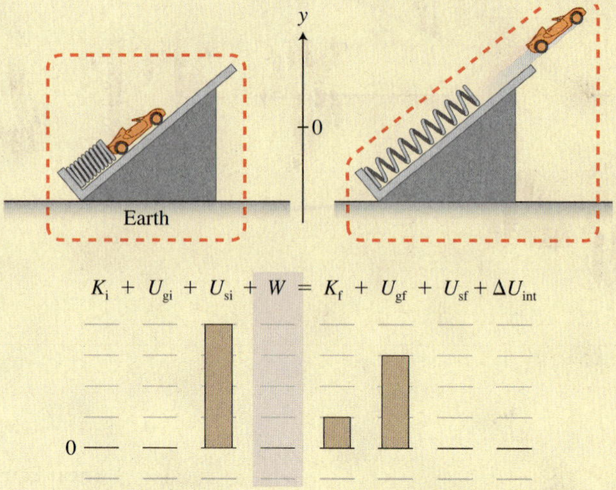  | $U_i + W = U_f$ Eq. (6.3) <br> $(K_i + U_{gi} + U_{si}) + W$ <br> $= (K_f + U_{gf} + U_{sf} + \Delta U_{int})$ |
| **Collisions** <br> ■ **Elastic:** momentum and kinetic energy of the system are constant—no changes in internal energy. <br> ■ **Inelastic:** momentum is constant but not kinetic energy—internal energy increases and kinetic energy decreases. <br> ■ **Totally inelastic:** an inelastic collision in which the colliding objects stick together. (Section 6.7) | | All collisions: <br> $\Sigma\vec{p}_i = \Sigma\vec{p}_f$ <br> Elastic collisions only: <br> $\Sigma K_i = \Sigma K_f$ <br> For other collisions, <br> $\Delta U_{int} > 0$ |
| **Power** ($P$) is the rate of energy conversion, or rate of work done on or by a system during a process. (Section 6.8) | 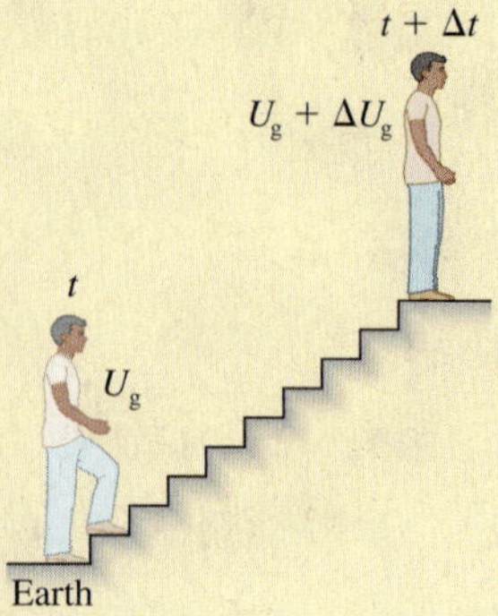  | $P = \left\lvert\dfrac{\Delta U}{\Delta t}\right\rvert$ or $P = \left\lvert\dfrac{W}{\Delta t}\right\rvert$ Eq. (6.9 or 6.10) |

MP For instructor-assigned homework, go to MasteringPhysics.

# Questions

## Multiple Choice Questions

1. In which of the following is positive work done by a person on a suitcase?
   (a) The person holds a heavy suitcase.
   (b) The person lifts a heavy suitcase.
   (c) The person stands on a moving walkway carrying a heavy suitcase.
   (d) All of the above (e) None of the above
2. Which answer best represents the system's change in energy for the following process? The system includes Earth, two carts, and a compressed spring between the carts. The spring is released, and in the final state, one cart is moving up a frictionless ramp until it stops. The other cart is moving in the opposite direction on a horizontal frictionless track.
   (a) Kinetic energy to gravitational potential energy
   (b) Elastic potential energy to gravitational potential energy
   (c) Elastic potential energy to kinetic energy and gravitational potential energy
3. Choose a process and system that match the following energy description: The kinetic energy of an object becomes gravitational potential energy of the system.
   (a) A pendulum bob released from a certain height swings to the lowest position; the system is the pendulum bob and the string.
   (b) A pendulum bob released from a certain height swings to the lowest position; the system is the pendulum bob and Earth.
   (c) A pendulum moves from the bottom of its swing to the top; the system is the pendulum bob and Earth.
   (d) A pendulum bob moves from the bottom of its swing to the top; the system is the pendulum bob and the string.
4. Three processes are described below. Choose one process in which there is work done on the system. The spring, Earth, and the cart are part of the system.
   (a) A relaxed spring rests upright on a tabletop. You slowly compress the spring. You then release the spring and it flies up several meters to its highest point.
   (b) A cart at the top of a smooth inclined surface coasts at increasing speed to the bottom (ignore friction).
   (c) A cart at the top of a smooth inclined surface slides at increasing speed to the bottom where it runs into and compresses a spring (ignore friction).
5. Choose which statement describes a process in which an external force does negative work on the system. The person is not part of the system.
   (a) A person slowly lifts a box from the floor to a tabletop.
   (b) A person slowly lowers a box from a tabletop to the floor.
   (c) A person carries a bag of groceries horizontally from one location to another.
   (d) A person holds a heavy suitcase.
6. Which example(s) below involve zero physics work? Choose all that apply.
   (a) A person holds a child.
   (b) A person pushes a car stuck in the snow but the car does not move.
   (c) A rope supports a swinging chandelier.
   (d) A person uses a self-propelled lawn mower on a level lawn.
   (e) A person pulls a sled uphill.
7. Estimate the change in gravitational potential energy when you rise from bed to a standing position.
   (a) No change (0 J) (b) About 250 J
   (c) About 2500 J (d) About 25 J
8. What does it mean if object 1 does +10 J of work on object 2?
   (a) Object 1 exerts a 10-N force on object 2 in the direction of its 1-m displacement.
   (b) Object 1 exerts a 1-N force on object 2 in the direction of its 10-m displacement.
   (c) Object 1 exerts a 10-N force on object 2 at a 60° angle relative to its 2-m displacement.
   (d) All of the above (e) None of the above
9. What does it mean if the gravitational potential energy of an Earth-apple system is 10 J?
   (a) Someone did 10 J of work lifting the apple.
   (b) Earth did negative work when someone was lifting the apple.
   (c) A 100-g apple is 10 m above the ground.
   (d) Parts (a) and (c) are both correct.
   (e) We need more information about the coordinate system and the object.
10. Imagine that you stretch a spring 3 cm, and then another 3 cm. Do you do more, less, or the same amount of work stretching it the second 3 cm compared with the first?
    (a) The same work (b) Less work (c) More work
11. Two small spheres of putty, A and B, hang from the ceiling on massless strings of equal length. Sphere A is raised to the side so its string is horizontal to the ground. It is released, swings down, and collides with sphere B (initially at rest). The spheres stick together and swing upward along a circular path to a maximum height on the other side. Which of the following principles must be used to determine this final height?
    I. The work-energy equation
    II. The impulse-momentum equation
    (a) I only (b) II only (c) Both I and II
    (d) Either I or II but not both (e) Neither I nor II
12. Two identical stones, A and B, are thrown from a cliff from the same height and with the same initial speed. Stone A is thrown vertically upward, and stone B is thrown vertically downward. Which of the following statements best explains which stone has a larger speed just before it hits the ground, assuming no effects of air friction?
    (a) Both stones have the same speed; they have the same change in $U_g$ and the same $K_i$.
    (b) A, because it travels a longer path
    (c) A, because it takes a longer time interval
    (d) A, because it travels a longer path and takes a longer time interval
    (e) B, because no work is done against gravity

## Conceptual Questions

13. Is energy a physical phenomenon, a model, or a physical quantity? Explain your answer.
14. Your friend thinks that the escape speed should be greater for more massive objects than for less massive objects. Provide a physics-based argument for his opinion. Then provide a counterargument for why the escape speed is independent of the mass of the object.
15. Suggest how you can measure the following quantities: work done by the force of friction, the power of a motor, the kinetic energy of a moving car, and the elastic potential energy of a stretched spring.
16. How can satellites stay in orbit without any jet propulsion system? Explain using work-energy ideas.
17. Why does the Moon have no atmosphere, but Earth does?
18. What will happen to Earth if our Sun becomes a black hole?
19. In the equation $U_g = mgy$, the gravitational potential energy is directly proportional to the distance of the object from a planet. In the equation $U_g = -G\frac{m_p m}{r}$, it is inversely proportional. How can you reconcile those two equations?

# Problems

Below, **BIO** indicates a problem with a biological or medical focus. Problems labeled **EST** ask you to estimate the answer to a quantitative problem rather than derive a specific answer. Problems marked with ✎ require you to make a drawing or graph as part of your solution. Asterisks indicate the level of difficulty of the problem. Problems with no * are considered to be the least difficult. A single * marks moderately difficult problems. Two ** indicate more difficult problems.

### 6.1 Work and energy

1. Jay fills a wagon with sand (about 20 kg) and pulls it with a rope 30 m along the beach. He holds the rope 25° above the horizontal. The rope exerts a 20-N tension force on the wagon. How much work does the rope do on the wagon?
2. You have a 15-kg suitcase and (a) slowly lift it 0.80 m upward, (b) hold it at rest to test whether you will be able to move the suitcase without help in the airport, and then (c) lower it 0.80 m. What work did you do in each case? What assumptions did you make to solve this problem?
3. * You use a rope to slowly pull a sled and its passenger 50 m up a 20° incline, exerting a 150-N force on the rope. (a) How much work will you do if you pull parallel to the hill? (b) How much work will you do if you exert the same magnitude force while slowly lowering the sled back down the hill and pulling parallel to the hill? (c) How much work did Earth do on the sled for the trip in part (b)?
4. A rope attached to a truck pulls a 180-kg motorcycle at 9.0 m/s. The rope exerts a 400-N force on the motorcycle at an angle of 15° above the horizontal. (a) What is the work that the rope does in pulling the motorcycle 300 m? (b) How will your answer change if the speed is 12 m/s? (c) How will your answer change if the truck accelerates?
5. You lift a 25-kg child 0.80 m, slowly carry him 10 m to the playroom, and finally set him back down 0.80 m onto the playroom floor. What work do you do on the child for each part of the trip and for the whole trip? List your assumptions.
6. A truck runs into a pile of sand, moving 0.80 m as it slows to a stop. The magnitude of the work that the sand does on the truck is $6.0 \times 10^5$ J. (a) Determine the average force that the sand exerts on the truck. (b) Did the sand do positive or negative work? (c) How does the average force change if the stopping distance is doubled? Indicate any assumptions you made.

### 6.2 and 6.3 Energy is a conserved quantity and Quantifying gravitational potential and kinetic energies

7. A 5.0-kg rabbit and a 12-kg Irish setter have the same kinetic energy. If the setter is running at speed 4.0 m/s, how fast is the rabbit running?
8. * **EST** Estimate your average kinetic energy when walking to physics class. What assumptions did you make?
9. * A pickup truck (2268 kg) and a compact car (1100 kg) have the same momentum. (a) What is the ratio of their kinetic energies? (b) If the same horizontal net force were exerted on both vehicles, pushing them from rest over the same distance, what is the ratio of their final kinetic energies?
10. * When does the kinetic energy of a car change more: when the car accelerates from 0 to 10 m/s or from 30 m/s to 40 m/s? Explain.
11. * When exiting the highway, a 1100-kg car is traveling at 22 m/s. The car's kinetic energy decreases by $1.4 \times 10^5$ J. The exit's speed limit is 35 mi/h. Did the driver reduce the car's speed enough? Explain.
12. * You are on vacation in San Francisco and decide to take a cable car to see the city. A 5200-kg cable car goes 360 m up a hill inclined 12° above the horizontal. What can you learn about the energy of the system from this information? Answer quantitatively. What assumptions did you need to make?
13. * **Flea jump** A $5.4 \times 10^{-7}$-kg flea pushes off a surface by extending its rear legs for a distance of slightly more than 2.0 mm, consequently jumping to a height of 40 cm. What physical quantities can you determine using this information? Make a list and determine three of them.
14. * **Roller coaster ride** A roller coaster car drops a maximum vertical distance of 35.4 m. (a) Determine the maximum speed of the car at the bottom of that drop. (b) Describe any assumptions you made. (c) Will a car with twice the mass have more or less speed at the bottom? Explain.
15. * **BIO EST Heart pumps blood** The heart does about 1 J of work while pumping blood into the aorta during each heartbeat. (a) Estimate the work done by the heart in pumping blood during a lifetime. (b) If all of that work was used to lift a person, to what height could an average person be lifted? Indicate any assumptions you used for each part of the problem.

16. * **Wind energy** Air circulates across Earth in regular patterns. A tropical air current called the Hadley cell carries about $2 \times 10^{11}$ kg of air per second past a cross section of Earth's atmosphere while moving toward the equator. The average air speed is about 1.5 m/s. (a) What is the kinetic energy of the air that passes the cross section each second? (b) About $1 \times 10^{20}$ J of energy was consumed in the United States in 2005. What is the ratio of the kinetic energy of the air that passes toward the equator each second and the energy consumed in the United States each second?
17. * BIO **Bone break** The tibia bone in the lower leg of an adult human will break if the compressive force on it exceeds about $4 \times 10^5$ N (we assume that the ankle is pushing up). Suppose that you step off a chair that is 0.40 m above the floor. If landing stiff-legged on the surface below, what minimum stopping distance do you need to avoid breaking your tibias? Indicate any assumptions you made in your answer to this question.
18. * EST BIO **Climbing Mt. Everest** In 1953 Sir Edmund Hillary and Tenzing Norgay made the first successful ascent of Mt. Everest. How many slices of bread did each climber have to eat to compensate for the increase of the gravitational potential energy of the system climbers-Earth? (One piece of bread releases about $1.0 \times 10^6$ J of energy in the body.) Indicate all of the assumptions used. Note: The body is an inefficient energy converter—see the reading passage at the end of this section.

## 6.4 Quantifying elastic potential energy

19. * A door spring is difficult to stretch. (a) What maximum force do you need to exert on a relaxed spring with a $1.2 \times 10^4$-N/m spring constant to stretch it 6.0 cm from its equilibrium position? (b) How much does the elastic potential energy of the spring change? (c) Determine its change in elastic potential energy as it returns from the 6.0-cm stretch position to a 3.0-cm stretch position. (d) Determine its elastic potential energy change as it moves from the 3.0-cm stretch position back to its equilibrium position.
20. * You compress a spring by a certain distance. Then you decide to compress it further so that its elastic potential energy increases by another 50%. What was the percent increase in the spring's compression distance?
21. * A moving car has 40,000 J of kinetic energy while moving at a speed of 7.0 m/s. A spring-loaded automobile bumper compresses 0.30 m when the car hits a wall and stops. What can you learn about the bumper's spring using this information? Answer quantitatively and list the assumptions that you made.
22. * The force required to stretch a slingshot by different amounts is shown in the graph in **Figure P6.22**. (a) What is the spring constant of the sling? (b) How much work does a child need to do to stretch the sling 15 cm from equilibrium?

**Figure P6.22**

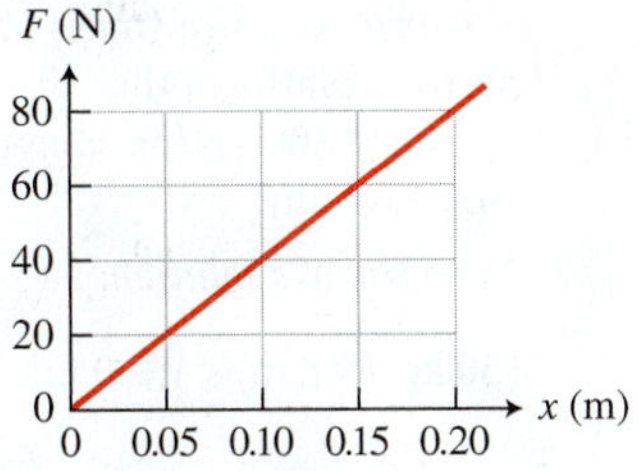


23. ** **Inverse bungee jump** The Ejection Seat at Lake Biwa Amusement Park in Japan (**Figure P6.23**) is an inverse bungee system. A seat with passengers of total mass 160 kg is connected to elastic cables on the sides. The seat is pulled down 12 m, stretching the cables. When released, the stretched cables launch the passengers upward above the towers to a height of about 30 m above their starting position at the ground. What can you learn about the mechanical properties of the cables using this information? Answer quantitatively. Assume that the cables are vertical.

**Figure P6.23**

## 6.5 Friction and energy conversion

24. * Jim is driving a 2268-kg pickup truck at 20 m/s and releases his foot from the accelerator pedal. The car eventually stops due to an effective friction force that the road, air, and other things exert on the car. The friction force has an average magnitude of 800 N. (a) Make a list of the physical quantities you can determine using this information and determine three of them. Specify the system and the initial and final states. (b) Would a heavier car travel farther before stopping or stop sooner? Identify the assumptions in your answer.
25. * A 1100-kg car traveling at 24 m/s coasts through some wet mud in which the net horizontal resistive force exerted on the car from all causes (mostly the force exerted by the mud) is $1.7 \times 10^4$ N. Determine the car's speed as it leaves the 18-m-long patch of mud.
26. * After falling 18 m, a 0.057-kg tennis ball has a speed of 12 m/s (the ball's initial speed is zero). Determine the average resistive force of the air in opposing the ball's motion. Solve the problem for the ball as your system and then repeat for a ball-Earth system. Are the answers the same or different?
27. * A water slide of length $l$ has a vertical drop of $h$. Abby's mass is $m$. An average friction force of magnitude $f$ opposes her motion. She starts down the slide at initial speed $v_i$. Use work-energy ideas to develop an expression for her speed at the bottom of the slide. Then evaluate your result using unit analysis and limiting case analysis.
28. ** You are pulling a crate on a rug, exerting a constant force on the crate $\vec{F}_{\text{Y on C}}$ at an angle $\theta$ above the horizontal. The crate moves at constant speed. Represent this process using a motion diagram, a force diagram, a momentum bar chart, and an energy bar chart. Specify your choice of system for each representation. Make a list of physical quantities you can determine using this information.
29. ** A 900-kg car initially at rest rolls 50 m down a hill inclined at an angle of 5.0°. A 400-N effective friction force opposes its motion. How fast is the car moving at the bottom? What distance will it travel on a similar horizontal surface at the bottom of the hill? Will the distance decrease or increase if the car's mass is 1800 kg?
30. * A car skids 18 m on a level road while trying to stop before hitting a stopped car in front of it. The two cars barely touch. The coefficient of kinetic friction between the first car and the road is 0.80. A policewoman gives the driver a ticket for exceeding the 35 mi/h speed limit. Can you defend the driver in court? Explain.

## 6.6 Skills for analyzing processes using the work-energy principle

31. * In a popular new hockey game, the players use small launchers with springs to move the 0.0030-kg puck. Each spring has a 120-N/m spring constant and can be compressed up to 0.020 m. What can you determine about the motion of the puck using this information? Make a list of quantities and determine their values.

32. * A 500-m-long ski slope drops at an angle of 6.4° relative to the horizontal. (a) Determine the change in gravitational potential energy of a 60-kg skier-Earth system when the skier goes down this slope. (b) If 20% of the gravitational potential energy change is converted into kinetic energy, how fast is the skier traveling at the bottom of the slope?

33. * A Frisbee gets stuck in a tree. You want to get it out by throwing a 1.0-kg rock straight up at the Frisbee. If the rock's speed as it reaches the Frisbee is 4.0 m/s, what was its speed as it left your hand 2.8 m below the Frisbee? Specify the system and the initial and final states.

34. A driver loses control of a car, drives off an embankment, and lands in a canyon 6.0 m below. What was the car's speed just before touching the ground if it was traveling on the level surface at 12 m/s before the driver lost control?

35. * You are pulling a box so it moves at increasing speed. Compare the work you need to do to accelerate it from 0 m/s to speed $v$ to the work needed to accelerate it from speed $v$ to the speed of $2v$. Discuss whether your answer makes sense. How many different situations do you need to consider?

36. * A cable lowers a 1200-kg elevator so that the elevator's speed increases from zero to 4.0 m/s in a vertical distance of 6.0 m. What is the force that the cable exerts on the elevator while lowering it? Specify the system, its initial and final states, and any assumptions you made. Then change the system and solve the problem again. Do the answers match?

37. ** **EST Hit by a hailstone** A 0.040-kg hailstone the size of a golf ball (4.3 cm in diameter) is falling at about 16 m/s when it reaches Earth's surface. Estimate the force that the hailstone exerts on your head—a head-on collision. Indicate any assumptions used in your estimate. Note that the cheekbone will break if something exerts a 900-N or larger force on the bone for more than 6 ms. Is this hailstone likely to break a bone?

38. * **BIO Froghopper jump** Froghoppers may be the insect jumping champs. These 6-mm-long bugs can spring 70 cm into the air, about the same distance as the flea. But the froghopper is 60 times more massive than a flea, at 12 mg. The froghopper pushes off for about 4 mm. What average force does it exert on the surface? Compare this to the gravitational force that Earth exerts on the bug.

39. * **Bar chart Jeopardy 1** Invent in words and with a sketch a process that is consistent with the qualitative work-energy bar chart shown in **Figure P6.39**. Then apply in symbols the generalized work-energy principle for that process.

**Figure P6.39**

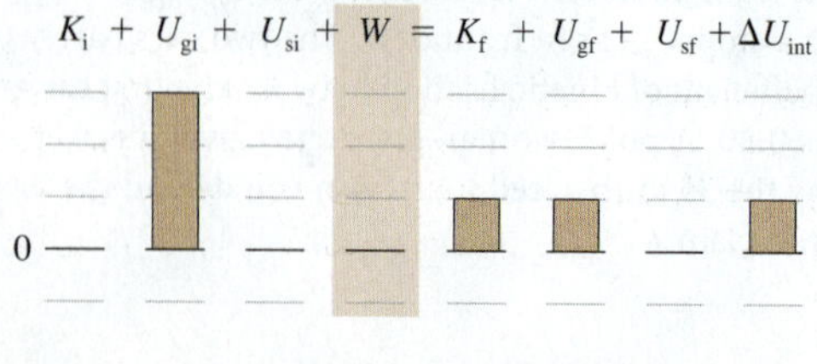


40. * **Bar chart Jeopardy 2** Invent in words and with a sketch a process that is consistent with the qualitative work-energy bar chart shown in **Figure P6.40**. Then apply in symbols the generalized work-energy principle for that process.

**Figure P6.40**

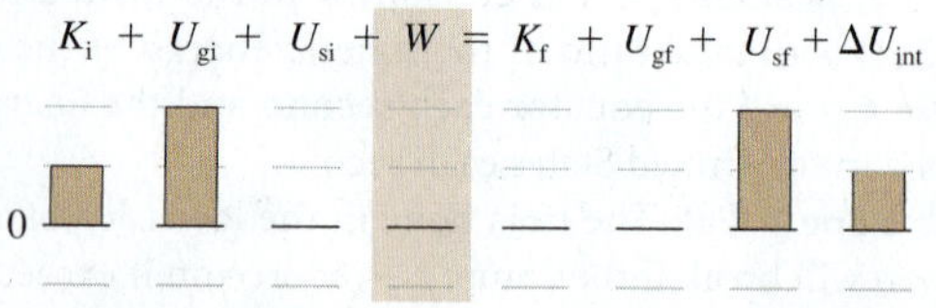


41. * **Equation Jeopardy 1** Construct a qualitative work-energy bar chart for a process that is consistent with the equation below. Then invent in words and with a sketch a process that is consistent with both the equation and the bar chart.

$$(1/2)(400\ \text{N/m})(0.20\ \text{m})^2 = (1/2)(0.50\ \text{kg})v^2 + (0.50\ \text{kg})(9.8\ \text{m/s}^2)(0.80\ \text{m})$$

42. * **Equation Jeopardy 2** Construct a qualitative work-energy bar chart for a process that is consistent with the equation below. Then invent in words and with a sketch a process that is consistent with both the equation and the bar chart.

$$(120\ \text{kg})(9.8\ \text{m/s}^2)(100\ \text{m})\sin 53^\circ = (1/2)(120\ \text{kg})(20\ \text{m/s})^2 + f_k(100\ \text{m})$$

43. * **Evaluation 1** Your friend provides a solution to the following problem. Evaluate his solution. Constructively identify any mistakes he made and correct the solution. Explain possible reasons for the mistakes.

*The problem:* A 400-kg motorcycle, including the driver, travels up a 10-m-long ramp inclined 30° above the paved horizontal surface holding the ramp. The cycle leaves the ramp at speed 20 m/s. Determine the cycle's speed just before it lands on the paved surface.

*Your friend's solution:*

$$(1/2)(400\ \text{kg})(20\ \text{m/s}) = (400\ \text{kg})(9.8\ \text{m/s}^2)(10\ \text{m}) + (1/2)(400\ \text{kg})v^2$$
$$= -13.2\ \text{m/s}$$

44. * **Evaluation 2** Your friend provides a solution to the following problem. Evaluate her solution. Constructively identify any mistakes she made and correct the solution. Explain possible reasons for the mistakes.

*The problem:* Jim (mass 50 kg) steps off a ledge that is 2.0 m above a platform that sits on top of a relaxed spring of force constant 8000 N/m. How far will the spring compress while stopping Jim?

*Your friend's solution:*

$$(50\ \text{kg})(9.8\ \text{m/s}^2)(2.0\ \text{m}) = (1/2)(8000\ \text{N/m})x$$
$$x = 0.25\ \text{m}$$

45. * A puck of mass $m$ moving at speed $v_i$ on a horizontal, frictionless surface is stopped in a distance $\Delta x$ because a hockey stick exerts an opposing force of magnitude $F$ on it. (a) Using the work-energy method, show that $F = mv_i^2/2\Delta x$. (b) If the stopping distance $\Delta x$ increases by 50%, by what percent does

the average force needed to stop the puck change, assuming that $m$ and $v_i$ are unchanged? Justify your result.

46. ** A rope exerts an 18-N force while lowering a 20-kg crate down a plane inclined at 20° (the rope is parallel to the plane). A 24-N friction force opposes the motion. The crate starts at rest and moves 10 m down the plane. Make a list of the physical quantities you can determine using this information and determine three of them. Specify the system and the initial and final states of the process.

## 6.7 Collisions: Putting it all together

47. ** You fire an 80-g arrow so that it is moving at 80 m/s when it hits and embeds in a 10-kg block resting on ice. (a) What is the velocity of the block and arrow just after the collision? (b) How far will the block slide on the ice before stopping? A 7.2-N friction force opposes its motion. Specify the system and the initial and final states for (a) and (b).

48. ** You fire a 50-g arrow that moves at an unknown speed. It hits and embeds in a 350-g block that slides on an air track. At the end, the block runs into and compresses a 4000-N/m spring 0.10 m. How fast was the arrow traveling? Indicate the assumptions that you made and discuss how they affect the result.

49. ** To confirm the results of Problem 48, you try a new experiment. The 50-g arrow is launched in an identical manner so that it hits and embeds in a 3.50-kg block. The block hangs from strings. After the arrow joins the block, they swing up so that they are 0.50 m higher than the block's starting point. How fast was the arrow moving before it joined the block?

50. ** A 1060-kg car moving west at 16 m/s collides with and locks onto a 1830-kg stationary car. (a) Determine the velocity of the cars just after the collision. (b) After the collision, the road exerts a $1.2 \times 10^4$-N friction force on the car tires. How far do the cars skid before stopping? Specify the system and the initial and final states of the process.

51. ** Jay rides his 2.0-kg skateboard. He is moving at speed 5.8 m/s when he pushes off the board and continues to move forward in the air at 5.4 m/s. The board now goes forward at 13 m/s. Determine Jay's mass and the change in the internal energy of the system during this process.

52. ** A 36-kg child is moving on a 2.0-kg skateboard at speed 6.0 m/s when she comes to a ledge that is 1.2 m above the surface below. Just before reaching the ledge, she pushes off the board. The board leaves the ledge moving horizontally and lands 8.0 m horizontally from the edge of the ledge. Make a list of the physical quantities describing the motion of the child after leaving the ledge and determine two of them. Describe any assumptions you made.

53. ** BIO **Falcons** While perched on an elevated site, a peregrine falcon spots a flying pigeon. The falcon dives, reaching a speed of 90 m/s (200 mi/h). The falcon hits its prey with its feet, stunning or killing it, then swoops back around to catch it in mid-air. Assume that the falcon has a mass of 0.60 kg and hits a 0.20-kg pigeon almost head-on. The falcon's speed after the collision is 60 m/s in the same direction. (a) Determine the final speed of the pigeon immediately after the hit. (b) Determine the internal energy produced by the collision. (c) Why does the falcon strike its prey with its feet and not head-on?

54. * When you play billiards, can you predict the velocities of the billiard balls after a collision if you know the velocity of a moving ball before the collision? Assume that the collision is head-on and elastic and that rotational motion can be ignored.

55. * A block of mass $m_1$ moving at speed $v$ toward the west on a frictionless surface has an elastic head-on collision with a second, stationary block of mass $m_2$. Determine expressions for the final velocity of each block.

56. ** A 4.0-kg block moving at 2.0 m/s toward the west on a frictionless surface has an elastic head-on collision with a second 1.0-kg block traveling east at 3.0 m/s. Determine the final velocity of each block. (b) Determine the kinetic energy of each block before and after the collision. Note: The block with the least initial kinetic energy actually gains energy and the one with the most loses an equal amount. This is analogous to what happens when cool air comes into contact with warm air. The cool air warms (its molecules speed up) and the warm air cools (its molecules slow down).

## 6.8 Power

57. (a) What is the power involved in lifting a 1.0-kg object 1.0 m in 1.0 s? (b) While lifting a 10-kg object 1.0 m in 0.50? (c) While lifting the 10-kg object 2.0 m in 1.0 s? (d) While lifting a 20-kg object 1.0 m in 1.0 s?

58. * A fire engine must lift 30 kg of water a vertical distance of 20 m each second. What is the amount of power needed for the water pump for this fire hose?

59. * BIO **Internal energy change while biking** You set your stationary bike on a high 80-N friction-like resistive force and cycle for 30 min at a speed of 8.0 m/s. Your body is 10% efficient at converting chemical energy in your body into mechanical work. (a) What is your internal chemical energy change? (b) How long must you bike to convert $3.0 \times 10^5$ J of chemical potential while staying at this speed? (This amount of energy equals the energy released by the body after eating three slices of bread.)

60. * BIO **Tree evaporation** A large tree can lose 500 kg of water a day. (a) How much work does the tree need to do to lift the water 8.0 m? (b) If the loss of water occurs over a 12-h period, what is the average power in watts needed to provide this increase in gravitational energy in the water-Earth system?

61. * **Climbing Mt. Mitchell** An 82-kg hiker climbs to the summit of Mount Mitchell in western North Carolina. During one 2.0-h period, the climber's vertical elevation increases 540 m. Determine (a) the change in gravitational potential energy of the climber-Earth system and (b) the power of the process needed to increase the gravitational energy.

62. * BIO EST **Sears stair climb** The fastest time for the Sears Tower (now Willis Tower) stair climb (103 flights, or 2232 steps) is about 20 min. (a) Estimate the mechanical power in watts for a top climber. Indicate any assumptions you made. (b) If the body is 20% efficient at converting chemical energy into mechanical energy, approximately how many joules and kilocalories of chemical energy does the body expend during the stair climb? Note: 1 food calorie = 1 kilocalorie = 4186 J.

63. * BIO EST **Exercising so you can eat ice cream** You curl a 5.5-kg (12 lb) barbell that is hanging straight down in your hand up to your shoulder. (a) Estimate the work that your hand does in lifting the barbell. (b) Estimate the average mechanical power of the lifting process. Indicate any assumptions used in making the estimate. (c) Assuming the efficiency described at the end of Problem 62, how many times would you have to lift the barbell in order to burn enough calories to

use up the energy absorbed by eating a 300-food-calorie dish of ice cream? (Problem 62 provides the joule equivalent of a food calorie.) List the assumptions that you made.

64. * * BIO **Salmon move upstream** In the past, salmon would swim more than 1130 km (700 mi) to spawn at the headwaters of the Salmon River in central Idaho. The trip took about 22 days, and the fish consumed energy at a rate of 2.0 W for each kilogram of body mass. (a) What is the total energy used by a 3.0-kg salmon while making this 22-day trip? (b) About 80% of this energy is released by burning fat and the other 20% by burning protein. How many grams of fat are burned? One gram of fat releases $3.8 \times 10^4$ J of energy. (c) If the salmon is about 15% fat at the beginning of the trip, how many grams of fat does it have at the end of the trip?

65. * EST Estimate the maximum horsepower of the process of raising your body mass as fast as possible up a flight of 20 stair steps. Justify any numbers used in your estimate. The only energy change you should consider is the change in gravitational potential energy of the system you-Earth.

66. * A 1600-kg car smashes into a shed and stops. The force that the shed exerts on the car as a function of a position at the car's center is shown in **Figure P6.66**. How fast was the car traveling just before hitting the shed?

**Figure P6.66**

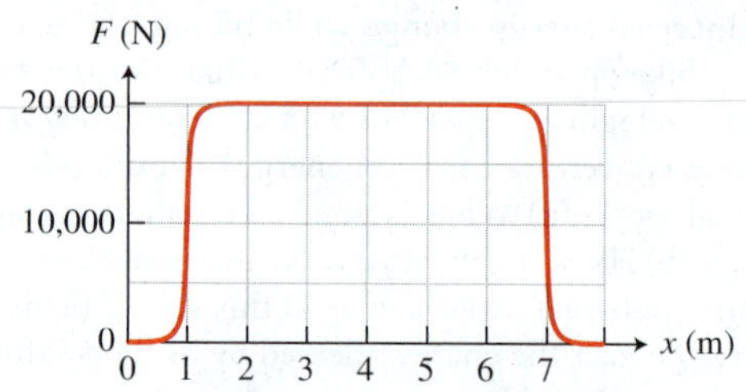


67. * * Suppose the car in Problem 66 was moving at twice the speed and was stopped in the same manner and in the same distance, only now by a more solidly constructed shed. Draw a new graph for the force that this new shed exerted on the car as a function of the car's position. Include the appropriate numbers on the force axis and on the distance axis.

## 6.9 Improving our model of gravitational potential energy

68. At what distance from Earth is the gravitational potential energy of a spaceship-Earth system reduced to half the energy of the system before the launch?

69. * **Possible escape of different air molecule types** (a) Determine the ratio of escape speeds from Earth for a hydrogen molecule ($H_2$) and for an oxygen molecule ($O_2$). The mass of the oxygen is approximately 16 times that of the hydrogen. (b) In the atmosphere, the average random kinetic energy of hydrogen molecules and oxygen molecules is the same. Determine the ratio of the average speeds of the hydrogen and the oxygen molecules. (c) Based on these two results, give one reason why our atmosphere lacks hydrogen but retains oxygen.

70. Determine the escape speed for a rocket to leave Earth's Moon.

71. Determine the escape speed for a rocket to leave the solar system.

72. If the Sun were to become a black hole, how much would it increase the gravitational potential energy of the Sun-Earth system?

73. * A satellite moves in elliptical orbit around Earth, which is one of the foci of the elliptical orbit. (a) The satellite is moving faster when it is closer to Earth. Explain why. (b) If the satellite moves faster when it is closer to Earth, is the energy of the satellite-Earth system constant? Explain.

74. * Determine the maximum radius Earth's Moon would have to have in order for it to be a black hole.

## General Problems

75. * * EST You wish to try bungee jumping, but want to make sure it is safe. The brochure provided at the ticket office says that the cord holding the jumper is initially 12 m long and has a spring constant of 160 N/m. The tower from which you plan to jump appears to be 10 floors high. Should you try the bungee jump? Explain your answer.

76. * Pose a problem involving work-energy ideas with real numbers. Then solve the problem choosing two different systems and discuss whether the answers were different.

77. * * Your dormitory has a nice balcony that looks over a pond in the grass below. You attach a 16-m-long rope to the limb of a tall tree beside the pond and pull it to the balcony so that it makes a 53° angle from the vertical. You hold the rope while standing on the balcony and then swing down. (a) How fast are you moving at the lowest point in your swing? Specify the system and the initial and final states. List the assumptions that you made. (b) How strong should the rope be to withstand your adventure (use your own mass for calculations)?

78. * * **Bungee jump at Squaw Valley** At the Squaw Valley ski area, 2500 m above sea level (**Figure P6.78**), a bungee jumper falls over a 152-m cliff above Lake Tahoe. Choose numbers for quantities, and make and solve a problem related to this bungee system.

**Figure P6.78**

79. * * **Six Flags roller coaster** A loop-the-loop on the Six Flags Shockwave roller coaster has a 10-m radius (**Figure P6.79**). The car is moving at 24 m/s at the bottom of the loop. Determine the force exerted by the seat of the car on an 80-kg rider when passing inverted at the top of the loop.

**Figure P6.79**

80. * * **Designing a ride** You are asked to help design a new type of loop-the-loop ride. Instead of rolling down a long hill to generate the speed to go around the loop, the 300-kg cart starts at rest (with two passengers) on a track at the same level as the bottom of the 10-m radius loop. The cart is pressed against a compressed spring that, when released, launches the cart along the track around the loop. Choose a spring of the appropriate spring constant to launch the cart so that the downward force exerted by the track on the cart as it passes the top of the loop is 0.2 times the force that Earth exerts on the cart. The spring is initially compressed 6.0 m.

81. ** BIO EST **Impact extinction** 65 million years ago over 50% of all species became extinct, ending the reign of dinosaurs and opening the way for mammals to become the dominant land vertebrates. A theory for this extinction, with considerable supporting evidence, is that a 10-km-wide $1.8 \times 10^{15}$-kg asteroid traveling at speed 11 km/s crashed into Earth. Use this information and any other information or assumptions of your choosing to (a) estimate the change in velocity of Earth due to the impact; (b) estimate the average force that Earth exerted on the asteroid while stopping it; and (c) estimate the internal energy produced by the collision (a bar chart for the process might help). By comparison, the atomic bombs dropped on Japan during World War II were each equivalent to 15,000 tons of TNT (1 ton of TNT releases $4.2 \times 10^9$ J of energy).

82. ** Newton's cradle is a toy that consists of several metal balls touching each other and suspended on strings (**Figure P6.82**). When you pull one ball to the side and let it strike the next ball, only one ball swings out on the other side. When you use two balls to hit the others, two balls swing out. Can you account for this effect using your knowledge about elastic collisions?

**Figure P6.82**

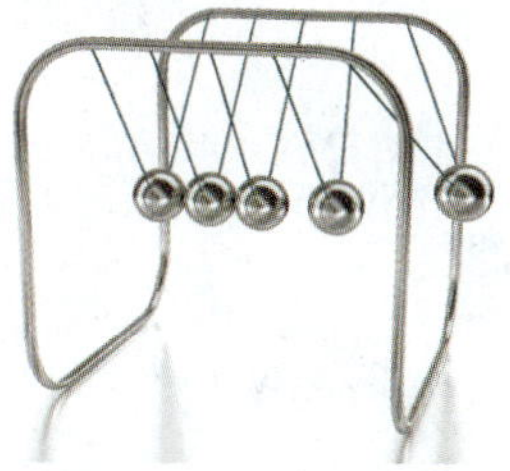

83. ** **Design of looping roller coaster** You are an engineer helping to design a roller coaster that carries passengers down a steep track and around a vertical loop. The coaster's speed must be great enough when at the top of the loop so that the rider stays in contact with the cart and the cart stays in contact with the track. Riders can withstand acceleration no more than a few "$g$'s", where one "$g$" is 9.8 m/s². What are some reasonable values for the physical quantities you can use in the design of the ride? For example, one consideration is the height at which the cart and rider should start so that they can safely make it around the loop and the radius of the loop.

## Reading Passage Problems

BIO **Metabolic Rate** Energy for our activities is provided by the chemical energy of the foods we eat. The absolute value of the rate of conversion of this chemical energy into other forms of energy ($\Delta E/\Delta t$) is called the metabolic rate. The metabolic rate depends on many factors—a person's weight, physical activity, the efficiency of bodily processes, and the fat-muscle ratio. **Table 6.8** lists the metabolic rates of people under several different conditions and in several different units of measure: 1 kcal = 1000 calories = 4186 J. Dieticians call a kcal simply a Cal. A piece of bread provides about 70 kcal of metabolic energy.

In 1 hour of heavy exercise a 68-kg person metabolizes 600 kcal − 90 kcal = 510 kcal more energy than when at rest. Typically, reducing kilocalorie intake by 3500 kcal (either by burning it in exercise or not consuming it in the first place) results in a loss of 0.45 kg of body mass (the mass is lost through exhaling carbon dioxide—the product of metabolism).

84. Why is the metabolic rate different for different people?
(a) They have different masses.
(b) They have different body function efficiencies.
(c) They have different levels of physical activity.
(d) All of the above

85. A 50-kg mountain climber moves 30 m up a vertical slope. If the muscles in her body convert chemical energy into gravitational potential energy with an efficiency of no more than 5%, what is the chemical energy used to climb the slope?
(a) 7 kcal (b) 3000 J
(c) 70 kcal (d) 300,000 J

86. If 10% of a 50-kg rock climber's total energy expenditure goes into the gravitational energy change when climbing a 100-m vertical slope, what is the climber's average metabolic rate during the climb if it takes her 10 min to complete the climb?
(a) 8 W (b) 80 W (c) 500 W
(d) 700 W (e) 800 W

87. EST A 68-kg person wishes to lose 4.5 kg in 2 months. Estimate the time that this person should spend in moderate exercise each day to achieve this goal (without altering her food consumption).
(a) 0.4 h (b) 0.9 h (c) 1.4 h
(d) 1.9 h (e) 2.4 h

88. EST A 68-kg person walks at 5 km per hour for 1 hour a day for 1 year. Estimate the *extra* number of kilocalories of energy used because of the walking.
(a) 40,000 kcal (b) 47,000 kcal
(c) 88,000 kcal (d) 150,000 kcal

89. Suppose that a 90-kg person walks for 1 hour each day for a year, expending 50,000 extra kilocalories of metabolic energy (in addition to his normal resting metabolic energy use). What approximately is the person's mass at the end of the year, assuming his food consumption does not change?
(a) 57 kg (b) 61 kg (c) 64 kg
(d) 66 kg (e) 67 kg

**Table 6.8 Energy usage rate during various activities.**

| Type of activity | $\Delta E/\Delta t$ (watts) | $\Delta E/\Delta t$ (kcal/h) | $\Delta E/\Delta t$ (kcal/day) |
|---|---|---|---|
| 45-kg person at rest | 80 | 70 | 1600 |
| 68-kg person at rest | 100 | 90 | 2100 |
| 90-kg person at rest | 120 | 110 | 2600 |
| 68-kg person walking 3 mph | 280 | 240 | 5800 |
| 68-kg person moderate exercise | 470 | 400 | 10,000 |
| 68-kg person heavy exercise | 700 | 600 | 14,000 |

**Figure 6.14** A kangaroo hopping.

Notice change in foot orientation when taking off and landing.

**Figure 6.15** The kangaroo stores elastic potential energy in its muscle and tendon when landing and uses some of that energy to make the next hop.

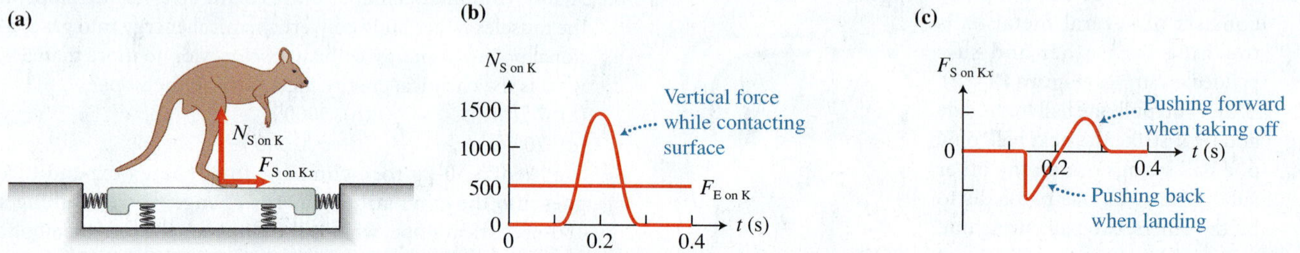


**BIO Kangaroo hopping** Hopping is an efficient method of locomotion for the kangaroo (see **Figure 6.14**). When the kangaroo is in the air, the Earth-kangaroo system has a combination of gravitational potential energy and kinetic energy. When the kangaroo lands, its Achilles tendons and the attached muscles stretch—a form of elastic potential energy. This elastic potential energy is used along with additional muscle tension to launch the kangaroo off the ground for the next hop. In the red kangaroo, more than 50% of the total energy used during each hop is recovered elastic potential energy. This is so efficient that the kangaroo's metabolic rate actually decreases slightly as its hopping speed increases from 8 km/h to 25 km/h.

The horizontal and vertical force components exerted by a firm surface on a kangaroo's feet while it hops are shown in **Figure 6.15a**. The vertical force $N_{\text{S on K}}$ (Figure 6.15b) varies: when the kangaroo is not touching the surface S, the force is zero; when it is pushing off, the force is about three times the gravitational force that Earth exerts on the kangaroo. The surface exerts a backward horizontal force ($F_{\text{S on K}\,x}$) on the kangaroo's foot while it lands and a forward horizontal force as it pushes off for the next hop (Figure 6.15c), similar to what happens to a human foot when landing in front of the body and when pushing off for another step when behind the body.

90. Why is hopping an energy-efficient mode of transportation for a kangaroo?
    (a) There is less resistance since there is less contact with the ground.
    (b) The elastic energy stored in muscles and tendons when landing is returned to help with the next hop.
    (c) The kangaroo has long feet that cushion the landing.
    (d) The kangaroo's long feet help launch the kangaroo.
91. Why does the horizontal force exerted by the ground on the kangaroo change direction as the kangaroo lands and then hops forward?
    (a) The backward force when it lands prevents it from slipping, and the forward force when taking off helps propel it forward.
    (b) One horizontal force is needed to help stop the kangaroo's fall and the other to help launch its upward vertical hop.
    (c) Both forces oppose the kangaroo's motion, but one looks like it is forward because the kangaroo is moving fast.
    (d) The kangaroo is not an inertial reference frame, and the forward force is not real.
    (e) All of the above
92. Which answer below is closest to the vertical impulse that the ground exerts on the kangaroo while it takes off?
    (a) Zero (b) $+50\ \text{N}\cdot\text{s}$ (c) $+150\ \text{N}\cdot\text{s}$
    (d) $-50\ \text{N}\cdot\text{s}$ (e) $-150\ \text{N}\cdot\text{s}$
93. Which answer below is closest to the vertical impulse due to the gravitational force exerted on the kangaroo by Earth during the short time interval while it takes off?
    (a) 0 (b) $+50\ \text{N}\cdot\text{s}$ (c) $+150\ \text{N}\cdot\text{s}$
    (d) $-50\ \text{N}\cdot\text{s}$ (e) $-150\ \text{N}\cdot\text{s}$
94. Suppose the net vertical impulse on the 50-kg kangaroo due to all external forces was $+100\ \text{N}\cdot\text{s}$. Which answer below is closest to its vertical component of velocity when it leaves the ground?
    (a) +2.0 m/s (b) +3.0 m/s (c) +4.0 m/s
    (d) +8.0 m/s (e) +10 m/s
95. Which answer below is closest to the vertical height above the ground that the kangaroo reaches if it leaves the ground traveling with a vertical component of velocity of 2.5 m/s?
    (a) 0.2 m (b) 0.3 m (c) 0.4 m
    (d) 0.6 m (e) 0.8 m

# 7 Extended Bodies at Rest

Why is it best to lift heavy objects with your knees bent and the object near your body?

Why are doorknobs located on the side of the door opposite the hinges?

Why must your biceps muscle exert about seven times more force on your forearm to lift a barbell than the force that Earth exerts on it?

Back pain is a major health problem. In 2009 in the United States, the medical costs related to back pain were over $80 billion a year, about the same as the yearly cost of treating cancer. Eighty percent of people will suffer back pain at some point, usually beginning between the ages of 30 and 50. Back pain often results from incorrect lifting, which compresses the disks in the lower back, causing nerves to be pinched. Understanding the physics principles underlying lifting can help us develop techniques that minimize this compression and prevent injuries.

**Be sure you know how to:**

- Define the point-like model for an object (Section 1.2).
- Draw a force diagram for a system (Section 2.1).
- Use the component form of Newton's second law (Section 3.5).

**So far in this book** we have primarily been modeling objects as point-like with no internal structure. This method is appropriate when the shapes of objects do not affect the consequences of their interactions with each other—for example, a car screeching to a halt, an elevator moving up a shaft, or an apple falling into a pile of leaves. However, objects in general and the human body in particular are extraordinarily complex, with many internal parts that rotate and move relative to one another. To study the body and other complex structures, we need to develop a new way of modeling objects and of analyzing their interactions.

# 7.1 Extended and rigid bodies

In earlier chapters we focused on situations in which real objects that have nonzero dimensions could be reasonably modeled as point-like. Although any real object is extended in three dimensions, we assumed that an object's size and internal structure were not important for understanding the phenomena. When we analyzed the motion of such extended objects as cars, we neglected their parts (the turning wheels, the moving engine parts, and so forth) that moved with respect to each other internally even though the car as a whole moved in a straight line. We also neglected the size of the car. When the car made a turn, we considered the radius of the turn the same for all parts of the car. Put another way, when we model an object as a point-like, we ignore the fact that different parts of the same object move differently. Such motion—when an object moves as a whole from one location to another, without turning—is called **translational motion.**

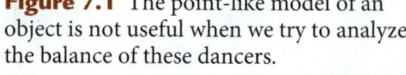

**Active Learning Guide ›**

In this chapter we will be analyzing more complex objects when they are at rest. We'd like, for example, to understand how the contemporary dancers in **Figure 7.1** manage to maintain their unusual balance. How large is the force exerted by the woman's hand on the man's foot? Where should each force be exerted and how large should their magnitudes be in order for the dancers to remain stable? We certainly cannot answer these questions by modeling the man and woman, who are extended objects, as point-like objects. We need a new model for extended objects and a new method for analyzing the forces that objects exert on each other. Our first task is to develop this new model for objects.

## Rigid bodies

Notice that at the instant shown in the photo the various parts of the dancers' bodies are not moving with respect to each other. They are acting as a single rigid object. This observation motivates a new model for an extended object, in which the size of the object is not zero (it is not a point-like object), but

**Figure 7.1** The point-like model of an object is not useful when we try to analyze the balance of these dancers.

parts of the object do not move with respect to each other. In physics, this model is called a **rigid body**.

> **Rigid body** A rigid body is a model of a real extended object. When we model an extended object as a rigid body, we assume that the object has a nonzero size but the distances between all parts of the object remain the same (the size and shape of the object do not change).

Many bones in your body can be reasonably modeled as rigid bodies, as can many everyday objects—buildings, bridges, streetlights, and utility poles. In this chapter we investigate what conditions are necessary for a rigid body to remain at rest.

## Center of mass

Let's start with some simple experiments. Place a piece of thin, flat cardboard on a very smooth table. If we consider the cardboard to be a point-like object, on a force diagram the upward normal force that the table exerts on the cardboard will balance the downward gravitational force that Earth exerts on the cardboard (**Figure 7.2a**)—the cardboard will not accelerate. Now, we place the cardboard on a very small surface—like the eraser of a pencil. The cardboard tilts and falls off (Figure 7.2b). Does this result mean that the eraser cannot exert the same upward force on the cardboard that the table did, or is there some other explanation? The model of the point-like object cannot explain the tilting, since point-like objects do not tilt. Perhaps we need to model the cardboard as a rigid body. Before we do this, let us learn a little more about rigid bodies in Observational Experiment **Table 7.1**.

**Figure 7.2** The cardboard is stable in (a) but not in (b). The place where the supporting force is exerted matters.

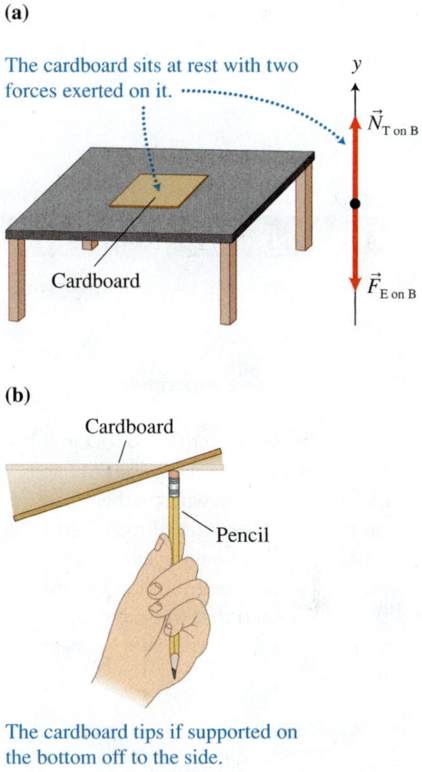
(a)
The cardboard sits at rest with two forces exerted on it.
y
$\vec{N}_{\text{T on B}}$
Cardboard
$\vec{F}_{\text{E on B}}$
(b)
Cardboard
Pencil
The cardboard tips if supported on the bottom off to the side.

**OBSERVATIONAL EXPERIMENT TABLE**

### 7.1 Pushing a board so that it moves without turning.

VIDEO 7.1

| Observational experiment | Analysis |
|---|---|
| We push with a pencil eraser at a point on the edge of a heart-shaped piece of flat cardboard so that the cardboard moves on a smooth surface *without* turning. We then exert the force on a different point and push the cardboard in a new direction so that it again moves without turning. We also push against the cardboard at points where it turns as it moves. | We draw lines across the top of the cardboard from the places and in the directions that the cardboard did not turn as it moved. |

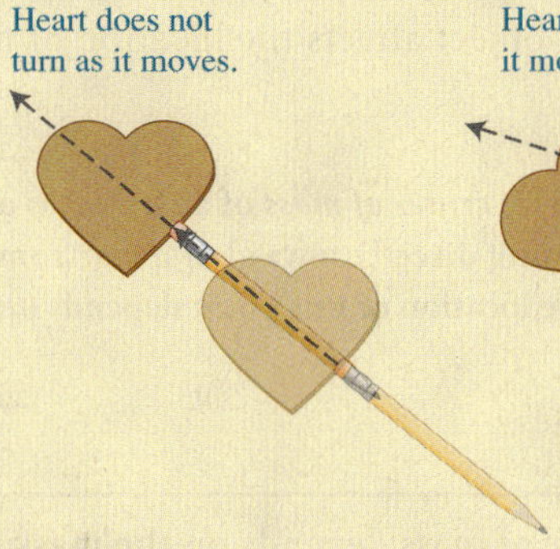
Heart does not turn as it moves.
Heart turns as it moves.

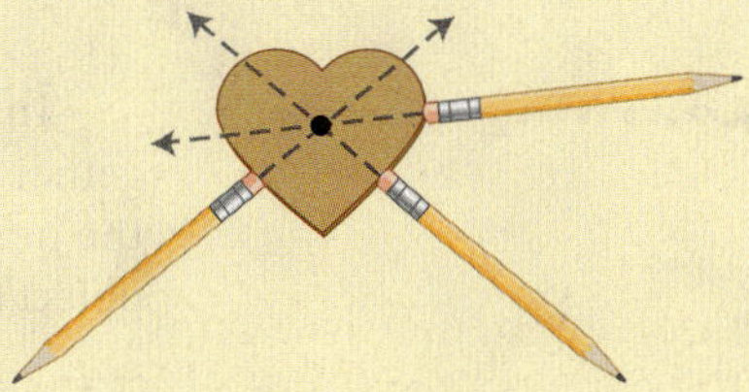

**Pattern**

1. All of the lines along which we had to push to move the cardboard without turning pass through a common point on the cardboard.
2. Pushing at the same locations in other directions causes the cardboard to turn as it moves.

**Active Learning Guide ›**

The analysis we did in Observational Experiment Table 7.1 indicates that a rigid body possesses a special point. If a force exerted on that object points directly toward or away from that point, the object will not turn. We call this point the object's **center of mass**. It appears that the center of mass is located approximately at the geometric center of the board (the geometric center is a point that in some sense is at the center of the object). In Testing Experiment **Table 7.2** we will test this idea that the center of mass of a rigid body is the point that we must push toward (or pull away from) in order to make an object move only translationally (as a whole, without turning).

**TESTING EXPERIMENT TABLE**

**7.2 Where is the center of mass?**

VIDEO 7.2

| Testing experiment | Prediction | Outcome |
|---|---|---|
| Place a heavy object (the dark circle) on the heart-shaped cardboard used in Table 7.1(somewhere other than on the board's center of mass) and repeat the experiments, each time pushing at different locations along the edges, such that in each experiment the cardboard moves *without* turning.   | If the idea is correct, and we push the cardboard so that in each experiment it moves but does not turn, then all of the lines along which the forces are exerted in all experiments should cross at one point: the new center of mass. Because we have added the heavy object, we expect it to be at a different location than in Table 7.1. | The lines along which the forces are exerted all cross at one point. This point is located somewhere between the old center of mass and the position of the added object. 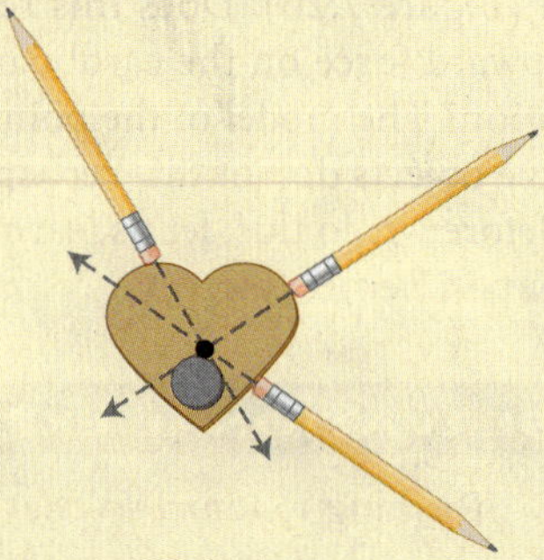 |

**Conclusion**

The outcome is consistent with the prediction—it supports the idea of a single center of mass in a rigid body or group of bodies.

**Active Learning Guide ›**

In Table 7.2, the center of mass of the heart and the added object together was different from the center of mass we found in Table 7.1 using only the heart. Thus, the mass distribution of an object affects the location of the object's center of mass.

**Center of mass (qualitative definition)** The *center of mass* of an object is a point where a force exerted on the object pointing directly toward or away from that point, will not cause the object to turn. The location of this point depends on the mass distribution of the object.

**TIP** Although the location of the center of mass depends on the mass distribution of the object, the mass of the object is not necessarily evenly distributed around the center of mass. We will learn more about the properties of the center of mass; we just want to caution you from taking the name of this point literally.

## Where is the gravitational force exerted on a rigid body?

At the beginning of the chapter we found that we could not balance cardboard on the eraser of a pencil. Why did the cardboard fall off? Imagine drawing the forces exerted on the cardboard. The cardboard interacts with two objects: the eraser and Earth. The eraser exerts an upward normal force on the cardboard at the point where it touches it. Earth exerts a downward gravitational force on every part of the cardboard. Is it possible to simplify the situation and to find one location at which we can assume that the entire gravitational force is exerted on the cardboard?

Let us go back to the heart-shaped cardboard from Observational Experiment Table 7.1. If we place the eraser exactly under the cardboard's center of mass, the cardboard does not tip over and fall (**Figure 7.3**). If the cardboard does not tip, it means that all forces exerted on it, including the force exerted by Earth and the force exerted by the eraser, pass through the center of mass. The normal force exerted by the eraser passes through the place where it contacts the cardboard—below the center of mass.

Earth exerts a small force on each small part of the board, but we can assume that the total force is exerted exactly at the center of mass. That is why sometimes the object's center of mass is called the object's **center of gravity.**

When we model something as a point-like object, we model it as if all of the object's mass is located at its center of mass. Likewise, we can apply what we know about translational motion for point-like objects to rigid bodies, as long as we apply the rules to their centers of mass.

**Review Question 7.1** You have an oval framed painting. How do you determine where you should insert a single nail into the frame so that the painting is correctly oriented in both vertical and horizontal directions?

**Figure 7.3** The eraser head can support the heart-shaped cardboard if the eraser is placed under the heart's center of mass.

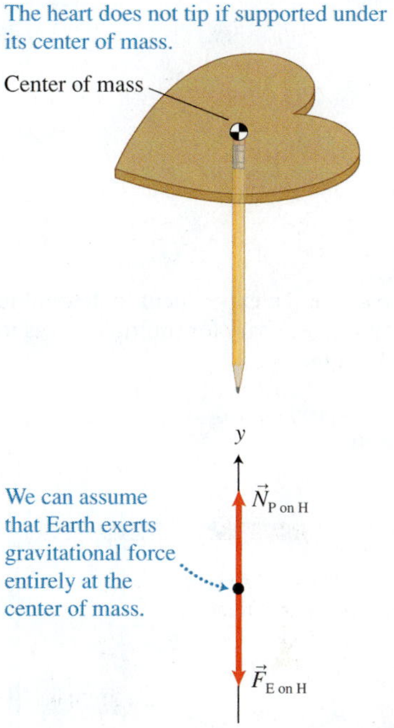

The heart does not tip if supported under its center of mass.
Center of mass
y
$\vec{N}_{\text{P on H}}$
We can assume that Earth exerts gravitational force entirely at the center of mass.
$\vec{F}_{\text{E on H}}$

**TIP** When multiple forces are exerted on a rigid body, the center of mass of the rigid body accelerates translationally according to Newton's second law

$$\vec{a} = \frac{\Sigma \vec{F}}{m}$$

# 7.2 Torque: A new physical quantity

We learned in the previous section that the turning effect of an individual force depends on where and in which direction the force is exerted on an object. The translational acceleration of the object's center of mass is still determined by Newton's second law, independently of where the force is exerted. In this section we will learn about the turning ability of a force that an object exerts on a rigid body.

## Axis of rotation

When objects turn around an axis, physicists say that they undergo **rotational motion**. The axis may be a fixed physical axis, such as the hinge of a door, or it may not, as in the case of a spinning top. In this chapter we will focus on the conditions under which objects that could potentially rotate do not do so.

Consider a door. When you push on the doorknob perpendicular to the door's surface ($\vec{F}_2$ in **Figure 7.4**), it rotates easily about the door hinges. We call the imaginary line passing through the hinges the **axis of rotation**. You know from experience that pushing a door at or near the axis of rotation ($\vec{F}_1$ in Figure 7.4) is not as effective as pushing the doorknob. You also know that the harder you push near the knob, the more rapidly the door starts moving. Lastly, pushing on the outside edge of the door toward the axis of rotation ($\vec{F}_3$ in Figure 7.4) does not move the door at all.

**Figure 7.4** Different forces have different effects in turning a door about its axis of rotation.

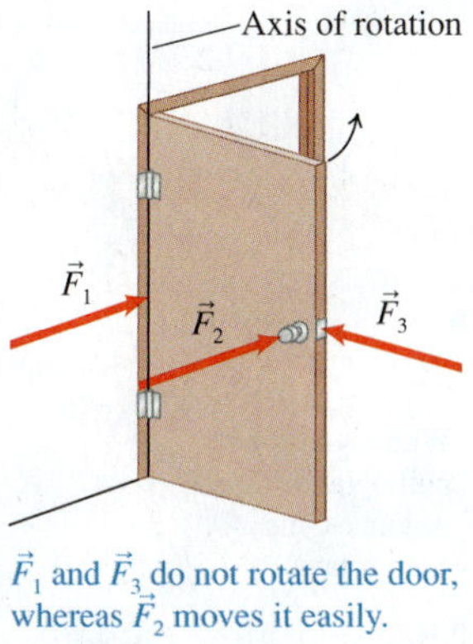

Axis of rotation
$\vec{F}_1$
$\vec{F}_2$
$\vec{F}_3$
$\vec{F}_1$ and $\vec{F}_3$ do not rotate the door, whereas $\vec{F}_2$ moves it easily.

These observations suggest that three factors affect the turning ability of a force: (1) the place where the force is exerted, (2) the magnitude of the force, and (3) the direction in which the force is exerted. Next, let's construct a quantitative expression for this turning ability.

## The role of position on the turning ability of a force

In order to construct a physical quantity that characterizes the turning ability of a force, we need to perform experiments where we exert measured forces at measured positions on a rigid body. We will take a 0.10-kg meter stick and suspend it at its center of mass from spring scale 2 (**Figure 7.5**); $\vec{F}_2$ is the force exerted on the meter stick by scale 2 at the point of suspension. Spring scale 1 pulls perpendicularly on the stick, exerting a downward force $\vec{F}_1$ at different locations on the left side, and scale 3 exerts a downward perpendicular force $\vec{F}_3$ at different locations on the right side. Earth exerts a 1.0-N force $\vec{F}_{\text{E on M}}$ on its center of mass, which is the point of suspension.

**Figure 7.5** An experiment to determine a condition necessary for multiple forces to balance a meter stick.

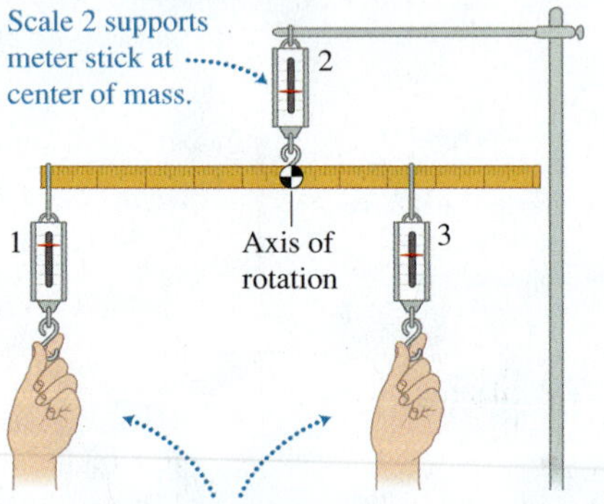

Scale 2 supports meter stick at center of mass.
2
1
Axis of rotation
3
We pull down on scales 1 and 3, exerting different forces at different places.

When either scale 1 or scale 3 pulls alone on the stick, the stick rotates (when scale 1 pulls on the stick, it rotates counterclockwise; when 3 pulls, it rotates clockwise). If we pull equally on scales 1 and 3 ($\vec{F}_1 = \vec{F}_3$) but the scales are located at different distances from the axis of rotation, the stick rotates (**Figure 7.6a**). We can use trial and error to find the combinations where the scales can be placed and pulled such that the stick does not rotate. For example, if scale 1 is twice as far from the axis of rotation and pulls half as hard as scale 3, the stick remains in equilibrium (Figure 7.6b). Similarly, if scale 1 is three times farther from the axis of rotation and exerts one-third of the force compared to scale 3, the stick remains in equilibrium (Figure 7.6c). When the stick does not rotate and does not move translationally, it is in a state of **static equilibrium.**

> **Static equilibrium** **An object is said to be in static equilibrium when it *remains* at rest (does not undergo either translational or rotational motion) with respect to a particular observer in an inertial reference frame.**

**Active Learning Guide ›**

If we explore more situations in which the stick is not rotating, we find that when we have two springs pulling perpendicular to the stick, the stick remains in static equilibrium when the product of the magnitude of the force

**Figure 7.6** (a) The meter stick does not balance even though equal downward forces are exerted on each side. (b) and (c) A greater force on one side nearer the pivot point balances a smaller force on the other side farther from the pivot point.

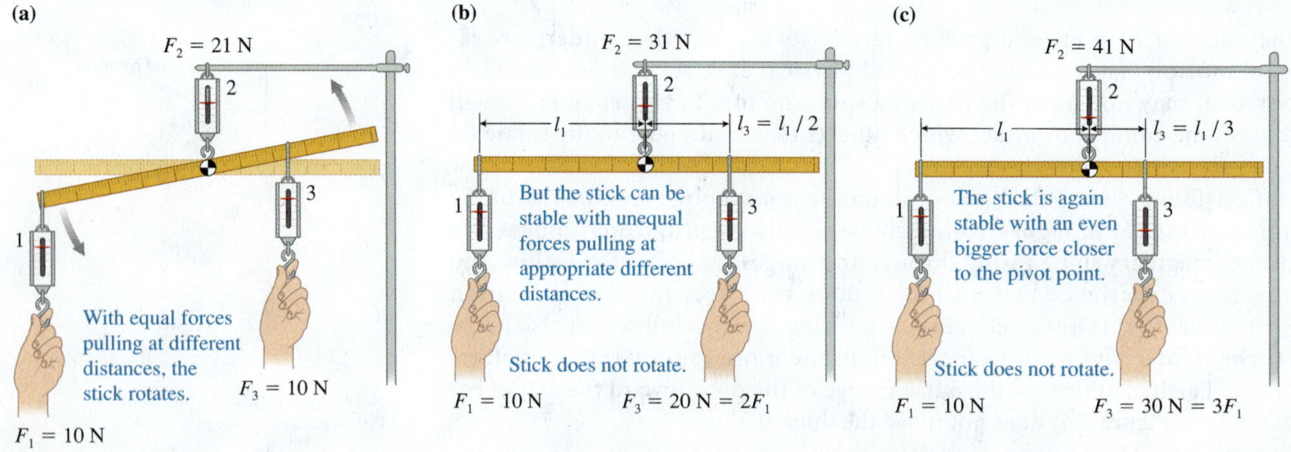

(a)
$F_2 = 21$ N
2
3
1
With equal forces pulling at different distances, the stick rotates.
$F_3 = 10$ N
$F_1 = 10$ N
(b)
$F_2 = 31$ N
2
$l_1$
$l_3 = l_1 / 2$
1
3
But the stick can be stable with unequal forces pulling at appropriate different distances.
Stick does not rotate.
$F_1 = 10$ N
$F_3 = 20 \text{ N} = 2F_1$
(c)
$F_2 = 41$ N
2
$l_1$
$l_3 = l_1 / 3$
1
3
The stick is again stable with an even bigger force closer to the pivot point.
Stick does not rotate.
$F_1 = 10$ N
$F_3 = 30 \text{ N} = 3F_1$

exerted by the scale on the stick ($F_1$ or $F_2$ in our experiment) and the distance between where the force is exerted and the axis of rotation ($l_1$ or $l_2$ in our experiment) is the same for both forces:

$$F_1 l_1 = F_2 l_2$$

In other words, the turning ability of the force on the left cancels the turning ability of the force on the right, and the object is in static equilibrium.

## The role of magnitude on the turning ability of a force

In addition, we notice that independently of whether the stick rotated or not, the reading of scale 2, $F_2$, supporting the stick always equals the sum of the readings of scales 1 and 3, $F_1 + F_3$, plus the magnitude of the force exerted by Earth on the stick $F_{\text{E on S}}$. In other words, in all cases the sum of the forces exerted on the stick was zero ($\Sigma F_y = 0$). This finding is consistent with what we know from Newton's laws—an object does not accelerate translationally if the sum of the forces exerted on it is zero. If it is originally at rest and does not accelerate translationally, then it remains at rest.

However, as we have seen from the experiment with the cardboard on the eraser, this sum-of-forces-equals-zero rule does not guarantee the rotational stability of rigid bodies (see Figure 7.2). Even when the sum of the forces exerted on the cardboard was zero, it could still start turning.

Another simple experiment helps illustrate this idea. Take a book with a glossy cover, place it on a smooth table (to minimize friction), and push it, exerting the same magnitude, oppositely directed force on each of two corners (as shown in **Figure 7.7a**). The force diagram in Figure 7.7b shows that the net force exerted on the book is zero—there is no translational acceleration. However, the book starts turning. The forces exerted by Earth and the table on the book pass through the book's center of mass; thus they do not cause turning. But the forces that you exert on the corners of the book do cause it to turn. Notice that these forces are of the same magnitude and are exerted at the same distance from the center of mass. You can imagine that there is an invisible axis of rotation passing through the center of mass perpendicular to the desk's surface. You would think that the turning effect caused by each force around this imaginary axis is the same; thus the two turning effects should cancel, as they did for scales 1 and 3 in the experiment described earlier. However, this does not happen. The fact that the book turns tells us that not only are the magnitude and placement of the force on the object important to describe the turning ability of the force, but the direction in which this force causes turning (for example, clockwise or counterclockwise around an axis of rotation) is also important.

By convention, physicists call counterclockwise turning about an axis of rotation positive and clockwise turning negative. So far, this is what we know about the new quantity that characterizes the turning ability of a force:

(a) It is equal to the product of the magnitude of the force and the distance the force is exerted from the axis of rotation.

(b) It is positive when the force tends to turn the object counterclockwise and negative when the force tends to turn the object clockwise.

(c) When one force tends to rotate an object counterclockwise and the other force tends to rotate an object clockwise, their effects cancel if $(F_{\text{counterclockwise}} l_1) + (-F_{\text{clockwise}} l_2) = 0$. In this case the object does not rotate.

Let's apply what we have devised so far to check whether this new quantity is useful for explaining other situations. Here is another simple experiment that you can do at home. Place a full milk carton or something of similar mass into a grocery bag. The bag with the milk carton by itself is not too heavy and

**Figure 7.7** The turning effect of a force must depend on more than $F$ and $l$.

(a)

The equal and opposite forces that you exert on the book at its corners cause the book to rotate.

(b)

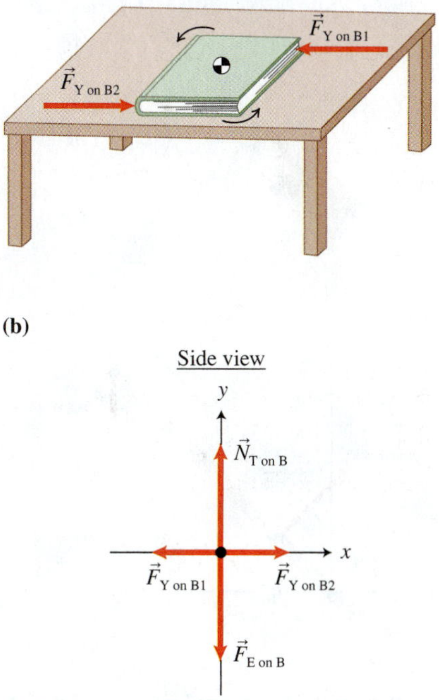

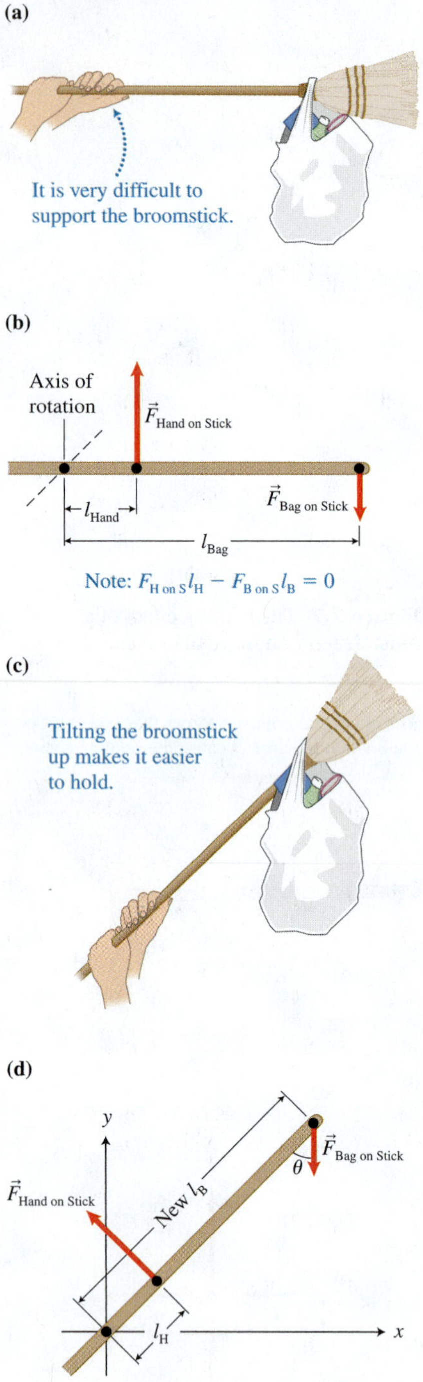


**Figure 7.8** Holding a bag at the end of a stick is more difficult when the stick is horizontal than when the stick is tilted up.

**Active Learning Guide ›**

you can easily lift it with one hand. Now hang the grocery bag from the end of a broomstick (**Figure 7.8a**). Try to support it by holding only the handle end of the broomstick with your hands close together. It is very difficult. Why?

The broomstick, the object of interest, can turn around an axis through the hand that is closest to you. The bag exerts a force on the broomstick far from this axis of rotation. This means that $l_{\text{Bag}}$, a quantity that we must find in order to determine the turning ability of the force exerted by the bag on the broomstick, is very large (Figure 7.8b). Your other hand, which is very close to the axis of rotation (distance $l_{\text{Hand}}$), must balance the effect of the bag. But since $l_{\text{Hand}}$ is so small, the force your hand exerts must be very large. The outcome of this experiment agrees with what we learned so far about the turning ability of a force.

Holding the broomstick perpendicular to your body is quite difficult. However, if you hold the broomstick at an angle above the horizontal (Figure 7.8c), you find that the bag becomes easier to support. Why? Perhaps this has to do with the direction the force is exerted relative to the broomstick (see Figure 7.8d).

## The role of angle on the turning ability of a force

Our current mathematical model of the physical quantity that characterizes the turning ability of the force ($\pm Fl$) takes into account the direction in which an exerted force can potentially rotate an object (clockwise or counterclockwise) but does not take into account the actual direction of the force. However, our experiment with the broomstick indicates that the angle at which we exert a force relative to the broomstick affects the turning ability of the force. We know from experience that pushing on a door on its outside edge directly toward the hinges does not cause it to rotate. The direction of the push must matter. How can we improve our model for the physical quantity to take the direction of the force into account?

To investigate this question we can change our experiment with the meter stick slightly by making scale 1 pull on the stick at an angle other than 90° (**Figure 7.9**). Scale 2, on the far right end of the meter stick, 0.50 m from the suspension point, will exert a constant force of 10.0 N downward at a 90° angle. Scale 1 on the far left end of the stick will pull at different angles $\theta$ so that the meter stick remains horizontal. The results are shown in **Table 7.3**. In all cases the stick is horizontal—therefore, the turning ability of the force on the right is balanced by the turning ability of the force on the left.

Using the data in the table, we see the effects of the magnitude and the angle of force $\vec{F}_1$ on its turning ability: the smaller the angle between the direction of the force and the stick, the larger the magnitude of the force that is necessary to produce the same turning ability. Thus we find that there are four factors that affect the turning ability of a force: (1) the direction (counterclockwise or clockwise) that the force can potentially rotate the object; (2) the magnitude of the force $F$, (3) the distance $l$ of the point of application of the force from the axis of rotation, and (4) the angle $\theta$ that the force makes relative to a line from the axis of rotation to the point of application of the

**Table 7.3 Magnitude, location, and direction of force and its turning ability.**

| Magnitude of $\vec{F}_1$ | Distance to the axis of rotation | Angle $\theta$ between $\vec{F}_1$ and the stick | Turning ability produced by $\vec{F}_3$ |
|---|---|---|---|
| 10.0 N | 0.50 m | 90° | $-(10.0\text{ N})(0.50\text{ m}) = -5.0\text{ N}\cdot\text{m}$ |
| 12.6 N | 0.50 m | 53° | $-(10.0\text{ N})(0.50\text{ m}) = -5.0\text{ N}\cdot\text{m}$ |
| 14.2 N | 0.50 m | 45° | $-(10.0\text{ N})(0.50\text{ m}) = -5.0\text{ N}\cdot\text{m}$ |
| 20.0 N | 0.50 m | 30° | $-(10.0\text{ N})(0.50\text{ m}) = -5.0\text{ N}\cdot\text{m}$ |

force. If we combine these four factors, the physical quantity characterizing the turning ability of a force takes a form such as:

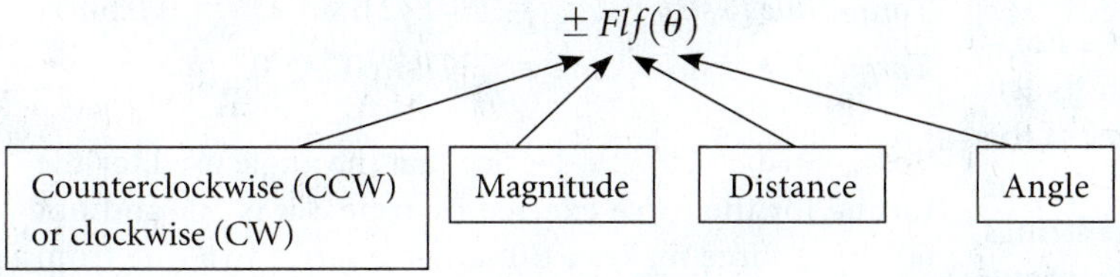

$\pm Flf(\theta)$
Counterclockwise (CCW) or clockwise (CW)
Magnitude
Distance
Angle

where $f(\theta)$ is some function of the angle $\theta$.

Consider the last row of Table 7.3. The force exerted by scale 1 is 0.50 m from the axis of rotation ($l = 0.50\text{ m}$), and the scale exerts a 20-N force $F_1$ on the stick at a 30° angle relative to the stick. This force produces the counterclockwise $+5.0\text{ N}\cdot\text{m}$ effect needed to balance the $-5.0\text{ N}\cdot\text{m}$ clockwise effect of scale 2. What value would $f(\theta)$ have to be to get this rotational effect?

$$+5.0\text{ N}\cdot\text{m} = (20.0\text{ N})(0.5\text{ m})\,f(30°)$$

or $f(30°)$ must be 0.50. Recall that $\sin 30° = 0.50$. Maybe the function $f(\theta)$ is the sine function, that is, $\tau = lF\sin\theta$. Is this consistent with the other rows in Table 7.3?

$$+(10.0\text{ N})(0.5\text{ m})(\sin 90°) = +5.0\text{ N}\cdot\text{m}\,(1.00) = +5.0\text{ N}\cdot\text{m}$$
$$+(12.6\text{ N})(0.5\text{ m})(\sin 53°) = +6.3\text{ N}\cdot\text{m}\,(0.80) = +5.0\text{ N}\cdot\text{m}$$
$$+(14.2\text{ N})(0.5\text{ m})(\sin 45°) = +7.1\text{ N}\cdot\text{m}\,(0.71) = +5.0\text{ N}\cdot\text{m}$$

This expression ($\tau = Fl\sin\theta$) is the mathematical definition of the new physical quantity that characterizes the ability of a force to turn (rotate) a rigid body. This physical quantity is called a **torque.** The symbol for torque is $\tau$, the Greek letter tau.

**Torque $\tau$ produced by a force** **The torque produced by a force exerted on a rigid body about a chosen axis of rotation is**

$$\tau = \pm Fl\sin\theta \qquad (7.1)$$

**where $F$ is the magnitude of the force, $l$ is the magnitude of the distance between the point where the force is exerted on the object and the axis of rotation, and $\theta$ is the angle that the force makes relative to a line connecting the axis of rotation to the point where the force is exerted (see Figure 7.10).**

Figure 7.10 illustrates the method for calculating the turning ability (torque) due to a particular force. In this case, we are calculating the torque due to the force that the slanted rope exerts on the end of a beam that supports a load hanging from the beam. The torque is positive if the force has a counterclockwise turning ability about the axis of rotation, and negative if the force has a clockwise turning ability. The SI unit for torque is newton · meter, N · m (the British system unit is lb · ft).

**TIP** Notice that the units of torque (N · m) are the same as the units of energy (N · m = J). Torque and energy are very different quantities. We will always refer to the unit of torque as newton · meter (N · m) and the unit of energy as joule (J).

**Figure 7.9** An experiment to determine the angle dependence of the turning ability caused by a force.

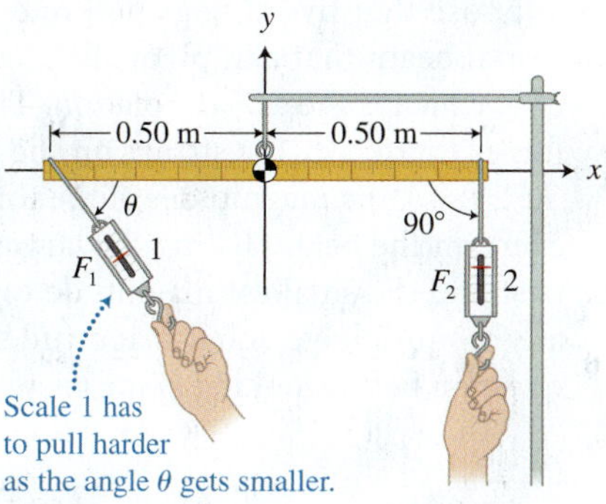

y
x
0.50 m
0.50 m
θ
90°
$F_1$
1
$F_2$
2
Scale 1 has to pull harder as the angle θ gets smaller.

**Figure 7.10** A method to determine the torque (turning ability) produced by a force.

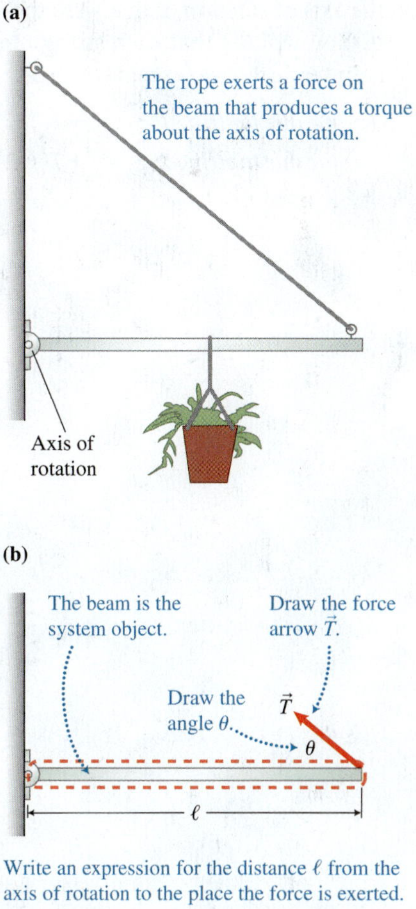

(a)
The rope exerts a force on the beam that produces a torque about the axis of rotation.
Axis of rotation
(b)
The beam is the system object.
Draw the force arrow $\vec{T}$.
Draw the angle θ.
$\vec{T}$
θ
ℓ
Write an expression for the distance ℓ from the axis of rotation to the place the force is exerted.

**QUANTITATIVE EXERCISE 7.1 Rank the magnitudes of the torques**

Suppose that five strings pull one at a time on a horizontal beam that can pivot about a pin through its left end, which is the axis of rotation. The magnitudes of the forces exerted by the strings on the beam are either $T$ or $T/2$. Rank the magnitudes of the torques that the strings exert on the beam, listing the largest magnitude torque first and the smallest magnitude torque last. Indicate if any torques have equal magnitudes. Try to answer the question before looking at the answer below.

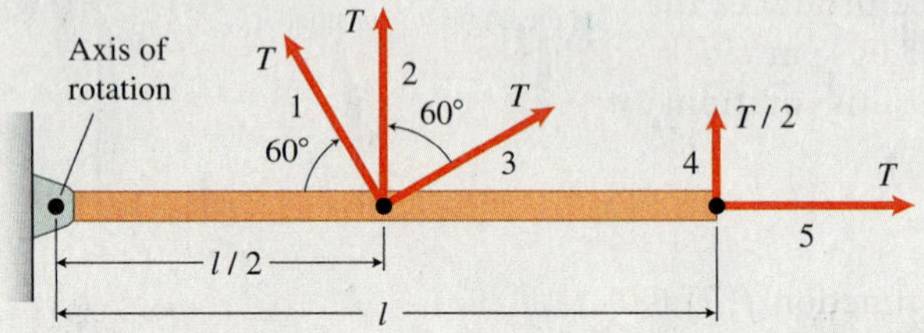


**Represent mathematically** A mathematical expression for the torque produced by each force is shown below. To understand why each torque is positive, imagine in what direction each string would turn the beam about the axis of rotation, if that were the only force exerted on it. You will see that each string tends to turn the beam counterclockwise (except string 5).

*Torque due to string 1:* $\tau_1 = +T(l/2)\sin 60° = +0.43\ Tl.$

*Torque due to string 2:* $\tau_2 = +T(l/2)\sin 90° = +0.50\ Tl.$

*Torque due to string 3:* $\tau_3 = +T(l/2)\sin 150° = +0.25\ Tl.$

*Torque due to string 4:* $\tau_4 = +(T/2)l\sin 90° = +0.50\ Tl.$

*Torque due to string 5:* $\tau_5 = +Tl\sin 0° = 0.$

**Solve and evaluate** Notice that the angle used for the torque for the force exerted by rope 3 was 30° and not 60°—the force makes a 30° angle relative to a line from the pivot point to the place where the string exerts the force on the beam. String 5 exerts a force parallel to the line from the pivot point to the place where it is exerted on the beam; as a result, torque 5 is zero. The rank order of the torques is $\tau_2 = \tau_4 > \tau_1 > \tau_3 > \tau_5$.

**Try it yourself:** Determine the torque caused by the cable pulling horizontally on the inclined drawbridge shown below. The force that the cable exerts on the bridge is 5000 N, the bridge length is 8.0 m, and the bridge makes an angle of 50° relative to the vertical support for the cable system.

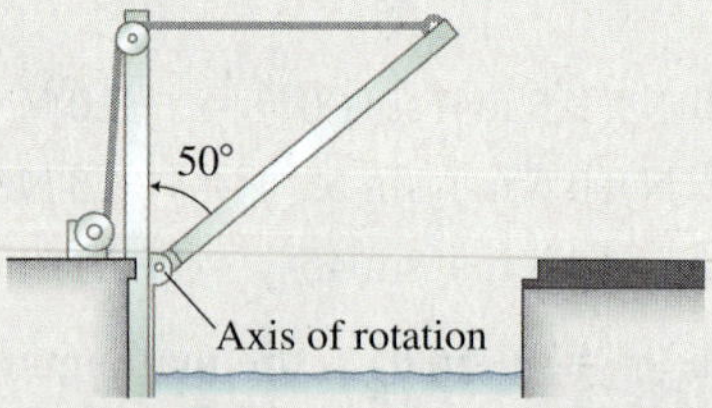


*Answer:* $\tau = +(5000\ \text{N})(8.0\ \text{m})\sin 40° = +26{,}000\ \text{N}\cdot\text{m}$. Note that we did not use 50° in our calculation. Why?

**TIP** To decide the sign of the torque that a particular force exerts on a rigid body about a particular axis, pretend that a pencil is the rigid body. Hold it with two fingers at a place that represents the axis of rotation (**Figure 7.11**) and exert a force on the pencil representing the force whose torque sign you wish to determine. Does that force cause the pencil to turn counterclockwise (a + torque) or clockwise (a − torque) about the axis of rotation?

**Figure 7.11** A method to determine the sign of a torque.

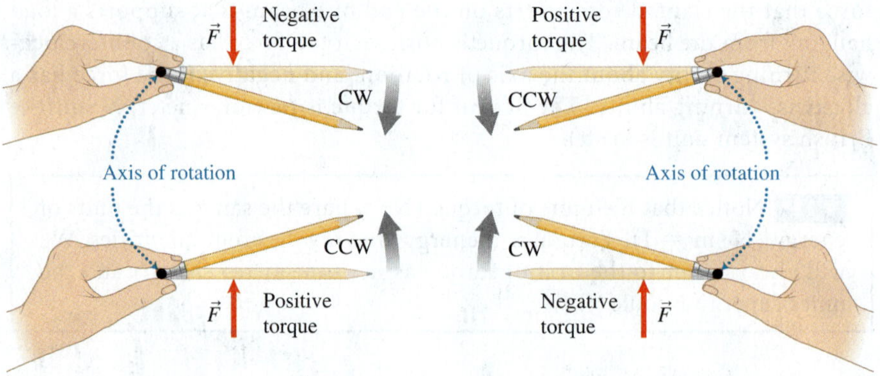

**EXAMPLE 7.2 A painter on a ladder**

A 75-kg painter stands on a 6.0-m-long 20-kg ladder tilted at 53° relative to the ground. He stands with his feet 2.4 m up the ladder. Determine the torque produced by the normal force exerted by the painter on the ladder for two choices of axis of rotation: (a) an axis parallel to the base of the ladder where it touches the ground and (b) an axis parallel to the top ends of the ladder where it touches the wall of the house.

**Sketch and translate** A sketch of the situation with the known information is shown below. The ladder is our system.

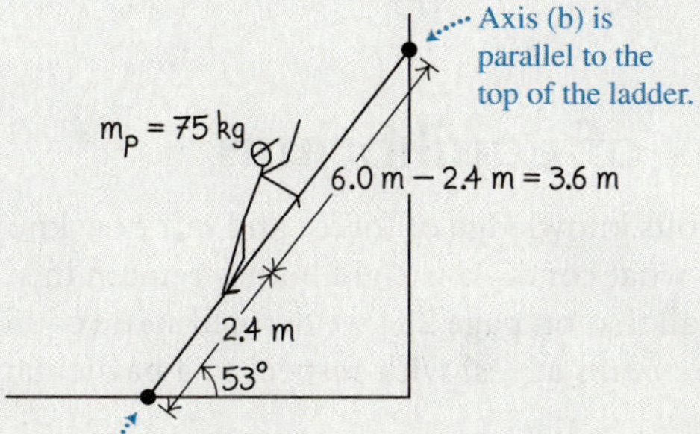
Find the torque that the painter exerts on the ladder about two different axes of rotation.
Axis (b) is parallel to the top of the ladder.
$m_P$ = 75 kg
6.0 m − 2.4 m = 3.6 m
2.4 m
53°
Axis (a) is parallel to the base of the ladder.

**Simplify and diagram** Four objects interact with the ladder: Earth, the painter's feet, the wall, and the ground. Our interest in this problem is only in the torque that the painter's feet exert on the ladder. Thus, we diagram for the ladder and the downward normal force $\vec{N}_{\text{P on L}}$ that the painter's feet exert on the ladder. The diagrams are for the two different axes of rotation.

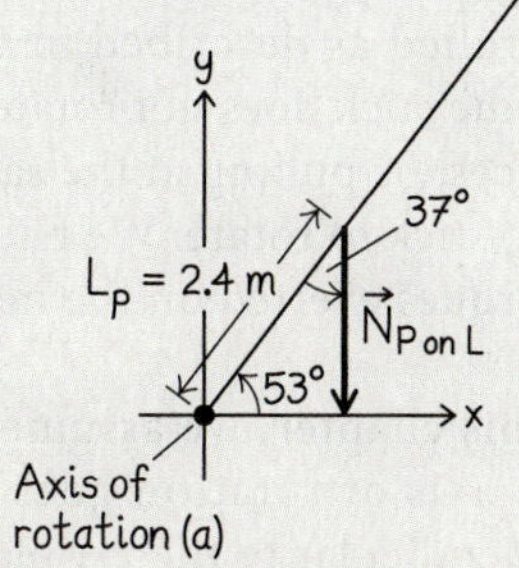
$\vec{N}_{\text{P on L}}$ causes a clockwise rotation about the axis of rotation at the bottom of the ladder (negative torque).
y
37°
$L_P$ = 2.4 m
$\vec{N}_{\text{P on L}}$
53°
x
Axis of rotation (a)

**Represent mathematically** We use Eq. (7.1) for the torque caused by the force that the painter's feet exert on the ladder (for simplicity, we write the magnitude of the force as $N$ with no subscripts): $\tau = \pm Nl\sin\theta$.

**Solve and evaluate** (a) For the first calculation, we choose the axis of rotation at the place where the feet of the ladder touch the ground. The torque produced by the normal force exerted by the painter's feet on the ladder is

$$\begin{aligned}\tau &= -Nl\sin\theta = -(m_{\text{P}}g)l\sin\theta \\ &= -(75\text{ kg})[(2.4\text{ m})(9.8\text{ N/kg})]\sin 37^\circ \\ &= -1100\text{ N}\cdot\text{m}\end{aligned}$$

Note that the normal force exerted by the painter's feet on the ladder has the same magnitude as the force $mg$ that Earth exerts on the painter, but it is not the same force, as it is exerted on a different object and is a normal force and not a gravitational force. That normal force tends to rotate the ladder clockwise about the axis of rotation (a negative torque). The force makes a 37° angle relative to a line from the axis of rotation to the place where the force is exerted.

(b) We now choose the axis of rotation parallel to the wall at the top of the ladder where it touches the wall. The torque produced by the normal force exerted by the painter's feet on the ladder is

$$\begin{aligned}\tau &= +Fl\sin\theta = +(m_{\text{P}}g)l\sin\theta \\ &= +(75\text{ kg})[(3.6\text{ m})(9.8\text{ N/kg})]\sin 143^\circ \\ &= +1600\text{ N}\cdot\text{m}\end{aligned}$$

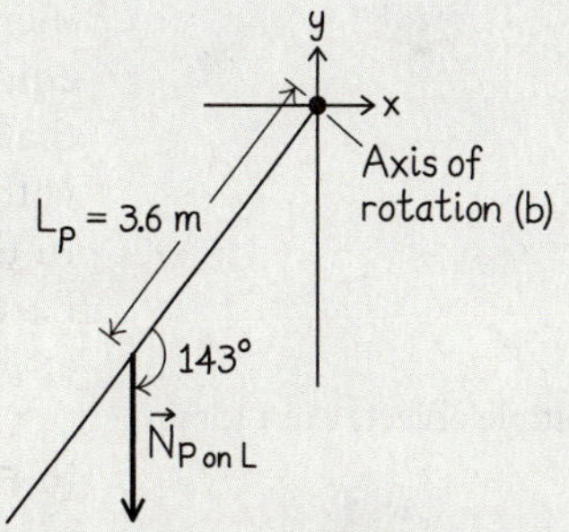
$\vec{N}_{\text{P on L}}$ causes a counterclockwise rotation about the axis of rotation at the top of the ladder (a positive torque).
y
x
Axis of rotation (b)
$L_P$ = 3.6 m
143°
$\vec{N}_{\text{P on L}}$

When we choose the axis of rotation at the top of the ladder, the downward force exerted by the painter's feet on the ladder tends to rotate the ladder counterclockwise about the axis at the top (a positive torque). This force makes a 143° angle with respect to a line from the axis of rotation to the place where the force is exerted.

Note that the torque depends on where we place the axis of rotation. You cannot do torque calculations without carefully defining the axis of rotation.

**Try it yourself:** Write an expression for the torque produced by the upward normal force exerted by the floor on the ladder about the same two axes: (a) at the base of ladder and (b) at the top of the ladder.

*(continued)*

*Answer:* (a) The torque will be 0 at the bottom, because the normal force passes through the axis. (b) About the axis at the top of the ladder, the torque is $\tau = -(6.0\text{ m})N_{\text{Floor on Ladder}} \sin 37°$. Note that the floor tends to push the ladder clockwise about an axis through the top of the ladder. The normal force makes a 37° angle relative to a line from the top axis to the place where the force is exerted.

**Review Question 7.2** Give an example of a situation in which (a) a torque produced by a force is zero with respect to one choice of axis of rotation but not zero with respect to another; (b) a force is not exerted at the axis of rotation, but the torque produced by it is zero anyway; and (c) several forces produce nonzero torques on an object, but the object does not rotate.

## 7.3 Conditions of equilibrium

We can combine our previous knowledge of forces and our new knowledge of torque to determine under what conditions rigid bodies remain in static equilibrium, that is, at rest. Recall that on page 234 we defined static equilibrium as a state in which an object *remains* at rest with respect to a particular observer in an inertial reference frame.

It is possible for an object to be at rest briefly. For example, a ball thrown upward stops for an instant at the top of its flight, but it does not remain at rest. Thus, the word *remains* is important in the expression "remains at rest"—the object has to stay where it is. The words "with respect to an observer in an inertial reference frame" are also an important part of the definition of static equilibrium. Recall from the chapter on Newtonian mechanics (Chapter 2) that if an observer is not in an inertial reference frame, an object can accelerate with respect to the observer even if the sum of the forces exerted on it is zero. In this chapter we will only consider observers who are at rest with respect to Earth, since that is the most common point of view for observing real-life situations involving static equilibrium.

**Figure 7.12** Multiple objects exert forces on a meter stick.

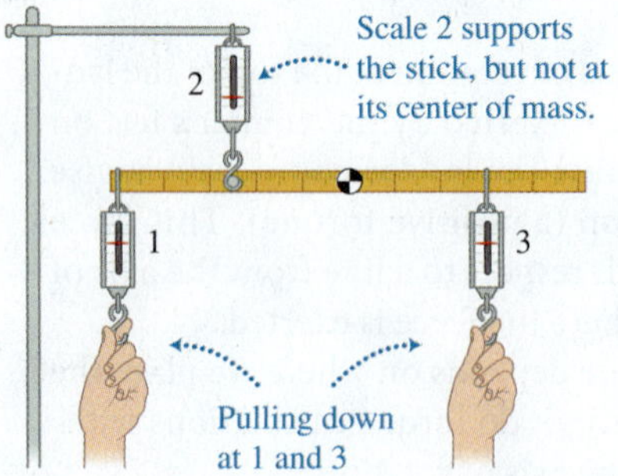
Scale 2 supports the stick, but not at its center of mass.
2
1
3
Pulling down at 1 and 3

We again suspend the same 0.1-kg meter stick from spring scale 2, as shown in **Figure 7.12**. However, the suspension point is no longer at the center of mass of the meter stick. You and your friend again pull on the stick at different positions with spring scales 1 and 3. When pulled as described in Observational Experiment **Table 7.4** on the next page, the stick does not rotate. Pulling at other positions while exerting the same forces, or pulling at the same positions while exerting different forces, causes the stick to rotate. We need to find a pattern in the combinations of forces and torques exerted on the meter stick that keep the stick in static equilibrium.

For most situations that we analyze in this chapter, we assume that the objects rotate in the $x$-$y$ plane and that the axis of rotation goes through the origin of the coordinate system and is perpendicular to the $x$-$y$ plane.

**OBSERVATIONAL EXPERIMENT TABLE**

## 7.4 Meter stick in static equilibrium.

VIDEO 7.4

| Observational experiment | Analysis |
| --- | --- |

**Experiment 1.** Three spring scales and Earth exert forces on a meter stick at locations shown below. Examine the forces and torques exerted on the stick. Choose the axis of rotation at the place where the string from scale 2 supports the stick. This choice determines the distances in the torque equations for each force.

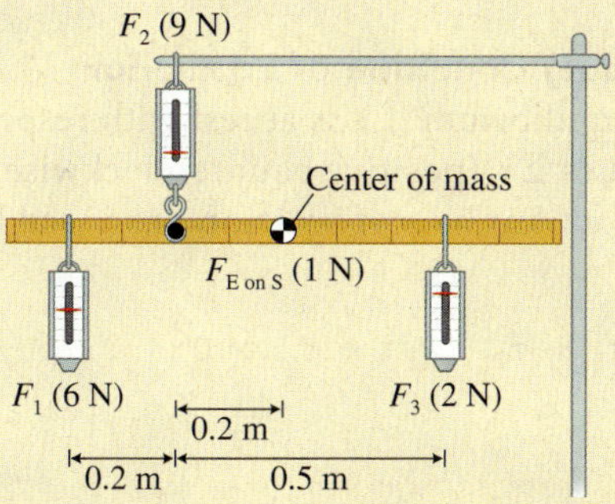


**Analysis:**

$\Sigma F_y = (-6.0\text{ N}) + 9.0\text{ N} + (-1.0\text{ N}) + (-2.0\text{ N}) = 0$

*Counterclockwise* torques:

$\tau_1 = (F_1)(l_1) = (6.0\text{ N})(0.20\text{ m}) = 1.2\text{ N}\cdot\text{m}$

*Clockwise* torques:

$\tau_2 = (F_2)(l_2) = (9.0\text{ N})(0) = 0$

$\tau_E = -(F_{\text{E on S}})(l_{\text{CM}}) = -(1.0\text{ N})(0.2\text{ m}) = -0.2\text{ N}\cdot\text{m}$

$\tau_3 = -(F_3)(l_3) = -(2.0\text{ N})(0.50\text{ m}) = -1.0\text{ N}\cdot\text{m}$

$\Sigma\tau = \tau_1 + \tau_2 + \tau_E + \tau_3$

$= +1.2\text{ N}\cdot\text{m} + 0 - 0.2\text{ N}\cdot\text{m} - 1.0\text{ N}\cdot\text{m} = 0$

**Experiment 2.** Three spring scales and Earth exert forces on a meter stick at locations shown below. Examine the forces and torques exerted on the stick. Choose the axis of rotation at the place where the string from scale 2 supports the stick.

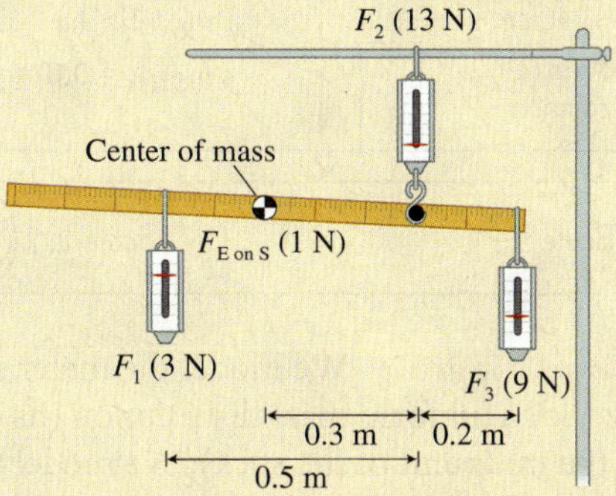


**Analysis:**

$\Sigma F_y = -3.0\text{ N} + (-1.0\text{ N}) + 13.\text{ N} + (-9.0\text{ N}) = 0$

*Counterclockwise*:

$\tau_1 = (F_1)(l_1) = (3.0\text{ N})(0.50\text{ m}) = 1.5\text{ N}\cdot\text{m}$

$\tau_E = (F_{\text{E on S}})(l_{\text{CM}}) = (1.0\text{ N})(0.3\text{ m}) = 0.3\text{ N}\cdot\text{m}$

*Clockwise*:

$\tau_2 = (F_2)(l_2) = (13\text{ N})(0) = 0$

$\tau_3 = -(F_3)(l_3) = -(9.0\text{ N})(0.20\text{ m}) = -1.8\text{ N}\cdot\text{m}$

$\Sigma\tau = \tau_1 + \tau_E + \tau_2 + \tau_3$

$= +1.5\text{ N}\cdot\text{m} + 0.3\text{ N}\cdot\text{m} + 0 - 1.8\text{ N}\cdot\text{m} = 0$

**Patterns**

- In both cases the net force exerted on the meter stick in the vertical direction is zero: $\Sigma F_y = 0$.
- In both cases the sum of the torques exerted on the meter stick equals zero: $\Sigma\tau = 0$.

The first pattern in both experiments is familiar to us. It is simply Newton's second law applied to the vertical axis of the meter stick for the case of zero translational acceleration. Because the sum of the vertical forces exerted on the meter stick is zero, there is no vertical acceleration. We had no horizontal forces, so the meter stick could not accelerate horizontally. The second pattern shows that in the experiments presented in the table in both cases the net torque is zero and the meter stick does not start turning. We will learn in the next chapter that these are examples of rigid bodies with zero rotational acceleration.

**TIP** Notice that you cannot determine the torque produced by a force without specifying the point at which the force is exerted on the object relative to the axis of rotation.

‹ Active Learning Guide

We can now state the conditions of static equilibrium mathematically as follows.

**Condition 1. Translational (Force) Condition of Static Equilibrium** An object modeled as a rigid body is in translational static equilibrium with respect to a particular observer if it is at rest with respect to that observer and the components of the sum of the forces exerted on it in the perpendicular $x$- and $y$-directions are zero:

$$\Sigma F_{\text{on O}\,x} = F_{1\,\text{on O}\,x} + F_{2\,\text{on O}\,x} + \cdots + F_{n\,\text{on O}\,x} = 0 \tag{7.2x}$$

$$\Sigma F_{\text{on O}\,y} = F_{1\,\text{on O}\,y} + F_{2\,\text{on O}\,y} + \cdots + F_{n\,\text{on O}\,y} = 0 \tag{7.2y}$$

The subscript $n$ indicates the number of forces exerted by external objects on the rigid body.

**TIP** Remember that all the gravitational forces exerted by Earth on the different parts of the rigid body can be combined into a single gravitational force being exerted on the center of mass of the rigid body.

**Condition 2. Rotational (Torque) Condition of Equilibrium** A rigid body is in turning or rotational static equilibrium if it is at rest with respect to the observer and the sum of the torques $\Sigma\tau$ (positive counterclockwise torques and negative clockwise torques) about any axis of rotation produced by the forces exerted on the object is zero:

$$\Sigma\tau = \tau_1 + \tau_2 + \cdots + \tau_n = 0 \tag{7.3}$$

## EXAMPLE 7.3 Testing the conditions of static equilibrium

Place the ends of a standard meter stick on two scales, as shown below. The scales each read 0.50 N. From this, we infer that the mass of the meter stick is about 0.10 kg (the gravitational force that Earth exerts on the meter stick would be $(0.10\,\text{kg})(9.8\,\text{N/kg}) = 1.0\,\text{N}$). Predict what each scale will read if you place a 5.0-kg brick 40 cm to the right of the left scale.

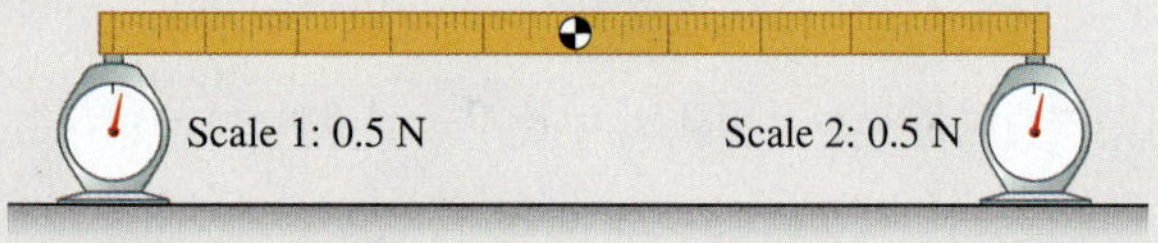


**Sketch and translate** A labeled sketch of the situation is shown below. We choose the stick as the system of interest and use a standard $x$-$y$ coordinate system. We choose the axis of rotation at the place where the left scale touches the stick. By doing this, we are making the torque produced by the normal force exerted by the left scale on the stick zero—that force is exerted exactly at the axis of rotation. With this choice, we remove one of the unknown quantities from the torque condition of equilibrium and will be able to use that condition to find the force exerted by the right scale on the meter stick.

Predict the scale readings after a 5.0-kg brick is placed at the 40-cm mark.

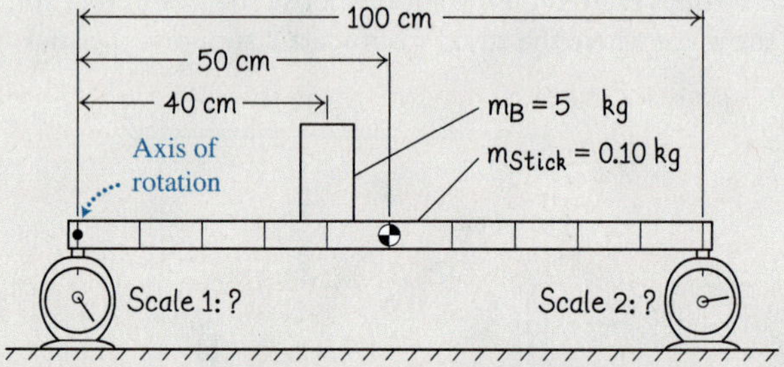


**Simplify and diagram** We model the meter stick as a rigid body with a uniform mass distribution (its center of mass is at the midpoint of the stick). We model the brick as a point-like object, and assume that the scales push up on the stick at the exact ends of the stick. We then draw a force diagram showing the forces exerted on the stick by Earth, the brick, and each of the scales. As noted, the left end of the stick has been chosen as the axis of rotation.

Analyzing the situation with the axis of rotation on the left side of the meter stick

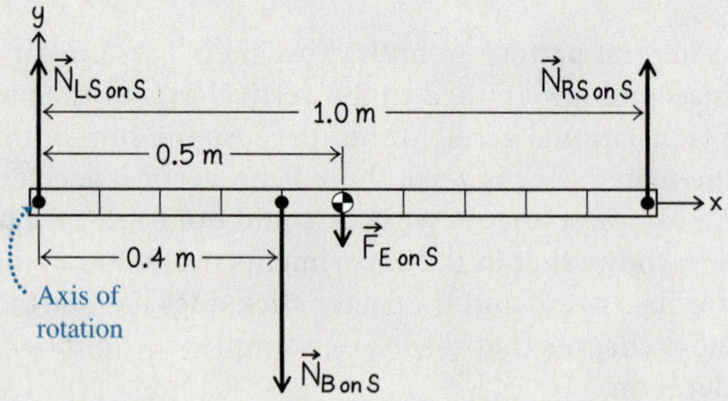

**Represent mathematically** According to the conditions of static equilibrium, the sum of the forces exerted on the meter stick should equal zero, as should the sum of the torques around the axis of rotation. The gravitational force exerted by Earth and the normal force exerted by the brick on the stick both have clockwise turning ability around the axis of rotation and produce negative torques. The force exerted by the right scale on the stick has counterclockwise turning ability and produces a positive torque. The force exerted by the left scale on the stick produces zero torque since it is exerted at the axis of rotation. The two conditions of equilibrium are then the following:

Translational (force) condition ($\Sigma F_y = 0$):

$$(-F_{\text{E on S}}) + (-N_{\text{B on S}}) + N_{\text{RS on S}} + N_{\text{LS on S}} = 0$$

Rotational (torque) condition ($\Sigma\tau = 0$):

$$-F_{\text{E on S}}(0.50\text{ m}) - N_{\text{B on S}}(0.40\text{ m}) + N_{\text{RS on S}}(1.00\text{ m}) = 0$$

Since none of the forces have $x$-components, we didn't apply the $x$-component form of the force condition of equilibrium.

Earth exerts a downward gravitational force on the brick of magnitude:

$$F_{\text{E on B}} = m_B g = (5.0\text{ kg})(9.8\text{ N/kg}) \approx 50\text{ N}$$

Thus, the stick must exert a balancing 50-N upward normal force on the brick. According to Newton's third law ($\vec{N}_{\text{B on S}} = -\vec{N}_{\text{S on B}}$), the brick must exert a downward 50-N normal force $\vec{N}_{\text{B on S}}$ on the stick.

**Solve and evaluate** We have two equations with two unknowns ($N_{\text{RS on S}}$ and $N_{\text{LS on S}}$). We first use the torque equilibrium condition to determine the magnitude of the force exerted by the right scale on the stick:

$$-[(0.10\text{ kg})(10\text{ N/kg})](0.50\text{ m})$$
$$-[(5.0\text{ kg})(10\text{ N/kg})](0.40\text{ m}) + N_{\text{RS on S}}(1.00\text{ m}) = 0$$

or

$$N_{\text{RS on S}} = 20.5\text{ N}$$

We can use this result along with the force equilibrium condition equation to determine the magnitude of the force exerted by the left scale on the stick.

$$-(0.10\text{ kg})(10\text{ N/kg}) - (5.0\text{ kg})(10\text{ N/kg})$$
$$+ 20.5\text{ N} + N_{\text{LS on S}} = 0$$

or

$$N_{\text{LS on S}} = 30.5\text{ N}$$

These predictions make sense because the sum of these two upward forces equals the sum of the two downward forces that Earth and the brick exert on the meter stick. Also, the force on the left end is greater because the brick is positioned closer to it, which sounds very reasonable. Performing this experiment, we find that the outcome matches the predictions.

**Using a different axis of rotation** Remember that we had the freedom to choose whatever axis of rotation we wanted. Let's try it again with the axis of rotation at 40 cm from the left side, the location of the brick. See the force diagram below. The force condition of equilibrium will not change since it does not depend on the choice of the axis of rotation:

$$(-F_{\text{E on S}}) + (-N_{\text{B on S}}) + N_{\text{RS on S}} + N_{\text{LS on S}} = 0$$

The axis of rotation is now at the 0.4-m position.

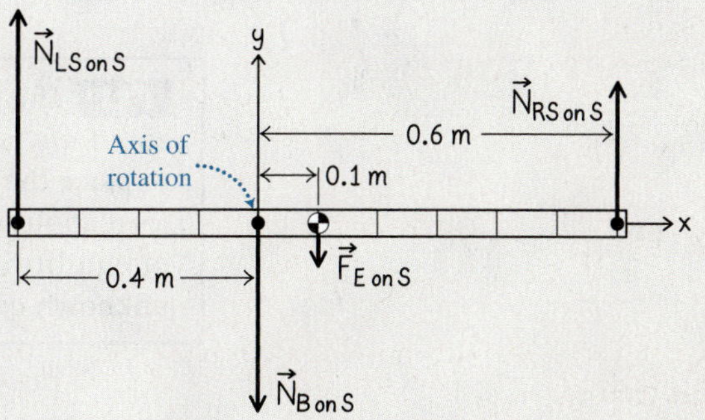


The torque condition will change:

$$[-F_{\text{E on S}}(0.10\text{ m})] + [-N_{\text{LS on S}}(0.40\text{ m})]$$
$$+ N_{\text{RS on S}}(0.60\text{ m}) = 0$$

Now we have two unknowns in each of the two equations and have to solve them simultaneously to determine the unknowns. This will be somewhat harder than when we chose the axis of rotation at the left end of the stick. Let's solve the force condition equation for $N_{\text{RS on S}}$ and substitute the result into the torque condition equation:

$$N_{\text{RS on S}} = F_{\text{E on S}} + N_{\text{B on S}} - N_{\text{LS on S}}$$
$$\Rightarrow -F_{\text{E on S}}(0.10\text{ m}) - N_{\text{LS on S}}(0.40\text{ m})$$
$$+ (F_{\text{E on S}} + N_{\text{B on S}} - N_{\text{LS on S}})(0.60\text{ m}) = 0$$

Combining the terms with $N_{\text{LS on S}}$ on one side, we get

$$N_{\text{LS on S}}(0.40\text{ m} + 0.60\text{ m}) = F_{\text{E on S}}(0.50\text{ m}) + N_{\text{B on S}}(0.60\text{ m})$$
$$= (1.0\text{ N})(0.50\text{ m}) + (50\text{ N})(0.60\text{ m}) = 30.5\text{ N}\cdot\text{m}$$

or $N_{\text{LS on S}} = 30.5$ N. Substituting back into the force condition equation, we find that

$$N_{\text{RS on S}} = 1.0\text{ N} + 50.0\text{ N} - 30.5\text{ N} = 20.5\text{ N}$$

These are the same results we obtained from the original choice of the axis of rotation. The choice of the axis

*(continued)*

of rotation does not affect the results. This makes sense, in the same way that choosing a coordinate system does not affect the outcome of an experiment. The concepts of axes of rotation and coordinate systems are mental constructs and should not affect the outcome of actual experiments.

**Try it yourself:** A uniform meter stick with a 50-g object on it is positioned as shown below. The stick extends 30 cm over the edge of the table. If you push the stick so that it extends slightly further over the edge, it tips over. Use this result to determine the mass of the meter stick.

The meter stick tips if it is extended more than 30 cm over the edge.

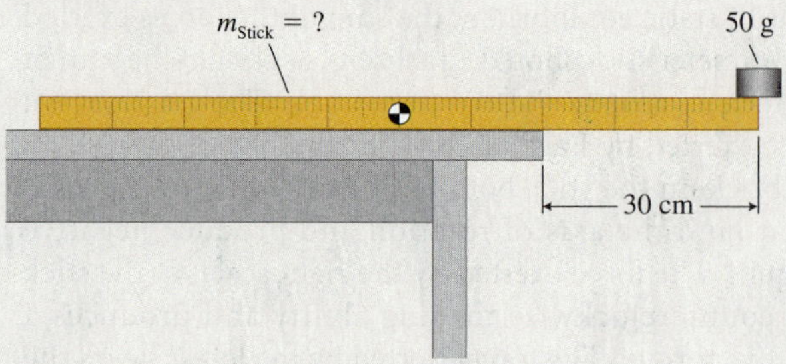


*Answer:* 75 g.

**TIP** If a rigid body is in static equilibrium, the sum of the torques about *any* axis of rotation is zero. It is often helpful in problem solving to place the axis at the place on the rigid body where the force you know least about is exerted. Then that force drops out of the second condition of equilibrium and you can use that equation to solve for some other unknown quantity.

**Review Question 7.3** How do we choose the location of the axis of rotation when we are applying the conditions of equilibrium for a rigid body?

## 7.4 Center of mass

Many extended bodies are not rigid—the human body is a good example. A high jumper crossing the bar is often bent into an inverted U shape (**Figure 7.13**). Why does this shape allow her to jump higher? At the moment shown in the photo, her legs, arms, and head are below the bar as the trunk of the body passes over the top. As each part of the body passes over the bar, the rest of the body is at a lower elevation so that her center of mass is always slightly below the bar. The high jumper does not have to jump as high because she is able to reorganize her body's shape so that her center of mass passes *under* or at least not significantly over the bar.

**Figure 7.13** Where is the jumper's center of mass?

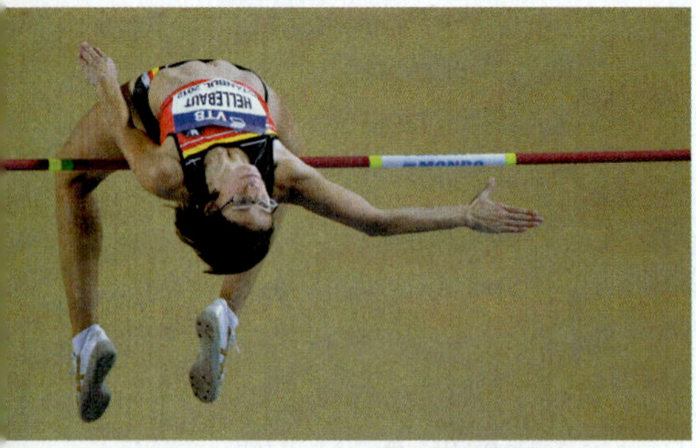

Without realizing it, we change the position of our center of mass with respect to other parts of the body quite often. Try this experiment. Sit on a chair with your back straight and your feet on the floor in front of the chair (see **Figure 7.14a**). Without using your hands, try to stand up; you cannot. No matter how hard you try, you cannot raise yourself to standing from the chair if your back is vertical.

Why can't you stand? The center of mass of an average person when sitting upright is near the front of the abdomen. Consider what a force diagram for the experiment would look like (Figure 7.14b), assuming somehow you managed to lift yourself slightly off the chair—again keeping the back straight. Earth exerts a downward force at your center of mass, and the floor exerts an upward normal force on your feet. The torques caused by these two forces about any axis of rotation causes you to rotate back onto the chair.

To stand, you must tilt forward in the clockwise direction and move your feet back under the chair (Figure 7.14c). This shifts your center of mass

**Figure 7.14** Getting out of a chair without using your hands.

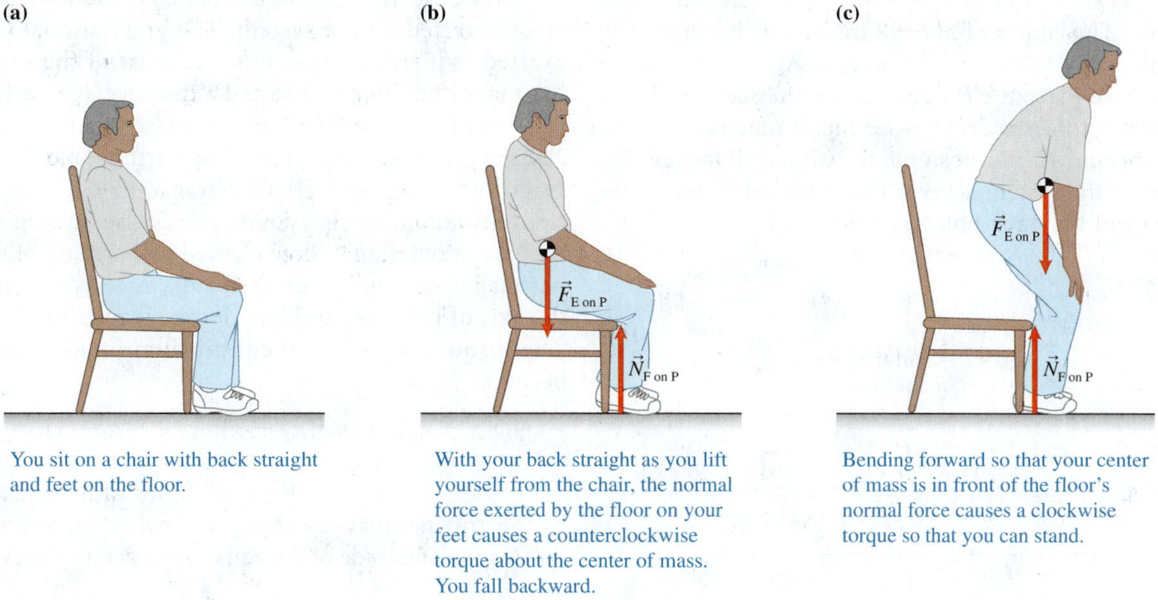
(a)
(b)
(c)
$\vec{F}_{\text{E on P}}$
$\vec{N}_{\text{F on P}}$
$\vec{F}_{\text{E on P}}$
$\vec{N}_{\text{F on P}}$
You sit on a chair with back straight and feet on the floor.
With your back straight as you lift yourself from the chair, the normal force exerted by the floor on your feet causes a counterclockwise torque about the center of mass. You fall backward.
Bending forward so that your center of mass is in front of the floor's normal force causes a clockwise torque so that you can stand.

forward so that the downward gravitational force exerted on you by Earth is in front of the upward normal force exerted by the floor on your feet. Now, the torques caused by these two forces allow you to rotate forward and stand without touching the chair seat.

## Calculating center of mass

How do we know that the center of mass of a sitting person is near the abdomen? In Section 7.1 we determined the location of an object's center of mass by investigating the directions along which one needs to push the object so it does not turn while being pushed on a flat smooth surface. When pushing in this way at different locations on the object, we found that lines drawn along the directions of these pushing forces all intersected at one point: the center of mass. This is a difficult and rather impractical way to find the center of mass of something like a human. Another method that we investigated consisted of balancing the object on a pointed support. This is also not very practical with respect to humans. Is there a way to predict where an object's center of mass is without pushing or balancing it? Our goal here is to develop a theoretical method that will allow us to determine the location of the center of mass of a complex object, such as the uniform seesaw with two apples shown in **Figure 7.15**. In the next example, we start with an object that consists of three other objects: two people of different masses and a uniform seesaw whose supporting fulcrum (point of support) can be moved (Figure 7.15). To determine the location of the center of mass of a system involving two people and a uniform beam, we find a place for the fulcrum to support the seesaw and the two people so that the system remains in static equilibrium.

**Figure 7.15** Where is the center of mass of this seesaw?

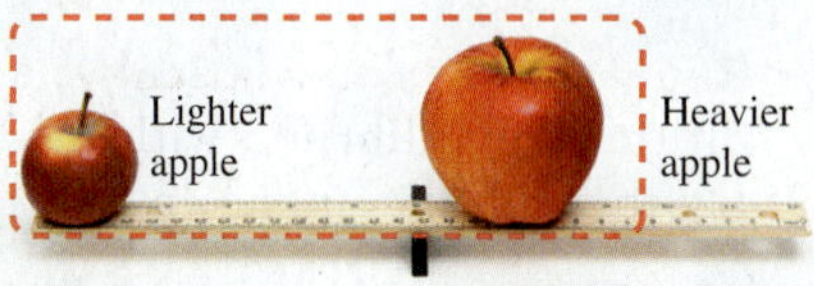
Lighter apple
Heavier apple

**EXAMPLE 7.4 Supporting a seesaw with two people**

Find an expression for the position of the center of mass of a system that consists of a uniform seesaw of mass $m_1$ and two people of masses $m_2$ and $m_3$ sitting at the ends of the seesaw beam ($m_2 > m_3$).

**Sketch and translate** The figure on the next page shows a labeled sketch of the situation. The two people are represented as blocks. We choose the seesaw and two blocks as the system and construct a mathematical equation that lets us calculate the center of mass of that system. We place the $x$-axis along the seesaw with its

*(continued)*

origin at some arbitrary position on the left side of the seesaw. The center of mass of the seesaw beam is at $x_1$ and the two blocks rest at $x_2$ and $x_3$. At what *position* $x$ should we place the fulcrum under the seesaw so that the system does not rotate—so that it remains in static equilibrium? At this position, the sum of all torques exerted on the system is zero. This position is the center of mass of the three-object system.

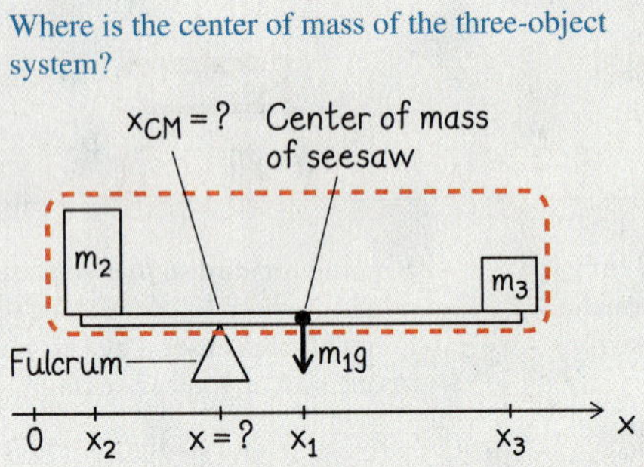


**Simplify and diagram** We model the seesaw as a rigid body and model each of the people blocks as point-like objects. Assume that the fulcrum does not exert a friction force on the seesaw. As you can see in the force diagram, we know the locations of all forces except the fulcrum force. We will calculate the unknown position $x$ of the fulcrum so the seesaw with two people on it balances—so that it satisfies the second condition of equilibrium.

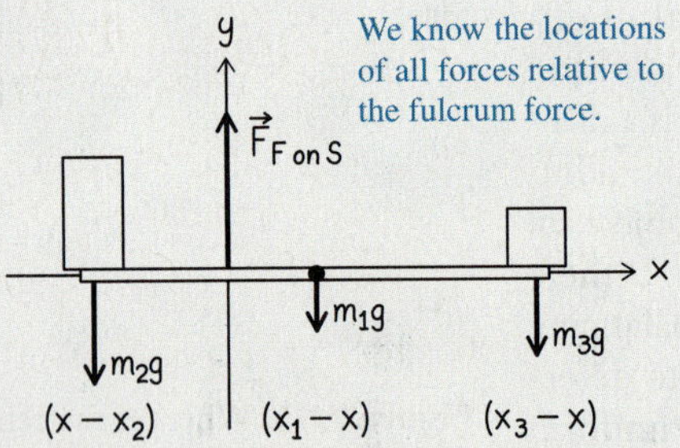


**Represent mathematically** Apply the torque condition of equilibrium with the axis of rotation going through the unknown fulcrum position $x$. Then determine the torques around this axis produced by the forces exerted on the system. The gravitational force exerted by Earth on the center of mass of the seesaw has magnitude $m_1g$, is exerted a distance $x_1 - x$ from the axis of rotation, and has clockwise turning ability. The gravitational force exerted by Earth on block 2 has magnitude $m_2g$, is exerted a distance $x - x_2$ from the axis of rotation, and has counterclockwise turning ability. The gravitational force exerted by Earth on block 3 has magnitude $m_3g$, is exerted a distance $x_3 - x$ from the axis of rotation, and has clockwise turning ability. The torque condition of equilibrium for the system becomes

$$m_2g(x - x_2) - m_1g(x_1 - x) - m_3g(x_3 - x) = 0$$

**Solve and evaluate** Divide all terms of the equation by the gravitational constant $g$ and collect all terms involving $x$ on one side of the equation to get

$$m_2x + m_1x + m_3x = m_2x_2 + m_1x_1 + m_3x_3$$

or

$$x(m_2 + m_1 + m_3) = m_2x_2 + m_1x_1 + m_3x_3$$

Divide both sides of the equation by $(m_2 + m_1 + m_3)$ to obtain an expression for the location of the center of mass of the three-object system:

$$x = \frac{m_1x_1 + m_2x_2 + m_3x_3}{m_1 + m_2 + m_3}$$

Let's evaluate this result. The units of $x$ are meters. Next check some limiting cases to see if the result makes sense. Imagine that there are no people sitting on the seesaw ($m_2 = m_3 = 0$). In this case, $x = \dfrac{m_1x_1}{m_1} = x_1$. The center of mass of the seesaw-only system is at the center of mass of the seesaw $x_1$, as it should be since we assumed its mass was uniformly distributed. Finally, if we increase the mass of one of the people on the seesaw, the location of the center of mass moves closer to that person.

**Try it yourself:** Where is the center of mass of a 3.0-kg, 2.0-m-long uniform beam with a 0.5-kg object on the right end and a 1.5-kg object on the left?

*Answer:* 0.8 m from the left end of the beam.

In Example 7.4 we arrived at an expression for the location of the center of mass of a three-object system where all objects were located along one straight line. We can apply the same method to a system whose masses are distributed in a two-dimensional $x$-$y$ plane. For such a two-dimensional system, we get the following:

**Center of mass (quantitative definition)** **If we consider an object as consisting of parts 1, 2, 3, . . . $n$ whose centers of masses are located at the coordinates $(x_1, y_1)$; $(x_2, y_2)$; $(x_3, y_3)$; . . . $(x_n, y_n)$, then the center of mass of this whole object is at the following coordinates:**

$$x_{cm} = \frac{m_1x_1 + m_2x_2 + m_3x_3 + \ldots + m_nx_n}{m_1 + m_2 + m_3 + \ldots + m_n}$$

$$y_{cm} = \frac{m_1y_1 + m_2y_2 + m_3y_3 + \ldots + m_ny_n}{m_1 + m_2 + m_3 + \ldots + m_n} \qquad (7.4)$$

Using the above equation for an object with a continuous mass distribution is difficult and involves calculus. For example, suppose you wanted to determine the $x_{cm}$ and $y_{cm}$ center-of-mass locations of the leg depicted in **Figure 7.16**. You would need to subdivide the leg into many tiny sections and insert the mass and position of each section into Eq. (7.4), and then add all the terms in the numerator and denominator together to determine the center of mass of the leg. For example, section 7 would contribute a term $m_7x_7$ in the numerator of the $x_{cm}$ equation and a term $m_7$ in the denominator. Usually, you will be given the location of the center of mass of such continuous mass distributions.

Knowledge of the center of mass helps you answer many questions: Why if you walk with a heavy backpack do you fall more easily? Why does a baseball bat have an elongated, uneven shape? Why do ships carry heavy loads in the bottom rather than near the top of the ship? Or why does a person lifting a barbell tilt backward (as in **Figure 7.17**)?

**Figure 7.16** Finding the center of mass of a continuous mass distribution.

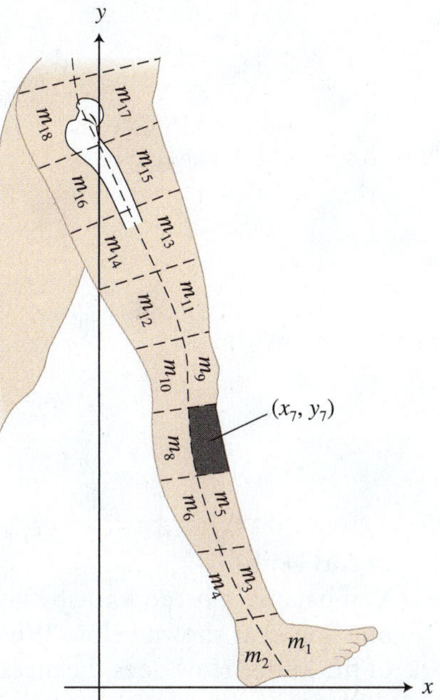

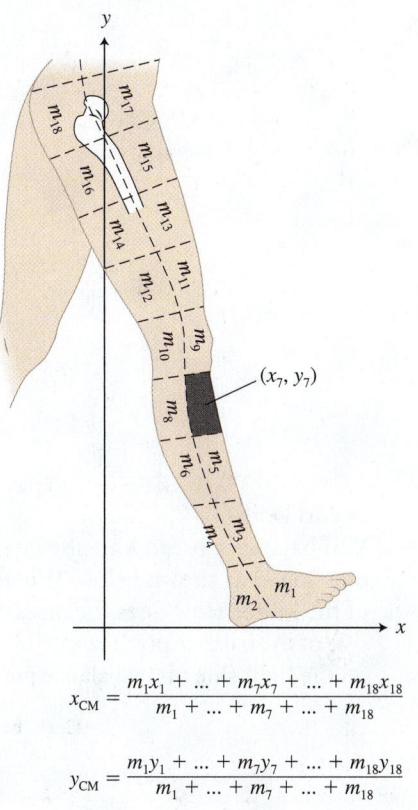

$$x_{CM} = \frac{m_1x_1 + \ldots + m_7x_7 + \ldots + m_{18}x_{18}}{m_1 + \ldots + m_7 + \ldots + m_{18}}$$

$$y_{CM} = \frac{m_1y_1 + \ldots + m_7y_7 + \ldots + m_{18}y_{18}}{m_1 + \ldots + m_7 + \ldots + m_{18}}$$

**Figure 7.17** Why is the weight lifter tilted backward?

## Mass distribution and center of mass

The term "center of mass" is deceiving. It might make you think that the center of mass of an object is located at a place where there is an equal amount of mass on each side. However, this is not the case. Consider again a uniform seesaw (see **Figure 7.18a**). Imagine that the mass of the seesaw is 10 kg, the length is 4.0 m, the mass of the person on the left end ($m_1$) is 70 kg, and the mass of the person on the right end ($m_3$) is 30 kg. Where is the center of mass of this two-person seesaw system and how much mass is on the left side and the right side of the center of mass?

To find the center of mass, we need a coordinate system with an origin. The origin can be anywhere. We put it at the location of the more massive person on the left side. The center of mass is then

$$x_{cm} = \frac{m_1x_1 + m_2x_2 + m_3x_3}{m_1 + m_2 + m_3} = \frac{(70\text{ kg}\cdot 0\text{ m}) + (10\text{ kg}\cdot 2.0\text{ m}) + (30\text{ kg}\cdot 4.0\text{ m})}{70\text{ kg} + 10\text{ kg} + 30\text{ kg}}$$

$$= 1.3\text{ m}$$

**Figure 7.18** The masses on each side of a system's center of mass are unequal.

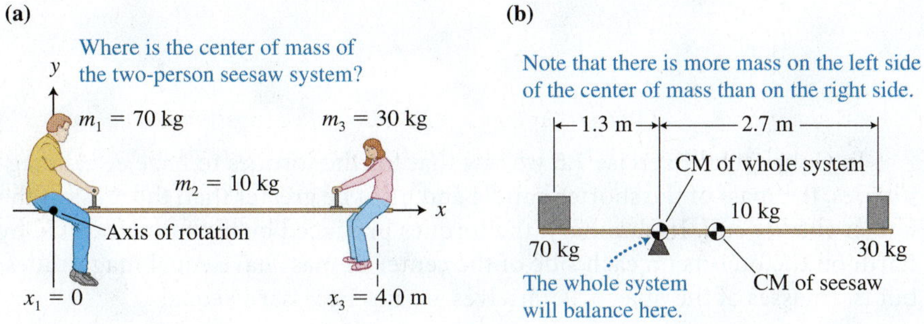

The seesaw would balance about a fulcrum located 1.3 m from $m_1$ and 2.7 m from $m_2$ (Figure 7.18b). We see that the masses are not equal. The mass on the left side of the center of mass is much greater than on the right side—70 kg versus 40 kg. The larger mass on the left is a shorter distance from the center of mass than the smaller masses on the right, which on average are farther from the center of mass. In other words, the masses on the right and on the left of the center of mass are not equal! However, the product of mass and distance on each side balances out, causing torques of equal magnitude. We could rename the center of mass as "the center of torque" to reflect the essence of the concept, but since this is not the term used in physics, we will continue to use the term center of mass.

Let's use another example to test this idea that the mass on the left and on the right side of the center of mass is not necessarily equal.

### CONCEPTUAL EXERCISE 7.5 Balancing a bread knife

You balance a bread knife by laying the flat side across one finger, as shown below. Where is the center of mass of the knife? How does the mass of the knife on the left side of the balance point compare to the mass of the knife on the right side of the balance point?

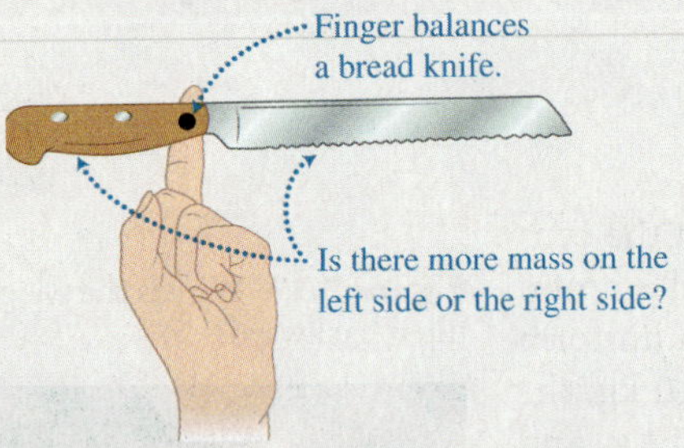

Finger balances a bread knife.
Is there more mass on the left side or the right side?

**Sketch and translate** We choose the knife as the system and orient a vertical $y$-axis upward.

**Simplify and diagram** Two objects exert forces on the knife: the finger exerts an upward normal force and Earth exerts a downward gravitational force. Since the knife is rotationally stable, these two forces must both pass through the axis of rotation at the location of the finger. If not, there would be an unbalanced torque exerted on the knife and it would tip. You can assume that the gravitational force is exerted at the knife's center of mass. Thus, the center of mass must be directly above the finger.

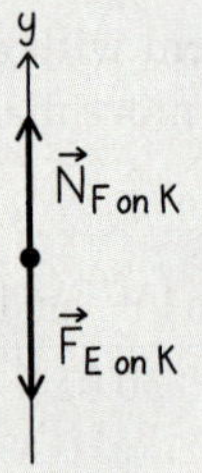


To determine how the mass on the right side of the balance point compares to the mass on the left side, we model the knife as two small spheres of mass $m_1$ and mass $m_2$, connected by a massless rod. The spheres are located at the center of mass of the respective sides of the knife. When the knife is balanced, there is less distance between sphere 1 and the balance point than between sphere 2 and the balance point. Thus, the mass of the handle end must be greater than the mass of the cutting end.

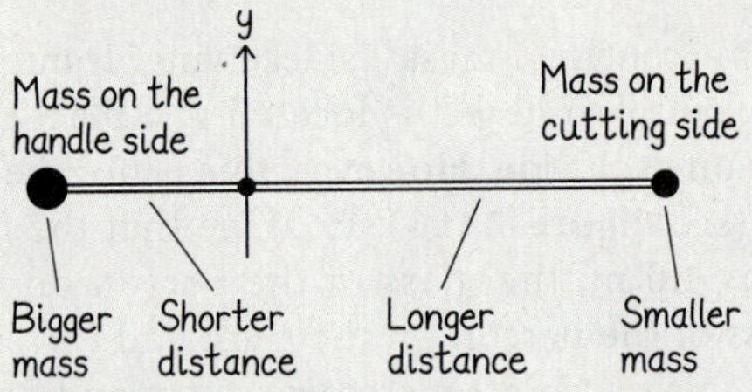


**Try it yourself:** A barbell has a 10-kg plate on one end and a 5-kg plate on the other end. Where is the center of mass of the barbell, which is also the balance point for the barbell? Ignore the mass of the 1.0-m-long rod that connects the plates on the ends.

*Answer:* The center of mass is 0.33 m from the 10-kg end and 0.67 m from the 5-kg end.

In Conceptual Exercise 7.5 we saw that for the torques to have equal magnitudes, the mass of the shorter handle end must be greater than the mass of the longer cutting end. In summary, the torques produced by the forces exerted by Earth on the objects on each side of the center of mass have equal magnitudes, but the masses of the objects themselves are not necessarily equal.

**Review Question 7.4** How could you find the center of mass of a person?

## 7.5 Skills for analyzing situations using equilibrium conditions

We often use the equations of equilibrium to determine one or two unknown forces if all other forces exerted on an object of interest are known. Consider the muscles of your arm when you lift a heavy ball or push down on a desktop (**Figure 7.19**). When you hold a ball in your hand, your biceps muscle tenses and pulls up on your forearm in front of the elbow joint. When you push down with your hand on a desk, your triceps muscle tenses and pulls up on a protrusion of the forearm behind the elbow joint. The equations of equilibrium allow you to estimate these muscle tension forces—see the next example, which describes a general method for analyzing static equilibrium problems. The right side of the table applies the general strategies to the specific problem provided.

◀ Active Learning Guide

**Figure 7.19** Muscles in the upper arm lift and push down on the forearm.

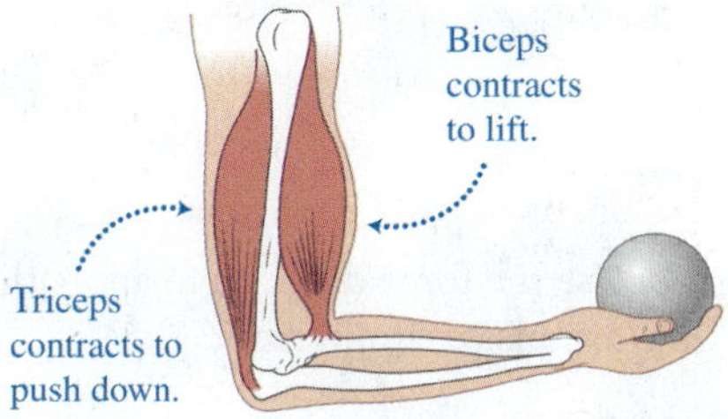


**PROBLEM-SOLVING STRATEGY Applying Static Equilibrium Conditions**

**EXAMPLE 7.6 Use the biceps muscle to lift**

Imagine that you hold a 6.0-kg lead ball in your hand with your arm bent. The ball is 0.35 m from the elbow joint. The biceps muscle attaches to the forearm 0.050 m from the elbow joint and exerts a force on the forearm that allows it to support the ball. The center of mass of the 12-N forearm is 0.16 m from the elbow joint. Estimate the magnitude of (a) the force that the biceps muscle exerts on the forearm and (b) the force that the upper arm exerts on the forearm at the elbow.

**Sketch and translate**

- Construct a labeled sketch of the situation. Include coordinate axes and choose an axis of rotation.
- Choose a system for analysis.

We choose the axis of rotation to be where the upper arm bone (the humerus) presses on the forearm at the elbow joint. This will eliminate from the torque equilibrium equation the unknown force that the upper arm exerts on the forearm.

We choose the system of interest to be the forearm and hand.

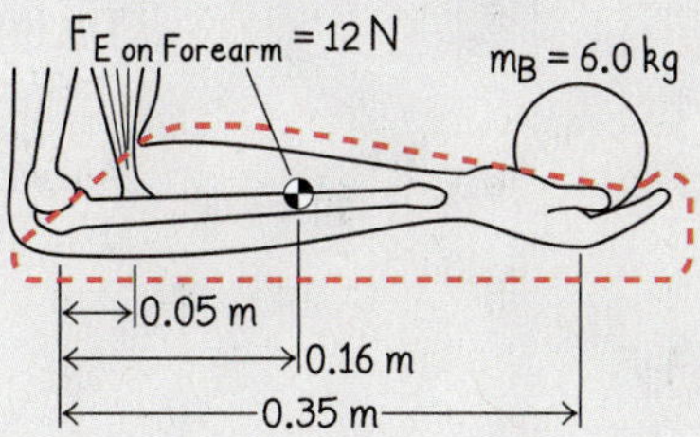


*(continued)*

**Simplify and diagram**

- Decide whether you will model the system as a rigid body or as a point-like object.
- Construct a force diagram for the system. Include the chosen coordinate system and the axis of rotation (the origin of the coordinate system).

Model the system as a rigid body and draw a force diagram for the forearm and hand.

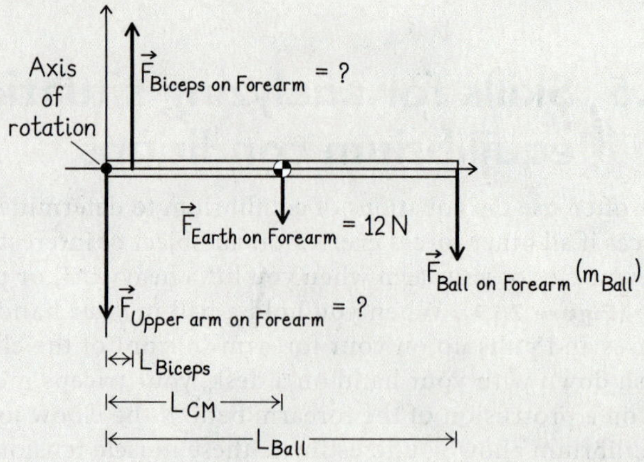


**Represent mathematically**

- Use the force diagram to apply the conditions of equilibrium.

$$\Sigma\tau = 0$$

$$(F_{\text{UA on FA}})(0) + (F_{\text{Biceps on FA}})(L_{\text{Biceps}} \sin 90°) + (-F_{\text{E on FA}}(L_{\text{CM}} \sin 90°)) + (-F_{\text{Ball on FA}}(L_{\text{Ball}} \sin 90°)) = 0$$

$$\Sigma F_y = 0$$

$$(-F_{\text{UA on FA}}) + F_{\text{Biceps on FA}} + (-F_{\text{E on FA}}) + (-F_{\text{Ball on FA}}) = 0$$

**Solve and evaluate**

- Solve the equations for the quantities of interest.
- Evaluate the results. Check to see if their magnitudes are reasonable and if they have the correct signs and units. Also see if they have the expected values in limiting cases.

Substitute $\sin 90° = 1.0$ and rearrange the torque equation to find $F_{\text{Biceps on FA}}$.

$$\begin{aligned} F_{\text{Biceps on FA}} &= [(F_{\text{E on FA}})(L_{\text{CM}}) + (F_{\text{Ball on FA}})(L_{\text{Ball}})]/L_{\text{Biceps}} \\ &= [(12\text{ N})(0.16\text{ m}) + (59\text{ N})(0.35\text{ m})]/(0.050\text{ m}) \\ &= 450\text{ N} \end{aligned}$$

Use the force equation to find $F_{\text{UA on FA}}$:

$$\begin{aligned} F_{\text{UA on FA}} &= F_{\text{Biceps on FA}} - F_{\text{E on FA}} - F_{\text{Ball on FA}} \\ &= 450\text{ N} - 12\text{ N} - 59\text{ N} = 380\text{ N} \end{aligned}$$

The 450-N force exerted by the biceps on the forearm is much greater than the 59-N force exerted by the ball on the forearm. This difference occurs because the force exerted by the biceps is applied much closer to the axis of rotation than the force exerted by the lead ball.

If the center of mass of the forearm were farther from the elbow, the biceps would have to exert an even larger force.

**Try it yourself:** How would the force exerted by the biceps on the forearm change if the biceps were attached to the forearm farther from the elbow?

*Answer:* A longer $L_{\text{Biceps}}$ in the torque equilibrium equation would mean that the biceps muscle would need to exert a smaller force on the forearm when lifting something.

## Applying the conditions of equilibrium in more complex situations

In the previous example the forces exerted on the system were exerted at right angles to the system. What if this is not the case? Consider the next example.

**EXAMPLE 7.7 Lifting a drawbridge**

A drawbridge across the mouth of an inlet on the coastal highway is lifted by a cable to allow sailboats to enter the inlet. You are driving across the 16-m-long drawbridge when the bridge attendant accidentally activates the bridge. You abruptly stop the car 4.0 m from the end of the bridge. The cable makes a 53° angle with the horizontal bridge. The mass of your car is 1000 kg and the mass of the bridge is 4000 kg. Estimate the tension force that the cable exerts on the bridge as it slowly starts to lift the bridge.

**Sketch and translate** We sketch the situation below and choose the bridge as the system. We place the axis of rotation where the drawbridge connects by a hinge to the roadway at the left side of the bridge—a good choice, as we have no information about that force.

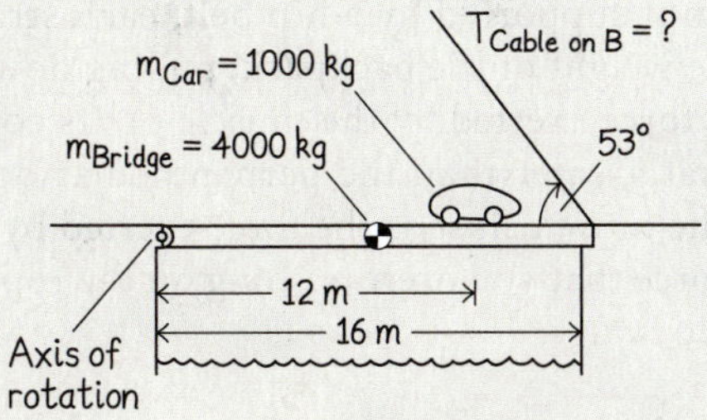


**Simplify and diagram** We model the car as a point-like object and the bridge as a rigid body with uniform mass distribution. The latter assumption means that Earth exerts a gravitational force on the center of the bridge. The bridge has just started to rise, so it is still approximately horizontal. Since it moves very slowly, we will assume that it is in static equilibrium. As we can see from the force diagram, four objects exert forces on the bridge. (1) The hinges on the left side exert a force $\vec{F}_{\text{H on B}}$ that is unknown in magnitude and direction. (2) Earth exerts a $(4000\text{ kg})(9.8\text{ N/kg}) = 39{,}200\text{ N}$ gravitational force $\vec{F}_{\text{E on B}}$ on the center of the bridge. (3) The car pushes down on the bridge, exerting a $(1000\text{ kg})(9.8\text{ N/kg}) = 9800\text{ N}$ force $\vec{F}_{\text{Car on B}}$ 4.0 m from the right side of the bridge. (4) The cable exerts an unknown force $\vec{T}_{\text{Cable on B}}$ on the right edge of the bridge at a 53° angle above the horizontal.

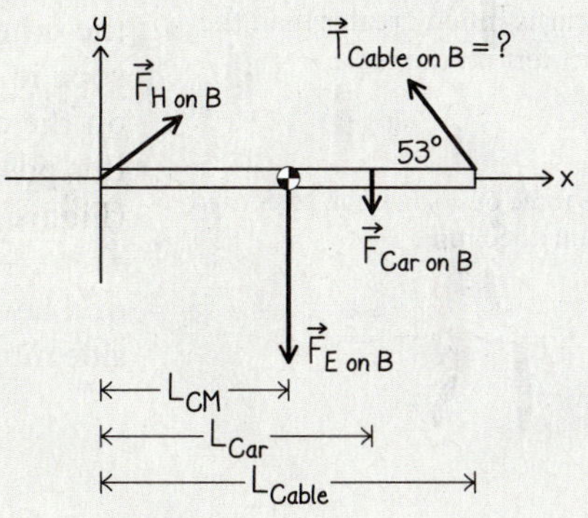


**Represent mathematically** Since four objects exert forces on the bridge, the torque condition of equilibrium will include four torques produced by these forces:

$$F_{\text{H on B}}(0) + (-F_{\text{E on B}}L_{\text{CM}}\sin 90°) + (-F_{\text{Car on B}}\,L_{\text{Car}}\sin 90°) + T_{\text{Cable on B}}\,L_{\text{Cable}}\sin 53° = 0.$$

Substitute $\sin 90° = 1.0$ and $\sin 53° = 0.80$ and rearrange the above to find an expression to determine the unknown tension force that the cable exerts on the bridge:

$$T_{\text{Cable on B}} = \frac{F_{\text{E on B}}L_{\text{CM}} + F_{\text{Car on B}}L_{\text{Car}}}{L_{\text{Cable}}\sin 53°}$$

**Solve and evaluate** Substitute the following values into the above equation: $L_{\text{CM}} = 8.0\text{ m}$, $L_{\text{Car}} = 12.0\text{ m}$, $L_{\text{Cable}} = 16.0\text{ m}$, $F_{\text{E on B}} = 3.92\times 10^4\text{ N}$, and $F_{\text{Car on B}} = 9800\text{ N}$. This yields

$$T_{\text{Cable on B}} = \frac{(3.92\times 10^4\text{ N})(8.0\text{ m}) + (9800\text{ N})(12.0\text{ m})}{(16\text{ m})\sin 53°}$$
$$= 34{,}000\text{ N}$$

The unit is correct. The value of 34,000 N is reasonable given that the bridge holds the 9800-N car near the free end and that Earth exerts a force of 39,000 N on the bridge in its middle at its center of mass.

**Try it yourself:** What force would the cable have to exert on the bridge if your car was 4.0 m from the hinged end of the bridge instead of 4.0 m from the free end?

*Answer:* 28,000 N. This makes sense since the car is exerting a smaller torque when it is closer to the axis of rotation.

## "Magnifying" a force

Our knowledge of equilibrium conditions allows us to understand how one can increase or decrease the turning ability of a force by exerting the force in a different location or in a different direction. Consider a situation when you need to get a car out of a rut in snow or mud. You know from experience that it is

**Figure 7.20** The force that the rope exerts on the car is much greater than the force that you exert on the rope.

**(a)**

To exert a large force on a stuck car, push from the side on a tautly tied rope.

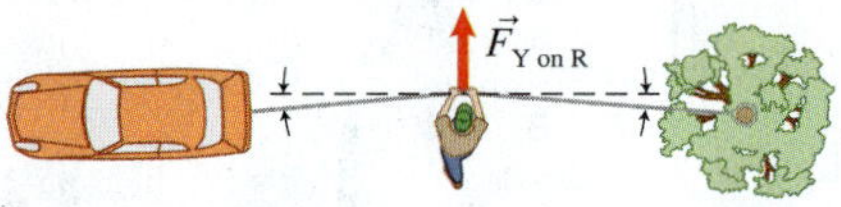


**(b)**

The force you exert on the rope is much less than the tension force exerted by the rope.

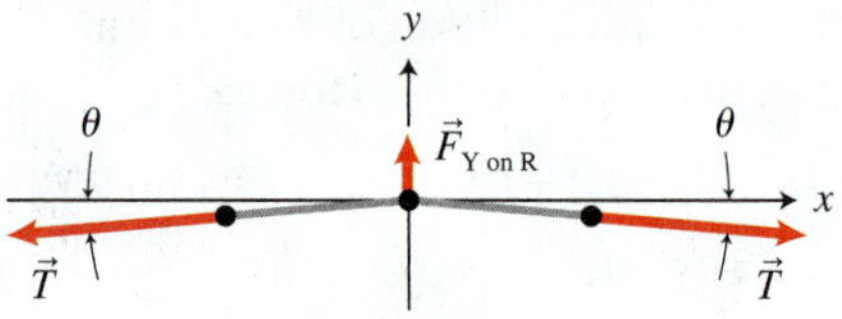


not easy. One way to do it is tie a rope to the front of the car and *tightly* wrap the other end around a tree (see **Figure 7.20a**). Then, push the middle of the rope in a direction perpendicular to the rope. The force that the rope exerts on the car is much greater than the force that you exert on the rope. How is this possible? Draw a force diagram for the short section of rope that you push (Figure 7.20b). For that small section of rope to remain stationary, the forces exerted on it must balance. Let's apply the *y*-component form of Newton's second law. The sum of the *y*-components of the tension forces pulling on each side must balance the force you exert on the rope:

$$-2T\sin\theta + (+F_{\text{Y on R}}) = 0$$

$$\Rightarrow T = \frac{F_{\text{Y on R}}}{2\sin\theta}$$

If the angle $\theta$ of the rope's deflection is small (that's why you need the rope to be tight at the beginning), then $\sin\theta$ is also small and the rope tension $T$ will be large.

Magnifying a force can have advantages or disadvantages, depending on the situation. For example, when you wear a backpack, the shoulder straps often rest on the trapezius muscles, which run across the top of your shoulders to your neck. If the backpack is not supported by a hip belt, each strap has to support approximately half the weight of the backpack, pulling down on the trapezius muscle. The tension force exerted by the muscle on its connection points at each end is somewhat greater than the perpendicular downward push that the strap exerts on the muscle, just as the force exerted by the rope on the car is greater than the force that you exert pushing on the rope. Carrying a heavy backpack can lead to injury.

**EXAMPLE 7.8 The impact of carrying a heavy backpack**

Assume that you are carrying a backpack with several books in it for a combined mass of 10 kg (Earth exerts about a 100-N gravitational force on it). This means that each of the two straps pulling on the trapezius muscle exerts a force of about 50 N on the muscle. This causes the muscle to deflect about 6° from the horizontal on each side of the strap. Estimate the force that the trapezius muscle exerts on its connecting points on the neck and shoulder (similar to the connections of the rope to the tree and the car in the previous example).

**Sketch and translate** Our sketch of the situation is shown below. Choose the section of muscle under the strap as the system of interest.

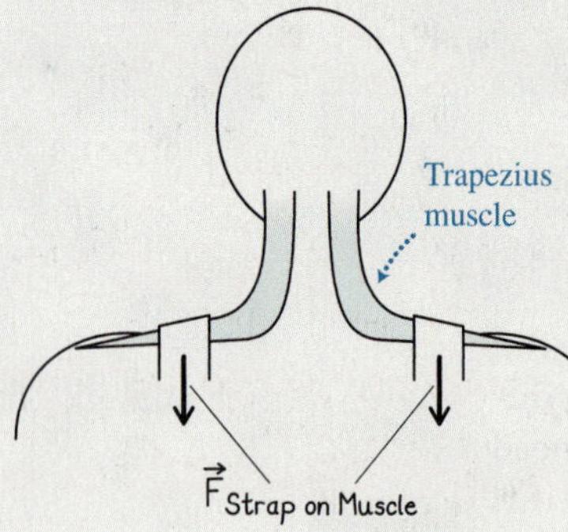


**Simplify and diagram** The force diagram shows the pull of the muscle tissue at an angle of 6° above the horizontal and the 50-N downward force exerted by the strap on that section of muscle.

$\vec{T}$ is large, so its *y*-components balance $\vec{F}_{\text{Strap}}$.

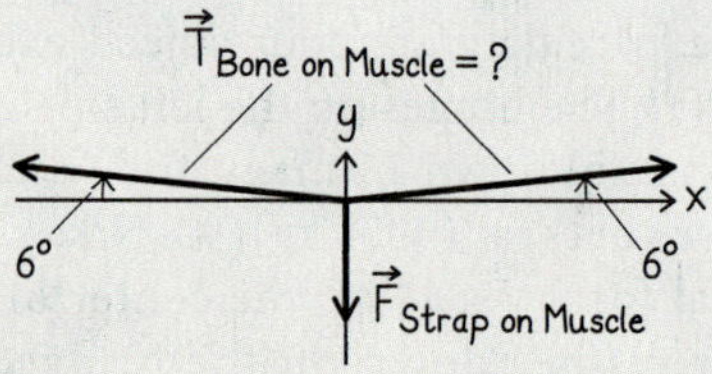


**Represent mathematically** We can apply the same equation we used to get the car out of the rut:

$$T_{\text{Bone on Muscle}} = \frac{F_{\text{Strap on Muscle}}}{2\sin\theta}$$

**Solve and evaluate** Substituting the known information in the above equation, we get

$$T_{\text{Bone on Muscle}} = \frac{50\text{ N}}{2\sin 6^\circ} = 240\text{ N (or 54 lb)}$$

This force is almost 2.5 times larger than the force exerted by Earth on the backpack.

**Try it yourself:** A 70-kg tightrope walker stands in the middle of a tightrope that deflects upward 5.0° on each side of where he stands. Determine the force that each half of the rope exerts on a short section of rope beneath the walker's feet.

*Answer:* 3900 N.

**Review Question 7.5** Earth exerts a 100-N force on a person's backpack. This force results in the person's trapezius muscle exerting a 240-N force on the bones it attaches to. How is this possible?

# 7.6 Stability of equilibrium

Often, objects can remain in equilibrium for a long time interval—you can sit comfortably for a long time on a living room couch without tipping. But sometimes equilibrium is achieved for only a short time interval—think of sitting on a chair and tilting it backward too far onto its rear legs.

## Equilibrium and tipping objects

You have probably observed that it is easier to balance and avoid falling while standing in a moving bus or subway train if you spread your feet apart in the direction of motion. By doing this you are increasing the **area of support**, the area of contact between the object and the surface it is supported by. To understand area of support, in Observational Experiment **Table 7.5** we model a person riding the subway as a rigid body and consider the torques produced by the forces exerted by the floor on her feet.

**OBSERVATIONAL EXPERIMENT TABLE**

**7.5 Stability of equilibrium.**

| Observational experiment | Analysis |
|---|---|
| *Stationary train with respect to the observer on the platform:*<br>(a) Person A stands with feet together.<br>(b) Person B stands with feet apart.<br>You, the observer, are on the ground watching the train. | *Stationary train:*<br>The forces and torques balance about an axis of rotation between their feet.<br>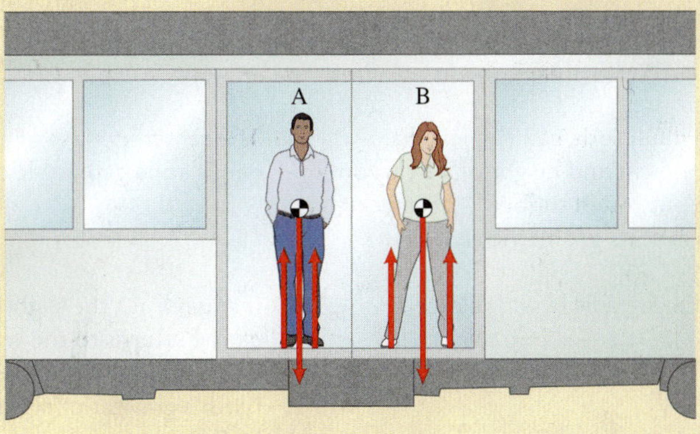 |



(*continued*)

| Observational experiment | Analysis |
|---|---|
| *Train accelerates to right with respect to the observer on the platform:*<br>Both people tend to remain stationary with respect to the platform (Newton's first law) and the train moves out from under them. They both tilt to the left. Consider the torques around an axis of rotation through the left foot of each person (as seen by the platform observer). | *Accelerating train:*<br>(a) The force exerted by Earth on A has counterclockwise turning ability and causes A to tip over.<br>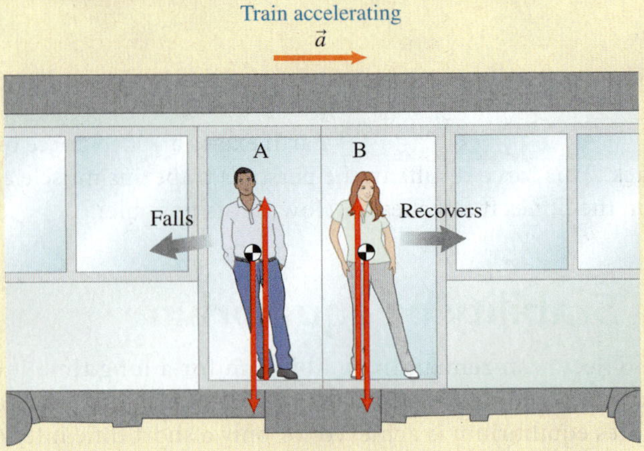<br><br>(b) The force exerted by Earth on B has clockwise turning ability and causes B to recover from the tilt without falling. |

**Patterns**

*Standing on stationary train:* The net torque exerted on both people about any point is zero.
*Standing on accelerating train:*
**PERSON A:** With the feet together, the gravitational force exerts a counterclockwise torque and the person falls. The gravitational force points outside the area of support provided by the feet.
**PERSON B:** With the feet apart, the gravitational force exerts a clockwise torque and the person recovers. Note that the gravitational force points between the feet.

The patterns we observed in Observational Experiment Table 7.5 lead us to a tentative rule about tipping:

> For an object in static equilibrium, if a vertical line passing through the object's center of mass is within the object's area of support, the object does not tip. If the line is not within the area of support, the object tips.

If this is a general rule, then we can use it to predict the angle at which an object with a known center of mass will tip over (see Testing Experiment **Table 7.6**).

**TESTING EXPERIMENT TABLE**

**7.6 Testing our tentative rule about tipping.**

VIDEO 7.6

| Testing experiment | Prediction | Outcome |
|---|---|---|
| Place a full soda can (typical diameter of 6 cm and height of 12 cm) on a flat but rough surface. Its center of mass is at its geometric center. | 1. The center of mass of a full soda can is at its geometric center.<br>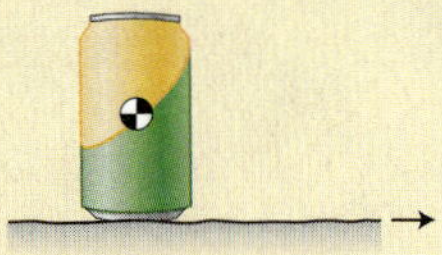 | |
| Tilt the can a little and release it. | 2. If you release the slightly tilted can, it returns to the vertical position because of the torque due to Earth's gravitational force exerted by Earth.<br>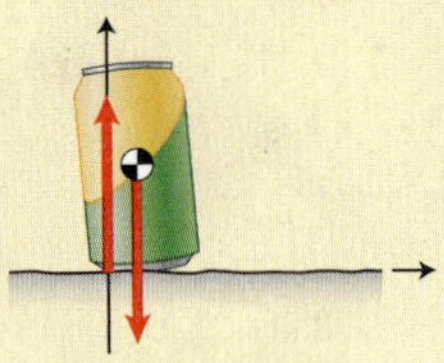 | The can returns to the vertical position. |

| | | |
|---|---|---|
| Tilt the can at larger and larger angles. Predict the angle at which the can will tip. | 3. If you tilt the can more and more, eventually the line defined by the gravitational force passes over the support point for the can at its bottom edge. Tilting the can more than this critical angle causes it to tip over. 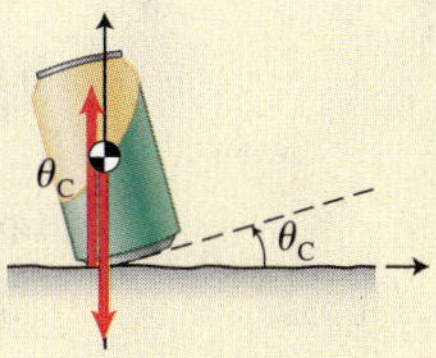  4. *Prediction:* For a can with a diameter of 6 cm and a height of 12 cm, this angle will be $\theta_C = \tan^{-1}(6\text{ cm}/12\text{ cm}) = 27°$ | The outcome matches the prediction. |
| Repeat the experiment, but this time use an open can of corn syrup with the same mass as the soda. The can will only be about half full. Predict the angle at which the can will tip. | The center of mass should now be about one-fourth of the way from the bottom (if we neglect the mass of the can). The critical tilt angle should now be $\theta_C = \tan^{-1}(6\text{ cm}/6\text{ cm}) = 45°$  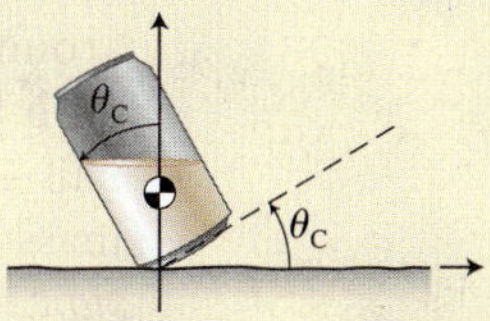  | The outcome matches the prediction within experimental uncertainty (about 1°). |

**Conclusion**

The outcomes are consistent with the predictions. We've increased our confidence in the tipping rule: For an object to be in static equilibrium, the line defined by the gravitational force exerted by Earth must pass within the object's area of support. If it is not within the area of support, the object tips over.

We now know that an object will tip if it is tilted so that the gravitational force passes beyond its area of support. If the area of support is large or if the center of mass is closer to the ground, more tipping is possible without the object falling over—it is more stable. This idea is regularly used in building construction. Tall towers (like the Eiffel Tower) have a wide bottom and a narrower top. The Leaning Tower of Pisa does not tip because a vertical line through its center of mass passes within the area of support (**Figure 7.21**). The same rule explains why you need to keep your feet apart when standing on a subway train entering or leaving a station (Observational Experiment Table 7.5).

**The equilibrium of a system is stable against tipping if the vertical line through its center of mass passes through the system's area of support.**

**Figure 7.21** Because $F_{\text{E on Tower}}$ passes through the base of the tower, the Leaning Tower of Pisa does not tip over.

## Equilibrium and rotating objects

Objects that can rotate around a fixed axis can also have either stable or unstable equilibria. Consider a ruler with several holes in it. If you hang the ruler on a nail using a hole near one end, it hangs as shown in **Figure 7.22a**. If you pull the bottom of the ruler to the side and release it, the ruler swings back and forth with decreasing maximum displacement from the equilibrium position, but eventually hangs straight down. This equilibrium position is called *stable* because the ruler always tries to return to that position if free to rotate. However, if you turn the ruler 180° so that the axis of rotation is at the bottom of the ruler (see Figure 7.22b), it can stay in this position only if very carefully balanced. If disturbed, the ruler swings down and never returns. In this case the ruler is in *unstable* equilibrium.

**Figure 7.22** Stable, unstable, and neutral equilibrium.

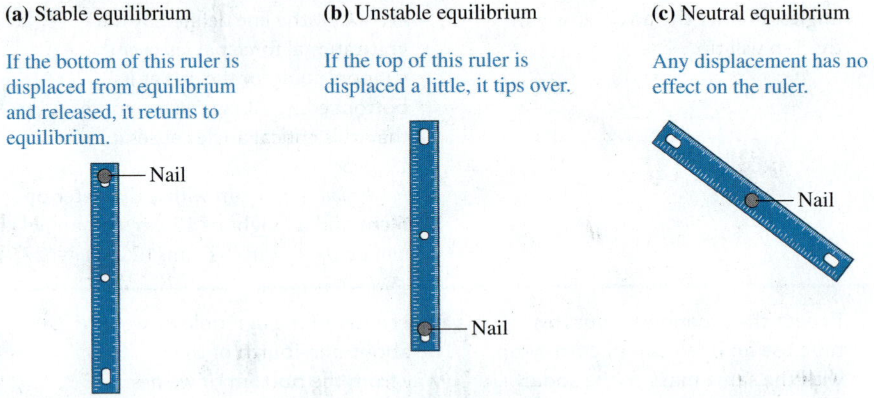

(a) Stable equilibrium
If the bottom of this ruler is displaced from equilibrium and released, it returns to equilibrium.
Nail
(b) Unstable equilibrium
If the top of this ruler is displaced a little, it tips over.
Nail
(c) Neutral equilibrium
Any displacement has no effect on the ruler.
Nail

The stability or lack of stability can be understood by considering the torque around the axis of rotation due to the force exerted by Earth on the object. In both positions a and b, the net force exerted on the ruler by Earth and by the nail (a normal force) is zero. The difference is that in the first case the center of mass is *below the axis of rotation*; in the second case it is *above the axis of rotation*. Torques produced by gravitational forces tend to lower the center of mass of objects. In other words, if it is possible for the object to rotate so that its center of mass becomes lower, it will tend to do so.

If we hang the ruler using the center hole (Figure 7.22c), it remains in whatever position we leave it. Both the normal force exerted by the nail and the gravitational force exerted by Earth produce zero torques. This is called *neutral* equilibrium.

**The equilibrium of a system is stable against rotation if the center of mass of the rotating object is below the axis of rotation.**

## CONCEPTUAL EXERCISE 7.9 Balancing a pencil

Is it possible to balance the pointed tip of a pencil on your finger?

**Sketch and translate** The figure to the right shows a sketch of the situation.

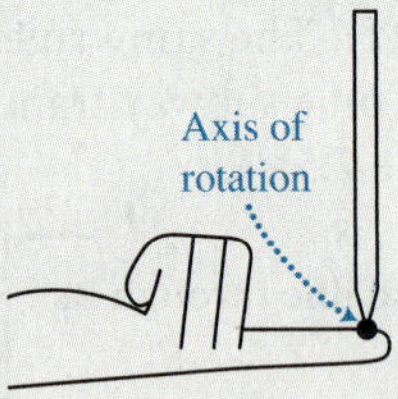


**Simplify and diagram** The forces exerted on the pencil are shown below. The tip of the pencil is the axis of rotation. When the pencil is tilted by only a small angle, the line defined by the gravitational force exerted by Earth on the pencil is not within the area of support of the pencil (which is just the pencil tip). This equilibrium is unstable. Having the center of mass *above* the axis of rotation leads to this instability. To make it stable, we need to lower the center of mass to below the axis of rotation. We can do it by attaching a pocketknife to the pencil, as shown below. Notice that the massive part of the knife is below the tip of the pencil and so is the center of mass of the system. Then when the pencil tilts, the torque due to the gravitational force brings it back to the equilibrium position.

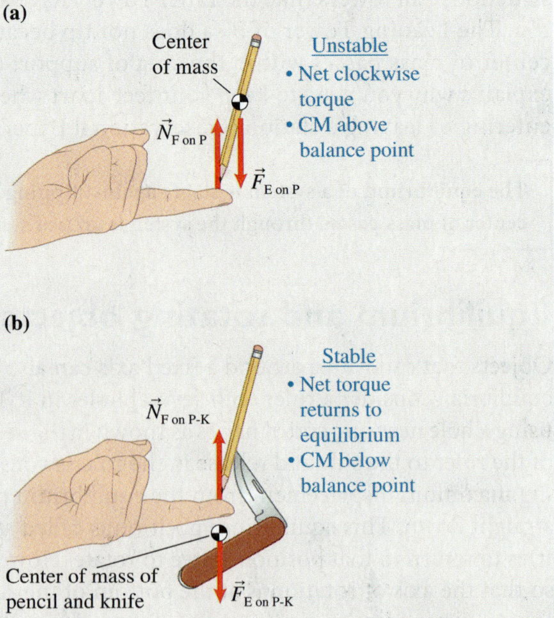

> **TIP** Always try to understand new situations in terms of ideas we have already discussed. The trick of balancing the pencil can be understood with the rules we expressed above: (1) the equilibrium is most stable when the center of mass of the system is in the lowest possible position or, equivalently, (2) when the gravitational potential energy of the system has the smallest value.

The rules we have learned about equilibrium and stability have many applications, including circus tricks. Think about where the center of mass is located for the bicycle and the two people shown in **Figure 7.23**. Another application involves vending machines.

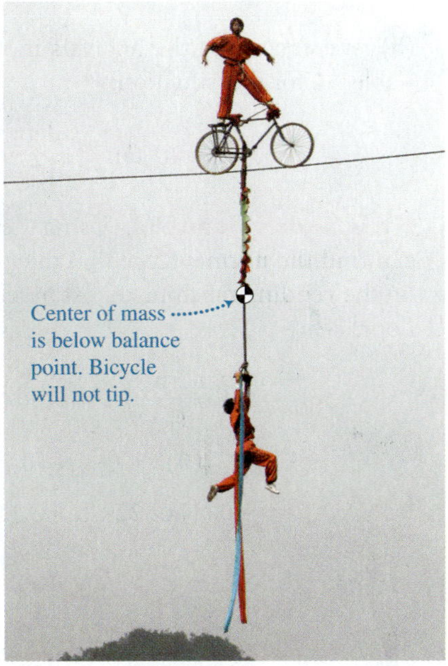


**Figure 7.23** Balancing a bicycle on a high wire may not be as dangerous as it looks.

### EXAMPLE 7.10 Tipping a vending machine

According to the U.S. Consumer Product Safety Commission, tipped vending machines caused 37 fatalities between 1978 and 1995 (2.2 deaths per year). Why is tipping vending machines so dangerous? A typical vending machine is 1.83 m high, 0.84 m deep, and 0.94 m wide and has a mass of 374 kg. It is supported by a leg on each of the four corners of its base.

(a) Determine the horizontal pushing force you need to exert on its front surface 1.5 m above the floor in order to just lift its front feet off the surface (so that it will be supported completely by its back two feet). (b) At what critical angle would it fall forward?

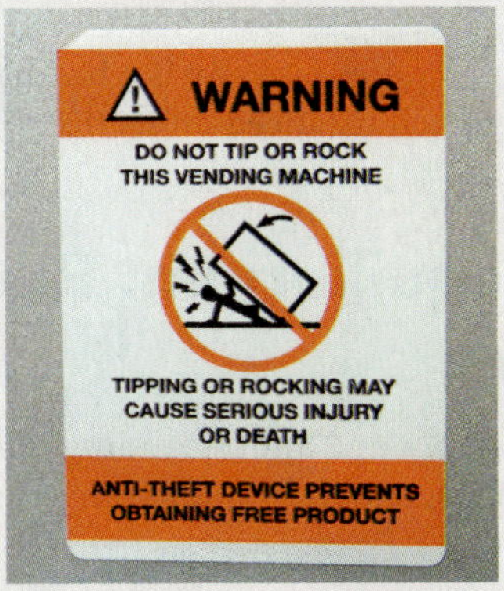


**Sketch and translate** (a) See the sketch of the situation below. The axis of rotation will go through the back support legs of the vending machine.

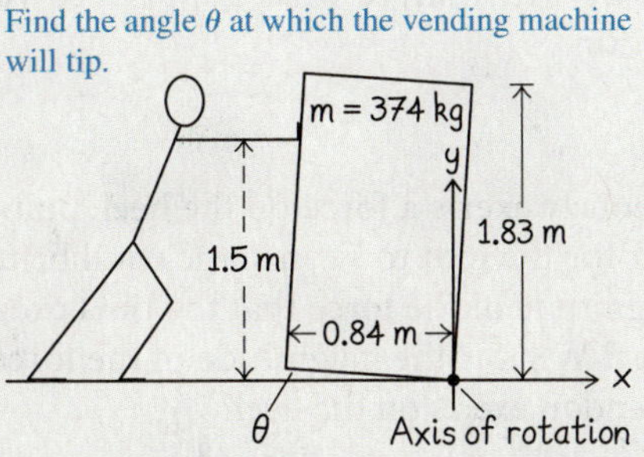


**Simplify and diagram** Model the vending machine as a rigid body. Three objects exert forces on the vending machine (shown in the side view force diagram below): the person exerting normal force $\vec{N}_{\text{P on M}}$ on the machine, Earth exerting gravitational force $\vec{F}_{\text{E on M}}$, and the floor exerting normal force $\vec{N}_{\text{F on M}}$ and static friction force $\vec{f}_{\text{s F on M}}$ on the machine.

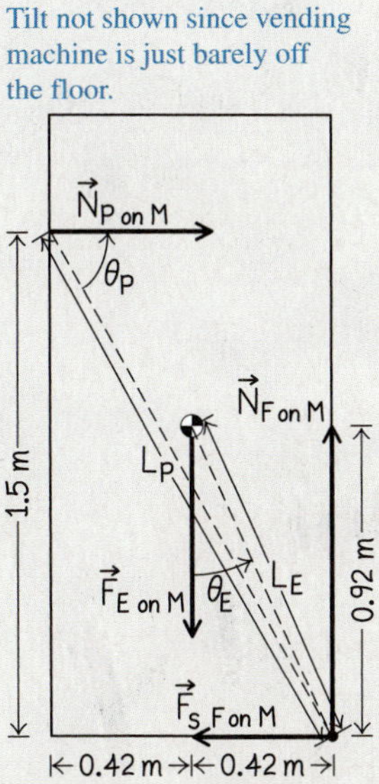


**Represent mathematically** (a) We use the torque condition of equilibrium to analyze the force needed to tilt the vending machine until the front legs are just barely off the floor. The normal force exerted by the person has a clockwise turning ability while the gravitational force has counterclockwise turning ability. The force exerted by the floor on the back legs does not produce a torque, since it is exerted at the axis of rotation.

$$-N_{\text{P on M}}L_{\text{P}}\sin\theta_{\text{P}} + F_{\text{E on M}}L_{\text{E}}\sin\theta_{\text{E}} + N_{\text{F on M}}(0) + f_{\text{s F on M}}(0) = 0$$

*(continued)*

(b) We can apply the analysis in Testing Experiment Table 7.6 for this situation:

$$\theta_C = \tan^{-1}\left(\frac{\text{depth}}{\text{height}}\right)$$

**Solve and evaluate** (a) Using the torque equation, we can find the normal force that the person needs to exert on the vending machine to just barely lift its front off the floor:

$$-N_{\text{P on M}}L_P\left(\frac{1.5\text{ m}}{L_P}\right) + (3700\text{ N})L_E\left(\frac{0.42\text{ m}}{L_E}\right) + N_{\text{F on M}}(0) + f_{s\text{ F on M}}(0) = 0$$

or $N_{\text{P on M}} = 1000\text{ N}$ or 220 lb.

(b) We find that the critical tipping angle is

$$\theta_{\text{tipping}} = \tan^{-1}\left(\frac{\text{depth}}{\text{height}}\right) = \tan^{-1}\left(\frac{0.42\text{ m}}{0.92\text{ m}}\right) = 25°$$

Both answers seem reasonable.

**Try it yourself:** Determine how hard you need to push against the vending machine to keep it tilted at a 25° angle above the horizontal.

*Answer:* 0 N. A vertical line through the center of mass passes through the axis of rotation, so that the net torque about that axis is zero. However, this equilibrium is unstable and represents a dangerous situation.

The chance of being injured by a tipped vending machine is small since a large force must be exerted on it to tilt it up, and it must be tilted at a fairly large angle before it reaches an unstable equilibrium. A more common danger is falling bookcases in regions subject to earthquakes. The base of a typical 2.5-m-tall bookcase is less than 0.3 m deep. The shelves above the base that are filled with books are the same size as the base. If tilted by just

$$\theta_C = \tan^{-1}\left(\frac{\text{depth}}{\text{height}}\right) \approx \tan^{-1}\left(\frac{0.15\text{ m}}{1.25\text{ m}}\right) \approx 7°$$

the bookcase can tip over. In earthquake-prone regions, people often attach a bracket to the top back of the bookcase and then anchor it to the wall.

**Review Question 7.6** Why is a ball hanging by a thread in stable equilibrium, while a pencil balanced on its tip is in unstable equilibrium?

## 7.7 Static equilibrium: Putting it all together

In this section we'll apply our understanding of static equilibrium to analyze three common situations: standing on your toes, lifting a heavy object, and safely climbing a ladder.

### Standing on your toes

Most injuries to the Achilles tendon occur during abrupt movement, such as jumps and lunges. However, we will analyze what happens to your Achilles tendon in a less stressed situation.

**EXAMPLE 7.11 Standing with slightly elevated heel**

Suppose you stand on your toes with your heel slightly off the ground. In order to do this, the larger of the two lower leg bones (the tibia) exerts a force on the ankle joint where it contacts the foot. The Achilles tendon simultaneously exerts a force on the heel, pulling up on it in order for the foot to be in static equilibrium. What is the magnitude of the force that the tibia exerts on the ankle joint? What is the magnitude of the force that the Achilles tendon exerts on the heel?

**Sketch and translate** First we sketch the foot with the Achilles tendon and the tibia. We choose the foot as the system of interest. Three forces are exerted on the foot: the tibia is pushing down on the foot at the ankle joint; the floor is pushing up on the ball of the foot and the toes; and the Achilles tendon is pulling up on the heel. We choose the axis of rotation as the place where the tibia presses against the foot.

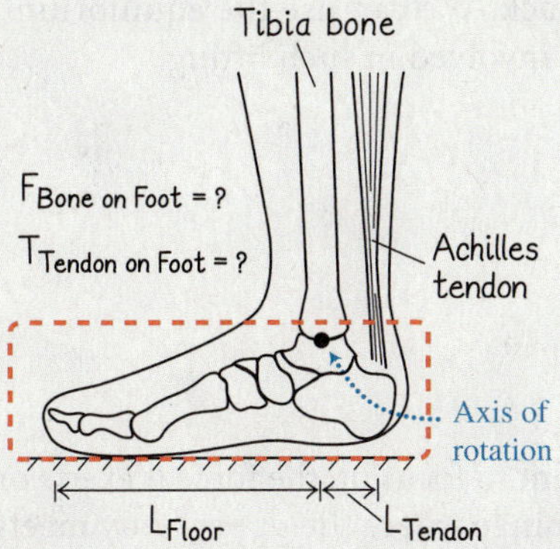


**Simplify and diagram** Model the foot as a very light rigid body. The problem says that the foot is barely off the ground, so we will neglect the angle between the foot and the ground and consider the foot horizontal. The gravitational force exerted on the foot by Earth is quite small compared with the other forces that are being exerted on it, so we will ignore it. A force diagram for the foot is shown below. When you are standing on the ball and toes of both feet, the floor exerts an upward force on each foot equal to half the magnitude of the gravitational force that Earth exerts on your entire body: $F_{\text{Floor on Foot}} = \dfrac{m_{\text{Body}}g}{2}$. The Achilles tendon pulls up on the heel of the foot, exerting a force $T_{\text{Tendon on Foot}}$. The tibia bone in the lower leg pushes down on the ankle joint exerting a force $F_{\text{Tibia on Foot}}$.

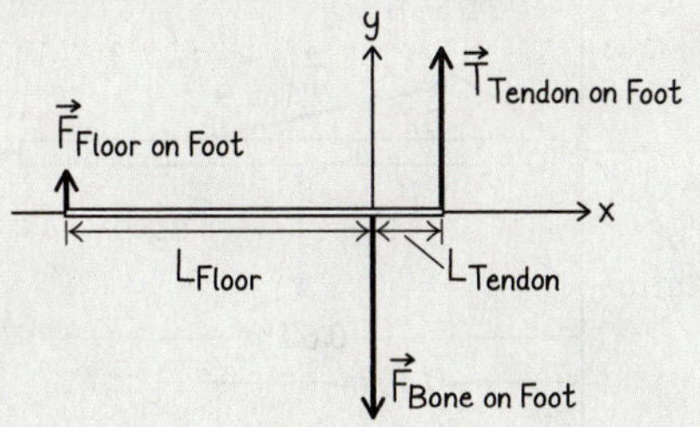


**Represent mathematically** Let's apply the conditions of equilibrium to this system. Note that the distance from the toes to the joint $L_{\text{Toe}}$ is somewhat longer than the distance from the joint to the Achilles tendon attachment point $L_{\text{Tendon}}$. The torque condition of equilibrium becomes

$$+[T_{\text{Tendon on Foot}}(L_{\text{Tendon}})] + F_{\text{Bone on Foot}}(0) - \frac{m_{\text{Body}}g}{2}(L_{\text{Toe}}) = 0$$

$$\Rightarrow T_{\text{Tendon on Foot}} = \frac{m_{\text{Body}}g}{2}\left(\frac{L_{\text{Toe}}}{L_{\text{Tendon}}}\right)$$

Now, apply the $y$-scalar component of the force condition of equilibrium:

$$\Sigma F_y = T_{\text{Tendon on Foot}} + (-F_{\text{Bone on Foot}}) + \frac{m_{\text{Body}}g}{2} = 0$$

$$\Rightarrow F_{\text{Bone on Foot}} = T_{\text{Tendon on Foot}} + \frac{m_{\text{Body}}g}{2}$$

**Solve and evaluate** The distance from the place where the bone contacts the foot to where the floor contacts the foot is about 5 times longer than the distance from the bone to where the tendon contacts the foot. Consequently, the force that the Achilles tendon exerts on the foot is about

$$T_{\text{Tendon on Foot}} = \frac{m_{\text{Body}}g}{2}\left(\frac{L_{\text{Toe}}}{L_{\text{Tendon}}}\right) = \frac{m_{\text{Body}}g}{2}(5) = \frac{5}{2}m_{\text{Body}}g$$

or two and a half times the gravitational force that Earth exerts on the body. Using $g = 10\ \text{N/kg}$ for a 70-kg person, this force will be about 1750 N. That's a very large force for something as simple as standing with your heel slightly elevated! The force exerted on the joint by the leg bone would be

$$F_{\text{Bone on Foot}} = T_{\text{Tendon on Foot}} + \left(\frac{m_{\text{Body}}g}{2}\right)$$

$$= 1750\ \text{N} + 350\ \text{N} = 2100\ \text{N}$$

This force is three times the weight of the person! The forces are much greater when moving. Thus, every time you lift your foot to walk, run, or jump, the tendon tension and joint compression are several times greater than the gravitational force that Earth exerts on your entire body.

**Try it yourself:** Estimate the increase in the magnitude of the force exerted by the Achilles tendon on the foot of the person in this example if his mass were 90 kg instead of 70 kg.

*Answer:* An increase of at least 500 N (110 lb).

**Figure 7.24** A bad way to lift. Lifting in this position causes considerable back muscle tension and disk compression in the lower back.

## Lifting from a bent position

Back problems often originate with improper lifting techniques—a person bends over at the waist and reaches to the ground to pick up a box or a barbell. In **Figure 7.24**, the barbell pulls down on the woman's arms far from the axis of rotation of her upper body about her hip area. This downward pull causes a large clockwise torque on her upper body. To prevent her from tipping over, her back muscles must exert a huge force on the backbone, thus producing an opposing counterclockwise torque. This force exerted by the back muscles compresses the disks that separate vertebrae and can lead to damage of the disks, especially in the lower back. We can use the equilibrium equations to estimate the forces and torques involved in such lifting.

**Active Learning Guide ›**

### EXAMPLE 7.12 Lifting incorrectly from a bent position

Estimate the magnitude of the force that the back muscle in the woman's back in Figure 7.24 exerts on her backbone and the force that her backbone exerts on the disks in her lower back when she lifts an 18-kg barbell. The woman's mass is 55 kg. Model the woman's upper body as a rigid body.

- The back muscle attaches two-thirds of the way from the bottom of her $l = 0.60$-m-long backbone and makes a 12° angle relative to the horizontal backbone.
- The mass of her upper body is $M = 33$ kg centered at the middle of the backbone and has uniform mass distribution. The axis of rotation is at the left end of the backbone and represents one of the disks in the lower back.

**Sketch and translate** The figure below is our mechanical model of a person lifting a barbell. We want to estimate the magnitudes of the force $T_{\text{M on B}}$ that the back muscle exerts on the backbone and the force $F_{\text{D on B}}$ that the disk in the lower back exerts on the backbone. The force that the disk exerts on the bone is equal in magnitude to the force exerted by the backbone on the disk. The upper body (including the backbone) is the system of interest, but we consider the back muscle to be external to the system since we want to focus on the force it exerts on the backbone. The hinge where the upper body meets the lower body is the axis of rotation.

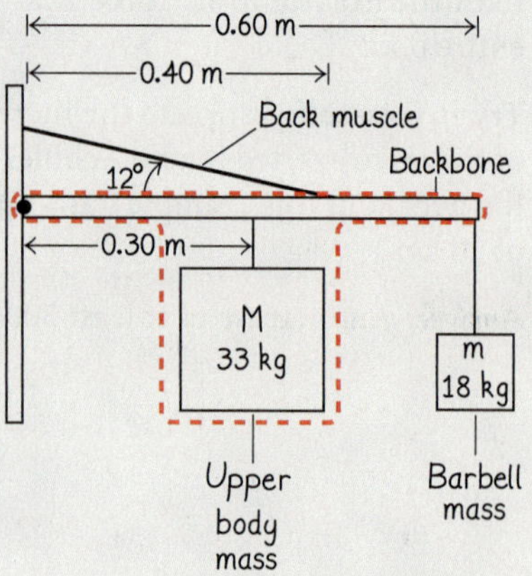


**Simplify and diagram** We next draw a force diagram for the upper body. The gravitational force that Earth exerts on the upper body $F_{\text{E on B}}$ at its center of mass is $Mg = (33\text{ kg})(9.8\text{ N/kg}) = 323\text{ N}$ (73 lb). The barbell exerts a force on the upper body equal to $mg = (18\text{ kg})(9.8\text{ N/kg}) = 176\text{ N}$ (40 lb). Because of our choice of axis of rotation, the force exerted by the disk on the upper body $F_{\text{D on B}}$ will not produce a torque. The gravitational force exerted by Earth on the upper body and the force that the barbell exerts on the upper body have clockwise turning ability, while the tension force exerted by the back muscles on the upper body has counterclockwise turning ability.

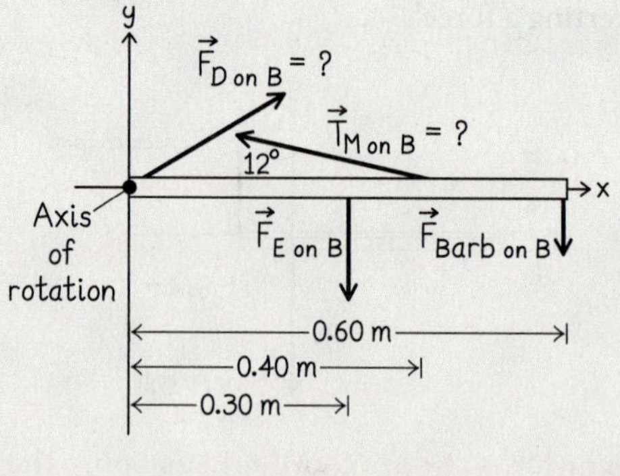


**Represent mathematically** The torque condition of equilibrium for the upper body is

$$\Sigma\tau = +(F_{\text{D on B}})(0) + [-(Mg)(l/2)\sin 90°] + (T_{\text{M on B}})(2l/3)\sin 12° + [-(F_{\text{Barb on B}})(l)\sin 90°] = 0$$

The $x$- and $y$-component forms of the force condition of equilibrium for the backbone are

$$\Sigma F_x = F_{\text{D on B}x} + (-T_{\text{M on B}} \cos 12^\circ) = 0$$
$$\Sigma F_y = F_{\text{D on B}y} + T_{\text{M on B}} \sin 12^\circ + (-mg) + (-Mg) = 0$$

where $F_{\text{D on B}x}$ and $F_{\text{D on B}y}$ are the scalar components of the force that the disk exerts on the upper body.

**Solve and evaluate** We can solve the torque equation immediately to determine the magnitude of the force that the back muscle exerts on the backbone:

$$T_{\text{M on B}} = \frac{(Mg)(l/2)(\sin 90^\circ) + (mg)(l)\sin 90^\circ}{(2l/3)(\sin 12^\circ)}$$

Note that the backbone length $l$ in the numerator and denominator of all of the terms in this equation cancels out. Thus,

$$T_{\text{M on B}} = \frac{(Mg)(1/2)(\sin 90^\circ) + (mg)(1)\sin 90^\circ}{(2/3)(\sin 12^\circ)}$$
$$= \frac{(323\text{ N})(0.50)(1.0) + (176\text{ N})(1)(1.0)}{(0.667\text{ m})(0.208)}$$
$$= 2440\text{ N }(550\text{ lb})$$

We then find $F_{\text{D on B}x}$ from the $x$-component force equation:

$$F_{\text{D on B}x} = +T_{\text{M on B}} \cos 12^\circ$$
$$= +(2440\text{ N})\cos 12^\circ$$
$$= +2390\text{ N}$$

and $F_{\text{D on B}y}$ from the $y$-component force equation:

$$F_{\text{D on B}y} = +Mg + mg - T_{\text{M on B}} \sin 12^\circ$$
$$= +323\text{ N} + 176\text{ N} - (2440\text{ N})(\sin 12^\circ)$$
$$= -8\text{ N}$$

Thus, the magnitude of $F_{\text{D on B}}$ is

$$F_{\text{D on B}} = \sqrt{(2390\text{ N})^2 + (-8\text{ N})^2} = 2390\text{ N }(540\text{ lb})$$

The direction of $\vec{F}_{\text{D on B}}$ can be determined using trigonometry:

$$\tan\theta = \frac{F_{\text{D on B}y}}{F_{\text{D on B}x}} = \frac{-8\text{ N}}{2390\text{ N}} = -0.0033$$

or $\theta = 0.19^\circ$ below the horizontal. We've found that the back muscles exert a force more than four times the gravitational force that Earth exerts on the person and that the disks of the lower back are compressed by a comparable force.

**Try it yourself:** Suppose that a college football lineman stands on top of a 1-inch-diameter circular disk. How many 275-lb linemen, one on top of the other, would exert the same compression force on the disk as that exerted on the woman's disk when she lifts the 40-lb barbell?

*Answer:* The magnitude of the force exerted on the vertebral disk is equivalent to two linemen ($2 \times 275$ lb $=$ 550 lb) standing on the 1-inch-diameter disk.

To lift correctly, keep your back more vertical with the barbell close to your body, as in **Figure 7.25**. Bend your knees and lift with your legs. With this orientation, the back muscle exerts one-third of the force that is exerted when lifting incorrectly. The disks in the lower back undergo one-half the compression they would experience from lifting incorrectly.

**Figure 7.25** A better way to lift things.

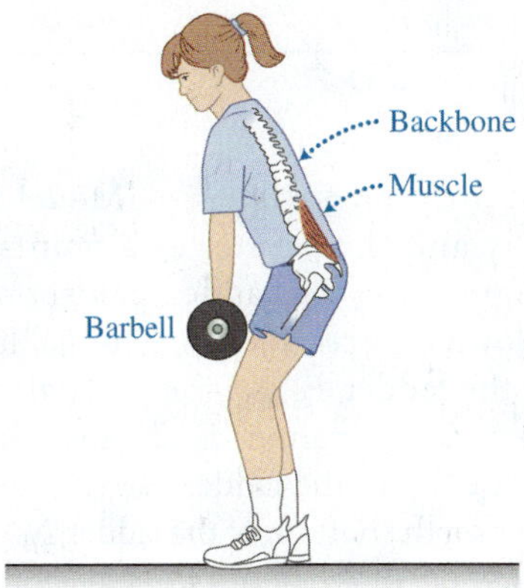


## Using a ladder safely

Each year about 25 people per 100,000 experience serious falls from ladders. Why are ladders sometimes dangerous? Consider the physics of one aspect of ladder use.

**EXAMPLE 7.13 Don't let your ladder slip**

Any time you have to climb a ladder, you want the ladder to remain in static equilibrium. At what angle should a 60-kg painter place his ladder against the wall in order to climb two-thirds of the way up the ladder and have the ladder remain in static equilibrium? The ladder's mass is 10 kg and its length is 6.0 m. The exterior wall of the house is very smooth, meaning that it exerts a negligible friction force on the ladder. The coefficient of static friction between the floor and the ladder is 0.50.

**Sketch and translate** We've sketched the situation below. If the ladder is tilted at too large an angle from the vertical, everyday experience indicates that it will slide down the wall. The static friction force exerted by the floor on the ladder is what is preventing the ladder from sliding, so one way to look at the situation is to ask, What is the maximum angle relative to the wall that the ladder can have before the static friction force is insufficient to keep the ladder in static equilibrium? We choose the ladder and the painter together as the system of interest.

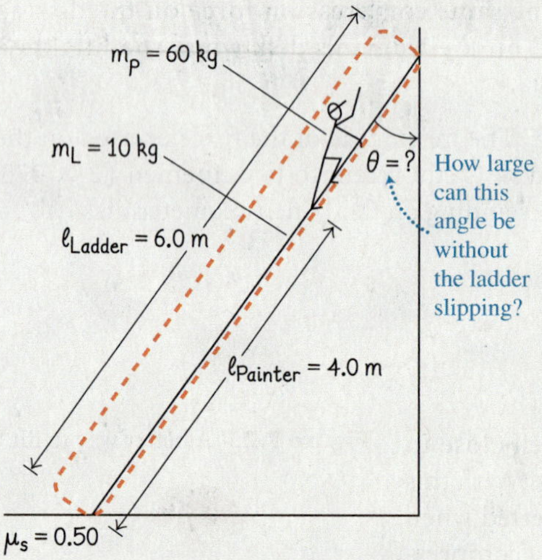


**Simplify and diagram** Model the ladder as a rigid body and the painter as a point-like object. A force diagram for the ladder-painter system includes the following forces: the gravitational force exerted by Earth on the ladder $\vec{F}_{\text{E on L}}$, the gravitational force that Earth exerts on the painter $\vec{F}_{\text{E on P}}$, the normal force of the wall on the top of the ladder $\vec{N}_{\text{W on L}}$, the normal force of the floor on the bottom of the ladder $\vec{N}_{\text{F on L}}$, and the static friction force of the floor on the bottom of the ladder $\vec{f}_{\text{sF on L}}$. We place the axis of rotation where the ladder touches the floor. Rather than determine the center of mass of the system, we have kept the gravitational forces exerted by Earth on the ladder and painter separate.

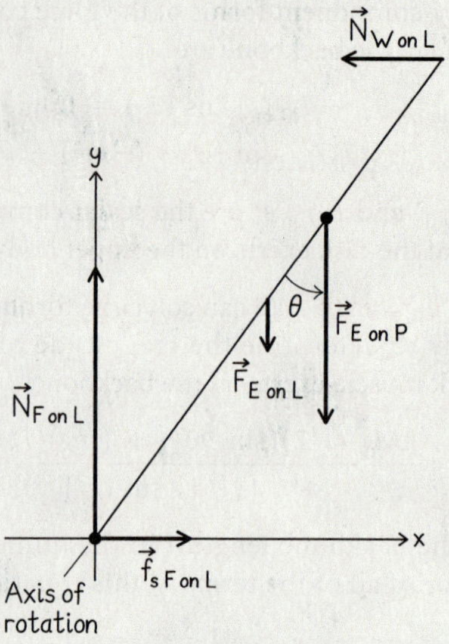


**Represent mathematically** Use the force diagram to apply the conditions of equilibrium. Since the forces in this situation point along more than one axis, we use both components of the force condition of equilibrium.

$y$-component force equation: $\Sigma F_y = N_{\text{F on L}} + (-m_{\text{Ladder}}g) + (-m_{\text{Painter}}g) = 0$

$x$-component force equation: $\Sigma F_x = (-N_{\text{W on L}}) + (f_{\text{s F on L}}) = 0$

We can insert the expression for the maximum static friction force ($f_{\text{s Max F on L}} = \mu_s N_{\text{F on L}}$) into the $x$-component force equation to get

$$N_{\text{W on L}} = \mu_s N_{\text{F on L}}$$

From the $y$-component equation, we get

$$N_{\text{F on L}} = m_{\text{Ladder}}g + m_{\text{Painter}}g$$

Combining the last two equations, we get

$$N_{\text{W on L}} = \mu_s(m_{\text{Ladder}} + m_{\text{Painter}})g$$

The torque equation is

$$\left[-F_{\text{E on L}}\left(\frac{l_{\text{Ladder}}}{2}\right)\sin\theta\right] + \left[-F_{\text{E on P}}\left(\frac{2l_{\text{Ladder}}}{3}\right)\sin\theta\right] + [N_{\text{W on L}}l_{\text{Ladder}}\sin(90° - \theta)] = 0$$

$$\Rightarrow -m_{\text{Ladder}}g\frac{l_{\text{Ladder}}}{2}\sin\theta - m_{\text{Painter}}g\frac{2l_{\text{Ladder}}}{3}\sin\theta + \mu_s(m_{\text{Ladder}} + m_{\text{Painter}})gl_{\text{Ladder}}\cos\theta = 0$$

**Solve and evaluate** We can cancel $gl_{\text{Ladder}}$ out of each term of the torque equation and then solve what remains for $\theta$:

$$\Rightarrow \frac{m_{\text{Ladder}}}{2}\sin\theta + \frac{2m_{\text{Painter}}}{3}\sin\theta - \mu_s(m_{\text{Ladder}} + m_{\text{Painter}})\cos\theta = 0$$

$$\Rightarrow \left(\frac{m_{\text{Ladder}}}{2} + \frac{2m_{\text{Painter}}}{3}\right)\frac{\sin\theta}{\cos\theta} - \mu_s(m_{\text{Ladder}} + m_{\text{Painter}}) = 0$$

$$\Rightarrow \left(\frac{m_{\text{Ladder}}}{2} + \frac{2m_{\text{Painter}}}{3}\right)\tan\theta = \mu_s(m_{\text{Ladder}} + m_{\text{Painter}})$$

$$\Rightarrow \tan\theta = \frac{\mu_s(m_{\text{Ladder}} + m_{\text{Painter}})}{\dfrac{m_{\text{Ladder}}}{2} + \dfrac{2m_{\text{Painter}}}{3}} = \frac{6\mu_s(1 + m_{\text{Painter}}/m_{\text{Ladder}})}{3 + 4m_{\text{Painter}}/m_{\text{Ladder}}} = \frac{6(0.50)(1+6)}{3+4(6)} = \frac{7}{9}$$

$$\Rightarrow \theta = 38^\circ$$

This seems reasonable and similar to what we observe in real life when someone uses a ladder. Some ladders come with a warning not to have the angle exceed 15°. As a limiting case, if the coefficient of static friction were zero, then our result says the angle would also have to be zero. This makes sense; without friction between the ground and the ladder, the ladder would need to be perfectly vertical or it would slip.

**Try it yourself:** If the exterior wall of the house were not smooth, how would the above method need to be changed?

***Answer:*** The wall will exert an upward friction force on the ladder. We would have to take this force into account for both force and torque conditions of equilibrium.

**Review Question 7.7** When lifting an object, the muscles in the body have to exert significantly larger forces than Earth exerts on the object being lifted. Use your understanding of torque and static equilibrium to explain this phenomenon.

# Summary

| Words | Pictorial and physical representations | Mathematical representation |
|---|---|---|
| **Center of mass** The gravitational force that the Earth exerts on an object can be considered to be exerted entirely on the object's center of mass. An external force pointing directly toward or away from the center of mass of a free object will not cause the object to turn or rotate. (Sections 7.1 and 7.4) | 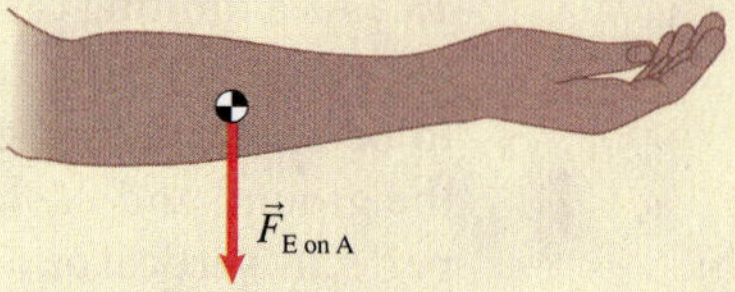  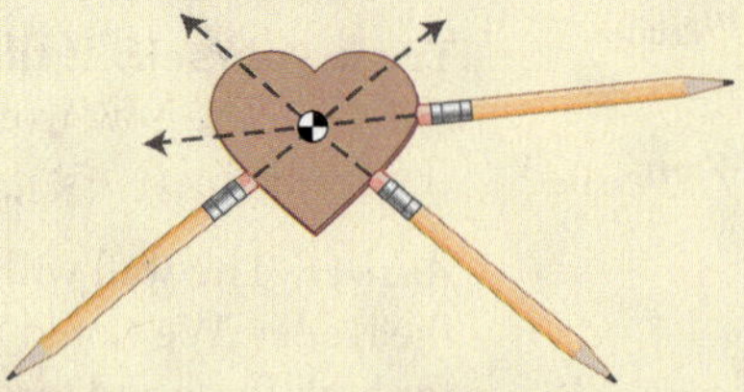 | $x_{cm} = \dfrac{m_1x_1 + m_2x_2 + m_3x_3 + \cdots + m_nx_n}{m_1 + m_2 + m_3 + \cdots + m_n}$ $y_{cm} = \dfrac{m_1y_1 + m_2y_2 + m_3y_3 + \cdots + m_ny_n}{m_1 + m_2 + m_3 + \cdots + m_n}$ Eq. (7.4) |
| A **torque** $\tau$ around an *axis of rotation* is a physical quantity characterizing the turning ability of a force with respect to a particular axis of rotation. The torque is positive if the force tends to turn the object counterclockwise and negative if it tends to turn the object clockwise about the axis of rotation. (Section 7.2) | 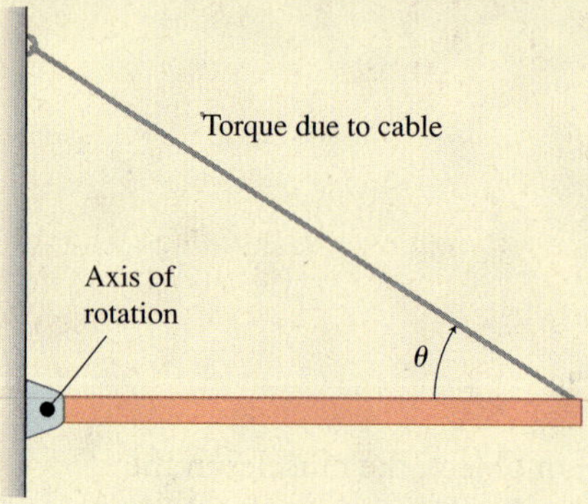  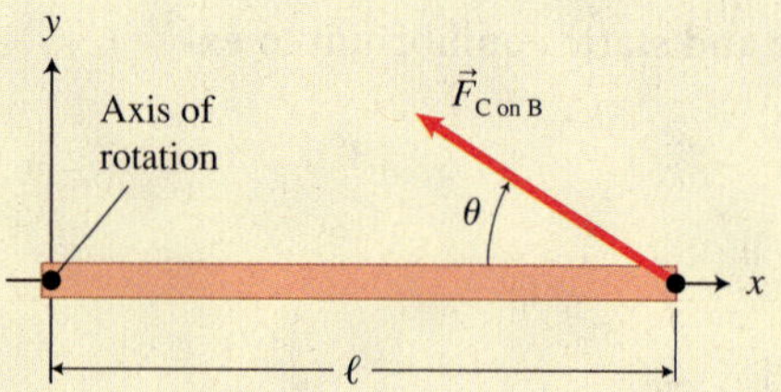  | $\tau = \pm Fl \sin\theta$ Eq. (7.1) |
| **Static equilibrium** is a state in which a rigid body is at rest and remains at rest both translationally and rotationally. (Section 7.3) | 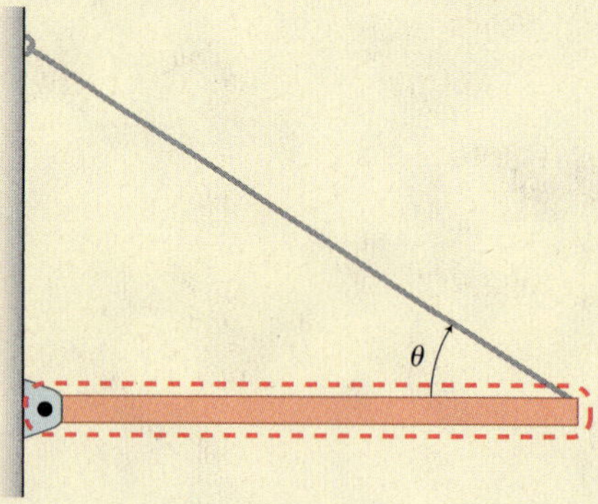  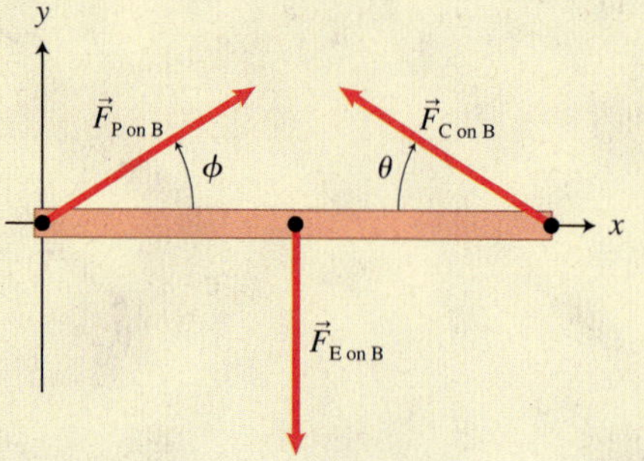  | *Translational (force) condition* $\Sigma F_x = F_{1\ on\ Bx} + \cdots + F_{n\ on\ Bx} = 0$ Eq. (7.2x) $\Sigma F_y = F_{1\ on\ By} + \cdots + F_{n\ on\ By} = 0$ Eq. (7.2y) *Rotational (torque) condition* $\Sigma\tau = \tau_1 + \tau_2 + \cdots + \tau_n = 0$ Eq. (7.3) |

The **equilibrium** of a system is stable against tipping line through its center of mass is within the system's area of support.

The **equilibrium** of a system is stable against rotating of mass of the rotating object is below the axis of rotation. (Section 7.6)

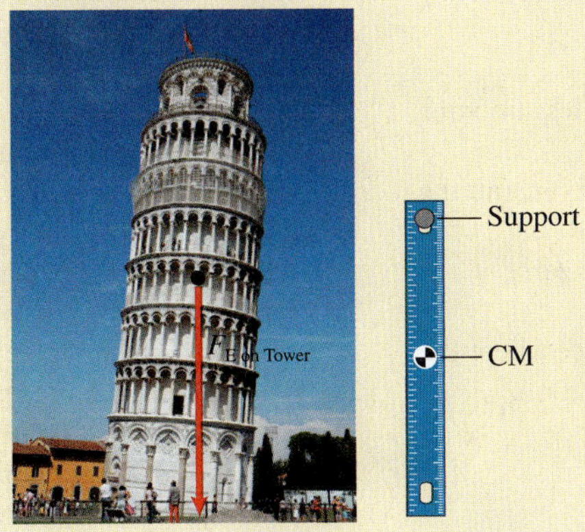


MP® For instructor-assigned homework, go to **MasteringPhysics.**

# Questions

## Multiple Choice Questions

1. A falling leaf usually flutters while falling. However, we have learned that the force that Earth exerts on an object is exerted at its center of mass and thus should not cause rotational motion. How can you resolve this contradiction?
   (a) A leaf is not a rigid body and the rule does not apply.
   (b) There are other forces exerted on the leaf as it falls besides the force exerted by Earth.
   (c) Some forces were not taken into account when we defined the center of mass.
2. You have an irregularly shaped object. To find its center of mass you can do which of the following?
   (a) Find a point where you can put a fulcrum to balance it.
   (b) Push the object in different directions and find the point of intersection of the lines of action of the forces that do not rotate the object.
   (c) Separate the object into several regularly shaped objects whose center of masses you know, and use the mathematical definition of the center of mass to find it.
   (d) All of the above
   (e) a and b only
3. A hammock is tied with ropes between two trees. A person lies in it. Under what circumstances are its ropes more likely to break?
   (a) If stretched tightly between the trees
   (b) If stretched loosely between the trees
   (c) The ropes always have equal likelihood to break.
4. Where is the center of mass of a donut?
   (a) In the center of the hole
   (b) Uniformly distributed throughout the donut
   (c) Cannot be found
5. Which of the following objects is at rest but not in static equilibrium?
   (a) An object thrown upward at the top of its flight
   (b) A person sitting on a chair
   (c) A person in an elevator moving down at constant speed
6. Which of the following objects is not at rest but in equilibrium?
   (a) An object thrown upward at the top of its flight
   (b) A person sitting on a chair
   (c) A person in an elevator moving down at constant speed
7. A physics textbook lies on top of a chemistry book, which rests on a table. Which force diagram below best describes the forces exerted by other objects on the chemistry book (**Figure Q7.7**)?

**Figure Q7.7**

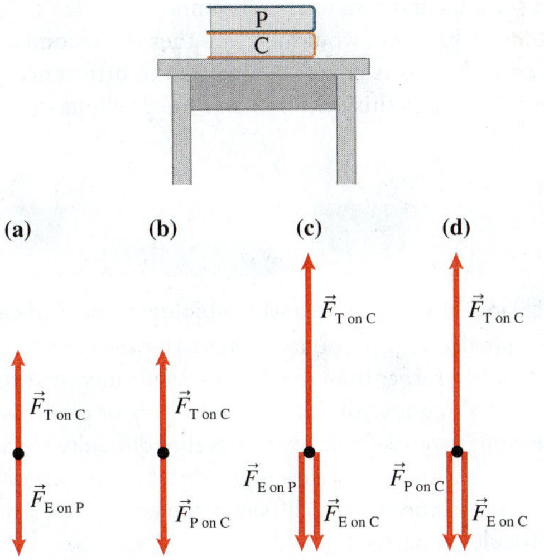


8. What does it mean if the torque of a force is positive?
   (a) The object exerting the force is on the right side of the axis of rotation.
   (b) The object exerting the force is on the left side of the axis of rotation.
   (c) The force points up.
   (d) The force points down.
   (e) None of these choices is necessarily correct.

9. Why do you tilt your body forward when hiking with a heavy backpack?
   (a) The backpack pushes you down.
   (b) Bending forward makes the backpack press less on your back.
   (c) Bending forward moves the center of mass of you and the backpack above your feet.
10. What does it mean if the torque of a 10-N force is zero?
    (a) The force is exerted at the axis of rotation.
    (b) A line parallel to and passing through the place where the force is exerted passes through the axis of rotation.
    (c) Both a and b are correct.

## Conceptual Questions

11. Is it possible for an object not to be in equilibrium when the net force exerted on it by other objects is zero? Give an example.
12. Explain the meaning of torque so that a friend not taking physics can understand.
13. ✎ Something is wrong with the orientation of the ropes shown in **Figure Q7.13**. Use the first condition of equilibrium for the hanging pulley to help explain this error, and then redraw the sketch as you would expect to see it.

**Figure Q7.13**

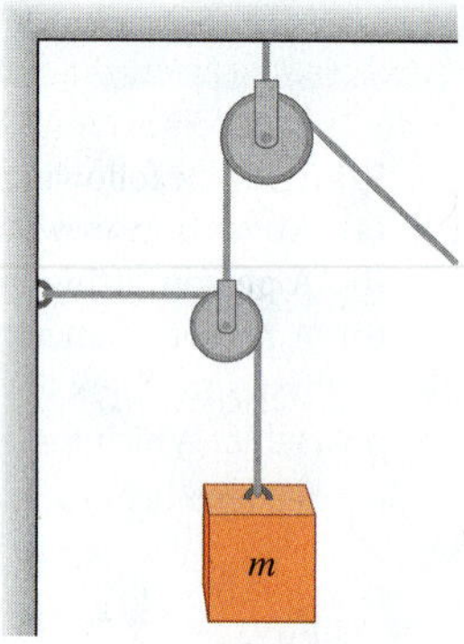


14. What are the two conditions of equilibrium? How do you know that these conditions create a state of equilibrium?
15. Describe an observational experiment that you will conduct so that a friend can discover the second condition of equilibrium.
16. Describe how you would test whether the second condition of equilibrium is correct. What is the difference between your answer to this question and the previous question?
17. Give three examples of situations in which an object is starting to turn even though the sum of the forces exerted on the object is zero.
18. Give an example of an object that is accelerating translationally even though the sum of the torques of the forces exerted on it about an axis of rotation is zero.
19. The force that the body muscles exert on bones that are used to lift various objects is usually five to ten times greater than the gravitational force that Earth exerts on the object being lifted. Explain and give an example.
20. When a person jumps off a boat, what happens to the location of the center of mass of the person-boat system?
21. ✎ A ladder leans against a wall. Construct a force diagram showing the direction of all forces exerted on the ladder. Identify two interacting objects for each force.
22. ✎ Using a crowbar, a person can remove a nail by exerting little force, whereas pulling directly on the nail requires a large force to remove it (you probably can't). Why? Draw a sketch to support your answer.
23. Is it more difficult to do a sit-up with your hands stretched in front of you or with them behind your head? Explain.
24. Sit on a chair with your feet straight down at the front of the chair. Keeping your back perpendicular to the floor, try to stand up without leaning forward. Explain why it is impossible to do it.
25. Can you balance the tip of a wooden ruler vertically on a fingertip? Why is it so difficult? Design a method to balance the ruler on your fingertip. Describe any extra material(s) you will use.
26. Try to balance a sharp wooden pencil on your fingertip, point down. [*Hint:* A small pocketknife might help by lowering the center of mass of the system.]
27. Design a device that you can use to successfully walk on a tight rope.
28. Why do tightrope walkers carry long, heavy bars?
29. Explain why it is easier to keep your balance while jumping on two feet than while hopping on one.
30. A carpenter's trick to keep nails from bending when they are pounded into a hard material is to grip the center of the nail with pliers. Why does this help?
31. Why does a skier crouch when going downhill?

# Problems

Below, **BIO** indicates a problem with a biological or medical focus. Problems labeled **EST** ask you to estimate the answer to a quantitative problem rather than develop a specific answer. Problems marked with ✎ require you to make a drawing or graph as part of your solution. Asterisks indicate the level of difficulty of the problem. Problems with no * are considered to be the least difficult. A single * marks moderately difficult problems. Two ** indicate more difficult problems.

### 7.2 Torque: A new physical quantity

1. Determine the torques about the axis of rotation $P$ produced by each of the four forces shown in **Figure P7.1**. All forces have magnitudes of 120 N and are exerted a distance of 2.0 m from $P$ on some unshown object O.

**Figure P7.1**

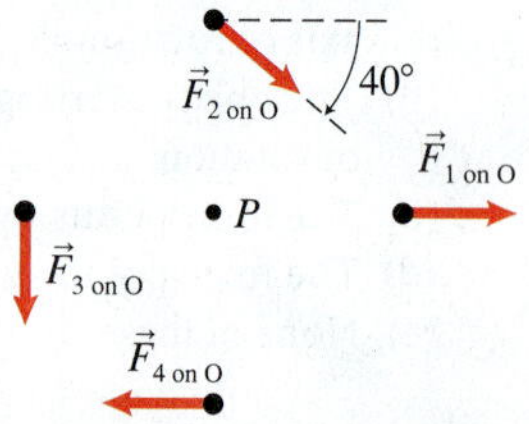


2. * **EST** Your hand holds a liter of milk while your arm is bent at the elbow in a 90° angle. Estimate the torque caused by the milk on your arm about the elbow joint. Indicate all numbers used in your calculations. This is an estimate, and your answer may differ by 10 to 50% from the answers of others.
3. * **EST Body torque** You hold a 4.0-kg computer. Estimate the torques exerted on your forearm about the elbow joint caused by the downward force exerted by the computer on the forearm and the upward 340-N force exerted by the biceps muscle on the forearm. Ignore the mass of the arm. Indicate any assumptions you make.
4. Three 200-N forces are exerted on the beam shown in **Figure P7.4**. (a) Determine the torques about the axis of rotation on the left produced by forces $\vec{F}_{1\text{ on B}}$ and $\vec{F}_{2\text{ on B}}$. (b) At what distance from the axis of rotation must $\vec{F}_{3\text{ on B}}$ be exerted to cause a torque that balances those produced by $\vec{F}_{1\text{ on B}}$ and $\vec{F}_{2\text{ on B}}$?

Figure P7.4

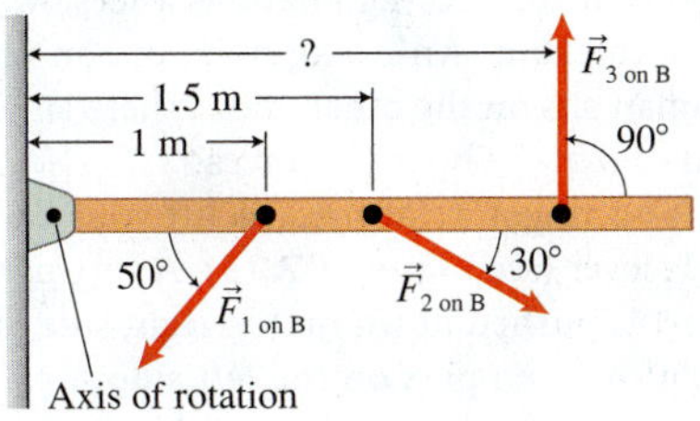


5. *A 2.0-m-long, 15-kg ladder is resting against a house wall, making a 30° angle with the vertical wall. The coefficient of static friction between the ladder feet and the ground is 0.40, and between the top of the ladder and the wall the coefficient is 0.00. Make a list of the physical quantities you can determine or estimate using this information and calculate them.

### 7.3 Conditions of equilibrium

6. Three friends tie three ropes in a knot and pull on the ropes in different directions. Adrienne (rope 1) exerts a 20-N force in the positive $x$-direction, and Jim (rope 2) exerts a 40-N force at an angle 53° above the negative $x$-axis. Luis (rope 3) exerts a force that balances the first two so that the knot does not move. (a) Construct a force diagram for the knot. (b) Use equilibrium conditions to write equations that can be used to determine $F_{\text{L on K}\,x}$ and $F_{\text{L on K}\,y}$. (c) Use equilibrium conditions to write equations that can be used to determine the magnitude and direction of $\vec{F}_{\text{L on K}}$.
7. Adrienne from Problem 6 now exerts a 100-N force $\vec{F}_{\text{A on K}}$ that points 30° below the positive $x$-axis and Jim exerts a 150-N force in the negative $y$-direction. How hard and in what direction does Luis now have to pull the knot so that it remains in equilibrium?
8. *Kate joins Jim, Luis, and Adrienne in the rope-pulling exercise described in the previous two problems. This time, they tie four ropes to a ring. The three friends each pull on one rope, exerting the following forces: $\vec{T}_{1 \text{ on R}}$ (50 N in the positive $y$-direction), $\vec{T}_{2 \text{ on R}}$ (20 N, 25° above the negative $x$-axis), and $\vec{T}_{3 \text{ on R}}$ (70 N, 70° below the negative $x$-axis). Kate pulls rope 4, exerting a force $\vec{T}_{4 \text{ on R}}$ so that the ring remains in equilibrium. (a) Construct a force diagram for the ring. (b) Use the first condition of equilibrium to write two equations that can be used to determine $T_{4 \text{ on R}\,x}$ and $T_{4 \text{ on R}\,y}$ (c) Solve these equations and determine the magnitude and direction of $\vec{T}_{4 \text{ on R}}$.
9. You hang a light in front of your house using an elaborate system to keep the 1.2-kg light in static equilibrium (see **Figure P7.9**). What are the magnitudes of the forces that the ropes must exert on the knot connecting the three ropes if $\theta_2 = 37°$ and $\theta_3 = 0°$?

Figure P7.9

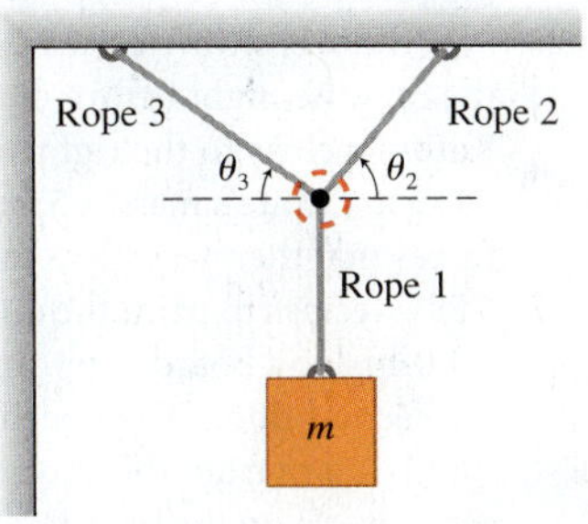


10. *Find the values of the forces the ropes exert on the knot if you replace the light in Problem 7.9 with a heavier 12-kg object and the ropes make angles of $\theta_2 = 63°$ and $\theta_3 = 45°$ (see Figure P7.9).
11. Redraw Figure P7.9 with $\theta_2 = 50°$ and $\theta_3 = 0°$. Rope 2 is found to exert a 100-N force on the knot. Determine $m$ and the magnitudes of the forces that the other two ropes exert on the knot.
12. *Determine the masses $m_1$ and $m_2$ of the two objects shown in **Figure P7.12** if the force exerted by the horizontal cable on the knot is 64 N.

Figure P7.12

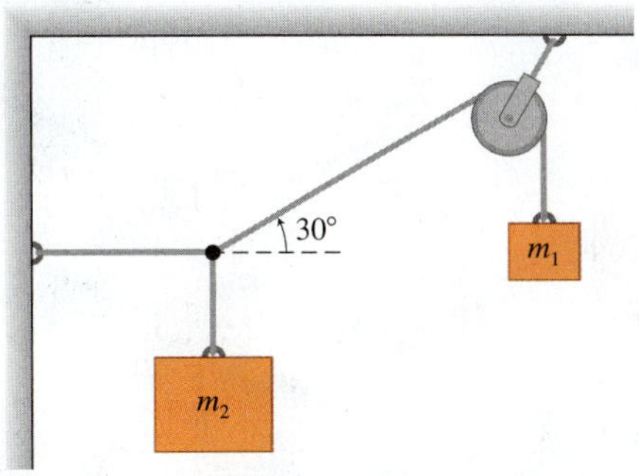


13. ***Lifting an engine*** You work in a machine shop and need to move a huge 640-kg engine up and to the left in order to slide a cart under it. You use the system shown in **Figure P7.13**. How hard and in what direction do you need to pull on rope 2 if the angle between rope 1 and the horizontal is $\theta_1 = 60°$?

Figure P7.13

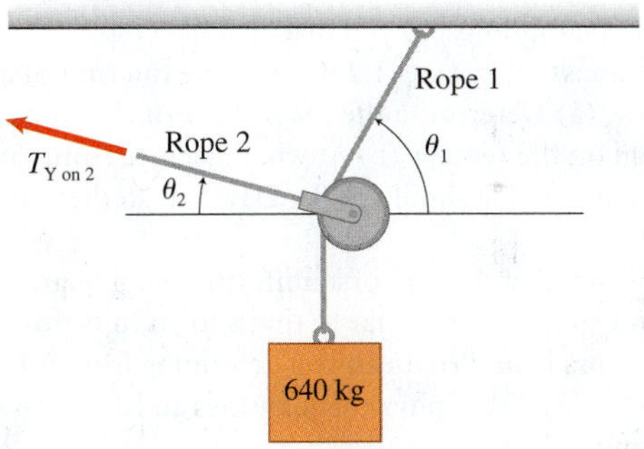


14. ***More lifting*** You exert a 630-N force on rope 2 in the previous problem (Figure P7.13). Write the two equations ($x$ and $y$) for the first condition of equilibrium using the pulley as the object of interest for a force diagram. Calculate $\theta_1$ and $\theta_2$. You may need to use the identity $(\sin\theta)^2 + (\cos\theta)^2 = 1$.
15. ***Even more lifting*** A pulley system shown in **Figure P7.15** will allow you to lift heavy objects in the machine shop by exerting a relatively small force. (a) Construct a force diagram for each pulley. (b) Use the equations of equilibrium and the force diagrams to determine $T_1$, $T_2$, $T_3$, and $T_4$.

Figure P7.15

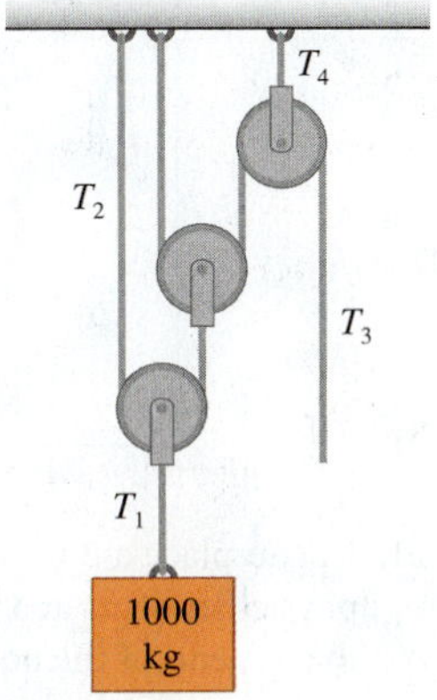


16. **Tightrope walking** A tightrope walker wonders if her rope is safe. Her mass is 60 kg and the length of the rope is about 20 m. The rope will break if its tension

exceeds 6700 N. What is the smallest angle at which the rope can bend up from the horizontal on either side of her to avoid breaking?

17. **Lifting patients** An apparatus to lift hospital patients sitting at the sides of their beds is shown in **Figure P7.17**. At what angle above the horizontal does the rope going under the pulley bend while supporting the 78-kg person hanging from the pulley?

**Figure P7.17**

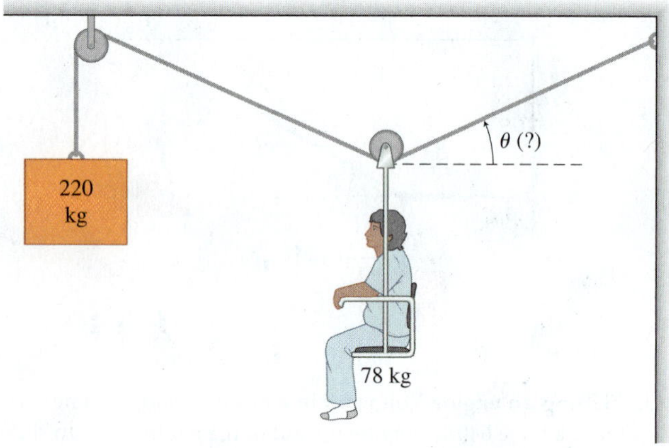


18. A mutineer on Captain Bligh's ship is made to "walk the plank." The plank, which extends 3.0 m beyond its support, will break if subjected to a torque greater than 3300 N·m. Will the sailor break the plank before stepping off its end? Explain. What assumptions did you make?

19. Brett (mass 70 kg) sits 1.2 m from the fulcrum of a uniform seesaw. (a) Determine the magnitude of the torque exerted by him on the seesaw. (b) At what distance from the fulcrum on the other side should 54-kg Dawn sit so that the seesaw is horizontal?

20. * You stand at the end of a uniform diving board a distance $d$ from support 2 (similar to that shown in **Figure P7.20**). Your mass is $m$. What can you determine from this information? Make a list of physical quantities and show how you will determine them.

**Figure P7.20**

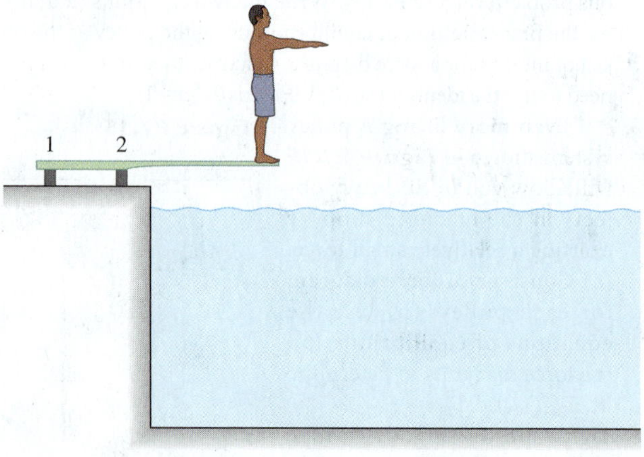


21. * You place a 3.0-m-long board across a chair to seat three physics students at a party at your house. If 70-kg Dan sits on the left end of the board and 50-kg Tahreen on the right end of the board, where should 54-kg Komila sit to keep the board stable? What assumptions did you make?

22. After dinner (see Problem 7.21), two guests decide to use the same 3.0-m-long 5.0-kg board as a seesaw, using a small bench as a fulcrum. An 82-kg man sits on one end and a 64-kg woman sits on the other end. Where should the bench be located so that the board balances?

23. **Car jack** You've got a flat tire. To lift your car, you make a homemade lever (see **Figure P7.23**). A very light 1.6-m-long handle part is pushed down on the right side of the fulcrum and a 0.050-m-long part on the left side supports the back of the car. How hard must you push down on the handle so that the lever exerts an 8000-N force to lift the back of the car?

**Figure P7.23**

24. * **Mobile** You are building a toy mobile, copying the design shown in **Figure P7.24**. Object A has a 1.0-kg mass. What should be the mass of object B? The numbers in Figure P7.24 indicate the relative lengths of the rods on each side of their supporting cords.

**Figure P7.24**

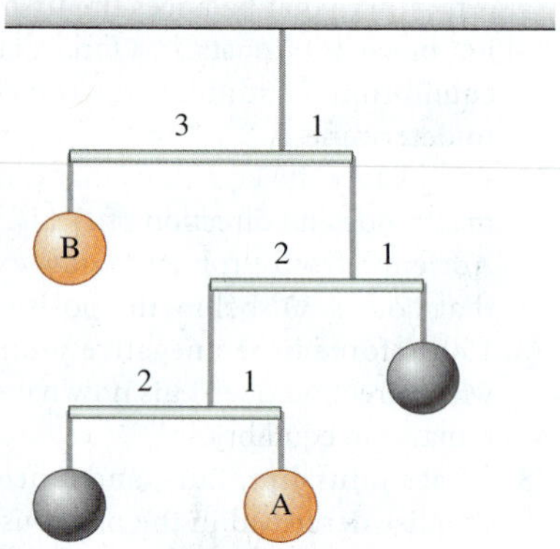


25. * **Another mobile** You are building a toy mobile similar to that shown in Figure P7.24 but with different dimensions and replacing the objects with cups. The bottom rod is 20 cm long, the middle rod is 15 cm long, and the top rod is 8 cm long. You put one penny in the bottom left cup, three pennies in the bottom right cup, eleven pennies in the middle right cup, and five pennies in the top left cup. (a) Draw a force diagram for each rod. (b) Determine the cord attachment points and lengths on each side for each rod. (c) What assumptions did you make in order to solve the problem?

26. ***EST** Kate is sitting on a 1.5-m-wide porch swing. Because of the rain the night before, the left side of the swing is wet, and Kate sits close to the right side. The mass of the swing seat is 10 kg and Kate's mass is 55 kg. Estimate the magnitudes of the forces that the two supporting cables exert on the swing.

27. *Ray decides to paint the outside of his uncle's house. He uses a 4.0-m-long board supported by cables at each end to paint the second floor. The board has a mass of 21 kg. Ray (70 kg) stands 1.0 m from the left cable. What are the forces that each cable exerts on the board?

28. *The fulcrum of a uniform 20-kg seesaw that is 4.0 m long is located 2.5 m from one end. A 30-kg child sits on the long end. Determine the mass a person at the other end would have to be in order to balance the seesaw.

29. *A 2.0-m-long uniform beam of mass 8.0 kg supports a 12.0-kg bag of vegetables at one end and a 6.0-kg bag of fruit at the other end. At what distance from the vegetables should the beam rest on your shoulder to balance? What assumptions did you make?

30. *A uniform beam of length $l$ and mass $m$ supports a bag of mass $m_1$ at the left end, another bag of mass $m_2$ at the right end, and a third bag $m_3$ at a distance $l_3$ from the left end ($l_3 < 0.5l$). At what distance from the left end should you support the beam so that it balances?

## 7.4 Center of mass

31. *An 80-kg person stands at one end of a 130-kg boat. He then walks to the other end of the boat so that the boat moves 80 cm with respect to the bottom of the lake. (a) What is the length of the boat? (b) How much did the center of mass of the person-boat system move when the person walked from one end to the other?

32. EST Two people (50 kg and 75 kg) holding hands stand on rollerblades 1.0 m apart. (a) Estimate the location of their center of mass. (b) The two people push off each other and roll apart so the distance between them is now 4.0 m. Estimate the new location of the center of mass. What assumptions did you make?

33. *You have a disk of radius $R$ with a circular hole of radius $r$ cut a distance $a$ from the center of the disk. Where is the disk's center of mass?

34. *A person whose height is 1.88 m is lying on a light board placed on two scales so that scale 1 is under the person's head and scale 2 is under the person's feet. Scale 1 reads 48.3 kg and scale 2 reads 39.3 kg. Where is the center of mass of the person?

35. **EST Estimate the location of the center of mass of the person described in the previous problem when he bends over and touches the floor with his hands.

36. *A seesaw has a mass of 30 kg, a length of 3.0 m, and fulcrum beneath its midpoint. It is balanced when a 60-kg person sits on one end and a 75-kg person sits on the other end. Locate the center of mass of the seesaw. Where is the center of mass of a uniform seesaw that is 3.0 m long and has a mass of 30 kg if two people of masses 60 kg and 75 kg sit on its ends?

37. **Find the center of mass of an L-shaped object. The vertical leg has a mass of $m_a$ of length $a$ and the horizontal leg has a mass of $m_b$ of length $b$. Both legs have the same width $w$, which is much smaller than $a$ or $b$.

38. **You have a 10-kg table with each leg of mass 1.0 kg—total mass 14 kg. If you place a 5.0-kg pot of soup in the back right corner of the table, where is the table's center of mass?

## 7.5 Skills for analyzing situations using equilibrium conditions

39. ***Using biceps to hold a child*** A man is holding a 16-kg child using both hands with his elbows bent in a 90° angle. The biceps muscle provides the positive torque he needs to support the child. Determine the force that each of his biceps muscles must exert on the forearm in order to hold the child safely in this position. Ignore the triceps muscle and the mass of the arm.

40. *BIO **Using triceps to push a table** A man pushes on a table exerting a 20-N downward force with his hand. Determine the force that his triceps muscle must exert on his forearm in order to balance the upward force that the table exerts on his hand. Ignore the biceps muscle and the mass of the arm. If you did not ignore the mass of the arm, would the force you determined be smaller or larger? Explain.

41. *BIO **Using biceps to hold a barbell** Find the force that the biceps muscle shown in Figure P7.26 exerts on the forearm when you lift a 16-kg barbell with your hand. Also determine the force that the bone in the upper arm (the humerus) exerts on the bone in the forearm at the elbow joint. The mass of the forearm is about 5.0 kg and its center of mass is 16 cm from the elbow joint. Ignore the triceps muscle.

42. **Leg support** A person's broken leg is kept in place by the apparatus shown in **Figure P7.42**. If the rope pulling on the leg exerts a 120-N force on it, how massive should be the block hanging from the rope that passes over the pulley?

**Figure P7.42**

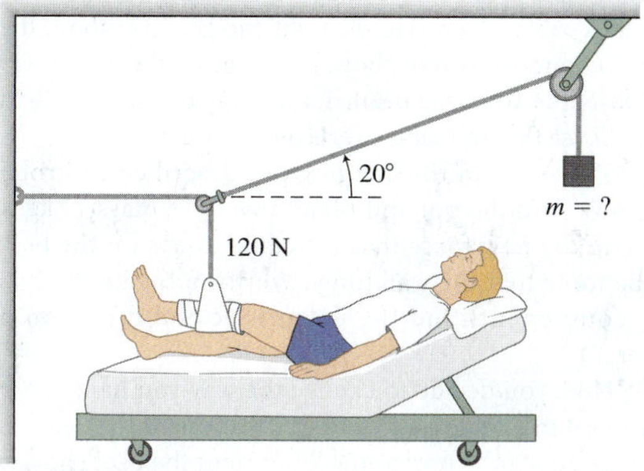


43. *BIO **Hamstring** You are exercising your hamstring muscle (the large muscle in the back of the thigh). You use an elastic cord attached to a hook on the wall while keeping your leg in a bent position (**Figure P7.43**). Determine the magnitude of the tension force $\vec{T}_{\text{H on L}}$ exerted by the hamstring muscles on the leg and the magnitude of compression force $\vec{F}_{\text{F on B}}$ at the knee joint that the femur exerts on the calf bone. The cord exerts a 20-lb force $\vec{F}_{\text{C on F}}$ on the foot.

**Figure P7.43**

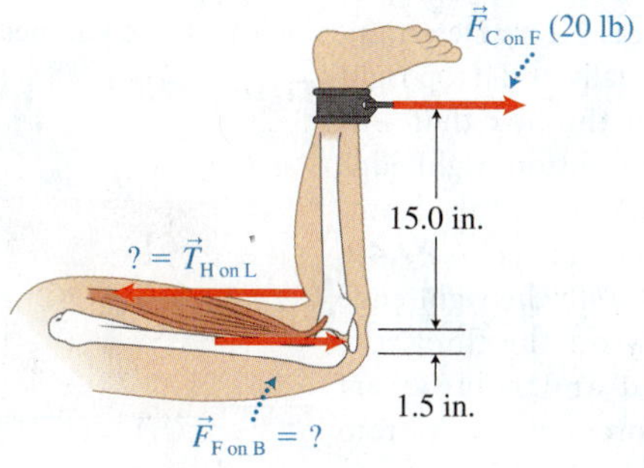


44. *You decide to hang a new 10-kg flowerpot using the arrangement shown in **Figure P7.44**. Can you use a slanted rope attached from the wall to the end of the beam if that rope breaks when the tension exceeds 170 N? The mass of the beam is not known but it looks light.

**Figure P7.44**

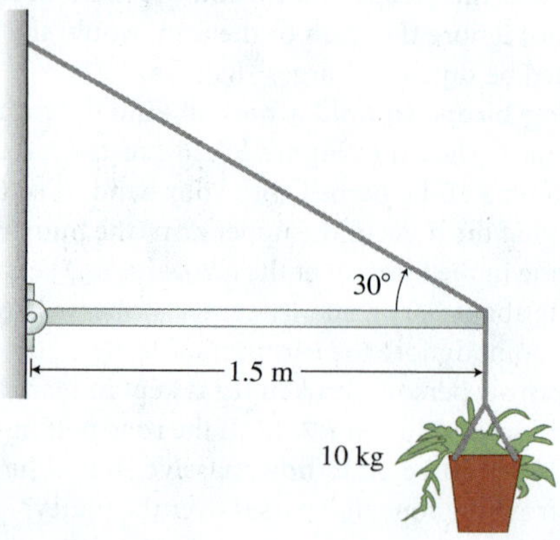


45. *You decide to hang another plant from a 1.5-m-long 2.0-kg horizontal beam that is attached by a hinge to the wall on the left. A cable attached to the right end goes 37° above the beam to a connecting point above the hinge on the wall. You hang a 100-N pot from the beam 1.4 m away from the wall. What is the force that the cable exerts on the beam?

46. **The plant in the hanging pot described in Problem 45 grows, and the pot and plant now have mass 12 kg. Determine the new force that the cable exerts on the beam and the force that the wall hinge exerts on the beam (its $x$- and $y$-components and the magnitude and direction of that force).

47. **Now you decide to change the way you hang the pot described in Problems 45 and 46. You orient the beam at a 37° angle above the horizontal and orient the cable horizontally from the wall to the end of the beam. The beam still holds the 2.0-kg pot and plant hanging 0.1 m from its end. Now determine the force that the cable exerts on the beam and the force that the wall hinge exerts on the beam (its $x$- and $y$-components and the magnitude and direction of that force).

48. ** **Diving board** The diving board shown in Figure P7.20 has a mass of 28 kg and its center of mass is at the board's geometrical center. Determine the forces that support posts 1 and 2 (separated by 1.4 m) exert on the board when a 60-kg person stands on the end of the board 2.8 m from support post 2.

49. **A uniform cubical 200-kg box sits on the floor with its bottom left edge pressing against a ridge. The length $L$ of a side of the box is 1.2 m. Determine the least force you need to exert horizontally at the top right edge of the box that will cause its bottom right edge to be slightly off the floor, as shown in **Figure P7.49**. (Note: With the right edge slightly off the floor, the ground and ridge exert their forces on the bottom left edge of the box.)

**Figure P7.49**

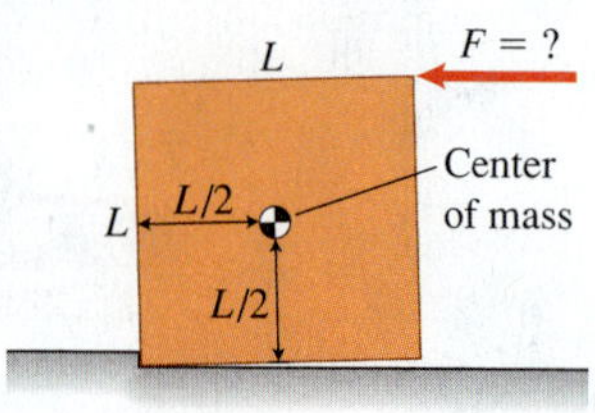


50. *If the force $F$ shown in Figure P7.49 is 840 N and the bottom right edge of the box is slightly off the ground, what is the mass of the cubical box of side 1.2 m?

51. We know from the second condition of equilibrium that if two different magnitude forces are exerted on the same object, their rotational effects can cancel if their torques are the same magnitude but opposite sign. For example, you can lift a heavy boulder by exerting a force much smaller than the weight of the boulder. Design an experiment where you can lift a 100-kg rock by exerting a downward 100-N push (it is much easier to push than to pull).

52. *What mechanical work must you do to lift a log that is 3.0 m long and has a mass of 100 kg from the horizontal to a vertical position? [Hint: Use the work-energy principle.]

## 7.6 Stability of equilibrium

53. * A 70-g meter stick has a 30-g piece of modeling clay attached to the end. Where should you drill a hole in the meter stick so that you can hang the stick horizontally in equilibrium on a nail in the wall? Draw a picture to help explain your decision.

54. **You are trying to tilt a very tall refrigerator (2.0 m high, 1.0 m deep, 1.4 m wide, and 100 kg) so that your friend can put a blanket underneath to slide it out of the kitchen. Determine the force that you need to exert on the front of the refrigerator at the start of its tipping. You push horizontally 1.4 m above the floor.

55. *You have an Atwood machine with two blocks each of mass $m$ attached to the ends of a string of length $l$. The string passes over a frictionless pulley down to the blocks hanging on each side. While pulling down on one block, you release it. Both blocks continue to move at constant speed, one up and the other down. Is the system still in equilibrium? Find the vertical component of the center of mass of the two-block system. Indicate all of your assumptions and the coordinate system used.

56. *EST You stand sideways in a moving train. Estimate how far apart you should keep your feet so that when the train accelerates at 2.0 m/s$^2$ you can still stand without holding anything. List all your assumptions.

## 7.7 Static equilibrium: Putting it all together

57. **BIO **Lift with bent legs** You injure your back at work lifting a 420-N radiator. To understand how it happened, you model your back as a weightless beam (**Figure P7.57**), analogous

**Figure P7.57**

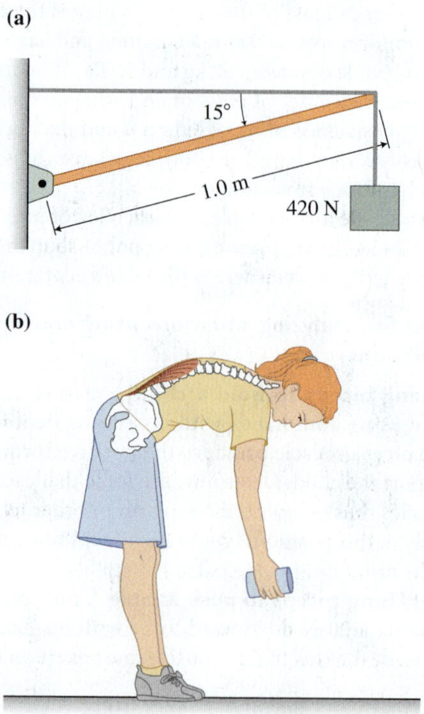

to the backbone of a person in a bent position when lifting an object. (a) Determine the tension force that the horizontal cable exerts on the beam (which is analogous to the force the back muscle exerts on the backbone) and the force that the wall exerts on the beam at the hinge (which is analogous to the force that a disk in the lower back exerts on the backbone). (b) Why do doctors recommend lifting objects with the legs bent?

58. * *Determine the tension force that the horizontal cable exerts on the beam in Figure P7.57, but with the beam tilted at 30° rather than 15°. The cable remains horizontal.

59. * *Determine the tension force that the horizontal cable exerts on the beam in **Figure P7.59** if the horizontal cable is moved down (compared to Figure P7.57) so that its right end attaches to the beam 0.30 m from its top right end. The cable remains horizontal and the beam is tilted at 15°.

**Figure P7.59**

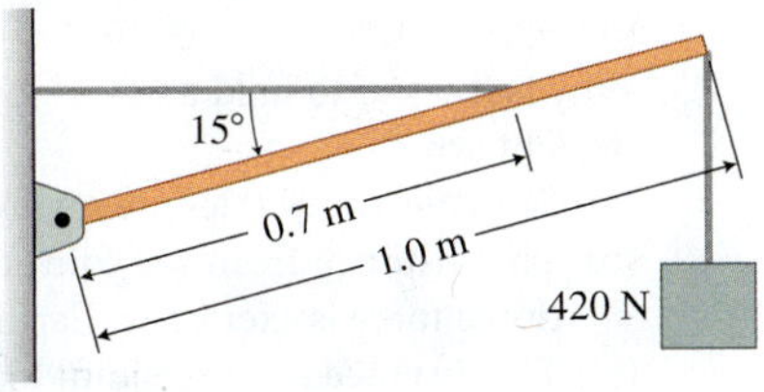


60. *BIO **Barbell lift I** A woman lifts a 3.6-kg barbell in each hand with her arm in a horizontal position at the side of her body and holds it there for 3 s (see **Figure P7.60**). What force does the deltoid muscle in her shoulder exert on the humerus bone while holding the barbell? The deltoid attaches 13 cm from the shoulder joint and makes a 13° angle with the humerus. The barbell in her hand is 0.55 m from the shoulder joint, and the center of mass of her 4.0-kg arm is 0.24 m from the joint.

**Figure P7.60**

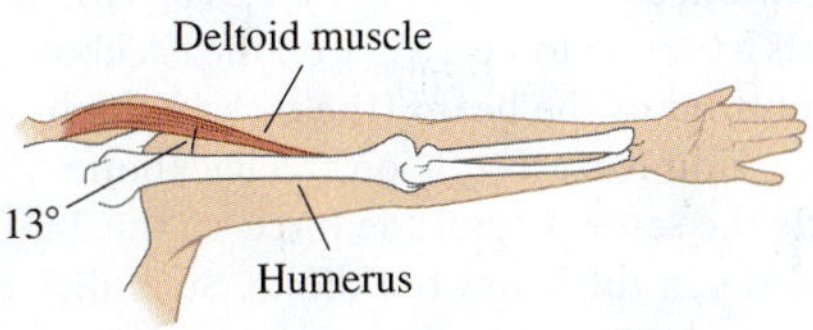


61. * *BIO **Barbell lift II** Repeat the previous problem with a 7.2-kg barbell. Determine both the force that the deltoid exerts on the humerus and the force that the lifter's shoulder joint exerts on her humerus.

62. *BIO **Facemask penalty** The head of a football running back (see **Figure P7.62**) can be considered as a lever with the vertebra at the bottom of the skull as a fulcrum (the axis of rotation). The center of mass is about 0.025 m in front of the axis of rotation. The torque caused by the force that Earth exerts on the 8.0-kg head/helmet is balanced by the torque caused by the downward forces exerted by a complex muscle system in the neck. That muscle system includes the trapezius and levator scapulae muscles, among others (effectively 0.057 m from the axis of rotation). (a) Determine the magnitude of the force exerted by the neck muscle system pulling down to balance the torque caused by the force exerted by Earth on the head. (b) If an opposing player exerts a downward 180-N (40-lb) force on the facemask, what muscle force would these neck muscles now need to exert to keep the head in equilibrium?

**Figure P7.62**

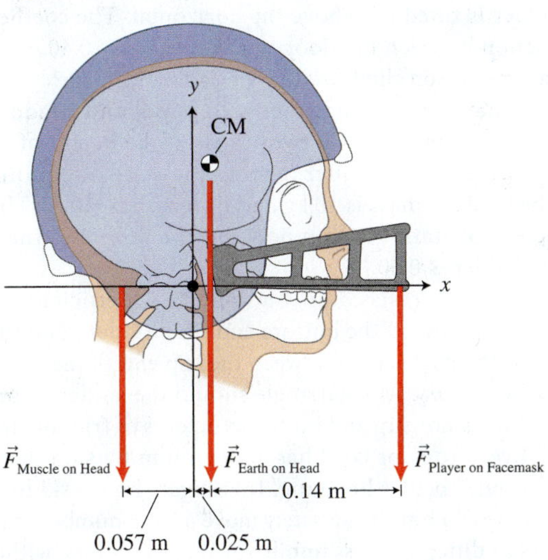


## General Problems

63. * *Design two experiments to determine the mass of a ruler, using different methods. Your available materials are the ruler, a spring, and a set of objects of standard mass: 50 g, 100 g, and 200 g. One of the methods should involve your knowledge of static equilibrium. After you design and perform the experiment, decide whether the two methods give you the same or different results.

64. *A board of mass $m$ and length $l$ is placed on a horizontal tabletop. The coefficient of static friction between the board and the table is $\mu$. How far from the edge of the tabletop can one extend the board before it falls off?

65. * ***Tightrope walker** A 60-kg 1.6-m-tall tightrope walker stands on a tightrope. (a) In his hands he holds a 10-kg 2.0-m-long horizontal bar. At each end of the bar are two 5.0-kg balls hanging from 0.50-m-long strings. How much does this apparatus lower his center of mass? (b) How long should the strings be so that the center of mass of the walker-bar-hanging ball system is at the level of the rope? Indicate all assumptions made for each part of the problem.

66. * ***Lecturing on a beam** A 70-kg professor sits on a 20-kg beam while lecturing (**Figure P7.66**). A rope attached to the end of the beam passes over a pulley and down to a harness that wraps around the professor (the professor is supported partly by the beam and partly by the harness). The professor sits in the middle of the beam. Determine the force that the rope exerts on the harness and professor and the force that the beam exerts on the professor.

**Figure P7.66**

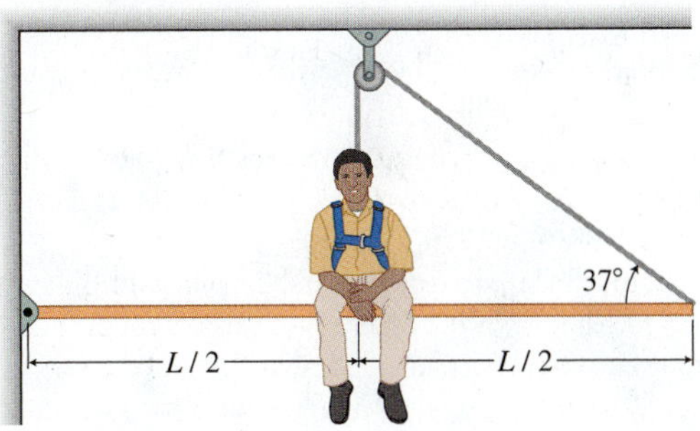

67. **A 70-kg person stands on a 6.0-m-long 50-kg ladder. The ladder is tilted 60° above the horizontal. The coefficient of friction between the floor and the ladder is 0.40. How high can the person climb without the ladder slipping?

68. **What is a safe angle between a wall and a ladder for a 60-kg painter to climb two-thirds of the height of the ladder without the ladder leaving the state of equilibrium? The ladder's mass is 10 kg and its length is 6.0 m. The coefficient of static friction between the floor and the feet of the ladder is 0.50.

69. **A ladder rests against a wall. The coefficient of static friction between the bottom end of the ladder and the floor is $\mu_1$; the coefficient between the top end of the ladder and the wall is $\mu_2$. At what angle should the ladder be oriented so it does not slip and both coefficients of friction are 0.50?

70. **Every rope or cord has a maximum tension that it can withstand before breaking. Investigate how a ski lift works and explain how it can safely move a large number of passengers of different mass uphill during peak hours, without the cord that carries the chairs breaking.

## Reading Passage Problems

BIO **Muscles work in pairs** Skeletal muscles produce movements by pulling on tendons, which in turn pull on bones. Usually, a muscle is attached to two bones via a tendon on each end of the muscle. When the muscle contracts, it moves one bone toward the other. The other bone remains in nearly the original position. The point where a muscle tendon is attached to the stationary bone is called the *origin*. The point where the other muscle tendon is attached to the movable bone is called the *insertion*. The origin is like the part of a door spring that is attached to the doorframe. The insertion is similar to the part of the spring that is attached to the movable door.

During movement, bones act as levers and joints act as axes of rotation for these levers. Most movements require several skeletal muscles working in groups, because a muscle can only exert a pull and not a push. In addition, most skeletal muscles are arranged in opposing pairs at joints. Muscles that bring two limbs together are called flexor muscles (such as the biceps muscle in the upper arm in **Figure 7.26**). Those that cause the limb to extend outward are called extensor muscles (such as the triceps muscle in the upper arm). The flexor muscle is used when you hold a heavy object in your hand; the extensor muscle can be used, for example, to extend your arm when you throw a ball.

**Figure 7.26** Muscles often come in flexor-extensor pairs.

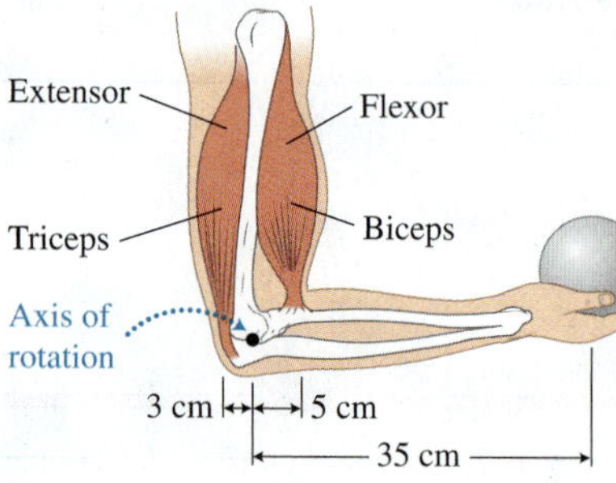


71. You hold a 10-lb ball in your hand with your forearm horizontal, forming a 90° angle with the upper arm (Figure 7.26). Which type of muscle produces the torque that allows you to hold the bell?
(a) Flexor muscle in the upper arm
(b) Extensor muscle in the upper arm.
(c) Flexor muscle in the forearm
(d) Extensor muscle in the forearm

72. In Figure 7.26, how far in centimeters from the axis of rotation are the forces that the ball exerts on the hand, that the biceps exerts on your forearm, and that the upper arm exerts on your forearm at the elbow joint?
(a) 0, 5, 35 (b) 35, 5, 0 (c) 35, 5, 3
(d) 35, 5, −3 (e) 30, 5, 0

73. Why is it easier to hold a heavy object using a bent arm than a straight arm?
(a) More flexor muscles are involved.
(b) The distance from the joint to the place where gravitational force is exerted by Earth on the object is smaller.
(c) The distance from the joint to the place where force is exerted by the object on the hand is smaller.
(d) There are two possible axes of rotation instead of one.

74. Why are muscles arranged in pairs at joints?
(a) Two muscles can produce a bigger torque than one.
(b) One can produce a positive torque and the other a negative torque.
(c) One muscle can pull on the bone and the other can push.
(d) Both a and b are true.

BIO **Improper lifting and the back** A careful study of human anatomy allows medical researchers to use the conditions of equilibrium to estimate the internal forces that body parts exert on each other while a person lifts in a bent position (**Figure 7.27a**). Suppose an 800-N (180-lb) person lifts a 220-N (50-lb) barbell in a bent position, as shown in Figure 7.27b. The cable (the back muscle) exerts a tension force $\vec{T}_{\text{M on B}}$ on the backbone and the support at the bottom of the beam (the disk in the lower back) exerts a compression force $\vec{F}_{\text{D on B}}$ on the backbone. The backbone in turn exerts the same magnitude force on the 2.5-cm-diameter fluid-filled disks in the lower backbone. Such disk compression can cause serious back problems. A force diagram of this situation is shown in Figure 7.27c. The magnitude of the gravitational force $\vec{F}_{\text{E on B}}$ that Earth exerts on the center of mass of the upper stomach-chest region is 300 N. Earth exerts a 380-N force on the head, arms, and 220-N barbell held in the hands. Using the conditions of equilibrium, we estimate that the back muscle exerts a 3400-N (760-lb) force $\vec{T}_{\text{M on B}}$ on the backbone and that the disk in the lower back exerts a 3700-N (830-lb) force $\vec{F}_{\text{D on B}}$ on the backbone. This is like supporting a grand piano on the 2.5-cm-diameter disk.

75. Rank in order the magnitudes of the distances of the four forces exerted on the backbone with respect to the joint (see Figure 7.27c), with the largest distance listed first.
(a) $1 > 3 > 2 > 4$ (b) $4 > 2 = 3 > 1$
(c) $4 > 3 > 2 > 1$ (d) $2 > 3 > 1 > 4$
(e) $1 > 2 > 3 > 4$

76. Rank in order the magnitudes of the torques caused by the four forces exerted on the backbone (see Figure 7.27c), with the largest torque listed first.
(a) $1 > 2 > 3 > 4$ (b) $2 = 3 > 1 > 4$
(c) $3 > 2 > 1 > 4$ (d) $2 > 1 > 3 > 4$
(e) $1 = 2 = 3 = 4$

**Figure 7.27** Analysis of a person's backbone when lifting from a bent position.

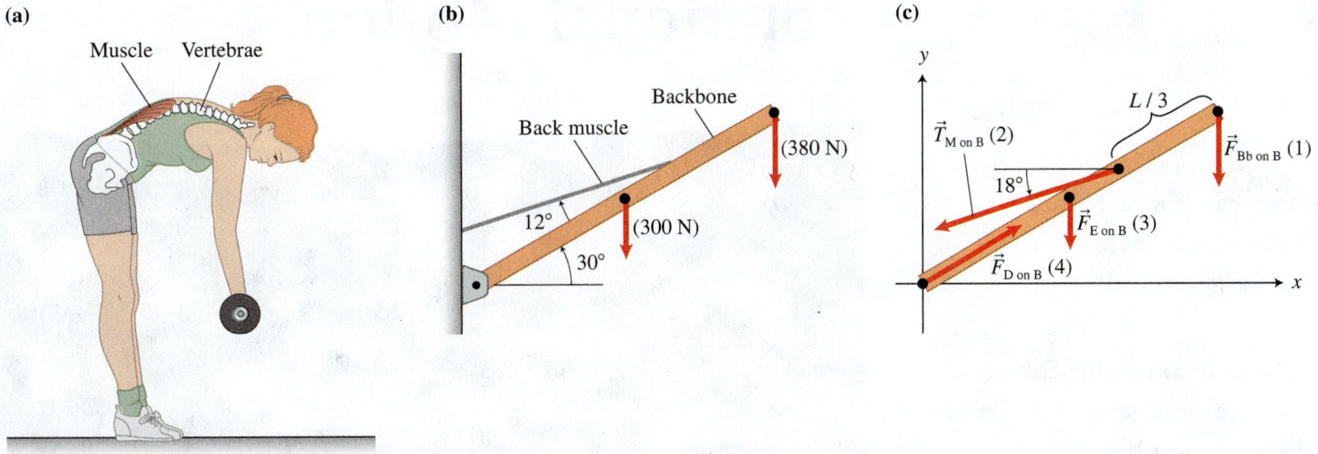


77. What are the signs of the torques caused by forces 1, 2, 3, and 4, respectively, about the origin of the coordinate system shown in Figure 7.27c?
(a) +, +, +, + (b) −, +, −, 0
(c) +, −, +, 0 (d) −, −, −, 0
(e) +, −, +, −

78. Which expression below best describes the torque caused by force $F_4 = F_{DonB}$, the force that the disk in the lower back exerts on the backbone of length $L$?
(a) 0 (b) $F_4(2L/3)\sin 12°$
(c) $F_4(L)\cos 30°$ (d) $-F_4(2L/3)\sin 12°$
(e) $-F_4(L)\cos 30°$

79. Which expression below best describes the torque caused by force $F_3 = F_{E\ on\ B}$, the force that Earth exerts on the upper body at its center of mass for the backbone of length $L$?
(a) 0 (b) $F_3(2L/3)\sin 12°$
(c) $F_3(L/2)\cos 30°$ (d) $-F_3(2L/3)\sin 12°$
(e) $-F_3(L/2)\cos 30°$

80. Which expression below best describes the torque caused by force $F_2 = T_{M\ on\ B}$ exerted by the muscle on the backbone?
(a) 0 (b) $F_2(2L/3)\sin 12°$
(c) $F_2(L)\cos 30°$ (d) $-F_2(2L/3)\sin 12°$
(e) $-F_2(L)\cos 30°$

# 8 Rotational Motion

How can a star rotate 1000 times faster than a merry-go-round?

Why is it more difficult to balance on a stopped bike than on a moving bike?

How is the Moon slowing Earth's rate of rotation?

**Be sure you know how to:**

- Draw a force diagram for a system (Section 2.1).
- Determine the torque produced by a force (Section 7.2).
- Apply conditions of static equilibrium for a rigid body (Section 7.3).

In 1967, a group of astrophysicists from Cambridge University in England was looking for quasars using an enormous radio telescope. Jocelyn Bell, a physics graduate student, operated the radio telescope and analyzed the nearly 30 meters of printed radio telescope data that were collected daily. Bell noticed a series of regular radio pulses in the midst of a lot of receiver noise. It looked like somebody was sending a radio message, turning the signal on and off every 1.33 seconds. This is an incredibly small time for astronomical objects. At first, Bell's advisor Anthony Hewish believed that they had found signals from extraterrestrial life. This idea received considerable support from the scientific community, although it eventually proved to be an incorrect assertion. The group had, in fact, discovered a new class of astronomical objects, called pulsars, which emit radio signals every second or so. The study of rotational motion explains how pulsars can emit signals so rapidly. In later chapters we will learn the mechanism behind them.

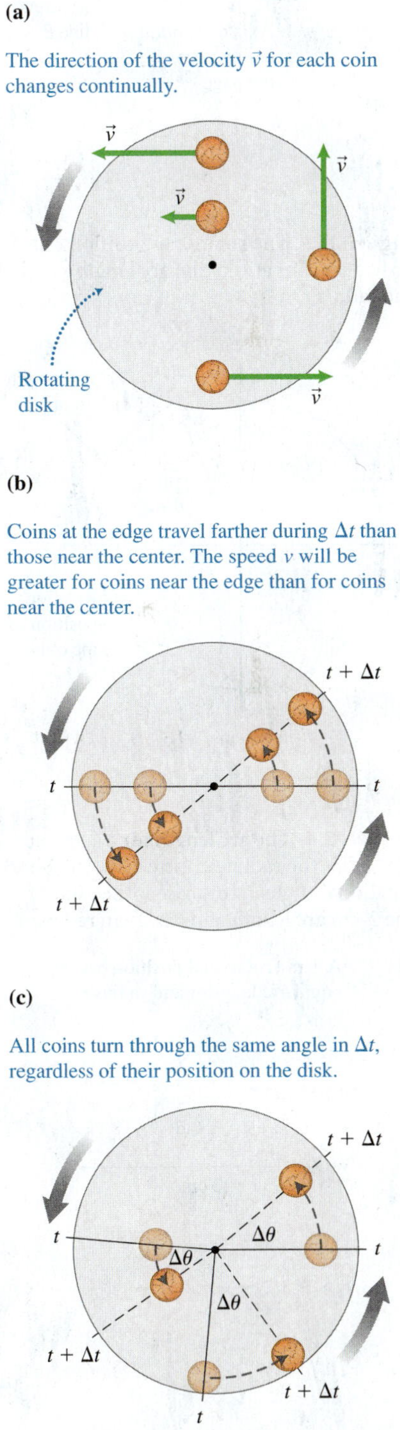

(a)
The direction of the velocity $\vec{v}$ for each coin changes continually.
$\vec{v}$
Rotating disk
(b)
Coins at the edge travel farther during $\Delta t$ than those near the center. The speed $v$ will be greater for coins near the edge than for coins near the center.
$t + \Delta t$
$t$
(c)
All coins turn through the same angle in $\Delta t$, regardless of their position on the disk.
$\Delta\theta$
$t + \Delta t$
$t$


**Figure 8.1** Top views comparing the velocities of coins traveling on a rotating disk.

**In the last chapter,** we learned about the torque that a force produces on a rigid body. However, we only analyzed rigid bodies that were in static equilibrium—they remained at rest. In many cases, however, objects do not remain at rest when torques are exerted—they rotate. For example, the human leg rotates slightly around the hip joint while a person walks, and a car tire rotates around the axle as the car moves. In this chapter, we will learn how to describe, explain, and predict such motions.

## 8.1 Rotational kinematics

In order to understand the motion of rotating rigid bodies, we will follow the same strategy that we used for linear motion. We start by investigating how to describe rotational motion and then develop rules that explain how forces and torques cause objects to rotate in the way they do. Ultimately, this investigation will enable us to understand many aspects of the natural and human-made world, from why a bicycle is so stable when moving to how the gravitational pull of the Moon slows the rotation of Earth.

One of the simplest examples of a rotating rigid body is a disk. Suppose you stand at a lab bench with a rotating disk on top of it. You wish to describe the counterclockwise motion of the disk quantitatively. This is trickier than it might seem at first. When we investigated the motion of point-like objects, we did not have to specify which part of the object we were describing, since the object was located at a single point. With a rigid body, there are infinitely many points to choose from. For example, imagine that you place small coins at different locations on the disk, as shown in **Figure 8.1a**. As the disk turns, you observe that the direction of the velocity of each coin changes continually (see the coins on the outer edge of the disk in Figure 8.1a). In addition, a coin that sits closer to the edge moves faster and covers a longer distance during a particular time interval than a coin closer to the center (Figure 8.1b). This means that different parts of the disk move not only in different directions, but also at different speeds relative to you.

On the other hand, there are similarities between the motions of different points on a rotating rigid body. Perhaps we can find physical quantities that have the same value regardless of which point we consider. In Figure 8.1c, we see that during a particular time interval, all coins at the different points on the rotating disk turn through the same angle. Perhaps we should describe the rotational position of a rigid body using an angle.

### Rotational (angular) position $\theta$

Consider again a disk that rotates on a lab bench about a fixed point. The axis of rotation passes through the center of and is perpendicular to the disk

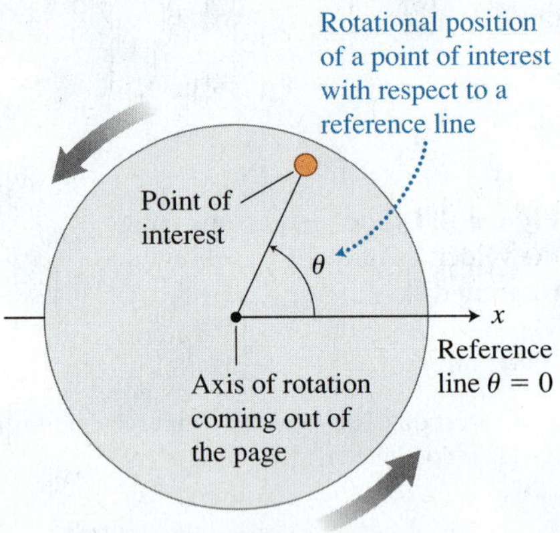


**Figure 8.2** The rotational position (also called the angular position) of a point on a rotating disk. The units of rotational position can be either degrees or radians.

(**Figure 8.2**). A fixed line *perpendicular* to the axis of rotation (like the positive $x$-axis in Figure 8.2) is used as a reference line. We can draw another line on the disk from the axis of rotation to a point of interest, for example, to a coin sitting on the rotating disk. The angle $\theta$ in the counterclockwise direction between the reference line and the line to the point of interest is the **rotational position** (or **angular position**) of the point of interest. The observer is stationary beside the lab bench and looking down on the disk.

> **Rotational position $\theta$** **The rotational position $\theta$ of a point on a rotating object (sometimes called the angular position) is defined as an angle in the counterclockwise direction between a reference line (usually the positive $x$-axis) and a line drawn from the axis of rotation to that point. The units of rotational position can be either degrees or radians.**

## Units of rotational position

Two units are commonly used to indicate the rotational position $\theta$ of a point on a rotating object: degrees and radians. The degree (°) is the most familiar. There are 360° in a circle. If a point on the turning object is at the top of the circle, its position is 90° from a horizontal, positive $x$-axis. When at the bottom of the circle, its position is 270° or, equivalently, −90°.

The unit for rotational position that is most useful in physics is the **radian**. It is defined in terms of the two lengths shown in **Figure 8.3**. The arc length $s$ is the distance in the counterclockwise direction along the circumference of the circle from the positive $x$-axis to the position of a point on the circumference of the rotating object. The other length is the radius $r$ of the circle. The angle $\theta$ in units of radians (rad) is the ratio of $s$ and $r$:

$$\theta\,(\text{in radians}) = \frac{s}{r} \tag{8.1}$$

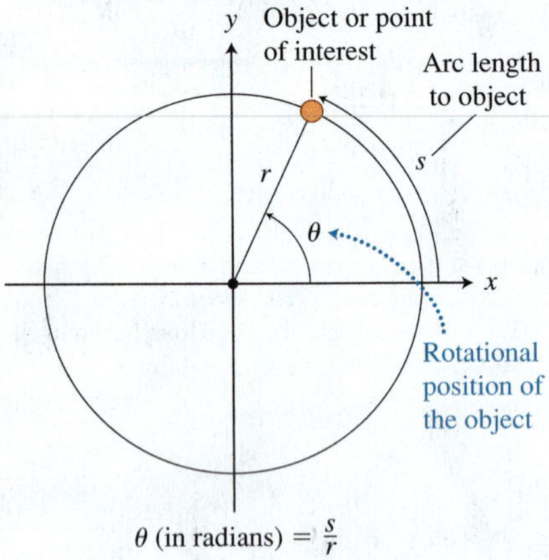


**Figure 8.3** The rotational position $\theta$ in radians is the ratio of the arc length $s$ and the radius $r$.

Note that the radian unit has no dimensions; it is the ratio of two lengths. We can multiply by the radian unit or remove the radian unit from an equation with no consequence. If we put the unit **rad** in the equation, it is usually because it is a reminder that we are using radians for angles.

> **TIP** From Eq. (8.1) we see that the arc length for a 1-rad angle equals the radius of the circle. For example, the 1-rad angle shown in **Figure 8.4** is the ratio of the 2-cm arc length and the 2-cm radius and is simply 1. If you use a calculator to work with radians, make sure it is in the radian mode.

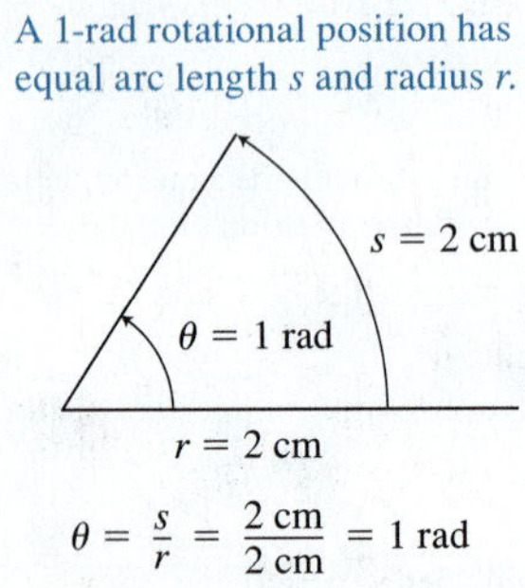


**Figure 8.4** The arc length for a 1-rad angle equals the radius of the circle. The 1-rad rotational angle in this case is the ratio of the 2-cm arc length and the 2-cm radius.

One complete rotation around a circle corresponds to a change in arc length of $2\pi r$ (the circumference of the circle) and a change in rotational position of

$$\theta\,(\text{one complete rotation}) = \frac{s}{r} = \frac{2\pi r}{r} = 2\pi$$

Thus, there are $2\pi$ radians in one circle. We can now relate the two rotational position units:

$$360° = 2\pi\text{ rad}$$

We can use this equation to convert between degrees and radians.

We can use Eq. (8.1) to find the arc length $s$ if the radius and rotational position $\theta$ are known:

$$s = r\theta\ (\text{for }\theta\text{ in radians only})$$

For example, if a car travels 2.0 rad (that is, from $\theta_0 = 0$ rad to $\theta = 2.0$ rad) around a highway curve of radius 100 m, the car travels a distance along the arc equal to

$$s = r(\theta - \theta_0) = (100\text{ m})(2.0\text{ rad} - 0) = 200\text{ m}$$

We dropped the radian unit in the answer because angles measured in radians are dimensionless.

**TIP** You cannot calculate arc length using $s = r\theta$ when $\theta$ is measured in degrees. You must first convert $\theta$ to radians.

**QUANTITATIVE EXERCISE 8.1 An old-fashioned watch**

Your analog watch with hour and minute hands reads 3:30. What is the rotational position in radians of each of these hands? Use a reference line from the axis of rotation through the 12:00 position. Assume (contrary to reality) that the hour hand points directly at "3".

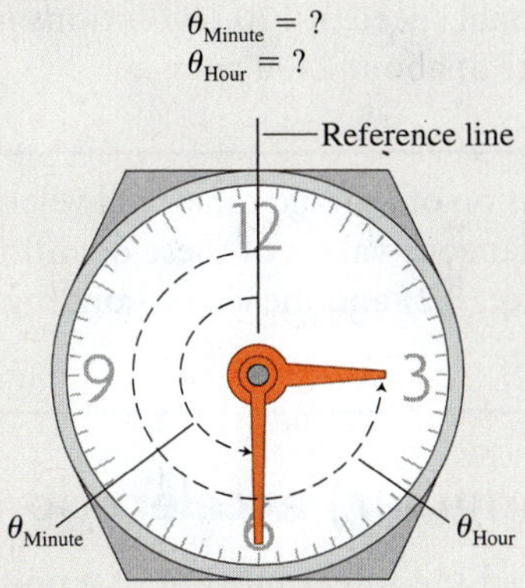


**Represent mathematically and solve** The rotational position of the hour hand is the angle in the counterclockwise direction from the reference line to the 3:00 hour hand. The rotational position of the hour hand is three-quarters of the way around the clock from 12:00 going in the counterclockwise direction. So $\theta_{Hour} = +(3/4)2\pi = +3\pi/2$ radians. The rotational position of the minute hand is the angle in the counterclockwise direction from the reference line to the 6:00 minute hand. The positive counterclockwise position of the minute hand is $\theta_{Minute} = +(1/2)2\pi = +\pi$ radians.

**Try it yourself:** Your watch reads 6:15. What is the rotational position in radians from a 12:00 reference line to each hand of the clock? What assumption did you make?

*Answer:* $+\pi$ for the hour hand (assumption: it is still pointing at 6:00) and $+3\pi/2$ for the minute hand.

## Rotational (angular) velocity $\omega$

When we were investigating the motion of a point-like object along a single axis, we defined the translational velocity of that object as the rate of change of its linear position. Thus, it seems natural to define the **rotational (angular) velocity** $\omega$ of a rigid body as the rate of change of each point's rotational position. Because all points on the rigid body rotate through the same angle in the same period of time (see Figure 8.1c), each point on the rigid body has the same rotational velocity. This means we can just refer to the rotational velocity of the rigid body itself, rather than to any specific point within it.

**Rotational velocity $\omega$** The average rotational velocity (sometimes called angular velocity) of a turning rigid body is the ratio of its change in rotational position $\Delta\theta$ and the time interval $\Delta t$ needed for that change (see **Figure 8.5**):

$$\omega = \frac{\Delta\theta}{\Delta t} \quad (8.2)$$

The sign of $\omega$ (omega) is positive for counterclockwise turning and negative for clockwise turning, as seen looking along the axis of rotation. *Rotational (angular) speed* is the magnitude of the rotational velocity. The most common units for rotational velocity and speed are radians per second (rad/s) and revolutions per minute (rpm).

**TIP** Rotational velocity is the same for all points of a rotating rigid body. It is independent of the distance of a chosen point on the rigid body from the axis of rotation.

**Figure 8.5** Each point on a rigid body has the same rotational velocity $\omega$.

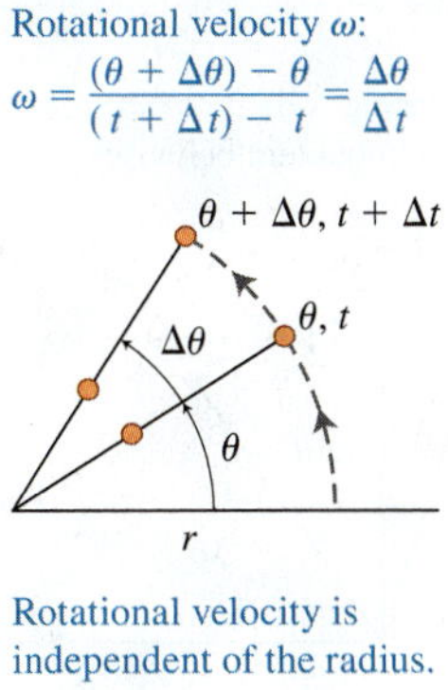

To distinguish the rotational velocity from the familiar velocity that characterizes the linear motion of an object, the latter is called linear velocity. When an object rotates, each point of the object has linear velocity. If you examine Figure 8.1a you see that the linear velocity vectors are tangent to the circle. Thus the linear velocity of a point on a rotating object is sometimes called **tangential velocity**.

The **revolution** is a familiar unit from everyday life. One revolution (rev) corresponds to a complete rotation about a circle and equals 360°. The revolution is not a unit of rotational position. It is a unit of *change* in rotational position $\Delta\theta$. Revolutions are usually used to indicate change in rotational position per unit time. For example, a motor that makes 120 complete turns in 1 min is said to have a rotational speed of 120 revolutions per minute (120 rpm). Automobile engines rotate at about 2400 rpm.

**TIP** The definition of average rotational velocity or rotational speed becomes the instantaneous values of these quantities if you consider a small time interval in Eq. (8.2) and the corresponding small change in the rotational position.

**Figure 8.6** Three rotational motion diagrams and the corresponding signs of the rotational accelerations.

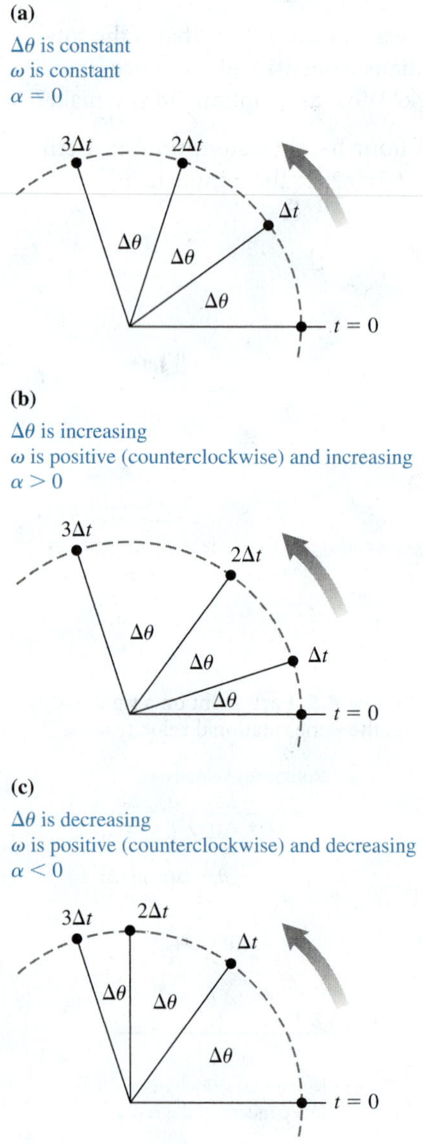


## Rotational (angular) acceleration $\alpha$

When we investigated the linear motion of a point-like object along a single axis, we developed the physical quantity acceleration to describe the object's change in velocity. This was translational acceleration, as it described the changing velocity of the object while moving from one position to another. We could apply the same translational acceleration idea to the center of mass of a rigid body that is moving as a whole from one position to another. But usually we are interested in the rate of change of the rigid body's rotational velocity, that is, its **rotational acceleration**. In other words, when the rotation rate of a rigid body increases or decreases, it has a nonzero rotational acceleration.

**Rotational acceleration $\alpha$** **The average rotational acceleration $\alpha$ (alpha) of a rotating rigid body (sometimes called angular acceleration) is its change in rotational velocity $\Delta\omega$ during a time interval $\Delta t$ divided by that time interval:**

$$\alpha = \frac{\Delta\omega}{\Delta t} \tag{8.3}$$

**The unit of rotational acceleration is (rad/s)/s = rad/s$^2$.**

**Figure 8.6** shows motion diagrams for three different types of rotational motion. Let's consider these rotational motion diagrams and try to develop a rule for how the sign of the rotational acceleration relates to the rotational velocity for a counterclockwise-turning disk. Note that when the disk's rotational velocity is constant (the lengths of arcs between the dots are the same), its rotational acceleration is zero (Figure 8.6a). When its counterclockwise rotational velocity (positive) is increasing (note that in Figure 8.6b the arcs between the dots increase in length), its rotational acceleration has the same sign (positive) as the rotational velocity. If the disk's counterclockwise rotational velocity (positive) is decreasing (note that in Figure 8.6c the arcs between the dots are shrinking), its rotational acceleration has the opposite sign (negative). Similarly, when a disk is rotating clockwise (negative rotational velocity) faster and faster, its rotational acceleration has the same sign (negative). If the disk's clockwise rotational velocity (negative) is decreasing, its rotational acceleration has the opposite sign (positive).

What could we conclude about signs of Earth's rotational velocity and acceleration if we were looking down on Earth from above the North Pole? The rotational velocity would have a positive sign (turning counterclockwise), and because the rotational velocity is constant, the rotational acceleration would be zero. (We will learn later that Earth's rotational velocity is not exactly constant.)

**TIP** The sign of the rotational acceleration is the same as the sign of the rotational velocity when the object rotates increasingly faster. The rotational acceleration has the opposite sign if the object is rotating increasingly slower.

## Relating translational and rotational quantities

Are there mathematical connections between physical quantities describing the rotational motion of a rigid body and the translational motion of different points on the body? Recall that the rotational position $\theta$ of a point on a turning object depends on the radial distance $r$ of that point from the axis of rotation and the length $s$ measured along the arc connecting that point to the reference axis (see Figure 8.3):

$$s = r\theta \tag{8.1}$$

If the angle changes by $\Delta\theta$, the distance of the point of the object along the arc changes by $\Delta s$, so that

$$\Delta s = r\,\Delta\theta$$

A similar relation exists between the speed of a point on a turning object and the rotational velocity of the turning object. An analogous relation also exists between the magnitude of acceleration of a point on a turning object and the rotational acceleration of the turning object. Suppose, for example, that a point on the object changes rotational position by $\Delta\theta$ in a time interval $\Delta t$. Its rotational velocity is $\omega = \Delta\theta/\Delta t$. The change in arc length is $\Delta s$ along its circular path, and its tangential speed (the speed of the object tangent to the circle, sometimes called linear speed) is $v_t = \Delta s/\Delta t$. Substituting for $\Delta s$, we get

$$v_t = \frac{\Delta s}{\Delta t} = \frac{r\,\Delta\theta}{\Delta t} = r\left(\frac{\Delta\theta}{\Delta t}\right) = r\omega \tag{8.4}$$

Notice that while the rotational speed of all points of the same rigid body is the same, the tangential (linear) speed of different points increases as their distance from the axis of rotation increases. A similar relationship can be derived that relates that point's acceleration $a_t$ tangent to the circle and its rotational acceleration $\alpha$:

$$a_t = \frac{\Delta v_t}{\Delta t} = \frac{r\,\Delta\omega}{\Delta t} = r\left(\frac{\Delta\omega}{\Delta t}\right) = r\alpha \tag{8.5}$$

The signs of the rotational position and velocity are positive for counterclockwise turning, and the signs of the translational position and velocity are also positive for counterclockwise motion.

**TIP** You get the familiar translational quantities for motion along the circular path by multiplying the corresponding angular rotational quantities by the radius $r$ of the circle.

To visualize this relationship, imagine five people (the point objects in **Figure 8.7**) holding on to a long stick that can rotate horizontally about a vertical pole to which it is attached on one end. These people hold on to the stick as it completes a full circle. The person closest to the pole moves the slowest, the next person moves a little faster, and the one at the free end has to almost run to keep the stick in his hands. At a particular time, all of them have the same rotational position$\theta$ and the same rotational velocity $\omega$. However, the linear distances and speeds are larger for the people farther from the axis of rotation (larger values of $r$).

**Figure 8.7** A top-view diagram of five people (represented by dots) holding on to a stick that rotates about a fixed pole. Note that the speed $v$ of each person depends on the person's distance from the axis of rotation.

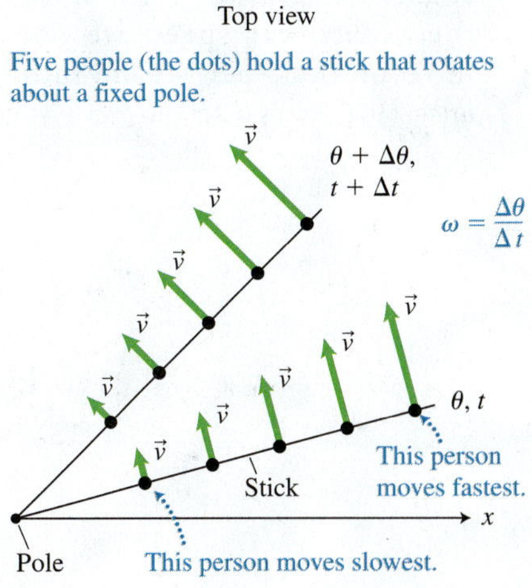

Top view
Five people (the dots) hold a stick that rotates about a fixed pole.
θ + Δθ, t + Δt
ω = Δθ/Δt
θ, t
This person moves fastest.
Stick
Pole
x
This person moves slowest.

People sometimes play crack the whip while ice skating. The skaters hold hands and skate in a line. Then the person on one end of the line stops and the others continue moving but are pulled toward the stopped person. The line now moves in a circle with the farthest person going very fast, like the end of a whip.

Black holes are an extreme case of rotational motion. Black holes form when some stars at the end of their lives collapse, forming small, very dense objects. If the star was spinning when it was young, it continues to spin when it becomes a black hole, only much faster. The matter near the outer edge of the black hole is usually extremely hot gas that orbits the black hole with a tangential speed near the speed of light. The figure below is an artist's rendition of what such a black hole might look like, if we could see it.

**QUANTITATIVE EXERCISE 8.2 Orbiting a black hole**

Black hole GRS 1915+105 in the constellation Aquila (the Eagle) is about 35,000 light-years from Earth. It was formed when the core of a star with about 14 solar masses (mass of 14 times the mass of our Sun) collapsed. The boundary of the black hole, called the event horizon, is a sphere with radius about 25 km. Surrounding the black hole is a stable gaseous cloud with an innermost 30-km-radius stable circular orbit. This cloud moves in a circle about the black hole about 970 times per second. Determine the tangential speed of matter in this innermost stable orbit.

**Represent mathematically** We model the cloud as a rotating disk with radius $r = 30{,}000$ m; to find the rotational speed we convert the rotations in revolutions per second into radians per second: $\omega = (970 \text{ rev/s})(2\pi \text{ rad/rev}) = 6100 \text{ rad/s}$. We need to find the tangential speed $v_t$ of the particles of gas in orbit around the black hole, which is related to the radius $r$ of the circular orbit and the rotational speed $\omega$ of the matter:

$$v_t = r\omega$$

**Solve and evaluate** The speed of the matter in the innermost stable orbit will be

$$v_t = r\omega = (30{,}000 \text{ m})(6100 \text{ rad/s}) = 1.8 \times 10^8 \text{ m/s}$$

This is slightly more than half the speed of light! Actually, the physics we have developed is only moderately applicable in this environment of extreme gravitational forces and high speeds. Our answer is about 20% high compared to a more sophisticated analysis done using Einstein's theory of general relativity. Nevertheless, using ideas from uniform circular motion, we estimate that the radial acceleration of matter moving in that circular orbit is about $v^2/r \approx 10^{11}\, g$. It's not a good place to visit.

**Try it yourself:** You ride a carnival merry-go-round and are 4.0 m from its center. A motion detector held by your friend next to the merry-go-round indicates that you are traveling at a tangential speed of 5.0 m/s. What are the rotational speed and time interval needed to complete one revolution on the merry-go-round?

*Answer:* $\omega = 1.3 \text{ rad/s}$; $T = 5.0$ s.

## Rotational motion at constant acceleration

Earlier, we developed equations that related the physical quantities $t$, $x$, $v$, and $a$, which we used to describe the translational motion of a point-like object along a single axis with constant acceleration. Similar equations relate the rotational kinematics quantities $t$, $\theta$, $\omega$, and $\alpha$, assuming the rotational acceleration is constant. We're not going to develop them based on observations in the way we did for the equations of translational motion, since the process will be

**Table 8.1** Equations of kinematics for translational motion with constant acceleration and the analogous equations for rotational motion with constant rotational acceleration.

| Translational motion | Rotational motion | |
|---|---|---|
| $v_x = v_{0x} + a_x t$ | $\omega = \omega_0 + \alpha t$ | (8.6) |
| $x = x_0 + v_{0x}t + \frac{1}{2}a_x t^2$ | $\theta = \theta_0 + \omega_0 t + \frac{1}{2}\alpha t^2$ | (8.7) |
| $2a_x(x - x_0) = v_x^2 - v_{0x}^2$ | $2\alpha(\theta - \theta_0) = \omega^2 - \omega_0^2$ | (8.8) |

nearly the same. Instead, we will rely on the connections we have seen between the translational and rotational quantities. The analogous rotational motion equations are provided in **Table 8.1** along with the corresponding translational motion equations. Because the quantities that describe motion depend on the choice of the reference frame, always note the location of the observer in a particular situation.

For rotational motion, $\theta_0$ is an object's rotational position at time $t_0 = 0$; $\omega_0$ is the object's rotational velocity at time $t_0 = 0$; $\theta$ and $\omega$ are the rotational position and rotational velocity at some later time $t$; and $\alpha$ is the object's constant rotational acceleration during the time interval from time zero to time $t$. The sign of the rotational position is positive for counterclockwise $\theta$ and negative for clockwise $\theta$ from the reference axis. The sign of the rotational velocity $\omega$ depends on whether the object is rotating counterclockwise $(+)$ or clockwise $(-)$. The sign of the rotational acceleration $\alpha$ depends on how the rotational velocity is changing; $\alpha$ has the same sign as $\omega$ if the magnitude of $\omega$ is increasing and the opposite sign of $\omega$ if $\omega$'s magnitude is decreasing.

**Review Question 8.1** Visualize an ice skater rotating faster and faster in a clockwise direction. What are the signs of rotational velocity and rotational acceleration? Then the skater starts slowing down. What are the signs of rotational velocity and acceleration now?

## 8.2 Torque and rotational acceleration

What causes a rigid body to have a particular rotational acceleration? When we investigated translational motion we learned that the acceleration of a point-like object was determined by its interactions with other objects, that is, forces that objects in the environment exerted on it. Perhaps there is an analogous way to think about what causes rotational acceleration. In the last chapter, we learned that the net torque produced by forces exerted on a system had to equal zero for the object to remain in static equilibrium, to not start rotating. What happens when the net torque isn't zero? Let's investigate this. In Observational Experiment **Table 8.2**, we perform four experiments with a bicycle that rests upside down so its front tire can rotate freely (**Figure 8.8**).

**Figure 8.8** The front tire of an inverted bicycle, used for the experiments in Observational Experiment Table 8.2.

**OBSERVATIONAL EXPERIMENT TABLE**

## 8.2 Turning effects of forces exerted on a bicycle tire.

VIDEO 8.2

| Observational experiment | Analysis |
|---|---|
| **Experiment 1:** Your bike sits upside down. You push on the front tire toward the axle. The tire does not turn. 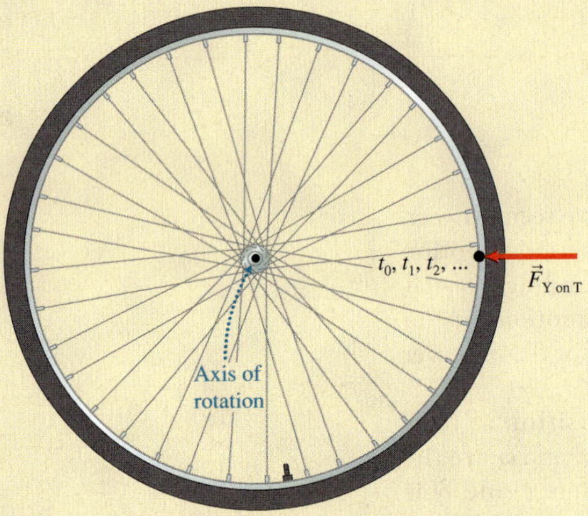  | A force is exerted on the tire, but the torque produced by the force about the axis of rotation (the axle) is zero. It has no effect on the tire's rotation. |
| **Experiment 2:** You push lightly and continuously on the outside of the tire in a counterclockwise (ccw) direction tangent to the tire. As you continue to push, the tire rotates ccw faster and faster. 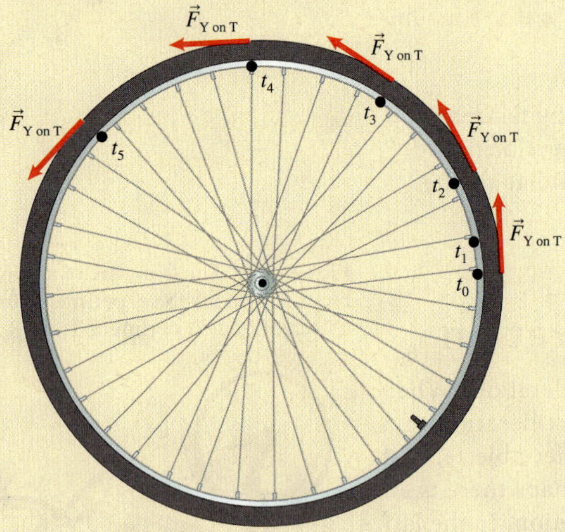  | Your pushing causes a force and a ccw torque. The tire has increasing positive rotational velocity and a positive rotational acceleration. |

**Experiment 3:** You release the spinning tire and watch it. The tire continues rotating ccw at a constant rate.

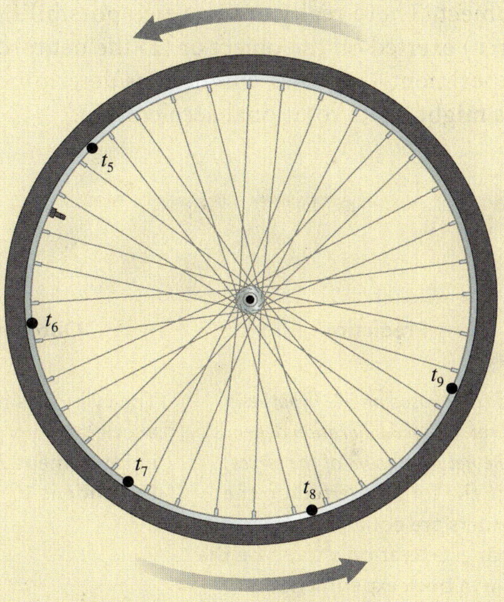


When the force and torque are zero, there is no change in rotation rate. The rotational velocity is constant and the rotational acceleration is zero.

**Experiment 4:** With the tire still rotating ccw fast, you gently and continuously push clockwise (cw) against the tire. The rotational speed decreases.

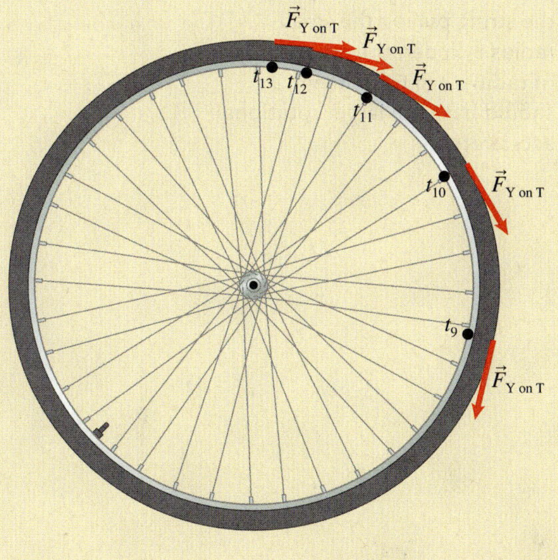


Pushing opposite the rotation causes a force and a cw torque and a decreasing ccw rotation rate. There is a clockwise (negative) rotational acceleration.

**Patterns**

- An external force that produces a *zero torque* on the tire does not change the tire's rotation rate. If the tire is at rest, it remains at rest.
- When there are no external forces exerting torques on a tire, its rotational velocity remains constant.
- An external force that produces a *nonzero torque* on the tire:
  - causes the tire to turn faster and faster if in the direction the tire is turning; and
  - causes the tire to turn slower and slower if opposite the direction it is turning.

The experiments in Table 8.2 indicate that a zero torque has no effect on the rotational motion of a rigid body. However, a nonzero torque does cause a change. If the torque is in the same direction as the direction of rotation of the rigid body, its rotational speed increases. If the torque is in the opposite direction, the rigid body's rotational speed decreases. These patterns are similar to the patterns that we found for translational motion.

Our goal is to determine what physical quantities cause rotational acceleration of an extended object. There really are just two possibilities: (1) the sum of the forces (net force) exerted on the object or (2) the net torque caused by the forces. Testing Experiment **Table 8.3** will help us determine which (if either) of these quantities might affect rotational acceleration.

**TESTING EXPERIMENT TABLE**

**8.3 Testing two hypotheses explaining rotational acceleration.**

| Testing experiment | Prediction | Outcome |
|---|---|---|
| You have a cylinder that can rotate on an axle. Half of the cylinder's thickness has a large radius ($R$) and the other half has a smaller radius $r$. The axle is fixed. When the cylinder rotates, both parts rotate.<br>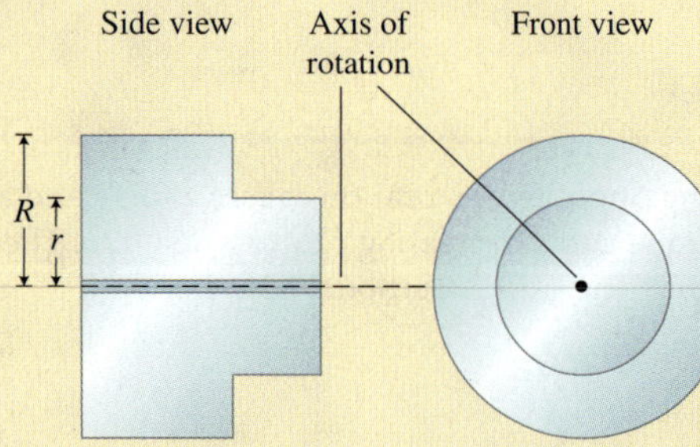<br><br>**Experiment 1.** Wrap a string around the part of the cylinder with the small radius $r$ and pull the string, exerting force $F$ on the string.<br>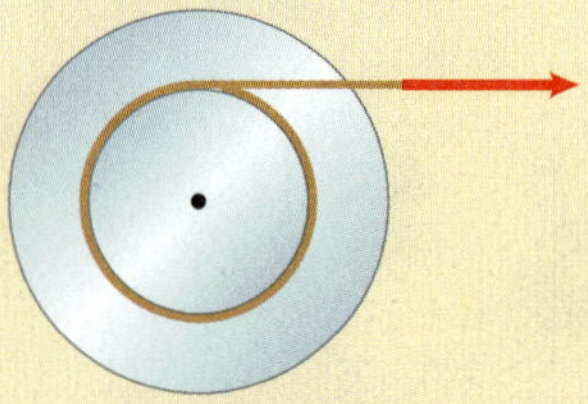<br>**Experiment 2.** Wrap a string around the part of the cylinder with the large radius $R$ and pull the string, exerting the same force $F$ on the string.<br>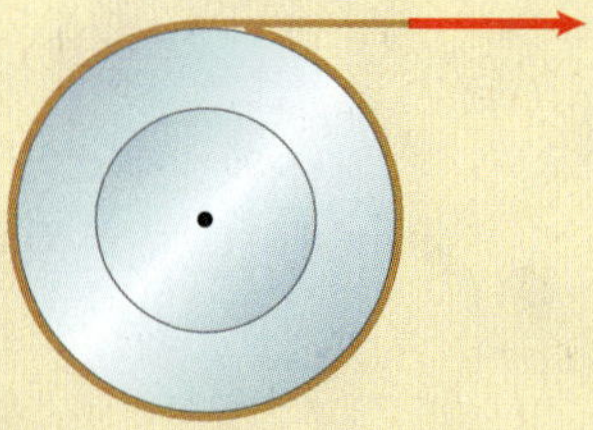<br>In which experiment will the cylinder have a greater rotational acceleration? | *Prediction based on the hypothesis that rotational acceleration depends on the net force (sum of the forces):* Since the forces exerted on the cylinders are equal, the rotational acceleration should be the same in both experiments.<br><br>*Prediction based on the hypothesis that rotational acceleration depends on the net torque:* Since the string pulling the large-radius cylinder produces a greater torque than the string pulling the small-radius cylinder, the cylinder of radius $R$ in Experiment 2 should have the larger rotational acceleration. | The cylinder's rotational acceleration is greater in Experiment 2 than in Experiment 1. |

**Conclusions**

- The outcome *does not* match the prediction based on the hypothesis that the *net force* exerted on an object causes its rotational acceleration.
- The outcome *does* match the prediction based on the hypothesis that the *net torque* exerted on an object causes its rotational acceleration.

Notice that the force exerted on the wheel in the first experiment in Table 8.2 produced a nonzero force and a zero torque on the wheel, and the wheel did not turn. Therefore, that experiment already disproves the hypothesis that the net force exerted on a rigid body causes its rotational acceleration. Additionally, the experiments in Table 8.3 disprove the force hypothesis and support the following provisional rule:

**Changes in rotational velocity** **Rotational acceleration depends on net torque. The greater the net torque, the greater the rotational acceleration.**

Notice how this is similar to what we learned when studying translational motion. A nonzero net force (sum of the forces) needs to be exerted on an object to cause its velocity to change. The greater the net force, the greater the translational acceleration of the object.

Remember that in all of the experiments we have performed so far in this chapter, rotational motion was all that was possible. Each rigid body was rotating about a fixed axis through its center of mass. If the rigid body were not held fixed, then a change in both translational and rotational motion could occur. The translational acceleration of the center of mass of such an object is determined by Newton's second law $\vec{a} = \Sigma\vec{F}/m$. The rotational acceleration around its center of mass will be determined by the ideas we will investigate over the next several sections.

**Review Question 8.2** How do we know that rotational acceleration of an object depends on the torque, not the net force exerted on it?

## 8.3 Rotational inertia

We have found that a nonzero net torque will cause an object's rotational velocity to change. What other quantities might affect rotational acceleration? We know that the translational acceleration of an object is directly proportional to the net force exerted on it and inversely proportional to its mass. It seems that mass should also somehow affect an object's rotational acceleration. However, a simple experiment allows us to see that the mass alone is not the answer. Lay a broom on a hard smooth floor. First try to increase the broom's rotational speed about a vertical axis by turning it with one hand holding the broomstick far from the broom head. Then, do it again, holding the broom in the middle nearer the broom head. It is much easier to increase the rotational speed holding it in the middle. You are turning the same mass, but the location of that mass from the axis of rotation seems to make a difference. We explore this idea in Observational Experiment **Table 8.4**.

**OBSERVATIONAL EXPERIMENT TABLE**

**8.4 Effect of mass distribution on rotational acceleration.**

| Observational experiment | Analysis |
| --- | --- |
| Two cylinders of the same radius and the same mass can rotate around an axis passing through their centers. One cylinder is solid and made of wood, and the other is hollow and made of iron. You wrap a string around each cylinder and pull equally hard on the strings wrapped around each cylinder. Both cylinders start accelerating, but the solid cylinder has a greater rotational acceleration than the hollow cylinder. 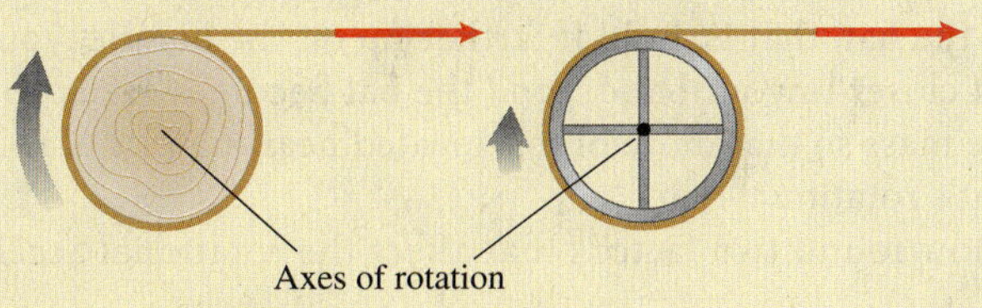 | If we assume that the force and the torque exerted on the cylinders are the same, then the only difference is the distribution of mass of the cylinders. Most of the mass of the hollow cylinder is located far from the axis of rotation. The mass of the solid cylinder is distributed more uniformly, some of it closer to the axis of rotation. |



**Pattern**

The same torque produces smaller rotational acceleration if the mass of the rigid body is distributed farther from the axis of rotation.

**Figure 8.9** Rotational acceleration decreases if the mass is farther from the axis of rotation, as in (b).

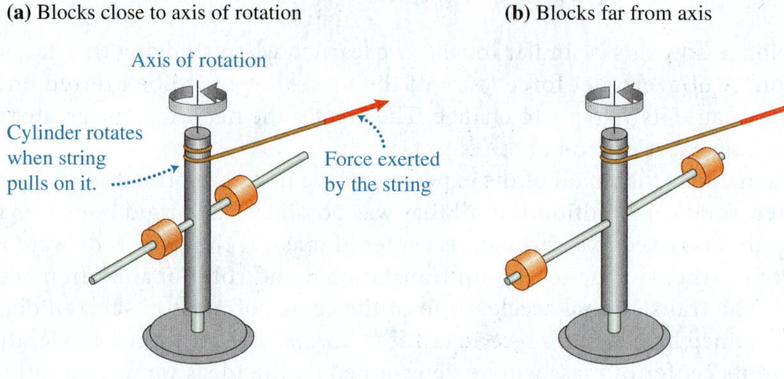


The pattern found in Table 8.4 is consistent with our simple broom rotation experiment. *If* it is correct and we change the distribution of the mass of a rigid body while keeping the total mass the same, *then* the rotational acceleration of the object due to a particular torque should decrease when the mass is moved farther away from the axis of rotation. Let's use this statement as a prediction of the outcome of a new experiment.

We use an apparatus that consists of metal blocks that can be moved on a metal rod so that they are different distances from the axis of rotation—see **Figure 8.9**. If we pull each string exerting the same force (represented by the solid red arrow), will the rotational acceleration be greater for the arrangement shown in Figure 8.9a or for that in Figure 8.9b? According to the above explanation, we predict that the rotational acceleration will be greater for arrangement (a) than arrangement (b) since the mass is nearer the axis of rotation in (a) than in (b). When we try the experiment, we find that the rate of rotation for (a) does change faster than for (b), consistent with the pattern we found in Table 8.4.

Thus, another important factor that affects the rotational acceleration of a rigid body is the *distribution* of mass of the rotating object. The closer the mass of the object to the axis of rotation, the easier it is to change its rotational motion. We call the physical quantity characterizing the location of the mass relative to the axis of rotation the **rotational inertia** (also known as the **moment of inertia**) of the object. Rotational inertia $I$ depends on both the total mass of the object and the distribution of that mass about its axis of rotation. For objects of the same mass, the more mass that is located near the axis of rotation, the smaller the object's rotational inertia will be. Likewise, the more mass that is located farther away from the axis of rotation, the greater the object's rotational inertia will be. For objects of different mass but the same mass distribution, the more massive object has more rotational inertia. The higher the rotational inertia of an object, the harder it is to change its rotational motion. In summary, this quantity is the rotational equivalent of mass.

A common example of the effect of mass distribution on the change of the rotational velocity of an object involves swinging a baseball bat. When you "choke up" on the bat, that is, hold it farther up on the handle, you move the mass of the bat closer to your hands, and the bat becomes easier to swing. By distributing the mass so that more of it is located near the axis of rotation, you decrease the bat's rotational inertia.

We have now found two factors that affect the rotational acceleration of an object:

- The rotational inertia of the object
- The net torque produced by forces exerted on the object

In the next two sections we will investigate quantitatively how these two factors affect the rotational acceleration of an object.

**Review Question 8.3** A solid wooden ball and a smaller solid metal ball have equal mass (the metal ball is smaller because it is much denser than wood). Both can rotate on an axis going through their centers. You exert a force on each that produces the same torque about the axis of rotation. Which sphere's rotational motion will change the least? Explain.

## 8.4 Newton's second law for rotational motion

Let's see if we can construct a quantitative relationship between rotational acceleration, net torque, and rotational inertia. We start with a simple example of rotational motion: a small block attached to a light stick that can move on a smooth surface in a circular path (**Figure 8.10**). The axis of rotation passes through a pin at the other end of the stick. After we analyze this case, we will generalize the result to the rotation of an extended rigid body. You push the block with your finger, exerting a small force $\vec{F}_{\text{F on B}}$ on the block tangent to the circular path. This push causes a torque, which in turn causes the block and stick's rotational velocity about the pin to increase.

**Figure 8.10** A top view of an experiment to relate torque and rotational acceleration.

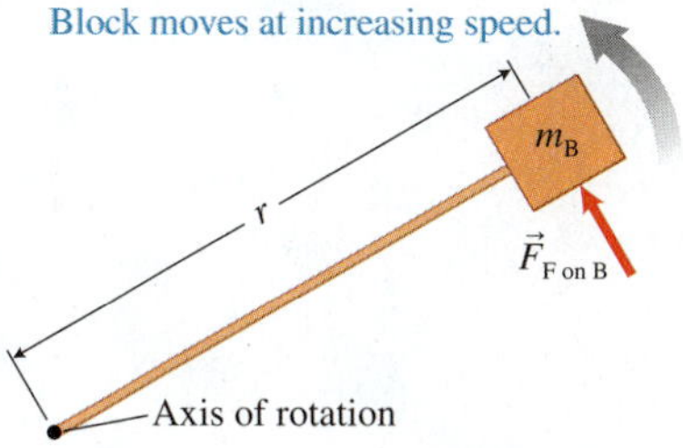
Block moves at increasing speed.
$m_B$
$r$
$\vec{F}_{\text{F on B}}$
Axis of rotation
Your finger (not shown) pushes the block, causing its rotational acceleration.

The torque produced by the force $\vec{F}_{\text{F on B}}$ is

$$\tau = rF_{\text{F on B}} \sin\theta = rF_{\text{F on B}} \sin 90° = rF_{\text{F on B}}$$

Since the block is small, we can reasonably model it as a point-like object. This allows us to apply Newton's second law. Since the mass of the block is much larger than the mass of the stick, we assume that the stick has no mass. The finger exerts a force of constant magnitude pushing lightly in a direction tangent to the block's circular path. Thus, the tangential component of Newton's second law for the block is

$$a_t = \frac{F_{\text{F on B}}}{m_B}$$

There is a mathematical way to get the torque produced by the pushing force. Rearrange the above equation to get

$$m_B a_t = F_{\text{F on B}}$$

Then, multiplying both sides of the equation by $r$, the radius of the circular path:

$$m_B \, r \, a_t = rF_{\text{F on B}}$$

Recall from Eq. (8.5) that $a_t = r\alpha$. Thus,

$$m_B \, r \, (r\alpha) = rF_{\text{F on B}}$$

The right side of this equation equals the torque $\tau$ caused by $\vec{F}_{\text{F on B}}$:

$$(m_B r^2)\, \alpha = \tau$$

$$\Rightarrow \alpha = \frac{\tau}{m_B r^2}$$

Examine the above equation and compare it to Newton's second law for the same object—the block not only acquires translational acceleration $a_t$ due to the force exerted on it by the finger, but also acquires rotational acceleration around the axis caused by the torque produced by that same force. This rotational acceleration is directly proportional to the torque produced by the force and inversely proportional to the mass of the block times the square of the distance between the block and the axis of rotation. The latter makes sense—we found experimentally that the farther the mass of the object is from the axis of rotation, the harder it is to change its rotational velocity. Thus the denominator in the equation above is an excellent candidate for the rotational inertia of the block about the pin (the axis of rotation in this situation).

In the above thought experiment, there was just a single force exerted on the object producing a torque about the axis of rotation. More generally, there could be several forces producing torques. It's reasonable that we should add the torques produced by all forces exerted on the object to determine its rotational acceleration:

$$\alpha = \frac{1}{m_B r^2}\Sigma\tau = \frac{1}{m_B r^2}(\tau_1 + \tau_2 + \ldots) \tag{8.9}$$

where $\tau_1, \tau_2, \ldots$ are the torques produced by forces $\vec{F}_{1 \text{ on O}}, \vec{F}_{2 \text{ on O}}, \ldots$ exerted on the object.

## Analogy between translational motion and rotational motion

Notice how similar Eq. (8.9)

$$\alpha = \frac{1}{m_B r^2}\Sigma\tau$$

is to Newton's second law for translational motion

$$\vec{a} = \frac{1}{m}\Sigma\vec{F}$$

When the same forces are exerted on a point-like object, we can describe its motion using two acceleration-type quantities—translational and rotational acceleration. The translational acceleration is determined by Newton's second law, and the rotational acceleration is determined by Eq. (8.9), called **Newton's second law for rotational motion.** There is a strong analogy between each of the three quantities in the two equations (see **Table 8.5**):

For translational motion, mass is a measure of an object's translational inertia—the tendency for its motion to not change. For the rotational motion of a point-like object, the object's mass times the square of its distance $r$ from the axis of rotation ($mr^2$) is a measure of the object's rotational inertia—the tendency for its rotational motion to not change. In summary, the quantity $mr^2$ is the rotational inertia $I$ of a point-like object of mass $m$ around the axis that is the distance $r$ from the location of the object.

For translational motion, the net force $\Sigma\vec{F}$ exerted on an object of interest by other objects causes that object's velocity to change—it has a translational acceleration ($\vec{a} = \Delta\vec{v}/\Delta t$). For rotational motion, the net torque $\Sigma\tau$ produced by forces exerted on the object causes its rotational velocity to change—it has a rotational acceleration ($\alpha = \Delta\omega/\Delta t$).

**Table 8.5** **Analogy between translational and rotational quantities in Newton's second law.**

| | Translational motion | Rotational motion |
|---|---|---|
| Inertia of a point-like object | $m$ | $mr^2$ |
| Cause of acceleration | $\Sigma\vec{F}$ | $\Sigma\tau$ |
| Acceleration | $\vec{a}$ | $\alpha$ |

**EXAMPLE 8.3 Pushing a rollerblader**

A 60-kg rollerblader holds a 4.0-m-long rope that is loosely tied around a metal pole. You push the rollerblader, exerting a 40-N force on her, which causes her to move increasingly fast in a counterclockwise circle around the pole. The surface she skates on is smooth, and the wheels of her rollerblades are well oiled. Determine the tangential and rotational acceleration of the rollerblader.

**Sketch and translate** We sketch the situation as shown below. We choose the rollerblader as the system object of interest.

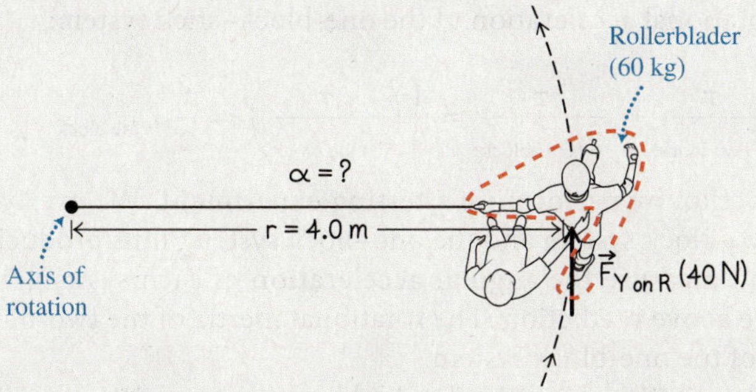


**Simplify and diagram** Since the size of the rollerblader is small compared to the length of the rope, we can model her as a point-like object. The figure below shows a force diagram for the rollerblader (viewed from above). Her tangential acceleration has no vertical component along a vertical axis (which would extend up and out of the page in the figure). The upward normal force $\vec{N}_{\text{F on R}}$ that the floor exerts on her balances the downward gravitational force $\vec{F}_{\text{E on R}}$ exerted by Earth on her (these forces are not shown in the figure). The tension force exerted by the rope on the rollerblader $\vec{T}_{\text{Rope on R}}$ points directly toward the axis of rotation, so that force produces no torque.

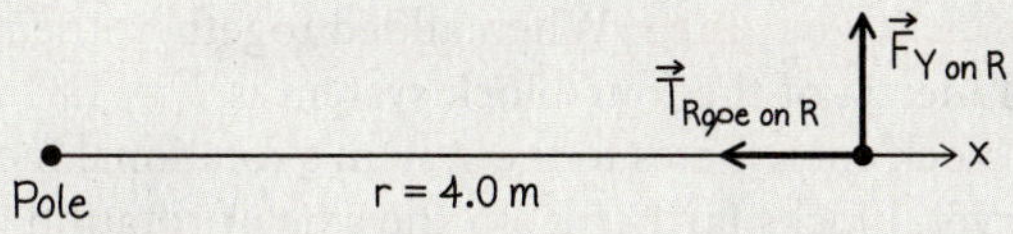


**Represent mathematically** From the force diagram we conclude that your pushing $\vec{F}_{\text{Y on R}}$ on the rollerblader is the only force that produces a nonzero torque $\tau = rF_{\text{Y on R}} \sin 90° = rF_{\text{Y on R}}$, where $r$ is the radius of the rollerblader's circular path.

Use Newton's second law in the tangential direction to determine the rollerblader's tangential acceleration:

$$a_t = \frac{1}{m_R}\Sigma F_t = \frac{F_{\text{Y on R}}}{m_R}$$

Use Newton's second law for rotational motion [Eq. (8.9)] to determine the rollerblader's rotational acceleration:

$$\alpha = \frac{1}{m_R r^2}\Sigma\tau = \frac{1}{m_R r^2}\tau_{\text{Y on R}} = \frac{1}{m_R r^2}(rF_{\text{Y on R}}) = \frac{F_{\text{Y on R}}}{m_R r}$$

**Solve and evaluate** For the tangential acceleration:

$$a_t = \frac{F_{\text{Y on R}}}{m_R} = \frac{40\text{ N}}{60\text{ kg}} = 0.67\text{ m/s}^2$$

For the rotational acceleration:

$$\alpha = \frac{F_{\text{Y on R}}}{m_R r} = \frac{40\text{ N}}{(60\text{ kg})(4.0\text{ m})} = 0.17\text{ rad/s}^2$$

Let's check the units; note that $\frac{\text{N}}{\text{kg m}} = \frac{\text{kg}\cdot\text{m}}{\text{s}^2\cdot\text{kg m}} = \frac{1}{\text{s}^2}$. Remember that the radian is not an actual unit. It is dimensionless. It is just a reminder that this is the angle unit appropriate for these calculations. So the units are correct, and the magnitudes for both results are reasonable. The rollerblader would have a rotational velocity of 0.17 rad/s after 1 s, 0.34 rad/s after 2 s, and so forth—the rotational velocity increases 0.17 rad/s each second.

**Try it yourself:** Suppose you exerted the same force on your friend, but the friend is holding an 8.0-m-long rope instead of a 4.0-m-long rope. How will this affect the rotational acceleration?

*Answer:* The tangential acceleration will not change, but the rotational acceleration will be half as large. The torque will be doubled because the distance from the axis of rotation to the point where the force is applied will be doubled. But the rotational inertia $(mr^2)$ in the denominator will be quadrupled because of the $r^2$. The combination will be a reduction of the rotational acceleration by half—0.08 rad/s$^2$.

## Newton's second law for rotational motion applied to rigid bodies

We know that the mass of an object composed of many small objects with masses $m_1$, $m_2$, $m_3$, etc. is the sum of the masses of its parts: $m = m_1 + m_2 + m_3 + \ldots$. Mass is a scalar quantity and therefore is always positive. The rotational inertia

**Figure 8.11** Top views comparing the effect of rotational inertia on rotational acceleration.

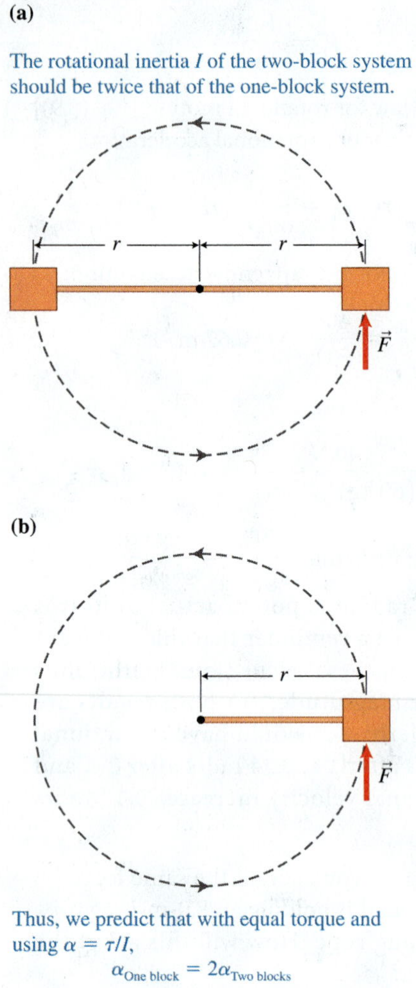


of a point-like object with respect to some axis of rotation is a scalar quantity. Thus, it is reasonable to think that the same rule applies to rigid bodies: the rotational inertia of a rigid body about some axis of rotation is the sum of the rotational inertias of the individual point-like objects that make up the rigid body.

(a)
The rotational inertia $I$ of the two-block system should be twice that of the one-block system.
$r$
$r$
$\vec{F}$
(b)
$r$
$\vec{F}$
Thus, we predict that with equal torque and using $\alpha = \tau/I$,
$\alpha_{\text{One block}} = 2\alpha_{\text{Two blocks}}$

To test this idea, we can use it to calculate the rotational inertia of a lightweight stick with a block attached to each end, as shown in **Figure 8.11a**. The axis of rotation is at the middle of the stick. If our reasoning is correct, the rotational inertia of this two-block rigid body should be twice the rotational inertia of a single block at the end of a stick that is half the length (Figure 8.11b): $I_{\text{Two block}} = 2I_{\text{One block}}$.

If this is correct, and we exert the same torque ($\tau = rF\sin 90° = rF$) on both rigid bodies, the rotational acceleration of the two-block–stick system should be half the rotational acceleration of the one-block–stick system:

$$\alpha_{\text{Two block}} = \frac{\tau}{I_{\text{Two block}}} = \frac{\tau}{2I_{\text{One block}}} = \frac{1}{2}\left(\frac{\tau}{I_{\text{One block}}}\right) = \frac{1}{2}\alpha_{\text{One block}}$$

We check this prediction by performing a testing experiment. We exert the same force on the two-block system and the one-block system, thus producing the same torque, and measure the angular acceleration of each system. The outcome matches the above prediction. The rotational inertia of the two-block system is twice that of the one-block system.

It appears that the rotational inertia of a rigid body that consists of several point-like parts located at different distances from the axis of rotation is the sum of the $mr^2$ terms for each part:

$$I = m_1r_1^2 + m_2r_2^2 + \ldots$$

Let's apply this idea.

### QUANTITATIVE EXERCISE 8.4

Use what you learned about rotational inertial to write an expression for the rotational inertia of the rigid body shown below. Each of the four blocks has mass $m$. They are connected with lightweight sticks of equal length $L/4$.

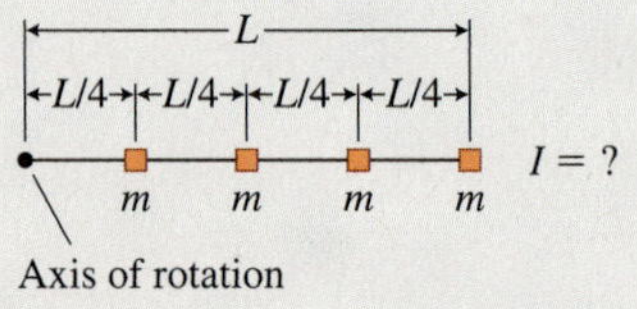


**Represent mathematically** Each block of mass $m$ contributes differently to the rotational inertia of the system. The farther the block from the axis of rotation, the greater its contribution to the rotational inertia of the system of blocks. We can add the rotational inertia of each block of mass $m$ about the axis of rotation:

$$I = m(L/4)^2 + m(2L/4)^2 + m(3L/4)^2 + m(4L/4)^2$$

**Solve and evaluate** When added together, the rotational inertia of the four-block system is $I = 1.88\ mL^2$. Each block contributes to the system's rotational inertia. However, blocks farther from the axis of rotation contribute much more than those near the axis. In fact, the block at the right side of the rod contributes more to the rod's rotational inertia ($mL^2$) than the other three blocks combined ($0.88\ mL^2$).

**Try it yourself:** Calculate the rotational inertia of the same system altered so that the axis of rotation passes perpendicular through the rod halfway between the two central blocks.

*Answer:* $0.313\ mL^2$.

## Calculating rotational inertia

We can calculate the rotational inertia of a rigid body about a specific axis of rotation in about the same way we determined the rotational inertia of the four-block system in Quantitative Exercise 8.4, by adding the rotational inertias of each part of the entire system. However, for most rigid bodies, the parts are not separate objects but are instead parts in a continuous distribution of mass—like in a door or baseball bat. In such a case we break the continuous distribution of mass into a very large number of very small pieces and add the rotational inertias for all of the pieces. Consider the person's leg shown in **Figure 8.12**, which we model as a rigid body if none of the joints bend. For example, mass element 7 contributes an amount $m_7 r_7^2$ to the rotational inertia of the leg. The rotational inertia of the whole leg is then

$$I = m_1 r_1^2 + m_2 r_2^2 + \ldots + m_7 r_7^2 + \ldots + m_{18} r_{18}^2 \qquad (8.10)$$

All $r$'s in the above are from the same axis of rotation. The rotational inertia would be different if we chose a different axis of rotation. There are other ways to do the summation process in Eq. (8.10); often it is done using integral calculus, and sometimes $I$ is determined experimentally.

**Table 8.6** gives the rotational inertias of some common objects for specific axes of rotation. Notice the coefficients in front of the $mR^2$ and $mL^2$ expressions for the objects of different shapes. The value of the coefficient is determined by

**Figure 8.12** Add the $mr^2$ of all the small parts to find the rotational inertia $I$ of the leg.

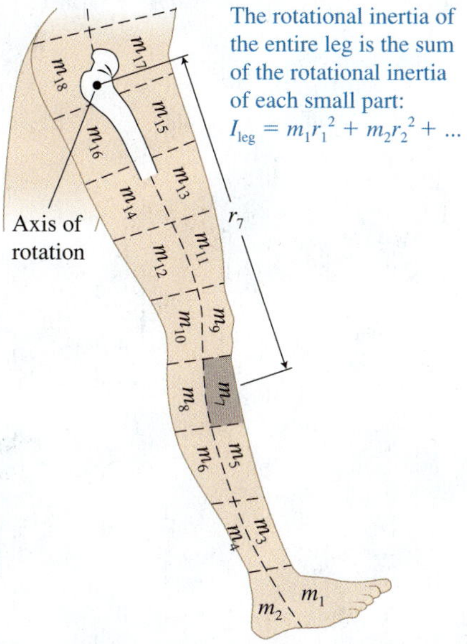


**Table 8.6 Expressions for the rotational inertia of standard shape objects.**

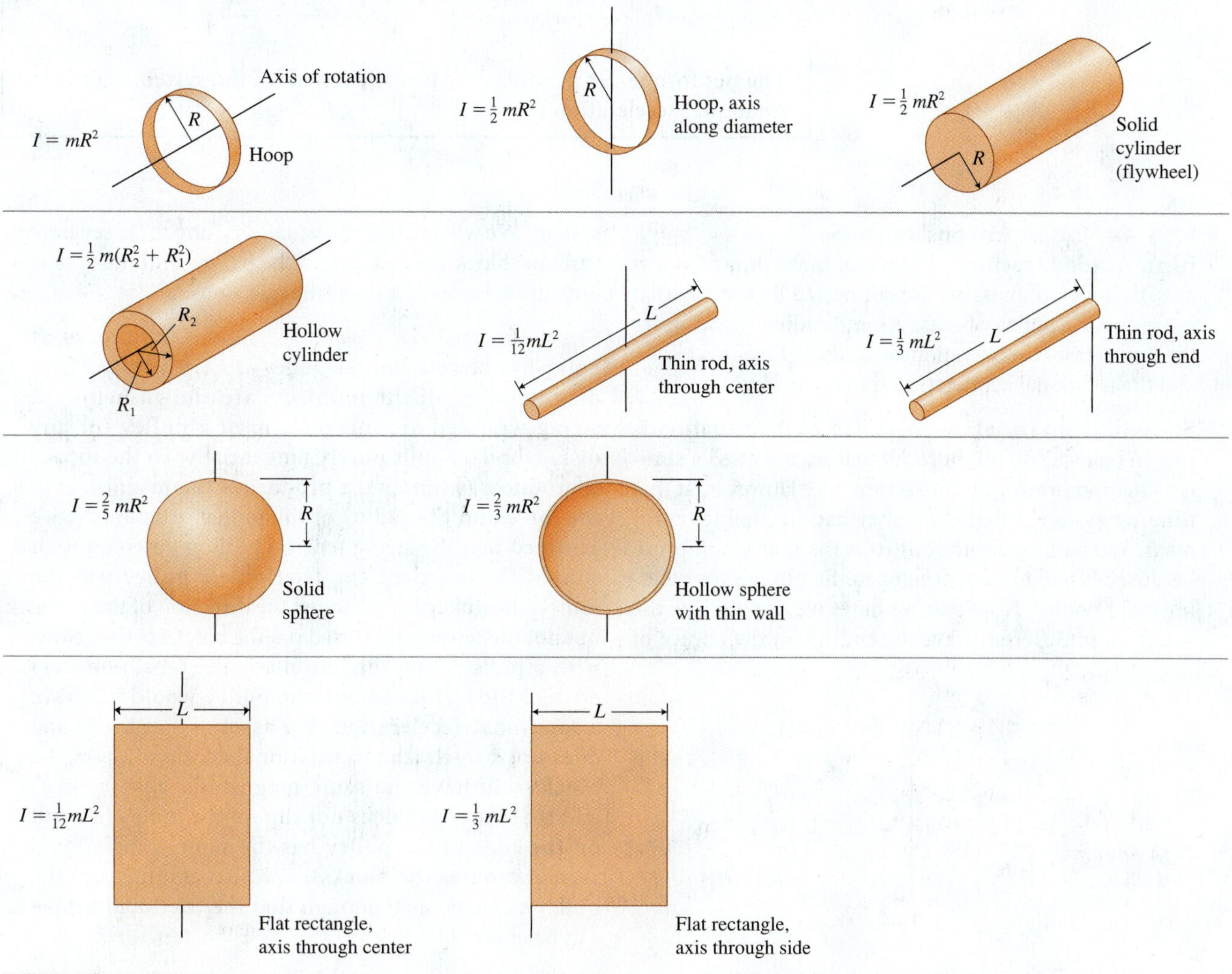

the mass distribution inside the object and the location of the axis of rotation. The closer the mass is to the axis of rotation, the less effect it has on the rotational inertial of the object. The same object has different rotational inertias for different axes of rotation.

We can now rewrite the rotational form of Newton's second law in terms of the rotational inertia of the rigid body.

**Rotational form of Newton's second law** **One or more objects exert forces on a rigid body with rotational inertia $I$ that can rotate about some axis. The sum of the torques $\Sigma\tau$ due to these forces about that axis causes the object to have a rotational acceleration $\alpha$:**

$$\alpha = \frac{1}{I}\Sigma\tau \tag{8.11}$$

**TIP** By writing Newton's second law in the form

$$\vec{a}_S = \frac{1}{m_S}\Sigma\vec{F}_{\text{on S}} = \frac{\vec{F}_{O_1 \text{ on S}} + \vec{F}_{O_2 \text{ on S}} + \ldots + \vec{F}_{O_n \text{ on S}}}{m_S}$$

we see the cause-effect relationship between the net force $\Sigma\vec{F}_{\text{on S}}$ exerted on the system and the system's resulting translational acceleration $\vec{a}_S$. The same idea is seen in Eq. (8.11), only applied to the rotational acceleration:

$$\alpha = \frac{1}{I}\Sigma\tau = \frac{\tau_1 + \tau_2 + \ldots + \tau_n}{I}$$

The net torque $\Sigma\tau$ produced by forces exerted on the system causes its rotational acceleration $\alpha$.

**EXAMPLE 8.5** **Atwood machine**

In an Atwood machine, a block of mass $m_1$ and a less massive block of mass $m_2$ are connected by a string that passes over a pulley of mass $M$ and radius $R$. What are the translational accelerations $a_1$ and $a_2$ of the two blocks and the rotational acceleration $\alpha$ of the pulley?

**Sketch and translate** A sketch of the situation is shown below. You might recall that we analyzed a similar situation previously (in Section 3.3). However, at that time we assumed that the pulley had negligible (zero) mass. We had no choice but to make that assumption because we had not yet developed the physics for rotating rigid bodies. Now that we have, we can analyze the situation in different ways depending on the choice of system. We will analyze the situation using three separate systems: block 1, block 2, and the pulley, and then combine the analyses to answer the questions.

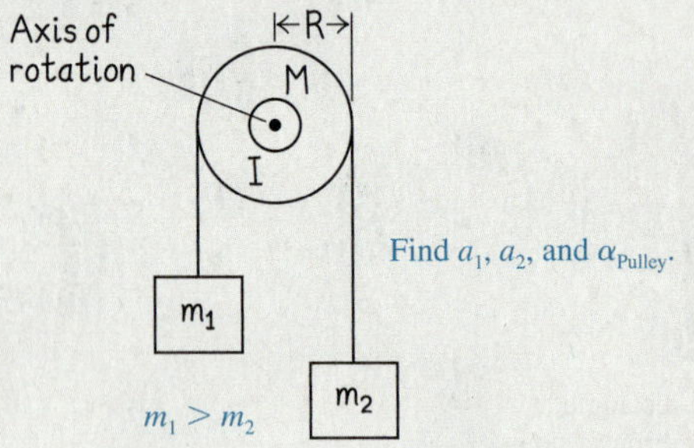


**Simplify and diagram** We model the blocks as point-like objects and the pulley as a rigid body. Force diagrams for all three objects are shown below. A string wrapped around the rim of a pulley (or any disk/cylinder) pulls purely tangentially, so the torque it produces is simply the product of the magnitude of the force and the radius of the pulley. Previously, we assumed that the string tension pulling down on each side of the massless and frictionless pulley was the same—the pulley just changed the direction of the string but not the tension it exerted on the blocks below. Now, with a pulley with nonzero mass, the tension differs on each side. If it did not, the pulley would not have a rotational acceleration. We assume that the string does not stretch (the translational acceleration of the blocks will have the same magnitude $a_1 = a_2 = a$) and that the string does not slip on the pulley (a point on the edge of the pulley has the same translational acceleration as the blocks). We also assume that the pulley's axle is oiled enough that the frictional torque can be ignored.

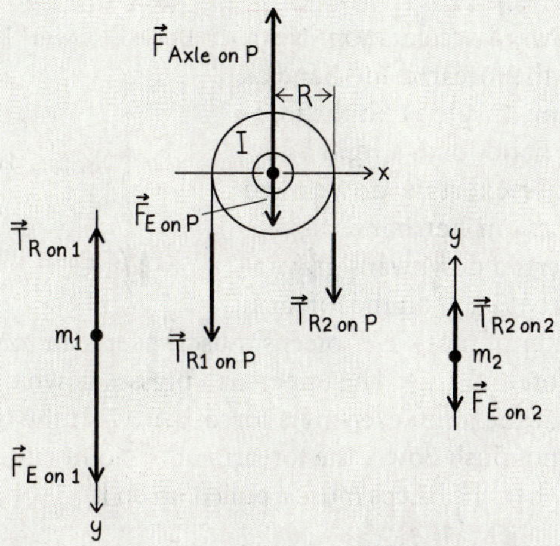


The translational acceleration of the hanging objects is due to the difference between the gravitational force that Earth exerts on them and the tension force that the string exerts on them. The rotational acceleration of the pulley is due to a nonzero net torque produced by the two tension forces exerted on the pulley.

We consider the pulley to be similar to a solid cylinder. Then according to Table 8.6, its rotational inertia around the axis that passes through its center is $I = \frac{1}{2}MR^2$, where $R$ is the radius of the pulley and $M$ is its mass.

**Represent mathematically** The force diagrams help us apply Newton's second law in component form for the two blocks and the rotational form for the pulley. The coordinate systems used in each case are shown. We choose the coordinate systems so the translational accelerations of both blocks are positive.

Block on the left: $\quad m_1a = +m_1g + (-T_{\text{R1 on 1}})$

Block on the right: $\quad m_2a = -m_2g + T_{\text{R2 on 2}}$

The pulley: $T_{\text{R1 on P}}R + (-T_{\text{R2 on P}}R) = I\alpha = \left(\frac{MR^2}{2}\right)\left(\frac{a}{R}\right)$

$$\Rightarrow (T_{\text{R1 on 1}})R - (T_{\text{R2 on 2}})R = \frac{MRa}{2}$$

$$\Rightarrow T_{\text{R1 on 1}} - T_{\text{R2 on 2}} = \frac{Ma}{2}$$

We now have three equations with three unknowns—the two tension forces exerted by the rope on the pulley and the magnitude of the acceleration $a$ of the blocks. We can write expressions for $T_{\text{R1 on 1}}$ and $T_{\text{R2 on 2}}$ using the first two equations. We then have

$$T_{\text{R1 on 1}} = m_1g - m_1a$$
$$T_{\text{R2 on 2}} = m_2g + m_2a$$

After substituting these expressions for the rope forces into the pulley equation, we get

$$(m_1g - m_1a) - (m_2g + m_2a) = Ma/2$$

**Solve and evaluate** This equation can be rearranged to get an expression for the translational acceleration of the blocks:

$$a = \frac{m_1 - m_2}{\frac{1}{2}M + m_1 + m_2}g$$

Notice that if we neglect the mass of the pulley, the acceleration becomes

$$a = \frac{m_1 - m_2}{m_1 + m_2}g,$$

a larger acceleration than with the pulley (and consistent with the result we got in Chapter 3). Thus, the massive pulley decreases the acceleration of the blocks, which makes sense. If the pulley mass $M$ is much heavier than the masses of the hanging objects $m_1$ and $m_2$, the acceleration becomes very small—it is almost like the blocks are hanging from a fixed massive object and not moving at all. We can find the rotational acceleration of the pulley by dividing the translational acceleration by the radius of the pulley:

$$\alpha = \frac{a}{R} = \left(\frac{m_1 - m_2}{\frac{1}{2}M + m_1 + m_2}\right)\frac{g}{R}$$

**Try it yourself:** Determine the translational acceleration of the blocks and the rotational acceleration of the pulley for the following given information: $m_1 = 1.2$ kg, $m_2 = 0.8$ kg, $M = 1.0$ kg, and $R = 0.20$ m.

*Answer:* $a = 1.6\ \text{m/s}^2$ and $\alpha = 7.8\ \text{rad/s}^2$.

Notice the calculations of the translational acceleration of the blocks in Example 8.5. The acceleration is equal to the sum of the forces divided by an effective total mass of the system.

$$a = \frac{m_1 - m_2}{\frac{1}{2}M + m_1 + m_2}g$$

Here you see that the pulley only contributes half of its mass due to its mass distribution. We can say that the pulley's "effective mass" is $M/2$, because of how its mass is distributed.

## EXAMPLE 8.6 Throwing a bottle

A woman tosses a 0.80-kg soft drink bottle vertically upward to a friend on a balcony above. At the beginning of the toss, her forearm rotates upward from the horizontal so that the hand exerts a 20-N upward force on the bottle. Determine the force that her biceps exerts on her forearm during this initial instant of the throw. The mass of her forearm is 1.5 kg and its rotational inertia about the elbow joint is $0.061\ \text{kg}\cdot\text{m}^2$. The attachment point of the biceps muscle is 5.0 cm from the elbow joint, the hand is 35 cm away from the elbow, and the center of mass of the forearm/hand is 16 cm from the elbow.

**Sketch and translate** A sketch of the situation is shown below. There is no information given about the kinematics of the process (for example, no way to directly determine the rotational acceleration of her arm). How can we use the rotational form of Newton's second law to determine the unknown force that the biceps muscle exerts on the woman's forearm during the throw? We do know the force her hand exerts on the bottle and the bottle's mass. So, in the first part of the problem, we can first use the translational form of Newton's second law with the bottle as the system to find the bottle's vertical acceleration at the beginning of the throw. We can then use this acceleration to find the angular acceleration of the arm and then finally use the rotational form of Newton's second law to find the force that the biceps muscle exerts on her arm during the throw. For this second part of the problem, we choose the lower arm and hand as the system of interest. The axis of rotation is at the elbow joint between the upper arm and the forearm.

Forearm tossing a bottle

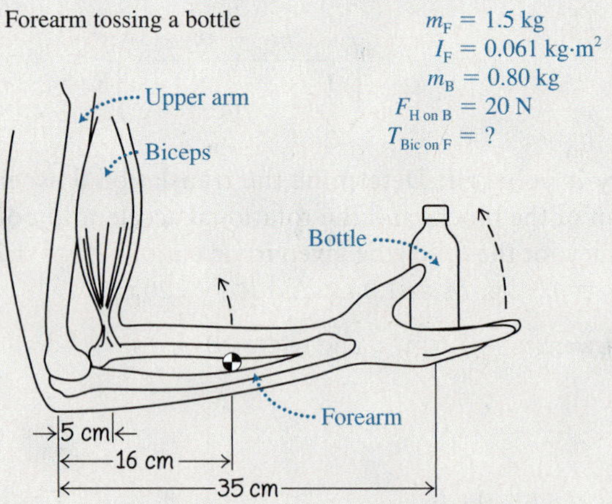


**Simplify and diagram** The figure at top right is a force diagram for the bottle as a system. Earth exerts a downward 7.8-N gravitational force $\vec{F}_{\text{E on B}}$ on the bottle and the woman's hand exerts an upward 20-N normal force $\vec{N}_{\text{H on B}}$ on the bottle. Since these forces do not cancel, the bottle has an initial upward acceleration. Next, consider the forearm and hand as the system. Assume that the forearm and hand form a rigid body. The bottle exerts a downward 20-N force on her hand $\vec{N}_{\text{B on H}}$. Earth exerts a downward gravitational force $\vec{F}_{\text{E on F}}$ on the forearm at its center of mass. Her biceps muscle exerts an upward tension force $\vec{T}_{\text{Bic on F}}$. The upper arm presses down on the forearm at the joint, exerting a force $\vec{F}_{\text{UA on F}}$. If the upper arm did not push down, the forearm at the joint would fly upward when the biceps muscle pulled up on it.

Bottle is system

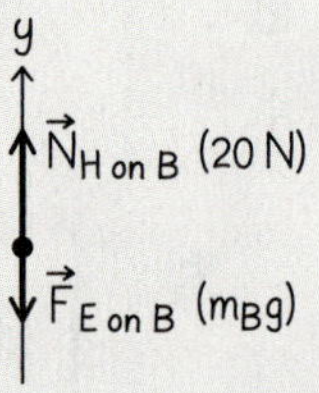


Forearm is system

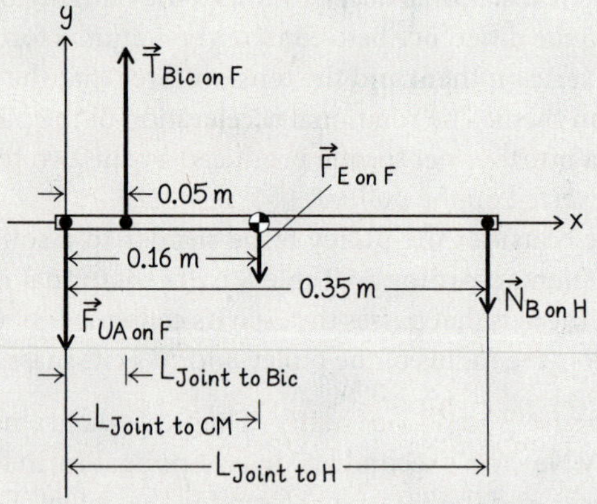


**Represent mathematically** We first analyze the bottle's motion to determine its translational acceleration; then determine the rotational acceleration of the forearm and hand system; and finally apply the rotational form of Newton's second law to find the force that the biceps needs to exert on the system to cause this rotational acceleration. Consider the initial instant of the bottle's upward trip. The $y$-component form of Newton's second law applied to the bottle can be used to determine the vertical acceleration $a_{\text{B}\,y}$ for the bottle:

$$a_{\text{B}\,y} = \frac{N_{\text{H on B}\,y} + F_{\text{E on B}\,y}}{m_\text{B}} = \frac{N_{\text{H on B}} + (-m_\text{B}g)}{m_\text{B}}$$

The rotational acceleration of the forearm/hand system at that instant is related to the vertical acceleration of the bottle:

$$\alpha_\text{F} = \frac{a_{\text{B}\,y}}{r}$$

where $r$ is the distance from the axis of rotation to the hand. The magnitude of the force that the biceps muscle exerts on the forearm ($T_{\text{Bic on F}}$) can be determined using the rotational form of Newton's second law applied to the forearm/hand. Notice that here the system consists of different parts joined together.

$$T_{\text{Bic on F}}L_{\text{Joint to Bic}} + (-F_{\text{E on F}}L_{\text{Joint to CM}}) + (-N_{\text{B on H}}L_{\text{Joint to H}}) = I_{\text{F}}\alpha_{\text{F}}$$

**Solve and evaluate** We now use the known values of the quantities to solve the problem:

$$a_{\text{B}\,y} = \frac{N_{\text{H on B}} - m_{\text{B}}g}{m_{\text{B}}} = \frac{20\ \text{N} - (0.80\ \text{kg})(9.8\ \text{N/kg})}{(0.80\ \text{kg})} = 15.2\ \text{m/s}^2$$

$$\alpha_{\text{F}} = \frac{a_{\text{B}\,y}}{r} = \frac{(15.2\ \text{m/s}^2)}{(0.35\ \text{m})} = +43.4\ \text{rad/s}^2$$

$$T_{\text{Bic on F}}(0.05\ \text{m}) - [(1.5\ \text{kg})(9.8\ \text{N/kg})](0.16\ \text{m}) - (20\ \text{N})(0.35\ \text{m}) = (0.061\ \text{kg}\cdot\text{m}^2)(43.4\ \text{rad/s}^2)$$

Solving the above equation, we find that $T_{\text{Bic on F}} = 240\ \text{N} = 54\ \text{lb}$, a reasonable magnitude for this force.

**Try it yourself:** Determine the force that the woman's biceps exerts on her forearm during the initial instant of a vertical toss of a 100-g rubber ball if she is exerting a 10-N force on the ball.

*Answer:* 430 N.

**Review Question 8.4** How is Newton's second law for rotational motion similar to Newton's second law for translational motion?

## 8.5 Rotational momentum

Earlier in this textbook (Chapters 5 and 6), we constructed powerful principles for momentum and energy that allowed us to analyze complex processes that involved translational motion. Is it possible to find analogous principles for the rotational (angular) momentum and rotational energy of extended bodies? Consider the rotational inertia involved in the Observational Experiment **Table 8.7** following experiments.

**OBSERVATIONAL EXPERIMENT TABLE**

**8.7 Observations concerning rotational motion.**

| Observational experiment | Analysis |
|---|---|
| (a) A figure skater initially spins slowly with a leg and two arms extended. Then she pulls her leg and arms close to her body and her spinning rate increases dramatically.   | *Initial situation:* Large rotational inertia $I$ and small rotational speed $\omega$.<br>*Final situation:* Smaller rotational inertia $I$ and larger rotational speed $\omega$. |

*(continued)*

| Observational experiment | Analysis |
| --- | --- |
| (b) A man sitting on a chair that can spin with little friction initially holds barbells far from his body and spins slowly. When he pulls the barbells close to his body, the spinning rate increases dramatically. 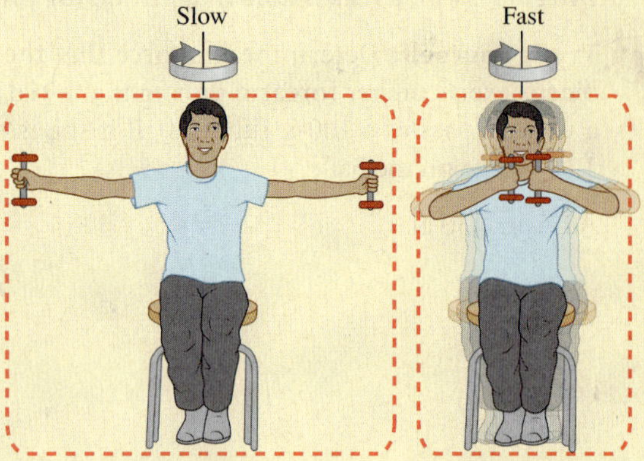  | *Initial situation:* Large rotational inertia $I$ and small rotational speed $\omega$.<br>*Final situation:* Smaller rotational inertia $I$ and larger rotational speed $\omega$. |

**Pattern**

- There are no external forces exerted on either person—no torques.
- As the mass distribution of the system moves closer to the axis of rotation, the system's rotational inertia $I$ decreases and the system's rotational speed $\omega$ increases (even though the net torque on the system is zero).

For each experiment in Table 8.7, the rotational inertia $I$ of the spinning person decreased (the mass moved closer to the axis of rotation). Simultaneously, the rotational speed $\omega$ of the person increased. When $I$ increases and $\omega$ decreases (or vice versa), $I\omega$ remains constant. We propose tentatively that when the rotational inertia $I$ of an extended body in an isolated system decreases, its rotational speed $\omega$ increases, and vice versa.

Let's test this idea qualitatively in Testing Experiment **Table 8.8**.

**TESTING EXPERIMENT TABLE**

### 8.8 Testing the idea that $\omega$ increases if $I$ decreases.

| Testing experiment | Prediction | Outcome |
| --- | --- | --- |
| A puck is tied to a string that passes through a hole in the center of an air table. As the puck moves, the string is pulled down through the table, decreasing the radius of the circular path of the puck. 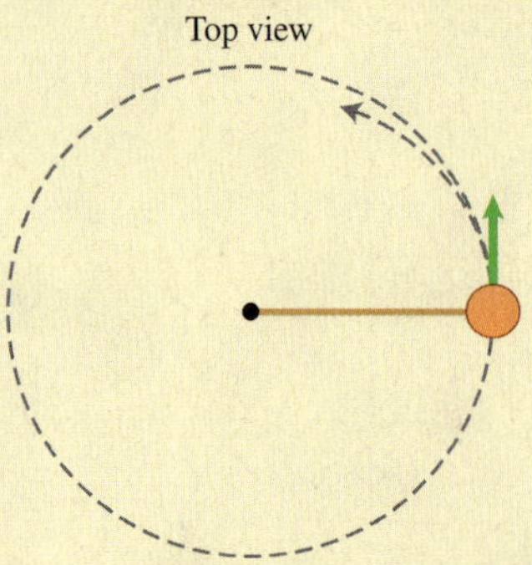  | As $r$ decreases, the puck's rotational inertia $I = mr^2$ decreases. According to the proposed rule, the puck's rotational speed $\omega$ should increase—it should take less time to complete one rotation. | We observe that the rotational speed increases—it takes less time to complete one rotation around the post. |

**Conclusion**

This result supports the idea that $\omega$ increases if the rotational inertia $I$ of an isolated system decreases.

## Rotational momentum is constant for an isolated system

Note that $I$ is the rotational analogue of the mass $m$ of a point-like object and $\omega$ is the rotational analogue of the translational velocity $\vec{v}$. The linear momentum of an object is the product of its mass $m$ and its velocity $\vec{v}$. Let's propose that a turning object's rotational momentum $L$ (analogous to linear momentum $\vec{p} = m\vec{v}$) is defined as

$$L = I\omega$$

In the chapter on linear momentum (Chapter 5) we derived a relationship [Eq. (5.4)] between the net force exerted on an object and the change in its linear momentum:

$$\Sigma\vec{F}(t_f - t_i) = \vec{p}_f - \vec{p}_i \quad (5.4)$$

where $\vec{p} = m\vec{v}$. Torque $\tau$ is analogous to force $\vec{F}$. Thus, using the analogy between rotational and translational motion, we write

$$\Sigma\tau(t_f - t_i) = L_f - L_i$$

where $L = I\omega$. If a system with one rotating body is isolated, then the external torque exerted on the object is zero. In such a case, the rotational momentum of the object does not change ($0 = L_f - L_i$), and the object's rotational momentum is constant ($L_f = L_i$), or

$$I_i\omega_i = I_f\omega_f$$

Note that this is consistent with our tentative qualitative rule. If the final value of one quantity ($I$ or $\omega$) increases for an isolated system, then the other quantity must decrease. Now, let's try a quantitative test of this proposed rule—similar to the qualitative test done in Testing Experiment Table 8.8.

**EXAMPLE 8.7 Puck on a string**

Attach a 100-g puck to a string and let the puck glide in a counterclockwise circle (positive rotational momentum) on a horizontal, frictionless air table. The other end of the string passes through a hole at the center of the table. You pull down on the string so that the puck moves along a circular path of radius 0.40 m. It completes one revolution in 4.0 s. If you pull harder on the string so the radius of the circle slowly decreases to 0.20 m, what is the new period of revolution?

**Sketch and translate** A sketch of the situation is shown below. We choose the puck as the system and place the axis of rotation at the center of the table where the string passes through the hole.

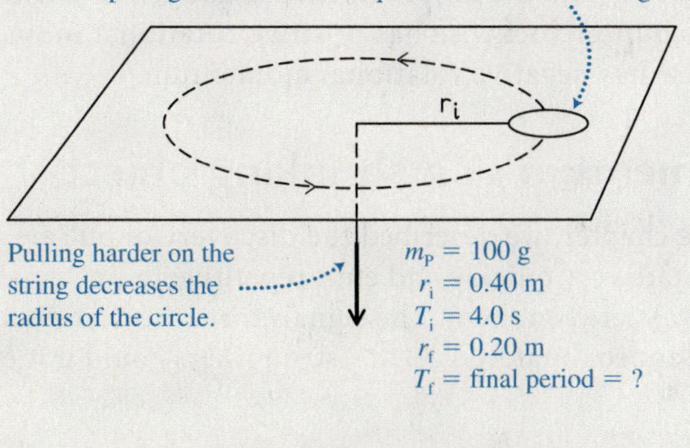


**Simplify and diagram** We model the puck as a point-like object. The force diagram to the right shows the situation viewed from the side of the air table looking from the back of the puck along its direction of motion. The tension force exerted by the string on the puck $\vec{F}_{S\text{ on }P}$ passes through the axis of rotation and therefore produces zero torque. The upward normal force that the air table exerts on the puck $\vec{N}_{T\text{ on }P}$ balances the downward gravitational force $\vec{F}_{E\text{ on }P}$ of Earth on the puck, so they have zero net effect on the rotational motion of the puck. Also, the net torque due to these two forces is zero. So the net torque produced by all forces exerted on the puck is zero. According to the rule we are testing, this means that the rotational momentum of the puck should be constant, even though the radius of its circular path is decreasing.

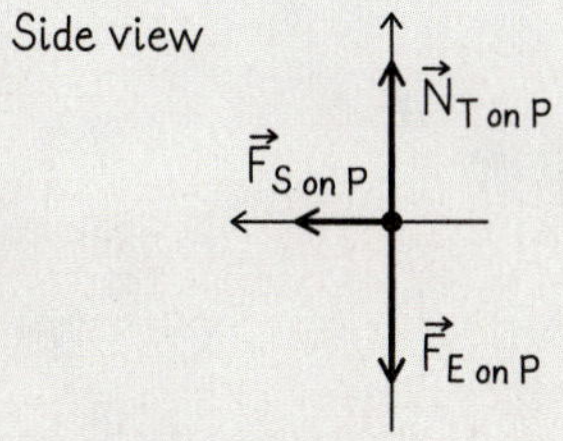


**Represent mathematically** Since the net external torque on the system is zero, the initial and final rotational momenta of the puck should be equal ($L_f = L_i$). Thus,

$$I_f\omega_f = I_i\omega_i$$

*(continued)*

The puck travels once around the circle ($2\pi$ rad) in a time interval of one period $T$. Thus, the puck's rotational speed is

$$\omega = \frac{2\pi}{T}$$

The rotational inertia of the puck around the axis of rotation is

$$I = mr^2$$

Combining these last three equations, we get

$$(mr_i^2)\left(\frac{2\pi}{T_i}\right) = (mr_f^2)\left(\frac{2\pi}{T_f}\right)$$

Dividing each side by common terms, we get

$$\frac{r_f^2}{T_f} = \frac{r_i^2}{T_i}$$

Now multiply both sides of the equation by $T_f T_i$ and divide by $r_i^2$ to obtain an expression for $T_f$:

$$T_f = \frac{r_f^2 T_i}{r_i^2}$$

**Solve and evaluate** Insert the known quantities to get

$$T_f = \frac{r_f^2 T_i}{r_i^2} = \frac{(0.20\ \text{m})^2(4.0\ \text{s})}{(0.40\ \text{m})^2} = 1.0\ \text{s}$$

Remember that this is a testing experiment. When the experiment is performed, we find that the time interval with the reduced radius is very close to 1.0 s. The outcome of the experiment is consistent with the prediction.

**Try it yourself:** An 80-kg roller skater holds a rope that loops around a thick metal post, causing him to skate in a circular path. It takes him 8.0 s to complete one rotation around the pole. He starts pulling himself inward along the rope so that the radius of his motion decreases from 2.0 m to 1.0 m. Determine his rotational speed once he has pulled himself inward to a radius of 1.0 m.

*Answer:* 3.14 rad/s.

**Rotational momentum and rotational impulse** We now have a quantitative relation between rotational momentum $L = I\omega$ and rotational impulse.

$$L_i + \Sigma\tau\Delta t = L_f \qquad (8.12)$$

**The initial rotational momentum of a turning object plus the product of the net external torque exerted on the object and the time interval during which it is exerted equals the final rotational momentum of the object.**

**If the net torque that external objects exert on the turning object is zero, or if the torques add to zero, then the rotational momentum $L$ of the turning object remains constant:**

$$L_f = L_i \text{ or } I_f\omega_f = I_i\omega_i \qquad (8.13)$$

**TIP** Rotational momentum is sometimes called angular momentum.

To explain most of the applications of torque and rotational momentum in this book, we account for their directions using positive or negative signs. A torque is positive if it tends to rotate the object counterclockwise and negative if it tends to rotate the object clockwise about the axis of rotation. A body rotating counterclockwise has positive rotational momentum and if rotating clockwise has negative rotational momentum.

## Rotational momentum of a shrinking object

At the beginning of the chapter, we described the discovery of pulsars, astronomical objects that rotate very quickly and emit repetitive radio signals with a very small time interval between them. The signals from the first discovered pulsar had a period of approximately 1.33 s. Astronomers could not explain at first how pulsars could rotate so rapidly. Most stars, including our Sun, rotate very much the way Earth does, usually taking several days to complete one rotation (about a month for our Sun).

However, as a star's core collapses and its mass moves closer to the axis of rotation, its rotational velocity increases because its rotational momentum is constant (assuming the star does not interact with any other objects). How much does a star need to shrink so that its period of rotation becomes seconds instead of days?

**EXAMPLE 8.8 A pulsar**

Imagine that our Sun ran out of nuclear fuel and collapsed. What would its radius have to be in order for its period of rotation to be the same as the pulsar described above? The Sun's current period of rotation is 25 days.

**Sketch and translate** First, sketch the process. The Sun is the system. We can convert the present period of rotation of the Sun into seconds ($T_i = 25\text{ days} = 2.16 \times 10^6\text{ s}$). Its mass is $m = 2.0 \times 10^{30}\text{ kg}$ and its radius is $R_i = 0.70 \times 10^9\text{ m}$. After the Sun collapses, its period of rotation will be $T_f = 1.33\text{ s}$. What will be its radius?

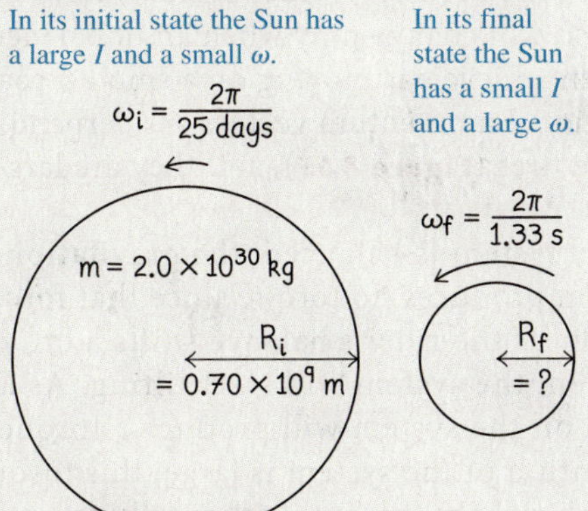
In its initial state the Sun has a large I and a small ω.
ωi = 2π / 25 days
m = 2.0 × 10^30 kg
Ri = 0.70 × 10^9 m
In its final state the Sun has a small I and a large ω.
ωf = 2π / 1.33 s
Rf = ?

**Simplify and diagram** Assume that the Sun is a sphere with its mass distributed uniformly. Assume also that it does not lose any mass as it collapses.

**Represent mathematically** Now, apply the principle of rotational momentum conservation (Eq. 8.14) to the Sun's collapse:

$$I_i\omega_i = I_f\omega_f$$

From Table 8.6 we find that the rotational inertia of a sphere rotating around an axis passing through its center is

$$I = \left(\frac{2}{5}\right)mR^2$$

The rotational velocity of an object is

$$\omega = \frac{\Delta\theta}{\Delta t} = \frac{2\pi}{T}$$

where $T$ is the period for one rotation. Combining the above three equations, we get

$$\left(\frac{2}{5}mR_i^2\right)\frac{2\pi}{T_i} = \left(\frac{2}{5}mR_f^2\right)\frac{2\pi}{T_f}$$

Dividing by the $2/5$, $2\pi$, and $m$ on each side of the equation, we get

$$\frac{R_i^2}{T_i} = \frac{R_f^2}{T_f}$$

**Solve and evaluate** Multiply both sides of the above by $T_f$ and take the square root:

$$R_f = \sqrt{\frac{R_i^2 T_f}{T_i}}$$

$$= \sqrt{\frac{(0.70 \times 10^9\text{ m})^2(1.33\text{ s})}{2.16 \times 10^6\text{ s}}}$$

$$= 5.5 \times 10^5\text{ m}$$

$$= 550\text{ km}$$

Although this is much smaller than the radius of Earth, models of stellar evolution actually do predict that the Sun's core will eventually shrink to this size and possibly smaller.

**Try it yourself:** When massive stars explode, the collapse can shrink their radii to about 10 km. What would be the period of rotation of such a star if it originally had a mass twice the mass of the Sun, a radius that was 1.3 times the Sun's radius, and the same initial period of rotation as the Sun (25 days)?

*Answer:* $2.6 \times 10^{-4}$ s.

**Figure 8.13** Using the right-hand rule to determine the direction of an object's rotational momentum $\vec{L}$.

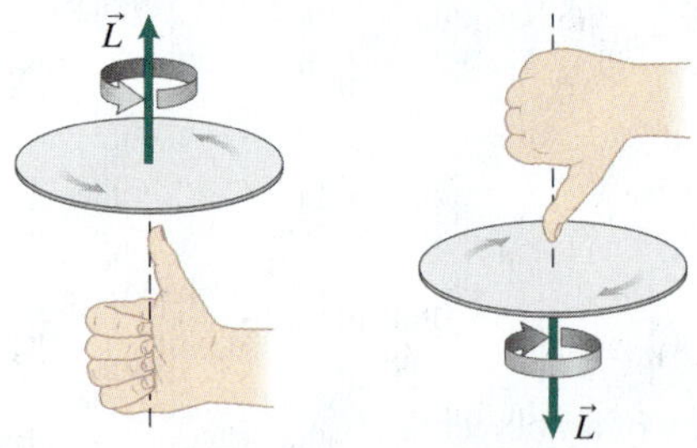


**Figure 8.14** A bicycle that is not moving is in unstable equilibrium.

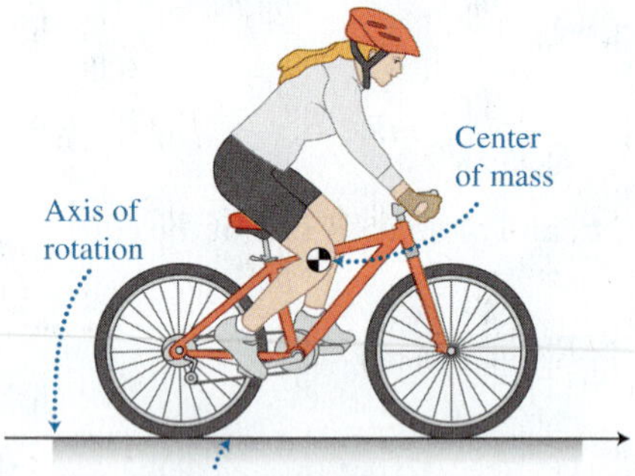


**Figure 8.15** Rotating bicycle tires have rotational momentum that stabilizes the bicycle.

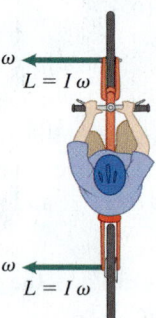


## Vector nature of torque, rotational velocity, and rotational momentum

We have not yet considered the vector nature of torque, rotational velocity, and rotational momentum. However several important applications depend on an understanding of the vector nature of these quantities. The vector direction of both rotational velocity and rotational momentum can be determined using **a right-hand rule.**

**Right-hand rule for determining the direction of rotational velocity and rotational momentum** Curl the four fingers of your right hand in the direction of rotation of the turning object. Your thumb, held perpendicular to the fingers, then points in the direction of both the object's rotational velocity and rotational momentum (**Figure 8.13**). To determine the **vector direction** of the torque that a force produces on an object (as opposed to the clockwise/counterclockwise way of describing it) about an axis of rotation, first imagine that the object is at rest and that the torque you are interested in is the only torque exerted on the object. Next, curl the fingers of your right hand in the direction that the torque would make the object rotate. Your thumb, held perpendicular to the fingers, shows the direction of this torque.

**Bicycling** We can use the vector nature of torque and rotational momentum to understand why a bicycle is much more stable when moving fast—especially if the bicycle has massive tires. Consider an axis of rotation parallel to the ground that passes through the two contact points of the tires with the ground. This axis of rotation is below the center of mass of the stationary bike—an unstable equilibrium (**Figure 8.14**). When the bicycle is moving quickly, the rotating tires (and therefore the bicycle + rider system) have considerable rotational momentum, which will change only when an unbalanced torque is exerted on the system. When a bicycle is moving on a smooth road, the rotational velocity and the rotational momentum vectors are perpendicular to the plane of rotation of the bike tires (**Figure 8.15**), and they are large due to the rapid rotation of the tires.

When the bike + rider system is balanced, the gravitational force exerted by Earth on the system produces no torque since that force points directly at the axis of rotation. If the rider's balance shifts a bit, or the wind blows, or the road is uneven, the system will start tilting. As a result, the gravitational force exerted on the system will produce a torque. However, since the rotational momentum of the system is large, this torque does not change its direction by much right away, but it takes only several tenths of a second for the torque to change the rotational momentum significantly. This is enough time for an experienced rider to make corrections to rebalance the system. The faster the person is riding the bike, the greater the rotational momentum of the system and the more easily she/he can keep the system balanced.

**Gyroscopes** Guidance systems for spaceships rely on the constancy of rotational momentum in isolated systems to help them maintain their chosen course. Once the ship is pointed in the desired direction, one or more heavy gyroscopes starts rotating. The gyroscope is a wheel whose axis of rotation keeps the ship oriented in the chosen direction. The gyroscope is similar to the rotating bicycle tires that help keep a rider upright without tipping or changing direction. Gyroscopes are also used in cameras to prevent them from vibrating or moving while the camera lens is open.

**Review Question 8.5** After a playground merry-go-round is set in motion, its rotational speed decreases noticeably if another person jumps on it. However, if a person riding the merry-go-round steps off, the rotational speed seems not to change at all. Explain.

## 8.6 Rotational kinetic energy

We are familiar with the kinetic energy $(1/2)mv^2$ of a single particle moving along a straight line or in a circle. It would be useful to calculate the kinetic energy of a rotating body—like Earth. Doing so would allow us to use the work-energy approach to solving problems involving rotation. Let's start by deriving an expression for the rotational kinetic energy of a single particle of mass $m$ moving in a circle of radius $r$ at speed $v$. According to the kinematics in Section 8.1, its linear speed $v$ and rotational speed $\omega$ are related:

$$v = r\omega$$

Thus, the kinetic energy of this particle moving in a circle can be written as

$$K_{\text{rotational}} = \frac{1}{2}mv^2 = \frac{1}{2}m(r\omega)^2 = \frac{1}{2}(mr^2)\omega^2 = \frac{1}{2}I\omega^2$$

where $I = mr^2$ is the rotational inertia of a particle moving a distance $r$ from the center of its circular path. The expression for the translational kinetic energy $(1/2)mv^2$ of a particle is similar to the rotational version, which involves the product of a mass-like term $I$ and the square of a speed-like term $\omega$. Can we use the expression $\frac{1}{2}I\omega^2$ for the rotational kinetic energy of a rotating rigid body?

To test this idea, consider a solid sphere of known radius $R$ and mass $m$ that can rotate freely on an axis. We wrap a string around the sphere and pull the string with a force probe exerting a constant force of a known magnitude so that the sphere starting at rest completes 5.0 revolutions (**Figure 8.16a**). After we stop pulling, we measure the rotational speed $\omega$ of the sphere. But before measuring it, we predict its value using this expression for rotational kinetic energy. If we choose the sphere as the system, the string is the only external object that exerts a force that causes a nonzero torque on the sphere. This string force does work, which changes the sphere's kinetic energy from zero to some new value (Figure 8.16b). Thus, the initial rotational kinetic energy of the sphere (zero) plus the work done by the string on the sphere during these five turns equals the final rotational kinetic energy of the sphere:

$$K_i + W = K_f$$

The string pulls parallel to the displacement of the edge of the sphere during the entire time. We can use the expression for the rotational kinetic energy under test to predict the magnitude of the final rotational speed:

$$K_i + W = K_f$$

$$0 + F_{\text{String on Sphere}}(5 \cdot 2\pi R)\cos 0 = \frac{1}{2}I\omega^2$$

where $5 \cdot 2\pi R$ is the distance the string is pulled—five circumferences of the sphere. The above leads to a prediction of the final rotational speed:

$$\omega_f = \sqrt{\frac{F_{\text{String on Sphere}} \cdot 20\pi R}{I}}$$

**Figure 8.16** A string pulls a solid sphere.

(a)

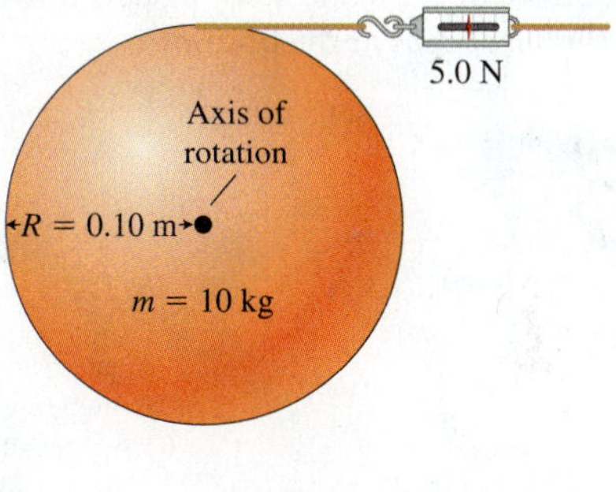


(b)

Work done by the string causes the rotational kinetic energy of the sphere to increase.

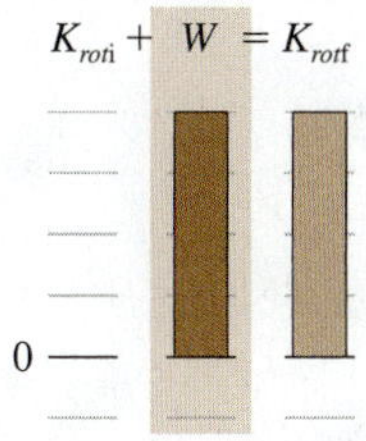

From Table 8.6 we know that the rotational inertia of a solid sphere of radius $R$ and mass $m$ is $I = (2/5)mR^2$ (the axis passes through the center of the sphere) where $m = 10$ kg and $R = 0.10$ m. Thus, the rotational inertia is

$$I = (2/5)mR^2 = (2/5)(10\text{ kg})(0.10\text{ m})^2 = 0.040\text{ kg}\cdot\text{m}^2$$

We pull the string so that it exerts a 5.0-N force on the edge of the sphere. Thus, the sphere's final speed should be

$$\omega_{\text{f}} = \sqrt{\frac{(5.0\text{ N})(20\,\pi)(0.10\text{ m})}{(0.040\text{ kg}\cdot\text{m}^2)}} = \left(28\,\frac{\text{rad}}{\text{s}}\right)\left(\frac{1\text{ rev}}{2\pi\text{ rad}}\right) = 4.5\,\frac{\text{rev}}{\text{s}}$$

When we measure the final angular velocity, it is about 4.5 rev/s.

Let's test our idea for the mathematical expression of rotational kinetic energy in Testing Experiment **Table 8.9**.

## TESTING EXPERIMENT TABLE

### 8.9 Testing the expression for rotational kinetic energy.

| Testing experiment | Prediction | Outcome |
|---|---|---|
| A solid cylinder and a hoop of the same radius and mass start rolling at the top of an inclined plane. Which object reaches the bottom of the plane first? What is the ratio of their speeds at the bottom?<br>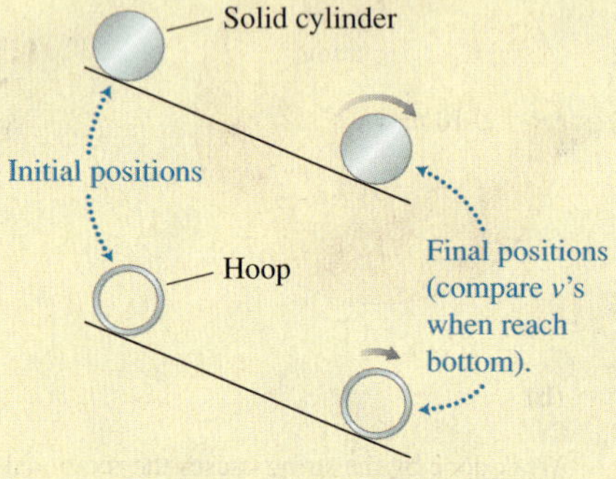 | Both Earth-object systems start with the same gravitational potential energy. As they roll, both acquire translational and rotational kinetic energies. We can represent the process in a work-energy bar chart.<br>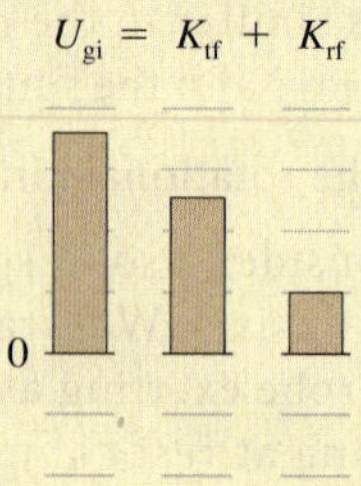<br>The bar chart helps us construct a mathematical description.<br>$$mgy_{\text{i}} = \frac{1}{2}mv_{\text{f}}^2 + \frac{1}{2}I\omega_{\text{f}}^2$$<br>The rotational speed $\omega$ and translational speed $v$ are related:<br>$$\omega = v/r$$<br>Substitute in the energy equation and then rearrange:<br>$$mgy_{\text{i}} = \frac{1}{2}\left(m + \frac{I}{r^2}\right)v_{\text{f}}^2$$<br>$$v_{\text{f}} = \sqrt{\frac{2mgy_{\text{i}}}{\left(m + \frac{I}{r^2}\right)}}$$<br>At the bottom of the inclined plane, the object with greater rotational inertia $I$ will have the least translational speed $v_{\text{f}}$.<br>$I_{\text{cylinder}} = (1/2)mr^2$ and $I_{\text{hoop}} = mr^2$. Thus, $I_{\text{hoop}} > I_{\text{cylinder}}$. Thus, the hoop should be moving more slowly. | The speeds, measured with a motion detector, are consistent with the prediction. |

*Calculations:* Insert the expressions for the rotational inertia of the solid cylinder and of the hoop to find their final speeds:

Cylinder $v_f = \sqrt{\frac{4}{3}gy_i}$

Hoop $v_f = \sqrt{gy_i}$

The ratio of their final speeds is $\sqrt{\frac{4}{3}}$, with the solid cylinder moving faster.

**Conclusion**

The outcome of the experiment supports the expression for the rotational kinetic energy of a rigid body:

$$K_{\text{rotational}} = \frac{1}{2}I\omega^2$$

This is the second testing experiment involving rotational kinetic energy in which the outcome matched the prediction. Given this support for our prediction and the lack of counterevidence, we will use this mathematical expression for a rigid body's rotational kinetic energy.

**Rotational kinetic energy** **The rotational kinetic energy of an object with rotational inertia $I$ turning with rotational speed $\omega$ is**

$$K_{\text{rotational}} = \frac{1}{2}I\omega^2 \quad (8.14)$$

**TIP** When you encounter a new physical quantity, always check whether its units make sense. In this particular case the units for $I$ are $\text{kg}\cdot\text{m}^2$ and the units for $\omega^2$ are $1/\text{s}^2$. Thus, the unit for kinetic energy is $\text{kg}\cdot\text{m}^2/\text{s}^2 = (\text{kg}\cdot\text{m}/\text{s}^2)\text{m} = \text{N}\cdot\text{m} = \text{J}$, the correct unit for energy.

## Flywheels for storing and providing energy

You stop your car at a stoplight. Before stopping, the car had considerable kinetic energy; after stopping, the kinetic energy is zero. It has been converted to internal energy due to friction in the brake pads. Unfortunately, this thermal energy cannot easily be converted back into a form that is useful. Is there a way to convert that translational kinetic energy into some other form of energy that would help the car regain translational kinetic energy when the light turns green?

Efforts are under way to use the rotational kinetic energy of flywheels (rotating disks) for this purpose. Instead of rubbing a brake pad against the wheel and slowing it down, the braking system would, through a system of gears or through an electric generator, convert the car's translational kinetic energy into the rotational kinetic energy of a flywheel. As the car's translational speed decreases, the flywheel's rotational speed increases. This rotational kinetic energy could then be used to help the car start moving, rather than relying entirely on the chemical potential energy of gasoline.

**EXAMPLE 8.9 Flywheel rotational speed**

A 1600-kg car traveling at 20 m/s approaches a stop sign. If it could transfer all of its translational kinetic energy to a 0.20-m-radius, 20-kg flywheel while stopping, what rotational speed would the flywheel acquire?

**Sketch and translate** A sketch of the situation is shown below. The system of interest will be the car, including the flywheel.

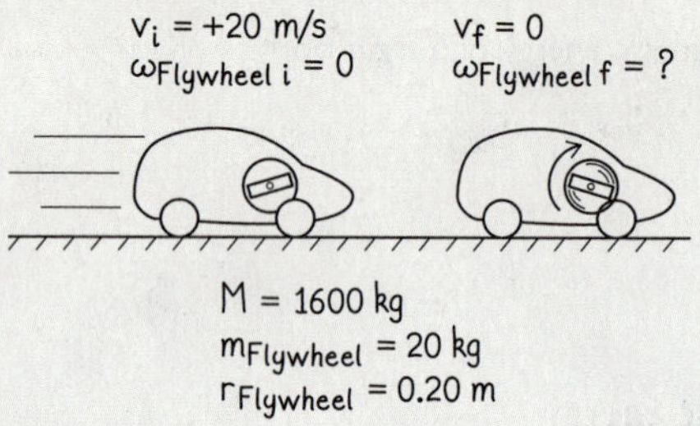


**Simplify and diagram** The process is represented in the figure below with a bar chart. The initial energy of the system is the car's translational kinetic energy; the final energy is the flywheel's rotational kinetic energy. Assume that the flywheel is a solid disk with rotational inertia of $(1/2)mr^2$ (see Table 8.6).

Braking converts the car's initial kinetic energy $K_{ti}$ into the flywheel's final rotational kinetic energy $K_{rf}$—saved for future use.

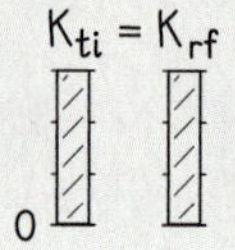


**Represent mathematically** Use the bar chart to help construct an energy conservation equation:

$$K_{\text{translational i}} = K_{\text{rotational f}}$$

$$\frac{1}{2}Mv_i^2 = \frac{1}{2}I\omega_f^2$$

Multiplying both sides of the equation by 2 and dividing by $I$, we get

$$\omega_f^2 = \frac{Mv_i^2}{I}$$

The rotational inertia of the disk (a solid cylinder) is $I_{\text{cylinder}} = (1/2)mr^2$. Thus,

$$\omega_f^2 = \frac{Mv_i^2}{I} = \frac{Mv_i^2}{\frac{1}{2}mr^2} = \frac{2Mv_i^2}{mr^2}$$

**Solve and evaluate** To find the rotational speed, take the square root of both sides of the above equation:

$$\omega_f = \frac{v_i}{r}\sqrt{\frac{2M}{m}} = \frac{20\text{ m/s}}{0.20\text{ m}}\sqrt{\frac{2(1600\text{ kg})}{(20\text{ kg})}} = 1300\text{ rad/s}$$

$$= 200\text{ rev/s} = 12{,}000\text{ rpm}$$

**Try it yourself:** Could you store more energy in a rotating hoop or in a rotating solid cylinder, assuming they have the same mass, radius, and rotational speed?

*Answer:* The hoop has a greater rotational inertia and therefore would have a greater rotational kinetic energy at the same rotational speed.

## Rolling versus sliding

Our knowledge of rotational kinetic energy helps explain a very simple but rather mysterious experiment that you can perform at home. For this experiment, you need two identical plastic water bottles. Fill one of them with snow. Pack the snow tightly. Fill the other one with water so that the mass of the water-filled bottle is the same as the mass of the snow-filled bottle. Place one of them at the top of an inclined plane (**Figures 8.17a** and b) and let it roll down. Then repeat the experiment with the other bottle. You observe an interesting effect. When the snow-filled bottle rolls down, it rotates and the solid snow inside rotates with the bottle. The bottle with water rotates, too, but the water inside does not rotate (Figure 8.17b). Thus, in effect the water slides down the incline and does not roll. Which bottle will win a race down the inclined plane if they travel side by side? Think about your prediction before you read on.

When the snow-filled bottle rolls down, it rotates as a solid cylinder, acquiring rotational kinetic energy in addition to translational kinetic energy.

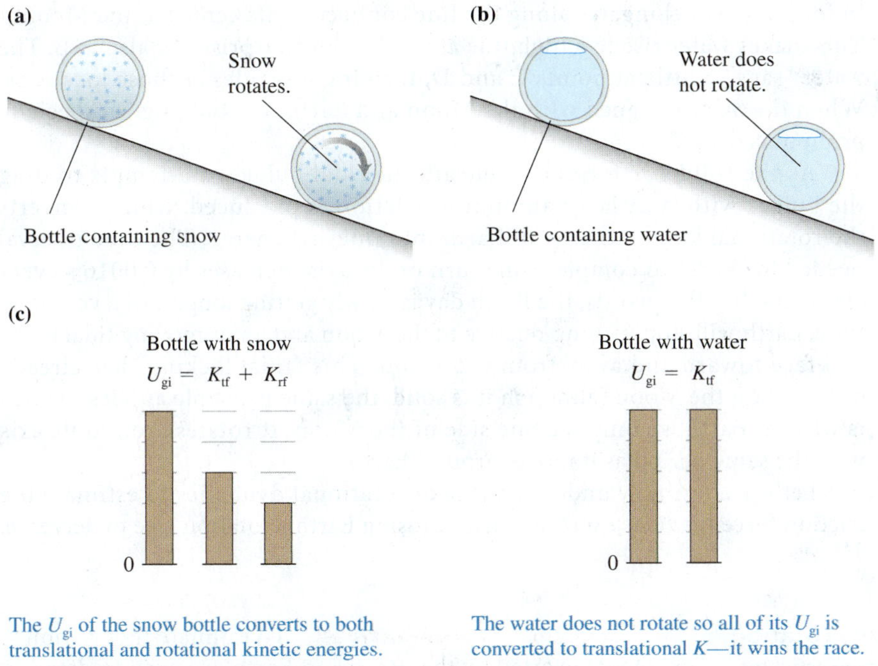
(a)
Snow rotates.
Bottle containing snow
(b)
Water does not rotate.
Bottle containing water
(c)
Bottle with snow
$U_{gi} = K_{tf} + K_{rf}$
0
Bottle with water
$U_{gi} = K_{tf}$
0
The $U_{gi}$ of the snow bottle converts to both translational and rotational kinetic energies.
The water does not rotate so all of its $U_{gi}$ is converted to translational $K$—it wins the race.

**Figure 8.17** Bottles with solid snow and liquid water race down identical inclines.

In the case of the water-filled bottle, only the bottle rotates. The water inside just translates. Rolling is a combination of translation and rotation, whereas sliding involves only translation. The energy bar charts help us understand the energy transformations during the process (Figure 8.17c). The water-filled bottle has no rotational kinetic energy and a larger translational kinetic energy at the end of the plane. It should reach the bottom first. It's a fun experiment—try it and see if the outcome matches the prediction!

**Review Question 8.6** Will a can of watery chicken noodle soup roll slower or faster down an inclined plane than an equal-mass can of thick, sticky English clam chowder?

## 8.7 Rotational motion: Putting it all together

We can use our knowledge of rotational motion to analyze a variety of phenomena that are part of our world. In this section, we consider two examples—the effect of the tides on the period of Earth's rotation (the time interval for 1 day) and the motion of bowling (also called pitching) in the sport of cricket.

### Tides and Earth's day

The level of the ocean rises and falls by an average of 1 m twice each day, a phenomenon known as the tides. Many scientists, including Galileo, tried to explain this phenomenon and suspected that the Moon was a part of the answer. Isaac Newton was the first to explain how the motion of the Moon actually creates tides. He noted that at any moment, different parts of Earth's surface are at different distances from the Moon and that the distance from a given location on Earth to the Moon varied as Earth rotated. As illustrated in **Figure 8.18**, point A is closer to the Moon than the center of Earth or point B are, and therefore the gravitational force exerted by the Moon on point A is greater than the gravitational force exerted on point B. Due to the difference

**Figure 8.18** The ocean bulges on both sides of Earth along a line toward the Moon.

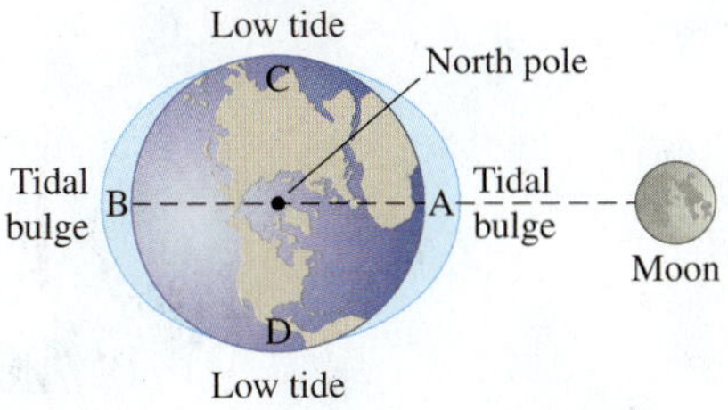
Low tide
C
North pole
Tidal bulge
B
A
Tidal bulge
Moon
D
Low tide

in forces, Earth elongates along the line connecting its center to the Moon's. This makes water rise to a high tide at point A and surprisingly also at B. The water "sags" a little at points C and D, forming low tides at those locations. When the Sun is aligned with the Moon and Earth, the bulging is especially pronounced.

As the solid Earth rotates beneath the tidal bulges, it attempts to drag the bulges with it. A large amount of friction is produced, which converts the rotational kinetic energy of Earth into internal energy. The time interval needed for Earth to complete one turn on its axis increases by 0.0016 s every 100 years. In other words, the Earth day is slowly getting longer. In a very long time, Earth will stop turning relative to the Moon and an unmoving tidal bulge will face toward and away from the Moon. This "tidal locking" has already occurred on the Moon (although it is solid, the same principle applies), which is why on Earth we only see one side of the Moon. It rotates around its axis with the same period as it moves around Earth.

Let's use our new understanding of rotational dynamics to estimate the friction force the tides exert on Earth, causing Earth's rotation rate to decrease.

**EXAMPLE 8.10 Tides slow Earth's rotation**

Estimate the effective tidal friction force exerted by ocean water on Earth that causes a 0.0016-s increase in Earth's rotation time every 100 years.

**Sketch and translate** The situation is already sketched in Figure 8.18. The solid Earth will be the system; the water covering most of its surface is considered an external object for this estimate. Earth rotates counterclockwise, taking 24 hours for one revolution (one period), as seen looking down on the North Pole. Remember that when an object rotates counterclockwise, its angular velocity is considered to be positive. Since Earth's rotation is gradually slowing, the tidal friction force is producing a negative torque, opposite the positive sign of the rotational velocity. We need to find the magnitude of the force that would increase the time of one revolution by 0.0016 s in 100 years.

**Simplify and diagram** Assume that the solid Earth is a sphere covered uniformly with water on its surface. Assume also that the frictional force exerted by the water on Earth is constant in magnitude and exerted at the equator. The force exerted by the tidal bulge on Earth produces a torque that opposes Earth's rotation.

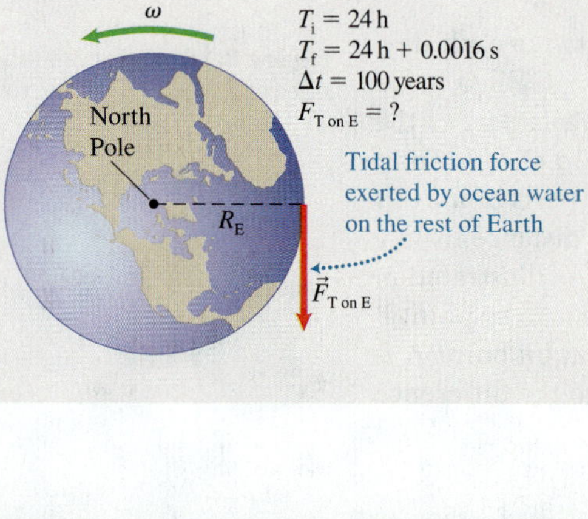


**Represent mathematically** To estimate the friction force exerted by the oceans on Earth, we need to determine the torque produced by that force. We can use the rotational form of Newton's second law if we can determine the rotational inertia of Earth and its rotational acceleration. The rotational acceleration of Earth is

$$\alpha = \frac{\Delta\omega}{\Delta t} = \frac{\omega_f - \omega_i}{\Delta t} = \frac{\left(+\frac{2\pi}{T_f}\right) - \left(+\frac{2\pi}{T_i}\right)}{\Delta t}$$

$$= \frac{2\pi}{\Delta t}\left(\frac{1}{T_f} - \frac{1}{T_i}\right)$$

where $T_i = 24\,\text{h}\left(\frac{3600\,\text{s}}{1\,\text{h}}\right) = 86{,}400\,\text{s}$, $T_f = T_i + \Delta T = 86{,}400\,\text{s} + 0.0016\,\text{s}$, and $\Delta t = 100\,\text{years}\left(\frac{365\,\text{days}}{1\,\text{year}}\right)\left(\frac{86{,}400\,\text{s}}{1\,\text{day}}\right) = 3.15 \times 10^9\,\text{s}$.

Because $T_f$ and $T_i$ are so close, your calculator will likely evaluate the rotational acceleration of Earth to be zero. To deal with this, we put the equation into another form.

$$\alpha = \frac{2\pi}{\Delta t}\left(\frac{1}{T_f} - \frac{1}{T_i}\right) = \frac{2\pi}{\Delta t}\left(\frac{T_i - T_f}{T_i T_f}\right)$$

$$= \frac{2\pi}{\Delta t}\left(\frac{-\Delta T}{T_i(T_i + \Delta T)}\right) = \frac{2\pi}{\Delta t}\left(\frac{-\Delta T}{T_i^2 + T_i\Delta T}\right)$$

Now look at the two terms in the denominator, $T_i^2$ and $T_i\Delta T$. Because $\Delta T$ is so small, the second term is much less than the first term. This means the second term can be dropped without affecting the result in any significant way. Thus,

$$\alpha = -\frac{2\pi\Delta T}{\Delta t T_i^2}$$

We can use the rotational form of Newton's second law to get an alternative expression for Earth's rotational acceleration:

$$\alpha = \frac{1}{I}\Sigma\tau = \frac{1}{\frac{2}{5}mR_E^2}(F_{\text{T on E}}R_E \sin 90°)$$

The $R_E$ in the numerator is the distance from the axis of rotation to the surface where the friction force is exerted, the radius of Earth. The $R_E$ in the denominator is also the radius of Earth.

$$\alpha = \frac{1}{\frac{2}{5}mR_E^2}(F_{\text{T on E}}R_E \sin 90°) = \frac{5F_{\text{T on E}}}{2mR_E}$$

Setting the two expressions for the magnitude of the rotational acceleration equal to each other, we get

$$\frac{2\pi\Delta T}{\Delta t T_i^2} = \frac{5F_{\text{T on E}}}{2mR_E}$$

**Solve and evaluate** Solve the previous equation for the force of the tides on Earth:

$$F_{\text{T on E}} = \frac{4\pi mR_E\Delta T}{5\Delta t T_i^2}$$
$$= \frac{4\pi(5.97 \times 10^{24}\,\text{kg})(6.38 \times 10^6\,\text{m})(0.0016\,\text{s})}{5(3.15 \times 10^9\,\text{s})(86{,}400\,\text{s})^2}$$
$$= 6.5 \times 10^9\,\text{N}$$

The magnitude of this friction force seems big, and it is. But when exerted on an object of such large mass ($5.97 \times 10^{24}$ kg), the effect is extremely tiny. By comparison, the gravitational force that the Sun exerts on Earth is $3.5 \times 10^{22}$ N.

**Try it yourself:** Estimate the time interval in years that it will take for Earth's rotation to change from 24 hours to 27 days.

*Answer:* Using Eq. (8.6), we find $t = (\omega - \omega_0)/\alpha \approx -\omega_0/\alpha = (-7 \times 10^{-5}\,\text{rad/s})/(-4 \times 10^{-22}\,\text{rad/s}^2) = 2 \times 10^{17}$ s, almost 10 billion years.

## Cricket bowling

In the game of cricket, a bowler (comparable to the pitcher in baseball) delivers the ball toward the batsman in an overhand motion with an almost straight arm, swinging the arm in a vertical circle (see **Figure 8.19**, below). The record speed for a cricket ball pitch is about the same as the top speed for a baseball pitch—about 44 m/s (100 mph).

**EXAMPLE 8.11 Cricket ball pitch**

Estimate the average force that the bowler's hand exerts on the cricket ball (mass 0.156 kg) during the pitch. The bowler's body is moving forward at about 4 m/s and the ball leaves his hand at 40 m/s relative to the bowler's torso and at 44 m/s relative to the ground.

**Sketch and translate** The situation is shown in Figure 8.19. The ball will be the system of interest. We analyze the situation from the point of view of the bowler. Just before starting the pitch, the speed of the ball with respect to the bowler is 0 m/s. Just as the ball leaves the bowler's hand, it travels horizontally at 40 m/s with respect to the bowler. What is the average force that the bowler exerts on the ball to cause this velocity change?

**Simplify and diagram** Assume that the bowler's arm (estimated length 1.0 m) completes one full circle ($2\pi$ rad) during a pitch. We model the ball as a point-like object and assume the force exerted by the bowler is the only force producing a significant torque on it. The axis of rotation is the bowler's shoulder.

**Figure 8.19** A cricket bowler's arm pushes the ball around a circular path during a pitch.

(continued)

**Represent mathematically** To estimate the force exerted on the ball by the bowler's arm, we use the rotational form of Newton's second law, which will require determining the ball's rotational inertia and its rotational acceleration. Since the ball is being modeled as a point-like object, its rotational inertia is $I = mr^2$. The rotational form of Newton's second law becomes

$$\alpha = \frac{1}{I}\Sigma\tau = \frac{1}{mr^2}(F_{\text{Bowler on Ball}}r)$$

$$\Rightarrow F_{\text{Bowler on Ball}} = mr\alpha$$

Next, we use rotational kinematics to determine the rotational acceleration of the ball:

$$F_{\text{Bowler on Ball}} = mr\alpha = mr\left(\frac{\omega_f^2 - \omega_i^2}{2(\theta_f - \theta_i)}\right)$$

$$= mr\left(\frac{\left(\frac{v}{r}\right)^2 - 0}{2(2\pi)}\right) = \frac{mv^2}{4\pi r}$$

**Solve and evaluate** We can now substitute the known information in the above to determine the force that the bowler's hand exerts on the ball:

$$F_{\text{Bowler on Ball}} = \frac{(0.156\text{ kg})(40\text{m/s})^2}{4\pi(1.0\text{ m})} = 20\text{ N}$$

The answer has the correct units and has a reasonable magnitude, although perhaps smaller than you might have expected. Remember that this is the force that the bowler's hand exerts on the ball. The muscles in his shoulder exert a much greater force on the arm during this throwing motion. The mass of a human arm is about 3–4 kg, whereas the mass of the cricket ball is 0.156 kg. Thus, most of the bowler's effort goes into swinging his own arm. The ball is just along for the ride!

Let's check some limiting cases. If the final velocity of the ball were zero, then the above equation indicates that the bowler would need to exert zero force on the ball. This makes sense. If the bowler's arm were longer, then according to the above equation, he would need to exert less force on the ball; it gets up to the desired speed more easily at the end of a long arm.

**Try it yourself:** If you could exert a 20-N push on a 0.156-kg ball along a straight line that is $2\pi(1.0\text{ m})$ long and the ball started at rest, how fast would it be traveling at the end of this push? Explain.

*Answer:* 40 m/s, the same speed as the cricket ball in the last example.

**Review Question 8.7** How can you explain the increasing length of a day on Earth?

# Summary

**Words** | **Pictorial and physical representations** | **Mathematical representation**

**Rotational kinematics** The rotational motion of a rigid body can be described using quantities similar to those for translational motion—rotational position $\theta$, rotational velocity $\omega$, and rotational acceleration $\alpha$. (Section 8.1)

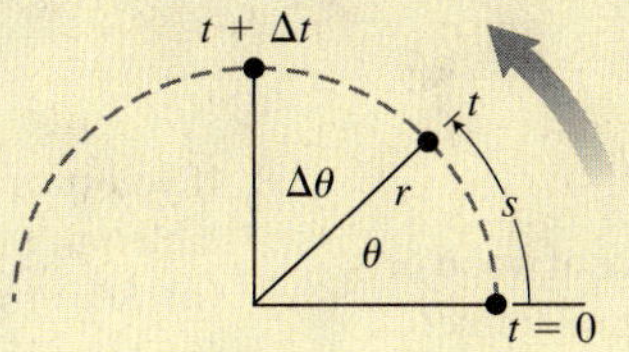


- Rotational position (in radians)

$$\theta = s/r \qquad \text{Eq. (8.1)}$$

- Rotational velocity (in rad/s)

$$\omega = \Delta\theta/\Delta t \qquad \text{Eq. (8.2)}$$

- Rotational acceleration (in rad/s$^2$)

$$\alpha = \Delta\omega/\Delta t \qquad \text{Eq. (8.3)}$$

**Rotational inertial** is the physical quantity equal to the sum of the $mr^2$ terms for each part of an object and depends on the distribution of mass relative to an axis of rotation. (Section 8.3)

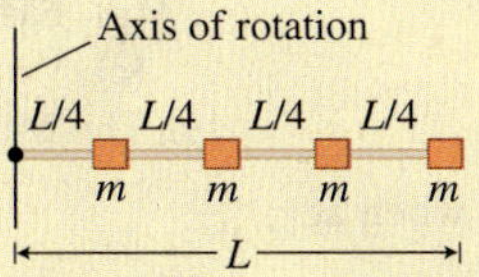


$$I = \Sigma mr^2 \qquad \text{Eq. (8.10)}$$

**Rotational dynamics** A rigid body's rotational acceleration equals the net torque produced by forces exerted on the body divided by its rotational inertia. (Section 8.4)

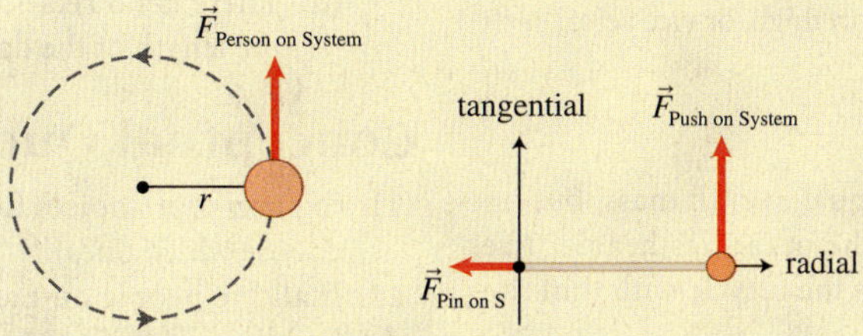


$$\alpha = \frac{\Sigma\tau}{I} \qquad \text{Eq. (8.11)}$$

**Rotational momentum** $L$ is the product of the rotational inertia $I$ of an object and its rotational velocity $\omega$, positive for counterclockwise rotation and negative for clockwise rotation. For an isolated system (zero net torque exerted on it), the rotational momentum of the system is constant. (Section 8.6)

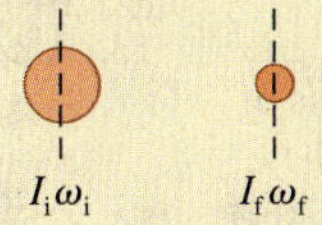


$$L = I\omega \qquad \text{Eq. (8.13)}$$

For isolated system,

$$I_i\omega_i = I_f\omega_f \qquad \text{Eq. (8.14)}$$

**Rotational kinetic energy** $K_{\text{rotational}}$ of a rigid body is energy due to the rotation of the object about a particular axis. This is another form of kinetic energy that is included in the work-energy principle. (Section 8.7)

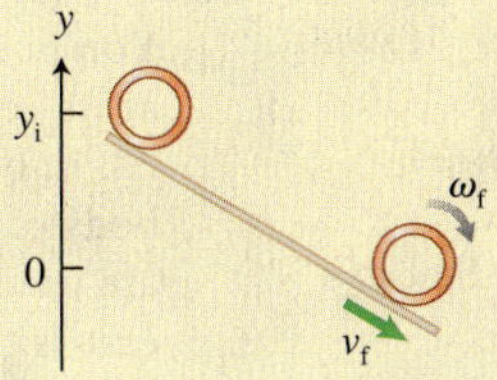


$$K_{\text{rotational}} = \frac{1}{2}I\omega^2 \qquad \text{Eq. (8.15)}$$

MP® For instructor-assigned homework, go to **MasteringPhysics.**

# Questions

## Multiple Choice Questions

1. Is it easier to open a door that is made of a solid piece of wood or a door of the same mass made of light fiber with a steel frame?
   (a) Wooden door
   (b) Fiber door with a steel frame
   (c) The same difficulty
   (d) Not enough information to answer
2. You push a child on a swing. Why doesn't the child continue in a vertical loop over the top of the swing?
   (a) The torque of the force that Earth exerts on the child pulls him back.
   (b) The swing does not have enough kinetic energy when at the bottom.
   (c) The swing does not have enough rotational momentum.
   (d) All of the above are correct.
3. In terms of the torque needed to rotate your leg as you run, would it be better to have a long calf and short thigh, or vice versa?
   (a) Long calf and short thigh
   (b) Short calf and long thigh
   (c) Does not matter
4. Suppose that two bicycles have equal overall mass, but one has thin lightweight tires while the other has heavier tires made of the same material. Why is the bicycle with thin tires easier to accelerate?
   (a) Thin tires have less area of contact with the road.
   (b) With thin tires, less mass is distributed at the rims.
   (c) With thin tires, you don't have to raise the large mass of the tire at the bottom to the top.
5. When riding a 10-speed bicycle up a hill, a cyclist shifts the chain to a larger-diameter gear attached to the back wheel. Why is this gear preferred to a smaller gear?
   (a) The torque exerted by the chain on the gear is larger.
   (b) The force exerted by the chain on the gear is larger.
   (c) You pedal more frequently to travel the same distance.
   (d) Both a and c are correct.
6. A meter stick is supported horizontally at each end by your fingers. A heavy object rests on one end of the stick. If you remove your fingers under the end that holds the object, that end of the meter stick falls faster than the object. Why?
   (a) The object is heavier than the meter stick.
   (b) The rotational inertia of the object is larger than that of a meter stick.
   (c) The acceleration of certain parts of the meter stick can be greater than $9.8\ \text{m/s}^2$.
   (d) Both b and c are correct.
7. You have a raw egg and a cooked egg. If you exert the same torque on each of them for the same period of time, which one spins faster in the end?
   (a) The raw egg
   (b) The cooked egg
   (c) They spin at the same speed.
   (d) The whole process is random and unpredictable.
8. If you turn on a coffee grinding machine sitting on a smooth tabletop, what do you expect it to do?
   (a) Start rotating in the same direction as the blades rotate
   (b) Start rotating in the direction opposite the blade rotation
   (c) Grind the coffee without any rotation of the machine
9. If you could accelerate from zero to 4 m/s when running anywhere on Earth's surface, where and in which direction would you run to increase the length of the day most?
   (a) East-West at the poles (b) East-West at the equator
   (c) West-East at the poles (d) West-East at the equator
10. The Mississippi River carries sediment from higher latitudes toward the equator. How does this affect the length of the day?
    (a) Increases the day
    (b) Decreases the day
    (c) Does not affect the day
    (d) There is no relation between the mass distribution and the length of the day.

## Conceptual Questions

11. Explain your choices for Questions 1–10 (your instructor will choose which ones).
12. If all the people on Earth took elevators to the tops of high buildings in their communities, estimate the effect of this on the length of the day. Explain.
13. A spinning raw egg, if stopped momentarily and then released by the fingers, will resume spinning. Explain. Will this happen with a hard-boiled egg? Explain.
14. Compare the magnitude of Earth's rotational momentum about its axis to that of the Moon about Earth. The tides exert a torque on Earth and Moon so that eventually they rotate with the same period. The object with the greatest rotational momentum will experience the smallest percent change in the period of rotation. Will Earth's solar day increase more than the moon's period of rotation decreases? Explain.
15. You lay a pencil on a smooth desk (ignore sliding friction). You push the pencil, exerting a constant force first directly at its center of mass and then close to the tip of the pencil. In both cases, the force is exerted perpendicular to the body of the pencil. If the forces that you exert on the pencil are exactly the same in magnitude and direction, in which case is the translational acceleration of the pencil greater in magnitude?
16. If you watch the dive of an Olympic diver, you note that she continues to rotate after leaving the board. However, her center of mass follows a parabolic curve. Explain why.
17. Explain why you do not tip over when riding a bicycle but do tip when stationary at a stoplight.
18. Sometimes a door is not attached properly and it will open by itself or close by itself. But it will never do both. Why?

# Problems

Below, **BIO** indicates a problem with a biological or medical focus. Problems labeled **EST** ask you to estimate the answer to a quantitative problem rather than develop a specific answer. Problems marked with ✎ require you to make a drawing or graph as part of your solution. Asterisks indicate the level of difficulty of the problem. Problems with no * are considered to be the least difficult. A single * marks moderately difficult problems. Two ** indicate more difficult problems.

## 8.1 Rotational kinematics

1. The sweeping second hand on your wall clock is 20 cm long. What is (a) the rotational speed of the second hand, (b) the translational speed of the tip of the second hand, and (c) the rotational acceleration of the second hand?
2. You find an old record player in your attic. The turntable has two readings: 33 rpm and 45 rpm. What do they mean? Express these quantities in different units.
3. * Consider again the turntable described in the last problem. Determine the magnitudes of the rotational acceleration in each of the following situations. Indicate the assumptions you made for each case. (a) When on and rotating at 33 rpm, it is turned off and slows and stops in 60 s. (b) When off and you push the play button, the turntable attains a speed of 33 rpm in 15 s. (c) You switch the turntable from 33 rpm to 45 rpm, and it takes about 2.0 s for the speed to change. (d) In the situation in part (c), what is the magnitude of the average tangential acceleration of a point on the turntable that is 15 cm from the axis of rotation?
4. You step on the gas pedal in your car, and the car engine's rotational speed changes from 1200 rpm to 3000 rpm in 3.0 s. What is the engine's average rotational acceleration?
5. You pull your car into your driveway and stop. The drive shaft of your car engine, initially rotating at 2400 rpm, slows with a constant rotational acceleration of magnitude 30 $\text{rad/s}^2$. How long does it take for the drive shaft to stop turning?
6. An old wheat-grinding wheel in a museum actually works. The sign on the wall says that the wheel has a rotational acceleration of 190 $\text{rad/s}^2$ as its spinning rotational speed increases from zero to 1800 rpm. How long does it take the wheel to attain this rotational speed?
7. **Centrifuge** A centrifuge at the same museum is used to separate seeds of different sizes. The average rotational acceleration of the centrifuge according to a sign is 30 $\text{rad/s}^2$. If starting at rest, what is the rotational velocity of the centrifuge after 10 s?
8. * **Potter's wheel** A fly sits on a potter's wheel 0.30 m from its axle. The wheel's rotational speed decreases from 4.0 rad/s to 2.0 rad/s in 5.0 s. Determine (a) the wheel's average rotational acceleration, (b) the angle through which the fly turns during the 5.0 s, and (c) the distance traveled by the fly during that time interval.
9. * During your tennis serve, your racket and arm move in an approximately rigid arc with the top of the racket 1.5 m from your shoulder joint. The top accelerates from rest to a speed of 20 m/s in a time interval of 0.10 s. Determine (a) the magnitude of the average tangential acceleration of the top of the racket and (b) the magnitude of the rotational acceleration of your arm and racket.
10. * An ant clings to the outside edge of the tire of an exercise bicycle. When you start pedaling, the ant's speed increases from zero to 10 m/s in 2.5 s. The wheel's rotational acceleration is 13 $\text{rad/s}^2$. Determine everything you can about the motion of the wheel and the ant.
11. * The speedometer on a bicycle indicates that you travel 60 m while your speed increases from 0 to 10 m/s. The radius of the wheel is 0.30 m. Determine three physical quantities relevant to this motion.
12. * You peddle your bicycle so that its wheel's rotational speed changes from 5.0 rad/s to 8.0 rad/s in 2.0 s. Determine (a) the wheel's average rotational acceleration, (b) the angle through which it turns during the 2.0 s, and (c) the distance that a point 0.60 m from the axle travels.
13. **Mileage gauge** The odometer on an automobile actually counts axle turns and converts the number of turns to miles based on knowledge that the diameter of the tires is 0.62 m. How many turns does the axle make when traveling 10 miles?
14. **Speedometer** The speedometer on an automobile measures the rotational speed of the axle and converts that to a linear speed of the car, assuming the car has 0.62-m-diameter tires. What is the rotational speed of the axle when the car is traveling at 20 m/s (45 mph)?
15. * **Ferris wheel** A Ferris wheel starts at rest, acquires a rotational velocity of $\omega$ rad/s after completing one revolution and continues to accelerate. Write an expression for (a) the magnitude of the wheel's rotational acceleration (assumed constant), (b) the time interval needed for the first revolution, (c) the time interval required for the second revolution, and (d) the distance a person travels in two revolutions if he is seated a distance $l$ from the axis of rotation.
16. * You push a disk-shaped platform on its edge 2.0 m from the axle. The platform starts at rest and has a rotational acceleration of 0.30 $\text{rad/s}^2$. Determine the distance you must run while pushing the platform to increase its speed at the edge to 7.0 m/s.
17. * **EST** Estimate what Earth's rotational acceleration would be in $\text{rad/s}^2$ if the length of a day increased from 24 h to 48 h during the next 100 years.

## 8.4 Newton's second law for rotational motion

18. A turntable turning at rotational speed 33 rpm stops in 50 s when turned off. The turntable's rotational inertia is $1.0 \times 10^{-2}\ \text{kg} \cdot \text{m}^2$. How large is the resistive torque that slows the turntable?
19. A 0.30-kg ball is attached at the end of a 0.90-m-long stick. The ball and stick rotate in a horizontal circle. Because of air resistance and to keep the ball moving at constant speed, a continual push must be exerted on the stick, causing a 0.036-N · m torque. Determine the magnitude of the resistive force that the air exerts on the ball opposing its motion. What assumptions did you make?
20. * **Centrifuge** A centrifuge with a 0.40-kg · $\text{m}^2$ rotational inertia has a rotational acceleration of 100 $\text{rad/s}^2$ when the power is turned on. (a) Determine the minimum torque that the motor supplies. (b) What time interval is needed for the centrifuge's rotational velocity to increase from zero to 5000 rad/s?
21. * **Airplane turbine** What is the average torque needed to accelerate the turbine of a jet engine from rest to a rotational velocity of 160 rad/s in 25 s? The turbine's rotating parts have a 32-kg · $\text{m}^2$ rotational inertia.

22. * The solid two-part pulley in **Figure P8.22** initially rotates counterclockwise. Two ropes pull on the pulley as shown. The inner part has a radius of $1.5a$, and the outer part has a radius of $2.0a$. (a) Construct a force diagram for the pulley with the origin of the coordinate system at the center of the pulley. (b) Determine the torque produced by each force (including the sign) and the resultant torque exerted on the pulley. (c) Based on the results of part (b), decide on the signs of the rotational velocity and the rotational acceleration.

**Figure P8.22**

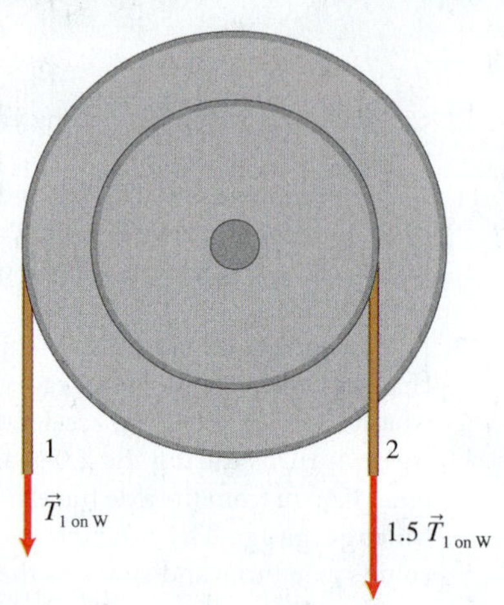


23. * The flywheel shown in Figure P8.22 is initially rotating clockwise. Determine the relative force that the rope on the right needs to exert on the wheel compared to the force that the left rope exerts on the wheel in order for the wheel's rotational velocity to (a) remain constant, (b) increase in magnitude, and (c) decrease in magnitude. The outer radius is $2.0a$ compared to $1.5a$ for the inner radius.

24. * The flywheel shown in Figure P8.22 is initially rotating in the clockwise direction. The force that the rope on the right exerts on it is $1.5T$ and the force that the rope on the left exerts on it is $T$. Determine the ratio of the maximum radius of the inner circle compared to that of the outer circle in order for the wheel's rotational speed to decrease.

25. * A pulley such as that shown in **Figure P8.25** has rotational inertia $10\ \text{kg}\cdot\text{m}^2$. Three ropes wind around different parts of the pulley and exert forces $T_{1\text{ on W}} = 80\ \text{N}$, $T_{2\text{ on W}} = 100\ \text{N}$, and $T_{3\text{ on W}} = 50\ \text{N}$. Determine (a) the rotational acceleration of the pulley and (b) its rotational velocity after 4.0 s. It starts at rest.

**Figure P8.25**

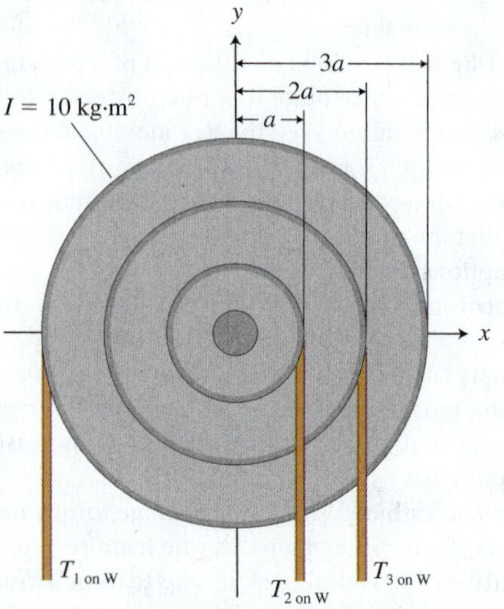


26. * **Equation Jeopardy 1** The equation below describes a rotational dynamics situation. Draw a sketch of a situation that is consistent with the equation and construct a word problem for which the equation might be a solution. There are many possibilities.

$$-(2.2\ \text{N})(0.12\ \text{m}) = [(1.0\ \text{kg})(0.12\ \text{m})^2]\alpha$$

27. * **Equation Jeopardy 2** The equation below describes a rotational dynamics situation. Draw a sketch of a situation that is consistent with the equation and construct a word problem for which the equation might be a solution. There are many possibilities.

$$-(2.0\ \text{N})(0.12\ \text{m}) + (6.0\ \text{N})(0.06\ \text{m}) = [(1.0\ \text{kg})(0.12\ \text{m})^2]\alpha$$

28. Determine the rotational inertia of the four balls shown in **Figure P8.28** about an axis perpendicular to the paper and passing through point A. The mass of each ball is $m$. Ignore the mass of the rods to which the balls are attached.

**Figure P8.28**

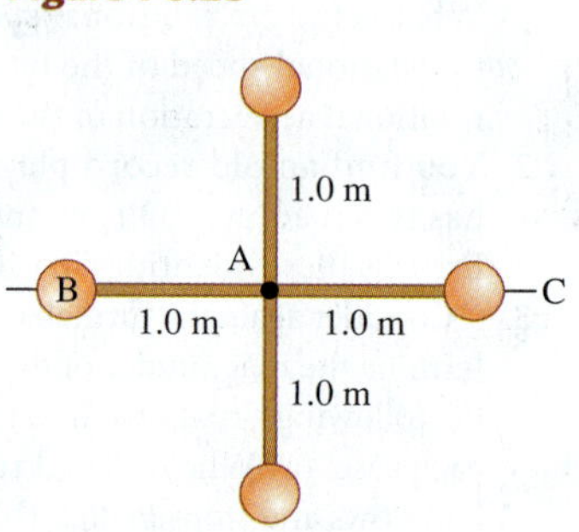


29. Repeat the previous problem for an axis perpendicular to the paper through point B.

30. Repeat the previous problem for axis BC, which passes through two of the balls.

31. **Merry-go-round** A mechanic needs to replace the motor for a merry-go-round. What torque specifications must the new motor satisfy if the merry-go-round should accelerate from rest to 1.5 rad/s in 8.0 s? You can consider the merry-go-round to be a uniform disk of radius 5.0 m and mass 25,000 kg.

32. * A small 0.80-kg train propelled by a fan engine starts at rest and goes around a circular track with a 0.80-m radius. The fan air exerts a 2.0-N force on the train. Determine (a) the rotational acceleration of the train and (b) the time interval needed for it to acquire a speed of 3.0 m/s. Indicate any assumptions you made.

33. * The train from the previous problem is moving along the rails at a constant rotational speed of 5.4 rad/s (the fan has stopped). Determine the time interval that is needed to stop the train if the wheels lock and the rails exert a 1.8-N friction force on the train.

34. * **Motor** You wish to buy a motor that will be used to lift a 20-kg bundle of shingles from the ground to the roof of a house. The shingles are to have a $1.5\text{-m/s}^2$ upward acceleration at the start of the lift. The very light pulley on the motor has a radius of 0.12 m. Determine the minimum torque that the motor must be able to provide.

35. * A thin cord is wrapped around a grindstone of radius 0.30 m and mass 25 kg supported by bearings that produce negligible friction torque. The cord exerts a steady 20-N tension force on the grindstone, causing it to accelerate from rest to 60 rad/s in 12 s. Determine the rotational inertia of the grindstone.

36. ** A string wraps around a 6.0-kg wheel of radius 0.20 m. The wheel is mounted on a frictionless horizontal axle at the top of an inclined plane tilted 37° below the horizontal.

The free end of the string is attached to a 2.0-kg block that slides down the incline without friction. The block's acceleration while sliding down the incline is 2.0 m/s$^2$. (a) Draw separate force diagrams for the wheel and for the block. (b) Apply Newton's second law (either the translational form or the rotational form) for the wheel and for the block. (c) Determine the rotational inertia for the wheel about its axis of rotation.

37. * Elena, a black belt in tae kwon do, is experienced in breaking boards with her fist. A high-speed video indicates that her forearm is moving with a rotational speed of 40 rad/s when it reaches the board. The board breaks in 0.0040 s and her arm is moving at 20 rad/s just after breaking the board. Her fist is 0.32 m from her elbow joint and the rotational inertia of her forearm is 0.050 kg · m$^2$. Determine the average force that the board exerts on her fist while breaking the board (equal in magnitude to the force that her fist exerts on the board). Ignore the gravitational force that Earth exerts on her arm and the force that her triceps muscle exerts on her arm during the break.

38. ** **Like a yo-yo** Sam wraps a string around the outside of a 0.040-m-radius 0.20-kg solid cylinder and uses it like a yo-yo (**Figure P8.38**). When released, the cylinder accelerates downward at (2/3) $g$. (a) Draw a force diagram for the cylinder and apply the translational form of Newton's second law to the cylinder in order to determine the force that the string exerts on the cylinder. (b) Determine the rotational inertia of the solid cylinder. (c) Apply the rotational form of Newton's second law and determine the cylinder's rotational acceleration. (d) Is your answer to part (c) consistent with the application of $a = r\alpha$, which relates the cylinder's linear acceleration and its rotational acceleration? Explain.

**Figure P8.38**

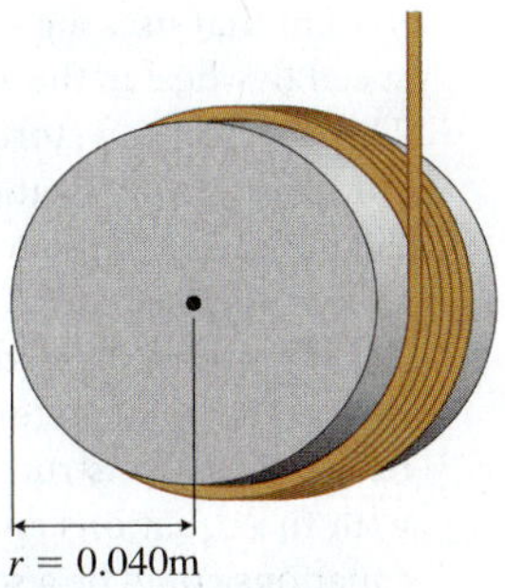


39. ** **Fire escape** A unique fire escape for a three-story house is shown in **Figure P8.39**. A 30-kg child grabs a rope wrapped around a heavy flywheel outside a bedroom window. The flywheel is a 0.40-m-radius uniform disk with a mass of 120 kg. (a) Make a force diagram for the child as he moves downward at increasing speed and another for the flywheel as it turns faster and faster. (b) Use Newton's second law for translational motion and the child force diagram to obtain an expression relating the force that the rope exerts on him and his acceleration. (c) Use Newton's second law for rotational motion and the flywheel force diagram to obtain an expression relating the force the rope exerts on the flywheel and the rotational acceleration of the flywheel. (d) The child's acceleration $a$ and the flywheel's rotational acceleration $\alpha$ are related by the equation $a = r\alpha$, where $r$ is the flywheel's radius. Combine this with your equations in parts (b) and (c) to determine the child's acceleration and the force that the rope exerts on the wheel and on the child.

**Figure P8.39**

40. ** An Atwood machine is shown in Example 8.5. Use $m_1 = 0.20$ kg, $m_2 = 0.16$ kg, $M = 0.50$ kg, and $R = 0.10$ m. (a) Construct separate force diagrams for block 1, for block 2, and for the solid cylindrical pulley. (b) Determine the rotational inertia of the pulley. (c) Use the force diagrams for blocks 1 and 2 and Newton's second law to write expressions relating the unknown accelerations of the blocks. (d) Use the pulley force diagram and the rotational form of Newton's second law to write an expression for the rotational acceleration of the pulley. (e) Noting that $a = R\alpha$ for the pulley, use the three equations from parts (c) and (d) to determine the magnitude of the acceleration of the hanging blocks.

41. ** A physics problem involves a massive pulley, a bucket filled with sand, a toy truck, and an incline (see **Figure P8.41**). You push lightly on the truck so it moves down the incline. When you stop pushing, it moves down the incline at constant speed and the bucket moves up at constant speed. (a) Construct separate force diagrams for the pulley, the bucket, and the truck. (b) Use the truck force diagram and the bucket force diagram to help write expressions in terms of quantities shown in the figure for the forces $T_{1\text{ on Truck}}$ and $T_{2\text{ on Bucket}}$ that the rope exerts on the truck and that the rope exerts on the bucket. (c) Use the rotational form of Newton's second law to determine if the tension force $T_{1\text{ on Pulley}}$ that the rope on the right side exerts on the pulley is the same, greater than, or less than the force $T_{2\text{ on Pulley}}$ that the rope exerts on the left side.

**Figure P8.41**

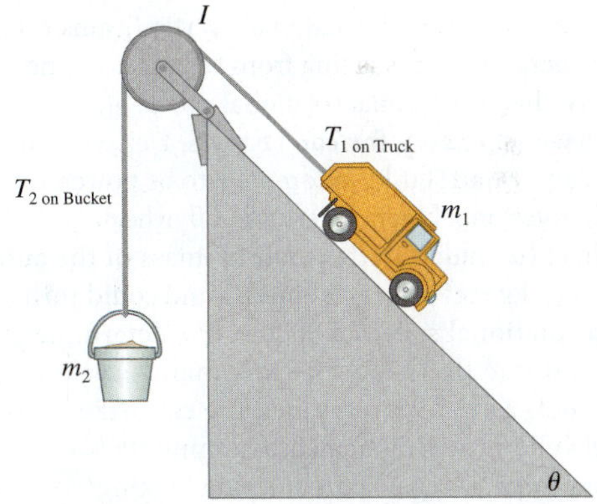


## 8.5 Rotational momentum

42. (a) Determine the rotational momentum of a 10-kg disk-shaped flywheel of radius 9.0 cm rotating with a rotational speed of 320 rad/s. (b) With what magnitude rotational speed must a 10-kg solid sphere of 9.0 cm radius rotate to have the same rotational momentum as the flywheel?

43. **Ballet** A ballet student with her arms and a leg extended spins with an initial rotational speed of 1.0 rev/s. As she draws her arms and leg in toward her body, her rotational inertia becomes 0.80 kg · m$^2$ and her rotational velocity is 4.0 rev/s. Determine her initial rotational inertia.

44. * A 0.20-kg block moves at the end of a 0.50-m string along a circular path on a frictionless air table. The block's initial rotational speed is 2.0 rad/s. As the block moves in the circle, the string is pulled down through a hole in the air table at the

axis of rotation. Determine the rotational speed and tangential speed of the block when the string is 0.20 m from the axis.

45. * **Equation Jeopardy 3** The equation below describes a process. Draw a sketch representing the initial and final states of the process and construct a word problem for which the equation could be a solution.

$$\left(\frac{2}{5}mR^2\right)\left(\frac{2\pi}{30\text{ days}}\right) = \left[\frac{2}{5}m\left(\frac{R}{100}\right)^2\right]\left(\frac{2\pi}{T_f}\right)$$

46. * A student sits motionless on a stool that can turn friction-free about its vertical axis (total rotational inertia $I$). The student is handed a spinning bicycle wheel, with rotational inertia $I_{wheel}$, that is spinning about a vertical axis with a counterclockwise rotational velocity $\omega_0$. The student then turns the bicycle wheel over (that is, through 180°). Estimate, in terms of $\omega_0$, the final rotational velocity acquired by the student.

47. * **Neutron star** An extremely dense neutron star with mass equal to that of the Sun has a radius of about 10 km—about the size of Manhattan Island. These stars are thought to rotate once about their axis every 0.03 to 4 s, depending on their size and mass. Suppose that the neutron star described in the first sentence rotates once every 0.040 s. If its volume then expanded to occupy a uniform sphere of radius $1.4 \times 10^8$ m (most of the Sun's mass is in a sphere of this size) with no change in mass or rotational momentum, what time interval would be required for one rotation? By comparison, the Sun rotates once about its axis each month.

## 8.6 Rotational kinetic energy

48. Determine the change in rotational kinetic energy when the rotational velocity of the turntable of a stereo system increases from 0 to 33 rpm. Its rotational inertia is $6.0 \times 10^{-3}$ kg · m².

49. A grinding wheel with rotational inertia $I$ gains rotational kinetic energy $K$ after starting from rest. Determine an expression for the wheel's final rotational speed.

50. ** **Flywheel energy for car** The U.S. Department of Energy had plans for a 1500-kg automobile to be powered completely by the rotational kinetic energy of a flywheel. (a) If the 300-kg flywheel (included in the 1500-kg mass of the automobile) had a 6.0-kg · m² rotational inertia and could turn at a maximum rotational speed of 3600 rad/s, determine the energy stored in the flywheel. (b) How many accelerations from a speed of zero to 15 m/s could the car make before the flywheel's energy was dissipated, assuming 100% energy transfer and no flywheel regeneration during braking?

51. * The rotational speed of a flywheel increases by 40%. By what percent does its rotational kinetic energy increase? Explain your answer.

52. ** **Rotating student** A student sitting on a chair on a circular platform of negligible mass rotates freely on an air table at initial rotational speed 2.0 rad/s. The student's arms are initially extended with 6.0-kg dumbbells in each hand. As the student pulls her arms in toward her body, the dumbbells move from a distance of 0.80 m to 0.10 m from the axis of rotation. The initial rotational inertia of the student's body (not including the dumbbells) with arms extended is 6.0 kg · m², and her final rotational inertia is 5.0 kg · m². (a) Determine the student's final rotational speed. (b) Determine the change of kinetic energy of the system consisting of the student together with the two dumbbells. (c) Determine the change in the kinetic energy of the system consisting of the two dumbbells alone without the student. (d) Determine the change of kinetic energy of the system consisting of student alone without the dumbbells. (e) Compare the kinetic energy changes in parts (b) through (d).

53. * A turntable whose rotational inertia is $1.0 \times 10^{-3}$ kg · m² rotates on a frictionless air cushion at a rotational speed of 2.0 rev/s. A 1.0-g beetle falls to the center of the turntable and then walks 0.15 m to its edge. (a) Determine the rotational speed of the turntable with the beetle at the edge. (b) Determine the kinetic energy change of the system consisting of the turntable and the beetle. (c) Account for this energy change.

54. * Repeat the previous problem, only assume that the beetle initially falls on the edge of the turntable and stays there.

55. * **Water turbine** A Verdant Power water turbine (a "windmill" in water) turns in the East River near New York City. Its propeller is 2.5 m in radius and spins at 32 rpm when in water that is moving at 2.0 m/s. The rotational inertia of the propeller is approximately 3.0 kg · m². Determine the kinetic energy of the turbine and the electric energy in joules that it could provide in 1 day if it is 100% efficient at converting its kinetic energy into electric energy.

56. * **Flywheel energy** Engineers at the University of Texas at Austin are developing an Advanced Locomotive Propulsion System that uses a gas turbine and perhaps the largest high-speed flywheel in the world in terms of the energy it can store. The flywheel can store $4.8 \times 10^8$ J of energy when operating at its maximum rotational speed of 15,000 rpm. At that rate, the perimeter of the rotor moves at approximately 1,000 m/s. Determine the radius of the flywheel and its rotational inertia.

57. * **Equation Jeopardy 4** The equations below represent the initial and final states of a process (plus some ancillary information). Construct a sketch of a process that is consistent with the equations and write a word problem for which the equations could be a solution.

$$(80\text{ kg})(9.8\text{ N/kg})(16\text{ m}) = \frac{1}{2}(80\text{ kg})v_f^2 + \frac{1}{2}(240\text{ kg}\cdot\text{m}^2)\omega_f^2$$

$$v_f = (0.40\text{ m})\omega_f$$

58. ** A bug of a known mass $m$ stands at a distance $d$ cm from the axis of a spinning disk (mass $m_d$ and radius $r_d$) that is rotating at $f_i$ revolutions per second. After the bug walks out to the edge of the disk and stands there, the disk rotates at $f_f$ revolutions per second. (a) Use the information above to write an expression for the rotational inertia of the disk. (b) Determine the change of kinetic energy in going from the initial to the final situation for the total bug-disk system.

59. ** **Merry-go-round** A 40-kg boy running at 4.0 m/s jumps tangentially onto a small stationary circular merry-go-round of radius 2.0 m and rotational inertia 80 kg · m² pivoting on a frictionless bearing on its central shaft. (a) Determine the rotational velocity of the merry-go-round after the boy jumps on it. (b) Find the change in kinetic energy of the system consisting of the boy and the merry-go-round. (c) Find the change in the boy's kinetic energy. (d) Find the change in the kinetic energy of the merry-go-round. (e) Compare the kinetic energy changes in parts (b) through (d).

60. ** Repeat the previous problem with the merry-go-round initially rotating at 1.0 rad/s in the same direction that the boy is running.

61. ** Repeat the previous problem with the merry-go-round initially rotating at 1.0 rad/s opposite the direction that the boy was running before he jumped on it.

62. * **Another merry-go-round** A carnival merry-go-round has a large disk-shaped platform of mass 120 kg that can rotate

about a center axle. A 60-kg student stands at rest at the edge of the platform 4.0 m from its center. The platform is also at rest. The student starts running clockwise around the edge of the platform and attains a speed of 2.0 m/s relative to the ground. (a) Determine the rotational velocity of the platform. (b) Determine the change of kinetic energy of the system consisting of the platform and the student.

63. ** A rough-surfaced turntable mounted on frictionless bearings initially rotates at 1.8 rev/s about its vertical axis. The rotational inertia of the turntable is $0.020\ \text{kg} \cdot \text{m}^2$. A 200-g lump of putty is dropped onto the turntable from 0.0050 m above the turntable and at a distance of 0.15 m from its axis of rotation. The putty adheres to the surface of the turntable. (a) Find the initial kinetic energy of the turntable. (b) What is the final rotational speed of the system (the lump of putty and turntable)? (c) What is the final linear speed of the lump of putty? Find the change in kinetic energy of (d) the turntable, (e) the putty, and (f) the putty-turntable combination. How do you account for your answers?

## 8.7 Rotational dynamics: Putting it all together

64. * **Stopping Earth's rotation** Suppose that Superman wants to stop Earth so it does not rotate. He exerts a force on Earth $\vec{F}_{\text{S on E}}$ at Earth's equator tangent to its surface for a time interval of 1 year. What magnitude force must he exert to stop Earth's rotation? Indicate any assumptions you make when completing your estimate.

65. * BIO **Triceps and darts** Your upper arm is horizontal and your forearm is vertical with a 0.010-kg dart in your hand (**Figure P8.65**). When your triceps muscle contracts, your forearm initially swings forward with a rotational acceleration of $35\ \text{rad/s}^2$. Determine the force that your triceps muscle exerts on your forearm during this initial part of the throw. The rotational inertia of your forearm is $0.12\ \text{kg} \cdot \text{m}^2$ and the dart is 0.38 m from your elbow joint. You triceps muscle attaches 0.03 m from your elbow joint.

**Figure P8.65**

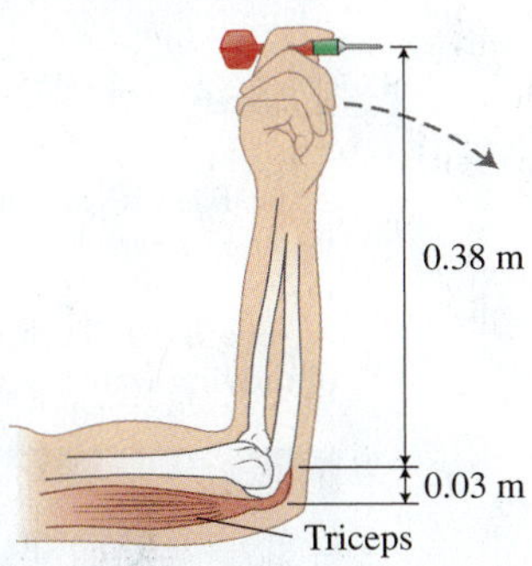


66. * BIO **Bowling** At the start of your throw of a 2.7-kg bowling ball, your arm is straight behind you and horizontal (**Figure P8.66**). Determine the rotational acceleration of your arm if the muscle is relaxed. Your arm is 0.64 m long, has a rotational inertia of $0.48\ \text{kg} \cdot \text{m}^2$, and has a mass of 3.5 kg with its center of mass 0.28 m from your shoulder joint.

**Figure P8.66**

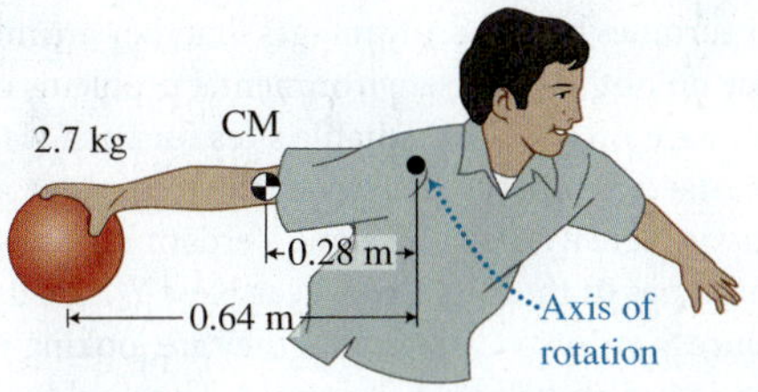


67. ** BIO **Leg lift** You are doing one-leg leg lifts (**Figure P8.67**) and decide to estimate the force that your iliopsoas muscle exerts on your upper leg bone (the femur) when being lifted (the lifting involves a variety of muscles). The mass of your entire leg is 15 kg, its center of mass is 0.45 m from the hip joint, and its rotational inertia is $4.0\ \text{kg} \cdot \text{m}^2$, and you estimate that the rotational acceleration of the leg being lifted is $35\ \text{rad/s}^2$. For calculation purposes assume that the iliopsoas attaches to the femur 0.10 m from the hip joint. Also assume that the femur is oriented 15° above the horizontal and that the muscle is horizontal. Estimate the force that the muscle exerts on the femur.

**Figure P8.67**

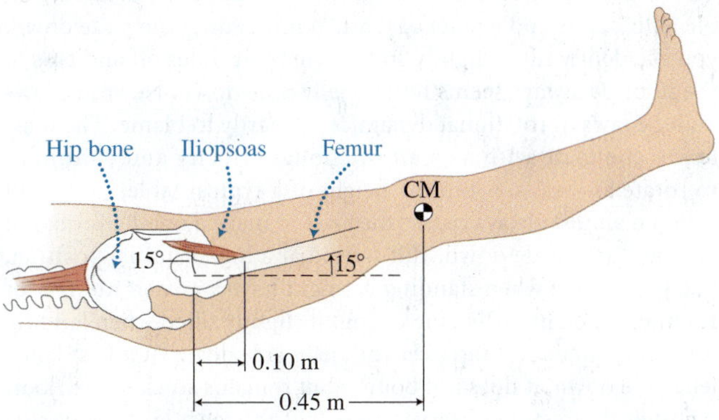


# General Problems

68. * BIO EST **Punting a football** Estimate the tangential acceleration of the foot and the rotational acceleration of the leg of a football punter during the time interval that the leg starts to swing forward in an arc until the instant just before the foot hits the ball. Indicate any assumptions that you make and be sure that your method is clear.

69. * EST Estimate the average rotational acceleration of a car tire as you leave an intersection after a light turns green. Discuss the choice of numbers used in your estimate.

70. ** EST **Door on fingers** Estimate the average force that a car door exerts on a person's fingers if the door is closed when the fingers are in the door opening. Justify all assumptions you make.

71. ** **Yo-yo trick** A yo-yo rests on a horizontal table. The yo-yo is free to roll but friction prevents it from sliding. When the string exerts one of the following tension forces on the yo-yo (shown in **Figure P8.71**), which way does the yo-yo roll? Try the problem for each force: (a) $\vec{T}_{\text{A S on Y}}$; (b) $\vec{T}_{\text{B S on Y}}$; and (c) $\vec{T}_{\text{C S on Y}}$. [*Hint:* Think about torques about a pivot point where the yo-yo touches the table.]

**Figure P8.71**

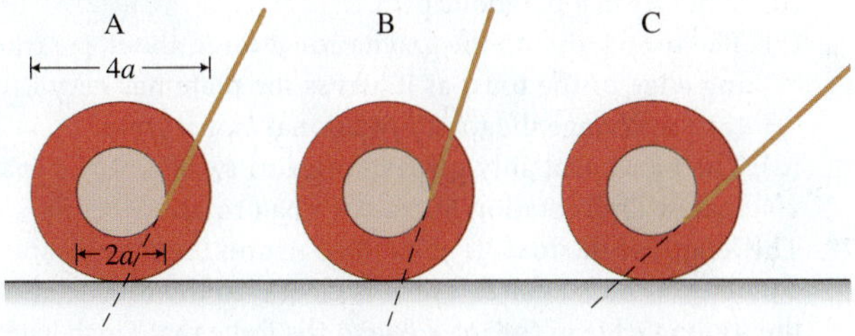


72. ** EST **Running to change time interval of day** At present, the motion of people on Earth is fairly random; the number moving east equals the number moving west, etc. Assume that we could get all of Earth's inhabitants lined up along the

land at the equator. If they all started running as fast as possible toward the west, estimate the change in the length of a day. Indicate any assumptions you made.

73. * EST **White dwarf** A star the size of our Sun runs out of nuclear fuel and, without losing mass, collapses to a white dwarf star the size of our Earth. If the star initially rotates at the same rate as our Sun, which is once every 25 days, determine the rotation rate of the white dwarf. Indicate any assumptions you make.

## Reading Passage Problems

**Toast lands jelly-side down** You are preparing the breakfast table with coffee and a plate of toast. While setting the plate down, you accidently tilt it slightly so that the toast slides off and falls to the floor. It always seems to land jelly-side down. Newton's laws and the laws of rotational dynamics are partly to blame. The toast leaves the plate with a small rotational velocity and continues to rotate as it falls. From the height of a typical table, this small rotation almost always causes the toast to make a one-half rotation and land jelly-side down. If it started at a higher initial position, like your elbow when standing, it might have a greater number of rotations and land jelly-side up; but if slightly tilted when landing it might bounce and flip over with jelly-side down. If it first lands jelly-side down, it does not bounce but remains stuck to the floor. Another interesting observation is that the toast lands jelly-side down only if it slides slowly off the plate. If it is shoved from the table with high velocity, it can land either way.

74. What is the force that provides the torque that causes the toast to rotate?
    (a) The normal force exerted by the plate on the trailing edge of the toast
    (b) The force due to air resistance exerted by the air on the toast
    (c) The gravitational force exerted by Earth on the toast when partly off the plate
    (d) The centripetal force of the toast's rotation
    (e) The answer depends on the choice of axis.
75. The toast is more likely to fall on the jelly side if it makes how many revolutions?
    (a) 0.5 revolutions (b) 0.8 revolutions
    (c) 0.9 revolutions (d) No revolutions
76. What does the number of revolutions that the toast sliding off the plate will make before it touches the floor depend on?
    (a) The amount of jelly
    (b) The height of its starting position
    (c) The length of the toast
77. Why does toast have a better chance of landing jelly-side up if it is quickly shoved off the plate or table?
    (a) It falls faster than if slowly slipping from the plate and does not have time to rotate.
    (b) It moves in a parabolic path.
    (c) The torque due to the gravitational force about the trailing edge of the toast as it leaves the plate has very little time to change the toast's rotational momentum.
    (d) The hand probably gives it an extra twist and the toast makes a full rotation instead of a half rotation.
78. The length of the toast is about 0.10 m and the mass is about 0.050 kg. Which answer below is closest to the torque about the trailing edge of the toast due to the force that Earth exerts on the toast when its trailing edge is just barely on the plate and the rest is off the plate?
    (a) Zero (b) $0.0025\ \text{N}\cdot\text{m}$
    (c) $0.005\ \text{N}\cdot\text{m}$ (d) $0.025\ \text{N}\cdot\text{m}$
    (e) $0.05\ \text{N}\cdot\text{m}$

**Tidal energy** Tides are now used to generate electric power in two ways. In the first, huge dams can be built across the mouth of a river where it exits to the ocean. As the ocean tide moves in and out of this tidal basin or estuary, the water flows through tunnels in the dam (see **Figure 8.20**). This flowing water turns turbines in the tunnels that run electric generators. Unfortunately, this technique works best with large increases in tides—a 5-m difference between high and low tide. Such differences are found at only a small number of places. Currently, France is the only country that successfully uses this power source. A tidal basin plant in France, the La Rance station, makes 240 megawatts of power—enough energy to power 240,000 homes. Damming tidal basins can have negative environmental effects because of reduced tidal flow and silt buildup. Another disadvantage is that they can only generate electricity when the tide is flowing in or out, for about 10 hours each day.

**Figure 8.20** Dams built across tidal basins can generate electric power.

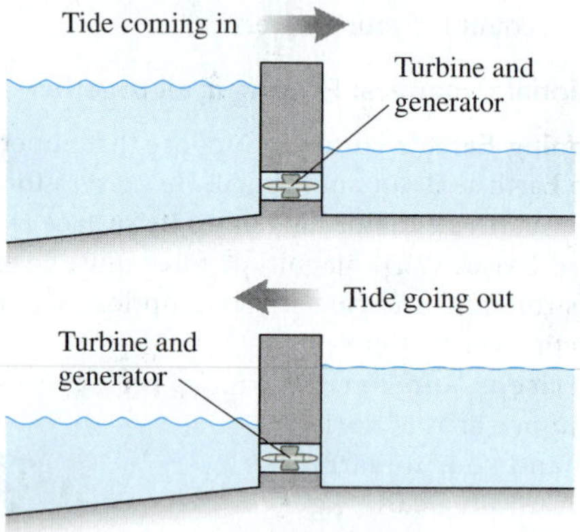


As the tide rises and falls, water passes through the turbine, which runs a generator.

**Figure 8.21** Turbines harness the energy of moving water.

A second method for collecting energy from the tidal flow (as well as all water flow) is to place turbines directly in the water—like windmills in moving water instead of in moving air (see **Figure 8.21**). These water turbines have the advantages that they are much cheaper to build, they do not have the environmental problems of a tidal basin, and there are many more suitable sites for such water flow energy farms. Also, the energy density of flowing water is about 800 times the energy density of dry air flow. Verdant Power is developing turbine prototypes in the East River near New York City and in the Saint Lawrence Seaway in Canada, and they are looking at other sites in the Puget Sound and all over the world. The worldwide potential for hydroelectric power is about 25 terawatts $= 25 \times 10^{12}\ \text{J/s}$—enough to supply the world's energy needs.

79. If the La Rance tidal basin station in France could produce power 24 hours a day, which answer below is closest to the daily amount of energy in joules that it could produce?
(a) 240 J (b) $240 \times 10^6$ J (c) $6 \times 10^9$ J
(d) $2.5 \times 10^{10}$ J (e) $2 \times 10^{13}$ J

80. Suppose a tidal basin is 5 m above the ocean at low tide and that the area of the basin is $4 \times 10^7$ sq m (about 4 miles by 4 miles). Which answer below is closest to the gravitational potential energy change if the water is released from the tidal basin to the low-tide ocean level? The density of water is 1000 kg/m$^3$. [Hint: The level does not change by 5 m for all of the water.]
(a) $5 \times 10^8$ J (b) $5 \times 10^{11}$ J (c) $1 \times 10^{12}$ J
(d) $5 \times 10^{12}$ J (e) $1 \times 10^{13}$ J

81. The La Rance tidal basin can only produce electricity when what is occurring?
(a) Water is moving into the estuary from the ocean.
(b) Water is moving into the ocean from the estuary.
(c) Water is moving in either direction.
(d) The moon is full.
(e) The moon is full and directly overhead.

82. Why do water turbines seem more promising than tidal basins for producing electric energy?
(a) Turbines are less expensive to build.
(b) Turbines have less impact on the environment.
(c) There are many more locations for turbines than for tidal basins.
(d) Turbines can operate 24 hours/day versus for only 10 hours/day for tidal basins.
(e) All of the above

83. Why do water turbines have an advantage over air turbines (windmills)?
(a) Air moves faster than water.
(b) The energy density of moving water is much greater than that of moving air.
(c) Water turbines can float from one place to another, whereas air turbines are fixed.
(d) All of the above
(e) None of the above

84. Which of the following is a correct statement about water turbines?
(a) Water turbines can operate only in moving tidal water.
(b) Water turbines can produce only a small amount of electricity.
(c) Water turbines have not had a proof of concept.
(d) Water turbines cause significant ocean warming.
(e) None of the above are correct statements.

# 9 Gases

Why does a plastic bottle left in a car overnight look crushed on a chilly morning?

How hard is air pushing on your body?

How long will the Sun shine?

**Be sure you know how to:**

- Draw force diagrams (Section 2.1).
- Use Newton's second and third laws to analyze interactions of objects (Section 2.8).
- Use the impulse-momentum principle (Section 5.3).

When you inflate the tires of your bicycle in a warm basement in winter, they tend to look a bit flat when you take the bike outside. The same happens to a basketball—you need to pump it up before playing outside on a cold day. A plastic bottle half full of water left in a car looks crushed on a chilly morning. What do all those phenomena have in common, and how do we explain them?

**When we studied energy (in Chapters 6 and 8),** we found that in many processes some of the mechanical energy of a system transforms into internal energy, resulting in a change in the temperature of the interacting objects. One of the goals of this chapter is to investigate the connection between temperature and internal energy. It turns out that the key to this connection lies in understanding the internal structure of matter.

## 9.1 Structure of matter

When we look at objects that surround us, we do not see their internal structure—water in a cup looks homogeneous, and the cup itself looks like one piece of material. However, they must be made up of *something*. What are the building blocks of matter? To begin to answer this question, consider a simple observational experiment.

Imagine that you dip a cotton ball in rubbing alcohol and wipe it across a piece of paper (**Figure 9.1**). The wet alcohol strip disappears gradually, with the edges of the strip disappearing first. You observe the same behavior when you wipe water or acetone on the paper, except that the water strip disappears more slowly and the acetone strip disappears more quickly. This phenomenon is the same one we observe with wet clothes and puddles as they dry.

**Figure 9.1** A disappearing moist strip on a piece of paper.

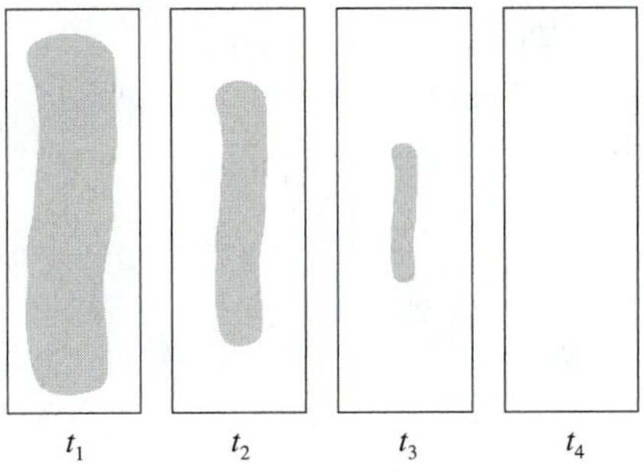


Since the alcohol disappeared gradually, it is reasonable to suggest that it is made of "pieces" too small to be seen. If the alcohol were composed of one piece, it would be gone all at once. However, the model of small pieces does not explain *how* the disappearance occurs. Let's try to construct some possible *mechanisms* that explain how the alcohol disappears (these mechanisms would also be applicable to the drying of water or acetone). Three of the many possible mechanisms are described below:

**‹ Active Learning Guide**

**Mechanism 1.** The little pieces of liquid move to the *inside* of the paper and are still there, even though the paper looks dry.

**Mechanism 2.** The air surrounding the paper somehow pulls the liquid pieces out of the paper.

**Mechanism 3.** The pieces of liquid are moving—they bump into each other and slowly bump each other out of the paper one by one.

All of these mechanisms seem viable. Testing Experiment **Table 9.1** will help us decide if we can rule out any of them.

**‹ Active Learning Guide**

**TESTING EXPERIMENT TABLE**

**9.1 Testing various mechanisms for the drying of wet objects.**

VIDEO 9.1

| Testing experiment 1 | Prediction based on each mechanism | Outcome |
|---|---|---|
| Weigh a dry strip of paper on a sensitive scale. Record its mass. Then moisten the paper with alcohol and record the mass of the wet paper. What happens to the mass of the strip when it dries? 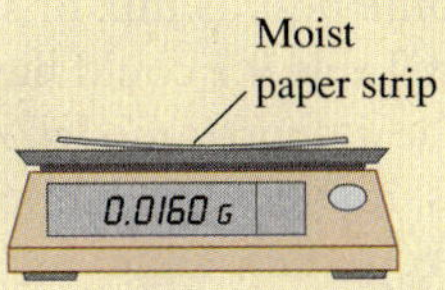  | **Mechanism 1.** If the alcohol pieces go "inside" the paper, then the mass of the paper after it dries should be greater than it was before it was made wet and exactly the same as it was just after it was made wet.<br>**Mechanism 2.** If the air pieces absorb and carry away the alcohol pieces, then after the paper dries, it should have the same mass as it had before it was made wet.<br>**Mechanism 3.** If the moving alcohol pieces bump each other out of the paper, then after the paper dries it should have the same mass as it had before it was made wet. | After the paper dries, its mass is the same as before it was made wet.<br><br>*(continued)* |

**Conclusion 1**

The alcohol pieces left the paper. Mechanism 1 is disproved by this experiment, but Mechanisms 2 and 3 are consistent with it. This leaves us with only two possible mechanisms.

| Testing experiment 2 | Prediction based on each mechanism | Outcome |
|---|---|---|
| Moisten two strips of paper. Place one inside a sealed glass jar attached to a vacuum pump. Place the other strip outside the jar. Pump the air from the jar. 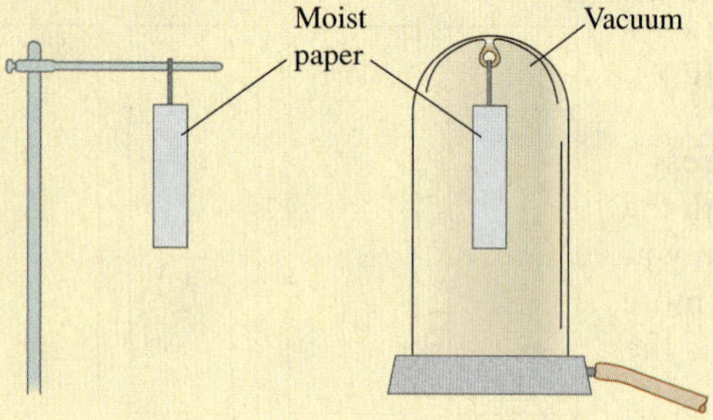  | **Mechanism 2.** If the air pulls the alcohol pieces out of the strip, then the strip inside the vacuum jar with little air should dry more slowly than the strip outside the jar.<br><br>**Mechanism 3.** If the moving alcohol pieces bump each other out of the paper, then the strip inside the vacuum jar should not dry more slowly than the strip outside the jar. | The paper inside the evacuated jar dries faster. |

**Conclusion 2**

Mechanism 2 is disproved by this experiment. Mechanism 3 is not disproved by it.* Now we have only one mechanism that has not been disproved.

| Testing experiment 3 | Prediction | Outcome |
|---|---|---|
| Add a droplet of colored alcohol to a glass of clear alcohol. What happens to the food coloring? 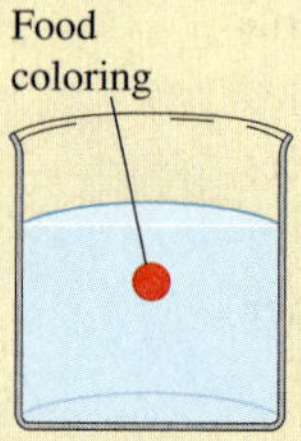  | Mechanism 3. If the small pieces of clear alcohol are moving and bumping each other, then they should bump the colored pieces and cause them to spread. | The color slowly spreads throughout the clear alcohol. 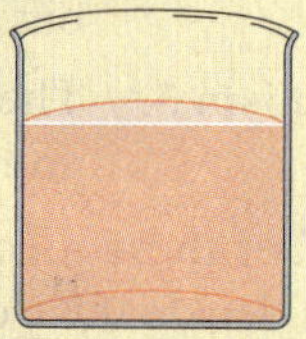 |

**Conclusion 3**

Mechanism 3 is supported by this experiment.

*Note that while Mechanism 3 was not disproved in Experiment 2, the mechanism does not explain why the strip in the vacuum jar actually dried faster.

Based on these experiments, it is reasonable to assume that alcohol and other liquids are composed of smaller objects, called **particles,** which move randomly in all directions. These particles need empty space between them so that particles of other materials can move between them, as happened in Experiment 3 in Table 9.1. This model of the internal structure of alcohol can be used to explain many other phenomena that we encounter—the way some liquids mix or smells spread. In fact, experiments such as those described in Table 9.1 could have led the Greek philosopher Democritus (460–370 B.C.) to the *atomistic* model. *Atomos* in Greek means *indivisible.* According to Democritus, matter was composed of small, indivisible pieces with different shapes and properties. Different substances were made of different combinations of these pieces. Democritus also suggested that the pieces were separated by tiny regions of completely empty space. Democritus proclaimed, "There is nothing in the world but atoms and empty space." We see that Democritus's views are in line with our reasoning.

In 1827, a Scottish botanist named Robert Brown used a microscope to observe the random movement of pollen granules in a droplet of water. To explain his observations, Brown used Mechanism 3—that water itself is composed of particles smaller than the granules and these particles move randomly between frequent collisions (**Figure 9.2a**). The water particles randomly hit the pollen granules from all directions and caused random changes in the position of a granule (Figure 9.2b). This experiment supported the model of water composed of invisible particles that were in continual random motion.

In 1905, Albert Einstein constructed a quantitative model to describe the phenomenon observed by Brown, called *Brownian motion.* He predicted the average distance that a granule of a certain size would move in a given time interval. Einstein's model predicted that the distance depended on the temperature of the liquid in a specific way. Later experiments by Jean Perrin were consistent with Einstein's predictions. Physicists consider Perrin's experiments strong support for the *particle structure of matter.*

We now know that **atoms** are the smallest objects that still retain the chemical properties of a particular element (hydrogen, oxygen, carbon, iron, gold, etc.). Atoms of these various elements combine to build solids, liquids, and gases. A **molecule** is a certain combination of atoms that bond together in a particular arrangement. Molecules may consist of two atoms (such as oxygen, $O_2$, and nitrogen, $N_2$), three atoms (such as water—two hydrogen atoms and one oxygen atom, $H_2O$), or many atoms. Protein molecules can consist of thousands of atoms. Atoms are composed of fundamental particles—protons, neutrons, and electrons. We will discuss these fundamental particles later in the book (Chapters 27 and 28). For now, we will use the word *particle* to indicate an object approximately the size of a molecule or smaller.

The particle model explains how we can smell things even when we are not near them. Suppose we open a bottle of perfume while standing in the middle of a room. Several minutes later, people all over the room can smell the perfume. According to the particle model, the little particles of perfume leave the bottle and gradually disperse, eventually arriving at our nostrils. Since everyone in the room eventually smells the perfume, the perfume particles must move in all directions. However, it takes time to smell the perfume if you are far away from the bottle. Why is that? Perfume particles leaving the bottle move quickly but collide with air particles along the way and reach your nose only after many collisions.

If air is composed of tiny particles, and those particles have mass, then air must have mass. We can test this hypothesis with another simple experiment. Take two rigid metal cylinders—one that has had the air evacuated with a vacuum pump, and the other filled with air. Then weigh them. If the hypothesis that air has mass is correct, the evacuated jar should weigh less than the unevacuated jar. The cylinder filled with air is indeed heavier than the evacuated cylinder (**Figure 9.3**). However, the difference in masses is small (8 g difference for cylinders of about $6 \times 10^{-3}\,m^3$ (1.5 gallons).

## Gases, liquids, and solids

We know from experience that gases are easy to compress, while liquids and solids are almost incompressible (**Figure 9.4**). The particle model helps us explain this difference: we assume that matter in all states is composed of small particles, but the amount of empty space between the particles is different in solids, liquids, and gases. In solids and liquids, the particles are closely packed (almost no empty space between them), while in gases the particles are packed more loosely (lots of empty space).

Gases tend to occupy whatever volume is available. If you take the air that fills a small cylinder and move it to a much larger cylinder, the air also fills the

**Figure 9.2** (a) Water particles moving randomly. (b) Their collisions with a pollen granule cause its random motion.

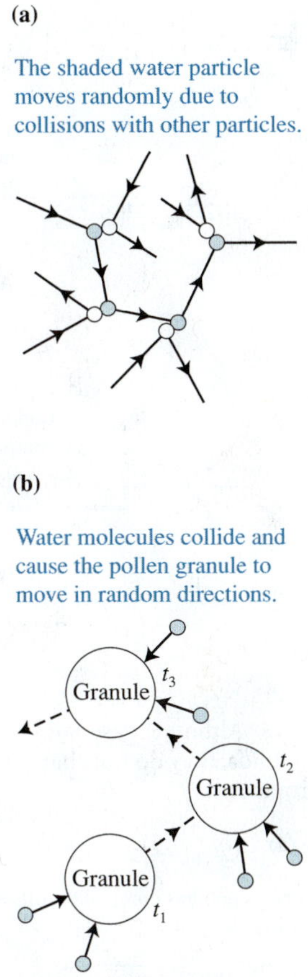
(a)
The shaded water particle moves randomly due to collisions with other particles.
(b)
Water molecules collide and cause the pollen granule to move in random directions.
Granule
$t_3$
Granule
$t_2$
Granule
$t_1$

**< Active Learning Guide**

**Figure 9.3** Air has mass.

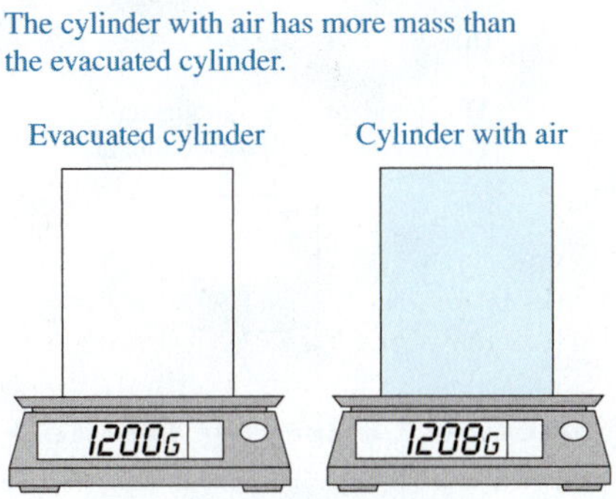
The cylinder with air has more mass than the evacuated cylinder.
Evacuated cylinder
Cylinder with air
1200G
1208G

**Figure 9.4** A gas can be compressed, but a liquid cannot.

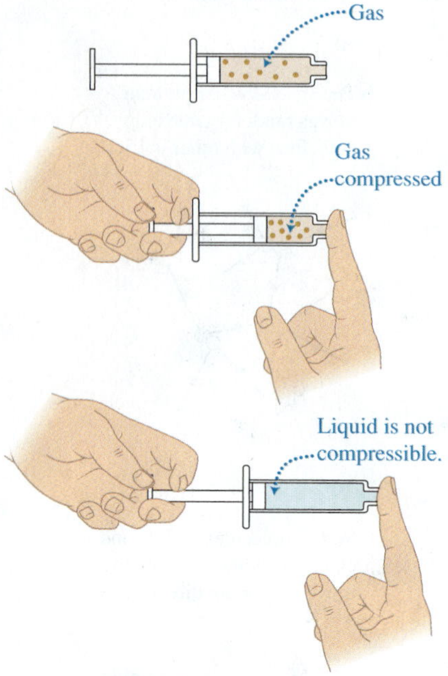
Gas
Gas
compressed
Liquid is not
compressible.

**Figure 9.5** Although gases and liquids are both fluids, they do not share all the same properties.

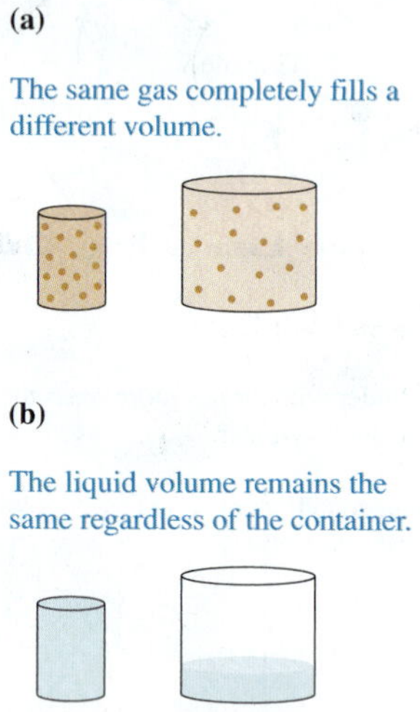
(a)
The same gas completely fills a different volume.
(b)
The liquid volume remains the same regardless of the container.

large cylinder (**Figure 9.5a**). In contrast, if we move the liquid filling a small container to a much larger container, the liquid volume remains the same independent of the container's shape (Figure 9.5b). Solids maintain not only their volume but also their shape.

In order for particles to form a substance, the particles must somehow be attracted to each other. We will learn later (in Chapters 14 and 27) that although the nature of this attractive force between particles is different from their gravitational attraction for each other, there are some similarities between the two types of attraction. For instance, in both cases the particles are attracted to each other at a distance, without any direct contact. The behavior of gases indicates that particles in gases must be so far from each other that such attractive forces are very small or even negligible. In solids the particles are close together, and these new attractive forces between them are strong. The attractive forces between particles in a liquid are stronger than for gas. In solids they are even stronger.

## Ideal gas model

From the observed behavior of gases and our explanations based on the particle nature of matter, we can conclude that gases will be the easiest to study, as we can neglect the very small forces between the particles. We start by constructing a simplified model of a gas as a system.

In this model, we assume that the average distance between the particles (molecules or atoms) is much larger than the size of the particles. We assume that the gas particles are point-like objects that only interact with each other and with any surfaces in their containers during collisions (**Figure 9.6**), similar to the impulsive interactions of a billiard ball with a side of a pool table. In our model the particles do not attract each other at a distance the way they do in solids and liquids. We also assume that the motions of these particles obey Newton's laws. Thus, between collisions the particles move in straight lines at constant velocity and only change their velocities during collisions. Altogether, these simplifying assumptions make up the **ideal gas model.**

**Ideal gas model** **A model of a system in which gas particles are considered point-like and only interact with each other and the walls of their container through collisions. This model also assumes that the particles and their interactions are accurately described using Newton's laws.**

*Ideal* in this context does not mean perfect; it means *simplified*. This is a simplified model with certain assumptions. Whether or not this model can be used to represent a real gas remains to be seen. Only testing experiments can resolve that issue.

How useful is this new model of a gas? As with our previous models of objects (point-like objects, rigid bodies), we will use it to describe and explain known phenomena and then to predict new phenomena. However, so far our model is only qualitative. We need to devise physical quantities to represent the features and behavior of the model. Only then can we use it to develop descriptions, explanations, and predictions of new phenomena involving gases. This process will then allow us to construct a mathematical description of an ideal gas and predict its behavior.

**Active Learning Guide ›**

**Figure 9.6** Ideal gas model. Gas particles are modeled as point-like objects that interact only when colliding with each other or with the walls of their containers.

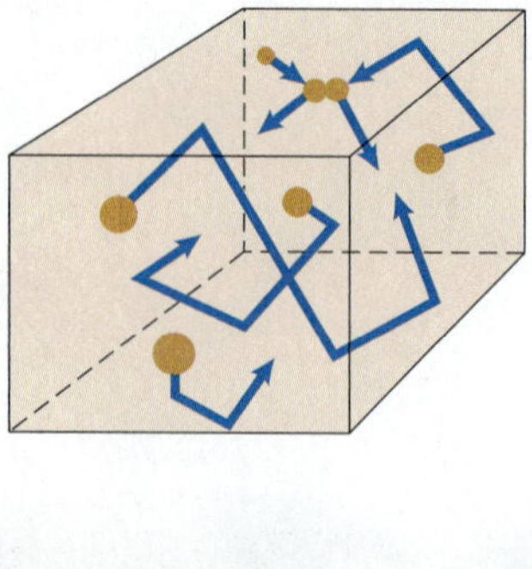

**Review Question 9.1** Use the particle model to explain how moist objects dry out.

# 9.2 Pressure, density, and the mass of particles

In this section we will identify several new physical quantities that are useful for describing the properties of the model of an ideal gas. These quantities are also applicable to the properties of real gases, liquids, and solids.

## Pressure

Imagine you are holding an air-filled balloon. Try to crush it a little bit. You feel the balloon resisting the crushing, as if something inside it pushes back on your fingers. How can we explain this "resistance" of the air inside the balloon? As we will learn later in this section, the particles of matter (atoms and molecules) are rather small. As air particles move randomly in space, they eventually collide with the solid surfaces of any objects in that space. In each of these collisions, the particle exerts an impulsive force on the object—like a tennis ball hitting a practice wall (see **Figure 9.7a**). However, when a huge number of particles bombard a solid surface at a constant rate, these collisions collectively exert an approximately constant force on the object (Figure 9.7b). This impulsive force must be what we feel when we are trying to squeeze the balloon. Notice that we have now constructed a *model of a process* that explains how the motion of the particles of gas inside the balloon accounts for the observational evidence—the apparent resistance of the balloon to squeezing. This ideal process that we imagined serves as a mechanism for what we observe.

We can test this process model with a simple experiment. Consider that when you inflate a balloon, its surface expands outward. Our model actually predicts this outcome because as we blow, we add air particles to the interior of the balloon; thus there are more particles inside colliding with the walls. This greater collision rate results in a larger outward average force on each part of the balloon's surface, causing it to expand outward.

On the other hand, there is also air outside the balloon, and particles of this air hit the outside of the balloon walls. In Testing Experiment **Table 9.2**, we investigate what will happen if we decrease the number of particles outside the balloon.

**Figure 9.7** Impulsive forces during collisions cause an approximately constant force against a wall.

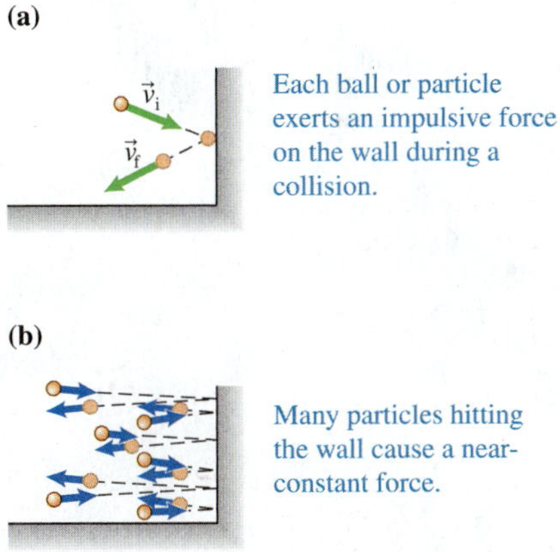


**TESTING EXPERIMENT TABLE**

**9.2 Testing the model of moving gas particles pushing on the surface.**

| Testing experiment | Prediction based on a model of gas particle motion | Outcome |
|---|---|---|
| Place a partially inflated balloon inside a vacuum jar. Seal the jar. 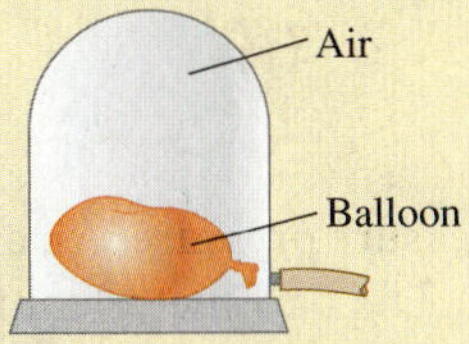  What happens to the balloon's shape when you start pumping air out of the jar? | As we remove air particles from outside the balloon, the collisions of particles on the outside of the balloon are less frequent and exert less force on each part of the balloon's outer surface. The collision rate of the particles inside the balloon does not decrease. Therefore, the balloon should expand. | As air outside of the balloon is removed, the balloon expands. 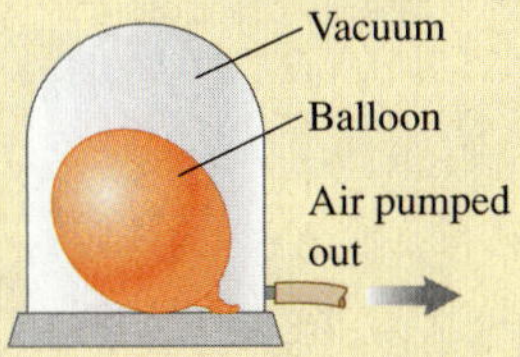  |

**Conclusion**

The model of air consisting of moving particles colliding with objects exposed to the air has not been disproved. The results of this experiment support the model.

In normal situations, an extremely large number of gas particles collide each second with the surface that is in contact with the gas. For example, about $10^{23}$ particles of air collide each second with each square centimeter of your skin. Making a force diagram for the skin by including these individual particle collision forces is not practical. A different approach is needed.

As we have discussed, although each particle collision is impulsive, the forces are so small and so frequent that the force exerted by the gas on the walls of the container can be modeled as a single constant force. The force also depends on the area of that surface—the bigger the area, the more particles push on it during collisions. Thus, instead of using force to describe gas processes, we use a new force-like physical quantity called **pressure.** As we will see (in Chapters 10 and 11), this quantity is useful for describing the behavior of liquids and solids as well.

**Pressure *P*** **Pressure is a physical quantity equal to the magnitude of the perpendicular component of the force $F_\perp$ that a gas, liquid, or solid exerts on a surface divided by the contact area $A$ over which that force is exerted:**

$$P = \frac{F_\perp}{A} \quad (9.1)$$

The SI unit of pressure is the pascal (Pa), where $1\ \text{Pa} = 1\ \text{N/m}^2$. In British units pressure is measured in pounds per square inch (psi). You will find a complete list of pressure units on the inside front cover of the text.

Pressure is easy to visualize when you think about two solid surfaces that contact each other. Compare the magnitude of force you exert on soft snow when wearing street shoes versus the magnitude of force that you exert on the snow when wearing snowshoes. The magnitude of force that you exert on the snow is the same, but the corresponding pressure is not. The snowshoes decrease the pressure you exert on the snow, since the force is spread over a much larger area. Similarly, when you decrease the area over which the force is spread, the pressure increases. Scissors and knives increase pressure on a surface because they decrease the area over which they exert a force.

Consider the concept of pressure inside a gas or liquid. The air particles hitting one wall in a room collectively exert an average force on the wall, and therefore a pressure. However, the air particles in the center of the room are not interacting with the wall at all, at least not for the moment. Yet, the air there has a pressure as well. If a table were placed in the center of the room, the air particles would exert an average force, and therefore a pressure, on the top, bottom, and sides of the table. Thus, air has a pressure whether or not a solid object is present.

## Measuring pressure

Many instruments are used to measure pressure. An aneroid barometer is used to measure gas pressure directly. The barometer contains a small aneroid cell (**Figure 9.8**). Inside, the cell has almost no air. A lever is attached to the cell's moveable wall. As the outside air pressure on this wall changes, the cell thickness changes, and the lever causes a pointer needle on the aneroid barometer to change, indicating the outside air pressure. Measurements show that the pressure of the atmospheric air at sea level is on average $10^5\ \text{N/m}^2$, or $10^5$ Pa. This atmospheric pressure defines yet another unit of pressure, called an atmosphere: $1.0\ \text{atm} = 1.0 \times 10^5\ \text{Pa} = 1.0 \times 10^5\ \text{N/m}^2$.

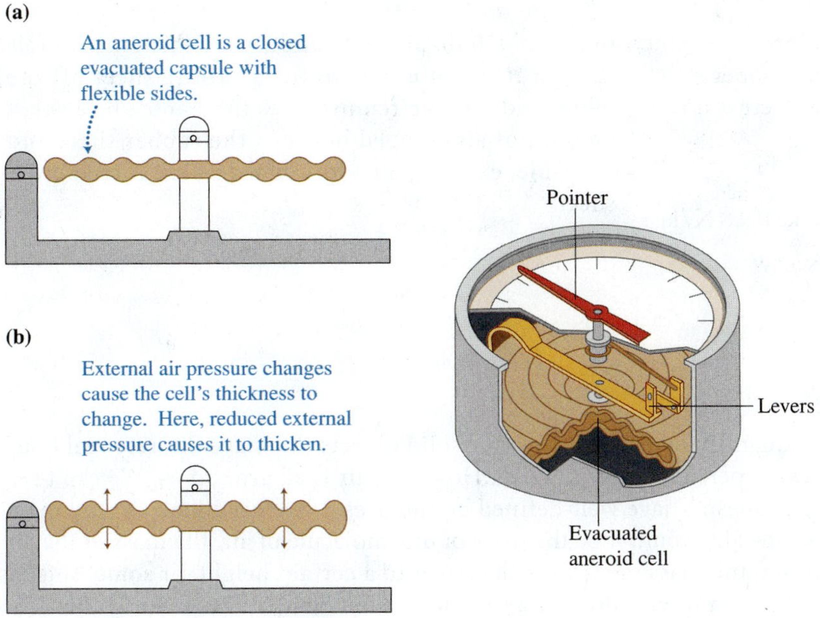


An aneroid cell is a closed evacuated capsule with flexible sides.

External air pressure changes cause the cell's thickness to change. Here, reduced external pressure causes it to thicken.

**Figure 9.8** An aneroid barometer consists of an evacuated aneroid cell that gets thicker or thinner depending on outside air pressure.

Another way to measure gas pressure is to compare the pressure of a gas in contact with a gauge to the atmospheric pressure. For example, when you use a tire gauge to measure the air pressure in a car tire, you are comparing the air pressure inside the tire to that of the atmosphere outside the tire. The pressure in the tire is called **gauge pressure.** If the pressure in a container is 1.0 atm, its gauge pressure is zero, because there is no difference between the pressure inside the container and outside of it.

**Gauge pressure $P_{gauge}$** **Gauge pressure is the difference between the pressure in some container and the atmospheric pressure outside the container:**

$$P_{gauge} = P - P_{atm} \tag{9.2}$$

**where $P_{atm} = 1.0\text{ atm} = 1.0 \times 10^5\text{ N/m}^2$.**

We are continually immersed in a fluid—the gaseous atmosphere. What is the magnitude of the force that the air exerts on one side of your body?

**QUANTITATIVE EXERCISE 9.1 The force that air exerts on your body**

Estimate the total force that the air exerts on the front side of your body, assuming that the pressure of the atmosphere is constant.

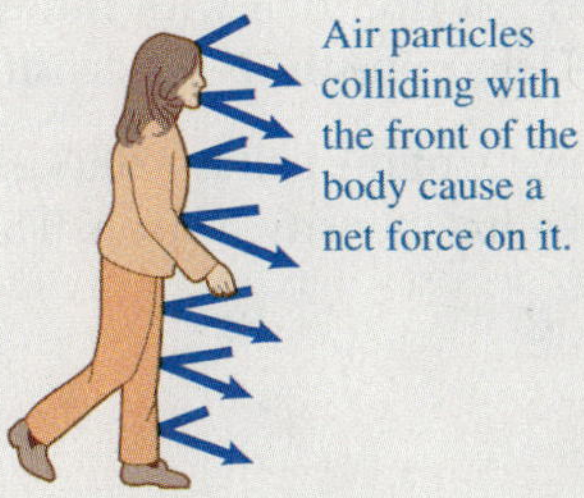

Air particles colliding with the front of the body cause a net force on it.

**Represent mathematically** To calculate the total force exerted by the air on the front of your body, multiply your front surface area by $10^5\text{ N/m}^2$, the atmospheric air pressure. To estimate the surface area of the front of your body, model your body as a rectangular box—your height times your width. Assume a height of 1.8 m and a width of 0.3 m. Use Eq. (9.1) to estimate the force:

$$F_{\text{A on F}} = P \cdot A$$

**Solve and evaluate** $F_{\text{A on F}} = P \cdot A = (10^5\text{ N/m}^2)[(1.8\text{ m})(0.3\text{ m})] \approx 5 \times 10^4\text{ N}$. That's $5 \times 10^4\text{ N} \times 0.22\text{ lb/N} = 1.1 \times 10^4\text{ lb}$, or 10,000 lb. So why aren't you immediately thrown backward? [*Hint:* Think about the force that the air behind you is exerting on you as well.]

(*continued*)

**Try it yourself:** What is the minimum force that you must exert to lift a 10 cm × 20 cm, 10-kg rubber sheet that is stuck flat to a tabletop? Assume that there is no air below the sheet.

*Answer:* $F_{\text{Y on S}} = (P_{\text{atm}} \cdot A) + mg = (10^5\ \text{N/m}^2)[(0.10\ \text{m})(0.20\ \text{m})] + (0.10\ \text{kg})(9.8\ \text{N/kg}) = 2000\ \text{N}$, or about 450 lb. Since it is not really this difficult (although it is not easy!) to lift a rubber sheet off the table, under real circumstances there must be a small amount of air trapped between the rubber sheet and the table, exerting an additional upward force on the sheet.

## Density

The quantity *mass* helps describe solid objects that have discrete real boundaries—a person, a car, or a ball. However, air is all around us. We can't see it, and it doesn't have well-defined boundaries. If we used the quantity mass to describe air, would it be the mass of one molecule of air, the mass of the air in a room, the mass of air over the street to a certain height, or something else? It's difficult to visualize air as a macroscopic object. For gases, a much more useful physical quantity is the mass of one unit of volume—**density.**

Density measures the mass of one cubic meter of a substance. For example, at sea level and 0°C the mass of $1.0\ \text{m}^3$ of air is 1.3 kg. We say that the density of air is $1.3\ \text{kg/m}^3$. If we had $2.0\ \text{m}^3$ of air at sea level, its mass would be 2.6 kg. Its density is still $1.3\ \text{kg/m}^3$, since

$$\frac{2.6\ \text{kg}}{2.0\ \text{m}^3} = 1.3\ \text{kg/m}^3.$$

**TIP** Density is different from mass. Air in a room has a particular mass and density. If you divide the room into two equal parts using a screen, the mass of air in each part will be half the total mass, but the density in each part will remain the same.

**Density $\rho$** The density $\rho$ (lowercase Greek letter "rho") of a substance or of an object equals the ratio of the mass $m$ of a volume $V$ of the substance (for example, air or water) divided by that volume $V$:

$$\rho = \frac{m}{V} \tag{9.3}$$

The unit of density is $\text{kg/m}^3$.

**QUANTITATIVE EXERCISE 9.2 The density of a person**

Estimate the density of a person.

**Represent mathematically** Assume the following about the person: mass is 80 kg; dimensions are 1.8 m tall, 0.3 m wide, and 0.1 m thick; and volume is $V \approx 1.8\ \text{m} \times 0.3\ \text{m} \times 0.1\ \text{m} = 0.054\ \text{m}^3$.

The person's density is given by Eq. (9.3):

$$\rho = \frac{m}{V}$$

**Solve and evaluate** Substitute the person's mass and volume into the above to get

$$\rho = \frac{80\ \text{kg}}{0.054\ \text{m}^3} = 1500\ \text{kg/m}^3$$

This is the correct unit for density. In a later chapter, we will decide if this is a reasonable estimate for the density of a human.

**Try it yourself:** An iron ball with radius 5.0 cm has a mass of 2.0 kg. Determine the ball's density.

*Answer:* The density of the ball is $3800\ \text{kg/m}^3$, much less than the $7860\text{-kg/m}^3$ density of iron. The ball must be hollow.

## Mass and size of particles

In 1811, an Italian scientist named Avogadro proposed that equal volumes of different types of gas, when at the same temperature and pressure, contain the same number of gas particles. Using Avogadro's hypothesis, scientists could determine the relative masses of different types of particles by comparing the masses of equal volumes of the gases. Presently, scientists use Avogadro's number $N_A$ to indicate the number of atoms or molecules present in 22.4 L ($22.4 \times 10^3\ \text{cm}^3$) of any gas at 0° C and standard atmospheric pressure.

The mass in grams of any substance that has exactly Avogadro's number of particles is equal to the atomic mass. For example, the atomic mass of molecular hydrogen is 2, that of molecular oxygen is 32, and that of lead is 207; therefore, 2 g of molecular hydrogen, 32 g of molecular oxygen, and 207 g of lead all have the same number of particles—exactly $N_A = 6.02 \times 10^{23}$. This number of particles is called a **mole** (**Figure 9.9**).

**Figure 9.9** A mole ($6 \times 10^{23}$ particles) of helium (4 g of helium in the balloon), water (18 g), and salt (58 g).

**Avogadro's number and the mole** Avogadro's number $N_A = 6.02 \times 10^{23}$ particles is called a mole. The number of particles in a mole is the same for all substances and is the number of particles whose total mass equals the atomic mass of that substance.

The mass of one mole of particles of any substance is called **molar mass.** One mole of hydrogen ($H_2$) has a mass of 2.0 g, one mole of oxygen ($O_2$) has a mass of 32 g, and one mole of lead has a mass of 207 g.

We can now easily determine the mass of a single gas particle of any substance (for example, hydrogen, oxygen, or nitrogen) by dividing the molar mass of the substance by $6.02 \times 10^{23}$, the number of particles in one mole of the substance:

$$m_{\text{particle}} = \frac{m_{\text{mole}}}{N_A}$$

We find that air (typically 70% $N_2$ with $m_{N_2} = 28$ g, 29% $O_2$ with $m_{O_2} = 32$ g, and small percentages of other gases) has a molar mass of about 29 g. Thus, the mass per air particle is approximately

**‹ Active Learning Guide**

$$m_{\text{air particle}} = \frac{29 \times 10^{-3}\ \text{kg}}{6.02 \times 10^{23}\ \text{air particles}} = 4.8 \times 10^{-26}\ \text{kg/air particle}$$

In addition to a particle's mass, its size is also important. The size of particles was estimated much later than the mass—in the 1860s by Josef Loschmidt. Loschmidt found the linear size of the particles that made up gases, liquids, and solids to be about $d \approx 10^{-9}\ \text{m} = 10^{-7}\ \text{cm} = 1\text{nm}$. Contemporary methods indicate that nitrogen and oxygen are about 0.3 nm.

**EXAMPLE 9.3 The average distance between air particles**

What is the average separation between nearby gas particles in the air, and how does it compare to the size of the particles themselves?

**Sketch and translate** Sketch the situation as a mole of air divided into cubes as shown at the right, with one particle located at the center of each cube. The diameter of an individual particle is $d$; the average distance between them is $D$.

We visualize the mole of gas divided into equal-sized cubes of volume $D^3$, each containing one particle of diameter $d$.

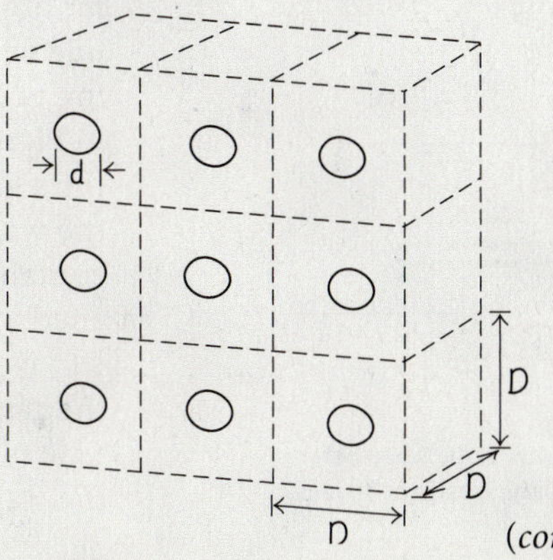


*(continued)*

**Simplify and diagram** Assume standard conditions, with one mole of gas particles occupying a volume of $22.4 \times 10^3\ \text{cm}^3$. Now, imagine this volume divided equally between the particles (the cubes in the figure) and each particle being contained in a cube whose volume is $D^3$. We can think of the distance between the particles as the side of this cube $D$.

**Represent mathematically** We determine the volume of the cube corresponding to each particle and then take the cube root of that volume to estimate the average distance between them.

$$V_{\text{per particle}} = \frac{V_{\text{one mole}}}{N_{\text{particles in a mole}}}$$

$$D = \sqrt[3]{V_{\text{per particle}}} = \sqrt[3]{\frac{V_{\text{one mole}}}{N_{\text{particles in a mole}}}}$$

**Solve and evaluate**

$$D = \sqrt[3]{\frac{V_{\text{one mole}}}{N_{\text{particles in a mole}}}} = \sqrt[3]{\frac{22.4 \times 10^3\ \text{cm}^3}{6.02 \times 10^{23}\ \text{particles}}}$$
$$= \sqrt[3]{3.72 \times 10^{-20}\ \text{cm}^3/\text{particle}}$$
$$= \sqrt[3]{37.2 \times 10^{-21}\ \text{cm}^3/\text{particles}}$$
$$= 3.34 \times 10^{-7}\ \text{cm}$$

Recall that the average diameter of a single particle of air is $3 \times 10^{-8}$ cm. Thus, the approximate distance between the particles of air is on average $\frac{3.34 \times 10^{-7}\ \text{cm}}{3 \times 10^{-8}\ \text{cm}} \approx 10$ times the size of the particles ($D/d = 10$). This is a lot of empty space. A macroscopic analogy would be 20 people lying down and spread uniformly over the area of a football field.

**Try it yourself:** Estimate the distance between water molecules in liquid water and compare this distance to their dimensions. The density of water is $1.0 \times 10^3\ \text{kg/m}^3$. The molar mass of water is $18 \times 10^{-3}$ kg/mole.

*Answer:* The volume occupied by one mole of water is about

$$V_{\text{one mole}} \approx \frac{(18\ \text{g})(10^{-3}\ \text{kg/g})}{1000\ \text{kg/m}^3} = 18 \times 10^{-6}\ \text{m}^3$$

The volume occupied by one molecule is

$$V_{\text{one molecule}} \approx \frac{18 \times 10^{-6}\ \text{m}^3}{6 \times 10^{23}} = 3 \times 10^{-29}\ \text{m}^3$$

The distance between particles is $D_{\text{between water molecules}} \approx 3 \times 10^{-8}$ cm. This is just a little larger than the size of the molecules, which means there is very little space between the water molecules.

**Review Question 9.2** The distance between air particles is very small—about $3 \times 10^{-7}$ cm. How can we say that there is considerable empty space in air?

## 9.3 Quantitative analysis of ideal gas

We can use the quantities pressure, density, and the mole to construct a mathematical description of an ideal gas that will allow us to make predictions about new phenomena.

To start, we make a few more simplifying assumptions. First, in addition to modeling the particles (atoms or molecules) as point-like objects whose motion is governed by Newton's laws, assume that the particles do not collide with each other—they only collide with the walls of the container, exerting pressure on the walls (in other words, they move like the model depicted in Figure 9.6 but with no particle collisions). This is a reasonable assumption for a gas of low density. Second, assume that the collisions of particles with the walls are elastic. This makes sense, as the pressure of the gas in a closed container remains constant, which would not happen if the particles' kinetic energy decreased during inelastic collisions.

**Figure 9.10** A gas particle bouncing back and forth between the walls of a container.

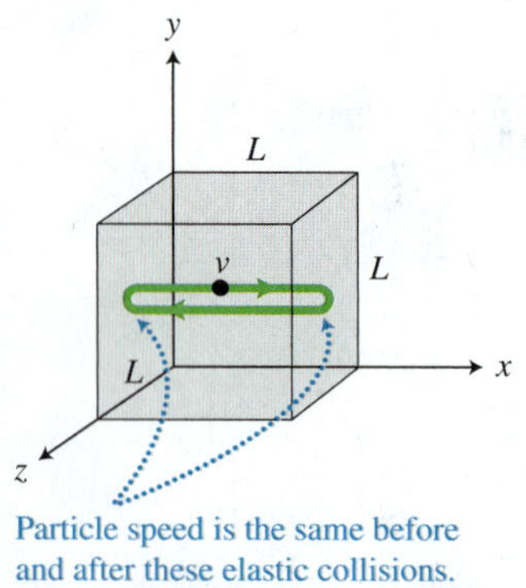


Now, let's construct a mathematical description of an ideal gas. Imagine the gas inside a cubic container with sides of length $L$ (see **Figure 9.10**). A particle moves at velocity $\vec{v}$ with respect to a vertical wall. When it hits the wall, the wall exerts a force on the particle that causes it to reverse direction. Since the kinetic energy of the particle is the same before and after the collision, the same is true for its speed. The wall exerts a force on the particle, and the particle in turn exerts an equal-magnitude and oppositely directed force on the wall (Newton's third law).

Let's use impulse-momentum ideas to analyze the particle–wall collision. During the collision with the right wall, it exerts a normal force on the particle in the negative $x$-direction. Before the collision, the particle has a positive $x$-component of velocity $v_{x\text{i}}$. After the collision, the particle has a negative $x$-component of velocity $v_{x\text{f}}$. The impulse-momentum equation gives

$$mv_{x\text{i}} + F_{\text{W on P }x}\Delta t = mv_{x\text{f}}$$

Because the particle's speed is the same before and after the collision, $v_{x\text{f}} = -v_{x\text{i}}$. In addition, because of Newton's third law, $F_{\text{W on P }x} = -F_{\text{P on W }x}$. Thus,

$$mv_{x\text{i}} + (-F_{\text{P on W }x}\Delta t) = -mv_{x\text{i}}$$
$$2mv_{x\text{i}} = F_{\text{P on W }x}\Delta t$$

In the above equation, $F_{\text{P on W }x}$ is the impulsive force that the particle exerts on the wall during the very short time interval $\Delta t$ that the particle is actually touching and colliding with the wall. How can we determine an average effect of these impulsive collisions in order to determine the pressure of the gas on the wall? We rewrite the right side of the equation as the product of the *average force* exerted by the particle on the wall from one collision to the next, multiplied by the time that *passes between collisions*. Note that this average force is much smaller than the impulsive force, since most of the time the particle is flying through the container and is not in contact with the wall (**Figure 9.11**). However, the time interval between collisions is longer than the impulsive time interval. The product of the big impulsive force and short time interval equals the product of the small average force and the long time interval between collisions.

Looking at the $x$-component of the motion of the particle, we see that the time interval between collisions with the wall is $\Delta t_{\text{between collisions}} = (2L/v_{x\text{i}})$ since the particle must travel a distance $2L$ in the $x$-direction before colliding with the wall again. So, our equation becomes

$$2mv_{x\text{i}} = F_{\text{P on W }x}\Delta t_{\text{collision}} = \overline{F_{\text{P on W }x}\Delta t_{\text{between collisions}}} = \overline{F_{\text{P on W }x}}\frac{2L}{v_{x\text{i}}}$$

The bar above the force in the last two expressions indicates that it is the *average force* exerted over the time interval between collisions. Multiplying by $v_{x\text{i}}$ and dividing by 2 gives

$$mv_{x\text{i}}^2 = \overline{F_{\text{P on W }x}L}$$

To relate this microscopic relationship to macroscopic quantities, such as the pressure of the gas and its volume, multiply both sides of the equation by $N$, the number of particles of gas in the container. Because the particles do not all move with the same speed, we also replace the quantity $v_{x\text{i}}^2$ (the square of the $x$-component of the velocity of an individual particle) with its average value for all the particles:

$$Nm\overline{v_x^2} = \left(N\frac{\overline{F_{\text{P on W }x}}}{L^2}\right)L^3$$

We have also multiplied the right side by $L^2/L^2$. The term in parentheses is the pressure exerted by the gas on the wall (the force exerted by all $N$ particles divided by the wall area $L^2$.) The $L^3$ outside the bracket is the volume occupied by the gas. Thus, the above equation becomes

$$Nm\overline{v_x^2} = PV$$

**Figure 9.11** A method to find the average force of particles colliding with the container wall.

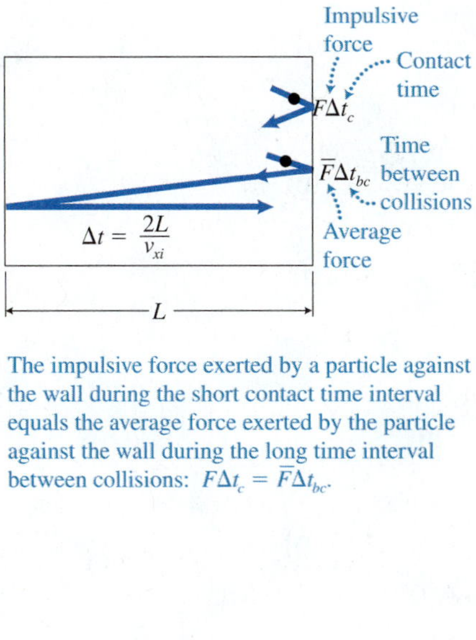


The impulsive force exerted by a particle against the wall during the short contact time interval equals the average force exerted by the particle against the wall during the long time interval between collisions: $F\Delta t_c = \overline{F}\Delta t_{bc}$.

**Figure 9.12** The average velocity squared $\overline{v^2}$ is the sum of the $x$, $y$, and $z$ average velocity squared components.

(a)

(b)

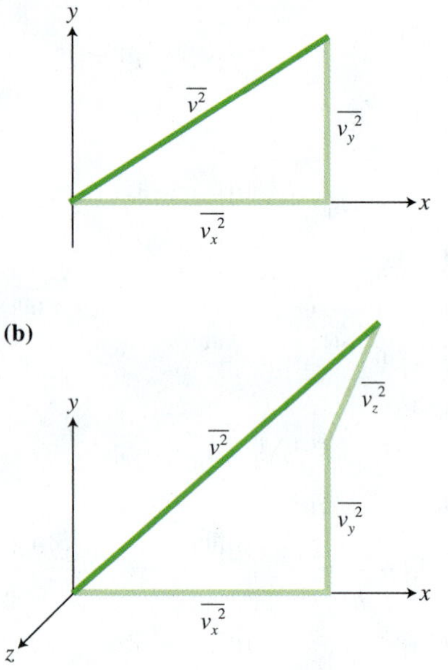


Note that this equation refers only to the motion of the particles along the $x$ -direction. The particles also move in the $y$- and $z$-directions. Those equations are

$$Nm\overline{v_y^2} = PV$$
$$Nm\overline{v_z^2} = PV$$

To make an equation that simultaneously takes into account the motion of the particles in all three directions, we just add these three equations:

$$Nm(\overline{v_x^2} + \overline{v_y^2} + \overline{v_z^2}) = 3PV$$

The quantity $(\overline{v_x^2} + \overline{v_y^2} + \overline{v_z^2})$ is the sum of the averages of the squares of the speeds of the particles in the three directions. Think of the meaning of this sum. We are adding squared $x$-, $y$-, and $z$-components of some vector. If we worked in two dimensions, the sum of the squared $x$- and $y$-components of some vector would be the square of the magnitude of that vector (the Pythagorean theorem—**Figure 9.12a**). In our three dimensional situation, $(\overline{v_x^2} + \overline{v_y^2} + \overline{v_z^2})$, this sum must be the average of the squared speed $(\overline{v_x^2} + \overline{v_y^2} + \overline{v_z^2}) = \overline{v^2}$, as depicted in Figure 9.12b. Using this new average of the squared speed, we can write

$$Nm\overline{v^2} = 3PV$$

The left side of the equation is the average kinetic energy of the particles if we multiply by ½. After multiplying by ½ and rearranging the equation, we get

$$PV = \frac{2}{3}N\left(\frac{1}{2}m\overline{v^2}\right) = \frac{2}{3}N\overline{K}$$

Dividing by $V$, we get

$$P = \frac{2}{3}\left[\frac{N\left(\frac{1}{2}m\overline{v^2}\right)}{V}\right] \qquad (9.4)$$

**TIP** Every time you derive a new equation, ask yourself whether it makes sense. In Eq. (9.4), the pressure is proportional to the average squared speed of the particles. This means that doubling the speed of all the particles quadruples the pressure. Try to explain this dependence before you read on.

When the particles have high speed, they (1) hit the walls of the container more frequently and (2) exert a greater force during the collisions. Both factors lead to a greater pressure. Thus, it is the speed squared and not just the speed of the particles that affects the pressure.

Suppose the number of particles $N$ and the particle average speed squared $\overline{v^2}$ remains the same, but the volume of the box decreases. Then the pressure must increase. This seems reasonable. The smaller the container, the more frequently particles collide with the walls, and the greater the pressure. A unit analysis with a little manipulation indicates that both sides of Eq. (9.4) have the units $kg/(m \cdot s^2)$, so the equation checks out from the point of view of dimensional analysis.

**QUANTITATIVE EXERCISE 9.4 How fast do they move?**

Estimate the average speed of air particles at standard conditions, when the air is at atmospheric pressure ($1.0 \times 10^5\ N/m^2$), and one mole of the air particles ($6.02 \times 10^{23}$ molecules) occupies 22.4 L or $22.4 \times 10^{-3}\ m^3$. Although air is composed of many types of particles, we will assume that the air particles have an average mass $m_{air} = 4.8 \times 10^{-26}$ kg/particle.

**Represent mathematically** We can estimate the average speed using Eq. (9.4):

$$P = \frac{2}{3}\left[\frac{N\left(\frac{1}{2}m\overline{v^2}\right)}{V}\right]$$

**Solve and evaluate** Multiply both sides by 3, divide by $Nm_p$, and take the square root to get

$$\overline{v} \approx \sqrt{\frac{3PV}{Nm_p}}$$

Inserting the appropriate values gives

$$\overline{v} \approx \sqrt{\frac{3PV}{Nm_p}} = \sqrt{\frac{3(1.0 \times 10^5\ \text{N/m}^2)(22.4 \times 10^{-3}\ \text{m}^3)}{(6.02 \times 10^{23})(4.8 \times 10^{-26}\ \text{kg})}}$$
$$= 480\ \text{m/s}$$

This number seems too high. We know that it takes several minutes for the smell of perfume to propagate across a room. How could it take so long if the perfume molecules move at hundreds of meters per second (about 1000 mph)? We will find out in the next section.

**Try it yourself:** The pressure in a diver's full oxygen tank is about $4 \times 10^7\ \text{N/m}^2$. What happens to the average speed of the particles when some of the oxygen is used up and the pressure in the tank drops to half this value? What assumptions did you make?

*Answer:* If we assume that the only change is in the number of particles in the tank, then their average speed should stay the same. However, when a gas expands (as this one does when the valve to the tank is opened), the gas inside the tank has to push outward against the environment in order to leave the tank. This requires energy, which lowers the average kinetic energy (and therefore the speed) of the particles inside.

## Time interval between collisions

We estimated that the average speed of air particles is $v = 480\ \text{m/s}$, or about 1000 mph. Then why does it take 5 to 10 minutes for the smell of perfume to travel across a room? Remember that we estimated that the average distance between particles in a gas at normal conditions is about $D = 3.3 \times 10^{-7}$ cm. While deriving our mathematical descriptions of the ideal gas model, we assumed that the particles do not collide with each other. Perhaps the gas particles actually do collide with neighboring particles and change direction due to each collision, thus making little progress in crossing the room. More detailed estimates show that particles collide about $10^9$ times a second under typical atmospheric conditions. They change direction at each collision, and even though they are moving very fast, their migration from one place to another is very slow.

## What's next?

So far, we have said nothing about the temperature of the gas. Could there be a connection between the temperature of a gas and the average kinetic energy of its particles? We consider this idea next.

**Review Question 9.3** In the expression $PV = \frac{1}{3}N(m_p\overline{v^2})$, pressure is proportional to the particle mass times the average squared speed of the particles. Thus, doubling the average speed of particles leads to quadrupling the pressure. Explain why this makes sense.

# 9.4 Temperature

As with other physical quantities, temperature can be measured. A common way to measure temperature is with a liquid thermometer. A liquid thermometer consists of a narrow tube connected to a bulb at the bottom (**Figure 9.13**). The bulb and part of the tube are filled with a liquid that expands predictably when heated and shrinks when cooled. To calibrate a thermometer, one marks the height of the liquid at the freezing and boiling conditions of water and then divides this interval by a set number of degrees. On the Celsius scale,

**Temperature conversions**

$T_F = (9/5)T_C + 32°$

$T_C = (5/9)T_F - 32°$

$T_K = T_C + 273.15°$

**Figure 9.13** Thermometers calibrated for (a) Celsius scale, (b) Fahrenheit scale, and (c) absolute (Kelvin) scale.

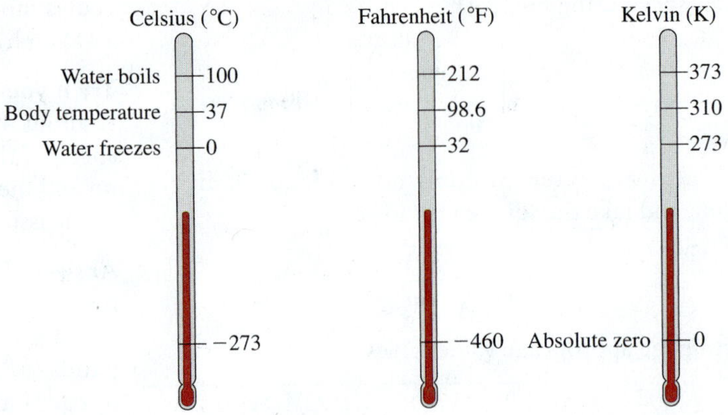


100 C° separates the boiling point and freezing point of water. On the Fahrenheit scale, 180 F° separates these same points (212 °F for boiling and 32 °F for freezing).

## What does temperature really quantify?

We measure the temperature of an object indirectly by measuring the changing volume of a liquid that contacts the object. But what is actually different about an object at higher temperature compared to an object at lower temperature? Consider the experiments in Observational Experiment **Table 9.3**.

**OBSERVATIONAL EXPERIMENT TABLE**

**9.3 Connecting properties of matter to its temperature.**

VIDEO 9.3

| Observational experiment | Analysis |
|---|---|
| A beach ball is soft when in a very cold garage. When taken into a warm house, it becomes firmer and bouncier. | Assuming that the air pressure inside and outside the house is the same, the particles of air inside the ball must exert a higher pressure when the gas inside the ball gets warmer. |
| A balloon filled with air shrinks if placed in ice water. | Assuming that the air pressure in the room stays the same, the particles of air inside the balloon must exert a lower pressure when the gas inside the balloon gets colder. |

**Pattern**

Changing the temperature of the gas seems to change its pressure.

**Active Learning Guide ›**

How can we explain this pattern? In the experiment with the ball taken into a warm room, we can hypothesize that the ball expands because the impulses of the particles against the inside walls are larger when the gas is warm then when it is cold. This would happen if the particles were moving faster. If so, they would also collide with the walls more frequently. In the balloon experiment the particles seem to exert a smaller impulse on the walls of the balloon when the gas is cooler. This would happen if the particles were moving slower. If so, they would also collide less frequently. Based on this reasoning, we can hypothesize that the temperature of a gas is related to the speed of the random motion of its particles. Is the temperature of the gas related to any other properties of the gas (the pressure or volume of the gas, or perhaps how many particles comprise the gas and how massive they are)?

Let's do more observational experiments, this time conducted in a physics laboratory. We place three different gases in three containers of different but known volumes. A pressure gauge measures the pressure inside each container. The number $N$ of particles (atoms or molecules) in each container is determined by measuring the mass of the gas $m_{gas}$ in each container and then calculating

$$N = \frac{m_{gas}}{m_{molar\ mass}} N_A,$$

where $N_A$ is Avogadro's number. Each container, with known $V$, $P$, and $N$ for each gas, is placed first in an ice water bath and then in boiling water, as depicted in **Figure 9.14**. Notice that the volume, pressure, and the number of particles in each container are different, but the temperature of the matter in the three containers is the same. Collected data show the following pattern: independently of the type of gas in a container, the ratio $PV/N$ is identical for all of the gases in the containers when they are at the same temperature:

$$\frac{P_N V_N}{N_N} = \frac{P_O V_O}{N_O} = \frac{P_{He} V_{He}}{N_{He}}$$

The ratio for the gases in the containers at $T_1 = 0\,°C$ is smaller than that for the gases in the containers at $T_2 = 100\,°C$.

From these experiments we can conclude that if you have any amount of a particular type of gas and know its pressure, volume, and the number of particles, then the ratio $PV/N$ only depends on the temperature of the gas. Maybe it equals the temperature? Consider the units of this quantity:

$$\frac{(N/m^2)m^3}{\text{particle}} = \frac{N \cdot m}{\text{particle}} = J/\text{particle}$$

The joule is a unit of energy, not temperature! Perhaps gas particles at the same temperature have the same average energy per particle. Remember that in the ideal gas model, the particles do not have any potential energy between them. This suggests that temperature is related to the average kinetic energy per particle of the gas.

Can we mathematically relate the energy per particle of the gas molecules to the temperature of the gas? The simplest relationship is a direct proportionality to the temperature:

$$\frac{PV}{N} = kT \tag{9.5}$$

where $k$ is a proportionality constant whose value we need to determine. Notice that this relationship immediately leads to a difficulty. The kinetic energy per particle is always a positive number. But in the Celsius and Fahrenheit scales, temperatures can have negative values. So the particle energy cannot be proportional to temperature if measured using either of those scales.

**Figure 9.14** The ratio $PV/N$ seems to depend only on the gas temperature.

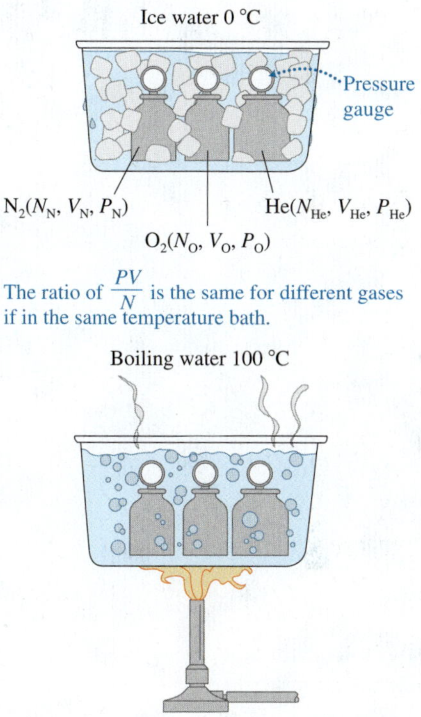


## Absolute (Kelvin) temperature scale and the ideal gas law

We need a scale in which the zero point is the lowest possible temperature. That way, all temperatures will be positive. This lowest possible temperature can be found by applying Eq. (9.5) to measurements with a container of gas at two different reference temperatures—the freezing and boiling temperatures of water. The data in **Table 9.4** were collected when a constant-volume metal container

**Table 9.4** ***PV/N* for one mole of gas in a 22.4-L container at two different temperatures.**

| Conditions in the bath | Pressure | Volume | $\frac{PV}{N} = kT$ |
|---|---|---|---|
| Ice water ($T$) | $1.013 \times 10^5\ \text{N/m}^2$ | $22.42 \times 10^{-3}\ \text{m}^3$ | $3.773 \times 10^{-21}\ \text{J}$ |
| Boiling water ($T + 100$) | $1.384 \times 10^5\ \text{N/m}^2$ | $22.42 \times 10^{-3}\ \text{m}^3$ | $5.154 \times 10^{-21}\ \text{J}$ |

with 1 mol of nitrogen ($N = N_A = 6.02 \times 10^{23}$ particles) was placed in baths at two different temperatures. If we assume that the ratio $PV/N$ is proportional to the absolute temperature of the gas, we can find the coefficient of proportionality.

We now have two equations with two unknowns: the constant $k$ and the water temperature $T$ on the new scale.

$$3.773 \times 10^{-21}\ \text{J} = kT$$
$$5.154 \times 10^{-21}\ \text{J} = k(T + 100)$$

We subtract the first equation from the second to get $k = 1.38 \times 10^{-23}$ J/degree. The freezing temperature $T$ of the water is then

$$T = \frac{3.773 \times 10^{-21}\ \text{J}}{k} = \frac{3.773 \times 10^{-21}\ \text{J}}{1.381 \times 10^{-23}\ \text{J/degree}} = 273.2 \text{ degrees.}$$

On this new scale, water freezes at $T = 273.2$ degrees above the lowest possible temperature. The lowest possible temperature on the new scale is 0 and on the Celsius scale should be $-273.2$ °C. Considerable modern research has refined this value to $-273.15$ °C. This temperature scale is called the absolute temperature scale or the Kelvin scale (because it was invented by William Thomson (Lord Kelvin) in 1848). Temperatures are described in kelvin (see the right scale in Figure 9.13). Temperature intervals on the Kelvin scale are the same as on the Celsius scale: a change in temperature of 1 C° is equivalent to a change in temperature of 1 K. Celsius temperatures are related to kelvin as follows:

$$T_K = T_C + 273.15 \tag{9.6}$$

In other words, a temperature of 273.15 K is equivalent to 0 °C.

The constant $k$ that we determined using the data in Table 9.4 is called Boltzmann's constant after the German physicist Ludwig Boltzmann (1844–1906):

$$k = 1.38 \times 10^{-23}\ \text{J/K}$$

**TIP** Note that Eq. (9.7) implies that when the absolute temperature of the ideal gas is zero, its pressure must be zero.

We can now rewrite Eq. (9.5) using the absolute temperature $T$ and the value of the constant $k$:

$$PV = NkT \tag{9.7}$$

Equation (9.7) is called **the ideal gas law**. It is also commonly used in a slightly different form. Rather than referring to the number of particles that comprise the gas (typically an extremely large number), we refer to the number $n$ of moles of the gas. Since one mole has Avogadro's number of particles, $N = nN_A$. Substituting this into Eq. (9.7), we get

$$PV = nN_AkT$$

The product of the two constants $N_A k$ is another constant called the **universal gas constant** $R$. Equation (9.7) then becomes

$$PV = nRT$$

where $R = N_A k = \left(6.02 \times 10^{23} \frac{\text{particles}}{\text{mole}}\right)\left(1.38 \times 10^{-23} \frac{\text{J}}{\text{K}}\right) = 8.3 \frac{\text{J}}{\text{K} \cdot \text{mole}}$

This is the more common form of the ideal gas law.

**Ideal gas law** For an ideal gas, the quantities pressure $P$, volume $V$, number of particles $N$, temperature $T$ (in kelvins), and Boltzmann's constant $k = 1.38 \times 10^{-23}$ J/K are related in the following way:

$$PV = NkT \tag{9.7}$$

The law can also be written in terms of the number of moles of particles $n$, and the universal gas constant $R = 8.3 \frac{\text{J}}{\text{K} \cdot \text{mole}}$:

$$PV = nRT \tag{9.8}$$

## Temperature and particle motion

Let us look back at what we have done so far. First, we found the relation between the pressure and volume of an ideal gas and the average kinetic energy of the particles that comprise it, $PV = \frac{2}{3}N\overline{K}$ [Eq. (9.4)]. This was a reasonable finding: the faster the particles move inside the gas, the more often and the harder they hit the walls. The particle mass and the number of them per unit volume $N/V$ also affect the pressure.

Next, we found that the product of the pressure and volume of a gas is related to the temperature of the gas, $PV = NkT$ [Eq. (9.7)]. We can now connect the average kinetic energy of the gas particles to the absolute temperature of the gas. Rearrange Eqs. (9.4) and (9.5) so they each have $PV/N$ on the left side. Insert the average kinetic energy $\overline{K} = \frac{1}{2}m\overline{v^2}$ in the right side of Eq. (9.4) and then set the right sides of the two equations equal to each other to get

$$\overline{K} = \frac{3}{2}kT \tag{9.9}$$

The temperature of a gas is an indication of the average random translational kinetic energy of the particles in the ideal gas. Note that temperature is an indication of not only the particle's speed but also its kinetic energy—the mass of the particle also matters. One implication of this discovery is that when you have a mixture of particles of different gases in one container (for example, in air there are nitrogen molecules, oxygen molecules, carbon dioxide molecules, etc.), the lighter molecules move faster than the heavier ones, though each species of particles has the same average kinetic energy (since each species will have the same temperature once the gas has mixed together thoroughly).

Before we move to an example, let's think more about temperature. Imagine that you have two metal containers with identical gases that have been sitting in the same room for a long time. One container is large and the other one is small. Which one has a higher temperature? Since the average kinetic energy per particle is the same in each container, the temperatures of the two gases are the same. However, the total kinetic energy of the particles in the large container is larger because it contains more particles.

Imagine another scenario: You have two containers with the same type of gas. In one container the gas is hot and in the other it is cold. What will happen if you mix those two gases together? The faster moving particles of the hot gas will collide with the slower moving particles of the cold gas. If we use the laws of momentum and kinetic energy conservation (assuming that collisions are elastic), we find that following a collision, the faster moving particle on average is moving slower than before the collision, and the slower particle is on average moving faster than before. Eventually, the particles of the two gases have the same average kinetic energy and therefore the same temperature. Physicists say that the gases are in *thermal equilibrium*.

**TIP** Only when temperature is measured in kelvins can Eq. (9.9) be used to calculate the average kinetic energy of the particles.

**QUANTITATIVE EXERCISE 9.5 Speed of air particles**

Estimate the average speed of air particles in a typical room. Air consists of particles whose average molar mass is $29 \times 10^{-3}\ \text{kg/mole}$.

**Represent mathematically** The temperature in an average room is about 20 °C or 293 K. The average kinetic energy of each particle at temperature $T$ is

$$\bar{K} = \frac{1}{2}m_p\overline{v^2} = \frac{3}{2}kT$$

The mass $m_p$ of one particle is the mass $M$ of a mole of that type of particle divided by the number of particles in one mole $N_A$:

$$m_p = \frac{M}{N_A}$$

**Solve and evaluate** Combining the last two equations, we find the square root of the average speed squared, called the **root-mean-square speed** (the rms speed) of the air particles. We'll use this as our estimate of the average speed of the particles:

$$\sqrt{\overline{v^2}} = \sqrt{\frac{3kT}{m_p}} = \sqrt{\frac{3kTN_A}{M}}$$

Inserting the appropriate values gives

$$\sqrt{\overline{v^2}} = \sqrt{\frac{3(1.38 \times 10^{-23}\ \text{J/K})(293\ \text{K})(6.02 \times 10^{23}\ \text{particles/mole})}{29 \times 10^{-3}\ \text{kg/mole}}}$$

$$= 500\ \text{m/s}$$

This speed is close to the speed calculated in Quantitative Exercise 9.4. Let's check the units:

$$\sqrt{\overline{v^2}} = \sqrt{\frac{\left(\frac{\text{J}}{\text{K}}\right)(\text{K})\left(\frac{1}{\text{mole}}\right)}{\left(\frac{\text{kg}}{\text{mole}}\right)}}$$

$$= \sqrt{\frac{\text{J}}{\text{kg}}} = \sqrt{\frac{\text{N}\cdot\text{m}}{\text{kg}}}$$

$$= \sqrt{\frac{\left(\text{kg}\cdot\frac{\text{m}}{\text{s}^2}\right)\cdot\text{m}}{\text{kg}}} = \frac{\text{m}}{\text{s}}$$

The units are consistent.

**Try it yourself:** What happens to the average speed of the molecules in the room when the temperature drops by one-half?

*Answer:* It depends on which scale we use to measure the temperature. If we use the Celsius scale, the temperature would be 10 °C, (283 K on the Kelvin scale) and the rms speed would be about 490 m/s. If we use the Kelvin scale, the temperature would be 147 K and the rms speed would be 350 m/s.

**Review Question 9.4** If there is a mixture of different molecules in a container, which ones have a higher average speed, the more massive molecules or the less massive molecules?

## 9.5 Testing the ideal gas law

In order to determine if the ideal gas law describes the behavior of real gases, we will use the law to predict the outcomes of some testing experiments. If the predictions match the outcomes, we gain confidence in the ideal gas law. In the experiments below we will keep one of the variables ($T$, $V$, or $P$) constant and predict the relation between the two other variables. Processes in which $T$, $V$, or $P$ are constant are called **isoprocesses.** The three types of isoprocesses we will investigate are **isothermal** ($T =$ constant), **isochoric** ($V =$ constant) and **isobaric** ($P =$ constant).

‹ Active Learning Guide

**TESTING EXPERIMENT TABLE**

**9.5 Does the ideal gas law apply to real gases?**

| Testing experiment | Prediction | Outcome |
|---|---|---|
| **Experiment 1:** Isothermal process. $n$ moles of gas are in a variable volume $V$ container that is held in an ice bath at constant 0 °C (273 K) temperature $T$. How does the pressure of the gas change as we change the volume of the container? We push the piston slowly so that the temperature of the gas is always the same as the ice bath.<br>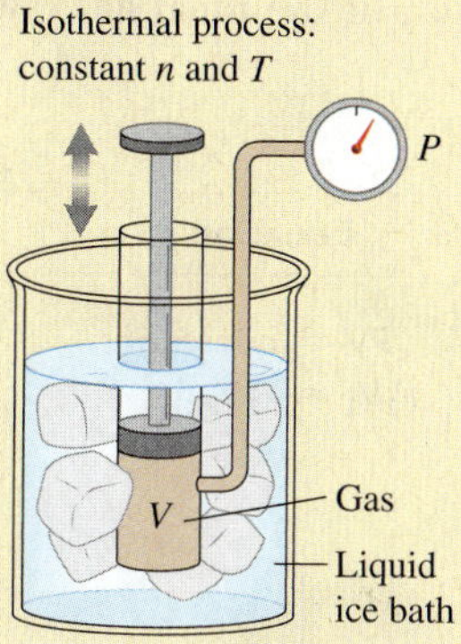<br> | According to the ideal gas law $PV = nRT$, during a constant temperature process, the product of $PV$ should remain constant. We predict that as the volume decreases, the pressure will increase so that the product remains constant. | Data collected:<br>$V\,(m^3)$ — $P\,(N/m^2)$<br>$3.0 \times 10^{-4}$ — $2.0 \times 10^5$<br>$6.0 \times 10^{-4}$ — $1.0 \times 10^5$<br>$9.0 \times 10^{-4}$ — $0.67 \times 10^5$<br>The product of volume and pressure remains constant in all experiments. |
| **Experiment 2:** Isochoric process. $n$ moles of gas and the gas volume $V$ are kept constant. The container is placed in different-temperature baths. How does the gas pressure change as the temperature changes?<br><br> | According to the ideal gas law $PV = nRT$, during a constant volume process, the ratio $\frac{P}{T} = \frac{nR}{V}$ should remain constant. We predict that the pressure should increase in proportion to the temperature. | Data collected:<br>$T\,(K)$ — $P\,(N/m^2)$<br>300 — $1.0 \times 10^5$<br>400 — $1.3 \times 10^5$<br>500 — $1.7 \times 10^5$<br>The ratio of pressure and temperature is constant in all experiments. |
| **Experiment 3:** Isobaric process. $n$ moles of gas and the gas pressure $P$ are held constant, as a piston in the gas container can move freely up and down keeping the pressure constant. How does the gas volume change as the temperature changes?<br>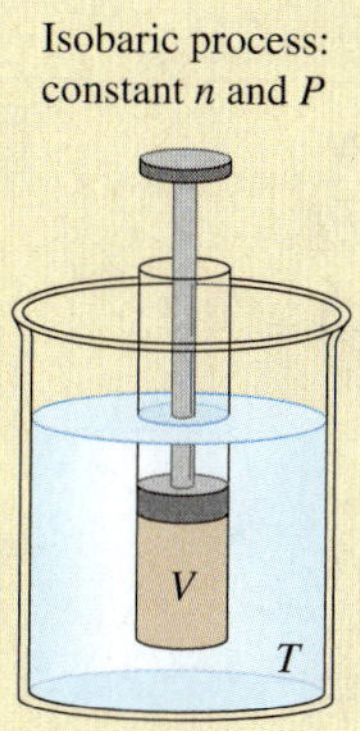<br> | According to the ideal gas law $PV = nRT$, during a constant pressure process, the ratio $\frac{V}{T} = \frac{nR}{P}$ should remain constant. We predict that the volume should increase in proportion to the temperature. | Data collected:<br>$T\,(K)$ — $V\,(m^3)$<br>300 — $3.0 \times 10^{-4}$<br>400 — $4.0 \times 10^{-4}$<br>500 — $5.0 \times 10^{-4}$<br>The ratio of volume and temperature remains constant in all experiments. |

(*continued*)

**Conclusion**

The outcomes of all three experiments are consistent with the predictions.

- In the first experiment, the product of pressure and volume remains constant, as predicted.
- In the second experiment, the pressure increases in direct proportion to the temperature, as predicted.
- In the third experiment, the volume increases in direct proportion to the temperature, as predicted.

The outcomes of the experiments in Testing Experiment **Table 9.5** were consistent with predictions based on the ideal gas law, giving us increased confidence that the law applies to real gases (however, we cannot say that we proved it). A summary of gas processes (some of which are not isoprocesses) is provided in **Table 9.6**.

## Reflection on the process of construction of knowledge

Let's pause here and reflect on the process through which we arrived at the mathematical version of the ideal gas law. The first step was to construct a

**Table 9.6** A summary of ideal gas law processes.

| Name | Constant quantities | Changing quantities | Equation | Graphical representation |
|---|---|---|---|---|
| Isothermal | $N$ or $n$, $T$ | $P$, $V$ | $PV = \text{constant}$<br>$P_1V_1 = P_2V_2$ | 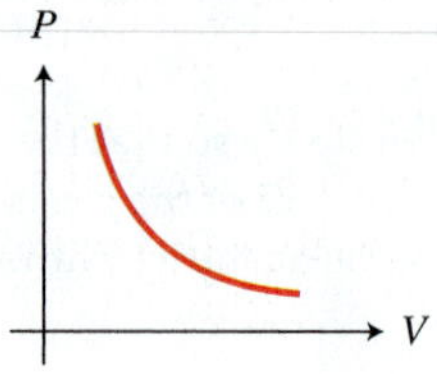 |
| Isobaric | $N$ or $n$, $P$ | $V$, $T$ | $\frac{V}{T} = \text{constant}$<br>$\frac{V_1}{T_1} = \frac{V_2}{T_2}$ | 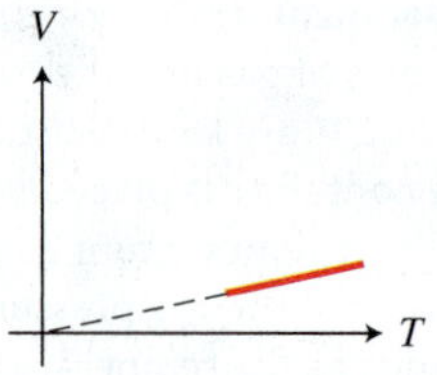 |
| Isochoric | $N$ or $n$, $V$ | $P$, $T$ | $\frac{P}{T} = \text{constant}$<br>$\frac{P_1}{T_1} = \frac{P_2}{T_2}$ | 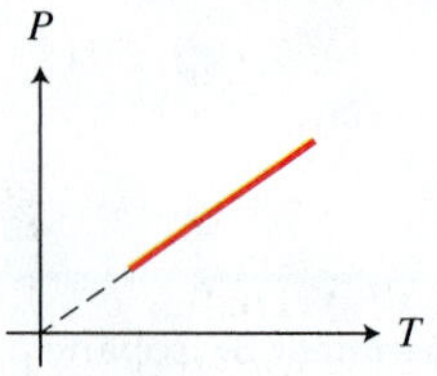 |
| [No name] | $N$ or $n$ | $P$, $V$, $T$ | $\frac{PV}{T} = \text{constant}$<br>$\frac{P_1V_1}{T_1} = \frac{P_2V_2}{T_2}$ | |
| [No name] | | $P$, $V$, $T$, $N$ or $n$ | $\frac{PV}{NT} = k$<br>$\frac{PV}{nT} = R$ | |

simplified model of a system that could represent a real gas—the ideal gas model. This involved making assumptions about the internal structure of gases. This model was based on some observations and also on the knowledge of particle motion and interactions developed earlier—Newton's laws of motion. We used this model to devise a mathematical description of the behavior of gases, the ideal gas law. We then tested its applicability to real gases by using it to predict how macroscopic quantities describing the gas (temperature, pressure, volume, and the amount of gas) would change during specific processes (isothermal, isobaric, and isochoric) and used the ideal gas law to construct equations that described those processes. These predictions were consistent with the outcomes of the new testing experiments.

The process of constructing the knowledge described above looks relatively smooth and straightforward—observe, simplify, explain, test. However, in real physics, knowledge construction is not that simple and straightforward. For example, the isoprocesses mentioned above were known in physics long before the ideal gas model was constructed. They were discovered in the 17th and 18th centuries and carried the names of the people who discovered them through patterns found in observational experiments. The relation $PV =$ constant for constant temperature processes is called **Boyle's law** and was discovered experimentally by Robert Boyle in 1662. The relation $V/T =$ constant for constant pressure processes is called **Charles's law** and was discovered by Jacques Charles in 1787, though the work was published by Joseph Gay-Lussac only in 1802. Gay-Lussac also discovered the relation $P/T =$ constant, now called **Gay-Lussac's law**. When these relations were discovered empirically, there was no explanation for why gases behaved in these ways. The explanations arrived much later via the ideal gas model. The real process of knowledge construction is often more complicated and nonlinear than how it is presented in a textbook. The skills you are learning by constructing knowledge through experimentation will prepare you for those more complicated situations.

## Applications of the ideal gas law

The ideal gas law has numerous everyday applications. Try to explain the following phenomena using the ideal gas law. Clearly state your assumptions.

- A sealed, half-full bottle of water shrinks when placed in the refrigerator.
- A container full of nitrogen explodes if the temperature rises too high.
- A bubble of gas expands as it rises from the bottom of a lake.
- Air rushes into your lungs when your diaphragm, a dome-shaped membrane, contracts.

Below we consider in greater detail several of these phenomena.

**Breathing** During inhalation, our lungs absorb oxygen from the air, and during exhalation, they release carbon dioxide, a metabolic waste product. Yet the lungs have no muscle to push air in or out. The muscle that makes inhaling and exhaling possible, the diaphragm, is not part of the lungs. The diaphragm is a large dome-shaped muscle that separates the rib cage from the abdominal cavity (see **Figure 9.15**).

The diaphragm works like a bellows. As the diaphragm contracts and moves down from the base of the ribs, the volume of the chest cavity and lungs increases. If we assume that for a brief instant both the temperature and the number of particles are constant (an isothermal process),

$$P = \frac{nRT}{V},$$

**Figure 9.15** The diaphragm is a large dome-shaped muscle that separates the rib cage from the abdominal cavity. Relaxing or contracting the diaphragm changes the volume of the lungs and chest cavity.

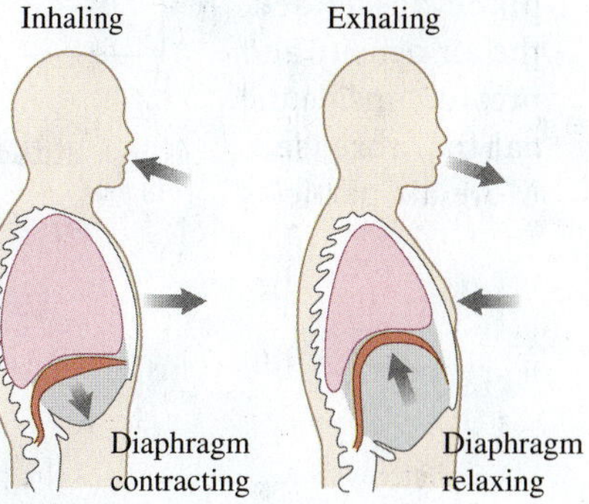


Diaphragm contraction and relaxation cause air to enter and leave the lungs.

then the pressure in the cavity and in the lungs will decrease. The pressure of the outside air is greater than that of the inside. Because of this, the outside air at normal atmospheric pressure enters the mouth or nostrils and fills the lungs with fresh new air (an increase $\Delta n > 0$ of the amount of air). To exhale, the opposite occurs. The diaphragm relaxes and rises, decreasing the volume of the chest cavity and the lungs, thus increasing the pressure inside (again assuming that for a brief instant both the temperature and the number of particles are constant),

$$P = \frac{nRT}{V}$$

Because the inside pressure is higher than the outside pressure, air is then forced out of the lungs.

**Your Water Bottle on an Airplane** The behavior of an empty plastic water bottle on an airplane is another example of an isothermal process.

**CONCEPTUAL EXERCISE 9.6 A shrunk bottle** On a flight from New York to Los Angeles you drink most of a bottle of water and then store it in your seat pocket. Just before landing, you take it out again as the flight attendant is collecting trash. The bottle has changed shape; it looks like someone crushed it. How can we explain this shape change?

**Sketch and translate** The bottle's volume decreased, though the temperature in the cabin didn't change much during the flight.

**Simplify and diagram** If we assume the bottle was perfectly sealed, and the temperature of the gas inside the bottle remained constant, then we can model the process as an isothermal process, represented graphically in the figure below. Although the cabin is pressurized, at higher elevations air pressure and air density inside the cabin are slightly less than at lower elevations. You closed the bottle when it was filled with low-density air at high elevation. As the plane descended, the air density and pressure inside the cabin increased. More air particles were hitting the outside walls of the empty bottle than were hitting the inside. The higher pressure from outside partially crushed the bottle. If you open the cap after landing, the bottle will pop back to its original shape.

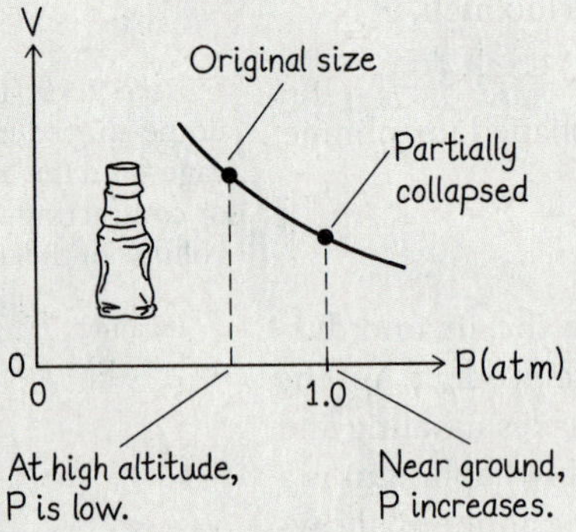


**Try it yourself:** A process is represented on a graph in the figure below. Describe the process in words. Assume the mass to be constant.

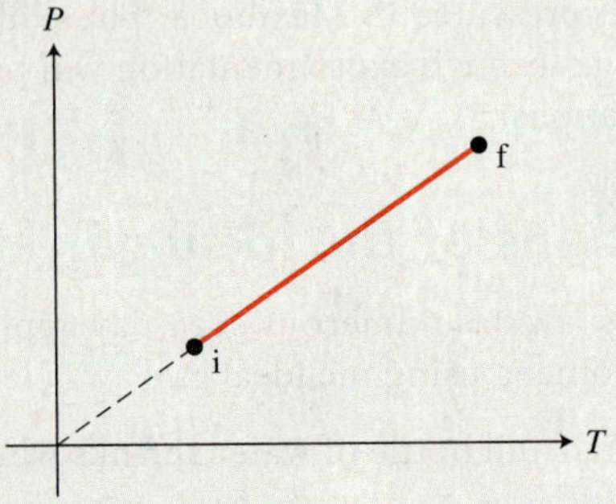


*Answer:* The graph represents an isochoric process. The graph line, if extended, passes through the origin; thus the pressure is directly proportional to temperature. The gas was in a sealed container. The gas container was first placed in a water bath at low temperature and then transferred to a water bath at high temperature. The volume of the gas is constant. The pressure increases as the particles move faster and collide with the walls of the container more often.

## Comparing different gases

How does the type of gas that is in a container affect the pressure of the gas? Consider the next conceptual exercise.

**CONCEPTUAL EXERCISE 9.7 Analyzing two types of gas**

You have two containers, each with pistons of equal mass that can move up and down depending on the pressure of the gas below. Each container has a nozzle that allows you to add gases to the containers. Container 1 holds nitrogen (molar mass $M_1$); we label its volume $V_1$. Container 2 ($V_2$) holds helium (molar mass $M_2 < M_1$). The volumes are the same ($V_1 = V_2$), and the containers sit in the same room. Determine everything you can about the situation.

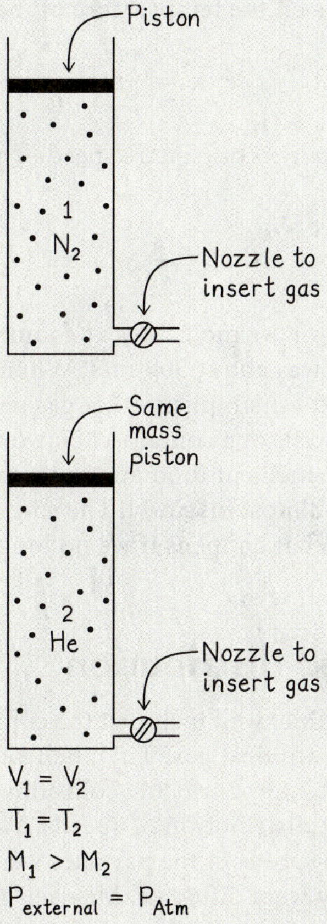


**Sketch and translate** We sketch the situation in the figure to the right. Quantities that we can try to compare between the two containers are the following:

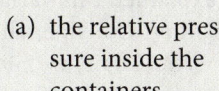

(a) the relative pressure inside the containers
(b) the average kinetic energy of a particle of each type of gas
(c) the mass of individual gas particles
(d) the rms speeds of each particle
(e) the number of particles in each container
(f) the mass of the gas in each container
(g) the density of the gas in the containers.

**Simplify and diagram** Assume that the ideal gas model accurately describes the gases.

(a) Consider the pressures inside the containers by analyzing the identical moveable pistons above the gases. Force diagrams for each piston are shown at right. Earth exerts a downward force $\vec{F}_{\text{E on P}}$, on the piston, the atmospheric gas above a piston pushes down on the piston $\vec{F}_{\text{Atm on P}}$, and the gases inside the containers push up on their pistons $\vec{F}_{\text{N}_2\text{ on P}}$ or $\vec{F}_{\text{He on P}}$. As the pistons are not accelerating, the net force exerted on each is zero. As the downward forces exerted by Earth and by the outside atmospheric air on each piston are the same, the upward forces exerted by the gases in each container on the piston must also be the same. Since the surface area of each piston is the same, the pressure of the gas inside each container must be the same, $P_{\text{N}_2} = P_{\text{He}}$.

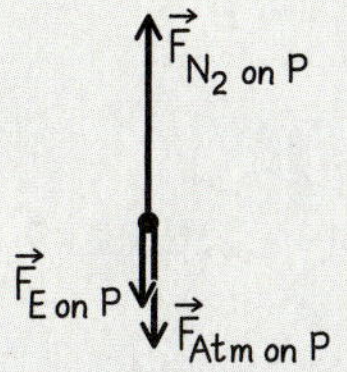


The downward forces on each piston are the same, so the upward force is also the same.

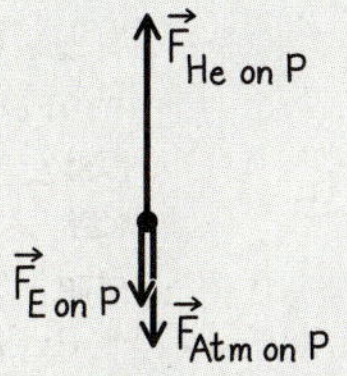


(b) Since the temperature of a gas depends only on the average kinetic energy of the particles, and the temperature of the two gases is the same, the average kinetic energy of each particle is the same:

$$\frac{3}{2}kT = \frac{1}{2}m_{\text{N}_2\text{ in 1}}\overline{v_{\text{N}_2\text{ in 1}}^2} = \frac{1}{2}m_{\text{He in 2}}\overline{v_{\text{He in 2}}^2}.$$

(c) and (d) Since the $N_2$ particles in gas 1 have a higher molar mass than the He particles in gas 2, the $N_2$ molecular mass $m = M_1/N_A$ is also the higher mass ($m_{\text{N}_2\text{ in 1}} > m_{\text{He in 2}}$). Using this result and that from (b), we find that $\overline{v_{\text{N}_2\text{ in 1}}^2} < \overline{v_{\text{He in 2}}^2}$. The more massive particles move slower.

(e) The pressure, volume, and temperature are the same for each gas. Thus, according to the ideal gas law, both gases have the same number of particles and the same number of moles of gas in their containers ($PV/kT = N$; $PV/RT = n$).

(f) and (g) Since the mass of a nitrogen molecule is greater than the mass of a helium atom, and there are equal numbers of particles in each container, the total mass of the gas in the nitrogen container must be greater than the total mass of the gas in the helium container ($M_{\text{N}_2\text{ in 1}} > M_{\text{He in 2}}$). Since the volumes of the containers are equal, the density $\rho = M/V$ of the gas in the nitrogen container must be greater than that in the helium container ($\rho_{\text{N}_2\text{ in 1}} > \rho_{\text{He in 2}}$).

**Try it yourself:** Why are the gases at the same temperature?

*Answer:* Both containers sit in the same room temperature environment.

**Review Question 9.5** What is the difference between the following two equations: $PV = 1/3\, N m\overline{v^2}$ and $PV = nRT$?

## 9.6 Speed distribution of particles

In our previous analysis, we found that the average molecular kinetic energy of gas particles depends on the temperature of the gas

$$\overline{K} = \frac{1}{2}m\overline{v^2} = \frac{3}{2}kT \tag{9.9}$$

Consequently, the root-mean-square speed of a gas atom or molecule (the rms speed) is

$$v_{\text{rms}} = \sqrt{\overline{v^2}} = \sqrt{\frac{3kT}{m}} \tag{9.10}$$

We found that for air molecules at room temperature, the average root-mean-square speed was about 500 m/s. When we derived relationships such as these, we assumed for simplicity that gas particles do not collide with each other, just with the walls of a container. However, we know that if gas particles did not collide, the smells of food and perfume would spread at hundreds of meters per second—almost instantly. The smells spread slowly, so the particles must be colliding. What happens if we no longer ignore collisions of particles with each other?

### Maxwell speed distribution

In 1860 James Clerk Maxwell included the collisions of the particles in his calculations involving an ideal gas. This inclusion led to the following prediction: at a particular temperature, the collisions of gas particles with each other cause a very specific distribution of speeds. When we were deriving Eq. (9.4) we assumed that the speeds of the particles were different, but we did not have any idea of why they were different. Maxwell's work explained this variability of speeds by the collisions of the particles with each other. Consider the yellow $^{20}$Ne line (neon atoms) in **Figure 9.16**. On the vertical axis, we plot the percentage of particles that have a particular speed. According to Maxwell, a certain percentage of the particles should have speeds around 100 m/s, more around 200 m/s, the most around 500 m/s, and fewer at higher speeds. Very few particles have extremely low or extremely high speeds. Most should have intermediate speeds. The most probable speed is at the highest point on the curve in Figure 9.16. Surprisingly, the root-mean-square speed of the particles at a particular temperature is the same as that predicted by the model with no collisions—Eq. (9.10).

**Figure 9.16** The Maxwell particle speed distributions at a particular temperature for four gases.

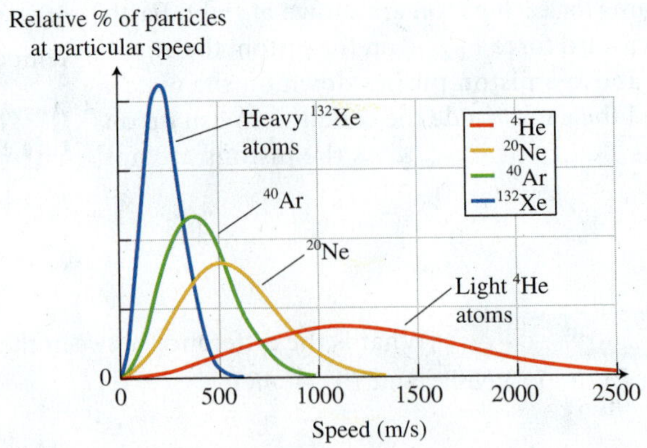

To test Maxwell's ideas, one needs to actually measure the speeds of atoms and molecules at a particular temperature and then compare the distribution of speeds to the calculated distribution curve shown in Figure 9.16. The task—measuring the speeds of objects that are $10^{-10}$ m in diameter—seems almost impossible. However, the problem was tackled and successfully solved by German physicist Otto Stern in 1920, many years after the development of Eq. (9.10). Stern's experiments led to a whole field of study called molecular beam spectroscopy.

An apparatus such as that shown in **Figure 9.17** is used in molecular beam spectroscopy. A gas is heated to some predetermined temperature. A small fraction of the rapidly moving gas particles leaves the container through slit A. Some of these particles pass through slits B and C, forming a narrow beam of particles that hits a rapidly rotating drum with a slit D. The particles can only enter the drum as slit D passes along the line from slits A to C. The particles that enter the drum travel across it to the other side, where they are detected by a sensitive film that produces a mark (a dot) when hit by a particle. Fast-moving particles hit the film almost directly across from the slit, whereas slow-moving particles hit the film somewhat later, as the drum has rotated farther. After the drum completes one rotation, a new group of particles enters the drum. Thus, even if the beam has only a few particles hitting slit D per rotation, after many rotations a denser pattern develops on the film.

The density of the number of particles hitting a particular part of the film indicates the relative speed of those particles. Thus, you can make a graph of the darkening of the film (the relative number of particles hitting a part of the film) versus the position on the film (the speed of particles hitting that part of the film). The pattern can be used to calculate the average particle speed squared. The experiment can be repeated multiple times with the gas at a different temperature each time. The determined average speed squared and gas temperature are consistent with Eq. (9.10). The measured speed distribution patterns in Stern's experiments matched the Maxwell predicted distributions perfectly (Figure 9.16). Finally, note that all of the speed distributions in Figure 9.16 were for different types of gas particles at the same temperature. Note that the less massive particles moved at higher speeds—in complete agreement with experimental results. These results provided strong support for the kinetic theory of gases.

Therefore, we can say that the ideal gas model is a productive model for describing gases. A combination of the model, the ideal gas law, and all of the testable predictions and testing experiments is called **kinetic molecular theory**—a theory that describes and explains the behavior of gases based on their particle structure.

**Figure 9.17** An apparatus to measure the speed distribution of particles in a gas.

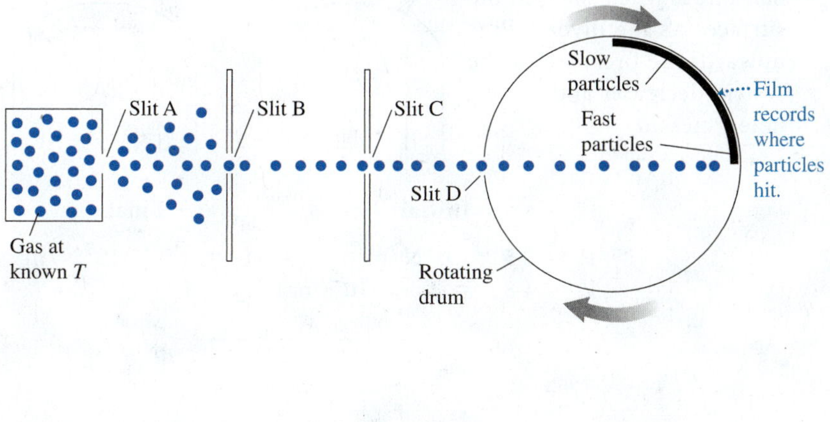

## Limitations of the ideal gas law

Isoprocesses were discovered empirically with experiments long before the model of ideal gas was constructed. However, the fact that the ideal gas law, derived from the ideal gas model, predicts those empirical results is evidence that the law itself and the ideal gas model on which the law is based are quite reliable. Nevertheless, the model and the gas law have their limitations.

For real gases such as air, the measurements of pressure and volume at the conditions of normal pressure ($1.0 \times 10^5$, as elsewhere) and temperature (room temperature) are consistent with predictions of the ideal gas law. But at very high pressures or very low temperatures, real measurements differ from those predictions (for example, the model of ideal gas does not predict that one can turn a gas into a liquid). The ideal gas law describes gases accurately only over certain temperature and pressure ranges.

**Review Question 9.6** What is the role of Stern's molecular beam experiment in the development of kinetic molecular theory?

# 9.7 Skills for analyzing processes using the ideal gas law

In this section, we adapt our problem-solving strategy to analyze gas processes. A general strategy for analyzing such processes is described on the left side of the table in Example 9.8 and illustrated on the right side for a specific process.

**PROBLEM-SOLVING STRATEGY** **Applying the Ideal Gas Law**

**EXAMPLE 9.8 Scuba diver returns to the surface**

The pressure of the air in a scuba diver's lungs when she is 15 m under the water surface is $2.5 \times 10^5\ \mathrm{N/m^2}$ (equal to the pressure of the water at that depth), and the air occupies a volume of 4.8 L. Determine the volume of the air in the diver's lungs when she reaches the surface, where the pressure is $1.0 \times 10^5\ \mathrm{N/m^2}$.

**Sketch and translate**

- Sketch the process. Choose a system and characterize the initial and final states of the system.
- Describe the process (a word description of the changes in the system between the initial and final states) in terms of macroscopic quantities (pressure, volume, temperature, moles of gas).

The gas inside the diver's lungs is the system. The initial state of the system is when the diver is underwater; the final state is when she is at the surface. As the diver swims upward, the pressure of the system decreases and its volume increases.

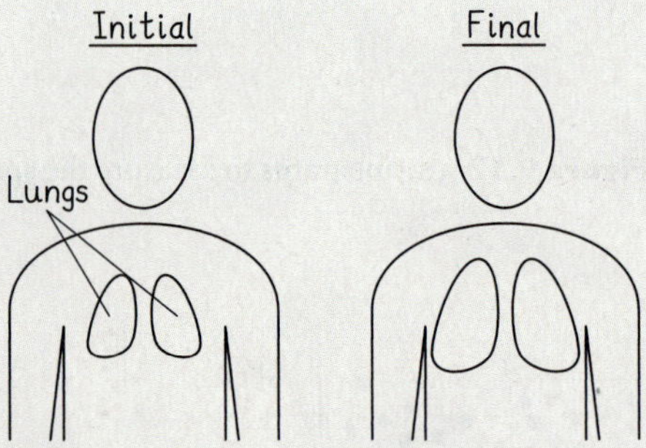


**Initial**

$P_i = 2.5 \times 10^5\ \mathrm{N/m^2}$

$V_i = 4.8\ \mathrm{L} = 4.8 \times 10^{-3}\ \mathrm{m^3}$

**Final**

$P_f = 1.0 \times 10^5\ \mathrm{N/m^2}$

$V_f = ?$

**Simplify and diagram**

- Decide if the system can be modeled as an ideal gas.
- Decide which macroscopic quantities remain constant and which do not.
- If helpful, draw $P$ vs. $V$, $P$ vs. $T$, and/or $V$ vs. $T$ graphs to represent the process.

Assume that the gas inside the lungs can be modeled as an ideal gas. Assume that the diver does not exhale, which means that the moles of gas remain constant. Assume the temperature of the gas is constant at body temperature. A pressure-versus-volume graph for the process is shown below.

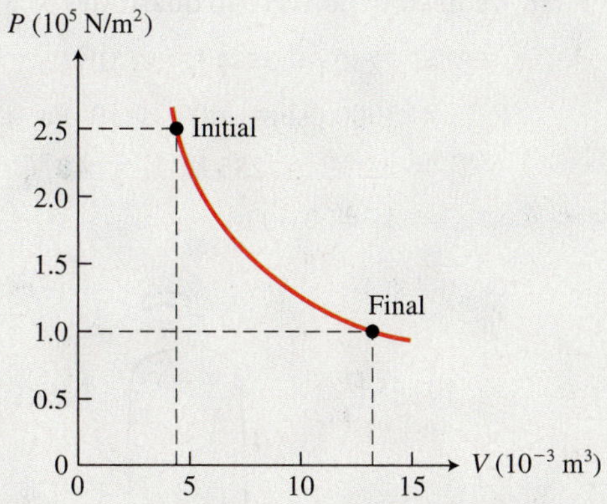


**Represent mathematically**

Use your sketch or the graphs to help construct a mathematical description.

$$P_i V_i = P_f V_f$$

$$\Rightarrow V_f = \frac{P_i V_i}{P_f}$$

**Solve and evaluate**

- Solve for the unknowns.
- Evaluate the answer: is it reasonable? (For example, evaluate the magnitude of the answer, its units, and how the solution behaves in limiting cases.)

$$V_f = \frac{P_i V_i}{P_f} = \frac{(2.5 \times 10^5\ \text{N/m}^2)(4.8 \times 10^{-3}\ \text{m}^3)}{(1.0 \times 10^5\ \text{N/m}^2)} = 12 \times 10^{-3}\ \text{m}^3$$

We found the lung volume to be 12 L, much larger than seems possible. Is the answer realistic? Remember that we assumed that the mass of the gas inside remains constant—the person does not exhale. As we see from the solution, it is important for divers to exhale as they are ascending and the gas expands. If not, they can suffer severe internal damage.

**Try it yourself:** How many moles of gas are in the diver's lungs and how many should she exhale so the final volume is only 6 L instead of 12? Assume that the gas temperature is 37 °C.

*Answer:* 0.47 moles; 0.23 moles.

**TIP** In some gas processes, notice that the graph lines for isobaric processes (constant pressure) pass through the origin in $V$-versus-$T$ graphs and those for isochoric (constant volume) processes pass through the origin in $P$-versus-$T$ graphs.

**EXAMPLE 9.9 Will the container burst?**

A 100-L oxygen tank filled at night was left outside in the sunshine the next day. When it was filled, the pressure was 2250 psi (pounds per square inch is another unit for pressure, where 1 psi $= 6.89 \times 10^3$ Pa) and the temperature was 12 °C. Is it dangerous to leave the container outside when the temperature is 40 °C, if the warning on the container says "pressure not to exceed 3000 psi?" Explain the process microscopically.

**Sketch and translate** Label a diagram with known quantities for the initial and final states, as shown in the figure below. The gas inside the tank is the system. The initial state is when it was filled at night. The final state is

(*continued*)

during the hot day. We need to find the final pressure $P_f$. First, we need to convert all quantities to SI units.

$$P_i = 2250 \text{ psi} = 1.55 \times 10^7 \text{ Pa}$$
$$P_{max} = 3000 \text{ psi} = 2.07 \times 10^7 \text{ Pa}$$
$$T_i = 12\,°\text{C} = 285 \text{ K};\ T_f = 40\,°\text{C} = 313 \text{ K}$$
$$V_i = V_f = 0.10 \text{ m}^3$$

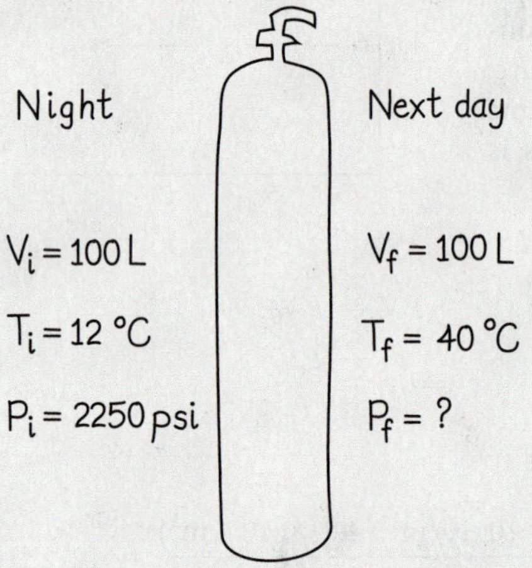


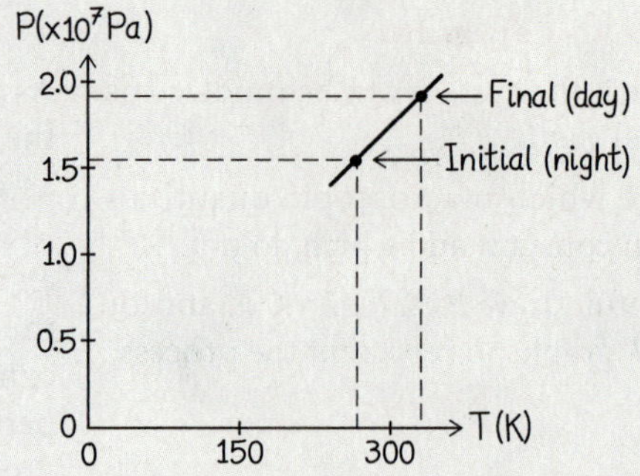


**Simplify and diagram** Can we model the gas in the container as an ideal gas? The pressure starts at $P_i = 1.55 \times 10^7$ Pa and is not to exceed $P_{max} = 2.07 \times 10^7$ Pa, which is more than 200 times atmospheric pressure. The particles of the gas are much closer together than at atmospheric pressure; thus the ideal gas model might not be applicable. However, we do not have another mathematical model to use, so we will use the ideal gas law, keeping in mind that our estimate of the final pressure might not be reasonable. The tank's volume remains constant, so we can use the mathematical description of an isochoric process.

**Represent mathematically** For an isochoric process,

$$\frac{P_i}{T_i} = \frac{P_f}{T_f}$$

**Solve and evaluate** Solving for $P_f$ and inserting the appropriate values gives

$$P_f = \frac{P_i T_f}{T_i} = \frac{(1.55 \times 10^7 \text{ Pa})(313 \text{ K})}{285 \text{ K}} = 1.70 \times 10^7 \text{ Pa}$$

We are still under the limit, but getting close. Taking into account that we are not sure that the ideal gas law applies to the gas inside, we should remember to put the tank in the shade!

**Try it yourself:** How many moles of gas are in the tank when its temperature is 12 °C and when it is 40 °C?

*Answer:* 650 moles; $n$ does not change—it's a sealed container.

**TIP** Notice that if you continue the graph line in the above Example, it passes through the origin, as the pressure is proportional to the absolute temperature.

**Review Question 9.7** Why is it helpful to know whether the mass of the gas is constant during a particular process?

## 9.8 Thermal energy, the Sun, and diffusion: Putting it all together

The ideal gas model applies to many real-world situations involving gases, provided that the gas particles are far from each other compared to their own sizes. Let's consider some of these situations.

## Thermal energy of air

Gas consisting of many atoms and molecules at normal temperature possesses thermal energy—the random kinetic energy of the constituent particles. How does this thermal energy compare to the gravitational potential energy of the system consisting of Earth and its atmosphere? To answer these questions we could estimate the thermal energy of a cup of air. A cup of air at 27 °C contains about $10^{22}$ molecules. Each molecule has some kinetic energy; the thermal energy of these molecules is the sum of their individual average kinetic energies:

$$U_{\text{thermal}} = N\left(\frac{3}{2}kT\right)$$

Or, using the provided information,

$$U_{\text{thermal}} = N\left(\frac{3}{2}\right)kT = (10^{22})\frac{3}{2}(1.38 \times 10^{-23}\,\text{J/K})[(273 + 27)\text{K}] = 60\,\text{J}$$

How high would we need to lift the air molecules in the cup so that the gravitational potential energy of the gas-Earth system equals the thermal energy of the gas? If we assume that the zero level of gravitational potential energy is at Earth's surface, then the gravitational potential energy of the system once the gas has been lifted to a height $h$ is $U_g = m_{\text{total}}gh$, where $m_{\text{total}}$ is the total mass of the gas. Thus,

$$U_{\text{thermal}} = N\left(\frac{3}{2}kT\right) = m_{\text{total}}gh$$

$$\Rightarrow h = \frac{N\left(\frac{3}{2}kT\right)}{m_{\text{total}}g}$$

The mass of the air molecules in the cup is

$$m_{\text{total}} = Nm_{\text{single particle}} = N\frac{M_{\text{mole}}}{N_A}$$

(the mass of one particle is its molar mass divided by the number of particles $N_A$ in one mole). Inserting the numbers, we get

$$h = \frac{U_{\text{thermal}}}{m_{\text{total}}g} = \frac{U_{\text{thermal}}}{N\frac{M_{\text{mole}}}{N_A}g} = \frac{60\,\text{J}}{\left[(10^{22})\frac{(29 \times 10^{-3}\,\text{kg})}{(6.02 \times 10^{23})}\right](9.8\,\text{N/kg})} = 13{,}000\,\text{m}$$

The thermal energy present in that cup of air is equivalent to the work needed to lift that air 13 km! This gives you an idea of the very significant amount of thermal energy in the atmosphere.

## How long will the Sun shine?

The technique that we used to determine the thermal energy of gases can be used to help analyze many kinds of systems, such as automobile engines, Earth's atmosphere, and the Sun. The mass of the Sun is about $2 \times 10^{30}$ kg, its radius is about $7 \times 10^{8}$ m, and the temperature of the Sun's surface is about $6 \times 10^{3}$ K. Its core temperature is about $10^{7}$ K. Every second, the Sun radiates about $4 \times 10^{26}$ J as visible light and other forms of radiation. How long will our Sun shine if its particles possess *only* thermal energy?

We first need to determine how much thermal energy the Sun possesses. Then we can divide this amount of energy by the amount it loses every second from the emission of visible light and other forms of radiation to find the number of seconds the Sun can shine if its thermal energy is the only source of this radiative energy. Although in reality there are some complications with modeling the material of the Sun as an ideal gas, we will do so in this case for simplicity.

To estimate the thermal energy of the Sun, we need to know the number of particles and the average temperature of these particles. The Sun consists mostly of hydrogen atoms. To find the number of hydrogen atoms in the Sun, we estimate the number of moles of hydrogen gas and then multiply by the number of particles in one mole (note that 1 mole of hydrogen has a mass of $1\text{ g} = 10^{-3}\text{ kg}$):

$$
\begin{aligned}
N &= \frac{m_{\text{Sun}}}{m_{\text{hydrogen atom}}} = \frac{m_{\text{Sun}}}{M_{\text{molar mass hydrogen}}/N_{\text{A}}} \\
&= \frac{(2 \times 10^{30}\text{ kg})}{(10^{-3}\text{ kg/mole})/(6 \times 10^{23}\text{ particles/mole})} \\
&= \frac{(2 \times 10^{30}\text{ kg})}{(1.7 \times 10^{-27}\text{ kg/particle})} \\
&= 12 \times 10^{56}\text{ particles}
\end{aligned}
$$

At the very high temperatures within the Sun, each hydrogen atom separates into two smaller particles called an electron and a proton—subjects of later study. This separation of hydrogen atoms doubles the number of particles to $24 \times 10^{56}$.

These particles do not spread out into space or collapse in toward the center of the Sun because two competing forces remain in balance. All parts of the Sun exert a gravitational force on all other parts. If we select a small volume inside the Sun as the system of interest and add the forces that all other particles of the Sun exert on the system (see **Figure 9.18**), the net gravitational force points toward the center of the Sun. There is another force exerted on the system. It is the pressure force exerted by other particles on the system. This force points outward and balances the gravitational force. As long as these forces balance each other, the Sun is in equilibrium and does not expand or collapse.

**Figure 9.18** Gas in the Sun is in equilibrium due to two forces.

This small volume of gas (the system) is kept in equilibrium by the gravitational force and pressure force exerted on it by other particles in the Sun.

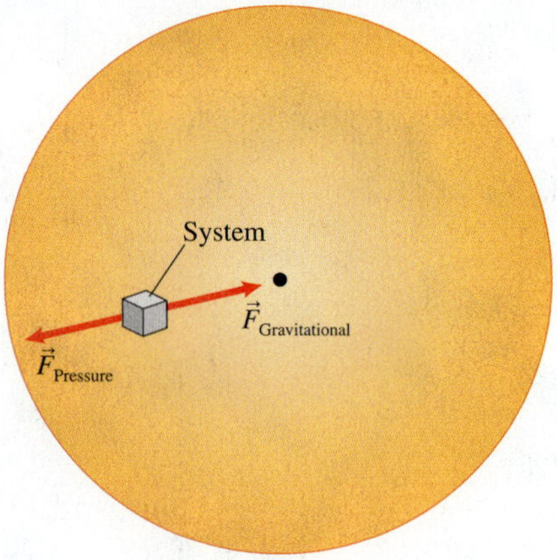


Now consider the total thermal energy of the Sun's particles for two extreme cases. A lower bound for the Sun's thermal energy assumes its temperature throughout equals its surface temperature. An upper bound for the Sun's thermal energy assumes its temperature throughout equals its core temperature.

$$
\begin{aligned}
U_{\text{thermal min}} &= \frac{3}{2}NkT_{\text{min}} \\
&= \frac{3}{2}(24 \times 10^{56}\text{ particles})(1.38 \times 10^{-23}\text{ J/K})(6 \times 10^{3}\text{ K}) \\
&= 3 \times 10^{38}\text{ J} \\
U_{\text{thermal max}} &= \frac{3}{2}NkT_{\text{max}} \\
&= \frac{3}{2}(24 \times 10^{56}\text{ particles})(1.38 \times 10^{-23}\text{ J/K})(10^{7}\text{ K}) \\
&= 5 \times 10^{41}\text{ J}
\end{aligned}
$$

Both numbers are so large that they imply that the Sun will shine for a long time. The life expectancy of the Sun in seconds (based on the thermal energy alone) equals the total thermal energy divided by the energy radiated per second, $R = 4 \times 10^{26}$ J/s. We can convert our result to years, noting that there are about $3 \times 10^7$ s in 1 year.

$$\Delta t_{\text{min}} = \frac{E_{\text{min}}}{R} = \frac{3 \times 10^{38}\,\text{J}}{4 \times 10^{26}\,\text{J/s}} = 7.5 \times 10^{11}\,\text{s} = \frac{7.5 \times 10^{11}\,\text{s}}{3 \times 10^{7}\,\text{s/year}} = 2 \times 10^{4}\,\text{years}$$

$$\Delta t_{\text{max}} = \frac{E_{\text{max}}}{R} = \frac{5 \times 10^{41}\,\text{J}}{4 \times 10^{26}\,\text{J/s}} = 1.2 \times 10^{15}\,\text{s} = \frac{1.2 \times 10^{15}\,\text{s}}{3 \times 10^{7}\,\text{s/year}} = 4 \times 10^{7}\,\text{years}$$

The maximum possible lifetime for the Sun if it simply converts its thermal energy into light and other forms of radiation is $4 \times 10^7$ years, or 40 million years. Yet we know that the geological age of Earth is $4.5 \times 10^9$ years—4.5 billion years! Earth cannot be older than the Sun; thus this simple estimate suggests that either the Sun's material cannot be modeled as an ideal gas or some source of energy other than the thermal energy of the Sun's particles has been supporting its existence for a time interval equal to the age of Earth. As the ideal gas model explains many phenomena occurring on the Sun, it means that there must be another energy source within the Sun that far exceeds the thermal energy present. (We will learn about it in Chapter 28.)

## Diffusion

Molecules spread from regions of high concentration to regions of low concentration due to their random motion. This process is called **diffusion.** The explanation of diffusion follows from the ideal gas model. Diffusion plays an important role in biological processes. Oxygen is carried by the hemoglobin in the blood from the heart to the tiny capillary vessels spread throughout the body. Oxygen diffuses from the oxygen-rich blood inside the capillaries to the oxygen-poor cells near the capillaries. Some of these cells may be muscle fibers. Since a muscle fiber needs oxygen to twitch, the action of muscles may be limited by the rate of oxygen diffusion into the fiber. Thus, diffusion limits the rate of some processes in our bodies.

**Review Question 9.8** How do we know that the Sun's thermal energy is not the main source of the light energy it produces?

# Summary

**Words** | **Pictorial and physical representations** | **Mathematical representation**

**Pressure** $P$ The perpendicular component of the force $F$ that another object or that a gas or liquid exerts perpendicular to a surface of area $A$ divided by that area. (Section 9.2)

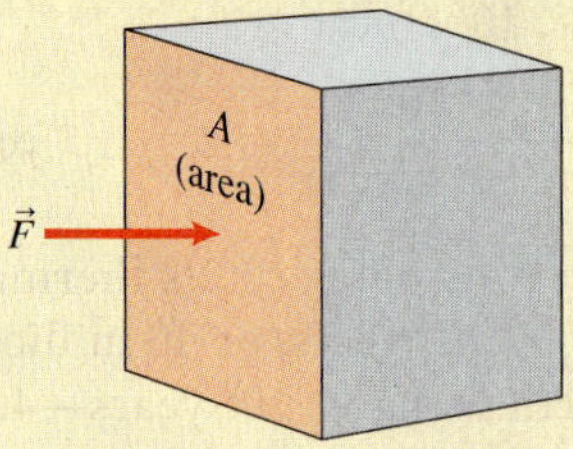


$$P = \frac{F_{\perp}}{A} \quad \text{Eq. (9.1)}$$

**Density** $\rho$ The mass $m$ of a substance divided by the volume $V$ that the substance occupies. (Section 9.2)

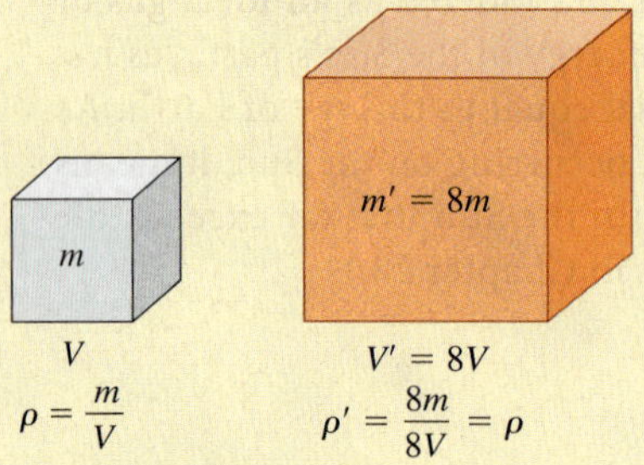


$$\rho = \frac{m}{V} \quad \text{Eq. (9.3)}$$

**Moles** $n$ **and Avogadro's number** $N_A$ A mole of any type of particle equals Avogadro's number of that type of particle. (Section 9.3)

$N_A = 6.02 \times 10^{23}$ particles/mole

**Temperature** $T$ **and temperature scales** Temperature measures how hot or cool a substance is. When measured in kelvins, the temperature is directly proportional to the average random kinetic energy of a particle in that gas. (Section 9.4)

The average kinetic energy per atom or molecule in a gas is

$$\bar{K} = \frac{3}{2}kT$$

$$T_F = (9/5)T_C + 32°$$

$$T_C = (5/9)(T_F - 32°)$$

$$T_K = T_C + 273.15° \quad \text{Eq. (9.6)}$$

**Thermal energy** $U_{thermal}$ The random kinetic energy of *all* atoms and molecules in a system. (Section 9.8)

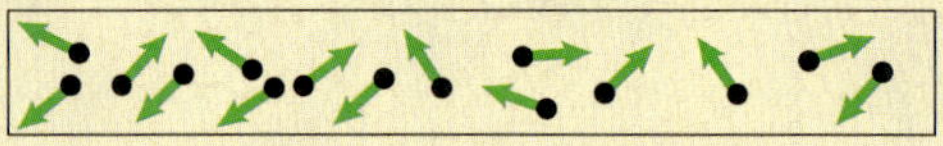

$$U_{thermal} = N\left(\frac{3}{2}kT\right)$$

**Ideal gas model and ideal gas law** The ideal gas model is a simplified model of gas in which atoms/molecules are considered to be point-like objects that obey Newton's laws. They only interact with each other and with the walls of the container during collisions exerting pressure. The ideal gas law relates the macroscopic quantities of such a gas. (Sections 9.1, 9.3, and 9.4)

$$P = \frac{2}{3}\left[\frac{N\left(\frac{1}{2}m\overline{v^2}\right)}{V}\right] \quad \text{Eq. (9.4)}$$

$$PV = NkT \quad \text{Eq. (9.7)}$$

$$PV = nRT \quad \text{Eq. (9.8)}$$

<table>
<tr><th colspan="4">Gas Processes (Examples)</th></tr>
<tr>
<td>Constant pressure (isobaric) process<br>A container with a frictionless plunger is filled with gas. The air outside is at constant pressure, thus the pressure inside the container is constant. (Section 9.5)</td>
<td>Microscopic The molecules inside collide with container walls at different speeds and varying frequency. If gas warms, the particles collide harder and more often, thus causing the gas to expand.</td>
<td>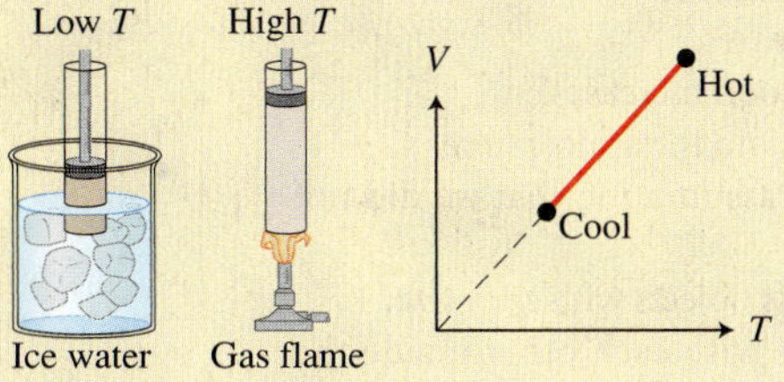
</td>
<td>Assume that $N$, $n$, and $P$ are constant. Then,<br>$$\frac{V_1}{T_1} = \frac{V_2}{T_2}$$</td>
</tr>
<tr>
<td>Constant volume (isochoric) process<br>A closed oxygen tank sits outside on a sunny summer day. Its volume is constant. (Section 9.5)</td>
<td>Microscopic As gas warms, the molecules inside move faster and collide with walls more often, thus exerting greater pressure on walls.</td>
<td>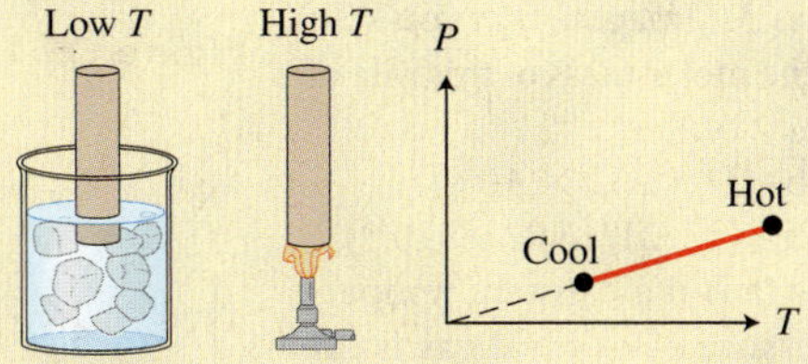
</td>
<td>Assume that $N$, $n$, and $V$ are constant. Then,<br>$$\frac{P_1}{T_1} = \frac{P_2}{T_2}$$</td>
</tr>
<tr>
<td>Constant temperature (isothermal) process<br>A closed plastic bottle shrinks as an airplane descends. The temperature inside the bottle is always equal to the temperature outside. (Section 9.5)</td>
<td>Microscopic As pressure increases, the collisions of air molecules against the outside of the bottle become more frequent, causing the bottle volume to decrease.</td>
<td>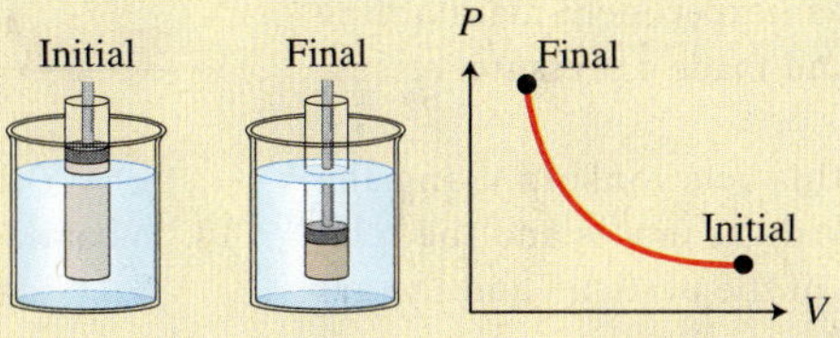
</td>
<td>Assume that $N$, $n$, and $T$ are constant. Then,<br>$$P_1V_1 = P_2V_2$$</td>
</tr>
</table>

MP® For instructor-assigned homework, go to **MasteringPhysics.**

# Questions

## Multiple-Choice Questions

1. What experimental evidence rejects the explanation that wet clothes become dry because the air absorbs the water?
   (a) The clothes dry faster if you blow air across them.
   (b) They do not dry if you put the wet clothes in a plastic bag.
   (c) The clothes dry faster under a vacuum jar with the air pumped out.
2. What is the difference between the words particle, molecule, and atom?
   (a) A particle is bigger than a molecule or an atom.
   (b) Particles can be microscopic and macroscopic, while atoms and molecules are only microscopic.
   (c) Molecules are made of atoms; both can be called particles.
   (d) All are correct.
   (e) Both b and c are correct.
3. You have a basketball filled with gas. Which method below changes its volume because of a mass change of the gas inside?
   (a) Put it into a refrigerator. (b) Squeeze it.
   (c) Pump more gas into it. (d) Hold it under water.
   (e) Leave it in the sunshine.
4. Choose the quantities describing the air inside a bike tire that do *not* change when you pump the tire.
   (a) Mass (b) Volume
   (c) Density (d) Pressure
   (e) Particle mass (f) Particle concentration
5. Which answer below does *not* explain the decrease in size of a basketball after you take it outside on a cold day?
   (a) The pressure inside the ball decreases.
   (b) The temperature of the gas inside the ball decreases.
   (c) The volume of the ball decreases.
   (d) The number of gas particles inside the ball decreases.

6. What causes balloons filled with air or helium to deflate as time passes?
   (a) The elasticity of the rubber decreases.
   (b) The temperature inside decreases.
   (c) The pressure outside the balloon increases.
   (d) The gas from inside diffuses into the atmosphere.
7. From the list below, choose the assumption that we did not use in deriving the ideal gas law.
   (a) Gas particles can be treated as objects with zero size.
   (b) The particles do not collide with each other inside the container.
   (c) The particles collide partially inelastically with the walls of the container.
   (d) The particles obey Newton's laws.
8. You have a mole of oranges of mass $M$. Imagine that you split each orange in half. What will be the molar mass of this pile of half oranges in kilograms?
   (a) $M/2$ (b) $M/(6.02 \times 10^{23})$
   (c) $2M/(6.02 \times 10^{23})$ (d) $(M/2)/(6.02 \times 10^{23})$
9. How did physicists come to know that at a constant temperature and constant mass, the pressure of an ideal gas is inversely proportional to its volume?
   (a) They could have conducted an experiment maintaining the gas as described above and made a pressure-versus-volume graph.
   (b) They could have derived this relationship using the equations describing the ideal gas model and the relationship between the speed of the particles and the gas temperature.
   (c) Both a and b are correct.
10. A gas is in a sealed container with a heavy top that is free to move. When the gas is heated, the top moves up, causing the volume of the gas to increase. Which equation below best describes this process?
   (a) $V_{\text{initial}}/T_{\text{initial}} = V_{\text{final}}/T_{\text{final}}$
   (b) $P_{\text{initial}}V_{\text{initial}} = P_{\text{final}}V_{\text{final}}$
   (c) $P_{\text{initial}}V_{\text{initial}} = nRT_{\text{final}}$
   (d) $P_{\text{initial}}V_{\text{initial}} = NkT_{\text{final}}$
11. A gas is in a sealed container with a heavy top that is free to move. With constant external pressure pushing down on the top, the top moves up, causing the volume of the gas to increase. Which graph below best represents this process (**Figure Q9.11**)?

**Figure Q9.11**

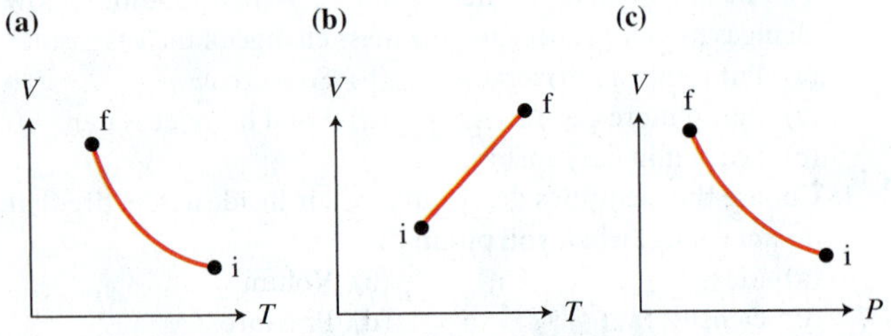


12. A completely closed rigid container of gas is taken from the oven and placed in ice water. Which graph at top right does not represent this process (**Figure Q9.12**)?

**Figure Q9.12**

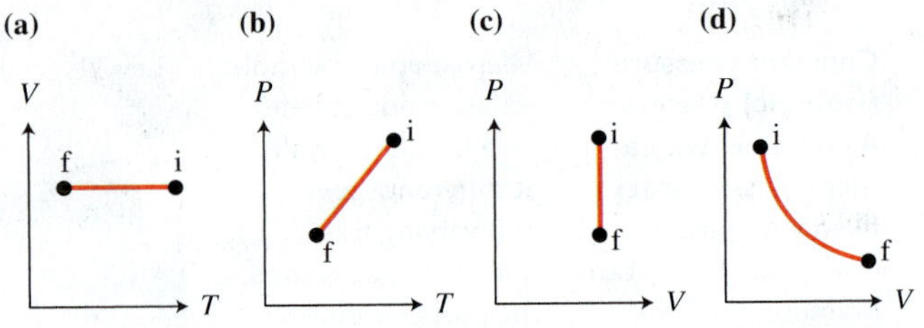


   (e) None of them do.
13. Which graph below does not represents a process described by the equation $V_{\text{initial}}/T_{\text{initial}} = V_{\text{final}}/T_{\text{final}}$ (**Figure Q9.13**)?

**Figure Q9.13**

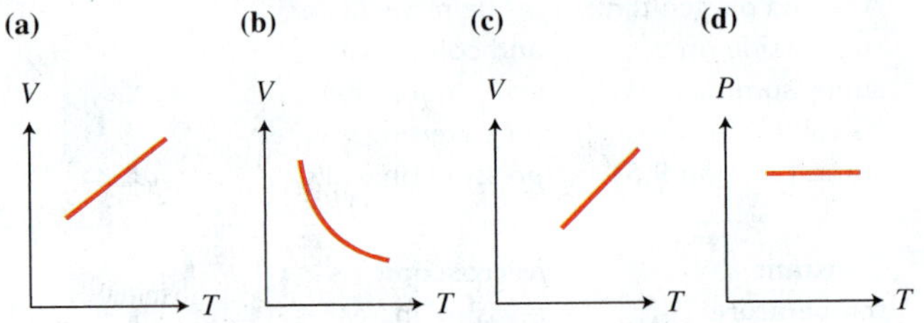


   (e) (a) and (b)
14. What does contracting and relaxing the diaphragm allow?
   (a) Greater $O_2$ absorption by humans and mammals compared to other species
   (b) Greater pressure difference between air in the lungs and outside air
   (c) The only way to facilitate air intake into and out of the lungs
   (d) a and b
   (e) a, b, and c
15. When is air is inhaled into the lungs?
   (a) When lung muscles cause them to expand
   (b) When the diaphragm contracts
   (c) When the chest expands
   (d) b and c
   (e) a, b, and c
16. What happens if the volume of the lungs increases?
   (a) The air pressure in the lungs increases.
   (b) The air pressure in the lungs decreases.
   (c) The amount $n$ of air in the lungs increases.
   (d) a and c can occur simultaneously.
   (e) b and c can occur simultaneously.

## Conceptual Questions

17. (a) How do we know if a real gas can be described as an ideal gas? (b) How can you decide if air in the physics classroom can be described using the ideal gas model?
18. Imagine that you are an astronaut on a space station. What happens if you open a perfume bottle in outer space?
19. Why does it hurt to walk barefoot on gravel?
20. In the magic trick in which a person lies on a bed of nails, why doesn't the person get hurt by the nails?

21. Why does food cook faster in a pressure cooker than in an open pot?
22. What does it mean if the density of a gas is $1.29 \text{ kg/m}^3$?
23. How many oranges would you have if you had two moles of oranges?
24. Imagine that you have an unknown gas. What experiments do you need to do and what real equipment do you need to determine the mass of one molecule of this gas?
25. One mole of chicken feathers is spread uniformly over the surface of Earth. *Estimate* the thickness of this layer of feathers. Justify any assumptions you made in your calculations. The radius of Earth is about 6400 km.
26. How many molecules are there in 1 g of air at normal conditions? If these molecules were distributed uniformly on Earth's surface, *estimate* the number that would be under your feet right now. The radius of Earth is about 6400 km.
27. Describe how temperature and one degree are defined on the Celsius scale.
28. Describe how temperature and one degree are defined on the Kelvin scale.
29. Why does sugar dissolve faster in hot tea than in cold water?
30. (a) Describe experiments that were used to test the predictions of the molecular kinetic theory. (b) What experiments revealed its limitations?
31. Give three examples of diffusion that are important for human life.
32. Why do very light gases such as hydrogen not exist in Earth's atmosphere but do exist in the atmospheres of giant planets such as Uranus and Saturn?
33. Why does the Moon have no atmosphere?

# Problems

Below, **BIO** indicates a problem with a biological or medical focus. Problems labeled **EST** ask you to estimate the answer to a quantitative problem rather than develop a specific answer. Problems marked with ✎ require you to make a drawing or graph as part of your solution. Asterisks indicate the level of difficulty of the problem. Problems with no * are considered to be the least difficult. A single * marks moderately difficult problems. Two ** indicate more difficult problems.

## 9.2 Pressure, density, and the mass of particles

1. * **EST** A water molecule in a glass of water has $10^{14}$ collisions with other molecules every second. Estimate the number of years a college football player would have to play (24 hours a day) to have the same number of collisions. Explain all of your assumptions.
2. What are the molar masses of molecular and atomic hydrogen, helium, oxygen, and nitrogen? What are their molecular masses?
3. **EST** Estimate the number of hydrogen atoms in the Sun. The mass of the Sun is $2 \times 10^{30}$ kg. About 70% of it by mass is hydrogen, 30% is helium, and there is a negligible amount of other elements.
4. * The average particle density in the Milky Way galaxy is about one particle per cubic centimeter. Express this number in SI units ($\text{kg/m}^3$). Indicate any assumptions you made.
5. * (a) What is the concentration (number per cubic meter) of the molecules in air at normal conditions? (b) What is the average distance between molecules compared to the dimensions of the molecules? (c) Can you consider air to be an ideal gas? Explain your answer.
6. * **EST** Estimate the number of collisions that one molecule of air in the physics classroom experiences every second. List all the assumptions that you made.
7. What is the mass of a water molecule in kilograms? What is the mass of an average air molecule in kilograms?
8. You find that the average gauge pressure in your car tires is about 35 psi. How many newtons per square meter is it? What is gauge pressure?
9. **BIO** **Forced vital capacity** Physicians use a machine called a spirometer to measure the maximum amount of air a person can exhale (called the forced vital capacity). Suppose you can exhale 4.8 L. How many kilograms of air do you exhale? What assumptions did you make to answer the question? How do these assumptions affect the result?
10. A container is at rest with respect to a desk. Inside the container a particle is moving horizontally at a speed $v$ with respect to the desk. It collides with a vertical wall of the container elastically and rebounds. Qualitatively, determine the direction and the speed of the particle if the wall is (a) at rest with respect to the desk; (b) moving in the same direction as the particle at a speed smaller than the particle's; and (c) moving in the direction opposite to the motion of the particle at a smaller speed.

## 9.3 Quantitative analysis of ideal gas

11. * **Hitting tennis balls against a wall** A 0.058-kg tennis ball, traveling at 25 m/s, hits a wall, rebounds with the same speed in the opposite direction, and is hit again by another player, causing the ball to return to the wall at the same speed. The ball returns to the wall once every 0.60 s. (a) Determine the force that the ball exerts on the wall averaged over the time between collisions. State the assumptions that you made. (b) If 10 people are practicing against a wall with an area of $30 \text{ m}^2$, what is the average pressure of the 10 tennis balls against the wall?
12. * Friends throw snowballs at the wall of a $3.0 \text{ m} \times 6.0 \text{ m}$ barn. The snowballs have mass 0.10 kg and hit the wall moving at an average speed of 6.0 m/s. They do not rebound. Determine the average pressure exerted by the snowballs on the wall if 40 snowballs hit the wall each second. Which problem, this or the previous problem, resembles the actions of the molecules of an ideal gas hitting the walls of their container?
13. * A ball moving at a speed of 3.0 m/s with respect to the ground hits a stationary wall at a 30° angle with respect to the surface of the wall. Determine the direction and the magnitude of the velocity of the ball after it rebounds. Explain carefully what physics principles you used to find the answer. What assumptions did you make? How will the answer change if one or more of them are not valid?
14. **Oxygen tank for mountains** Consider an oxygen tank for a mountain climbing trip. The mass of one molecule of

oxygen is $5.3 \times 10^{-26}$ kg. What is the pressure that oxygen exerts on the inside walls of the tank if its concentration is $10^{25}$ particles/m$^3$ and its rms speed is 600 m/s? What assumptions did you make?

15. You have five molecules with the following speeds: 300 m/s, 400 m/s, 500 m/s, 450 m/s, and 550 m/s. What is their average speed?
16. What is the rms speed of the molecules in the previous problem? Is it different from the average speed?
17. Two gases in different containers have the same concentration and same rms speed. The mass of a molecule of the first gas is twice the mass of a molecule of the second gas. What can you say about their pressures? Explain.
18. * You are hiking up a mountain. About halfway up you pass through a cloud and become moist from cloud water. How can this water be at such a high elevation? To answer this question, compare the molar masses and densities of dry air and humid air. Explain. List all of your assumptions.
19. * BIO **Breathing** You are breathing heavily while hiking up the mountain. To inhale, you expand your diaphragm and lungs. Explain, using your knowledge of gas pressure, why this mechanical movement leads to the air flowing into your nose or mouth. Support your reasoning with diagrams if necessary.
20. * **Oxygen tank for mountain climbing** An oxygen container that one can use in the mountains has a 90-min oxygen supply at a speed of 6 L/min. Determine everything you can about the gas in the container. Make reasonable assumptions.

## 9.4 Temperature

21. You are cooking dinner in the mountains. At 7000 feet, water boils at 92.3 °C. Convert the boiling temperature to °F and suggest two possible reasons why the boiling temperature is lower at this elevation than at sea level.
22. Your temperature, when taken orally, is 98.6 °F. When taken under your arm, it's 36.6 °C. Are these results consistent? Explain.
23. On top of Mount Everest, the temperature is −19 °C in July. Being a physicist, you determine by how many degrees Celsius one needs to change the air temperature to double the average kinetic energy of its molecules. Explain your reasoning.
24. Air consists of many different molecules, for example, $N_2$, $O_2$, $H_2O$, and $CO_2$. Which molecules are the fastest on average? The slowest on average? Explain.
25. What is the average kinetic energy of a particle of air at standard conditions?
26. Air is a mixture of molecules of different types. Compare the rms speeds of the molecules of $N_2$, $O_2$, and $CO_2$ at standard conditions. What assumptions did you make?
27. * How many moles of air are in a regular 1-L water bottle when you finish drinking the water? What assumptions did you make? How do these assumptions affect your result?
28. * At approximately what temperature does the average random kinetic energy of a $N_2$ molecule in an ideal gas equal the macroscopic translational kinetic energy of a copper atom in a penny that is dropped from the height of 1.0 m?
29. ** A molecule moving at speed $v_1$ collides head-on with a molecule of the same mass moving at speed $v_2$. Compute the speeds of the molecules after the collision. What assumptions did you make? How does the answer to this problem explain why the mixing of hot and cold gases causes the cold gas to become warmer and the hot gas to become cooler?
30. **Balloon flight** For a balloon ride, the balloon must be inflated with helium to a volume of 1500 m$^3$ at sea level. The balloon will rise to an altitude of about 12 km, where the temperature is about −52 °C and the pressure is about 20 kPa. How much helium should be put into the balloon? What assumptions did you make?
31. * BIO **Ears pop** The middle ear has a volume of about 6.0 cm$^3$ when at a pressure of $1.0 \times 10^5$ N/m$^2$ (1.0 atm). Determine the volume of that same air when the air pressure is $0.83 \times 10^5$ N/m$^2$, as it is at an elevation of 1500 m above sea level (assume the air temperature remains constant). If the volume of the middle ear remains constant, some air will have to leave as the elevation increases. That is why ears "pop."
32. * Even the best vacuum pumps cannot lower the pressure in a container below $10^{-15}$ atm. How many molecules of air are left in each cubic centimeter in this "vacuum?" Assume that the temperature is 273 K.
33. **Pressure in interstellar space** The concentration of particles (assume neutral hydrogen atoms) in interstellar gas is 1 particle/cm$^3$, and the average temperature is about 10 K. What is the pressure of the interstellar gas? How does it compare to the best vacuum that can be achieved on Earth (see the previous problem)?

## 9.5 Testing the ideal gas law; 9.6 Speed distribution of particles; 9.7 Skills for analyzing processes using the ideal gas law

34. * Describe experiments to determine if each of the three gas isoprocess laws works. The experiments should be ones that you could actually carry out.
35. * The following data were collected for the temperature and volume of a gas. Can this gas be described by the ideal gas model? Explain how you know.

| Temperature (°C) | Volume (ml) |
|---|---|
| 11 | 95.0 |
| 25 | 100.0 |
| 47 | 107.5 |
| 73 | 116.0 |
| 159 | 145.1 |
| 233 | 170.0 |
| 258 | 177.9 |

36. * Describe a mechanical model of Stern's molecular beam experiment.
37. * Explain the microscopic mechanisms for the relation of macroscopic variables for an isothermal process, an isobaric process, and an isochoric process.
38. * **Scuba diving** The pressure of the air in a diver's lungs when he is 20 m under the water surface is $3.0 \times 10^5$ N/m$^2$, and the air occupies a volume of 4.8 L. How many moles of air should he exhale while moving to the surface, where the pressure is $1.0 \times 10^5$ N/m$^2$?
39. * BIO EST **Alveoli surface area** Estimate the size of the surface area of a single alveolus in the lungs.
40. * When surrounded by air at a pressure of $1.0 \times 10^5$ N/m$^2$, a basketball has a radius of 0.12 m. Compare its volume at this condition with the volume that it would have if you take it 15 m below the water surface where the pressure is $2.5 \times 10^5$ N/m$^2$. What assumptions did you make?
41. ** You have gas in a container with a movable piston. The walls of the container are thin enough so that its temperature stays the same as the temperature of the surrounding medium. You have baths of water of different temperatures, different objects that you can place on top of the piston, etc.

(a) Describe how you could make the gas undergo an isothermal process so that the pressure inside increases by 10%, then undergo an isobaric process so that the new volume decreases by 20%, and finally undergo an isochoric process so that the temperature increases by 15%. (b) Represent all processes in $P$-versus-$T$, $V$-versus-$T$, and $P$-versus-$V$ graphs. (c) What are the new pressure, volume, and temperature of the gas?

42. * **Bubbles** While snorkeling, you see air bubbles leaving a crevice at the bottom of a reef. One of the bubbles has a radius of 0.060 m. As the bubble rises, the pressure inside it decreases by 50%. Now what is the bubble's radius? What assumptions did you make to solve the problem?

43. ** **Diving bell** A cylindrical diving bell, open at the bottom and closed at the top, is 4 m tall. Scientists fill the bell with air at the pressure of $1.0 \times 10^5\ \text{N/m}^2$. The pressure increases by $1.0 \times 10^5\ \text{N/m}^2$ for each 10 m that the bell is lowered below the surface of the water. If the bell is lowered 30 m below the ocean surface, how many meters of air space are left inside the bell? Why doesn't water enter the entire bell as it goes under water? Draw several sketches for this problem.

44. * **Mount Everest** (a) Determine the number of molecules per unit volume in the atmosphere at the top of Mount Everest. The pressure is $0.31 \times 10^5\ \text{N/m}^2$, and the temperature is $-30\ ^\circ\text{C}$. (b) Determine the number of molecules per unit volume at sea level, where the pressure is $1.0 \times 10^5\ \text{N/m}^2$ and the temperature is $20\ ^\circ\text{C}$.

45. * EST **Breathing on Mount Everest** Using the information from Problem 44, estimate how frequently you need to breathe on top of Mount Everest to inhale the same amount of oxygen as you do at sea level. The pressure is about one-third the pressure at sea level.

46. **Capping beer** You would like to make homemade beer, but you are concerned about storing it. Your beer is capped into a bottle at a temperature of $27\ ^\circ\text{C}$ and a pressure of $1.2 \times 10^5\ \text{N/m}^2$. The cap will pop off if the pressure inside the bottle exceeds $1.5 \times 10^5\ \text{N/m}^2$. At what maximum temperature can you store the beer so the gas inside the bottle does not pop the cap? List the assumptions that you made.

47. * **Car tire** With a tire gauge, you measure the pressure in a car tire as $2.1 \times 10^5\ \text{N/m}^2$. How can this be if you know that absolute pressure in the tire is three times higher than atmospheric? The tire looks okay. What's the deal?

48. * **Car tire dilemma** Imagine a car tire that contains 5.1 moles of air when at a gauge pressure of $2.1 \times 10^5\ \text{N/m}^2$ (the pressure above atmospheric pressure) and a temperature of $27\ ^\circ\text{C}$. The temperature increases to $37\ ^\circ\text{C}$, the volume decreases to 0.8 times the original volume, and the gauge pressure decreases to $1.6 \times 10^5\ \text{N/m}^2$. Can these measurements be correct if the tire did not leak? If it did leak, then how many moles of air are left in the tire?

49. There is a limit to how much gas can pass through a pipeline, because the pipes can only tolerate so much pressure on the walls. To increase the amount of gas going through the pipeline, engineers decide to cool the gas (to reduce its pressure). Suggest how much they should lower the temperature of the gas if they want to increase the mass per unit time by 1.5 times.

50. Explain how you know that the volume of one mole of gas at standard conditions is 22.4 L.

51. * At what pressure is the density of $-50\ ^\circ\text{C}$ nitrogen gas ($N_2$) equal to 0.10 times the density of water?

52. * In the morning, the gauge pressure in your car tires is 35 psi. During the day, the air temperature increases from $20\ ^\circ\text{C}$ to $30\ ^\circ\text{C}$ and the pressure increases to 36.5 psi. By how much did the volume of one of the tires increase? What assumptions did you make?

53. **Equation Jeopardy 1** The equation below describes a process. Construct a word problem for a process that is consistent with the equations (there are many possibilities). Provide as much detailed information as possible about your proposed process.

$$\frac{1.2 \times 10^5\ \text{N/m}^2}{293\ \text{K}} = \frac{2.0 \times 10^5\ \text{N/m}^2}{T}$$

54. * **Equation Jeopardy 2** The equation below describes a process. Construct a word problem for a process that is consistent with the equations (there are many possibilities). Provide as much detailed information as possible about your proposed process.

$$\Delta n = \frac{(0.67 \times 10^5\ \text{N/m}^2)(0.60 \times 10^{-6}\ \text{m}^3)}{(8.3\ \text{J/mole}\cdot\text{K})(303\ \text{K})} - \frac{(1.00 \times 10^5\ \text{N/m}^2)(0.60 \times 10^{-6}\ \text{m}^3)}{(8.3\ \text{J/mole}\cdot\text{K})(310\ \text{K})}$$

55. ** The $P$-versus-$T$ graph in **Figure P9.55** describes a cyclic process comprising four hypothetical parts. (a) What happens to the pressure of the gas in each part? (b) What happens to the temperature of the gas in each part? (c) What happens to the volume of the gas in each part? (d) Explain each part microscopically. (e) Use the information from (a)–(c) to represent the same parts in $P$-versus-$V$ and $V$-versus-$T$ graphs. [*Hint:* It helps to align the $P$-versus-$V$ graph *beside* the $P$-versus-$T$ graph using the same $P$ values on the ordinate (vertical axes) and to place the $V$-versus-$T$ graph *below* the $P$-versus-$T$ graph using the same $T$ values on the abscissa (horizontal axes). This helps keep the same scale for the variables.]

**Figure P9.55**

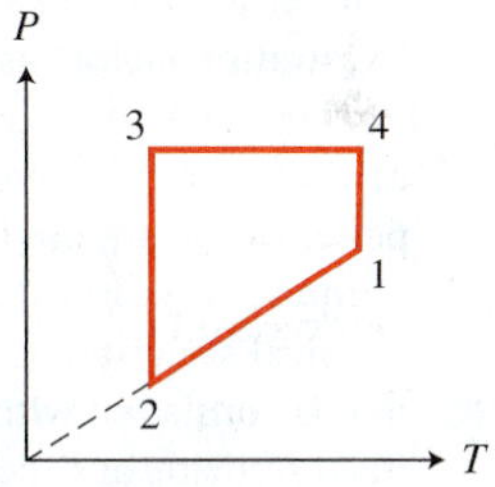


56. ** The $V$-versus-$T$ graph in **Figure P9.56** describes a cyclic process comprising four hypothetical parts. (a) What happens to the pressure of the gas in each parts? (b) What happens to the temperature of the gas in each parts? (c) What happens to the volume of the gas in each parts? (d) Explain each part microscopically. (e) Use the information from (a)–(c) to represent the same process in a $P$-versus-$T$ graph (below the $V$-versus-$T$ graph) and a $P$-versus-$V$ graph (beside the $P$-versus-$T$ graph). See the hint in the previous problem about the graph alignments.

**Figure P9.56**

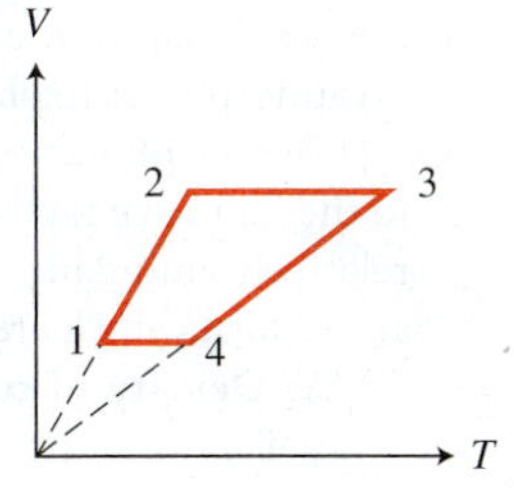


## 9.8 Thermal energy, the Sun, and diffusion: Putting it all together

57. ** EST **Sun's life expectancy** (a) Estimate the average kinetic energy of the particles in the Sun. Assume that it is made of atomic hydrogen and that its average temperature is 100,000 K. The mass of the Sun is $2 \times 10^{30}$ kg. (b) For how long would the Sun shine using this energy if it radiates $4 \times 10^{26}$ W/s? Is your answer reasonable? Explain.

58. * The temperature of the Sun's atmosphere near the surface is about 6000 K, and the concentration of atoms is about $10^{15}$ particles/$m^3$. What is the average pressure and density of its atmosphere? What assumptions did you make to solve the problem?

59. ** A gas that can be described by the ideal gas model is contained in a cylinder of volume $V$. The temperature of the gas is $T$. The particle mass is $m$, and the molar mass is $M$. Write an expression for the total thermal energy of the gas. Now, imagine that the exact same gas has been placed in a container of volume $2V$. What happens to its pressure? What happens to its temperature? What happens to its thermal energy?

60. * BIO EST **Breathing and metabolism** We need about 0.7 L of oxygen per minute to maintain our resting metabolism and about 2 L when standing and walking. Estimate the number of breaths per minute for a person to satisfy this need when resting and when standing and walking. What assumptions did you make? Remember that oxygen is about 21% of the air.

61. ** In 1896, Lord Rayleigh showed that a mixture of two gases of different atomic masses could be separated by allowing some of it to diffuse through a porous membrane into an evacuated space. Rayleigh proposed that the molecules of lighter gas diffuse through the membrane faster, leaving the heavier gas behind in the original space. He described this "separation factor" as equal to $\sqrt{m_2/m_1}$, where $m_1$ is the molecular weight of a lighter gas and $m_2$ is the molecular weight of a heavier gas. Give a reason why the separation factor depends on the square root of the ratio of the molecular masses.

62. * **Equation Jeopardy 3** The three equations below describe a physical situation. Construct a word problem for a situation that is consistent with the equations (there are many possibilities). Provide as much detailed information as possible about the situation.

$$m = (1.3\ \text{kg/m}^3)(3.0\ \text{m} \times 5.0\ \text{m} \times 2.0\ \text{m})$$

$$N = m/[(29 \times 10^{-3}\ \text{kg})/(6.0 \times 10^{23}\ \text{particles})]$$

$$U_{\text{thermal}} = N(3/2)(1.38 \times 10^{-23}\ \text{J/K})(273\ \text{K})$$

## General Problems

63. * **No $H_2$ in Earth atmosphere** Explain why Earth has almost no free hydrogen in its atmosphere.

64. * **No atmosphere on the Moon** Why does the Moon have no atmosphere? Explain.

65. * **Different planet compositions** Explain why planets closer to the Sun have low concentrations of light elements, but are relatively abundant in giant planets such as Jupiter, Uranus, and Saturn, which are far from the Sun.

66. ** EST **Density of our galaxy** Estimate the average density of particles in our galaxy, assuming that the most abundant element is atomic hydrogen. There are about $10^9$ stars in the galaxy and the size of the galaxy is about $10^5$ light-years. A light-year is the distance light travels in 1 year moving at a speed of $3 \times 10^8$ m/s. What do you need to assume about the stars in order to answer this question?

67. * BIO **Breathing** Observe yourself breathing and count the number of times you inhale per second. During each breath you probably inhale about 0.50 L of air. How many oxygen molecules do you inhale if you are at sea level?

68. ** **Car engine** During a compression stroke of a cylinder in a diesel engine, the air pressure in the cylinder increases from $1.0 \times 10^5\ \text{N/m}^2$ to $50 \times 10^5\ \text{N/m}^2$, and the temperature increases from 26 °C to 517 °C. Using this information, how would you convince your friends that knowledge about ideal gases can help explain how hot gases burned in the car engine affect the motion of the car?

69. ** EST How can the pressure of air in your house stay constant during the day if the temperature rises? Estimate the volume of your house and the number of moles of air that leave the house during the daytime. Assume that nighttime temperature and daytime temperature differ by about 10 °C. List all other assumptions that are necessary to answer the question.

70. ** **Tell-all problem** Tell everything you can about the process described by the pressure-versus-volume graph shown in **Figure P9.70**.

**Figure P9.70**

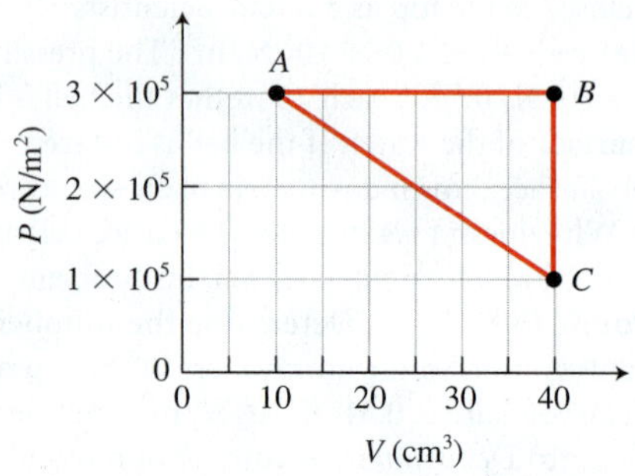


71. ** Two massless, frictionless pistons are inside a horizontal tube opened at both ends. A 10-cm-long thread connects the pistons. The cross-sectional area of the tube is 20 $cm^2$. The pressure and temperature of gas between the pistons and the outside air are the same and are equal to $P = 1.0 \times 10^5\ \text{N/m}^2$ and $T = 24$ °C. At what temperature will the thread break if it breaks when the tension reaches 30 N?

72. * A closed cylindrical container is divided into two parts by a light, movable, frictionless piston. The container's total length is 100 cm. Where is the piston located when one side is filled with nitrogen ($N_2$) and the other side with the same mass of hydrogen ($H_2$) at the same temperature?

## Reading Passage Problems

BIO **Vascular wall tension and aortic blowout** The walls of blood vessels contain varying amounts of elastic fibers that allow the vessels to expand and contract as the pressure and amount of fluid inside vary (these fibers are more prevalent in the aorta and large arteries than in the small arterioles and capillaries). These fibers in the cylindrical walls produce a wall tension $T$, defined as

$$T = \frac{F}{L}$$

where $L$ is the length of an imaginary cut parallel to the axis of the vessel and $F$ is the magnitude of the force that each side of the cut must exert on the other side to hold the two sides together.

Three forces are exerted on a short section of wall fiber—the system. (1) The fluid inside pushes outward, due to fluid pressure from inside $P_{\text{inside fluid pushing out}} = P_{\text{out}}$; (2) fluid outside the vessel pushes inward, due to fluid pressure from outside $P_{\text{outside fluid pushing in}} = P_{\text{in}}$; and (3) the wall next to the system exerts wall tension $T$ on each side of that wall. The pressure difference across the wall $\Delta P = P_{\text{out}} - P_{\text{in}}$, the wall tension $T$, and the radius $R$ of the cylindrical vessel are related by Laplace's law:

$$\Delta P = \frac{T}{R} \quad \text{or} \quad T = \Delta P \cdot R$$

The inward gauge pressure $P_{in}$ of tissue surrounding the vessels is approximately zero. Thus, the pressure difference $\Delta P = P_{out} - P_{in} = P_{vessel} - 0 = P_{vessel}$ is the gauge pressure in the blood vessel. We can now estimate the wall tension for different types of vessels. The tension in the aorta is approximately

$$T = \Delta P \cdot R \approx (100 \text{ mm Hg})\left(\frac{133 \text{ N/m}^2}{1 \text{ mm Hg}}\right)(1.3 \times 10^{-2} \text{ m})$$
$$= 170 \text{ N/m}$$

Using similar reasoning, the wall tension in the low-pressure, very small radius capillaries is about 0.016 N/m—about 0.0003 times the tension needed to tear a facial tissue. Because such little tension is needed to hold a capillary together, its wall can be very thin, allowing easy diffusion of various molecules across the wall.

The walls of a healthy aorta can easily provide the tension needed to support the increased blood pressure when it fills with blood from the heart during each heartbeat. However, aging and various medical conditions may weaken the aortic wall in a short section, and increased blood pressure can cause it to stretch. The weakened wall bulges outward; this is called an aortic aneurism. The increased radius causes increased tension, which can increase bulging. This cycle can result in a rupture to the aorta: an aortic blowout.

73. Why is the wall tension in capillaries so small?
   (a) There are so many capillaries.
   (b) Their radii are so small.
   (c) The outward pressure of the blood inside is so small.
   (d) b and c
   (e) a, b, and c
74. Which answer below is closest to the wall tension in a typical arteriole of radius 0.15 mm and 60 mm Hg blood pressure?
   (a) 0.001 N/m (b) 0.01 N/m
   (c) 0.1 N/m (d) 1 N/m
   (e) 10 N/m
75. According to Laplace's law, elevated blood pressure in an artery should cause the wall tension in the artery to do what?
   (a) Increase (b) Remain unchanged
   (c) Decrease (d) Impossible to decide
76. As a person ages, the fibers in arteries become less elastic and the wall tension increases. According to Laplace's law, this will cause the blood pressure to do what?
   (a) Increase (b) Remain unchanged
   (c) Decrease (d) Impossible to decide
77. Aortic blowout occurs when part of the wall of the aorta becomes weakened. What does this cause?
   (a) A bulge and increased radius of the aorta when the blood pressure inside increases
   (b) An increased radius of the aorta, which causes increased tension in the wall
   (c) An increased tension in the aorta, which causes the radius to increase
   (d) a and b
   (e) a, b, and c

**BIO Portable hyperbaric chamber** In 1997, a hiking expedition was stranded for 38 days on the Tibetan side of Mount Everest at altitudes over 5200 m. For 30 days the party was stranded above 6500 m in severe weather conditions. While at altitudes over 8000 m, 10 climbers suffered acute mountain sickness. A 37-year-old climber with acute pulmonary edema (buildup of fluid in the lungs) was treated with a portable Gamow bag (see **Figure 9.19**), named after its inventor, Igor Gamow. The Gamow bag is a windowed cylindrical portable hyperbaric chamber constructed of nonpermeable nylon that requires constant pressurization with a foot pump attached to the bag. The climber enters the bag, and a person outside pumps air into the bag so that the air pressure inside is somewhat higher than the outside pressure.

The bag and pump have a 6.76-kg mass. The volume of the inflated bag is 0.476 $m^3$. The maximal bag pressure is $0.14 \times 10^5 \text{ N/m}^2$ above the air pressure at the site where it is used. In the 1997 climb, with the temperature at $-20$ °C, the bag was filled in about 2 min with 10–20 pumps per minute. This raised the pressure in the bag to $0.58 \times 10^5 \text{ N/m}^2$ (equivalent to an elevation of 4400 m) instead of the actual outside pressure of $0.43 \times 10^5 \text{ N/m}^2$ at the 6450-m elevation. The treatment lasted for 2 h, with the person inhaling about 15 times/min at about 0.5 L/inhalation, and was successful—the pulmonary edema disappeared.

**Figure 9.19** A Gamow bag, used to help climbers with acute altitude sickness.

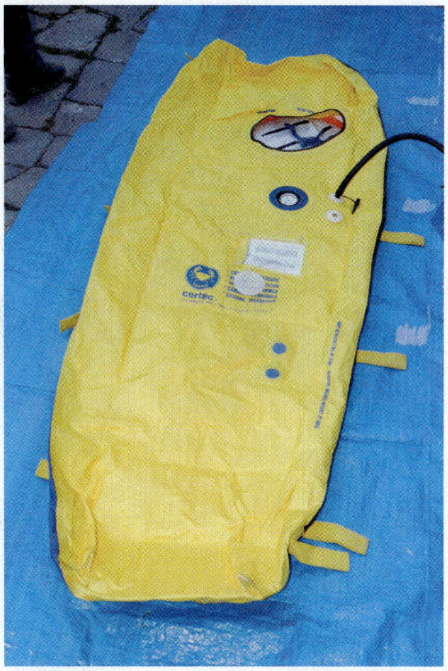

78. What is closest to the volume of the Gamow bag?
   (a) 50 L (b) 100 L (c) 200 L (d) 500 L (e) 1000 L
79. What is closest to the temperature at the 6450-m elevation on the day described in the problem?
   (a) 37 K (b) 253 K (c) 20 K (d) $-20$ K (e) 273 K
80. What is closest to the number $n$ of gram-moles of air in the filled bag when at 4400 m?
   (a) 3 g · moles (b) 10 g · moles
   (c) 13 g · moles (d) 110 g · moles
   (e) 170 g · moles
81. What is closest to the number $n$ of gram-moles of air in the bag if at the 6450-m pressure?
   (a) 3 g · moles (b) 10 g · moles
   (c) 13 g · moles (d) 110 g · moles
   (e) 170 g · moles
82. Estimate the fraction of the air in the Gamow bag that an occupant would inhale in 1 h, assuming no replacement of the air.
   (a) 0.01 (b) 0.1 (c) 0.3 (d) 0.5 (e) 1.0

# 10 Static Fluids

**How can a hot air balloon travel for hours in the sky?**

**Why is it dangerous for scuba divers to ascend to the surface quickly from a deep-sea dive?**

**Why does a 15-g nail sink in water and a cargo ship float?**

**Be sure you know how to:**

- Draw a force diagram for a system of interest (Section 1.6).
- Apply Newton's second law in component form (Section 3.5).
- Define pressure (Section 9.2).

The first balloon flight occurred in 1731 in Russia, in a balloon filled with smoke. Fifty years later the brothers Montgolfier used heated air to fill a balloon in France. That ride carried a sheep, a duck, and a chicken in the balloon's basket. Both smoke and hot air are less dense than cold air. By carefully balancing the density of the balloon and its contents with the density of air, the pilot can control the force that the outside cold air exerts on the balloon. What do pressure, volume, mass, and temperature have to do with this force?

**In the previous chapter,** we constructed the ideal gas model and used it to explain the behavior of gases. That model was built on the knowledge of the particle nature of matter and a Newtonian analysis of the motion of those particles. The temperature of the gas and the pressure it exerted on surfaces played an important role in the phenomena we analyzed. We ignored the effect of the gravitational force exerted by Earth on the gas particles. This simplification was reasonable, since in most of the processes we analyzed the gases had little mass and occupied a relatively small region of space, like in a piston. In this chapter, our interest expands to include phenomena in which the force exerted by Earth plays an important role. We will confine the discussion to static fluids—fluids that are not moving.

## 10.1 Density

We are familiar with the concept of density (from Chapter 9). To find the density of an object or a substance, determine its mass and volume and then calculate the ratio of the mass and volume:

$$\rho = \frac{m}{V} \tag{10.1}$$

Archimedes (Greek, 287–212 B.C.) discovered how to determine the density of an object of irregular shape. First determine its mass using a scale. Then determine its volume by submerging it in a graduated cylinder with water (**Figure 10.1**). Finally, divide the mass in kilograms by the volume in cubic meters to find the density in kilograms per cubic meter. Using this method we find that the density of an iron nail is $7800\ \text{kg/m}^3$—relatively large. A gold coin has an even larger density—$19{,}320\ \text{kg/m}^3$. The universe, though, contains much denser objects. For example, the rapidly spinning neutron star known as a pulsar (discussed in Chapter 8) has a density of approximately $10^{18}\ \text{kg/m}^3$. **Table 10.1** lists the densities of various solids, liquids, and gases.

**Figure 10.1** Measuring the density of an irregularly shaped object.

1. Measure mass of object.
2. Place the object in water in a graduated cylinder.
3. Measure the volume change of the water. Volume change of water = volume of object.
4. Density = $\rho = m/V$

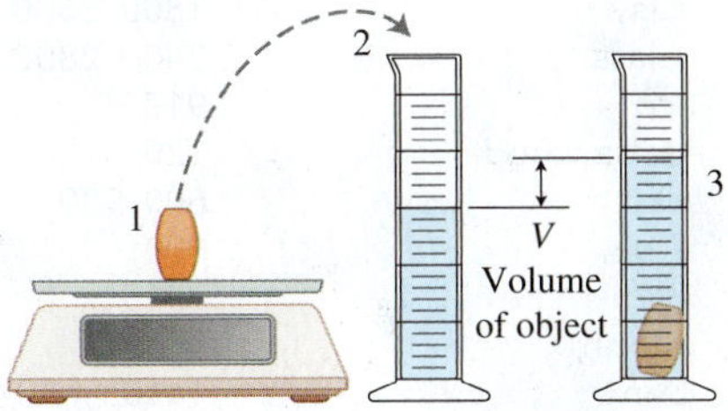


**QUANTITATIVE EXERCISE 10.1**

**Ping-pong balls with different densities**

Saturn has the lowest density of all the planets in the solar system ($M_{\text{Saturn}} = 5.7 \times 10^{26}\ \text{kg}$ and $V_{\text{Saturn}} = 8.3 \times 10^{23}\ \text{m}^3$). The average density of a neutron star is $10^{18}\ \text{kg/m}^3$. Compare the mass of a ping-pong ball filled with material from Saturn with that of the same ball filled with material from a neutron star. An empty ping-pong ball has a 0.037-m diameter (0.020-m radius) and a 2.7-g mass.

**Represent mathematically** To find the mass of a ping-pong ball filled with a particular material, we add the mass of the ball alone and the calculated mass of the material inside:

$$m_{\text{filled ball}} = m_{\text{ball}} + m_{\text{material}}$$

where

$$m_{\text{material}} = \rho_{\text{material}} V_{\text{ball}}$$

*(continued)*

The density of the neutron star is given, and the density of Saturn can be found using the operational definition $\rho_{\text{Saturn}} = \frac{m_{\text{Saturn}}}{V_{\text{Saturn}}}$. The interior of the ping-pong ball is a sphere of volume

$$V_{\text{sphere}} = \frac{4}{3}\pi R^3$$

Assume that the plastic shell of the ball has negligible volume. The mass of either filled ball is

$$m_{\text{filled ball}} = m_{\text{ball}} + m_{\text{material}} = m_{\text{ball}} + \rho_{\text{material}} V_{\text{ball}}$$
$$= m_{\text{ball}} + \rho_{\text{material}}(4/3)\pi R_{\text{ball}}^3$$

**Solve and evaluate** For the neutron-star-filled ball:

$$m_{\text{neutron star ball}} = (0.003\text{ kg}) + (10^{18}\text{ kg/m}^3)\frac{4}{3}\pi(0.020\text{ m})^3$$
$$= 3.4 \times 10^{13}\text{ kg}$$

For the Saturn-filled ball:

$$m_{\text{Saturn ball}} = m_{\text{ball}} + \left(\frac{m_{\text{Saturn}}}{V_{\text{Saturn}}}\right)\left(\frac{4}{3}\pi R_{\text{ball}}^3\right)$$
$$= 0.003\text{ kg} + \left(\frac{5.7 \times 10^{26}\text{ kg}}{8.3 \times 10^{23}\text{ m}^3}\right)\left(\frac{4}{3}\pi(0.020\text{ m})^3\right)$$
$$= 0.003\text{ kg} + 0.023\text{ kg} = 0.026\text{ kg}$$

The material from Saturn has less mass than an equal volume of water. The ball filled with the material from a neutron star has a mass of more than a billion tons!

**Try it yourself:** The mass of the ping-pong ball filled with soil from Earth's surface is 0.050 kg. What is the density of the soil?

*Answer:* $1400\text{ kg/m}^3$.

**Table 10.1** **Densities of various solids, liquids, and gases.**

| Solids | | Liquids | | Gases | |
|---|---|---|---|---|---|
| **Substance** | **Density (kg/m$^3$)** | **Substance** | **Density (kg/m$^3$)** | **Substance** | **Density (kg/m$^3$)** |
| Aluminum | 2700 | Acetone | 791 | Dry air 0° C | 1.29 |
| Copper | 8920 | Ethyl alcohol | 789 | 10° C | 1.25 |
| Gold | 19,300 | Methyl alcohol | 791 | 20 °C | 1.21 |
| Iron | 7860 | Gasoline | 726 | 30 °C | 1.16 |
| Lead | 11,300 | Mercury | 13,600 | Helium | 0.178 |
| Platinum | 21,450 | Milk | 1028–1035 | Hydrogen | 0.090 |
| Silver | 10,500 | Seawater | 1025 | Oxygen[2] | 1.43 |
| Bone | 1700–2000 | Water 0 °C | 999.8 | | |
| Brick | 1400–2200 | 3.98 °C | 1000.00 | | |
| Cement | 2700–3000 | 20 °C | 998.2 | | |
| Clay | 1800–2600 | Blood plasma | 1030 | | |
| Glass | 2400–2800 | Blood whole[1] | 1050 | | |
| Ice | 917 | | | | |
| Balsa wood | 120 | | | | |
| Oak | 600–900 | | | | |
| Pine | 500 | | | | |
| Planet Earth | 5515 | | | | |
| Moon | 3340 | | | | |
| Sun | 1410 | | | | |
| Universe (average) | $10^{-26}$ | | | | |
| Pulsar | $10^{11}$–$10^{18}$ | | | | |

[1]Densities of liquids are at 0 °C unless otherwise noted.
[2]Densities of gases are at 0 °C and 1 atm unless otherwise noted.

## Density and floating

Understanding density allows us to pose questions about phenomena that we observe almost every day. For example, why does oil form a film on water? If you pour oil into water or water into oil, they form layers (see **Figure 10.2a**). Independently of which fluid is poured first—the layer of oil is always on top of the water. The density of oil is less that the density of water. If you pour corn syrup and water into a container, the corn syrup forms a layer below the water (Figure 10.2b); the density of corn syrup is $1200\text{ kg/m}^3$, greater than the density of water. Why, when mixed together, is the lower density substance always on top of the higher density substance?

Similar phenomena occur with gases. Helium-filled balloons accelerate upward in air while air-filled balloons accelerate (slowly) downward. The mass of helium atoms is much smaller than the mass of any other molecules in the air. (Recall that at the same pressure and temperature, atoms and molecules of gas have the same concentration; because helium atoms have much lower mass, their density is lower.) The air-filled balloon must be denser than air. The rubber with which the skin of any balloon is made is denser than air. We can disregard the slight compression of the gas by the balloon, because even though it increases the density of the gas, the effect is the same for both the air and helium in the balloons.

**Figure 10.2** Less dense matter floats on denser matter.

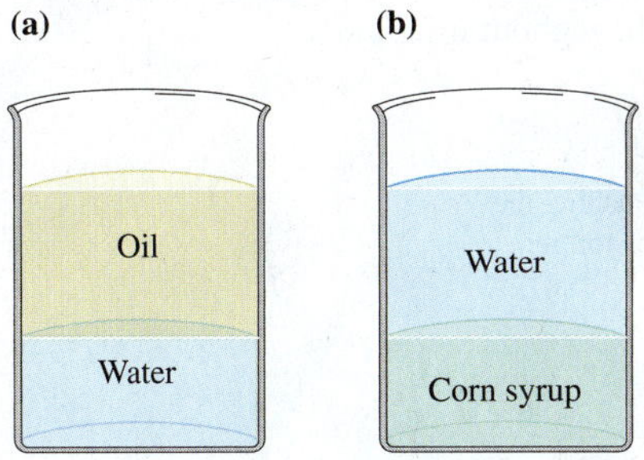


## Solid water floats in liquid water

The solid form of a particular substance is almost always denser than the liquid form of the substance, with one very significant exception: liquid water and solid ice. Since ice floats on liquid water, we can assume that the density of ice is less than that of water. This is in fact true: the density of water changes slightly with temperature and is the highest at 4 °C: 1000 $kg/m^3$. The density of ice is 917 $kg/m^3$. Ice has a lower density because in forming the crystal structure of ice, water molecules spread apart. The fact that water expands when it forms ice is important for life on Earth (see the second Reading Passage at the end of this chapter). Fish and plants living in lakes survive cold winters in liquid water under a shield of ice and snow at the surface. Water absorbed in the cracks of rocks freezes and expands in the winter, cracking the rock. Over the years, this process of liquid water absorption, freezing, and cracking eventually converts the rock into soil.

Why do denser forms of matter sink in less dense forms of matter? We learned (in Chapter 9) that the quantity *pressure* describes the forces that fluids exert on each other and on the solid objects they contact. Let us investigate whether pressure explains, for example, why a nail sinks in water or why a hot air balloon rises.

◀ Active Learning Guide

**Review Question 10.1** How would you determine the density of an irregularly shaped object?

# 10.2 Pressure exerted by a fluid

We know that as gas particles collide with the walls of the container in which they reside, they exert pressure. In fact, if you place any object inside a gas, the gas particles exert the same pressure on the object as the gas exerts on the walls of the container. Do liquids behave in a similar way? In the last chapter we learned that the particles in a liquid are in continual random motion, somewhat similar to particles in gases.

Let's conduct a simple observational experiment. Take a plastic water bottle and poke several small holes at the same height along its perimeter. Close the holes with tacks, fill the bottle with water, open the cap, remove the tacks, and observe what happens (**Figure 10.3**). Identically-shaped parabolic streams of water shoot out of the holes. The behavior of the water when the tacks are removed is analogous to a person leaning on a door that is suddenly opened from the other side—the person falls through the door. Evidently, the water inside must push out perpendicular to the wall of the bottle, just as gas pushes out perpendicular to the wall of a balloon. Due to their similar behaviors, liquids and gases are often studied together and are collectively referred to as fluids. In addition, since the four streams are identically shaped, the pressure at all points at the same depth in the fluid is the same.

**Figure 10.3** Arcs of water leaving holes at the same level in a bottle.

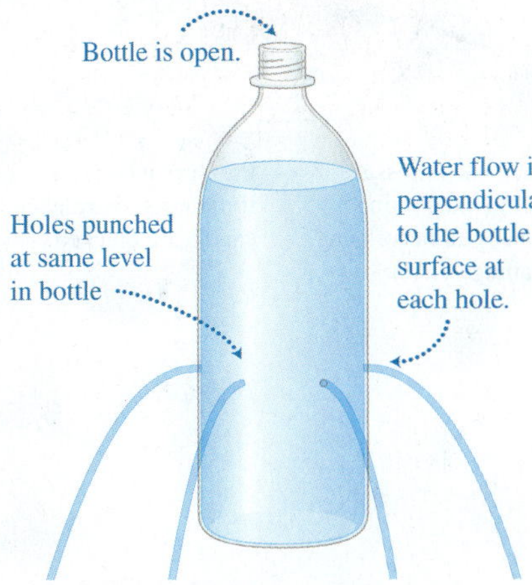

**Figure 10.4** Pascal's first law: Increasing the pressure of a fluid at one location causes a uniform pressure increase throughout the fluid.

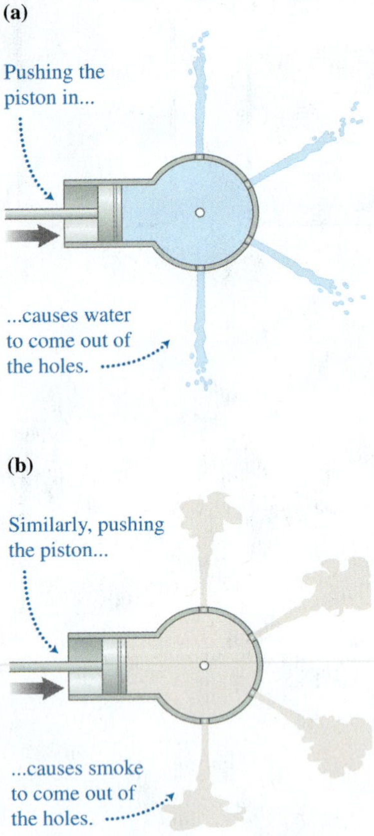


**Active Learning Guide ›**

**Figure 10.5** Glaucoma is an increase in intraocular pressure, caused by blockage of the ducts that normally drain aqueous humor from the eye.

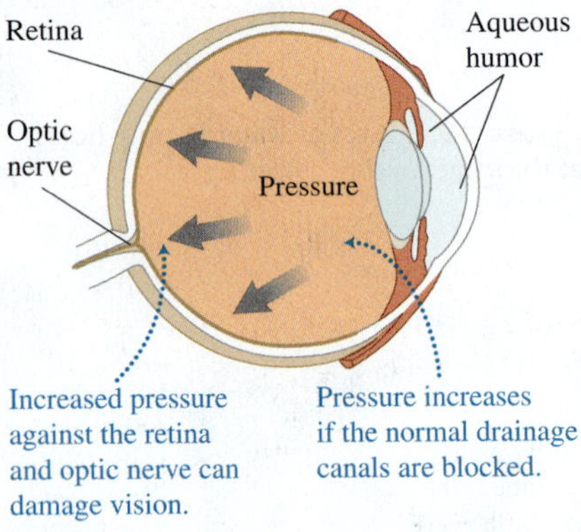

Retina
Aqueous humor
Optic nerve
Pressure
Increased pressure against the retina and optic nerve can damage vision.
Pressure increases if the normal drainage canals are blocked.

## Pascal's first law

Many practical applications involve situations in which an external object (for example, a piston) exerts a force on a particular part of a fluid. What happens to the pressure at other places inside the fluid? To investigate how the pressure at different points in a fluid are related, we use a special instrument that consists of a round glass bulb with holes in it connected to a glass tube with a piston on the other side (**Figure 10.4a**). When we fill the apparatus with water and push the piston, water comes out of all of the holes, not only those that align with the piston. When we fill the apparatus with smoke and push the piston, we get the same result (Figure 10.4b). The liquid and the gas behave similarly.

How can we explain this observation? Pushing the piston in one direction caused a greater pressure in the fluid close to the piston. It seems that almost immediately the pressure throughout the fluid increased as well, as the fluid was pushed out of *all* of the holes in the bulb in the same way. This phenomenon was first discovered by French scientist Blaise Pascal in 1653 and is called Pascal's first law.

**Pascal's first law An increase in the pressure of a static, enclosed fluid at one place in the fluid causes a uniform increase in pressure throughout the fluid.**

The above experiment describes Pascal's first law macroscopically. We can also explain Pascal's first law at a microscopic level. Particles inside a container move randomly in all directions. When we push harder on one of the surfaces of the container, the fluid compresses near that surface. The molecules near that surface collide more frequently with their neighbors farther from the surface. They in turn collide more frequently with their neighbors. The extra pressure exerted at the one surface quickly spreads and soon there is increased pressure throughout the fluid.

## Glaucoma

Pascal's first law can help us understand a common eye problem—glaucoma. A clear fluid called aqueous humor fills two chambers in the front of the eye (**Figure 10.5**). In a healthy eye, new fluid is continually secreted into these chambers while old fluid drains from the chambers through sinus canals. A person with glaucoma has closed drainage canals. The buildup of fluid causes increased pressure throughout the eye, including at the retina and optic nerve, which can lead to blindness. Ophthalmologists diagnose glaucoma by measuring the pressure at the front of the eye. The eye pressure of a person with glaucoma is about $3000\ \mathrm{N/m^2}$ above atmospheric pressure.

## Hydraulic lift

One of the technical applications of Pascal's first law is a hydraulic press, a form of simple machine that converts small forces into larger forces, or vice versa. Automobile mechanics use hydraulic presses to lift cars, and dentists and barbers use them to raise and lower their clients' chairs. The hydraulic brakes of an automobile are also a form of hydraulic press. Most of these devices work on the simple principle illustrated in **Figure 10.6**, although the actual devices are usually more complicated in construction.

In Figure 10.6, a downward force $\vec{F}_{1\text{ on L}}$ is exerted by piston 1 (with small area $A_1$) on the liquid. This piston compresses a liquid (usually oil) in the lift. The pressure in the fluid just under piston 1 is

$$P_1 = \frac{F_{1\text{ on L}}}{A_1}$$

Because the pressure changes uniformly throughout the liquid, the pressure under piston 2 is also $P = F_{1\text{ on L}}/A_1$, assuming the pistons are at the same elevation. Since piston 2 has a greater area $A_2$ than piston 1, the liquid exerts a greater upward force on piston 2 than the downward force on piston 1:

$$F_{\text{L on 2}} = PA_2 = \left(\frac{F_{1\text{ on L}}}{A_1}\right)A_2 = \left(\frac{A_2}{A_1}\right)F_{1\text{ on L}} \quad (10.2)$$

Since $A_2$ is greater than $A_1$, the lift provides a significantly greater upward force $F_{\text{L on 2}}$ on piston 2 than the downward push of the smaller piston 1 on the liquid $F_{1\text{ on L}}$.

**Figure 10.6** Schematic of a hydraulic lift.

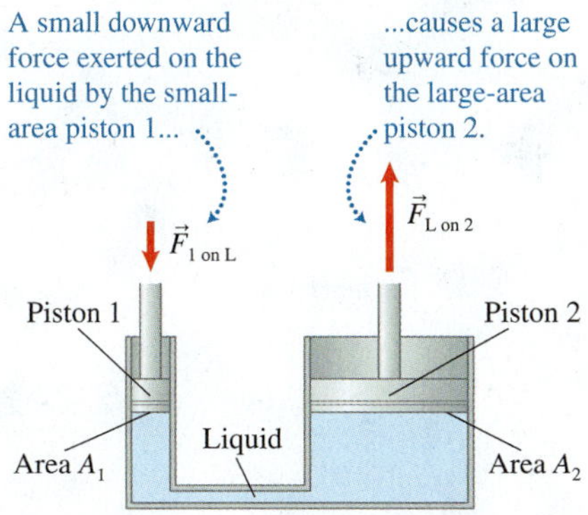


**‹ Active Learning Guide**

**EXAMPLE 10.2 Lifting a car with one hand**

A hydraulic lift similar to the press described above has a small piston with surface area 0.0020 m² and a larger piston with surface area 0.20 m². Piston 2 and the car placed on piston 2 have a combined mass of 1800 kg. What is the minimal force that piston 1 needs to exert on the fluid to slowly lift the car?

**Sketch and translate** The situation is similar to that shown in Figure 10.6. We need to find $F_{1\text{ on L}}$ so the fluid exerts a force great enough to support the mass of the car and piston 2. The hydraulic lift Eq. (10.2) should then allow us to determine $F_{1\text{ on L}}$.

**Simplify and diagram** Assume that the levels of the two pistons are the same and that the car is being lifted at constant velocity. Use the force diagram for the car and piston 2 (see below diagram) and Newton's second law to determine $F_{\text{L on 2}}$. Note that the force that the liquid exerts on the large piston 2 $F_{\text{L on 2}}$ is equal in magnitude to the force that piston 2 and the car exert on the liquid $F_{2\text{ on L}}$, which equals the downward gravitational force that Earth exerts on the car and piston: $F_{2\text{ on L}} = m_{\text{Car+piston}}g$.

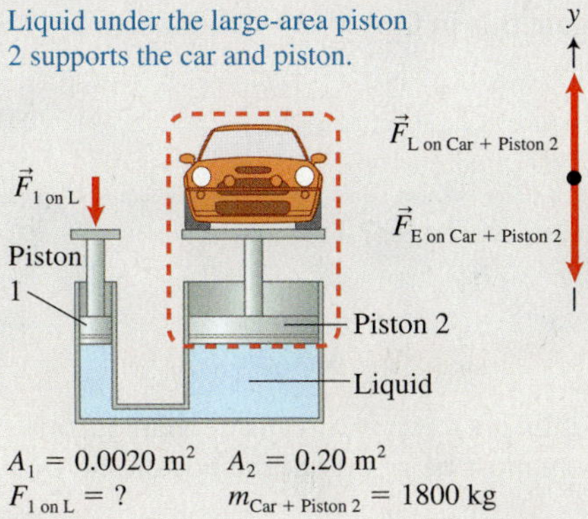


**Represent mathematically** We rewrite the hydraulic lift Eq. (10.2) to determine the unknown force:

$$F_{1\text{ on L}} = \left(\frac{A_1}{A_2}\right)F_{2\text{ on L}} = \left(\frac{A_1}{A_2}\right)m_{\text{Car+piston}}g$$

**Solve and evaluate**

$$F_{1\text{ on L}} = \frac{(0.0020\text{ m}^2)}{(0.20\text{ m}^2)}[(1800\text{ kg})(9.8\text{N/kg})] = 180\text{ N}.$$

That is the force equal to lifting an object of mass 18 kg, which is entirely possible for a person. The units are also consistent.

**Try it yourself:** If you needed to lift the car about 0.10 m above the ground, what distance would you have to push down on the small piston? Specify the assumptions you made.

*Answer:* 10 m. Since liquids are essentially incompressible, the total volume of the liquid won't change significantly. The downward pushing distance (10 m) times the area of piston 1 equals the upward lifting distance (0.10 m) times the area of piston 2—the volume changes are equal. The small piston must push down significantly farther than the large piston will rise.

**Review Question 10.2** If you poke many small holes in a closed toothpaste tube and squeeze it, the paste comes out equally from all holes. Why?

## 10.3 Pressure variation with depth

Pascal's first law states that an increase in the pressure in one part of an enclosed fluid results in an increase at all other parts of the fluid. Does that mean that the pressure is the same throughout a fluid—for example, in a vertical column of fluid? To test this hypothesis, consider another experiment with a water bottle. This time, poke holes vertically along one side of the bottle. Place tacks in the holes, and fill the bottle with water. Leave the cap off. If the pressure is the same throughout the liquid, when the tacks are removed the water should come out of each hole in the same arcs, as shown in **Figure 10.7a**.

**Figure 10.7** Water seems to be pushed harder from holes deeper in the water.

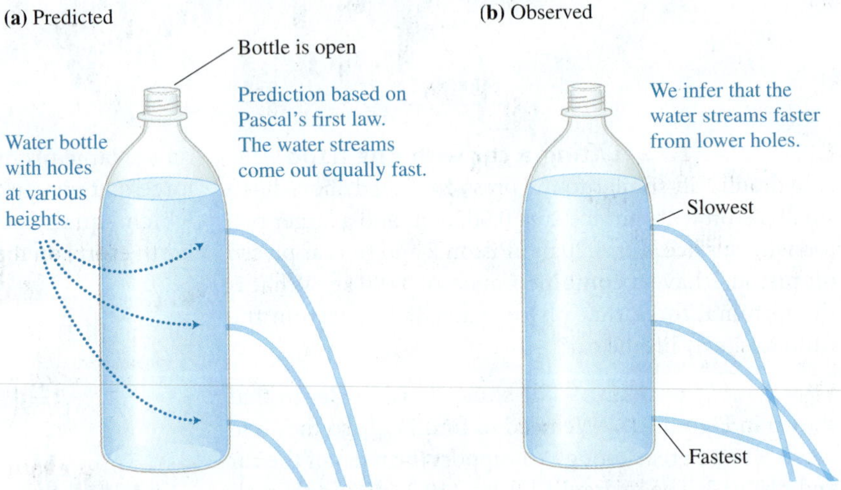


**Active Learning Guide ›**

However, when the tacks are removed, we observe that the water squirts out in a wider arc at the bottom than at the top (Figure 10.7b). Should we abandon Pascal's first law now because the prediction based on it did not match the outcome of the experiment? In such cases, scientists do not immediately throw out the principle but first examine the additional assumptions that were used to make the prediction. In our first experiment with the water bottle, we did not consider the impact of poking holes at different heights. Maybe this was an important factor in the experiment. Let's investigate this in Observational Experiment **Table 10.2**.

**OBSERVATIONAL EXPERIMENT TABLE**

**10.2 How does the location of the holes affect the streams leaving the holes?**

| Observational experiment | Analysis |
|---|---|
| **Experiment 1.** Place two tacks on each side of a plastic bottle, one hole above the other, and fill the bottle with water above the top tack. Remove the tacks. Water comes out on the left and right and the stream from the lower holes shoots farther. 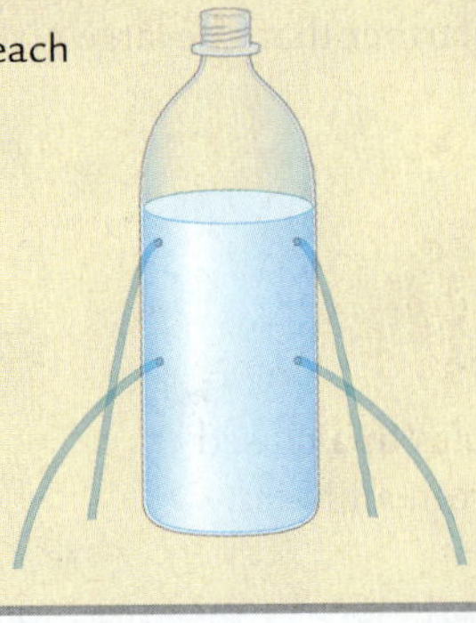 | There must be greater pressure inside than outside. The pressure must be greater at the bottom holes than at the top holes. |

| Observational experiment | Analysis |
|---|---|
| **Experiment 2.** Repeat Experiment 1 but this time fill the bottle with water to the same distance above the bottom tack as it was filled above the top tack in Experiment 1. Remove the tacks. The stream comes out the bottom holes with the same arc as it came out of the top holes in Experiment 1. 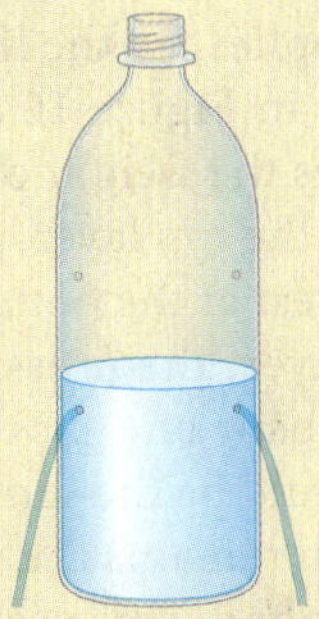 | The total water depth seems not to matter, just the height of the water above the hole. |
| **Experiment 3.** Repeat Experiment 1 using a thinner bottle with the water level initially the same distance above the top tack as it was in Experiment 1. Remove the tacks. The water streams are identical to those in Experiment 1. 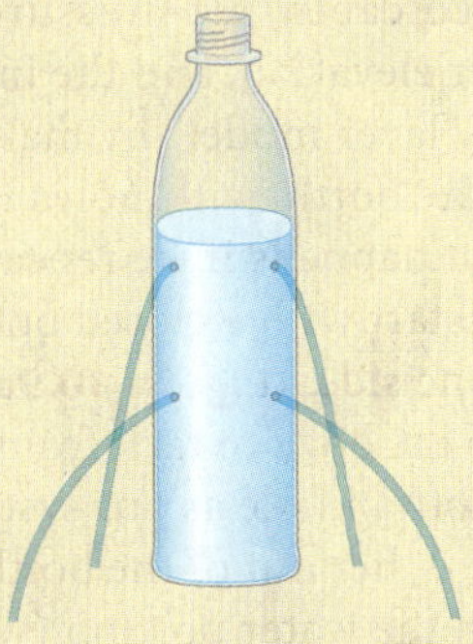 | Because the water comes out in exactly the same arc in a bigger bottle and in a smaller bottle when the water level above the top tack is at the same height, we can conclude that the mass of the water in the bottle does not affect the pressure. |

**Patterns**

The stream shape at a particular level:

- Depends on the height of the water above the hole.
- Is the same in different directions at the same level.
- Does not depend on the amount of liquid (volume or mass) above the hole (just the height of the water above the hole).
- Does not depend on the amount (mass or volume) or depth of the water below the hole.

**Figure 10.8** Pressure increases with depth.

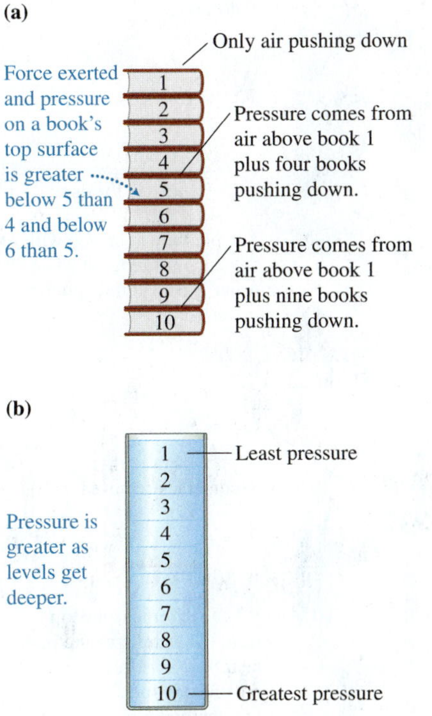


From the patterns above we reason that the pressure of the liquid at the hole depends only on the *height* of the liquid above the hole, and not on the *mass* of the liquid above. We also see that the pressure at a given depth is the same in all directions. This is consistent with the experience you have when you dive below the surface of the water in a swimming pool, lake, or ocean. The pressure on your ears depends only on how deep you are below the surface. Pascal's first law fails to explain this pressure variation at different depths below the surface. Now we need to understand why the pressure varies with depth and to devise a rule to describe this variation quantitatively.

## Why does pressure vary at different levels?

To explain the variation of pressure with depth we can use an analogy of stacking ten books on a table (**Figure 10.8a**). Imagine that each book is a layer of water in a cylindrical tube (see Figure 10.8b). Consider the pressure (force per unit area) from above on the top surface of each book. The only force exerted on the top surface of the top book is due to air pushing down from above. However, there are in effect two forces exerted on the top surface of the second book: the force that the top book exerts on it (equal in magnitude to the weight of the book) plus the force exerted by the air on the top book. The top surface of the bottom book in the stack must balance the force exerted by the nine books above it (equal in magnitude to the weight of nine books) plus the

**Figure 10.9** Water coming from holes on the side of a bottle behaves differently if the bottle is closed at the top.

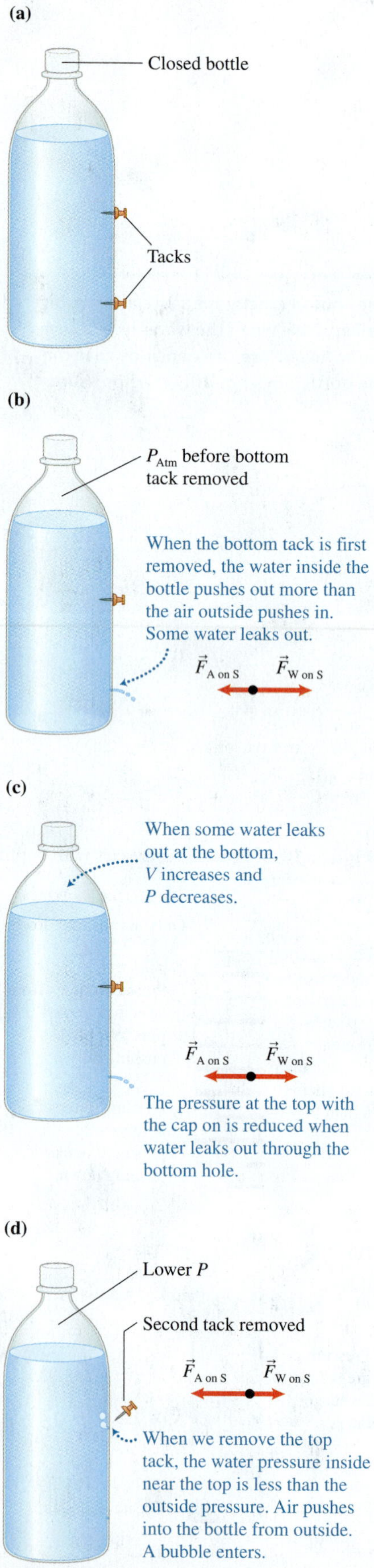
(a)
Closed bottle
Tacks
(b)
$P_{Atm}$ before bottom tack removed
When the bottom tack is first removed, the water inside the bottle pushes out more than the air outside pushes in. Some water leaks out.
$\vec{F}_{A\text{ on }S}$
$\vec{F}_{W\text{ on }S}$
(c)
When some water leaks out at the bottom, $V$ increases and $P$ decreases.
$\vec{F}_{A\text{ on }S}$
$\vec{F}_{W\text{ on }S}$
The pressure at the top with the cap on is reduced when water leaks out through the bottom hole.
(d)
Lower $P$
Second tack removed
$\vec{F}_{A\text{ on }S}$
$\vec{F}_{W\text{ on }S}$
When we remove the top tack, the water pressure inside near the top is less than the outside pressure. Air pushes into the bottle from outside. A bubble enters.

pressure force exerted by the air on the top book. So the pressure increases on the top surface of each book in the stack as we go lower in the stack.

Similar reasoning applies for the liquid-filled tube divided into a number of imaginary thin layers in Figure 10.8b. Air pushes down on the top layer. The second layer balances the weight of the top layer plus the force exerted by the air pushing down on the top layer, and so on. The pressure is lowest at the top of the fluid and greatest at the bottom.

Note that, at each layer, the pressure is the same in all directions. If we could take a pressure sensor and place it inside the container of water, the readings of the sensor would be the same independent of the orientation of the sensor as long as its depth remains the same.

This helps us understand why a sealed empty water bottle on an airplane collapses as the plane descends from a higher elevation to a lower elevation. Even in a pressurized cabin, air pressure at the lower elevation is slightly greater than at higher elevation, and the increased pressure crushes the bottle.

We can test this "layer model" by making one small change in the experiment that involves the bottle with holes in it. So far we have kept the cap of the bottle open. What happens if we repeat the experiment, but this time with the cap closed? Let's start with a closed bottle full of water with two tacks, one above the other on one side. (**Figure 10.9a**) What does the model predict will happen if we remove the bottom tack from the hole? If we consider a tiny portion of water at the bottom hole as our system of interest, then the acceleration of that system can be either out of the bottle, into the bottle, or zero. There are two forces exerted on the water portion (we will call it the system $S$); the force exerted by the outside air pushing inward $\vec{F}_{A\text{ on }S}$ and the force of the inside water pushing outward $\vec{F}_{W\text{ on }S}$ (Figure 10.9b). If one of the forces is smaller, then the water will either accelerate outward and we will see the bottle leaking, or it will accelerate inward and we will see bubbles of air coming in.

What does the model predict will happen? Closing the bottle traps a little bit of air above the water. The air is at atmospheric pressure since the bottle was originally open. Inside the water at the position of the bottom tack, in addition to the pressure due to air in the bottle above pushing down on the water, there is additional pressure from the water pressing downward. At the level of the lower tack, the force that the water exerts on the water at the hole (the system) should be greater than the force exerted by the air outside the hole on that same water. If you remove that tack, the water should accelerate outward.

However, when water comes out of the hole, the volume available to the air at the top of the closed bottle increases. Because the bottle is closed, there is no way for air to enter the bottle. As a result, the air expands to fill the larger volume and the air pressure decreases. This pressure decrease spreads throughout the fluid, decreasing the magnitude of the force exerted by the water on the system. As water drains out, eventually the forces exerted by the water and the air on the system balance (Figure 10.9c). Thus, the model predicts that some water will come out of the bottle when we remove one tack, but then the leaking will soon stop. This is exactly what happens when we perform the experiment.

Now, what happens if you remove the higher tack while the bottom hole remains open? Try to predict the outcome before continuing. After you make your prediction, look at Figure 10.9d. Remember that water flow from the bottom tack stops when the inside pressure at the lower tack equals the outside air pressure pushing in. Thus the pressure in the water at the level of the higher opening will be less than the air pressure outside. Thus, the outside air will exert a greater inward force on water at the higher opening than the outward force that water in the bottle exerts on that same portion. Air flows into the top hole, forming bubbles, which float to the top of the closed bottle. The water pressure increases and more water squirts out of the lower hole, reducing pressure at the top of the bottle again. More air comes in the upper hole, and another squirt of water comes out of the bottom hole. The bottle drains in small squirts from the bottom hole. Is this what you predicted?

## How can we quantify pressure change with depth?

We know that pressure increases with depth, but does the pressure depend on the depth linearly, quadratically, or in some other way? Consider the shaded cylinder C of water shown in **Figure 10.10a** as our system of interest. The walls on opposite sides of the cylinder push inward, exerting equal-magnitude and oppositely directed forces—the forces exerted by the sides cancel. What about the forces exerted by the water above and below? If the pressure at elevation $y_2$ is $P_2$ and the cross-sectional area of the cylinder is $A$, then the fluid above pushes down, exerting a force of magnitude $F_{\text{fluid above on C}} = P_2A$ (Figure 10.10b). Similarly, fluid from below the shaded section of fluid at elevation $y_1$ exerts on the cylinder an upward force of magnitude $F_{\text{fluid below on C}} = P_1A$. Earth exerts a third force on the shaded cylinder $\vec{F}_{\text{E on C}}$ equal in magnitude to $m_\text{C}g$, where $m_\text{C}$ is the mass of the liquid in the cylinder. Since the liquid is not accelerating, these three forces add to zero. Choosing the $y$-axis pointing up, we have

$$\Sigma F_y = (-F_{\text{fluid above on C}}) + F_{\text{fluid below on C}} + (-m_\text{C}g) = 0$$

Substituting the earlier expressions for the forces, we have

$$(-P_2A) + P_1A + (-m_\text{C}g) = 0$$

The mass of the fluid in the shaded cylinder is the product of the fluid's density and the volume of the cylinder:

$$m_\text{C} = \rho_{\text{fluid}}V = \rho_{\text{fluid}}[A(y_2 - y_1)]$$

Substituting this expression for the mass in the above expression for the forces, we get

$$-P_2A + P_1A - \rho_{\text{fluid}}[A(y_2 - y_1)]g = 0$$

Divide by the common $A$ in all of the terms and rearrange to get

$$P_1 = P_2 + \rho_{\text{fluid}}(y_2 - y_1)g$$

This is Pascal's second law. As we see, pressure varies linearly with depth.

**Pascal's second law—variation of pressure with depth** The pressure $P_1$ in a static fluid at position $y_1$ can be determined in terms of the pressure $P_2$ at position $y_2$ as follows:

$$P_1 = P_2 + \rho_{\text{fluid}}(y_2 - y_1)g, \qquad (10.3)$$

where $\rho_{\text{fluid}}$ is the fluid density, assumed constant throughout the fluid, and $g = 9.8\ \text{N/kg}$. The positive $y$-direction is up.

**TIP** When using Pascal's second law [Eq. (10.3)], picture the situation and be sure to include a vertical $y$-axis that points upward and has a defined origin, or zero point. Then choose the two points of interest and identify their vertical $y$-positions relative to the axis. This lets you relate the pressures at those two points.

In order to test whether the pressure of a fluid depends on the depth and not on the mass of the fluid, Blaise Pascal filled a barrel with water and inserted a long, narrow vertical tube into the water from above. He then sealed the barrel (see **Figure 10.11**). He predicted that when he filled the tube with water, the barrel would burst even though the mass of water in the thin tube was small, because the pressure of the water in the barrel would depend on the height, not the mass, of the water column. The barrel burst, matching Pascal's prediction: thus supporting the idea that the height of the liquid above, not the amount of water above, determined the pressure.

**Figure 10.10** Using Newton's second law to determine how fluid pressure changes with the depth in the fluid.

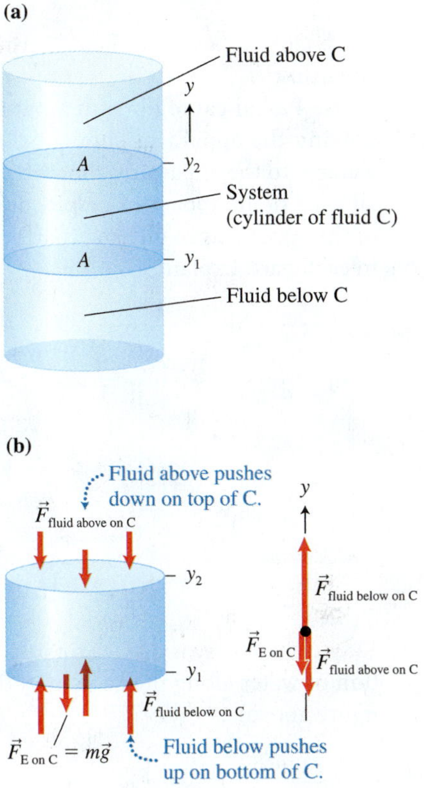


**‹ Active Learning Guide**

**Figure 10.11** Pascal tests his second law.

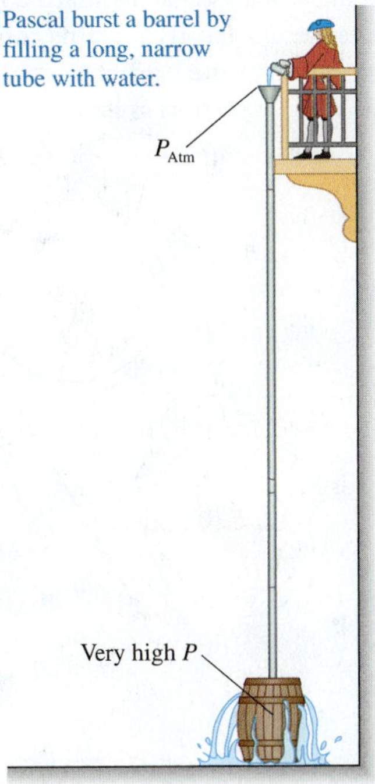

Equation (10.3) applies to any static fluid, whether liquid or gas. This explains why air pressure is greater at the bottom of a mountain than at the top.

**CONCEPTUAL EXERCISE 10.3 Pascal's paradox**

Blaise Pascal came up with a paradoxical situation involving the apparatus shown below. When he poured water into the apparatus, the water level was the same in all parts of the apparatus despite differences in the shapes of the apparatus in different parts and the mass of water in each part. Explain.

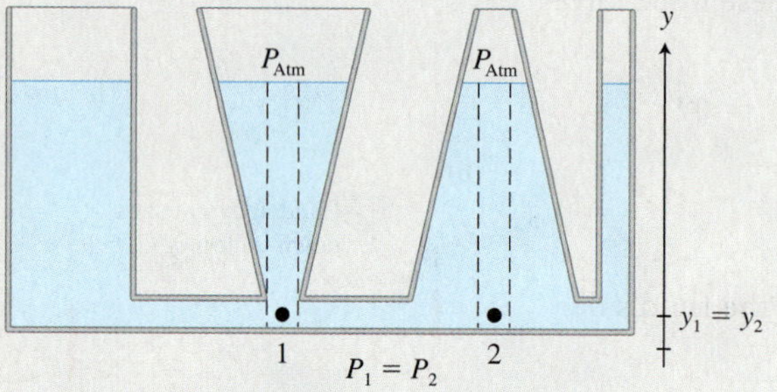


**Sketch and translate** Consider the horizontal portion of water along the bottom of the apparatus (see the figure above).

**Simplify and diagram** Since this water is stationary, the pressure at all points along the bottom must be the same. Consider in particular the water at positions 1 and 2. These pressures can be equal only if the columns of water above points 1 and 2 are the same height. Thus, each part of Pascal's unusually shaped device fills to the same level.

**Try it yourself:** Use your understanding of pressure to explain why our ears pop when we climb a mountain or during takeoff in an airplane.

*Answer:* When we begin climbing a mountain, the air pressure inside $P_{\text{inside}}$ the eardrum and the air pressure $P_1$ of outside air pushing in on the eardrum are equal. At the top of the mountain, the air pressure $P_2$ outside is less than the air pressure $P_{\text{inside}}$ inside. We can equalize these pressures by venting a little air from the middle ear. That's the popping sensation. The same thing happens when you are in an ascending airplane.

**QUANTITATIVE EXERCISE 10.4 Pop your ears**

If your ears did not pop, then what would be the net force exerted by the inside and outside air on your eardrum at the top of a 1000-m-high mountain? You start your hike from sea level. The area of your eardrum is $0.50\text{ cm}^2$. The density of air at sea level at standard conditions is $1.3\text{ kg/m}^3$. Assume the air density remains constant during the hike. The situation at the start at $y_1 = 0$ and at the end of the hike at $y_2 = 1000$ m is sketched below.

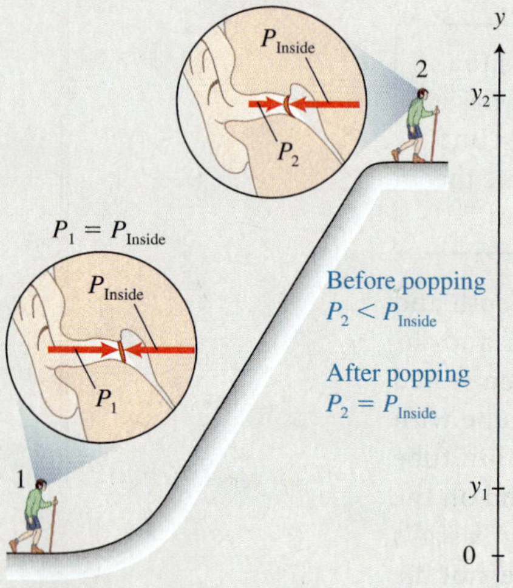


**Represent mathematically** We use an upward pointing vertical $y$-axis with the origin at sea level. Assume that the air pressure inside the eardrum remains constant at its sea level value $P_{\text{inside}} = P_1$. The air pressure difference between the top of the mountain and sea level would be

$$P_2 - P_1 = \rho(y_1 - y_2)g$$

When at 1000 m above sea level, the outside air exerts a lower pressure $P_2$ on the eardrum. So the net force exerted by the air on the drum is the pressure difference of air pushing out $P_{\text{inside}} = P_1$ and air pushing in $P_2$ times the area of the eardrum: $F_{\text{net air on drum}} = (P_1 - P_2)A$. This pressure difference can be determined using Pascal's second law.

**Solve and evaluate**

$$\begin{aligned} P_1 - P_2 &= \rho(y_2 - y_1)g \\ &= (1.3\text{ kg/m}^3)(1000\text{ m} - 0)(9.8\text{ N/kg}) \\ &= +0.13 \times 10^5\text{ N/m}^2 \end{aligned}$$

This is 0.13 atm!

$$\begin{aligned} F_{\text{net air on drum}} &= (P_1 - P_2)A \\ &= (0.13 \times 10^5\text{ N/m}^2)(0.5\text{ cm}^2) \\ &\quad \times (1\text{ m}/100\text{ cm})^2 \\ &= +0.6\text{ N} \end{aligned}$$

The net force is exerted outward and is about half the gravitational force that Earth exerts on an apple. No wonder it can hurt until you get some air out of that middle ear!

**Try it yourself:** Determine the difference in water pressure on your ear when you are 1.0 m underwater compared to when you are at the surface. The density of water is $1000\ kg/m^3$.

**Answer:** $+9800\ N/m^2$ greater pressure when under the water, or 0.1 atm.

**Review Question 10.3** Pascal's first law says that an increase in pressure in one part of an enclosed liquid results in an increase in pressure throughout all parts of that liquid. Why then does the pressure differ at different heights?

# 10.4 Measuring atmospheric pressure

We can now use Pascal's second law to develop a method for measuring atmospheric air pressure.

## Measuring atmospheric pressure

During Galileo's time (1600s), suction pumps were used to lift drinking water from wells and to remove water from flooded mines. The suction pump was like a long syringe. The pump consisted of a piston in a long cylinder that pulled up water (**Figure10.12a**). Such pumps could lift water a maximum of 10.3 m. Why 10.3 meters?

Evangelista Torricelli (1608–1647), one of Galileo's students, hypothesized that the pressure of the air in the atmosphere could explain the limit to how far water could be lifted. Torricelli did not know of Pascal's second law, which was published in the year that Torricelli died. However, it is possible that Torricelli's work influenced Pascal. Let's analyze the situation shown in Figure 10.12b.

Consider the pressure at three places: point 1, at the water surface in the pool outside the cylinder; point 2, at the same elevation only inside the cylinder; and point 3, in the cylinder 10.3 m above the pool water level. The pressure at point 1 is atmospheric pressure. The pressure at point 2, according to Pascal's second law, is also atmospheric pressure, since it is at the same level as point 1. To get the water to the 10.3-m maximum height, the region above the water surface inside the cylinder and under the piston must be at the least possible pressure—essentially a vacuum. Thus, we assume that the pressure at point 3 is zero.

Now we will use Eq. (10.3) with $y_3 - y_2 = 10.3$ m to predict the pressure $P_2$:

$$P_2 = P_3 + \rho_{\text{water}}(y_3 - y_2)g = 0 + (1.0 \times 10^3\ \text{kg/m}^3)(10.3\ \text{m} - 0)(9.8\ \text{N/kg})$$
$$= 1.01 \times 10^5\ \text{N/m}^2$$

This number is exactly the value of the atmospheric pressure that we encountered in our discussion of gases (in Chapter 9). The atmospheric pressure pushing down on the water outside the tube can push water up the tube a maximum of 10.3 m if there is a vacuum (absence of any matter) above the water in the tube.

At Torricelli's time the value of normal atmospheric pressure was unknown, so the huge number that came out of this analysis surprised Torricelli. Not believing the result, he tested it using a different liquid—mercury. Mercury is 14 times denser than water ($\rho_{\text{Hg}} = 13{,}600\ \text{kg/m}^3$); hence, the column of mercury should rise only 1/14 times as high in an evacuated tube. However,

**Figure 10.12** A piston pulled up a cylinder causes water to rise to a maximum height of 10.3 m.

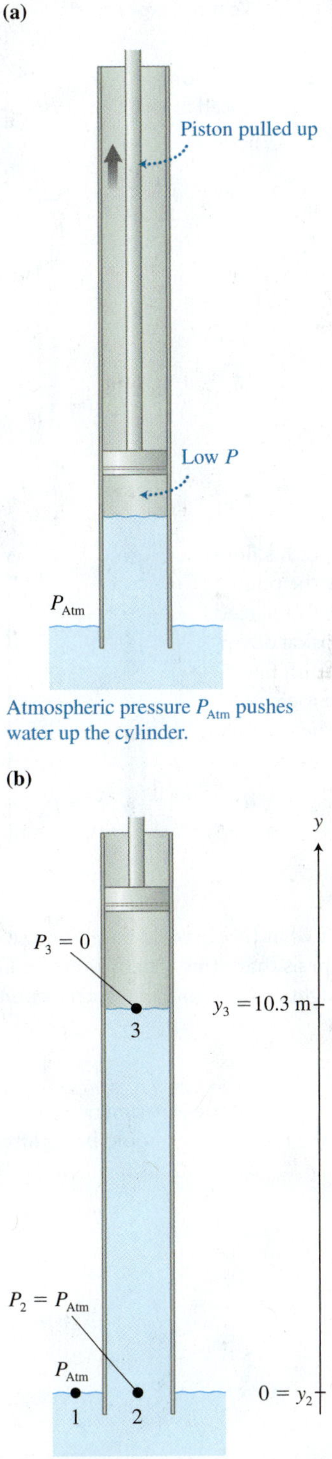

instead of using a piston to lift mercury, Torricelli devised a method that guaranteed that the pressure at the top of the column was about zero. Consider Testing Experiment **Table 10.3**.

## TESTING EXPERIMENT TABLE

### 10.3 Testing Torricelli's hypothesis using mercury.

| Testing experiment | Prediction based on Torricelli's hypothesis that atmospheric pressure limits the height of the liquid in a suction pump | Outcome |
|---|---|---|
| ■ Torricelli filled a long glass tube closed at one end with mercury. 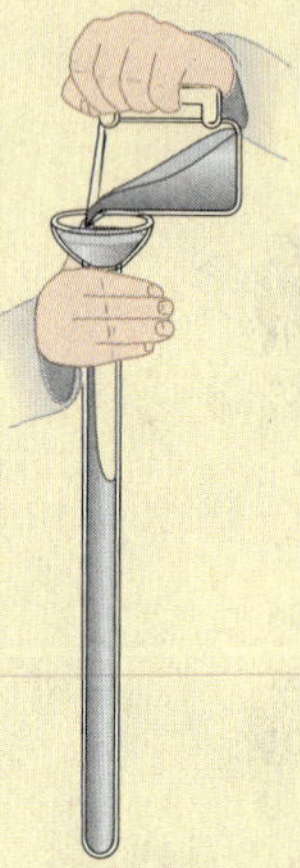 | Mercury should start leaking from the tube into the dish. When it leaks, it leaves an empty evacuated space at the top of the tube. It will leak until the height of the mercury column left in the tube produces the same pressure as the atmosphere at the bottom of the column at position 1. The height of the mercury in the tube should be $y_2 - y_1 = \frac{P_1 - P_2}{\rho_{mercury} g}$ $= \frac{(1.01 \times 10^5\,N/m^2 - 0)}{(13.6 \times 10^3\,kg/m^3)(9.8\,N/kg)}$ $= 0.76\,m$ | Torricelli observed some mercury leaking from the tube and then the process stopped. He measured the height of the remaining mercury to be 0.76 m = 760 mm, in agreement with the prediction. |
| ■ He put his finger over the open end and placed it upside down in a dish filled with mercury. He then removed his finger. 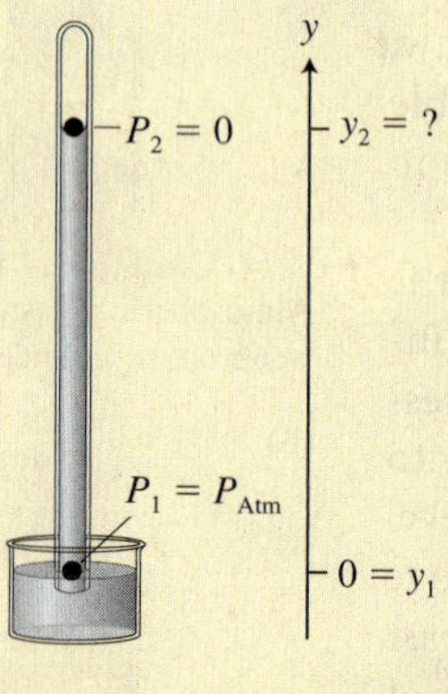  Predict what he observed based on the hypothesis that atmospheric pressure limits the height of the liquid in a suction pump. | | |

**Conclusion**

The outcome of the experiment was consistent with the prediction based on Torricelli's hypothesis that atmospheric pressure limits the height of the liquid being lifted in a suction pump. Thus the hypothesis is supported by evidence.

Torricelli also used his understanding of pressure and fluids to predict that in the mountains, where the atmospheric pressure is lower, the height of the mercury column should be lower. Experiments have shown that the mercury level indeed decreases at higher elevation. These experiments supported the explanation that atmospheric air pushes the liquids upward into the tubes.

Torricelli's apparatus with the mercury tube became a useful device for measuring atmospheric pressure—called a barometer. However, since mercury is toxic, the Torricelli device has since been replaced by the aneroid barometer (described in Chapter 9).

**TIP** We now understand why pressure is often measured and reported in mm Hg and why atmospheric pressure is 760 mm Hg. The atmospheric pressure (101,000 $N/m^2$) can push mercury of density (13,600 $kg/m^3$) 760 mm up a column.

## Diving bell

Our understanding of atmospheric pressure allows us to explain many simple experiments that lead to important practical applications. For example, have you ever submerged a transparent container upside down under water? If you do, you will see that at first little water enters the container; then as the inverted container is pushed deeper into the water, more water enters the container. One practical application of this phenomenon is a diving bell—a large bottomless chamber lowered under water with people and equipment inside. Divers use the diving bell to take a break and refill on oxygen.

‹ Active Learning Guide

**EXAMPLE 10.5 Diving bell**

The bottom of a 4.0-m-tall cylindrical diving bell is at an unknown depth underwater. The pressure of the air inside the bell is 2.0 atm (it was 1 atm before the bell entered the water). The average density of ocean water is slightly higher than fresh water, $\rho_{\text{ocean water}} = 1027\ kg/m^3$. How high is the water inside the bell and how deep is the bottom of the bell under the water?

**Sketch and translate** We sketch the situation below. We want to find the height $h$ of the water in the bell and the depth $d$ of the bottom of the bell under the water.

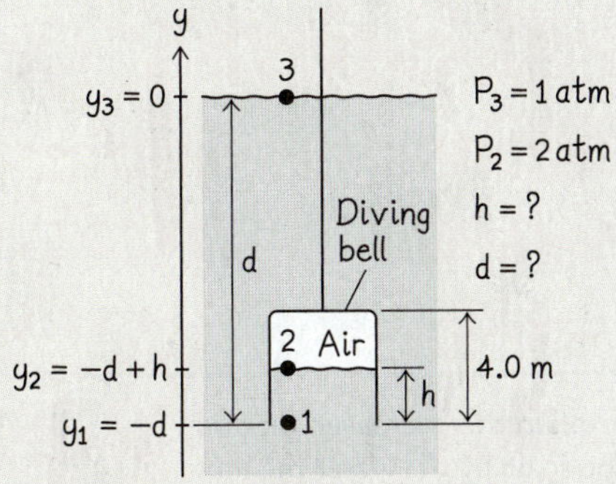


**Simplify and diagram** Assume that Boyle's law ($PV$ is constant for a constant temperature gas) applies to the air inside the bell and use it to relate the state of the air inside the bell before it is submerged to its state after being submerged. This will let us find the ratio of its volumes before and after, and from that we can determine the height of the water inside the bell. Once we have that result, we can use Pascal's second law to determine how far the bottom of the bell is under the water.

**Represent mathematically** Use Boyle's law for the process of submerging the bell, where the initial state is just before the bell starts to be submerged in the water and the final state is when submerged to some unknown depth. We have:

$$P_i V_i = P_f V_f$$

Next, use Pascal's second law to determine the depth of the diving bell. The water surface $y_3 = 0$ will be the origin, with the positive direction pointing upward. The bottom of the bell will be at $y_1 = -d$. Compare the pressure at the ocean surface $P_3$ to the pressure at the water surface inside the bell $P_2$.

$$P_2 = P_3 + \rho(y_3 - y_2)g$$

**Solve and evaluate** The air volume inside the diving bell is half of what it was before entering the water.

$$V_f = \frac{P_i}{P_f}V_i = \frac{1.0\text{ atm}}{2.0\text{ atm}}V_i = \frac{1}{2}V_i$$

Thus, the bell is half full of water, which means the height of the water level inside is

$$h = \frac{1}{2}(4.0\text{ m}) = 2.0\text{ m}$$

*(continued)*

Rearrange Pascal's second law to solve for the position $y_2 = -d + h$ of the water level inside the bell:

$$y_2 = y_3 - \frac{P_2 - P_3}{\rho g}$$
$$= 0 - \frac{(2.02 \times 10^5 \text{ Pa} - 1.01 \times 10^5 \text{ Pa})}{(1027 \text{ kg/m}^3)(9.8 \text{ N/kg})}$$
$$= 0 - 10.0 \text{ m}$$

But $y_2 = -d + h = -d + 2.0$ m. Set these two expressions for $y_2$ equal to each other:

$$y_2 = -d + 2.0 \text{ m} = 0 - 10.0 \text{ m}$$

Thus, $d = 12.0$ m. The position of the bottom of the bell is 12.0 m below the ocean's surface.

**Try it yourself:** Suppose the air pressure in the bell is 3.0 atm. Now how high is water in the bell and how deep is the bottom of the bell?

*Answer:* The water is 2.7 m high in the bell (1.3 m of air at top) and the bottom is 23 m below the water surface.

**Review Question 10.4** What does it mean if atmospheric pressure is 760 mm of mercury?

## 10.5 Buoyant force

**Active Learning Guide ›**

Pascal's first law tells us that pressure changes in one part of a fluid result in pressure changes in other parts. Pascal's second law describes how the pressure in a fluid varies depending on the depth in the fluid. Do these laws explain why some objects float and others sink? Consider Observational Experiment **Table 10.4.**

**OBSERVATIONAL EXPERIMENT TABLE**

**10.4 Effect of depth of submersion on a steel block suspended in water.**

VIDEO 10.4

| Observational experiment | Analysis |
|---|---|
| **Experiment 1.** Hang a 1.0-kg block from a spring scale. The force that the scale exerts on the block balances the downward force that Earth exerts on the block ($mg = (1.0 \text{ kg})(9.8 \text{ N/kg}) = 9.8$ N). 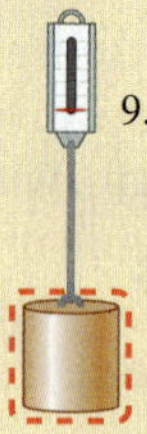  | Force diagram for the block B: 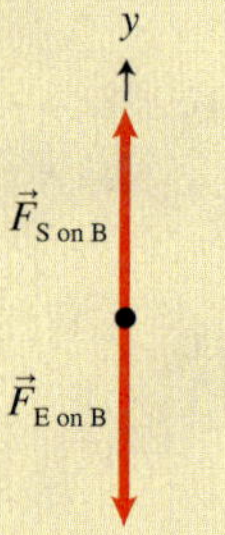  |
| **Experiment 2.** Lower the block into a container of water, so it is partially submerged. The water level rises. The reading of the scale decreases. 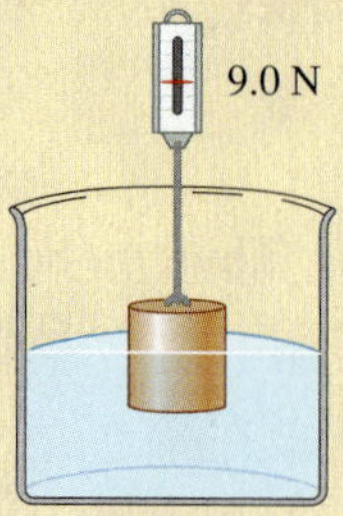  | We explain the decreased reading on the scale by the water pushing upward a little on the block. 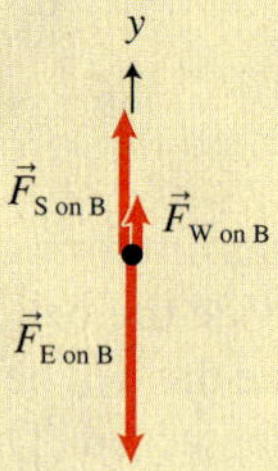  |

<table>
<tr><th>Observational experiment</th><th>Analysis</th></tr>
<tr><td>Experiment 3. Lower the same block into the container of water only to the point where the block is completely submerged. As the water level rises, the reading of the scale decreases.<br>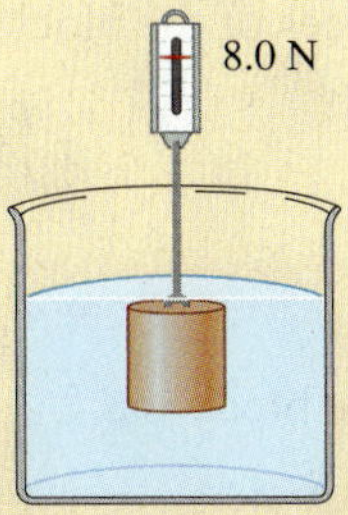<br></td><td>The upward force exerted by the water increases.<br>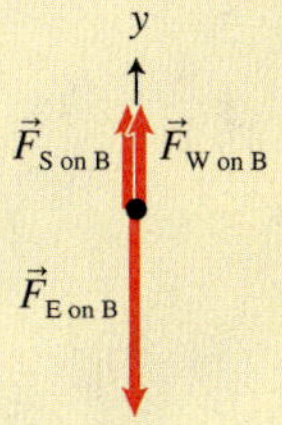<br></td></tr>
<tr><td>Experiment 4. Lower the block into the container of water so that the block is completely submerged near the bottom. The water level and the reading of the scale do not change.<br>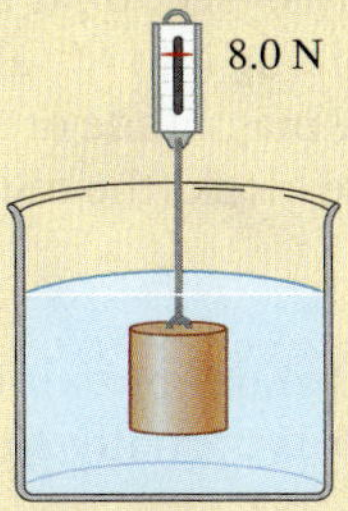<br></td><td>The upward force exerted by the water does not change once it is completely submerged.<br>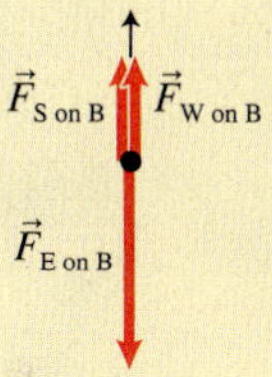<br></td></tr>
<tr><th colspan="2">Patterns</th></tr>
<tr><td colspan="2">We notice two effects:<br>1. The level of the water in the container rises as more of the block is submerged in the water.<br>2. The scale reading decreases as more of the block is submerged. The water exerts an upward force on the block. The magnitude of this force depends on how much of the block is submerged. After it is totally submerged, the force does not change, even though the depth of submersion changes.</td></tr>
</table>

In Table 10.4, after the block is completely submerged, the scale reads 8.0 N instead of 9.8 N. Evidently, the water exerts a 1.8-N upward force on the block. What is the mechanism responsible for this force?

## The magnitude of the force of fluid on a submerged object

Consider *only* the fluid forces exerted on the block shown in **Figure 10.13a**. The fluid pushes inward on the block from all sides, including the top and the bottom. The forces exerted by the fluid on the vertical sides of the block cancel, since the pressure at a specific depth is the same magnitude in all directions.

What about the fluid pushing down on the top and up on the bottom of the block? The pressure is greater at elevation $y_1$ at the bottom of the block than at elevation $y_2$ at the top surface of the block. Consequently, the force exerted by the fluid pushing up on the bottom is greater than the force exerted by the fluid pushing down on the top of the block. Arrows in Figure 10.13b represent the forces that the fluid exerts on the top and bottom of the block. The vector sum of these two fluid forces always points up and is called a buoyant force $\vec{F}_{\text{F on O}}$ (fluid on object).

To calculate the magnitude of the upward buoyant force $F_{\text{F on O}}$ exerted by the fluid on the block, we use Eq. (10.3) to determine the upward pressure $P_1$ of the fluid on the bottom surface of the block compared to the downward pressure $P_2$ of the fluid on the top surface (see Figure 10.13):

$$P_1 = P_2 + \rho_{\text{fluid}}(y_2 - y_1)g$$

**Figure 10.13** A fluid exerts an upward buoyant force on the block.

(a)

The force $\vec{F}_{\text{FB on B}}$ exerted by the fluid below on the block is greater than the force $\vec{F}_{\text{FA on B}}$ exerted by the fluid above.

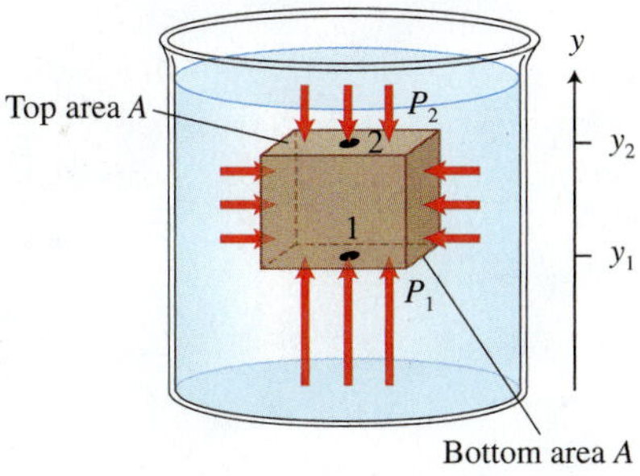


(b)

The upward force of the fluid on the bottom surface is greater than the downward force of the fluid on the top surface.

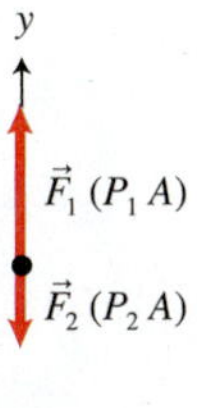

**Active Learning Guide ›**

The magnitudes of the forces exerted by the fluid on the top and on the bottom of the object are the products of the pressure $P$ and the area $A$ of the top and bottom surfaces of the block:

$$P_1A = P_2A + \rho_{\text{fluid}}(y_2 - y_1)Ag$$

or

$$F_1 = F_2 + \rho_{\text{fluid}}(y_2 - y_1)Ag$$

The volume of the block is

$$V = A(y_2 - y_1)$$

where $A$ is the cross-sectional area of the block and $(y_2 - y_1)$ is its height. Substitute this volume into the above force equation and rearrange it to get an expression for the magnitude of the total upward **buoyant force** $F_{\text{F on O}}$ that the fluid exerts on the block (the object of interest):

$$F_{\text{F on O}} = F_1 - F_2 = \rho_{\text{fluid}}\,gV$$

Note that for a totally submerged block, $V$ is the volume of the block. However, when it is partially submerged, $V$ is the volume of the submerged part.

We can now understand the results of the experiments in Table 10.4. When submerging the block further into the water, the scale reading decreased because the buoyant force was increasing. The scale reading stopped changing after the block was completely under the water. Once completely underwater, the submerged volume did not change; the upward buoyant force exerted by the fluid on the block remained constant. This equation is known as Archimedes' principle.

**Active Learning Guide ›**

**Archimedes' principle—the buoyant force** A stationary fluid exerts an upward buoyant force on an object that is totally or partially submerged in the fluid. The magnitude of the force is the product of the fluid density $\rho_{\text{fluid}}$, the volume $V_{\text{fluid}}$ of the fluid that is displaced by the object, and the gravitational constant $g$:

$$F_{\text{F on O}} = \rho_{\text{fluid}} V_{\text{fluid}}\, g \quad (10.4)$$

**TIP** (1) If an object is completely submerged in the fluid (as in the case of the block), the volume used in Eq. (10.4) is just the volume of the object. However, if the object floats, the volume in the equation then equals the volume of space taken up by the object below the fluid's surface. (2) The derivation of Eq. (10.4) was for a solid cube, but the result applies to objects of any shape, though calculus is needed to establish that.

**Review Question 10.5** Why does a fluid exert an upward force on an object submerged in it?

## 10.6 Skills for analyzing static fluid processes

In this section we adapt our problem-solving strategy to analyze processes involving static fluids. A general strategy is described on the left side of the table in Example 10.6 and illustrated on the right side for that specific problem.

## PROBLEM-SOLVING STRATEGY Analyzing Static Fluid Processes

### EXAMPLE 10.6 Buoyant force exerted by air on a human

Suppose your mass is 70.0 kg and your density is $970\ \text{kg/m}^3$. If you could stand on a scale in a vacuum chamber on Earth's surface, the reading of the scale would be $mg = (70.0\ \text{kg})(9.80\ \text{N/kg}) = 686\ \text{N}$. What will the scale read when you are completely submerged in air of density $1.29\ \text{kg/m}^3$?

### Sketch and translate

- Make a labeled sketch of the situation and choose the system of interest.
- Include all known information in the sketch and indicate the unknown(s) you wish to determine.

You are the system object.

The scale reads 686 N when in a vacuum. Your density is $970\ \text{kg/m}^3$ and the density of air is $1.29\ \text{kg/m}^3$. What does it read when you are submerged in air?

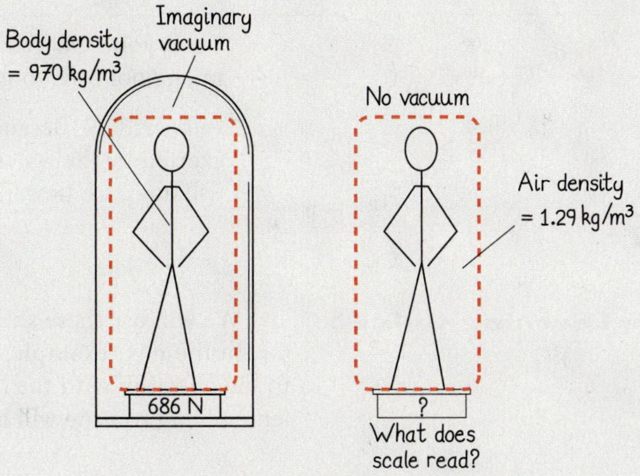


### Simplify and diagram

- Indicate any assumptions you are making.
- Identify objects outside the system that interact with the system object(s).
- Construct a force diagram for the system, including a vertical coordinate axis. The buoyant force is just one of the forces included in the diagram.

Assume that the air density is uniform.

Three objects exert forces on you. Earth exerts a downward gravitational force $F_{\text{E on Y}} = mg = 686\ \text{N}$. The air exerts an upward buoyant force $F_{\text{A on Y}} = \rho_{\text{air}}\, g V_{\text{you}}$. The scale exerts an unknown upward normal force of magnitude $N_{\text{S on Y}}$.

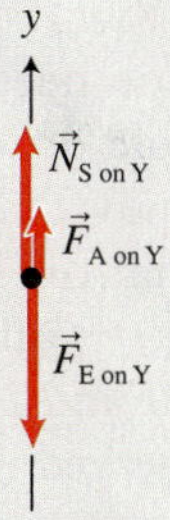


### Represent mathematically

- Use the force diagram to help apply Newton's second law in component form.
- Use the expression for the buoyant force and the definitions of pressure and density if needed.

The $y$-component form of Newton's second law for your body with zero acceleration is (assuming the upward direction as positive)

$$0 = N_{\text{S on Y}} + F_{\text{A on Y}} + (-F_{\text{E on Y}})$$

or

$$N_{\text{S on Y}} = +F_{\text{E on Y}} - F_{\text{A on Y}}$$

The buoyant force that the air exerts on your body has magnitude $F_{\text{A on Y}} = \rho_{\text{air}} V_{\text{you}} g$. The volume of your body is $V_{\text{you}} = (m/\rho_{\text{body}})$. The magnitude of the buoyant force that the air exerts on you is

$$F_{\text{A on Y}} = \rho_{\text{air}}\left(\frac{m}{\rho_{\text{body}}}\right)g = mg\left(\frac{\rho_{\text{air}}}{\rho_{\text{body}}}\right)$$

*(continued)*

Thus the reading of the scale should be:

$$N_{\text{S on Y}} = mg - mg\left(\frac{\rho_{\text{air}}}{\rho_{\text{body}}}\right)$$

**Solve and evaluate**

- Insert the known information and solve for the desired unknown.
- Evaluate the final result in terms of units, reasonable magnitude, and whether the answer makes sense in limiting cases.

$$N_{\text{S on Y}} = +686\text{ N} - (686\text{ N})\frac{1.29\text{ kg/m}^3}{970\text{ kg/m}^3} = 685\text{ N}$$

According to Newton's third law, the force that you exert on the scale $N_{\text{Y on S}}$ is equal in magnitude to the force the scale exerts on you.

The reading of the scale is actually 0.1% less when you step on the scale in air—not a big deal. We can usually neglect air's buoyant force. Notice that the atmospheric air pushes up on objects, and not down.

**Try it yourself:** What will the scale read if you weigh yourself in a swimming pool with your body completely submerged?

*Answer:* 0 N. Because you are less dense than water, the buoyant force exerted by the water on you will completely support you and the scale will not push upward on you at all.

**Active Learning Guide ›** We will use these strategies to analyze several more situations in the chapter. In the next example, it might not be obvious how to arrive at the answer to the question with the information provided. Following the suggested problem-solving routine will help you arrive at a solution.

**EXAMPLE 10.7 Is the crown made of gold?**

You need to determine if a crown is made from pure gold or some less valuable metal. From Table 10.1 you know that the density of gold is 19,300 kg/m$^3$. You find that the force that a string attached to a spring scale exerts on the crown is 25.0 N when the crown hangs in air and 22.6 N when the crown hangs completely submerged in water.

**Sketch and translate** We draw a sketch of the situation and label the givens. If you could measure the mass $m_C$ of the crown and its volume $V_C$, you could calculate the density $\rho_C = m_C/V_C$ of the crown—it should be 19,300 kg/m$^3$. You can determine the mass of the crown easily from the measurement of the scale when the crown hangs in air. But how can you determine the volume from the given information? Crowns have irregular shapes, and it would be difficult to determine its volume by simple measurements and calculations.

**Simplify and diagram** Let's just follow the recommended strategy and see what happens. First, we draw force diagrams for the crown hanging in air and again when hanging in water. When the crown is in air, the upward force exerted by the string attached to the spring scale balances the downward force exerted by Earth. We ignore the buoyant force that air exerts on the crown when hanging in air, since it will be very small in magnitude compared with the other forces exerted on the crown. When the crown is in water, the upward force exerted by the string (the force measured by the scale) and the upward buoyant force that the water exerts on the crown combine to balance the downward gravitational force that Earth exerts on the crown.

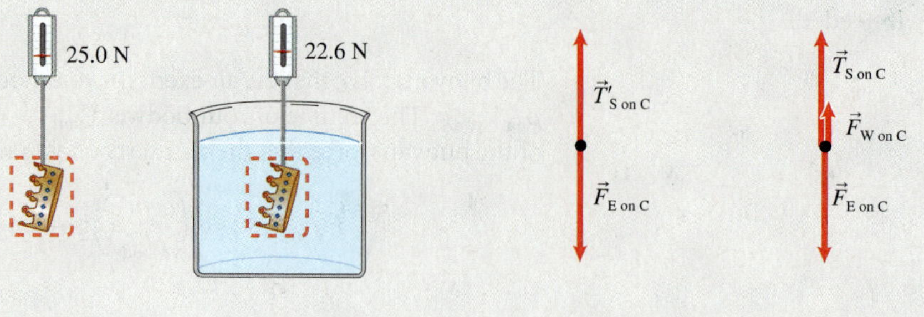

**Represent mathematically** Since the crown is in equilibrium, the forces exerted on it must add to zero in both cases. When the crown is hanging in air, the vertical component form of Newton's second law is

$$0 = \Sigma F_y = T'_{\text{S on C}} + (-F_{\text{E on C}})$$

where $T'_{\text{S on C}}$ is the 25.0-N string tension force exerted on the crown when it is suspended in air. Thus,

$$F_{\text{E on C}} = m_{\text{C}}g = 25.0\ \text{N}$$

or

$$m_{\text{C}} = \frac{25.0\ \text{N}}{9.8\ \text{N/kg}} = 2.55\ \text{kg}$$

The vertical component form of Newton's second law when the crown hangs in water becomes (the upward direction is positive):

$$\Sigma F_y = T_{\text{S on C}} + F_{\text{W on C}} + (-F_{\text{E on C}}) = 0$$

where $T_{\text{S on C}} = 22.6\ \text{N}$ is the magnitude of the string tension force, and the buoyant force that the water exerts on the crown is $F_{\text{W on C}} = \rho_{\text{W}} V_{\text{C}} g$. Substituting in the above, we get

$$T_{\text{S on C}} + \rho_{\text{W}} V_{\text{C}} g - m_{\text{C}} g = 0$$

**Solve and evaluate** We see now that the last equation can be used to determine the volume of the crown:

$$V_{\text{C}} = \frac{m_{\text{C}}g - T_{\text{S on C}}}{\rho_{\text{W}}\, g}$$

$$= \frac{25.0\ \text{N} - 22.6\ \text{N}}{(1000\ \text{kg/m}^3)(9.8\ \text{N/kg})} = 0.000245\ \text{m}^3$$

We now know the crown mass and volume and can calculate its density:

$$\rho = \frac{m}{V} = \frac{2.55\ \text{kg}}{0.000245\ \text{m}^3} = 10{,}400\ \text{kg/m}^3$$

Oops! Since $10{,}400\ \text{kg/m}^3$ is much less than the $19{,}300\ \text{kg/m}^3$ density of gold, the crown is not made of pure gold. The goldsmith must have combined the gold with some less expensive metal.

**Try it yourself:** What is the density of the crown if the scale reads 0 when submerged in water?

*Answer:* $1000\ \text{kg/m}^3$.

**Review Question 10.6** Two objects have the same volume, but one is heavier than the other. When they are completely submerged in oil, on which one does the oil exert a greater buoyant force?

‹ Active Learning Guide

## 10.7 Buoyancy: Putting it all together

As we learned in Section 10.1, whether an object floats or sinks depends on its density relative to the density of the fluid. The reason for this lies in the interactions of the object with the fluid and Earth. Specifically, the magnitude of the buoyant force is $F_{\text{F on O}} = \rho_{\text{fluid}} V_{\text{submerged part}}\, g$, and the magnitude of the force exerted by Earth is $F_{\text{E on O}} = \rho_{\text{object}} V_{\text{object}}\, g$. These forces are exerted in the opposite directions. The relative magnitudes of the forces and consequently the relative magnitudes of the densities of the fluid and the object determine what happens to the object when placed in the fluid.

- If the object's density is less than that of the fluid $\rho_{\text{object}} < \rho_{\text{fluid}}$, then $\rho_{\text{object}} V_{\text{object}}\, g < \rho_{\text{fluid}} V_{\text{object}}\, g$; the object floats *partially* submerged since the buoyant force can balance the gravitational force with less than the entire object below the surface of the fluid.
- If the densities are the same $\rho_{\text{object}} = \rho_{\text{fluid}}$, then $\rho_{\text{object}} V_{\text{object}}\, g = \rho_{\text{fluid}} V_{\text{object}}\, g$; the sum of the forces exerted on the object is zero and it remains wherever it is placed totally submerged at any depth in the fluid.
- If the object is denser than the fluid $\rho_{\text{object}} > \rho_{\text{fluid}}$, then $\rho_{\text{object}} V_{\text{object}}\, g > \rho_{\text{fluid}} V_{\text{object}}\, g$; the magnitude of the gravitational force is always greater than the magnitude of the buoyant force. The object sinks at increasing speed until it reaches the bottom of the container.

These cases show that by changing the density of an object relative to the density of the fluid, the object can be made to float or sink in the same fluid. In this section we investigate this phenomenon and its many practical applications.

## How do submarines manage to sink and then rise in the water?

A submarine's density increases when water fills its compartments. With enough water in the compartments, the submarine's density is greater than that of the water outside, and it sinks. When the water is pumped out, leaving behind empty compartments, the submarine's density decreases. With enough air in the compartments, its density is less than that of the outside water, and the submarine rises toward the surface.

## Building a stable ship

For years ships were made of wood. In the middle of the 17th century, people decided to try building metal ships. Many thought that this idea was absurd: Iron is denser than water and an iron boat would certainly sink. In 1787 British engineer John Wilkinson succeeded in building the first iron ship that did not sink. Since the middle of the 19th century, large ships have been made primarily of steel, which is less dense that iron but much denser than water. These ships can float because part of the volume of a ship is filled with air, which reduces the average density of the ship to a density lower than that of water.

### EXAMPLE 10.8 Should we take this trip?

An empty life raft of cross-sectional area 2.0 m × 3.0 m has its top edge 0.36 m above the waterline. How many 75-kg passengers can the raft hold before water starts to flow over its edges? The raft is in seawater of density 1025 kg/m$^3$.

**Sketch and translate** We make a sketch of the unloaded raft. As people get on the raft, it sinks deeper into the water, and the upward buoyant force increases until the raft reaches a maximum submerged volume, when the maximum number of people are on board. The maximum submerged volume is $V_{\text{submerged}} = 2.0\text{ m} \times 3.0\text{ m} \times 0.36\text{ m} = 2.16\text{ m}^3$. We need to determine the maximum buoyant force the seawater can exert on the raft and then decide how to convert this into the number of passengers the raft can hold. The raft and the passengers are our system of interest.

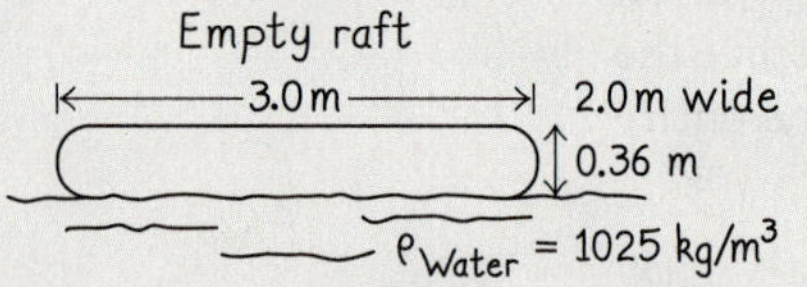


Empty raft

**Simplify and diagram** We have no information about the mass of the raft; thus we will assume it is negligible. We draw a sketch of the filled raft and a force diagram for the raft with passengers (the system). The vertical axis points up. There are two forces exerted on the system: the upward force exerted by the water $\vec{F}_{\text{W on S}}$ of magnitude $\rho_{\text{water}} g V_{\text{submerged}}$ and the downward force exerted by Earth $\vec{F}_{\text{E on S}}$. The magnitude of the force Earth exerts on $N$ people is $N_{\text{people}} m_{\text{person}} g$. As the system is in equilibrium, the net force exerted on it is zero.

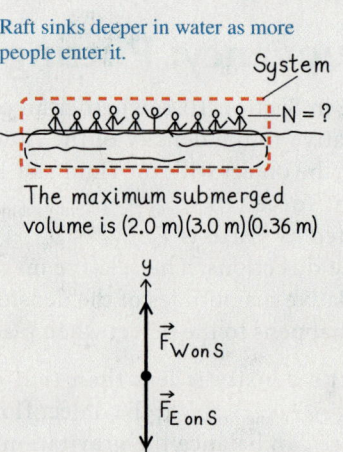

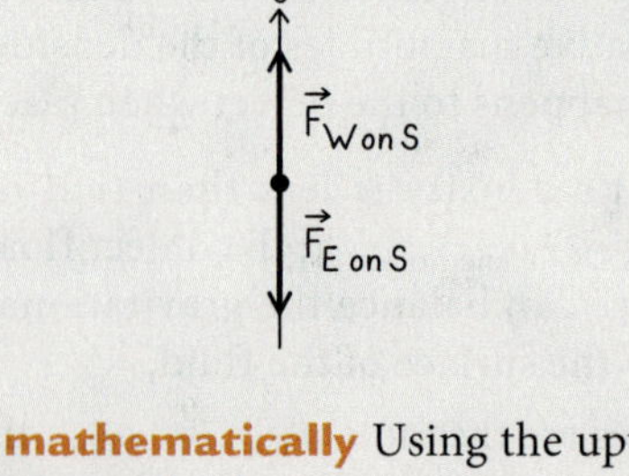


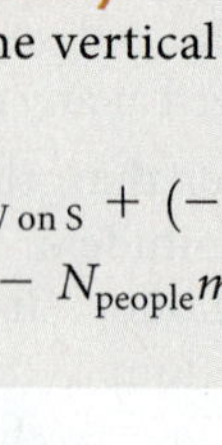


Raft sinks deeper in water as more people enter it.

**Represent mathematically** Using the upward direction as positive, apply the vertical component form of Newton's second law:

$$\Sigma F_y = F_{\text{W on S}} + (-F_{\text{E on S}}) = 0$$

$$\rho_{\text{water}} g V_{\text{submerged}} - N_{\text{people}} m_{\text{person}}\, g = 0$$

Assuming that all people have the same mass, we find the number of people:

$$N_{\text{people}} = \frac{\rho_{\text{water}} V_{\text{submerged}} g}{m_{\text{person}} g} = \frac{\rho_{\text{water}} V_{\text{submerged}}}{m_{\text{person}}}$$

**Solve and evaluate**

$$N_{\text{people}} = \frac{\rho_{\text{water}} V_{\text{submerged}}}{m_{\text{person}}}$$

$$= \frac{(1025\ \text{kg/m}^3)(0.36\ \text{m} \times 2.0\ \text{m} \times 3.0\ \text{m})}{75\ \text{kg}}$$

$$= 29.5$$

The raft can precariously hold 29 passengers, which is a reasonable number. The number is inversely proportional to the mass of a person. This makes sense—the heavier the people, the fewer of them the raft should hold. The units, dimensionless, also make sense. We assumed that the raft has negligible mass. If we take the mass into account, the number of people will be smaller.

**Try it yourself:** Suppose that 10 people of average mass 80 kg entered the raft. Now how far would the water line be from the top of the raft?

*Answer:* The raft would sink 0.13 m into the water, and the water line would be 0.23 m below the top edge of the raft.

Making a ship or a raft float is only part of the challenge of building watercraft. Another problem is to maintain stable equilibrium for the ship, allowing it to right itself if it tilts to one side due to wind or rough seas. Refresh your knowledge of stable equilibrium (Section 7.6) before you read on.

Consider a floating bottle partially filled with sand. Earth exerts a gravitational force at the center of mass of the bottle (**Figure 10.14a**). The buoyant force exerted by the water on the bottle is effectively exerted at the geometrical center of the part of the bottle that is underwater, which equals the center of mass of the displaced water. If this point is above the center of mass of the bottle, then any slight tipping causes these forces to produce a torque that attempts to return the bottle to an upright position (see Figure 10.14b). However, if the geometrical center of the part of the bottle that is underwater is above the center of mass of the bottle, slight tipping causes the gravitational force to produce a torque that enhances the tipping—unstable equilibrium (Figure 10.14c).

So to build a good ship, make the average density of the ship less than the density of water, and make the ship's center of mass lower than the center of mass of the fluid the ship displaces. This is why ships have their cargo stored at the bottom.

**‹ Active Learning Guide**

**Figure 10.14** Making a bottle float with stable equilibrium.

(a)

Bottle partially filled with sand floats upright if the center of mass is below the geometrical center of bottle.

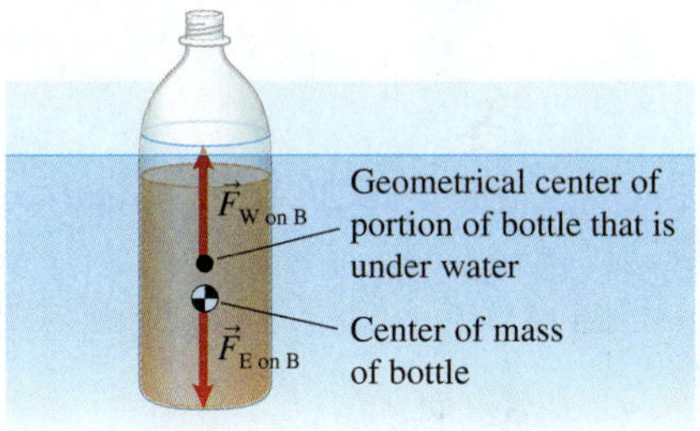


(b)

If the bottle is tipped, torques due to forces exerted on the bottle return it to the upright position.

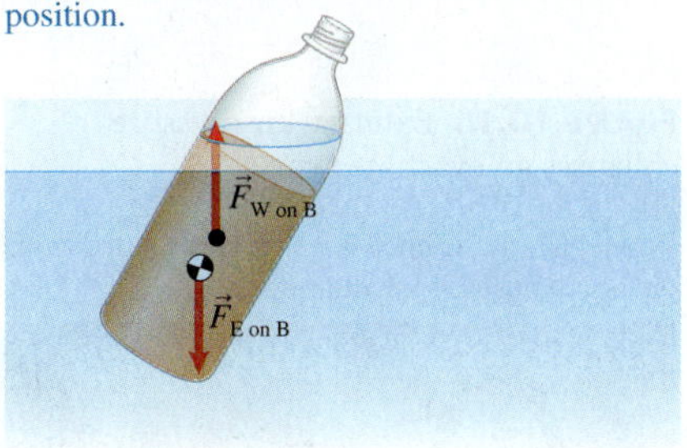


(c)

If this bottle tips slightly, torques due to forces exerted on the bottle cause it to overturn.

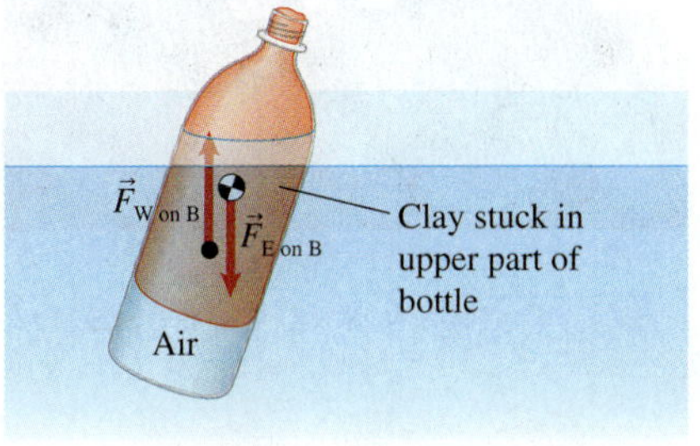


## Ballooning

Balloons used for transportation are filled with hot air. Why hot air? The density of 100 °C hot air is 0.73 times the density of 0 °C air. Thus, balloonists can adjust the average density of the balloon (the balloon's material, people, equipment, etc.) to match the density of air so that the balloon can float at any location in the atmosphere (up to certain limits). A burner under the opening of the balloon regulates the temperature of the air inside the balloon and hence its volume and density. This allows control over the buoyant force that the outside cold air exerts on the balloon. The same approach that we used in Example 10.8 allows us to predict that a balloon with radius of 5.0 m and a mass of 20.0 kg filled with hot air at the temperature of 100 °C can carry about 160 kg (three slim 53-kg people or two medium mass 80-kg people). This is not a heavy load.

At one time, hydrogen was used in closed balloons instead of air. The density of hydrogen is 1/14 times the density of air. Unfortunately, hydrogen can burn explosively in the presence of oxygen. The hydrogen-filled Hindenburg, a German

**Table 10.5 The pressure of air and the pressure due to the oxygen in the air (called partial pressure) at different elevations.**

| Location | Elevation (m) | $P_{air}$ (atm) | $P_{oxygen}$ (atm) |
|---|---|---|---|
| Sea level | 0 | 1.0 | 0.21 |
| Mount Washington | 1917 | 0.93 | 0.18 |
| Pikes Peak | 4300 | 0.59 | 0.12 |
| Mount McKinley | 6190 | 0.47 | 0.10 |
| Mount Everest | 8848 | 0.34 | 0.07 |
| Jet travel | 12,000 | 0.23 | 0.05 |

**Figure 10.15** The Hindenburg explosion.

airship, caught fire and exploded in 1937, killing 36 people (see **Figure 10.15**). Balloonists then turned to helium—an inert gas that does not interact readily with other types of atoms.

## Effects of altitude on humans

As we know, the pressure that a liquid exerts on the walls of a container and on any object submerged in it increases with depth. Water pressure is lower near the surface than deep underwater. Similarly, the pressure of atmospheric air decreases as one moves higher and higher in altitude—see **Table 10.5**.

Climbers and balloonists have to guard against altitude sickness, caused by the low pressure and lack of oxygen. Below 3000 m, there is little effect of altitude on performance. Between 3000 m and 4600 m, climbers experience compensated hypoxia—increased heart and breathing rates. Between 4600 m and 6000 m, manifest hypoxia sets in. Heart and breathing rates increase dramatically, and cognitive and sensory function and muscle control decline. Climbers may feel lethargy and euphoria and even experience hallucinations. Between 6000 m and 8000 m, climbers undergo critical hypoxia, characterized by rapid loss of muscular control, loss of consciousness, and possibly death.

These symptoms were exhibited clearly in the attempt on April 15, 1875 by three French balloon pioneers to set an altitude record. They carried bags of oxygen with them, but slowly lost the mental awareness needed to use the bags as their elevation increased. Instruments indicate that the balloon reached a maximum elevation of 8600 m twice. During the second time, two of the balloonists died. The third lost consciousness but survived.

**Figure 10.16** External air pressure collapses an evacuated can.

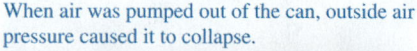

When air was pumped out of the can, outside air pressure caused it to collapse.

## Scuba diving

The sport of scuba diving depends on an understanding of fluid pressure and buoyant force to avoid cases of excess internal pressure, oxygen overload, and decompression sickness.

Air exerts a 50,000-N force (over 5 tons) on the surface of your body. Fortunately, fluids inside the body push outward and balance the force exerted by the outside air. For example, the pressure inside your lungs is approximately atmospheric pressure. What would happen if the fluid pressure on the inside remained constant while the pressure on the outside doubled or tripled? Would you be crushed, the way a can or barrel is crushed by outside air pressure when the air pressure inside the can is much lower than the pressure outside (**Figure 10.16**)? Scuba divers face this problem.

We can determine the pressure $P_d$ at a depth $d$ in the water compared to the atmospheric pressure at the water's surface using Pascal's second law [Eq. (10.3)]. We've defined the zero point to be at the depth $d$:

$$\begin{aligned} P_2 &= P_1 + \rho(y_1 - y_2)g \\ &= P_{\text{surface}} + \rho_{\text{water}}(d - 0)g \\ &= (1 \times 10^5\ \text{Pa}) + (1 \times 10^3\ \text{kg/m}^3)d\,(10\ \text{N/kg}) \\ &= (1 \times 10^5\ \text{Pa}) + (1 \times 10^4\ \text{Pa/m})d \end{aligned}$$

At depth $d = 10$ m, the water pressure will be $2 \times 10^5\ \text{N/m}^2$, or 2 atm. At 40 m below the water surface, the pressure is about $5 \times 10^5\ \text{N/m}^2$, or 5 atm. This would surely be a problem for a scuba diver if the internal pressure were only 1 atm!

To avoid this problem, divers breathe compressed air. While moving slowly downward, a diver adjusts the pressure outlet from the compressed air tank in order to accumulate gas from the cylinder into her lungs and subsequently into other body parts, increasing the internal pressure to balance the increasing external pressure. If a diver returns too quickly to the surface, the great gas pressure in the lungs can force bubbles of gas into the bloodstream. These bubbles can behave like blood clots, blocking blood flow to the brain and possibly causing death. Blood vessels could rupture if the pressure difference between the inside and outside of the vessel is too great. Thus, a diver rises to the surface slowly so that pressure changes gradually and bubbles of gas do not form. This gradual process is called **decompression.**

When humans travel to dangerous environments (mountaintops, the deep sea, or outer space), physics intersects with human physiology. A careful understanding of gases, liquids, and the effects of changing pressures on the human body is needed to allow humans to survive in these places. As we explore the universe, we will need to learn to adapt to ever more challenging environments.

**Review Question 10.7** A ship's waterline marks the maximum safe depth of the ship in the water when it has a full cargo. An empty ship is at the dock with its waterline somewhat above the water level. How could you estimate its maximum cargo?

# Summary

| Words | Pictorial and physical representations | Mathematical representation |
| --- | --- | --- |
| **Pressure *P*** The ratio of the force *F* that a fluid exerts perpendicular to a surface of area *A* divided by that area. The pressure is caused by fluid particles colliding elastically with objects in contact with the fluid. (Section 10.2) |  | $P = \dfrac{F_\perp}{A}$ Eq. (9.1) |
| **Density $\rho$** The ratio of the mass *m* of a substance divided by the volume of that substance. (Section 10.1) | 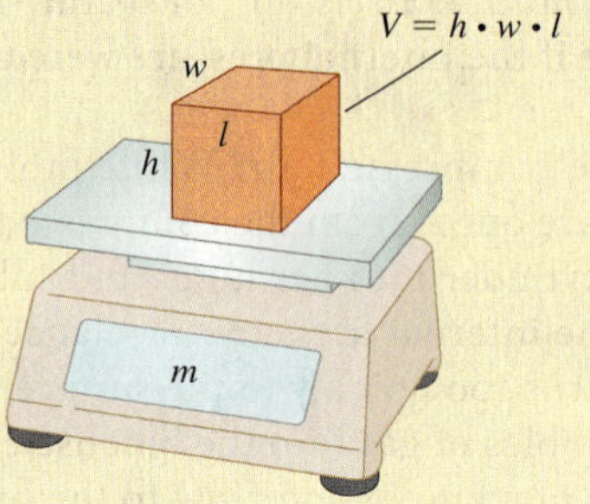 | $\rho = \dfrac{m}{V}$ Eq. (10.1) |
| **Pascal's first law—hydraulic press** An increase in the pressure in one part of an enclosed fluid increases the pressure throughout the fluid. In a hydraulic press, a small force $F_1$ exerted on a small piston of area $A_1$ can cause a large force $F_2$ to be exerted on a large piston of area $A_2$. (Section 10.2) | 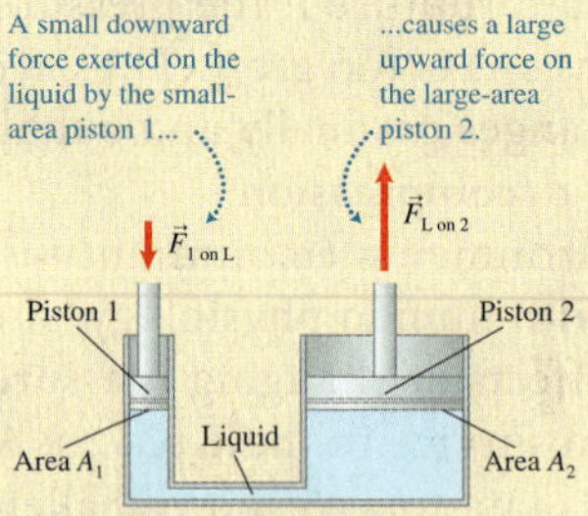 | For a hydraulic press: $F_{\text{L on 2}} = PA_2 = (A_2/A_1)\,F_{1\text{ on L}}$ Eq. (10.2) |
| **Pascal's second law—variation of pressure with depth** On a vertical upward-pointing *y*-axis, the pressure of a fluid $P_2$ at position $y_2$ depends on the pressure $P_1$ at position $y_1$ and on the density of the fluid. (Section 10.3) | 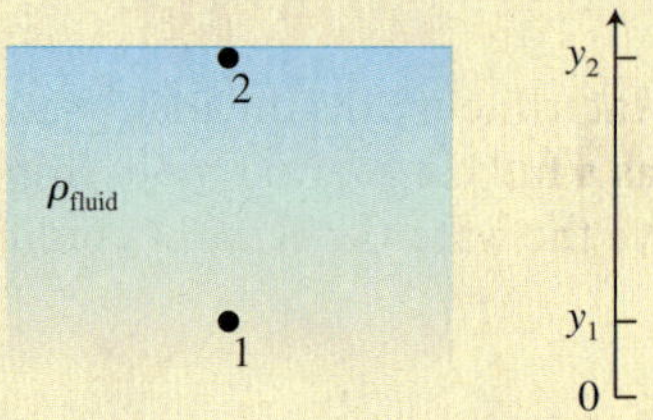 | $P_1 = P_2 + \rho_{\text{fluid}}(y_2 - y_1)g$ Eq. (10.3) |
| **Buoyant force** A fluid exerts an upward-pointing buoyant force on an object totally or partially immersed in the fluid. The force depends on the density of the fluid and on the volume of the fluid displaced. (Section 10.5) | 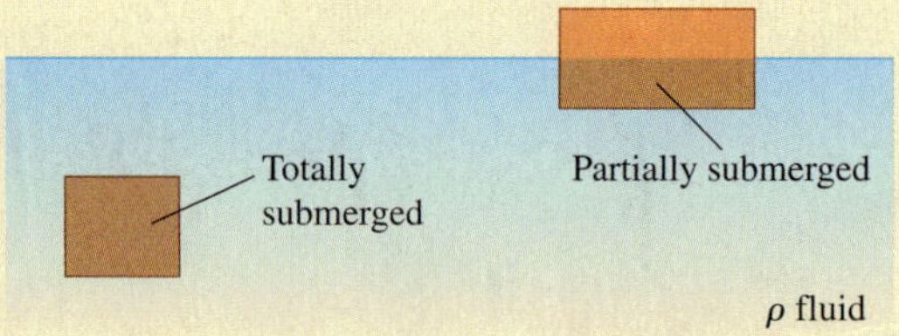 | $F_{\text{F on O}} = \rho_{\text{fluid}} V_{\text{fluid displaced}}$ Eq. (10.4) |
| **Newton's second law** Use the standard problem-solving strategies with this law (sketches, force diagrams, math descriptions) to find some unknown quantity. The problems often involve the buoyant force, pressure, and density. (Section 10.6) | 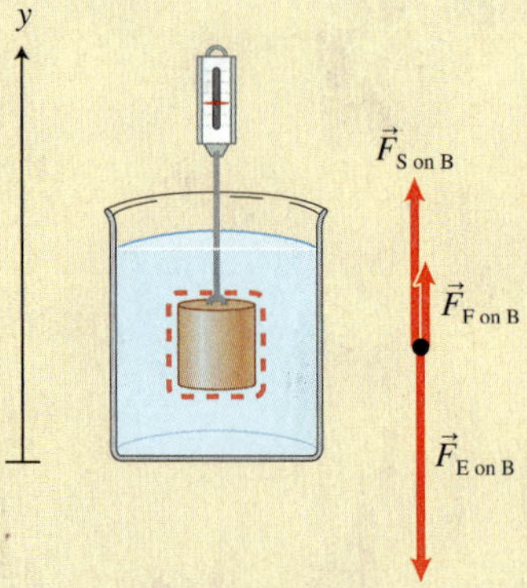 | $a_y = \Sigma \dfrac{F_y}{m}$ Eq. (2.6) |

MP® For instructor-assigned homework, go to MasteringPhysics.

# Questions

## Multiple Choice Questions

1. What does it mean if the pressure exerted by a fluid equals $10 \text{ N/m}^2$?
   (a) The fluid exerts a force of 10 N.
   (b) The surface area of the object is $1 \text{ m}^2$.
   (c) The ratio of the magnitude of the force component perpendicular to the surface area is $10 \text{ N/m}^2$.
   (d) All three choices are correct.
2. Rank in increasing order the pressure that the *italicized objects* exert on the surface.
   I. A person standing with *bare feet* on the floor
   II. A person in *skis* standing on snow
   III. A person in *rollerblades* standing on a road
   IV. A person in *ice skates* standing on ice
   (a) I, II, III, IV (b) IV, III, I, II (c) IV, III, II, I (d) III, II, IV, I (e) II, I, III, IV
3. Choose a device that reduces the pressure caused by a force.
   (a) Scissors (b) Knife (c) Snowshoes (d) Nail (e) Syringe
4. What does it mean if the density of a material equals $2000 \text{ kg/m}^3$?
   (a) The mass of the material is 2000 kg.
   (b) The volume of the material is $1 \text{ m}^3$.
   (c) The ratio of the mass of any amount of this material to the volume is equal to $2000 \text{ kg/m}^3$.
5. Is a material with a density of $10 \text{ kg/m}^3$ more or less dense than a material that causes a pressure of $10 \text{ N/m}^2$?
   (a) More (b) Less (c) The same
   (d) Not enough information to answer
6. If you hold a sheet of paper in your hands by the ends so it is oriented horizontally, what is the net force that the atmosphere exerts on the paper?
   (a) Downward (b) Upward
   (c) Zero—the up and down pressure are equal and cancel
7. If you hold a cylinder vertically, what is the net force exerted by the atmospheric pressure on it?
   (a) Downward (b) Upward (c) Zero
8. How do we know that a fluid exerts an upward force on an object submerged in the fluid?
   (a) Fluid pushes on the object in all directions.
   (b) The reading of a scale supporting the object when submerged in the fluid is less than when not in the fluid.
   (c) The fluid pressure on the bottom of the object is greater than the pressure on the top.
   (d) Both (b) and (c) are correct.
9. When you suspend an object from a spring scale, it reads 15 N. Then you place the same object and scale under a vacuum jar and pump out the air. What happens to the reading of the scale?
   (a) It increases slightly.
   (b) It decreases slightly.
   (c) It says the same.
   (d) Don't have enough information to answer.
10. Why can't a suction pump lift water higher than 10.3 m?
    (a) Because it does not have the strength to pull up higher
    (b) Because the atmospheric pressure is equal to the pressure created by a 10.3-m-high column of water
    (c) Because suction pumps are outdated lifting devices
    (d) Because most suction cups have an opening to the bulb that is too narrow
11. If Torricelli had a wider tube in his mercury barometer, what would the height of the mercury column in the tube do?
    (a) Decrease (b) Increase (c) Stay the same
12. If Torricelli took his mercury barometer to the mountains, what would the height of the mercury column in the tube do?
    (a) Decrease (b) Increase (c) Stay the same
13. Two identical beakers with the same amount of water sit on the arms of an equal arm balance. A wooden block floats in one of them. What does the scale indicate?
    (a) The beaker with the block is heavier.
    (b) The beaker without the block is heavier.
    (c) The scale shows that both beakers weigh the same.
14. What would happen to the level of water in the oceans if all icebergs presently floating in the oceans melted?
    (a) The level would rise.
    (b) The level would stay the same.
    (c) The level would drop.
15. A piece of steel and a bag of feathers are suspended from two spring scales in a vacuum. Each scale reads 100 N. What happens when you repeat the experiment outside under normal conditions?
    (a) The scale with feathers reads more than the scale with steel.
    (b) The scale with feathers reads less that the scale with steel.
    (c) The scales have the same reading, but the reading is less than the reading in a vacuum.
16. A metal boat floats in a pool. What happens to the level of the water in the pool if the boat sinks?
    (a) It rises. (b) It falls. (c) It stays the same.
17. A helium balloon floats in a car with the windows closed. In which direction will the balloon move with respect to the ground when the car accelerates forward from a stop sign?
    (a) Opposite the direction of the car's acceleration
    (b) It will not move.
    (c) In the same direction as the car's acceleration
18. Will a boat float higher or lower in salt water than in fresh water?
    (a) Higher (b) Lower (c) The same
19. Three blocks are floating in oil as shown in **Figure Q10.19**. Which block has a higher density?
    (a) A (b) B (c) C
    (d) All blocks have the same density.
20. Three blocks are floating in oil as shown in Figure Q10.19. On which block does the oil exert a greater buoyant force?
    (a) A (b) B (c) C
    (d) The oil exerts the same force on all of them.

**Figure Q10.19**

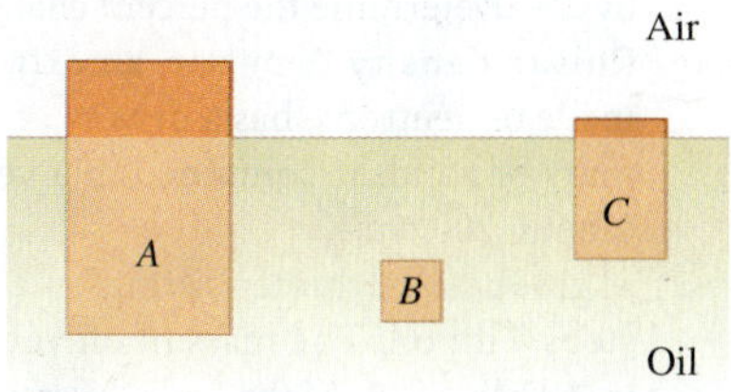

## Conceptual Questions

21. Describe a method to measure the density of a liquid.
22. How can you determine the density of air?
23. Design an experiment to determine whether air has mass.
24. Does air exert an upward force or a downward force on an object submerged in the air? How can you test your answer experimentally?
25. What causes the pressure that air exerts on a surface that is in the air?
26. Why, when you fill a teapot with water, is the water always at the same level in the teapot and in the spout?
27. What experimental evidence supports Pascal's first law?
28. Fill a plastic cup to the very top with water. Put a piece of paper on top of the cup so that the paper covers the cup at the edges and is not much bigger that the surface of the cup. Turn the cup and paper upside down (practice over the sink first) and hold the bottom of the cup (now on the top). Why doesn't the water fall out of the cup?
29. Why does a fluid exert an upward force on an object submerged in the fluid?
30. Describe how you could predict whether an object will float or sink in a particular liquid without putting it into the liquid.
31. Why can you lift objects while in water that are too heavy to lift when in the air?
32. When placed in a lake, an object either floats on the surface or sinks. It does not float at some intermediate location between the surface and the bottom of the lake. However, a weather balloon floats at some intermediate distance between Earth's surface and the top of its atmosphere. Explain.
33. A flat piece of aluminum foil sinks when placed under water. Take the same piece and shape it so that it floats in the water. Explain why the method worked.
34. Ice floats in water in a beaker. Will the level of the water in the beaker change when the ice melts? Explain.
35. The density of ice at 0 °C is less than the density of water at 0 °C. Why is this very important for the existence of life on Earth?
36. How would you determine the density of an irregular-shaped unknown object if (a) it sinks in water and (b) it floats in water? List all the steps and explain the reasoning behind them.
37. Why do people sink in fresh water and in most seawater (if they do not make an effort to stay afloat) but do not sink in the Dead Sea?

# Problems

Below, **BIO** indicates a problem with a biological or medical focus. Problems labeled **EST** ask you to estimate the answer to a quantitative problem rather than develop a specific answer. Problems marked with ✎ require you to make a drawing or graph as part of your solution. Asterisks indicate the level of difficulty of the problem. Problems with no * are considered to be the least difficult. A single * marks moderately difficult problems. Two ** indicate more difficult problems.

### 10.1 Density

1. **BIO Water in human body** About two-thirds of your body mass consists of water. Determine the volume of water in a 70-kg person.
2. * Determine the average density of Earth. What data did you use? What assumptions did you make?
3. * **EST Height of atmosphere** Use data for the normal pressure and the density of air near Earth's surface to estimate the height of the atmosphere, assuming it has *uniform* density. Indicate any additional assumptions you made. Are you on the low or high side of the real number?
4. **EST** A single-level home has a floor area of 200 $m^2$ with ceilings that are 2.6 m high. Estimate the mass of the air in the house.
5. * **BIO** A diet decreases a person's mass by 5%. Exercise creates muscle and reduces fat, thus increasing the person's density by 2%. Determine the percent change in the person's volume.
6. **Pulsar density** A pulsar, an extremely dense rotating star made of neutrons, has a density of $10^{18}$ kg/$m^3$. Determine the mass of a pulsar contained in a volume the size of your fist (about 200 $cm^3$).
7. A graduated cylinder sitting on a platform scale is filled in steps with oil. The mass of oil versus its volume is reported in **Table 10.6**. Make a mass-versus-volume graph for the oil and from the graph determine its density.

**Table 10.6**

| Oil volume ($m^3$) | Oil mass (kg) |
|---|---|
| $5.00 \times 10^{-5}$ | $4.4 \times 10^{-2}$ |
| $10.0 \times 10^{-5}$ | $9.0 \times 10^{-2}$ |
| $15.0 \times 10^{-5}$ | $13.5 \times 10^{-2}$ |
| $20.0 \times 10^{-5}$ | $18.2 \times 10^{-2}$ |
| $25.0 \times 10^{-5}$ | $22.4 \times 10^{-2}$ |

8. * Use the graph lines in **Figure P10.8** to determine the densities of the three liquids in SI units. If you place them in one container, how will they position themselves? How does

**Figure P10.8**

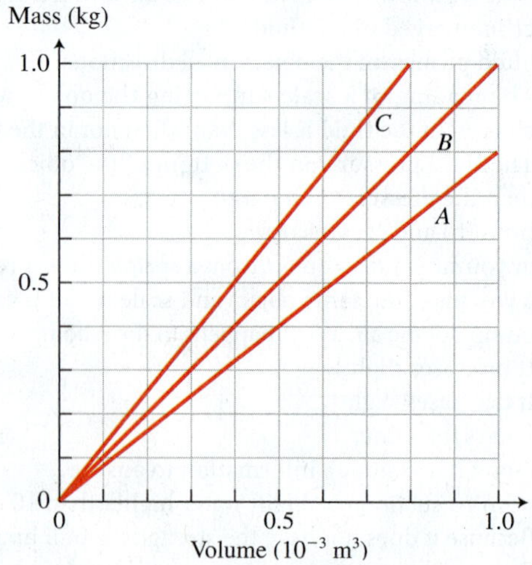

the density of each liquid change as its volume increases? As its mass decreases? Compare the masses of the three liquids when they occupy the same volume. Compare the volumes of the three liquids when they have the same mass.

9. * Imagine that you have gelatin cut into three cubes: the side of cube A is $a$ cm long, the side of cube B is double the side of A, and the side of cube C is three times the side of A. Compare the following properties of the cubes: (a) density, (b) volume, (c) surface area, (d) cross-sectional area, and (e) mass.

10. Determine the density of the material whose mass-versus-volume graph line is shown in **Figure P10.10**. If you double the mass of this substance, what will happen to its density? What substance might this be?

**Figure P10.10**

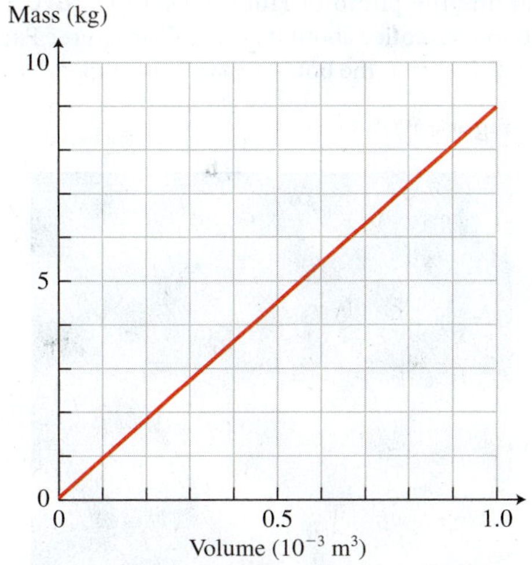


11. An object made of material A has a mass of 90 kg and a volume of 0.45 $m^3$. If you cut the object in half, what would be the density of each half? If you cut the object into three pieces, what would be the density of each piece? What assumptions did you make?

12. You have a steel ball that has a mass of 6.0 kg and a volume of $3.0 \times 10^{-3}$ $m^3$. How can this be?

13. A material is made of molecules of mass $2.0 \times 10^{-26}$ kg. There are $2.3 \times 10^{29}$ of these molecules in a 2.0-$m^3$ volume. What is the density of the material?

14. You compress all the molecules described in Problem 13 into 1.0 $m^3$. Now what is the density of the material? What type of material could possibly behave this way?

15. * Bowling balls are heavy. However, some bowling balls float in water. Use available resources to find the dimensions of a bowling ball and explain why some balls float while others do not.

16. * **EST** Estimate the average density of a glass full of water and then the glass when the water is poured out (do not forget the air that now fills the glass instead of water).

### 10.2 Pressure exerted by a fluid

17. * Anita holds her physics textbook and complains that it is too heavy. Andrew says that her hand should exert no force on the book because the atmosphere pushes up on it and balances the downward pull of Earth on the book (the book's weight). Jim disagrees. He says that the atmosphere presses down on things and that is why they feel heavy. Who is correct? Approximately how large is the force that the atmosphere exerts on the bottom of the book? Why does this force not balance the force exerted by Earth on the book?

18. **EST Force exerted by the air on your state** Estimate the force exerted by Earth's atmosphere on the state where you are taking your physics course.

19. * The air pressure in the tires of a 980-kg car is $3.0 \times 10^5$ N/$m^2$. Determine the average area of contact of each tire with the road.

20. * **EST** Estimate the pressure that you exert on the floor while wearing hiking boots. Now estimate the pressure under each heel if you change into high-heeled shoes. Indicate any assumptions you made.

21. **Hydraulic car lift** You are designing a hydraulic lift for a machine shop. The average mass of a car it needs to lift is about 1500 kg. What should be the specifications on the dimension of the pistons if you wish to exert a force on a smaller piston of not more than 500 N? How far down will you need to push the piston in order to lift the car 30 cm?

22. **Venus pressure and underwater pressure** Atmospheric pressure on Venus is $9.0 \times 10^6$ N/$m^2$. How deep underwater on Earth would you have to go to feel the same pressure?

23. **EST Force of air on forehead** Estimate the force that air exerts on your forehead. Describe the assumptions you made.

24. * A cylindrical iron plunger is held against the ceiling, and the air is pumped from inside it. A 72-kg person hangs by a rope from the plunger (**Figure P10.24**). List the quantities that you can determine about the situation and determine them. Make assumptions if necessary.

**Figure P10.24**

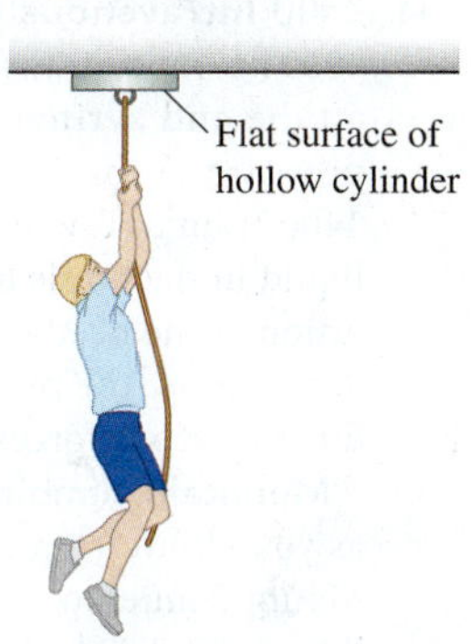


25. You have a rubber pad with a handle attached to it (**Figure P10.25**). If you press the pad firmly on a smooth table, it is impossible to lift it off the table. Why? What force would you need to exert on the handle to lift it? The surface area of the pad is 0.023 $m^2$.

26. * You vacuum up a small piece of paper on the floor. Draw a force diagram for the paper just as it is being lifted up into the vacuum cleaner.

**Figure P10.25**

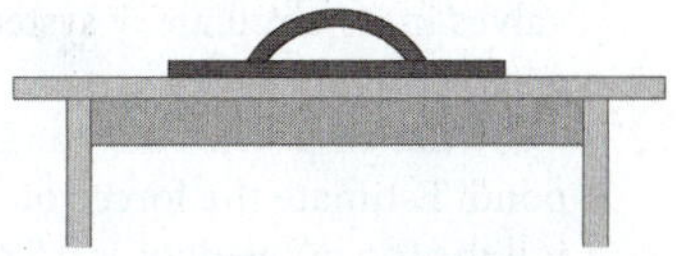

27. **EST Toy bow and arrow** A child's toy arrow has a suction cup on one end. When the arrow hits the wall, it sticks. Draw a force diagram for the arrow stuck on the wall and estimate the magnitudes of the forces exerted on it when it is in equilibrium. The mass of the arrow is about 10 g. Why are the words "suction cup" not appropriate?

### 10.3 Pressure variation with depth

28. **Pressure on the Titanic** The Titanic rests 4 km (2.5 miles) below the surface of the ocean. What physical quantities can you determine using this information?

29. You have three reservoirs (**Figure P10.29**). Rank the pressures at the bottom of each and explain your rankings. Then rank the net force that the water exerts on the bottom of each reservoir. Explain your rankings.

**Figure P10.29**

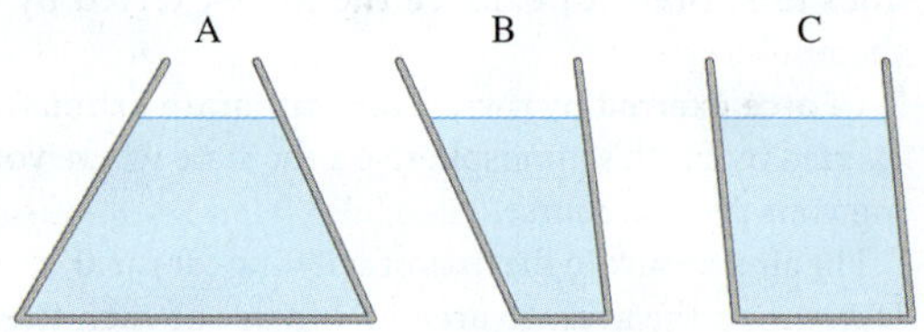


30. * A bucket filled to the top with water has a piece of ice floating in it. (a) Will the pressure on the bottom of the bucket change when the ice melts? Explain. (b)Will the level of water rise when the ice melts? Explain.
31. **Water reservoir and faucet** The pressure at the top of the water in a city's gravity-fed reservoir is $1.0 \times 10^5\ N/m^2$. Determine the pressure at the faucet of a home 42 m below the reservoir.
32. **Dutch boy saves Holland** An old story tells of a Dutch boy who used his fist to plug a 2.0-cm-diameter hole in a dike that was 3.0 m below sea level, thus preventing the flooding of part of Holland. What physical quantities can you determine from this information? Determine them.
33. EST BIO **Blood pressure** Estimate the pressure of the blood in your brain and in your feet when standing, relative to the average pressure of the blood in your heart of $1.3 \times 10^4$-$N/m^2$ above atmospheric pressure.
34. * BIO **Intravenous feeding** A glucose solution of density $1050\ kg/m^3$ is transferred from a collapsible bag through a tube and syringe into the vein of a person's arm. The pressure in the arm exceeds the atmospheric pressure by $1400\ N/m^2$. How high above the arm must the top of the liquid in the bottle be so that the pressure in the glucose solution at the needle exceeds the pressure of the blood in the arm? Ignore the pressure drop across the needle and tubing due to viscous forces.
35. * **Mountain climbing** Determine the change in air pressure as you climb from elevation of 1650 m at the timberline of Mount Rainier to its 4392-m summit, assuming an average air density of $0.82\ kg/m^3$. Will the real change be more or less than the one you calculated? Explain.
36. EST BIO **Giraffe raises head** Estimate the pressure change of the blood in the brain of a giraffe when it lifts its head from the grass to eat a leaf on an overhead tree. Without special valves in its circulatory system, the giraffe could easily faint when lifting its head.
37. * EST **Car in pond** Your car slides off an embankment into a pond. Estimate the force you must exert on the door to open it if the top of the door is 0.5 m below the surface. Describe in detail how you might escape without opening the door.
38. **Drinking through a straw** You are drinking water through a straw in an open glass. Select a small volume of water in the straw as a system and draw a force diagram for the water inside this volume that explains why the water goes up the straw.
39. * **More straw drinking** While you are drinking through the straw, the pressure in your mouth is 30 mm Hg below atmospheric pressure. What is the maximum length of a straw in an open glass that you can use to drink a fruit drink of density $1200\ kg/m^3$?
40. * Your office has a $0.020\ m^3$ cylindrical container of drinking water. The radius of the container is about 14 cm. When the container is full, what is the pressure that the water exerts on the sides of the container halfway down from the top? All the way down?
41. * EST BIO **Eardrum** Estimate the net force on your $0.5\text{-cm}^2$ eardrum that air exerts on the inside and the outside after you drive from Denver, Colorado (elevation 1609 m) to the top of Pikes Peak (elevation 4301 m). Assume that the air pressure inside and out are balanced when you leave Denver and that the average density of the air is $0.80\ kg/m^3$. What other assumptions did you make?
42. BIO **Eardrum again** You now go snorkeling. What is the pressure on your eardrum when you are 2.4 m under the water, assuming the pressure was equalized before the dive?
43. Water and oil are poured into opposite sides of an open U-shaped tube. The oil and water meet at the exact center of the U at the bottom of the tube. If the column of oil of density $900\ kg/m^3$ is 16 cm high on one side, how high is the water on the other side?
44. * Examine the photo of Hoover Dam (**Figure P10.44**). What do you notice about its vertical structure? Explain why a dam is thicker at the bottom than at the top.

**Figure P10.44**

45. * A test tube of length $L$ and cross-sectional area $A$ is submerged in water with the open end down so that the edge of the tube is a distance $h$ below the surface. The water goes up into the tube so its height inside the tube is $l$. Describe how you can use this information to decide whether the air that was initially in the tube obeys Boyle's law. List your assumptions.

**Figure P10.48**

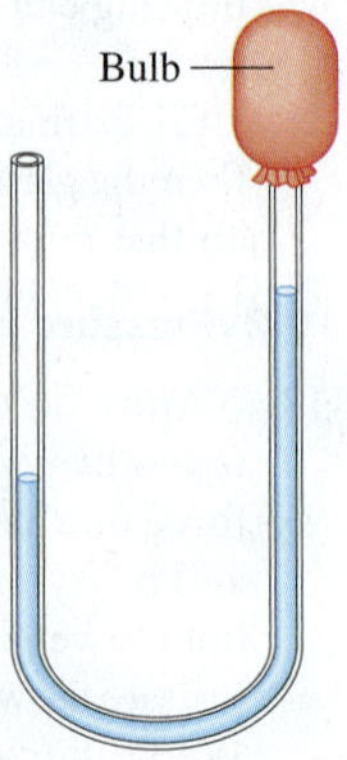


### 10.4 Measuring atmospheric pressure

46. The reading of a barometer in your room is 780 mm Hg. What does this mean? What is the pressure in pascals?
47. How long would Torricelli's barometer have had to be if he had used oil of density $950\ kg/m^3$ instead of mercury?
48. Sometimes gas pressure is measured with a device called a *liquid manometer* (**Figure P10.48**). Explain how this

instrument can be used to measure the pressure of gas in a bulb attached to one of the tubes.

49. You use a liquid manometer with water to measure the pressure inside a rubber bulb. Before you squeeze the bulb, the water is at the same level in both legs of the tube. After you squeeze the bulb, the water in the opposite leg rises 20 cm with respect to the leg connected to the bulb (**Figure P10.49**). What is the pressure in the bulb? What assumptions did you make? How will the answer change if the assumptions are not valid?

**Figure P10.49**

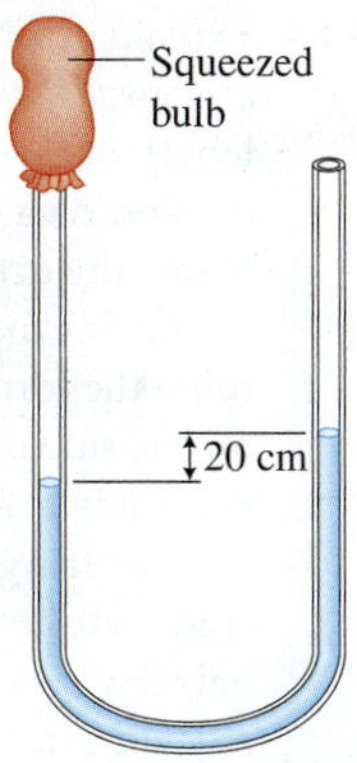


50. * In a mercury-filled manometer (**Figure P10.50**), the open end is inserted into a container of gas and the closed end of the tube is evacuated. The difference in the height of the mercury is 80 mm. The radius of the connecting tube is 0.50 cm. (a) Determine the pressure inside the container in newtons per square meter. (b) An identical manometer has a connecting tube that is twice as wide. If the difference in the height of the mercury is the same, then what is the pressure in the container?

**Figure P10.50**

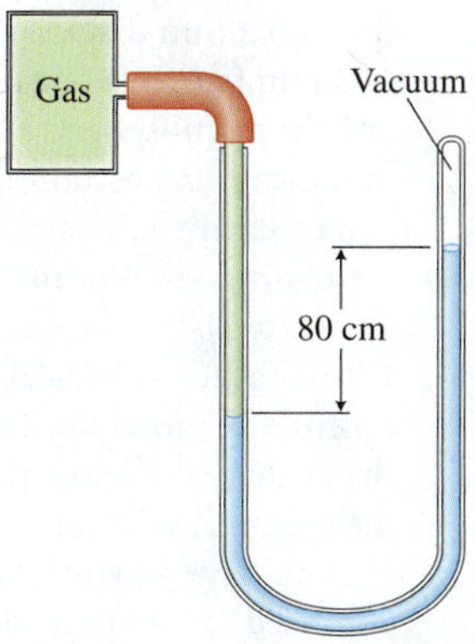


51. A liquid manometer contains two liquids. Liquid A on the left side is $h$ meters higher ($h << 1.0$ m) than the liquid on the right side. What do you know about those two liquids? What assumptions did you make? If the assumptions were not true, how would the answer be different?

52. Examine the reading of the manometer that you use to measure the pressure inside car tires. What are the units? Does the manometer measure the absolute pressure of the air inside the tires or gauge pressure? How do you know?

53. * Marjory thinks that the mass of a liquid above a certain level should affect the pressure at this level. Describe how you will test her idea.

## 10.5 Buoyant force

54. Draw a force diagram for an object that is floating at the surface of a liquid.

55. Draw a cubic object that is completely submerged in a fluid but not resting on the bottom of the container. Then draw arrows to represent the forces exerted by the fluid on the top, sides, and bottom of the object. Make the arrows the correct relative lengths. What is the direction of the total force exerted by the fluid on the object?

56. Draw a force diagram for a helium balloon that you just released. Then draw a force diagram for an air-filled balloon that you just released.

57. You are holding a brick that is completely submerged in water. Draw a force diagram for the brick. Why does it feel lighter in water than when you hold it in the air?

58. * This textbook says that the upward force that a fluid exerts on a submerged object is equal in magnitude to the product of the density of the fluid, the gravitational constant $g$, and the volume of the submerged part of the object. Where did this equation come from?

59. * **Design** This textbook says that the upward force that a fluid exerts on a submerged object is equal in magnitude to the product of the density of the fluid, the gravitational constant $g$, and the volume of the submerged part of the object. Design an experiment to test this expression, including a prediction about the outcome of the experiment.

60. * You have four objects at rest, each of the same volume. Object A is partially submerged, and objects B, C, and D are totally submerged in the same container of liquid, as shown in **Figure P10.60**. Draw a force diagram for each object. Rank the densities of the objects from least to greatest and indicate whether any objects have the same density.

**Figure P10.60**

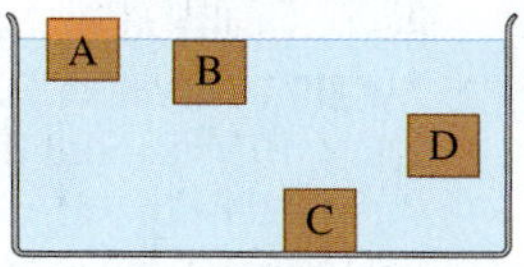


61. * **Does air affect what a scale reads?** A 60-kg woman with a density of 980 kg/m$^3$ stands on a bathroom scale. Determine the reduction of the scale reading due to air.

62. * When analyzing a sample of ore, a geologist finds that it weighs 2.00 N in air and 1.13 N when immersed in water. What is the density of the ore? What assumptions did you make to answer the question? If the assumptions are not correct, how would the answer be different?

63. * A pin through a hole in the middle supports a meter stick. Two identical blocks hang from strings at an equal distance from the center so the stick is balanced. What happens to the stick if one block is submerged in water of density 1000 kg/m$^3$ and the other block in kerosene of density 850 kg/m$^3$?

64. ** A meter stick is supported by a pin through a hole in the middle. (a) Two blocks made of the same material but different sizes hang from strings at different positions in such a way that the stick balances. What happens when the blocks hang entirely submerged in beakers of water? (b) Next you hang two blocks of different masses but the same volume at different positions so the stick balances. What happens when these blocks hang completely submerged in beakers of water? Support your answer for each part using force diagrams with arrows drawn with the correct relative lengths.

## 10.6 Skills for analyzing static fluid processes

65. **Goose on a lake** A 3.6-kg goose floats on a lake with 40% of its body below the 1000-kg/m$^3$ water level. Determine the density of the goose.

66. * **Floating in seawater** A person of density of $\rho_1$ floats in seawater of density $\rho_2$. What fraction of the person's body is submerged? Explain.

67. * **Floating in seawater** A person of density of 980 kg/m$^3$ floats in seawater of density 1025 kg/m$^3$. What can you determine using this information? Determine it.

68. **(a) Determine the force that a vertical string exerts on a 0.80-kg rock of density of 3300 kg/m$^3$ when it is fully submerged in water of density 1000 kg/m$^3$. (b) If the force exerted by the string supporting the rock increases by 12% when the rock is submerged in a different fluid, what is that fluid's density? (c) If the density of another rock of the same volume is 12% greater, what happens to the buoyant force the water exerts on it?

69. * **Snorkeling** A 60-kg snorkeler (including snorkel, mask, and other gear) displaces 0.058 $m^3$ of water when 1.2 m under the surface. Determine the magnitude of the buoyant force exerted by the 1025-$kg/m^3$ seawater on the person. Will the person sink or drift upward?
70. * A helium balloon of volume 0.12 $m^3$ has a total mass (the helium plus the balloon) of 0.12 kg. Determine the buoyant force exerted on the balloon by the air if the air has density 1.13 $kg/m^3$. Determine the initial acceleration of the balloon when released.
71. ** A bucket filled to the top with water has a piece of ice floating in it. Will the pressure on the bottom change when the ice melts? Justify your answer.
72. * BIO **Protein sinks in water** A protein molecule of mass $1.1 \times 10^{-22}$ kg and density $1.3 \times 10^3\ kg/m^3$ is placed in a vertical tube of water of density 1000 $kg/m^3$. (a) Draw a motion diagram and a force diagram at the moment immediately after the molecule is released. (b) Determine the initial acceleration of the protein.
73. * How can you determine if a steel ball of known radius is hollow? List the equipment that you will need for the experiment, and describe the procedure and calculations. Can you determine how big the hollow part is if present in the ball?
74. ** **Crown composition** A crown is made of gold and silver. The scale reads its mass as 3.0 kg when in air and 2.75 kg when in water. Determine the masses of the gold and the silver in the crown. The density of gold is 19,300 $kg/m^3$ and that of silver is 10,500 $kg/m^3$.

### 10.7 Buoyancy: Putting it all together

75. * **Wood raft** Logs of density 600 $kg/m^3$ are used to build a raft. What is the weight of the maximum load that can be supported by a raft built from 300 kg of logs?
76. ** A cylinder has radius $R$. How high should a column of liquid be so that the magnitude of the force averaged over the side wall surface area that the liquid exerts on the wall equals the magnitude of the force that the liquid exerts on the bottom surface of the cylinder? Explain.
77. **Standing on a log** A log is $L$ long and $d$ in diameter. What is the mass of a person who can stand on the log without getting her feet wet?
78. * **Ferryboat** A ferryboat is 12 m long and 8 m wide. Two cars, each of mass 1600 kg, ride on the boat for transport across the lake. How much farther does the boat sink into the water?
79. EST **Iceberg** Estimate the fraction of the volume of an iceberg that is underwater.
80. * **Life preserver** A life preserver is manufactured to support a 70-kg person with 20% of his volume out of the water. If the density of the life preserver is 100 $kg/m^3$ and it is completely submerged, what must its volume be? List your assumptions.

## General Problems

81. * Compare the density of water at 0 °C to the density of ice at 0 °C. Suggest possible explanations in terms of the molecular arrangements inside the liquid and solid forms of water that would account for the difference. If necessary, use extra resources to help answer the question.
82. * **Collapsing star** The radius of a collapsing star destined to become a pulsar decreases by 10% while at the same time 12% of its mass escapes. Determine the percent change in its density.
83. ** EST BIO **Syringe pressure** You are getting a flu shot. Estimate the average pressure of the fluid entering your arm during the shot. Indicate any assumptions you made. Compare this to the pressure of your shoes on the ground when standing.
84. Explain qualitatively and quantitatively how we drink through a straw. Make sure you can account for the water going up the length of the straw.
85. * **Deep dive** The *Trieste* research submarine traveled 10.9 km below the ocean surface while exploring the Mariana Trench in the South Pacific, the deepest place in the ocean. Determine the force needed to prevent a 0.10-m-diameter window on the side of the submarine from imploding. The density of the water is 1025 $kg/m^3$.
86. ** **Bursting a wine barrel** Pascal placed a long 0.20-cm -radius tube in a wine barrel of radius 0.24 m. He sealed the barrel where the tube entered it. When he added wine of density 1050 $kg/m^3$ to the tube so the column of wine was 8.0 m high, the cover of the barrel burst off the top of the barrel. What was the net force that caused the cover to come off?
87. * BIO **Lowest pressure in lungs** Experimentally determine the maximum distance you can suck water up a straw. Use this number to determine the pressure in your lungs above or below atmospheric pressure while you are sucking. Be sure to indicate any assumptions you made and show clearly how you reached your conclusion.
88. * **Landing on Venus** Atmospheric pressure on Venus is $92 \times 10^5\ N/m^2$. Suppose that NASA is planning to land a 1.0-m-radius spherical research vehicle on Venus. (a) Determine the force on each square centimeter of its surface. (b) What is the buoyant force exerted by the atmosphere on the spherical vehicle?
89. ** You have an empty water bottle. Predict how much mass you need to add to it to make it float half-submerged. Then add the calculated mass and explain any discrepancy that you found. How did you make your prediction?
90. ** BIO **Flexible bladder helps fish sink or rise** A 1.0-kg fish of density 1025 $kg/m^3$ is in water of the same density. The fish's bladder contains 10 $cm^3$ of air. The bladder compresses to 4 $cm^3$, reducing its volume by 6 $cm^3$. Now what is the density of the fish? Will it sink or rise? Explain.
91. * **Plane lands on Nimitz aircraft carrier** When a 27,000-kg fighter airplane lands on the deck of the aircraft carrier Nimitz, the carrier sinks 0.25 cm deeper into the water. Determine the cross-sectional area of the carrier.
92. ** To determine the density of an object and an unknown liquid, it is first weighed in air, then in water, and then in an unknown liquid. The readings of the scale are $T_1$, $T_2$, and $T_3$, respectively. Suggest a method of using these data to determine the density of the object and of the liquid. Decide what additional equipment and measurements you would need to make to test whether the results of the first method are correct.
93. ** Two upward-moving balloons carry equal loads. The first balloon has an upward acceleration of $(g/3)$. The second balloon moves up at constant speed. The density of the gas inside both balloons is one-third the density of air. The volume of the first balloon is $V_1$. What is the volume of the second balloon? The masses of the balloons are the same.
94. Derive an equation for determining the unknown density of a liquid by measuring the magnitude of a force $T_{S\ on\ O}$ that a string needs to exert on a hanging object of unknown mass $m$ and density $\rho$ to support it when the object is submerged in the liquid.

## Reading Passage Problems

BIO **Free diving** So-called "no-limits" free divers slide to deep water on a weighted sled that moves from a boat down a vinyl-coated steel cable to the bottom of a dive site. The diver reaches depths where a soda can would implode. After reaching the target depth, the diver releases the sled and an air bag opens and brings the diver quickly back to the surface. The divers have no external oxygen supply—just lungs full of air at the start of the dive. In August 2002, Tanya Streeter of the Cayman Islands held the women's no-limits free dive record at 160 m. In 2005 Patrick Musimu set the men's record with a 209.6-m free dive in the Red Sea just off the Egyptian coast (the record was later broken by Herbert Nitsch of Austria).

Musimu's 2005 dive took 3 minutes 28 seconds. He began the dive with his 9-L lungs full of air. By the time he passed the 200-m mark, Musimu's lungs had contracted to the size of a tennis ball. His body transferred blood from his limbs to essential organs such as the heart, lungs, and brain. This "blood shift" occurs when mammals submerge in water. Blood plasma fills the chest cavity, especially the lungs. Without this adaptation, the lungs would shrink and press against the chest walls, causing permanent damage. When he reached his target, Musimu released the weighted segment of the specialized sled that had taken him down and opened an airbag, which began his return to the surface at an average speed of 3–4 m/s.

95. The pressure of the water when Musimu was 209.6 m below the surface was closest to which of the following?
(a) 2 atm (b) 3 atm (c) 21 atm
(d) 22 atm (e) 200 atm

96. Assuming Musimu weighs 670 N (150 lb) and is 1.6 m tall, 0.30 m wide, and 0.15 m thick, which answer below is closest to the magnitude of the force that the deep water exerted on one side of his body?
(a) 0 (b) 670 N (130 lb)
(c) 15,000 N (3000 lb) (d) $10^5$ N (20,000 lb)
(e) $10^6$ N (200,000 lb)

97. Musimu's training allows him to hold up to 9 L = 9000 cm$^3$ of air when in a 1 atm environment. Which answer below is closest to the volume of that air if at pressure 22 atm?
(a) 100 cm$^3$ (b) 200 cm$^3$ (c) 400 cm$^3$
(d) 9000 cm$^3$ (e) $2 \times 10^5$ cm$^3$

98. As Musimu descends, the buoyant force that the water exerts on him
(a) remains approximately constant.
(b) increases a lot because the pressure is so much greater.
(c) decreases significantly because his body is being compressed and made much smaller.
(d) is zero for the entire dive.
(e) There is not enough information to answer the question.

99. Why don't his lungs, heart, and chest completely collapse?
(a) The return balloon helps counteract the external pressure.
(b) There is no external force pushing directly on the organs.
(c) The sled that helps him descend protects the front of his body.
(d) Blood plasma moves from his extremities to his chest and the organs in it.
(e) The air originally in the lungs is transferred to the vital organs.

100. Using the dimensions in Question 96, which answer below is closest to the buoyant force that the water exerts on Musimu (without his sled or his return balloon)? Assume that the density of water is 1000 kg/m$^3$.
(a) 200 N (b) 400 N (c) 700 N
(d) 1000 N (e) $2 \times 10^6$ N

**Lakes freeze from top down** We all know that ice cubes float in a glass of water. Why? Virtually every substance contracts when it solidifies—the solid is denser than the liquid. If this happened to water, ice cubes would sink to the bottom of a glass, and ice sheets would sink to the bottom of a lake. Fortunately, this doesn't happen. Liquid water expands by 9% when it freezes into solid ice at 0 °C, from a liquid density of 1000 kg/m$^3$ to a solid density of 917 kg/m$^3$. Consequently, in the winter when the water in a lake freezes, the solid ice stays at the top, forming an ice sheet. Snow covering the icy surface forms a protective blanket that insulates the ice and water below and helps to keep the lake from completely freezing into a solid chunk of ice. Fish and lake plants below the ice survive during the winter.

The expansion of water when it freezes has another important environmental benefit: the so-called freeze-thaw effect on sedimentary rocks. Water is absorbed into cracks in these rocks and then freezes in cold weather. The solid ice expands and cracks the rock—like a wood cutter splitting logs. This continual process of liquid water absorption, freezing, and cracking releases mineral and nitrogen deposits into the soil and can eventually break the rock down into soil.

101. When is water denser?
(a) When liquid at 0 °C.
(b) When solid ice at 0 °C.
(c) Water is always 1000 kg/m$^3$.
(d) When it is near room temperature.

102. Why does water freeze from the top down?
(a) The denser water at 0 °C sinks below the ice.
(b) The less dense ice at 0 °C rises above the liquid water at 0 °C.
(c) The solid ice is denser than the liquid, just like for metals.
(d) a and b
(e) a, b, and c

103. Using Newton's second law, expressions for buoyant force and other forces, and the densities of liquid and solid water at 0 °C, find the fraction of an iceberg or an ice cube that is under liquid water.
(a) 0.84 (b) 0.88 (c) 0.92
(d) 0.96 (e) 1.00

104. A swimming pool at 0 °C has a very large chunk of ice floating in it—like an iceberg in the ocean. When the ice melts, what happens to the level of the water at the edge of the pool?
(a) It rises. (b) It stays the same.
(c) It drops. (d) It depends on the size of the chunk.

105. Which of the following are benefits of the decrease in the density of water when it freezes?
(a) Fish and plants can survive winters without being frozen.
(b) Over time, soil is formed from sedimentary rocks.
(c) Water pipes when frozen in the winter do not burst.
(d) Two of the above three.
(e) All of the first three.

# 11 Fluids in Motion

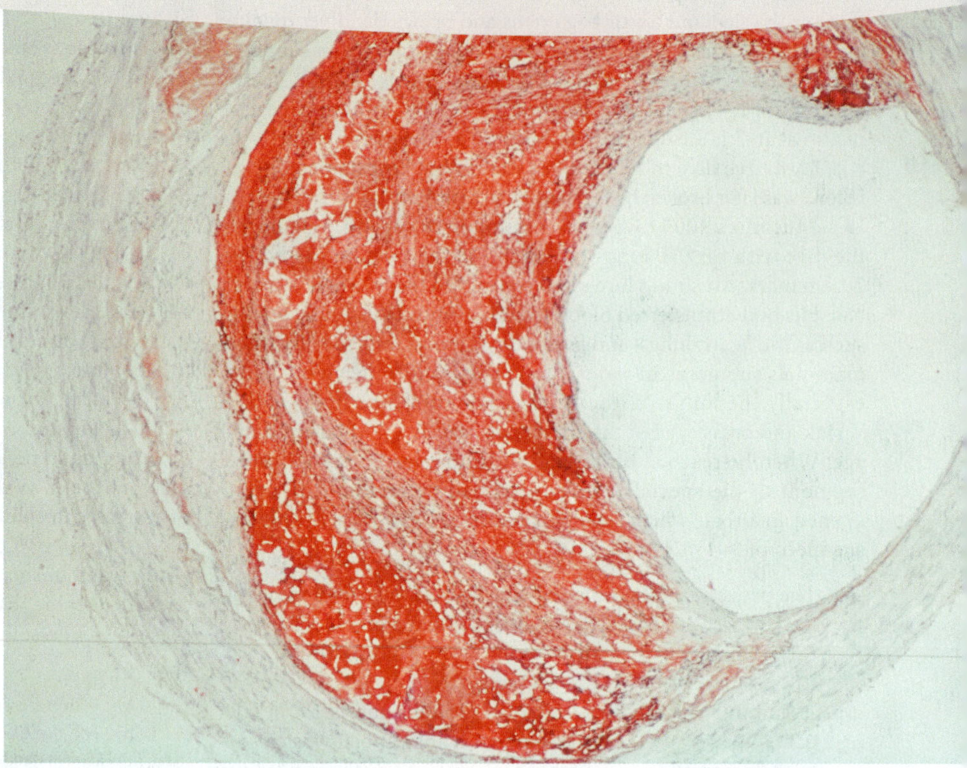

How does blood flow dislodge plaque from an artery?

Why can a strong wind cause the roof to blow off a house?

Why do people snore?

**Be sure you know how to:**

- Draw work-energy bar charts (Section 6.2).
- Apply the concept of pressure to explain the behavior of liquids (Section 10.2).
- Draw force diagrams and apply Newton's second law (Section 3.2).

Plaque (fatty deposits that accumulate on the walls of an artery) grows larger as cholesterol-laden blood flows by. As plaque on the artery wall grows, blood flows past at higher speed. The fast-flowing blood pulls on the plaque like a suction cup. If the blood is moving fast enough, it can completely remove the deposit from the wall. The plaque floats downstream until it reaches another narrow opening, where it may become lodged and completely stop blood flow. If this stoppage occurs in an artery in the heart, it can cause a heart attack by preventing blood from flowing to the cardiac muscles responsible for pumping the blood. In this chapter we will learn why blood flows faster through an artery clogged with plaque and why the fast-moving stream of blood tends to pull the plaque off the artery wall.

**In the previous chapter,** we investigated the behavior of static fluids. We learned that a fluid's pressure increases with depth beneath the surface of the fluid, that the physics of static fluids explains how a hydraulic lift works, and that a fluid exerts a buoyant force on objects partially or completely submerged in it. In all of these situations, however, the fluid was at rest. What happens when a gas or liquid moves across a surface—for example, when air moves across the roof of a house or when blood moves through a blood vessel? In this chapter, we will investigate and explain phenomena involving moving fluids—fluid dynamics.

**Figure 11.1** Air blowing across the top of the ball causes it to float in the air.

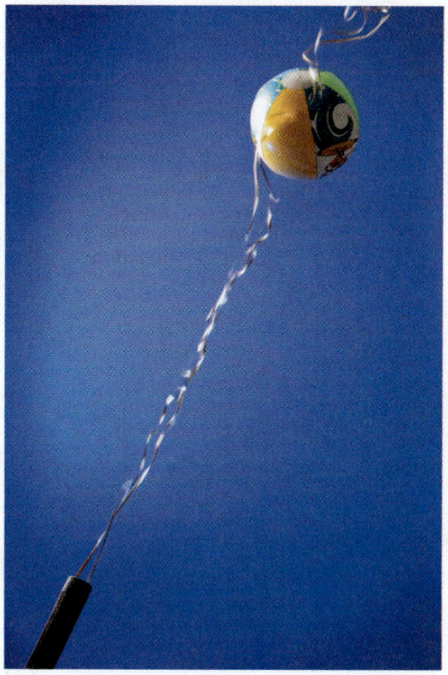

## 11.1 Fluids moving across surfaces—Qualitative analysis

How does air blowing over the top of a beach ball, shown in **Figure 11.1**, lift and support the ball? To answer questions like this, we must compare the forces that stationary air exerts on a surface to the forces exerted on the surface by moving air. In Observational Experiment **Table 11.1**, we apply this analysis to three systems. We know the directions of the forces that external objects exert on these systems and can thus deduce the direction of the net force due to air pressure exerted on different sides of these objects.

**OBSERVATIONAL EXPERIMENT TABLE**

**11.1 Does fluid pressure against a surface depend on the motion of the fluid across the surface?**

VIDEO 11.1

| Observational experiment | Analysis |
| --- | --- |
| **Experiment 1.** We blow air through a straw between two lightbulbs hanging from strings. The bulbs move closer to each other. 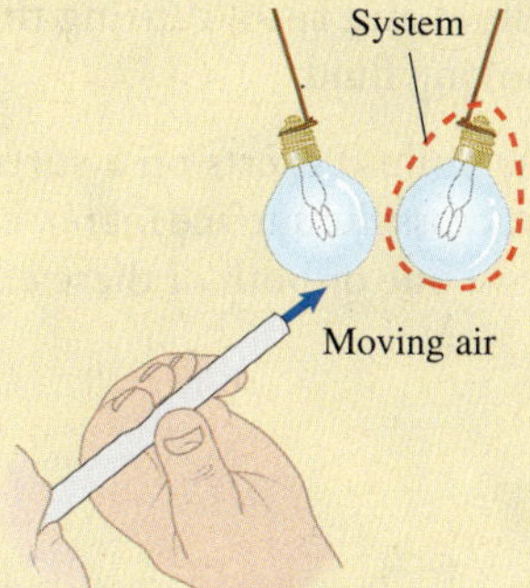  | Choose the right bulb as the system. After it comes to equilibrium, the forces exerted on the bulb are balanced. The vertical component of the force exerted by the string balances the downward force exerted by Earth. The horizontal component of the force exerted toward the right by the string must balance the net force exerted by the air. Thus, the force exerted toward the left by the non-moving air on the right side of the bulb must be greater than the force exerted toward the right by the air moving on the left side of the bulb. The pressure of moving air must be lower than the pressure of nonmoving air. 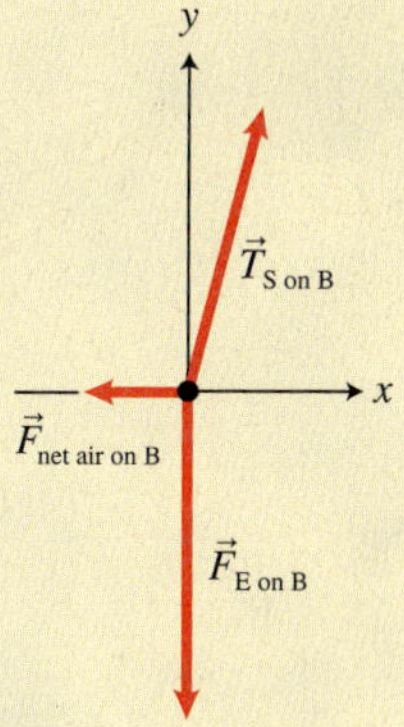  |

*(continued)*

| Observational experiment | Analysis |
|---|---|
| **Experiment 2.** As a car with a canvas-covered trailer moves fast, the canvas bulges upward as though something is pushing up from below. 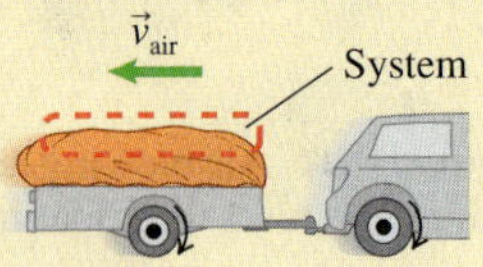 | Choose part of the canvas as the system. Earth exerts a downward gravitational force on the canvas. In order for the canvas to bulge, the stationary air below the canvas and the moving air above must exert a net upward force. This can only happen if the air pressure pushing the canvas up is greater than the air pressure pushing the canvas down. Therefore, we can conclude that the air pressure on the side of moving air is less that on the side of stationary (with respect to the canvas) air.  |
| **Experiment 3.** We blow a stream of air across the top of a large air-filled balloon. The moving air allows it to float with no other support. 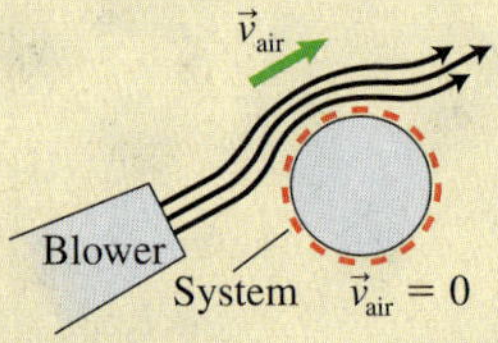 | Choose the floating balloon as the system. Earth exerts a downward gravitational force. To remain stationary, the net force caused by the air must point upward toward the moving-air side of the balloon. This can only happen if the air pressure pushing the balloon up is greater than the air pressure pushing the balloon down. Therefore, we can conclude that the air pressure on top of the balloon (the side of the moving air) is less than that on the bottom of the balloon (the side of the stationary air) 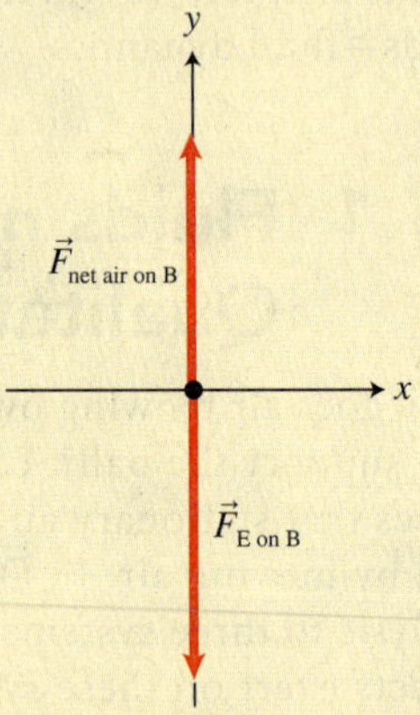 |

**Pattern**

Based on the analysis of each experiment using the force diagrams, we can infer that moving air exerts a smaller force on the system and consequently has a smaller pressure then stationary air.

In each of the three experiments in Table 11.1 the air exerted a net force on the object in the direction of the side past which the air was moving. This means that stationary air exerted a greater pressure on the object than the moving air. Two explanations might account for the observed pattern.

**Explanation 1: Temperature.** The pressure on one side of an object decreases because a moving fluid (such as air) is warmer than a stationary fluid; the warm fluid has a lower density than the cold fluid and rises, causing the pressure to decrease on the side with the warm, moving fluid.

**Explanation 2: Fluid speed.** The pressure that a fluid exerts on a surface decreases as the speed with which the fluid moves across the surface increases.

Let's devise an experiment that can disprove one or both of these explanations. See Testing Experiment **Table 11.2**.

**TESTING EXPERIMENT TABLE**

**11.2 Testing the two explanations for why moving fluid exerts a lower pressure on a surface than stationary fluid.**

VIDEO 11.2

| Testing experiment | Prediction | Outcome |
|---|---|---|
| **Experiment 1.** What happens to a piece of paper held at the corners when you blow hard across the *top* surface of the paper? 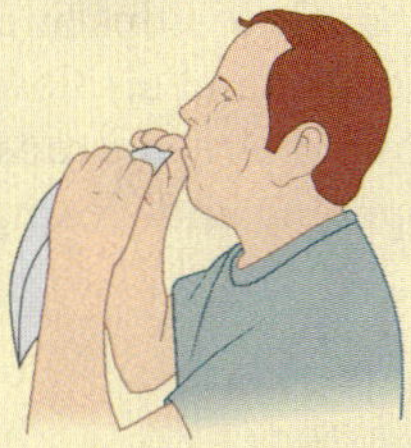 | ***Prediction based on Explanation 1:*** The moving air originating inside your body is warmer than the external air below the paper. The warm air above the paper rises, reducing the pressure on the top surface. The paper will rise.<br><br>***Prediction based on Explanation 2:*** There is less pressure on the top surface where the air is moving compared to the pressure from the stationary air below the paper. The greater pressure from below will cause the paper to rise. | The paper rises when blowing air across the top. 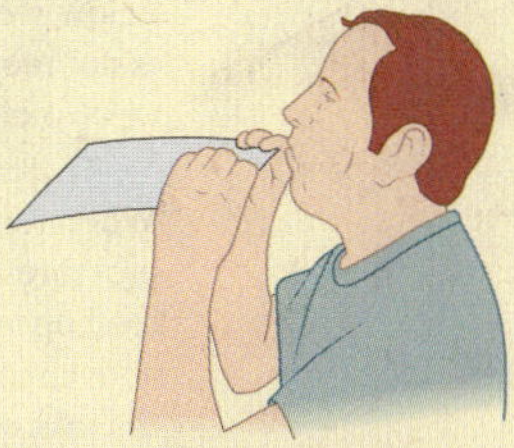 |
| **Experiment 2.** What happens if you vigorously blow air through a straw *under* a card folded in an inverted-U shape? 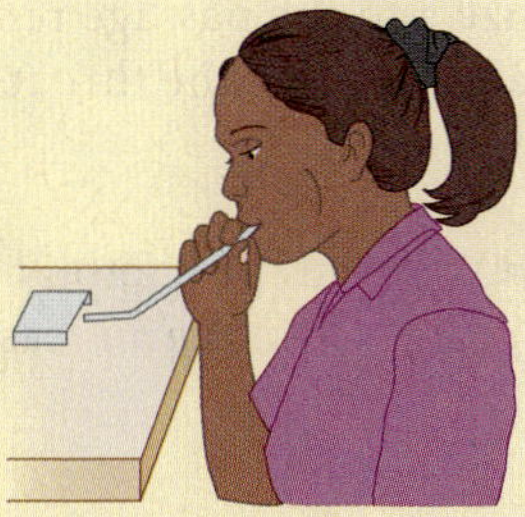 | ***Prediction based on Explanation 1:*** The card should bend up because the warm air from your breath is less dense than surrounding air and tends to rise.<br><br>***Prediction based on Explanation 2:*** The air moving under the card exerts less pressure on the bottom surface than the stationary air above the card—the card should bend down. | The card bends down. 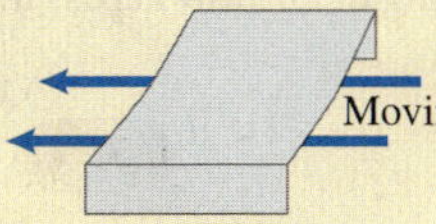  |

**Conclusion**

- In Experiment 1, both explanations correctly predicted the outcome. Thus, neither explanation can be rejected based on this experiment.
- In Experiment 2, the outcome matches the prediction based on Explanation 2 and is opposite the prediction based on Explanation 1. We can reject Explanation 1 based on the outcome of this experiment.

Our testing experiments did not reject Explanation 2. In addition, if we repeat Experiment 1 in Table 11.1, carefully observing what happens when we increase the speed at which air is moving, we notice that the angle of the threads supporting the bulbs increases. This means that the difference in the pressure of stationary and moving air increases with the increase of air speed. At this point, without evidence to the contrary, we can assume that as a fluid's speed increases, the pressure that the moving fluid exerts on the surface decreases. Explanation 2 is a qualitative version of a rule formulated in 1738 by Daniel Bernoulli and named in his honor.

**‹Active Learning Guide**

**Bernoulli's principle** **The pressure that a fluid exerts on a surface *decreases* as the speed with which the fluid moves across the surface *increases*.**

In short, the magnitude of the speed of the fluid has a significant effect on the pressure—the greater the speed, the lower the pressure. Bernoulli's principle has important fluid-flow implications in biological systems—for example, in the flow of blood through blood vessels. The blood pressure against the wall of a vessel depends on how fast the blood is moving—pressure is lower when the blood is moving faster. Let's look at two other applications of Bernoulli's principle.

## Clarinet reed

A clarinet is a woodwind instrument. The musician induces sound by blowing air into the clarinet's mouthpiece, where air moves across the top of a reed (see **Figure 11.2**). Air pressure across the top of the reed decreases relative to the pressure below the reed, where the air is stationary. In response to the drop in pressure, the flexible reed rises and closes the mouthpiece of the clarinet. Once the mouthpiece is closed, the airflow stops and the pressure above and below the reed equalizes. The reed opens and air flows again across the top of the reed. Once again, pressure above the reed drops, the reed rises, and the mouthpiece closes. The rhythmic opening and closing of the reed initiates the sound heard from a clarinet.

**Figure 11.2** A clarinet reed closes when air starts moving across the top and opens when air flow over the reed stops.

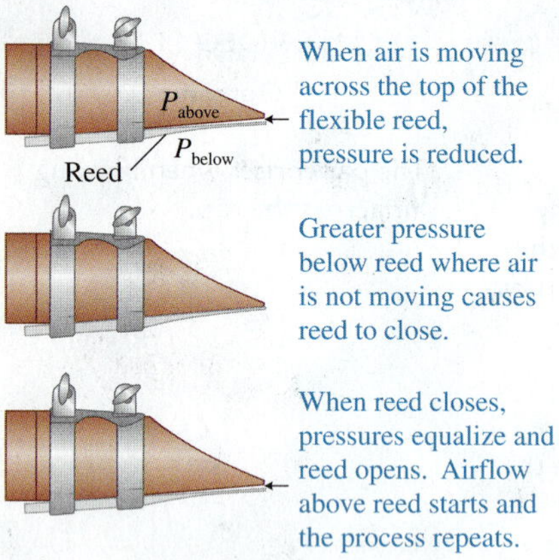


## Snoring

A snoring sound occurs when air moving through the narrow opening above the soft palate at the back of the roof of the mouth has lower pressure than nonmoving air below the palate (**Figure 11.3**). The normal air pressure below the soft palate, where the air is not moving, pushes the palate closed. When airflow stops, the pressures equalize and the passage reopens. The rhythmic opening and closing of the soft palate against the throat leads to the snoring sound (like the vibration of a clarinet reed).

**Figure 11.3** Snoring occurs when the soft palate opens and closes due to the starting and stopping of air flow across it.

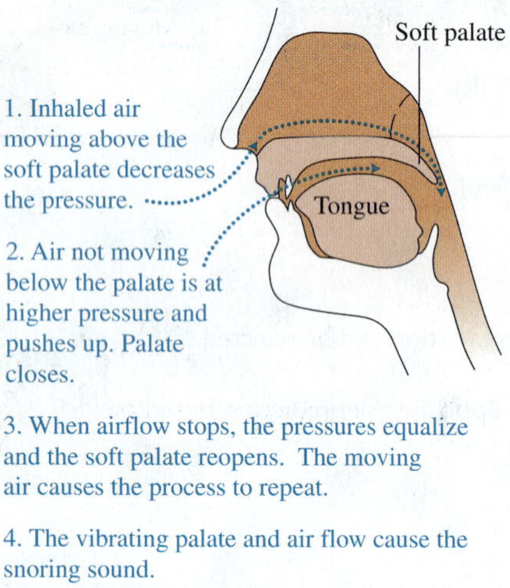

Soft palate
Tongue
1. Inhaled air moving above the soft palate decreases the pressure.
2. Air not moving below the palate is at higher pressure and pushes up. Palate closes.
3. When airflow stops, the pressures equalize and the soft palate reopens. The moving air causes the process to repeat.
4. The vibrating palate and air flow cause the snoring sound.

**Review Question 11.1** Why did the lightbulbs in the first observational experiment come together when the air was moving between them?

# 11.2 Flow rate and fluid speed

We have found qualitatively that the pressure of a fluid against a surface is lower when the fluid is moving across the surface than when it is static. In order to apply this idea quantitatively later, we need first to quantify the rate of fluid flow.

Perhaps you have taken a shower in which little water flowed from the showerhead. In physics, we would say the water flow rate was low. The **flow rate $Q$** is an important consideration in designing showerheads. A smaller flow rate will save water, but a larger flow rate will get you cleaner. Flow rate is defined as the volume $V$ of fluid that moves through a cross section of a pipe divided by the time interval $\Delta t$ during which it moved (see **Figure 11.4**):

$$Q = \frac{V}{\Delta t} \tag{11.1}$$

The SI unit of flow rate is $m^3/s$, but you may also see it as $ft^3/s$, $ft^3/min$, gallons/min, or any unit of volume divided by any unit of time interval. Notice that flow rate in $m^3/s$ is different from the speed of the fluid $v$ in m/s.

**Figure 11.4** Flow rate is the volume of fluid that passes a cross section of a vessel in a given time interval.

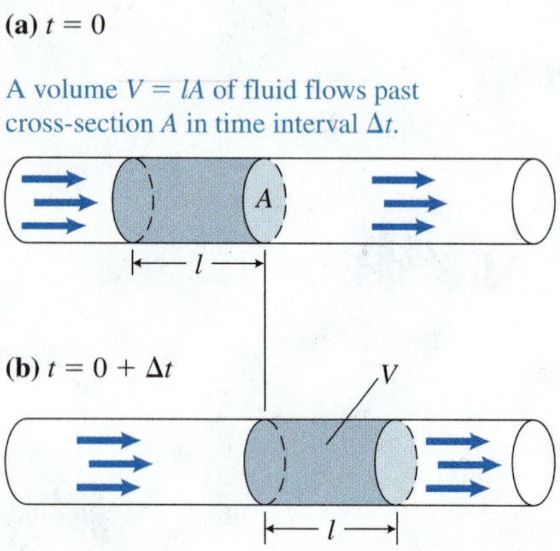


**TIP** The symbols $V$, $t$, and $Q$ are also used in other aspects of physics. For example, a lowercase $v$ denotes speed, the capital letter $T$ is used for temperature, and in future chapters we will use $Q$ for two other unrelated quantities. Because these symbols are often used to indicate different quantities, it is important when working with equations to try to visualize their meaning with concrete images (for example, the volume of water flowing out of a faucet during 1 s).

You probably feel intuitively that the flow rate should be related to the speed of the moving fluid. To explore the relationship, consider Figure 11.4a. Over a certain time interval $\Delta t$ the darkened volume of fluid passes a cross section of area $A$ at some position along the pipe. Thus, after a time $\Delta t$, the back part of this fluid volume has in effect moved forward to the position shown in Figure 11.4b. The volume $V$ of fluid in the darkened portion of the cylinder is the product of its length $l$ and the cross-sectional area $A$ of the pipe:

$$V = lA$$

Thus, the fluid flow rate is

$$Q = \frac{V}{\Delta t} = \frac{lA}{\Delta t} = \left(\frac{l}{\Delta t}\right)A$$

However, $l$ is also the distance the fluid moves in a time interval $\Delta t$. Thus, $\frac{l}{\Delta t}$ is the average fluid speed $v$. Substituting $v = \frac{l}{\Delta t}$ into the above equation, we find that

$$Q = \frac{V}{\Delta t} = vA \quad (11.2)$$

The flow rate is equivalent to the average fluid speed multiplied by the cross-sectional area of the vessel.

◂Active Learning Guide

**QUANTITATIVE EXERCISE 11.1 Speed of blood flow in aorta**

The heart pumps blood at an average flow rate of $80\ \text{cm}^3/\text{s}$ into the aorta, which has a diameter of 1.5 cm. Determine the average speed of blood flow in the aorta.

**Represent mathematically** The flow rate can be determined by rearranging Eq. (11.2):

$$v = \frac{Q}{A}$$

where the cross-sectional area of the aorta is

$$A = \pi r^2 = \pi\left(\frac{d}{2}\right)^2$$

**Solve and evaluate** Combining the above two equations, we find that the average speed of blood flow in the aorta is

$$v = \frac{Q}{A} = \frac{Q}{\pi(d/2)^2} = \frac{(80\ \text{cm}^3/\text{s})}{\pi(1.5\ \text{cm}/2)^2} = 45\ \text{cm/s}$$

The unit is correct. The magnitude is reasonable—about half a meter each second.

**Try it yourself:** Determine the average speed of blood flow if the diameter is reduced from 1.5 cm to 1.0 cm—with the same flow rate.

*Answer:* 100 cm/s.

Notice that blood speed more than doubles when the aorta diameter decreases by 33%. Vessel diameter has a very significant effect on the flow rate of fluid through a vessel, including those in biological systems. The narrower the blood vessel, the faster the blood flows, increasing the risk of dislodging plaque. Likewise, the narrower the airway from the nose to the mouth, the faster the air moves and the more likely you are to snore. These effects depend on the speed of the fluid, like blood or air in different parts of a vessel or pipe in which the diameter changes from one section to another. Let's see if we can relate the speed of the fluid in one part of a vessel to another.

## Continuity equation

Frequently, the radius of a single pipe varies from one part of the pipe to another. How does this affect the flow rate and the speed of the moving fluid in the different parts of the pipe? Consider blood flowing through the artery

**Figure 11.5** The flow speed $v_2 > v_1$ depends on the cross-sectional area of pipe carrying the fluid.

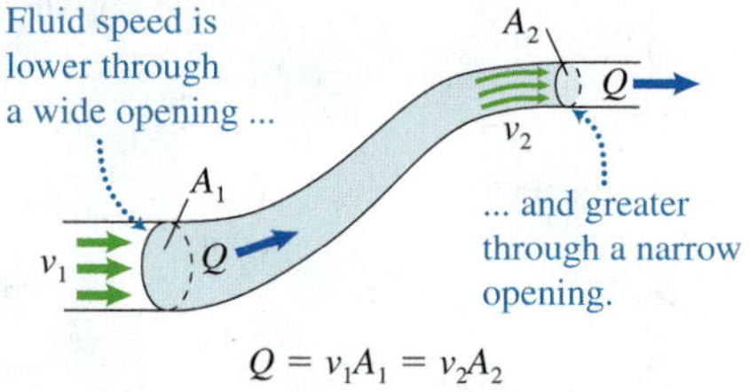


shown in **Figure 11.5**. Is it possible for more fluid to flow through cross section 1 (labeled $A_1$ on the figure) than through cross section 2 ($A_2$)? In such a case, more fluid would be entering the region between 1 and 2 than was leaving that region. This scenario might occur if the region between 1 and 2 were expanding. It could also occur if the fluid were compressed so that it took up less space between 1 and 2. However, let's assume that the vessel does not change its shape and remember that liquids (as opposed to gases) are almost incompressible. This means that the amount of fluid entering at 1 must equal the amount of fluid leaving at 2. What must change is the speed of the fluid as it travels through the narrower part of the vessel. In a narrow section of the vessel ($A_2$) the speed will be greater in order to keep the flow rate constant. Thus, the flow rate past cross section 1 will equal that past cross section 2:

$$Q_1 = v_1A_1 = v_2A_2 = Q_2 \tag{11.3}$$

where $v_1$ is the average speed of the fluid passing cross section $A_1$ and $v_2$ is the average speed of the fluid passing cross section $A_2$. Equation (11.3) is called the **continuity equation** and is used to relate the cross-sectional area and average speed of fluid flow in different parts of a rigid vessel carrying an incompressible fluid.

### QUANTITATIVE EXERCISE 11.2 Blood flow speed

Blood normally flows at an average speed of about 10 cm/s in a large artery with a radius of about 0.30 cm. Assume that the radius of a small section of the artery is reduced by one-half because of *atherosclerosis*, a thickening of the arterial walls. Determine the speed of the blood as it passes through the constriction—point 2 in the figure below.

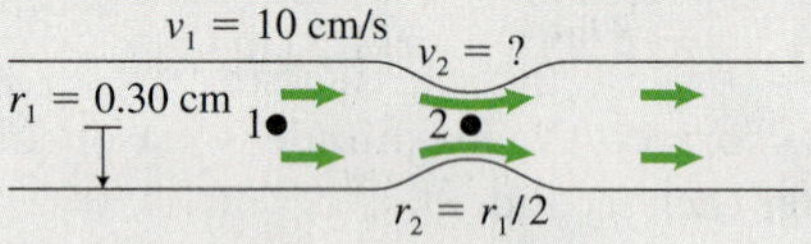


**Represent mathematically** Point 1 indicates the wider part of the artery and point 2 is the constricted part. Assume that the cross-sectional areas are circular ($A = \pi r^2$). We can rearrange the continuity equation to find the average speed of blood flow past point 2 in terms of its speed past 1 and the two cross-sectional areas:

$$v_2 = \left(\frac{A_1}{A_2}\right)v_1$$

**Solve and evaluate** To find $v_2$ in terms of $v_1$, insert expressions for the two areas:

$$v_2 = \left(\frac{A_1}{A_2}\right)v_1 = \left(\frac{\pi r_1^2}{\pi r_2^2}\right)v_1 = \left(\frac{r_1}{r_2}\right)^2 v_1 = \left(\frac{r_1}{0.5r_1}\right)^2 v_1$$
$$= 4v_1$$

Thus, $v_2$ will be 40 cm/s if $v_1$ is 10 cm/s. The increase in speed will have a significant effect on the pressure that the blood exerts on the atherosclerotic material (called plaque) on the vessel wall.

**Try it yourself:** Suppose the radius of this small section of artery is reduced to one-third its normal value because of atherosclerosis. What now is the speed of blood flow past this constriction?

*Answer:* $9v_1$, or 90 cm/s.

**Figure 11.6** Why is $v_2 > v_1$?

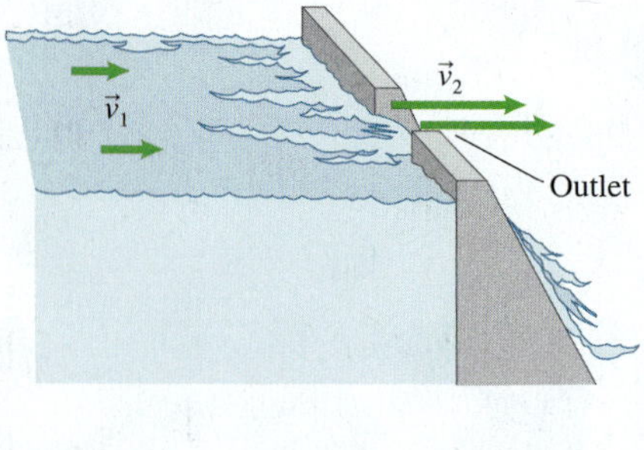


**Review Question 10.2** Why does water in a river flow more slowly just before a dam than it does while passing through the outlet of the dam (**Figure 11.6**)?

## 11.3 Causes and types of fluid flow

We've learned qualitatively that the fluid pressure against a surface decreases as the fluid speed increases. We have also learned that flow rate of a fluid through a pipe depends on the speed of the fluid and the cross-sectional area of the pipe. The next question is: What causes fluids to flow?

Fluid flow is caused by differences in pressure. When the pressure in one region of the fluid is lower than in another region, the fluid tends to flow from the higher pressure region toward the lower pressure region. For example, large masses of air in Earth's atmosphere move from regions of high pressure into regions of low pressure. Blood flows through the circulatory system from the arterial side, which has a gauge pressure of about 100 mm Hg, to the venous side, which has a gauge pressure of about 5 mm Hg.

What kinds of flow can occur? We're familiar with different kinds of fluid flow. There is the smooth flow that we see in a wide river and the more turbulent flow that we see as water rushes and swirls through a narrow channel. Studies of fluid flow in wind tunnels indicate that there are two primary kinds of flow: streamline (or laminar) flow and turbulent flow. In **streamline flow**, every particle of fluid that passes a particular point follows the same path as particles that preceded it. This is the smooth flow that we see in a wide river (see the fluid flowing smoothly in the wide tube in **Figure 11.7**). **Turbulent flow**, on the other hand, is characterized by agitated, disorderly motion. Instead of following a given path, the fluid forms whirlpool patterns called eddies, which come and go randomly, or sometimes become semi-stable (see **Figure 11.8**). Turbulent flow occurs when a fluid moves around objects and through pipes at high speed.

**Figure 11.7** Streamline flow.

Streamline flow of water through a tube

**Figure 11.8** Turbulent flow.

Turbulent flow of water through a narrow part of a tube

The force exerted by the fluid on objects is called the **drag force** and is somewhat greater during turbulent flow than during streamline flow. Designing a car so that air moves over it with streamline flow reduces the drag force that the air exerts on the car and improves gasoline mileage. Placing a curved dome above the cab of a truck deflects air up and over the trailer, reducing turbulent flow and increasing gas mileage by more than 10% (see **Figure 11.9**).

**Review Question 11.3** Is it easier for the heart to pump blood if the flow of the blood through the blood vessels is streamline or if it is turbulent? Explain.

## 11.4 Bernoulli's equation

Earlier in this chapter we developed the qualitative version of Bernoulli's principle: the pressure of a fluid against a surface decreases as the speed of the fluid across the surface increases. A quantitative version of Bernoulli's principle relates the properties (pressure, speed) of a fluid at one position to its properties at another position.

**Figure 11.9** A dome over the cab of a truck reduces the turbulent flow of air and increases the truck's fuel efficiency.

**‹Active Learning Guide**

To derive the quantitative version of Bernoulli's principle, called Bernoulli's equation, we again use the case of a fluid flowing through a pipe, as shown in **Figure 11.10a**. We assume that (1) the fluid is incompressible, (2) the fluid flows without friction, and (3) the flow is streamline. We can apply the work-energy equation to describe the behavior of the fluid as it moves a short distance along the vessel. Consider the system to be composed of the shaded volume of fluid pictured in Figures 11.10a and b and Earth.

Figure 11.10a shows us the initial state of the system. As the volume of fluid flows to the right, the fluid behind it at position 1 exerts a force of magnitude $F_1 = P_1A_1$ to the right where $P_1$ is the fluid pressure against the left side of the volume, and $A_1$ is the cross-sectional area of the left side of the volume. Simultaneously, the fluid ahead of the system at position 2 exerts a force in the opposite direction of magnitude $F_2 = P_2A_2$, where $P_2$ is the fluid pressure against the right side of the volume and $A_2$ is the cross-sectional area of the right side of the volume.

In Figure 11.10b, the shaded volume of fluid has moved to the right. Because the pipe is narrower at 2, the right side of the fluid at position 2 moves a greater distance than the left side of the same volume of fluid in the wider part

**Figure 11.10** Applying the work-energy equation to fluid flow.

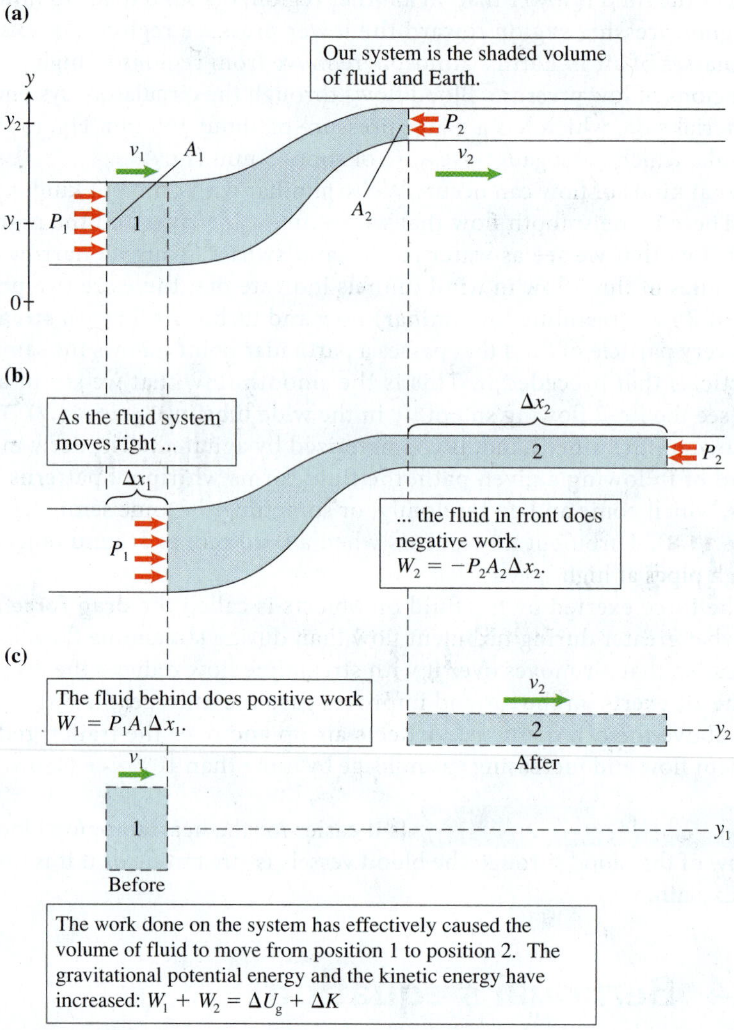
(a)
Our system is the shaded volume of fluid and Earth.
$y$
$y_2$
$y_1$
0
$v_1$
$A_1$
$P_1$
1
$P_2$
$v_2$
$A_2$
(b)
As the fluid system moves right ...
$\Delta x_1$
$\Delta x_2$
2
$P_1$
$P_2$
... the fluid in front does negative work. $W_2 = -P_2A_2\Delta x_2$.
(c)
The fluid behind does positive work $W_1 = P_1A_1\Delta x_1$.
$v_1$
1
Before
$v_2$
2
After
$y_2$
$y_1$
The work done on the system has effectively caused the volume of fluid to move from position 1 to position 2. The gravitational potential energy and the kinetic energy have increased: $W_1 + W_2 = \Delta U_g + \Delta K$

of the pipe at position 1. The net effect of the movement of the fluid a short distance to the right is summarized in Figure 11.10c. The volume of fluid initially at position 1 moving at speed $v_1$ has now been transferred, in effect, to position 2 where it moves at speed $v_2$. The volume of fluid stays constant, since we assume the fluid is incompressible. The fluid is now moving faster through the narrow tube at position 2 than it was earlier when moving through the wider tube at position 1, an increase in kinetic energy. The fluid at position 2 is at a higher elevation than when at position 1, thus the gravitational potential energy of the system increases. The energies changed as a result of the work done by the forces exerted by the fluid behind and ahead of the shaded volume. We can represent this quantitatively using the generalized work-energy equation [Eq. (6.3)].

$$(K_1 + U_{g1}) + W = (K_2 + U_{g2})$$

If we move the terms with the subscript 1 to the right side of the equation, we have

$$W = (K_2 - K_1) + (U_{g2} - U_{g1})$$

or

$$W = \Delta K + \Delta U_g \tag{11.4}$$

Let us now write expressions for each of the terms in the above equation.

**Work Done** Two forces are doing work on the system. The fluid behind the system exerts a force $F_1$ to the right on the left side of the system over a distance $\Delta x_1$. The fluid ahead of the system exerts a force $F_2$ to the left on the right side of the system over a distance $\Delta x_2$. (Figures 11.10a and b show fluid pressures; the corresponding forces have magnitudes $F_1 = P_1A_1$ and $F_2 = P_2A_2$). The force $F_1$ does positive work since it points in the direction of the motion of the system. The force $F_2$ does negative work since it points in the direction opposite the motion of the system. The total work done on the system is

$$W = F_1\Delta x_1 \cos 0° + F_2\Delta x_2 \cos 180°$$
$$= P_1A_1\Delta x_1 - P_2A_2\Delta x_2$$

The volume of fluid $V$ that has moved from the left to the right is $V = A_1\Delta x_1 = A_2\Delta x_2$ since the fluid is incompressible. The preceding expression for work becomes

$$W = P_1V - P_2V = (P_1 - P_2)V$$

**Change in Kinetic Energy** The mass $m$ of the system (the moving volume of fluid) is related to its density $\rho$ and volume $V$:

$$m = \rho V$$

As the system moves from 1 to 2, its speed changes from $v_1$ to $v_2$. Thus, the kinetic energy change of the mass $m$ of fluid shown in Figure 11.10c is

$$\Delta K = \frac{1}{2}mv_2^2 - \frac{1}{2}mv_1^2 = \frac{1}{2}\rho V v_2^2 - \frac{1}{2}\rho V v_1^2$$

**Change in Gravitational Potential Energy** The gravitational potential energy of the system has also changed, because the system has moved from elevation $y_1$ to elevation $y_2$. The change in gravitational potential energy is then

$$\Delta U_g = mg(y_2 - y_1) = \rho Vg(y_2 - y_1)$$

We can now substitute the above three expressions into Eq. (11.4) to get

$$(P_1 - P_2)V = \left(\frac{1}{2}\rho V v_2^2 - \frac{1}{2}\rho V v_1^2\right) + \rho V g(y_2 - y_1)$$

If the common $V$ is canceled from each term, we find that

$$P_1 - P_2 = \frac{1}{2}\rho(v_2^2 - v_1^2) + \rho g(y_2 - y_1)$$

By dividing by $V$ in that last step we have changed the units of each term in the equation from energy (measured in joules) to energy density (measured in joules per cubic meter). Energy density is the same as energy per unit of volume of the fluid and appears on the right side of the above equation. The left hand side represents the amount of work done on the fluid per unit volume of fluid.

**Bernoulli's equation relates the pressures, speeds, and elevations of two points along a single streamline in a fluid:**

$$P_1 - P_2 = \frac{1}{2}\rho(v_2^2 - v_1^2) + \rho g(y_2 - y_1) \quad (11.5)$$

**The equation can be rearranged into an alternate form:**

$$\frac{1}{2}\rho v_1^2 + \rho g y_1 + P_1 = P_2 + \frac{1}{2}\rho v_2^2 + \rho g y_2 \quad (11.6)$$

**The sum of the kinetic and gravitational potential energy densities and the pressure at position 1 equals the sum of the same three quantities at position 2.**

Bernoulli's equation describes the flow of a frictionless, nonturbulent, incompressible fluid. The equation is the quantitative equivalent of Bernoulli's principle, which we developed in Section 11.1.

## Using Bernoulli bar charts to understand fluid flow

Bernoulli's equation looks fairly complex and might be difficult to use for visualizing fluid dynamics processes. However, since Bernoulli's equation is based on the work-energy principle, we can represent such processes using energy bar charts similar to the ones used in Chapter 6 (here the bars represent pressures and energy densities). The following Reasoning Skill box describes how to construct a fluid dynamics bar chart for a process. The procedure is illustrated for the following process: A fire truck pumps water through a big hose up to a smaller hose on the ledge of a building. Water sprays out of the smaller hose onto a fire in the building. Compare the pressure in the hose just after leaving the pump to the pressure at the exit of the small hose.

**REASONING SKILL** Constructing a bar chart for a moving fluid

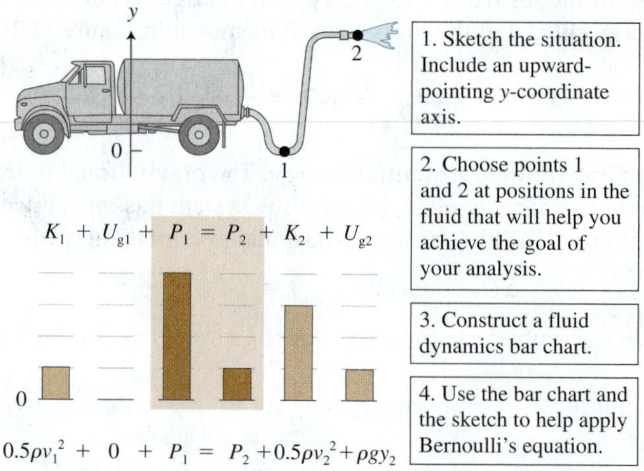

y
2
0
1
1. Sketch the situation. Include an upward-pointing y-coordinate axis.
2. Choose points 1 and 2 at positions in the fluid that will help you achieve the goal of your analysis.
$K_1 + U_{g1} + P_1 = P_2 + K_2 + U_{g2}$
3. Construct a fluid dynamics bar chart.
0
4. Use the bar chart and the sketch to help apply Bernoulli's equation.
$0.5\rho v_1^2 + 0 + P_1 = P_2 + 0.5\rho v_2^2 + \rho g y_2$


To start, first draw a sketch of the process. Then, choose positions 1 and 2 at appropriate locations in order to help answer the question. One of the positions might be a place where you want to determine the pressure in the fluid and the other position a place where the pressure is known. For the water pump-hose process, it is useful to choose position 1 at the exit of the pump (the location of the unknown pressure) and position 2 at the exit of the water from the small hose (at known atmospheric pressure).

Next, represent this process by placing bars of appropriate relative lengths on the chart (the absolute lengths are not known). It is often easiest to start by analyzing the gravitational potential energy density. Use a vertical $y$-axis with a well-defined origin to keep track of the gravitational potential energy densities. For the fire hose process, choose position 1 as the origin of the vertical coordinate system. The gravitational potential energy density at position 1 is then zero. The exit of the water from the small hose is at higher elevation; thus, there is a positive gravitational potential energy density bar for position 2 (in the bar chart we arbitrarily assign it one positive unit of energy density).

Next, consider the kinetic energy. The water flows from a wider hose at position 1 to a narrower hose at position 2. Thus the kinetic energy density at 2 is greater than at 1—thus the longer bar at position 2. We arbitrarily assume that the kinetic energy density bar for position 1 is one unit and for position 2 is three units.

Notice now that the total length of the bars on the right side of the chart is much higher than on the left side (the difference is three units). To account for the difference we need to consider the change in pressure. Since the fluid pressures at 1 and 2 are analogous to the work done on the system in an ordinary work-energy bar chart, $P_1$ and $P_2$ appear in the shaded box in the center of the bar chart, where work is represented. The difference in the pressure heights should account for the total difference in the energy densities. Thus, we draw the bar for $P_1$ three units higher than for $P_2$. The bar chart is now complete. We can use it to write a mathematical description for the process and solve for any unknown quantity.

**Review Question 11.4** Compare and contrast work-energy bar charts and Bernoulli bar charts.

## 11.5 Skills for analyzing processes using Bernoulli's equation

In this section, we adapt our problem-solving strategy to analyze processes involving moving fluids. In this case, we describe and illustrate a strategy for finding the speed of water as it leaves a bottle. The general strategy is on the left side of the table and the specific process is on the right.

### PROBLEM-SOLVING STRATEGY Applying Bernoulli's Equation

**EXAMPLE 11.3 Removing a tack from a water bottle**

What is the speed with which water flows from a hole punched in the side of an open plastic bottle? The hole is 10 cm below the water surface.

#### Sketch and translate

- Sketch the situation. Include an upward-pointing $y$-coordinate axis. Choose an origin for the axis.
- Choose points 1 and 2 at positions in the fluid where you know the pressure/speed/position or which involve the quantity you are trying to determine.
- Choose a system.

- Choose the origin of the vertical $y$-axis to be the location of the hole.
- Choose position 1 to be the place where the water leaves the hole and position 2 to be a place where the pressure, elevation, and water speed are known—at the water surface $y_2 = 0.10\text{ m}$ and $v_2 = 0$. The pressure in Bernoulli's equation at both positions 1 and 2 is atmospheric pressure, since both positions are exposed to the atmosphere ($P_1 = P_2 = P_{atm}$).
- Choose Earth and the water as the system.

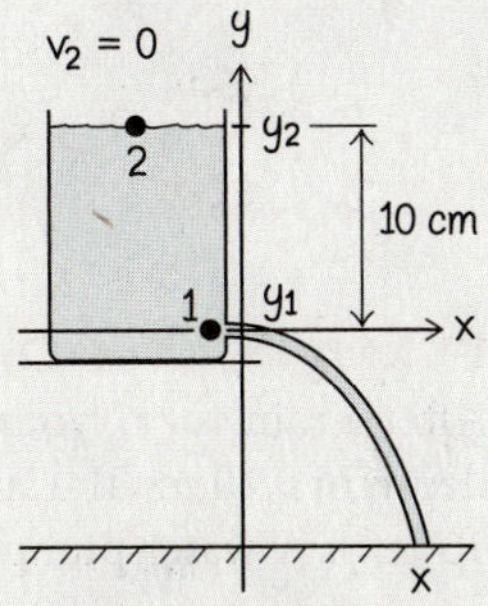


#### Simplify and diagram

- Identify any assumptions you are making. For example, can we assume flow without friction?
- Construct a Bernoulli bar chart.

- Assume that the fluid flows without friction.
- Assume that $y_2$ and $y_1$ stay constant during the process, since the elevation of the surface decreases slowly compared to the speed of the water as it leaves the tiny hole.
- Draw a bar chart that represents the process.

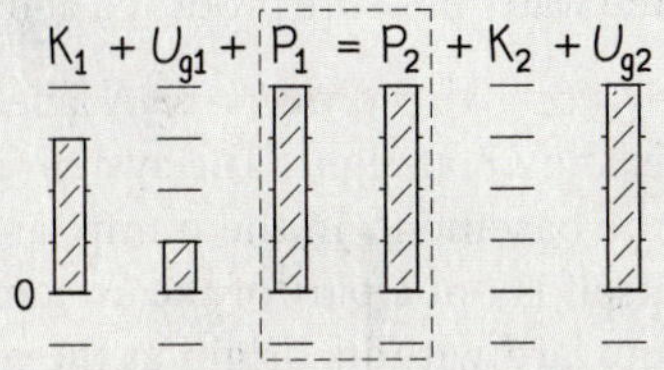


*(continued)*

**Represent mathematically**

- Use the sketch and bar chart to help apply Bernoulli's equation.
- You may need to combine Bernoulli's equation with other equations, such as the equation of continuity $Q = v_1A_1 = v_2A_2$ and the definition of pressure $P = \frac{F}{A}$.

- We see from the sketch and the bar chart that the speed of the fluid at position 2 is zero (zero kinetic energy density) and that the elevation is zero at position 1 (zero gravitational potential energy density). Also, the pressure is atmospheric at both 1 and 2. Thus,

$$(1/2)\rho(0)^2 + \rho g y_2 + P_{atm} = P_{atm} + (1/2)\rho v_1^2 + \rho g(0)$$
$$\Rightarrow \rho g y_2 = (1/2)\rho v_1^2$$

**Solve and evaluate**

- Solve the equations for an unknown quantity.
- Evaluate the results to see if they are reasonable (the magnitude of the answer, its units, how the answer changes in limiting cases, and so forth).

- Solve for $v_1$

$$v_1 = \sqrt{2gy_2}$$

Substituting for $g$ and $y_2$, and $y_2$, we find that

$$v_1 = \sqrt{2(9.8\ \text{m/s}^2)(0.10\ \text{m})} = 1.4\ \text{m/s}$$

- The unit m/s is the correct unit for speed. The magnitude seems reasonable for water streaming from a bottle (if we obtained 120 m/s it would be unreasonably high).

**Active Learning Guide ›**

**Try it yourself:** In the above situation the water streams out of the bottle onto the floor a certain horizontal distance away from the bottle. The floor is 1.0 m below the hole. Predict this horizontal distance using your knowledge of projectile motion. [*Hint:* Use Eqs. (3.7) and (3.8).]

*Answer:* The equations yield a result of 0.63 m. However, if we were to actually perform this experiment with a tack-sized hole, the water would land short of our prediction because there is friction between a small hole and the water. In order to make the water land 0.63 m from the bottle, we must increase the diameter of the hole to about 3 mm. We discuss the effect of friction on fluid flow later in the chapter.

**EXAMPLE 11.4 Drying your basement**

After a rainstorm, your basement is filled with water to a depth of 0.10 m. The surface area of the basement floor is 150 $m^2$. A water pump with a short 1.2-cm-radius outlet pipe connects to a 0.90-cm-radius hose that in turn goes out the basement window to the ground outside, 2.4 m above the pump outlet pipe. The pump can remove water from the basement at a rate of 6.8 $m^3$/hour (1800 gal/h). Determine the time interval needed to remove the water and determine the water pressure produced at the pump outlet pipe.

**Sketch and translate** A labeled sketch of the situation is shown at right. The system is Earth and the water in the basement, in the pump, and in the hose. The pump itself is not a part of the system. A vertical $y$-axis points upward with its origin at the pump outlet pipe. We can calculate the volume of water in the basement and use this volume and the flow rate $Q = 6.8\ \text{m}^3/\text{h}$ to determine the time interval needed to remove the water.

The system is the shaded water and Earth.

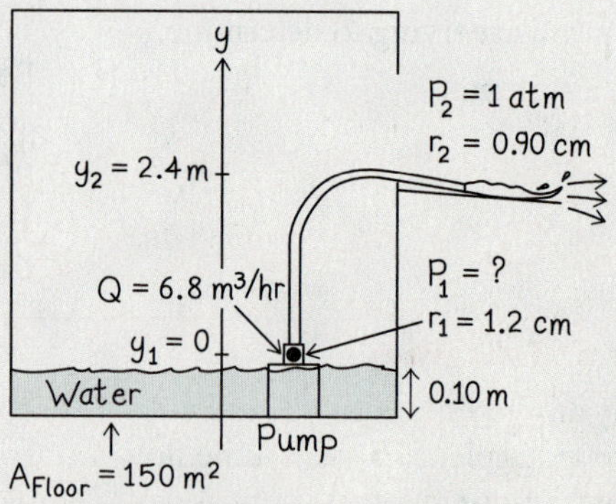


We use Bernoulli's equation to determine the pressure at the pump outlet pipe. Point 1 is at the outlet pipe in the basement at vertical position $y_1 = 0$ (the place where we want to determine the pressure). Position 2 is the exit point of the hose on the ground outside at vertical position $y_2 = +2.4$ m (we know that pressure $P_2 = P_{atm}$, since the fluid is exposed to the air). Since we know the fluid flow rate $Q$, we use $Q = Av$ to determine the speed of the fluid at the pump outlet pipe and at the exit of the hose (since the fluid flow rate must be the same at both positions).

**Simplify and diagram** Assume that the water is incompressible and that the water flows without friction or turbulence. Then draw a fluid dynamics bar chart to represent the process. We include the following energy density changes.

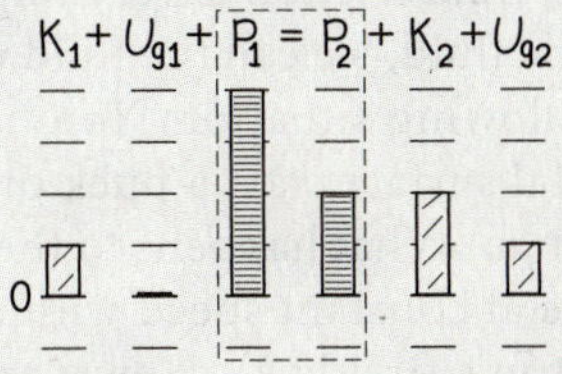


**Kinetic energy density** The fluid moves faster through the narrower hose at position 2 than through the wider pipe at position 1. Thus, the kinetic energy density $K_2$ is greater than $K_1$.

**Gravitational potential energy density** Position 2 is at a higher elevation than position 1. Thus, $U_{g2}$ is greater than $U_{g1}$, which we know is zero, having chosen the origin of the vertical $y$-axis at position 1.

**Pressure** To balance the initial-final energy densities, the pressure $P_1$ at the outlet of the pump in the basement must be greater than the atmospheric pressure $P_2 = P_{atm}$ of the air as water leaves the hose at position 2.

**Represent mathematically** The volume $V$ of water in the basement is the product of its area $A = 150\ \text{m}^2$ and its depth $d = 0.10$ m; thus, $V = A \cdot d = (150\ \text{m}^2)(0.10\ \text{m}) = 15\ \text{m}^3$. The time interval for pumping all the water out is

$$\Delta t = \frac{V}{Q}$$

The bar chart helps us apply Bernoulli's equation [Eq. (11.5) or (11.6)]. There is nonzero term in the equation for each nonzero bar in the chart. We get

$$\frac{1}{2}\rho v_1^2 + P_1 = P_2 + \frac{1}{2}\rho v_2^2 + \rho g y_2$$

Substitute for the pressure at point 2 and rearrange the equation to find $P_1$:

$$P_1 = P_{atm} + \frac{1}{2}\rho(v_2^2 - v_1^2) + \rho g y_2$$

The speeds $v_1$ and $v_2$ are determined using Eq. (11.2): $v_1 = Q/A_1$ and $v_2 = Q/A_2$.

**Solve and evaluate** The flow rate is 6.8 m$^3$/h. The time interval needed to pump all of the water out of the basement will be

$$\Delta t = \frac{V}{Q} = \frac{15\ \text{m}^3}{6.8\ \text{m}^3/\text{h}} = 2.2\ \text{h}$$

The unit for time interval is correct, and the magnitude is reasonable—usually pumping water from the basement takes hours, not seconds or years.

Next, use Bernoulli's equation to determine the pressure at the pump outlet. First use the expression for flow rate Eq. (11.2) to determine the speed of the water at positions 1 and 2. The flow rate is $Q = 6.8\ \text{m}^3/\text{h}(1\text{h}/3600\ \text{s}) = 0.00188\ \text{m}^3/\text{s}$. Thus, the water speeds at positions 1 and 2 are

$$v_1 = \frac{Q}{A_1} = \frac{Q}{\pi r_1^2} = \frac{0.00188\ \text{m}^3/\text{s}}{\pi(0.012\ \text{m})^2} = 4.2\ \text{m/s}$$

$$v_2 = \frac{Q}{A_2} = \frac{Q}{\pi r_2^2} = \frac{0.00188\ \text{m}^3/\text{s}}{\pi(0.0090\ \text{m})^2} = 7.4\ \text{m/s}$$

The pressure $P_1$ is determined using Eq. (11.5):

$$\begin{aligned} P_1 &= P_{atm} + \frac{1}{2}\rho(v_2^2 - v_1^2) + \rho g y_2 \\ &= (1.0 \times 10^5\ \text{N/m}^2) + \frac{1}{2}(1000\ \text{kg/m}^3)[(7.4\ \text{m/s})^2 \\ &\quad - (4.2\ \text{m/s})^2] + (1000\ \text{kg/m}^3)(9.8\ \text{N/kg})(2.4\ \text{m}) \\ &= (1.0 \times 10^5\ \text{N/m}^2) + (0.19 \times 10^5\ \text{N/m}^2) \\ &\quad + (0.24 \times 10^5\ \text{N/m}^2) \\ &= 1.4 \times 10^5\ \text{N/m}^2 \end{aligned}$$

Note that $1\ \text{kg/m} \cdot \text{s}^2 = 1\ \text{N/m}^2$. This would be a gauge pressure of $0.4 \times 10^5\ \text{N/m}^2$ or 0.4 atm, certainly possible with a pump from most hardware stores.

**Try it yourself:** Assume that the hose from the pump to the outside has a radius of 0.50 cm instead of 0.90 cm. Now what is the speed of the water exiting the hose and the pressure at the pump outlet pipe?

*Answer:* 24 m/s (a hose with a smaller radius produces a somewhat higher water speed) and $4.0 \times 10^5\ \text{N/m}^2$ (we need a huge increase in pressure because of the smaller radius hose).

**Review Question 11.5** In Example 11.3 we said that the pressure was the same at two levels when we drew the bar chart. Doesn't the pressure in a fluid increase with depth?

# 11.6 Viscous fluid flow

In our previous discussions and examples in this chapter, we assumed that fluids flow without friction. That is, we assumed no interaction either between the fluid and the walls of the pipes they flow in, or between the layers of the fluid. However, in Example 11.3 we found that this assumption was not reasonable. In fact, for many processes, such as the transport of blood in the small vessels in our bodies, fluid friction is very important. When we cannot neglect this friction inside the fluid, we call the fluid **viscous.**

Consider the following situation. You have an object that can slide on a frictionless horizontal surface, say, a puck on smooth ice. You push the puck abruptly and then let go. What happens to the puck? Once in motion, the puck will continue to slide at constant speed with respect to the ice even if nothing else pushes it (Newton's first law). However, if there is friction between the contacting surfaces (there is a little sand in the ice), then the puck starts slowing down; for it to continue moving at constant speed, someone or something has to push it forward to balance the opposing friction force.

By analogy, if a fluid flows through a horizontal tube without friction, we would expect it to continue to flow at a constant rate with no additional forward pressure. But if friction is present, there must be greater pressure at the back of the fluid than at the front of the fluid. If this is the case, the force exerted on any volume of the fluid due to the forward pressure is greater than the force exerted on the same volume of the fluid due to the pressure in the opposite direction.

## Factors that affect fluid flow rate

What factors affect the flow rate in the vessel with friction? What is the functional dependence of those factors? Let's think about the physical properties of the fluid and the vessel that can affect the flow rate. The following quantities might be important.

**Pressure Difference** The flow rate should depend on how hard the fluid is pushed forward, that is, on the difference between the fluid pressure pushing forward from behind and the fluid pressure pushing back from in front of the fluid, or $(P_1 - P_2)$.

**Radius of the Tube** The radius $r$ of the tube carrying the fluid should affect the flow rate. From everyday experience we know that it is more difficult to push (a greater pressure difference) fluid through a tube of tiny radius than through a tube with a large radius.

**Length of the Tube** The length $l$ of the tube might also affect the ease of fluid flow. A long tube offers more resistance to flow than a shorter tube.

**Fluid Type** Water flows much more easily than molasses does. Thus some property of a fluid that characterizes its "thickness" or "stickiness" should affect the flow.

Let's design an experiment to investigate exactly how the first three of these four factors ($P$, $r$, and $l$) affect the fluid flow rate $Q$. As shown in **Figure 11.11**, a pump that produces an adjustable pressure $P_1$ causes fluid to flow through tubes of different radii $r$ and lengths $l$. We collect the fluid exiting the tube and measure the flow rate $Q$, which is the volume $V$ of fluid leaving the tube in a certain time interval $\Delta t$ divided by that time interval. The results of the experiments are reported in **Table 11.3**.

How is the flow rate affected by each of the three factors?

*Pressure difference* Looking at the first three rows of the table, we notice that the flow rate is proportional to the pressure difference ($Q \propto P_1 - P_2$).

**Figure 11.11** How do $P_1 - P_2$, $r$, and $l$ affect the flow rate $Q$?

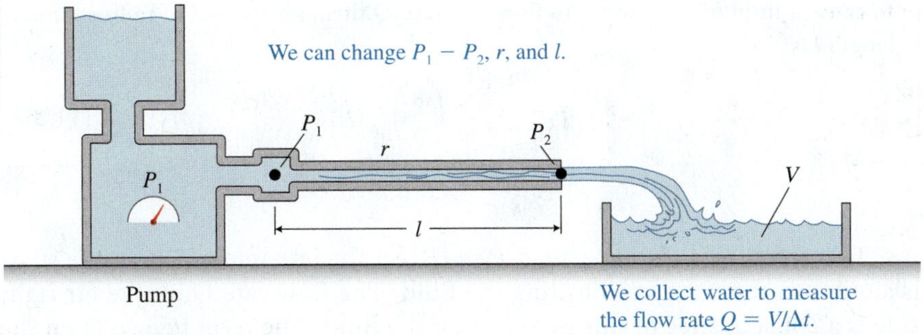
We can change $P_1 - P_2$, $r$, and $l$.
$P_1$
$P_2$
$r$
$P_1$
$V$
$l$
Pump
We collect water to measure the flow rate $Q = V/\Delta t$.

**Table 11.3 Different quantities affect the flow rate $Q$ of fluid through a tube. The data are reported in relative units.**

| $P_1 - P_2$ (Pressure difference) | $r$ (Radius) | $l$ (Length) | $Q$ (Flow rate) |
|---|---|---|---|
| 1 | 1 | 1 | 1 |
| 2 | 1 | 1 | 2 |
| 3 | 1 | 1 | 3 |
| 1 | 2 | 1 | 16 |
| 1 | 3 | 1 | 81 |
| 1 | 1 | 2 | 0.5 |
| 1 | 1 | 3 | 0.33 |

*Radius of the tube* Looking at rows 1, 4, and 5, we notice that the flow rate increases rapidly as the radius increases. Doubling the radius causes the flow rate to increase by a factor of 16 ($2^4$). Tripling the radius causes the flow rate to increase by a factor of 81 ($3^4$). The flow rate is proportional to the fourth power of the radius of the tube ($Q \propto r^4$).

*Length of the tube* Looking at rows 1, 6, and 7, we notice that the flow rate decreases as the length of the tube increases. It is proportional to the inverse of the length ($Q \propto 1/l$).

These three relationships can be combined in a single equation:

$$Q \propto \frac{r^4(P_1 - P_2)}{l}$$

## Viscosity and Poiseuille's law

In this experiment we did not investigate the fourth factor: the type of fluid. Under the same conditions, water flows faster than oil, which flows faster than molasses. If we use the same pressure difference to push different fluids through the same tube, we find that the fluids have different flow rates. The quantity by which we measure this effect on flow rate is called the **viscosity** $\eta$ of the fluid. The flow rate is inversely proportional to viscosity:

$$Q \propto \frac{1}{\eta}$$

In 1840, using an experiment similar to that described above, French physician and physiologist Jean Louis Marie Poiseuille established a relationship between these physical quantities. However, instead of writing the flow rate in terms of the other four quantities, he wrote an expression for the pressure difference needed to cause a particular flow rate.

**Poiseuille's law** **The forward-backward pressure difference $P_1 - P_2$ needed to cause a fluid of viscosity $\eta$ to flow at a rate $Q$ through a vessel of radius $r$ and length $l$ is**

$$P_1 - P_2 = \left(\frac{8}{\pi}\right)\frac{\eta l}{r^4}Q \qquad (11.8)$$

**TIP** Notice that the pressure difference needed to cause a particular flow rate is proportional to the inverse of the fourth power of the radius of the vessel. If the radius of a vessel carrying fluid is reduced by a factor of 0.5, the pressure difference needed to cause the same flow rate must increase by $(1/0.5)^4 = 16$. We need 16 times the pressure difference to cause the same flow rate.

The pressure difference term $P_1 - P_2$ on the left side of Poiseuille's law is similar to the net force pushing the fluid. The flow rate $Q$ on the far right side is a consequence of this net push on the fluid. The term before $Q$ on the right side (the $\left(\frac{8}{\pi}\right)\frac{\eta l}{r^4}$ term) can be thought of as the resistance of the fluid to flow—the resistance is greater if the fluid has greater viscosity $\eta$, is greater for a longer vessel (greater $l$), and is far more resistive if the vessel through which the fluid flows has a smaller radius ($1/r^4$ is much greater for small $r$). This idea has many applications relative to the circulatory system—see the example a little later in this section.

From Poiseuille's law we can determine the unit for viscosity. To do this, we express the viscosity using other quantities in Eq. (11.8):

$$P_1 - P_2 = \left(\frac{8}{\pi}\right)\frac{\eta l}{r^4}Q \Rightarrow \eta = \frac{(P_1 - P_2)r^4\pi}{8Ql}$$

We use the latter equation to find the units for the viscosity. Remember that the units of pressure are

$$\text{Pa} = \frac{\text{N}}{\text{m}^2} = \frac{\text{kg}\cdot\text{m}}{\text{s}^2\cdot\text{m}^2} = \frac{\text{kg}}{\text{s}^2\cdot\text{m}}$$

and the units for flow rate are $\text{m}^3/\text{s}$. Using these units we get

$$\eta = \frac{(\text{kg})(\text{m}^4)(\text{s})}{(\text{s}^2\cdot\text{m})(\text{m}^3)(\text{m})} = \frac{\text{kg}}{\text{s}\cdot\text{m}}.$$

We can also rewrite the last combination of units as $\text{N}\cdot\text{s/m}^2$. A list of viscosities of several fluids appears in **Table 11.4** using $\text{N}\cdot\text{s/m}^2$ for the units of $\eta$.

**Table 11.4** Viscosities of some liquids and gases.

| Substance | Viscosity $\eta$ $(\text{N}\cdot\text{s/m}^2)$ |
|---|---|
| Air (30 °C) | $1.9 \times 10^{-5}$ |
| Water vapor (30 °C) | $1.25 \times 10^{-5}$ |
| Water (0 °C) | $1.8 \times 10^{-3}$ |
| Water (20 °C) | $1.0 \times 10^{-3}$ |
| Water (40 °C) | $0.66 \times 10^{-3}$ |
| Water (80 °C) | $0.36 \times 10^{-3}$ |
| Blood, whole (37 °C) | $4 \times 10^{-3}$ |
| Oil, SAE No. 10 | 0.20 |

**QUANTITATIVE EXERCISE 11.5 Blood flow through a narrow artery**

Because of plaque buildup, the radius of an artery in a person's heart decreases by 40%. Determine the ratio of the present flow rate to the original flow rate if the pressure across the artery, its length, and the viscosity of blood are unchanged.

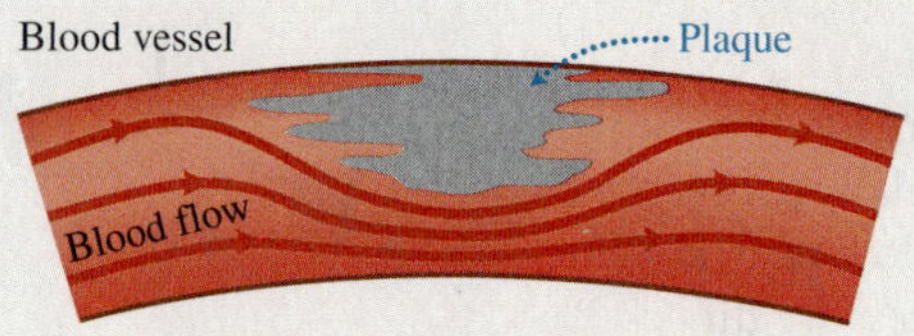


**Represent mathematically** In this exercise, we are interested in the change in the flow rate and not in the change in pressure. Consequently, we rearrange Poiseuille's law for the flow rate in terms of the other quantities:

$$Q = \left(\frac{\pi}{8}\right)\left(\frac{\Delta P}{\eta l}\right)r^4$$

If the radius decreases by 40%, the new radius is $100\% - 40\% = 60\%$ of the original. Thus the radius $r$ of the vessel at the present time is related to the radius $r_0$ years earlier by the equation $r = 0.60r_0$.

**Solve and evaluate** The ratio of the flow rates is

$$\frac{Q}{Q_0} = \frac{\left(\frac{\pi}{8}\right)\left(\frac{\Delta P}{\eta l}\right)r^4}{\left(\frac{\pi}{8}\right)\left(\frac{\Delta P}{\eta l}\right)r_0^{\,4}} = \frac{r^4}{r_0^{\,4}} = \left(\frac{r}{r_0}\right)^4 = (0.60)^4 = 0.13$$

The flow rate is only 13% of the original flow rate! To compensate for such a dramatically reduced flow rate, the person's blood pressure will increase.

**Try it yourself:** Determine the reduction in flow rate, assuming a constant pressure difference, if the radius of the vessel is reduced 90% (to 0.10 times its original value). This is not an unusual reduction for people with high blood pressure.

*Answer:* $Q/Q_0 = 0.0001$, or 0.01% of its original value!

**Limitations of Poiseuille's Law: Reynolds Number** Poiseuille's law describes the flow of a fluid accurately only when the flow is streamline, or laminar. Experiments indicate that to determine when the flow is laminar or turbulent, one needs to determine what is called the Reynolds number $R_e$:

$$R_e = \frac{2\bar{v}r\rho}{\eta} \qquad (11.9)$$

where $\bar{v}$ is the average speed of the fluid, $\rho$ is its density, $\eta$ is the viscosity, and $r$ is the radius of the vessel that carries the fluid. Experiments show that if the Reynolds number is less than 2000, the fluid flow is laminar; if it is more than 3000, the flow is turbulent; and between 2000 and 3000 the flow is unstable and can be either laminar or turbulent.

**Review Question 11.6** Describe some of the physics-related effects on the cardiovascular system of medication that lowers the viscosity of blood).

## 11.7 Applying fluid dynamics: Putting it all together

We can explain many phenomena using ideas from fluid dynamics. In this section, we'll analyze a roof being ripped off a house during a high-speed wind and the dislodging of plaque in an artery.

### Blowing the roof off a house

You've no doubt seen images of roofs being blown from houses during tornadoes or hurricanes. How does that happen? On a windy day, the air inside the house is not moving, whereas outside the air is moving very rapidly. The air pressure inside the house is therefore greater than the air pressure outside, creating a net pressure against the roof and windows that pushes outward. If the net pressure becomes great enough, the roof and/or the windows will blow outward off of the house. In the following example, we do a quantitative estimate of the net force exerted by the inside and outside air on a roof.

**EXAMPLE 11.6 Effect of high-speed air moving across the roof of a house**

During a storm, air is moving at speed 45 m/s (100 mi/h) across the top of the 200-$m^2$ flat roof of a house. Estimate the net force exerted by the air pushing up on the inside of the roof and the outside air pushing down on the outside of the roof. Indicate any assumptions made in your estimate.

**Sketch and translate** The situation is shown below. We need to determine the pressure just above and below the roof.

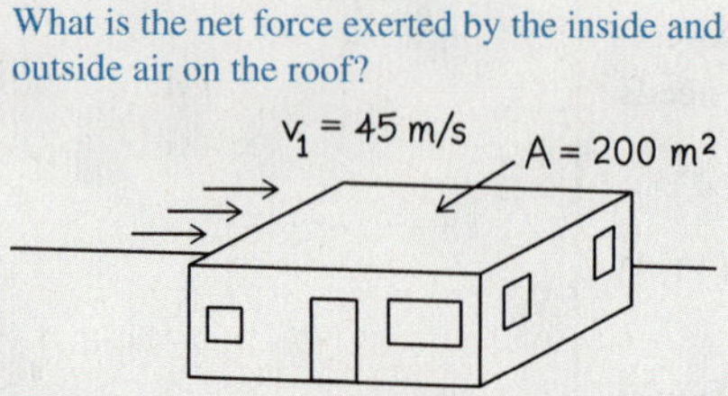


**Simplify and diagram** A force diagram for the roof is shown below. The air above the house exerts a downward force on the roof $F_{1 \text{ on R}} = P_1 A$, where $P_1$ is the air pressure above the house and $A$ is the area of the roof. The air inside the house pushes up, exerting a force on the roof $F_{2 \text{ on R}} = P_{\text{atm}} A$, where $P_{\text{atm}}$ is the assumed atmospheric pressure of the stationary air inside the house. We assume the air is incompressible and flows without friction or turbulence and that the roof is fairly thin so that the air has approximately the same gravitational potential energy density at points 1 and 2.

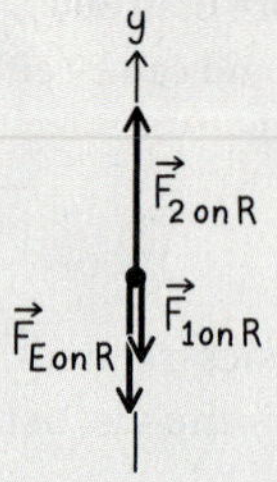


**Represent mathematically** With the $y$-axis oriented upward, the net force exerted by the air on the roof is

$$F_{\text{net Air}} = F_{2 \text{ on R}} - F_{1 \text{ on R}} = P_{\text{atm}}A - P_1 A$$
$$= (P_{\text{atm}} - P_1)A$$

We use Bernoulli's equation to find this pressure difference.

$$P_2 + \frac{1}{2}\rho v_2^2 + \rho g y_2 = \frac{1}{2}\rho v_1^2 + \rho g y_1 + P_1$$

$$\Rightarrow P_{\text{atm}} + 0 + \rho g y_2 = \frac{1}{2}\rho v_1^2 + \rho g y_1 + P_1$$

$$\Rightarrow P_{\text{atm}} - P_1 = \frac{1}{2}\rho v_1^2 + (\rho g y_1 - \rho g y_2) = \frac{1}{2}\rho v_1^2 + 0$$

We can now determine the net force exerted by the air on the roof.

$$F_{\text{net Air}} = (P_{\text{atm}} - P_1)A = \frac{1}{2}\rho v_1^2 A$$

**Solve and evaluate**

$$F_{\text{net Air}} = \frac{1}{2}\rho v_1^2 A$$
$$= \frac{1}{2}(1.3\ \text{kg/m}^3)(45\ \text{m/s})^2(200\ \text{m}^2)$$
$$= 2.6 \times 10^5\ \text{N}$$

The result is an upward net force that is enough to lift more than ten cars of combined mass 30,000 kg.

We were a little lax in applying Bernoulli's equation in Example 11.6. The equation relates the properties of a fluid at two points along the same streamline. A streamline does not flow between just below the roof and just above the roof. We could, with some more complex reasoning, use the equation correctly by considering two streamlines that start far from the house at the same pressure. One ends up in the house under its roof with the air barely moving. The other passes just above the roof with the air moving fast. We would get the same result in a somewhat more cumbersome manner.

**Try it yourself:** A 2.0 m $\times$ 2.0 m canvas covers a trailer. The trailer moves at 29 m/s (65 mi/h). Determine the net force exerted on the canvas by the air above and below it.

*Answer:* An upward force of 2000 N (about 500 lb). No wonder the canvas covering a truck trailer moving on a highway balloons outward.

**TIP** Remember: Hurricane winds do not tear the roofs from houses. The air inside the house actually pushes the roof upward from the house. Similarly, when you drink through a straw, the atmospheric air pushes down on the liquid in the glass, which then moves up into the lower pressure air in the straw.

## Dislodging plaque

The physical principles of a roof being lifted from a house also explain how plaque can become dislodged from the inner wall of an artery. Consider the figure in Example 11.5. The plaque may block a considerable portion of the area where blood normally flows. Suppose the radius of the vessel opening is one-third its normal value because of the plaque. Then the area available for blood flow, proportional to $r^2$, is about one-ninth the normal value. The speed of flow in the narrowed portion of the artery will be about 9 times greater than in the unblocked part of the vessel. The kinetic energy density term in Bernoulli's equation is proportional to $v^2$ and therefore is 81 times greater in the constricted area than in the open part of the vessel.

Notice that in Bernoulli's equation, the sum of the gravitational potential energy density, the kinetic energy density, and the pressure at one location should equal the sum of the same three terms at some other location along a streamline in the blood. As blood speeds by the plaque, its kinetic energy density is 81 times greater, and consequently its pressure is much less than the pressure in the open vessel just before and just after the plaque. This pressure differential could cause the plaque to be pulled off the wall and tumble downstream, causing a blood clot (a process called thrombosis). Let's estimate the net force that the blood exerts on the plaque.

**EXAMPLE 11.7 A clogged artery**

Blood flows through the unobstructed part of a blood vessel at a speed of 0.50 m/s. The blood then flows past a plaque that constricts the cross-sectional area to one-ninth the normal value. The surface area of the plaque parallel to the direction of blood flow is about $0.60\ \text{cm}^2 = 6.0 \times 10^{-5}\ \text{m}^2$. Estimate the net force that the fluid exerts on the plaque.

**Sketch and translate** A simplified sketch of the situation is shown below. Point 1 is above the plaque in stationary blood pooled in a channel where the plaque attaches to the artery wall. Point 2 is in the bloodstream below the plaque, where blood flows rapidly past it. The net force will depend on the differences in pressure at points 1 and 2. Thus, we need to first find the pressure difference. Since points 1 and 2 are not on the same streamline, we cannot automatically use Bernoulli's equation. But since the streamlines that do go through points 1 and 2 were side by side before they reached the plaque, each streamline will have the same $P + (1/2)\rho v^2 + \rho g y$ value. This means we can equate the $P + (1/2)\rho v^2 + \rho g y$ values at points 1 and 2.

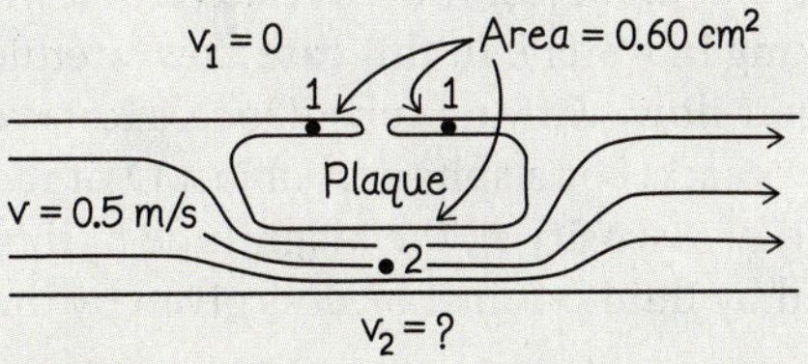


**Simplify and diagram** Assume for simplicity that the blood is nonviscous and flows with laminar flow without turbulence. Assume also that the vertical distance between points 1 and 2 is small ($y_1 - y_2 \approx 0$) and that the area of the stationary blood above the plaque is the same as the area where the blood moves below the plaque. A bar chart represents the process. The blood pressure at position 2 is less than at position 1 because the blood flows at high speed through the constricted artery, whereas it sits at rest in the channels at position 1($v_1 = 0$).

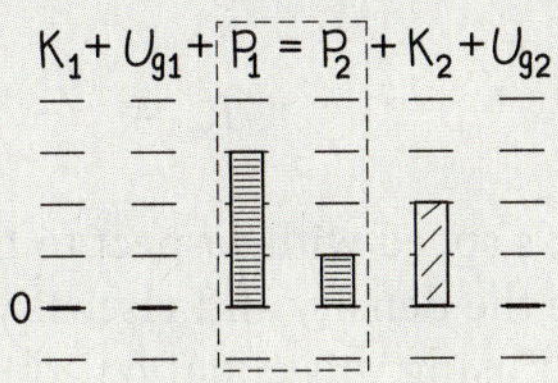


**Represent mathematically** Compare the two points using Bernoulli's equation:

$$P_1 - P_2 = \frac{1}{2}\rho\,(v_2^2 - v_1^2) + \rho g\,(y_2 - y_1)$$

$$= \frac{1}{2}\rho\,(v_2^2 - 0) + 0 = \frac{1}{2}\rho\, v_2^2$$

Since the cross-sectional area of the vessel at the location of the plaque is one-ninth its normal area (the radius is one-third the normal value), the blood must be flowing at nine times its normal speed. The net force exerted by the blood on the plaque downward and perpendicular to the direction the blood flows will be

$$F_{\text{net blood on P}\,y} = F_{\text{blood 1 on P}} + (-F_{\text{blood 2 on P}})$$

$$= P_1 A - P_2 A = (P_1 - P_2)A$$

$$\Rightarrow F_{\text{net blood on P}\,y} = \frac{1}{2}\rho\, v_2^2 A$$

**Solve and evaluate**

$$F_{\text{net blood on P}\,y}$$

$$= \frac{1}{2}(1050\ \text{kg/m}^3)\,(9 \times 0.5\ \text{m/s})^2(6.0 \times 10^{-5}\ \text{m}^2)$$

$$= 0.64\ \text{N} \approx 0.6\ \text{N}.$$

This is about the weight of one-half of an apple pulling on this tiny plaque. In addition, an "impact" force caused by blood hitting the plaque's upstream side contributes to the risk of breaking the plaque off the side wall of the vessel. The loose plaque can then tumble downstream and block blood flow in a smaller vessel in the heart (causing a heart attack) or in the brain (causing a stroke).

**Try it yourself:** Air of density $1.3\ \text{kg/m}^3$ moves at speed 10 m/s across the top surface of a clarinet reed that has an area of $3\ \text{cm}^2$. The air below the reed is not moving and is at atmospheric pressure. Determine the net force exerted on the reed by the air above and below it.

*Answer:* 0.02 N upward, toward the inside of the mouthpiece.

**Figure 11.12** Measuring blood pressure

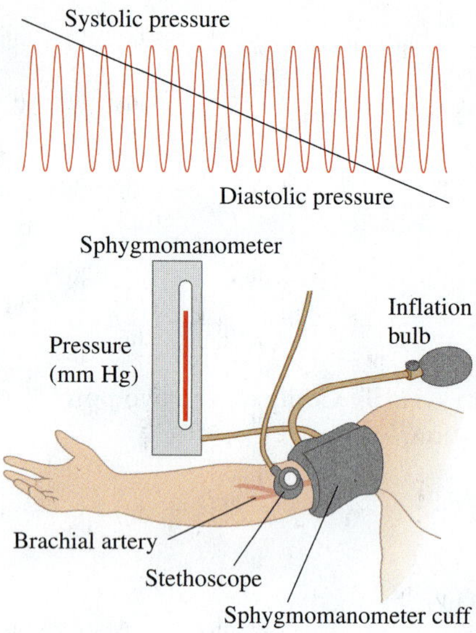


## Measuring blood pressure

To measure a person's blood pressure, a nurse uses a device called a sphygmomanometer (see **Figure 11.12**). She places a cuff around the upper arm of a patient at about the level of the heart and places a stethoscope on the inside of the elbow above the brachial artery in the arm. The nurse then increases the gauge pressure in the cuff to about 180 mm Hg by pumping air into the cuff. The expanded cuff pushes on the brachial artery and stops blood flow in the arm. Then the nurse slowly releases the air from the cuff, decreasing the pressure of the air in it. When the pressure in the cuff is equal to the systolic pressure (120 mm Hg if the systolic blood pressure is normal), blood starts to squeeze through the artery past the cuff. The flow is intermittent and turbulent and causes a sound heard with the stethoscope. This turbulent sound continues until the cuff pressure decreases below the diastolic pressure (80 mm Hg for normal diastolic blood pressure). At that point the artery is continually open and blood flow is laminar and makes no sound. The systolic and diastolic pressure numbers together make up the blood pressure measurement.

**Review Question 11.7** Why does a roof lift up off a house during a high-speed wind?

# 11.8 Drag force

So far in this chapter all of our analyses have focused on a moving fluid—air flowing past a roof or blood flowing through an artery. In Section 11.6 we were concerned with the friction-like resistance that occurs as these fluids move through a tube. Now we focus on solid objects moving through a fluid—for example, a swimmer moving through water, a skydiver falling through the air, or a car traveling through air. As you know from experience, the fluid in these and in other cases exerts a resistive **drag force** on the object moving through the fluid. So far we have been neglecting this force in our mechanics problems. Now we will not only learn how to calculate this force but also learn whether our assumptions were reasonable: for example, is the resistive drag force insignificant when people and objects fall from small and large heights?

**Laminar Drag Force** Imagine that an object moves relatively slowly through a fluid (for example, a rock sinking in water). In this case the water flows around the object in streamline laminar flow, with no turbulence. However, the fluid does exert a drag force on the object. For a spherical object O of radius $r$ moving at speed $v$ through a liquid of viscosity $\eta$, the magnitude of this nonturbulent drag force $F_{\text{D F on O}}$ exerted by fluid on the object is given by the equation

$$F_{\text{D F on O}} = 6\pi\eta r v \tag{11.10}$$

This equation is called **Stokes's law**. Notice that the drag force is proportional to the speed of the object relative to the fluid and to the radius of the object.

**Turbulent Drag Force** A rock falls through air much faster than through water. In this case the motion of the air past the falling rock is turbulent and Eq. (11.10) does not apply. A different Reynolds number can be used to decide whether the flow of fluid past an object is laminar or turbulent:

$$R_e = \frac{v l \rho}{\eta} \tag{11.11}$$

where $v$ is the object's speed with respect to the fluid, $l$ is the length of the object, and $\rho$ and $\eta$ are the density and viscosity of the fluid. When the Reynolds number is calculated using this equation, the threshold value for the laminar flow is 1. If the Reynolds number is more than 1, the flow is turbulent and we cannot use Stokes's law. In this case, a new equation for drag force applies:

$$F_{\text{D F on O}} \approx \frac{1}{2} C_{\text{D}} \rho A v^2 \qquad (11.12)$$

where $\rho$ is the density of fluid, $A$ is the cross-sectional area of the object as seen along its line of motion, and $C_{\text{D}}$ is a dimensionless number called the *drag coefficient*. The drag coefficient depends on the shape of the object (the lower the number, the smaller the drag force and the more streamline the flow past the object).

## Drag force exerted on a moving vehicle

Does Stokes's law apply to moving cars? At 60 mi/h (about 30 m/s), for a car about 2 m wide in air with density 1.3 kg/m$^3$ and viscosity $2 \times 10^{-5}\ \text{N} \cdot \text{s/m}^2$, the estimated Reynolds number will be

$$R_{\text{e}} = \frac{vl\rho}{\eta} = \frac{(30\ \text{m/s})(2\ \text{m})(1.3\ \text{kg/m}^3)}{(2 \times 10^{-5}\ \text{N} \cdot \text{s/m}^2)} \approx 4 \times 10^6.$$

This is much more than 1. We need to use Eq. (11.12) for the drag force.

**QUANTITATIVE EXERCISE 11.8 Drag force exerted on a car**

Estimate the magnitude of the drag force that air exerts on a compact car traveling at 27 m/s (60 mi/h). The drag coefficient $C_{\text{D}}$ is approximately 0.5 for a well-designed car, and the air density is 1.3 kg/m$^3$.

**Represent mathematically** The flow of air past the car is turbulent, and we use Eq. (11.12) to estimate the drag force that the air exerts on the car:

$$F_{\text{D F on O}} = \frac{1}{2} C_{\text{D}} \rho A v^2$$

**Solve and evaluate** We estimate that the cross-sectional area of a car is 2 m$^2$. Thus,

$$F_{\text{D F on O}} = \frac{1}{2}(0.5)(1.3\ \text{kg/m}^3)(2\ \text{m}^2)(27\ \text{m/s})^2 = 470\ \text{N}$$

or a force of about 100 lb. Designing cars to minimize drag force improves fuel economy.

**Try it yourself:** Estimate the drag force (without using a calculator) if the car's speed is 13 m/s, half the value in the quantitative exercise.

*Answer:* About one-fourth the answer above, or about 120 N.

## Drag force exerted on a falling person

We can estimate whether the drag force exerted on a person falling from a building is significant. For example, assume that a 70-kg person accidentally falls from a second-floor window. The force exerted on him by Earth is about 700 N. The buoyant force exerted on the person by the air is about

$$F_{\text{B A on P}} = \rho_{\text{air}} g V_{\text{body}} \approx (1.3\ \text{kg/m}^3)(9.8\ \text{N/kg})\frac{(70\ \text{kg})}{(1000\ \text{kg/m}^3)} \approx 0.9\ \text{N}$$

where the person's volume is his mass divided by his density. We see that the buoyant force is very small. To estimate the drag force, assume that the drag coefficient is 1, that the person's cross-sectional area is about $(1.6\ \text{m})(0.3\ \text{m}) = 0.5\ \text{m}^2$, and that the air density is 1.3 kg/m$^3$. The speed after falling freely a distance of 2 m is about 6 m/s ($v_{\text{f}}^2 = 2g(h_{\text{f}} - h_{\text{i}})$). Thus, our estimate for the drag force exerted by air on the person after falling 2 m is about

$$F_{\text{D F on P}} = \frac{1}{2}(1)(1.3\ \text{kg/m}^3)(0.5\ \text{m}^2)(6\ \text{m/s})^2 \approx 12\ \text{N}$$

**TIP** Note that if the speed of a car increases by two times, the drag force exerted on it quadruples. Thus, because of air drag, when you increase your driving speed, you reduce your gas mileage.

Thus, for a person falling from small heights, the drag force is significant but small. What about falls from higher heights?

## Terminal speed

As a skydiver falls through the air, her speed increases and the drag force that the air exerts on the diver also increases. Eventually, the diver's speed becomes so great that the resistive drag force that the air exerts on the diver equals the downward gravitational force that Earth exerts on the diver. The net force exerted on the diver is balanced, so the diver moves downward at a constant speed, known as **terminal speed**. Let's estimate the terminal speed for a skydiver.

**EXAMPLE 11.9 Terminal speed of skydiver**

Estimate the terminal speed of a 60-kg skydiver falling through air of density $1.3\ \text{kg/m}^3$, assuming a drag coefficient $C_D = 0.6$.

**Sketch and translate** The situation is sketched below. When the diver is moving at terminal speed, the forces that the air exerts on the diver and that Earth exerts on the diver balance—the net force is zero. We choose the diver as the system of interest with vertical $y$-axis pointing upward.

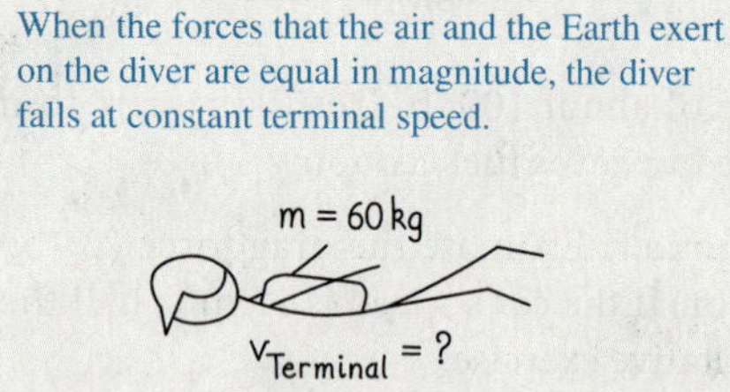

When the forces that the air and the Earth exert on the diver are equal in magnitude, the diver falls at constant terminal speed.
m = 60 kg
$v_{\text{Terminal}}$ = ?


**Simplify and diagram** A force diagram for the diver is shown to the right. Assume that the buoyant force that the air exerts on the diver is negligible in comparison to the other forces exerted on her and that the drag force involves turbulent airflow past the diver.

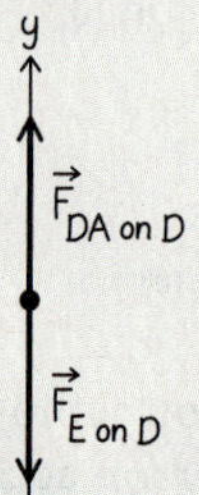


**Represent mathematically** Use the force diagram to help apply Newton's second law for the diver:

$$ma_y = \Sigma F_y$$
$$\Rightarrow 0 = F_{\text{D A on D}} + (-F_{\text{E on D}})$$
$$\Rightarrow 0 = \frac{1}{2} C_D \rho A v_{\text{terminal}}^2 - mg$$

**Solve and evaluate** Solving for the diver's terminal speed gives

$$v_{\text{terminal}} = \sqrt{\frac{2mg}{C_D \rho A}}$$

All of the quantities in the above expression are known except the cross-sectional area of the diver along her line of motion. If we assume that she is 1.5 m tall and 0.3 m wide, her cross-sectional area is about $0.5\ \text{m}^2$. We find that her terminal speed is

$$v_{\text{terminal}} = \sqrt{\frac{2(60\ \text{kg})(9.8\ \text{N/kg})}{(0.6)(1.3\ \text{kg/m}^3)(0.5\ \text{m}^2)}} = 55\ \text{m/s}$$

The unit is correct. The magnitude seems reasonable—about 120 mi/h.

**Try it yourself:** Suppose she pulled her legs near her chest so she was more in the shape of a ball. How qualitatively would that affect her terminal speed? Explain.

*Answer:* Her cross-sectional area would be smaller, and according to the above equation, her terminal speed would be greater.

**Review Question 11.8** When a skydiver falls downward at constant terminal speed, shouldn't the resistive drag force that the air exerts on the skydiver be a little less than the downward gravitational force that Earth exerts on the diver? If they are equal, shouldn't the diver stop falling? Explain.

# Summary

| Words | Pictorial and physical representations | Mathematical representation |
|---|---|---|
| **Flow rate** The flow rate $Q$ of a fluid is the volume $V$ of fluid that passes a cross section in a tube divided by the time interval $\Delta t$ needed for that volume to pass. The flow rate also equals the product of the average speed $v$ of the fluid and the cross-sectional area A of the vessel. (Section 11.2) | 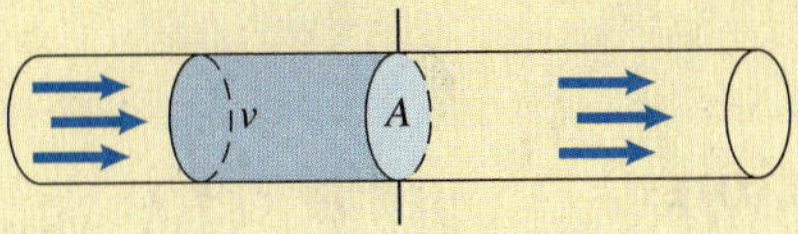 | *Flow rate Q:* $Q = \dfrac{V}{\Delta t} = vA$ Eq. (11.2) |
| **Continuity equation** If fluid does not accumulate, the flow rate into a region (position 1) must equal the flow rate out of the region (position 2). At position 1 the fluid has speed $v_1$ and the tube has cross-sectional area $A_1$. At position 2 the fluid has speed $v_2$ and the tube has cross-sectional area $A_2$. (Section 11.2) | 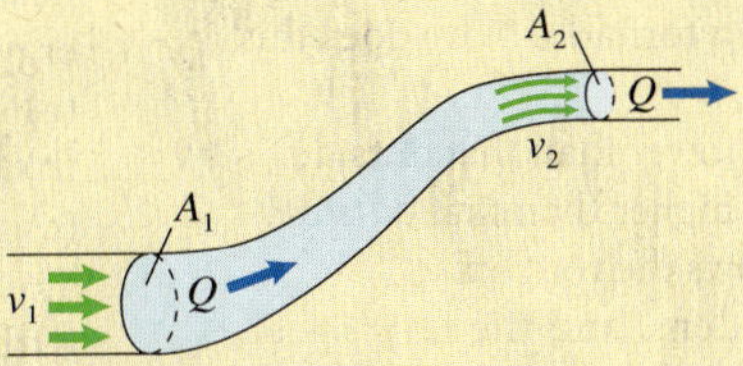 | *Continuity equation:* $Q = v_1A_1 = v_2A_2$ Eq. (11.3) |
| **Bernoulli's equation** For a fluid flowing without friction or turbulence, the sum of the kinetic energy density $(1/2)\rho v^2$, the gravitational potential energy density $\rho g y$, and pressure $P$ of the fluid is a constant. (Section 11.4) | 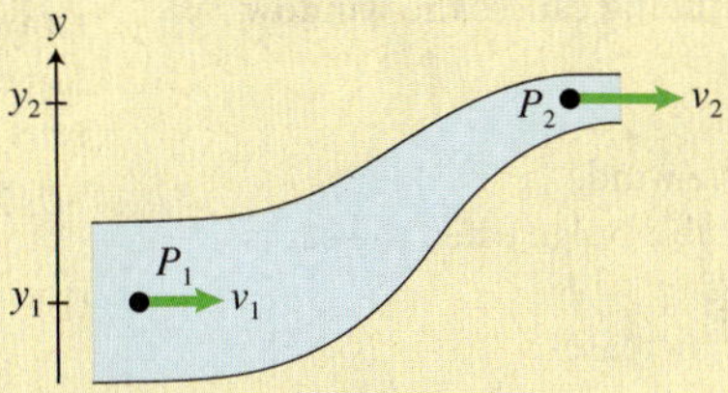 | *Bernoulli's equation:* $\frac{1}{2}\rho v_1^2 + \rho g y_1 + P_1 = P_2 + \frac{1}{2}\rho v_2^2 + \rho g y_2$ Eq. (11.6) |
| **Poiseuille's law** For viscous fluid flow, the pressure drop $(P_1 - P_2)$ across a fluid of viscosity $\eta$ flowing in a tube depends on the length $l$ of the tube, its radius $r$, and the fluid flow rate $Q$. (Section 11.6) | 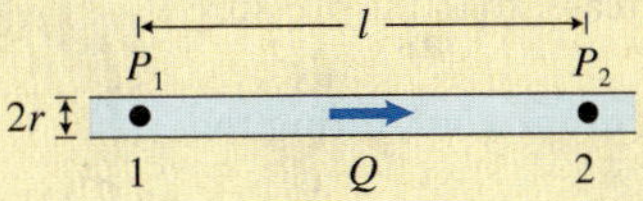 | *Poiseuille's law:* $P_1 - P_2 = \dfrac{8}{\pi}\dfrac{\eta l}{r^4}Q$ Eq. (11.8) |
| **Laminar drag force** When a spherical object (like a balloon falling in air) moves slowly through a fluid, the fluid exerts a resistive drag force on the object that is proportional to the object's speed $v$. **Stokes's law** describes the force. (Section 11.8) |  | *Laminar drag force (Stokes's law):* $F_D = 6\pi\eta r v$ Eq. (11.10) |
| **Turbulent drag force** For an object moving at faster speed through a fluid (like a tilted airplane wing), turbulence occurs and the resistive drag force is proportional to the square of the speed. (Section 11.8) |  | *Turbulent drag force:* $F_D = \frac{1}{2}C_D\rho A v^2$ Eq. (11.12) |

MP For instructor-assigned homework, go to MasteringPhysics.

# Questions

## Multiple Choice Questions

1. Two empty soda cans stand on a smooth tabletop. If you blow horizontally between the cans, what will they do?
   (a) Move further apart.
   (b) Move closer together.
   (c) Stay where they are.
2. A roof is blown off a house during a tornado. Why does this happen?
   (a) The air pressure in the house is lower than that outside.
   (b) The air pressure in the house is higher than that outside.
   (c) The wind is so strong that it blows the roof off.
3. A river flows downstream and widens, and the flow speed slows. As a result, the pressure of the water against a dock downstream compared to upstream will be
   (a) higher. (b) lower. (c) the same.
4. Air blowing past the window of a building causes the window to be blown
   (a) out. (b) in.
   (c) It depends on the direction of the wind.
5. Why does the closed top of a convertible bulge when the car is riding along a highway?
   (a) The volume of air inside the car increases.
   (b) The air pressure is greater outside the car than inside.
   (c) The air pressure inside the car is greater than the pressure outside.
   (d) The air blows into the front part of the roof, lifting the back part.
6. How does Bernoulli's principle *help* explain air going up the chimney of a house?
   (a) Air blowing across the top of the chimney reduces the pressure above the chimney.
   (b) The air above the chimney attracts the ashes.
   (c) The hot ashes seek the cooler outside air.
   (d) The gravitational potential energy is lower above the chimney.
7. Why does cutting the end of an envelope and blowing air past the cut end cause the envelope to bulge open?
   (a) The blown air goes in and increases the pressure inside.
   (b) The blown air blowing past the outside exerts less pressure than air inside.
   (c) The person blowing tends to squish the edges of the envelope.
8. As a river approaches a dam, the width of the river increases and the speed of the flowing water decreases. What can explain this effect?
   (a) Bernoulli's equation
   (b) The continuity equation
   (c) Poiseuille's law
9. What is an incompressible fluid?
   (a) A law of physics
   (b) A physical quantity
   (c) A model of an object
10. What is viscous flow?
    (a) A physical phenomenon
    (b) A law of physics
    (c) A physical quantity
11. The heart does about 1 J of work pumping blood during one heartbeat. What is the immediate first and main type of energy that increases due to the heart's work?
    (a) Kinetic energy
    (b) Thermal energy
    (c) Elastic potential energy
12. You have a glass of water with a straw in it (see **Figure Q11.12**). The end of the straw is about 1 cm above the water. You now blow hard through a second straw so that air moves across the top of the first straw (do not blow into it). Predict what happens.
    (a) The water level in the straw goes down.
    (b) The water level in the straw goes up and possibly sprays out of the straw.
    (c) The water level in the straw is not affected by the air moving above it.

**Figure Q11.12**

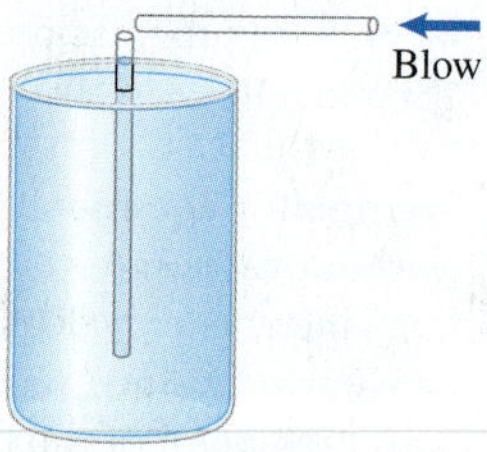


## Conceptual Questions

13. A hair dryer blowing air over a ping-pong ball will support it, as shown in **Figure Q11.13**. Construct a force diagram for the ball. Explain in terms of forces how the ball can remain in equilibrium.

**Figure Q11.13**

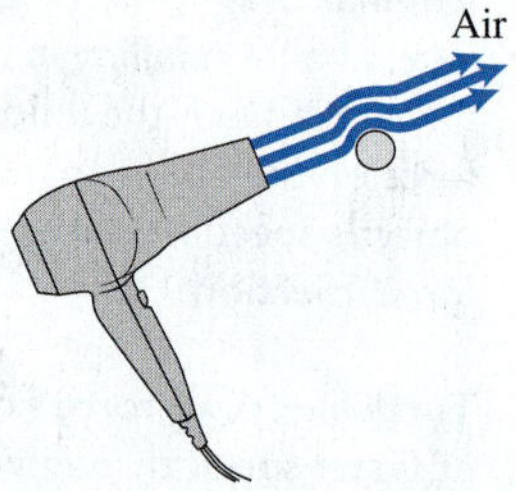


14. You have two identical large jugs with small holes on the side near the bottom. One jug is filled with water and the other with liquid mercury. The liquid in each jug, sitting on a table, squirts out the side hole into a container on the floor. Which container, the one catching the water or the one catching the mercury, must be closer to the table in order to catch the fluid? Or should they be placed at the same distance? Which jug will empty first, or do they empty at the same time? Explain.
15. Why does much of the pressure drop in the circulatory system occur across the arterioles (small vessels carrying blood to the capillaries) and capillaries as opposed to across the much larger diameter arteries?
16. If you partly close the end of a hose with your thumb, the water squirts out farther. Give at least one explanation for why this phenomenon occurs.

# Problems

Below, BIO indicates a problem with a biological or medical focus. Problems labeled EST ask you to estimate the answer to a quantitative problem rather than develop a specific answer. Problems marked with ✎ require you to make a drawing or graph as part of your solution. Asterisks indicate the level of difficulty of the problem. Problems with no * are considered to be the least difficult. A single * marks moderately difficult problems. Two ** indicate more difficult problems. Unless stated otherwise, assume in these problems that atmospheric pressure is $1.01 \times 10^5\ \text{N/m}^2$ and that the densities of water and air are $1000\ \text{kg/m}^3$ and $1.3\ \text{kg/m}^3$, respectively.

### 11.1 and 11.2 Fluids moving across surfaces—qualitative analysis and Flow rate and fluid speed

1. **Watering plants** You water flowers outside your house. (a) Determine the flow rate of water moving at an average speed of 32 cm/s through a garden hose of radius 1.2 cm. (b) Determine the speed of the water in a second hose of radius 1.0 cm that is connected to the first hose.
2. **Irrigation canal** You live near an irrigation canal that is filled to the top with water. (a) It has a rectangular cross section of 5.0-m width and 1.2-m depth. If water flows at a speed of 0.80 m/s, what is its flow rate? (b) If the width of the stream is reduced to 3.0 m and the depth to 1.0 m as the water passes a flow-control gate, what is the speed of the water past the gate?
3. **Fire hose** During a fire, a firefighter holds a hose through which $0.070\ \text{m}^3$ of water flows each second. The water leaves the nozzle at an average speed of 25 m/s. What information about the hose can you determine using these data?
4. The main waterline for a neighborhood delivers water at a maximum flow rate of $0.010\ \text{m}^3/\text{s}$. If the speed of this water is 0.30 m/s, what is the pipe's radius?
5. * BIO **Blood flow in capillaries** The flow rate of blood in the aorta is $80\ \text{cm}^3/\text{s}$. Beyond the aorta, this blood eventually travels through about $6 \times 10^9$ capillaries, each of radius $8.0 \times 10^{-4}$ cm. What is the speed of the blood in the capillaries?
6. * **Irrigating a field** It takes a farmer 2.0 h to irrigate a field using a 4.0-cm-diameter pipe that comes from an irrigation canal. How long would the job take if he used a 6.0-cm pipe? What assumption did you make? If this assumption is not correct, how will your answer change?

### 11.4 Bernoulli's equation

7. ✎ Represent the process sketched in **Figure P11.7** using a qualitative Bernoulli bar chart and an equation (include only terms that are not zero).

**Figure P11.7**

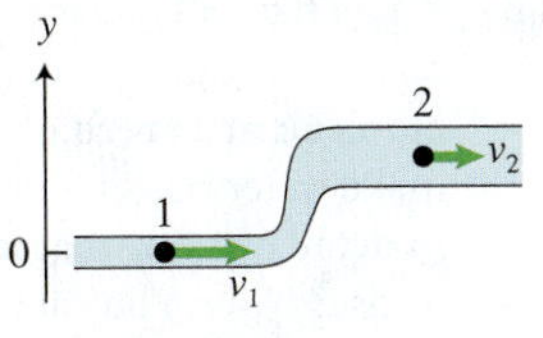


8. ✎ Represent the process sketched in **Figure P11.8** using a qualitative Bernoulli bar chart and an equation (include only terms that are not zero).

**Figure P11.8**

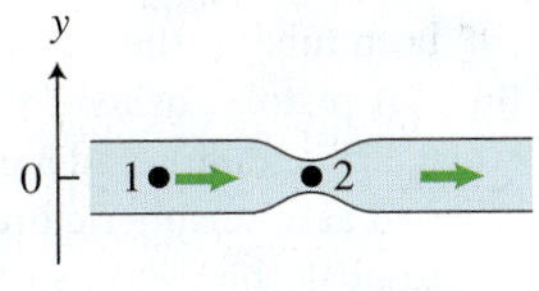


9. ✎ **Fluid flow problem** Write a symbolic equation (include only terms that are not zero) and draw a sketch of a situation that could be represented by the qualitative Bernoulli bar chart shown in **Figure P11.9** (there are many possibilities).
10. Repeat Problem 9 using the bar chart in **Figure P11.10**.
11. Repeat Problem 9 using the bar chart in **Figure P11.11**.
12. Repeat Problem 9 using the bar chart in **Figure P11.12**.

**Figure P11.9**

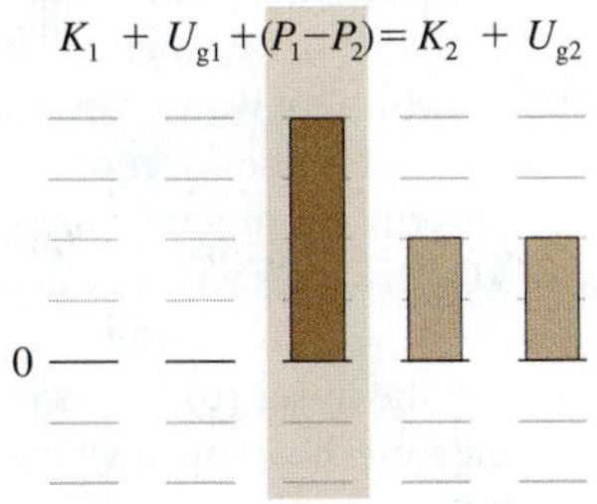


**Figure P11.10**

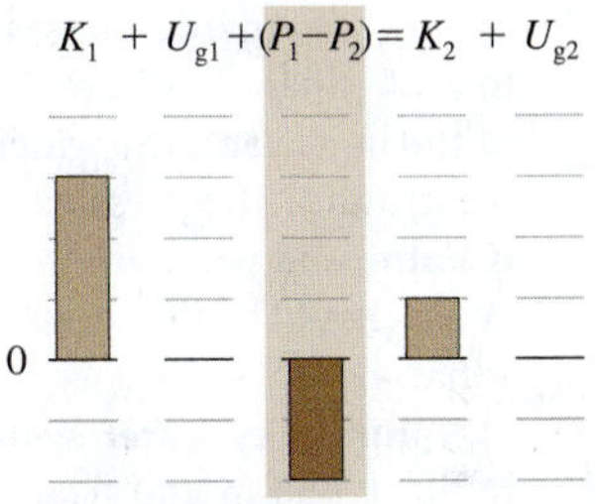


**Figure P11.11**

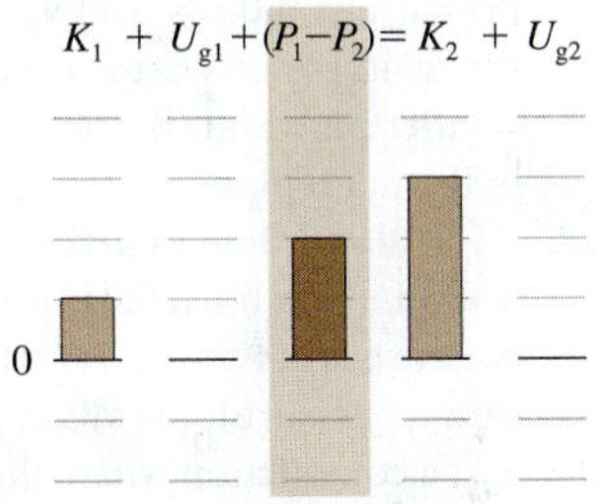


**Figure P11.12**

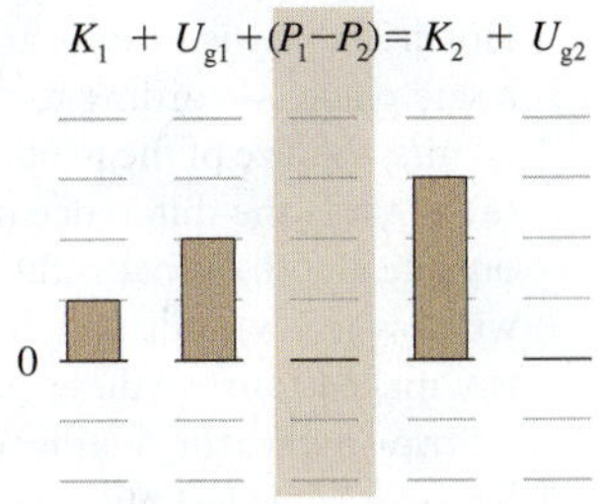


13. ✎ An application of Bernoulli's equation is shown below. Construct a qualitative Bernoulli bar chart that is consistent with the equation and draw a sketch of a situation that could be represented by the equation (there are many possibilities).
$\rho g y_2 = 0.5 \rho v_1^{\,2}$
14. Repeat Problem 13 using the equation below. The size of the symbols represents the relative magnitudes of the physical quantities at two points.
$0.5\rho v_1^2 + (P_1 - P_2) = 0.5\rho v_2^2$
15. Repeat Problem 13 using the equation below. The size of the symbols represents the relative magnitudes of the physical quantities at two points.
$0.5\rho v_1^2 + (P_1 - P_2) = 0.5\rho v_2^2 + \rho g y_2$
16. * ✎ **Wine flow from barrel** While visiting a winery, you observe wine shooting out of a hole in the bottom of a barrel. The top of the barrel is open. The hole is 0.80 m below the top surface of the wine. Represent this process in multiple ways (a sketch, a bar chart, and an equation) and apply Bernoulli's equation to a point at the top surface of the wine and another point at the hole in the barrel.
17. ✎ **Water flow in city water system** Water is pumped at high speed from a reservoir into a large-diameter pipe. This pipe connects to a smaller diameter pipe. There is no change in elevation. Represent the water flow from the large pipe to the smaller pipe in multiple ways—a sketch, a bar chart, and an equation.

## 11.5 Skills for analyzing processes using Bernoulli's equation

18. * The pressure of water flowing through a 0.060-m-radius pipe at a speed of 1.8 m/s is $2.2 \times 10^5\ N/m^2$. What is (a) the flow rate of the water and (b) the pressure in the water after it goes up a 5.0-m-high hill and flows in a 0.050-m-radius pipe?
19. * **Siphoning water** You want to siphon rainwater and melted snow from the cover of an above-ground swimming pool. The cover is 1.4 m above the ground. You have a plastic hose of 1.0-cm radius with one end in the water on the pool cover and the other end on the ground. (a) At what speed does water exit the hose? (b) If you want to empty the pool cover in half the time, how much wider should the hose be? (c) How much faster does the water flow through this wider pipe?
20. * **Cleaning skylights** You are going to wash the skylights in your kitchen. The skylights are 8.0 m above the ground. You connect two garden hoses together—a 0.80-cm-radius hose to a 1.0-cm-radius hose. The smaller hose is held on the roof of the house and the wider hose is attached to the faucet on the ground. The pressure at the opening of the smaller hose is 1 atm, and you want the water to have the speed of 6.0 m/s. What should be the pressure at ground level in the large hose? What should be the speed?
21. ** **Community water system** A large city waterline pipe has radius 0.060 m and feeds ten smaller pipes, each of radius 0.020 m, that carry water to homes. The flow rate of water in each of the smaller pipes is to be $6.0 \times 10^{-3}\ m^3/s$, and the pressure is $4.00 \times 10^5\ N/m^2$. The homes are 10.0 m above the main pipe. What is the average speed of the water in (a) a smaller pipe and (b) the main pipe? (c) What is the pressure in the main pipe?
22. * BIO **Blood flow in artery** Blood flows at an average speed of 0.40 m/s in a horizontal artery of radius 1.0 cm. The average pressure is $1.4 \times 10^4\ N/m^2$ above atmospheric pressure (the gauge pressure). (a) What is the average speed of the blood past a constriction where the radius of the opening is 0.30 cm? (b) What is the gauge pressure of the blood as it moves past the constriction?
23. * **Window in wind** You are on the 48th floor of the Windom Hotel. The day is stormy, and you wonder whether the hotel is a safe place. According to the weather report, the air speed is 20 m/s; the size of the window in your room is 1.0 m × 2.0 m. (a) What is the difference in pressure between the inside and outside air? (b) What is the net force that the air exerts on the window (magnitude and direction)? (c) What assumption did you make to answer these questions?
24. * **Straw aspirator** A straw extends out of a glass of water by a height $h$. How fast must air blow across the top of the straw to draw water to the top of the straw?
25. * **Gate for irrigation system** You observe water at rest behind an irrigation dam. The water is 1.2 m above the bottom of a gate that, when lifted, allows water to flow under the gate. Determine the height $h$ from the bottom of the dam that the gate should be lifted to allow a water flow rate of $1.0 \times 10^{-2}\ m^3/s$. The gate is 0.50 m wide.

## 11.6 Viscous fluid flow

26. * A 5.0-cm-radius horizontal water pipe is 500 m long. Water at 20 °C flows at a rate of $1.0 \times 10^{-2}\ m^3/s$. (a) Determine the pressure drop due to viscous friction from the beginning to the end of the pipe. (b) What radius pipe must you use if you want to keep the pressure difference constant and double the flow rate?
27. **Fire hose** A volunteer firefighter uses a 5.0-cm-diameter fire hose that is 60 m long. The water moves through the hose at 12 m/s. The temperature outside is 20 °C. What is the pressure drop due to viscous friction across the hose?
28. **Another fire hose** The pump for a fire hose can develop a maximum pressure of $6.0 \times 10^5\ N/m^2$. A horizontal hose that is 50 m long is to carry water of viscosity $1.0 \times 10^{-3}\ N \cdot s/m^2$ at a flow rate of $1.0\ m^3/s$. What is the minimum radius for the hose?
29. * **Solar collector water system** Water flows in a solar collector through a copper tube of radius $R$ and length $l$. The average temperature of the water is $T$ °C and the flow rate is $Q\ cm^3/s$. Explain how you would determine the viscous pressure drop along the tube, assuming the water does not change elevation.
30. * BIO **Blood flow through capillaries** Your heart pumps blood at a flow rate of about $80\ cm^3/s$. The blood flows through approximately $9 \times 10^9$ capillaries, each of radius $4 \times 10^{-4}$ cm and 0.1 cm long. Determine the viscous friction pressure drop across a capillary, assuming a blood viscosity of $4 \times 10^{-3}\ N \cdot s/m^2$.

## 11.7 Applying fluid dynamics: Putting it all together

31. * BIO **Flutter in blood vessel** A person has a 5200-N/m² gauge pressure of blood flowing at 0.50 m/s inside a 1.0-cm-radius main artery. The gauge pressure outside the artery is $3200\ N/m^2$. When using his stethoscope, a physician hears a fluttering sound farther along the artery. The sound is a sign that the artery is vibrating open and closed, which indicates that there must be a constriction in the artery that has reduced its radius and subsequently reduced the internal blood pressure to less than the external 3200-N/m² pressure. What is the maximum artery radius at this construction?
32. * BIO **Effect of smoking on arteriole radius** The average radius of a smoker's arterioles, the small vessels carrying blood to the capillaries, is 5% smaller than those of a nonsmoker. (a) Determine the percent change in flow rate if the pressure across the arterioles remains constant. (b) Determine the percent change in pressure if the flow rate remains constant.
33. * **Roof of house in wind** The mass of the roof of a house is $2.1 \times 10^4$ kg and the area of the roof is $160\ m^2$. At what speed must air move across the roof of the house so that the roof is lifted off the walls? Indicate any assumptions you made.
34. * You have a U-shaped tube open at both ends. You pour water into the tube so that it is partially filled. You have a fan that blows air at a speed of 10 m/s. (a) How can you use the fan to make water rise on one side of the tube? Explain your strategy in detail. (b) To what maximum height can you get the water to rise? Note: You cannot touch the water yourself.
35. * Determine the ratio of the flow rate through capillary tubes A and B (that is, $Q_A/Q_B$). The length of A is twice that of B, and the radius of A is one-half that of B. The pressure across both tubes is the same.
36. * A piston pushes 20 °C water through a horizontal tube of 0.20-cm radius and 3.0-m length. One end of the tube is open and at atmospheric pressure. (a) Determine the force needed to push the piston so that the flow rate is $100\ cm^3/s$. (b) Repeat the problem using SAE 10 oil instead of water.

37. * Engineers use a venturi meter to measure the speed of a fluid traveling through a pipe (see **Figure P11.37**). Positions 1 and 2 are in pipes with surface areas $A_1$ and $A_2$, with $A_1$ greater than $A_2$, and are at the same vertical height. How can you determine the relative speeds at positions 1 and 2 and the pressure difference between positions 1 and 2?

**Figure P11.37**

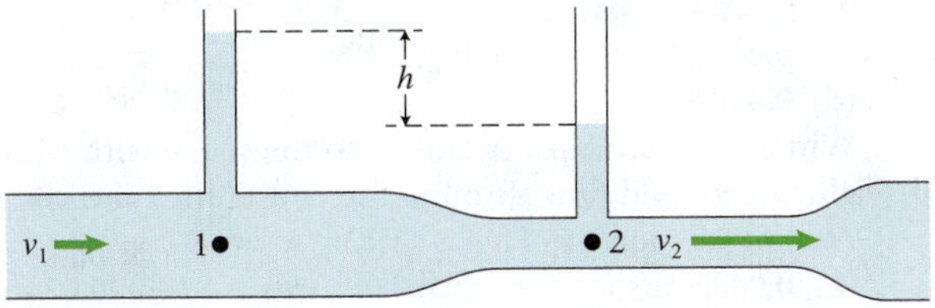


38. * How can you use the venturi meter system (see Problem 37) to determine whether viscous fluid needs an additional pressure difference to flow at the same speed as a nonviscous fluid?

### 11.8 Drag force

39. **Car drag** A 2300-kg car has a drag coefficient of 0.60 and an effective frontal area of 2.8 $m^2$. Determine the air drag force on the car when traveling at (a) 24 m/s (55 mi/h) and (b) 31 m/s (70 mi/h).

40. * EST **Air drag when biking** Estimate the drag force opposing your motion when you ride a bicycle at 8 m/s.

41. BIO **Drag on red blood cell** Determine the drag force on a red blood cell with a radius of $1.0 \times 10^{-5}$ m and moving through 20 °C water at speed $1.0 \times 10^{-5}$ m/s. (Assume laminar flow.)

42. * BIO EST **Protein terminal speed** A protein of radius $3.0 \times 10^{-9}$ m falls through a tube of water with viscosity $\eta = 1.0 \times 10^{-3}\ \text{N} \cdot \text{s/m}^2$. Earth exerts a constant downward $3.0 \times 10^{-22}$-N force on the protein. (a) Use Stokes's law and the information provided to estimate the terminal speed of the protein. Assume no buoyant force is exerted on the protein. (b) How many hours would be required for the protein to fall 0.10 m?

43. * Earth exerts a constant downward force of $7.5 \times 10^{-13}$ N on a clay particle. The particles settle 0.10 m in 820 min. Determine the radius of a clay particle. Assume no buoyant force is exerted on the clay particle. The viscosity of water is $1.0 \times 10^{-3}\ \text{N} \cdot \text{s/m}^2$.

44. * A sphere falls through a fluid. Earth exerts a constant downward 0.50-N force on the sphere. The fluid exerts an opposing drag force on the fluid given by $F_D = 2v$, where $F_D$ is in newtons if $v$ is in meters per second. Determine the terminal speed of the sphere.

45. * **Terminal speed of balloon** A balloon of mass $m$ drifts down through the air. The air exerts a resistive drag force on the balloon described by the equation $F_D = 0.03v^2$ where $F_D$ is in newtons if $v$ is in meters per second. What is the terminal speed of the balloon?

## General Problems

46. ** EST A cooler filled with water has a hole of radius 0.40 cm at the bottom. The hole is originally closed with a plug. The cooler is about 1.0 m tall, and the bottom has area 0.4 m × 0.6 m. Determine the initial flow rate of water after removing the plug. Estimate how long it will take to empty the cooler. What assumptions did you make? If they are not valid, will the real time be greater or smaller than the estimate?

47. ** Design an elevator-like device that can lift you to your dorm room by blowing air across the top surface of the elevator. Be sure to provide the details in your design and indicate any difficulties you might encounter.

48. ** BIO **Pressure needed for intravenous needle** A glucose solution of viscosity $2.2 \times 10^{-3}\ \text{N} \cdot \text{s/m}^2$ and density $1030\ \text{kg/m}^3$ flows from an elevated open bag into a vein. The needle into the vein has a radius of 0.20 mm and is 3.0 cm long. All other tubes leading to the needle have much larger radii, and viscous forces in them can be ignored. The pressure in the vein is $1000\ \text{N/m}^2$ above atmospheric pressure. (a) Determine the pressure relative to atmospheric pressure needed at the entrance of the needle to maintain a flow rate of $0.10\ \text{cm}^3/\text{s}$. (b) To what elevation should the bag containing the glucose be raised to maintain this pressure at the needle?

49. ** **Viscous friction with Bernoulli** We can include the effect of viscous friction in Bernoulli's equation by adding a term for the thermal energy generated by the viscous retarding force exerted on the fluid. Show that the term to be added to Eq. (11.5) for flow in a vessel of uniform cross-sectional area $A$ is

$$\frac{\Delta U_{\text{Th}}}{V} = \frac{4\pi\eta l v}{A}$$

where $v$ is the average speed of the fluid of viscosity $\eta$ along the center of a pipe whose length is $l$.

50. ** (a) Show that the work $W$ done per unit time $\Delta t$ by viscous friction in a fluid with a flow rate $Q$ across which there is a pressure drop $\Delta P$ is

$$\frac{W}{\Delta t} = \Delta P Q = Q^2 R = \frac{\Delta P^2}{R}$$

where $R = 8\eta l/\pi r^4$ is called the *flow resistance* of the fluid moving through a vessel of radius $r$. (b) By what percentage must the work per unit time increase if the radius of a vessel decreases by 10% and all other quantities including the flow rate remain constant (the pressure does not remain constant)?

51. ** EST BIO **Thermal energy in body due to viscous friction** Estimate the thermal energy generated per second in a normal body due to the viscous friction force in blood as it moves through the circulatory system.

52. ** BIO **Essential hypertension** Suppose your uncle has hypertension that causes the radii of his 40,000 arterioles to decrease by 20%. Each arteriole initially was 0.010 mm in radius and 1.0 cm long. By what factor does the resistance $R = 8\eta l/\pi r^4$ to blood flow through an arteriole change because of these decreased radii? The pressure drop across all of the arterioles is about 60 mm Hg. If the flow rate remains the same, what now is the pressure drop change across the arteriole part of the circulatory system?

53. * **Parachutist** A parachutist weighing 80 kg, including the parachute, falls with the parachute open at a constant 8.5-m/s speed toward Earth. The drag coefficient $C_D = 0.50$. What is the area of the parachute?

54. A 0.20-m-radius balloon falls at terminal speed 0.40 m/s. If the drag coefficient is 0.50, what is the mass of the balloon?

55. ** **Terminal speed of skier** A skier going down a slope of angle $\theta$ below the horizontal is opposed by a turbulent drag

force that the air exerts on the skier and by a kinetic friction force that the snow exerts on the skier. Show that the terminal speed is

$$v_T = \left[\frac{2mg(\sin\theta - \mu\cos\theta)}{C_D\,\rho A}\right]^{1/2}$$

where $\mu$ is the coefficient of kinetic friction between the skis and the snow, $\rho$ is the density of air, $A$ is the skier's frontal area, and $C_D$ is the drag coefficient.

56. ** A grain of sand of radius 0.15 mm and density 2300 $kg/m^3$ is placed in a 20 °C lake. Determine the terminal speed of the sand as it sinks into the lake. Do not forget to include the buoyant force that the water exerts on the grain.

57. ** EST **Comet crash** On June 30, 1908, a monstrous comet fragment of mass greater than $10^9$ kg is thought to have devastated a 2000-$km^2$ area of remote Siberia (this impact was called the Tunguska event). Estimate the terminal speed of such a comet in air of density 0.70 $kg/m^3$. State all of your assumptions.

## Reading Passage Problems

BIO **Intravenous (IV) feeding** A patient in the hospital needs fluid from a glucose nutrient bag. The glucose solution travels from the bag down a tube and then through a needle inserted into a vein in the patient's arm (**Figure 11.13a**). Your study of fluid dynamics makes you think that the bag seems a little low above the arm and the narrow needle seems long. You wonder if the glucose is actually making it into the patient's arm. What height should the bag (open at the top) be above the arm so that the glucose solution (density 1000 $kg/m^3$ and viscosity $1.0 \times 10^{-3}\ N\cdot s/m^2$) drains from the open bag down the 0.6-m-long, $2.0 \times 10^{-3}$-m radius tube and then through the 0.020-m-long, $4.0 \times 10^{-4}$-m radius needle and into the vein? The gauge pressure in the vein in the arm is +930 $N/m^2$ (or 7 mm Hg). The nurse says the flow rate should be $0.20 \times 10^{-6}\ m^3/s$ (0.2 $cm^3/s$).

**Figure 11.13** (a) A glucose solution flowing from an open container into a vein. (b) The analysis of the needle in this system.

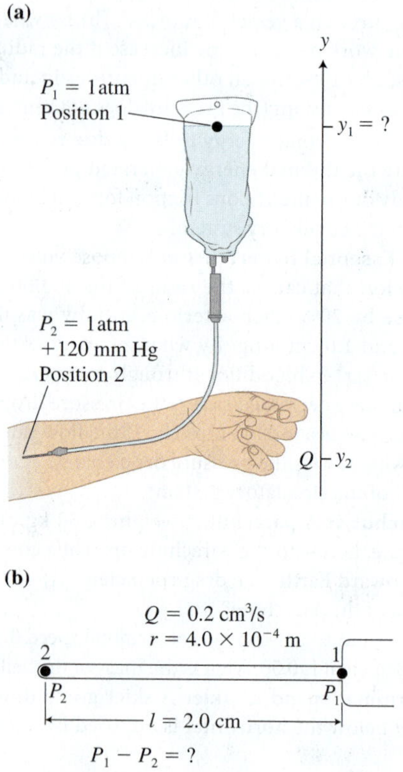


58. Which answer below is closest to the speed with which the glucose should flow out of the end of the needle at position 2 in Figure 11.13b?
(a) 0.0004 m/s (b) 0.004 m/s (c) 0.04 m/s
(d) 0.4 m/s (e) 4 m/s

59. Which answer below is closest to the speed with which the glucose should flow through the end of the tube just to the right of position 1 in Figure 11.13b?
(a) 0.0002 m/s (b) 0.002 m/s (c) 0.02 m/s
(d) 0.2 m/s (e) 2 m/s

60. Assume that there is no resistive friction pressure drop across the needle (as could be determined using Poiseuille's law). Use the Bernoulli equation and the results from Problems 58 and 59 to determine which answer below is closest to the change in pressure between positions 1 and 2 ($P_1 - P_2$) in Figure 11.13b.
(a) 8 $N/m^2$ (b) 80 $N/m^2$ (c) 800 $N/m^2$
(d) 8000 $N/m^2$ (e) 80,000 $N/m^2$

61. Now, in addition to the Bernoulli pressure change from position 1 to position 2 calculated in Problem 60, there may be a Poiseuille resistive friction pressure change across the needle from position 1 to position 2. Which answer below is closest to that pressure change?
(a) 0.4 $N/m^2$ (b) 4 $N/m^2$ (c) 40 $N/m^2$
(d) 400 $N/m^2$ (e) 4000 $N/m^2$

62. The blood pressure in the vein at position 2 in Figure 11.13b at the exit of the needle into the blood is 930 $N/m^2$. Use this value and the results of Problems 60 and 61 to determine which answer below is closest to the gauge pressure at position 1 in the tube carrying the glucose to the needle.
(a) 1010 $N/m^2$ (b) 1410 $N/m^2$ (c) 1980 $N/m^2$
(d) 2800 $N/m^2$ (e) 4620 $N/m^2$

63. Suppose that there is no Poiseuille resistive friction pressure decrease from the top of the glucose solution in the open bag (position 1 in Figure 11.13a) through the tube and down to position 2 near the entrance to the needle. Which answer below is closest to the minimum height of the top of the bag in order for the glucose to flow down from the tube and through the needle into the blood? Remember that the pressure at position 1 is atmospheric pressure, which is a zero gauge pressure.
(a) 0.04 m (b) 0.08 m (c) 0.14 m
(d) 0.27 m (e) 0.60 m

64. Suppose there is a Poiseuille resistive friction pressure decrease from the top of the glucose solution (position 1 in Figure 11.13a) through the tube and down to position 2 near the entrance to the needle. How will this affect the placement of the bag relative to the arm?
(a) The bag will need to be higher.
(b) The bag can remain the same height above the arm.
(c) The bag can be placed lower relative to the arm
(d) Too little information is provided to answer the question.

BIO **The human circulatory system** In the human circulatory system, depicted in **Figure 11.14**, the heart's left ventricle pumps about 80 $cm^3$ of blood into the aorta. The blood then moves into

**Figure 11.14** A schematic representation of the circulatory system including the pressure variation across different types of vessels.

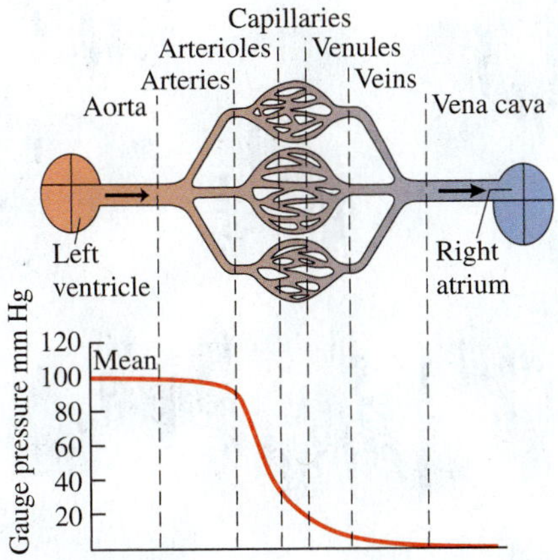


a larger and larger number of smaller radius vessels (aorta, arteries, arterioles, and capillaries). After the capillaries, which deliver nutrients to the body cells and absorb waste products, the vessels begin to combine into a smaller number of larger radius vessels (venules, small veins, large veins, and finally the vena cava). The vena cava returns blood to the heart (see **Table 11.5**).

The flow rate $Q$ of blood through the arteries equals the flow rate through the arterioles, which equals the flow rate through the capillaries, and so forth. The average blood gauge pressure in the aorta is about 100 mm Hg. The pressure drops as blood passes through the different groups of vessels and is approximately 0 mm Hg when it returns to the heart at the vena cava.

A working definition of the resistance $R$ to flow by a group of vessels is the ratio of the gauge pressure drop $\Delta P$ across those vessels divided by the flow rate $Q$ through the vessels:

$$R = \frac{\Delta P}{Q} \tag{11.13}$$

The gauge pressure drop across the whole system is (100 mm Hg – 0), and the total resistance is

$$R_{\text{total}} = \frac{\Delta P_{\text{total}}}{Q} = \frac{100 \text{ mm Hg}}{80 \text{ cm}^3/\text{s}} = 1.25 \frac{\text{mm Hg}}{\text{cm}^3/\text{s}}$$

**Table 11.5 A rough description of the different types of vessels in the circulatory system.**

| Vessel type | Number of vessels | Approximate radius (mm) |
|---|---|---|
| Aorta | 1 | 5 |
| Large arteries | 40 | 2 |
| Smaller arteries | 2400 | 0.4 |
| Arterioles | 40,000,000 | 0.01 |
| Capillaries | 1,200,000,000 | 0.004 |
| Venules | 80,000,000 | 0.02 |
| Small veins | 2400 | 1 |
| Large veins | 40 | 3 |
| Vena cava | 1 | 6 |

The gauge pressure drop across the whole system is the sum of the drops across each type of vessel:

$$\Delta P_{\text{total}} = \Delta P_{\text{aorta}} + \Delta P_{\text{arteries}} + \Delta P_{\text{arterioles}} + \Delta P_{\text{capillaries}} + \cdots + \Delta P_{\text{vena cava}}$$

Now rearrange and insert Eq. (11.13) into the above for the pressure drop across each part:

$$QR_{\text{total}} = QR_{\text{aorta}} + QR_{\text{arteries}} + QR_{\text{arterioles}} + QR_{\text{capillaries}} + \cdots + QR_{\text{vena cava}}$$

Canceling the common flow rate through each group of vessels, we have an expression for the total resistance of the circulatory system:

$$R_{\text{total}} = R_{\text{aorta}} + R_{\text{arteries}} + R_{\text{arterioles}} + R_{\text{capillaries}} + \cdots + R_{\text{vena cava}}$$

The measured gauge pressure drop across the arterioles is about 50 mm Hg, and the arteriole resistance is

$$R_{\text{arterioles}} = \frac{\Delta P_{\text{arterioles}}}{Q} = \frac{50 \text{ mm Hg}}{80 \text{ cm}^3/\text{s}} = 0.62 \frac{\text{mm Hg}}{\text{cm}^3/\text{s}}$$

or about 50% of the total resistance. The next most resistive group of vessels is the capillaries, at about 25% of the total resistance. These percentages vary significantly from person to person. A person with essential hypertension has arterioles and capillaries that are reduced in radius. The resistance to blood flow increases dramatically (the resistance has a $1/r^4$ dependence). The blood pressure has to be greater (for example, double the normal value) in order to produce reasonable flow to the body cells. Even with the increased pressure, the flow rate may still be lower than normal.

65. The capillaries typically produce about 25% of the resistance to blood flow. Which pressure drop below is closest to the pressure drop across the group of capillaries?
(a) 5 mm Hg (b) 15 mm Hg (c) 25 mm Hg
(d) 35 mm Hg (e) 45 mm Hg

66. We found that the arteriole resistance to fluid flow was about 0.62 mm Hg/(cm$^3$/s). By what factor would you expect the resistance of all the arterioles to change if the radius of each arteriole decreased by 0.8?
(a) 1.3 (b) 1.6 (c) 2.4
(d) 0.4 (e) 0.6

67. Why is the resistance to fluid flow through unobstructed arteries relatively small compared to resistance to fluid flow through the arterioles and capillaries?
(a) The arteries are nearer the heart.
(b) There are a relatively small number of arteries.
(c) The artery radii are relatively large.
(d) b and c
(e) a, b, and c

68. The huge number of capillaries and venules is needed to
(a) provide nutrients (such as $O_2$) and remove waste products from all of the body cells.
(b) distribute water uniformly throughout the body.
(c) reduce the resistance of the circulatory system.
(d) b and c
(e) a, b, and c

69. Which number below best represents the ratio of the resistance of a single capillary to the resistance of a single arteriole, assuming they are equally long?
(a) 40 (b) 6 (c) 2.5
(d) 0.4 (e) 0.026

# 12 First Law of Thermodynamics

How is the $CO_2$ concentration in the atmosphere related to the melting of glaciers?

How does sweating protect our bodies from overheating?

Why does the thermal energy of a spoon increase when you place it in a cup of hot tea even though no external objects do work on it?

**Be sure you know how to:**

- Identify a system and decide on the initial and final states of a process (Section 2.1).
- Draw an energy bar chart and use it to help apply the work-energy equation (Section 6.2).
- Apply your understanding of molecular motion to explain gas processes (Section 9.6).

Earth's average temperature is increasing. The 1980s were the warmest decade in recorded history. However, each decade since has beaten that record. Before the Industrial Revolution of the 1800s, the concentration of atmospheric $CO_2$ ranged from 200 ppm (parts per million) during relatively cold periods to 280 ppm during relatively warm periods. The $CO_2$ concentration is now over 370 ppm and is increasing at a rate of about 1.6 ppm/year. What is the connection between the concentration of $CO_2$ in Earth's atmosphere and the changes in its average global temperature? See one consequence in **Figure 12.1**.

**In this chapter,** we combine our understanding of the random motion of atoms and molecules (kinetic molecular theory) with the ideas of work and energy to describe and explain such phenomena as how clouds form, what is causing global climate change, and why athletes get overheated on hot summer days. In physics, these questions are part of the study of **thermodynamics**.

When we studied work and energy (Chapter 6), we learned that a system's energy could change when external objects do work on it. However in real life we know many examples of processes that can change the internal energy of a system without external objects exerting forces and doing work. Imagine for example that you put a room-temperature spoon into a cup of hot coffee. Our understanding of work-energy processes does not explain why the spoon gets hot. However, we can use thermodynamics to analyze such processes.

**Figure 12.1** Warming Earth causes glaciers to melt. Photos taken in 1941 and 2004 show the melting of Alaska's Muir glacier.

## 12.1 Internal energy and work in gas processes

The ideal gas model describes a gas consisting of atoms and molecules that are assumed to be identical point objects—particles. These particles do not interact with each other at a distance but do obey Newton's laws when colliding with each other and with the walls of their container. We determined (in Chapter 9) an expression for the energy of the particles in a container of such a gas.

### Thermal energy of ideal gas

Imagine a bottle filled with air. The air is our system. What energy does this system possess? It does not have any gravitational potential energy (Earth is not in the system) and it has no organized kinetic energy. However, the gas consists of individual particles that are moving randomly.

Earlier (in Chapter 9) we reasoned that each gas particle has some average kinetic energy

$$\overline{K}_{\text{particle}} = \frac{3}{2}kT_{\text{K}}$$

due to its random motion. $T$ is the absolute (kelvin) temperature of the gas. Consequently, the $N$ molecules in the container have a total random kinetic energy (called **thermal energy**):

$$U_{\text{thermal}} = N\left(\frac{3}{2}kT_{\text{K}}\right)$$

Do gas molecules possess any potential energy? We assumed that ideal gas particles do not interact at a distance; thus, the system has no potential energy due to particle interactions. Therefore, in an ideal gas the total internal energy of the gas particles equals its thermal energy.

**‹ Active Learning Guide**

When we studied ideal gases (Chapter 9), we expressed the thermal energy of the gas in different ways. The number $N$ of particles can also be written as the number of moles $n$ of the gas times Avogadro's number of particles $N_{\text{A}}$ in one mole ($N = nN_{\text{A}}$). Also, recall that $N_{\text{A}}k = R$, the universal gas constant and $T = T_{\text{K}}$ is the temperature in kelvins. Thus we can rewrite the equation as

$$U_{\text{thermal}} = N\left(\frac{3}{2}kT\right) = nN_{\text{A}}\left(\frac{3}{2}kT\right) = n\left(\frac{3}{2}N_{\text{A}}kT\right) = \frac{3}{2}nRT \quad (12.1)$$

Note that the thermal energy of this gas depends only on its absolute temperature and on the number of moles of gas. It does not depend on the volume it occupies.

The thermal energy of a gas is called a **state function.** A state function is a property of a system that describes the system at a particular time. It is possible that we will need several state functions to completely describe the system.

> **TIP** We distinguish between the thermal energy
>
> $$U_{\text{thermal}} = N\left(\frac{3}{2}kT\right)$$
>
> of all $N$ particles in a container of gas and the average random kinetic energy
>
> $$\overline{K} = \frac{3}{2}kT$$
>
> of a single particle. Imagine a small and a large container of the same gas, both at the same temperature. The gas in the larger container has more thermal energy even though the average kinetic energy of individual molecules in both containers is the same.

**EXAMPLE 12.1 Thermal energy of the gas in a room**

Construct an expression that could be used to estimate the thermal energy of air at pressure $P$ in a room of volume $V$. Use the expression to estimate the thermal energy of the air filling a small bedroom.

**Sketch and translate** We have an expression for the thermal energy of a container of gas in terms of its absolute temperature $T$ and the number $n$ of moles of particles (Eq. 12.1). We need to connect these quantities to the gas pressure $P$ and the room volume $V$. The ideal gas law seems promising for relating these quantities.

**Simplify and diagram** Assume that the air in the room is an ideal gas, that the air pressure is $P_{\text{atm}} = 1.0 \times 10^5\ \text{N/m}^2$, and that the dimensions of a small bedroom are $4\ \text{m} \times 3\ \text{m} \times 3\ \text{m} = 36\ \text{m}^3$.

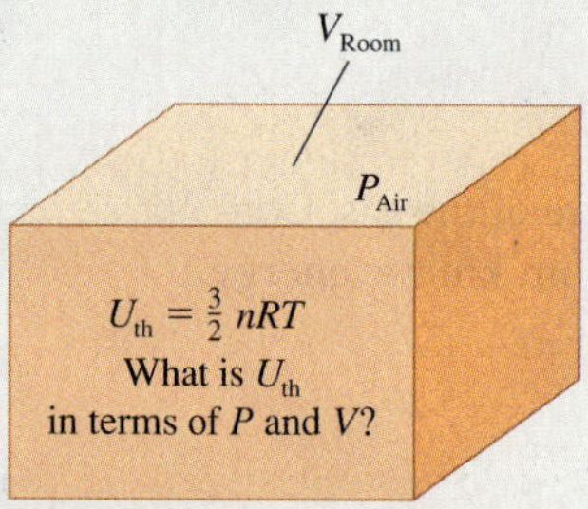


**Represent mathematically** The thermal energy of the gas is given by Eq. (12.1): $U_{\text{th}} = (3/2)nRT$. Using the ideal gas law, $PV = nRT$, we find that

$$U_{\text{th}} = \frac{3}{2}nRT = \frac{3}{2}PV$$

We can use this expression to estimate the thermal energy of the gas.

**Solve and evaluate** The thermal energy of the gas in a small bedroom is approximately:

$$U_{\text{th}} = \frac{3}{2}PV = \frac{3}{2}(10^5\ \text{N/m}^2)(36\ \text{m}^3)$$
$$= 54 \times 10^5\ \text{N}\cdot\text{m} \approx 5 \times 10^6\ \text{J}$$

The result is equivalent to the energy given off by a 100-J/s (100-watt) lightbulb during a 50,000-s (14-hour) time interval—a significant amount of energy! Notice that when estimating, we keep only one significant figure.

**Try it yourself:** What is the thermal energy of the air in an empty 1-L soft drink bottle? State your assumptions.

*Answer:* 150 J. We assumed that the gas pressure inside the bottle equaled atmospheric pressure and that the air inside the bottle could be modeled as an ideal gas.

## Work done on a gas

**Figure 12.2** The piston does work on the gas, changing its thermal energy.

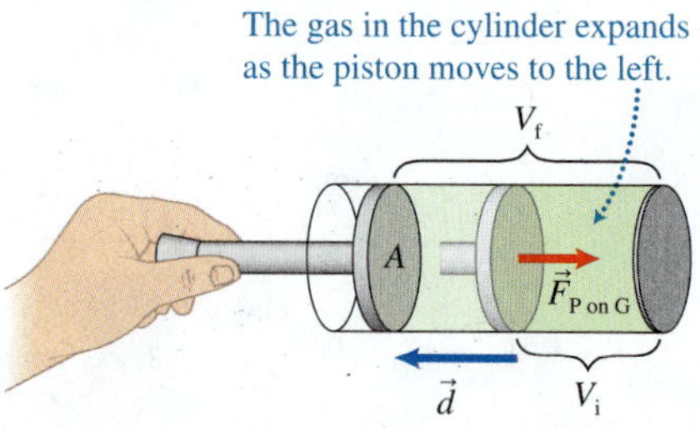


Gases play an important role in many common mechanical devices. For example, the explosion of a mixture of air and gasoline in the cylinder of a car's engine increases the gas pressure against the piston in that cylinder. The gas at high pressure pushes the piston outward, which causes the driveshaft to rotate and causes the car wheels to turn. How much work does the gas in the cylinder do during such an expansion?

Imagine that you have a gas (the system) at high pressure in a cylinder with a movable piston (**Figure 12.2**). You hold the piston and allow it to move slowly outward (to the left in Figure 12.2). The gas pushes to the left on the piston and the piston in turn pushes toward the right on the gas, exerting a force on the gas $\vec{F}_{\text{P on G}}$. The gas inside the piston expands slowly. What work is done by the force that the piston exerts on the gas? Assume that the piston moves outward a distance $d$, allowing the gas to expand from its initial volume $V_i$ to its final larger volume $V_f$.

Recall that work is the product of the magnitude of the force exerted on an object, the object's displacement, and the cosine of the angle between the directions of the force and the displacement ($W = Fd\cos\theta$). This equation only works for a *constant* magnitude force. In our case, as the high-pressure gas expands, there is a decrease in the frequency of collisions of particles with the piston, and the pressure decreases. Thus, the piston exerts a *changing* force on the expanding gas.

To determine the work done by the piston on the gas, we could use calculus. However, it is also possible to determine an expression for the work done on the changing volume of gas by breaking the big volume change into many small increments and adding the work done during each small change. Let's start with a process in which the piston moves outward in small steps each of distance $\Delta x$ starting at position $x_i$ and ending at position $x_f$ (**Figure 12.3a**). For each step the gas volume change is small, the gas pressure changes little, and the force that the piston exerts on the gas remains approximately constant. The gas pushes outward on the piston whether the piston is causing an expansion or contraction of the gas. The piston in turn exerts a force on the gas in the opposite direction (toward the right in Figure 12.3a). If the gas is expanding, then the piston in Figure 12.3a is moving left, and the work done by the piston on the gas for the $\Delta x$ expansion is

$$W_{\text{Piston on Gas}} = F_{\text{P on G}}\Delta x\cos(180°) = -F_{\text{P on G}}\Delta x$$

The force that the piston exerts on the gas and the piston's displacement point in opposite directions—hence the minus sign.

The change in the volume of the gas when the piston moves out a distance $\Delta x$ is $\Delta V = A\Delta x$ where $A$ is the cross-sectional area of the piston. Rearranging the above, we find that $\Delta x = (\Delta V)/A$. Recall also the definition of pressure: $P = (F/A)$ or $F = P \cdot A$. Insert these expressions for $F$ and $\Delta x$ into the above expression for work to get

$$W_{\text{Piston on Gas}} = -F_{\text{P on G}}\Delta x = -(P \cdot A)\left(\frac{\Delta V}{A}\right) = -P\Delta V \qquad (12.2)$$

As the gas has expanded and has a larger volume, $\Delta V$ is positive. Since the piston is pushing opposite the expansion, it does negative work on the gas and the energy of the gas decreases. If instead the piston exerted a force that caused the gas to compress, the force would do positive work on the system ($\Delta V$ would be negative and it would cancel the negative sign in Eq. (12.2); the gas would have increased energy).

Remember also that this is the work done *on the gas* by the external force that the piston exerts on the gas during a small change in the gas volume. This work done by the piston on the gas during a small expansion of the gas is the negative of the area under the narrow dark-shaded column of the $P$-versus-$V$ graph line in Figure 12.3b. This area is the height $P$ of the narrow column times its width $\Delta V$.

We split the large change in gas volume for the whole process into $n$ tiny volume changes. For each small volume change, assume that the pressures are approximately constant (Figure 12.3b). Then add the work done during each of these volume changes to get the total work done by the piston on the gas:

$$W_{\text{Piston on Gas}} = (-P_1\Delta V_1) + (-P_2\Delta V_2) + \cdots + (-P_n\Delta V_n)$$

Each term is the negative of the area of a narrow column in Figure 12.3b. Thus, the work done during the expansion from volume $V_i$ to volume $V_f$ is the negative area under the entire graph line between those initial and final volumes.

**Figure 12.3** Determine the work done on a changing volume of gas.

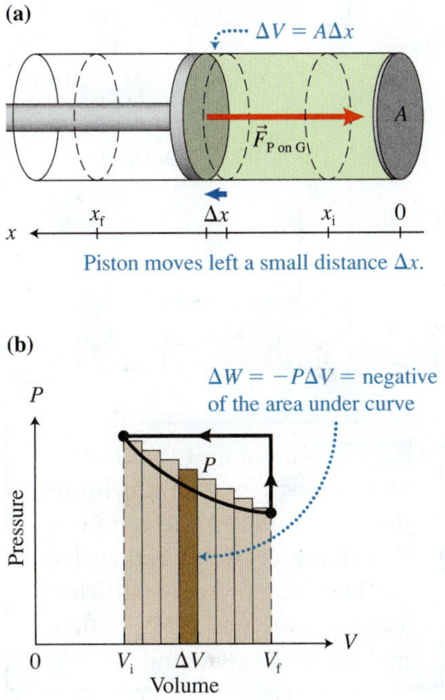

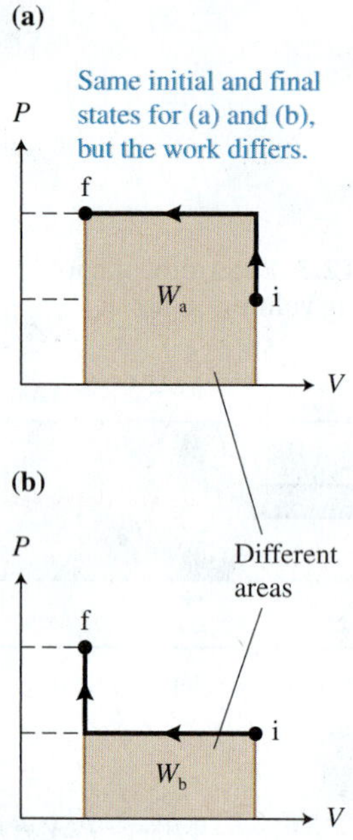


**Figure 12.4** Work depends on the way in which the system changes from the initial to the final state.

> **TIP** Notice that the thermal energy of the gas in a cylinder depends only on the temperature of the gas when in that state (and on the number of particles). Temperature and internal thermal energy are state functions.

## Work depends on the process between the initial and final states

Many different processes can take a gas from an initial pressure $P_i$ and volume $V_i$ to a final pressure $P_f$ and volume $V_f$ (different curves or lines on a $P$-versus-$V$ graph). The amount of work done on the gas can differ in each case (see the examples in **Figures 12.4a** and b). Note that in Figure 12.4a, to go from the initial to the final state, the gas first has a constant volume and increasing pressure and then a constant pressure and decreasing volume. On the other hand, in (b), the gas first has a constant-pressure volume decrease and then a constant-volume pressure increase to get to the same final state. Note that the work done for each process, represented by the shaded area under the graph lines, is different—less for the process in (b) than the process in (a), even though the initial and final states are the same. The work done on or by a gas depends on how the pressure and volume change in going from the initial to the final state. The work depends on the process that occurs in changing state and is *not* a state function. It does not describe a single state or the difference between the two states of the system.

## Work done *by* a gas *on* its environment

The gas mixture in the cylinders of your automobile engine is part of a motor. In this case we are not interested in the work done by the environment (the piston) on the gas (the system) but are instead interested in the work done by that gas on the environment. When the gas in the cylinder expands, it does positive work on its environment. The environment in turn does the same magnitude negative work on the gas. The work that the piston does on the gas is equal in magnitude and opposite in sign to the work that the gas does on the piston—an example of Newton's third law. Thus, we can define work involving gas changes as follows:

> **Work done on or by a gas** **The work done by the environment on a gas $W_{\text{Environment on Gas}}$ when a gas changes from volume $V_i$ to $V_f$ is the negative of the area under the $P$-versus-$V$ graph for that process. The work is negative if the gas volume increases and positive if its volume decreases. The work done by the gas on the environment $W_{\text{Gas on Environment}}$ has the same magnitude but the opposite sign. If the pressure is constant, the work done by the environment on the gas is**
>
> $$W_{\text{Environment on Gas}}(\text{constant pressure process}) = -P(V_f - V_i) = -P\Delta V \quad (12.2)$$
>
> **and the work done by the gas on the environment is**
>
> $$W_{\text{Gas on Environment}}(\text{constant pressure process}) = +P(V_f - V_i) = +P\Delta V \quad (12.3)$$

**Review Question 12.1** Imagine that a balloon expands when brought from a cold garage into a warm room. The room is at atmospheric pressure and the change in the volume of the balloon is $\Delta V$. What is the work that the air in the room does on the balloon during the process?

# 12.2 Two ways to change the energy of a system

**Active Learning Guide ›**

Throughout this text, we have stressed the importance of defining a system—a carefully identified object or group of objects on which we focus our attention. Everything outside the system is the environment. A system is characterized by physical quantities, such as the system's internal energy, mass, volume, temperature, and number of particles. The interaction of the environment with the system is defined in terms of the work done by the environment on the system. We devised the generalized work-energy equation [Eq. (6.3)] that summarizes these ideas:

$$U_\text{i} + W_\text{Environment on System} = U_\text{f}$$

Using this equation, we find that the total initial energy $U_\text{i}$ of the system plus the work done on the system by the environment $W_\text{Environment on System}$ equals the total final energy $U_\text{f}$ of the system. The energy of the system can be in many different forms, but the only way the total energy of the system can change is if the environment does work on the system.

$$(K_\text{i} + U_\text{gi} + U_\text{si} + \cdots) + W_\text{E on S} = (K_\text{f} + U_\text{gf} + U_\text{sf} + \cdots + \Delta U_\text{internal})$$

Are there situations where these ideas are inadequate? Consider Observational Experiment **Table 12.1**. To analyze these experiments we will use a rewritten version of the generalized work-energy equation:

$$W_\text{Environment on System} = U_\text{f} - U_\text{i}$$

**OBSERVATIONAL EXPERIMENT TABLE**

**12.1 Using the work-energy principle to analyze two processes.**

| Observational experiment | Analysis |
|---|---|
| **Experiment 1.** Place a fixed amount of gas in a fixed-volume container. A flame below causes the gas to warm from temperature $T_\text{i}$ to $T_\text{f}$. 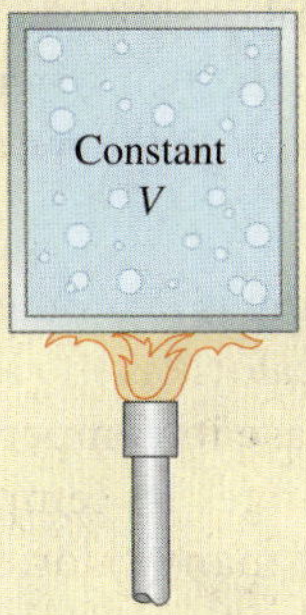  | ■ The temperature of the gas increased; hence its thermal energy increased: $\Delta U_\text{th} > 0$. ■ The work done on the gas was zero, because the gas volume did not change: $W = 0$. ■ Thus, $0 = \Delta U_\text{th}$. *According to the generalized work-energy equation, this process cannot occur—the left side is zero and the right side is a positive number.* |
| **Experiment 2.** Place a fixed amount of gas in a cylindrical enclosure with a piston that can move freely up or down in the cylinder, thus keeping the pressure exerted on the gas constant. A flame below causes the gas to expand and warm from temperature $T_\text{i}$ to $T_\text{f}$. It takes longer to reach $T_\text{f}$ than it did in Experiment 1. 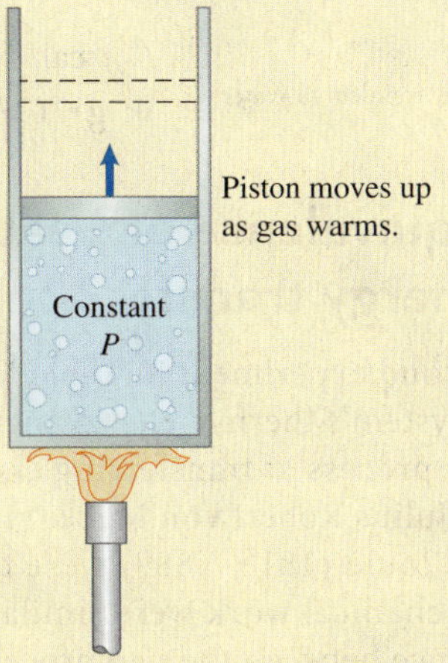  | ■ The temperature of the gas increased; hence its thermal energy increased: $\Delta U_\text{th} > 0$. ■ The piston pushed down on the gas opposite the direction of its expansion; thus negative work was done on the gas ($W_\text{Piston on Gas} < 0$). ■ Thus, $-\lvert W\rvert = +\lvert U_\text{th}\rvert$. *According to the generalized work-energy equation, this process cannot occur—the left side is negative and the right side is positive.* |

**Pattern**

Analyzing both experiments using the work-energy principle, we find that according to this principle the processes cannot possibly occur. However, since we observe these processes in real life, they must be possible. We conclude that either the principle is incorrect or it needs modification.

The generalized work-energy principle is unable to account for the energy changes in the two experiments in Table 12.1. What modification in the principle would help us explain what happened in those two experiments?

## Particle motion explains the change in thermal energy of the gas

To reconcile the work-energy principle with the observational experiments described in Table 12.1, we need to assume that energy can be transferred to the gas by means other than work. As we know, the **temperature** of a substance is related to the random motion and vibration of the particles of which the substance is made. Consider Experiment 1. In a hot flame the particles are moving very quickly in random directions. They collide with the slower randomly moving particles in the cool cylinder containing the gas. The particles in that cool cylinder gain random kinetic energy and move faster and/or vibrate more. The particles have greater random motion, which corresponds to a higher temperature, so the cylinder gets warmer. Thus the energy of the hot flame was transferred to the cooler cylinder without any work done. The same mechanism transfers energy from the cylinder walls to the cool gas inside it. Overall, the energy from the hot flame is transferred to the cool gas. This new way of transferring energy from an object at one temperature (the hot flame) to an object at a different temperature (the cool gas) is called **heating**.

**TIP** Note that we use the term *heating* rather than *heat.* We do this deliberately. Heating does not mean the actual increase of the temperature of the system; it means the process by which energy is transferred from a hot object to a cooler object (without any work being done).

**Heating** **$Q$ is a physical quantity that characterizes a process of transferring energy from the environment to a system, which is at a different temperature. The environment does no work on the system.**

The SI unit of heating is the joule. However, another common unit of heating is the calorie. One calorie is the amount of energy that must be transferred to 1 g of water to increase its temperature by 1 degree C. If you have a known mass of water that changes its temperature by a known number of degrees, you can determine how many calories of energy were transferred to this water through the process of heating. To do this you must multiply one calorie by the mass and the temperature difference; the answer will be in units of calories:

$$Q_{\text{Environment to Water}} = \left(1 \frac{\text{cal}}{\text{g} \cdot {}^\circ\text{C}}\right)(m_{\text{water}})(\Delta T)$$

## Testing the equivalence of heating and work as a means of energy transfer

Recall that in the second experiment in Table 12.1 negative work was done on the system, but the system's thermal energy increased. How do we explain this effect using our new process of transferring energy by heating? German physician and physicist Julius Robert von Mayer (1814–1878) and English brewer and physicist James Joule (1818–1889) were the first scientists to speculate that heating and mechanical work were similar processes in terms of energy transfer. The difference between the two processes, they hypothesized, is that for the energy to be transferred to a system through heating, there must be some temperature difference between the system and the environment. This is not the case with mechanical work. Both men performed experiments to test this idea.

First we will describe James Joule's testing experiment in Testing Experiment **Table 12.2**.

**TESTING EXPERIMENT TABLE**

**12.2 Testing the equivalence of work and heating as a means for energy transfer.**

| Testing experiment | Prediction based on the idea under test | Outcome |
|---|---|---|
| Joule placed a paddle wheel in water. The wheel was connected with a string passing over pulleys tied to a heavy block. When the block went down, the paddle wheel rotated in the water. 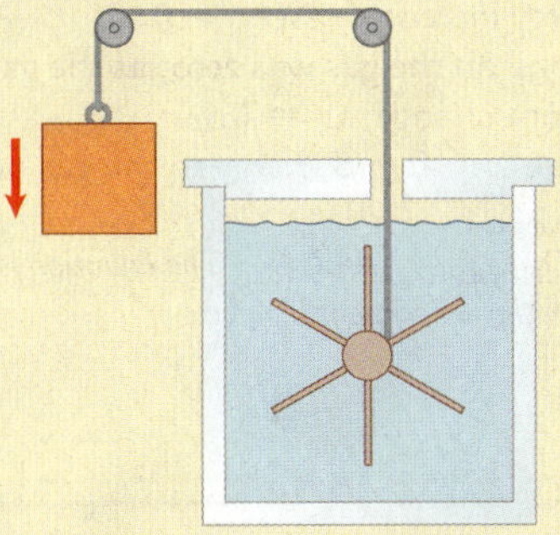 | Our system will be Earth, block, paddle wheel, and water. The work $W$ done in lifting a block of mass $m$ a height $h$ equals the gain in the gravitational potential energy of the system $\Delta U_g = m_{\text{block}}gh$ measured in joules. When the block is released and returns to its starting position, the paddle wheel turning in the water should cause the water to gain the same amount of thermal energy. We can predict the temperature change of the water assuming that the gravitational potential energy of the system is converted into the internal energy of the water: $m_{\text{block}}gh = \left(1\frac{\text{cal}}{\text{g}\cdot{}^\circ\text{C}}\right)(m_{\text{water}})(\Delta T)$ or $\Delta T = \dfrac{m_{\text{block}}gh}{\left(1\frac{\text{cal}}{\text{g}\cdot{}^\circ\text{C}}\right)(m_{\text{water}})}.$ To validate the assumption and minimize the transfer of energy to the container, pulleys, strings, and surrounding air, Joule covered the container in blankets, used light and smooth pulleys, and made other adjustments. | Joule lifted the block multiple times to get a bigger temperature increase and repeated the experiment many times. In all experiments the observed change in temperature of the water agreed with the predicted change. |
| **Conclusion** | | |
| The amount that the system's energy changes when external objects do mechanical work on it is the same as when external objects transferred the same amount of energy by heating. | | |

Based on Table 12.2 we can conclude that the thermal energy of a system can be changed by two means: mechanical work and heating. Mechanical work and heating are both ways of transferring energy between the environment and the system. Joule found that 1.0 J of work done by the person lifting the block and transferred to the kinetic energy of the paddle wheel inside the liquid caused the same effect as the transfer of 0.24 cal through the process of heating to the same liquid. In independent experiments others observed the same effect:

$$1.0\text{ J} = 0.24\text{ cal}$$
$$1.0\text{ cal} = 4.2\text{ J}$$

Consequently, the amount of energy that we need to transfer to 1 kg of water to change its temperature by 1 °C is 1000 cal or 4200 J.

**Review Question 12.2** Suppose that heating with an electric burner transfers 400 cal of energy to a pan of water. How many joules of mechanical energy would produce the same change in the water's thermal energy?

## 12.3 First law of thermodynamics

Now that we have established the equivalence of heating and work as mechanisms through which energy can be transferred between an environment and a system, we can incorporate the concept of heating into the generalized work-energy principle. Let's try now to explain the outcomes of the experiments in Table 12.1 by including a heating term in the equation:

$$W_{\text{Environment on System}} + Q_{\text{Environment to System}} = \Delta U_{\text{System}}$$

**OBSERVATIONAL EXPERIMENT TABLE**

**12.3 Using heating to explain the experiments performed in Table 12.1.**

| Observational experiment | Analysis |
|---|---|
| **Experiment 1.** Place a fixed amount of gas in a fixed-volume container. A flame below causes the gas to warm from temperature $T_i$ to $T_f$. 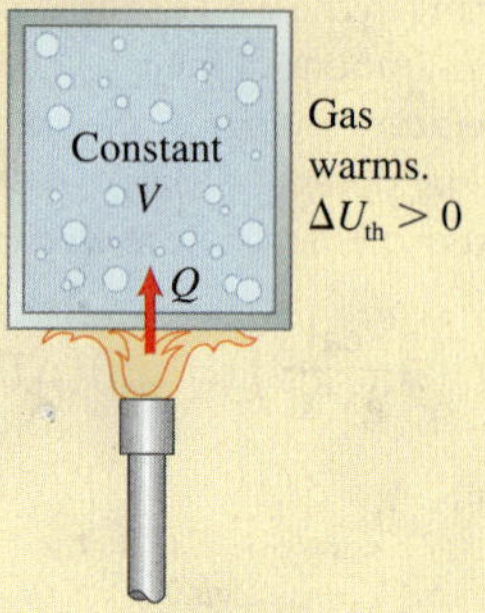  | ■ The temperature of the gas increased; hence its thermal energy increased: $\Delta U_{th} > 0$. ■ The work done on the gas was zero, as the gas volume did not change: $W = 0$. ■ The hot flame transferred energy to the gas by heating: $Q_{\text{Flame to Gas}} > 0$. ■ Thus, $0 + Q_{\text{Flame to Gas}} = \Delta U_{th}$. *The equation is consistent with the experimental outcome.* |
| **Experiment 2.** Place a fixed amount of gas in a cylindrical enclosure with a piston that can move freely up or down in the cylinder, thus keeping the pressure exerted on the gas constant. A flame below causes the gas to expand and warm from temperature $T_i$ to $T_f$. It takes longer to reach the final temperature than in Experiment 1. 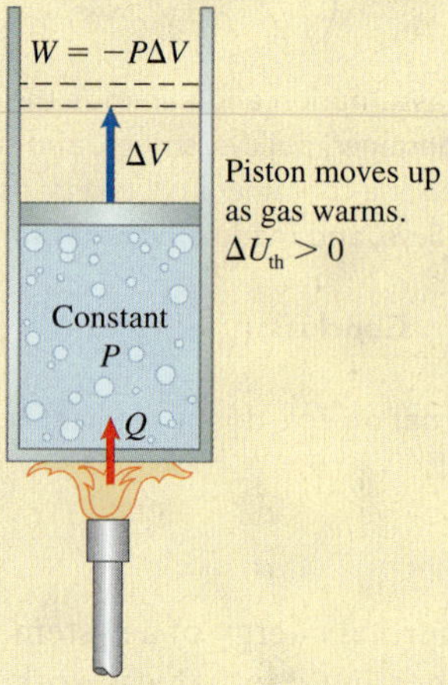  | ■ The temperature of the gas increased; hence the thermal energy increased: $\Delta U_{th} > 0$. ■ The piston pushed down on the gas opposite the direction of its expansion, thus negative work was done on the gas ($W_{\text{Piston on Gas}} < 0$). ■ The hot flame transferred more energy to the gas by heating than in Experiment 1—the gas was exposed to the flame for a longer period of time. $Q_{\text{Flame to Gas}} > 0$. ■ Thus, $-\lvert W \rvert + Q_{\text{Flame to Gas}} = +\Delta U_{th}$. *The equation is consistent with the experimental outcome.* |

**Pattern**

The new work-heating-energy change principle explains the outcomes of both experiments.

The work-heating-energy equation succeeds in explaining many phenomena that the old work-energy principle could not—such as the change of the temperature of water (the system) placed on a hot electric stove (the environment) and the change in the temperature of a hot hard-boiled egg (the system) placed in a bowl of cold water (the environment). In each of these cases, the energy of the system changes with no work done on it. But in all cases the system contacts a part of the environment that is *at a different temperature* than the system.

## A note about temperature, thermal energy, and heating

We've learned that energy can only be transferred spontaneously through heating when the temperatures of the system and the environment are different. This is different than saying that their thermal energies must be different. Imagine a small spoon at room temperature placed in a bathtub filled with water at room temperature. Although the water in the bathtub has far more thermal energy than the spoon does, there is no energy transfer from the water to the spoon.

The spoon and bathtub example highlights the difference between three physical quantities: temperature, thermal energy, and heating. Temperature is the physical quantity that quantifies the *average* random kinetic energy of the individual particles that comprise the object. Temperature does not depend on the number of particles in the object. Thermal energy is a physical quantity that quantifies the *total* random kinetic energy of all the particles. Adding more particles at the same temperature to the system increases the system's thermal energy. Heating is the physical quantity that quantifies the *process* through which some *amount* of thermal energy is transferred between the system and the environment when they are at different temperatures. If two systems have the same temperatures but different thermal energies, there is no transfer of energy between them through heating.

**TIP** When the temperature of a system is lower than that of the environment, "positive heating" occurs—the system gains energy from the environment; when the temperature of the system is higher than the environment's, "negative heating" occurs—the system loses energy to the environment.

## Quantitative analysis of a process involving heating

The physician Robert Mayer approached the idea of work-heating-energy from a different angle. His reasoning is used in the following example.

### EXAMPLE 12.2 Quantitative analysis of heating a gas

It was known in Mayer's times that 1.0 g of air requires 0.17 cal of energy transferred through heating at *constant volume* for its temperature to increase by 1.0 °C and 0.24 cal of energy transferred through heating at *constant pressure* for the temperature to change by 1.0 °C. Are these data consistent with the work-heating-energy equation and the equivalence of work and heating?

**Sketch and translate** A sketch for each experiment is shown below. The system is the gas. In the first experiment an electric heater warms the gas (transfer of energy though heating) while the gas remains at constant volume. In the second experiment the heater again warms the gas but this time at constant pressure $P$. As the gas warms at constant pressure, it expands and pushes outward on a piston.

(a) Experiment 1. Constant volume heating

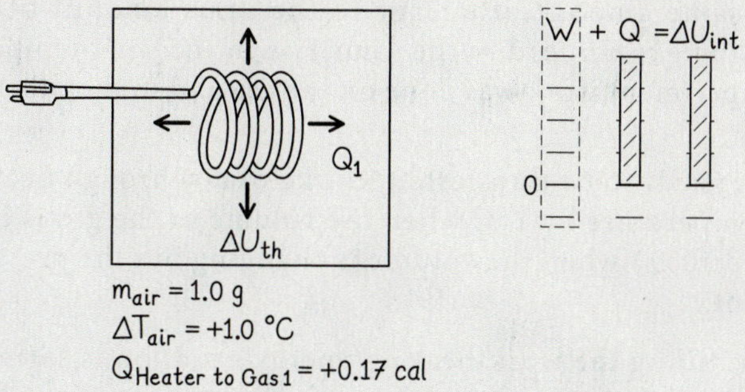


(b) Experiment 2. Constant pressure heating

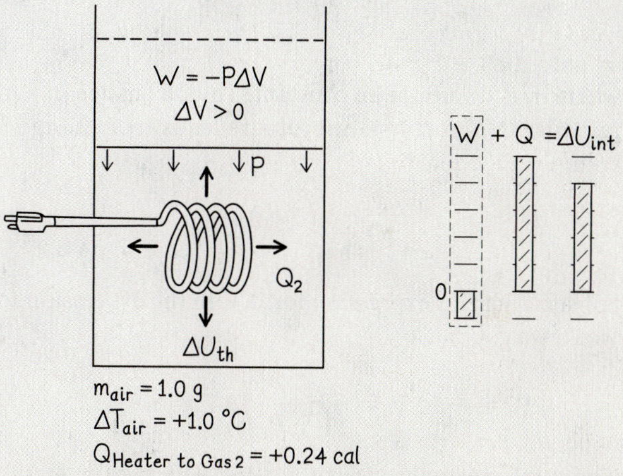


**Simplify and diagram** We assume that the gas in each experiment is an ideal gas and that the piston moves without friction. We can represent each experiment with an energy bar chart. Notice the $Q$ bar next to the work bar; it indicates the energy transferred through heating to the system. The temperature increases by 1.0 °C in each experiment—the same internal energy change in both cases. In the second experiment, the piston does negative work on the gas (because the force exerted by the piston on the gas points opposite the direction of expansion of the gas). To balance the bar chart in the second experiment, the heating must be greater than in the first experiment.

**Represent mathematically** In the first experiment, the heating of the gas leads to a change in its internal energy:

$$0 + Q_{\text{Heater to Gas 1}} = \Delta U_{\text{Thermal Gas 1}}$$

*(continued)*

In the second experiment, in addition to heating the gas, the environment does negative work on the gas:

$$W_{\text{Environment on Gas 2}} + Q_{\text{Heater to Gas 2}} = \Delta U_{\text{Thermal Gas 2}}$$

Since both gases 1 and 2 undergo the same temperature change, the internal energy changes are equal $(\Delta U_{\text{Thermal Gas 1}} = \Delta U_{\text{Thermal Gas 2}})$. Consequently,

$$Q_{\text{Heater to Gas 1}} = W_{\text{Environment on Gas 2}} + Q_{\text{Heater to Gas 2}}$$

or

$$W_{\text{Environment on Gas 2}} = Q_{\text{Heater to Gas 1}} - Q_{\text{Heater to Gas 2}}$$

We could determine the work done by an independent method involving the definition of work and the ideal gas law. The work done by the piston on the gas in experiment 2 was $W_{\text{Environment on Gas 2}} = -P\Delta V$, where $P$ was the constant pressure of the gas and $\Delta V$ was the volume change of the gas ($W_{\text{Environment on Gas 2}} < 0$, since pressure is always a positive number and the gas expands, thus its volume change is positive).

To find work, we need to determine $\Delta V$. Using the ideal gas law for a constant pressure process, we can write

$$V = \left(\frac{nR}{P}\right)T,$$

where $n$, $R$, and $P$ are all constant. Thus, a change in temperature $\Delta T$ at constant pressure leads to a change in volume $\Delta V$ equal to

$$\Delta V = \left(\frac{nR}{P}\right)\Delta T$$

Substituting this expression for $\Delta V$ in the expression for work, we get:

$$W = -P\Delta V = -P\left(\frac{nR}{P}\right)\Delta T = -nR\Delta T$$

The number of moles of gas is $n = m/M$, where $m = 1.0$ g is the mass of gas and $M = 29.0$ g/mole is the molecular mass of air.

$$n = \frac{1.0\text{ g}}{29.0\text{ g}/1.0\text{ mole}} = 0.0345\text{ moles}$$

**Solve and evaluate** We can now calculate the work $W$ done by the environment on the gas in two ways. The first method gives us

$$W_{\text{Environment on Gas 2}} = Q_{\text{Heater to Gas 1}} - Q_{\text{Heater to Gas 2}}$$
$$= 0.17\text{ cal} - 0.24\text{ cal} = -0.07\text{ cal}$$

The second method gives us

$$W = -nR\Delta T = -(0.0345\text{ mole})\left(\frac{8.3\text{ J}}{\text{mole}\cdot{}^\circ\text{C}}\right)(1.0\,^\circ\text{C})$$
$$= -0.29\text{ J} = -0.29\text{ J}\left(\frac{0.24\text{ cal}}{1.0\text{ J}}\right) = -0.07\text{ cal}$$

The two independent methods of calculating the work produce consistent results. This strengthens our confidence in the new work-heating-energy equation.

**Try it yourself:** If in the experiment Mayer had 1.0 mole of gas, what would be the ratio of the heating needed to raise the temperature of this gas by one degree at constant pressure and the heating needed to raise the temperature of the same mole of gas by one degree at constant volume?

*Answer:* 5/3. The thermal energy increase in each gas is $(3/2)nR\Delta T$ and the negative work done on the constant pressure gas is $nR\Delta T$. Thus,

$$\frac{Q_{\text{Env. to Gas 2 (constant pressure)}}}{Q_{\text{Env. to Gas 1 (constant volume)}}} = \frac{\Delta U_{\text{th}} - (-P\Delta V)}{\Delta U_{\text{th}}}$$
$$= \frac{(3/2)nR\Delta T + nR\Delta T}{(3/2)nR\Delta T} = \frac{5}{3}$$

This ratio is different from what Mayer would have calculated, 0.24 cal/0.17 cal = 1.4 = 7/5. This is because the ideal gas law assumes molecules are point-like and have no internal structure. This assumption is only marginally reasonable for air in this situation.

Notice two important points here.

1. Providing the same amount of energy to the same amount of gas through heating *might not* lead to the same rise in the gas's temperature if in one experiment work was done on the gas and in the other no work was done.
2. The energy that needs to be transferred to 1 kg of air through heating to change its temperature by 1 °C when the volume of the gas is constant is 710 J and 1000 J when the volume is changing but the pressure remains constant.

What we have been calling the work-heating-energy equation is formally known as the **first law of thermodynamics**, a fundamental principle of physics.

**First law of thermodynamics** **Consider a process involving a system of interest. The system's internal energy change $(U_f - U_i) = \Delta U_{\text{System}}$ is equal to the amount of work $W$ done on the system plus the amount of energy $Q$ transferred to the system through the process of heating:**

$$W_{\text{Environment on System}} + Q_{\text{Environment to System}} = \Delta U_{\text{System}} \quad (12.4a)$$

We can rewrite the first law by including all types of energy in a system:

$$(U_{gi} + K_i + U_{si}) + W + Q = (U_{gf} + K_f + U_{sf} + \Delta U_{int}) \qquad (12.4b)$$

Think of how you can warm your right hand (the system) after coming from a trip outside on a cool windy day. You can press your other hand against it and rub them together—transferring energy to your right hand by doing work on it (notice that both of your hands are at the same temperature). You can also submerge your right hand in warmer water—transferring energy to it through heating.

**TIP** Heating, like work, is a physical quantity characterizing a process during which energy is transferred between a system and its environment. Both heating and work lead to a change in the energy of a system. Remember, heating and work are not things that reside within the system.

**Review Question 12.3** How are the first law of thermodynamics and the work-energy equation similar? How are they different?

## 12.4 Specific heat

We need to transfer 1000 cal or 4200 J to 1 kg of water to change its temperature by 1 °C. Do all materials need the same amount of energy to change their temperature by 1 °C per unit mass? Consider the experiments in Observational Experiment **Table 12.4**.

**OBSERVATIONAL EXPERIMENT TABLE**

**12.4 Energy transferred to different materials to achieve the same temperature change.**

| Observational experiment | Analysis |
|---|---|
| **Experiment 1.** A closed Styrofoam container holds 1.0 kg of water at 20 °C and a thermometer. A large pan of boiling water (at 100 °C) holds a 1.0 kg iron block. You transfer the block very quickly to the container with 20 °C water. You observe the temperature rising and then stabilizing at around 28 °C. | Although the masses of the water and iron block were the same, the water temperature increased by 8 °C whereas the iron block's temperature decreased by 100 °C − 28 °C = 72 °C. We assume that the water received 4200 J of energy for each 1 °C temperature change, or a total of $8(4200\ \text{J}) = 3.4 \times 10^4\ \text{J}$. The iron must have lost the same amount of energy. Thus the energy loss per 1 °C for the 1.0-kg iron block was $\frac{3.4 \times 10^4\ \text{J}}{(1.0\ \text{kg})(72\ ^\circ\text{C})} = \frac{470\ \text{J}}{\text{kg} \cdot {}^\circ\text{C}}$ |
| **Experiment 2.** You repeat Experiment 1, replacing the iron block with a 1.0 kg aluminum block. The final temperature is 34 °C. | The masses of the water and aluminum block were the same (1.0 kg), but the water temperature increased by 14 °C whereas the aluminum block's temperature decreased by 100 °C − 34 °C = 66 °C. We assume that the water received 4200 J of energy for each 1 °C temperature change, or a total of $14(4200\ \text{J}) = 5.88 \times 10^4\ \text{J}$. The aluminum must have lost the same amount of energy. Thus the energy loss per 1 °C for the 1.0 kg aluminum block was $\frac{5.88 \times 10^4\ \text{J}}{(1.0\ \text{kg})(66\ ^\circ\text{C})} = \frac{890\ \text{J}}{\text{kg} \cdot {}^\circ\text{C}}$ |

**Pattern**

The same amount of energy added to or removed from objects of the same mass by heating or cooling leads to different changes of temperature of these objects. Alternatively, different amounts of energy are gained or lost for each 1 °C temperature change, depending on the type of material.

To explain the patterns in Table 12.4 we hypothesize that there is a physical quantity that characterizes how much additional energy 1 kg of a substance needs to change its temperature by 1°C, and this physical quantity is different for

**Table 12.5** **Specific heats ($c$) of various solid and liquid substances.**

| Solid Substances | $c$ (J/kg · °C) | Liquid Substances | $c$ (J/kg · °C) |
|---|---|---|---|
| Iron, steel | 450 | Water | 4180 |
| Copper | 390 | Methanol | 2510 |
| Aluminum | 900 | Ethanol | 2430 |
| Silica glass | 840 | Benzene | 1730 |
| Sodium chloride | 860 | Ethylene glycol | 2420 |
| Lead | 130 | | |
| Ice | 2090 | | |
| Wood | ~1700 | | |
| Sand | ~800 | | |
| Brick | ~800 | | |
| Concrete | ~900 | | |
| Human body | ~3500 | | |

different substances. As we found, the amount of energy is different for water, iron, and aluminum and is about

$$4200 \frac{\text{J}}{\text{kg} \cdot {}^\circ\text{C}},\ 470 \frac{\text{J}}{\text{kg} \cdot {}^\circ\text{C}},\ \text{and } 890 \frac{\text{J}}{\text{kg} \cdot {}^\circ\text{C}},$$

respectively. These values are the values of a physical quantity, called the **specific heat** of each substance.

> **Specific heat $c$ is the physical quantity equal to the amount of energy that needs to be added to 1 kg of a substance to increase its temperature by 1 °C. The symbol for specific heat is $c$ and the units are $\frac{\text{J}}{\text{kg} \cdot {}^\circ\text{C}}$. This energy is added through heating or work or both.**

**TIP** Every time you encounter a new physical quantity, think of its meaning. For example, what does it mean that the specific heat of ethanol is 2430 J/kg · °C? It means that we would have to transfer 2430 joules of energy to 1 kg of ethanol to change its temperature by 1 °C.

**Table 12.5** lists specific heat capacities for several solid and liquid materials.

Water has the largest specific heat in Table 12.5. Notice that the specific heats of sand, bricks, and concrete are about one-fifth that of water. These materials have much greater temperature changes than water when equal masses of these materials absorb the same energy. This is one reason why the sand on a beach or the concrete beside a swimming pool feels so much hotter than the adjacent water on a sunny day. The sand and concrete have bigger temperature changes than the water when exposed to the same sunlight.

The physical quantity specific heat allows us to construct an equation that can be used to determine the total amount of energy $\Delta U$ that needs to be transferred to a mass $m$ of some type of material to cause its temperature to change by $\Delta T$. If the amount of energy equal to the specific heat $c$ in joules (J) needs to be transferred to 1 kg of material to change its temperature by 1 °C, then to change the energy of $m$ kg of the same material we need to transfer $c \cdot m$ J. If we now want to change the temperature by $\Delta T$ degrees instead of 1 degree, we need to transfer $c \cdot m\Delta T$ joules. Note that the unit of the product of these three quantities is the unit of energy:

$$\frac{\text{J}}{\text{kg} \cdot \text{C}^\circ} \times \text{kg} \times {}^\circ\text{C} = \text{J}$$

Therefore, we conclude that an amount of energy $\Delta U$ must be added to a substance of mass $m$ and specific heat $c$ to cause its temperature to change by $\Delta T$:

$$\Delta U = cm\Delta T \tag{12.5}$$

The change in the energy of the system can occur both due to the process of heating and due to external objects doing work on the system. This leads to an interesting consequence. Imagine that you are heating a gas in a rigid container. No work is done on the gas since it is not expanding, so the change in thermal energy of the gas is exactly equal to the heating. However, if the container is very flexible and there is atmospheric air outside, that atmospheric air does negative work on the system, reducing the change in thermal energy. To achieve the same change in temperature as in the previous process and thus the same change in thermal energy of the gas, one needs to transfer more energy through heating. This is exactly what happened in Example 12.2. Recall the statement of the problem: It was known in Mayer's times that 1.0 g of air requires 0.17 cal of energy transferred through heating at *constant volume* for its temperature to increase by 1.0 °C and 0.24 cal of energy transferred through heating at *constant pressure* for the temperature to change by 1.0 °C. In the first case there is no work done on the system, and thus there is less energy transferred through heating needed to change its temperature by one degree. If the work done on the gas is negative and its magnitude is exactly equal to the positive energy transferred through heating, the temperature of the gas and its thermal energy do not change. However, in most applications, we deal with solids and liquids whose volume changes very little. In this case the thermal energy changes only due to the process of heating, and thus $Q + W = Q + 0 = \Delta U = cm\Delta T$.

Specific heat is a useful quantity. For example, we can use it to determine how much thermal energy a car engine would need to gain by heating in order to warm it to the boiling temperature of the water in its cooling system. Normally, a radiator cooling system prevents the engine from getting too hot, but what if the radiator isn't functioning?

**QUANTITATIVE EXERCISE 12.3 Heating a car engine**

A car has a 164-kg mostly aluminum engine that when operating at highway speeds converts chemical potential energy into thermal energy at a rate of $P = 1.8 \times 10^5\ \text{J/s}$ (240 horsepower). Suppose the cooling system shuts down and all of the thermal energy continually generated by the engine remains in the engine. Determine the time interval for the engine's temperature to change from $T_i = 20\ °\text{C}$ to $T_f = 100\ °\text{C}$, the boiling temperature of water.

**Represent mathematically** Assume that the energy is transferred from the burning fuel to the engine through the mechanism of heating, which produces the temperature change in the engine:

$$Q_{\text{Fuel to Engine}} = c_{\text{Engine}} m_{\text{Engine}} (T_{\text{Engine, f}} - T_{\text{Engine, i}})$$

The burning fuel converts chemical energy into thermal energy at a rate of

$$P = \frac{\Delta U_{\text{thermal}}}{\Delta t} = \frac{\Delta Q}{\Delta t} = \frac{Q_{\text{Fuel to Engine}}}{\Delta t}$$

Thus, the time interval $\Delta t$ needed for the burning fuel to warm the engine is

$$\Delta t = \frac{Q_{\text{Fuel to Engine}}}{P}$$

**Solve and evaluate** The engine is made mostly of aluminum with specific heat $c_{\text{aluminum}} = 900\ \text{J/kg} \cdot °\text{C}$. Thus,

$$Q_{\text{Fuel to Engine}} = \left(900 \frac{\text{J}}{\text{kg} \cdot °\text{C}}\right)(164\ \text{kg})(100\ °\text{C} - 20\ °\text{C})$$
$$= 1.2 \times 10^7\ \text{J}$$

The engine will reach boiling temperature in

$$\Delta t = \frac{Q_{\text{Fuel to Engine}}}{P}$$
$$= \frac{(1.2 \times 10^7\ \text{J})}{(1.8 \times 10^5\ \text{J/s})} = 67\ \text{s}$$

The unit is correct. The engine warms quickly. An operating cooling system is important!

**Try it yourself:** You hold two spoons of identical shape and volume in a flame. One spoon is aluminum and one is steel. Why does the temperature of the aluminum spoon change more quickly than that of the steel spoon, even though the specific heat of aluminum is about twice that of steel?

*Answer:* The density of aluminum is about one-third of the density of steel; hence the steel spoon is more massive.

## Sharing energy through the process of heating when objects are in contact

**Active Learning Guide ›**

If you have a fever, bathing in cool bath water lowers your temperature and raises the water temperature. If you are pulled from a cold lake, to warm up you can bathe in hot bath water—the hot water warms your body and your cool body cools the water. These changes in the thermal energies of two objects can be explained using the idea of energy transfer: the hot object loses thermal energy to the cold object through the process of heating, and the cold object gains energy through the process of heating from the hot object. We can summarize the above discussion as follows:

$$Q_{\text{Hot to Cold}} + Q_{\text{Cold to Hot}} = 0$$

It is important to note that when a cold object and a hot object are in contact, the energy transfers from the hot object to the cold object until their temperatures become the same. At that point the average random kinetic energies of their particles are the same and there is no more energy transfer. The objects are said to be in **thermal equilibrium**.

**TIP** When object 1 at high temperature is in contact with object 2 at lower temperature, the process of energy transfer from 1 to 2 is still called heating, despite the fact that the temperature of object 1 goes down.

**EXAMPLE 12.4 Warming the body of a hypothermic person**

The kayak of a 70-kg man tips on a spring day and he falls into a cold stream. When he is rescued from the cold water, his body temperature is 33 °C (91.4 °F). You place him in 50 kg of warm bath water at temperature 41 °C (105.8 °F). What is the final temperature of the man and the water?

**Sketch and translate** Draw a sketch of the situation. Since the initial bath water temperature is higher than the man's initial temperature, the decrease in the thermal energy of the water equals the increase in the thermal energy of the man. The specific heats of water and of humans are listed in Table 12.5.

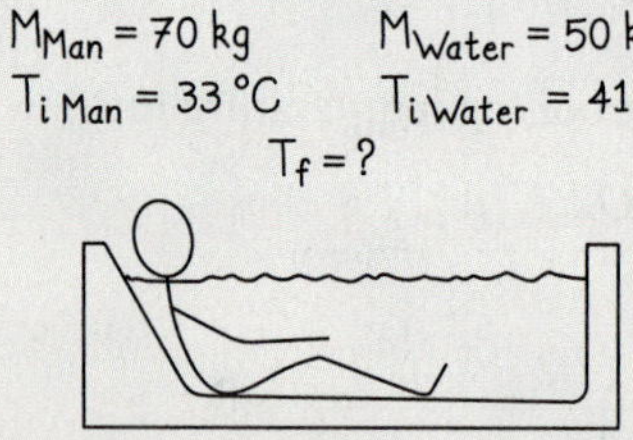


**Simplify and diagram** Assume that no energy is transferred from the water and man to the surroundings. In addition, assume that chemical energy converted into thermal energy in the man's body is small and can be ignored during the short time interval in the water.

**Represent mathematically** The thermal energy change of the man equals the energy transferred from the water by heating to the man: $Q_{\text{Water to Man}} = m_{\text{Man}}c_{\text{Man}}(T_{\text{f Man}} - T_{\text{i Man}})$. The thermal energy change of the water equals the energy transferred by heating from the man to the water: $Q_{\text{Man to Water}} = m_{\text{Water}}c_{\text{Water}}(T_{\text{f Water}} - T_{\text{i Water}})$. We also know that $T_{\text{f Water}} = T_{\text{f Man}} = T_{\text{f}}$ and that $Q_{\text{Man to Water}} + Q_{\text{Water to Man}} = 0$. Putting this all together, we get

$$m_{\text{Water}}c_{\text{Water}}(T_{\text{f}} - T_{\text{i Water}}) + m_{\text{Man}}c_{\text{Man}}(T_{\text{f}} - T_{\text{i Man}}) = 0$$

Notice that the first expression on the left side of the above equation will be negative, as the water's final temperature is less than its initial temperature. The above can now be solved for the final temperature (try to do it yourself before you look at the expression shown below).

**Solve and evaluate** Solving for the final temperature of the system, we get

$$T_{\text{f}} = \frac{m_{\text{Water}}c_{\text{Water}}T_{\text{i Water}} + m_{\text{Man}}c_{\text{Man}}T_{\text{i Man}}}{m_{\text{Water}}c_{\text{Water}} + m_{\text{Man}}c_{\text{Man}}}$$

$$= \frac{(50\text{ kg})(4180\text{ J/kg}\cdot{}^{\circ}\text{C})(41\,{}^{\circ}\text{C}) + (70\text{ kg})(3470\text{ J/kg}\cdot{}^{\circ}\text{C})(33\,{}^{\circ}\text{C})}{(50\text{ kg})(4180\text{ J/kg}\cdot{}^{\circ}\text{C}) + (70\text{ kg})(3470\text{ J/kg}\cdot{}^{\circ}\text{C})}$$

$$= 36.7\,{}^{\circ}\text{C}$$

The answer is reasonable: the final temperature is 3.7 °C above the initial temperature of the man and 4.3 °C below the water's initial temperature. The final temperature is 0.3 °C below the normal 37 °C body temperature.

**Try it yourself:** Fill a Styrofoam cup with 100 g of water at 90 °C and then place a 30-g aluminum spoon at 25 °C into it. The water cools to 86 °C and the spoon warms to 86 °C, the same temperature as the water. Use these results to estimate the specific heat of aluminum. What assumptions did you make?

*Answer:* 900 J/kg·°C. We assumed that the spoon is completely submerged in the water and no energy is transferred through heating to the cup or any other part of the environment.

**Review Question 12.4** What experimental evidence indicates that the specific heat of aluminum or iron is much smaller that the specific heat of water?

# 12.5 Applying the first law of thermodynamics to gas processes

◂ Active Learning Guide

Let us now apply the first law of thermodynamics to several processes involving gases. We will represent the processes using words, graphs, bar charts, and equations and explain what happens using a microscopic approach (molecules and their motion) and a first law of thermodynamics approach (energy transfer to or from the system and its changes). Let's begin with a simple process.

## Effect of a moving piston on the temperature of an enclosed gas

Suppose a gas is enclosed in a cylinder with a piston on the right end that can move, allowing the gas to expand or contract (see **Figure 12.5a**). How does the average speed of the gas particles and therefore the gas temperature change if the gas expands?

Consider a situation in which one particle collides with the piston. The particle will bounce off the piston, but because the piston is moving away from the particle, its speed will not on average be as great after the collision as it was before the collision, and the gas temperature will decrease slightly. Consider an extreme example of such a collision. Suppose the particle is moving at velocity $+10\ \text{m/s}$ with respect to Earth when it collides with a piston that is moving outward in the same direction as the particle, only at $+9\ \text{m/s}$ with respect to Earth (Figure 12.5b). Then with respect to the piston, the particle is moving towards the piston at $+1\ \text{m/s}$. Since the collision is elastic, the particle rebounds at $-1\ \text{m/s}$ (Figure 12.5c) with respect to the piston. With respect to Earth, then, the particle should be moving to the right at 8 m/s, because $(-1\ \text{m/s}) + (9\ \text{m/s}) = 8\ \text{m/s}$. Its speed has decreased by 20%—from 10 m/s to 8 m/s with respect to Earth, and it is therefore a much "cooler" particle.

In a real situation, the piston moves much more slowly than the particles and their speeds decrease less dramatically. But they do move somewhat more slowly after colliding with the expanding piston. As a result, the temperature of the gas decreases as the piston moves outward. Similarly, if the piston moves inward, the average speed of the particles after colliding with the piston will be greater than before the collision. In this case, the temperature of the gas increases as the piston moves inward.

Now that we have an understanding of this, let's move on to more complex processes.

**Figure 12.5** Piston motion affects speed of gas particles. (a) The piston is moving right. (b) Particle speed of a gas particle before it hits a piston that is moving away. (c) The gas particle speed is lower after it hits the piston.

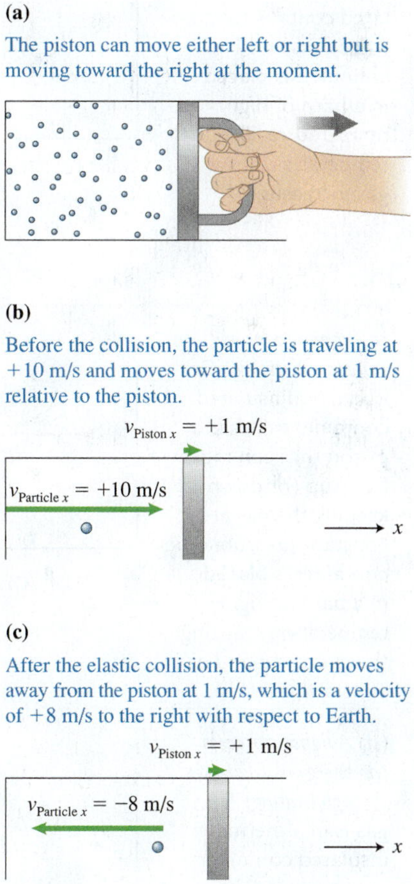
(a)
The piston can move either left or right but is moving toward the right at the moment.
(b)
Before the collision, the particle is traveling at +10 m/s and moves toward the piston at 1 m/s relative to the piston.
$v_{\text{Piston }x} = +1$ m/s
$v_{\text{Particle }x} = +10$ m/s
x
(c)
After the elastic collision, the particle moves away from the piston at 1 m/s, which is a velocity of +8 m/s to the right with respect to Earth.
$v_{\text{Piston }x} = +1$ m/s
$v_{\text{Particle }x} = -8$ m/s
x

## Analyzing gas processes

In **Table 12.6**, we analyze various gas processes using the first law of thermodynamics using particle-based and energy-based explanations. In all of these processes the number of moles $n$ of gas is assumed to be constant. Recall what we have learned about isoprocesses (in Chapter 9)—processes occurring in a gas of constant mass in which one of the following physical quantities also remained constant: $P$, $T$, or $V$. In addition, we will analyze a new type of process called an **adiabatic** process. In an adiabatic process no energy is transferred through heating ($Q = 0$).

**Table 12.6** **Application of the first law to gas processes. In each process, the gas starts at position 1 (the initial state) on the *P*-versus-*V* graph and moves to position 2 (the final state).**

| Process | Graph | Particle-level explanation | Explanation and bar chart using the first law of thermodynamics |
|---|---|---|---|
| (a) *Isothermal process (constant temperature)* The gas is in a non-insulated container with a piston, all submerged in a large bath of water at constant temperature, which is the same as the temperature of the gas. Someone pulls the piston out very slowly. | 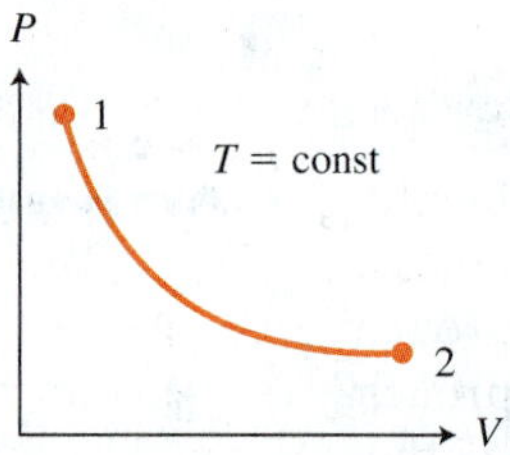 | As the piston is pulled out, the gas molecules collide elastically with the piston and rebound at a slower speed than before the collision, which causes a temperature decrease. But the container is in contact with the water bath, which provides heating so the gas remains at constant temperature. | ■ As the temperature of the gas stays constant, $\Delta U_{int} = 0$.<br>■ The external work done on the gas is negative ($W < 0$).<br>■ Transfer of energy through heating ($Q > 0$) is needed to make the internal energy change zero.<br>■ $-\lvert W \rvert + Q = 0$<br>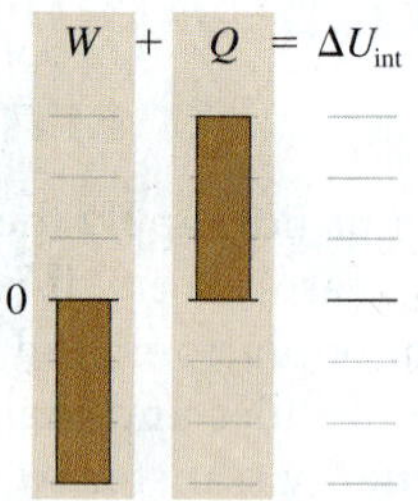 |
| (b) *Isochoric process (constant volume)* The gas is in a noninsulated container with a fixed piston. The container is placed in a bath of higher temperature. The temperature of the gas increases. | 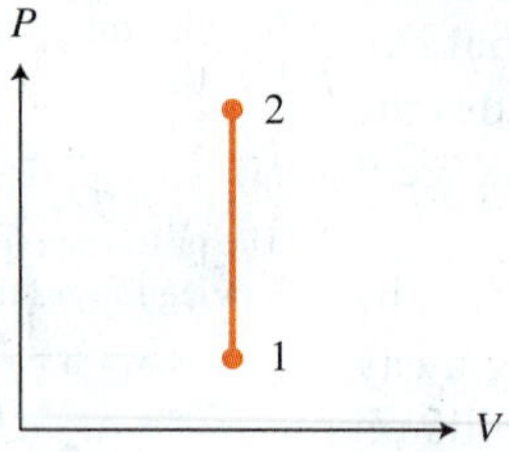 | As the gas warms, its molecules move faster and faster, colliding with the walls of the container more often and more violently. The gas pressure increases. | ■ Because the volume is constant, there is no work done on the gas ($W = 0$).<br>■ Energy is transferred to the gas through heating ($Q > 0$).<br>■ Transfer of energy through heating leads to an increase in the thermal energy of the gas ($\Delta U_{int} > 0$).<br>■ $0 + Q = +\Delta U_{int}$<br>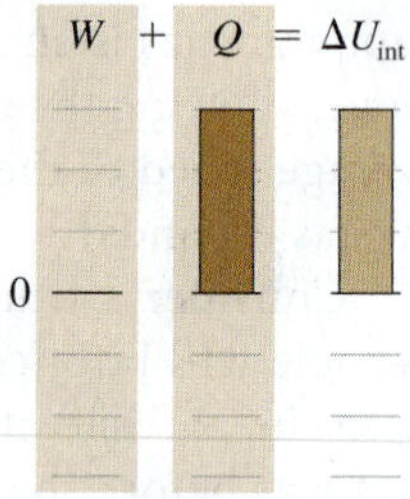 |
| (c) *Isobaric process (constant pressure)* Gas is in a noninsulated container that has a piston that can move freely up (or down), keeping the gas at constant pressure. The container is placed in a bath at higher temperature, causing the gas to expand at constant pressure. | 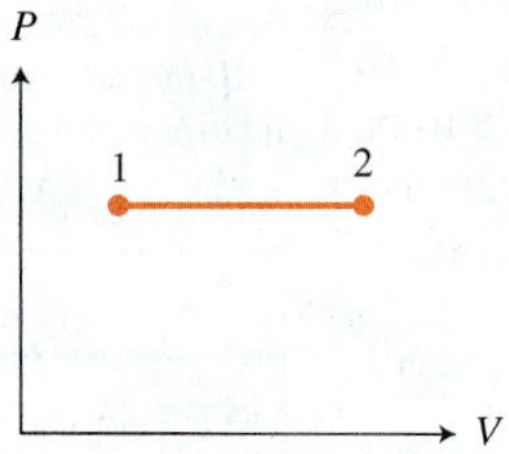 | The hot temperature bath warms the gas by heating. But the gas warms less because the piston moves outward, slowing the particles that hit it and also causing negative work to be done by the environment on the gas. | ■ The environment does negative work on the gas ($W < 0$).<br>■ Energy is transferred through heating ($Q > 0$) and is greater than the negative work.<br>■ The thermal energy of the gas increases ($\Delta U_{int} > 0$).<br>■ $-\lvert W \rvert + Q = +\Delta U_{int}$<br>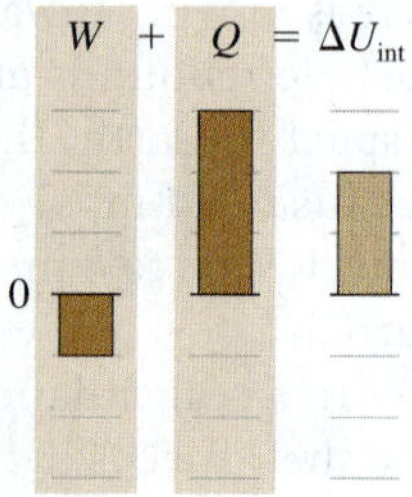 |
| (d) *Adiabatic process (no energy transferred through heating)* The gas is in a thermally insulated container or is compressed very quickly. | 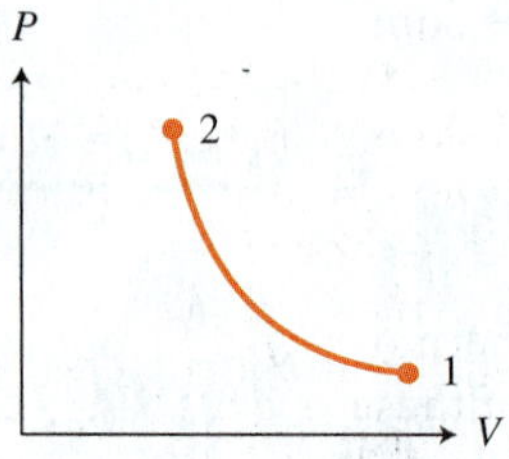 | As the gas is being compressed, the molecules collide with a piston moving inward; the speed of the particles reflected off the incoming piston is greater than before, and the gas's temperature increases. | ■ The environment does positive work on the gas ($W > 0$).<br>■ Since the process happens very quickly or the gas is in an insulated container, we assume that there is no transfer of energy through heating ($Q = 0$).<br>■ The internal thermal energy of the gas increases ($\Delta U_{int} > 0$).<br>■ $W + 0 = +\Delta U_{int}$<br>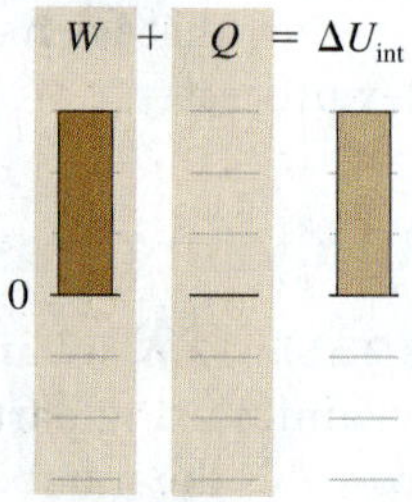 |

## Skills for solving gas problems using the first law of thermodynamics

Everyday life offers many examples of isoprocesses and adiabatic processes, where one quantity $T$, $P$, $V$, or $Q$ is constant (in an adiabatic process $Q = \text{const} = 0$). A useful strategy for analyzing such processes using the first law of thermodynamics is described on the left side of the table in Example 12.5 and illustrated for the problem in that example.

### PROBLEM-SOLVING STRATEGY Using thermodynamics for gas law processes

**EXAMPLE 12.5 Hot air balloon**

A burner heats 1.0 m$^3$ of air inside a small hot air balloon. Initially, the air is at 37 °C and atmospheric pressure. Determine the amount of energy that needs to be transferred to the air through heating (in joules) to make it expand from 1.0 m$^3$ to 1.2 m$^3$.

**Sketch and translate**

- Make a labeled sketch of the process.
- Choose a system of interest and the initial and final states of the process.

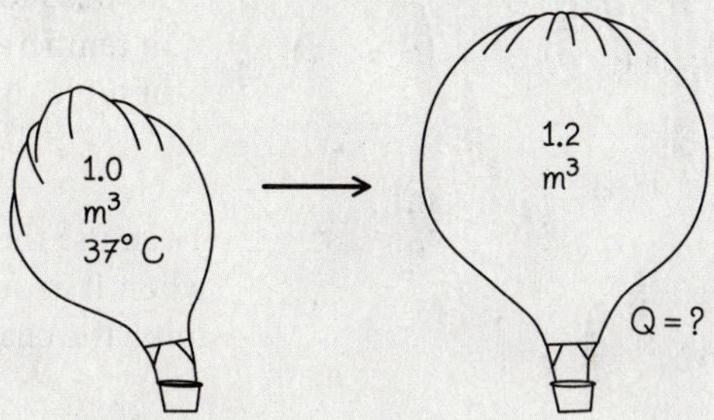


$$V_i = 1.0\ \text{m}^3 \qquad V_f = 1.2\ \text{m}^3$$
$$P = \text{constant} = 1.0 \times 10^5\ \text{N/m}^2$$
$$T_i = 310\ \text{K} \qquad T_f = ?$$
$$Q_{\text{Environment to System}} = ?$$

The system is the gas inside the balloon. The initial state is before the flame was turned on; the final state is after the balloon has partially expanded.

**Simplify and diagram**

- Decide whether you can model the system as an ideal gas.
- Decide whether the gas undergoes one of the isoprocesses.
- Determine whether you can ignore any interactions of the environment with the system.
- Draw the following representations if appropriate: a work-heating-energy bar chart and/or a $P$-versus-$V$ graph.

- We can model the air as an ideal gas since conditions are close to normal atmospheric conditions.
- Assume that the gas expands slowly so that it remains at constant atmospheric pressure—an isobaric process. This also assumes that the skin of the balloon does not exert a significant additional inward pressure on the gas. Assume that the number of moles of air in the balloon remains constant.

***P*-versus-*V* graph**

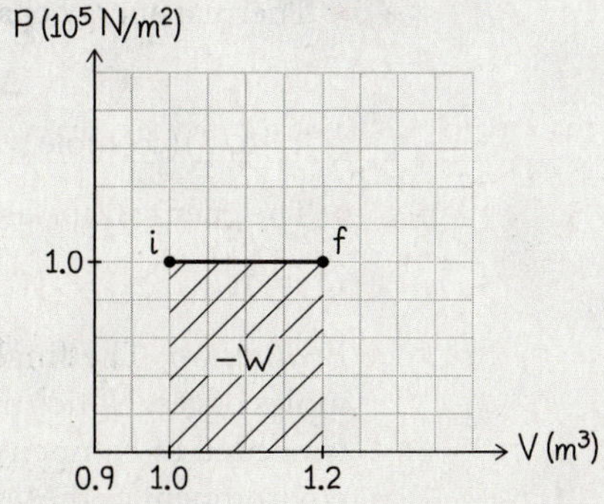


***Q*-*W*-Δ*U* bar chart**

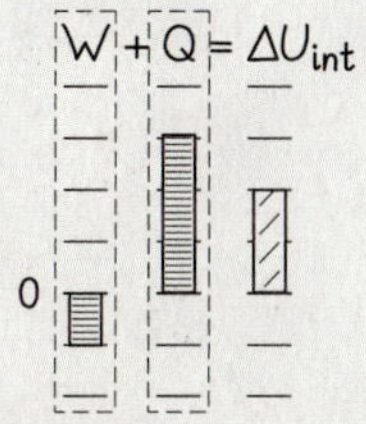


*(continued)*

## Represent mathematically

- Use the bar chart to help apply the first law of thermodynamics:

$$U_{gi} + K_i + U_{si} + W + Q = U_{gf} + K_f + U_{sf} + \Delta U_{int}.$$

Decide if the mechanical energy of the system is constant (kinetic, gravitational potential, or elastic).

- If the system can be modeled as an ideal gas, use, if needed, gas law equations such as $PV = nRT = NkT$.

- For this process, the initial and final gravitational potential and (nonrandom) kinetic energies of the gas are equal, and there are no springs in the system. Thus the first law equation becomes

$$W + Q = \Delta U_{int}$$

We need to determine the amount of energy transferred by heating:

$$Q = \Delta U_{int} - W.$$

The gas expands; thus the environment pushes in the direction opposite to the motion of the balloon walls and does negative work (area under $P$-versus-$V$ graph line with a negative sign):

$$W = -P\Delta V$$

- To find the change in internal energy, we need to use the expression for the internal energy of the ideal gas $U_{int} = \frac{3}{2}nRT$. To find the change, we need to determine the number of moles of gas and then the change in temperature. We can use the ideal gas law to determine the number of moles of air initially in the 1.0 $m^3$ of air:

$$n = P_i V_i / RT_i$$

Then use the ideal gas law to determine the final temperature of the air when it expands to the final volume: $T_f = P_f V_f / nR$. From this, determine the change in thermal energy of the air:

$$\Delta U_{int} = (3/2)nR(T_f - T_i)$$

- We can now put everything together into the equation

$$Q = \Delta U_{int} - W$$

## Solve and evaluate

- Solve the equations for the unknown quantity.
- Evaluate the results to see if they are reasonable (the magnitude of the answer, its units, and how the solution changes in limiting cases).

The work done on the gas is:

$$W = -P\Delta V = -(1.0 \times 10^5\ \text{N/m}^2)(1.2\ \text{m}^3 - 1.0\ \text{m}^3) = -0.2 \times 10^5\ \text{N}\cdot\text{m} = -2.0 \times 10^4\ \text{J}$$

- The number of moles of air is

$$n = P_i V_i / RT_i = (1 \times 10^5\ \text{N/m}^2)(1\ \text{m}^3)/(8.3\ \text{J/mole}\cdot\text{K})(310\ \text{K}) = 39\ \text{moles}$$

- The final temperature of the air is

$$T_f = P_f V_f / nR = (1 \times 10^5\ \text{N/m}^2)(1.2\ \text{m}^3)/(39\ \text{mole})(8.3\ \text{J/mole}\cdot\text{K}) = 371\ \text{K}$$

- Thermal energy change of the gas is

$$\Delta U_{thermal} = (3/2)nR(T_f - T_i) = (3/2)(39\ \text{mole})(8.3\ \text{J/mole}\cdot\text{K})[(371 - 310)\text{K}] = +3.0 \times 10^4\ \text{J}$$

- The energy transferred to the gas through heating is:

$$Q = \Delta U_{thermal} - W = +3.0 \times 10^4\ \text{J} - (-2.0 \times 10^4\ \text{J}) = +5.0 \times 10^4\ \text{J}$$

*Evaluation:* The final temperature is reasonable, as is the number of moles of gas. Notice that almost half of the energy transferred by heating is needed to compensate for negative work done on the system by the environment as the gas system expands.

**Try it yourself:** Air rushing out of the open valve of a pumped bicycle tire feels cool. Why?

*Answer:* This is an approximately adiabatic process ($Q = 0$) since it happens so rapidly. The gas is expanding, so the environment does negative work on the gas ($W < 0$). Thus, the internal energy change should also be negative ($W + 0 = \Delta U_{\text{int}}$). The gas cools as it expands adiabatically.

In the next example we analyze an isochoric process.

‹ Active Learning Guide

## EXAMPLE 12.6 Temperature control in a domed stadium

You are the heating-cooling contractor for a new 68,000-person sports dome construction project. Estimate the air temperature change in the building caused by the metabolic processes of the spectators during a football game, assuming that the building is perfectly insulated and has no air conditioning. Identify any other assumptions that you make.

**Sketch and translate** A sketch of the situation is shown below. Choose the air in the dome as the system. If we assume that the volume of the air inside the dome remains constant and that no air escapes the dome, then no work is done on the air by the environment ($W = -P\Delta V = 0$ and $0 + Q = \Delta U_{\text{th}}$). The change in the internal energy (thermal energy) of the gas equals the energy transferred by heating from the spectators to the air in the dome (see the bar chart below):

$$Q_{\text{People to Air}} = \Delta U_{\text{th Air}}$$

(a)

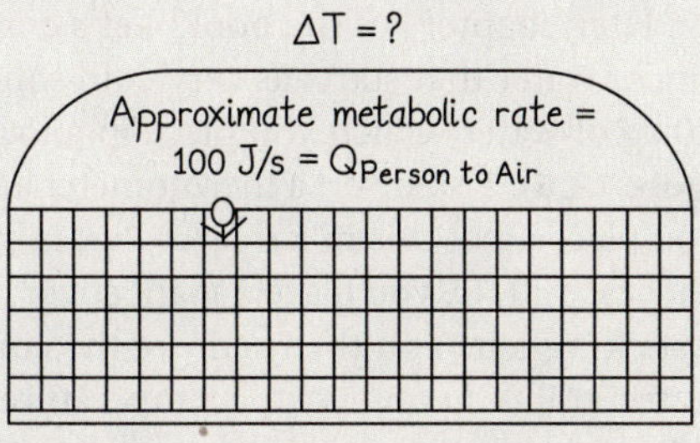


(b)

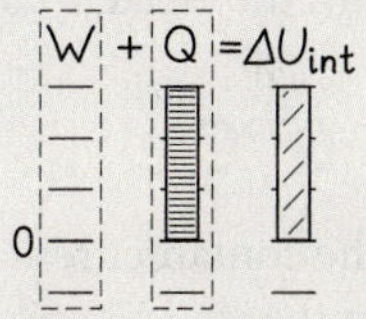


We can use Eq. (12.5) to estimate the air temperature change $\Delta T$ caused by this thermal energy change during the game:

$$\Delta T = \frac{\Delta U_{\text{th Air}}}{c_{\text{Air}} m_{\text{Air}}} = \frac{Q_{\text{People to Air}}}{c_{\text{Air}} m_{\text{Air}}}$$

To complete this task, we need to estimate $Q_{\text{People to Air}}$ and the mass of the air inside the dome $m_{\text{Air}}$. We also need to know the specific heat $c_{\text{Air}}$ of the air at constant volume.

**Simplify and diagram** The stadium holds $N = 68{,}000$ spectators, each with a resting metabolic rate $MR_{\text{Person}}$ of about 100 J/s (about 100 W). Assume the game lasts about 3 hours (notice that we do not worry about significant figures here, as we are doing an estimate). Also assume that there is no energy transfer between the stadium and the environment.

**Represent mathematically** The energy that an average person adds to the air during the game is

$$Q_{\text{Person to Air}} = MR_{\text{Person}} \Delta t$$

where $MR_{\text{Person}} = 100\ \text{W} = 100\ \text{J/s}$ is the metabolic rate (power output) of a resting person. Since about $N$ people are at the game, the total energy added to the air through heating during the game is approximately

$$Q_{\text{People to Air}} = NQ_{\text{Person to Air}} = N \cdot MR_{\text{Person}} \Delta t$$

To find the temperature change we need to know the mass of the air and its specific heat. The mass of the air will be the density of air times the stadium volume:

$$m_{\text{Air}} = \rho_{\text{Air}} V_{\text{Stadium}}$$

**Solve and evaluate** The total energy added through the process of heating to the air by the people during the game is

$$Q_{\text{People to Air}} = N \cdot MR_{\text{Person}} \Delta t$$
$$= (68{,}000\ \text{persons})[(100\ \text{J/s})/\text{person}]\left(3\ \text{h}\frac{3600\ \text{s}}{1\ \text{h}}\right)$$
$$\approx 7 \times 10^{10}\ \text{J}$$

What is the mass of the air in the dome? The approximate inside volume $V_{\text{Stadium}}$ of such a stadium is $(140\ \text{m} \times 130\ \text{m} \times 40\ \text{m}) \approx 7 \times 10^5\ \text{m}^3$. The density of air at normal conditions is $\rho_{\text{Air}} = 1.3\ \text{kg/m}^3$ and the specific heat at constant volume is about

*(continued)*

$\approx 700\ \text{J/kg}\cdot{}^\circ\text{C}$. Thus, our estimate of the mass of the air inside the stadium is

$$m_{\text{Air}} = \rho_{\text{Air}} V_{\text{Stadium}} = (1.3\ \text{kg/m}^3)(7 \times 10^5\ \text{m}^3) = 9 \times 10^5\ \text{kg} \approx 10^6\ \text{kg}$$

We can now estimate the expected change in temperature:

$$\Delta T = \frac{Q_{\text{People to Air}}}{c_{\text{Air}} m_{\text{Air}}} \approx \frac{(7 \times 10^{10}\ \text{J})}{(700\ \text{J/kg}\cdot{}^\circ\text{C})(10^6\ \text{kg})} \approx 100\ {}^\circ\text{C}$$

Wow! If the air starts at 20 °C (room temperature), the final temperature would be about 120 °C. This example illustrates the importance of having a cooling system that can remove energy from the stadium at the same rate that it is being added by the people. Next time you attend a crowded party, note the temperature in the room. It becomes warmer due to the 100 J/s heating effect of each person.

**Try it yourself:** If the volume of a tent is about 9 m$^3$, estimate how much the air in the tent warms during an 8-hour night while two people sleep in it. What assumptions did you make? How reasonable are they?

*Answer:* 700 °C! We assumed that the air in the tent exchanges energy only with the two sleeping people (100 J/s each)—the result is unreasonable. A large amount of energy must transfer to the environment through the walls of the tent.

**Figure 12.6** An experiment to study phase changes and temperature changes due to heating.

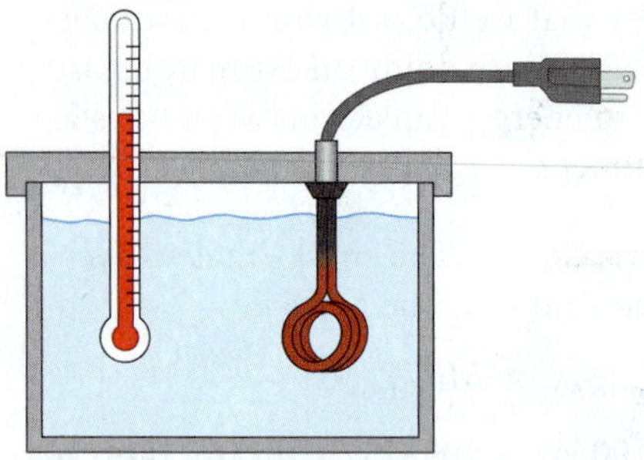

**Active Learning Guide ›**

**Review Question 12.5** Describe two situations in which the same amount of energy is supplied through heating to the same mass of gas at the same initial conditions, but the gas ends up at a different final temperature in each situation.

## 12.6 Changing state

We have found that providing energy to a system through heating or by doing work causes its temperature to change. Are there any processes in which the thermal energy of a system changes but the temperature of the system does not? Suppose you warm an ice cube in a Styrofoam cup so that some of it melts and becomes liquid water. Is the melted water warmer than the solid ice cube that remains? The transformation of ice into water is a **phase change**—a process during which a substance changes from one state to another. We are most familiar with the gas, liquid, and solid states. There are other states, such as plasma. Plasma is what fills neon lights and is the main state of matter in stars. We will learn about plasma in later chapters of the book. Let's consider an experiment involving the heating of water that starts as very cold solid ice.

Assume that you have 1.0 kg of water, which you place in a foam container with the top initially open (**Figure 12.6**). You insert a thermometer and an electric heating coil into the water. You place all of this in a freezer that is set at −40 °C. After the water freezes and reaches −40 °C, you place a foam cover over the container with the thermometer sticking out. You then remove the container from the freezer and turn on the electric coil. You observe the temperature change with time. The graph in **Figure 12.7** shows the relationship between the temperature and energy transferred to the container from the electric coil.

**Figure 12.7** Effect of heating on ice, liquid water, and steam.

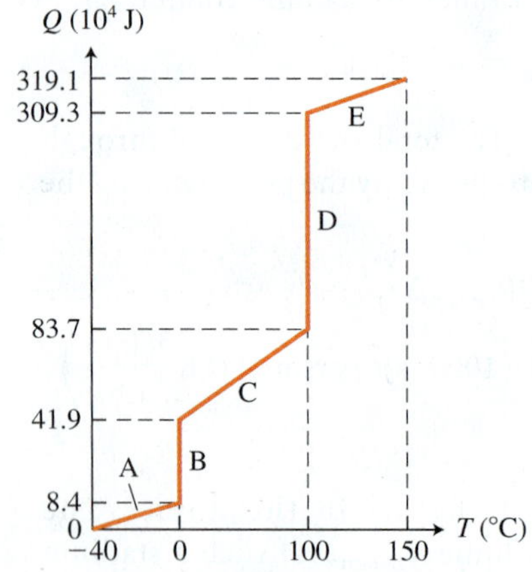


A Warming ice
B Melting ice
C Warming water
D Boiling water
E Warming steam

### Warming from −40 °C to 0 °C

Between −40 °C and 0 °C, the contents of the container remain frozen (A in Figure 12.7). The energy transferred to the ice from the electric coil causes the ice to warm and eventually reach its melting temperature at 0 °C. Because you know the power output of the coil, you learn that it takes 2090 J to warm 1.0 kg of ice by 1.0 °C. In other words, it takes $8.4 \times 10^4$ J to warm 1.0 kg of ice from −40 °C to 0 °C. This means that the specific heat of ice is 2090 J/kg · °C. It takes less energy to warm ice by 1.0 °C than it does to warm the same amount of liquid water. So far the transfer of energy to the ice led to a change of its temperature—no surprises here.

## Melting and freezing

At 0 °C the graph changes suddenly, becoming vertical (B in Figure 12.7). We observe that when the ice reaches 0 °C, it starts to melt, but the thermometer reading does not change. According to the graph, it takes $3.35 \times 10^5$ J of energy to completely melt 1.0 kg of ice. While this energy is added, the temperature of the system does not change. Why?

Remember that in a solid, neighboring molecules are close to each other and are attracted and bonded to each other—that is why solids keep their shape. Their potential energy of interaction is negative, meaning they are bound to one another much like the Earth-Moon system is bound by negative gravitational potential energy. The molecules vibrate and therefore have kinetic energy of random motion (though they do not fly around freely as they would in a gas). As the temperature of the solid rises toward its melting temperature, the vibrations of the neighboring water molecules become larger in amplitude and less ordered. When the solid reaches its melting temperature, the vibrations of neighboring molecules become so disruptive that the molecules begin to separate from one another and are no longer bound (**Figure 12.8**). There is still considerable interaction between neighboring molecules, but the interaction is less than in the rigid solid. Therefore, when melting or freezing occurs, it is the potential energy of particle interactions that changes. When we transfer energy to the solid material at the melting temperature, all of this energy goes into changing the potential energy of particle interactions, not the kinetic energy—thus the temperature does not change.

**Figure 12.8** Melting. Depiction of structural change as ice in (a) melts to form liquid water in (b).

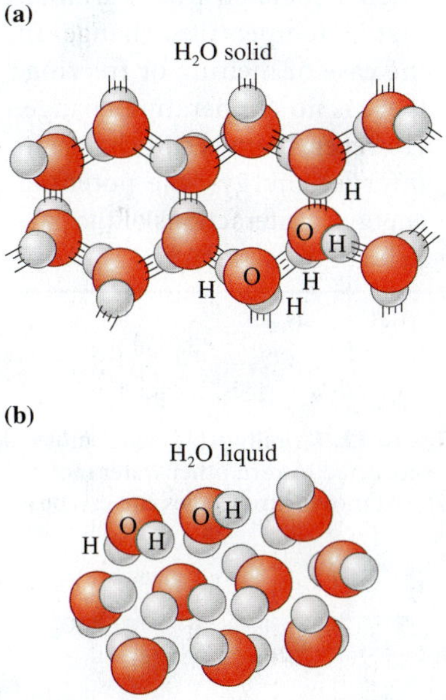


The energy needed to melt 1.0 kg of ice into water is called the **heat of fusion** of water and is given the symbol $L_f$ ($L_{f\,water} = 3.35 \times 10^5$ J/kg). (The heat of fusion is sometimes called the *latent heat of fusion* to underscore that this energy does not lead to temperature change; the change is in a hidden or latent form.) When a liquid freezes, it releases the same amount of energy (in other words, the reaction occurs in the opposite direction). The heat of fusion differs for different materials; their melting-freezing temperatures also differ (see **Table 12.7**). The unit of heat of fusion is joules per kilogram.

**Energy to melt or freeze** **The energy in joules needed to melt a mass $m$ of a solid at its melting temperature, or the energy released when a mass $m$ of the liquid freezes at that same temperature, is**

$$\Delta U_{int} = \pm mL_f \qquad (12.6)$$

**$L_f$ is the heat of fusion of the substance (see Table 12.7). The plus sign is used when the substance melts and the minus sign when it freezes.**

**Table 12.7** **Heats of fusion and vaporization.**

| Substance | Melting temperature (°C) | $L_f$, Heat of fusion at the melting temperature (J/kg) | Boiling temperature (°C) | $L_v$, Heat of vaporization at the boiling temperature (J/kg) |
|---|---|---|---|---|
| Water | 0 | $3.35 \times 10^5$ | 100 | $2.256 \times 10^6$ |
| Ethanol | −114 | $1.01 \times 10^5$ | 78 | $0.837 \times 10^6$ |
| Hydrogen ($H_2$) | −259 | $0.586 \times 10^5$ | −253 | $0.446 \times 10^6$ |
| Oxygen ($O_2$) | −218 | $0.138 \times 10^5$ | −183 | $0.213 \times 10^6$ |
| Nitrogen ($N_2$) | −210 | $0.257 \times 10^5$ | −196 | $0.199 \times 10^6$ |
| Aluminum | 660 | $3.999 \times 10^5$ | 2520 | $10.9 \times 10^6$ |
| Copper | 1085 | $2.035 \times 10^5$ | 2562 | $4.73 \times 10^6$ |
| Iron | 1538 | $2.473 \times 10^5$ | 2861 | $6.09 \times 10^6$ |

> **TIP** The term heat of fusion is confusing, as the word *heat* is often associated with warming, that is, temperature change. In the case of melting or freezing, there is no temperature change. However, there is still change in internal energy—the potential energy of interactions of the particles in the system.

## Warming liquid water from 0 °C to 100 °C

Once the ice has completely melted (C in Figure 12.7), further transfer of energy by heating causes the liquid water to warm. The graph in this region has a slope of 4180 J/°C and indicates that the specific heat of liquid water is 4180 J/kg · °C. With continued heating, the water temperature continues to rise.

## Boiling and condensation

At 100 °C, the graph again changes abruptly (D in Figure 12.7). We observe that the water is boiling and steam is coming out. The temperature of the water remains at the boiling temperature until all of the liquid water boils into vapor. From the graph we see that when the water reaches 100 °C, we must add $2.256 \times 10^6$ J of energy to completely convert 1.0 kg of liquid water to gaseous water vapor (steam). Once again, energy has been added to the water, but its temperature does not change. Where does the energy go? For a molecule to leave the surface of a liquid, it must have enough kinetic energy to break away from the neighboring molecules, which are exerting attractive forces on it (**Figure 12.9a**). A fast-moving water molecule can escape, while a slow-moving one cannot (Figure 12.9b). Escape of the fast-moving molecules leaves the remaining slow-moving "cooler" molecules behind. A continuous supply of energy through heating when the substance is at the boiling temperature causes a continual number of water molecules to gain the energy needed to leave the water surface but does not change the water temperature until all molecules have become water vapor. The energy transferred to the liquid leads to a change in the potential energy component of the internal energy.

**Figure 12.9** Boiling. (a) Water molecules are attracted toward other water molecules. (b) Fast moving molecules can escape.

(a)

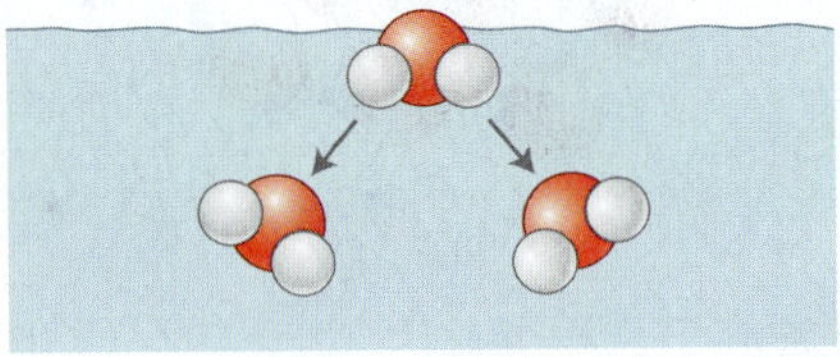

A water molecule at the surface of the water is prevented from leaving the water by its attraction to neighboring molecules.

(b)

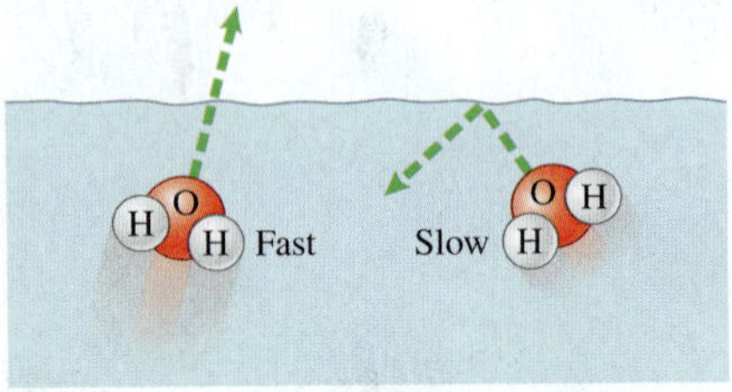


A fast-moving water molecule can escape from a liquid when it hits the surface. A slow-moving one cannot escape.

Liquids can evaporate at any temperature, not just at the boiling temperature. At any temperature some molecules have enough kinetic energy to escape the liquid. However, as they leave, the temperature of the liquid left behind decreases, since the average kinetic energy of the molecules left behind decreases. This is how we cool ourselves by evaporation on hot days.

The **heat of vaporization** of a substance $L_v$ (sometimes called the *latent heat of vaporization*) is the energy needed to transform 1 kg of a liquid at its boiling temperature into its gaseous state at that same temperature. For water, $L_{v\,water} = 2.256 \times 10^6$ J/kg. When 1.0 kg of the gas turns into liquid (condenses) at its boiling temperature, it releases the same amount of energy. The heat of vaporization differs for different substances; their boiling temperatures also differ (see Table 12.7). The unit of heat of vaporization is joules per kilogram.

> **Energy to boil or condense** The energy in joules needed to vaporize a mass $m$ of a liquid at its boiling temperature, or the energy released when a mass $m$ of a gas condenses at that same temperature is
>
> $$\Delta U_{int} = \pm mL_v \tag{12.7}$$
>
> $L_v$ is the heat of vaporization of the substance (see Table 12.7). The plus sign is used when the substance vaporizes and the minus sign when it condenses.

The heat of fusion, melting temperature, heat of vaporization, and boiling temperature depend on the type of substance, as you can see in Table 12.7.

Notice that the values of the heats of fusion and vaporization are much larger than the specific heat of these substances. This means that much more energy is needed to change the *state* of a substance than to change the *temperature* of the substance by one degree. Also notice that the heats of vaporization are significantly larger than the heats of fusion; more energy is required to boil the same mass of the same substance than to melt it. Finally, notice that the values are provided for boiling temperatures. Liquids evaporate at temperatures below the boiling temperature; however, the energy needed to keep them at the same temperature during this evaporation is greater. For example, the heat of vaporization at the boiling temperature for water is $2.256 \times 10^6(\text{J/kg})$ whereas at body temperature (when sweat evaporates) the heat of vaporization is $L_{\text{v water}} \approx 2.4 \times 10^6\ \text{J/kg}$.

## Warming gaseous water vapor above 100 °C

We cannot perform the next part of the experiment (E in Figure 12.7) with the apparatus shown in Figure 12.6 since the water vapor would escape. But the experiment could be performed if the container were airtight. In that case, if we continued to add energy to 1.0 kg of water vapor, we would find that its temperature increased 1 °C for each approximately 2000 J of additional energy added to it. This tells us that the specific heat of water vapor is approximately $c_{\text{water vapor}} = 2000\ \text{J/kg}\cdot{}^\circ\text{C}$.

## Reversing the process

We now put the 1.0 kg of hot water vapor back in our −40 °C freezer and slowly remove thermal energy to reverse the process shown graphically in Figure 12.7. We find that the water first cools to its boiling temperature (100 °C for water). It then releases considerable energy as it condenses to form liquid water. Next, it cools to 0 °C as more energy is removed. The liquid water again releases considerable energy as it freezes. The ice finally cools to a final temperature of −40 °C.

**QUANTITATIVE EXERCISE 12.7 Sweating in the Grand Canyon**

Eugenia carries a 10-kg backpack on a hike down into the Grand Canyon and then back up. Sweat evaporates off her skin at a rate of 0.28 g/s. At what rate is thermal energy transferred from inside her body to the surface of her skin in order to provide the energy needed to vaporize the sweat on her skin at this rate? At body temperature the heat of vaporization for sweat is about $L_{\text{v Sweat}} \approx 2.4 \times 10^6\ \text{J/kg}$.

**Represent mathematically** The sweat on the skin will be the system. The energy needed to evaporate a small mass $\Delta m$ of the sweat that is added to the sweat from inside Eugenia's body is $Q_{\text{Eugenia to Sweat}} = +\Delta m \cdot L_{\text{v}}$. Dividing each side by a small time interval $\Delta t$, we get

$$\frac{Q_{\text{Eugenia to Sweat}}}{\Delta t} = \frac{\Delta m_{\text{Sweat}} L_{\text{v Sweat}}}{\Delta t} = \left(\frac{\Delta m_{\text{Sweat}}}{\Delta t}\right) L_{\text{v Sweat}}$$

where the rate of sweating is $\dfrac{\Delta m_{\text{Sweat}}}{\Delta t} = 0.28\ \text{g/s} = 0.28 \times 10^{-3}\ \text{kg/s}$.

**Solve and evaluate** We find the rate of energy transfer through evaporative heating is

$$\frac{Q_{\text{Eugenia to Sweat}}}{\Delta t} = \left(\frac{\Delta m_{\text{Sweat}}}{\Delta t}\right) L_{\text{v Sweat}}$$
$$= (0.28 \times 10^{-3}\ \text{kg/s})(2.4 \times 10^6\ \text{J/kg})$$
$$= 670\ \text{J/s} = 670\ \text{W}$$

This is the rate that energy is being transferred from inside her body to the moisture on her skin. Another way to think about this is that Eugenia's body is being cooled at a rate of 670 W!

The evaporation of sweat is a natural cooling mechanism that protects us from overheating. Increasing the rate of evaporation from the skin increases the body's cooling rate. A sick child's fever is often reduced by rubbing the skin with alcohol, since alcohol evaporates much faster than water.

**Try it yourself:** After swimming in a pool, you come out with 300 g of water absorbed in your swimsuit. How much energy will this water absorb from the environment to evaporate? How can you evaluate your answer?

*Answer:* $7.2 \times 10^5$ J. If we assume that the swimsuit dries only using the energy generated by your body at a rate of 100 J/s, it should take about 7200 s to dry, which is about 2 hours—this is a reasonable time. When the Sun is out, a swimsuit dries somewhat faster, so the absorption of energy from the Sun's radiation must be significant.

We can often solve problems in multiple ways. We could solve the following problem by treating the two objects (hot coffee and an ice cube) as two different systems. However, this time we will treat them as a single system. The hot object loses thermal energy as its temperature decreases, and the cooler object gains an equal amount of internal energy (potential and thermal) as it melts and warms.

**EXAMPLE 12.8 Cooling hot coffee**

You add 10 g of ice at temperature 0 °C to 200 g of coffee at 90 °C. Once the ice and coffee reach equilibrium, what is their temperature? Indicate any assumptions you made.

**Sketch and translate** A sketch of the process is shown below. We choose the coffee and ice as the system. The ice gains thermal energy and melts at 0 °C. Then the melted ice, now liquid water, warms to the final equilibrium temperature. Simultaneously, the coffee loses an equal amount of thermal energy as it cools until the mixture is at a single final equilibrium temperature.

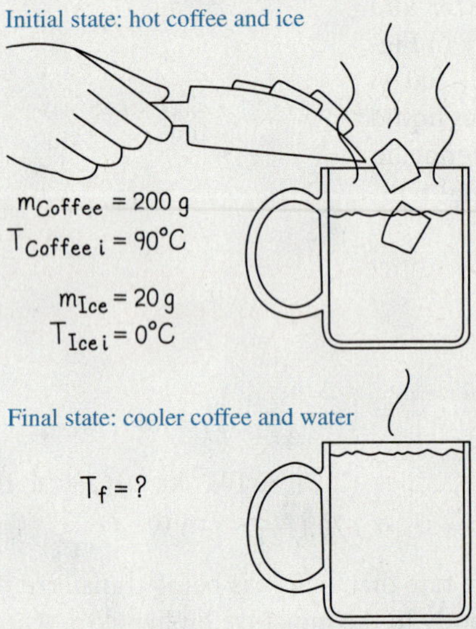


**Simplify and diagram** The coffee and ice are the system. Assume that the coffee-ice cube system is isolated; there is no energy transferred between the system and the environment, including the cup. Assume also that the coffee has the same specific heat as water, which is reasonable since coffee is mostly water. The process is represented qualitatively in a temperature-versus-time ($T$-versus-$t$) graph using two lines for the ice and one line for the coffee. Notice that after some time, they reach the same thermal equilibrium temperature.

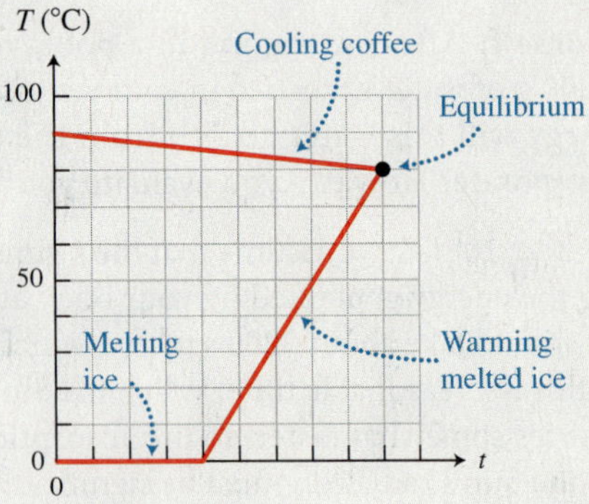


**Represent mathematically** The energy of the system is constant, so the internal energy change of the ice plus the thermal energy change of the coffee must add to zero.

$$\Delta U_{\text{int}} = \Delta U_{\text{int Ice}} + \Delta U_{\text{th Coffee}} = 0$$

The ice first melts, then warms. The coffee cools.

$$(m_{\text{Ice}}L_{\text{Ice}} + m_{\text{Ice}}c_{\text{Water}}(T_{\text{f}} - T_{\text{Ice i}})) + m_{\text{Coffee}}c_{\text{Coffee}}(T_{\text{f}} - T_{\text{Coffee i}}) = 0$$

Note that the system reaches equilibrium, so $T_{\text{Ice f}} = T_{\text{Coffee f}} = T_{\text{f}}$. Also, the initial temperature of the melted ice is $T_{\text{Ice i}} = 0\,°\text{C}$.

**Solve and evaluate** Solve for the final temperature $T_{\text{f}}$:

$$T_{\text{f}} = \frac{m_{\text{Coffee}}c_{\text{Coffee}}T_{\text{Coffee i}} + m_{\text{Ice}}c_{\text{Water}}(0\,°\text{C}) - m_{\text{Ice}}L_{\text{Ice}}}{m_{\text{Ice}}c_{\text{Water}} + m_{\text{Coffee}}c_{\text{Coffee}}}$$

$$= \frac{(0.2\text{ kg})(4180\text{ J/kg}\cdot°\text{C})(90\,°\text{C}) + 0 - (0.01\text{ kg})(3.35 \times 10^5\text{ J/kg})}{(0.01\text{ kg})(4180\text{ J/kg}\cdot°\text{C}) + (0.2\text{ kg})(4180\text{ J/kg}\cdot°\text{C})}$$

$$= 82\,°\text{C}$$

This is a reasonable temperature considering the small amount of ice used.

**Try it yourself:** You pour 200 g water at 5.0 °C into a Styrofoam container and add 500 g of ice at −40 °C. What is the result? Describe the steps in solving the problem and determine the numerical answer.

*Answer:* Four possible outcomes are (a) all ice melts and the resulting water is above 0 °C; (b) all water freezes and the resulting ice is below 0 °C; (c) some but not all of the water freezes and the resulting mixture is at 0 °C; (d) some but not all ice melts and the mixture is at 0 °C. To decide which outcome is correct, first determine how much energy the water will release to cool to 0 °C and how much energy the ice will need to warm to 0 °C. Calculations show that the water will release $4.2 \times 10^3$ J and the ice will require $42 \times 10^3$ J. Thus cases (b) and (c) are possible. Final answer: the mixture will be at 0 °C and the amount of water that freezes is 110 g.

**Review Question 12.6** Why are the units for specific heat and heat of fusion/vaporization different?

## 12.7 Heating mechanisms

So far we have described the process of heating as a way for energy to be transferred into or out of a system. However, we haven't discussed the energy transfer mechanisms by which heating occurs. These mechanisms underlie and explain many phenomena. For example, how can we reduce energy losses from our homes during the winter? Why is the "dry heat" in Arizona less uncomfortable than a hot, humid day in Mississippi? Why should we be concerned about climate change? These questions are related to energy transfer mechanisms and, in particular, to the rate at which these transfers occur.

We define heating rate $H$ as the energy transfer by heating $Q$ from one object to another divided by the time interval $\Delta t$ needed for that heating energy transfer:

$$\text{Heating rate} = H = \frac{Q}{\Delta t}$$

This transfer can occur through several mechanisms that we discuss below.

Why is the handle hot, even though it has not touched the flame?

### Conduction

You are boiling water for morning coffee at your campsite. You have two cups: one cup is Styrofoam and the other aluminum. Which cup would you prefer to hold just after filling it with hot water? The water inside both cups is at the same temperature. Experience tells us that the Styrofoam cup will feel more comfortable while the aluminum cup may burn our fingers.

Later, you cook scrambled eggs in a frying pan on the camp stove. You start to remove the pan from the stove, but the handle is very hot, even though the handle hasn't touched the stove's hot burner. How can you explain these observations?

Recall that the temperature of a substance is related to the average random kinetic energy of the particles of which it is made. The high-energy, high-speed hot water molecules hit the inside wall of the aluminum cup or the Styrofoam cup, making the particles on the inside wall of each cup vibrate faster. These more energetic particles in the wall then bump into neighboring particles in the wall, which in turn bump into their neighbors, which are closer to the outside wall. Thus, the thermal energy transfers from particle to particle from the inside wall to the outside wall. Your fingers touch the outside wall. When its surface is hot, the high-energy particles on the surface bump the particles on the surface of your skin, causing them to move faster. The high-energy vibrations of the particles on your skin can produce a burning sensation. The process by which thermal energy is transferred through physical contact is called **conductive heating** or **cooling.**

Why does the aluminum cup of hot water feel hotter than the Styrofoam cup filled with equally hot water? Aluminum transfers thermal energy from particle to particle quickly. Styrofoam transfers thermal energy more slowly. We say that aluminum is a good thermal energy conductor and Styrofoam is a poor thermal energy conductor. A quantity called **thermal conductivity** $K$ characterizes the rate at which a particular material transfers thermal energy (see **Table 12.8**). The larger the thermal conductivity, the faster thermal energy transfers from molecule to molecule in that material. You want to use a material with low thermal conductivity to make a cup for holding hot

**Table 12.8** Thermal conductivity of materials (at 25 °C).

| Material | $K$ (W/m · °C) |
|---|---|
| ***Metals*** | |
| Aluminum | 250 |
| Brass | 109 |
| Copper | 400 |
| Stainless steel | 16 |
| ***Other solids*** | |
| Soil | 0.6–4 |
| Brick | 1.31 |
| Concrete, stone | 1.7 |
| Attic insulation | 0.04 |
| Glass, window | 0.9 |
| Ice (at 0 °C) | 2.18 |
| Styrofoam | 0.03 |
| Wood | 0.04–0.14 |
| Fat | 0.2 |
| Muscle | 0.4 |
| Bone | 0.4 |
| ***Other materials*** | |
| Air | 0.024 |
| Water | 0.58 |

beverages. Likewise, you want to use a material with low thermal conductivity to insulate your home—to prevent thermal energy from leaving the house during the winter or to prevent it from entering during the summer.

**Conductive heating/cooling** **The rate of conductive heating $H_{\text{Conduction}}$ from the hot side of a material to the cool side is proportional to the temperature difference between the two sides $(T_{\text{Hot}} - T_{\text{Cool}})$, the thermal conductivity $K$ of the material separating the sides, and the cross-sectional area $A$ across which the thermal energy flows. The rate is inversely proportional to the distance $l$ that the thermal energy travels.**

$$H_{\text{Conduction}} = \frac{Q_{\text{Hot to Cool}}}{\Delta t} = \frac{KA(T_{\text{Hot}} - T_{\text{Cool}})}{l} \tag{12.8}$$

## CONCEPTUAL EXERCISE 12.9 Snow on the roof

In the morning on a cloudy winter day your house with an attached garage has a light snow covering it. What might you expect to see 24 hours later, assuming that the temperature stays low and there is no new snow?

**Sketch and translate** The melting on different parts of the roof depends on the conductive thermal energy transfer through the roof from inside the house to the outside.

**Simplify and diagram** Assume that the energy delivered to the roof comes from inside the house and not from the Sun. Assume also that the insulation and roofing material is the same in all parts of the house ($K$ is the same for the roof over all parts of the house). Now divide the house into two parts: the living quarters and the garage. Thermal energy transfer occurs from the warmer regions inside the house through the roof to cooler regions outside of the house. The conductive heating transfer rate depends on the temperature difference $(T_{\text{Hot inside}} - T_{\text{Cold outside}})$. The temperature difference is greater from in the living area to the outside than it is from the inside of the garage to the outside. Thus, the thermal energy transfer rate should be greater through the roof above the living quarters than it is through the roof over the garage. Because of this, there will be more snow melting above the living area than above the garage.

The photo below confirms this reasoning. The faster melting over the living area tells us that energy is being transferred out of the house more quickly than out of the garage.

Can we reduce the amount of energy we are releasing to the environment (and therefore keep more of the energy—and warmth—in the house)? To reduce energy transfer, we could slow the heating transfer rate by using lower thermal conductivity $K$ insulation or thicker insulation. When we use thicker insulation [thereby increasing length $l$ in Eq. (12.8)], the thermal energy has to travel a longer distance across the interface.

**Try it yourself:** Your skin temperature is about 35 °C. Imagine that you touch the wooden top of a desk in a room where the temperature is 20 °C. The desk feels comfortable. Then you touch the steel leg of the desk and it feels cold. Does it mean that the legs of the desk are at a lower temperature than the wooden top?

*Answer:* The legs and the desktop are at the same temperature—room temperature. When you touch each of them with your hand, which is somewhat warmer than these objects, thermal energy transfers from your hand to the objects. The larger thermal conductivity of steel makes this transfer occur at a higher rate, which is why the steel feels cooler to your hand.

The concept of thermal conductivity explains why a thermos keeps drinks hot or cold. The thermos depicted in **Figure 12.10** consists of two metal cups separated by a layer of low-density air (as close to a vacuum as we can get). Because the particles of air in the nearly evacuated space between the metal cups are very far from each other, they slowly transfer energy by conduction, resulting in a very small thermal conductivity. (In a perfect vacuum $K$ would be zero, because there would be no particles to interact with each other.) Thus, if a liquid in the inner cup is cold, it stays cold; if it is warm, it stays warm.

Another application of the concept of thermal conductivity is fur that allows animals to get through cold winters. The fur consists of hair strands separated by air; the strands work almost like a thermos in preventing the conductive energy transfer from a warm body to a cold environment.

**Figure 12.10** How a thermos bottle works. A thermos consists of an inner chamber surrounded by a vacuum.

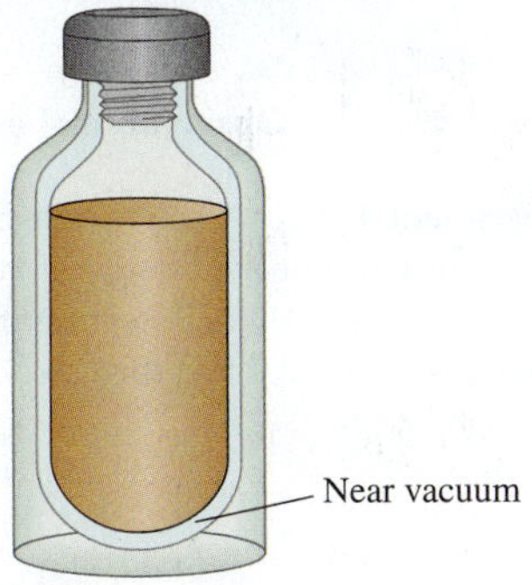


## Evaporation

At the beginning of this section we asked, Why are the hot, dry summers in Arizona less uncomfortable than hot, humid summers elsewhere in the United States? The answer is that we more effectively transfer energy by evaporation in a dry climate. On a humid day water not only evaporates from the skin but also condenses on the skin from the gaseous water vapor in the air. When gaseous water vapor converts to liquid water (condenses), energy is released and returned to your body, raising your temperature. To stay cool, you want the rate of evaporation to be somewhat greater than the rate of condensation. In a "dry" climate with low humidity, not much water vapor condenses on your skin. You evaporate more water than condenses on you, and there is a net cooling effect. **Figure 12.11** indicates some typical evaporation rates under different circumstances.

**Figure 12.11** Evaporation rates for different conditions.

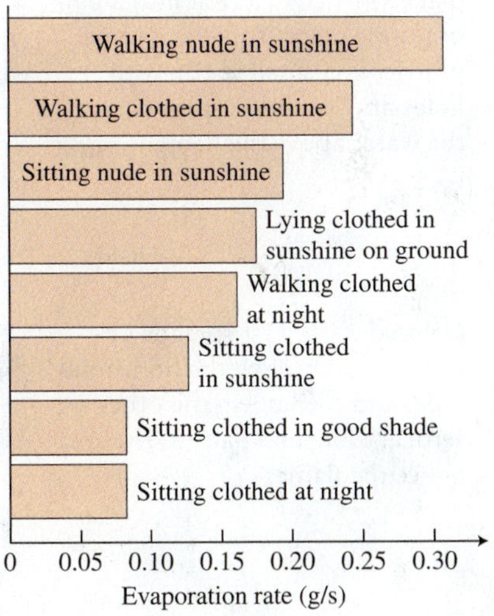


**Evaporative heating/cooling** Evaporative heating occurs when a gas condenses on a surface. Evaporative cooling occurs when a liquid evaporates from a surface. The evaporative energy transfer rate $H_{\text{Evaporation}}$ is proportional to the rate at which liquid mass evaporates or condenses $\Delta m/\Delta t$ on the surface and to the heat of vaporization $L_v$ of the liquid at the temperature of the liquid:

$$H_{\text{Evaporation}} = \pm\left(\frac{\Delta m}{\Delta t}\right)L_v \qquad (12.9)$$

We use the positive sign for condensation (energy transferred to the system) and the negative sign for evaporation (energy transferred out of the system).

## Convection

Conduction as the heating mechanism works well in materials in which particles interact with each other. In gases, for example, the particles are so far apart that these interactions are almost nonexistent. In liquids the particles interact but much less than in a solid. How does thermal energy transfer in those materials? Consider the experiments in Observational Experiment **Table 12.9** on the next page.

## OBSERVATIONAL EXPERIMENT TABLE

### 12.9 Convective heating/cooling.

**Observational experiment**

**Experiment 1.** You light a candle and hold your hand near one side of it. Your hand does not feel much warmer than when it is far away from the candle.

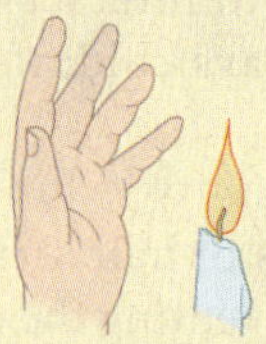

However, if you place your hand above the candle, you can only hold it there for a short time interval before you risk burning yourself.

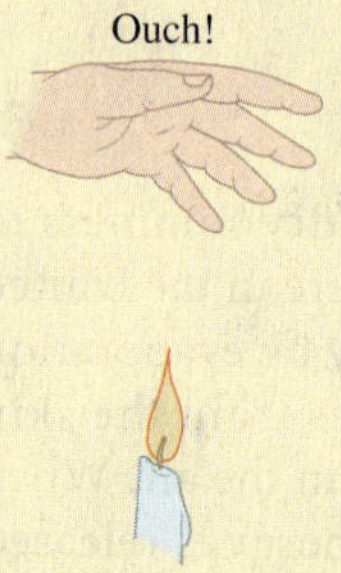


**Analysis**

The air near the flame quickly warms and expands. Since its density is now less than the density of the surrounding air, it rises and this warmed air contacts the skin of your hand directly above the candle.

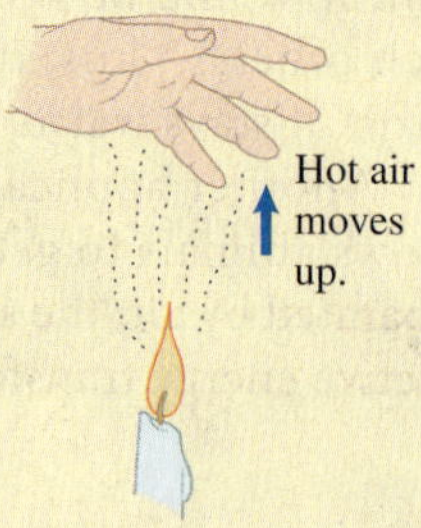


**Observational experiment**

**Experiment 2.** You have the apparatus shown, placed over a flame. Potassium permanganate ($KMnO_4$) is a powdery substance of dark purple color that when placed in water turns it into a pink or purple color. You place some powder in a small spoon with holes and lower the spoon into the water above the flame.

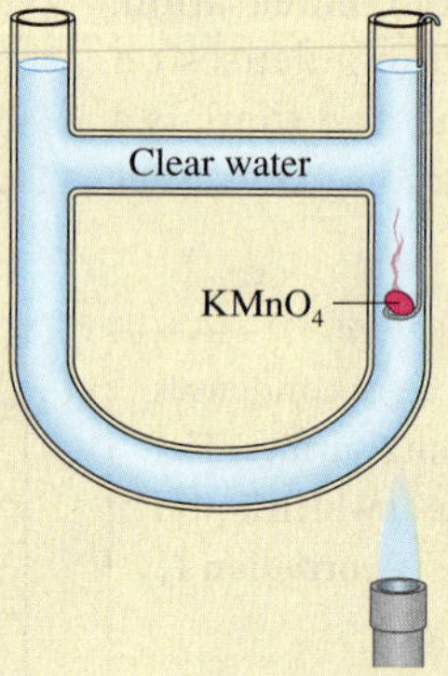


Colored water first spreads upward, then through the horizontal pipe, and then down the other vertical tube, and back to the area above the flame.

**Analysis**

Thermal conduction warms the water next to the flame. This warm water expands and rises. The warm water that moved up starts a *flow* of warm liquid that moves around the apparatus. As the pink water moves horizontally, it cools and sinks, replacing the water heated by the flame. The pink water is then heated again and the process repeats.

**Pattern**

In both processes the warmed fluid becomes less dense and rises. As the warm fluid rises, it is replaced by cooler fluid.

The processes described in Table 12.9 are called **natural convective heating**.

**Natural convective heating/cooling** **Natural convective heating/cooling is the process of energy transfer by a fluid moving in a vertical direction away from/toward the center of Earth. Fluids such as air or water rise because their density decreases as their temperature increases and sink because their density increases as their temperature decreases.**

While convection can occur as a natural process, it can also be induced. For example, in a forced-air heating system, air in the furnace is warmed by a gas or electric heater. The warmed air is forced by a fan through ducts to different rooms in the house. Other ducts return cooler air back to the furnace to be warmed. In cars, burning gasoline makes the engine very hot. To cool it, a pump moves fluid (water or a coolant) from the radiator through hoses into and through the engine (**Figure 12.12**). Thermal energy is transmitted by conduction to the coolant from the engine. The energy is then carried away by the fluid (forced convective thermal energy transfer) through a radiator hose and back to the radiator, where outside air moves past the fins in the radiator and cools the fluid again. This process is called forced convective heating (or cooling). The forced-air heating system in a house and the cooling system in cars are examples of **forced convection.**

**Figure 12.12** A car's engine is cooled mostly by forced convection.

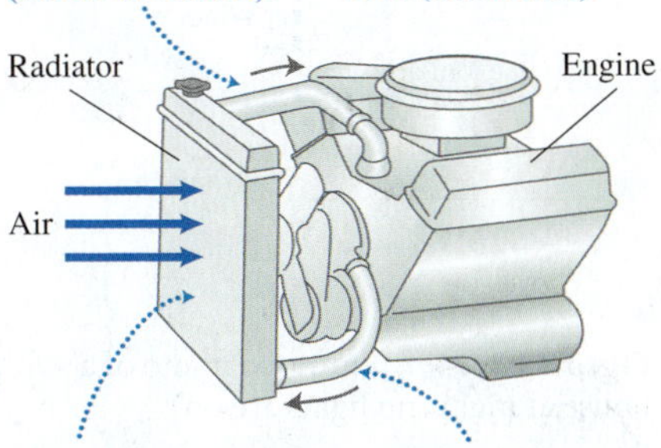
1. Cool water travels from radiator to engine (forced convection).
2. Water passes through hot engine cooling the engine and warming water (conduction).
Radiator
Engine
Air
4. Air passes through radiator causing hot water to cool (convection).
3. Hot water travels from engine to radiator (forced convection).

**Forced convective heating/cooling** **In forced convective heating, a hot fluid is forced to flow past a cooler object, transferring thermal energy to the object. The fluid moves away with less thermal energy. In forced convective cooling, a cool fluid moves past a warmer object (for example, a cold winter wind moving across your warm skin). The object transfers thermal energy to the cooler fluid. The now-warmed fluid carries the thermal energy away from the object.**

Calculating the rate of convective energy transfer is a complex subject and depends on many factors. Often, the calculations are estimates and involve the use of tables of data assembled by energy transfer specialists. There are many practical examples of natural and forced convective heating and cooling.

A *convection oven* has advantages over a traditional oven. In a traditional oven, a thin layer of air surrounds the food. This air is at a temperature between that of the food and the hotter air in the rest of the oven. Thus, cooking occurs at a lower temperature than the temperature of air in the oven. In a convection oven, a fan (maybe more than one) at the back of the oven moves heated air around the cooking compartment. This causes fresh hot air to continually move past the food. Cooking time is reduced by about 20% and cooking temperature can be 10–20 °C lower than in a traditional oven. You get the same result with less energy use.

## Radiation

Consider another phenomenon. You are sitting by a campfire on a cool evening. The air above the flames warms and rises with the smoke from the fire, an example of natural convective heating. You sit on a log at the side of the fire and are also warmed by it. The warming is not the result of the convection from the fire—the warm air is moving upward and does not move across your skin. You notice that the part of your body facing the flame is warmed, while the part of your body facing away from the flame remains cool. Why is that?

Consider a similar process. Outside Earth's atmosphere is an approximate vacuum with low concentrations of atomic particles. The particles are very far from each other and therefore essentially do not transfer energy by conduction. The Earth has remained warm enough for habitation because the Sun irradiates Earth in much the same way that a campfire warms the part of your body facing it. The Earth is warmer on the side facing the Sun and cooler on the side facing away from it. Both the Sun and the campfire emit infrared radiation and some light that is the source of this thermal energy.

> **Radiative heating/cooling** **All objects, hot and cool, emit different forms of radiation that can travel through a vacuum. The hotter the object, the greater the rate of radiation emission. Objects in the path of this radiation may absorb it, gaining energy.**

**Figure 12.13** A passive solar heating system.

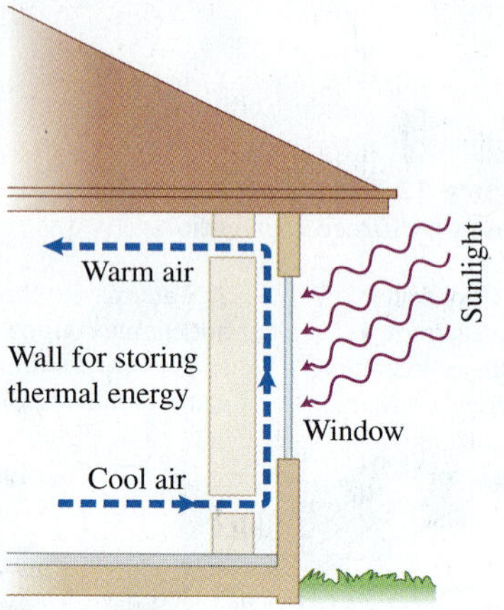


*Passive solar heating* is an important application of radiative heating (and other heating mechanisms). Passive solar buildings are constructed so that they have walls with large heat capacities for storing radiative energy (sunlight) that passes through windows during the daytime (**Figure 12.13**). The light energy absorbed by the wall transforms its energy into the thermal energy of the surface atoms and molecules in the wall. These warmer atoms and molecules transfer their energy by conduction to other atoms and molecules deeper in the wall. A well-designed wall has large mass $m$ with a high specific heat $c$. As a result, the wall's temperature change

$$\Delta T = \frac{Q}{m \cdot c}$$

will be small during the day, because both $m$ and $c$ are large. The large amount of energy transferred to the wall by heating during the day is released slowly at night when it is needed. Energy is transferred to the building at night by natural convective heating as cool air moves up past the warm wall and into the building as well as from radiative heating from the wall into the building.

Radiation is also used for diagnostic methods in what is called *thermography*. One application is breast thermography, which detects differences in infrared radiation from warmer and cooler surfaces. Thermography detects "hot spots" caused by the increased metabolic activity in cancer cells and combines advanced digital technology with ultrasensitive infrared cameras for safe and early screening of breast cancer.

**Figure 12.14** An infrared photo of a house at night (no lights are on).

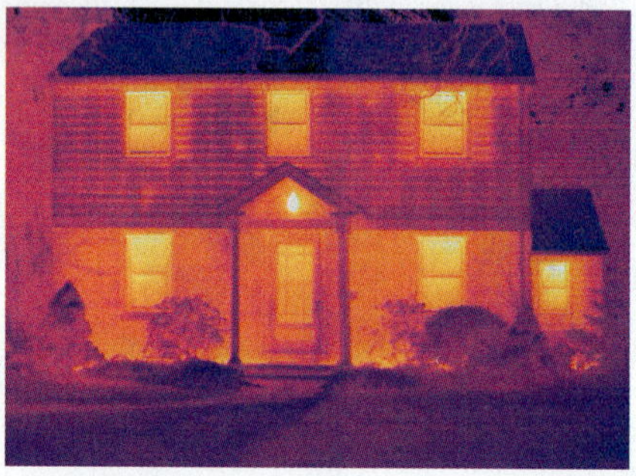

Thermography was also employed to take the picture in **Figure 12.14** of a house at night. The parts of the house that are radiating energy at the fastest rate are yellow. The next fastest radiating part is orange, then pink, purple, violet, and black. None of these regions are sources of visible light. All lights in the house are off. This radiation is being emitted simply because the house is not at zero absolute temperature.

**Review Question 12.7** What are the most effective ways of transferring energy through solids, liquids, gases, and vacuum?

**Figure 12.15** Earth's average temperature and $CO_2$ concentration for the past 400,000 years.

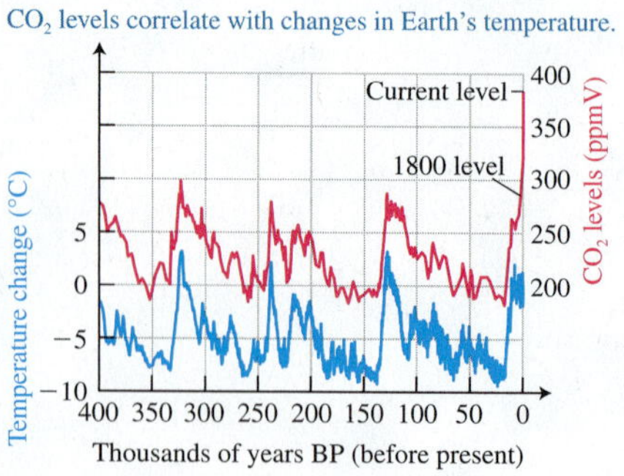


## 12.8 Climate change and controlling body temperature: Putting it all together

The first law of thermodynamics and the various heating mechanisms help us understand two crucial phenomena—global climate change and body temperature control.

### The greenhouse effect and climate change

**Figure 12.15** shows a strong correlation between the $CO_2$ concentration in Earth's atmosphere and the average Earth temperature during the last 400,000 years. During the coldest periods, when much of Earth was covered with glaciers, the $CO_2$ levels were lowest. During warm periods, some parts of Earth became much dryer while other parts experienced flooding. During these warm periods, the $CO_2$ levels were highest. Glaciers melted, leading to a rising ocean and coastal flooding. More frequent and severe weather events occurred (such as hurricanes, typhoons, droughts, and blizzards). Scientific evidence indicates that the concentration of $CO_2$ in our atmosphere plays a big part in these dramatic climate changes.

The cause of the increase in atmospheric $CO_2$ is a complex subject. Part of the increase is due to human activities, such as the increasing use of fossil fuels and the removal of forests and plant life, which absorb $CO_2$.

We can qualitatively investigate the role of $CO_2$ concentration in the Earth's changing temperature by applying the first law of thermodynamics to the system pictured in **Figure 12.16**. The system inside the dashed lines consists of Earth and its atmosphere. The Earth/atmosphere system gains energy through radiative heating from the Sun ($Q_{\text{Sun to Earth}} > 0$) and loses energy through radiation that it emits into space ($Q_{\text{Space to Earth}} < 0$):

$$Q_{\text{Sun to Earth}} + Q_{\text{Space to Earth}} = \Delta U_{\text{System}}$$

If we divide both sides of this equation by an arbitrary time interval $\Delta t$ during which heating occurs, we get an equation that involves the heating rates and the rate of the system's energy change:

$$\frac{Q_{\text{Sun to Earth}}}{\Delta t} + \frac{Q_{\text{Space to Earth}}}{\Delta t} = \frac{\Delta U_{\text{System}}}{\Delta t}$$

or

$$H_{\text{Sun to Earth}} + H_{\text{Space to Earth}} = \frac{\Delta U_{\text{System}}}{\Delta t}$$

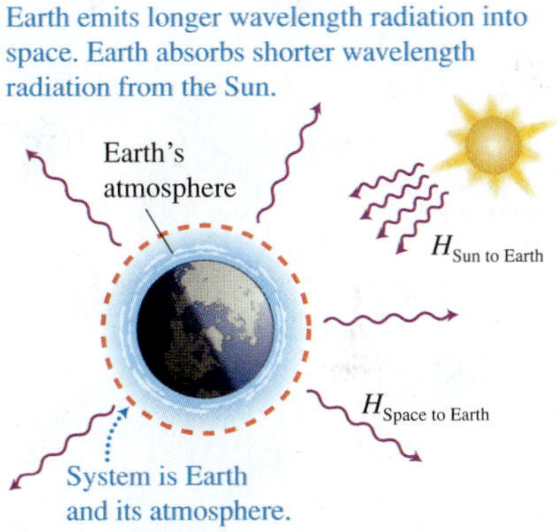
Earth emits longer wavelength radiation into space. Earth absorbs shorter wavelength radiation from the Sun.
Earth's atmosphere
$H_{\text{Sun to Earth}}$
$H_{\text{Space to Earth}}$
System is Earth and its atmosphere.

**Figure 12.16** The Earth's temperature (including the atmosphere) depends on the balance of incoming and outgoing radiation.

The terms on the left are the radiative heating rates from the Sun to Earth ($H_{\text{Sun to Earth}} > 0$) and from Earth to space ($H_{\text{Space to Earth}} < 0$). If $H_{\text{Sun to Earth}} + H_{\text{Space to Earth}} = 0$, then the internal energy of the system (Earth and its atmosphere) remains constant. If $H_{\text{Sun to Earth}} + H_{\text{Space to Earth}} > 0$, the system's energy increases. If $H_{\text{Sun to Earth}} + H_{\text{Space to Earth}} < 0$, its energy decreases.

Which type of radiation is emitted by an object depends on the temperature of the object. Hot objects such as the Sun emit more energy as short-wavelength infrared radiation and visible light. Cooler objects such as Earth emit longer wavelength infrared radiation and very little short-wavelength infrared radiation and visible light. Carbon dioxide is a strong absorber of long-wavelength infrared radiation, but not a strong absorber of short-wavelength infrared radiation and visible light. This means that carbon dioxide does not absorb much of the radiation Earth receives from the Sun, but it does absorb considerable radiation emitted by Earth. Since the magnitude of $H_{\text{Space to Earth}}$ decreases as $CO_2$ concentration increases, then $H_{\text{Sun to Earth}} + H_{\text{Space to Earth}} > 0$ and the thermal energy of the Earth/atmosphere system increases, which manifests as a global temperature increase. Our simple model based on the first law of thermodynamics and the mechanisms of energy transfer predicts global warming if $CO_2$ concentration is increasing.

The concentration of $CO_2$ in Earth's atmosphere is now at the highest level in the last 400,000 years (Figure 12.15). The hottest years since Earth temperature data were first recorded in 1861 have all occurred since 1990, with each decade getting slightly warmer than the previous decades.

## Controlling body temperature

A healthy human body maintains a central core temperature of about 37.0 °C (98.6 °F). To do so, the body must shed excess thermal energy that is converted from chemical energy due to metabolic processes. This is especially challenging during vigorous exercise, when the body's core temperature could increase by as much as 6 °C in only 15 min, which could lead to convulsions or even brain damage.

Energy transfer is the key to maintaining a consistent temperature. We can analyze how the human body uses heating mechanisms of energy transfer by applying the first law of thermodynamics ($W_{\text{Environment on System}} + Q_{\text{Environment to System}} = \Delta U_{\text{System}}$) to the body (the system) during a short time interval $\Delta t$. Divide the quantities in the first law by the time interval during which the changes occurred:

$$\frac{W_{\text{Environment on System}}}{\Delta t} + \frac{Q_{\text{Environment to System}}}{\Delta t} = \frac{\Delta U_{\text{System}}}{\Delta t}$$

The second term in the above equation is the heating energy transfer rate ($Q_{\text{Environment to System}}/\Delta t$) from the environment to the system. We can separate $Q$ into four separate terms, which represent the rates of conduction, convection, radiation, and evaporation (represented by $H$ symbols with subscripts):

$$\frac{W_{\text{Environment on System}}}{\Delta t} + (H_{\text{Conduction}} + H_{\text{Convection}} + H_{\text{Radiation}} + H_{\text{Evaporation}})$$
$$= \frac{\Delta U_{\text{Thermal}}}{\Delta t} + \frac{\Delta U_{\text{Chemical}}}{\Delta t}$$

We also substituted the two types of internal energy change that can occur in a person's body: the rate of thermal energy change and the rate of chemical potential energy change. This latter term is the metabolic rate, the rate of energy released by the chemical reactions occurring in the body.

Now, let's apply the above equation to a real process. Suppose that a person runs at moderate speed on an indoor air-conditioned track. **Table 12.10** shows quantities that are representative of a real process, although the quantitative methods used to determine the numbers have not been developed in this book.

Consider each value in Table 12.10. Air exerts a drag force that opposes the runner's motion, doing work on the runner at a rate of $-100$ J/s. The cool air in the indoor facility passes across the runner's skin and causes a $-300$ J/s convective cooling rate as the skin loses thermal energy and the air gains it. The walls of the indoor facility are cooler than the runner's skin temperature, and consequently the runner absorbs energy transmitted by radiative heating from the walls at a lower rate than the body emits energy by radiative cooling from the skin (a net $-75$ J/s radiative energy transfer rate). Finally, the runner's evaporative energy transfer rate equals $-325$ J/s. Adding all these, we get $-800$ J/s. To keep the body temperature constant, the runner's metabolic rate must be 800 J/s. If the person's metabolic rate is more than 800 J/s, the person would need additional cooling in order to maintain a constant body core temperature.

**Review Question 12.8** How does the concentration of $CO_2$ in the atmosphere contribute to the rise of the atmosphere's temperature?

**Table 12.10 Work, heating, and energy changes in the body of a runner (moderate speed in an air-conditioned building).**

| Type of energy change | Work and heating rates (watts = J/s) |
|---|---|
| Drag force exerted by air (work) | −100 |
| Conduction | ≈0 |
| Thermal energy transfer due to convection | −300 |
| Transfer of energy by radiative cooling | −75 |
| Cooling due to evaporation of sweat | −325 |
| | **Rate of Internal Energy Change (watts = J/s)** |
| Thermal | ≈0 |
| Chemical (metabolic rate) | −800 |

# Summary

**Words** | **Pictorial and physical representations** | **Mathematical representation**

**Thermal energy $U_{\text{thermal}}$** The random kinetic energy of the atoms and molecules in a substance. (Section 12.1)

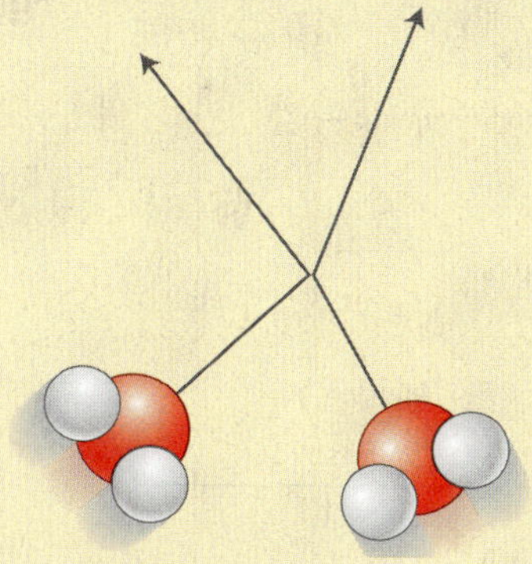

$$U_{\text{thermal}} = \frac{3}{2}nRT = \frac{3}{2}NkT_K$$

Eq. (12.1)

(Ideal gas, $T$ in kelvins)

**Work $W$** The process of energy transfer from the environment to the system due to mechanical motion. For gases, instead of force and distance to describe work, we use the pressure of the environment against the system and the volume change of the system. For gases the work equals the negative of the area under a $P$-versus-$V$ graph). (Section 12.1)

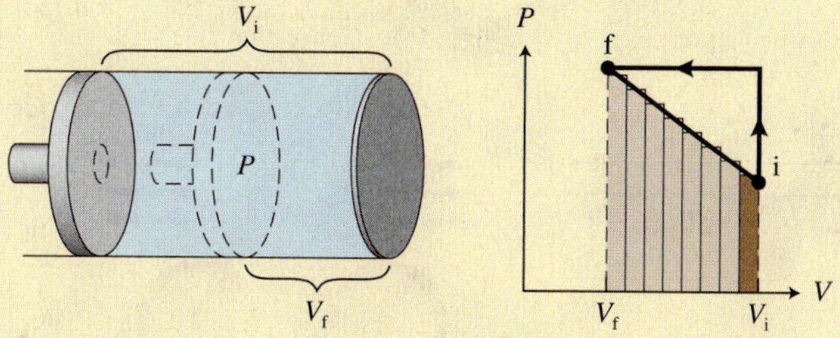


$$W_{\text{Environment on System}} = -P\Delta V$$

Eq. (12.2)

for constant $P$

**Heating $Q$** Process of energy transfer from the environment to the system due to differences in their temperatures. $Q > 0$ if energy is transferred into the system; $Q < 0$ if energy is transferred out of the system. (Section 12.2)

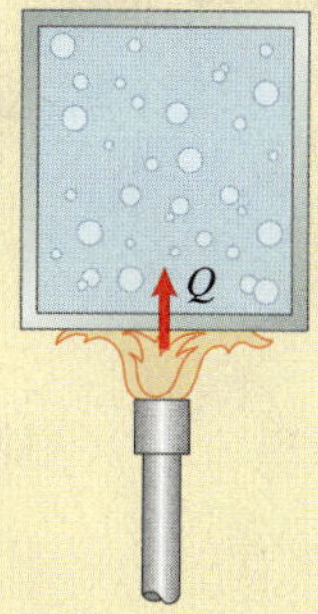


**First law of thermodynamics** A system's energy can change $\Delta U_{\text{int}} = (U_f - U_i)$ during a process due to the work $W$ done on the system by the environment and/or due to the heating $Q$ of the system by the environment. (Section 12.3)

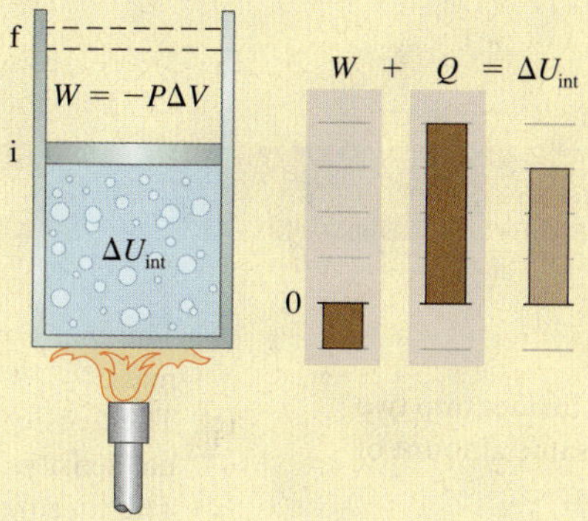


$$W_{\text{Environment on System}} + Q_{\text{Environment to System}} = \Delta U_{\text{System}}$$

Eq. (12.4)

*(continued)*

| Words | Pictorial and physical representations | Mathematical representation |
|---|---|---|

**Consequences of heating** (no work done on the system)

- ***Temperature change*** The system's temperature changes when there is no phase change occurring.
- ***Phase changes*** System melts ($\Delta U > 0$) or freezes ($\Delta U < 0$) at constant temperature. System boils ($\Delta U > 0$) or condenses ($\Delta U < 0$) at constant temperature. (Sections 12.4 and 12.6)

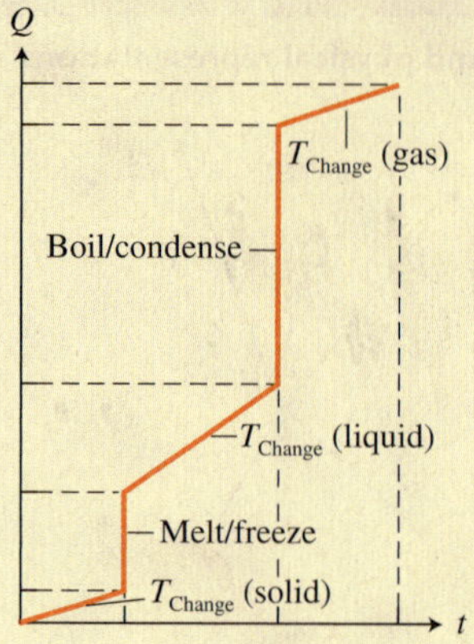


$\Delta U = cm\Delta T$ Eq. (12.5)

$Q = \pm mL_f$ Eq. (12.6)

$Q = \pm mL_v$ Eq. (12.7)

**Thermodynamic energy transfer mechanisms**

- ***Conduction*** $H_{Conduction}$ Energy transfer from particle to particle via contact.
- ***Convection*** $H_{Convection}$ Energy transfer by particles moving from one place to another.
- ***Evaporation*** $H_{Evaporation}$ Energy transfer due to evaporation/condensation on surface.
- ***Radiation*** $H_{Radiation}$ Energy transfer via absorption or emission of radiation. (Section 12.7)

Conduction

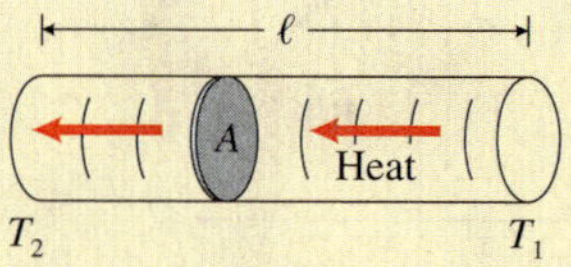


Convection

Evaporation

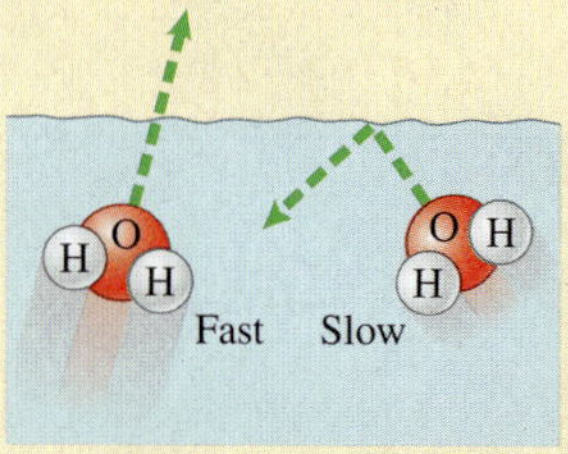


Radiation

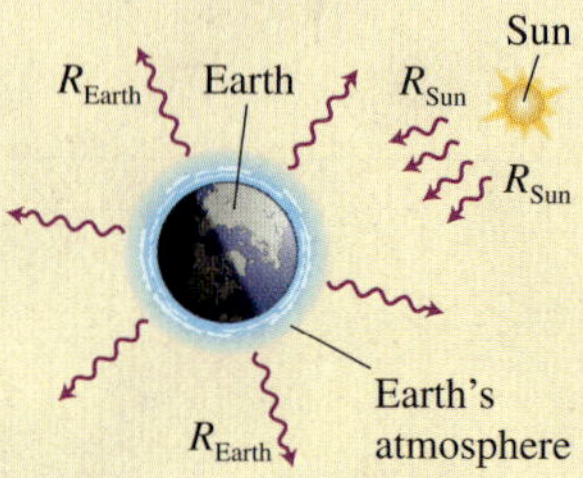


$H = Q/\Delta t$

Conduction:

$H_{Conduction} = KA(T_{Hot} - T_{Cold})/\ell$ Eq. (12.8)

Convection: *No equation*

Evaporation:

$H_{Evaporation} = (\Delta m/\Delta t)L_v$ Eq. (12.9)

Radiation: *No equation*

MP® For instructor-assigned homework, go to **MasteringPhysics.**

# Questions

## Multiple Choice Questions

1. An ideal gas in a container is separated with a divider into two identical half-size containers, each with the same amount of gas. Which of the following statements are correct?
   (a) The temperature of the gas in each container is half the previous temperature.
   (b) The mass of gas in each container is half the original mass.
   (c) The density of the gas in each container is half the original density.
   (d) The thermal energy of gas in each container is half the original thermal energy.

2. A container of gas has a movable piston, which you hold with your hand. You remove your hand and simultaneously place the container in a vacuum. Which of the following statements are correct? (More than one statement may be correct.)
   (a) The gas volume increases.
   (b) The internal energy of the gas does not change.
   (c) The temperature of the gas decreases.
   (d) The internal energy of the gas decreases.
3. A container of gas has a movable piston, which you hold with your hand. You remove your hand. Which of the following statements are always correct at the moment at which you release the piston? (More than one statement may be correct.)
   (a) The volume of the gas will increase.
   (b) The internal energy of the gas will not change.
   (c) The temperature of the gas will decrease.
   (d) The internal energy of the gas will decrease.
   (e) None of the above are always true.
4. Which of the following are possible means to change the energy of a system?
   (a) Do work on the system.
   (b) Heat the system.
   (c) Transfer energy without doing any work or heating.
5. Which gas has the greatest specific heat?
   (a) A gas in a constant volume container
   (b) The same gas in a constant pressure container with a movable piston
   (c) The specific heat is the same for both containers.
   (d) There is not enough information to answer the question.
6. The specific heat of water is 4200 J/°C · kg. Which of the following statements is true?
   (a) One kilogram of water stores 4200 J of energy when its temperature is 1 °C.
   (b) If you add 4200 J to water, its temperature will increase by 1 °C.
   (c) If you add 42,000 J to 1 kg of water, its temperature will increase by 10 °C.
7. How much heat is stored in 10 kg of water at 95 °C?
   (a) 42,000 J
   (b) 399,000 J
   (c) Zero; heat is not energy but is a process for transferring thermal energy.
8. We define the specific heat of a material as the energy that must be transferred to 1.0 kg of that material in order to cause it to warm 1.0 °C. What happens to the specific heat if we transfer twice that much energy?
   (a) The specific heat doubles.
   (b) The specific heat halves.
   (c) The specific heat does not change.
9. Dublin, Ireland and Edmonton, Alberta, Canada are at the same latitude (53.5°). The January and July average temperatures in Dublin are 5 °C and 16 °C, respectively; in Edmonton they are −12 °C and 21 °C, respectively. Why is the temperature variation so much greater in Edmonton than in Dublin?
   (a) Dublin and Edmonton are in different hemispheres.
   (b) Dublin is a bigger city.
   (c) Dublin is on the ocean.
10. Why do people sometimes wear fur coats in the winter in northern cities?
   (a) Fur reflects the sunlight.
   (b) Fur has low thermal conductivity.
   (c) Wearing a fur coat raises your temperature.

## Conceptual Questions

11. Match each heating mechanism (left column) with a corresponding phenomenon (right column). There will be more than one match for some mechanisms.

| Mechanism | Phenomenon |
|---|---|
| a. Conduction | 1. You feel cold when wearing wet clothes. |
| b. Convection | 2. A mother rubs a sick child with alcohol. |
| c. Evaporation | 3. A farmer tills the soil around a fruit tree in preparation for a cold winter. |
| d. Radiation | 4. You wear white clothes in the summer. |
| | 5. Ice chests are made of Styrofoam. |
| | 6. An electric heating coil is inserted at the bottom of a water container, not at the top. |

12. Your friend says, "Heat rises." Do you agree or disagree? If you disagree, what can you tell your friend to convince him of your opinion?
13. Suggest practical ways for determining the specific heats of different liquids and solids.
14. Suggest practical ways to measure heats of melting and evaporation for ice and water.
15. A solar thermal storage tank holds 2000 kg of water. Approximately what mass of rocks would store the same amount of thermal energy (assuming the same temperature change)?
16. A farmer's fruit storage cellar is unheated. To prevent the fruit from freezing, the farmer places a barrel of water in the cellar. Explain why this helps prevent the fruit from freezing.
17. Why does an egg take the same time interval to cook in water that is just barely boiling as in water that boils vigorously?
18. Why does food cook faster in a pressure cooker than in an open kettle?
19. A potato into which several nails have been pushed bakes faster than a similar potato with no nails. Explain. List the physics ideas and principles that you used to answer this question.
20. Explain why double-paned windows help reduce winter energy losses in a home.
21. The water in a paper cup can be boiled by placing the cup directly over the flame from a candle or from a Bunsen burner. Explain why this is possible without burning a hole in the cup.
22. Provide two reasons why blowing across hot soup or coffee helps lower its temperature. How can you test your explanations?
23. Placing a moistened finger in the wind can help identify the wind's direction. Explain why.
24. Joggers often accumulate large amounts of water on their skin when running with the wind. When running against the wind, their skin may seem almost dry. Give two explanations. How can you test your explanations?
25. Why does covering a keg of beer with wet towels on a warm day help keep the beer cool?
26. Explain why dogs can cool themselves by panting.
27. Some houses are heated by circulating hot oil or water through baseboards. Such a heating system is always positioned close to the floor. Why?
28. If on a hot summer day you place one bare foot on a hot concrete swimming pool deck and the other bare foot on an

adjacent rug at the same temperature as the concrete, the concrete feels hotter. Why?

29. A woman has a cup of hot coffee and a small container of room-temperature milk, which she plans to add to the coffee. If the woman must wait 10 minutes before drinking the coffee, and she wants it to be as hot as possible at that time, should she add the milk and wait 10 minutes or wait 10 minutes and add the milk? Explain.
30. Look carefully at your surroundings for one or more days. Make four recommendations of ways to reduce loss of energy by heat transfer. What physical ideas and principles did you use to make these recommendations?

# Problems

Below, **BIO** indicates a problem with a biological or medical focus. Problems labeled **EST** ask you to estimate the answer to a quantitative problem rather than develop a specific answer. Problems marked with ✎ require you to make a drawing or graph as part of your solution. Asterisks indicate the level of difficulty of the problem. Problems with no * are considered to be the least difficult. A single * marks moderately difficult problems. Two ** indicate more difficult problems.

## 12.1 Internal energy and work in gas processes

1. * **EST** Estimate the thermal energy of the air in your bedroom. List all of the assumptions you make.
2. A helium-filled balloon has a volume of $0.010\ \text{m}^3$. The temperature in the room and in the balloon is 20 °C. What are the average speed and the average kinetic energy of a particle of helium inside the balloon? What is the thermal energy of the helium?
3. * Imagine that the helium balloon from the previous problem was placed in an evacuated container of volume $0.020\ \text{m}^3$ and that the balloon popped when it touched a sharp edge on the inside of the container. (a) How much work is done on the helium gas? (b) What happens to the temperature of the helium, its density, the gas pressure, the average kinetic energy of each particle, and the thermal energy of the helium gas? Provide quantitative answers.
4. * You accidentally release a helium-filled balloon that rises in the atmosphere. As it rises, the temperature of the helium inside decreases from 20 °C to 10 °C. What happens to the average speed of helium atoms in the balloon and the thermal energy of the helium inside the balloon? Describe the assumptions you made.
5. * Air in a cylinder with a piston and initially at 20 °C expands at constant atmospheric pressure. (a) What is the work that the piston does on the gas if the air expands from $0.030\ \text{m}^3$ to $0.043\ \text{m}^3$? (b) How many moles of gas are in the container? (c) Suppose that the work leads to a corresponding change in thermal energy (there is no heating). What is the final temperature of the gas?
6. * In an empty rubber raft the pressure is approximately constant. You push on a large air pump that pushes $1.0\ \text{L}\ (1.0 \times 10^{-3}\ \text{m}^3)$ of air into the raft. You exert a 20-N force while pushing the pump handle 0.02 m. (a) Determine the work done on the gas. (b) If all of the work is converted to thermal energy of the 1.0 L of gas, what is the temperature increase of the gas?

## 12.2–12.4 Two ways to change the energy of a system; First law of thermodynamics; Specific heat

7. * **EST BIO Body temperature change** A drop in temperature of the human body core from 37 °C to about 31 °C can be fatal. Estimate the thermal energy that must be removed from a human body to cause this temperature change.
8. * **BIO Temperature change of a person** A 50-kg person consumes about 2000 kcal of food in one day. If 10% of this food energy is converted to thermal energy that does not leave the body, what is the person's temperature change? What assumptions did you make?
9. Determine the amount of thermal energy provided by heating to raise the temperature of (a) 0.50 kg of water by 10 °C, (b) 0.50 kg of ethanol by 10 °C, and (c) 0.50 kg of iron by 10 °C.
10. **EST** Estimate the time interval required for a 600-kg cast iron car engine to warm from 30 °C to 1500 °C (approximately the melting temperature of iron) if burning fuel in the engine as it idles produces thermal energy at a rate of 8000 J/s and none of the energy escapes the car engine.
11. * A lead bullet of mass $m$ traveling at $v_i$ penetrates a wooden block and stops. (a) Represent the process with a bar chart. What system did you choose? (b) Assuming that 50% of the initial kinetic energy of the bullet is converted into thermal energy in the bullet, write an expression that would allow you to determine the block's temperature increase. (c) List all of the physics ideas that you used to solve this problem.
12. * **BIO Exercising warms body** A 50-kg woman repeatedly lifts a 20-kg barbell 0.80 m from her chest to an extended position above her head. (a) If her body retains 10 J of thermal energy for each joule of work done while lifting, how many times must she lift the barbell to warm her body 0.50 °C? (b) State any assumptions you used. (c) List all of the physics ideas that you used to solve this problem.
13. * You add 25 g of milk at 10 °C to 200 g of coffee (essentially water) at 70 °C. The coffee is in a Styrofoam cup. If the specific heat of milk is 3800 J/kg · °C, by how much will the coffee temperature decrease when the milk is added? Indicate any assumptions you made.
14. * You add 20 °C water to 0.20 kg of 40 °C soup. After a little mixing, the water and soup mixture is at 34 °C. The specific heat of the soup is 3800 J/kg · °C. Determine everything you can using this information.
15. **BIO Cooling a hot child** A 30-kg child has a temperature of 39.0 °C (102.2 °F). How much thermal energy must be removed from the child's body by some heating process to lower his temperature to the normal 37.0 °C (98.6 °F) body temperature?
16. * **Impact of extinction-causing meteorite** Scientists have proposed that 65 million years ago in what is now the Yucatan peninsula of Mexico, a $1.2 \times 10^{16}$-kg meteorite moving at speed 11 km/s collided with Earth, and the resulting harsh conditions led to the extinction of many species, including the dinosaurs. (a) Calculate the kinetic energy of the meteorite before the collision. (b) If 20% of this energy was converted to

thermal energy in the meteorite, which had a specific heat of $900 \mathrm{~J/kg \cdot {}^\circ C}$, by how much did its temperature increase?

17. * You pour 250 g of tea into a Styrofoam cup, initially at 80 °C, and stir in a little sugar using a 100-g aluminum 20 °C spoon and leave the spoon in the cup. What is the highest possible temperature of the spoon when you finally take it out of the cup? What is the temperature of the tea at that time? What assumptions did you make to answer the questions?
18. ** A 500-g aluminum container holds 300 g of water. The water and aluminum are initially at 40 °C. A 200-g iron block at 0 °C is added to the water. What can you determine using this information? State any assumptions you used. List all physics ideas that you used to solve this problem.
19. * A 150-g insulated aluminum container holds 250 g of water initially at 20 °C. A 200-g metal block at 60 °C is added to the water, resulting in a final temperature of 22.8 °C. What type of metal is the block? What assumptions did you make to answer the question?

## 12.5 Applying the first law of thermodynamics to gas processes

In the problems in this section, clearly describe your system and indicate important objects in its environment. Label the work as $W_{\text{Environment on System}}$, heating as $Q_{\text{Environment to System}}$, and the change in internal energy as $\Delta U_{\text{int}}$.

20. ** Gas in a container with a movable piston initially at volume $V_1$, pressure $P_1$, and a very high temperature $T_1$ expands at constant pressure until its temperature and volume became $T_2$ and $V_2$. (a) Describe the process using the concepts of work, heating, and internal energy. (b) Draw a bar chart representing the process. (c) Calculate the work that the environment did on the gas. (d) Explain the process from a microscopic point of view. (e) Represent the process using $P$-versus-$V$, $P$-versus-$T$, and $V$-versus-$T$ graphs. (f) Repeat steps (a)–(e) for a situation in which the gas started with the same initial state but expanded at constant temperature instead of constant pressure.
21. ** Gas in a closed container undergoes a cyclic process from state 1 to state 2 and then back to state 1 (**Figure P12.21**). Describe the processes 1-2 and 2-1 qualitatively using the concepts of work, heating, and internal energy. (a) What happened to the thermal energy of the gas as it went from 1 to 2 and then from 2 to 1? What is the net change in the internal energy after the gas returned to state 1? (b) On the $P$-versus-$V$ graph, show the magnitude of the work that was done on the gas by the environment during process 1-2 and during process 2-1. (c) Was the total work done on the gas positive, negative, or zero during the entire process 1-2-1? (d) Discuss the heating of the gas during process 1-2 and then 2-1. Was the total heating of the gas positive, negative, or zero during the whole process 1-2-1?

**Figure P12.21**

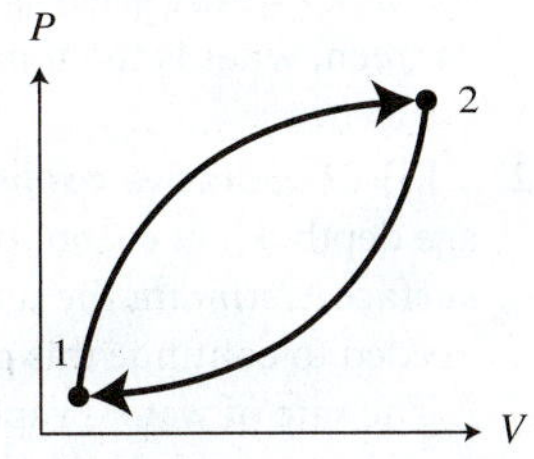


22. * **Jeopardy problem** A gas process is described mathematically as follows: $100 \mathrm{~J} + (-P)(0.001 \mathrm{~m^3}) = 0$. Pose a problem for which this description could be the answer. Describe the process macroscopically and microscopically.
23. * **Jeopardy problem** A gas process is described mathematically as follows: $Q + 120 \mathrm{~J} = 50 \mathrm{~J}$. Pose a problem for which this description could be the answer. Describe the process macroscopically and microscopically.
24. Use the first law of thermodynamics to devise a mathematical description of a process in which gas is being heated (positive heating) but its temperature does not change. Represent the process with a bar chart.
25. * Use the first law of thermodynamics to devise a mathematical description of a process in which gas is being cooled (negative heating) but its temperature increases. Represent the process with a bar chart.
26. * Derive an expression for the amount of energy that must be provided through heating for one mole of gas of molar mass $M$ to have a temperature increase of 1 K. Consider different processes and decide whether the amount of heating is independent of the process.
27. * You are making a table for specific heats of gases. Compared to the specific heats of solid and liquid substances, what additional information do you need to provide when you are listing the values of specific heat for each gas (oxygen, hydrogen, etc.)? Explain your answer.
28. ** One of the experiments that Joule used to test the idea of energy conservation was measurement of the temperature of a gas during the process of gas expansion into a vessel from which all air was evacuated. (a) Why did he choose this experiment? (b) What outcome would he predict based on the first law of thermodynamics? (c) Draw a picture of the experiment and provide macroscopic and microscopic reasoning for your answer.
29. EST **Temperature change in Carrier Dome** On March 5, 2006, a new college basketball attendance record of 33,633 was set in Syracuse University's Carrier Dome in the last regular-season game against Villanova. The volume of air in the dome is about $1.5 \times 10^6 \mathrm{~m^3}$. Estimate the temperature change of the air in 2 h, if all the seats in the dome are filled and each person transfers his or her metabolic thermal energy to the air in the dome at a rate of 100 W (100 J/s). Assume that no thermal energy leaves the air through the walls, floor, or ceiling of the dome.

## 12.6 Changing state

30. Determine the energy needed to change a 0.50-kg block of ice at 0 °C into water at 20 °C.
31. * When $1.4 \times 10^5 \mathrm{~J}$ of energy is removed from 0.60 kg of water initially at 20 °C, will all the water freeze? If not, how much remains unfrozen?
32. An electric heater warms ice at a rate of $H$. (a) What do you need to do to determine the mass of ice that melts in $\Delta t$ min? (b) What assumptions did you use? How will your answer change if you use different assumptions?
33. * Determine the number of grams of ice at 0 °C that must be added to a cup with 250 g of tea at 40 °C to cool the tea to 35 °C.
34. An ice-making machine removes thermal energy from ice-cold water at a rate of $H$. Determine the time interval needed to form $m$ kg of ice at ice melting temperature.
35. **Preventing freezing in canning cellar** A tub containing 50 kg of water is placed in a farmer's canning cellar, initially at 10 °C. On a cold evening the cellar loses thermal energy through the walls at a rate of 1200 J/s. Without the tub of water, the fruit would freeze in 4 h (the fruit freezes at −1 °C

because the sugar in the fruit lowers the freezing temperature). By what time interval does the presence of the water delay the freezing of the fruit?

36. * **Passive solar energy storage material** A Dow Chemical product called TESC-81 (primarily a salt, calcium chloride hexahydrate) is used as an energy-storage material for solar applications. Energy from the Sun raises the temperature of the solid material, causing it to melt at 27 °C (81 °F). At night the energy is released as the salt cools and returns to the solid state. (a) Determine the energy required to raise the temperature of 1.0 kg of solid TESC-81 from 20 °C to the liquid state at 27 °C. (b) How warm would 1.0 kg of water become if it started at 20 °C and absorbed the same energy? (c) Discuss the desirability of TESC-81 as a thermal energy-storage material compared to water. For TESC-81, $c\,(\text{solid}) = 1900\ \text{J/kg}\cdot{}^\circ\text{C}$ and $L_f = 1.7 \times 10^5\ \text{J/kg}$.

37. How much energy is required to convert (a) 0.10 kg of water at 100 °C to steam at 100 °C and (b) 0.10 kg of liquid ethanol at 78 °C to ethanol vapor at 78 °C?

38. **Cooling with alcohol rub** During a back rub, 80 g of ethanol (rubbing alcohol) is converted from a liquid to a gas. Determine the thermal energy removed from a person's body by this conversion. Indicate any assumptions you made.

39. **Energy in a lightning flash** A lightning flash releases about $10^{10}$ J of electrical energy. If all this energy is added to 50 kg of water (the amount of water in a 165-lb person) at 37 °C, what are the final state and temperature of the water?

40. A kettle containing 0.75 kg of boiling water absorbs thermal energy from a gas stove at a rate of 600 J/s. What time interval is required for the water to boil away, leaving a charred kettle?

41. **Cooling nuclear power plant** A nuclear power plant generates waste thermal energy at a rate of $1000\ \text{MW} = 1000 \times 10^6\ \text{W}$. If this energy is transferred by hot water passing through tubes in the water in an evaporative cooling tower, how much water must evaporate to cool the plant (a) per second and (b) per day?

## 12.7 Heating mechanisms

In the problems in this section, clearly identify the system and objects in the environment that interact with the system. Label the work as $W_{\text{Environment on System}}$, heating as $Q_{\text{Environment to System}}$, and the change in internal energy as $\Delta U_{\text{int}}$.

42. ** EST **Energy changes when it rains** Estimate the energy that is released or absorbed as water condenses and falls to Earth. Use the following information. Clouds are formed when moisture in the gaseous state in the air condenses. A rainstorm follows, dropping 2 cm of rain over an area 2 km × 2 km. Note that the mass of 1 $\text{m}^3$ of water is 1000 kg.

43. * **Insulating a house** You insulate your house using insulation rated as R-12, which will conduct 1/12 Btu/h of thermal energy through each square foot of surface if there is a 1 °F temperature difference across the material (R-12 insulation is said to have a thermal resistance $R$ of $12\ \text{h}\cdot\text{ft}^2\cdot{}^\circ\text{F/Btu}$). The conduction rate through a material of area $A$, across which there is a temperature difference $T_2 - T_1$, is

$$H_{\text{Conduction}} = (1/R)A\,(T_2 - T_1)$$

Use this information to determine the conductive energy flow rate across (a) an 8.0 ft × 16.0 ft R-15 wall; (b) a 3.0 ft × 7.0 ft R-4 door; and (c) a 3.0 ft × 4.0 ft R-1.5 window. Assume that the inside temperature is 68 °F and the outside temperature is 20 °F. (d) Convert each answer from Btu/h to J/s = W.

44. ** **Igloo thermal energy conduction** A typical snow igloo (thermal conductivity about 1/10 of ice) is shaped like a hemisphere of radius 1.5 m with 0.36-m thick walls. What is the conductive heating rate through the walls if the inside temperature is 10 °C and the outside temperature is −10 °C? [*Hint:* Use the conductive heating equation from the previous problem, but replace $1/R$ by the thermal conductivity $K$.]

45. After a vigorous workout, you stand in shorts in a 20 °C room in front of a fan that blows air past you. What are the signs (+ or −) of the convective and radiative heating rates? Will you perspire to keep cool? Explain your answer.

46. To cool hot soup, you blow across the top of a bowl of soup. Assuming that the soup is the system, what are the signs (+ or −) of the different heating mechanisms and the effect of this energy transfer on the soup?

47. While blowing across the bowl of soup in the previous problem, you wonder how efficiently the soup can cool by itself through evaporation. You notice that the bowl of hot soup loses 0.40 g of water by evaporation in 1 min. What is the average evaporative heating rate of the soup during that minute? Assume that soup is primarily water.

48. EST ** **Solar collector** You wish to install a solar panel that will run at least five lightbulbs, a TV, and a microwave. How large should the panel be if the average sunlight incident on a photoelectric solar collector on a roof for the 8-h time interval is 700 $\text{W/m}^2$ and the radiant energy is converted to electricity with an efficiency of 20%?

49. BIO **Marathon** You are training for a marathon. While training, you lose energy by evaporation at a rate of 380 W. How much water mass do you lose while running for 3.5 h?

50. **Cooling beer keg** A keg of beer is covered with a wet towel. Imagine that the keg gains energy from its surroundings at a rate of 20 W. (a) At what rate in grams per second must water evaporate from a towel placed over the keg to cool the keg at the same rate that energy is being absorbed? (b) How much water in grams is evaporated in 2.0 h?

51. * A canteen is covered with wet canvas. If 15 g of water evaporates from the canvas and if 50% of the thermal energy used to evaporate the water is supplied by the 400 g of water in the canteen, what is the temperature change of the water in the canteen?

52. * EST **Evaporative cooling** Each year a layer of water of average depth 0.8 m evaporates from each square meter of Earth's surface. *Estimate* the average energy transfer rate in watts needed to continue this process.

53. * The rate of water evaporation from a fish bowl is 0.050 g/s and the natural thermal energy transfer rate to the bowl by conduction, convection, and radiation is +36 W. What power electric heater must you buy to keep the temperature in the fish bowl constant?

54. BIO **Tree leaf** A tree leaf of mass of 0.80 g and specific heat of $3700\ \text{J/kg}\cdot{}^\circ\text{C}$ absorbs energy from the sunlight at a rate of 2.8 J/s. If this energy is not removed from the leaf, how much does the temperature of the leaf change in 1 min? [*Note:* Do not be surprised if your answer is large. A leaf clearly needs other heat transfer mechanisms to control its temperature.]

55. **Warming a spaceship** Your friend says that natural convection would not work on a spaceship orbiting the Earth. Do you agree or disagree with her statement? Explain.

56. * **Passive solar energy storage** Solar energy entering the windows of your house is absorbed and stored by a concrete wall of mass $m$. The wall's temperature increases by 10 °C during the sunlight hours. What mass of water, in terms of $m$, would have the same temperature increase if it absorbed an equal amount of energy? What assumptions did you make?

### 12.8 Climate change and controlling body temperature: Putting it all together

57. Which is lighter: dry or wet air? Explain your answer.
58. ** If you drop a burning candle, it stops burning almost instantly. Suggest two explanations for this phenomenon and then propose testing experiments to rule out those explanations.
59. BIO **Losing liquid while running** While running, you need to transfer 320 J/s of thermal energy from your body to the moisture on your skin in order to remain at the same temperature. What mass of perspiration must you evaporate each second? Indicate any assumptions you made.
60. * BIO **Running a marathon** When you run a marathon, the opposing force that air exerts on you does −150 J of work each second. You convert 1000 J of internal chemical energy to thermal energy each second. (a) How much thermal energy must be removed from your body per second to keep your temperature constant? (b) If 50% of this energy loss is caused by evaporation, how much water do you lose per second? (c) How much water do you lose in 3 h?
61. ** **Global climate change** Assume that because of increasing $CO_2$ concentration in the atmosphere, the net radiation energy transfer rate for Earth and its atmosphere is $+0.002 \times 1350\ \mathrm{W/m^2}$, corresponding to a 0.2% decrease in radiation leaving Earth. (a) Determine the extra thermal energy added to Earth and its atmosphere in 10 years. [*Note:* The radiation falls on an area approximately equal to $\pi r_{\text{Earth}}^2$ where $r_{\text{Earth}} = 6.4 \times 10^6$ m.] (b) If 30% of this energy is used to melt the polar ice caps, how many kilograms of ice will melt in 10 years? (c) How many cubic meters of ice will melt? (d) By how much will the level of the oceans rise in 10 years?
62. * **Standard house 1** On an average winter day (3 °C or 38 °F) in a typical house, energy already in the house is lost at the following rates: (i) 2.1 kW is lost through partially insulated walls and the roof by conduction; (ii) 0.3 kW is lost through the floor by conduction; and (iii) 1.9 kW is lost by conduction through the windows. Additional heating is also needed at the following rates: (iv) 2.3 kW to heat the air infiltrating the house through cracks, flues, and other openings and (v) 1.1 kW to humidify the incoming air (because warm air must contain more water vapor than cold air for people to be comfortable). What is the total rate at which energy is lost from this house?
63. * **Standard house 2** On the same day in the same house described in the previous problem, some thermal energy is supplied by heating in the following amounts: (i) sunlight through windows, 0.5 kW; (ii) thermal energy given off by the inhabitants, 0.2 kW; and (iii) thermal energy from appliances, 1.2 kW. How many kilowatts must be supplied to this standard house by the heating system to keep its temperature constant?
64. * **Standard house 3** Suppose that the following design changes are made to the house described in the previous two problems: (i) additional insulation of walls, roof, and floors, cutting thermal losses by 60%; (ii) tightly fitting double-glazed windows with selective coatings to reduce the passage of infrared light, cutting conduction losses by 70%; and (iii) elimination of cracks, closing of flues, and so on, cutting infiltration losses by 70%. What is the total rate at which energy is lost from this house?
65. * **Standard house 4** After further improvements (shifting windows from the north to the south sides and replacing outmoded appliances), thermal energy is supplied to the house described above at the following rates: (i) sunlight through windows, 1.0 kW; (ii) people's warmth, 0.2 kW; and (iii) appliance warmth, 0.8 kW. How many kilowatts must be supplied by the heating system of this house to keep it at constant temperature?

## General Problems

66. ** EST BIO **Metabolism warms bedroom** Because of its metabolic processes, your body continually emits thermal energy. Suppose that the air in your bedroom absorbs all of this thermal energy during the time you sleep at night. Estimate the temperature change you expect in this air. Indicate any assumptions you make.
67. * EST You have an 850-W electric kettle. Estimate the least amount of time you have to boil water before 10 guests arrive for a tea break. State clearly all numbers that you use in your estimate.
68. * EST **House ventilation** For purposes of ventilation, the inside air in a home should be replaced with outside air once every 2 hours. This air infiltration occurs naturally by leakage through tiny cracks around doors and windows, even in well-caulked and weather-stripped homes. (a) Estimate the mass of air lost every 2 h and each second. (b) Estimate the energy per second needed to warm outside air leaking into the house during a winter night. State your assumptions.
69. BIO **Frostbite** When exposed to very cold temperatures, the human body maintains core body temperature by reducing blood circulation to the skin and extremities, and skin and extremity temperatures drop. This can eventually lead to frostbite. Explain why this helps conserve thermal energy.
70. ** EST **Heating an event center with metabolic energy** Estimate the temperature change in some enclosure on your campus during an athletic event. Assume that there is no thermal energy transfer into or out of the building during the event. Indicate any other assumptions you make.
71. ** EST **Lightning warms body** A lightning flash releases about $10^{10}$ J of electrical energy. Quantitatively estimate the effect on your body if you absorbed 10% of the energy. State clearly any assumptions you made.

## Reading Passage Problems

**Cloud formation** Air consists mostly of nitrogen ($N_2$), with a molecular mass of 28, and oxygen ($O_2$), with a molecular mass of 32. A water molecule ($H_2O$) has molecular mass 18. According to the ideal gas law ($N/V = P/kT$), dry air at a particular pressure and temperature has the same particle density (number of particles per unit volume) as humid air at the same pressure and temperature. Consequently, humid air, whose low-mass water molecules replace more massive nitrogen and oxygen molecules, is less dense than dry air—the humid air rises. Atmospheric pressure decreases with elevation since there is less air above that is pushing down. At about 5000 m above the Earth's surface, the pressure is about

0.5 atm. Assuming an ideal gas, the gas volume $V$ increases as pressure $P$ decreases.

What happens to the air temperature as humid air rises? Air is a poor thermal energy conductor, and there is little heating from neighboring air ($Q \approx 0$, an adiabatic expansion process). As the rising gas expands, the neighboring environmental gas pushes in the opposite direction of the increasing volume of this rising gas. The environment does negative work on the rising air system ($W_{\text{by Environment on System}} < 0$). According to the first law of thermodynamics ($W + Q = \Delta U_{\text{System}}$): $-|W| + 0 = \Delta U_{\text{System}}$. The system's internal thermal energy decreases with a corresponding temperature decrease—about $-10\,°\text{C}$ for each 1000-m increase in altitude.

When the humid air reaches its dew point temperature, it starts to condense into water droplets (cloud formation). When condensation occurs, energy is released. There is a competition between decreasing thermal energy as the air expands and increasing thermal energy as the water vapor condenses. The air now cools at a lower rate of about $-5\,°\text{C}$ for each 1000-m increase in elevation.

72. If no condensation occurred, how high would 40 °C humid air have to rise before its temperature decreased to 10 °C?
(a) 1000 m (b) 2000 m (c) 3000 m
(d) 6000 m (e) 10,000 m
73. When rising humid air starts to condense, why does its temperature change less rapidly with increasing elevation?
(a) Its density increases, making it more difficult to change temperature.
(b) Thermal energy released during condensation causes less thermal energy change.
(c) The gas expands less, causing less negative work.
(d) The temperature of the surrounding air is changing less.
74. After crossing a mountain top on a warm sunny day, what should cool dry air do?
(a) Sink, because it is denser than warmer air below
(b) Warm, because the surrounding gas does positive work in causing it to contract
(c) Not change, as the surrounding air transfers little thermal energy by heating ($Q \approx 0$)
(d) a and b are correct.
75. Why does humid air rise in dry air?
(a) Water is attracted to the clouds above.
(b) A water molecule has lower mass than other air molecules.
(c) 1 $m^3$ of humid air has fewer molecules than 1 $m^3$ of dry air at the same $T$ and $P$.
(d) b and c are correct.
(e) None of the above is correct.
76. If 1 $m^3$ of dry air rises 5000 m and has no temperature change, what would its volume be?
(a) 0.3 $m^3$ (b) 0.5 $m^3$ (c) 1.5 $m^3$
(d) 2.0 $m^3$ (e) Too little information to answer
77. **EST** The magnitude of the thermal energy released from the water molecules to the air if 1.0 g of water vapor condensed to 1.0 g of liquid water is closest to which of the following?
(a) 300 J (b) 500 J (c) 1000 J
(d) 2000 J (e) $2 \times 10^6$ J

**Meteorite impact** The great Arizona crater was created by the impact of a meteorite of estimated $5 \times 10^8$ kg mass. The meteorite's speed before impact was about 10,000 m/s. Large amounts of rock found near the crater appeared to have melted on impact and then solidified as it cooled, indicating that the temperature of the rock during impact reached at least 1700 °C, the melting temperature of the rock. Is this possible?

Consider Earth and the meteorite as the system. The initial state of the process is the meteorite moving fast just before hitting Earth's surface. The final state is several minutes after the collision. What types of energy transformation occurred? The meteorite had kinetic energy before impact. In the collision, the meteorite dug a hole in Earth, forming the crater. The displaced soil was raised a distance approximately equal to the diameter of the meteorite. This inelastic collision produced considerable internal energy—thermal energy of the meteorite and of Earth's surface matter at the collision site. If the temperature change was high enough, the meteorite and/or parts of Earth may have undergone one or more phase changes. We summarize the process as follows:

$$K_i = \Delta U_{gf} + \Delta U_{\text{thermal}}$$

In the following questions, estimate different energies and energy changes. In addition to the information already given, you can use the following: 3300 kg/$m^3$ meteorite density; 840 J/kg·°C specific heat for the solid meteorite; $2.7 \times 10^5$ J/kg heat of fusion, the same as that of iron; 1000 J/kg·°C specific heat for the liquid meteorite, the same as that of iron; and $6.4 \times 10^6$ J/kg heat of vaporization, the same as for iron.

78. **EST** The initial kinetic energy of the meteorite was closest to
(a) $2 \times 10^{13}$ J (b) $3 \times 10^{15}$ J
(c) $3 \times 10^{16}$ J (d) $5 \times 10^{16}$ J
79. **EST** The radius of the meteorite was closest to
(a) 10 m (b) 30 m (c) 50 m
(d) 100 m (e) 1000 m
80. **EST** The gravitational potential energy change was closest to
(a) $10^7$ J (b) $10^9$ J (c) $10^{11}$ J
(d) $10^{13}$ J (e) $10^{15}$ J
81. **EST** The energy needed to warm the solid meteorite to its melting temperature at 1700 °C is closest to
(a) $2 \times 10^{13}$ J (b) $5 \times 10^{14}$ J
(c) $7 \times 10^{14}$ J (d) $3 \times 10^{15}$ J
82. **EST** The energy needed to melt the solid meteorite at its 1700 °C melting temperature is closest to
(a) $2 \times 10^{13}$ J (b) $5 \times 10^{14}$ J
(c) $7 \times 10^{14}$ J (d) $3 \times 10^{15}$ J
83. **EST** The energy needed to vaporize the melted meteorite at its 2600 °C boiling temperature is closest to
(a) $2 \times 10^{13}$ J (b) $5 \times 10^{14}$ J
(c) $7 \times 10^{14}$ J (d) $3 \times 10^{15}$ J
84. Is the initial kinetic energy enough to vaporize the meteorite?
(a) Yes
(b) No
(c) Too little information to answer

# Second Law of Thermodynamics 13

Why is the statement "we need to conserve energy" incorrect in terms of physics?

How does the inside of a refrigerator stay cold?

You add a pinch of salt to soup and it spreads out evenly in the soup. Why doesn't the salt spontaneously come back together into crystals?

In a gasoline-powered car, the energy of the fuel is converted into kinetic energy when the car moves and into internal thermal energy exhausted to the air. We worry about the increasing cost of a decreasing supply of fossil fuels. Does it mean we will run out of energy? We know from physics that energy does not disappear—it is converted from one form to another. If we consider the whole universe as our system, the amount of energy in the universe is constant no matter what we do. So why can't we just reuse the energy of fuel extracted from the ground? So far we have treated all forms of energy as equal—one form can be converted to another, but the total never changes. However, as we learn in this chapter, this is not entirely true.

**Be sure you know how to:**

- Identify a system and the initial and final states of a well-defined process (Sections 6.1 and 6.2).
- Determine the energy transferred through heating to, the work done on, and the internal energy change of a system (Sections 12.1 and 12.3).
- Apply the first law of thermodynamics to a process (Section 12.5).

**Our discussions of energy** until now have been based on the work-heating-energy principle. If a part of the universe chosen as the system gains energy because of work done on it by external forces or because of energy transfer through heating, then the environment surrounding the system loses an equal amount of energy, or vice versa. In short, the total energy of a system plus its environment is constant. However, there is something more subtle occurring. Some processes, despite being consistent with energy conservation, never occur. A ball that you hold above the floor will fall, bounce a few times, and then stop. If we choose the system to be Earth, the ball, and the floor, then we would say that the gravitational potential energy of the system is converted to internal energy after the ball stops bouncing (both the ball and the floor have gotten a little warmer). However, you never observe a ball that has been resting on the floor spontaneously lose thermal energy (cool down) and gain the equal amount of kinetic energy needed to jump up into the air. Such a process does not violate energy conservation, so why doesn't it occur? That question is the subject of this chapter.

**Figure 13.1** The warmer water at the bottom of the waterfall never cools and goes back to the top of the waterfall.

## 13.1 Irreversible processes

As we have already learned, the total energy of a system and its environment is constant. Energy can be transferred between a system and its environment through the mechanisms of mechanical work or by heating. There can also be changes in the forms of energy within a system. For example, the gravitational potential energy of the water-Earth system at the top of a waterfall converts into kinetic energy as the water falls and then to thermal energy when it hits the rocks at the bottom of the falls (**Figure 13.1**). But the warmer water at the bottom never cools and flows back up in the reverse direction. Are our ideas about energy incomplete?

Let's think more deeply about kinetic energy. Imagine a boulder falling off a cliff. All of the particles within the boulder follow more or less the same path. The same is true of the particles in a moving car or in a flying airplane. Thus we can say that the kinetic energy of these objects is organized. What about the motion of water molecules in a glass of water? The molecules also have kinetic energy; however, their motion is random rather than organized—"less organized" than the kinetic energy of a falling boulder. Now that we've made this distinction, consider the experiments in Testing Experiment **Table 13.1**.

**TESTING EXPERIMENT TABLE**

**13.1 Use the first law of thermodynamics to predict what might happen in the following processes and reverse processes.**

VIDEO 13.1

| Testing experiment | Prediction based on the first law of thermodynamics (energy transfer through work and heating) | Outcome |
|---|---|---|
| **Experiment 1a.** A pendulum bob is raised to the side and released. What happens to the bob? | **1a.** We predict that as the bob moves back and forth, the size of the swing will decrease due to air resistance and friction in the bearing at the top of the string. Eventually, the bob will stop, and the bearing and the air will be slightly warmer. | The predicted outcome for Experiment 1a always occurs. |

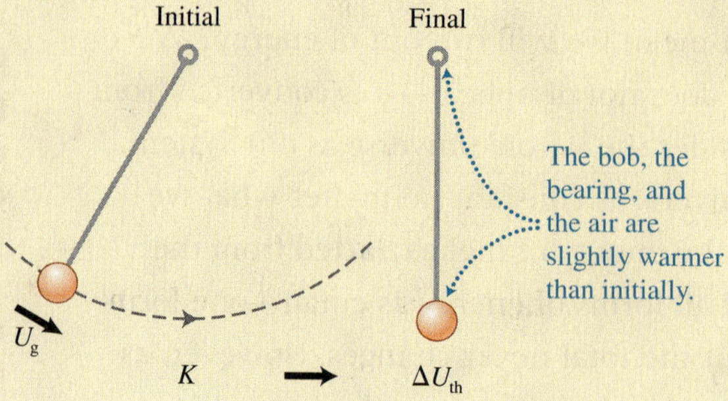

<table>
<tr>
<td>Experiment 1b Reverse process. A pendulum bob hangs straight down at rest. What could happen to the bob that is consistent with the first law of thermodynamics?</td>
<td>1b. The bob, air, and bearing at top could cool and the bob would convert the decreasing thermal energy to kinetic energy and start swinging.<br>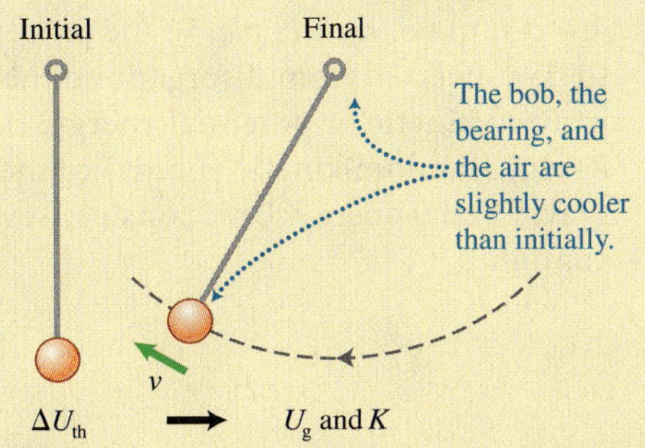<br></td>
<td>The predicted outcome for Experiment 1b never occurs.</td>
</tr>
<tr>
<td>Experiment 2a. A car is coasting on a horizontal road with its motor off. What will happen to the car?</td>
<td>2a. We predict that the car will gradually slow down and stop. The initial kinetic energy of the car will convert into internal thermal energy due to friction. The car tires and road will become warmer.<br>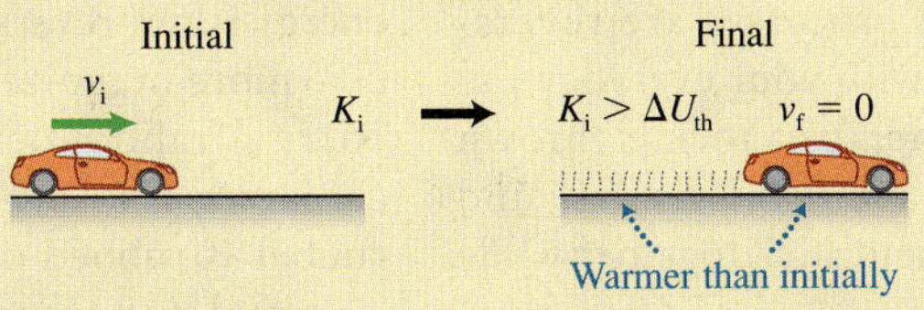<br></td>
<td>The predicted outcome of Experiment 2a occurs.</td>
</tr>
<tr>
<td>Experiment 2b Reverse process. A car with its motor off is at rest on a horizontal road. What might happen to the car that is consistent with the first law of thermodynamics?</td>
<td>2b. The car tires and road could cool and the decreased thermal energy would be converted into the car's organized kinetic energy as it gradually begins moving.<br>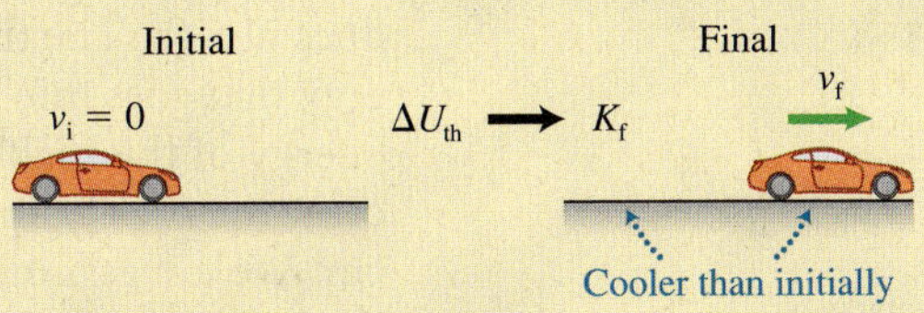<br></td>
<td>The predicted outcome of Experiment 2b never occurs.</td>
</tr>
<tr>
<td>Experiment 3a. An ice cube is sitting on a tabletop. What will eventually happen to the ice cube?</td>
<td>3a. We predict that the ice cube will gain thermal energy from the tabletop and air, causing the ice to melt and leaving a puddle of water on the tabletop. The tabletop and surrounding air will get a little cooler.<br>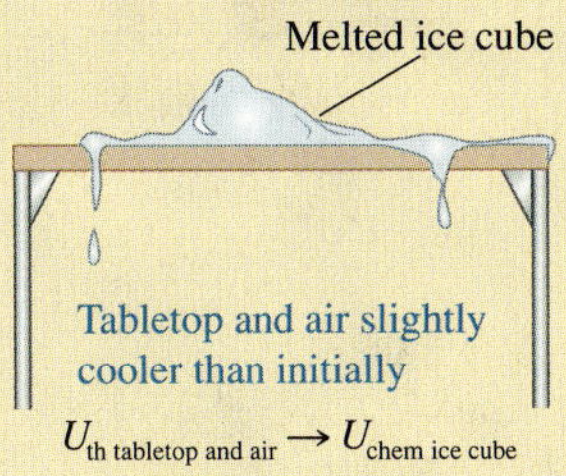<br></td>
<td>The predicted outcome of Experiment 3a occurs.</td>
</tr>
<tr>
<td>Experiment 3b Reverse process. A small puddle of cold water sits on a tabletop. What might happen to this puddle over time?</td>
<td>3b. The puddle could transfer thermal energy to the tabletop and air, causing the puddle's temperature to decrease below freezing so it turns into ice.<br>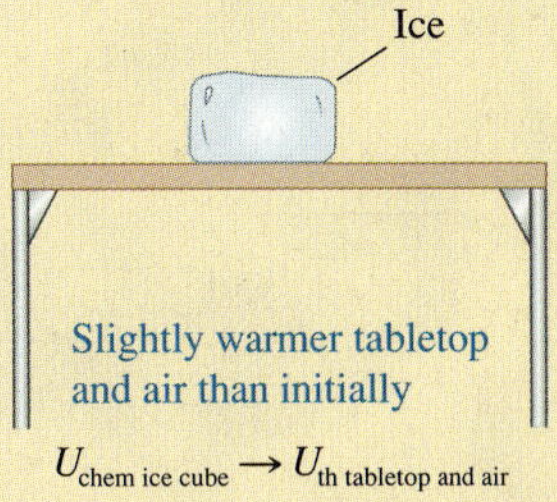<br></td>
<td>The predicted outcome of Experiment 3b never occurs.</td>
</tr>
</table>

**Conclusion**

- In the experiments in which the outcome matches the prediction (consistent with the first law of thermodynamics), the organized energy converts into less organized energy.
- In the experiments in which the outcomes do not match the predictions (also consistent with the first law of thermodynamics), the unorganized energy would have to convert into more organized energy. Such processes never occur.

**Active Learning Guide ›**

We conclude based on the testing experiments in Table 13.1 that some processes, although allowed by the first law of thermodynamics, do not occur in nature. Processes that occur in one direction but never occur in the opposite direction are said to be *irreversible*. In the processes that never occur, energy would have to be converted from disorganized thermal energy into an equal amount of organized kinetic or potential energy. It seems that an isolated system's thermal energy (random kinetic energy) cannot be converted into organized kinetic or gravitational energy. Let's consider several other situations to see if this pattern continues.

### CONCEPTUAL EXERCISE 13.1 Irreversible processes?

Use the ideas of "organized" and "random" energy to decide which of the following processes are reversible. Then compare the results with your everyday experience. (a) A car skids to a stop; (b) a rock is thrown upward and reaches a height $h$ above the ground; and (c) two fast-moving cars collide and stick together while stopping.

**Sketch and translate** The initial and final states of the processes and the choices of systems are sketched below.

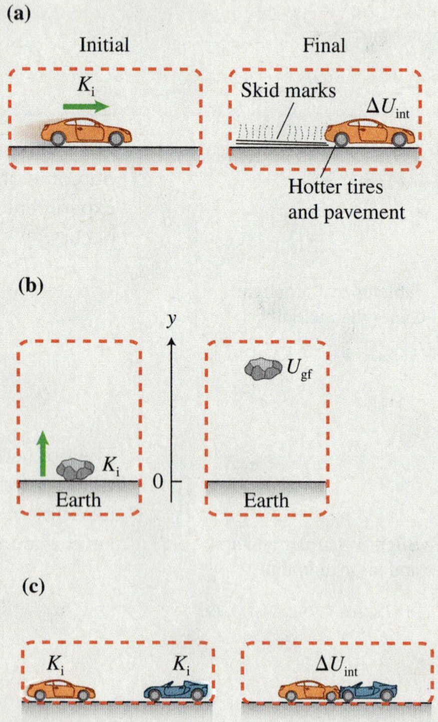


**Simplify and diagram** According to the pattern we observed in Table 13.1, a process in which a system goes from more organized energy to less organized random energy is not reversible. In (a), the kinetic energy of the car (more organized energy) converts into the thermal energy (random energy) of the tires and road as the car is stopping and into internal potential energy as the rubber is rubbed off the tires while skidding to a stop. Thermal and internal potential energy are less organized forms of energy; thus the reverse process (the hot road and tires and the skid marks coming back together to get the car moving) should not occur, and definitely does not. (b) The thrown rock had organized kinetic energy that goes into the organized gravitational potential energy of the rock-Earth system; thus the reverse process (the rock falling back down) should be able to occur. We are assuming that air resistance is small, since the increase in thermal energy of the air would be a conversion from more organized to less organized energy. (c) Before the collision, the cars have organized kinetic energy. After the collision, the cars are deformed (a change in the internal energy of the molecular structure of the cars) and the temperatures of the cars and the road have increased. Both the deformation and the temperature increase reflect more random, disorganized forms of energy. Thus, the process is not reversible (the warmed road and tires and deformed cars will not spontaneously be converted back into two undamaged moving cars).

**Try it yourself:** You place a droplet of food coloring in a glass of water. Describe the process and decide whether it is reversible or not based on the idea of organized energy.

*Answer:* The food coloring spreads throughout the water. We never see it spontaneously reform into its original organized droplet. The process is irreversible.

**Reversible and irreversible processes** Many processes in nature occur in only one direction—they are called *irreversible processes.* In the allowed direction, energy converts from more organized forms to less organized forms. In the reverse, unallowed processes, energy would have to convert from a less organized form to a more organized form. However, the unallowed reverse process does not contradict energy conservation.

## Usefulness of different types of energy in doing work

You may have noticed that in all processes investigated so far in this chapter, no work was done on the system by the environment. Let's investigate a situation in which we wish to do some work—to compress a spring (the system object) a small distance. Imagine that a cart of mass $m$ (not in the system) moves at a speed $v$. The cart has enough kinetic energy (if considered as a separate system) so that if it hits the spring, it will compress the spring a reasonable amount (**Figure 13.2a**). If the cart has a mass of 10 kg and moves at a speed of 5 m/s before hitting the spring, the kinetic energy of the cart would be 125 J. Some of this energy is transferred to the spring during the collision so that the spring gains elastic potential energy.

Imagine a second process. A container filled with 1 kg of water is warmed from room temperature (20 °C) to boiling temperature. This requires

$$\Delta U = mc\Delta T = (1\text{ kg})(4186\text{ J/kg}\cdot{}^\circ\text{C})(100\,^\circ\text{C} - 20\,^\circ\text{C}) \approx 330{,}000\text{ J}$$

of energy transferred through heating—far more than the energy needed to compress the spring by the same amount as the moving cart compressed it. You place the spring in the water. The hot water does not compress the spring at all. The thermal energy of the spring (disorganized energy) increases as it warms, but its elastic potential energy (organized energy) does not change. Apparently, disorganized forms of energy (such as thermal energy) do not have the ability to do work nearly as well as organized forms of energy.

◀ **Active Learning Guide**

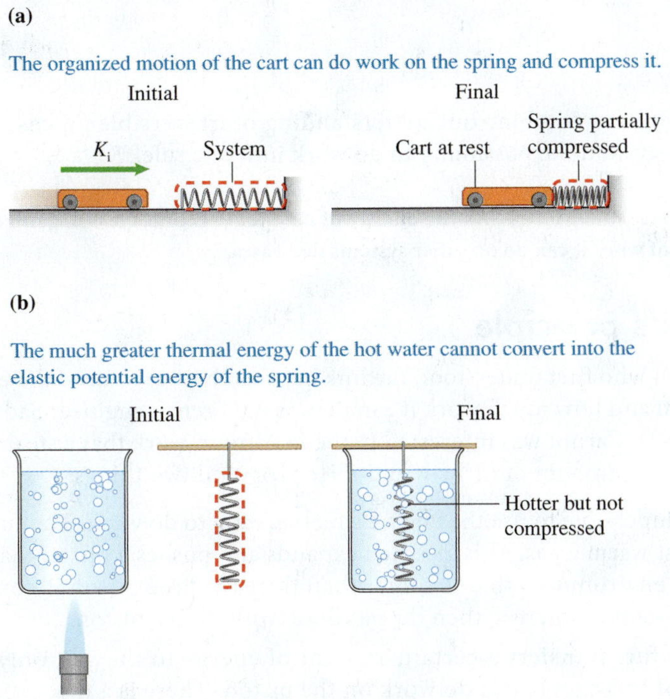
(a)
The organized motion of the cart can do work on the spring and compress it.
Initial
Final
$K_i$
System
Cart at rest
Spring partially compressed
(b)
The much greater thermal energy of the hot water cannot convert into the elastic potential energy of the spring.
Initial
Final
Hotter but not compressed

**Figure 13.2** The organized energy in (a) can do work, whereas the disorganized energy in (b) cannot.

**CONCEPTUAL EXERCISE 13.2 What type of energy is best for doing work?**

Below we describe three systems that have the same amount of energy. Imagine that each of these systems can interact with another system of interest and transfer energy to that other system by doing work on it. Arrange these systems in the approximate order of how much work they can do on another system, listing the best system for doing work first.

1. 1000 J of chemical potential energy stored in the molecules of gasoline
2. 1000 J of thermal energy in the air particles of the room where you now reside
3. 1000 J of gravitational potential energy between Earth and a 50-kg barbell held 2.0 m above the ground

**Sketch and translate** The three systems are sketched below.

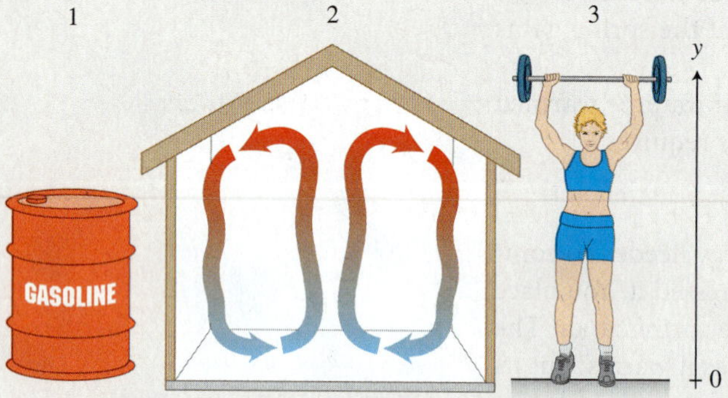


**Simplify and diagram** Possible energy conversion processes for each of the above are described at right.

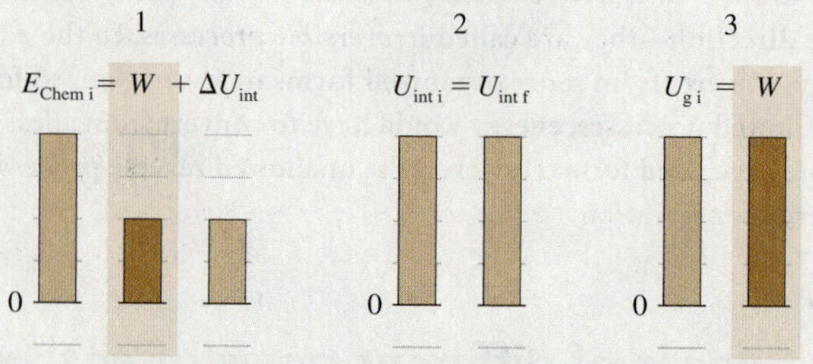


1. The chemical potential energy in the gasoline can do some work (for example, to run a small motor that partially compresses a spring). But the gasoline burning also increases the thermal energy in its surroundings.
2. The thermal energy in the air cannot run a motor or cause a spring to compress if they are at the same temperature as the air in the room. The thermal energy in the air does not change and is not a useful form of energy for doing work.
3. The gravitational potential energy between the barbell and Earth is useful, and nearly all of it is available for doing work on a spring or turning a paddle in water.

The bar charts represent these possible energy conversion processes. In summary, the ranking for the usefulness of the energy is $3 > 1 > 2$.

**Try it yourself:** Which system is more useful for doing work on another system: a block moving upward with 10 J of kinetic energy or a block-Earth system in which the block is lifted so that the system has 10 J of gravitational potential energy relative to the Earth's surface? Ignore air resistance.

*Answer:* They are equally useful. Both systems have the same amount of organized energy (10 J).

Now we can combine our understanding of irreversible processes, organized energy, and the possibility to do work into one rule.

**Active Learning Guide ›**

**Usefulness of energy** **As the energy of a system becomes less organized, the amount of work it can do on other systems decreases.**

## Carnot's principle

The person who first understood the link between how organized the energy is in a system and how much work it can do was the French engineer Sadi Carnot (1796–1832). Carnot was interested in the maximum work that a steam engine with a certain amount of fuel could do. He proposed two things:

1. It is impossible to use the engine's fuel directly to do work. The fuel needs to first warm a gas. This gas then expands and pushes a piston that moves in an environment that is cooler than the gas. If the piston is considered the system of interest, then the gas does work on the piston.
2. If the fuel transfers a certain amount of energy to the gas, only a fraction of this energy can do work on the piston. There is a theoretical limit

on the work that can be done. The remaining energy is transferred to the environment, where it manifests as thermal energy. Additionally, all real-world engines fail to reach the theoretical limit when they operate.

Carnot devised a rule similar to the one we just did: that the energy in a closed system evolves from more useful to less useful forms for doing work. He also said that energy in a less useful form cannot be converted entirely into the same amount of energy in a more useful form.

**Carnot's principle** **It is impossible to build an engine that uses all of its thermal energy to do mechanical work on its environment.**

## Another idea about the direction of energy conversion

Observational Experiment **Table 13.2** indicates another way to think about the preferred direction for energy transfer during irreversible processes.

**OBSERVATIONAL EXPERIMENT TABLE**

**13.2 Direction of energy transfer.**

VIDEO 13.2

| Observational experiment | Analysis |
|---|---|
| **Experiment 1.** You place a room temperature spoon in a cup of hot tea. The spoon gets warmer and the tea gets cooler until they reach the same temperature. | Energy is transferred from the hot tea to the cool spoon until they are at the same temperature. |
| **Experiment 2.** The air conditioning in a building fails on a very hot day. The temperature in the building increases until it reaches a similar temperature to the outside temperature. | Energy is transferred through the walls from the hot outside of the building to the cool inside until the inside and outside are at the same temperature. |

**Pattern**

In both experiments, energy is transferred from a high-temperature region to a low-temperature region.

In the experiments in Table 13.2, you do not observe the reverse of the processes, in which a cool object becomes cooler and a hot object hotter. Although the first law of thermodynamics allows this, such processes do not happen. This leads us to another formulation concerning the direction of energy conversion in isolated systems.

**Direction of thermal energy transfer** **In an isolated system, energy always transfers from a warmer region to a cooler region.**

## The second law of thermodynamics

Let's review the boxed statements in this section. We have determined that as the energy of a system becomes less organized, the amount of work it can do on other systems decreases. Carnot's principle tells us that it is impossible to build an engine that uses all of its thermal energy to do mechanical work on its environment. Finally, we have concluded that in an isolated system, energy always transfers from a warmer region to a cooler region. All of these statements are different versions of **the second law of thermodynamics**. The second law describes what processes can and cannot occur in an isolated system. While the first law of thermodynamics is a principle about energy conservation, the second law is a principle of *energy transfer* between two regions at different temperatures and *energy conversion* from one form of energy to another. In order to make a more quantitative version of the second law, we must construct a new physical quantity that describes the degree of organization of a system's energy.

**Review Question 13.1** When you burn gasoline in your car, the car can gain organized kinetic energy as it moves faster and faster. Does this contradict the second law of thermodynamics?

## 13.2 Statistical approach to irreversible processes

**Active Learning Guide ›**

One way to understand the preference of systems for certain forms of energy is to do a statistical analysis of the probability that a system is in one state versus another. Consider a very simple thought experiment with a box holding identical atoms that are labeled so we can tell them apart. We drop a thin divider down a slot in the middle of the box, thus dividing it in half. Let's start with four atoms in the box. At regular time intervals we insert the divider, count the number of atoms on each side, and then remove the divider. The five possible distributions are shown in **Figure 13.3**. Each distribution is characterized using the symbol $i$, which has integer values 0 through 4 indicating the number of atoms in the left half. For example, the $i = 0$ distribution has zero atoms on the left and four on the right; the $i = 1$ distribution has one on the left and three on the right; and so forth. Each of these five distributions is called a **macrostate** of the system of four atoms. Each macrostate has a particular number of atoms $n_l$ on the left and $n_r$ on the right. In thermodynamics we call the variable quantities $n_l$ and $n_r$ *state variables*; they distinguish one particular macrostate of a system from another.

**Figure 13.3** The possible states of four atoms found on each side of a divider after its insertion in the box.

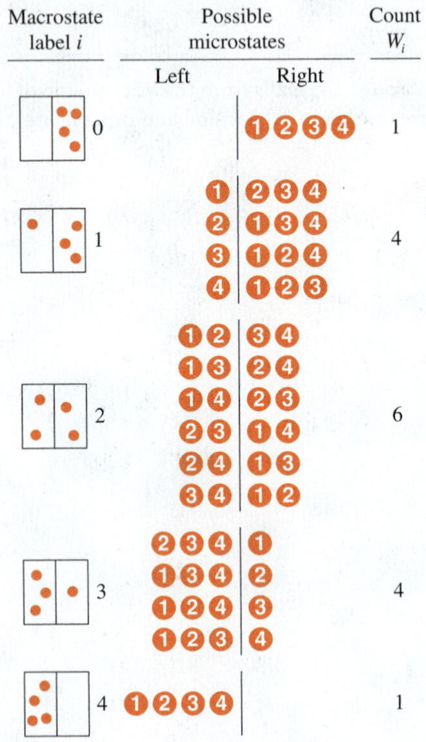


Earlier, we said that each atom is labeled so we can tell them apart. Real atoms don't have labels, but this is a thought experiment so we can construct the situation however we like. Our goal now is to determine the different unique ways for each macrostate to exist. For example, there are four different ways to get one atom on the left and three on the right—each of the four atoms could be alone on the left with the other three on the right. We call this number of ways of getting a particular macrostate the count $W_i$ of that state (note that here we use the letter symbol $W$ for the number of possible microstates, although it is used elsewhere to designate work). Macrostate 0 with zero atoms on the left and four on the right can occur in only one way—if all four atoms are on the right. For macrostate 0, $W_0 = 1$. As we said above, macrostate 1 can occur in four different ways, $W_1 = 4$. Column 3 in Figure 13.3 lists the number of ways of getting the different macrostates. Each of these ways is called a **microstate** of that particular macrostate. All microstates are equally probable. The probability of a macrostate occurring depends on the number of microstates it has. For example, macrostate 2 has six unique ways to arrange four atoms to get two on each side.

In general, for any number of atoms we can determine the number $W_i$ of microstates of the $i$th macrostate by using the equation

$$W_i = \frac{n!}{(n_l!)(n_r!)} \tag{13.1}$$

where $n$ is the total number of atoms and $n_l$ and $n_r$ are the number of atoms on the left and on the right, respectively.

We can check that Eq. (13.1) works for the four-atom example, whose results are shown in Figure 13.3. Consider the number of microstates for macrostate 2:

$$W_2 = \frac{4 \times 3 \times 2 \times 1}{(2 \times 1)(2 \times 1)} = 6$$

The equation works. You can try the equation for the other macrostates.

If we add all the numbers in the third column of Figure 13.3, we find that there are 16 microstates, the number of unique ways in which the four atoms can be arranged in the two halves of the box. The *probability* $P_i$ of a particular macrostate occurring is the number of microstates for that macrostate (the count) divided by the total number of microstates. For example, the probability of finding zero atoms in the left box is 1 in 16; the probability of finding one atom in the left half is 4 in 16; and the probability of observing two atoms in each half is 6 in 16.

For the box with four atoms, there is a small chance (1 in 16) that you will observe all atoms on one side of the box. With a larger number of atoms, the situation is quite different. With 10 atoms, for instance, the state with 10 on the left and 0 on the right is 252 times less likely to occur than the macrostate with 5 on each side. That is,

$$\frac{W_0}{W_5} = \frac{1}{252}$$

With 100 atoms, the state with 100 on the left and zero on the right is $1/10^{29}$ as likely to occur as the state with 50 on each side. If we increase the number of atoms to one million, states with equal numbers of atoms on each side are overwhelmingly more probable than states with an unequal number on each side. For example, for one million atoms there is almost a $10^{87}$ times better chance of observing an equal number on each side than of observing 51% of the atoms on the left and 49% on the right. This 50-50 distribution is the state with the highest probability—the state that is overwhelmingly most likely to occur. This most probable state is the state with the greatest number of microstates and has the greatest randomness especially compared to the state with all particles on the same side.

**TIP** The exclamation marks (!) in Eq. (13.1) indicate the factorial function. It means that we multiply the number before the exclamation mark by all the positive integers equal to and smaller than the number. Thus, $8! = 8 \times 7 \times 6 \times 5 \times 4 \times 3 \times 2 \times 1$, and $2! = 2 \times 1$. The notation 8! is called eight factorial, and $n!$ is called $n$ factorial. Zero factorial is defined as one ($0! = 1$).

**QUANTITATIVE EXERCISE 13.3 Five atoms**

Suppose the box with the divider contains five atoms. (a) Identify the six macrostates that can occur. (b) Determine the count of the 0 on left, 5 on right macrostate and that of the 3 on left, 2 on right macrostate.

**Represent mathematically** Use Eq. (13.1) with $n = 5$ to determine the count for the different macrostates:

$$W_i = \frac{5 \times 4 \times 3 \times 2 \times 1}{(n_l!)(n_r!)}.$$

**Solve and evaluate** (a) The macrostates $(n_l, n_r)$ are (0,5), (1,4), (2,3), (3,2), (4,1), and (5,0).

(b) The count for the $i = 0$ macrostate [the (0,5) state] is

$$W_0 = \frac{5 \times 4 \times 3 \times 2 \times 1}{(0!)(5 \times 4 \times 3 \times 2 \times 1)} = 1$$

The count for the $i = 3$ macrostate [the (3,2) state] is

$$W_3 = \frac{5 \times 4 \times 3 \times 2 \times 1}{(3 \times 2 \times 1)(2 \times 1)} = 10$$

**Try it yourself:** Determine the count for a container with seven atoms with 2 on the left side and 5 on the right side.

*Answer:* 21.

## Connecting atoms in a box to physics

In general, an isolated system with a large numbers of particles will evolve toward states that have a higher probability of occurring. The highest probability state is the one in which the particles are evenly distributed—in other words, the state with the highest level of randomness. This state of maximum probability is often called the *equilibrium state* of the system. An isolated system in the most probable state is said to have reached equilibrium.

The same rules apply in the real world. The state of a system with particles moving randomly (random thermal energy) is much more probable than the state with the particles all moving together in the same direction. To characterize this randomness or disorder in the system, physicists use a physical quantity called **entropy,** a measure of the probability of a particular macrostate.

**Entropy** **is a physical quantity that quantifies the probability of a state occurring and of the degree of disorder or disorganization of a system when in that state. In statistical thermodynamics, the entropy $S_i$ of the $i$th macrostate is defined in terms of the count $W_i$ of that macrostate:**

$$\text{Entropy of } i\text{th macrostate} = S_i = k \ln W_i \qquad (13.2)$$

**in which $\ln W_i$ is the natural logarithm of the count $W_i$ and $k$ is a proportionality constant (Boltzmann's constant) with a value $k = 1.38 \times 10^{-23}$ J/K (joules/kelvin).***

The entropy of the five macrostates for the distribution of four atoms on the left or right side of a box is $S_0 = 0$, $S_1 = 1.4k$, $S_2 = 1.8k$, $S_3 = 1.4k$, and $S_4 = 0$. Try to use Eq. (13.2) to confirm these results. (The use of logarithms is reviewed briefly in the appendix.)

At the beginning of this section we set out to construct a physical quantity that described the degree of organization of a system's energy. We learned that isolated systems spontaneously evolve toward states whose energy is more disorganized. We now have a way to quantify this statement: isolated systems evolve toward states with higher entropy.

## Entropy of an expanding gas—statistical approach

**Active Learning Guide >**

Ultimately, we would like a quantitative way to understand the concept of entropy in terms of macroscopic quantities such as temperature $T$ and heating $Q$ rather than in terms of counts of microstates. In working toward this, let's first examine from a statistical point of view a familiar process: a gas expanding at constant temperature.

### EXAMPLE 13.4 Entropy change of an expanding gas

Suppose you have a gas composed of just six atoms, each moving randomly and with the same amount of energy. The atoms start grouped together in one tiny box (situation 1). The box then doubles in size (situation 2), triples in size (situation 3), and finally reaches six times the original size (situation 4). The gas has expanded its volume to six times its original value. Determine the count and the entropy of the maximum probability state for each of the four situations.

**Sketch and translate** The situations are pictured at right. For a gas with six atoms and a box that is twice as big, the maximum probability state has three atoms in each half of the box. With a box that is three times

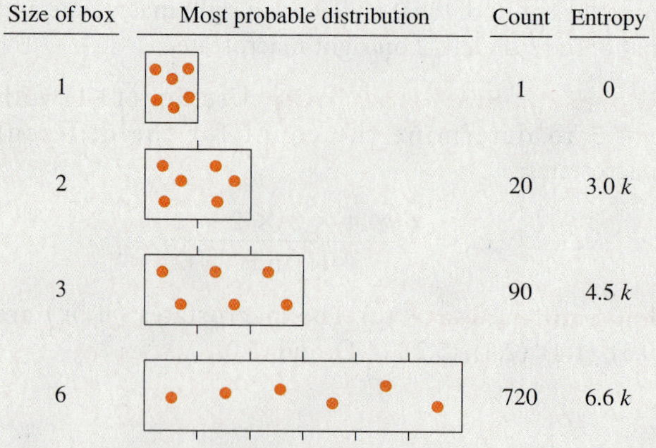

Size of box
Most probable distribution
Count
Entropy
1
1
0
2
20
3.0 k
3
90
4.5 k
6
720
6.6 k

As the box gets bigger, the count of the most probable distribution increases dramatically.

*A J/K is sometimes called an entropy unit (eu). Thus, 1 J/K = 1 eu.

bigger, the maximum probability state has two atoms in each third of the box. The maximum probability state with a six times bigger box has one atom in each sixth of the box.

**Simplify and diagram** We assume that the atoms are labeled. Then for situation 2, the distribution 1, 2, 3 on side 1 and 4, 5, 6 on side 2 differs from 1, 2, 4 on side 1 and 3, 5, 6 on side 2—these are two distinct microstates.

**Represent mathematically** The count of the number of microstates with $n$ atoms in $N$ distinct parts of a container is

$$W_N = \frac{n!}{n_1!n_2! \cdots n_N!}$$

**Solve and evaluate** The count for each maximum probability macrostate, as calculated using the above equation, is reported in the figure above. For example, for situation 3 the count is

$$W_3 = \frac{6!}{2!2!2!} = 90$$

You can check the numbers for the other maximum probability macrostates. The entropy of each state is determined using

$$S_n = k \ln W_n$$

and is reported in the figure on the previous page. Note that as the number of possible locations for the atoms increases, the maximum probability count and entropy increases dramatically. As the gas expands, its entropy increases. From an energy point of view, we would say that the energy of the gas evolved from being localized in a smaller region and potentially being more useful for doing work to being distributed evenly over a larger region and being much less useful.

**Try it yourself:** Four atoms start in a small box. They then expand into a box that is twice as large and finally into a box that is four times as large. Determine the count of the maximum probability macrostate for each situation. Label the first box as a single location, the second box as having two places the atoms can reside, and the largest box as having four places the atoms can reside.

*Answer:* 1; 6; 24.

This last example illustrates quantitatively the second law of thermodynamics expressed using entropy.

**Second law of thermodynamics** **Spontaneous processes in an isolated system tend to proceed in the direction of increasing entropy.**

Remember, entropy is a measure of the probability of a particular macrostate. High-entropy states are more disorganized than lower entropy states. These high-probability disorganized states with higher entropy are less able to do work on their environment than less probable organized states with lower entropy.

**Review Question 13.2** Explain why turning on an extra lightbulb in your apartment does not prevent energy from being conserved but still affects the entropy.

## 13.3 Connecting the statistical and macroscopic approaches to irreversible processes

So far, we have related entropy to the number of microstates (the count) for a particular macrostate. The greater the number of microstates, the greater the disorder of the system when in that macrostate and the greater the entropy of the system. In systems with large entropy, the energy is very disorganized (for example, there is more random thermal energy) and the system can do less work on its environment.

These ideas are all based on our observations of real-world phenomena and on the statistical analysis of simple systems. We can actually calculate the entropy of these simple systems (involving 10 or even 100 particles). But real macroscopic systems involve vastly larger numbers of particles ($10^{23}$ or more, for example). It would be difficult to count the number of microstates in a macrostate of such a large system. We need another method to determine the entropy in real-world macroscopic systems with large numbers of particles. Can we express entropy using macroscopic physical quantities such as temperature?

## Macroscopic definition of entropy change

Consider a macroscopic process during which a gas (modeled as an ideal gas) is heated by a Bunsen burner flame and expands at constant temperature. When the gas in a system expands, it pushes outward on the environment. Therefore, the environment (in this case, a piston) exerts a force on the system that does negative work (**Figure 13.4a**). The bar chart in Figure 13.4b depicts this process. The system's internal energy does not change since this process is isothermal. Since the environment is doing negative work on the system, the energy provided to the gas by the environment through heating must transfer energy so that the system's thermal energy remains constant.

**Figure 13.4** Isothermal expansion. (a) The gas expands isothermally. (b) The negative work done by the piston on the gas is balanced by the positive heating.

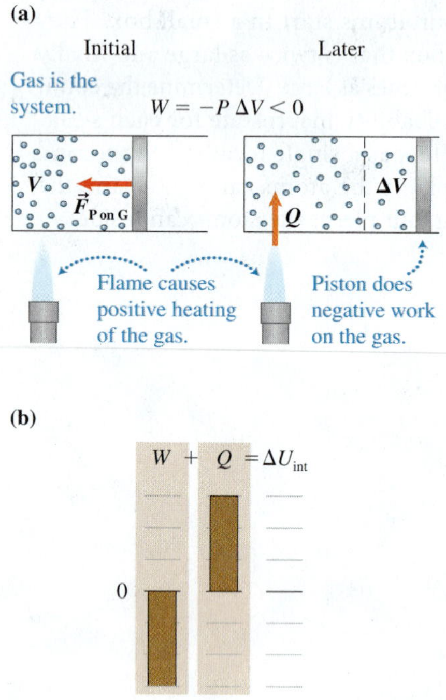


The volume of the gas increases and the gas particles now have more places to reside than before—the entropy of the system should increase even though the internal energy does not change. We can understand this connection through a limiting case analysis. Imagine that at the beginning, the volume of the gas is almost zero—all particles are crammed together and the number of microstates of that macrostate of the system is 1. Now, if the gas expands, the particles have more room and can occupy this room in many different configurations. Thus, the count of microstates and therefore the entropy of the system increases.

We can use the bar chart in Figure 13.4b to apply the first law of thermodynamics to this isothermal process:

$$+Q_{\text{Env to Sys}} + W_{\text{Env on Sys}} = \Delta U_{\text{int}} = 0$$

or

$$Q_{\text{Env to Sys}} = -W_{\text{Env on Sys}}$$

We know that the work done by the environment on the gas is $W_{\text{Env on Sys}} = -P\,\Delta V$. Thus,

$$Q_{\text{Env to Sys}} = +P\Delta V$$

As the gas expands at constant temperature, its volume increases and its pressure decreases. We use the ideal gas law $P = nRT/V$ to substitute for $P$ in the above:

$$Q_{\text{Env to Sys}} = +P\Delta V = \frac{nRT}{V}\Delta V$$

Solving for $\Delta V/V$, we get

$$\frac{\Delta V}{V} = \left(\frac{1}{nR}\right)\frac{Q_{\text{Env to Sys}}}{T}$$

Remember that this is an isothermal process in which positive energy transfer through heating is required to balance the negative work done by the environment as the gas expands. Because of the increasing volume, more space is available for the same number of gas particles at the same temperature. We can say that the entropy increases because

$$\frac{\Delta V}{V} > 0$$

It appears that either $\Delta V/V$ or $Q/T$ could be used as macroscopic measures of the increase in disorder of the system—measures of the system's entropy change. The units of $Q/T$ are J/K, the same as the units of entropy from the microscopic entropy definition using Eq. (13.2): $S_i = k \ln W_i$. Note that the natural logarithm of any quantity is unitless, and $k$ (Boltzmann's constant) has units J/K. Thus, the unit of statistical entropy is the same as the unit of $Q/T$. It seems reasonable to define the entropy change of a macroscopic system as $Q/T$.

Often, when energy is transferred between an environment and a system through heating, the temperature of the system changes. What temperature should be used to determine the system's entropy change? To deal with this, let's define the entropy change $\Delta S$ as a small amount of heating $Q$ when the system is at a particular temperature $T$. Since this is a small amount of energy, the temperature of the system will not change significantly during the process. Alternatively, we can use the average temperature of the process to find the change in entropy. Here's the summary.

**Macroscopic definition of entropy change** **The entropy change $\Delta S_{Sys}$ of a system during a process equals the ratio of the energy transferred to the system from the environment through heating $Q$ divided by the system's temperature $T_{Sys}$ while the process is occurring:**

$$\Delta S_{Sys} = \frac{Q_{Env\ to\ Sys}}{T_{Sys}} \tag{13.3}$$

**$T_{Sys}$ is either the constant temperature of the system or the average temperature if the temperature changes.**

**TIP** When you use Eq. (13.3), the temperature must have units of kelvin.

Consider a very simple example. You leave a 1-L water bottle outside on a sunny day. Let's estimate the change in the entropy of the water in the bottle. Assume that the mass of the water in the bottle is about 1 kg, and the temperature inside changes from 20 °C to 30 °C. To change the temperature of water, the environment must have transferred energy to it through heating:

$$\Delta U_{Sys} = Q_{Env\ to\ Sys} = mc\Delta T = \left(4200\frac{\text{J}}{\text{kg}\cdot{}^\circ\text{C}}\right)(1\ \text{kg})(10\ {}^\circ\text{C}) = 42{,}000\ \text{J}$$

The average temperature is 25 °C. Thus the entropy change is

$$\Delta S_{Sys} = \frac{Q_{Env\ to\ Sys}}{T_{Sys}} = \frac{42{,}000\ \text{J}}{(25 + 273)\ \text{K}} = 140\frac{\text{J}}{\text{K}}$$

Let's apply this definition to a process in which objects at different temperatures reach thermal equilibrium.

## Entropy change of a system and its environment

Consider a process in which a warm object in the environment at temperature $T_{Env}$ transfers a small amount of energy through heating $Q > 0$ to a system that is at a lower temperature $T_{Sys}$. What was the net entropy change for the system plus the environment, known as the entropy change of the universe: $\Delta S_{Universe} = \Delta S_{Sys} + \Delta S_{Env}$?

$$\Delta S_{Sys} = \frac{Q_{Env\ to\ Sys}}{T_{Sys}} = \frac{Q}{T_{Sys}}$$

and

$$\Delta S_{Env} = \frac{Q_{Sys\ to\ Env}}{T_{Env}} = \frac{-Q}{T_{Env}}$$

or

$$\Delta S_{Universe} = \frac{Q}{T_{Sys}} + \frac{(-Q)}{T_{Env}} = Q\left(\frac{1}{T_{Sys}} - \frac{1}{T_{Env}}\right) = Q\left(\frac{T_{Env} - T_{Sys}}{T_{Sys}T_{Env}}\right) > 0$$

The heating occurred because the environment was at a higher temperature than the system $T_{Env} > T_{Sys}$. But this also means that the magnitude of the system's entropy change was greater than the magnitude of the environment's entropy change. Thus, the entropy change of the system and its environment is positive; the entropy of the system and environment increased. This process occurred when objects of different temperature came in contact and achieved thermal equilibrium. This leads to yet another version of the second law of thermodynamics.

Active Learning Guide ›

**Second law of thermodynamics** **During any process that involves the transfer of energy through heating, the net change in entropy of the system and its environment is always greater than zero.**

## EXAMPLE 13.5 Mixing hot and cool water

We add 0.50 kg of water at 70 °C to 0.50 kg of water at 30 °C. Estimate the net entropy change of all the water once the system has come to equilibrium.

**Sketch and translate** The process is sketched in the figure below. The warm water cools and the cool water warms.

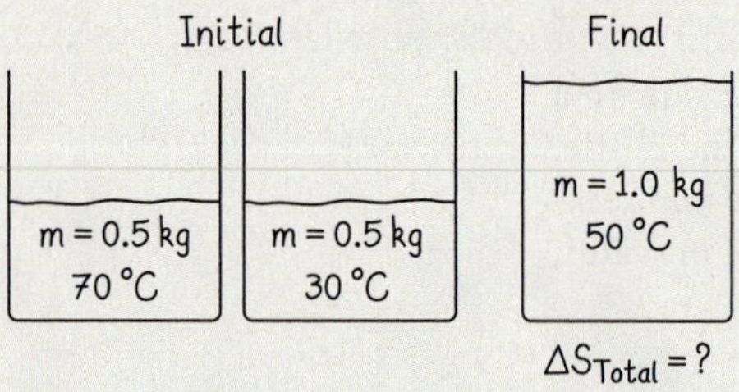


Equal amounts of warm and cool water combine to form intermediate temperature water.

**Simplify and diagram** Consider this as a two-system process; the warm water (the warm system) transfers energy to the cool water (the cool system) through heating. Assume that the combined system is isolated from its environment. Notice that in this process, the temperature of both systems continually changes. Therefore, we use the average temperature of each system as they undergo the cooling or warming process. Since there are equal amounts of water, the equilibrium temperature will be 50 °C. The average temperature of the warm system is then

$$\frac{70\,°\text{C} + 50\,°\text{C}}{2} = 60\,°\text{C}$$

while the average temperature of the cool system is

$$\frac{30\,°\text{C} + 50\,°\text{C}}{2} = 40\,°\text{C}$$

**Represent mathematically** The thermal energy transferred from the warm water to the cool water is determined using Eq. (12.5):

$$Q_{\text{Warm to Cool}} = mc_{\text{water}}(50\,°\text{C} - 30\,°\text{C}).$$

Similarly, the negative heating of the warm water is

$$Q_{\text{Cool to Warm}} = mc_{\text{water}}(50\,°\text{C} - 70\,°\text{C})$$

We now use Eq. (13.3) to add the entropy changes of the warm water and of the cool water:

$$\Delta S_{\text{Total}} = \Delta S_{\text{Cool}} + \Delta S_{\text{Warm}} \approx \frac{Q_{\text{Warm to Cool}}}{T_{\text{Cool}}} + \frac{Q_{\text{Cool to Warm}}}{T_{\text{Warm}}}$$

**Solve and evaluate** The specific heat of water is $c = 4180\ \text{J/kg}\cdot°\text{C}$. The magnitude of the energy transferred through heating is

$$\begin{aligned} Q_{\text{Warm to Cool}} &= mc_{\text{water}}(50\,°\text{C} - 30\,°\text{C}) \\ &= (0.50\ \text{kg})\left(\frac{4180\ \text{J}}{\text{kg}\cdot°\text{C}}\right)(20\,°\text{C}) = 41{,}800\ \text{J} \end{aligned}$$

The cool to warm water heating is −41,800 J. We can now make an estimate of the total entropy change:

$$\begin{aligned} \Delta S_{\text{Total}} &= \Delta S_{\text{Cool}} + \Delta S_{\text{Warm}} \\ &\approx \frac{+41{,}800\ \text{J}}{(273 + 40)\text{K}} + \frac{-41{,}800\ \text{J}}{(273 + 60)\text{K}} \\ &= +134\ \text{J/K} - 126\ \text{J/K} = +8\ \text{J/K} \end{aligned}$$

Since the net entropy of the combined system increases, this process can occur spontaneously (the second law of thermodynamics.)

Suppose you started with all the water at 50 °C and waited for half the water to cool to 30 °C while the other half warmed to 70 °C (the reverse of the above process). This process is consistent with energy conservation (the first law of thermodynamics). The entropy change for such a process would be

$$\begin{aligned} \Delta S_{\text{Total}} &= \Delta S_{\text{Cool}} + \Delta S_{\text{Warm}} \\ &\approx \frac{-41{,}800\ \text{J}}{(273 + 40)\text{K}} + \frac{+\,41{,}800\ \text{J}}{(273 + 60)\text{K}} \\ &= -134\ \text{J/K} + 126\ \text{J/K} = -8\ \text{J/K} \end{aligned}$$

The entropy of the system would decrease. This process will not occur spontaneously (the second law of thermodynamics.)

**Try it yourself:** Determine the entropy change of a 0.10-kg chunk of ice at 0 °C that melts and becomes a 0.10-kg puddle of liquid water at 0 °C. The heat of fusion of water is $3.35 \times 10^5$ J/kg.

*Answer:* +120 J/K.

## Entropy and complex organic species

Living beings develop from less complex forms to more complex forms. Does this contradict what we have learned about entropy and the second law of thermodynamics? What about when an organism grows? Clearly, its organization and complexity increase.

The growth of an organism does not violate the second law of thermodynamics. The key reason is that the second law applies only to an isolated system; a growing organism is definitely not an isolated system. It interacts with the environment. If we consider the environment as part of the system, then the analysis is quite different.

Plant tissues, for example, absorb energy from the Sun and carbon in the air and water from the soil (**Figure 13.5**), initiating complicated processes that lead to the eventual formation of carbohydrates, some amino acids, fatty acids, and thousands of other compounds by a myriad of other reactions. The formation of these complex useful compounds seems like a decrease in entropy and a violation of the second law of thermodynamics. In each reaction in the complex set of reactions, two or more atoms or molecules combine to form a more complex, lower entropy molecule. However, when these reactions occur, they release energy through heating into the environment. The corresponding entropy increase of the environment is over two times greater than the corresponding decrease due to the formation of the larger molecule. Photosynthesis produces a net increase in the entropy of the Earth system.

**Figure 13.5** The flower is not an isolated system.

What about humans? A person grows and develops into a complex, highly ordered adult by consuming and converting complex molecules with low entropy, such as carbohydrates, fats, and proteins, into simpler molecules with higher entropy. Each year a healthy adult consumes about 500 kg, or a half-ton, of this low-entropy food. During consumption over the course of a year a human transfers roughly two billion joules of energy to the environment, increasing its thermal energy, a form of energy that has little ability to do work. The entropy of the human may decrease, becoming more complex and organized, but the entropy of the environment increases significantly, more than enough to compensate.

In industrialized countries that burn low-entropy fuels such as petroleum and coal to meet their energy needs, the entropy increase of the environment is almost 100 times greater per person than in nonindustrialized countries. The scientific basis of the "energy crisis" rests with the second law of thermodynamics and the increase in entropy of the universe as a result of the conversion of useful energy to much less useful forms. Once we convert the organized energy of petroleum to less organized forms of internal energy, the entropy of the universe has increased and the potential of using that energy for useful purposes is lost forever.

**Review Question 13.3** The internal energy of a system is a state function, but work and heating are not. Is entropy a state function? Explain.

# 13.4 Thermodynamic engines and pumps

We have learned that over time an isolated system's energy converts into less useful forms—for example, mechanical energy converts into thermal energy. Is it possible, though, to use the thermal energy of one system to do work on another system?

Suppose that a system has two "reservoirs" at different temperatures. A reservoir could be a pond of cool water, a flame that converts water into steam, or the burning fuel in the cylinder of a car engine. If the hot and

**Figure 13.6** Thermodynamic engine. (a) A schematic illustration of the primary components of a thermodynamic engine. (b) A bar chart for the net change during a thermodynamic engine cycle. (c) An electric power plant is a thermodynamic engine.

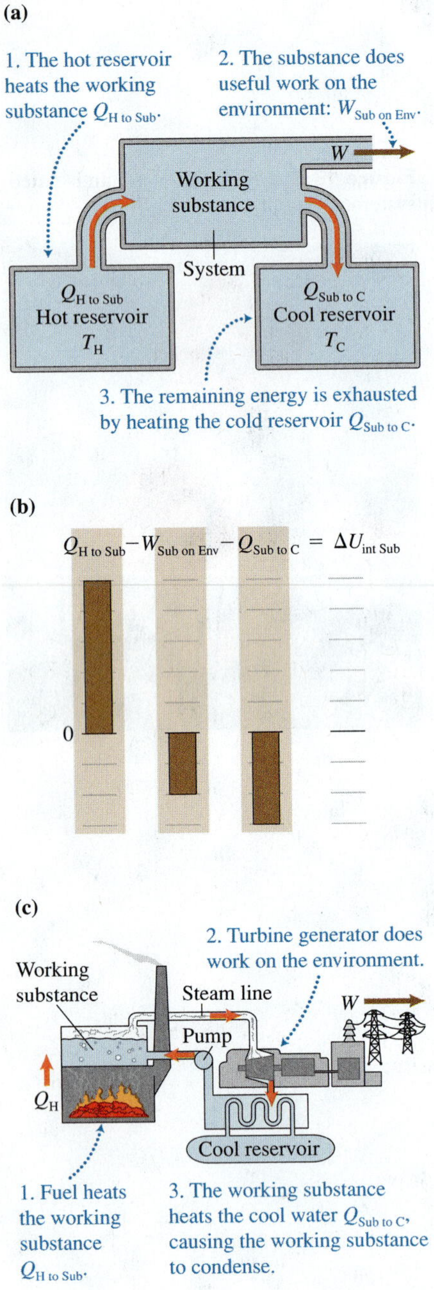

(a)
1. The hot reservoir heats the working substance $Q_{H\ to\ Sub}$.
2. The substance does useful work on the environment: $W_{Sub\ on\ Env}$.
W
Working substance
System
$Q_{H\ to\ Sub}$
Hot reservoir
$T_H$
$Q_{Sub\ to\ C}$
Cool reservoir
$T_C$
3. The remaining energy is exhausted by heating the cold reservoir $Q_{Sub\ to\ C}$.
(b)
$Q_{H\ to\ Sub} - W_{Sub\ on\ Env} - Q_{Sub\ to\ C} = \Delta U_{int\ Sub}$
0
(c)
2. Turbine generator does work on the environment.
Working substance
Steam line
W
Pump
$Q_H$
Cool reservoir
1. Fuel heats the working substance $Q_{H\ to\ Sub}$.
3. The working substance heats the cool water $Q_{Sub\ to\ C}$, causing the working substance to condense.

cold reservoirs are in contact, thermal energy flows spontaneously from the warmer reservoir to the cooler one. We can harness the energy of this device in order to work on another system. Such device is called a **thermodynamic engine.** Even more surprising, we can create a device that makes a cool object even colder, such as a refrigerator.

We will begin by examining how a thermodynamic engine works. We will then take a look at how a refrigerator works.

## Thermodynamic engine

A thermodynamic engine (**Figure 13.6a**) has three main parts: (1) an input reservoir with hot material at temperature $T_H$, (2) some kind of working substance that can expand and contract and do useful mechanical work, and (3) an output reservoir with cool material at a lower temperature $T_C$. The working substance travels through the engine and undergoes thermodynamic processes that cause some desired effect—for example, doing useful work such as turning an electric generator. At the end of each cycle, the working substance returns to its starting state. It then repeats the sequence.

The hot reservoir is maintained at temperature $T_H$ usually by burning some kind of fuel (for example, coal, natural gas, or nuclear fuel in a power plant, or the gasoline in a car engine). The working substance is usually a gas that can expand significantly to do work. The cold reservoir may be a pool of cold water that continually flows to transfer energy away from the working substance.

In the first step of the cycle, the hot reservoir transfers thermal energy to the working substance by heating $Q_{H\ to\ Sub}$. In the second step, the working substance, which is now hot and has expanded, pushes a piston or turns the turbine of a generator. Thus, the working substance does positive work on the environment $W_{Sub\ on\ Env}$. According to the second law of thermodynamics, it is impossible to use 100% of the energy transferred by heating from the hot reservoir into work done by the working substance. What happens to the rest of the energy?

In the third step, the energy is transferred from the substance to the cold reservoir $Q_{Sub\ to\ C}$. Since the working substance is the system, this would be negative heating of the substance by the cold reservoir $Q_{C\ to\ Sub}$. The working substance, now cooler, contracts. A compressor also helps in this contraction, doing some work on the substance but somewhat less work than the substance did on the environment in step 2. In other words, during step 3 the environment does positive work on the working substance; however, the net work done by the environment on the working substance during the whole process is negative.

In summary, we have positive heating of the substance by the hot reservoir ($Q_{H\ to\ Sub}$), the positive net work done by the working substance on the environment ($W_{Sub\ on\ Env}$), and the positive heating of the cold reservoir by the substance ($Q_{Sub\ to\ C}$). The bar chart in Figure 13.6b represents the process. Since the working substance returns to the same state at the end of each loop around the engine, the net internal energy change of the working substance is zero. Thus, we can put this into an equation form using the first law of thermodynamics as applied to the working substance:

$$Q_{H\ to\ Sub} + W_{Env\ on\ Sub} + Q_{C\ to\ Sub} = 0$$

The second and third terms in this equation are negative (see the bar chart). We can switch their subscripts and the signs of those two terms:

$$Q_{H\ to\ Sub} - W_{Sub\ on\ Env} - Q_{Sub\ to\ C} = 0$$

Thus, the net positive work that the substance does on the environment is

$$W_{\text{Sub on Env}} = Q_{\text{H to Sub}} - Q_{\text{Sub to C}} \tag{13.4}$$

All of the quantities in the above equation have positive values. A schematic model of the thermodynamic engine in an electric power plant is shown in Figure 13.6c.

The efficiency of a thermodynamic engine is the work done by the working substance on the environment divided by the energy transferred from the hot reservoir to the working substance:

$$e = \frac{W_{\text{Sub on Env}}}{Q_{\text{H to Sub}}} \tag{13.5}$$

**EXAMPLE 13.6 Efficiency of a thermodynamic engine**

In a car engine, gasoline is ignited by a spark plug to generate an explosion in a cylinder. Each explosion transfers 700 J of energy from the burning fuel to the gas in the cylinder. During the expansion (and later contraction) of the gas, the gas does a net 200 J of mechanical work on the environment (on a piston that pushes outward and helps propel the car). (a) How much energy is transferred through heating to the car's cooling system during this cycle? (b) What is the efficiency of the engine?

**Sketch and translate** The working substance is the gas (mostly air) inside the cylinder. The hot reservoir is the burning fuel in the cylinder that is ignited by the spark plug. The cooling system is the cold reservoir. One cycle of the process can be represented as shown in the figure below. $Q_{\text{H to Sub}} = +700\text{ J}$ (the energy transferred from the burning fuel to the gas inside the cylinder) and $W_{\text{Sub to Env}} = +200\text{ J}$ (the hot air in the cylinder expands and contracts doing positive work against the piston in the cylinder, which ultimately causes the wheels to turn.) We need to find the energy transferred through heating to the cooling system of the car and the efficiency of the engine.

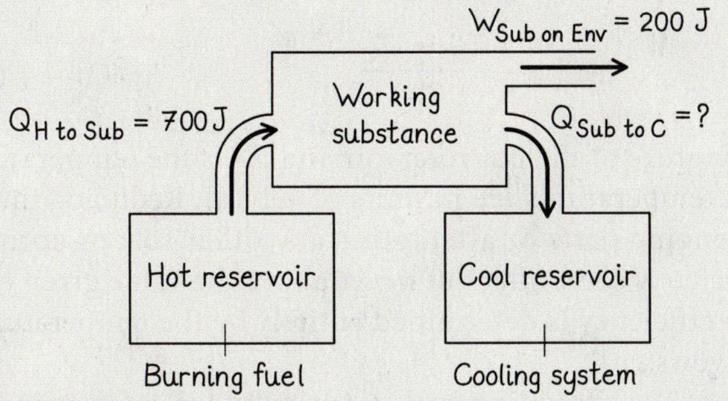


**Simplify and diagram** We model the car's engine as a thermodynamic engine and represent the process with a bar chart in the figure at right.

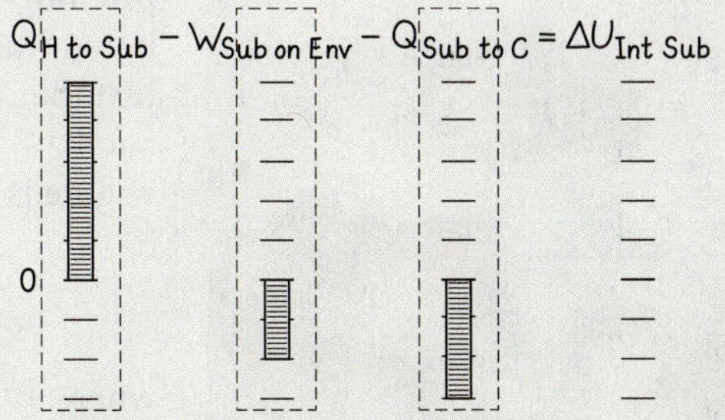


**Represent mathematically** Rearrange Eq. (13.4) to determine how much energy is transferred through heating to the car's cooling system:

$$W_{\text{Sub on Env}} = Q_{\text{H to Sub}} - Q_{\text{Sub to C}}$$
$$\Rightarrow W_{\text{Sub}} = Q_{\text{H}} - Q_{\text{C}}$$
$$\Rightarrow Q_{\text{C}} = Q_{\text{H}} - W_{\text{Sub}}$$

Then use Eq. (13.5) to determine the efficiency:

$$e = \frac{W_{\text{Sub}}}{Q_{\text{H}}}$$

**Solve and evaluate** Using the above equations, and noting that the substance does positive work on the environment (the environment does negative work on the system), we find that

$$Q_{\text{C}} = Q_{\text{H}} - W_{\text{Sub}} = +700\text{ J} - 200\text{ J} = +500\text{ J}$$

Over 70% of the energy that the substance receives from the hot reservoir is moved to the cold reservoir. The engine's efficiency is

$$e = \frac{W_{\text{Sub on Env}}}{Q_{\text{H to Sub}}} = \frac{200\text{ J}}{700\text{ J}} = 0.29$$

The result agrees with what you know from your everyday experience when you stand near a parked car when the engine is running. The car's exhaust system pumps hot gas into the atmosphere, carrying away energy that has served

*(continued)*

no useful purpose. The hood and radiator grill of the car is also warm, releasing even more energy to the environment.

**Try it yourself:** Another car engine has the same efficiency as the car in this example. If the engine does work on its environment at a rate of 40 hp $(1\text{ hp} = 746\text{ J/s})$, how much work does the engine do on its environment each second, and how much energy through heating must be transferred from the burning fuel to the gas in the cylinders each second?

*Answer:* $W_{\text{Sub on Env}} = (40\text{ hp})(1.0\text{ s})\left(\frac{746\text{ J/s}}{1\text{ hp}}\right) = 30{,}000\text{ J}$

and

$$Q_{\text{H to Sub}} = \frac{W_{\text{Sub on Env}}}{e} = \frac{30{,}000\text{ J}}{0.29} = 1.0 \times 10^5\text{ J}.$$

During this second, there might be as many as 100 explosions in the cylinders of the engine.

Diesel engines are more efficient than standard gasoline engines.

## Maximum efficiency of a thermodynamic engine

The 29% efficiency of the automobile engine in the last example is a little depressing—most of the chemical potential energy stored in the gasoline ends up being transferred to the environment rather than being used to do work. Is it just that car engines are poorly designed, or is there a more fundamental reason why they are so inefficient? Recall that the efficiency of a thermodynamic engine is

$$e = \frac{W_{\text{Sub on Env}}}{Q_{\text{H to Sub}}}$$

where $W_{\text{Sub on Env}}$ is the useful work done by the working substance in the engine on the environment and $Q_{\text{H to Sub}}$ is the energy transferred to the engine by the hot reservoir.

According to the first law of thermodynamics, the net work done by the engine on its environment is the energy transferred from the hot reservoir to the working substance through heating minus the energy transferred from the working substance to the cold reservoir:

$$W_{\text{Sub on Env}} = Q_{\text{H to Sub}} - Q_{\text{Sub to C}}$$

Thus, the efficiency of a thermodynamic engine can be written as

$$e = \frac{W_{\text{Sub on Env}}}{Q_{\text{H to Sub}}} = \frac{Q_{\text{H to Sub}} - Q_{\text{Sub to C}}}{Q_{\text{H to Sub}}}$$

In England, when thermodynamic engines were being developed to pump water out of mines, scientists and engineers wondered if they could improve the efficiency of the engines. Sadi Carnot, using ideas of entropy and the second law of thermodynamics, showed that the maximum efficiency that a thermodynamic engine could possibly have was

$$e_{\text{max}} = \frac{T_{\text{H}} - T_{\text{C}}}{T_{\text{H}}} \tag{13.6}$$

where $T_{\text{H}}$ is the temperature of the hot reservoir and $T_{\text{C}}$ is the temperature of the cold reservoir (all temperatures are in units of kelvin). Reducing internal friction between the engine parts or attempting any other sort of optimization may improve efficiency somewhat but never above the value given by Eq. (13.6). The maximum efficiency is determined entirely by the temperatures of the hot and cold reservoirs.

The temperature of the exploding fuel in the cylinders of cars is about 200 °C, while the temperature of the exhaust system of the car is about 60 °C. Thus, the maximum efficiency of such an engine is

$$e_{\text{max}} = \frac{T_{\text{H}} - T_{\text{C}}}{T_{\text{H}}} = \frac{(200 + 273)\text{K} - (60 + 273)\text{K}}{(200 + 273)\text{K}} = \frac{473\text{ K} - 333\text{ K}}{473\text{ K}} \approx 0.30$$

We know that gas mileage in cars can vary significantly in efficiency. The above number is for a relatively low-efficiency car—the number can vary from below 0.3 to above 0.4.

The equation on the previous page indicates ways we might increase the maximum efficiency of an engine.

1. Increase the temperature of the hot reservoir. For example, a pebble bed nuclear power plant reactor under development can operate at a temperature up to 1600 °C with efficiency up to 50% (supposedly more safely than traditional nuclear reactors). The "pebble bed" name comes from the shape of the fuel elements, which are tennis ball-sized pebbles. Diesel fuel is burned at a higher temperature than gasoline, making diesel car engines more efficient than gasoline engines.
2. Reduce the temperature of the cold reservoir. Doing so requires a more advanced cooling system. For example, turbocharged and supercharged car engines have a device known as an intercooler, a special radiator through which the compressed air passes and is cooled before it enters the cylinder.

## Thermodynamic pumps (refrigerators or air conditioners)

What happens if we reverse the thermodynamic engine process so that the work done by the environment on the gas is positive? There is then less heating into the working substance than heating out of the working substance. We investigate this process in Testing Experiment **Table 13.3**.

**TESTING EXPERIMENT TABLE**

**13.3 Reversing the operation of a thermodynamic engine.**

| Testing experiment | Prediction | Outcome |
|---|---|---|
| We make a machine that has the same cold and hot reservoirs as the thermodynamic engine but performs all the operations in reverse order. The environment will do positive work on the working substance, and there will be net heating out of the working substance. 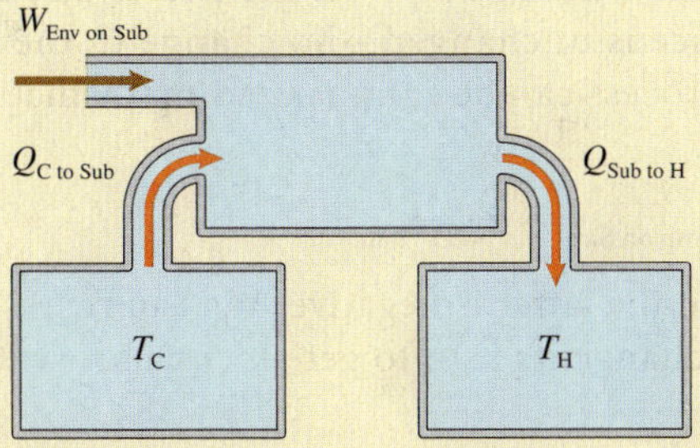  | If we reverse the energy bar chart used to analyze thermodynamic engines, we conclude that the cold reservoir gets cooler and the hot reservoir gets warmer. 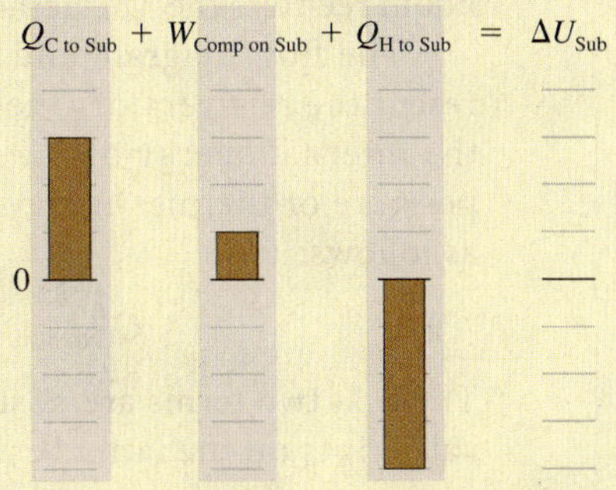  | We can build a refrigerator—it makes the cold food inside the refrigerator cooler and the air outside the refrigerator warmer. |

**Conclusion**

It is possible to create a machine that will transfer thermal energy from a cooler reservoir to a hotter reservoir if the environment does positive work on the working substance. This machine cools the working substance and warms the environment.

**Figure 13.7** A schematic drawing of a refrigerator.

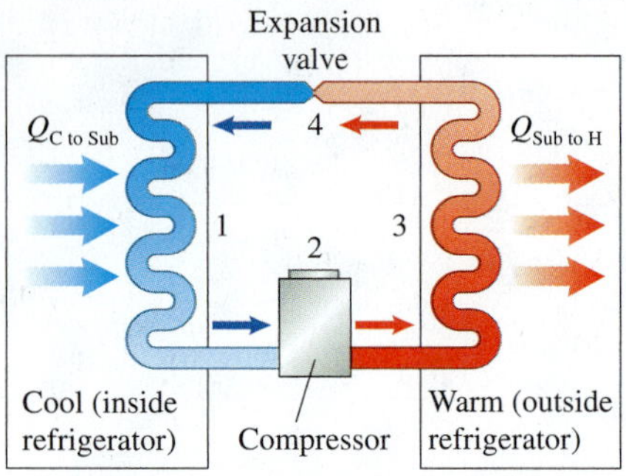


By reversing the thermodynamic engine process, we have invented a machine called a **thermodynamic pump**. A refrigerator is a type of thermodynamic pump. So is an air conditioner. As you can see from the schematic drawing in **Figure 13.7**, in a thermodynamic pump a working substance called a refrigerant takes thermal energy from a "cold reservoir" (position 1 in Figure 13.7). The substance moves through the pipes in the direction of the small blue and red arrows and delivers some of the thermal energy to a hot reservoir outside the pump (position 3 in the figure). The net effect is that thermal energy is transferred from the cool reservoir to the substance ($Q_{\text{C to Sub}}$), which transfers some of the thermal energy to the warmer air outside the pump ($Q_{\text{Sub to H}}$).

The substance can take energy from the cold reservoir that is at lower temperature and deliver it to the hot reservoir at higher temperature because it is not an isolated system. In addition to the two reservoirs at low and high temperatures, the substance is connected to a motor (a compressor) that does positive work on it ($W_{\text{Comp on Sub}}$).

Let's consider in detail how a refrigerator works. The refrigerant is typically a fluid with a boiling point below 0 °C, such as a hydrofluorocarbon (HFC). The boiling temperature of HFCs is −29.9 °C; below this temperature, most of the substance is in liquid form; above this temperature, it is a gas. The food compartment is the "cold reservoir," which transfers thermal energy to the substance, which runs through coils in the appliance. As the substance's temperature increases, it partially evaporates ($Q_{\text{C to Ref}}$). As the substance flows through the inside of the refrigerator, more of it evaporates. Energy transfers from the food to the substance, leading to a decrease in the food temperature. As the substance leaves the interior of the refrigerator, it is warmer and much more of it is in the gaseous state than when it entered.

The refrigerant then passes through a compressor (position 2) that does positive work on the substance ($W_{\text{Comp on Sub}}$), adiabatically compressing it to higher internal energy and temperature ($Q = 0$, $\Delta V < 0$, and $P|\Delta V| + 0 = \Delta U_{\text{Sub}}$). The substance is now hotter than the environment outside the refrigerator (considered the "hot reservoir"). The substance passes through a condenser coil (position 3) on the outside of the refrigerator, where it transfers thermal energy to the outside environment ($Q_{\text{Sub to H}}$), which is the hot reservoir. At this point, the substance temperature decreases, and it condenses in part back from a gas to a liquid.

The condensed substance passes through a pressure-lowering device, the expansion valve (position 4). There it expands adiabatically ($Q = 0$, $\Delta V > 0$, and $-P\Delta V + 0 = \Delta U_{\text{Sub}}$), causing its internal energy to decrease. At this point, even more of the substance condenses into a cool liquid, and this liquid reenters the refrigerator for another trip around the cycle.

The flow diagram and bar chart in Testing Experiment Table 13.3 represent the refrigerator (the thermodynamic pump). At the end of one cycle, the internal energy of the substance is unchanged—no change in the temperature or thermal energy. The process can be summarized mathematically as follows:

$$Q_{\text{C to Sub}} + W_{\text{Comp on Sub}} + Q_{\text{H to Sub}} = 0$$

The first two terms are positive and the latter is negative. We can reverse the subscripts on the latter term and change its sign to get an equation with all positive terms:

$$Q_{\text{C to Sub}} + W_{\text{Comp on Sub}} - Q_{\text{Sub to H}} = 0 \qquad (13.7)$$

Air conditioners operate in precisely the same way. The efficiencies of refrigerators and air conditioners are rated in terms of a performance coefficient $K$:

$$K_{\text{Ref}} = \frac{Q_{\text{C to Sub}}}{W_{\text{Comp on Sub}}} = \frac{Q_{\text{C to Sub}}}{Q_{\text{Sub to H}} - Q_{\text{C to Sub}}} \qquad (13.8)$$

The second law of thermodynamics limits this performance coefficient:

$$K_{\text{Ref max}} \leq \frac{T_C}{T_H - T_C} \qquad (13.9)$$

The temperatures must be in kelvin (K).

## Using a thermodynamic pump to warm a house

A thermodynamic pump can also be used to warm a home in the winter. The only difference is that the outside of the house (cold in the winter) plays the role of the inside of the refrigerator. Energy is transferred through heating from the cooler air outside (the cold reservoir at temperature $T_C$) to the warmer air inside (the warm reservoir at temperature $T_H$). A compressor plugged into a wall outlet does work on the working substance (a refrigerant) to make the transfer possible. A sketch of this operation is shown in **Figure 13.8a**. The process is represented by a flow diagram in Figure 13.8b and by a bar chart in Figure 13.8c. The system is the working substance (like the refrigerant). The goal of this device is to warm something (the house) rather than cool something. The compressor does the work that makes possible the transfer of energy through heating to the house. Thus, the performance coefficient $K_{\text{Home pump}}$ is the ratio of the energy provided through heating to the house and the work that must be done for this energy transfer to occur:

$$K_{\text{Home Pump}} = \frac{Q_{\text{Sub to H}}}{W_{\text{Comp on Sub}}} \leq \frac{T_H}{T_H - T_C} \qquad (13.10)$$

Again, the temperatures must be in kelvin (K). From Eq. (13.10), you see that the performance coefficient is always larger than 1.

**TIP** The term "thermodynamic pump" does not mean that any sort of physical material is being moved around. It is energy that is being "pumped" from a cooler region to a warmer region. This transfer is possible because of the work being done on the working substance by the environment—by the compressor.

**Figure 13.8** Thermodynamic pump. (a) A thermodynamic pump on the outside wall of a house. (b) A schematic for the pump and (c) a bar chart for a cycle of the pump.

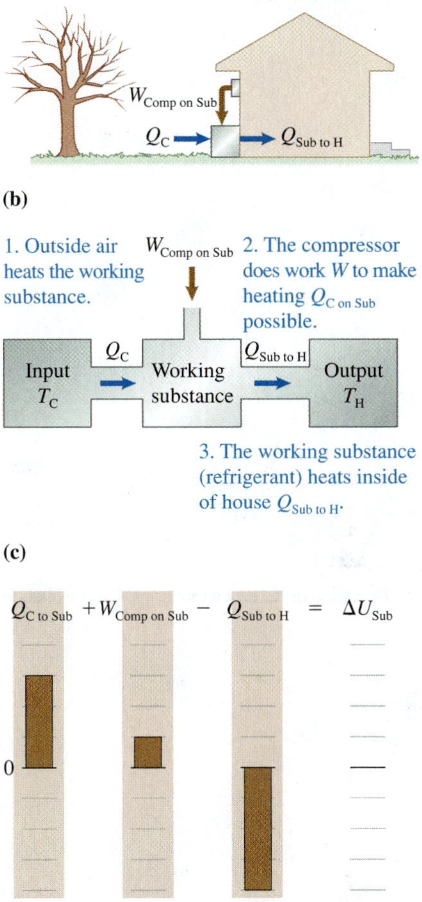


### QUANTITATIVE EXERCISE 13.7 Warming a house in the winter

A thermodynamic pump transfers 14,000 J of energy through heating into a house. The air temperature outside is −10 °C and inside is 20 °C. Determine the maximum performance coefficient of the thermodynamic pump and the minimum work that the pump's compressor must do on the working substance to cause this energy transfer.

**Represent mathematically** Energy $Q_{\text{Sub to W}} = +14{,}000\ \text{J}$ is transferred from the working substance through heating to the house. Our goal is to determine the maximum performance coefficient of the pump and the work $W_{\text{Env on Sub}}$ that the pump's compressor must do if the pump is operating at maximum efficiency. The maximum performance coefficient is given by Eq. (13.10):

$$K_{\text{Pump max}} = \frac{T_H}{T_H - T_C}$$

We can now determine the minimum work that must be done using the definition of the performance coefficient.

$$K_{\text{Pump}} = \frac{Q_{\text{Sub to H}}}{W_{\text{Comp on Sub}}} \Rightarrow W_{\text{Comp on Sub}} = \frac{Q_{\text{Sub to H}}}{K_{\text{Pump max}}}$$

**Solve and evaluate**

$$K_{\text{Pump max}} = \frac{T_H}{T_H - T_C} = \frac{(273 + 20)\text{K}}{(273 + 20)\text{K} - (273 - 10)\text{K}} = 9.8$$

$$W_{\text{Comp on Sub}} = \frac{Q_{\text{Sub to H}}}{K_{\text{Pump max}}} = \frac{14{,}000\ \text{J}}{9.8} = 1430\ \text{J}$$

The units are correct for the respective quantities and the numbers are reasonable. Unfortunately, thermodynamic pumps do not usually work at maximum performance, but they are still efficient ways to warm up a house in the winter and cool it in the summer (when behaving as an air conditioner).

**Try it yourself:** A refrigerator has a performance coefficient of 4. How much work must the compressor do to transfer 4000 J of energy from inside the refrigerator to the outside air?

*Answer:* 1000 J.

**Review Question 13.4** Why is there no contradiction between the second law of thermodynamics and the operation of refrigerators and air conditioners that transfer thermal energy from cooler objects to warmer objects?

## 13.5 Automobile efficiency and power plants: Putting it all together

People often talk about energy conservation in terms of saving energy. As we know now, the problem is not about running out of energy on Earth. Energy cannot be destroyed or used up. From a sustainability perspective, "energy conservation" means making efficient use of useful (low-entropy) forms of energy. In this section we consider two areas in which significant amounts of energy have been converted from useful forms to less useful forms with less then desired efficiency.

### Automobile efficiency

Transportation accounts for about 28% of U.S. energy use, and almost all of this demand is satisfied by burning some form of petroleum product (such as unleaded gasoline). Some of the chemical potential energy in the fuel input to cars is eventually transformed into less useful thermal energy in the engine and the drivetrain and through various forms of friction or unproductive use such as idling at a stop light (see **Figure 13.9**). The conversion of useful energy into less useful forms is known as **energy degradation**. Reducing energy degradation improves automobile efficiency.

**Active Learning Guide ›**

**Energy degradation in the engine (62%)** The maximum efficiency of gasoline-powered internal combustion engines depends on the temperatures of the combustion chamber and the exhaust system, but is not very high (30–40% maximum). Thus, 60–70% of the initial chemical energy of the gasoline is converted to thermal energy. Additional energy degradation is caused by engine friction (pistons rubbing against cylinders and friction in all other moving parts of the engine) and energy used to pump air into and out of the engine. Diesel engines operate at higher pressure and temperature of the burning fuel and as a result are more efficient than gasoline engines—up to 50% efficiency.

**Drivetrain losses (6%)** Energy is lost in the transmission and in the moving parts of the drivetrain. Technologies under development such as automated manual transmissions and continuously variable transmissions can reduce those losses.

**Other friction-like losses (13%)** Energy is converted to thermal energy due to air drag resistance (a form of friction), rolling resistance (another form of friction that involves the continual flexing of the tires as they rotate), and friction

**Figure 13.9** Energy degradation in an automobile.

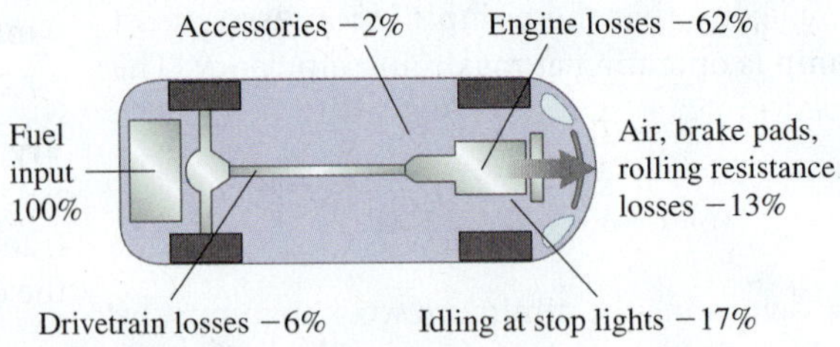

produced in the disk brakes when stopping the car. Any time you use your brakes to stop the car, the car's kinetic energy is converted to useless thermal energy. Newer braking systems convert a car's kinetic energy while stopping into the electrical potential energy of a charged battery or rotational kinetic energy of a turning flywheel. This stored energy can be reused to get the car moving again.

**Energy degradation from idling (17%)** Significant energy degradation occurs when a car is idling at stoplights or when it is parked with its engine running. Technologies such as integrated starter/generator systems help reduce these transformations by automatically turning the engine off when the vehicle comes to a stop and restarting it instantaneously when the accelerator is pressed. In some hybrid cars, the gasoline engine turns off completely when the car stops and the only working motor is electric.

**Accessories (2%)** Additional energy is used to run accessories such as the air conditioner, power steering, and windshield wipers.

A final important efficiency factor for automobiles involves energy used to gain speed. A vehicle's driveline must provide enough energy for the wheels to turn fast enough so the road surface can cause the vehicle to accelerate (the car's kinetic energy increases). The acceleration is directly related to the mass of the car. The less mass a vehicle has, the less work must be done to increase the car's kinetic energy. A smaller car will typically have less mass than a larger one, and manual transmissions have less mass than automatic transmissions.

### QUANTITATIVE EXERCISE 13.8 Temperature difference in engine

Note in Figure 13.9 that 62% of useful fuel energy input to that particular car's internal combustion engine is transformed into less useful forms—the engine of that particular car is only about 38% efficient before other degradations that occur, such as braking. If this is the maximum possible efficiency of this engine, what is the temperature difference between the burning fuel in the engine cylinders and the temperature of the air when expelled from the engine?

**Represent mathematically** The maximum efficiency of a thermodynamic engine is

$$e_{\text{max}} = \frac{T_{\text{H}} - T_{\text{C}}}{T_{\text{H}}}$$

Because of the $T_{\text{H}}$ in the denominator, we cannot solve directly for the temperature difference $T_{\text{H}} - T_{\text{C}}$. Instead, we make a rough estimate of $T_{\text{C}}$ and then determine $T_{\text{H}}$. We'll estimate $T_{\text{C}}$ by estimating the temperature of the exhaust pipe that leaves the car engine. The exhaust pipe is part of the car's cooling system, and we assume it to have a temperature $T_{\text{C}} = 80\ ^\circ\text{C}$. Multiplying both sides of the equation for the efficiency by $T_{\text{H}}$, we get

$$T_{\text{H}}e_{\text{max}} = T_{\text{H}} - T_{\text{C}}$$

$$\Rightarrow T_{\text{C}} = T_{\text{H}} - T_{\text{H}}e_{\text{max}} = T_{\text{H}}(1 - e_{\text{max}})$$

$$\Rightarrow T_{\text{H}} = \frac{T_{\text{C}}}{1 - e_{\text{max}}}$$

**Solve and evaluate** The maximum efficiency of this particular engine is 0.38. The temperature of the burning fuel in the cylinder is then

$$T_{\text{H}} = \frac{(273 + 80)\text{K}}{1 - 0.38} = 570\ \text{K}$$

Remember that kelvin (K) temperatures must be used in efficiency equations. The above temperature in degrees Celsius is $570\ \text{K} - 273\ \text{K} = 300\ ^\circ\text{C}$. However, gasoline automatically ignites at 260 °C. With a spark from a spark plug, it would ignite at a lower temperature. A real engine is probably not operating as hot as calculated and consequently is not as efficient. But our outcome seems reasonable as an estimate.

The result is depressing. The efficiencies of typical internal combustion engines are limited by thermodynamics to less than 0.4. To improve automobile efficiency, we need to change to a different type of engine—diesel, hybrid, electric, hydrogen, low mass, or perhaps something even more advanced.

**Try it yourself:** What temperature changes would result in increased efficiency of a thermodynamic engine?

*Answer:* Increase the temperature difference between the hot and cold reservoirs. Reducing the temperature of the cold reservoir also helps.

## Efficiency of a coal power plant

A power plant is another example of a thermodynamic engine. The Tennessee Valley Authority's (TVA) Kingston Fossil Plant is a typical coal-fired power plant that burns about 14,000 tons (1 ton is 1000 kg) of coal per day in a boiler (the plant's high-temperature reservoir). Here are the steps in the plant's operation:

1. The boiler transfers thermal energy to water (the working substance), converting it to steam at a temperature of about 540 °C.
2. The steam, at very high pressure, flows through a turbine, which spins generators to produce electric energy (the working substance does work on the environment in this step).
3. After leaving the generator, the steam is cooled in a condenser, which is in contact with cold river water (the cold temperature reservoir). The condenser converts the steam back into water at about 32–38 °C and at lower pressure. The low temperature and pressure on the output side of the turbine help move the high-pressure steam at the input to the turbine through the turbine.
4. After the condenser, the cooled water is returned to the boiler and the cycle begins again.

The Kingston plant generates power at an average rate of about $9.1 \times 10^8$ watts, enough electric energy to supply 540,000 homes.

**EXAMPLE 13.9 Energy from coal**
Using the information in the previous discussion of the Kingston coal plant, (a) estimate its maximum possible efficiency, (b) determine the energy transfer rate through heating from the boiler to the working substance, (c) determine the plant's actual efficiency. Note that 1 kg of coal releases roughly $24 \times 10^6$ J of energy when burned.

**Sketch and translate** We already have a sketch of the situation in Figure 13.6c. Information about this particular plant is provided in the text above. We need to find (a) the maximum efficiency $e_{max}$, (b) the rate

$$\frac{Q_{\text{H to Sub}}}{\Delta t}$$

that energy transfers through heating from the hot reservoir to the working substance, (c) the actual efficiency $e$, (d) and the rate

$$\frac{Q_{\text{Sub to C}}}{\Delta t}$$

that energy transfers through heating from the working substance to the cool reservoir. The working substance is the water that is boiling and condensing, the hot reservoir is the boiler (at 540 °C), and the cool reservoir is the river water (at 35 °C.)

**Simplify and diagram** We can model the plant as a thermodynamic engine, such as the one shown in Figure 13.6c. Assume that the rate

$$\frac{W_{\text{Sub on Env}}}{\Delta t}$$

by which work is done by the working substance on the environment equals the electric power generated by the Kingston plant. Also assume that all the chemical potential energy released by the burning coal is transferred to the working substance.

**Represent mathematically** The maximum efficiency of a thermodynamic engine is

$$e_{max} = \frac{T_H - T_C}{T_H}$$

The first law of thermodynamics written in the form of a rate equation and applied to this process is

$$\left(\frac{Q_{\text{H to Sub}}}{\Delta t}\right) - \left(\frac{W_{\text{Sub on Env}}}{\Delta t}\right) - \left(\frac{Q_{\text{Sub to C}}}{\Delta t}\right) = 0$$

All of the terms in parentheses are positive.

**Solve and evaluate** (a) The maximum thermodynamic efficiency of the plant depends on the temperatures of the hot and cold reservoirs and is

$$e_{\text{max}} = \frac{T_{\text{H}} - T_{\text{C}}}{T_{\text{H}}} = \frac{(540 + 273)\text{K} - (35 + 273)\text{K}}{(540 + 273)\text{K}} = 0.62$$

(b) We know that the average rate of electric energy produced by the TVA plant is about

$$\frac{W_{\text{Sub on Env}}}{\Delta t} = 9.1 \times 10^8 \text{ watts} = 9.1 \times 10^8 \text{ J/s}$$

The mass of coal burned per second is determined from the tons of coal burned each day:

$$\frac{\Delta m}{\Delta t} = \frac{(14{,}000 \text{ tons/day})(1000 \text{ kg/ton})}{(24 \text{ h/1 day})(3600 \text{ s/1 h})} = 162 \text{ kg/s}$$

Thus, the rate of thermal energy transfer by burning coal (the hot reservoir) to the working substance is

$$\frac{Q_{\text{H to Sub}}}{\Delta t} = (162 \text{ kg/s})(24 \times 10^6 \text{ J/kg}) = 3.9 \times 10^9 \text{ J/s}$$

(c) The actual efficiency is the useful output divided by the input:

$$\frac{W_{\text{Sub on Env}}/\Delta t}{Q_{\text{H to Sub}}/\Delta t} = \frac{9.1 \times 10^8 \text{ J/s}}{3.9 \times 10^9 \text{ J/s}} = 0.23$$

The plant is only operating at half the maximum possible efficiency allowed by the second law of thermodynamics.

**Try it yourself:** Determine the energy transfer rate from the working substance to the river water in the above power plant.

*Answer:* $\left(\frac{Q_{\text{Sub to C}}}{\Delta t}\right) = \left(\frac{Q_{\text{H to Sub}}}{\Delta t}\right) - \left(\frac{W_{\text{Sub on Env}}}{\Delta t}\right)$
$= 3.9 \times 10^9 \text{ J/s} - 9.1 \times 10^8 \text{ J/s} = 3.0 \times 10^9 \text{ J/s}$

The plant is transferring more energy into the environment as disorganized thermal energy than organized electric energy.

**Review Question 13.5** A thermodynamic engine does useful work equal to only 30% or so of the energy supplied to the engine. The rest is transferred to the environment as useless thermal energy. What are the reasons for this inefficiency?

# Summary

| Words | Pictorial and physical representations | Mathematical representation |
|---|---|---|

**Statistical definition of entropy** The entropy $S$ of an energy state is a measure of the probability of that state occurring. The greater the entropy, the less useful the energy in the system is for doing work. The entropy depends on the count $W_i$ of a state. (Section 13.2)

| Microstate arrangements | | Count $W_i$ | Entropy $S_i$ |
|---|---|---|---|
| 1 | 2 3 4 | 4 | 1.4 k |
| 2 | 1 3 4 | | |
| 3 | 1 2 4 | | |
| 4 | 1 2 3 | | |

$$W_i = \frac{n!}{n_l! n_r!} \quad \text{Eq. (13.1)}$$

$$S_i = k \ln W_i \quad \text{Eq. (13.2)}$$

where

$$k = 1.38 \times 10^{23}\ \text{J/K}$$

**Thermodynamic definition of entropy** A system's entropy changes by $\Delta S_{\text{Sys}}$ when energy is transferred through heating $Q_{\text{Env to Sys}}$ from the environment while the system is at a temperature $T_{\text{Sys}}$. (Section 13.3)

$$\Delta S_{\text{Sys}} = \frac{Q_{\text{Env to Sys}}}{T_{\text{Sys}}} \quad \text{Eq. (13.3)}$$

where $T_{\text{Sys}}$ is in kelvins.

**Second law of thermodynamics** The entropy of any system plus its environment increases during any irreversible process. The increase in entropy causes energy to be transferred into more probable forms that have less ability to do useful work. (Sections 13.2; 13.3)

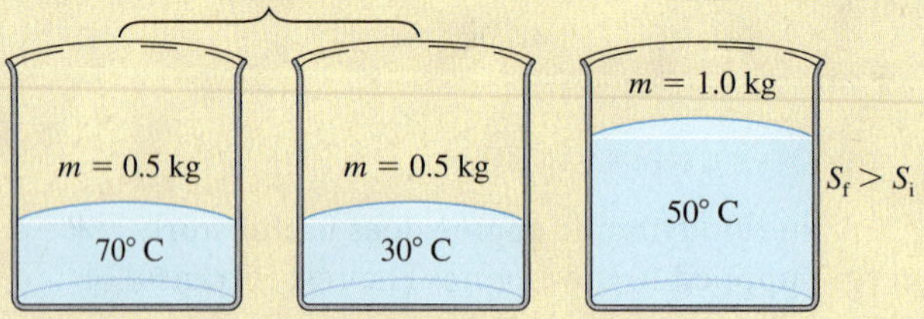


During any irreversible process

$$\Delta S_{\text{Sys}} + \Delta S_{\text{Env}} > 0.$$

**Thermodynamic engines and pumps** Energy transferred through heating $Q_{\text{H to Sub}}$ from a hot reservoir at temperature $T_{\text{H}}$ to a working substance allows the substance to do work $W_{\text{Sub on Env}}$ on the environment. Energy is then transferred through heating from the working substance to a cold reservoir $Q_{\text{Sub to C}}$ at temperature $T_{\text{C}}$. During one cycle of operation, the working substance's internal energy does not change $\Delta U_{\text{Sub}} = 0$. The maximum efficiency of this process is $e_{\text{max}}$. A thermodynamic pump is the reverse of this process. (Section 13.6)

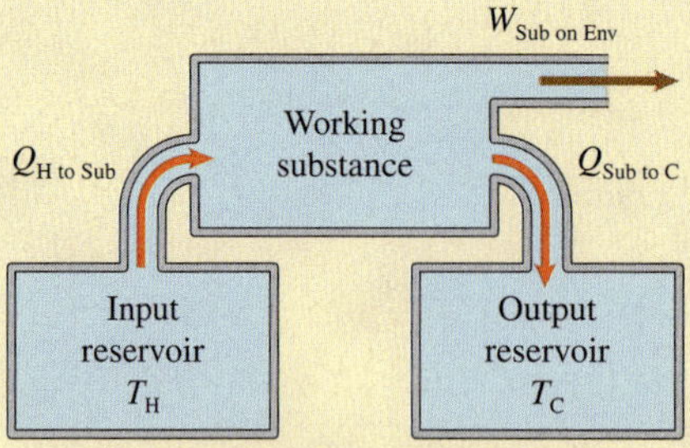


Thermodynamic engine

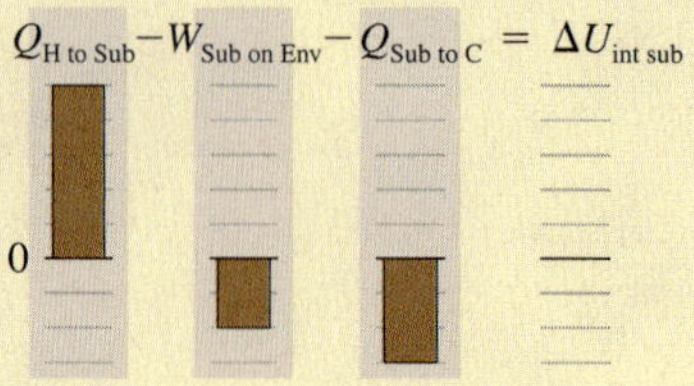


$$Q_{\text{H to Sub}} - W_{\text{Sub on Env}} - Q_{\text{Sub to C}} = 0 \quad \text{Eq. (13.4)}$$

All quantities are positive.

$$e = \frac{W_{\text{Sub on Env}}}{Q_{\text{H to Sub}}} \quad \text{Eq. (13.5)}$$

$$e_{\text{max}} = \frac{T_{\text{H}} - T_{\text{C}}}{T_{\text{H}}} \quad \text{Eq. (13.6)}$$

where $T$ is in kelvins.

$$Q_{\text{C to Sub}} + W_{\text{Env on Sub}} - Q_{\text{Sub to H}} = 0$$

*Refrigeration:*

$$K = \frac{Q_{\text{C to Sub}}}{W_{\text{Env on Sub}}} \leq \frac{T_{\text{C}}}{T_{\text{H}} - T_{\text{C}}} \quad \text{Eq. (13.8; 13.9)}$$

*Home heating:*

$$K = \frac{Q_{\text{Sub to H}}}{W_{\text{Env on Sub}}} \leq \frac{T_{\text{H}}}{T_{\text{H}} - T_{\text{C}}} \quad \text{Eq. (13.10)}$$

MP® For instructor-assigned homework, go to **MasteringPhysics.**

# Questions

## Multiple Choice Questions

1. Which of the following processes is reversible?
   (a) The room temperature collision of two atoms
   (b) The collision of two cars
   (c) The collision of a pool cue with a pool ball
   (d) The collision of two billiard balls
2. In physics the collision of billiard balls is usually considered to be elastic. What *observable* reason could you have for not choosing the collision of billiard balls as an example of a reversible process?
   (a) You hear sound produced by the collision.
   (b) There is change in the internal energy of the billiard balls.
   (c) All macroscopic processes are irreversible.
3. Why can't all of the thermal energy in a system be converted into mechanical work done on an environment?
   (a) The environment would do equal negative work on the system.
   (b) Energy would not be conserved.
   (c) The disordered energy in the system cannot be completely converted into ordered energy that can do work.
4. Which form of energy listed below is the most useful?
   (a) 1000 J of thermal energy in very hot gas in a cylinder closed by a moveable piston
   (b) 1000 J of thermal energy in cold gas in a cylinder closed by a moveable piston
   (c) 1000 J of thermal energy in water at 15 °C in the rocks at the bottom of a waterfall
   (d) 1000 J of thermal energy in the water in a boiler at 700 °C
   (e) 1000 J of chemical energy stored in the chemical bonds of a piece of wood
5. When driving a car (the system), what object does work on the car to make it move at constant speed in the presence of air resistance and rolling friction?
   (a) The engine
   (b) The surface of the road
   (c) The fuel in the car's tank
6. The law of energy conservation says that energy is always conserved; it can only be transferred from one system to another or converted from one form to another. Why then do environmentalists argue for energy conservation?
   (a) Because they are talking about isolated systems
   (b) Because the environmentalists mean the energy that is useful for doing work
   (c) Because when a car burns fuel, its energy is used up
7. Choose the best slogan(s) from the physics point of view for protecting energy resources. Give the reasons for your choice.
   (a) Conserve energy.
   (b) Reduce entropy increase.
   (c) Conserve useful energy.
8. When the engine in a car overheats, you can help lower the temperature of the engine by doing what?
   (a) Turning the air conditioner on
   (b) Turning off all fanning mechanisms
   (c) Turning on the heater
9. If you leave a refrigerator door open for a long time, the temperature in the room does what?
   (a) Increases (b) Decreases (c) Stays the same
10. Entropy can be calculated using which of the following expressions?
   (a) $Q/T$ (b) $10^6$ (c) $\Delta Q/T$
   (d) None of these
11. In what way does human life contradict the second law of thermodynamics?
   (a) The law does not apply to living organisms.
   (b) The human body is an isolated system.
   (c) When a human body interacts with the environment, the entropy of the body decreases.
   (d) There is no contradiction.
12. When a drop of ink enters a glass of water and spreads out, the entropy of the ink-water system does what?
   (a) Increases (b) Decreases (c) Stays the same
13. Choose the best reason for why the following statement is incorrect: When you clean your room and decrease its disorder, you violate the second law of thermodynamics.
   (a) The room is an isolated system. Entropy does not change.
   (b) Rooms are macroscopic objects; the second law applies only to molecules.
   (c) The room is not an isolated system. Your work leads to considerable entropy increase.

## Conceptual Questions

14. Describe five everyday examples of processes that involve increases in entropy. Be sure to state all parts of the system and environment involved in these entropy increases.
15. A cup of hot coffee sits on your desk. Assume that some of the thermal energy in the coffee can be converted to gravitational potential energy of the coffee-Earth system—enough so that the coffee rises up above the rim of the cup. Estimate the drop in temperature needed to accomplish this energy conversion. Why does this conversion not occur?
16. In terms of the statistical definition of entropy, why is there a good chance that the entropy due to the distribution of five atoms in a box can decrease, whereas the chance of an entropy decrease for redistribution of $10^6$ atoms in a box is negligible?
17. The entropy of the molecules that form leaves on a tree decreases in the spring of each year. Is this a contradiction of the second law of thermodynamics? Explain.
18. Give three examples of processes, other than those described in this book, in which potentially useful forms of energy are converted to less useful forms of energy.

# Problems

Below, **BIO** indicates a problem with a biological or medical focus. Problems labeled **EST** ask you to estimate the answer to a quantitative problem rather than develop a specific answer. Problems marked with require you to make a drawing or graph as part of your solution. Asterisks indicate the level of difficulty of the problem. Problems with no * are considered to be the least difficult. A single * marks moderately difficult problems. Two ** indicate more difficult problems.

## 13.1 Irreversible processes

1. * **BIO** **Types of energy and reversibility of a process** Describe the types of energy that change, the work done on the system, and the energy transferred through heating during the following processes. Indicate whether a reverse process can occur. (a) Water at the top of Niagara Falls cascades onto the blades of an electric generator near the bottom of the falls, rotating the blades and generating an electric current that causes a lightbulb to glow. The water, generator, lightbulb, and Earth are the system. (b) Each second, your body converts 100 J of metabolic energy (converting complex molecules from food) to thermal energy transferred to the air surrounding your body. The system is your body and the surrounding air. (c) The hot gas in a cylinder pushes a piston, which causes the blades of an electric generator to turn, which in turn causes a lightbulb to glow briefly. The system is the original hot gas (which cools while pushing the piston), the generator, and the lightbulb (which first glows and then stops glowing and cools down).
2. * For the following processes, choose the initial and final states and describe the process using the physical quantities internal energy, work, and heating. Explain why the process is irreversible. (a) A large foam ball is moving vertically up at speed $v$ and reaches a maximum height $h'$ somewhat less than $\sqrt{v^2/2g}$. The ball, Earth, and air are the system. (b) Two cups of water, one cold and the other hot, are mixed in an insulated bowl. The mixture reaches an intermediate temperature. The water in the cups is the system.
3. **Hourglass** An hourglass starts with all of the sand in the top bulb. During the next hour, the sand slowly leaks into the bottom bulb. Describe the energy changes in a system that includes the glass, sand, and Earth. Is this a reversible or irreversible process? Explain.
4. **Car hits tree** Your car slides on ice and runs into a tree, causing the front of the car to become slightly hotter and crumpled. The car and ice are the system. Indicate what object does the work on the system. Indicate whether heating occurs. Identify the types of energy that change. Are these quantities positive or negative? Explain.
5. **EST** **BIO** **Human metabolism** A 60-kg person consumes about 2000 kcal of food in one day. If 10% of this food energy is converted to thermal energy and cannot leave the body, estimate the temperature change of the person. [*Note:* 1 kcal = 4180 J.] Is this a reversible or irreversible process?

## 13.2 Statistical approach to irreversible processes

6. * (a) Identify all of the macrostate distributions for five atoms located in a box with two halves. (b) Determine the number of microstates for each macrostate. (c) Determine the entropy of each state.
7. * Repeat the previous problem for a system with six atoms.
8. * Determine the ratio of the number of microstates (count) of a system of eight atoms when in a macrostate with four atoms on the left side of a container and four on the right side and when in a macrostate with seven atoms on the left and one on the right. Which state has the greatest entropy? Explain.
9. * **Person lost on island** The probability that a lost person wandering about an island will be on the north part is one-half and on the south part is also one-half. (a) Determine the probability that three lost people wandering about independently will all be on the south half. (b) Repeat part (a) for the probability of six lost people all being on the south half of the island.
10. * **Parachutists landing on island** Parachutists have an equal chance of landing on the south half of a small island or the north half. If eight parachutists jump at one time, what is the ratio of the probability that six land on the north half and two on the south half to the probability that four land on each half?
11. * Determine the ratio of the counts of a system of 20 atoms when in a macrostate with 10 atoms on the left half of a box and 10 on the right half and when in a macrostate with 18 atoms on the left and 2 on the right. (b) Do the same for a system with 10 coins for the states with 5 coins on the left and 5 on the right compared to 9 coins on the left and 1 on the right. (c) When you compare your answers to parts (a) and (b), what do you infer about a similar ratio for a system with 10,000 atoms?
12. * Nine numbered balls are dropped randomly into three boxes. The numbers of balls falling into each box are labeled $n_1$, $n_2$, and $n_3$. (a) Identify five of the many possible arrangements or macrostates of the balls. (b) Determine the ratio of the count for the equal distribution ($n_1 = 3$, $n_2 = 3$, and $n_3 = 3$) and for the 0, 0, 9 distribution. (c) Determine the ratio of the count for the equal distribution and for the 2, 3, 4 distribution. [*Note:* The count is given by

$$W = \frac{n!}{n_1!\,n_2!\,n_3!},$$

where $n$ is the total number of balls.]
13. ** **Rolling dice** Two dice are rolled. Macrostates of these dice are distinguished by the total number for each roll (that is, 2, 3, 4, . . . , 12). (a) Determine the number of microstates for each macrostate. For example, there are three microstates for macrostate 4: (2, 2), (3, 1), and (1, 3). (b) What is the macrostate with greatest entropy? (c) What is the macrostate with least entropy? Explain.
14. * (a) Apply your knowledge of probability to explain why a drop of food coloring in a glass of clear water spreads out so that all of the water has an even color after some time. (b) Discuss whether after the food coloring spreads evenly in a glass of clear water it could condense back to an original droplet.
15. Explain using your knowledge of probability why a gas always occupies the entire volume of its container.

## 13.3 Connecting the statistical and macroscopic approaches to irreversible processes

16. * **EST** Estimate the total change in entropy of two containers of water. One container holds 0.1 kg of water at 70 °C and is warmed to 90 °C by heating from contact with the other container. The other container, also holding 0.1 kg of water, cools from 30 °C to 10 °C. Is this energy transfer process allowed by the first law of thermodynamics? By the second?

17. * **EST** (a) You add 0.1 kg of water at 0 °C to 0.3 kg of iced tea at 70 °C. Determine the final temperature of the mixture after it reaches equilibrium. The specific heat of iced tea is the same as water. (b) Estimate the entropy change of this system during this process. Is it allowed by the second law of thermodynamics? Explain.

18. * **Entropy change of a house** A house at 20 °C transfers $1.0 \times 10^5$ J of thermal energy to the outside air, which has a temperature of −15 °C. Determine the entropy change of the house-outside air system. Is this process allowed by the second law of thermodynamics?

19. * **Barrel of water in cellar in winter** A barrel containing 200 kg of water sits in a cellar during the winter. On a cold day, the water freezes, releasing thermal energy to the room. This energy passes from the cellar to the outside air, which has a temperature of −20 °C. Determine the entropy change for this process if the cellar remains at 0 °C.

20. * **EST** (a) Determine the final temperature when 0.1 kg of water at 10 °C is added to 0.3 kg of soup at 50 °C. What assumptions did you make? (b) Estimate the entropy change of this water-soup system during the process. Does the second law of thermodynamics allow this process?

21. * A 5.0-kg block slides on a level surface and stops because of friction. Its initial speed is 10 m/s and the temperature of the surface is 20 °C. Determine the entropy change of the block, which is the system in this process.

22. ** A 5.0-kg block slides from an initial speed of 8.0 m/s to a final speed of zero. It travels 12 m down a plane inclined at 15° with the horizontal. Determine the entropy change of the block-inclined plane-Earth system for this process if originally the block and the inclined plane were at 27 °C. Why did the block stop?

## 13.4 Thermodynamic engines and pumps

23. **Maximum efficiencies** Determine the maximum efficiencies of the thermodynamic engines described below. (a) Burning coal heats the gas in the turbine of an electric power plant to 700 K. After turning the blades of the generator, the gas is cooled in cooling towers to 350 K. (b) An inventor claims to have a thermodynamic engine that attaches to a car's exhaust system. The temperature of the exhaust gas is 90 °C and the temperature of the output of this proposed heat engine is 20 °C. (c) Near Bermuda, ocean water is about 24 °C at the surface and about 10 °C at a depth of 800 m.

24. * **BIO** **Efficiency of woman walking** A 60-kg woman walking on level ground at 1 m/s metabolizes energy at a rate of 230 W. When she walks up a 5° incline at the same speed, her metabolic rate increases to 370 W. Determine her efficiency at converting chemical energy into gravitational potential energy.

25. * **Nuclear power plant** A nuclear power plant operates between a high-temperature heat reservoir at 560 °C and a low-temperature stream at 20 °C. (a) Determine the maximum possible efficiency of this thermodynamic engine. (b) Determine the heating rate (J/s) from the high-temperature heat reservoir to the power plant so that it produces 1000 MW of power (work/time).

**Figure P13.26**

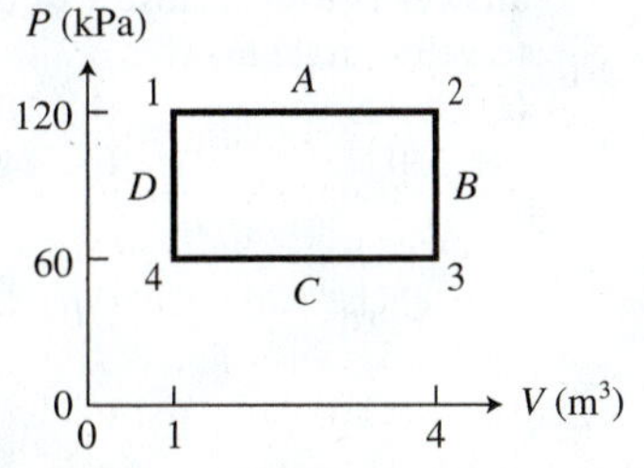


26. ** A cyclic process involving 1 mole of ideal gas is shown in **Figure P13.26.** (a) Determine the work done on the gas by the environment during each step (*A*, *B*, *C*, and *D*) of the cycle. (b) Determine the net work done on the gas $W_{\text{Env on Gas}}$. (c) Find the work that the gas does on the environment; it is equal to $W_{\text{Gas on Env}} = -W_{\text{Env on Gas}}$. (d) Use the ideal gas law to determine the temperature of the gas at each corner of the process (1, 2, 3, and 4). (e) Use the temperatures found in (d) to determine the thermal energy of the gas at each corner of the process and the change in internal thermal energy during each step of the process. (f) Use the results recorded in the previous parts and the first law of thermodynamics to determine the heating during each step of the process. (g) Determine the efficiency of the process.

**Figure P13.27**

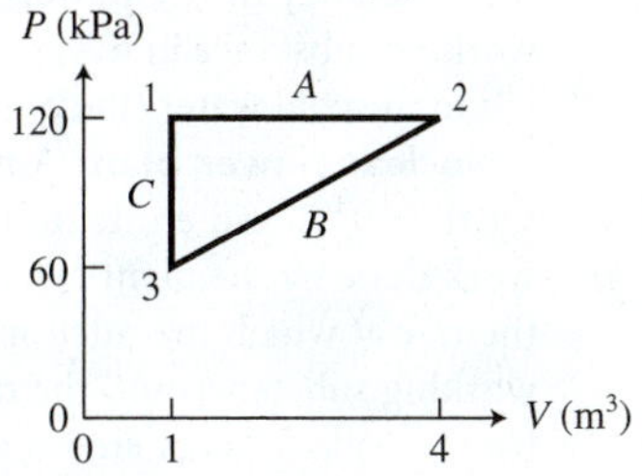


27. ** A cyclic process involving 1 mole of ideal gas is shown in **Figure P13.27** (a) Determine the work done on the gas by the environment during each step (*A*, *B*, and *C*) of the cycle. (b) Determine the net work done on the gas $W_{\text{Env on Gas}}$. (c) Find the work that the gas does on the environment equal to $W_{\text{Gas on Env}} = -W_{\text{Env on Gas}}$. (d) Use the ideal gas law to determine the temperature of the gas at each corner of the process (1, 2, and 3). (e) Use the temperatures found in (d) to determine the thermal energy of the gas at each corner of the process and the change in internal thermal energy during each step of the process. (f) Use the results recorded in the previous parts and the first law of thermodynamics to determine the heating during each step of the process. (g) Determine the efficiency of the process.

28. * **Home thermodynamic pump** A heat pump collects thermal energy from outside air at 5 °C and delivers it into a house at 40 °C. (a) Determine the maximum coefficient of performance (the maximum coefficient is determined in the same way as for a thermodynamic engine). (b) If the motor of the heating pump uses 1000 J of electrical energy to do work during a certain time interval, how much thermal energy is delivered into the house through heating, assuming the heating pump works at the maximum coefficient of performance? (c) Repeat (b) for a coefficient of performance of 2.0.

29. ** **Ice-making machine** An ice-making machine needs to convert 0.20 kg of water at 0 °C to 0.20 kg of ice at 0 °C. The room temperature surrounding the ice machine is 20 °C. (a) How much thermal energy must be removed from the water? (b) Determine the minimum work needed to extract this energy by the ice-making machine. (c) How much energy is deposited in the room?

## 13.5 Automobile efficiency and power plants: Putting it all together

30. * **Automobile engine** An automobile engine has a power output for doing work of 150 kW (about 200 hp). The efficiency of the engine is 0.32. Determine the heating input per second to the engine by burning gasoline and the heating rate of the engine to the environment.

31. **Diesel car engine** A diesel engine in a car does 1000 J of work due to 2800 J of heating caused by the combustion of diesel fuel in its cylinders. Determine the efficiency of the engine and the thermal energy emitted by the engine to the environment.

32. ** **Gas used in car's engine** During each cycle of an automobile's gasoline engine operation, the gasoline burned

provides 12,000 J of energy through heating to the engine and the engine does 3600 J of work. Gasoline provides $4.4 \times 10^7$ J of energy for each kilogram of gasoline burned. (a) Determine the efficiency of the engine. (b) Determine the thermal energy exhausted from the engine during each cycle. (c) Determine the mass of the gasoline burned during each cycle. (d) If the engine has 80 cycles/s, how much gasoline does the engine use in 1 h in kilograms and in gallons? The density of gasoline is $737\ \text{kg/m}^3$.

33. **Nuclear power plant** A nuclear power plant does useful work generating electric energy at a rate of 500 MW. The energy transfer rate to the electric generator from the high-temperature nuclear fuel (the hot reservoir) is 1200 MW. Determine the efficiency of the power plant and the rate at which the working substance in the plant transfers energy through heating to the cold water (the cool reservoir).

34. * **Nuclear power plant** A nuclear power plant warms water to 500 °C and emits it at 100 °C. You want to get useful work done by the plant at a rate of $1.0 \times 10^9$ J/s. Determine the rate at which the nuclear fuel must provide energy to the working substance and the rate of thermal energy exhausted from the plant to the environment.

## General Problems

35. ** **EST BIO Body efficiency—experiment design** Describe an experiment that can be performed to estimate the average efficiency of a human body.

36. ** The following equations represent the four parts (A, B, C, and D) of a cyclic process with a gas. In this case we consider the work done by the system on the environment and consequently $Q - W = \Delta U_{\text{int}}$.

$$W = (3.0 \times 10^5\ \text{N/m}^2)(0.020\ \text{m}^3 - 0.010\ \text{m}^3) + 0 + (1.0 \times 10^5\ \text{N/m}^2)(0.010\ \text{m}^3 - 0.020\ \text{m}^3) + 0$$

$$\Delta U_{\text{int}} = (3/2)(1.0\ \text{mole})(8.3\ \text{J/mole} \cdot \text{K})[(700\ \text{K} - 360\ \text{K}) + (480\ \text{K} - 700\ \text{K}) + (240\ \text{K} - 480\ \text{K}) + (360\ \text{K} - 240\ \text{K})]$$

$$Q = Q_A + Q_B + Q_C + Q_D$$

(a) Draw a $P$-versus-$V$ graph for the process with labeled axes (including a scale).
(b) Determine the net change in the internal energy during the entire cycle.
(c) Determine the heating of the system for each of the four parts of the process.

## Reading Passage Problem

**Fuel used to counter air resistance** The resistive drag force that the air exerts on an automobile is

$$F_{\text{Air on Car}} = (1/2)CA\rho v^2$$

where $C$ is a drag coefficient that depends on how streamlined the car is, $\rho$ is the density of the air, $A$ is the cross-sectional area of the car along its line of motion, and $v$ is the car's speed relative to the air. Consider the effect of air resistance on fuel consumption in a typical car during a 160 km (100 mi) trip while traveling at 22 m/s (50 mi/h). The resistive force exerted by the air on the car is about 200 N. Thermal energy produced due to this resistive force during the trip is

$$\Delta U_{\text{thermal}} = F_{\text{Air on Car}}d = (200\ \text{N})(1.6 \times 10^5\ \text{m}) = 3.2 \times 10^7\ \text{J}$$

One liter of gas releases about $3.25 \times 10^7$ J of energy. So the gasoline energy that is equivalent to the resistive thermal energy produced during this trip is

$$(3.2 \times 10^7\ \text{J})/(3.25 \times 10^7\ \text{J/L}) = 0.98\ \text{L} = 0.26\ \text{gallons}$$

In summary, you use about one-quarter gallon of gasoline to counter air resistance during the 160 km trip. Unfortunately, the car is only about 13% efficient (see Figure 13.9), where

$$\text{Efficiency} = \frac{\text{useful energy output}}{\text{energy input}} = 0.13$$

Consequently, your useful output of 0.98 liters requires an energy input of (0.98 L)/0.13 = 7.5 L of gas to overcome this air resistance. This is about half of the gasoline a typical car consumes when driving at a steady speed for 160 km. The other half is needed to overcome rolling resistance.

37. Assuming a 200-N drag force when traveling at 22 m/s through air of density $1.3\ \text{kg/m}^3$, what is the closest value to the product $CA$ in the drag force equation for the vehicle?
(a) $0.50\ \text{m}^2$ (b) $0.62\ \text{m}^2$ (c) $0.86\ \text{m}^2$
(d) $1.1\ \text{m}^2$ (e) $1.5\ \text{m}^2$

38. At 22 m/s the magnitude of the resistive force that air exerts on the car is about 200 N. Which answer below is closest to the magnitude of the drag force when the car is traveling at 31 m/s?
(a) 100 N (b) 140 N (c) 280 N
(d) 400 N (e) 520 N

39. The amount of fuel used to counter air resistance should do what?
(a) Increase in proportion to the speed squared
(b) Increase in proportion to the speed
(c) Be the same independent of the speed
(d) Decrease in proportion to the inverse of the speed
(e) Decrease in proportion to the inverse of the speed squared

40. The 200-N resistive force of the air in this problem is closest to which answer below?
(a) 800 lb (b) 90 lb (c) 60 lb
(d) 45 lb (e) 30 lb

41. Why is the resistive force of the air on a Hummer H3 traveling at 22 m/s about 600 N instead of 200 N?
(a) The Hummer has a bulky, less streamlined shape.
(b) The Hummer has a greater cross-sectional area along the line of motion.
(c) The Hummer has greater mass.
(d) a and b
(e) a, b, and c

42. The value of $CA$ for a Ford Escape Hybrid is $1.08\ \text{m}^2$. Which answer below is closest to the drag force on this car when traveling at 22 m/s?
(a) 130 N (b) 180 N (c) 270 N
(d) 350 N (e) 440 N

# Electric Charge, Force, and Energy

# 14

How does a photocopier work?

Why is your hair attracted to a plastic comb after combing?

How is a metal soda can different from a plastic bottle on a microscopic level?

When you make a photocopy, you place a document on the glass surface of the copy machine, close the lid, and push the start button. Seconds later, a copy appears on a new sheet of paper. The toner inside the copier reproduced the original image on the new sheet. How does the toner stick to the paper, and how does the toner "know" where to stick?

**Be sure you know how to:**

- Identify a system and construct a force diagram for it (Section 2.1).
- Identify a system and construct an energy bar chart for it (Section 6.2).
- Convert a force diagram and an energy bar chart into a mathematical statement (Sections 2.3 and Section 6.6).

**Figure 14.1** Rubbing different materials causes an interaction between them.

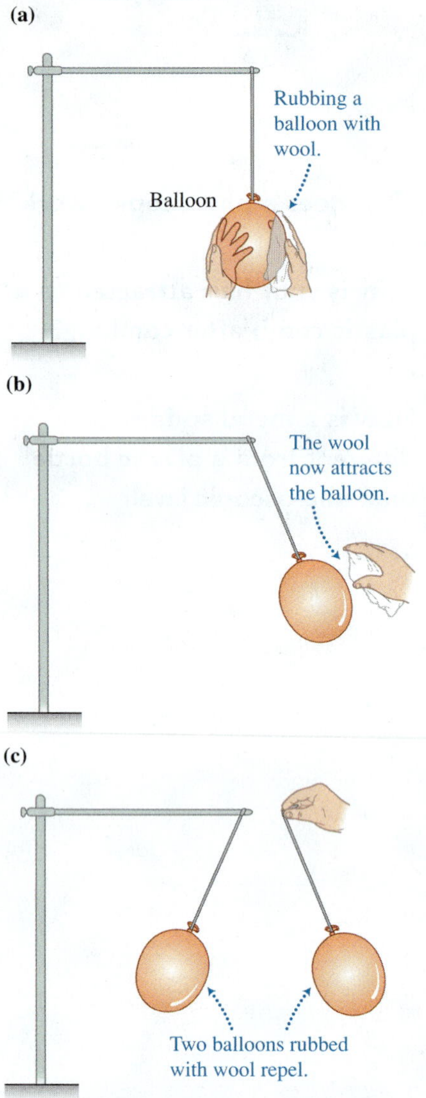


The drum inside a copy machine attracts small particles of toner powder through the same mechanism by which a rubbed balloon sticks to your hair. Then a strong beam of light illuminates the original paper. This light reflects off blank parts of the paper but does not reflect off dark regions containing text or pictures. The light reflected from blank regions of the paper hits the drum and causes the powder to be removed. Powder is left behind in places corresponding to the text and pictures. The drum then turns and presses against the copy paper, transferring the desired image. Increasing the temperature bakes the powder so that it sticks to the new page.

**In Chapters 9–13,** we learned that all objects are made of tiny particles. What holds these particles together? For that matter, what holds the particles that make up the paper in this textbook together? The interaction responsible for holding the particles together is the same interaction that makes the toner stick to a copier drum and a balloon stick to your hair, as well as many other phenomena that we observe in our everyday world.

## 14.1 Electrostatic interactions

If you rub a balloon with a wool sweater (**Figure 14.1a**) and then bring the sweater near the balloon, the sweater attracts the balloon (Figure 14.1b). If you bring a second balloon rubbed the same way near the first balloon, they repel each other (Figure 14.1c). Let's investigate this phenomenon in detail, in Observational Experiment **Table 14.1**.

**OBSERVATIONAL EXPERIMENT TABLE**

**14.1 Experimenting with rubbed objects.**

VIDEO 14.1

| Observational experiment | Analysis |
|---|---|
| **Experiment 1.** Hang two small balloons from two thin strings. Rub each balloon with felt. The balloons repel. Bringing the balloons closer to each other increases the angle between the threads. 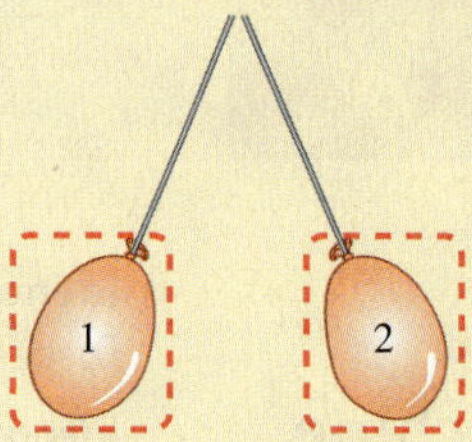  | The balloons must exert a repulsive force on each other. This unknown force increases in magnitude as the balloons get closer to each other. 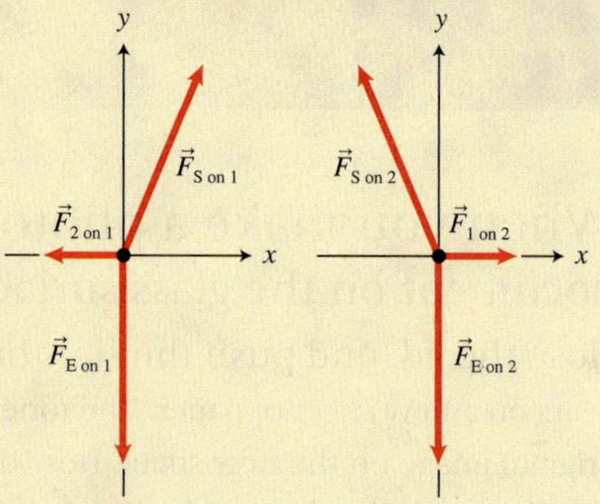  |

**Experiment 2.** Now, bring the felt used to rub the balloons close to one of the rubbed balloons. The balloon is attracted to the felt. The attraction strengthens as the felt is held closer.

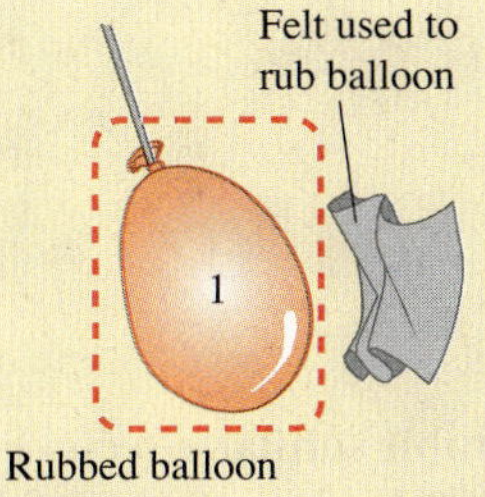


The felt exerts an unknown attractive force on the rubbed balloon. The force strengthens as the felt gets closer to the balloon.

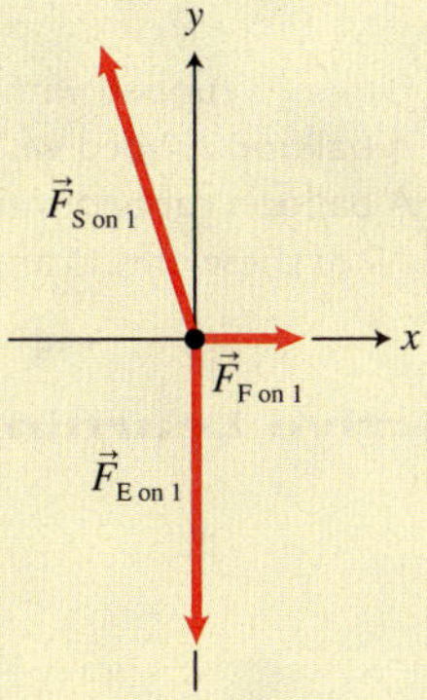


**Experiment 3.** Repeat Experiment 1, only this time rub two small balloons with plastic wrap. The balloons repel. Bringing the balloons closer to each other increases the angle between the threads.

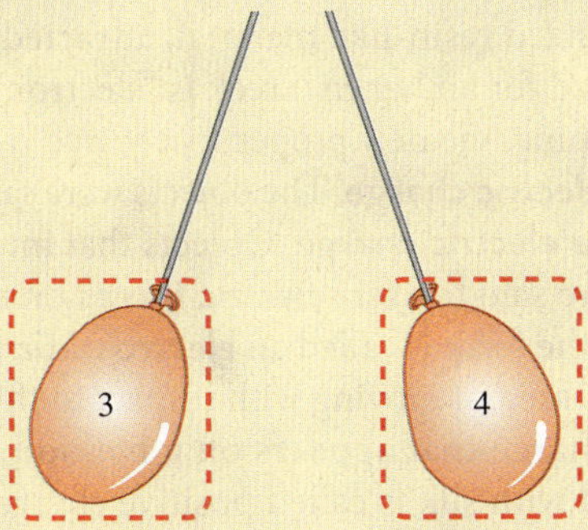


Each balloon exerts an unknown repulsive force on the other balloon. This force increases in magnitude as the balloons get closer to each other.

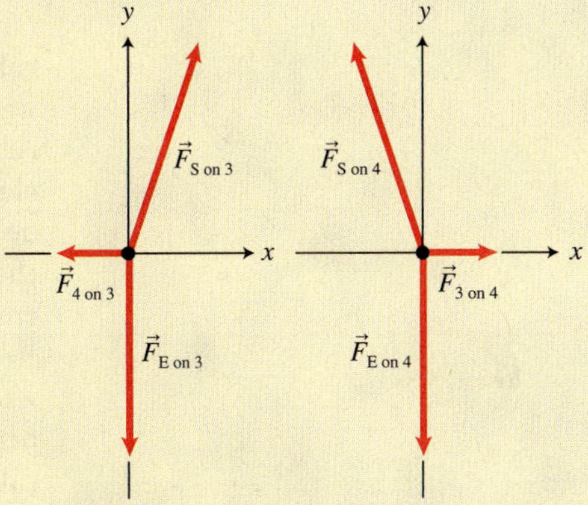


**Experiment 4.** Bring the plastic wrap close to one of the balloons that was rubbed with plastic wrap. The balloon is attracted to the plastic. The attraction is stronger as the plastic wrap is held closer.

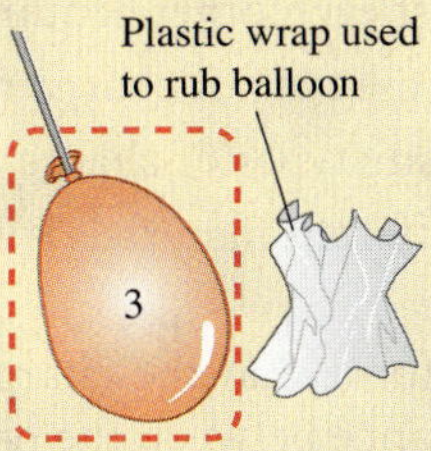


The plastic wrap exerts an unknown attractive force on the rubbed balloon. The force is stronger the closer the plastic wrap is to the balloon.

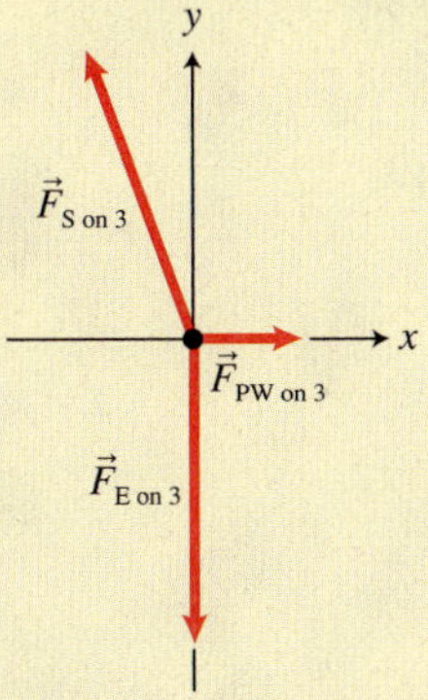


**Experiment 5.** Bring a balloon rubbed with felt near a balloon rubbed with plastic wrap. The balloons are attracted to each other. The attraction strengthens as the balloons are held closer.

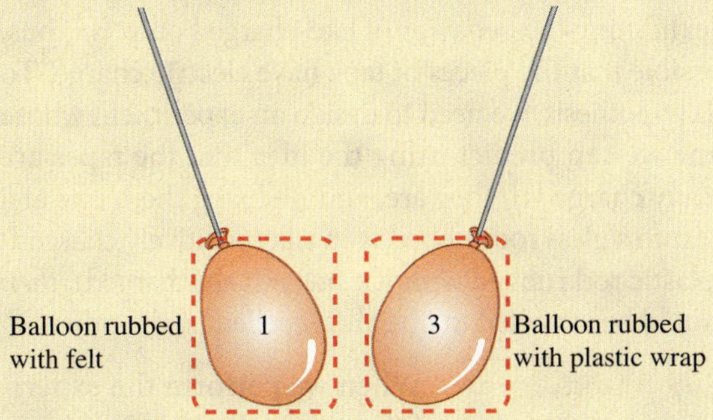


An unknown force pulls each balloon toward the other. This force increases in magnitude as the balloons get closer to each other.

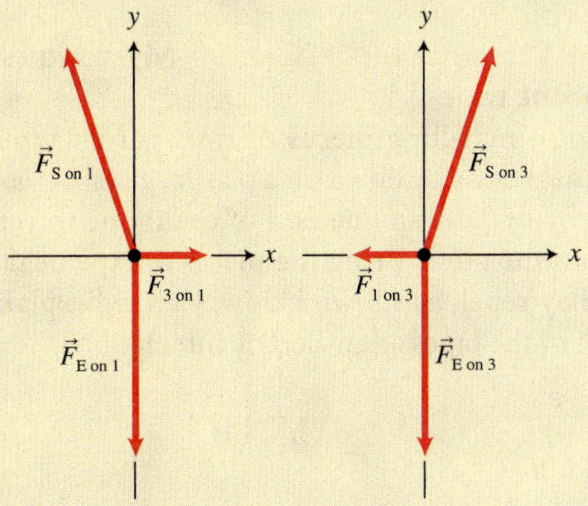


(*continued*)

**Patterns**

- Balloons rubbed with the same material repel each other.
- A balloon rubbed with a particular material is attracted to that material.
- A balloon rubbed with felt attracts a balloon rubbed with plastic wrap.
- All of these effects are stronger the closer the objects are to each other.

**Active Learning Guide ›**

Many other objects exhibit similar effects. For example, you can use plastic rods instead of balloons and rub them with felt (like Experiment 1), or glass rods and rub them with silk (like Experiment 3). In all of these experiments, we find that more vigorous rubbing increases the magnitude of the forces that the objects exert on each other.

Ancient people observed similar effects with different materials. The Greeks noticed that amber, a fossilized resin-like material, attracted particles after it was rubbed with wool. The word for amber in Greek is "electron"; thus this attraction was called electrical. Over time, the new property acquired by the materials due to rubbing came to be called **electric charge**. The objects were said to acquire **positive electric charge** or **negative electric charge**. Objects that interact with each other because they are charged are said to exert **electric forces** on each other. When the charged objects are at rest, the force is called an **electrostatic force;** the word *static* emphasizes that the objects are not moving with respect to the observer.

By convention, the charge that appears on a balloon rubbed with plastic wrap or a glass rod rubbed with silk is called positive charge; the charge that appears on a balloon rubbed with wool or felt or on a plastic rod rubbed with felt is called negative charge. With these ideas we can reinterpret the patterns that we found in the experiments in Observational Experiment Table 14.1:

1. Materials rubbed against each other acquire electric charge.
2. Two objects with the same type of charge repel each other.
3. Two objects with opposite types of charge attract each other.
4. Two objects made of different materials rubbed against each other acquire opposite charges.
5. The magnitude of the force that the charged objects exert on each other increases when the distance between the objects decreases.
6. Sometimes more vigorous rubbing leads to a greater force exerted by the rubbed objects on each other.

Remember, we have *not* observed electric charge in any of these experiments. We *have* observed attraction and repulsion and used the concept of electric charge to explain these observations. The model of two electric charges is sufficient to explain all of the experiments that we have done so far.

**CONCEPTUAL EXERCISE 14.1 Mysterious transparent tape**

Take two 9-inch-long pieces of transparent tape and place them sticky side down on a plastic, glass, or wooden tabletop. Now pull on one end of each tape to remove them from the table. Bring the pieces of tape near each other. They repel, as shown below. Can we explain the repulsion of the tapes as an electric interaction?

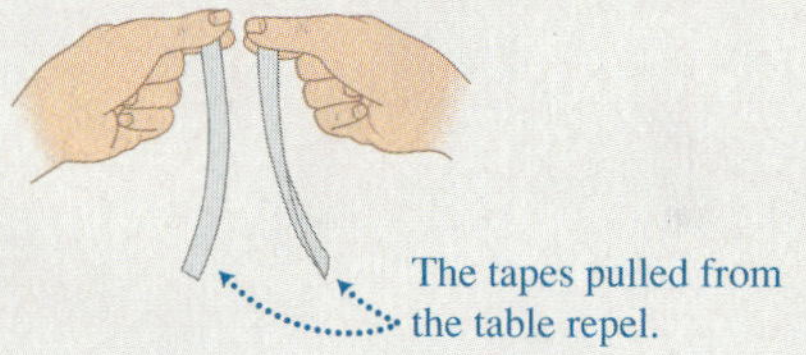
The tapes pulled from
the table repel.

**Sketch and translate** The pieces of tape repel each other, exhibiting the behavior of like-charged objects. Thus, it is possible that the pieces of tape have electric charge. To test this hypothesis, we need to design an experiment whose outcome we can predict using the idea that the tapes are electrically charged. *If* they are, and we bring them one at a time next to a glass rod rubbed with silk (positively charged) and a plastic rod rubbed with felt (negatively charged), *then* they should be attracted to one and repelled by the other.

**Simplify and diagram** When we perform the experiment, we see that a piece of tape is repelled by the plastic rod rubbed with felt and attracted to the glass rod rubbed with silk. Does it mean that the tapes are electrically charged? We cannot say for sure, but our experiment does not rule out the hypothesis that the pieces of tape are electrically charged.

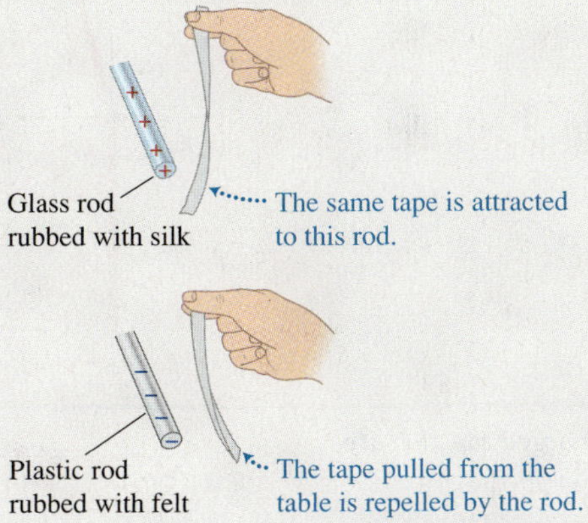


**Try it yourself:** Take two pieces of tape and fold the end of each piece over about one-half centimeter to make a "handle." Place one piece of tape on top of a table and then the second piece on top of the first. Then pull them together off the table and separate them by pulling the "handles" away from each other. The pieces of tape attract each other. Design an experiment to determine the sign of each of their charges.

*Answer:* According to our charge sign convention, a plastic rod rubbed with felt is negatively charged and a glass rod rubbed with silk is positively charged. Hold both rods near the two pieces of tape to observe repulsion or attraction—this will tell you the signs of the charges of the tape pieces.

## Charged objects attract uncharged objects

So far, we have learned that objects rubbed with different materials can have opposite charges and attract each other electrically. However, everyday observations show that uncharged lightweight objects, like small bits of paper that have not been rubbed against anything, are attracted to charged objects. When we bring a plastic comb that has been rubbed with felt or a synthetic sweater near small bits of paper, the bits of paper are attracted to the comb (**Figure 14.2**). Small pieces of aluminum foil are attracted to the comb even more readily than paper. Why are uncharged objects attracted to charged objects? We explore this question in Observational Experiment **Table 14.2.**

**Figure 14.2** Uncharged paper pieces are attracted to a charged comb.

**OBSERVATIONAL EXPERIMENT TABLE**

**14.2 Interactions of charged and uncharged objects.**

VIDEO 14.2

| Observational experiment | Analysis |
|---|---|
| **Experiment 1.** Hang a balloon from a string next to a wall (neither the balloon nor the wall are charged). The balloon hangs straight down. Next, hang a negatively charged balloon near the same wall. The negatively charged balloon is attracted to the wall.   | The uncharged wall exerts an attractive force on the negatively charged balloon. 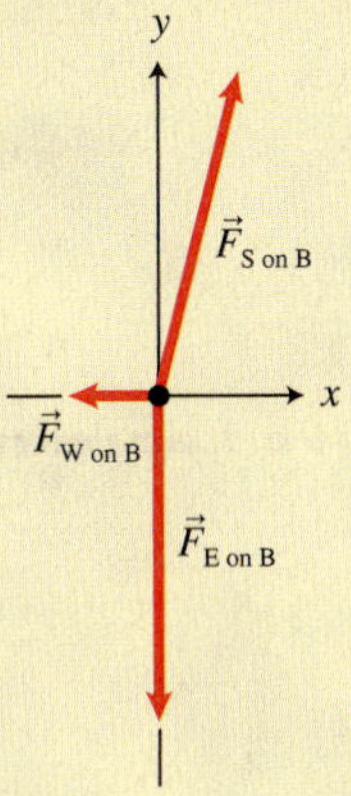  |

*(continued)*

| Observational experiment | Analysis |
|---|---|
| **Experiment 2.** Hang a positively charged balloon near the uncharged wall. The balloon is attracted to the wall. 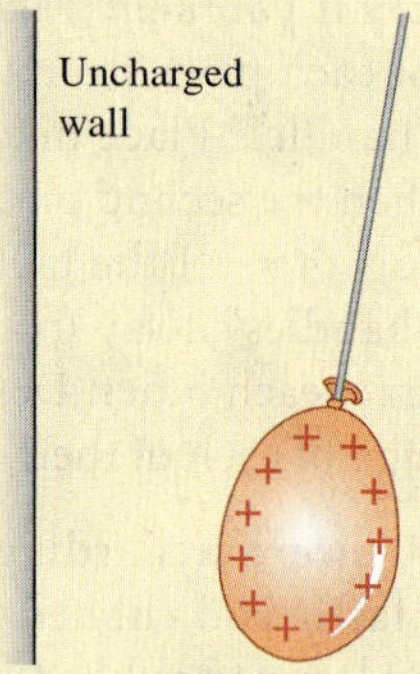  | The uncharged wall exerts an attractive force on the positively charged balloon. 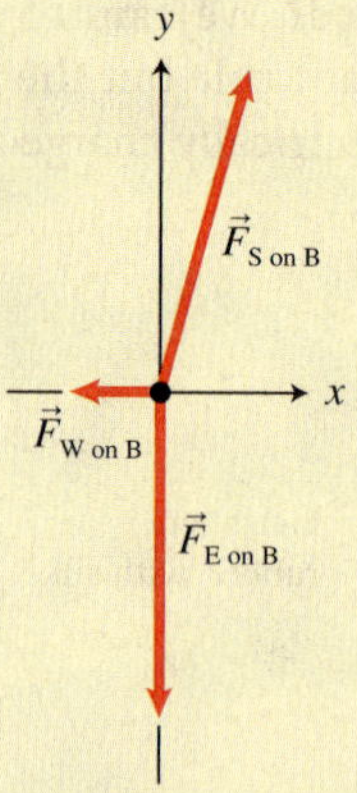  |
| **Experiment 3.** Repeat Experiment 1, only this time bring the negatively charged balloon near an uncharged metal plate. The balloon has a stronger attraction to the plate than to the wall. 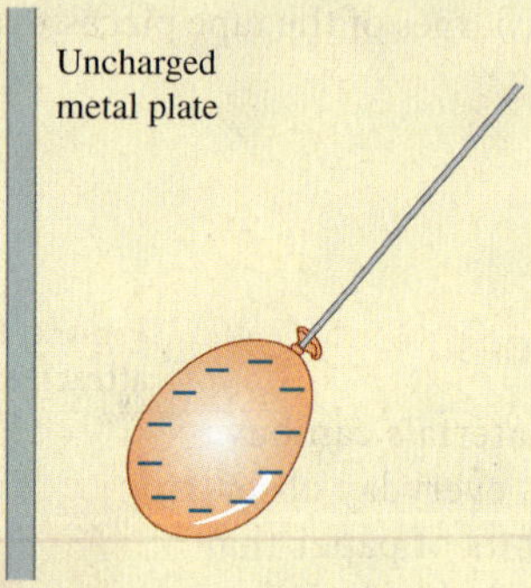  | The uncharged metal plate exerts a larger magnitude force on the negatively charged balloon than the wall in Experiment 1 does. 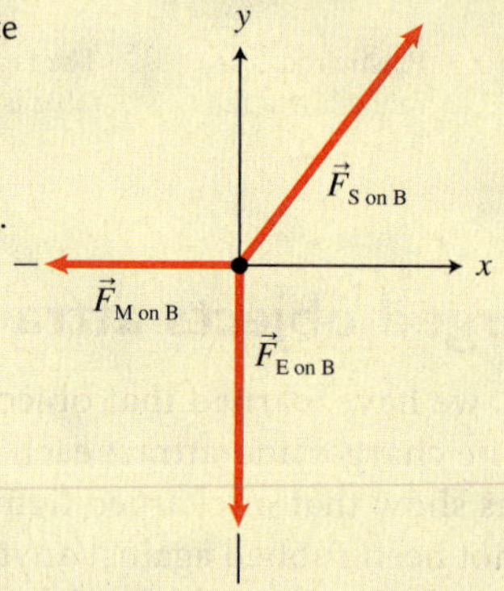  |
| **Experiment 4.** Repeat Experiment 2, only this time bring the positively charged balloon near an uncharged metal plate. The balloon has a stronger attraction to the plate than to the wall. 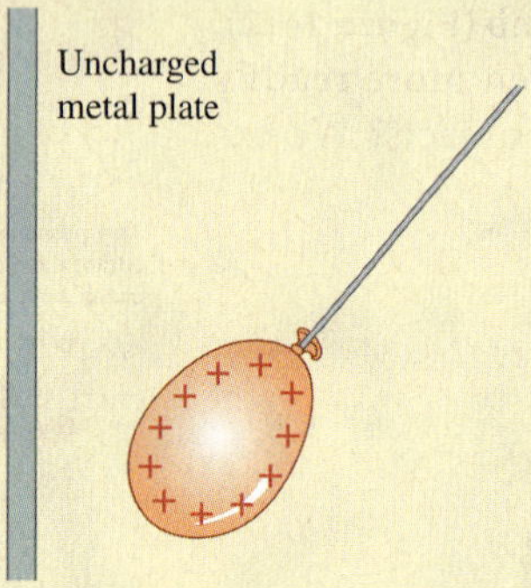  | The uncharged metal plate exerts a larger magnitude force on the positively charged balloon than the wall in Experiment 2 does. 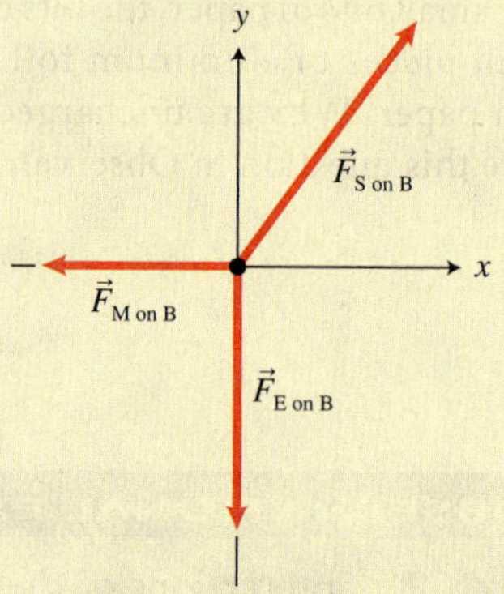  |
| **Pattern** | |
| The positively and negatively charged balloons were attracted to uncharged materials (the wall and the metal). The intensity of interaction was much stronger with the metal plate than with the nonmetal wall. | |

We see from the experiments in Table 14.2 that both positively and negatively charged objects are attracted to uncharged objects. We add a new statement to our conceptual model of the electric interaction:

**Active Learning Guide ›**

**Uncharged objects are attracted to both positively and negatively charged objects; uncharged metal objects are attracted more strongly than nonmetal objects.**

In the next section, we develop conceptual explanations for these and other observations.

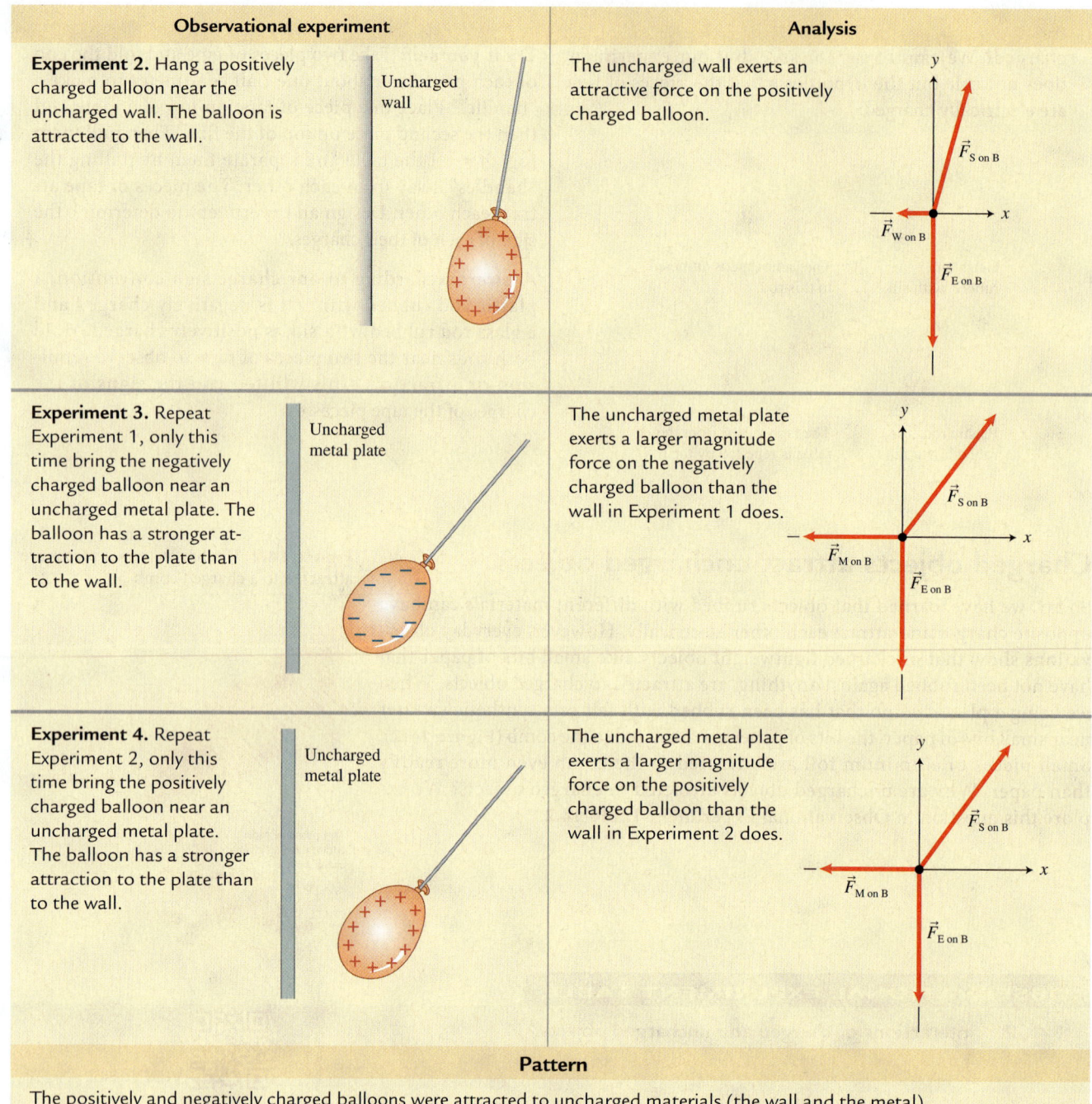

**Review Question 14.1** To decide whether an object is electrically charged, we need to observe its repulsion from some other object, not its attraction. Why is the attraction insufficient?

## 14.2 Explanations for electrostatic interactions

Electrically charged objects interact with each other: they repel or attract. They also attract uncharged objects. What is the mechanism behind this interaction?

### Is the electrical interaction just a magnetic interaction?

Do you recall observing any other interaction that involved repulsion? If you've ever played with magnets, you probably know that magnets have poles (north and south). Like poles repel, while unlike poles attract. Both poles attract some objects that are not magnets but contain iron—nails, paper clips, etc. Electrically charged objects also attract and repel each other and attract uncharged objects. Perhaps the electric interaction is actually the magnetic interaction, just described using different terminology. We test this hypothesis in Testing Experiment **Table 14.3**.

‹ **Active Learning Guide**

**TESTING EXPERIMENT TABLE**

**14.3 Testing the electric = magnetic interaction hypothesis.**

VIDEO 14.3

| Testing experiment | Prediction | Outcome |
|---|---|---|
| **Experiment 1.** Bring a negatively charged plastic rod near the north pole of a magnet. | If the electric interaction is the same as the magnetic interaction, then the negatively charged rod should either attract or repel the north pole of the magnet. | The negatively charged rod attracts the north pole. 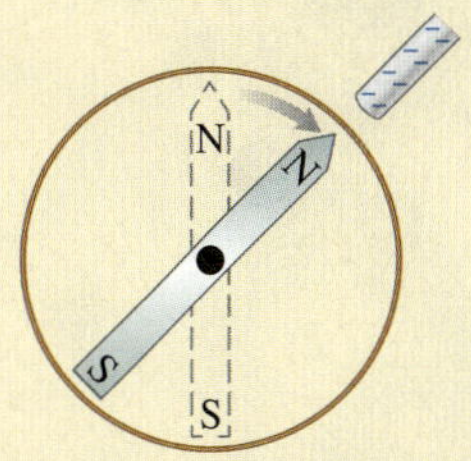  |
| **Experiment 2.** Bring the negatively charged plastic rod near the south pole of a magnet. | If electric interaction is the same as the magnetic interaction, then the negatively charged rod should repel the south pole of the magnet. | The negatively charged rod also attracts the south pole. 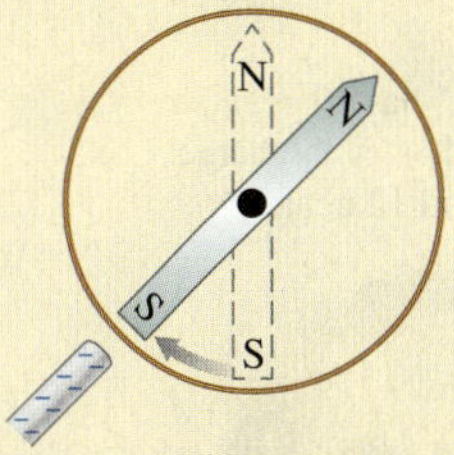  |

**Conclusion**

The outcome of the second experiment is inconsistent with the prediction. However, both outcomes can be explained if we assume that the magnet behaves similarly to an uncharged metal object with both poles attracting negatively and positively charged objects.

The experiments in Table 14.3 show that the same charged rod attracts both the north and the south poles of a magnet. This disproves our hypothesis that electric interactions are the same as the interactions of magnets. We need a different mechanism to explain electrostatic interactions.

## Fluid models of electric charge

Benjamin Franklin (1706–1790) proposed that a weightless electric fluid is present in all objects. According to Franklin, too much electric fluid in an object makes it positively charged and too little makes it negatively charged. If you rub two different types of material against each other, the electric fluid may move from one material to the other. The material that loses some electric fluid becomes negatively charged and the material that gains some electric fluid becomes positively charged. The fluid can also move inside the uncharged object when another charged object is present, making its sides positive or negative. This movement of the fluid inside explains why an uncharged object is attracted to a charged object.

In another model there are two types of electric fluids—positive and negative. In an object that is not rubbed, the two fluids are in balance. After rubbing, one object loses some positive fluid and becomes negative while the other one gains some positive fluid and becomes positive. Objects with an excess of the same fluid repel each other, and objects with an excess of different fluids attract. The fluids can move inside objects, making their sides charged even when the object overall is uncharged. Both of these fluid models account for all of our observations so far.

**Figure 14.3** A simplified model of Millikan and Fletcher's oil drop experiment.

The force exerted by the charged plates P on the drop D balances the gravitational force exerted by Earth on the drop. As a result, the drop falls at constant velocity.

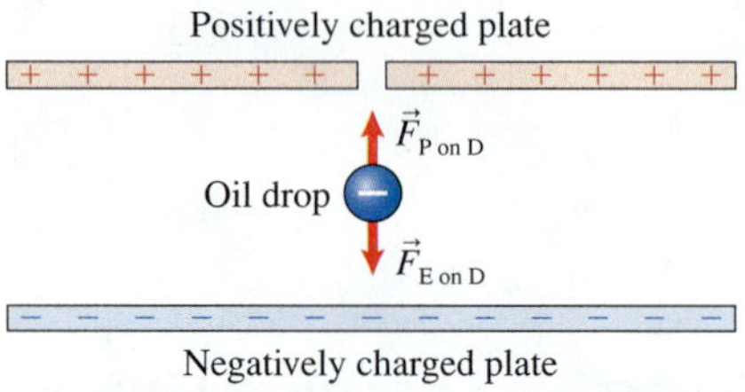


**Figure 14.4** A simplified model of (a) a carbon atom and (b) a sodium ion. Protons (+) and neutrons (which have no charge) are massive particles in the nucleus. Electrons (−) have very small masses compared to protons.

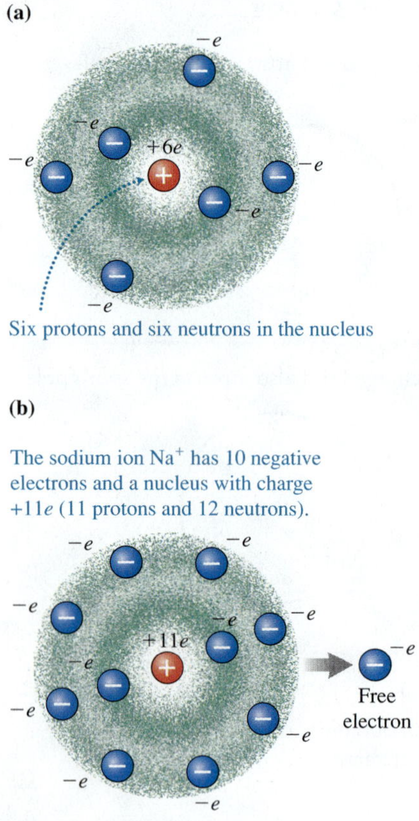


## Particle model of electric charge

In 1897, J.J. Thomson, explaining the results of his experiments with cathode rays, proposed that electric charge was not associated with a weightless fluid but was carried by the particles that comprise matter. In 1909, Robert Millikan and his graduate student Harvey Fletcher tested this hypothesis. They placed tiny droplets of oil between two oppositely charged horizontal metal plates. As the tiny droplets of oil moved between the plates, they picked up ions or electrons from the air, which had been ionized with X-rays. The drops drifted up or down, depending on the charge on the plates. Millikan and Fletcher adjusted the electric charge on the plates so that the drop velocity was constant because of the balance of three forces: viscous drag, the electrical force, and Earth's gravitational force. If the magnitude of the charge of the droplets changed, the researchers would adjust the charge of the plates to maintain the steady speed of the droplets (**Figure 14.3**).

Thousands of measurements demonstrated that the electric charge of each oil droplet changed only by multiples of some discrete unit of electric charge. This pattern indicated to Millikan and Fletcher that electric charge was carried by microscopic particles. Removing or adding these particles to an object changed the charge of the object. Each charged particle had a charge equal to some smallest indivisible amount of electric charge. This particle, the electron, also had a very small mass, several thousand times less than the mass of an average atom.

## Contemporary model

Today we know that the explanation of charging by rubbing is actually very complicated. Atoms, the basic units of matter, consist of three types of particles: negatively charged electrons, positively charged protons (the same magnitude charge as electrons), and uncharged neutrons (**Figure 14.4a**). It is possible for an electron to leave one object and move to a different object. Such a transfer occurs because the atoms of the material that gains electrons hold them more tightly than the atoms of the material that loses them. An atom that has lost one electron has a net charge of $+e$ (it has one more proton than it has electrons) and is called a **positive ion** (a positively charged sodium ion is shown in Figure 14.4b). An atom that gains an extra electron has a net charge of $-e$ and is called a **negative ion**.

We now understand why rubbed objects acquire opposite charges. Two objects start as **neutral**—the total electric charge of each is zero. During rubbing, one object gains electrons and becomes negatively charged. The other loses an equal number of electrons and with this deficiency of electrons becomes positively charged. Sometimes when you rub two objects against each other, no transfer of electrons occurs. When the electrons in both materials are bound equally strongly to their respective atoms, no transfer occurs during rubbing.

We can represent this process using integers. A neutral object can be represented with a zero. A zero can be made up of a sum of positive and negative numbers—for example, $+10 + (-10) = 0$. When we rub object 1 with a second object 2 that pulls an electron from object 1, the negative number of object 1 becomes smaller while its positive number stays the same: $+10 + (-10) - (-1) = +10 + (-9) = +1$. The object is positively charged.

## CONCEPTUAL EXERCISE 14.2 Shirt and sweater attract

You pull your sweater and shirt off together and then pull them apart. You notice that they attract each other—a phenomenon called "static" in everyday life. Explain the mechanism behind this attraction and suggest an experiment to test your explanation.

The sleeves attract after they are pulled apart.

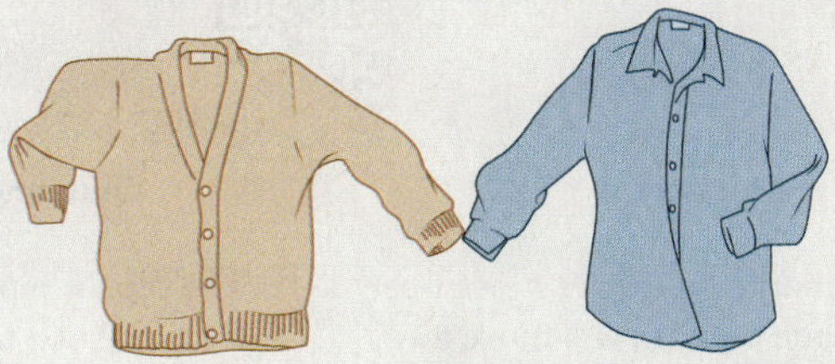

**Sketch and translate** Electric attraction between the sweater and shirt is a possible explanation. We need to explain how pulling the sweater and shirt apart leads to their opposite charges.

**Simplify and diagram** According to the particle model of electric charge, some of the electrons from one garment transfer to the other garment when they rub against each other. Before rubbing, both objects were neutral. If some negative electrons transfer from one object to the other, the former object now has a net positive charge while the latter has a net negative charge. If this hypothesis is correct, then a transparent tape strip pulled off a table should be repelled by one of the garments and attracted to the other. When we perform the experiment, shown below, the outcome matches this prediction.

Negatively charged tape is attracted to the sweater . . .

. . . and repelled from the shirt.

**Try it yourself:** You suspect that some objects after rubbing acquire a third type of charge. How can you test this hypothesis?

*Answer:* Imagine that rubbing a balloon with some new type of material (spandex, for example) gives the balloon a third type of charge. Then according to our pattern—two balloons rubbed with spandex should repel—and they do. But they should also attract both the balloon rubbed with felt and the balloon rubbed with plastic (if we assume that all differently charged objects attract each other). However, this never happens. If the balloon rubbed with spandex attracts the balloon rubbed with felt, it always repels the balloon rubbed with plastic, and vice versa. No experiment ever performed has required the existence of a third type of electric charge in order to explain its outcome.

**Review Question 14.2** The model of charging by rubbing involved the transfer of negatively charged electrons. How then can a neutral object become positively charged?

## 14.3 Conductors and nonconductors (dielectrics)

We found in Observational Experiment Table 14.2 that an electrically charged balloon was attracted to an uncharged wall and was more strongly attracted to an uncharged sheet of metal. The former is a so-called nonconductor (dielectric) and the latter a conductor. The meaning of these terms will be apparent by the end of this section.

**Figure 14.5** A metal bar is attracted to charged objects.

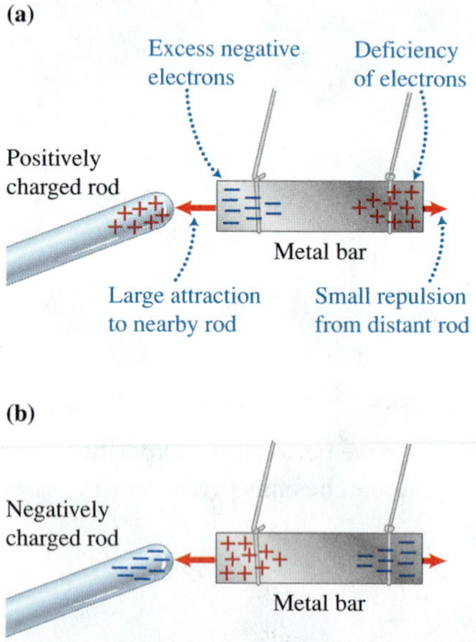


### Conductors

Suppose that in metals, some electrons are not bound to their respective atoms but can move freely throughout the metal. When we bring a positively charged rod next to a metal bar (without touching it), these **free electrons** in the metal bar should move closer to the positively charged rod, leaving the other side of the metal bar with a deficiency of electrons and therefore positively charged (**Figure 14.5a**). The negatively charged side of the bar is closer to the positively charged rod and thus should be attracted to it more strongly than the more distant positively charged side of the bar is repelled by it. When a negatively charged object is brought near an uncharged metal bar, the free electrons are repelled, resulting in similar charge separation (but with opposite orientation) and attraction (Figure 14.5b). This is why the metal compass needle in Table 14.3 was attracted to both positively and negatively charged rods. The attraction had nothing to do with magnetic poles—it was all about the movement of free electrons in the conductive compass needle. In Testing Experiment **Table 14.4**, we test the idea that electrons can move freely in metals.

**TESTING EXPERIMENT TABLE**

**14.4 Are there free electrons in a soft drink can?**

VIDEO 14.4

| Testing experiment | Prediction | Outcome |
|---|---|---|
| Place an empty metal soft drink can lengthwise on a cardboard box (the can is not charged). Hang a 10-cm-long piece of negatively charged transparent tape (charged by pulling it from a tabletop) from a wooden dowel so that its bottom end is near one end of the can but does not touch the can. The can attracts the tape. What will happen if you bring a strongly negatively charged plastic rod (rubbed with felt) near the other end of the can without touching it? 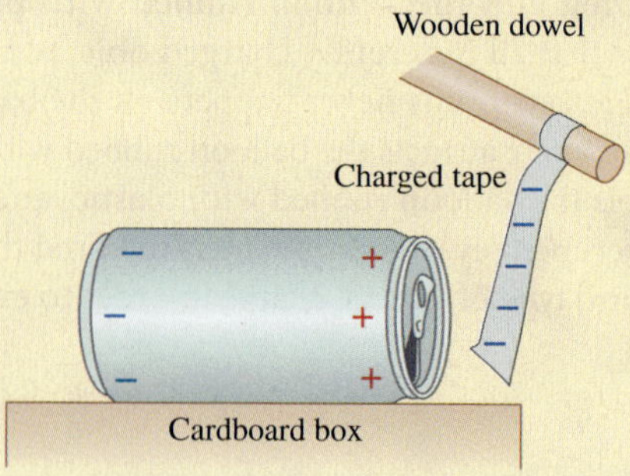  | If negatively charged particles move freely in the can, some of them will move away from the negatively charged plastic rod toward the end of the can near the negatively charged tape. The tape should be repelled from this end of the can. 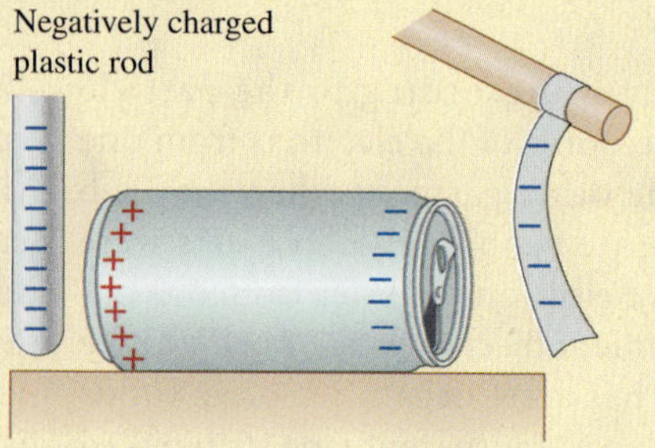  | The tape is repelled from the end of the can. |

**Conclusion**

The outcome is consistent with the prediction. This experiment supports our hypothesis that there are free electrons inside metals. We can repeat the experiment, bringing a positively charged glass rod (rubbed with silk) near the left side of the can. Using the same logic, we predict that the negatively charged tape on the other end should now be attracted to the can. This is exactly what happens. These two experiments give us more confidence in our free electron model of metals.

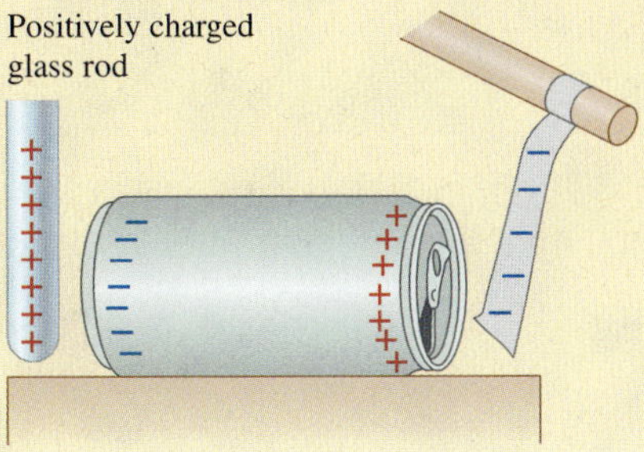


**Figure 14.6** A neutral foil ball interacts with a negatively charged plastic rod.

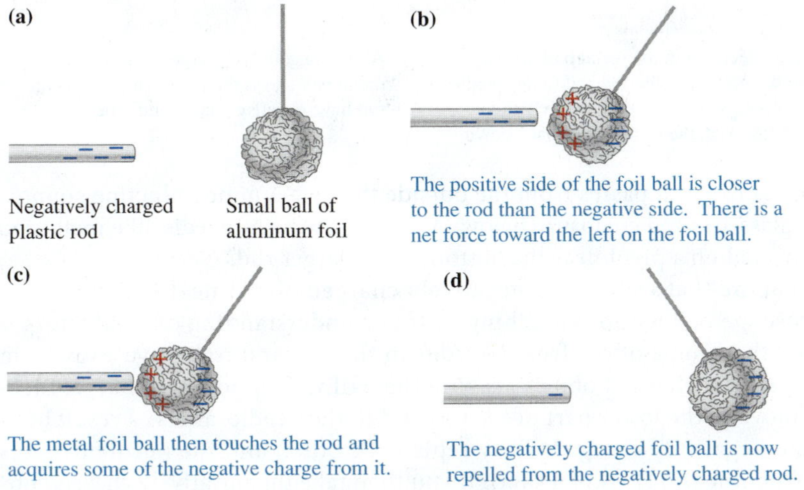

(a)
Negatively charged plastic rod
Small ball of aluminum foil
(b)
The positive side of the foil ball is closer to the rod than the negative side. There is a net force toward the left on the foil ball.
(c)
The metal foil ball then touches the rod and acquires some of the negative charge from it.
(d)
The negatively charged foil ball is now repelled from the negatively charged rod.


If we repeat these experiments with other metals, we obtain similar results. The results support our model of the freely moving charged particles in metals. We call metals electric **conductors**, materials in which some of the charged particles can move. In metals the moving charges are negatively charged free electrons. Some nonmetallic materials, such as graphite and silicon, can be conductors under certain conditions. In liquid conductors, such as nondistilled water or human blood, the moving charged particles can be positively or negatively charged ions. In hot gases, the moving charged particles can be both free electrons and ions.

Consider another experiment that depends on free electron motion in metals. An uncharged ball of aluminum foil hangs from a thin string (**Figure 14.6a**). When a negatively charged plastic rod is brought near the metal foil, free electrons move away from the negative rod, leaving the ball with positively and negatively charged sides (Figure 14.6b). The negative rod exerts a greater attractive force on the nearer positive charges on the ball than it exerts a repulsive force on the more distant negative charges. The foil ball touches the negatively charged rod and acquires some of its excess electrons (Figure 14.6c). Now both the rod and the foil ball are negatively charged and they repel (Figure 14.6d). The ball now swings away from the rod.

**‹ Active Learning Guide**

**Figure 14.7** An electroscope.

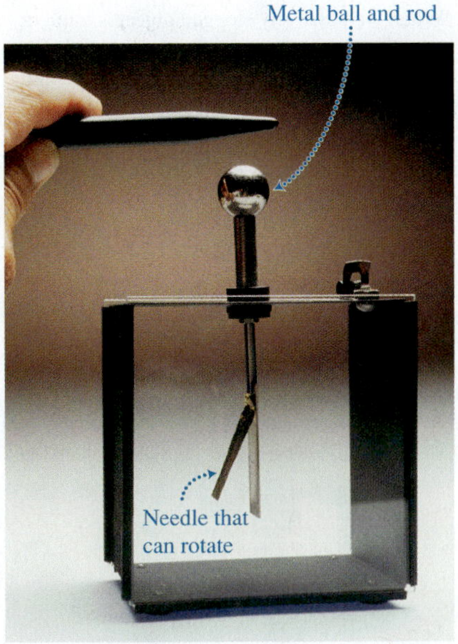


## The electroscope

Electroscopes, such as that illustrated in **Figure 14.7**, are useful tools for studying electrostatic interactions. They rely on the movement of free electrons in metal conductors. An electroscope consists of a metal ball attached

**Figure 14.8** Charging an electroscope.

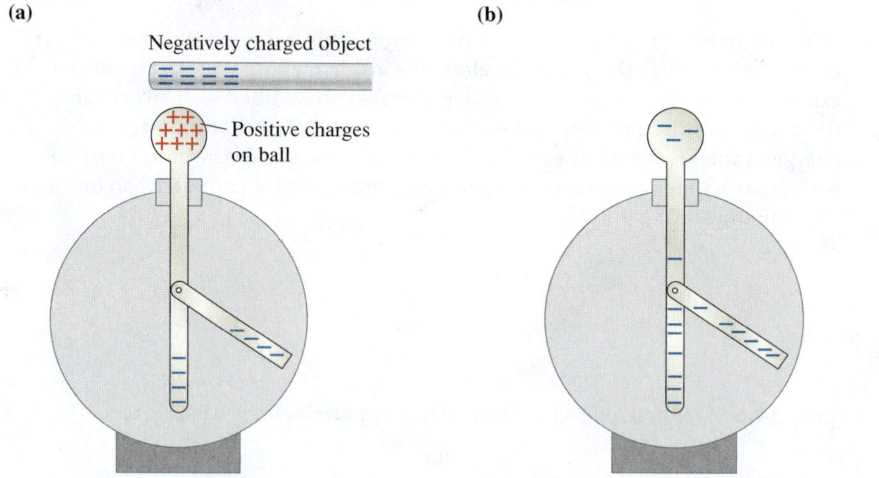


Some free electrons in the metal ball and rod move to the bottom of the rod and to the needle — they repel. When the negatively charged object is removed, the needle goes back down.

After a negatively charged object touches the metal ball and is then removed, the needle and rod have negative charge and repel.

to a metal rod that passes from the outside through a nonconducting support into a glass-fronted enclosure below. A very lightweight needle-like metal rod is connected on a pivot near the bottom of the larger rod.

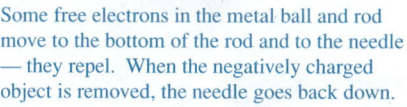

Imagine that we bring a negatively charged object near the top of the electroscope ball without touching it. If our understanding of conductors is correct, then some of the free electrons in the ball and rod move away from the negatively charged object, leaving the ball with a positive charge. Electrons move to the lower part of the rod and to the needle, and as a result both become negatively charged. They repel each other, and the needle deflects away from the rod (**Figure 14.8a**). If you then take the negatively charged object away from the top of the electroscope, the needle returns to its original vertical position.

Now, we repeat the experiment, only this time, we touch the top of the electroscope with the charged rod. We observe the same deflection again; only this time when the rod is removed, the needle stays deflected (Figure 14.8b). Touching the negatively charged object to the ball on the top of the electroscope transfers electrons to the electroscope. If we then remove the negatively charged object from the electroscope, the electroscope ball, rod, and needle are now negatively charged throughout, and again the rod and needle repel each other. When we repeat the experiments with a positively charged object instead, we observe the same outcomes.

If instead of just touching the top of the electroscope with the charged rod, we carefully rub the charged rod against the top of the electroscope as if we are scrubbing the charges from the rod, we see that the angle of the deflection increases—the angle of deflection is related to the magnitude of the electric charge of the electroscope.

**Figure 14.9** A simplified model of how a charged rod polarizes atoms in an insulating material.

(a)

Neutral atoms with charge uniformly distributed

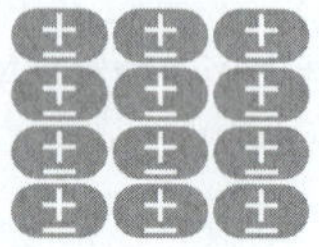

(b)

When the positively charged object is brought near the dielectric material, its atoms and molecules become polarized.

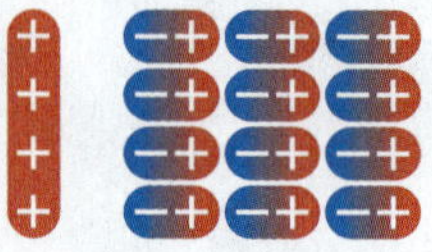

There is now a weak attraction between the charged object and the dielectric material.

(c)

A similar thing happens if we bring a negatively charged object near a non-conducting material.

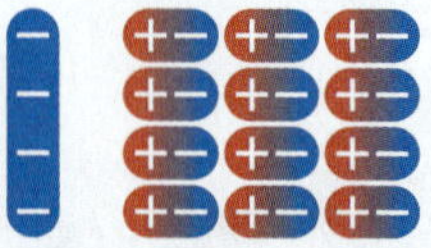

## Dielectrics

The presence of electrically charged particles that can freely move inside conductors explains how a charged object attracts a neutral metal object. How can we explain the attraction of a neutral nonmetal object and a charged object, such as observed in Table 14.2?

Plastic, glass, and other nonmetal materials do not have free electrons or any other charged particles that are free to move inside. Such materials are called **insulators** or **dielectrics.** All the electrons are tightly bound to their atoms/molecules (**Figure 14.9a**). If a charged object (let's make it positive) is

**Active Learning Guide ›**

brought nearby, it exerts forces on the negatively charged electrons and on the positively charged nuclei inside the atoms of the dielectric material. Negatively charged electrons are pulled slightly closer to the charged bar, and the positively charged nuclei are pushed slightly away. This causes a slight atomic-scale separation of the negative and positive charges in the material, a process called **polarization.** When the atoms are in this polarized state, they are called **electric dipoles.** An electric dipole is any object that is overall electrically neutral but has its negative and positive charges separated. Polarization leads to a small accumulation of charge on the surface of the object (depicted in Figure 14.9b). A similar polarization occurs if you bring a negatively charged object near a nonconducting material (Figure 14.9c).

Due to this polarization, electrically neutral pieces of paper are attracted to a charged comb that has been run through your hair, and an electrically neutral wall attracts a charged balloon. Although both conductors and dielectrics are attracted to charged objects when uncharged, there is a big difference between them. If you perform an experiment similar to the one in Testing Experiment Table 14.4 but replace the metal can with a plastic bottle, the tape is not affected by a charged bar on the other side of the plastic bottle. There are no free charged particles in the bottle.

Let's summarize what we have learned about the electric properties of materials.

**Materials can be divided into two groups: conductors and insulators (dielectrics). Both materials are composed of oppositely charged particles with a total charge of zero. In conductors, some of the charged particles can move freely. In dielectrics, charges can only be redistributed slightly (set off), a process called polarization.**

We can test the difference between conductors and dielectrics. Imagine that you have two aluminum cups (conductors) and two plastic cups (dielectrics). The aluminum cups are touching each other, as are the plastic cups. You bring a negatively charged rod next to each pair of cups (**Figures 14.10a** and b). While holding the rod nearby, you separate the cups by pushing them apart with a plastic ruler. If our understanding of conductors and insulators is correct, the metal cups will be oppositely charged; the cup closer to the rod will be positively charged and the cup farther from the rod will be negatively charged (Figure 14.10c). The plastic cups should remain uncharged (Figure 14.10d). We can check this prediction using a strip of transparent tape pulled off a table. The strip of tape is attracted to one of the metal cups and repelled from the other. However, the charged tape is attracted slightly to both plastic cups (each plastic cup becomes slightly polarized and then attracts the negatively charged tape). Since the outcome of the experiment matched the prediction, we gain confidence in our model.

**TIP** Electric charge is completely defined by the effects it produces. You cannot see electric charge; it does not correspond to any amount of mass, color, length, width, or any other physical quantity.

## Is the human body a conductor or a dielectric?

In the previous experiment you used a plastic ruler to separate the metal cups. If you had touched the metal cups with your hand instead of the ruler, the charged tape would have been attracted to both. This is because the human body is a conductor. As your hand approaches a positively charged cup, the free electrons on the surface of your hand and arm move toward it (**Figure 14.11a**). When you touch the positively charged metal cup, electrons from the surface of your hand travel to the cup until the cup has a net charge of approximately

**Figure 14.10** The effect of a charged rod on metal and plastic cups.

(a)

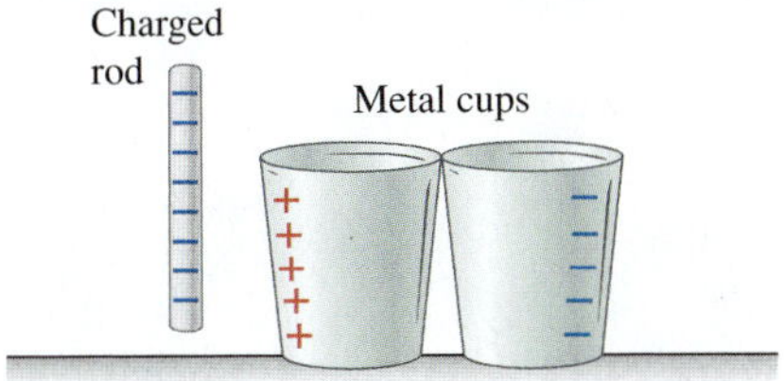


The rod polarizes the charge on the cups so that there are excess electrons on the right cup and a deficiency of electrons on the left cup.

(b)

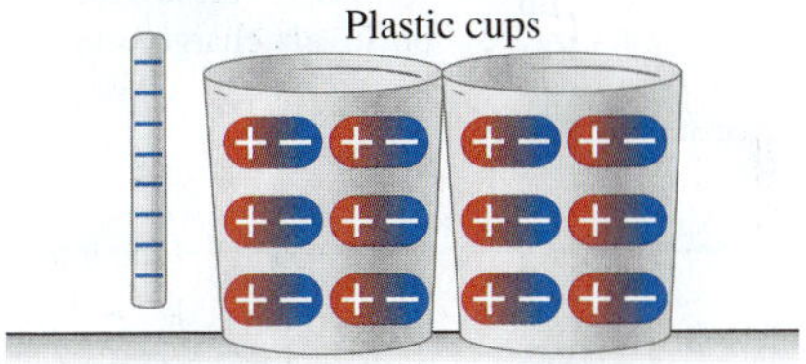


Atoms become polarized but there is no net charge separation.

(c)

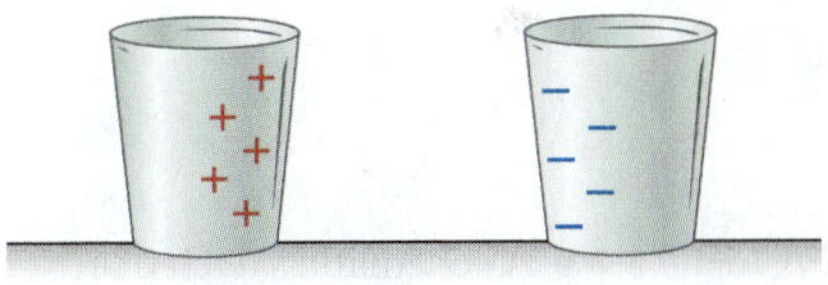

When separated, the metal cups are oppositely charged.

(d)

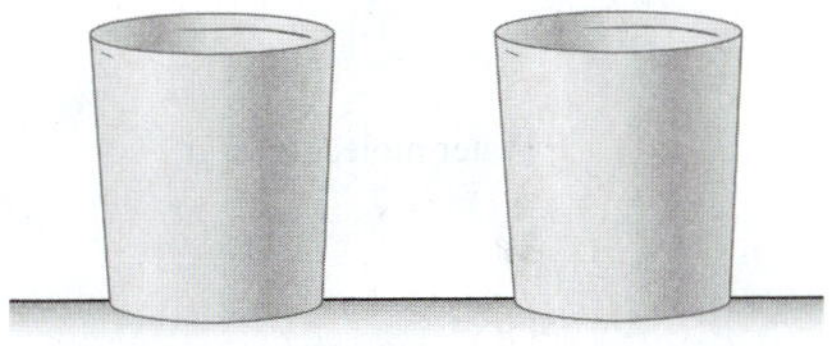

When separated, the plastic cups are uncharged.

**Figure 14.11** Free electrons in the body, an electrical conductor, move on the fingers toward a positively charged cup.

(a)

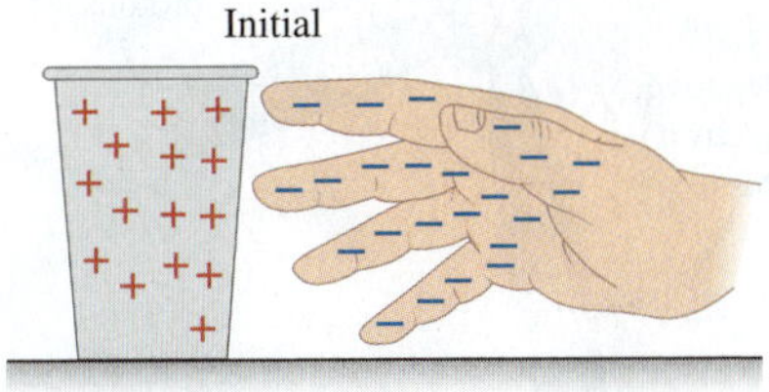


If your hand touches the cup, free electrons on the hand transfer to the cup until the net charge on the cup is zero.

(b)

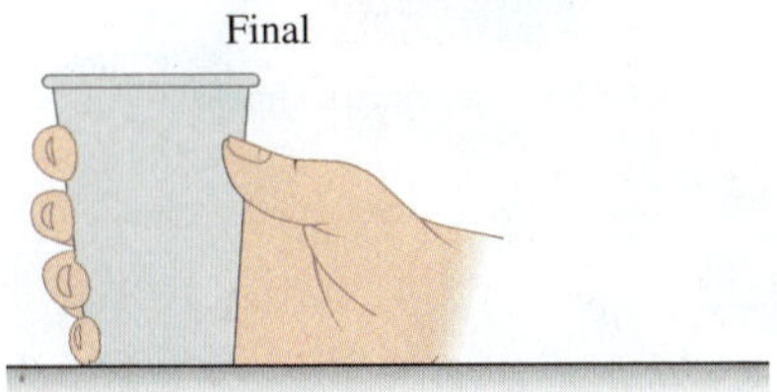

zero (Figure 14.11b). A similar process happens when you touch a negatively charged cup, only this time the electrons transfer from the cup to your skin. Sometimes when you touch a charged object, you feel a shock and might even see a spark. We will explain these phenomena later in the book.

## Grounding

Instead of touching the charged metal cup with your hand, you can safely discharge it (make its net charge effectively zero) by connecting a wire to both the cup and a metal pipe that goes into the ground (**Figure 14.12**). This process is called **grounding**. To explain why a grounded metal cup discharges, let's assume that Earth is a large conductor. If the cup is charged positively, negative electrons in the ground are attracted toward the cup and travel from the ground to the cup, causing it to become neutral. For a negatively charged cup, electrons in the cup travel through the grounding wire into the ground. Grounding electric appliances is extremely important in order to avoid electric shocks.

**Figure 14.12** A wire that is "grounded" on one end to a pipe in the ground can discharge the positively charged cup.

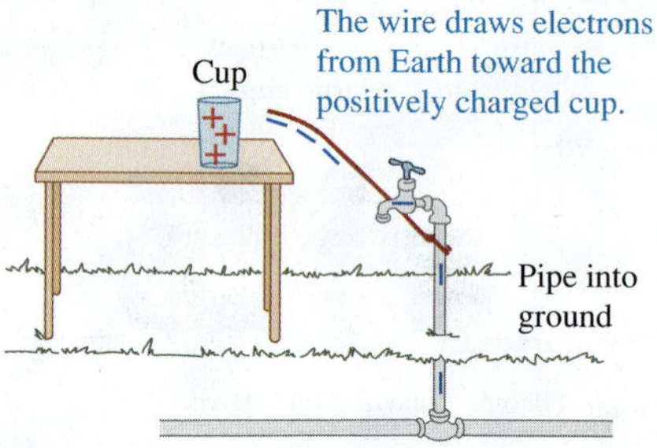


## Discharging by moisture in the air

If you leave a negatively or positively charged object alone for a long time interval, the object can become neutral without being grounded. This can happen for two reasons. First, the air has a few electrically charged particles, which are attracted to any object that has an opposite charge. The second reason is the presence of water vapor in the air. The electric charge in water molecules is distributed in such a way that the molecule is a natural electric dipole even without the polarizing influence of an external charged object (**Figure 14.13a**). The hydrogen side of the molecule is slightly positive while the oxygen side is slightly negative (Figure 14.13b), thus producing an electric dipole (Figure 14.13c). When water molecules are near a charged object (in this case negative), they will be attracted to it. The positive sides of the molecules contact the object and remove electrons, discharging it. That is why on a rainy or humid day, most of the experiments described so far will not work at all—you will have a hard time charging objects by rubbing and keeping them charged.

**Figure 14.13** A schematic sketch of a water molecule.

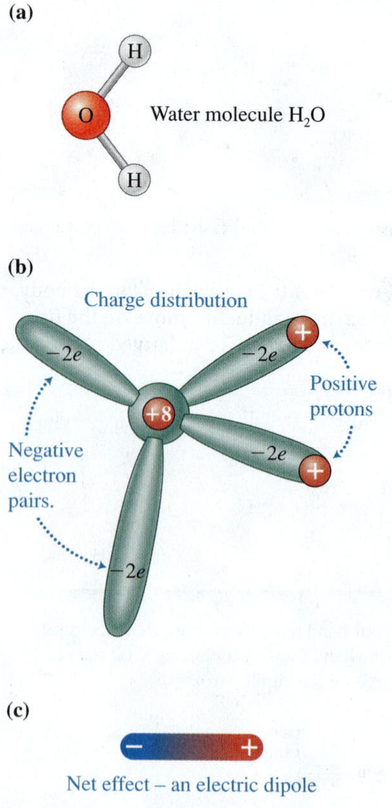


## Properties of electric charge

Think back to our explanation of charging by rubbing. When a plastic rod is rubbed by felt, the rod becomes negatively charged. The felt becomes positively charged. However, rubbing did not create the charges—it just transferred some electrons from one object to the other. If we assume that each transferred electron has the same electric charge independent of the presence of other electrons, then the total electric charge of an isolated system should be constant. The charge of a nonisolated system changes only if charged objects are transferred in or out of the system. Thus, similar to mass, momentum, and energy, *electric charge is a conserved quantity*; it changes in a predictable way in a nonisolated system and is constant in isolated systems.

Another property of electric charge can be deduced from the Millikan-Fletcher experiments. Electric charge cannot be divided infinitely—the smallest amount of charge that an object can have is the charge of one electron. This property is called charge **quantization.**

**Electric charge** Electric charge (symbol $q$ or $Q$) is a property of objects that participate in electrostatic interactions. Electric charge is quantized—you can only change an object's charge by increments, not continuously. Electric charge is conserved. The unit for electric charge is the coulomb (C). The smallest increment of charge is that of one electron $-e = -1.6 \times 10^{-19}$ C.

Our next goal is to develop a quantitative description of the forces that charged objects exert on each other.

**Review Question 14.3** One cannot charge a held metal object by rubbing. Why?

## 14.4 Coulomb's force law

We observed in the experiments so far that the nearer rubbed objects are to each other, the greater the forces they exert on each other. The forces also increase when the magnitude of the charges of the interacting objects increases. So the force that a charged object exerts on another must depend on the distance $r$ between them and on their electric charges $q_1$ and $q_2$.

Charles Coulomb determined the relationship between these quantities in 1785. The experimental apparatus he used is called a torsion balance (**Figure 14.14**). Coulomb hung a light glass rod from a thin wire. At the ends of the rod he attached identical small metal spheres. He then charged a third metal sphere and touched one of the spheres attached to the rod. The two spheres consequently had the same electric charge, and they repelled. The wire twisted until the torque exerted by the wire on the glass rod balanced the torque exerted by one charged sphere on the other. Coulomb measured the angle of this twist and used it to determine the electric force exerted by one sphere on the other. He measured the distance between the repelling spheres and varied it to see how the force of repulsion depended on the distance between them.

**Figure 14.14** Coulomb's torsion balance, an apparatus for measuring the electric force between charged objects.

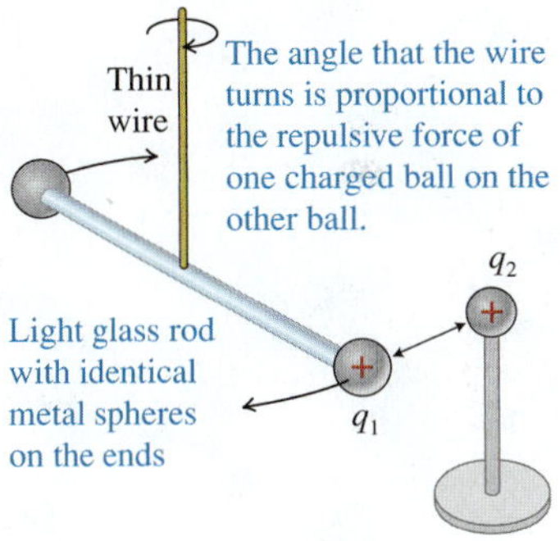


In Coulomb's time scientists did not know how to directly measure electric charge, and the unit for electric charge did not exist. So Coulomb used relative charges rather than absolute charges. Imagine two identical metal spheres—one charged and one uncharged. If you bring them in contact, what would be the charge on each? Coulomb reasoned that since the spheres were identical, the charge would distribute equally between them. Using the same method, he could achieve fractions of charge by touching a charged sphere to identical uncharged spheres many times. This way he could have charges of $q$, $\frac{1}{2}q$, $\frac{1}{4}q$, and so forth on the original sphere. He needed to have spheres of different charges to find out how the force exerted by one on the other depended on the magnitude of their charges. **Table 14.5** shows a simplified version of data that Coulomb might have collected. It indicates how the repulsive force that one metal sphere exerted on another depended on their separation and on their charges. (In reality, Coulomb's data did not follow such clear patterns and included experimental uncertainty.)

To find patterns in the data presented in Table 14.5, let's focus on one variable at a time. In Experiments 1 and 2, the only quantity that changes is

**Table 14.5** Coulomb's data.

| Experiment | Charge $q_1$ | Charge $q_2$ | Distance $r$ | Force $F_{q_1 \text{ on } q_2}$ |
|---|---|---|---|---|
| 1 | 1 (unit) | 1 (unit) | 1 (unit) | 1 (unit) |
| 2 | ½ | 1 | 1 | ½ |
| 3 | ¼ | 1 | 1 | ¼ |
| 4 | 1 | ½ | 1 | ½ |
| 5 | 1 | ¼ | 1 | ¼ |
| 6 | ½ | ½ | 1 | ¼ |
| 7 | ¼ | ¼ | 1 | 1/16 |
| 8 | 1 | 1 | 2 | ¼ |
| 9 | 1 | 1 | 3 | 1/9 |
| 10 | 1 | 1 | 4 | 1/16 |

the magnitude of the charge $q_1$. When the charge is halved, the force exerted by $q_1$ on $q_2$ is also halved. Looking at Experiment 3, we see that dividing the charge by four reduced the force to one-fourth. The force seemed to be directly proportional to the magnitude of the charge $q_1$. Experiments 4 and 5 show the same pattern for the magnitude of the charge $q_2$. If a quantity is directly proportional to two independent quantities, it has to be proportional to their product. Thus, we infer from the table that $F_{q_1 \text{ on } q_2} \propto q_1 q_2$. All 10 experiments are consistent with this conclusion.

From the data we also see that doubling the distance between the objects decreased the force by a factor of 4, and tripling the distance decreased the force by a factor of 9. It appears that the force is inversely proportional to the square of the separation between the objects

**Active Learning Guide ›**

$$F_{q_1 \text{ on } q_2} \propto \frac{1}{r^2}$$

We can now combine these patterns into one mathematical expression, called Coulomb's law.

**Coulomb's law** **The magnitude of the electric force that point-like object 1 with electric charge $q_1$ exerts on point-like object 2 with electric charge $q_2$ when they are separated by a distance $r$ is given by the expression**

$$F_{q_1 \text{ on } q_2} = k\frac{|q_1||q_2|}{r^2} \qquad (14.1)$$

**where $k$ is a proportionality constant that depends on the system of units used. In the SI system,**

$$k = 9.0 \times 10^9 \frac{\text{N} \cdot \text{m}^2}{\text{C}^2}$$

**When using the above to determine the magnitude of the electric force, we do not include the signs of the electric charges $q_1$ and $q_2$.**

**The force that object 1 exerts on object 2 points away from object 1 if they have same sign charges, and toward object 1 if they have opposite sign charges.**

**TIP** If the interacting objects are large conductors and near each other, the electrons will redistribute on their surfaces, making the effective distance smaller than the distance between the centers of the objects (and difficult to estimate). Thus, we only use Coulomb's law for point-like objects. These objects can be either conductors or dielectrics.

Notice that, mathematically, the expression for the electric force that two charged objects exert on each other

$$F_{q_1 \text{ on } q_2} = k\frac{|q_1||q_2|}{r^2}$$

is analogous to the expression for the gravitational force that any two objects with mass exert on each other

$$F_{m_1 \text{ on } m_2} = G\frac{m_1 m_2}{r^2}$$

Note that the gravitational force depends on the masses of the objects and the electric force depends on the charges of the objects. Also, the gravitational force is always an attractive force, whereas the electric force can be attractive or repulsive. The proportionality constants also have very different values.

## Comparing the magnitude of the electric force to the gravitational force

Let's compare the electric force to the gravitational force exerted by the proton on the electron in a hydrogen atom. A proton has a charge of $+1.6 \times 10^{-19}$ C and mass of $1.7 \times 10^{-27}$ kg. An electron has a charge of $-1.6 \times 10^{-19}$ C and mass of $9.1 \times 10^{-31}$ kg. They are separated in the atom by about $10^{-10}$ m.

Electric force:

$$F_{q_1 \text{ on } q_2} = k\frac{q_1 q_2}{r^2} = (9.0 \times 10^9\,\text{N}\cdot\text{m}^2/\text{C}^2)\frac{(1.6 \times 10^{-19}\,\text{C})^2}{(10^{-10}\,\text{m})^2} = 2.3 \times 10^{-8}\,\text{N}$$

Gravitational force:

$$F_{m_1 \text{ on } m_2} = G\frac{m_1 m_2}{r^2} = (6.67 \times 10^{-11}\,\text{N}\cdot\text{m}^2/\text{kg}^2)\frac{(1.7 \times 10^{-27}\,\text{kg})(9.1 \times 10^{-31}\,\text{kg})}{(10^{-10}\,\text{m})^2}$$
$$= 1.0 \times 10^{-47}\,\text{N}$$

If we divide the electric force by the gravitational force, the result is about $2 \times 10^{39}$. The electric force that the proton exerts on the electron is about $2 \times 10^{39}$ times greater than the gravitational force that the proton exerts on the electron!

What about the gravitational force exerted by Earth on the electron?

$$F_{\text{E on } m_2} = G\frac{m_\text{E} m_2}{r^2} = (6.67 \times 10^{-11}\,\text{N}\cdot\text{m}^2/\text{kg}^2)\frac{(6.0 \times 10^{24}\,\text{kg})(9.1 \times 10^{-31}\,\text{kg})}{(6.4 \times 10^6\,\text{m})^2}$$
$$= 8.9 \times 10^{-30}\,\text{N}$$

The gravitational force exerted by Earth on the electron is 18 orders of magnitude larger than the gravitational force exerted by the proton on the electron, but still about 22 orders of magnitude weaker than the electric force exerted by the proton on the electron. That is why physicists can confidently ignore gravitational forces when dealing with atomic size particles.

We have learned that the electric force that electrons and nuclei in atoms exert on each other is what holds atoms together. Although the atoms themselves are electrically neutral and should not interact with each other via electric forces, they do in fact interact. This interaction occurs because the electrons of one or more atoms can interact with the nuclei of other atoms. These forces are responsible for the existence of groups of bound atoms—molecules. In addition, the electric forces that the charged parts of electrically neutral molecules exert on each other are responsible for molecules forming liquids and solids at sufficiently low temperatures. Many forces that we encounter in everyday life—tension forces, friction forces, normal forces, buoyant forces, etc.—can be explained using Coulomb's law. You can use Coulomb's law to help explain why a liquid exerts an upward buoyant force on a submerged object, or why carpet exerts a friction force on your feet.

**TIP** Since Coulomb's law contains the product of the two charges, the electric force that object 1 exerts on object 2 is exactly the same in magnitude as the electric force that object 2 exerts on object 1. Coulomb's law is consistent with Newton's third law.

**‹ Active Learning Guide**

**CONCEPTUAL EXERCISE 14.3 Interactions of charged objects**

Two equal-mass small aluminum foil balls A and B have electric charges $+q$ (on A) and $+3q$ (on B). (a) Compare the magnitude of the electric force that the foil ball with the smaller charge exerts on the ball with the larger charge ($F_{\text{A on B}}$) to the force that the larger charged ball exerts on the smaller charged ball ($F_{\text{B on A}}$). Justify your answer. (b) If the balls are suspended by equal-length strings from the same point, will one string hang at a greater angle from the vertical than the other? Justify your answer.

**Sketch and translate** A labeled sketch of the situation is shown at right. We choose each ball as a separate system of interest. Ball A interacts with Earth, the string, and ball B. Ball B interacts with Earth, the string, and ball A.

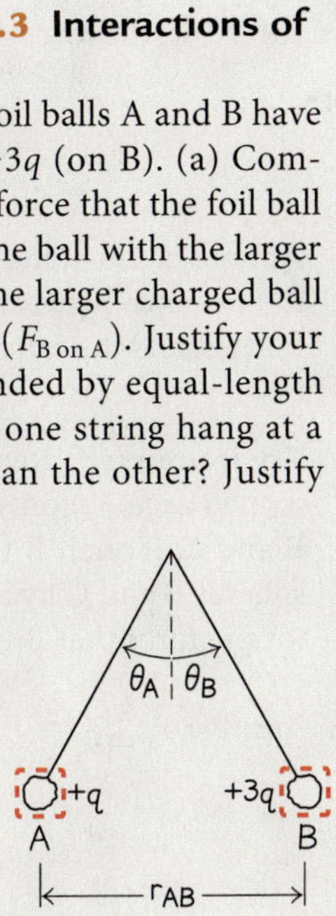


**Simplify and diagram** We will model the balls as point-like objects (see the force diagrams below).

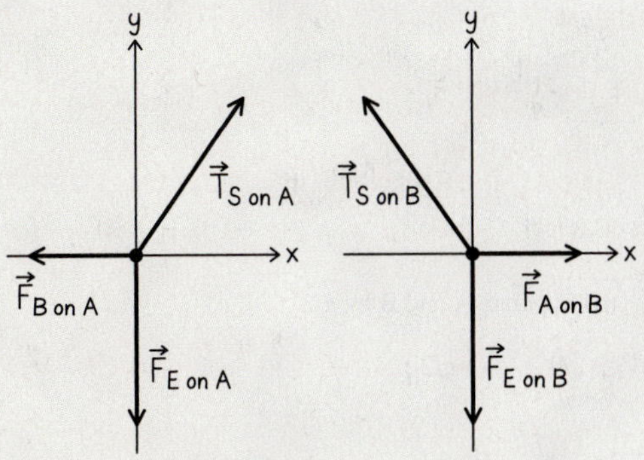


(a) The electric forces that the balls exert on each other have the same magnitude, according to Newton's third law and Coulomb's law ($F_{\text{A on B}} = kq_\text{A}q_\text{B}/r_{\text{AB}}^2 = kq_\text{B}q_\text{A}/r_{\text{AB}}^2 = F_{\text{B on A}}$). (b) Since the balls have the same

*(continued)*

mass, Earth exerts the same force on each. Thus the forces exerted by the strings must also have the same magnitude since the net force exerted on each ball equals zero. Therefore, the angles of the strings relative to the vertical should also be the same.

**Try it yourself:** Assume that ball A has twice the mass of B and the charges on the balls are the same as in the above exercise. Qualitatively compare the angles that the strings make with the vertical.

*Answer:* The angle made by the string for ball B is twice the angle that the string for ball A makes with the vertical.

## QUANTITATIVE EXERCISE 14.4 Coulomb's law and proportional reasoning

Three identical metal spheres A, B, and C are on separate insulating stands (part a in the figure below). You charge sphere A with charge $+q$. Then you touch sphere B to sphere A and separate them by distance $d$ (part b in the figure). (a) Write an expression for the electric force that the spheres exert on each other. What assumptions did you make? (b) You now take sphere C, touch it to sphere A (part c in the figure), then remove sphere C. You then separate spheres A and B by a distance $d/2$ (part d in the figure). Write an expression for the magnitude of the force that spheres A and B exert on each other.

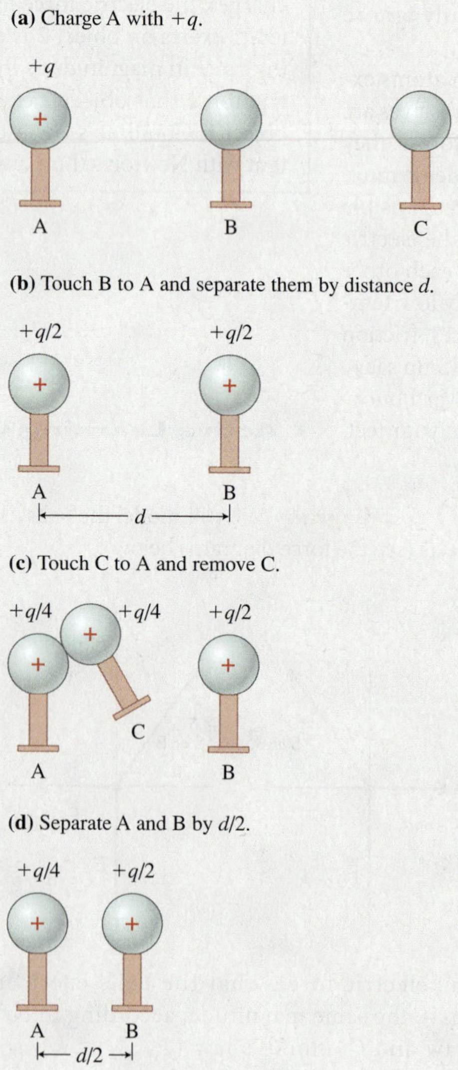

(a) Charge A with +q.
+q
A
B
C
(b) Touch B to A and separate them by distance d.
+q/2
+q/2
A
B
d
(c) Touch C to A and remove C.
+q/4
+q/4
+q/2
C
A
B
(d) Separate A and B by d/2.
+q/4
+q/2
A
B
d/2

**Represent mathematically** Assume that we can model the spheres as point-like objects: (a) Assume that after sphere A touches B, each of them has a charge of $+q/2$.

(b) After C touches A, the charge on A halves again, so that A now has charge $+q/4$. The distance between A and B is $d/2$. We can use Coulomb's law to write an expression for the force that each charge exerts on the other for each part of the problem.

**Solve and evaluate** (a) The magnitude of the electric force that A and B exert on each other is

$$F_{\text{A on B}} = F_{\text{B on A}} = k\frac{(q/2)(q/2)}{d^2} = k\frac{q^2/4}{d^2} = k\frac{q^2}{4d^2}$$

(b) Now, the force that A and B exert on each other is

$$F_{\text{A on B}} = F_{\text{B on A}} = k\frac{(q/4)(q/2)}{(d/2)^2} = k\frac{q^2/8}{d^2/4} = k\frac{q^2}{2d^2}$$

Although one of the charges got smaller, the decrease in the distance between A and B more than compensated for this, and the force became twice as great as in the first case.

**Try it yourself:** Repeat the previous exercise, starting with charge $+q$ on sphere A and no charge on spheres B and C. Touch B to A. Then touch C to B. Separate spheres B and C by a distance $d/2$. Write an expression for the force that B exerts on C.

*Answer:* $F_{\text{B on C}} = F_{\text{C on B}} = k\dfrac{(q/4)^2}{(d/2)^2} = k\dfrac{q^2}{4d^2}$

**EXAMPLE 14.5 Net electric force**

The metal spheres on the insulating stands from the previous exercise have the following electric charges: $q_A = +2.0 \times 10^{-9}\,\text{C}$, $q_B = +2.0 \times 10^{-9}\,\text{C}$, and $q_C = -4.0 \times 10^{-9}\,\text{C}$. The spheres are placed at the corners of an equilateral triangle whose sides have length $d = 1.0\,\text{m}$ with C at the top of the triangle. What is the magnitude of the total (net) electric force that spheres A and B exert on C?

**Sketch and translate** A labeled sketch of the situation is shown below. All three angles of the triangle are 60°.

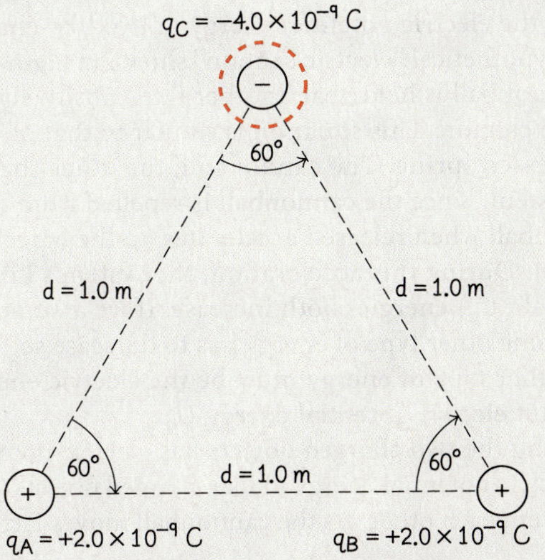


**Simplify and diagram** Assume that the distance between the spheres is much bigger than their radii so that they can be modeled as point-like objects. Choose sphere C as the system of interest and draw a force diagram for it. We are only interested in the magnitude of the net electric force exerted on C. Thus, we include in the diagram only the electric forces exerted on C by A and by B. We see that the vector sum of $\vec{F}_{A\text{ on }C}$ and $\vec{F}_{B\text{ on }C}$ points straight down.

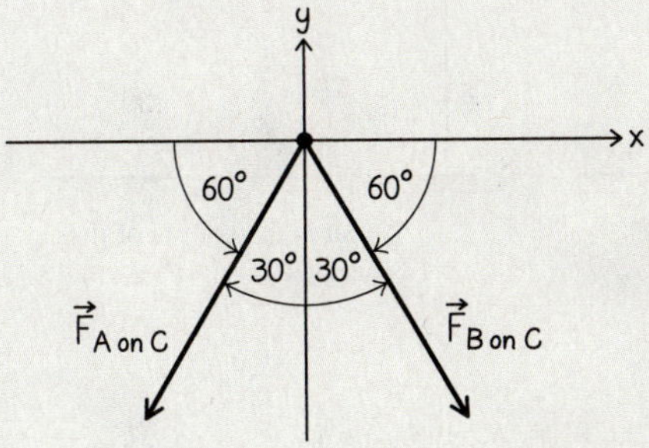


**Represent mathematically** Each of the spheres A and B exerts a force on C of magnitude

$$F_{A\text{ on }C} = F_{B\text{ on }C} = k\frac{|q_A||q_C|}{d^2}$$

where $d$ is the length of a side of the triangle. Each force vector points along the line connecting the respective spheres. The $x$-components of the two force vectors cancel each other and the resultant force has only a $y$-component. Using trigonometry, we find that the net $y$-component force that A and B exert on C is

$$F_{\text{Net on }C\,y} = (-F_{A\text{ on }C}\sin 60°) + (-F_{B\text{ on }C}\sin 60°)$$

$$= -2F_{A\text{ on }C}\sin 60° = -2\left(k\frac{|q_A||q_C|}{d^2}\right)\sin 60°$$

**Solve and evaluate** Inserting the appropriate values gives

$$F_{\text{Net on }C\,y} = -2\left((9.0 \times 10^9\,\text{N}\cdot\text{m}^2/\text{C}^2) \times \frac{(2.0 \times 10^{-9}\,\text{C})(4.0 \times 10^{-9}\,\text{C})}{1.0\,\text{m}^2}\right)\sin 60°$$

$$= -1.2 \times 10^{-7}\,\text{N}$$

The force has magnitude $1.2 \times 10^{-7}\,\text{N}$ and points straight down. This is a very small force. If the spheres had relatively small masses (a few hundred grams) and were mounted on stands on which the surface exerted a regular friction force, we would probably not observe any effects of these electric forces.

**Try it yourself:** What charges should you place at the corners of an equilateral triangle with a horizontal base so that the net force on the top charge is horizontally to the left?

*Answer:* The bottom left could be $-q$, the bottom right $+q$, and the top center sphere could have any positive charge.

**Review Question 14.4** Two charged objects (1 and 2) with charges $q_1 = q$ and $q_2 = 28q$ are placed $r$ meters away from each other. What is the ratio of the electric force that object 1 exerts on 2 to the force that 2 exerts on 1? Explain your answer.

## 14.5 Electric potential energy

In the previous section we learned to describe the interactions of electric charges in terms of forces. In this section we will learn to describe these interactions in terms of energy.

**Active Learning Guide >**

When you have a system of two objects exerting gravitational forces on each other (an apple-Earth system, for example), the system possesses gravitational potential energy. Since Coulomb's law is mathematically very similar to the law of universal gravitation, it is reasonable to suggest that a system of electrically charged objects also possesses some sort of electric energy. Let's investigate this idea.

### Electric potential energy: A qualitative analysis

Let's begin by looking at the electric potential energy of two like-charged objects that are part of the hypothetical "electric cannon" shown in **Figure 14.15a**. A positively charged cannonball is held near another fixed positively charged object in the barrel of the cannon. This situation is similar to that of an object pressed against a compressed spring. The cannonball, the other charged object, and Earth are the system. Since the cannonball is repelled from the fixed-charge object, the cannonball when released accelerates up the barrel and out of its end (Figure 14.15b). During this acceleration, the system's kinetic ($K$) and gravitational potential ($U_g$) energies both increase. If we assume that the system is isolated, then some other type of energy has to decrease so that these two can increase. This other type of energy must be the electric energy suggested above. We will call it **electric potential energy** $U_q$.

The system comprising the two charged objects has electric potential energy. It seems that the electric potential energy of like-charged objects decreases as they get farther apart from each other. As the cannonball moves farther from

**Figure 14.15** The electric potential energy of two like-charged objects.

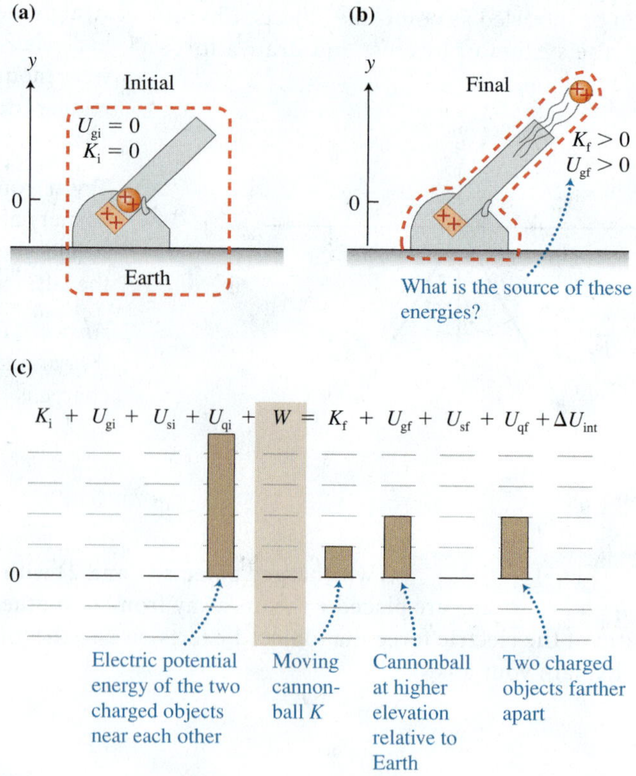

the fixed-charge object, the electric potential energy of the system decreases as it is converted into kinetic and gravitational potential energy.

We can represent this process using a bar chart (Figure 14.15c). The initial state is the moment at which the cannonball is near the other fixed-charge object; the final state is when the cannonball is moving at the end of the barrel. For this process we assign the zero level for gravitational potential energy to be at the initial position of the cannonball. We must also make a zero-level assignment for electric potential energy. We say that when two charged objects are so far apart that they essentially do not interact, then they have zero electric potential energy.

Let us consider a situation in which two oppositely charged objects interact. In this case we will use as an example a hypothetical "electric nutcracker" (**Figure 14.16a**). The system consists of the two oppositely charged blocks, one of which can slide without friction. When the negatively charged block is released and moves nearer the nut, the kinetic energy of the system increases (Figure 14.16b). Thus electric potential energy must decrease. Assuming the zero level is when the two oppositely charged blocks are infinitely far apart, we conclude that the electric potential energy of a pair of unlike charges is less than zero. The bar chart in Figure 14.16c represents this process. The initial electric potential energy of the system is negative. As the objects come closer together, the kinetic energy of the system increases and its electric potential energy decreases, becoming even more negative. Our next step is to devise a mathematical expression for electric potential energy.

## Electric potential energy: a quantitative analysis

To derive an expression for the change in electric potential energy of two electric charges whose separation changes, we use the generalized work-energy principle:

$$W = \Delta U_{\text{system}} \tag{6.3}$$

where $W$ is the work done on the system by objects in the environment and $\Delta U_{\text{system}}$ is the resulting change in the system's energy. To derive an expression for $\Delta U_q$, we choose a hypothetical system that has the following feature:

**Figure 14.16** The electric potential energy of two oppositely charged objects.

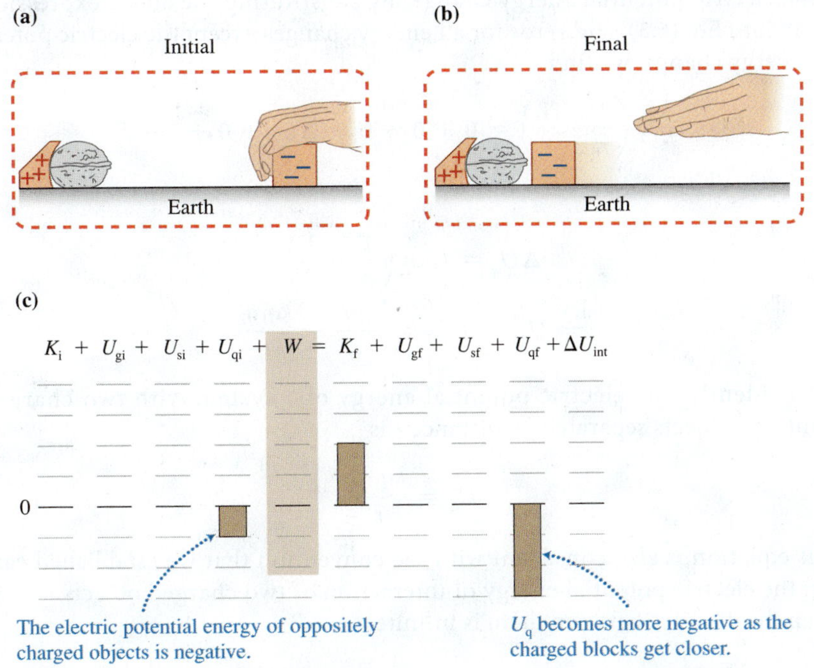

(a)
Initial
Earth
(b)
Final
Earth
(c)
$K_i + U_{gi} + U_{si} + U_{qi} + W = K_f + U_{gf} + U_{sf} + U_{qf} + \Delta U_{int}$
0
The electric potential energy of oppositely charged objects is negative.
$U_q$ becomes more negative as the charged blocks get closer.

**Figure 14.17** The work done in pushing two like-charged objects closer together.

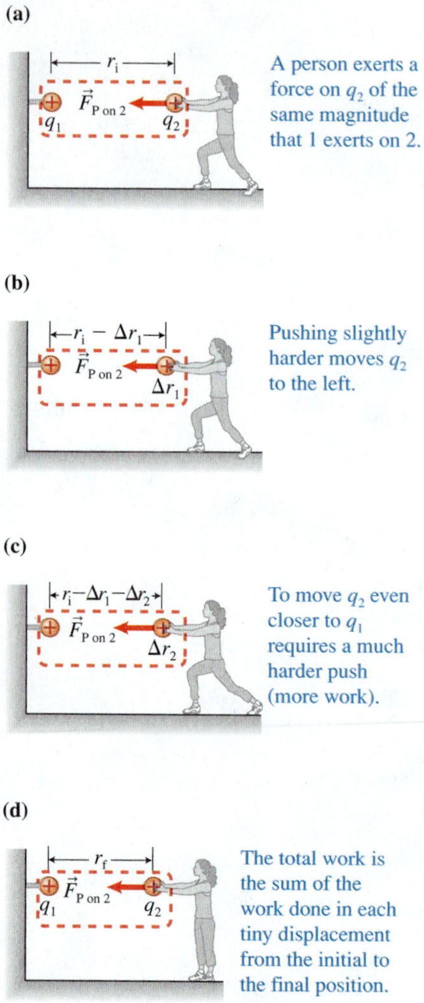
(a)
$r_i$
$\vec{F}_{P\text{ on }2}$
$q_1$
$q_2$
A person exerts a force on $q_2$ of the same magnitude that 1 exerts on 2.
(b)
$r_i - \Delta r_1$
$\vec{F}_{P\text{ on }2}$
$\Delta r_1$
Pushing slightly harder moves $q_2$ to the left.
(c)
$r_i - \Delta r_1 - \Delta r_2$
$\vec{F}_{P\text{ on }2}$
$\Delta r_2$
To move $q_2$ even closer to $q_1$ requires a much harder push (more work).
(d)
$r_f$
$\vec{F}_{P\text{ on }2}$
$q_1$
$q_2$
The total work is the sum of the work done in each tiny displacement from the initial to the final position.

when work is done on this system, only its electric potential energy changes. The system shown in **Figure 14.17** consists only of the point-like charged objects 1 and 2 with like charges $q_1$ and $q_2$, initially separated by a distance $r_i$. The like charges exert a repulsive force on each other of magnitude

$$F_{q_1 \text{ on } q_2} = k\frac{|q_1||q_2|}{r^2} \tag{14.1}$$

Imagine that we wish to prevent object 2 from flying away from 1 by exerting a force on 2 toward 1 of the same magnitude that the charged object 1 exerts on 2 (Figure 14.17a). If we push object 2 just a tiny bit harder, object 2 can be displaced a small distance $\Delta r_1$ toward object 1 (Figure 14.17b). We use Eq. (6.1) to calculate the work $\Delta W_1$ (here the symbol delta $\Delta$ indicates a small amount of work and not the change in work) done by the pushing force during this small displacement (the force and displacement are in the same direction; thus the cosine of the angle between them is 1):

$$\Delta W_1 = F_{P\text{ on }2}\Delta r_1 = \frac{k|q_1||q_2|}{r_i^{\,2}}\Delta r_1$$

This calculation is only approximate, though, because the force needed to push object 2 increases as its separation from object 1 decreases. If $\Delta r_1$ is small, the equation is a good approximation for the work done during that small displacement of object 2. If we wish to push object 2 through the next small displacement $\Delta r_2$ (Figure 14.17c), we must exert a larger force, since object 2 is now closer to object 1. For each step closer, the amount of work needed to move object 2 closer to object 1 increases.

The total work done in moving the charge from an initial separation $r_i$ to a final separation $r_f$ (Figure 14.17d) equals the sum of the work for each small step. This type of infinitesimal addition is done easily using calculus. The result of such a calculation is

$$W = \Delta W_1 + \Delta W_2 + \Delta W_3 + \cdots = \frac{kq_1q_2}{r_f} - \frac{kq_1q_2}{r_i} = kq_1q_2\left(\frac{1}{r_f} - \frac{1}{r_i}\right)$$

At each step of the process represented in Figure 14.17, there is no acceleration and therefore no change in kinetic energy or any other kind of energy. Thus we can reason that the only energy change of the system due to this work is the electric potential energy change. By substituting the above expression for $W$ into Eq. (6.3), and zeros for all energy changes except the electric potential energy change, we find

$$kq_1q_2\left(\frac{1}{r_f} - \frac{1}{r_i}\right) = 0 + 0 + 0 + \Delta U_q + 0 + \cdots$$

Therefore,

$$\Delta U_q = kq_1q_2\left(\frac{1}{r_f} - \frac{1}{r_i}\right)$$

$$\Rightarrow U_{qf} - U_{qi} = \frac{kq_1q_2}{r_f} - \frac{kq_1q_2}{r_i}$$

Evidently, the electric potential energy of a system with two charged point-like objects separated by distance $r$ is

$$U_q = \frac{kq_1q_2}{r}$$

This equation is also consistent with the convention that we established earlier: the electric potential energy of interaction of two charged objects is zero when the distance between them is infinite.

**Active Learning Guide ›**

The change in **electric potential energy** $\Delta U_q$ of a system of two charged objects $q_1$ and $q_2$ when they are moved from an initial separation $r_i$ to a final separation $r_f$ is

$$\Delta U_q = U_{q\,f} - U_{q\,i} = kq_1q_2\left(\frac{1}{r_f} - \frac{1}{r_i}\right) \quad (14.2)$$

Electric potential energy is measured in units of joules. Equation (14.2) is valid for both positively and negatively charged objects, provided the signs of the charges are included.

**TIP** Two points are worth noticing: (a) electric potential energy is proportional to $1/r$ and not to $1/r^2$ (as is the force in Coulomb's law), and (b) mathematically the expression for the electric potential energy is similar to the expression for the gravitational potential energy of a system with two objects with mass,

$$U_g = -G\frac{m_1m_2}{r}$$

**EXAMPLE 14.6 Changes in electric potential energy**

Two oppositely charged objects (with positive charge $+q$ and negative charge $-q$) are separated by distance $r_i$. Will the electric potential energy of the system decrease or increase if you pull the objects farther apart? Explain.

**Sketch and translate** The initial state was the two oppositely charged objects (the system) a distance $r_i$ apart. You pull the $-q$ charged object so that it is farther from the positively charged object—the final state—as shown below.

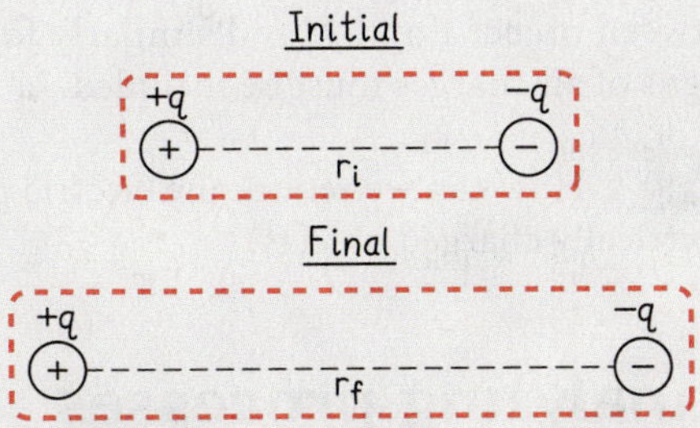


Because the direction of the force that you exert on the $-q$ charged object is the same as the direction of its displacement, positive work is being done on the system, increasing the electric potential energy of the system. If we pulled $-q$ infinitely far from $+q$, the final electric potential energy of the system would be zero. Thus, the initial electric potential energy must be negative. To understand this, imagine that you add a positive number to an unknown number and get zero: $x + 5 = 0$. What is the unknown number? $x = -5$ as $(-5) + 5 = 0$. The initial energy was negative.

**Simplify and diagram** We model the two objects as point-like particles, and we assume that only the electric potential energy changes. A bar chart representing the process is shown at the right.

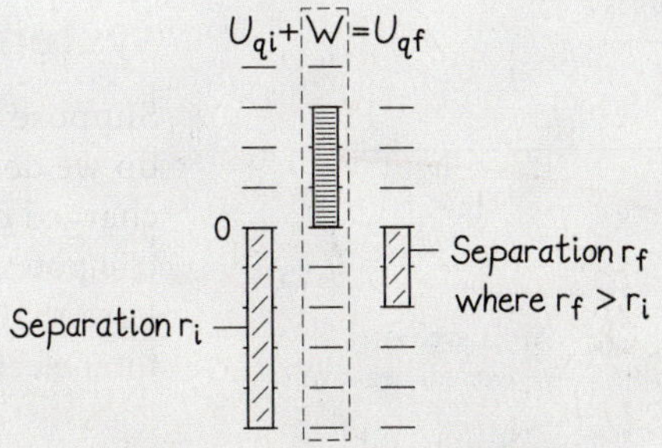


**Represent mathematically** We now use the bar chart and the generalized work-energy principle to represent the process mathematically:

$$U_i + W = U_f$$

$$\Rightarrow \frac{k(+q)(-q)}{r_i} + W = \frac{k(+q)(-q)}{r_f}$$

**Solve and evaluate** Notice that if $r_i$ is less than $r_f$, the electric potential energy term on the left side is more negative than that on the right side—consistent with the bar chart. This means that the electric potential energy of the system is negative and changes from a larger negative value to a smaller negative value as the charges get farther apart—an increase in energy. This outcome is consistent with our conceptual understanding of the process.

**Try it yourself:** In a hydrogen atom a proton and an electron have the same magnitude charge $1.6 \times 10^{-19}$ C but opposite signs of charge. The distance between them is $0.53 \times 10^{-10}$ m. What is the change in electric potential energy if the electron is moved far away from the nucleus?

*Answer:* $4.3 \times 10^{-18}$ J.

**Figure 14.18** The electric potential energy-versus-distance graphs for (a) like charged objects and (b) unlike charged objects.

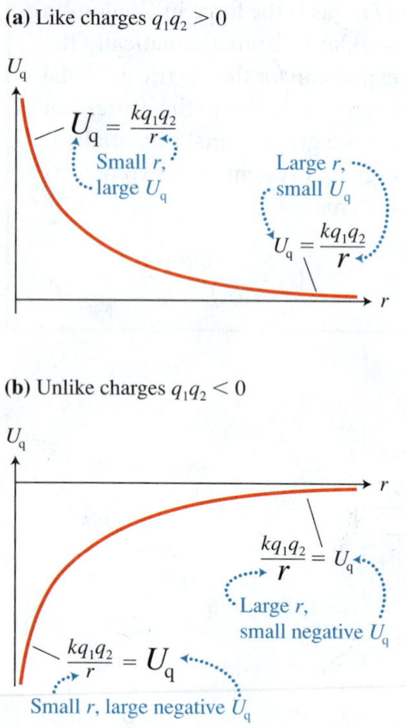


## Graphing the electric potential energy versus distance

So far we have represented electric potential energy with bar charts and with mathematics. Using a graph is another way to conceptualize this abstract physical quantity. Consider first how the electric potential energy changes as the separation $r$ between two like-charged objects is varied. Because of the $1/r$ dependence, the electric potential energy approaches positive infinity when the separation approaches zero (see **Figure 14.18a**). The electric potential energy becomes less positive and approaches zero as the like charges are moved far apart. For a system with two unlike-charged objects, the electric potential energy approaches negative infinity when their separation approaches zero and becomes less negative (increases) and approaches zero as they are moved far apart (Figure 14.18b).

## Electric potential energy of multiple charge systems

Suppose a system has several interacting electrically charged objects. How do we determine the electric potential energy of such a system? Each pair of charged objects has an associated electric potential energy, and the total electric potential energy of the system is the sum of the energies of all pairs. Suppose that the system has three charged objects labeled 1, 2, and 3. Then the total electric potential energy is

$$U_q = U_{q\,12} + U_{q\,13} + U_{q\,23} = \frac{kq_1q_2}{r_{12}} + \frac{kq_1q_3}{r_{13}} + \frac{kq_2q_3}{r_{23}} \tag{14.3}$$

where $r_{12}$ is the distance between object 1 and 2, and similarly for the other terms. Remember that the signs of all charges must be included.

**Review Question 14.5** How can we reduce the electric potential energy of a system of two electrically charged objects?

# 14.6 Skills for analyzing processes involving electric force and electric potential energy

While solving problems with electrically charged objects, you can use a problem-solving strategy similar to that used in the dynamics and energy chapters. The process is described on the left side of the table below and illustrated on the right side in the solution for Example 14.7.

**PROBLEM-SOLVING STRATEGY Analyzing processes involving electric force and electric potential energy**

**EXAMPLE 14.7 Where to put the charge?**

Two electrically charged objects with charges $q_1 = +1.0 \times 10^{-9}\,\text{C}$ and $q_2 = +2.0 \times 10^{-9}\,\text{C}$ are separated by 1.0 m. Where should you place a third electrically charged object so that the net electric force exerted on it by the first two objects is zero?

**Sketch and translate**

- Sketch the process described in the problem statement. Label the physical quantities.
- Choose an appropriate system.

We do not yet know where to place the third object. It could be in any of the three regions shown, but it has to be on the line connecting objects 1 and 2 so that the forces that 1 and 2 exert on this third object are parallel and pointing in opposite directions. We choose the object of unknown electric charge $q$ as the system. The system interacts with objects 1 and 2.

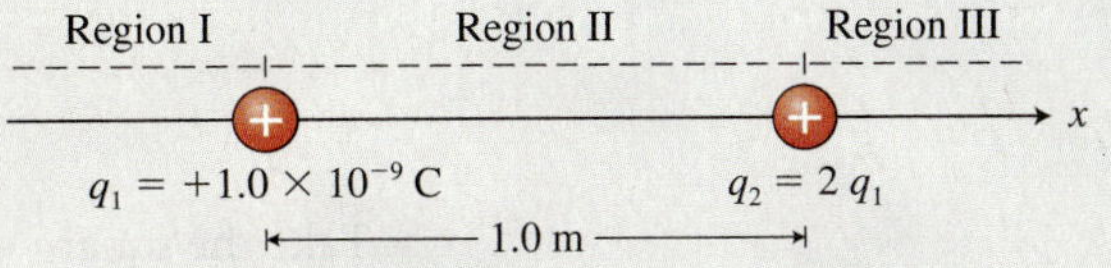


**Simplify and diagram**

- Decide whether you can consider the charged objects to be point-like. Decide what other interactions you will consider and what interactions you can ignore.
- Construct a force diagram for the system. Choose appropriate coordinate axes. If you are using the work-energy principle, construct an energy bar chart. Decide where the zeros for potential energies are.

Consider all objects to be smaller than the distances between them and, thus, point-like.

Draw force diagrams for the charged object $q$ in the three possible regions. The net electric force exerted on $q$ is the vector sum of the two forces exerted by $q_1$ and $q_2$.

(I) If $q$ is positioned to the left of both charged objects, the forces due to $q_1$ and $q_2$ both point left and cannot add to zero.

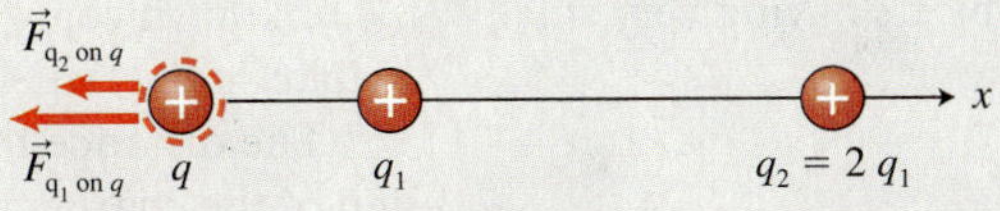


(III) If $q$ is positioned to the right of the two charges, both forces point right and cannot add to zero.

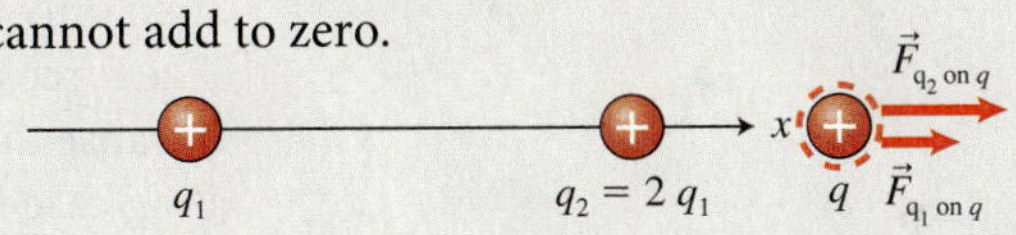


(II) If $q$ is between the charges, the forces point in opposite directions and can add to zero. If $q$ is closer to the smaller charge, the reduced distance will compensate for the smaller $q_1$.

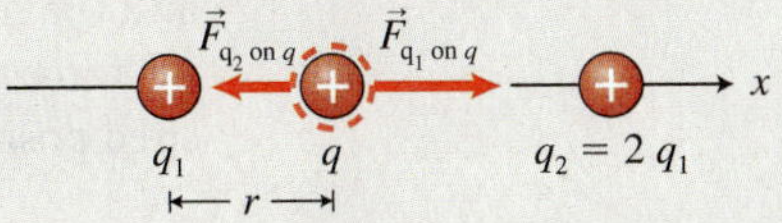


**Represent mathematically**

- Use the force diagram to apply the component form of Newton's second law to the process (or use the energy bar chart to apply the generalized work-energy equation).
- If necessary, use kinematics equations to describe the motion of the object.

Both forces are horizontal (their $y$-components are zero). Thus, we use only the horizontal $x$-component form of Newton's second law. With the positive $x$-direction toward the right and taking the signs of the components into account, we get

$$F_{q_1 \text{ on } q} + (-F_{q_2 \text{ on } q}) = 0$$

*(continued)*

**PROBLEM-SOLVING STRATEGY** ***(continued)***

Using Coulomb's law for electric force, the fact that the total distance between the charges is 1.0 m, and that $q_2 = 2q_1$ and labeling the distance between $q_1$ and $q$ as $r$, the above equation becomes

$$\frac{kq_1q}{r^2} = \frac{k2q_1q}{[(1.0\text{ m}) - r]^2}$$

Divide both sides of the equation by $kqq_1$ to get

$$\frac{1}{r^2} = \frac{2}{[(1.0\text{ m}) - r]^2}$$

Take the square root of both sides of the equation:

$$\frac{1}{r} = \frac{\sqrt{2}}{(1.0\text{ m}) - r}$$

**Solve and evaluate**

- Rearrange the equation and solve for the unknown quantity. Verify that your answer is reasonable with respect to sign, unit, and magnitude. Also make sure the equation applies for limiting cases, such as objects having very small or very large charge.

Rearranging the last equation, we get

$$\sqrt{2}r = (1.0\text{ m}) - r \quad \text{or} \quad r = 0.41\text{ m}$$

The location where a net electric force of zero will be exerted on $q$ is a distance 0.41 m from charged object 1. The result looks reasonable, as the unknown charge $q$ should be closer to the smaller magnitude charge $q_1$ than to the larger charge $q_2$.

The distance that we found does not depend on the magnitude or sign of the charge $q$. The net force exerted on $q$ by the two other objects will be zero regardless of the sign of $q$.

One limiting case is if one of the charges $q_1$ or $q_2$ is zero. If $q_1 = 0$, then only $q_2$ could exert a force on $q$ and the net force could *not* be zero. The force equation becomes $0 + (-F_{q_2 \text{ on } q}) = 0$, which has no solution because the right and left sides are never equal to each other. So this limiting case is consistent with our mathematics.

**Try it yourself:** Suppose we placed $q$ slightly closer to $q_1$ than the position calculated in this example. What happens to $q$?

*Answer:* The magnitude of the force that $q_1$ exerts on $q$ increases slightly and the magnitude of the force that $q_2$ exerts on $q$ decreases slightly. The net force exerted on $q$ is no longer zero—$q$ will accelerate away from the desired position if it is negative and toward the desired position if it is positive.

In the next example we apply the steps of the problem-solving strategy to a process whose analysis requires an energy approach.

**EXAMPLE 14.8 Radon decay in lungs**

Suppose that a radon atom in the air in a home is inhaled into the lungs. While in the lungs, the nucleus of the radon atom undergoes radioactive decay, emitting an $\alpha$ (alpha) particle, which is composed of two protons and two neutrons. (These high-energy alpha particles can damage lung tissue—a reason for keeping radon concentration low in homes.) During this process, the radon nucleus turns into a polonium nucleus with charge $+84e$ and mass $3.6 \times 10^{-25}$ kg. The $\alpha$ particle has charge $+2e$ and mass $6.6 \times 10^{-27}$ kg. Suppose the two particles are initially separated by $1.0 \times 10^{-15}$ m and are at rest. How fast is the $\alpha$ particle moving when it is very far from the polonium nucleus? (Note: 1 mm would be very far for such a process since even 1 mm is much larger than the size of a nucleus.)

**Sketch and translate** We draw a sketch showing the initial and final states of the system, choosing the product polonium nucleus and the escaping $\alpha$ particle as the system. We need to determine the speed of the $\alpha$ particle when it is infinitely far from the polonium nucleus.

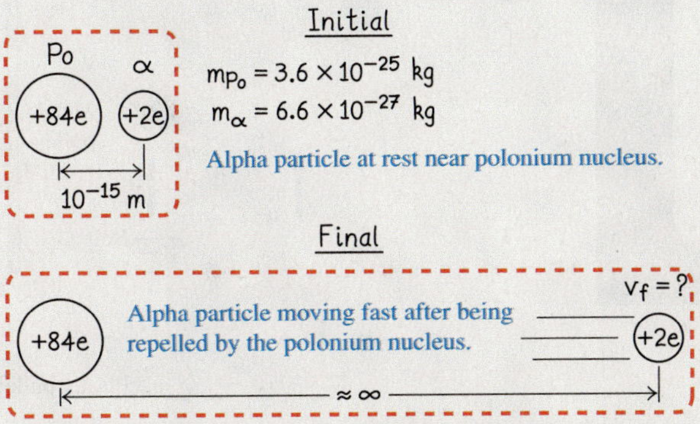


**Simplify and diagram** Model the nuclei and $\alpha$ particle as point-like objects and neglect the gravitational attraction between them. Assume that they are at rest at the start of the process and that we can ignore the final kinetic energy of the massive polonium nucleus. A work-energy bar chart for the process is shown at the right. In the initial state, the system has electric potential energy; in the final state, the $\alpha$ particle has kinetic energy, and since the particles are comparatively far apart we assume the system has zero electric potential energy.

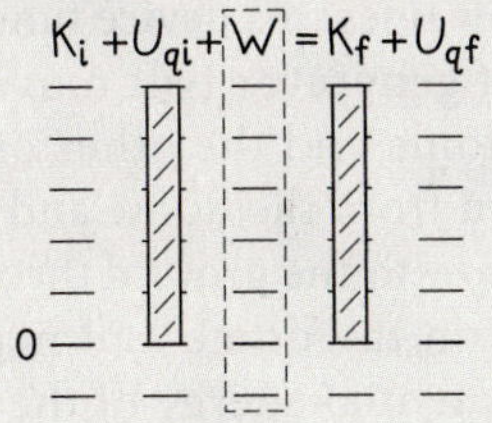


**Represent mathematically** Each nonzero bar in the bar chart turns into a nonzero term in the generalized work-energy equation.

$$\frac{k(+84e)(+2e)}{r_i} = \frac{1}{2}m_\alpha v_f^2$$

Rearranging the above, we get an expression for the final speed of the $\alpha$ particle:

$$v_f = \sqrt{\frac{2k(84e)(2e)}{m_\alpha r_i}}$$

**Solve and evaluate** Substituting the appropriate values:

$$v_f = \sqrt{\frac{2k(84e)(2e)}{m_\alpha r_i}}$$

$$= \sqrt{\frac{2(9.0 \times 10^9\ \text{N}\cdot\text{m}^2/\text{C}^2)(84.0 \times 1.6 \times 10^{-19}\ \text{C})(2.0 \times 1.6 \times 10^{-19}\ \text{C})}{(6.6 \times 10^{-27}\ \text{kg})(1.0 \times 10^{-15}\ \text{m})}}$$

$$= 1.1 \times 10^8\ \text{m/s}$$

We estimate that the final alpha particle speed will be about $10^8$ m/s.

**Try it yourself:** In the above example we assumed that the massive polonium nucleus remained at rest. Use impulse-momentum ideas to determine how fast the polonium nucleus was actually moving following the decay process. Compare the kinetic energies of the nucleus and the $\alpha$ particle.

*Answer:* $v_{polonium} = 0.0183\, v_{alpha}$. The kinetic energies, which depend on the speed squared, are related by $K_{polonium} = 0.0183\, K_{alpha}$. We were justified in ignoring the final kinetic energy of the polonium in this example.

**Review Question 14.6** How would our reasoning in Example 14.8 change if we chose the $\alpha$ particle alone as the system instead of both the nucleus and the escaping $\alpha$ particle?

## 14.7 Charge separation and photocopying: Putting it all together

We can separate charge by rubbing together two objects that are made of different materials. However, the magnitude of the resulting charge is quite low. In humid weather the separated charges on these objects are quickly neutralized by interactions with the air, making charge-related investigations difficult. To study the behavior of charged objects more easily we need charges of large magnitudes. Specialized equipment such as Van de Graaff generators (**Figure 14.19**) and Wimshurst machines are commonly used for such studies in physics courses.

**Figure 14.19** A charged Van de Graaff generator.

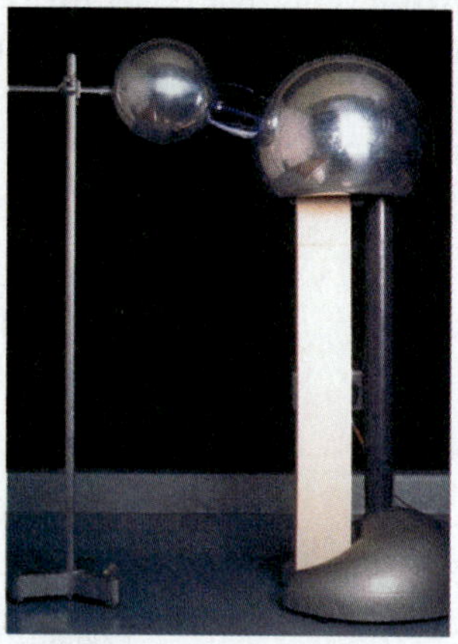

**Figure 14.20** How a Van de Graaff generator works.

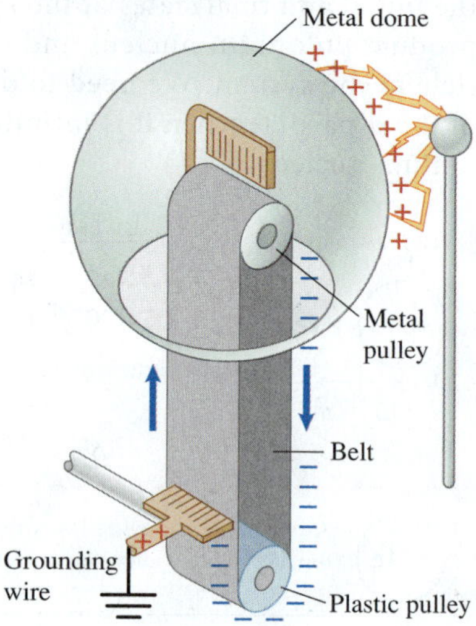


## Van de Graaff generator

A Van de Graaff generator is made of a plastic cylinder with a motorized moving belt inside and a hollow metal dome at the top. When the belt is moving, you can hear a cracking sound and see sparks around the dome. Some generators have an attachment—a metal ball on an insulating stand. The sparks between the dome and this ball or any nearby metal objects can be large, implying large charge separation. How does this device function?

A Van de Graaff generator belt moves on two rollers (one on the top inside of the metal dome and the other on the bottom—see **Figure 14.20**). Electrons are removed from the dome and then transferred by the belt to the bottom, where they flow to the ground through a grounding wire. This charge separation (positive on the dome and negative on the ground) greatly increases the electric potential energy of the dome/ground system. The charge accumulating on the dome is significant—on the order of $10^{-5}$ C.

### CONCEPTUAL EXERCISE 14.9 Making your hair stand on end.

The woman shown below has her hands on the dome of a Van de Graaff generator and her feet on an electrically insulated footstool. Why is her hair standing on end?

**Sketch and translate** Hair can stand up when the individual strands are repelled from each other and from the body. For this to happen, the hair and body need to have the same sign of electric charge.

**Simplify and diagram** If we assume that the human body is a fairly good conductor of electric charge, the woman's body is acting like an extension of the dome. Electrons flow from her to the dome, leaving her and each strand of her hair positively charged. This makes them repel from her body and from the other strands—they stand on end.

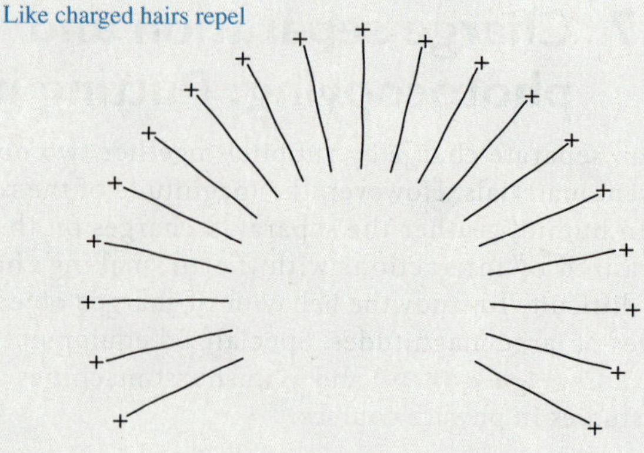

**Try it yourself:** You place a stack of aluminum pie plates on top of an uncharged Van de Graaff generator and then turn on the generator. Predict what happens to the plates and explain your prediction.

*Answer:* If we assume that electrons transfer from the plates to the positively charged dome, the now positively charged plates repel one another and the dome. They fly up off the dome.

Now imagine that you are holding a grounded metal ball about 10 cm from the dome of a charged Van de Graaff generator. What is actually happening in the air that results in a flash and a sharp cracking sound? The explanation involves what are known as **cosmic rays**, high-energy particles that continually rain down on Earth's atmosphere from space. These cosmic rays ionize atoms, producing free electrons and positively charged ions. High-energy particles produced during naturally occurring radioactive decay can do the same. The positive charge on the dome of a Van de Graaff generator attracts the free electrons already present in the air, causing them to accelerate toward the dome, colliding with other atoms and causing some of them to ionize. When these free electrons recombine with atoms, light is produced and we see a spark.

## EXAMPLE 14.10 What causes the sparking from a Van de Graaff generator?

The energy needed to remove an electron from a hydrogen atom is about $2 \times 10^{-18}$ J (about the same for other atoms too). The average distance a free electron in air will travel between collisions, called the mean free path, is about $10^{-6}$ m. The dome has a 0.15-m radius and a $+10^{-5}$-C charge. Could a free electron in the air gain enough kinetic energy to ionize an atom and cause a spark as it travels that short distance toward the charged dome?

**Sketch and translate** The situation is sketched below. We need to compare the kinetic energy that the electron gains between collisions to the energy of ionization of the atom.

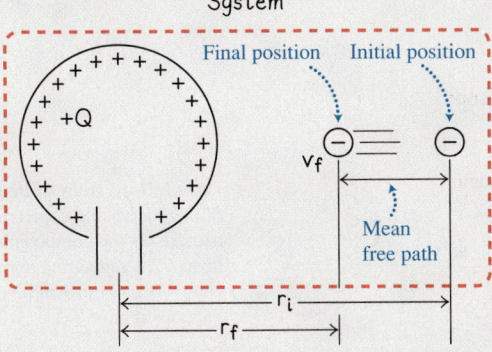

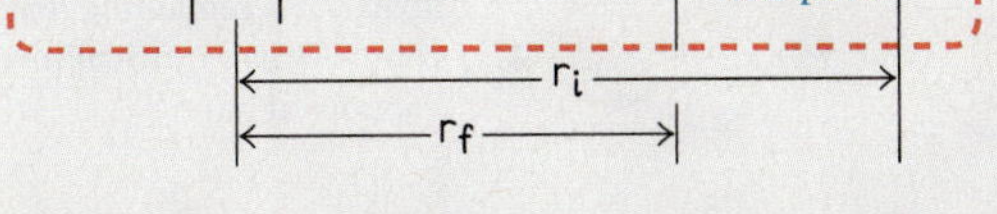


**Simplify and diagram** Assume that the electron is near the surface of the dome. The system will be the electron and the dome. We model both as point-like objects. This is reasonable for the electron and reasonable for the dome as well, but only because it is spherical in shape. (We will discuss that in more detail when we discuss the electric field in the next chapter.) An energy bar chart representing the process is shown below. The zero level of electric potential energy is at infinity. The electron-dome system starts with negative electric potential energy and zero kinetic energy. As the electron accelerates toward the dome, its kinetic energy increases and electric potential energy decreases.

**Represent mathematically** Use the bar chart to help apply the generalized work-energy equation to the process:

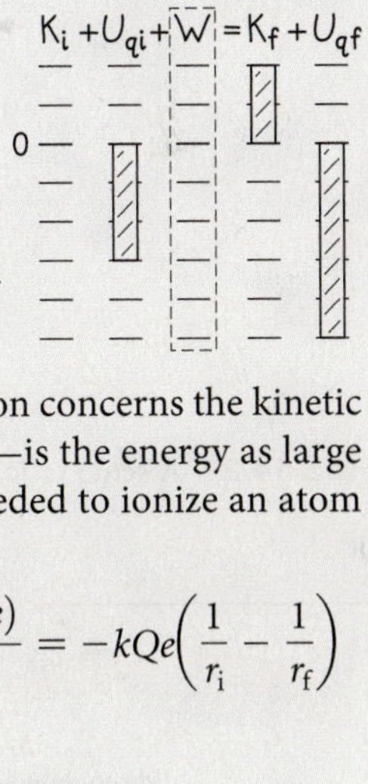


$$\frac{k(Q)(-e)}{r_i} = \frac{1}{2}mv_f^2 + \frac{k(Q)(-e)}{r_f}$$

**Solve and evaluate** Our question concerns the kinetic energy $(1/2)mv_f^2$ of the electron—is the energy as large as or does it exceed the energy needed to ionize an atom $\Delta U \approx 2 \times 10^{-18}$ J?

$$\frac{1}{2}mv_f^2 = \frac{k(Q)(-e)}{r_i} - \frac{k(Q)(-e)}{r_f} = -kQe\left(\frac{1}{r_i} - \frac{1}{r_f}\right)$$

$$= kQe\left(\frac{r_i - r_f}{r_i r_f}\right)$$

$$= (9 \times 10^9\ \text{N} \cdot \text{m}^2/\text{C}^2)(10^{-5}\ \text{C})$$

$$\times (1.6 \times 10^{-19}\ \text{C})\left(\frac{10^{-6}\ \text{m}}{(0.15\ \text{m})^2}\right)$$

$$\approx (10^{10})(10^{-5})(10^{-19})\left(\frac{10^{-6}}{10^{-2}}\right)(\text{N} \cdot \text{m})$$

$$\approx 10^{-18}\ \text{N} \cdot \text{m} = 10^{-18}\ \text{J}$$

*(continued)*

This is comparable to the energy needed to ionize an atom. Note that some electrons will travel farther than the average distance and therefore will have even more kinetic energy.

**Try it yourself:** Why are we ignoring the positive ions in the air and considering only the acceleration of the free electrons due to the positive charge on the dome?

*Answer:* The positive ions are much more massive than the electrons and do not accelerate to sufficient speeds to have the kinetic energy needed to ionize other atoms.

**Figure 14.21** A Wimshurst machine.

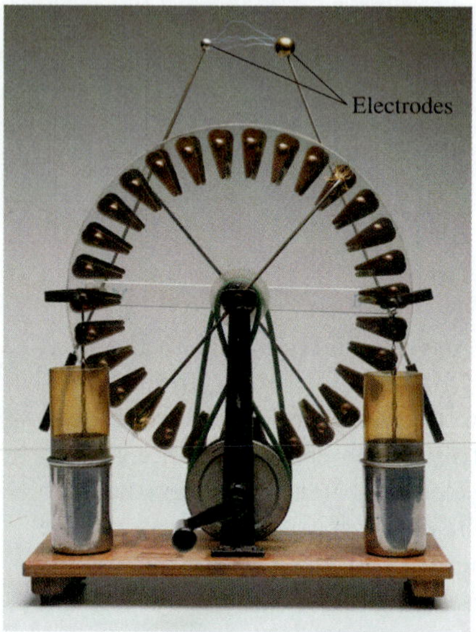


## Wimshurst machine

The Wimshurst machine (**Figure 14.21**) is another device that can produce large charge separations. The device, invented in the 1880s, consists of two plastic disks that rotate in opposite directions. Charge separation in the Wimshurst machine occurs by a complex process that we will not explain here. As a result of this process, one of the electrodes becomes positively charged and the other becomes negatively charged. When the two electrodes are brought near one another, we see sparks as long as 5–10 cm. The Wimshurst machine not only allows us to create large charges on the electrodes but also allows us to study the creation of the spark and its dependence on the distance between the electrodes—one goal of the next chapter.

## A photocopier

The photocopying process depends on the electrical attraction of oppositely charged particles—like how a balloon rubbed on a wool sweater attracts tiny bits of paper or sugar off a tabletop. In a copy machine the charged balloon is a drum that gets electrically charged with an image of the page being copied, and the paper bits are dark toner particles that stick to the places where the drum is charged. The toner then transfers to paper in the exact location of the letters or pictures on the copied page.

Let's analyze this process in more detail. The drum is covered with a special photoconductive material. When we turn on the photocopier, the drum becomes positively charged beneath the photoconductive material and negatively charged on the outside of the photoconductive material (**Figure 14.22a**).

**Figure 14.22** A schematic representation of the process of making a photocopy.

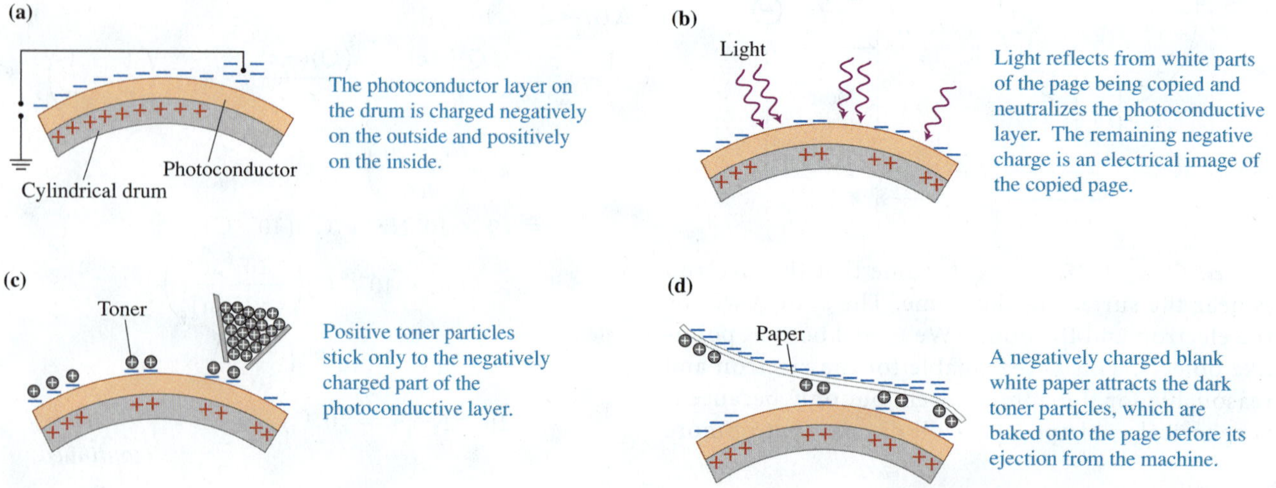

(a)
Photoconductor
Cylindrical drum
The photoconductor layer on the drum is charged negatively on the outside and positively on the inside.
(b)
Light
Light reflects from white parts of the page being copied and neutralizes the photoconductive layer. The remaining negative charge is an electrical image of the copied page.
(c)
Toner
Positive toner particles stick only to the negatively charged part of the photoconductive layer.
(d)
Paper
A negatively charged blank white paper attracts the dark toner particles, which are baked onto the page before its ejection from the machine.

During copying, a strong source of light moves across the page from under the glass cover. This light reflects off white regions of the page but does not reflect from dark regions. The reflected light reaches the photoconductive material where electrons on the surface of the photoconductive material absorb the light and move inside the photoconductive material to neutralize the positive charge. The drum surface is now neutral in places where light was reflected from the original page being copied. The drum remains negatively charged in areas below the dark text of the page that was being copied. Thus, the negative electrons on the top of the drum form an electrical image of the text being copied (Figure 14.22b).

The next step is covering the drum with the toner, which is positively charged. The toner sticks to the negatively charged "electrical image" on the drum (Figure 14.22c). Then a blank negatively charged paper wraps around the drum and pulls off the toner (Figure 14.22d). The drum and the paper are heated and pressed together to make this transfer more effective and to bake the toner on the surface of the paper. Finally, the new copy is ejected and a rubber material wipes the drum clean of remaining toner and illuminates it with light to remove all remaining charge.

**Review Question 14.7** In a Van de Graaff generator, where does the energy emitted as light in a spark come from?

# Summary

| Words | Pictorial and physical representations | Mathematical representation |
|---|---|---|

**Electric charge** is a physical quantity that characterizes how charged objects participate in electrostatic interactions.

(a) Charge comes in two types—positive and negative.
(b) Like charged objects repel and unlike charged objects attract.
(c) The smallest charge is the charge of an electron ($-e = -1.6 \times 10^{-19}$ C).
(d) Electric charge is constant in an isolated system. (Sections 14.1 and 14.2)

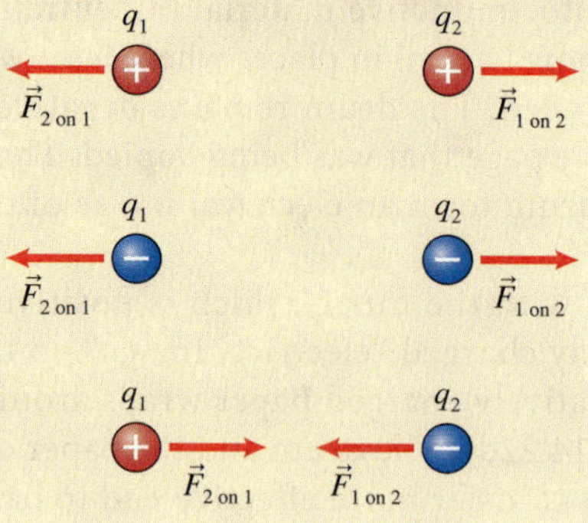


---

**Conductors** are materials in which electrically charged particles can move freely. (Section 14.3)

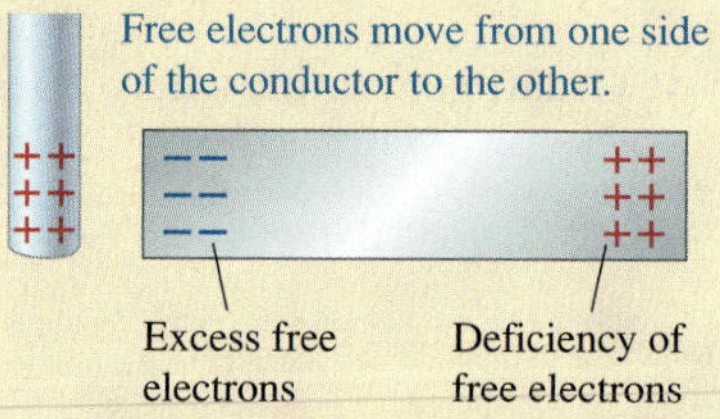


---

**Dielectrics** (insulators) are materials in which electrically charged particles cannot move freely. However, the electric charges in the atoms and molecules in the material can rearrange slightly (a process called polarization), allowing them to participate in electric interactions. (Section 14.3)

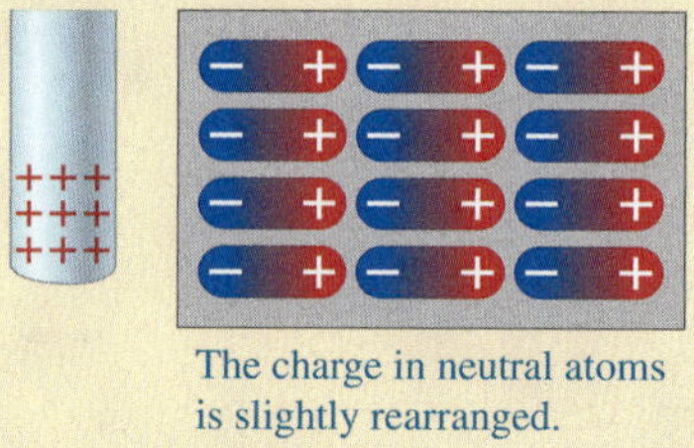

The charge in neutral atoms is slightly rearranged.

---

**Coulomb's law** is used to determine the magnitude of the electric force that point-like objects with charges $q_1$ and $q_2$ exert on each other when separated by a distance $r$. (Section 14.4)

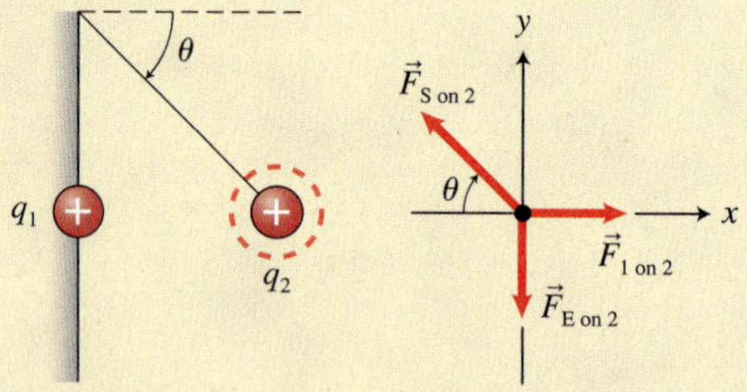


$$|F_{q_1 \text{ on } q_2}| = |F_{q_2 \text{ on } q_1}| = \frac{k|q_1||q_2|}{r^2}$$ Eq. (14.1)

$$k = 9.0 \times 10^9 \frac{\text{N} \cdot \text{m}^2}{\text{C}^2}$$

---

The **electric potential energy** $U_q$ of a system of point-like objects with charges $q_1$ and $q_2$ separated by a distance $r$ is positive for like charges and negative for unlike charges (include the signs of the charges when using this equation.) The zero level of electric potential energy is usually chosen to be when they are infinitely far apart. (Section 14.5)

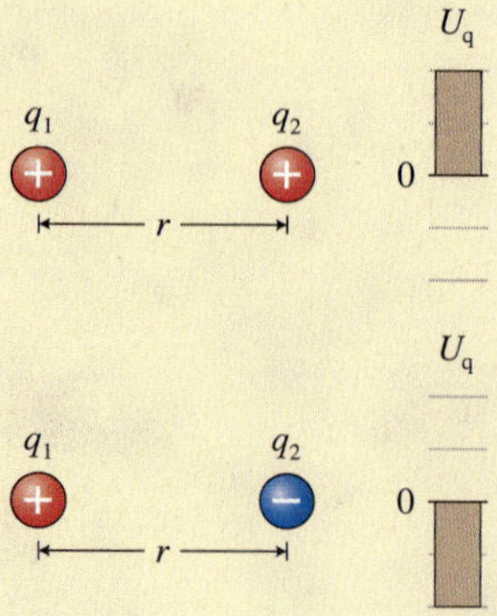


$$U_q = \frac{kq_1q_2}{r}$$ Eq. (14.2)

MP® For instructor-assigned homework, go to MasteringPhysics.

# Questions

## Multiple Choice Questions

1. Which of the following occur when two objects are rubbed against each other?
   (a) Both acquire the same sign charge.
   (b) They acquire opposite sign charge.
   (c) No new charges are created.
   (d) Both b and c are correct.
2. With which statements do you disagree?
   (a) If two objects repel, they have like charges.
   (b) If two objects attract, they have opposite charges.
   (c) If two objects attract, one of them can be charged and the other one can be neutral.
   (d) None of them
3. Which explanation agrees with the contemporary model of electric charge?
   (a) If an object is charged negatively by rubbing, it has lost positively charged particles.
   (b) If an object is charged positively by rubbing, it has lost negatively charged particles.
   (c) If an object is charged positively by rubbing, it has acquired positively charged particles.
   (d) If an object is charged negatively by rubbing, it has acquired negatively charged particles.
   (e) a and c
   (f) b and d
4. When an object gets charged by rubbing, where does the electric charge originate?
   (a) It comes from the air surrounding the object.
   (b) It is created by the rubbing energy.
   (c) The process of rubbing leads to a redistribution of charge, not the creation of charge.
5. Choose all of the quantities that are constant in an isolated system.
   (a) Mass (b) Velocity
   (c) Electric charge (d) Momentum
   (e) Mechanical energy (f) Total energy
   (g) Temperature (h) Entropy
6. Identically charged objects A and B are separated by a distance $d$. You measure the force $F$ that A exerts on B. If you now separate the objects by distance $1.4d$, what will be the new force that A exerts on B?
   (a) $1.4\,F$ (b) $2.0\,F$ (c) $\frac{1}{2}F$
   (d) $\frac{1}{1.4}F$ (e) $F$
7. When separated by distance $d$, identically charged objects A and B exert a force of magnitude $F$ on each other. If you reduce the charge of A to one-half its original value, and the charge of B to one-fourth, and reduce the distance between the objects by half, what will be the new force that they exert on each other?
   (a) $2F$ (b) $\frac{1}{2}F$ (c) $4F$
   (d) $\frac{F}{4}$ (e) $F$
8. Balloon A has charge $q$, and identical mass balloon B has charge $10q$. You hang them from threads near each other. Choose all of the statements with which you agree.
   (a) The force that A exerts on B is 1/10 the force that B exerts on A.
   (b) The force that A exerts on B is 10 times the force that B exerts on A.
   (c) A and B exert the same magnitude forces on each other.
   (d) The angle between the thread supporting A and the vertical is less than the angle between the thread supporting B and the vertical.
9. Imagine that two charged objects are the system of interest. When the objects are infinitely far from each other, the electric potential energy of the system is zero. When the objects are close to each other, the electric potential energy is positive. Which of the following statements is(are) incorrect?
   (a) Both objects are positively charged.
   (b) Both objects are negatively charged.
   (c) One object is negatively charged and the other one is positively charged.
10. Two objects with charges $+q$ and $-2q$ are separated by a distance $r$. You slowly move one of the objects closer to the other so that the distance between them decreases by half. Considering the two objects as the system, which statements are incorrect?
    (a) The electric potential energy of the system doubles.
    (b) The electric potential energy of the system decreases by half.
    (c) The magnitude of the electric potential energy doubles.
    (d) The magnitude of the electric potential energy decreases by half.
11. Charged objects A and B are separated by a distance $d$. Choose object A alone as the system. You slowly push the other charged object B closer to A, decreasing the distance between them to $0.3d$. Based on the given information, choose all of the correct statements.
    (a) The electric potential energy of the system increases.
    (b) The electric potential energy of the system decreases.
    (c) The electric potential energy of the system does not change.
    (d) You do positive work on the system.
    (e) You do negative work on the system.
    (f) You do zero work on the system.
12. Two small objects, each with charge $+5.0 \times 10^{-9}$ C, are separated by a distance of 50 cm. What is the magnitude and the direction of the electric force that they exert on an identical third object that is 50 cm from each of them?
    (a) $1.6 \times 10^{-6}$ N, perpendicular to the line connecting the charges
    (b) $1.6 \times 10^{-6}$ N, parallel to the line connecting the charges
    (c) $1.8 \times 10^{-6}$ N, perpendicular to the line connecting the charges
    (d) $1.8 \times 10^{-6}$ N, parallel to the line connecting the charges

## Conceptual Questions

13. Describe the differences between the electric force and the gravitational force and some experiments that can be explained through these differences.
14. Describe the difference between charged objects and magnets and some experiments that can be explained through these differences.
15. At one time it was thought that electric charge was a weightless fluid. An excess of this fluid resulted in a positive charge; a deficiency resulted in a negative charge. Describe an experiment for which this hypothesis provides a satisfactory explanation.
16. How do you know that there are only two types of charge, and not three or more types?
17. Describe one currently accepted explanation for how electric charge is transferred from one object to another.
18. An object acquires a positive electric charge due to rubbing. Does its mass increase, decrease, or remain the same? Explain. How is its mass affected if it acquires a negative electric charge? Explain.
19. List everything that you know about electric charges.
20. What experimental evidence supports the idea that conducting materials have freely moving electrically charged particles inside them?
21. You have an aluminum pie pan with pieces of aluminum foil attached to it. Predict what will happen if you touch the pan with a plastic rod rubbed with wool.
22. You have a charged metal ball. How can you reduce its charge by half?
23. You have a foam rod rubbed with felt and a small aluminum foil ball attached to a thread. Describe what happens when you slowly approach the ball with the rod and then touch the ball. Explain why this happens.
24. A positively charged metal ball A is placed near metal ball B. Measurements demonstrated that the force between them is zero. Is ball B charged? Explain.
25. Prove that if the charge on B in the previous question is positive but of small magnitude, the balls will be attracted to each other.
26. ✎ Two metal balls of the same radius are placed a small distance apart. Will they interact with the same magnitude force when they have like charges as when they have opposite charges? Explain your answer. Include a sketch showing the charge distribution.
27. Describe the experiments that were first used to determine a quantitative expression for the magnitude of the force that one electrically charged object exerts on another electrically charged object. What are the limitations of this expression?
28. The electrical force that one electric charge exerts on another is proportional to the product of their charges—that is, $F_{1\text{ on }2} \propto q_1q_2$. How would Coulomb's observations be different if the force were proportional to the sum of their charges $(q_1 + q_2)$?
29. Why isn't Coulomb's law valid for large conducting or dielectric objects, even if they are spherically symmetrical?
30. How is electric potential energy similar to spring potential energy? How is it different?

## Problems

Below, **BIO** indicates a problem with a biological or medical focus. Problems labeled **EST** ask you to estimate the answer to a quantitative problem rather than develop a specific answer. Problems marked with ✎ require you to make a drawing or graph as part of your solution. Asterisks indicate the level of difficulty of the problem. Problems with no * are considered to be the least difficult. A single * marks moderately difficult problems. Two ** indicate more difficult problems.

In some of these problems you may need to know that the mass of an electron is $9.1 \times 10^{-31}$ kg and the charge is $-1.6 \times 10^{-19}$ C.

### 14.1–14.3 Electrostatic interactions; Explanations for electrostatic interactions; Conductors and nonconductors (dielectrics)

1. **BIO Ventricular defibrillation** During ventricular fibrillation, the muscle fibers of the heart's ventricles undergo uncoordinated rapid contractions, resulting in little or no blood circulation. To restore the heart's normal rhythm, a defibrillator sends an abrupt jolt of about −0.20 C of electric charge through the chest into the heart. How many electrons pass through the body during this defibrillation?
2. * You rub two 4.0-mg balloons with a wool sweater. The balloons hang from 0.50-m-long very light strings. When you attach the strings together at the top, the balloons hang away from each other each string making an angle of 37° with the vertical. (a) Represent the situation with the force diagram for each balloon and determine the magnitudes of the forces on the diagram. (b) What can you say about the magnitudes of the forces that the balloons exert on each other? Explain. (c) Will the relative magnitudes change if the charge on one balloon is two times larger than on the other? How do you know?
3. * ✎ Two balloons of different mass hang from strings near each other. You charge them about the same amount by rubbing each balloon with wool. Draw a force diagram for each of the balloons. Compare the angles of the threads with the vertical. How do your answers depend on whether the balloons have the same or different magnitude charge?
4. * ✎ **Lightning** A cloud has a large positive charge. Assume that this is the only cloud in the sky over Earth and that Earth is a good electrical conductor. Draw a sketch showing electric charge distribution on Earth due to the cloud's electric charge. Explain why a person's hair might stand on end before a lightning strike.

### 14.4 Coulomb's force law

5. **EST** (a) Earth has an excess of 60 electrons on each square centimeter of surface. Determine the electric charge in coulombs on each square centimeter of surface. (b) If, as you walk across a rug, about $10^{-22}$ kg of electrons transfer to your body, estimate the number of electrons and the total charge in coulombs on your body.

6. You have a small object with 0.0020 C of electric charge and another small object with 0.0060 C of charge. Compare the magnitude of the electric force that the 0.0020-C object exerts on the 0.0060-C object to the force that the 0.0060-C object exerts on the 0.0020-C object. Explain your answer.
7. Determine the electrical force that two protons in the nucleus of a helium atom exert on each other when separated by $2.0 \times 10^{-15}$ m.
8. * Two charged objects exert a 4.0-N force on each other when separated by 1.0 m. (a) What can you determine using this information? (b) You then perform four experiments: you double the separation; you reduce the separation by one-half; you reduce the magnitude of one charge by one-half; and you double both charges. What quantitative information about the interaction of the objects can you determine for each of the experiments?
9. * Two identical small metal spheres are separated by $1.0 \times 10^5$ m and exert an 18.9-N repulsive force on each other. A wire is connected between the spheres and then removed. The spheres now repel each other, exerting a 22.5-N force. (a) Explain why the force that these two objects exert on each other changed. (b) Determine everything you can about the situation before the wire was connected and after it was removed.
10. * Determine the number of electrons that must be transferred from Earth to the Moon so that the electrical attraction between them is equal in magnitude to their present gravitational attraction. What is the mass of this number of electrons?
11. BIO **Ions on cell walls** The membrane of a body cell has a positive ion of charge $+e$ on the outside wall and a negative ion of charge $-e$ on the inside wall. Determine the magnitude of the electrical force between these ions if the membrane thickness is $0.80 \times 10^{-9}$ m. Ignore the effect of the material in which the ions are located.
12. Sodium chloride (table salt) consists of sodium ions of charge $+e$ arranged in a crystal lattice with an equal number of chlorine ions of charge $-e$. The mass of each sodium ion is $3.82 \times 10^{-26}$ kg and that of each chlorine ion $5.89 \times 10^{-26}$ kg. Suppose that the sodium ions could be separated into one pile and the chlorine ions into another. What mass of salt would be needed to get 1.00 C of charge into the sodium ion pile and $-1.00$ C into the chlorine ion pile?
13. * **Electrical attraction** Two friends each contain about $4 \times 10^{28}$ electrons and an equal number of protons. What will happen in terms of their interaction if 1% of one friend's electrons are transferred to the other, who is about 100 m away? What other physical quantities can you determine using this information?
14. * **Hydrogen atom** In a simplified model of a hydrogen atom, the electron moves around the proton nucleus in a circular orbit of radius $0.53 \times 10^{-10}$ m. Use this information to determine at least four physical quantities related to this information.
15. * Three 1.0-C charged objects are equally spaced on a straight line. The separation of each object from its neighbor is 100 m. Find the force exerted on the center object if (a) all charges are positive, (b) all charges are negative, and (c) the rightmost charge is negative and the other two are positive.
16. * Two objects with charges $q$ and $4q$ are separated by 1.0 m. (a) Determine the sign, magnitude, and position of a third charged object that causes all three objects to remain in equilibrium. (b) Is the equilibrium stable or unstable? How do you know?
17. * **Salt crystal** Four ions ($Na^+$, $Cl^-$, $Na^+$, and $Cl^-$) in a row are each separated from their nearest neighbor by $3.0 \times 10^{-10}$ m. The charge of a sodium ion is $+e$ and that of a chlorine ion is $-e$. Determine the electric force exerted on the chlorine ion at the right end of the row due to the other three ions.
18. * A +1.0-C charged object and a +2.0-C charged object are separated by 100 m. Where should a $-1.0 \times 10^{-3}$-C charged object be located on a line between the positively charged objects so that the net electrical force exerted on the negatively charged object is zero?
19. ** BIO **Bee pollination** Bees acquire an electric charge in flight from friction with the air, which causes pollen to cling to them. The pollen is then attracted to the stigma of the next flower (**Figure P14.19**). Suppose the bee's body has a charge of $-10^{-9}$ C and is about $3 \times 10^{-3}$m from the front edge of a spherical granule of pollen of diameter $5 \times 10^{-5}$ m. Charged particles in the pollen become polarized with $+10^{-11}$ C on the front edge and $-10^{-11}$ C on the backside of the pollen $(3 \times 10^{-3} + 5 \times 10^{-5})$ m from the bee. What useful physical quantities can you determine?

**Figure P14.19**

20. * A triangle with equal sides of length $1.0 \times 10^3$ m has $-2.0$-C charged objects at each corner. Determine the electrical force (magnitude and direction) exerted on the object at the top corner due to the two objects at the base of the triangle.
21. * You have a small metal sphere hanging from a thread and another sphere attached to a plastic handle. You charge one of the spheres and then touch it with the second sphere. Draw a force-versus-time graph for the force that one sphere exerts on the other as you move the sphere on the handle slowly and steadily away from the hanging sphere. What information do you need in order to determine the magnitudes of the quantities on the graph?
22. * After the experiment in Problem 21, you touch the hanging sphere with your hand. Then you touch that hanging sphere with the sphere on the handle. Draw a force-versus-time graph for the force that one sphere exerts on the other as you move the sphere on the handle slowly and steadily away from the hanging sphere. How is this graph different from the graph in the previous problem?
23. ** Coulomb's law is formulated for point-like charged objects. Imagine that you have two point-like charged objects $q_1$ and $q_2$ of unspecified sign separated by a distance $d$ and you also have two large spheres of radius $R$ with the same charges $q_1$ and $q_2$. The distance between the centers of the spheres is $d$. Compare the electric force that the point-like objects exert on each other to the force that the two spheres exert on each other. Consider all possibilities.

### 14.5 Electric potential energy

24. (a) Determine the change in electric potential energy of a system of two charged objects when a $-1.5$-C charged object and a $-4.0$-C charged object move from an initial separation of 500 km to a final separation of 100 km. (b) What other quantities can you calculate using this information?
25. You have a system of two positively charged objects separated by some arbitrary finite distance. (a) What is the sign of

their potential energy? (Remember that charges that are infinitely far from each other have zero potential energy.) (b) What can you do to decrease this energy? (c) Draw an energy bar chart for this process of decreasing the energy.

26. You have a system of two negatively charged objects separated by some arbitrary finite distance. (a) What is the sign of their potential energy? (Remember that charges that are infinitely far from each other have zero potential energy.) (b) What can you do to decrease this energy? (c) Draw an energy bar chart for this process of decreasing the energy.

27. Repeat (a)—(c) of Problem 26 for a system with a negatively charged object and a positively charged object separated by some arbitrary finite distance.

28. * BIO **Heart's dipole charge** At one instant of time, a person's heart has a dipolar charge on it, with a positive charge on the bottom and a negative charge on the top of the heart. Will the electric potential energy of a sodium ion (charge $+e$) on the left side of the heart (as seen by another person) and at a level halfway between the heart's charges increase, decrease, or remain the same if the sodium ion moves up? What will happen to the potential energy of a chlorine ion (charge $-e$) that starts at the same place and moves up? Justify each answer.

29. The metal sphere on the top of a Van de Graaff generator has a relatively large positive charge. In which direction must a positively charged ion in the air move relative to the sphere in order for the electrical energy the ion-generator system to decrease? Justify your answer.

30. ** EST An electron is 0.10 cm from an object with electric charge of $+3.0 \times 10^{-3}$ C. (a) Determine the magnitude of the electrical force $F_{\text{O on } e}$ that the object exerts on the electron. (b) The electron is pulled so that it moves to a distance of 0.11 cm from the charged object. Determine the magnitude of the electrical force $F_{\text{O on } e}'$ exerted by the charged object on the electron when at this distance. (c) Estimate the work done by the average force pulling the electron

$$\left(\frac{F_{\text{O on } e} + F_{\text{O on } e}'}{2}\right)\Delta x$$

(d) Compare this number to the change in electric potential energy of the electron–charged-object system as the electron moves away from the object. Why should the numbers be approximately equal?

31. * (a) An object with charge $q_4 = +3.0 \times 10^{-5}$ C is moved to position C from infinity (**Figure P14.31**). $q_1 = q_2 = q_3 = +10.0 \times 10^{-4}$ C. Determine as many work-energy quantities as you can that characterize this process. Make sure you specify the system. (b) Repeat your calculations, but for $q_1 = q_3 = -0.01$ C and $q_2 = +0.01$ C.

**Figure P14.31**

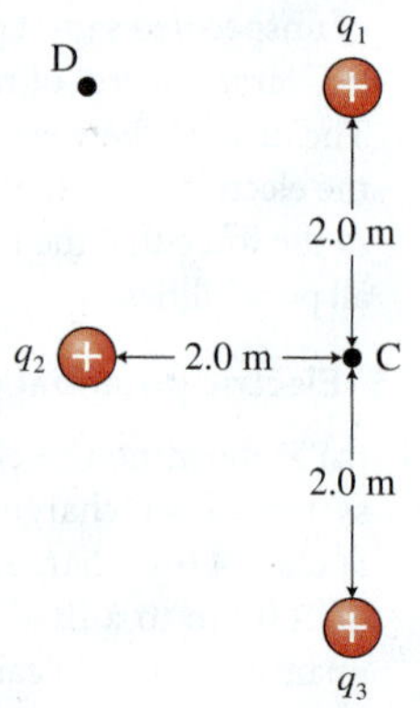


32. * An object with charge $+2.0 \times 10^{-5}$ C is moved from position C to position D in Figure P14.31. $q_1 = q_3 = +10.0 \times 10^{-5}$ C and $q_2 = -20.0 \times 10^{-5}$ C. All four charged objects are the system. What work-energy-related quantities can you determine for the process?

33. ** A stationary block has a charge of $+6.0 \times 10^{-4}$ C. A 0.80-kg cart with a charge of $+4.0 \times 10^{-4}$ C is initially at rest and separated by 4.0 m from the block. The cart is released and moves along a frictionless surface to a distance of 10.0 m from the block. Determine as many values of the physical quantities describing the motion of the cart as you can.

### 14.6 Skills for analyzing processes involving electric force and electric potential energy

34. ** An electric cannon, such as shown in Figure 14.15, consists of a 10-kg metal ball with charge $+2.0 \times 10^{-4}$ C compressed into a plastic barrel so that it is 0.10 m from an equally charged object at the closed end of the barrel. The barrel is oriented at 37° with respect to the horizontal. When the ball is released, it shoots 3.0 m along the barrel from its starting position because of the repulsive force between the two charged objects. Determine three physical quantities that describe the motion of the ball after it leaves the barrel.

35. * **Equation Jeopardy 1** The solution to a problem is represented by the following equation:

$$(9.0 \times 10^9\ \text{N m}^2/\text{C}^2)(0.020\ \text{C})(0.050\ \text{C})/(100\ \text{m})^2 + (9.0 \times 10^9\ \text{N m}^2/\text{C}^2)(0.010\ \text{C})(0.050\ \text{C})/(50\ \text{m})^2 = (10\ \text{kg})a_x$$

Sketch a situation that the equation might represent and formulate a problem for which it is a solution (there are multiple possibilities).

36. * **Equation Jeopardy 2** The solution to a problem is represented by the following equation:

$$(9.0 \times 10^9\ \text{N m}^2/\text{C}^2)(0.020\ \text{C})(0.050\ \text{C})/(5\ \text{m}) = (1/2)(10\ \text{kg})v_x^2 + (9.0 \times 10^9\ \text{N m}^2/\text{C}^2)(0.020\ \text{C})(0.050\ \text{C})/(20\ \text{m})$$

Sketch a situation that the equation might represent and formulate a problem for which it is a solution (there are multiple possibilities).

37. * **Equation Jeopardy 3** The solution to a problem is represented by the following equation:

$$(9.0 \times 10^9\ \text{N m}^2/\text{C}^2)(0.020\ \text{C})(0.050\ \text{C})/(100\ \text{m})^2 - (9.0 \times 10^9\ \text{N m}^2/\text{C}^2)(0.010\ \text{C})(0.050\ \text{C})/(50\ \text{m})^2 = (10\ \text{kg})a_x$$

Sketch a situation that the equation might represent and formulate a problem for which it is a solution (there are multiple possibilities).

38. * **Equation Jeopardy 4** The solution to a problem is represented by the following equation:

$$(9.0 \times 10^9\ \text{N m}^2/\text{C}^2)(-0.020\ \text{C})(0.050\ \text{C})/(20\ \text{m}) = (1/2)(10\ \text{kg})v_x{}^2 + (9.0 \times 10^9\ \text{N m}^2/\text{C}^2)(-0.020\ \text{C})(0.050\ \text{C})/(5\ \text{m})$$

Sketch a situation that the equation might represent and formulate a problem for which it is a solution (there are multiple possibilities).

39. * **Evaluate the solution** A student was given the following problem: A 0.20-kg cannonball with a $+1.0 \times 10^{-4}$ C charge starts at rest 0.40 m from a fixed $+2.0 \times 10^{-4}$ C charge. When the cannonball is released, it flies vertically upward. How fast is it moving when 10 m above the fixed charge? The student gave this solution:

$$(9.0 \times 10^9\ \text{N m}^2/\text{C}^2)(+2.0 \times 10^{-4}\ \text{C})(+1.0 \times 10^{-4}\ \text{C})/(0.40\ \text{m})^2 = (1/2)(10\ \text{kg})v_y{}^2 + (0.20\ \text{kg})(9.8\ \text{m/s}^2)(10\ \text{m})$$

or $v_y = 15$ m/s. Evaluate the solution to this problem and correct any errors you find.

40. * Construct separate force diagrams for each charged object shown in **Figure P14.40**. Use two-letter subscripts identifying each force.

**Figure P14.40**

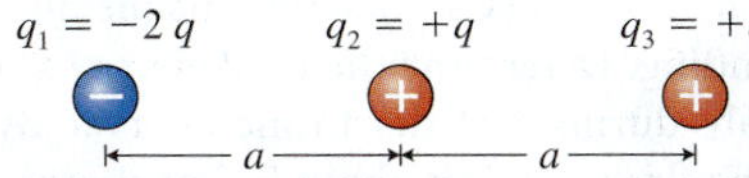


41. * The six objects shown in **Figure P14.41** have equal-magnitude electric charge. Adjacent objects are separated by distance $a$. Write an expression in terms of $q$ and $a$ for the force that the five objects on the right exert on the positive charge on the left.

**Figure P14.41**

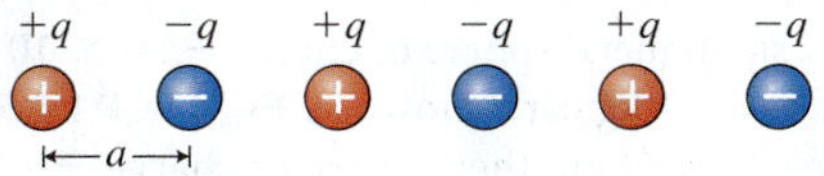


42. * A small metal ball with positive charge $+q$ and mass $m$ is attached to a very light string, as shown in **Figure P14.42**. A larger metal ball with negative charge $-Q$ is securely held on a plastic rod to the ceiling. Write an expression for the magnitude of the force $T$ that the string exerts on the ball. Define any other quantities used in your expression.

**Figure P14.42**

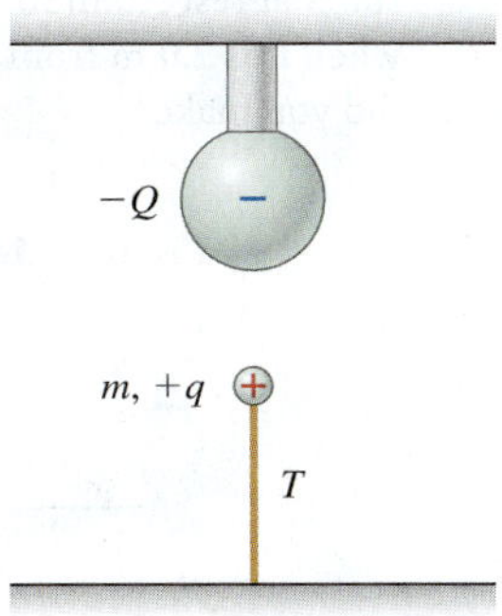


43. * Four objects each with charge $+1.0 \times 10^{-4}$ C are located at the corners of a square whose sides are 2.0 m long. Determine the values of two physical quantities describing the situation but not provided in the givens.

44. ** Two 5.0-g aluminum foil balls hang from 1.0-m-long threads that are suspended from the same point at the top. The charge on each ball is $+5.0 \times 10^{-6}$ C. Make a list of the physical quantities that you can determine using this information. Determine the values of two of those physical quantities.

45. * A 0.060-kg ball with charge $+3.0 \times 10^{-6}$ C hangs from a 0.50-m-long string at an angle of 53° below the horizontal. The string is attached at the top on the wall above a second charged object, as shown in **Figure P14.45**. (a) Determine the charge on the second object. (b) Make a list of other physical quantities that you can determine using the information in the problem. Describe how you can determine two of those quantities.

**Figure P14.45**

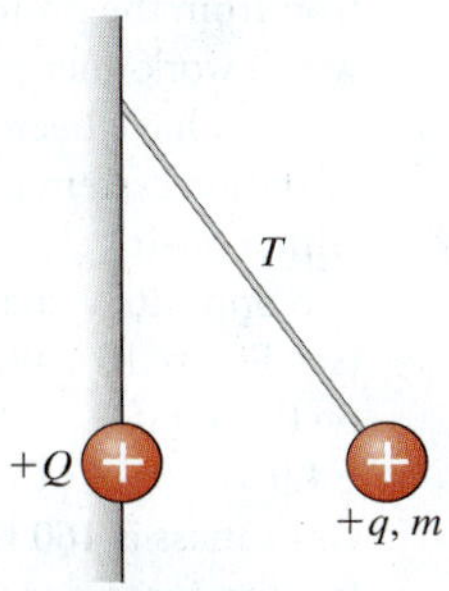


46. * A 0.40-kg cart with charge $+4.0 \times 10^{-5}$ C starts at rest on a horizontal frictionless surface 0.50 m from a fixed object with charge $+2.0 \times 10^{-4}$ C. When the cart is released, it moves away from the fixed object. (a) How fast is the cart moving when very far (infinity) from the fixed charge? (b) When 2.0 m from the fixed object?

47. * **Electric cannon** The cannonball shown in **Figure P14.47** has charge $q = +8.0 \times 10^{-5}$ C and mass 0.10 kg. The fixed object in the cannon has a charge of $Q = +9.0 \times 10^{-4}$ C. The ball starts at $d = 0.10$ m from the fixed charged object. (a) Determine the maximum height of the cannonball after its release. (b) Show how you can determine two other physical quantities relevant to the process.

**Figure P14.47**

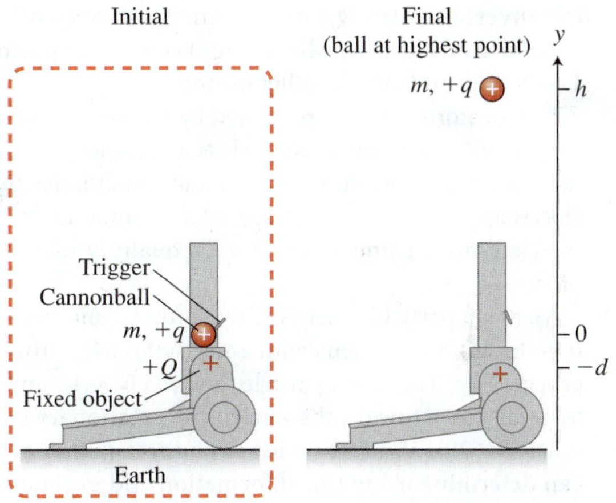


48. * **Electric shuttle train** A shuttle train moves from one station to another. It is powered by equal-magnitude, opposite-sign charges on top of each station (see **Figure P14.48**). If a 7000-N friction force opposes the train's motion, how great must the dipole charge be to induce an initial acceleration of $1.0 \text{ m/s}^2$? The train's mass is $2.0 \times 10^4$ kg, and it has a positive charge of $+3.0 \times 10^{-2}$ C on its roof.

**Figure P14.48**

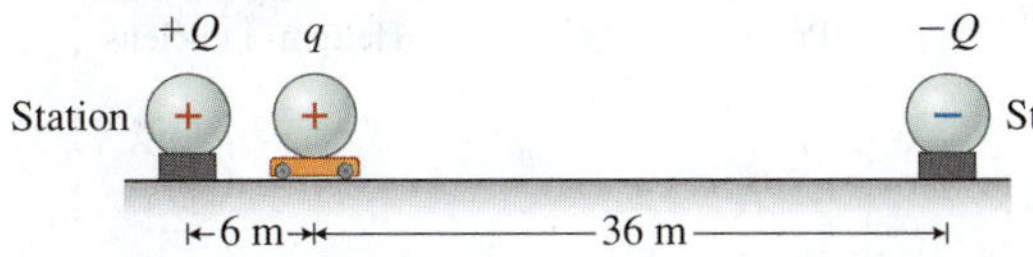


49. * A small metal sphere with electric charge $+q$ is brought to a distance $d$ from a metal sphere with charge $Q$ sitting on the top of a nonconducting support. The mass of the small sphere is $m$. What are the direction and magnitude of the electric force exerted by the large sphere on the small sphere? Make sure you analyze different possibilities.

## 14.7 Charge separation and photocopying: Putting it all together

50. * The dome of a Van de Graaff generator has a radius of 0.10 m. An electron in a hydrogen atom is located at a distance 0.20 m from the center of the dome. Determine the magnitude of the positive charge needed on the dome to exert a force on the electron that is equal to the force exerted by the atom's nucleus on the electron. Is this a reasonable amount of charge on the dome? Explain.

51. ** A Van de Graaff generator is placed in rarefied air at 0.4 times the density of air at atmospheric pressure. The average distance that free electrons move between collisions (mean free path) in that air is $(1/0.4) \times 10^{-6}$ m. Determine the positive charge needed on the generator dome so that a free electron located 0.20 m from the center of the dome will gain at the end of the mean free path length the $2.0 \times 10^{-18}$ J of kinetic energy needed to ionize a hydrogen atom during a collision.

52. * Two protons each of mass $1.67 \times 10^{-27}$ kg and charge $+e$ are initially at rest and separated by $1.0 \times 10^{-14}$ m (approximately the radius of a nucleus). When released, the protons fly apart. (a) Determine the change in their electric potential energy when they are $1.0 \times 10^{-10}$ m apart (approximately the radius of an atom). (b) If the electric potential energy change is converted entirely into the kinetic energy of the protons (shared equally), what is the speed of one proton when $1.0 \times 10^{-10}$ m from the other proton?

53. * Two protons, initially separated by a very large distance ($r_i$ is infinity), move directly toward each other with the same initial speed $v_i$. (a) Determine their initial speeds if the distance of closest approach when their speeds are zero is $4.0 \times 10^{-14}$ m. (b) Determine some other physical quantity relevant to the process.

54. * An alpha particle consists of two protons and two neutrons together in one nucleus with a mass of $6.64 \times 10^{-27}$ kg and charge $+2e$. The alpha particle flies at $3.0 \times 10^{7}$ m/s from a large distance toward the nucleus of a stationary gold atom (charge $+79e$). (a) Make a list of physical quantities that you can determine using this information and explain how you will calculate one of them. (b) Determine the distance of the alpha particle from the gold nucleus when it stops.

55. * Determine the speed that the proton shown in **Figure P14.55** must be moving in order to get within $1.0 \times 10^{-15}$ m of the helium-3 nucleus that has two protons and one neutron. Assume that the helium nucleus is attached to a massive molecule and does not move.

**Figure P14.55**

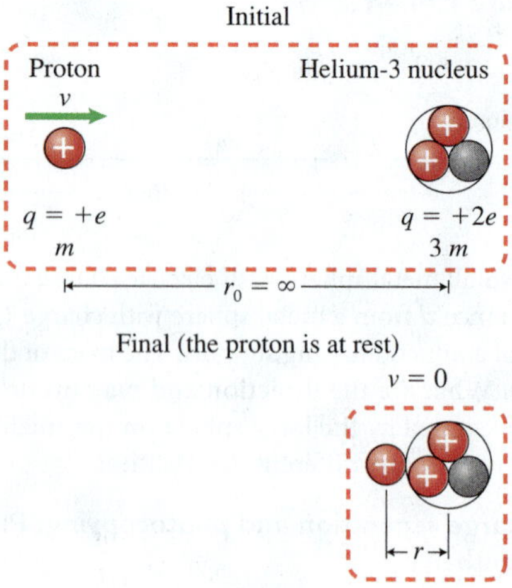


## General Problems

56. ** Suppose that Earth and the Moon initially have zero charge. Then 1000 kg of electrons are transferred from Earth to the Moon. Determine the radius of a stable moon orbit when both the electrical and gravitational forces of attraction are exerted on the Moon and it completes one rotation about Earth in 27.5 days.

57. * BIO **Calcium ion synapse transfer** Children have about $10^{16}$ synapses that can transfer signals between neurons in the brain and between neurons and muscle cells. Suppose that these synapses simultaneously transmit a signal, sending 1000 calcium ions ($Ca^{2+}$) across the membrane at each synaptic ending. Determine the total electric charge transfer in coulombs during that short time interval. By comparison, a lightning flash involves about 5 C of charge transfer. (Note: This is a fictional scenario. All human neurons do not simultaneously produce signals.)

58. * BIO **DNA stretch** The DNA molecule is spiral shaped, like a spring. A 10-$\mu$m-long DNA molecule has an effective spring constant of $10^{-8}$ N/m. Suppose a positive ion of charge $+e$ is attached on one end of the molecule and a positive ion of charge $+e$ is attached on the other end. What distance is the DNA stretched because of the electrical repulsion of these two ions?

59. ** A small metal sphere of charge $-2.0 \times 10^{-4}$ C sits on the roof of a 5.0-kg cart, shown in **Figure P14.59.** The sphere moves toward another sphere of charge $+1.5 \times 10^{-3}$ C, located on the wall. As it moves, the cart pulls a cable that passes over a pulley and lifts an object with mass 10 kg. The cart starts at rest 5.0 m from the wall charge. (a) What is its speed when it is 2.0 m from the wall charge? (b) What assumptions did you make?

**Figure P14.59**

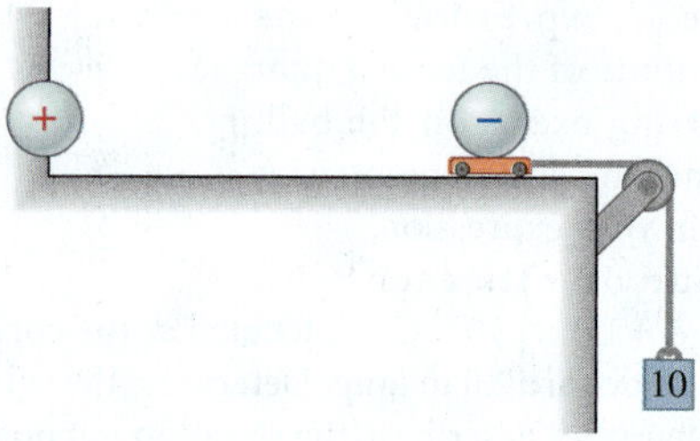


60. ** **Energy in the hydrogen atom** (a) Write an expression for the total energy of the electron-proton system in a hydrogen atom. (b) Use circular motion dynamics and any other ideas you may need to show that the positive kinetic energy of the electron is half the magnitude of the negative electric potential energy of the proton-electron system. (c) Based on these results, do you need to do positive or negative work to remove the electron from the proton? Explain. Represent the removal process with a work-energy bar chart.

61. ** You have been asked to analyze the feasibility of a transportation system that is operated using electric charges. A cart with a positively charged ball on the top is to move between two oppositely charged balls at opposite ends of a straight rail (see Figure P14.48). Suppose the charge on the stationary ball on the left is $+4.0 \times 10^{-4}$ C and the charge on the right ball is $-4.0 \times 10^{-4}$ C; the charge on the cart is $+2.0 \times 10^{-4}$ C. The cart's mass is 160 kg (including a passenger). A 60-N effective friction force opposes the motion of the cart. Determine the cart's acceleration when at the position shown in the figure. (Note that the acceleration changes as the cart's position relative to the stationary charges changes.) Is this a feasible way to transport the cart along the rail?

62. * **Electric nutcracker** Suppose the electric nutcracker shown in Figure 14.16 has a stationary $+5.0 \times 10^{-5}$-C positive charge initially separated 0.40 m from a $-2.0 \times 10^{-5}$-C negatively charged 0.50-kg block. When the block is released, it accelerates toward the positive charge and the nut. Determine its speed when it is 0.10 m from the negatively charged block—just before it hits the nut. Evaluate the feasibility of this nutcracker device.

63. ** **EST Shocking your friend** (a) You shuffle across the rug and then place your finger near your friend's nose, causing a small spark that transfers about $10^{-9}$ C of charge from you to your friend. Determine the number of electrons transferred. (b) Estimate the fraction of electrons in your body that were transferred to the friend. Note that the electron mass is about 1/20,000 the mass of the atom—the nuclei are much more massive—and the mass of an average atom in the body is about $2 \times 10^{-26}$ kg. (Note: Don't get hung up on minutia—this is a rough estimate.)

## Reading Passage Problems

**Static cling** You pull your clothes from the dryer and find that they stick together. You remove your coat and find that your skirt or pants sticks to your legs. Static cling like this can occur for two reasons. First, different types of atoms have greater or lesser affinity for additional electrons. When two different materials are rubbed together, the atoms with a greater electron affinity attract electrons from the material with a lesser electron affinity. Because of this, the substance that gains electrons becomes negatively charged and the substance that loses electrons becomes positively charged. The differently charged materials attract and stick together.

Second, static cling can occur between charged and uncharged objects. For instance, you may notice that a sock removed from the dryer is attracted toward an uncharged sweater you are wearing. Or sometimes your skirt or pants sticks to your leg. This happens because the molecules in a charged piece of clothing cause the electric charge inside the molecules of the nearby uncharged objects to slightly redistribute (to become polarized) so that the unlike charge of the molecule moves closer to the charged object and is attracted more than the same molecular charge of the same sign, which is slightly farther away (see Figures 14.9b and c).

Some people use fabric softeners to prevent static cling. These products coat cloth fibers with a thin layer of electrically conductive molecules, thus preventing buildup of static electricity. You can also use a metal clothes hanger to remove electric charge from already clean charged clothes. Brush the hanger on the inside of the garment from top to bottom.

Shoes scuffing on different surfaces can also cause electric charge transfer. For that reason, hospital personnel wear special shoes in hospital operating rooms to avoid sparking that might ignite flammable gases in the room.

64. You rub a balloon against your wool sweater and then place the balloon on the wall—it sticks. Why?
    (a) The balloon and wall have opposite electric charges.
    (b) The molecules on the wall redistribute their charge so that the charge opposite that on the balloon is nearest the balloon.
    (c) Electric charge in Earth is pulled to the part of the wall nearest the balloon.
    (d) a and c are correct.
    (e) a, b, and c are correct.

65. As you unload the clothes dryer, you find a sock clinging to a shirt. As you pull the sock off the shirt, you do positive work on the sock. What is the main form of energy increase?
    (a) Gravitational (b) Elastic
    (c) Electric potential (d) Thermal
    (e) c and d

66. Table salt, $Na^+Cl^-$, is made of ionized sodium and chlorine atoms. Which atom is most attractive for an excess electron?
    (a) Na (b) Cl
    (c) They are equally attractive.

67. You add fabric softener to your next load of wash. Your clothes do not cling after they emerge from the dryer. What do the water softener molecules do?
    (a) Carry away all the excess electrons.
    (b) Cause sparks in the dryer that discharge the excess charge on the clothes.
    (c) Form a protective coating that prevents the charge from joining the clothes.
    (d) Make the cloth fluffy so the charge comes off naturally.
    (e) Join the cloth molecules and make a conductive layer that prevents excess charge from accumulating on the clothes.

68. You put many strips of aluminum foil in the dryer along with your clothes. Which answer below best represents the condition of the clothes after leaving the dryer?
    (a) The clothes are uncharged because the excess charge is on the aluminum strips.
    (b) The clothes are uncharged because they transfer excess charge to the strips, which transfer it to the metal dryer walls.
    (c) The clothes are charged but the strips are not because they are conductive.
    (d) The clothes are charged because the strips are not connected to anything.
    (e) None of these answers is reasonable.

69. You remove electric charge from your clean slacks by rubbing a metal clothes hanger down the inside of the slacks. Which answer below represents the best explanation for why this works?
    (a) The charge travels from the cloth to the metal to your hand through your body to the ground.
    (b) The metal hanger absorbs all the charge.
    (c) The metal causes tiny sparks that send the charge into the air.
    (d) The metal provides charge to the cloth that neutralizes its charge.
    (e) None of these answers is reasonable.

**Electrostatic exploration** Geologists sometimes analyze the distribution of materials under Earth's surface, materials such as iron, water, oil, or dry soil. The process they use is called electrostatic exploration. Electrodes are placed in the ground about 800 m apart. An electric generator is connected to the electrodes and causes them to become oppositely charged. The opposite-sign charges on the electrodes cause electrically charged ions in the matter below the surface to move. The moving ions are detected by equipment on the surface. This helps the geologist decide what type of matter is below the surface. What causes this motion? To help answer this question, determine the net electric force exerted on ions at different places below the surface.

70. Which arrow in **Figure P14.70** best represents the force exerted on a positive ion at position A?
(a) I (b) II (c) III
(d) IV (e) V (f) VI

**Figure P14.70**

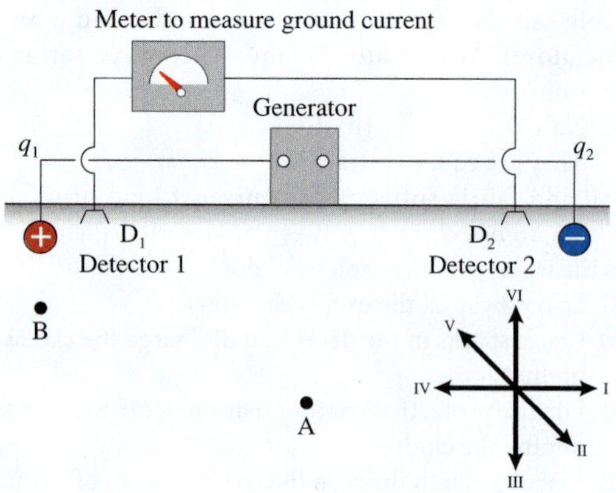


71. Which arrow in Figure P14.70 best represents the force exerted on a negative ion at position A?
(a) I (b) II (c) III
(d) IV (e) V (f) VI

72. Which arrow in Figure P14.70 best represents the force exerted on a positive ion at position B?
(a) I (b) II (c) III
(d) IV (e) V (f) VI

73. Which arrow in Figure P14.70 best represents the force exerted on a negative ion at position B?
(a) I (b) II (c) III
(d) IV (e) V (f) VI

74. Based on the analysis in the first four questions, which charge detector in Figure P14.70 will be collecting negative electrons?
(a) $D_1$ (b) $D_2$
(c) Both $D_1$ and $D_2$ (d) Neither $D_1$ nor $D_2$

75. The resistance to the motion of negative electrons through different types of materials is, in decreasing order, dry soil, moist soil, underground water, oil, and iron ore. How can this knowledge and the measurement of the charge reaching the detector per unit time help identify what type of material is under Earth's surface?
(a) More electrons will reach the detector from iron than from dry soil.
(b) The electrons reaching the detector will carry water with them if water is under the surface.
(c) The electrons reaching the detector are affected very little by what's under the surface.
(d) The detector will get more electric charge if the resistance to flow is less, and resistance is related to the type of material.
(e) a and d

76. Electrode 1 has charge $q_1 = +1.0 \times 10^{-5}$ C and electrode 2 has charge $q_2 = -1.0 \times 10^{-5}$ C. The electrodes are separated by 800 m and a free electron is located 300 m under the ground halfway between electrode 1 and electrode 2. What is the magnitude of the electric force that the electrodes exert on an electron at this position closest to?
(a) $1.1 \times 10^{-19}$ N (b) $1.4 \times 10^{-7}$ N (c) $2.8 \times 10^{-7}$ N
(d) $9.2 \times 10^{-20}$ N (e) $2.8 \times 10^{-14}$ N

# The Electric Field 15

**Why is it safe to sit in a car during a lightning storm?**

**How does a lightning rod protect a building?**

**How does electrocardiography work?**

Imagine that you are in a car during a thunderstorm. Lightning—an electric discharge strikes the road just in front of you. Are you and your car in danger? Many people think that cars are safe because the rubber tires isolate them from the ground. We will learn in this chapter why it is not the rubber tires but the metal body of the car that protects passengers from the effects of the lightning. We will also learn why you are safer in the car during a thunderstorm than under than a tall tree.

**Previously (in Chapter 14)** we learned how to describe electrostatic interactions in two ways: (1) with a force exerted by one charged object on another and (2) with the electric potential energy possessed by systems of charged objects. We

**Be sure you know how to:**

- Find the force that one charged object exerts on another charged object (Section 14.4).
- Determine the electric potential energy of a system (Section 14.5).
- Explain the differences in the internal structure of electric conductors and dielectrics (Section 14.3).

**Figure 15.1** The electric field model for interactions between electrically charged objects.

(a)

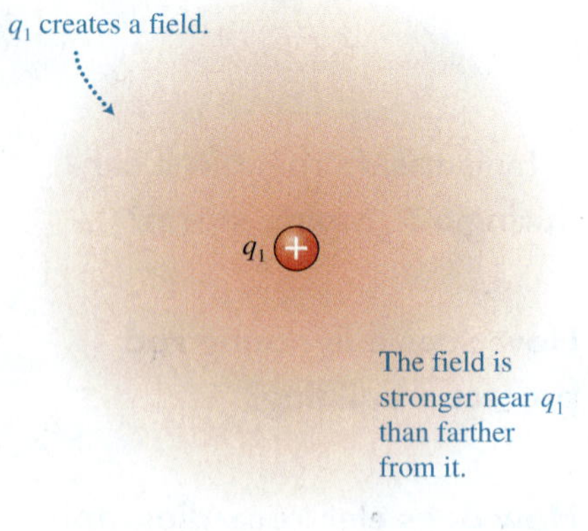
$q_1$ creates a field.
$q_1$
The field is stronger near $q_1$ than farther from it.

(b)

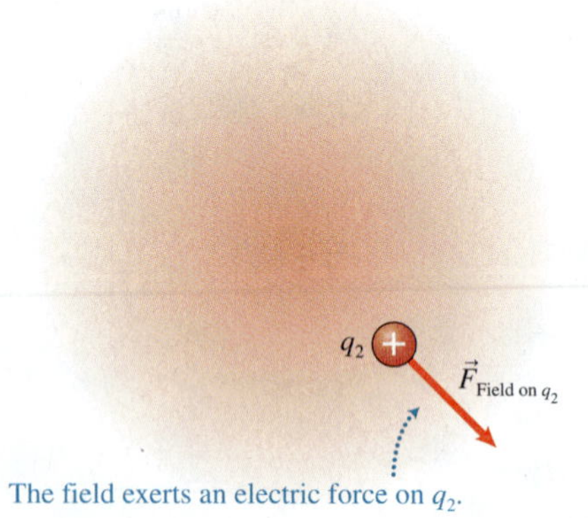
$q_2$
$\vec{F}_{\text{Field on } q_2}$
The field exerts an electric force on $q_2$.

**Figure 15.2** The field approach for the gravitational force.

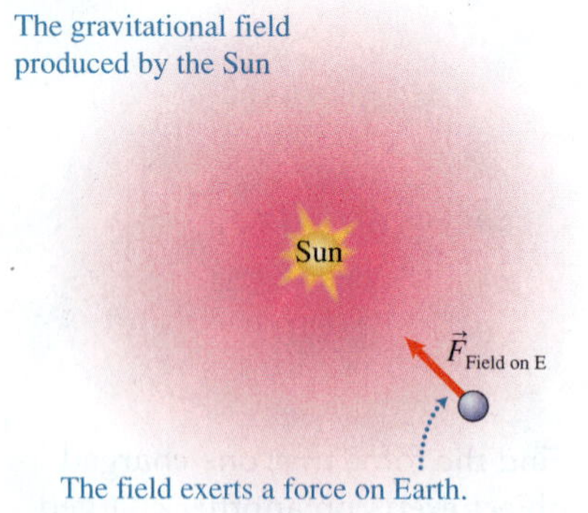
The gravitational field produced by the Sun
Sun
$\vec{F}_{\text{Field on E}}$
The field exerts a force on Earth.

also discovered that charged objects can exert forces on each other without being in direct contact. This is only the second interaction we have encountered with this property, the gravitational interaction being the first. How does one charged object "know" about the presence of another when they are not in direct contact? This chapter helps us answer this question and develop a new approach for analyzing electric processes.

## 15.1 A model of the mechanism for electrostatic interactions

When we pull two pieces of transparent tape off a table and place them near each other, they repel. How does an electrically charged object, such as a piece of tape, affect the other piece of tape without touching it? You could ask the same question about gravitational interactions—how does the Sun pull on Earth even though they are millions of kilometers apart?

Historically, there have been two answers to this question. The first model was called *interaction at a distance*. Isaac Newton supported this model for the gravitational interaction, and Charles Coulomb supported it for the electrostatic interaction. In this model, both interactions just happen. For example, if you move one of the two interacting charges, the other charge instantaneously "senses" that movement and responds accordingly.

The second model for electric interactions, suggested by Michael Faraday, involved an agent that acts as an intermediary between the charges. It was some sort of *electric disturbance* in the region surrounding them. Physicists call this electric disturbance an **electric field**. According to this **electric field model**, a charged object $q_1$ electrically disturbs the region surrounding itself (**Figure 15.1a**). If you place a second charged object $q_2$ in this region, the electric field due to object 1 exerts a force on object 2 (Figure 15.1b). The same is true for object 2—it creates its own field that exerts a force on object 1.

We can apply the field model to the gravitational interaction. For example, how does the Sun pull on Earth, located about 150 million km away? According to the field model, the Sun gravitationally disturbs the region around it in an invisible way. This disturbance is called a **gravitational field** (**Figure 15.2**). The Sun's gravitational field exerts a force on Earth that causes it to travel in an almost circular orbit around the Sun.

An analogy may help you visualize the idea of a field. Imagine a very large horizontal elastic sheet with a grid on it that has been pulled tight. The sheet represents a gravitational field. Now, imagine a massive object such as the Sun resting on the sheet. The Sun causes the sheet to bend, more so closer to the Sun than farther away from it. This bending represents the field due to the Sun. Now suppose a much smaller object such as a comet passes by the Sun. The comet is so small that it does not disturb the elastic sheet field. But the Sun's field disturbs the comet's path as it passes—less when the comet is farther away (the gravitational force is less) and more when closer (the gravitational force is greater). The Sun's distortion of the elastic gravitational sheet acts as an intermediary, allowing the comet and the Sun to interact with each other even though they are not in contact. Similarly, the electric field acts as an intermediary between charged objects, allowing the objects to interact without direct contact with each other.

The elastic sheet analogy allows us to make a prediction based on the field model: the interaction between the objects changes at a finite speed. When you move one of the two interacting charges (object 1), the other one does not "sense" that movement instantly—it takes time for the change in the field to reach object 2. In the model of interaction at a distance, any change in the position of object 1 leads to the immediate change in the force exerted on object 2. We will return to this question when we study electromagnetic waves (in Chapter 24).

## Gravitational field due to a single object with mass

Let's develop a gravitational field approach for the gravitational force that one object with mass exerts on another. In the region near Earth, Earth's contribution to the gravitational field is the dominant one. Imagine Earth and one of three small objects A, B, or C. We call these objects *test objects*, because we use them as detectors to probe the field (**Figure 15.3a**). We place the test objects one at a time at the same location near Earth. The masses of these test objects are $m_A = m$, $m_B = 2m$, and $m_C = 3m$. Consider the gravitational force that Earth exerts on each test object (Figure 15.3b).

$$F_{\text{E on A}} = G\frac{m_E}{r^2}m_A = m_A\left(G\frac{m_E}{r^2}\right)$$

$$F_{\text{E on B}} = G\frac{m_E}{r^2}m_B = m_B\left(G\frac{m_E}{r^2}\right) = 2F_{\text{E on A}}$$

$$F_{\text{E on C}} = G\frac{m_E}{r^2}m_C = m_C\left(G\frac{m_E}{r^2}\right) = 3F_{\text{E on A}}$$

The directions of these forces are all toward the center of Earth. However, the magnitudes of the forces differ: they are proportional to the masses of the test objects. Despite the differences in magnitude, however, the ratio of the magnitude of the force exerted on each object and the mass of that object is identical for all three objects:

$$\frac{F_{\text{E on A}}}{m_A} = \frac{F_{\text{E on B}}}{m_B} = \frac{F_{\text{E on C}}}{m_C} = G\frac{m_E}{r^2}$$

Consider the objects to be near Earth's surface. When we substitute the values of $G$, $m_E$, and Earth's radius $r_E$, we find that

$$G\frac{m_E}{r_E^2} = \left(6.67 \times 10^{-11}\,\frac{\text{N}\cdot\text{m}^2}{\text{kg}^2}\right)\frac{(5.97 \times 10^{24}\,\text{kg})}{(6.37 \times 10^6\,\text{m})^2} = 9.8\ \text{N/kg}$$

Since this value does not depend on the mass of any test object, we speculate that this value might be a mathematical description of the "strength" of Earth's gravitational field at a particular location. Since the gravitational force has direction, we say that the gravitational field close to Earth's surface at a particular location has a magnitude of 9.8 N/kg and points directly toward the center of Earth. Until now, we have called this quantity free-fall acceleration. Now we characterize the gravitational field using the quantity **$\vec{g}$ field.** We define the $\vec{g}$ field at any location as the gravitational force exerted by the field on a test object at that location, divided by the mass of that object:

$$\vec{g} = \frac{\vec{F}_{\text{Field on Object}}}{m_{\text{Object}}} \tag{15.1}$$

The magnitude of the field at any location does not depend on the test object; it depends on the mass of the object(s) creating the field (in this case, Earth) and the location where the field is measured: $g = G\,(m_E/r_E^2)$.

**Figure 15.3** Earth exerts a gravitational force on each of the three test objects placed at the same location.

(a)

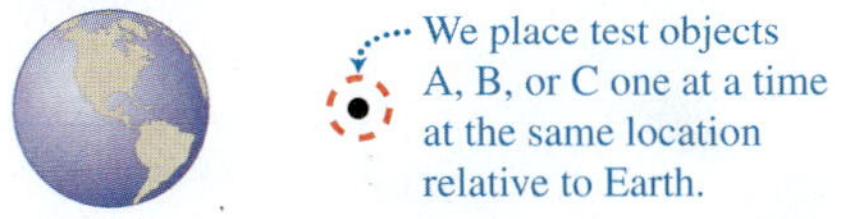
We place test objects A, B, or C one at a time at the same location relative to Earth.

(b)

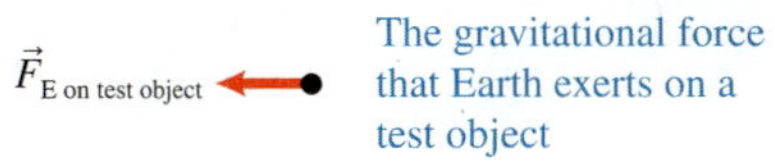
$\vec{F}_{\text{E on test object}}$
The gravitational force that Earth exerts on a test object

## Electric field due to a single point-like charged object

Let's use a similar approach to construct a physical quantity for the "strength" of the electric field. Imagine an object with positive electric charge $Q$ (we use capital $Q$ to denote the object whose field we are investigating), and one of three test objects K, L, or M placed at a distance $r$ from the center of

**Figure 15.4** An object with charge $Q$ exerts an electric force on one of three charged objects placed at the same location.

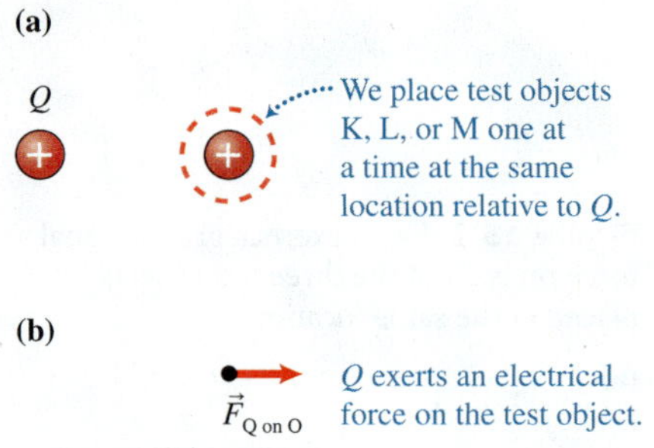
(a)
Q
We place test objects K, L, or M one at a time at the same location relative to Q.
(b)
$\vec{F}_{\text{Q on O}}$
Q exerts an electrical force on the test object.

object Q (**Figure 15.4a**). Objects K, L, and M have small positive charges $q_K = q$, $q_L = 2q$, and $q_M = 3q$ (here we use lowercase $q$ to denote that these are test objects). Use Coulomb's law to compare the magnitudes of the electric forces that Q exerts on K, L, or M (Figure 15.4b):

$$F_{\text{Q on K}} = k\frac{Qq_K}{r^2} = q_K\left(k\frac{Q}{r^2}\right)$$

$$F_{\text{Q on L}} = k\frac{Qq_L}{r^2} = q_L\left(k\frac{Q}{r^2}\right) = 2F_{\text{Q on K}}$$

$$F_{\text{Q on M}} = k\frac{Qq_M}{r^2} = q_M\left(k\frac{Q}{r^2}\right) = 3F_{\text{Q on K}}$$

Notice that Q exerts electric forces proportional to the electric charge of the test objects K, L, or M. However, the ratio of the magnitude of the electric force exerted on objects K, L, or M and the electric charge of each object equals the same value:

$$\frac{F_{\text{Q on K}}}{q_K} = \frac{F_{\text{Q on L}}}{q_L} = \frac{F_{\text{Q on M}}}{q_M} = k\frac{Q}{r^2}$$

Since these ratios all have the same value, the ratio $(F_{\text{Q } on \text{ K}}/q_k)$ could be a mathematical description of the strength of the electric field produced by the charge $Q$ at that location.

We can now think of the objects K, L, and M as probes of the electric field—called **test charges.** We use these test charges to examine the electric field produced by some object Q—the **source charge.** If we agree that a test charge is always positive and relatively small in magnitude so that it has little effect on the source charge, we can define the physical quantity that characterizes the electric field at a location as a vector with magnitude $(F_{\text{Q on } q_{\text{test}}}/q_{\text{test}})$ that points in the direction of the electric force that is exerted on the positive test charge placed at that location. This physical quantity is called the $\vec{E}$ field.

**The $\vec{E}$ field is the physical quantity that characterizes the electric field at a location. To determine the $\vec{E}$ field at a specific location, place an object with a small positive test charge $q_{\text{test}}$ at that location and measure the electric force exerted on that object. The $\vec{E}$ field at that location equals the ratio**

$$\vec{E} = \frac{\vec{F}_{\text{Q on } q_{\text{test}}}}{q_{\text{test}}} \tag{15.2}$$

**and points in the direction of the electric force exerted on the positive test charge. The $\vec{E}$ field is independent of the test charge used to determine the field. The unit of the electric field is newtons per coulomb (N/C).**

Using the above operational definition of the $\vec{E}$ field at a point, we found that the magnitude of the $\vec{E}$ field produced by a single source object with charge Q was

$$E = \frac{F_{\text{Q on } q_{\text{test}}}}{q_{\text{test}}} = k\frac{|Q|q_{\text{test}}}{r^2 q_{\text{test}}} = k\frac{|Q|}{r^2}$$

We can interpret this field as follows:

$$E = k\frac{|Q_{\text{Source}}|}{r^2} \tag{15.3}$$

Effect (arrow to $E$) Cause (arrow to $|Q_{\text{Source}}|$)

The object with electric charge $Q$ is the cause, and the $\vec{E}$ field produced by it is the effect. Note also that the $\vec{E}$ field vector at any location points *away* from the object creating the field if $Q$ is positive and *toward* it if $Q$ is negative.

## $\vec{E}$ field due to multiple charged objects

How do we determine the magnitude and direction of the $\vec{E}$ field at a particular point caused by more than one charged object? We start by considering the field created by two charged objects in Observational Experiment **Table 15.1**.

### OBSERVATIONAL EXPERIMENT TABLE

**15.1** $\vec{E}$ **field due to two charged objects.**

VIDEO 15.1

| Observational experiment | Analysis |
|---|---|
| **Experiment 1.** Two identical metal spheres are on insulating stands with equal positive charge. You charge a small aluminum foil ball positively and hang it from an insulated thread between the two spheres. When hanging closest to the left sphere, the foil ball is repelled away, toward the right sphere. When hanging nearer the right sphere, it is repelled away, toward the left sphere. At the exact middle between the spheres, the foil ball hangs straight down. 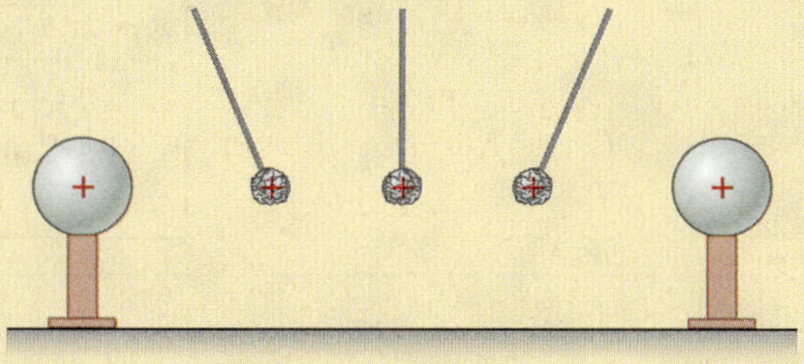 | Regard the foil ball as a test charge to determine the direction of the $\vec{E}$ field at different places between the spheres. The $\vec{E}$ field points right when nearest the left sphere, is zero in the exact middle between the spheres, and points left when nearest the right sphere. 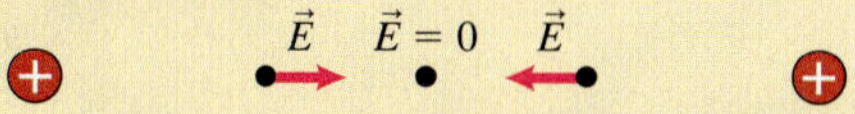  |
| **Experiment 2.** Repeat Experiment 1, only this time with the left sphere charged negatively and the right sphere charged positively. When the foil ball hangs at any place between the spheres, the ball is attracted toward the left negative sphere and repelled from the right sphere. 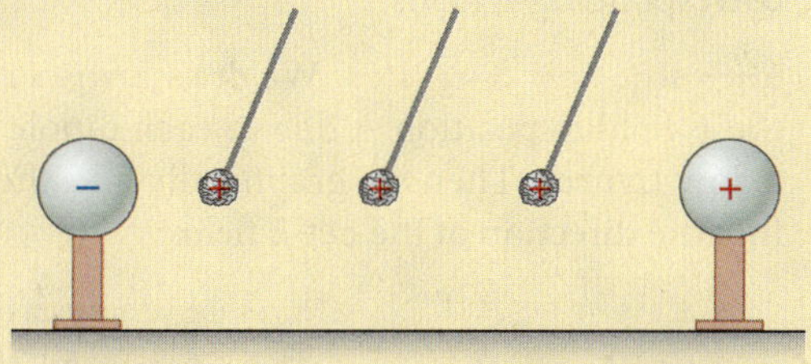 | The foil ball indicates that the $\vec{E}$ field points left at all points between the spheres. 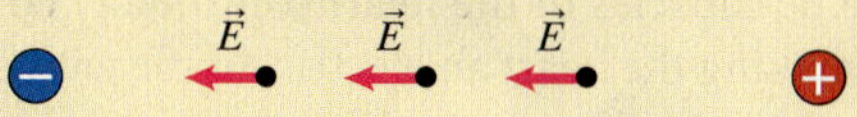  |

**Patterns**

- In Experiment 1, the zero field at the exact middle must have been due to the vector addition of equal magnitude but oppositely directed $\vec{E}$ fields caused by the two spheres.
- In Experiment 1, when on either side of the middle, the nearer sphere caused a greater repulsion than the farther sphere. The opposing $\vec{E}$ fields only partially canceled each other.
- In Experiment 2, both spheres caused an $\vec{E}$ field to the left, which added to form a stronger $\vec{E}$ field to the left.

In summary, the $\vec{E}$ fields due to individual charged objects add as vector quantities.

It appears that when multiple charged objects are present, each object creates its own contribution to the $\vec{E}$ field. At a particular location, the field is the *vector* sum of the $\vec{E}$ field contributions due to all of these charges. This summation effect is an example of the superposition principle. The fact that we add the $\vec{E}$ field contributions means that they do not affect each other, but they all affect a test charge placed at a particular location.

**Superposition principle** **The $\vec{E}$ field at a point of interest is the *vector* sum of the individual contributions to the $\vec{E}$ field of each charged object:**

$$\vec{E} = \vec{E}_1 + \vec{E}_2 + \vec{E}_3 + \cdots \tag{15.4}$$

The Reasoning Skill box illustrates step by step how to use the superposition principle and graphical vector addition to estimate the direction of the $\vec{E}$ field at a point of interest.

**REASONING SKILL 1** Estimating the $\vec{E}$ field at a position of interest

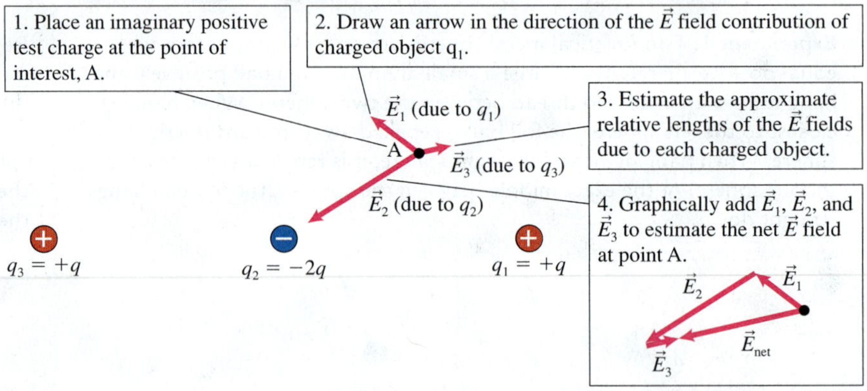

1. Place an imaginary positive test charge at the point of interest, A.
2. Draw an arrow in the direction of the $\vec{E}$ field contribution of charged object $q_1$.
3. Estimate the approximate relative lengths of the $\vec{E}$ fields due to each charged object.
4. Graphically add $\vec{E}_1$, $\vec{E}_2$, and $\vec{E}_3$ to estimate the net $\vec{E}$ field at point A.
$\vec{E}_1$ (due to $q_1$)
A
$\vec{E}_3$ (due to $q_3$)
$\vec{E}_2$ (due to $q_2$)
$q_3 = +q$
$q_2 = -2q$
$q_1 = +q$
$\vec{E}_2$
$\vec{E}_1$
$\vec{E}_{net}$
$\vec{E}_3$

## CONCEPTUAL EXERCISE 15.1 The $\vec{E}$ field in body tissue due to the heart dipole charge

The muscles of the heart continually contract and relax, making the heart an electric dipole with equal-magnitude positive and negative electric charges, such as shown in the figure below. Estimate the direction of the $\vec{E}$ field at position A in the body tissue to the left side of the midpoint between the dipole charges.

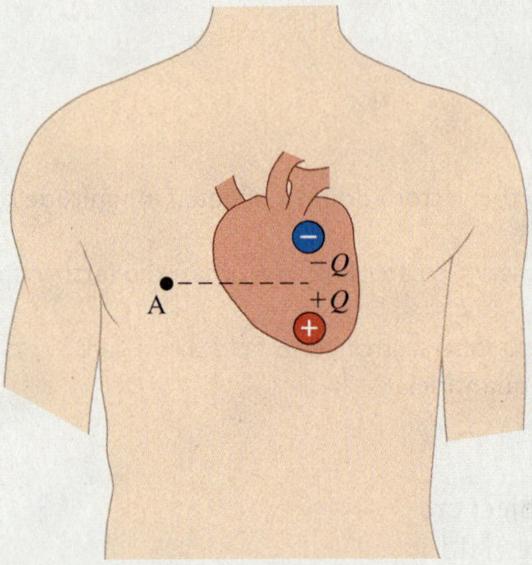

$-Q$
$+Q$
A

**Sketch and translate** Imagine placing a positive test charge at position A to estimate the direction of the forces exerted on it by each dipole charge. We will use the directions of those forces to find the direction of the corresponding $\vec{E}$ fields.

**Simplify and diagram** We draw arrows representing the $\vec{E}$ field at position A due to each dipole charge (see below figure). Then we graphically add the $\vec{E}$ fields to find the direction of the net $\vec{E}$ field.

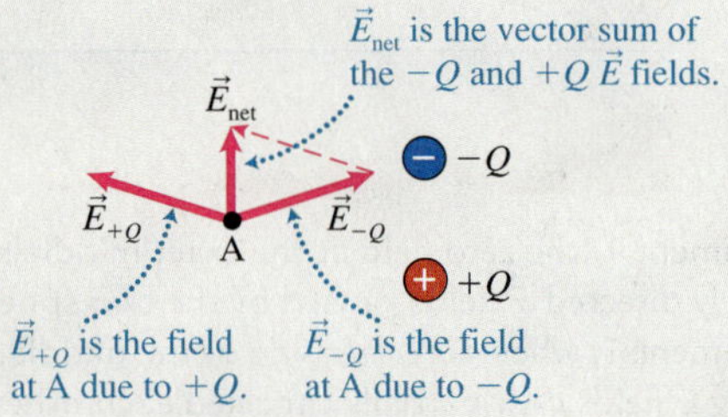

$\vec{E}_{net}$ is the vector sum of the $-Q$ and $+Q$ $\vec{E}$ fields.
$\vec{E}_{net}$
$-Q$
$\vec{E}_{+Q}$
$\vec{E}_{-Q}$
A
$+Q$
$\vec{E}_{+Q}$ is the field at A due to $+Q$.
$\vec{E}_{-Q}$ is the field at A due to $-Q$.

**Try it yourself:** Repeat the above for a position that is in the body tissue several centimeters to the right side of the center of the heart dipole.

*Answer:* The net $\vec{E}$ field on the right side also points up in the same direction as the $\vec{E}$ field on the left side of the heart dipole.

## $\vec{E}$ field lines

We have just learned how to estimate the $\vec{E}$ field at a single location in the vicinity of several charged objects. To represent the overall shape of the $\vec{E}$ field, we use electric field lines, or $\vec{E}$ *field lines.* Imagine two source charges ($+Q$ and $-Q$) organized as shown in **Figure 15.5a.** We start by using the superposition principle and graphical vector addition to find the net electric field $\vec{E}_{net}$ at position A due to both charges (see Figure 15.5a). Next, we use the same method to draw the $\vec{E}_{net}$ vector for a position just to the right of A (Figure 15.5b). The field to the right of A points toward the right and downward a little, since the negative charge $-Q$ is slightly closer and more attractive to a positive test charge at that point than the slightly more distant positive charge $+Q$ is repulsive to it.

Next, we draw a third $\vec{E}$ field vector for a position near the tip of the second $\vec{E}$ field vector. We continue drawing the $\vec{E}_{net}$ vectors for adjacent positions along the direction of the previous $\vec{E}_{net}$ vector, eventually getting a series of $\vec{E}_{net}$ vectors that seem to follow one after the other from A to the negative charge $-Q$, as shown in Figure 15.5c. The $\vec{E}_{net}$ vectors from the positive charge to A have a similar shape and are also shown in Figure 15.5c. Now, draw a single line that passes through each position tangent to these vectors. The line looks as shown in Figure 15.5d. If you repeat the process for a series of positions farther and closer to a line between the two charges, you get a series of lines such as shown in Figure 15.5e—the $\vec{E}$ field lines for the source charges shown in Figure 15.5a.

Using a similar process, you could construct the $\vec{E}$ field lines for the single positively charged object shown in **Figure 15.6a**; for the single negatively charged object shown in Figure 15.6b; and for the two equal-magnitude positively charged objects shown in Figure 15.6c. Note that at locations closer to the charged objects, the lines are closer together than when the locations are farther away. This means that the number of lines per unit area (the density of the lines) is larger where the $\vec{E}$ field is stronger. Similarly, if you have two source charges, then at the same relative location the magnitude of the $\vec{E}$ field is larger next to a bigger charge. Thus the number of lines emanating from the charge represents the magnitude of that charge. If you have two charged objects next to each other with the magnitude of one charge twice the magnitude of the other, and you draw six lines coming out of the first object, then the second object should have 12 lines coming out of it.

> **TIP** In this book all figures are two-dimensional, but in real life the situations are three-dimensional. The $\vec{E}$ field vectors and $\vec{E}$ field lines are really in a three-dimensional space.

**Figure 15.5** Constructing $\vec{E}$ field lines that represent the $\vec{E}$ field in the space near two source charges.

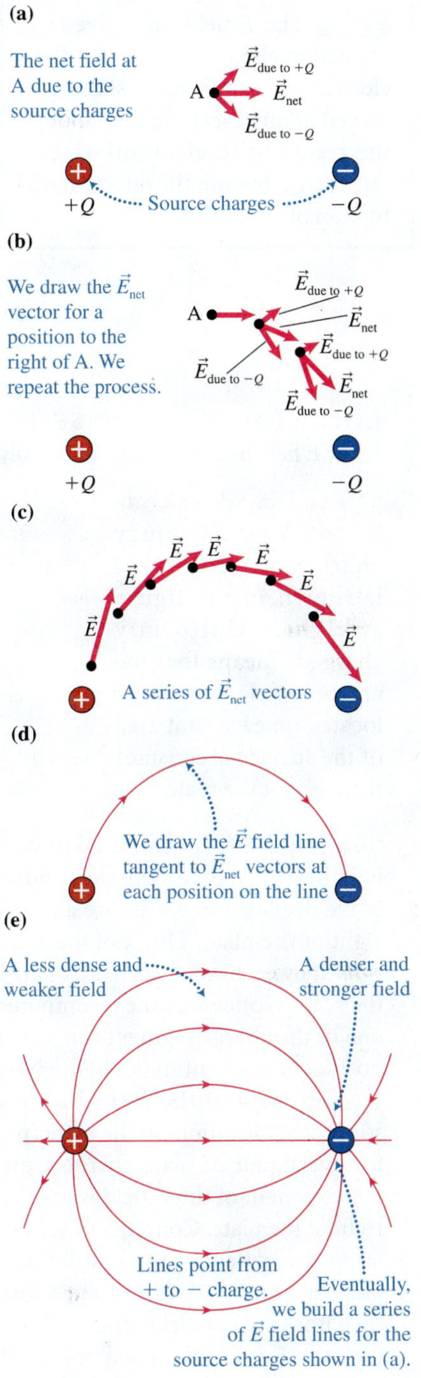


**Figure 15.6** $\vec{E}$-field lines for: (a) a single positive charge, (b) a single negative charge, and (c) two positive charges.

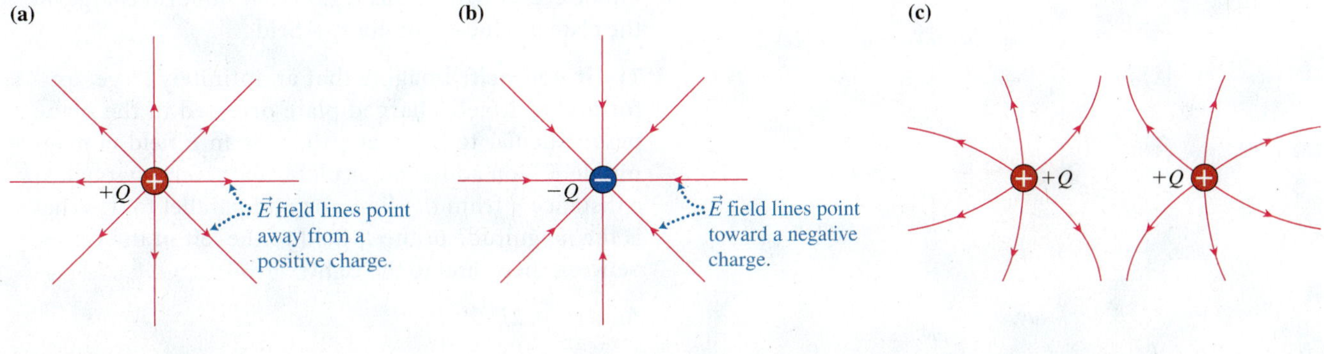

> **TIP** The $\vec{E}$ field lines give information about the force that the electric field exerts on a test charge placed at any location, and about the resulting acceleration of the test charge, but not the direction of motion of the test charge.

**Summary—$\vec{E}$ field lines:**

- $\vec{E}$ field lines are drawn so that the $\vec{E}$ field vectors at positions on those lines are tangent to the lines.
- $\vec{E}$ field lines begin on positively charged objects and end on negatively charged objects.
- The number of $\vec{E}$ field lines beginning or ending on a charged object is proportional to the magnitude of that object's charge. Therefore, on an electric field diagram you will see twice as many $\vec{E}$ field lines leaving a charge of $+2q$ as $+1q$.
- The density of the $\vec{E}$ field lines in a region is proportional to the magnitude of the $\vec{E}$ field in that region.

## CONCEPTUAL EXERCISE 15.2 Uniform $\vec{E}$ field

Draw $\vec{E}$ field lines for a large uniformly charged plate of glass.

**Sketch and translate** A uniformly charged plate of glass is shown in the figure at right. "Uniformly charged" means that the amount of electric charge located on each unit area of the surface is constant throughout the plate.

The distribution of charge on an infinitely large uniformly charged plate of glass

**Simplify and diagram** Assume that the plate of glass is infinitely large in both perpendicular directions in the plane of its surface. Choose a position of interest to the right of the plate. Think of the $\vec{E}$ field at this position as caused by each small segment of positive source charge on the plate. Notice that the $y$-component of the $\vec{E}$ field from one of the charge segments on the plate (for example, position 1 segment in the figure below) is canceled by the $y$-component of the $\vec{E}$ field from a charge segment at some other location on the plate (position 2). This occurs for every pair of plate charge segments. Therefore, the $y$-component of the $\vec{E}$ field is zero at every position to the right of the plate. Consequently, the $\vec{E}$ field is perpendicular to the plate (see the first figure at the right) at every point to the right of the plate (close or far away). The $\vec{E}$ field lines are parallel to the field at every position. Therefore, the $\vec{E}$ field lines must point away from the plate and be perpendicular to it (see the second figure at the right).

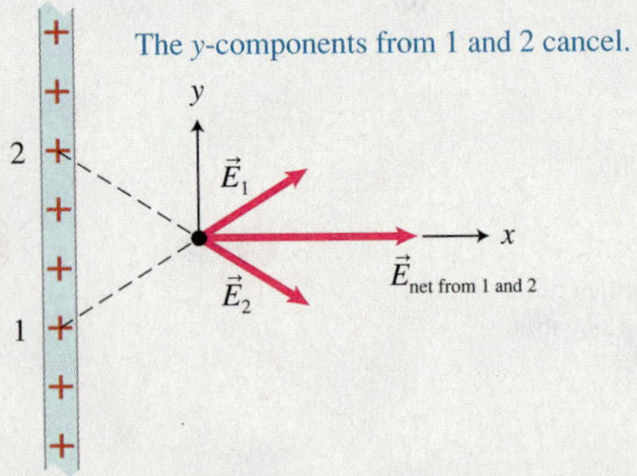

The $y$-components from 1 and 2 cancel.
$y$
$x$
2
1
$\vec{E}_1$
$\vec{E}_2$
$\vec{E}_{\text{net from 1 and 2}}$

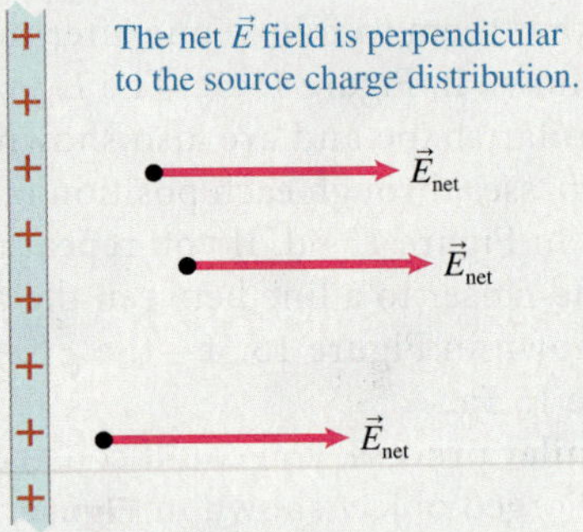

The net $\vec{E}$ field is perpendicular to the source charge distribution.
$\vec{E}_{\text{net}}$
$\vec{E}_{\text{net}}$
$\vec{E}_{\text{net}}$

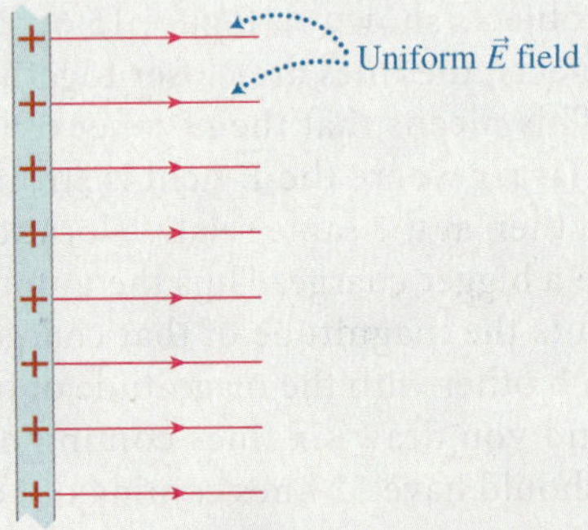

Uniform $\vec{E}$ field

The plate is infinitely large, so the $\vec{E}$ lines should look the same at any point on the right side of the plate. If the $\vec{E}$ field decreased in magnitude as you moved farther from the plate, the lines would have to spread farther apart. But they don't. The field lines must have a uniform density—the same separation between the $\vec{E}$ field lines everywhere. This means the $\vec{E}$ field has the same magnitude everywhere in this region. The uniform charge on the plate produces a uniform $\vec{E}$ field.

**Try it yourself:** Imagine that an infinitely large, uniformly positively charged plate oriented in the plane perpendicular to the page produces an $\vec{E}$ field of magnitude $E$. You add another plate, negatively charged, at a distance $d$ from the first one and parallel to it. What is the magnitude of the $\vec{E}$ field to the left of the plates, between them, and to the right?

*Answer:* 0, $2E$, 0.

The analysis of the uniform $\vec{E}$ field in the last conceptual exercise will be very useful later in the chapter, especially when we study capacitors.

**Review Question 15.1** How do you estimate the direction of the $\vec{E}$ field at a point located near two point-like charged objects?

## 15.2 Skills for determining $\vec{E}$ fields and analyzing processes with $\vec{E}$ fields

You will encounter two main types of problems involving $\vec{E}$ fields. In the first type, you will know the charge distribution and must determine the $\vec{E}$ field at one or more points caused by these source charges. In the second type, you will need to analyze a process involving a charged object that is stationary or moving in a given $\vec{E}$ field. Let's start by describing a strategy for solving the first type of problem.

### Determining the $\vec{E}$ field produced by given source charges

The Reasoning Skill box below illustrates a general strategy for determining the $\vec{E}$ field at a particular position caused by a given group of source charges.

**REASONING SKILL 2** Determine the $\vec{E}$ field

Follow the steps shown below to determine the magnitude and direction of the $\vec{E}$ field at position I.

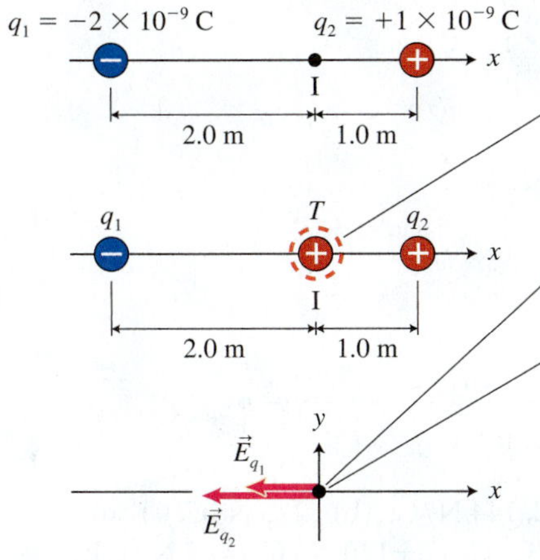


1. *System* Place a small, positively charged test object (the system) at the position where you want to determine the $\vec{E}$ field.

2. *$\vec{E}$ field diagram* Construct an $\vec{E}$ field diagram for the point of interest. Choose coordinate axes.

3. *$\vec{E}$ field components* The magnitude of the $\vec{E}$ field contribution of a pointlike charged object is

$$E = k\frac{|Q|}{r^2}.$$

Use this, any appropriate trig functions, and the appropriate sign to determine the components of the $\vec{E}$ field contributions from each charged object. Include these components in the step 4 equations.

$E_x = -k|q_1|/(2.0\text{ m})^2 - k|q_2|/(1.0\text{ m})^2$
$= -(9.0 \times 10^9\ \text{N·m}^2/\text{C}^2)(2 \times 10^{-9}\ \text{C})/(2.0\text{ m})^2$
$-(9.0 \times 10^9\ \text{N·m}^2/\text{C}^2)(1 \times 10^{-9}\ \text{C})/(1.0\text{ m})^2$
$= -4.5\ \text{N/C} - 9.0\ \text{N/C}$
$= -13.5\ \text{N/C}$ or $-14$ N/C (two significant digits)

For this problem, $E_y = 0$ since neither contribution has a $y$ component. Thus $E$ has 14 N/C magnitude and points in the negative $x$ direction. Note: Choose $E_x$ signs based on the arrow directions in the diagram (not on the signs of the charges).

4. *Superposition principle*
4a. Add $x$ components.
4b. Add $y$ components.
4c. Use the resulting $x$ and $y$ components of the vector to determine its magnitude and direction. The magnitude is:

$$E = \sqrt{E_x^2 + E_y^2}.$$

The angle $\theta$ that the electric field makes with the positive or negative $x$ axis is determined using:

$$\tan\theta = \frac{E_y}{E_x}.$$

## EXAMPLE 15.3 $\vec{E}$ field due to multiple charges

Two small metal spheres attached to insulating stands reside on a table a distance $d$ apart. The left sphere has positive charge $+q$ and the right has negative charge $-q$. Determine the magnitude and direction for the $\vec{E}$ field at a distance $d$ above the center of the line connecting the spheres.

**Sketch and translate** The sketch in the accompanying figure shows the source charges $+q$ and $-q$ that are producing the $\vec{E}$ field and the place (the point of interest) where we want to determine the $\vec{E}$ field—both its magnitude and direction. To estimate the direction of the vectors, imagine placing a positively charged test object $q_T$ at that point. (Remember, the fields exist without the test charge.)

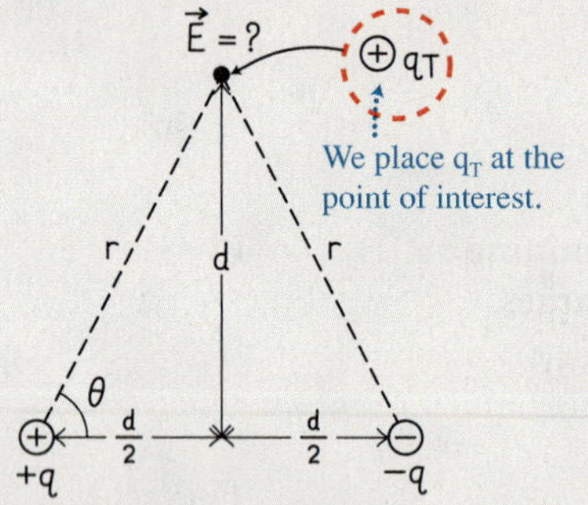


**Simplify and diagram** Next, construct an $\vec{E}$ field diagram for the point of interest, showing an $\vec{E}$ field arrow for each source charge that contributes to the $\vec{E}$ field at that point (see figure at right). Include coordinate axes in the diagram.

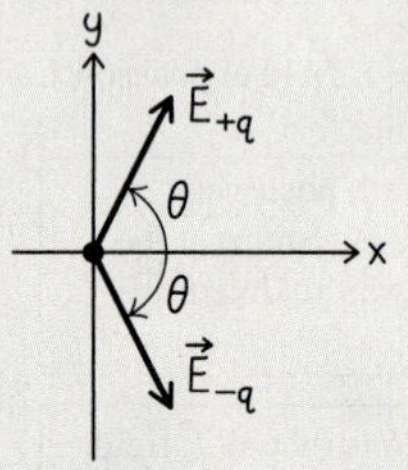


**Represent mathematically** Use Eq. (15.3) to determine the magnitude of the $\vec{E}$ field contributions from both $+q$ and $-q$: $E_{+q} = E_{-q} = k(q/r^2)$. Use the Pythagorean theorem to determine the distance $r$ from $+q$ and $-q$ to the point of interest:

$$r^2 = \left(\frac{d}{2}\right)^2 + d^2 = \frac{5d^2}{4} \text{ or } r = \frac{\sqrt{5}d}{2}$$

Thus, the magnitude of the $\vec{E}$ field contributions from both $+q$ and $-q$ is

$$E_{+q} = E_{-q} = k\frac{q}{\left(\frac{\sqrt{5}d}{2}\right)^2} = \frac{4kq}{5d^2}$$

The angle $\theta$ that the $\vec{E}$ field vector makes relative to the positive $x$-axis can be determined using the triangle in the first figure:

$$\tan\theta = \frac{\text{opposite side}}{\text{adjacent side}} = \frac{d}{d/2} = 2.0 \text{ or } \theta = 63.4°$$

Now determine the $x$- and $y$-components of the $\vec{E}$ field.

$$E_x = E_{+qx} + E_{-qx} = E_{+q}\cos\theta + E_{-q}\cos\theta$$
$$= 2E_{+q}\cos\theta = 2\left(\frac{4kq}{5d^2}\right)\cos 63.4° = 0.716\frac{kq}{d^2}$$
$$E_y = E_{+qy} + E_{-qy} = +E_{+q}\sin 63.4° + (-E_{-q})\sin 63.4° = 0$$

The $y$-components of the two $\vec{E}$ field contributions add to zero.

**Solve and evaluate** The magnitude of the $\vec{E}$ field is

$$E = \sqrt{E_x^2 + E_y^2} = \sqrt{\left(\frac{0.716\,kq}{d^2}\right)^2 + 0^2} = 0.72\frac{kq}{d^2}$$

and it points in the positive $x$-direction.

**Try it yourself:** Determine the $\vec{E}$ field at point I in the figure below. (a) Determine the magnitudes of the $\vec{E}$ field contributions at point I from charges $+q$ and $-q$. (b) Determine the $x$- and $y$-components of those $\vec{E}$ field contributions. (c) Determine the net $x$- and $y$-components of the $\vec{E}$ field at point I. (d) Finally, determine the magnitude and the direction of the $\vec{E}$ field at point I.

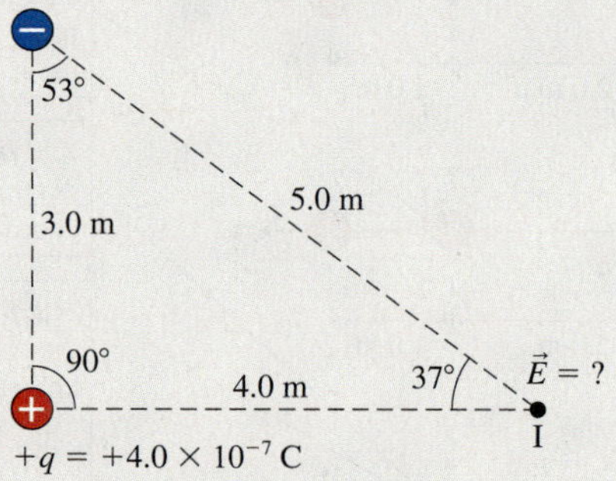


*Answer:* (a) 225 N/C, 144 N/C; (b) (225 N/C, 0) and (−115 N/C, +87 N/C); (c) (+110 N/C, +87 N/C); (d) 140 N/C, 38° above $+x$-axis (toward the right).

## Analyzing a process involving a known $\vec{E}$ field

If the $\vec{E}$ field at some location is known, we can use Eq. (15.2) to determine the electric force that the field exerts on an object with charge $q$ if at that location:

$$\vec{F}_{\vec{E}\text{ on }q} = q\vec{E} \tag{15.5}$$

We can now combine our understanding of electric field and force with our knowledge of Newton's laws to solve problems.

---

**PROBLEM-SOLVING STRATEGY Incorporating the $\vec{E}$ field into Newton's second law**

**EXAMPLE 15.4 A charged object in a known $\vec{E}$ field**

A spring made of a dielectric material with spring constant 50 N/m hangs from a large uniformly charged horizontal plate. The uniform $\vec{E}$ field produced by the plate has magnitude $2.0 \times 10^5$ N/C and points down. A 0.20-kg ball with charge $+4.0 \times 10^{-5}$ C hangs at the end of the spring. Determine the distance the spring is stretched from its equilibrium length. Assume that $g = 10$ N/kg.

### Sketch and translate

- Draw a labeled sketch of the situation. Include the symbols for the known and unknown quantities that you plan to use.
- Choose the system.

The situation is sketched at the right. The charged ball is the system.

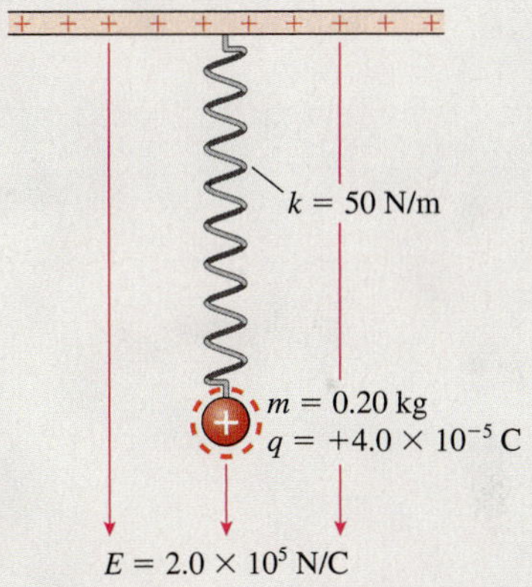


### Simplify and diagram

- Determine the $\vec{E}$ field produced by the environment. Is it produced by point-like charges (making it nonuniform) or by large charged plates (making it uniform)?
- Construct a force diagram for the system if necessary. Do not forget the axes.
- State any assumptions you have made.

The field is uniform. The upward direction is positive.
Assume that the spring has no mass.
A force diagram for the ball is shown at the right.

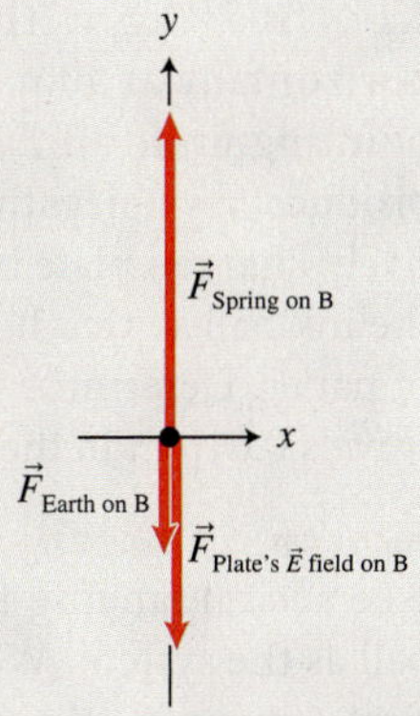


*(continued)*

**Represent mathematically**

- Use the force diagram to help apply Newton's second law in component form. Use component addition to determine the net force along each coordinate axis. Determine the acceleration of the system if needed.
- If necessary, use kinematics equations to describe the motion of the system.

The ball is not accelerating, so the forces exerted on the ball balance. None of the forces have $x$-components:

$$F_{\text{S on B}\,y} + F_{\text{E on B}\,y} + F_{\text{Plate on B}\,y} = 0$$
$$\Rightarrow k\Delta y + (-mg) + (-qE) = 0$$
$$\Rightarrow \Delta y = \frac{mg + qE}{k}$$

**Solve and evaluate**

- Combine the above equations and complete the solution.
- Check the direction, magnitude, and units and decide whether the result makes sense in limiting cases.

$$\Delta y = \frac{(0.20\text{ kg})(10\text{ N/kg}) + (4.0 \times 10^{-5}\text{ C})(2.0 \times 10^{5}\text{ N/C})}{50\text{ N/m}}$$
$$= 0.20\text{ m}$$

The units for the stretch distance are meters, as they should be. Check using limiting cases. As both the gravitational force and the electric force point in the same direction, eliminating either of them should reduce the distance that the spring stretches. Note in the equation above that $\Delta y$ decreases if $m = 0$ or if $q = 0$. Also, a stiff spring (larger $k$) should stretch less if we have the same mass ball and the same electric charge—also consistent with our result.

**Try it yourself:** If the plate were negatively charged, producing the same magnitude electric field but now pointing up, what would the spring stretch or compression be?

*Answer:* $-0.12$ m (the spring is compressed 0.12 m).

**EXAMPLE 15.5 Electric field deflects a tiny ink ball**

Inside an inkjet printer a tiny ball of black ink of mass $1.1 \times 10^{-11}$ kg with charge $-6.7 \times 10^{-12}$ C moves horizontally at 40 m/s. The ink ball enters an upward-pointing uniform $\vec{E}$ field of magnitude $1.0 \times 10^{4}$ N/C produced by a negatively charged plate above and a positively charged plate below. The plates are used to deflect the ink ball so that it lands at a particular spot on a piece of paper. Determine the deflection of the ink ball after it travels 0.010 m in the $\vec{E}$ field.

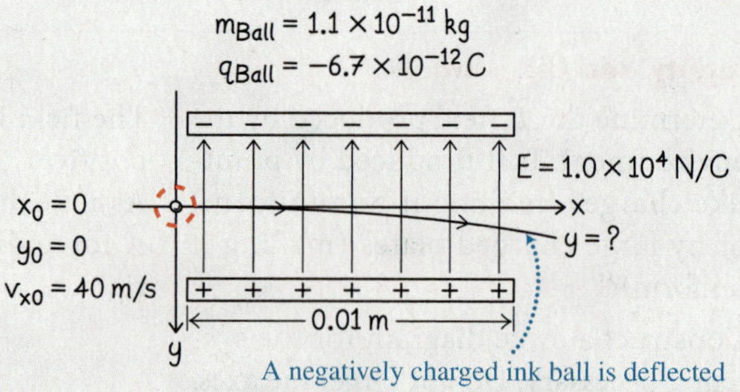

A negatively charged ink ball is deflected downward by the electric field.

**Sketch and translate** After sketching the situation (see accompanying figure), we choose the charged ink ball as the system. We break the problem into two parts: first determine the acceleration of the ball due to the forces being exerted on it and then use kinematics to determine its vertical displacement as it passes through the region with the electric field.

**Simplify and diagram** Assume that the ink ball is a point-like object and that the plates produce a uniform $\vec{E}$ field. A force diagram for the ball is shown in the figure on the next page. Earth and the $\vec{E}$ field exert downward forces on the negatively charged ball.

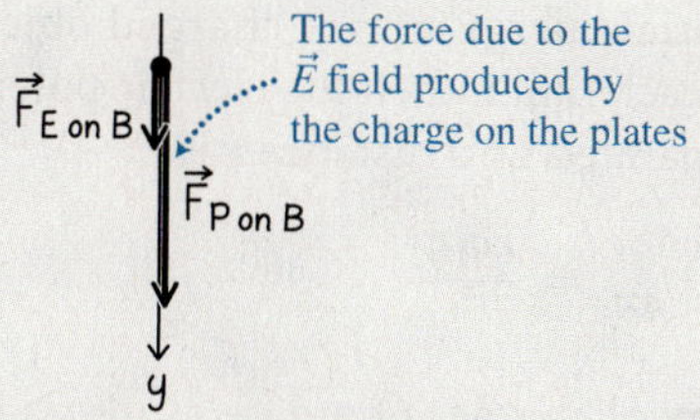


**Represent mathematically** The forces exerted on the ink ball do not have *x*-components, so we use only the *y*-component form of Newton's second law to determine the *y*-component of the acceleration (we choose the positive *y*-direction pointing down):

$$a_y = \frac{1}{m}\sum F_y = \frac{1}{m}(F_{\text{P on B}\,y} + F_{\text{E on B}\,y})$$

$$= \frac{1}{m}(qE_y + mg_y) = \frac{qE_y}{m} + g_y$$

The ink ball does not accelerate in the *x*-direction, so the *x*-component of its velocity remains constant at 40 m/s during the time interval it takes to traverse the region with nonzero $\vec{E}$ field. This time interval is $\Delta t = \Delta x / v_x$. During this time interval the ball's displacement in the *y*-direction is

$$y - y_0 = v_{0y}\Delta t + \frac{1}{2}a_y \Delta t^2$$

**Solve and evaluate** Substituting for $a_y$, $\Delta t$, and $v_{0y} = 0$ gives

$$y - y_0 = 0 + \frac{1}{2}\left(\frac{qE_y}{m} + g_y\right)\left(\frac{\Delta x}{v_x}\right)^2$$

We can now insert the known information ($E_y = -1.0 \times 10^4\,\text{N/C}$, $q = -6.7 \times 10^{-12}\,\text{C}$, $m = 1.1 \times 10^{-11}\,\text{kg}$, and $g_y = +10\,\text{N/kg}$) into the above to determine the ink ball deflection:

$$y - y_0 = \frac{1}{2}\left(\frac{(-6.7 \times 10^{-12}\,\text{C})(-1.0 \times 10^4\,\text{N/C})}{1.1 \times 10^{-11}\,\text{kg}} + 10\,\text{N/kg}\right)\left(\frac{0.010\,\text{m}}{40\,\text{m/s}}\right)^2 = 1.9 \times 10^{-4}\,\text{m} = 0.19\,\text{mm}$$

The distance unit is correct, and the value seems reasonable given the size of a letter on a printed page. To get a bigger deflection, we can use a larger electric field.

**Try it yourself:** What should the mass of the ink ball be so that if you reverse the direction of the $\vec{E}$ field, the ball will have zero acceleration?

*Answer:* $6.7 \times 10^{-9}$ kg.

**Review Question 15.2** You have an object with charge $+Q$ on the left and a second object with charge $-2Q$ at a distance *d* on the right. Sketch the $\vec{E}$ field vector at the same distance *d* directly above the midpoint between the charged objects.

# 15.3 The *V* field

So far we have been studying electric fields using a force approach. However, in many problems in mechanics and electrostatics, an energy approach allows us to focus on the initial and final stages of a process without knowing what happened in between. Can we describe electric fields using the concepts of work and energy? In order to do this, we need to describe the electric field not as the force-related $\vec{E}$ field, but instead as an energy-related field.

## Electric potential due to a single charged object

To construct this new way of describing the electric field, we will think of the electric interaction in terms of the electric potential energy of a system with a point-like object with electric charge Q (a source charge) and charged objects K, L, or M with different small electric charges $q_K$, $q_L$, or $q_M$ (test charges) placed

**Figure 15.7** Consider the electric potential energy $U_q$ of a system with source charge Q and test charge $q = q_K, q_L$, or $q_M$.

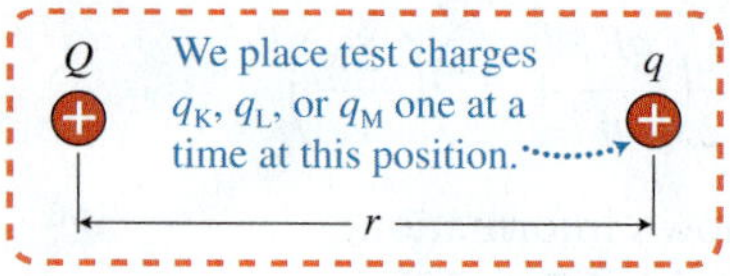
Q
We place test charges
$q_K, q_L$, or $q_M$ one at a
time at this position.
q
r

one at a time at a distance $r$ from Q (**Figure 15.7**). Recall (from Chapter 14) that if we consider the electric potential energy of two charged objects to be zero when they are infinitely far from each other, then the electric potential energy of two point-like charged objects separated by a distance $r$ is

$$U_{q_1 q_2} = \frac{kq_1 q_2}{r}$$

Thus, the electric potential energy of charges Q and $q_K$, $q_L$, or $q_M$ is

$$U_{Qqk} = \frac{kQq_k}{r} = q_K\left(\frac{kQ}{r}\right)$$

$$U_{QqL} = \frac{kQq_L}{r} = q_L\left(\frac{kQ}{r}\right)$$

$$U_{QqM} = \frac{kQq_M}{r} = q_M\left(\frac{kQ}{r}\right)$$

If we divide the electric potential energy of the Q-$q_K$, Q-$q_L$, or Q-$q_M$ charge pair by the charges $q_K$, $q_L$, or $q_M$ (that is, $U_{qQ}/q$), we find that all three ratios equal the same quantity:

$$\frac{U_{Qq_K}}{q_K} = \frac{U_{Qq_L}}{q_L} = \frac{U_{Qq_M}}{q_M} = \frac{kQ}{r}$$

Since these ratios all have the same value ($kQ/r$), this ratio could be a mathematical representation of the electric energy field caused by charge Q at a distance $r$ from Q. We call this physical quantity the ***V* field** or **electric potential** due to Q at a distance $r$ from Q.

***V* field (or electric potential) due to a single charge** **To determine the $V$ field due to a single source charge at a specific location, place a test charge $q_{test}$ at that location and determine the electric potential energy $U_{Q\,q\text{test}}$ of a system consisting of the test charge and the source charge that creates the field. The $V$ field at that location equals the ratio**

$$V = \frac{U_{Qq_{\text{test}}}}{q_{\text{test}}} = \frac{kQ}{r} \qquad (15.6)$$

**The unit of electric potential is joule/coulomb (J/C) and is called the volt (V).**

**TIP**

- Both the $V$ field and the $\vec{E}$ field at a specific location are independent of the test charge and characterize the properties of space at that location.
- Unlike the $\vec{E}$ field, which is a vector quantity, the $V$ field (electric potential) is a scalar quantity. It can have a positive or a negative value depending on the sign of the electric charge Q of the object that creates the $V$ field at a particular location.

## The superposition principle and the *V* field due to multiple charges

Now that we can determine mathematically the $V$ field produced by a single point-like charged object [Eq. (15.6)], we can use the superposition principle to determine the *V* field at a specific location produced by several charged objects. Using the same superposition principle idea as was used for the $\vec{E}$ field, we have

$$V = V_1 + V_2 + V_3 + \ldots = \frac{kQ_1}{r_1} + \frac{kQ_2}{r_2} + \frac{kQ_3}{r_3} + \ldots \tag{15.7}$$

where $Q_1, Q_2, Q_3, \ldots$ are the source charges (including their signs) creating the field and $r_1, r_2, r_3, \ldots$ are the distances between the source charges and the location where we are determining the *V* field. Because the *V* field is a scalar field rather than a vector field (like the $\vec{E}$ field is), it is much easier to apply the superposition principle.

**QUANTITATIVE EXERCISE 15.6 Electric potential due to heart dipole**

Suppose that the heart dipole charges $-Q$ and $+Q$ are separated by distance $d$ as in the figure below. Write an expression for the *V* field due to both charges at point A, a distance $d$ to the right of the $+Q$ charge.

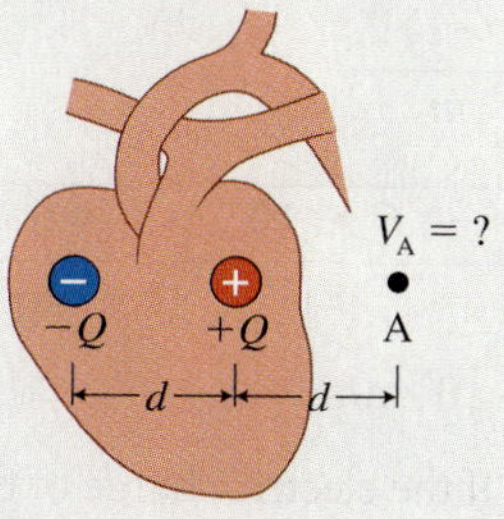


**Represent mathematically** Use Eq. (15.7) to write an expression for the *V* field:

$$V = V_{+Q} + V_{-Q} = \frac{k(+Q)}{r_{+Q}} + \frac{k(-Q)}{r_{-Q}}$$

**Solve and evaluate** Inserting the distances, we get

$$V = \frac{k(+Q)}{d} + \frac{k(-Q)}{2d} = +\frac{kQ}{2d}$$

**Try it yourself:** Write an expression for the *V* field at a distance $d$ to the left of $-Q$.

*Answer:* $V = \dfrac{k(-Q)}{d} + \dfrac{k(+Q)}{2d} = -\dfrac{kQ}{2d}.$

## Finding the electric potential energy when the *V* field is known

If we know the electric potential at a specific location, we can rearrange the definition of the *V* field [Eq. (15.6)] to determine the electric potential energy of a system that includes the source charges and another charge $q$ that is in the vicinity of the source charges:

$$U_{\text{Source Q and } q} = qV_{\text{Due to source Q}} \tag{15.8}$$

For example, we could use this equation to determine the electric potential energy of the heart dipole in Quantitative Exercise 15.6 and a sodium ion in the tissue near the dipole.

Since the value of the electric potential depends on the choice of zero level (as does electric potential energy), we often use the difference in electric potential between two points, called the **potential difference**, to analyze processes using electric potential.

> **Potential difference** $\Delta V$ **between two points A and B is equal to the difference in the values of electric potential at those points:** $\Delta V = V_B - V_A$.

Potential difference is a very useful quantity—if we know the potential difference between two points A and B, we can predict the direction of acceleration of a charged object. A positively charged object accelerates from regions of higher electric potential toward regions of lower potential (like an object falling to lower elevation in Earth's gravitational field). A negatively charged particle tends to do the opposite, accelerating from regions of lower potential toward regions of higher potential. Consider the next example.

**EXAMPLE 15.7 X-ray machine**

Inside an X-ray machine is a wire (called a filament) that, when hot, ejects electrons. Imagine one of those electrons, now located outside the wire. It starts at rest and accelerates through a region where the $V$ field increases by 40,000 V. The electron stops abruptly when it hits a piece of tungsten at the other side of the region, producing X-rays (we will learn more about that when we discuss quantum optics). How fast is the electron moving just before it reaches the tungsten?

**Sketch and translate** The situation is sketched in the figure. The system consists of the electron and the electric field of the tube. We choose the zero level of the $V$ field to be at the initial position of the electron, just outside the filament.

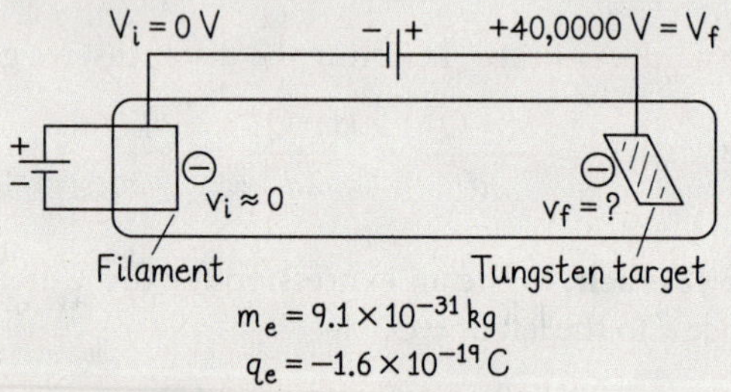


**Simplify and diagram** Assume that the gravitational force that Earth exerts on the electron is not significant. A bar chart for the process is shown at right. In the initial state, the system has neither electric potential energy (the electric potential is zero there) nor kinetic energy (the electron is not moving). In the final state, the system has positive kinetic energy and negative electric potential energy. The electric potential energy is determined using $U_q = qV$ with $q < 0$ and $V_f > 0$.

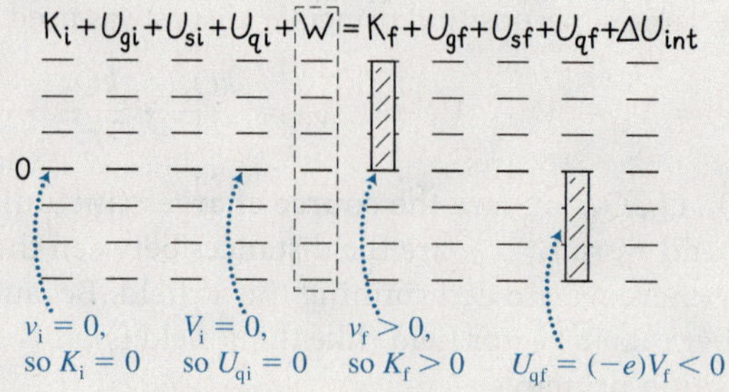


**Represent mathematically** Each nonzero bar in the bar chart turns into a term in the generalized work-energy equation:

$$0 = \frac{1}{2} m_e v_f^2 + (-e) V_f$$

**Solve and evaluate**

$$v_f = \sqrt{-\frac{2(-e)V_f}{m_e}}$$

$$= \sqrt{\frac{-2(-1.6 \times 10^{-19}\,\text{C})(4.0 \times 10^4\,\text{V})}{9.11 \times 10^{-31}\,\text{kg}}}$$

$$= 1.2 \times 10^8\ \text{m/s}$$

*Limiting case check:* If the electric charge of the electron were zero, it would not participate in the electric interaction at all and its final speed should be zero; it is. If the electric potential at the final location were the same as at the initial location, the final speed of the electron should be zero; and it is.

**Try it yourself:** A 0.10-kg cart with a charge of $+6.0 \times 10^{-5}$ C rests on a 37° incline. Through what potential difference must the cart move along the incline so that it travels 2.0 m up the incline before it stops? The cart starts at rest. Assume that $g = 10$ N/kg.

*Answer:* The potential difference needed is −20,000 V.

**Figure 15.8** Equipotential surfaces produced by an object with charge $+Q$.

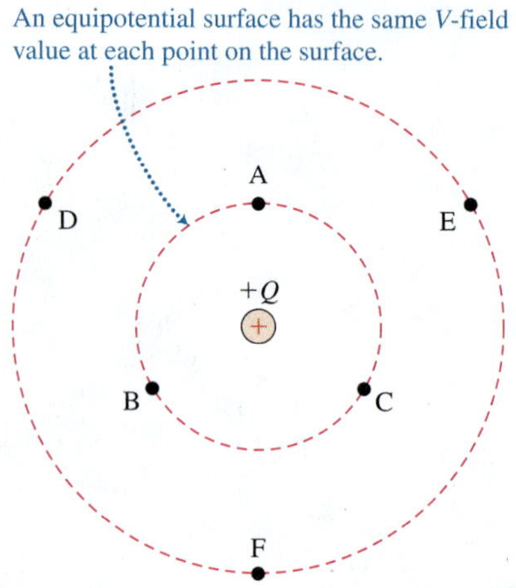


## Equipotential surfaces—representing the $V$ field

$\vec{E}$ field lines help us visualize the $\vec{E}$ field. We will now develop a new way to visualize the $V$ field at different points surrounding one or more charged objects. We start by considering a positively charged point-like charged object $+Q$ (shown in **Figure 15.8**). The values of the $V$ field at points A, B, and C are all the same. The values of the $V$ field at D, E, and F are all the same but different than the values at A, B, and C. If we imagine the surface to which all the points of equal $V$ field value belong (for example, points A, B, and C), the surface will be a sphere surrounding the source charge $+Q$. These surfaces of constant electric potential $V$ are called **equipotential surfaces**. The surfaces are spheres (they look like circles on a two-dimensional page) because the $V$ field produced by a point-like charged object has the same value at all points equidistant from the charged object. The electric potential energy of a system with the source charge $+Q$ and a test charge $+q$ is constant if the test charge remains on an equipotential surface. No work is needed to move the test charge $+q$ to some other place on that equipotential surface.

The $\vec{E}$ field lines for a single positively charged source charge point outward along radial lines. The spherical equipotential surfaces are perpendicular to the $\vec{E}$ field lines at every point (**Figure 15.9a**). This is a general feature of every charge distribution; the $\vec{E}$ field lines and the equipotential surfaces are perpendicular to each other at every point in the region surrounding the charges (see the $\vec{E}$ field lines and the $V$ field surfaces for the dipole and two equal-magnitude positive charge distributions in Figures 15.9b and c).

The relative distances between neighboring equipotential surfaces are smaller where the $\vec{E}$ field is greater—the $V$ field changes faster where the $\vec{E}$ field is stronger. Consider the graph of $V$ field-versus-distance from $+Q$ ($V$-versus-$r$ graph) shown in **Figure 15.10.** The closer to $+Q$, the more rapidly the $V$ field changes with changing distance $r$. Thus in Figure 15.9a the adjacent equipotential surfaces vary by the same amount of potential from one surface to the next and are closer together when nearer $+Q$ and further apart when farther away from $+Q$. The density of the $\vec{E}$ field lines is greatest near the $+Q$ charge and less dense when far from it.

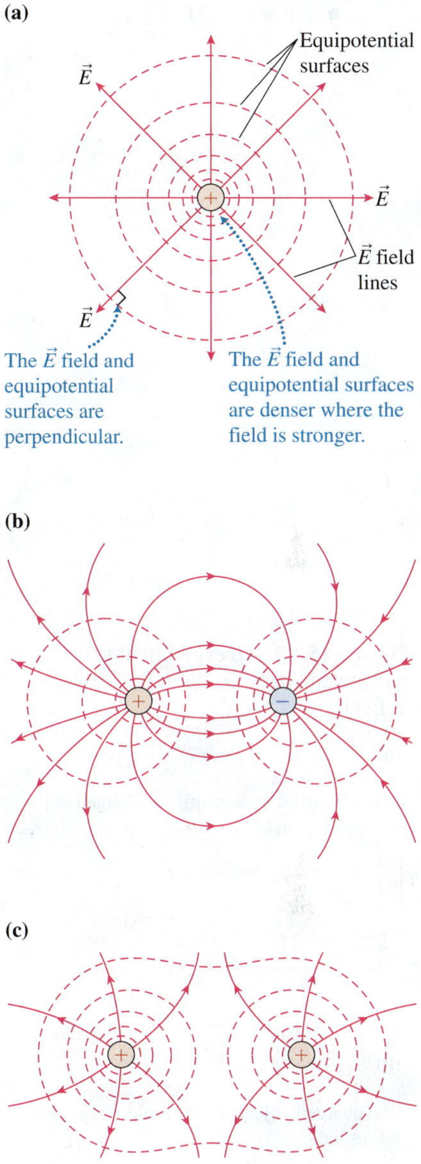
(a)
Equipotential surfaces
$\vec{E}$
$\vec{E}$
$\vec{E}$ field lines
$\vec{E}$
The $\vec{E}$ field and equipotential surfaces are perpendicular.
The $\vec{E}$ field and equipotential surfaces are denser where the field is stronger.
(b)
(c)

**Figure 15.9** The equipotential surfaces and $\vec{E}$ field lines produced by (a) a positively charged point object, (b) an electric dipole, and (c) two equal-magnitude positively charged objects.

## Contour maps—an analogy for equipotential surfaces

An analogy might help you visualize equipotential surfaces. If you were hiking in the mountains you would move up and down between different elevations. A contour map displays elevation with lines for each regular elevation increase above sea level (**Figure 15.11**). As you hike up or downhill, you move from one corresponding contour line to another on the map. Your elevation change also changes the gravitational potential energy of the you-Earth system. However, at all points along a single contour line on the map, that gravitational potential energy is the same, $U_g = mgy$, where $y$ is the elevation of that contour. Just as Earth's gravitational field can be represented by contour lines on a flat map, the electric field can be represented by lines indicating equipotential surfaces. Where the contour lines on a map are closer together, the elevation

**Figure 15.10** The $V$ field of a $+Q$-charged object changes rapidly when near the object.

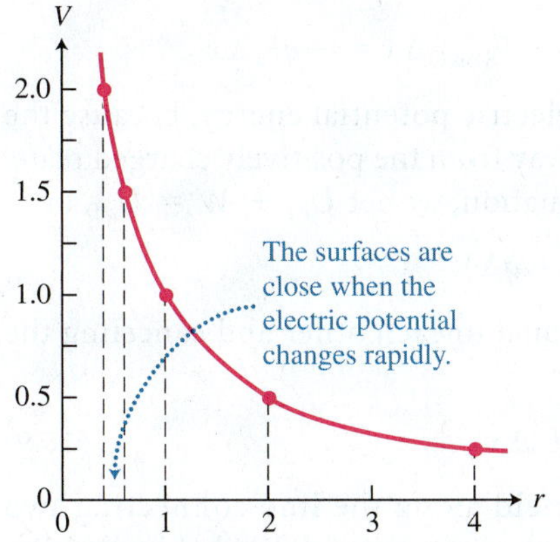
V
2.0
1.5
1.0
0.5
0
0
1
2
3
4
r
The surfaces are close when the electric potential changes rapidly.

**Figure 15.11** Constant elevation contour lines are closest together where the hill is steepest. Constant electric potential surfaces are closest where the $\vec{E}$ field is strongest.

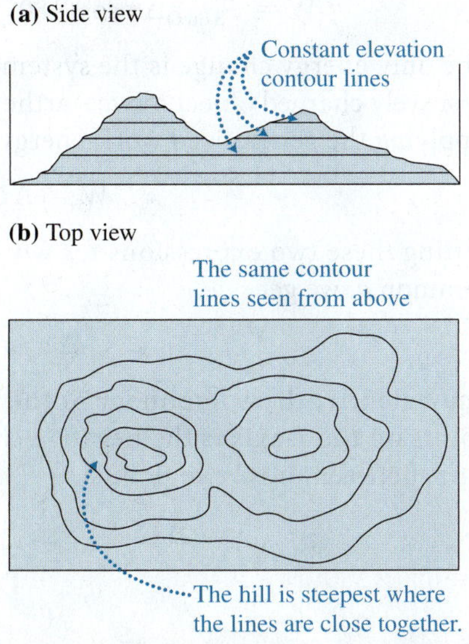
(a) Side view
Constant elevation contour lines
(b) Top view
The same contour lines seen from above
The hill is steepest where the lines are close together.

**Figure 15.12**

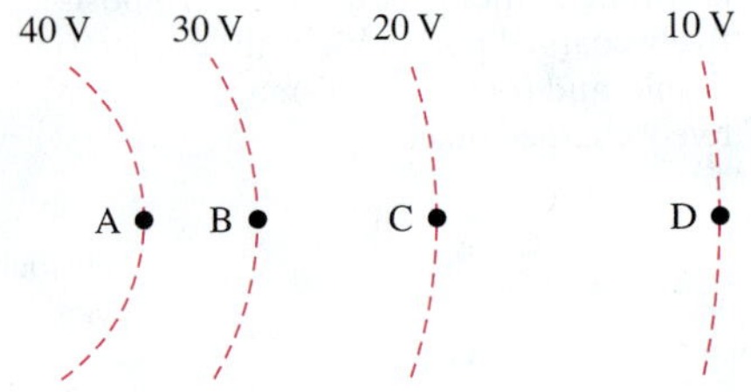
40 V
30 V
20 V
10 V
A
B
C
D

changes faster with the position; where the equipotential surfaces are closer together, the electric field is stronger.

Positively charged objects accelerate from regions of higher $V$ field toward regions of lower $V$ field. Objects with mass tend to move in the "downhill" direction—a sled going down a hill. For negatively charged objects, the analogy breaks down, since there are no objects with negative mass. Nonetheless, negatively charged objects tend to accelerate in the "uphill" direction, from regions of low $V$ field to regions of high $V$ field.

**Review Question 15.3** Compare the work needed to move an object with charge $q$ from point A to point B to the work needed to move the same object from point C to point D in the electric potential region depicted in **Figure 15.12.** Explain.

## 15.4 Relating the $\vec{E}$ field and the $V$ field

We know qualitatively that the $V$ field varies most rapidly with position where the $\vec{E}$ field is strongest. We can also examine this idea quantitatively.

### Deriving a relation between the $\vec{E}$ field and $\Delta V$

Consider the uniform $\vec{E}$ field produced by an electrically charged infinitely large glass plate. We attach a small object with charge $+q$ to the end of a very thin wooden stick and place the charged object and stick in the electric field produced by the plate (**Figure 15.13a**). The electric field exerts a force on the charged object in the positive $x$-direction $F_{\vec{E}\text{ on O}\,x} = +qE_x$. The stick exerts a force on the charged object with an $x$-component $F_{\text{S on O}\,x} = -F_{\text{S on O}}$ pointing toward the plate (Figure 15.13b) in the negative $x$-direction. If the charged object does not move or moves slowly with zero acceleration, the $x$-component form of Newton's second law applied to the charged object is $\sum F_x = F_{\vec{E}\text{ on O}\,x} + F_{\text{S on O}\,x} = +qE_x - F_{\text{S on O}} = 0$, or

$$F_{\text{S on O}} = qE_x$$

**Figure 15.13** An experiment to derive a relationship between the $\vec{E}$ field and the $V$ field.

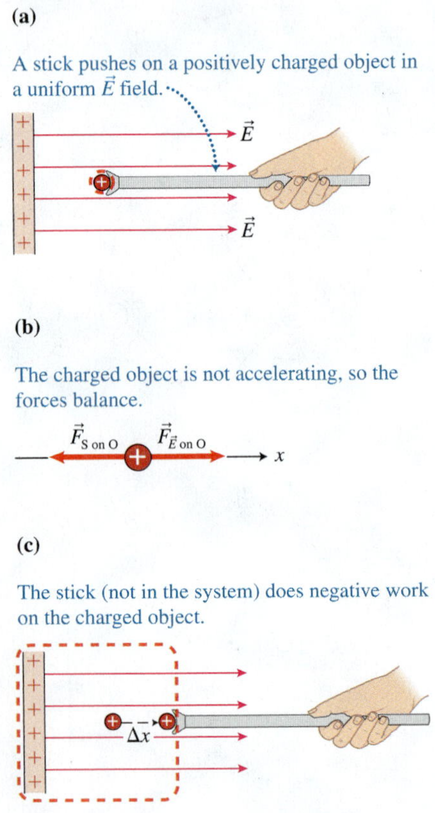
(a)
A stick pushes on a positively charged object in a uniform $\vec{E}$ field.
$\vec{E}$
$\vec{E}$
(b)
The charged object is not accelerating, so the forces balance.
$\vec{F}_{\text{S on O}}$
$\vec{F}_{\vec{E}\text{ on O}}$
x
(c)
The stick (not in the system) does negative work on the charged object.
Δx

Now, let's do a work-energy analysis for a process in which the charged object (still attached to the stick) is moved slowly a small distance $\Delta x$ farther from the plate (Figure 15.13c). For our system we choose the charged object and the electric field, but not the stick, which is part of the environment and does work on the system. The stick exerts a force on the charged object opposite its displacement and does negative work on the system:

$$W = F_{\text{S on O}}\Delta x \cos(180°) = -F_{\text{S on O}}\Delta x = -qE_x\Delta x$$

The only energy change is the system's electric potential energy, because the positively charged object moves farther away from the positively charged plate. Applying the generalized work-energy equation, we get $U_{qi} + W = U_{qf}$, or

$$W = \Delta U_q = q\Delta V$$

Setting these two expressions for work equal to each other and canceling the common $q$, we get

$$\Delta V = -E_x\Delta x \qquad (15.9)$$

Equivalently, the component of the $\vec{E}$ field along the line connecting two points on the $x$-axis is the negative change of the $V$ field divided by the distance between those two points:

$$E_x = -\frac{\Delta V}{\Delta x} \qquad (15.10)$$

The magnitude of the $\vec{E}$ field component in a particular direction indicates how fast the $V$ field (electric potential) changes in that direction. The $\vec{E}$ field points in the direction of decreasing $V$ field: hence the minus sign in Eq. (15.10). Similar equations apply for other directions if the situation is in two or three dimensions. Although we derived Eqs. (15.9) and (15.10) using the example of a uniform $\vec{E}$ field, the equations represent a general result that relates the component of the $\vec{E}$ field in the chosen direction to the rate of change of the $V$ field in that direction. The relation between $\vec{E}$ field and $V$ field tells us two things: (1) in a region where the $V$ field is constant, the $\vec{E}$ field is zero, and (2) if you have two points at different potential, the closer those points are the stronger the $\vec{E}$ field between them will be. This explains, for example, why there is a spark between you and your bedroom door when your hand is close to the handle but not when it is far away.

**CONCEPTUAL EXERCISE 15.8** Can you think of locations relative to charge distributions where (a) the $V$ field at a particular location is zero but the $\vec{E}$ field is not, and (b) the $\vec{E}$ field is zero but the $V$ field is not zero?

**Sketch and translate** See the examples in the first two figures shown here. Consider positions that are at the exact center between the charged objects.

**Simplify and diagram** (a) Note that at the position between the charged objects in figure (a), the electric field points toward the right (see figure (c)) but the potential is zero:

$$V = \frac{k(+Q)}{d/2} + \frac{k(-Q)}{d/2} = 0$$

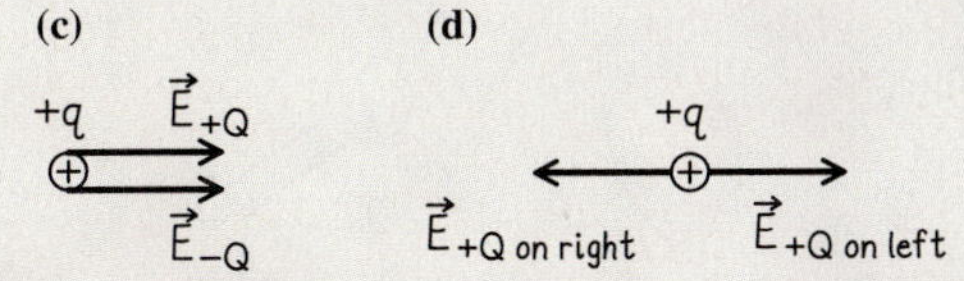


(b) Note that at the point between the charges in figure (b) the electric field is zero (see figure (d)) but the potential is

$$V = \frac{k(+Q)}{d/2} + \frac{k(+Q)}{d/2} = \frac{4k(+Q)}{d}$$

**Try it yourself:** Suppose four equal positively charged objects are at the corners of a square. Is there any place in the plane of the square where either the $V$ field at a particular location is zero but the $\vec{E}$ field is not, or the $\vec{E}$ field is zero but the $V$ field is not? Explain.

*Answer:* At the center of the square, the electric $\vec{E}$ field is zero, but the electric $V$ field is positive.

## Testing the relation between the $\vec{E}$ field and $\Delta V$

We derived Eq. (15.10) using our knowledge of work-energy, the $\vec{E}$ field, and the $V$ field. How can we test this relationship experimentally? Air (a dielectric under normal conditions) can turn into a conductor if free electrons in the air are accelerated to such speeds that they can knock an electron out of an air molecule on the next collision (see Example 14.10). The two electrons then move on and knock additional electrons out of other air molecules, creating a cascade, called **dielectric breakdown**. When these electrons recombine with molecules that have lost an electron, light is produced—a spark. Experiments indicate that dielectric breakdown occurs in dry air when the magnitude of the $\vec{E}$ field exceeds about $3 \times 10^6\,\text{N/C} = 3 \times 10^6\,\text{V/m}$.

To conduct the testing experiment for Eq. (15.10), we use a Van de Graaff generator (Testing Experiment **Table 15.2**). This generator has a 30-cm-radius sphere dome, which when fully charged has an electric potential of 450,000 V with respect to Earth, which we consider to be at zero potential.

**TESTING EXPERIMENT TABLE**

**15.2 Testing the relationship between the $\vec{E}$ field and the $V$ field.**

VIDEO 15.2

| Testing experiment | Prediction | Outcome |
|---|---|---|
| Using a Van de Graaff generator and a grounded metal sphere on a wooden handle, charge the Van de Graaff generator so that its potential is 450,000 V and do not charge the second sphere. The two spheres have a constant potential difference $\Delta V = (450{,}000\text{ V} - 0) = 450{,}000\text{ V}$. Now slowly move the sphere on the handle closer to the generator sphere. Predict the distance that will be between them when you see a spark. 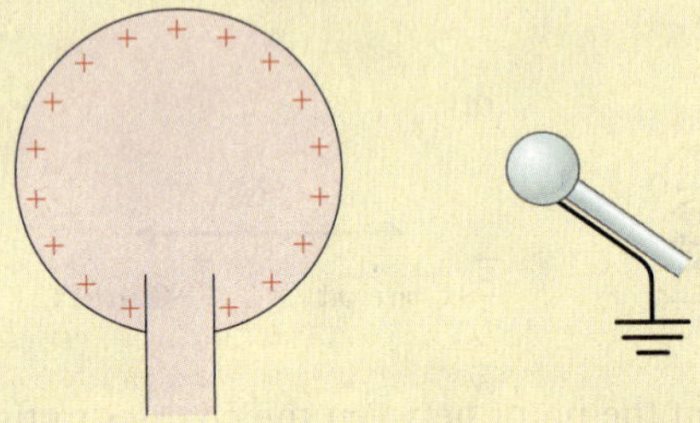 | Note that $\Delta V = (450{,}000\text{ V} - 0)$ between the spheres. If Eq. (15.10) is correct, and we place the spheres 0.15 m apart, the magnitude of the $\vec{E}$ field should be $\frac{\Delta V}{\Delta x} = \frac{450{,}000\text{ V}}{0.15\text{ m}} = 3.0 \times 10^6\text{ V/m}$, enough to cause dielectric breakdown. Thus, we predict a spark when the spheres are about 0.15 m = 15 cm apart. | We see a spark when the distance between them is about 15 cm. |

**Conclusion**

The outcome of the experiment matches the prediction. This provides experimental support for the relation between potential difference $\Delta V$ and the $\vec{E}$ field described by Eq. (15.10).

Imagine that you shuffle across a rug and then reach for a metal doorknob. Just before you touch the doorknob, a spark jumps between you and the knob. You estimate that the distance the spark jumps is about 1 cm. The value for the dielectric breakdown of air allows us to estimate the magnitude of the potential difference between your body and the doorknob:

$$\Delta V = E\Delta x = (3 \times 10^6\text{ V/m})(0.01\text{ m}) = 3 \times 10^4\text{ V} = 30{,}000\text{ V}$$

This is a huge potential difference. Fortunately, the potential difference is not dangerous; the amount of electric charge that flows is what is dangerous—and it is very small in this case.

**Review Question 15.4** Why does a spark jump between your fingers and a doorknob only when your hand is close to the doorknob?

## 15.5 Conductors in electric fields

We can use the concepts of the $\vec{E}$ field and the $V$ field to understand more precisely the behavior of electrically conducting materials and applications such as shielding and grounding.

### Electric field of a charged conductor

Imagine a metal sphere of radius $R$ that we touch with a negatively charged plastic rod (**Figure 15.14a**). Some of the excess electrons on the rod move to the metal sphere (Figure 15.14b). The electrons transferred to the sphere

**Figure 15.14** Charging a metal sphere.

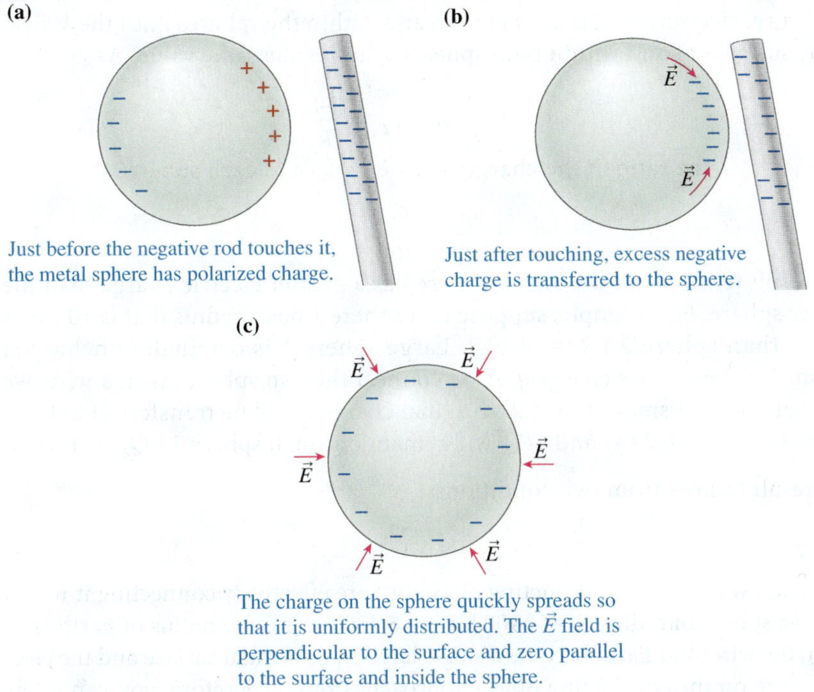
(a)
Just before the negative rod touches it, the metal sphere has polarized charge.
(b)
Just after touching, excess negative charge is transferred to the sphere.
(c)
The charge on the sphere quickly spreads so that it is uniformly distributed. The $\vec{E}$ field is perpendicular to the surface and zero parallel to the surface and inside the sphere.

**Figure 15.15** The $\vec{E}$ field lines and equipotential surfaces for: (a) a point charge, and (b) a charged metal sphere.

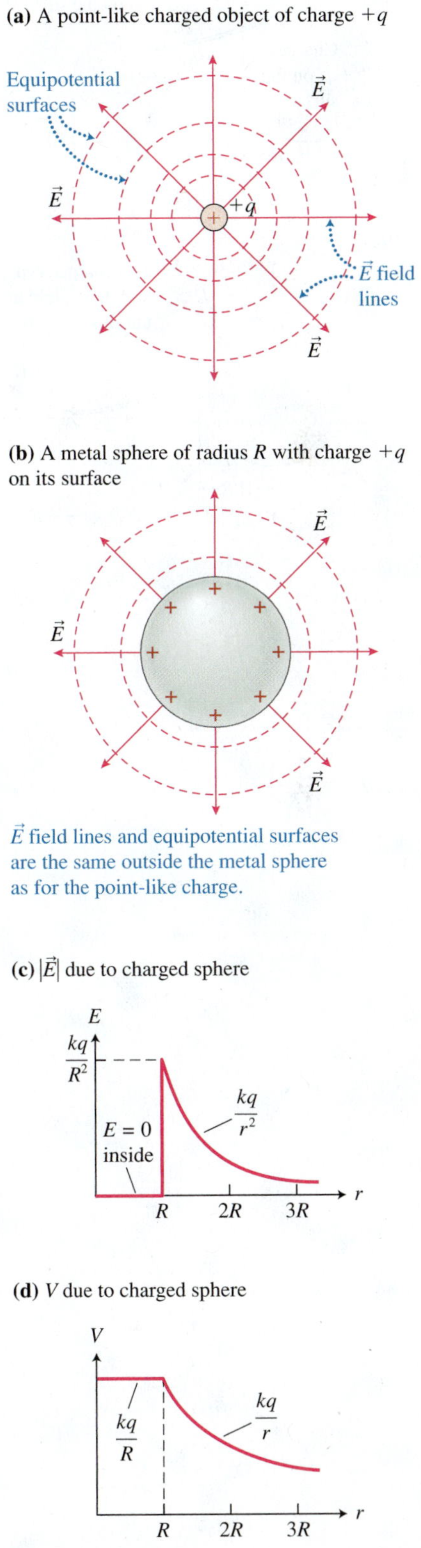


create an electric field in the metal that causes free electrons that are already in the metal to accelerate away from that spot. As a result of repulsion, the free electrons in the conductor quickly redistribute until equilibrium is reached, at which point the $\vec{E}$ field inside the conductor and parallel to its surface becomes zero (Figure 15.14c). If it were not zero, the electrons would continue to move. Since the $\vec{E}$ field is zero within and parallel to the surface of the sphere, this also means that the $V$ field has the same value at all points on the sphere. If there were an electric field within or parallel to the surface of the sphere, it would exert a force on free electrons inside that would cause them to move.

Outside the charged conductor the field is not zero. **Figures 15.15a** and b illustrate the $\vec{E}$ field lines and equipotential surfaces for a point-like charged object with charge $+q$ and for a metal sphere of radius $R$ with charge $+q$ on its surface. We won't do the calculation here since it requires calculus, but the magnitude of the $\vec{E}$ field and the value of the $V$ field outside the sphere are the same as that produced by a point-like charged object:

$$E = k\frac{|q|}{r^2} \text{ and } V = k\frac{q}{r}$$

where $r$ is the distance from the center of the sphere ($r \geq R$). Figures 15.15c and d show graphs of $E(r)$ and $V(r)$ for the charged metal sphere. Inside the sphere $E = 0$ and $V = k(q/R)$, the same value as on the surface. To summarize: when a conductor is charged, all electric charges reside on the surface and there is no electric field inside it. This equation also explains another important application—grounding.

## Grounding

Grounding discharges an object made of conducting material by connecting it to Earth. Here we will explain how grounding works by applying our new understanding of conductors and electric potential.

**Figure 15.16** (a) When you connect the different size spheres by a metal wire, they are at the same electric potential. (b) The charge/area and the $\vec{E}$ field are greater on the small sphere.

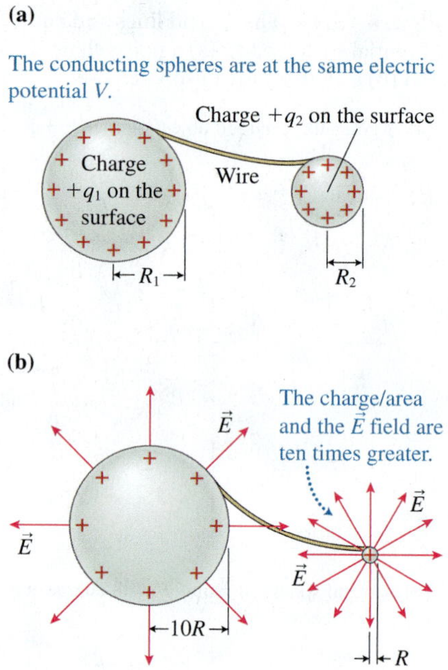
(a)
The conducting spheres are at the same electric potential V.
Charge $+q_2$ on the surface
Charge $+q_1$ on the surface
Wire
$R_1$
$R_2$
(b)
The charge/area and the $\vec{E}$ field are ten times greater.
$\vec{E}$
$10R$
$R$

Imagine that we have two conducting metal spheres of radii $R_1$ and $R_2$ with charges $q_1$ and $q_2$, respectively (**Figure 15.16a**). If we connect them with a long metal wire, electrons will move between and within the spheres until the $V$ field on the surfaces of and within both spheres achieves the same value. As

$$V_1 = \frac{kq_1}{R_1} \text{ and } V_2 = \frac{kq_2}{R_2}$$

and $V_1 = V_2$, the ratio of the charges on the surface of each sphere is

$$\frac{q_1}{q_2} = \frac{R_1}{R_2}$$

This result tells us that the bigger sphere has a greater electric charge than the smaller sphere. For example, suppose that sphere 1 has a radius that is 10 times greater than sphere 2 ($R_1 = 10R_2$). Large sphere 1 is originally uncharged and small sphere 2 has charge $q$. If we connect the two spheres with a wire, we find that 10/11 of small sphere 2's original charge $q$ will be transferred to large sphere 1 ($Q_1 = 10/11q$) and 1/11 will remain on small sphere 2 ($Q_2 = 1/11q$).

This result follows from two conditions:

$$\frac{Q_1}{Q_2} = \frac{R_1}{R_2} \text{ and } Q_1 + Q_2 = q$$

When we ground a conducting object, we are effectively connecting it with a wire to a sphere of radius $R = 6400$ km $= 6{,}400{,}000$ m, the radius of Earth. The object, the wire, and Earth become a single large equipotential surface and the electric charge on the conducting object approaches zero. Therefore, you can safely touch the grounded conducting object (which might be a toaster or a high-definition television) while standing on the ground without experiencing an electric shock.

Another important consequence of connecting different size conductors is that the $\vec{E}$ field lines will be denser on the small sphere and the $\vec{E}$ field will be stronger. In the above example with spheres of radii $R_1 = 10R_2$, the large sphere had a greater charge $Q_1 = (10/11)q$ and the small sphere had charge $Q_2 = (1/11)q$ The charge per unit surface area for the large sphere is

$$\frac{Q_1}{A_1} = \frac{10q}{11 \cdot 4\pi(10R_2)^2} = \frac{q}{11 \cdot 4\pi \cdot 10R_2^2},$$

while for the smaller sphere, the charge per unit surface area is

$$\frac{Q_2}{A_2} = \frac{q}{11 \cdot 4\pi R_2^2}$$

The charge per unit surface area is 10 times greater on the small sphere (Figure 15.16b). The $\vec{E}$ field lines will be ten times denser on the small sphere and the electric field will be 10 times stronger. As a result, on a pointed surface, such as a lightning rod, the electric field will be strongest at the tip.

## Uncharged conductor in an electric field—shielding

In our chapter-opening story we asked why you would be safe in a car during a lightning storm. Now we can answer that question. Imagine that we take a noncharged hollow conducting object and place it in a region with a uniform nonzero $\vec{E}$ field whose value is $\vec{E}_0$ (**Figure 15.17a**). The free electrons inside the object redistribute due to electric forces, producing their own contribution $\vec{E}_1$ to the $\vec{E}$ field (Figure 15.17b). This continues until the $\vec{E}$ field within the conducting object is reduced to zero (Figure 15.17c):

$$\vec{E} = \vec{E}_0 + \vec{E}_1 = 0$$

The $\vec{E}$ field produced by the environment $\vec{E}_0$ is cancelled by the contribution produced by the conductor. The net $\vec{E}$ field outside the conducting object is

**Figure 15.17** A hollow conducting object in an external $\vec{E}_0$ field.

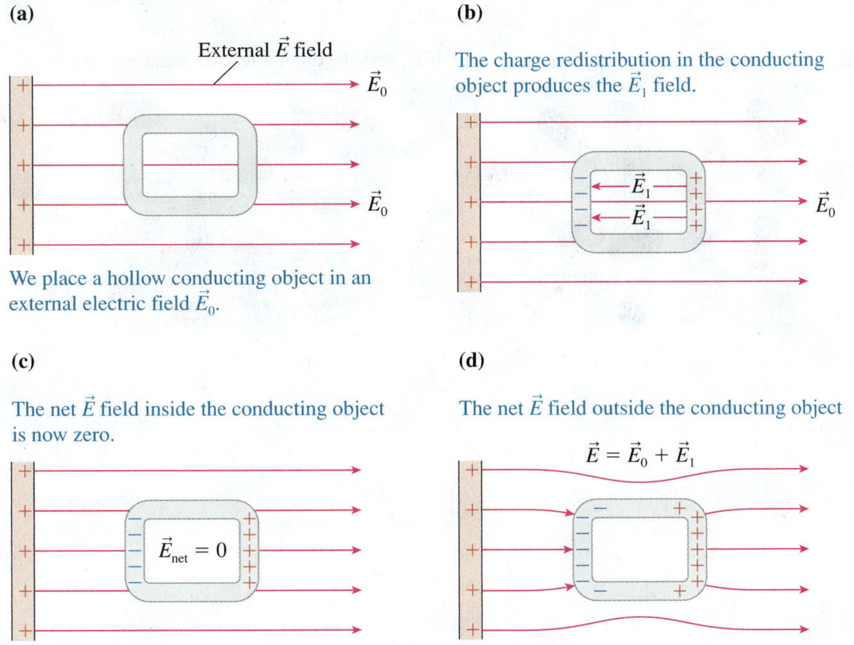

(a)
External $\vec{E}$ field
$\vec{E}_0$
We place a hollow conducting object in an external electric field $\vec{E}_0$.
(b)
The charge redistribution in the conducting object produces the $\vec{E}_1$ field.
$\vec{E}_1$
$\vec{E}_0$
(c)
The net $\vec{E}$ field inside the conducting object is now zero.
$\vec{E}_{net} = 0$
(d)
The net $\vec{E}$ field outside the conducting object
$\vec{E} = \vec{E}_0 + \vec{E}_1$

**Figure 15.18** A passenger is safe in a car struck by lightning.

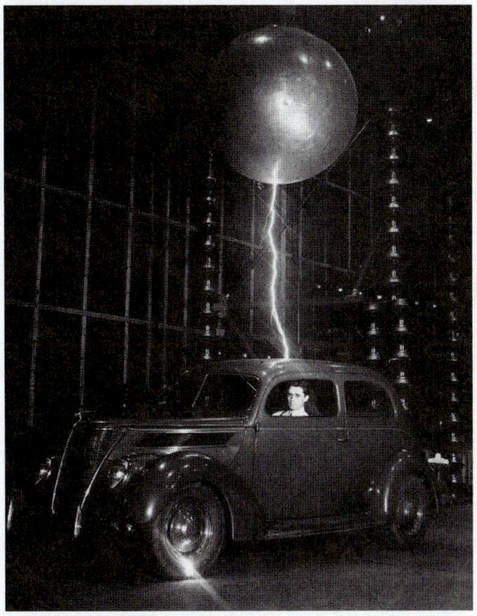

now a superposition of two $\vec{E}$ fields (Figure 15.17d). Because the conductor effectively "destroys" the external $\vec{E}$ field inside itself by creating the oppositely directed $\vec{E}_1$ field, the interior is protected from the external field. This effect is called **shielding**.

The hollow conducting object we just described might well be a car. Note that because the $\vec{E}$ field inside the conductor is zero, a person inside the car during a lightning storm is completely unaffected by the outside electric processes as long as she/he does not touch the surface of the car (**Figure 15.18**).

**Review Question 15.5** In this section you read that the $\vec{E}$ field inside the conductor is always zero. Why is this true?

## 15.6 Dielectric materials in an electric field

In this section we investigate the behavior of dielectric (nonconducting) materials. Recall from Chapter 14 that dielectric materials are attracted to both positively and negatively charged objects. How does our concept of the electric field contribute to the explanation of this phenomenon?

All atoms are composed of positively charged nuclei surrounded by negatively charged electrons (**Figure 15.19a**). If an atom in a dielectric material resides in a region with an external electric field $\vec{E}_0$, the field exerts a force on the atom's positive nucleus in the direction of the field $\vec{E}_0$ and a force in the opposite direction on the atom's negatively charged electrons. The nucleus and the electrons are displaced slightly in opposite directions until the force that the field exerts on each of them is balanced by the force they exert on each other due to their attraction (Figure 15.19b). Now the centers of the positive and negative charges are spatially separated. Such a system is familiar to us as an electric dipole (Figure 15.19c). We can represent an electric dipole using a

**Figure 15.19** The electric field polarizes an atom.

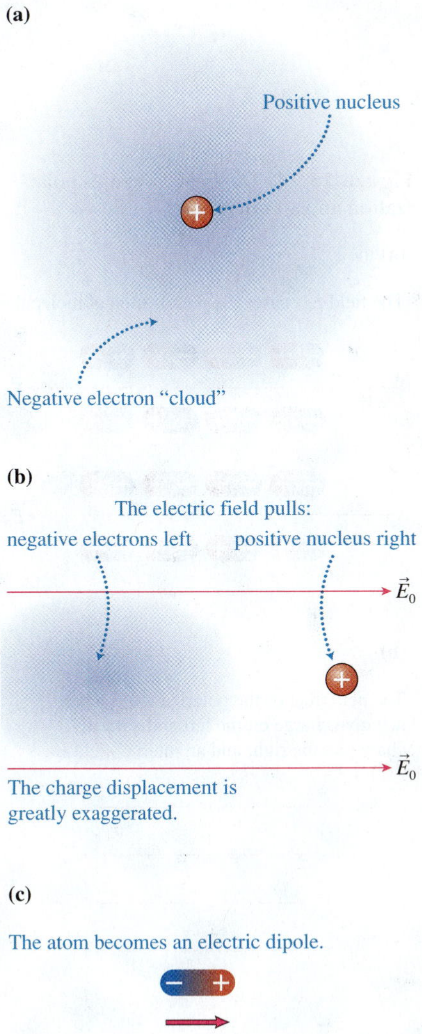

(a)
Positive nucleus
Negative electron "cloud"
(b)
The electric field pulls:
negative electrons left
positive nucleus right
$\vec{E}_0$
$\vec{E}_0$
The charge displacement is greatly exaggerated.
(c)
The atom becomes an electric dipole.

**Figure 15.20** Polar water molecules in an external electric field.

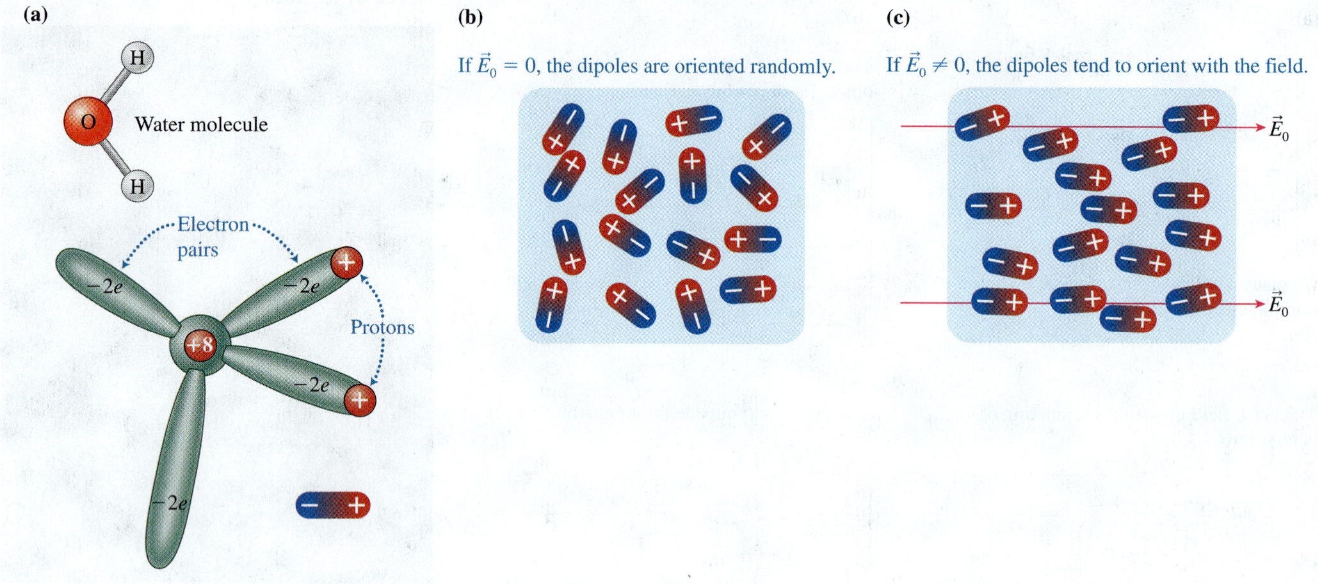


Electron pairs

If $\vec{E}_0 \neq 0$, the dipoles tend to orient with the field.

**Figure 15.21** The field $\vec{E}_0$ causes polarization and an internal field $\vec{E}_1$.

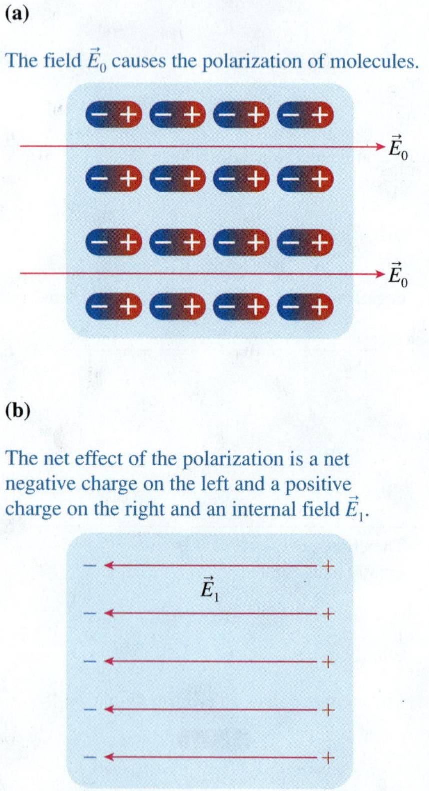


vector $\vec{p}$ that points from the negative charge center of the system to its positive charge center.

(a)
The field $\vec{E}_0$ causes the polarization of molecules.
$\vec{E}_0$
$\vec{E}_0$
(b)
The net effect of the polarization is a net negative charge on the left and a positive charge on the right and an internal field $\vec{E}_1$.
$\vec{E}_1$

However, some molecules, such as water, are natural electric dipoles even when the external $\vec{E}$ field is zero (**Figure 15.20a**). Without an external electric field, water molecules bump into each other and the dipoles are oriented randomly (Figure 15.20b). When the external $\vec{E}$ field is nonzero, the electric field exerts forces in opposite directions on the ends of the molecules, producing a torque that tends to rotate the molecules so that their electric dipoles align with the $\vec{E}$ field (Figure 15.20c).

When a dielectric material is placed in a nonzero $\vec{E}$ field, the atomic and molecular electric dipoles that result form a layer of positive electric charge on one surface of the material and a negative layer on the other surface (**Figure 15.21a**). These layers produce an additional contribution to the $\vec{E}$ field, an $\vec{E}_1$ field inside the dielectric material that points in the opposite direction to the external field $\vec{E}_0$ (Figure 15.21b). The net electric field inside is $\vec{E} = \vec{E}_0 + \vec{E}_1$. Since $\vec{E}_1$ points opposite $\vec{E}_0$, $E < E_0$. Thus, dielectric materials reduce the magnitude of the $\vec{E}$ field inside the material, similar to what happens in conductors. However, unlike conductors, the $\vec{E}$ field within does not decrease to zero. Thus, a dielectric material cannot completely shield its interior from an external electric field.

The ability of a dielectric material to decrease the $\vec{E}$ field varies from material to material. Physicists use a physical quantity to characterize this ability—the **dielectric constant** $\kappa$ (Greek letter kappa). The dielectric constant of a material determines how much the material reduces the external $\vec{E}$ field (the field produced by the environment) within itself. The larger the value of $\kappa$, the more the $\vec{E}_0$ external field is reduced in the material.

The dielectric constant of a material is defined as follows:

$$\kappa = \frac{E_0}{E} \tag{15.11}$$

where $E_0$ is the magnitude of the electric field produced by the environment and $E = E_0 - E_1$ is the magnitude of the electric field within the dielectric material. You can see from Eq. (15.11) that the dielectric constant is dimensionless. **Table 15.3** lists sample dielectric constants for many types of materials.

**Table 15.3** Dielectric constants for different types of materials.*

| Type of material | Dielectric constant ($\kappa$) |
|---|---|
| Vacuum | 1.0000 |
| Dry air | 1.0006 |
| Wax | 2.25 |
| Mica | 2–7 |
| Glass | 4–7 |
| Paper | 3–6 |
| Axon membrane | 8 |
| Body tissue | 8 |
| Ethanol | 26 |
| Methanol | 31 |
| Water | 81 |

*At 20° C.

Using Eq. (15.11), we can calculate the magnitude of the electric force that one charged object (object 1) exerts on another charged object (object 2) when both are inside a material with a dielectric constant $\kappa$:

$$F_{1 \text{ on } 2 \text{ in dielectric}} = q_2 E_{1 \text{ in dielectric}} = q_2 \frac{E_{1 \text{ in vacuum}}}{\kappa} = \frac{F_{1 \text{ on } 2 \text{ in vacuum}}}{\kappa}$$

The force that object 1 exerts on object 2 is reduced by $\kappa$ compared with the force it would exert in a vacuum. Inside the dielectric material, Coulomb's law is now written as

$$F_{1 \text{ on } 2} = k \frac{|q_1||q_2|}{\kappa r^2} \quad (15.12)$$

Note in **Figure 15.22** that the positive ends of the dipole water molecules group around a negative ion and in effect reduce its net charge. The same thing happens to a positive ion in water. There is a reduction in the force that charged objects exert on each other when in that dielectric material. We can interpret Eq. (15.12) in this way:

$$F_{1 \text{ on } 2} = k \frac{\left(\frac{q_1 q_2}{\kappa}\right)}{r^2}$$

The charge product $q_1 q_2$ in Coulomb's law has been reduced by a factor of $1/\kappa$.

**Figure 15.22** The force between positive and negative ions is reduced by the polar water molecules.

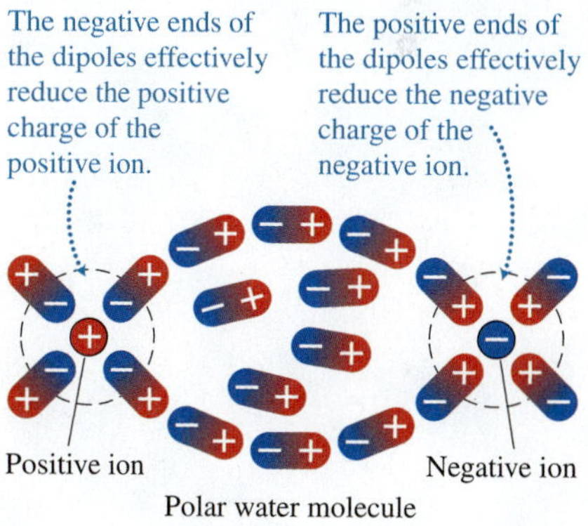

The negative ends of the dipoles effectively reduce the positive charge of the positive ion.
The positive ends of the dipoles effectively reduce the negative charge of the negative ion.
Positive ion
Negative ion
Polar water molecule

## Salt dissolves in blood but not in air

The presence of dielectric materials that have polar molecules (like water) has dramatic effects on processes occurring inside the human body. One important function of water is to dissolve compounds, making their elements available for physiological use. For instance, the nervous system uses sodium ions in the transmission of information. Solid salt (NaCl) dissolves into sodium ($Na^+$) and chlorine ($Cl^-$) ions in the blood, which is made up primarily of water.

Table salt comes in the form of an *ionic crystal.* Because a chlorine atom is more attractive to electrons than a sodium atom, an electron is transferred from a sodium atom (leaving it with electric charge $+1.6 \times 10^{-19}$ C) to the chlorine atom (giving it electric charge $-1.6 \times 10^{-19}$ C). The electric force that the oppositely charged sodium and chlorine ions exert on each other holds the salt crystal together. The distance between the two ions is about the same as the characteristic size of a molecule (about $10^{-10}$m). The magnitude of the electrical force that the ions exert on each other is then

$$F_{\text{Na on Cl}} = F_{\text{Cl on Na}} = k\frac{|q_{\text{Na}}||q_{\text{Cl}}|}{r^2}$$
$$= (9 \times 10^9\,\text{N}\cdot\text{m}^2/\text{C}^2)\frac{(1.6 \times 10^{-19}\,\text{C})^2}{(10^{-10}\,\text{m})^2} \approx 2 \times 10^{-8}\,\text{N}$$

The electric potential energy of a sodium-chlorine ion pair is about

$$U_q = k\frac{q_{\text{Na}}q_{\text{Cl}}}{r}$$
$$= (9 \times 10^9\,\text{N}\cdot\text{m}^2/\text{C}^2)\frac{(+1.6 \times 10^{-19}\,\text{C})(-1.6 \times 10^{-19}\,\text{C})}{(10^{-10}\,\text{m})} \approx -2 \times 10^{-18}\,\text{J}$$

The energy is negative because of the opposite charges of the ions. To separate the ions, approximately $+2 \times 10^{-18}$ J must be added to the system.

Atoms, molecules, and ions also have positive kinetic energy due to their random thermal motion. A rough estimate of this energy is $K = (3/2)k'T$, an expression we learned (in Chapter 9) for the kinetic energy of the particles in an ideal gas, where $k'$ is Boltzmann's constant (do not confuse it with the $k$ used in Coulomb's law). The average positive kinetic energy at room temperature (300 K) is approximately

$$K = \frac{3}{2}k'T = \frac{3}{2}(1.4 \times 10^{-23}\,\text{J/K})(300\,\text{K}) \approx 6 \times 10^{-21}\,\text{J}$$

This positive kinetic energy of the individual ions is much smaller than the negative electric potential energy holding them together. Thus, the total energy of the system is negative and the ions remain bound together when salt is in air.

When the salt is placed in water or blood, however, two changes occur. First, in water, there are many more collisions between molecules than there are in air. Although most of the collisions will not break an ion free from the crystal, some might. Remember, $K = (3/2)k'T$ is the *average* kinetic energy of the molecules. A few molecules will have enough random kinetic energy to knock an ion free from the crystal during a collision. Second, any ions that do become separated from the crystal by collisions do not exert nearly as strong an attractive force on each other because of the dielectric effect of the blood. If we assume that blood is mostly water, the attractive force that the two ions exert on each other decreases by a factor of 81, the dielectric constant of water:

$$F_{\text{Na on Cl in water}} = k\frac{e^2}{\kappa r^2}$$
$$= (9 \times 10^9\,\text{N}\cdot\text{m}^2/\text{C}^2)\frac{(1.6 \times 10^{-19}\,\text{C})^2}{81(10^{-10}\,\text{m})^2} \approx 3 \times 10^{-10}\,\text{N}$$

The electric potential energy is also smaller by a factor of 81:

$$U_q = k\frac{-e^2}{\kappa r}$$
$$= (9 \times 10^9\,\text{N}\cdot\text{m}^2/\text{C}^2)\frac{-(1.6 \times 10^{-19}\,\text{C})^2}{81(10^{-10}\,\text{m})} \approx -3 \times 10^{-20}\,\text{J}$$

This energy is almost the same as the average kinetic energy of the random motion of the water molecules. This means that the random kinetic energy of the water molecules is sufficient to keep the sodium and chlorine ions from recombining. This allows the nervous system to use the freed sodium ions to transmit information.

**TIP** You can think of a conductor as a medium with a dielectric constant of infinity.

**Review Question 15.6** What are the differences between conducting and dielectric materials when they are placed in a region with a nonzero $\vec{E}$ field? Explain these differences.

## 15.7 Capacitors

We have learned about practical applications of conductors in electric fields, such as grounding and shielding. Another important application involving electric fields and conductors is storing energy in the form of electric potential energy. When positively and negatively charged objects are separated, the system's electric potential energy increases. How can this charge separation be maintained so that the electric potential energy can be stored for useful purposes? This is accomplished with a device known as a *capacitor*.

A **capacitor** consists of two conducting surfaces separated by a nonconducting material. Although a variety of configurations are possible, the simplest are parallel plate capacitors made of two metal plates separated by air, rubber, paper, or some other dielectric material (**Figure 15.23a**). The role of a capacitor is to store electric potential energy.

To charge a capacitor, one usually connects its plates to the terminals of a battery. The positive battery terminal is connected by a conducting wire to one plate, for example the left capacitor plate. Then an $\vec{E}$ field produced by the battery causes electrons to flow from the left capacitor plate toward the positive battery terminal (Figure 15.23b). Electrons likewise flow from the negative battery terminal to the right capacitor plate. This charge transfer continues until the capacitor plate, conducting wire, and battery terminal on each side are at the same potential (Figure 15.23c). Effectively, the battery removes electrons from one plate and deposits them on the other. The plate with a deficiency of electrons is positively charged with charge $+q$ and the plate with excess electrons is negatively charged with charge $-q$. The total charge of the capacitor is zero.

If we consider the capacitor plates to be large flat conductors, the charges should distribute evenly on the plates. Each charged plate produces a uniform $\vec{E}$ field (Figure 15.23d). When the plates are close to each other, the fields from each plate overlap between the plates and outside the plates. Outside the plates, the $\vec{E}$ field from the positively charged plate points opposite the $\vec{E}$ field from the negatively charged plate (Figure 15.23e) and they cancel each other. Between the plates the $\vec{E}$ fields are in the same direction and add. Thus, the net $\vec{E}$ field is strong between the plates and zero outside (Figure 15.23f).

According to Eq. (15.10), the magnitude of the $\vec{E}$ field between the plates relates to the potential difference $\Delta V$ from one plate to the other and the distance $d$ separating them:

$$E = \left|-\frac{\Delta V}{\Delta x}\right| = \left|\frac{\Delta V}{d}\right|$$

Thus, if the potential difference across the plates doubles (for example, you connect the capacitor to a 12-V battery instead of a 6-V battery), the magnitude of the $\vec{E}$ field between the plates also doubles. Recall that $\vec{E}$ field lines are created by charges on the plates. Thus to double the $\vec{E}$ field, the charge on the plates has to double. We conclude that the magnitude of the charge $q$ on the plates ($+q$ on one plate and $-q$ on the other) should be proportional to the potential difference $V$ across the plates: $q \propto \Delta V$. Or, if we use a proportionality constant,

$$q = C|\Delta V| \tag{15.13}$$

The proportionality constant $C$ in this equation is called the **capacitance** of the capacitor. In the above, $q$ is the magnitude of the charge on each plate. The unit of capacitance is 1 coulomb/volt = 1 farad (1 C/V = 1 F) in honor of Michael Faraday, whose experiments helped establish the atomic nature of electric charge.

**Figure 15.23** Charging a capacitor.

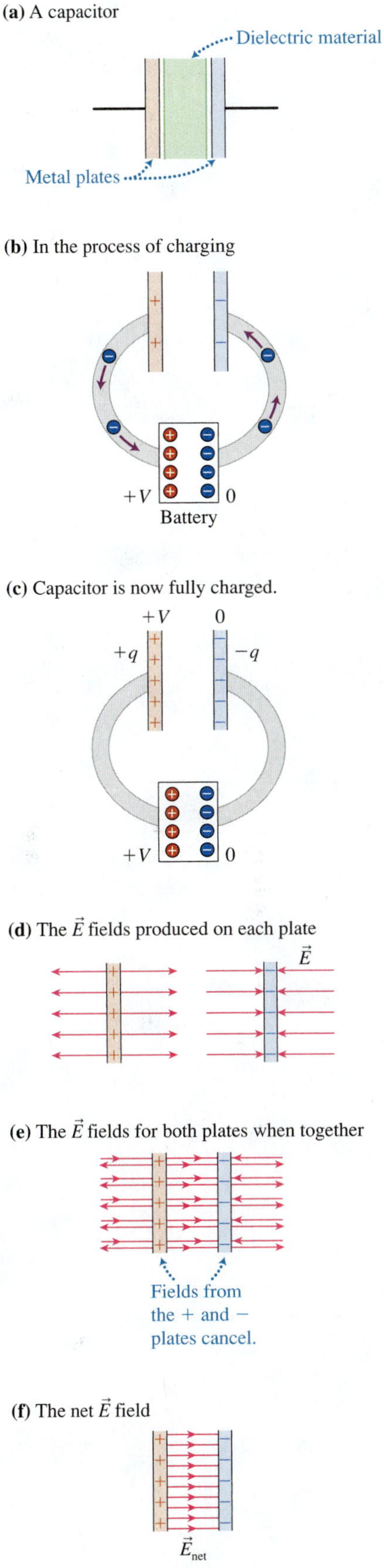

## Capacitance of a capacitor

What properties of capacitors determine their capacitance? It might seem that Eq. (15.13) answers this question. However, the capacitance of a capacitor does not change when the charge on the plates or the potential difference across them is changed. These two quantities are proportional to each other ($q \propto \Delta V$); to double the charge on the plates we need to double the potential difference across the same capacitor. But the ratio $C = q/|\Delta V|$ remains the same. This ratio is an operational definition of capacitance.

So then what does affect the capacitance of a particular capacitor? Imagine two capacitors whose plates have different surface areas and are connected to the same potential difference source. The capacitor with the larger surface area $A$ plates should be able to maintain more charge separation because there is more room for the charge to spread out (**Figure 15.24a**).

**Figure 15.24** Capacitance depends on $A$, $d$, and $\kappa$.

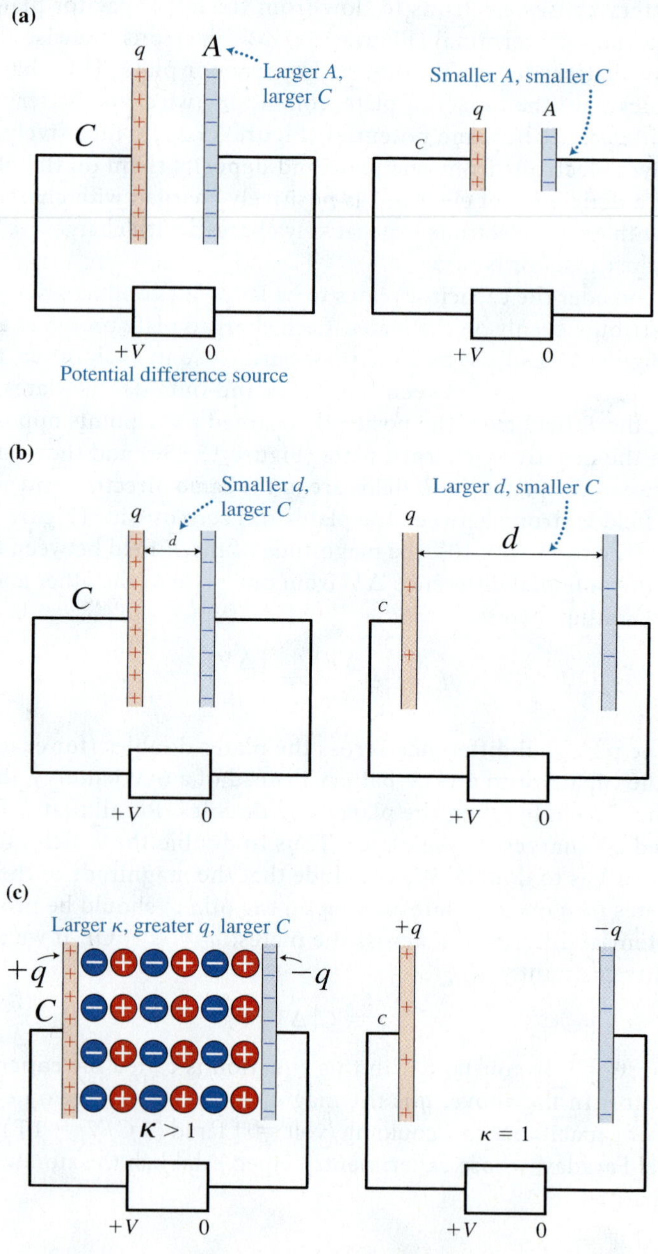

Now imagine two capacitors with the same surface area and the same potential difference across the plates but with different distances $d$ between the plates (Figure 15.24b). Since $|\Delta V| = Ed$ and $\Delta V$ is constant, a larger distance $d$ between the plates leads to a smaller magnitude $\vec{E}$ field between the plates. But since the magnitude of this $\vec{E}$ field is proportional to the amount of electric charge on the plates, a larger plate separation leads to a smaller magnitude of electric charge on the plates ($+q$ on one and $-q$ on the other). This means the capacitance of the capacitor decreases with increasing $d$.

In addition to surface area and distance between the plates, capacitance is affected by the dielectric constant $\kappa$ of the material between the plates. Material between the plates with a large dielectric constant becomes polarized by the electric field between the plates (Figure 15.24c). The positive sides of the polar molecules near the negatively charged plate attract more negative charge onto the negative plate, and vice versa for the positive plate. Thus, more charge moves onto the capacitor plates for capacitors whose plates are separated by material of high dielectric constant. The capacitance increases in proportion to the dielectric constant $\kappa$ of the material.

Thus we conclude that the capacitance of a particular capacitor should increase if the surface area $A$ of the plates increases, decrease if the distance $d$ between them is increased, and increase if the dielectric constant $\kappa$ of the material between them increases. A careful derivation (which we will not go through) provides the following result for a parallel plate capacitor:

$$C_{\text{Parallel plate capacitor}} = \frac{\kappa A}{4\pi k d} \tag{15.14}$$

where $k = 9.0 \times 10^9\,\text{N}\cdot\text{m}^2/\text{C}^2$. Let's check the units for capacitance:

$$\frac{(\text{m}^2)}{\left(\dfrac{\text{N}\cdot\text{m}^2}{\text{C}^2}\right)\text{m}} = \frac{\text{C}^2}{\text{N}\cdot\text{m}} = \frac{\text{C}^2}{\text{J}} = \frac{\text{C}}{\text{J}}\text{C} = \frac{\text{C}}{\text{V}} = \text{F}$$

The units check.

**TIP** There are two similar looking symbols in Eq. (15.14). $\kappa$ is the dielectric constant of the material between the plates, and $k$ is the constant in Coulomb's law.

**QUANTITATIVE EXERCISE 15.9 Capacitance of a textbook**

Estimate the capacitance of your physics textbook, assuming that the front and back covers (area $A = 0.050\,\text{m}^2$, separation $d = 0.040\,\text{m}$) are made of a conducting material. The dielectric constant of paper is approximately 6.0. Second, determine what the potential difference must be across the covers in order for the textbook to have a charge separation of $10^{-6}$ C (one plate has charge $+10^{-6}$ C and the other has charge $-10^{-6}$ C).

**Represent mathematically** The capacitance of a parallel plate capacitor is

$$C = \frac{\kappa A}{4\pi k d}$$

The potential difference $\Delta V$ needed to produce a charge separation of $q = 10^{-6}$ C is

$$\Delta V = \frac{q}{C}$$

**Solve and evaluate**

$$C = \frac{\kappa A}{4\pi k d} = \frac{(6.0)(0.050\,\text{m}^2)}{4\pi(9.0 \times 10^9\,\text{N}\cdot\text{m}^2/\text{C}^2)(0.040\,\text{m})}$$

$$= 7.0 \times 10^{-11}\,\text{F} = 70\,\text{pF}$$

(Here, $p$ stands for "pico", the metric prefix for $10^{-12}$.)

$$\Delta V = \frac{q}{C} = \frac{10^{-6}\,\text{C}}{7.0 \times 10^{-11}\,\text{F}} = 1.4 \times 10^4\,\text{V} = 14\,\text{kV}$$

This is a rather small capacitance, but reasonable given the large plate separation and the low dielectric value. The farad is a very large unit, and capacitors with picofarad or nanofarad capacitances are quite common.

**Try it yourself:** Approximately how long must the book cover be to have a 1-F capacitance?

*Answer:* If we assume the same plate separation, plate width, and dielectric constant of the paper, then the covers should be about $3 \times 10^9$ m long, or about 3 million km long! One farad is a large capacitance.

The previous example shows that making a capacitor with a large capacitance requires a large surface area. It is possible to make capacitors of much larger capacitance by making multiple very thin layers of specially designed materials. These are called supercapacitors, and they can reach thousands of farads in capacitance.

## Body cells as capacitors

Capacitors are used in all devices that require some instant action—such as camera flashes or computer keyboards. They are also present in the tuning circuits of radios, music amplifiers, etc. Biological capacitors are found inside our bodies. Cells, including nerve cells, have capacitor-like properties (see Example 15.10). The conducting "plates" are the fluids on either side of a moderately non-conducting cell membrane. In this membrane, chemical processes cause ions to be "pumped" across the membrane. As a result, the membrane's inner surface becomes slightly negatively charged while the outer surface becomes slightly positively charged.

**EXAMPLE 15.10 Capacitance of and charge on body cells**

Estimate (a) the capacitance $C$ of a single cell and (b) the charge separation $q$ of all of the membranes of the human body's $10^{13}$ cells. Assume that each cell has a surface area of $A = 1.8 \times 10^{-9}\,\text{m}^2$, a membrane thickness of $d = 8.0 \times 10^{-9}\,\text{m}$, a $\Delta V = 0.070$-V potential difference across the membrane wall, and a membrane dielectric constant $\kappa = 8.0$.

**Sketch and translate** A body cell is sketched in the figure below. The information needed to answer the questions is given in the problem statement.

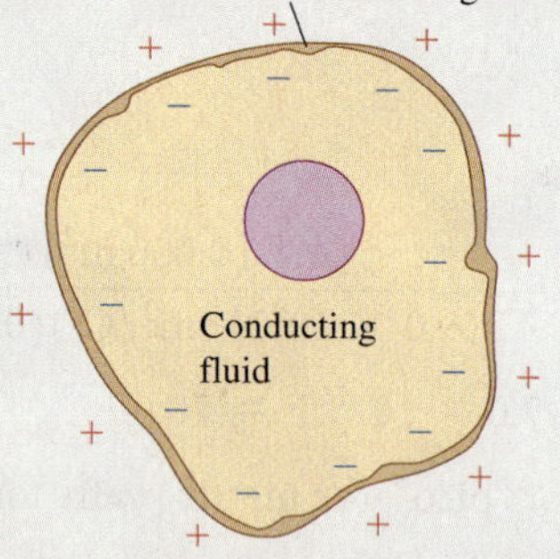


**Simplify and diagram** The thickness of the membrane $d = 8.0 \times 10^{-9}\,\text{m}$ is much less than the dimensions of the cell (roughly the square root of the surface area $\sqrt{A} = 4.2 \times 10^{-5}\,\text{m}$). Thus, up close the cell membrane appears almost flat, just as Earth appears flat to people on its surface. Thus, we can use the expression for the capacitance of a parallel plate capacitor to make our estimate.

**Represent mathematically** The capacitance of one cell is

$$C = \frac{\kappa A}{4\pi k d}$$

The total charge separation $q$ on the $10^{13}$ cells (biological capacitors) in the body is then

$$q_{\text{total}} = (10^{13})\,C|\Delta V|$$

**Solve and evaluate** The capacitance of one cell is

$$C = \frac{(8.0)(1.8 \times 10^{-9}\,\text{m}^2)}{4\pi(9.0 \times 10^{9}\,\text{N}\cdot\text{m}^2/\text{C}^2)(8.0 \times 10^{-9}\,\text{m})}$$
$$= 1.6 \times 10^{-11}\,\text{F}$$

The total capacitance of all $10^{13}$ cells is then

$$C_{\text{total}} = 10^{13}(1.6 \times 10^{-11}\,\text{F}) = 160\,\text{F}$$

The total charge separated by the membranes of all of the cells is approximately

$$q = C_{\text{total}}\Delta V = (160\,\text{F})(0.070\,\text{V}) \approx 10\,\text{C}$$

Although these calculations are approximate, it is clear that the separation of electric charge (−10 C total on the inside walls of cell membranes and +10 C on the outside walls) is an important part of our metabolic processes. This 10-C electric charge separation is huge—about the same as the charge transferred during a lightning flash. Fortunately, our bodies' cells do not all discharge simultaneously in one surge.

**Try it yourself:** Suppose you doubled the membrane thickness of all body cells. How would the potential difference across them have to change in order for the cells to maintain the same charge separation?

*Answer:* Doubling the membrane thicknesses would reduce the capacitance by half. Thus, you would need to double the potential difference across the membranes to maintain the same charge separation.

## Energy of a charged capacitor

A capacitor is essentially a system of positively and negatively charged objects that have been separated from each other so that the system has electric potential energy. To determine electric potential energy in a charged capacitor, we start with an uncharged capacitor and then calculate the amount of work that must be done on the capacitor to move electrons from one plate to the other.

This process is sketched in **Figure 15.25** in the form of a thought experiment. It's not possible for us to manually grab electrons and move them from one capacitor plate to the other. Usually, this work is performed by an external source such as a battery. However, in this thought experiment, we will move increments of charge $-\Delta q$ from one plate to the other. Note that the charge is negative because these charge increments are composed of electrons. After the first charge increment has been moved, the plate from which it was taken has a charge of $+\Delta q$ and the plate to which it is taken has a charge of $-\Delta q$.

Only a very small amount of work must be done to move this first $-\Delta q$ since the left plate is uncharged while moving it (Figure 15.25b). The next $-\Delta q$ is more difficult to move (Figure 15.25c) since the left plate now has a charge of $-\Delta q$ that repels the $-\Delta q$ we are trying to move there. Additionally, the $+2\Delta q$ charge on the right plate pulls back on it. The more charge increments we move from one plate to the other, the more difficult it becomes to move the next one. Eventually, with increasing effort, we will have transferred a total (negative) charge of $-q = (-\Delta q) + (-\Delta q) + \cdots$ to the left plate, leaving the right plate with a charge of $+q$. We can represent the whole process of charging the capacitor with a work-energy bar chart (**Figure 15.26**). The external object (us moving the charge increments) does work on the system (the capacitor) and the electric potential energy of the system increases.

The electric potential energy increase can be calculated using Eq. (15.8):

$$\Delta U_q = q\Delta V_{\text{average}}$$

where $\Delta V_{\text{average}}$ is the average potential difference from one plate to the other during the process of charging. Since the initial potential difference is zero and the final is $\Delta V$, the average potential difference between the plates is

$$\Delta V_{\text{average}} = \frac{0 + \Delta V}{2} = \frac{\Delta V}{2}$$

Substituting this expression for $\Delta V_{\text{average}}$ into Eq. (15.8), we get

$$U_q = q\Delta V_{\text{average}} = q\frac{\Delta V}{2}$$

(We dropped the $\Delta$, as $U_{qi} = 0$.) The above expression can be written in three different ways using Eq. (15.13), $q = C|\Delta V|$:

$$U_q = \frac{1}{2}q|\Delta V| = \frac{1}{2}C|\Delta V|^2 = \frac{q^2}{2C} \qquad (15.15)$$

Note that the process of charging a capacitor is similar to stretching a spring; at the beginning a smaller force is needed to stretch the spring by a certain amount compared to the much greater force needed when the spring is already stretched.

**Figure 15.25** The energy stored while charging a capacitor.

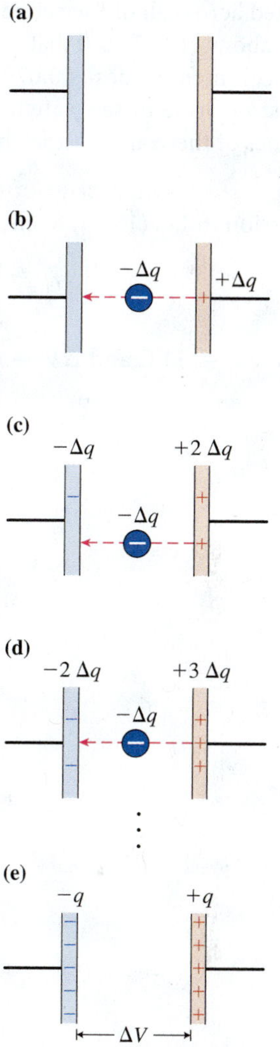


**Figure 15.26** A bar chart representing the work done to charge a capacitor.

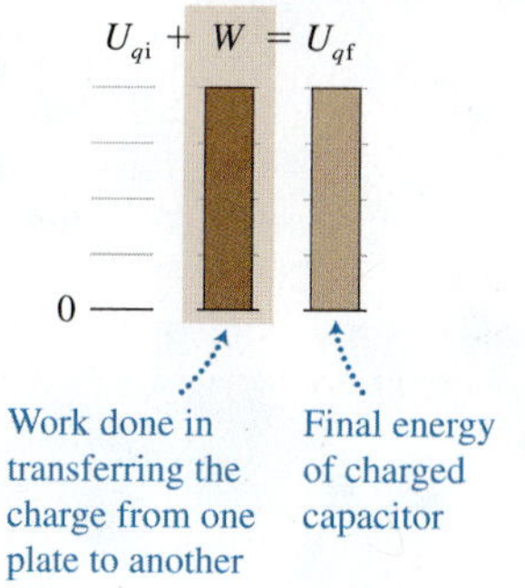

**QUANTITATIVE EXERCISE 15.11 Energy needed to charge human body cells**

In Example 15.10, we estimated that the total charge separated across all of the cell membranes in a human body was about 11 C. Recall that the potential difference across the cell membranes is 0.070 V. Estimate the work that must be done to separate the charges across the membranes of these approximately $10^{13}$ cells.

**Represent mathematically** We can use the first expression in Eq. (15.15) to answer this question:

$$U_q = \frac{1}{2}q\Delta V$$

where $q = 11\text{ C}$ and $\Delta V = 0.070\text{ V}$.

**Solve and evaluate**

$$U_q = \frac{1}{2}q\Delta V = \frac{1}{2}(11\text{ C})(0.070\text{ V}) = 0.40\text{ J}$$

These cells are continually being charged and discharged as part of the body's metabolic processes.

**Try it yourself:** Estimate the energy needed to charge body cells each day assuming that each cell discharges once per second.

*Answer:* $35{,}000\text{ J} = 8\text{ kcal}$, or about one-tenth the energy provided by a piece of bread. This is not much. However, some cells might discharge up to 400 times per second, though some might discharge fairly infrequently. The result is a rough estimate.

## Energy density of electric field

We have found that the energy stored in a charged capacitor is

$$U_q = \frac{1}{2}C|\Delta V|^2$$

Where is this energy stored? The difference between a charged and an uncharged capacitor is the presence of the electric field in the region between the plates. We hypothesize that it is the electric field that possesses this energy.

What physical quantities affect the amount of this stored energy? It takes more energy to charge a larger capacitor (for example, a capacitor with larger plates) that has the same electric field between the plates. To have a measure of energy independent of the capacitor volume, we will use the physical quantity of **energy density**. This energy density quantifies the electric potential energy stored in the electric field per cubic meter of volume. The $\vec{E}$ field energy density $u_E$ is

$$u_E = \frac{U_q}{V}$$

where $U_q$ is the electric potential energy stored in the electric field in that region, and $V$ is the volume of the region.

**TIP** Notice that here the letter $V$ stands for volume, not for the electric potential. Sometimes similar or identical symbols are used for different physical quantities. When working with mathematical expressions, always ask yourself: "What does each of these symbols represent?"

Let us rewrite the above equation for the electric field energy density in terms of the $\vec{E}$ field, using our knowledge of the electric energy of a parallel plate capacitor:

$$u_E = \frac{U_q}{V} = \frac{\left(\frac{1}{2}C|\Delta V|^2\right)}{V} = \frac{\left(\frac{1}{2}\left(\frac{\kappa A}{4\pi k d}\right)(Ed)^2\right)}{V}$$

We used Eqs. (15.14) and (15.9) to substitute for $C$ and $\Delta V$. $A$ is the plate area, $d$ is the plate separation, $E$ is the magnitude of the $\vec{E}$ field between the plates, and $\kappa$ is the dielectric constant of the material between the plates. Simplifying this, we have

$$u_E = \frac{\kappa A(Ed)^2}{8\pi k d V} = \frac{\kappa}{8\pi k} E^2 \frac{Ad}{V}$$

Since the volume of the region between the plates is $V = Ad$, this equation simplifies to

$$u_E = \frac{\kappa}{8\pi k} E^2 \quad (15.16)$$

The energy density depends only on the magnitude of the $\vec{E}$ field, the properties of the dielectric, and a few mathematical and physical constants. We'll use this idea in a later chapter on electromagnetic waves.

**TIP** Although we derived the expression for energy density for a parallel plate capacitor, it can be applied for any electric field.

**Review Question 15.7** A parallel plate capacitor has two oppositely charged plates, each producing its own electric field. Why is there no electric field outside the capacitor?

# 15.8 Electrocardiography and lightning: Putting it all together

Let's apply what we have learned about the $\vec{E}$ and $V$ fields to understand how electrocardiography works and how lightning occurs.

## Electrocardiography

The human heart has four chambers. During a 1-s heartbeat, each chamber pumps in a specific sequence. For example, the lower right chamber (the right ventricle) pumps blood through the lungs, where carbon dioxide is exchanged for oxygen, and the lower left chamber (the left ventricle) pumps freshly oxygenated blood into the aorta for a new trip around the circulatory system to provide oxygen for the body cells.

An electric charge separation occurs when muscle cells in the heart contract during the pumping process. For example, when the left ventricle pumps blood into the aorta, many left ventricle muscle cells contract. As each muscle cell contracts, positive and negative charges separate (this process is represented schematically in **Figure 15.27**). The simultaneous contraction of the large number of cells in the heart muscle of the left ventricle results in a relatively large charge separation—an electric dipole. On the other hand, the number of simultaneously contracting muscle cells is much smaller when the left atrium pumps blood into the left ventricle. Thus, during the 1-s cycle of a heartbeat, there is an electric dipole with continually changing magnitude and orientation that depends at each instant on the number and the orientation of the muscle cells that are contracting at that instant.

This electric dipole produces a $V$ field that extends outside the heart to the body tissue. A device called an electrocardiogram (ECG) detects that $V$ field. Data from an ECG can identify abnormalities, such as enlarged left or right ventricles. How does an ECG work? Electrodes, in the form of pads, are placed on the skin. These pads transmit data to the ECG about the potential difference between the various pairs of pads. The ECG uses these data to reconstruct the changing electric dipole on the heart.

**Figure 15.27** An electric dipole is produced by each heart muscle cell contraction.

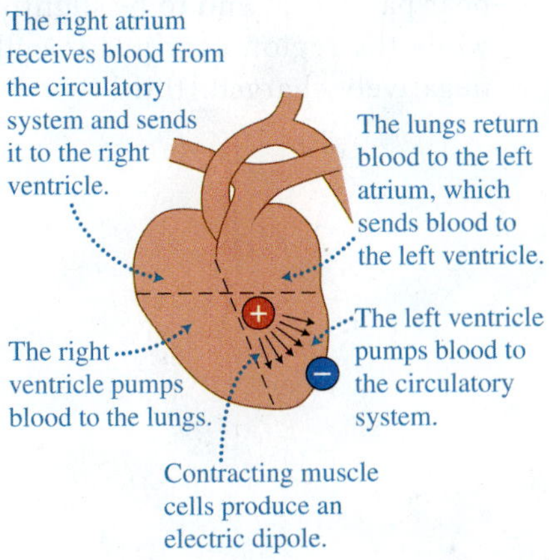

The right atrium receives blood from the circulatory system and sends it to the right ventricle.
The lungs return blood to the left atrium, which sends blood to the left ventricle.
The left ventricle pumps blood to the circulatory system.
The right ventricle pumps blood to the lungs.
Contracting muscle cells produce an electric dipole.
+
−

## CONCEPTUAL EXERCISE 15.12 Heart's electric dipole and ECG

The first figure shows a simplified electric dipole charge distribution on a heart at one instant during a heartbeat and two ECG pads on opposite shoulders of the person's body. What do we expect these pads to measure at that particular instant? (a) Draw $\vec{E}$ field vectors produced by the heart's dipole charge representing the electric field at the location of the dot in the figure. (b) Determine the direction of the forces exerted by the electric field on a positive sodium ion $Na^+$ and on a negative chlorine ion $Cl^-$ in the body tissue at that location.

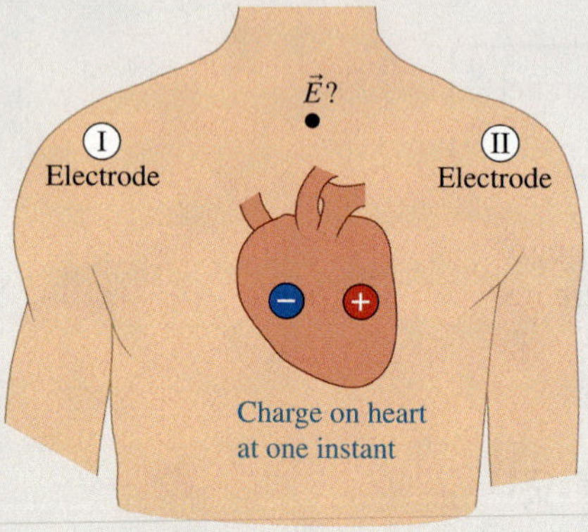


**Sketch and translate** The situation is sketched in the figure at top right. Each charge of the heart's electric dipole produces a contribution to the $\vec{E}$ field at each point in space, including the location of interest. To find the $\vec{E}$ field at that point, we graphically add the $\vec{E}$ field contributions produced by each dipole charge. The electric field exerts a force on ions located there. The force points in the same direction as the $\vec{E}$ field for positive ions and in the opposite direction for negative ions.

**Simplify and diagram** The figure at top right shows the $\vec{E}$ field contribution from each heart dipole charge along with the graphical addition of these two contributions to estimate the direction of the $\vec{E}$ field. The $\vec{E}$ field exerts a force on positive sodium ions toward ECG pad I and on negative chlorine ions toward ECG pad II (second figure at right). As a result, the region near pad I will tend to be slightly positively charged, while the region near pad II will tend to be slightly negatively charged (third figure at right). Thus, pad I will measure a slightly higher $V$ field than pad II. The device to which the pads are connected then detects this difference and a physician uses the data to infer the health of the heart.

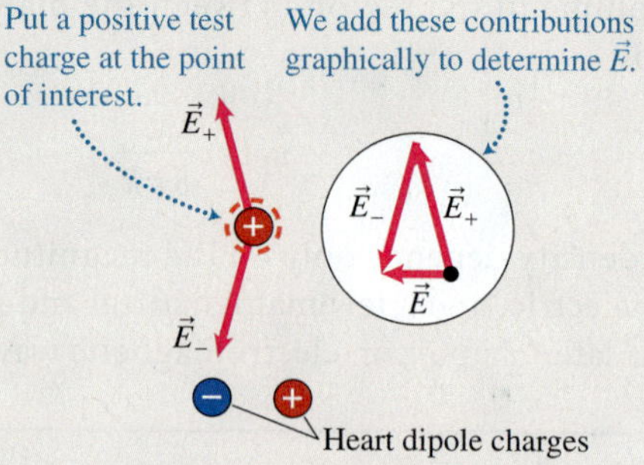


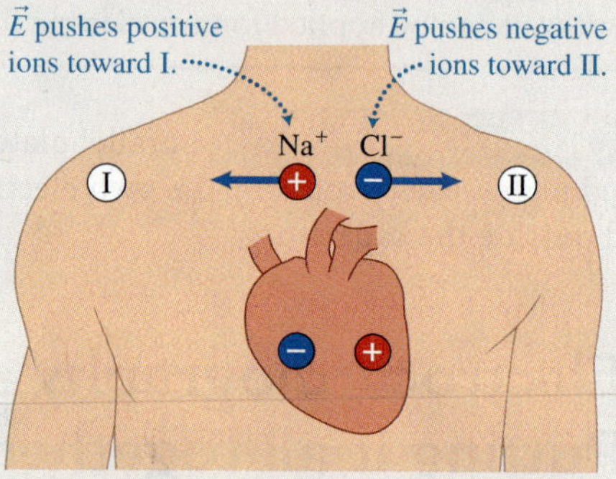


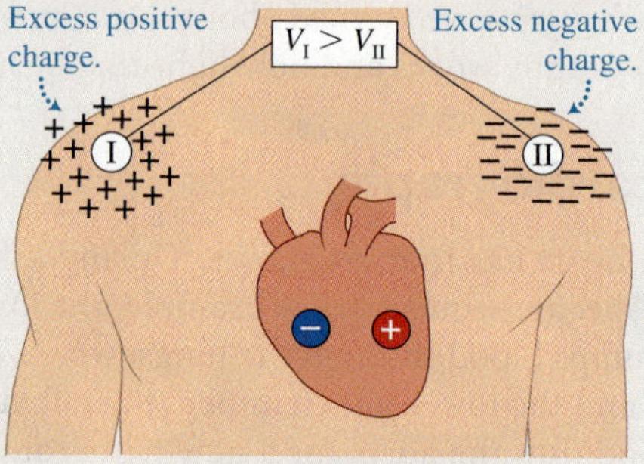


**Try it yourself:** What change(s) in the last example would cause the potential difference between leads I and II to be qualitatively less?

*Answer:* The potential difference would be reduced if fewer muscle cells were contracting (this would reduce the dipole charge on the heart) or if the orientation of the dipole was different (for example, the dipole was oriented more vertically).

A physician looks at the varying potential difference between different pairs of electrode pads and from this can learn about the heart's changing electric dipole. This dipole in turn indicates the number of muscle cells contracting at different times. A person with left ventricular hypertrophy (an enlarged left ventricle) will have a larger than normal dipole and potential difference during contraction of the left ventricle. **Figure 15.28** compares a normal ECG to one showing left ventricular hypertrophy.

## Lightning

When the $\vec{E}$ field in air or in some other material is very large, free electrons accelerate and quickly acquire enough kinetic energy to ionize atoms and molecules in their path when colliding with them. This produces additional free electrons, which in turn accelerate, collide with, and ionize yet more atoms and molecules. This dielectric breakdown, an intense cascade of electrons flowing opposite the $\vec{E}$ field direction, occurs suddenly and often produces a spark or a bigger flash—lightning.

Although there are many types of lightning, the commonly seen type is cloud-to-ground lightning. In cloud-to-ground lightning, water droplets and ice particles in the cloud rub against water droplets in the ascending air, causing some particles to become negatively or positively charged (**Figure 15.29a**). The air with positively charged particles is less dense and accumulates at the top of the cloud (in the P region) while the air with negatively charged particles is denser and remains at the bottom (in the N region). Electrons in the ground directly below the cloud are repelled by the cloud and move away, leaving the ground positively charged under the cloud.

Electrons leap downward from the N region of the cloud (Figure 15.29b) in a 15- to 30-m-long "stepped leader." The electrons make successive 15- to 30-m jumps toward the ground at intervals of one-millionth of a second (Figure 15.29c), leaving a trail of negative ions. As the electrons approach a high point on the ground (like the building in Figure 15.29d), the $\vec{E}$ field in the air above that point becomes so large that positive ions (called the positive streamer) are produced and pulled in an upward direction from the ground and building to meet the stepped leader. Electrons and positive ions recombine to produce a flash of light. After this breakdown, negative charge in the stepped leader farther above the ground can now rush down through the region of ionized air (Figure 15.29e). This intense flow of electrons originating farther above the ground causes more flashing light. The flash, which appears to move up, and the large downward electron flow make up what is called the return stroke. Eventually, a large number of electrons originally in the cloud

**Figure 15.28** Electrocardiogram signals across a pair of electrode pads for: (a) a healthy heartbeat, and (b) for an enlarged left ventricle heartbeat.

(a) Normal heart

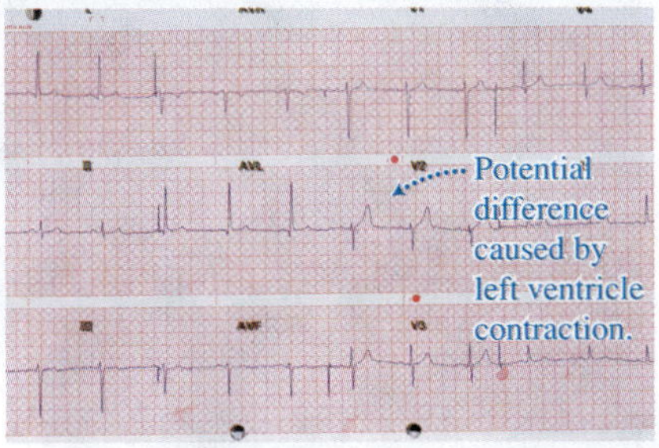
Potential difference caused by left ventricle contraction.

(b) Left ventricular hypertrophy

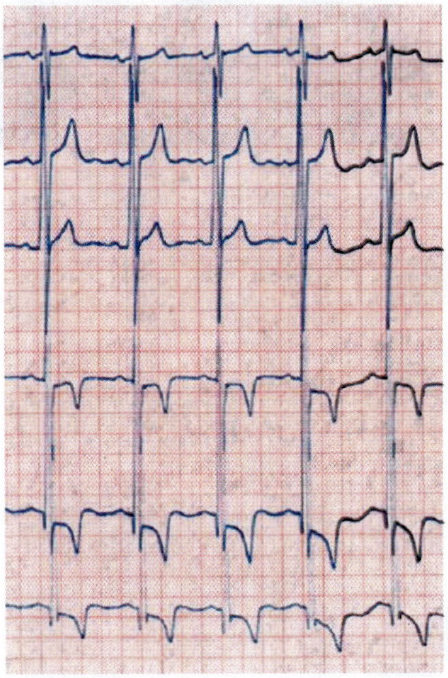

**Figure 15.29** How cloud-to-ground lightning forms.

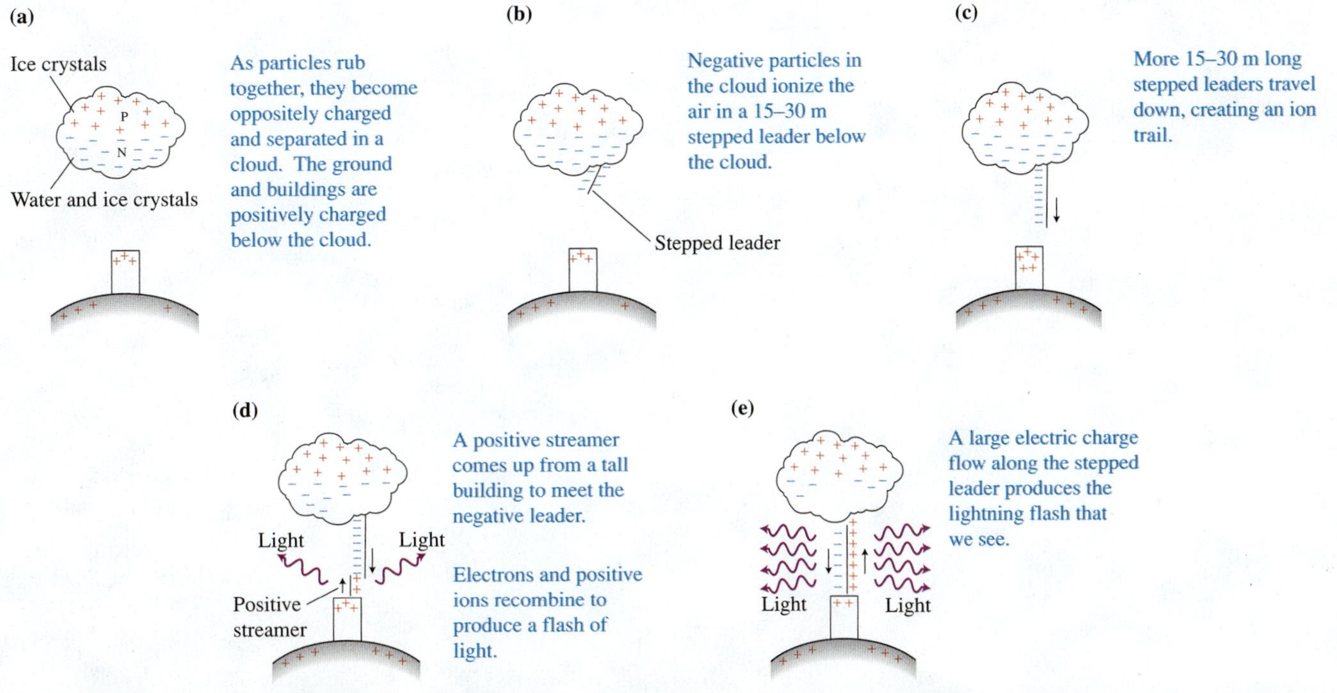
(a)
Ice crystals
P
N
Water and ice crystals
As particles rub together, they become oppositely charged and separated in a cloud. The ground and buildings are positively charged below the cloud.
(b)
Negative particles in the cloud ionize the air in a 15–30 m stepped leader below the cloud.
Stepped leader
(c)
More 15–30 m long stepped leaders travel down, creating an ion trail.
(d)
A positive streamer comes up from a tall building to meet the negative leader.
Light
Light
Positive streamer
Electrons and positive ions recombine to produce a flash of light.
(e)
A large electric charge flow along the stepped leader produces the lightning flash that we see.
Light
Light

will have made their way to the ground. About $-5$ C of electrons move down during the stepped leader and another $-20$ to $-30$ C of electrons move down during the return stroke. The air inside the streamers becomes hot and expands abruptly—we hear sound.

**Figure 15.30** Lightning striking a lightning rod high on a building.

## Lightning rods

A lightning rod is a tall, pointy metal object on the top of a building that extends from Earth's surface. Earth's surface and the rod form an equipotential surface. The radius of curvature of the tip of the rod is much smaller than other nearby objects. Consequently, as we discussed in the section on grounding, the surface charge density and the $\vec{E}$ field around the rod are much stronger than at other points on the building or ground. Dielectric breakdown occurs between the cloud and the lightning rod. Drawing lightning to the rod and away from the building prevents damage to the building and its inhabitants (**Figure 15.30**).

If you are caught outside during a lightning storm, do not make a lightning rod of yourself or stand beneath something that will act like one (such as a tree). If you are in an open space, find a ravine, valley, or depression in the ground. Crouch so that you do not project above the surrounding landscape. Do not lie on the ground. A large potential difference may develop between your head and feet, causing an undesirable flow of electric charge through your body. When crouched, keep your feet close together.

**Review Question 15.8** If you were to put a pointy metal object on the roof of a house to serve as a lightning rod, what do you need to do to make sure the metal rod works as a lightning rod?

# Summary

| Words | Pictorial and physical representations | Mathematical representation |
|---|---|---|
| **$\vec{E}$ field** The $\vec{E}$ field is a force-like quantity characterizing the electric field at a specific location. A source charge creates an electric field that exerts a force on a test charge placed in that field at that location. The $\vec{E}$ field at a the chosen location is equal to the electric force exerted by a source charge on a positively charged test object divided by the test object's charge (Section 5.1). | 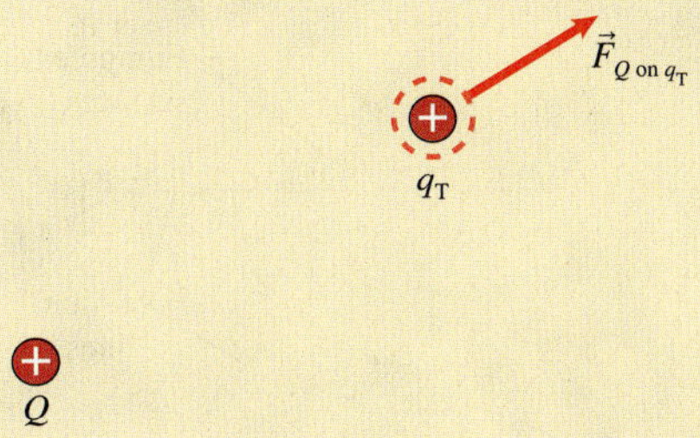  | $\vec{E} = \dfrac{\vec{F}_{Q \text{ on } q_{test}}}{q_{test}}$ Eq. (15.2)<br>$\vec{F}_{Q \text{ on } q_{test}} = q_{test}\vec{E}$ Eq. (15.5) |
| **Superposition principle** The $\vec{E}$ field at a location is the *vector* sum of the contributions of *n* source charges to the $\vec{E}$ field at that location (Section 5.1). | 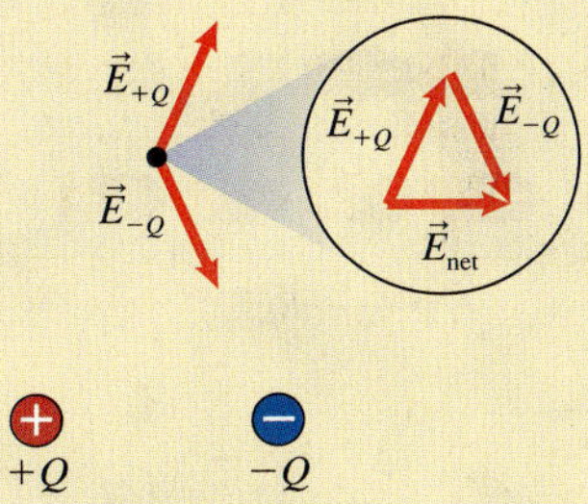  | $\vec{E} = \vec{E}_1 + \vec{E}_2 + \vec{E}_3 + \cdots$<br>Eq. (15.4) |
| ***V* field** (electric potential) is an energy-like quantity characterizing the electric field at a point. The *V* field (electric potential) at a position of interest is equal to the electric potential energy $U_q$ of a small positively charged test charge $q_T$ and the source charge divided by the test object's charge. Include the signs of charges when calculating $U_q$. (Section 5.3) | | $V = \dfrac{U_{Qq_{test}}}{q_{test}} = \dfrac{kQ}{r}$ Eq. (15.6)<br>For one source charge:<br>$V = \dfrac{kQ_1}{r_1}$<br>$U_{\text{Source } Q \text{ and } q} = qV_{\text{Due to source } Q}$<br>Eq. (15.8) |
| **Superposition principle for *V* field** The *V* field at a specific location is the algebraic sum of the contributions of *n* source charges to the *V* field at that location. (Section 5.4) | 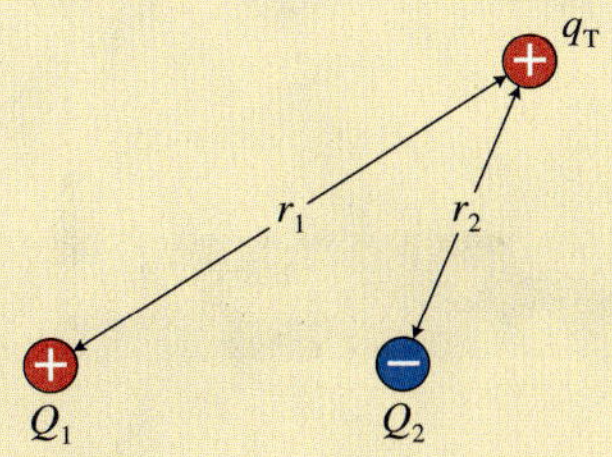  | $V = V_1 + V_2 + V_3 + \cdots$<br>Eq. (15.7) |
| **Relationship of $\vec{E}$ field and *V* field** The greater the magnitude of the $\vec{E}$ field at a particular location, the faster the *V* field changes with position near that location. The $\vec{E}$ field points from higher *V* field values to lower *V* field values. (Section 5.4) | 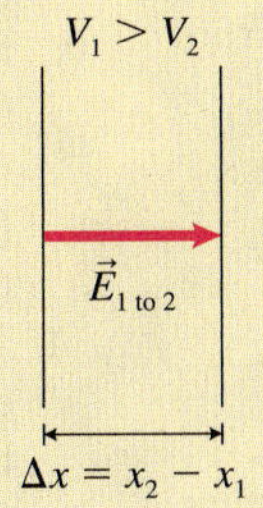  | $\Delta V = -E_x \Delta x$ Eq. (15.9)<br>or<br>$E_x = -\dfrac{\Delta V}{\Delta x}$ Eq. (15.10) |

*(continued)*

| Words | Pictorial and physical representations | Mathematical representation |
|---|---|---|
| **Representing the $\vec{E}$ field and the $V$ field** $\vec{E}$ field lines and equipotential surfaces represent the field. $\vec{E}$ field lines are perpendicular to equipotential surfaces (Section 5.4). | 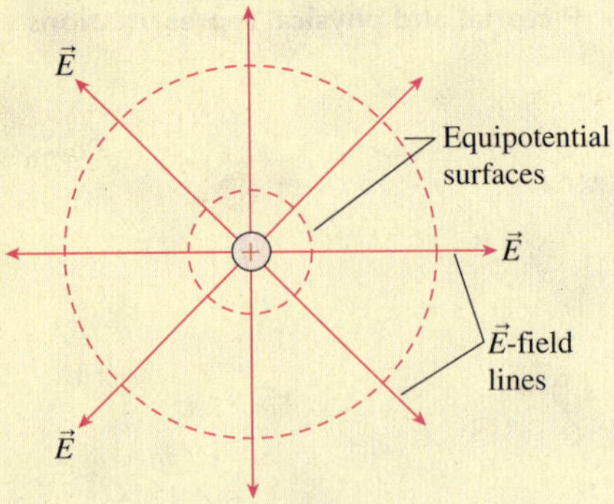 | |
| **Dielectric constant** $\kappa$ A dielectric material partially cancels the $\vec{E}$ field within it produced by the environment. This reduces the magnitude of the force exerted by charged objects on each other within the material (Section 5.6). | 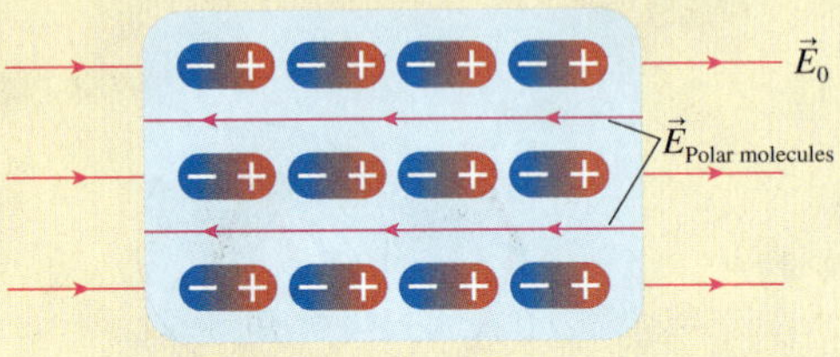 $\vec{E} = \vec{E}_0 - \vec{E}_{\text{Polar molecules}}$ | $\kappa = \frac{E_0}{E}$ Eq. (15.11)<br>$F_{1\text{ on }2} = k\frac{\lvert q_1 \rvert \lvert q_2 \rvert}{\kappa r^2}$ Eq. (15.12) |
| **Capacitors** A capacitor is a device that (when charged) stores electric potential energy. It has two conducting surfaces separated by a nonconducting material. To charge a capacitor, one connects its plates to the points at different electric potential (like the terminals of a battery). The charge separation stores electric potential energy in the capacitor. (Section 5.7) | 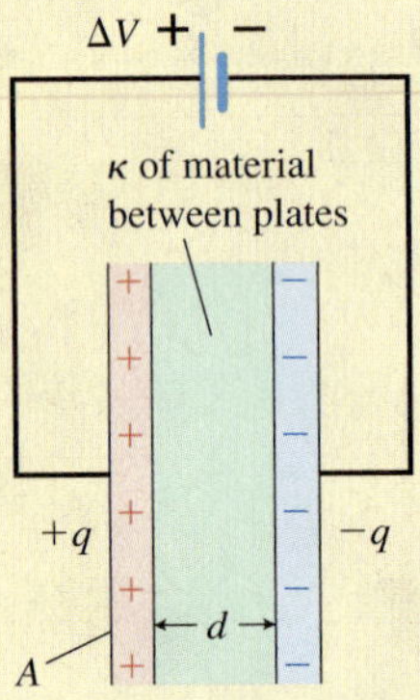 | $q = C\lvert \Delta V \rvert$ Eq. (15.13)<br>$C = \frac{\kappa A}{4\pi k d}$ Eq. (15.14)<br>for parallel plate capacitors<br>$U_q = \frac{1}{2}q\lvert \Delta V \rvert = \frac{1}{2}C\lvert \Delta V \rvert^2 = \frac{1}{2}\frac{q^2}{C}$ Eq. (15.15) |

MP® For instructor-assigned homework, go to **MasteringPhysics.**

# Questions

## Multiple Choice Questions

1. What does the $\vec{E}$ field at point A, which is a distance $d$ from the source charge, depend on?
   (a) the magnitude of the test charge
   (b) the sign of the test charge
   (c) the magnitude and the sign of the source charge
   (d) the magnitude and sign of the test charge
2. Why can you shield an object from an external electric field using a conductor?
   (a) There are freely moving electric charges inside conductors.
   (b) There are positive and negative charges inside conductors.
   (c) Both a and b are essential.
3. If you place an object made of a conducting material in an external electric field, what will be the magnitude of the $\vec{E}$ field inside the material?
   (a) less than outside
   (b) more than outside
   (c) zero
   (d) It depends on the size of the material.
4. Two identical positive charges are located at a distance $d$ from each other. Where are both the $\vec{E}$ field and the electric potential zero?
   (a) exactly between the charges
   (b) at a distance $d$ from both charges
   (c) Both a and b are correct.
   (d) None of these choices is correct.

## Conceptual Questions

5. How do we use the model of the electric field to explain the interaction between charges?
6. Describe a procedure to determine the $\vec{E}$ field at some point.
7. What does it mean if the $\vec{E}$ field at a certain point is 5 N/C and points north?
8. A very small positive charge is placed at one point in space. There is no electric force exerted on it. (a) What is the value of the $\vec{E}$ field at that point? (b) Does this mean there are no electric charges nearby? Explain. (c) Suggest one charge distribution (two or more charges) that would produce a zero $\vec{E}$ field at a point.
9. How do we create an $\vec{E}$ field with parallel lines and uniform density (equally spaced)?
10. Draw a sketch of the $\vec{E}$ field lines caused by (a) two positive charges of the same magnitude, (b) two negative charges of the same magnitude, and (c) two positive charges of different magnitudes.
11. Draw a sketch of the $\vec{E}$ field lines caused by (a) a positive and a negative charge of the same magnitude, (b) a positive and a negative charge of different magnitudes, and (c) a positive charge between two negative charges.
12. Jim thinks that $\vec{E}$ field lines are the paths that test charges would follow if placed in the $\vec{E}$ field. Do you agree or disagree with his statement? If you disagree, give a counter example.
13. Can $\vec{E}$ field lines cross? Explain why or why not.
14. An electron moving horizontally from left to right across the page enters a uniform $\vec{E}$ field that points toward the top of the page. Draw vectors indicating the direction of the electric force exerted by the field on the electron and its acceleration. Draw a sketch indicating the path of the electron in this field. What motion in the gravitational field is similar to the motion of the electron?
15. (a) What does it mean if the electric potential at a certain point in space is 10 V? (b) What does it mean if the potential difference between two points is 10 V?
16. Explain how grounding works.
17. Explain how shielding works.
18. Explain the difference between the microscopic structures of polar and of nonpolar dielectric materials. Give an example of each.
19. Explain why in Coulomb's law we divide the force exerted by one charged object on the other by the value of the dielectric constant if the charged objects are submerged in a dielectric material.
20. What does it mean if the dielectric constant $\kappa$ of water is 81? What is $\kappa$ of a metal paper clip?
21. Describe the relation between the quantities $\vec{E}$ field and $V$ field.
22. If the $V$ field in a region is constant, what is the $\vec{E}$ field in this region? If the $\vec{E}$ field is constant in the region, what does the $V$ field look like? Explain your answers.
23. Why are uncharged pieces of a dielectric material attracted to charged objects?
24. Draw equipotential surfaces and label them in order of decreasing potential for (a) one positive charge, (b) one negative charge, (c) two identical positive point charges at a distance $d$ from each other, and (d) a negatively charged infinitely large metal plate.
25. Show a charge arrangement and a point in space where the potential produced by the charges is zero but the $\vec{E}$ field is not zero.
26. (a) Explain what happens if you place a conductor in an external $\vec{E}$ field. How can you test your explanation? (b) What happens if you cut the conductor in half (in the direction perpendicular to the external $\vec{E}$ field vectors) while keeping it in the external $\vec{E}$ field? How can you test your answer?
27. (a) Explain what happens if you place a dielectric material in an external $\vec{E}$ field? (b) What happens if you cut a dielectric in half (in the direction perpendicular to the external $\vec{E}$ field vectors) while keeping it in the external $\vec{E}$ field? How can you test your answer?
28. Explain why the charge on an electrical conducting material is located on its surface.

# Problems

Below, **BIO** indicates a problem with a biological or medical focus. Problems labeled **EST** ask you to estimate the answer to a quantitative problem rather than develop a specific answer. Problems marked with a pencil icon require you to make a drawing or graph as part of your solution. Asterisks indicate the level of difficulty of the problem. Problems with no * are considered to be the least difficult. A single * marks moderately difficult problems. Two ** indicate more difficult problems.

In some of these problems you may need to know that the mass of an electron is $9.11 \times 10^{-31}$ kg, the mass of a proton is $1.67 \times 10^{-27}$ kg, the electric charge of the electron is $-1.6 \times 10^{-19}$ C, and the charge of a proton is $1.6 \times 10^{-19}$ C.

### 15.1 A model of the mechanism for electrostatic interactions

1. * (a) Construct a graph of the magnitude of the $\vec{E}$ field-versus-position for the $\vec{E}$ field created by a point-like object with charge $+Q$. (b) Using the same set of axes, draw a graph for the field produced by an object of charge $+2Q$. (c) Using the same set of axes, draw a graph for the field produced by an object of charge $-2Q$.
2. A uranium nucleus has 92 protons. (a) Determine the magnitude of the $\vec{E}$ field at a distance of $0.58 \times 10^{-12}$ m from the nucleus (about the radius of the innermost electron orbit around the nucleus). (b) What is the magnitude of the force exerted on an electron by this $\vec{E}$ field? (c) What assumptions did you make? If the assumptions are not valid, will the magnitude in part (b) be an overestimate or underestimate of the actual value?
3. The electron and the proton in a hydrogen atom are about $10^{-10}$ m from each other. What quantities related to the $\vec{E}$ field can you determine using this information?
4. * Use the superposition principle to draw $\vec{E}$ field lines for the two objects whose charges are given. Consider all objects to be point-like and choose the distance you want between them: (a) $+q$ and $+q$; (b) $+q$ and $+3q$; (c) $+q$ and $-q$.

5. * Use the superposition principle to draw $\vec{E}$ field lines for the following objects whose charges are given. Consider all objects to be point-like and choose the distance you want between them: (a) $+q$ and $-3q$; (b) $+q$, $+q$, and $-3q$.
6. * $\vec{E}$ field lines for a field created by an arrangement of charged objects are shown in **Figure P15.6.** (a) Where are these objects located, and what are the signs of their electric charge? (b) What else can you determine using the information? Give two examples.

**Figure P15.6**

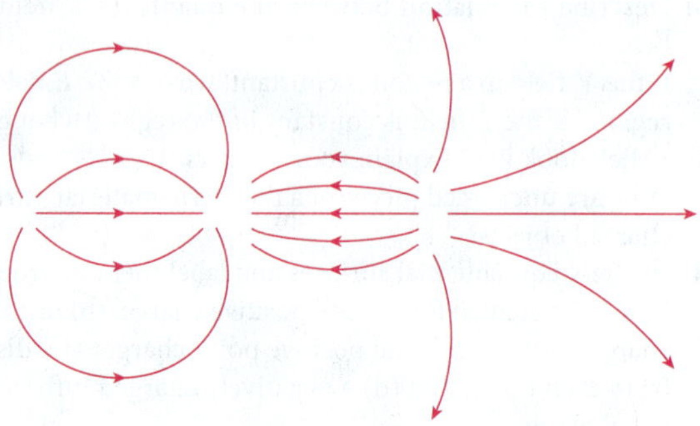

## 15.2 Skills for determining $\vec{E}$ fields and analyzing processes with $\vec{E}$ fields

7. * Two objects with charges $+4.0 \times 10^{-9}$ C and $+3.0 \times 10^{-9}$ C are 50 cm from each other. Find a location where the $\vec{E}$ field due to these two charged objects is zero.
8. * A $+4.0 \times 10^{-9}$-C charged object is 6.0 cm along a horizontal line toward the right of a $-3.0 \times 10^{-9}$-C charged object. Determine the $\vec{E}$ field at a point 4.0 cm to the left of the negative charge.
9. ** A $+4.0 \times 10^{-9}$-C charged object is 4.0 cm along a horizontal line toward the right of a $-3.0 \times 10^{-9}$-C charged object. Determine the $\vec{E}$ field at a point 3.0 cm directly above the negative charge.
10. ** A distance $d$ separates two objects, each with charge $+q$. Determine an expression for the $\vec{E}$ field at a point that is a distance $d$ from both charges.
11. * A point-like charged object with a charge $+q$ is placed in an external horizontal uniform electric field $\vec{E}_0$ that points to the right. Determine an expression for the net $\vec{E}$ field at a distance $d$ from the charge (a) to the right of it and along a horizontal line going through the charge $+q$ and (b) to the left of it along that same line.
12. * A 3.0-g aluminum foil ball with a charge of $+4.0 \times 10^{-9}$ C is suspended on a string in a uniform horizontal $\vec{E}$ field. The string makes an angle of 30°with the vertical. What information about the $\vec{E}$ field can you determine for this situation?
13. ** (a) If the string in the previous problem is cut, how long will it take the ball to fall to the floor 1.5 m below the level of the ball? (b) How far will the ball travel in the horizontal direction while falling? Indicate all the assumptions that you made.
14. * EST **Using Earth's $\vec{E}$ field for flight** Earth has an electric charge on its surface that produces a 150-N/C $\vec{E}$ field, which points down toward the center of Earth. Estimate the electric charge that a person would need so that the electric force that the field exerts on the person will support the person in the air above Earth. Indicate all of your assumptions. Is this a reasonable idea? Explain.
15. * An electron moving with a speed $v_0$ enters a region where an $\vec{E}$ field points in the same direction as the electron's velocity. What will happen to the electron in this field? Answer the question qualitatively and quantitatively (using symbols).
16. * A 1.0-g aluminum foil ball with a charge of $1.0 \times 10^{-9}$ C hangs freely from a 1.0-m-long thread. What happens to the ball when a horizontal 5000-N/C $\vec{E}$ field is turned on? Answer the question as fully as possible.
17. A 0.50-g oil droplet with charge $+5.0 \times 10^{-9}$ C is in a vertical $\vec{E}$ field. In what direction should the $\vec{E}$ field point and what magnitude should it have so that the droplet moves at constant speed?
18. * An electron is ejected into a horizontal uniform $\vec{E}$ field at a parallel horizontal velocity of 1000 m/s. Describe everything you can about the motion of the electron.
19. * **Equation Jeopardy 1** The equations below describe one or more physical processes. Solve the equations for the unknowns and write a problem statement for which the equations are a satisfactory solution.

$$-(0.10\text{ kg})(9.8\text{ N/kg}) + T\sin 53^\circ = 0$$
$$+(-1.0 \times 10^{-6}\text{ C})E_x + T\cos 53^\circ = 0$$

20. * **Equation Jeopardy 2** The equations below describe one or more physical processes. Solve the equations for the unknowns and write a problem statement for which the equations are a satisfactory solution.

$$(+1.0 \times 10^{-5}\text{ C})(-4.0 \times 10^4\text{ N/C}) = (0.20\text{ kg})a_x$$
$$0 = (8.0\text{ m/s}) + a_x t$$

## 15.3–15.4 The $V$ field and Relating the $\vec{E}$ field and the $V$ field

21. During a lightning flash, $-15$ C of charge moves through a potential difference of $8.0 \times 10^7$ V. Determine the change in electric potential energy of the field-charge system.
22. * (a) Sketch a $V$-versus-position graph for the electric potential created by a point-like object with charge $+Q$. (b) Using the same set of axes, draw a graph for an object with charge $+2Q$ and for an object with charge $-2Q$.
23. A horizontal distance $d$ separates two objects each with charge $+q$. Determine the value of the electric potential at a point that is located at a distance $d$ from each charge.
24. * Two objects with charges $-q$ and $+q$ are separated by a distance $d$. Determine an expression for the $V$ field at a point that is located at a distance $d$ from each charge.
25. * Four objects with the same charge $-q$ are placed at the corners of a square of side $d$. Determine the values of the $\vec{E}$ field and $V$ field in the center of the square.
26. **Spark jumps to nose** An electric spark jumps from a person's finger to your nose and through your body. While passing through the air, the spark travels across a potential difference of $2.0 \times 10^4$ V and releases $3.0 \times 10^{-7}$ J of electric potential energy. What is the charge in coulombs, and how many electrons flow?
27. Two $-3.0 \times 10^{-6}$-C charged point-like objects are separated by 0.20 m. Determine the potential (assuming zero volts at infinity) at a point (a) halfway between the objects and (b) 0.20 m to the side of one of the objects (and 0.40 m from the other) along a line joining them.
28. * The potential difference from the cathode (negative electrode) to the screen of an old television set is +22,000 V. An electron leaves the cathode with an initial speed of zero.

Determine everything you can about the motion of the electron in the TV set using this information.

29. Imagine a 10,000-kg shuttle bus that carries a +15-C charged sphere on its roof. Through what potential difference must the bus travel to acquire a speed of 10 m/s, assuming no friction force exerted on the bus?

30. BIO **Electric field in body cell** The electric potential difference across the membrane of a body cell is +0.070 V (higher on the outside than on the inside). The cell membrane is $8.0 \times 10^{-9}$ m thick. Determine the magnitude and direction of the $\vec{E}$ field through the cell membrane. Describe any assumptions you made.

31. * BIO EST **Energy used to charge nerve cells** A nerve cell is shaped like a cylinder. The membrane wall of the cylinder has a +0.07-V potential difference from the inside to the outside of the wall. To help maintain this potential difference, sodium ions are pumped from inside the cell to the outside. For a typical cell, $10^9$ ions are pumped each second. (a) Determine the change in chemical energy each second required to produce this increase in electric potential energy. (b) If there are roughly $7 \times 10^{11}$ of these cells in the body, how much chemical energy is used in pumping sodium ions each second? (c) Estimate the fraction of a person's metabolic rate used to pump these ions.

32. * **Equation Jeopardy 3** The equation below describes one or more physical processes. Solve the equation for the unknown and write a problem statement for which the equation is a satisfactory solution.

$$0 = (1/2)(100\text{ kg})(6.0\text{ m/s})^2 + (2.0 \times 10^{-4}\text{ C})\Delta V$$

33. * **Equation Jeopardy 4** The equation below describes one or more physical situations. Solve the equation for the unknown and write a problem statement for which the equation is a satisfactory solution.

$$(9.0 \times 10^9\text{ Nm}^2/\text{C}^2)(+2.0 \times 10^{-5}\text{ C})/(2000\text{ m}) + (9.0 \times 10^9\text{ Nm}^2/\text{C}^2)(-2.0 \times 10^{-5}\text{ C})/(1000\text{ m}) = \Delta V$$

## 15.5–15.6 Conductors and dielectrics in electric fields

34. * A metal sphere has no charge on it. A positively charged object is brought near, but does not touch the sphere. Show that this object can exert a force on the sphere even though the sphere has no net charge. How can you test your answer experimentally?

35. * A Van de Graaff generator of radius 0.10 m has a charge of about $-1.0 \times 10^{-6}$ C on it. The Van de Graaff generator is then turned off and grounded. How many excess electrons remain on its dome?

36. * A metal ball of radius $R_1$ has a charge $\frac{1}{2}Q$. Later it is connected to a metal ball of radius $R_2$. What is the fraction of the charge $Q$ that remains on the first ball?

37. * (a) Draw a microscopic representation of what happens to the charge distribution in a conductor placed in an external uniform electric field. For simplicity, make the conductor of some regular shape. How can you test your explanation? (b) What happens if you cut the conductor in half (in the direction perpendicular to the direction of the $\vec{E}$ field) while keeping it in the external field? How can you test your answer? (c) Repeat steps (a) and (b) for a dielectric.

38. ** A positively charged metal ball A is placed near metal ball B. Measurements demonstrate that the force that they exert on each other is zero. Is ball B charged? Explain.

39. * Positively charged metal sphere A is placed near metal sphere B. B has a very small positive charge. Explain why the spheres could attract each other. Draw sketches of the spheres and their charge distribution to support your answer.

40. * Two small metal spheres A and B have different electric potentials (A has a higher potential). Describe in words and mathematically what happens if you connect them with a wire. What assumptions did you make?

41. * Two metal spheres of the same radius are placed close to each other. (a) Will they exert the same magnitude force on each other when they have like charges as when they have opposite charges? Explain. (b) What assumptions did you make? (c) How will you set up an experiment to test your answer?

42. An electric dipole such as a water molecule is in a uniform $\vec{E}$ field. (a) Will the force exerted by the field cause the dipole to have a linear acceleration along a line in the direction of $\vec{E}$? Explain. (b) Will the field exert a torque on the dipole? Explain.

43. * BIO **Electric field of a fish** An African fish called the aba has a charge $q = +1.0 \times 10^{-7}$ C at its head and an equal magnitude negative charge $-q$ at its tail (see **Figure P15.43**). Determine the magnitude and direction of the electric field at position A and the force exerted on a hydroxide ion (charge $-e$) at that point. The fish and ion are in water.

**Figure P15.43**

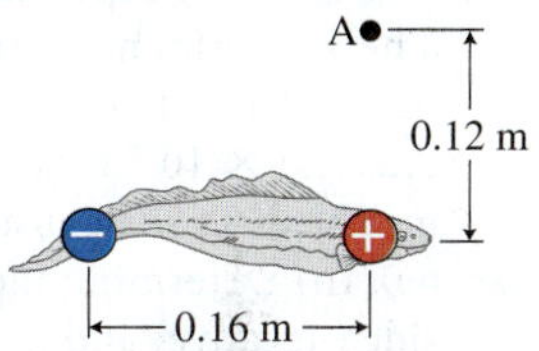


44. BIO **Body cell membrane electric field** (a) Determine the average magnitude of the $\vec{E}$ field across a body cell membrane. A 0.07-V potential difference exists from one side to the other and the membrane is $7.5 \times 10^{-9}$ m thick. (b) Determine the magnitude of the electrical force on a sodium ion (charge $+e$) in the membrane. Assume that the dielectric constant is 1.0 (it is actually somewhat larger).

45. ** **Earth's electric field** Earth has an electric charge of approximately $-5.7 \times 10^5$ C distributed relatively uniformly on its surface. Determine as many quantities as possible about the electrical properties of the space around Earth. Use any additional information that you need.

46. * **Geological exploration** Determine the $\vec{E}$ field at position A in **Figure P15.46** due to the dipole charges produced by a geologist's electrodes. The dielectric constant of the soil is 8.0 and the dipole charge $q = 4.0 \times 10^{-3}$ C. Determine the force exerted by this field on a sodium ion (charge $+e$) at position A.

**Figure P15.46**

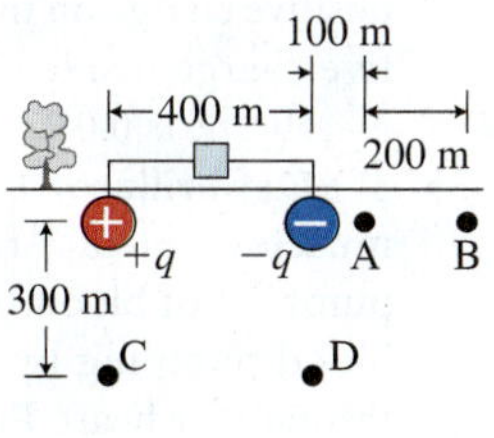


## 15.7 Capacitors

47. You have a parallel-plate capacitor. (a) Determine the average $\vec{E}$ field between the plates if a 120-V potential difference exists across the plates. Their separation is 0.50 cm. (b) A spark will jump if the magnitude of the $\vec{E}$ field exceeds $3.0 \times 10^6$ V/m when air separates the plates. What is the closest the plates can be placed to each other without sparking?

48. ** A capacitor of capacitance $C$ with a vacuum between the plates is connected to a source of potential difference $\Delta V$. (a) Write an expression for the charge on each of the plates and for the total energy stored by the capacitor. (b) You then fill

the capacitor with a dielectric material of dielectric constant $\kappa$ while the capacitor remains connected to the same potential difference. What are the new charge on the plates and the new energy stored by the capacitor? (c) Where did the energy change come from?

49. ** A capacitor of capacitance $C$ with a vacuum between the plates is connected to a source of potential difference $\Delta V$. (a) Write an expression for the charge on each of the plates and for the total energy stored by the capacitor. (b) The capacitor is then disconnected from the potential difference source and you fill it with a dielectric that has a dielectric constant $\kappa$. What are the new charge on the plates and the new energy? (c) Where did the energy change come from or where did the energy go?

50. How does the capacitance of a parallel plate capacitor change if you double the magnitude of the charge on its plates? If you triple the potential difference across the plates? What assumptions did you make?

51. * BIO EST **Axon capacitance** The long thin cylindrical axon of a nerve carries nerve impulses. The axon can be as long as 1 m. (a) Estimate the capacitance of a 1.0-m-long axon of radius $4.0 \times 10^{-6}$ m with a membrane thickness of $8.0 \times 10^{-9}$ m. The dielectric constant of the membrane material is about 6.0. (b) Determine the magnitude of the charge on the inside (negative) and outside (positive) of the membrane wall if there is a 0.070-V potential difference across the wall. (c) Determine the energy stored in this axon capacitor when charged.

52. ** **Sphere capacitance** A metal sphere of radius $R$ has an electric charge $+q$ on it. Determine an expression for the electric potential $V$ on the sphere's surface. Use the definition of capacitance to show that the capacitance of this isolated sphere is $R/k$, where $k$ is the constant used in Coulomb's law.

53. ** You have a capacitor of capacitance $C$ with plate separation $d$ and filled with a dielectric with a dielectric constant $\kappa$. Design three different problems that you can solve using this information. Then solve the problems.

54. * BIO EST **Capacitance of red blood cell** Assume that a red blood cell is spherical with a radius of $4 \times 10^{-6}$ m and with wall thickness of $9 \times 10^{-8}$ m. The dielectric constant of the membrane is about 5. Assuming the cell is a parallel plate capacitor, estimate the capacitance of the cell and determine the positive charge on the outside and the equal-magnitude negative charge inside when the potential difference across the membrane is 0.080 V.

55. BIO **Defibrillator** During ventricular fibrillation the heart muscles contract randomly, preventing the coordinated pumping of blood. A defibrillator can often restore normal blood pumping by discharging the charge on a capacitor through the heart. Paddles are held against the patient's chest, and a $6 \times 10^{-6}$-F charged capacitor is discharged in several milliseconds. If the capacitor energy is 250 J, what potential difference was used to charge the capacitor?

### 15.8 Electrocardiography and lightning: Putting it all together

56. * The dielectric strength of air is $3 \times 10^6$ V/m. As you walk across a synthetic rug, your body accumulates electric charge, causing a potential difference of 6000 V between your body and a doorknob. What can you can determine using this information?

57. * **Charged cloud causes electric field on Earth** The electric charge of clouds is a complex subject. Consider the simplified model shown in **Figure P15.57**. A positive charge is near the top of the cloud and a negative charge is near the bottom. Determine the direction of the $\vec{E}$ field on Earth at point P below the cloud and explain it so that a classmate can understand why there is positive charge on the ground directly below the cloud.

**Figure P15.57**

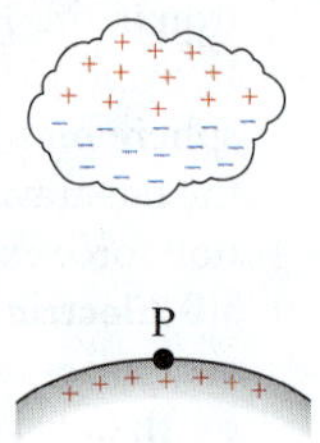


58. * BIO **Heart's dipole charge** The heart has a dipole charge distribution with a charge of $+1.0 \times 10^{-7}$ C that is 6.0 cm above a charge of $-1.0 \times 10^{-7}$ C. Determine the $\vec{E}$ field (magnitude and direction) caused by the heart's dipole at a distance of 8.0 cm directly above the heart's positive charge. All charges are located in body tissue of dielectric constant 7.0. What is the force exerted on a sodium ion (charge $+1.6 \times 10^{-19}$ C) at that point?

## General Problems

59. ** A speck of dust of mass 0.10 μg floats between two oppositely charged horizontal metal plates separated by 0.10 m. The potential difference between the plates is 200 V. The speck is not falling at first, but when a beam of ultraviolet light hits it, it starts accelerating downward. However, increasing the potential difference to 250 V reduces the downward acceleration of the speck to zero. Explain the phenomenon qualitatively and quantitatively using symbols.

60. ** A 0.050-kg cart has a spherical dome that is charged to $+1.0 \times 10^{-5}$ C. The cart is at the top of a 2.0-m-long inclined plane that makes a 30° angle with the horizontal. The cart and the plane are placed in an upward-pointing vertical uniform 4000-N/C $\vec{E}$ field. How long will it take the cart to reach the bottom of the plane? What assumptions did you make?

61. ** BIO **Can a shark detect an axon $\vec{E}$ field?** A nerve signal is transmitted along the long, thin axon of a neuron in a small fish. The transmission occurs as sodium ions $Na^+$ transfer like tipping dominos across the axon membrane from outside to inside. Each short section of axon gets an excess of about $6 \times 10^8$ sodium ions/mm. Determine the $\vec{E}$ field 4.0 cm from the axon produced by the excess sodium ions on the inside of the axon and an equal number of negative ions on the outside of a 1-mm length of axon. The ions are separated by the $8 \times 10^{-9}$-m-thick axon membrane. Will a shark that is able to detect fields as small as $10^{-6}$ N/C be able to detect that axon field? Explain.

62. * EST **Lightning warms water** A lightning flash occurs when $-40$ C of charge moves from a cloud to Earth through a potential difference of $4.0 \times 10^8$ V. Estimate how much water can boil as a result of energy released during the process. Describe the assumptions you made.

63. * In a hot water heater, water warms when electric potential energy is converted into thermal energy. (a) Determine the energy needed to warm 180 kg of water by 10 °C. (b) If $-10.0$ C of electric charge passes through a $+120$-V potential difference in the heating coils each second, determine the time needed to warm the water by 10 °C.

64. ** BIO **Electrophoresis** Electrophoresis is used to separate biological molecules of different dimensions and electric charge (the molecules can have different charge depending on the pH

of the solution). A particular molecule of radius $R$ with charge $q$ is in a viscous solution that has an $\vec{E}$ field across it. The field exerts an electric force on the molecule and the viscous solution exerts an opposing drag force $F_{\text{drag}} = DRv$, where $D$ is a constant drag coefficient that depends on the shape and other features of the molecule and the solution, and $v$ is the molecule's speed in the solution. When the molecule gets up to speed, the electric force exerted on it by the field is equal in magnitude and opposite in direction to the drag force. Show that during time interval $\Delta t$, the molecule will travel a distance $\Delta x = \dfrac{qE\,\Delta t}{DR}$. Describe any assumptions you made.

65. ** BIO **Energy stored in axon electric field** An axon has a surface area of $5 \times 10^{-6}\ \text{m}^2$ and the membrane is $8 \times 10^{-9}$ m thick. The dielectric constant of the membrane is 6. (a) Determine the capacitance of the axon considered as a parallel plate capacitor. (b) If the potential difference across the membrane wall is 0.080 V, determine the magnitude of the charge on each wall. (c) Determine the energy needed to charge that axon capacitor. (d) Determine the magnitude of the $\vec{E}$ field across the membrane due to the opposite sign charges across the membrane walls. (e) Calculate the energy density of that field. (f) Multiply the volume of space occupied by that field by the volume of the membrane to get the total energy stored in the field. How do the answers to (c) and (f) compare?

## Reading Passage Problems

BIO **Electric discharge by eels** In several aquatic animals, such as the South American electric eel, electric organs produce 600-V potential difference pulses to ward off predators as well as to stun prey. **Figure 15.31** illustrates the key component that produces this electric shock—an **electrocyte**. The interior of an inactive electrocyte (Figure 15.31a) has an excess of negatively charged ions. The exterior has an excess of positively charged sodium ions ($Na^+$) on the left side and positively charged potassium ions ($K^+$) on the right. There is a −90-mV electric potential difference from outside the cell membrane to inside the electrocyte cell membrane, but zero potential from one exterior side to the other exterior side. How then does an eel produce the 600-V potential difference necessary to stun an intruder?

The eel's long trunk and tail contain many electrocytes placed one after the other in columns (Figures 15.31b and c). Each electrocyte contains several types of ion channels, which when activated by a nerve impulse allow sodium ions to pass through channels *on the left flat side* of each electrocyte from outside the cell to the inside. This causes the electric potential across that cell membrane to change from −90 mV to +50 mV. The electric potential from the left external side to the right external side of an electrocyte is now about 100 mV (Figure 15.31c). With about 6000 electrocytes placed one after the other (in series), the eel is able to produce an electric impulse of over 600 V (6000 electrocytes in series times 0.10 V per electrocyte). The discharge lasts about 2–3 ms.

**Figure 15.31** An electric eel's shocking system.

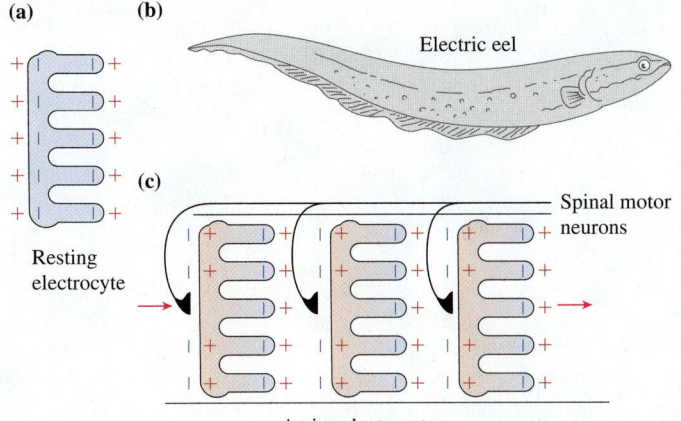

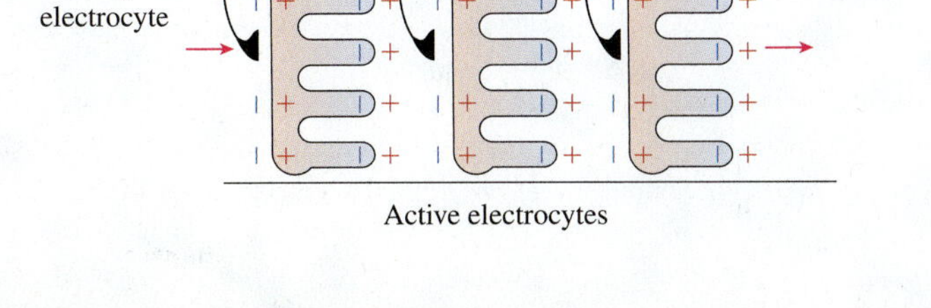


66. Suppose you have a 1.0-F capacitor (very large capacitance) with a 0.10-V potential difference across the capacitor. What is the magnitude of the electric charge on each plate of the capacitor?
(a) 0.05 C (b) 0.10 C (c) 0.20 C (d) 1.0 C (e) 10 C

67. Suppose you place two of these 1.0-F capacitors with the charge calculated in the previous question as shown in **Figure P15.67a**. What is the net potential difference across the two capacitors (from one dot to the other)?
(a) 0.05 V (b) 0.10 V (c) 0.20 V (d) None of these
(e) Not enough information

**Figure P15.67**

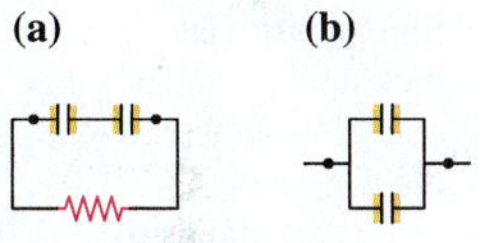


68. If both capacitors are discharged simultaneously, how much electric charge goes through the wire shown in pink in Figure P15.67a?
(a) 0.05 C (b) 0.10 C (c) 0.20 C (d) 1.0 C (e) 10 C

69. Suppose you place the two capacitors with the 0.10-V potential difference across each capacitor as shown in Figure P15.67b. What is the net potential difference across the two capacitors?
(a) 0.05 V (b) 0.10 V (c) 0.20 V (d) None of these
(e) Not enough information

70. Look at the electrocyte shown in Figure 15.31c. What causes the 0.10-V potential difference from the outer left to the outer right side of the cell?
(a) The membrane is thicker on the left than on the right.
(b) The ion distribution across the left membrane is different than across the right membrane.
(c) The left and right membranes have different capacitances.
(d) b and c
(e) a, b, and c

71. Suppose that one cell of the electrocyte is regarded as a small capacitor with a 0.10-V potential difference across it. How should we arrange 10 cells to get a 1.0-V potential difference across them?
(a) In series , as in Figure P15.67a
(b) In parallel, as in Figure P15.67b
(c) Randomly so that they do not cancel each other
(d) Not enough information

**Electrostatic precipitator (ESP)** Electrostatic precipitators are a common form of air-cleaning device. ESPs are used to remove particle emissions from smoke moving up smokestacks in coal and oil-fired electricity-generating plants and pollutants from the boilers in oil refineries. You can buy portable ESPs or whole-house ESPs that connect to the cold-air return on the furnace. These devices remove about 95% of dirt and 85% of microscopic particles from the air.

A basic electrostatic precipitator contains a negatively charged horizontal metal grid (made of thin wires) and a stack of large, flat, vertically oriented metal collecting plates, with the plates typically spaced about 1 cm apart (only two plates are shown in **Figure 15.32**). Air flows across the charged grid of wires and then passes between the stack of plates. A large negative potential difference (tens of thousands of volts) is applied between the wires and the plates, creating sparks that ionize particles in the air around the thin wires. Negatively charged smoke particles flow upward between the plates. The charged particles are attracted to and stick to the oppositely charged plates and are thus removed from the moving gas.

**Figure 15.32** An electrostatic precipitator.

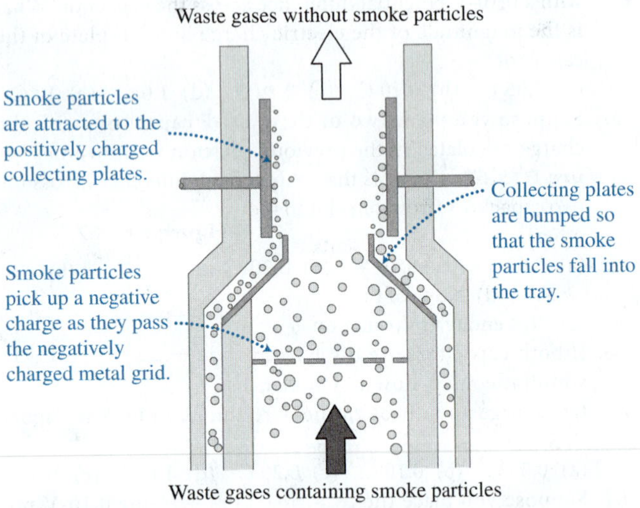

Waste gases without smoke particles
Smoke particles are attracted to the positively charged collecting plates.
Collecting plates are bumped so that the smoke particles fall into the tray.
Smoke particles pick up a negative charge as they pass the negatively charged metal grid.
Waste gases containing smoke particles

72. An $\vec{E}$ field of approximately $3 \times 10^6$ V/m is needed to ionize dry air molecules. If the potential difference between the thin wires and metal plates is 60,000 V, what minimum distance must the wires be from the plates in order to cause dry air to ionize?
(a) 1.0 cm (b) 2.0 cm (c) 4.0 cm (d) 50 cm (e) 50 m

73. Why are the smoke particles attracted to the closely spaced plates?
(a) The particles are negatively charged.
(b) The particles are positively charged.
(c) The plates are positively charged.
(d) The plates are negatively charged.
(e) a and c are correct.
(f) b and c are correct.

74. Suppose a $2.0 \times 10^{-6}$-kg dust particle with charge $-4.0 \times 10^{-9}$ C is moving vertically up a chimney at speed 6.0 m/s when it enters the +2000-N/C $\vec{E}$ field pointing away from a metal collection plate of an electrostatic precipitator. If the particle is 2.0 cm from the plate at that instant, which answer below is closest to the magnitude of its horizontal acceleration toward the plate?
(a) $2.0 \times 10^{-5}$ m/s$^2$ (b) $1.2 \times 10^{-4}$ m/s$^2$
(c) $4.0 \times 10^{-4}$ m/s$^2$ (d) 4.0 m/s$^2$
(e) 24 m/s$^2$

75. Suppose everything is the same as in the previous problem. Which answer below is closest to the vertical distance the dust particle will move before hitting the plate?
(a) 0.01 m (b) 0.06 m (c) 0.1 m (d) 0.6 m (e) 1.0 m

76. What is the purpose of giving a negative charge to the particles collected by the precipitator?
(a) The particles are then attracted to the positively charged plates while moving upward.
(b) The particles are then repelled from the positively charged plates while moving upward.
(c) The extra mass slows the upward movement of the particles.
(d) The negative charge cancels the positive charge the particles have when entering the precipitator.
(e) a and c are correct.

77. Suppose the average particle moving with air up a chimney has a mass of $4.0 \times 10^{-6}$ kg and charge of $-6.0 \times 10^{-8}$ C. The particles move with the air at speed 6.0 m/s and average distance of 3.0 cm from a vertical plate. The plate is 0.60 m tall. Which answer below is closest to the minimum horizontal electric field that the plate must have so that the particles will be collected before moving past the top of the plate?
(a) 100 N/C (b) 200 N/C
(c) 400 N/C (d) 1000 N/C
(e) 4000 N/C

# DC Circuits 16

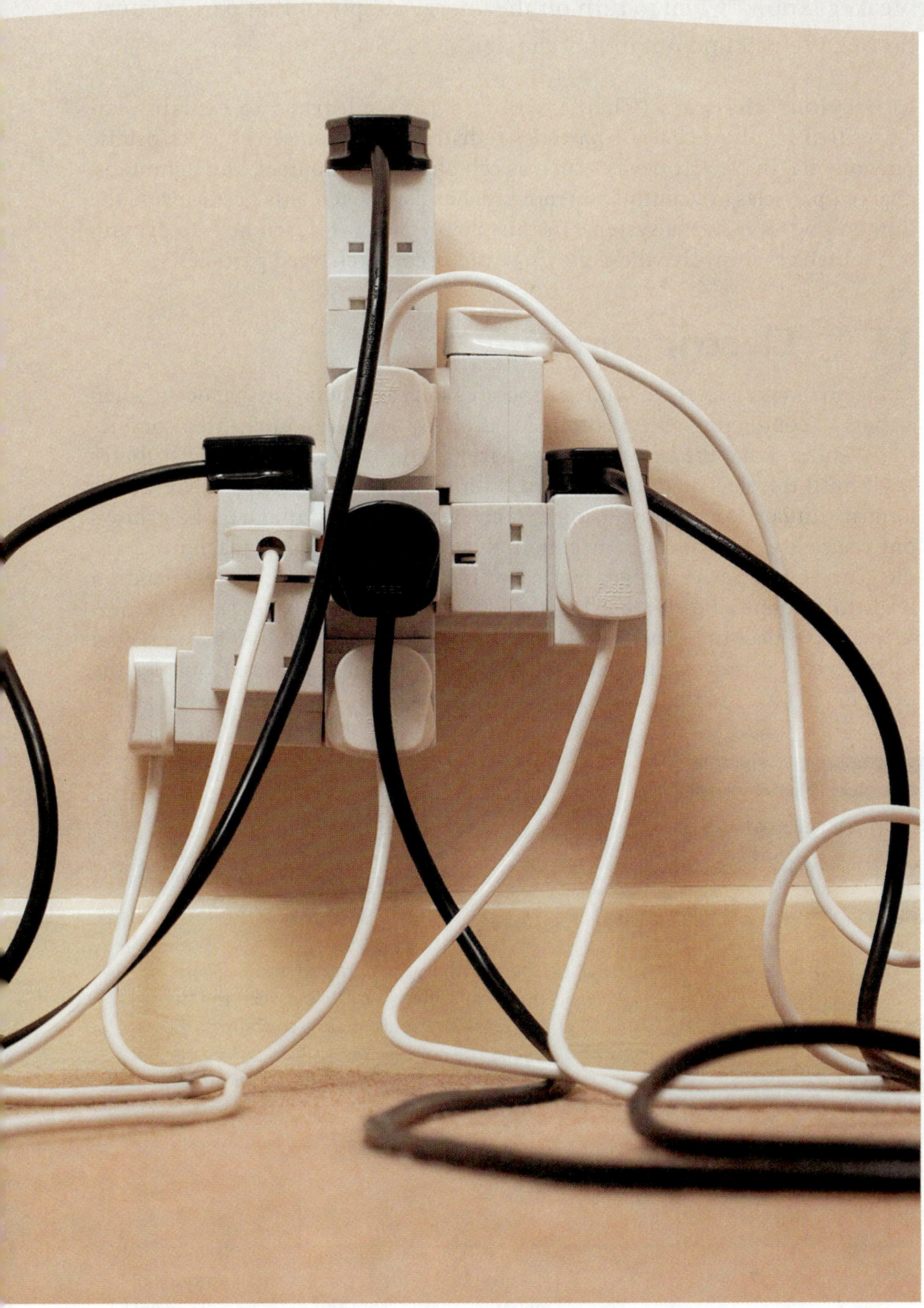

**Why are modern buildings equipped with electrical circuit breakers?**

**How can you use an electric circuit to model the human circulatory system?**

**Why is it dangerous to use a hair dryer while taking a bath?**

On a cold Sunday morning you turn on a space heater and the toaster oven. Then you start the washing machine and the dishwasher. Finally, you turn on the coffee grinder, and suddenly the power in that part of the house shuts off. What happened?

**Be sure you know how to:**

- Apply the concept of the electric field to explain electric interactions (Section 15.1).
- Define the $V$ field (electric potential) and electric potential difference $\Delta V$ (Section 15.3).
- Explain the differences in internal structure between conducting materials and nonconducting materials (Section 14.3).

When you turn on too many appliances that run on a single circuit, a device called a circuit breaker disconnects that part of the house from the external supply of electric energy. The disconnection prevents wires in the walls from overheating and catching fire. How does a circuit breaker "know" when to turn off the energy supply to that part of your house? We will find out in this chapter.

**In previous chapters** (Chapters 14 and 15) we learned to explain processes that involved charged particles redistributing themselves: electrostatic phenomena. In electric devices such as cell phones, computers, and lightbulbs, charged particles are continually moving. Similar movements occur inside the human body's nervous system. In this chapter we will learn how to explain phenomena that involve these moving, microscopic, charged particles.

## 16.1 Electric current

In the previous chapter we studied processes such as grounding that occur when a charged conducting object is connected with a wire to an uncharged conductor. We found that the excess electric charge on the charged object redistributes itself until the electric potential $V$ at both conductors becomes equal. Let us look at similar experiments using a Wimshurst machine (Figure 14.20 shows one example of a Wimshurst machine). Recall that cranking the machine's handle generates opposite charges in the two metal spheres. The charge separation leads to a potential difference between the spheres (**Figure 16.1**). In the new Wimshurst machine experiments, we'll try to identify common features that might relate to transferring electric charge for useful purposes. Consider the experiments in Observational Experiment **Table 16.1**.

**Figure 16.1** The charged Wimshurst spheres produce an $\vec{E}$ field.

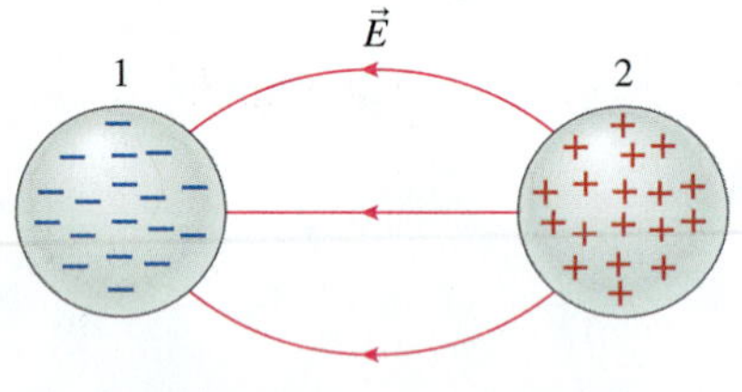


**OBSERVATIONAL EXPERIMENT TABLE**

**16.1 Electric potential difference and charge transfer.**

VIDEO 16.1

| Observational experiment | | Analysis |
|---|---|---|
| **Experiment 1.** Crank the handle of a Wimshurst machine and then bring the oppositely charged spheres of the machine close to each other (about 5 cm apart). You see a big spark.<br>After the spark, when you again bring the spheres close, no more sparking occurs. | 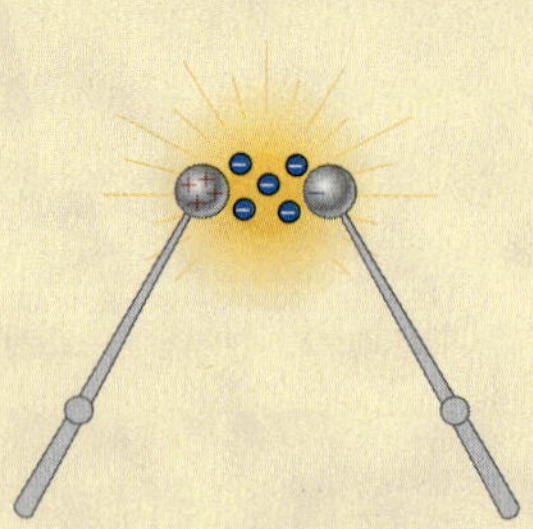 | There is charge separation and nonzero potential difference $\Delta V$ between the two spheres. The spark means that the air between the spheres becomes a conductor, leading to the rapid discharge and the production of light. |
| **Experiment 2.** Crank the handle of the Wimshurst machine and hang a light aluminum foil ball from an insulating thread between the oppositely charged spheres. The ball swings back and forth from one sphere to the other for a few minutes and then stops.<br>If you remove the foil ball and bring the spheres near each other, no spark occurs. | 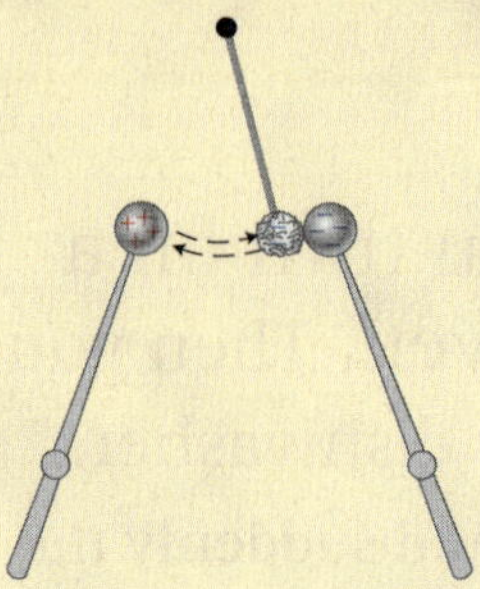 | The foil ball must acquire a small amount of negative charge from the negative sphere each time it touches it. The ball carries the negative charge to the positive sphere and deposits it there and then returns to the negative sphere to repeat the process. This continues until the spheres are discharged and the potential difference between them is zero. The original electric potential energy is converted to mechanical energy and internal energy. |

| Observational experiment | Analysis |
|---|---|
| **Experiment 3.** Crank the handle of the Wimshurst machine and connect the leads of a neon bulb between the charged spheres. There is a flash of light from the bulb. <br><br> If you remove the bulb and bring the spheres close, no spark occurs. <br> 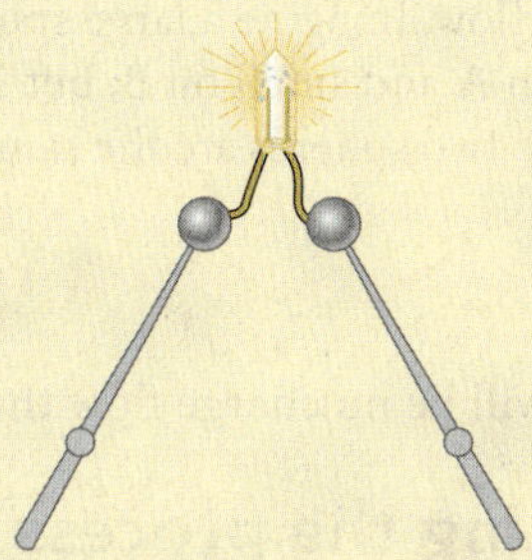 | Before the bulb touches them, the spheres are charged and at different potentials. The bulb and its leads provide a conduction path to discharge the spheres. The discharge causes a flash of light from the bulb. |

**Patterns**

- In all three experiments, the Wimshurst machine started with negative charge on one sphere and positive charge on the other sphere. There was a potential difference $\Delta V$ between the spheres.
- This charge separation and potential difference led to a flow of charge from one sphere to the other.
- The charge flow involved different observable consequences: a spark of light, the vibrating ball of foil, or the flash of the bulb.
- After the charge flow, the Wimshurst spheres were discharged and the potential difference between them was zero. No more sparking or movement could occur.

The observable events in Table 16.1 were only able to occur because of

1. An initial charge separation and resulting potential difference $\Delta V$ between the oppositely charged spheres. After the discharges occurred, resulting in equipotential spheres, nothing further could happen.
2. The presence of a charge conduction pathway (the foil ball, the leads of the neon bulb, and the conducting neon inside). No flash or movement occurred if the charged spheres were far apart and the electric field was too weak between the spheres to ionize the air.
3. A process that converted the electric potential energy of the spheres into some other form of energy: the mechanical energy of the foil ball, or the flash of light from the neon bulb, or the flash from the air itself.

Each of these three processes occurred abruptly. If we want to convert electrical energy to light or mechanical energy for a substantial time interval, we need to learn how to keep the processes described in Table 16.1 happening continuously.

## Fluid flow and charge flow

A fluid flow analogy may help us better understand the electric potential difference and conduction pathways of these electrical processes. You have two containers with water—A and B. A is almost full and B is almost empty. You connect a hose between the two containers (**Figure 16.2a**). Water starts flowing from A to B until the levels are the same (Figure 16.2b), at which point the water flow stops. The amount of water in container A is analogous to the excess positive charge on Wimshurst sphere 2, and the difference in water pressure in the hose is analogous to the potential difference between the spheres. The hose provides a pathway for water to flow until the pressure at both ends of the hose is the same; the electric charge flows until the electric potential at each sphere is the same.

Notice that it is the pressure difference and not the difference in the amount of water in each container that makes the water flow. Imagine a large container A full of water with the water level the same as in small container B—the water will

**Figure 16.2** A water flow analogy for electric charge flow.

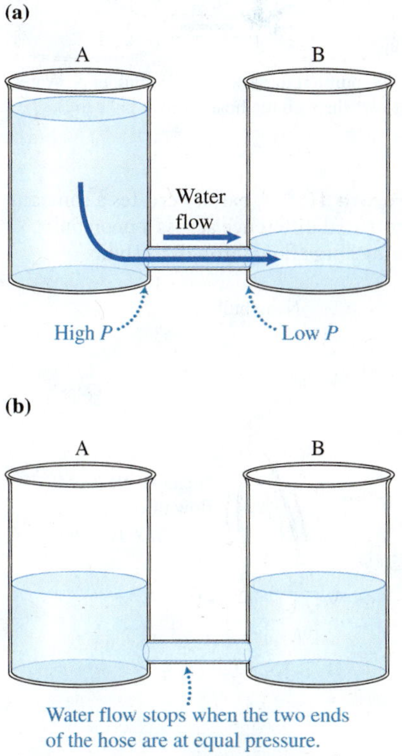

**Figure 16.3** There is no: (a) water flow if $P_A = P_B$ or (b) charge flow if $V_A = V_B$.

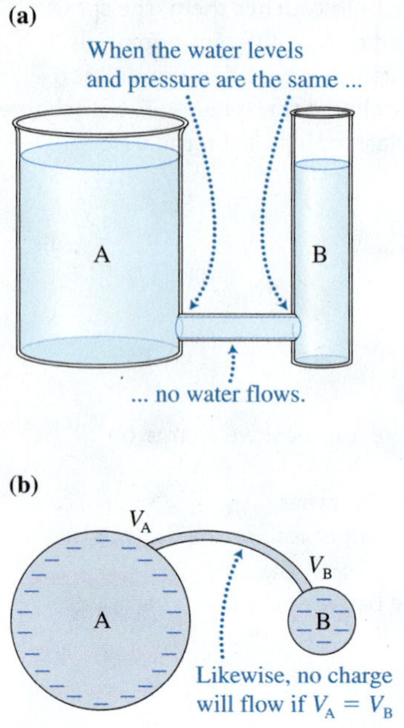
(a)
When the water levels and pressure are the same ...
A
B
... no water flows.
(b)
$V_A$
$V_B$
A
B
Likewise, no charge will flow if $V_A = V_B$

**Figure 16.4** The pump creates a constant pressure difference from B to A, causing a water flow in the hose from A to B.

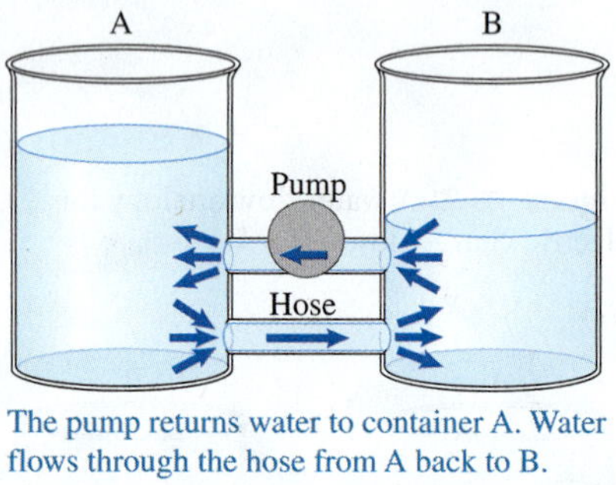
A
B
Pump
Hose
The pump returns water to container A. Water flows through the hose from A back to B.

**Figure 16.5** A battery creates a constant potential difference across a neon bulb and charge flow through the bulb.

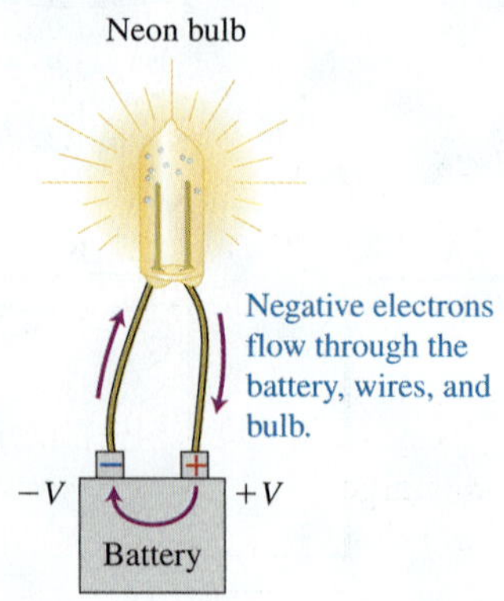
Neon bulb
Negative electrons flow through the battery, wires, and bulb.
$-V$
$+V$
Battery

not flow (**Figure 16.3a**). Similarly, it is not the total charge difference between the spheres of the Wimshurst machine but the potential difference that makes the charge flow. Imagine a large sphere A and a small sphere B. Suppose the charge is large on A and small on B; but if the $V$ fields (electric potentials) on the surfaces of these two spheres are the same

$$V = k\frac{q}{R}$$

there will be no charge flow through the wire connecting them (Figure 16.3b).

## Making the process continuous

We can make fluid flow continuously by connecting the containers with another hose attached to a pump that moves water from B back to A (**Figure 16.4**). Pumping water from B back to A maintains a pressure difference between the ends of the hose and results in a continuous flow from A to B.

To achieve a steady flow of electric charge, we need a device that can maintain a steady potential difference between, for example, the leads of the neon bulb. Let's attach wires from the positive and negative terminals of a battery to the leads of the neon bulb (**Figure 16.5**). The bulb does in fact have a steady glow. Thus, it appears that the battery produces a steady potential difference across its terminals, which in turn causes a steady flow of electric charge through the neon bulb. The battery is equivalent to the pump in the water analogy.

## Electric current

The flow of charged particles moving through a wire between two locations that are at different electric potentials is a physical phenomenon called **electric current**. When the charged particles always move in the same direction, the current is called **direct.** A system of devices such as a battery, wires, and a bulb that allows for the continuous flow of charge is called an **electric circuit.** An electric circuit that has a direct current in it is called a **DC circuit.** In most electric circuits, the moving charged particles are free electrons in the wires and circuit elements. However, as we will learn later in the chapter, in some circuits those moving particles can be positively or negatively charged ions.

According to a simplified model of the internal structure of a wire, the wire consists of a **crystal lattice** composed of positive metal ions and **free electrons**. Free electrons (lost by the atoms that became ions) move within the lattice structure (**Figure 16.6a**). The ions constitute the mass of the material and are relatively stationary except for minor vibrations due to their thermal motion. In this model, the electrons move chaotically inside the wire like a cloud of mosquitoes on a day when the air is still—each mosquito moves, but the cloud remains at rest above your head. When the wire is placed in an external electric field, the electrons accelerate opposite the direction of the $\vec{E}$ field. They slow

**Figure 16.6** (a) Free electrons drift randomly, moving fast between collisions but with no preferred direction. (b) When an electric field is present, the electrons continue this random motion but now drift toward the higher potential and opposite the direction of the $\vec{E}$ field.

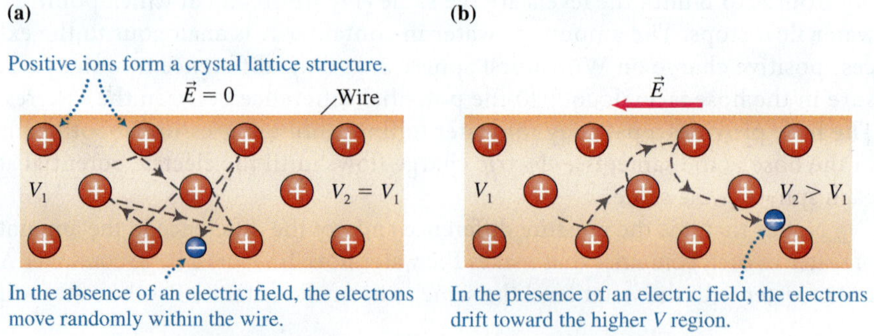
(a)
Positive ions form a crystal lattice structure.
$\vec{E} = 0$
Wire
$V_1$
$V_2 = V_1$
In the absence of an electric field, the electrons move randomly within the wire.
(b)
$\vec{E}$
$V_1$
$V_2 > V_1$
In the presence of an electric field, the electrons drift toward the higher $V$ region.

down when they "collide" with the ions and then accelerate again in the same direction (Figure 16.6b). This motion, called **drift,** occurs in the same direction for all of the electrons—just as a cloud of mosquitoes collectively moves in a single direction on a windy day. The collisions of electrons with ions make the ions vibrate faster, warming the wire. You can test this phenomenon by connecting two terminals of a battery with a wire for a brief moment—both the battery and the wire get hot!

Free electrons in a wire drift toward the direction of higher *V* field (toward the positive terminal of a battery)—see **Figure 16.7a**. However, traditionally, the direction of electric current in a circuit is defined as opposite the direction of the electrons' drifting motion. Imagine that positively charged particles are flowing from the higher electric potential terminal (labeled "+" on the battery) through the wires and electrical devices in the circuit to the terminal at lower electric potential (labeled "−"), then back through the battery to the positive terminal, where the cycle repeats (Figure 16.7b).

This definition of the direction of the electric current as a flow of positively charged particles is a historical quirk. When scientists first studied electric processes, they did not know about electrons and instead thought that an invisible positively charged electric fluid flowed around the circuit. For most situations involving electric circuits, it does not matter if we model the process in terms of electrons flowing one way or positive charges flowing the other way, so this historical convention is not a problem. Physicists use the term "electric current" to describe the drifting of electrically charged particles as well as the physical quantity that characterizes how much charge is moved through a conductor per unit time.

**Electric current** **The magnitude of the physical quantity of electric current *I* in a wire equals the magnitude of the electric charge *q* that passes through a cross section of the wire divided by the time interval $\Delta t$ needed for that amount of charge to pass:**

$$I = \frac{|q|}{\Delta t} \tag{16.1}$$

**The unit of current is the ampere A, equivalent to one coulomb per second C/s. A current of 1 A (one ampere, or amp) means that 1 C of charge passes through a cross section of the wire every second. The direction of the current is in the direction positive charges would move.**

**In SI units, the unit for electric current is a fundamental unit. The unit of electric charge (1 coulomb) is defined in terms of the ampere as $1\text{ C} = (1\text{A})(1\text{ s})$. In the next chapter, we study the phenomenon that can be used to define the ampere.**

**Figure 16.7** The electric current is defined as opposite the direction that electrons move.

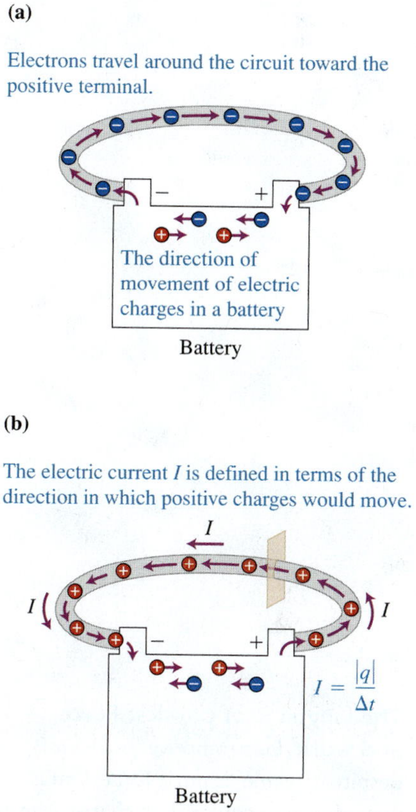
(a)
Electrons travel around the circuit toward the positive terminal.
−
+
The direction of movement of electric charges in a battery
Battery
(b)
The electric current I is defined in terms of the direction in which positive charges would move.
I
I
I
−
+
I = |q|/Δt
Battery

## QUANTITATIVE EXERCISE 16.1 Electric current calculation

Each second, $1.0 \times 10^{17}$ electrons pass from right to left past a cross section of a wire connecting the two terminals of a battery. Determine the magnitude and direction of the electric current in the wire.

**Represent mathematically** The electric charge of an electron is $-1.6 \times 10^{-19}$ C. Since we know how many electrons ($N$) pass a cross section of the wire in 1 s, we can find the total electric charge passing that cross section each second and therefore the current:

$$I = \frac{|q|}{\Delta t} = \frac{|-eN|}{\Delta t}$$

**Solve and evaluate**

$$I = \frac{|-eN|}{\Delta t} = \frac{|-(1.6 \times 10^{-19}\text{ C})(1.0 \times 10^{17})|}{1\text{ s}}$$
$$= 1.6 \times 10^{-2}\text{ A} = 16\text{ mA}$$

The direction of the current is from left to right, opposite the direction of electron flow.

**Try it yourself:** If there is a 2.0-A current through a wire for 20 min, what is the total charge that moved through the wire during this time interval?

*Answer:* 2400 C.

**Review Question 16.1** What condition(s) are needed for electric charge to continuously travel from one place to another?

## 16.2 Batteries and emf

We have found that a battery creates a constant potential difference that can cause a neon bulb to glow continuously, analogous to how a water pump can cause water to flow continuously. If the current is a moving positive charge, the battery does work on the charge to move it within the battery from the lower electric potential negative terminal to the higher electric potential positive terminal. Once there, the charge can then travel around the external circuit (the bulb and connecting wires) and cause the bulb to glow continuously. Batteries maintain a potential difference by means of a nonelectrostatic chemical process.

The work done by a battery per unit charge moved inside the battery from one terminal to the other is the battery's **emf** $\varepsilon$. When first conceived, the term for this work was electromotive force, but now it is just called emf. (See The Language of Physics in the margin.)

If you connect a neon bulb to the battery terminals, the bulb will glow and get a little warmer as work done by the battery results in the production of light and an increase in the thermal energy of the external circuit (bulb and connecting wires). If instead of a neon bulb, you were to connect a toaster to the battery, the work would be converted into thermal energy and light energy in the toaster—the toaster's coils would glow red and get hot.

**The language of physics: Force and work** Electromotive force (emf), despite its name, is not a force. Emf is work done per coulomb of charge. The term *electromotive force* was coined by Alessandro Volta (1745–1827), who invented the battery. At that time the terms "force," "energy, " and "power" were used somewhat interchangeably. The linguistic distinction between force and energy had not yet been made clear. For example, kinetic energy was called "live force" and potential energy was called "dead force."

**Emf $\varepsilon$ The emf $\varepsilon$ equals the work $W$ done by a battery per coulomb of electric charge $q$ that is moved through the battery from one terminal to the other in order to maintain a potential difference at the battery terminals:**

$$\varepsilon = \frac{W}{q} \tag{16.2}$$

The unit of emf is $\mathrm{J/C} = \mathrm{V}$, the same as the unit of electric potential difference. The emf of standard AAA, AA, C, and D batteries is 1.5 V. The small rectangular batteries used in smoke detectors have an emf of 9.0 V. The physical size of the battery is not related to the emf but to its storage capacity—the total charge it can move before it must be replaced or recharged. The bigger the size, the larger the product of $q = I\Delta t$, and the longer the time interval during which the battery works (assuming constant current).

**CONCEPTUAL EXERCISE 16.2 Graphing electric potential in a circuit**

You connect a 9.0-V battery to a small motor. Describe the changes in electric potential in the circuit with a graph.

**Sketch and translate** A sketch of a motor connected to the battery is shown at right. The battery has two terminals, + and −. The emf of the battery is 9.0 V.

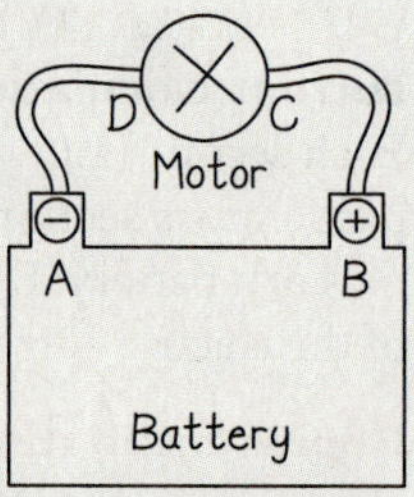


**Simplify and diagram** To represent the changes in electric potential in the circuit, use a $V$-versus-location graph for an imaginary positively charged particle moving through the circuit (see figure on the next page). Assume that the electric potential of the − terminal at A is zero and that of the + terminal at B is 9.0 V. The battery does work on the charged particle-circuit system, increasing its electric potential energy. This work is similar to a person lifting a ball up a hill. When the ball reaches the top of the hill and starts rolling down, the potential energy of the system begins to decrease. Similarly, after leaving the battery at B, the imaginary positively charged particle travels along a wire to the motor at C. We don't know yet the decrease in potential in the connecting wire, but will learn later that it is very small. There is also a small decrease in potential as the charge travels along the wire from D back to the battery at A. Thus, most of the potential decrease should occur across the motor. The potential decreases back to zero at A.

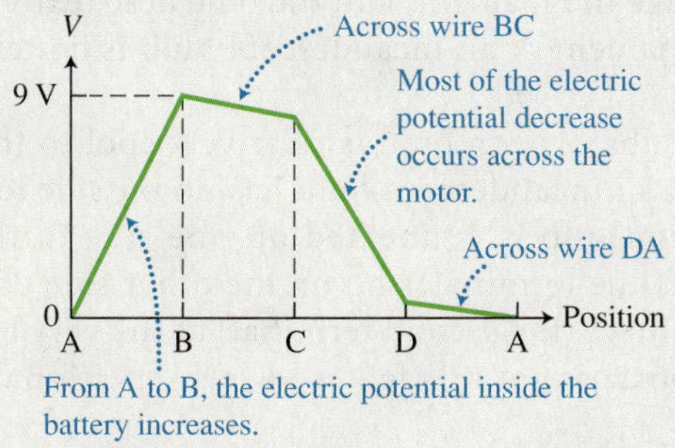
V
9 V
0
Across wire BC
Most of the electric potential decrease occurs across the motor.
Across wire DA
Position
A B C D A
From A to B, the electric potential inside the battery increases.

**Try it yourself:** How would the sketch and graph differ if there were two identical small motors attached one after the other to the battery?

*Answer:* It's the same battery, so there is still a 9.0-V potential increase from its negative to its positive terminal. While passing through the two-motor circuit, about half of the decrease in potential (4.5 V) is across the first motor and the other 4.5-V decrease is across the second motor (see figure below).

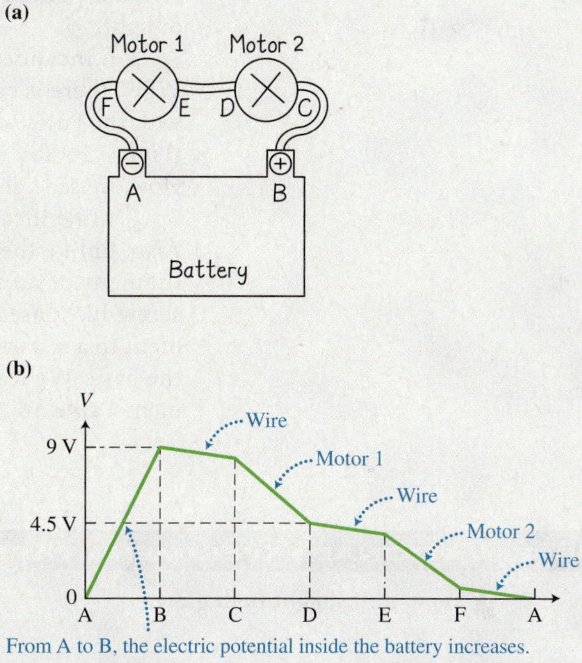
(a)
Motor 1
Motor 2
F
E
D
C
A
B
Battery
(b)
V
9 V
4.5 V
0
Wire
Motor 1
Wire
Motor 2
Wire
A B C D E F A
From A to B, the electric potential inside the battery increases.

**Review Question 16.2** Describe the changes in electric potential through a circuit and relate them to the motion of charged particles.

# 16.3 Making and representing simple circuits

An electric circuit is a system of devices such as a battery, wires, and a bulb (or some other elements, such as a motor) that allows for the continuous flow of charge. However, just having those elements next to each other does not make a circuit—the electrical connections among them are important. In this section we investigate how to make a circuit. However, first we need to learn more about lightbulbs—common indicators of current in a circuit.

## Neon and incandescent bulbs

All lightbulbs have a mechanism that converts electrical energy into light and thermal energy when electric charge flows through them. Here we will consider two types of bulbs—neon and incandescent—that use different mechanisms for this energy conversion.

A neon bulb consists of a glass bulb filled with low-pressure neon gas. Two thin wires (called **electrodes**) extend from the bottom of the bulb (**Figure 16.8a**). The leads outside the bulbs are the extensions of the electrodes. If the potential difference between the electrodes is high enough, the gas inside undergoes dielectric breakdown and becomes a conductor connecting the two electrodes. As a result, ions and free electrons are now present in the space

**Figure 16.8** Comparing a neon bulb and an incandescent bulb.

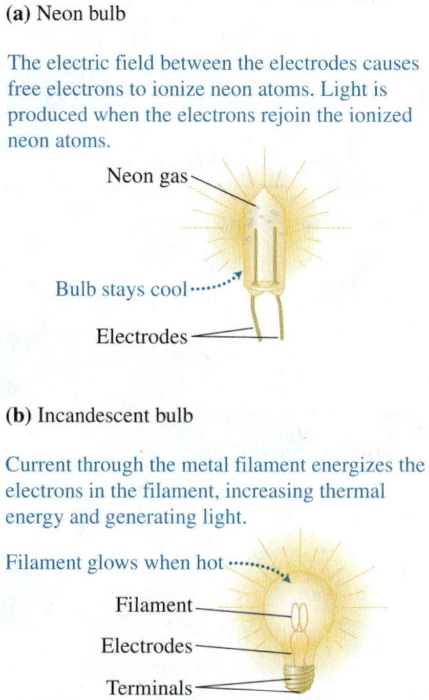
(a) Neon bulb
The electric field between the electrodes causes free electrons to ionize neon atoms. Light is produced when the electrons rejoin the ionized neon atoms.
Neon gas
Bulb stays cool
Electrodes
(b) Incandescent bulb
Current through the metal filament energizes the electrons in the filament, increasing thermal energy and generating light.
Filament glows when hot
Filament
Electrodes
Terminals

between the electrodes. As the electric field associated with the potential difference causes them to travel in opposite directions, some electrons will recombine with ions and emit light. Even if the potential difference across the terminals of the neon bulb only lasts for a very short time, we still see a flash of light.

An incandescent bulb has a metal filament inside of it instead of a gas. When there is current through the filament, the interactions of free electrons with the lattice of atoms make the filament, and thus the bulb, extremely hot (Figure 16.8b). When the filament of an incandescent bulb is hot enough to glow, we see light.

Unlike incandescent bulbs, a neon bulb is relatively cool to the touch. Also, unlike the neon bulb, an incandescent bulb has no outside leads. The filament of an incandescent bulb is connected on one side to the metal screw-like base of the bulb (one terminal) and on the other side of the filament to a separate metal contact (the second terminal) at the very bottom of the base. We learn the importance of this feature in Observational Experiment **Table 16.2**.

**OBSERVATIONAL EXPERIMENT TABLE**

**16.2 Making a flashlight bulb glow.**

VIDEO 16.2

| Observational experiment | Analysis |
|---|---|
| Connection 1—No light is produced. 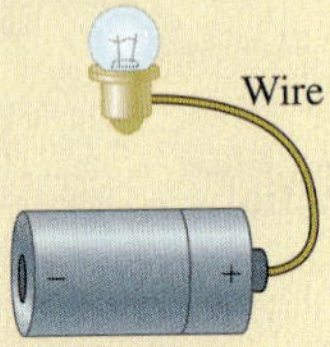  | A wire connects one bulb terminal to one battery terminal. |
| Connection 2—no light 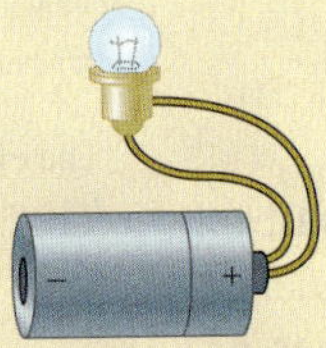 | Both bulb terminals are connected to the same battery terminal with a wire. |
| Connection 3—no light 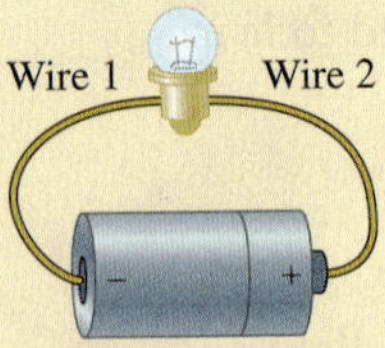  | One bulb terminal is connected to both battery terminals with wires. |
| Connection 4—light  | The two bulb terminals are connected to the two battery terminals with wires. |

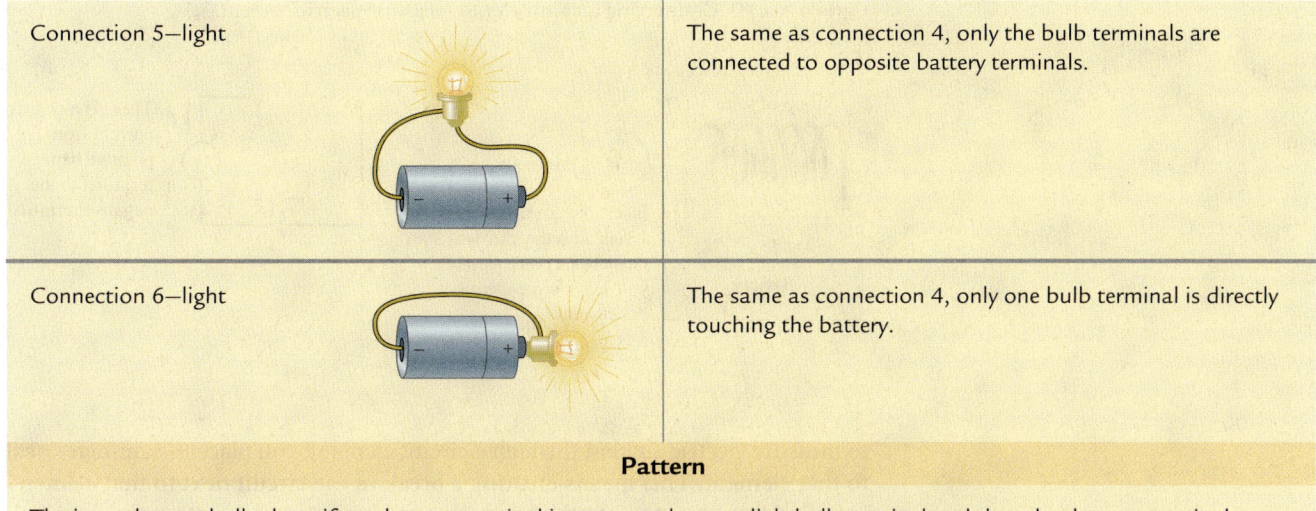

| | |
|---|---|
| Connection 5—light  | The same as connection 4, only the bulb terminals are connected to opposite battery terminals. |
| Connection 6—light 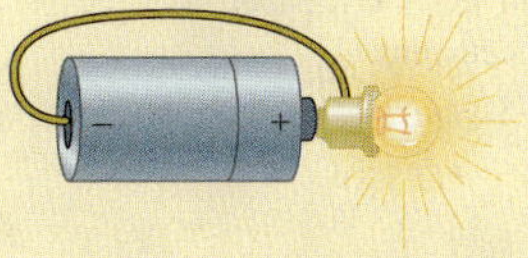 | The same as connection 4, only one bulb terminal is directly touching the battery. |

**Pattern**

The incandescent bulb glows if one battery terminal is connected to one lightbulb terminal and the other battery terminal is connected to the other bulb terminal either with a metal wire or by direct contact. This arrangement is an example of a **complete circuit**.

In Table 16.2 we identified conditions necessary to have a circuit with an electric current in it. To check whether any circuit you build is complete, trace a path of an imaginary positive charge moving from the positive terminal of the battery to the negative terminal (either in a real circuit or in the circuit you drew on paper). The path has to pass along conducting material at every location. Check the circuits in which the bulb did not light in the table above using this rule.

## Symbols for the elements in electric circuits

It is inconvenient to represent electric circuits by drawing realistic likenesses of actual batteries and bulbs. Instead, scientists and engineers use simplified symbols for the elements in an electric circuit. Diagrams with these symbols are called **circuit diagrams**. The symbols for some common circuit elements are shown in **Figure 16.9**. Notice that the positive terminal of a battery is represented with a longer line than the negative terminal. Straight lines represent connecting wires. Bulbs are represented by circles with a loop in the middle (representing the filament). A sawtooth line represents any resistive device, an element in a circuit that converts electric energy into some other type of energy. We will discuss resistive devices—often called **resistors**—later in the chapter. Lightbulbs and toasters are all examples of resistors. Switches—devices that allow us to insert a dielectric (such as air) into a conductive path of the circuit to stop the current—have an open and closed position. Devices that measure current through a circuit element or the potential difference across it are called ammeters and voltmeters, respectively. Their symbols are circles with the letters A and V inside.

**Figure 16.9** The symbols used in circuit diagrams.

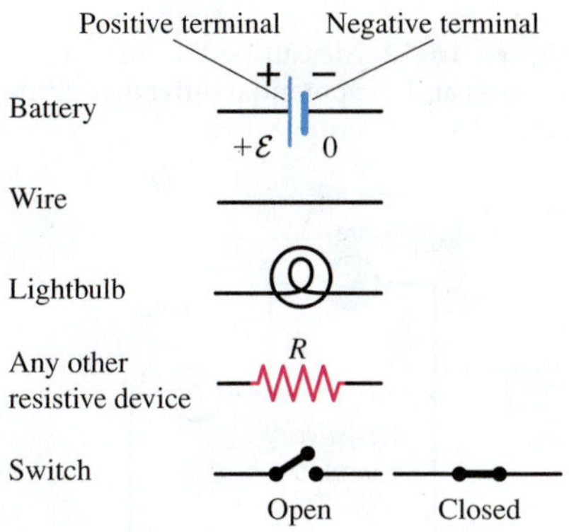


## Ammeters

Recall our water analogy. An **ammeter** acts like a water flow meter. If you wish to measure how much water flows through a cross section of a pipe each second (the *flow rate*), you insert a flow meter into the pipe—just before or after the cross section that interests you. The water must pass through the flow meter. Similarly,

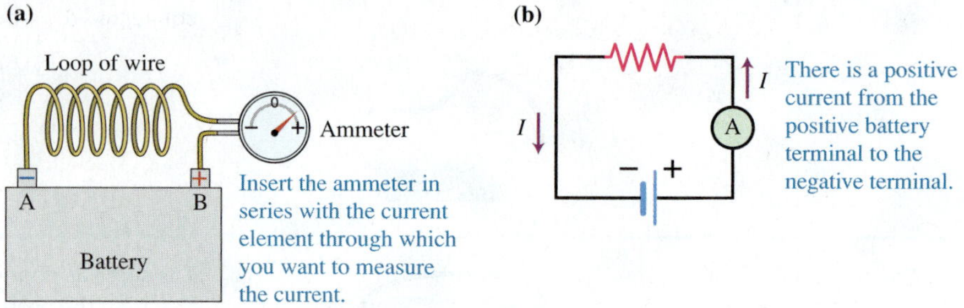
(a)
Loop of wire
Ammeter
A
B
Battery
Insert the ammeter in series with the current element through which you want to measure the current.
(b)
I
I
A
There is a positive current from the positive battery terminal to the negative terminal.

**Figure 16.10** Connecting an ammeter to measure electric current.

to measure electric current through a circuit element, you place the ammeter next to that element. This means creating a break in the circuit next to that element, inserting the ammeter into the break, and reconnecting the wires to the terminals of the ammeter (**Figure 16.10a**). The circuit diagram in Figure 16.10b shows the arrangement. Ammeters have positive and negative terminals, which determine the sign of the ammeter reading. It is positive when the current is in the positive to negative terminal direction and negative when the current is in the negative to positive terminal direction.

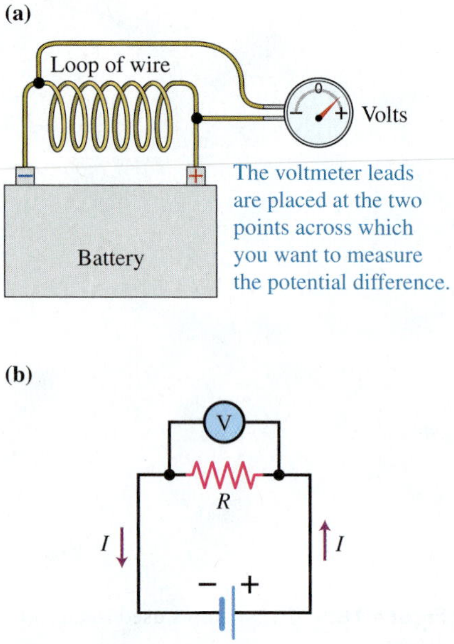
(a)
Loop of wire
Volts
Battery
The voltmeter leads are placed at the two points across which you want to measure the potential difference.
(b)
V
R
I
I

**Figure 16.11** Connecting a voltmeter to measure potential difference.

## Voltmeters

A voltmeter measures the electric potential difference between two points in a circuit. Using a voltmeter to measure potential difference is analogous to using a pressure meter to measure the water pressure difference between two points in a water system. To measure the pressure difference you would submerge one terminal of the pressure meter into the water at the first point and the other terminal into the water at the second point. To measure the electric potential difference between two points in a circuit, you place a voltmeter so that its terminals are connected to those two points (**Figure 16.11a**). A circuit diagram for that connection is shown in Figure 16.11b. The voltmeter has a positive reading when the terminal labeled with a + sign is connected to the point with a higher electric potential than the terminal labeled with the − sign.

A multimeter is a device that combines the functions of an ammeter with a voltmeter. When you use a multimeter be sure you check the settings—accidentally measuring potential difference when the multimeter is on the ammeter setting will harm the device.

**Review Question 16.3** Explain the meaning of the term "a complete circuit."

# 16.4 Ohm's law

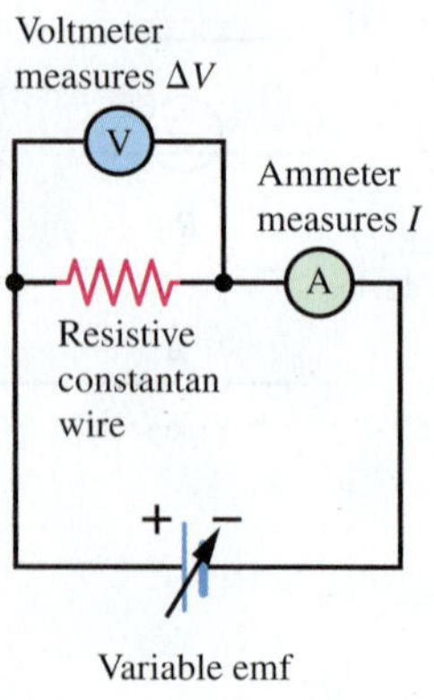
Voltmeter measures $\Delta V$
V
Ammeter measures $I$
A
Resistive constantan wire
Variable emf

**Figure 16.12** Measuring the current through and the potential difference across a resistive constantan wire.

Is there a relationship between the potential difference across a circuit element and the electric current through it? To determine such a relationship, we conduct an experiment (Observational Experiment **Table 16.3**). We use different length wires made of constantan (an alloy of nickel and copper) and connect the ends of one of the wires to a variable emf indicated by the symbol ↗ (**Figure 16.12**). A variable emf can change both in magnitude and in direction. We build two circuits, each with a different length constantan wire (the resistive device in this case). We then vary the potential difference across each and record the resulting current through it. We focus only on the current through and potential difference across the constantan wires, not the connecting wires.

**OBSERVATIONAL EXPERIMENT TABLE**

## 16.3 Developing a relationship between current through and potential difference across a resistive element.

**Observational experiment**

The electric circuit is shown in Figure 16.12. By changing the setting of the variable emf, we vary the potential difference $\Delta V$ across the ends of the constantan resistive wire and measure the current $I$ through it. Then we repeat the experiment using a constantan wire of different length. Finally, we graph the results.

**Analysis**

| Wire 1 potential difference $\Delta V$ (volts) | Wire 1 current $I$ (amps) | Wire 2 potential difference $\Delta V$ (volts) | Wire 2 current $I$ (amps) |
|---|---|---|---|
| 0 | 0.00 | 0 | 0.00 |
| 1.0 | 0.37 | 1.0 | 0.19 |
| 1.5 | 0.56 | 1.5 | 0.28 |
| 2.0 | 0.74 | 2.0 | 0.37 |
| 2.5 | 0.94 | 2.5 | 0.47 |
| −1.0 | −0.37 | −1.0 | −0.19 |
| −1.5 | −0.56 | −1.5 | −0.28 |
| −2.0 | −0.74 | −2.0 | −0.37 |
| −2.5 | −0.94 | −2.5 | −0.47 |

The graphs $I$ versus $\Delta V$ for both resistive wires are straight lines that pass through the origin.

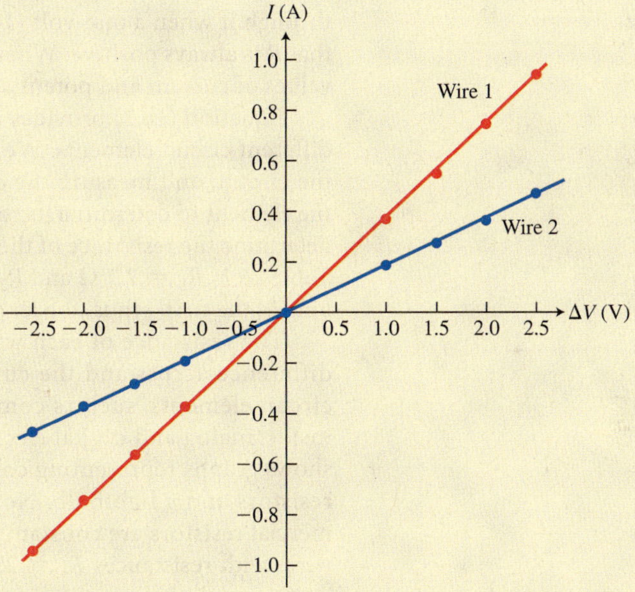


**Pattern**

The graphs show that the current through both resistive wires is directly proportional to the potential difference across them. When the potential difference reverses direction, so does the current. The slopes of the lines are different for different resistive wires.

The relationship we observed in Table 16.3 is reasonable given our conceptual understanding. The larger the potential difference across the wire, the larger the $\vec{E}$ field inside. This means a larger electric force is exerted on the electrons, resulting in larger accelerations of the electrons between the collisions with the ions in the lattice and thus a larger drift velocity and a larger current through the wire. Since the pattern is the same in both directions, it seems the electric properties of the metal are independent of the direction of current.

What is the meaning of the slope of the line for each wire, a constant called $G$? The constant $G$ for wire 1 is twice as large as that for wire 2. At the same time we notice that for the same potential difference across each wire, there is always twice as much current through 1 as through 2. Thus $G$ seems to quantify how easily the charged particles in the circuit flow (the less they interact with the ions, the less their kinetic energy is converted into the thermal energy of the lattice). The quantity $G$ is called **conductance.** Circuit elements with higher conductance require a smaller potential difference across them to have the same current through them.

To characterize how a circuit element resists a current, we use another quantity, called **resistance** $R$:

$$R = \frac{1}{G}$$

We can write the relationship between current and potential difference using resistance $R$:

$$I = \frac{\Delta V}{R} \tag{16.3}$$

where $R$—resistance—is the physical quantity that characterizes the degree to which an object resists a current. The unit of the physical quantity resistance $R$ is called the **ohm** (Greek omega, $\Omega$), where 1 ohm $=$ 1 volt/ampere ($1\ \Omega = 1\ \text{V/A}$). A 1.0-ohm resistor has a one-ampere (1-A) electric current passing through it when a one-volt (1-V) potential difference is placed across it. Note that $R$ is always positive. When we write Eq. (16.3), we usually use the absolute values of current and potential difference.

Equation (16.3) provides us with a method to determine the resistances of different circuit elements. We connect each circuit element into a circuit, close the circuit, and measure the current through and potential difference across the element to determine its resistance $R = \Delta V/I$. Using the above method we determine the resistance of the two constantan wires used in the experiments in Table 16.3: $R_1 = 2.7\ \Omega$ and $R_2 = 5.3\ \Omega$. Physically, the second wire is twice as long as the first—thus, the length of the wire must affect its resistance.

The resistance of each wire remains constant even though the potential difference across and the current through it change. Is this true for other circuit elements, such as commercial resistors that you see in many circuits inside analog and digital electronic equipment and lightbulbs? **Figure 16.13** shows graphs representing collected data for three elements: two commercial resistors and a lightbulb. Notice that the slopes of the graphs for the commercial resistors are constant. They behave in a similar fashion to constantan wires with resistances $R_1 = 25.0\ \Omega$ and $R_2 = 50.0\ \Omega$, independent of the current through them.

**Figure 16.13** $I$-vs-$\Delta V$ graphs for ohmic and non-ohmic circuit elements.

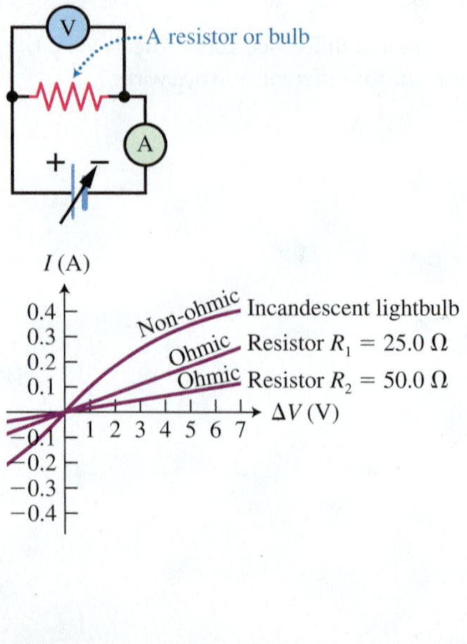


Unlike for the commercial resistors, however, the slope of the graph for the lightbulb decreases as the potential difference increases. It takes more potential difference to cause the same increase in current. In other words, the resistance of the lightbulb increases with increasing potential difference and current. This relationship must be taken into account when building a circuit, because the resistance of such a circuit element depends on the current through it. Note that for all of these elements, including the lightbulb, reversing the polarity of the battery does not change the shape of the curve.

Equation (16.3) is called Ohm's law in honor of 19th-century physicist Georg Ohm, who discovered it experimentally. It is a cause-effect relationship that predicts the value of the current through a resistive device when the potential difference across it and its resistance are known. If the resistance of a circuit element does not depend on the potential difference across it (such as constantan wires or all commercial resistors), the element is called **ohmic.** Circuit elements that cannot be modeled as ohmic devices (such as a lightbulb) are called **non-ohmic**. You will learn more about non-ohmic devices at the end of this chapter.

**Ohm's law** **The current $I$ through a circuit element (other than a battery) can be determined by dividing the potential difference $\Delta V$ across the circuit element by its resistance $R$:**

$$I = \frac{\Delta V}{R} \tag{16.3}$$

**TIP** Note that $\Delta V$ is the potential difference between one side of the device and the other (hence the $\Delta$), whereas current $I$ is the flow rate of charge through the device.

For ohmic devices, resistance $R$ is constant and independent of the magnitude or direction of potential difference across it. For non-ohmic devices, the resistance $R$ is variable, and its value depends on magnitude and sometimes on the direction of the potential difference across it.

Constantan wire, nichrome wire (a nickel-chromium alloy), commercial resistors that are used in many circuit boards, many parts of the human body, some geological formations, and salt water are ohmic devices, but neon bulbs and transistors (a fundamental circuit element in computer chips) are not. In **Figure 16.14** you see an $I$-versus-$\Delta V$ graph for another non-ohmic device, a diode. A diode is a circuit element that consists of two different kinds of materials that cause it to be asymmetrical with respect to the potential difference across it. When one lead of a diode is at lower potential than the other lead, the diode's resistance is very low and a large current results (the right side of the graph in Figure 16.14). However, if the potential difference is reversed, the diode's resistance is very high and only a tiny current results. In other words, the diode's resistance is strongly dependent on the orientation of the potential difference across it. Diodes are useful in circuits in which you need the resistance to be different for each direction of current. Using a diode in a circuit allows you to achieve one-directional current even if the potential difference is not.

**Figure 16.14** An $I$-vs-$\Delta V$ graph of a non-ohmic diode.

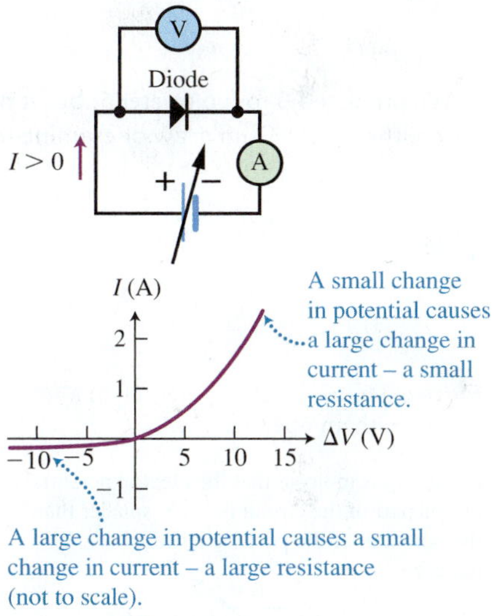


Let's test Ohm's law on a slightly more complicated circuit (**Figure 16.15**). A battery and switch are connected to two resistors $R_1 = 25\ \Omega$ and $R_2 = 50\ \Omega$. There are four ammeters in the circuit and five voltmeters. When the switch is closed, the meters have the readings as shown in the figure. The current through all the elements is the same. The same current enters the 25-$\Omega$ resistor as leaves it and that same current enters the 50-$\Omega$ resistor and then leaves it. Current is not used up!

Voltmeter 2 across a connecting wire and 5 across the closed switch both have zero readings. This is only possible if the resistance of the connecting wire and the closed switch is very small, or almost zero: $\Delta V = IR \approx I \cdot 0 = 0$. Now let's open the switch and use Ohm's law to make predictions about the meter readings.

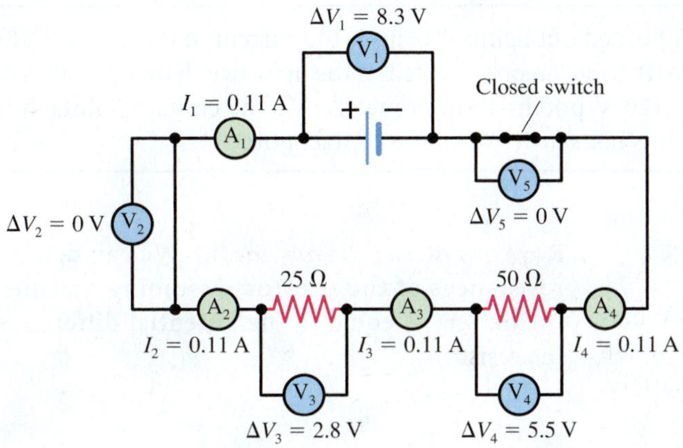


**Figure 16.15** Examine the ammeter and voltmeter readings.

## TESTING EXPERIMENT TABLE

### 16.4 Applying Ohm's law to an open circuit.

VIDEO 16.4

**Testing experiment**

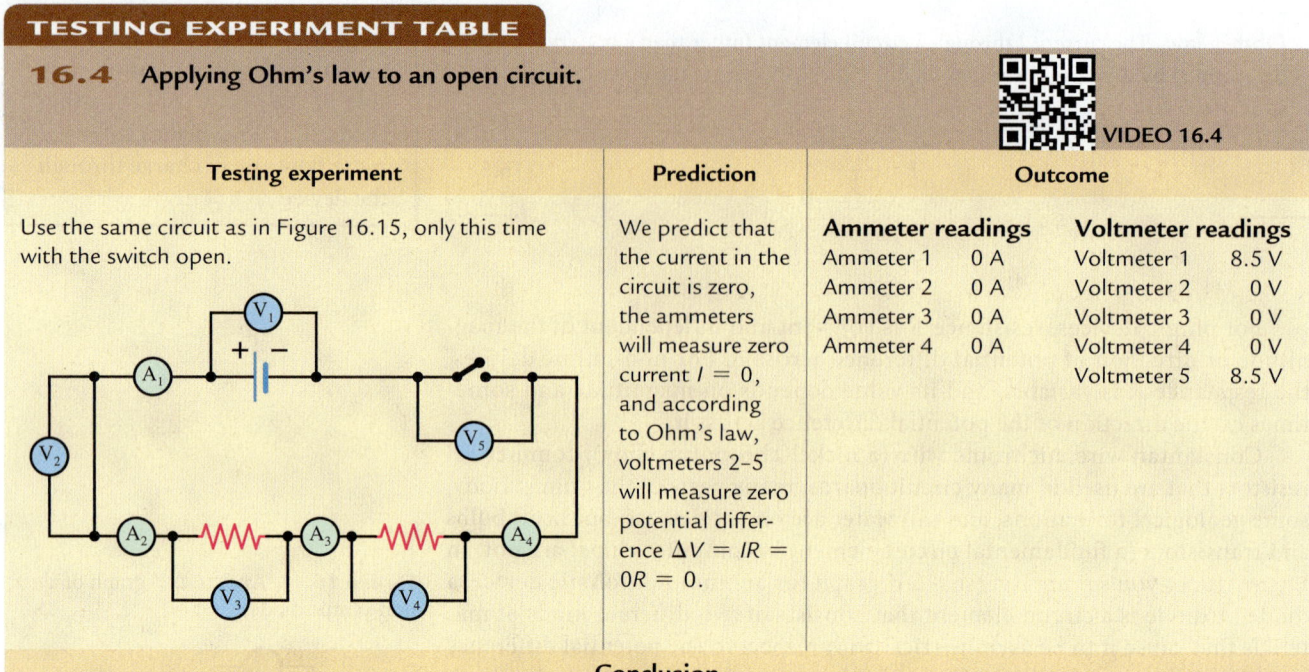

Use the same circuit as in Figure 16.15, only this time with the switch open.

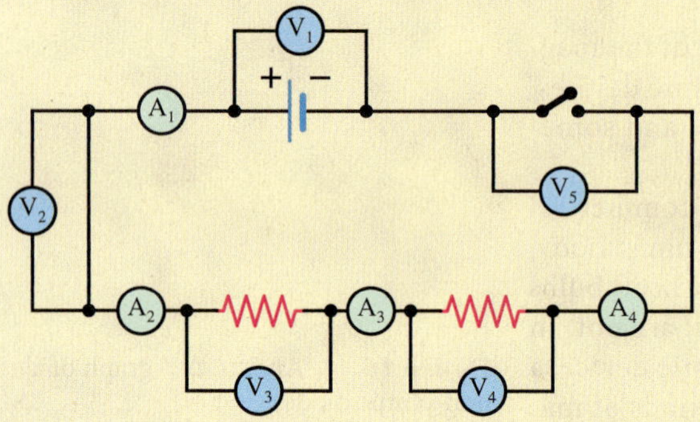


**Prediction**

We predict that the current in the circuit is zero, the ammeters will measure zero current $I = 0$, and according to Ohm's law, voltmeters 2–5 will measure zero potential difference $\Delta V = IR = 0R = 0$.

**Outcome**

| Ammeter readings | | Voltmeter readings | |
|---|---|---|---|
| Ammeter 1 | 0 A | Voltmeter 1 | 8.5 V |
| Ammeter 2 | 0 A | Voltmeter 2 | 0 V |
| Ammeter 3 | 0 A | Voltmeter 3 | 0 V |
| Ammeter 4 | 0 A | Voltmeter 4 | 0 V |
| | | Voltmeter 5 | 8.5 V |

**Conclusion**

We predicted 0 for voltmeter 5, but it measured 8.5 V. The outcome of the experiment does not match the prediction. We need to either revise Ohm's law or examine how we apply it.

When a mismatch between the prediction and the outcome happens, we need to evaluate the reasoning that led to the prediction. When the switch is open, there is no current in the circuit, so on first glance Eq. (16.3) written as $\Delta V = IR$ predicts a zero potential difference. Let us look at the battery first. Even when there is no current in the circuit, the battery has potential difference across the terminals equal to its emf. Thus, the non-zero reading of voltmeter 1 makes sense. But why would there be potential difference across the switch when there is no current in the circuit? When the switch is open, it stops being a conductor; its resistance becomes infinite. We cannot apply relation $\Delta V = IR$, as we do not know the result of multiplying zero by infinity. Therefore, when we made a prediction that did not match the outcome, it was not Ohm's law that failed us but how we applied it. How do we explain the existence of the potential difference across the switch? When the switch is open, the whole part of the circuit connected to the positive terminal of the battery is at a potential of 8.5 V. The part of the circuit connected to the negative battery terminal is at a potential of 0.0 V (**Figure 16.16**).

**Figure 16.16** Electric potential in a circuit with an open switch.

Minus signs indicate that the electric potential of this part of the circuit is 8.5 V smaller than the potential of the positive terminal of the battery.

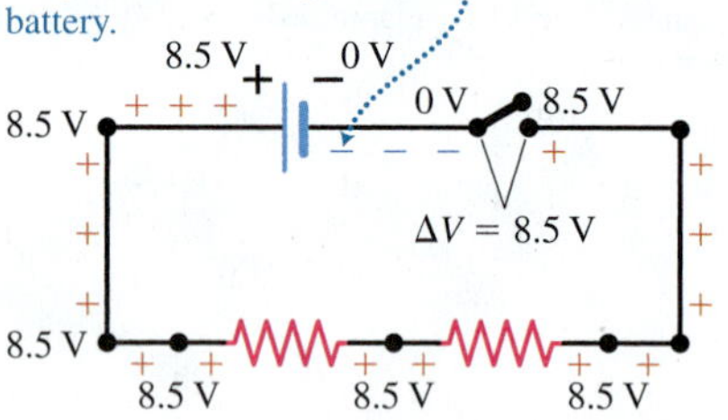


> **TIP** A burned out lightbulb causes the current in the line with the bulb to be zero. It is like an open switch. If the light switch on the wall is on, there may be a 120-V potential difference across the contact points in the bulb canister. It is not safe to touch the contact points!

**QUANTITATIVE EXERCISE 16.3 Applying Ohm's law**

When commercial resistor 1 is connected to a 9-V battery, the current through the resistor is 0.1 A. When commercial resistor 2 is connected to the same battery, the current through it is 0.3 A. What can you learn about the resistors? What assumptions did you make?

**Represent mathematically** We can determine the resistances of the resistors assuming that the emf of the battery is equal to the potential difference across the resistors:

$$R = \frac{\Delta V}{I} = \frac{\varepsilon}{I}$$

**Solve and evaluate** For resistor 1, $R = \frac{\varepsilon}{I} = \frac{9.0\ \text{V}}{0.1\ \text{A}} = 90\ \Omega$; for resistor 2, $R = \frac{\varepsilon}{I} = \frac{9.0\ \text{V}}{0.3\ \text{A}} = 30\ \Omega$. We assumed that the potential difference across the resistors is equal to the emf of the battery. To evaluate the result we need to use a voltmeter to measure the potential difference across the resistors to make sure they are both equal to the emf of the battery.

**Try it yourself:** You connect resistor 1 from the exercise above to an unknown battery. The current through the resistor is 0.2 A. What information about the situation does this number provide? What assumptions did you make?

*Answer:* If we assume that the resistance of the commercial resistors is independent of the potential difference across them, we can deduce that the emf of the battery is about 18 V.

**Review Question 16.4** Why does it make sense that the current is the same through all the elements in the circuit in Figure 16.15?

# 16.5 Qualitative analysis of circuits

So far most of our analyses have applied to circuits with one element connected to a battery. Real-world circuits contain many elements. In this section we will learn about common arrangements of circuit elements and analyze the effect of these arrangements on the electric current in the circuits.

**Figure 16.17** The bulbs are in series.

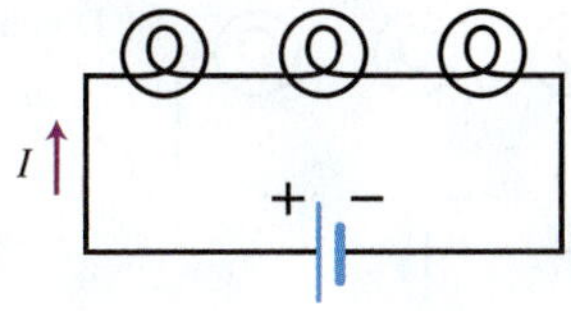


## Circuits in series

In Observational Experiment **Table 16.5** we use lightbulbs as circuit elements. We start by analyzing bulbs arranged in **series**, one after the other as shown in **Figure 16.17**.

**OBSERVATIONAL EXPERIMENT TABLE**

**16.5 Multiple bulbs arranged in series.**

VIDEO 16.5

| Observational experiment | Analysis |
|---|---|
| **Experiment 1.** One bulb connected to a battery is bright. 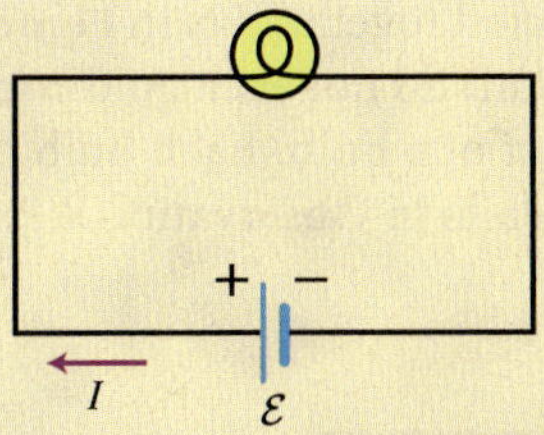 | The lightbulb is lit when the circuit is complete. |
| **Experiment 2.** Two bulbs connected one after the other (in series) to the battery from Experiment 1 are each dimmer than the bulb in 1 but have the same brightness as each other. The bulbs are of equal brightness but dimmer than in experiment 1. 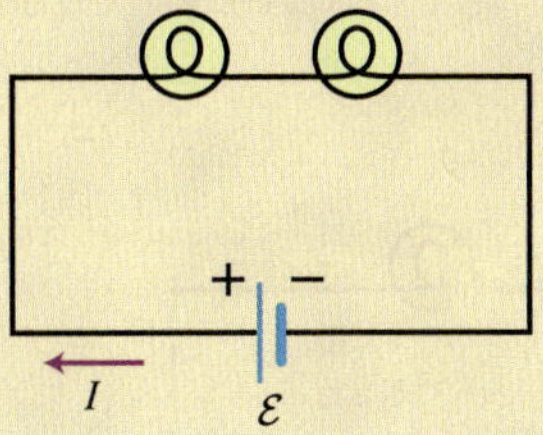 | ■ There must be less current through the bulbs than in Experiment 1. ■ Each bulb must have the same lower current through it (since they are equally bright). |

(continued)

| Observational experiment | Analysis |
|---|---|
| **Experiment 3.** Three bulbs connected in series to the same battery are each dimmer than the bulbs in Experiments 1 and 2 but have the same brightness as each other. 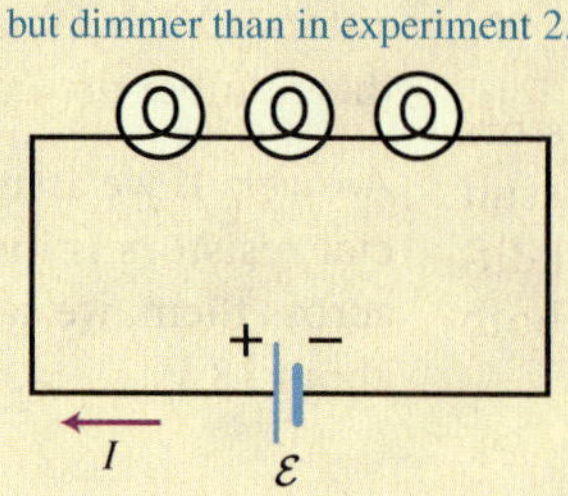 | ■ There must be less current through the bulbs than in Experiments 1 and 2. ■ Each bulb must have the same even lower current through it (since they are equally bright.) |



**Patterns**

- The brightness of all identical bulbs arranged in series is the same.
- Adding more bulbs arranged in series decreases the brightness of all bulbs.

**Figure 16.18** The effect of series bulbs on current.

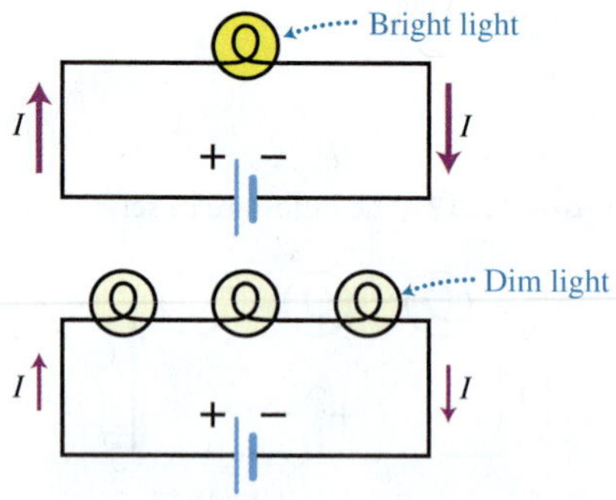



**Figure 16.19** Bulbs in parallel.

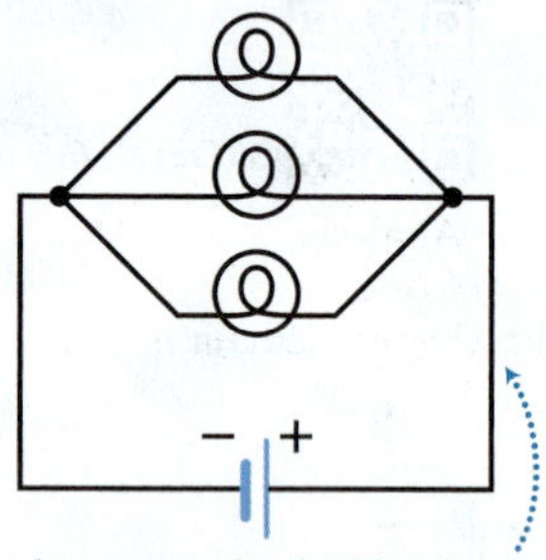

− +
An electron moving in this wire goes through only one of the bulbs on its way to the other side of the circuit.

The first pattern in Table 16.5 makes sense if we assume that the magnitude of the current through a bulb affects how bright the bulb is. Electric current is the flow of charged electrons. Since none of those electrons can escape the circuit, the current must be the same in the first bulb as in the last bulb in the series circuit—it is not used up. This reasoning is consistent with the results of the experiment in Figure 16.15 that showed us that in a similar circuit with resistors instead of the bulbs, the current was the same through all of the resistors.

The second pattern we observed in Table 16.5 indicated that adding more bulbs in series reduced the brightness of each bulb. Evidently, there is less current through series bulbs when there are more of them (**Figure 16.18**). Less current means the amount of charge passing through a cross-sectional area of any element of a circuit per unit time is smaller. Adding more bulbs must increase the resistance of the circuit, reducing the electric current *everywhere* in the circuit (similar to how putting a filter in a hose slows the flow of water through the whole water system). A battery is not a source of constant current.

## Circuits in parallel

We can also arrange bulbs in **parallel**: side by side with the terminals on each side connected together, as in **Figure 16.19**. The charged particles flowing in a parallel circuit do not each go through all the bulbs; each charged particle goes through just one bulb. Each bulb has its own current. We investigate simple parallel circuits in Observational Experiment **Table 16.6**.

**OBSERVATIONAL EXPERIMENT TABLE**

**16.6 Multiple bulbs arranged in parallel.**

VIDEO 16.6

| Observational experiment | Analysis |
|---|---|
| **Experiment 1.** One bulb connected to a battery is bright. 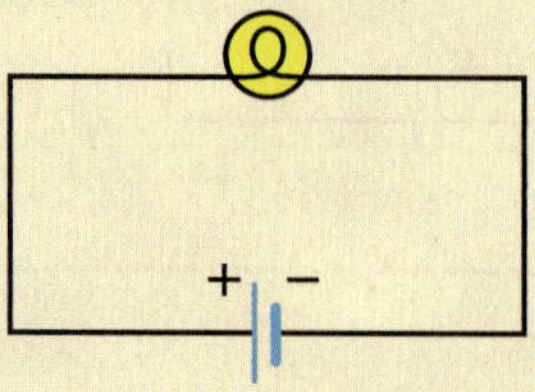 | The bulb is lit when the circuit is complete. |

| Observational experiment | | Analysis |
|---|---|---|
| **Experiment 2.** Two identical bulbs connected in parallel to the same battery are as bright as the bulb in Experiment 1 and the same brightness as each other. | 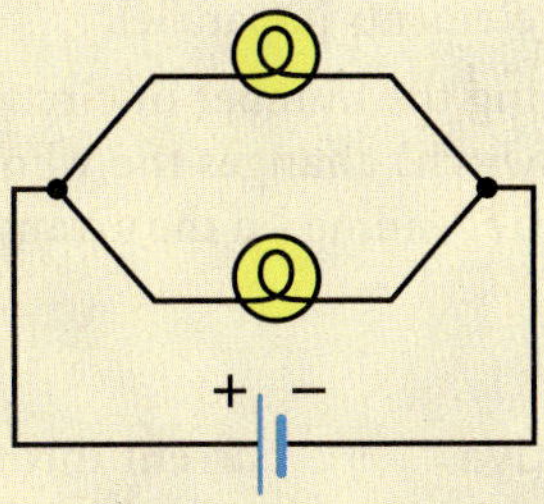 | Since both bulbs have the same brightness, they must have an equal current through them. |
| **Experiment 3.** Three identical bulbs connected in parallel to the same battery are as bright as the bulbs in Experiments 1 and 2 and the same brightness as each other. | 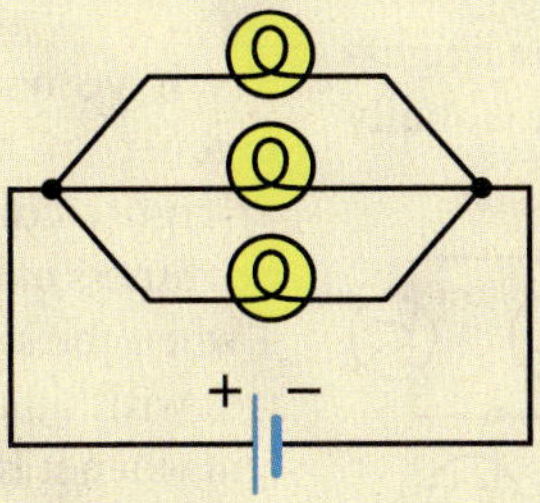 | There must be an equal current through each bulb in every experiment, even when additional bulbs are added. |
| **Pattern** | | |
| The brightness appears to be the same in all of the identical bulbs in parallel and is not affected by the presence of the other bulbs. | | |

How can we explain the pattern in Table 16.6? In order for each identical bulb to be equally bright and have the same brightness as the bulb in Experiment 1, the current must be the same through each of them. In other words, when bulbs are arranged in parallel, they behave as though they were connected to the battery all by themselves and are not affected by the presence of the other bulbs (**Figures 16.20a** and b). Since the bulbs are identical, they have identical resistance (same resistance $R$ for the same current $I$). Therefore, the potential difference across them should be the same ($I = \Delta V/R$) and equal to the potential difference across the battery. The battery must drive a current through the bulbs simultaneously—the total current driven by the battery must double for two parallel bulbs and triple for three parallel bulbs (Figure 16.20b). We see again that the battery is *not* a constant current source—it sends a smaller current through bulbs arranged in series and a larger total current through bulbs in parallel. This also means that multiple bulbs connected in parallel must have less collective resistance than a single bulb because as external resistance decreases, current through the battery increases.

Conventionally, circuits such as in Figure 16.20b are represented differently—see Figure 16.20c.

We can now summarize our qualitative investigations of electric circuits:

- When circuit elements are connected in series, the current through each element is the same; when circuit elements are connected in parallel, the battery is connected across each parallel element and the potential difference across each element is the same.
- Adding more circuit elements in series *decreases* the total current through all the elements and increases the effective total resistance of the elements.
- Adding more elements in parallel *increases* the total current through the parallel elements and reduces the effective total resistance of the elements.

**Figure 16.20** Different ways of representing three bulbs connected in parallel to the battery. The bulbs connected in parallel to the battery have the same potential difference across them as a single bulb.

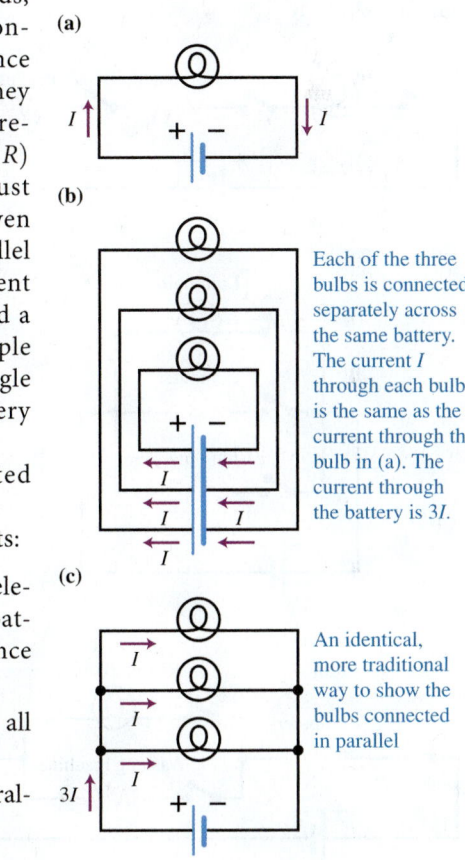

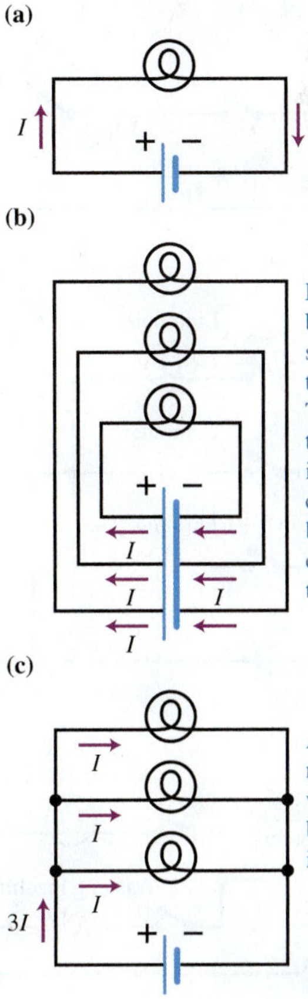


Each of the three bulbs is connected separately across the same battery. The current $I$ through each bulb is the same as the current through the bulb in (a). The current through the battery is $3I$.

An identical, more traditional way to show the bulbs connected in parallel

- The battery is *not* a source of constant current: it drives a smaller current through circuit elements arranged in series and a larger total current through circuit elements in parallel.
- Changing the number of circuit elements (for example, by opening or closing a switch) changes the whole circuit. Total current may decrease or increase depending on the arrangement of circuit elements.

**CONCEPTUAL EXERCISE 16.4 Qualitative reasoning—rating bulb brightness**

The circuit shown in the figure has four identical bulbs. Rank the bulbs in terms of brightness, listing the brightest bulb first. Indicate whether any bulbs are equally bright. Explain your reasoning.

**Sketch and translate** The situation is sketched in the adjacent figure.

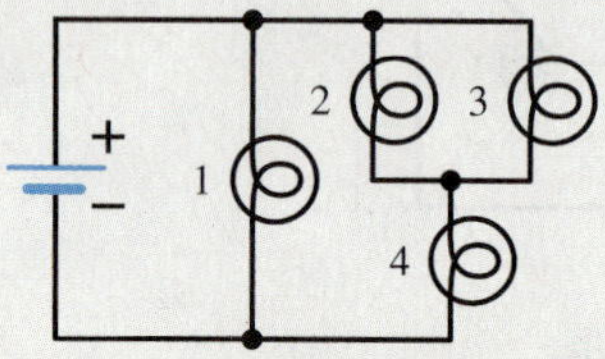


**Simplify and diagram** Bulbs 2, 3, and 4 definitely offer more resistance to charge flow than the single bulb 1. Thus we expect that bulb 1 will be the brightest. Bulbs 2 and 3 in parallel will have the same current and should be equally bright. The sum of the currents in bulbs 2 and 3 equals the current through bulb 4—thus 4 will be brighter than 2 and 3. Thus, we rate the bulbs in terms of brightness as $1 > 4 > 2 = 3$.

**Try it yourself:** In the circuit in the figure at right, compare the brightness of bulbs 1 and 4 when the switch in series with bulb 3 is open and when it is closed.

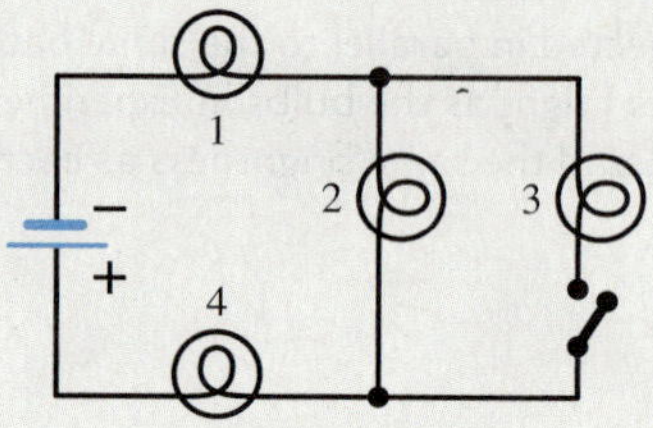


*Answer:* When the switch is open, bulbs 1, 2, and 4 are in series. They are equally bright. When the switch is closed, bulbs 2 and 3 are in parallel and there is less resistance to current through that part of the circuit. Thus, with reduced resistance, the current will be greater through bulbs 1 and 4—they will be brighter.

**Figure 16.21** The possible ways to wire household appliances.

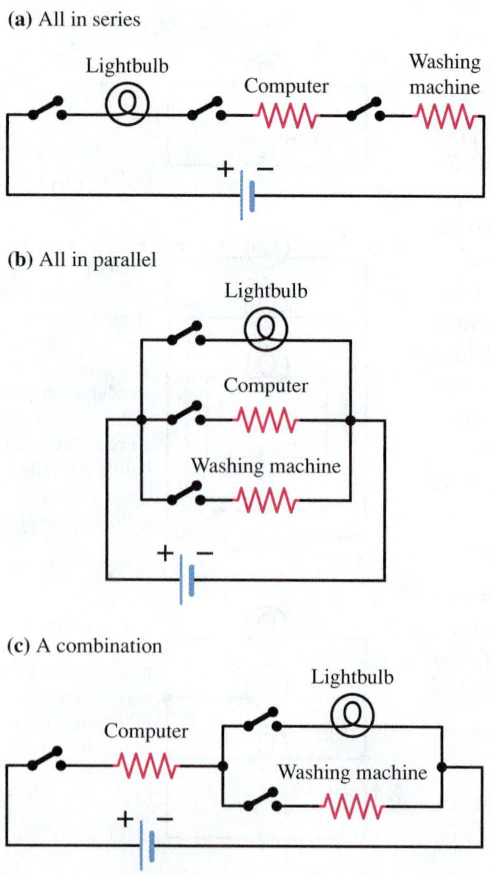


## How is your house wired?

Although the current in the electric circuit in our homes is different in many ways from the current we study in this chapter, there are many similarities that allow us to apply what we are learning in this book to the electrical wiring in a home.

You turn on a lightbulb, a computer, and a washing machine. Are they connected in series or in parallel? There are three possible arrangements (**Figure 16.21**): (a) all the devices are in series, (b) all are in parallel, or (c) some combination of the two. If they are all in series, when you turn one of the devices off, the circuit will open. There will be no current through the circuit and all of the devices will be off. From your everyday experience, you know that does not happen. The devices must not be in series. If they are all in parallel, as in (b), then turning any one device on or off will not affect the others. This is what we observe, so parallel wiring is a possibility. In (c) the devices are arranged neither all in series nor all in parallel. Still, turning off one device also turns off others. This is not what happens in your house, so this sort of arrangement is not possible. The only choice that is consistent with the way devices in your house behave is (b), parallel.

**Review Question 16.5** What experimental evidence supports the hypothesis that in your house the appliances are connected in parallel and not in series?

## 16.6 Joule's law

So far, we have mostly used identical lightbulbs in all of our experiments. Identical bulbs in series in a closed circuit have the same brightness. However, if we connect the two bulbs shown in **Figure 16.22a** in series to a battery, we observe that bulb A is brighter than B. If we reverse the order of the bulbs, bulb A is still the brighter one (Figure 16.22b). We know that in a series circuit, there is the same current through each bulb. Thus, it is not just the current through a bulb that determines its brightness. In this section we will learn what other quantities affect a bulb's brightness.

**Figure 16.22** The bulb's brightness depends on something other than the current through the bulb.

(a)

(b)

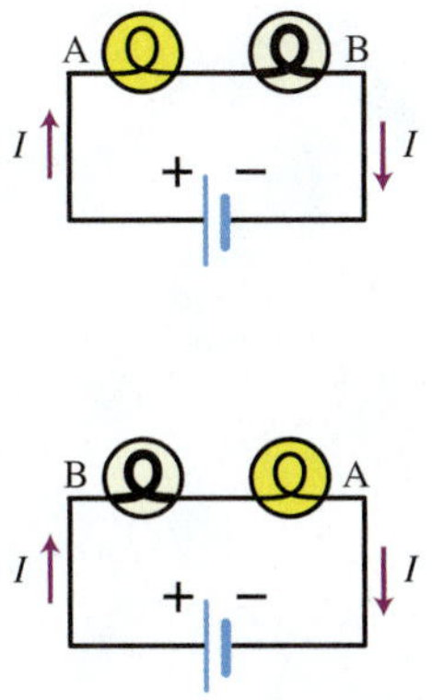


Bulb A is brighter than bulb B in both circuits.

### Electric power

An incandescent lightbulb filament is hot and glows when there is current through it (the free electrons drift through the lattice and interact with ions, making the ions vibrate faster). Consider the bulb and connecting wires as the system and the battery as an external object. The battery does work, increasing the electric potential energy of the system. When there is current through a circuit element, the electric potential energy is continuously converted into other forms of energy (thermal energy, light energy, mechanical energy, etc.). Earlier (in Chapter 6) we learned that power is the physical quantity for the rate of energy conversion. In this case it is the rate at which the thermal energy of a resistive device increases:

$$P = \left|\frac{\Delta U_{\text{thermal}}}{\Delta t}\right|$$

The amount by which $U_{\text{thermal}}$ increases equals the amount that the electric potential energy of the system $U_q$ decreases. Therefore:

$$P = \left|\frac{-\Delta U_q}{\Delta t}\right|$$

As negatively charged electrons (total electric charge $-Q$) travel through a resistive device, they travel from the side that is at low electric potential to the side that is at high electric potential. This movement of electrons results in a change in the electric potential energy of the system $\Delta U_q = q\Delta V = -Q\Delta V$. Therefore:

$$P = \left|\frac{Q\Delta V}{\Delta t}\right| = \frac{Q}{\Delta t}|\Delta V| = I|\Delta V|$$

The above has the correct power units (watts; $1\text{ W} = 1\text{ J/s}$).

$$\text{A}\cdot\text{V} = \frac{\text{C}}{\text{s}}\cdot\frac{\text{J}}{\text{C}} = \frac{\text{J}}{\text{s}} = \text{W}$$

James Joule determined this expression in experiments using a current-carrying wire to warm water. Later, Heinrich Lenz and Alexandre-Edmund Becquerel repeated the experiments using alcohol instead of water and achieved similar results. Now the expression for the rate of electric potential energy conversion into thermal energy is called **Joule's law**.

**Joule's law** **The rate $P$ at which electric potential energy is converted into thermal energy $\Delta U_{\text{thermal}}$ in a resistive device equals the magnitude of the potential difference $\Delta V$ across the device multiplied by the current $I$ through the device:**

$$P = \left|\frac{\Delta U_{\text{thermal}}}{\Delta t}\right| = I|\Delta V| \qquad (16.4)$$

**The language of physics: Electricity: charge, current, potential difference, and power**

The word "electricity" is used in many ways. People say that electricity flows, that electricity is used to heat buildings, and that electricity comes out of a plug in the wall. When people say that electricity flows, they are referring to the flow of charged particles in wires (electric current); when they talk about heating with electricity, they mean electric potential energy being transformed into thermal energy that heats the environment; and when they talk about wall sockets, they usually refer to a 120-V effective potential difference across any appliance plugged into the socket. When you use the word electricity, make sure you understand which of these ideas you are referring to.

Because potential difference and current are related through Ohm's law $I = \Delta V/R$, Joule's law can be written in two alternate forms:

$$P = I|\Delta V| = I(IR) = I^2R$$
$$P = I|\Delta V| = \left(\frac{\Delta V}{R}\right)|\Delta V| = \frac{(\Delta V)^2}{R} \quad (16.5)$$

Examine the different forms of Eq. (16.5). The top version says that the power is proportional to the resistance of the circuit element; the bottom version says that it is inversely proportional. Do these contradict each other? As the current through and potential difference across the element are not independent from each other, there is no contradiction. It is usually most convenient to use the $I^2R$ expression when comparing power in elements connected in series and $(\Delta V)^2/R$ when elements are connected in parallel.

If we assume that the brightness of a bulb is directly related to its power, we can use Joule's law to explain the experiment at the beginning of this section. The lightbulbs in Figure 16.22 had different brightnesses even though the current through them was the same. Based on Eq. (16.4), $P = I|\Delta V|$, the potential difference across bulb A had to be greater than across bulb B. This could only happen if bulb A had a greater resistance than bulb B, $R_A > R_B$, as $I_A = I_B$ thus $P_A = I^2R_A > P_B = I^2R_B$. More resistance means a higher rate of conversion of electric energy to thermal energy! We test this explanation in Testing Experiment **Table 16.7**.

**TESTING EXPERIMENT TABLE**

**16.7 Predicting the brightness of different bulbs.**

VIDEO 16.7

| Testing experiment | Prediction | Outcome |
|---|---|---|
| Connect bulb A and bulb B to separate identical power supplies with a switch in series in each circuit. Do not close the switches. 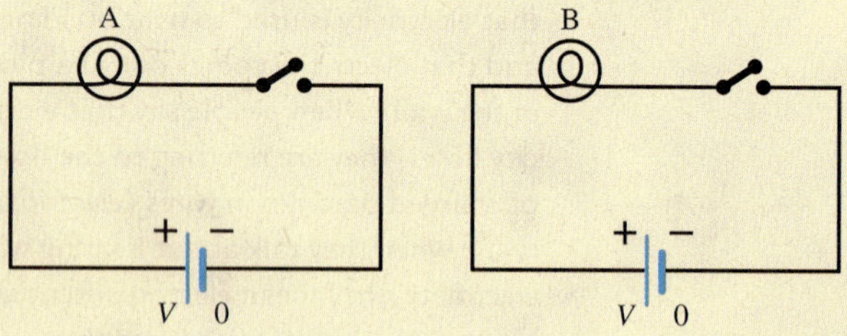  | The bulbs are now connected across the same potential difference $\Delta V_A = \Delta V_B$. If the power is $P = I|\Delta V|$, and $R_A > R_B$, then $I_A < I_B$ and $P_A < P_B$. When we close the switches, lightbulb A should be dimmer than B, but both should be brighter than in the series circuit in Figure 16.22 since the potential difference across each bulb is now higher. | When we close the switches, we observe that bulb A is dimmer than B and both are brighter than when in series in the experiment shown in Figure 16.22. |

**Conclusion**

The bulb's brightness depends on the potential difference across it and on the current through it.

The conclusion in the above table seems to contradict everyday experience. We buy household lightbulbs according to their labeled "wattage"—for example, a 60-W lightbulb—and not their resistance. What is the difference between a 60-W lightbulb and a 100-W lightbulb? The labeling of household lightbulbs assumes that they are used in circuits where the potential difference across them is 120 V. The resistance of a 60-W bulb with a 120-V potential difference[1] across it is

$$P = \frac{(\Delta V)^2}{R} \Rightarrow R = \frac{(\Delta V)^2}{P} = \frac{(120\text{ V})^2}{60\text{ W}} = 240\ \Omega$$

The resistance of a 100-W bulb under the same conditions turns out to be 144 Ω. This makes sense. A bulb with a lower resistance will have a larger current through it, and since $P = I|\Delta V|$, a larger current results in a larger power output. If you put the same bulbs in series, the 60-W bulb will be brighter than the 100-W bulb. In series, the current is the same in each bulb and the power used is $P = I^2 R$. Thus, the larger resistance 60-W bulb will use more power and will be brighter than the 100-W bulb.

## Paying for electric energy

Electric power companies charge their customers according to the amount of electric potential energy that is transformed into other energy forms in the devices that the customer uses. Utility companies do not use the joule. Instead, they use an energy unit called the **kilowatt-hour**. A kilowatt-hour (kW · h) is the electric potential energy that a 1000-W device transforms to other energy forms in a time interval of one hour ($1\text{ kW}\cdot\text{h} = 3.6 \times 10^6\text{ J}$).

**Review Question 16.6** If you place a 100-W bulb and a 60-W bulb in a series circuit, which one will be brighter? Explain.

# 16.7 Kirchhoff's rules

Most electric circuits consist of combinations of several resistive elements, switches, one or more power supplies that may have a variable potential difference output, and a variety of other circuit elements. For example, the electric circuit shown in **Figure 16.23** represents a very simple model of the wiring in a home. In this section we develop techniques to determine the electric current through each circuit element in a complicated circuit. However, we start with a very simple circuit.

**Figure 16.23** A simplified model of a home wiring system.

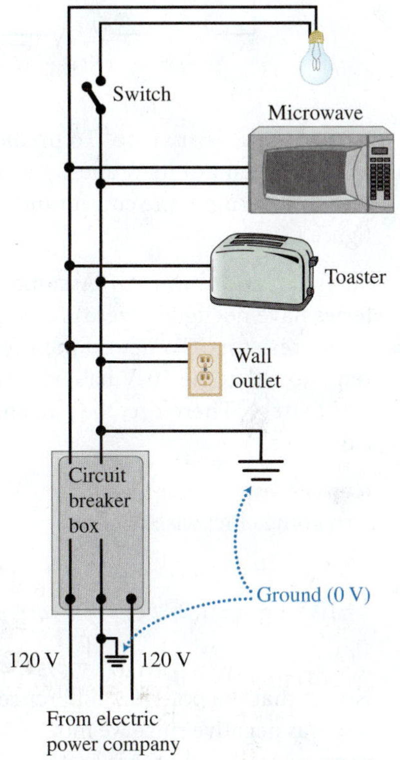


## Kirchhoff's loop rule

The current (the direction of movement of positively charged particles) in the circuit shown in **Figure 16.24a** is counterclockwise. Let us trace the change in electric potential as we move with these charges around the circuit (Figure 16.24b). The sum of the changes in potential for each step along the trip must add to zero, as the potential when we return to the starting position must be the same as when we left that position (the potential cannot have multiple values at one location):

*Changes* in electric potential

| Across battery | | Across connecting wire | | Across resistor | | Across connecting wire | | |
|---|---|---|---|---|---|---|---|---|
| $+\varepsilon$ | $+$ | 0 | $+$ | $(-IR)$ | $+$ | 0 | $=$ | 0 |

A pattern (that we shall call the loop pattern) accounts for our reasoning:

> The algebraic sum of the changes in electric potential around any closed circuit loop is zero.

Let us test this pattern.

**Figure 16.24** The changing electric potential around an electric circuit.

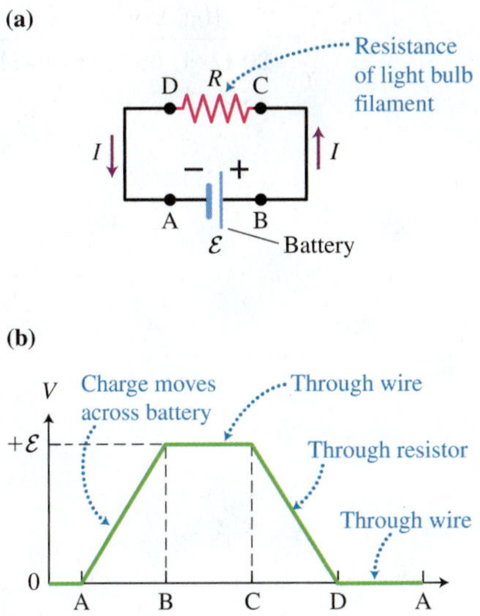


[1]Note that the potential difference across house appliances is not constant in time; it varies 60 times per second. When we use Joule's law derived for constant potential difference, we assume it also applies for variable potential difference and current (averaged in a special way).

**EXAMPLE 16.5 Testing the loop pattern**

Use the loop pattern to predict the potential difference between point A and point B in the electric circuit shown in the figure.

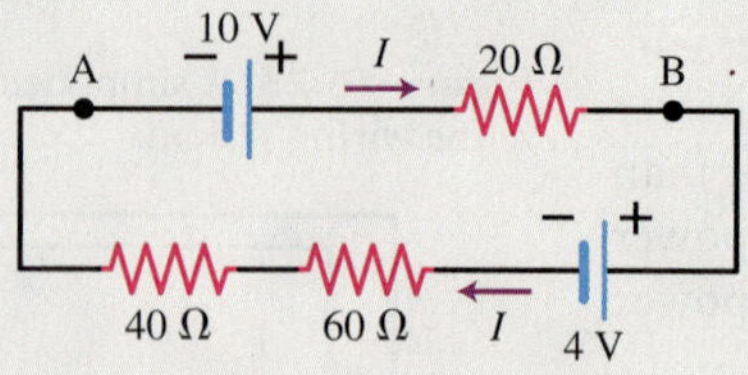


**Sketch and translate** To predict the potential difference between points A and B, we first use the loop pattern to determine the current in the circuit shown in the figure.

**Simplify and diagram** Assume that the wires and batteries have negligible (zero) resistance compared to that of the resistors. To determine the direction of the current, note that the 10-V battery has a larger emf than the 4-V battery. Therefore, the current in the circuit will be clockwise.

**Represent mathematically** We apply the loop pattern going clockwise from A:

$$\Delta V_{10\text{ V}} + \Delta V_{20\,\Omega} + \Delta V_{4\text{ V}} + \Delta V_{60\,\Omega} + \Delta V_{40\,\Omega} = 0$$

$$(+10\text{ V}) + [-I(20\,\Omega)] + (-4\text{ V}) + [-I(60\,\Omega)] + [-I(40\,\Omega)] = 0$$

Notice that the potential difference across the 4.0-V battery was negative since we moved from the positive to the negative terminal of that battery. Also, the potential differences across the resistors were negative as we moved across these resistors in the clockwise downstream direction of the electric current.

**Solve and evaluate** We can now determine the current:

$$I = \frac{10.0\text{ V} - 4.0\text{ V}}{20\,\Omega + 60\,\Omega + 40\,\Omega} = 0.050\text{ A}$$

To determine the potential difference from A to B, we follow the circuit along any continuous path from A to B and add the potential differences across each element in the circuit along that path. There are two possible paths. We first move clockwise from A through the 10.0-V battery and 20-Ω resistor:

$$\Delta V_{AB} = \Delta V_{10\text{ V}} + \Delta V_{20\,\Omega} = +10\text{ V} - (0.050\text{ A})(20\,\Omega) = +9.0\text{ V}$$

The *V* field value at B is 9.0 V higher than at A.

We can also calculate the potential change from A to B by moving counterclockwise upstream through the 40-Ω and 60-Ω resistors (opposite the current) and through the 4.0-V battery from its negative to positive terminal.

$$\Delta V_{AB} = \Delta V_{40\,\Omega} + \Delta V_{60\,\Omega} + \Delta V_{4\text{ V}} = +(0.050\text{ A})(40\,\Omega) + (0.050\text{ A})(60\,\Omega) + 4.0\text{ V} = +9.0\text{ V}$$

The potential difference from A to B is the same no matter which path we take. When we take a voltmeter and connect its terminals to A and B (its negatively marked terminal to A and its positively marked terminal to B), we can see if the outcome and prediction are consistent. The voltmeter reads about 9 V, supporting the loop pattern.

**Try it yourself:** Suppose that in the previous circuit, there was an open switch inserted in the vertical wire on the left side before position A. Now, apply the loop pattern to determine the potential difference across the switch.

*Answer:* $\Delta V_{\text{switch}} = -6\text{ V}$. The top of the switch closer to the 10-V battery will be at a lower potential. The presence of potential difference across the switch when it is open is consistent with our experimental finding in Table 16.4.

**TIP** To indicate the signs of the electric potential changes across the resistors in the previous example, we chose and indicated in the circuit drawing the assumed direction of the current. The potential change across a resistor if moving in the direction of the current is $\Delta V = -IR$, and if moving across the resistor opposite the current is $\Delta V = +IR$. We *must* indicate a current direction in the circuit diagram and keep the signs of $I$ and $\Delta V$ consistent while making a trip around the circuit.

This "loop pattern" was first developed in 1845–1846 by German physicist Gustav Robert Kirchhoff (1824–1887) and is known as **Kirchhoff's loop rule**.

**Kirchhoff's loop rule** The sum of the electric potential differences $\Delta V$ across the circuit elements that make up a closed path (called a loop) in a circuit is zero.

$$\sum_{\text{Loop}} \Delta V = 0 \quad (16.6)$$

**EXAMPLE 16.6 Internal resistance of a battery**

You buy a 9.0-V battery. To check whether it really produces a 9.0-V potential difference across its terminals, you use a voltmeter and find that it reads 9.0 V. Satisfied, you build the circuit shown in the figure. An ammeter (not shown) indicates a 1.5-A current through the 5.0-Ω resistor. A voltmeter reads 7.5 V across the battery and −7.5 V across the resistor. (a) Explain why the measurements are unexpected. (b) What are possible reasons for the discrepancy between the expected and actual results?

**Sketch and translate** The figure shows your circuit diagram. We know the emf of the battery $\varepsilon = 9.0\text{ V}$ but the potential difference across it is only 7.5 V. The current through the resistor $I = 1.5\text{ A}$, but the application of Ohm's law indicates that the current should be $I = \frac{\varepsilon}{R} = \frac{9.0\text{ V}}{5.0\ \Omega} = 1.8\text{ A}$. We need to explain those readings.

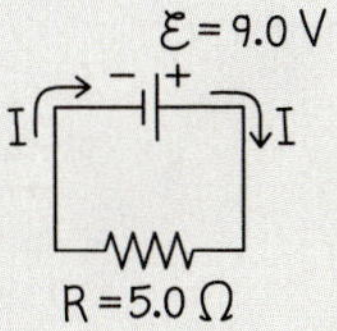


**Simplify and diagram** The current could be smaller than predicted if there is some other resistor connected to the battery. Perhaps the battery has its own **internal resistance**. We actually measure the battery potential as shown in the figure at the right.

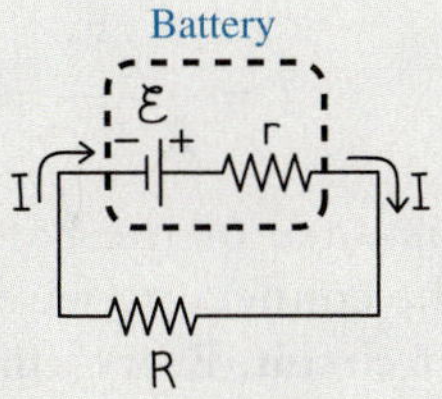


**Represent mathematically** Using this hypothetical resistance, we can write the loop rule as $+\varepsilon - Ir - IR = 0$ and use it to find the internal resistance of the battery $r = \frac{\varepsilon - IR}{I}$.

**Solve and evaluate** The internal resistance of the battery is

$$r = \frac{\varepsilon - IR}{I} = \frac{9.0\text{ V} - (1.5\text{ A})(5.0\ \Omega)}{1.5\text{ A}} = 1.0\ \Omega$$

This explains why the current is less than 1.8 A:

$$I = \frac{\varepsilon}{R + 1.0\ \Omega} = \frac{9.0\text{ V}}{5.0\ \Omega + 1.0\ \Omega} = 1.5\text{ A}$$

The idea of internal resistance is consistent with observations that batteries feel warm to the touch after they have been operating for a while.

We can also explain why the voltmeter reading is less in the closed circuit than in the open circuit. When there is current in the circuit, the loop rule is

$$\varepsilon - Ir = IR$$
$$\Delta V_{\text{batt}} = IR = (1.5\text{ A})(5.0\ \Omega) = 7.5\text{ V}$$

where $\Delta V_{\text{batt}} = \varepsilon - Ir$ is the potential difference across the battery. Notice that whenever there is a current through the battery, the potential difference across it is less than its emf; how much less depends on the current through the battery and on its internal resistance. If there is no circuit attached to the battery (no current), then the potential difference across it equals its emf. If $r \ll R$, then $\Delta V_{\text{batt}} \approx \varepsilon$.

**Try it yourself:** What would be the current through the 5.0-Ω resistor and the potential difference across the battery if the battery's internal resistance was 0.1 Ω instead of 1.0 Ω?

*Answer:* 1.76 A and 8.8 V.

## Kirchhoff's junction rule

We developed the loop rule for a circuit with only one loop. However, many electric circuits have several loops. In multiloop circuits, the charged particles moving along a wire can arrive at a junction where several wires meet. The charged particles comprising the current cannot vanish or be created out of nothing (electric charge is a conserved quantity). Thus, the sum of the currents into a junction must equal the sum of the currents out of a junction.

**TIP** If $r \ll R$, then $\Delta V_{\text{batt}} \approx \varepsilon$.

**Figure 16.25** Kirchhoff's junction rule for electric currents.

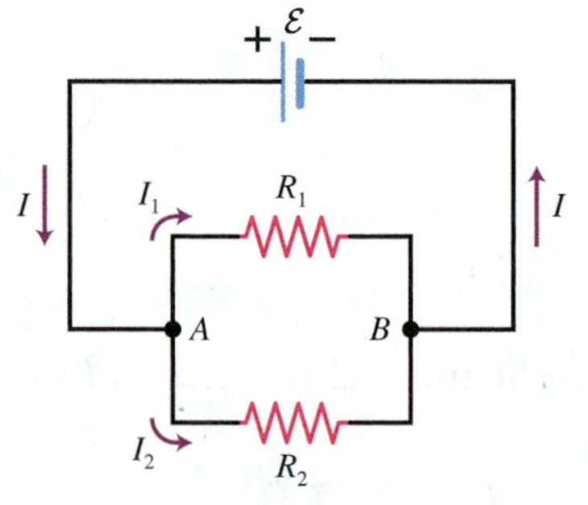


### Sign conventions

When using the loop rule for a closed circuit, follow these sign conventions:

- The potential difference across a resistor is $-IR$ when the loop traverses the resistor in the direction of the current through it and $+IR$ when the loop traverses the resistor in the direction opposite the current through it.
- The potential difference across a battery (assuming zero internal resistance) is $+\varepsilon$ when the loop traverses the battery from its negative terminal to its positive terminal and $-\varepsilon$ when the loop traverses the battery from its positive terminal to its negative terminal. The potential difference instead is $\pm(\varepsilon - Ir)$ if the battery's internal resistance cannot be neglected.
- Assuming the resistance of a connecting wire is zero, all points along that wire are at the same potential.

**Figure 16.26** A short circuit across a battery can cause a very large current through it.

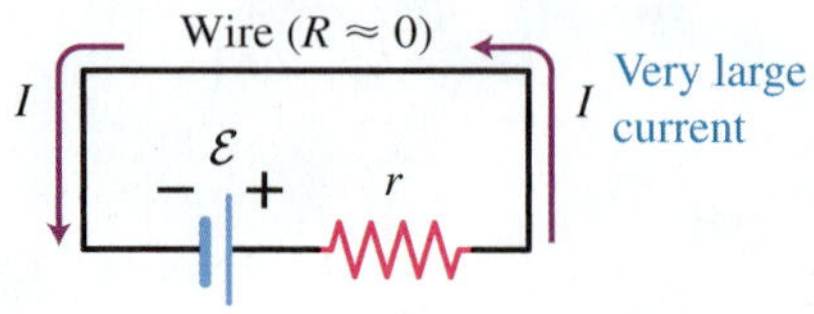


Consider the multiloop circuit shown in **Figure 16.25**. The battery at the top causes a counterclockwise current $I$ in the circuit. The current entering junction A equals the sum of the currents leaving A. Expressed mathematically:

$$I = I_1 + I_2$$

When the currents $I_1$ (through $R_1$) and $I_2$ (through $R_2$) meet at junction B, the two currents recombine to form a single larger current $I$ that leaves junction B:

$$I_1 + I_2 = I$$

For the circuit in Figure 16.25, the current $I$ moving upward through the wire on the right side of the circuit equals the downward current $I$ through the wire on the left side of the circuit; the current leaving the positive terminal of the battery equals the current entering the negative terminal. The current is not used up in a circuit. We can summarize this splitting and joining of currents at junctions as **Kirchhoff's junction rule.**

**Kirchhoff's junction rule** **The total rate at which electric charge enters a junction equals the total rate at which electric charge leaves the junction:**

**Sum of currents into junction = Sum of currents out**

**In symbols:**

$$\sum_{\text{In}} I = \sum_{\text{Out}} I \qquad (16.7)$$

## Short circuit

Imagine that you accidentally connect the terminals of a battery with a connecting wire of approximately zero resistance $R = 0$ (**Figure 16.26**). The wire and the battery become very hot. Notice that when the external resistance is $R = 0$, the current through the battery is

$$I = \frac{\varepsilon}{R + r} = \frac{\varepsilon}{0 + r} = \frac{\varepsilon}{r}$$

If the internal resistance of the battery is small, the current becomes large; it can burn the connecting wire and destroy the battery. This situation is called a **short circuit.** Every time you build an electric circuit, before you turn it on check that you did not connect a single wire directly to the two battery terminals without any other resistive devices in series with it.

**Review Question 16.7** Where is the electric potential higher: where the current "enters" a resistor or where it "leaves" it? Where is the electric potential higher: where the current "enters" a conducting wire or where it "leaves" it?

# 16.8 Series and parallel resistors

In the previous sections we learned about resistors in series and parallel. Here we learn how to describe such resistor combinations mathematically.

## Resistors in series

Suppose that you arrange three resistors ($R_1$, $R_2$, and $R_3$) in series with a battery of emf $\varepsilon$. Can we replace these three resistors with a single resistor without changing the current through the battery (**Figure 16.27**)? Use Kirchhoff's loop rule and imagine traveling counterclockwise around the circuit starting at the negative terminal of the battery:

$$+\varepsilon - IR_1 - IR_2 - IR_3 = 0$$

$$\Rightarrow I = \frac{\varepsilon}{R_1 + R_2 + R_3}$$

The current through each element in this circuit is the same since there are no junctions. It appears that a single resistor $R = R_1 + R_2 + R_3$ would have the same effect on the current in the circuit as the three separate resistors $R_1$, $R_2$, and $R_3$.

**Figure 16.27** A single equivalent resistor replaces the three series resistors.

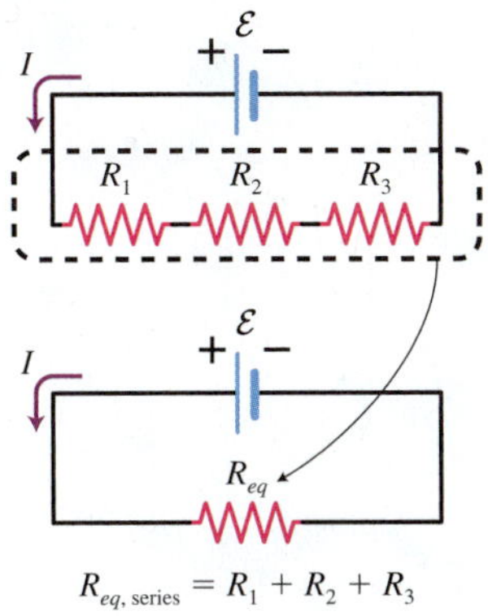


**Series resistance** When resistors are connected in series,

- **The current through each resistor is the same: $I_1 = I_2 = I_3 = \ldots$**
- **The potential difference across each resistor is $\Delta V_1 = IR_1$, $\Delta V_2 = IR_2, \ldots$**
- **The potential difference across the entire series arrangement of resistors is $\Delta V = \Delta V_1 + \Delta V_2 + \Delta V_3 + \ldots$**
- **The equivalent resistance of the resistors arranged in series is the sum of the resistances of the individual resistors:**

$$R_{eq,\,\text{series}} = R_1 + R_2 + R_3 + \ldots \quad (16.8)$$

## Resistors in parallel

When resistors are arranged in parallel, one side of each resistor is connected to one common junction and the other side of each resistor is connected to another common junction (**Figure 16.28**). We know from Kirchhoff's junction rule that the current $I$ into the left junction equals the sum of the currents through each resistor:

$$I = I_1 + I_2 + I_3$$

The potential difference between the junctions A and B at the sides of this group of parallel resistors is the same regardless of the path taken between those two points:

$$\Delta V_1 = \Delta V_2 = \Delta V_3 = \Delta V$$

We can use Ohm's law to determine the current through each resistor:

$$I_1 = \frac{\Delta V}{R_1},\; I_2 = \frac{\Delta V}{R_2},\; I_3 = \frac{\Delta V}{R_3}$$

Inserting the above into Kirchhoff's junction rule, we get

$$I = I_1 + I_2 + I_3 = \frac{\Delta V}{R_1} + \frac{\Delta V}{R_2} + \frac{\Delta V}{R_3}$$

Now, suppose that we replace these three resistors by a single equivalent parallel resistor $R_{eq,\,\text{parallel}}$ that causes the same current $I$ in the circuit. The current will be

$$I = \frac{\Delta V}{R_{eq,\,\text{parallel}}}$$

Combining these two results, we get

$$\frac{\Delta V}{R_{eq,\,\text{parallel}}} = \frac{\Delta V}{R_1} + \frac{\Delta V}{R_2} + \frac{\Delta V}{R_3}$$

$$\Rightarrow \frac{1}{R_{eq,\,\text{parallel}}} = \frac{1}{R_1} + \frac{1}{R_2} + \frac{1}{R_3}$$

$$\Rightarrow R_{eq,\,\text{parallel}} = \left(\frac{1}{R_1} + \frac{1}{R_2} + \frac{1}{R_3}\right)^{-1}$$

**Figure 16.28** A single equivalent resistor replaces the three parallel resistors.

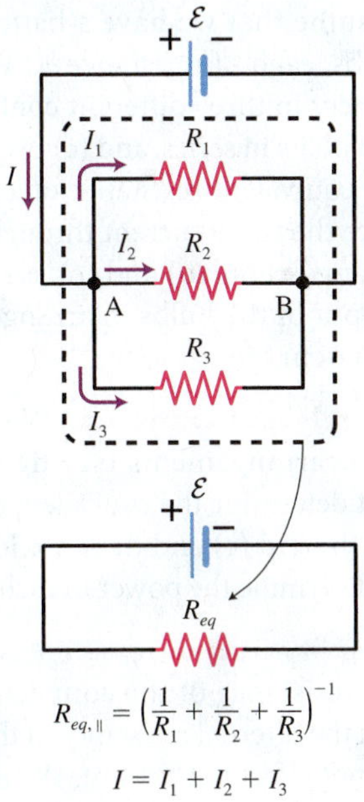

**Parallel resistance** When resistors are connected in parallel,

- The potential difference across each individual resistor is the same:

$$\Delta V_1 = \Delta V_2 = \Delta V_3 = \Delta V$$

- The sum of the currents through them equals the total current through the parallel arrangement:

$$I = I_1 + I_2 + I_3 = \frac{\Delta V}{R_1} + \frac{\Delta V}{R_2} + \frac{\Delta V}{R_3}$$

- The equivalent resistance of resistors in parallel is

$$R_{eq,\,\text{parallel}} = \left(\frac{1}{R_1} + \frac{1}{R_2} + \frac{1}{R_3} + \cdots\right)^{-1} \tag{16.9}$$

Note that when you connect resistors in series, the equivalent resistance is larger than the largest resistance; when you connect resistors in parallel, the equivalent resistance is smaller than the smallest resistance.

**TIP** Do not forget the $(-1)$ exponent when you calculate the equivalent resistance of resistors in parallel.

## Quantitative reasoning about electric circuits

In Section 16.5 we learned to reason qualitatively about currents in electric circuits. Here we analyze similar circuits quantitatively.

### EXAMPLE 16.7 Brightness of bulbs connected in parallel and in series

Assume that we have a battery and two identical lightbulbs, each of resistance $R$. We connect the bulbs to the battery in three different configurations: (a) one bulb, (b) two bulbs in series, and (c) two bulbs in parallel. Determine the equivalent resistance of a single resistor that will produce the same current through the battery as produced in arrangements (b) and (c). Next, compare the total power output of the bulbs in arrangements (b) and (c) with the output in (a).

**Sketch and translate** We draw the three arrangements (see figures). We first determine the equivalent resistance for (b) and (c) and then use Joule's law to determine the power in each case.

(a)

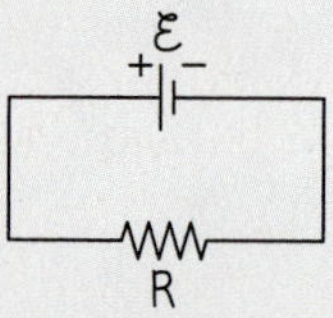


(b)

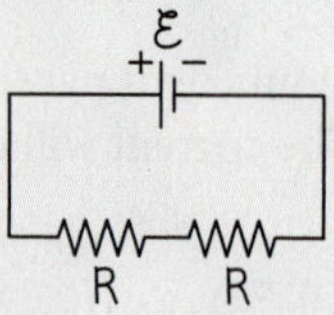


(c)

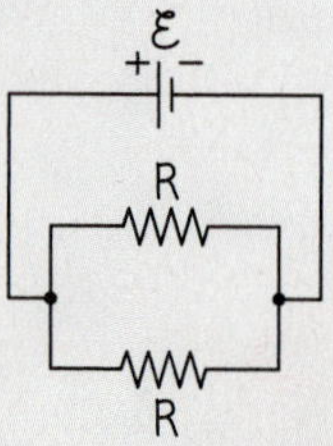


**Simplify and diagram** Assume that the resistance of the connecting wires and the internal resistance of the battery do not affect the current. We model the identical bulbs as identical ohmic resistors, thus ignoring that their resistance depends on the current through them.

**Represent mathematically** The two identical bulbs in series have an equivalent series resistance:

$$R_{eq,\,\text{series}} = R + R = 2R$$

The two identical bulbs in parallel have an equivalent parallel resistance:

$$R_{eq,\,\text{parallel}} = \left(\frac{1}{R} + \frac{1}{R}\right)^{-1} = \left(\frac{2}{R}\right)^{-1} = \frac{R}{2}$$

The current in each simplified circuit (with the multiple resistors replaced by a single equivalent resistor) equals the battery emf divided by the circuit equivalent resistance:

$$I = \frac{\varepsilon}{R_{eq}}$$

The power output of the equivalent resistor (the rate at which electric potential energy is transformed into thermal energy) is

$$P = I^2 R_{eq}$$

**Solve and evaluate** (a) Circuit (a) has resistance $R$. The current in this circuit is $I_{\text{One bulb}} = \varepsilon/R$. The power output of this circuit is

$$P_{\text{One bulb}} = I^2_{\text{One bulb}} R = \left(\frac{\varepsilon}{R}\right)^2 R = \frac{\varepsilon^2}{R}$$

(b) The equivalent resistance of circuit (b) is $R_{\text{Series}} = 2R$. The current in this circuit is

$$I_{\text{Series}} = \frac{\varepsilon}{R_{eq,\,\text{series}}} = \frac{\varepsilon}{2R} = \frac{1}{2} I_{\text{One bulb}}$$

The power output of this circuit is

$$P_{\text{Series}} = I^2_{\text{Series}} R_{eq,\,\text{series}} = \left(\frac{\varepsilon}{R_{eq,\,\text{series}}}\right)^2 R_{eq,\,\text{series}} = \left(\frac{\varepsilon}{2R}\right)^2 2R$$

$$= \frac{1}{2}\frac{\varepsilon^2}{R} = \frac{1}{2} P_{\text{One bulb}}$$

(c) The equivalent resistance of circuit (c) is $R_{eq,\text{ parallel}} = R/2$. The current in this circuit is

$$I_{\text{Parallel}} = \frac{\varepsilon}{R_{eq,\text{ parallel}}} = \frac{\varepsilon}{R/2} = 2I_{\text{One bulb}}$$

The power output of this circuit is

$$P_{\text{Parallel}} = I^2_{\text{Parallel}} R_{eq,\text{ parallel}} = \left(\frac{\varepsilon}{R_{eq,\text{ parallel}}}\right)^2 R_{eq,\text{ parallel}}$$
$$= \left(\frac{\varepsilon}{R/2}\right)^2 R/2 = 2\frac{\varepsilon^2}{R} = 2P_{\text{One bulb}}$$

Notice that the total power output of the circuit doubles when you go from one bulb to two parallel bulbs. This means you will be getting twice as much light. On the other hand, when you go from one bulb to two series bulbs, the total power output halves. Each of the two series bulbs is one-quarter as bright as the single bulb.

**Try it yourself:** Two equally bright identical bulbs in series are shown in the first figure. What happens to the brightness of the bulbs and the power of the circuit if you place a wire (called a shorting wire) in parallel with bulb B (see the second figure)? Explain.

*Answer:* Bulb A becomes much brighter and bulb B goes out. The shorting wire has zero resistance; thus all current is through the shorting wire, not through B. The resistance of the new circuit is half the original resistance. Thus, the current in the circuit doubles and the brightness of the remaining bulb quadruples.

**(a)**

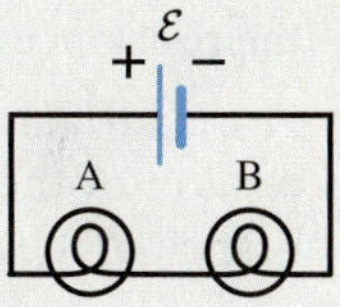


**(b)**

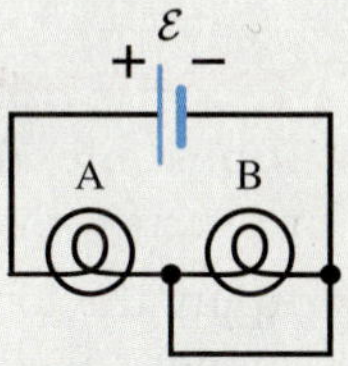


**Review Question 16.8** Rank the four identical bulbs shown in **Figure 16.29** from brightest to dimmest. Explain your reasoning.

**Figure 16.29** Rank the brightness of the bulbs.

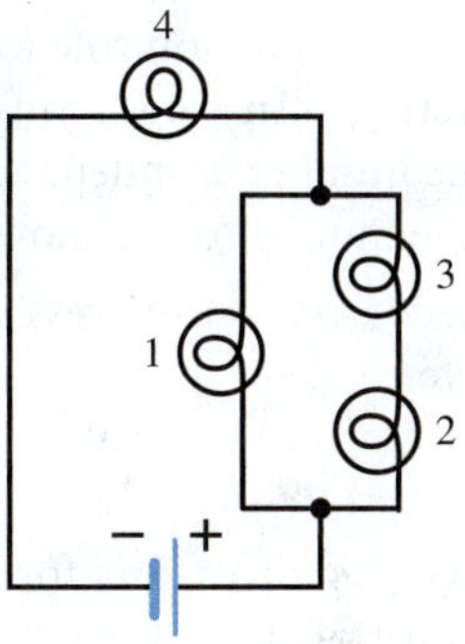


## 16.9 Skills for solving circuit problems

The left side of the table below describes a general strategy for quantitatively solving circuit problems. The right side illustrates its application to the following problem.

**PROBLEM-SOLVING STRATEGY Applying Kirchhoff's rules**

**EXAMPLE 16.8 Solving circuit problems**

Determine the currents in each branch of the circuit shown in the figure. The 9.0-V battery has a 2.0-Ω internal resistance and the 3.0-V battery has negligible internal resistance.

**Sketch and translate**

- Draw the electric circuit described in the problem statement and label all the known quantities.
- Decide which resistors are in series with each other and which are in parallel.

- We first draw the circuit with known information (see the figure).
- We cannot simplify the circuit using the series or parallel equivalent resistance equations because there are batteries in two of the branches and the parallel equivalent resistance rule applies for branches with resistors only.
- Since there are three unknown currents, $I_1$, $I_2$, and $I_3$, we need three independent equations to solve for the currents. We use the loop rule twice and the junction rule once.

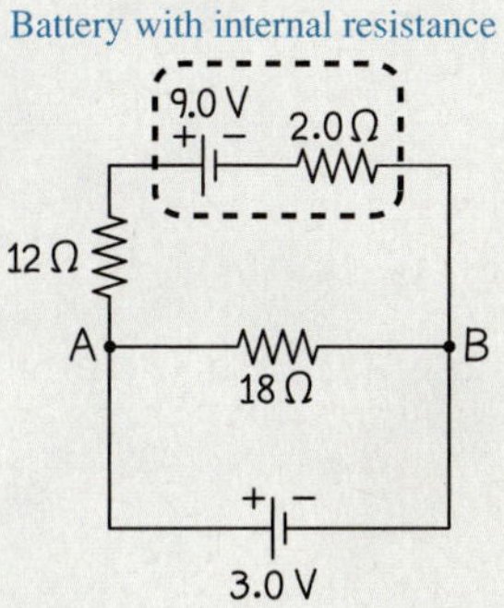


*(continued)*

### Simplify and diagram

- Decide whether you can neglect the internal resistance of the battery and/or the resistance of the connecting wires.
- Draw an arrow representing the direction of the electric current in each branch of the circuit.

- The text of the problem does not provide information about the resistances of the wires, so we will assume their resistance is approximately zero.
- Add an arrow and a label for the current in each branch of the circuit. Each arrow points in the direction that we think is the direction of the current in that branch (see the figure at the right).

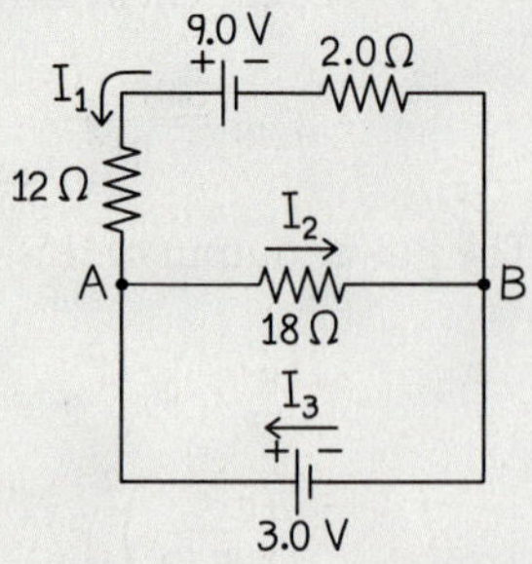


### Represent mathematically

- If possible, replace combinations of resistors with equivalent resistors.
- Apply the loop rule for the potential changes as you move around one or more different loops of the circuit. Each additional loop you choose must include at least one branch of the circuit that you have not yet included.
- Once you have included branches, apply the junction rule for one or more junctions. In total, you will need the same number of independent equations as the number of unknown currents.
- If necessary, use expressions for electric power.

Apply the loop rule for two different loops, First, start at B and move clockwise around the bottom loop:

$$\Delta V_{3\,\text{V}} + \Delta V_{18\,\Omega} = 0$$
$$+3.0\text{ V} - I_2(18\ \Omega) = 0$$

Next, start at B and move counterclockwise around the outside loop:

$$\Delta V_{2\,\Omega} + \Delta V_{9\,\text{V}} + \Delta V_{12\,\Omega} + \Delta V_{3\,\text{V}} = 0$$
$$-I_1(2\ \Omega) + 9.0\text{ V} - I_1(12\ \Omega) - 3.0\text{ V} = 0$$

Notice that we cross the 3.0-V battery from the positive to the negative side, a decrease in potential.

Apply the junction rule to junction A:

$$I_1 + I_3 = I_2$$

### Solve and evaluate

- Solve the equations for the unknown quantities. Check whether the directions of the current and the magnitude of the quantities make sense.

Using the first loop:

$$I_2 = \frac{3.0\text{ V}}{18\ \Omega} = 0.17\text{ A}$$

Using the second loop:

$$I_1 = \frac{9.0\text{ V} - 3.0\text{ V}}{2\ \Omega + 12\ \Omega} = 0.43\text{ A}$$

Using the junction A:

$$I_3 = 0.17\text{ A} - 0.43\text{ A} = -0.26\text{ A}$$

Since $I_3$ turned out negative, it means our initial choice for its direction was incorrect. The magnitude is still correct.

**Try it yourself:** Determine the potential difference from A to B in two ways: (1) through the 18-Ω resistor and (2) through the 12-Ω resistor, the 2-Ω resistor, and the 9-V battery.

*Answer:* Both paths give us −3.0 V.

In the next example we will use the idea of equivalent resistance to simplify a circuit and determine the current through the battery.

**EXAMPLE 16.9 Using equivalent resistors to solve a circuit problem**

Determine the total current $I$ through the battery in the figure below.

**Sketch and translate** To find the current $I$, first simplify the circuit by combining the resistors into a single equivalent resistor.

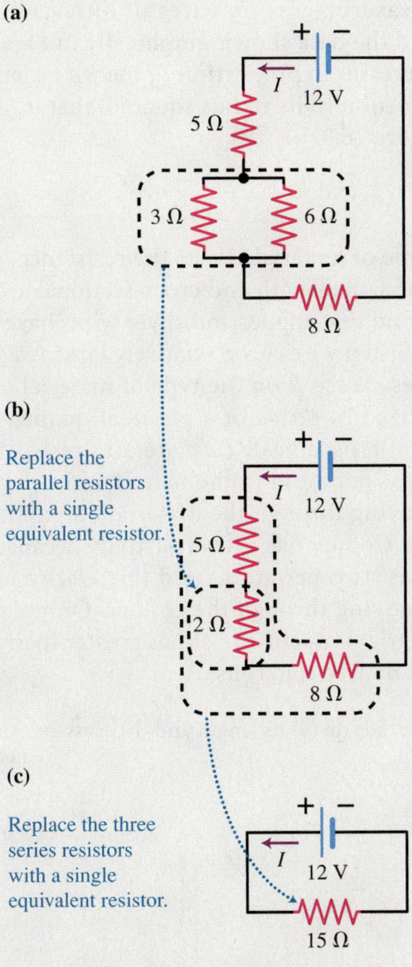


**Simplify and diagram** Assume that the battery and the connecting wires have negligible resistance. The circuit is a combination of series and parallel parts, so simplify the circuit in steps. First, replace the parallel 3.0-Ω and 6.0-Ω resistors in (a) by the 2.0 Ω equivalent resistor in (b). Then replace the three resistors in series in (b) with the one equivalent resistor in (c). The new circuit is shown in (c). Once we have a circuit with a single resistor and a single battery, we can determine the current through the battery.

**Represent mathematically** First we combined the two resistors in parallel:

$$R_{eq,\ \text{parallel}} = \left(\frac{1}{3.0\ \Omega} + \frac{1}{6.0\ \Omega}\right)^{-1}$$

We then added this combined resistor to the other resistors in series to determine the equivalent resistance of the entire circuit:

$$R_{eq} = 5.0\ \Omega + \left(\frac{1}{3.0\ \Omega} + \frac{1}{6.0\ \Omega}\right)^{-1} + 8.0\ \Omega$$

Our simplified circuit has the single equivalent resistor attached to the 12-V battery. The current $I$ is

$$I = \frac{\varepsilon}{R_{eq}}$$

**Solve and evaluate** Completing these calculations gives

$$R_{eq,\ \text{parallel}} = \left(\frac{1}{3.0\ \Omega} + \frac{1}{6.0\ \Omega}\right)^{-1} = \left(\frac{2}{6.0\ \Omega} + \frac{1}{6.0\ \Omega}\right)^{-1}$$

$$= \left(\frac{3}{6.0\ \Omega}\right)^{-1} = 2.0\ \Omega$$

$$R_{eq} = 5.0\ \Omega + 2.0\ \Omega + 8.0\ \Omega = 15.0\ \Omega$$

$$I = \frac{12.0\ \text{V}}{15.0\ \Omega} = 0.8\ \text{A}$$

This value for the current is reasonable given the resistances and battery emf.

**Try it yourself:** Find the equivalent resistance of the combination of resistors shown in the figure at right.

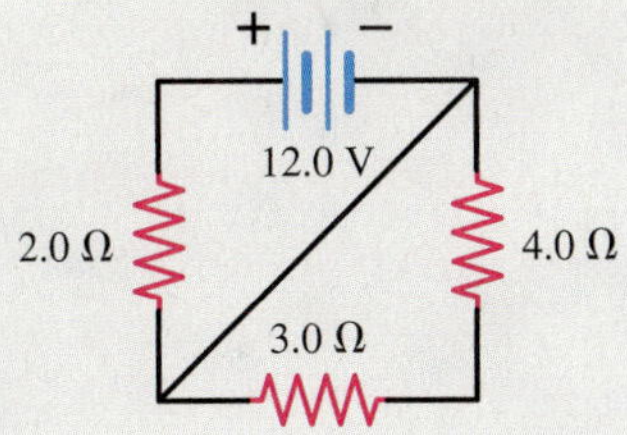


*Answer:* 2 Ω.

**Review Question 16.9** What does it mean when you get a negative value for the current in the circuit when using Kirchhoff's rules?

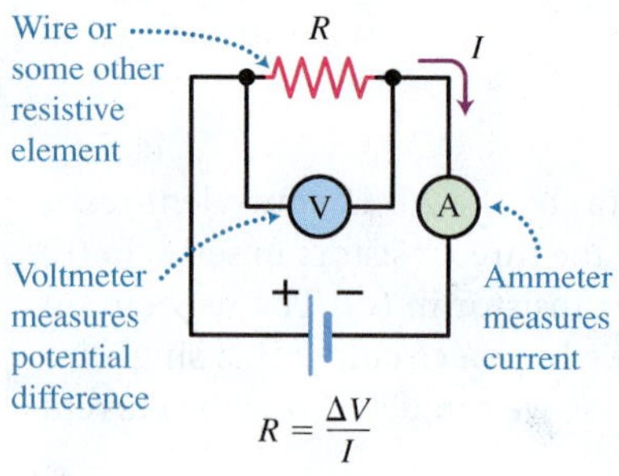


**Figure 16.30** A circuit used to measure resistance.

## 16.10 Properties of resistors

What properties of a resistor affect the value of its resistance? Consider again our water flow analogy, in which a hose connects two containers of water. A wider hose should allow more water to pass through a cross section of the hose each second, and a longer hose should reduce the amount of water passing through per unit time (because of friction exerted by the walls of the hose on the water). By analogy, we suspect that the length and the cross-sectional area of a wire will affect its electrical resistance. Also, if a resistor is made of a dielectric material, then the current through it will be nearly zero, meaning it has a huge resistance. So, it seems that resistance must also depend on the internal properties of the material of which it is made. The experimental setup in **Figure 16.30** will help us decide how the resistance of a circuit element depends on its geometric properties (its length $L$ and cross-sectional area $A = \pi r^2$). We make measurements on wires of different lengths and radii. The measurements yield the data shown graphically in **Figure 16.31**. It appears that the resistance increases in proportion to the wire's length (Figure 16.31a) and inversely in proportion to its radius squared, that is, to the cross-sectional area of the wire (Figure 16.31b).

$$R \propto L, \text{ and } R \propto \frac{1}{r^2} \propto \frac{1}{A}$$

To investigate how the type of material affects the resistance, we measure the resistance of wires with the same length and cross-sectional area but made from different materials. We find that copper and silver wires have small resistance, and constantan and tungsten wires have relatively large resistance.

The dependence of the resistance $R$ on the type of material of which the resistor is made is characterized in terms of a physical quantity called the **resistivity** $\rho$ of that material. Microscopically, a material's resistivity depends on the number of free electrons per atom in the material, the degree of "difficulty" the electrons have moving through the material due to their interactions with it, and other factors. Copper has a low resistivity because of its large concentration of free electrons (two per atom) and the relative lack of difficulty they experience while moving through the copper. On the other hand, the resistivity of glass (a dielectric) is about $10^{20}$ times greater than that of copper because it contains almost no free electrons.

**Figure 16.31** Wire resistance depends on (a) its length and (b) cross-sectional area.

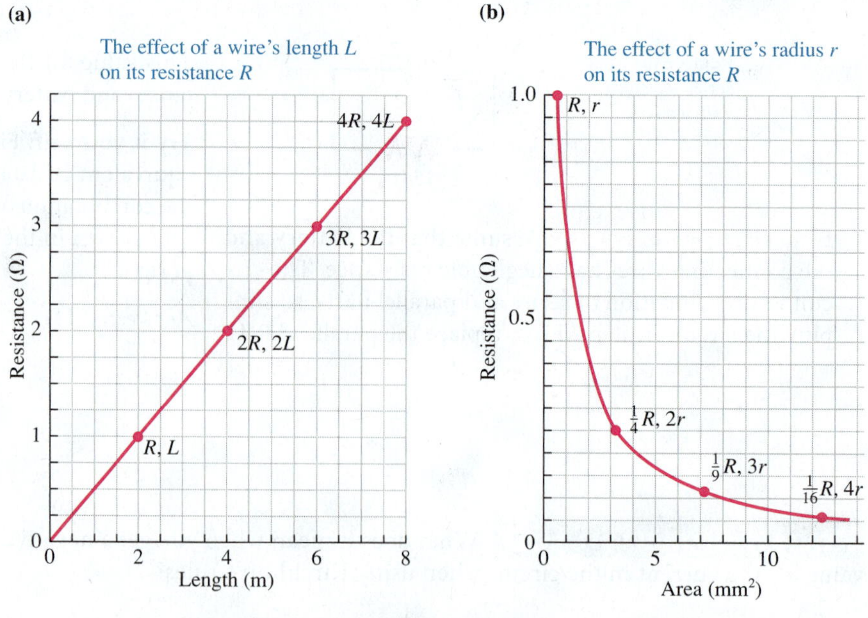

Alternatively, materials are sometimes characterized in terms of another physical quantity called **conductivity** $\sigma$. The conductivity of a material is the reciprocal of its resistivity ($\sigma = 1/\rho$).

We can assemble an equation for the resistance of a resistive circuit element by combining these three dependencies.

**Electrical resistance** The electrical resistance $R$ of a resistive circuit element depends on its geometric structure (its length $L$ and cross-sectional area $A$) and on the resistivity $\rho$ of the material of which it is made:

$$R = \rho\frac{L}{A} \quad (16.10)$$

Eq. (16.10) helps us understand why when you connect resistors in series, the total resistance of the circuit increases, and when you connect the resistors in parallel, the total resistance decreases. Connecting the resistors in series is similar to increasing the length of the resistor ($R \propto L$), and connecting the resistors in parallel is similar to increasing the cross-sectional area of the resistor $R \propto 1/A$.

**TIP** **Table 16.8** lists resistivity for materials at a particular temperature. As we know, the resistance of a lightbulb is greater when it is hot than when cold; its resistivity changes with temperature. The resistivity of some other materials, such as constantan, does not change much with temperature.

In the previous sections we disregarded the resistance of connecting wires in circuits. Since wires are long and thin, according to Eq. (16.10) they could have quite high resistance. Let's investigate the resistance of such wires.

**TIP** The Greek letter $\rho$ is used to designate resistivity as well as the density of a substance. Do not confuse the two!

**Table 16.8 The resistivity $\rho$ of different materials.**

| Material | Resistivity ($\Omega \cdot$ m, at 20 °C) |
|---|---|
| ***Metals*** | |
| Silver | $1.6 \times 10^{-8}$ |
| Copper | $1.7 \times 10^{-8}$ |
| Aluminum | $2.8 \times 10^{-8}$ |
| Tungsten | $5.2 \times 10^{-8}$ |
| Constantan | $50 \times 10^{-8}$ |
| ***Dielectrics*** | |
| Ordinary glass | $9 \times 10^{11}$ |
| Hard rubber | $1 \times 10^{16}$ |
| Shellac | $1 \times 10^{14}$ |
| Dry wood | $10^{14}$–$10^{16}$ |
| ***Human tissue*** | |
| Blood | 1.5 |
| Lung tissue | 20 |
| Fat | 25 |
| Trunk | 5 |
| Skin | 5000–50,000 |
| ***Geological materials*** | |
| Igneous rocks | $10^{2}$–$10^{7}$ |
| Sedimentary rocks | $1$–$10^{5}$ |
| Ground water | $\approx 10$ |

**QUANTITATIVE EXERCISE 16.10 Resistance of connecting wires**

The connecting wires that we use in electric circuit experiments are usually made of copper. What is the resistance of a 10-cm-long piece of copper connecting wire that has a diameter of 2.0 mm $= 2.0 \times 10^{-3}$ m?

**Represent mathematically** The resistance of a wire is $R = \rho\frac{L}{A}$. The cross-sectional area of a round wire is $A = \pi r^2$. Therefore, the resistance is $R = \rho\frac{L}{\pi r^2} = \rho\frac{L}{\pi(d/2)^2}$.

**Solve and evaluate** Insert the value for the resistivity of copper from Table 16.8 and the wire dimensions into the above:

$$R = \rho\frac{L}{\pi(d/2)^2} = (1.7 \times 10^{-8}\ \Omega \cdot \mathrm{m})\frac{0.1\ \mathrm{m}}{\pi(1.0 \times 10^{-3}\ \mathrm{m})^2}$$
$$= 5.4 \times 10^{-4}\ \Omega$$

This is a tiny amount of resistance. If the current through this wire was 1 A, the potential difference across it would be very small:

$$\Delta V = IR = (1.0\ \mathrm{A})(5.4 \times 10^{-4}\ \Omega) = 5.4 \times 10^{-4}\ \mathrm{V}$$

Typical lightbulb filaments have resistances of 100–300 Ω and will therefore have a potential difference of 100–300 V across them when the current through them is 1 A. Connecting wires are deliberately made of material with low resistivity so that the potential difference across them is negligible.

**Try it yourself:** A power line from the power plant to your home has the same cross-sectional area as in this example but is 50 km long. What is the resistance of the power line and the potential difference across it if the current through it is 1 A?

*Answer:* 270 Ω, 270 V. However, power lines are actually designed with much greater cross-sectional areas so that their resistance is kept low.

## Microscopic model of resistivity

Recall the microscopic structure of a metal: a crystal structure of metal atoms whose outer electrons are free to roam throughout the bulk of the metal. The free electrons moving chaotically inside the metal are similar to ideal gas particles. This model suggests a microscopic explanation for resistance. As the electrons drift through the wire, perhaps they collide with the metal ions. These collisions increase the internal energy of the metal ions, causing them to vibrate more vigorously. These vibrating ions become even more of an obstacle for the electrons. That explains why a hotter metal has a higher resistivity than a cooler one and thus explains why the resistance of the lightbulb filament increases with temperature. Paul Drude first suggested this model in 1900, now known as the Drude model. The Drude model does not correctly predict the details of many properties of metals (quantum mechanical models do much better). However, the Drude model is still a concrete and convenient way to understand qualitatively the reasons for electrical resistance and the mechanism of transforming electrical energy to thermal energy.

## Superconductivity

In the early 1900s Dutch physicist Kamerlingh Onnes was attempting to liquefy helium, the only gas that had not yet been liquefied at that time. In 1908 he achieved the desired result—the helium condensed from a gas to a liquid at 4.2 K. Onnes continued experimenting with helium and focused his attention on measuring the electric resistivity of materials at this low temperature. Many physicists, including Lord Kelvin himself, thought that at very low temperatures, electrons in metals should completely stop moving (as the average kinetic energy of the particles is proportional to temperature), thus making the resistivity infinite. Others, including Onnes, hypothesized that the resistivity decreased with temperature and would eventually become zero at zero K (**Figure 16.32a**).

**Figure 16.32** Superconductivity: (a) Predictions about low temperature resistivity. (b) The measured resistivity for mercury.

(a)

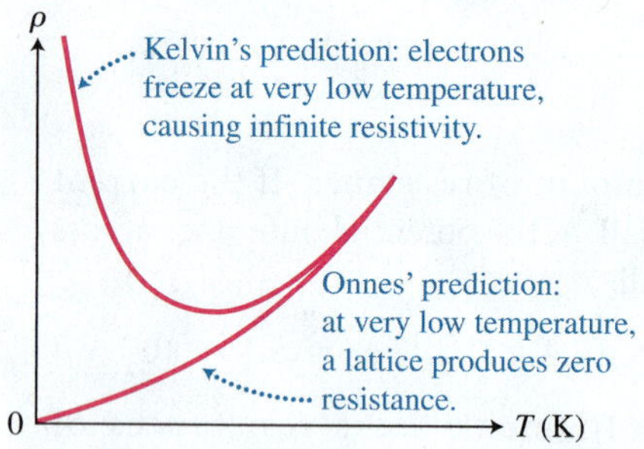
ρ
Kelvin's prediction: electrons freeze at very low temperature, causing infinite resistivity.
Onnes' prediction: at very low temperature, a lattice produces zero resistance.
0
T (K)

(b)

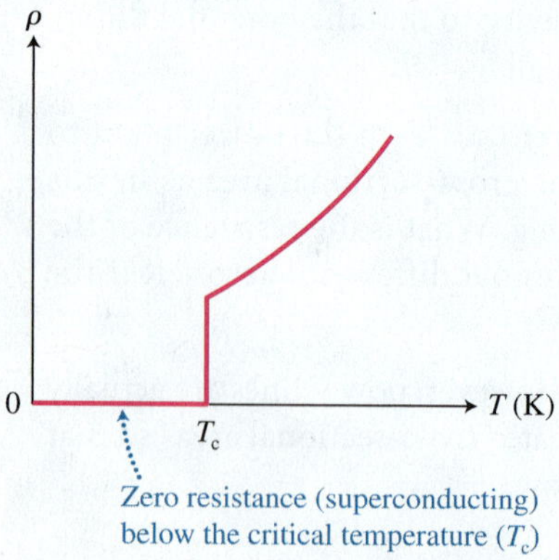
ρ
0
T (K)
$T_c$
Zero resistance (superconducting) below the critical temperature ($T_c$)

In 1911 Onnes made an electric circuit by connecting the terminals of a battery to a sample of mercury. He measured the current through it and the potential difference across it. When Onnes submerged the mercury sample in liquid helium at 4.2 K and repeated his measurement, he found that there was still current through the sample but the potential difference across it dropped to zero—this meant that the resistance of the sample was zero. Basically, he observed that the resistivity of mercury became zero not at 0 K, but at 4.2 K (Figure 16.32b). Onnes had discovered what became known as **superconductivity**. Superconductivity was a surprising result that was not predicted and wasn't successfully explained until about 50 years later, using quantum mechanics. At very low temperatures, the electrons as a whole no longer transfer energy to the metal lattice ions, allowing the electrons to move through the lattice with exactly zero resistance.

In the meantime, Onnes and other physicists continued experimenting with different metals and found that many of them, including aluminum and tin, become superconductors at very low temperatures, but that others, including gold and silver, do not.

Unfortunately, the temperatures at which superconductivity was achieved were very low, never having a critical temperature above 20 K. This made the practical applications of superconductivity almost impossible. In addition, the existing theory predicted that superconductivity could not occur above about 30 K.

In 1986 a seemingly improbably event occurred. Bednorz and Müller observed superconductivity in a material at 35 K. The material they used was not even a metal. It was a complex copper-based ceramic compound. With this as guidance, physicists manufactured similar compounds and raised the high-temperature superconductivity record to 95 K. In 1993, a complex ceramic compound was found to have a critical temperature of 138 K. In 2008 a new family of iron-based high-temperature superconductors was discovered. While there is a long way to go, the goal of discovering a room-temperature superconductor no longer seems completely impossible.

The applications of superconductors are numerous. One is transportation by magnetic levitation, the unique property of superconductors to float suspended above magnets. Superconductors can also be used to make superstrong electromagnets. Using superconductors to transmit information will bypass limitations on data rates caused by electrical resistance. Finally, superconductors can be used as a basis for supersensitive thermometers.

## Semiconductors

The electrical resistance of metal conductors increases with temperature. However, the resistivity of another type of conducting material (called a **semiconductor**) has the opposite dependence: it decreases with increasing temperature. Silicon is an example. When cooled to near 0 K, silicon has practically infinite resistivity. As its temperature increases, its resistivity decreases. Unlike in metals, the outer electrons of silicon atoms are not free—they are tightly bound to the atoms. When the temperature of a silicon crystal increases, some of the electrons do become free, leaving behind a positively charged ion in the crystal lattice. The location of this missing electron is called a **hole** (**Figure 16.33**). If charged objects outside the silicon crystal are used to produce a nonzero $\vec{E}$ field within the silicon, these newly freed conduction electrons accelerate in a direction opposite the $\vec{E}$ field, similar to the way they do in metals. In addition, the electrons of atoms next to the holes can move into the hole. The hole then moves to the atom where the electron came from. In effect, these positively charged holes behave similarly to positively charged particles, accelerating in the direction of the $\vec{E}$ field. Because the number of conduction electrons and holes increase as the temperature of the semiconductor increases, the resistivity of a semiconductor decreases with increasing temperature.

Semiconductors are used in nearly all modern technology, including consumer electronics, medical electronic products, and solar power. We will learn later in the book how solar energy is converted into electric energy.

**Figure 16.33** Conduction in a semiconductor.

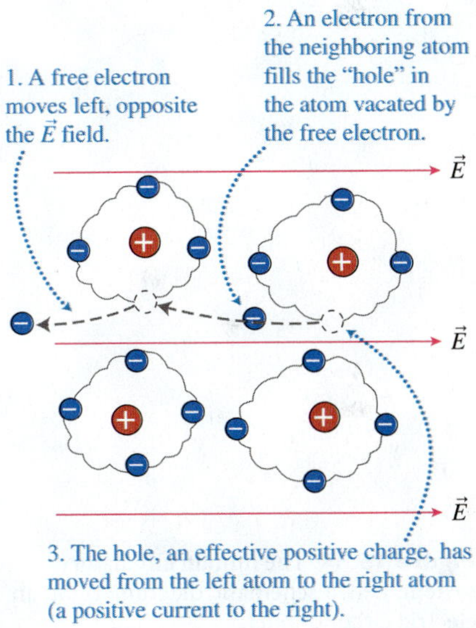

2. An electron from the neighboring atom fills the "hole" in the atom vacated by the free electron.
1. A free electron moves left, opposite the $\vec{E}$ field.
$\vec{E}$
$\vec{E}$
$\vec{E}$
3. The hole, an effective positive charge, has moved from the left atom to the right atom (a positive current to the right).

## Electrolytes

Electrolytes are substances containing free ions that make the substance electrically conductive. Several electrolytes are critical in human physiology. For example, free sodium $Na^+$ and chlorine $Cl^-$ ions in the body's tissues move in opposite directions forming currents ($I^+$ and $I^-$) in response to electric fields produced by electric dipoles in the heart. Both types of ions contribute to the current ($I_{total} = I^+ + I^-$). In nerve cells, $Na^+$, $Cl^-$, and other ion's move across nerve membranes, transmitting electrical signals from the brain to other parts of the body—to activate muscles, for example.

**Review Question 16.10** Why does the resistance of a lightbulb increase as the current through it increases?

# 16.11 Human circulatory system and circuit breakers: Putting it all together

In this section we consider two applications of electric circuits: (1) using an electric circuit to model the human circulatory system and (2) circuit breakers and fuses, a common feature in homes and buildings.

## Modeling the human circulatory system as an electric circuit

Many biological processes can be modeled using ideas of electric circuits. **Figure 16.34** shows a schematic representation of the circulatory system and an electric circuit diagram model of that system. The analogy between the elements of the circulatory system and the elements of the electric circuit includes the following:

- The pressure difference produced by the heart is analogous to the emf produced by the battery.
- The flowing blood is analogous to the flowing electric charge (the electric current).
- The blood vessels, which resist blood flow, are analogous to the electric resistors that resist the electric current moving through them.

In the electric model of the circulatory system, the resistor connected to the right side of the battery is analogous to the aorta, the single vessel through which all blood flows as it leaves the heart. There is normally a small pressure drop across this vessel analogous to a small negative potential difference across the aorta resistor.

Most of the resistance to blood flow occurs in the smaller arterial vessels (called arterioles) and the capillaries, the tiny vessels that deliver nutrients to body cells and carry waste products away. The venules (the small veins that connect the capillaries to the veins) and the vena cava (the large vein that leads directly into the heart) have flexible walls and low resistance and correspond to resistors with low resistance, and therefore have a small potential difference across them.

In the next example we use an electric circuit model to analyze the effect of various resistance changes on the work the heart must do to pump the blood. We will focus on the part of the human circulatory system that includes the heart, aorta, the small vessels in the kidneys, and the vena cava (see Figure 16.34). The relative resistance of each group of resistors for a normal heart and kidney circulatory system is given in the figure in Exercise 16.11. For example, the resistance of the aorta is about 10 Ω, whereas the net relative resistance of all of the parallel arterioles in the kidney is about 50 Ω. Note that each arteriole has very large resistance because of its small radius, but there are so many in parallel that the total arteriole resistance is reasonable (50 Ω). Now what does the heart (the battery) have to do to keep the blood flow constant if the resistance of any of the vessels changes?

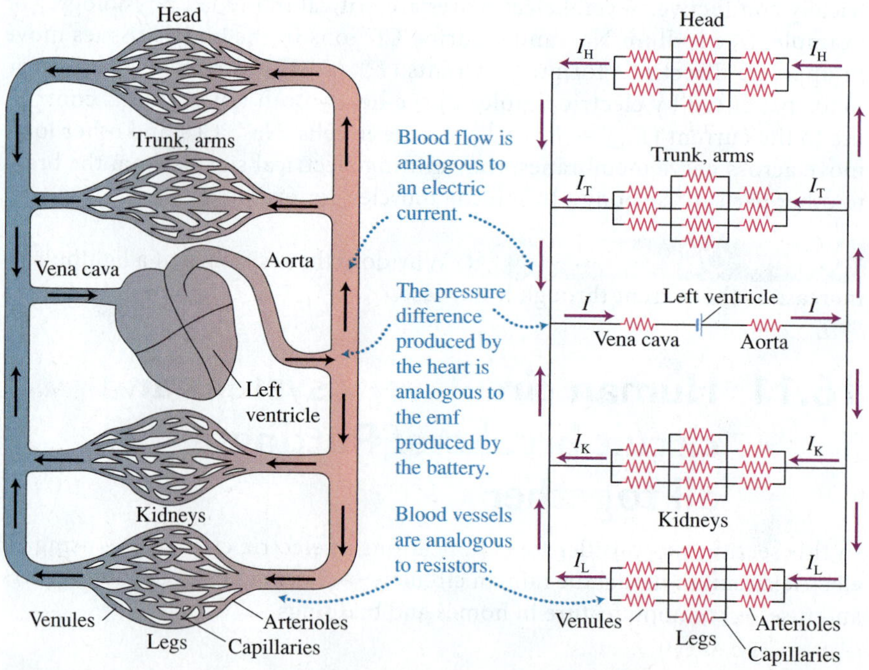


**Figure 16.34** The human circulatory system: Left, a schematic diagram; right, an electric circuit model.

**QUANTITATIVE EXERCISE 16.11 The effect of resistance in blood vessels on blood pressure**

Assuming that the heart is a 100-V battery, (a) determine the electric current (blood flow) through the circulatory system shown in Figure 16.34. (b) Now assume that the aorta has considerable plaque buildup and its diameter is reduced so that its resistance increases by a factor of 5. (Recall from Chapter 11 that the resistance is proportional to $\frac{1}{r^4}$—a fivefold increase in resistance is not rare.) What battery emf is needed to get the same current (blood flow) as in (a)? (c) Suppose the diameters of the arterioles are reduced so that their resistance increases by a factor of 2—a condition called essential hypertension. In this case, what battery emf is needed to get the same current (blood flow) as in (a)?

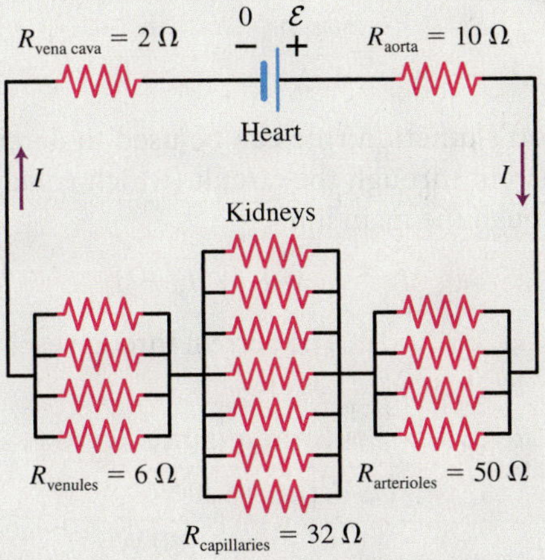


**Represent mathematically** (a) The current can be determined using Kirchhoff's loop rule:

$$\varepsilon_{\text{battery}} - IR_{\text{aorta}} - IR_{\text{arterioles}} - IR_{\text{capillaries}} - IR_{\text{venules}} - IR_{\text{vena cava}} = 0$$

where for the "normal" circulatory system $\varepsilon_{\text{battery}} = 100\text{ V}$, $R_{\text{aorta}} = 10\ \Omega$, $R_{\text{arterioles}} = 50\ \Omega$, $R_{\text{capillaries}} = 32\ \Omega$, $R_{\text{venules}} = 6\ \Omega$, and $R_{\text{vena cava}} = 2\ \Omega$.

(b) For hypertension due to increased aorta resistance, $R'_{\text{aorta}} = 5(10\ \Omega) = 50\ \Omega$. What is $\varepsilon'_{\text{battery}}$ so the current is the same as in (a)?

(c) For essential hypertension due to increased arteriole resistance, $R''_{\text{arterioles}} = 2(50\ \Omega) = 100\ \Omega$. What does $\varepsilon''_{\text{battery}}$ have to be so that the current is the same as in (a)?

**Solve and evaluate** (a) Rearrange the Kirchhoff's law equation and substitute the known information to get the current:

$$I = \frac{\varepsilon_{\text{battery}}}{R_{\text{aorta}} + R_{\text{arterioles}} + R_{\text{capillaries}} + R_{\text{venules}} + R_{\text{vena cava}}}$$

$$= \frac{100\text{ V}}{10\ \Omega + 50\ \Omega + 32\ \Omega + 6\ \Omega + 2\ \Omega} = 1.0\text{ A}$$

(b) We can now substitute the new aorta resistance and determine the battery emf needed to get a 1.0-A current with the higher resistance aorta:

$$\begin{aligned}\varepsilon'_{\text{battery}} &= I(R'_{\text{aorta}} + R_{\text{arterioles}} + R_{\text{capillaries}} + R_{\text{venules}} \\ &\quad + R_{\text{vena cava}}) \\ &= (1.0\text{ A})(50\ \Omega + 50\ \Omega + 32\ \Omega + 6\ \Omega + 2\ \Omega) \\ &= (1.0\text{ A})(140\ \Omega) = 140\text{ V}\end{aligned}$$

In order to maintain the same flow of blood, the system with a partially blocked aorta must operate at a much higher pressure (analogous to emf). Hence, this person has high blood pressure.

(c) We repeat the above, only this time with an increased arteriole resistance:

$$\begin{aligned}\varepsilon''_{\text{battery}} &= I(R_{\text{aorta}} + R''_{\text{arterioles}} + R_{\text{capillaries}} + R_{\text{venules}} \\ &\quad + R_{\text{vena cava}}) \\ &= (1.0\text{ A})(10\ \Omega + 100\ \Omega + 32\ \Omega + 6\ \Omega + 2\ \Omega) \\ &= (1.0\text{ A})(150\ \Omega) = 150\text{ V}\end{aligned}$$

The increased resistance of the arterioles is causing an increase in blood pressure.

**Try it yourself:** Determine what battery emf would be needed to maintain blood flow if the person in the previous exercise had both increased aortic and arteriole resistance.

***Answer:*** 190 V, nearly twice the normal emf.

## Circuit breakers and fuses

Although the current through appliances connected to wall sockets (called **alternating current,** or AC) differs in some ways from the current we studied in this chapter (**direct current,** or DC), many of the same physics ideas apply to both. As noted earlier in the chapter, household appliances are connected in parallel. When we connect more appliances in parallel, the total resistance of the circuit decreases. As a result, the total current in the main line (the connection to the outside power grid) of the circuit increases. Remember that the main line provides a constant effective potential difference of 120 V. When the current in the main line becomes too large, the line can become hot enough (due to its own resistance) that the insulation around the wire melts; that hot material can set adjacent objects on fire. To prevent this from occurring, the electrical systems in buildings are installed with circuit interrupters: either **fuses** or **circuit breakers**.

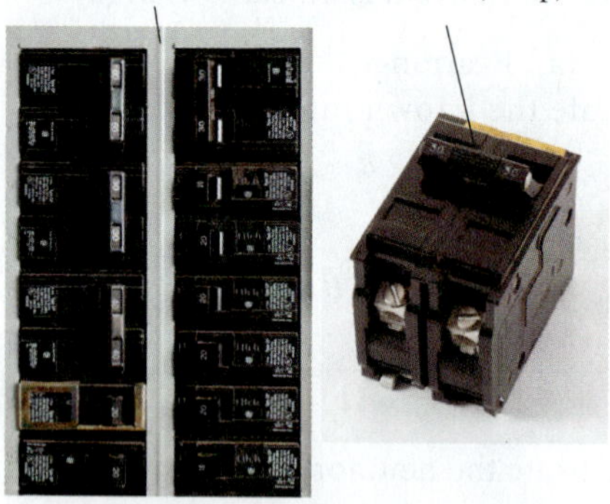


**Figure 16.35** A circuit breaker panel box and a 30-A breaker switch.

A fuse is a piece of wire made of an alloy of lead and tin. This alloy melts at a relatively low temperature. Thus, when the current through the fuse gets too large to be safe, the increase in thermal energy of the fuse melts the wire. The melted fuse is a gap in the circuit, which causes the current to quickly drop to zero. In order to reconnect the circuit, the fuse must be replaced.

Modern electrical wiring uses circuit breakers rather than fuses (**Figure 16.35**). There are different kinds of circuit breakers, but all of them are based on the same principle—when the current in a circuit increases to a critical value, the circuit breaker opens a switch, cutting off the current through the circuit. Circuit breakers are far more convenient than fuses because they are not self-destructive. A circuit breaker that has been tripped can quickly be reset instead of having to be replaced.

## QUANTITATIVE EXERCISE 16.12 Current caused by parallel appliances

A 1380-W electric heater, a 180-W computer, a 120-W lightbulb, and a 1200-W microwave are all plugged into wall sockets in the same part of a house (see figure). When the current through the main line exceeds 20 A, the circuit breaker opens the circuit. Will the circuit breaker open the circuit if all these appliances are in operation at the same time?

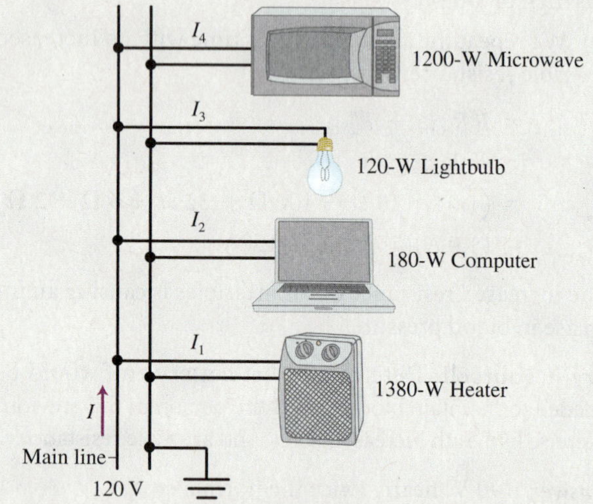


**Represent mathematically** The connecting wires and the circuit breaker have negligible resistance. Consequently, there is no electric potential difference across them, and as a result the potential difference across each appliance is 120 V. We can use one form of Joule's law to determine the current in each appliance:

$$I = \frac{P}{|\Delta V|} = \frac{P}{120\text{ V}}$$

Kirchhoff's junction rule can be used to determine the total current through the circuit (which equals the current through the main line):

$$I = I_1 + I_2 + I_3 + I_4$$

**Solve and evaluate** The current through each appliance is

$$\text{Heater: } I_1 = \frac{P_1}{|\Delta V_1|} = \frac{1380\text{ W}}{120\text{ V}} = 11.5\text{ A}$$

$$\text{Computer: } I_2 = \frac{P_2}{|\Delta V_2|} = \frac{180\text{ W}}{120\text{ V}} = 1.5\text{ A}$$

$$\text{Light: } I_3 = \frac{P_3}{|\Delta V_3|} = \frac{120\text{ W}}{120\text{ V}} = 1.0\text{ A}$$

$$\text{Microwave: } I_4 = \frac{P_4}{|\Delta V_4|} = \frac{1200\text{ W}}{120\text{ V}} = 10.0\text{ A}$$

The total current through the main line $I$ is

$$I = I_1 + I_2 + I_3 + I_4$$
$$= 11.5\text{ A} + 1.5\text{ A} + 1.0\text{ A} + 10.0\text{ A} = 24.0\text{ A}$$

Since the current $I$ exceeds 20 A, the circuit breaker will open and cause the current to stop.

**Try it yourself:** Determine the resistance of each of the above appliances.

*Answer:* Heater, 10 Ω; computer, 80 Ω; light, 120 Ω; microwave, 12 Ω.

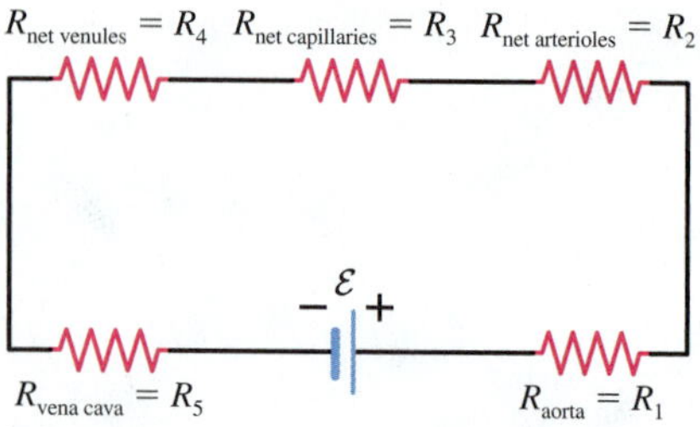


**Figure 16.36** A simplified circuit model of the circulator system.

**Review Question 16.11** If the net resistance of all the arterioles $R_2$ in the simplified circulatory circuit in **Figure 16.36** were to double, conceptually what would need to happen to keep the blood flow the same?

# Summary

| Words | Pictorial and physical representations | Mathematical representation |
|---|---|---|
| **Electric current** Electric current $I$ equals the electric charge $Q$ that passes through a cross section of a circuit element divided by the time interval $\Delta t$ for that amount of charge to pass. (Section 16.1) | 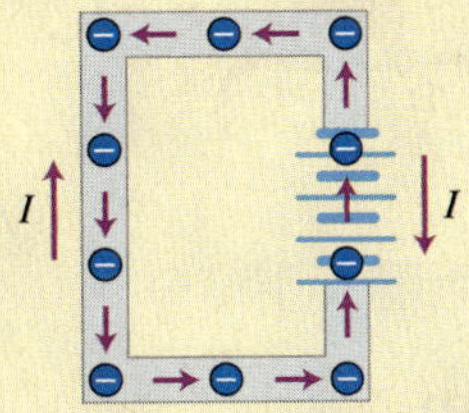 | $I = \frac{\lvert Q \rvert}{\Delta t}$ Eq. (16.1)<br>Unit: 1 A (ampere) = 1 C/s |
| **Emf** The work done per unit charge by a power source (such as a battery) to move charge from one terminal to other. A battery that is not connected to anything has a potential difference $\varepsilon$ across its terminals. When there is current in the circuit, the potential difference across the battery may be less than the emf due to internal resistance of the battery. (Sections 16.2 and 16.7) | 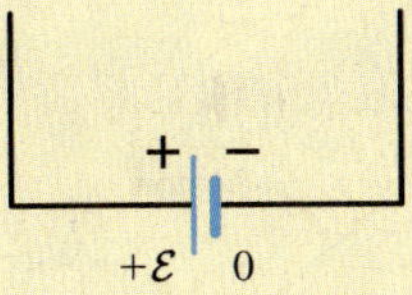 | $\varepsilon = \frac{W}{q}$ Eq. (16.2)<br>Unit: 1 V (volt) = 1 J/C |
| **Electrical resistance** The electrical resistance $R$ of a wire depends on its geometrical structure: its length $L$, cross-sectional area $A$, and the resistivity of the material $\rho$. (Section 16.10) | 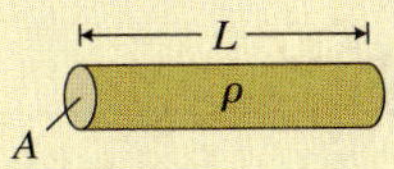 | $R = \rho \frac{L}{A}$ Eq. (16.10) |
| **Ohm's law** The current $I$ through a circuit element equals the potential difference $\Delta V$ across it divided by its resistance $R$. For ohmic devices, the resistance is independent of the current through the resistor. (Section 16.4)<br>**Joule's law** determines the rate of conversion of electric potential energy into other forms of energy. (Section 16.6) | 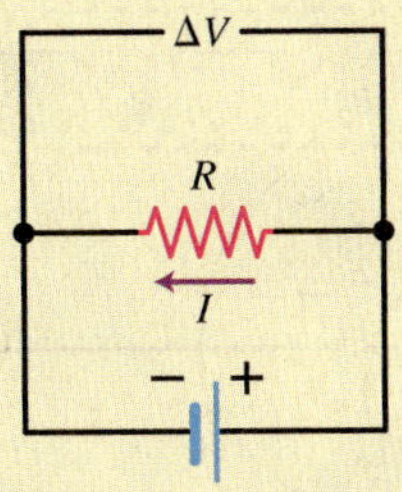 | $I = \frac{\Delta V}{R}$ Eq. (16.3)<br>$P = \left\lvert \frac{\Delta U}{\Delta t} \right\rvert = I \lvert \Delta V \rvert$ Eq. (16.4)<br>$= I^2 R = (\Delta V)^2/R$ Eq. (16.5)<br>Unit: 1 W (watt) = 1 J/s |
| **Kirchhoff's junction rule** The algebraic sum of all currents into a junction in a circuit equals the algebraic sum of the currents out of the junction. (Section 16.7) | 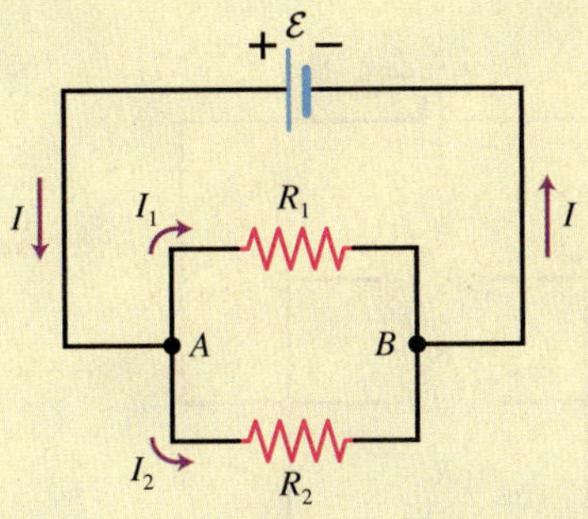<br>$I = I_1 + I_2$ | $\sum_{\text{in}} I = \sum_{\text{out}} I$ Eq. (16.7) |

*(continued)*

| Words | Pictorial and physical representations | Mathematical representation |
|---|---|---|
| **Kirchhoff's loop rule** The algebraic sum of the potential differences across circuit elements around any closed circuit loop is zero. For resistors, the potential difference is $-IR$ if moving across the resistor in the direction of the current and $+IR$ if moving across the resistor opposite the direction of current. For batteries (assuming no internal resistance) the potential difference is $+\varepsilon$ if moving from the negative terminal to the positive terminal and $-\varepsilon$ if moving from positive to negative. (Section 16.7) | 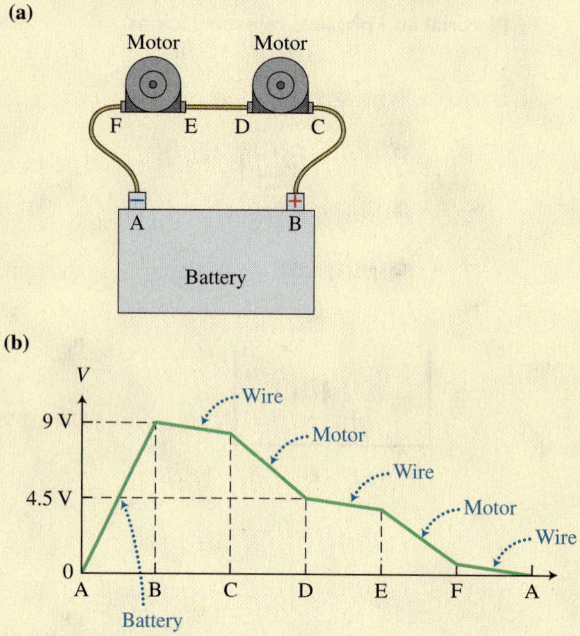 | $\sum_{\text{loop}} \Delta V = 0$ Eq. (16.6)<br>*Battery:* $\Delta V = \pm\varepsilon$, assuming no internal resistance<br>*Resistor:* $\Delta V = \pm IR$ (including the internal resistance of the battery $Ir$) |
| **Series resistance** When resistors are connected in series, the current through each resistor is the same. The potential difference across each resistor can differ. The equivalent resistance is the sum of individual resistances. (Section 16.8) | 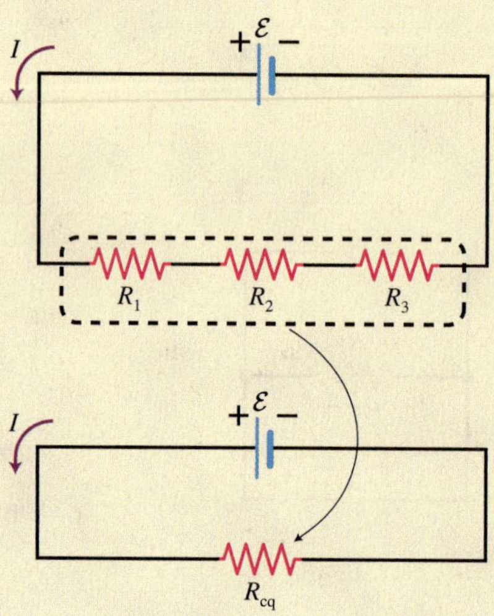 | $I_1 = I_2 = I_3 = \ldots$<br>$R_{\text{eq, series}} = R_1 + R_2 + R_3 + \ldots$<br>Eq. (16.8) |
| **Parallel resistance** The potential difference across each resistor is the same. The total current through the parallel resistors is the sum of the currents through each. The total equivalent resistance of resistors in parallel is smaller than the resistance of each individual resistor. (Section 16.8) | 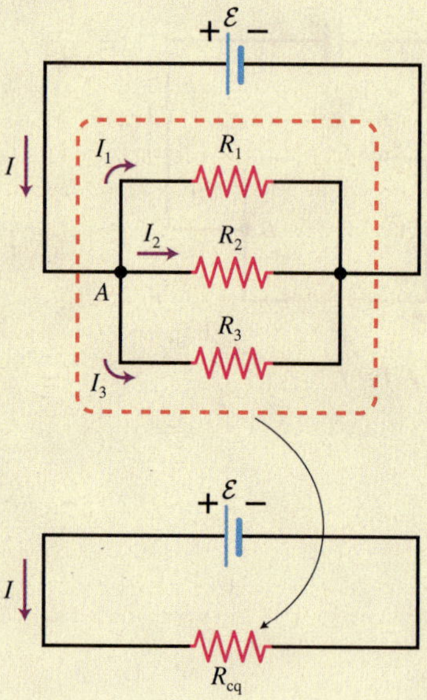 | $\Delta V_1 = \Delta V_2 = \Delta V_3 = \ldots$<br>$I = I_1 + I_2 + I_3 + \ldots$<br>$R_{\text{eq, parallel}} = \left(\frac{1}{R_1} + \frac{1}{R_2} + \frac{1}{R_3} + \ldots\right)^{-1}$<br>Eq. (16.9) |

MP® For instructor-assigned homework, go to MasteringPhysics.

# Questions

## Multiple Choice Questions

1. A single bulb is connected to the battery shown in **Figure Q16.1**. What can you say about the current at different points in the circuit?
   (a) The current is greatest at point A.
   (b) The current is greatest at point D.
   (c) The current is the same at all points in the circuit.
   (d) There is too little information to answer.

**Figure Q16.1**

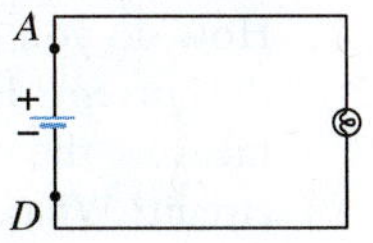


2. Two identical bulbs are connected in parallel across the battery shown in **Figure Q16.2**. There is an open switch next to bulb 2. Initially, bulb 1 is bright (strong current) and bulb 2 is dark (no current). How do the electric currents change when the switch is closed?
   (a) The current through the battery is the same but now splits between 1 and 2.
   (b) The current in 1 is still bigger than that in 2.
   (c) The current in 1 does not change and is the same as in 2.
   (d) The current through the battery doubles.
   (e) Both c and d are correct.

**Figure Q16.2**

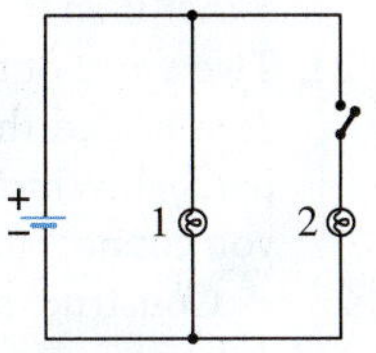


3. Compare the potential difference across bulbs 1 and 2 in Figure Q16.2 after the switch is closed. Which statement is correct?
   (a) The potential difference is greater across bulb 1.
   (b) The potential difference is greater across bulb 2.
   (c) The potential difference is the same across bulbs 1 and 2.
   (d) The potential difference across the battery doubles.
   (e) None of the above.
4. Two identical bulbs are in series as shown in **Figure Q16.4**. Which statement is correct?
   (a) The current through the battery is twice as great as when one bulb was present.
   (b) The current through the battery is half as great as when one bulb was present.
   (c) The current is greater in bulb 1 than in bulb 2.
   (d) The current is greater in bulb 2 than in bulb 1.
   (e) The currents in bulbs 1 and 2 are identical.
   (f) b and e are correct.
   (g) a and c are correct.

**Figure Q16.4**

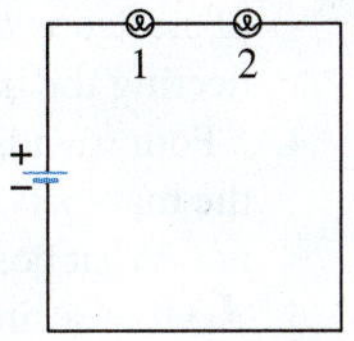


5. Which statement below about the potential difference across the battery and across the identical bulbs 1 and 2 shown in Figure Q16.4 is correct?
   (a) The potential difference is greater across bulb 1 than across bulb 2.
   (b) The potential difference is greater across bulb 2 than across bulb 1.
   (c) The potential difference is the same across bulbs 1 and 2.
   (d) The potential difference across the battery is twice what it was with only one bulb present.
   (e) c and d are correct.
6. Three circuits with identical bulbs and emf sources are shown in **Figure Q16.6**. Rank the circuits in terms of the ammeter readings, with the largest ammeter reading listed first.
   (a) A > B > C (b) A > C > B (c) B > A > C
   (d) B > C > A (e) C > A > B

**Figure Q16.6**

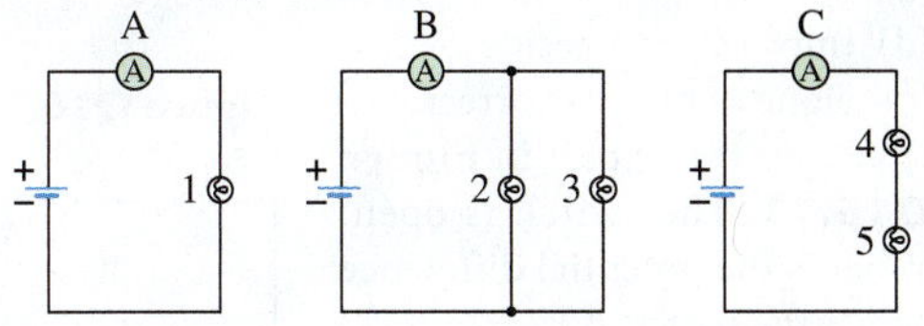


7. Rank in order the potential differences across each of the five bulbs in the three circuits in Figure Q16.6, listing the largest potential difference first. Indicate equal potential differences with equal signs. Assume that there is zero potential difference across the ammeters.
   (a) $1 > 2 = 3 > 4 = 5$
   (b) $2 > 3 > 1 > 4 > 5$
   (c) $2 = 3 > 1 > 4 = 5$
   (d) $1 = 2 = 3 > 4 = 5$
   (e) None of these
8. Rank in order the five identical bulbs in the circuit shown in **Figure Q16.8**, listing the brightest bulb first. Indicate equally bright bulbs with an equal sign.
   (a) $1 > 5 > 2 > 3 = 4$
   (b) $1 = 5 = 2 > 3 = 4$
   (c) $1 > 5 > 2 = 3 = 4$
   (d) $1 = 2 = 3 = 4 = 5$
   (e) None of these is correct.

**Figure Q16.8**

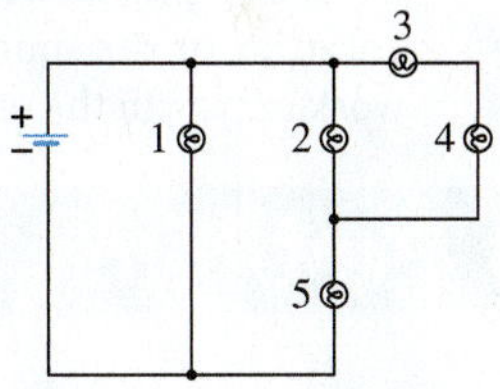


9. Four identical bulbs are shown in the circuit in **Figure Q16.9**. With the switch open, bulbs 1, 2, and 3 are equally bright and bulb 4 is dark. How does the brightness of bulbs 1, 2, and 3 change when the switch is closed?
   (a) 1 and 2 stay the same and 3 gets brighter.
   (b) 1, 2, and 3 stay the same.
   (c) 1 and 3 get brighter and 2 gets dimmer.
   (d) 1, 2, and 3 all get brighter.
   (e) None of these is correct.

**Figure Q16.9**

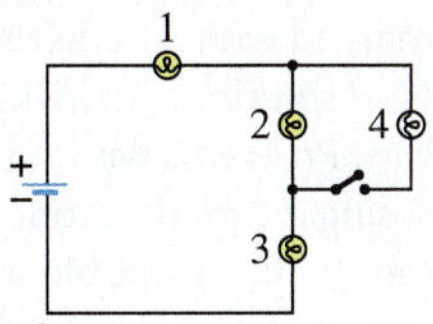


10. Four identical bulbs are shown in the circuit in **Figure Q16.10** with the switch open. How does the brightness of bulbs 1, 3, and 4 change when the switch is closed?

**Figure Q16.10**

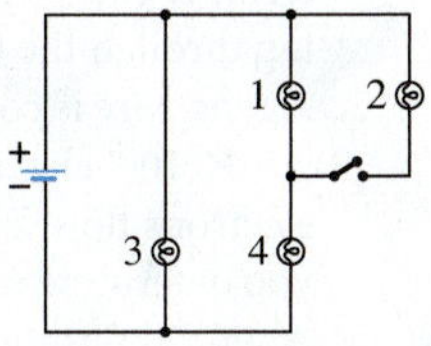

(a) 1 gets dimmer, 4 gets brighter, and 3 stays the same.
(b) 1 stays the same, 4 gets brighter, and 3 gets dimmer.
(c) 1 stays the same, 4 gets brighter, and 3 stays the same.
(d) 1, 3, and 4 all stay the same.
(e) None of these is correct.

11. Consider the circuit shown in **Figure Q16.11**. How is the current through resistor 1 related to the currents through the other two resistors?
(a) $I_1 = I_2$
(b) $I_1 = I_3$
(c) $I_1 = I_2 = I_3$
(d) $I_1 = I_2 + I_3$
(e) $I_1 = I_2 - I_3$

**Figure Q16.11**

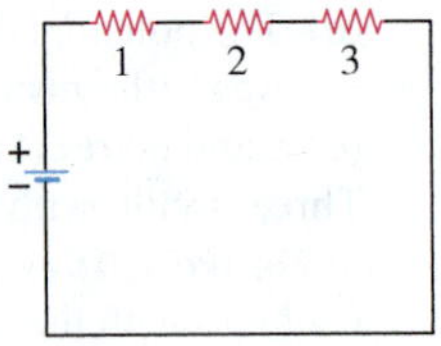


12. Consider the circuit shown in Figure Q16.11. How is the current through resistor 2 related to the current through resistor 3?
(a) $I_2 = I_3$ (b) $I_2 > I_3$ (c) $I_2 < I_3$
(d) Impossible to decide
(e) None of these is correct.

13. Consider the circuit in **Figure Q16.13**. The switch is open. What is the potential difference across the switch $\Delta V_{\text{A to B}}$?
(a) Zero (b) 9.0 V
(c) 4.5 V (d) −4.5 V

**Figure Q16.13**

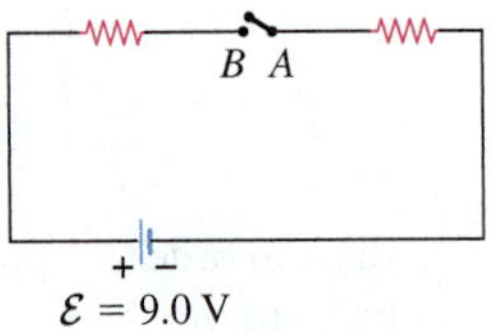


## Conceptual Questions

14. What is the role of a battery in an electric circuit? Describe a mechanical analogy.
15. Compare and contrast the physical quantities emf and potential difference.
16. **Birds on high power lines** Why can birds perch on a 100,000-V power line with no adverse effects?
17. **Preventing electric shock** When a person is repairing electrical equipment and can encounter high potential differences, he or she should keep one hand in a back pocket and work only with the other hand. Why?
18. (a) How can you measure electric current through a resistor? (b) How can you measure potential difference across a resistor? (c) How can you determine the resistance of a resistor?
19. (a) What does it mean if the current through a resistor is 3 A? (b) What does it mean if the potential difference across the resistor is 3 V? (c) What does it mean if the resistance of the resistor is 3 Ω?
20. Why do resistors become warm and sometimes hot when there is electric current through them?
21. At one time aluminum rather than copper wires were used to carry electric current through homes. Which wire must have the larger radius if they are the same length and need to have the same electrical resistance? Explain.
22. How do you connect an ammeter in a circuit to measure the current? How do you connect a voltmeter in a circuit to measure the potential difference between two points in the circuit? What implications do these connections have for the resistances of these measuring instruments?
23. Why do we connect resistive elements in a home in parallel rather than in series?
24. Two wires of the same length and cross section, one made of copper and the other made of aluminum, are connected in parallel. Which one will be hotter after 30 s? What happens if you connect them in series? Explain.
25. Construct an electric circuit that is analogous to your circulatory system. Indicate the corresponding parts of the two systems.
26. Most modern Christmas tree lights are connected in parallel. (a) Describe one advantage of lights connected in parallel. (b) Suppose you connect these lights in series. Will they be brighter or darker? Explain. What modification would make the lights equally bright? Explain.
27. How would you convince a person who missed a class on Ohm's law that the law is valid?
28. Why would a physics professor argue that Ohm's law should not be called a law but a relationship?
29. Use the laws of energy and charge conservation to explain Kirchhoff's rules.

# Problems

Below, **BIO** indicates a problem with a biological or medical focus. Problems labeled **EST** ask you to estimate the answer to a quantitative problem rather than develop a specific answer. Problems marked with a pencil require you to make a drawing or graph as part of your solution. Asterisks indicate the level of difficulty of the problem. Problems with no * are considered to be the least difficult. A single * marks moderately difficult problems. Two ** indicate more difficult problems.

### 16.1–16.3 Electric current, Batteries and emf, and Making and representing simple circuits

1. A bulb in a table lamp has a current of 0.50 A through it. Determine two physical quantities related to the electrons passing through the wires leading to the bulb.
2. A long wire is connected to the terminals of a battery. In 8.0 s, $9.6 \times 10^{20}$ electrons pass a cross section along the wire. The electrons flow from left to right. What physical quantities can you determine using this information?
3. A typical flashlight battery will produce a 0.50-A current for about 3 h before losing its charge. Determine the total number of electrons that have moved past a cross section of wire connecting the battery and lightbulb.
4. * Four friends each have a battery, a bulb, and one wire. They use the materials to try to light their bulb. Two succeed and two do not. What possible circuit arrangements could they have made?
5. Draw a circuit that has a battery, a lightbulb, and connecting wires. Draw a schematic of a water flow system that allows a continual circulation of water and compare the function of each element of the circuit to each element of the water flow system.
6. Add another battery to the circuit described in Problem 5. In how many different ways can you do this? Draw analogous changes in the water flow system.
7. Add another lightbulb to the circuit with one battery described in Problem 5. In how many different ways can you connect the second bulb? What do you need to add to the water flow system to make it similar to the circuit?
8. * A 9.0-V battery is connected to a resistor so that there is a 0.50-A current through the resistor. For how long should the battery be connected in order to do 200 J of work while separating charges?

### 16.4–16.5 Ohm's law and Qualitative analysis of circuits

9\. A graph of the electric potential versus location in a series circuit with 1.0 A of current is shown in **Figure P16.9**. Draw a circuit in which such change could occur.

**Figure P16.9**

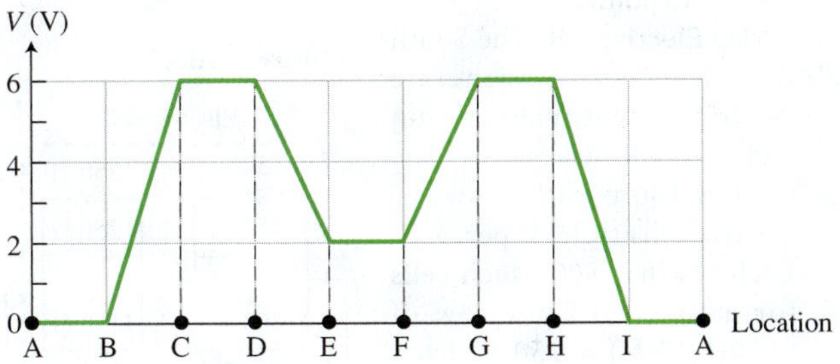


10\. Sketch a potential-versus-location graph for the circuit shown in **Figure P16.10**. Start at A and move clockwise around the circuit.

**Figure P16.10**

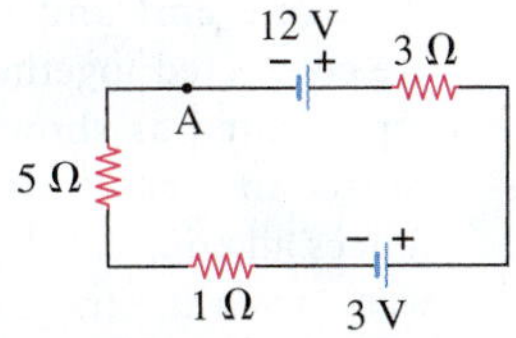


11\. BIO **Electric currents in the body** A person accidentally touches a 120-V electric line with one hand while touching a ground wire with the other hand. Determine the current through the body when the hands are dry (100,000-Ω resistance) and when wet (5000-Ω resistance). If the current exceeds about 10 mA, muscular contractions may prevent the person from releasing the wires—a dangerous situation. Is the person in danger with dry hands? With wet hands? Explain.

12\. (a) An automobile light has a 1.0-A current when it is connected to a 12-V battery. What can you determine about the light using this information? (b) What potential difference is needed to produce a current of 5.0 mA through a 2.0-Ω resistor?

13\. * If a long wire is connected to the terminals of a 12-V battery, $6.4 \times 10^{19}$ electrons pass a cross section of the wire each second. Make a list of the physical quantities whose values you can determine using this information and determine three of them.

14\. Determine the current through a 2.5-Ω flashlight filament when connected across two 1.5-V batteries in series (a potential difference of 3.0 V).

15\. * You have a circuit with a 50-Ω, a 100-Ω, and a 150-Ω resistor connected in series. (a) Rank the current through them from highest to lowest. (b) Rank the potential difference across them from highest to lowest. Explain your rankings.

16\. * You have a circuit with a 50-Ω, a 100-Ω, and a 150-Ω resistor connected in parallel to the same battery. (a) Rank the current through them from highest to lowest. (b) Rank the potential difference across them from highest to lowest.

### 16.6 Joule's law

17\. * You connect a 50-Ω resistor to a 9-V battery whose internal resistance is 1 Ω. (a) Determine the electric power dissipated by the resistor. (b) Will a lightbulb with a 10-Ω resistance dissipate more or less power if connected to the same battery?

18\. ** EST **Making tea** You use an electric teapot to make tea. It takes about 2 min to boil 0.5 L of water. (a) Estimate the power of the heater. What are your assumptions? (b) Estimate the current through the heater. State your assumptions.

19\. * If a long wire is connected to the terminals of a 12-V battery, $6.4 \times 10^{19}$ electrons pass a cross section of the wire each second. What is the rate of work being done by the battery?

20\. ** Three friends are arguing with each other. Aidan says that a battery will send the same current to any circuit you connect to it. Cathy says that the battery will produce the same potential difference at its terminals independent of the circuit connected to it. Eugenia says that the battery will always produce the same electric power independent of the circuit connected to it. Describe experiments that you can perform to convince all of them that their opinions are wrong in at least one situation.

21\. * You have a 40-W lightbulb and a 100-W bulb. What do these readings mean? Can we say that the 100-W lightbulb is always brighter? Explain and give examples.

22\. * Does a 60-W lightbulb have more or less resistance than a 100-W bulb? Explain.

### 16.7–16.8 Kirchhoff's rules and Series and parallel resistors

23\. * (a) Write two loop rule equations and one junction rule equation for the circuit in **Figure P16.23**. (b) Use these equations to determine the current in each branch of the circuit for the case in which $R_1 = 0\ \Omega$, $R_2 = 18\ \Omega$, $R_3 = 9\ \Omega$, and $\varepsilon = 60$ V.

**Figure P16.23**

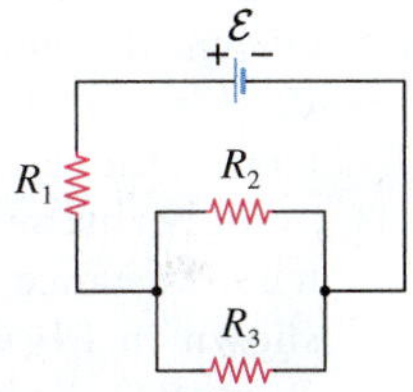


24\. * Repeat the previous problem for the case in which $R_1 = 50\ \Omega$, $R_2 = 30\ \Omega$, $R_3 = 15\ \Omega$, and $\varepsilon = 120$ V.

25\. ** The current through resistor $R_1$ in Figure P16.23 is 2.0 A. Determine the currents through resistors $R_2$ and $R_3$ and the emf of the battery. $R_1 = 4\ \Omega$, $R_2 = 10\ \Omega$, and $R_3 = 40\ \Omega$.

26\. * Use Kirchhoff's rules to prove in general that the currents $I_2$ and $I_3$ through resistors $R_2$ and $R_3$ shown in Figure P16.23 satisfy the relation $I_2/I_3 = R_3/R_2$.

27\. ** (a) Write Kirchhoff's loop rule for the circuit shown in **Figure P16.27** for the case in which $\varepsilon_1 = 20$ V, $\varepsilon_2 = 8$ V, $R_1 = 30\ \Omega$, $R_2 = 20\ \Omega$, and $R_3 = 10\ \Omega$. (b) Determine the current in the circuit. (c) Using this value of current, start at position A and move clockwise around the circuit, calculating the electric potential change across each element in the circuit (be sure to indicate the sign of each change). (d) Add these potential changes around the whole circuit. Explain your result.

**Figure P16.27**

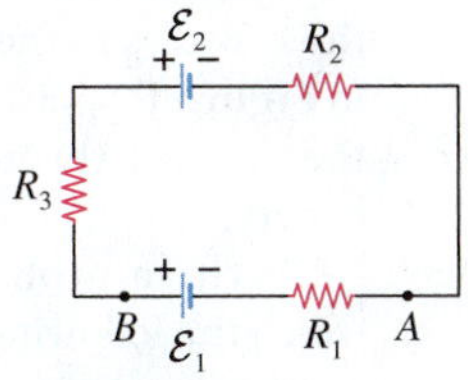


28\. ** Repeat parts (a) and (b) of the previous problem for the case in which $\varepsilon_1 = 12$ V, $\varepsilon_2 = 3$ V, $R_1 = R_2 = 1\ \Omega$, and $R_3 = 16\ \Omega$. (c) Determine the potential difference from A to B. (d) Draw a potential-versus-position graph starting at any location in the circuit and returning to the same point.

29\. ** (a) Determine the value of $\varepsilon_1$ so that there is a clockwise current of 1.0 A in the circuit shown in Figure P16.27, where $\varepsilon_2 = 12$ V, $R_1 = 2\ \Omega$, $R_2 = 1\ \Omega$, and $R_3 = 12\ \Omega$. (b) Determine the potential difference from B to A. (c) Draw a graph representing the potential at different points of the circuit.

30\. ** (a) Write the loop rule for two different loops in the circuit shown in **Figure P16.30** and the junction rule for point A. Solve the equations to find the current in each loop when $\varepsilon_1 = 3$ V, $\varepsilon_2 = 6$ V, $R_1 = 10\ \Omega$, $R_2 = 20\ \Omega$, and $R_3 = 30\ \Omega$. (b) Determine the potential difference from A to B. Check your answer by taking a different path from A to B. (c) Sketch a potential-versus-position graph for each of the three loops, assuming the potential is zero at B.

Figure P16.30

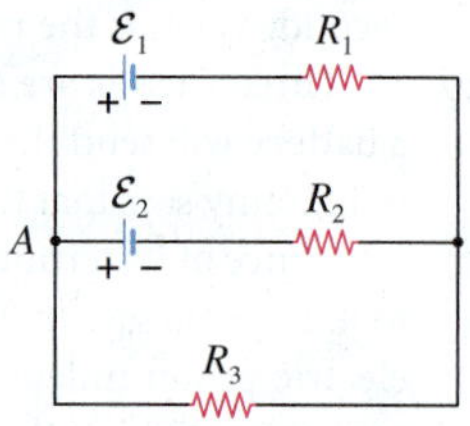


31. ** Determine the value of $R_2$, shown in Figure P16.30, so that the current through $R_3$ equals twice that through $R_2$. The values of other circuit elements are $\varepsilon_1 = 12\text{ V}, \varepsilon_2 = 15\text{ V}, R_1 = 15\ \Omega$, and $R_3 = 30\ \Omega$.

32. * Determine (a) the equivalent resistance of resistors $R_1$, $R_2$, and $R_3$ in **Figure P16.32** for $R_1 = 28\ \Omega, R_2 = 30\ \Omega$, and $R_3 = 20\ \Omega$ and (b) the current through the battery if $\varepsilon = 10$ V.

Figure P16.32

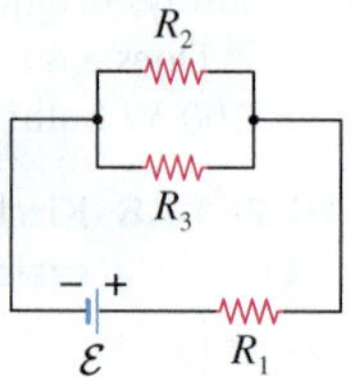


33. * (a) Determine the equivalent resistance of resistors $R_1$, $R_2$, and $R_3$ in Figure P16.32 for $R_1 = 50\ \Omega, R_2 = 30\ \Omega$, and $R_3 = 15\ \Omega$. (b) Determine the current through $R_1$ if $\varepsilon = 120$ V. (c) Use Kirchhoff's loop rule and your result from part (b) to determine the current through $R_2$ and through $R_3$.

34. * Determine the equivalent resistance of the resistors shown in **Figure P16.34** if $R_1 = 60\ \Omega, R_2 = 30\ \Omega, R_3 = 20\ \Omega, R_4 = 20\ \Omega, R_5 = 60\ \Omega, R_6 = 20\ \Omega$, and $R_7 = 10\ \Omega$.

Figure P16.34

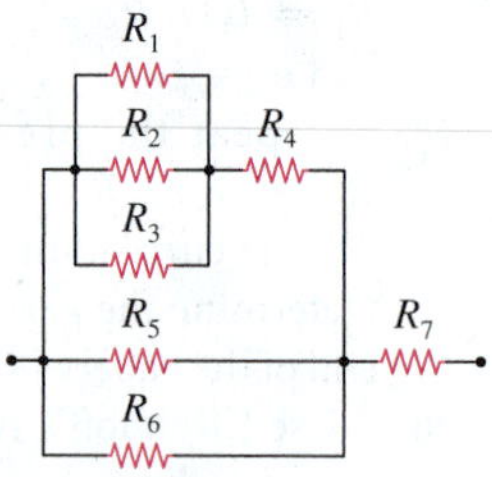


35. * Determine the equivalent resistance of the resistors in Figure P16.34 if $R_1 = R_2 = 20\ \Omega$, $R_3 = 10\ \Omega, R_4 = 25\ \Omega, R_5 = 30\ \Omega, R_6 = 10\ \Omega$, and $R_7 = 50\ \Omega$.

36. * Determine (a) the equivalent resistance of the resistors in the circuit in **Figure P16.36** and (b) the current through the battery.

Figure P16.36

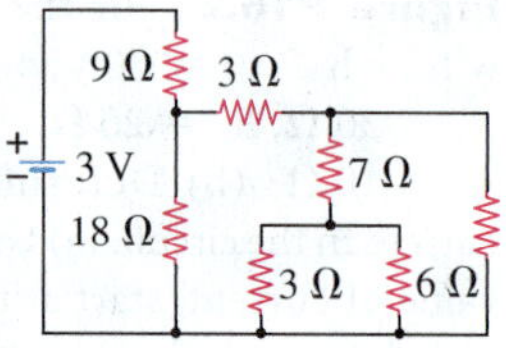


37. ** Write a problem for which the following mathematical statement can be a solution:

$$1\text{ A} = \frac{\varepsilon - (1\text{ A})(3\Omega)}{R}$$

## 16.9 Skills for solving circuit problems

38. * **Home wiring** A simplified electrical circuit for a home is shown in **Figure P16.38**. (a) Determine the currents through the circuit breaker, the lightbulb, the microwave oven, and the toaster. (b) Determine the electric power used by each appliance.

Figure P16.38

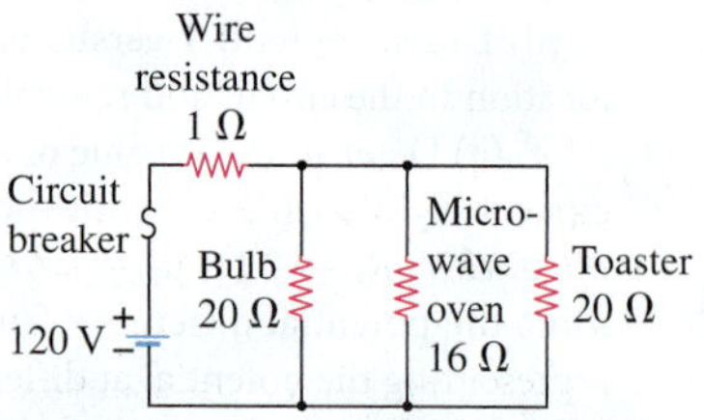


39. ** (a) Write Kirchhoff's rules for two loops and one junction in the circuit shown in **Figure P16.39**. (b) Solve the equations for the current in each branch of the circuit when $\varepsilon_1 = 10\text{ V}, \varepsilon_2 = 2\text{ V}, R_1 = 50\ \Omega$, $R_2 = 200\ \Omega$, and $R_3 = 20\ \Omega$. (c) Determine the potential difference across resistor $R_3$ from point B to point A.

Figure P16.39

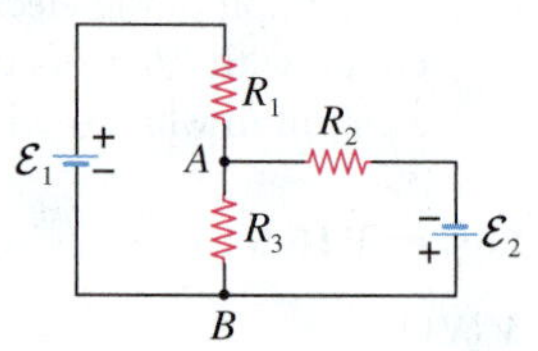


40. ** BIO **Electric eel** The South American eel can generate electric current that can stun and even kill nearby fish. The eel has 140 parallel rows of electric cells (0.15 V per cell). Each row has 5000 such cells for a total emf per row of $5000(0.15\text{ V}) = 750$ V. Each row of cells also has about 1250 Ω of internal resistance. Because each row has the same emf and the rows are connected together on each side, the eel's circuit can be represented as shown in **Figure P16.40**—a 750-V emf source in series with 140 1250-Ω parallel internal resistances all connected across an external resistance (the seawater from the front of the eel to its back) of about 800 Ω. Can the eel produce enough current to be dangerous to a person? Explain your answer. Identify all assumptions that you made.

Figure P16.40

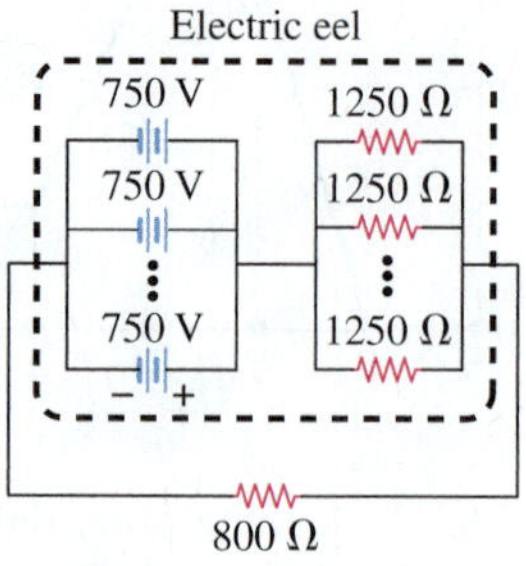


41. * **Home wiring** A 120-V electrical line in a home is connected to a 60-W lightbulb, a 180-W television set, a 300-W desktop computer, a 1050-W toaster, and a 240-W refrigerator. How much current is flowing in the line?

42. ** Determine (a) the equivalent resistance, (b) the current through the battery, and (c) the power supplied by the battery for the circuit shown in **Figure P16.42**. (d) Use Kirchhoff's loop rule to determine the currents through the 60-Ω, 40-Ω, and 30-Ω resistors.

Figure P16.42

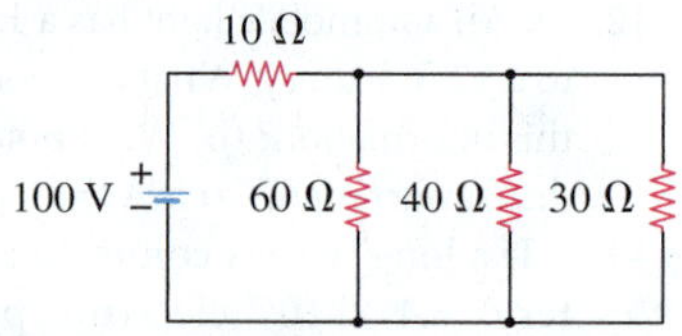


43. * **Tree lights** Nine tree lights are connected in parallel across a 120-V potential difference. The cord to the wall socket carries a current of 0.36 A. (a) Determine the resistance of one of the bulbs. (b) What would the current be if the bulbs were connected in series?

44. ** Two lightbulbs use 30 W and 60 W, respectively, when connected in parallel to a 120-V source. How much power does each bulb use when connected in series across the 120-V source, assuming that their resistances remain the same?

45. * Three identical resistors, when connected in series, transform electrical energy into thermal energy at a rate of 15 W (5 W per resistor). Determine the power consumed by the resistors when connected in parallel to the same potential difference.

46. ** **Impedance matching** A battery has an emf of 12 V and an internal resistance of 3 Ω. (a) Determine the power delivered to a resistor $R$ connected to the battery terminals for values of $R$ equal to 1, 2, 3, 4, 5, and 6 Ω. (b) Plot on a graph the calculated values of $P$ versus the different values of $R$. Connect the points by a smooth curve. Confirm that the maximum power is delivered when $R$ has the same resistance as the internal resistance of the power source (3 Ω in this example).

### 16.10 Properties of resistors

47. * A 100-m-long copper wire of radius 0.12 mm is connected across a 1.5-V battery. Make a list of the physical quantities that you can determine using this information and determine the values of three of them.

48. * **BMT subway rail resistance** The BMT subway line in New York City stretches roughly 30 km from the Bronx to Brooklyn. The electrified rail on which it runs has a cross section of about 40 $cm^2$ and is made of steel with a resistivity of $10 \times 10^{-8}\ \Omega \cdot m$. Make a list of all the physical quantities describing different properties of the rail that you can determine using this information. Then calculate the values of one quantity related to the electrical properties and one quantity not related to the electrical properties.

49. * **Thermometer** A platinum resistance thermometer consists of a coil of 0.10-mm-diameter platinum wire wrapped in a coil. Determine the length of wire needed so that the coil's resistance at 20 °C is 25 Ω. The resistivity of platinum at this temperature is $1.0 \times 10^{-7}\ \Omega \cdot m$.

50. As the potential difference in volts across a thin platinum wire increases, the current in amperes changes as follows: $(\Delta V, I) = (0, 0), (1.0, 0.112), (3.0, 0.337)$, and $(6.1, 0.675)$. Plot a graph of potential difference as a function of current and indicate whether the platinum wire satisfies Ohm's law. Explain how you made your decision.

51. * BIO **Respiration detector** A respiration detector monitors a person's breathing. One type consists of a flexible hose filled with conductive salt water (resistivity of 5.0 Ω · m). Electrodes at the ends of the tube measure the resistance of the fluid in the tube. The tube is wrapped around a person's chest. When the person inhales and exhales, the tube stretches and contracts and its resistance changes. Determine the factor by which the resistance of the fluid changes when the hose is stretched so that its length increases by a factor of 1.1. The water volume remains constant.

52. * A wire whose resistance is $R$ is stretched so that its length is tripled while its volume remains unchanged. Determine the resistance of the stretched wire.

53. ** **Ratio reasoning** Determine the ratio of the resistances of two wires that are identical except that (a) wire A is twice as long as wire B, (b) wire A has twice the radius of wire B, and (c) wire A is made of copper and wire B is made of aluminum. Be sure to show clearly how you arrive at each answer.

54. ** **Electronics detective** You need to determine the mass and length of the wire inside a particular electronic device. You cannot take it out. Devise a method to do this by using a battery, ammeter, voltmeter, and micrometer (a device that measures small distances.)

55. ** A battery produces a 2.0-A current when connected to an unknown resistor of resistance $R$. When a 10-Ω resistor is connected in series with $R$, the current drops to 1.2 A. (a) Determine the emf of the battery and the resistance $R$. (b) What assumptions did you make? Do your assumptions make each of the values you determined greater than or less than their actual values? Explain.

56. BIO **Resistance of human nerve cell** Some human nerve cells have a long, thin cylindrical cable (the axon) from their inputs to their outputs. Consider an axon of radius $5 \times 10^{-6}$ m and length 0.6 m. The resistivity of the fluid inside the axon is 0.5 Ω · m. Determine the resistance of the fluid in this axon.

### 16.11 Human circulatory system and circuit breakers: Putting it all together

57. * A fuse for one line in your home's electrical system melts if the electric current through it is greater than 30 A. Will the fuse melt if the following appliances are all connected in parallel to that line: a 13-Ω toaster, an 18-Ω dishwasher, a 24-Ω refrigerator, and a 15-Ω heater? The potential difference across the devices is 120 V.

58. * BIO **Circulatory system** (a) Use Kirchhoff's rules to determine the current in each branch of the circuit shown in **Figure P16.58** for the case in which $\varepsilon = 90$ V, $R_1 = 10\ \Omega$, $R_2 = 30\ \Omega$, and $R_3 = 60\ \Omega$. (b) Repeat the current calculations if $R_1$ increases to 30 Ω, while $R_2$ and $R_3$ remain unchanged. (c) Discuss briefly how this electrical circuit is analogous to the circulatory system of a person and the effect of a decrease in the size of a main artery on the flow of blood to other parts of the body.

**Figure P16.58**

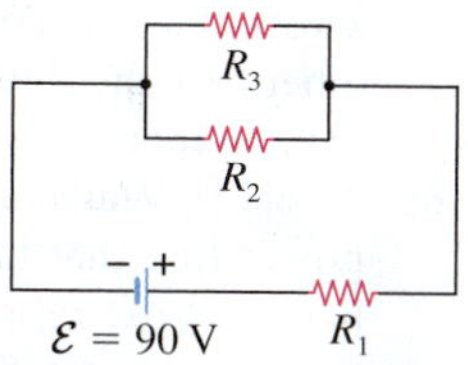


59. * BIO **Circulatory system** Suppose the electric circuit shown in Figure P16.58 models the circulatory system with 10,000 arterioles in parallel, each of resistance 100 Ω. (a) Determine the net resistance of the arterioles. (b) If the resistance of each arteriole increased by 50% because of narrowing, what now would be the net resistance of the 10,000 arterioles? (c) By how much would the potential difference across these arteriole resistors change if the flow rate (current) is to remain the same?

60. * BIO **Arteriosclerosis** Recall that the resistance of a vessel to fluid flow is inversely proportional to the radius of the vessel to the fourth power. (a) By what factor would a resistor representing the aorta shown in Figure P16.58 change if the aorta radius of a person was 0.20 times what it had been years earlier? (b) If the pressure drop across the aorta had originally been 5 mm Hg with normal blood flow, what would the pressure drop now be in order to have the same flow rate? Note that the blood pressure will have to increase to maintain the same flow rate.

## General Problems

61. ** **People current** Suppose that all the people on the Earth moved at the same speed around a circular track. Approximately how many times per second would each person pass the starting line if that "people current" equaled the number of electrons passing a cross section in a wire when there is a 1.0-A current through it?

62. ** EST **Bird on a power line** A bird stands on a 20,000-V bare copper power line. A 10-A current passes through the line. Estimate the current through the bird. Be sure to discuss all of your assumptions and indicate clearly the method you use in making the estimate.

63. ** A 5.0-A current caused by moving electrons flows through a wire. (a) Determine the number of electrons that flow past a cross section each second. (b) The same number of water molecules moves along a stream. Determine the volume of water that moves each second under a bridge that passes over the stream.

64. ** BIO **Current across membrane wall of axon** An axon cable that connects the input and output ends of a human nerve cell is $5 \times 10^{-6}$ m in radius and 0.5 m long. The thickness of the membrane surrounding the fluid in the axon is $6 \times 10^{-9}$ m, its resistivity is $1.6 \times 10^{7}\ \Omega \cdot m$, and there is a

0.070-V potential difference across the membrane. (a) Determine the current through the membrane. (b) If the current is due to $Na^+$ ions, how many of these ions leak across the membrane wall per second? (c) How does the current change if the membrane is twice as thick?

65. * BIO EST **Lifting forearm by electric current** You are asked to evaluate an idea for making it possible for a stroke patient to contract the biceps muscle in order to lift the forearm. The muscle will contract if a current of about 20 mA flows through the upper arm. Estimate the potential difference across electrodes, one inserted near the top of the arm and the other near the elbow, that will produce this current. Indicate any assumptions you made.

66. ** BIO EST **Warm arms** You accidentally touch a 10,000-V power line that has fallen during a storm. Your wet skin has a 1000-Ω resistance. Your body is about 70% water. Estimate the rate of thermal energy production in the body and the time interval needed to start the water boiling.

67. ** You have a switch, a power supply, a 75-W bulb, and a 15-W bulb. Build a circuit satisfying the following condition: when the switch is on, the 75-W bulb is on. When the switch is off, the 75-W bulb no longer glows, but the 15-W bulb does.

68. ** **Wiring a staircase** Devise an electric circuit that will allow you to turn a stairway lightbulb on and off from the bottom and from the top of the stairway.

69. * **Wiring high-fidelity speakers** Your high-fidelity amplifier has one output for a speaker of resistance 8 Ω. How can you arrange two 8-Ω speakers, one 4-Ω speaker, and one 12-Ω speaker so that the amplifier powers all speakers and their equivalent resistance when connected together in this way is 8 Ω?

70. ** **Shoe safety** According to the Occupational Health and Safety Administration (OSHA) standards, electrostatic dissipative shoes, which reduce the accumulation of static electricity, should be worn around sensitive computer equipment so the machines do not receive static shocks. To prevent potentially dangerous sparks, they should also be worn where an explosive atmosphere, such as flammable vapors, may be present—for example, in grain elevators and hospital operating rooms. The current through the shoes can be tested using the system shown in **Figure P16.70**. The emf is 50 V, the resistor $R$ is 1.2 MΩ, and the voltmeter reads 3.6 V. Is the current through the person less than 150 $\mu$A (the maximum safe level)? Indicate any assumptions you made.

**Figure P16.70**

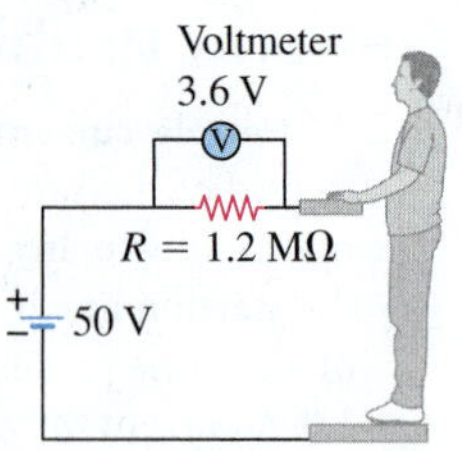


71. ** **Gas or electric heating?** What information should you collect to decide whether to use electricity or gas to heat your house in the winter? Search relevant information and write a report that will convince your roommates that one way is better than the other.

72. ** EST **Electric water heater** An electric hot water heater has a 7000-W resistive heating element. (a) Determine the energy needed to warm the 120 kg of water in the heater from 20 °C to 60 °C. (b) What time interval is needed to warm the water? (c) Estimate the electric energy cost of warming the water at recent rates.

73. ** BIO EST **The hands and arms as a conductor** While doing laundry you reach to turn on the light above the washing machine. Unfortunately, you touch an exposed 120-V wire in the broken switch box. Your other hand is supporting you on the top of the grounded washing machine. Assume that the resistance is 100,000 Ω across the dry skin of each hand and that the resistivity of tissue inside your arms and body is 5 Ω · m. (a) Estimate the total resistance of the tissue inside your arms from one side to the other (ignore the resistance across your chest). (b) Estimate the total resistance across the skin of one hand through the tissue in your arms and then across the skin of your other hand. (c) Are you in danger?

## Reading Passage Problems

BIO **Signals in nerve cells stimulate muscles** The input end of a human nerve cell is connected to an output end by a long, thin, cylindrical axon. A signal at the input end is caused by a stretch sensor, a temperature sensor, contact with another cell or nerve, or some other stimulus. At the output end, the nerve signal can stimulate a muscle cell to perform a function (to contract, provide information to the brain, etc.).

The axon of a so-called unmyelinated human nerve cell has a radius of $5 \times 10^{-6}$ m and a membrane that is $6 \times 10^{-9}$ m thick. The membrane has a resistivity of about $1.6 \times 10^{7}\ \Omega \cdot \mathrm{m}$. The fluid inside the axon has resistivity of about 0.5 Ω · m. The membrane wall has proteins that pump three sodium ions ($Na^+$) out of the axon for each two potassium ions ($K^+$) pumped into the axon. In the resting axon, the concentration of these ions results in a net positive charge on the outside of the membrane compared to negative charge on the inside. Because of the unequal charge distribution, there is a −70 mV potential inside compared to outside the axon.

When an external source stimulates the input end of the nerve cell so the potential inside reaches about −50 mV, gates or channels in the membrane walls near that input open and sodium ions rush into the axon. This stimulates neighboring gates to swing open and sodium ions rush into the axon farther along. This disturbance quickly travels along the axon—a nerve impulse. The potential across the inside of the membrane changes in 0.5 ms from −70 mV to +30 mV relative to the outside. Immediately after this depolarization, potassium ion gates open and positively charged potassium ions rush out of the axon, repolarizing the axon. Sodium and potassium ion pumps then return the axon and its membrane to their original configuration.

74. Which answer below is closest to the resistance of the fluid inside a 0.5-m-long unmyelinated axon?
(a) $10^9$ Ω (b) $10^6$ Ω (c) $10^3$ Ω
(d) $10^{-3}$ Ω (e) $10^{-6}$ Ω

75. An electric signal transmitted in a metal wire involves electrons moving parallel to the wire. What does an electric signal in an axon involve?
(a) Sodium ions moving across the axon membrane perpendicular to its length
(b) Potassium ions moving across the axon membrane perpendicular to its length
(c) Sodium and potassium ions moving inside the axon parallel to its length
(d) Only sodium ions moving inside the axon parallel to its length
(e) a and b

76. Which answer below is closest to the magnitude of the $\vec{E}$ field in the resting axon membrane?
(a) 10 V/m (b) $10^3$ V/m (c) $10^5$ V/m
(d) $10^7$ V/m (e) $10^9$ V/m

77. The charge density on the axon membrane walls (positive outside and negative inside) is $1.0 \times 10^{-4}\ \mathrm{C/m^2}$. Suppose the membrane of a 1.0-m-long axon discharged completely

in 0.02 s. What answer below is closest to the electric current across the membrane wall of one such hypothetical axon discharge?
(a) 0.2 A (b) 0.02 A (c) $2 \times 10^{-4}$ A
(d) $2 \times 10^{-5}$ A (e) $2 \times 10^{-7}$ A

78. The horizontal 4-Ω resistors in the two circuits in **Figure P16.78** represent the resistance of a small horizontal length of fluid inside an axon. The 15-Ω resistors represent the resistance across the axon membrane for a tiny length of axon. The 10-Ω resistor to the right side of each "ladder" is the effective resistance in front of the axon sections under consideration. Determine the resistance between B and G of the small length of axon shown in Figure P16.78a and between A and G of the slightly longer length axon shown in Figure P16.78b.
(a) 4.2 Ω, 8.4 Ω (b) 10 Ω, 10 Ω (c) 29 Ω, 48 Ω
(d) 10 Ω, 12 Ω (e) 4.2 Ω, 12 Ω

**Figure P16.78**

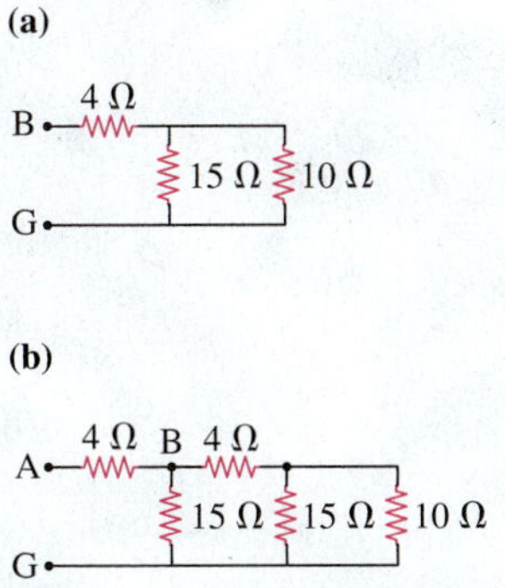


79. Suppose nerve impulses travel at 100 m/s in the axons of nerve cells from your fingers to your brain and then back again to your fingers in order to stimulate muscles that lift your fingers off the hot burner of a stove. Which answer below is closest to the time interval needed for the nerve signal transmission along the axons?
(a) 0.01 s (b) 0.001 s (c) 0.2 s
(d) 0.02 s (e) 0.002 s

**BIO Effect of electric current on human body** Nerve impulses are initiated at the input end of a nerve cell, travel along a relatively long axon (cable), and then cause an effect at the output end of the cell—for example, the initiation of a muscle contraction in a muscle cell. The nerve impulse is initiated by a stimulus that lowers the potential difference from outside the cell to inside from its normal −70 mV to about −50 mV.

A potential difference across two parts of the body (for example, the 120-V potential difference from a wall socket from one hand to the other or from the hands to the feet) can initiate an electric current in the body that stimulates nerve endings and triggers nerve signals that cause muscular contraction. Even worse, the current in the body can upset the rhythmic electrical operation of the heart. The heart muscles might be stimulated randomly in what is called ventricular fibrillation—a random contraction of the ventricles, which can be deadly. A rough guide to the effects of electric current on the body at different current levels is provided in **Figure 16.37**. Under dry conditions, human skin has high electrical resistance. Wet skin dramatically lowers the body's resistance and makes electrocution more likely to occur.

**Figure 16.37** The effects of electric current on the human body.

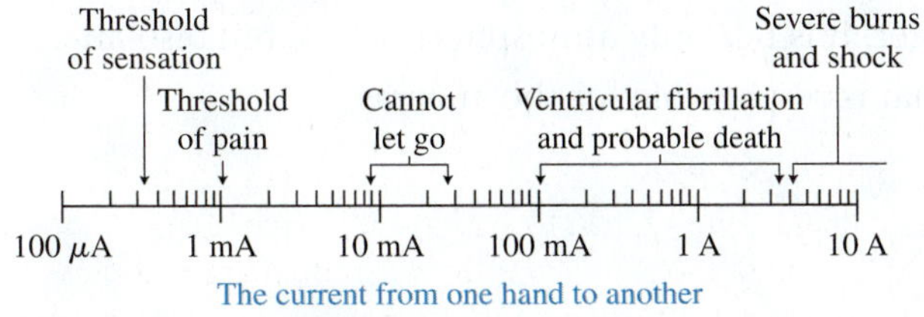


80. The electrical resistance across dry skin is about 100,000 Ω. Suppose a person with dry skin puts one hand on a 120-V power cord from a home wall socket while the other hand is touching a metal object at 0 V (at what is called ground). Which condition described below is most likely to occur?
(a) No sensation (b) Threshold of pain
(c) Cannot let go (d) Ventricular fibrillation
(e) Severe burns and shock

81. The electrical resistance across wet skin is about 1000 Ω. Suppose a person with wet skin puts one hand on a 120-V power cord from a home wall socket while the other hand is touching a metal object at 0 V (at what is called ground). Which condition described below is most likely to occur?
(a) No sensation (b) Threshold of pain
(c) Cannot let go (d) Ventricular fibrillation
(e) Severe burns and shock

82. Suppose the electrical resistance across your wet skin is about 1000 Ω. Which answer below is closest to the least potential difference from one hand that will cause slight pain?
(a) 0.1 V (b) 1 V (c) 10 V
(d) 100 V (e) 1000 V

83. Suppose the electrical resistance across your wet skin is about 1000 Ω. Which answer below is closest to the least potential difference from one hand to the other that will cause ventricular fibrillation?
(a) 0.1 V (b) 1 V (c) 10 V
(d) 100 V (e) 1000 V

84. When muscular contraction caused by electrical stimulation prevents a person from releasing contact with the potential difference sources, lower current can be extremely dangerous. For example, a 100-mA current for 3 s causes about the same effect as a 900-mA current for 0.03 s. In which of these situations is the least electric charge transferred through the body?
(a) 100 mA for 3 s (b) 900 mA for 0.03 s
(c) Too little information to decide

85. Why is it dangerous to place a hair dryer, radio, or other electric appliance that is plugged into a wall socket near a bathtub?
(a) The water provides a conductive path for current, which heats the metal cover on the appliance and can cause burns.
(b) If the appliance is accidentally knocked into the tub while a person is bathing, large currents could pass through the person's low-resistance body because of the 120-V potential difference that powers the appliance.
(c) There is no potential for danger, because electric appliances are grounded.
(d) a and b

86. Occasionally, the electric circuit that produces a coordinated pumping of blood from the four chambers of the heart becomes disturbed. Ventricular fibrillation can occur—random muscle contractions that produce little or no blood pumping. To stop the fibrillation, two defibrillator pads are placed on the chest and a large current (about 14 amps) is sent through the heart, restarting its normal rhythmic pattern. The current lasts 10 ms and transfers 140 J of electric energy to the body. Which answer below is closest to the potential difference between the defibrillator pads?
(a) 100 V (b) 400 V (c) 1000 V
(d) 5000 V (e) 10,000 V

# 17 Magnetism

What causes the northern lights?

How does Earth protect us from the solar wind and cosmic rays?

Are we really walking northward when we follow a compass needle?

**Be sure you know how to:**

- Use the electric field concept to explain how electrically charged objects exert forces on each other (15.1).
- Find the direction of the electric current in a circuit (16.1).
- Apply Newton's second law to a particle moving in a circle (4.4).

The Aurora Borealis—the northern lights—illuminates the autumn Arctic sky with a flickering greenish glow that is streaked with red and sometimes blue or violet.

A comparable phenomenon—the Aurora Australis—appears in the autumn sky above southern Australia, southern South America, and the Antarctic. The color of the auroras comes from the chemical composition of particles in the atmosphere (oxygen and nitrogen). But why do we see auroras near Earth's poles and only rarely anywhere else? Doesn't Earth's atmosphere surround the planet more or less uniformly? As we will learn in this chapter, it is not only atmospheric gases, but also another phenomenon that is responsible for the auroras.

**In previous chapters** we learned that charged objects attract and repel each other—similar to the way magnets do. However, electrically charged objects do not exhibit magnetic properties. Are electricity and magnetism unrelated phenomena or are they connected in some way? Until 1820 physicists believed that they were completely different. Since then, scientists have made many discoveries about the relationships between electricity and magnetism. These discoveries have made possible electromagnets, electric motors, magnetic flow meters, magnetic resonance imaging (MRI), and many other devices. We begin this chapter by learning about the connections between electricity and magnetism and how scientists discovered them.

**Figure 17.1** Like magnetic poles repel and unlike poles attract.

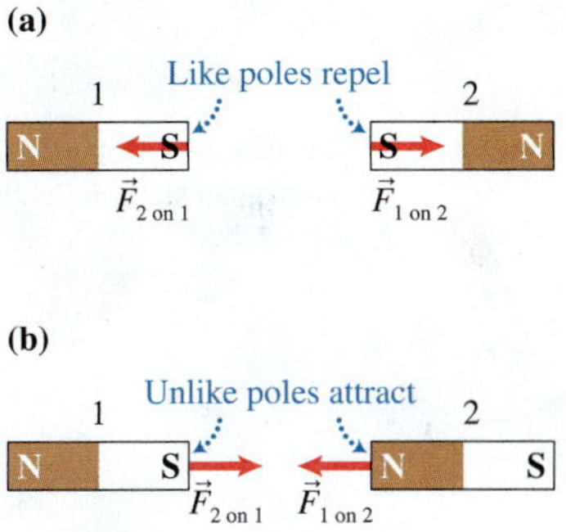


# 17.1 The magnetic interaction

Magnets become familiar through toys, refrigerator magnets, and countless household applications. Very early we discover that, when brought near each other, magnets can both attract and repel. The two sides of a magnet are called **poles**—the **north pole** and the **south pole.** If you bring the like poles of two magnets (north to north and south to south) near each other, they repel. If you bring opposite poles near each other, they attract (**Figures 17.1a** and b). Additionally, both sides of a magnet attract certain other objects, such as those made from steel or iron, even if those objects aren't magnets themselves.

Magnets always have two poles. If you break a magnet into two pieces, each piece still has two poles—a north pole and a south pole (**Figure 17.2**). If you break one of those pieces again, each smaller piece has two poles, and so on. Unlike electrically charged objects that can be either negatively or positively charged, there are no magnets with a single pole (a so-called monopole).

**Figure 17.2** Breaking a magnet into pieces produces more magnets, each with a north and south pole.

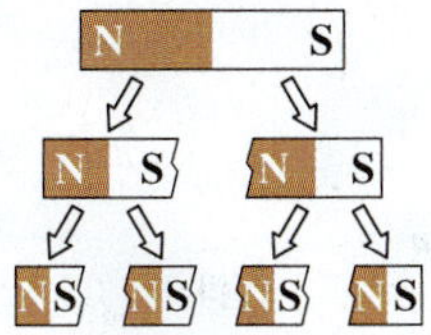


The labels "north" and "south" originated in 11th-century China. People noticed that if they put a tiny magnet on a low-friction pivot (or attached it to a piece of wood floating in water), one end always pointed toward geographic north. This property of a magnet resulted in the names for the two ends: the north pole points toward geographical north; the south pole points toward geographical south. The device became known as a **compass**.

Since the north pole of a compass is attracted to the south pole of another magnet, and since compasses that are not near other magnets always point approximately toward Earth's geographic north pole, we can infer that Earth itself acts as a giant magnet with its magnetic south pole close to its geographic north pole and its magnetic north pole close to its geographic south pole (**Figure 17.3**).

**Figure 17.3** The magnetic and geographic poles of Earth.

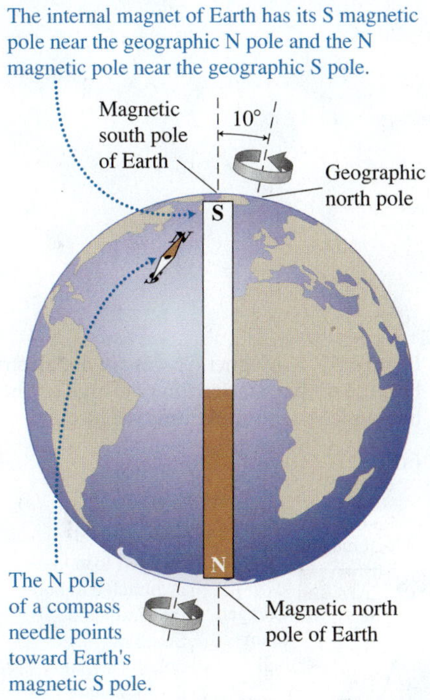


## The magnetic interaction depends on separation

Suppose you lay a compass on one side of a wooden table and place another magnet on the other side with one of its poles facing the compass, as shown in

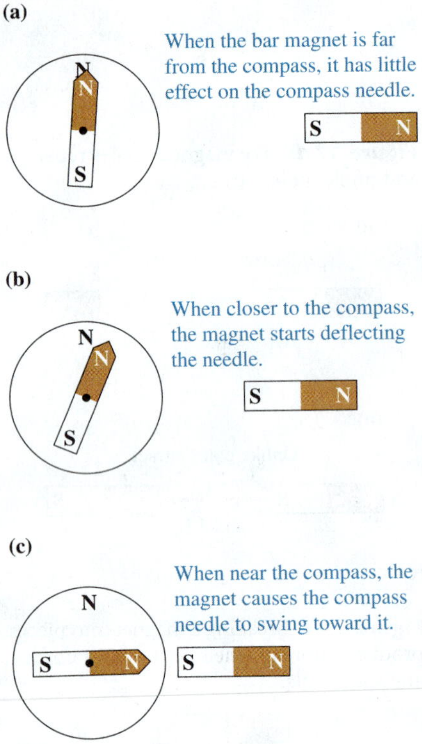
(a)
When the bar magnet is far from the compass, it has little effect on the compass needle.
N
N
S
S
N
(b)
When closer to the compass, the magnet starts deflecting the needle.
N
N
S
S
N
(c)
When near the compass, the magnet causes the compass needle to swing toward it.
N
S
N
S
N

**Figure 17.4** The strength of the interaction between magnets depends on their separation.

**Figure 17.4**. If the compass and the magnet are far apart, the compass needle will turn very little from its orientation to Earth's magnetic pole (Figure 17.4a). If you move the magnet a little closer to the compass, the compass needle will turn a little from its original orientation (Figure 17.4b). If you place the magnet very close to the compass, the compass needle will swing abruptly around so that one pole of the compass points toward the opposite pole of the magnet (Figure 17.4c). The interaction between the magnets increases in strength as their separation decreases.

## The magnetic and electrical interactions are different

Both magnets and electrically charged objects attract and repel, and the intensity of their interactions depends on their separation. However, these are different interactions. Recall our test of the electric and magnetic interactions (Testing Experiment Table 14.3), which showed that electrically charged objects *do not* interact with magnets in the same way that magnets interact with magnets. For example, a positively charged polystyrene foam ball is attracted to *both* ends of a metal magnet (**Figure 17.5a**). This happens because the free electrons inside the metal magnet redistribute themselves in the presence of the charged foam ball so that the ball is electrically attracted to the magnet (Figure 17.5b). The charged ball has a *slight* attraction to both ends of a magnet made of nonconducting material such as ceramic (similar to the attraction it would have to a dielectric material) (Figure 17.5c). Magnetic poles *are not* electric charges.

**Review Question 17.1** How do we know that magnetic poles are not electric charges?

# 17.2 Magnetic field

The electric field explains how electrically charged objects interact without contact. Since magnets also interact without contact, it is reasonable to suggest the existence of some kind of field, in this case a **magnetic field,** as the mechanism behind magnetic interactions. Thus we can assume that a magnet produces a magnetic field—a magnetic disturbance with which other objects with magnetic properties (another magnet, anything made of iron, etc.) interact. In this chapter we will learn how to describe this field qualitatively and quantitatively and how to predict its effects on objects that possess their own magnetic properties.

## Direction of the magnetic field

To describe the magnetic field, we use a vector physical quantity called the **$\vec{B}$ field.** The $\vec{B}$ field has a magnitude and direction. We can use a compass to

**Figure 17.5** Magnetic poles are not electric charges. (a) Both poles of a metal magnet attract a positively charged foam ball because of (b) polarization of charge on the magnet. (c) There is a slight interaction between the foam ball and the magnetic poles of a nonconducting ceramic magnet.

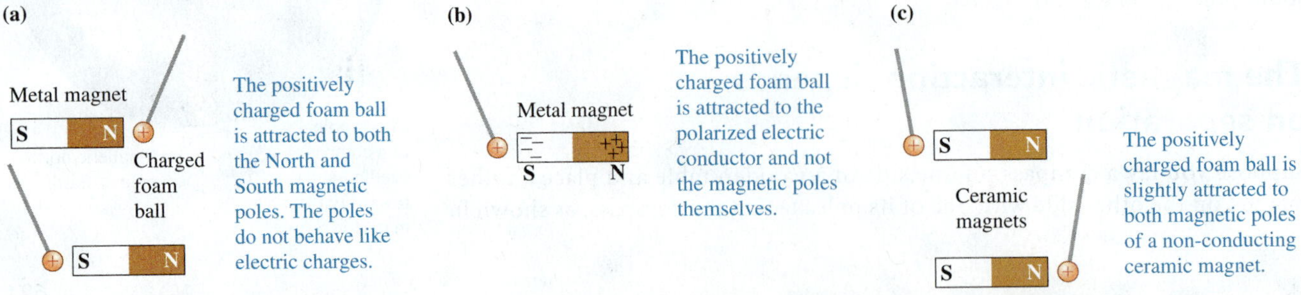
(a)
Metal magnet
S
N
Charged foam ball
S
N
The positively charged foam ball is attracted to both the North and South magnetic poles. The poles do not behave like electric charges.
(b)
Metal magnet
S
N
The positively charged foam ball is attracted to the polarized electric conductor and not the magnetic poles themselves.
(c)
S
N
Ceramic magnets
S
N
The positively charged foam ball is slightly attracted to both magnetic poles of a non-conducting ceramic magnet.

**Figure 17.6** Compasses indicate the $\vec{B}$ field directions at various locations near (a) a bar magnet and (b) a horseshoe magnet.

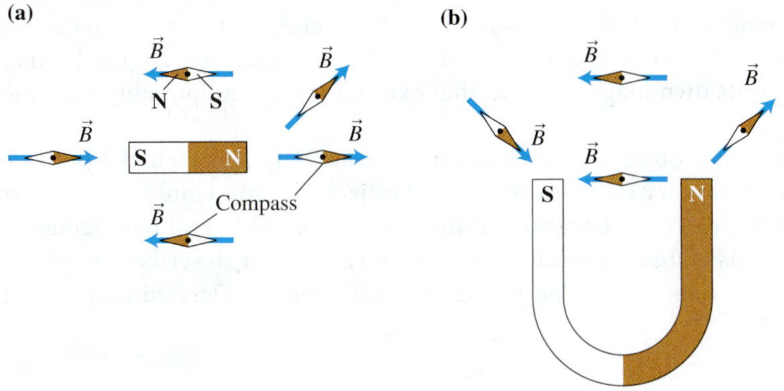


detect the direction of the $\vec{B}$ field at a particular location, just as we can use a test charge for the $\vec{E}$ field. The direction of the $\vec{B}$ field is defined as the direction that the north pole of the compass needle points. We can also determine the direction of the $\vec{B}$ field by drawing a vector from the south pole to the north pole of the compass inside the compass. **Figure 17.6a** shows the direction of the $\vec{B}$ field at several points near a bar magnet. Figure 17.6b shows the $\vec{B}$ field vectors near a horseshoe magnet.

**Figure 17.7** Compasses indicate $\vec{B}$ field lines.

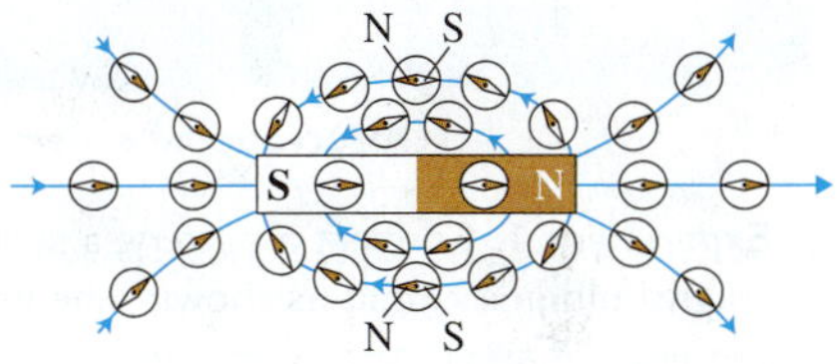


## Representing the magnetic field: field lines

To represent the magnetic field in a region, we will draw $\vec{B}$ field lines (similar to $\vec{E}$ field lines). We can visualize the lines by placing multiple compasses around a bar magnet, as shown in **Figure 17.7**. Following the direction of the south-north vector for each compass as we move from compass to compass, we find that the compasses are tangent to lines surrounding the magnet—the blue lines in Figure 17.7. Instead of compasses, we can use hundreds of tiny iron filings (which act like tiny compasses) sprinkled on a thin piece of transparent plastic placed on top of the magnet. The filings form a pattern that looks similar to the lines formed by the compasses (**Figure 17.8**). Note that the filings that are directly on top of the magnet align with the south-north axis of the magnet. These lines are so-called **$\vec{B}$ field lines**. They are used to represent the $\vec{B}$ field produced by the magnet in the whole region, not just at a point as one $\vec{B}$ field vector indicates. Similar to $\vec{E}$ field lines, $\vec{B}$ field lines represent not only the direction of the $\vec{B}$ field, but also its magnitude. The lines are closer together in regions where the $\vec{B}$ field magnitude is greater—see the pattern shown in **Figure 17.9**.

**Figure 17.8** Iron filings indicate $\vec{B}$ field lines.

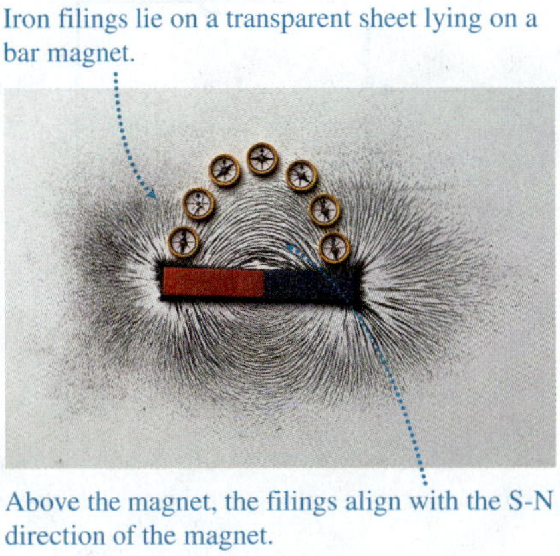

Iron filings lie on a transparent sheet lying on a bar magnet.
Above the magnet, the filings align with the S-N direction of the magnet.


**TIP** Unlike $\vec{E}$ field lines, which begin on positively charged objects and end on negatively charged objects, $\vec{B}$ field lines do not have beginnings or ends. Notice in Figure 17.9 that the lines form complete closed loops.

**Figure 17.9** $\vec{B}$ field lines.

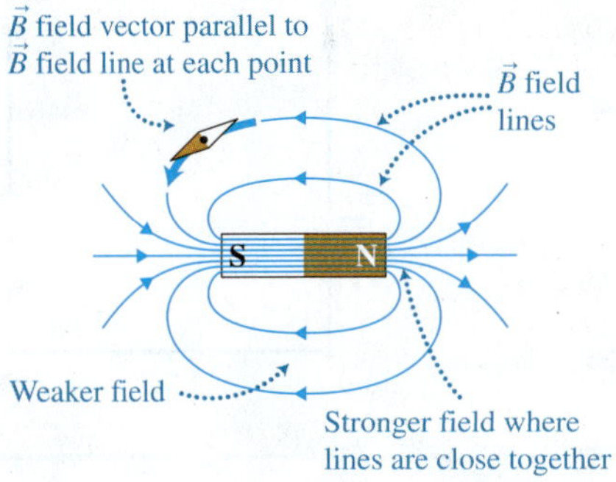


**The magnetic $\vec{B}$ field and its representation by $\vec{B}$ field lines** The direction of the magnetic $\vec{B}$ field at a particular location is defined as the direction that the north pole of a compass needle points when at that location. The $\vec{B}$ field vector at a location is tangent to the direction of the $\vec{B}$ field line at that location. Both point in the same direction. The density of lines in a region represents the magnitude of the $\vec{B}$ field in that region—where the $\vec{B}$ field is stronger, the lines are closer together.

## Do other objects produce a magnetic field?

We can now explain the interaction between two magnets (let's call them magnet A and magnet B) in the following way. Magnet A creates a magnetic field around itself, which exerts a force (and possibly a torque) on magnet B. Magnet B creates its own magnetic field that exerts a force (and possibly a torque) on magnet A.

Do any other objects besides magnets create magnetic fields? We found earlier that stationary electrically charged objects do not. Could electric currents (moving electric charges) produce magnetic fields and be affected by magnetic fields? Observational Experiment **Table 17.1** describes simplified versions of two experiments performed in 1820 by Hans Oersted that investigated this idea.

**OBSERVATIONAL EXPERIMENT TABLE**

**17.1 Do electric currents produce magnetic fields?**

VIDEO 17.1

| Observational experiment | Analysis |
|---|---|
| **Experiment 1.** Connect a battery, a switch, some wires, and a lightbulb in a circuit as shown. The bulb indicates the presence of an electric current in the circuit. With the switch open (the bulb is off), place compasses beneath, on top of, and to the sides (not shown) of one of the wires. Notice the direction the compasses point. 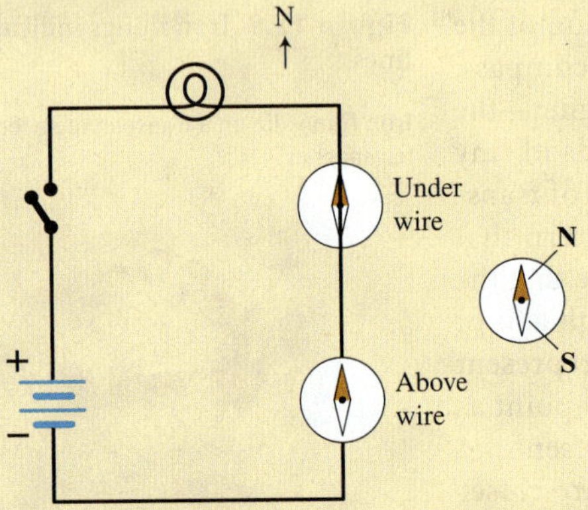  | Without current in the circuit, the needles point toward geographic north (magnetic south). Thus, the wire does not possess any magnetic properties. |
| Close the switch (the bulb is on). Notice the directions that the compasses above and beneath the wire now point. 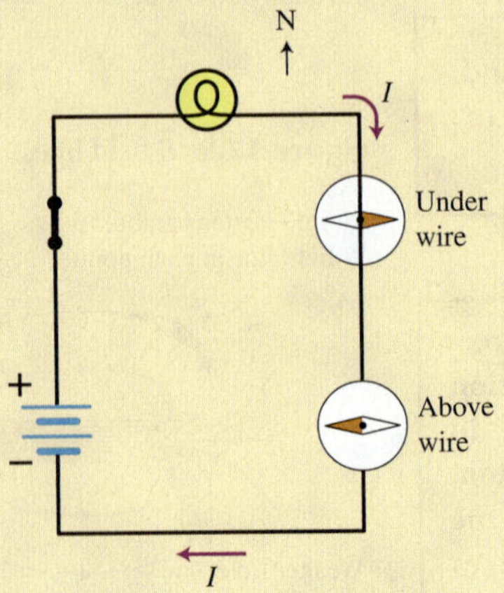  | The current in the wire affects the orientation of the compasses placed nearby. Therefore, the current-carrying wire might possess magnetic properties. |

**Experiment 2.** Reverse the direction of the current. The compasses on top of and beneath the wire point as shown.

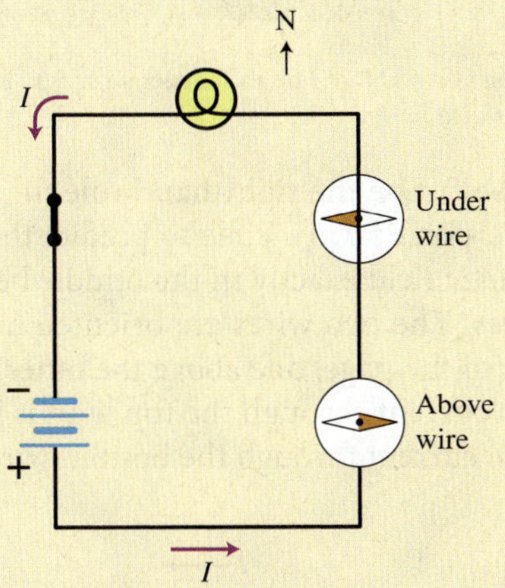


The change in the orientation of the compass indicates that magnetic properties of the wire depend on the direction of the current through it.

**Pattern**

Electric current affects the orientation of a compass. The orientation depends on the direction of the current. With the current in the wire in Experiment 1, the compasses orient in the direction of the circles around the wire shown at right: toward the right when under the wire, up on the right side of the wire, toward the left on top of the wire, and down on the left side of the wire. The circles clearly represent the effect of the current on the compasses.

In Experiment 2, the compass orientations are reversed. Circles in the opposite direction would represent the effect of this current on the compasses.

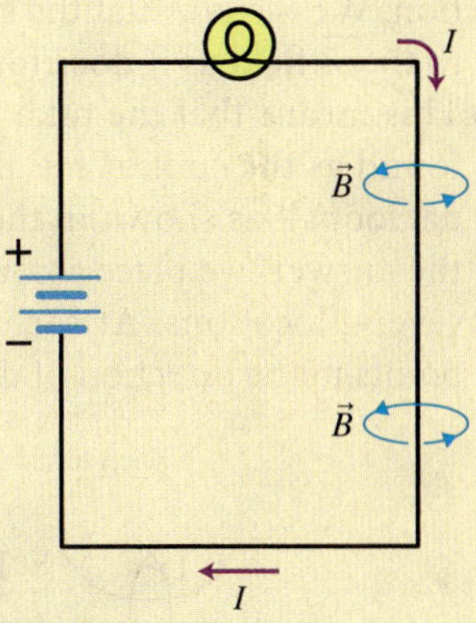


We conclude that an electric current does produce a magnetic field. The direction of the $\vec{B}$ field depends on the direction of the current. The $\vec{B}$ field lines form closed circles around the current. Their direction changes when the direction of the current changes.

Since the current is made of electrically charged particles that are in collective motion with respect to the compass, this means that charged objects in motion produce a magnetic field and stationary charged objects do not. Through the experiments in Table 17.1 we have deduced a pattern in the direction of the $\vec{B}$ field vectors and $\vec{B}$ field lines around a current-carrying wire. Imagine grasping the wire with your right hand, with your thumb pointing in the direction of the current. When you do so, your fingers wrap around the current in the direction of the $\vec{B}$ field lines (**Figure 17.10**). This method for determining the shape of the $\vec{B}$ field produced by the electric current in a wire is called the **right-hand rule** for the $\vec{B}$ field.

> **Right-hand rule for the $\vec{B}$ field** To determine the direction of the field lines produced by a current, grasp the wire with your right hand, with your thumb pointing in the direction of the current. Your four fingers wrap around the current in the direction of the $\vec{B}$ field lines. At each point on a $\vec{B}$ field line, the $\vec{B}$ field vector is tangent to the line and points in the direction of the line.

**Figure 17.10** The right-hand rule for the direction of the $\vec{B}$ field produced by an electric current.

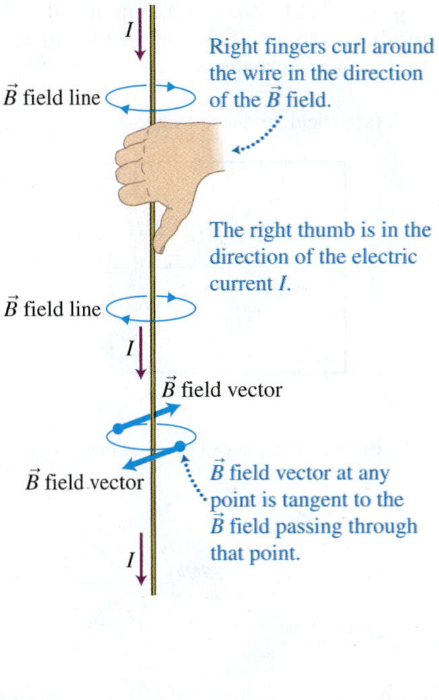

## CONCEPTUAL EXERCISE 17.1 Magnetic field of a solenoid

Draw the magnetic field lines of a solenoid connected to a battery.

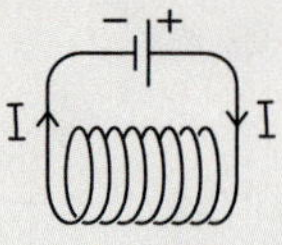

I
I
Solenoid

**Sketch and translate** A **solenoid** is a wire with a large number of loops in a cylindrical shape. For simplicity in this exercise we will use a solenoid with eight loops. To draw the $\vec{B}$ field lines produced by a current in the solenoid, we use the right-hand rule for the $\vec{B}$ field lines produced by each loop. We then combine them for a complete picture of the $\vec{B}$ field produced by the solenoid.

**Simplify and diagram** Consider three adjacent loops near the center of the solenoid. The $\vec{B}$ field lines inside each loop point in the same direction (see the figure below). Therefore the $\vec{B}$ field vectors also point in the same direction. We assume that the superposition principle applies to the $\vec{B}$ field as it does for the $\vec{E}$ field (see Chapter 15). This means that the total $\vec{B}$ field at any location can be found as the sum of the $\vec{B}$ field vectors due to individual loops—as shown in the figure at top right. To check the answer, we place a compass close to the solenoid at several locations. At each location the compass needle points in the direction of the $\vec{B}$ field.

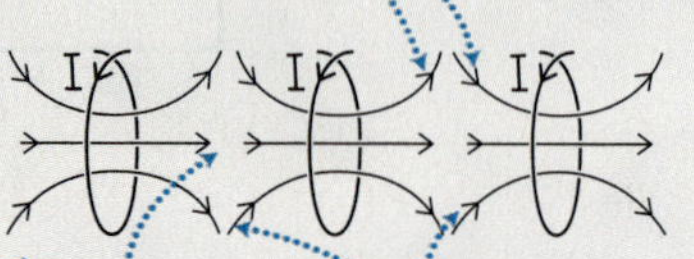

Each turn produces a $\vec{B}$ field.
I
I
I
The $\vec{B}$ fields from adjacent loops add here ...
... and cancel here.

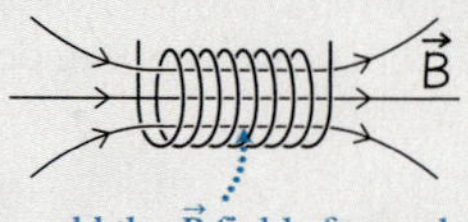

$\vec{B}$
When you add the $\vec{B}$ fields for each turn, you get a uniform field inside.

**Try it yourself:** Use the right-hand rule for the $\vec{B}$ field and the superposition principle to predict the direction of the magnetic field exactly in the middle between two straight wires. The two wires are oriented horizontally in the plane of the page, one above the other (see figure below). The current through the top wire is toward the right and the current through the bottom wire is toward the left.

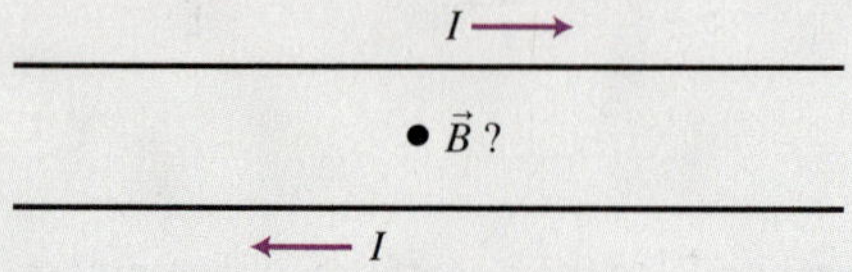

$I$
$\vec{B}$ ?
$I$

*Answer:* The $\vec{B}$ field contribution of each wire at the location of interest points into the page. Therefore, the $\vec{B}$ field at that location points into the page.

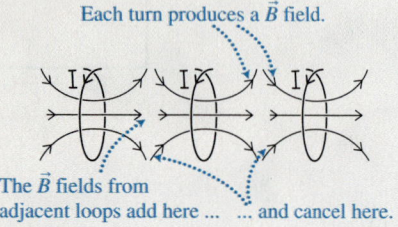

**Figure 17.11** A current loop or coil produces a $\vec{B}$ field that is almost identical to that of a bar magnet.

(a) $\vec{B}$ field produced by a loop

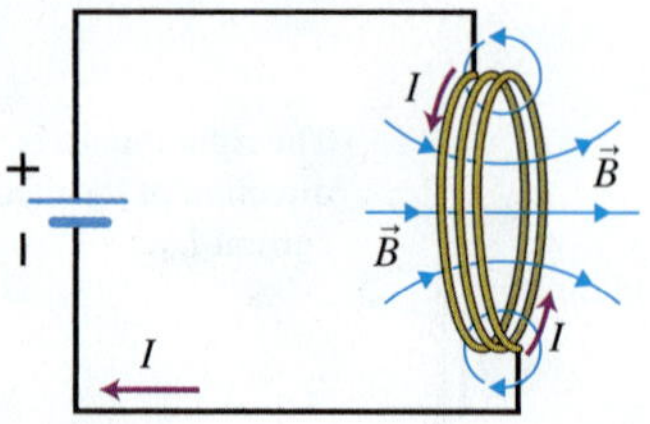

$I$
+
$\vec{B}$
$\vec{B}$
$I$
$I$

(b) $\vec{B}$ field produced by a bar magnet

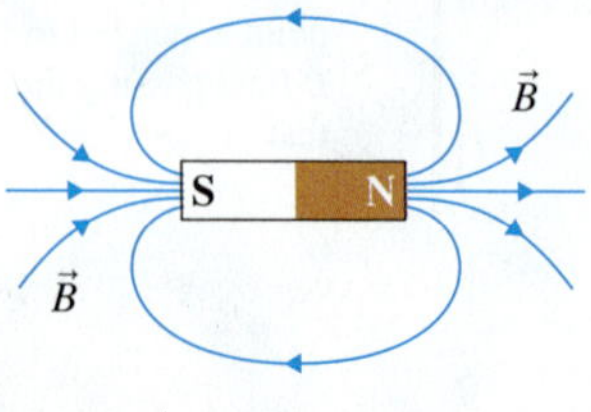

$\vec{B}$
S
N
$\vec{B}$

Notice that the $\vec{B}$ field produced by the current in a loop or a coil (a wire that is bent into multiple loops) (**Figure 17.11a**) and that produced by a bar magnet (Figure 17.11b) are very similar. The $\vec{B}$ field in the region to the right of the plane of the coil (or loop) looks just like the $\vec{B}$ field around the north pole region of the bar magnet. Likewise, the region to the left of the plane of the coil looks just like the $\vec{B}$ field in the south pole region of the bar magnet. Wire coils with current act magnetically in a very similar way to bar magnets and are known as **electromagnets**. We will discuss electromagnets further in Section 17.8.

**Review Question 17.2** What is the direction of the $\vec{B}$ field at a point halfway between two parallel wires with currents in the same direction?

## 17.3 Magnetic force exerted by the magnetic field on a current-carrying wire

If a current-carrying wire is similar in some ways to a magnet, then a magnetic field should exert a magnetic force on a current-carrying wire similar to the force it exerts on another magnet. When we place a bar magnet near a circuit with long connecting wires, we find that the magnet sometimes pulls on the wire and sometimes does not—the effect depends on the relative directions of the $\vec{B}$ field and the current in the wire.

## The magnetic field exerts a force on a current-carrying wire

To investigate how the relative directions of the $\vec{B}$ field and current-carrying wire affect their interactions, we conduct the experiments described in Observational Experiment **Table 17.2**. We use a horseshoe magnet because between the poles of such a magnet the $\vec{B}$ field lines are almost parallel to each other (see **Figure 17.12**).

**Figure 17.12** Between the poles of a horseshoe magnet, the $\vec{B}$ field lines are almost parallel to each other

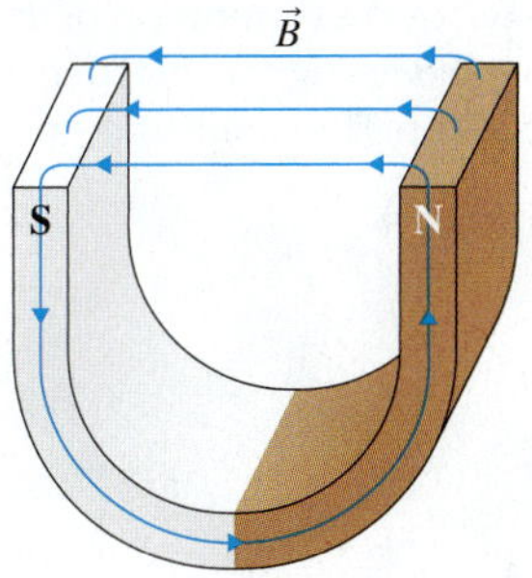


**OBSERVATIONAL EXPERIMENT TABLE**

**17.2 Direction of magnetic force.**

VIDEO 17.2

| Observational experiment | Analysis |
|---|---|
| **Experiment 1.** Hang a horizontal straight wire between the poles of a horseshoe magnet using conducting support wires connected to a battery. Orient the wire parallel to the $\vec{B}$ field lines. Attach a battery and open switch in series to the wire. Turn the current on by closing the switch. The wire does not move. If you reverse the current, the wire still does not move. 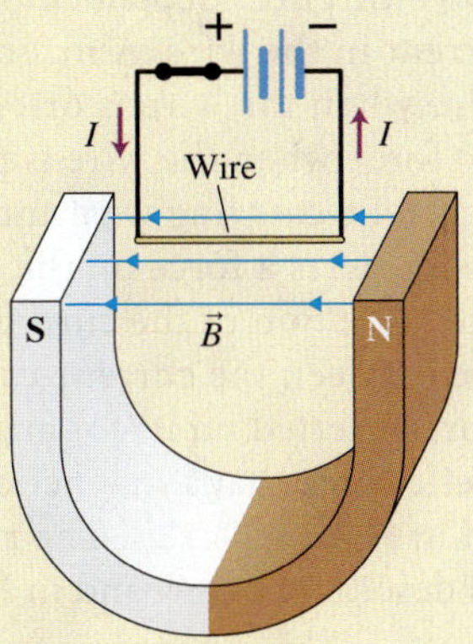  | The $\vec{B}$ field lines for both experiments are parallel to the direction of the current. The magnet does not exert a force on the current-carrying wire. |
| **Experiment 2.** Orient the wire perpendicular to the $\vec{B}$ field lines and turn the current on by closing the switch. The wire bends down. 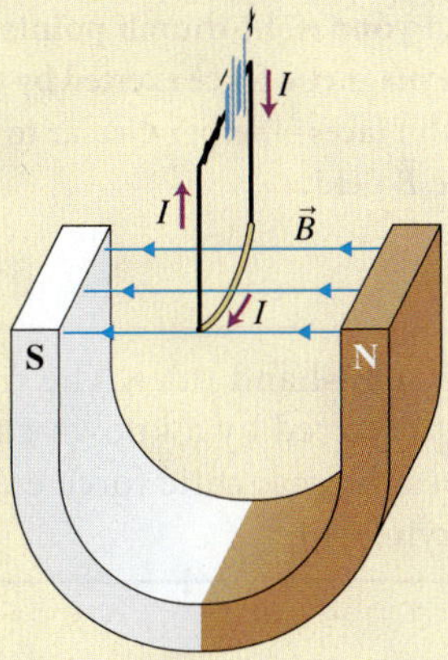  | The field lines are perpendicular to the direction of the current in the wire. The force exerted on the wire by the magnetic field is downward and perpendicular to both the current and the $\vec{B}$ field. |

(*continued*)

| Observational experiment | Analysis |
|---|---|
| **Experiment 3.** Repeat Experiment 2, but reverse the orientation of the battery, reversing the direction of the current. The wire bends up. 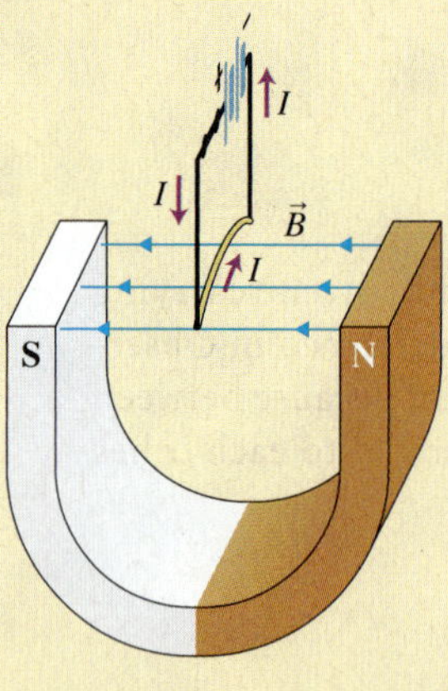 | The field lines are perpendicular to the direction of the current in the wire. The force exerted on the wire by the magnetic field is upward and perpendicular to both the current and the $\vec{B}$ field. |



**Pattern**

- The magnetic field does not exert a force on the current-carrying wire if the $\vec{B}$ field is parallel to the wire.
- When the $\vec{B}$ field is perpendicular to the direction of the current, the magnetic field exerts a force on the current-carrying wire that is perpendicular to both the direction of the wire and the $\vec{B}$ field. The direction of this force depends on the direction of the current-carrying wire and the direction of the $\vec{B}$ field.

**Figure 17.13** A method to determine the direction of the force that a magnetic field exerts on a current-carrying wire.

(a) Magnetic force

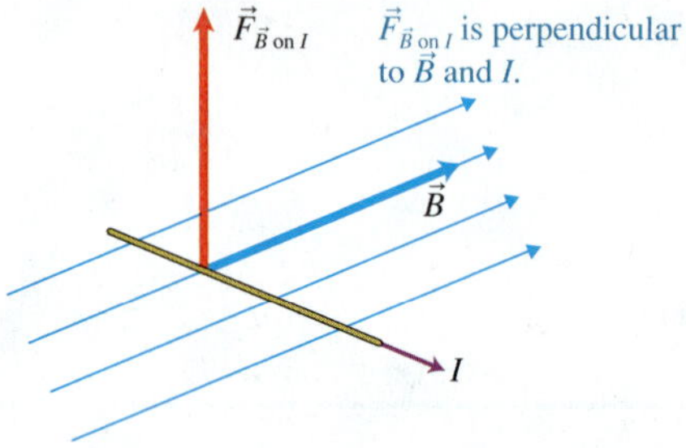


(b) Right-hand rule for magnetic force

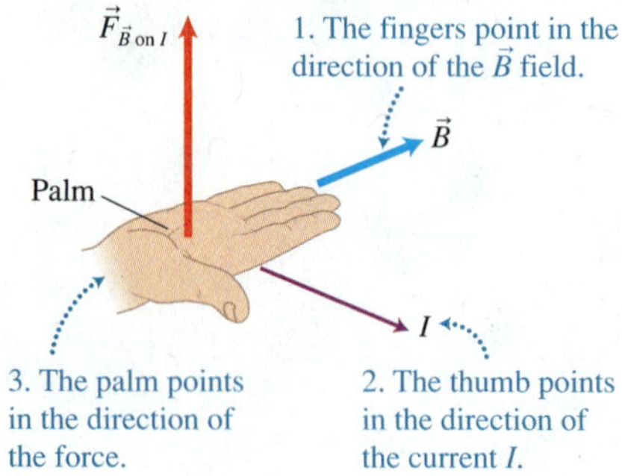


In Table 17.2 we found that the magnet sometimes exerted a force on the wire and sometimes did not—the effect depended on the relative orientations of the $\vec{B}$ field and the current in the wire. A magnetic field does exert a force on a current-carrying wire when the wire is oriented perpendicular to the $\vec{B}$ field and does not exert a force when the wire is parallel. Other experiments indicate that when the current-carrying wire and the $\vec{B}$ field form any nonzero angle, the magnetic field exerts a force on the current-carrying wire that is perpendicular to both the direction of the current and the direction of the $\vec{B}$ field. The larger the angle between the current-carrying wire and the $\vec{B}$ field, the larger the magnetic force exerted on it (for angles up to 90°).

The direction of the magnetic force that a magnetic field exerts on a current-carrying wire is illustrated in **Figure 17.13a**. The right-hand rule for determining the force direction is described below and in Figure 17.13b.

**Right-hand rule for the magnetic force** **Hold your right hand flat with your thumb extended from the four fingers. Orient your hand so that your fingers point in the direction of the $\vec{B}$ field and your right thumb points along the direction of the current. The direction of the magnetic force exerted by the magnetic field on the current is the direction your palm faces—perpendicular to both the direction of the current and the direction of the $\vec{B}$ field.**

**TIP** Do not confuse the two right-hand rules. The right-hand rule for the field describes the $\vec{B}$ field produced by a known current. The right-hand rule for the force describes the magnetic force exerted by a known magnetic field on a current-carrying wire.

## Forces that current-carrying wires exert on each other

If a current-carrying straight wire produces a magnetic field, the field should exert a force on a second current-carrying straight wire placed nearby. Similarly, the magnetic field produced by the second wire's current should exert a force on the first wire's current. According to Newton's third law, the forces that these wires exert on each other should point in opposite directions and have the same magnitudes. Let us test this reasoning with the experiments described in Testing Experiment **Table 17.3**.

**TESTING EXPERIMENT TABLE**

### 17.3 Do two current-carrying wires interact as magnets?

| Testing experiment | Prediction | Outcome |
| --- | --- | --- |
| Two vertical strips (A and D) of aluminum foil are next to each other, each connected by wires at their ends to the terminals of their own batteries. 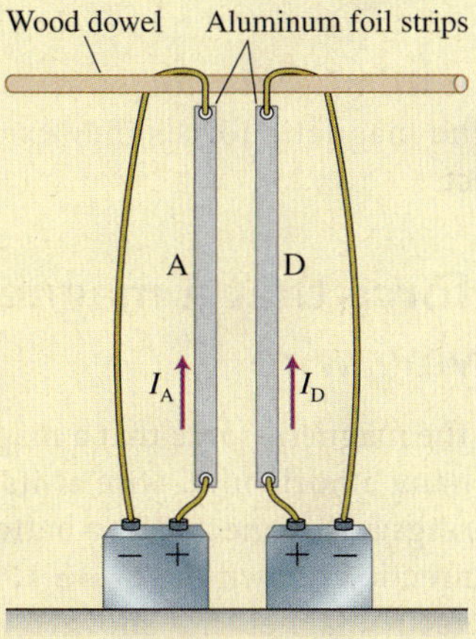  Predict what happens when the currents in the strips are in the same direction. | Choose strip D as the system. Using the right-hand rule for the $\vec{B}$ field, we find that the $\vec{B}_A$ field produced by strip A at D points into the page. Using the right-hand rule for force, we find that this field exerts a force on strip D $\vec{F}_{A\text{ on }D}$ that points to the left. 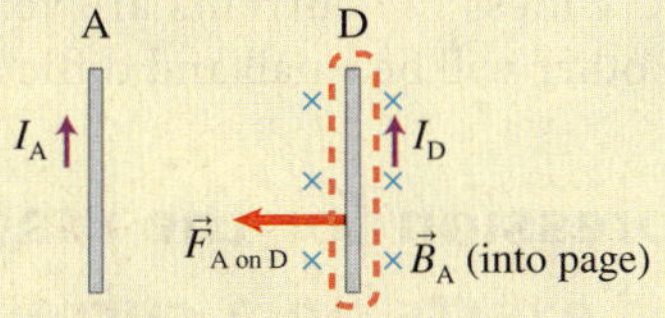  Repeat the procedure for strip A as the system. The $\vec{B}$ field $\vec{B}_D$ produced by strip D at A points out of the page. This field exerts a force $\vec{F}_{D\text{ on }A}$ that points to the right. 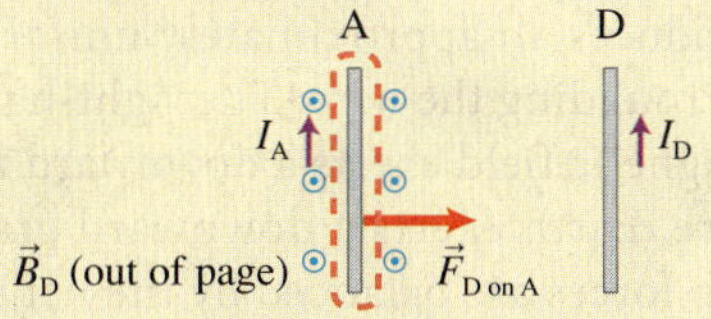  The strips should bend toward each other. | When we do the experiment, the strips bend toward each other. 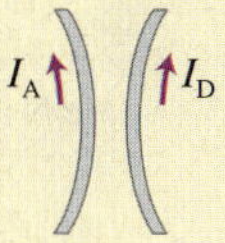  |
| Predict what happens when the currents in the strips are in the opposite directions. | We repeat the same process to make our prediction. When the current in D is in the downward direction instead of up, the same analysis shows that the strips should repel. | When the currents are in the opposite directions, they repel. 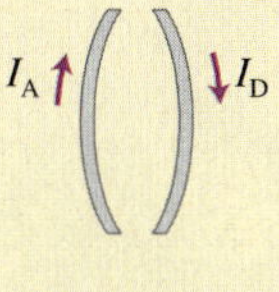  |

**Conclusion**

We found that two current-carrying wires exert forces on each other whose direction can be predicted using both right-hand rules. The results are consistent with Newton's third law—the forces that the strips exert on each other point in opposite directions.

**Figure 17.14** Two coils attract when the currents in them are as shown.

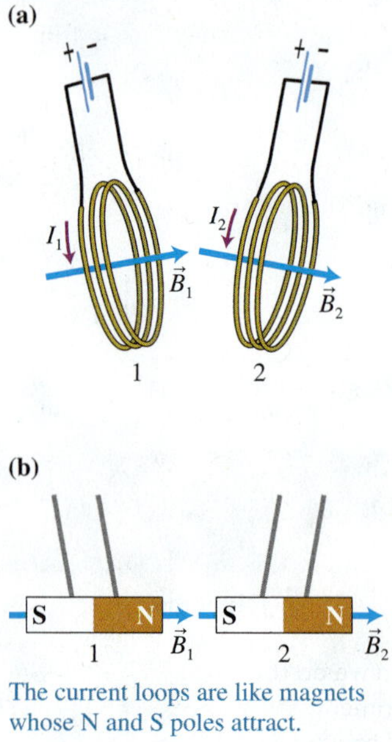


We can use the same method to predict what happens to the two current-carrying coils of wire when the current is as shown in **Figure 17.14a**. The coils are like bar magnets with their poles in the same direction. The north pole of the left coil attracts the south pole of the right coil, and vice versa (see Figure 17.14b). The coils should attract—and they do!

**TIP** In Table 17.3 the wires attract and repel each other via magnetic forces. The wires do not interact via electric forces, because the net electric charge of each wire is zero.

André-Marie Ampère in 1820 conducted numerous such experiments and found that the magnitude of the force that one current-carrying wire exerts on another parallel current was directly proportional to the magnitude of each current and inversely proportional to the distance between the wires. This relationship became the basis for the unit of the electric current in the SI system—the **ampere** (**Figure 17.15**).

**Ampere (A)** **When two 1.0-m-long parallel wires are separated by 1.0 m (as in Figure 17.15) and there are equal-magnitude electric currents through the wires so that they exert a force of $2.0 \times 10^{-7}$ N on each other, the current $I$ in each wire is defined to have a magnitude of 1.0 A.**

**Figure 17.15** Ampere's experiment for the unit of electric current.

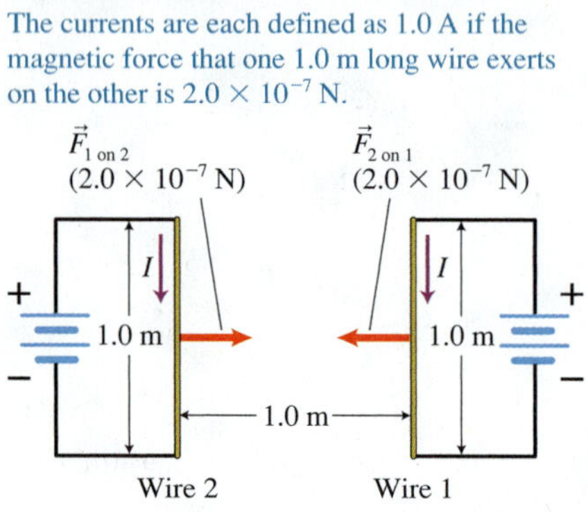


From this definition it is apparent that the magnetic forces are rather weak. Unless the currents are very large, the magnetic forces they exert on each other will be small and difficult to detect.

## Expression for the magnetic force that a magnetic field exerts on a current-carrying wire

To determine a mathematical expression for the magnetic force that a magnetic field exerts on a current-carrying wire, we hang a horizontal wire at its ends from conducting springs. The tops of the springs are connected to a battery so that the current through the wire is in the direction shown in **Figure 17.16a**. We place the wire between the poles of an electromagnet (not shown), which produces an approximately uniform horizontal variable $\vec{B}$ field in the region surrounding the wire. The right-hand rule for magnetic force indicates that the magnetic field exerts a downward magnetic force $\vec{F}_{\vec{B}\text{ on W}}$ on the current in the wire. Earth exerts a downward gravitational force on the wire $\vec{F}_{\text{E on W}}$. These two forces are balanced by the net upward force of the two springs on the wire $\vec{F}_{\text{S on W}}$, as shown in the force diagram in Figure 17.16b. Knowing the spring constant of the springs and the mass of the wire, we can use the stretch of the springs to deduce the magnitude of the magnetic force exerted on different length wires when different currents are in them. The collected data are shown in **Table 17.4**.

Notice that the magnitude of the magnetic force exerted on the wire is proportional to the current $I$ through the wire (the first three rows), to the length $L$ of the wire (the second three rows), and to the sine of the angle $\theta$ between the direction of the current and the direction of the $\vec{B}$ field (the last four rows). We can represent this mathematically:

$$F_{\vec{B}\text{ on W}} \propto IL\sin\theta \tag{17.1}$$

When one quantity is proportional to another quantity, the ratio of the two is constant: $\frac{F_{\vec{B}\text{ on W}}}{IL\sin\theta} = \text{constant}$. If you place the same current-carrying

**Table 17.4** **Magnitude of the magnetic force.**

| Current $I$ in the wire (A) | Length $L$ of wire (m) | Orientation angle $\theta$ between the wire and the $\vec{B}$ field | Magnitude of magnetic force $F$ exerted on the wire (N) |
|---|---|---|---|
| $I$ | $L$ | 90° | $F$ |
| $2I$ | $L$ | 90° | $2F$ |
| $3I$ | $L$ | 90° | $3F$ |
| $I$ | $L$ | 90° | $F$ |
| $I$ | $2L$ | 90° | $2F$ |
| $I$ | $3L$ | 90° | $3F$ |
| $I$ | $L$ | 0° | 0 |
| $I$ | $L$ | 30° | $0.5F$ |
| $I$ | $L$ | 60° | $0.87F$ |
| $I$ | $L$ | 90° | $F$ |

**Figure 17.16** By varying the $\vec{B}$ field, we can develop an expression for the force that the magnetic field exerts on the current-carrying wire.

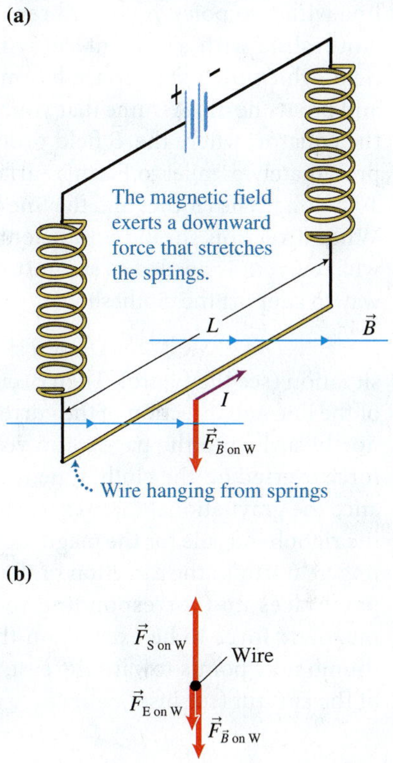


wire in the magnetic field of an electromagnet in which you can increase the magnetic field by increasing the electric current in the magnet's coil windings, you find that the proportionality constant increases as the $\vec{B}$ field increases. It appears that the proportionality constant is the magnitude of the $\vec{B}$ field.

We can use the above mathematical relation to define the magnitude of the $\vec{B}$ field in a particular region as the ratio of the magnitude of the maximum force that the field exerts on a current-carrying wire to the product of the length $L$ of the wire and the current $I$ in the wire. This maximum force is exerted when the wire is perpendicular to the direction of the $\vec{B}$ field.

$$B = \frac{F_{\vec{B}\text{ on W max}}}{IL} \tag{17.2}$$

Equation (17.2) allows us to define a unit for the $\vec{B}$ field, known as the **tesla, T** (named in honor of the Serbian inventor Nikola Tesla [1856–1943]). In a particular region (assuming a uniform field), the $\vec{B}$ field of magnitude 1 T exerts a force of 1 N on a 1-m-long wire with a 1-A current through it when the wire is oriented perpendicular to the $\vec{B}$ field: $1\text{ T} = 1\text{ N/A}\cdot\text{m}$. A 1-T $\vec{B}$ field is very strong. By comparison, the average value of the $\vec{B}$ field produced by Earth at the surface is $5 \times 10^{-5}$ T. Good quality bar magnets produce a $\vec{B}$ field near their poles of about 0.04 T in magnitude.

We can now use the definition of the magnitude of the $\vec{B}$ field to rewrite the expression for the magnetic force exerted by the field on a current.

**Magnetic force exerted on a current** **The magnitude of the magnetic force $F_{\vec{B}\text{ on W}}$ that a uniform magnetic field $\vec{B}$ exerts on a straight current-carrying wire of length $L$ with current $I$ is**

$$F_{\vec{B}\text{ on W}} = ILB\sin\theta \tag{17.3}$$

**where $\theta$ is the angle between the directions of the $\vec{B}$ field and the current $I$. The direction of this magnetic force is given by the right-hand rule for the magnetic force.**

## EXAMPLE 17.2 Magnetic field supports a clothesline

You wonder if instead of supporting your clothesline with two poles you could replace the poles and the clothesline with a current-carrying wire in Earth's $\vec{B}$ field, which near the surface has magnitude $5 \times 10^{-5}\,\text{T}$ and points north. Assume that your house is located near the equator, where the $\vec{B}$ field produced by Earth is approximately parallel to Earth's surface. The clothesline is 10 m long. The clothes and the line have a mass of 2.0 kg. What direction should you orient the clothesline and what current is needed to support it? Is this a promising way to support the clothesline?

**Sketch and translate** Draw a sketch representing the situation (see the figure). Then decide on the orientation of the line and direction of the current. The $\vec{B}$ field points northward (into the page), and you want the magnetic force exerted on the clothesline to point upward to balance the gravitational force exerted on it by Earth. Using the right hand rule for the magnetic force, you point your fingers north in the direction of the magnetic field. Your palm faces up (corresponding to the desired upward magnetic force to be exerted on the clothesline). Your thumb now points toward the east, the needed direction of the current (see inset).

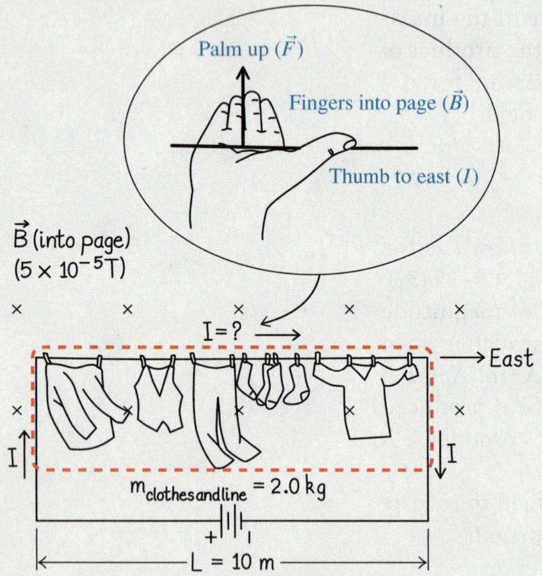


**Simplify and diagram** Assume that the $\vec{B}$ field produced by Earth is uniform in the region of the clothesline and neglect the mass of connecting (vertical) wires. Draw a force diagram for the clothesline + clothes system. Earth exerts a downward gravitational force ($\vec{F}_{\text{E on Cl}}$); Earth's magnetic field exerts an upward magnetic force ($\vec{F}_{\vec{B}\text{ on Cl}}$). Choose the upward direction as positive.

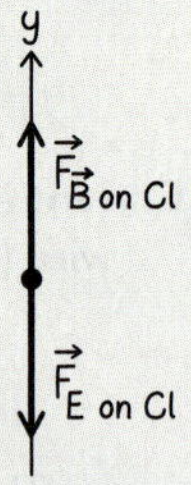


**Represent mathematically** For the system to remain at rest (zero acceleration), the $y$-components of the forces exerted on it must add to zero:

$$\sum F_y = F_{\text{E on Cl}\,y} + F_{\vec{B}\text{ on Cl}\,y}$$
$$= (-F_{\text{E on Cl}}) + F_{\vec{B}\text{ on Cl}} = 0$$

or

$$-m_{\text{Cl}}g + ILB\sin\theta = 0$$

**Solve and evaluate** We can solve the above for the current $I$ and insert the known quantities to get

$$I = \frac{m_{\text{Cl}}g}{LB\sin\theta} = \frac{(2.0\,\text{kg})(9.8\,\text{N/kg})}{(10.0\,\text{m})(5 \times 10^{-5}\,\text{T})(\sin 90°)}$$
$$= 3.9 \times 10^{4}\,\text{A}$$

This is a serious problem. Home wiring only supports about 20 A before circuit breakers shut off the current for safety reasons. There is no way to safely run a current of 39,000 A through the electrical system of a home.

**Try it yourself:** A 2.0-m-long wire has a 10-A current through it. The wire is oriented south to north and located near the equator. Earth's $\vec{B}$ field has a $5.0 \times 10^{-5}$-T magnitude in the vicinity of the wire. What is the magnetic force exerted by the field on the wire?

*Answer:* The wire is parallel to the $\vec{B}$ field. Thus, $\sin\theta = \sin(0°) = 0$ and the magnetic force exerted on the wire is zero.

## Summary of the differences between gravitational, electric, and magnetic forces

- The gravitational and electric forces exerted on objects do not depend on the direction of motion of those objects, whereas the magnetic force does. If the direction of the electric current is parallel or antiparallel (pointing exactly in the opposite direction) to the $\vec{B}$ field, no magnetic force is exerted on it.

**Figure 17.17** The directions of different forces relative to their fields.

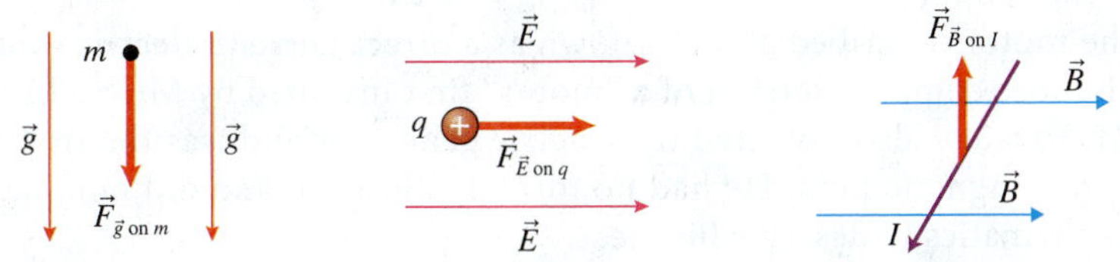


The gravitational force is parallel to the $\vec{g}$ field.

The electrical force is parallel to the $\vec{E}$ field.

The magnetic force is perpendicular to the magnetic field and to the electric current.

- While the forces exerted by the gravitational and the electric fields are always in the direction of the $\vec{g}$ or $\vec{E}$ field (or opposite that direction in the case of a negatively charged object), the force exerted by the magnetic field on a current-carrying wire is perpendicular to both the $\vec{B}$ field and the direction of the electric current (**Figure 17.17**).

## The direct current electric motor

A **motor** is a device that converts electric energy into mechanical energy, specifically rotational or translational kinetic energy. A simple motor consists of a rectangular current-carrying coil placed between the poles of a large electromagnet (**Figure 17.18a**). The coil rotates around an axle. How does this device convert electric energy into the rotational energy of the coil?

Let us start with the rectangular coil oriented so that the plane of the coil is parallel to the $\vec{B}$ field. The currents through sides 1 and 3 of the coil are perpendicular to the $\vec{B}$ field, so the field exerts a force on them. The currents through sides 2 and 4 are parallel to the $\vec{B}$ field: the force exerted on them is zero.

According to the right-hand rule for magnetic force, the magnetic field exerts an upward force on wire 1 of the coil and a downward force on wire 3 of the coil (Figure 17.18b). These forces each produce a torque around the axle that causes the coil to start rotating clockwise.

As the coil turns, the orientations of the currents relative to the $\vec{B}$ field change, as do the magnetic forces exerted by the field on these sides. As the coil reaches an orientation with its surface perpendicular to the $\vec{B}$ field, the magnetic field exerts forces on each side of the coil that tend to stretch it but are not able to turn it (Figure 17.18c). The coil turns past this orientation, reaching the one shown in Figure 17.18d. Using the right-hand rule for magnetic forces again, we find that the torques produced by the magnetic forces exerted on sides 1 and 3 cause the coil to accelerate in the counterclockwise direction, slowing down and reversing the direction of its rotation. This could be a serious problem unless we can reverse the direction of the current when the plane of the coil passes the perpendicular to the $\vec{B}$ field orientation (shown in Figure 17.18c) so that the net magnetic torque remains in the clockwise direction. Consequently, for the net torque produced by the magnetic force exerted on the coil to always remain clockwise, the current through the coil must change direction each time the coil passes the vertical orientation.

This reversal of the current is made possible using a device known as a **commutator** (**Figure 17.19**). A commutator consists of two semicircular rings that are attached to the rotating coil. Sliding contacts (metal brushes) connect the battery to the commutator rings. When the brushes change from one ring

**Figure 17.18** A magnetic field exerts a torque on a current-carrying coil.

(a)

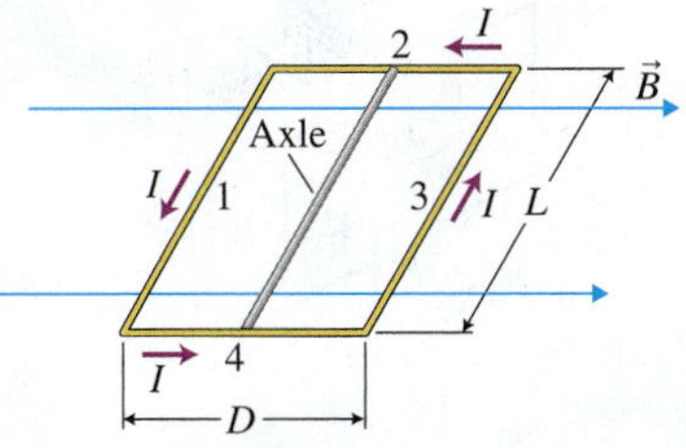


(b)

The magnetic forces exerted on sides 1 and 3 cause a clockwise torque on the coil.

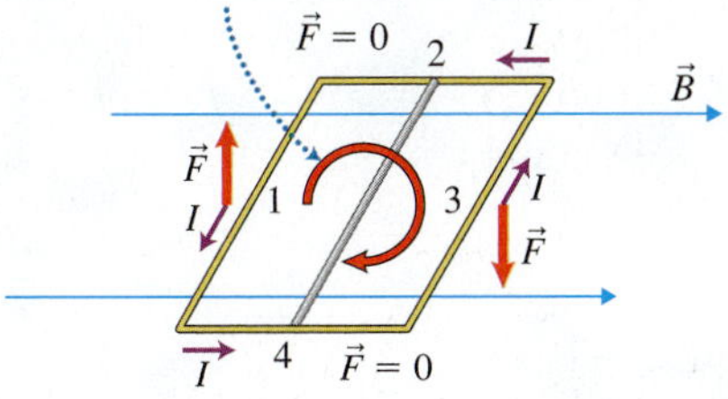


(c)

The magnetic forces due to the magnetic field stretch the coil but cause zero torque.

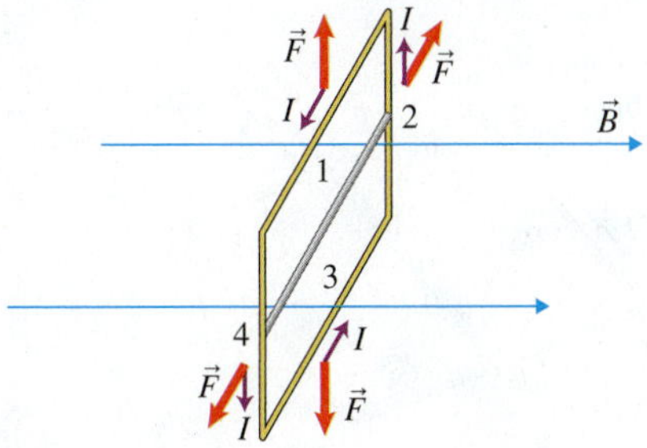


(d)

The magnetic field now exerts forces on sides 1 and 3 that cause a counterclockwise torque.

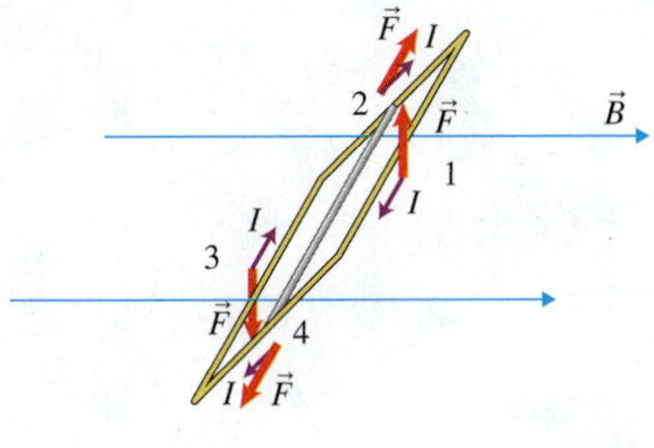

**Figure 17.19** A commutator ring allows a motor's steady magnetic field to continue exerting a torque on the coil in one direction.

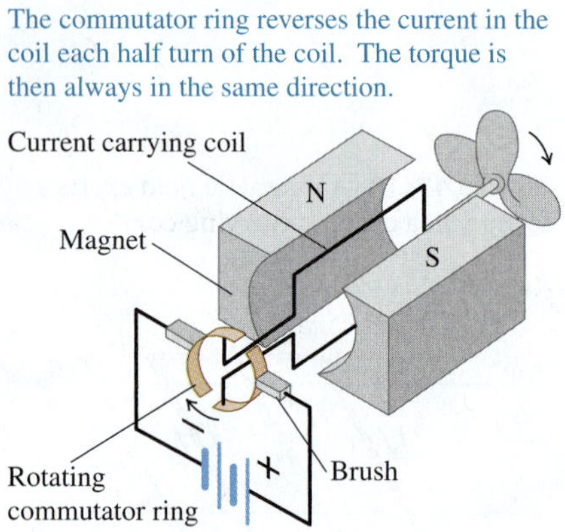
The commutator ring reverses the current in the coil each half turn of the coil. The torque is then always in the same direction.
Current carrying coil
N
Magnet
S
Rotating commutator ring
Brush

to the other, the direction of the current is reversed, as is the direction of the torque exerted on the coil.

The motor described above, known as a **direct current electric motor,** is a much more complex version of a "motor" first invented by Michael Faraday in 1821. Faraday also invented the electric generator and was the first to describe the magnetic field. He had no formal education and did not use complex mathematics to describe his ideas.

## Torque exerted on a current-carrying loop

We see that the torque produced by the magnetic forces exerted on a loop depends on the orientation of the loop relative to the $\vec{B}$ field. The magnitude of this torque depends on how far from the loop's rotation axis these magnetic forces are exerted. The torque that the magnetic force exerts on sides 1 and 3 of the loop in Figure 17.18b is directly proportional to the distance $D/2$ from the axis of rotation to where the force is exerted—shown in Figure 17.18a. As there are two equal-magnitude torques exerted on the loop with turning ability in the same direction, the total torque is proportional to $D$. The magnitude of the magnetic force exerted on wire 1 and on wire 3 depends on the length $L$ of that side of the loop (Figure 17.18a). Therefore, the total torque should be proportional to the product of $D$ and $L$, which is the area $A$ of the loop. In addition, if we have a coil with $N$ loops, the torque will be $N$ times the torque exerted on each loop.

We arrive at an expression for the magnitude of the torque that magnetic forces exert on a current-carrying coil:

$$|\tau_{\vec{B}\text{ on Coil}}| = NBAI\sin\theta \tag{17.4}$$

where $N$ is the number of turns in the coil, $B$ is the magnitude of the $\vec{B}$ field in the region of the coil (assumed uniform), $A$ is the area of the coil, $I$ is the electric current through the coil, and $\theta$ is the angle between a vector perpendicular to the coil's surface (called the **normal vector**) and the direction of the $\vec{B}$ field.

**Figure 17.20** The normal vector $\vec{n}$ indicates the orientation of a loop or coil's magnetic dipoles $\vec{p}_m$ relative to a $\vec{B}$ field.

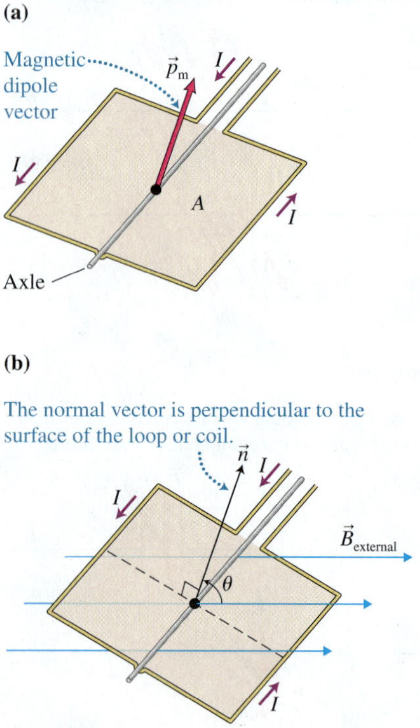
(a)
Magnetic dipole vector
$\vec{p}_m$
I
A
Axle
(b)
The normal vector is perpendicular to the surface of the loop or coil.
$\vec{n}$
I
$\vec{B}_{external}$
θ

## Magnetic dipole moment

The product of the current $I$ and area $A$ in Eq. (17.4) is called the magnitude of the **magnetic dipole moment** $\vec{p}_m$ of the loop ($|\vec{p}_m| = IA$). The direction of the dipole moment vector is perpendicular to the surface of the loop and in the direction of the $\vec{B}$ field produced by current at the center of the loop (**Figure 17.20a**). The bigger the dipole moment of a loop, the greater the torque that an external magnetic field exerts on it:

$$|\tau_{\vec{B}\text{ on Coil}}| = NBAI\sin\theta = NBp_m\sin\theta$$

where $\theta$ is the angle between the normal vector perpendicular to the surface of the current loop (same direction as the dipole moment vector) and the external $\vec{B}$ field vector (see Figure 17.20b).

## Using a coil in a magnetic field to measure current—an ammeter

We can use Eq. (17.4) as the basis for a method to measure the current through a wire. Place a coil of wire between the poles of a horseshoe magnet whose $\vec{B}$ field is known. Orient the coil so that its normal vector is perpendicular to the $\vec{B}$ field. Connect this coil in series with the wire you want to measure the current through. When the current is on, the magnetic field exerts forces on the current through the coil that result in a magnetic torque exerted on the coil. This torque tends to align the magnetic dipole moment of the coil with the external field. If we attach springs to the turning sides of the coil, the springs exert torques that oppose the magnetic torque. The coil then turns until the spring torques balance the magnetic torque (**Figure 17.21**). The current through the coil can then be determined by Eq. (17.4):

$$I = \frac{|\tau_{\vec{B}\text{ on Coil}}|}{NAB\sin(90° - \theta)} = \frac{|\tau_{\text{Springs on Coil}}|}{NAB\sin(90° - \theta)}$$

The torque exerted by the springs will be proportional to their stretch distance, so all quantities on the right-hand side of the equation are measurable. This allows us to measure the current $I$ through the coil. This method is used in analog ammeters.

**Review Question 17.3** Equation (17.2) defines the magnitude of the $\vec{B}$ field at a point as $B = F_{\vec{B}\text{ on W max}}/IL$. Explain the meaning of every symbol in this equation. Do any quantities on the right side of the equation cause the $\vec{B}$ field to have a bigger or smaller magnitude? Explain.

**Figure 17.21** A device for measuring electric current.

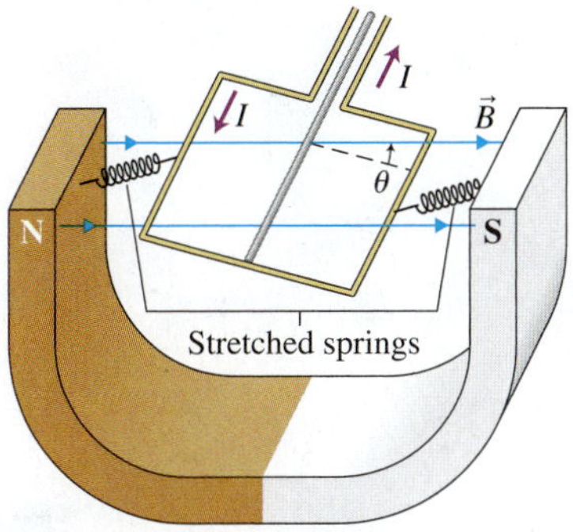


# 17.4 Magnetic force exerted on a single moving charged particle

We have learned that the magnetic field exerts a force on a current-carrying wire. The current through a wire is the result of the collective motion of a huge number of electrically charged particles—electrons. Thus, it is reasonable to conclude that the magnetic field also exerts a force on each individual electron. In fact, it is because of those forces that physicists were able to discover electrons. In addition, if the magnetic field exerts a force on moving electrons, it is reasonable to conclude that it also exerts a force on other moving charged particles, such as protons and helium nuclei. Those particles bombard Earth from the Sun and distant stars. The magnetic force exerted on individual particles protects us from the harmful effects of those particles.

## Direction of the force that the magnetic field exerts on a moving charged particle

An **oscilloscope** (**Figure 17.22a**) demonstrates how a magnetic field affects moving electrons. An oscilloscope consists of two electrodes—a **cathode** and a hollow **anode**—in an evacuated glass enclosure. The cathode is connected to the negative terminal of a power source and the anode to the positive terminal. A current caused by a separate power source keeps the cathode hot. Due to its high temperature, the cathode emits electrons (more on this in Chapter 26). The potential difference between the hot cathode and the anode causes the electrons to accelerate toward the anode, through the hole in the anode, and onto a screen. The screen is treated with a material that glows green when hit by electrons. We can use the location of the green dot on the screen to infer the path of the electrons within the oscilloscope.

If a magnetic field due to a hand-held bar magnet (not shown) exerts a force on individual electrons similar to the force it exerts on an electric current in a wire, then we should be able to use the right-hand rule for the magnetic force to predict the direction that the electrons will be deflected. The rule was formulated in terms of electric current, which by convention is the direction that positively charged particles move. The magnetic force exerted on negative electrons moving toward the screen should be in the opposite direction to the one given by the right-hand rule for the magnetic force. To test this reasoning, we orient the magnet so that its $\vec{B}$ field points perpendicular to the electron path shown in Figure 17.22a (into the page, as

**Figure 17.22** An electron beam in an oscilloscope is deflected by a magnetic field.

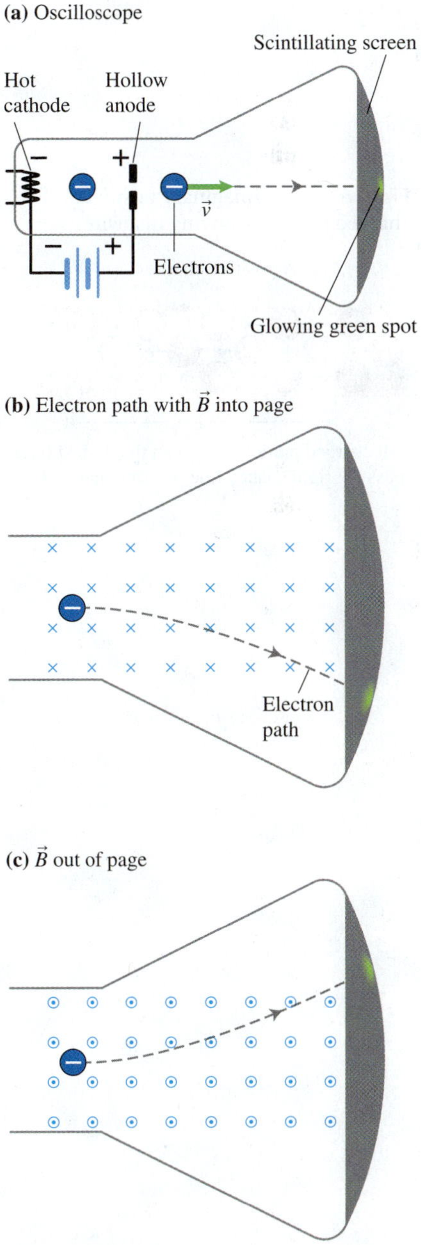

in Figure 17.22b). The right-hand rule for the magnetic force predicts that positively charged particles should be deflected upward while the negatively charged electrons inside the oscilloscope should deflect downward. If we reverse the direction of the magnetic field, the electrons should be deflected upward (Figure 17.22c). When the experiment is performed, the outcome is consistent with the prediction. This result supports the idea that the magnetic force exerted by the magnetic field on individual charged objects is similar to the one it exerts on currents. Let's construct a quantitative relationship for the magnitude of the magnetic force exerted by the magnetic field on an individual charged particle.

## Magnitude of the force that a magnetic field exerts on a moving charged particle

We know that a magnetic field exerts on a current-carrying wire a force of magnitude

$$F_{\vec{B}\text{ on W}} = ILB\sin\theta$$

We use this to develop an expression for the magnitude of the force that the magnetic field exerts on a single charged object with charge $q$ moving at speed $v$ (the current $I$ in the wire consists of a large number of moving charged particles—**Figure 17.23**). Although we know that electrons move in the wire when current is present, for simplicity we use positively charged particles in this calculation.

**Figure 17.23** Imaginary positively charged particles moving in a wire.

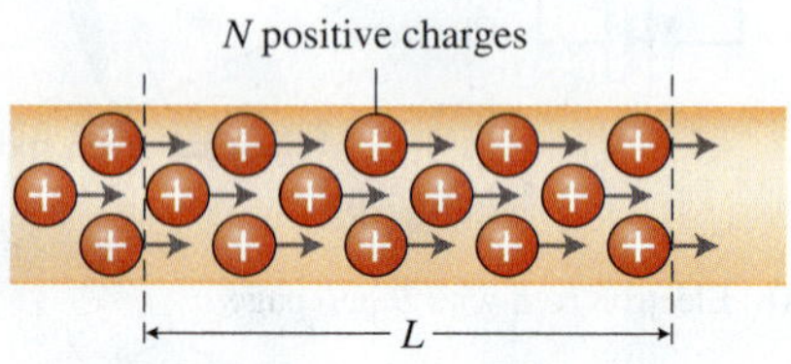
N positive charges
L
All charged particles between the dashed lines pass the right dashed line in a time interval Δt.

Imagine that between the two dashed lines there are $N$ moving charged particles. In a time interval $\Delta t$, all of them pass through the dashed line on the right. Thus, the electric current through this wire is

$$I = \frac{Nq}{\Delta t}$$

The speed of the charged particles is $v = L/\Delta t$ since a particle at the left dashed line takes a time interval $\Delta t$ to reach the right dashed line. Rearrange this for $\Delta t = L/v$ and substitute in the above:

$$I = \frac{Nqv}{L}$$

Inserting this into $F_{\vec{B}\text{ on W}} = ILB\sin\theta$, we get

$$F_{\vec{B}\text{ on W}} = \left(\frac{Nqv}{L}\right)LB\sin\theta = N(qvB\sin\theta)$$

This is the magnitude of the force exerted by the magnetic field on all $N$ moving charged particles. The magnitude of the force exerted by the field on a single charged particle is then

$$F_{\vec{B}\text{ on }q} = |q|vB\sin\theta$$

**Magnetic force exerted by the magnetic field on an individual charged particle** **The magnitude of the magnetic force that the magnetic field of magnitude $B$ exerts on a particle with electric charge $q$ moving at speed $v$ is**

$$F_{\vec{B}\text{ on }q} = |q|vB\sin\theta \qquad (17.5)$$

**where $\theta$ is the angle between the direction of the velocity of the particle and the direction of the $\vec{B}$ field. The direction of this force is determined by the right-hand rule for the magnetic force. If the particle is negatively charged, the force points opposite the direction given by the right-hand rule.**

**Figure 17.24** The right-hand rule for the magnetic force exerted by a magnetic field on a charged particle.

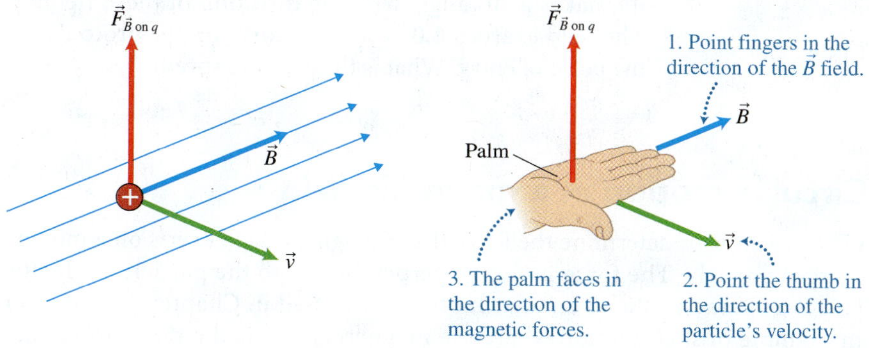
(a)
(b)
$\vec{F}_{\vec{B}\text{ on }q}$
$\vec{B}$
$\vec{v}$
$\vec{F}_{\vec{B}\text{ on }q}$
1. Point fingers in the direction of the $\vec{B}$ field.
$\vec{B}$
Palm
$\vec{v}$
3. The palm faces in the direction of the magnetic forces.
2. Point the thumb in the direction of the particle's velocity.

The direction of the magnetic force that a magnetic field exerts on a moving charged particle is illustrated in **Figure 17.24a**. The right-hand rule for determining the force direction is described below (and in Figure 17.24b).

**Right-hand rule for the direction of the magnetic force exerted on a moving charged particle** **Hold your right hand flat with your thumb perpendicular to your fingers and pointing in the direction of the object's velocity. Point your fingers in the direction of the $\vec{B}$ field. The direction of the magnetic force exerted by the magnetic field on a positively charged particle is in the direction your palm faces—perpendicular to both the velocity and the $\vec{B}$ field. The force exerted by the magnetic field on a negatively charged particle is in the opposite direction.**

**TIP** Note that if the velocity and the $\vec{B}$ field are parallel, $\sin 0° = 0$ and the force is zero. The force is maximum when the object's velocity and the $\vec{B}$ field are perpendicular.

**QUANTITATIVE EXERCISE 17.3 Particles in a magnetic field**

Each of the lettered dots shown in the figure represents a small object with an electric charge $+2.0 \times 10^{-6}$ C moving at speed $3.0 \times 10^{7}$ m/s in the directions shown. Determine the magnetic force (magnitude and direction) that a 0.10-T $\vec{B}$ field exerts on each object. The $\vec{B}$ field points in the positive $y$-direction.

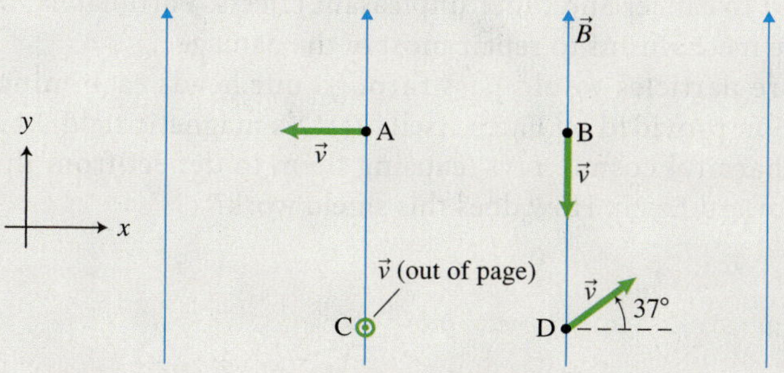
$\vec{B}$
y
x
A
$\vec{v}$
B
$\vec{v}$
$\vec{v}$ (out of page)
C
$\vec{v}$
37°
D

**Represent mathematically** First, use the right-hand rule for the magnetic force to determine the directions of the magnetic force exerted on each object. (a) For object A, point the fingers of your right hand toward the top of the page in the direction of $\vec{B}$. Then orient your hand so that your thumb points to the left in the direction of $\vec{v}$. Your palm points into the page. This is the direction of the magnetic force exerted on object A. (b) Object B moves in a direction opposite to $\vec{B}$ ($\theta = 180°$); thus, the magnetic force is zero. (c) For C, point your fingers toward the top of the page and the thumb out of the page. Your palm then faces left, so the magnetic force exerted on object C points in the negative $x$-direction. (d) For object D, point your fingers toward the top of the page and your thumb parallel to the page pointing 37° above rightward. Your palm faces out of the page, the direction of the force exerted on D.

**Solve and evaluate** Use Eq. (17.5) to determine the magnitude of each force:

$$F_{\vec{B}\text{ on A}} = (2.0 \times 10^{-6}\,\text{C})(3.0 \times 10^{7}\,\text{m/s})(0.10\,\text{T})\sin(90°) = 6.0\,\text{N}$$

$$F_{\vec{B}\text{ on B}} = (2.0 \times 10^{-6}\,\text{C})(3.0 \times 10^{7}\,\text{m/s})(0.10\,\text{T})\sin(180°) = 0$$

$$F_{\vec{B}\text{ on C}} = (2.0 \times 10^{-6}\,\text{C})(3.0 \times 10^{7}\,\text{m/s})(0.10\,\text{T})\sin(90°) = 6.0\,\text{N}$$

$$F_{\vec{B}\text{ on D}} = (2.0 \times 10^{-6}\,\text{C})(3.0 \times 10^{7}\,\text{m/s})(0.10\,\text{T})\sin(53°) = 4.8\,\text{N}$$

*(continued)*

**Try it yourself:** The equation below represents the solution to a problem. Devise a possible problem that is consistent with the equation:

$$(1.6 \times 10^{-19}\,\text{C})v(0.50 \times 10^{-5}\,\text{T})\sin(30^\circ) = 1.0 \times 10^{-18}\,\text{N}$$

*Answer:* One possible problem: A proton enters Earth's magnetic field far above the surface. The proton's velocity makes a 30° angle with the direction of the $\vec{B}$ field. The field exerts a $1.0 \times 10^{-18}$ N force on the proton at the point of entry. What is the proton's speed?

## Circular motion in a magnetic field

**Figure 17.25** A charged particle moving perpendicular to a magnetic field moves in a circle.

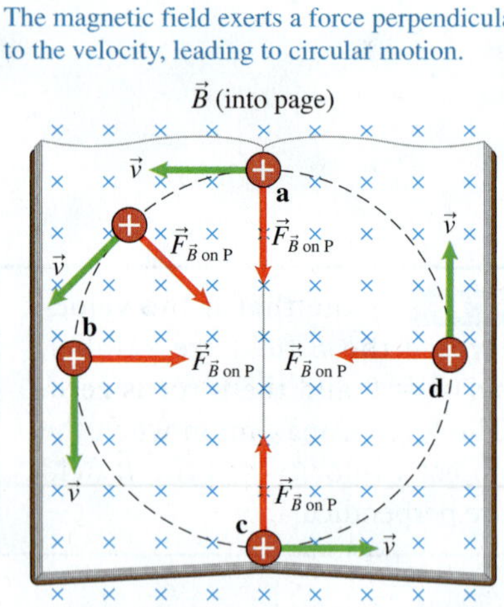

The magnetic field exerts a force perpendicular to the velocity, leading to circular motion.
$\vec{B}$ (into page)
$\vec{v}$
a
b
c
d
$\vec{F}_{B \text{ on } P}$

We know how to determine the force that a magnetic field exerts on a moving charged particle. The force is always perpendicular to the particle's velocity. This is a characteristic of circular motion (discussed in Chapter 4). Consider an example of such motion where the magnetic force is the radial force that causes constant speed circular motion.

Imagine that a positively charged particle moves to the left across the top of your open textbook (position **a** in **Figure 17.25**) when a uniform magnetic field pointing into the page turns on. Using the right-hand rule for the magnetic force, we find that the field exerts a magnetic force on the particle that points downward. The particle continues to move forward, but the direction of its velocity changes and now points slightly downward. The force exerted by the magnetic field always points perpendicular to the particle's velocity. This deflects the particle further. Once it has made a quarter turn and reaches position **b**, the particle is moving toward the bottom of the page, and the magnetic field exerts a force toward the right. This pattern persists—the magnetic force exerted on the particle always points toward the center of the particle's path. Thus, in a uniform $\vec{B}$ field a charged particle that initially moves perpendicular to the $\vec{B}$ field will move along a circular path in a plane perpendicular to the field.

## Cosmic rays

The deflection of charged particles in a magnetic field has an important everyday application. Charged particles zoom past and *through* us every minute, bombarding Earth and its inhabitants. These particles, called **cosmic rays**, are usually electrons, protons, and other elementary particles produced by various astrophysical processes including those occurring in the Sun and sources outside the solar system. Every minute about 20 of these fast-moving charged particles pass through a person's head. These particles can cause genetic mutations, which can lead to cancer and other unpleasant effects. Fortunately, our bodies have multiple mechanisms to repair most of the damage.

Thousands more particles would pass through our heads each minute without the protection provided by Earth itself. Earth's magnetic field serves as a shield against harmful cosmic rays, causing them to deflect from their original trajectory toward Earth. How does this shield work?

### EXAMPLE 17.4 Motion of protons in Earth's magnetic field

Determine the path of a cosmic ray proton flying into Earth's atmosphere above the equator at a speed of about $10^7$ m/s and perpendicular to Earth's magnetic field. The average magnitude of Earth's $\vec{B}$ field in this region is approximately $5 \times 10^{-5}$ T. The mass $m$ of a proton is approximately $10^{-27}$ kg.

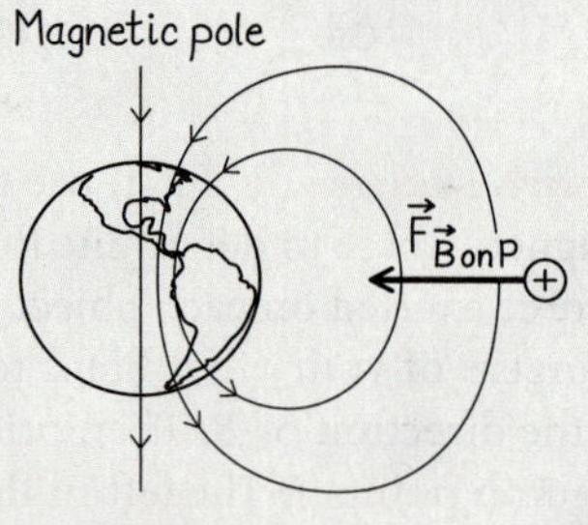

Magnetic pole
$\vec{F}_{BonP}$

**Sketch and translate** The situation is sketched in the figure.

**Simplify and diagram** Consider a short distance that the proton travels high above Earth and assume that in this region the $\vec{B}$ field vectors are parallel to Earth's surface and have a constant magnitude of $5 \times 10^{-5}$ T. We neglect the gravitational force that Earth exerts on the proton since it is extremely small in comparison to the magnetic force exerted on it. A force diagram for the proton is shown above.

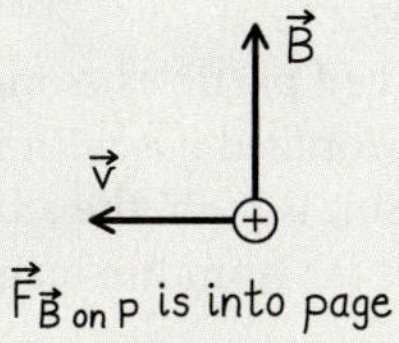


**Represent mathematically** When the velocity of a charged particle is perpendicular to the $\vec{B}$ field, it moves in a circular path at constant speed. We can use the radial $r$ component form of Newton's second law to relate the magnetic force exerted on the proton to its resulting motion. The force exerted by the magnetic field points toward the center of the proton's circular path (see the figure below):

$$a_r = \frac{v^2}{r} = \frac{1}{m}\sum F_r = \frac{1}{m}(F_{\vec{B} \text{ on P } r}) = \frac{1}{m}(|q|vB\sin\theta)$$

$$= \frac{|q|vB\sin(90°)}{m} = \frac{|q|vB}{m}$$

This equation can be used to determine the radius $r$ of the proton's circular path and to determine the period $T$ of its motion, noting that

$$v = \frac{2\pi r}{T}$$

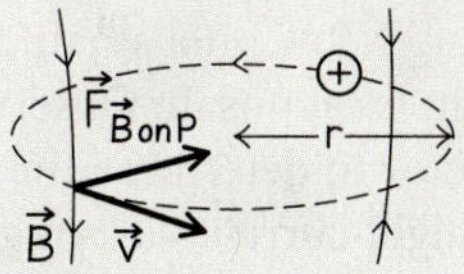


A cosmic ray proton (⊕) enters Earth's magnetic field high above Earth's surface. The magnetic force causes it to move in a circle.

**Solve and evaluate** We used Newton's second law to develop an expression describing the proton's circular motion:

$$\frac{v^2}{r} = \frac{|q|vB}{m}$$

Multiply both sides by the product of $r$ and $m$ and then rearrange to get an expression for $r$:

$$\frac{v^2 mr}{r} = \frac{|q|vBmr}{m}$$

$$\Rightarrow mv^2 = |q|vBr$$

$$\Rightarrow r = \frac{mv}{|q|B}$$

$$\approx \frac{(10^{-27}\text{ kg})(10^7\text{ m/s})}{|1.6 \times 10^{-19}\text{ C}|(5 \times 10^{-5}\text{ T})} \approx 10^3\text{ m}$$

The period $T$ of the proton's motion is then

$$T = \frac{2\pi r}{v} \approx \frac{2\pi(10^3\text{ m})}{10^7\text{ m/s}} \approx 10^{-3}\text{ s}$$

Something interesting happens if we combine the equations for $r$ and $T$:

$$T = \frac{2\pi r}{v} = \frac{2\pi}{v}\left(\frac{mv}{|q|B}\right) = \frac{2\pi m}{|q|B}$$

We find that the period of the proton's motion depends only on the mass and charge of the proton and the magnitude of the $\vec{B}$ field. It does not depend on the speed of the proton.

**Try it yourself:** What happens to the motion of a proton that enters Earth's magnetic field parallel to the $\vec{B}$ field?

*Answer:* The motion of the proton will not be affected by the magnetic field.

Earth's magnetic field acts as a shield, protecting life on the surface from cosmic ray particles. A cosmic ray proton entering the magnetic field of Earth at one-third light speed, or $1 \times 10^8$ m/s, will move in a helical path with a radius of 1 or more km. Earth's magnetic field extends several tens of thousands of kilometers above the surface, so life on Earth is well protected. The Earth's atmosphere itself serves as an additional shield, absorbing those protons that enter it. However, astronauts traveling outside of Earth's atmosphere have neither the atmospheric shield nor the magnetic shield. They are exposed to higher levels of cosmic radiation than is present on Earth's surface. This exposure is an important consideration in planning future long missions to Mars and other planets.

**TIP** Do not confuse the period $T$ with the unit for the magnetic field, the tesla T.

## The auroras

Charged particles moving in Earth's magnetic field actually follow helical paths around the $\vec{B}$ field lines. The charged particles entering the field travel in a helical path that follows the $\vec{B}$ field lines and enter the Earth's atmosphere near the poles (**Figure 17.26**). These particles collide with molecules in the atmosphere, ionizing the molecules. When the electrons recombine with the ionized molecules, the excess energy is radiated as light. We see this light in the upper atmosphere

**Figure 17.26** Paths of particles entering Earth's magnetic field from space.

Particles entering slightly less than perpendicular to Earth's magnetic field follow helical paths toward the poles, causing auroras.

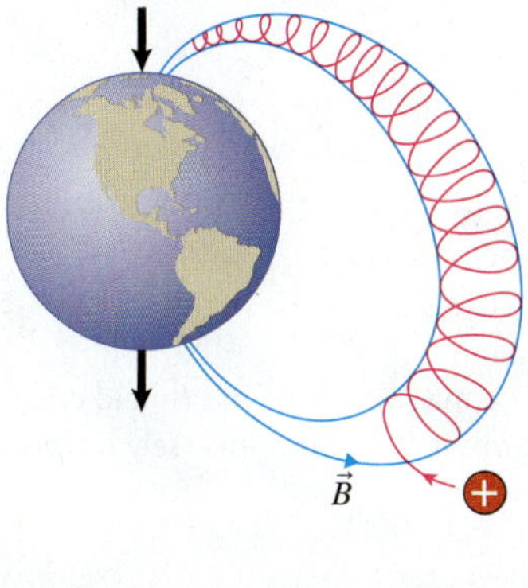

**Figure 17.27** An aurora.

in the region of the magnetic poles—the auroras mentioned at the beginning of the chapter (**Figure 17.27**).

Often these charged particles come from solar flares caused by the interactions of the Sun's hot ionized gas with its magnetic field. Therefore, when magnetic activity on the Sun is high, the auroras become more intense. On occasion, the auroras are visible far from the magnetic pole regions, sometimes even quite close to the equator.

**Review Question 17.4** If the magnetic force is always perpendicular to the velocity of a charged particle, can it do any work on the particle? Explain your answer.

## 17.5 Magnetic fields produced by electric currents

So far we have qualitatively investigated the $\vec{B}$ fields produced by current-carrying straight wires, loops, and solenoids. To analyze problems such as the danger of magnetic fields due to high-power transmission lines, we need to know how to quantitatively determine the magnitude and direction of a $\vec{B}$ field produced by a particular current configuration (the source)—the subject of this section.

### The $\vec{B}$ field produced by an electric current in a long straight wire

An electric current in a long straight wire produces a magnetic field whose $\vec{B}$ field lines circle around the wire (**Figure 17.28**). What is the magnitude of that field at various locations near the wire? To answer this question we need to place some kind of a detector (test object) of magnetic field at different locations and investigate the effects of the field on this object. In Section 17.3 we learned that a magnetic field will produce a maximum torque $|\tau_{\vec{B}\text{ on C max}}| = NBAI$ on a small current-carrying coil used to probe a magnetic field at different locations. In Observational Experiment **Table 17.5**, we use such a coil as a detector to determine the relative strength of the magnetic field produced by a long straight current-carrying wire.

**Figure 17.28** The magnetic field caused by a long current-carrying wire.

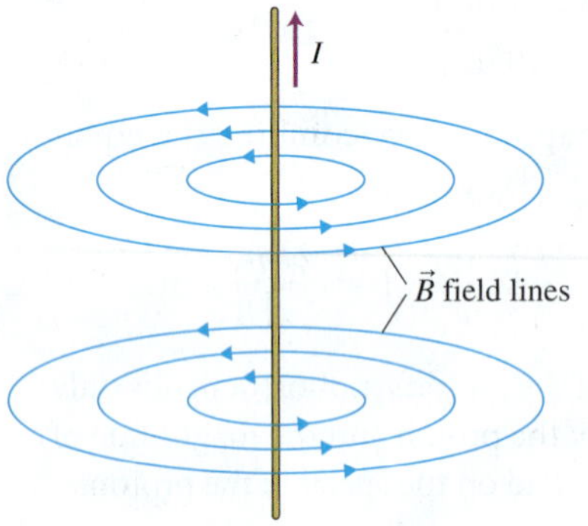


**OBSERVATIONAL EXPERIMENT TABLE**

**17.5 $\vec{B}$ field around a straight current-carrying wire.**

| Observational experiment | Analysis |
|---|---|
| Use a small light coil connected to a battery as a magnetic field detector. Use the maximum torque exerted on the detector to determine the magnitude of the $\vec{B}$ field surrounding a long straight wire with different currents $I$ and at different distances $r$ from the wire (the source). | Look for a pattern in how $B$ depends first on $I$ and then on $r$. |

| $I_{source}$ (the straight wire) | $r$ (distance between wire and detector) | $B$ (produced by straight wire) |
|---|---|---|
| $I_{wire}$ | $r$ | $B$ |
| $2I_{wire}$ | $r$ | $2B$ |
| $3I_{wire}$ | $r$ | $3B$ |
| $I_{wire}$ | $2r$ | $B/2$ |
| $I_{wire}$ | $3r$ | $B/3$ |
| $I_{wire}$ | $r/2$ | $2B$ |

**Pattern**

The magnitude of the $\vec{B}$ field created by a long straight current-carrying wire is directly proportional to the magnitude of the current $I_{source}$ and inversely proportional to the distance $r$ between the wire and the location where the field is measured.

We can express the pattern identified in Table 17.5 mathematically as follows:

$$B_{\text{straight wire}} \propto \frac{I_{\text{wire}}}{r}$$

Since we can experimentally measure all three of the quantities appearing in this relationship, it's possible to determine the constant of proportionality that will turn it into an equation. Traditionally this constant is written as $\mu_0/2\pi$, where $\mu_0 = 4\pi \times 10^{-7}\ \text{T}\cdot\text{m/A}$. We now have an expression for the magnitude of the $\vec{B}$ field at a perpendicular distance $r$ from a long straight current-carrying wire:

$$B_{\text{straight wire}} = \frac{\mu_0}{2\pi}\frac{I_{\text{wire}}}{r} \tag{17.6}$$

Note that the farther you move from the current-carrying wire (larger $r$), the smaller the magnitude of the $\vec{B}$ field; the greater the current (larger $I_{\text{wire}}$), the larger the magnitude of the $\vec{B}$ field.

## Magnetic permeability

With this value of $\mu_0$, the $\vec{B}$ field magnitude produced 0.10 m from a long straight wire with a current of 10.0 A is $2.0 \times 10^{-5}$ T. The constant $\mu_0$ is known as the **magnetic permeability**. When using Eq. (17.6), we assume that the region where we measure the $\vec{B}$ field is a vacuum (as opposed to a medium such as water or oil). The magnetic permeability of air is approximately equal to $\mu_0$ as well. However, the magnetic permeability of iron is about 1000 times greater than $\mu_0$. In other words, the $\vec{B}$ field within iron is 1000 times greater than if only air were there. In general, Eq. (17.6) is written as

$$B_{\text{straight wire}} = \frac{\mu}{2\pi}\frac{I}{r}$$

where $\mu$ is the magnetic permeability of the substance present at the location of interest.

**Figure 17.29** summarizes the information about magnetic fields created by different configurations of current-carrying wires. We will use the magnitude of the $\vec{B}$ field at the center of a single current-carrying loop next to analyze the magnetic field produced by the motion of an electron in a hydrogen atom.

## $\vec{B}$ field due to electron motion in an atom

In an early 20th-century model of the hydrogen atom, the electron was thought to move rapidly in a tiny circular path around the nucleus of the atom (similar to the figure in Example 17.5). Although the model itself is outdated, it allows us to make reasonable predictions about magnetic properties of individual atoms and objects made from them. In this model the electron motion is like a circular electric current $I$ that produces a $\vec{B}$ field. The $\vec{B}$ field at the center of a current-carrying loop of radius $r$ is

$$B = \frac{\mu_0 I}{2r}$$

The electron's circular motion also produces a magnetic dipole moment of magnitude $p_m = IA$ equal to the product of the electron current $I$ and the area $A$ of

**Figure 17.29** Expressions for the $\vec{B}$ field produced by (a) a straight wire, (b) a loop or coil, and (c) a solenoid.

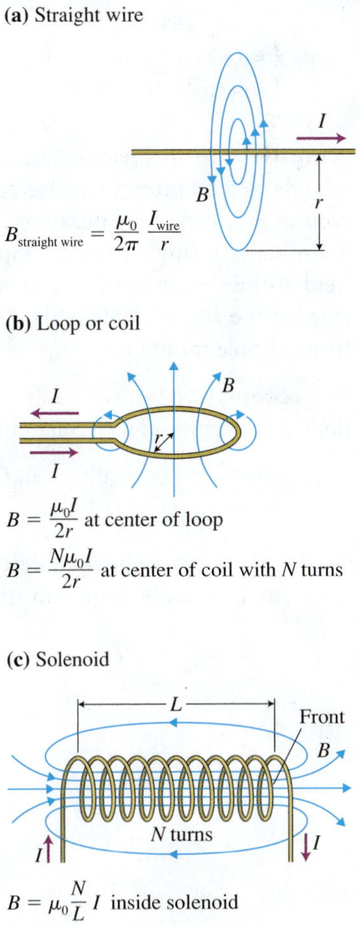

the loop. If these dipole moments are significant in magnitude, they could potentially help explain magnetic properties of materials. In the next example we make a rough estimate of the hydrogen atom's electric current $I$ and the $\vec{B}$ field and dipole moment produced by this current.

**EXAMPLE 17.5 Magnetic field produced by electron in a hydrogen atom**

In the above-mentioned early 20th-century model of the hydrogen atom, the electron was thought to move in a circle of radius $0.53 \times 10^{-10}$ m, orbiting once around the nucleus every $1.5 \times 10^{-16}$ s. Determine the magnitude of the $\vec{B}$ field produced by the electron at the center of its circular orbit and its dipole moment.

**Sketch and translate** A sketch of the moving electron is shown in the figure along with the known information.

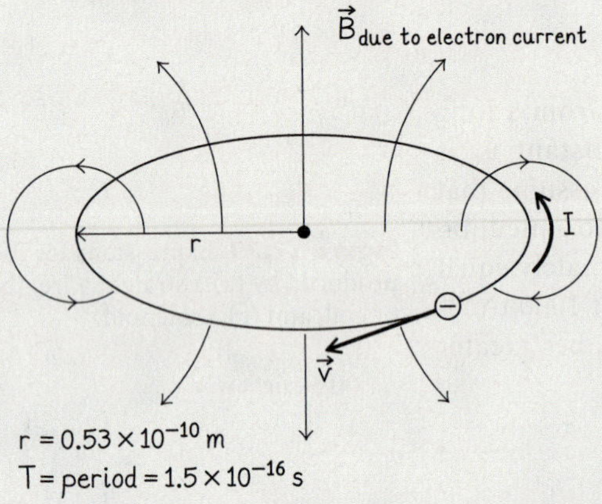


**Simplify and diagram** The electron's motion corresponds to a counterclockwise current (opposite the direction of travel of the negatively charged electron). This is similar to a single current loop. The direction of the $\vec{B}$ field at the center can be determined by the right-hand rule for the $\vec{B}$ field. This is also the direction of the electron's dipole moment.

**Represent mathematically** The magnitude of the $\vec{B}$ field at the center of a circular current $I$ is

$$B = \frac{\mu_0 I}{2r}$$

To determine $B$ we have to determine the electric current due to the electron's motion. Electric current is $I = q/\Delta t$, where $q$ is the magnitude of the total electric charge that passes a cross section of a wire in time $\Delta t$. The cross section in this case is a single point along the electron's orbit. The single electron passes that point once every $1.5 \times 10^{-16}$ s.

**Solve and evaluate** Using these ideas, we can calculate the magnitude of the current:

$$I = \frac{q}{\Delta t} = \frac{1.6 \times 10^{-19}\,\text{C}}{1.5 \times 10^{-16}\,\text{s}} = 1.1 \times 10^{-3}\,\text{A}$$

The magnitude of the $\vec{B}$ field at the center of the loop is then

$$B = \frac{\mu_0 I}{2r} = \frac{(4\pi \times 10^{-7}\,\text{T}\cdot\text{m/A})(1.1 \times 10^{-3}\,\text{A})}{2(0.53 \times 10^{-10}\,\text{m})} = 13\,\text{T}$$

This is a huge $\vec{B}$ field, especially compared to the $10^{-5}$-T $\vec{B}$ field of Earth. The magnitude of the dipole moment of an electron is

$$p_\text{m} = IA = (1.1 \times 10^{-3}\,\text{A})[\pi(0.53 \times 10^{-10}\,\text{m})^2]$$
$$= 9.7 \times 10^{-24}\,\text{A}\cdot\text{m}^2$$

This is a tiny quantity compared to the magnetic moment of a macroscopic current-carrying loop. However, one object can easily contain a trillion trillion electrons.

**Try it yourself:** We found in the preceding example that a moving electron in an atom produces a very strong magnetic field. Suggest an explanation for why all materials are not strong magnets.

*Answer:* One possible reason is that the electron orbits of the individual atoms are oriented randomly and therefore so are their dipole moments. If this is correct, their dipole moments would cancel each other. Another reason could be that most atoms have more than one electron and magnetic moments produced by each of these electrons add to zero so that the $\vec{B}$ field produced by each individual atom is zero.

**Review Question 17.5** The definition of a 1-A current states that two parallel 1.0-m-long current-carrying wires separated by 1.0 m with a current of 1 A in each exert an attractive force on each other of magnitude $2.0 \times 10^{-7}$ N. Where does this value of $2.0 \times 10^{-7}$ N originate?

# 17.6 Skills for analyzing magnetic processes

Problems involving magnetic interactions are often of two main types: (1) determine the magnetic force exerted on a current or on an individual moving charged object by the magnetic field or (2) determine the $\vec{B}$ field produced by a known source such as an electric current. In this section we will develop skills needed to solve such problems. The general procedure is described on the left side of the skill box and illustrated with a solution to the following example on the right side.

## PROBLEM-SOLVING STRATEGY Applying our knowledge of the magnetic field

### EXAMPLE 17.6 Magnetic force problem

A horizontal metal wire of mass 5.0 g and length 0.20 m is supported at its ends by two very light conducting threads. The wire hangs in a 49-mT magnetic field, which points perpendicular to the wire and out of the page. The maximum force each thread can exert on the wire before breaking is 39 mN. What is the minimum current through the wire that will cause the threads to break?

**Sketch and translate**

- Sketch the process described in the problem.
- Choose the system of interest.
- Show the direction of the $\vec{B}$ field and the direction of the electric current (or the velocity of a charged particle) if known.
- Decide whether the problem asks to find a $\vec{B}$ field produced by an electric current or to find a magnetic force exerted by an external field on a moving charged particle or wire with electric current.

A sketch of the situation is shown along with the known information and the unknown quantity. We do not yet know the direction of the current.

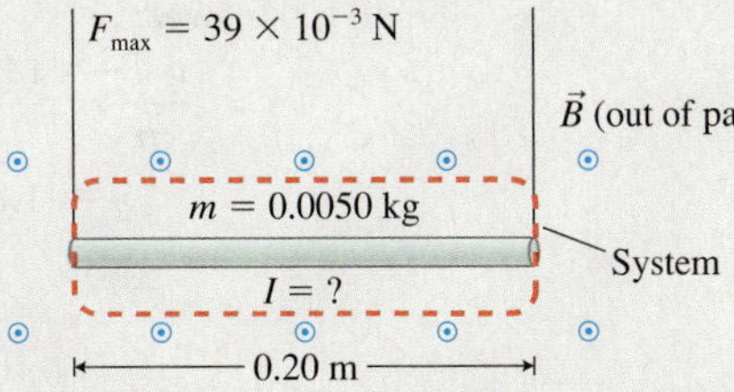


The horizontal wire is the system. What downward force exerted by the magnetic field, when added to the gravitational force exerted by Earth, will be enough to break the threads? Then what current will produce this downward magnetic force?

This is a problem about finding the force exerted by the external field on an object.

**Simplify and diagram**

- Decide whether the $\vec{B}$ field can be considered uniform in the region of interest.
- Draw a force diagram for the system (the object in the field region) if necessary. Use the right-hand rule for the magnetic force to find an unknown force, current, velocity, or field direction if needed.
- Use the right-hand rule for the $\vec{B}$ field if the problem is about the field of a known source.

- Nothing specific is mentioned about the $\vec{B}$ field, so we consider it uniform in the vicinity of the wire.
- Construct a force diagram for the wire. The wire interacts with the threads ($\vec{F}_{T\,on\,W}$), Earth ($\vec{F}_{E\,on\,W}$), and the magnetic field ($\vec{F}_{\vec{B}\,on\,W}$). With the $\vec{B}$ field pointing out of the page, the magnetic force points down in the desired direction if the wire current is toward the right. We choose the $y$-axis as pointing down.

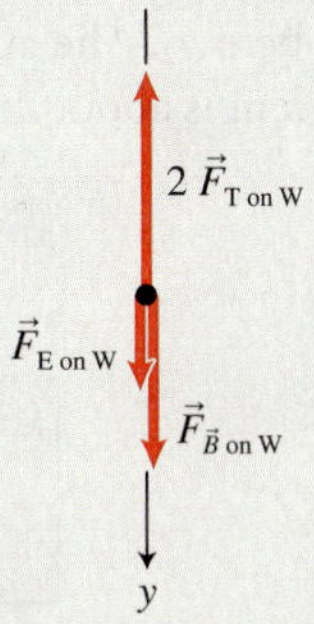


*(continued)*

**Represent mathematically**

- Describe the situation mathematically using the expressions for magnetic force exerted on a current or charged particle and the expressions for the $\vec{B}$ field produced by currents.
- If necessary, use Newton's second law in component form and/or kinematics.

The wire is in equilibrium ($\Sigma F_y = 0$). In component form,

$$F_{\text{E on W}\,y} + 2F_{\text{T on W}\,y} + F_{\vec{B}\text{ on W}\,y} = 0$$
$$\Rightarrow mg + 2(-F_{\text{T on W}}) + ILB = 0$$

Move the terms that do not contain $I$ to the right side of the equation: $ILB = 2F_{\text{T on W}} - mg$. Then divide both sides by $LB$:

$$I = \frac{2F_{\text{T on W}} - mg}{LB}$$

**Solve and evaluate**

- Use the mathematical representation of the process to determine the unknown quantity.
- Evaluate the results—units, magnitude, and limiting cases—to make sure they are reasonable.

Inserting the appropriate values:

$$I = \frac{2(39 \times 10^{-3}\,\text{N}) - (5.0 \times 10^{-3}\,\text{kg})(9.8\,\text{N/kg})}{(0.2\,\text{m})(49 \times 10^{-3}\,\text{T})} = 3.0\,\text{A}$$

This is a large current for a thin wire but is not completely unreasonable.

Looking at a limiting case: if the wire mass is such that $mg = 2F_{\text{T on W}}$, then the required current is zero—the gravitational force that Earth exerts would be enough to break the wire.

**Try it yourself:** An electron enters a $1.0 \times 10^{-2}$-T $\vec{B}$ field perpendicular to the $\vec{B}$ field lines. It then completes a semicircular path of radius $1.0 \times 10^{-3}$ m and leaves the $\vec{B}$ field region traveling in the opposite direction. Its mass is $9.11 \times 10^{-31}$ kg. What is the speed of the electron?

*Answer:*

$$v = \frac{qrB}{m} = \frac{(1.6 \times 10^{-19}\,\text{C})(1.0 \times 10^{-3}\,\text{m})(1.0 \times 10^{-2}\,\text{T})}{(9.11 \times 10^{-31}\,\text{kg})}$$
$$= 1.8 \times 10^{6}\,\text{m/s}$$

## EXAMPLE 17.7 Determine the $\vec{B}$ field

Determine the $\vec{B}$ field 5.0 cm from a long straight wire that is connected in series with a 5.0-Ω resister and a 9.0-V battery.

**Sketch and translate** Make a sketch of the situation including the electric circuit connected to the wire (see figure below). The current in the circuit is clockwise. The problem is about finding the field produced by a known source.

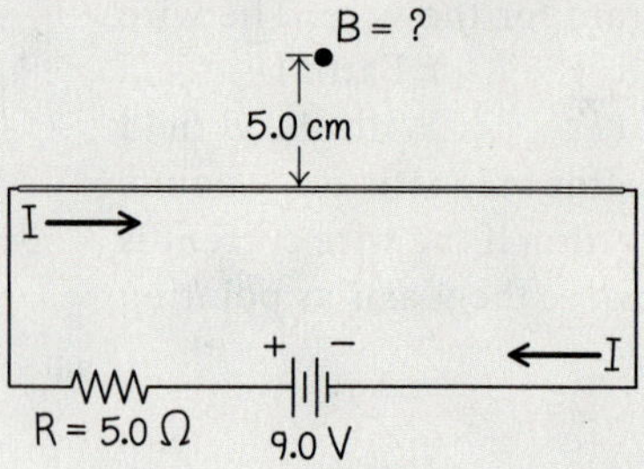


**Simplify and diagram** Assume that the only contribution to the $\vec{B}$ field at the point of interest comes from the long wire. The other wires in the circuit contribute as well, but since they are somewhat farther away by comparison, we will neglect their contributions. Assume also that the other connecting wires and the battery have zero resistance. Using the right-hand rule for the $\vec{B}$ field, we find that below the wire the field points into the paper, and above the wire it points out of the paper (see figure below).

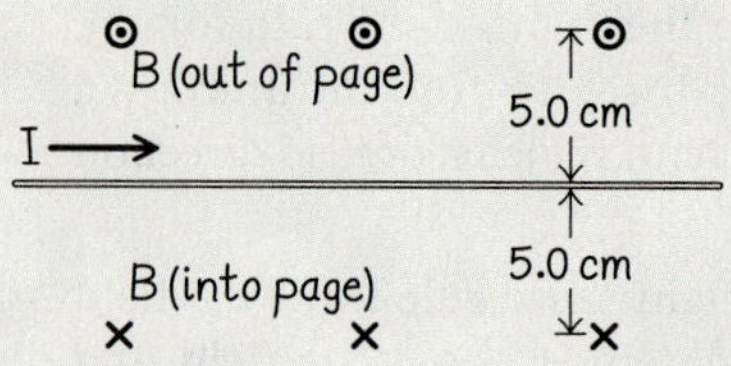

**Represent mathematically** The magnitude of the $\vec{B}$ field produced by a long straight current-carrying wire is given by Eq. (17.6):

$$B_{\text{straight wire}} = \frac{\mu_0}{2\pi}\frac{I}{r}$$

Using Ohm's law and the loop rule (traversing the loop clockwise), we can determine the current through the wire:

$$\Delta V_{\text{batt}} + \Delta V_{\text{R}} = 0$$
$$\Rightarrow \varepsilon + (-IR) = 0$$

**Solve and evaluate** Combining the above equations gives

$$B_{\text{straight wire}} = \frac{\mu_0}{2\pi}\frac{I}{r} = \frac{\mu_0}{2\pi}\frac{(\varepsilon/R)}{r} = \frac{\mu_0}{2\pi}\frac{\varepsilon}{rR}$$

$$= \frac{4\pi \times 10^{-7}\,\text{T}\cdot\text{m/A}}{2\pi}\,\frac{9.0\,\text{V}}{(0.050\,\text{m})(5.0\,\Omega)}$$

$$= 7.2 \times 10^{-6}\,\text{T}$$

We can do a limiting case analysis for the case of a discharged battery. If $\varepsilon = 0$, then there should be no current and the $\vec{B}$ field should be zero, consistent with the equation.

**Try it yourself:** Estimate the magnitude of the $\vec{B}$ field produced by an electron beam in an oscilloscope at a point 1.0 m to the side of the beam. Assume that $10^{10}$ electrons hit the screen every second and that they move at a speed of $10^7\,\text{m/s}$.

*Answer:* $3.2 \times 10^{-16}\,\text{T}$.

## Intensity modulated radiation therapy

Intensity modulated radiation therapy (IMRT) is a powerful cancer-fighting technology in which a high dose of X-rays is directed at a tumor, leaving most of the surrounding tissue untouched. The IMRT machine accelerates electrons to the desired kinetic energy, then uses a magnetic field to bend them 90° into a tungsten alloy target, resulting in the production of X-rays. Movable metal leaves then shape the X-ray beam to match the shape of the tumor. The accelerator and radiation device rotate around the patient with the leaves continually changing to match the 3-D shape of the tumor (**Figure 17.30**). IMRT works well for prostate cancer as well as for tumors of the head, neck, and other delicate areas. In the next example, we consider the magnetic field that bends the electron beam inside the IMRT.

**Figure 17.30** An intensity modulated radiation therapy (IMRT) machine.

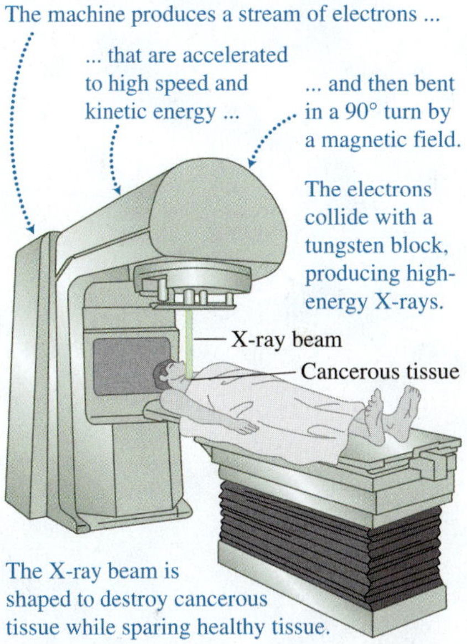

The machine produces a stream of electrons ...
... that are accelerated to high speed and kinetic energy ...
... and then bent in a 90° turn by a magnetic field.
The electrons collide with a tungsten block, producing high-energy X-rays.
X-ray beam
Cancerous tissue
The X-ray beam is shaped to destroy cancerous tissue while sparing healthy tissue.

**EXAMPLE 17.8 Magnetic field that bends the electron beam in an IMRT device**

Estimate the magnitude of the $\vec{B}$ field needed for the IMRT machine. For the estimate, assume that the electrons are moving at $2 \times 10^8\,\text{m/s}$, the mass of the electrons is $9 \times 10^{-31}\,\text{kg}$, and the radius of the turn is $5\,\text{cm} = 0.05\,\text{m}$.

**Sketch and translate** The bending process is sketched in the figure to the right.

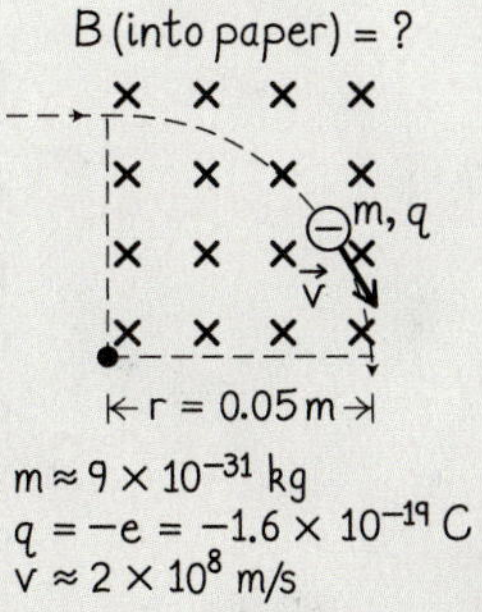


*(continued)*

**Simplify and diagram** This is an estimate and we assume that the physics principles we will use apply without modification to these electrons traveling at a significant fraction of light speed. A force diagram for the electron part of the way around the 90° curved arc is shown in the figure above. We neglect the gravitational force that Earth exerts on the electron. The magnetic force points in the radial direction perpendicular to the electron's velocity at each point along its path.

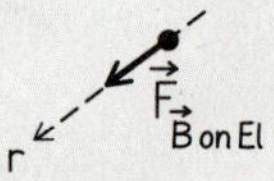


**Represent mathematically** Apply Newton's second law to the radial direction. The magnitude of the acceleration is determined by the magnitude of the sum of the forces exerted on the electron in the radial direction and by the mass of the electron:

$$m_{\text{El}}a_r = m_{\text{El}}\frac{v^2}{r} = \Sigma F_r$$

The only force with a nonzero radial component is the magnetic force. The magnitude of that force is $F_{\vec{B}\text{ on El}} = evB$, where $e$ is the magnitude of the electron's electric charge. We get

$$m_{\text{El}}\frac{v^2}{r} = evB$$

**Solve and evaluate** Divide each side of the equation by $ev$ to determine the magnitude of the magnetic field $B$:

$$B = \frac{m_{\text{El}}v}{er} = \frac{(9 \times 10^{-31}\text{ kg})(2 \times 10^8\text{ m/s})}{(1.6 \times 10^{-19}\text{ C})(0.05\text{ m})}$$

$$= 0.023\text{ T} \approx 0.02\text{ T}$$

This is an easily attained magnetic field. Thus, there should be no difficulty bending the electron beam by 90°.

**Try it yourself:** Suppose that this is a beam of protons instead of electrons. How would this affect the required magnetic field?

*Answer:* Protons are about 2000 times more massive than electrons. Thus, the magnetic field would need to be about 2000 times greater.

**Review Question 17.6** What is the difference between the right-hand rule for the magnetic force and the right-hand rule for the $\vec{B}$ field?

## 17.7 Flow speed, electric generator, and mass spectrometer: Putting it all together

So far in this chapter, we have only considered applications involving the magnetic field and the magnetic forces and torques it can exert. In this section, we will examine some applications that involve a combination of magnetic and electric phenomena, including the magnetohydrodynamic generation of electric power and the measurement of the speed at which blood flows. Both of these applications involve electrically charged objects moving in a region that has both nonzero $\vec{B}$ field and $\vec{E}$ fields perpendicular to each other. We will also investigate how our knowledge of magnetic fields helps us determine the masses of ions using a mass spectrometer.

### Ions moving through perpendicular $\vec{B}$ field and $\vec{E}$ field

Imagine that you have a $\vec{B}$ field in the region between two conducting plates that points into the page (see **Figure 17.31a**). Initially the plates are not charged. Both positively and negatively charged particles (ions) move downward between the plates perpendicular to the direction of the $\vec{B}$ field. According to the right-hand rule for magnetic force, the field exerts a magnetic force on the positively charged particles toward the right (Figure 17.31b). The trajectory of these particles becomes circular, and they collide with and are collected by the plate on the right. The same thing happens to the negatively charged particles, except that the magnetic force exerted by the magnetic field on them points to the left, so they collect on the left plate (Figure 17.31c.) This device separates positively and negatively charged particles.

As the plates become oppositely charged, they produce an electric field with an $\vec{E}$ field vector that points to the left in the region between the plates (Figure 17.31d). This field now exerts an electric force on the positive ions that points to the left and on the negative ions that points to the right. In both cases, this electric force points in the opposite direction to the magnetic force. The increasing $\vec{E}$ field due to the accumulation of electric charge on the plates quickly becomes large enough so that it prevents further accumulation of charged particles on the plates. When the electric and magnetic forces exerted on the moving charged particles balance, the ions travel with constant velocity downward despite the presence of both a $\vec{B}$ field and an $\vec{E}$ field (Figure 17.31e). Mathematically, for a positively charged particle:

$$F_{\vec{B}\text{ on }q\,x} + F_{\vec{E}\text{ on }q\,x} = 0$$
$$\Rightarrow (+F_{\vec{B}\text{ on }q}) + (-F_{\vec{E}\text{ on }q}) = 0$$
$$\Rightarrow |q|vB\sin(90°) - |q|E = 0$$
$$\Rightarrow vB - E = 0 \quad \text{or} \quad E = vB \tag{17.7}$$

We have learned (in Chapter 15) that the $\vec{E}$ field can be expressed in terms of the magnitude of the potential difference $\Delta V$ across the plates ($E = \Delta V/d$), where $d$ is the distance between the plates. Thus, the potential difference across the plates will be

$$\Delta V = Ed = vBd$$

Figure 17.31 Charge separation of moving charged particles.

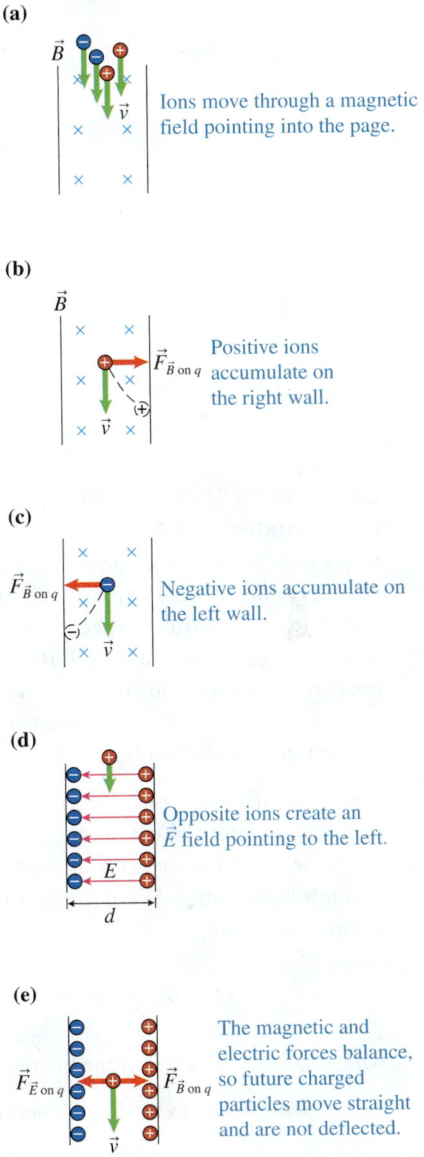

(a)
Ions move through a magnetic field pointing into the page.
(b)
Positive ions accumulate on the right wall.
(c)
Negative ions accumulate on the left wall.
(d)
Opposite ions create an $\vec{E}$ field pointing to the left.
(e)
The magnetic and electric forces balance, so future charged particles move straight and are not deflected.


## Magnetohydrodynamic generator

The process described above has multiple practical applications, one of which is magnetohydrodynamic (MHD) power generation. An MHD generator converts the random kinetic energy of high-temperature charged particles into electric potential energy. In an MHD generator, pulverized coal (gaseous fuel) enters a combustion chamber where it is burned at high temperature (1000–2000 °C) and pressure. When alkali metals, such as potassium, are injected into the burning gas, free electrons and positively charged ions form. The electrons and ions pass from the combustion chamber through a nozzle and into a magnetic field region such as depicted in **Figure 17.32**. The negative electrons and positive ions accumulate on opposite plates beside the pathway of the moving gas, producing a potential difference between the plates. The plates, like the terminals of a battery, serve as a supplemental power source for an attached circuit (shown as the bulb in Figure 17.32). Because they are continually recharged by the hot moving ionized gas, the MHD plates maintain a constant potential difference, resulting in a current through the attached circuit.

MHD generators are used at some older coal-fired power plants to improve the efficiency of power generation from 40% to about 60%. New technologies have replaced MHD generators in newly built coal-fired electric power plants. However, MHD generators are valuable tools for upgrading older coal plants for efficiency. Scientists also use the MHD model to explain the northern lights by considering Earth as a giant MHD generator in which the charged particles come from the Sun.

Figure 17.32 Magnetohydrodynamic power generation.

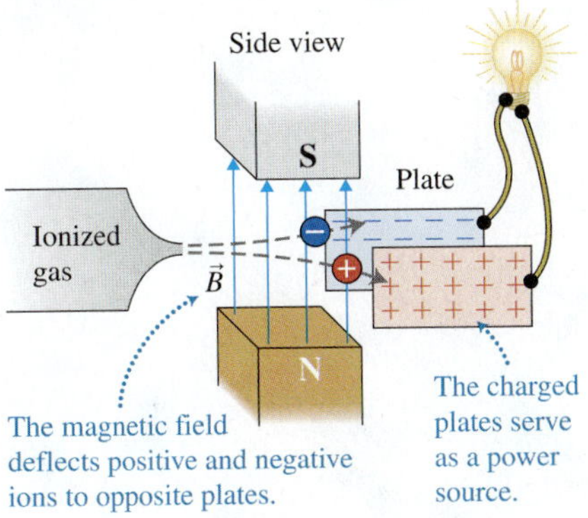


## Magnetic flow meter

A magnetic flow meter with a $\vec{B}$ field perpendicular to flowing charged particles is widely used to monitor water flow, in such applications as municipal and industrial water use, wastewater flow, the flow of cooling water for power plants, and well water use rates. The magnetic flow meter has no rotor to get stuck in sand or debris and no bearings to wear out and is virtually maintenance-free. Commercial flow meters can measure flow rates from 4800 to 12,000 L/min.

The magnetic flow meter works only for fluids with moving ions, which includes most fluids. A magnetic field is oriented perpendicular to the vessel through which the fluid flows. Oppositely charged ions in the fluid are pushed by the magnetic field to opposite walls, producing a potential difference $\Delta V$ across the walls of the vessel. By measuring $\Delta V$, the magnitude of the $\vec{B}$ field, and the diameter $d$ of the vessel, the equation $v = \Delta V/Bd$ can be used to determine the fluid's speed $v$. With this information we can determine the fluid's volume flow rate (a topic that you learned in Chapter 11):

$$Q = Av = (\pi r^2)\frac{\Delta V}{Bd} = \frac{\pi d^2}{4}\frac{\Delta V}{Bd} = \frac{\pi d \Delta V}{4B}$$

**QUANTITATIVE EXERCISE 17.9 Blood speed meter**

Is the general magnetic flow meter idea feasible for measuring blood speed in an artery? Estimate the potential difference you would expect to measure as blood in an artery passes through a 0.10-T $\vec{B}$ field region. Assume the heart pumps about 80 $\text{cm}^3$ of blood each second (the approximate volume for each heart beat) and the diameter of the artery is about 1.0 cm.

**Represent mathematically** Use $\Delta V = vBd$ to estimate the potential difference $\Delta V$ expected across opposite walls of the artery and then decide if this is large enough to measure. We need first to estimate the speed of the blood using

$$Q = Av = (\pi r^2)v = \frac{\pi d^2}{4}v$$

where $Q$ is the blood volume flow rate.

**Solve and evaluate** The speed of the blood in this artery is

$$v = \frac{Q}{(\pi d^2)/4} = \frac{4Q}{\pi d^2} = \frac{4(80\ \text{cm}^3/\text{s})\left(\frac{1\ \text{m}}{100\ \text{cm}}\right)^3}{\pi(1.0\ \text{cm})^2\left(\frac{1\ \text{m}}{100\ \text{cm}}\right)^2}$$

$$= 1\text{m/s}$$

This speed will result in a potential difference across the walls of the artery of

$$\Delta V = vBd$$

$$= (1\text{m/s})(0.10\ \text{T})(1.0\ \text{cm})\left(\frac{1\ \text{m}}{100\ \text{cm}}\right)$$

$$= 1 \times 10^{-3}\ \text{V}$$

The potential difference is easily measured, so this is in fact a practical way to measure blood flow in major arteries.

**Try it yourself:** The flow rate of water from a 5.0-cm-diameter irrigation pipe is 200 gallons/min (1 gallon = $3.79 \times 10^{-3}\text{m}^3$). The water passes through a 0.1-T magnetic field. Determine the average speed of the water in the pipe and the potential difference across the flow meter in that pipe.

*Answer:* 6.4 m/s and 0.032 V.

## Mass spectrometer

In Section 17.5 we learned that when a charged particle's velocity is perpendicular to the direction of the $\vec{B}$ field, the particle moves in a circular path. The radius $r$ of the circle depends on the particle's mass:

$$r = \frac{mv}{|q|B}$$

This relationship is the foundation for the mass spectrometer. The mass spectrometer helps determine the mass of ions, molecules, and even elementary particles such as protons and electrons. It also can determine the relative concentration of atoms of the same chemical element that have slightly different masses (known as isotopes, which we will learn about in Chapter 28).

A mass spectrometer produces ions through heating, collisions, or some other mechanism. An electric field accelerates the ions into a device called a velocity selector that has perpendicular electric and magnetic fields ($\vec{E}$ and $\vec{B}_1$ fields). These fields allow only those ions moving at a predetermined speed to pass through [see Eq. (17.7)]. The ions then enter a region with a different uniform $\vec{B}_2$ field, traveling perpendicular to the field lines. As a result, the ions move in circular paths. The radius of the circle is measured by observing the place where the ions strike a detector after moving halfway around a circle (**Figure 17.33**). The ion mass is then determined using the method in the exercise below.

Mass spectrometry has many scientific uses. For example, researchers can use the mass spectrometer to measure the concentration of two isotopes of oxygen in glacial ice: oxygen-16 and oxygen-18. The oxygen-16/18 isotope ratio changes over geologic time due to global climate conditions, and therefore can be used as a way to determine the age of plant and animal remains found in the ice.

**Figure 17.33** A mass spectrometer.

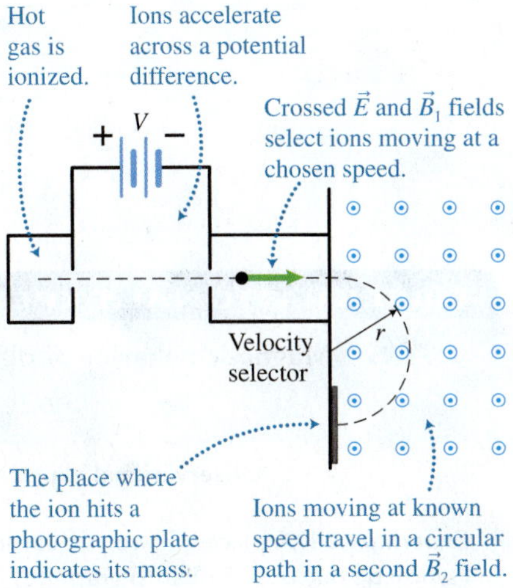


## QUANTITATIVE EXERCISE 17.10 Mass spectrometer

An atom or molecule with a single electron removed is traveling at $1.0 \times 10^6\,\text{m/s}$ when it enters a mass spectrometer's 0.50-T uniform $\vec{B}$ field region. Its electric charge is $+1.6 \times 10^{-19}\,\text{C}$. It moves in a circle of radius 0.20 m until it hits the detector (see figure). Determine (a) the magnitude of the magnetic force that the magnetic field exerts on the ion and (b) the mass of the ion.

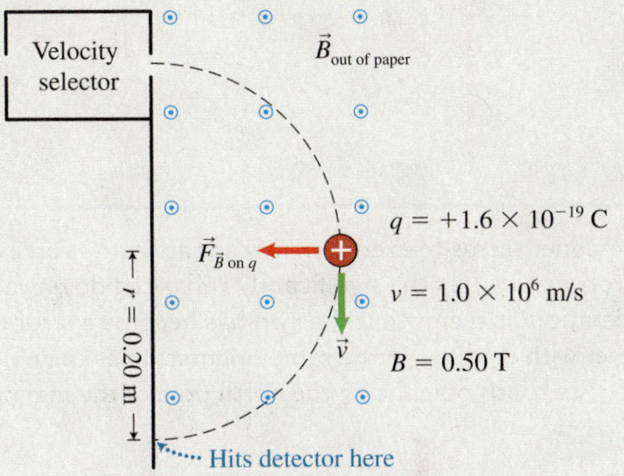


**Represent mathematically** The magnitude of the force that the magnetic field exerts on the moving ion is determined using Eq. (17.5):

$$F_{\vec{B}\text{ on }q} = |q|vB\sin\theta$$

This force perpendicular to the ion's velocity causes its radial acceleration:

$$ma_r = |q|vB$$

$$\Rightarrow m\frac{v^2}{r} = |q|vB$$

The mass of the ion is

$$m = \frac{|q|rB}{v}$$

**Solve and evaluate** Now insert the appropriate values to determine the two unknowns:

$$F_{\vec{B}\text{ on }q} = |q|vB\sin\theta$$
$$= (+1.6 \times 10^{-19}\,\text{C})(1.0 \times 10^6\,\text{m/s})(0.50\,\text{T})\sin(90°)$$
$$= 8.0 \times 10^{-14}\,\text{N}$$

and

$$m = \frac{|q|Br}{v} = \frac{(+1.6 \times 10^{-19}\,\text{C})(0.50\,\text{T})(0.20\,\text{m})}{1.0 \times 10^6\,\text{m/s}}$$
$$= 1.6 \times 10^{-26}\,\text{kg}$$

The magnitude of the force exerted on the ion is small but reasonable since it is the force exerted on a single ion. The mass of the ion is also reasonable (the mass of a single proton is $1.67 \times 10^{-27}$ kg).

**Try it yourself:** What would be the radius of the circular path of a mass $1.7 \times 10^{-26}$-kg particle?

*Answer:* 0.21 m.

**Review Question 17.7** How does an MHD generator produce a constant potential difference?

## 17.8 Magnetic properties of materials

Materials, even among metals, have widely varying magnetic properties. For example, magnets strongly attract objects made from iron, such as paper clips, but do not exert an observable magnetic force on objects made from aluminum, such as soda cans. Iron has the ability to greatly amplify the $\vec{B}$ field surrounding it. How can we explain this? See Observational Experiment **Table 17.6.**

**OBSERVATIONAL EXPERIMENT TABLE**

**17.6 Magnetic behavior of different materials.**

| Observational experiment | Analysis |
|---|---|
| **Experiment 1.** A piece of pyrolytic carbon (synthetic material similar to graphite) is placed above a strong magnet; the carbon levitates above the magnet. 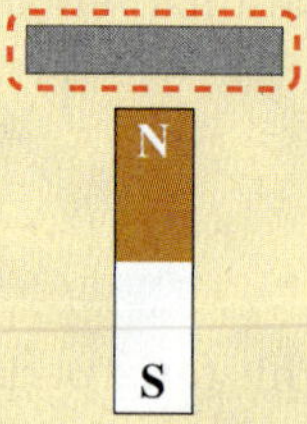  | The magnet must exert a repulsive force on the pyrolytic carbon that balances the gravitational force that Earth exerts on it. It seems that the carbon has become a strong magnet with its poles oppositely aligned to the magnet below it (its north pole facing the north pole of the magnet). 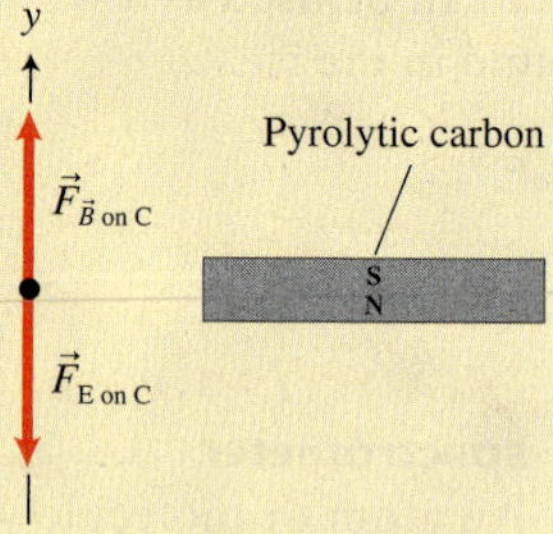  |
| **Experiment 2.** A piece of aluminum is placed above the same magnet; it falls downward with an acceleration slightly greater than 9.8 $m/s^2$. 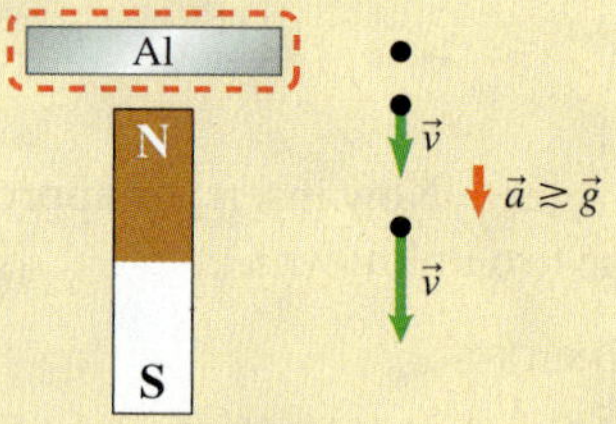  | The magnet seems to exert a small attractive force on the aluminum that causes the enhanced downward acceleration. It seems that the aluminum has become a weak magnet with its poles in the same orientation as the magnet below (its south pole facing the north pole of the magnet). 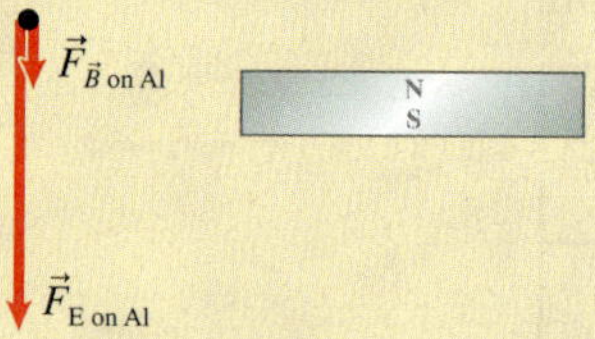  |
| **Experiment 3.** A piece of iron is placed above the same magnet; it falls downward with an acceleration significantly greater than 9.8 $m/s^2$. 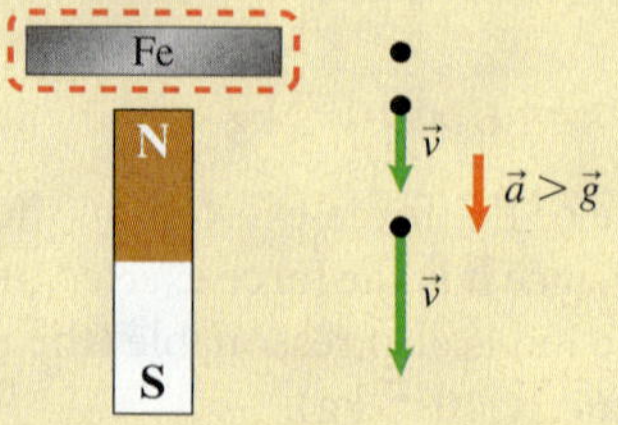  | The magnet seems to exert a significant attractive force on the iron that causes the significantly enhanced downward acceleration. It seems that the iron has become a strong magnet with its poles in the same orientation as the magnet below (its south pole facing the north pole of the magnet). 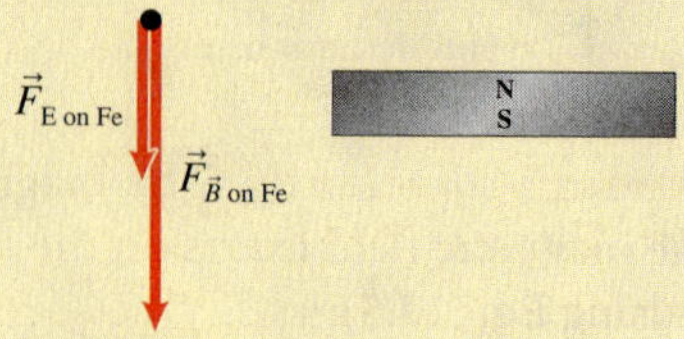  |

**Pattern**

All of the materials have magnetic properties. Some become "magnetized" opposite the direction of the other magnet and some in the same direction as the external magnet. The amount of this "magnetization" also varies in strength.

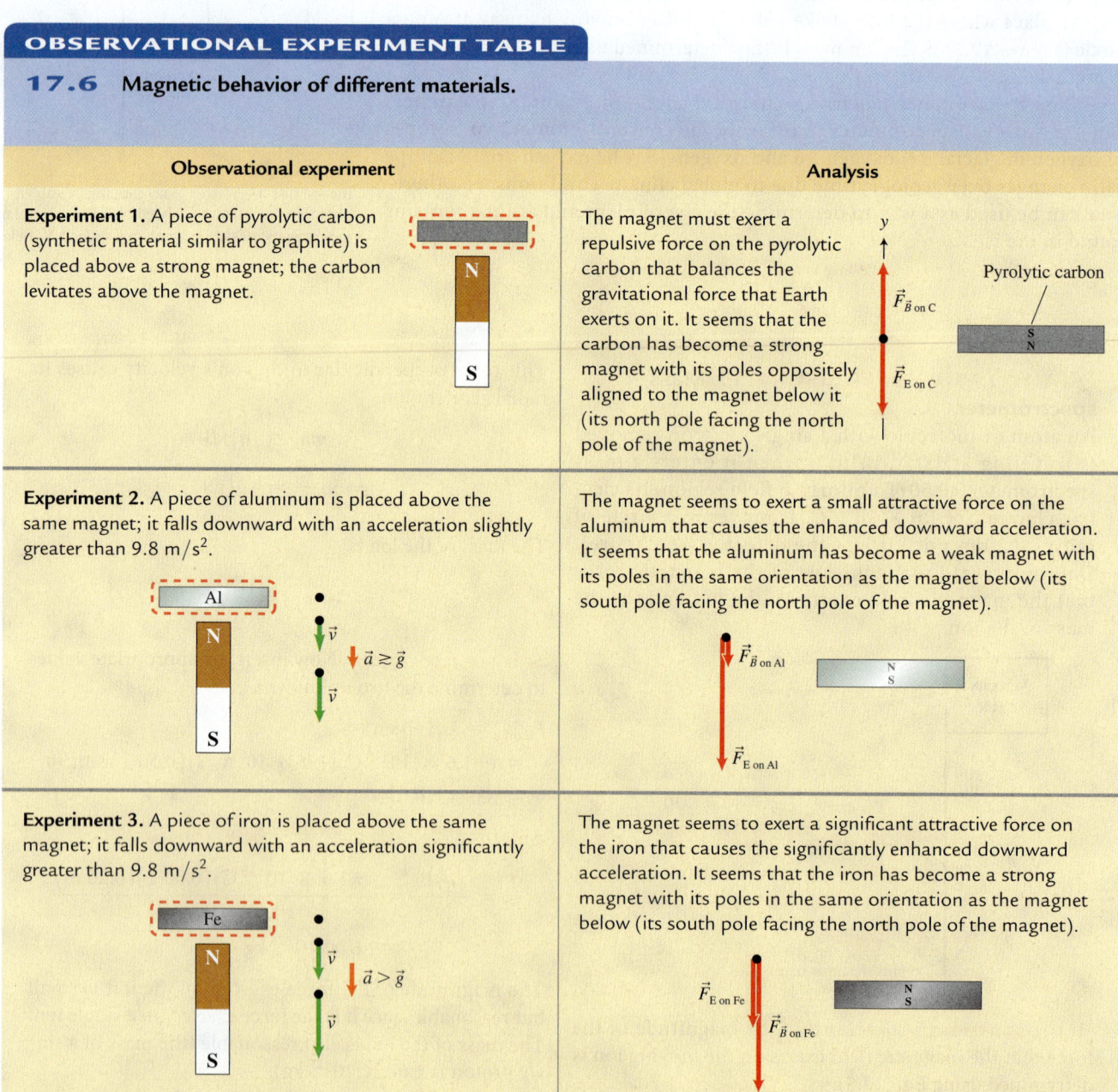

It turns out that *all* materials belong to one of the three types of materials encountered in the table: the materials that are repelled by magnets are called **diamagnetic** (like pyrolytic carbon or water); those that are weakly attracted are called **paramagnetic** (like aluminum); and those that are strongly attracted are called **ferromagnetic** (like iron). Because aluminum is paramagnetic, we did not observe a significant interaction of the soda can and the magnet. Ferromagnetic materials also retain their magnetization to a certain degree even after the external magnetic field that magnetized them is removed. Let's look at the mechanisms behind diamagnetism, paramagnetism, and ferromagnetism.

## Magnetic properties of atoms

To explain these three magnetic properties of materials, we have to understand the magnetic behavior of individual atoms. The model of an atom that we use has a point-like positively charged nucleus at its center and point-like negatively charged electrons moving at high speeds in circular orbits around the nucleus. Due to this motion, each electron has a magnetic dipole moment (**Figure 17.34a**) and acts like a tiny bar magnet (Figure 17.34b). In addition to this electron orbital magnetic moment, the electron itself is like a tiny magnet, which also contributes to the total magnetic moment produced by the atom.

In atoms with more than one electron, the magnetic moments produced by the individual electrons often cancel each other. This happens because the electrons tend to pair up and orient themselves in opposite directions.

**Figure 17.34** Electron motion in an atom effectively produces a tiny "bar magnet."

(a)

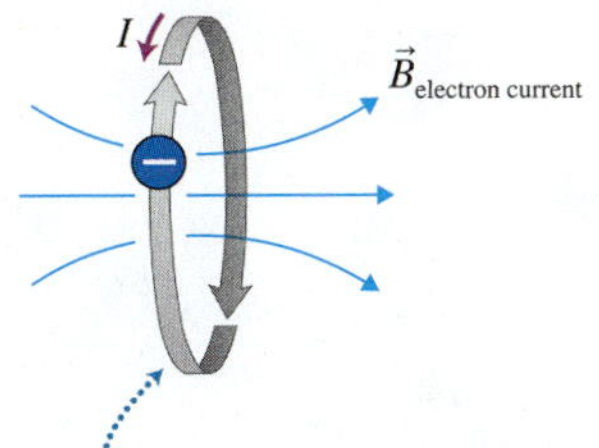
$I$
$\vec{B}_{\text{electron current}}$
The electron motion produces a circular electric current.

(b)

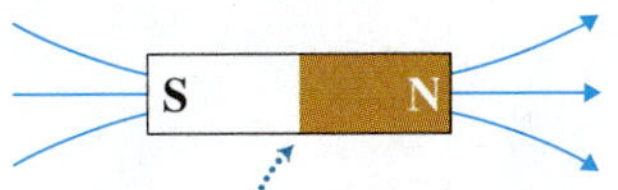
S
N
The electron current "magnet"

## Diamagnetic materials

In diamagnetic materials such as water, graphite, bismuth, and pyrolytic carbon, the magnetic moments produced by individual electrons in the atoms cancel each other, making the total field produced by the atom zero. When diamagnetic materials are placed in a region with a nonzero external $\vec{B}_{ex}$ field, the motion of the electrons in the individual atoms changes slightly. Remember, the orbiting electrons behave like tiny currents, and therefore the external magnetic field will exert forces on them. The electrons whose magnetic moments align with the external field slow down a little and their magnetic moment decreases; the electrons whose magnetic moments are opposite to the external field speed up, so their magnetic moment increases. The result is that the net $\vec{B}_{atom}$ produced by the electrons in each atom is no longer exactly zero; it now points in the direction opposite the external field $\vec{B}_{ex}$, causing the diamagnetic object to be repelled by the magnet.

## Paramagnetic materials

In most atoms, the orbital magnetic moments of the electrons and the intrinsic moments of the electrons themselves usually cancel. If they don't cancel exactly, the atom will have a magnetic moment similar to that of a small bar magnet (**Figure 17.35a**). These paramagnetic materials include aluminum, sodium, and oxygen.

When a paramagnetic material is placed in a $\vec{B}_{ex}$ field, the atoms behave like tiny (but weak) compasses and tend to align with that $\vec{B}_{ex}$ field. The magnetic moments of individual atoms no longer add to zero but instead contribute a small field that adds to the $\vec{B}_{ex}$ field (Figure 17.35b). In most materials this paramagnetic effect produces only a slight enhancement of the external $\vec{B}_{ex}$ field (about $10^{-5}$ times greater). The random motion of the particles causes the magnetic moments of the atoms to remain mostly randomly oriented.

**Figure 17.35** Paramagnetism.

(a)

$\vec{B}$ field of randomly oriented atomic "magnets"

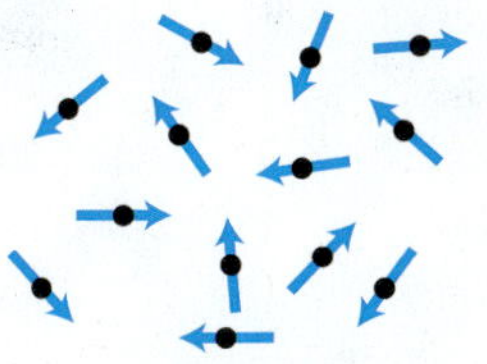

(b)

In an external $\vec{B}$ field, the atoms make a slight contribution to the $\vec{B}$ field.

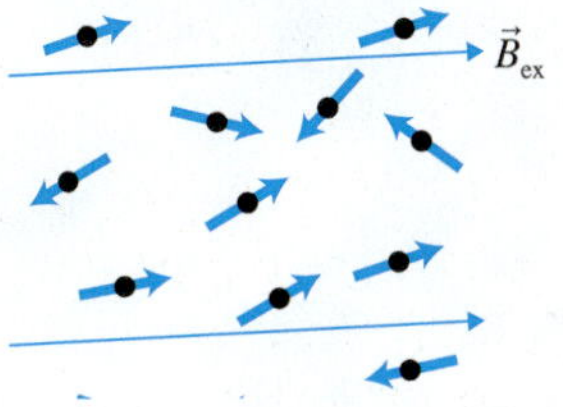
$\vec{B}_{ex}$

**Figure 17.36** Ferromagnetism.

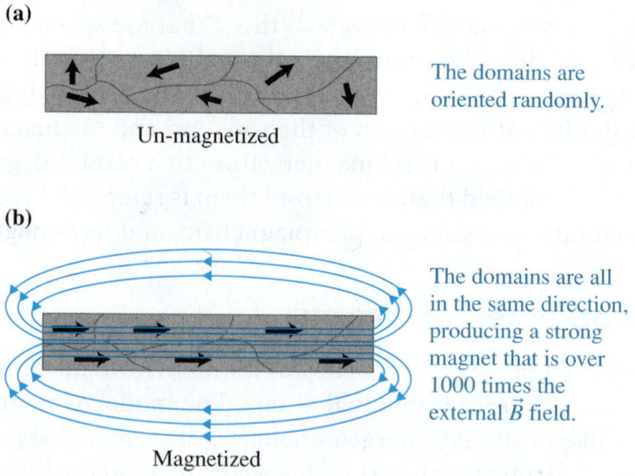
(a)
The domains are oriented randomly.
Un-magnetized
(b)
The domains are all in the same direction, producing a strong magnet that is over 1000 times the external $\vec{B}$ field.
Magnetized

## Ferromagnetic materials

Ferromagnetic materials, such as iron, nickel, and cobalt, have individual atoms with magnetic moments, just like in paramagnetic materials. However, the "magnetization" effect in an external magnetic field is thousands of times stronger in ferromagnetic materials.

Even when not in an external magnetic field, neighboring atoms in ferromagnetic materials tend to line up in small, localized regions called **domains.** Each domain may include $10^{15}$ to $10^{16}$ atoms and occupy a space less than a millimeter on a side. In a piece of iron that is un-magnetized, like a nail, the domains are oriented randomly, and the magnetic moments of the domains add to zero (**Figure 17.36a**). If an un-magnetized piece of iron is placed in a region with a nonzero $\vec{B}_{ex}$ field, the domains with magnetic moments oriented in the direction of the $\vec{B}_{ex}$ field increase in size, while those oriented in other directions decrease in size. When the external magnetic field is removed, the magnetic moments of the domains remain aligned and the iron now produces its own strong $\vec{B}$ field (Figure 17.36b). This is how steel nails become magnets after being placed in external magnetic fields and also how permanent magnets are created. This alignment of domains also explains why each piece of a broken magnet is still a complete magnet. Each piece has its domains aligned before splitting. Thus, each piece has its own north and south pole.

Understanding ferromagnetism helps explain how a device called an **electromagnet** works. Picture a current through a solenoid (a wire that has been shaped into a series of coils) that produces a $\vec{B}$ field that resembles the field of a bar magnet. Now, insert an iron bar into the solenoid. The $\vec{B}$ field produced by the wire causes the magnetic domains within the iron to line up. The now-magnetized iron produces its own contribution to the $\vec{B}$ field, a contribution that is up to several thousands of times stronger than the $\vec{B}$ field produced by the current. This is an electromagnet.

The permanent magnetization of ferromagnetic materials has many practical applications. Hard disk drives use them to store data. Airport metal detectors, transformers, electric motors, loudspeakers, electric generators, and permanent magnets all depend on the magnetization of ferromagnetic materials.

**Review Question 17.8** Why is there a difference in the behavior of paramagnetic and diamagnetic materials when they are placed in a region with nonzero $\vec{B}_{ex}$ field?

# Summary

| Words | Pictorial and physical representations | Mathematical representation |
|---|---|---|
| **Magnets** always have a north pole and a south pole. A magnetic field is produced by magnets or electric charges moving relative to the observer (Section 17.1). | 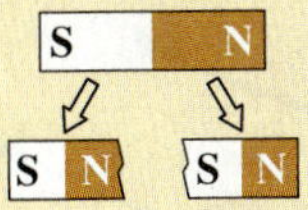 | |
| **A magnetic field** is represented by $\vec{B}$ field vectors and $\vec{B}$ field lines. The $\vec{B}$ field vector at a point is tangent to the direction of the $\vec{B}$ field line at that point. The separation of lines in a region represents the magnitude of the $\vec{B}$ field in that region—the closer the lines, the stronger the $\vec{B}$ field (Sections 17.1, 17.3, 17.5). | 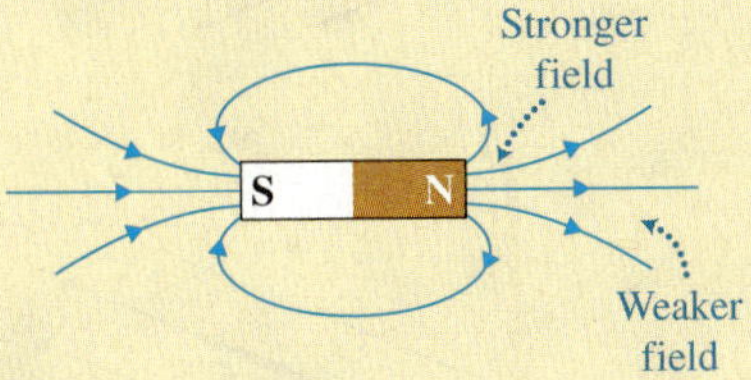 | The magnitude of a magnetic field $\vec{B}$ can be found as<br>$B = \dfrac{F_{\vec{B}\text{ on I max}}}{IL}$ Eq. (17.2)<br>The fields produced by a long wire and at the center of a coil are:<br>$B_{\text{straight wire}} = \dfrac{\mu_0 I}{2\pi r}$ Eq. (17.6)<br>$B_{\text{center of coil}} = \dfrac{N\mu_0 I}{2r}$ |
| **Right-hand rule for the $\vec{B}$ field:** To find the orientation of the $\vec{B}$ field produced by a current, grasp the wire with your right hand so that your thumb points in the direction of the current. Your four fingers will wrap around the wire in the direction of the $\vec{B}$ field lines (Section 17.2). | 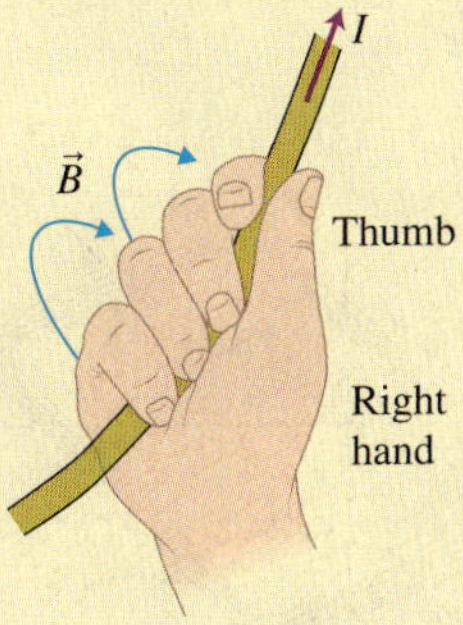 | |
| **Right-hand rule for the magnetic force:** Point the fingers of your open right hand in the direction of the magnetic field. Orient your hand so that your thumb points along the direction of motion of the electric current or charged particle. If the particle is positively charged (or is a current) the magnetic force exerted on it is in the direction your palm is facing. If the particle is negatively charged, it points in the opposite direction (Section 17.3). | 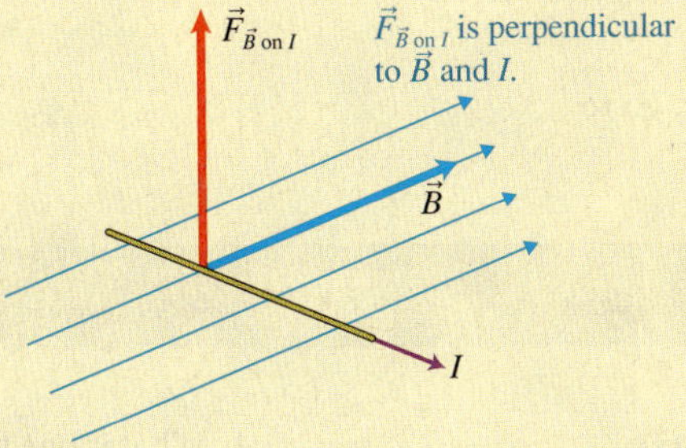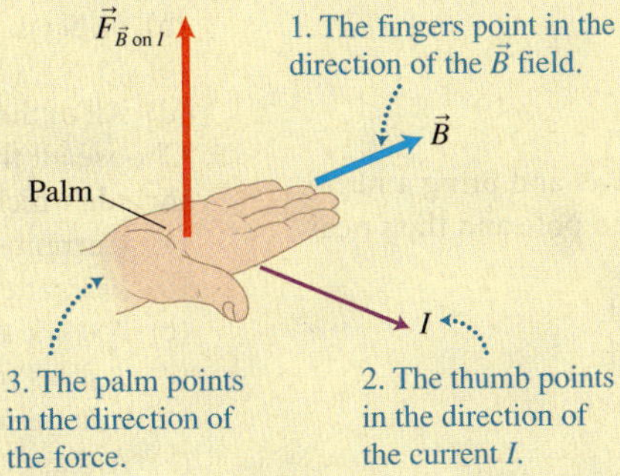 | |

(*continued*)

| Words | Pictorial and physical representations | Mathematical representation |
|---|---|---|
| **Magnitude of magnetic force** depends on the object on which the magnetic field exerts a force (either a current-carrying wire or an individual moving charged particle), the magnitude of the $\vec{B}$ field vector, the relative orientation of the $\vec{B}$ field vector, and the direction of motion of charged objects. (Section 17.3, 17.5). | 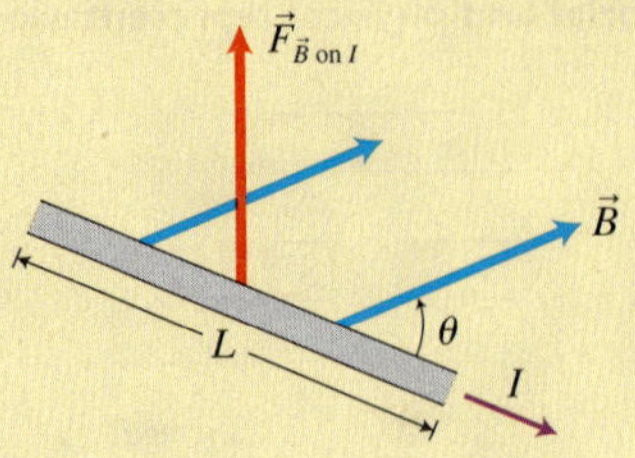  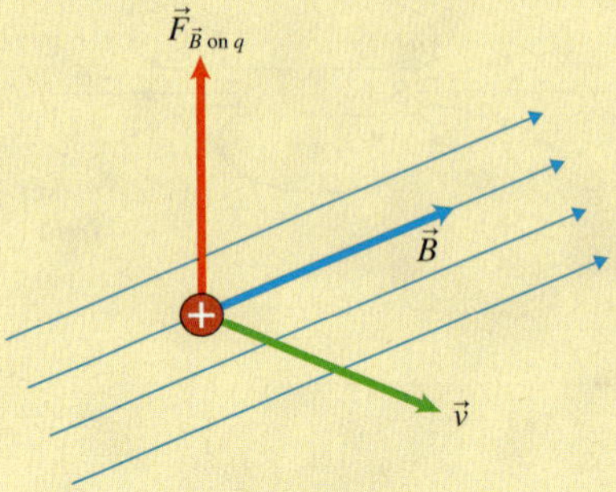  | $F_{\vec{B}\text{ on }I} = ILB\sin\theta$ Eq. (17.3)<br>where $\theta$ is the angle between $I$ and $\vec{B}$.<br>$F_{\vec{B}\text{ on }q} = \lvert q \rvert vB\sin\theta$ Eq. (17.5)<br>where $\theta$ is the angle between $\vec{v}$ and $\vec{B}$. |
| **Magnetic torque exerted on current-carrying coil:** A magnetic field exerts forces on the current passing through the wires of a coil, resulting in a torque on the coil (Section 17.3). | 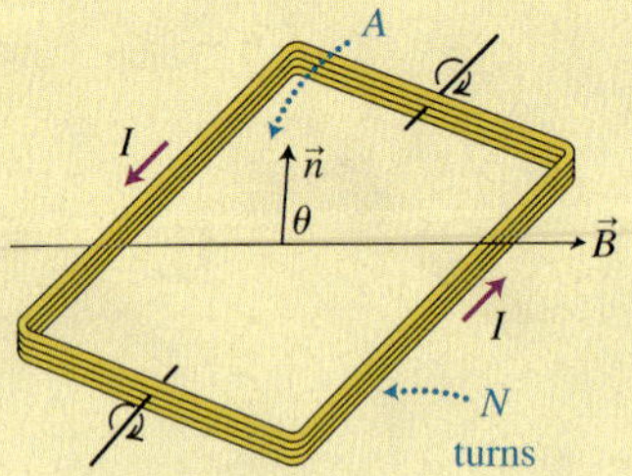  | $\lvert \tau_{\vec{B}\text{ on C}} \rvert = NBAI\sin\theta$ Eq. (17.4) |

MP® For instructor-assigned homework, go to MasteringPhysics.

# Questions

## Multiple Choice Questions

1. Magnets are similar to
   (a) individual electric charges
   (b) electric dipoles
   (c) current-carrying wire loops
2. You place a metal bar magnet on a swivel and bring a negatively charged plastic rod near the north pole and then near the south pole. What do you observe?
   (a) The north pole turns toward the rod.
   (b) The south pole turns toward the rod.
   (c) The poles do not interact with the rod.
   (d) Both poles are attracted to the rod.
3. An electron moves left to right in the plane of the page when it enters a magnetic field going into the page. The acceleration of the electron is
   (a) upward
   (b) downward
   (c) in the direction of motion
   (d) opposite to the direction of motion
   (e) into the page
   (f) out of the page
4. What is one tesla?
   (a) $1\text{ N}/(1\text{ m}\cdot 1\text{ A})$
   (b) $(1\text{ N}\cdot\text{m})/(1\text{ A}\cdot 1\text{ m}^2)$
   (c) $(1\text{ N})/(1\text{ C}\cdot 1\text{ m/s})$
   (d) All of the above
5. Choose all that apply. Objects that produce magnetic fields include which of the following?
   (a) Current-carrying wires
   (b) Permanent magnets
   (c) A compass
   (d) A glass rod rubbed with silk sitting on the table observed by a person standing on the floor
   (e) A glass rod rubbed with silk when placed on a moving truck and observed by a person standing on the ground
   (f) Current in a solenoid
6. What is one difference between magnetic and electric field lines?
   (a) Magnetic lines start on the poles and electric field lines start on electric charges.
   (b) Magnetic lines do not start or end anywhere, whereas electric field lines do have a beginning and end.
   (c) Magnetic field lines are shorter than electric field lines.

7. You use a current-carrying wire to produce a magnetic field. If you double the current in the wire, what happens to the magnitude of the $\vec{B}$ field it produces?
   (a) It doubles.
   (b) It does not change.
   (c) It becomes half as big.
8. If you double the current in a wire that you use as the detector of an external magnetic field, what happens to the force that the magnetic field exerts on it?
   (a) It doubles.
   (b) It does not change.
   (c) It becomes half as big.
9. If you triple the speed of a particle entering a magnetic field, what happens to the radius of the helix that it makes? Explain the reasoning that can lead a student to choose each of the answers. You do not need to agree with that reasoning.
   (a) It triples.
   (b) It becomes one-third as big.
   (c) It increases by 9 times.
   (d) It will not change if the magnetic field does not change.
10. Choose all of the units that are fundamental SI units.
   (a) Coulomb (b) Volt (c) Ampere
   (d) Newton (e) Meter (f) Second
   (g) Gram (h) Kilogram

## Conceptual Questions

11. In 1911 physicists measured a magnetic field around a beam of electrons. Draw $\vec{B}$ field lines for this field.
12. How do you know that electric currents create magnetic fields?
13. How can you determine if there is a magnetic field in a certain region?
14. You have a magnet on which the poles are not marked. How can you determine which pole is north and which is south if you have (a) another magnet and (b) a current-carrying coil? Describe what you will do so a reader can repeat the experiment and get the same results.
15. List all of the ways to detect a magnetic field in a particular region. Explain how they work and how you can use them to determine the magnitude and the direction of the $\vec{B}$ field.
16. List all of the ways that you could produce a magnetic field.
17. What is the difference between the right-hand rule for the magnetic force and the right-hand rule for the $\vec{B}$ field? Provide examples of problems in which you would need to use each of them.
18. A current-carrying wire is placed in a magnetic field, as shown in **Figure Q17.18**. In which direction does the magnetic field exert a force on the wire?

**Figure Q17.18**

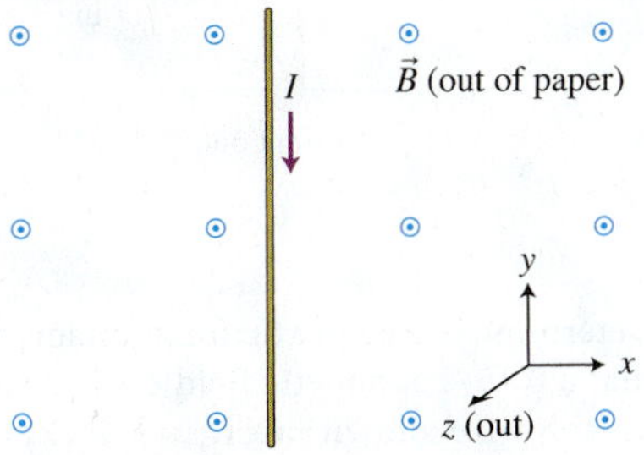


19. An electron flies through the magnetic field shown in **Figure Q17.19**. In which direction does the magnetic field exert a force on it? What effect does the field have on the magnitude of the speed of the electron? Describe the path of the electron. Describe what would be different if a proton flew in the same direction through the same field.

**Figure Q17.19**

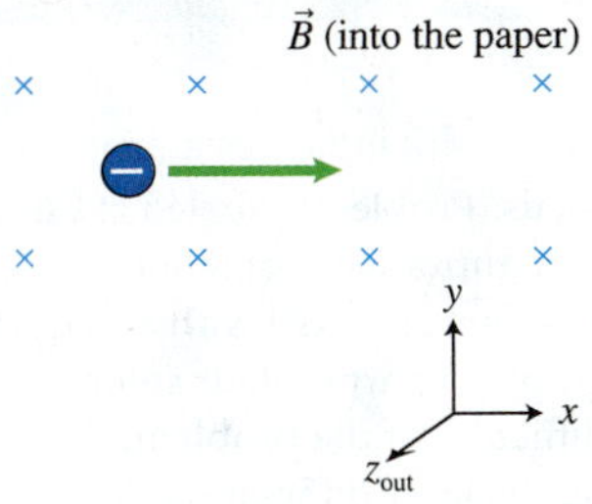


20. Find the direction of the magnetic field whose effect on a charged object is as shown in **Figure Q17.20**.

**Figure Q17.20**

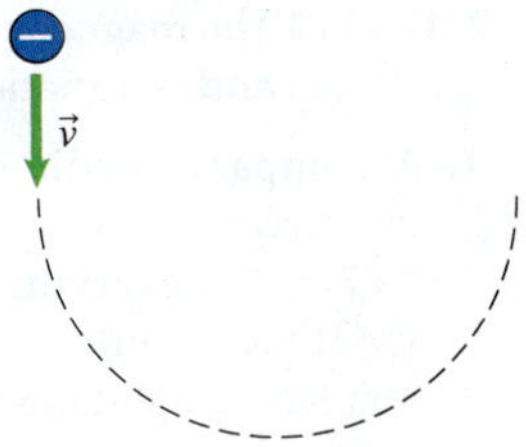


21. A beam of electrons is not deflected as it moves through a region of space in which a magnetic field exists. Give two explanations why the electrons might move along a straight path.
22. A beam of electrons moving toward the east is deflected upward by a magnetic field. Determine the direction in which the magnetic field $\vec{B}$ points. Repeat for an $\vec{E}$ field. Describe the effect of both fields on the magnitude of the speed of the electron.
23. Why are residents of northern Canada shielded less from cosmic rays than are residents of Mexico?
24. A U-shaped wire with a current in it hangs with the bottom of the U between the poles of an electromagnet (see **Figure Q17.24**). When the field in the magnet is increased, does the U swing toward the right or toward the left? Explain.

**Figure Q17.24**

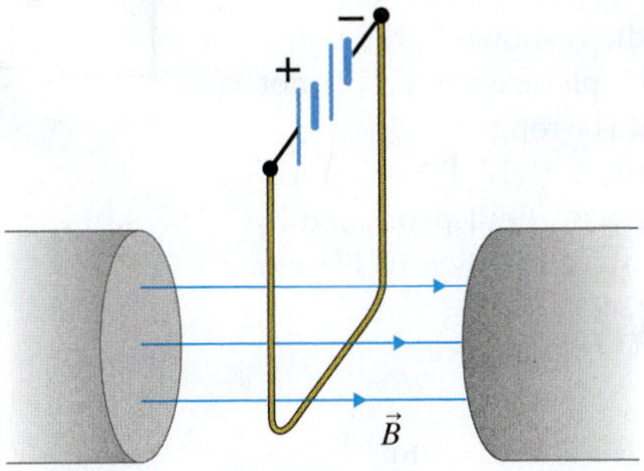


25. An electron enters a solenoid at a small angle relative to the magnetic field inside. Describe the electron's motion.
26. Two parallel wires carry electric current in the same direction. Does the moving charge in one wire cause a magnetic force to be exerted on the moving charge in the other wire? If so, in what direction is the force relative to the wires? Explain. Repeat for currents moving in opposite directions.
27. Why is the magnetic field so much greater at the end of a solenoid that has an iron core than at the end of a similar solenoid with the same current through it but with an air core?
28. Give examples of objects whose motion will be affected by a magnetic field but will not be affected by an electric field. Explain how you know.
29. Give examples of objects whose motion will be affected by an electric field but will not be affected by a magnetic field. Explain how you know.
30. Give examples of objects whose motion will be affected by both a magnetic field and an electric field. Explain how you know.

# Problems

Below, **BIO** indicates a problem with a biological or medical focus. Problems labeled **EST** ask you to estimate the answer to a quantitative problem rather than develop a specific answer. Problems marked with ✎ require you to make a drawing or graph as part of your solution. Asterisks indicate the level of difficulty of the problem. Problems with no * are considered to be the least difficult. A single * marks moderately difficult problems. Two ** indicate more difficult problems.

## 17.1–17.2 The magnetic interaction and Magnetic field

1. A compass needle deflects in the direction shown in **Figure P17.1**. Say everything you can about the circuit.

**Figure P17.1**

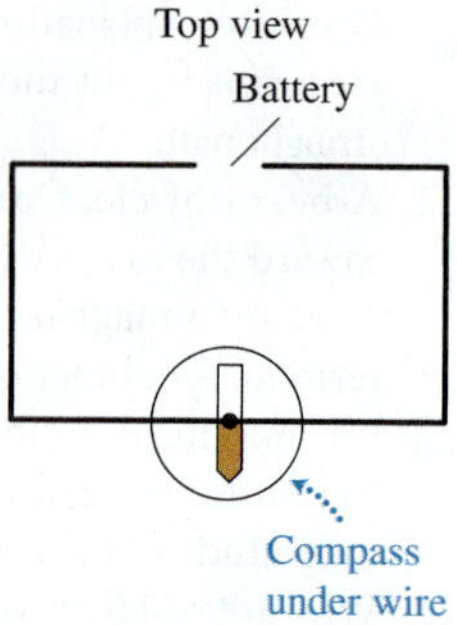


2. You have a lightbulb with a constant current through it. (a) What happens if a compass is placed under the constant current two-wire cable to the bulb? (b) What happens if you separate the wires and place the compass under one of the separated wires? Explain your answers for both parts.
3. The current through a circuit is shown in **Figure P17.3**. The deflection of a compass needle is shown in the figure. Is the picture correct? If not, what is wrong?

**Figure P17.3**

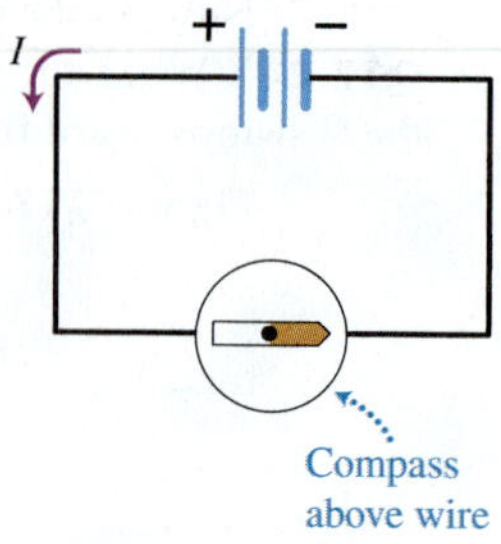


4. Draw $\vec{B}$ field lines for the magnetic field produced by the objects shown in **Figure P17.4**.

**Figure P17.4**

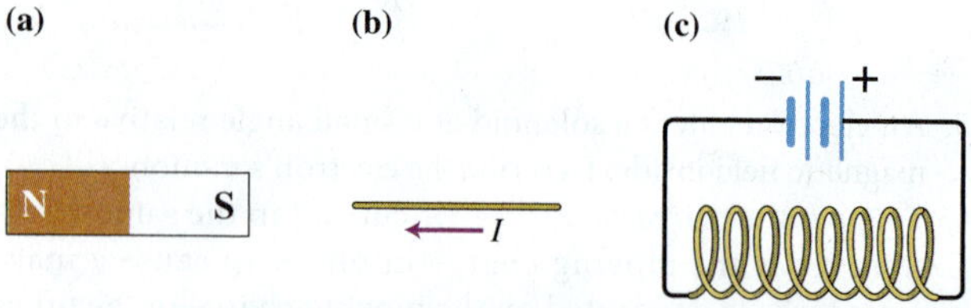


5. * A magnetic field is said to exist in a region of space. Describe three experiments you could do to confirm this statement.

## 17.3 Magnetic force exerted by the magnetic field on a current-carrying wire

6. * In Houston, Earth's $\vec{B}$ field has a magnitude of $5.2 \times 10^{-5}$ T and points in a direction 57° below a horizontal line pointing north. Determine the magnitude and direction of the magnetic force exerted by the magnetic field on a 10-m-long vertical wire carrying a 12-A current straight upward.
7. * A 15-g 10-cm-long wire is suspended horizontally between the poles of a horseshoe magnet. When the 0.50-A current in the wire is turned on, the wire jumps up and out of the magnet. What can you learn about the magnet using this information?
8. ** A metal rod of length $l$ and mass $m$ is suspended in a magnetic field from two light wires that are connected to a battery with emf $\varepsilon$. When a vertical magnetic field $\vec{B}$ is turned on, the supporting wires make an angle $\theta$ with the vertical. What can you learn about the circuit using this information? Make a list of physical quantities and explain how to determine three of them.
9. * A metal rod is connected to a battery through two stiff metal wires that hold the rod horizontally. The rod is between the poles of a horseshoe magnet that is sitting on a mass-measuring platform scale, which reads 100 g. Draw the magnetic poles of the magnet and the battery connected to the metal rod so that when you turn the current in the circuit on (a) the reading of the scale supporting the magnet increases; (b) the reading decreases.
10. * After you turned on the current in the circuit described in the previous problem, the scale supporting the magnet read 106 g. The rod is 7.0 cm long and the current through it is 1.0 A. What can you learn about the magnet using this information?
11. * **Equation Jeopardy** Describe a problem for which the following equation is a solution:

$$0.70\text{ A} = \frac{3.0\text{ N}}{(0.20\text{ m})B}$$

12. ** A square coil with 30 turns has sides that are 16 cm long. When it is placed in a 0.30-T magnetic field, a maximum torque of 0.60 N · m is exerted on the coil. What can you learn about the coil from this information?
13. ** A 5.0-A current runs through a 10-turn $(0.12\text{ m})^2$-area coil. The coil is in a 0.15-T magnetic field. What is the direction of the magnetic moment of the coil? Determine the torque exerted on the coil (the magnitude and the direction it tends to turn the coil if initially at rest) for each orientation shown in **Figure P17.13**.

**Figure P17.13**

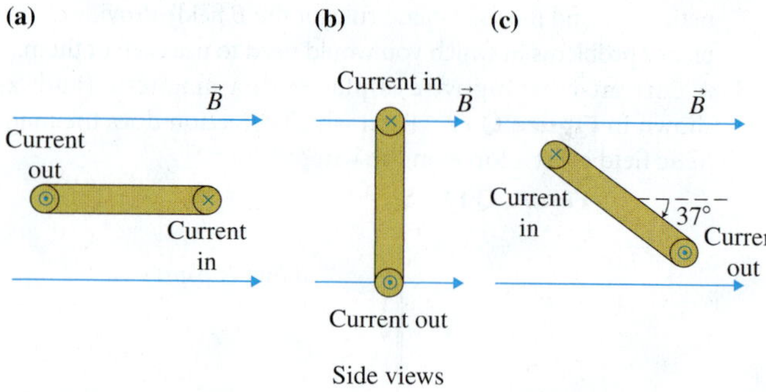


14. * (a) Determine the magnetic force (magnitude and direction) that a 0.15-T magnetic field exerts on segments a, b, and c of the wire shown in **Figure P17.14**. Segments a and c are 2.0 cm long, and segment b is 10.0 cm long. A 3.0-A current flows through the wire. (b) Determine the net torque caused by forces $\vec{F}_{\vec{B}\text{ on }a}$ and $\vec{F}_{\vec{B}\text{ on }c}$ and the torque caused by force $\vec{F}_{\vec{B}\text{ on }b}$.

**Figure P17.14**

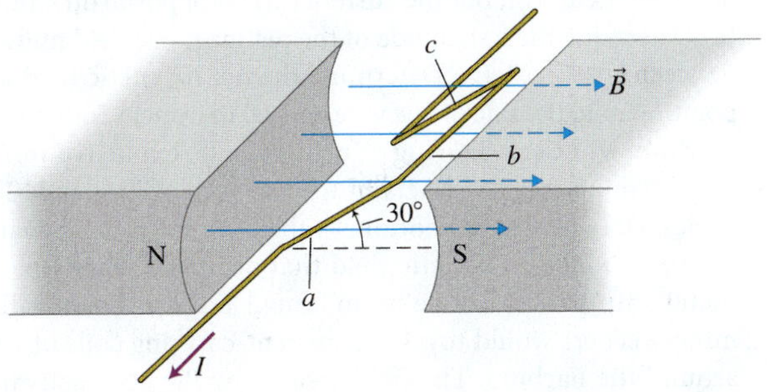


15. * A 500-turn coil of wire is hinged to the top of a table, as shown in **Figure P17.15**. Each side of the movable part of the coil has a length of 0.50 m. (a) In which direction should a magnetic field point to help lift the free end of the coil off the table? (b) Determine the torque caused by a 0.70-T field pointing in the direction described in part (a) when there is a 0.80-A current through the wire and when the coil is parallel to the table.

**Figure P17.15**

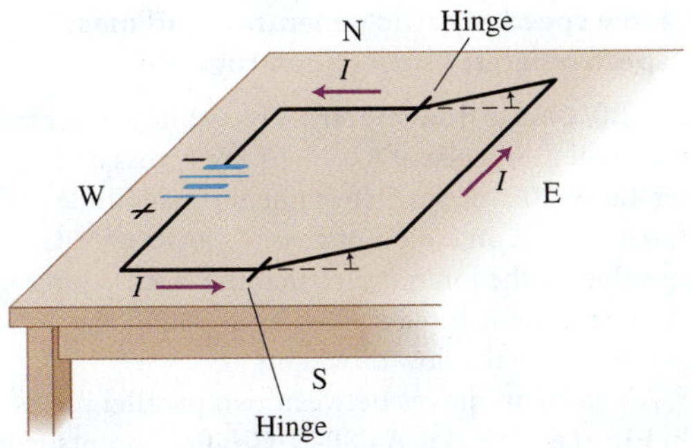


16. * **Electric motor 1** An electric motor has a square armature with 500 turns. Each side of the coil is 12 cm long and carries a current of 4.0 A. The magnetic field inside the motor is 0.60-T. Choose two orientations of the coil for which you can calculate force-related and torque-related quantities. Draw pictures to show the relative orientations of the coil and the magnetic field vectors.

17. ** A 20-turn 5.0 cm by 5.0 cm coil hangs with the plane of the coil parallel to a 0.12-T magnetic field. (a) Determine the torque on the coil when there is a 0.50-A current through it. (b) The support for the coil exerts an opposing torque, which increases as the coil is deflected. The opposing torque is calculated using the equation $\tau = -0.016\theta$ (in units of N · m), with the angle $\theta$ in units of radians (**Figure P17.17**). At approximately what angle does the magnetic torque balance the torque caused by the supporting wire?

**Figure P17.17**

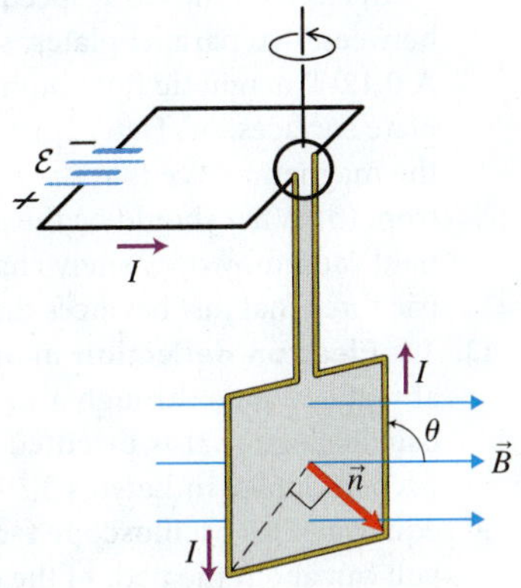


18. * **Electric motor 2** An electric motor has a circular armature with 250 turns of radius 5.0 cm through which an 8.0-A current flows. The maximum torque exerted on the armature as it turns in a magnetic field is 1.9 N · m. Show the orientation of the armature coil in the magnetic field when this torque is exerted on it and determine the magnitude of the field.

## 17.4 Magnetic force exerted on a single moving charged particle

19. Each of the lettered dots a–d shown in **Figure P17.19** represents a $+1.0 \times 10^{-5}$-C charged particle moving at speed $2.0 \times 10^7$ m/s. A uniform 0.50-T magnetic field points in the positive $z$-direction. Determine the magnitude of the magnetic force that the field exerts on each particle and indicate carefully, in a drawing, the direction of the force.

**Figure P17.19**

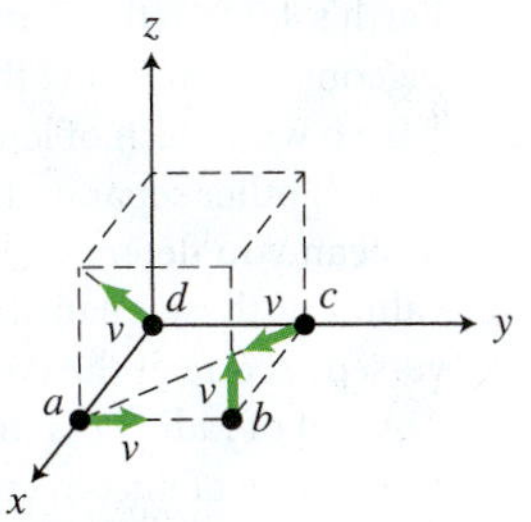


20. **Duck gets a lift** A duck accumulates a positive charge of $3.0 \times 10^{-8}$ C while flying north at speed 18 m/s. Earth's magnetic field at the duck's location has a magnitude of $5.3 \times 10^{-5}$ T and points in a direction 62° below a horizontal line pointing north. Determine the magnitude and direction of the force exerted by Earth's magnetic field on the duck.

21. An electron of mass $9.1 \times 10^{-31}$ kg moves horizontally toward the north at $3.0 \times 10^7$ m/s. Determine the magnitude and direction of a $\vec{B}$ field that will exert a magnetic force that balances the gravitational force that Earth exerts on the electron.

22. A 1000-kg car moves west along the equator. At this location Earth's magnetic field is $3.5 \times 10^{-5}$ T and points north parallel to Earth's surface. If the car carries a charge of $-2.0 \times 10^{-3}$ C, how fast must it move so that the magnetic force balances 0.010% of Earth's gravitational force exerted on the car?

23. * BIO **Magnetic force exerted by Earth on ions in the body** A hydroxide ion ($OH^-$) in a glass of water has an average speed of about 600 m/s. (a) Determine the electrical force between the hydroxide ion (charge $-e$) and a positive ion (charge $+e$) that is $1.0 \times 10^{-8}$ m away (about the separation of 30 atoms). (b) Determine the maximum magnetic force that Earth's $3.0 \times 10^{-5}$-T magnetic field can exert on the ion. (c) On the basis of these two calculations, does it seem likely that Earth's magnetic field has much effect on the biochemistry of the body?

24. ** EST **Design a magnetic shield** An alien from a planet in the galaxy M31 (Andromeda) has a ray gun that shoots protons at a speed of $2.0 \times 10^5$ m/s. Design a magnetic shield that will deflect the protons away from your body. Using rough estimates, show that your magnetic shield will, in fact, protect you. Indicate the orientation of your protective device's magnetic field relative to the direction of the proton beam and the direction in which the ions are deflected.

25. * EST An electron beam moves toward an oscilloscope screen. Estimate its vertical displacement caused by Earth's magnetic field.

26. ** An electron and a proton, moving side by side at the same speed, enter a 0.020-T magnetic field. The electron moves in a circular path of radius 7.0 mm. Describe the motion of the proton qualitatively and quantitatively.

## 17.5 Magnetic fields produced by electric currents

27. An east-west electric power line carries a 500-A current toward the east. The line is 10 m above Earth's surface. Determine the magnitude and direction of the magnetic field at Earth's surface directly under the wire. What assumptions did you make?

28. **Pigeons** A solenoid of radius 1.0 m with 750 turns and a length of 5.0 m surrounds a pigeon cage. What current must be in the solenoid so that the solenoid field just cancels Earth's $4.2 \times 10^{-5}$-T magnetic field (used occasionally by the pigeons to determine the direction they travel)?

29. * Two wires each of length $L$ with current $I$ are placed parallel to each other separated by a distance $d$. What physical quantities can you determine using this information? How will the values of these quantities change if one of the currents is reversed? If one of the currents is doubled?

30. * A coil of radius $r$ is made of $N$ circular loops. It is connected to a battery of known emf and internal resistance. The coil produces a magnetic field whose magnitude at the center of the coil is $\vec{B}$. Make a list of physical quantities you can determine using this information and show how to determine two of them.

31. ** **EST** You have a 9-V battery, a set of 10 50-Ω resistors, and a long (2-m) connecting wire. Estimate the maximum magnitude of the $\vec{B}$ field that you can produce with this equipment. Make sure you explain how you arrive at your estimate.

## 17.6 Skills for analyzing magnetic processes

32. * **Equation Jeopardy 1** The equation below describes a process involving magnetism. Solve for the unknown quantity and draw a sketch that represents a possible process described by the equation.

$$(1.6 \times 10^{-19}\,\text{C})(2.0 \times 10^{7}\,\text{m/s})(3.0 \times 10^{-5}\,\text{T}) = (1.7 \times 10^{-27}\,\text{kg})(2.0 \times 10^{7}\,\text{m/s})^2/r$$

33. * **Equation Jeopardy 2** The equation below describes a process involving magnetism. Solve for the unknown quantity and draw a sketch that represents a possible process described by the equation.

$$2T - (0.020\,\text{kg})(9.8\,\text{N/kg}) + (10\,\text{A})(0.10\,\text{m})(0.10\,\text{T}) = 0$$

34. * **Equation Jeopardy 3** The equation below describes a process involving magnetism. Solve for the unknown quantity and draw a sketch that represents a possible process described by the equation.

$$100(2.0\,\text{A})(4.0 \times 10^{-2}\,\text{m}^2)(0.20\,\text{T}) - m(9.8\,\text{N/kg})(0.10\,\text{m}) = 0$$

35. ** **EST** The magnitude of the $\vec{B}$ field inside a solenoid is given by the equation $B = \mu_0 I\,(N/L)$, where $N$ is the number of turns and $L$ is the length of the solenoid. (a) Describe an experiment that can help you test this relation. (b) Explain whether this equation is an operational definition of the magnitude of the $\vec{B}$ field magnitude or a cause-effect relationship. (c) Powerful industrial solenoids produce $\vec{B}$ field magnitudes of about 30 T. Estimate the relevant physical quantities for such solenoids.

36. ** **Electron current and magnetic field in H atom** In a simplified model of the hydrogen atom, an electron moves with a speed of $1.09 \times 10^{6}$ m/s in a circular orbit with a radius of $2.12 \times 10^{-10}$ m. (a) Determine the time interval for one trip around the circle. (b) Determine the current corresponding to the electron's motion. (c) Determine the magnetic field at the center of the circular orbit and the magnetic moment of the atom.

37. * Two long, parallel wires are separated by 2.0 m. Each wire has a 30-A current, but the currents are in opposite directions. (a) Determine the magnitude of the net magnetic field midway between the wires. (b) Determine the net magnetic field at a point 1.0 m to the side of one wire and 3.0 m from the other wire.

38. * **Minesweepers** During World War II, explosive mines were dropped by the Nazis in the harbors of England. The mines, which lay at the bottom of the harbors, were activated by the changing magnetic field that occurred when a large metal ship passed above them. Small English boats called minesweepers would tow long, current-carrying coils of wire around the harbors. The field created by the coils activated the mines, causing them to explode under the coils rather than under ships. (a) Determine the current that must flow in one long, straight wire to create a 0.0050-T magnetic field at a depth of 20 m under the water. The magnetic permeability of water is about the same as that of air. (b) How might the field be created using a smaller current?

39. * You have a compass, a 9-V battery, and a 50-Ω resistor. You wish to show your friends how Oersted's experiment works. Will you be able to do this with the available equipment? Explain how you made your decision.

## 17.7 Flow speed, electric generator, and mass spectrometer: Putting it all together

40. ** **BIO Blood flow meter** A blood flow meter measures a potential difference of $8.0 \times 10^{-5}$ V across a vessel of diameter $4.0 \times 10^{-3}$ m. (a) Determine the magnitude of the electric force exerted on an ion of charge $1.6 \times 10^{-19}$ C. (b) At what speed must the ion move so that the electric force is balanced by the magnetic force exerted by a 0.040-T field oriented perpendicular to the flow direction?

41. * An electron moves between two parallel plates, as shown in **Figure P17.41**. A 480-N/C $\vec{E}$ field points from the upper plate toward the lower plate. A 0.12-T magnetic field points into the paper. (a) Determine the electric force (magnitude and direction) that the electric field exerts on an electron between the plates. (b) At what speed and in which direction (right or left) must the electron move so that the magnetic field exerts an opposing force that balances the electric force? (If narrow slits are placed at the entrance and exit of the plates, the device allows charged particles of only one speed to pass through both slits. Such a device is called a *velocity selector*.)

**Figure P17.41**

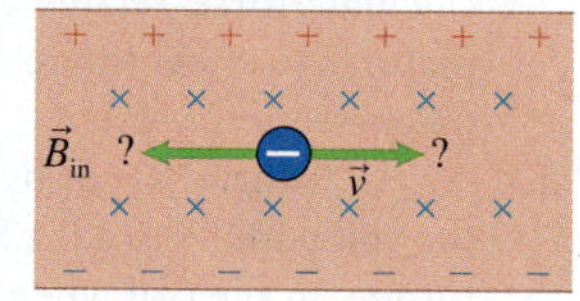


42. * An electron moves at speed $8.0 \times 10^{6}$ m/s toward the right between two parallel plates, such as shown in Figure P17.41. A 0.12-T magnetic field points into the paper parallel to the plate surfaces. (a) Determine the magnitude and direction of the magnetic force that the magnetic field exerts on the electron. (b) What should be the magnitude and direction of an $\vec{E}$ field caused by oppositely charged plates to produce an electric force that just balances the magnetic force?

43. ** **Electron deflection in an oscilloscope** Electrons start at rest and pass through a +28-kV potential difference in an oscilloscope that is oriented so that the electron is moving perpendicular to Earth's $3.5 \times 10^{-5}$-T magnetic field at the equator. The oscilloscope faces east. Determine everything you can about the path of the electron.

44. ** **Mass spectrometer** A mass spectrometer has a velocity selector that allows ions traveling at only one speed to pass with no deflection through slits at the ends. While moving through the velocity selector, the ions pass through a 60,000-N/C $\vec{E}$ field and a 0.0500-T $\vec{B}$ field. The quantities $\vec{v}$, $\vec{E}$, and $\vec{B}$ are mutually perpendicular. (a) Determine the speed of the ions that are not deflected. (b) After leaving the velocity selector, the ions continue to move in the 0.0500-T magnetic field. Determine the radius of curvature of a singly charged lithium ion, whose mass is $1.16 \times 10^{-26}$ kg.

45. ** **Mass spectrometer 2** One type of mass spectrometer accelerates ions of charge $q$, mass $m$, and initial speed zero through a potential difference $\Delta V$. The ions then enter a magnetic field where they move in a circular path of radius $r$. How is the mass of the ions related to these other quantities?

46. * An ion with charge $1.6 \times 10^{-19}$ C moves at speed $1.0 \times 10^{6}$ m/s into and perpendicular to the 0.30-T magnetic field of a mass spectrometer. After entering the field, the ion moves in a circular path of radius 0.31 m. What physical quantities describing the field and the ion can you determine using this information? Determine them.

## General Problems

47. * A box has either an electric field or a magnetic field inside. Describe experiments that you might perform to determine which field is present and its orientation.

48. ** A wire, shown in **Figure P17.48**, moves downward perpendicular to a magnetic field. (a) In what direction will the electrons in the wire move? (b) After the electrons are forced in one direction, leaving positive charges behind, an $\vec{E}$ field and potential difference $\Delta V$ develop from one end of the wire to the other. The electrons no longer move, since the electric and magnetic forces exerted on them balance. Show that this happens when the potential difference from one end of the wire to the other is $\Delta V = vLB$, where $v$ is the speed of the wire, $L$ is its length, and $B$ is the magnitude of the magnetic field. Describe two real-world situations where this phenomenon might occur.

**Figure P17.48**

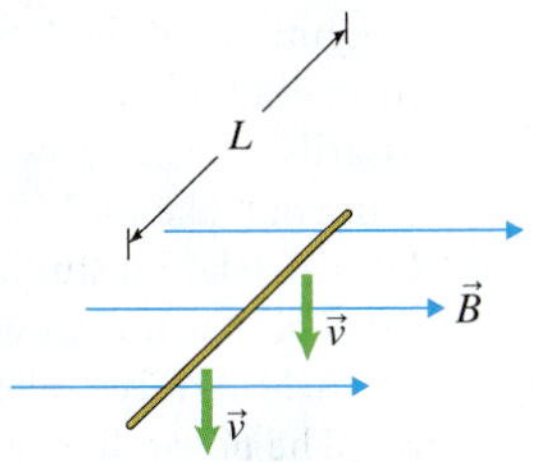


49. * **Charge separation on Boeing 747** Determine the potential difference developed across the wings of a Boeing 747 airliner when traveling north at 240 m/s. The wingspan is 60 m. Earth's $5.0 \times 10^{-5}$-T magnetic field points in a direction 60° below a horizontal line directed north. In solving this problem, follow the outline developed in the previous problem. Note that the magnetic field is not perpendicular to the plane's velocity.

50. ** **EST** Particles in cosmic rays are mostly protons, which have energies up to about $10^{20}$ eV, where 1 eV is the change in the kinetic energy of an electron moving through a potential difference of 1 V. Use this information to estimate the path of these particles in Earth's atmosphere. State the assumptions that you are making in your estimate.

## Reading Passage Problems

**BIO Magnetic resonance imaging (MRI)** In magnetic resonance imaging (MRI), a patient lies in a strong 1- to 2-T magnetic field $\vec{B}$ oriented parallel to the body. This field is produced by a large superconducting solenoid. The MRI measurements depend on the magnetic dipole moment $\vec{\mu}$ of a proton, the nucleus of a hydrogen atom. The proton magnetic dipoles can have only two orientations: either with the field or against the field. The energy needed to reverse this orientation ("flip" the protons) from with the field to against the field is exactly $\Delta U = 2\mu B$ (like the energy needed to turn a compass needle from north to south).

The pulse of a radio frequency probe field irradiates the patient's body in the region to be imaged. If this probe field is tuned correctly so that its energy equals the $\Delta U = 2\mu B$ needed to reverse the orientation of the protons from with the external $B$ field to against it, a reasonable number of protons will flip. When the protons return to their initial orientation and a lower energy state, they emit this same radio frequency radiation in different directions. This radiation is detected and provides a measure of the concentration of protons in the region irradiated by the probe field. The proton concentration differs in fat, muscle, and bone tissue, and in healthy and diseased tissue. Thus, the probe signal makes an image of the tissue type in each local region.

The MRI image of an internal body part is made by adjusting an auxiliary magnetic field, which varies the external $B$ field over the region being examined so that the probe field energy equals the flipping energy $\Delta U = 2\mu B$ in only a small area of the body. A measurement is made at that point. The external magnetic field is then adjusted to flip protons in a neighboring small area of the body. Continual shifts in the magnetic field and detection of proton concentrations at different tiny locations produce a map of proton concentration in the body. The MRI image of the lower back in **Figure 17.37** indicates an L45 disk that has partially collapsed—it has lost water and because it contains fewer protons produces a darker MRI image.

**Figure 17.37** A magnetic resonance image (MRI) of the lower back showing a partially collapsed disk.

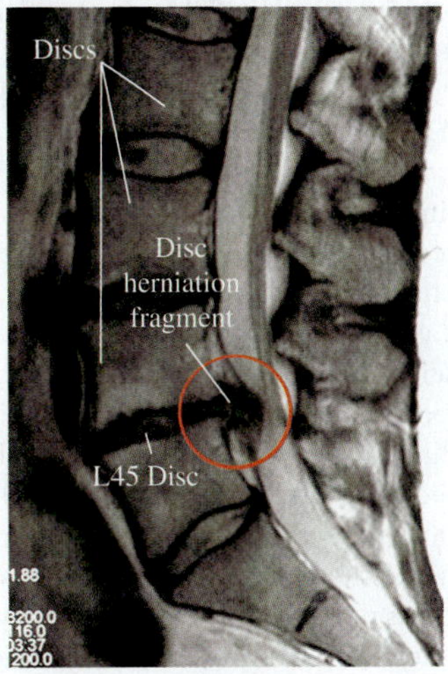

Discs
Disc herniation fragment
L45 Disc
1.88
The L45 disc is darker than the others. It has less water and fewer protons —a sign of a collapsed disc.

51. The solenoid magnetic field in an MRI apparatus is greater than Earth's $5.0 \times 10^{-5}$ T magnetic field by about how many times?
(a) 40 (b) 400 (c) 4000 (d) 40,000 (e) 400,000

52. The energy of the probe field causes protons to flip when in a 1.50000-T magnetic field. A $\pm 0.001$-T variation in the magnetic field causes a mismatch with the radio frequency flipping field and no flipping at a distance of 0.002 m from where the

magnetic field is matched to the flipping field. Which answer below is closest to the change in the $B$ field per unit distance?
(a) 0.08 T/m (b) 0.8 T/m (c) 8 T/m
(d) $0.003\ \text{T}\cdot\text{m}$ (e) 0.000003 T/m

53. Using your answer to Problem 52, determine the quantity closest to the amount that the auxiliary magnetic field causes the magnetic field to vary over a 0.20-m region of the body being mapped by MRI.
(a) 4 T (b) 2 T (c) 0.002 T (d) 0.02 T (e) 0.2 T

54. The MRI apparatus is able to look at proton concentration (and hence hydrogen concentration) at one tiny part of the body by doing what?
(a) Aiming the probe field at the whole body
(b) Varying the probe field frequency so that only protons in one place are flipped
(c) Varying the $B$ field over the body so $\Delta U = 2\,\mu B$ matches the probe field energy at one small location
(d) Placing a small hole in a body shield so the probe field reaches only one part of body
(e) All of the above

55. Which answer below is closest to the energy of the probe field needed to flip protons? The magnetic dipole moment of a proton is $1.41 \times 10^{-26}$ J/T.
(a) 4 J (b) $4 \times 10^{-10}$ J (c) $4 \times 10^{-25}$ J
(d) $4 \times 10^{-26}$ J (e) $4 \times 10^{-27}$ J

56. Why might a herniated disk projecting slightly out from between two vertebrae look different in an MRI image than a nonherniated disk?
(a) The vertebrae adjacent to a herniated disk are closer than vertebrae beside a nonherniated disk.
(b) There is a different concentration of hydrogen atoms in bone and in disks.
(c) Protons in the herniation produce an image that can be seen.
(d) b and c
(e) a, b, and c

**BIO Power lines—do their magnetic fields pose a risk?** In 1979 an epidemiologic study alleged a correlation between childhood leukemia in Denver neighborhoods and nearby high-voltage electric power lines. Since then there have been many studies exploring health risks caused by magnetic and electric fields from high-voltage power lines. Epidemiologic and animal studies that hinted at problems are not supported by independent tests at other laboratories. No mechanism linking power lines and cancer has been found. In recent years, large-scale scientific studies have had negative results, and scientific and medical societies have issued official statements that power lines are not a health risk.

Power lines produce both electric and magnetic fields. The interior of the human body is an electrical conductor, and as a result, electric fields are greatly reduced in magnitude within the body. The electric fields inside the body from power lines are much smaller than electric fields normally existing in the body.

However, magnetic fields are not reduced in the body. Earth's magnetic field, approximately $5 \times 10^{-5}$ T, is very small and not regarded as a health threat. Thus, it is interesting to compare Earth's magnetic field to fields produced by high power lines. The magnetic field $B$ produced at a distance $r$ from a straight wire with an electric current $I$ is

$$B = (2 \times 10^{-7}\ \text{T}\cdot\text{m/A})I/r$$

The magnetic field from a high-voltage power line located 40 m above the ground carrying a 100-A current is much smaller than Earth's $\vec{B}$ field.

Wires that provide electric power for household appliances also produce electric and magnetic fields. The current in the wire for a 500-W space heater is about 5 A. With the wire located several meters from your body, the magnetic field of such an appliance is somewhat smaller than Earth's magnetic field. By comparison, laboratory mice lived for several generations in 0.0010-T magnetic fields (20 times Earth's magnetic field) without any adverse effects.

During the last three decades, electric power use has increased the magnitudes of the $\vec{B}$ field created by power lines to which Americans are exposed by roughly a factor of 20. Yet during that same time interval, leukemia rates have slowly declined. It seems unlikely that magnetic fields produced by home appliances or high-voltage power lines are a hazard.

57. Which answer below is closest to the ratio of the power line $\vec{B}$ field on the ground 50 m below a 100-A current and Earth's $\vec{B}$ field at the same location?
(a) 0.001 (b) 0.01 (c) 0.1 (d) 10 (e) 100

58. A 550-W toaster oven is connected to a 110-V wall outlet. Which answer below is closest to the electric current in one of the wires from the wall outlet to the oven?
(a) 0.2 A (b) 50,000 A (c) 1 A (d) 5 A (e) 20 A

59. Which answer below is closest to the ratio of the magnetic field produced 0.4 m from the cable for the toaster oven in the last problem and Earth's magnetic field? (This is assuming the current flows in only one direction—which it does not.)
(a) 0.001 (b) 0.003 (c) 0.05 (d) 0.5 (e) 5

60. Leukemia rates have declined in recent years, whereas the magnitudes of the $\vec{B}$ fields created by power lines in our environment have increased significantly. Why does this not necessarily rule out power line magnetic fields as a contributing cause of leukemia?
(a) Correlation studies have no cause-effect relationship.
(b) There is no known mechanism for power line magnetic field-induced deaths.
(c) The power line magnetic fields cannot penetrate clothing and skin.
(d) Perhaps the power line magnetic field-induced cancers have increased from 0.001 to 0.025 of the cases and other causes have decreased.
(e) a, b, and d
(f) All of the above

61. Why would scientists be concerned about the relationship between magnetic fields and human health but not so much about electric fields and human health?
(a) The human body is a conductor.
(b) Many molecules are dipoles.
(c) There are moving electrically charged particles inside the body.
(d) Magnetic fields are more dangerous than electric fields.
(e) Electric fields are reflected from clothing but magnetic fields are not.

62. The 1979 epidemiologic study is a(n)
(a) testing experiment.
(b) observational experiment.
(c) hypothesis and a testing experiment.
(d) observational experiment and a hypothesis.

63. The mice experiment in the study of the effects of magnetic fields on health is a(n)
(a) observational experiment. (b) testing experiment.
(c) assumption. (d) proof.

# Electromagnetic Induction 18

How might electrical stimulation of the brain help treat certain diseases?

How does a microphone work?

How does a vending machine sort coins?

Because human cells, including nerve cells, contain ions that can move, they are electrically conductive. For some time, scientists have believed that the electrical nature of the brain could be used to help treat certain diseases. However, the skull is a fairly good electrical insulator. Until recently, the only options for electrical stimulation of the brain were either to apply a very high potential difference across points on the skull or to surgically implant electrodes into the brain. Now there is a promising new way to study and alter the electric activity of the brain noninvasively. This technology, called transcranial magnetic stimulation (TMS), may help treat mood disorders, Parkinson's disease, and Huntington's disease. TMS studies have also

**Be sure you know how to:**

- Find the direction of the $\vec{B}$ field produced by an electric current (Section 17.2).
- Find the direction of the magnetic force exerted on moving electric charges (Sections 17.3 and 17.4).
- Explain how an electric field produces a current in a wire and how that current relates to the resistance of the wire (Sections 16.1 and 16.4).

helped medical researchers understand the processes involved in neural repair, learning, and memory.

TMS treatment is fairly simple: a clinician places a small coil of wire on or near the patient's scalp. The changing current through this coil produces an abrupt electric current in the brain directly under the coil, even though there is no electrical connection between the outside coil and the brain. How is this possible?

**In the last chapter,** we learned that an electric current through a wire produces a magnetic field. Could the reverse happen? Could a magnetic field produce a current? It took scientists many years to answer this question. In this chapter, we will investigate the conditions under which this can happen.

## 18.1 Inducing an electric current

In the chapter on circuits (Chapter 16), we learned that an electric current results when a battery or some other device produces an electric field in a wire. The field in turn exerts an electric force on the free electrons in the wire connected to the battery. As a result, the electrons move in a coordinated manner around the circuit—an electric current.

**Active Learning Guide›**

In this section we will learn how to produce a current in a circuit without a battery—a process called **inducing** a current. We start by analyzing some simple experiments in Observational Experiment **Table 18.1**. The experiments involve a bar magnet and a coil. The coil is not connected to a battery but is connected to a galvanometer that detects both the presence of a current and its direction. A galvanometer works like an ammeter, but it is usually not calibrated in specific units. See if you can find any patterns in the outcomes of the experiments.

**OBSERVATIONAL EXPERIMENT TABLE**

**18.1 Inducing an electric current using a magnet.**

VIDEO 18.1

| Observational experiment | Analysis |
|---|---|
| **Experiment 1.** You hold a magnet motionless in front of a coil. | The galvanometer reads zero. There is no current through the coil.  |

| | | |
|---|---|---|
| **Experiment 2.** You move the magnet toward the coil or move the coil toward the magnet. | The galvanometer needle moves to the right, indicating a current through the coil. | 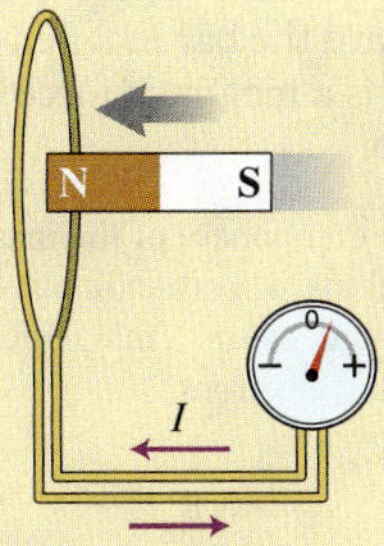 |
| **Experiment 3.** You move the magnet away from the coil or move the coil away from the magnet. | The galvanometer needle moves to the left, indicating a current through the coil but opposite the direction in the last experiment. | 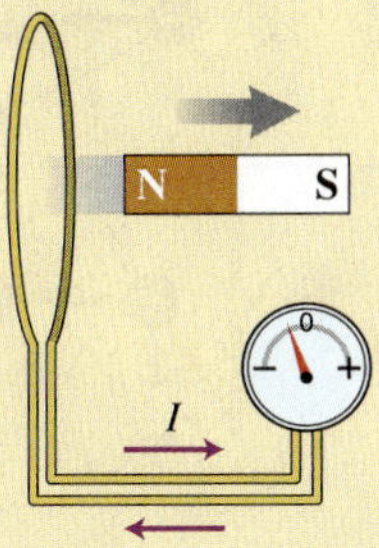 |
| **Experiment 4.** You turn the magnet 90° so that the poles are now perpendicular to their previous position. | The galvanometer registers a current while the magnet is turning. | 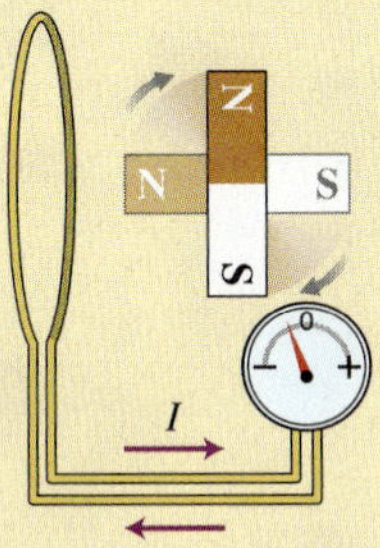 |
| **Experiment 5.** You collapse the sides of the coil together so its opening becomes very small.<br><br>You pull open the sides of the collapsed coil so the area becomes large again. | In both cases, the galvanometer registers a current while the coil's area is changing, but the direction is different in each case. |  |

**Patterns**

Although no battery was used, an electric current was induced in the coil when the magnet and coil moved toward or away from each other. Current was also induced when the coil's orientation relative to the magnet or the area of the coil changed.

In Table 18.1 there was no battery, yet the galvanometer registered an electric current through a coil. For the current to exist, there must be some source of emf. What produced the emf in the experiments in Table 18.1?

Recall from our study of magnetism (in Chapter 17) that a magnetic field can exert a force on moving electrically charged particles. The force exists only if the magnetic field or a component of the field is perpendicular to the velocity of the electrically charged particles. Let's consider again Experiment 2 in Table 18.1 in which the coil moves toward the magnet; for simplicity, we use a square loop made of conducting wire (**Figure 18.1**). Inside the wire there are positively charged ions that make up the lattice of the metal (and cannot leave their

**Figure 18.1** When the metal loop moves toward the bar magnet, the magnetic field exerts a magnetic force on electrons in the loop.

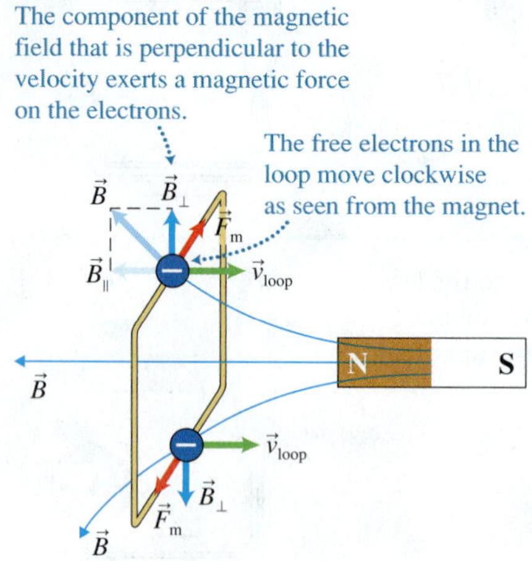
The component of the magnetic field that is perpendicular to the velocity exerts a magnetic force on the electrons.
The free electrons in the loop move clockwise as seen from the magnet.
$\vec{B}$
$\vec{B}_\perp$
$\vec{F}_m$
$\vec{B}_\parallel$
$\vec{v}_{loop}$
N
S
$\vec{B}$
$\vec{v}_{loop}$
$\vec{B}_\perp$
$\vec{F}_m$
$\vec{B}$

**Figure 18.2** Changing $\vec{B}$ or $A$ causes an induced current.

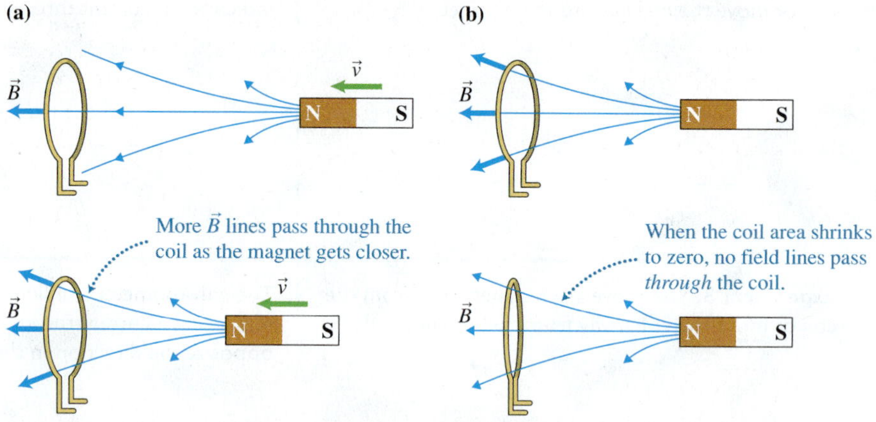
(a)
(b)
$\vec{B}$
$\vec{v}$
N
S
More $\vec{B}$ lines pass through the coil as the magnet gets closer.
When the coil area shrinks to zero, no field lines pass *through* the coil.

locations) and negatively charged free electrons. The $\vec{B}$ field produced by the bar magnet points away from the magnet and spreads out as shown in the figure.

Notice that at the top and bottom sections of the loop, a component of the $\vec{B}$ field is perpendicular to the velocity of the loop as it moves toward the right. As a result, the magnetic field exerts a force on each electron in the wire. This force causes the electrons in the wire to accelerate clockwise around the loop as viewed from the magnet (use the right-hand rule for the magnetic force on the negatively charged electrons, as discussed in Section 17.4). Note also that electrons in the vertical section of the loop closest to us accelerate upward, while the electrons in the vertical section farthest from us accelerate downward. The overall result is that due to the relative motion between the loop and the bar magnet, the electrons start moving around the loop in a coordinated fashion—we have an induced electric current.

Thus, it seems that the currents produced in Table 18.1 can be explained using the concept of the magnetic force we have previously developed (in Chapter 17). Does magnetic force also explain how transcranial magnetic stimulation (TMS) works? It does not. In the TMS procedure, the coil is not moving relative to the brain, whereas the magnetic force-based explanation requires motion of the loop relative to the magnetic field.

Perhaps there is another explanation. Let's examine Table 18.1 from a different perspective, one that focuses on the $\vec{B}$ field itself rather than on any sort of motion. When the magnet or coil moved or rotated with respect to the magnetic field, the number of $\vec{B}$ field lines going through the area of the coil increased or decreased (**Figure 18.2a**). The number of $\vec{B}$ field lines through the coil also changed when the area of the coil changed (Figure 18.2b). Thus, an alternative explanation for the pattern we observed in the experiments is that when the number of $\vec{B}$ field lines through the coil's area changes, there is a corresponding emf produced in the coil, which leads to the induced current.

We can summarize these two explanations for the induced current as follows:

*Explanation 1:* The induced current is caused by the magnetic force exerted on moving electrically charged objects (for example, the free electrons in conducting wires that are moving relative to the magnet).

*Explanation 2:* Any process that changes the number of $\vec{B}$ field lines through a coil's area induces a current in the coil. The mechanism explaining how it happens is unclear at this point.

Let's test these explanations with an experiment involving a change in the number of $\vec{B}$ field lines through a coil's area, but with no relative motion. Recall that we can create a magnetic field using either a wire that carries current or a permanent magnet (Chapter 17). The current-carrying coil in **Figure 18.3a** has a $\vec{B}$ field that resembles the $\vec{B}$ field of the bar magnet in Figure 18.3b.

**Figure 18.3** A current-carrying loop and a bar magnet create magnetic fields that have the same distribution of $\vec{B}$ field lines.

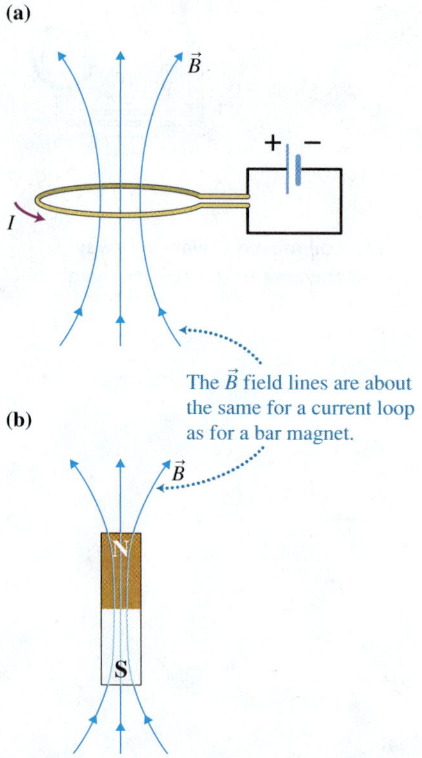
(a)
$\vec{B}$
+ −
$I$
The $\vec{B}$ field lines are about the same for a current loop as for a bar magnet.
(b)
$\vec{B}$
N
S

Testing Experiment **Table 18.2** uses two coils, such as those shown in **Figure 18.4**. Coil 1 on the bottom is connected to a battery and has a switch to turn the current through coil 1 on and off. When the switch is open, there is no current in coil 1. When the switch is closed, the current in coil 1 produces a magnetic field whose $\vec{B}$ field lines pass through coil 2's area. For each of the experiments we will use the two explanations to predict whether or not there should be an induced electric current in coil 2.

**Figure 18.4** Coil 1 is positioned directly under coil 2. Can a magnetic field from coil 1 induce a current in coil 2?

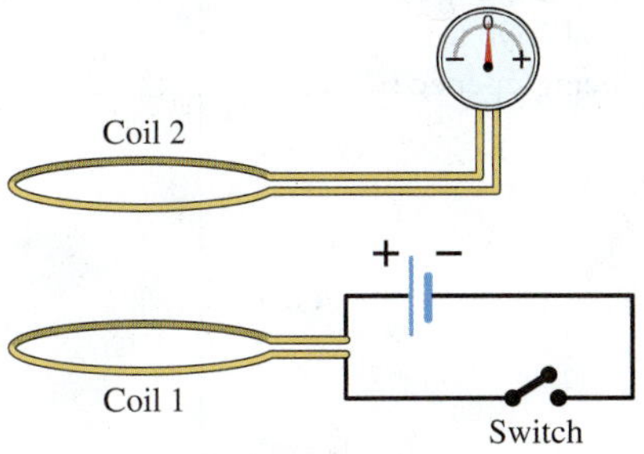


**TESTING EXPERIMENT TABLE**

**‹Active Learning Guide**

### 18.2 Testing the explanations for induced current.

VIDEO 18.2

| Testing experiment | Will a current be induced in coil 2? Prediction based on Explanation 1: Induced current is due to magnetic force exerted on moving charged particles. | Will a current be induced in coil 2? Prediction based on Explanation 2: Induced current is due to changing number of $\vec{B}$ field lines through coil. | Outcome |
|---|---|---|---|
| **Experiment 1.** The switch in the circuit for coil 1 is open. There is no current in coil 1. Is there any current in coil 2? | There is no current in coil 1, thus there is no magnetic field at coil 2. Neither coil is moving. No current will be induced in coil 2. | There is no current in coil 1; therefore, there is no change in the number of $\vec{B}$ field lines through coil 2's area. No current will be induced in coil 2. | 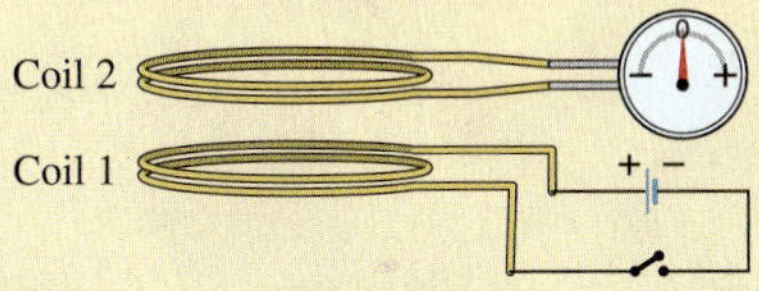  The galvanometer registers no current in coil 2. |
| **Experiment 2.** You close the switch in the circuit for coil 1. While the switch is being closed, the current in coil 1 increases rapidly from zero to a steady final value. Is there any current in coil 2 while the switch is being closed? | Neither coil is moving, thus no current will be induced in coil 2. | The increasing current in coil 1 produces an increasing $\vec{B}$ field. This changes the number of $\vec{B}$ field lines through coil 2's area. A brief current should be induced in coil 2. | 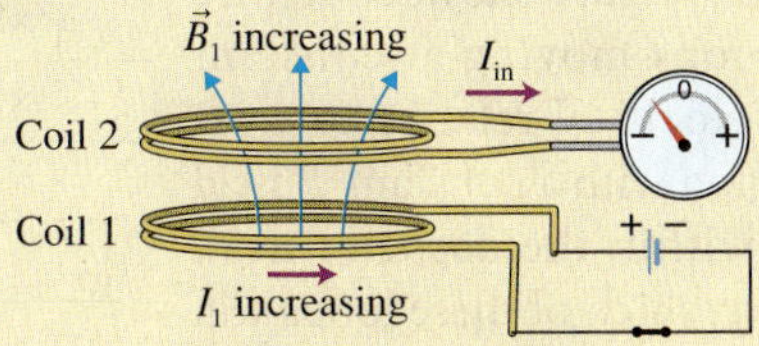  Just as the switch closes, the galvanometer needle briefly moves to the left and then returns to vertical, indicating a brief induced current in coil 2. |
| **Experiment 3.** You keep the switch in the circuit for coil 1 closed. The current in coil 1 has a steady value. Is there current in coil 2? | Neither coil is moving. Thus no current will be induced in coil 2. | There is a steady current in coil 1, which results in a steady $\vec{B}$ field. Thus, the number of $\vec{B}$ field lines through coil 2's area is not changing. Therefore, no current will be induced in coil 2. | 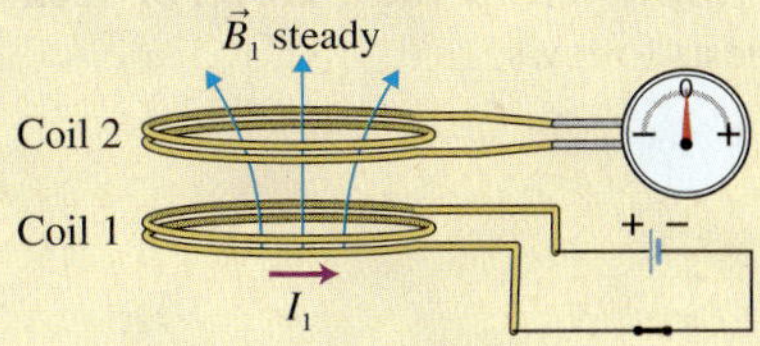  The galvanometer registers no current in coil 2. |

(continued)

| | | | |
|---|---|---|---|
| **Experiment 4.** You open the switch again. Is there any current in coil 2 while the switch is being opened? | Neither coil is moving. Thus no current will be induced in coil 2. | The decreasing current in coil 1 produces a decreasing $\vec{B}$ field. This changes the number of $\vec{B}$ field lines through coil 2's area, which should induce a brief current in coil 2. | 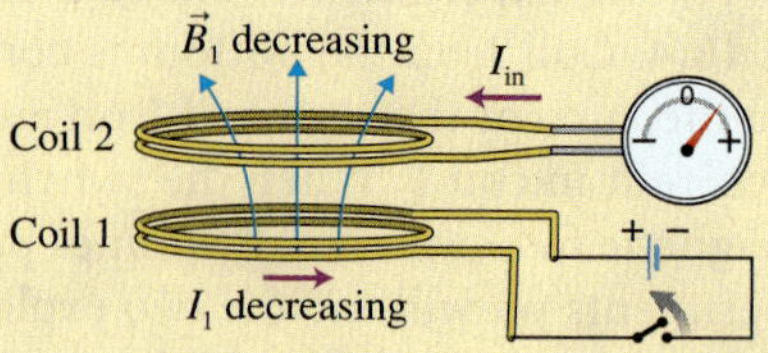<br>Just as the switch opens, the galvanometer needle briefly moves to the right (opposite the direction in experiment 2), then returns to the vertical, indicating a brief induced current in coil 2. |

**Conclusion**

The predictions based on Explanation 2 matched the outcomes in all four experiments. The predictions based on Explanation 1 did not match the outcomes in two of the four experiments.

Motion is not necessary to have an induced current. In contrast, when the number of $\vec{B}$ field lines through a coil's area changes, there is an induced current in that coil. Explanation 1 has been found not to be generally valid, but Explanation 2 has been found to be generally valid.

We have learned that it is possible to have a current in a closed loop of wire without using a battery. This phenomenon of inducing a current using a changing $\vec{B}$ field is called **electromagnetic induction**.

Electromagnetic induction explains how transcranial magnetic stimulation (TMS) works. When the physician closes the switch in the circuit with the coil that rests on the outside of the skull, the increasing current in the coil produces a changing $\vec{B}$ field within the brain, which in turn induces current in the brain's electrically conductive tissue. A current is briefly and noninvasively induced in a small region of the brain.

**CONCEPTUAL EXERCISE 18.1 Moving loops in a steady magnetic field**

The figure shows four wire loops moving at constant velocity $\vec{v}$ relative to a region with a constant $\vec{B}$ field (within the dashed lines). In which of these loops will electric currents be induced?

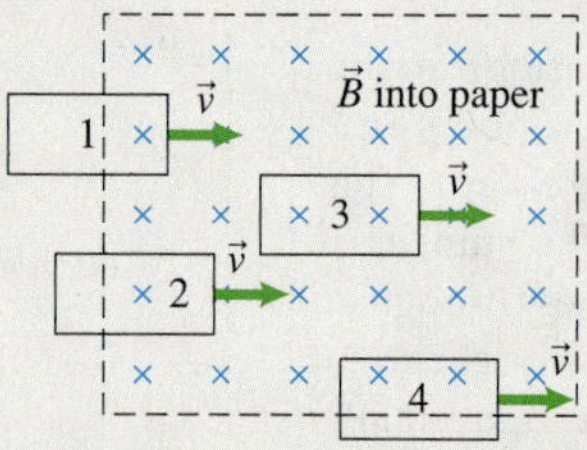


**Sketch and translate** An electric current will be induced in a loop whenever the number of $\vec{B}$ field lines through the loop's area changes.

**Simplify and diagram** For each case, current should be as follows:

1. The loop is moving into the field region, so the number of $\vec{B}$ field lines through its area is increasing. A current will be induced in the loop.
2. A current will be induced in the loop for the same reasons as (1).
3. The loop is completely within the field region, thus the number of $\vec{B}$ field lines is not changing. As a result, no current is induced.
4. No current is induced for the same reason as in (3).

**Try it yourself:** What happens to loops 3 and 4 as they are leaving the magnetic field region?

*Answer:* The number of $\vec{B}$ field lines through each loop will change and a brief current will be induced as the loops leave the $\vec{B}$ field region.

## Discovery of electromagnetic induction

We observed electromagnetic induction with relative ease. However, the first observation of this phenomenon in 1831 was more difficult. Following Hans Oersted's discovery in 1820 that an electric current produces a magnetic force on a compass needle, scientists wondered if the reverse might occur—could an object with magnetic properties produce an electric current? In 1821, British experimentalist Michael Faraday began investigating the possibility. Two difficulties lay before him—one conceptual and one technical. The conceptual difficulty was that although a steady current always produces a magnetic field, a steady magnetic field does not always produce an electric current. The technical difficulty was that the galvanometers of the time could not detect the weak currents induced in a single loop. Coils of wire (multiple loops of wire) were needed to amplify the magnetic effect. However, individual wires cannot be in contact with each other; they must be insulated from touching each other. At the time, no process existed for producing insulated wires; thus there was no way to make coils.

The technical problem was solved when American physicist Joseph Henry devised and published a method for insulating wires by wrapping them in silk. Henry was the first to observe a current being induced in a coil. He did not immediately publish his discovery, however. In 1831, upon learning about Henry's work, Faraday used Henry's insulation method to induce current in coils in his laboratory. Since then, scientists and engineers have devised many practical devices based on electromagnetic induction, such as electric generators, magnetic credit card readers, the transformers needed for modern electric power grids, and the pick-up coils for string and percussion instruments.

**Figure 18.5** A schematic of a dynamic microphone. The sound wave pushes a diaphragm to which a coil of wire is attached. The diaphragm/wire system moves with respect to the magnet.

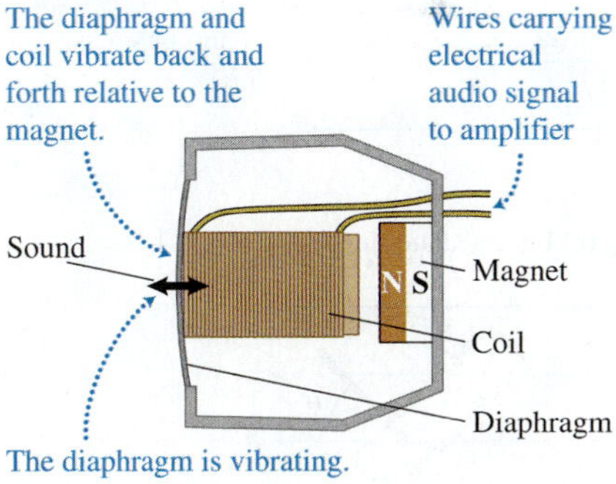


## Applying electromagnetic induction: Dynamic microphones and seismometers

A dynamic microphone that converts sound vibrations into electrical oscillations employs the principle of electromagnetic induction (**Figure 18.5**). When a sound wave, such as a singer's voice, strikes the diaphragm inside the microphone, the diaphragm oscillates. This oscillation moves a coil of wire attached to the diaphragm alternately closer and farther from a magnet in the microphone, corresponding to locations with stronger and weaker magnetic field. This changing magnetic field through the coil induces a changing current in the coil. The changes in the current mirror the sound waves that led to the production of this current; thus this current can then be used to store the details of the sound electronically.

A seismometer operates by the same principle. A seismometer is a sensor that detects seismic waves during an earthquake. The seismometer has a massive base with a magnet that vibrates as seismic waves pass. At the top, a coil attached to a block hangs at the end of a spring (**Figure 18.6**). The spring acts as a shock absorber that reduces the vibrations of the hanging block and coil relative to the base. The motion of the base relative to the coil induces a current through the coil, which produces a signal that is recorded on a seismograph.

**Figure 18.6** A simple seismometer. The base with the magnet attached vibrates when the seismic wave passes. The vibrating magnet induces a current in the coil above.

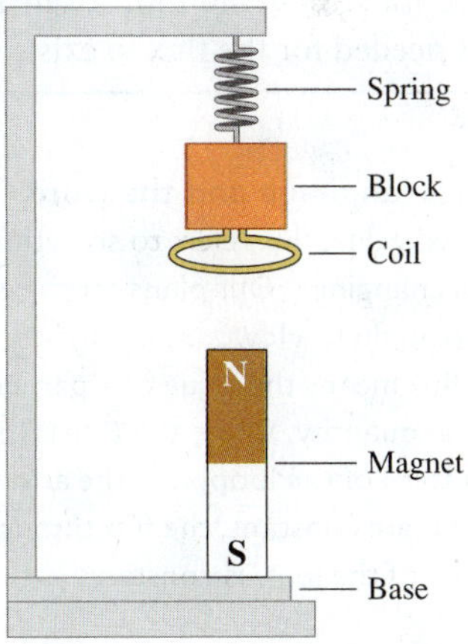


**Review Question 18.1** Your friend thinks that relative motion of a coil and a magnet is absolutely necessary to induce current in a coil that is not connected to a battery. Support your friend's point of view with a physics argument. Then provide a counterargument and describe an experiment you could perform to disprove your friend's idea.

# 18.2 Magnetic flux

In the last section we found that an electric current is induced in a coil when the number of $\vec{B}$ field lines through the coil's area changes. This occurred when

- the strength of the $\vec{B}$ field in the vicinity of the coil changed, or
- the area $A$ of the coil changed, or
- the orientation of the $\vec{B}$ field relative to the coil changed.

**Figure 18.7** Flux depends on the angle between the normal vector to the loop surface and the direction of the $\vec{B}$ field.

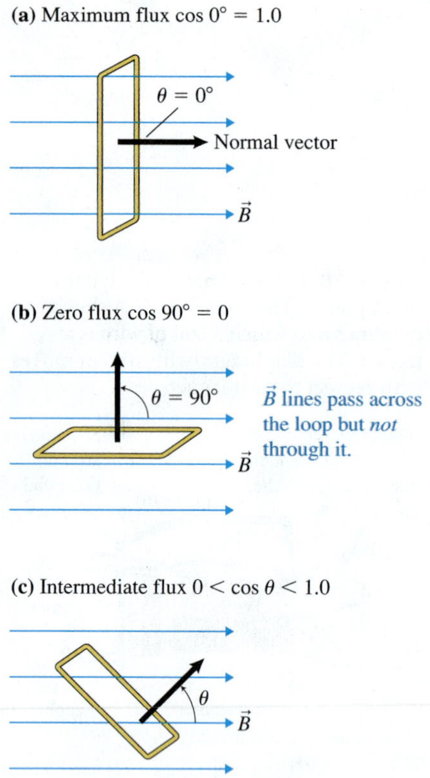


In this section we will construct a physical quantity for the number of $\vec{B}$ field lines through a coil's area. Based on the analysis in the last section, changes in that quantity should cause an induced electric current in the coil. Physicists call this physical quantity **magnetic flux** $\Phi$.

We have already defined the magnetic flux qualitatively as the number of $\vec{B}$ field lines passing through a particular two-dimensional area $A$. The greater the magnitude of the $\vec{B}$ field passing through the area, the greater the number of field lines through the area. Additionally, if the area itself is larger, the number of field lines through the area is greater. If we double one or the other, the number of lines through the area should double. This suggests that the magnetic flux is proportional to the magnitude of the $\vec{B}$ field passing through the area and to the size of the area itself. Mathematically,

$$\Phi \propto BA$$

How do we include in the above the dependence on the orientation of the loop relative to the $\vec{B}$ field lines? Imagine a rigid loop of wire in a region with uniform $\vec{B}$ field. If the plane of the loop is perpendicular to the $\vec{B}$ field lines, then a maximum number of field lines pass through the area (**Figure 18.7a**). If the plane of the loop is parallel to the $\vec{B}$ field, zero field lines pass through the area (Figure 18.7b). In between these two extremes, the magnetic flux takes on intermediary values (Figure 18.7c).

To describe this relative orientation we use a line perpendicular to the plane of the loop—the black normal vector shown in Figure 18.7. The angle $\theta$ between this vector and the $\vec{B}$ field lines quantifies this orientation. Since the cosine of an angle is at a maximum when the angle is zero and a minimum when the angle is 90°, the magnetic flux through the area is also proportional to $\cos\theta$. This leads to a precise definition for the magnetic flux through an area.

**Magnetic flux $\Phi$** **The magnetic flux $\phi$ through a region of area $A$ is**

$$\phi = BA\cos\theta \qquad (18.1)$$

**where $B$ is the magnitude of the uniform magnetic field throughout the area and $\theta$ is the angle between the direction of the $\vec{B}$ field and a normal vector perpendicular to the area. The SI unit of magnetic flux is the unit of the magnetic field (the tesla T) times the unit of area ($m^2$), or $T \cdot m^2$. This unit is also known as the weber (Wb).**

Equation (18.1) assumes that the magnetic field throughout the area is uniform and that the area is flat. For situations in which this is not the case, you first split the area into small subareas within which the $\vec{B}$ field is approximately uniform; then add together the magnetic fluxes through each. This book will not address such cases.

In Section 18.1, we proposed that when the number of magnetic field lines through a wire loop changes, a current is induced in the loop. We can now refine that idea and say that *current is induced when there is a change in the magnetic flux through the loop's area.* In other words, if the magnetic flux throughout the loop's area is steady, no current will be induced.

**TIP** Often we measure magnetic flux through a particular area, such as the area inside a wire loop. But the loop itself is not needed for the flux to exist.

### Physics language and the word flux

In everyday life, flux refers to something that is changing. "Our plans for the weekend are in flux." However, in physics, the word flux means the value of a particular physical quantity. When the $\vec{B}$ field, the orientation of the loop, and the area of the loop are constant, the flux through the area of the loop is constant.

**QUANTITATIVE EXERCISE 18.2 Flux through a book cover**

A book is positioned in a uniform 0.20-T $\vec{B}$ field that points from left to right in the plane of the page, shown in the figures on the next page. (For simplicity, we depict the book as a rectangular loop.) Each side of the book's cover measures 0.10 m. Determine the magnetic flux through the cover when (a) the cover is in the plane of the page (figure a), (b) the cover is perpendicular to the plane of the page and the normal vector makes a 60° angle with the $\vec{B}$ field (figure b), and (c) the book's cover area is perpendicular to the plane of the page and the normal vector points toward the top of the page (figure c).

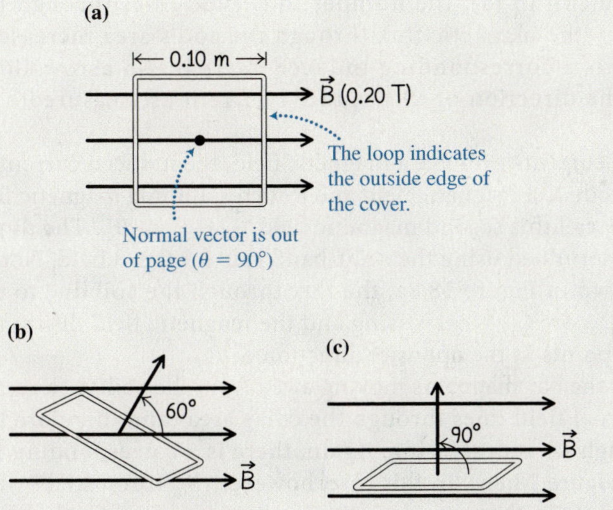
(a)
0.10 m
$\vec{B}$ (0.20 T)
The loop indicates the outside edge of the cover.
Normal vector is out of page ($\theta = 90°$)
(b)
60°
$\vec{B}$
(c)
90°
$\vec{B}$

**Represent mathematically** The magnetic flux through an area is determined by Eq. (18.1) $\Phi = BA\cos\theta$. The angle in each of the three situations is (a) $\theta = 90°$, (b) $\theta = 60°$, and (c) $\theta = 90°$.

**Solve and evaluate** The magnetic flux through the book cover in each case is

(a) $\Phi = (0.20\text{ T})(0.10\text{ m})^2\cos(90°) = 0$

(b) $\Phi = (0.20\text{ T})(0.10\text{ m})^2\cos(60°) = 1.0 \times 10^{-3}\text{ T}\cdot\text{m}^2$

(c) $\Phi = (0.20\text{ T})(0.10\text{ m})^2\cos(90°) = 0$

We can evaluate these results by comparing the calculated fluxes to the number of $\vec{B}$ field lines through the book cover's area. Note that for the orientation of the book in (a) and (c), the $\vec{B}$ field lines are parallel to the book's area and therefore do not go through it. Those positions are consistent with our mathematical result. The orientation for the book in (b) is such that some $\vec{B}$ field lines do pass through the book, which is also consistent with the nonzero mathematical result.

**Try it yourself:** A circular ring of radius 0.60 m is placed in a 0.20-T uniform $\vec{B}$ field that points toward the top of the page. Determine the magnetic flux through the ring's area when (a) the plane of the ring is perpendicular to the surface of the page and its normal vector points to the right and (b) the plane of the ring is perpendicular to the surface of the page and its normal vector points toward the top of the page.

*Answer:* (a) 0; (b) $0.23\text{ T}\cdot\text{m}^2$.

**Review Question 18.2** You have a bar magnet and a gold ring. How should you position the ring relative to the magnet so that the magnetic flux through the circular area inside the ring is zero?

‹Active Learning Guide

## 18.3 Direction of the induced current

Recall that in the experiments in Section 18.1, the galvanometer registered current in one direction for some of the experiments and in the opposite direction for others. We discovered the conditions under which a current can be induced in a wire loop, but can we also explain the direction of this current? That is the goal of this section.

**Figure 18.8** shows the results of two experiments in which the number of $\vec{B}$ field lines through a wire coil's area is changing. As the bar magnet

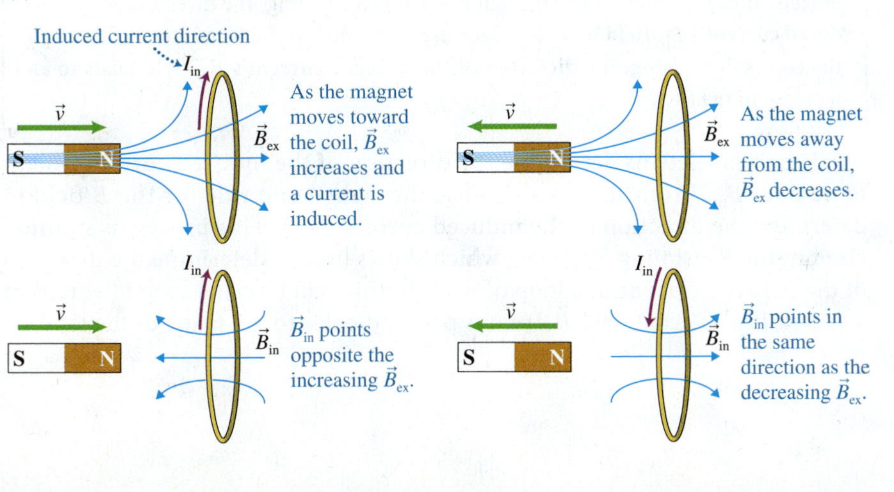
(a)
Induced current direction
$I_{in}$
$\vec{v}$
S
N
$\vec{B}_{ex}$
As the magnet moves toward the coil, $\vec{B}_{ex}$ increases and a current is induced.
$I_{in}$
$\vec{v}$
S
N
$\vec{B}_{in}$
$\vec{B}_{in}$ points opposite the increasing $\vec{B}_{ex}$.
(b)
$\vec{v}$
S
N
$\vec{B}_{ex}$
As the magnet moves away from the coil, $\vec{B}_{ex}$ decreases.
$I_{in}$
$\vec{v}$
S
N
$\vec{B}_{in}$
$\vec{B}_{in}$ points in the same direction as the decreasing $\vec{B}_{ex}$.

**Figure 18.8** How is the changing $\vec{B}_{ex}$ related to the direction of $I_{in}$ and $\vec{B}_{in}$? The number of the $\vec{B}_{ex}$ field lines passing though the coil increases in (a) and decreases in (b).

**Active Learning Guide>**

moves closer to the coil in (a), the number of $\vec{B}$ field lines through the coil's area increases (the magnetic flux through the coil's area increases). As expected, there is a corresponding induced current. An arrow along the coil indicates the direction of this induced current as measured by a galvanometer.

Because electric currents produce a magnetic field, the induced current in the coil must also produce a magnetic field and a corresponding magnetic flux through the coil. We call this second magnetic field $\vec{B}_{\text{induced}}$ or $\vec{B}_{\text{in}}$. The direction of $\vec{B}_{\text{in}}$ can be determined using the right-hand rule for the $\vec{B}$ field. Notice that in the case shown in Figure 18.8a, the flux through the coil due to the magnet (called $\vec{B}_{\text{external}}$ or $\vec{B}_{\text{ex}}$) is increasing and the magnetic field due to the induced current $\vec{B}_{\text{in}}$ points in the opposite direction of $\vec{B}_{\text{ex}}$.

In Figure 18.8b, the bar magnet is moving away from the coil. As a result, the number of external field lines through the coil's area (and therefore the magnetic flux through it) is decreasing. Again, there is a corresponding induced current (see Figure 18.8b). In this case, however, $\vec{B}_{\text{in}}$ (produced by the induced current) points in the same direction as $\vec{B}_{\text{ex}}$ (produced by the magnet). Can we find a pattern in these data?

Notice that in both cases $\vec{B}_{\text{in}}$ points in the direction that diminishes the change in the external flux through the coil. In the first experiment the flux through the coil was increasing. In that situation, $\vec{B}_{\text{in}}$ pointed in the opposite direction to $\vec{B}_{\text{ex}}$, as if to resist the increase. In the second experiment, the external flux through the coil was decreasing and $\vec{B}_{\text{in}}$ pointed in the same direction as $\vec{B}_{\text{ex}}$, as if to resist the decrease. In both situations, the $\vec{B}_{\text{in}}$ due to the induced current resisted the *change* in the external flux through the coil. Put another way, $\vec{B}_{\text{in}}$ points in whatever direction is necessary to try to keep the magnetic flux through the coil's area constant.

Consider what would happen if the reverse occurred. Suppose an increasing external magnetic flux through the loop led to an induced magnetic field in the same direction as the external field. In that case, the magnetic flux due to the induced field would augment rather than reduce the total flux through the loop. This would cause an even a greater induced current, which would cause yet a greater induced magnetic field and a steeper increase in magnetic flux. In other words, just by lightly pushing a bar magnet toward a loop of wire, you would cause a runaway induced current that would continually increase until the wire melted. Of course, this would violate the conservation of energy. If such a scenario were possible, we could heat water by simply moving a bar magnet over a coil in a large tank of water.

This pattern concerning the direction of the induced current was first developed in 1833 by the Russian physicist Heinrich Lenz.

**Lenz's law** **The direction of the induced current in a coil is such that its $\vec{B}_{\text{in}}$ field opposes the change in the magnetic flux through the coil's area produced by other objects. If the magnetic flux through the coil is increasing, the direction of the induced current's $\vec{B}_{\text{in}}$ field leads to a decrease in the flux. If the magnetic flux through the coil is decreasing, the direction of the induced current's $\vec{B}_{\text{in}}$ field leads to an increase in the flux.**

Lenz's law lets us determine the direction of the induced current's magnetic field $\vec{B}_{\text{in}}$. From there we can use the right-hand rule for the $\vec{B}$ field to determine the direction of the induced current itself. This process is summarized in the Reasoning Skill box, which shows how to determine the direction of the induced current in a loop of wire. In this skill box, the loop of wire is in a decreasing external field $\vec{B}_{\text{ex}}$ that is perpendicular to the plane of the loop.

**REASONING SKILL** Determine the direction of an induced current

The magnetic flux through a loop or coil can change because of a change in the external magnetic field, a change in the area of a loop or coil, or a change in its orientation. Because of a flux change, a current is induced in a direction that can be determined as shown below.

1. Determine the initial external magnetic flux $\Phi_{ex\,i}$ through the coil (represented below by the number and direction of the external magnetic field lines).

2. Determine the final external magnetic flux $\Phi_{ex\,f}$ through the coil (represented below by the smaller number of external magnetic field lines).

$\vec{B}_{ex\,i}$
$\vec{B}_{ex\,f}$

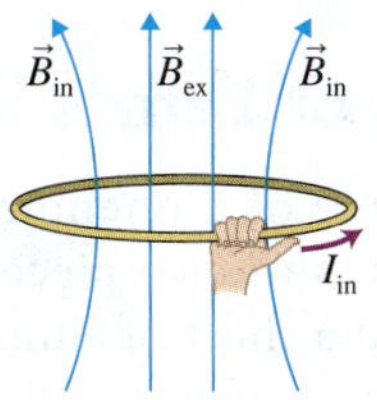

$\vec{B}_{in}$
$\vec{B}_{ex}$
$\vec{B}_{in}$
$I_{in}$

3. The induced flux opposes the *change* in the external flux (a decrease in upward external flux in this case; $\vec{B}_{in}$ points up so that the net upward flux does not decrease).

4. Use the right-hand rule for the magnetic field to determine the direction of the induced current that will produce the $\vec{B}_{in}$ and the induced flux.

**CONCEPTUAL EXERCISE 18.3 Practice using Lenz's law**

A loop in the plane of the page is being pulled to the right at constant velocity out of a region of a uniform magnetic field $\vec{B}_{ex}$. The field is perpendicular to the loop and points out of the paper in the region inside the rectangular dashed line (see the first figure below). Determine the direction of the induced electric current in the loop when it is halfway out of the field, as shown in the second figure below.

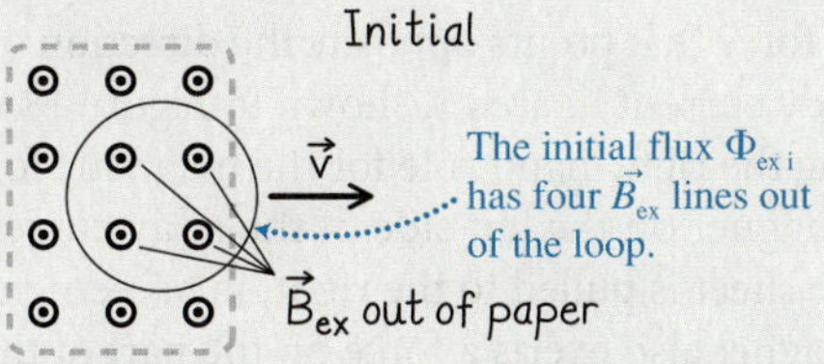

Initial
$\vec{v}$
The initial flux $\Phi_{ex\,i}$ has four $\vec{B}_{ex}$ lines out of the loop.
$\vec{B}_{ex}$ out of paper

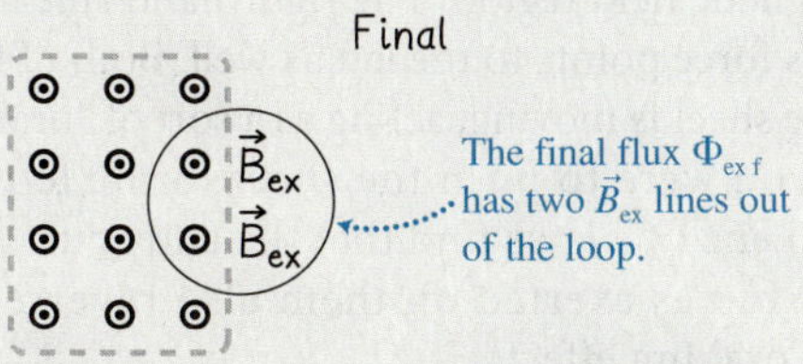

Final
$\vec{B}_{ex}$
$\vec{B}_{ex}$
The final flux $\Phi_{ex\,f}$ has two $\vec{B}_{ex}$ lines out of the loop.

**Sketch and translate** The number of $\vec{B}_{ex}$ lines passing through the coil's area decreases by half during the time interval from when the coil is completely in the field to when it is halfway out. Thus, the external magnetic flux through the loop's area has decreased.

**Simplify and diagram** Thus, according to Lenz's law the induced magnetic flux and induced magnetic field $\vec{B}_{in}$ should point out of the paper, as shown below, thus keeping the flux through the loop closer to the initial flux. The direction of the induced current that causes this induced magnetic field is determined using the right-hand rule for the $\vec{B}$ field; the induced electric current through the loop is counterclockwise.

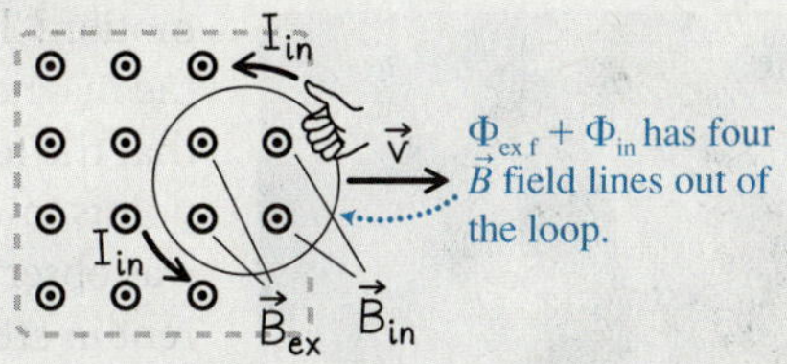

$I_{in}$
$\vec{v}$
$\Phi_{ex\,f} + \Phi_{in}$ has four $\vec{B}$ field lines out of the loop.
$I_{in}$
$\vec{B}_{ex}$
$\vec{B}_{in}$

**Try it yourself:** Notice that the induced current in the left side of the loop is still in the external magnetic field $\vec{B}_{ex}$ (as shown in the figure above). Determine the direction of the force that the external magnetic field $\vec{B}_{ex}$ exerts on the induced current in the left side of the loop.

*Answer:* To the left, opposite the direction of the loop's velocity.

**Figure 18.9** Inducing an eddy current by pulling a metal sheet through a magnetic field. (a) Area 1 is leaving the field and area 2 is entering the field. (b) Changing flux through both areas induces eddy currents. The external magnetic field exerts a force on the eddy currents opposite the direction of motion of the sheet.

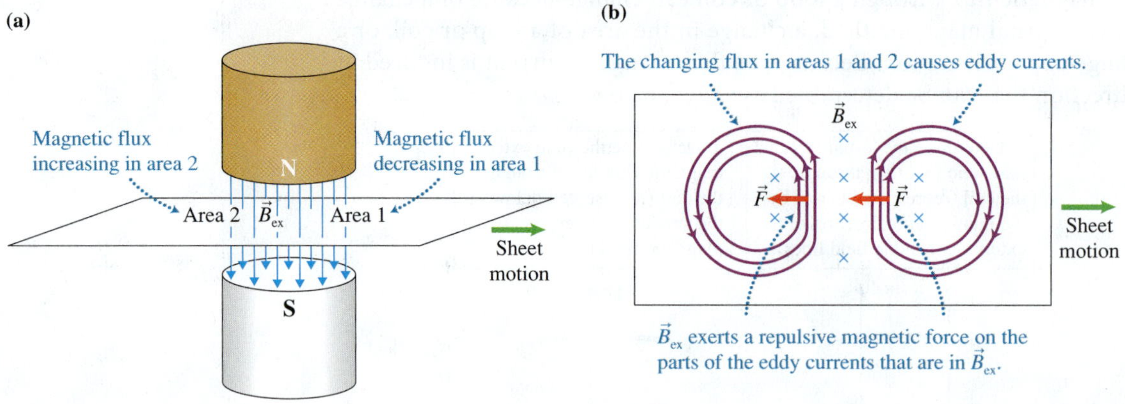
(a)
Magnetic flux increasing in area 2
Magnetic flux decreasing in area 1
N
Area 2
$\vec{B}_{ex}$
Area 1
Sheet motion
S
(b)
The changing flux in areas 1 and 2 causes eddy currents.
$\vec{B}_{ex}$
$\vec{F}$
$\vec{F}$
Sheet motion
$\vec{B}_{ex}$ exerts a repulsive magnetic force on the parts of the eddy currents that are in $\vec{B}_{ex}$.

## Eddy currents: an application of Lenz's law

Conceptual Exercise 18.3 shows an example of a phenomenon called an **eddy current.** An eddy current usually occurs when a piece of metal moves through a magnetic field. If you were to hold a sheet of aluminum or copper between the poles of a strong horseshoe magnet, you would find that neither of the poles attracts the sheet. However, when you move the sheet out from between the poles of the magnet, especially if you pull it quickly, you encounter resistance. This force is similar to the force exerted on the loop leaving the magnetic field region in the Try It Yourself part of Conceptual Exercise 18.3.

Let's examine this phenomenon. **Figure 18.9a** shows a metal sheet between the poles of an electromagnet. Pulling the sheet to the right decreases the external magnetic flux through area 1. This is similar to the situation in Conceptual Exercise 18.3, although in that exercise the external $\vec{B}$ field was out of the page instead of down. This decrease in flux induces an eddy current in the metal sheet around area 1, which circles or curls clockwise in this region (see Figure 18.9b) and according to Lenz's law produces an induced field that points in the same direction as the external $\vec{B}$ field.

At the same time, area 2 of the sheet is entering the magnetic field region, so the magnetic flux through that area is increasing. This change in magnetic flux induces a counterclockwise eddy current.

So, how do we explain the force that points opposite the direction of the sheet's motion? The left side of the eddy current in area 1, shown in Figure 18.9b, is still in the magnetic field region. Using the right-hand rule for the magnetic force, we find that the force exerted by the magnet on the left side of the eddy current in area 1 points toward the left when the sheet is pulled to the right, in agreement with what was observed. Similarly, the magnet also exerts a force on the part of the eddy current in area 2 that is in the magnetic field region. The right-hand rule for the magnetic force determines that this force points to the left as well. Both of these forces point opposite the direction the sheet is moving, acting as a sort of "braking" force.

What would happen if you were to push the sheet to the left instead of pulling it to the right? Using Lenz's law, we find that the eddy currents reverse direction, and the magnetic forces exerted on them also reverse direction, again resulting in a magnetic braking effect.

**Figure 18.10** Magnetic braking is used to stop the Tower of Terror II ride at the Gold Coast in Australia. The electromagnet induces eddy currents in the wheels of the car. Magnetic braking stops the car (originally traveling at 161 km/h) in about 200 m.

Many technological applications rely on the phenomenon of resistance to the motion of a nonmagnetic metal material through a magnetic field. For instance, the braking system used in some cars, trains, and amusement park car rides consists of strong electromagnets with poles on either side of the turning metal wheels of the vehicle (**Figure 18.10**). When the electromagnet is turned on, magnetic forces are exerted by the electromagnet on the eddy currents in the wheels and oppose their motion. As the turning rate of the wheels decreases, the eddy currents decrease, and therefore the braking forces decrease.

Coin sorters in vending machines also rely on magnetic braking. The coins roll down a track and through a magnetic field, which induces eddy currents that slow each coin to a speed that is based on the coin's metal type and size. The coins exit the field region at different speeds and fly like projectiles into a bin that is specific for each type of coin. Sensors detect where the coins land to determine how much the customer paid.

**Review Question 18.3** What difficulty would occur if the $\vec{B}$ field produced by the induced current enhanced the change in the external field rather than opposed the change? Give a specific example.

## 18.4 Faraday's law of electromagnetic induction

In the first two sections of this chapter, we found that when the magnetic flux through a coil's area changes, there is an induced electric current in the coil. The flux depends on the magnitude of the $\vec{B}$ field, the area of the coil, and the orientation of the coil relative to the $\vec{B}$ field. Our next goal is to construct a quantitative version of this idea that will allow us to predict the magnitude of the induced current through a particular coil. We begin with Observational Experiment **Table 18.3**, in which we will determine what factors affect the magnitude of the induced current produced by the flux change.

**‹Active Learning Guide**

**OBSERVATIONAL EXPERIMENT TABLE**

**18.3 Factors affecting the magnitude of induced current.**

VIDEO 18.3

| Observational experiment | Analysis |
|---|---|
| **Experiment 1.** Rapidly move a magnet toward a coil and observe the galvanometer needle. Repeat the process, only move the magnet slowly. | 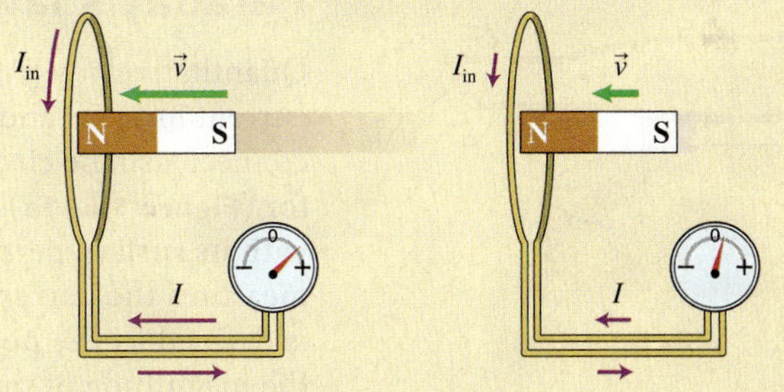  The galvanometer registers a larger induced current when the magnet moves rapidly toward the coil compared with when it moves slowly. |
| **Experiment 2.** Rotate a magnet rapidly in front of a stationary coil. Repeat the process, only rotate the magnet slowly. | 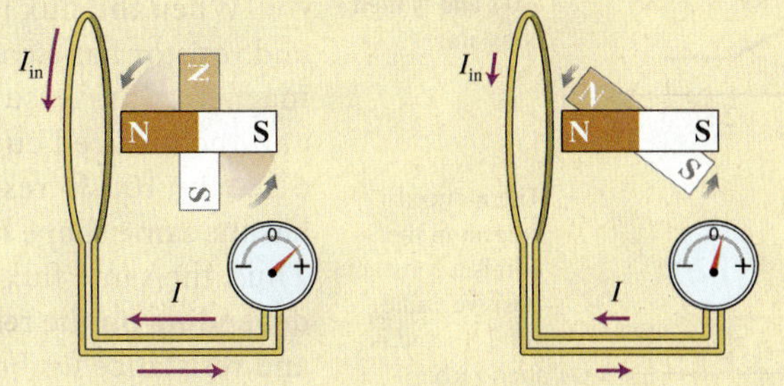  The galvanometer registers a larger induced current when the magnet rotates rapidly compared with when it rotates slowly. |

*(continued)*

| Observational experiment | Analysis |
|---|---|
| **Experiment 3.** Use two coils, each with a different number of turns. Move the magnet toward the coil with the greater number of turns. Then move the magnet at the same speed toward the coil with fewer turns. | 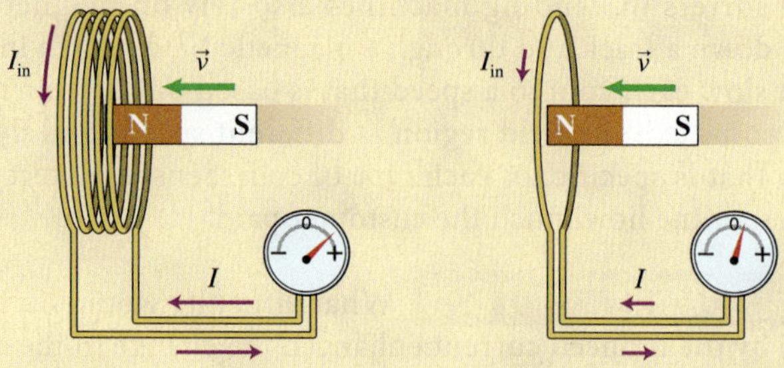  The galvanometer registers a larger induced current in the coil with a larger number of turns. |

**Patterns**

- The speed at which the magnet moved or rotated affected the induced current. The shorter the time interval for the change of the flux through the coil, the greater the induced current.
- The induced current is greater in a coil with a larger number of turns than in a coil with a smaller number of turns.

**Figure 18.11** An experiment to quantitatively relate induced emf ($\varepsilon_{in}$) and induced current ($I_{in}$) to flux change. (a) The $\vec{B}_{ex}$ of the electromagnet decreases at a constant rate; (b) although the flux changes with time, the current induced in the loop is constant. When the flux is constant, the current is zero.

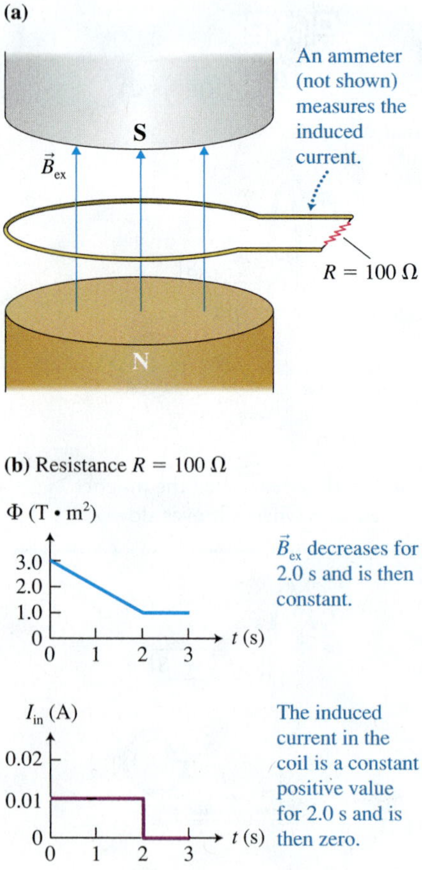


In the experiments in Observational Experiment Table 18.3, we found that the induced current was greater if the same change in magnetic flux through a coil occurred in a shorter time interval $\Delta t$. Additionally, the induced current through the coil was greater in a coil with a larger number of turns $N$.

> **TIP** Notice that even though we were investigating the effects of two quantities on the magnitude of the induced current, we changed only one of them at a time so that we could investigate those effects separately.

## Faraday's law of electromagnetic induction

Quantitatively, what is the relationship between the magnitude of the induced current in a coil and the change in magnetic flux through that coil's area? Let's connect a single circular loop with zero resistance in series with a 100-$\Omega$ resistor (**Figure 18.11a**). The loop is placed between the poles of an electromagnet with its surface perpendicular to the magnetic field. An ammeter (not shown) measures the current in the loop and resistor. The upward magnetic field (up is defined as the positive direction) decreases steadily for 2.0 s (we decrease the magnitude of the $\vec{B}$ field by decreasing the current in the coils of the electromagnet; note that this is not the induced current), after which the field and flux remain constant at a smaller positive value. The magnetic flux-versus-time and the induced current-versus-time graphs are shown in Figure 18.11b.

When the flux is *changing* at a constant rate, the current through the loop and resistor has a constant value. For example, for the first 2.0 s the *slope* of the magnetic flux-versus-time graph has a constant negative value of $-1.0\ \text{T}\cdot\text{m}^2/\text{s}$, and the induced current has a constant positive value of +0.010 A. If we replace the 100-$\Omega$ resistor with a 50-$\Omega$ resistor, the current-versus-time graph has the same shape but its magnitude during the first 2.0 s doubles to +0.020 A. Thus, the same flux change produces different size currents in the same loop depending on the resistance of the circuit. However, the product of the current and resistance for both situations has the same value:

$$(0.010\ \text{A})(100\ \Omega) = (0.020\ \text{A})(50\ \Omega) = 1.0\ \text{A}\cdot\Omega = 1.0\ \text{V}$$

It almost seems that there must be a 1.0-V battery in series with the resistor, but there isn't.

The changing flux through the loop caused the induced current. However, the unit of the *slope* of the flux-versus-time graph is not the ampere, but the volt:

**‹Active Learning Guide**

$$\frac{\Delta\Phi}{\Delta t} = \frac{-2.0\,\text{T}\cdot\text{m}^2}{2.0\,\text{s}} = -1.0\left(\frac{\text{N}}{\text{C}\dfrac{\text{m}}{\text{s}}}\right)\frac{\text{m}^2}{\text{s}} = -1.0\,\frac{\text{N}\cdot\text{m}}{\text{C}} = -1.0\,\frac{\text{J}}{\text{C}} = -1.0\,\text{V}$$

(We used the expression for the magnetic force $F_{\text{magnetic}} = qvB$ or $B = F_{\text{magnetic}}/qv$ to convert the magnetic field in tesla (T) to other units.) Thus the induced current is a consequence of the emf induced in the coil. The changing flux acts as a "battery" that produces a 1.0-V emf that causes the electric current. You could apply Kirchhoff's loop rule to the circuit shown in Figure 18.11a. There is a +1.0-V potential change across the loop and a −1.0-V potential change across the resistor.

The magnitude of the emf in the loop depends only on the rate of change of flux through the loop (the slope of the flux-versus-time graph $\varepsilon_{\text{in}} = |\Delta\Phi/\Delta t|$). When we repeat this experiment for a coil with $N$ loops, we find that the magnitude of the induced emf increases in proportion to the number $N$ of loops in the coil. Thus,

$$\varepsilon_{\text{in}} = N\left|\frac{\Delta\Phi}{\Delta t}\right| \tag{18.2}$$

This expression is known as **Faraday's law of electromagnetic induction**. However, the mathematical expression was actually developed by James Clerk Maxwell.

**Faraday's law of electromagnetic induction** **The average magnitude of the induced emf $\varepsilon_{\text{in}}$ in a coil with $N$ loops is the magnitude of the ratio of the magnetic flux change through the loop $\Delta\Phi$ to the time interval $\Delta t$ during which that flux change occurred multiplied by the number $N$ of loops:**

$$\varepsilon_{\text{in}} = N\left|\frac{\Delta\Phi}{\Delta t}\right| = N\left|\frac{\Delta[BA\cos\theta]}{\Delta t}\right| \tag{18.3}$$

The direction of the current induced by this emf is determined using Lenz's law. A minus sign is often placed in front of the $N$ to indicate that the induced emf opposes the change in magnetic flux. However, we will not use this notation.

Faraday realized that the phenomenon of electromagnetic induction has the same effect on devices attached to a coil as a battery does. This idea enabled him to build the first primitive electric generator that produced an induced current in a coil.

## QUANTITATIVE EXERCISE 18.4

### Hand-powered computer

To power computers in locations not reached by any power grid, engineers have developed hand cranks that rotate a 100-turn coil in a strong magnetic field. In a quarter-turn of the crank of one of these devices, the magnetic flux through the coil's area changes from 0 to $0.10\,\text{T}\cdot\text{m}^2$ in 0.50 s. What is the average magnitude of the emf induced in the coil during this quarter-turn?

**Represent mathematically** Faraday's law [Eq. (18.3)] can be used to determine the average magnitude of the emf produced in the coil:

$$\varepsilon_{\text{in}} = N\left|\frac{\Delta\Phi}{\Delta t}\right| = N\left|\frac{\Phi_{\text{f}} - \Phi_{\text{i}}}{t_{\text{f}} - t_{\text{i}}}\right|$$

*(continued)*

**Solve and evaluate** Inserting the appropriate values, we find:

$$\varepsilon_{in} = (100)\left|\frac{(0.10\ \text{T}\cdot\text{m}^2) - 0}{0.50\ \text{s} - 0}\right| = 20\ \text{V}$$

The magnitude of this emf is about what is required by laptop computers. However, the computer only works while the coil is turning. Laptops typically have a power requirement of about 50 watts (50 joules each second), which means considerable strength and endurance would be needed to keep the coil rotating.

**Try it yourself:** Determine the average emf produced in the coil if you turn the coil one quarter-turn in 1.0 s instead of in 0.50 s.

*Answer:* 10 V.

**Review Question 18.4** Why do we write the law of electromagnetic induction in terms of emf rather than in terms of induced current?

## 18.5 Skills for analyzing processes involving electromagnetic induction

**Active Learning Guide›**

Faraday's law enables us to design and understand practical applications of electromagnetic induction. For example, to design an automobile ignition system that uses spark plugs, engineers must estimate how quickly the magnetic field through a coil must be reduced to zero to produce a large enough emf to ignite a spark plug. An engineer designing an electric generator will be interested in the rate at which the generator coil must turn relative to the $\vec{B}$ field to produce the desired induced emf. The general strategy for analyzing questions like these is described and illustrated in Example 18.5.

### PROBLEM-SOLVING STRATEGY Problems involving electromagnetic induction

#### EXAMPLE 18.5 Determine the $\vec{B}$ field produced by an electromagnet

To determine the $\vec{B}$ field produced by an electromagnet, you use a 30-turn circular coil of radius 0.10 m (30-Ω resistance) that rests between the poles of the magnet and is connected to an ammeter. When the electromagnet is switched off, the $\vec{B}$ field decreases to zero in 1.5 s. During this 1.5 s the ammeter measures a constant current of 180 mA. How can you use this information to determine the initial $\vec{B}$ field produced by the electromagnet?

**Sketch and translate**

- Create a labeled sketch of the process described in the problem. Show the initial and final situations to indicate the change in magnetic flux.
- Determine which physical quantity is changing ($\vec{B}$, $A$, or $\theta$), thus causing the magnetic flux to change.

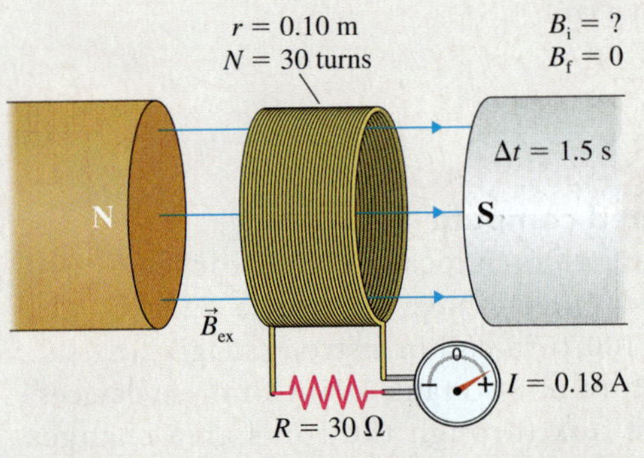


The changing quantity (from time 0.0 s to 1.5 s) is the magnitude of the $\vec{B}$ field produced by the electromagnet. Due to this change, the flux through the coil's area changes; thus there is an induced current in the coil. We can use the law of electromagnetic induction to find the magnitude of the initial $\vec{B}$ field produced by the electromagnet.

**Simplify and diagram**

- Decide what assumptions you are making: Does the flux change at a constant rate? Is the magnetic field uniform?
- If useful, draw a graph of the flux and the corresponding induced emf-versus-clock reading.
- If needed, use Lenz's law to determine the direction of the induced current.

Assume that

- The current in the electromagnet changes at a constant rate, thus the flux through the coil does also.
- The $\vec{B}$ field in the vicinity of the coil is uniform.
- The $\vec{B}$ field is perpendicular to the coil's surface.

**Represent mathematically**

- Apply Faraday's law and indicate the quantity ($\vec{B}$, $A$, or $\cos\theta$) that causes a changing magnetic flux.
- If needed, use Ohm's law and Kirchhoff's loop rule to determine the induced current.

$$\varepsilon_{\text{in}} = N\left|\frac{\Phi_f - \Phi_i}{t_f - t_i}\right| = N\left|\frac{0 - B_i A\cos\theta}{\Delta t}\right|$$

The number of turns $N$ and the angle $\theta$ remain constant. The magnitude of the magnetic field changes.

Substitute $\varepsilon_{\text{in}} = IR$ and $A = \pi r^2$ and solve for $B_i$:

$$B_i = \frac{IR\Delta t}{N\pi r^2 \cos\theta}$$

**Solve and evaluate**

- Use the mathematical representation to solve for the unknown quantity.
- Evaluate the results—units, magnitude, and limiting cases.

The plane of the coil is perpendicular to the magnetic field lines, so $\theta = 0$ and $\cos 0° = 1$. Inserting the appropriate quantities:

$$B_i = \frac{IR\Delta t}{N\pi r^2} = \frac{0.18\text{ A} \times 30\ \Omega \times 1.5\text{ s}}{30 \times \pi \times (0.1\text{ m})^2} = 8.6\text{ T}$$

This is a very strong $\vec{B}$ field but possible with modern superconducting electromagnets. Let's check the units:

$$\frac{\text{A}\cdot\Omega\cdot\text{s}}{\text{m}^2} = \frac{\text{V}\cdot\text{s}}{\text{m}^2} = \frac{\text{J}\cdot\text{s}}{\text{C}\cdot\text{m}^2} = \frac{\text{N}\cdot\text{m}\cdot\text{s}}{\text{C}\cdot\text{m}^2} = \frac{\text{N}}{\text{C}\cdot(\text{m/s})} = \text{T}$$

The units match. As a limiting case, a coil with fewer turns would require a larger $\vec{B}$ field to induce the same current. Also, if the resistance of the circuit is larger, the same $\vec{B}$ field change induces a smaller current.

**Try it yourself:** Determine the current in the loop if the plane of the loop is parallel to the magnetic field. Everything else is the same.

*Answer:* Zero.

In the following three sections, we consider practical applications of electromagnetic induction that involve a change in (1) the magnitude of the $\vec{B}$ field, (2) the area of the loop or coil, or (3) the orientation of the coil relative to the $\vec{B}$ field. All of these processes involve the same basic idea: a changing magnetic flux through the area of a coil or single loop is accompanied by an induced emf around the coil or loop. In turn, this emf induces an electric current in the coil or loop.

**Review Question 18.5** In the last example, why did we assume that the $\vec{B}$ field was uniform?

## 18.6 Changing $\vec{B}$ field magnitude and induced emf

We have learned that a current is induced in a coil (or single loop, or electrically conductive region in the case of eddy currents) when the magnetic flux through the coil's area changes. In this section, we consider examples where the flux change is due to a change in the magnitude of the $\vec{B}$ field throughout the coil's area. This is the case in transcranial magnetic stimulation (TMS), which we investigated qualitatively earlier in this chapter.

**EXAMPLE 18.6 Transcranial magnetic stimulation**

The magnitude of the $\vec{B}$ field from a TMS coil increases from 0 T to 0.2 T in 0.002 s. The $\vec{B}$ field lines pass through the scalp into a small region of the brain, inducing a small circular current in the conductive brain tissue in the plane perpendicular to the field lines. Assume that the radius of the circular current in the brain is 0.0030 m and that the tissue in this circular region has an equivalent resistance of 0.010 Ω. What are the direction and magnitude of the induced electric current around this circular region of brain tissue?

**Sketch and translate** We first sketch the situation: a small coil on the top of the scalp and a small circular disk region inside the brain tissue through which the changing magnetic field passes (see the figure below). The change in magnetic flux through this disk is caused by the increasing $\vec{B}$ field produced by the TMS coil (called $\vec{B}_{ex}$). This change in flux causes an induced emf, which produces an induced current in the brain tissue. The direction of this current can be determined using Lenz's law.

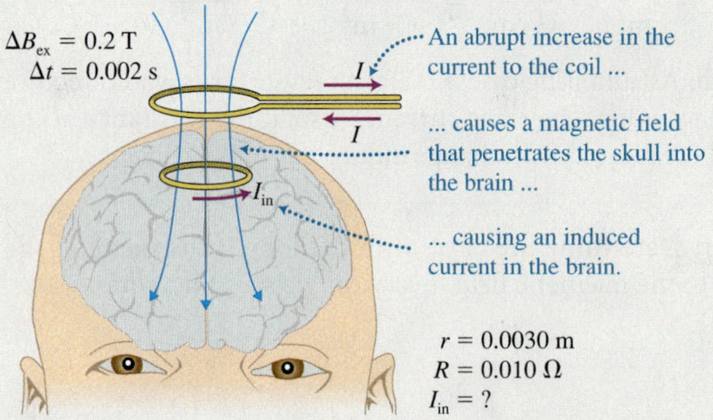
$\Delta B_{ex} = 0.2$ T
$\Delta t = 0.002$ s
An abrupt increase in the current to the coil ...
... causes a magnetic field that penetrates the skull into the brain ...
... causing an induced current in the brain.
$I$
$I$
$I_{in}$
$r = 0.0030$ m
$R = 0.010\ \Omega$
$I_{in} = ?$

**Simplify and diagram** Assume that $\vec{B}_{ex}$ throughout the disk of brain tissue is uniform and increases at a constant rate. Model the disk-like region as a single-turn coil. Viewed from above, $\vec{B}_{ex}$ points into the page (shown as X's in the figure top right). Since the number of $\vec{B}$ field lines is increasing into the page, the $\vec{B}_{in}$ field produced by the induced current will point out of the page (shown as dots). Using the right-hand rule for the $\vec{B}_{in}$ field, we find that the direction of the induced current is counterclockwise.

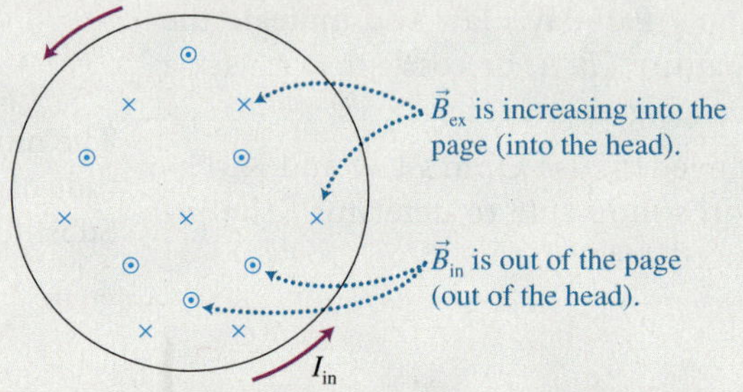
Top view of induced current
$\vec{B}_{ex}$ is increasing into the page (into the head).
$\vec{B}_{in}$ is out of the page (out of the head).
$I_{in}$

**Represent mathematically** To find the magnitude of the induced emf, use Faraday's law:

$$\varepsilon_{in} = N\left|\frac{\Phi_f - \Phi_i}{t_f - t_i}\right|$$

where the magnetic flux through the loop at a specific clock reading is $\Phi = B_{ex}A\cos\theta$. The area $A$ of the loop and the orientation angle $\theta$ between the loop's normal vector and the $\vec{B}_{ex}$ field are constant, so

$$\varepsilon_{in} = N\left|\frac{B_{ex\,f}A\cos\theta - B_{ex\,i}A\cos\theta}{t_f - t_i}\right| = NA\cos\theta\left|\frac{B_{ex\,f} - B_{ex\,i}}{t_f - t_i}\right|$$

Using our understanding of electric circuits, we relate this induced emf to the resulting induced current:

$$I_{in} = \frac{\varepsilon_{in}}{R}$$

**Solve and evaluate** Combine these two equations and solve for the induced current:

$$I_{in} = \frac{1}{R}\varepsilon_{in} = \frac{1}{R}\left(NA\cos\theta\left|\frac{B_{ex\,f} - B_{ex\,i}}{t_f - t_i}\right|\right)$$

$$= \frac{N\pi r^2\cos\theta}{R}\left|\frac{B_{ex\,f} - B_{ex\,i}}{t_f - t_i}\right|$$

$$= \frac{(1)\pi(0.0030\text{ m})^2(1)}{0.010\ \Omega}\left|\frac{0.2\text{ T} - 0}{0.002\text{ s} - 0}\right| = 0.28\text{ A}$$

Note that the normal vector to the loop's area is parallel to the magnetic field; therefore, $\cos\theta = \cos(0°) = 1$.

This is a significant current and could affect brain function in that region of the brain.

**Try it yourself:** A circular coil of radius 0.020 m with 200 turns lies so that its area is parallel to this page. A bar magnet above the coil is oriented perpendicular to the coil's area, its north pole facing toward the coil. You quickly (in 0.050 s) move the bar magnet sideways away from the coil to a location far away. During this 0.050 s, the magnitude of the $\vec{B}$ field throughout the coil's area changes from 0.40 T to nearly 0 T. Determine the average magnitude of the induced emf around the coil while the bar magnet is being moved away.

*Answer:* 2.0 V.

**Review Question 18.6** How can a coil of wire, a battery, and a switch placed outside a patient's head produce electric currents in the brain?

## 18.7 Changing area and induced emf

When located in a region with nonzero $\vec{B}$ field, a change in a coil's area also results in a change in the magnetic flux through that area. There is then a corresponding induced emf producing an induced electric current circling the area.

### EXAMPLE 18.7 Lighting a bulb

A 10-Ω lightbulb is connected between the ends of two parallel conducting rails that are separated by 1.2 m, as shown in the figure below. A metal rod is pulled along the rails so that it moves to the right at a constant speed of 6.0 m/s. The two rails, the lightbulb, its connecting wires, and the rod make a complete rectangular loop circuit. A uniform 0.20-T magnitude $B_{ex}$ field points downward, perpendicular to the loop's area. Determine the direction of the induced current in the loop, the magnitude of the induced emf, the magnitude of the current in the lightbulb, and the power output of the lightbulb.

(a)

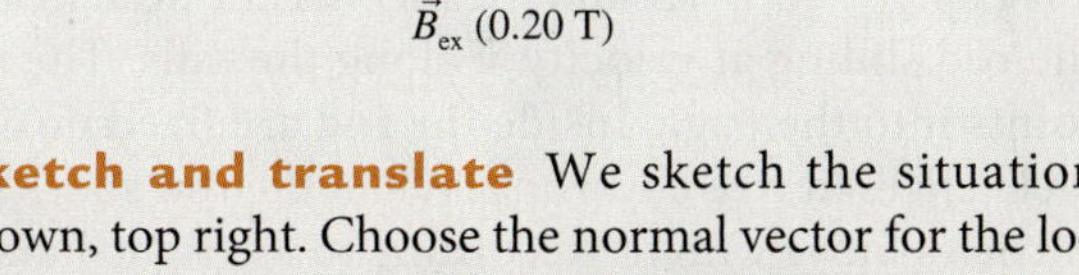

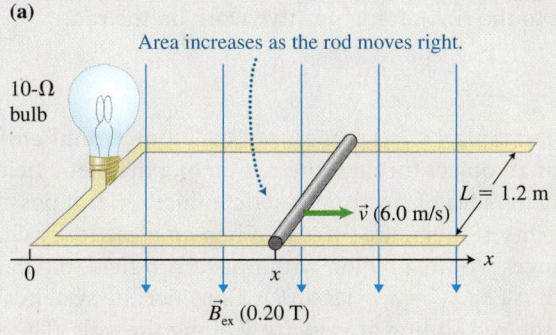


**Sketch and translate** We sketch the situation as shown, top right. Choose the normal vector for the loop's area to point upward. The loop's area increases as the rod moves away from the bulb. Because of this, the magnitude of the magnetic flux through the loop's area is increasing as the rod moves to the right.

(b)

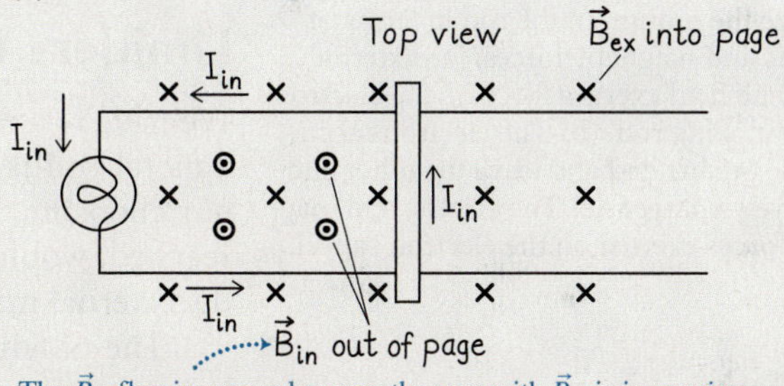


The $\vec{B}_{ex}$ flux increases because the area with $\vec{B}_{ex}$ is increasing.

**Simplify and diagram** Assume that the rails, rod, and connecting wires have zero resistance. The induced field $\vec{B}_{in}$ due to the loop's induced current should point upward, resisting the change in the downward increasing magnetic flux through the loop's area. Using the right-hand rule for the $\vec{B}_{in}$ field, we find that the direction of the induced current is counterclockwise.

**Represent mathematically** To find the magnitude of the induced emf, use Faraday's law:

$$\varepsilon_{in} = N\left|\frac{\Phi_f - \Phi_i}{t_f - t_i}\right|$$

The angle between the loop's normal line and the $B_{ex}$ field is 180°. The magnitude of the $B_{ex}$ field is constant. Therefore:

$$\varepsilon_{in} = (1)\left|\frac{B_{ex}A_f\cos(180°) - B_{ex}A_i\cos(180°)}{t_f - t_i}\right|$$

$$= B_{ex}\left|\frac{A_f - A_i}{t_f - t_i}\right|$$

*(continued)*

The area $A$ of the loop at a particular clock reading equals the length $L$ of the sliding rod times the $x$-coordinate of the sliding rod (the origin of the $x$-axis is placed at the bulb.) Therefore:

$$\varepsilon_{in} = B_{ex}\left|\frac{Lx_f - Lx_i}{t_f - t_i}\right| = B_{ex}L\left|\frac{x_f - x_i}{t_f - t_i}\right| = B_{ex}Lv$$

The quantity inside the absolute value is the $x$-component of the rod's velocity, the absolute value of which is the rod's speed $v$. Using Ohm's law, the induced current depends on the induced emf and the bulb resistance:

$$I_{in} = \frac{\varepsilon_{in}}{R}$$

The power output of the lightbulb will be

$$P = I_{in}^2 R$$

**Solve and evaluate** The magnitude of the induced emf around the loop is

$$\varepsilon_{in} = B_{ex}Lv = (0.20\ \mathrm{T})(1.2\ \mathrm{m})(6.0\ \mathrm{m/s}) = 1.44\ \mathrm{V}$$

The current in the bulb is

$$I_{in} = \frac{1.44\ \mathrm{V}}{10\ \Omega} = 0.14\ \mathrm{A}$$

The lamp should glow, but just barely, since its power output is only

$$P = I_{in}^2 R = (0.14\ \mathrm{A})^2(10\ \Omega) = 0.21\ \mathrm{W}$$

**Try it yourself:** Suppose the rod in the last example moves at the same speed but in the opposite direction so that the loop's area decreases. Determine the magnitude of the induced emf, the magnitude of current in the bulb, and the direction of the current.

*Answer:* 1.4 V, 0.14 A, and clockwise.

**Figure 18.12** An alternative way to analyze the motion of the rod in terms of electric and magnetic forces. (a) External magnetic field exerts a force on the electrons in the moving rod. (b) The electrons accumulate on one end and leave the other end positively charged (c) The electric and magnetic forces exerted on the electron cancel.

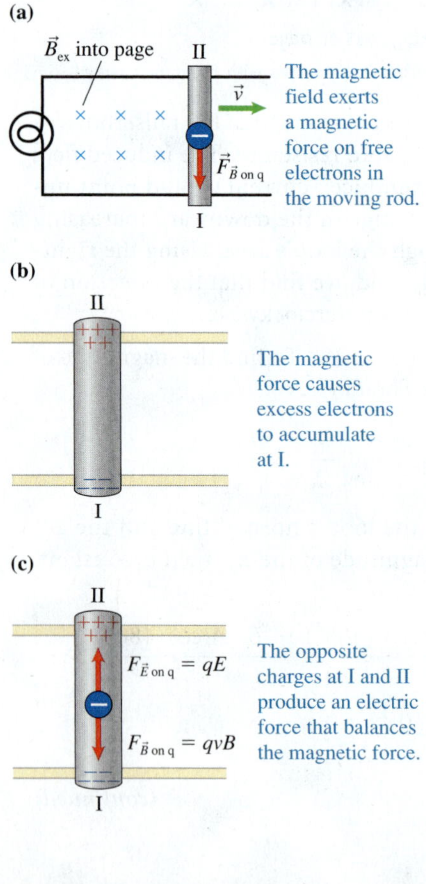
(a)
$\vec{B}_{ex}$ into page
II
I
$\vec{v}$
$\vec{F}_{B\ on\ q}$
The magnetic field exerts a magnetic force on free electrons in the moving rod.
(b)
II
I
The magnetic force causes excess electrons to accumulate at I.
(c)
II
I
$F_{\vec{E}\ on\ q} = qE$
$F_{\vec{B}\ on\ q} = qvB$
The opposite charges at I and II produce an electric force that balances the magnetic force.

## Limitless electric energy?

We have just found that the induced emf depended on the speed with which a metal rod is pulled along the rails. By accelerating the rod to a speed of our choosing, we could induce whatever emf we desired. Once this speed is reached, would this method of obtaining the emf by moving a conductor in the external magnetic field maintain the current indefinitely?

The moving rod resulted in an emf and an induced current in the rails, bulb, and rod. However, if we use the right-hand rule for magnetic force, we find that the magnetic field exerts a force on the induced current in the rod toward the left, causing it to slow down. From an energy perspective, the kinetic energy of the rod is being transformed into light and thermal energy in the bulb. In order to keep the rod moving at constant speed, some other object must exert a force on it to the right doing positive work on the rod.

## Motional emf

The emf produced in Example 18.7 is sometimes called **motional emf**; it is caused by the motion of an object through the region of a magnetic field. We explained this emf using the idea of electromagnetic induction. Is it possible to understand it just in terms of magnetic forces? When an electrically charged object with charge $q$ moves within a region with nonzero $\vec{B}$ field, the field exerts a magnetic force on it ($F_{\vec{B}\ on\ q} = |q| vB\sin\theta$). Consider the system shown in **Figure 18.12a** with the rod sliding at velocity $\vec{v}$ along the rails. The external magnetic field $\vec{B}_{ex}$ points into the page. Inside the rod are fixed positively charged ions and negatively charged free electrons. When the rod slides to the right, all of its charged particles move with it. According to the right-hand rule for the magnetic force, the external magnetic field exerts a magnetic force on the electrons toward end I. The positive charges cannot move inside the rod, but the free electrons can. The electrons accumulate at end I, leaving end II with a deficiency of electrons (a net positive charge). The ends of the rod become charged, as shown in Figure 18.12b.

**‹Active Learning Guide**

These separated charges create an electric field $\vec{E}$ in the rod that exerts a force of magnitude $F_{\vec{E}\,\text{on}\,q} = qE$ on other electrons in the rod; the electric field exerts a force on negative electrons toward II (Figure 18.12c) opposite the direction of the magnetic force. When the magnitude of the electric force equals the magnitude of the magnetic force, the accumulation of opposite electric charge at the ends of the rod ceases. Then,

$$\underbrace{qvB}_{\text{Magnetic force}} = \underbrace{qE}_{\text{Electric force}} \quad \text{or} \quad E = vB$$

An electric potential difference is produced between points I and II that depends on the magnitude of the electric field $\vec{E}$ in the rod and the distance $L$ between ends I and II:

$$\varepsilon_{\text{motional emf}} = |\Delta V_{\text{I}-\text{II}}| = EL = vBL \qquad (18.4)$$

The above expression for motional emf is the same expression we derived in Example 18.7 using Faraday's law. Thus, for problems involving conducting objects moving in a magnetic field, we can use either Faraday's law or the motional emf expression to determine the emf produced—either method will provide the same result.

**Review Question 18.7** Suppose the rod in Example 18.7 was one-third the length and the magnetic field was four-fifths the magnitude. How fast would the rod need to move to produce the same emf? Would the current induced in this case be the same as for Example 18.7? Explain.

## 18.8 Changing orientation and induced emf

In the previous two sections, we investigated processes where emf was induced when the magnitude of the $\vec{B}$ field changed or when the area of a loop within the $\vec{B}$ field region changed. In this section we investigate what happens when the orientation of a loop changes relative to the direction of the $\vec{B}$ field. This process has many practical applications, the most important being the electric generator.

### The electric generator

Worldwide, we convert an average of 310 J per person of electric potential energy into other less useful energy forms every second. Electric generators make this electric potential energy available by converting mechanical energy (such as water rushing through a hydroelectric dam) into electric potential energy.

To understand how an electric generator works, consider a very simple device that consists of a loop of wire attached to a turbine (a propeller-like object that can rotate). The loop is positioned between the poles of an electromagnet that produces a steady uniform $\vec{B}$ field. A Bunsen burner next to the turbine heats a flask of water (**Figure 18.13**). The water is converted to steam, which strikes the blades of the turbine, causing the turbine to rotate. The loop of wire attached to the turbine rotates in the $\vec{B}$ field region. When the loop's surface is perpendicular to the $\vec{B}$ field, the magnetic flux through the loop's area is at a maximum. One quarter turn later, the $\vec{B}$ field lines are parallel to the loop's area and the flux through it is zero. After another quarter turn, the flux is again at its maximum magnitude, but negative in value since the

**Figure 18.13** A homemade version of an electric generator. The steam turns the turbine, which turns the wire loop with respect to the magnet.

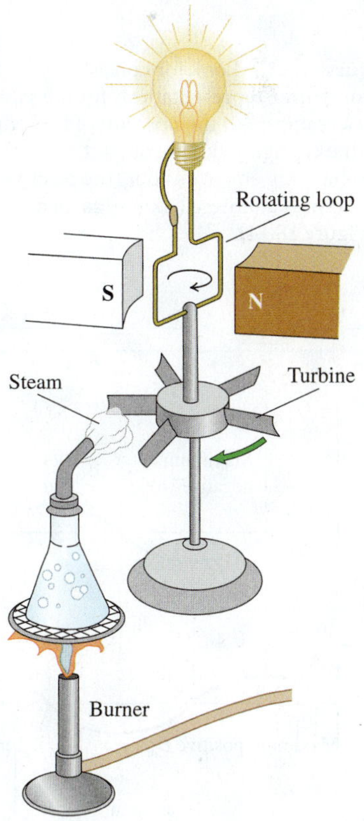

orientation of the loop's area is opposite what it was originally. This changing magnetic flux through the loop's area causes a corresponding induced emf, which produces a current that changes direction each time the loop rotates one half turn. Current that periodically changes direction in this way is known as **alternating current** (AC).

A coal-fired power plant is based on this process. Coal is burned to heat water, converting it to steam. The high-pressure steam pushes against turbine blades, causing the turbine and an attached wire coil to rotate in a strong $\vec{B}$ field. The resulting emf drives the electric power grid.

## Emf of a generator

How can we determine an expression for the emf produced by an electric generator? Consider the changing magnetic flux through a loop's area as it rotates with constant rotational speed $\omega$ in a constant uniform $\vec{B}_{ex}$ field (**Figure 18.14**). If there is an angle $\theta$ between the loop's normal vector and the $\vec{B}_{ex}$ field, then the flux through the loop's area is

**Figure 18.14** The magnetic flux through the loop changes as the angle between the normal vector to the loop and the $\vec{B}$ field lines changes.

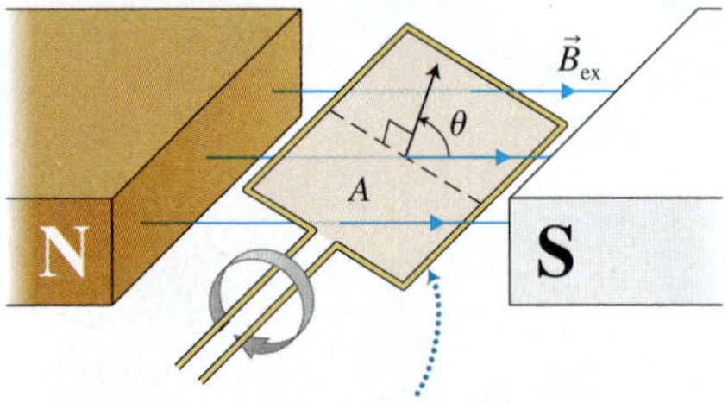


$$\Phi = B_{ex}A\cos\theta$$

Since the loop is rotating, $\theta$ is continuously changing. The loop is rotating with zero rotational acceleration ($\alpha = 0$); we can describe the motion with rotational kinematics (see Chapter 8):

$$\theta = \theta_0 + \omega t + \frac{1}{2}\alpha t^2 = \theta_0 + \omega t + \frac{1}{2}(0)t^2 = \theta_0 + \omega t$$

If we define the initial orientation $\theta_0$ to be zero, then

$$\theta = \omega t$$

where $\omega$ is the constant rotational speed of the loop. This means that the magnetic flux $\Phi$ through the loop's area as a function of time $t$ is

$$\Phi = B_{ex}A\cos(\omega t)$$

For a side view of the rotating loop see **Figure 18.15a**. Figure 18.15b shows a graph of the magnetic flux through the loop's area as a function of time.

According to Faraday's law Eq. (18.2), the induced emf around a coil (a multi-turn loop) is

$$\varepsilon_{in} = N\left|\frac{\Delta\Phi}{\Delta t}\right| = N\left|\frac{\Phi_f - \Phi_i}{t_f - t_i}\right|$$

**Figure 18.15** Look for consistency among three representations for the same clock reading: (a) the position of the loop, (b) the changing flux through it, and (c) the changing emf around it (the positive direction is counterclockwise as seen in Figure 18.14).

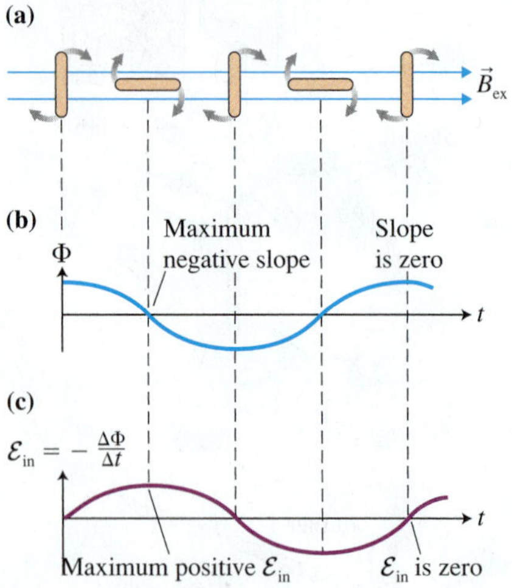


Since $\Phi$ is continually changing, we should use calculus to write the above equation. However, in this text, we will simply show the result:

$$\varepsilon_{in} = NB_{ex}A\omega\sin(\omega t) \tag{18.5}$$

where $N$ is the number of turns in a coil rotating between the poles of the magnet.

Figure 18.15c shows a graph of the induced emf as a function of clock reading. Comparing Figures 18.15b and c, you will see a pattern. The value of $\varepsilon_{in}$ at a particular clock reading equals the negative value of the slope of the $\Phi$-versus-$t$ graph at that same clock reading. This makes sense, since the induced emf is related to the rate of change of the magnetic flux through the loop's area. Slopes represent exactly that, rates of change.

Electric power plants in the United States produce an emf with frequency $f$ equal to 60 Hz (Hz is a unit of frequency; 60 Hz means the emf undergoes 60 full cycles in 1 second). This corresponds to a generator coil with a rotational speed

$$\omega = 2\pi f = 2\pi(60\text{ Hz}) = 120\pi\text{ rad/s}$$

These power plants can produce a peak (maximum) emf as high as 20 kV. The peak emf produced by a generator occurs when $\sin(\omega t) = 1$ and when $\sin(\omega t) = -1$. At those times,

$$\mathcal{E}_{\text{in max}} = NB_{\text{ex}}A\omega. \quad (18.6)$$

**QUANTITATIVE EXERCISE 18.8 Bicycle light generator**

The label on the Schmidt E6 bicycle dynamo headlight indicates that the light has a power output of 3 W and a peak emf of 6 V. The generator (also called a dynamo) for the lightbulb has a cylindrical hub that rubs against the edge of the bike tire, causing a coil inside the generator to rotate, as shown below. When the bicycle is traveling

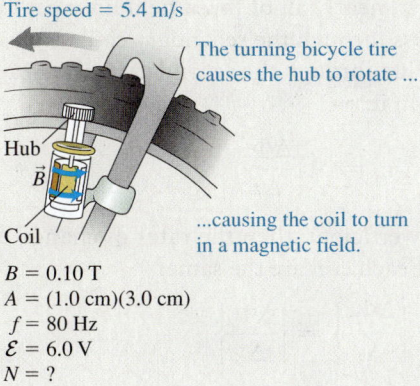

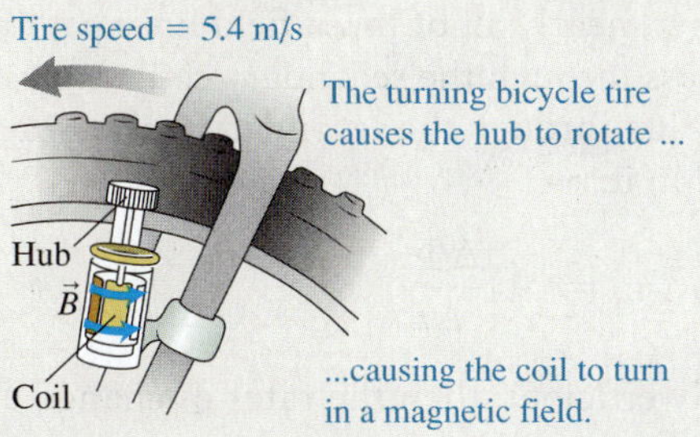


$B = 0.10$ T
$A = (1.0\text{ cm})(3.0\text{ cm})$
$f = 80$ Hz
$\mathcal{E} = 6.0$ V
$N = ?$

at a speed of 5.4 m/s, the coil rotates with frequency of 80 Hz (80 revolutions per second). The $\vec{B}$ field in the vicinity of the coil is uniform and has a magnitude of 0.10 T. The coil is a rectangle with dimensions 1.0 cm × 3.0 cm. Without taking the light apart, determine how many turns there are in the generator coil.

**Represent mathematically** The number of turns in the coil is related to the maximum emf the generator can produce [Eq. (18.6)]:

$$\mathcal{E}_{\text{in max}} = NBA\omega = NBA(2\pi f)$$

**Solve and evaluate** Solving for $N$ and inserting the appropriate values:

$$N = \frac{\mathcal{E}_{\text{in max}}}{2\pi f BA} = \frac{6.0\text{ V}}{2\pi(80\text{ Hz})(0.10\text{ T})(0.01\text{ m} \times 0.03\text{ m})} = 400$$

A generator coil with this number of turns is reasonable. Let's check limiting cases. If the magnetic field, the coil area, or the frequency is larger, then fewer turns are needed for the peak emf to be 6.0 V, which is reasonable.

**Try it yourself:** While riding your bike up a hill, you pedal harder; however, your bike speed reduces from 5.4 m/s to 2.7 m/s. How would these conditions affect the emf produced by the bicycle light generator in the last example?

*Answer:* The peak emf would be 3.0 V, since the loop turning frequency would decrease to half the previous value.

**Review Question 18.8** How does the law of electromagnetic induction explain why there is an induced emf in a rotating generator coil?

## 18.9 Transformers: Putting it all together

Another useful application of electromagnetic induction is the **transformer**, a device that increases or decreases the maximum value of an alternating emf.

A transformer consists of two coils, each wrapped around an iron core (ferromagnetic) (**Figure 18.16**). The core confines the magnetic field produced by the electric current in one coil so that it passes through the second coil instead of spreading outside. An alternating emf across the primary coil is converted into a larger or smaller alternating emf across the secondary coil, depending on the number of loops in each coil.

Transformers are used in many electronic devices. They are also essential for transmitting electric energy from a power plant to your house. The rate of this electric energy transmission is proportional to the product of the emf across the power lines and the electric current in the lines. If the emf is low, considerable electric current is needed to transmit a considerable amount of energy. However, due to the electrical resistance of the power lines much of the electric energy is converted into thermal energy. The rate of this conversion is $P = I^2R$. To reduce

**Figure 18.16** A transformer changes the input/output emf depending on the ratio of the turns.

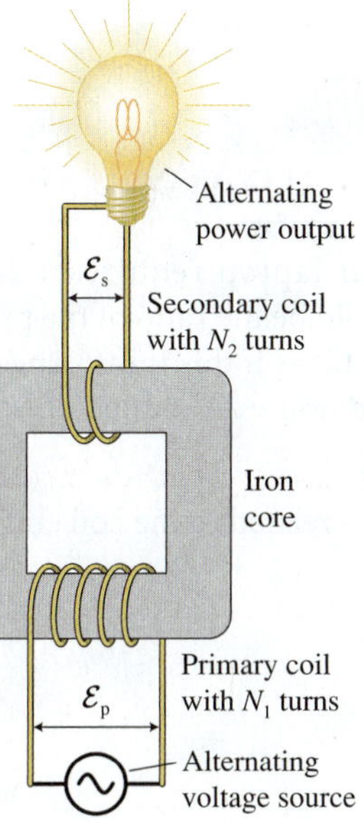

the $I^2R$ losses, the transmission of electric energy is done at high peak emf (about 20,000 V) and low current. Transformers then reduce this peak emf to about 170 V for use in your home. How does a transformer change the peak emf?

Suppose there is an alternating current in the **primary coil**, the coil connected to an external power supply. The **secondary coil** is connected to an electrical device, but this device requires an emf that is different from what the external power supply produces. The alternating current in the primary coil produces a $\vec{B}$ field within the transformer core. Since the current is continuously changing, the magnetic flux through the primary coil's area is also continuously changing. Thus, an emf is induced in the primary coil:

$$\varepsilon_{\mathrm{P}} = N_{\mathrm{P}}\left|\frac{\Delta\Phi_{\mathrm{P}}}{\Delta t}\right|$$

where the p subscript refers to the primary coil.

In an efficient transformer, nearly all of the $\vec{B}$ field lines passing through the primary coil's area also pass through the secondary coil's area. As a result, there is a changing magnetic flux through the secondary coil's area as well as a corresponding emf produced in it:

$$\varepsilon_{\mathrm{s}} = N_{\mathrm{s}}\left|\frac{\Delta\Phi_{\mathrm{s}}}{\Delta t}\right|$$

If the transformer is perfectly efficient, then the rates of change of the magnetic flux through one turn of each coil are the same:

$$\left|\frac{\Delta\Phi_{\mathrm{P}}}{\Delta t}\right| = \left|\frac{\Delta\Phi_{\mathrm{s}}}{\Delta t}\right|$$

Using the results from Faraday's law:

$$\frac{\varepsilon_{\mathrm{P}}}{N_{\mathrm{P}}} = \frac{\varepsilon_{\mathrm{s}}}{N_{\mathrm{s}}}$$

$$\Rightarrow \varepsilon_{\mathrm{s}} = \frac{N_{\mathrm{s}}}{N_{\mathrm{P}}}\varepsilon_{\mathrm{P}} \tag{18.7}$$

We see that the emf in the secondary coil can be substantially larger or smaller than the emf in the primary coil depending on the number of turns in each coil. Engineers use this result to design transformers for specific purposes. For example, a **step-down transformer** can convert the 120-V alternating emf from a wall socket to a 9-V alternating emf, which is then converted to DC to power a laptop computer.

**QUANTITATIVE EXERCISE 18.9 Transformer for laptop**

Your laptop requires a 24-V emf to function. What should be the ratio of the primary coil turns to secondary coil turns if this transformer is to be plugged into a standard house AC outlet (effectively a 120-Vemf)?

**Represent mathematically** The ratio we are looking for is related to the coil emfs by Eq. (18.7):

$$\varepsilon_{\mathrm{s}} = \frac{N_{\mathrm{s}}}{N_{\mathrm{P}}}\varepsilon_{\mathrm{P}}$$

**Solve and evaluate** Solving for the ratio and inserting the appropriate values:

$$\frac{N_{\mathrm{P}}}{N_{\mathrm{s}}} = \frac{\varepsilon_{\mathrm{P}}}{\varepsilon_{\mathrm{s}}} = \frac{120\ \mathrm{V}}{24\ \mathrm{V}} = 5$$

The primary coil needs to have five times the number of turns as the secondary coil. This is a step-down transformer, since the resulting secondary coil peak emf is lower than the primary coil peak emf.

**Try it yourself:** If the primary coil had 200 turns, how many turns should the secondary coil have to reduce the peak emf from 120 V to 6 V?

***Answer:*** 10 turns.

**Figure 18.17** How a transformer converts the emf from a 12-V car battery to a 20,000-V spark.

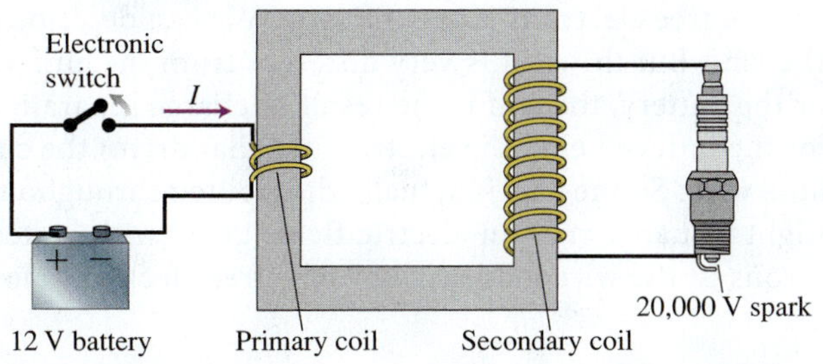

1. Current in the primary circuit produces a magnetic field in the magnetic core.
2. Opening the switch causes an abrupt decrease in the primary current and in the core magnetic field ...
Electronic switch
I
20,000 V spark
12 V battery
Primary coil
Secondary coil
3. ... which causes a large induced potential difference in the secondary coil and dielectric breakdown (spark) across the spark plug gap.

Some transformers are designed to increase rather than decrease emf. In a car that uses spark plugs for ignition, a transformer converts the 12-V potential difference of the car battery to the 20,000-V potential difference needed to produce a spark in the engine's cylinder (**Figure 18.17**). The battery supplies a steady current in the transformer. An electronic switching system in the circuit can open the circuit, stopping the current in the primary coil in a fraction of a millisecond. This causes an abrupt change in the magnetic flux through the primary coil of the transformer, which leads to an induced emf in the secondary coil. The secondary coil is attached to a spark plug that has a gap between two conducting electrodes. When the potential difference across the gap becomes sufficiently high, the air between the electrodes ionizes. When the ionized atoms recombine with electrons, the energy is released in the form of light—a spark that ignites the gasoline.

The induced emf in the secondary coil is much greater than the 12 V in the primary coil for three reasons. First, the secondary coil has many more turns than the primary coil ($N_s \gg N_p$). Second, the magnetic flux through the primary coil's area decreases very quickly (the $\Delta t$ in the denominator of Faraday's law is very small), resulting in a large induced emf $\varepsilon_p$ to which $\varepsilon_s$ is proportional [see Eq. (18.2)]. Third, the ferromagnetic core (usually iron) passing through the two coils significantly increases the $\vec{B}$ field within it (Section 17.8). For these three reasons, it is possible for $\varepsilon_s \gg \varepsilon_p$; the 12-V car battery can produce a 20,000-V potential difference across the electrodes of a spark plug.

**Review Question 18.9** How does a transformer achieve different induced peak emfs across its primary and secondary coils?

## 18.10 Mechanisms explaining electromagnetic induction

Faraday's law *describes* how a changing magnetic flux through a wire loop is related to an induced emf, but it does not *explain* how the emf comes about. In this section we will explain the origin of the induced emf.

**Figure 18.18** A changing $\vec{B}_{ex}$ creates an electric field $\vec{E}$ that induces an electric current.

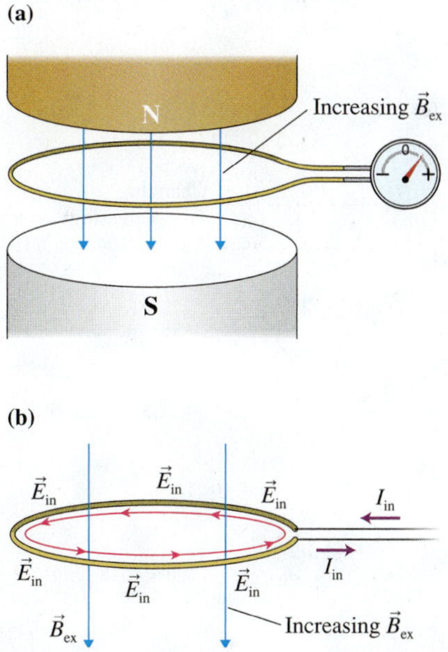


### A changing $\vec{B}_{ex}$ field has a corresponding $\vec{E}$ field

We know that a changing magnetic flux induces an electric current in a stationary loop (**Figure 18.18**). Because the loop is not moving, there is no net magnetic force exerted on the free electrons in the wire. Thus, an electric field must be present. The electric field that drives the current exists throughout the wire. This electric field is not produced by charge separation, but by a changing magnetic field.

If we were to represent it with $\vec{E}$ field lines, those lines would have no beginning or end—they would form closed loops. This electric field essentially "pushes" the free electrons along the loop. We can describe it quantitatively with the emf. But this emf is very different from the emf produced by a battery. For the battery, the emf is the result of charge separation across its terminals. For the induced emf, the electric field that drives the current is everywhere in the wire. So the emf is actually distributed throughout the entire loop. You might visualize it as an electric field "gear" with its teeth hooked into the electrons in the wire loop, pushing the free electrons along the wire at every point.

**Figure 18.19** A summary of some of the main ideas of electricity and magnetism.

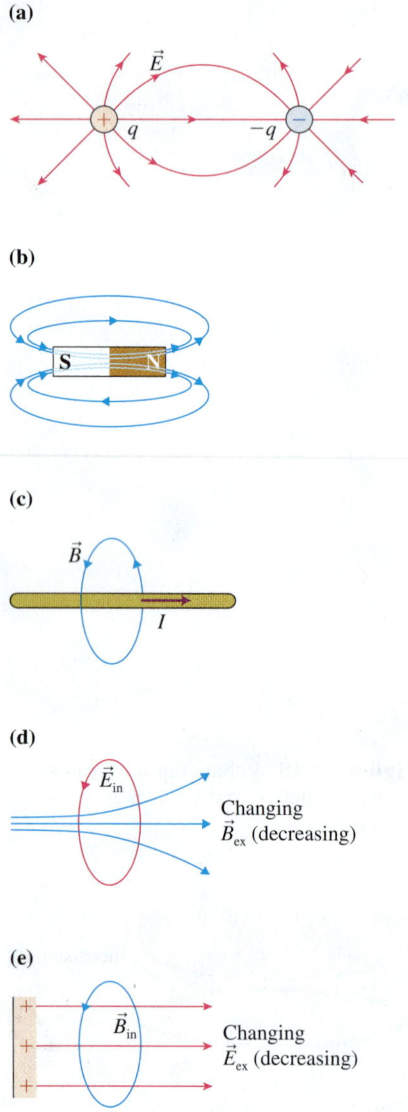


## What do we now know about electricity and magnetism?

We have learned a great deal about electric and magnetic phenomena. We learned about electrically charged objects that interact via electrostatic (Coulomb) forces. Stationary electrically charged objects produce electric fields, and electric field lines start on positive charges and end on negative charges (**Figure 18.19a**). In our study of magnetism (Chapter 17), we learned that moving electrically charged objects and permanent magnets interact via magnetic forces and produce magnetic fields. Magnetic field lines do not have beginnings or ends (Figure 18.19b), as there are no individual magnetic charges (magnetic monopoles).

When we studied electric circuits (Chapter 16) we learned that electric fields cause electrically charged particles inside metal wires to move in a coordinated way—electric currents. Later we learned that electric currents produce magnetic fields (Figure 18.19c). In this chapter, we learned about the phenomenon of electromagnetic induction and its explanation: a changing magnetic field is always accompanied by a corresponding electric field (Figure 18.19d). However, this new electric field is not produced by electric charges, and its field lines do not have beginnings or ends.

Except for the lack of magnetic charges, there is symmetry between electric and magnetic fields. This symmetry leads us to pose the following question: If in a region where the magnetic field is changing there is a corresponding electric field, is it possible that in a region where the electric field is changing there could be a corresponding magnetic field (Figure 18.19e)? This hypothesis, suggested in 1862 by James Clerk Maxwell, led to a unified theory of electricity and magnetism, a subject we will investigate in our chapter on electromagnetic waves (Chapter 24).

**Review Question 18.10** Explain how (a) an electric current is produced when part of a single wire loop moves through a magnetic field and how (b) an electric current is produced when an external magnetic flux changes through a closed loop of wire.

# Summary

| Words | Pictorial and physical representations | Mathematical representation |
|---|---|---|
| The **magnetic flux** through a loop's area depends on the size of the area, the $\vec{B}$ field magnitude, and the orientation of the loop relative to the $\vec{B}$ field. (Sections 18.2, 18.4) | 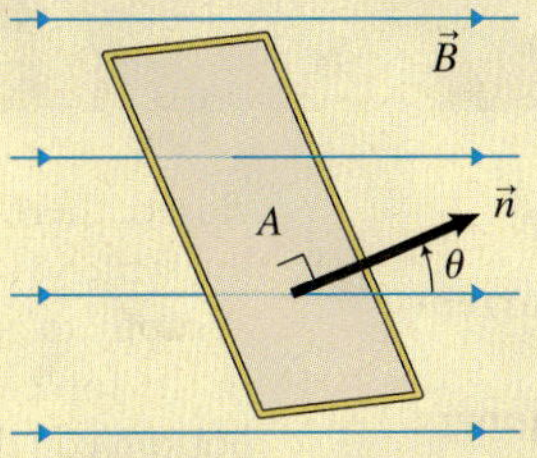 | $\Phi = BA\cos\theta$ Eq. (18.1)<br>when B is constant through the loop area |
| **Electromagnetic induction** In a region with a changing magnetic field, there is a corresponding induced electric field. When a wire loop or coil is placed in this region, the magnetic flux through that loop changes and an electric current is induced in the loop. (Section 18.4) | 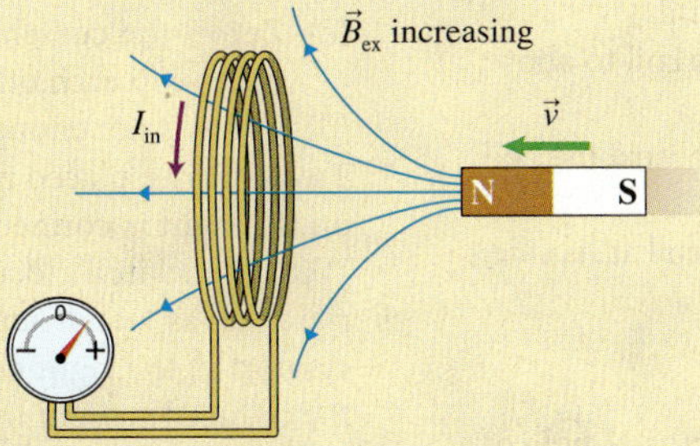<br>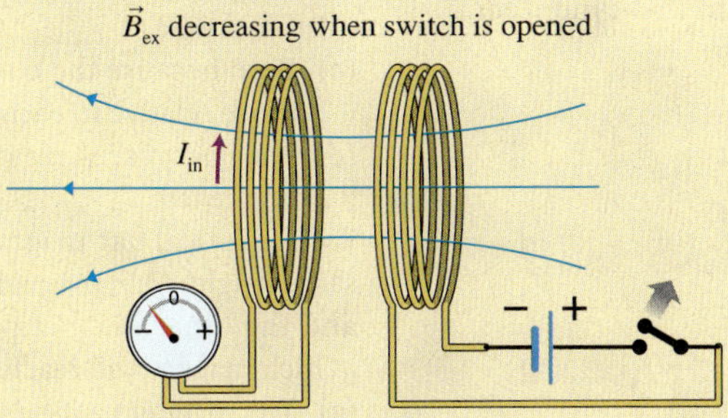 | $\varepsilon_{in} = N\left\lvert\frac{\Delta\Phi}{\Delta t}\right\rvert = N\left\lvert\frac{\Phi_f - \Phi_i}{t_f - t_i}\right\rvert$<br>Eq. (18.2) |
| **Lenz's law** When current is induced, its direction is such that the $\vec{B}$ field it produces opposes the change in the magnetic flux through the loop. (Section 18.3) | 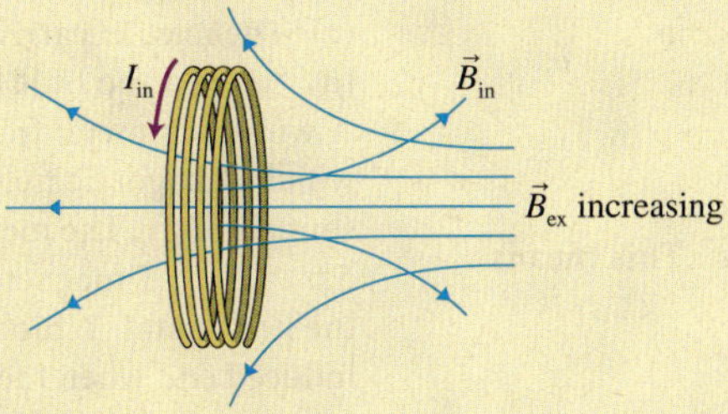 | |
| An **electric generator** produces an emf by rotating a coil within a region of strong $\vec{B}$ field, an important application of electromagnetic induction. (Section 18.8) | 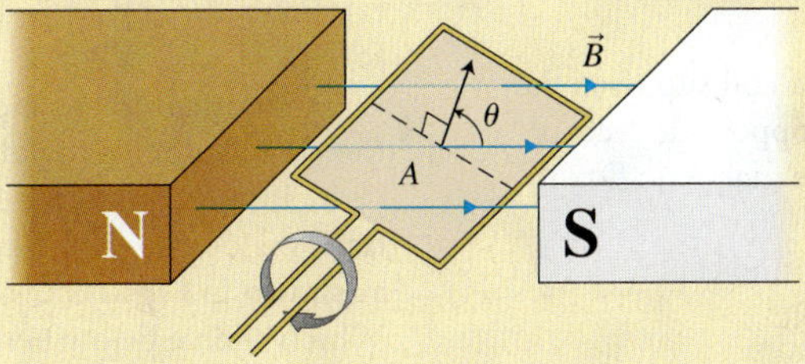 | $\varepsilon_{in} = NB_{ex}A\omega\sin(\omega t)$ Eq. (18.5) |
| **Transformers** are electrical devices used to increase or decrease the peak value of an alternating emf. (Section 18.9) | 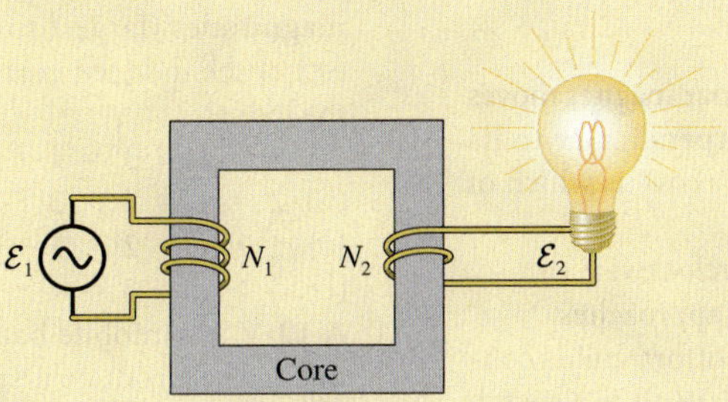 | $\varepsilon_s = \frac{N_s}{N_p}\varepsilon_p$ Eq. (18.7) |

# Questions

## Multiple Choice Questions

1. In which of the following experiments is electric current induced?
   I. The north pole of a strong bar magnet is held stationary in front of a coil.
   II. A strong magnet is rotating in front of the coil.
   III. A strong magnet moves from in front of the coil to above it.
   IV. The north pole of a strong magnet moves toward the coil from the side.
   V. A strong magnet rotates in the plane of the coil at its side.
   (a) I, II, IV, and V (b) III and IV
   (c) II, III, and V (d) II and III
   (e) II and IV
2. If you move the coil in **Figure Q18.2** toward the N pole of the large electromagnet, would an electric current be induced?

**Figure Q18.2**

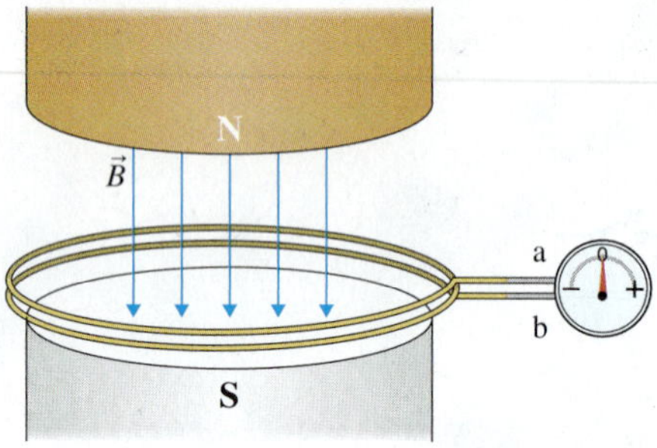


   (a) Yes (b) No
   (c) It depends on the current in the electromagnet.
3. The magnetic flux through a loop is 10 $T \cdot m^2$. This means that:
   (a) The magnetic field is parallel to the loop.
   (b) No current is induced.
   (c) Since the flux is not zero, current is induced.
   (d) Both (a) and (c) could be correct.
4. Your friend says that the emf induced in a coil supports the changing flux through the coil rather than opposes it. According to your friend, what happens when the magnetic flux increases slightly?
   (a) The induced current will increase continuously.
   (b) The coil will get hot and eventually melt.
   (c) The induced emf will get larger.
   (d) None of the above.
   (e) (a), (b), and (c) will occur.
5. A metal ring lies on a table. The S pole of a bar magnet moves down toward the ring from above and perpendicular to its surface. The induced current as seen from above is which of the following?
   (a) Clockwise (b) Counterclockwise
   (c) Zero—it only changes when the N pole approaches
6. One coil is placed on top of another. The bottom coil is connected in series to a battery and a switch. With the switch closed, there is a clockwise current in the bottom coil. When the switch is opened, the current in the bottom coil decreases abruptly to zero. What is the direction of the induced current in the top coil, as seen from above while the current in the bottom coil decreases?
   (a) Clockwise (b) Counterclockwise
   (c) Zero—the current is induced only when the coils move relative to each other
   (d) There is not enough information to answer this question.
7. Two coils are placed next to each other on the table. The coil on the right is connected in series to a battery and a switch. With the switch closed, there is a clockwise current in the right coil as seen from above. When the switch is opened, the current in the right coil decreases abruptly to zero. What is the induced current in the coil on the left as seen from above while the current in the right coil decreases?
   (a) Clockwise (b) Counterclockwise
   (c) Zero because the current is only present when the coils move relative to each other
   (d) Zero because there is no magnetic field through the coil on the left
8. Two identical bar magnets are dropped vertically from the same height. One magnet passes through an open metal ring, and the other magnet passes through a closed metal ring. Which magnet will reach the ground first?
   (a) The magnet passing through the closed ring
   (b) The magnet passing through the open ring
   (c) The magnets arrive at the ground at the same time.
   (d) There is too little information to answer this question.
9. A window's metal frame is essentially a metal loop through which a magnetic field can change when the window swings shut abruptly. The metal frame is 1.0 m $\times$ 0.50 m and Earth's $5.0 \times 10^{-5}$ T magnetic field makes an angle of 53° relative to the horizontal. Which answer below is closest to the average induced emf when the window swings shut 90° in 0.20 s from initially parallel to the field?
   (a) $9.9 \times 10^{-5}$ V (b) $7.5 \times 10^{-5}$ V
   (c) $9.9 \times 10^{-2}$ V (d) $12.5 \times 10^{-4}$ V
10. Four identical loops move at the same velocity toward the right in a uniform magnetic field inside the dashed lines, as shown in **Figure Q18.10**. Which choice below best represents the ranking of the magnitudes (largest to smallest) of the induced currents in the loops?

**Figure Q18.10**

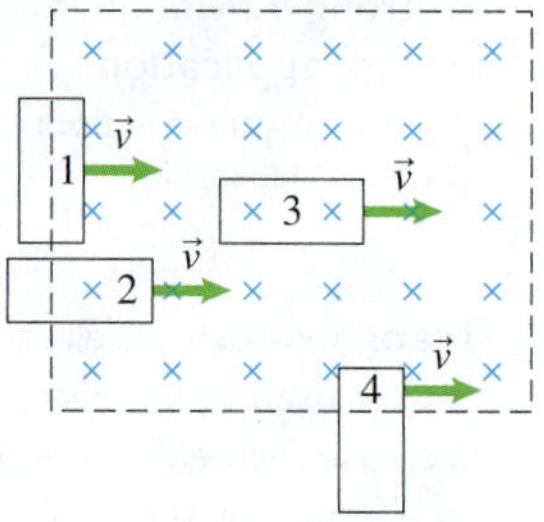


   (a) $1 = 2 > 3 = 4$
   (b) $2 = 1 = 4 > 3$
   (c) $3 > 4 > 2 > 1$
   (d) $1 > 2 > 3 = 4$ (e) $1 = 2 = 3 = 4$
11. A 12-V automobile battery provides the thousands of volts needed to produce a spark across the gap of a spark plug. Which of the following mechanisms is involved?

(a) The secondary loop connected to the spark plug has many more turns than the primary loop attached to the battery.
(b) The current in the primary coil is reduced quickly.
(c) An iron core increases the magnetic flux through the primary and secondary coils.
(d) All three mechanisms are involved.
(e) Only mechanisms (a) and (b) are involved.

12. A respiration detector consists of a coil placed on a person's chest and another placed on the person's back. There is a constant current in one coil. What causes an induced current to be produced in the other coil?
(a) The person's heart beats.
(b) The person's breathing causes the coil separation to change.
(c) The person moves.
(d) All three of the above occur.
13. A transformer has a small number of turns in the primary coil and a large number in the secondary. The electric power input to the primary coil is $P = I\Delta V$. The secondary coil with more turns will have a greater emf (in volts) across it than the primary coil. If we connect the secondary coil across a lightbulb, we get which of the following?
(a) Extra power because of the higher emf
(b) The same power because of higher emf and lower current
(c) Less power because the current is less
(d) There is too little information to answer this question.

## Conceptual Questions

14. A bar magnet falling with the north pole facing down passes through a coil held vertically. Sketch flux-versus-time and emf-versus-time graphs for the magnet approaching the coil and passing through it. What assumptions did you make?
15. An induction cooktop has a smooth surface. When on high, the surface does not feel warm, yet it can quickly cook soup in a metal bowl. However, it cannot cook soup in a ceramic or glass bowl. Explain how the cooktop works.
16. Use your knowledge of Lenz's law to find the direction of the induced current in a coil when a magnet is falling through it. How many possible answers can you give?
17. Why does the magnetic field of the induced current oppose the change in the external magnetic field?
18. Describe three common applications of electromagnetic induction.
19. Two rectangular loops A and B are near each other. Loop A has a battery and a switch. Loop B has no battery. Imagine that a current starts increasing in loop A. Will there be a current in loop B? Samir argues that there will be current. Ariana argues that there will be no current. Provide experimental evidence to support the claims of both students.
20. An apnea monitor can prevent sudden infant death syndrome by sounding an alarm when a sleeping infant stops breathing. One coil carrying an alternating electric current is placed on the chest of an infant and a second coil is placed on the infant's back. Explain how the apnea monitor detects the cessation of breathing.
21. ✎ A simple metal detector has a coil with an alternating current in it. The current produces an alternating magnetic field. If a piece of metal is near the coil, eddy currents are induced in the metal. These induced eddy currents produce induced magnetic fields that are detected by a magnetic field detection device. Draw a series of sketches representing this process, including the appropriate directions of the magnetic fields at one instant of time, and indicate two applications for this device.
22. ✎ Construct flux-versus-time and emf-versus-time graphs that explain how an electric generator works.
23. How is it possible to get a 2000-V emf from a 12-volt battery?

# Problems

Below, **BIO** indicates a problem with a biological or medical focus. Problems labeled **EST** ask you to estimate the answer to a quantitative problem rather than derive a specific answer. Problems marked with ✎ require you to make a drawing or graph as part of your solution. Asterisks indicate the level of difficulty of the problem. Problems with no * are considered to be the least difficult. A single * marks moderately difficult problems. Two ** indicate more difficult problems.

### 18.1 Inducing an electric current

1. * You and your friend are performing experiments in a physics lab. Your friend claims that in general, something has to move in order to induce a current in a coil that has no battery. What experiments can you perform to support her idea? What experiments can you perform to reject it?
2. * Your friend insists that a strong magnetic field is required to induce a current in a coil that has no battery. Describe one experiment that she and you can perform to observe that a strong magnetic field helps induce an electric current and two experiments where no current is induced even with a strong magnetic field. What should you conclude about your friend's idea?
3. You decide to use a metal ring as an indicator of induced current. If there is a current, the ring will feel warm in your hand. You place the ring around a solenoid, as shown in **Figure P18.3**. (a) Will the ring feel warm if there is constant nonzero current in the solenoid? (b) Will the ring feel warm if the current in the solenoid is alternating? Explain your answers.

**Figure P18.3**

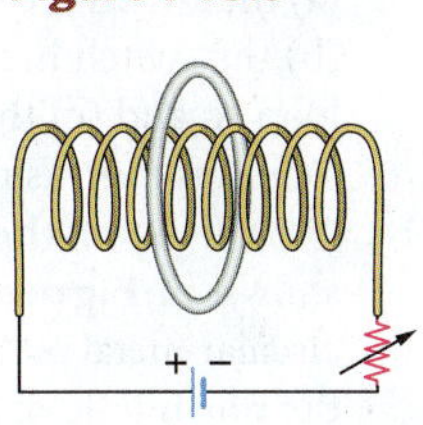


4. * To check whether a lightbulb permanently attached to a coil is still good, you place the coil next to another coil that is attached to a battery, as shown in **Figure P18.4**. Explain how or whether each of the following actions can help you determine if the lightbulb is ok. (a) Close the switch in circuit A. (b) Keep the switch in circuit A closed. (c) Open the switch in circuit A.

**Figure P18.4**

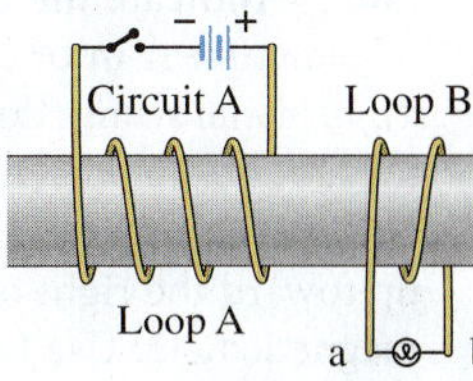


5. * **Flashlight without batteries** A flashlight that operates without batteries is lying on your desk. The light illuminates only when you continuously squeeze the flashlight's handle. You also notice that paper clips tend to stick to the outside

of the flashlight. What physical mechanism might control the operation of the flashlight?

6. You need to invent a practical application for a coil of wire that detects the vibrations or movements of a nearby magnet. Describe your invention. (The application should not repeat any described in this book.)

7. * **Detect burglars entering windows** Describe how you will design a device that uses electromagnetic induction to detect a burglar opening a window in your ground floor apartment. Include drawings and a word description.

8. * A coil connected to an ammeter can detect alternating currents in other circuits. Explain how this system might work. Could you use it to eavesdrop on a telephone conversation being transmitted through a wire?

## 18.2 Magnetic flux

9. The $\vec{B}$ field in a region has a magnitude of 0.40 T and points in the positive $z$-direction, as shown in **Figure P18.9**. Determine the magnetic flux through (a) surface abcd, (b) surface bcef, and (c) surface adef.

**Figure P18.9**

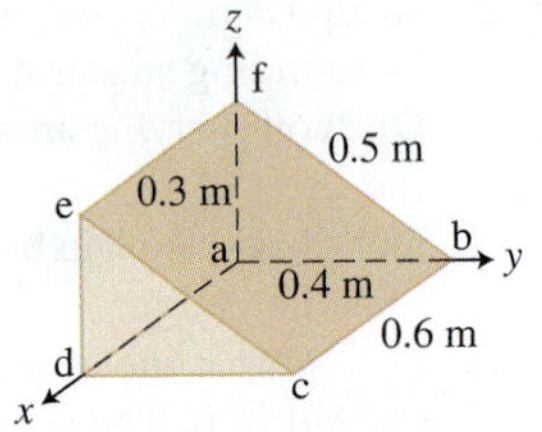


10. **EST** How do you position a bicycle tire so that the magnetic flux through it due to Earth's magnetic field is as large as possible? Estimate this maximum flux. What assumptions did you make?

11. **EST** Estimate the magnetic flux through your head when the $\vec{B}$ field of a 1.4-T MRI machine passes through your head.

12. **EST** Estimate the magnetic flux through the south- and west-facing windows of a house in British Columbia, where Earth's $\vec{B}$ field has a magnitude of $5.8 \times 10^{-5}$ T and points roughly north with a downward inclination of 72°. Explain how you made the estimates.

## 18.3 Direction of the induced current

13. * You perform experiments using an apparatus that has two insulated wires wrapped around a cardboard tube (Figure P18.4). Determine the direction of the current in the bulb when (a) the switch is closing and the current in loop A is increasing, (b) the switch has just closed and there is a steady current in loop A, and (c) the switch has just opened and the current in loop A is decreasing.

14. * You have the apparatus shown in **Figure P18.14**. A circular metal plate swings past the north pole of a permanent magnet. The metal consists of a series of rings of increasing radius. Indicate the direction of the current in one ring (a) as the metal swings down from the left into the magnetic field and (b) as the metal swings up toward the right out of the magnetic field. Use Lenz's law to justify your answers.

**Figure P18.14**

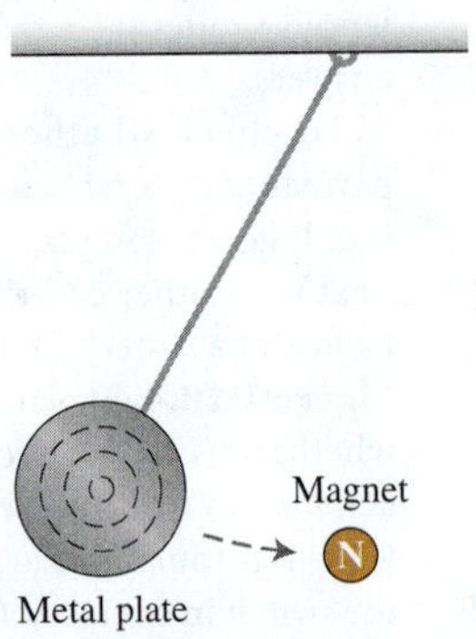


15. * You suggest that eddy currents can stop the motion of a steel disk that vibrates while hanging from a spring. Explain how you can do this without touching the disk.

16. * Your friend thinks that an induced magnetic field is *always* opposite the changing external field that induces an electric current. Provide a detailed description of a situation in which this idea would violate energy conservation.

## 18.4–18.7 Faraday's law of electromagnetic induction; Skills for analyzing processes involving electromagnetic induction; Changing $\vec{B}$ field magnitude and induced emf; Changing area and induced emf

17. The magnetic flux through three different coils is changing as shown in **Figure P18.17**. For each situation, draw a corresponding graph showing qualitatively how the induced emf changes with time.

**Figure P18.17**

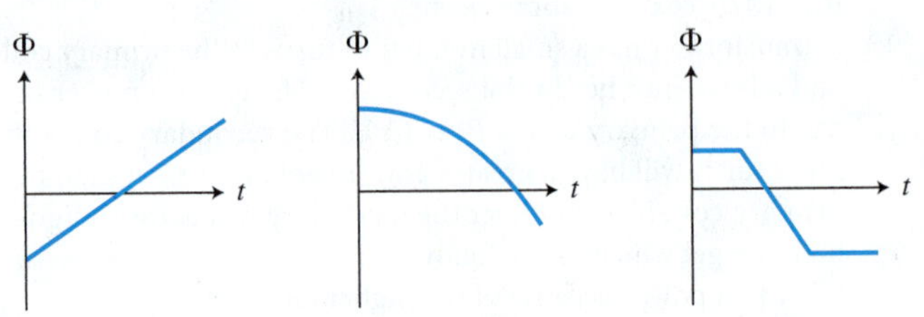


18. The magnetic flux through three different coils is changing as shown in **Figure P18.18**. For each situation, draw a corresponding graph showing quantitatively how the induced emf changes with time.

**Figure P18.18**

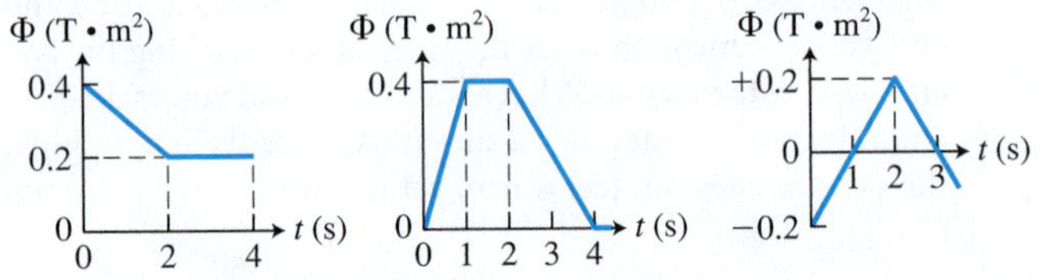


19. A magnetic field passing through two identical coils decreases from a magnitude of $B_{ex}$ to zero in the time interval $\Delta t$. The first coil has twice the number of turns as the second. (a) Compare the emfs induced in the coils. (b) How can you change the experiment so that the emfs produced in them are the same?

20. **BIO** **Stimulating the brain** In transcranial magnetic stimulation (TMS) an abrupt decrease in the electric current in a small coil placed on the scalp produces an abrupt decrease in the magnetic field inside the brain. Suppose the magnitude of the $\vec{B}$ field changes from 0.80 T to 0 T in 0.080 s. Determine the induced emf around a small circle of brain tissue of radius $1.2 \times 10^{-3}$ m. The $\vec{B}$ field is perpendicular to the surface area of the circle of brain tissue.

21. * To measure a magnetic field produced by an electromagnet, you use a circular coil of radius 0.30 m with 25 loops (resistance of 25 Ω) that rests between the poles of the magnet and is connected to an ammeter. While the current in the electromagnet is reduced to zero in 1.5 s, the ammeter in the coil shows a steady reading of 180 mA. Draw a picture of the experimental setup and determine everything you can about the electromagnet.

22. You want to use the idea of electromagnetic induction to make the bulb in your small flashlight glow; it glows when the potential difference across it is 1.5 V. You have a small bar magnet and a coil with 100 turns, each with area $3.0 \times 10^{-4}$ m². The magnitude of the $\vec{B}$ field at the front of

the bar magnet's north pole is 0.040 T and reaches 0 T when it is about 4 cm away from the pole. Can you make the bulb light? Explain.

23. * BIO **Breathing monitor** An apnea monitor for adults consists of a flexible coil that wraps around the chest (**Figure P18.23**). When the patient inhales, the chest expands, as does the coil. Earth's $\vec{B}$ field of $5.0 \times 10^{-5}$ T passes through the coil at a 53° angle relative to a line perpendicular to the coil. Determine the average induced emf in such a coil during one inhalation if the 300-turn coil area increases by 42 $cm^2$ during 2.0 s.

**Figure P18.23**

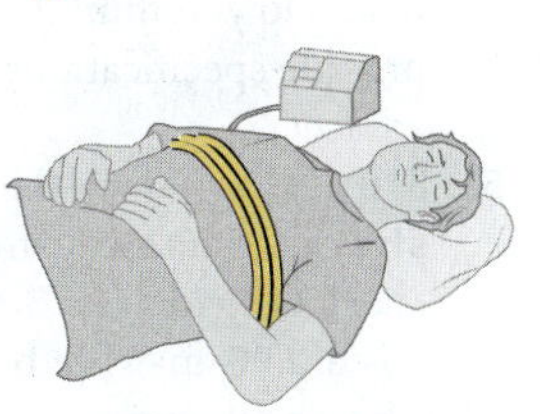

24. * A bar magnet induces a current in an $N$-turn coil as the magnet moves closer to it (**Figure P18.24**). The coil's radius is $R$ cm, and the average induced emf across the bulb during the time interval is $\varepsilon$ V. (a) Make a list of the physical quantities that you can determine using this information; (b) Is the direction of the induced current from lead a to b, or from b to a? Explain.

**Figure P18.24**

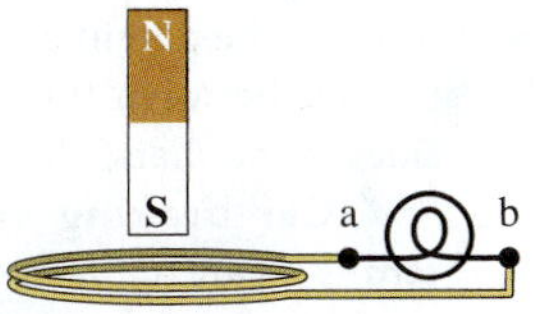


25. * You have a coil of wire with 10 turns each of 1.5 cm radius. You place the plane of the coil perpendicular to a 0.40-T $\vec{B}$ field produced by the poles of an electromagnet (Figure Q18.2). (a) Find the magnitude of the average induced emf in the coil when the magnet is turned off and the field decreases to 0 T in 2.4 s. (b) Is the direction of the induced current in the galvanometer from lead a to b, or from b to a? Explain.
26. * An experimental apparatus has two parallel horizontal metal rails separated by 1.0 m. A 2.0-Ω resistor is connected from the left end of one rail to the left end of the other. A metal axle with metal wheels is pulled toward the right along the rails at a speed of 20 m/s. Earth's uniform $5.0 \times 10^{-5}$-T $\vec{B}$ field points down at an angle of 53° below the horizontal. Make a list of the physical quantities you can determine using this information and determine two of them.
27. * Two horizontal metal rails are separated by 1.5 m and connected at their ends by a 3.0-Ω resistor to form a long, thin U shape. A metal axle with metal wheels on each side rolls along the rails at a speed of 25 m/s. Earth's $\vec{B}$ field has a magnitude of $5.0 \times 10^{-5}$ T and tilts downward 68° below the horizontal. Make a list of the physical quantities you can determine using this information and determine two of them.
28. A Boeing 747 with a 65-m wingspan is cruising northward at 250 m/s toward Alaska. The $\vec{B}$ field at this location is $5.0 \times 10^{-5}$ T and points 60° below its direction of travel. Determine the potential difference between the tips of its wings.
29. A circular loop of radius 9.0 cm is placed perpendicular to a uniform 0.35-T $\vec{B}$ field. You collapse the loop into a long, thin shape in 0.10 s. What is the average induced emf while the loop is being reshaped? What assumptions did you make?
30. * EST BIO **Magnetic field and brain cells** Suppose a power line produces a $6.0 \times 10^{-4}$-T peak magnetic field 60 times each second at the location of a neuron brain cell of radius $6.0 \times 10^{-6}$ m. Estimate the maximum magnitude of the induced emf around the perimeter of this cell during one-half cycle of magnetic field change.
31. * You need to test Faraday's law. You have a 12-turn rectangular coil that measures 0.20 m × 0.40 m and an electromagnet that produces a 0.25-T magnetic field in a well-defined region that is larger than the area of the coil. You also have a stopwatch, an ammeter, a voltmeter, and a motion detector. (a) Describe an experiment you will design to test Faraday's law. (b) How will you calculate the measurable outcome of this experiment using the materials available? (c) Describe how you can test Lenz's law with this equipment.
32. * You build a coil of radius $r$ (m) and place it in a uniform $\vec{B}$ field oriented perpendicular to the coil's surface. What is the total electric charge that passes through the coil's wire loops if the $\vec{B}$ field decreases at a constant rate to zero? The resistance of the coil's wire is $R$ (Ω).
33. * **Equation Jeopardy 1** Invent a problem for which the following equation might be a solution.

$$0.01\ \text{V} = (100)\frac{(A)\cos 0°(0.12\ \text{T} - 0)}{(1.2\ \text{s} - 0)}$$

34. * **Equation Jeopardy 2** Invent a problem for which the following equation might be a solution.

$$0.01\text{V} = 100\frac{\pi(0.10\ \text{m})^2(0.12\ \text{T})(\cos 0° - \cos 90°)}{(t - 0)}$$

35. * **Equation Jeopardy 3** Invent a problem for which the following equation might be a solution.

$$\varepsilon = -(35)\frac{(0.12\ \text{T})(\cos 0°)[\ (0)^2 - \pi(0.10\ \text{m})^2\ ]}{(3.0\ \text{s} - 0)}$$

## 18.8 Changing orientation and induced emf

36. * EST **Generator for space station** Astronauts on a space station decide to use Earth's magnetic field to generate electric current. Earth's $\vec{B}$ field in this region has the magnitude of $3.0 \times 10^{-7}$ T. They have a coil that rotates 90° in 1.2 s. The area inside the coil measures 5000 $m^2$. Estimate the number of loops needed in the coil so that during that 90° turn it produces an average induced emf of about 120 V. Indicate any assumptions you made. Is this a feasible way to produce electric energy?
37. * EST The surface of the coil of wire discussed in Problem 25 is initially oriented perpendicular to the field, as shown in Figure Q18.2. (a) Estimate the magnitude of the average induced emf if the coil is rotated 90° in 0.050 s in the 0.40-T field. The coil's surface is now parallel to the $\vec{B}$ field. (b) Determine the magnitude of the average induced emf if the coil is rotated another 90° in 0.020 s.
38. * A toy electric generator has a 20-turn circular coil with each turn of radius 1.8 cm. The coil resides in a 1.0-T magnitude $\vec{B}$ field. It also has a lightbulb that lights if the potential difference across it is about 1 V. You start rotating the coil, which is initially perpendicular to the $\vec{B}$ field. (a) Determine the time interval needed for a 90° rotation that will produce an average induced emf of 1.0 V. (b) Use a proportion technique to show that the same emf can be produced if the time interval for one rotation is reduced by one-fourth while the radius of the coil is reduced by one-half.
39. A generator has a 450-turn coil that is 10 cm long and 12 cm wide. The coil rotates at 8.0 rotations per second in a 0.10-T magnitude $\vec{B}$ field. Determine the generator's peak voltage.
40. * You need to make a generator for your bicycle light that will provide an alternating emf whose peak value is 4.2 V. The

generator coil has 55 turns and rotates in a 0.040-T magnitude $\vec{B}$ field. If the coil rotates at 400 revolutions per second, what must the area of the coil be to develop this emf? Describe any problems with this design (if there are any).

41. * **Evaluating a claim** A British bicycle light company advertises flashing bicycle lights that require no batteries and produce no resistance to riding. A magnet attached to a spoke on the bicycle tire moves past a generator coil on the bicycle frame, inducing an emf that causes a light to flash. The magnet and coil never touch. Does this lighting system really produce no resistance to riding? Justify your answer.

42. * The alternator in an automobile produces an emf with a maximum value of 12 V when the engine is idling at 1000 revolutions per minute (rpm). What is the maximum emf when the engine of the moving car turns at 3000 rpm?

43. * A generator has a 100-turn coil that rotates in a 0.30-T magnitude $\vec{B}$ field at a frequency of 80 Hz (80 rotations per second) causing a peak emf of 38 V. (a) Determine the area of each loop of the coil. (b) Write an expression for the emf as a function of time (assuming the emf is zero at time zero). (c) Determine the emf at 0.0140 s.

44. * A 10-Hz generator produces a peak emf of 40 V. (a) Write an expression for the emf as a function of time. Indicate your assumptions. (b) Determine the emf at the following times: 0.025 s, 0.050 s, 0.075 s, and 0.100 s. (c) Plot these emf-versus-time data on a graph and connect the points with a smooth curve. What is the shape of the curve?

### 18.9 Transformers: Putting it all together

45. You need to build a transformer that can step the emf up from 120 V to 12,000 V to operate a neon sign for a restaurant. What will be the ratio of the secondary to primary turns of this transformer?

46. Your home's electric doorbell operates on 10 V. Should you use a step-up or step-down transformer in order to convert the home's 120 V to 10 V? Determine the ratio of the secondary to primary turns needed for the bell's transformer.

47. A 9.0-V battery and switch are connected in series across the primary coil of a transformer. The secondary coil is connected to a lightbulb that operates on 120 V. Draw the circuit. Describe in detail how you can get the bulb to light—not necessarily continuously.

48. * You are fixing a transformer for a toy truck that uses an 8.0-V emf to run it. The primary coil of the transformer is broken; the secondary coil has 30 turns. The primary coil is connected to a 120-V wall outlet. (a) How many turns should you have in the primary coil? (b) If you then connect this primary coil to a 240-V source, what emf would be across the secondary coil?

### 18.10 Mechanisms explaining electromagnetic induction

49. ** A wire loop has a radius of 10 cm. A changing external magnetic field causes an average 0.60-N/C electric field in the wire. (a) Determine the work that the electric field does in pushing 1.0 C of electric charge around the loop. (b) Determine the induced emf caused by the changing magnetic field. (c) You measure a 0.10-A electric current. What is the electrical resistance of the loop?

## General Problems

50. * **Ice skater's flashing belt** You are hired to advise the coach of the Olympic ice-skating team concerning an idea for a costume for one of the skaters. They want to put a flat coil of wire on the front of the skater's torso and connect the ends of the coil to lightbulbs on the skater's belt. They hope that the bulbs will light when the skater spins in Earth's magnetic field. Do you think that the system will work? If so, could you provide specifications for the device and justification for your advice?

51. BIO **Hammerhead shark** A hammerhead shark (**Figure P18.51**) has a 0.90-m-wide head. The shark swims north at 1.8 m/s. Earth's $\vec{B}$ field at this location is $5.0 \times 10^{-5}$ T and points 30° below the direction of the shark's travel. Determine the potential difference between the two sides of the shark's head.

**Figure P18.51**

52. * **Car braking system** You are an inventor and want to develop a braking system that not only stops the car but also converts the original kinetic energy to some other useful energy. One of your ideas is to connect a rotor coil (the rotating coil of the generator) to the turning axle of the car. When you press on the brake pedal, a switch turns on a steady electric current to a stationary coil (an electromagnet called the stator) that produces a steady magnetic field in which the rotor turns. You now have a generator that produces an alternating current and an induced emf—electric power. Make a simple drawing of the rotor and stator at one instant and determine the direction of the magnetic force exerted on the rotor. Does this force help brake the car? Explain.

53. ** Your professor asks you to help design an electromagnetic induction sparker (a device that produces sparks). Include drawings and word descriptions for how it might work, details of its construction, and a description of possible problems.

54. ** In a new lab experiment, two parallel vertical metal rods are separated by 1.0 m. A 2.0-Ω resistor is connected from the top of one rod to the top of the other. A 0.20-kg horizontal metal bar falls between the rods and makes contact at its ends with the rods. A $\vec{B}$ field of $0.50 \times 10^{-4}$ T points horizontally between the rods. The bar should eventually reach a terminal falling velocity (constant speed) when the magnetic force of the magnetic field on the induced current in the bar balances the downward force due to the gravitational pull of the Earth. (a) Develop in symbols an expression for the current through the bar as it falls. (b) Determine in symbols an expression for the magnetic force exerted on the falling bar (and determine the direction of that force). Remember that an electric current passes through it, and the bar is falling in the magnetic field. (c) Determine the final constant speed of the falling bar (d) Is this process realistic? Explain.

55. ** EST **Designing a sparker** Your friend decides to use a device that converts some mechanical energy into the production of a spark to ignite lighter fluid. Use the information and the questions below to decide whether his sparker will work. The sparker has a coil connected across a very short gap (0.1 mm) between the ends of the wire in the coil. (a) Estimate the potential difference needed across this gap to cause

dielectric breakdown (a spark) to ignite the fumes from the wick. Dielectric breakdown occurs when the magnitude of the $\vec{E}$ field is $3 \times 10^6$ V/m or greater. (b) Estimate, based on mechanical properties, the shortest time interval that you think a person can push a small magnet from several centimeters away to the surface of a coil. (c) As the magnet is pushed toward the coil, the field in the coil increases and causes an induced emf. If the magnetic flux inside one loop increases by $10^{-6}\ \mathrm{T \cdot m^2}$ as the magnet moves forward, how many coil turns are needed to produce the emf to cause a spark? Is this a reasonable lighter system?

56. ** You have a 12-V battery, some wire, a switch, and a separate coil of wire. (a) Design a circuit that will produce an emf around the coil even though it is not connected to the battery. (b) Show, using appropriate equations, why your system will work. (c) Describe one application for your circuit.

57. * **Design a burglar alarm** You decide to build your own burglar alarm. Your window frames are wood, so you decide to fit the sides with metal sliders and the bottom with metal strips. These changes will turn the window area into a metal loop whose size changes as the window opens. Your idea is that an electric current will be induced in this loop as the window opens in Earth's magnetic field. The current can set off an alarm if a burglar enters. How feasible is your idea?

58. * You want to build a generator for a multi-day canoe trip. You have a fairly large permanent magnet, some wire, and a lightbulb. Design a generator and provide detailed specifications for it. (Ideas for the design could include cranking a handle or placing a paddle wheel that turns a coil in a nearby stream.)

59. ** **Free energy from power line** While on a camping trip, you decide to get some free electric energy. A power line is 12 m above the ground and carries an alternating current. You place a 0.50 m $\times$ 3.0 m coil with 100 turns below the wire so it lies with the 3.0-m side on the ground. The coil is connected to a lightbulb. (a) Will the lightbulb glow? (b) Indicate in a drawing the orientation of the coil relative to the power line so that a maximum changing flux passes through the coil. (c) If the current in the power line decreases from 200 A to 0 A in 1/240 s, what is the average emf induced in the coil? [Hint: Determine the $\vec{B}$ field produced by the long straight power line (see Chapter 17)]. Describe any assumptions that you make.

60. * EST A sparker used to ignite lighter fluid in a barbeque grill is shown in **Figure P18.60**. You compress a knob at the end of the sparker. This compresses a spring, which when released moves a magnet at the end of the knob quickly into a 200-turn coil. The change in flux through the coil induces an emf that causes a spark across the 0.10-mm gap at the end of the sparker. (a) Estimate the time interval needed for the change in flux in order to produce this spark. Indicate any assumptions you made. (b) Is this a realistic process? Explain.

**Figure P18.60**
Note: The size of the gap is not to scale.

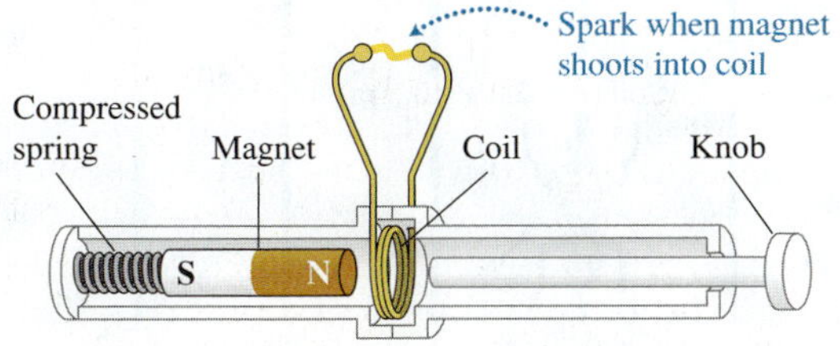


61. * BIO EST **MRI power failure** Jose needs an MRI (magnetic resonance imaging) scan. During the exam, Jose lies in a region of a very strong 1.5-T magnetic field that points down into his chest from above. A sudden power failure causes the power supply for the magnet to shut down, reducing the magnetic field from 1.5 T to 0 T in 0.50 s. Consequently, the $\vec{B}$ field through Jose's 0.3-m by 0.4-m chest decreases. The conductive fluid tissue inside his body along the edge of his chest is a loop, with the chest as the area inside this loop. (a) Estimate the induced emf around this conducting loop as the $\vec{B}$ field decreases. (b) If the resistance of his body tissue around this loop is 5 $\Omega$, what is the induced current passing around his body? (c) What is the direction of the current?

62. * **Magstripe reader** A magstripe reader used to read a credit card number or a card key for a hotel room has a tiny coil that detects a changing magnetic field as tiny bar magnets pass by the coil. Calculate the magnitude of the induced emf in a magstripe card reader coil. Assume that the magstripe magnetic field changes at a constant rate of 500 mT/ms as the region between two tiny magnets on the stripe passes the coil. The reader coil is 2.0 mm in diameter and has 5000 turns.

63. Show that when a metal rod $L$ meters long moves at speed $v$ perpendicular to $\vec{B}$ field lines, the magnetic force exerted by the field on the electrically charged particles in the rod produces a potential difference between the ends of the rod equal to the product $BLv$.

64. ** EST **The Tower of Terror ride** Figure 18.10 shows a Tower of Terror vehicle near the vertical end of its ride. (a) Is its 161-km/h speed what you would expect of an object after a 115-m fall? Explain. (b) Estimate the time interval for the free-fall part of its trip. (c) Estimate the average acceleration of the vehicle while stopping due to its magnetic braking.

## Reading Passage Problems

BIO **Magnetic induction tomography (MIT)** Magnetic induction tomography is an imaging method used in mineral, natural gas, oil, and groundwater exploration; as an archaeological tool; and for medical imaging. MIT has also been used to measure topsoil depth in agricultural soils. Topsoil depth is information that many farmers need: for instance, corn yield is much higher in soil that has a deep topsoil layer above the underlying, impermeable claypan. Using a trailer attached to a tractor, a farmer can map an 80,000-m$^2$ (about 20-acre) field for topsoil depth in about 1 hour.

**Figure 18.20** shows how MIT works. A time-varying electric current in a source coil (Figure 18.20a) induces a changing magnetic field that passes into the region to be imaged—in this case, the soil (Figure 18.20b). This changing magnetic field induces a weak induced electric current in topsoil and a stronger induced current in the more conductive claypan soil at the same depth. (Figure 18.20c; the current direction here is drawn as though the source current and source fields are increasing). This changing induced electric current in turn produces its own induced magnetic field (Figure 18.20d). The induced magnetic field passes out of the region being mapped to a detector coil (Figure 18.20e) near the source coil. The nature of the signal at the detector (its magnitude and phase) provides information about the region being mapped. A strong signal returned to the detector coil indicates a claypan layer near the surface; a weak signal returns if the clay layer is deeper below the surface.

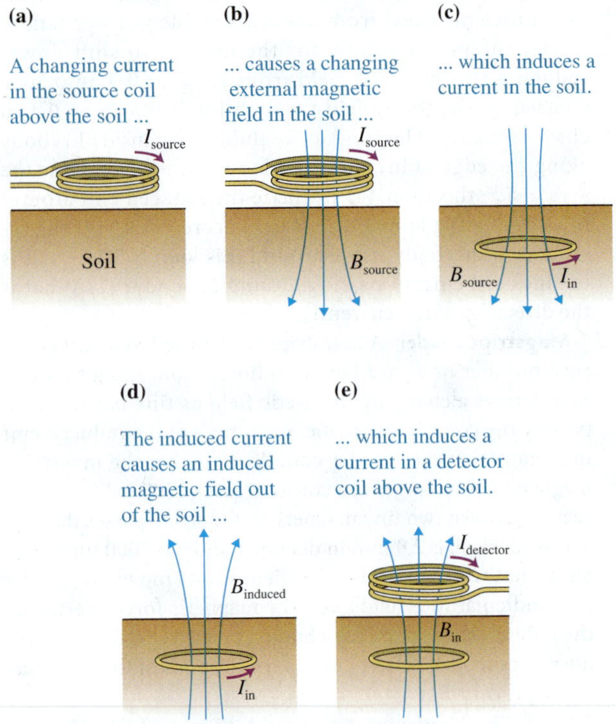
(a)
A changing current in the source coil above the soil ...
$I_{source}$
Soil
(b)
... causes a changing external magnetic field in the soil ...
$I_{source}$
$B_{source}$
(c)
... which induces a current in the soil.
$B_{source}$
$I_{in}$
(d)
The induced current causes an induced magnetic field out of the soil ...
$B_{induced}$
$I_{in}$
(e)
... which induces a current in a detector coil above the soil.
$I_{detector}$
$B_{in}$

**Figure 18.20** Magnetic induction tomography. Changing current in the source coil produces induced current in the soil. The induced magnetic field can then be detected.

65. Why is the detected signal from an MIT apparatus greater if a moist conductive layer is near the surface?
   (a) The signal is reflected better from the top of a nearby conductive layer.
   (b) The induced current is greater if the soil is moist and conductive.
   (c) The induced magnetic field from the induced current is bigger if its source is near the detection coil.
   (d) All three of the above reasons
   (e) b and c
66. All other conditions being equal, why is the induced current greater in claypan soil than in topsoil?
   (a) Claypan soil has a higher concentration of magnetic ions compared to topsoil.
   (b) Claypan soil is partly metallic in composition.
   (c) Claypan soil has greater density than loose topsoil.
   (d) The clay is closely packed, moist, and a better electrical conductor than loose, dry topsoil.
67. Why is MIT used to search noninvasively for mineral deposits (iron, copper, zinc)?
   (a) The minerals are good conductors of electricity and produce strong induced currents and strong returning magnetic fields.
   (b) The minerals absorb the incident magnetic field indicating their presence by a lack of returning signal.
   (c) The minerals produce their own returning magnetic fields.
   (d) The minerals attract the incoming magnetic field and reflect it directly above the minerals.
68. Which of the statements below about magnetic induction tomography (MIT) and transcranial magnetic stimulation (TMS), studied in Section 18.6, are true?
   (a) Both MIT and TMS have source currents in coils, source magnetic fields, and induced currents.
   (b) MIT detects the induced magnetic field produced by the induced current, and TMS does not.
   (c) MIT provides information directly about the imaged area, whereas TMS disrupts some brain activity, and the disruption is measured in some other way.
   (d) a and c only (e) a, b, and c
69. Describe all the changes that would occur if the source current were in the direction shown in Figure 18.20 but decreasing instead of increasing.
   (a) The induced current would be in the opposite direction.
   (b) The induced magnetic field would be in the opposite direction.
   (c) The detected current would be in the opposite direction.
   (d) a and b (e) a, b, and c

**BIO Measuring the motion of flying insects** Studying the motion of flying animals, particularly small insects, is difficult. One method researchers use involves attaching a tiny coil with miniature electronics to the neck of an insect and another coil to its thorax (**Figure 18.21**). They place the insect in a strong magnetic field and observe the changing orientations and induced emfs of the two coils in the field as the insect flies. Suppose that a 50-turn coil of radius $2.0 \times 10^{-3}$ m is attached to a tsetse fly that is flying in a $4.0 \times 10^{-3}$-T magnetic field. The tsetse fly makes a 90° turn in 0.020 s. Consider the average magnitude of the induced emf that occurs due to the turn of the tsetse fly and its coil.

70. Which of the quantities $B_{ex}$, $A$, or $\theta$ is changing as the fly turns?
   (a) $B_{ex}$ (b) $A$ (c) $\theta$
   (d) All of them (e) None of them
71. Which answer is closest to the magnitude of the flux change?
   (a) $1 \times 10^{-4}\ \text{T} \cdot \text{m}^2$ (b) $2 \times 10^{-6}\ \text{T} \cdot \text{m}^2$
   (c) $3 \times 10^{-7}\ \text{T} \cdot \text{m}^2$ (d) $5 \times 10^{-8}\ \text{T} \cdot \text{m}^2$
72. Which answer is closest to the induced emf on the tsetse fly coil during the 90° turn?
   (a) $6 \times 10^{-2}$ V (b) $1 \times 10^{-4}$ V
   (c) $4 \times 10^{-6}$ V (d) $2 \times 10^{-7}$ V
73. Which of the following could double the emf produced when the fly turns 90°?
   (a) Double the number of turns in the coil.
   (b) Double the coil's area.
   (c) Double the magnitude of the external magnetic field.
   (d) Get the tsetse fly to take twice as long to turn.
   (e) a, b, and c

**Figure 18.21** The coil changes its orientation with respect to the external magnetic field as the fly makes a turn.

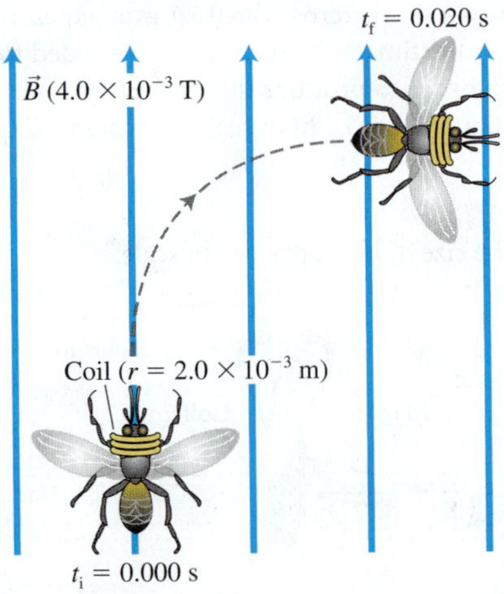
$t_f = 0.020$ s
$\vec{B}$ $(4.0 \times 10^{-3}$ T)
Coil ($r = 2.0 \times 10^{-3}$ m)
$t_i = 0.000$ s

# 19 Vibrational Motion

**Why should soldiers not march in step when they go over a bridge?**

**Why do you need to "pump" your legs when you begin swinging on a park swing?**

**How can you carry a full cup of coffee without splashing it?**

On a cold day in January, 1905, a Russian cavalry squadron of 50 soldiers was crossing a bridge over the Fontanka River in St. Petersburg when it collapsed. The stone bridge had stood for 79 years, with many more people simultaneously walking or riding horses across it than there were soldiers in that squadron. However, unlike ordinary crowds, the Russian soldiers were marching in step as they crossed the bridge. Why did their marching cause the bridge to collapse? We will find out in this chapter.

**Be sure you know how to:**

- Apply Hooke's law to analyze forces exerted by springs (Section 6.4).
- Use radians to describe angles (Section 8.1).
- Draw an energy bar chart for a process and convert it into an equation (Section 6.2 and 6.6).

We have studied linear motion—objects moving in straight lines at either constant velocity or constant acceleration. We have also studied objects moving at constant speed in a circle. In the case of circular motion, the direction of the velocity and acceleration of the objects changes, although the magnitude of both usually remains constant. In this chapter we encounter a new type of motion during which both direction and speed change. You experience this motion every day—when you walk, when you talk, and when you listen to music.

## 19.1 Observations of vibrational motion

**Figure 19.1** Vibrational motion. Your arms and legs undergo repetitive back-and-forth motion relative to the body while you walk.

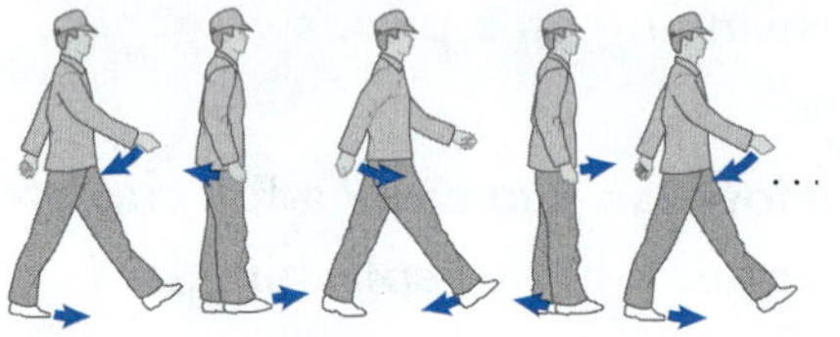

When you walk, your arms and legs swing back and forth. These motions repeat themselves; your arm or your leg passes all of the same points multiple times along its path relative to your body (**Figure 19.1**). The back-and-forth motion of an object that passes through the same positions, first moving in one direction and then in the opposite direction, is an important feature of **vibrational motion**. Investigating this type of motion is the goal of this chapter.

Let's start with some observational experiments using a cart on a smooth surface attached to a spring (Observational Experiment **Table 19.1**). Recall that a spring exerts a force on an object that stretches or compresses the spring. The force is variable in magnitude (the force is directly proportional to the stretch or compression of the spring) and points in the direction opposite to the direction in which the spring is stretched or compressed.

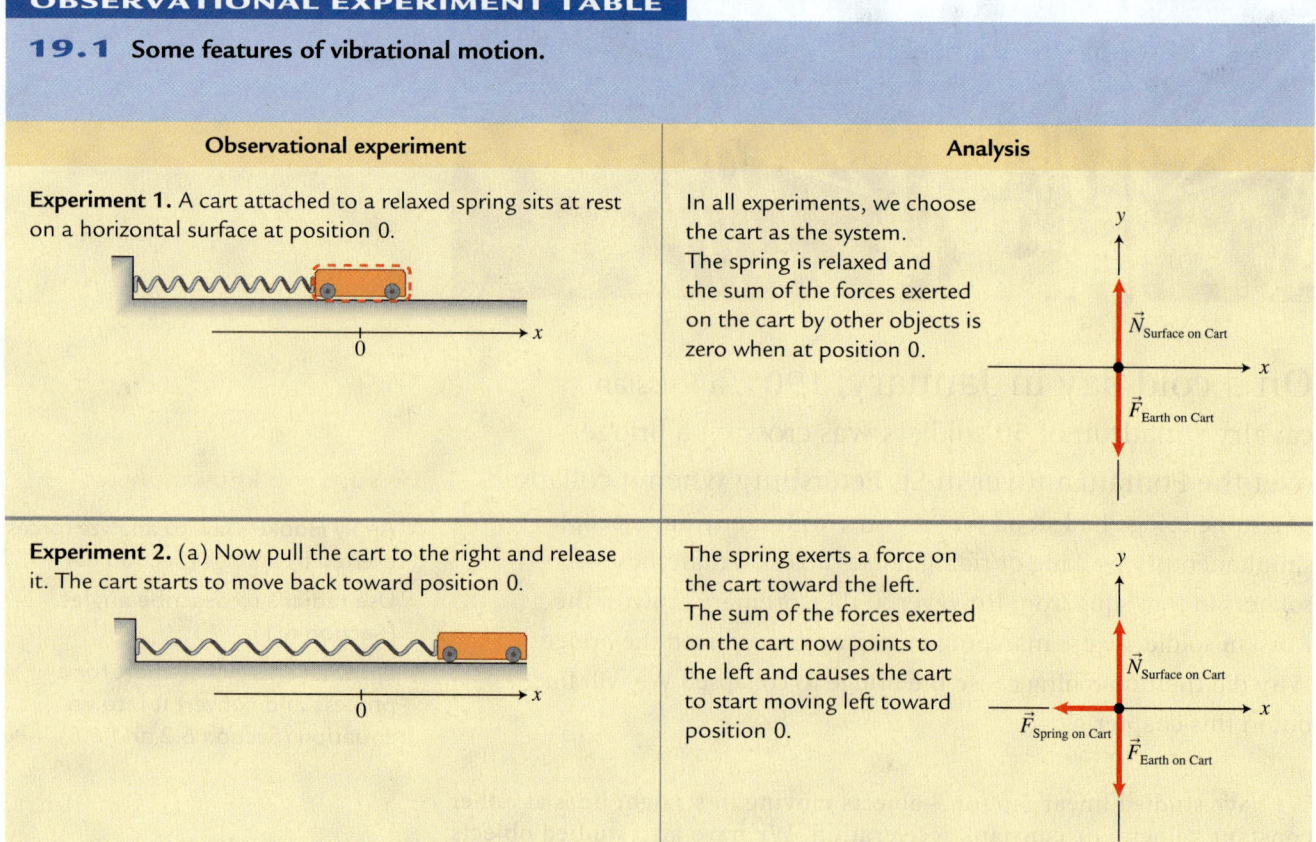

**Active Learning Guide›**

**OBSERVATIONAL EXPERIMENT TABLE**

**19.1 Some features of vibrational motion.**

| Observational experiment | Analysis |
|---|---|
| **Experiment 1.** A cart attached to a relaxed spring sits at rest on a horizontal surface at position 0. 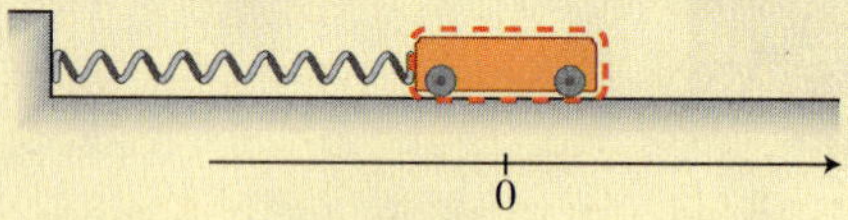  | In all experiments, we choose the cart as the system. The spring is relaxed and the sum of the forces exerted on the cart by other objects is zero when at position 0. 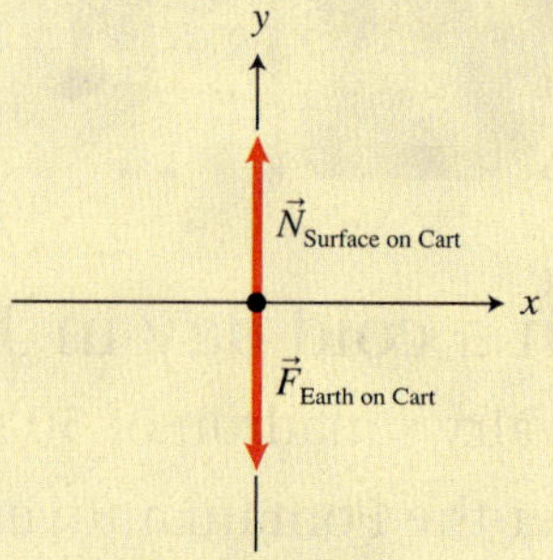  |
| **Experiment 2.** (a) Now pull the cart to the right and release it. The cart starts to move back toward position 0. 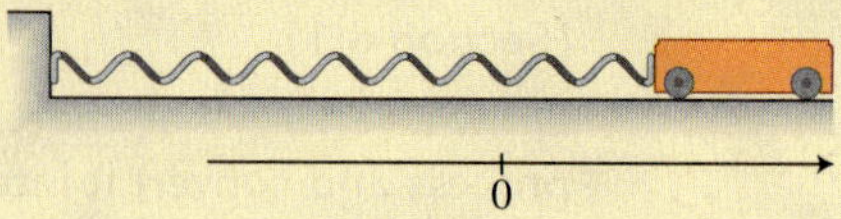  | The spring exerts a force on the cart toward the left. The sum of the forces exerted on the cart now points to the left and causes the cart to start moving left toward position 0. 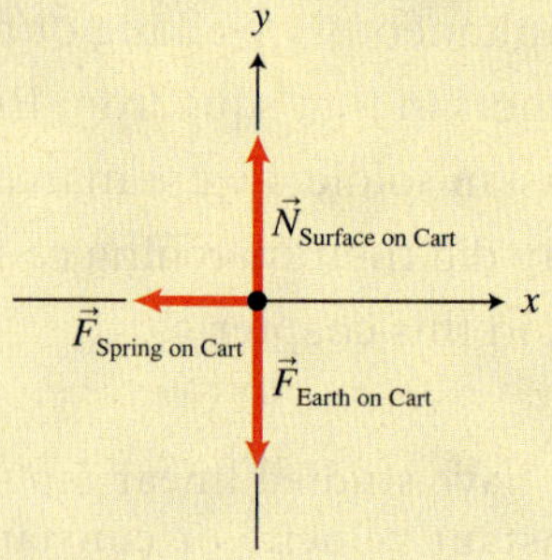  |

(b) The cart is moving fast when it reaches position 0 and overshoots that position.

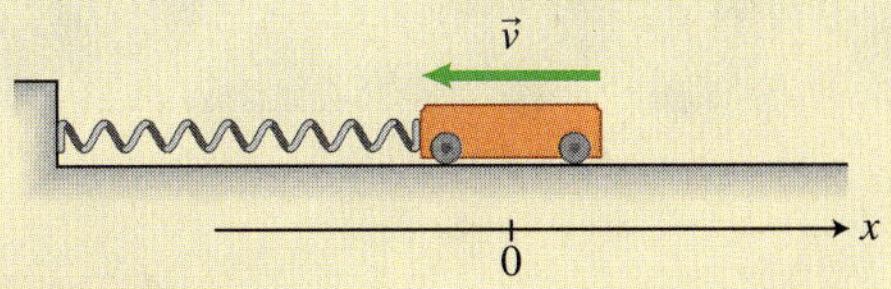


As the cart passes position 0, the sum of the forces that other objects exert on the cart is again zero. But since it is moving, it continues moving.

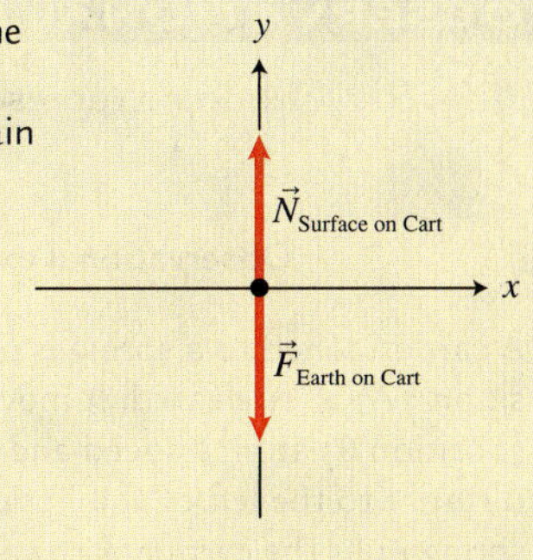

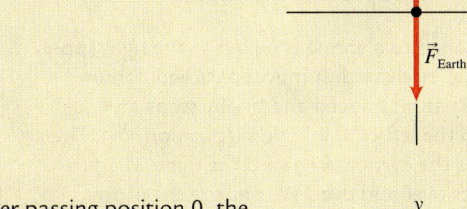

(c) The cart now slows down and eventually stops to the left of position 0, then starts moving back to the right toward position 0.

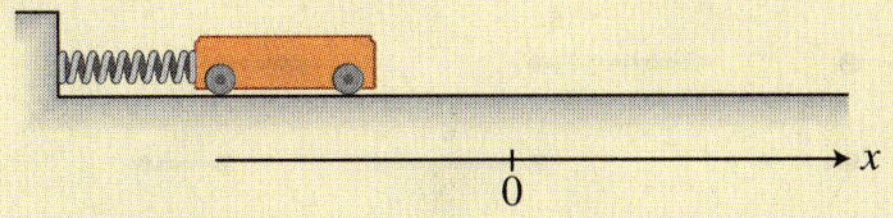


After passing position 0, the spring exerts a force to the right toward position 0; the sum of the forces points to the right, causing the cart to slow down, stop, and move back toward position 0.

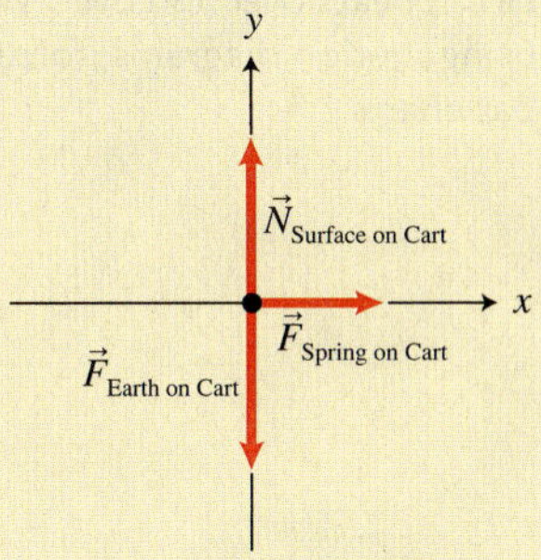


(d) The cart overshoots again and eventually stops where it started, on the right side of 0. The motion then repeats itself.

**Patterns**

The system is at rest in a particular position when not vibrating. When the cart is at rest in this position, the sum of the forces that other objects exert on it is zero. When the cart is displaced from this position and released, the cart moves back and forth, passing through that position in two opposite directions. If displaced from the rest position, a force is exerted on the vibrating object that tends to return it to that position.

In Table 19.1, the cart attached to the spring vibrates back and forth through what is called its **equilibrium position**—the place where it resides when not disturbed. When the cart is displaced on either side of the equilibrium position, the spring exerts a so-called **restoring force** on the cart that tends to return it to that equilibrium position. However, the cart is moving when it reaches the equilibrium position and therefore overshoots, moving to the other side before turning back toward the equilibrium position.

**‹Active Learning Guide**

**Equilibrium position** (or just **equilibrium**) **The position at which a vibrating object resides when not disturbed. When resting at this position, the sum of the forces that other objects exert on it is zero. During vibrational motion the object passes back and forth through this position from two opposite directions.**

**Restoring force** **When an object is displaced from equilibrium, some other object exerts a force with a component that always points opposite the direction of the vibrating object's displacement from equilibrium. This force tends to restore the vibrating object back toward equilibrium.**

We have analyzed the vibrational motion of the cart-spring system using sketches and force diagrams. In Observational Experiment **Table 19.2**, we also use motion diagrams and energy bar charts to see if there are other common characteristics of vibrating systems.

**OBSERVATIONAL EXPERIMENT TABLE**

## 19.2 Multiple representation analysis of a cart on a spring.

**Observational experiment**

A cart attached to a spring is pulled to the right (position +A) and released. It moves past equilibrium (position 0) at high speed and briefly stops an equal distance to the left of equilibrium (position −A). The spring pulls the cart back toward the right. The process repeats over and over. We analyze the process using motion diagrams, force diagrams, and energy bar charts.

**Analysis**

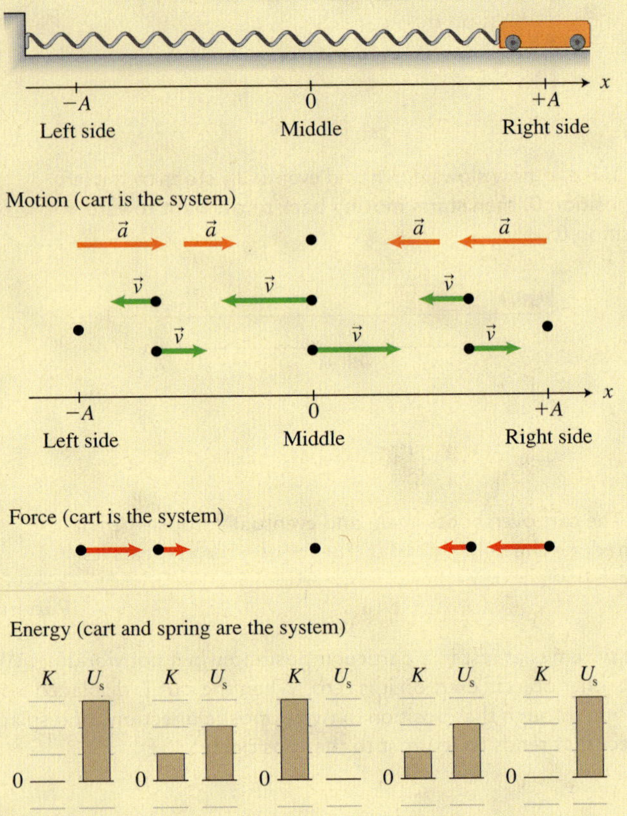


**Pattern**

***Restoring force*** The restoring force is zero as the vibrating object passes through the equilibrium position and has maximum magnitude when at the extreme positions on the left and right.

***Potential and kinetic energy*** The energy of the vibrating system (the cart and spring) varies between maximum potential energy when at the extreme positions to maximum kinetic energy as the object passes equilibrium. In between, the energy is a combination of kinetic and potential energy.

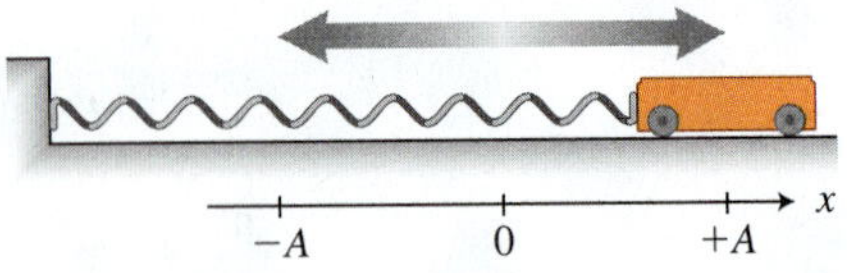


**Figure 19.2** The cart vibrates between $+A$ and $-A$. The distance from the equilibrium position to $A$ is called the amplitude of the vibration.

In Table 19.2, the acceleration and the force that the spring exerted on the cart were greatest at its extreme positions—at $x = \pm A$. This distance from the equilibrium position to an extreme position of the motion is called the **amplitude** of the vibration (**Figure 19.2**).

> **Amplitude** **The amplitude $A$ of a vibration is the magnitude of the maximum displacement of the vibrating object from its equilibrium position.**

The patterns described in the observational experiment tables are general features of vibrational motion. We summarize them below:

1. During vibrational motion, an object passes through the same positions, first moving in one direction and then in the opposite direction.
2. The object passes one position (the equilibrium position) at high speed, first in one direction, then the other. When it overshoots the equilibrium

position, a component of a restoring force exerted on it by some other object points back toward equilibrium.

3. A system composed of the vibrating object and the object exerting the restoring force has maximum potential energy when the vibrating object is at extreme positions ($x = \pm A$). The system has maximum kinetic energy when the vibrating object is passing through the equilibrium position ($x = 0$).

Many objects undergo vibrational motion: the strings of musical instruments, a drumhead, or the branches of trees on a windy day (see **Figure 19.3**). Vibrations are not always desirable. Tall buildings must be constructed so that they do not sway excessively during earthquakes or on very windy days, and cars must have shock absorbers to dampen the up-and-down vibrations that occur on a bumpy road. In other cases, vibrations are a desirable result: feel the surface of a speaker when music is playing.

**‹Active Learning Guide**

**Figure 19.3** Tall trees swaying in the wind undergo vibrational motion.

**Review Question 19.1** Brian places a book on a desk and pushes it right to left on the desk. He says the book undergoes vibrational motion. Explain what features of the motion might make Brian say this. Then explain why this motion does not fit our definition of vibrational motion.

## 19.2 Period and frequency

Amplitude is one physical quantity we can use to describe vibrational motion. Another important quantity is the time interval for one complete vibration. Consider the cart attached to a spring in Figure 19.2. The time interval for one complete vibration from $x = +A$ to $-A$ and back to $+A$ is called the **period** $T$ of the vibration.

We are familiar with this quantity from our study of circular motion. For example, one day is the time interval (the period) for Earth to make one complete rotation on its axis—the time interval for the appearance of the Sun above the horizon on two consecutive mornings.

**Period** **The period $T$ of a vibrating object is the time interval needed for the object to make one complete vibration—from the clock reading when it passes through a position while moving in a certain direction until the next clock reading when it passes through that *same* position moving in the *same* direction. The unit of period is the second.**

The number of vibrations each second is called the **frequency** $f$ of the vibration. If an object makes 4 vibrations each second, the frequency is 4 vib/s and the period is 0.25 s. The most commonly used frequency unit is the hertz (Hz), where 1 hertz equals 1 vibration per second: 1 Hz = 1 vib/s (or 1 cycle/s). The words vibration and cycle indicate what is being counted but are not themselves units. These words (cycle and vibration) can be removed from the units in equations involving frequency. Thus, an appropriate unit for frequency is 1/s or $s^{-1}$.

**Frequency** **The frequency $f$ of vibrational motion is the number of complete vibrations of the system during one second. Frequency is related to period:**

$$f = \frac{1}{T} \tag{19.1}$$

**The unit for frequency is the hertz (Hz), where 1 Hz = 1 vib/s = 1 $s^{-1}$.**

**QUANTITATIVE EXERCISE 19.1 The vibration of a building**

The Empire State Building sways back and forth in high wind at a vibration frequency of 0.125 Hz. What is its period of vibration?

**Represent mathematically** Since we know the vibration frequency, we can determine the period $T = 1/f$.

**Solve and evaluate** The period is

$$T = \frac{1}{f} = \frac{1}{0.125\ \text{Hz}} = \frac{1}{0.125\ \text{vib/s}} = 8.0\ \text{s/vib} = 8.0\ \text{s}$$

The period is quite long! Observations of the tower show that the amplitude of vibration at the top of the building in a high wind is less than 4 cm. The long period of vibration and relatively small amplitude means this vibration is unlikely to damage the building or affect its occupants.

**Try it yourself:** You sit on a rocking chair. It takes 12 s to complete six rocks—all the way back and all the way forward again. Determine the period and the frequency of the rocking.

*Answer:* 2.0 s period and 0.5 rocks/s $= 0.5\ \text{s}^{-1} = 0.5$ Hz frequency.

**Review Question 19.2** Can we say that the period of vibration depends on the frequency or that the frequency depends on the period? Explain your answer.

## 19.3 Kinematics of vibrational motion

**Active Learning Guide>**

To help describe vibrational motion quantitatively, let's look at the data from an experiment in which a motion detector collects position-versus-time, velocity-versus-time, and acceleration-versus-time data for a cart vibrating on a spring. Graphs of the data are shown in **Figure 19.4**.

The position of the object changes periodically, as does the slope of the graph. Note that the period for all three graphs is 4.0 s. Note also the relationship between the position and acceleration graphs. The acceleration has its maximum negative value when the position has its maximum positive value and vice versa. This makes sense. When the cart is at the rightmost positive $x$-position, the spring is exerting its maximum leftward negative $x$-force on the cart, causing maximum leftward negative acceleration.

**Figure 19.4** Kinematics graphs for vibrational motion.

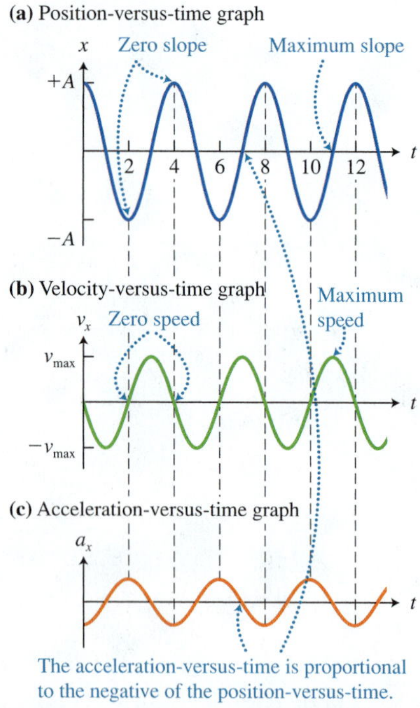


### Consistency of motion diagram and graphs

Is a motion diagram for vibrational motion consistent with the graphs in Figure 19.4? We can rotate the motion diagram in Table 19.2 so that it is oriented vertically beside a position-versus-time graph (**Figure 19.5a**) with the first five points numbered 0–4. We can move the $\vec{v}$ arrows from the motion diagram down to the velocity-versus-time graph (Figure 19.5b) and the $\vec{a}$ arrows from the diagram down to the acceleration-versus-time graph (Figure 19.5c). The graphs and the motion diagram are consistent with each other.

### Consistency of graphs with each other

Notice also that the three graphs in Figure 19.4 are consistent with each other. Recall that velocity is the rate of change of position $v_x = (\Delta x/\Delta t)$—the slope of the position-versus-time graph. The velocity has its maximum positive value when the position-versus-time graph has its maximum positive slope. The slope of the position-versus-time graph is zero when $x = A$, as is the velocity. Note that at other clock readings, the slope of the position-versus-time graph has the same sign and relative magnitude as the velocity-versus-time graph.

Recall that acceleration is the rate of change of velocity $a_x = (\Delta v_x/\Delta t)$. The acceleration-versus-time graph is also consistent with the velocity-versus-time graph. For example, the acceleration is maximum when the velocity-versus-time graph has its maximum slope at $t = 2$ s and 6 s. The acceleration is zero when the slope of the velocity-versus-time graph is zero at $t = 3$ s.

**Figure 19.5** The motion diagram and kinematics graphs are consistent with each other.

**(a)** Comparing a motion diagram to its corresponding position-versus-time graph

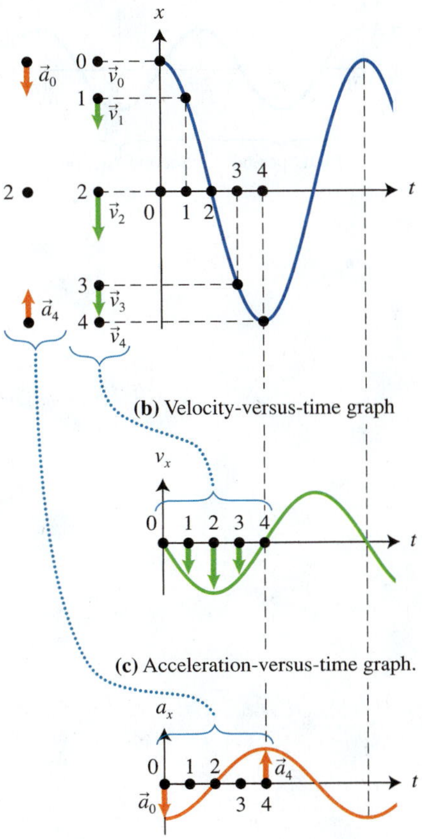


## Mathematical description of position as a function of time

Our goal now is to devise a periodic mathematical function whose graphical representation is similar to the graphs shown in Figure 19.4. Sinusoidal functions (in which the value of the function is proportional to the sin or cos of an angle) are the simplest periodic functions. The meanings of sine and cosine can be best understood using a unit circle (**Figure 19.6**) whose **radius vector** has a length of 1 unit and makes an angle $\theta$ with the $x$-axis (Figure 19.6a). We can then resolve the radius vector into its $x$- and $y$-components (Figure 19.6b):

$$r_x = r\cos\theta = (1)\cos\theta = \cos\theta$$
$$r_y = r\sin\theta = (1)\sin\theta = \sin\theta$$

Imagine now that the radius vector rotates counterclockwise around the origin of the coordinate system. The angle $\theta$ changes continuously from 0 to 360° or from 0 to $2\pi$ rad. Consequently, the $x$-component of the radius vector changes from +1 to 0 to −1 to 0 and then back to +1; the $y$-component changes from 0 to +1 to 0 to −1 and then back to 0 (Figure 19.6c). If the radius vector continues rotating, the functions $\cos\theta$ and $\sin\theta$ become periodic functions with period $2\pi$ rad or 360°.

If instead of a unit circle, you have a circle of radius $A$, the rotating radius vector will now have components:

$$R_x = A\cos\theta$$
$$R_y = A\sin\theta$$

**Figure 19.6** (a–c) A unit circle used to define the cos and sin functions and (d) a circle of radius $A$ used to define sinusoidal functions.

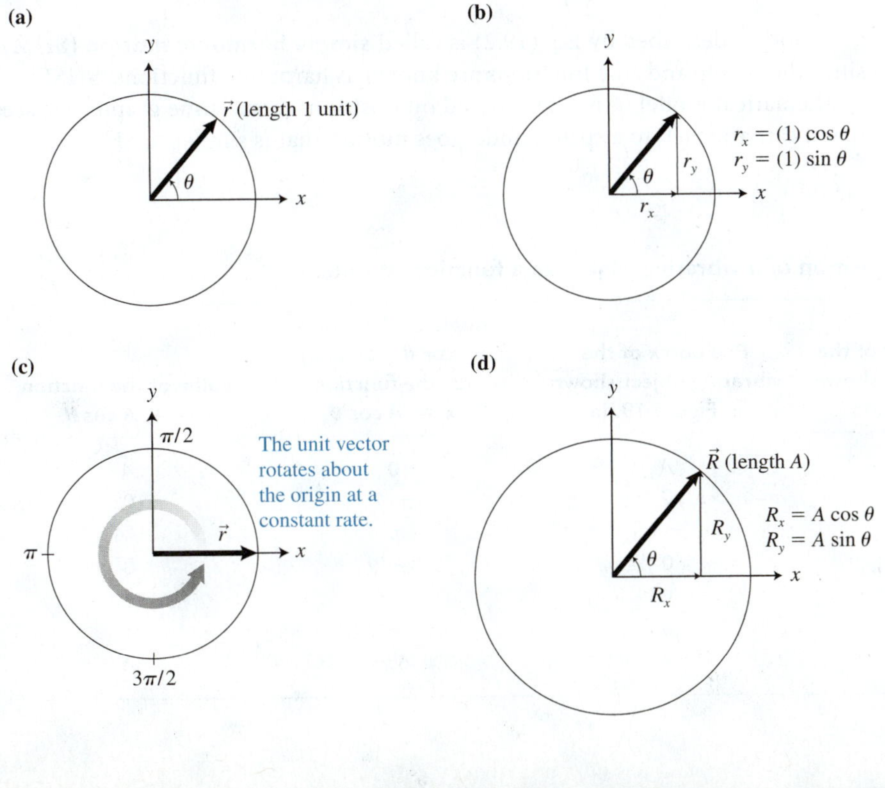

**Figure 19.7** A graph of an $A \cos \theta$ function.

This graph has the same shape as the position-versus-time graph in Figure 19.4a.

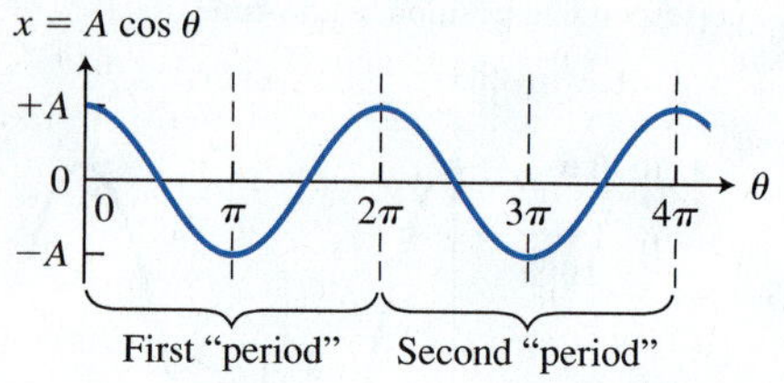


as in Figure 19.6d. In these functions $A$ is the maximum value or amplitude of the function, and $\theta$ is an angle that changes by $2\pi$ radians ($2\pi \text{ rad} = 360°$) during each period of the function. We can think of $R_x$ and $R_y$ as the functions $x = A\cos\theta$ and $y = A\sin\theta$. A graph of $x = A\cos\theta$, shown in **Figure 19.7**, looks very similar to the position-versus-time graph produced by the motion detector shown in Figure 19.4a. However, in Figure 19.7 the independent variable is the angle $\theta$ and not the time $t$. We need a sinusoidal function of time to describe the motion of a vibrating object.

In **Table 19.3**, we compare the changes in position of the vibrating object as a function of clock reading shown in Figure 19.4a and the changes in position as a function of angle of the sinusoidal function shown in Figure 19.6d.

The function $x = A\cos\theta$ matches the actual vibrational motion well. If we compare more times, the match remains strong. Now, we just need to replace $\theta$ with an appropriate function of time. Let's take the ratio of the angle and clock reading at any of the clock readings listed in the table. For example,

$$\frac{\theta}{t} = \frac{2\pi}{T}$$

or

$$\theta = \left(\frac{2\pi}{T}t\right)$$

We can write a periodic function $x(t)$ to represent the position-versus-time graph:

$$x = A\cos\left(\frac{2\pi}{T}t\right) \tag{19.2}$$

Notice that the above function has $x = +A$ at $t = 0$. If an object is at $x = 0$ at $t = 0$, you can either adjust the cos function by adding $-(\pi/2)$ or use the sine function to describe the motion:

$$x = A\sin\left(\frac{2\pi}{T}t\right)$$

since $\sin\theta = \cos\left(\theta - \frac{\pi}{2}\right)$

Motion described by Eq. (19.2) is called **simple harmonic motion** (SHM) since the cosine and sine functions are known as harmonic functions. SHM is a mathematical model of motion. Based on position-versus-time graphs, we see that a cart attached to a spring undergoes motion that is similar to SHM.

**Table 19.3 Position of a vibrating object as a function of time.**

| Clock reading $t$ of the vibrating object shown in Figure 19.4a | Position $x$ of the vibrating object shown in Figure 19.4a | Angle of the radius vector $\theta$ (radians) for the function $x = A\cos\theta$ | Value of the function $x = A\cos\theta$ |
|---|---|---|---|
| 0 (0 s) | $A$ | 0 | $A$ |
| $T/4$ (1 s) | 0 | $\pi/2$ | 0 |
| $T/2$ (2 s) | $-A$ | $\pi$ | $-A$ |
| $3T/4$ (3 s) | 0 | $3\pi/2$ | 0 |
| $T$ (4 s) | $A$ | $2\pi$ | $A$ |
| $2T$ (8 s) | $A$ | $4\pi$ | $A$ |
| $3T$ (12 s) | $A$ | $6\pi$ | $A$ |

**QUANTITATIVE EXERCISE 19.2 Describing the motion of two objects**

The graphical descriptions of the vibrational motions of two objects are shown in the figure. Write mathematical descriptions (equations) for these motions.

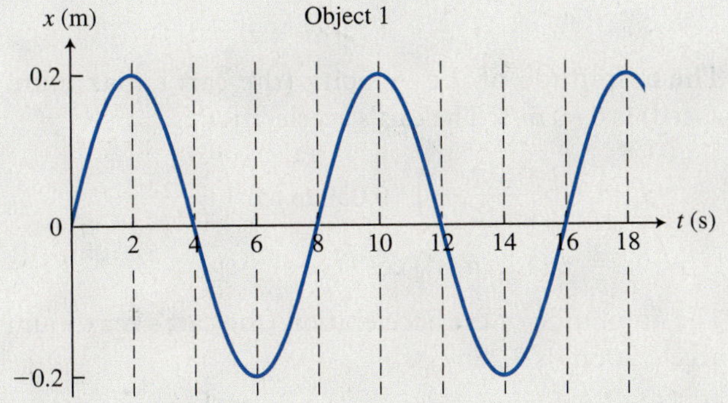


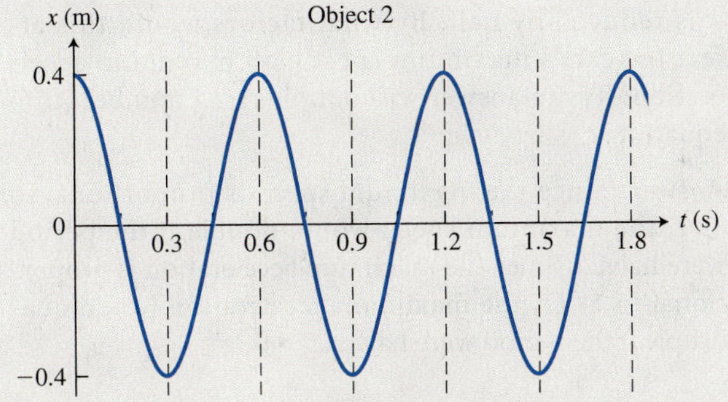


**Represent mathematically** (a) Vibrating object 1 has its maximum displacement of $A = 0.2$ m and a period $T = 8$ s. At time 0 the object is at zero and then its position increases. Thus, the function that describes its motion is a sine function:

$$x = A\sin\left(\frac{2\pi}{T}t\right)$$

(b) Vibrating object 2 has a maximum displacement $A = 0.4$ m at time zero and has period $T = 0.6$ s. Thus, the function

$$x = A\cos\left(\frac{2\pi}{T}t\right)$$

describes its motion.

**Solve and evaluate** The mathematical description of vibrating object 1 is

$$x = (0.2\text{ m})\sin\left(\frac{2\pi}{8\text{ s}}t\right)$$

The mathematical description of vibrating object 2 is

$$x = (0.4\text{ m})\cos\left(\frac{2\pi}{0.6\text{ s}}t\right)$$

**Try it yourself:** Write a mathematical description for the displacement of an object that has amplitude 0.10 m and frequency of 5.0 Hz, and starts vibrating from position $x = -A$.

*Answer:* $x = -(0.10\text{ m})\cos\left(\frac{2\pi}{0.20\text{ s}}t\right)$. Note that the period is $T = \frac{1}{f} = \frac{1}{5.0\text{ s}^{-1}} = 0.20\text{ s}$.

## Mathematical description of velocity and acceleration as a function of time

We see in Figure 19.4 that the position-, velocity-, and acceleration-versus-time graphs are all sinusoidal. They have the same period but have different amplitudes and reach their respective maxima and minima at different clock readings. We discussed the qualitative relationships among the three graphs earlier in this section. What are the mathematical descriptions of the three functions? When you have a function of time, you can find its rate of change by taking the derivative of the function. Those operations are studied in calculus; we will not show them in this book but will provide you with the final results. If the position function is given by Eq. (19.2)

$$x = A\cos\left(\frac{2\pi}{T}t\right)$$

then the velocity and acceleration functions are

$$v_x = -\left(\frac{2\pi}{T}\right)A\sin\left(\frac{2\pi}{T}t\right) \quad (19.3)$$

$$a_x = -\left(\frac{2\pi}{T}\right)^2 A\cos\left(\frac{2\pi}{T}t\right) \quad (19.4)$$

where $A$ is the amplitude of the vibration and $T$ is the period of the vibration. Notice that the velocity function has amplitude $(2\pi/T)A$ while the acceleration function has amplitude $(2\pi/T)^2A$. Note also that the function for acceleration looks exactly the same as the function for position multiplied by $-(2\pi/T)^2$. We can say that the acceleration of a vibrating object is directly proportional to its displacement from equilibrium. Specifically, $a_x = -(2\pi/T)^2x$. Finally, we see that the clock readings when the velocity and acceleration reach their respective maxima and minima match those in Figure 19.4.

**QUANTITATIVE EXERCISE 19.3 Equation Jeopardy**

The position of a vibrating cart attached to a spring is described by the equation

$$x = (0.050\text{ m})\cos((12\text{ s}^{-1})t)$$

What can you determine about the motion?

**Represent mathematically** Compare the above equation with the position-versus-time equation for simple harmonic motion: $x = A\cos\left(\frac{2\pi}{T}t\right)$. We see that the amplitude $A$ of the vibration is 0.050 m and $\frac{2\pi}{T} = 12\text{ s}^{-1}$. The period is then $T = (2\pi/12)\text{ s} = (\pi/6)\text{ s}$. The frequency of vibration is $f = 1/T = (6/\pi)\text{ s}^{-1} = (6/\pi)\text{ Hz}$. The object has its maximum positive displacement at time zero.

**Solve and evaluate** The velocity of the vibrating cart is

$$v_x = -\frac{2\pi}{(\pi/6)\text{ s}}(0.050\text{ m})\sin\left(\frac{2\pi}{(\pi/6)\text{ s}}t\right)$$
$$= -(0.60\text{ m/s})\sin((12\text{ s}^{-1})t)$$

The amplitude of the velocity (the cart's maximum speed) is 0.60 m/s. The cart's acceleration is

$$a_x = -\left(\frac{2\pi}{(\pi/6)\text{ s}}\right)^2(0.050\text{ m})\cos\left(\frac{2\pi}{(\pi/6)\text{ s}}t\right)$$
$$= -(7.2\text{ m/s}^2)\cos((12\text{ s}^{-1})t)$$

The amplitude of the acceleration (the cart's maximum acceleration) is $7.2\text{ m/s}^2$.

**Try it yourself:** Suppose the amplitude of vibration in this example remained the same, but the period was reduced by half. By what factors would this affect the cart's maximum speed and maximum acceleration? Try to answer without plugging numbers into equations.

*Answer:* Since the maximum speed is proportional to $1/T$, the maximum speed would double if the period were halved. Since the maximum acceleration is proportional to $1/T^2$, the maximum acceleration would quadruple if the period were halved.

**Review Question 19.3** The velocity of an object attached to a spring changes as

$$v_x = -(0.84\text{ m/s})\sin\left(\left(4.19\frac{1}{\text{s}}\right)t\right)$$

Say all you can about its motion.

**Figure 19.8** Dynamics of simple harmonic motion.

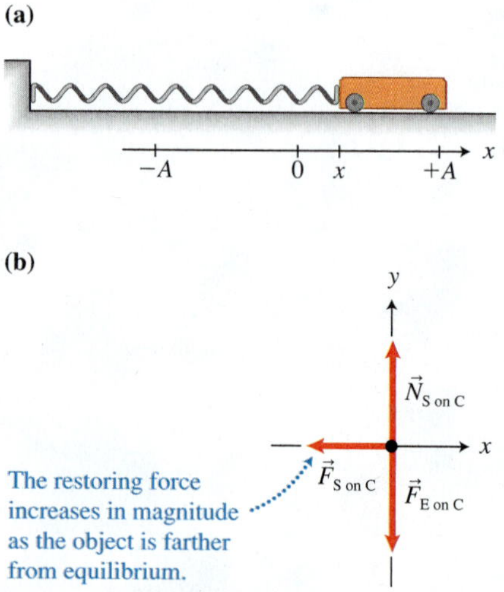


## 19.4 The dynamics of simple harmonic motion

So far, we have described simple harmonic motion (SHM) using sketches, motion diagrams, kinematics graphs, and kinematics equations. Now we will investigate the mechanism behind SHM.

### Forces and acceleration

Consider again a cart on a frictionless surface attached to a spring (**Figure 19.8a**). We use a light (massless) spring that when stretched exerts a force on the cart. According to Hooke's law, that force is $F_{\text{S on C}x} = -kx$. Position $x$ is the displacement of the end of the spring from its unstretched position to where it is attached to the cart. The other end of the spring is fixed. When the cart is at position $x$ relative to the unstretched equilibrium position, the spring exerts a restoring force on the cart (the object) that pulls it toward the equilibrium position (Figure 19.8b). The force component along the $x$-direction is

$$F_{\text{S on C}\,x} = -kx \tag{19.5}$$

The spring is the only object that exerts a force with a nonzero $x$-component on the cart. Thus, the $x$-component form of Newton's second law $\left(a_x = \dfrac{\Sigma F_x}{m}\right)$ becomes

$$a_x = \frac{-kx}{m} = -\frac{k}{m}x \tag{19.6}$$

The above expression shows that the cart's acceleration $a_x$ is proportional to the negative of its displacement $x$ from the equilibrium position. This is exactly how the displacement and acceleration of an object undergoing SHM are related [Eq. (19.4)]. If the cart is on the positive side of the equilibrium position, the spring pulls it in the negative direction so that its acceleration component is negative. If the cart is on the negative side of the equilibrium position, the spring pushes it in the positive direction so its acceleration component is positive. Note that the farther the cart is from equilibrium, the greater the magnitude of the force that the spring exerts on the cart and the greater the magnitude of the cart's resulting acceleration.

‹Active Learning Guide

We now understand *why* the acceleration-versus-time graph in Figure 19.4c is the same shape as the position-versus-time graph in Figure 19.4a with the sign of $a$ being opposite the sign of $x$ at all times; the restoring force exerted on the vibrating object is proportional to the displacement of the object from equilibrium but opposite in direction. Whenever this relation holds for a system, that system's motion can be described mathematically as simple harmonic motion.

Stop for a moment and think how unusual Eq. (19.6) is. When we studied linear motion at constant acceleration (Chapter 1), the acceleration of a moving object was the same at any location and did not depend on the position. When we studied constant speed circular motion (Chapter 4), the direction of acceleration changed with changing position along the circle but the magnitude remained the same. In SHM the acceleration of an object changes in time in both magnitude and direction but is synchronized with the displacement of the object. Thus the sum of the forces exerted on an object moving in a circle at constant speed is constant in magnitude and only changes direction (it always points toward the center of the circle). The sum of the forces exerted on an object undergoing SHM not only changes direction during the motion but also changes magnitude.

## Period of vibrations of a cart attached to a spring

We now understand that Eq. (19.6) explains the relationship between the position and the acceleration of the cart. Can it also explain the frequency or period of vibration of a spring-cart system? What properties of a vibrating system affect the period of the vibration? Intuitively, you probably think that the period depends on the amplitude—the larger the amplitude, the longer it should take the cart to go through a full cycle. This is easy to check by simply starting the vibration with various amplitudes. We find that the period is essentially constant even when the amplitude is doubled or tripled (provided the amplitude is not too large). This seems surprising, but on reflection is reasonable. The farther we pull the cart from its equilibrium position, the greater the elastic force exerted on it by the spring, the greater the acceleration of the cart, and the faster it moves toward the equilibrium position. Thus it moves faster but over a longer distance—the two effects compensate. The period is independent of the amplitude.

Two other factors could affect the period of the vibration. The stiffness of the spring should make a difference, because a stiffer spring pulls harder on the cart, increasing the acceleration of the cart. The mass of the cart should also affect the period, as a more massive cart is harder to accelerate. We can use Newton's second law [Eq. (19.6)] and the kinematics equations that describe position and acceleration to derive an expression for the period of vibration of a cart with mass $m$ at the end of a spring of spring constant $k$. We start with Eq. (19.6):

$$a_x = -\frac{k}{m}x$$

Substitute the expressions for $a_x$ and $x$ in the above [Eqs. (19.2) and (19.4)] to get

$$-\left(\frac{2\pi}{T}\right)^2 A\cos\left(\frac{2\pi}{T}t\right) = -\frac{k}{m}A\cos\left(\frac{2\pi}{T}t\right)$$

Divide both sides of the equation by $-A\cos\left(\frac{2\pi}{T}t\right)$ and solve for $T$:

$$\left(\frac{2\pi}{T}\right)^2 = \frac{k}{m}$$

$$\Rightarrow 4\pi^2\frac{m}{k} = T^2$$

Thus, the period is

$$T = 2\pi\sqrt{\frac{m}{k}} \qquad (19.7)$$

Notice that in this expression for period, there is no dependency on the amplitude. We can also check the units: $\sqrt{\frac{\text{kg}}{\text{N/m}}} = \sqrt{\frac{(\text{kg} \times \text{m})\text{s}^2}{\text{kg} \times \text{m}}} = \text{s}$, the correct unit for period. The frequency $f$ of vibration of an object attached to a spring is $\frac{1}{T}$, or

$$f = \frac{1}{2\pi}\sqrt{\frac{k}{m}} \qquad (19.8)$$

In our derivation we assumed that the spring obeys Hooke's law, that the spring has zero mass, and that the cart is a point-like object. We also neglected friction. In real life most springs obey Hooke's law if you do not stretch them too far, but it is impossible to find a massless spring or to eliminate frictional effects entirely. However, even with these assumptions, the above equation agrees closely with experiments in which friction is minimal and the mass of the spring is small compared with the mass of the attached object.

**QUANTITATIVE EXERCISE 19.4**

**Vibrating chair**

You want to design a bouncy seat for your nephew using a bungee cord (see figure at right). The chair should allow a 12-kg baby (including the mass of the seat) to bounce up and down naturally at a frequency of about 0.40 Hz. Estimate the spring constant the cord should have.

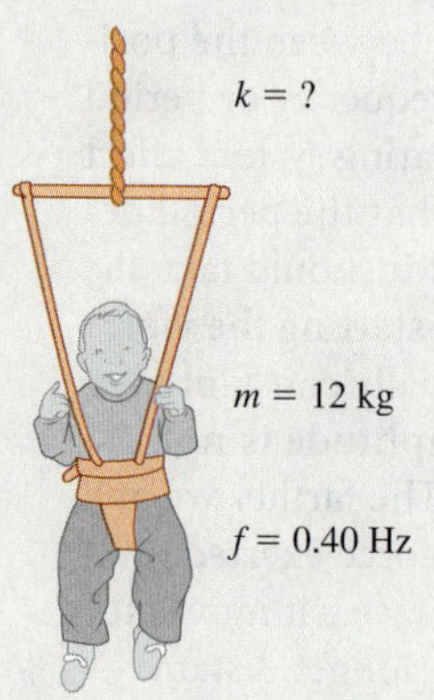


**Represent mathematically** Model the cord as a linear spring and the child+seat as a point-like object. We know the requested vibration frequency and the proposed mass of the child and seat. We can use Eq. (19.8) to find $k$:

$$f = \frac{1}{2\pi}\sqrt{\frac{k}{m}}$$

**Solve and evaluate** Square both sides of the above equation to get $f^2 = (1/4\pi^2)(k/m)$. Then, multiplying both sides by $4\pi^2 m$, we have

$$k = 4\pi^2 mf^2 = 4\pi^2(12\text{ kg})(0.4\text{ s}^{-1})^2$$

$$\simeq 80\text{ kg/s}^2 = 80\text{ N/m}$$

**Try it yourself:** If the total mass of the child+seat were increased to 24 kg without changing the spring constant, what would be the frequency of vibration? Describe any additional difficulties that will result from this change.

*Answer:* $f = 0.28\ \text{s}^{-1}$ or $T = 3.5$ s. The cord will stretch twice as far—it may not fit in a doorway.

**Review Question 19.4** What will happen to the period of a cart on a spring if you attach a second spring next to the first one, making two of them pull on the cart together?

## 19.5 Energy of vibrational systems

So far we have used the force approach to analyze the motion of vibrational systems quantitatively. Let us investigate how the energy approach can help us understand the process better. Consider again the vibration of a cart at the end of a spring (our system). The cart starts one vibration cycle at time $t = 0$ at position $x = +A$.

As the cart-spring system vibrates, the energy of the system continuously changes from all elastic when at $x = +A$ to all kinetic when passing through the equilibrium position at $x = 0$ to all elastic again when stopping for an instant at the far left of the vibration at $x = -A$ (see the far right column of **Table 19.4**). The cycle repeats on the way back to its starting position. Thus, the energy cycles between elastic and kinetic twice during each period. Assuming that no external forces do work on the system, the system's total energy is constant. Its value equals the maximum elastic potential energy when at $x = \pm A$, the maximum kinetic energy when at $x = 0$, and some combination of the two at other positions.

$$U = \frac{1}{2}kA^2 = \frac{1}{2}mv_{\text{max}}^2 = \frac{1}{2}mv^2 + \frac{1}{2}kx^2 \qquad (19.9)$$

**Table 19.4 Variation of energy during one vibration.**

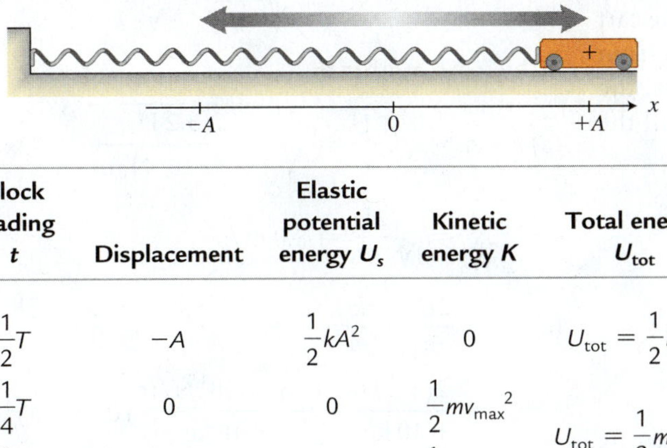


| Clock reading $t$ | Displacement | Elastic potential energy $U_s$ | Kinetic energy $K$ | Total energy $U_{\text{tot}}$ |
|---|---|---|---|---|
| $\frac{1}{2}T$ | $-A$ | $\frac{1}{2}kA^2$ | 0 | $U_{\text{tot}} = \frac{1}{2}kA^2$ |
| $\frac{1}{4}T$ | 0 | 0 | $\frac{1}{2}mv_{\text{max}}^2$ | $U_{\text{tot}} = \frac{1}{2}mv_{\text{max}}^2$ |
| $\frac{3}{4}T$ | 0 | 0 | $\frac{1}{2}mv_{\text{max}}^2$ | |
| 0 | $A$ | $\frac{1}{2}kA^2$ | 0 | $U_{\text{tot}} = \frac{1}{2}kA^2$ |
| $T$ | $A$ | $\frac{1}{2}kA^2$ | 0 | |

The second equality in Eq. (19.9) can be rearranged to give a relationship between the amplitude of the vibration and the cart's maximum speed:

$$v_{max} = \sqrt{\frac{k}{m}}A$$

**Active Learning Guide>**

This is exactly what follows from the kinematics equations describing simple harmonic motion [Eqs. (19.2) and (19.3)]. It also makes sense conceptually: When the mass of the cart is large, it should move slowly—having the mass in the denominator makes sense. If the spring is stiff, the cart will move faster—having the spring constant in the numerator also makes sense.

> **TIP** In the above discussion we neglected the interactions of the system with the surface of the track and with the air. These would both do negative work on the system and gradually decrease its energy, eventually bringing the vibrating system to rest.

**EXAMPLE 19.5 Vibrating cart at end of spring**

A spring with a $1.6 \times 10^4\ \text{N/m}$ spring constant and a 0.10-kg cart at its end has a total vibrational energy of 3.2 J. (a) Determine the amplitude of the vibration. (b) Determine the cart's maximum speed. (c) Determine the cart's speed when it is displaced 0.010 m from equilibrium. (d) What would be the amplitude of the vibration if the energy of the system was doubled?

**Sketch and translate** Use the sketch of the system in Table 19.4 to visualize the situation. The cart+spring is the system. We need to find the amplitude of vibration $A$, the maximum cart speed $v_{max}$, the cart speed $v$ when $x = 0.010$ m, and the amplitude $A'$ when the system energy is doubled ($U' = 2U$).

**Simplify and diagram** Assume that the spring is massless and neglect interactions between the system and the air and surface.

**Represent mathematically** (a) The speed of the cart is zero when it is at its maximum displacement from equilibrium—when $x = \pm A$. At these positions, the cart stops for an instant to change direction so all the energy of the system is elastic potential energy:

$$U = \frac{1}{2}m\,0^2 + \frac{1}{2}kA^2 = \frac{1}{2}kA^2$$

Solving for $A$ we find that

$$A = \sqrt{\frac{2U}{k}}$$

(b) The maximum speed occurs when the cart passes its equilibrium position (when $x = 0$). The energy of the system at that moment is all kinetic:

$$U = \frac{1}{2}mv_{max}{}^2 + \frac{1}{2}k\,0^2$$

or

$$v_{max} = \sqrt{\frac{2U}{m}}$$

(c) The system's energy when the cart is displaced 0.010 m from equilibrium is a combination of kinetic energy and elastic potential energy:

$$U = \frac{1}{2}mv^2 + \frac{1}{2}kx^2$$

or

$$v^2 = \frac{2U}{m} - \frac{k}{m}x^2$$

(d) When the energy is doubled so that $U' = 2U$, the vibrational amplitude of the cart is

$$A' = \sqrt{\frac{2U'}{k}} = \sqrt{\frac{4U}{k}}$$

**Solve and evaluate**

(a) $$A = \sqrt{\frac{2U}{k}} = \sqrt{\frac{2(3.2\ \text{J})}{1.6 \times 10^4\ \text{N/m}}} = 0.020\ \text{m}$$

(b) $$v_{max} = \sqrt{\frac{2U}{m}} = 8.0\ \text{m/s}$$

(c) $$v^2 = \frac{2U}{m} - \frac{k}{m}x^2$$

$$= \frac{2(3.2\ \text{J})}{0.10\ \text{kg}} - \frac{1.6 \times 10^4\ \text{N/m}}{0.10\ \text{kg}}(0.010\ \text{m})^2$$

$$= 48\ \text{m}^2/\text{s}^2$$

or $v = 6.9$ m/s.

(d) When the energy is doubled,

$$A' = \sqrt{\frac{2U'}{k}} = \sqrt{\frac{4U}{k}} = \sqrt{\frac{4(3.2\text{ J})}{1.6 \times 10^4\text{ N/m}}} = 0.028\text{ m}$$

Notice that the amplitude does not double when its energy doubles since $U$ is proportional to $A^2$ and not to $A$.

**Try it yourself:** What is the maximum stretch of the spring if the maximum speed of the cart is 5.0 m/s, its mass is 70 g, and the spring constant is 30 N/m?

*Answer:* 0.24 m.

**Review Question 19.5** The period of vibration of a cart on a spring is 2.0 s. The elastic potential energy fluctuates in a repetitive pattern. What is the period of the elastic potential energy variation of the system? What is the period of the total energy of the system?

## 19.6 The simple pendulum

**‹Active Learning Guide**

Springs are common in many objects that we use every day; they are an important component of mattresses, doors, shock absorbers, and office chairs. However, we seldom observe the springs themselves—most of them are hidden. One vibrating system in which the motion is very apparent is a pendulum. Grandfather clocks and metronomes use pendulums, of course. Another version of a pendulum is the oscillating movement of our arms and legs when we walk. In fact, much of animal locomotion involves complicated pendulum-like motions (**Figure 19.9**). For now, let's consider a simplified model of a pendulum system that has a compact object (called a *bob*) at the end of a comparatively long and massless string and undergoes small amplitude vibrations (**Figure 19.10**). This idealized system is called a **simple pendulum**.

**Figure 19.9** While walking, a giraffe's legs behave like complex pendulums, especially the front legs, which have three sections swinging in different directions.

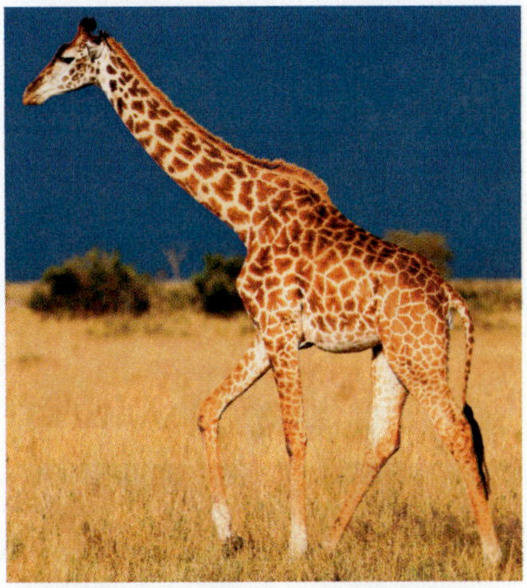

When the bob shown in **Figure 19.11a** hangs straight down at rest, the sum of the forces exerted on it is zero. This is the equilibrium position for the bob. If you pull the bob to the right side and release it, it swings to the left, overshooting the equilibrium position (Figure 19.11b), and then stops briefly at about the same distance to the left as it started on the right (Figure 19.11c). The bob then swings back toward the equilibrium position, overshoots, and stops about where it started.

Two objects interact with the bob of the pendulum. The string S exerts a force in the radial direction that is always perpendicular to the path of the bob. Earth exerts a downward gravitational force that we can resolve into a radial component $r$ along the string and a tangential component $t$ perpendicular to the string and the radius of the circle. The tangential component of the force exerted by Earth on the bob points toward the equilibrium position—toward the left when the bob is on the right side of equilibrium (Figure 19.11b) and toward the right when on the left side (Figure 19.11c). This component of the gravitational force is what is always trying to restore the pendulum bob to its equilibrium position. Hence, the tangential component of the gravitational force exerted by Earth on the bob is the restoring force. Thus, the ideas of equilibrium position and restoring force also apply to the vibrational motion of a pendulum.

**Figure 19.10** A simple pendulum.

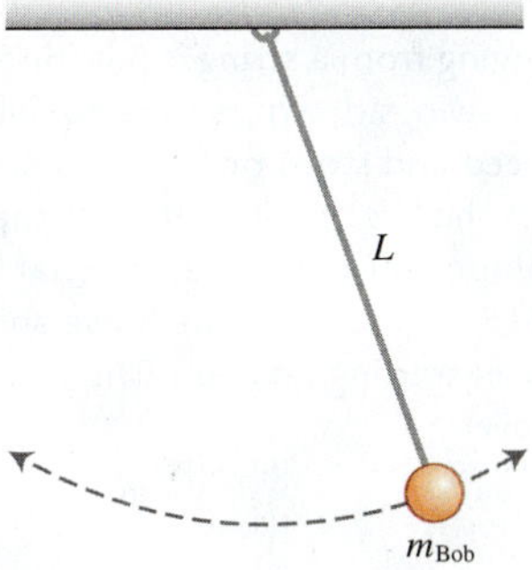


Consider the pendulum in Observational Experiment **Table 19.5**, on the next page.

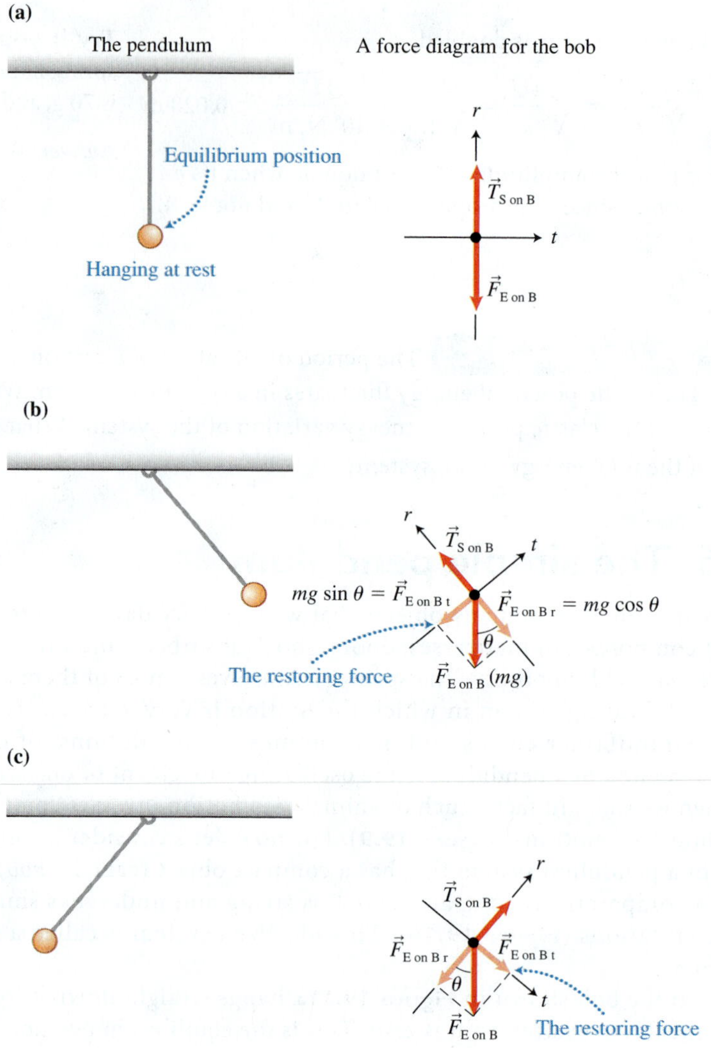


**Figure 19.11** Analyzing the forces exerted on a pendulum bob during pendulum motion.

**OBSERVATIONAL EXPERIMENT TABLE**

**19.5 Multiple representation analysis of the pendulum.**

VIDEO 19.5

| Observational experiment | Analysis |
|---|---|
| A bob hanging from a string is pulled to the right and released. It swings down past the equilibrium position at high speed and stops on the other side the same distance to the left that it started on the right. The tangential component of the gravitational force pulls the bob back toward equilibrium. It overshoots and then returns to its starting position. The process repeats over and over. | 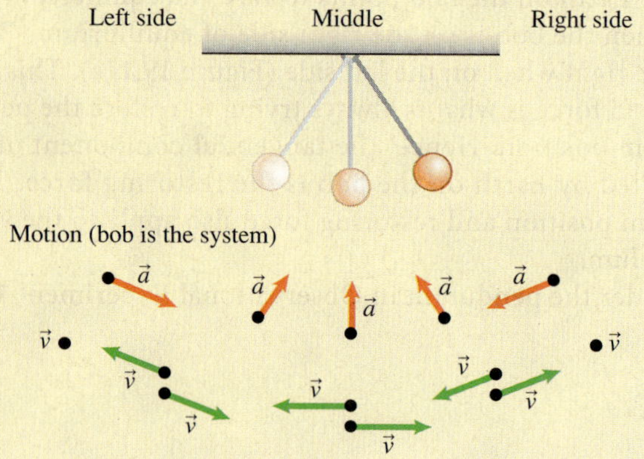  |

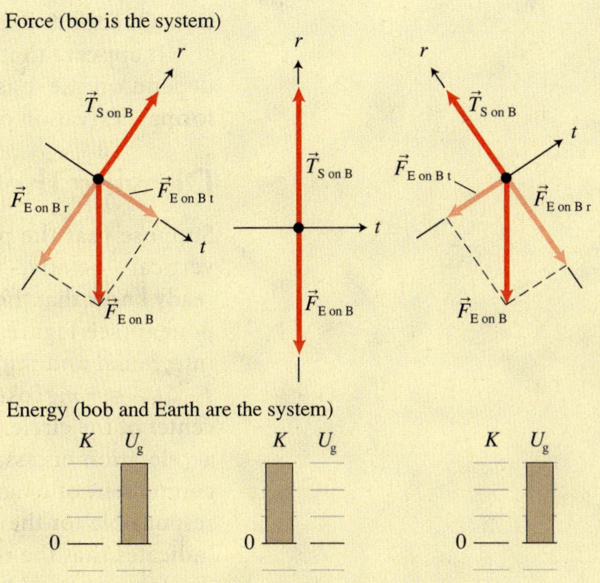


**Pattern**

***Restoring force*** The restoring force is zero as the vibrating object passes through the equilibrium position and has maximum magnitude when at the extreme positions on the left and right sides.

***Potential and kinetic energy*** The energy of the vibrating system varies between maximum potential energy when at the extreme positions to maximum kinetic energy as the object passes equilibrium. In between, the energy is a combination of kinetic and potential energy.

We find that the motion of the pendulum has the same patterns as the motion of the cart on a spring: it passes the equilibrium position from two different directions, there is a restoring force exerted on the bob, and the system's energy oscillates between maximum potential and maximum kinetic.

## Experimental investigation of the period $T$ of a simple pendulum

To investigate how the properties of a simple pendulum affect the period, we record its period for different vibration amplitudes $A$, bob masses $m$, and string lengths $L$ (changing them one at a time). The results appear in **Table 19.6**. Notice that for the pendulum the amplitude of the angular displacement is

**Table 19.6 Effect of bob mass, string length, and amplitude on pendulum period.**

| Bob mass $m$ (kg) | String length $L$ (m) | Amplitude $\theta$ (°) | Period $T$ (s) |
|---|---|---|---|
| 1.0 | 1.0 | 10.0 | 2.0 |
| **2.0** | 1.0 | 10.0 | 2.0 |
| **3.0** | 1.0 | 10.0 | 2.0 |
| 1.0 | 1.0 | 10.0 | 2.0 |
| 1.0 | **2.0** | 10.0 | 2.8 |
| 1.0 | **3.0** | 10.0 | 3.5 |
| 1.0 | **4.0** | 10.0 | 4.0 |
| 1.0 | 1.0 | **15.0** | 2.0 |
| 1.0 | 1.0 | **20.0** | 2.0 |
| 1.0 | 1.0 | **25.0** | 2.0 |

measured in terms of the angle $\theta$ that the string makes with the vertical direction. Quantities that have been changed are boldfaced.

It appears that the period depends only on the string length. It does not depend on the mass of the bob or the amplitude of vibrations. Let's make a formal derivation of the period and see if it agrees with the above.

## Deriving the period $T$ of a simple pendulum

Suppose that the pendulum string has an angular displacement $\theta$ from the vertical—positive to the right and negative to the left (Figure 19.11). We already know that the string exerts a force on the bob that only has a radial component (see Figures 19.11b and c). The force exerted by Earth can be resolved into radial and tangential components. The radial component of magnitude $F_{\text{E on B}\,r} = -mg\cos\theta$ is along the radial axis and points outward, away from the center of the circle. The two radial components provide the bob with the radial acceleration necessary to keep the ball following a circular path. The tangential component of magnitude $F_{\text{E on B}\,t} = -mg\sin\theta$ serves as a restoring force (it is responsible for the change in the bob's speed along the arc). The negative sign indicates that the restoring force points opposite the direction of the angular displacement and back toward the equilibrium position.

Recall that for small angles, the value of $\theta$ in radian units is approximately equal to $\sin\theta$ (you can check this using your calculator). For example, when $\theta = 0.3491$ radians (equal to 20°), $\sin\theta = 0.3420$. $\theta$ and $\sin\theta$ differ by only 2%, even for this moderately large angle. Because $\theta \approx \sin\theta$ for small angles, the restoring force can be written as $F_{\text{E on B}\,t} = -mg\,\theta$, where $\theta$ must be in radians.

The angular displacement $\theta$ is related to the linear position $x$ along the arc of the swinging ball (the solid line on the arc in **Figure 19.12**):

**Figure 19.12** The linear displacement $x$ of the bob is related to the string's angular displacement $\theta$ and length $L$.

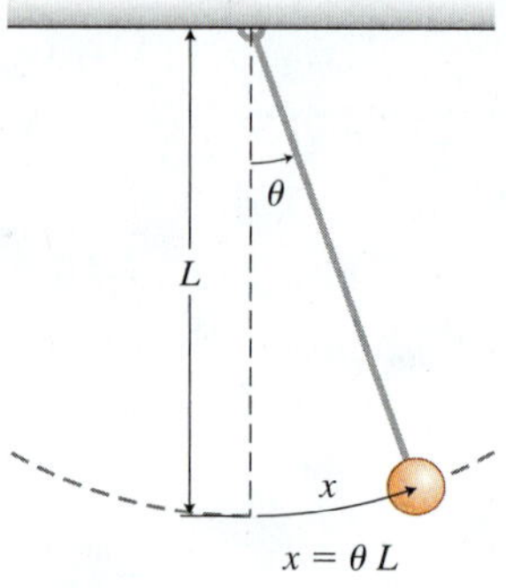


$$x = L\theta$$

where $L$ is the length of the pendulum string. Thus, we find that the tangential component of the restoring force is

$$F_{\text{E on B}t} = -mg\,\theta = -mg\left(\frac{x}{L}\right) = -\left(\frac{mg}{L}\right)x \tag{19.10}$$

Equation (19.10) for the restoring force exerted on a vibrating pendulum bob is very similar to the restoring force exerted on a mass at the end of a spring ($F_x = -kx$); the only difference is that the spring constant $k$ for a spring has been replaced by $mg/L$. Since the restoring force is directly proportional to the bob's displacement from equilibrium, the bob is in simple harmonic motion. Consequently, the period of a pendulum is given by Eq. (19.7) with $mg/L$ replacing $k$:

$$T = 2\pi\sqrt{\frac{m}{k}} = 2\pi\sqrt{\frac{m}{mg/L}} = 2\pi\sqrt{\frac{L}{g}}$$

Thus, the period of a pendulum is

$$T = 2\pi\sqrt{\frac{L}{g}} \tag{19.11}$$

Notice that the period of a simple pendulum depends only on its length and on the magnitude of the gravitational constant. It does not depend on the mass of the object hanging at its end or on the amplitude of vibration. This agrees with the patterns we discovered in Table 19.6. The restoring gravitational force is proportional to the bob's mass while the bob's acceleration $a_t = \Sigma F_t/m = -mg\theta/m$ is inversely proportional to its mass. Since mass cancels in the equation, the pendulum's motion does not depend on the bob's mass.

**EXAMPLE 19.6 Leg swinging period**

Estimate the number of steps that is "natural" for a leg to take per second while walking: in other words, the natural swinging frequency of the leg. We will treat the leg as a simple pendulum.[1] The calculations in this example are approximate.

**Sketch and translate** Imagine a leg that is about 1 m long swinging like a pendulum from the hip. To estimate the frequency, the only thing we need to know is the length of the pendulum.

**Simplify and diagram** Although the leg swings as a pendulum, it does not resemble a simple pendulum since its mass is concentrated not at the foot but can be thought to be concentrated at the center of mass of the leg. Therefore, our result will be only an estimate. We'll assume that the center of mass of the leg is at the knee and that the knee is at the midpoint of the leg.

**Represent mathematically** If a leg is 1 m long, then $L = 0.5$ m and its natural swinging frequency is calculated using Eq. (19.11).

$$f = \frac{1}{T} = \frac{1}{2\pi}\sqrt{\frac{g}{L}}$$

**Solve and evaluate** Inserting the estimated length in the above, we get

$$f \approx \frac{1}{2\pi}\sqrt{\frac{9.8\ \mathrm{m/s^2}}{0.5\ \mathrm{m}}} = 0.7\ \mathrm{s}^{-1}$$

The period of one step is the inverse of $f$, or 1.4 s.

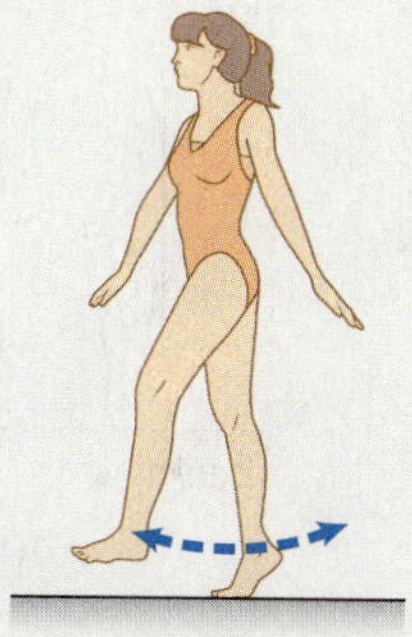

According to our analysis, it is natural to take one step with your right leg (or with your left leg) approximately each 1.4 s. To check if this estimate is close to the real value, you can stand on the toes of one foot and swing your other leg freely back and forth at what seems like its natural unforced frequency (see figure). Count the number of swings in 60 s. When you divide 60 s by this number, you have the period for one swing.

Longer legs have longer periods of vibration. Giraffes take relatively few steps per second because of their long legs (a low walking frequency). A small dog like a Chihuahua has short legs and a high walking frequency. Each leg of the dog in the photo below takes about two steps for each step taken by the man.

**Try it yourself:** What should happen to your walking frequency when you walk wearing ankle weights? Why?

*Answer:* The natural swinging frequency should decrease as the center of mass of the leg is lower and the pendulum becomes longer.

[1]In actuality, the leg is not a simple pendulum but is an example of what is called a physical pendulum—an extended body that swings back and forth about a pivot point. The swinging period of a physical pendulum is $f = (1/2\pi)\sqrt{mgL/I}$, where $L$ is the distance from the pivot point to its center of mass, $m$ is its mass, and $I$ is the rotational inertia of the pendulum about its pivot point.

## Energy analysis of a pendulum

We can also analyze the motion of a pendulum using the energy approach. We will do so in Conceptual Exercise 19.7.

**CONCEPTUAL EXERCISE 19.7 Energy of a swing at different locations**

A child sits on a swing that hangs straight down and is at rest. Draw energy bar charts for the child+swing+Earth system; (a) as a person pulls the child back in preparation for the first swing; (b) at the moment the person releases the swing while it is at its elevated position; (c) as the swing passes the equilibrium position; (d) when it reaches half the maximum height on the other side; and (e) as it passes the equilibrium position moving in the opposite direction.

**Sketch and translate** Start with the sketches of the process at the required instants (see the left sides of the accompanying figures).

*(continued)*

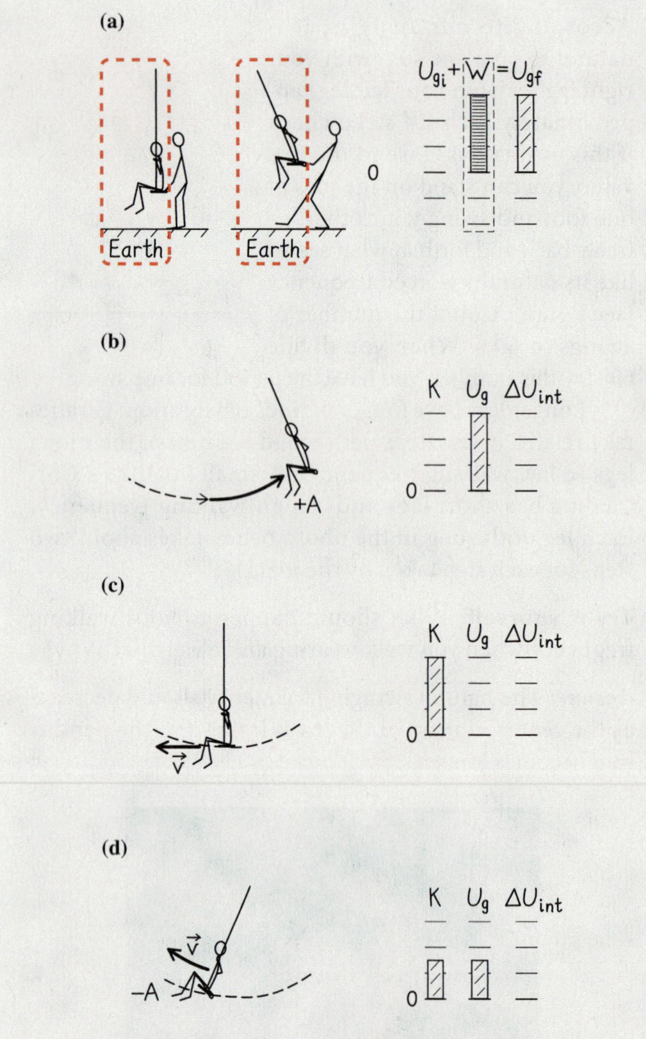


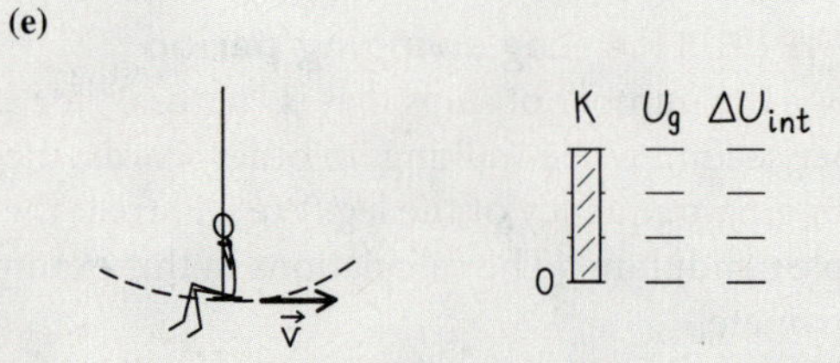


**Simplify and diagram** We ignore air resistance and friction at the ropes' pivot points. Choose the zero level of gravitational potential energy at the position of the seat when the swing is vertical. The ropes are always perpendicular to the child's motion and consequently do zero work on the system. The only external object that does work on the system is the person who raises the swing to its starting position. We can now represent the process with bar charts (see the right sides of the figures). Note that the first chart shows work done on the system. Then energy switches between all gravitational potential energy at $x = \pm A$ and all kinetic energy when passing through the equilibrium position at $x = 0$ to half gravitational and half kinetic at position (d).

**Try it yourself:** Suppose that you do not ignore air resistance and friction and that after several swings the child returns to only half the original maximum height on the right side. Construct a bar chart for that moment.

*Answer:* We include the air and the pivots in the system. Half of the original gravitational potential energy has been converted to internal energy due to air resistance and friction at the ropes' pivots, as shown below.

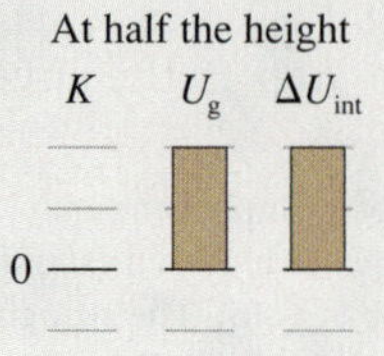


**Figure 19.13** An astronaut Body Mass Measurement Device.

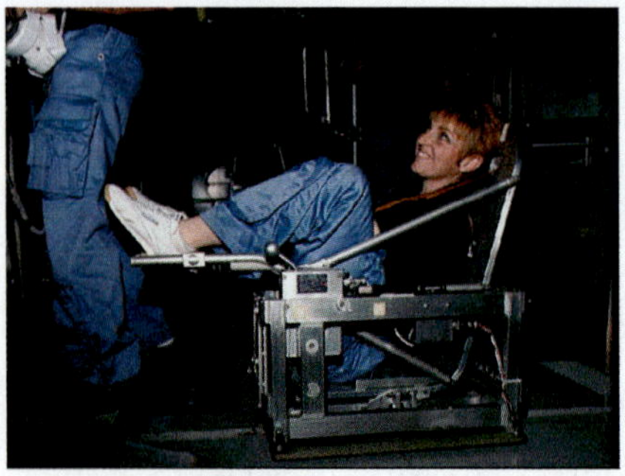

**Review Question 19.6** Your grandfather's pendulum clock is slow. How should you adjust the position of the bob to correct for this?

## 19.7 Skills for analyzing processes involving vibrational motion

Astronauts in space can lose significant body mass because their muscles do not get sufficient use and begin to atrophy. Since traditional scales do not function in orbit (both the scale and the astronaut are in free fall), NASA developed the Body Mass Measurement Device (BMMD) to monitor the astronauts' mass. The BMMD consists of a chair on which the astronaut sits (**Figure 19.13**). A spring holds the chair in place. When the chair is pulled back and released, it vibrates at a frequency that depends on the mass of the chair plus astronaut.

Let's use the BMMD as an example to describe a general strategy for analyzing processes involving vibrational motion.

**Active Learning Guide›**

## PROBLEM-SOLVING STRATEGY Applying Our Knowledge of Vibrational Motion

### EXAMPLE 19.8 Astronaut mass

The BMMD chair (mass 32 kg) has a vibrational period of 1.2 s when empty. When an astronaut sits on the chair, the period changes to 2.1 s. Determine (a) the effective spring constant of the chair's spring, (b) the mass of the astronaut, and (c) the maximum vibrational speed of the astronaut if the amplitude of vibration is 0.10 m.

### Sketch and translate

- Sketch the process described in the problem statement. Label physical quantities.
- Choose an object (or objects) as the system of interest. Depending on the process you might need to analyze several systems.

We choose the chair and the spring as the system for (a) and the astronaut, chair, and spring for (b) and (c).

$$m_{\text{Chair}} = 32\text{ kg}$$
$$T_{\text{Chair}} = 1.2\text{ s}$$
$$T_{\text{Ast+Chair}} = 2.1\text{ s}$$
$$k = ?$$
$$v_{\max} = ?$$

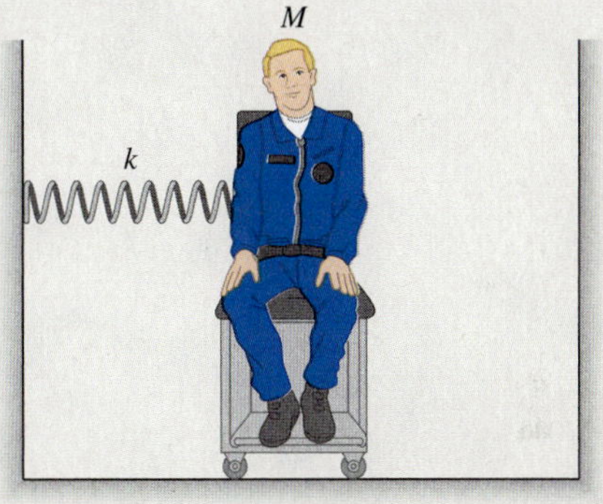


### Simplify and diagram

- Identify and evaluate assumptions and approximations.
- Represent the process with force diagrams and/or bar charts if needed.

Assume that the chair's spring obeys Hooke's law and that the spring's mass is much smaller than the mass of the chair plus astronaut. We determine the effective spring constant as though there were one spring pushing and pulling the chair. For (c) we represent the process using a bar chart starting from the state when the chair is pulled the farthest from equilibrium and ending as the chair and person pass equilibrium.

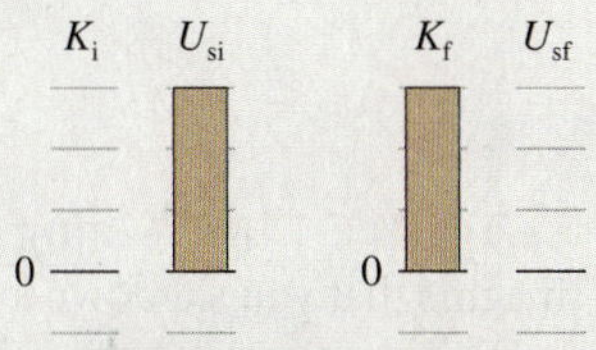


### Represent mathematically

- If necessary, use kinematics equations to describe the changing motion of the object.
- If necessary, use force diagrams to apply the component form of Newton's second law to the problem or use bar charts to apply work-energy principles.
- If necessary, use the expressions for the period of an object attached to a spring or to a pendulum.

(a) Use $T_{\text{Chair}} = 2\pi\sqrt{\dfrac{m}{k}}$ to determine the spring constant with the chair empty.

(b) Use $T_{\text{Ast+Chair}} = 2\pi\sqrt{\dfrac{m + M}{k}}$ to determine the astronaut mass.

(c) Determine the energy in the spring-astronaut-chair system when at the extreme position $x = +A$: $U = \dfrac{1}{2}kA^2$. Then use this energy to determine the maximum speed as the cart passes through equilibrium using $U = \dfrac{1}{2}kA^2 = \dfrac{1}{2}(m + M)v_{\max}^2$.

*(continued)*

**Solve and evaluate**

- Solve for the unknowns. Evaluate the solution—is it reasonable? Consider the magnitude of the answer, its units, limiting cases, etc.

(a) Square $T_{\text{Chair}} = 2\pi\sqrt{\dfrac{m}{k}}$ or $T^2_{\text{Chair}} = 4\pi^2\dfrac{m}{k}$. Rearrange to get

$$k = 4\pi^2\frac{m}{T^2_{\text{Chair}}} = 4\pi^2\frac{32\text{ kg}}{(1.2\text{ s})^2} = 877\text{ N/m}$$

(b) Square $T_{\text{Ast+Chair}} = 2\pi\sqrt{\dfrac{m+M}{k}}$ or $T^2_{\text{Ast+Chair}} = 4\pi^2\dfrac{m+M}{k}$.

Rearrange to get

$$m + M = \frac{T^2_{\text{Ast+Chair}}\,k}{4\pi^2} \quad\text{or}\quad M = \frac{(2.1\text{ s})^2(877\text{ N/m})}{4\pi^2} - 32\text{ kg} = 66\text{ kg}$$

(c) From $U = \frac{1}{2}kA^2 = \frac{1}{2}(m+M)v_{\text{max}}^{\;2}$, $kA^2 = (m+M)v_{\text{max}}^{\;2}$.

$$\text{Thus } v_{\text{max}} = A\sqrt{\frac{k}{m+M}} = (0.10\text{ m})\sqrt{\frac{(877\text{ N/m})}{32\text{ kg} + 66\text{ kg}}} = 0.30\text{ m/s.}$$

$$\text{Check the units: } \text{m}\sqrt{\frac{\text{N/m}}{\text{kg}}} = \sqrt{\frac{\text{m}^2(\text{N/m})}{\text{kg}}} = \sqrt{\frac{\text{m}(\text{kg}\times\text{m})}{\text{kg}\times\text{s}^2}} = \text{m/s.}$$

The units are correct. The magnitudes are reasonable—a person can have a mass of 66 kg and the speed of 0.3 m/s is a realistic speed for such a process.

**Try it yourself:** What would be the period of vibration of the machine if used by an 80-kg astronaut?

*Answer:* About 2.25 s.

## EXAMPLE 19.9 Skiing and vibrations

Imagine that you ski down a slope wearing a Velcro ski vest and then continue skiing on a horizontal surface at the bottom of the hill. There you run into a padded, Velcro-covered cart, which is also on skis (see the figure below). A 1280-N/m spring is attached to the other end of the cart and also to a wall. The spring compresses after your 60-kg body hits and sticks to the 20-kg cart. Your speed is 16 m/s just before you hit the cart. (a) What is your maximum speed after joining with the cart? (b) What is the maximum compression of the spring? (c) What is the period of the vibrational motion? (d) What is your maximum acceleration and where does it occur?

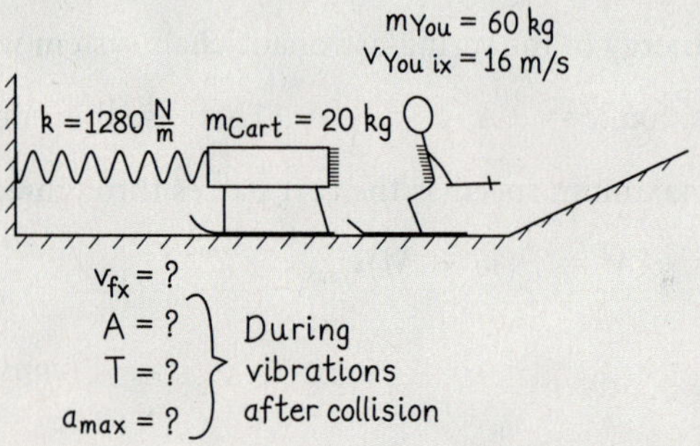


**Sketch and translate** Draw a labeled sketch (see the first figure at right). (a) The collision with the cart is a completely inelastic collision because you stick to the cart. If you and the cart are the system, then during the moment of the collision, the momentum of the system is constant (the same before and after the collision). We can use impulse-momentum ideas to determine your speed immediately after the collision. (b) Now choose a different system: you, the cart, and the spring. The kinetic energy of you and the cart is converted into the spring's elastic potential energy as the spring compresses. (c) This is now a spring-object system that vibrates—you can determine the period of the vibration and the frequency. (d) The maximum acceleration is when the spring is compressed or stretched the most—when $x = \pm A$. We can determine the magnitude of the maximum acceleration using Newton's second law.

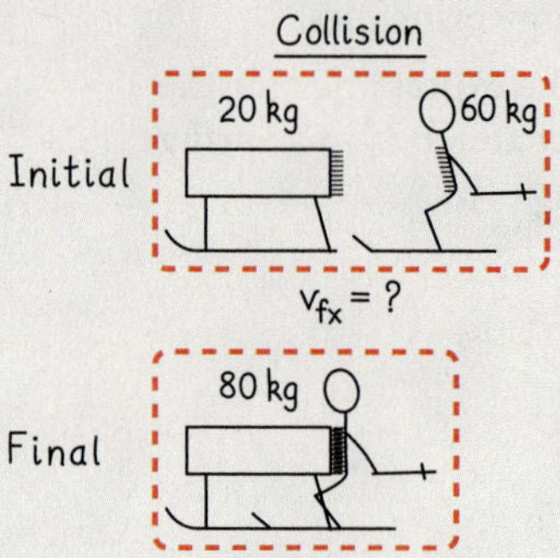

**Simplify and diagram** Assume that the snowy surface is perfectly smooth (zero friction), and that the spring has zero mass and obeys Hooke's law. (a) A momentum bar chart represents your collision with the cart (see the first figure at the right). There are no horizontal forces exerted on the you+cart system along the $x$-direction during the instant of collision. Thus the $x$-component of momentum is constant. (b) Immediately after the collision, you and the cart, now joined, compress the spring. An energy bar chart represents this process (see the second figure above).

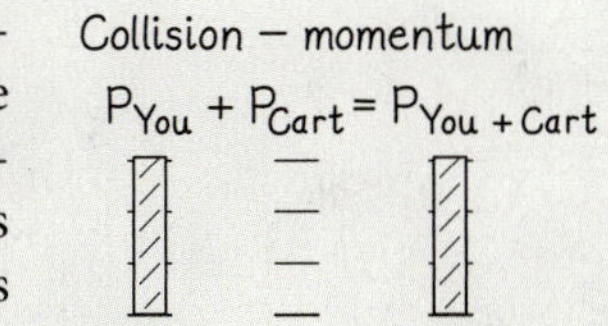


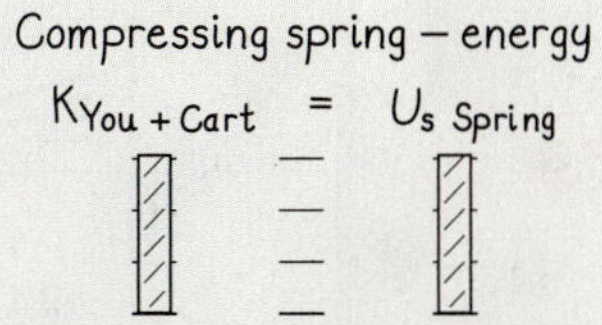


**Represent mathematically** (a) Use momentum constancy for the collision with the cart to find the speed of you+cart after the collision:

$$m_{You}v_{You\,i\,x} + m_{Cart}v_{Cart\,i\,x} = m_{You+Cart}v_{f\,x}$$

(b) Use energy constancy to determine the maximum compression distance:

$$\frac{1}{2}m_{You+Cart}v_{f\,x}^{\ 2} = \frac{1}{2}kA^2$$

(c) Determine the period using

$$T = 2\pi\sqrt{\frac{m_{You+Cart}}{k}}$$

(d) Use Newton's second law to determine the maximum acceleration during the vibrations after the collision:

$$a_x = \frac{-kx}{m_{You+Cart}}$$

$a$ is maximum when the spring has its maximum compression (when $x = -A$).

**Solve and evaluate** (a) The velocity of you and the cart immediately after the collision is

$$v_{f\,x} = \frac{m_{You}v_{You\,i\,x} + m_{Cart}v_{Cart\,i\,x}}{m_{You+Cart}} = \frac{(60\text{ kg})(16\text{ m/s}) + 0}{(60\text{ kg} + 20\text{ kg})}$$
$$= 12\text{ m/s}$$

The speed seems reasonable, and the sign matches the original direction of motion.

(b) Using the equation $\frac{1}{2}m_{You+Cart}\,v_f^{\,2} = \frac{1}{2}k\,A^2$, we can cancel the (1/2), divide both sides of the equation by $k$, and take the square root. We find that the maximum displacement of the cart from its equilibrium position is

$$A = \sqrt{\frac{m_{You+Cart}}{k}}\,v_f = \sqrt{\frac{80\text{ kg}}{1280\text{ N/m}}}(12\text{ m/s}) = 3.0\text{ m}$$

The compression seems a little high but the skier was moving with high speed; the units are fine.

(c) The period for the you+cart vibration is

$$T = 2\pi\sqrt{\frac{m_{You+Cart}}{k}} = 2\pi\sqrt{\frac{80\text{ kg}}{1280\text{ N/m}}} = 1.6\text{ s}$$

(d) The cart started at its equilibrium position ($x = 0$) and was moving at 12 m/s after you collided with it. This is its maximum speed. The amplitude of the vibration was 3.0 m. Thus, the maximum acceleration of the you+cart system when at the $x = -A$ position is

$$a_{max} = \frac{-k(-A)}{(m_{You+Cart})} = \frac{(1280\text{ N/m})(3.0\text{ m})}{(60\text{ kg} + 20\text{ kg})} = 48\text{ m/s}^2.$$

The maximum acceleration of you and the cart is five times free-fall acceleration. The units for the result are the units of acceleration: $\frac{(\text{N/m}) \times (\text{m})}{(\text{kg})} = \frac{\text{N}}{\text{kg}} = \frac{\text{kg} \times \text{m/s}^2}{\text{kg}} = \text{m/s}^2$.

**Try it yourself:** (a) How would the period of vibration change if you hit the cushion at half the velocity in the example? (b) What would be the period of vibration for a person whose mass is half of yours and whose speed before the collision with the cart is 0.8 times your speed?

*Answer:* (a) The period is independent of the amplitude. (b) The period will be 1.24 s.

**Review Question 19.7** Why was it important to assume that the springs obey Hooke's law in Examples 19.8 and 19.9?

## 19.8 Including friction in vibrational motion

In the preceding sections, we have mostly ignored the effect of friction on vibrating objects. Without it, though, a car would continue vibrating on its suspension system for miles after crossing a bump on a road and tall buildings would continue to sway even after the wind had died down.

**Figure 19.14** Damped and undamped oscillators.

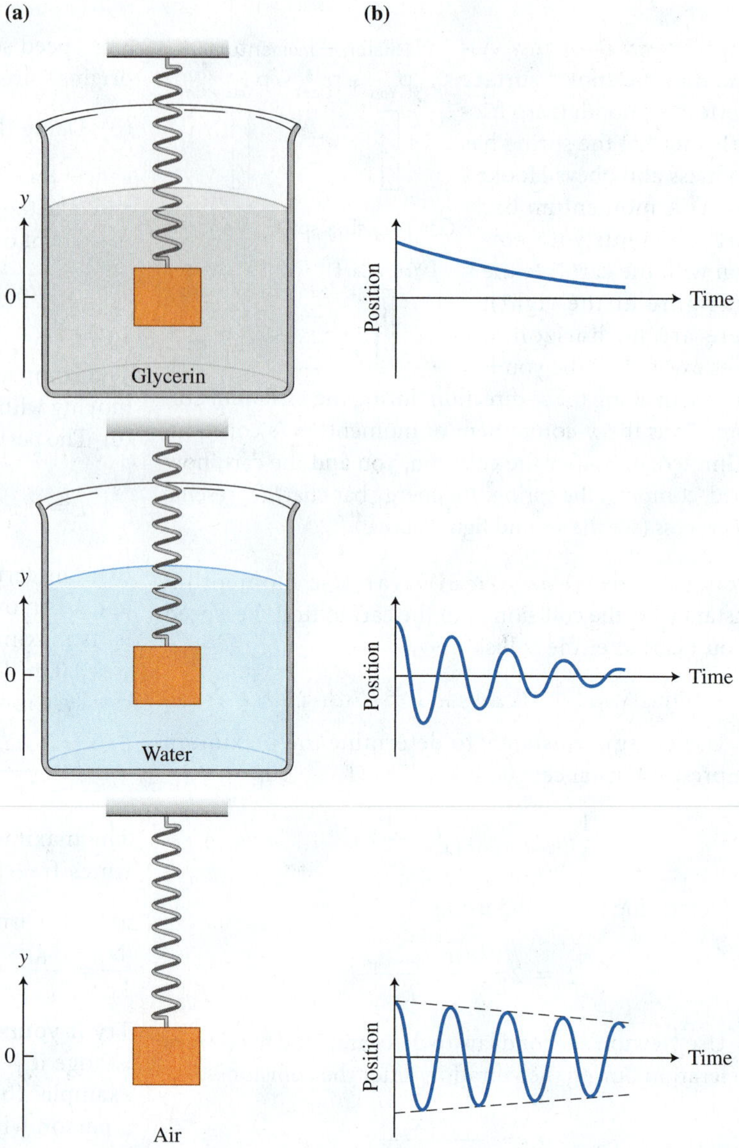


**Figure 19.15** A shock absorber for a mountain bike.

You can observe the effects of friction on a simple system. Fill one container with glycerin (a thick, viscous liquid) and another with water. Hang three identical blocks from three identical springs (**Figure 19.14a**). Hang one block in the glycerin, the second in the water, and the third in air. Start the blocks vibrating by moving the top of each spring up and down for a short time interval and then holding them still. The block in the air continues to vibrate with almost constant period and with very slowly reducing amplitude. The block in water makes several vibrations with significantly decreasing amplitude before stopping. (Its frequency also decreases, but this is difficult to observe.) The block in glycerin does not complete even one full vibration. Graphs of the position versus time for the three blocks are shown in Figure 19.14b. This phenomenon of decreasing vibration amplitude and increasing period is called **damping**.

Damping is a useful aspect of the design of vehicles and bridges. For example, a mountain bike has shock absorbers to decrease the force exerted by the surface on the tire when the tire hits a bump or a hole in the trail. The shock absorbers consist of springs and oil-filled dampers in the front fork of the bicycle and underneath the seat. The spring is wrapped around a piston filled with oil (**Figure 19.15**). During a collision, the spring compresses and increases the stopping distance, thus smoothing the effect of the collision on the rider. When the spring compresses, the piston pushes oil through a small hole. This process converts the spring's elastic potential energy into internal energy in the oil. Instead of continuing to bounce up and down, the spring returns quickly to its noncompressed position.

Damping is not desirable in all cases. Spiders that spin webs depend on vibrations to locate and restrain prey. An insect striking the web will produce vibrations that allow the spider to locate the insect. The further these vibrations travel before damping out, the more likely the spider will locate its prey (**Figure 19.16**).

**Figure 19.16** Vibrations in the spiderweb help the spider locate the captured prey.

Consider position-versus-time graphs for vibrational systems with friction. We distinguish between a system that is weakly damped and one that is overdamped or critically damped. A **weakly damped** system continues to vibrate for many periods. The vibration amplitude slowly decreases (**Figure 19.17a**) while the period increases slightly. In an **overdamped system**, the damping is so great that all of the elastic energy of the stretched vibrating system is converted to internal energy as the vibrating object makes its first move toward the equilibrium position (Figure 19.17b). When overdamped, the vibrating system takes a long time to return to the equilibrium position, if it ever does (some doors have an overdamped vibrating system with a spring and oil-based damper that slowly returns the door to its closed position after opening). In a **critically damped system,** the vibrating object returns to equilibrium in the shortest possible time but does not overshoot the equilibrium position (Figure 19.17c). The suspension system of a car should be critically damped so that it returns to equilibrium quickly and avoids continued vibrations.

**Figure 19.17** Position-versus-time graphs of weakly damped, overdamped, and critically damped oscillators.

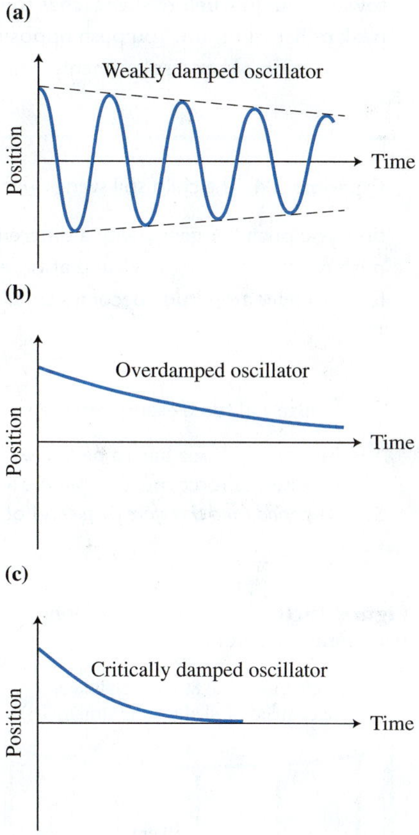


**Review Question 19.8** What features of damped vibrational motion make it different from SHM?

## 19.9 Vibrational motion with an external driving force

All real vibrations are damped and eventually stop unless energy is added to the system to compensate. In Observational Experiment **Table 19.7**, we investigate how external interactions of the environment with a vibrating system could lead to continuous vibrations.

**OBSERVATIONAL EXPERIMENT TABLE**

**19.7 A forced vibration.**

| Observational experiment | Analysis |
|---|---|
| **Experiment 1.** A young child sits on a swing. You exert a steady force on the child and swing until the swing cable is at such an angle that the swing stops (this angle depends on the construction of the swing). There are no vibrations. 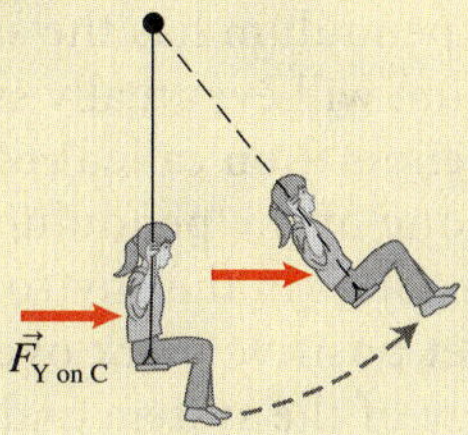 | The swing, child, and Earth are the system (but not you). You exerted a constant force that did positive work on the swing. But no vibrations occurred during this time. |



(*continued*)

| Observational experiment | Analysis |
|---|---|

**Experiment 2.** A young child swings with a period $T = 2\pi\sqrt{\frac{L}{g}}$. You push her gently and briefly every time just after she reaches the peak of her vibration. Each brief push is in the same direction as her displacement. The pushes have the same period as her swinging.

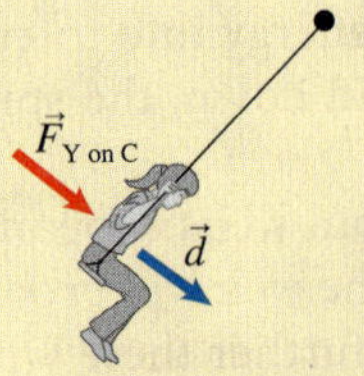


You exerted a variable force that did positive work on the swing. The energy of the system increased. The amplitude of the swing vibrations increases.

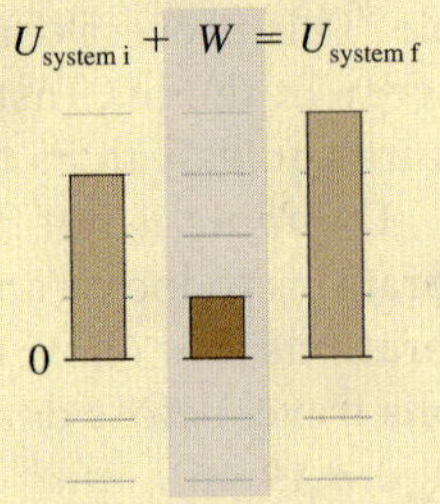


**Experiment 3.** The child still swings with period $T = 2\pi\sqrt{\frac{L}{g}}$. This time you push her gently and briefly as she swings back toward you, just before she reaches the peak of her vibration. You push opposite the direction of her displacement.

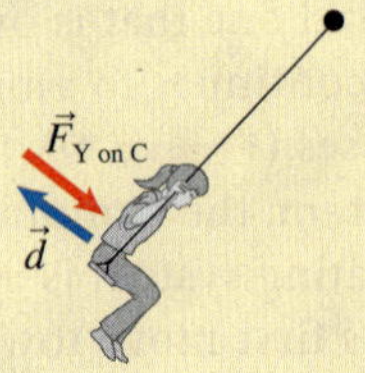


You did negative work on the swing. The energy of the system decreased. The amplitude of the swing vibrations decreases.

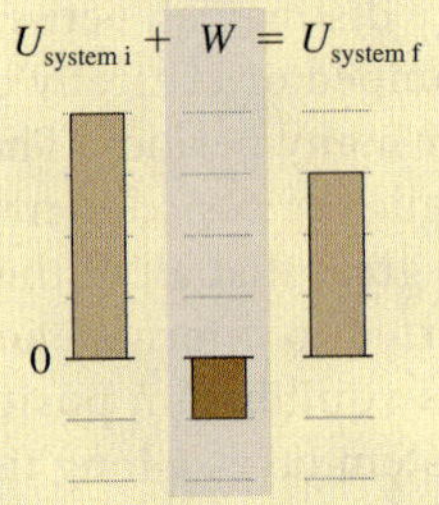


**Experiment 4.** The child still swings with period $T = 2\pi\sqrt{\frac{L}{g}}$. This time you push her gently with a different period. Sometimes you push as she is moving back and at other times as she is moving forward. Her amplitude becomes small and stays small.

You do positive work sometimes and negative work at other times (the net work over time is zero). Friction and air resistance cause the swinging amplitude to decrease.

**Pattern**

For an external force exerted on the swinging system to cause the amplitude to increase:

1. The external force has to be a *variable force*.
2. The external force must do *positive work* on the vibrating system.
3. The *period (and therefore frequency)* of the external force *must match* that of the vibrating system.

**Figure 19.18** Connected 80-cm-long pendulum oscillators.

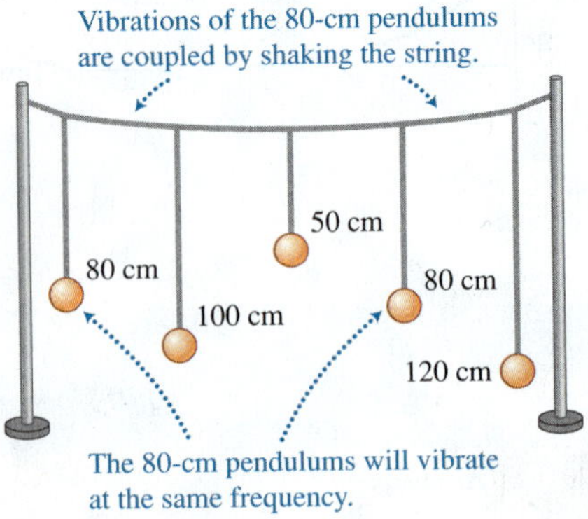


To test these patterns, we use them to predict the results of the following experiment. A string is tied between two ring stands—the string sags a little (**Figure 19.18**). Five pendulums of different lengths are attached to the string (80 cm, 100 cm, 50 cm, 80 cm, and 120 cm). What happens if we pull the 80-cm-long pendulum on the left in the direction perpendicular to the orientation of the string and let it go?

The swinging 80-cm-long pendulum causes the horizontal string to move back and forth at the natural frequency of the 80-cm-long pendulum:

$$f = \frac{1}{2\pi}\sqrt{\frac{9.8\text{ m/s}^2}{0.80\text{ m}}} = 0.56\text{ s}^{-1} = 0.56\text{ Hz}$$

This 0.56-Hz motion of the vertical string causes it to pull the horizontal string back and forth at that frequency, which in turn tugs on the other vertical strings and thus does work on the other four pendulums. Since only the other 80-cm pendulum has the same natural frequency, we predict that only that pendulum will eventually swing with large amplitude. This is what we observe, a phenomenon called **resonance**—the increase in the amplitude of vibrations of one of the pendulums. Resonance occurs in vibrating systems when some object in the environment exerts a force on it that varies in time and does net positive work over time. This work leads to an increase in the total energy of the system and therefore to the increase of the amplitude of vibrations. This increase in energy occurs when the frequency of the external force is either the same as the natural frequency of the system without damping or close to the natural frequency for the systems with damping. The child on the swing exhibited another example of resonance, when the frequency of the variable force exerted on her was close to the natural frequency of the swing.

As we continue to observe the pendulums, an even more unusual thing happens. As time passes, the amplitude of the original 80-cm pendulum decreases to zero as the amplitude of the second 80-cm pendulum reaches a maximum. Then the first pendulum begins vibrating again as the second pendulum's amplitude decreases. Energy is being gradually transferred from one 80-cm-long pendulum to the other. This is an example of **energy transfer through resonance**.

Many everyday phenomena can be explained using the concept of energy transfer through resonance. We have found that transfer of energy from one object to another depends on two conditions:

1. the natural frequencies of the objects must be close, and
2. a mechanism must exist that allows one object to do positive work on the other.

We can see an example of energy transfer through resonance in our chapter-opening story about the Russian soldiers marching across a bridge. When the frequency of their steps approached the natural vibration frequency of the bridge, the marching initiated large-amplitude vibrations in the bridge, which led to its collapse. Any small vibrations that the bridge had were amplified by the positive work done by the soldiers on the bridge. As the amplitude of the vibrations increased, parts of the bridge twisted and pulled apart and eventually the bridge collapsed.

Resonance caused the collapse of the Tacoma Narrows Bridge in Washington only four months after its completion. Unbeknownst to the engineers who designed it, vibrations near one of the bridge's resonant frequencies would increase in amplitude rather than damp out. On November 7, 1940, a strong wind combined with the bridge's natural twisting vibration caused an amplitude vibration of over 3 m at the sides of the bridge (**Figure 19.19a**). It collapsed after about an hour (Figure 19.19b).

Vocalists have been known to break wine glasses by producing sounds of particular frequencies. To do so, they must match the natural frequency of the wine glass (about 560 Hz). You are probably familiar with another example of energy transfer through resonance: spilling a beverage while carrying it across a room.

**Figure 19.19** The collapse of the Tacoma Narrows bridge.

(a)

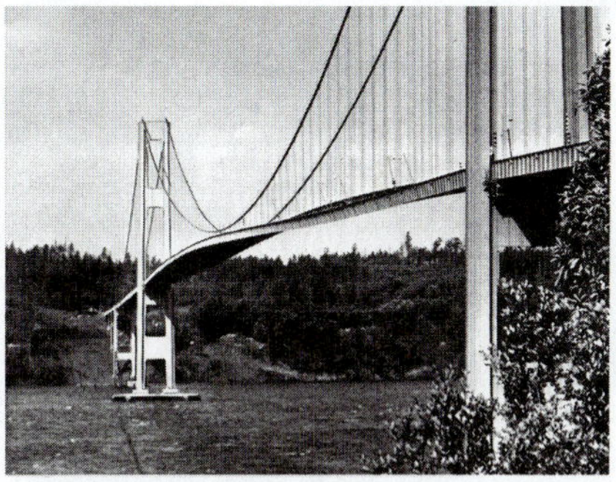

(b)

**CONCEPTUAL EXERCISE 19.10 Carrying a glass of juice**

Snehal is crossing a room while carrying a glass of cranberry juice. Estimate the stepping frequency that might cause the juice to splash onto the carpet. The natural frequency of vibrations of the juice sloshing in the nearly full glass is about 3 vibrations/s.

**Sketch and translate** Snehal's arm is a vibrating object that interacts with the glass of juice as well as with his legs.

**Simplify and diagram** Assume that Snehal's hand and arm oscillate at the same frequency at which his feet hit the ground. To avoid spilling the cranberry juice, he should try to walk so that his feet do not hit the ground 3 times per second (or 1.5 steps/s for each foot). That is a rapid pace, so it is unlikely that Snehal will cause resonant vibrations in the glass.

**Try It Yourself:** Do a follow-up investigation. First, determine the natural sloshing frequency in some container of liquid. Then walk with each of your legs swinging at the same frequency to see if the liquid spills. Think about what other factors affect the possibility of a coupled resonance between your stepping and the sloshing vibrations of liquid in a container. List the possible factors and then do experiments to determine the effect of changing those factors.

**Review Question 19.9** Describe the phenomenon of energy transfer by resonance and explain why it happens.

## 19.10 Vibrational motion in everyday life: Putting it all together

We can apply our knowledge of vibrational motion to explain many microscopic and macroscopic processes. In this section we investigate the microscopic vibrations of atoms in carbon dioxide molecules and the physics of bungee jumping.

### Radiation absorption by molecules

We can use what we know about vibrational motion to understand some aspects of global climate change. To maintain a consistent climate, Earth and its atmosphere as a system rely on a balanced exchange of energy in and out of the system. The energy absorbed by the Earth-atmosphere system, which enters the system in the form of visible light and infrared radiation from the Sun, must be balanced by the energy transferred out of that system. The energy is transferred out when vibrating molecules on Earth emit lower frequency infrared radiation that travels away from Earth. The radiation from the Sun is at a much higher temperature than that emitted by Earth.

Some of the radiation emitted by molecular vibration on Earth is absorbed by $CO_2$ molecules in the atmosphere—a form of resonant energy transfer. This absorption causes the vibrational energy of these atmospheric molecules to increase. The excited $CO_2$ molecules re-emit the infrared radiation, much of which is reabsorbed by molecules on Earth. Thus, less energy is transferred out of the system than is transferred in—the planet warms. This **greenhouse effect** is similar to what happens to a person sitting in a car on a hot summer day. The energy from the Sun in the form of sunlight enters the car and is converted into additional thermal energy of the air in the car. The hot air in the car emits lower frequency infrared radiation, which is absorbed and reflected by the car's windows. Thus, energy does not radiate out of the car—the temperature inside the car rises.

Let's look briefly at the $CO_2$ vibrations that are involved in this energy absorption.

**EXAMPLE 19.11 $CO_2$ vibration**

Within molecules, atoms interact with each other via attractive and repulsive electric forces. These forces maintain a fairly consistent distance between the atoms. When the atoms get close to each other they repel; when they are pulled away they attract. Thus we can model the bond between the atoms as a spring. One of the vibrations in a $CO_2$ molecule involves an asymmetric stretching and compression of the CO bonds between the carbon and oxygen atoms (see the figure at right). This bond has an effective spring constant of 1400 N/m. The vibration of the molecule has a natural frequency of $7.0 \times 10^{13}$ Hz. (a) Determine the effective mass of the $CO_2$ molecule when vibrating in this way. (b) When a $CO_2$ molecule in the atmosphere absorbs infrared energy emitted by Earth, the energy of the vibration is $4.7 \times 10^{-20}$ J. Estimate the amplitude of the vibration.

One CO bond is compressed while the other stretches, and the process then reverses at high frequency.

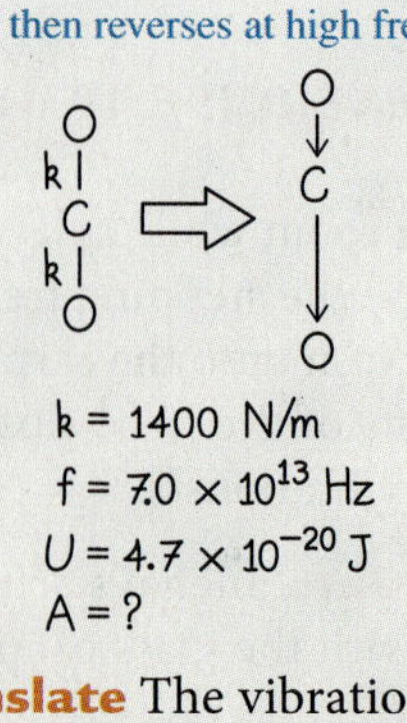


**Sketch and translate** The vibration is depicted in the figure above. We will use the expression for the vibration frequency of a cart with mass $m$ and spring with spring constant $k$ to estimate an effective mass of the vibrating molecule. We can use $1/2\,kA^2$ for the energy of vibration to estimate the amplitude.

**Simplify and diagram** Assume that the molecule can be modeled as a system undergoing SHM with a mass $m$ at the end of a spring of spring constant $k$. We are also assuming that the physics ideas we are using apply to atomic-scale objects (which turns out not to be a very good assumption).

**Represent mathematically** The frequency of vibration of a mass at the end of a spring is

$$f = \frac{1}{2\pi}\left(\frac{k}{m}\right)^{1/2}$$

The energy of vibration is

$$U_{\text{total}} = \frac{1}{2}kA^2$$

**Solve and evaluate** The effective mass of the vibrating system is

$$m = \frac{k}{4\pi^2 f^2} = \frac{(1400\ \text{N/m})}{4\pi^2(7.0 \times 10^{13}\ \text{s}^{-1})^2} = 7.2 \times 10^{-27}\ \text{kg}$$

The amplitude of vibration is

$$A = \left(\frac{2U_{\text{total}}}{k}\right)^{1/2} = \left(\frac{2(4.7 \times 10^{-20}\ \text{J})}{1400\ \text{N/m}}\right)^{1/2} = 8.2 \times 10^{-12}\ \text{m}$$

Are these numbers reasonable? The mass $7.2 \times 10^{-27}$ kg is approximately the mass of an oxygen atom. The amplitude is a few percent of the typical distance between atoms in a molecule. Both answers thus seem reasonable.

**Try it yourself:** Check the expressions used for each answer to see if they have the correct units.

*Answer:* $m = \dfrac{k}{4\pi^2 f^2}$ has units $\dfrac{(\text{N/m})}{(\text{s}^{-1})^2} = \left(\dfrac{\text{kg}\cdot\text{m}}{\text{s}^2}\right)\left(\dfrac{1}{\text{m}}\right)\left(\dfrac{1}{\text{s}^{-2}}\right) = \text{kg}$, the correct unit for mass. $A = \left(\dfrac{2U_{\text{total}}}{k}\right)^{1/2}$ has units $\left(\dfrac{\text{J}}{\text{N/m}}\right)^{1/2} = \left(\dfrac{\text{N}\cdot\text{m}}{1}\cdot\dfrac{\text{m}}{\text{N}}\right)^{1/2} = (\text{m}^2)^{1/2} = \text{m}$, the correct unit for amplitude.

## Vibrating at the end of a bungee cord

‹Active Learning Guide

When we studied work and energy (in Chapter 6), we used energy principles to examine the first modern bungee jump by four members of the Oxford University Dangerous Sport Club. They each fell from the 76-m-high Clifton Suspension Bridge in Bristol, England while attached to a rubber bungee cord. Now let's consider a different use for a bungee cord. In a so-called inverse bungee jump, the stretched cord launches the person upward for the start of an exciting vibrational ride.

**EXAMPLE 19.12 Inverse bungee jump**

A bungee cord with a 49-N/m spring constant will be used for an inverse bungee jump by a 70-kg person. The person starts by hanging stationary at the end of a cord that stretches an unknown distance $A$ (the unstretched cord is 32 m long). The person hanging at the end of the cord is then pulled down an additional distance $A$ so the cord is now stretched a distance $2A$. Finally, the person is released and vibrates up and down over a distance $2A$. Determine the amplitude $A$ of the vibration, the maximum speed of the jumper during each vibration, and the period of one vibration.

**Sketch and translate** We start by sketching the situation (see the sketch at the right). The person vibrates around an equilibrium position, the place where he hangs at the end of the cord when not vibrating. The gravitational force that Earth exerts on the person is constant during the entire vibration. However, the force that the cord exerts on the person continually changes—increasing as the cord stretches downward. We will use two systems for analysis: for the energy analysis the jumper, the cord, and Earth are the system. For the force analysis, the jumper will be the system.

**Simplify and diagram** Assume that the cord stretches linearly (that is, it does not exceed its elastic limit). Also, ignore air resistance and the mass of the bungee cord. The $y$-axis points upward with the origin at the equilibrium position where the jumper hangs at rest. Force

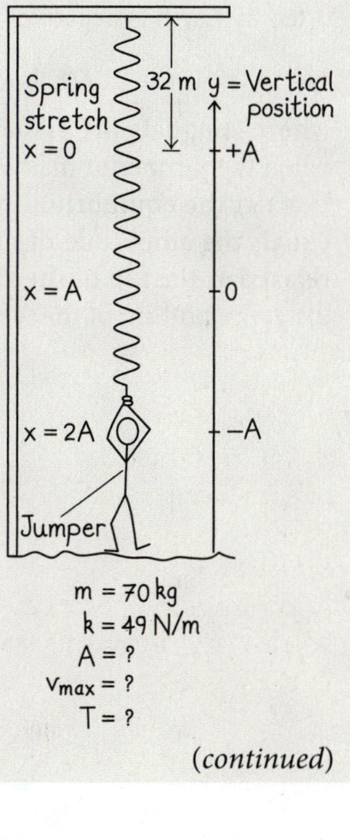


*(continued)*

diagrams for the jumper as a system when at the top of the vibration, while passing through equilibrium, and at the bottom of the vibration are shown at the right. Bar charts represent the energy of the jumper-cord-Earth system when the person is at these three positions. Assume that the system is isolated and thus its energy is constant. When the person is at the highest point in the vibration (a distance $A$ above the equilibrium position), the cord is not stretched and the system has only gravitational potential energy. When the person is passing through equilibrium, he has maximum speed and kinetic energy. Also, the system has elastic potential energy as the cord is now stretched a distance $A$. At the bottom of the vibration, the system has considerable elastic potential energy as the cord is stretched a distance $2A$; there is also negative gravitational potential energy in the system as the person is below the origin of the vertical y-axis.

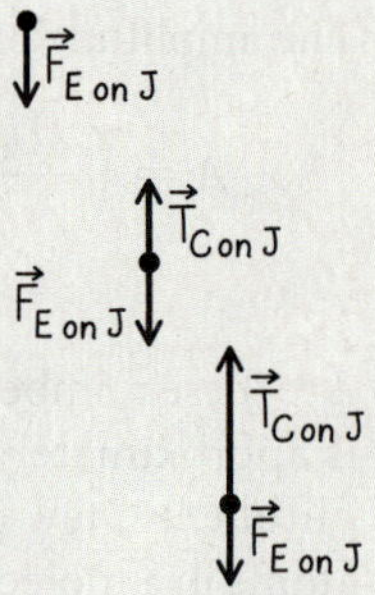


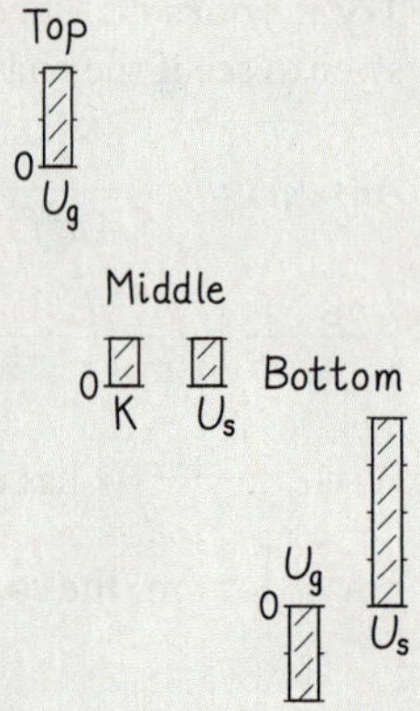


**Represent mathematically** We apply the vertical component form of Newton's second law to the jumper to determine how far the cord is stretched when hanging at the equilibrium position (see the second force diagram in the figure above). As the positive $y$-direction is up, the component of the elastic force exerted on the jumper is positive and the component of the gravitational force exerted by Earth is negative:

$$kA + (-mg) = 0$$

where $A$ equals the distance that the cord is stretched when the jumper of mass $m$ hangs at rest (zero acceleration) at the equilibrium position. This distance $A$ also equals the amplitude of the vibration—the cord will be relaxed at the top of the vibration and will be stretched $2A$ at the bottom of the vibration.

At the top of the vibration, the system's energy will all be gravitational potential energy $+mgA$ and at the bottom it will be elastic and gravitational $\frac{1}{2}k(2A)^2 - mgA$. These energies are equal, and they equal the maximum kinetic energy plus elastic energy when the jumper passes through the equilibrium position $\frac{1}{2}mv_{max}{}^2 + \frac{1}{2}kA^2$. Thus,

$$+mgA = \frac{1}{2}mv_{max}{}^2 + \frac{1}{2}kA^2 = \frac{1}{2}k(2A)^2 - mgA$$

The period of the vibration will be

$$T = 2\pi\sqrt{\frac{m}{k}}$$

**Solve and evaluate** The cord will be stretched $A$ when the jumper is at equilibrium:

$$A = \frac{mg}{k} = \frac{(70\text{ kg})(9.8\text{ N/kg})}{(49\text{ N/m})} = 14\text{ m}$$

We can determine the jumper's maximum speed using the left equality in the energy equation above:

$$v_{max} = \left(\frac{2mgA - kA^2}{m}\right)^{1/2}$$

$$= \left(\frac{2(70\text{ kg})(9.8\text{ m/s}^2)(14\text{ m}) - (49\text{ N/m})(14\text{ m})^2}{70\text{ kg}}\right)^{1/2}$$

$$= 11.7\text{ m/s} \approx 12\text{ m/s or 26 mph}$$

The period for one down-up-down vibration will be

$$T = 2\pi\left(\frac{m}{k}\right)^{1/2} = 2\pi\left(\frac{70\text{ kg}}{49\text{ N/m}}\right)^{1/2} = 7.5\text{ s}$$

All of the numbers above have the correct units and reasonable magnitudes.

**Try it yourself:** Suppose the bungee cord used for the inverse bungee jump had a spring constant of 40 N/m. Determine the amplitude of the ride, the maximum speed during a vibration, and the time interval for one vibration.

*Answer:* 17 m; 13 m/s; 8.3 s.

**Review Question 19.10** How are increased $CO_2$ in the atmosphere, vibrational motion, and global climate change related?

# Summary

| Words | Pictorial and physical representations | Mathematical representation |
|---|---|---|
| **Vibrational motion** is the repetitive movement of an object back and forth about an **equilibrium position.** This vibration is due to the **restoring force** exerted by another object that tends to return the first object to its equilibrium position. An object's maximum displacement from equilibrium is the **amplitude** $A$ of the vibration. (Sections 19.1 and 19.6) | 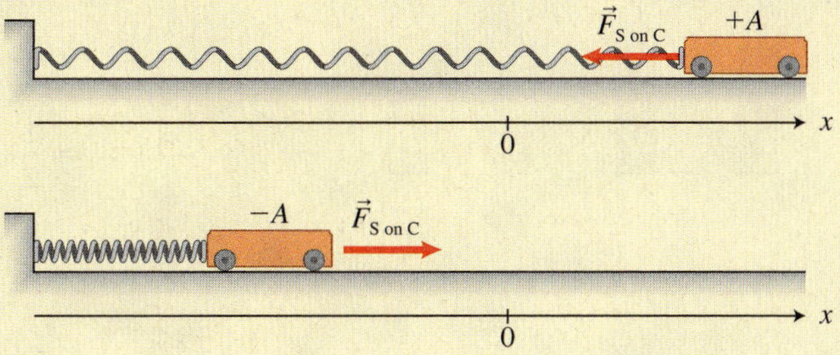 | Object at end of spring: $F_{\text{Restoring}\,x} = -kx$ Eq. (19.5) Pendulum: $F_{\text{Restoring}\,x} = -\left(\frac{mg}{L}\right)x$ Eq. (19.10) |
| **Period** $T$ is the time interval for one complete vibration, and **frequency** $f$ is the number of complete vibrations per second (in hertz). The frequency is the reciprocal of the period. (Section 19.2) | 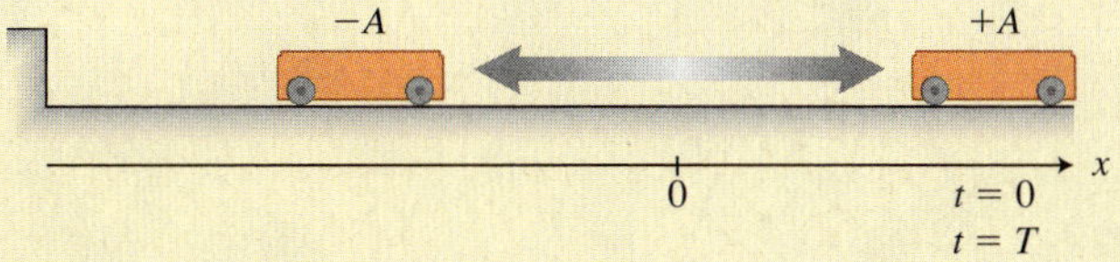 | $f = \frac{1}{T}$ Eq. (19.1) |
| Kinematics equations describe the changing position $x$, velocity $v_x$, and acceleration $a_x$ of a vibrating object as a function of time. (Section 19.3) | 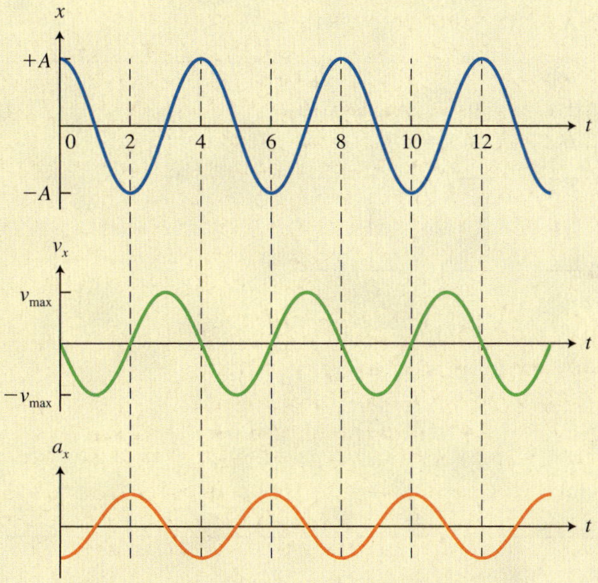 | $x = A\cos\left(\frac{2\pi}{T}t\right)$ Eq. (19.2) $v_x = -\left(\frac{2\pi}{T}\right)A\sin\left(\frac{2\pi}{T}t\right)$ Eq. (19.3) $a_x = -\left(\frac{2\pi}{T}\right)^2 A\cos\left(\frac{2\pi}{T}t\right)$ Eq. (19.4) |
| **Period** $T$ and **frequency** $f$ for an object vibrating at the end of a spring depend on the mass $m$ of the object and on the spring constant $k$ of the spring. (Section 19.4) | 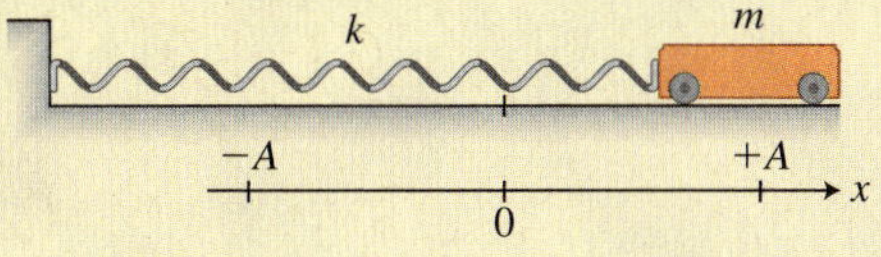 | $T = 2\pi\sqrt{\frac{m}{k}}$ Eq. (19.7) |

(*continued*)

| Words | Pictorial and physical representations | Mathematical representation |
|---|---|---|
| The energy of a spring-object system for an object vibrating horizontally at the end of a spring changes continuously from elastic potential energy when at the extreme positions to maximum kinetic energy when passing through the equilibrium position to a combination of energy types at other positions. (Section 19.5) | 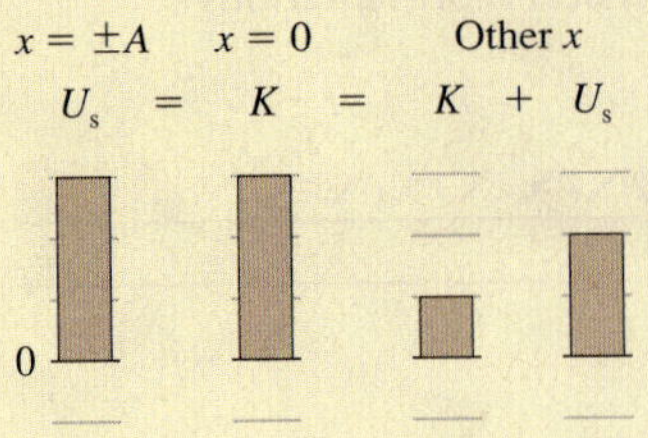  | $U = \frac{1}{2}kx^2 + \frac{1}{2}mv^2$ <br> $= \frac{1}{2}kA^2$ <br> $= \frac{1}{2}mv_{max}{}^2$ Eq. (19.9) |
| Period $T$ of a simple pendulum depends on the length $L$ of the string and the gravitational constant $g$. (Section 19.6) | 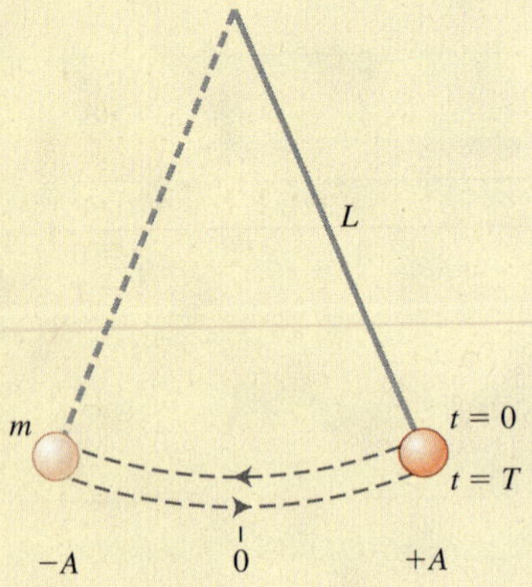  | $T = 2\pi\sqrt{\frac{L}{g}} = \frac{1}{f}$ Eq. (19.11) |
| Energy of a pendulum-Earth system converts continuously from gravitational potential energy when it is at the maximum height of a swing to kinetic energy when it is passing through the lowest point in the swing to a combination of energy types at other positions. (Section 19.6) | 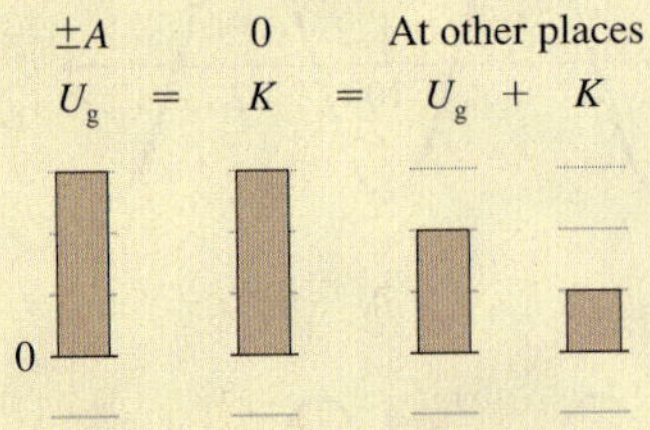  | $U = mgy + \frac{1}{2}mv^2$ <br> $= mgy_{max}$ <br> $= \frac{1}{2}mv_{max}{}^2$ Eq. (19.9) |
| **Resonant energy transfer** occurs if: <br> **1.** Two objects have the same or very similar natural frequencies. <br> **2.** Positive work is done by one on the other. (Section 19.9) | | |

# Questions

## Multiple Choice Questions

1. What are the features that make vibrational motion *different* from circular motion? Choose all that apply.
   (a) Vibrational motion is periodic.
   (b) Vibrational motion repeats itself.
   (c) During vibrational motion, there is a periodic change in the form of system energy.
   (d) Vibrational motion has a specific equilibrium point through which the system passes from different directions.
2. What does it mean if the amplitude of an object's vibration is constant and equal to 0.10 m?
   (a) The object's maximum distance from the starting point is 0.10 m.
   (b) The object's displacement from the equilibrium position is $\pm 0.10$ m.
   (c) The magnitude of the object's maximum displacement from the equilibrium position is 0.10 m.
3. What does it mean if the period of an object's vibration equals 0.60 s?
   (a) The motion lasts for 0.60 s.
   (b) The time interval needed for the object to move between the two farthest points from the equilibrium position is 0.60 s.
   (c) The time interval needed for the object to move through the same point in the same direction is 0.60 s.
4. What is the period of the kinetic or the potential energy change if the period of position change of an object attached to a spring is 2.0 s?
   (a) The energy does not change. (b) 4.0 s
   (c) 2.0 s (d) 1.0 s
5. A cart undergoing simple harmonic motion has a 2.0-Hz frequency and 3.0-cm amplitude. At time zero, the cart is at $x = +A$. Which of the three position-versus-time graphs shown in **Figure Q19.5** best represents the motion?

**Figure Q19.5**

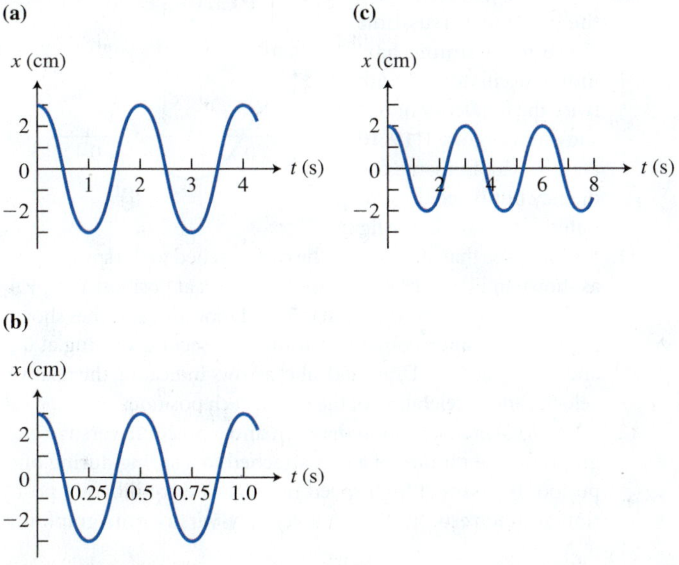


6. You have a simple harmonic oscillator. At what position(s) during simple harmonic motion is an oscillator's speed zero?
   (a) At the equilibrium position
   (b) At the maximum distances from equilibrium
   (c) At $t = 0$
   (d) Because the oscillator is constantly moving, there is no place where its speed is zero.
7. You have a simple harmonic oscillator. Where is its acceleration zero?
   (a) At the equilibrium position
   (b) At the maximum distances from equilibrium
   (c) At $t = 0$
   (d) Because the oscillator is constantly moving, there is no place where its acceleration is zero.
8. You have a simple harmonic oscillator. Where is the magnitude of its acceleration at its maximum?
   (a) At the equilibrium position
   (b) At the maximum distances from equilibrium
   (c) At $t = 0$
   (d) The acceleration is constant.
9. Why do you believe that an object attached to a light spring undergoes simple harmonic motion after it is displaced from the equilibrium position?
   (a) Because the position-versus-time graph is a sinusoidal-type function
   (b) Because the acceleration of the object is proportional to its displacement with a negative sign
   (c) Because the motion is periodic and has a constant period
   (d) a and c are correct.

## Conceptual Questions

10. (a) Give three common examples of vibrational motion with damping. Explain the reasons for the damping. (b) Give a common example of a forced vibration. (c) Give a common example of resonance. Do not use any examples already used in this chapter.
11. An object of known mass hangs at the end of a spring, causing it to stretch a distance $\Delta L$. How can you determine the frequency of vibration of this object if you know only the mass and the stretching distance?
12. Describe two different ways to estimate the spring constant of a rubber band using only a 100-g mass, a stopwatch, and a meter stick. Describe the assumptions that you need to make in each estimate.
13. You have a small metal ball attached to a 1.0-m string. Explain two different methods that you can use with this apparatus to measure free-fall acceleration at this location.
14. A pendulum clock is running too fast. Explain how you can fix the problem.
15. What simplifications were used to derive the formula for the period of vibration of an object at the end of a spring and of a simple pendulum?
16. A pendulum clock is moved from the Mississippi Delta region to the Rocky Mountains. Will the clock run faster or slower in the Rockies? Why? How can it be adjusted to compensate for this change? Describe your assumptions.

17. Oil is often found in a geological structure called a salt dome. The rock in the salt dome is less dense than surrounding rock and is pushed up so that it "floats" on surrounding rock. Explain how a sensitive pendulum can detect a salt dome.
18. By experiment and by calculation, estimate the natural period and frequency of vibration of your arm. Carefully explain the values of the numbers used used in each estimate.
19. A pendulum and a block hanging at the end of a spring are both carefully adjusted to make one vibration each second when on Earth's surface. Will the period of vibration of the pendulum or of the block on the spring be affected if they are placed on the Moon? Explain.
20. Will the frequency of vibration of a swing be greater or less when you stand on the swing than when you sit? Explain.
21. Estimate the time interval required for a Chihuahua to take a single step. Estimate the time interval required for a giraffe to take a single step.
22. The amplitude of vibration of a swing slowly decreases to zero if you do not pump your legs while swinging. Explain what happens to the energy of the system.
23. If you walk with your arms hanging down, they often oscillate forward and back at the same frequency as your stepping frequency. However, if your arms are bent at the elbow so that the forearms are horizontal, they often do not oscillate. Why?
24. You have a pendulum with a 1-m string. What is the period of vibration of the pendulum when it is placed on a space ship orbiting Earth? Explain.

# Problems

Below, **BIO** indicates a problem with a biological or medical focus. Problems labeled **EST** ask you to estimate the answer to a quantitative problem rather than develop a specific answer. Problems marked with require you to make a drawing or graph as part of your solution. Asterisks indicate the level of difficulty of the problem. Problems with no * are considered to be the least difficult. A single * marks moderately difficult problems. Two ** indicate more difficult problems.

## 19.1 Observations of vibrational motion

1. * You have a pendulum pulled to the side, a heavy beach ball pushed partly under water, and a medicine ball lifted some height above the ground. Each object is then released. (a) Draw a picture of the motion of each object and a motion diagram for each. (b) Indicate the equilibrium position. (c) Draw two or more force diagrams at key points in the motion. (d) Use (a) through (c) to reason whether the motions of these objects can be considered vibrational motions.
2. * Consider the three objects described in Problem 1. (a) Choose a system for each and construct energy bar charts for each motion. What initial and final states did you choose? (b) What assumptions did you make?
3. **Exercise stretch cord** You want to determine the spring constant of an exercise stretch cord. You pull the cord with a force probe that exerts a 50-N force on the cord, causing it to stretch 20 cm. (a) What is the spring constant of the cord? Describe your reasoning and assumptions. (b) How can you test your answer?
4. * **Two exercise cords** In order to increase resistance, you put two exercise cords together as shown in **Figure P19.4**. What is the effective spring constant of the two bands compared to one? Use data from Problem 3. Explain.

**Figure P19.4**

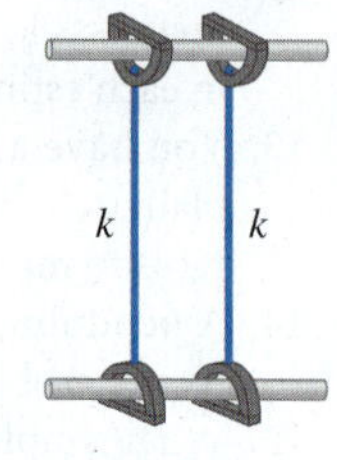


5. * You have a ball bearing and a bowl. You let the ball roll down from the top of the bowl; it moves up and down the bowl's wall for some time. (a) Is this vibrational motion? Explain why or why not. (b) Draw a force diagram for the ball for four different positions and indicate what force or force component is responsible for the acceleration toward the equilibrium position. (c) Draw energy bar charts for these four ball-Earth system positions.

## 19.2 Period and frequency

6. ** (a) Draw a sketch of an object attached to a vertical spring. Indicate the equilibrium position. (b) Show this object vibrating up and down and indicate the amplitude of vibrations on the sketch. (c) Describe an experiment you would perform to determine the period of vibration of the object. (d) Perform the experiment. Include experimental uncertainty in your result.
7. Draw a sketch of a pendulum. Indicate the equilibrium position. Show this object vibrating and indicate the amplitude of vibration on the sketch.
8. **Concert A and $O_2$ vibration** (a) Musicians in an orchestra tune their instruments to what is called "concert A," a frequency of 440 Hz. Determine the period for one vibration. (b) The atoms in an oxygen molecule complete one vibration in a time interval of $2.11 \times 10^{-14}$ s. Determine the frequency of vibration of $O_2$.
9. **BIO** **Hearing range** A doctor is checking your hearing. (a) If the period of the lowest-frequency sound you can hear is 0.050 s, then what is its frequency? (b) If the highest-frequency sound you can hear is 20,000 Hz, then what is its period?

## 19.3 Kinematics of vibrational motion

10. Draw a graph showing the position-versus-time curve for a simple harmonic oscillator (a) with twice the frequency of that shown in **Figure P19.10** and (b) with the same frequency but twice the amplitude as shown in the figure.

**Figure P19.10**

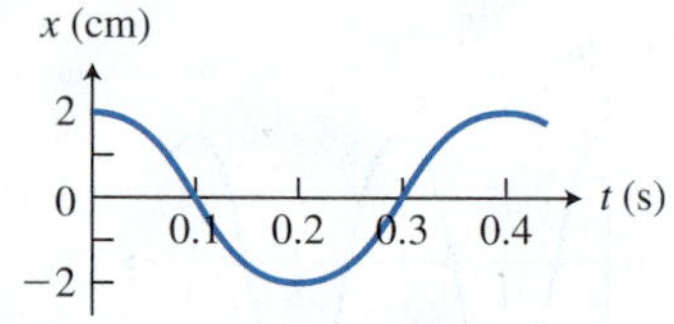


11. * Suppose that at time zero the cart attached to the spring such as shown in Figure 19.8 is released from rest at position $x = +A$ and that its period of vibration is 8.0 s. Draw nine sketches showing the cart's approximate position each second starting at 0 s and ending at 8.0 s. Draw and label arrows indicating the relative velocity and acceleration of the cart at each position.
12. ** (a) Sketch a motion diagram and a position-versus-time graph for the motion of a cart attached to a spring during one period. It passes at high speed through the equilibrium position at time zero. (b) Sketch a velocity-versus-time graph for

the cart. (c) Sketch an acceleration-versus-time graph for the cart. (d) Draw another representation of the process; explain how this representation allows you to learn more about the process than the kinematics graphs do.

13. ** The motion of a cart is described as $x = (0.17\text{ m})\sin(\pi s^{-1})t$. Say everything you can about this motion and represent it using graphs, motion diagrams, and energy bar charts.

14. ** Sketch an acceleration-versus-time graph for the simple harmonic motion of an object of your choice. Indicate the amplitude of the acceleration and the period of the vibration. Underneath the graph draw corresponding position-versus-time and velocity-versus-time graphs with the correct shape but not necessarily the correct amplitudes.

15. ** Devise a position-versus-time function that describes the simple harmonic motion of an object of your choice. Choose the amplitude and the period of vibration, and draw corresponding velocity-versus-time and acceleration-versus time graphs. Draw a motion diagram for one period of its motion.

16. * The position of a vibrating object changes as a function of time as $x = (0.2\text{ m})\cos(\pi s^{-1})t$. Say everything you can about this motion. Write an expression for the velocity and acceleration as functions of time. Draw graphs for each function.

17. * The velocity of a vibrating object changes as a function of time as $v = -(0.6\text{ m/s})\cos(2\pi)$. (a) Say everything you can about the motion. Draw a sketch of the situation when the motion starts. (b) Represent the motion with a position-versus-time graph and a work-energy bar chart.

18. * A cart at the end of a spring undergoes simple harmonic motion of amplitude 10 cm and frequency 5.0 Hz. (a) Determine the period of vibration. (b) Write an expression for the cart's position at different times, assuming it is at $x = -A$ when $t = 0$. Determine its position (c) at 0.050 s and (d) at 0.100 s.

## 19.4 The dynamics of simple harmonic motion

19. You exert a 100-N pull on the end of a spring. When you increase the force by 20% to 120 N, the spring's length increases 5.0 cm beyond its original stretched position. What is the spring constant of the spring and its original displacement?

20. **Metronome** You want to make a metronome for music practice. You use a 30-g object attached to a spring to serve as the time standard. What is the desired spring constant of the spring if the object needs to make 1.00 vibrations each second? What are the assumptions that you made?

21. Determine the frequency of vibration of the cart shown in **Figure P19.21**.

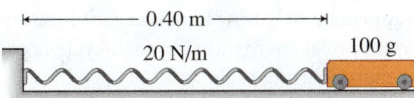

**Figure P19.21**

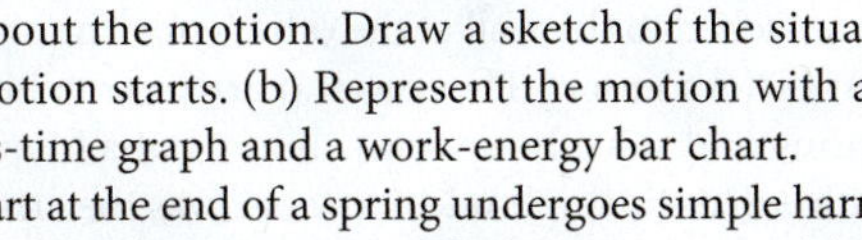


22. * A spring with a cart at its end vibrates at frequency 6.0 Hz. (a) Determine the period of vibration. (b) Determine the frequency if the cart's mass is doubled while the spring constant remains unchanged and (c) the frequency if the spring constant doubles while the cart's mass remains the same.

23. A cart with mass $m$ vibrating at the end of a spring has an extra block added to it when its displacement is $x = +A$. What should the block's mass be in order to reduce the frequency to half its initial value?

24. ** A 2.0-kg cart vibrates at the end of an 18-N/m spring. (a) Make a list of physical quantities you can determine about the vibrations and determine two of them. (b) If a second 18-N/m spring is attached beside the first one, what will be the period of the vibration?

25. * What were the main ideas that we used to derive the expression for the period of an object vibrating at the end of a spring? Explain the assumptions that we made. When is the expression for the period not valid for a spring of known $k$?

## 19.5 Energy of vibrational systems

26. * A spring with a spring constant of 1200 N/m has a 55-g ball at its end. (a) If the energy of the system is 6.0 J, what is the amplitude of vibration? (b) What is the maximum speed of the ball? (c) What is the speed when the ball is at a position $x = +A/2$? What assumptions did you make to solve the problem? How do you know if your answer makes sense?

27. ** A person exerts a 15-N force on a cart attached to a spring and holds the cart steady. The cart is displaced 0.060 m from its equilibrium position. When the person stops holding the cart, the system cart+spring undergoes simple harmonic motion. (a) Determine the spring constant of the spring. (b) Determine the energy of the system. (c) Write an expression $x(t)$ for the motion of the cart. (d) Draw as many graphical representations of the motion as you can.

28. * A spring with spring constant $2.5 \times 10^4$ N/m has a 1.4-kg cart at its end. (a) If its amplitude of vibration is 0.030 m, what is the total energy of the cart+spring system? (b) What is the maximum speed of the cart? (c) If the energy is tripled, what is the new amplitude? (d) What is the maximum speed of the cart? (e) What assumptions did you make to solve the problem? If the assumptions were not reasonable, how would the answers change?

29. * **Proportional reasoning** By what factor must we increase the amplitude of vibration of an object at the end of a spring in order to double its maximum speed during a vibration? Explain.

30. * **Proportional reasoning** By what factor must we increase the amplitude of vibration of an object at the end of a spring in order to double the total energy of the system? Explain. What will happen to the speed the object? What assumptions did you make?

31. * **Monkey trick at zoo** A monkey has a cart with a horizontal spring attached to it that she uses for different tricks. In one trick, the monkey sits on the vibrating cart. When the cart reaches its maximum displacement from equilibrium, the monkey picks up a 0.30-kg cantaloupe from a trainer. The mass of the monkey and the cart together is 3.0 kg. The spring constant is 660 N/m. The amplitude of horizontal vibrations is 0.24 m. Determine the ratio of the maximum speed of the monkey before and after she picks up the cantaloupe.

32. * A cart attached to a spring vibrates with amplitude $A$. (a) What fraction of the total energy of the cart-spring system is elastic potential energy and what fraction is kinetic energy when the cart is at position $x = A/2$? (b) At what position is the cart when its kinetic energy equals its elastic potential energy?

33. A 2.0-kg cart attached to a spring undergoes simple harmonic motion so that its displacement is described by $x = (0.20\text{ m})\sin[(2\pi/2.0\text{ s})t]$. Construct energy bar charts for the cart-spring system at times $t = 0$, $t = T/4$, $t = T/2$, $t = 3T/4$, and $t = T$.

34. * **Equation Jeopardy** The following expression describes a situation involving vibrational motion. Sketch a process and devise a problem for which the expression might be an answer.

$$\frac{1}{2}(20{,}000\ \text{N/m})(0.20\ \text{m})^2 = \frac{1}{2}(100\ \text{kg})v^2 + \frac{1}{2}(20{,}000\ \text{N/m})(0.10\ \text{m})^2$$

## 19.6 The simple pendulum

35. **Pendulum clock** Shawn wants to build a clock whose pendulum makes one swing back and forth each second. (a) What is the desired length of the rod (assumed to have negligible mass) holding the metal ball at its end? (b) Will the rod need to be shorter or longer if he includes the mass of the rod in the calculations? Explain.
36. Show that the expression for the frequency of a pendulum as a function of its length is dimensionally correct.
37. A pendulum swings with amplitude 0.020 m and period of 2.0 s. What is its maximum speed?
38. * **Proportional reasoning** You are designing a pendulum clock whose period can be adjusted by 10% by changing the length of the pendulum. By what percent must you be able to change the length to provide this flexibility in the period? Explain.
39. **Building demolition** A 500-kg ball at the end of a 30-m cable suspended from a crane is used to demolish an old building. If the ball has an initial angular displacement of 15° from the vertical, determine its speed at the bottom of the arc.
40. * You have a pendulum with a long string whose length you can vary but cannot measure, a small ball, and a stopwatch. Describe two experiments that you can design to determine the height of a ladder. Indicate the assumptions that you use in each method.
41. * **Variations in *g*** The frequency of a person's pendulum is 0.3204 Hz when at a location where $g$ is known to be exactly $9.800\ \text{m/s}^2$. Where might the same pendulum be when its frequency is 0.3196 Hz? What is $g$ at that location?
42. EST A graph of position versus time for an object undergoing simple harmonic motion is shown in **Figure P19.42**. Estimate from the graph the amplitude and period of the motion and determine the object's frequency. If the object is a pendulum, what is its length?

**Figure P19.42**

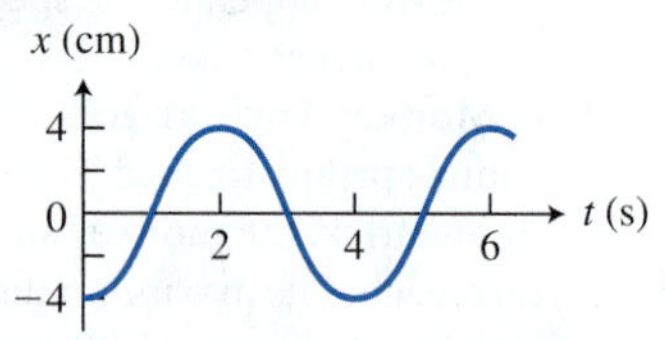


43. Determine the period of a 1.3-m-long pendulum on the Moon.

## 19.7 Skills for analyzing processes involving vibrational motion

44. * **Trampoline vibration** When a 60-kg boy sits at rest on a trampoline, it sags 0.10 m at the center. (a) What is the effective spring constant for the trampoline? (b) The trampoline is pulled downward an extra 0.050 m by a strap sewed under the center of the trampoline. When the strap is released, what are the energy and frequency of the trampoline? What assumptions did you make?
45. ** A 1.2-kg block sliding at 6.0 m/s on a frictionless surface runs into and sticks to a spring. The spring is compressed 0.10 m before stopping the block and starting its motion back in the opposite direction. What can you determine about the vibrations that start after the collision? Make a list of physical quantities and determine four of them.
46. * **Proportional reasoning** If you double the amplitude of vibration of an object at the end of a spring, how does this affect the values of $k$, $T$, $U$, and $v_{max}$?
47. ** EST **Willis Tower vibration** The mass of the Willis (formerly Sears) Tower in Chicago is about $2 \times 10^8$ kg. The tower sways back and forth at a frequency of about 0.10 Hz. (a) Estimate the effective spring constant for this swaying motion. Explain why you included a particular number of significant figures. (b) A gust of wind hitting the building exerts a force of about $4 \times 10^6$ N. By approximately how much is the top of the building displaced by the wind? Explain why you included a particular number of significant figures. State the assumptions that you made. Does your answer make sense?
48. * EST BIO **Annoying sound** Low-frequency vibrations (less than 5 Hz) are annoying to humans if the product of the amplitude and the frequency squared ($Af^2$) equals $0.5 \times 10^{-2}\ \text{m} \cdot \text{s}^{-2}$ (or more). Estimate the frequency and amplitude of a buzzer that produces these annoying vibrations if the device vibrates a total mass of 0.12 kg and has a vibrational energy of 0.012 J.
49. ** You shoot a 0.050-kg arrow into a 0.50-kg wooden cart that sits on a horizontal, frictionless surface at the end of a spring that is attached to the wall at the other end. The arrow hits the cart and sticks into it. The cart and arrow together compress the spring and start the system vibrating at a frequency of 2.0 Hz with a 0.20-m amplitude. How fast was the arrow moving? State any assumptions you made to solve the problem.
50. * **Pendulum on Mars** The frequency of a pendulum is 39% less when on Mars than when on Earth's surface. Use this fact to determine Mars's gravitational constant.

## 19.8 Including friction in vibrational motion

51. * This chapter stated that when damping is present, the period of vibration of a system is more than without damping. How can you test this assertion? Provide the details.
52. * You have a pendulum whose length is 1.3 m and bob mass is 0.20 kg. The amplitude of vibration of the pendulum is 0.07 m. (a) Determine the maximum energy of the bob-Earth system. Explain why you included a particular number of significant figures. (b) How much mechanical energy is converted to internal energy before the pendulum stops?
53. * You have a 0.10-kg cart on a spring. The spring constant of the spring is 20 N/m. The cart's initial vibration amplitude is 0.10 m. (a) Make a list of physical quantities you can determine using this information and determine three of them. (b) For approximately how long will the vibrations last if 10% of the mechanical energy during each cycle is converted into internal energy during each cycle?
54. * You have a spring that stretches 0.070 m when a 0.10-kg block is attached to and hangs from it. Imagine that you slowly pull down with a spring scale so the block is now below the equilibrium position where it was hanging at rest. The scale reading when you let go of the block is 3.0 N. (a) Where was the block when you let go? (b) Determine the work you did stretching the spring. (c) What was the energy of the spring-Earth system when you let go? (d) How far will the block rise after you release it? (e) The vibrations last for 50 cycles. Qualitatively represent the beginning of cycle #1 and beginning of cycle # 25 with an energy bar chart.

55. * Imagine that you have a cart on a spring that moves on a rough surface. (a) Represent the cart's motion with a motion diagram for one period. (b) Draw force diagrams for each quarter of a period. (c) Draw energy bar charts for each quarter of a period. (d) Draw position-, velocity-, and acceleration-versus-time graphs for each period. (e) On these graphs, use a dashed line to sketch graphs for the same cart on the same spring, assuming no friction between the cart and the surface.

### 19.9 Vibrational motion with an external driving force

56. EST **Twins on a swing** How frequently do you need to push a swing with twin brothers on it compared to when pushing the swing with one on it? What assumptions did you make?

57. * (a) Determine the maximum speed of a girl on a 3.0-m-long swing when the amplitude of vibration is 1.2 m. (b) What assumptions did you make? Is the child's vibrational motion damped? Explain. (c) Under what conditions is the motion a forced vibration? Explain.

58. You have a 0.20-kg block on a 10-N/m spring that oscillates up and down. How often do you need to push the block upward when it passes the bottom of its motion to increase the amplitude of its vibrations?

59. **Sloshing water** You carry a bucket with water that has a natural sloshing period of 1.7 s. At what walking speed will water splash if your step is 0.90 m long?

60. * **Feeling road vibrations in a car** If the average distance between bumps on a road is about 10 m and the natural frequency of the suspension system in the car is about 0.90 Hz, at what speed will you feel the bumps the most?

### 19.10 Vibrational motion in everyday life: Putting it all together

61. * EST **H atom vibration** A hydrogen atom of mass $1.67 \times 10^{-27}$ kg is attached to a very large protein by a bond that behaves much like a spring. (a) If the vibrational frequency of the hydrogen is $1.0 \times 10^{14}$ Hz, what is the "effective" spring constant of this spring-like bond? (b) If the total vibrational energy is $kT$ ($k$ is Boltzmann's constant and $T$ is the temperature in kelvins), approximately what is the classical amplitude of vibration at room temperature ($T = 300$ K)? By comparison, the diameter of a hydrogen atom is about $10^{-10}$ m. State any assumptions that you made.

62. **Child's bouncy chair** You are designing a bouncy chair for a neighbor's child. A 1.0-kg chair hangs from a spring that is attached to the ceiling. When you put the 10.0-kg child in the chair, the system vibrates with a period of 3.0 s. What would be the period if the child's mother (mass 50 kg) sits on the same chair? What assumptions did you make?

63. * You attach a block (mass $m$) to a spring (spring constant $k$) that oscillates in a vertical direction. Determine an expression for the period of its vibrations. Sketch a position-versus-time graph for the vibrations, assuming that the motion starts when the block is at its lowest maximum displacement from the equilibrium position. What assumptions did you make?

64. * You attach a 1.6-kg object to a spring, pull it down 0.12 m from the equilibrium position, and let it vibrate. You find that it takes 5.0 s for the object to complete 10 vibrations. Make a list of physical quantities that you can determine about the motion of the object and determine five of them.

## General Problems

65. * **Traveling through Earth** A hole is drilled through the center of Earth. The gravitational force exerted by Earth on an object of mass $m$ as it goes through the hole is $mg(r/R)$, where $r$ is the distance of the object from Earth's center and $R$ is the radius of Earth ($6.4 \times 10^6$ m). Will an object dropped into the hole execute simple harmonic motion? If yes, find the period of the motion. How does one-half this time compare with the time needed to fly in an airplane halfway around Earth? What assumptions did you make to solve the problem?

66. * EST Estimate the effective spring constant of the suspension system of a car. Describe your technique carefully. How can you test your answer to determine if it makes sense?

67. * Use dimensional analysis to show that the expressions for the periods of an object attached to a spring and for a simple pendulum are reasonable.

68. * BIO **Vibration amplitude in ear** The weakest sound a human ear can possibly hear makes the ear vibrate with the energy of about $10^{-19}$ J. If the spring constant of the ear is 20 N/m, what is the amplitude of vibration of the eardrum? What assumptions did you make? How does this compare to the size of an atom (about $0.5 \times 10^{-10}$ m)?

69. ** A 5.0-g bullet traveling horizontally at an unknown speed hits and embeds itself in a 0.195-kg block resting on a frictionless table. The block slides into and compresses a 180-N/m spring a distance of 0.10 m before stopping the block and bullet. Determine the initial speed of the bullet. State any assumptions that you made to solve the problem.

70. ** You have a pendulum of mass $m$ and length $L$ that is displaced an angle $\theta$ at the start of the swinging. (a) Determine an expression for the energy of the bob-Earth system. (b) Determine an expression for the maximum vertical height of the bob with respect to the equilibrium position. (c) Determine an expression for the maximum speed of the bob. (d) What assumptions did you make? (e) Discuss how the answer to each of the questions changes if relevant assumptions are not valid.

71. ** A pendulum clock works for many days before the swinging stops. Describe several mechanisms that allow an actual pendulum clock to continue working without decreasing amplitude. Write a report explaining how it works.

72. ** **Foucault's pendulum** In 1851, the French physicist Jean Foucault hung a large iron ball on a wire about 67 m (220 ft) long to show that Earth rotates. The pendulum appears to continuously change the plane in which it swings as time elapses. Determine the swinging frequency of this pendulum. Explain why the behavior of Foucault's pendulum provides supporting evidence for the hypothesis that Earth is a noninertial reference frame and helps reject the model of a geocentric universe (Earth at the center).

73. ** You push down on a raft floating on a lake. The raft sinks and then vibrates up and down for a short time. How does the period of its vibrations depend on the size and mass of the raft?

## Reading Passage Problems

BIO **Transportation to the other side of Earth** It has been suggested that an efficient way to travel from the United States to Australia would be to dig one or more holes through Earth and fall through one of the holes in a capsule. The gravitational force that Earth exerts on you and on the capsule when the capsule is at a distance $r$ ($r < R$ of Earth) equals that of the mass $M$ inside a sphere of radius $r$ with its center at Earth's center. At Earth's center, the force is zero. At some other distance $r$ from the center of Earth, the force would be

$$F_{\text{E on Y}} = \frac{GM_{\text{inside } r} m_{\text{Y}}}{r^2}$$

where $M_{\text{inside } r}$ equals the volume of a sphere of radius $r$ times the density of Earth. Substituting for $M_{\text{inside } r}$, you find that the force that Earth exerts on you and on the capsule is proportional to $r$ and provides a restoring force. You start at rest at Earth's surface and accelerate as you get closer to the center. At the center you are moving at maximum speed and Earth exerts no force on you. After you pass the center, Earth starts pulling back on you, causing your speed to decrease. With no friction or air resistance, you should stop on the other side of Earth. If you can find an expression for the restoring force that Earth exerts on you, you can substitute that in place of $k$ in the expression for period $T = 2\pi(m/k)^{1/2}$. The travel time will be half the period.

74. Which expression below represents the mass $m$ of Earth inside a sphere of radius $r$ smaller than the radius $R$ of Earth? Note that $\rho$ is the density of Earth, assumed uniform.
(a) $(4/3)\pi r^3 \rho$ (b) $G(4/3)\pi r^3 \rho$
(c) $GM_{\text{Earth}}/r^2$ (d) $G(4/3)\pi r\rho$
(e) $mg$
75. Which expression below represents the magnitude of the restoring force that Earth exerts on an object of mass $m$ when a distance $r$ from the center of Earth? Note that $\rho$ is the density of Earth, assumed uniform.
(a) $m(4/3)\pi r^3 \rho$ (b) $mG(4/3)\pi r^3 \rho$
(c) $GM_{\text{E}}m/r^2$ (d) $mG(4/3)\pi r\rho$
(e) $mgr$
76. Which expression below represents the period $T$ for one oscillation from Earth's surface through the center, to the other side, and then back again?
(a) $2\pi(3/G4\pi\rho)^{1/2}$ (b) $2\pi(G4\pi\rho/3)^{1/2}$
(c) $2\pi(3m/G4\pi\rho)^{1/2}$ (d) $2\pi(mG4\pi\rho/3)^{1/2}$
(e) $2\pi(m/k)^{1/2}$
77. During the trip, when will your acceleration be greatest?
(a) At the beginning and end of the trip
(b) At the end of the trip
(c) When passing through the center of Earth
(d) The same for entire the trip
78. During the trip, when will your speed be greatest?
(a) At the beginning and ends of the trip
(b) At the end of the trip
(c) When passing through the center of Earth
(d) The same for entire the trip
79. If the radius of Earth is $6.4 \times 10^6$ m, its mass is $6.0 \times 10^{24}$ kg, and $G = 6.67 \times 10^{-11}\,\text{N}\cdot\text{m}^2/\text{kg}^2$, which answer is closest to the time interval for one trip through Earth?
(a) 0.7 h (b) 1.4 h (c) 3.2 h
(d) 4.8 h (e) 6.3 h

**BIO Resonance vibration transfer and the ear** When you push a person on a swing, a series of small pushes timed to match the swinger's swinging frequency makes the person swing with larger amplitude. If timed differently, the pushing is ineffective. The board shown in **Figure 19.20** (from the Exploratorium in San Francisco) is made of rods of different length with identical balls on the ends of each rod. Each rod vibrates at a different natural frequency, the long rod on the left at lower frequency and the short rod on the right at higher frequency. If you shake the board at the high frequency at which the short rod vibrates, the short rod swings with large amplitude while the others swing a little. If you shake the board at the middle frequency at which the two center rods vibrate, the center rods undergo large-amplitude vibrations and the rods on each end do not vibrate. Imagine now that you have a fancy board with 15,000 rods, each of slightly different length, the shortest on the left and the longest on the right. Shaking the board at a particular frequency causes the rods in one small region of the board to vibrate at this frequency and has little effect on the others.

The inner ear (the cochlea) is a little like this fancy board. Sound reaching the tympanic membrane, or eardrum, is greatly amplified by three tiny bones in the middle ear—the hammer, anvil, and stirrup (**Figure 19.21**). These bones vibrate, pushing on the fluid in the inner ear and causing vibrations along its entire length. A basilar membrane with about 15,000 hair cells passes along the center of the inner ear. The basilar membrane is narrow and stiff near the entrance to the inner ear and wide and more flexible near the end. When a single-frequency vibration causes the fluid to vibrate, the membrane and the hair cells respond best at a single place—high frequencies near the oval window and low frequencies near the end of the basilar membrane. The bending of these hairs causes those

**Figure 19.20** Coupled oscillations of the shaking board and rods.

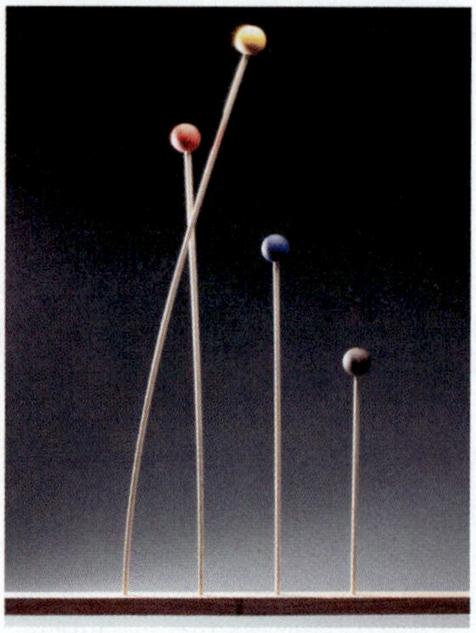

The short rod with a ball on the end responds to higher frequency shaking of the board, whereas the middle length rods respond to intermediate frequency shaking.

**Figure 19.21** Anatomy of the human ear.

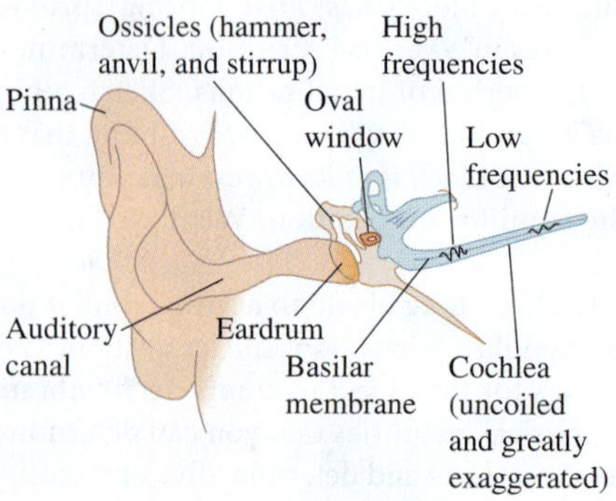

nerve cells to fire. Thus, we detect the frequency of the sound by the part of the membrane from which the nerve signal comes.

80. What causes vibrations in the fluid in the cochlea?
    (a) The eardrum pushes against the cochlea.
    (b) Sound waves push against the cochlea.
    (c) A bone in the middle ear pushes against the cochlea.
    (d) The fluid vibrates when something outside the ear vibrates.
    (e) The fluid does not vibrate—the hair cells vibrate.
81. How does the basilar membrane distinguish different frequency vibrations?
    (a) The hair cells have different lengths.
    (b) The dimensions and stiffness of the cochlea vary along its length.
    (c) The basilar membrane vibrates where the cochlea dimension and stiffness match the vibration frequency.
    (d) The cochlea fluid resonates in only one part of the cochlea.
    (e) All the hair cells bend back and forth at only the frequency of the vibration.
82. If you shake the board shown in Figure 19.20 at a frequency higher than the natural frequency of the rod on the right, then what happens?
    (a) None of the rods vibrate.
    (b) All of the rods vibrate.
    (c) The shortest rod vibrates.
    (d) The longest rod vibrates.
    (e) The middle rods vibrate.
83. If you were to shake the special board (the one that has 15,000 rods of varying length) at one particular frequency, then what would happen?
    (a) None of the rods would vibrate.
    (b) All of the rods would vibrate.
    (c) A small number of rods at one location would vibrate.
    (d) A disturbance would travel back and forth along the board.
84. You hang four pendulum bobs from strings connected to a wooden dowel. The strings are different lengths. How can you get the second longest pendulum bob to vibrate while the other three do not—without touching the pendulums?
    (a) Shake the dowel back and forth.
    (b) Shake the dowel back and forth at the resonant frequency of that pendulum.
    (c) Move the dowel sideways at any frequency.
    (d) Blow air on that bob.

# 20 Mechanical Waves

**How do bats "see" in the dark?**

**How can you transmit energy without transmitting matter?**

**Why does the pitch of a train whistle change as it passes you?**

**Be sure you know how to:**

- Describe simple harmonic motion verbally, graphically, and algebraically (Section 19.3).
- Determine the period and amplitude of vibrations using a graph (Section 19.3).
- Determine the power of a process (Section 6.8).

You have probably heard the expression "blind as a bat." It turns out that bats are not blind; many species have very sensitive eyes that can see even when there is very little light. However, the organs that bats use to avoid obstacles at night and find their prey are their ears. While flying, they emit sound that travels to the prey, reflects, and travels back to the bat. The bat's brain processes the delay time and automatically calculates the distance to the prey. Why don't we ever hear the sounds produced by bats?

**Previously (in Chapter 19),** we studied the vibrational motion of an object. This motion occurs when a system is disturbed from its equilibrium position, moves back toward equilibrium, and then overshoots. The cause of this process is a restoring force that continually returns the object toward its equilibrium position, where it overshoots again and the cycle repeats. So far we have looked only at the vibrational motion of one object, such as a pendulum. We have not considered what happens to the environment that is in contact with the vibrating object. In this chapter we focus on the effect that the vibrating object has on the medium that surrounds it. An example of such an effect is the propagation of sound.

# 20.1 Observations: Pulses and wave motion

Our everyday experience indicates that when an object vibrates, it also disturbs the medium surrounding it. We call the object that causes the vibration the **source.** For example, place a heavy metal cylinder attached to a spring so that the cylinder is partially submerged in a tub with water (**Figure 20.1**). Push down on the cylinder a little and release it. The vibrating cylinder (the source) sends ripples (waves) across the tub. In Observational Experiment **Table 20.1** we observe the simplest case of wave motion—when the source of waves vibrates only once, sending a short disturbance, or a **pulse**, into the surrounding medium.

**Figure 20.1** A vibration disturbs the surrounding medium.

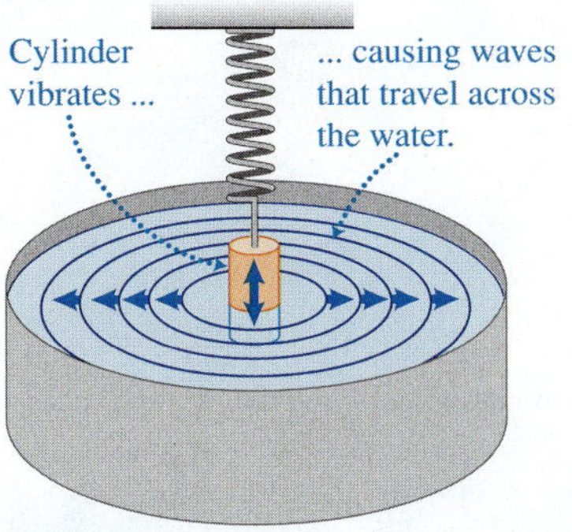

Cylinder vibrates ...
... causing waves that travel across the water.

**OBSERVATIONAL EXPERIMENT TABLE**

**20.1 Disturbances in different media.**

VIDEO 20.1

| Observational experiment | Analysis |
|---|---|
| **Experiment 1.** Tie a rope to a doorknob and then shake the end that you are holding one time, producing a quick disturbance—a pulse. The pulse travels along the rope horizontally. The part of the rope marked with a ribbon moves up and down when the pulse reaches it. 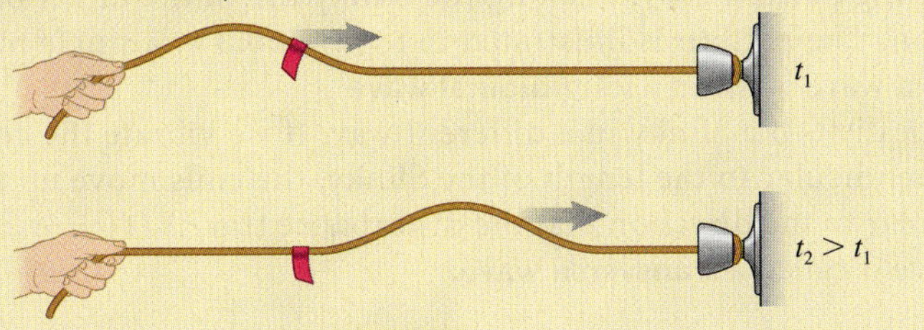 | The source of the pulse is your hand pulling the rope; the medium in which the pulse propagates is the rope itself. The pulse moves along the rope, but the rope fibers (see the little ribbon on the rope) return to their original position after the disturbance passes—they do not travel along the rope. It takes time for the pulse to travel to the other end of the rope. |
| **Experiment 2.** Place a beach ball and polystyrene foam pieces on the surface of a swimming pool. Push the ball up and down once. A pulse in the shape of a circular disturbance spreads outward. The foam pieces bob up and down but do not travel across the pool. 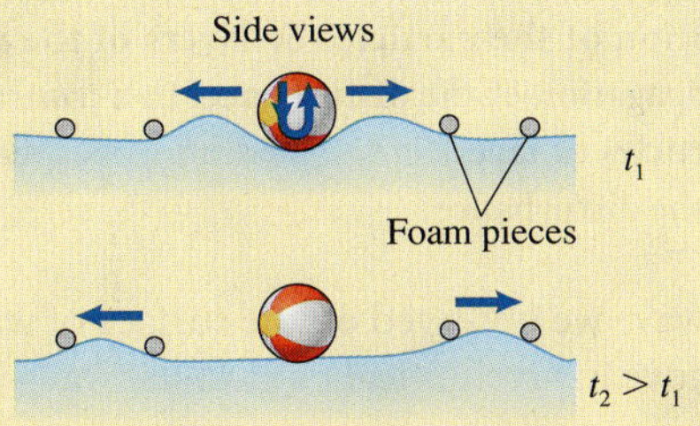 | The source of the pulse is the ball; the medium is the surface layer of water in the pool. The circular disturbance produced by the ball's up and down motion spreads; the water itself does not move across the pool, as the foam pieces on the water surface only move up and down. It takes time for the pulse to reach the far sides of the pool. |



**Patterns**

In the two experiments, we saw that

- a moving disturbance was created by a vibrating source—your hand or the ball.
- the disturbance moved through a material at an observable speed.
- the particles of the medium did not travel with the disturbance.

## Waves and wave fronts

In the beach ball experiment, when the vibrating ball (the source) moves down, the water (the medium) under the ball's bottom surface is pushed out to

‹Active Learning Guide

**Figure 20.2** (a) A wave front produced by a single beach ball vibration and (b) wave fronts produced by repetitive beach ball vibrations.

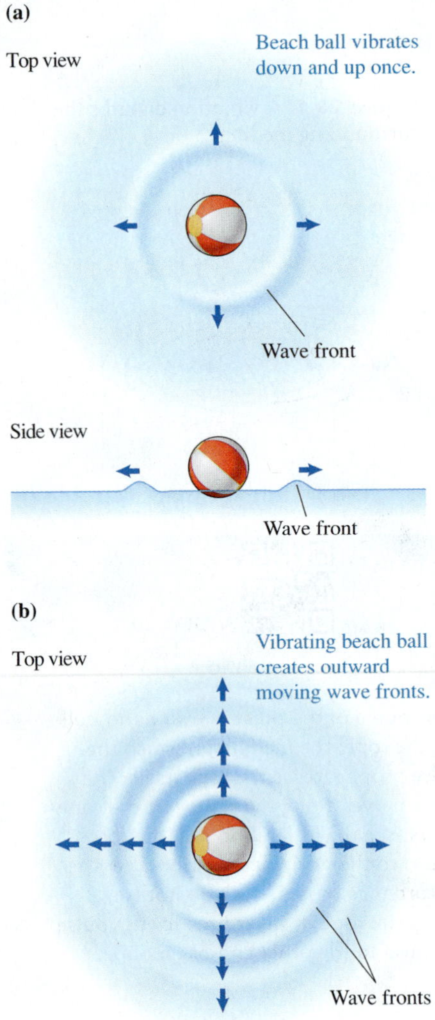
(a)
Top view
Beach ball vibrates down and up once.
Wave front
Side view
Wave front
(b)
Top view
Vibrating beach ball creates outward moving wave fronts.
Wave fronts

the side of the ball, forming a hump in the water at the sides of the ball. This hump falls back toward its equilibrium level and overshoots, pushing water slightly farther away upward into a hump. This disturbance propagates outward, but the water at a particular location mostly vibrates up and down, and not outward in the direction the wave travels. The vibrating source and the interactions of the neighboring sections of the medium result in a coordinated vibration—a wave. Similar phenomena occur when you shake the end of a long rope up and down. In this case, you are the source of the disturbance and the rope is the medium. We can now define wave motion:

> **Wave motion** **involves a disturbance produced by a vibrating object (a source). The disturbance moves, or propagates, through a medium and causes points in the medium to vibrate. When the disturbed medium is physical matter (solid, liquid, or gas), the wave is called a mechanical wave.**

The moving circular disturbance itself created by the beach ball on the water is called a **wave front** (see the top view in **Figure 20.2a**). Every point on the wave front has the same displacement from equilibrium at the same clock reading. If instead of pushing the ball one time, you push it down and up in a regular pattern, you will see a continuous group of humps and dips (waves consisting of many wave fronts) of circular shape moving outward on the surface of the water (Figure 20.2b).

## Two kinds of waves

The stretchy toy called a Slinky provides another good example of wave motion. If we (the source) push or pull the end coil of a stretched Slinky (the medium) several times, it will pull the coil attached to it. That coil pulls the next coil, which in turn pulls the next coil, and so on. We see a compressing and stretching pattern moving along the Slinky toy in the direction of wave propagation. The pattern is illustrated in **Figure 20.3** at a single moment in time. Such a wave is called a **longitudinal wave**.

We can disturb a Slinky in a different way. If we vibrate the coil up and down perpendicular to the length of the Slinky, the coils move up and down perpendicular to the direction that the disturbance travels (**Figure 20.4**). This type of wave is called a **transverse wave.**

**Active Learning Guide>**

> **Longitudinal and transverse waves** **In a *longitudinal wave* the vibrational motion of the particles or layers of the medium is parallel to the direction of propagation of the disturbance. In a *transverse wave* the vibrational motion of the particles or layers of the medium is perpendicular to the direction of propagation of the disturbance.**

The wave we observed on the surface of water is more complicated; the water layers move in an elliptical path so the wave is part transverse and part longitudinal.

**Figure 20.3** A longitudinal wave on a Slinky.

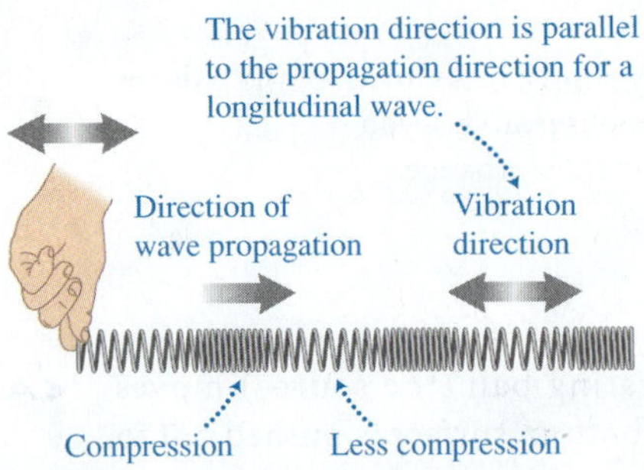
The vibration direction is parallel to the propagation direction for a longitudinal wave.
Direction of wave propagation
Vibration direction
Compression
Less compression

**Figure 20.4** A transverse wave on a Slinky.

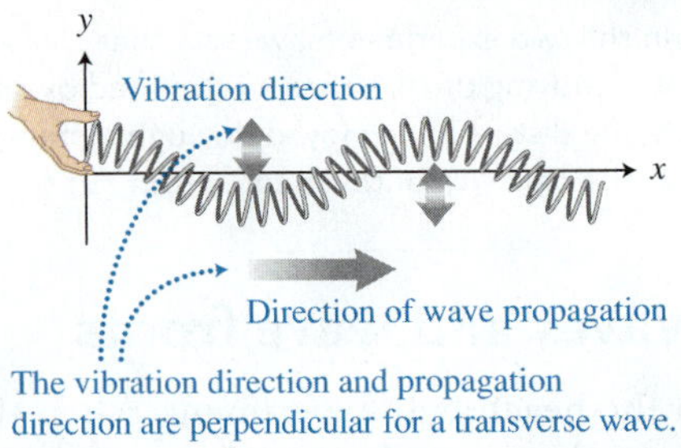
y
Vibration direction
x
Direction of wave propagation
The vibration direction and propagation direction are perpendicular for a transverse wave.

## Reflection of waves

We observe another important property of waves when watching waves on the surface of the water, on a Slinky, or on a rope. When a wave reaches the wall of the container or the end of the Slinky or rope, it reflects off the end and moves in the opposite direction. In fact, when a wave encounters any boundary between different media, some of the wave is reflected back. **Figure 20.5** shows this reflection for one pulse on a rope. The wave traveling to the right and reflected wave traveling to the left exist in the rope simultaneously, complicating the analysis. Thus, if we want to study simple wave motion and not worry about reflected waves, we need to have a very long medium.

**Figure 20.5** A pulse on a rope at different times.

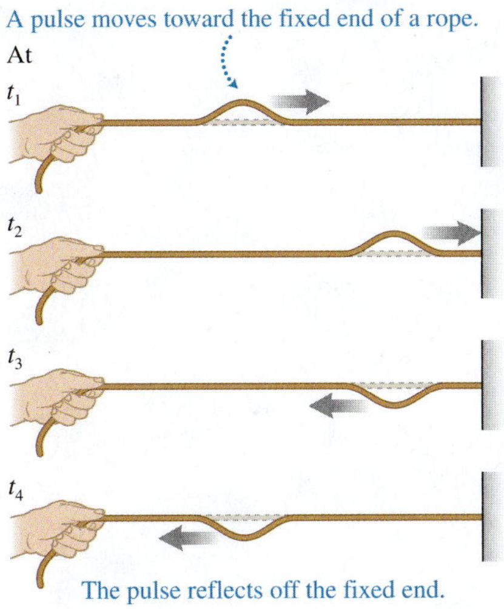


## Sound waves

So far we have considered mechanical waves propagating in ropes, Slinkies, and water. Sound waves are also created by vibrating objects and propagate in different media—air and other gases, water, and even solids. A simple source of sound waves is a tuning fork. If you strike a tuning fork, you hear sound. You can feel the vibrations of the prongs if you touch them. In fact, the moment you touch them, the sound stops, since you've stopped their vibrations. If you place the vibrating prongs in water you see ripples going out (**Figure 20.6**). What happens so that our ears hear a sound? The vibrating prongs of the tuning fork cause air pressure changes, and these pressure changes travel through the air from one place to another. This varying pressure wave is what we call sound. These pressure variations lead to longitudinal compressions and decompressions (expansions) of the medium, meaning that sound is a longitudinal wave. Sound propagates in water (it's how dolphins communicate) and even in solid objects (seismic waves in Earth). If a tuning fork vibrates in a vacuum, there will be no medium for the tuning fork prongs to vibrate against and consequently no way for pressure variations to occur. Thus sound cannot propagate in a vacuum. We will study sound waves in more detail later in the chapter.

**Figure 20.6** A vibrating tuning fork produces sound along with waves on the surface of the water.

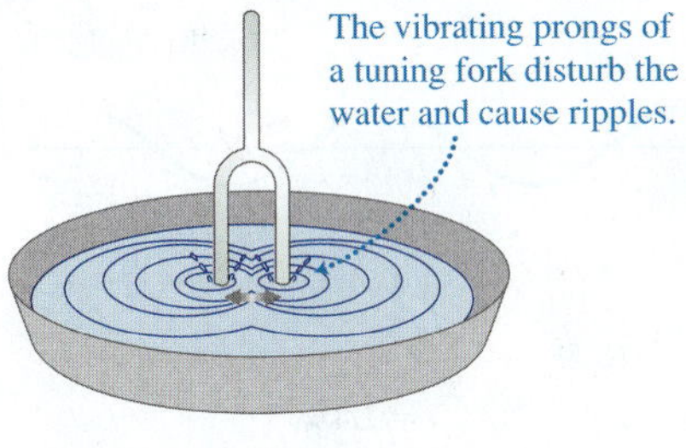

The vibrating prongs of a tuning fork disturb the water and cause ripples.

**Review Question 20.1** How do you produce a longitudinal wave on a Slinky? How do you produce a transverse wave?

# 20.2 Mathematical descriptions of a wave

To understand waves and predict their behavior, we need to describe them using physical quantities and find relationships between those physical quantities.

We will use the waves traveling along an infinitely long rope (so long that we can ignore reflected waves) and a wave source that vibrates with simple harmonic motion (SHM; discussed in Chapter 19). This motion can be created by a motor that vibrates the end of a rope up and down perpendicular to the rope (**Figure 20.7a**). The motor (the source) produces a transverse wave in the rope (the medium). At time zero, the source starts at $y = +A$. The source vibrates sinusoidally; thus the source displacement and the end of the rope displacement from equilibrium are described by a sinusoidal function of time (Figure 20.7b):

$$y = A\cos\left(\frac{2\pi}{T}t\right) \qquad (20.1)$$

where $y$ is the displacement of that end of the rope from its equilibrium position, $A$ is the amplitude of the vibration, and $T$ is the period of vibration. The left end of the rope moves up and down, pulling the next part of the rope,

**Figure 20.7** A wave produced on a rope by a vibrating source.

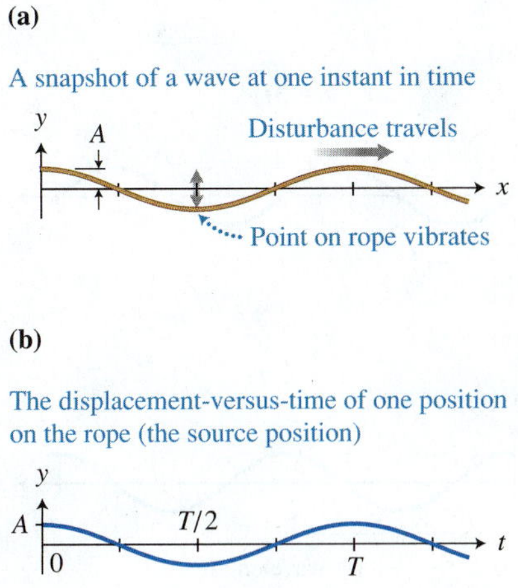

Active Learning Guide›

which pulls its neighboring part, which pulls its neighboring part, and so on. The wave propagates along the rope at a particular speed. When the disturbance reaches a point at a distance $x$ from the source, that point starts vibrating at the same period $T$ as the source. We now have four quantities that describe the wave started by this vibrating source. The first three are familiar to us from Chapter 19: they describe the motion of a vibrating object. The fourth quantity (speed), as it applies to waves, is new.

**Period** $T$ **in seconds is the time interval for one complete vibration of a point in the medium anywhere along the wave's path.**

**Frequency** $f$ **in Hz** $(s^{-1})$ **is the number of vibrations per second of a point in the medium as the wave passes.**

**Amplitude** $A$ **is the maximum distance of a point of the medium from its equilibrium position as the wave passes.**

**Speed** $v$ **in m/s is the distance a disturbance travels during a time interval, divided by that time interval.**

**Figure 20.8** The shape of a transverse wave at five consecutive times.

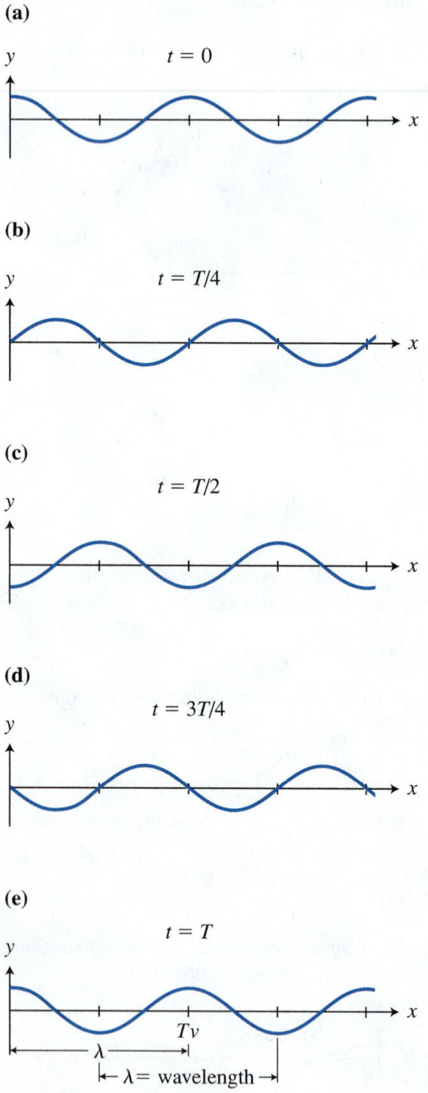


## Mathematical description of a traveling sinusoidal wave

We know how to describe the motion of the wave source when it executes simple harmonic motion. Now our goal is to construct a description of a traveling sinusoidal wave at different places along its path and at different times (clock readings). We assume that the mechanical energy of the source is not converted to the internal energy of the medium. In other words, no damping occurs—the amplitude of the vibration is the same at all points. We also assume that the speed of the wave propagation is constant throughout the medium. We choose the positive $x$-direction as the direction in which the disturbance travels at speed $v$. The left end of the rope at $x = 0$ is attached to the vibrating source. According to Eq. (20.1), the source oscillates up and down with a vertical displacement

$$y(0, t) = A\cos\left(\frac{2\pi}{T}t\right)$$

The zero in the parentheses indicates that the function $y(0, t)$ describes the motion of the medium at the position $x = 0$ and clock reading $t$. The shape of the rope at five different times is shown in **Figure 20.8**. If you look at any other position along the $x$-axis, the rope vibrates in the same way as it does at $x = 0$ but at a different instant in the vibration cycle. For example, at the first tick mark to the right of $x = 0$, you see that at time zero, the rope displacement is $-A$; at time $t = T/4$, its $y$-displacement is zero, then $+A$, then 0, and finally back to $-A$.

We can mathematically describe the disturbance $y(x, t)$ of a point of the rope at some positive position $x$ to the right of $x = 0$ at some arbitrary time $t$. Every point to the right of $x = 0$ vibrates with the same frequency, period, and amplitude as the rope at $x = 0$. But there is a time delay $\Delta t = x/v$ for the disturbance at $x = 0$ to reach the location $x$. The vibrational disturbance at position $x$ at time $t$ is the same as the disturbance at $x = 0$ was at the earlier time $t - x/v$. Thus, the vibration of the rope at position $x$ can be described as

$$y(x, t) = A\cos\left[\frac{2\pi}{T}\left(t - \frac{x}{v}\right)\right] \qquad (20.2)$$

The snapshot graph of the disturbance (Figure 20.8e) shows that the disturbance pattern repeats at a distance $Tv$. This distance separates neighboring locations on the rope that have the same displacement $y$ and the same slope.

This distance is called the **wavelength** of the wave and is represented by the Greek letter $\lambda$ (lambda):

$$\lambda = Tv = \frac{v}{f}$$

where $v$ is the speed of travel of the wave, $T$ is the period of vibration of each part of the rope, and $f$ is the frequency of vibration $(1/T)$.

**‹Active Learning Guide**

**Wavelength** $\lambda$ **equals the distance between two nearest points on a wave that at any clock reading have exactly the same displacement and shape (slope). It is also the distance between two consecutive wave fronts:**

$$\lambda = Tv = \frac{v}{f} \tag{20.3}$$

We can substitute this expression for wavelength into Eq. (20.2) and rearrange a little to get a function of the two variables $t$ and $x$ that allows us to predict the displacement $y$ of any point $x$ at any time $t$ in a medium through which the wave propagates:

$$y(x, t) = A\cos\left[\frac{2\pi}{T}\left(t - \frac{x}{v}\right)\right] = A\cos\left[2\pi\left(\frac{t}{T} - \frac{x}{Tv}\right)\right] = A\cos\left[2\pi\left(\frac{t}{T} - \frac{x}{\lambda}\right)\right]$$

**TIP** You can think of a wavelength as the distance that the wave propagates during one period.

**Mathematical description of a traveling sinusoidal wave** **The displacement from equilibrium $y$ of a point at location $x$ at time $t$ when a wave of period $T$ travels at speed $v$ in the positive $x$-direction through a medium is described by the function**

$$y = A\cos\left[2\pi\left(\frac{t}{T} - \frac{x}{\lambda}\right)\right] \tag{20.4}$$

**The wavelength $\lambda$ of this wave equals $\lambda = Tv$.**

**TIP** Notice that Eq. (20.4) is mathematically symmetric with respect to the period of the wave and its wavelength (the terms in which each appears have the same form). This is because there are two repetitive processes in a wave. If the location $x$ is fixed, that point in the medium vibrates in time with period $T$. If the time $t$ is fixed (as in a photograph), the space has a wavelike appearance with wavelength $\lambda$ (see Figure 20.8). We replaced the notation for the function $y(x, t)$ with just $y$ for simplicity.

If we write Eq. (20.4) as $y = A\cos\left[2\pi\frac{t}{T} - 2\pi\frac{x}{\lambda}\right]$, the part $2\pi(x/\lambda)$ describes the difference between the displacement of a point at a distance $x$ from the origin of the wave and the displacement of the point at the origin. The $2\pi(x/\lambda)$ is called a **phase**. Two points in a medium are said to be "in phase" if at every clock reading their displacements are exactly the same. This means that two points separated by a distance equal to an integer multiple of wavelengths (one wavelength, two wavelengths, etc.) are always in phase (**Figure 20.9**).

**Figure 20.9** Points a and b have the same phase at any clock reading.

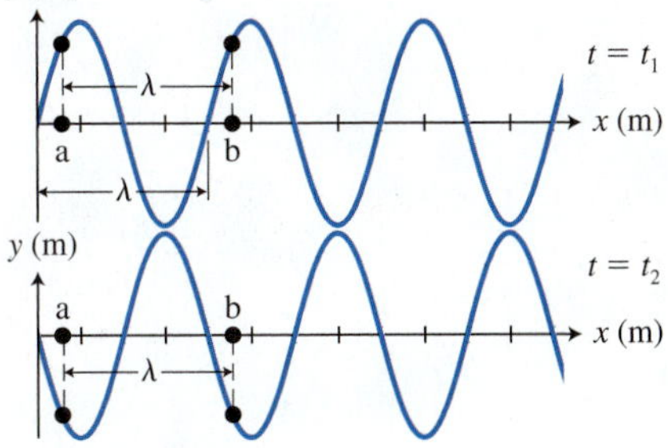

**CONCEPTUAL EXERCISE 20.1 Making sense of a mathematical description of a wave**

A wave on an infinitely long Slinky is described mathematically as

$$y = (0.10\ \text{m})\cos\left[2\pi\left(\frac{t}{0.70\ \text{s}} - \frac{x}{1.8\ \text{m}}\right)\right]$$

Say everything you can about this wave and represent it graphically.

**Sketch and translate** Compare the given equation with the wave equation [Eq. (20.4)]: $y = A\cos\left[2\pi\left(\frac{t}{T} - \frac{x}{\lambda}\right)\right]$. Therefore, $A = 0.10$ m, $T = 0.70$ s, and $\lambda = 1.8$ m. We can also say that at $t = 0$, the starting point with the coordinate $x = 0$ had a displacement of

$$y(0, 0) = (0.10\ \text{m})\cos\left[2\pi\left(\frac{0}{0.70\ \text{s}} - \frac{0}{1.8\ \text{m}}\right)\right] = 0.10\ \text{m}.$$

We can write a mathematical equation for the motion of the wave source at location $x = 0$: $y_{\text{source}} = (0.10\ \text{m})\cos\left[2\pi\left(\frac{t}{0.70\ \text{s}}\right)\right]$. Notice that at $x = 0$, $y$ is a function of time only, as we are looking at the disturbance at only one location. Rearranging the equation $\lambda = Tv$, we get $v = \lambda/T = (1.8\ \text{m})/(0.70\ \text{s}) = 2.6\ \text{m/s}$, a reasonable number.

**Simplify and diagram** Knowing the amplitude and the period, we can draw a displacement-versus-time (clock reading) graph for one particular location in the medium (we do it for $x = 0$ in the first figure below). Knowing the amplitude and the wavelength, we can draw a graph that represents an instantaneous picture of the whole Slinky at a particular time (clock reading) (shown for $t = 0$ in the second figure).

**(a)**

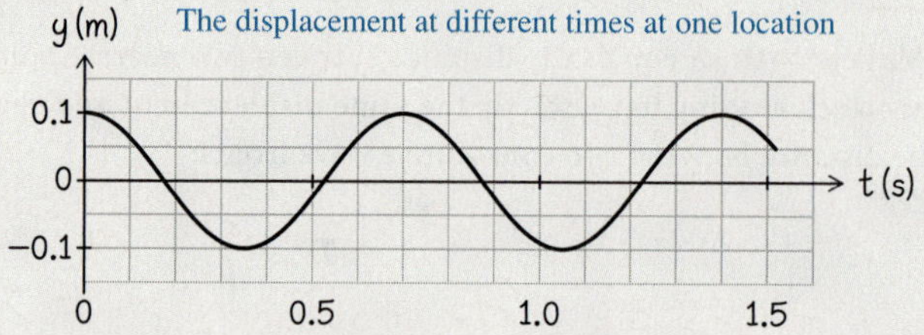


**(b)**

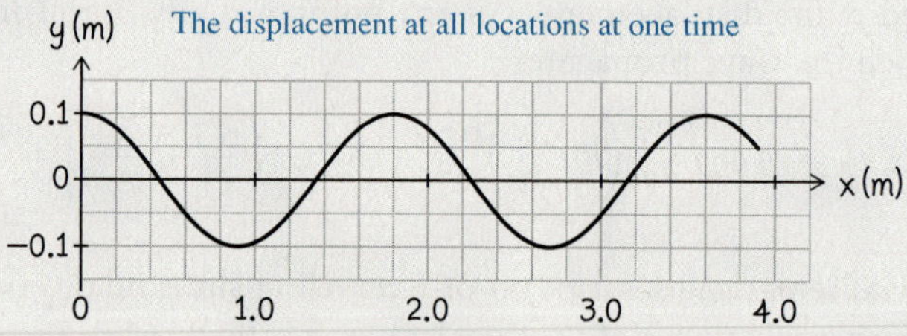


**Try it yourself:** A Slinky vibrates with amplitude 15 cm and period 0.50 s. The speed of the wave on the Slinky is 4.0 m/s. Write the wave equation for the Slinky.

*Answer:* $y = (0.15\ \text{m})\cos\left[2\pi\left(\frac{t}{0.50\ \text{s}} - \frac{x}{2.0\ \text{m}}\right)\right]$.

**TIP** Because Eq. (20.4) is a function of two variables, when we represent waves graphically we need two graphs, such as those in the figures in Exercise 20.1—one showing how one point of the medium changes its position with time ($y$ versus $t$; $x =$ constant) and the other one showing the displacements of multiple points of the medium at the same clock reading—a snapshot of the wave ($y$ versus $x$; $t =$ constant).

**Review Question 20.2** Compare and contrast what is meant by the speed of a vibrating object and the speed of a wave.

## 20.3 Dynamics of wave motion: Speed and the medium

We have defined the speed of a wave as the distance that a disturbance travels in the medium during a time interval divided by that time interval. This operational definition of speed tells us how to determine the speed but does not explain why a particular wave has a certain speed. Our goal in this section is to investigate what determines the wave speed. Consider the results reported in Observational Experiment **Table 20.2**.

OBSERVATIONAL EXPERIMENT TABLE

## 20.2 What affects wave speed?

VIDEO 20.2

| Observational experiment | Analysis |
|---|---|
| Two people are holding the ends of a Slinky. The person on the left starts a pulse on the Slinky. The person on the right measures the force pulling on the Slinky and the time interval $\Delta t$ for the pulse to travel the distance $d$. They vary the amplitude, frequency, and pulling force one at a time. | Assume that the speed of a pulse is the same as the speed of a wave and that the pulse does not slow down or speed up as it travels. The speed of the pulse is $v = \frac{\text{distance}}{\text{time interval}} = \frac{d}{\Delta t} = \frac{(4.0\text{ m})}{(1.0\text{ s})} = 4.0\text{ m/s}$ |

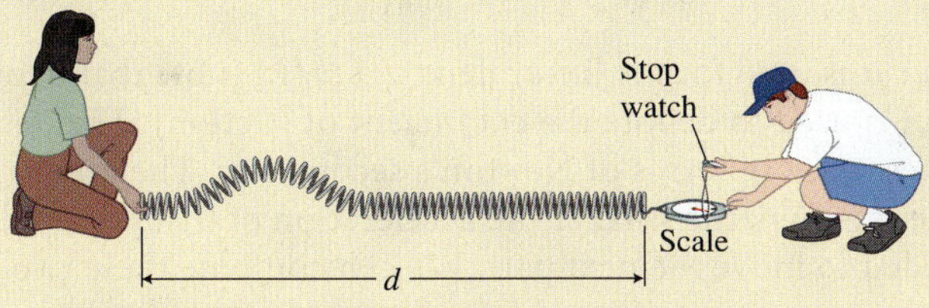


| Distance (m) | Time Interval (s) |
|---|---|
| 4.0 | 1.0 |

**Effect of amplitude** Change the amplitude of the pulses:

| Amplitude (m) | Distance (m) | Time Interval (s) | Speed (m/s) |
|---|---|---|---|
| 0.1 | 4.0 | 1.0 | 4.0 |
| 0.2 | 4.0 | 1.0 | 4.0 |
| 0.3 | 4.0 | 1.0 | 4.0 |

**Effect of frequency** Change the frequency of the pulses (we send several pulses one after another):

| Frequency (Hz) | Distance (m) | Time Interval (s) | Speed (m/s) |
|---|---|---|---|
| 2.0 | 4.0 | 1.0 | 4.0 |
| 4.0 | 4.0 | 1.0 | 4.0 |
| 6.0 | 4.0 | 1.0 | 4.0 |

**Effect of force** Change the force pulling on the end of the Slinky (the length of the Slinky also changes):

| Force (N) | Distance (m) | Time Interval (s) | Speed (m/s) |
|---|---|---|---|
| 2 | 4.0 | 1.0 | 4.0 |
| 4 | 8.0 | 1.0 | 8.0 |
| 6 | 12.0 | 1.0 | 12.0 |

**Patterns**

- The speed of a pulse *does not depend* on either the amplitude or frequency.
- The speed *does depend* on the magnitude of force exerted on the end.

The patterns that we found in Table 20.2 are surprising—the speed of the pulse does not depend on the amplitude of the pulse or on its frequency. The source of the waves does not affect the speed of the wave on the Slinky!

However, pulling harder on the end of the Slinky does increase the wave speed. In addition, as we stretch the Slinky, the coils spread farther apart, reducing its "linear density"—its mass per unit length. The force pulling on the Slinky and its linear density seem to affect the speed of a pulse or wave on the Slinky.

If we do similar experiments with stiff springs that stretch less than a Slinky, we find that the wave speed is proportional to the square root of the force pulling on the end of the spring. This applies to other objects as well, such as strings.

$$v \propto \sqrt{F_{\text{M on M}}}$$

The subscripts M and M indicate the pulling force that one part of the medium exerts on a neighboring part. This result makes sense—the more you pull on the spring, the more strongly adjacent parts interact with each other and therefore the faster they transfer the disturbance.

Other experiments indicate that speed is also affected by the mass per unit length of vibrating particles (the coils in the experiments in Observational Experiment Table 20.2). The speed of a wave is inversely proportional to the square root of the mass per unit length of the medium:

**Active Learning Guide>**

$$v \propto \frac{1}{\sqrt{\dfrac{m}{L}}} = \frac{1}{\sqrt{\mu}}$$

Here the Greek letter $\mu$ is used for the linear density $(m/L)$ (note that in other contexts this symbol is used to denote the coefficient of friction). The dependence on mass makes sense in terms of Newton's second law. The greater the mass of a part of the medium, the smaller the acceleration of that part and the longer the time needed to move the next part. We can combine these two factors into one equation:

$$v = \sqrt{\frac{F_{\text{M on M}}}{\mu}}$$

Let us check whether the units are correct:

$$v = \sqrt{\frac{\text{N}}{\text{kg/m}}} = \sqrt{\left(\frac{\text{kg}\cdot\text{m}}{\text{s}^2}\right)\left(\frac{\text{m}}{\text{kg}}\right)} = \frac{\text{m}}{\text{s}}$$

The units are correct. To make sure that the speed does in fact depend on the above properties, we can perform the experiment in Testing Experiment **Table 20.3**. We will take a rope of known mass and length, attach one end to a wall, and pull the other end with a spring scale, exerting a known force. Then we will pluck the rope once, sending a pulse. We will use the above expression for speed to predict the time interval for a pulse to travel from one end to the wall and back again.

**TESTING EXPERIMENT TABLE**

### 20.3 Testing the expression for wave speed.

| Testing experiment | Prediction | Outcome |
|---|---|---|
| Predict the time interval for a pulse to travel from one end of a 0.60-kg rope to the other end and back again. The rope is pulled by a spring scale. Since the pulse travels very fast, to minimize experimental uncertainty we will count 10 trips. 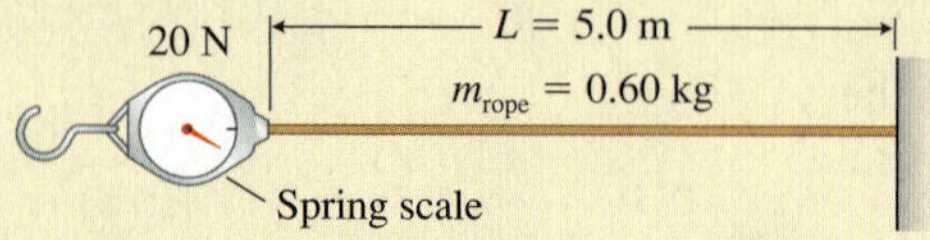  | The wave speed is $v = \sqrt{\dfrac{F_{\text{S on R}}}{m/L}} = \sqrt{\dfrac{20\text{ N}}{0.60\text{ kg}/5.0\text{ m}}} = 13\text{ m/s}$ The time interval should be $\Delta t = \dfrac{2L}{v} = 2(5.0\text{ m})/(13\text{ m/s}) = 0.77\text{ s}.$ We predict that 10 round trips will take 7.7 s. | We conduct the experiment several times and measure the time as 7.7 ± 0.5 s. The predicted outcome lies within this interval. |

**Conclusion**

The outcome is consistent with the prediction. We did not disprove that the speed of a wave depends on the properties of the medium.

**Wave speed** The speed $v$ of a wave on a string or in some other one-dimensional medium depends on how hard the string or medium is pulled $F_{\text{M on M}}$ (one part of the medium pulling on an adjacent part) and on the mass per unit length $\mu = m/L$ of the medium:

$$v = \sqrt{\frac{F_{\text{M on M}}}{\mu}} \quad (20.5)$$

**TIP** Note two mathematical expressions for the speed of the pulse:

$$v = d/\Delta t \text{ and } v = \sqrt{\frac{F_{\text{M on M}}}{\mu}}.$$

The first expression is an operational definition for speed, while the second is a cause-effect relationship.

We have just learned that the speed of waves in a medium in which no mechanical energy is converted into internal energy depends only on the properties of the medium. However, the period and frequency of a wave depend on the vibration source. Since the medium vibrates at the same frequency as the source, the wavelength of the wave depends on both the frequency of the source and the speed of the wave through the medium. If the vibration frequency is low and the wave speed high, the crest of the wave will travel a greater distance before the next crest leaves the source. Thus the wavelength is large:

$$\boldsymbol{\lambda} = \boldsymbol{v}/f$$

However, if the wave speed is low and/or the frequency high, the wavelength is small:

$$\lambda = v/\boldsymbol{f}$$

The wave speed also depends on the type of wave (or pulse). In the same medium, transverse waves (and pulses) travel more slowly than longitudinal waves (and pulses). The reason for the difference is in the types of deformation of the medium that occur when the waves propagate. The transverse waves that can only propagate in solids require shear deformation—the layers of the medium slide across each other—while the longitudinal waves require compressions and decompressions. The respective forces that parts of the medium exert on each other for the shear deformations are weaker than for the compressions/decompressions.

**TIP** The equation $\lambda = Tv = v/f$ is a cause-effect relationship for the wavelength.

**QUANTITATIVE EXERCISE 20.2 Force exerted on a violin G string**

A violin G string is 0.33 m long and has a mass of 2.2 g. Determine the force exerted on the ends of the string in order for a wave to travel at the speed of 128 m/s.

**Represent mathematically** We know that

$$v = \sqrt{\frac{F_{\text{S on S}}}{\mu}} = \sqrt{\frac{F_{\text{S on S}}}{m/L}}$$

Rearrange this to determine the magnitude of the tension force $F_{\text{S on S}}$ that adjacent parts of the string exert on each other and that the peg at the end of the violin exerts on the string.

**Solve and evaluate** Rearranging the above equation, we get

$$F_{\text{S on S}} = \frac{m}{L}v^2 = \frac{(2.2 \times 10^{-3}\text{ kg})}{(0.33\text{ m})}(128\text{ m/s})^2$$
$$= 110\text{ kg}\cdot\text{m/s}^2 = 110\text{ N}$$

This is a reasonable tension (about 20 lb), and the unit is correct.

**Try it yourself:** Would the tension be higher or lower in a 1.4-g violin A string if the waves traveled at a speed of 288 m/s in it?

*Answer:* Higher—about 350 N.

**Review Question 20.3** Why do the speeds of the waves on the A string and the lower frequency G string of a violin differ?

# 20.4 Energy, power, and intensity of waves

So far we have investigated waves that propagate in a one-dimensional medium such as a rope. If there is no damping of the vibrations as the wave propagates, then the amplitude of the vibration at any point in the one-dimensional medium is the same as that of the source. What happens if the wave propagates through a two-dimensional medium (like the surface of a small lake) or a three-dimensional medium (like the air surrounding a balloon that pops and produces sound)? Do the amplitudes of these waves remain constant as they propagate outward from the source?

**Figure 20.10** Wave crests are shown at one time on the surface of water.

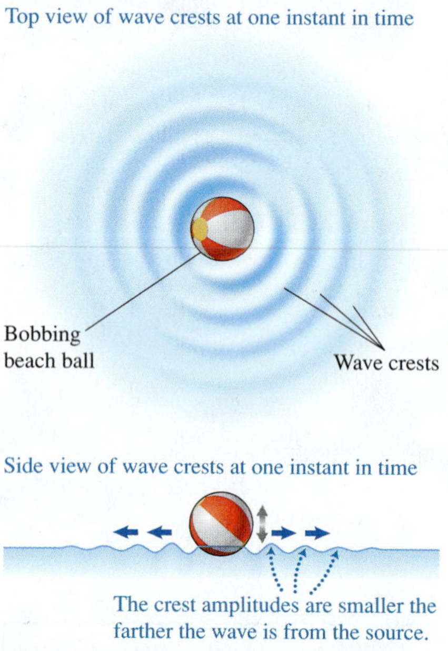
Top view of wave crests at one instant in time
Bobbing beach ball
Wave crests
Side view of wave crests at one instant in time
The crest amplitudes are smaller the farther the wave is from the source.

## Wave amplitude and energy in a two-dimensional medium

Consider **Figure 20.10**. A beach ball bobs up and down in water in simple harmonic motion, producing circular waves that travel outward across the water surface in all directions. The shaded circles represent the peaks of the waves. We observe that the amplitudes of the crests decrease as the waves move farther from the source. Why?

Suppose that the wave source uses 10 J of energy every second to produce a pulse, and that all of that energy is transferred to the pulse (one pulse per second). As the pulse propagates outward in all directions, more particles vibrate, since the circumference of the circular pulse increases as it gets farther from the source. However, the total energy of all vibrating particles in the pulse cannot exceed 10 J. Since there are more vibrating particles as the pulse moves outward, the vibrational energy per particle will have to decrease.

Now, let us do a quantitative analysis. Surround the source with two imaginary rings (the dashed lines in **Figure 20.11**), one at a distance $R$ from the source and the other at distance $2R$. If the energy output of the source per unit time is 10 J/s (its power output), then this energy first moves through the circumference $C_1 = 2\pi R$ and then through the second larger circumference $C_2 = 2 \cdot 2\pi R = 2C_1$. The circumference of the second ring is two times greater than the first, but the same energy per unit time moves through it. Consequently the energy per unit circumference length ($\text{energy}/C = \text{energy}/2\pi r$) passing the second ring is one-half that passing the first ring. We see that the energy per unit time and per unit circumference length is inversely proportional to the distance from the source (assuming that there is no damping).

**Figure 20.11** The wave crest amplitude decreases on a two-dimensional surface with distance from the source.

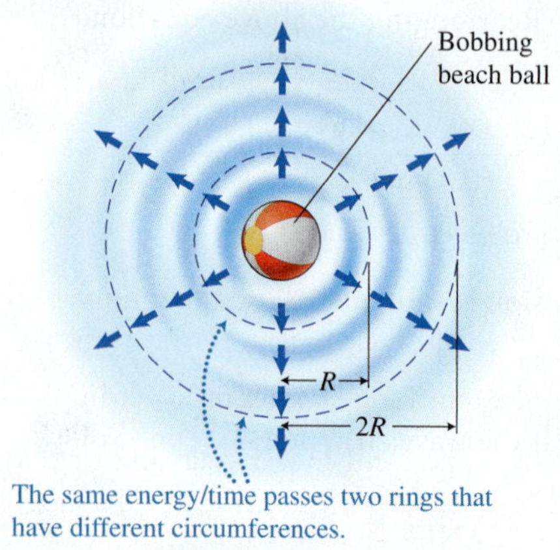
Snapshot of wave crests at one instant in time
Bobbing beach ball
R
2R
The same energy/time passes two rings that have different circumferences.

**Two-dimensional waves produced by a point source** The energy per unit circumference length and per unit time crossing a line perpendicular to the direction that the wave travels decreases as $1/r$, where $r$ is the distance from the point source of the wave.

Earlier (in Chapter 19) we learned that the amplitude of vibrations and the energy of the vibrating system are related to each other. Thus, if the energy per circumference length and per unit time of a two-dimensional wave decreases as the wave propagates, so should the amplitude. The following example investigates the changes in amplitude mathematically. What happens to the amplitude (not the energy) of two-dimensional waves as they move away from the source?

**EXAMPLE 20.3 Water wave in a pool**

You do a cannonball jump off a high board into a pool. The wave amplitude is 50 cm at your friend's location 3.0 m from where you enter the water. What is the wave amplitude at a second friend's location 5.0 m from where you enter the water?

**Sketch and translate** We consider the water wave to be a two-dimensional wave. The amplitude of the impulse at $r_1 = 3.0\text{ m}$ from the source is $A_{r1} = 50\text{ cm}$. We wish to find the amplitude of the impulse at a distance $r_2 = 5.0\text{ m}$ from the source.

**Simplify and diagram** Assume that there is no conversion of mechanical energy into internal energy (no damping), so the changing amplitude is only due to increased distance from the source.

**Represent mathematically** The energy per circumference length perpendicular to the direction of wave travel and per unit time decreases as $1/r$, where $r$ is the distance from the source. Thus,

$$\frac{U_{r2}}{U_{r1}} = \frac{(1/r_2)}{(1/r_1)} = \frac{r_1}{r_2}$$

We also know [from Chapter 19, Eq. (19.9)] that the energy of vibration is proportional to the square of the amplitude of the vibration: $U \propto A^2$. Thus:

$$\frac{U_{r2}}{U_{r1}} = \frac{A_{r2}^{\,2}}{A_{r1}^{\,2}} = \frac{r_1}{r_2}$$

Therefore,

$$A_{r2} = \left(\frac{r_1}{r_2}\right)^{1/2} A_{r1}$$

**Solve and evaluate** Thus,

$$A_{r2} = \left(\frac{r_1}{r_2}\right)^{1/2} A_{r1} = \left(\frac{3.0\text{ m}}{5.0\text{ m}}\right)^{1/2}(0.50\text{ m})$$

$$= (0.77)(0.50\text{ m}) = 0.39\text{ m}$$

This decreased amplitude farther from the source seems reasonable. The amplitude of a 2-D wave decreases as one over the square root of the distance from the source, $A \propto 1/\sqrt{r}$.

**Try it yourself** You set up longitudinal vibrations on a really long Slinky. If there is no friction or other form of energy conversion from mechanical energy, how does the amplitude at a distance of 2.0 m from the source compare to the amplitude 4.0 m from the source?

*Answer:* They are the same—the Slinky is a one-dimensional medium and the same vibrational energy passes every position on it.

## Wave amplitude and energy in a three-dimensional medium

Consider a three-dimensional medium. A fire alarm goes off and sound moves outward in all directions. Consider two imaginary spheres that surround the source (**Figure 20.12**), one at distance $R$ from the source and the other at distance $2R$. Assume that the energy output of the source per unit time is 10 J/s. This energy first moves through the surface area $A_1 = 4\pi R^2$ and then through the second larger area $A_2 = 4\pi(2R)^2 = 4A_1$. The area of the second sphere is four times the first, but the same energy per unit time moves through it. Consequently, the energy per unit area through the second sphere is one-fourth that through the first sphere. The energy per unit time per unit area is inversely proportional to the distance squared from the source. This conclusion is consistent (qualitatively) with our experience: we know that the farther we are from the alarm, the quieter the sound becomes.

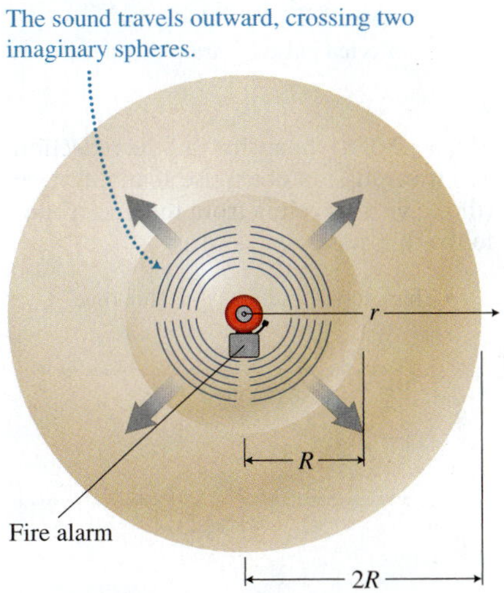
The sound travels outward, crossing two imaginary spheres.
r
R
Fire alarm
2R

**Figure 20.12** The sound decreases in amplitude and loudness at a greater distance from the source.

**Three-dimensional waves produced by a point source** The energy per unit area per unit time passing across a surface perpendicular to the direction that the wave travels decreases as $1/r^2$, where $r$ is the distance from the point source of the wave.

**TIP** The symbol for the area and the amplitude of vibrations is the same—$A$. Make sure you interpret the meaning of $A$ based on the context of the situation.

## Wave power and wave intensity

Earlier (in Chapter 6), we defined the power $P$ of a process as energy change per unit time interval ($P = \Delta U/\Delta t$). Since this power is spread over the area the waves travel through, a more useful quantity for describing the wave is **intensity**.

The intensity of a wave is defined as the energy per unit area per unit time interval that crosses perpendicular to an area in the medium through which it travels.

$$\text{Intensity} = \frac{\text{Energy}}{\text{time} \cdot \text{area}} = I = \frac{\Delta U}{\Delta t \cdot A} = \frac{P}{A} \qquad (20.6)$$

The unit of intensity $I$ is equivalent to joules per second per square meter ($\text{J/s} \cdot \text{m}^2$). Since J/s is a watt, the unit of intensity is then watts per square meter ($\text{W/m}^2$).

We know that the intensity of a wave propagating in a three-dimensional medium decreases in proportion to the inverse of the distance squared from the source ($I \propto 1/r^2$). We have also learned that the energy of a vibrating system depends on the amplitude of the vibration. For example, for an object attached to a spring:

$$U_{\text{s max}} = \frac{1}{2}kA^2$$

Thus, if the energy (the intensity for 3-D waves) on the left side of the above equation decreases by one-fourth (1/4) when the distance from the point source doubles, the amplitude on the right side decreases by one-half (since 1/2 squared is 1/4).

**Review Question 20.4** Explain what happens to the intensity of a wave as it propagates in a three-dimensional medium.

**Figure 20.13** Snapshots of the reflection and transmission of a pulse at an interface (the wave is traveling from the thin rope to the thick rope).

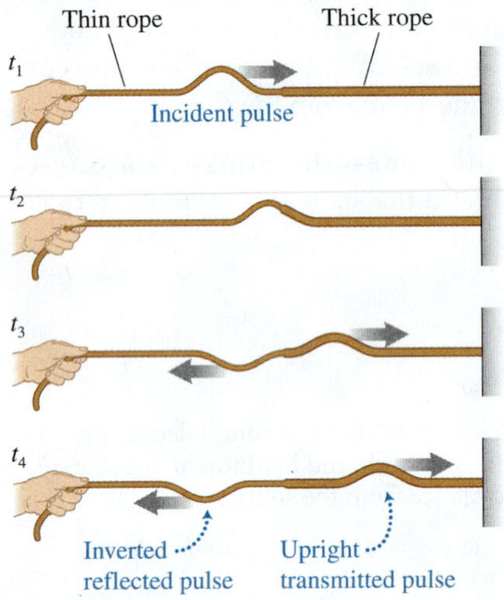


**Figure 20.14** Snapshots of the reflection and transmission of a pulse at an interface (the wave is traveling from the thick rope to the thin rope).

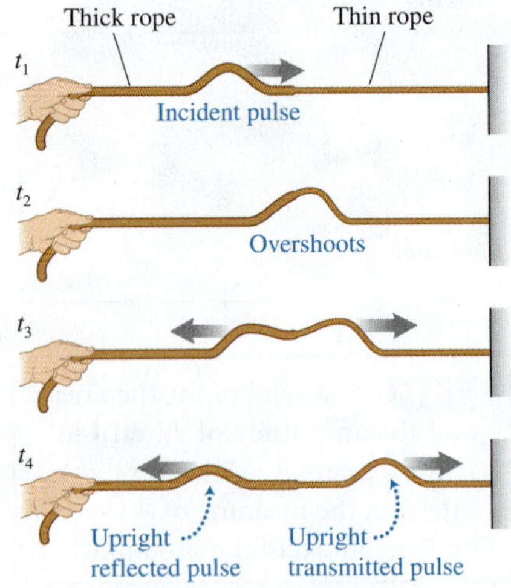


## 20.5 Reflection and impedance

If you hold one end of a rope whose other end is fixed (as in Figure 20.5) and shake it once, you create a transverse traveling incident pulse. We already know that when the pulse reaches the fixed end, the reflected pulse bounces back in the opposite direction. We also see that the reflected pulse is **inverted**—oriented downward as opposed to upward.

Until now, we have considered pulses and waves moving through a homogeneous medium, a medium with the same properties everywhere. What happens to a wave when there is an abrupt change from one medium to another?

Imagine holding a thin rope (small mass per unit length) that is attached to a thicker rope (larger mass per unit length) on the right. You send an upright transverse pulse to the right in the thin rope (**Figure 20.13**). A partially reflected inverted pulse (oriented like the pulse reflected off the fixed end) returns to your hand and a partially transmitted upright pulse travels in the thicker rope.

With the ropes connected in the opposite order (the thick rope on the left and thin rope on the right), you send an upright transverse pulse along the thick rope toward the right (**Figure 20.14**). Now, a partially reflected upright pulse returns to your hand and a partially transmitted upright pulse travels in the thin rope. Why?

In the first experiment, because of its larger mass per unit length, the second thick rope is harder to accelerate upward than a thin rope would be. A small-amplitude upward pulse is initiated in the thick rope. As the thick rope does not "give" as easily (it has a larger mass per unit length), it exerts a downward force, starting an inverted reflected pulse back toward the left in the thin rope. The same happens when the pulse reached the end of the rope—we can think of your hand holding the rope or a wall to which the rope is attached as an infinitely thick rope.

In the second experiment, when the upright pulse in the thick rope reaches the interface of the two ropes, the small rope restrains the thick rope less than the thick rope restrains itself. The thick rope initiates a large upright transmitted pulse in the thin rope (which "gives" more easily because of its smaller mass per unit length). The small rope overshoots its position relative to the thick rope and starts a small upright reflected pulse back in the thick rope.

Understanding the patterns that occur when waves travel from one medium to another allows us to better transmit information and protect ourselves from unwanted signals. In the first experiment a significant amount of the pulse energy traveled back from the thicker medium—thus less energy was transmitted forward. Similarly, your hard skull reflects most of the energy from sound waves traveling through the air. However, the structure of the ear is such that some of this energy can reach the inner ear and be detected.

The pattern of traveling waves in varying media also allows geologists to "see" under Earth's surface. Radio waves reflected from the bottom of the Antarctic ice sheet are characteristic of an ice-rock interface rather than an ice-water interface. Thus, geologists have determined that the ice sheet rests on rock rather than water. Geologists have also used the pattern of motion of seismic waves to determine that there is a sharp, distinctive boundary between the Earth's mantle and its outer core.

## Impedance

A physical quantity called **impedance** characterizes the degree to which waves are reflected and transmitted at the boundary between different media. When a wave travels through a medium, parts of the medium move with respect to each other, or the medium gets distorted. The impedance $Z$ of a medium is a measure of the difficulty a wave has in distorting the medium and depends on both the medium's elastic and inertial properties. The elastic property is related to the interactions between the particles in the medium (the stronger the interactions, the greater the impedance). The inertial property characterizes the density of the medium. Impedance is defined as of the square root of the product of the elastic and inertial properties of the medium:

$$\text{Impedance} = Z = \sqrt{(\text{Elastic property})(\text{Inertial property})} \qquad (20.7)$$

If two media connected together at an interface have different elastic and inertial properties (for example, two different linear density ropes woven together), a wave traveling in one medium is partially reflected and partially transmitted at the boundary between the media. If the product of the elastic and inertial properties of the two media is the same, a wave moves from one medium to the other as if there is no change. If the impedances of two media are very different, most of the wave energy is reflected back into the first medium and does not travel into the second medium. One application of this effect is the use of **ultrasound** (a compression wave with frequencies higher than 20,000 Hz), which takes advantage of the differing densities of internal structures to "see" inside the body, for example, to determine the shape of a fetus inside the mother's uterus.

Ultrasound is poorly transmitted from air outside the body to tissue inside the body. The impedance of tissue is much greater than that of air. As a result, most of the ultrasound energy is reflected at the air-body interface and does not travel inside. To overcome this problem, the area of the body to be scanned is covered with a gel that helps "match" the impedance between the emitter and the body surface. The emitter is then held directly against the gel-covered area. This matching allows the ultrasound wave to propagate into the body instead of being reflected off the body's surface.

Similarly, one part of a seismometer is a block that contacts the Earth and has the same impedance as Earth and vibrates with it during an earthquake. These devices are so sensitive (as no waves reflect off them) that they can detect a man jumping on the ground 1 km from the seismometer.

Many animals can detect seismic signals. Elephants have stiff cartilage and dense fat in their feet, which matches the impedance of Earth's surface. Elephants at times lean forward on their front feet, seemingly to detect vibrations in the ground—for example, prior to approaching a water hole. Reports describe Asian elephants as vigorously responding to earthquakes, even trumpeting at the approach of an earthquake before humans feel the quake.

Not all pulses and waves are reflected at the boundary between two media. Some of a wave's energy may be absorbed at the boundary between two media or in any part of a medium through which the wave travels. The coordinated vibrations of the atoms in the medium are turned into random kinetic energy, that is, into thermal energy. The rate of this conversion varies greatly. The slower the rate of conversion, the longer the wave lives. A pillow over your ears absorbs sound, especially the high-frequency sound from birds. Two pillows absorb even more.

**Review Question 20.5** Why is it impossible to create a traveling wave on a Slinky that only propagates in one direction?

## 20.6 Superposition principle and skills for analyzing wave processes

**Active Learning Guide›**

We have studied the behavior of a single wave traveling in a medium and initiated by a simple harmonic oscillator. Most periodic or repetitive disturbances of a medium are combinations of two or more waves of the same or different frequencies traveling through the same medium at the same time. The sound coming from a violin, for instance, is a combination of almost 20 sinusoidal waves, each of different frequency and amplitude. In Observational Experiment **Table 20.4** we investigate what happens when two or more waves simultaneously pass through a medium.

**OBSERVATIONAL EXPERIMENT TABLE**

**20.4 Adding two pulses passing through a medium.**

| Observational experiment | | Analysis |
|---|---|---|
| An upright pulse traveling right from the left side of a rope meets an inverted pulse traveling left from the right side of the rope. The rope looks undisturbed as the pulses pass in the middle. After passing, the pulses continue moving in the same direction with the same amplitude as if they had never met. | 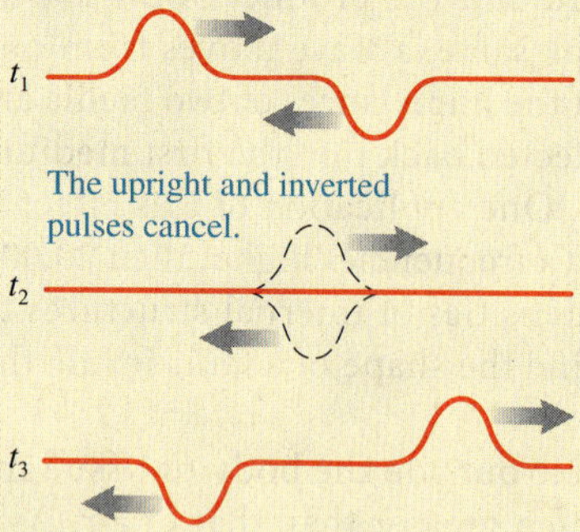  | The part of the rope where the pulses meet is pulled up by the pulse coming from the left and down by the pulse coming from the right, causing a net zero disturbance. |
| Two upright pulses travel from each end of the rope toward the center. When the pulses pass the center, the disturbance is twice that if only one pulse was present. After passing the center of the rope, the pulses continue moving in the same direction with the same amplitude as though they had never met. | 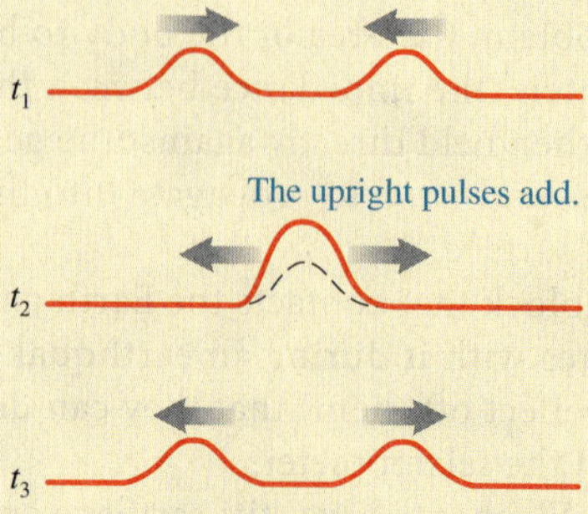  | The part of the rope where the pulses meet is pulled up by both the right- and left-moving pulses, resulting in a double amplitude disturbance. |

**Pattern**

When pulses meet, the resultant disturbance is the sum of the two.

It appears that each pulse pulls the rope separately and the final displacement of the rope at a particular time and location is just a combination of the two pulses at that time and location—adding them together:

$$y_{net} = y_1 + y_2 + \ldots \tag{20.8}$$

This looks similar to the other kinds of superposition we've encountered before (for forces, the $\vec{E}$ field, and the $\vec{B}$ field). Let's test this superposition principle for waves.

Imagine we have two vibrating sources in water—for example, you push beach balls A and B up and down at the same frequency, one in each hand (**Figure 20.15a**). They each send out sinusoidal waves. Consider point C in Figure 20.15a. The sources A and B vibrate with the same amplitude and period, each sending a wave toward point C. The source A wave will cause C to vibrate as

$$y_A = A\cos\left[\frac{2\pi}{T}\left(t - \frac{a}{v}\right)\right]$$

The source B wave will cause C to vibrate as

$$y_B = A\cos\left[\frac{2\pi}{T}\left(t - \frac{b}{v}\right)\right]$$

If distances $a$ and $b$ are equal, then the two vibrations arrive in the same phase at point C

$$2\pi\frac{a}{Tv} = 2\pi\frac{b}{Tv}$$

If one wave causes C to move up, the other one will also cause it to move up (the displacements from the A and B waves at C are shown in Figure 20.15b). *If* the idea of superposition is correct *and* point C is at the same distance from both sources, *then* the amplitude of vibration at C will be twice the amplitude caused by one wave source. There are many points at the same distance from both sources, and the vibrations at these points will have twice the amplitude caused by one wave.

More generally, the superposition principle predicts a similar effect for any point where the difference in distance from source A and from source B is an integer multiple of the wavelength (if $a - b = n\lambda$ for $n = 0, \pm 1, \pm 2, \ldots$). At any of these points, the resultant vibration amplitude due to both waves will be the sum of the amplitudes of the individual waves.

Suppose now that we choose a point D located so that the wave from A arrives there with a maximum positive displacement and the wave from B arrives with a maximum negative displacement (see **Figure 20.16a**). Then the A wave pushes point D up while the B wave pushes it down the same distance. The net displacement is almost zero. This situation happens only if one wave has to travel a distance that is one-half, or three-halves, or five-halves, etc. times the wavelength $\lambda$ longer than the other (Figure 20.16b). We conclude that at

**Figure 20.15** The superposition at point C of two waves that add.

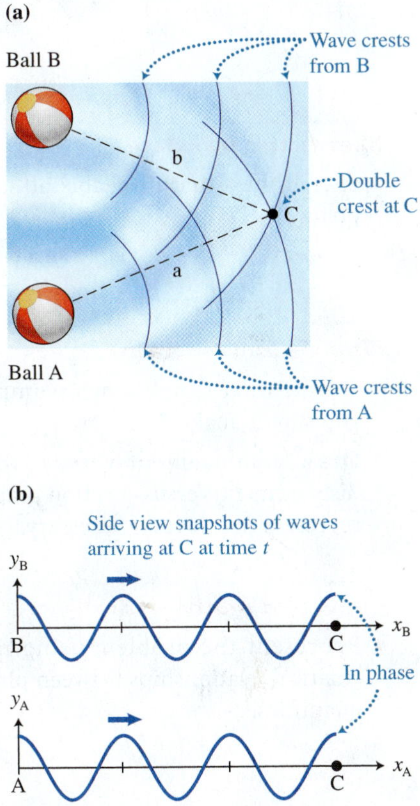


**Figure 20.16** The superposition at point D of two waves that cancel.

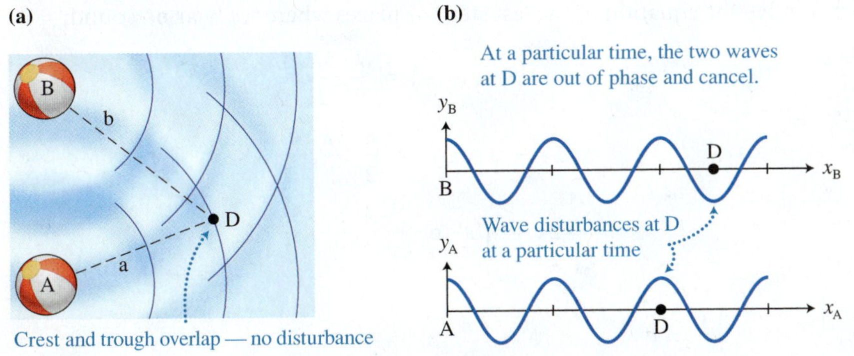

locations where $a - b = (n + 1/2)\lambda$ for $n = 0, \pm 1, \pm 2, \ldots$, there will be little vibration because the two waves will nearly cancel each other.

Let's test this canceling effect. Take two speakers at different locations and predict places between them where our ears will hear little sound. A procedure for analyzing wave interference processes is described on the left side of the table and applied to the problem in Example 20.4 on the right side.

## PROBLEM-SOLVING STRATEGY Applying our knowledge of wave motion

### EXAMPLE 20.4 Where is there no sound?

Two sound speakers separated by 100 m face each other and vibrate in unison at a frequency of 85 Hz. Determine three places on a line between the speakers where you cannot hear any sound.

**Sketch and translate**

- Sketch the situation. Label all known quantities.

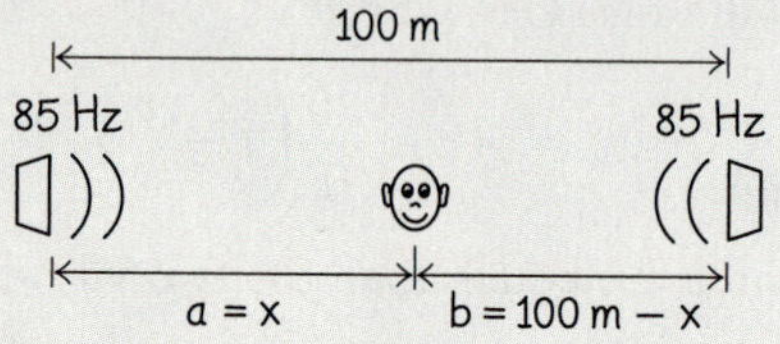


**Simplify and diagram**

- Decide what simplifying assumptions you should make.
- Draw displacement-versus-time or displacement-versus-position graphs to represent the waves if necessary.

Assume that the amplitudes of the waves are equal when they arrive at the point of interest. The person should hear no sound at places where the distance from the left speaker to the person is an odd multiple of a half wavelength farther than the distance from the right speaker. The displacement-versus-position graph for this situation is shown in Figure 20.16b.

**Represent mathematically**

- Represent the problem using mathematical relationships between physical quantities.

You should hear no sound if

$$a - b = \left(n + \frac{1}{2}\right)\lambda \text{ for } n = 0, 1, 2, \ldots .$$

Here $a$ is $x$ and $b$ is $100\text{ m} - x$:

$$x - (100\text{ m} - x) = \frac{\lambda}{2}, \frac{3\lambda}{2}, \frac{5\lambda}{2} \ldots$$

where

$$\lambda = \frac{v_{\text{sound}}}{f} = \frac{340\text{ m/s}}{85\text{ Hz}} = \frac{340\text{ m/s}}{85\text{ s}^{-1}} = 4.0\text{ m}$$

**Solve and evaluate**

- Solve the equations for the unknown quantity.
- Evaluate whether the results are reasonable (the magnitude of the answer, its units, how the solution behaves in limiting cases, and so forth).

Solve the equation in the last step for places where we hear no sound:

$$x - (100\text{ m} - x) = \frac{\lambda}{2}, \frac{3\lambda}{2}, \frac{5\lambda}{2} \ldots$$

$$\Rightarrow 2x - 100\text{ m} = \frac{\lambda}{2}, \frac{3\lambda}{2}, \frac{5\lambda}{2} \ldots$$

$$\Rightarrow 2x = 100\text{ m} + \frac{\lambda}{2}, \frac{3\lambda}{2}, \frac{5\lambda}{2} \ldots$$

$$\Rightarrow x = \left[50\text{ m} + \frac{\lambda}{4}\right], \left[50\text{ m} + \frac{3\lambda}{4}\right], \left[50\text{ m} + \frac{5\lambda}{4}\right] \ldots$$

Using $\lambda = 4.0$ m, we expect no sound at

$$x = 51\text{ m}, 53\text{ m}, 55\text{ m}, \ldots$$

Note that the first place is 51 m from the left speaker and 49 m from the right (one half wavelength different). The next place is 53 m − 47 m = 6 m or 1.5 wavelengths. The answers seem reasonable. When we conduct the experiment we find that at the predicted locations we hear almost no sound.

**Try it yourself:** For the situation described in the above example, where should you stand to hear sound with twice the amplitude?

*Answer:* The condition for these places is that the distances from the ear to each source should differ by an integer multiple (including zero) of the wavelength of the waves: $x - (100\text{ m} - x) = 0, \lambda, 2\lambda, \ldots$. This condition is satisfied at positions 50 m, 52 m, 54 m, ....

We can now assert the superposition principle for waves with greater confidence.

**Superposition principle for waves** **When multiple waves pass through the same medium at the same time, the net displacement at a particular time and location in the medium is the sum of the displacements that would be caused by each wave if it were alone in the medium at that time. Mathematically this statement can be written as**

$$y_{\text{net}} = y_1 + y_2 + \ldots \quad (20.8)$$

The process of two or more waves of the same frequency overlapping is called **interference**. Places where the waves add to create a larger disturbance are called locations of **constructive interference** (point C in Figure 20.15). Places where two waves add to produce a smaller disturbance are called locations of **destructive interference** (point D in Figure 20.16).

## Huygens' principle

When a beach ball bobs up and down in water, circular waves move outward. We can explain the formation of the waves using the superposition principle, first explained this way by Christiaan Huygens (1629–1695). It begins with a wave front represented by curved line AB moving at speed $v$ away from a point source $S$ (**Figure 20.17**). To determine the location of the wave front a short time interval $\Delta t$ later, one needs to draw a large number of circular arcs, each centered on a different point of wave front AB. Each arc represents a **wavelet** that moves away from a point on the original wave front. The addition of these wavelets produces a new wave front whose shape can be determined if one adds the disturbances due to all wavelets. The new wave front is determined by drawing a tangent to the front edges of the wavelets—line CD in Figure 20.17. To test Huygens' idea, consider two new experiments in Testing Experiment **Table 20.5**.

**Figure 20.17** Using Huygens' principle to create a new wave front.

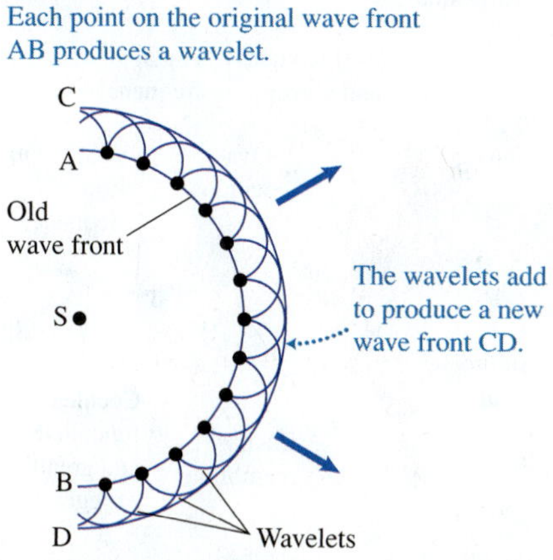
Each point on the original wave front AB produces a wavelet.
C
A
Old wave front
S
The wavelets add to produce a new wave front CD.
B
D
Wavelets

**TESTING EXPERIMENT TABLE**

**20.5 Testing Huygens' principle.**

VIDEO 20.5

| Testing experiment | Prediction | Outcome |
|---|---|---|
| Take a strip of cardboard with many closely spaced pushpins pushed through it from the bottom in a line. 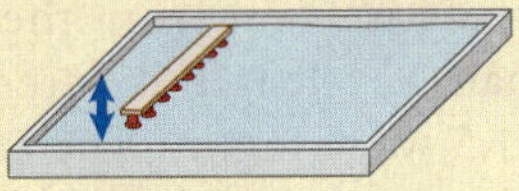 Place the cardboard just above the water so that when it vibrates, the pins move up and down, each creating a wavelet in the water. Then shake the cardboard to produce vibrations of the pins in the water. | When added, the wavelets produced by each pin should produce a straight wave front. 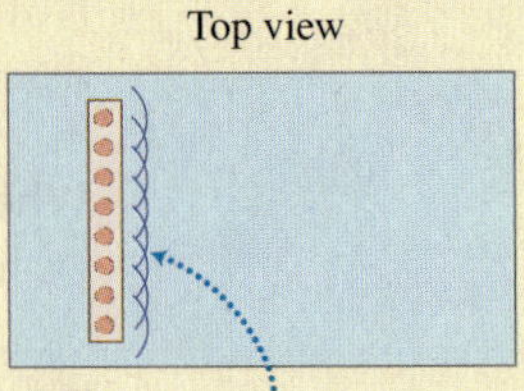  | We observe the predicted straight wave front. |
| Remove the pins from the cardboard and shake the board. When the board vibrates in a tray of water, it creates a wave with wave fronts that look like lines (plane waves). Place a barrier with a narrow slit opening in the path of the plane waves. 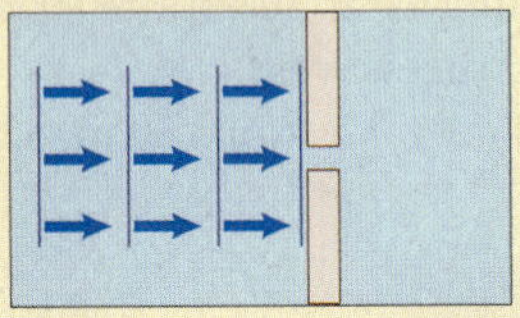  | The vibrations of water beyond the narrow slit should be circular wavelets. 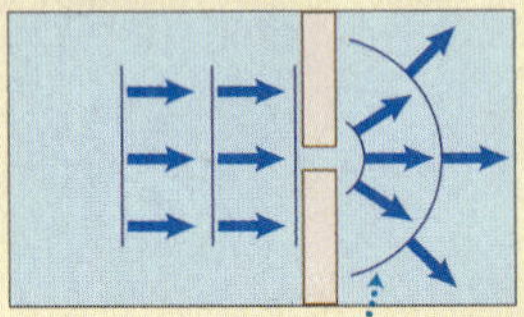  | We observe a circular wavelet spreading beyond the narrow slit. |

**Conclusion**

Predictions based on Huygens' principle match the outcomes of the testing experiments. These experiments do not disprove the principle but rather give us more confidence in the principle.

**Review Question 20.6** Your friend says that it is impossible for two waves to arrive at the same point and still have no vibrational motion at this point. How can you convince him that this is possible?

## 20.7 Sound

We hear sound when pressure variations at the eardrum cause it to vibrate (**Figure 20.18**). Amplification mechanisms in the outer ear and middle ear increase the pressure variations up to a factor of 180 and transform them into a pressure fluctuation in the fluid inside the inner ear. Nerve cells in the inner ear can distinguish sounds differing by just 3 Hz over a range of approximately 20 Hz to 20,000 Hz. Pressure waves in this frequency range are called **sound**. Higher or lower frequency pressure waves are called **ultrasound** or **infrasound,** respectively. Many animals, such as bats, dolphins, whales, dogs, and some fish, can sense ultrasonic or infrasonic frequencies.

**Figure 20.18** A simplified sketch of the human ear.

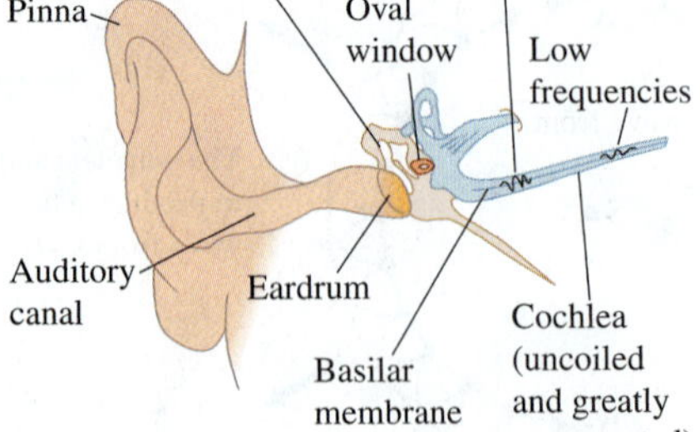


### Loudness and intensity

**Loudness** is determined primarily by the amplitude of the sound wave. The larger the amplitude, the louder the sound. However, human ears are not equally sensitive to all frequencies of sound. They are most sensitive to frequencies from 2000 to 5000 Hz and least sensitive to sounds at the lowest and highest frequencies: 20 Hz and 20,000 Hz. This means that equal-amplitude sound waves of different frequencies will not have the same perceived loudness to humans. A wave at 2000 Hz will seem much louder than a wave of the same amplitude but of 15,000-Hz frequency.

Sound waves are traveling pressure variations. Normal atmospheric pressure is $1.0 \times 10^5\ \mathrm{N/m^2}$. The pressure variation above and below atmospheric pressure of a barely audible sound is called the **threshold of audibility**. For the normal ear of a young person, this threshold is about $2 \times 10^{-5}\ \mathrm{N/m^2}$, or less than one-billionth of atmospheric pressure. Sound that creates pressure that is about $10^6$ times greater than this threshold, or about $20\ \mathrm{N/m^2}$, damages the sensitive cells of the inner ear as well as the auditory nerves. Instruments that measure the loudness of a sound do not measure the sound's pressure amplitude but instead measure its intensity: the energy per unit area per unit time interval, or $\mathrm{W/m^2}$ [Eq. (20.6)] delivered by the sound to the instrument.

**QUANTITATIVE EXERCISE 20.5 Sound energy at a construction site**

The intensity of sound at a construction site is $0.10\ \mathrm{W/m^2}$. If the area of the eardrum is $0.20\ \mathrm{cm^2}$, how much sound energy is absorbed by one ear in an 8-h workday?

**Represent mathematically** The intensity $I$ of the sound, the area $A$ of the eardrum, and the time interval $\Delta t$ of the exposure have been given. We wish to determine the sound energy $\Delta U$ absorbed by the ear. Equation (20.6) relates all of these quantities:

$$\text{Intensity} = I = \frac{\text{Energy}}{\text{time} \cdot \text{area}} = \frac{\Delta U}{\Delta t \cdot A}$$

To find the energy, multiply both sides by the time interval and the area:

$$\Delta U = I \cdot \Delta t \cdot A$$

**Solve and evaluate** In using the above, we need to convert the time interval into seconds and the area into $\mathrm{m^2}$.

$$\Delta U = (0.10\ \mathrm{W/m^2})\left[(8.0\ \mathrm{h})\left(\frac{3600\ \mathrm{s}}{1\ \mathrm{h}}\right)\right] \times \left[(0.20\ \mathrm{cm^2})\left(\frac{10^{-4}\ \mathrm{m^2}}{1\ \mathrm{cm^2}}\right)\right] = 0.058\ \mathrm{J}$$

This looks like a small number. However, our perception of sound energy depends on the power per unit area, not the total energy. Our environment includes sounds that differ in intensity from about $10^{-12}\ \mathrm{W/m^2}$ (barely audible) to nearly $1\ \mathrm{W/m^2}$ (a painful sound). The sound intensity in a typical classroom is about $10^{-7}\ \mathrm{W/m^2}$.

**Try it yourself:** How would the answer differ for a child with an eardrum that is half the area?

*Answer:* The sound energy would be half as big.

## Intensity level

Because of the wide variation in the range of sound intensities, a quantity called **intensity level** is commonly used to compare the intensity of one sound to the intensity of a reference sound. Intensity level is defined on a base 10 logarithmic[1] scale as follows:

$$\beta = 10 \log_{10} \frac{I}{I_0} \qquad (20.9)$$

where $I_0$ is a reference intensity to which other intensities $I$ are compared (for sound, $I_0 = 10^{-12}\ \mathrm{W/m^2}$, the smallest sound intensity typically detectable by humans). Since $I$ and $I_0$ in Eq. (20.9) have the same units, intensity level is a dimensionless quantity. Intensity level does have a dimensionless unit associated with it, called the decibel (dB). An increase of 10 dB is equivalent to a 10-fold increase in intensity. The unit serves as a reminder of what we are quantifying, much like the radian is a dimensionless reminder of one way of quantifying angles. The intensity levels $\beta$ of several familiar sounds are listed in **Table 20.6.**

[1] If $x = b^y$, then $y = \log_b(x)$. For common logarithms $b = 10$. Thus for $y = \log(x)$, $x = 10^y$.

**Table 20.6** **Intensities and intensity levels of common sounds.**

| Source of sound | Intensity ($W/m^2$) | Intensity level (dB) | Description |
|---|---|---|---|
| Large rocket engine (nearby) | $10^6$ | 180 | |
| Jet takeoff (nearby) | $10^3$ | 150 | |
| Pneumatic riveter; machine gun (nearby) | 10 | 130 | |
| Rock concert with amplifiers (2 m); jet takeoff (60 m) | 1 | 120 | Pain threshold |
| Construction noise (3 m) | $10^{-1}$ | 110 | |
| Moving subway train (nearby) | $10^{-2}$ | 100 | |
| Heavy truck (15 m) | $10^{-3}$ | 90 | Constant exposure |
| Niagara Falls (nearby) | $10^{-3}$ | 90 | endangers hearing |
| Noisy office with machines; inside an average factory | $10^{-4}$ | 80 | |
| Busy traffic | $10^{-5}$ | 70 | |
| Normal conversation (1 m) | $10^{-6}$ | 60 | |
| Quiet office | $10^{-7}$ | 50 | Quiet |
| Library | $10^{-8}$ | 40 | |
| Soft whisper (5 m) | $10^{-9}$ | 30 | |
| Rustling leaves | $10^{-10}$ | 20 | Barely audible |
| Normal breathing | $10^{-11}$ | 10 | |
| | $10^{-12}$ | 0 | Hearing threshold |

### QUANTITATIVE EXERCISE 20.6 Busy classroom

The sound in an average classroom has an intensity of $10^{-7}\ W/m^2$. (a) Determine the intensity level of that sound. (b) If the sound intensity doubles, what is the new sound intensity level?

**Represent mathematically** To calculate the intensity level $\beta$, we need to compare the given sound intensity $I$ with the reference intensity for sound $I_0 = 10^{-12}\ W/m^2$ and then take 10 times the log of that ratio (see the appendix for more information on logarithms):

$$\beta = 10 \log\left(\frac{I}{I_0}\right)$$

**Solve and evaluate** (a) The ratio of the sound intensity $I$ and the reference intensity $I_0$ is

$$\frac{I}{I_0} = \frac{10^{-7}\ W/m^2}{10^{-12}\ W/m^2} = 10^5$$

Since $\log 10^5$ is 5, the sound intensity level is

$$\beta = 10 \log 10^5 = 10(5) = 50\ dB$$

The classroom is pretty quiet.

(b) If $I$ is doubled, the ratio of $I$ and $I_0$ becomes

$$\frac{I}{I_0} = \frac{2 \times 10^{-7}\ W/m^2}{10^{-12}\ W/m^2} = 2 \times 10^5$$

The sound intensity level is then

$$\beta = 10 \log(2 \times 10^5) = 10(5.3) = 53\ dB$$

A doubling of intensity corresponds to an intensity level increase of just 3 dB.

**Try it yourself:** The sound intensity at a concert is $0.10\ J/s \cdot m^2$. What is the intensity level?

*Answer:* 110 dB.

**TIP** Remember that a 60-dB sound has 10 times the intensity of a 50-dB sound, which has 10 times the intensity of a 40-dB sound, and so on.

**Review Question 20.7** How is the loudness of sound related to the physical quantities that characterize wave motion?

# 20.8 Pitch, frequency, and complex sounds

Tuning forks of different sizes produce sounds of approximately the same intensity, but we hear each as having a different **pitch**—the perception of the frequency of a sound. The shorter the length of the tuning fork, the higher the frequency and the higher the pitch. As with loudness, pitch is not a physical quantity but a subjective impression. It does not have any units.

## Complex sounds, waveforms, and frequency spectra

There is more to sound than the sensations of its loudness and pitch. For example, an oboe and a violin playing concert A equally loudly at 440 Hz sound very different. What other property of sound causes this different sensation to our ears? Let's use a microphone and a computer to look at a pressure-versus-time graph of a sound wave produced by playing one note on a piano. Notice that the graph in **Figure 20.19** has negative values; thus it must be representing the pressure difference from atmospheric, or gauge, pressure. There is a characteristic frequency, but the wave is not sinusoidal. What causes this different shape?

To help answer this question, let's use two tuning forks with the second fork vibrating at twice the frequency of the first. When we strike each fork separately, the waves are sinusoidal (**Figures 20.20a** and b). When we hit the two forks simultaneously, the wave looks different (Figure 20.20c)—it is more like that recorded when playing the piano. It must be a superposition of the waves from the two tuning forks. This complex pattern is called a **waveform**—a combination of two or more waves of different frequencies and possibly different amplitudes.

It is possible, using a microphone and a device known as a spectrum analyzer, to decompose a complex waveform into its component frequencies. The amplitude of each component can be plotted versus the frequency of that component, a graph called the **frequency spectrum** of the waveform. A frequency spectrum of the waveform in Figure 20.20c is shown in **Figure 20.21**. The height of the line at each frequency is proportional to the amplitude of the wave at that frequency. In this case, the graph shows that the amplitude and frequency of wave 2 are greater than those of wave 1.

Most sounds we hear are complex waves consisting of sinusoidal waves of many different frequencies. "White noise"—the sound you hear if you tune a radio to a frequency that doesn't match a nearby radio station—includes waves of a broad range of frequencies. We seldom associate a pitch with white noise. However, if the complex sound is the result of the addition of a relatively large amplitude lowest frequency component (called the **fundamental**) with several higher frequency components that are whole-number multiples of the fundamental (called **harmonics**), then we usually identify the pitch of the sound as being that of the fundamental frequency. The waveforms and frequency spectra of complex waves from a piano and a violin while playing concert A at 440 Hz are shown in **Figure 20.22**. Notice that the violin A has many frequency components, which are all multiples of 440 Hz (for example, 880 Hz, 1320 Hz, 1760 Hz, 2200 Hz, and even higher frequencies). The violin's complex wave has a richer tone than the less complex piano sound at the same frequency and loudness. Different musical instruments have different frequency spectrums that contribute in part to the quality of the sound we associate with that instrument.

**Figure 20.19** A pressure-versus-time graph of sound from one piano note.

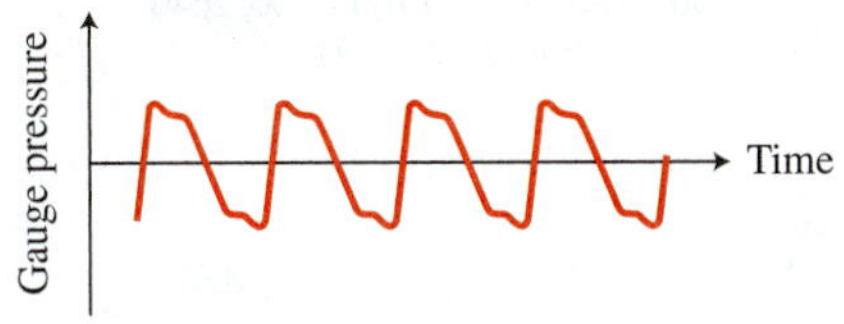


**Figure 20.20** Pressure-versus-time graphs produced by one or two tuning forks.

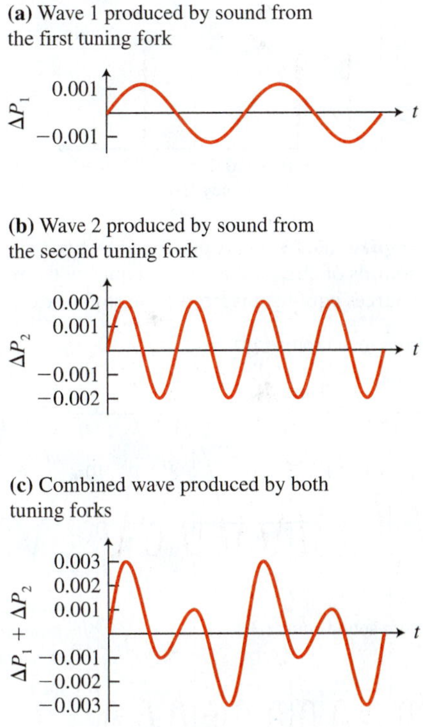


**Figure 20.21** A frequency spectrum produced by two tuning forks.

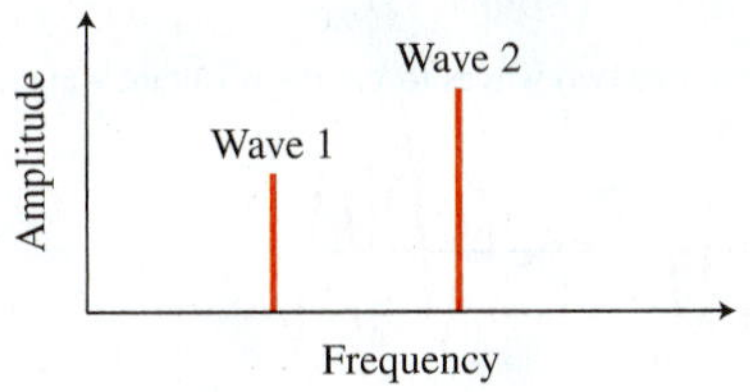

**Figure 20.22** Waveforms and frequency spectra for a piano and a violin playing concert A.

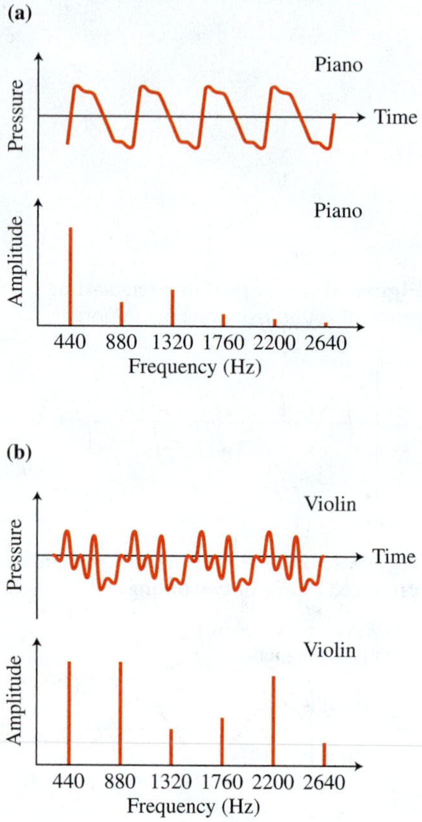


**Figure 20.23** Beats produced by two sounds of almost the same frequency from sources equidistant from a microphone.

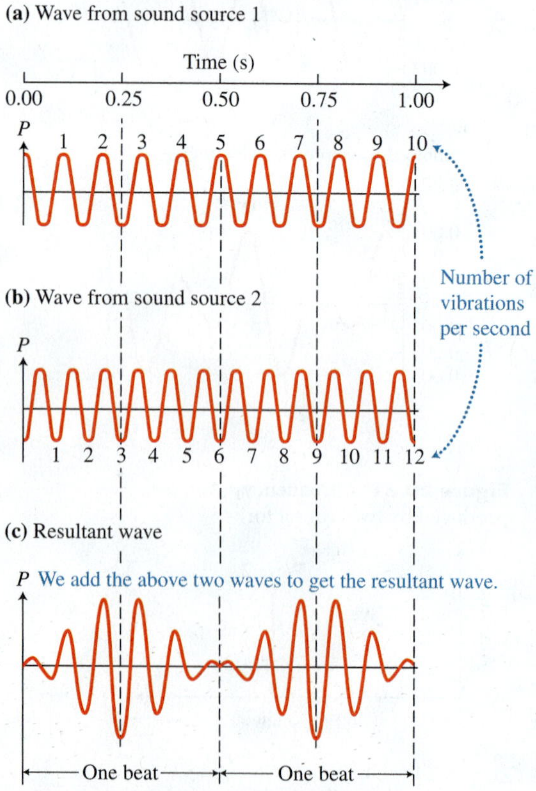


## Beat and beat frequencies

What happens when two sound waves of close but not identical frequencies overlap? Suppose that two sound sources are equidistant from a microphone that records the air pressure variations due to the two sound sources as a function of time (**Figure 20.23**). Source 1 vibrates at 10 Hz and source 2 vibrates at 12 Hz. Figures 20.23a and b show the pressure variations due to wave 1 and wave 2 during a 1.0-s interval. Note that in each case only one wave source is playing at a time. Consider the pressure-versus-time wave pattern we would get if both sound sources played at the same time.

We add the two waves shown in Figures 20.23a and b to see the pattern produced when both sounds play simultaneously (Figure 20.23c). Note in Figure 20.23c that at time 0.00 s the positive pressure caused by wave 1 cancels the negative pressure caused by wave 2. The net disturbance at time 0.00 s is zero and there is no sound. At 0.25 s, both waves have a maximum negative pressure variation, so the net disturbance is twice that of a single wave—the sound is loud. At 0.50 s, the two wave pressure variations again cancel each other, causing a zero pressure disturbance and no sound.

The pattern resulting from adding two waves of close frequencies shown in Figure 20.23c from 0.00 s to 0.50 s is called a **beat.** The vibration has a frequency of 11 Hz (the average of the two wave frequencies) but its amplitude is modulated from zero to approximately twice the normal amplitude and back to zero again at 0.50 s. A second beat is produced from 0.50 s to 1.00 s. The **beat frequency** is the number of beats per second—in this case 2 beats/s—the sound goes on and off twice each second. The beat frequency is just the difference in the two wave frequencies: $12\text{ Hz} - 10\text{ Hz} = 2\text{ Hz}$. This modulated vibration frequency is easily heard if the beat frequency is a few hertz or less.

**A beat is a wave that results from the superposition of two waves of about the same frequency. The beat (the net wave) has a frequency equal to the average of the two frequencies and has variable amplitude. The frequency with which the amplitude of the net wave changes is called the beat frequency $f_{beat}$; it equals the difference in the frequencies of the two waves:**

$$f_{beat} = |f_1 - f_2| \quad (20.10)$$

Beats are useful in precise frequency measurements. For example, a piano tuner can easily set the frequency of the middle C string on the piano to 262 Hz by listening to the beat frequency produced when the piano string and a 262-Hz tuning fork are sounded simultaneously. If the tuner hears a beat frequency of 3 Hz, then the piano string must be vibrating at either 259 Hz or 265 Hz—a difference of 3 Hz from the frequency of the tuning fork. The string is then tightened to increase the string frequency or loosened to lower the frequency until the beat frequency is reduced to zero (you do not hear beats anymore) and the string is vibrating at exactly 262 Hz.

**Review Question 20.8** What happens when two sinusoidal sounds, one at twice the frequency of the other, are produced simultaneously? How do we know?

# 20.9 Standing waves on strings

The superposition of waves is the basis for understanding how musical instruments produce sound. Let's consider stringed instruments first. Start by observing the vibration of a rope with one end attached to a fixed support (a wall) while the other end is held in your hand. You shake that end up and down. Depending on the frequency with which you shake your hand, the rope has either very small amplitude irregular vibrations or large amplitude sine-shaped vibrations. The latter appear not to be traveling waves but are instead coordinated vibrations, with parts of the rope vibrating up and down with large amplitude and other parts not vibrating at all. Such waves are called **standing waves.** Displacement-versus-position graphs for several of those waves for one instant are shown in **Figure 20.24** (page 758). Because stringed musical instruments have both ends of their strings fixed, the motion of the strings forms standing waves. How is the first standing wave vibration in Figure 20.24 produced? We will explore this question in Observational Experiment **Table 20.7**.

## OBSERVATIONAL EXPERIMENT TABLE

### 20.7 Producing the first standing wave vibration.

VIDEO 20.7

**Observational experiment**

You shake a rope that is fixed at one end at a relatively low frequency and observe the large-amplitude vibration.

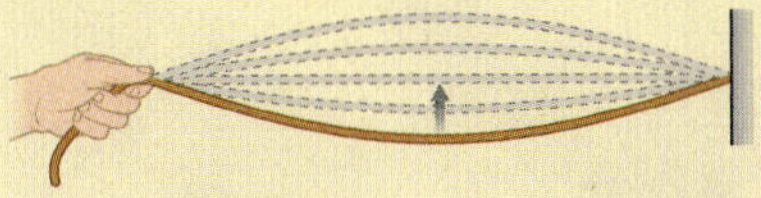

**Analysis**

A pulse starts moving to the right (b–d). When it reaches the fixed end, the upward pulse is reflected and inverted to a downward pulse traveling back toward your hand (e–g). When it reaches your hand, it once again reflects and inverts to an upward pulse (h). To make this pulse bigger, you begin another upward shake, timing the shake so that it adds to the previous upward pulse that has just been reflected (h).

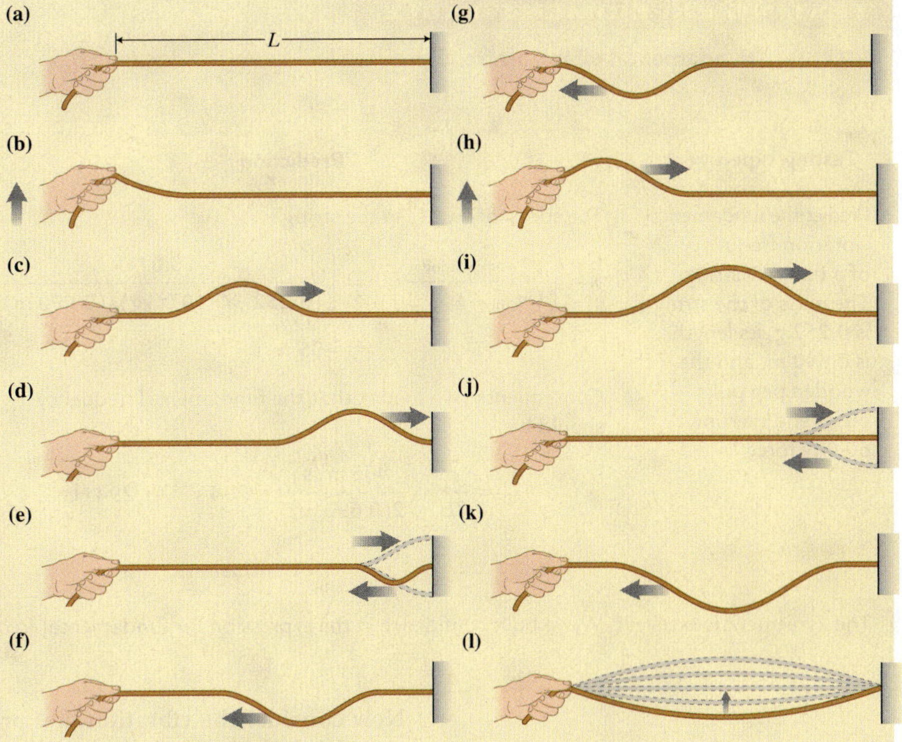


**Active Learning Guide>**

**Pattern**

- You must shake the rope upward each time a previous pulse returns. The amplitude of the disturbance traveling along the rope grows.
- You need one up and down shake during the time interval $(2L/v)$ that it takes for the pulse to travel down the string and back:

$$f_1 = \frac{1 \text{ vibration}}{\dfrac{2L}{v}} = \frac{v}{2L}$$

**Figure 20.24** Standing waves on a string.

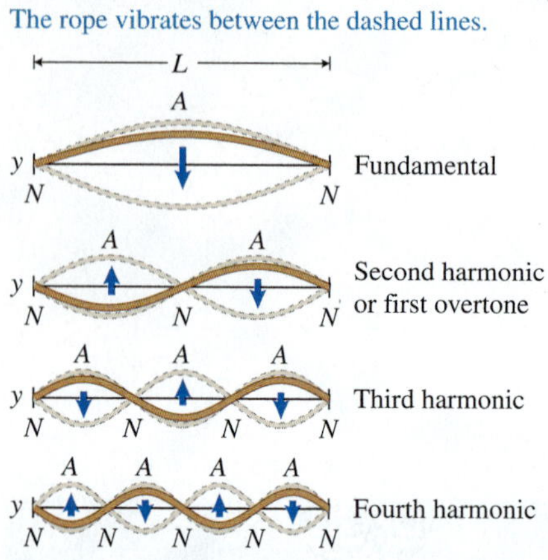


The subscript 1 in the last equation in Table 20.7 means the lowest standing wave frequency. The numerator in that equation is the number of vibrations, but *vibration* is not a unit (similar to the radian and the decibel). Thus, this lowest frequency vibration of the rope is one up and down shake per time interval $\tau = 2L/v$ or

$$f_1 = \frac{1}{\tau} = \frac{1}{\frac{2L}{v}} = \frac{v}{2L} \tag{20.11}$$

This frequency is called the **fundamental frequency.** If the shaking frequency is just a little greater or a little less than this, each new pulse will interfere destructively with previous pulses, and there will no longer be a time-independent pattern; the standing wave will disappear and multiple traveling waves will bump here and there, like splashing water. At frequency $f_1 = v/(2L)$, each new pulse constructively interferes with previous pulses and creates a large-amplitude vibration. The wavelength of this fundamental frequency vibration is

$$\lambda_1 = \frac{v}{f_1} = \frac{v}{\frac{v}{2L}} = 2L$$

The length of the rope equals one half of the wavelength of the wave that has this fundamental frequency. We will test these ideas in Testing Experiment **Table 20.8**.

## TESTING EXPERIMENT TABLE

### 20.8 Fundamental vibration frequency.

| Testing experiment | Prediction | Outcome |
|---|---|---|
| Predict the fundamental vibration frequency of a banjo D string. The mass of the string is 0.252 g, its length is 0.690 m, and the wooden peg pulls on its end exerting a 60.0-N force. | The speed of a wave in the string is $v = \left[\frac{F_{\text{Peg on String}}}{(m_{\text{string}}/L_{\text{string}})}\right]^{1/2} = \left[\frac{(60.0\text{ N})}{(0.252 \times 10^{-3}\text{ kg})/(0.690\text{ m})}\right]^{1/2} = 405\text{ m/s}$ Consequently, we predict that the fundamental frequency should be $f_1 = \frac{v}{2L} = \frac{405\text{ m/s}}{2(0.690\text{ m})} = 294\text{ s}^{-1} = 294\text{ Hz}$ | When we pluck a banjo D string and compare it to the frequency of a tuning fork, we find $f_1 = 294$ Hz. |

**Conclusion**

The agreement is excellent. We've built confidence in the expression for fundamental frequency.

Now consider the vibration that produces a standing wave on the rope at twice the frequency of the fundamental vibration (the second wave in Figure 20.24). You vibrate the rope so that your hand has two up and down shakes during the time interval that is needed for a pulse to make a round trip down the rope and back (**Figure 20.25**). The new pulse adds to a pulse started earlier that has completed a round trip back to your hand. Thus, the frequency $f_2$ of this second standing wave vibration is

$$f_2 = \text{second standing wave vibration} = \frac{2\text{ vibrations}}{\frac{2L}{v}} = 2\frac{v}{2L}$$

We can see a pattern now.

> **Standing wave frequencies on a string** The frequencies $f_n$ at which a string with fixed ends can vibrate are
>
> $$f_n = n\left(\frac{v}{2L}\right) \text{ for } n = 1, 2, 3, \ldots \qquad (20.12)$$
>
> where $v$ is the speed of the wave on the string and $L$ is the string length.

Since $\lambda = v/f$, we can write an expression for the wavelengths of these allowed standing wave vibrations:

$$\lambda_n = \frac{v}{f_n} = \frac{v}{n\left(\frac{v}{2L}\right)} = \frac{2L}{n} \text{ for } n = 1, 2, 3, \ldots \qquad (20.13)$$

Note that a rope or string will vibrate with large amplitude at any frequency for which a pulse produced by the wave source (for example, the vibrating hand) is reinforced by a new pulse after completing a round trip up the string and back again. Thus standing waves are caused by the superposition of waves moving in opposite directions.

How are standing waves different from traveling waves? When a traveling wave moves along the string, each point on the string vibrates between the maximum and minimum displacements (between $y = \pm A$). However different points reach the maximum displacements at different times (the phase of the wave is different at different locations). In a standing wave, only very specific points on the string, called **antinodes** (labeled A in Figure 20.24), reach the maximum displacement. Other points, called **nodes** (labeled N in Figure 20.24), do not vibrate at all. However, each point reaches its respective maximum displacement at the same time—the points vibrate in phase. When you play a musical instrument, you excite multiple standing wave vibrations. For example, when you bow a violin string, you simultaneously excite 10 or 20 standing wave vibrations.

**Figure 20.25** Producing the second standing wave vibration on a string or rope.

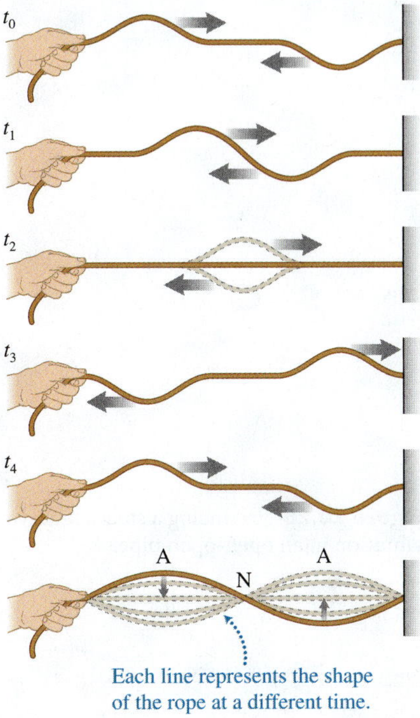

$t_0$
$t_1$
$t_2$
$t_3$
$t_4$
A
N
A
Each line represents the shape of the rope at a different time.

### EXAMPLE 20.7 Playing a banjo

A banjo D string is 0.69 m long and has a fundamental frequency of 294 Hz. Shortening the string (by holding the string down over a fret) causes it to vibrate at a higher fundamental frequency. Where should you press to play the note F-sharp, which has a fundamental frequency of 370 Hz?

**Sketch and translate** We know that when the length of the string is $L = 0.69$ m, the fundamental frequency is $f_1 = 294$ Hz. At what length of the string $L'$ is the fundamental frequency $f_1' = 370$ Hz?

**Simplify and diagram** Assume that the speed of the wave on the string is independent of where the finger pushes down on the string.

**Represent mathematically** The fundamental frequency $f_1$ relates to the string length $L$ by the equation $f_1 = v/(2L)$, or $L = v/(2f_1)$. We cannot solve immediately for the unknown length when the string is vibrating at 370 Hz since we do not know the speed $v$ of the wave on the string. But we can use this idea to write an expression for the speed of the wave for both cases (the speed is the same for both cases):

$$v = f_1(2L) = f_1'(2L')$$

**Solve and evaluate** Dividing both sides of the expression for the speed by $f_1'$ and by 2, we get an expression for $L'$:

$$L' = \frac{f_1 L}{f_1'} = \left(\frac{f_1}{f_1'}\right)L = \left(\frac{294\text{ Hz}}{370\text{ Hz}}\right)(0.69\text{ m}) = 0.55\text{ m}$$

The answer is reasonable, as it is less than 0.69 m. To find where to press on the string, we need to find the difference between the original length $L$ and the new length $L'$:

$$L - L' = 0.69\text{ m} - 0.55\text{ m} = 0.14\text{ m}$$

You should press 0.14 m from the end of the string.

**Try it yourself:** If tuned correctly, the fundamental frequency of the A string of a violin is 440 Hz. List the frequencies of several other standing wave vibrations of the A string.

*Answer:* The other standing wave frequencies are whole number multiples of the fundamental: 880 Hz, 1320 Hz, 1760 Hz, and so forth.

**Review Question 20.9** A horizontal string of length $L$ has one end passing over a pulley with a block hanging at the end. A motor attached to the other end of the string exerts a constant horizontal force on the string. When the motor is turned on, it vibrates the end of the string up and down at frequencies that increase slowly from zero to a high frequency. What do you observe?

## 20.10 Standing waves in air columns

Take a bottle, partially fill it with water, and blow across the top of the bottle. You hear sound with a pitch that depends on the amount of water in the bottle. The same phenomenon underlies how sound is made in musical instruments made of pipes or tubes, such as organs, clarinets, flutes, and trumpets. How can we explain this phenomenon?

### Standing waves in open-open pipes

**Figure 20.26** Producing a standing wave vibration in an open-open pipe.

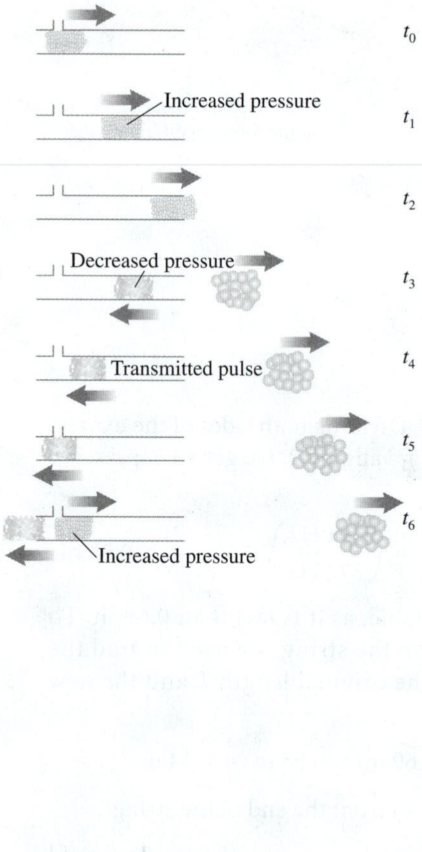


A flute is a wind instrument that is an example of an **open-open** pipe, so called because it is open on both ends. Blowing into the open-open pipe initiates pressure pulses at that end of it. Alternating increased and decreased pressure pulses move down the air column, compressing the air in the instrument in a repetitive fashion (**Figure 20.26** $t_0 \rightarrow t_2$). Near the opposite end, the flute is also open at the nearest open valve. The high-pressure pulse expands into open space outside of the valve, leaving behind a pulse of decreased pressure that reflects back into the tube. This change from high to low pressure is similar to the phase change of a pulse on a string reflected off a fixed end. This phase change does not occur when a pressure pulse reaches a closed end.

The pulse of decreased pressure moves back toward the other end of the pipe where it started ($t_3 \rightarrow t_5$) and is reflected at the open end on the left ($t_6$). The higher pressure outside the pipe causes the decreased pressure pulse to compress as air rushes in, becoming a high-pressure pulse. If a new high-pressure pulse is initiated at this time, it interferes constructively with the reflected pulse, and its amplitude increases. If this is done repeatedly, a large-amplitude standing wave vibration builds up in the open-open pipe. If, however, the new air pressure pulse is too early or too late, it interferes destructively with the reflected pulse, and the amplitude decreases.

The frequency of standing wave vibrations in pipes depends on the time interval needed for a pulse to travel down the pipe and back again. If the sound pulse travels at a speed $v$ along a pipe of length $L$, then the time interval $T$ needed to travel a distance $2L$ is $T = 2L/v$. To reinforce the vibration of air in the pipe, pressure pulses should occur once each time period $T$ or at a frequency $f = 1/T = v/(2L)$.

Just as we can make a string vibrate at multiples of the fundamental frequency, the pipe can also vibrate with large amplitude if excited by air pressure pulses at whole number multiples of the fundamental frequency.

**Standing wave vibration frequencies in open-open pipes** **The resonant frequencies $f_n$ of air in an open-open pipe of length $L$ in which sound travels at speed $v$ are**

$$f_n = n\left(\frac{v}{2L}\right) \text{ for } n = 1, 2, 3, \ldots \tag{20.14}$$

**where $v$ is the speed of sound in the medium that fills the pipe.**

## Standing waves in one-end closed pipes—an open-closed pipe

Next consider a pipe that is closed at one end and open at the other end, an **open-closed** pipe, such as a clarinet, trumpet, or the human throat (the closed end being the reed, mouthpiece, and vocal cords, respectively). Blowing into the reed or mouthpiece initiates a high-pressure pulse near the closed end of the pipe. After reflection off the open end, a low-pressure pulse returns to the closed end of the pipe and reflects. However, reflection off the closed end causes it to remain a low-pressure pulse. The pulse has to make another complete round trip in the pipe before a high-pressure pulse returns to the reed. Thus, the pulse has to make two round trips or travel a distance $4L$ before a new high-pressure pulse from the reed produces a fundamental frequency vibration. Thus, the fundamental frequency of an open-closed pipe is $f_1 = v/(4L)$ instead of $f_1 = v/(2L)$ for an open-open pipe. The fundamental frequency of an open-closed pipe clarinet is half the fundamental frequency of an open-open pipe flute of about the same length. Another difference is that the standing wave frequencies of open-closed pipes are not integer multiples of the fundamental but instead are odd whole number multiples, as the even whole numbers would cause waves in the open-closed pipe that interfere destructively.

‹Active Learning Guide

**Standing wave vibration frequencies in open-closed pipes** **The resonant frequencies $f_n$ of air in an open-closed pipe of length $L$ in which sound travels at speed $v$ are**

$$f_n = n\left(\frac{v}{4L}\right) \text{ for } n = 1, 3, 5, \ldots \qquad (20.15)$$

## Exciting the sound and changing the pitch of a wind instrument

We can apply our understanding of standing waves in pipes to explain how wind instruments produce sounds of a particular frequency. In clarinets and saxophones the sources of vibrations are the reeds. In trumpets, trombones, and French horns, the sources of vibrations are the vibrating lips and mouthpiece. The vibrations produced by the reeds and mouthpieces are not pure tones but are instead pressure pulses at a variety of frequencies. The pipes attached to the reeds and mouthpieces reinforce only those input frequency vibrations that are at resonant frequencies of the instruments.

A resonant frequency for a particular instrument can be changed by opening one or more valves or keys. Opening the valve makes a hole that serves as the open end of the pipe. Thus, the air column length in the pipe is changed, as is its fundamental frequency. Because usually more than one resonant frequency is excited at a time, sound from the instrument consists of multiple resonant frequencies. The combination of resonant frequencies contributes in part to the quality of the sound from the instrument.

## Testing our understanding of standing waves in pipes

*If* our reasoning is correct that sound in pipe-like instruments is due to standing waves inside of them, *then* changing the medium (which changes the wave speed) inside a pipe should change the fundamental frequency produced by the pipe even though the length of the pipe has not changed.

**CONCEPTUAL EXERCISE 20.8 Funny talk**

What should happen to a person's voice after inhaling helium and singing a note?

**Sketch and translate** The speed $v$ of sound in helium gas (930 m/s) is much greater than in air (340 m/s), and the fundamental frequency $f_1$ is proportional to the speed of sound in the medium.

**Simplify and diagram** Model the vocal chamber as an open-closed pipe with the vocal cords at the closed end and the mouth at the open end. The length of the chamber does not change during the experiment. Since the speed of sound in helium is almost three times higher than in air, the fundamental frequency of the sound should increase by a factor of 3 and we should hear a higher pitch. Inhaling helium before singing or talking creates a very high pitched sound—as predicted. However, without measuring the frequency we cannot say how much greater the frequency is.

**Try it yourself:** In our conceptual exercise with helium, we used gas with a greater speed of sound. What happens if we could inhale a gas that has a lower sound speed than in air, such as $CO_2$?

*Answer:* The pitch of one's voice would be lower. However, this gas is dangerous to inhale; nobody should perform this experiment. Thus this exercise is purely theoretical.

**Review Question 20.10** When we studied traveling waves, we decided that the wavelength of a wave depends on the frequency of vibration and the speed of the wave in the medium. Is this statement true for standing waves in pipes?

## 20.11 The Doppler effect: Putting it all together

**Active Learning Guide›**

When you hear the sound from the horn of a passing car, its pitch is noticeably higher than normal as it approaches and is noticeably lower than normal as it moves away. This phenomenon is an example of the **Doppler effect**. The Doppler effect occurs when a source of sound and an observer move with respect to each other and/or with respect to the medium in which the sound travels.

### Is there a pattern in these observed frequency shifts?

In Observational Experiment **Table 20.9**, we will investigate the Doppler effect, using a battery-powered whistle (the source) that generates a source frequency $f_S$ and a detector (the observer) that records the observed frequency $f_O$ using a spectrum analyzer, a special device that measures the frequency of sound. Put each on a cart that can move toward or away from the other. In all experiments the air in which the sound propagates is at rest with respect to Earth.

**OBSERVATIONAL EXPERIMENT TABLE**

**20.9 Effects of motion on observed frequency.**

| Observational experiment | Analysis |
|---|---|
| Place a battery-powered whistle (the source) on one cart and an observer with a spectrum analyzer on another cart. The carts can move toward or away from each other.<br>**(a) Source** stationary with respect to the ground<br>**Observer** moving toward source<br>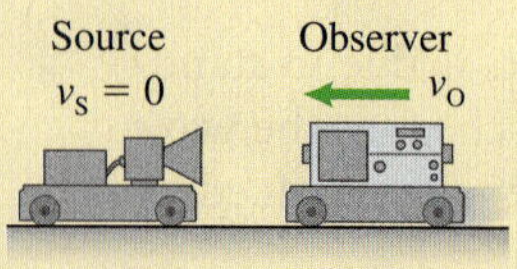<br> | The frequency spectrum for the signal emitted by the source ($f_S$) and heard by the observer ($f_O$)<br>$f_S < f_O$<br>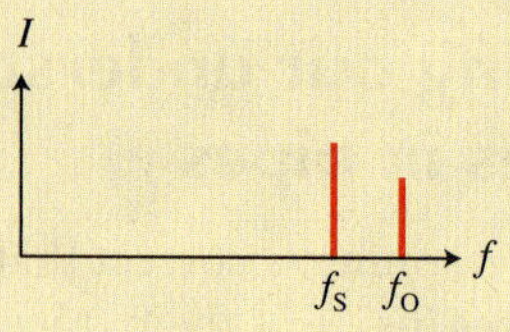<br> |

| Observational experiment | Analysis | |
|---|---|---|
| **(b) Source** moving toward observer<br>**Observer** stationary with respect to the ground<br>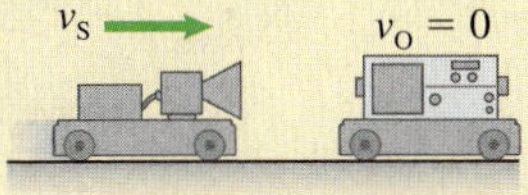  | The frequency spectrum:<br>$f_S < f_O$ | 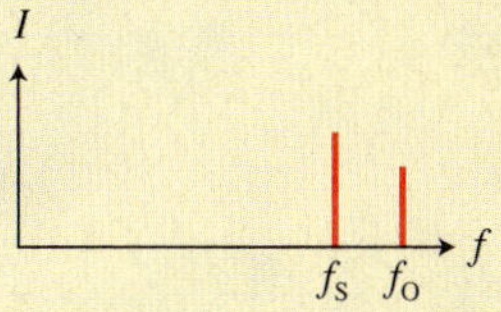  |
| **(c) Source** stationary with respect to the ground<br>**Observer** moving away from source<br>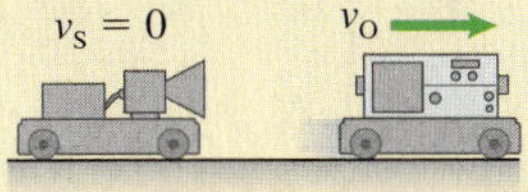  | The frequency spectrum:<br>$f_S > f_O$ | 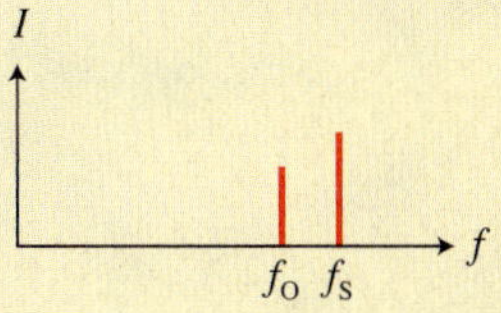  |
| **(d) Source** moving away from observer<br>**Observer** stationary<br>  | The frequency spectrum:<br>$f_S > f_O$ | 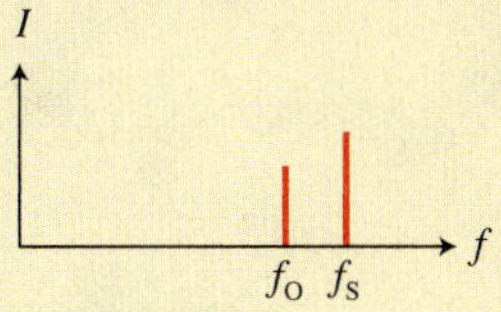  |
| **(e) Source** moving with respect to the ground<br>**Observer** moving at the same speed in the same direction<br>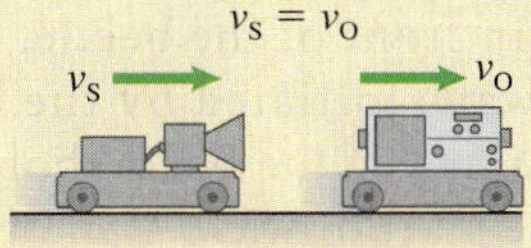  | The frequency spectrum:<br>$f_S = f_O$ | 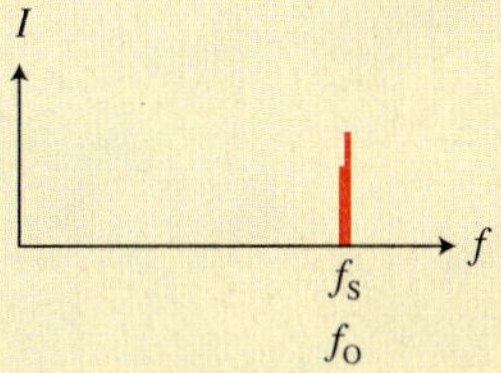  |

**Patterns**

- (a) and (b) Source and/or observer are moving *toward* each other: the observed frequency is *greater than* the emitted source frequency.
- (c) and (d) Source and/or observer are moving *away from* each other: the observed frequency is *less than* the source frequency.
- (e) There is no relative motion: the observed and emitted frequencies are the same.

Let's develop a quantitative explanation for the patterns observed in Table 20.9.

## Doppler effect for the source moving relative to the medium

We can use the waves created by a water beetle bobbing up and down at a constant frequency on the water to explain the Doppler effect. If the beetle bobs up and down in the same place, the pattern of wave crests appears as shown in **Figure 20.27a**. The crests move away from the source at a constant speed. The distance between adjacent crests is one wavelength and is the same toward observer A as toward observer B. The frequency of waves reaching both observers is also the same.

Now suppose the bobbing beetle moves to the right at a speed slower than the wave speed. Each new wave produced by the beetle originates from a point farther to the right (Figure 20.27b). As a result, B receives four wave crests in a shorter time interval than the source took to produce them, $f_O > f_S$. A receives four wave crests in a longer time than the source took to produce them, $f_O < f_S$.

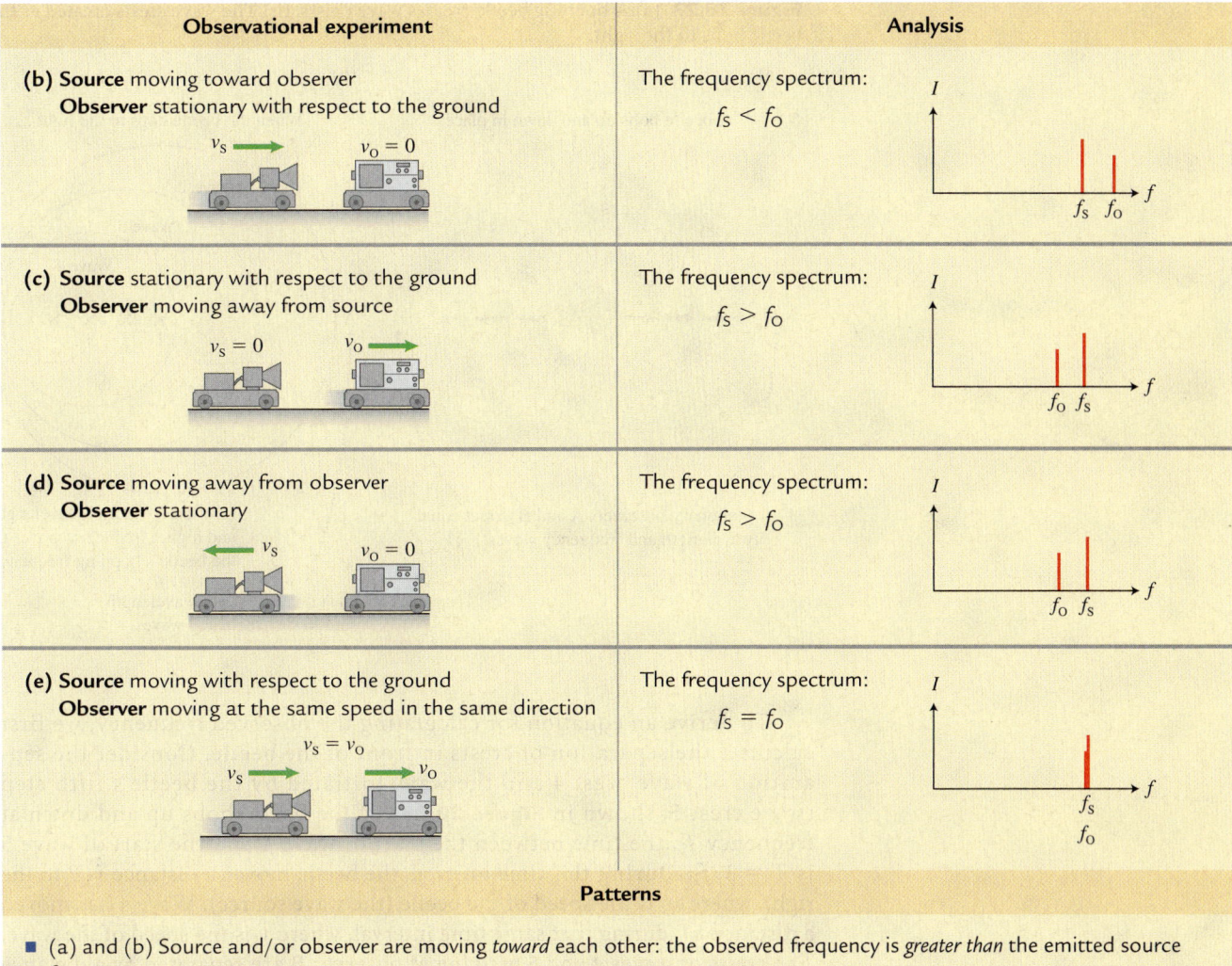

**Figure 20.27** (a) A bobbing beetle creates wave crests. (b) The wave crests created as the beetle hops to the right.

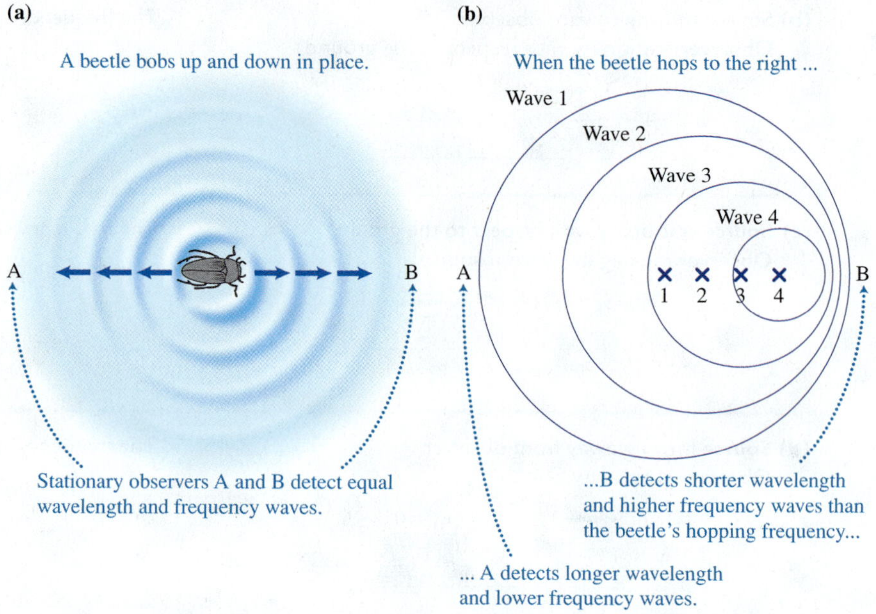
(a)
A beetle bobs up and down in place.
A
B
Stationary observers A and B detect equal wavelength and frequency waves.
(b)
When the beetle hops to the right ...
Wave 1
Wave 2
Wave 3
Wave 4
A
1
2
3
4
B
...B detects shorter wavelength and higher frequency waves than the beetle's hopping frequency...
... A detects longer wavelength and lower frequency waves.

To derive an equation for calculating the observed frequency, we first calculate the separation of crests in front of the beetle. Consider the separation of wave crest 4 and the wave initiated by the beetle's fifth step (wave crest 5, shown in **Figure 20.28**). If the beetle bobs up and down at frequency $f_S$, the time between the start of wave 4 and the start of wave 5 is $T = 1/f_S$. During this time interval, the beetle moves a distance $v_S T$ to the right, where $v_S$ is the speed of the beetle (the wave source). Wave 4 has moved a distance $vT$ during that same time interval, where $v$ is the speed of the wave. The crests of waves 4 and 5 arriving at observer B are separated by a distance $\lambda_B{}'$ (one wavelength), given by

$$\lambda_B{}' = vT - v_S T$$

Similarly, the beetle is moving away from A (see Figure 20.27) and the wavelength of the wave reaching A is

$$\lambda_A{}' = vT + v_S T$$

The waves are incident on observer B at a frequency

$$f_{OB} = \frac{\text{wave speed}}{\text{wavelength}} = \frac{v}{\lambda_B{}'} = \frac{v}{T(v - v_S)} = \frac{1}{T}\frac{v}{(v - v_S)} = f_S \frac{v}{v - v_S}$$

where $f_S = 1/T$ is the source frequency. The waves are incident on observer A at frequency

$$f_{OA} = \frac{\text{wave speed}}{\text{wavelength}} = \frac{v}{\lambda_A{}'} = \frac{v}{T(v + v_S)} = \frac{1}{T}\frac{v}{(v + v_S)} = f_S \frac{v}{v + v_S}$$

Combining these two equations, we get

$$f_O = f_S \frac{v}{v \mp v_S} \qquad (20.16)$$

We use the minus sign when the source is moving toward the observer and the plus sign when the source is moving away from the observer. This equation is consistent with the patterns discovered in Table 20.9.

**Figure 20.28** The wavelength is reduced in front of the hopping beetle.

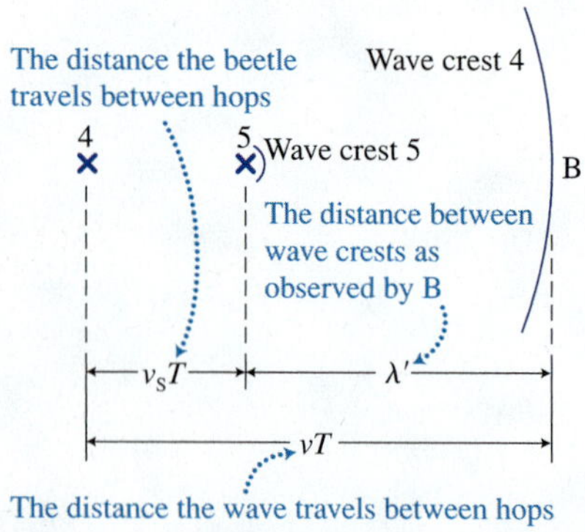
The distance the beetle travels between hops
Wave crest 4
4
5
Wave crest 5
B
The distance between wave crests as observed by B
$v_S T$
$\lambda'$
$vT$
The distance the wave travels between hops

## Doppler effect for the observer moving relative to the medium

The observed frequency also differs from the source frequency if the observer moves with respect to the medium and the source is stationary. If the observer moves toward the source at speed $v_O$, the wave fronts appear to move past the observer at speed $v + v_O$ and the observer encounters the fronts more frequently than if she had remained stationary. If the observer moves away from the source, the wave fronts appear to move past the observer at speed $v - v_O$ and she encounters the fronts less frequently.

Mathematically,

$$f_O = \frac{v \pm v_O}{\lambda} = \frac{v \pm v_O}{v/f_S} = f_S\frac{v \pm v_O}{v} \tag{20.17}$$

This equation can be used to calculate the frequency $f_O$ detected by an observer moving at speed $v_O$ toward (plus sign) or away from (minus sign) a stationary source emitting waves at frequency $f_S$.

## General equation for the Doppler effect for sound

We arrive at a general equation for the Doppler effect by combining Eqs. (20.16) and (20.17).

**Doppler effect for sound** **When a sound source and sound observer move relative to each other and/or the medium, the observed sound frequency $f_O$ differs from the source sound frequency $f_S$:**

$$f_O = f_S\left(\frac{v \pm v_O}{v \mp v_S}\right) \tag{20.18}$$

**where $v$ is the speed of waves through the medium, $v_O$ is the speed of the observer relative to the medium (use the plus sign if the observer moves toward the source and a minus sign if the observer moves away from the source), and $v_S$ is the speed of the source relative to the medium (use the minus sign if the source moves toward the observer and a plus sign if the source moves away from the observer).**

**TIP** Once you have chosen which signs to use in Eq. (20.18), check for consistency. When the source and observer are getting closer, $f_O > f_S$, and so on.

## Measuring the speed of blood flow in the body

The Doppler effect is used to measure the speed of blood flow in the body. Ultrasound waves ($f >$ 20,000 Hz) are directed into an artery (**Figure 20.29**). The waves are reflected by red blood cells back to a receiver. The frequency detected at the receiver $f_R$ relative to that emitted by the source $f_S$ indicates the blood cell's speed and the speed of the blood flow. This process has two parts: (1) the waves leave the source at frequency $f_S$ and strike the cell at frequency $f'$. (2) The waves reflect from the cell at frequency $f'$ and return to the receiver where they are detected at frequency $f_R$.

**Figure 20.29** Using the Doppler effect to measure blood flow speed.

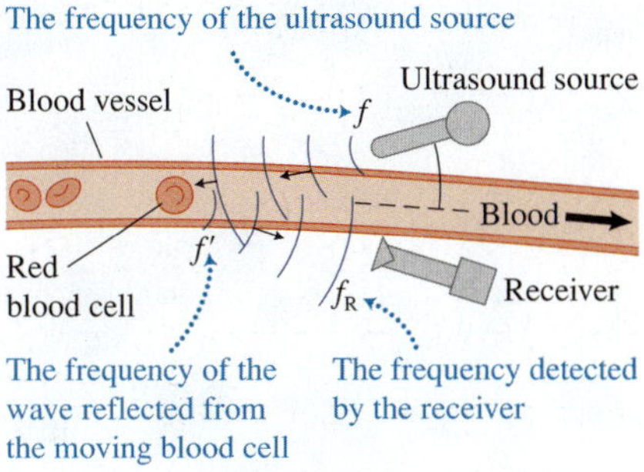


The frequency of the ultrasound $f'$ striking a red blood cell is related to the frequency $f_S$ emitted from the source by the equation

$$f' = f_S\left(\frac{v + 0}{v - v_b}\right) \tag{20.19}$$

where $v$ is the speed of sound in blood and $v_b$ is the speed of the blood. The red cells reflect this sound, acting like a source with frequency $f'$. The frequency detected at the receiver $f_R$ is related to $f'$ by the equation

$$f_R = f'\left(\frac{v + v_b}{v - 0}\right)$$

Substituting for $f'$ from Eq. (20.19), we find the frequency detected at the receiver $f_R$ relative to the frequency emitted by the sound source $f_S$:

$$f_R = f_S\left(\frac{v + 0}{v - v_b}\right)\left(\frac{v + v_b}{v - 0}\right) = f_S\left(\frac{v + v_b}{v - v_b}\right) \quad (20.20)$$

This result allows us to measure the speed of blood using the Doppler effect.

**EXAMPLE 20.9 Buzzer moves in circle**

A friend with a ball attached to a string stands on the floor and swings it in a horizontal circle. The ball has a 400-Hz buzzer in it. When the ball moves toward you on one side, you measure a frequency of 412 Hz. When the ball moves away on the other side of the circle, you measure a frequency of 389 Hz. Determine the speed of the ball.

**Sketch and translate** Draw a sketch of the situation. You detect a higher frequency when the ball moves toward you and lower frequency when it moves away, consistent with our understanding of the Doppler effect.

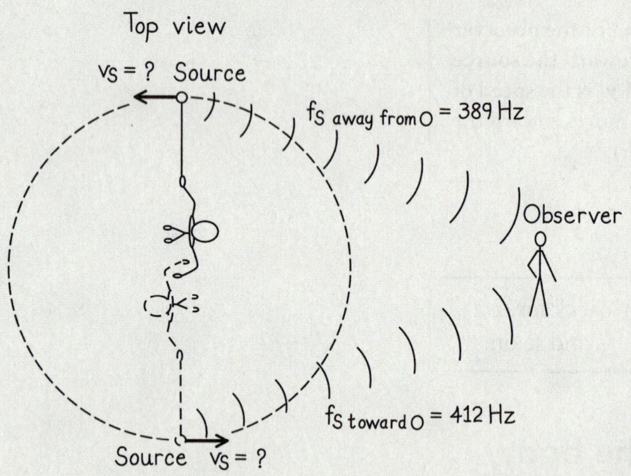


**Simplify and diagram** The speed of sound in air is constant and equals 340 m/s. The air is stationary with respect to the ground.

**Represent mathematically** We use Eq. (20.18) to find an expression for the ratio of the frequency when the source (the buzzer) is moving toward you (the observer) and when moving away. This ratio can then be rearranged to get an expression for the speed of the source (the buzzer):

$$\frac{f_{S\ \text{moving toward O}}}{f_{S\ \text{moving away from O}}} = \frac{f_S\dfrac{v + 0}{v - v_S}}{f_S\dfrac{v + 0}{v + v_S}} = \frac{v + v_S}{v - v_S}$$

where $v$ is the speed of sound in air and $v_S$ is the speed of the source.

**Solve and evaluate** Solving for $v_S$, we get

$$v_S = \left(\frac{f_{S\ \text{moving toward O}} - f_{S\ \text{moving away from O}}}{f_{S\ \text{moving toward O}} + f_{S\ \text{moving away from O}}}\right)v$$

$$= \left(\frac{412\ \text{Hz} - 389\ \text{Hz}}{412\ \text{Hz} + 389\ \text{Hz}}\right)(340\ \text{m/s}) = 9.8\ \text{m/s}$$

The ball was moving at 9.8 m/s—the correct unit and a reasonable magnitude.

**Try it yourself:** What frequency do you hear when the ball is moving perpendicular to your line of sight directly in front of you?

*Answer:* 400 Hz. At that instant, the ball is not moving toward or away from you—it moves in the perpendicular direction—there is no Doppler shift and you should hear the buzzer's frequency of 400 Hz.

**Review Question 20.11** An ambulance siren blares continuously as it approaches you and then moves away. How does the sound from the siren change and why?

# Summary

**Words** | **Pictorial and physical representations** | **Mathematical representation**

**Wave speed** $v$ depends on the properties of the medium (such as a force $F_{\text{M on M}}$ that one part of medium exerts on another part and a mass/length quantity $\mu = m/L$). The *wavelength* $\lambda$ is equal to the shortest distance between two points that vibrate in phase and depends on the frequency $f$ and the wave speed $v$. (Sections 20.2 and 20.3)

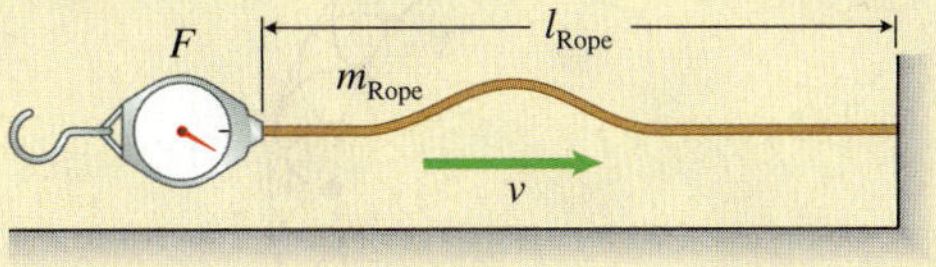


$$\lambda = vT = \frac{v}{f} \qquad \text{Eq. (20.3)}$$

$$v = \sqrt{\frac{F_{\text{M on M}}}{\mu}} \qquad \text{Eq. (20.5)}$$

**Traveling sinusoidal wave** The traveling wave expression allows us to determine the displacement from equilibrium $y$ of a point in the medium at location $x$ at time $t$ when a wave of wavelength $\lambda$ travels at speed $v$. In a traveling wave all points vibrate with the same amplitude but have different phases. (Section 20.2)

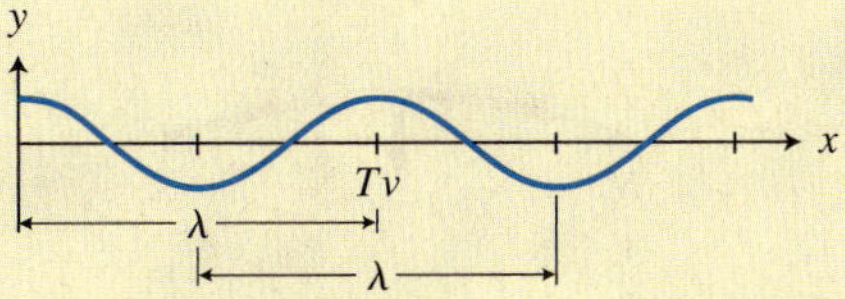


$$y = A\cos\left[2\pi\left(\frac{t}{T} - \frac{x}{\lambda}\right)\right] \qquad \text{Eq. (20.4)}$$

**Intensity** $I$ of a wave is the energy per unit area per time interval that crosses an area perpendicular to the direction of travel of the wave.

**Intensity level** $\beta$ is a logarithmic measure of the intensity $I$ relative to reference intensity $I_0$. (Sections 20.4 and 20.7)

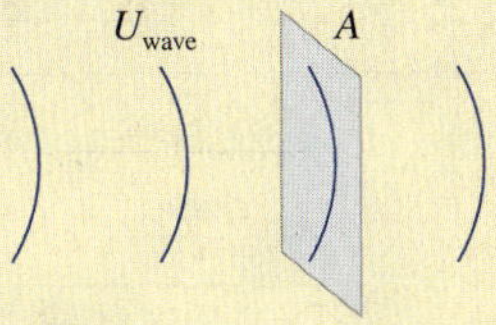


$$I = \frac{\Delta U}{\Delta t \cdot A} = \frac{P}{A} \qquad \text{Eq. (20.6)}$$

$$\beta = 10\log_{10}\left(\frac{I}{I_0}\right) \qquad \text{Eq. (20.9)}$$

where for sound $I_0 = 10^{-12}\,\text{J/m}^2\cdot\text{s}$

**Superposition principle** When two or more waves pass through the same medium at the same time, the net displacement of any point in the medium is the sum of the displacements that would be caused by each wave if alone in the medium. (Section 20.6)

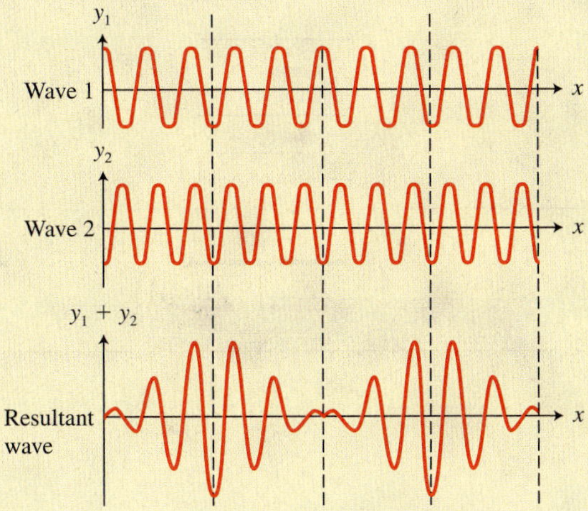


$$y_{\text{net}} = y_1 + y_2 + y_3 + \ldots \qquad \text{Eq. (20.8)}$$

(*continued*)

| Words | Pictorial and physical representations | Mathematical representation |
|---|---|---|
| **Huygens' principle** Every vibrating point of a medium produces a wavelet surrounded by a circular wave front. The new wave front is the superposition of all wavelets due to the vibrations of all points on the previous wave front. (Section 20.6) | 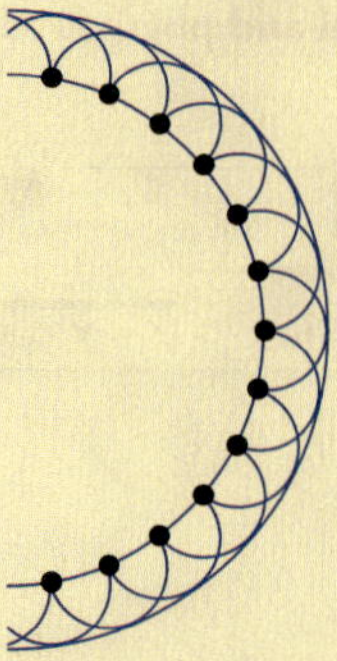 | |
| **Standing waves on strings** are the result of the constructive interference of reflected transverse waves of frequencies $f_n$ traveling at speed $v$ in both directions on a string of length $L$. The vibrating points in a standing wave have different amplitudes, but all vibrate in phase. (Section 20.9) | 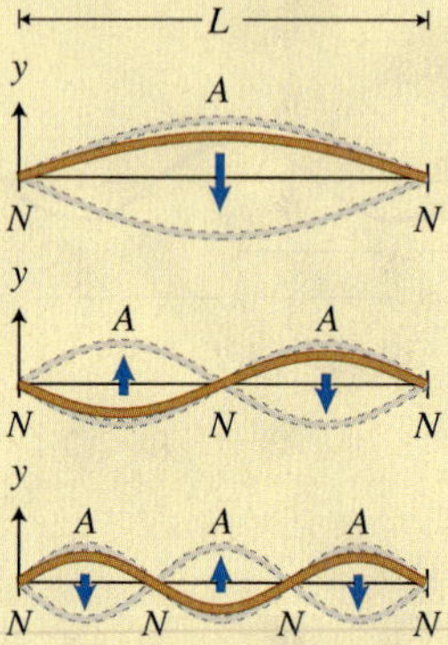  | $f_n = n\left(\frac{v}{2L}\right)$ for $n = 1, 2, 3, \ldots$ Eq. (20.12) |
| **Standing waves in open-open pipes** are similar to standing waves on strings except that they are longitudinal pressure waves of frequencies $f_n$ traveling at speed $v$ in a tube of length $L$ with two open ends. (Section 20.10) | 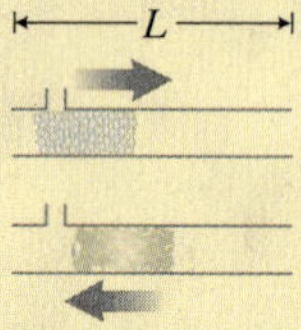  | $f_n = n\left(\frac{v}{2L}\right)$ for $n = 1, 2, 3, \ldots$ Eq. (20.14) |
| **Standing waves in open-closed pipes** are similar to standing waves in open pipes, but with a pipe having one closed end. (Section 20.10) | 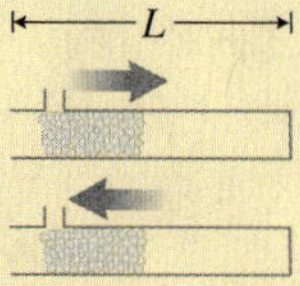  | $f_n = n\left(\frac{v}{4L}\right)$ for $n = 1, 3, 5, \ldots$ Eq. (20.15) |
| **Doppler effect for sound** When a sound source and sound observer move relative to each other, the frequency $f_O$ of the observed sound differs from the frequency $f_S$ of the source. $v_O$ and $v_S$ are the observer and source speeds (magnitudes of the velocities) and $v$ is the speed of the wave through the medium. (Section 20.11) | 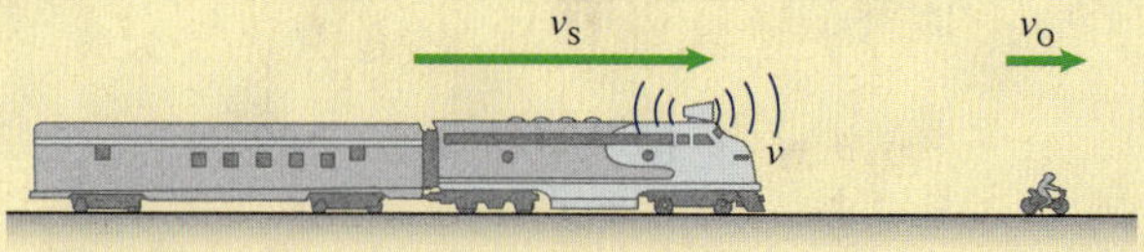  | $f_O = f_S\left(\frac{v \pm v_O}{v \mp v_S}\right)$ Eq. (20.18) Use + in front of $v_O$ if observer moves toward source and − if away; use − in front of $v_S$ if source moves toward observer and + if away. |

MP® For instructor-assigned homework, go to MasteringPhysics.

# Questions

## Multiple Choice Questions

1. Which statement(s) are applicable to the propagation of a pulse?
   (a) The particles in the pulse travel from one place to another.
   (b) The particles in the pulse do not travel from one place to another.
   (c) The particles in the medium interact with each other.
   (d) Energy travels but the medium does not.
2. When a bell rings, what happens to the vibrational energy initially present in the bell-Earth system?
   (a) It is lost.
   (b) It transfers to vibrational energy of nearby particles and then farther away from the bell.
   (c) It eventually transforms into internal energy of the surrounding medium and the bell.
   (d) Both b and c are applicable.
3. What does it mean if the period of a wave is 0.20 s?
   (a) The wave continues to travel for 0.20 s before stopping.
   (b) The wavelength is 0.20 s.
   (c) Each particle in the wave completes one vibration in 0.20 s.
4. What does it mean if the speed of a wave is 300 m/s?
   (a) Each particle moves at a speed of 300 m/s.
   (b) The wave disturbance travels 300 m during each second.
   (c) The wave disturbance travels 300 m for each wavelength.
5. What does it mean if the wavelength of a wave is 4.0 m?
   (a) The wave continues to travel for 4.0 m before stopping.
   (b) The period of the wave is 4.0 m.
   (c) The distance between adjacent identical locations in the disturbance is 4.0 m.
6. If you wish to graph one period of a wave on a string, which two variables should you have on the axes?
   (a) $v$ and $T$ (b) $y$ and $x$ (c) $y$ and $t$
   (d) $T$ and $\lambda$
7. If you wish to graph the disturbance pattern of a wave on a long string at one clock reading, which two variables should you have on the axes?
   (a) $y$ and $\lambda$ (b) $y$ and $x$ (c) $y$ and $t$
   (d) $T$ and $\lambda$
8. Which of the following choices explains why sound travels faster in water than in air?
   (a) The wavelength is shorter in water than in air.
   (b) The frequency is higher in water than in air.
   (c) The density of water is greater than the density of air.
   (d) Water molecules are heavier than air molecules.
   (e) There is a much stronger interaction between liquid water molecules than between air molecules.
9. Which mathematical expression represents a cause-effect relationship for a traveling wave:?
   (a) $f = \lambda v$ (b) $v = f/\lambda$ (c) $\lambda = v/f$
10. What is pitch?
    (a) A physical phenomenon (something that occurs in nature or in a laboratory)
    (b) A physical quantity
    (c) A unit
    (d) None of these is correct.
11. What is the frequency of a wave?
    (a) A physical phenomenon (b) A physical quantity
    (c) A unit (d) None of these is correct.
12. Two small spheres vibrate up and down on the surface of a large tank of water. We observe places in the water where the surface does not vibrate. What is this statement?
    (a) A hypothesis
    (b) A description of a physical phenomenon
    (c) A physics law (d) A physical quantity
13. What is Huygens' principle?
    (a) A law describing the behavior of waves
    (b) A hypothesized mechanism of wave propagation
    (c) A physical phenomenon involving waves
    (d) A physical quantity describing waves

## Conceptual Questions

14. Can a wave have a period of 2.0 s, a speed of 20 m/s, and a wavelength of 10 m?
15. You drop a rock in a pond and observe a wave propagating on the surface of the pond. How can you estimate the frequency, speed, and wavelength of the wave?
16. What physics ideas were necessary to construct Eq. (20.4)? What math ideas did you need?
17. How do you know that the wavelength of a wave can be calculated by multiplying the speed of the wave in a medium by the period of the vibrating source?
18. Invent an explanation for the difference in the speeds of longitudinal and transverse waves in the same medium. Then devise an experiment to test your explanation.
19. What conditions are necessary to create a sinusoidal wave in a certain medium? What is needed to create a wave of a different frequency? Different speed? Different wavelength?
20. Invent and describe an experiment to estimate the speed of sound in air. You can use only everyday items.
21. Explain why the speed of sound is greater in humid air than in dry air at the same pressure and temperature.
22. How can you use the time delay between a lightning flash and the sound of thunder to estimate the distance to the place where the flash occurred? Justify any assumptions you make in an example calculation.
23. Describe two useful types of information a geologist can obtain by detecting sound reflected from different materials under Earth's surface.
24. Two speakers hang from racks placed in an open field. When sound of the same frequency comes from both speakers, no sound is heard in some locations. The sound is loud in other places at about the same distance from the speakers. Explain.
25. Two identical sound waves are sent down a long hall. Does the resultant wave necessarily have amplitude twice that of the constituent waves? Explain. Under what conditions would we hear no sound?
26. Sound waves of all frequencies in the audio frequency range (20–20,000 Hz) travel at the same speed in air. Explain how some common experiences support this statement.
27. Explain why the properties of the medium affect the speed of sound in it.
28. How can you show that an object producing sound can be used to detect it?
29. Describe the common features and differences between traveling waves and standing waves.

30. Why do different guitar strings sound different when you pluck them, even though they are the same length?
31. Explain how a xylophone works.
32. A child riding on a merry-go-round blows a whistle. Describe the variation in sound frequency heard by a stationary observer some distance from the merry-go-round. Be sure to indicate in what direction the child is moving relative to the observer at different times in this frequency variation.

# Problems

Below, **BIO** indicates a problem with a biological or medical focus. Problems labeled **EST** ask you to estimate the answer to a quantitative problem rather than derive a specific answer. Problems marked with a pencil icon require you to make a drawing or graph as part of your solution. Asterisks indicate the level of difficulty of the problem. Problems with no * are considered to be the least difficult. A single * marks moderately difficult problems. Two ** indicate more difficult problems.

### 20.1 and 20.2 Observations: Pulses and wave motion and Mathematical descriptions of a wave

Assume that the speed of sound in air is 340 m/s for all of these problems unless stated otherwise.

1. * Imagine that you and your friend pull tight on the ends of a rope lying on a smooth floor. You shake your end of the rope, causing a transverse pulse. (a) Draw a displacement-versus-time graph representing the motion of your end of the rope. (b) Draw two displacement-versus-position graphs representing all of the points on the rope at two different clock readings.
2. ***Design** Design an experiment to determine whether longitudinal or transverse pulses have a higher speed on a Slinky. Describe the assumptions you make when determining the speed from the collected data. Discuss how experimental uncertainties will affect your decision.
3. Imagine that you are standing in a swimming pool with a beach ball. You push the ball up and down several times. Draw a picture representing wave fronts in the pool. Explain what you mean by the words *wave front.*
4. * Explain the meaning of each symbol in the equation

$$y = A\cos\left[\frac{2\pi}{T}\left(t - \frac{x}{v}\right)\right]$$

and summarize how we devised this equation.

5. * **Tell all** You have a sinusoidal wave that is described by the function

$$y = (0.2\text{ m})\sin\left[\frac{\pi}{2}\left(t + \frac{x}{20\text{ m/s}}\right)\right]$$

(a) Say everything you can about this wave. Pay attention to the positive sign in the equation. (b) Draw a $y$-versus-$t$ graph and a $y$-versus-$x$ graph for the wave.

6. (a) Draw a position-versus-time graph for one coil of an infinitely long Slinky when a longitudinal wave of frequency 2.0 Hz and speed 3.0 m/s propagates on the Slinky. (b) Draw a displacement-versus-position graph for one particular time for a piece of the Slinky. (c) What is the wavelength of the wave?
7. A longitudinal wave of amplitude 3.0 cm, frequency 2.0 Hz, and speed 3.0 m/s travels on an infinitely long Slinky. How far apart are the two nearest points on the Slinky that at one particular time both have the maximum displacements from their equilibrium positions? Explain your reasoning.
8. A boat is moving up and down in the ocean with a period of 1.7 s caused by a wave traveling at a speed of 4.0 m/s. What other physical quantities relevant to this wave can you determine using this information? Determine them.
9. * A large goose lands in a lake and bobs up and down for a short time. A fisherman notices that the first wave created by the goose reaches the shore in 8.0 s. The distance between two wave crests is 80 cm, and in 2.0s he sees four waves hit the shore. How can the fisherman use these observations to determine how far from the shore the goose landed?
10. * **Equation Jeopardy** The equation below describes the variation of pressure at different positions and times (relative to atmospheric pressure) caused by a sound wave:

$$\Delta P = (2.0\text{ N/m}^2)\cos\left[2\pi\left(\frac{t}{0.010\text{ s}} - \frac{x}{3.4\text{ m}}\right)\right]$$

Say everything you can about this wave.

11. **BIO Hearing** People can hear sounds ranging in frequency from about 20 Hz to 20,000 Hz. Determine the wavelengths of the sounds at these two extremes. What assumptions are you making? If these assumptions are not correct, how will your answer change?
12. **BIO Dolphin sonar** A dolphin has a sonar system that emits sounds with a frequency of $2.0 \times 10^5$ Hz. Determine everything you can about this sound wave. Remember that the sounds emitted by the dolphin travel in water.
13. **Antarctic ice** Radio waves travel at a speed of $1.7 \times 10^8$ m/s through ice. A radio wave pulse sent into the Antarctic ice reflects off the rock at the bottom and returns to the surface in $32.9 \times 10^{-6}$ s. How deep is the ice? What assumptions did you make to solve the problem?
14. **Lightning and thunder** You see a flash of lightning, and 2.4 s later you hear thunder coming from the same location. (a) Why is there this long delay? (b) How far away did the lightning flash occur? Justify any assumptions you make in your calculations.

### 20.3 Dynamics of wave motion: Speed and the medium

15. * A pulse travels at speed $v$ on a stretched rope. By what factor must you increase the force you exert on the rope to cause the speed to increase by a factor of 1.30?
16. * **Ratio reasoning** Two ropes have equal length and are stretched the same way. The speed of a pulse on rope 1 is 1.4 times the speed on rope 2. Determine the ratio of the masses of the two ropes $(m_1/m_2)$.
17. * **Telephone line** A telephone lineman is told to stretch the wire between two poles so the poles exert an 800-N force on the wire. As the lineman does not have a scale to measure forces, he decides to measure the speed of a pulse created in the wire when he hits it with a wrench. The pulse travels 60 m from one pole to the other and back again in 2.6 s. The 60-m wire has a mass of 15 kg. Should the wire be tightened or loosened? Explain. What assumptions did you make?

18. * **Violin strings** The speed of a wave on a violin A string is 288 m/s and on the G string is 128 m/s. Use this information to determine the ratio of mass per unit length of the strings. What assumptions did you make?
19. * **Design** Describe two experiments to determine the speed of propagation of a transverse wave on a rope. You have the following tools to use: a stopwatch, a meter stick, a mass-measuring scale, and a force-measuring device. Use whatever other items you need for your experiments.

### 20.4 Energy, power, and intensity of waves

20. ** Show using a sketch and mathematics that the intensity of a two-dimensional wave is inversely proportional to the distance from a source. Give three examples that will explain the meaning of "inversely proportional to the distance" in this case.
21. ** Show using a sketch and mathematics that the intensity of a three-dimensional wave is inversely proportional to the distance squared from the source. Give three examples that will explain what it means to be inversely proportional to the distance squared in this case.
22. * EST **Energy from the Sun** The Sun radiates energy at a rate of about $4 \times 10^{26}$ W. Estimate how much of the Sun's energy 1 $m^2$ of Earth's surface facing the Sun receives in 1 h. Earth is about $150 \times 10^{6}$ km from the Sun. What assumptions did you make?
23. * EST **More energy from the Sun** Use the data from the previous problem to estimate the energy coming to Earth from the Sun each second. The radius of Earth is about 6400 km. State your assumptions.
24. * Compare the intensity of a 100-W lightbulb while you are reading this book to the intensity of the Sun. What assumptions did you make?
25. ** At what distance from the Sun is its intensity the same as that of a 100-W lightbulb that is 1.0 m from you?
26. ** EST We can hear airplanes flying. Estimate the smallest power of the sound produced by the airplane engine that we can still hear. Explain your estimation method.

### 20.5 and 20.6 Reflection and impedance and Superposition principle and skills for analyzing wave processes

27. * Explain why the transverse pulse traveling on a rope held by two people reflects in the opposite orientation each time it reaches a person.
28. * **Design** Use your knowledge of waves to explain echoes. Use your explanation to devise a system to measure distances to objects that cannot be reached directly.
29. * BIO **Bat echo** A bat receives a reflected sound wave from a fly. If the reflected wave returns to the bat 0.042 s after it is sent, how far is the fly from the bat? Would you expect the reflected wave pulse to be in or out of phase with the incident wave pulse? Explain.
30. **Sound wave in Earth** A sound wave created by an explosion at Earth's surface is reflected by a discontinuity of some type under Earth's surface (**Figure P20.30**). Determine the distance from the surface to the discontinuity. Does the discontinuity have a greater or lesser impedance to sound than the surface above it? Explain. Sound travels at about 3000 m/s through the top layer of Earth.

**Figure P20.30**

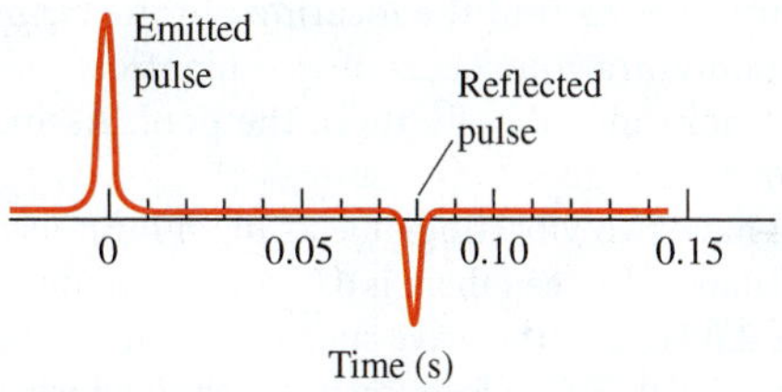


31. A 5.0-kg rope that is 20 m long is woven to an 8.0-kg rope that is 16 m long. The ropes are pulled taut and a pulse initiated in one is reflected at their interface. Draw a picture of what happens just after the pulse reaches the interface between the ropes.
32. * The pulses shown in **Figure P20.32** (shown at time zero) are moving toward each other at speeds of 10 m/s. Draw $y(x)$ graphs showing the resultant pulse at 0.10 s, 0.20 s, and 0.30 s.

**Figure P20.32**

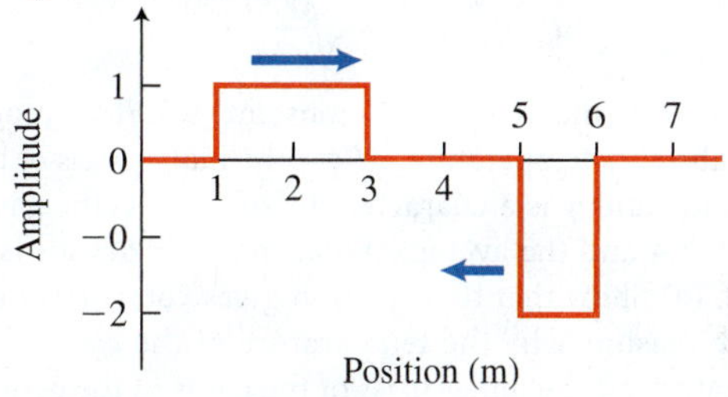


33. * Repeat the previous problem for the case where the pulse on the right is upright rather than inverted.
34. Find a resultant wave for the two waves shown in **Figure P20.34** at the instant they are traveling through the same medium.

**Figure P20.34**

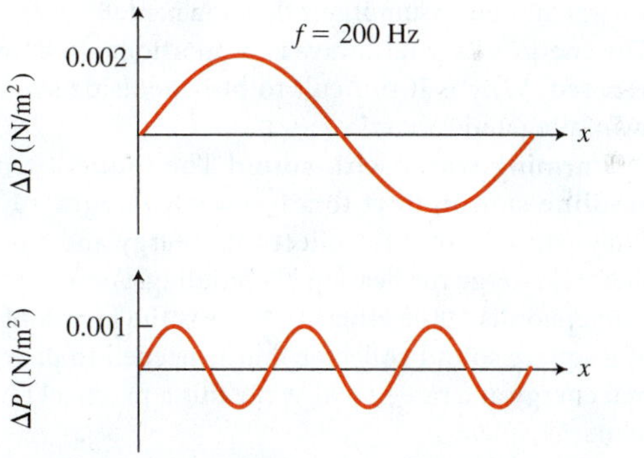


35. * Two waves shown in **Figure P20.35** at zero clock reading move toward each other at a speed of 10 m/s. Draw graphs of the resultant wave at times of 0.10 s, 0.20 s, and 0.30 s.

**Figure P20.35**

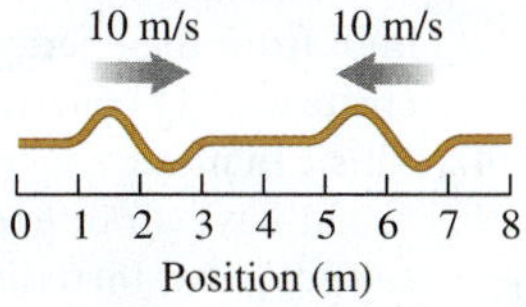


36. * Use Huygens' principle to find the shape of the wave fronts of a wave generated by the long edge of a flat piece of plastic floating horizontally and vibrating up and down in a swimming pool.
37. * Use Huygens' principle and a wave front representation of waves to show that if you place a screen with a small circular hole in the path of a wave with flat wave fronts, the wave fronts beyond the screen will be circular.

38. * Use Huygens' principle and a wave front representation of waves to find the locations in the swimming pool where no vibrations occur if two identical point-like objects synchronously vibrate in the pool. Assume the pool is infinitely large.

39. * You have two vibrating objects in an infinitely large pool. The distance between them is 6.0 m. Their frequency of vibration is 2.0 Hz and the wave speed is 4.0 m/s. The vibrations are sinusoidal. Find a location between them where the water does not vibrate and another location between them where the water vibrates with the largest amplitude.

## 20.7 and 20.8 Sound and Pitch, frequency, and complex sounds

40. **Design** Describe an experiment to convince a friend that sound is a wave.

41. ** The speed of sound in an ideal gas is given by the relationship

$$v = \sqrt{\frac{\gamma RT}{M}},$$

where $R =$ the universal gas constant $= 8.314\ \mathrm{J/mol\ K}$; $T =$ the absolute temperature; $M =$ the molar mass of the gas in kg/mol; and $\gamma$ is a characteristic of the specific gas. For air, $\gamma = 1.4$ and the average molar mass for dry air is 28.95 g/mol. (a) Show that the equation gives you correct units. (b) Give reasons why the temperature of the gas is in the numerator and the molar mass of the gas is in the denominator.

42. * Using the information from problem 41, calculate the speed of sound in the air. What assumptions are you making?

43. * **Speed of sound in summer and winter** Using the information in problem 41, estimate how much faster sound travels in summer than in winter. Explain how you arrived at your answer and the assumptions that you made.

44. The energy of a sound wave is proportional to its amplitude squared. Why is it difficult to hear poolside sounds when swimming underwater?

45. * **Warming water with sound** The sound intensity at a gasoline station next to a freeway averages $10^{-3}\ \mathrm{W/m^2}$. The owner decides to collect this energy and convert it to thermal energy for heating his building. Assuming that this conversion is 100% efficient, what is the length of one side of a square sound collector that is needed to provide thermal energy at a rate of 500 W? Is this a practical idea for the owner? Explain.

46. * **Supersonic jet** The sound intensity 5 km from the place where a supersonic jet takes off is $0.60\ \mathrm{W/m^2}$. Determine the area of a sound collector you would need to run a 40-W lightbulb from the energy collected. What might require you to create a larger collector?

47. **Music** In music a very soft sound called "pianississimo" (ppp) might have an intensity of about $10^{-8}\ \mathrm{W/m^2}$. A very loud sound called fortississimo (fff) might have an intensity of about $10^{-2}\ \mathrm{W/m^2}$. Convert these intensities to intensity levels (units of dB).

48. Two sounds differ by 1 dB. What is the difference in their intensities?

49. Calculate the change in intensity level when a sound intensity is increased by a factor of 8, by a factor of 80, and by a factor of 800.

## 20.9 and 20.10 Standing waves on strings and Standing waves in air columns

50. **Banjo string** A banjo D string is 0.69 m long and has a fundamental frequency of 294 Hz. (a) Determine the speed of a wave or pulse on the string. (b) Identify three other frequencies at which the string can vibrate.

51. **Banjo fret** How far from the end of the banjo string discussed in the previous problem must a fret be so that the string's fundamental frequency is 330 Hz when you hold it down at the fret?

52. * **Violin string** A 0.33-m-long violin string has a mass of 0.89 g. The peg exerts a 45 N force on it. What can you determine about the sound produced by that string?

53. * A person secures a 5.0-m-long rope of mass 0.40 kg at one end and pulls on the rope, exerting a 120-N force. The rope vibrates in three segments with nodes separating each segment. List the physical quantities you can determine using this information and determine three of them.

54. * **Dislodge grackle** A canary sits 10 m from the edge of a 30-m-long clothesline, and a grackle sits 5 m from the other end. The rope is pulled by two poles that each exerts a 200-N force on it. The mass per unit length is 0.10 kg/m. At what frequency must you vibrate the line in order to dislodge the grackle while allowing the canary to sit undisturbed?

55. * EST Estimate the fundamental frequency of vibration of a telephone line between adjacent poles near where you live. Explain how you arrived at your answer.

56. * **Ratio reasoning** Two wires on a piano are the same length and are pulled by pegs that exert the same force on the wires. Wire 1 vibrates at a fundamental frequency 1.5 times that of wire 2. Determine the ratio of their masses $(m_1/m_2)$.

57. * **Ratio reasoning** By what percent does the frequency of a piano string change if the force that the peg exerts on it increases by 10%?

58. (a) Determine the first three standing wave frequencies of a 40-cm-long open pipe. (b) Do the same for a 40-cm-long closed pipe.

59. * **Brooklyn-Battery Tunnel** The 2779-m Brooklyn-Battery Tunnel, connecting Brooklyn and Manhattan, is one of the world's longest underwater vehicular tunnels. (a) Determine its fundamental frequency of vibration. (b) What harmonic must be excited so that it resonates in the audio region at 20 Hz or greater?

60. * **Flute** A wooden flute, open at both ends, is 0.48 m long. (a) Determine its fundamental vibration frequency. (b) How far from one end should a finger hole be placed to produce a sound whose frequency is four-thirds that calculated in part (a)? Be sure to justify how you arrive at your answer.

61. * **Organ pipe** The lowest three standing wave vibration frequencies of an organ pipe are 120 Hz, 360 Hz, and 600 Hz. (a) Is the pipe open or closed, and what is its length? (b) Determine the frequencies of the first two harmonic vibrations on a pipe of the same length but of the other type than that described in part (a).

62. * EST **Music bottle** (a) Use the dimensions of a small soft drink bottle to estimate its fundamental resonant frequency when empty. (b) Determine the depth of water that must be added to increase its frequency by a factor of 4/3.

63. * The speed of sound can be measured using the apparatus shown in **Figure P20.63**. A 440-Hz tuning fork vibrating above a tube partially filled with water initiates sound waves in the tube. The air inside the tube vibrates at the same frequency as the tuning fork when the water in the tube is lowered 0.20 m and 0.60 m from the top of the tube. Use this information to determine the speed of sound in air.

**Figure P20.63**

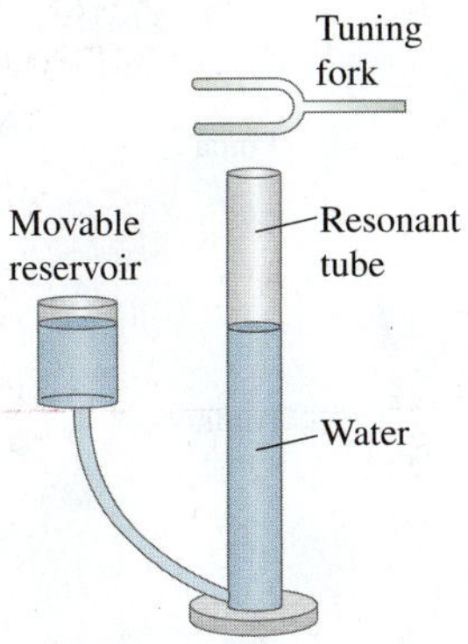


64. ** EST The fundamental frequency of a closed pipe, such as your vocal tract, is 240 Hz when filled with air (a mix of different molecules) that has molar mass of 29 g/mole. Using ideas from the kinetic theory of gases, estimate the vibration frequency when the vocal tract is filled with helium that has a molar mass of 4 g/mole.

### 20.11 The Doppler effect: Putting it all together

65. * BIO **Speed of blood** A source of ultrasound emits waves at a frequency of $2.00 \times 10^6$ Hz. The waves are reflected by red blood cells moving toward the source at a speed of 0.30 m/s. Determine the frequency of sound detected at a receiver next to the source. The speed of sound in the blood is 1500 m/s.

66. * **Car horn** A car horn vibrates at a frequency of 250 Hz. Determine the frequency a stationary observer hears as the car (a) approaches at a speed of 20 m/s and (b) departs at 20 m/s. If the car is stationary, what frequency is heard (c) by an observer approaching the car at 20 m/s and (d) by an observer departing from the car at 20 m/s?

67. * **Train whistle** A car drives at a speed of 25 m/s along a road parallel to a railroad track. A train traveling at 15 m/s sounds a horn that vibrates at 300 Hz. (a) If the train and car are moving toward each other, what frequency of sound is heard by a person in the car? (b) If the train and car are moving away from each other, what frequency of sound is heard in the car?

68. * **Circular motion sound source** A whistle with frequency 400 Hz moves at speed 20 m/s in a horizontal circle at the end of a rotating stick. Determine the highest and lowest frequencies heard by a person riding a bicycle at speed 10 m/s toward the whistle.

69. * BIO **Bat echo** A bat emits short pulses of sound at a frequency of $1.60 \times 10^5$ Hz. As the bat swoops toward a flat wall at speed 30 m/s, this sound is reflected from the wall back to the bat. (a) What is the frequency of sound incident on the wall? (b) Consider the wall as a sound source at the frequency calculated in part (a). What frequency of sound does the bat hear coming from the wall?

70. * **Hungry student** A hungry student working in a cafeteria decides to eat from plates of food that pass on a conveyor belt. The plates are separated by 3.0 m, and the belt moves at a speed of 9.0 m/min. (a) How many plates of food does the student eat per minute? (b) As the student's hunger is appeased, he moves with the belt at a speed of 6.0 m/min. Determine the number of plates of food that now reach the student each minute. (This change is similar to the Doppler shift for a moving listener.)

71. ** **Detecting speed of baseball** A Doppler speed meter operating at exactly $1.02 \times 10^5$ Hz emits sound waves and detects the same waves after they are reflected from a baseball thrown by the pitcher. The receiver "mixes" the reflected wave with a small amount of the emitted wave and measures a beat frequency of $0.30 \times 10^5$ Hz. How fast is the ball moving?

## General Problems

72. * Use Huygens' principle and a wave front representation of waves to show that a plane wave incident on a barrier traveling at an angle $\theta$ relative to a line perpendicular to the barrier (the so-called normal line) reflects, forming a wave front that travels at angle $\theta$ on the other side of the normal line.

73. * Use Huygens' principle and a wave front representation of waves to show that when a plane wave traveling in one medium hits a border with another different-speed medium that is not perpendicular to the direction of wave propagation, the wave changes its direction of propagation.

74. ** **Design** Describe an experiment that you can perform to measure the speed of sound in air using a graduated cylinder and a tuning fork that produces sound of a known frequency. Draw a picture. Carefully outline your experimental and mathematical procedure.

75. ** BIO EST **Reaction time** A basketball player's teammate shouts at her to catch a ball. Estimate the time required for the sound to travel a distance of 10 m, be detected by the ear, travel as a nerve impulse to the brain, be processed, then travel as a nerve signal for muscle action back to an arm, and finally cause a muscle contraction. Nerve impulses travel at a speed of about 120 m/s in humans. You will have to make reasonable assumptions about quantities not stated in the problem.

76. ** EST While camping, you record a thunderclap whose intensity is $10^{-2}\ \text{W/m}^2$. The clap reaches you 3.0 s after a flash of lightning. Estimate the total acoustical power generated by the bolt of lightning. Clearly state all assumptions that you made.

77. ** BIO **Blood speed** A red blood cell travels at speed 0.40 m/s in a large artery. A sound of frequency $2.00 \times 10^6$ Hz enters the blood opposite the direction of flow. (a) Determine the frequency of sound reflected from the cell and detected by a receiver. (b) If the emitted and received sounds are combined in the receiver, what beat frequency is measured?

## Reading Passage Problems

BIO **Bats and echolocation** If you shout "hello" near a vertical canyon wall, an echo "hello" returns after a short delay. If it returns in 2.0 s and sound travels through air at 340 m/s, then the total distance the signal traveled is 680 m. Therefore, the canyon wall must be half that distance from you (340 m). Bats use a similar process to hunt their prey. Bats travel and locate food at night by emitting ultrasound waves that bounce off prey and return to the bat. The time interval between signal emission and its return to the bat tells the bats which insects are closer than others. The amplitude of the reflected wave indicates the size of the prey—a large animal causes a large amplitude reflected wave. The bat can tell the left-right location of the object by the relative time delays of the return sound between the left and right ears. Rows of horizontal folded sound sensors in the bat's ears allow them to determine the up-down location of the prey. Finally, the bat can determine if the prey is moving toward it, moving away from it, or is stationary by the Doppler shift of the returning signal. The relationship between the source frequency of the bat and the frequency of the reflected sound is given by Eq. (20.20):

$$f_R = f_S\left(\frac{v + 0}{v - v_{P\text{ to }B}}\right)\left(\frac{v + v_{P\text{ to }B}}{v - 0}\right) = f_S\left(\frac{v + v_{P\text{ to }B}}{v - v_{P\text{ to }B}}\right)$$

78. You have a bat-like echolocation system on your car, which emits a 20,000-Hz frequency sound that returns to the car in 0.18 s after being reflected by another stationary car. Which answer below is closest to the distance from your car to the other car?
(a) 10 m (b) 20 m (c) 30 m
(d) 40 m (e) 50 m

79. If the car from Problem 78 is moving at 20 m/s toward the stationary car, which answer below is closest to the frequency of the reflected wave detected by the moving car from the stationary car?
(a) 18,000 Hz (b) 19,000 Hz (c) 20,000 Hz
(d) 21,000 Hz (e) 23,000 Hz

80. Which answer below is closest to the distance that the moving car in Problem 79 travels during the 0.18 s?
(a) 0.18 m (b) 0.36 m (c) 1.8 m
(d) 3.6 m (e) 5.4 m

81. Compare your answers to Problems 78 and 80. If you want a more accurate measure of the object's distance from you, you must consider the distance you travel during the time delay for the sound to get from you to the object and back to you. By approximately what percent does your change in position affect the distance calculated in Problem 78?
(a) 6% (b) 12% (c) 18%
(d) 24% (e) 30%

82. While your car from Problem 78 is stationary, you emit a 20,000-Hz signal and get a 22,000-Hz signal back from a reflecting object. What can you say about the object?
(a) It is moving away from you.
(b) It is stationary.
(c) It is moving toward you.

83. Your echolocation system has one transmitter in the middle of the front of the car and two detectors of reflected waves, one by each headlight (like the bat's ears). You detect a reflected signal in the left detector slightly before you detect the signal from the right detector. Where must the reflecting object be?
(a) To the left of center
(b) Straight in front of you
(c) To the right of center

**BIO Human ear** In the human ear (shown in **Figure 20.30**), the exterior ear flap, called the pinna, gathers and guides sound waves into the auditory canal. The sound waves create a pressure variation at the eardrum, a variation that is about two times as great as it would be without the pinna and auditory canal. This pressure variation causes the eardrum to vibrate. A large fraction of the vibrational energy is transmitted through the three small bones in the middle ear called the hammer, anvil, and stirrup. These three bones, collectively known as the ossicles, constitute a lever system that increases the pressure exerted on a small membrane of the inner ear known as the oval window. The pressure increase is possible for two reasons. (1) The hammer feels a small pressure variation over the large area of its contact with the eardrum, whereas the stirrup exerts a large pressure variation over the small area of its contact with the oval window. This difference in areas increases the pressure by a factor of 15 to 30. (2) The three bones act as a lever that increases the force that the stirrup exerts on the oval window, although the displacement resulting from the force is less than if the bones were not present. Because of these effects, pressure fluctuation of the stirrup against the oval window can be 180 times greater than the pressure fluctuation of a sound wave in air before it reaches the ear.

**Figure 20.30**

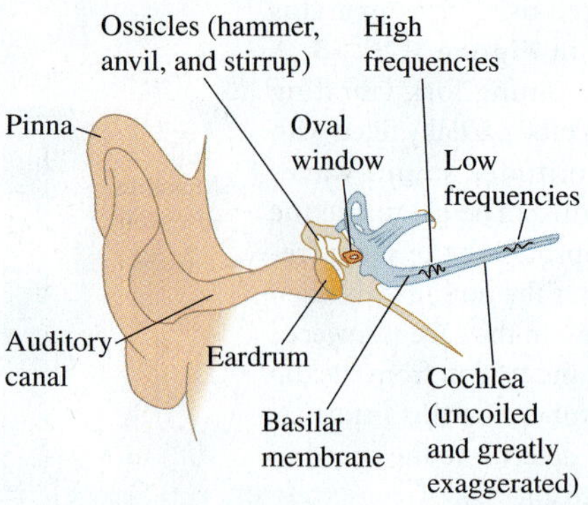


The increased pressure fluctuation against the oval window causes a fluctuating pressure in the fluid inside the cochlea of the inner ear. The fluctuating pressure is sensed by nerve cells along the basilar membrane. Nerves nearest the oval window respond to high-frequency sounds, whereas nerves farther from the window respond to low-frequency sounds. Thus, our ability to distinguish the frequency of a sound depends on the variation in sensitivity of different cells to different frequencies along the basilar membrane. The basilar membrane is only about 3 cm long, and yet a normal ear can distinguish sounds that differ in frequency by about 0.3%. This ability to distinguish sounds of slightly different frequency spans the human hearing range from about 20 Hz to 20,000 Hz.

84. The human ear can detect sound waves whose intensity is about $10^{-12}\text{ W/m}^2$. Where should you be with respect to a stereo speaker that produces sound of intensity $10^{-5}\text{ W/m}^2$ when 1 m from the speaker if you do not want to listen to the music?
(a) About 3 m away (b) About 30 m away
(c) About 3 km away
(d) It depends on the size of the speaker.

85. What amplifies the air pressure in the ear?
(a) Pinna (2 times) and ossicles (30 times)
(b) Pinna (2 times), auditory canal (3 times), and ossicles (30 times)
(c) Pinna and auditory canal (2 times), the area change from the eardrum to the oval window (30 times), and the lever action of the ossicles (3 times)

86. Where is the mechanism that allows the ear to distinguish between low-frequency and high-frequency sounds located?
(a) The fluid inside the cochlea of the inner ear
(b) In the oval window
(c) In the basilar membrane (d) both b and c

87. What frequency difference can the human ear distinguish near 1000 Hz?
(a) 100 Hz (b) 10 Hz (c) 3 Hz (d) 1 Hz

88. The threshold for pressure variation of a barely audible sound is about $2 \times 10^{-5}\text{ N/m}^2$. Which answer below is closest to the pressure variation of a barely audible sound in the cochlea of the inner ear?
(a) $1 \times 10^{-7}\text{ N/m}^2$ (b) $1 \times 10^{-6}\text{ N/m}^2$
(c) $2 \times 10^{-5}\text{ N/m}^2$ (d) $4 \times 10^{-3}\text{ N/m}^2$
(e) $4 \times 10^{-2}\text{ N/m}^2$

# Reflection and Refraction

# 21

How can you make something become invisible?

How does a periscope work?

What is fiber optics?

A magician breaks a glass vial with a hammer and places the pieces of broken glass in a beaker full of oil. Then she says "Abracadabra," places her hand in the beaker, and takes out an intact vial! Did the magician hide another intact vial in the oil prior to doing the trick? You should be able to see an intact vial in the transparent oil, shouldn't you? In this chapter you will learn how to explain this trick!

**In this chapter** we begin our investigation of light. Is light different from everything we have studied, or can we use the principles and tools we have

**Be sure you know how to:**

- Apply ideas of impulse-momentum to explain collisions (Section 5.4).
- Draw a wave front (Sections 20.1 and 20.6).
- Apply Huygens' principle (Section 20.6).

already developed to understand light and its useful applications? Many of the remaining chapters in this textbook involve using and improving the model of light we develop in this chapter as well as investigating many physical processes involving light.

## 21.1 Light sources, light propagation, and shadows

Some ancient people thought that humans emitted special invisible rays from their eyes. These rays reached out toward objects and wrapped around them to collect information about the objects. The rays then returned to the person's eyes with this information. If this ancient model were accurate, humans should be able to see in total darkness. However, a simple experiment disproves this model. If you sit for a while in a room with absolutely no light sources, no matter how long you wait, you will still see nothing. There must be some other explanation for how we see things. The experiments in Observational Experiment **Table 21.1** use a laser pointer as a source of light to explore the phenomenon of light and seeing.

**Active Learning Guide>**

**OBSERVATIONAL EXPERIMENT TABLE**

**21.1 How do we see objects?**

VIDEO 21.1

| Observational experiment | Analysis |
|---|---|
| **Experiment 1.** In a dark room, shine a laser pointer so that you see a spot of light on the wall. You do not see light going from the laser pointer to the wall. Place your hand along the straight line connecting the laser to the bright spot on the wall. The bright spot disappears from the wall and you see it on your hand. If you place your hand along a straight line connecting the spot on the wall and your eye, you do not see the spot. | The light has to travel to the wall along a straight line. When the light reaches the wall, it somehow bounces off and then travels to your eyes so that you see the bright spot on the wall. Everyone in the room also sees light from the bright spot. |

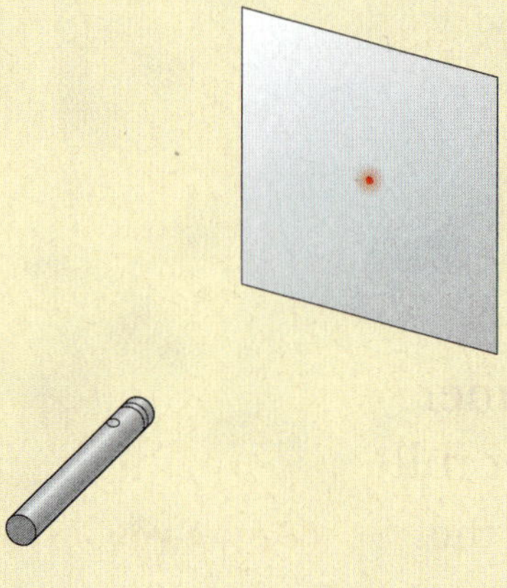

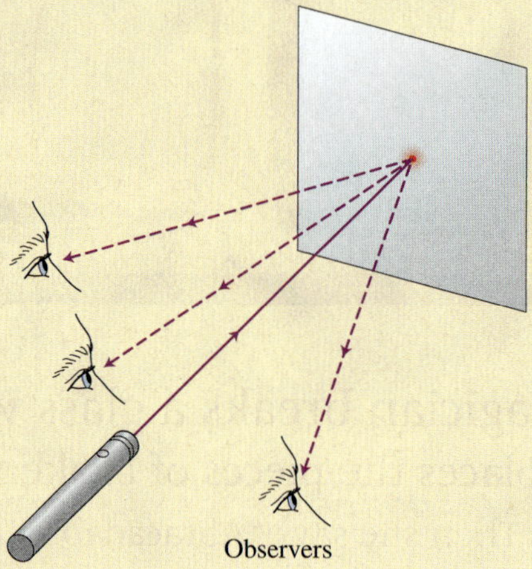

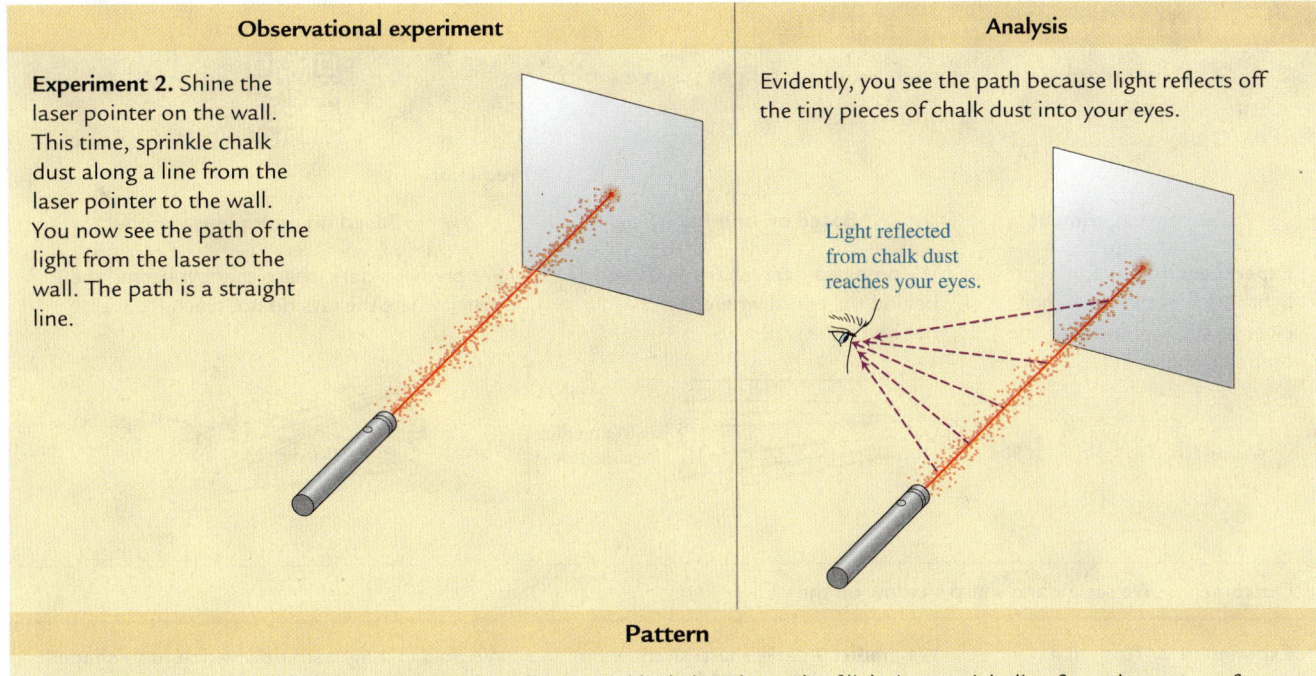

**Observational experiment**

**Experiment 2.** Shine the laser pointer on the wall. This time, sprinkle chalk dust along a line from the laser pointer to the wall. You now see the path of the light from the laser to the wall. The path is a straight line.

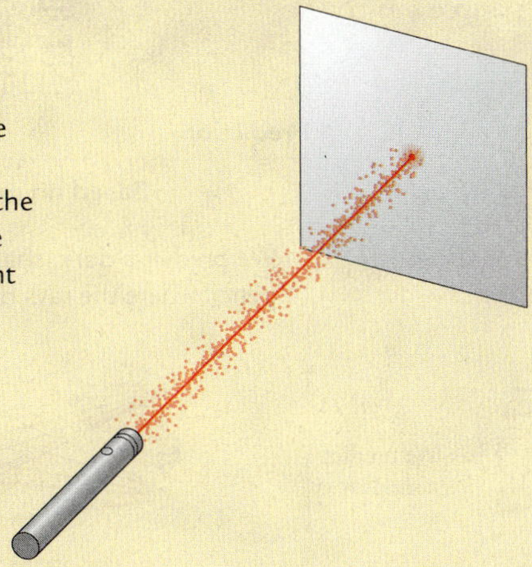

**Analysis**

Evidently, you see the path because light reflects off the tiny pieces of chalk dust into your eyes.

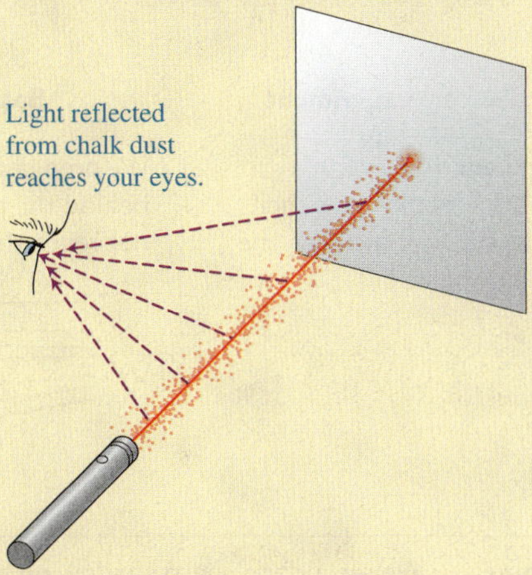
Light reflected
from chalk dust
reaches your eyes.

**Pattern**

We can see objects (even tiny ones such as dust) illuminated by light. The path of light is a straight line from the source of light to the object and then (assuming that the behavior of light does not change) another straight line of reflected light from the object to our eyes.

Table 21.1 indicates that to see something, we need a source of light and an object off which the light bounces (reflects) and then reaches the eyes of the observer. Light travels in a straight-line path between the source of the light and the object reflecting the light, then in another straight line between that object and our eyes.

How do we represent the light sent by a light source? How do objects illuminated by a light source reflect light into our eyes? We will start by investigating the first question.

**Figure 21.1** A lightbulb illuminates the ceiling of a room.

## Representing light coming from different sources

A laser pointer is useful for studying light propagation because the emitted light emerges as one narrow beam. However, most light sources, such as lightbulbs and candles, do not emit light as a single beam. These **extended light sources** consist of multiple points, each of which emits light. When we turn on a lightbulb in a dark room, the walls, the floor, and the ceiling of the room are illuminated (**Figure 21.1**). Obviously, the bulb sends light in all directions. But does one point of the shining bulb send light in one direction, or does each point send light in multiple directions? Both of these ideas can explain why the walls, floor, and ceiling are illuminated. Let's investigate those two possible models of how extended sources emit light. To do this we will represent the travel of light from one location to another with a **light ray,** drawn as a straight line and an arrow. Diagrams that include light rays are called **ray diagrams**.

*Model 1:* Each point of an extended light source emits light that can be represented by one outward-pointing ray (**Figure 21.2a**). Different points send rays in different directions.

*Model 2:* Each point on an extended light source emits light in multiple directions represented by multiple rays (Figure 21.2b).

**‹Active Learning Guide**

To help us determine which model better explains real phenomena, we use the models to make predictions about the outcome of three experiments (Testing Experiment **Table 21.2**). All of the experiments are conducted in an otherwise dark room.

**TESTING EXPERIMENT TABLE**

## 21.2 How many rays does each point on a light source emit?

VIDEO 21.2

| Testing experiment | Predictions: Based on one-ray model | Predictions: Based on multiple-ray model |
|---|---|---|
| **Experiment 1.** Turn on a lightbulb and place a pencil close to the wall between the bulb and the wall.  | We predict a dark, sharp shadow behind the pencil where the rays do not reach the wall. 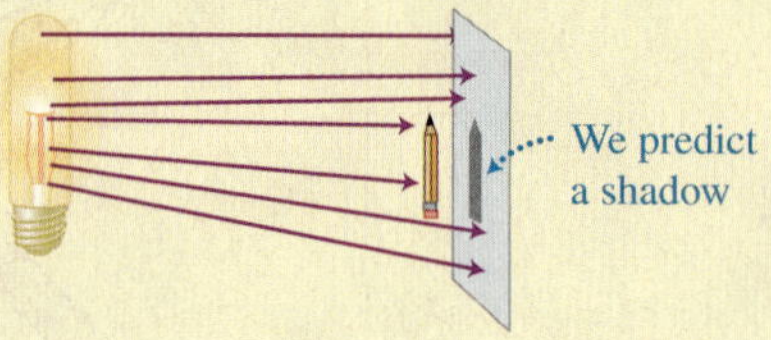  | We predict a dark, sharp shadow behind the pencil where the rays do not reach the wall. 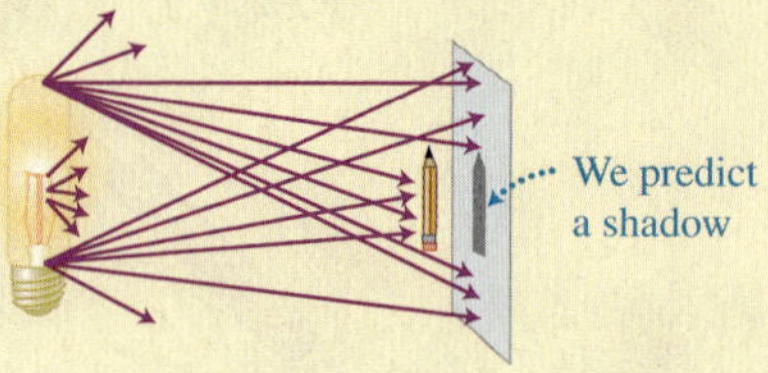  |
| **Outcome** | We see a dark, sharp shadow on the wall. | |
| **Experiment 2.** Turn on a lightbulb and place a pencil closer to the bulb between the bulb and the wall.  | We predict a dark, sharp shadow on the wall, as in Experiment 1. 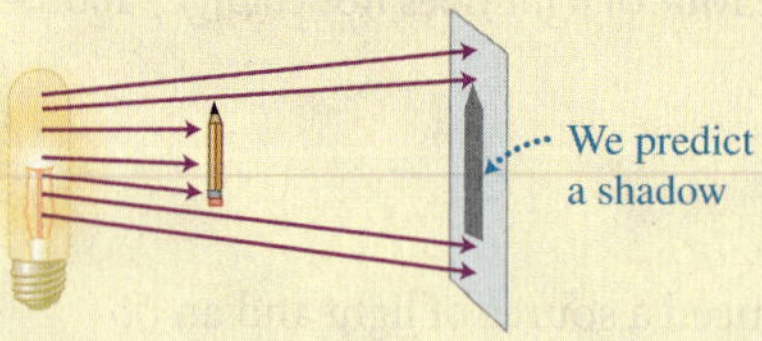  | We predict a light shadow with a fuzzy shaded region in the middle. 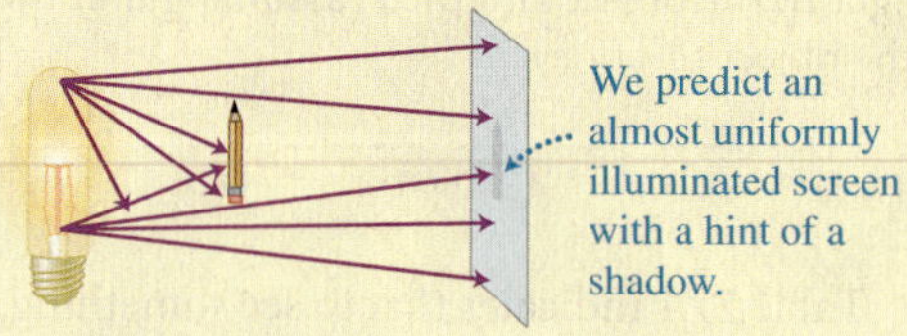  |
| **Outcome** | We see a fuzzy, light shadow (not as dark as in Experiment 1). | |
| **Experiment 3.** Cover the bulb with aluminum foil and poke a hole in the foil in the middle of the bulb facing the wall. Turn the bulb on. 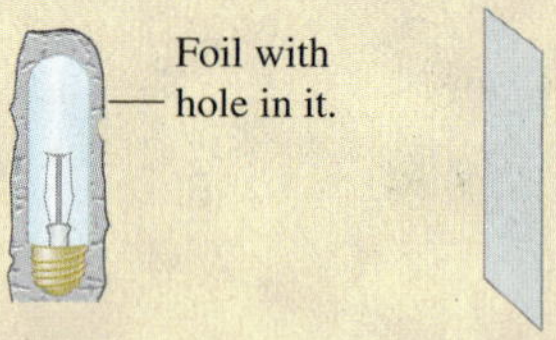  | We predict that we will see only a spot on the wall directly in front of the hole. 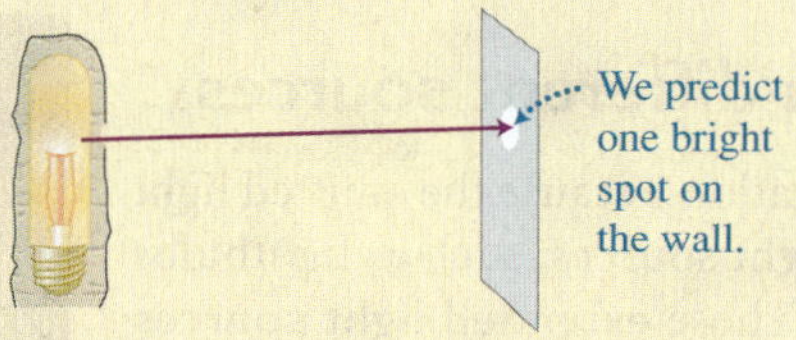  | We predict that the wall will be dimly lit. 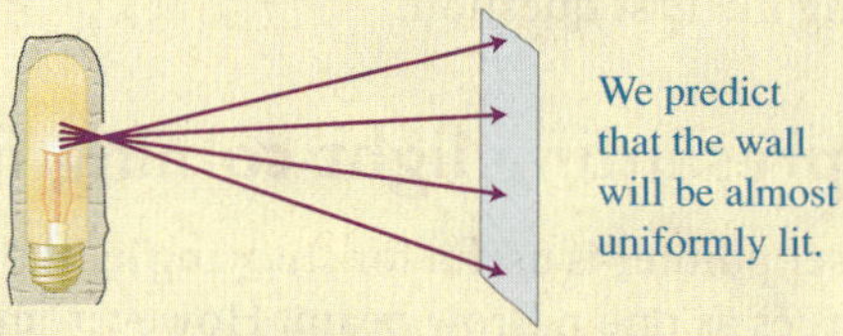  |
| **Outcome** | We observe that the wall is dimly lit. If we cover the first hole and poke a hole in a different place, the result remains the same. | |

### Conclusions

**Experiment 1.** Both models gave predictions that matched the outcome of this experiment. Neither is disproved.

**Experiments 2 and 3.** The predictions based on the one-ray model did not match the outcomes of the experiments; thus, the model is disproved. The predictions based on the multiple-ray model matched the outcomes of both experiments; thus, the model is supported (not proved).

**Figure 21.2** Two models of light emission from a bulb: (a) each point of the filament sends one ray; (b) each point of the filament sends multiple rays.

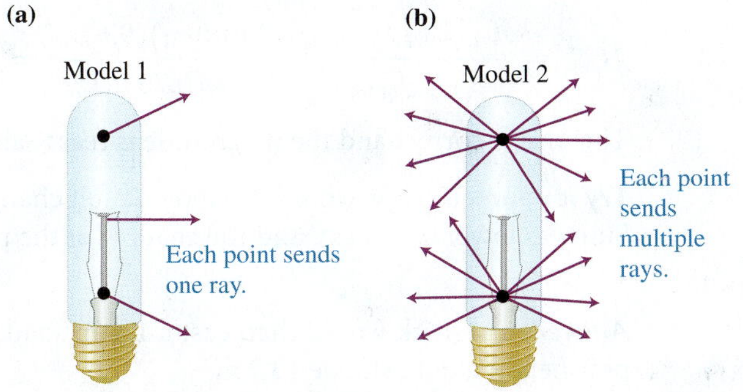


**Figure 21.3** Ray models of sunlight and laser light. (a) Only parallel rays of sunlight reach Earth; (b) parallel rays produce sharp shadows; (c) a laser beam can be modeled with one ray of light.

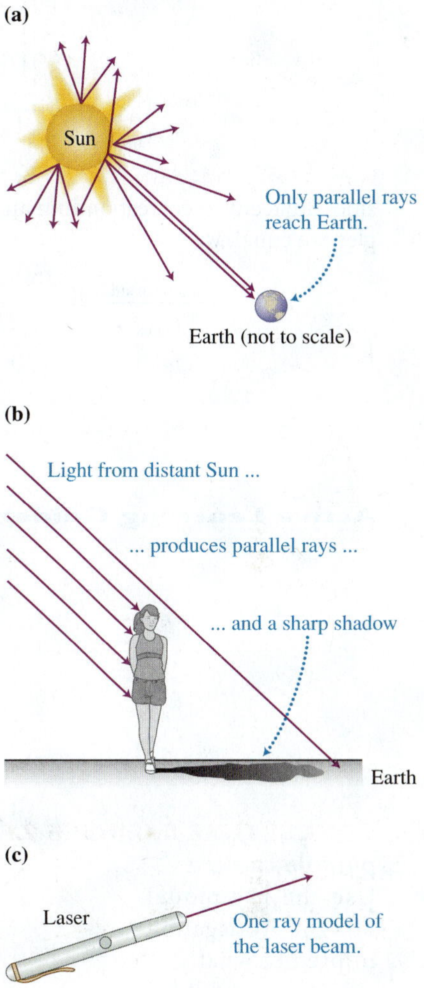


The testing experiments above disprove the one-ray model and support the idea that each point on an extended light source emits light in many different directions. This light can be represented by multiple rays diverging from that point.

**TIP** In Experiment 3 we made a small hole in the aluminum foil covering the bulb. Light coming through the hole traveled in all directions. Thus, light from a point-like light source must be represented using multiple rays—with the exception of laser light, which can be represented using one ray.

## Shadows and semi-shadows

The experiments in Table 21.2 revealed a new phenomenon—shadows. In Experiment 1 we found a sharp **shadow,** or **umbra,** behind the pencil on the wall. A shadow is a region behind the object where no light reaches. In Experiment 2 we found that there was no extremely dark shadow anywhere on the wall. What we observed is called a **semi-shadow,** also called a **penumbra**. A semi-shadow is a region where some light reaches and some does not. It appears as a fuzzy shadow.

The shadows of our bodies on a sunny day are very sharp—they aren't semi-shadows. Why? The Sun is so *far* from Earth that the only light from the Sun that reaches a small region on Earth is the light inside a narrow cylinder that extends from the Sun to Earth (**Figure 21.3a**). We can represent sunlight as a collection of parallel rays (Figure 21.3b). Similarly, we can represent the laser's very narrow beam of light with just one ray (Figure 21.3c).

**‹Active Learning Guide**

**EXAMPLE 21.1 Height of a streetlight pole**

On a sunny day, a streetlight pole casts a 9.6-m long shadow on the ground. You have a meter stick which when held vertical, casts a 0.70 m shadow. Use this information to determine the height of the pole.

**Sketch and translate** We first sketch the situation, as shown below. We represent the sunlight as parallel rays striking Earth's surface. The meter stick and the pole block the sunlight. The shadow of the stick is 0.70 m. The shadow of the pole is 9.6 m.

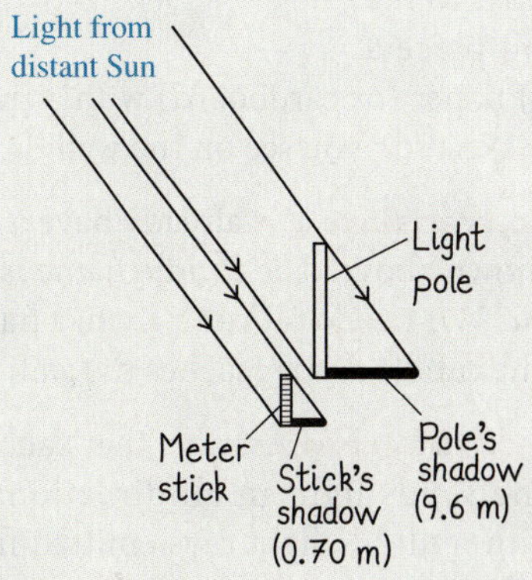


*(continued)*

**Simplify and diagram** The lines representing the shadows, the light rays, and the objects form two similar triangles, as shown.

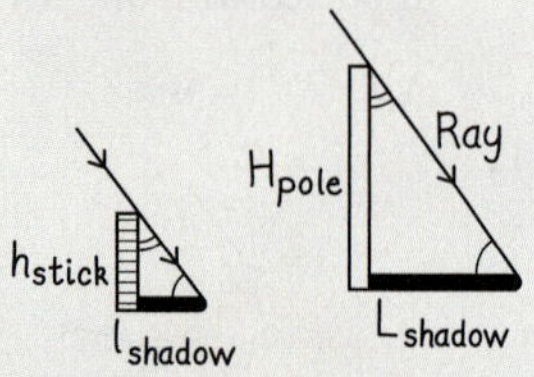


**Represent mathematically** Because the ratios of the sides adjacent to corresponding angles for the two triangles are equal, we have

$$\frac{l_{\text{shadow of stick}}}{h_{\text{stick}}} = \frac{L_{\text{shadow of pole}}}{H_{\text{pole}}}$$

**Solve and evaluate** Now, rearrange this equation to determine the height of the pole $H_{\text{pole}}$:

$$H_{\text{pole}} = \frac{h_{\text{stick}} L_{\text{shadow of pole}}}{l_{\text{shadow of stick}}} = \frac{(1.0\text{ m})(9.6\text{ m})}{(0.70\text{ m})} = 13.7\text{ m}$$

The unit is correct and the magnitude is reasonable.

**Try it yourself:** How would the above result change if the Sun was lower in the sky and the shadow of the pole was 19.2 m?

*Answer:* The stick would then cast a 1.4-m shadow. The pole height would still be 13.7 m.

## Pinhole camera

**Active Learning Guide›**

If you hold a candle flame about 1 m from a blank wall, you do not see a projection of the flame on the wall. Since each point on the flame is a point source of light and each point source emits light in all directions, the wall is illuminated by light coming from all of the points on the candle flame.

However, we can use a piece of cardboard with a very small hole in it to make a sharp projection of the flame on a wall in a dark room. Conceptual Exercise 21.2 explains how this projection is formed.

**CONCEPTUAL EXERCISE 21.2 Simulating a pinhole camera**

Use the ray model of light propagation to predict what you will see in the following experiment. You place a lit candle several meters from the wall in an otherwise dark room. Between the candle and the wall (and close to the candle), you place a piece of stiff paper (or cardboard) with a small hole in it (see figure). What do you see on the wall?

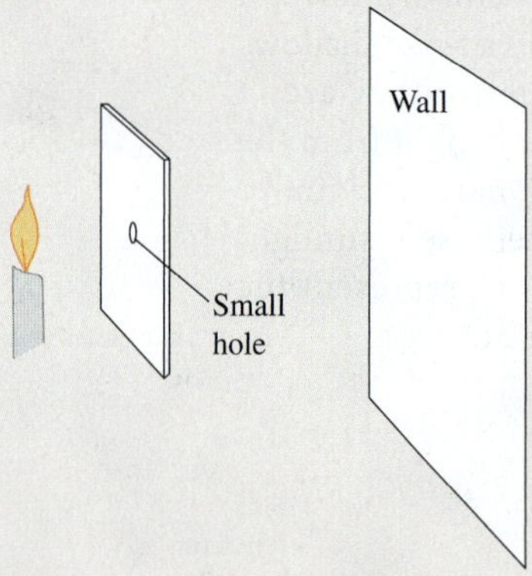


**Sketch and translate** We already have a sketch of the situation, shown above. The candle flame is an extended light source. We need to predict what happens when light from the candle flame reaches the wall.

**Simplify and diagram** Assume that each point of the candle flame sends light in all directions. Represent this light with multiple light rays emitted from only the top and the bottom of the flame. Most of these rays are blocked by the paper and do not reach the wall, as shown in the figure at the right. However, ray 1 from the bottom of the flame passes through the hole and reaches the wall. Ray 2 from the top of the flame reaches the wall but below where ray 1 did. Because of where these rays reach the wall, we predict that we will see an upside-down projection of the flame on the wall. That is what you actually see when you perform the experiment: the candle flame upside down on the wall.

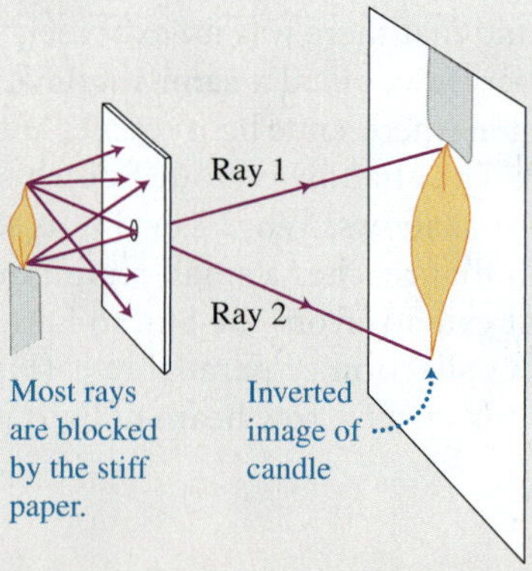


**Try it yourself:** Use the ray diagram to predict what you will see if you (a) move the candle closer to the hole; (b) move it farther from the hole; and (c) keep the candle and the stiff paper fixed relative to each other but move them together away from the wall.

*Answer:* (a) The upside-down candle on the wall gets bigger; (b) it gets smaller; (c) it gets bigger.

The cardboard with a small hole in it is the foundation of a pinhole camera, also called a **camera obscura**. Such a camera consists of a lightproof box with a very small hole in one wall and a photographic plate or film inside the box on the opposite wall. Before the invention of modern cameras that use lenses, pinhole cameras were used to make photographs. To photograph a person, you would shine intense light on the person for a long time. A small amount of light that reflected off the person would pass through the hole and form an inverted projection of the person on the film. In **Figure 21.4**, the outside world is projected through a pinhole onto a wall.

**Figure 21.4** A pinhole view of the outside world on a wall. Note that the buildings are projected upside down.

**Review Question 21.1** A light source placed in front of a screen with a small hole in it will be projected upside down on the wall behind the screen. Why is this true?

## 21.2 Reflection of light

In the last section we learned that people see objects because light is reflected off of objects into our eyes. In this section we investigate the phenomenon of reflection. We will start with a simple case of a narrow beam of light that we can represent with a single ray and a very good, smooth reflecting surface—a mirror.

In **Figure 21.5**, light from a laser pointer at A shines on a mirror supported vertically on a horizontal surface (a top view is shown). Light striking the mirror is called **incident** light. After reflection, the reflected laser light beam is as shown. Notice line OC in the figure—it is perpendicular to the surface of the mirror at the point where the incident light hits the mirror. This line is called a **normal line** (recall that in mathematics and physics "normal" means perpendicular). **Table 21.3** lists some angles related to the incident and reflected beams as measured by a protractor lying on the table.

The angle between the incident beam AO and the normal line CO is the **angle of incidence**. The angle between the reflected beam BO and the normal line CO is the **angle of reflection**. The data in the table show that the angles in columns two and three (the incident angle and the reflected angle, respectively) are always equal to each other. We can formulate this pattern as a mathematical rule for reflection:

*The angle of reflection = The angle of incidence*

**TIP** The angle of incidence and the angle of reflection are always the angles that the light beams form with the normal line and not the angles that they form with the surface of the mirror.

Let's use this result to predict the outcome of the new experiment in Conceptual Exercise 21.3.

**Table 21.3 Angles indicating relative directions of incident and reflected light beams from a mirror.**

| Angle AOB between the incident and reflected beams | Angle between the incident beam AO and the normal line CO | Angle between the reflected beam BO and the normal line CO |
|---|---|---|
| 0° | 0° | 0° |
| 40° | 20° | 20° |
| 60° | 30° | 30° |
| 90° | 45° | 45° |
| 120° | 60° | 60° |
| 160° | 80° | 80° |

### A Note about Language

Light propagates along straight lines, which we represent as *rays*. We use the term "a ray of light" as a model that describes and helps to predict light phenomena. When we talk about the actual light emitted from a source and traveling in space, we call it a "beam."

**‹Active Learning Guide**

**Figure 21.5** Light reflection from a mirror. Light beams and the protractor are in the horizontal plane and the mirror is positioned vertically (coming out of the page).

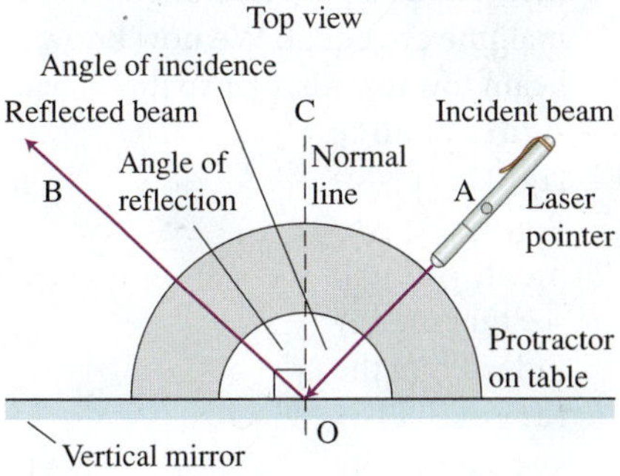

## CONCEPTUAL EXERCISE 21.3 Testing the reflection rule

Two mirrors stand on a table with their faces forming an angle greater than 90° (see the sketch below). Lay a laser pointer flat on the table and place the target in front of mirror 2. Use the rule of reflection being tested to predict how to aim a laser beam so that it hits mirror 1 and then mirror 2, and then finally passes directly over the target.

**Sketch and translate** The situation is already sketched in the first figure. Multiple laser orientations and positions will work. To guarantee that the beam will pass over the target, we'll work backward. Start by drawing the beam backward from the target to mirror 2, then to mirror 1, then finally to an appropriate position and orientation for the laser pointer.

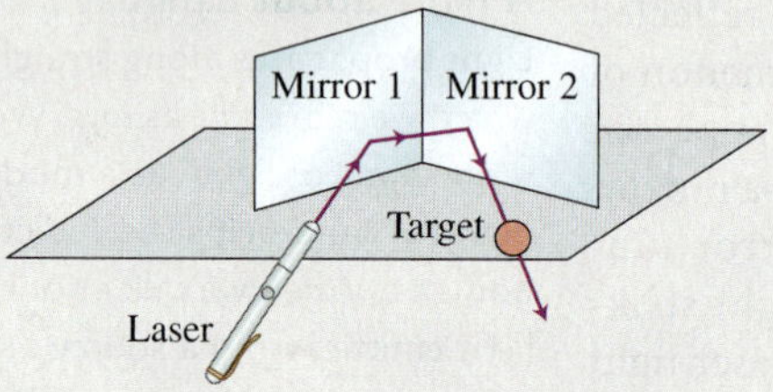


**Simplify and diagram** We draw top-view ray diagrams to represent the process in steps, representing each laser beam as one ray. The ray that hits mirror 2 and then the target is shown in the figure below. Note the orientation of the normal line in the figure and the equal angles of incidence and reflection. The reflection of that ray off mirror 1 is shown above, right. Again, the incident and reflected angles relative to mirror 1's normal line are equal. We now know how to direct the laser beam toward mirror 1 so that it will hit the target (second figure, above right). If you do the experiment, you will see that the ray indeed hits the target! This supports the reflection rule being tested.

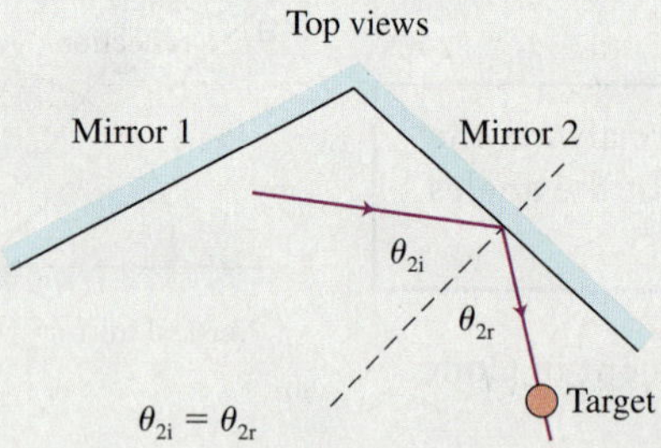


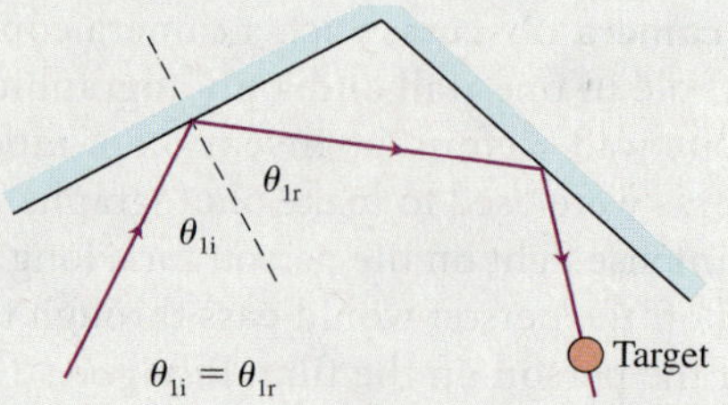


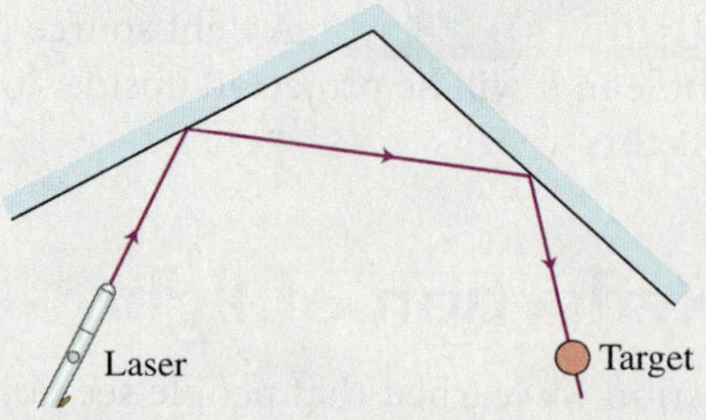


**Try it yourself:** Suppose the mirrors make a 90° angle relative to each other. Show that the ray reflected from one mirror and then the other is parallel to but in the opposite direction of the ray incident on the first mirror.The direction of the incident beam is perpendicular to the line where the mirrors intersect.

*Answer:* See the figure below. The incident ray reflects off mirror 1 and then mirror 2. The angle between the incident and reflected ray is equal to $2\theta_1 + 2\theta_2$. Note that the sum of the angles in a triangle equals 180° and the normal lines to the mirrors make a 90° angle with each other ($\theta_1 + \theta_2 + 90° = 180°$). Thus, $\theta_1 + \theta_2 = 90°$ and the angle between the incident and the reflected ray is $2(\theta_1 + \theta_2) = 2 \times 90° = 180°$.

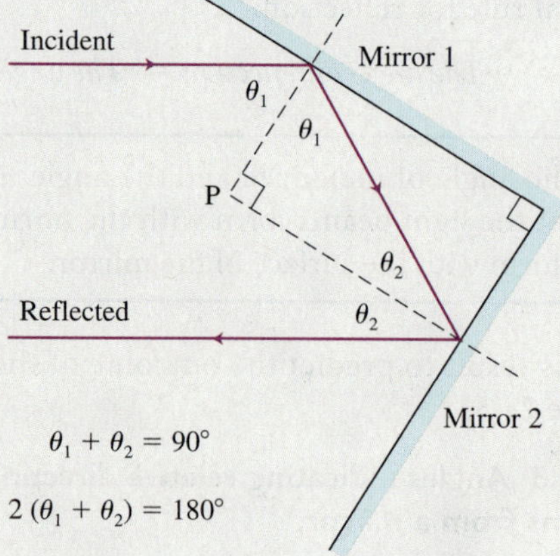


We can now formulate a relationship that is called the law of reflection.

**Law of reflection** **When a narrow beam of light, represented by one ray, shines on a smooth surface such as a mirror, the angle between the incident ray and the normal line perpendicular to the surface equals the angle between the reflected ray and the normal line (the angle of reflection equals the angle of incidence). The incident beam, reflected beam, and the normal line are in the same plane.**

$$\theta_{reflection} = \theta_{incidence} \tag{21.1}$$

## Specular and diffuse reflection

When we shine a laser beam on a wall, everyone in the room can see the bright spot. According to the law of reflection, the light beam should reflect at a particular angle determined by the angle of incidence. Does the fact that everyone sees the bright spot disprove the law of reflection?

Before we reject the law of reflection, we must carefully examine the assumptions we made when developing it; the reflecting surface is very smooth (like a mirror) and the light beam shining on it can be represented by one light ray. What if the reflecting surface is bumpy and the actual laser beam is wider than the bumps on the surface (**Figure 21.6a**)? In that case we cannot represent the reflected laser light as a single ray. The microscopic parts of the surface are oriented at different angles, and since the light from the laser hits many of these, the light is reflected in many directions. This explains why everyone in the room can see some of the light reflected from the wall. If instead of the wall we had a perfectly clean mirror (with no dust on it) and repeated the experiment in a dark room, only one person would be able to see the light reflected from the spot on the mirror illuminated by the laser.

The phenomenon in which a beam of light is reflected by a smooth surface (like a mirror) is called **specular reflection.** During specular reflection, a narrow incident beam represented by parallel rays is reflected as a narrow beam of light represented by parallel rays according to the law of reflection (**Figure 21.7a**).

The phenomenon in which light is reflected by a "bumpy" surface is called **diffuse reflection**. During diffuse reflection, different parts of a light beam (parallel incident rays) strike parts of the surface oriented at different angles with respect to the incident light. Thus the reflected beam becomes very wide (the reflected rays go in different directions) (Figure 21.7b).

We can now explain how we saw the path of the laser beam in the experiments in Table 21.1. Each tiny speck of dust reflects incident light in multiple directions. As there are myriads of dust particles, reflected rays reach our eyes and we can see "the path" of light. Understanding diffuse reflection helps us explain the phenomenon captured in the photo in **Figure 21.8.** Sunlight coming through the church windows reflects off the dust in the air and reaches different observers below.

In the next example, we use both diffuse and specular reflection to understand an everyday phenomenon.

**Figure 21.6** Reflection from a bumpy surface. Different spots on the surface are oriented at different angles with respect to incident light.

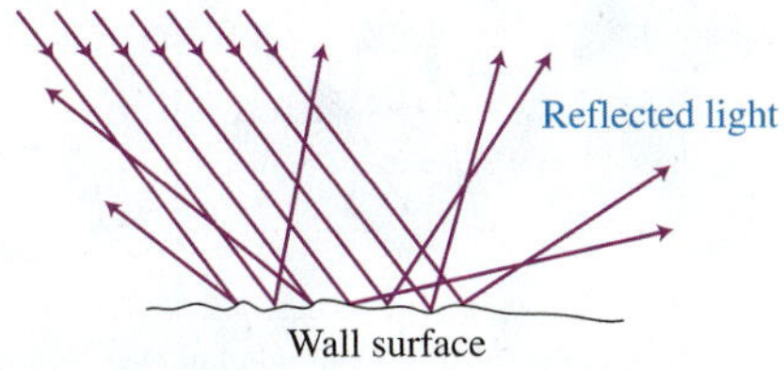


**Figure 21.7** (a) Specular light reflection from a smooth surface. Parallel incident rays remain parallel after reflection.
(b) Diffuse light reflection from a bumpy surface. Incident parallel rays diverge after reflection.

(a)

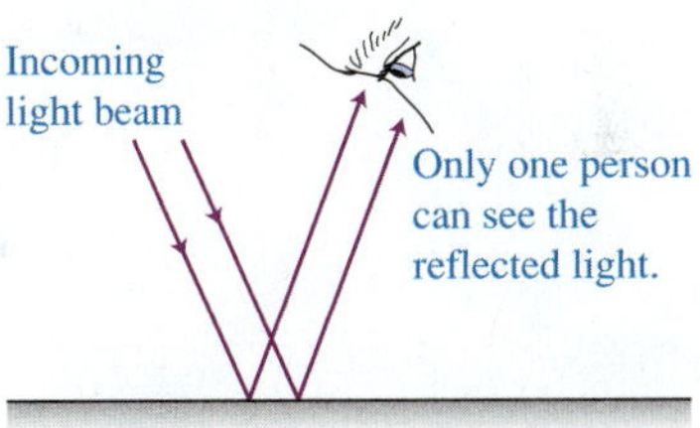


(b)

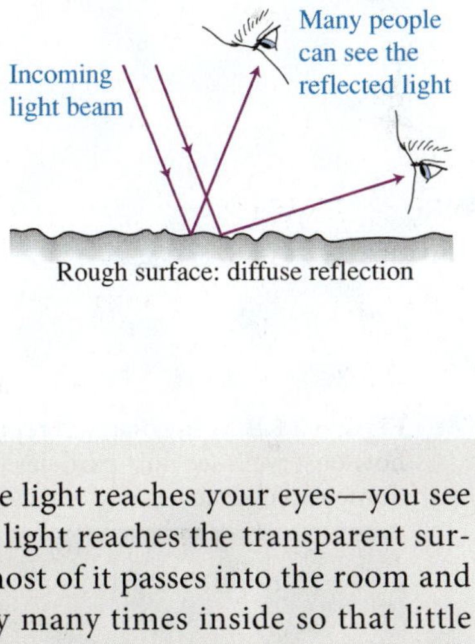


### CONCEPTUAL EXERCISE 21.4 Dark windows on a sunny day

On a sunny day, if you look at a house with its lights off, the uncovered windows look almost black but the outside walls do not. How can we explain this difference?

**Sketch and translate** If the windows look black, they must reflect very little sunlight to your eyes. Our goal is to explain why.

**Simplify and diagram** When light shines on the rough walls of the house, it reflects back diffusely at different angles and some of the light reaches your eyes—you see the wall easily. When light reaches the transparent surface of the window, most of it passes into the room and then reflects diffusely many times inside so that little comes back out. A small amount reflects off the smooth glass window as specular reflection (see the front view of the house and the top view inside the room in the figures on the next page). If you are not in the one correct location to see that reflected light, you see very little light coming from the window—it appears dark. But if you are standing in that one location, you see a bright reflection of the Sun in the window.

*(continued)*

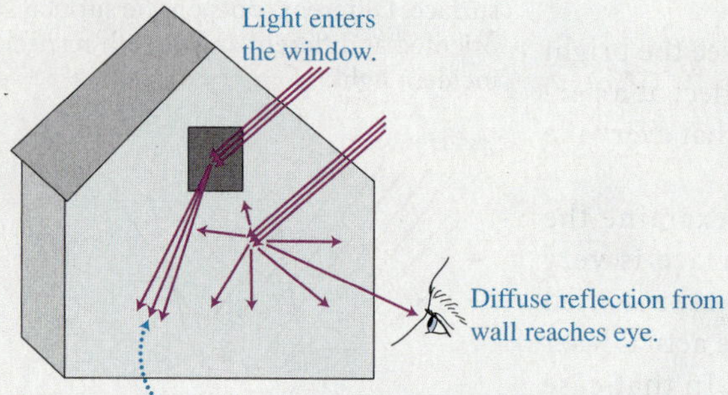
Light enters
the window.
Diffuse reflection from
wall reaches eye.
Specular reflection from window does not reach eye.

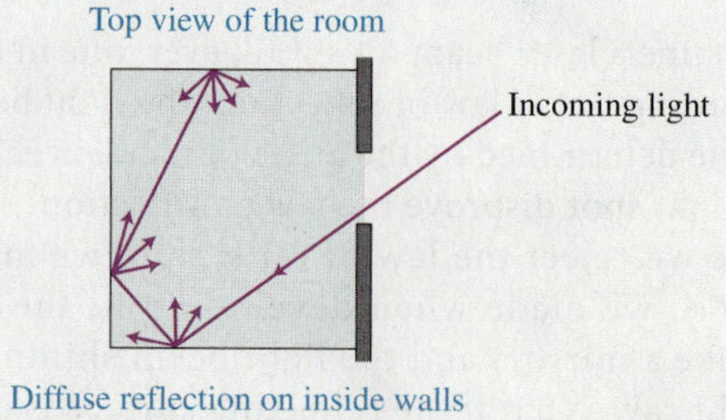
Top view of the room
Incoming light
Diffuse reflection on inside walls

**Try it yourself:** Why is the pupil of your eye dark?

*Answer:* The pupil is a hole in the eye—similar to a window in a house. Incident light enters the pupil and very little is reflected back—the pupil looks dark.

## Red eye effect

You are familiar with the so-called "red eye" effect that occurs occasionally when we take photographs with a flash (see **Figure 21.9**). The red eye effect is especially common at night or with low background lighting when the pupil is wide open. When the flash illuminates the open iris, light reflects from the red blood vessels in the retina on the back of the eye. Some of this reflected light passes back out of the pupil and makes the pupil appear red.

**Review Question 21.2** How can we test the law of reflection?

# 21.3 Refraction of light

At the shore of a lake, you see sunlight reflecting off the water's surface. But you also see rocks and sea plants under the surface. To see them, light must have entered the water, reflected off the rocks and plants, returned to the water surface, and then traveled from the surface to your eyes. If you have ever tried to use a stick to touch a rock under the surface of a pond or lake, you know that it is not easy. Although you carefully point the stick, you miss. It looks like you always extend the stick farther away from you than is needed to touch the rock. Why is that?

**Active Learning Guide›**

To answer this question, we will observe what happens when we shine a laser beam at different angles into water (see Observational Experiment **Table 21.4**).

**Figure 21.8** A ray diagram represents how observers see dust particles in the dome of a church.

**Figure 21.9** The baby's eyes in this photo have red circles in them. Light from the flash is reflected from the red blood vessels in the retina at the back surface of the inside of the eye.

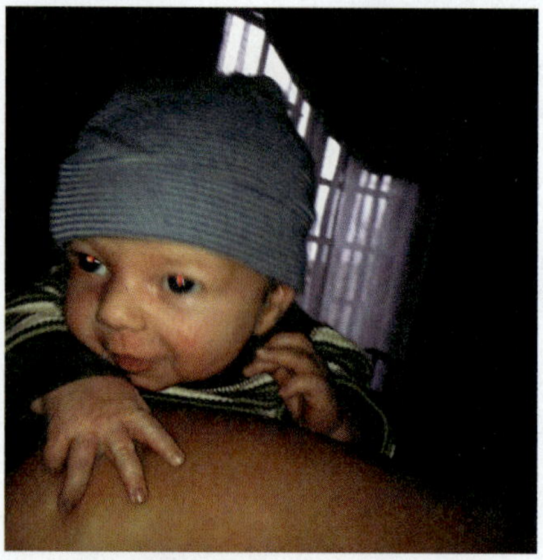

OBSERVATIONAL EXPERIMENT TABLE

## 21.4 The light path changes from air to water.

VIDEO 21.4

| Observational experiment | Analysis |
|---|---|
| **Experiment 1.** Shine a laser beam through the air straight down into a glass container filled with water with a few drops of milk added to make the beam visible. The container sits on a supporting ring so that light can leave through the bottom. We see red spots on the ceiling and floor of the room and on the bottom of the container. 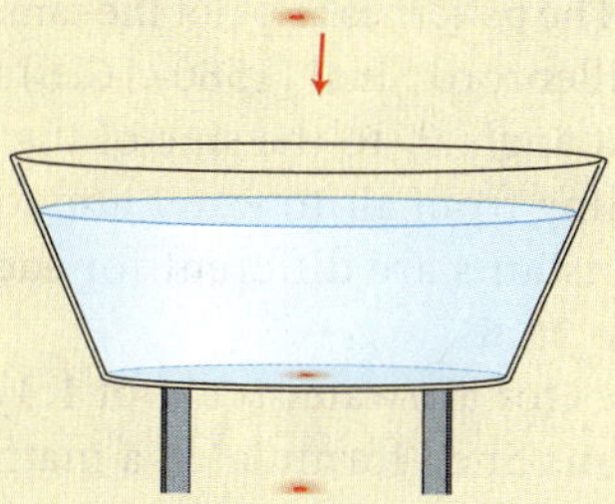 | Draw light rays to describe the appearance of the red spots. For simplicity we do not draw rays from the spots to our eyes. 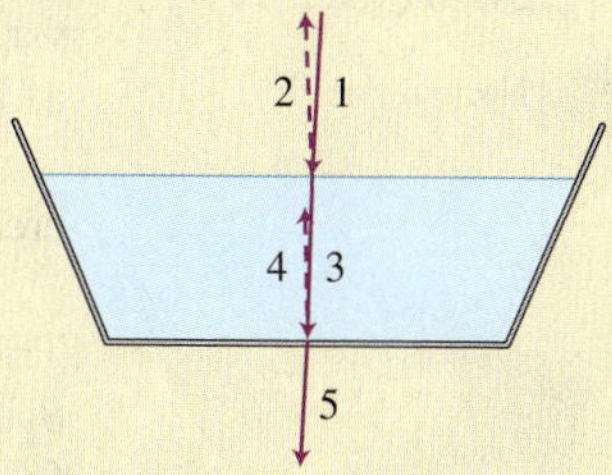  |
| **Experiment 2.** Shine the laser at an angle, and all four spots change their locations. 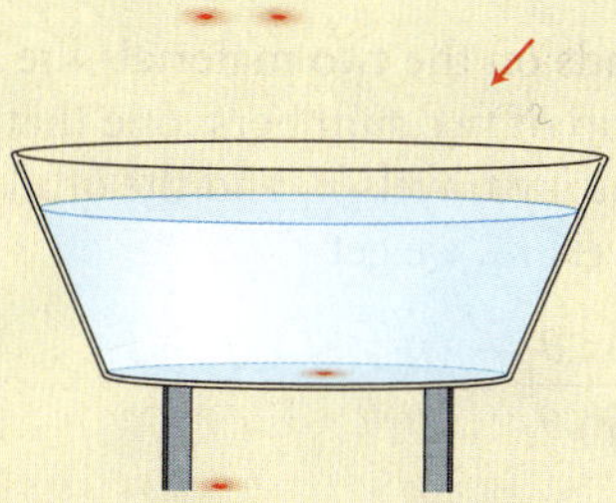 | The path of the ray changes as it enters water—it bends toward the normal, and then the path changes again when the ray emerges into the air—it bends away from the normal. 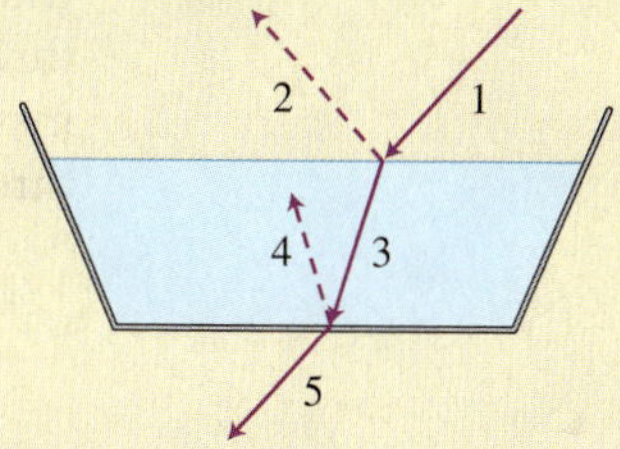  |
| **Experiment 3.** Increase the angle, and the spots change their locations even more. 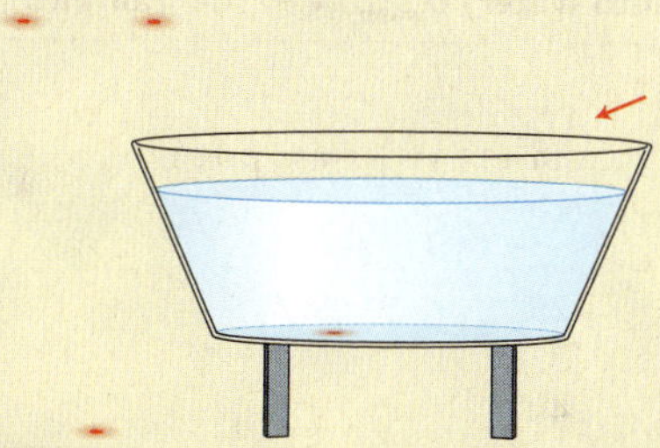 | 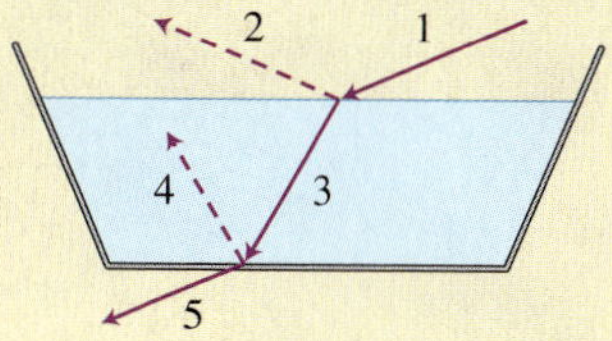  |

**Patterns**

When light shines at the air-water boundary at the top surface, the incident light beam represented by ray 1:

- Partially reflects back (ray 2) at the same angle as the angle of incidence; and
- Partially passes into the second medium (ray 3), bending at the interface toward the normal line.

Similar things happen to the light beam represented by ray 3. However, there are some differences. When ray 3 reaches the bottom water-air interface it:

- Partially reflects (ray 4) at the same angle as the angle of incidence; and
- Partially passes from the water into the air below the container (ray 5), bending at the interface away from the normal line.

Note in Table 21.4 that when the incident light represented by rays 1 and 3 is perpendicular to the boundary of the surfaces (Experiment 1), light reflects back along the same line (rays 2 and 4) and passes into the second medium without bending (ray 5). However, if the light is not perpendicular to the surface, it bends and travels in a different direction than in the previous medium. If we were to replace the water with a thick piece of transparent glass, we would observe a similar phenomenon. The change in the path of light when light travels from one medium to another is called **refraction**.

**Active Learning Guide>**

To develop a mathematical relationship between the angle of incidence and the **angle of refraction** (the angle the refracted beam makes with the normal line), we could do an experiment similar to the one we did when studying reflection. Such an experiment is shown in **Figure 21.10**. We could shine a laser beam at horizontal air-water and air-glass interfaces and record the angles of incidence and refraction. The results are shown below in **Table 21.5**.

**Figure 21.10** An experiment to measure the incident and refracted angles of light at an air-water (or air-glass) interface.

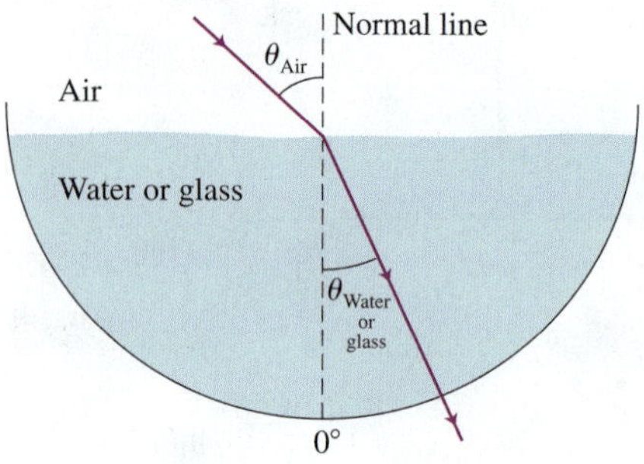


Notice that as the angle of incidence increases, the angle of refraction also increases but not in an obvious pattern. The pattern is also not the same for all materials.

In 1621 the Dutch scientist Willebrord Snell (1580–1626) found a pattern. The ratio of the sine of the incident angle $\theta_1$ to the sine of the refraction angle $\theta_2$ remains constant for light traveling from air to water and for light traveling from air to glass. However, the constants are different for each pair of media (see columns four and five of **Table 21.6**).

Notice that the ratio of the sines for air/water is about 1.3 and for air/glass is about 1.5. From these observations Snell formulated a mathematical model for refraction phenomena:

$$\frac{\sin \theta_1}{\sin \theta_2} = n_{1 \text{ to } 2} \tag{21.2}$$

where $n_{1 \text{ to } 2}$ is a number that depends on the two materials the light is traveling through. If we split $n_{1 \text{ to } 2}$ into a ratio of two numbers, one that depends on the material through which the incident ray travels $n_1$ and the other on the material through which the refracted ray travels $n_2$, we get

$$\frac{\sin \theta_1}{\sin \theta_2} = \frac{n_2}{n_1}$$

**Table 21.5 Angles of incidence and refraction between laser beams and the normal line.**

| Incident angle (ray in air) $\theta_{air}$ | Refraction angle (air into water) $\theta_{water}$ | Refraction angle (air into glass) $\theta_{glass}$ |
|---|---|---|
| 0° | 0° | 0° |
| 10° | 8° | 7° |
| 20° | 15° | 13° |
| 30° | 22° | 19° |
| 40° | 29° | 25° |
| 50° | 35° | 30° |
| 60° | 40° | 35° |

**Table 21.6 Pattern found by Snell for ratio of the sines of the incident and refraction angles.**

| Air $\sin \theta_1$ | Water $\sin \theta_2$ | Glass $\sin \theta_2$ | Air/water $(\sin \theta_1)/(\sin \theta_2)$ | Air/glass $(\sin \theta_1)/(\sin \theta_2)$ |
|---|---|---|---|---|
| 0.000 | 0.000 | 0.000 | | |
| 0.174 | 0.131 | 0.114 | 1.33 | 1.53 |
| 0.342 | 0.259 | 0.225 | 1.32 | 1.52 |
| 0.500 | 0.374 | 0.326 | 1.34 | 1.53 |
| 0.643 | 0.485 | 0.423 | 1.33 | 1.52 |
| 0.766 | 0.573 | 0.500 | 1.34 | 1.53 |
| 0.866 | 0.649 | 0.569 | 1.33 | 1.52 |

We call $n_1$ and $n_2$ the indexes of refraction of medium 1 and medium 2, respectively. Rearranging this, we get what has become known as **Snell's law** (or the law of refraction).

**Snell's law** **(the law of refraction) relates the refraction angle $\theta_2$ to the incident angle $\theta_1$ and the indexes of refraction of the incident medium $n_1$ and the refracted medium $n_2$:**

$$n_2 \sin \theta_2 = n_1 \sin \theta_1 \qquad (21.3)$$

**The angles of incidence and refraction are measured with respect to the normal line where the ray hits the interface between the two media. If the refracted ray is closer to the normal than the incident ray, then medium 2 is more optically dense (with a higher index of refraction) than medium 1. The incident ray, refracted ray, and the normal are in the same plane.**

If we define the index of refraction of air as 1.00, then the indexes of refraction of water and of the glass used in Tables 21.5 and 21.6 are 1.33 and 1.53, respectively. Notice that glass refracted the light ray more toward the normal line than did water; the glass is more optically dense than water. The refractive indexes of several materials are given in **Table 21.7**, using yellow light as an example.

**TIP** Light going from a lower to a higher index of refraction will bend toward the normal, but light going from a higher to a lower index of refraction will bend away from the normal.

### EXAMPLE 21.5 Concentration of glucose in blood

The refractive index of blood increases as the blood's glucose concentration increases (for normal glucose content it ranges from 1.34 to 1.36; it is higher than the 1.33 for water because of all the additional organic components in the blood). Therefore, measuring the index of refraction of blood can help determine its glucose concentration.

In a hypothetical process, a hemispherical container holds a small sample of blood (see figure). A narrow laser beam enters perpendicular to the bottom curved surface and into the sample. The light reaches the blood-air interface at a 40.0° angle relative to the normal line. The light leaves the blood and passes through the air to a row of tiny light detectors at the top indicating that the light beam left the blood at a 61.7° angle. Determine the refractive index of the blood.

**Sketch and translate** First, we sketch the situation (see the figure). We use Snell's law to determine the index of refraction of the blood $n_{1\text{ blood}}$. The incident angle is $\theta_{1\text{ blood}} = 40.0°$. The refraction angle is $\theta_{2\text{ air}} = 61.7°$. The index of refraction of the air is $n_{2\text{ air}} = 1.00$.

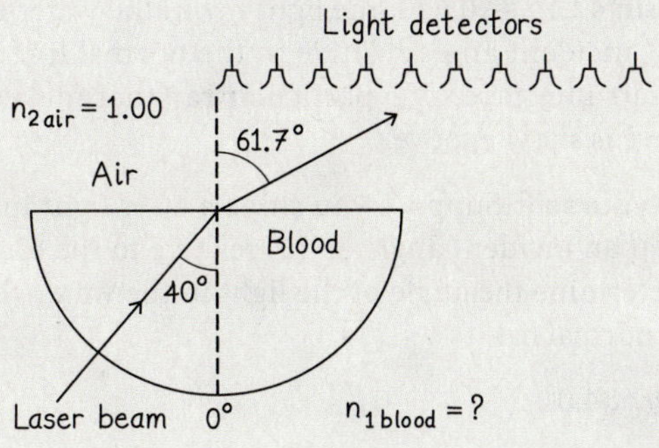


**Simplify and diagram** The diagram in the figure can be used as a ray diagram if we assume that the laser beam is narrow.

**Represent mathematically** Snell's law for this situation is

$$n_{1\text{ blood}} \sin \theta_{1\text{ blood}} = n_{2\text{ air}} \sin \theta_{2\text{ air}}$$

**Solve and evaluate** Dividing both sides of Snell's law equation by $\sin \theta_{1\text{ blood}}$, we have

$$n_{1\text{ blood}} = \frac{n_{2\text{ air}} \sin \theta_{2\text{ air}}}{\sin \theta_{1\text{ blood}}} = \frac{1.00 \cdot \sin 61.7°}{\sin 40.0°} = 1.37$$

This result is higher than the normal index of refraction for blood.

**Try it yourself:** Pure blood plasma has a higher index of refraction than blood. Suppose the refractive index of the patient's plasma is 1.43 instead of 1.37 for her blood. What then would the refraction angle be if the incident angle were still 40.0°?

*Answer:* 66.8°.

**Table 21.7 Refractive indexes of yellow light in various substances.**

| Material | Refractive index $n$ |
|---|---|
| Vacuum | 1.00000 |
| Air | 1.00029 |
| Carbon dioxide | 1.00045 |
| Water (20°C) | 1.333 |
| Ethyl alcohol | 1.362 |
| Glucose solution (25%) | 1.372 |
| Glucose solution (50%) | 1.420 |
| Glucose solution (75%) | 1.477 |
| Glass | 1.517–1.647 |
| Fluorite | 1.434 |
| Diamond | 2.417 |

At the beginning of this section we discussed the difficulty in touching an object under water with a stick. Refraction helps us understand why. See the following example.

**EXAMPLE 21.6 Retrieving a coin in a fountain**

Light from a coin at the bottom of a fountain reaches your eye at an angle of 27.0° below the horizontal. At what angle should you look to see the coin?

**Sketch and translate** We sketch the situation and label the known quantities and the unknowns, as shown below. You see the coin because sunlight reflecting from it enters your eye. The light traveling in the air after leaving the water makes an angle of 27.0° relative to the horizontal (making the refraction angle $\theta_{2\,\text{air}} = 63.0°$). The coin *appears* to be somewhere on this line. However, you know that light rays bend when they pass from water into air. Thus we need to find the angle through which the incident ray travels in water to determine the direction of the ray in water. The indexes of refraction of air and water are $n_{1\,\text{water}} = 1.33$ and $n_{2\,\text{air}} = 1.00$, respectively.

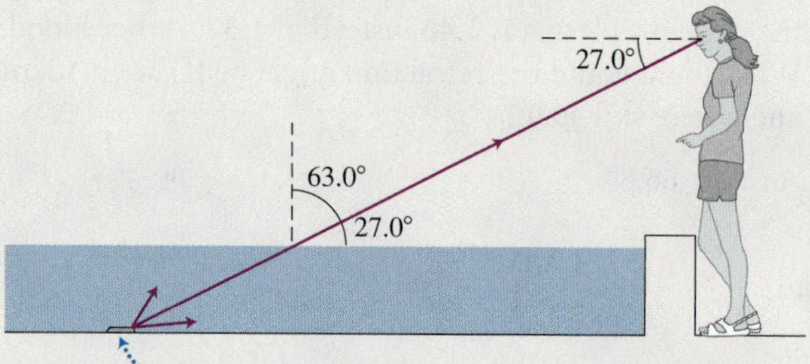


**Simplify and diagram** Assume that just one ray represents the light from the coin to your eye, as shown below. It changes direction when it goes from the water to the air.

**Represent mathematically** We want to determine the incident angle of the light ray when it reaches the water-air interface so that after refraction, the ray moves toward your eyes at a 27.0° angle below the horizontal (a 63.0° angle relative to the normal line). We use Snell's law to find the incident angle between the light ray moving in the water and the normal line:

$$n_{1\,\text{water}} \sin \theta_{1\,\text{water}} = n_{2\,\text{air}} \sin \theta_{2\,\text{air}}$$

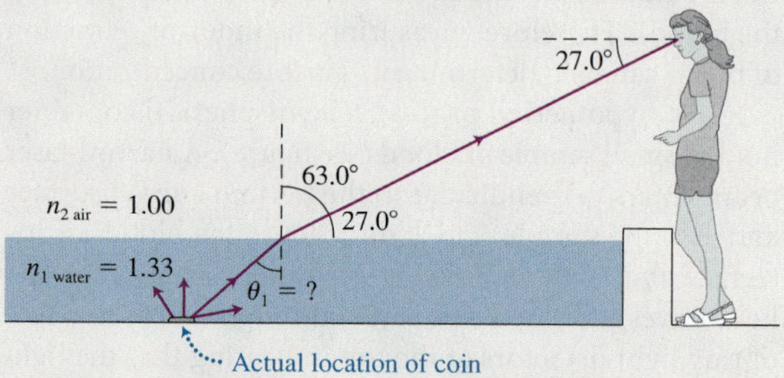


**Solve and evaluate** Dividing both sides of the equation by $n_{1\,\text{water}}$ and inserting the appropriate quantities, we have

$$\sin \theta_{1\,\text{water}} = \frac{n_{2\,\text{air}} \sin \theta_{2\,\text{air}}}{n_{1\,\text{water}}} = \frac{1.00 \sin 63.0°}{1.33} = 0.670$$

Since $\sin 42.1° = 0.670$, the light ray in the water makes a 42.1° incident angle relative to the normal line at the water-air interface. The particular ray that makes it to your eye is shown above.

**Try it yourself:** Suppose you shine a laser light into the water at an incident angle of 42° relative to the horizontal. Determine the angle of the light in the water relative to the normal line.

*Answer:* 34.0°.

**TIP** When you draw ray diagrams, remember:

a. Most objects do not emit light; they reflect light shining on them. For simplicity, we draw them as light-emitting objects, though really they are just sources of reflected light.
b. Each point of an object emits rays in all directions. We only draw the ones that are the most convenient for describing the situation.
c. No rays come from our eyes. We see objects because light from them reaches our eyes. When drawing a ray diagram, think of which rays will reach the eyes of the observer.

## Restoring a broken glass

Imagine that we place a piece of glass (instead of a coin) in water. Will you be able to see it? As long as light reflects off the glass and reaches your eye, you will. Since the index of refraction of glass is different from that of water, there is an **optical boundary** between them. When light traveling in water hits the glass, it refracts at the water-glass boundary and it also reflects some light back to the eyes of an observer, thus making the glass visible. Now, imagine that you have a liquid whose refractive index is exactly equal to that of the glass. A good example of such a liquid is vegetable oil. Light traveling through this oil will not reflect off the piece of glass submerged in it or refract at the surface of the glass because there is effectively no optical boundary between the two materials. Light will just continue traveling in straight lines as if the glass were not there. Thus a piece of glass becomes invisible in vegetable oil! This explains the magic trick described in the chapter opening story. (Note: If you decide to try the experiment, remember to fill the unbroken vial with oil before you put it in the oil bath; otherwise it will be full of air and the air will make it visible!)

In summary, in order for us to see things, they must either radiate light (like lightbulbs or fire) or reflect light (like planets, trees, and snowflakes). Reflection occurs off a transparent object only if the optical density of the reflecting object is different from that of the material that surrounds it.

**Review Question 21.3** Why is the expression "light travels in a straight line" not entirely accurate?

## 21.4 Total internal reflection

In two examples in the last section, light traveled from a more optically dense medium to a less optically dense medium, for example, from water to air. In such cases the light bent away from the normal line when entering the less dense medium. This behavior has some very important applications in the transmission of light by optical fibers.

Consider the situation represented by the ray diagram in **Figure 21.11a**. You perform a series of experiments in which an incident ray under water hits a water-air interface at an increasingly larger angle relative to the normal line. As the incident angle gets bigger, the angle in the air between the refracted ray and the normal line gets even bigger. At the so-called **critical angle of incidence** $\theta_c$ (Figure 21.11c), the refraction angle is 90°. The refracted ray travels along the water-air interface. Notice that when this happens, the incident angle is still less than 90°. At incident angles larger than the critical angle, $\theta_1 > \theta_c$, the light is totally reflected back into the water—none escapes into the air (Figure 21.11d).

**Figure 21.11** Total internal reflection occurs when $\theta_1 > \theta_c$, that is, when the angle of incidence is larger than the critical angle.

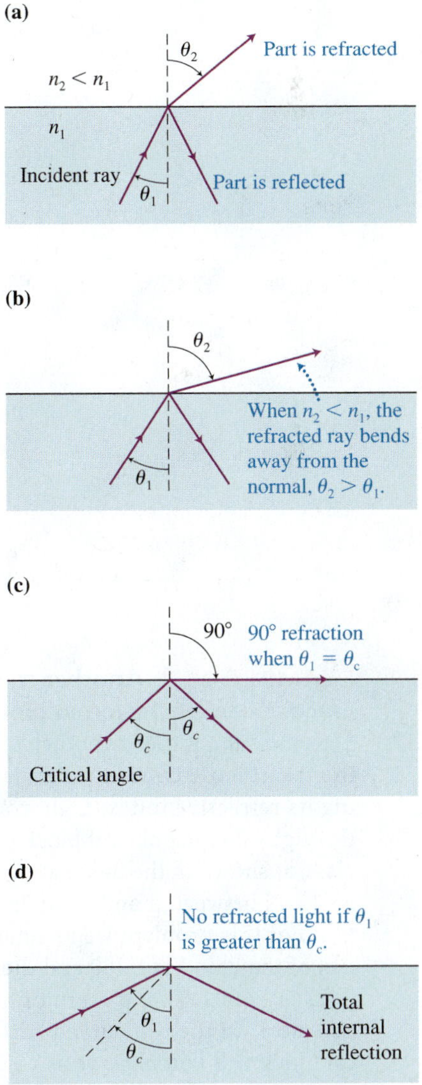


‹**Active Learning Guide**

This behavior occurs whenever light attempts to travel from a more optically dense medium 1 of refractive index $n_1$ to a less optically dense medium 2 of refractive index $n_2$ $(n_1 > n_2)$. We can use Snell's law to determine the critical angle in terms of the two indexes of refraction: $n_1 \sin\theta_1 = n_2 \sin\theta_2$ where $\theta_1 = \theta_c$ and $\theta_2 = 90°$. Remember that $\sin 90° = 1$. Inserting these values and solving for $\sin\theta_c$, we have

$$n_1 \sin\theta_c = n_2 \sin 90°$$

$$\sin\theta_c = \frac{n_2 \sin 90°}{n_1} = \frac{n_2(1)}{n_1} = \frac{n_2}{n_1}$$

If the incident angle $\theta_1$ is greater than $\theta_c$, there is no solution to Snell's law, as $\sin\theta_2$ would be greater than 1.00 (the maximum value of sine is 1.00). For $\theta_1 > \theta_c$, all of the light is reflected back into the water. There is no refracted ray. Remember that this applies only for situations when $n_1 > n_2$. This phenomenon is called **total internal reflection**.

**Total internal reflection** **When light travels from a more optically dense medium 1 of refractive index $n_1$ into a less optically dense medium 2 of refractive index $n_2$ $(n_1 > n_2)$, the refracted light beam in medium 2 bends away from the normal line. If the incident angle in medium 1 $\theta_1$ is greater than a critical angle $\theta_c$, *all* of the light is reflected back into the denser medium. The critical angle is determined by**

$$\sin\theta_c = \frac{n_2}{n_1} \quad (21.4)$$

**TIP** Total internal reflection occurs only when light travels from a medium with a higher refractive index to a medium with a lower one.

The refractive index is a fundamental physical property of a substance and is often used to identify an unknown substance, confirm its purity, or measure its concentration. Refractometers, instruments that measure refractive index, have medical and industrial applications. Veterinarians use portable refractometers to measure the total serum protein in blood and urine, and also to detect drug tampering in racehorses.

## EXAMPLE 21.7 Another method to measure glucose concentration in blood

Let's look at another hypothetical method for determining the glucose concentration in the blood by measuring its refractive index. Light travels in a narrow beam through a hemispherical block of high-refractive-index glass, as shown in the figure at right. A thin layer of blood is placed between it and another hemispherical block on top. For this problem we assume that the refractive index of the blood is 1.360 and the refractive index of the glass is 1.600. What pattern of light reaching tiny light detectors on the top curved surface of the hemispherical block will you observe as you move the light source clockwise around the edge of the curved surface of the bottom block?

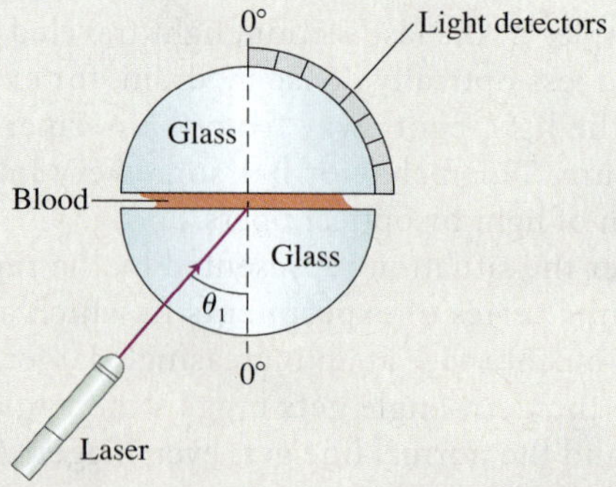


**Sketch and translate** We already have a sketch of the apparatus (in the figure above). Our goal is to predict what happens to the refracted beam if we vary the angle of incidence.

**Simplify and diagram** We assume that the light source is always oriented perpendicular to the curved surface on the bottom and points toward the same point on the glass-blood interface. We also assume that the blood layer is thin. Next we draw a ray diagram for four rays (as shown in the figure below).

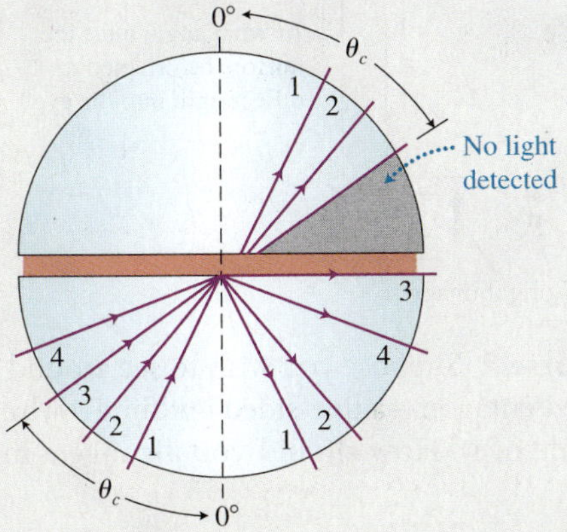


Ray 1 bends away from the normal line at the first glass-blood interface. It is partially reflected and partially refracted into the blood. On the top surface of the blood, it bends back toward the normal line as it moves into the hemispherical glass block above (part of it is reflected at the second interface—not shown). The net angular deflection of the ray is zero—ray 1 in the lower block is parallel to ray 1 in the top block. Ray 2 has a greater incident angle with the first glass-blood interface and also bends away from the normal line as it enters the blood. It is partially transmitted into the blood and partially reflected. On the top surface of the blood it bends back toward the normal line as it moves into the hemispherical glass block above—to the right of ray 1. Ray 3 has a greater incident angle and is refracted at 90° in the blood. Hence, the incident angle is the critical angle. Ray 4 has an even greater incident angle than the critical angle and is totally reflected back into the lower hemispherical block.

The detectors on the top surface stop detecting light when the incident angle is larger than the critical angle. Thus, you can detect the critical angle by the place where the light stops arriving at the top block.

**Represent mathematically** We can use Eq. (21.4) to determine the critical angle:

$$\sin \theta_c = \frac{n_2}{n_1}$$

**Solve and evaluate** Using the above with the glass index of refraction $n_1 = 1.600$ and the blood index of refraction $n_2 = 1.360$, we find that

$$\sin \theta_c = \frac{n_2}{n_1} = \frac{1.360}{1.600} = 0.850$$

A 58.2° angle has a sine equal to 0.850. When incident light reaches this angle, the detectors on the top hemisphere will stop detecting light. As long as the incident angle is smaller than 58.2°, the detectors on the top of the apparatus will detect light. Thus the apparatus allows us to measure the critical angle and use it to determine the index of refraction of the blood and glucose concentration.

**Try it yourself:** For a different sample of blood, you do not detect light at angles of 59.2° and larger. What is the refractive index of this new blood sample?

*Answer:* 1.374.

**Review Question 21.4** Why did we study total internal reflection after our investigation of refraction and not after reflection?

## 21.5 Skills for analyzing reflective and refractive processes

When you solve problems involving light, use ray diagrams to help in your reasoning and quantitative work. The diagrams will also help you evaluate the final answer. In this section we will work through three exercises: a conceptual exercise that does not involve calculations, an example that requires the use of all the problem-solving steps and, finally, an equation Jeopardy example that shows how to interpret the mathematical description of light phenomena using ray diagrams.

**TIP** When drawing ray diagrams, be sure to use a ruler and graph paper.

## CONCEPTUAL EXERCISE 21.8

### Build a periscope

A periscope is an instrument that allows you to observe your surroundings from a concealed location (for example, a submarine might have a periscope that allows sailors to look at other ships on the surface while under water). A simple periscope is a tube with two mirrors positioned parallel to each other at a certain angle relative to the horizontal direction. Determine that angle.

**Sketch and translate** Imagine an observer at a lower elevation than the object she wants to see. What arrangement of the mirrors allows light traveling horizontally from the object to reach the observer's eyes?

**Simplify and diagram** Horizontal light striking the first mirror reflects downward. For this light to reach the second mirror, it must travel vertically through the tube (see figure). If the mirror is positioned at a 45° angle relative to the horizontal, the reflected light (according to the law of reflection) will be 90° relative to the incident light, in other words, pointing down. At the lower mirror, also oriented at 45° relative to the horizontal, the light is reflected horizontally to the eye.

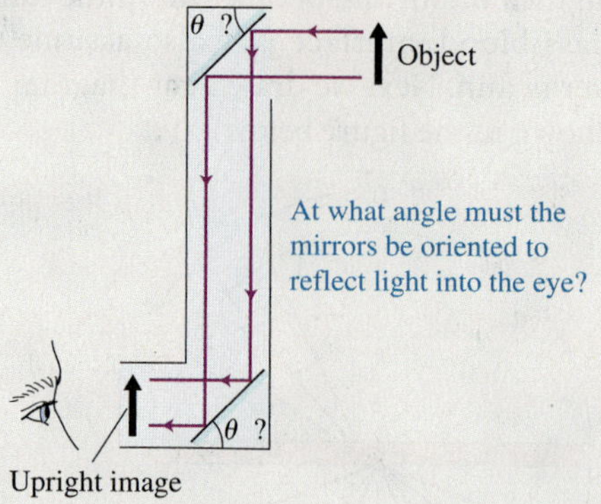


**Try it yourself:** Suppose you wish to see around the corner of the Pentagon—a five-sided building—while standing outside of it. How should you arrange a mirror or mirrors?

*Answer:* One mirror oriented at a 36.0° angle relative to your side of the building will do.

## PROBLEM-SOLVING STRATEGY Analyzing processes involving reflection and refraction

### EXAMPLE 21.9 Hiding mosquito fish

A mosquito fish hides from a kingfisher bird at the bottom of a shallow lake, 0.40 m below the surface. A leaf has blown onto the lake and floats above the mosquito fish. How big should the leaf be so the kingfisher cannot see its prey from any location above the water?

**Sketch and translate**

- Sketch the described situation or process.
- Indicate all the known and unknown quantities.
- Outline a solution for the problem as best you can.

- The situation is sketched below.

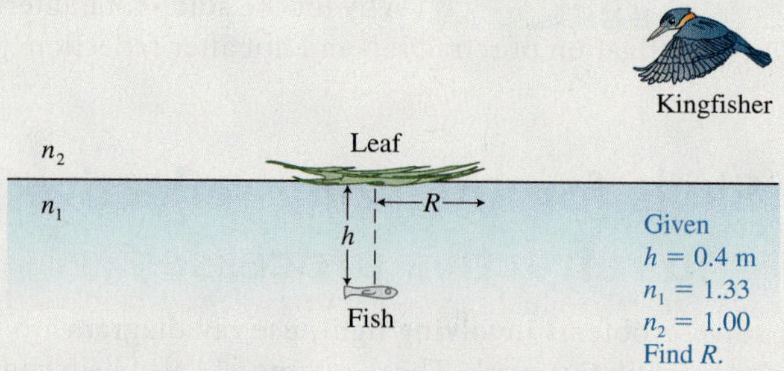


- We want to know what size leaf we need so that any light reflected from the fish and reaching the water surface undergoes total internal reflection and does not leave the water. Light incident at a smaller angle relative to the normal hits the leaf.

**Simplify and diagram**

- Indicate any assumptions you are making.
- Draw a ray diagram showing all relevant paths the light travels. Consider the light rays originating from the object and eventually reaching the observer. In the diagram you can represent the observer by an eye.

- Model the fish as a shining point particle and the leaf as circular.
- Draw a ray diagram. $\theta_c$ is the critical angle. For incident angles greater than $\theta_c$, there is no refracted ray—all light is totally reflected.

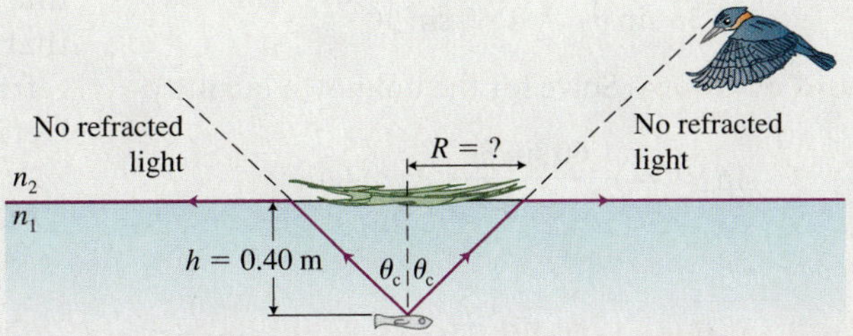


**Represent mathematically**

- Use the sketch and ray diagram to help construct a mathematical description using the law of reflection, or Snell's law for refraction, or the application of Snell's law for total internal reflection.

- Use Eq. (21.4) $\sin\theta_c = n_2/n_1$, where $n_1 = 1.33$ (the refractive index of water) and $n_2 = 1.00$ (the refractive index of air).
- Knowing the angle $\theta_c$ we can determine the radius of the leaf: $R = h\tan\theta_c$.

**Solve and evaluate**

- Solve the equations for the unknown quantities.
- Evaluate the results to see if they are reasonable (the magnitude of the answer, its units, how the answer changes in limiting cases, and so forth).

- We find that $\sin\theta_c = 1.00/1.33 = 0.752$. The angle whose sin is 0.752 equals 48.8°.
- Thus $\tan\theta_c = \tan 48.8° = 1.14$.
- We can now determine the radius of the leaf: $R = h\tan\theta_c = (0.40\text{ m})(1.14) = 0.46\text{ m}$.
- This is an unrealistically big leaf. The fish is not safe. Note that the ray diagram shows the large radius of the leaf compared to the depth of the fish.

**Try it yourself:** If the leaf radius is 0.20 m, how far below the leaf would the mosquito fish have to be so that it could not be seen by the kingfisher bird?

*Answer:* 0.18 m or less.

## Equation Jeopardy

‹Active Learning Guide

In the following example, you are asked to interpret the mathematical description of a process and construct a ray diagram that represents that process. Then you are asked to invent a word problem that the equation could be used to solve. The problem-solving procedure is reversed in a Jeopardy problem.

**EXAMPLE 21.10 Jeopardy problem**

The equation below describes a physical process. Invent a problem for which the equation would provide a solution.

$$1.33 \sin \theta_2 = 1.60 \sin 30°$$

**Solve and evaluate** Solve for the unknown quantity:

$$\sin \theta_2 = \frac{1.60 \sin 30°}{1.33} = 0.60$$

or $\theta_2 = 37°$.

**Represent mathematically** The equation appears to be an application of Snell's law.

**Simplify and diagram** Light in a medium with refractive index 1.60 (something like glass) is moving into a second medium with refractive index 1.33 (probably water). The incident angle is $\theta_1 = 30°$ (as shown in the figure). The light bends away from the normal since the refractive index decreases from glass to water.

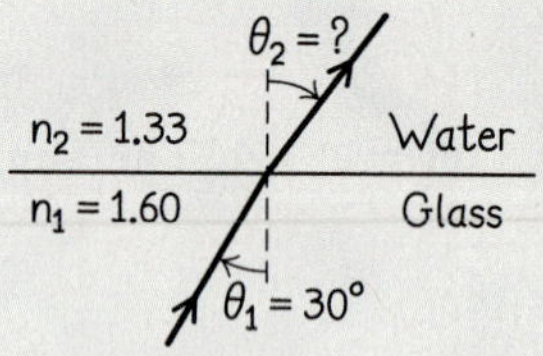


**Sketch and translate** One possible problem is as follows: A narrow beam of light moves up through the thick glass bottom of an aquarium and hits an interface with the water at a 30° angle relative to the normal line between the two media, as shown below. Assume that the aquarium is surrounded by air. The index of refraction of the glass is 1.60 and of the water is 1.33. Determine the angle of the refracted light in the water.

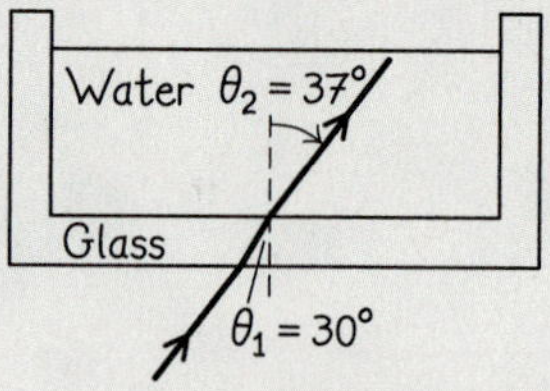


**Try it yourself:** Describe a process consistent with the equation $n_2 \sin 90° = 1.60 \sin 48°$.

*Answer:* Light in a glass-like material of refractive index 1.60 is incident on a different medium with refractive index 1.19. The light is refracted 90°. Thus, the critical angle is 48°.

**Review Question 21.5** What is the critical angle for total internal reflection for light going from water into glass of refractive index 1.56?

## 21.6 Fiber optics, prisms, mirages, and the color of the sky: Putting it all together

In this section we consider several applications of reflection, refraction, and total internal reflection: fiber optics, prisms, mirages, and the color of the sky.

### Fiber optics

Imagine light traveling inside a long, thin, flexible, glass cylinder (like a wire made of flexible glass): such a cylinder is called an **optical fiber**. Fiber optic filaments are used in telecommunications to transmit high-speed light-based data and in medicine to see inside the human body during surgery. The following example will help you understand the physics behind fiber optics.

**EXAMPLE 21.11 Trapping light inside glass**

Imagine that you have a long glass block of refractive index 1.56 surrounded by air, as shown in the figure below. Light traveling inside the block hits the top horizontal surface at a 41° angle. What happens next?

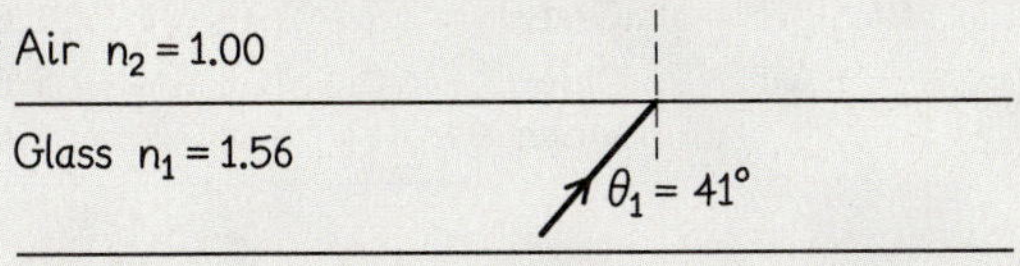


**Sketch and translate** The situation is sketched above. As light travels from glass into air, we need to compare the incident angle to the critical angle for total internal reflection.

**Simplify and diagram** Assume that the top and bottom surfaces of the block are parallel to each other. The light is hitting the top of the glass at a 41° angle. There are three possibilities: the light leaves the block, the light refracts so it moves along the surface, or the light is totally reflected back into the block. See the figure below.

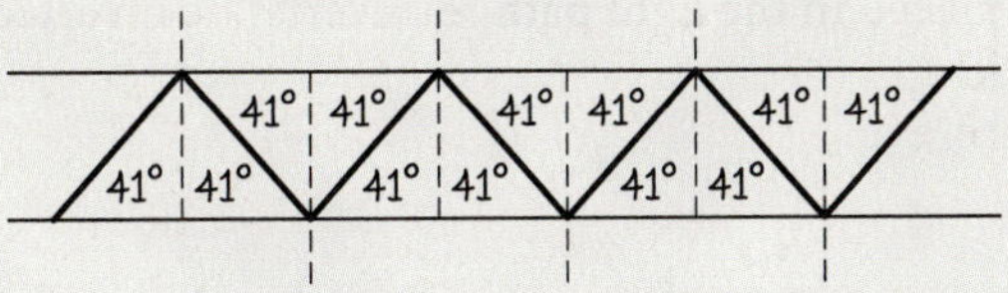


**Represent mathematically** The critical angle for the glass-air interface according to Eq. (21.4) is

$$\sin\theta_c = \frac{n_2}{n_1} = \frac{1.00}{1.56} = 0.64, \text{ or } \theta_c = 40°$$

If the incident angle of light in the glass is greater than this critical angle, all light is reflected at the glass-air interface. When it reaches the opposite side of the block it hits the bottom surface at the same angle and thus reflects back again.

**Solve and evaluate** We found above that $\theta_c = 40°$. The given 41° angle is slightly greater than the critical angle for total internal reflection. Thus the light is totally internally reflected during the first incidence on the upper surface. From there it moves down and to the right and hits the bottom surface at a 41° angle, where it is totally internally reflected again. The process of total internal reflection continues as the light travels the length of the block.

**Try it yourself:** What happens if the light hits the top surface at 45°? At 38°?

*Answer:* For 45° incidence, total internal reflection occurs (it occurs for all light incident at 40° or more). For 38° incidence (since it is less than 40°), some light refracts out of the top and bottom at each incidence causing the intensity of light within the block to diminish.

Within optical fibers, light is totally internally reflected, even when the fiber is bent, allowing rapid transmission of light and data. In surgery, fiber optics can guide the surgeon's hands (**Figure 21.12a**). A thin bundle of tiny glass fibers transmits light into the area being illuminated and then transmits the reflected light out for viewing (Figure 21.12b). A tiny tool can be inserted along with the fibers to, for example, clean cartilage or repair tendons. Only a small incision is needed to insert the fibers and the tool, reducing the amount of trauma to the joint and surrounding tissue.

**Figure 21.12** (a) A fiber optic view inside a knee during surgery. (b) The fiber optic endoscope used to look inside the body.

(a)

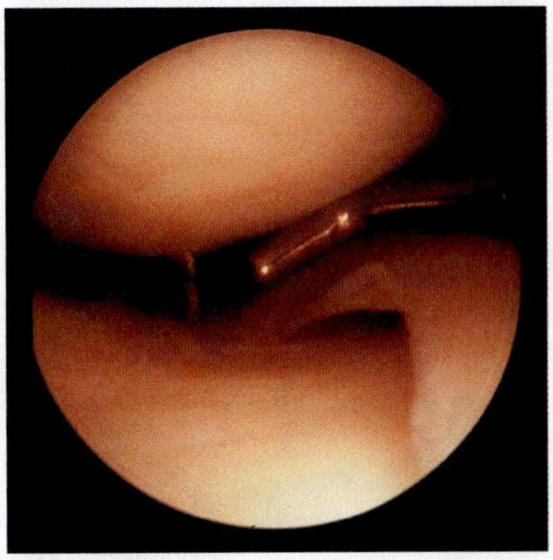

(b)

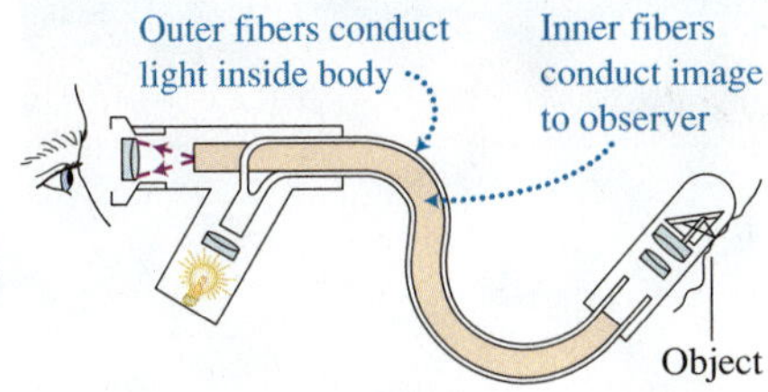


## Prisms

Centuries ago, Isaac Newton observed a thin beam of light coming into a room through a tiny hole in one of the shutters. By chance, this beam illuminated a prism sitting on a desk. To Newton's surprise, he saw a rainbow band of colored light on the opposite wall (**Figure 21.13a**). The band was much wider than the original beam, with violet light on the bottom and red on top. Was the prism creating these colors? To test this idea, he put an identical prism after the first one, but upside down (Figure 21.13b), and the colored band disappeared—the spot on the wall was white and small. He concluded that the prism did not create the colors; it somehow could separate the different colors out of white light and then recombine them back into white light. Newton further reasoned that the refractive index was greater for violet light and smaller for red, explaining the separation of colors (see **Table 21.8**).

Since the critical angle for a glass-air interface is less than 45°, glass prisms with 45-45-90° right angles are used in many optical instruments such as telescopes and binoculars to achieve total reflection of light through

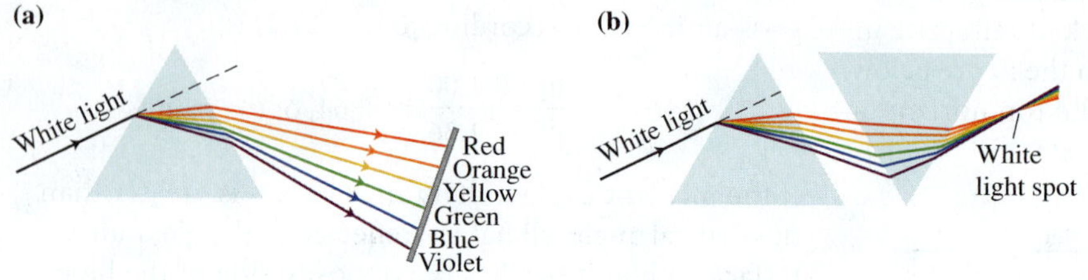


**Figure 21.13** (a) A prism separates white light into a light spectrum. (b) A second prism combines the spectrum back into white light.

**Table 21.8 Refractive indexes of glass for different colors.**

| Color of Light | $n_{glass}$ |
|---|---|
| Red | 1.613 |
| Yellow | 1.621 |
| Green | 1.628 |
| Blue | 1.636 |
| Violet | 1.661 |

**Figure 21.14** Reflecting prisms. (a) A prism creates an image oriented at 90° relative to the object; (b–c) two ways to produce an image oriented at 180° relative to the object.

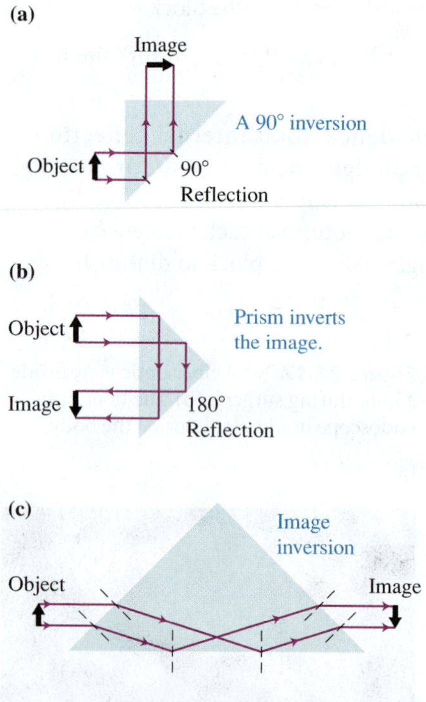


90° or 180° angles. Examples of the reflecting ability of prisms are shown in **Figure 21.14**. Prisms are better for reflection than mirrors for several reasons. First, prisms reflect almost 100% of the light incident on the prism via total internal reflection, whereas mirrors reflect somewhat less than 100%. Second, most mirrors are made of glass over a thin sheet of metal, which tarnishes and loses its reflective ability with age. Prisms retain their reflective ability. Finally, prisms can invert an image, that is, make it appear upside down (Figures 21.14b and c). This inversion may seem like a disadvantage; however, lenses in optical devices such as binoculars cause an object viewed through them to appear inverted. A prism placed in the light path re-inverts the inverted image so that it appears right side up.

## Mirages

One of the interesting consequences of the refraction of light is the formation of a **mirage**, such as the shimmering on roadways that looks like water (**Figure 21.15**). How can we explain this observation?

On a hot day, hot air may hover just above the pavement. The air above is cooler. This hot air is less dense and has a lower index of refraction than the cooler air above it. When light from the sky passes through air with a gradually changing index of refraction, its path gradually bends. This bending of light leads us to perceive that the source of light is at a different location than it actually is (see **Figure 21.16**).

For simplicity we will consider only one ray coming from point A in the sky (Figure 21.16a). Instead of a continuous variation in the refractive index as the light slants downward, we will assume for simplicity that it passes through several layers of different refractive index (represented as parallel layers). Its path changes according to Snell's law. At points 1 and 2, the light moving into layers of lower refractive index bends away from the normal line and its angle with the pavement decreases. At some point (point 3) the incident angle becomes so large that the ray undergoes total internal reflection and starts going up. After passing through several layers it enters the eye of the observer, whose brain perceives the ray as traveling along the straight line that originated at point B on the ground.

If we now consider more rays coming from a section of the sky (Figure 21.16b), the observer would see them originating in the vicinity of B. This location will look blue and will shimmer due to convection in the air above the road surface. The result looks like water on the road (though it is actually light from the sky).

**Figure 21.15** A mirage. We see wet pavement on a dry day.

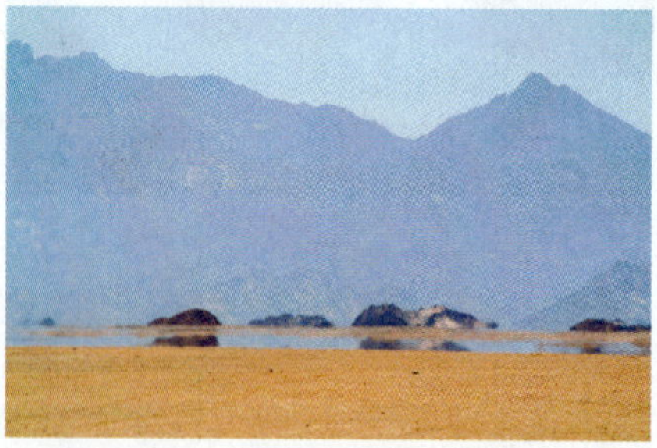

**Figure 21.16** The formation of a mirage. An observer perceives that light is coming from the pavement, when in fact the source of light is the sky.

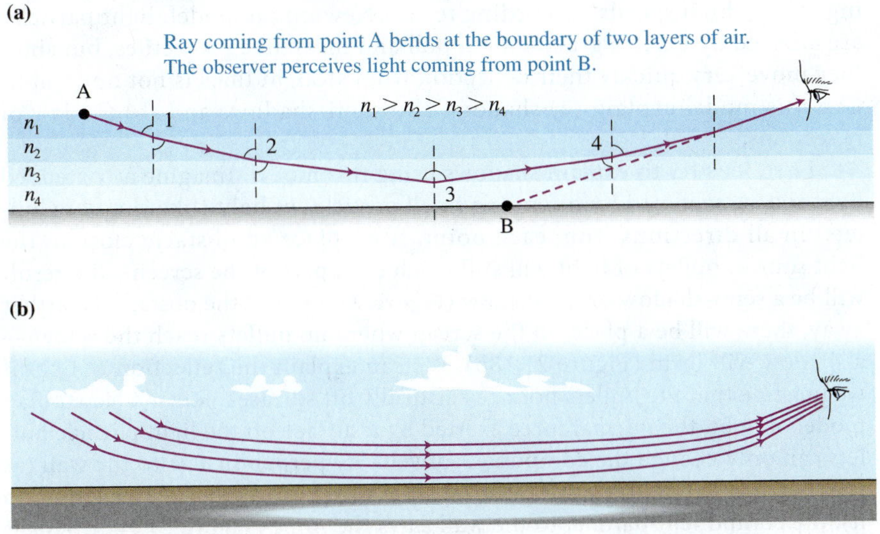


## Color of the sky

We learned in this chapter that we could represent light reaching Earth from the Sun as a beam of parallel rays. Sunlight contains all visible colors. Why then does the entire sky look blue?

Molecules, dust particles, and water droplets in the atmosphere that are along the path of light from the Sun to Earth reflect sunlight in all directions, similar to the chalk dust reflecting laser light in the first experiments in this chapter. However, if the atmosphere were reflecting all colors of light similar to the way chalk dust does, the atmosphere should be the same color as the Sun.

Instead, we see a sky that is primarily blue. Due to their sizes, atmospheric particles reflect blue light more efficiently than other colors (**Figure 21.17**). Thus, all colors other than blue pass through the atmosphere without changing direction as much. Because the blue light is reflected in all directions, even when we look at the sky away from the Sun, reflected blue light from the atmosphere still reaches our eyes. This explanation is supported by probes sent to other planets, where the atmospheres have different chemical and physical compositions than ours. The skies of Venus and Mars are different colors even though illuminated by the same sunlight as Earth.

If our atmosphere reflected all colors the same way, we would see the sky as white—this is in fact what we see when the sky is covered with clouds. The water droplets in clouds reflect all colors equally.

**Figure 21.17** Why the sky is blue. Blue light scatters off atmospheric particles and reaches our eyes from all directions. Light of other colors is not scattered as much.

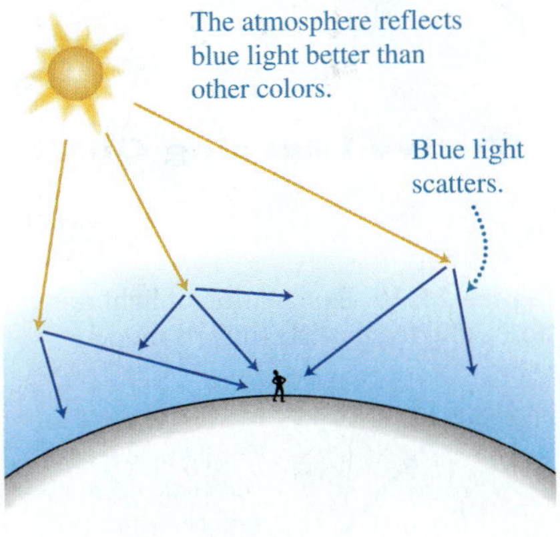

The atmosphere reflects blue light better than other colors.
Blue light scatters.

**Review Question 21.6** Why is the sky blue? Why are most clouds white?

# 21.7 Explanation of light phenomena: two models of light

We have developed a ray model of light that describes the way light behaves. What other models can we use to explain the way light behaves?

**Figure 21.18** Newton's bullet model of light.

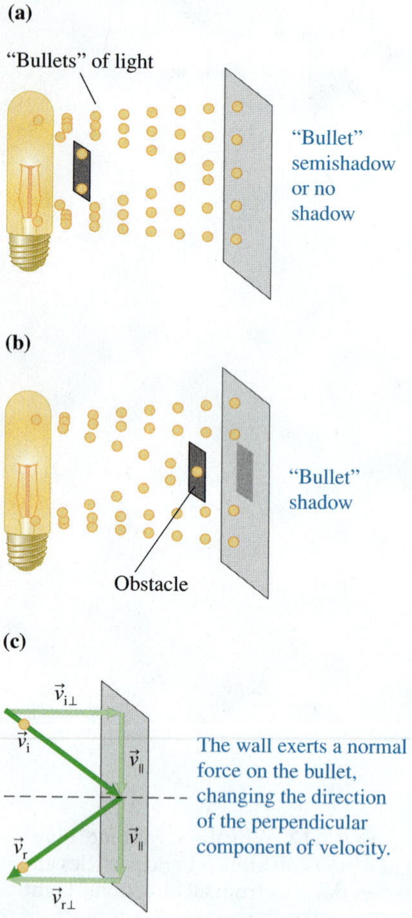
(a)
"Bullets" of light
"Bullet" semishadow or no shadow
(b)
"Bullet" shadow
Obstacle
(c)
$\vec{v}_{i\perp}$
$\vec{v}_i$
$\vec{v}_{\parallel}$
$\vec{v}_{\parallel}$
$\vec{v}_r$
$\vec{v}_{r\perp}$
The wall exerts a normal force on the bullet, changing the direction of the perpendicular component of velocity.

## Particle model

Isaac Newton modeled light as a stream of very small, low-mass particles moving at very high speeds. According to this Newtonian model, light particles are affected by Earth's gravitational pull and move like projectiles, but since they move very quickly their deflection from straight lines is not noticeable. Can this model explain (a) shadows and semi-shadows and (b) the law of reflection?

First, let's try to explain shadows using this model. Imagine an extended light source such as a bulb sending small particles of light (like shooting bullets) in all directions from each point. If we place an obstacle close to the light source, bullets of light will still reach each part of the screen—the result will be a semi-shadow or no shadow (**Figure 21.18a**). If the obstacle is farther away, there will be a place on the screen where no bullets reach the screen—a shadow will form (Figure 21.18b). We can explain the reflection of light if we imagine that the bullets bounce elastically off surfaces. For the Newtonian model of light, the normal force exerted by a surface on the light particle bullets can only change the component of velocity perpendicular to the wall (as the acceleration of an object is in the same direction as the force exerted on it); the component parallel to the wall stays the same (Figure 21.18c). This is consistent with the law of reflection. Newton's particle model of light does successfully explain light propagation and reflection.

Can we explain refraction using the particle model of light? In order for the model to be consistent with Snell's experiments on refraction, the light particles will have to speed up when they refract into a more optically dense medium (see **Figure 21.19**). Thus the particle model suggested that the denser medium exerts an attractive force on the light particles, causing their speeds to increase. Accurate measurements of the speed of light turned out to be very challenging, and for many years the speed of light could only be determined in air or in a vacuum (it was found to be 299,792,458 m/s—about $3 \times 10^8$ m/s). In addition, the particle model requires an additional interaction between the surface and the light particles that causes them to speed up when entering the denser medium. Scientists prefer explanations that are as simple as possible. Another model of light that did not require this additional interaction was proposed.

**Active Learning Guide›**

## Wave model of light

**Figure 21.19** Explanation of light refraction using the particle (bullet) model.

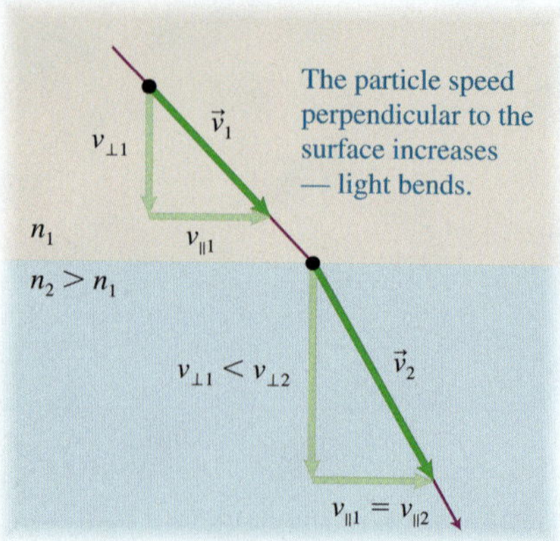
The particle speed perpendicular to the surface increases — light bends.
$v_{\perp 1}$
$\vec{v}_1$
$n_1$
$v_{\parallel 1}$
$n_2 > n_1$
$v_{\perp 1} < v_{\perp 2}$
$\vec{v}_2$
$v_{\parallel 1} = v_{\parallel 2}$

Simultaneously with Newton's development of the particle model of light, Christiaan Huygens was constructing a wave model of light. The motivation for the model could have come from the observations that light reflects off objects similar to the way sound reflects. Recall (from Chapter 20) that Huygens' wave propagation ideas involved disturbances of a medium caused by each point on a wave front producing a circular wavelet. Imagine that we have a wave with a wave front moving upward parallel to the page (see **Figure 21.20**). We choose six dots on this wave front, and from each dot we draw a wavelet originating from it (as shown in Figure 21.20). Each dot represents the source of a wavelet produced by the wave disturbance passing that point. According to Huygens' principle, each small wavelet disturbance produces its own semicircular disturbance that moves up the page in the direction the wave is traveling. Now, note places where the wavelets add together to form bigger waves. These places are where the wave front will be once it has moved a short distance up the page. We also draw an arrow (a ray) indicating the direction the wave is traveling perpendicular to the wave fronts and consistent with the ray model of light propagation. In this model, the ray is the line perpendicular to the wave front.

Waves travel at different speeds depending on the properties of the medium. For example, water waves travel more slowly in shallow water than in deeper

**Figure 21.20** Wavelets produce a new wave front.

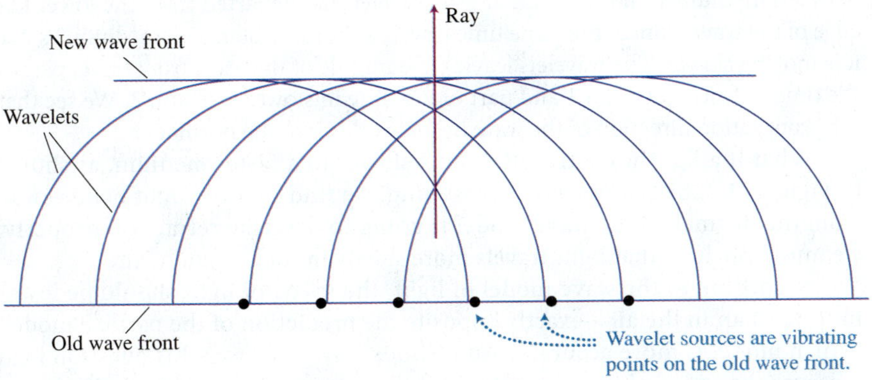


water. Sound travels more slowly in cold air than in warm. What does the wave model predict will happen when waves propagate through media with different wave speeds? Imagine that the speed of the wave shown in **Figure 21.21** is greater on the right side than on the left side. What happens to the wave front?

In the sketch in Figure 21.21, the six dots are part of a wave crest again moving toward the top of the page. Since the wave travels more slowly on the left side than on the right side, the wavelets originating from the left side have smaller radii than those on the right. New wave crests form at places where the wavelets add together to form a bigger wave disturbance than that caused by a single wavelet. A ray perpendicular to the new wave front indicates the wave's approximate path as it moves up the page. The wave bends toward the region where it moves more slowly. The wave changes direction when it travels in a medium with regions having different wave speeds.

## Wave model and refraction

Imagine a light wave moving in a less optically dense medium 1 and reaching an interface with a denser medium 2 at a nonzero angle of incidence (**Figure 21.22**). Using the wave model of light, we can now draw wave fronts for this wave as it travels from medium 1 into medium 2. During a certain time interval, the wavelet that

**Figure 21.21** The wave front changes direction (the old wave front is horizontal and the new wave front is not) as the wave travels at different speeds on the right and left side of the figure.

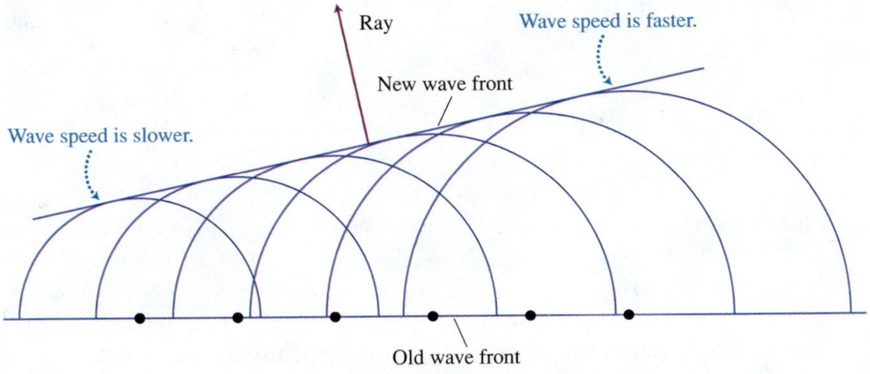

**Figure 21.22** Huygens' principle explains why the wave changes directions when it reaches the interface between two media.

**(a)**

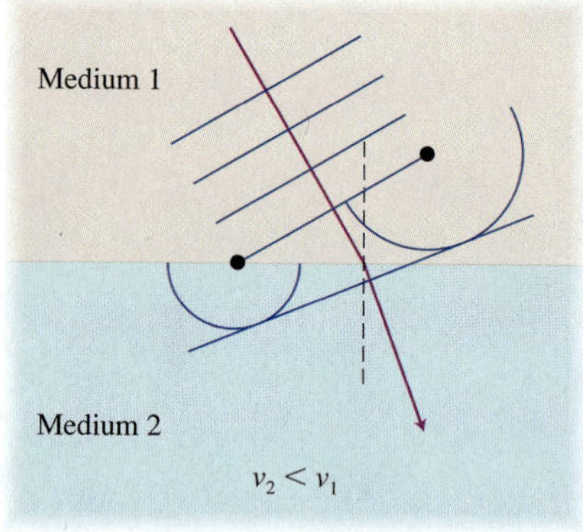


The wave travels farther on the right side in the less optically dense medium 1 than on the left side in denser medium 2.

**(b)**

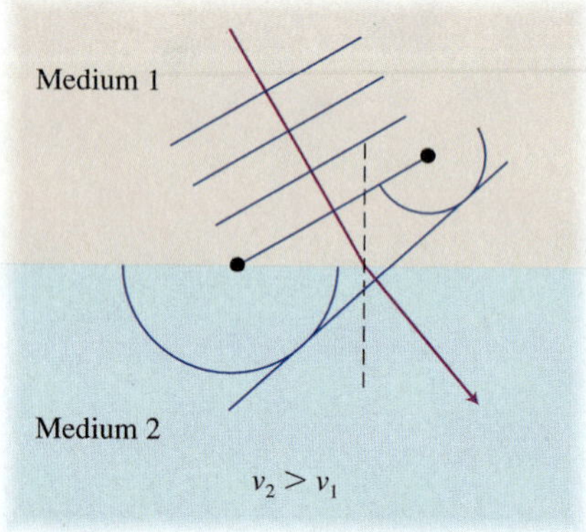


Medium 2 is less optically dense than medium 1, causing the wave to bend out rather than in.

earlier departed from the right edge of the wave front is just reaching the boundary between medium 1 and medium 2. The wavelet that departed from the lower left edge of the wave front at the same time travels a shorter distance in medium 2 since it is moving slower. The wavelets leaving the middle of the wave front travel part of the time in faster medium 1 and part of the time in slower medium 2. We see that the propagation direction of the wave becomes closer to the normal.

What if a light wave travels from a slower to a faster medium, as shown in Figure 21.22b? Using similar reasoning, we find that the light bends away from the normal. If the wave model is going to describe refraction properly, we must conclude that light travels more slowly in water than in air.

According to the wave model of light, the speed of light should be lower in water than in the air—exactly opposite the prediction of the particle model. Which model is more accurate? An obvious way to answer this question is to measure the speed of light in water. But this experiment is difficult. Thus, we are left with two models of light that both explain the reflection and refraction of light but lead to different predictions about its speed in water. This dilemma existed in physics for a long time due to the difficulty in measuring the speed of light in different media. Ultimately, the resolution came not from measurements of the speed of light but from overwhelming experimental support for the wave model. We will discuss the wave model in more detail in a later chapter (Chapter 23).

## Unanswered questions

Although we have learned a great deal about light, there are still questions that we have not answered. We do not know why different media bend light differently. We still do not know whether light propagates faster or slower in water and other media compared with air. We do not know how objects radiate light. Why do some stars look white while some look red? We also have not decided which model of light is better—the particle model or the wave model. We will investigate these questions in the coming chapters.

**Review Question 21.7** What is the difference between the two models of light and the predictions they make for the speed of light in water?

# Summary

| Words | Pictorial and physical representations | Mathematical representation |
|---|---|---|
| **Light sources and light rays** Lightbulbs, candles, and the Sun are examples of **extended sources** that emit light. Light from such sources illuminates other objects, which reflect the light. We see an object because incident light reflects off of it and reaches our eyes. We represent the travel of light with **rays** (drawn as lines and arrows). Each point of a shining object or a reflecting object sends rays in all directions. (Section 21.1) | 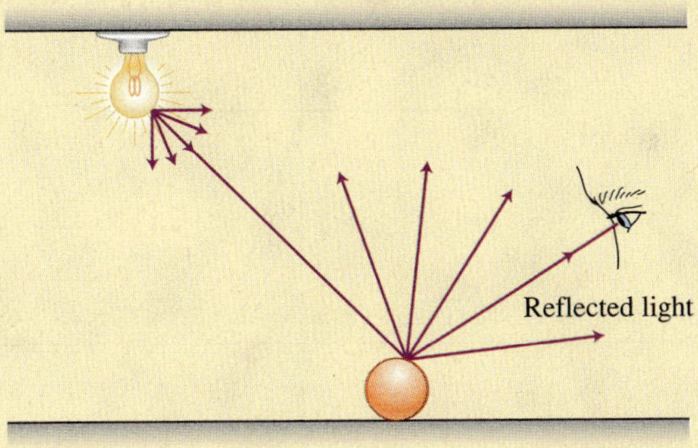 | |
| **Law of reflection** When a ray strikes a smooth surface such as a mirror, the angle between the incident ray and the normal line perpendicular to the surface equals the angle between the reflected ray and the normal line (the **angle of incidence** equals the **angle of reflection**). This phenomenon is called **specular reflection**. (Section 21.2) | 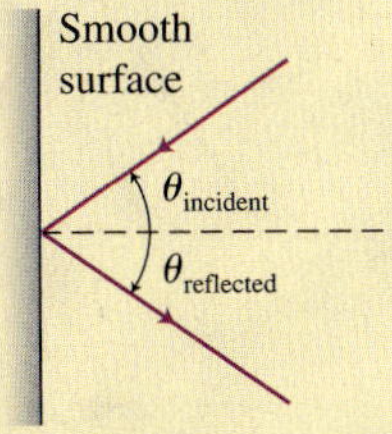 | $\theta_{\text{reflected}} = \theta_{\text{incident}}$ Eq. (21.1) |
| **Diffuse reflection** If light is incident on an irregular surface, the incident light is reflected in many different directions. This phenomenon is called **diffuse reflection.** (Section 21.2) | 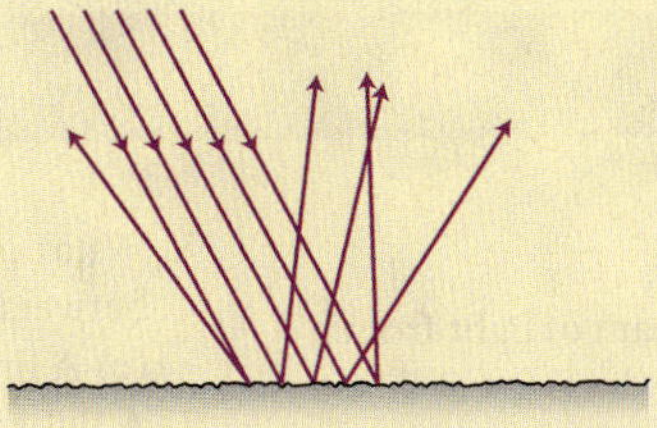 | |
| **Refraction** If the direction of travel of light changes as it moves from one medium to another, the light is said to refract (bend) as it moves between the media. (Section 21.3) | 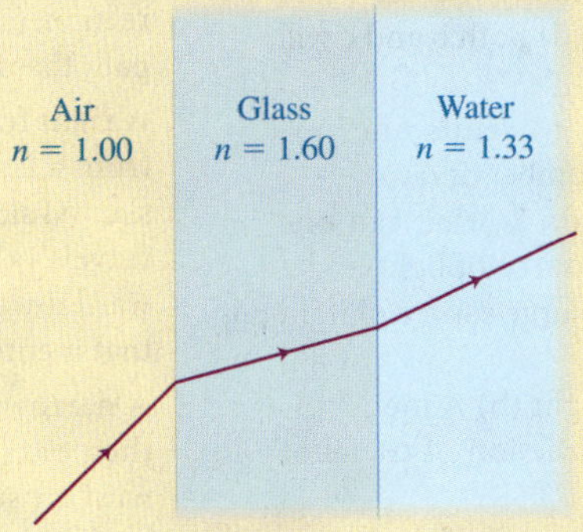 | |

*(continued)*

| Words | Pictorial and physical representations | Mathematical representation |
|---|---|---|
| **Snell's law** Light going from a lower to a higher index of refraction will bend toward the normal, but going from a higher to a lower index of refraction it will bend away from the normal. (Section 21.3) | 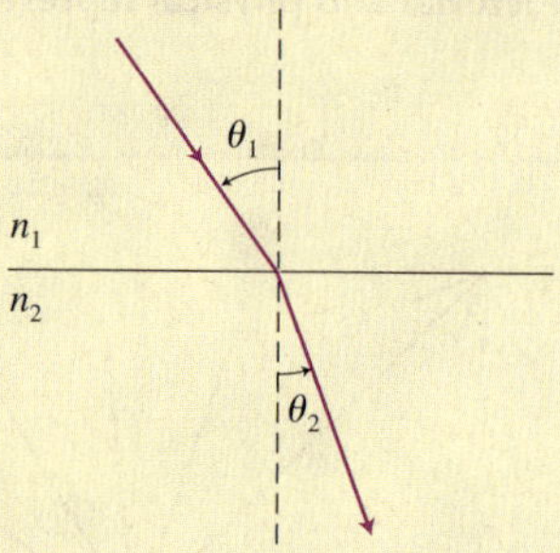 | $n_2 \sin\theta_2 = n_1 \sin\theta_1$ Eq. (21.3) |
| **Total internal reflection** If light tries to move from a more optically dense medium 1 of refractive index $n_1$ into a less optically dense medium 2 of refractive index $n_2$ ($n_1 > n_2$), the refracted light in medium 2 bends away from the normal line. If the incident angle in medium 1 $\theta_1$ is greater than a critical angle $\theta_c$, *all* of the light is reflected back into the denser medium. (Section 21.4) | 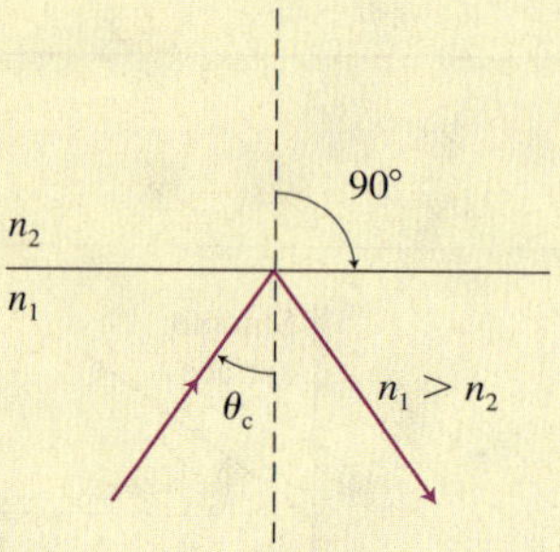 | $\sin\theta_c = \dfrac{n_2}{n_1}$ Eq. (21.4) |

MP For instructor-assigned homework, go to MasteringPhysics.

# Questions

## Multiple Choice Questions

1. How can you convince your friend that a beam of light from a laser pointer travels to a wall along a straight line? (a) Tell him that light rays travel in straight lines. (b) Sprinkle chalk dust along the laser beam. (c) Try to move along the laser beam blocking its light and note the light path. (d) Both b and c will work. (e) All answers (a)–(c) will work.
2. Each point of a light-emitting object (a) sends one ray. (b) sends two rays. (c) sends an infinite number of rays.
3. To test the model mentioned in Question 2, what can one use? (a) A pinhole camera (b) A small source of light and a large obstacle near a distant screen (c) A large source of light and a small obstacle close to the screen
4. What is a light ray? (a) A thin beam of light (b) A model invented by physicists to represent the direction of travel of light (c) A physical law
5. What is a semi-shadow? (a) A shadow that is half the size of the original shadow (b) A place where some light rays from an object reach and rays from other parts of the object do not. (c) A scientific term for a shadow
6. Why don't we see semi-shadows of objects on Earth produced by sunlight? (a) The Sun is too bright. (b) The Sun is very far away. (c) We live very far from the North Pole.
7. What is a normal line? (a) A line parallel to the boundary between two media (b) A vertical line separating two media (c) A line perpendicular to the boundary between two media
8. A light ray travels through air and then passes through a thin rectangular glass block. It exits (a) parallel to the original direction. (b) bent toward the normal line. (c) along the identical path that it entered the block. (d) bent away from the normal.
9. A right triangular prism sits on a base. A narrow light beam from a laser travels through air and then passes through the slanted side of the prism and out the vertical back side. It now travels (a) parallel to the original direction. (b) bent downward toward the base of the prism. (c) along the identical path that it entered the block. (d) bent upward away from the base.
10. A narrow light beam from a laser travels through water and then passes through a hollow right triangular prism filled with air submerged in the water. The light exits the vertical back side of the prism and is now traveling (a) parallel to the original direction. (b) bent down toward the base of the prism. (c) along the identical path that it entered the prism. (d) bent upward away from the base of the prism.
11. We can observe total internal reflection when light travels (a) from air to glass. (b) from water to glass. (c) from glass to water. (d) from air to water to glass.

## Conceptual Questions

12. What effects of light radiation and reflection are responsible for the fact that we can see objects in a room?
13. Why does the inside of a well look black?
14. What do you need to do to create different size shadows of the same stick?
15. Explain how a sundial works (a sundial is just a vertical stick in the ground).
16. You are trying to see what is happening in a room through a keyhole in a door. Where should you place your eye to see the most: closer or farther away from the hole? Explain.
17. If you stand near a light pole at night, the shadow of your foot (if slightly off the ground) is rather sharp while the shadow of your head looks fuzzy. Why?
18. In what cases can you see only a semi-shadow of an object and not a shadow?
19. Why when moving away from a lightbulb does your shadow become sharper and sharper?
20. Why can't you see a shadow of a pencil on the wall if you place it close to a lightbulb?
21. The visible diameters of the Moon and the Sun are almost the same. What should you see if the Moon were in the same line of sight as the Sun? Explain.
22. The shadow of the Moon on Earth is 200 km wide. Describe what you would see if you were in that area during the moment described in the previous question.
23. You stand at the side of a road. Why is light from a passing car's headlights easier to see in foggy weather than in clear weather?
24. Standing beside a river you can see the bottom of the river close to the bank but not in the middle of the river, even if the river is shallower there. Why?
25. ✎ You look at a fish underwater. Draw a ray diagram of the light that enters your eye from the fish.
26. Take a pencil and try to touch a penny on the bottom of a large pan of water while looking down into the pan at an angle. Explain why it is difficult to do this.
27. Explain why a submarine very far from a ship cannot send a light signal to it.
28. ✎ Will a beam of light experience total internal reflection going from glass into water or from water into glass? Explain (include drawings).
29. Explain how a prism turns a beam of white light into a rainbow.
30. On a hot sunny day, a highway sometimes looks wet. Why?
31. Why can't you see stars during the day?
32. What light phenomena can be explained using the particle model of light?
33. What phenomena can be explained using a wave model of light?
34. How is it possible that two different models can explain the same phenomenon? Give an example of another phenomenon that can be successfully explained by two different models (think of areas other than optics).

# Problems

Below, **BIO** indicates a problem with a biological or medical focus. Problems labeled **EST** ask you to estimate the answer to a quantitative problem rather than derive a specific answer. Problems marked with ✎ require you to make a drawing or graph as part of your solution. Asterisks indicate the level of difficulty of the problem. Problems with no * are considered to be the least difficult. A single * marks moderately difficult problems. Two ** indicate more difficult problems.

### 21.1 Light sources, light propagation, and shadows

1. **Tree height** You are 165 cm tall and stand under a tree. The tree shadow is 34 m long and your shadow is about twice your height. How tall is the tree?
2. * **Lighting for surgeon** A shadow from a surgeon's hand obstructs her view while operating. Make suggestions for an alternative light source that avoids this difficulty. Include one or more sketches for your proposed plan.
3. ✎ **Lunar eclipse** A lunar eclipse happens when the Moon, Earth, and Sun are aligned in that order (the Moon is in the shadow of Earth). Aristotle used this phenomenon to determine the shape of Earth. He proposed that Earth has a round shape. Draw a picture to describe his reasoning.
4. ✎* **Shadows during romantic dinner** You and a friend are having a romantic candlelight dinner. You notice that the light shadows of your hands on the wall look fuzzy. However, the shadow of a glass is very sharp and crisp. Where are you and your friend sitting with respect to the candle and the wall? Where is the glass? To answer these questions, draw ray diagrams assuming that the candle is an extended light source.
5. ✎* **Pinhole camera (camera obscura)** You want to make a pinhole camera with a blank wall as the screen and you (or a friend) as the object of interest. Draw a sketch showing the wall, the pinhole, the place you or your friend will sit, the best location for the Sun or some other light source, and the location of people viewing the wall. Will the image appear upright or inverted? Explain.
6. ✎* **Solar eclipse** Only observers in a very narrow region on Earth (about 200 km diameter) can see a total solar eclipse. In the region of such an eclipse, there is no sunlight and a person can see stars during daytime. Draw the arrangement of the Sun, the Moon, and Earth during a total solar eclipse.
7. * **Tree height 2** Your summer ecology research job involves documenting the growth of trees at an experimental site. One day you forget your tree-height-measuring instrument. How can you determine the height of trees without it? Provide a sketch for your method.
8. * **Sundial** The same day that you forgot the height-measuring instrument (see Problem 7), you also forgot your watch. You left your house at 7:00 a.m. and drove to the site in 2.0 hours. How can you build a sundial so that you can leave the site at 6:00 p.m. to make it to a concert? What assumptions did you make?

### 21.2 Reflection of light

9. ✎ You have a small mirror. While holding the mirror, you see a light spot on a wall at the same height as the mirror. At what angle are you holding the mirror if the Sun is 50° above the horizontal? Draw a ray diagram to answer the question. What assumptions did you make?

10. You see a well and wonder if there is water in it. How should you hold a small mirror to see the well's bottom? The Sun's rays are hitting the ground at an angle of 60° above the horizontal. Draw a ray diagram to answer the question. What assumptions did you make?

11. * **Design** Design an experiment that you can perform to test the law of reflection. Describe the instruments that you are going to use and their experimental uncertainty (half of the smallest division). How certain will you be of your results?

12. * Draw four separate mirrors: one horizontal, the next 30° above the horizontal, the third 90° above the horizontal (vertical), and the fourth at an angle of 120° relative to the first. For each mirror draw an incident ray hitting the mirror at its middle (the ray should not be perpendicular to the mirror) and then draw a reflected ray.

13. Design a mirror arrangement so that light from a laser pointer will travel in exactly the opposite direction after it reflects off the mirror(s), even if you change the direction the laser pointer is pointing.

14. Two mirrors are oriented at right angles. A narrow light beam strikes the horizontal mirror at an incident angle of 65°, reflects from it, and then hits the vertical mirror. Determine the angle of incidence at the vertical mirror and the direction of the light after leaving the vertical mirror. Include a sketch with your explanation.

15. * **Songbird reflection** Draw rays (maybe a side view) to help explain how the Sun shines on the back of the inverted bird in **Figure P21.15**. Specify where the observer is.

**Figure P21.15**

16. * You are driving along a street on a sunny day. Only a few apartment building windows appear bright; the rest are pitch black. Explain this difference and include a ray diagram to help with your explanation.

17. * **Making a car light** A lightbulb is placed somewhere on the radial axis of the bowl mirror shown in **Figure P21.17**. Carefully draw three different narrow light beams leaving the bulb that reflect from the bowl mirror and move away. Would this arrangement be useful for a headlight? Explain.

**Figure P21.17**

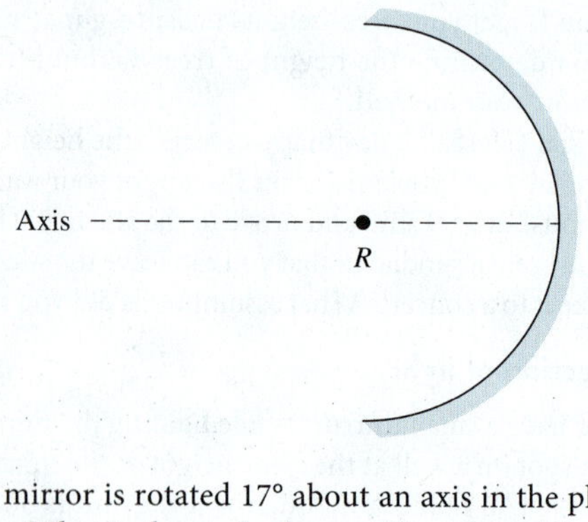


18. A flat mirror is rotated 17° about an axis in the plane of the mirror. What is the angle change of a reflected light beam if the direction of the incident beam does not change?

## 21.3 Refraction of light

19. (a) A laser beam passes from air into a 25% glucose solution at an incident angle of 35°. In what direction does light travel in the glucose solution? (b) The beam travels from ethyl alcohol to air at an incident angle of 12°. Determine the angle of the refracted beam in the air. (c) Draw pictures for (a) and (b) showing the interface between the media, the normal line, the incident rays, the reflected rays, the refracted rays, and the angles of these rays relative to the normal line.

20. A beam of light passes from glass with refractive index 1.58 into water with a refractive index 1.33. The angle of the refracted ray in water is 58.0°. Draw a sketch of the situation showing the interface between the media, the normal line, the incident ray, the reflected ray, the refracted ray, and the angles of these rays relative to the normal line.

21. A beam of light passes from air into a transparent petroleum product, cyclohexane, at an incident angle of 48°. The angle of refraction is 31°. What is the index of refraction of the cyclohexane?

22. * **Moving laser beam** An aquarium open at the top has 30-cm-deep water in it. You shine a laser pointer into the top opening so it is incident on the air-water interface at a 45° angle relative to the vertical. You see a bright spot where the beam hits the bottom of the aquarium. How much water (in terms of height) should you add to the tank so the bright spot on the bottom moves 5.0 cm?

23. * **Lifting light** You have a V-shaped transparent empty container such as shown in **Figure P21.23**. When you shine a laser pointer horizontally through the empty container, the beam goes straight through and makes a spot on the wall. (a) What happens to this spot if you fill the container with water just a little above the level at which the laser beam passes through the container? (b) What happens if you fill the container to the very top? Indicate any assumptions used and draw a ray diagram for each situation.

**Figure P21.23**

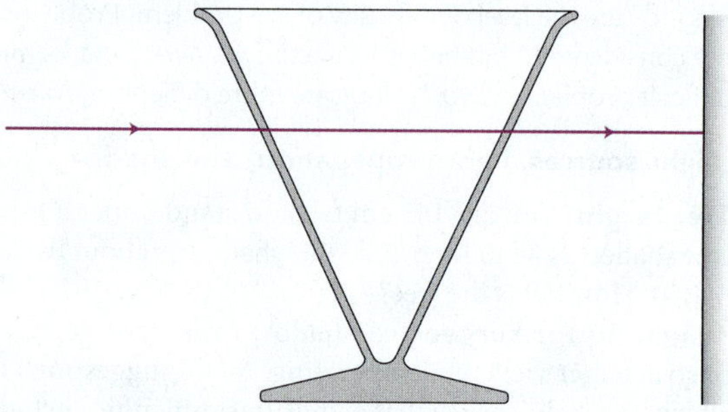

24. A light beam hits the interface between air and an unknown material at an angle of 43° relative to the normal. The reflected ray and the refracted ray make an angle of 108° with respect to each other. What is the index of refraction of the material?

25. * A light ray passes from air through a glass plate with refractive index 1.60 into water. The angle of the refracted ray in the water is 42.0°. Determine the angle of the incident ray at the air-glass interface.

26. **BIO Vitreous humor** Behind the lens of the eye is the vitreous humor, a jellylike substance that occupies most of the eyeball. The refractive index of the vitreous humor is 1.35 and that of the lens is 1.44. A narrow beam of light traveling in the lens

comes to the interface with the vitreous humor at a 23° angle. What is its direction relative to the interface when in the vitreous humor?

27. * You have a block of glass with an equilateral prism-shaped opening filled with air inside. Draw the path of a light ray that strikes the glass plate parallel to one of the sides of the prism and then passes through the prism.
28. * You watch a crab in an aquarium. Light traveling in air enters a sheet of glass at the side of the aquarium and then passes into the water. If the angle of incidence at the air-glass interface is 22°, what are the angles at which the light wave travels in the glass and in the water? Indicate any assumptions you made.
29. * Light moving up and toward the right in air enters the side of a cube of gelatin of refractive index 1.30 at an incident angle of 80°, Determine the angle at which the light leaves the top surface of the cube. How does the angle change if the refractive index of the gelatin is slightly greater? Explain.

## 21.4 Total internal reflection

30. * **Can your light be seen?** You swim under water at night and shine a laser pointer so that it hits the water-air interface at an incident angle of 52°. Will a friend see the light above the water? Explain.
31. Light is incident on the boundary between two media at an angle of 30°. If the refracted light makes an angle of 42°, what is the critical angle for light incident on the same boundary?
32. **Diamond total reflection** Determine the critical angle for light inside a diamond incident on an interface with air.
33. Determine the refractive index of a glucose solution for which the critical angle for light traveling in the solution incident on an interface with air is 42.5°. How would the critical angle change if the glucose concentration were slightly greater? Explain.
34. * You wish to use a prism to change the direction of a beam of light 90° with respect to its original direction. Describe the shape of the prism and its orientation with respect to the original beam to achieve this goal.
35. Light is incident on the boundary between two media at an angle of 34°. If the refracted light makes an angle of 47°, what is the critical angle for light incident on the same boundary?
36. * **Prism total reflection** What must be the minimum value of the refractive index of the prism shown in **Figure P21.36** in order that light is totally reflected where indicated? Will some of the light make it out of the top surface? Explain.

**Figure P21.36**

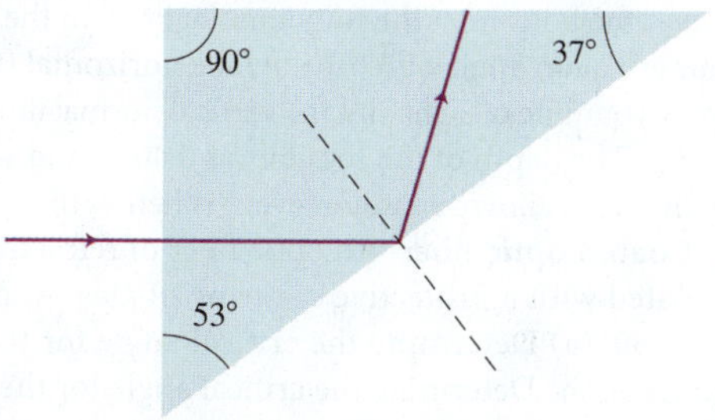


37. * **Gems and critical angles** In gemology, two of the most useful pieces of information concerning an unknown gem are the refractive index of the stone and its mass density. The refractive index is often determined using a critical angle measurement. Determine the refractive index of the gemstone shown in **Figure P21.37a**. The critical angle for the total reflection of light at the gem-air interface is 37.28°. A simple sketch of the experiment combined with the ray diagram is shown in Figure P21.37b. (The apparatus used to make these measurements is a complicated optical instrument.)

**Figure P21.37**

(a)

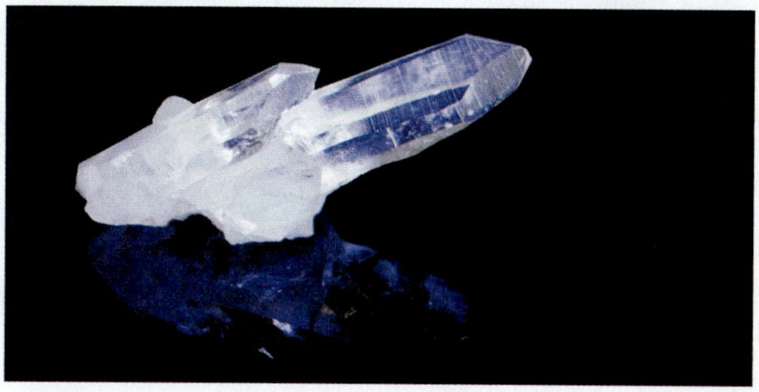

(b)

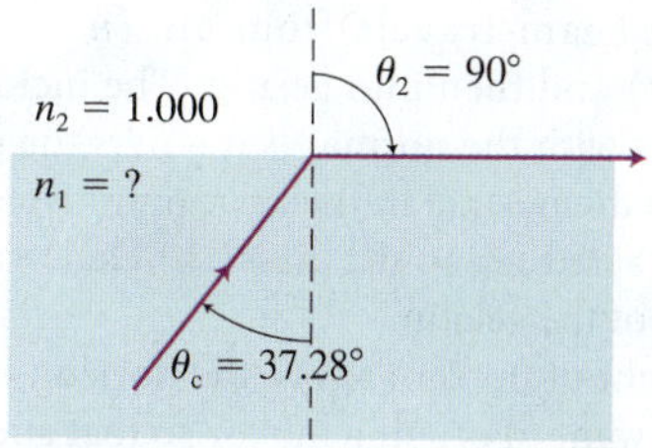


38. (a) The refractive index for the gem aquamarine is 1.57. Determine the critical angle for light traveling inside aquamarine when reaching an air interface. (b) You have a tourmaline gem and find that a laser beam in air incident on the air-tourmaline interface at a 50° angle has a refracted angle in the tourmaline of 28.2°. Determine the refractive index and critical angle of tourmaline.

## 21.5 Skills for analyzing reflective and refractive processes

39. * **Seeing bottom of pool** While you are sitting on a chair at the edge of a 1.2-m-deep swimming pool, your eyes are 1.5 m above the surface of the water. You see an acorn on the bottom of the pool at a 45° angle below the horizontal. What is the horizontal distance between you and the acorn?
40. * **Invisible in pool?** A swimming pool is 1.4 m deep and 12 m long. Is it possible for you to dive to the very bottom of the pool so people standing on the deck at the end of the pool do not see you? Explain.
41. * (a) Rays of light are incident on a glass-air interface. Determine the critical angle for total internal reflection ($n_{glass} = 1.58$). (b) If there is a thin, horizontal layer of water ($n_{water} = 1.33$) on the glass, will a ray incident on the glass-water interface at the critical angle determined in part (a) be able to leave the water (**Figure P21.41**)? Justify your answer.

**Figure P21.41**

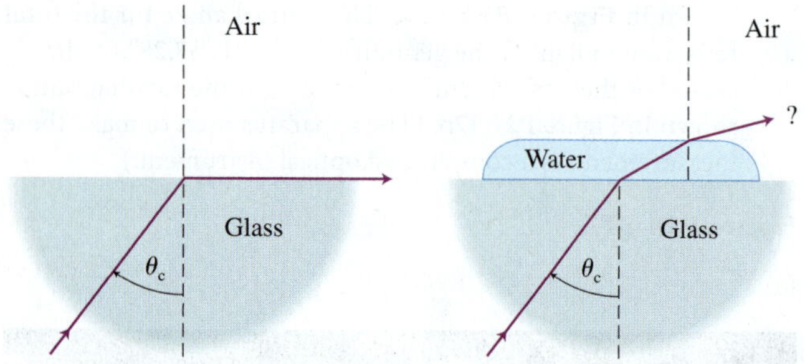


42. **Equation Jeopardy** Tell all that you can about a process described by the following equation: $1.33 \sin 30° = 1.00 \sin \theta_2$.
43. **Equation Jeopardy** Tell all that you can about a process described by the following equation: $1.33 \sin \theta_c = 1.00 \sin 90°$.
44. * **Equation Jeopardy** Tell all that you can about a process described by the following equation: $1.00 \sin 53° = 1.60 \sin \theta_2 = 1.33 \sin \theta_3$.
45. * When reaching a boundary between two media, an incident ray is partially reflected and partially refracted. At what angle of incidence is the reflected ray perpendicular to the incident ray? The indexes of refraction for the two media are known.
46. * A laser beam travels from air ($n = 1.00$) into glass ($n = 1.52$) and then into gelatin. The incident ray makes a 58.0° angle with the normal as it enters the glass and a 36.4° angle with the normal in the gelatin. (a) Determine the angle of the refracted ray in the glass. (b) Determine the index of refraction of the gelatin.
47. * The height of the Sun above the horizon is 25°. You sit on a raft and want to orient a mirror so that sunlight reflects off the mirror and travels at an angle of 45° in the water of refractive index 1.33. How should you orient the mirror?

### 21.6 Fiber optics, prisms, mirages, and the color of the sky: Putting it all together

48. * The prism shown in Figure P21.36 is immersed in water of refractive index 1.33. Determine the minimum value of the refractive index for the prism so that the light is totally internally reflected where shown. Will any light leave the top surface of the prism? Explain.
49. * **Light pipe** Rays of light enter the end of a light pipe from air at an angle $\theta_1$ (**Figure P21.49**). The refractive index of the pipe is 1.64. Determine the greatest angle $\theta_1$ for which the ray is totally reflected at the top surface of the glass-air interface inside the pipe.

**Figure P21.49**

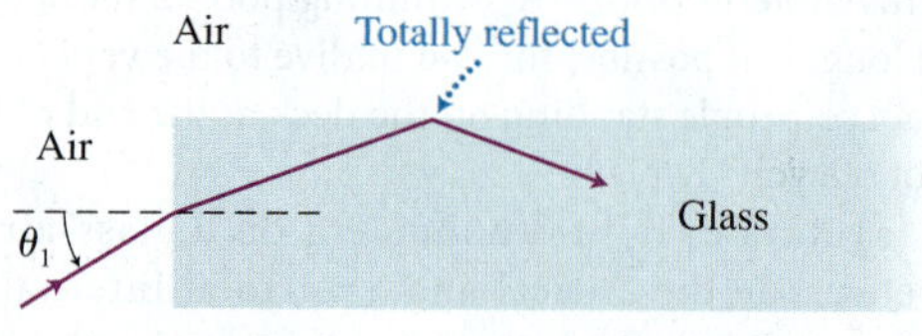


50. * A light ray is incident on a flat piece of glass. The angle of incidence is $\theta_1$ and the thickness of the glass is $d$. Determine the distance that the ray travels in the glass. At what angle will the ray emerge from the glass on the other side? Draw a ray diagram to help explain your solution.
51. * **Prism** You have a triangular prism made of glass of refractive index 1.60, with angles of 30°-90°-60°. The short side is oriented vertically. A horizontal ray hits the middle of the slanted side of the prism. Draw the path of a ray as it passes into and through the prism. Determine all angles for its trip through the prism.

## General Problems

52. * **Euclid's coin trick** In the third century B.C. Euclid performed the following experiment. He put a coin at the bottom of a mug with opaque sides. He placed his head above the mug at a position where he could *not* see the coin. Then without changing the position of his eyes, he added water to the mug and could see the coin. Repeat the experiment and explain how it works.
53. ** You have a candle and a large piece of paper with a triangular hole slightly larger than the candle flame cut in it. You place the paper between the candle and the wall. Draw ray diagrams to show what you see on the wall when the paper is placed (a) near the candle, (b) halfway between the candle and wall, and (c) near the wall.
54. ** **Design** You wish to investigate the properties of different transparent materials and decide to design a refractometer, a device used to measure the refractive index of an unknown liquid. A light beam in the liquid is refracted into a material whose refractive index is known and is less than that of the unknown liquid. The incident beam is adjusted for total internal reflection, and the equation for the angle of total internal reflection is used to determine the unknown index. Provide suggestions for a design for such a device and perform a sample calculation to show how it might work.
55. * You place a point-like source of light at the bottom of a container filled with vegetable oil of refractive index 1.60. At what height from the light source do you need to place a 0.30-cm circular cover so no light emerges from the liquid?
56. * There is a light pole on one bank of a small pond. You are standing up while fishing on the other bank. After reflection from the surface of the water, part of the light from the bulb at the top of the pole reaches your eyes. Use a ray diagram to help find a point on the surface of the water from where the reflected ray reaches your eyes. Determine an expression for the distance from this point on the water to the bottom of the light pole if the height of the pole is $H$, your height is $h$, and the distance between you and the light pole is $l$.
57. ** Imagine that while in Rome you are admiring the famous *Fontana dei Quattro Fiumi*, or the Fountain of the Four Rivers, in the center of Piazza Navona. You think about retrieving a coin at the bottom of the fountain. Light from the coin reaches your eye at an angle of 37° below the horizontal (the angle between your line of sight and the vertical normal line to the water is 53°). The depth of the fountain is 0.40 m and your height is 1.6 m (1.2 m above the water level). Where is the coin?
58. ** **Coated optic fiber** An optic fiber of refractive index 1.72 is coated with a protective covering of glass of refractive index 1.50. (a) Determine the critical angle for the fiber-glass interface. (b) Determine the critical angle for the glass-air interface. (c) Determine the critical angle for a fiber-air interface (no glass covering). (d) Suppose a ray hits the fiber-glass interface at the angle calculated in (c). What is the angle of refraction of that ray when it reaches the glass-air interface? Will it leave the optic fiber? Explain.

59. ** You put a mirror at the bottom of a 1.4-m-deep pool. A laser beam enters the water at 30° relative to the normal, hits the mirror, reflects, and comes back out of the water. How far from the water entry point will the beam come out of the water? Draw a ray diagram.

60. ** A scuba diver stands at the bottom of a lake that is 12 m deep. What is the distance to the closest points at the bottom of the lake that the diver can see due to light from these points being totally reflected by the water surface? The height of the diver is 1.8 m and $n_{water} = 1.33$.

## Reading Passage Problems

**Rainbows** How is a rainbow formed (**Figure 21.23**)? Recall that the index of refraction of a medium is slightly different for different colors. When white light from the Sun enters a spherical raindrop, as shown in **Figure 21.24**, the light is refracted or bent. After reflecting off the back surface of the drop, the light is refracted again as it leaves the front surface.

Each drop separates the colors of light. An observer on the ground with her back to the Sun sees at most one color of light coming from a particular drop (see **Figure 21.25**). If the observer sees red light from a drop (for example, the top drop in Figure 21.25), the violet light for that same drop is deflected above her head. However, if she sees violet light coming from a drop lower in the sky, the red light from that drop is deflected below her eyes onto the ground. She sees red light when her line of view makes an angle of 42° with the beam of sunlight and violet light when the angle is 40°. Other colors of light are seen at intermediate angles.

**Figure 21.23** When you see a rainbow in the sky, the colors always appear in the same order, with red on the outside of the rainbow's curve and violet inside.

**Figure 21.24** Different colors of light are refracted and reflected by different amounts by a spherical raindrop. The color violet is refracted more than red.

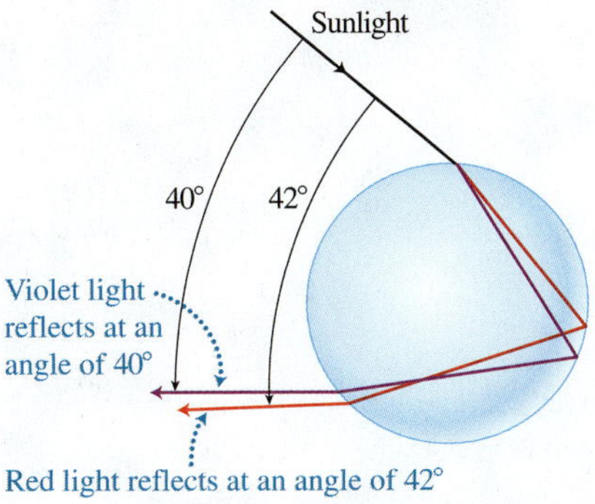


**Figure 21.25** An observer on the ground sees only one color of light from each raindrop. The drops that send the color red are above the drops that send violet.

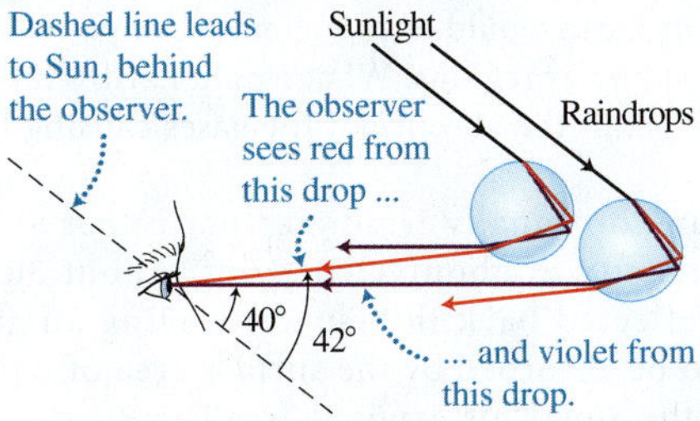


61. Why do you see rainbows only when you are between the Sun and the location of the rainbow?
(a) The sunlight must be refracted and reflected from raindrops and needs to move back and downward from its original direction.
(b) You need to intercept the light coming from the Sun.
(c) If you are looking toward the Sun, rainbows would block the sunlight.
(d) The sunlight is refracted by the raindrops and changes direction.

62. Suppose that light entered one side of a square water droplet of refractive index 1.33 at an incident angle of 20°. Some of the light reflects off the back surface and then goes back out on the same side as it entered. What is the closest angle to the angle of refraction of light leaving the square droplet?
(a) 15° (b) 18° (c) 20°
(d) 24° (e) 35°

63. Raindrops reflect different colors of light at different angles. Why do we see the parts of the rainbow as different colors of light rather than all colors coming from each?
(a) The net deflection of light seen from one drop depends on its refractive index.
(b) All of the raindrops reflecting a particular color have the same angular deflection relative to the direction of the Sun.
(c) Your eye sees only one color coming from a particular raindrop.
(d) The different colors are reflected and refracted at different angles.
(e) All of the above

64. Which choice is closest to the minimum angle of incidence of red light against the back wall of the water drop in Figure 21.24 so that it is totally reflected at that wall?
(a) 16° (b) 21° (c) 41°
(d) 45° (e) 49°

65. Repeat the previous problem, only for the violet light.
(a) 15° (b) 20° (c) 41°
(d) 45° (e) 49°

66. Why is violet light refracted more than red light?
(a) The violet light travels a shorter distance in the drop than the red light.
(b) The red light travels more slowly than the violet light.
(c) The refractive index of water for violet light is greater than that for red light.
(d) All of the above

**Earth energy balance** Gases in Earth's atmosphere, such as carbon dioxide and water vapor, act like a blanket that reduces the amount of energy that Earth radiates into space. This phenomenon is called the greenhouse effect. Without the greenhouse effect, most of Earth would have a climate comparable to that of the polar or subpolar regions. What would Earth's mean surface temperature be, in the absence of the gases causing the greenhouse effect?

The Sun continually irradiates our upper atmosphere with an intensity of about 1360 $W/m^2$. About 30% of this energy is reflected back into space, leaving an average of 950 $W/m^2$ to be absorbed by the surface area of Earth that is exposed to the Sun. This exposed area is $\pi R_{Earth}^2$, the Earth's circular cross section that intercepts the sunlight. All objects, including Earth, emit radiation at a rate proportional to the fourth power of the surface temperature ($T^4$) in kelvins (K). Thus, the radiation is emitted from Earth's surface, which as a sphere has the area $4\pi R_{Earth}^2$.

To maintain a constant temperature, Earth's radiation rate must equal its energy absorption rate from the Sun. A fairly simple calculation indicates that the two rates are equal when the average surface temperature of Earth is 255 K, or about 0 °F. However, Earth's mean surface temperature is much warmer than that—about 288 K. The calculations neglected the effect of greenhouse gases in the atmosphere. These gases absorb infrared radiation emitted by Earth and reflect some of it back to Earth. Thus, Earth emits less energy into space than it would without greenhouse gases. The energy absorbed from the Sun and the reduced energy emitted into space have caused the Earth to warm to its mean surface temperature of 288 K.

Over the past two centuries the concentration of carbon dioxide in our atmosphere has increased from a pre-industrial level of about 270 parts per million to 380 parts per million. This increase in carbon dioxide and other greenhouse gases has been caused by the burning of fossil fuels and the removal of forests, which absorb carbon dioxide. The carbon dioxide concentration in the atmosphere is expected to reach 600–700 parts per million by 2100. If that occurs, it will be warmer in 2100 than at any time in the last half million years.

67. Why is there essentially zero greenhouse effect on the Moon?
    (a) There is no photosynthesis on the Moon.
    (b) There is no carbon dioxide on the Moon.
    (c) There is no gaseous atmosphere on the Moon.
    (d) Only b and c are correct.
    (e) All of a, b, and c are correct.

68. The average Earth surface temperature without its atmosphere would be 255 K. Why?
    (a) At that temperature, the emission rate of radiation from Earth would just balance the absorption rate of radiation from the Sun.
    (b) The emission of Earth radiation is from a sphere of area $4\pi R_{Earth}^2$, whereas the absorption of radiation from the Sun is from an area $\pi R_{Earth}^2$.
    (c) Earth's cross section would be significantly reduced because of the lack of the atmosphere.
    (d) Answers a and b are correct.
    (e) Answers a, b, and c contribute about equally to this temperature calculation.

69. The Sun irradiates Earth's outer atmosphere at a rate that is closest to which of the following?
    (a) $1 \times 10^{16}$ J/s (b) $4 \times 10^{16}$ J/s
    (c) $2 \times 10^{17}$ J/s (d) $7 \times 10^{17}$ J/s
    (e) 1400 J/s

70. Because of the greenhouse effect, Earth's average surface temperature is 288 K instead of 255 K. Because of this higher temperature, Earth's surface emits radiation at a rate that is higher by a factor of approximately which of the following?
    (a) 1.13 (b) 1.28 (c) 1.63
    (d) 1.87 (e) 2.21

71. Because of the increased temperature of Earth's surface (288 K compared to 255 K) due to the greenhouse effect, its energy emission rate has increased significantly. Why hasn't Earth cooled down as a result?
    (a) It is cooling down.
    (b) The Sun absorbs much of Earth's extra radiation, and the Sun is warming and emitting more radiation.
    (c) There is a long delay between the change in temperature and the increased rate of radiation by Earth.
    (d) Earth's atmosphere absorbs some of the outgoing radiation and emits it back to Earth's surface.
    (e) All of the above are correct.

72. What will the increasing concentration of greenhouse gases do?
    (a) Increase the reflection rate of sunlight.
    (b) Increase the reflection of Earth's radiation back to the surface.
    (c) Increase the cross-sectional area of the Earth, causing more sunlight to be absorbed.
    (d) Shield Earth from dangerous ultraviolet radiation.
    (e) Provide extra nourishment for plants on Earth.

# Mirrors and Lenses

# 22

**How do eyeglasses correct your vision?**

**When you look in a mirror, where is the face you see?**

**In what ways is a cell phone camera similar to a human eye?**

Two of your roommates wear glasses. Komila needs them to drive and Jason needs them to read and write. When they buy glasses, Komila gets the glasses labeled as −2.0 and Jason gets the ones labeled as +1.5. What vision problems do Komila and Jason have, and what do those numerical labels mean? In this chapter you will learn how to answer these questions.

**In this chapter** we will apply what we have learned about reflection and refraction to mirrors and lenses. In the process we will learn how cameras, telescopes, microscopes, human vision, and corrective lenses work.

**Be sure you know how to:**

- Apply the properties of similar triangles (Math Review Appendix).
- Draw ray diagrams and normal lines (Sections 21.1, 21.2, and 21.3).
- Use the laws of reflection and refraction (Sections 21.2 and 21.3).

**Figure 22.1** Seeing light from an object.

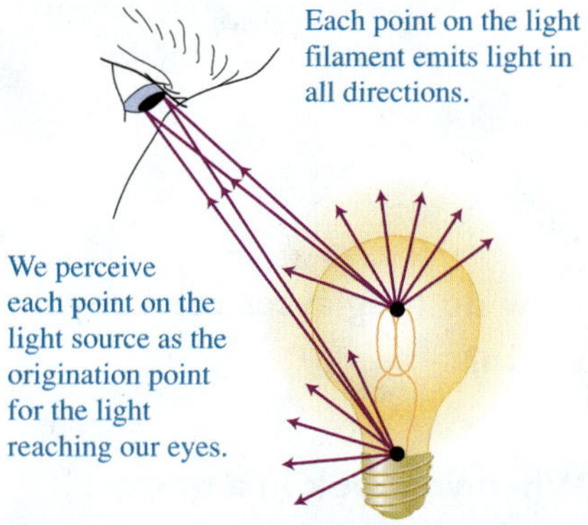

Each point on the light filament emits light in all directions.
We perceive each point on the light source as the origination point for the light reaching our eyes.

## 22.1 Plane mirrors

The simplest mirror is a **plane mirror**—a flat, reflective surface, often a metal film covered in glass. When you stand in front of a plane mirror, you see a reflection of yourself. How does the second "you" appear?

Recall the model that we created to describe how extended objects emit light. Each point on a luminous object sends light rays in all directions. Some of these rays enter our eyes and we see the object at the place where those rays originate (**Figure 22.1**). Now, suppose that a small shining object is in front of a mirror. What happens to the rays emitted by the object when they reach the mirror? Consider Observational Experiment **Table 22.1**.

**OBSERVATIONAL EXPERIMENT TABLE**

**22.1 Seeing a point object in a plane mirror.**

| Observational experiment | Analysis |
|---|---|
| We place a small lightbulb on a table about 20 cm in front of a plane mirror that is held perpendicular to the table, with one side resting on the table. Observers A, B, and C place rulers on the table and point them in the direction of the bulb that they see in the mirror. The solid lines on the figure indicate the orientations of their rulers. Note that the dashed extensions of the solid lines all intersect at one point behind the mirror. 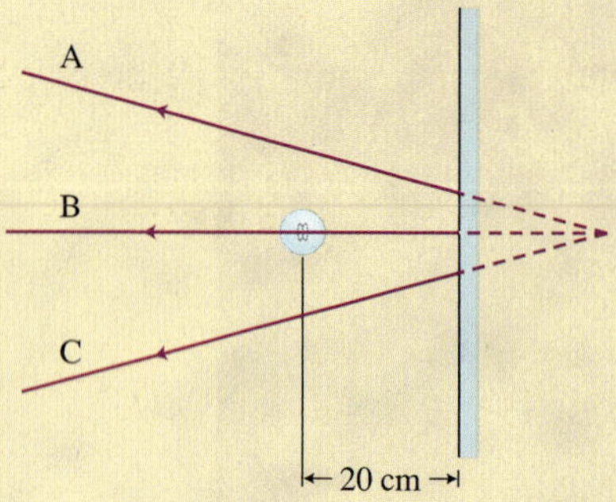  | For observer A to see an image of the lightbulb when looking at the mirror, one or more rays of light reflected from the mirror must reach her/his eyes. Rays between 1 and 2 do reach observer A's eyes. 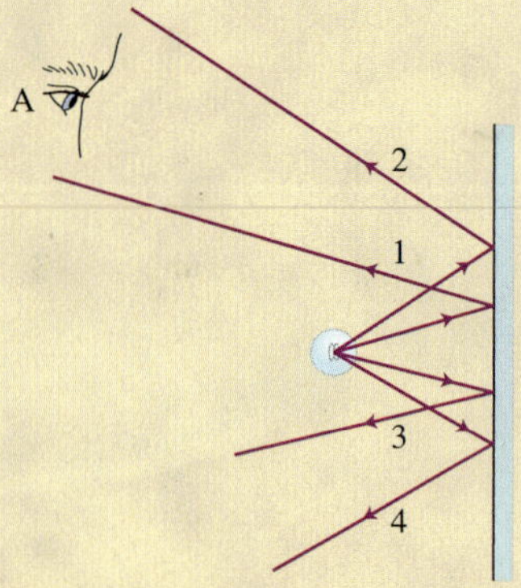  |

**Patterns**

- The rulers all point to the same spot behind the mirror. Rays between 1 and 2 reaching observer A seem to originate from that spot. This is the location of the perceived image of the bulb produced by the mirror.
- The ray diagram shows that the perceived image of the bulb is produced at the same distance behind the mirror as the bulb is in front of it.

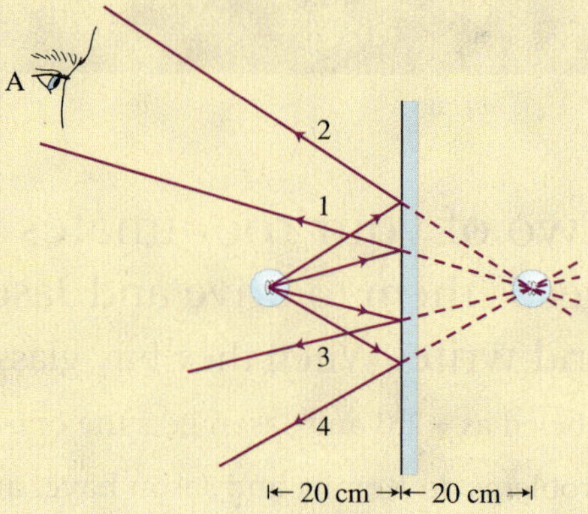


In Table 22.1 we found that the image of the real object seen in the mirror is located where light reflected from the mirror to the eye of the observer *seems* to originate. This perceived image is behind the mirror and not on the surface of the mirror. We also found, using the ray diagram, that the image is exactly the same distance behind the plane mirror as the object is in front of it. Let us test these findings in Testing Experiment **Table 22.2**.

## TESTING EXPERIMENT TABLE

### 22.2 Testing the image location of a plane mirror.

| Testing experiment | Prediction | Outcome |
|---|---|---|
| **Experiment 1.** We repeat the Table 22.1 experiment for observer A but cover the part of the mirror directly in front of the bulb. | If the image is due to the reflected rays, then even if we cover part of the mirror, some of the reflected rays will still reach our eyes. The location of the image should not change. 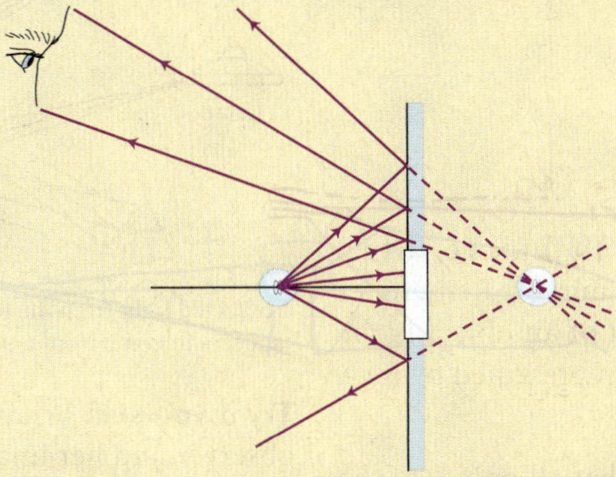 | We see the image of the bulb exactly where it was before. |
| **Experiment 2.** Replace the mirror with a clean sheet of glass (it reflects some of the light and allows some of it to pass through). Place a lit bulb 20 cm in front of the glass. Where must you place a second identical bulb, bulb 2, behind the glass so that it overlaps the image created by the light reflected off the glass sheet? | According to our assumptions, the image of the first bulb will be exactly the same distance behind the mirror as the bulb is in front. Thus, bulb 2 should be placed 20 cm behind the glass. 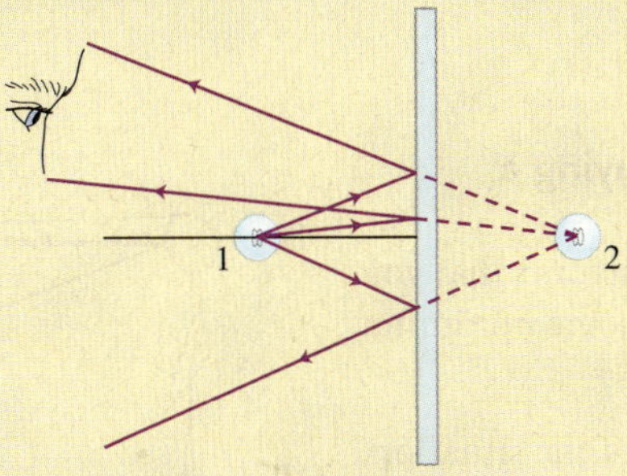 | The light from bulb 2 comes from the same location as the light from bulb 1 appears to come from after reflection. |

**Conclusions**

- **Experiment 1.** The position of the image formation is consistent with our previous experiment. It also disproves the common idea that the image forms on the surface of the mirror.
- **Experiment 2.** This experiment gives us confidence that the image is the same distance behind the mirror as the object is in front of it.

The image we see in a plane mirror is not real; there is no real light coming from behind the mirror. The reflected light reaching your eyes appears to originate from a point behind the mirror. What we see in the mirror is called a **virtual image**.

**Plane mirror virtual image** A plane mirror produces a virtual image that is the same distance behind the mirror as the object is in front of it. The reflected light seems to diverge from the image behind the mirror. But no light actually leaves that image—you see light reflected from the mirror.

**CONCEPTUAL EXERCISE 22.1 Where is the lamp?**

You place a lamp in front of a mirror and tilt it so that the top and the bottom of the lamp are at different distances from the mirror, at the position shown in the figure below. Where do you see the image of the lamp produced by the mirror?

**Sketch and translate** We need to find the virtual image of the lamp produced by the mirror. Remember that the virtual image of each point of an object is the place from which the reflected light, represented by rays, appears to diverge.

**Simplify and diagram** Assume that all points of the lamp (including the base) send out rays in all directions. For simplicity, we will consider the top and the bottom of the lamp and locate the images of each of these two end points. Assume that the images of all other points on the lamp are formed between the end point images. As you can see from the ray diagram below, the image of each point is behind the mirror, and each point is at the same distance behind the mirror as the point on the lamp is in front of the mirror.

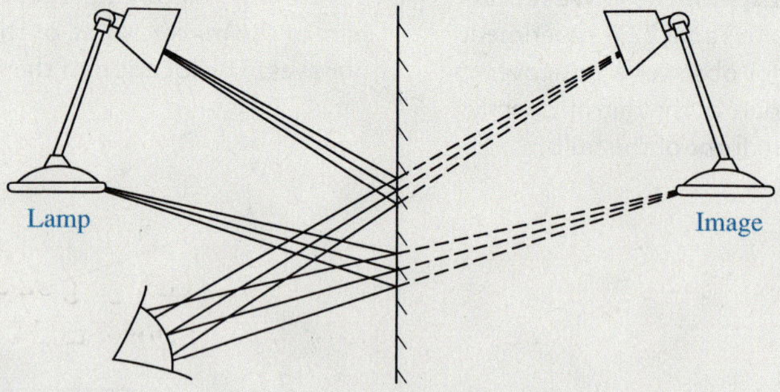


Reflected light from the top and bottom of the tilted lamp appears to come from a tilted image behind the mirror.

**Try it yourself:** What happens to the distance between an observer and her image in a plane mirror if the observer moves backward, doubling her distance from the mirror?

*Answer:* The distance between the observer and the image quadruples.

**CONCEPTUAL EXERCISE 22.2 Buying a mirror**

Your only requirement when buying a mirror is that you can see yourself from head to toe. What is the minimum length mirror you should buy?

**Sketch and Translate** Draw a sketch of the situation, as shown below. Let $h$ be your height. Express the mirror height $H$ as a fraction or a multiple of your height. Let $s$ be your distance from the mirror. Since the goal is to see your entire body in the mirror, rays from the top of your head and from the bottom of your feet must reflect off the mirror into your eyes.

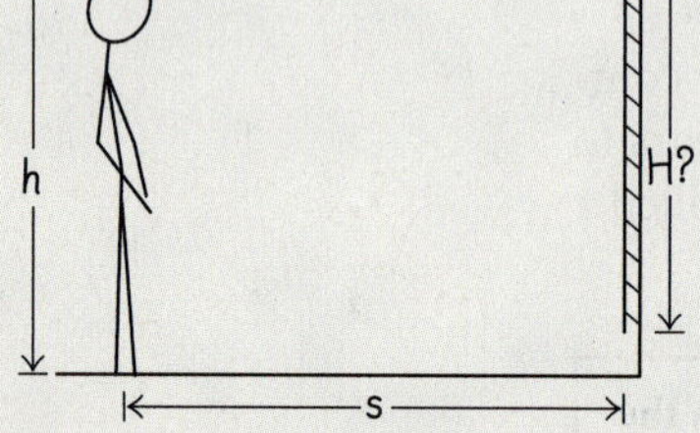


**Simplify and diagram** Assume that the mirror is mounted on a vertical wall and that all parts of your body are the same distance from the mirror. Also assume that you are a shining object of a particular height $h$. Finally, let $h_1$ be the vertical distance between your feet and eyes and $h_2$ be the vertical distance between the top of your head and your eyes. This means $h_1 + h_2 = h$.

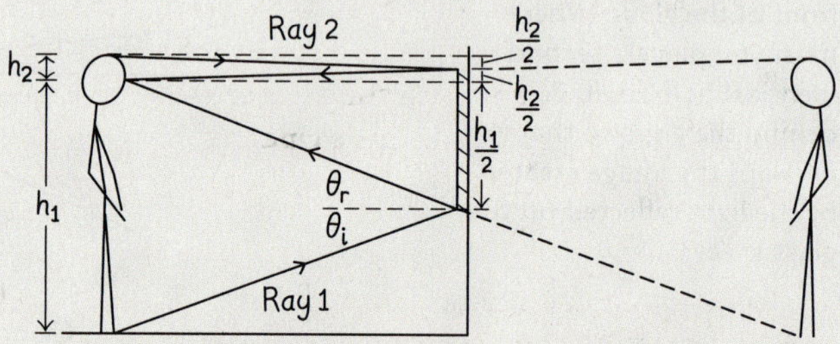


You will see your toes if ray 1 from your toes reaches your eyes after reflection. Ray 1 hits the mirror at a height $h_1/2$ above your toes. You will see the top of your head if ray 2 from the top of your head reaches your eyes after reflection. Ray 2 hits the mirror at a distance $h_2/2$ below a horizontal line from the top of your head to the mirror. From the diagram we see that you can cut off both the bottom $h_1/2$ and the top $h_2/2$ of the mirror and still see your whole body. Thus the length of the smallest mirror is $h_1/2 + h_2/2 = h/2$. Note that size of the mirror does not depend on the distance $s$ between you and the mirror—you can see your entire body in the mirror no matter how far you stand from it.

**Try it yourself:** You are 1.6 m tall and stand 2.0 m from a plane mirror. How tall is your image? How far from the mirror on the other side is it? What is your image height if you double your distance from the mirror?

*Answer:* 1.6 m; 2.0 m; the same size.

**Figure 22.2** Features of (a) concave and (b) convex mirrors.

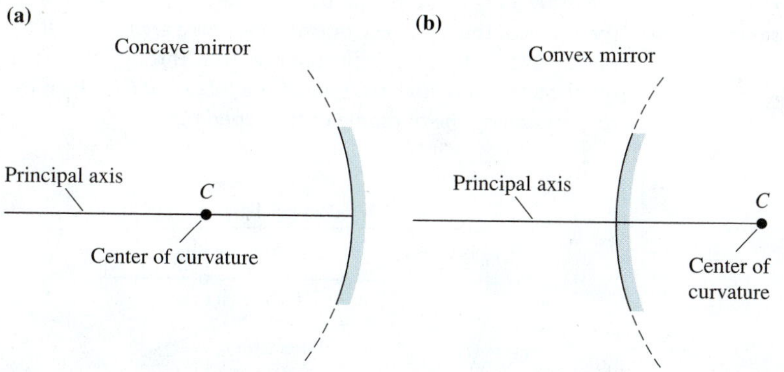


**Figure 22.3** A convex mirror provides visibility at blind spots and around corners.

**Review Question 22.1** A mirror is hanging on a vertical wall. Will you see more or less of your body if you step closer to the mirror?

## 22.2 Qualitative analysis of curved mirrors

Curved mirrors are cut from a spherically shaped piece of glass backed by a metal film. **Concave mirrors** are often used for magnification, in telescopes and cosmetic mirrors. The curve of a concave mirror bulges away from the light it reflects (see **Figure 22.2a**). In a **convex mirror,** the curve bulges toward the light it reflects (see Figure 22.2b). Convex mirrors are used as passenger-side rearview mirrors and to provide visibility at blind spots, such as hallway corners and driveway exits (**Figure 22.3**).

In both kinds of curved mirrors, the center of the sphere of radius $R$ from which the mirror is cut is called the **center of curvature** $C$ of the mirror. The imaginary horizontal line connecting the center of curvature with the center of the mirror's surface is called the **principal axis**.

**Figure 22.4** A concave mirror reflects a laser beam.

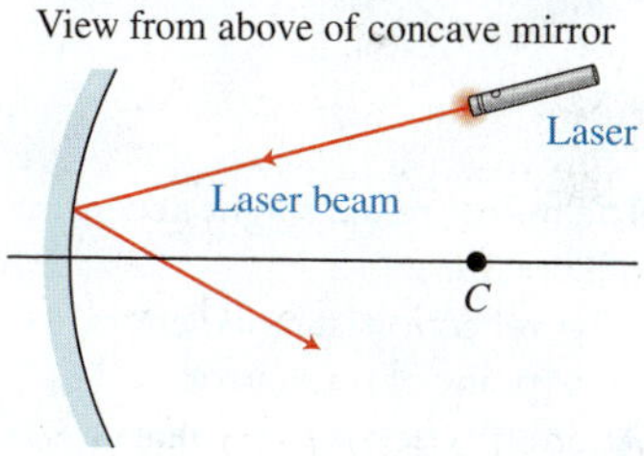


### Concave mirrors

In Observational Experiment **Table 22.3**, we cut a narrow band from a concave mirror and lay it on its edge on a piece of paper (**Figure 22.4**). The paper allows us to observe the paths of laser light incident and reflected from the mirror.

**OBSERVATIONAL EXPERIMENT TABLE**

**22.3 Using ray diagrams and the law of reflection to analyze the path of reflected laser light from a concave mirror.**

| Observational experiment | Analysis |
|---|---|
| **Experiment 1.** We shine a laser beam parallel to the plane of the page toward a concave mirror and trace the path of the incident and reflected light on the page. 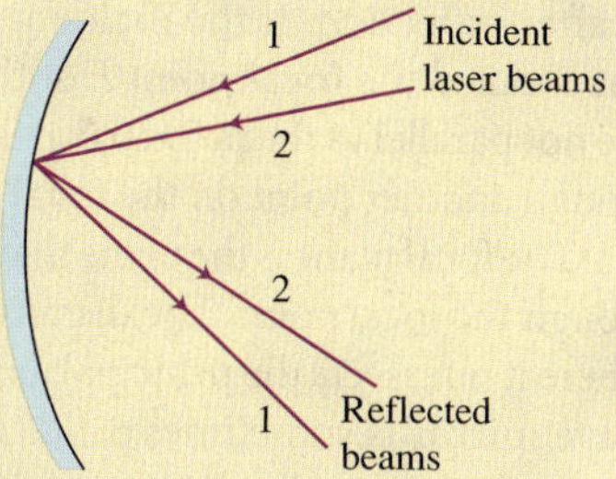  | We use the law of reflection to draw a ray diagram. The dashed line from the place where the ray hits the mirror to the center of curvature is a normal line perpendicular to the surface of the mirror. The law of reflection accounts for the path of the reflected ray. 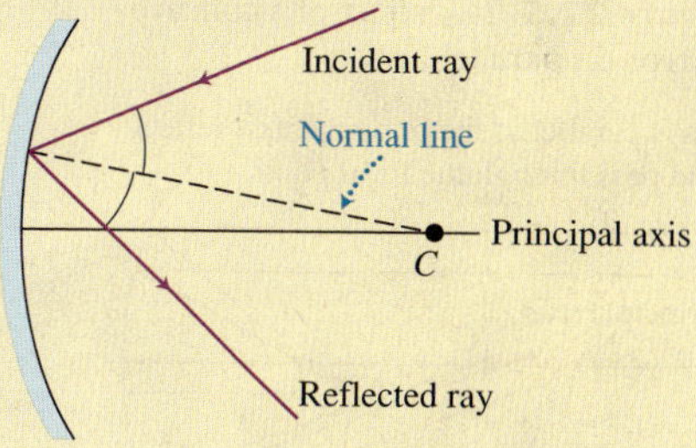  |

*(continued)*

**Experiment 2.** We send several beams parallel to the principal axis. Reflected beams all pass near the same point on the principal axis.

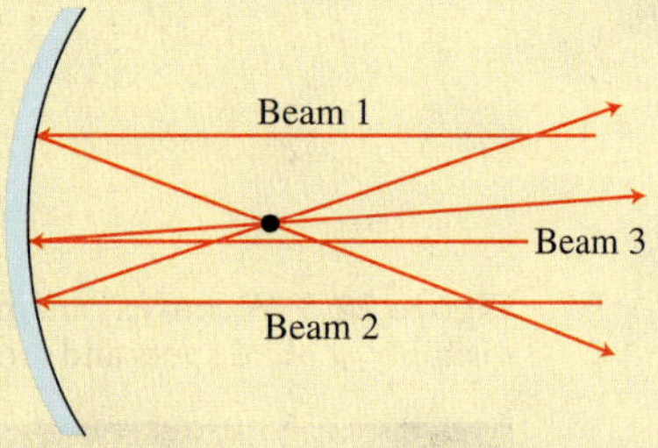


We draw a ray diagram and use the law of reflection to analyze the paths of the light. The normal lines here are radii of the mirror. We find that all reflected rays pass through the same point on the principal axis (called the *focal point F* of the mirror), matching the outcome of the experiment.

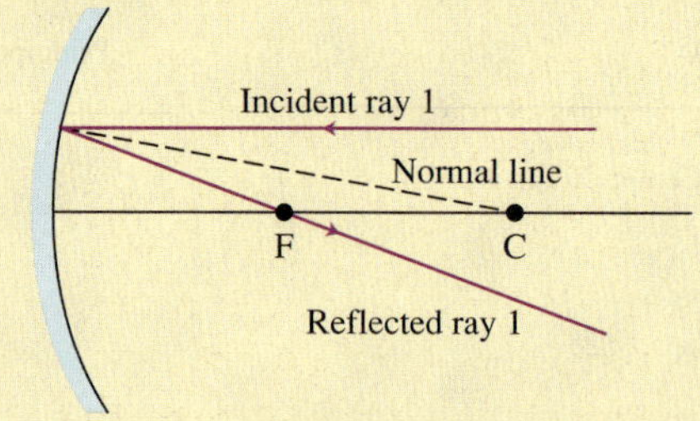


**Experiment 3.** This time send several beams parallel to each other but not parallel to the principal axis. One ray passes though the center of curvature C. All reflected rays pass near the same point.

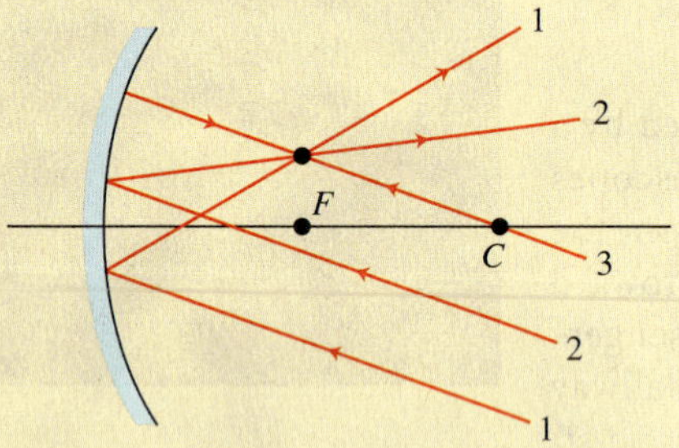


We use the law of reflection to draw a ray diagram to analyze the paths of the light. The ray passing through the center of curvature is perpendicular to the mirror and thus reflects back on itself. All other rays after reflection pass through the same point on a line perpendicular to the principal axis and through the focal point. This point is in the *focal plane* of the mirror. The ray diagram matches the outcome of the experiment.

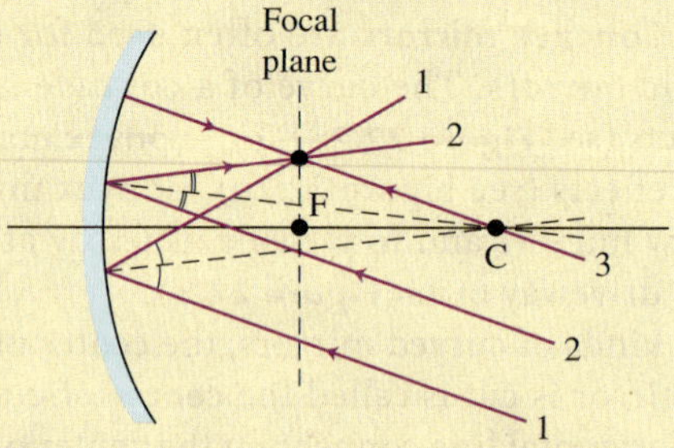


**Patterns**

- The mirror reflects light according to the law of reflection. The normal line at any point on the mirror goes through the center of curvature.
- After reflection, all incident rays traveling parallel to the principal axis pass through the same point on the principal axis—the focal point of the mirror.
- After reflection, all incident rays traveling parallel to each other but not parallel to the principal axis intersect at a common point. This point is on the line perpendicular to the principal axis through the focal point.

**Figure 22.5** The effect of a concave mirror on parallel rays.

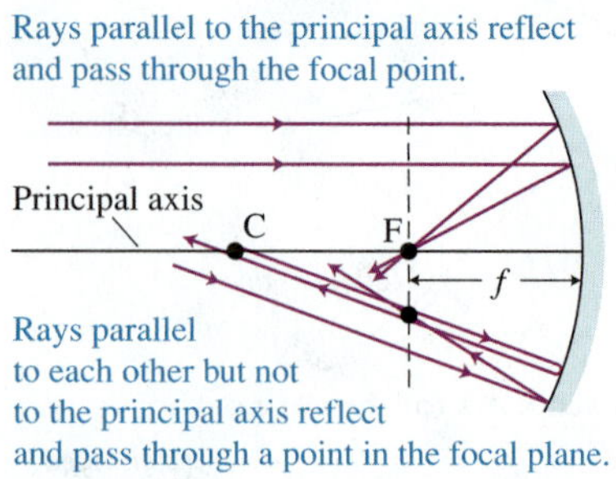


Table 22.3 shows us that a curved mirror reflects light in agreement with the law of reflection. But unlike a flat mirror, it causes parallel incident rays to pass through a single point after reflection. If the incident rays are also parallel to the principal axis, this point is called the **focal point** *F* of the mirror (see Experiment 2 in Table 22.3). If they are not parallel to the principal axis (but are still parallel to each other), they all pass through another point on the **focal plane** of the mirror (see Experiment 3 in Table 22.3). The focal plane is the plane that is perpendicular to the principal axis and passes though the focal point. The distance from the focal point to the surface of the mirror where it intersects the principal axis is called the **focal length** $f$ (**Figure 22.5**). A concave mirror is sometimes called a *converging mirror* because incident parallel rays converge after reflection from it. Since we can consider the rays of the Sun to be parallel, we can easily determine the location of the focal point and the focal plane of any converging mirror if we use the Sun as the light source.

If a mirror has a large radius of curvature compared to its size and the incident rays are close to the principal axis, the focal point is approximately halfway between the center of the mirror and the center of curvature ($f = R/2$). See the geometric proof in **Figure 22.6**.

**TIP** The verb *converge* does not mean that the rays reach the focal point and stop there. They continue traveling after passing through this point.

**Figure 22.6** The focal length $f$ of a curved mirror is half the radius of curvature $R$.

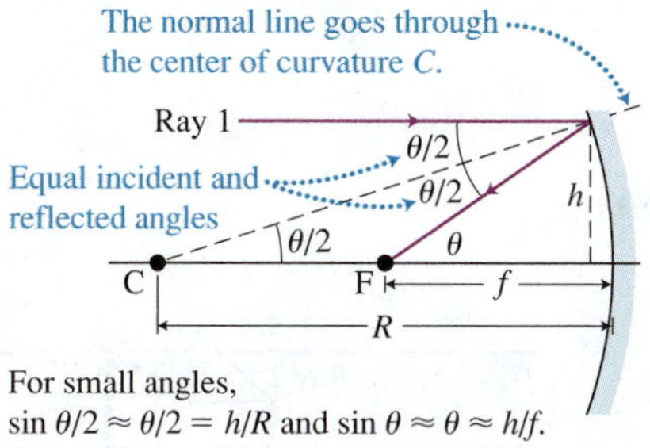


For small angles,
$\sin \theta/2 \approx \theta/2 = h/R$ and $\sin \theta \approx \theta \approx h/f$.
Thus, $2(\theta/2) = 2(h/R) = h/f$, or $f = R/2$.

## Using a ray diagram to locate the image formed by a concave mirror

We can use a ray diagram to locate the image of an object that is placed in front of a concave mirror. The method is illustrated in Reasoning Skill 22.1. The object in this case is a lightbulb that is placed beyond the center of curvature of a concave mirror. We will represent the object with an arrow that originates on the principal axis. For the time being we only locate the image of the top of the arrow and assume that the image of the base of the object (the bottom of the arrow) is also on the axis. We will validate this assumption later.

**REASONING SKILL 22.1** Constructing a ray diagram to locate the image of an object produced by a concave mirror

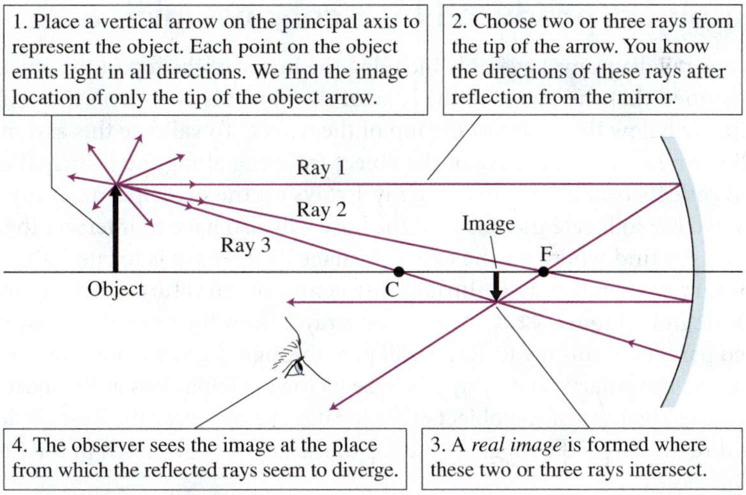

1. Place a vertical arrow on the principal axis to represent the object. Each point on the object emits light in all directions. We find the image location of only the tip of the object arrow.
2. Choose two or three rays from the tip of the arrow. You know the directions of these rays after reflection from the mirror.
Ray 1
Ray 2
Ray 3
Image
F
Object
C
4. The observer sees the image at the place from which the reflected rays seem to diverge.
3. A *real image* is formed where these two or three rays intersect.

Ray 1 travels from the tip of the object toward the mirror parallel to the principal axis. After reflection, it passes through the focal point $F$.
Ray 2 travels from the tip of the object through the focal point $F$. After reflection it travels parallel to the principal axis.
Ray 3 passes through the center of curvature $C$. It hits the mirror perpendicular to its surface and reflects back in the same direction.

The image formed is a **real inverted image**: it is **inverted,** meaning that the image points opposite the direction of the object; it is also a **real image**, meaning that the reflected rays actually pass through that image location and then reach the eyes of the observer. If you placed a small screen (a piece of paper, your hand, etc.) at that image location, you would see a sharp (focused), inverted image (an upside-down bulb) projected onto it. If you place the screen closer to the mirror or farther away than the image location, reflected light will reach it but the image will be fuzzy because the rays pass through different points on the screen.

**TIP** Remember that each point on an object emits an infinite number of rays that reflect off the mirror according to the law of reflection. Rays 1, 2, and 3 in the Reasoning Skill were chosen because we know how they reflect without needing to draw the normal line to the point of incidence and measure the angles of incidence and reflection.

**Figure 22.7** Using a ray diagram to locate the image of an object located near a concave mirror. Because the rays diverge after reflection, the virtual image appears behind the mirror.

(a)

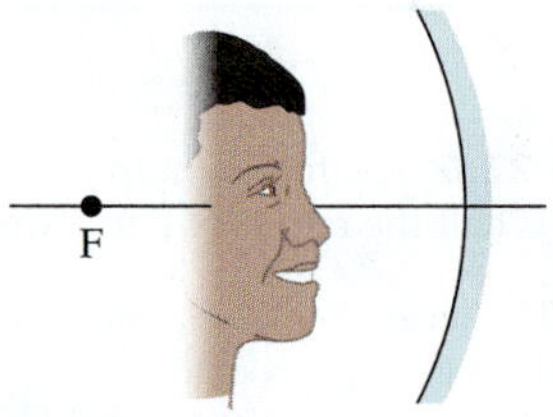


(b)

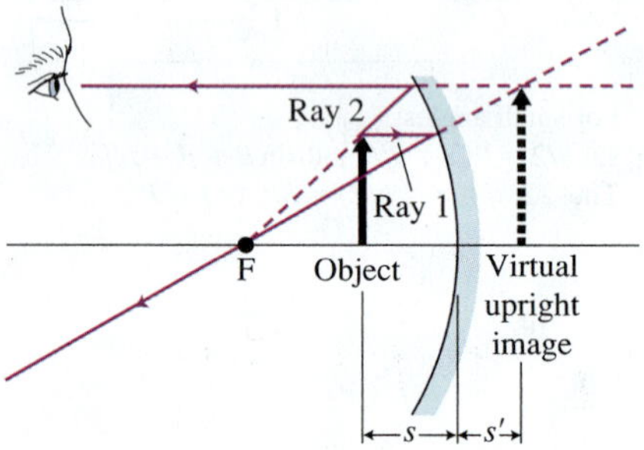


In Reasoning Skill 22.1 we used three special rays to locate the image of an object produced by a concave mirror. Ray 1 traveled from the tip of the object toward the mirror parallel to the principal axis. After reflection, ray 1 passed through the focal point. Ray 2 traveled from the tip of the object through the focal point. After reflection, it traveled parallel to the principal axis. Finally, ray 3 passed through the center of curvature, hit the mirror perpendicular to its surface, and reflected back in the same direction. We will use these three rays throughout our study of mirrors.

Not all reflected rays result in a real image. Consider the situation in **Figure 22.7a**, in which your face is closer to a concave mirror than the focal point *F*. We will use rays 1 and 2, as described above, to construct a ray diagram. In Figure 22.7b we see that reflected rays 1 and 2 diverge. For a person standing behind you and the mirror, the light seems to originate from a point behind the mirror. This point is the location of the **virtual image** produced by the mirror. The image is *virtual* (similar to the images formed by plane mirrors) because the light does not actually come from that point—it just appears to. The image is magnified (bigger than the object) and upright. You see such an image when you use a concave mirror to apply makeup or shave.

## Locating the image of the base of an object produced by a concave mirror

So far, our ray diagrams have only included the image of the top of the object. We have assumed that the image of the base of the object is always on the principal axis directly below the image of the top of the object. To validate this assumption we will use a ray from the base of the object traveling along the principal axis as ray 1. It reflects back on itself. Since ray 1 stays on the principal axis, any other ray that we use to locate the image of the base will also have to intersect the principal axis. To find where on the axis the image of the base is located, first draw an arbitrary imaginary ray C through the center of curvature. That ray reflects back through C (**Figure 22.8**). Now draw a ray 2 from the base of the object and oriented parallel to this ray C. Ray 2 will pass through the same point on the focal plane as the imaginary ray C. Ray 2 intersects the principal axis at the location of the image of the base of the object. This locates the image of the base. It does in fact lie directly above the image of the top of the arrow (not shown in the figure).

**Figure 22.8** Locating the base of the image produced by a curved mirror.

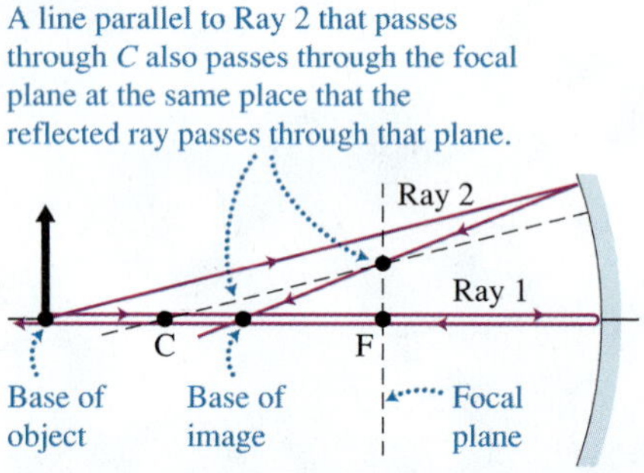


## Convex mirrors

Let's investigate convex mirrors. Consider Observational Experiment **Table 22.4**.

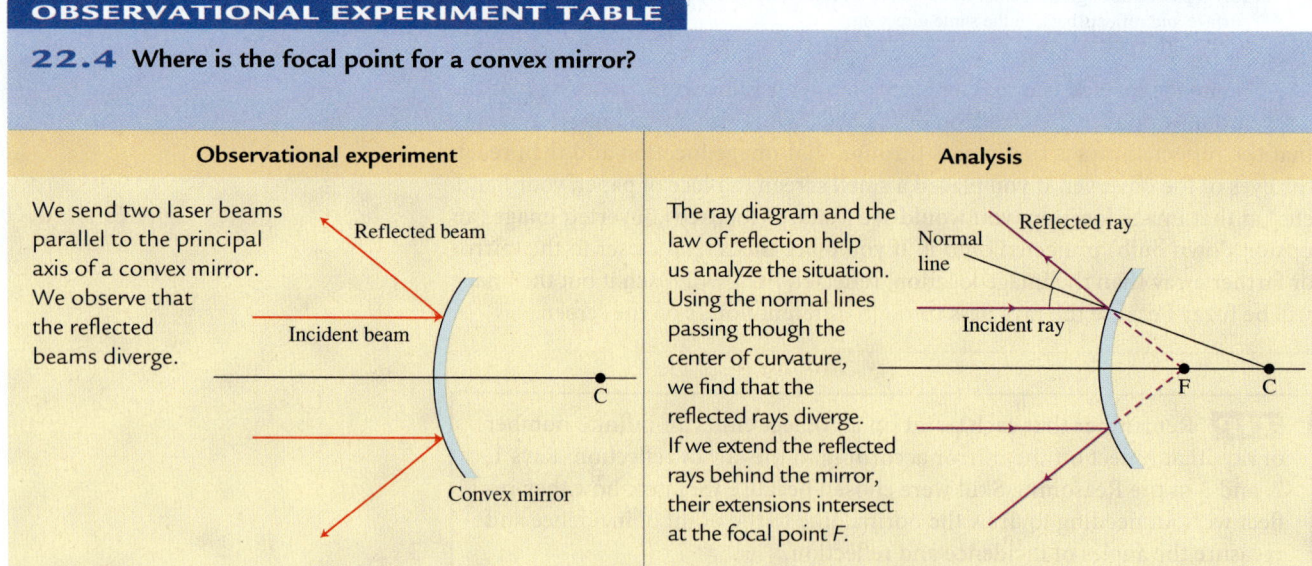

**OBSERVATIONAL EXPERIMENT TABLE**

**22.4 Where is the focal point for a convex mirror?**

| Observational experiment | Analysis |
|---|---|
| We send two laser beams parallel to the principal axis of a convex mirror. We observe that the reflected beams diverge. 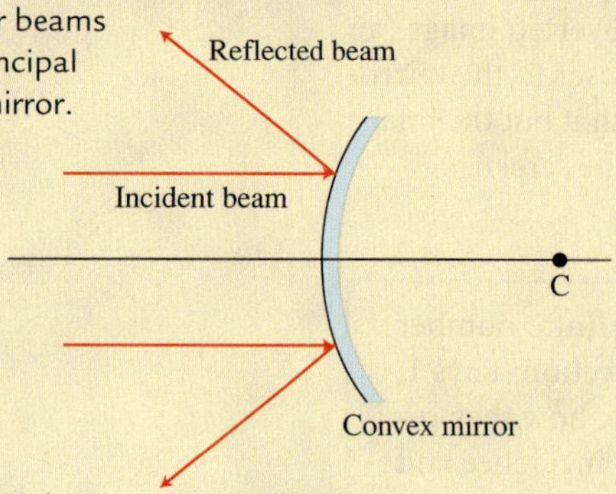 | The ray diagram and the law of reflection help us analyze the situation. Using the normal lines passing though the center of curvature, we find that the reflected rays diverge. If we extend the reflected rays behind the mirror, their extensions intersect at the focal point *F*. 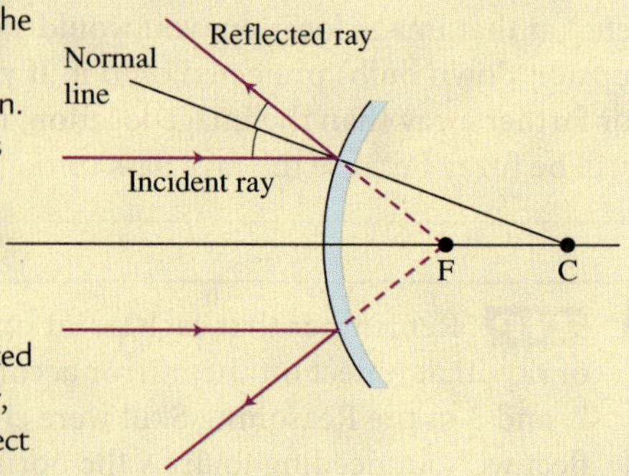 |

**Pattern**

- Light rays moving parallel to the principal axis of a convex mirror reflect and diverge away from each other.
- Lines drawn backward along the direction of the reflected light cross the axis at a focal point *F* behind the mirror.

Recall that for a concave mirror, rays parallel to the principal axis reflect through a focal point in *front* of the mirror. However, for a convex mirror, rays parallel to the principal axis diverge after reflection. They diverge from what is called a **virtual focal point** *F* behind the mirror. As with a concave mirror, the focal length $f$ equals half the radius of curvature $R$ of the mirror ($f = R/2$), as shown in **Figure 22.9**. Let's use a new set of ray diagrams to investigate images formed by convex mirrors.

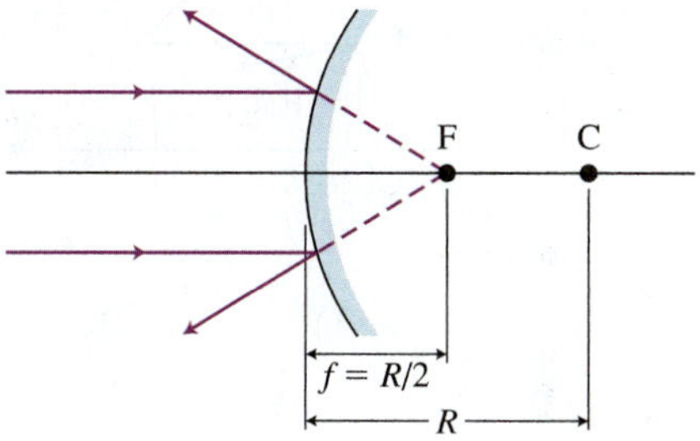


**Figure 22.9** A convex mirror reflects rays parallel to the principal axis. The continuations of the reflected rays pass through the focal point behind the mirror.

### REASONING SKILL 22.2 Constructing a ray diagram to locate the image produced by a convex mirror

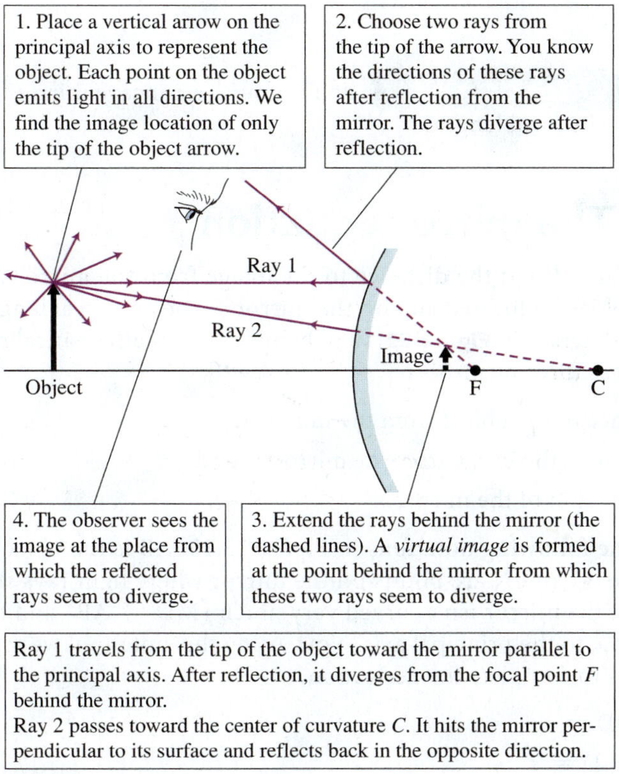

1. Place a vertical arrow on the principal axis to represent the object. Each point on the object emits light in all directions. We find the image location of only the tip of the object arrow.
2. Choose two rays from the tip of the arrow. You know the directions of these rays after reflection from the mirror. The rays diverge after reflection.
Ray 1
Ray 2
Image
Object
F
C
4. The observer sees the image at the place from which the reflected rays seem to diverge.
3. Extend the rays behind the mirror (the dashed lines). A *virtual image* is formed at the point behind the mirror from which these two rays seem to diverge.
Ray 1 travels from the tip of the object toward the mirror parallel to the principal axis. After reflection, it diverges from the focal point *F* behind the mirror.
Ray 2 passes toward the center of curvature *C*. It hits the mirror perpendicular to its surface and reflects back in the opposite direction.


In Reasoning Skill 22.2 the object was about $1.6R$ from the mirror. An upright virtual image was formed about $0.4R$ behind the mirror. What happens to the image when the object is moved closer to the mirror?

### CONCEPTUAL EXERCISE 22.3 Looking into a convex mirror

You hold a convex mirror $0.7R$ behind a pencil. Approximately where is the image of the pencil, and what are the properties of the image?

**Sketch and translate** Represent the pencil by an arrow and use a ray diagram to locate the image of the top of the pencil (the tip of the arrow).

*(continued)*

**Simplify and diagram** Draw a ray diagram. We use rays 1 and 2 described in Reasoning Skill 22.2 to locate the image. The image is virtual and upright about $0.3R$ behind the mirror and it is smaller than the object. This is what you would see if you looked at your reflection on the back of a shiny tablespoon. The images formed by convex mirrors are always upright, virtual (behind the mirror), and reduced in size compared to the objects.

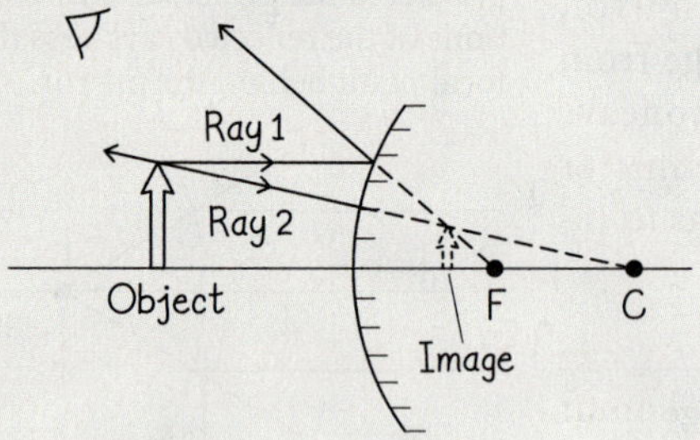


**Try it yourself:** In order to obtain the following types of images of an object, should you use a concave or convex mirror? Where should you place the object relative to the mirror? (a) Real, bigger than the object; (b) real, smaller than the object; (c) virtual, bigger than the object; (d) virtual, smaller than the object; (e) real and the same size as the object.

*Answers:* (a) Concave: the object should be between $R$ and $f$ from the mirror; (b) concave: the object should be farther than $R$ from the mirror; (c) concave: the object is closer than a focal length from the mirror; (d) convex: the object can be at any location; (e) concave: the object should be at $2f$ from the mirror—right at the center of curvature.

**Review Question 22.2** You've found a concave mirror. How can you estimate its focal length?

## 22.3 The mirror equation

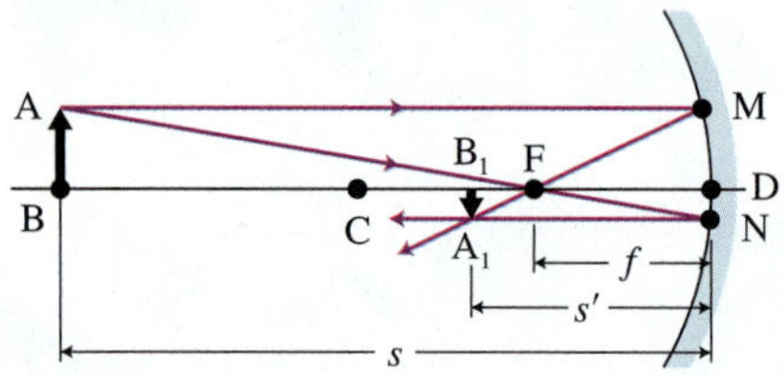


**Figure 22.10** A ray diagram to help develop the mirror equation.

We have learned that the distance to the image from the mirror depends on where the object is located and on the mirror's type and focal length. We will use the ray diagram in **Figure 22.10** to help derive a mathematical relationship between these three quantities:

- the distance of the object from the mirror $s$,
- the distance of the image from the mirror $s'$, and
- the focal length of the mirror $f$.

Note the following notations in Figure 22.10: AB is the object, and $A_1B_1$ is the image. M and N are points on the mirror where light rays strike it. We assume that the mirror isn't curved very much ($MD \approx AB$) and the rays are incident close to the principal axis. Now, using this notation, we make the following steps to complete the derivation.

- $AM \approx BD = s$
- $A_1N \approx B_1D = s'$
- $MD \approx AB$ and $ND \approx A_1B_1$ (if the mirror is not very curved).
- ABF and NDF are similar triangles. Thus,

$$\frac{ND}{DF} = \frac{A_1B_1}{DF} = \frac{AB}{BF} \text{ or } \frac{AB}{A_1B_1} = \frac{BF}{DF} = \frac{s-f}{f}$$

- $A_1B_1F$ and MDF are similar triangles. Thus,

$$\frac{A_1B_1}{B_1F} = \frac{MD}{DF} = \frac{AB}{DF} \text{ or } \frac{AB}{A_1B_1} = \frac{DF}{B_1F} = \frac{f}{s'-f}$$

- Setting the previous two $AB/A_1B_1$ ratios equal to each other, we find that

$$\frac{s-f}{f} = \frac{f}{s'-f} \tag{22.1}$$

After some algebra (see the Tip), we get a relationship called the **mirror equation**:

$$\frac{1}{s} + \frac{1}{s'} = \frac{1}{f} \quad (22.2)$$

**TIP** To get from Eq. (22.1) to (22.2), multiply both sides by the product of $f(s' - f)$. Then carry out the multiplication on the left side of the equation and add the quantities. You should have $ss' - sf - fs' = 0$. Now divide both sides of the equation by the product $ss'f$.

The mirror equation [Eq. (22.2)] allows us to predict the distance $s'$ of the image from the mirror given the distance $s$ of the object from the mirror and the mirror's focal length $f$. Before we test the equation experimentally, let us check to see if it is consistent with an extreme case. We know that a concave mirror causes rays parallel to the principal axis to pass through the focal point after reflection. If an object is extremely far away along the principal axis, we can assume that rays from the object reaching the mirror are parallel to the principal axis. *If* Eq. (22.2) is correct, and we use infinity for the object distance, *then* we should find that the image distance for that object equals the focal length distance. Using Eq. (22.2) we get

$$\frac{1}{\infty} + \frac{1}{s'} = 0 + \frac{1}{s'} = \frac{1}{f}$$

Therefore, the image is at a distance equal to the focal length away from the mirror. This result is consistent with the prediction based on the extreme case analysis.

**TIP** When you divide any number (other than infinity) by infinity, the result is zero.

**EXAMPLE 22.4 Where should the screen be?**

You place a candle 0.80 m from a concave mirror with a radius of curvature of 0.60 m. Where should you place a paper screen to see a sharp image of the candle?

**Sketch and translate** Draw a sketch of the situation, as shown below. Assemble all parts of the experiment on a meter stick so the distances can be easily measured. The known quantities are $s = 0.80$ m and $f = R/2 = 0.30$ m. To find where to place the screen, we need to determine $s'$.

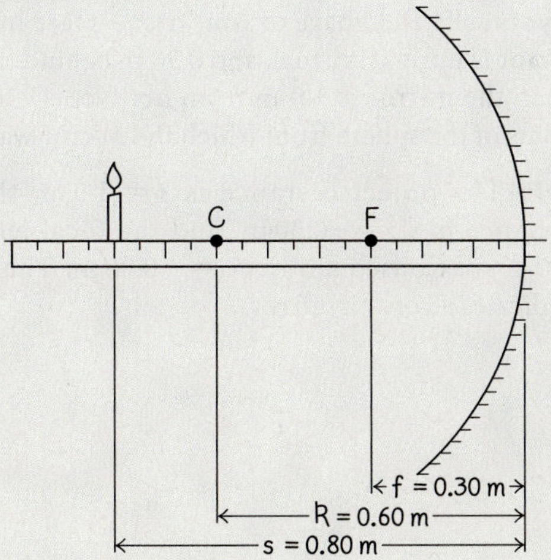


**Simplify and diagram** Model the candle as an arrow. We will use a ray diagram to locate the image of the tip of the arrow, as shown below. For this ray diagram, we will use rays 1 and 3 as described in Reasoning Skill 22.1. You can estimate the image distance from the diagram—it looks like the image distance is a little less than two-thirds the object distance.

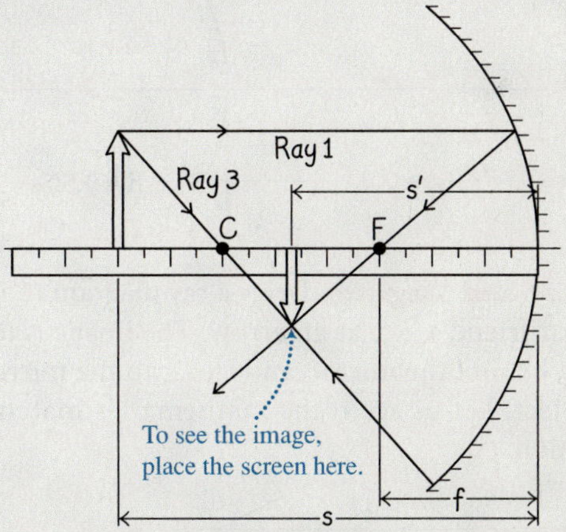


**Represent mathematically** We see from the ray diagram that the image is real, inverted, and closer to the mirror than the candle. Rearrange Eq. (22.2) to find $s'$:

$$\frac{1}{s'} = \frac{1}{f} - \frac{1}{s}$$

*(continued)*

**Solve and evaluate**

$$\frac{1}{s'} = \frac{1}{f} - \frac{1}{s} = \frac{1}{0.30\text{ m}} - \frac{1}{0.80\text{ m}}$$

$$= \frac{0.80\text{ m} - 0.30\text{ m}}{(0.30\text{ m})(0.80\text{ m})} = 2.08\text{ m}^{-1}$$

Thus, $s' = (1/2.08\text{ m})^{-1} = 0.48\text{ m}$. The image is closer to the mirror than the object, as predicted by our ray diagram. Thus, the predictions using mathematics and the ray diagram are consistent. When we place a screen where we predicted, we see an inverted real image of the flame. We will make the screen small and place it below the principal axis to avoid blocking light from the candle.

**Try it yourself:** Determine the distance of the image when the candle is 0.20 m from the mirror. Does the answer make sense to you?

*Answer:* $-0.60$ m. The image is virtual and behind the mirror. Evidently, the negative sign means that the image is virtual.

In the Try It Yourself part of Example 22.4, the mirror equation led to a negative value for the image distance. The image was virtual. We need to agree on some sign conventions when using Eq. (22.2).

- We put a negative sign in front of the distance $s'$ for a virtual image that appears to be behind a mirror.
- We put a negative sign in front of the focal length $f$ for a convex mirror.

Let us see if these sign conventions work when we use the mirror equation for a convex mirror.

**EXAMPLE 22.5 Test the mirror equation for a convex mirror**

A friend's face is 0.60 m from a convex 0.50-m-radius mirror. Where does the image of her face appear to you when you look at the mirror?

**Sketch and translate** The situation is sketched below. The known information is $s = 0.60\text{ m}$ and $f = -R/2 = -(0.50\text{ m})/2 = -0.25\text{ m}$. Note that for a convex mirror, the focal length is a negative number.

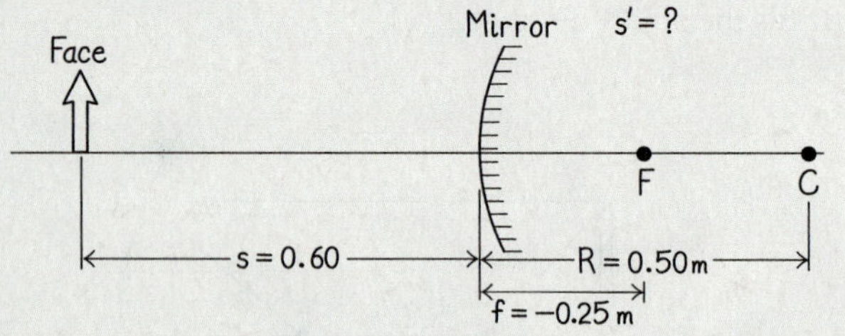


**Simplify and diagram** Draw a ray diagram representing your friend's face as an arrow. The image is upright, virtual, behind the mirror, and closer to the mirror than the object. Let us see if the mathematics matches this prediction.

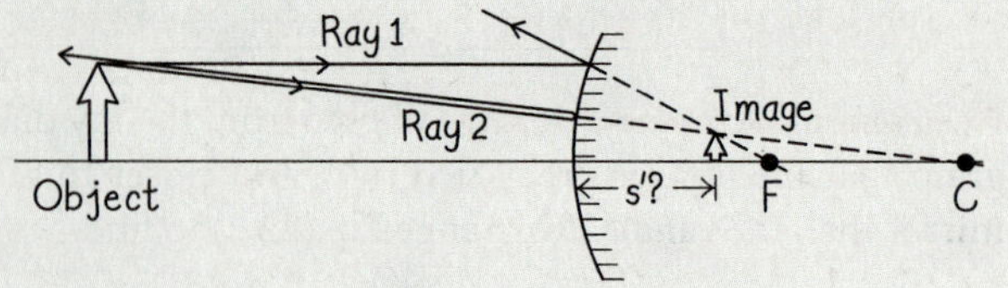


**Represent mathematically** Rearrange the mirror equation [Eq. (22.2)] to locate the image of your friend's face.

$$\frac{1}{s'} = \frac{1}{f} - \frac{1}{s}$$

**Solve and evaluate** Insert the known information:

$$\frac{1}{s'} = \frac{1}{-0.25\text{ m}} - \frac{1}{0.60\text{ m}} = -5.67\text{ m}^{-1}$$

The image distance is $s' = (1/-5.67\text{ m})^{-1} = -0.18\text{ m}$. We see that the image distance is negative and its magnitude is less than the magnitude of the object distance—consistent with the ray diagram.

**Try it yourself:** The image of your friend's face in a convex mirror is upright, virtual, and 0.30 m behind the mirror when the mirror is 1.0 m from her face. Determine the radius of the sphere from which the mirror was cut.

*Answer:* The object distance is $s = 1.0\text{ m}$, the image distance is $s' = -0.30\text{ m}$, and the focal length is $f = -0.43\text{ m}$. Consequently, $R = -0.86\text{ m}$. The minus sign indicates a convex mirror.

It appears that the mirror equation works equally well for concave and convex mirrors provided that appropriate sign conventions are used.

**Mirror equation** **The distance $s$ of an object from the surface of a mirror, the distance $s'$ of the image from the surface of the mirror, and the focal length $f$ of the mirror are related by the mirror equation:**

$$\frac{1}{s} + \frac{1}{s'} = \frac{1}{f} \tag{22.2}$$

**The focal length is half the radius of curvature: $f = R/2$. The following sign conventions apply for mirrors:**

- **The focal length is positive for concave mirrors and negative for convex mirrors.**
- **The image distance is positive for a real image and negative for a virtual image.**

## Magnification

You have probably noticed that the size of an image produced by a curved mirror is sometimes bigger and sometimes smaller than the size of the object. The change in the size of the image compared to the size of the object is a quantity called **linear magnification** $m$. Linear magnification is defined in terms of the image height $h'$ and the object height $h$, where the heights are the perpendicular sizes of the image and object relative to the principal axis:

$$\text{linear magnification} = \frac{\text{image height}}{\text{object height}} = m = \frac{h'}{h} \tag{22.3}$$

A height is positive if the image or object is upright, and negative if inverted.

It is often not possible to calculate magnification as defined above since the object and image heights are sometimes unknown. However, we can determine the magnification using the mirror equation and the object distance and image distance. Consider **Figure 22.11**, which will be used to derive a relation between $h'$ and $h$ and $s'$ and $s$. In Figure 22.11a we draw the image of the object using rays 1 and 2. Then in Figure 22.11b we draw a new ray that travels from the top of the object to the center of the mirror and then to the top of the inverted image. The reflected ray makes the same angle $\theta$ with the principal axis as does the incident ray. Thus,

$$\tan\theta = \frac{h}{s} = \left|\frac{h'}{s'}\right|$$

Therefore, the absolute value of the magnification is $|m| = |h'/h| = |s'/s|$. If the image is real, it will be inverted. Remember, the height of the inverted image is considered negative. Hence $h'/h$ will be negative, whereas $s'$ will be positive—the ratios have the opposite signs. To account for this, we add a negative sign. Thus, $m = (h'/h) = -(s'/s)$.

**Linear magnification** **$m$ is the ratio of the image height $h'$ and object height $h$ and can be determined using the negative ratio of the image distance $s'$ and object distance $s$:**

$$m = \frac{h'}{h} = -\frac{s'}{s} \tag{22.4}$$

**Remember that a height is positive for an upright image or object and negative for an inverted image or object. Also note that $m$ is a unitless physical quantity.**

**Figure 22.11** To determine the linear magnification, first (a) use rays 1 and 2 to find the image. Then (b) use ray 3 reflecting at the center of the mirror to help relate $h'$ and $h$ to $s'$ and $s$. Notice the similar triangles used to write the relation $\tan\theta = (h/s) = |h'/s'|$.

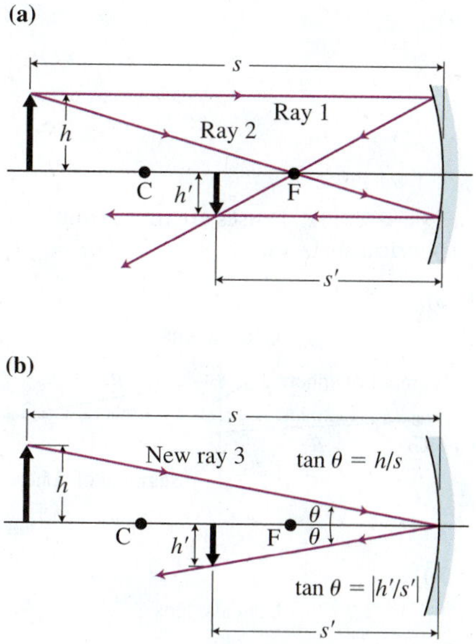

**EXAMPLE 22.6 Your face in a mirror**

You use a concave mirror with a radius of curvature of 0.32 m for putting on makeup or shaving. When your face is 0.08 m from the mirror, what are the image size and magnification of a 0.0030-m-diameter birthmark on your face?

**Sketch and translate** Sketch a ray diagram, such as the one shown below. The givens are $h = 0.30$ cm, $s = 0.08$ m, and $f = (R/2) = +0.16$ m. We need to determine the linear magnification $m$ and then the diameter $h'$ of the birthmark image.

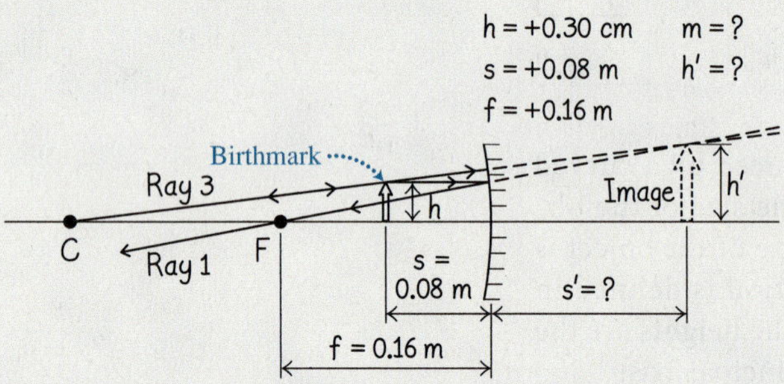


**Simplify and diagram** Represent the birthmark as an arrow. The reflected rays diverge. Extend the reflected rays back through the mirror to find where they intersect. This is the place where reflected light seems to come from and is the location of the image of the arrow's tip. The image is enlarged, upright, and virtual. We do not use ray 2 from Reasoning Skill 22.1 in this diagram because it does not hit the mirror.

**Represent mathematically** First use Eq. (22.2) to determine the image distance $s'$. Then use Eq. (22.4) to find the magnification $m$ and the image height $h'$.

**Solve and evaluate** Rearranging Eq. (22.2), we find

$$\frac{1}{s'} = \frac{1}{f} - \frac{1}{s} = \frac{1}{(0.16\text{ m})} - \frac{1}{(0.08\text{ m})} = -6.25\text{ m}^{-1}$$

or $s' = (1/-6.25\text{ m})^{-1} = -0.16$ m. Use Eq. (22.4) to determine the linear magnification:

$$m = -\frac{s'}{s} = -\frac{(-0.16\text{ m})}{(0.08\text{ m})} = +2.0$$

The magnification is also $m = (h'/h)$, so

$$h' = mh = (+2.0)(0.30\text{ cm}) = +0.60\text{ cm}$$

The birthmark image is upright and two times bigger than the object.

**Try it yourself:** You hold a 1.0-cm-tall coin 0.20 m from a concave mirror with focal length +0.60 m. Is the image of the coin upright or inverted, and what are its magnification and height?

*Answer:* The image is upright; $m = 1.5$ and $h' = 1.5$ cm.

**Review Question 22.3** You place a concave mirror on a stand so it is vertical and slowly move a candle closer to the mirror from a relatively large distance away. What does your friend need to do to make an image appear on a paper screen as you move the candle closer to the mirror?

**TIP** Notice that the physical size of the mirror does not enter into the mirror equation and does not affect the magnification. However, its focal length (and therefore its radius of curvature) does. If you cut the same mirror in half, the size of the image will not change.

**Figure 22.12** Lenses are made from spherical surfaces.

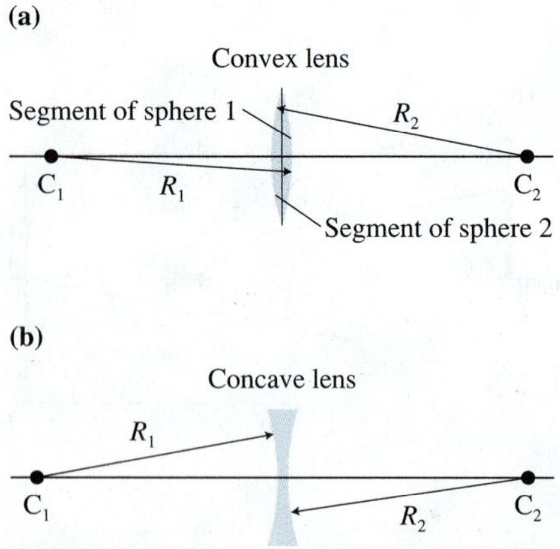


## 22.4 Qualitative analysis of lenses

Flat and curved mirrors allow us to create magnified images of objects using the reflection of light. We can also create images using a lens. A **lens** is a piece of glass or other transparent material with two curved surfaces that produces images of objects by changing the direction of light through refraction. Lenses, like curved mirrors, can be concave or convex. A convex lens (**Figure 22.12a**) is thicker in the middle than at its edges; it consists of two sections of spheres faced with the convex parts pointing out. A concave lens (Figure 22.12b) is thinner in the middle than at its edges; it consists of two sections of spheres faced with the convex part pointing inward. In this book we only investigate lenses in which the two surfaces are identical sections of spheres. However, there are practical uses for lenses with one curved surface and one flat, or with two differently curved surfaces. We will start with convex lenses, as they are easier to analyze.

## Convex lenses

In Observational Experiment **Table 22.5**, we show the paths of several parallel laser beams passing through a convex lens made of solid glass. Remember that the index of refraction of glass is greater than that of air. So when a ray of light moves from air to glass, it bends toward the normal line; when the ray leaves the lens (from glass to air), it bends away from the normal line. The overall result is that the ray bends toward the principal axis.

**OBSERVATIONAL EXPERIMENT TABLE**

**22.5 Laser beams passing through a convex lens.**

| Observational experiment | Analysis |
|---|---|
| **Experiment 1.** We shine three laser beams parallel to the principal axis of a convex glass lens. 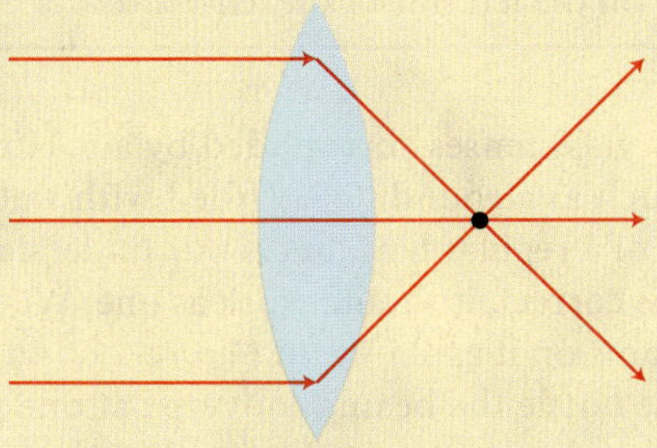 | After passing through the lens, the rays pass through the same point on the principal axis. We use the law of refraction to analyze the paths of the rays. Ray 2 is perpendicular to the boundary of the two media and thus does not bend. Rays 1 and 3 bend toward the normal at the air-glass interface and away from the normal at the glass-air interface. 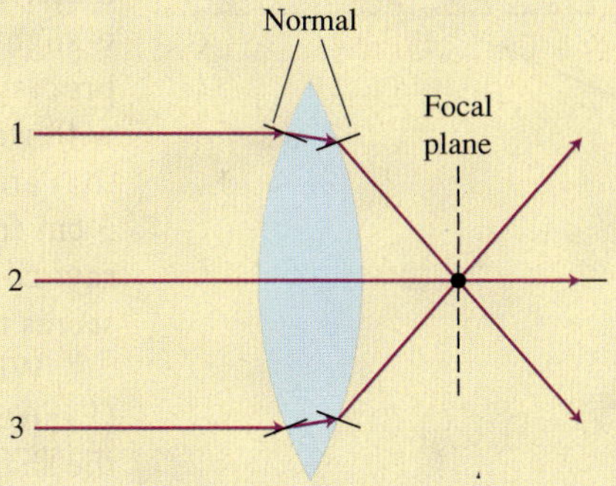  |
| **Experiment 2.** We shine three laser beams parallel to each other at an arbitrary angle relative to the principal axis of the lens. 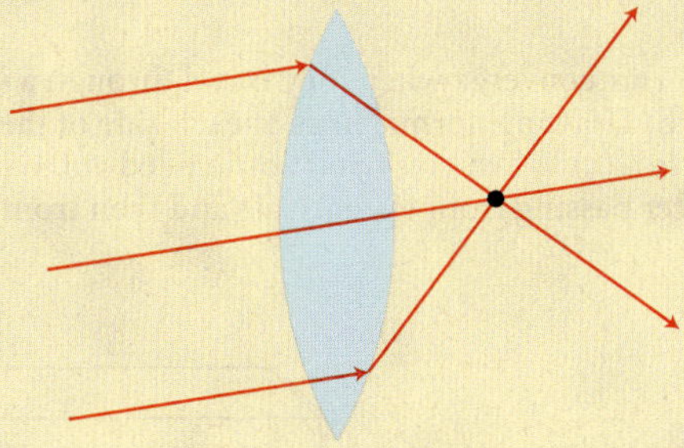 | The rays pass through a point on the plane perpendicular to the principal axis that passes through the focal point found above. We use the law of refraction to analyze the paths of the rays. Although ray 2 is not perpendicular to the boundary, it passes straight through the lens—a small bending toward the normal on the air-glass interface is cancelled by the small bending away from the normal at the glass-air interface 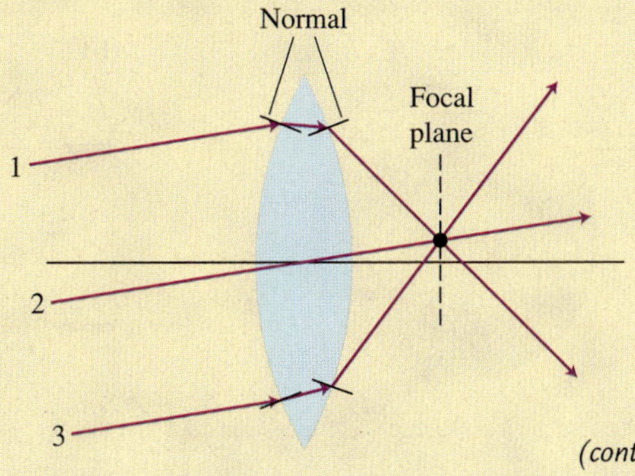  |

*(continued)*

**Patterns**

1. In all experiments the rays passing through the center of the lens do not bend.
2. The rays parallel to the principal axis pass through the same point after the lens. We call this point the focal point, similar to the focal point of a curved mirror.
3. The rays parallel to each other pass through a point on the plane perpendicular to the principal axis through the focal point. We call this plane the focal plane.

**Figure 22.13** Thick and thin convex lenses.

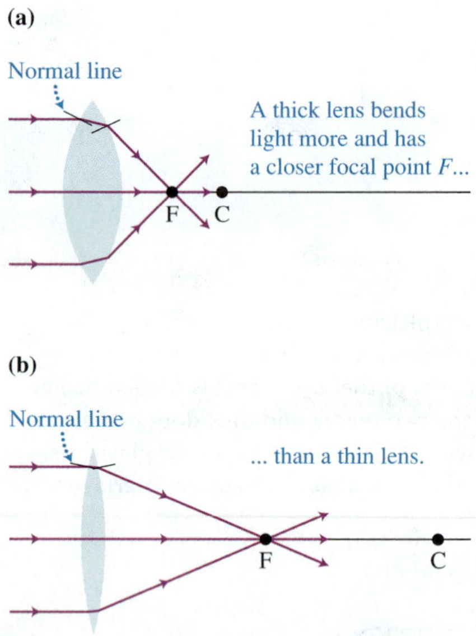


We have found that a convex lens made of glass is similar to a concave mirror where incident rays parallel to the principal axis intersect at a **focal point** after passing through the lens. Note in **Figure 22.13** that the rays converge more steeply when passing through a lens with highly curved surfaces than when passing through a lens with less curvature. The focal point is closer to a lens with more curvature than for a lens with less curvature.

**TIP** A lens with two symmetrical curved surfaces has two focal points at equal distances on each side of the lens. In all of our experiments we shine light from the left side of the lens. As a result, the focal point is on the right side of the lens. If we repeat the experiments, shining the light from the right, the rays will converge on the left side of the lens.

So far we have used only glass lenses surrounded by air. However, other materials can act as lenses, such as a round bottle filled with water. Its shape is slightly different from that of a regular lens, but if our understanding of the processes that involve lenses is correct, it should work as one. We fill the bottle with water and shine laser beams on it as shown in **Figure 22.14a**. We observe that after passing through the bottle the beams converge at one point (about 5 cm from the bottle) and then diverge again. Figure 22.14b shows why the rays converge (notice the direction of the normal at each interface). The bottle seems to work as a convex lens with a focal length of about 5 cm.

What happens if we place an empty bottle in a container filled with water (Figure 22.14c)? The laser beams diverge. This test shows us that it is not only the shape of the lens but also the refractive index of the material from which it is made and the index of the surrounding material that determine whether the lens is converging or diverging. In addition, the focal length of a lens depends not only on the radius of curvature of the surfaces but also on the refractive indexes of the material between the surfaces and the outside matter. The closer

**Figure 22.14** (a) Parallel rays in air converge when they travel through a spherically shaped bottle filled with water. (b) Drawing normal lines at each side of the bottle explains the paths of the rays in (a). (c) When you put a bottle filled with air into a tank of water, parallel rays diverge after passing from water to air and then from air to water.

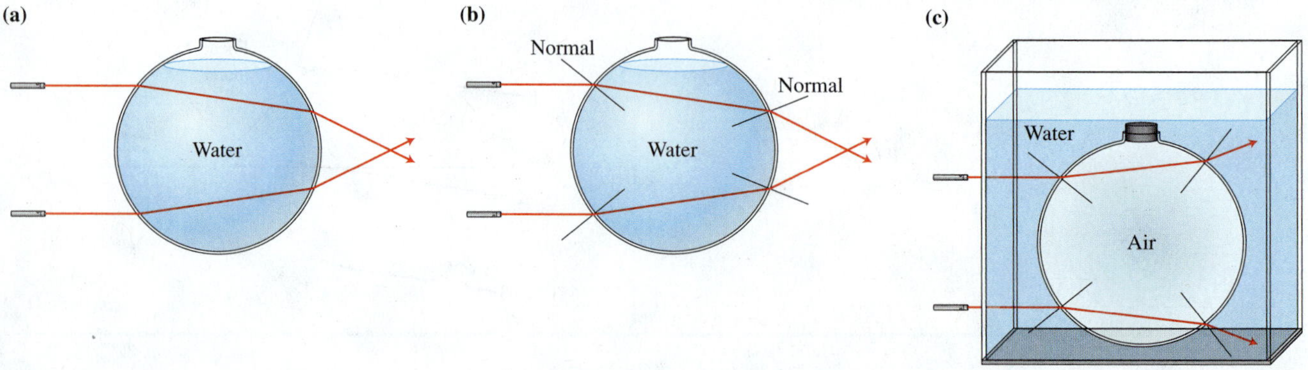

the refractive indexes are to each other, the less the lens bends light rays and therefore the longer its focal length.

## Concave lenses

We can perform an experiment comparable to Observational Experiment 22.5, sending parallel laser beams through a concave glass lens (**Figure 22.15**). Laser beams parallel to the principal axis diverge after passing through the concave lens. The dashed lines extended back reveal that these beams seem to diverge from a single point on the axis. This point is called the **virtual focal point**.

Lenses are used in magnifying glasses, cameras, eyeglasses, microscopes, telescopes, and many other devices. Some of these devices produce an image that is smaller than the object (camera), and some produce an image that is larger than the object (magnifying glass). Lens ray diagrams help us understand how these devices work. The method for constructing ray diagrams for lenses is summarized in Reasoning Skill 22.3.

**Figure 22.15** The focal point of a concave lens.

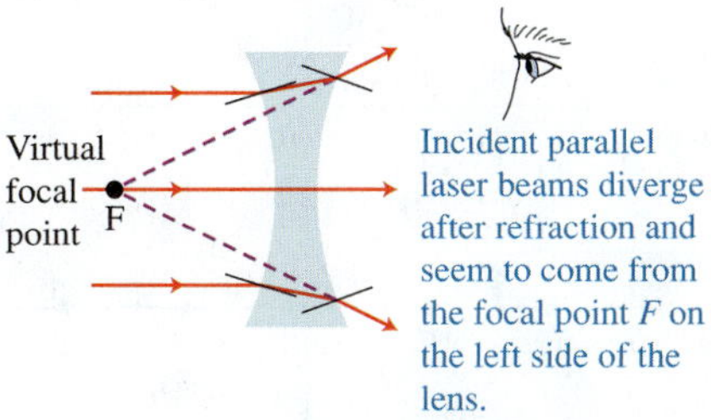


### REASONING SKILL 22.3 Constructing ray diagrams for single-lens situations

Convex lens

1. Draw the principal axis for the lens and a vertical line perpendicular to the axis representing the location of the lens. Label the focal points on each side of the lens.

2. Place an arrow at the object position with its base on the axis a distance $s$ from the center of the lens.

3. *Convex lenses* Draw two or three rays.
Ray 1 moves parallel to the principal axis and after refraction through the lens passes through the focal point on the right side of the lens.
Ray 2 passes directly through the middle of the lens and its direction does not change.
Ray 3 passes through the focal point on the left and refracts through the lens so that it moves parallel to the axis on the right.

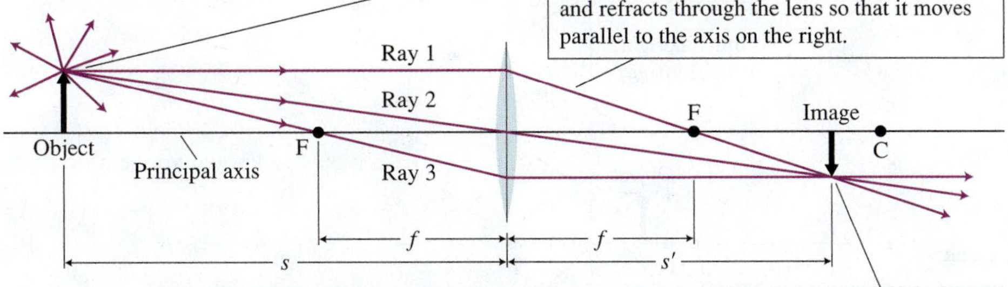


4. The place where the rays intersect on the right side at a distance $s'$ from the lens is the location of a real image of the object.

Concave lens

Steps 1 and 2 are the same as above.
3. *Concave lenses* Draw two or three rays.
Ray 1 moves parallel to the principal axis and is refracted away from the principal axis. Ray 1 appears to come from the focal point on the left side of the lens.
Ray 2 passes directly through the middle of the lens and for thin lenses its direction does not change.
Ray 3 is refracted before it reaches the focal point on the right. The ray moves parallel to the axis on the right.

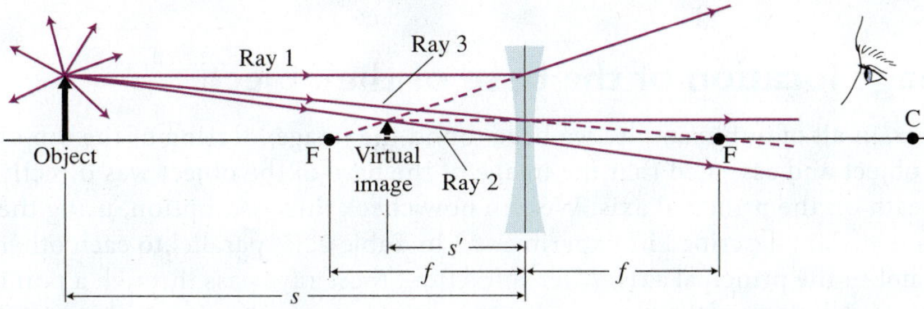


4. If the rays diverge after passing through the lens, the place from which they seem to diverge is on the left side at distance $s'$ from the lens at the location of a virtual image.

**Table 22.6 Ray diagrams for various lenses.**

| Situation | Ray diagram and description of the image | Application |
|---|---|---|
| $s > 2f$ 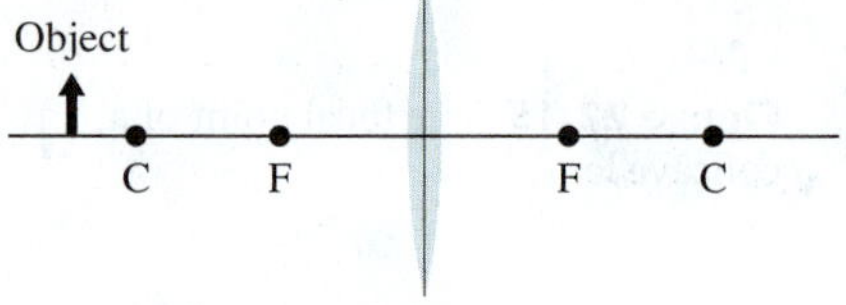  | 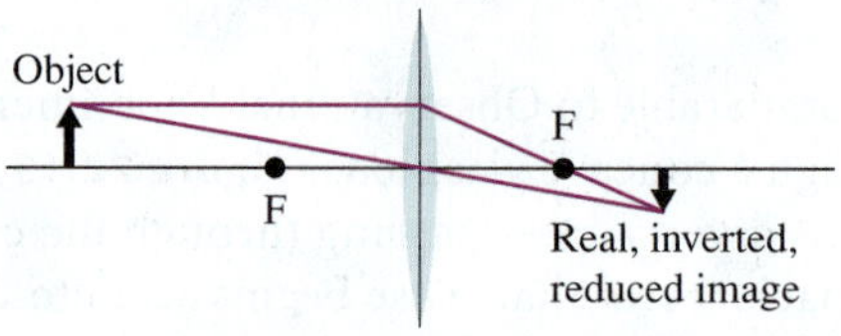  | Camera, human eye |
| $2f > s > f$ 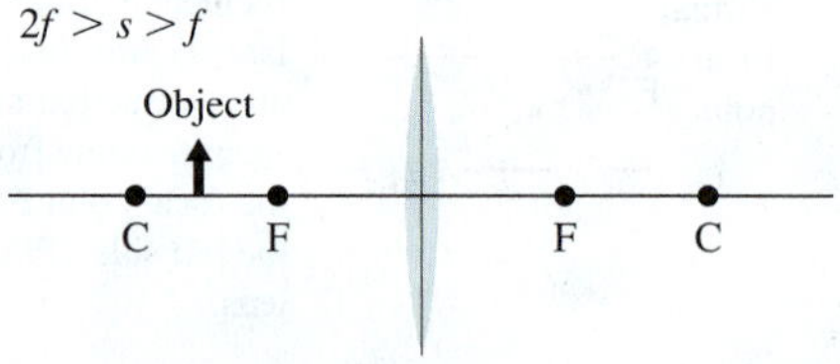  | 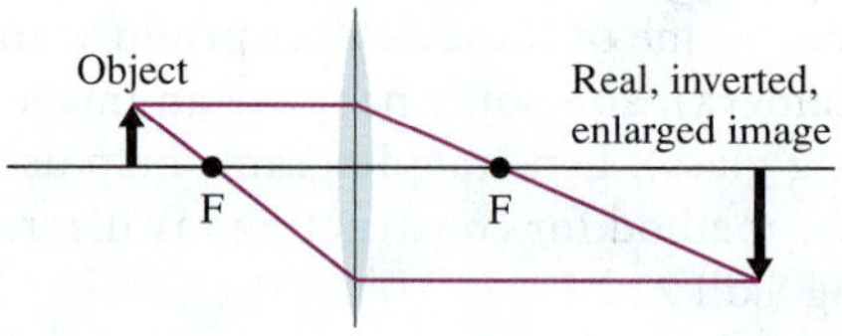  | Digital projector |
| $f > s$ 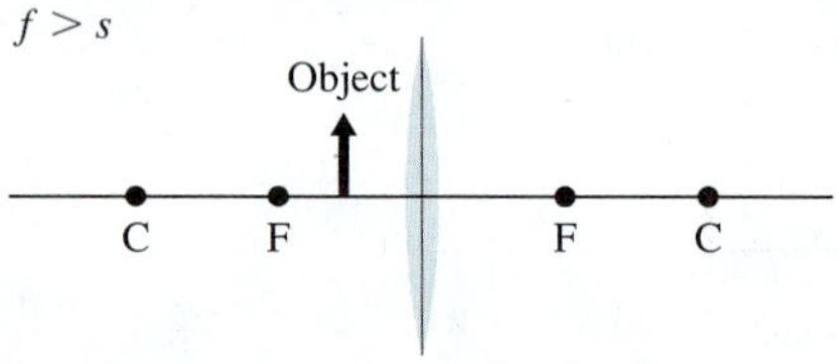  | 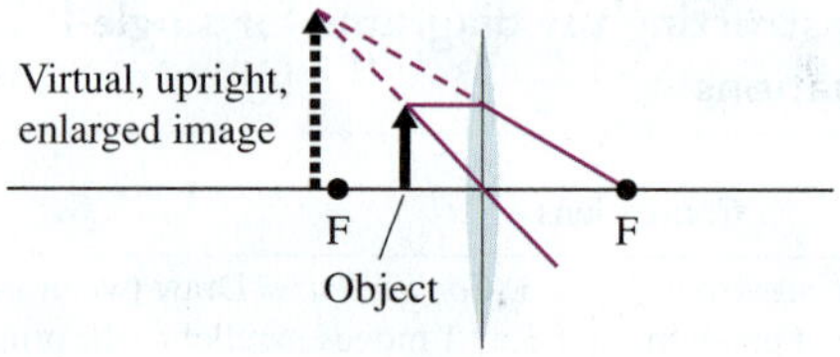  | Magnifying glass |
| $s > f$ 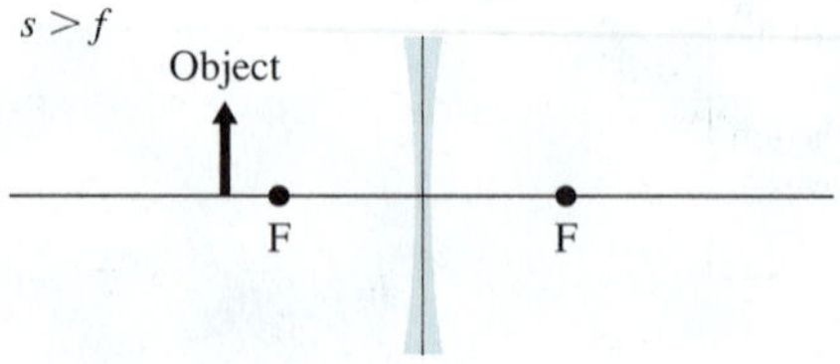  | 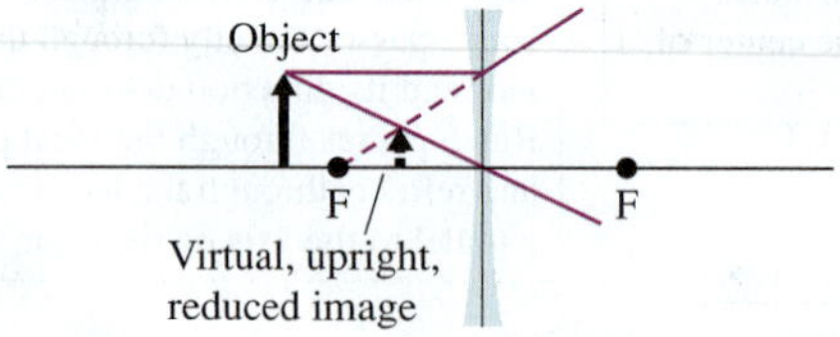  | Glasses for seeing distant objects |
| $f > s$ 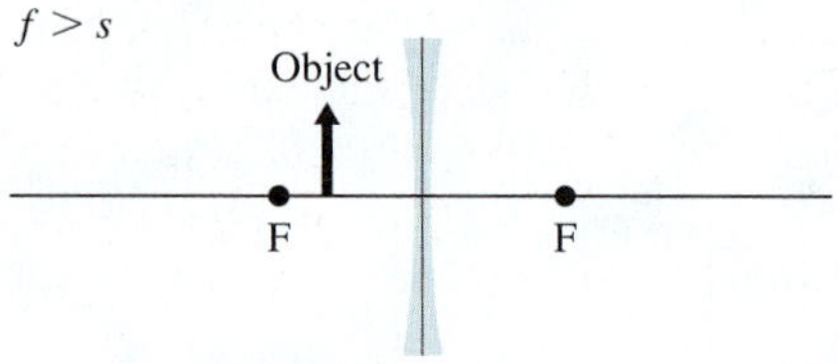  | 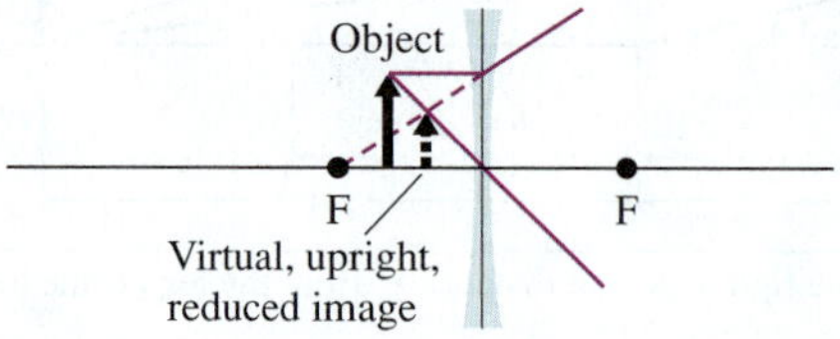  | No known applications |

**Figure 22.16** Parallel rays pass though the same point in the focal plane.

The path of each ray can be explained by refraction relative to the normal lines to each air-glass interface. The rays converge at the same distance from the lens as rays that are parallel to the principal axis.

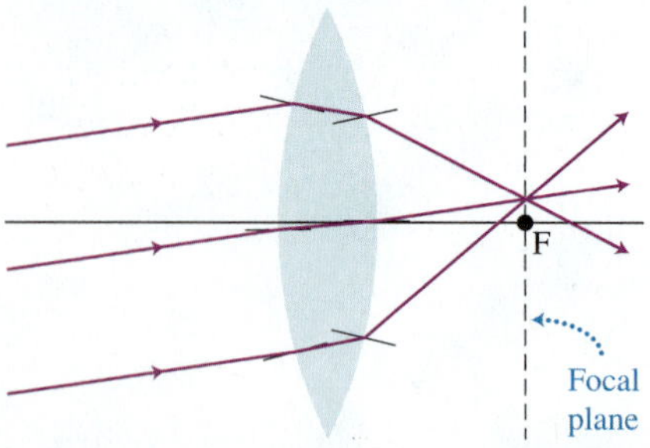


**Table 22.6** summarizes possible combinations of the locations of objects and images produced by different types of lenses. The right-hand column lists common applications for each lens type. For example, in a camera we need an image that is reduced and real. In a digital projector we need an image that is magnified and real. In a magnifying glass we want an upright enlarged virtual image.

## Image location of the base of the object

So far, in all of our examples we have found the image location of the top of the object and assumed that the image of the base of the object was directly beneath on the principal axis. We will now check this assumption, using the rays that were described in Experiment 2 in Table 22.5, parallel to each other but not to the principal axis. After refraction, these rays pass through a point on the focal plane of the lens—a plane perpendicular to the principal axis and passing through the focal point (**Figure 22.16**).

**Figure 22.17** Locating the image of the base of an object.

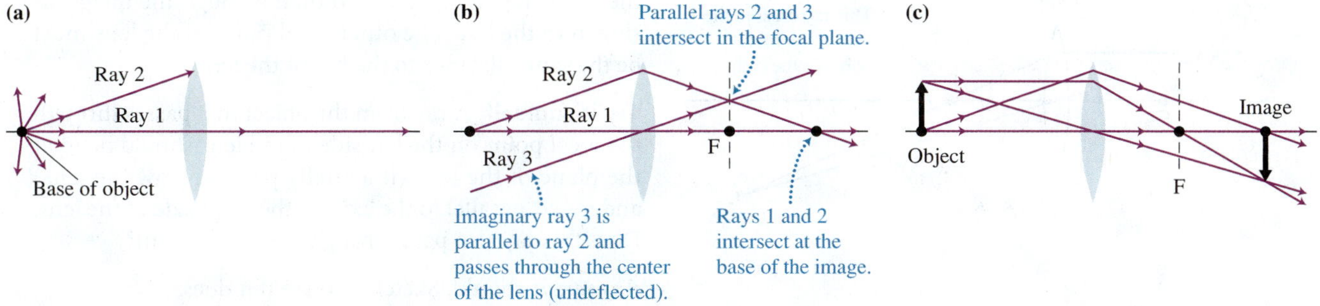


Now let's draw a ray diagram for the base of an object that lies on the principal axis (**Figure 22.17a**). An infinite number of rays leave the base of the object. Ray 1, parallel to the principal axis, passes through the center of the lens and does not bend. This means that the image of the base point should be on the principal axis—but where? We can draw an imaginary ray 3 (Figure 22.17b) parallel to ray 2 passing through the center of the lens (undeflected). After leaving the lens, *parallel* rays 2 and 3 pass through the same point in the focal plane.

The image of the object appears where bent ray 2 intersects ray 1 on the principal axis (Figure 22.17b). Now we can draw the image of the arrow using the top and the bottom points (Figure 22.17c).

**CONCEPTUAL EXERCISE 22.7 Lens Jeopardy**

The image of a shining point-like object S is produced by a lens (see figure below). The line is the principal axis of the lens. Where is the lens, what kind of lens is it, and where are its focal points?

**Sketch and translate** Let's first decide the location of the lens that produced the image S′. We know that the shining point S sends light in all directions. Rays that reach the lens bend and pass through the image point S′. We need to find rays with paths that will help us to find that lens and its focal points.

**Simplify and diagram** A ray that passes through the center of the lens does not bend. This ray also passes through the image point. Thus, we draw a straight line from the object to the image. The point where the line crosses the principal axis will be the location of the center of the lens, as shown below. The lens must be convex since the image is on the opposite side of the lens from the object.

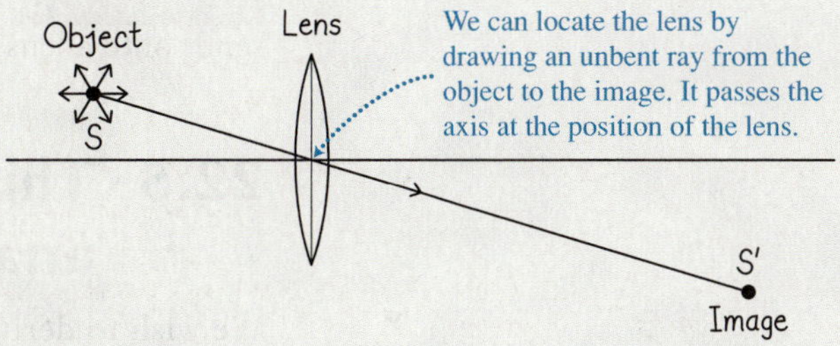

Object
S
Lens
We can locate the lens by drawing an unbent ray from the object to the image. It passes the axis at the position of the lens.
S′
Image

To locate a focal point, draw a ray from the object and parallel to the principal axis. This ray will refract through the lens and pass through the focal point on

*(continued)*

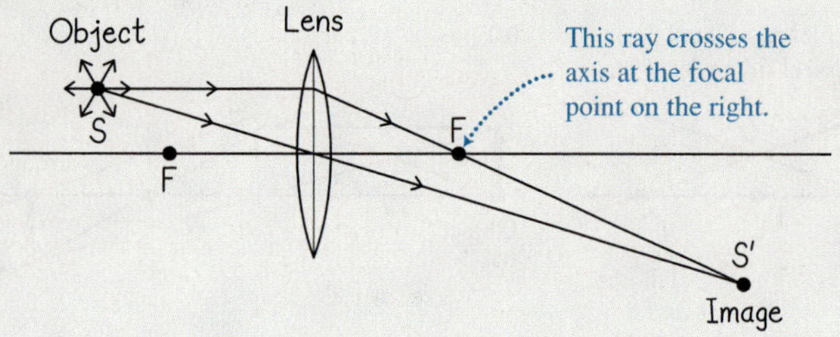


the right side of the lens and then through the image, as shown to the left. The other focal point of the lens must be the same distance to the left of the lens.

**Try it yourself:** A ray from the object that passes through the focal point on the left side of the lens should bend at the plane of the lens (it actually passes below the lens) and move parallel to the axis on the right side of the lens. Does this ray also pass through the image point?

*Answer:* It should. Sketch it to see if it does.

**CONCEPTUAL EXERCISE 22.8 A partially covered lens**

Imagine that you have an object, a lens, and a screen. You place an object at a position $s > 2f$ from the lens and cover the top half of the lens. Half of the object is above the principal axis, half below. What part of the image will you see on the screen: the top or the bottom?

**Sketch and translate** To answer the question, we need to draw a ray diagram and see what happens to the image with the top part of the lens covered.

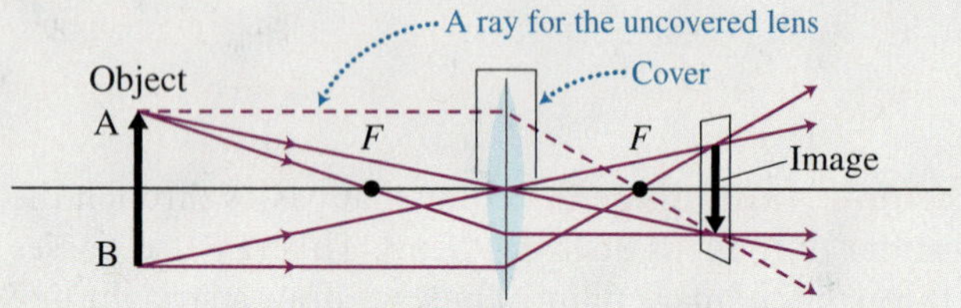


**Simplify and diagram** We represent the object with an arrow and find the image by examining rays radiated from points A and B (the top and the bottom of the object). With the top of the lens covered, the rays moving toward the top of the lens do not pass through, but rays moving toward the bottom half do. Remember that *all* rays from a point on the object that pass through the lens pass through the corresponding image point. Covering the top half of the lens still allows some light from all points on the object to reach the image location, so the entire image will form. However, the image brightness will decrease as less light passes through the lens.

**Try it yourself:** Imagine that you cover the central part of the lens. You have a shining object far away from the lens. How will the location of the image and its size change compared to the situation when not covered?

*Answer:* The size and location will not change, but the brightness will decrease.

**Review Question 22.4** How do we know how many rays an object sends onto a lens?

## 22.5 Thin lens equation and quantitative analysis of lenses

We wish to derive a relation between $s$, $s'$, and $f$ that can be used to make a quantitative prediction for the location and properties of the image formed by a lens. To do this we use the same technique that was used for curved mirrors. We use a convex lens in the derivation, but the results apply to concave lenses as well. As we did with mirrors, we will assume that the radii of curvature of the spheres from which the sides of the lens were cut are much larger than the size of the lens. Such a lens is called a **thin lens.** Both convex and concave lenses can be thin. Let's look at the geometry of an object at a known distance from a convex lens and its image. AB is the object and $A_1B_1$ is the image (**Figure 22.18**).

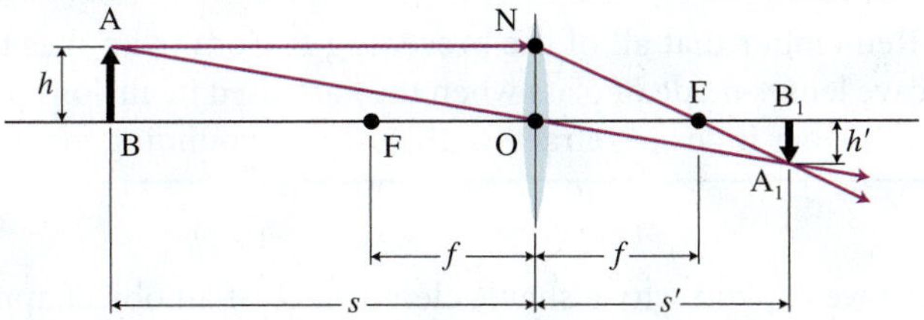


**Figure 22.18** We can use a ray diagram to develop a relation between the location of an object, the image, and the focal distance of a lens.

- We assume that the lens is very thin and that all of the rays are close to the principal axis. Triangles BAO and $B_1A_1O$ are similar triangles. Thus $(AB/s) = (A_1B_1/s')$. After rearranging we get

$$\frac{A_1B_1}{AB} = \frac{s'}{s}$$

- Triangles NOF and $A_1B_1F$ are similar. Thus $(NO/f) = (A_1B_1/(s' - f))$. Also, NO = AB. Thus,

$$\frac{AB}{f} = \frac{A_1B_1}{s' - f}$$

- Rearranging this last equation, we get

$$\frac{A_1B_1}{AB} = \frac{s' - f}{f}$$

- Setting the above two equations with $(A_1B_1/AB)$ on the left equal to each other, we get

$$\frac{s'}{s} = \frac{s' - f}{f} \tag{22.5}$$

Using algebra similar to what we used to derive the mirror equation, we get a relationship that is called the **thin lens equation**:

$$\frac{1}{s} + \frac{1}{s'} = \frac{1}{f} \tag{22.6}$$

Similar to the mirror equation, the thin lens equation will help us predict the location of the image when we know the location of the object and the focal length of the lens.

Let's see if this equation works for extreme cases. Imagine an object that is located at the focal point. Rays from this object will refract through the lens so that they are parallel. Since they neither converge nor diverge, no image will form. For an object at a focal point $(s = f)$:

$$\frac{1}{s} + \frac{1}{s'} = \frac{1}{f} \text{ or } \frac{1}{f} + \frac{1}{s'} = \frac{1}{f}$$

Therefore, $(1/s') = 0$ or $s' = \infty$. Saying that the image forms infinitely far from the lens is equivalent to saying that no image forms. The formation of an image by a lens is summarized below, including sign conventions for lenses.

**Thin lenses** **The distance $s$ of an object from a lens, the distance $s'$ of the image from the lens, and the focal length $f$ of the lens are related by the thin lens equation:**

$$\frac{1}{s} + \frac{1}{s'} = \frac{1}{f} \tag{22.6}$$

**Several sign conventions are important when using the thin lens equation.**

1. **The focal length $f$ is positive for convex lenses and negative for concave lenses.**
2. **The image distance $s'$ is positive for real images and negative for virtual images.**

**Figure 22.19** The lens produces a sharp image only at $s'$.

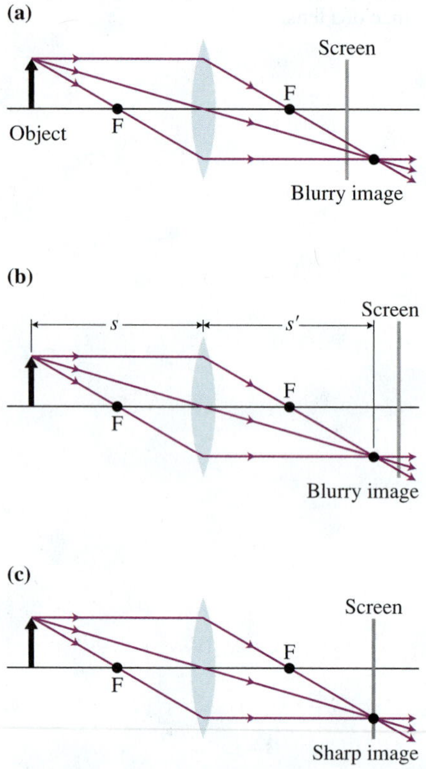


> **TIP** Remember that all of the preceding reasoning applies to convex and concave lenses made of glass when they are used in air (or made of any material of higher index of refraction than the surrounding medium).

How can we explain why a sharp, clear image of an object appears in only one place for a particular lens? If we place a screen closer to the lens than distance $s'$ (**Figure 22.19a**), then the rays leaving the tip of the arrow have not yet converged to a single point. As a result, the light from the tip of the arrow is spread out on the screen and the image will be blurry. If we place the screen at a distance beyond $s'$ (Figure 22.19b), where rays from the tip are diverging after having converged at $s'$, we again get a blurry image. Only at distance $s'$ from the lens do we see a sharp, clear image (Figure 22.19c). Being able to locate the position where a sharp image will appear is important for all applications involving optical instruments.

## Digital projectors

A digital projector creates an enlarged real image of a small object inside the projector, an image that can be seen on an external screen. How do these devices work? Here we focus on their optical components.

### EXAMPLE 22.9 Digital projector

An object (a small display of the desired image inside the projector) is 0.20 m from a +0.19-m focal length lens. Where should we place the external screen in order to get a focused image?

**Sketch and translate** See the labeled sketch of the situation, shown below. The goal is to determine the correct distance $s'$ of the screen from the projector lens.

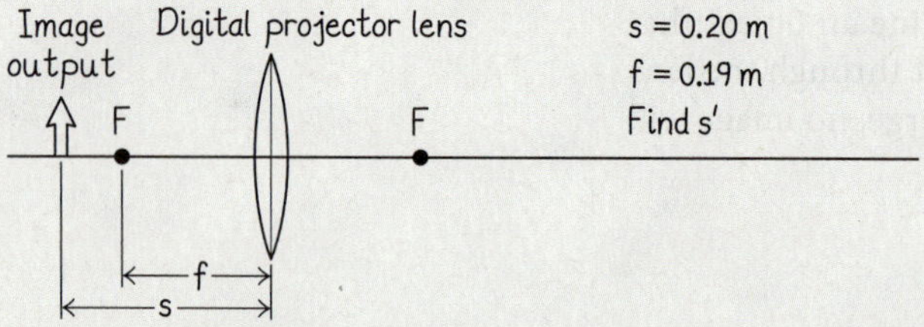


**Simplify and diagram** Draw a ray diagram for the situation, such as the one shown below. Notice that the image is enlarged, real, and inverted relative to the object. We must invert what is shown on the internal display to obtain an upright image.

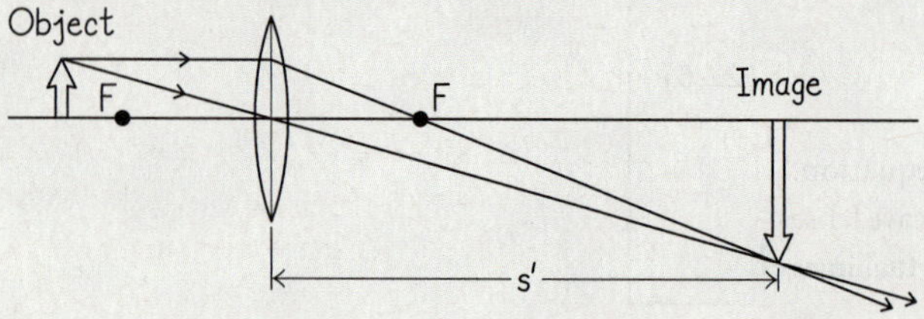


**Represent mathematically** We use the thin lens equation [Eq. (22.6)] to determine the unknown image distance $s'$:

$$\frac{1}{s} + \frac{1}{s'} = \frac{1}{f}$$

**Solve and evaluate** Solve for $s'$ and substitute the known quantities to get

$$\frac{1}{s'} = \frac{1}{f} - \frac{1}{s} = \frac{1}{(0.19\text{ m})} - \frac{1}{(0.20\text{ m})}$$

$$= \frac{0.20\text{ m} - 0.19\text{ m}}{(0.19\text{ m})(0.20\text{ m})} = 0.263\text{ m}^{-1}$$

Inverting, we get

$$s' = \frac{1}{(0.263\text{ m}^{-1})} = 3.8\text{ m}$$

The screen should be placed 3.8 m from the lens—a reasonable distance for a projector in a small conference room.

**Try it yourself:** Suppose the screen needs to be 4.8 m from the lens. What distance should the small display inside the projector be from the same lens to produce a focused image on the more distant screen?

*Answer:* $s = 0.198$ m (the lens and internal display need to be moved closer to each other).

**TIP** Forgetting to invert when solving for $s'$ is a common mistake.

## Linear magnification in lenses

Just as with mirrors, lenses can produce images that are bigger or smaller in size than the original objects. To characterize the relative sizes of the object and the image produced by a lens, we use the same physical quantity for magnification that we used for curved mirrors:

$$\text{linear magnification} = \frac{\text{image height}}{\text{object height}} = m = \frac{h'}{h} \qquad (22.3)$$

where the heights $h'$ and $h$ are positive if the image or object is upright and negative if inverted. As with mirrors, one can relate the linear magnification to the image distance $s'$ and the object distance $s$. Using the similar triangles in Figure 22.18, we can write

$$\text{linear magnification} = \frac{\text{image height}}{\text{object height}} = m = \frac{h'}{h} = -\frac{s'}{s} \qquad (22.4)$$

**EXAMPLE 22.10 Looking at an insect through a magnifying glass**

You use a convex lens of focal length +10.0 cm to look at a tiny insect on a book page. The lens is 5.0 cm from the paper. Where is the image of the insect? If the insect is 1.0 cm in size, how large is the image?

**Sketch and translate** Draw a labeled sketch of the situation, as shown.

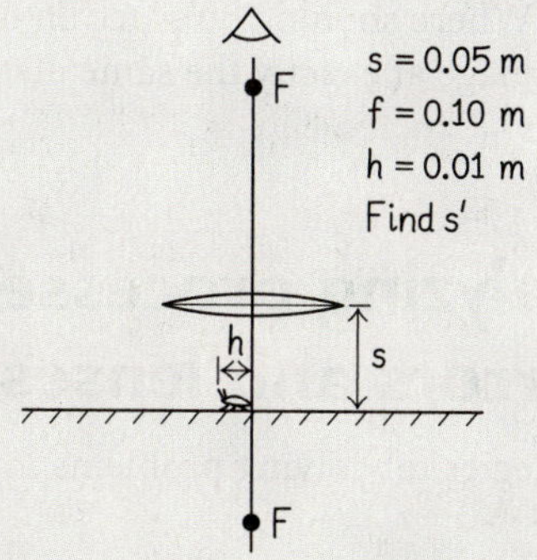


**Simplify and diagram** Assume that the lens is parallel to the paper looking down at the insect. For convenience we will draw the ray diagram horizontally, as shown below. From the ray diagram, we conclude that the image is virtual and is on the same side of the lens as the real insect. It is farther away, upright, and enlarged. This means that the image is *beneath* the book where the insect lies. This is where the light reaching your eye appears to originate.

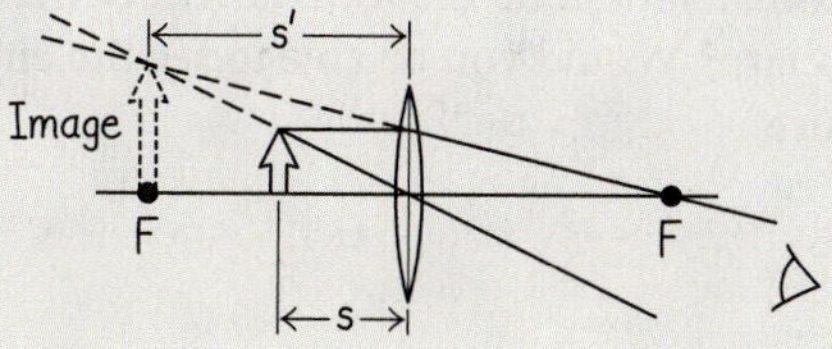


**Represent mathematically** Use the thin lens equation [Eq. (22.6)] to find the image distance $s'$:

$$\frac{1}{s} + \frac{1}{s'} = \frac{1}{f}$$

Then use the linear magnification equation [Eq. (22.4)] to determine the magnification and the height of the insect's image:

$$m = \frac{h'}{h} = -\frac{s'}{s}$$

**Solve and evaluate** Insert $f = 0.10$ m and $s = 0.05$ m into the lens equation to get

$$\frac{1}{s'} = \frac{1}{f} - \frac{1}{s} = \frac{1}{0.10\text{ m}} - \frac{1}{0.05\text{ m}} = -10\text{ m}^{-1}$$

Therefore, $s' = (1/-10\text{ m})^{-1} = -0.10$ m. The minus sign indicates that the image is virtual, consistent with the ray diagram. The linear magnification is

$$m = -\frac{s'}{s} = -\frac{(-0.10\text{ m})}{(+0.05\text{ m})} = +2.0$$

The image is upright (the plus sign) and twice the object size or $2.0(1.0\text{ cm}) = 2.0$ cm, consistent with the ray diagram.

**Try it yourself:** An image seen through a 10-cm focal length convex lens is exactly the same size as the object but inverted and on the opposite side of the lens. Where must the object and image be located?

*Answer:* The object must be $s = 2f = 20$ cm from the lens on one side, while the image must be $s' = 2f = 20$ cm from the lens on the other side.

**EXAMPLE 22.11 Find the image**

You place an object 20 cm to the left of a concave lens whose focal length is $-10$ cm (the negative sign indicates a concave lens). Where is the image located? Is it real or virtual?

**Sketch and translate** A labeled sketch of the situation is shown below.

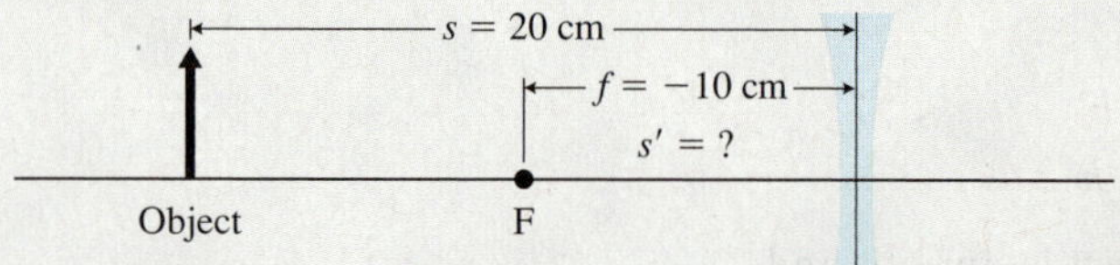


**Simplify and diagram** Draw a ray diagram for the situation (see below). The image is smaller than the object, virtual, and closer to the lens than the object. So, in your quantitative results, the image distance should be negative and the absolute value should be less than 10 cm.

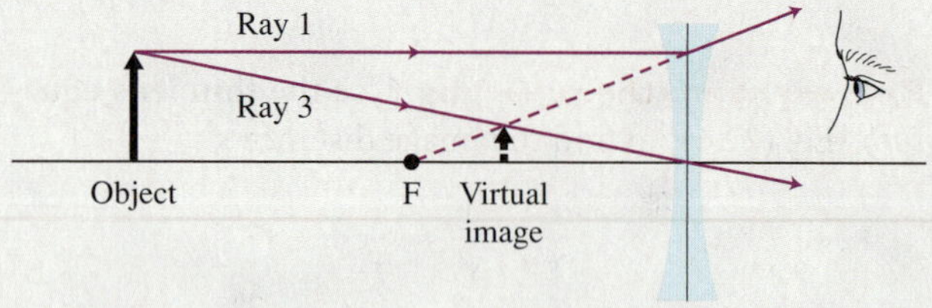


**Represent mathematically** We can use the thin lens equation [Eq. (22.6)] to determine the image location:

$$\frac{1}{s} + \frac{1}{s'} = \frac{1}{f}$$

**Solve and evaluate** Insert $f = -0.10$ m and $s = 0.20$ m into the above to get

$$\frac{1}{s'} = \frac{1}{f} - \frac{1}{s} = \frac{1}{-0.10\text{ m}} - \frac{1}{0.20\text{ m}} = -15\text{ m}^{-1}$$

or $s' = (1/-15\text{ m})^{-1} = -0.067$ m or $-6.7$ cm, consistent with the ray diagram.

**Try it yourself:** Where will the image be if you place an object exactly at the focal point of a concave lens?

*Answer:* The image will be at a distance $f/2$, on the same side of the lens as the object.

**Review Question 22.5** Where should you place an object with respect to a convex lens to have an image at exactly the same distance from the lens? What size will the image be?

## 22.6 Skills for analyzing processes involving mirrors and lenses

Example 22.12 illustrates the strategies for solving problems involving mirrors and lenses.

**PROBLEM-SOLVING STRATEGY Processes involving mirrors and lenses**

**Figure 22.20** A single-lens camera made in the late 1800s.

**EXAMPLE 22.12 A camera**

A camera made in the 1880s consisted of a single lens and a light-sensitive film placed at the image location (**Figure 22.20**). To focus light on the image, the photographer would change the image distance—the distance from the lens to the film. Imagine that the image distance is 20 cm and that the film is a 16 cm × 16 cm square. A 1.9-m-tall person stands 8.0 m from the camera. What should be the focal length of the camera's lens for the image to be sharp? Would you be able to see the entire body of the person in the picture?

**Sketch and translate**

- Sketch the situation in the problem statement.
- Include the known information and the desired unknown(s) in the sketch.

The situation is sketched below:

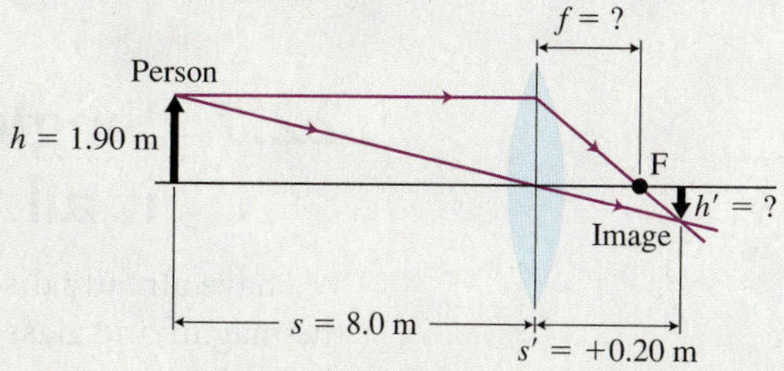


- Known quantities:

  object distance $s = 8.0$ m, image distance $s' = +0.20$ m, object size $h = 1.90$ m.

- Unknowns:

  focal length $f$ of the lens and the image height $h'$ compared to the 0.16-m film size.

**Simplify and diagram**

- Assume the lens/mirror is only slightly curved and the rays are incident near the principal axis.
- Draw a ray diagram representing the situation in the problem.

- A ray diagram with the person represented as a shining arrow is included in the sketch.
- The object distance is $s > 2f$ so the image is real, inverted, and smaller than the object.

**Represent mathematically**

- Use the picture and ray diagram to help construct a mathematical description of the situation.

- Use the lens equation to find the focal length:

$$\frac{1}{f} = \frac{1}{s} + \frac{1}{s'}$$

- Use the linear magnification equation [Eq. (22.4)] to find the image height:

$$h' = mh = -\left(\frac{s'}{s}\right)h$$

**Solve and evaluate**

- Solve the equations for an unknown quantity.
- Evaluate the results to see if they are reasonable (the magnitude of the answer, its units, how the answer changes in limiting cases, and so forth).

$$\frac{1}{f} = \frac{1}{s} + \frac{1}{s'} = \frac{1}{8.0\text{ m}} + \frac{1}{0.20\text{ m}}$$

$$= 0.125\text{ m}^{-1} + 5.0\text{ m}^{-1} = 5.125\text{ m}^{-1}$$

$$f = \frac{1}{5.125\text{ m}^{-1}} = 0.20\text{ m}$$

The image size is

$$h' = -\left(\frac{s'}{s}\right)h = -\left(\frac{0.20\text{ m}}{8.0\text{ m}}\right)(1.9\text{ m}) = -0.048\text{ m}$$

The inverted image size is 5 cm and will easily fit on the film.

**Try it yourself:** Suppose the 1.9-m-tall person stands 4.0 m from the +0.20-m focal length lens. Now where is the image formed relative to the lens, and what is the image height? Will you get a good picture? Explain.

*Answer:* $s' = 0.21$ m from the lens, $m = -0.053$ (the image is inverted), and $h' = -0.10$ m. The image distance is larger than 0.20 m; thus, the image will not be sharp on the screen that is 0.20 m away from the lens.

**Review Question 22.6** If we have a mathematical equation for lenses and mirrors, what is the purpose of drawing ray diagrams when solving lens and mirror problems?

## 22.7 Single-lens optical systems: Putting it all together

We have already discussed several applications for mirrors and lenses, from the magnifying glass to the digital projector. In this section we will investigate three single-lens optical systems: cameras, the human eye, and two kinds of corrective lenses.

### Photography and cameras

**Figure 22.21** The physics of a simple camera. The focal distance of the lens is fixed, but the image location can be varied.

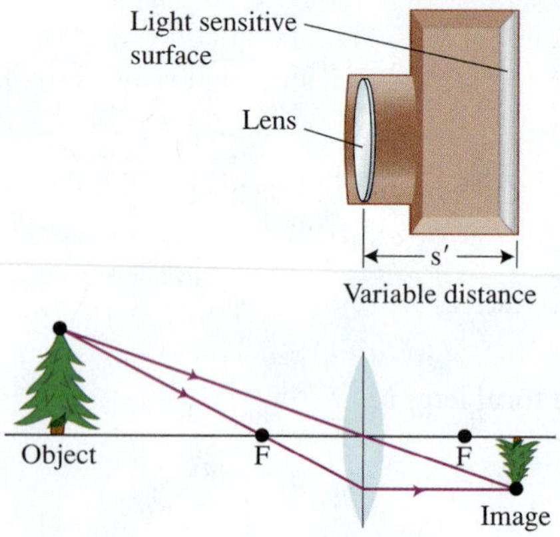


A simple camera (**Figure 22.21**) has a lens of fixed focal length. Light from an object enters the camera through the lens, which focuses the light on a surface that has light-sensitive properties. In analog cameras this surface is film; in digital cameras the film has been replaced by an image sensor that consists of a two-dimensional array of light-sensitive elements—pixels. There are two types of image sensors: CCD (charge-coupled devices) and CMOS (complementary metal oxide semiconductors). To produce sharp images of objects located at different distances from the lens, the lens is moved relative to the film or the image sensor. When the picture includes multiple objects at different distances, you have to choose the object to focus on when you take the photo. But what if you could take the picture and decide what object you wanted in focus later?

A new camera provides that capability using an idea called **light field photography**. In light field photography, the image sensor records *all* the light entering the camera, not just what would produce a sharp image on the focal plane. In fact, a light field camera doesn't have a conventional lens at all (**Figure 22.22**). Instead, a two-dimensional array of tiny micro-lenses is placed just in front of the image sensor. Behind each micro-lens is its own personal portion of the image sensor, its own grid of pixels. Since each micro-lens has its own grid of pixels, the camera records not only how much light of each color reaches each micro-lens but also the direction of each ray reaching that micro-lens. Using this camera, a photographer can choose an object to focus on after the picture has been taken, since the camera is effectively focusing on all objects at once.

**Figure 22.22** A light field camera allows you to choose which object to focus on after you have taken the picture.

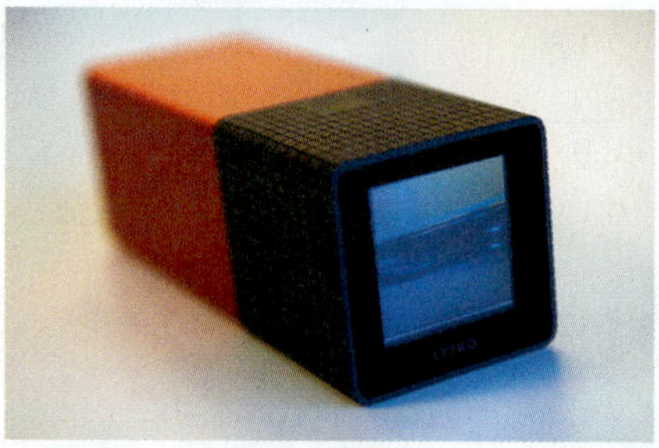

### Optics of the human eye

In many ways, the human eye resembles an expensive digital video camera. It is equipped with a built-in cleaning and lubricating system, an exposure meter, an automatic field finder, and about 130 million photosensitive elements that continuously send electric signals to the brain (the equivalent of a 130-megapixel digital camera CCD) in real time. Light from an object enters the **cornea** (**Figure 22.23**) and passes through a transparent lens. An **iris** in front of the lens widens or narrows, like the aperture on a camera that regulates the amount of light entering the device. The **retina** plays the role of the film.

**Figure 22.23** The human eye.

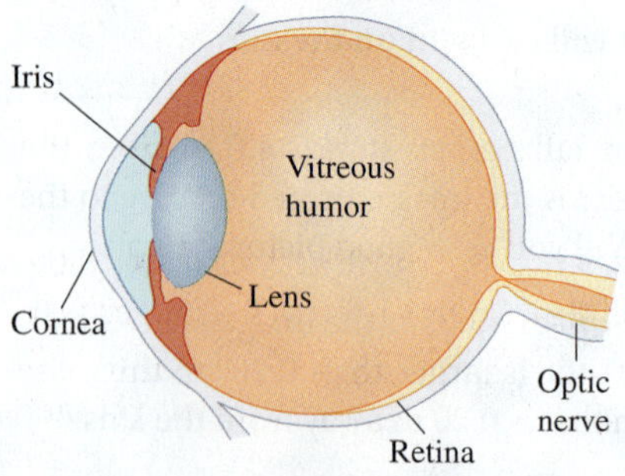


A normal human eye can produce sharp images of objects located anywhere from about 10–25 cm to hundreds of kilometers away. Unlike a camera, which moves a fixed focal length lens to accommodate different object distances, the eye has a fixed image distance of about 2.1 cm (the distance from the lens to the retina) and a variable focal length lens system.

**Figure 22.24** As the image distance is fixed in the eye, the lens changes shape to change the focal length, so that the image of any object is always on the retina.

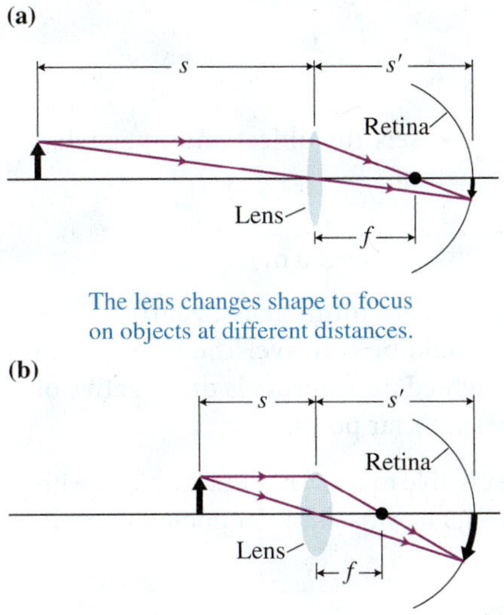
(a)
s
s′
Retina
Lens
f
The lens changes shape to focus on objects at different distances.
(b)
s
s′
Retina
Lens
f

The changing focal length of the eye's lens is illustrated in **Figure 22.24**. When the eye looks at distant objects, muscles around the lens of the eye relax, and the lens becomes less curved (Figure 22.24a), increasing the focal length and allowing an image to form on the retina. As the object moves closer, the corresponding image moves behind the retina, appearing blurry. To compensate, the eye muscles contract, increasing the curvature of the lens, reducing the focal length (Figure 22.24b) and moving the image back onto the retina. The contraction of these muscles causes your eyes to become tired after reading for many hours.

The **far point** is the most distant point at which the eye can form a sharp image of an object on the retina with the eye muscles relaxed. The **near point** of the eye is the nearest an object can be to the eye and still have a sharp image produced on the retina with the eye muscles tensed. For the normal human eye, the far point is effectively at infinity (we can focus on the Moon and on distant stars) and the near point is at approximately 25 cm from the eye.

When you swim with your eyes open, objects look blurry. If you wear a mask, however, you can see objects clearly. Why can't the eye produce sharp images on the retina underwater? What is the role of the mask or goggles? The answer is that the lens is not the only optical element in the eye. The moist, curved surface of the eye (the cornea) acts as a sort of lens as well. Under normal conditions, the cornea works with the rest of the eye to form a sharp image on the retina. Without the cornea, your eye cannot bend the light passing through it to make a sharp image on your retina.

In water, the cornea stops acting as a lens because its refractive index is about the same as the refractive index of water. Therefore, the eye cannot produce sharp images on the retina and everything looks blurry. When you wear a mask, you put air between the eye and the water, and your cornea can operate as a lens again.

**Figure 22.25** Correcting myopia (nearsightedness): (a) uncorrected; (b) corrected; (c) glasses help "move" the distant object to the eye's far point.

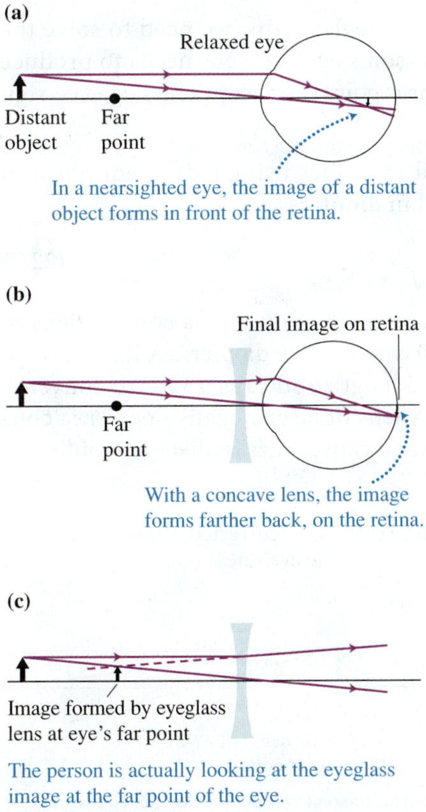
(a)
Relaxed eye
Distant object
Far point
In a nearsighted eye, the image of a distant object forms in front of the retina.
(b)
Final image on retina
Far point
With a concave lens, the image forms farther back, on the retina.
(c)
Image formed by eyeglass lens at eye's far point
The person is actually looking at the eyeglass image at the far point of the eye.

## Corrective lenses

Corrective lenses compensate for the inability of the eye to produce a sharp image on the retina. The two most common vision abnormalities corrected with lenses are **myopia** (nearsightedness) and **hyperopia** (farsightedness).

**Myopia** A person with myopia, or nearsightedness, can see close-up objects with clarity but not those that are distant. The far point of a nearsighted person may be only a few meters away rather than at infinity.

Myopia is caused by a larger-than-normal eyeball (greater than 2.5 cm in diameter) or a lens that is too curved. In such cases the image of a distant object is formed in front of the retina (**Figure 22.25a**) even when the eye muscles are relaxed. Placing a concave lens in front of the eye corrects nearsighted vision by causing rays from an object to diverge, so that when they pass through the eye lens, an image is formed farther back in the eye (Figure 22.25b). If an object is very distant (that is, $s = \infty$), then the focal length of the concave lens is chosen so that its virtual image is formed at the far point of the eye (Figure 22.25c). Light passing through the eyeglass lens appears to come from the image at the far point, not from the more distant object, which allows the eye to produce an image on the retina. To calculate the desired focal length of the eyeglass lens, we set $s = \infty$ and $s' = -(\text{far point distance})$. The negative sign accounts for the fact that the image produced by the eyeglass is virtual. We can then calculate the value of the desired focal length $f$ of the concave lens using the thin lens equation.

**Hyperopia** Unlike nearsighted people, hyperopic (farsighted) people are able to produce sharp images of distant objects on the retina but cannot do it for nearby objects, such as a book or a cell phone screen. Whereas the normal eye has a near point at about 25 cm, a farsighted person may have a near point several meters from the eye. If the object is closer than the person's near point, the image is

**EXAMPLE 22.13 Glasses for a nearsighted person**

Alex is nearsighted and his far point is 2.0 m from his eyes. What should be the focal length of his eyeglass lens so that he can focus on a very distant object ($s = \infty$)?

**Sketch and translate** A sketch of the eyeglass lens is shown in Figure 22.25b. The image of an object at infinity is to be formed at the far point 2.0 m in front of the lens.

**Simplify and diagram** Assume that the distance between the eyeglass lens and the eye lens is small compared to 2.0 m. The ray diagrams in Figure 22.25c represent the optics of the eyeglass lens. Without glasses the image of a distant object is formed before the light reaches his retina and appears blurred. When Alex wears glasses, the object at infinity ($s = \infty$) has an image 2.0 m to the left of the eyeglass lens ($s' = -2.0\text{ m}$).

**Represent mathematically** We can use the given values of $s$ and $s'$ in the lens equation [Eq. (22.6)] to determine the focal length of the eyeglass lenses:

$$\frac{1}{s} + \frac{1}{s'} = \frac{1}{f}$$

**Solve and evaluate** Insert the object and image distances in the above equation:

$$\frac{1}{f} = \frac{1}{\infty} + \frac{1}{(-2.0\text{ m})}$$

Therefore, $f = -2.0$ m. The number is negative. This means that the lens should be concave. The focal length of the lens for a nearsighted person equals the negative of the distance to the person's far point.

**Try it yourself:** If Alex is able to see very distant objects with glasses of focal length –1.5 m, what is his far point distance?

*Answer:* 1.5 m.

formed behind the retina (**Figure 22.26a**) and appears blurred. Farsightedness may occur if the diameter of a person's eyeball is smaller than normal or if the lens is unable to curve sufficiently when the surrounding muscles contract.

Convex lenses are used to correct farsighted vision. The eyeglass lens slightly converges light from a nearby object so that the image produced by the eye lens moves forward onto the retina (Figure 22.26b). The lens produces a virtual image of the object at or beyond the person's near point (Figure 22.26c).

To determine the focal length of the eyeglass lens, we need to solve the thin lens equation for $f$. A farsighted person's eyeglass lens needs to produce an image at their near point, $s' = -(\text{near point distance})$. The negative sign means that the image is virtual.

**TIP** Optical power $P$ is not related in any way to energy conversion rate $P$, which we learned about in earlier chapters.

**Active Learning Guide ›**

**Optical power** An optometrist prescribes glasses using a different physical quantity—the *optical power P*, measured in **diopters**:

$$P = \frac{1}{f} \quad (22.7)$$

where $f$ is measured in meters. For example, the power of a concave lens of focal length $f = -50$ cm is $P = 1/(-50\text{ cm}) = -2.0$ diopters. A lens of large positive power has a small positive focal length and causes rays to converge rapidly after passing through the lens. A lens of large negative power is concave. It has a small negative focal length and causes rays to diverge rapidly.

**Figure 22.26** Correcting hyperopia (farsightedness): (a) uncorrected; (b) corrected; (c) glasses help "move" the closely positioned object to the eye's near point.

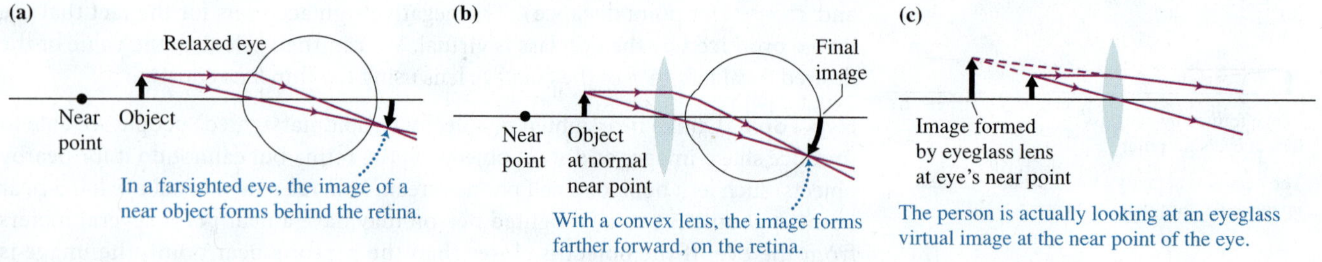
(a)
Relaxed eye
Near point
Object
In a farsighted eye, the image of a near object forms behind the retina.
(b)
Final image
Near point
Object at normal near point
With a convex lens, the image forms farther forward, on the retina.
(c)
Image formed by eyeglass lens at eye's near point
The person is actually looking at an eyeglass virtual image at the near point of the eye.

**EXAMPLE 22.14 Glasses for a farsighted person**

Eugenia is farsighted with a near point of 1.5 m. If you were an optometrist, what glasses would you prescribe for her so she can read a book held 25 cm from her eyes?

**Sketch and translate** Eugenia holds a book 0.25 m away (see Figure 22.26b) and wants her glasses to make her eyes think it is 1.5 m away at her near point. Thus $s = 0.25\text{ m}$ and $s' = -1.5\text{ m}$ for her glasses (on the same side of the lens as the book). We need to find their focal length $f$ and power $P$.

**Simplify and diagram** A ray diagram for the optics of the eyeglass system is shown in Figure 22.26c.

**Represent mathematically** Use Eq. (22.6) to find the focal length of the lenses:

$$\frac{1}{s} + \frac{1}{s'} = \frac{1}{f}$$

**Solve and evaluate** Insert the known object and image distances in the previous equation to determine the focal length of the desired glasses:

$$\frac{1}{f} = \frac{1}{0.25\text{ m}} + \frac{1}{-1.50\text{ m}}$$
$$= 4.0\text{ m}^{-1} - 0.67\text{ m}^{-1} = 3.33\text{ m}^{-1}$$

Thus,

$$f = \frac{1}{3.33\text{ m}^{-1}} = 0.30\text{ m}$$

and the power of the glasses is $P = (1/f) = 1/(0.30\text{ m}) = 3.3$ diopters.

**Try it yourself:** Eugenia went to a drugstore to buy glasses, but her only choice was 4.0-diopter glasses. Where will Eugenia need to hold the book when wearing her new glasses so its image is at her 1.5-m near point?

*Answer:* 21 cm.

**Review Question 22.7** What is the main difference between how a camera produces sharp images of objects at different distances and how the human eye does it?

# 22.8 Angular magnification and magnifying glasses

The linear magnification of an optical system compares only the heights of the image and the object, but the apparent size of an object as judged by the eye depends not only on its height but also on its distance from the eye. For example, a pencil held 25 cm from your eye appears longer than one held 100 cm away (**Figure 22.27**). In fact, the pencil may appear longer than a 100-story building if the building is several miles away, even though the pencil is actually much shorter.

The impression of an object's size is quantified by its **angular size** $\theta$, shown in Figure 22.27. The angular size depends on an object's height $h$ and its distance $r$ from the eye:

$$\theta = \frac{h}{r} \quad (22.8)$$

Remember that an angle $\theta$, in radians, is the ratio of an arc length ($h$ in Figure 22.27) and the radius of a circle ($r$ in Figure 22.27: $\theta = (h/r)$). We can use $h$ for the length of the straight line between the corners of the triangle only if $h$ is very close to the actual arc length. This is only true for small angles. Thus Eq. (22.8) is an approximation for small angle $\theta$.

If a person looks at an object through one or more lenses, he or she sees light that appears to come from the final image of the system of lenses. If the person's eye is close to the last lens (as in a typical telescope or microscope), then the angular size $\theta'$ of the image is

$$\theta' = \frac{h'}{s'} \quad (22.9)$$

**Figure 22.27** The angular size $\theta$ of an object.

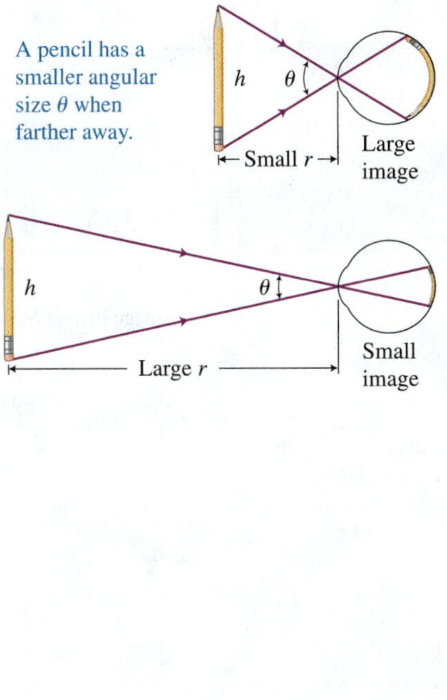

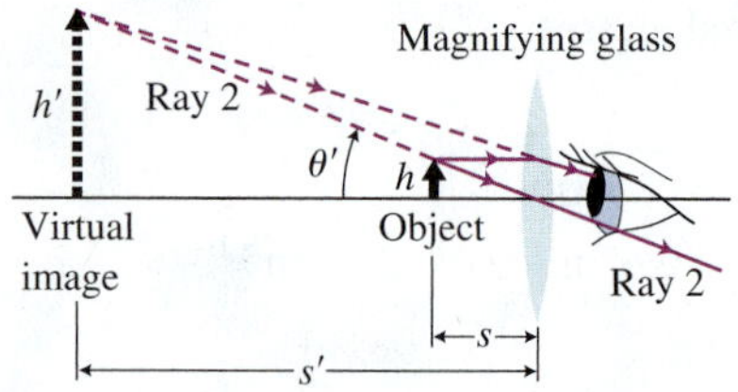


**Figure 22.28** A magnifying glass produces an enlarged virtual image of the closely positioned object.

where $h'$ is the height of the final image and $s'$ is the distance of the final image from the last lens in the system. The **angular magnification** $M$ of an optical system is defined as the ratio of $\theta'$ and $\theta$:

$$M = \frac{\theta'}{\theta} = \frac{\text{Angular size of the final image of optical system}}{\text{Angular size of object as seen by the unaided eye}} \qquad (22.10)$$

A magnifying glass consists of a single convex lens. To use the lens as a magnifying glass, we position the object between its focal point and the lens, as shown in **Figure 22.28**. A magnified virtual image is formed behind the object. The light appears to be coming from this enlarged image, which is farther away from the lens than the object. In other words, when you use a magnifying glass to look at small print on a page on a desk, the image you are viewing is located under the desk!

Note that in Figure 22.28 both the object and the image make the same angle $\theta'$ with the principal axis ($\theta' = (h'/s') = (h/s)$). The object has the same angular size when we look directly at it when it is a distance $s$ from our eye as when we look through the magnifying glass at the image a distance $s'$ from our eye. Using the magnifying glass lets you place the object closer than your near point (and hence have a larger angular size) and use the magnifying glass to focus it clearly on your retina.

To calculate the angular magnification of the magnifying glass, compare the angular size $\theta'$ of the image seen through the magnifying glass and the angular size $\theta$ of the object seen with the unaided eye when the object is as close as possible and still appears clear (at your near point). The angular size of the image viewed through the magnifying glass is $\theta' = (h'/s')$. By considering ray 2 in Figure 22.28, we find that $(h'/s') = (h/s)$. Thus,

$$\theta' = \frac{h}{s}$$

where $h$ is the actual height of the object and $s$ is its distance from the magnifying glass.

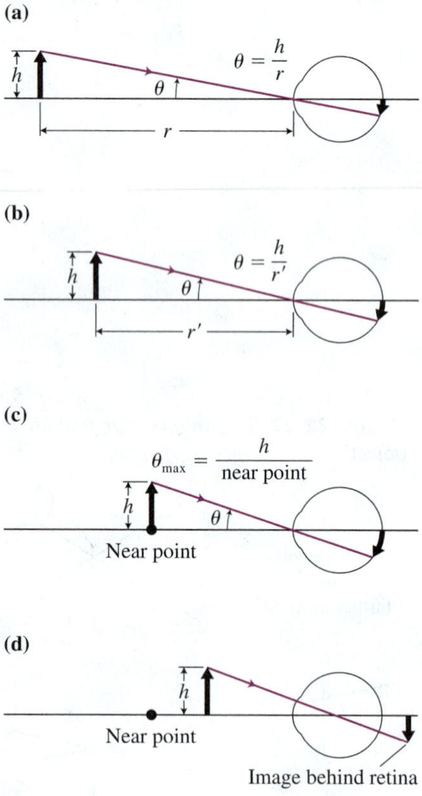


**Figure 22.29** Maximum angular size $\theta_{max}$ of an object whose image is on the retina. If we bring the object closer, the image is behind the retina and what we see is blurred.

Next consider the angular size of the object as seen with the unaided eye. The angular size $\theta$ of an object of height $h$ that we hold a distance $r$ from the eye is $\theta = (h/r)$ (**Figure 22.29a**). As we bring the object closer to the eye (Figure 22.29b and c), its angular size increases. If we bring the object closer than the near point, the image on the retina is blurred (Figure 22.29d). The *maximum* angular size $\theta_{max}$ of an object of height $h$, when viewed by the unaided eye and when focused on the retina, is

$$\theta_{max} = \frac{h}{\text{Near point distance}} \qquad (22.11)$$

The angular magnification of the image is the ratio of the angular size of the image $\theta'$ and the maximum angular size $\theta_{max}$ as seen with the unaided eye. Thus,

$$M = \frac{\theta'}{\theta_{max}} = \frac{h/s}{h/(\text{Near point distance})} = \frac{\text{Near point distance}}{s} \qquad (22.12)$$

**Review Question 22.8** If a person with normal vision and a farsighted person use the same magnifying glass, who will get the greater magnification? Why?

# 22.9 Telescopes and microscopes

Telescopes and microscopes use multiple lenses to produce images that are magnified more than is possible with a single lens. The technique for locating the final image and calculating its magnification requires careful attention to several details—provided below for a system with two lenses. This process applies to both telescopes and microscopes.

To find the final image of a two-lens system:

Step 1: Construct a sketch of the system of lenses (**Figure 22.30a**). The lenses are separated by a distance $d$ and have focal lengths $f_1$ and $f_2$. The object is a distance $s_1$ from lens 1.

Step 2: Now, use Eq. (22.6) to determine the location $s_1'$ of the image formed by the first lens (Figure 22.30b). If $s_1'$ is positive, the image is to the right of lens 1. If $s_1'$ is negative, the image is to the left of lens 1. In our example, $s_1'$ is positive.

Step 3: Next, note that the image of lens 1 is now the object for lens 2. The object distance $s_2$ (Figure 22.30c) is $s_2 = d - s_1'$ where $d$, the separation of the lenses, is a positive number. It is possible for $s_2$ to be negative if the image formed by lens 1 is to the right of lens 2.

Step 4: Finally, use the thin lens equation to determine the image distance $s_2'$ of the image formed by lens 2 (Figure 22.30d):

$$\frac{1}{s_2'} = \frac{1}{f_2} - \frac{1}{s_2}$$

This is the location of the final image relative to the second lens. If $s_2'$ is positive, the final image is real and located to the right of lens 2. If $s_2'$ is negative, the final image is virtual and to the left of lens 2. In our example, the final image is real and to the right of lens 2.

Step 5: The total linear magnification $m$ of the two-lens system equals the product of the linear magnifications of each lens: $m = m_1 m_2$. If $m$ is positive, the final image has the same orientation as the original object. If $m$ is negative, the final image is inverted relative to the original object. Techniques for calculating the total angular magnification $M$ are illustrated in the discussions and examples that follow.

**Figure 22.30** Analyzing a two-lens optical system. The image produced by lens 1 becomes the object for lens 2.

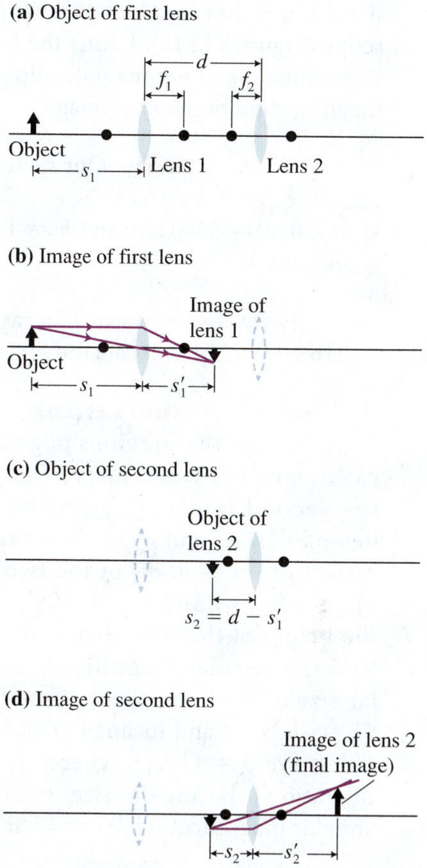


## Telescopes

Galileo is believed to have been the first to study the night sky and the Sun with a telescope. He discovered mountains and craters on the Moon, the moons of Jupiter, stars inside the Milky Way, and sunspots. His telescope consisted of two lenses—one convex and the other concave. Today, a more common version has two convex lenses separated by a distance slightly less than the sum of their focal lengths (**Figure 22.31a**). When you observe a distant object, the first lens produces a real image just beyond its own focal length (Figure 22.31b). The second lens is located so that the image from the first lens is just inside the focal point of the second lens. The image from the first lens becomes the object for the second lens. The second lens produces a magnified inverted virtual image (Figure 22.31c), which we observe by looking through the telescope.

**Figure 22.31** A telescope. (a) A telescope has two lenses separated by a specific distance—nearly the sum of the focal distances of the two lenses; (b) the first lens produces a real image smaller than the object; (c) the second lens produces an enlarged virtual image.

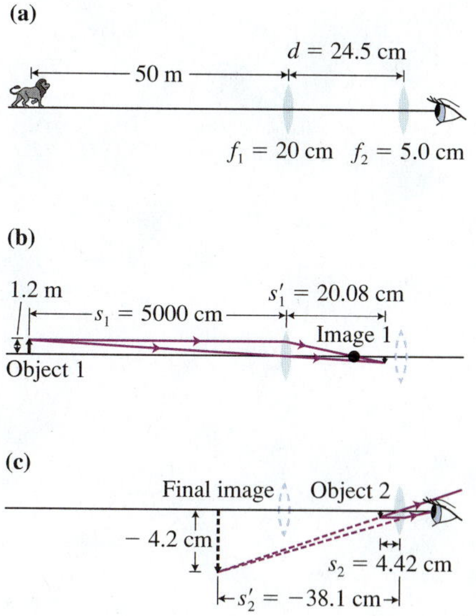

**EXAMPLE 22.15 Looking at a lion with a telescope**

A 1.2-m-tall lion stands 50 m from the first lens of a telescope (Figure 22.31a). Locate the final image of the lion. Determine the linear magnification, height, and angular magnification of the final image.

**Sketch and translate** Our sketch shows that the givens are $f_1 = +20.0\text{ cm}$, $f_2 = +5.0\text{ cm}$, $d = 24.5\text{ cm}$, $s_1 = 50\text{ m} = 5000\text{ cm}$, and $h = 1.2\text{ m}$. We need to find $s_2'$, $m$, and $M$.

**Simplify and diagram** The ray diagrams in Figure 22.31b and c help us locate the first and second images.

**Represent mathematically** Use the procedure described on the previous page to determine the locations of the first image, the second object, and the second image $s_1'$, $s_2$, and $s_2'$. Then use the values of $s_1$, $s_1'$, $s_2$ and $s_2'$ to determine the linear magnification $m = m_1 m_2$ of the two-lens system, where $m_1 = -(s_1'/s_1)$ and $m_2 = -(s_2'/s_2)$. Next, determine the height of the final image of the lion $h' = mh$. To find the angular magnification, compare the angular size of the lion as seen through the optical system $\theta' = (h_2'/s_2')$ and its angular size as seen with the unaided eye $\theta = (h/r_1)$, where $r_1 = s_1 + d$. Then take the ratio of the angular sizes in order to determine the angular magnification: $M = (\theta/\theta')$.

**Solve and evaluate** This procedure gives us results of $s_1' = 20.08\text{ cm}$, $s_2 = 4.42\text{ cm}$, and $s_2' = -38.1\text{ cm}$. Notice that $s_2'$ is negative, indicating a virtual image in agreement with the ray diagram. To determine the linear magnification:

$$m = m_1 m_2 = \left(-\frac{s_1'}{s_1}\right)\left(-\frac{s_2'}{s_2}\right)$$

$$= \frac{(20.08\text{ cm})(-38.1\text{cm})}{(5000\text{ cm})(4.42\text{ cm})} = -0.035$$

Notice that the linear magnification is less than 1; the final image is smaller than the original object. The image height is $h_2' = mh = (-0.035)(1.2\text{ m}) = -0.042\text{ m} = -4.2\text{ cm}$.

Finally, determine the angular magnification $M$:

$$\theta' = \frac{h_2'}{s_2'} = \frac{(-4.2\text{ cm})}{(-38.1\text{ cm})} = 0.11\text{ rad}$$

$$\theta = \frac{h}{s_1} = \frac{(1.2\text{ m})}{(50\text{ m})} = 0.024\text{ rad}$$

$$M = \frac{\theta'}{\theta} = \frac{0.11\text{ rad}}{0.024\text{ rad}} = 4.6$$

Although the image of the lion is smaller than the actual lion, its angular size is almost five times bigger when viewed through the telescope than when seeing it without a telescope from 50 m away. This telescope succeeds in making the lion appear bigger than it really is.

**Try it yourself:** How can the lion appear larger when its final image is smaller?

*Answer:* The final image was much closer than the lion (131 times closer) and thus looked bigger through the telescope, even though its final image was smaller (29 times smaller).

## The Compound Microscope

A compound microscope, like a telescope, has two convex lenses and passes light reflected from the object through two lenses to form the image. A microscope magnifies tiny nearby objects instead of large, distant ones.

Both lenses have relatively short focal lengths and are separated by 10–20 cm (**Figure 22.32a**). The object is placed just outside and very close to the focal point of the first lens (called the **objective** lens). The real, enlarged inverted image produced by the first lens is just inside the focal point of the second lens (called an **eyepiece**), as shown in Figure 22.32b. When an observer looks through the eyepiece, she sees an inverted, enlarged virtual image of the object (Figure 22.32c).

Consider the magnification of such a system. The image produced by the objective lens has linear magnification $m_1 = -(s_1'/s_1)$. The image produced by the objective lens becomes the object for the eyepiece lens. Since this object is located just inside the focal point of the eyepiece lens, the eyepiece lens acts as a magnifying glass used to view the real image produced by the objective lens (Figure 22.32c). The angular magnification of this magnifying glass is, according to Eq. (22.12), $M_2 = (\text{Near point distance}/s_2)$. The total angular magnification of the microscope is

$$M = m_1 M_2 = -\frac{s_1'}{s_1}\frac{\text{Near point distance}}{s_2}$$

**Figure 22.32** Compound microscope. (a) A microscope has two lenses separated by a small distance; (b) the first lens produces an enlarged real image of the closely positioned object; (c) the second lens produces an enlarged virtual image.

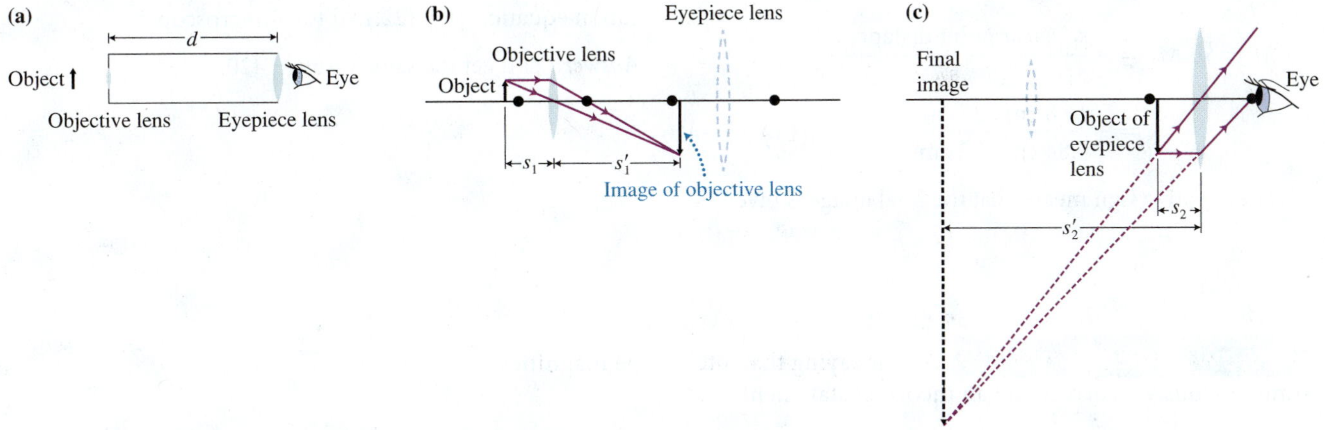


This expression for angular magnification can be rewritten in terms of the focal lengths of the lenses and their separation $d$. Consider **Figure 22.33**. Notice that $s_1 \approx f_1$; $s_1' \approx d - f_2$; and $s_2 \approx f_2$. By substituting these values in the previous equation, we have

$$M = \left(-\frac{d - f_2}{f_1}\right)\left(\frac{\text{Near point distance}}{f_2}\right) \quad (22.13)$$

This expression is easy to use because the magnification depends only on the focal lengths of the lenses, their separation, and the near point distance of the eye. The very best optical microscopes provide angular magnifications of about 1000.

**Figure 22.33** Developing an alternative expression for angular magnification.

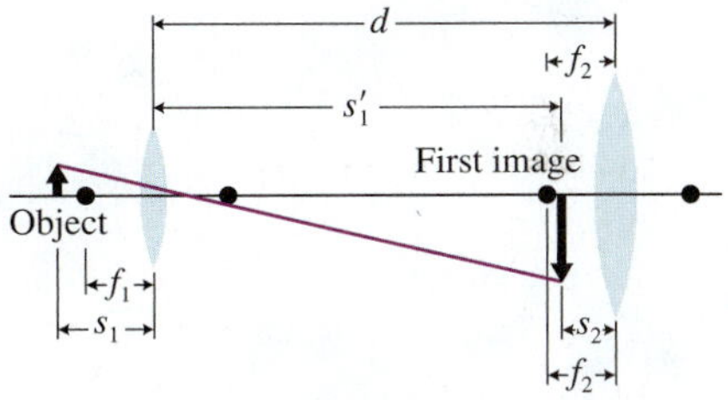


**EXAMPLE 22.16 Seeing a cell with a microscope**

A compound microscope has an objective lens of focal length 0.80 cm and an eyepiece of focal length 1.25 cm. The lenses are separated by 18.0 cm. If a red blood cell is located 0.84 cm in front of the objective lens, where is the final image of the cell, and what is its angular magnification? The viewer's near point is 25 cm.

**Sketch and translate** We have the following information: $f_1 = +0.80$ cm, $f_2 = +1.25$ cm, $d = 18.0$ cm, and $s_1 = +0.84$ cm. We need to find $s_2'$ and the angular magnification $M$ of the microscope. We can use Figure 22.32a as a sketch of the situation.

**Simplify and diagram** Assume that the red blood cell is brightly illuminated and can be modeled as a shining arrow. We use the ray diagrams in Figure 22.32b and c.

**Represent mathematically** First find the image produced by the first lens:

$$\frac{1}{s_1'} = \frac{1}{f_1} - \frac{1}{s_1}$$

This image becomes the object for the second lens. The object distance for the second lens is $s_2 = d - s_1'$. Use this distance to find the location of the image produced by the second lens (the final image):

$$\frac{1}{s_2'} = \frac{1}{f_2} - \frac{1}{s_2}$$

To find the angular magnification $M$ we can use

$$M = m_1 M_2 = -\frac{s_1'}{s_1}\frac{\text{Near point distance}}{s_2}$$

**Solve and evaluate**

$$\frac{1}{s_1'} = \frac{1}{f_1} - \frac{1}{s_1} = \frac{1}{0.80\text{ cm}} - \frac{1}{0.84\text{ cm}} = 0.0595\text{ cm}^{-1}$$

Thus, $s_1' = 1/(0.0595\text{ cm}^{-1}) = +16.8$ cm. The object distance for the second lens is $s_2 = d - s_1' = 18.0\text{ cm} - 16.8\text{ cm} = 1.2$ cm. The final image distance is

$$\frac{1}{s_2'} = \frac{1}{f_2} - \frac{1}{s_2} = \frac{1}{1.25\text{ cm}} - \frac{1}{1.20\text{ cm}} = -0.033\text{ cm}^{-1}$$

*(continued)*

or $s_2' = 1/(0.033\ \text{cm}^{-1}) = -30$ cm. The total angular magnification is

$$M = m_1 M_2 = -\frac{s_1'}{s_1}\frac{\text{Near point distance}}{s_2}$$

$$= -\frac{16.8\ \text{cm}}{0.84\ \text{cm}}\frac{25\ \text{cm}}{1.2\ \text{cm}} = -(20)(21) = -420$$

The negative sign means that the final image is inverted.

**Try it yourself:** How does the result using the magnifications for each lens compare to the approximate magnification equation [Eq. (22.13)] for a microscope?

*Answer:* You get the same result, $-420$.

**Review Question 22.9** Why is saying that a telescope magnifies simultaneously a correct and an incorrect statement?

# Summary

| Words | Pictorial and physical representations | Mathematical representation |
|---|---|---|

**Plane mirror:** A plane mirror produces a virtual image at the same distance behind the mirror as that of the object in front. A **virtual image** is at the position where the paths of the reflected rays seem to originate from behind the mirror. (Section 22.1)

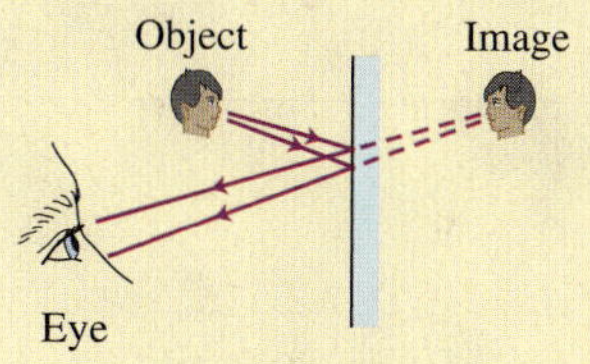


$$s = s'$$

**Concave and convex mirrors**: The distance *s* of an object from the surface of a concave or convex mirror, the distance *s'* of the image from the surface of the mirror, and the focal length *f* of the mirror are related by the mirror equation. The **focal point** *F* is the location through which all incident rays parallel to the principal axis pass after reflection (concave mirrors) or appear to have come from (convex mirrors). The *sign conventions* for these quantities are:

- The **focal length** *f* is positive for concave mirrors and negative for convex mirrors.
- The **image distance** *s'* is positive for real images and negative for virtual images. (Sections 22.2 and 22.3)

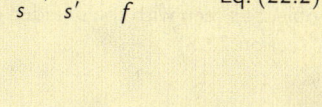

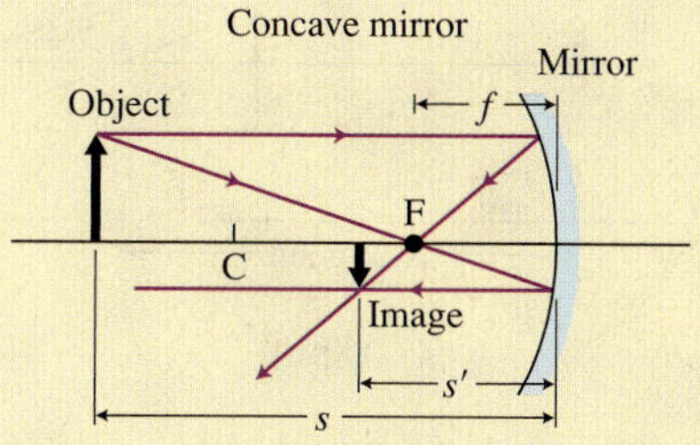


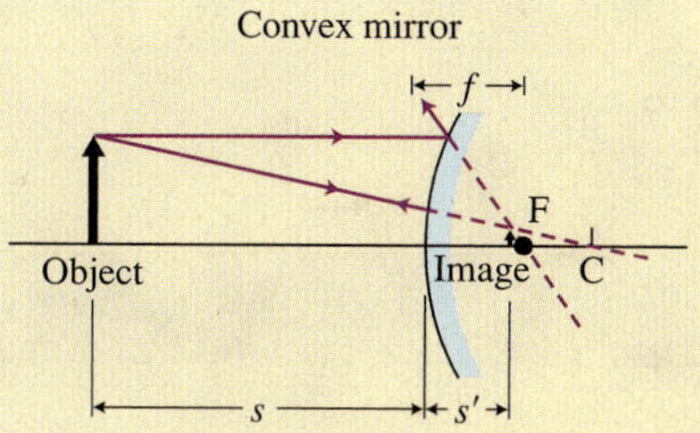


$$\frac{1}{s} + \frac{1}{s'} = \frac{1}{f} \qquad \text{Eq. (22.2)}$$

**Convex and concave lenses**: The distance *s* of an object from a lens, the distance *s'* of the image from the lens, and the focal length *f* of the lens are related by the thin lens equation. The *sign conventions* are:

- The **focal length** *f* is positive for convex lenses and negative for concave lenses.
- The **image distance** *s'* is positive for real images and negative for virtual images. (Sections 22.4 and 22.5)

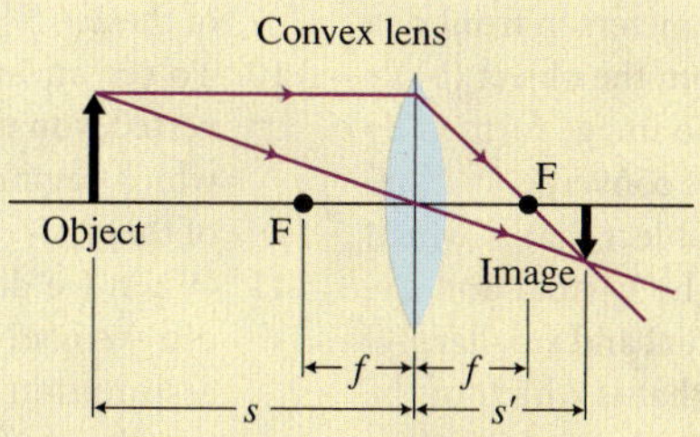


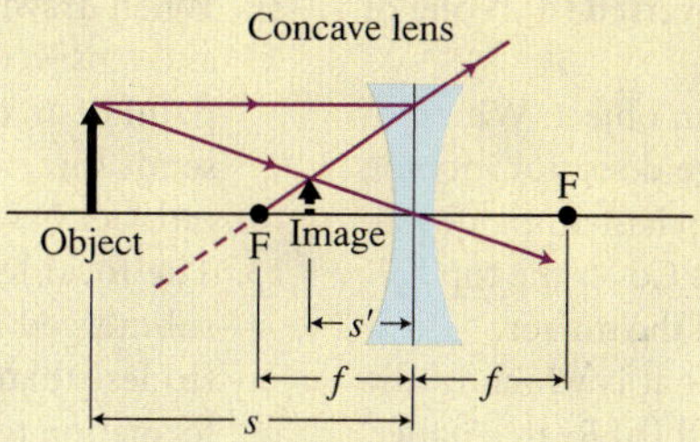


$$\frac{1}{s} + \frac{1}{s'} = \frac{1}{f} \qquad \text{Eq. (22.6)}$$

*(continued)*

**Linear magnification** $m$ is the ratio of the height $h'$ of the image and the height $h$ of the object. The heights are positive if upright and negative if inverted. The linear magnification can be determined using the negative ratio of the image distance $s'$ to the object distance $s$. (Sections 22.3, 22.5, 22.7, and 22.8)

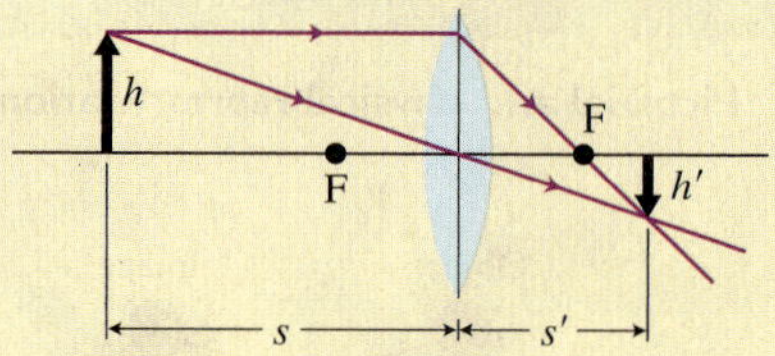


$$m = \frac{h'}{h} = -\frac{s'}{s} \quad \text{Eq. (22.4)}$$

**Angular magnification** $M$ is the ratio of the angular size $\theta'$ of an image as seen through a lens or lenses and the maximum angular size $\theta_{max}$ of the object as seen with the unaided eye. (Section 22.8)

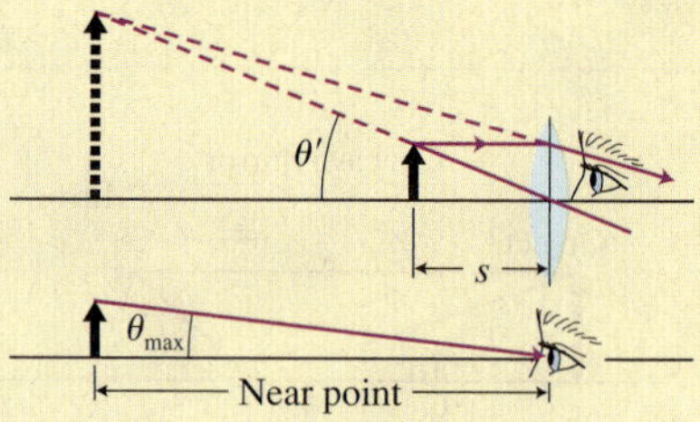


$$M = \frac{\theta'}{\theta_{max}} = \frac{\text{Near point distance}}{s} \quad \text{Eq. (22.12)}$$

MP® For instructor-assigned homework, go to **MasteringPhysics.**

# Questions

## Multiple Choice Questions

1. Where does the image of an object in a plane mirror appear? (a) In front of the mirror (b) On the surface of the mirror (c) Behind the mirror.
2. Where does the image of an object that is $s$ meters in front of a plane mirror appear? (a) $s$ meters away from the object (b) $2s$ meters away from the object (c) There is no image formed by the plane mirror as the reflected rays do not converge.
3. A plane mirror produces an image of an object that is which of the following? (a) Real and enlarged (b) Virtual and the same size (c) Virtual and enlarged (d) Virtual and smaller.
4. A concave mirror can produce an image that is which of the following? (a) Real, enlarged, and upright (b) Virtual, smaller, and inverted (c) Virtual, enlarged, and inverted (d) None of the answers is correct.
5. A concave mirror produces an image of an object. Which of these actions would ensure that the image does not include the top half of the object? (a) Cover the top half of the object. (b) Cover the bottom half of the object. (c) Cover the top half of the mirror. (d) Cover the bottom half of the mirror.
6. A convex mirror can produce an image that is which of the following? (a) Real, enlarged, and inverted (b) Real, smaller, and inverted (c) Real, smaller, and upright (d) Virtual, smaller, and upright (e) Virtual, enlarged, and upright.
7. A virtual image is the image produced (a) on a screen by reflected rays. (b) on a screen by refracted rays. (c) at a point where diverging rays appear to originate.
8. What is the relationship beween the focal length of a lens and the size of a glass sphere that the lens was cut from? (a) $f = R$ (b) $f = 2R$ (c) $f = R/2$ (d) $f$ and $R$ are not related.
9. To see an image of an object that is enlarged, real, and inverted, you need to place the object in front of a convex lens in which region? (a) $s > 2f$ (b) $2f > s > f$ (c) $f > s$ (d) None of these.
10. To see an image of an object that is enlarged, real, and inverted, you need to place the object in front of a concave lens in which region? (a) $s > 2f$ (b) $2f > s > f$ (c) $f > s$ (d) None of these.
11. When we derived the thin lens equation [Eq. (22.6)], what did we use? (a) A ray diagram (b) Similar triangles (c) An assumption that the curvature of the lens is much less than the distances $f$, $s$, and $s'$ (d) a, b, and c (e) a and b only.
12. When drawing images of objects produced by curved mirrors and lenses, which of the following ideas is required? (a) Each point of an object sends two rays. (b) Each point of an object sends three rays. (c) Each point of an object sends an infinite number of rays.
13. The focal length of a glass lens is 10 cm. When the lens is submerged in water, what is its new focal length? (a) 10 cm (b) less than 10 cm (c) more than 10 cm (d) Not enough information to answer.
14. A microbiologist uses a microscope to look at body cells because the microscope will do which of the following? (a) Magnify the linear size of the cells. (b) Magnify the angular size of the cells. (c) Bring the cells closer. (d) Make the cells brighter. (e) Both (a) and (b) are correct.
15. The human eye works in a similar way to which of the following? (a) A microscope (b) An overhead projector (c) A telescope (d) A camera.

## Conceptual Questions

16. When we draw a ray passing through the center of a lens, it does not bend independent of its direction. Explain why and state the implicit assumptions that we use to draw such rays.
17. Explain why your image in a plane mirror has its left and right sides reversed, while the images of your head and legs are not inverted.
18. You run toward a building with walls of a metallic, reflective material. Your speed relative to the ground is 1.5 m/s. How fast does your image appear to move toward you? Explain.
19. Explain how we derived the mirror equation.
20. Explain how we derived the thin lens equation.
21. Explain the difference between a real and a virtual image.
22. You stand in front of a mirror. You see that your head is enlarged and your legs and feet are shrunk. How can this be? Provide a ray diagram to support your answer.
23. A bubble of air is suspended underwater. Draw a ray diagram showing the approximate path followed by a light ray passing from the water (a) through the upper half of the bubble, (b) through its center, and (c) through its lower half. Does the bubble act like a converging or diverging lens? Explain.
24. A bubble of oil is suspended in water. Draw ray diagrams as in Question 23 and explain the difference between the diagrams in Questions 23 and 24.
25. BIO A person underwater cannot focus clearly on another object under the water. Yet if the person wears goggles, he or she can see underwater objects clearly. Explain.
26. The lens of a slide projector is adjusted so that a clear image of the slide is formed on a screen. If the screen is moved farther away, must the lens be moved toward or away from the slide? Explain.
27. BIO The retina has a blind spot at the place where the optic nerve leaves the retina. Design and describe a simple experiment that allows you to perceive the presence of this blind spot. Why doesn't the blind spot affect your normal vision?
28. BIO Find a farsighted person. Design an experiment to find out what lenses will help this person read a book holding it 25 cm away. Borrow this person's glasses and invent a simple experiment to measure the approximate focal lengths of the lenses of his or her glasses. Perform the experiment. Are the real glasses close to your "prescription?"
29. BIO Find a nearsighted person. Design an experiment to find out what lenses will help this person see objects that are far away. Borrow this person's glasses and invent a simple experiment to measure the approximate focal lengths of the lenses of his or her glasses. Perform the experiment. Are the real glasses close to your "prescription?"

# Problems

Below, BIO indicates a problem with a biological or medical focus. Problems labeled EST ask you to estimate the answer to a quantitative problem rather than develop a specific answer. Problems marked with require you to make a drawing or graph as part of your solution. Asterisks indicate the level of difficulty of the problem. Problems with no * are considered to be the least difficult. A single * marks moderately difficult problems. Two ** indicate more difficult problems.

### 22.1 Plane mirrors

1. You need to teach your friend how to draw rays to locate the images of objects produced by a plane mirror. Outline the steps that she needs to take.
2. Place a pencil in front of a plane mirror so that it is not parallel to the mirror. Draw an image that the mirror forms of the pencil and show using rays how the image is formed.
3. * Use geometry to prove that the virtual image of an object in a plane mirror is at exactly the same distance behind the mirror as the object is in front.
4. * You are 1.8 m tall. Where should you place the top of a mirror on the wall so you can see the top of your head? Where should you stand with respect to the wall?
5. What is the smallest size plane mirror a 190-cm-tall person should buy to see himself if he mounts the mirror on a vertical wall?
6. * Two people are standing in front of a plane mirror. Each of them claims that she sees her own image but not the image of the other person. Draw a ray diagram to find out if this is possible.
7. * **Test an idea** Describe an experiment that you can conduct to test that the image produced by a plane mirror is virtual.

### 22.2 and 22.3 Qualitative analysis of curved mirrors and The mirror equation

8. * Describe in detail an experiment to find the image of a candle produced in a curved mirror. Draw pictures of the experimental setup and a ray diagram. Show on the ray diagram where your eye is.
9. * Explain with a ray diagram how (a) a concave mirror and (b) a convex mirror produce images of objects. Make sure that you explain the choice of rays and the location of your eye.
10. * **Test an idea** Describe an experiment to test the idea that a convex mirror never produces a real image of an object.
11. * **Test an idea** Describe an experiment to test the mirror equation. What are you going to measure? What are you going to calculate?
12. * **Tablespoon mirror** You look at yourself in the back (convex shape) of a shiny steel tablespoon. Describe and explain what you see as you bring the spoon closer to your face.
13. * **Partially covered mirror** Your friend thinks that if he covers half of a concave mirror, he will see half of himself in it. Do you agree with his opinion? Support your opinion with a ray diagram. Why would he have such an opinion? Design an experiment to test whether he is correct.
14. * Use ray diagrams and the mirror equation to locate the position, orientation, and type of image of an object placed

in front of a concave mirror of focal length 20 cm. The object distance is (a) 200 cm, (b) 40 cm, and (c) 10 cm.

15. Repeat Problem 14 for a convex mirror of focal length −20 cm.

16. Use ray diagrams and the mirror equation to locate the images of the following objects: (a) an object that is 10 cm from a concave mirror of focal length 7 cm and (b) an object that is 10 cm from a convex mirror of focal length −7 cm.

17. * **Sinking ships** A legend says that Archimedes once saved his native town of Syracuse by burning the enemy's fleet with a mirror. Describe quantitatively the type of mirror that Archimedes could have used to burn ships that were 150 m away. Justify your answer.

18. * EST **Fortune-teller** A fortune-teller looks into a silver-surfaced crystal ball with a radius of 10 cm and focal length of −5 cm. (a) If her eye is 30 cm from the ball, where is the image of her eye? (b) Estimate the size of that image.

19. * You view yourself in a large convex mirror of −1.2-m focal length from a distance of 3.0 m. (a) Locate your image. (b) If you are 1.7 m tall, what is your image height?

20. * **Seeing the Moon in a mirror** The Moon's diameter is $3.5 \times 10^3$ km, and its distance from Earth is $3.8 \times 10^5$ km. Determine the position and size of the image formed by the Hale Telescope reflecting mirror, which has a focal length of +16.9 m.

21. * You view your face in a +20-cm focal length concave mirror. Where should your face be in order to form an image that is magnified by a factor of 1.5?

22. * **Buying a dental mirror** A dentist wants to purchase a small mirror that will produce an upright image of magnification +4.0 when placed 1.6 cm from a tooth. What mirror should she order? Say everything you can about the mirror.

23. * **Using a dental mirror** A dentist examines a tooth that is 1.0 cm in front of the dental mirror. An image is formed 8.0 cm behind the mirror. Say everything you can about the mirror and the image of the tooth.

## 22.4 Qualitative analysis of lenses

24. * You have a convex lens and a candle. Describe in detail an experiment that you will perform to find the image of the candle that this lens produces. Draw pictures of the experimental setup and a ray diagram. Show on the ray diagram where your eye is.

25. * Explain how to draw ray diagrams to locate images produced by objects in front of convex and concave lenses. Focus on the choice of rays and how you know where and what type of image is produced.

26. * Draw ray diagrams to show how a convex lens can produce (a) a real image that is smaller than the object, (b) a real image larger than the object, and (c) a virtual image.

27. * Use a ruler to draw ray diagrams to locate the images of the following objects: (a) an object that is 30 cm from a convex lens of +10-cm focal length, (b) an object that is 14 cm from the same lens, and (c) an object that is 5 cm from the same lens. (Choose a scale so that your drawing fills a significant portion of the width of a paper.) Measure the image locations on your drawings and indicate if they are real or virtual, upright or inverted.

28. * Repeat the procedure described in Problem 27 for the following lenses and objects: (a) an object that is 30 cm from a concave lens of −10-cm focal length, (b) an object that is 14 cm from the same lens, and (c) an object that is 5 cm from the same lens.

29. * Repeat the procedure described in Problem 27 for the following lenses and objects: (a) an object that is 7 cm from a convex lens of +10-cm focal length and (b) an object that is 7 cm from a concave lens of −10-cm focal length.

30. * Repeat the procedure in Problem 27 for the following lenses and objects: (a) an object that is 20 cm from a convex lens of +10-cm focal length, (b) an object that is 5 cm from the same lens, (c) an object that is 20 cm from a concave lens of −10-cm focal length, and (d) an object that is 5 cm from the lens in part (c).

31. * **Partially covering lens** Your friend thinks that if she covers one half of a convex lens she will only be able to see half of the object. Do you agree with her opinion? Why would she have such an opinion? Provide physics arguments. Design an experiment to test her idea.

## 22.5 and 22.6 Thin lens equation and quantitative analysis of lenses and Skills for analyzing processes involving mirrors and lenses

32. * Use ray diagrams to locate the images of the following objects: (a) an object that is 10 cm from a convex lens of +15-cm focal length and (b) an object that is 10 cm from a concave lens of −15-cm focal length. (c) Calculate the image locations for parts (a) and (b) using the thin lens equation. Check for consistency.

33. * Use ray diagrams to locate the images of the following objects: (a) an object that is 6.0 cm from a convex lens of +4.0-cm focal length and (b) an object that is 8.0 cm from a diverging lens of −4.0-cm focal length. (c) Calculate the image locations for parts (a) and (b) using the thin lens equation. Check for consistency.

34. Light passes through a narrow slit, and then through a lens and onto a screen. The slit is 20 cm from the lens. The screen, when adjusted for a sharp image of the slit, is 15 cm from the lens. What is the focal length of the lens?

35. * Describe two experiments that you can perform to determine the focal length of a glass convex lens. Is it a converging or a diverging lens? How do you know?

36. * BIO **Shaving/makeup mirror** You wish to order a mirror for shaving or makeup. The mirror should produce an image that is upright and magnified by a factor of 2.0 when held 15 cm from your face. What type and focal length mirror should you order?

37. * BIO **Face in concave mirror** You view your face in a concave mirror of focal length 20 cm and wish it to be magnified by a factor of 1.5. What should you do?

38. * **Clothing store mirror** A cylindrical mirror in a clothing store causes the customer's width to have a linear magnification of +0.85 when the customer stands 2.5 m from the mirror. Determine everything you can about the mirror.

39. * A large concave mirror of focal length 3.0 m stands 20 m in front of you. Describe the changing appearance of your image as you move from 20 m to 1.0 m from the mirror. Indicate distances from the mirror where the change in appearance is dramatic.

40. ***EST Mirrors for truck** Two mirrors (one plane and one convex) at the side of a truck are shown in **Figure P22.40**. Estimate the focal length of the curved mirror. Explain the reasoning behind your estimate.

**Figure P22.40**

## 22.7 Single-lens optical systems: Putting it all together

41. **Camera** You are using a camera with a lens of focal length 6.0 cm to take a picture of a painting located 3.0 m from the camera lens. Where should the film be positioned in relation to the lens in order to capture the image?

42. * **Camera** A camera with an 8.0-cm focal length lens is used to photograph a person who is 2.0 m tall. The height of the image on the film must be no greater than 3.5 cm. (a) Calculate the closest distance the person can stand to the lens. (b) For this object distance, how far should the film be located from the lens?

43. **BIO Slide projector** A slide projector produces real, inverted, enlarged images of the slides on a screen (slides are put upside down in the projector). If a slide is located 12.6 cm from the 12.0-cm focal length lens of a projector, (a) what should be the distance between the screen and the lens? (b) What is the height of the image of a person on the screen who is 2.0 cm tall on the slide?

44. **BIO Photo of carpenter ant** You take a picture of a carpenter ant with an old fashioned camera with a lens 18 cm from the film. At what places can the 4.0-cm focal length convex camera lens be located so you see a sharp image of the ant on the film?

45. **Photo of secret document** A secret agent uses a camera with a 5.0-cm focal length lens to photograph a document whose height is 10 cm. At what distance from the lens should the agent hold the document so that an image 2.5 cm high is produced on the screen? [Note: The real image is inverted.]

46. * **Photo of landscape** To photograph a landscape 2.0 km wide from a height of 5.0 km, Joe uses an aerial camera with a lens of 0.40-m focal length. What is the width of the image on the detector surface?

47. * Make a rough graph of image distance versus object distance for a convex lens of a known focal length as the object distance varies from infinity to zero.

48. * Make a rough graph of linear magnification versus object distance for a convex lens of 20-cm focal length as the object distance varies from infinity to zero. Indicate in which regions the image is real and in which regions it is virtual.

49. *Repeat Problem 48 for a concave lens of $-20$-cm focal length.

50. **BIO Eye** The image distance for the lens of a person's eye is 2.10 cm. Determine the focal length of the eye's lens system for an object (a) at infinity, (b) 500 cm from the eye, and (c) 25 cm from the eye.

51. **BIO Lens-retina distance** Assume that the eye accommodates to objects at different distances by altering the distance from the lens system to the retina. If the lens system has a focal length of 2.10 cm, what is the lens-retina distance needed to view objects at (a) infinity, (b) 300 cm, and (c) 25 cm?

52. **BIO Nearsighted and farsighted** (a) A woman can produce sharp images on her retina only of objects that are from 150 cm to 25 cm from her eyes. Indicate the type of vision problem she has and determine the focal length of eyeglass lenses that will correct her problem. (b) Repeat part (a) for a man who can produce sharp images on his retina only of objects that are 3.0 m or more from his eyes. He would like to be able to read a book held 30 cm from his eyes.

53. * **BIO Prescribe glasses** A man who can produce sharp images on his retina only of objects that lie from 80 cm to 240 cm from his eyes needs bifocal lenses. (a) Determine the desired focal length of the upper half of the glasses used to see distant objects. (b) Determine the focal length of the lower half used to read a paper held 25 cm from his eyes.

54. * **BIO Correcting vision** A woman who produces sharp images on her retina only of objects that lie from 100 to 300 cm from her eyes needs bifocal lenses. (a) Determine the desired power of the upper half of the glasses used to see distant objects. (b) Determine the power of the lower half used to read a book held 30 cm from her eyes.

55. * **BIO Where are the far and near points?** (a) A woman wears glasses of 50-cm focal length while reading. What eye defect is being corrected and what approximately are the near and far points of her unaided eye? (b) Repeat part (a) for a man whose glasses have a –350-cm focal length. He wears the glasses while driving a car.

56. * **BIO Age-related vision changes** A 35-year-old patent clerk needs glasses of 50-cm focal length to read patent applications that he holds 25 cm from his eyes. Five years later, he notices that while wearing the same glasses, he has to hold the patent applications 40 cm from his eyes to see them clearly. What should be the focal length of new glasses so that he can read again at 25 cm?

## 22.8 Angular magnification and magnifying glasses

57. * **BIO Looking at an aphid** You examine an aphid on a plant leaf with a magnifying glass of $+6.0$-cm focal length. You hold the glass so that the final virtual image is 40 cm from the lens. If you assume that your near point is at 30 cm, then what is the angular magnification? How will the magnification change if you are farsighted? Nearsighted?

58. * **Reading with a magnifying glass** You examine the fine print in a legal contract with a magnifying glass of focal length 5.0 cm. (a) How far from the lens should you hold the print to see a final virtual image 30 cm from the lens (at the eye's near point)? (b) Determine the angular magnification of the magnifying glass.

59. * **Seeing an image with a magnifying glass** A person has a near point of 150 cm. (a) What is the nearest distance at which she needs to hold a magnifying glass of 5.0-cm focal length from print on a page and still have an image formed beyond her near point? (b) Determine the angular magnification for an image at her near point.

60. * **Stamp collector** A stamp collector is viewing a stamp through a magnifying glass of 5.0-cm focal length. Determine the object distance for virtual images formed at (a) negative infinity, (b) $-200$ cm, and (c) $-25$ cm. (d) Determine the angular magnification in each case.

## 22.9 Telescopes and microscopes

61. ** You place a $+20$-cm focal length convex lens at a distance of 30 cm in front of another convex lens of focal length $+4.0$ cm. Then you place a candle 100 cm in front of the first

lens. Find (a) the location of the final image of the candle, (b) its orientation, and (c) whether it is real or virtual.

62. * You place a +25-cm focal length convex lens at a distance of 50 cm in front of a concave lens with a focal length of −40 cm. Then you place a small lightbulb (2 cm tall) 30 cm in front of the convex lens. Determine (a) the location of the final image, (b) its orientation, and (c) whether it is real or virtual.

63. * **EST** You place a candle 10 cm in front of a convex lens of focal length +4.0 cm. Then you place a second convex lens, also of focal length +4.0 cm, at a distance of 12 cm from the first lens. (a) Use a ray diagram to locate the final image (keep the scale). (b) Using measurements on your ray diagram, estimate the linear magnification of the object. Be sure to show your rays and/or estimation technique for each step. Do not use equations!

64. * **EST** Repeat Problem 63 for an object located 6.0 cm from a convex lens of focal length 3 cm separated by 11 cm from another convex lens of focal length 1 cm.

65. ** You measure the focal length of a concave lens by first forming a real image of a light source using a convex lens. The image is formed on a screen 20 cm from the lens. You then place the concave lens halfway between the convex lens and the screen. To obtain a sharp image, you need to move the screen 15 cm further away from the lenses. How does this experiment help you determine the focal length of the concave lens? What is the focal length?

66. ** **Telescope** A telescope consists of a +4.0-cm focal length objective lens and a +0.80-cm focal length eyepiece that are separated by 4.78 cm. Determine (a) the location and (b) the height of the final image for an object that is 1.0 m tall and is 100 m from the objective lens.

67. ** **Yerkes telescope** The world's largest telescope made only from lenses (with no mirrors) is located at the Yerkes Observatory near Chicago. Its objective lens is 1.0 m in diameter and has a focal length of +18.9 m. The eyepiece has a focal length of +7.5 cm. The objective lens and eyepiece are separated by 18.970 m. (a) What is the location of the final image of a Moon crater $3.8 \times 10^5$ km from Earth? (b) If the crater has a diameter of 2.0 km, what is the size of its final image? (c) Determine the angular magnification of the telescope by comparing the angular size of the image as seen through the telescope and the object as seen by the unaided eye.

68. ** **Telescope** A telescope consisting of a +3.0-cm objective lens and a +0.60-cm eyepiece is used to view an object that is 20 m from the objective lens. (a) What must be the distance between the objective lens and eyepiece to produce a final virtual image 100 cm to the left of the eyepiece? (b) What is the total angular magnification?

69. ** **Design a telescope** You are marooned on a tropical island. Design a telescope from a cardboard map tube and the lenses of your eyeglasses. One lens has a +1.0-m focal length and the other has a +0.30-cm focal length. The telescope should allow you to view an animal 100 m from the objective with the final image being formed 1.0 m from the eyepiece. Indicate the location of the lenses and the expected angular magnification.

70. ** **Microscope** A microscope has a +0.50-cm objective lens and a +3.0-cm eyepiece that is 20 cm from the objective lens. (a) Where should the object be located to form a final virtual image 100 cm to the left of the eyepiece? (b) What is the total angular magnification of the microscope, assuming a near point of 25 cm?

71. ** **BIO Dissecting microscope** A dissecting microscope is designed with a larger than normal distance between the object and the objective lens. The microscope has an objective lens of +5.0-cm focal length and an eyepiece of +2.0-cm focal length. The lenses are separated by 15 cm. The final virtual image is located 100 cm to the left of the eyepiece. (a) Determine the distance of the object from the objective lens. (b) Determine the total angular magnification.

72. ** **Microscope** A microscope has an objective lens of focal length +0.80 cm and an eyepiece of focal length +2.0 cm. An object is placed 0.90 cm in front of the objective lens. The final virtual image is 100 cm from the eyepiece at the position of minimum eyestrain. (a) Determine the separation of the lenses. (b) Determine the total angular magnification.

73. ** **Microscope** Determine the lens separation and object location for a microscope made from an objective lens of focal length +1.0 cm and an eyepiece of focal length +4.0 cm. Arrange the lenses so that a final virtual image is formed 100 cm to the left of the eyepiece and so that the angular magnification is −260 for a person with a near point of 25 cm.

## General Problems

74. * **Figure P22.74** shows three cases of the primary axis of a lens (the lens is not shown) and the location of a shining object and its image. In each case, find the location and the type of the lens (convex or concave) that could produce the image and find the focal points of the lens. Then draw a ray diagram to help justify each choice.

**Figure P22.74**

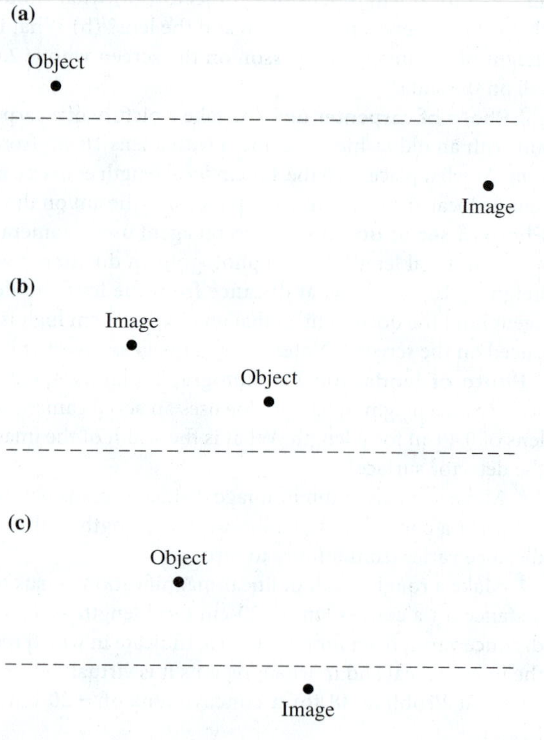


75. * **Jeopardy problem** The equations below describe a process that involves one lens. Determine the unknown quantities and write a word description of an optics situation that is consistent with the equations.

$$\frac{1}{4.0\text{ m}} + \frac{1}{s'} = \frac{1}{0.10\text{ m}}$$

$$h' = -\left(\frac{s'}{4.0\text{ m}}\right)(1.6\text{ m})$$

76. * **Jeopardy problem** The equations below describe a process involving more than one lens. Determine the unknown quantities and write a word description of an optics situation that is consistent with the equations.

$$\frac{1}{100\text{ m}} + \frac{1}{s_1{}'} = \frac{1}{0.10\text{ m}}$$
$$s_2 = (0.14\text{ m}) - s_1{}'$$
$$\frac{1}{s_2} + \frac{1}{s_2{}'} = \frac{1}{0.042\text{ m}}$$

77. * **Jeopardy problem** The equations below describe a process involving more than one lens. Determine the unknown quantities and write a word description of an optics situation that is consistent with the equations.

$$\frac{1}{0.14\text{ m}} + \frac{1}{s_1{}'} = \frac{1}{0.10\text{ m}}$$
$$s_2 = (0.395\text{ m}) - s_1{}'$$
$$\frac{1}{s_2} + \frac{1}{s_2{}'} = \frac{1}{0.050\text{ m}}$$

78. ** Design a system that will allow you to project the picture from your neighbors' TV set onto the wall of their living area that can be seen from your porch. They'd love to sit on your porch and watch TV while you barbeque. Design such a system and discuss its limitations.

79. ** **Two-lens camera** A two-lens camera (see **Figure P22.79**) has one lens with focal length +15.0 cm located 12 cm from the film and a second lens of focal length +13.0 cm a variable distance $d$ of 5.0 to 10.0 cm from the film. Determine the range of distances at which you can photograph objects and achieve sharp images on the film.

**Figure P22.79**

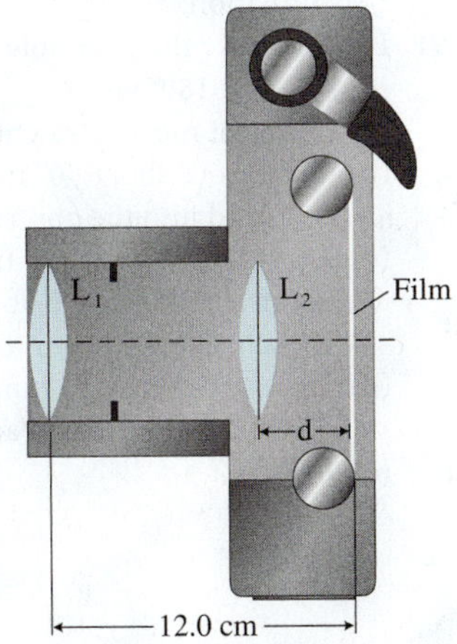

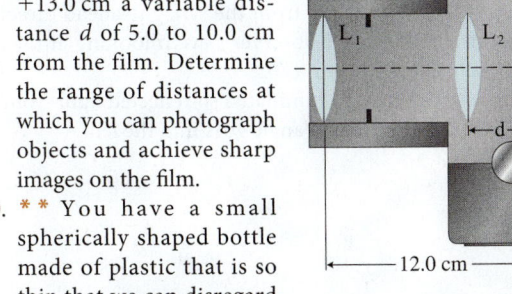

80. ** You have a small spherically shaped bottle made of plastic that is so thin that we can disregard its effects for optical processes. Show that the bottle works as a converging lens when filled with water, but when the same bottle is empty, closed, and submerged in water, it becomes a diverging lens. If you have access to the appropriate materials, test your ray diagram experimentally.

## Reading Passage Problems

BIO **Laser surgery for the eye** LASIK (laser-assisted *in situ* keratomileusis) is a surgical procedure intended to reduce a person's dependency on glasses or contact lenses. Laser eye surgery corrects common vision problems, such as myopia (nearsightedness), hyperopia (farsightedness), astigmatism (blurred vision resulting from corneal irregularities), or some combination of these. In myopia, the cornea, the clear covering at the front of the eye, is often too highly curved, causing rays from distant objects to form sharp images in front of the retina (**Figure 22.34**). LASIK refractive surgery can flatten the cornea so that images of distant objects form on the retina. A knife cuts a flap in the cornea with a hinge left at one end of this flap. The flap is folded back, exposing the middle section of the cornea. Pulses from a computer-controlled laser vaporize a portion of the tissue and the flap is replaced.

**Figure 22.34** LASIK surgery to correct nearsighted vision.

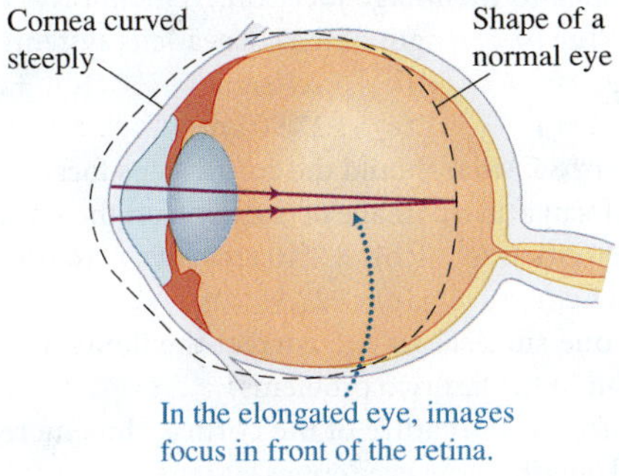


**Figure 22.35** Conductive keratoplasty (CK) surgery to correct farsighted vision.

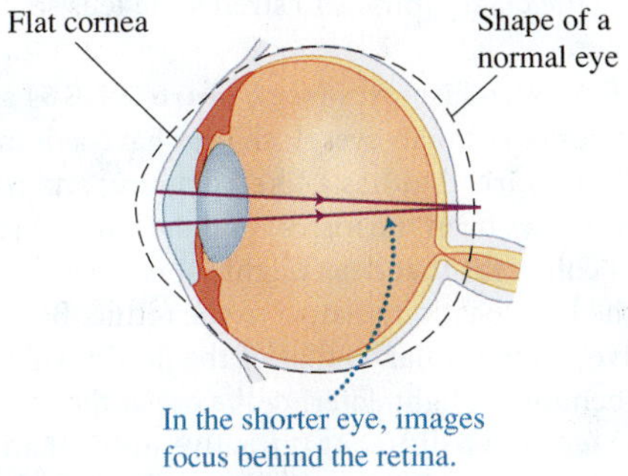


In farsighted people, an object held at the normal near point of the eye (about 25–50 cm from the eye) forms an image behind the retina (**Figure 22.35**). Increasing the curvature of the cornea causes rays from near objects to produce an image on the retina. Conductive keratoplasty (CK) can increase the curvature of the cornea to correct farsightedness. In CK, a tiny probe releases radio frequency energy in a circular pattern on the outside of the cornea. This circular pattern creates a constrictive band (like the tightening of a belt), increasing the overall curvature of the cornea.

81. Your eyeball is 2.10 cm from the cornea to the retina. You look at a sign on the freeway that is effectively an infinite distance away. The image for the sign is 1.96 cm from the cornea within the eye. Which answer below is closest to the focal length of the cornea-lens system?
(a) +1.96 cm (b) +2.10 cm (c) +2.26 cm
(d) +2.42 cm (e) +27.90 cm

82. In Problem 81, what should the cornea-lens focal length be in order to form a sharp image on the retina of the sign?
(a) +1.96 cm (b) +2.10 cm (c) +2.26 cm
(d) +2.42 cm (e) +27.90 cm

83. What is one surgical way to correct nearsighted vision?
(a) Reduce the curvature of the cornea, thus reducing the focal length of the eye's optics system.
(b) Increase the curvature of the cornea, thus increasing the focal length of the eye's optics system.
(c) Increase the iris opening to allow more light to enter.
(d) Use the ciliary muscle to increase the thickness of the lens.

84. Your eyeball is 2.10 cm from the cornea to the retina. A book held 30 cm from your eyes produces an image 2.26 cm from the entrance to the image location. Which answer below is closest to the focal length of the cornea-lens system?
(a) +1.96 cm (b) +2.10 cm (c) +2.26 cm
(d) +2.42 cm (e) +27.90 cm

85. In Problem 84, what should the cornea-lens focal length be in order to form a sharp image of the book on the retina?
(a) +1.96 cm (b) +2.10 cm (c) +2.26 cm
(d) +2.42 cm (e) +27.90 cm

86. What is one surgical way to correct the farsighted vision of the person in the last two problems?
(a) Reduce the curvature of the cornea, thus increasing the focal length of the eyes optics system.
(b) Increase the curvature of the cornea, thus reducing the focal length of the eyes optics system.
(c) Open the iris wider to allow more light to enter.
(d) Cause the ciliary muscle to stretch the lens so it becomes thinner.

BIO **Bulging fish eyes** Fish eyes (see **Figure 22.36**) are evolutionary precursors to human eyes. Fish eyes have a dome-shaped cornea covering an iris that has a fixed opening and a spherical lens, which protrudes from the iris. Behind the lens is a retina with many rod and cone cells that detect light. Muscles attached to the lens change the lens location relative to the retina. Because water has a refractive index similar to that of the fluid inside fish eyes, there is little bending of light entering the eye at the water-cornea interface. The lens has a different refractive index than the fluid surrounding the lens and bends and focuses light so that images are formed on the retina. The fish eye lens is spherical in shape and collects light from a 180° field of view (**Figure 22.37**).

**Figure 22.36** A fish eye.

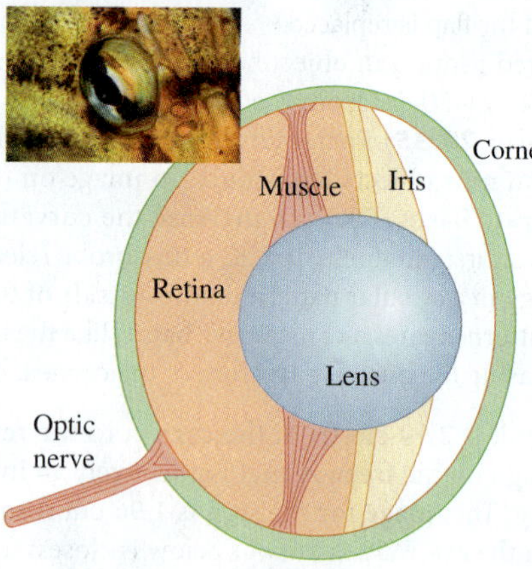


**Figure 22.37** A fisheye lens collects light from a 180° field of view.

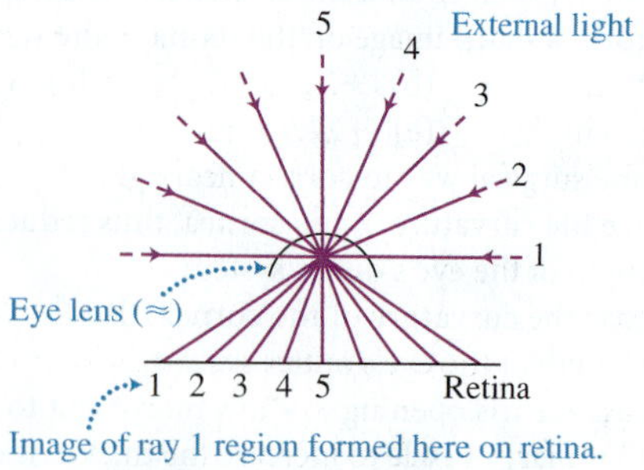


Camera manufacturers now make fisheye lenses that produce an image from a 180° spread in front of the lens. Fisheye lens applications include IMAX projection, hemispherical photography, and peepholes that provide 180° views outside doors.

87. Why is the fish eye in Figure 22.36 black?
(a) The lens is black.
(b) No light reflects out from behind the lens.
(c) The fluid behind the lens is black.
(d) The retina is black.

88. Which of the following is the same for the fish eye and the human eye?
(a) The muscles attached to the lens change their shape.
(b) The corneas help focus the light.
(c) The lenses have a spherical shape.
(d) The irises help focus the light.

89. Which ray in **Figure P22.89** best represents the path of the incident light through the fisheye lens?
(a) I (b) II
(c) III (d) IV
(e) V

**Figure P22.89**

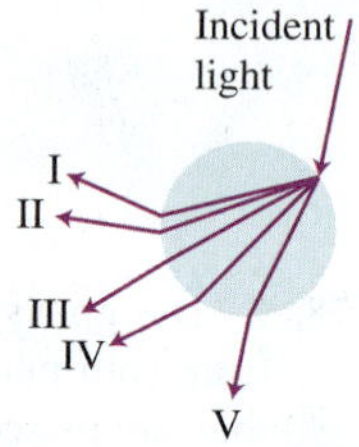


90. Which of the following statements regarding the fish eye and the human eye is not true?
(a) They both have a cornea.
(b) They both have an iris.
(c) They both have a lens.
(d) They both rely on the cornea and the lens for image formation.

91. How can the the peephole lens in a hotel door have an approximately 180° view?
(a) The light moves straight through the lens into eye.
(b) Light from about 90° from the straight-ahead direction is refracted into the observer's eye if very close to the lens.
(c) Light from about 90° from the straight-ahead direction is refracted into the observer's eye if looking at the lens from inside the room.
(d) The outside hall is illuminated so reflected light from the objects inside the 180°angle goes into the lens.
(e) b and d.

# Wave Optics

# 23

Why do we see colors in soap bubbles?

Does a particle or wave model best describe the behavior of light?

What limits our ability to see two closely positioned objects as separate?

Soap bubbles display a remarkable array of colors. Watch a bubble hanging on a bubble wand. You will see horizontal bands appear on the surface and repeatedly change colors. As time progresses, the bands of color on the top of the bubble widen until a dark band takes their place and the colors disappear. The dark band is a sign that the bubble's life is coming to an end—and then it pops. In this chapter you will learn why soap bubbles display such brilliant colors and why those colors disappear just before the bubble pops.

**Be sure you know how to:**

- Draw wave fronts for circular and plane waves (Section 20.1).
- Relate wave properties such as frequency, speed, and wavelength (Section 20.2).
- Apply Huygens' principle and the superposition principle (Section 20.6).

**In our chapter** on reflection and refraction (Chapter 21) we developed two models of light: a particle model in which a light beam is modeled as a stream of tiny "light bullets," and a wave model in which light is modeled as a propagating vibration—a wave. The particle model predicted that the speed of light in water should be greater than in air; the wave model predicted the opposite. Eventually, in 1850, Hippolyte Fizeau and Léon Foucault established that light travels more slowly in water than in air, finally disproving the particle model. However, even before Fizeau and Foucault, the wave model was gaining wide acceptance, because of experiments performed by Thomas Young in the early 1800s.

## 23.1 Young's double-slit experiment

In 1801, Englishman Thomas Young performed a testing experiment for the particle model of light. Testing Experiment **Table 23.1** describes a contemporary version of the experiment and the prediction of its outcome based on the particle model.

**TESTING EXPERIMENT TABLE**

**23.1 Recreating Young's experiment to test the particle model of light.**

VIDEO 23.1

| Testing experiment | Prediction based on the particle model of light | Outcome |
|---|---|---|
| We shine a laser beam on a glass plate covered with a black film. The film has two narrow slits that are very close to each other. Light passes through the slits and reaches the screen that is parallel to the plate with slits. | We will see two bright regions on the screen in the same shape as the slits. | We see *many* bright regions. |
| 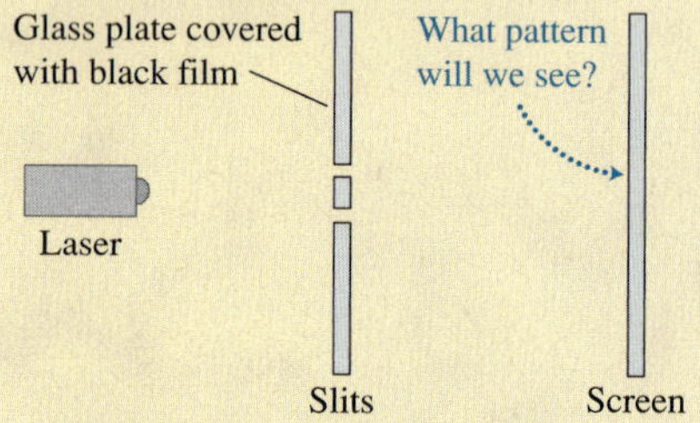  | 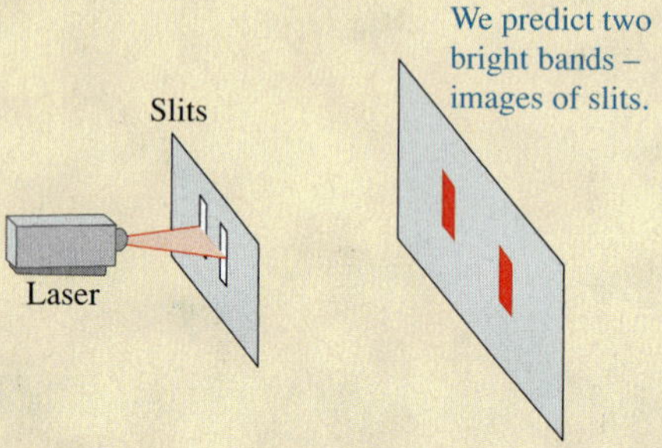  | 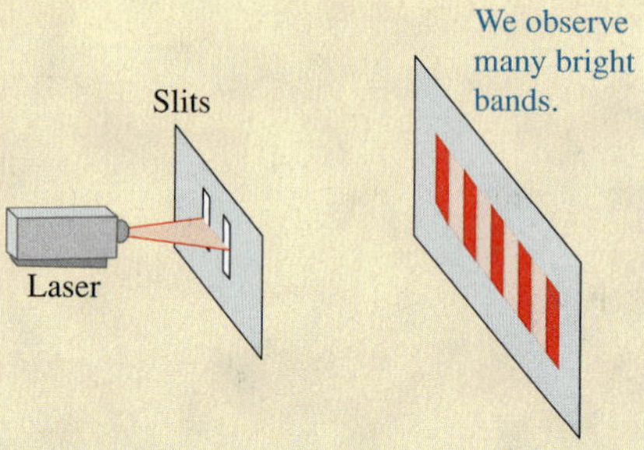  |

**Conclusion**

The prediction based on the particle model of light does not match the outcome, disproving the particle model.

In Table 23.1, light from the two slits produced a pattern of alternating bright regions on the screen (called **bands** or **fringes**). This outcome is inconsistent with the particle model of light.

Thomas Young explained the bright fringes using a wave model and Huygens' principle (see Chapters 20 and 21 for more on Huygens' principle). According to this principle, every point on a wave front is the source of a circular-shaped wavelet that moves in the forward direction at the same speed as the wave. A new wave front results from adding these wavelets together—the superposition (*interference*) of these wavelets.

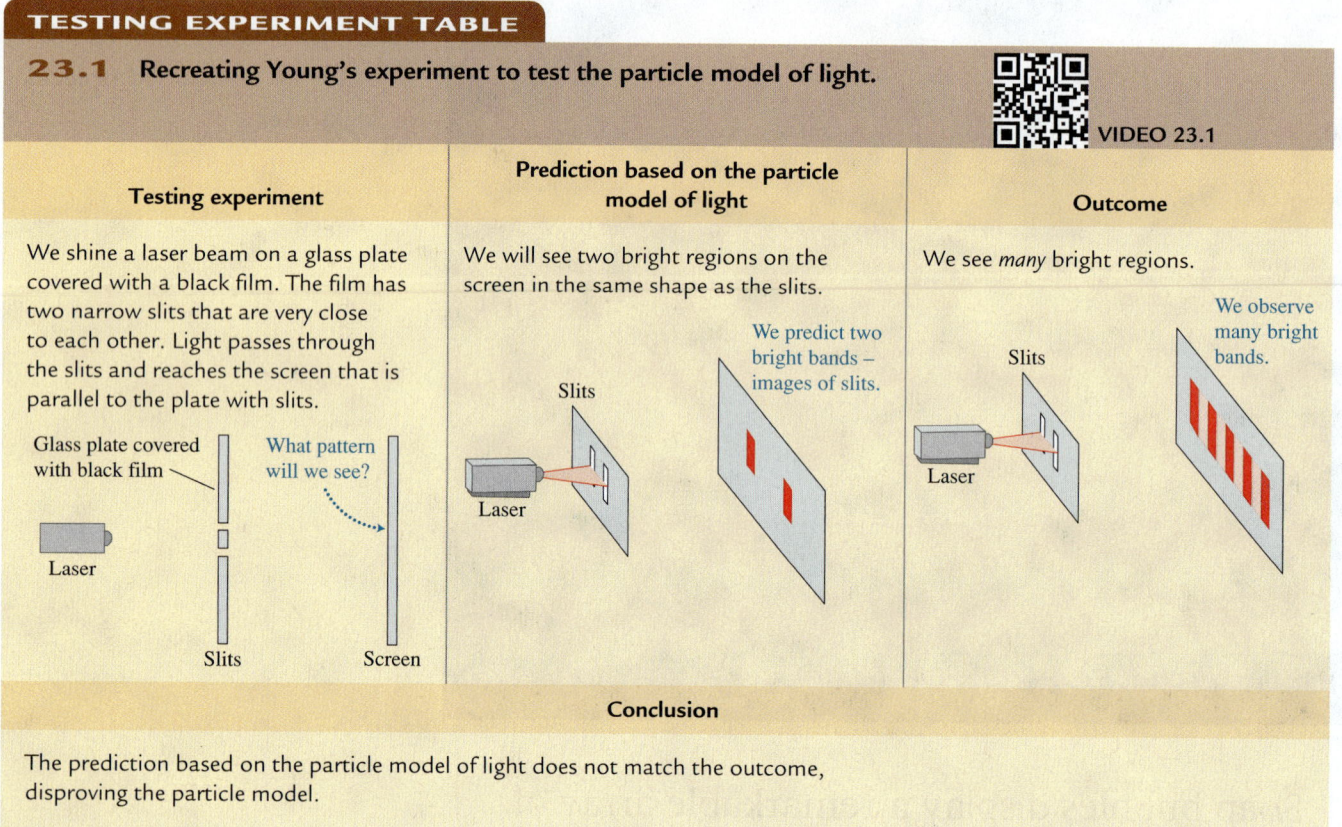

We can use the wave model of light to explain the outcome of the testing experiment. Imagine that the laser emits some kind of regular disturbance analogous to the sound wave produced by a tuning fork, only the disturbance is emitted in a single direction rather than spreading in all directions. We will agree that the most light will be present wherever the waves oscillate with the highest amplitude. Where the waves have zero amplitude there is no light. In other words, the amplitude of the light wave incident at a particular location on the screen determines how bright the screen is at that location.

The laser beam consists of plane wave fronts (**Figure 23.1a**) whose crests and troughs are incident on the two closely spaced slits and produce synchronized (in-phase) disturbances within each slit. According to Huygens' principle, the slits become sources of in-phase wavelets. The new wavelets generated from each slit spread outward and eventually irradiate a screen. There, the disturbances due to both wavelets combine according to the superposition principle. At a particular time, the wavelet disturbances produced by each slit appear as shown in Figure 23.1b. At some points the crests from the wavelets from each slit overlap (the positions of solid dots), resulting in double-sized crests. The troughs between the crests also overlap, resulting in double-sized troughs. This *constructive interference* results in a double-amplitude wave. These alternating double crests and double troughs arrive at the screen shown in Figure 23.1, causing a large-amplitude light wave—a bright band (or fringe). This occurs along line 0 and along the two lines labeled 1. It also occurs along several other lines not shown in the figure.

Along lines about halfway between these bright bands, you find that the crest of one wavelet overlaps with the trough of the other (the open circles in Figure 23.1c). The wavelets are out of phase along those lines and therefore cancel each other. This *destructive interference* means no light is present—a dark band (or fringe) is present.

Between the centers of the bright and dark bands, the wavelets are neither exactly in nor out of phase with each other and the wave amplitudes are between double and zero; the light has intermediate brightness. Since the pattern on the screen can be explained using the idea of interference of the wavelets from the two slits, it is called an **interference pattern**. The wave model of light explains the fringes observed by Thomas Young while the particle model does not.

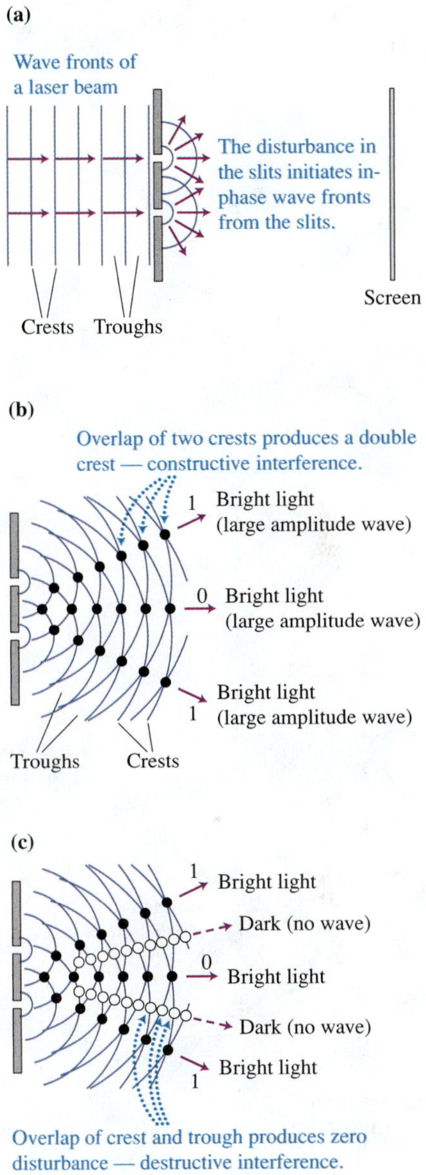


**Figure 23.1** (a) Double slits are the source of wavelets that can interfere. (b) When wave crests or troughs overlap, a large-amplitude wave occurs. (c) When a crest overlaps a trough, they cancel and there is no wave disturbance.

**Qualitative wave-based explanation of Young's experiment** The interference pattern of dark and bright bands produced by light passing through two closely spaced narrow slits can be explained by the addition or superposition of wavelets originating from the two slits. The length of the path that each wave travels to a particular spot on the screen determines the phase of the wave at that location. Locations where two waves arrive in phase are the brightest; locations where two waves arrive with different phases are darker; and locations where the waves arrive completely out of phase are dark.

## What is a light wave—What is being disturbed?

Imagine that we want to draw a graphical representation of a light wave—a disturbance-versus-position graph (similar to graphs we drew in Chapter 20 for mechanical waves). We know, for example, that a sound wave involves a moving disturbance of air pressure and density and that a wave on a Slinky involves a moving disturbance of the closeness of the Slinky coils. What is being disturbed when a light wave propagates?

In the early 1800s, physicists thought that a medium was needed for the propagation of any wave. They proposed the existence of a massless invisible medium for light called **ether**, which supposedly filled the entire universe and vibrated when light traveled through it. This vibrating ether model was accepted until the late 1800s. Later in this book (in Chapters 24 and 25) we will learn about a contemporary theory of light and the experiments that disproved the existence of ether. For now, we will just assume that light is a wave-like disturbance and that the nature of that disturbance is unknown. We will also assume that the disturbances from different waves can add and subtract from each other—as in the double-slit experiment.

## Quantitative analysis of the double-slit experiment

A mathematical description of the phenomenon observed by Young involves calculating the location of bright and dark bands on a screen for a particular light wavelength $\lambda$, slit separation $d$, and distance $L$ of the screen from the slits. As discussed earlier, we consider the slits to be sources of synchronized circular wavelets. At all locations where the waves arrive in phase, there will be constructive interference; where they arrive completely out of phase, there will be destructive interference.

Now consider the brightest locations on the screen, as shown in **Figure 23.2a**. The wavelets travel the same distance from the slits to point 0, and the waves arrive at this point in phase no matter how far the screen is from the slits. Figure 23.2b represents the wave disturbances from the two slits at one instant of time. This bright band or fringe in the center is called the **0th order maximum**.

Just to either side of this location the waves arrive slightly out of phase and the light is dimmer. The brightness changes slowly, from a maximum at the center where the waves arrive exactly in phase, to zero where waves arrive

**Figure 23.2** Double-slit wave interference. (a) A schematic of the bright and dark bands on a screen. (b) Producing the central 0th bright band. (c) Destructive interference between the 0th and 1st order bright bands. (d) Producing the 1st order bright band and (e) the 2nd order bright band.

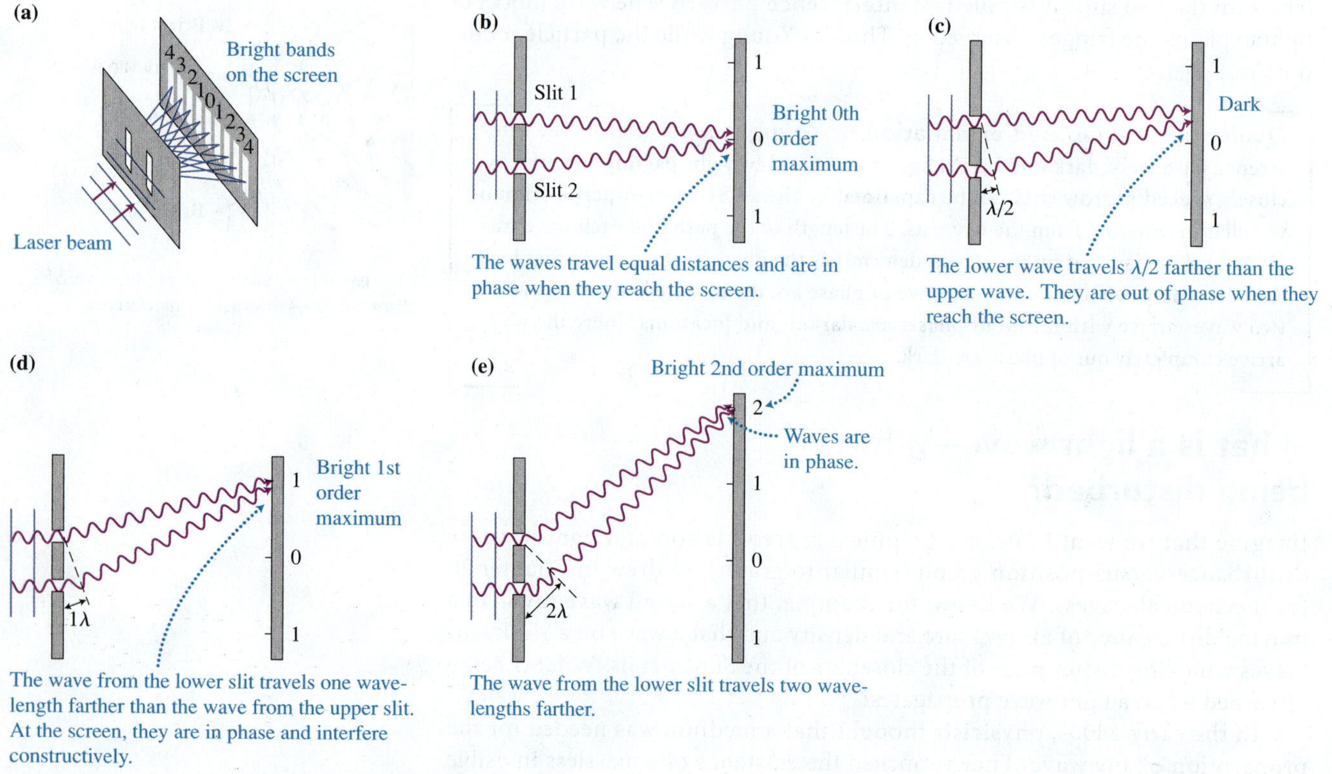
(a)
Bright bands on the screen
4 3 2 1 0 1 2 3 4
Laser beam
(b)
Slit 1
Slit 2
1
0
1
Bright 0th order maximum
The waves travel equal distances and are in phase when they reach the screen.
(c)
1
Dark
0
1
λ/2
The lower wave travels λ/2 farther than the upper wave. They are out of phase when they reach the screen.
(d)
Bright 1st order maximum
1
0
1
1λ
The wave from the lower slit travels one wavelength farther than the wave from the upper slit. At the screen, they are in phase and interfere constructively.
(e)
Bright 2nd order maximum
2
Waves are in phase.
1
0
1
2λ
The wave from the lower slit travels two wavelengths farther.

completely out of phase. At these two locations the difference in the path length traveled by the two waves is one-half wavelength (Figure 23.2c).

As we move away from the dark band on each side of the center, the screen becomes brighter again as the path length difference between the waves becomes larger. When the path length difference reaches one whole wavelength at location 1, we find another point of maximum brightness (Figure 23.2d). The wave from the lower slit travels exactly one wavelength farther than the distance traveled by the wave from the upper slit. This bright band is called the **1st order maximum**. A similar maximum appears at the other location 1 below location 0.

We can use similar reasoning to find the 2nd, 3rd, and 4th order maxima. The path length differences for the centers of these bright bands are $2\lambda$, $3\lambda$, and $4\lambda$, respectively (see the second bright band in Figure 23.2e). We can use this thinking to predict the locations of these bands using geometry.

## Mathematical location of the *m*th bright band

Consider the general case in **Figure 23.3**. The small shaded triangle next to the slits has a hypotenuse $d$ equal to the slit separation. The two red lines are the distances that the two waves travel to get to the center of the $m$th bright band on the screen (position $P_m$ in Figure 23.3). The path length difference between the waves equals an integer number of whole wavelengths—$m$ wavelengths in this case. The side of the small shaded triangle opposite angle $\theta$ is the path difference for the two waves reaching location $P_m$ on the screen. If there is an interference maximum at that location on the screen, then the path length difference $\Delta$ that the waves originating at the two slits travel to $P_m$ should equal an integer number of wavelengths:

$$\Delta = m\lambda, \text{ where } m = 0, \pm 1, \pm 2, \pm 3, \text{etc.}$$

Using trigonometry: $\sin\theta = (\Delta/d) = (m\lambda/d)$. The angle $\theta$ in the small triangle equals the angle $\theta_m$ between the horizontal and the $m$th bright band. Thus,

$$\sin\theta_m = \frac{m\lambda}{d} \quad (23.1)$$

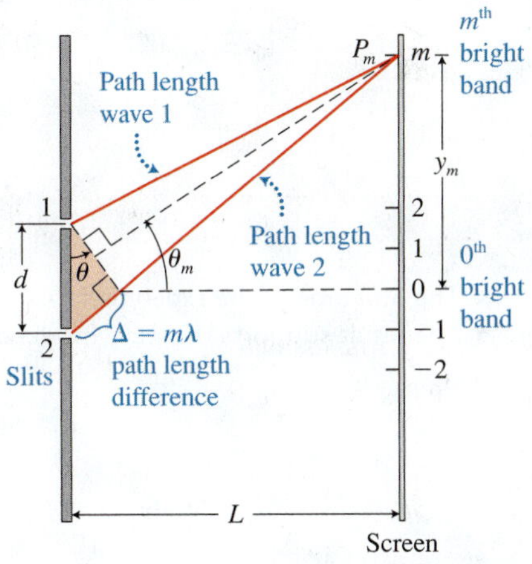


**Figure 23.3** The geometry for finding the position of the $m$th bright band on a screen.

This derivation has led us to a quantitative relation between the slit separation $d$, the angle $\theta_m$ at which the $m$th maximum occurs, and the wavelength $\lambda$ of the light. If we know the distance $L$ between the slits and the screen, we can find the distance $y_m$ between the 0th maximum on the screen and the $m$th maximum in Figure 23.3:

$$\tan\theta_m = \frac{y_m}{L} \quad (23.2)$$

If the angle $\theta_m$ is very small, the tangent of the angle equals the sine of the angle and also equals the angle itself if measured in radians: $\tan\theta_m \approx \theta_m = (y_m/L)$ and $\sin\theta_m \approx \theta_m = (m\lambda/d)$. Combining these two relationships, we find that for small angles relative to the horizontal, the bright bands on the screen are located at

$$y_m = \frac{m\lambda L}{d} \quad (23.3)$$

## Testing these expressions for bright band locations on a screen

Since it is not easy to independently measure the wavelength of light, we will instead test the inverse relationship between the location of the fringes $y_m$ and the slit separation $d$ (Testing Experiment **Table 23.2**).

**TESTING EXPERIMENT TABLE**

**23.2 Slit separation and locations of the bright spots.**

VIDEO 23.2

| Testing experiment | Prediction | Outcome |
|---|---|---|
| ■ We shine red laser light through slits with separation $d = 0.050$ mm onto a screen at $L = 2.0$ m from the slits. The first bright fringe on the screen from the center maximum is at $y_1 = 2.75$ cm.<br>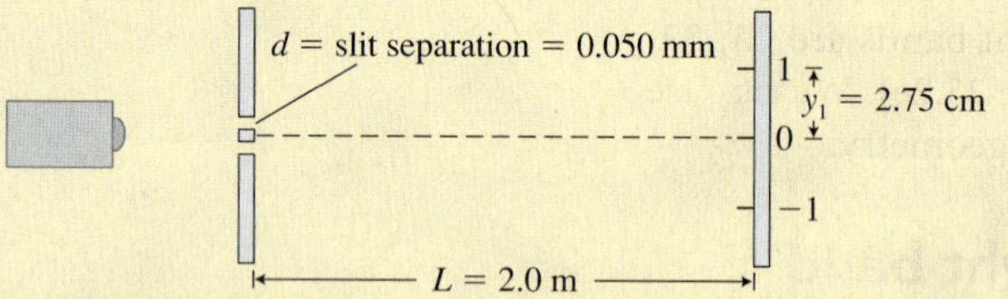<br> | We solve Eq. (23.3) for $\lambda$:<br>$$\lambda = \frac{y_m d}{mL}$$<br>$m$ is 1 since we are interested in the first bright fringe. Since the same laser is used for both slit separations, we can equate the right-hand sides of this equation for each case. | We measure a distance of $4.6 \pm 0.1$ cm between the center $m = 0$ maximum and the $m = 1$ maxima. |
| ■ Using the same laser, we predict the location of the first bright band on the screen if using slits of a different separation $d' = 0.030$ mm. | $$\frac{y_1 d}{(1)L} = \frac{y_1' d'}{(1)L}$$<br>Multiplying by $L$ and solving for $y_1'$, we have our prediction:<br>$$y_1' = \frac{y_1 d}{d'} = \frac{(2.75\text{ cm})(0.050\text{ mm})}{0.030\text{ mm}} = 4.6\text{ cm}$$ | |

**Conclusion**

- The outcome of the experiment matches the prediction from Eq. (23.3).
- The result supports the wave model of light.

Because the outcome of the experiment agreed with the prediction that followed from the equations we derived, we gain confidence in the wave model of light. Many other experiments have also agreed with predictions based on the wave model of light.

Now that we have some trust in Eq. (23.3), we can use it to determine the wavelength of the red laser light used in Table 23.2.

$$\lambda = \frac{y_m d}{mL} = \frac{(2.75\text{ cm})(0.050\text{ mm})}{(1)(2.0\text{ m})} = 6.9 \times 10^{-7}\text{ m}$$

To simplify the expression of such small numbers, physicists record wavelengths of light in nanometers; 1 nm $= 10^{-9}$ m. The wavelength for red light described in Table 23.2 is about 690 nm. We can now summarize our findings about double-slit interference.

**Double-slit interference** When light of wavelength $\lambda$ passes through two narrow slits separated by distance $d$, it forms a series of bright and dark bands on a screen beyond the slits. The angular deflection $\theta_m$ of the center of the $m$th bright band to the side of the central $m = 0$ bright band is determined using the equation

$$d \sin \theta_m = m\lambda \qquad (23.1)$$

where $m = 0, \pm 1, \pm 2, \pm 3$, etc. The distance $y_m$ on a screen from the center of the central bright band to the center of the $m$th bright band depends on the angular deflection $\theta_m$ and on the distance $L$ of the screen from the slits:

$$L \tan \theta_m = y_m \qquad (23.2)$$

**For small angles:**

$$y_m = \frac{m\lambda L}{d} \quad (23.3)$$

We see from Eq. (23.1) that for light of wavelength $\lambda$, the greater the slit separation $d$, the smaller the angular deflection of the bright spots $\theta_m$. Thus, to observe distinct bands, we need slits that are very close to each other. Also, the longer the wavelength, the larger the angular deflection for the same slit separation. Finally, using Eq. (23.2), we see that the greater the distance $L$ of the screen from the slits, the greater the distance $y_m$ of the $m$th bright band from the central bright band on the screen.

**TIP** We could have done the derivation for the minima (the locations where two waves arrive out of phase and cancel each other) instead of for the maxima and obtained a relation for the angles at which the dark bands occur. This happens when the path length difference from the two slits to the dark band equals an odd number of half wavelengths: $\Delta = (2m + 1)(\lambda/2)$ where $m = 0, \pm 1, \pm 2, \pm 3$, etc. and consequently

$$\sin\theta = \frac{(2m + 1)\lambda/2}{d}$$

Thomas Young used sunlight in his original experiments since lasers had not been invented yet. He found that the bands of light on the screen were colored like a rainbow, except for the central band, which was white. Each band that he observed had blue light on the inside edge (closest to the central maximum) and red light on the outside edge (**Figure 23.4**). We found that the angular deflection to the center of the $m$th bright band should be $\sin\theta_m = (m\lambda/d)$. If the wavelength $\lambda$ is longer, the angular deflection $\theta_m$ should be greater. Evidently, red light must have a longer wavelength than blue light, with green in between. We can check this idea.

**Figure 23.4** Pattern of white light produced by two slits on a screen.

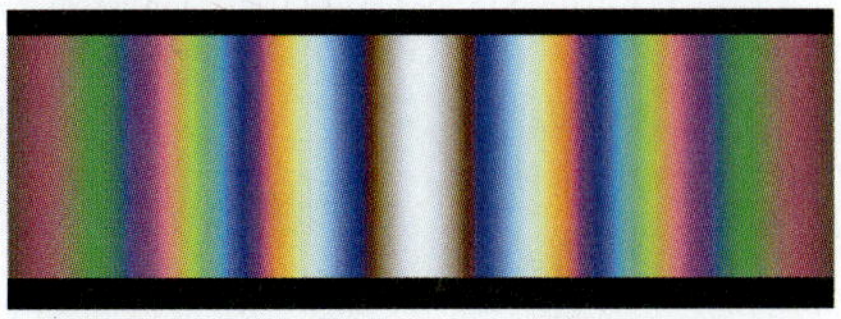

### QUANTITATIVE EXERCISE 23.1 Wavelength of green light

You repeat the experiment described in Table 23.2 using slits separated by $d = 0.050$ mm and a screen at $L = 2.0$ m from the slits. This time, you use a green laser instead of a red laser. The second maximum ($m = 2$) appears on the screen at $y_2 = 4.4$ cm from the center of the 0th order maximum. Determine the wavelength of this light.

**Represent mathematically** We use Eq. (23.3) $y_m = (m\lambda L/d)$ to determine the wavelength.

**Solve and evaluate** To determine $\lambda$, multiply both sides by $d$ and divide by $mL$ to get $(y_m d/mL) = \lambda$. Now substitute the known values:

$$\lambda = \frac{y_m d}{mL} = \frac{(4.4 \times 10^{-2}\,\text{m})(5.0 \times 10^{-5}\,\text{m})}{2 \times 2.0\,\text{m}}$$

$$= 5.5 \times 10^{-7}\,\text{m} = 550\,\text{nm}$$

The order of magnitude for the wavelength is the same as for red light, but the value is smaller, consistent with Young's observations.

**Try it yourself:** What is the angle between the direction of light going toward the central maximum and the direction to the first *dark* fringe?

*Answer:* $5.5 \times 10^{-3}$ rad.

**Review Question 23.1** How is it possible to get *multiple* bright bands of light on a screen after light of a particular wavelength passes through two narrow, closely spaced slits?

# 23.2 Index of refraction, light speed, and wave coherence

Let's use the wave model of light with Snell's law of refraction to determine how the refractive index of a medium depends on the speed of light in that medium.

## Relating the refractive index and the speed of light in a substance

In the chapter on reflection and refraction (Chapter 21) we learned that light bends toward the normal when it travels from a medium with lower refractive index to a medium with a higher index, for example, when it travels from water to glass. We can determine the change in the direction of light propagation using Snell's law: $n_2 \sin \theta_2 = n_1 \sin \theta_1$ (**Figure 23.5a**). Because $n_2 > n_1$, $\sin \theta_2 < \sin \theta_1$ and $\theta_2 < \theta_1$. If we are using the wave model to explain this change, we need to consider that the wave fronts that are perpendicular to the propagation of the wave also bend at the interface. For a wave to bend toward the normal, its speed must be smaller in medium 2 than in medium 1. We know that the frequency of the wave does not change as it passes from medium 1 to medium 2. Therefore, the faster moving light in medium 1 has a longer wavelength than the slower moving light in medium 2.

**Figure 23.5** (a) The wave slows as it moves from medium 1 to medium 2, causing the wavelength to decrease. (b) The angles of incidence and refraction are related to the wavelengths.

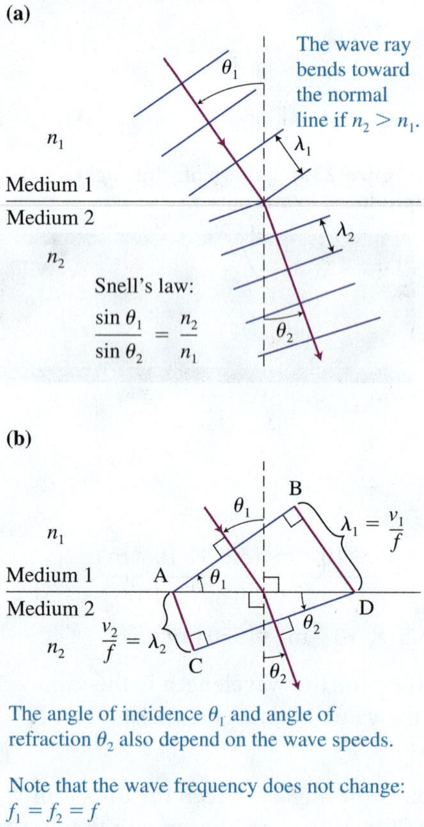
(a)
The wave ray bends toward the normal line if $n_2 > n_1$.
$\theta_1$
$\lambda_1$
$n_1$
Medium 1
Medium 2
$\lambda_2$
$n_2$
Snell's law:
$\frac{\sin \theta_1}{\sin \theta_2} = \frac{n_2}{n_1}$
$\theta_2$
(b)
B
$\theta_1$
$\lambda_1 = \frac{v_1}{f}$
$n_1$
Medium 1
A
$\theta_1$
D
Medium 2
$\theta_2$
$\frac{v_2}{f} = \lambda_2$
$n_2$
C
$\theta_2$
The angle of incidence $\theta_1$ and angle of refraction $\theta_2$ also depend on the wave speeds.
Note that the wave frequency does not change: $f_1 = f_2 = f$

Consider the right triangles DAB and ADC in Figure 23.5b. We know that the angle of incidence $\theta_1$ is equal to the angle DAB because the corresponding sides of $\theta_1$ and of DAB are perpendicular to each other. The angle of refraction $\theta_2$ is equal to the angle ADC for the same reason. The side opposite angle DAB (that is, DB) is one wavelength $\lambda_1$ of the light in medium 1. Similarly, the side opposite angle ADC (that is, AC) is one wavelength $\lambda_2$ of the light in medium 2. Thus, the sines of the angles are

$$\sin \text{DAB} = \sin \theta_1 = \frac{\lambda_1}{\text{DA}} \quad \text{and} \quad \sin \text{ADC} = \sin \theta_2 = \frac{\lambda_2}{\text{DA}}$$

Now, use the above to take the ratio of the sines of the angles $\theta_1$ and $\theta_2$:

$$\frac{\sin \theta_1}{\sin \theta_2} = \frac{\lambda_1/\text{DA}}{\lambda_2/\text{DA}} = \frac{\lambda_1}{\lambda_2}$$

We can use the relation between wavelength $\lambda$, frequency $f$, and speed $v$ of a light wave to rewrite the above in terms of the speeds of the light in the two media:

$$\lambda_1 = \frac{v_1}{f} \quad \text{and} \quad \lambda_2 = \frac{v_2}{f}$$

After substitution, we get

$$\frac{\sin \theta_1}{\sin \theta_2} = \frac{\lambda_1}{\lambda_2} = \frac{v_1/f}{v_2/f} = \frac{v_1}{v_2}$$

From Snell's law,

$$\frac{\sin \theta_1}{\sin \theta_2} = \frac{n_2}{n_1}$$

Thus, the ratio of the speeds of light in two media 1 and 2 should be equal to the inverse ratio of their indexes of refraction:

$$\frac{v_1}{v_2} = \frac{n_2}{n_1}$$

We know that the index of refraction of water is 1.33 and of air is 1.00. Thus the speed of light in water should equal the ratio of its speed in air $c$ ($3.0 \times 10^8$ m/s) divided by the $n = 1.33$ index of refraction of water:

$$v = \frac{c}{1.33} = \frac{(3.0 \times 10^8\ \text{m/s})}{1.33} = 2.26 \times 10^8\ \text{m/s}$$

This is exactly the speed that French physicists Hippolyte Fizeau and Leon Foucault obtained in their 1850 experiments to determine the speed of light in water.

Thus the wave model of light not only explains why light bends at the boundary of two media but also explains Snell's law by connecting the medium's index of refraction to the speed of light in that medium.

**Refractive index** **The refractive index $n$ of a medium equals the ratio of the speed of light $c$ in vacuum (or in air) to the speed of light $v$ in the medium:**

$$n = \frac{c}{v} \qquad (23.4)$$

## Does the refractive index depend on the color of the light?

We already know that when white light passes through a prism, it comes out as a set of colored bands (see **Figure 23.6a**). In other words, different colors of light are refracted differently by the prism. How does the wave model explain this? Snell's law of refraction [Eq. (21.3)] as applied to Figure 23.6b is

$$n_{\text{glass}} = \frac{n_{\text{air}} \sin \theta_{\text{air}}}{\sin \theta_{\text{glass}}}$$

Because the refracted angle in glass for blue light $\theta_{\text{glass}}$ is less than that for red light, the refractive index $n_{\text{glass}}$ for blue light must be greater than that for red light. The refractive index is related to the speed of light in the medium by Eq. (23.4). The different indexes of refraction for different colors mean that light of different colors or light waves of different frequencies travel at different speeds in the same medium. Only in a vacuum does light of all frequencies travel at the same speed.

**Figure 23.6** The refractive index $n$ of a medium is different for different wavelengths $\lambda$ of light.

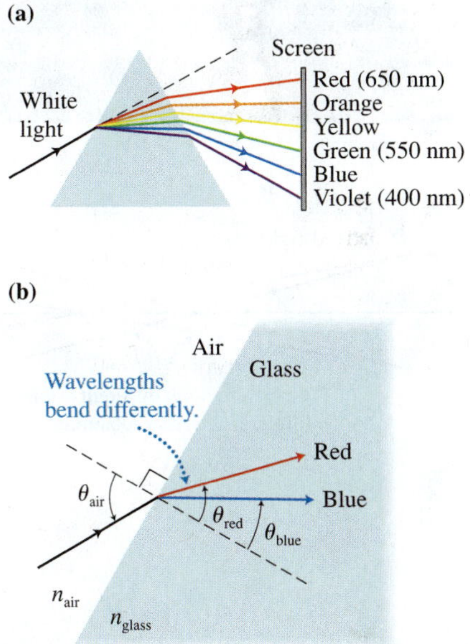


**QUANTITATIVE EXERCISE 23.2 Speed of light in a medium**

A laser pointer emits green light that has a 550-nm wavelength when in air. (a) What is the light's frequency? (b) What are the frequency, speed, and wavelength of the laser light as it travels through glass of refractive index 1.60?

**Represent mathematically** The frequency of a wave can be determined using its speed and wavelength $f = (c/\lambda)$. The speed of light in glass can be found using $n = (c/v)$.

**Solve and evaluate** (a) We are given the wavelength of the light when in air: $\lambda = 550 \times 10^{-9}$ m. The speed of the light when in air is $c = 3.0 \times 10^8$ m/s. Thus, the frequency of the light is

$$f = \frac{c}{\lambda} = \frac{3.0 \times 10^8\ \text{m/s}}{550 \times 10^{-9}\ \text{m}} = 5.5 \times 10^{14}\ \text{Hz}$$

(b) This light has the same $5.5 \times 10^{14}$-Hz frequency when in glass, because the frequency of a traveling wave is determined by its source, not by the medium. However, its speed and wavelength change:

$$v_{\text{in glass}} = \frac{c}{n_{\text{glass}}} = \frac{3.0 \times 10^8\ \text{m/s}}{1.60} = 1.9 \times 10^8\ \text{m/s}$$

$$\lambda_{\text{in glass}} = \frac{v_{\text{in glass}}}{f} = \frac{1.9 \times 10^8\ \text{m/s}}{5.5 \times 10^{14}\ \text{Hz}} = \frac{1.9 \times 10^8\ \text{m/s}}{5.5 \times 10^{14}\ \text{s}^{-1}}$$

$$= 3.5 \times 10^{-7}\ \text{m} = 350\ \text{nm}$$

**Try it yourself:** What is the wavelength of green light traveling through water of refractive index 1.33?

*Answer:* 410 nm.

**Figure 23.7** (a) Chromatic aberration is evident in this magnification of newsprint. (b) The purple and red lights form images at different places due to differences in their refractive indexes.

(a)

(b)

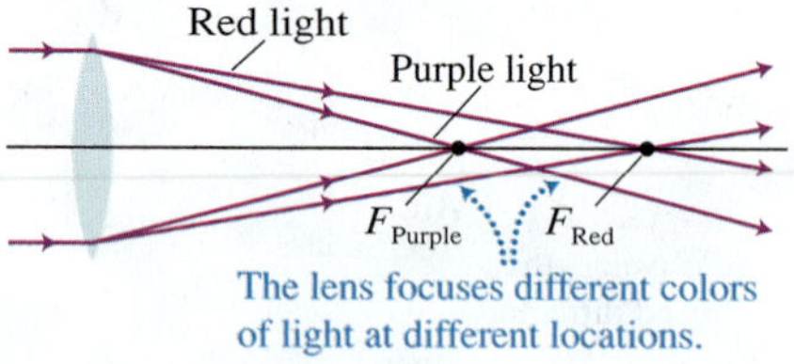
Red light
Purple light
$F_{Purple}$
$F_{Red}$
The lens focuses different colors of light at different locations.

**Figure 23.8** Two lightbulbs do not produce an interference pattern.

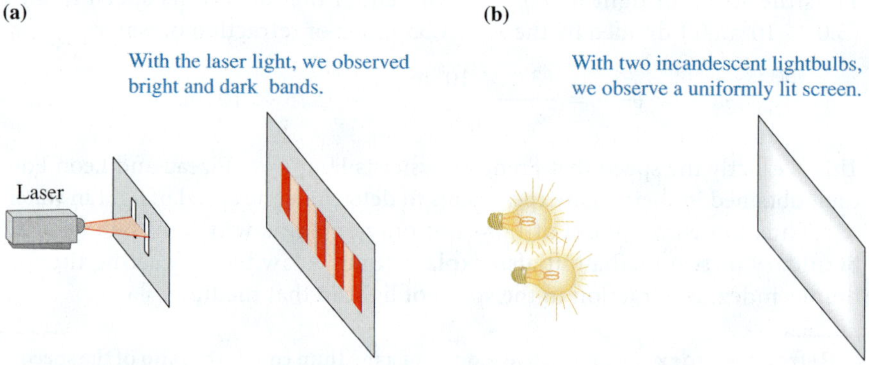
(a)
With the laser light, we observed bright and dark bands.
Laser
(b)
With two incandescent lightbulbs, we observe a uniformly lit screen.

**TIP** Note that when light travels from one medium to another its frequency does not change, but its wavelength does.

## Chromatic aberration in lenses—A practical problem in optical instruments

Photographs taken on a sunny day with a long exposure time have purple edges—an effect called *purple fringing*. Purple fringing occurs because the refractive index is wavelength dependent, resulting in a lens having a slightly smaller focal length for purple light than for red light ($F_{\text{purple}} < F_{\text{red}}$). Thus, the location of an image for the purple part of an object is slightly different from the image location for the red part of that object, an effect called **chromatic aberration**. Since the image locations are slightly different, their magnifications, which depend on image location, will also be slightly different. When the magnification of the purple image is slightly greater than the other colors, purple fringing occurs (**Figure 23.7**).

## Monochromatic and coherent waves

Our observations of double-slit interference (Table 23.1) involved laser light passing through two very narrow slits, each of which acted as a light source. If we replace the laser light with two small lightbulbs we do not observe an interference pattern. Instead, we see a uniformly illuminated screen (**Figure 23.8**). Why?

In order for two waves to interfere consistently over time, two conditions must be met. First, the waves must vibrate at the same constant frequency, producing what are called **monochromatic waves**. Second, the phases of the waves must be synchronized, meaning that as time passes they must arrive at a particular location consistently in phase, out of phase, or somewhere in between. The phase difference between the two waves must remain constant in time. When this occurs, the waves are called **coherent waves**. The synchronized vibrating beach balls shown in **Figure 23.9** produce coherent monochromatic waves and interference. For light, if the slits are close to each other, the same wave front excites coherent light wavelets from each slit and therefore the wavelets are coherent and monochromatic (**Figure 23.10**).

**Figure 23.9** The vibrating beach balls are monochromatic and coherent and produce a stable interference pattern.

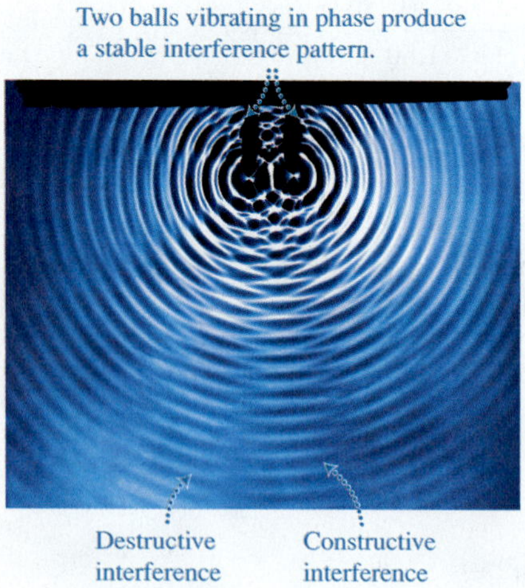
Two balls vibrating in phase produce a stable interference pattern.
Destructive interference
Constructive interference

Two lightbulbs do not produce an interference pattern on a wall in a room because each point on an extended light source sends light in all directions. If we add a second bulb, the waves reaching the wall come from even more light sources. The probability that the phase difference among all of them will remain constant in time is zero. Thus the waves add randomly and produce no interference pattern on the wall.

Now we can summarize.

**Coherent monochromatic waves** **Only waves of constant frequency (*monochromatic*) and having constant phase difference (*coherent*) can add to produce an interference pattern. In Young's double-slit experiment, the light from the slits is coherent because the disturbance at each slit is caused by the same passing wave front.**

**Figure 23.10** A laser produces coherent and monochromatic disturbances at the slits.

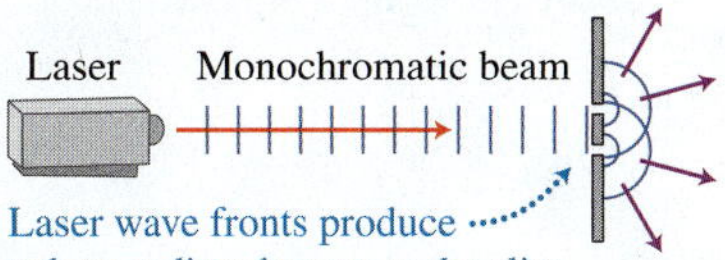
Laser
Monochromatic beam
Laser wave fronts produce coherent disturbances at the slits.

In his original experiment with sunlight, Young made a tiny hole in a window shutter for the light to pass through before it hit two slits. This hole acted as a single point-like source illuminating the two slits. Although he worked with white light, for each frequency of light the slits functioned as a pair of coherent monochromatic sources and therefore produced a steady interference pattern for that frequency (color). The resulting pattern on the screen had bands of different color at different locations.

**Review Question 23.2** The refractive index of glass is about 1.6. What does this number tell us about the speed of light in glass? What does it tell us about the frequency of the light? Its wavelength?

## 23.3 Gratings: An application of interference

Thomas Young's double-slit experiments helped establish the wave model of light. However, double slits had no practical applications in his lifetime. We now know that the interference of light passing through a large number of narrow closely spaced slits (called a **grating**) can be used to analyze the wavelengths of light emitted by biological materials, particles in Earth's atmosphere, and even stars and galaxies.

How does a grating work? When light of wavelength $\lambda$ passes through two narrow closely spaced slits, we observe the interference maxima at angular deflections determined by the equation $\sin\theta_m = (m\lambda/d)$, where $m = 0, \pm 1, \pm 2, \pm 3, \ldots$ If we were to use a meter that measures the intensity of light at different locations on the screen, we would find that the intensity of light reaching the screen changes with the meter's position on the screen. A simplified graph of this intensity as measured on a screen is shown in **Figure 23.11a**. (We do not show that the overall brightness changes across the screen.) For each peak there is a gradual change of brightness from the very maximum at the locations determined by the equation above to zero at the locations where $\sin\theta = ((2m + 1)\lambda/2)/d$.

If we use four slits instead of two, we again see bright spots on the screen at the same locations. But the intensity-versus-position graph is different. It now has three minima between the bright bands. The total amount of light reaching the screen increases by a factor of 2 and the bright bands are brighter and narrower (Figure 23.11b), and do not change position. If we increase to 16 slits, the bright bands still do not change positions (the horizontal axes in Figure 23.11a–c show the same locations of the bright bands) but become even brighter and narrower (notice the change in the readings on the vertical axis for the three graphs). There are many more minima between the bright bands, and we see almost no light on the screen between the bright bands (Figure 23.11c). This pattern is a due to the complicated interference of the coherent wavelets coming from many slits. Having such narrow bright bands allows us to distinguish more clearly between different frequencies of light passing through the slits.

A typical grating has hundreds of slits per millimeter, and the bright bands are very intense and narrow with almost complete darkness between them (similar to Figure 23.11c). The locations of the bright bands are determined using

**Figure 23.11** The bright interference bands sharpen with an increasing number of slits.

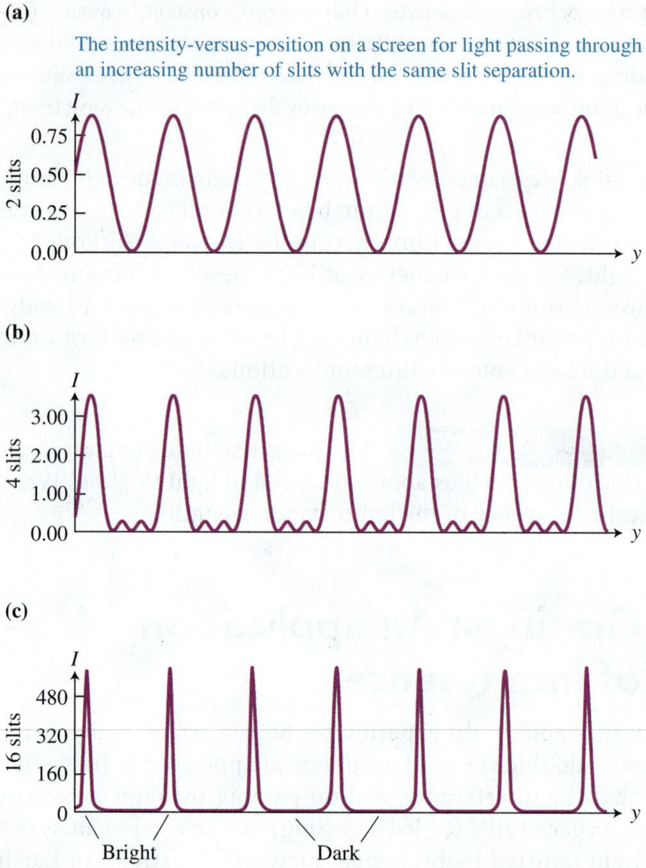


Eq. (23.3). The distance $d$ between the slits is usually dramatically less for gratings than for double slits. Consequently, the bright bands are seen much farther apart.

Thus, for a grating with spacing between adjacent slits $d$, the bright bands (interference maxima) occur at the angles determined from the expression

$$\sin\theta_m = \frac{m\lambda}{d}, \text{ where } m = 0, \pm 1, \pm 2, \pm 3, \ldots \quad (23.5)$$

**TIP** Gratings are usually labeled in terms of the number of slits per millimeter or per centimeter, for example, 100 slits/cm. In such a grating, the distance between the slits would be one hundredth of a centimeter. If a grating has 1000 slits/mm, the distance between the slits is one thousandth of a millimeter. You can find the distance between the slits by dividing 1 by the number of slits per unit length. The distance comes out in the respective units (cm or mm).

**QUANTITATIVE EXERCISE 23.3 Double-slit and grating interference**

You have a slide with two narrow slits separated by 0.10 mm and a grating with 100 slits/mm. Compare the angular deflections for the $m = 1$ maximum (see figure top left of next page) using the double slit and the grating for light of wavelength 670 nm. Describe the differences in the interference pattern that you observe.

**Represent mathematically** We need to find the angle $\theta_1$ for both cases. Use the equation $\sin\theta_m = (m\lambda/d)$ with $m = 1$.

**Solve and evaluate** To find the slit separation for the grating, divide 1 by the number of slits per mm:

$$d_{\text{grating}} = \frac{1}{100 \text{ slits/mm}} = 10^{-2} \text{ mm} = 10^{-5} \text{ m}.$$

For the double slit, $d_{\text{slits}} = 10^{-4}$ m.

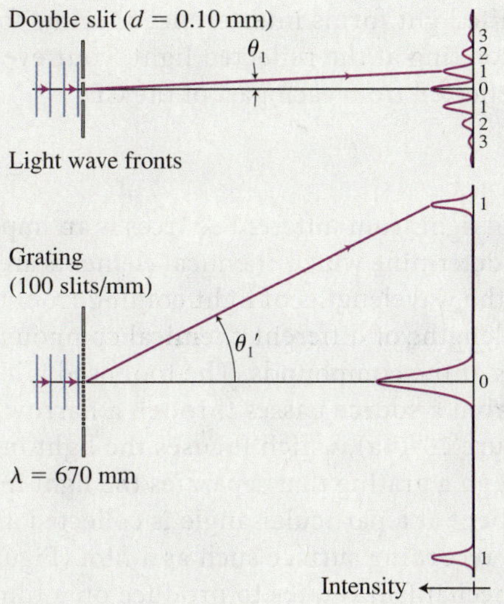


$$\text{Slits: } \sin\theta_1 = \frac{\lambda}{d_{\text{slits}}} = \frac{670 \times 10^{-9}\,\text{m}}{10^{-4}\,\text{m}} = 6.7 \times 10^{-3}$$

$$\text{Grating: } \sin\theta_1{}' = \frac{\lambda}{d_{\text{grating}}} = \frac{670 \times 10^{-9}\,\text{m}}{10^{-5}\,\text{m}} = 6.7 \times 10^{-2}$$

As expected, the bright regions are much farther apart from each other for the grating than for the double slit. The bright regions from the grating look like sharp dots, and those from the double slit look like fuzzy bands.

**Try it yourself:** What is the angle of deflection of the 1st order maximum for a slide with three very narrow slits, each separated from each other by 0.10 mm?

*Answer:* Adding or removing slits without changing their separation does not change the location of the maxima. With more slits, the spacing between bright bands is the same; the maxima are brighter and narrower.

**Figure 23.12** A white light diffraction pattern caused by a grating.

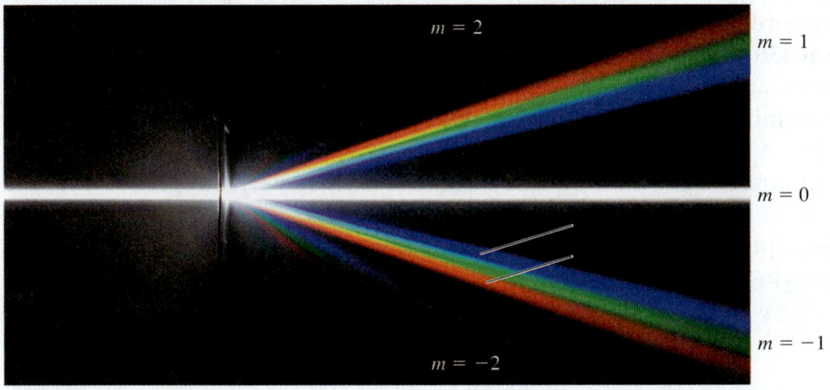


## White light incident on grating

What happens when white light shines on a grating? After passing through the grating, light of all wavelengths produces a bright band at the center (the $m = 0$ band); thus on a screen beyond the grating, you see a central maximum of white light. However, according to $\sin\theta_m = (m\lambda/d)$ different wavelengths of the white light are deflected at different angles for each $m$ (other than $m = 0$). The greater the wavelength, the larger the angle from the center of the $m = 0$ white band to the $m$th bright band. Consequently, each $m \neq 0$ band is a rainbow of colors on the screen, with the blue color always closer to the central maximum than the red, because light of smaller wavelength has a smaller deflection angle—$\sin\theta_m = (m\lambda/d)$ (**Figure 23.12**). If the slit separation $d$ is smaller, the colors are separated more. The colored band, called a **spectrum,** looks exactly like the spectrum produced when white light passes through a prism. However, the mechanisms are different. A spectrum produced by a prism is a result of the different speeds with which light waves of different frequencies travel in a medium. A spectrum produced by a grating is a result of the light of different wavelengths interfering constructively at different locations.

**Figure 23.13** White light reflection from a CD.

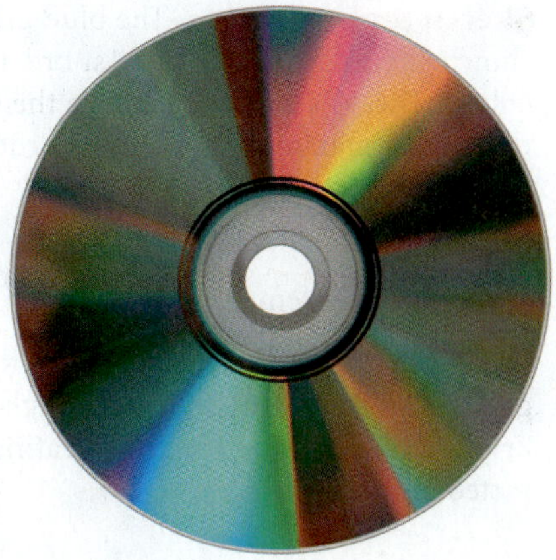

## CDs and DVDs—reflection gratings

The reflective surfaces of CDs and DVDs (**Figure 23.13**) consist of spirals of closely spaced grooves with many color bands. This makes a CD a type of

**Figure 23.14** Spectrometer. (a) A grating separates the light from a source into different angular deflections. (b) The detector focuses light of particular wavelengths on a film. (c) The spectrum produced by light from a mercury lamp.

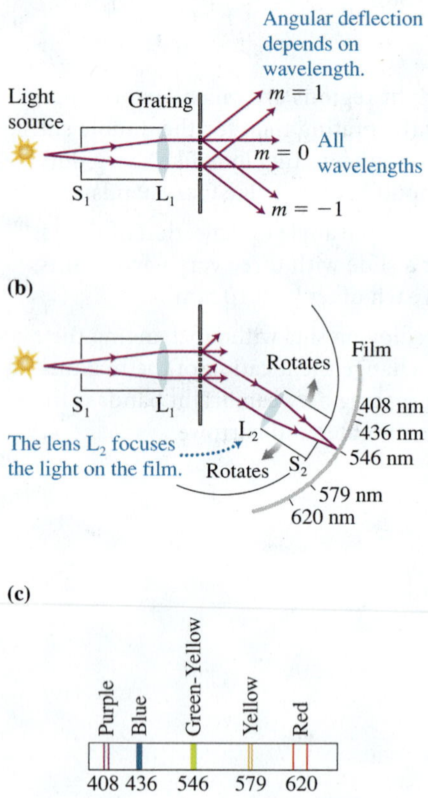


grating that reflects light instead of transmitting it. The grooves play the role of the slits: the reflected white light forms interference maxima for different colors at different angles. Looking at the reflected light, your eye intercepts only one color of the light reflected from each part of the CD.

## Spectrometer

Analyzing the wavelengths of light from different sources is an important tool in astronomy. Astronomers determine which chemical elements are present in distant stars by identifying the wavelengths of light coming from those stars. Chemists measure the wavelengths of different chemical compounds and are able to determine properties of the compounds. The tool used is a **spectrometer** (**Figure 23.14**). Light from a source passes through a narrow slit $S_1$ and then through a lens $L_1$ (Figure 23.14a), which focuses the light into a beam. This beam then passes through a grating that separates the light into its various wavelengths. The light bent at a particular angle is collected and focused by lens $L_2$ and slit $S_2$ onto a recording surface such as a film (Figure 23.14b). The detector lens and slit mechanism rotates to produce on a film the spectrum of wavelengths of light coming from the source (Figure 23.14c).

A spectrometer can be used to analyze light from any source. The Sun and incandescent lightbulbs emit a broad range of light frequencies—a continuous spectrum of light. However, hot, low-density gases do not. Instead, they emit light of specific wavelengths that indicates that specific types of atoms and molecules are present in the gas. For example, mercury vapor, whose spectrum was recorded in Figure 23.14c, emits a pair of purple lines near 408 nm, a blue line at 436 nm, a greenish-yellow line at 546 nm, a yellowish pair of lines near 579 nm, and two dim red lines near 620 nm. We will learn more about why atoms emit specific wavelengths of light in later chapters.

Spectra also provide information about the expansion of the universe. In the spectra of light from distant galaxies, the characteristic spectral lines are shifted to longer wavelengths (lower frequencies). This so-called red shift is an example of the Doppler effect for light and indicates that distant galaxies are moving away from us—supporting the idea that the universe is expanding.

### EXAMPLE 23.4 Hydrogen spectrum

You wish to determine the frequencies of light emitted by atoms in a hydrogen gas-discharge lamp. You use a spectrometer with a 4000 slits/cm grating. The light-sensitive detector of the spectrometer is 0.500 m from the grating, as shown. The image produced by the detector looks as shown at right. Determine the wavelengths of the hydrogen lines in the red and blue-green parts of the visible spectrum.

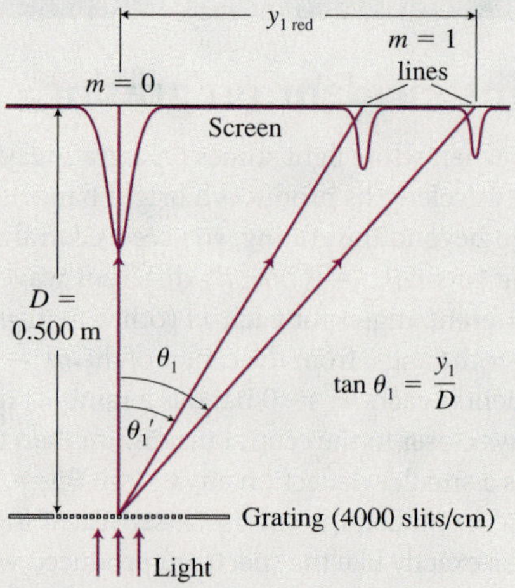


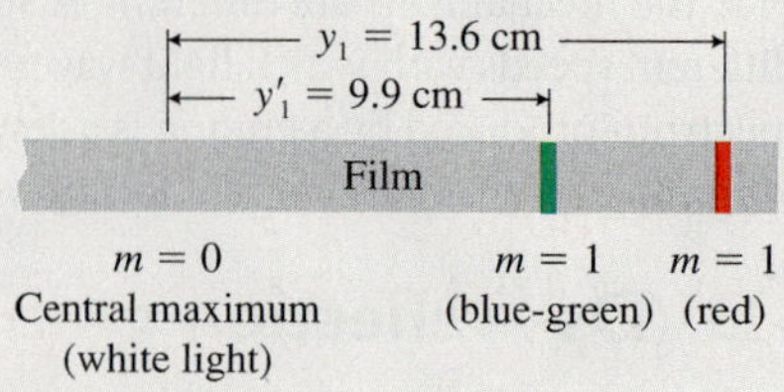


**Sketch and translate** The blue-green and red bands shown should be the $m = 1$ 1st order maxima for these colors. The $y_1$'s in the figure are the distances between the central maximum and the 1st order $m = 1$ maxima. The slit separation is

$$d = \frac{1}{4000 \text{ lines/cm}} = 2.50 \times 10^{-4} \text{ cm} = 2.5 \times 10^{-6} \text{ m}$$

**Simplify and diagram** Assume that the slits of the grating are coherent sources of light. The path of the bright bands of light from the grating to the film is depicted in the figure.

**Represent mathematically** Using the diagram above and geometry, we get for each color:

$$\text{Red:} \quad \tan\theta_1 = \frac{y_1}{D}$$

$$\text{Blue-green:} \quad \tan\theta_1{}' = \frac{y'_1}{D}$$

Use these relations to find the corresponding angles. Then use the value for $d$ and Eq. (23.3) $\lambda = d\sin\theta_1$ to find the wavelength of each line.

**Solve and evaluate**

$$\text{Red:} \quad \tan\theta_1 = \frac{y_1}{D} = \frac{0.136\text{ m}}{0.500\text{ m}} = 0.272 \text{ or } \theta_1 = 15.2°$$

$$\lambda = d\sin\theta_1 = (2.50 \times 10^{-6}\text{ m})\sin 15.2°$$
$$= 6.55 \times 10^{-7}\text{ m} = 655\text{ nm}$$

$$\text{Blue-green:} \quad \tan\theta_1{}' = \frac{y'_1}{D} = \frac{0.099\text{ m}}{0.500\text{ m}}$$
$$= 0.198 \text{ or } \theta_1 = 11.2°$$

$$\lambda' = d\sin\theta_1{}' = (2.50 \times 10^{-6}\text{ m})\sin 11.2°$$
$$= 4.86 \times 10^{-7}\text{ m} = 486\text{ nm}$$

The wavelength for red light is greater than that for blue-green, as expected. The orders of magnitude for the wavelengths are reasonable—hundreds of nanometers for the wavelengths of visible light.

**Try it yourself:** Suppose a different atom emits yellow light of 580 nm. Where would it fall on the above pattern?

*Answer:* About 12 cm from the $m = 0$ central maximum.

**Review Question 23.3** How does the location of the maxima change when we add more slits to an interference device (assuming that we make them the same distance from each other)? How does the width of each maximum change?

## 23.4 Thin-film interference

The beautiful, swirling colors that occur on the surfaces of soap bubbles also appear in the thin oil films that float on water, the shimmering patterns of some butterflies, the brilliant feathers of many hummingbirds, and the elegant peacock's tail (**Figure 23.15**). This array of color is not caused by pigment but by the interference of light waves reflected off of the boundaries of the surface. All of these examples—the bubbles, the oil, and the animals—are the result of thin-film interference.

**Figure 23.15** The colorful peacock tail feathers are caused by thin-film interference.

### Bright and dark bands due to reflected monochromatic light

**Figure 23.16a** shows a photo of a thin soap film with white light shining on it. Let's start by analyzing the pattern produced when we irradiate the soap film with red light only (Figure 23.16b). We see a regular pattern of bright and dark bands produced by the reflected red light, a little like a double-slit interference pattern. What causes these patterns?

We have learned that light is partially transmitted and partially reflected at interfaces between media with different refractive indexes. For a soap film and other thin films, light reflection occurs at both the front surface and the back surface (Figure 23.16c). These two reflected waves can add constructively to make brighter reflected light, destructively to make no reflected light, and anything in between.

Two factors affect the way in which the light reflected from the front surface combines with the light reflected from the back surface.

1. *Phase change upon reflection.* We know from our study of mechanical waves (Chapter 20) that a reflected wave changes phase by $180° = \pi$ rad if the second medium is denser than the first. For light, this would happen

**Figure 23.16** Soap bubble patterns caused by (a) white light and (b) red light. (c) Rays that interfere to cause the colors and fringes.

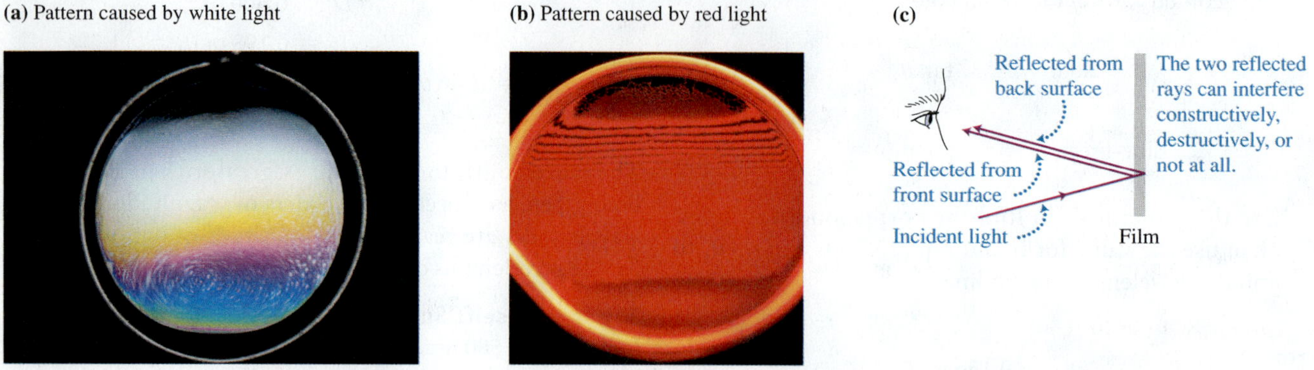


**Figure 23.17** (a) and (b) Phase changes for light that reflects off a soap bubble film. (c) There is also a phase change because of a path length difference of light reflected from the front and the back of the film.

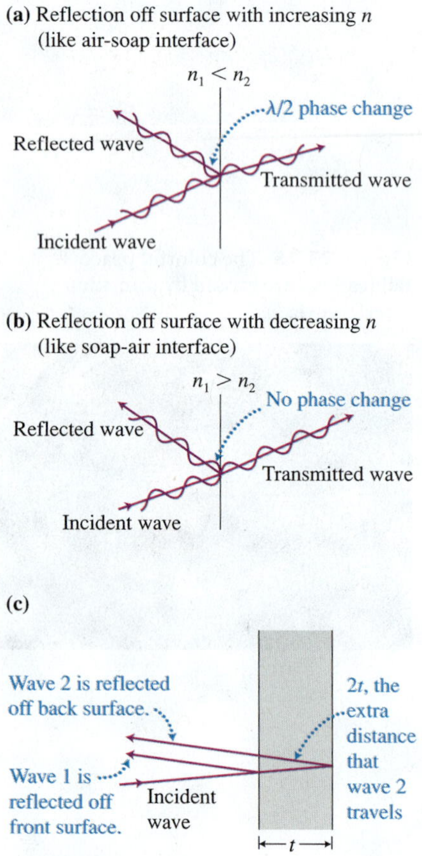


if the second medium had a greater refractive index than the first. However, the phase of the reflected wave remains the same if the second medium has a smaller refractive index than the first. Thus, when light traveling in air reflects off the front surface of the soap bubble film of refractive index 1.3, there is a phase change of $180° = \pi$ rad (**Figure 23.17a**). When light traveling inside the soap bubble film reflects off the back surface, there is no phase change (Figure 23.17b) since the refractive index of air is less than 1.3. So only the light reflecting off of the front of the bubble undergoes a phase change.

2. *Phase difference due to path length difference.* In Figure 23.17c we see that wave 2, which reflects from the back surface of the thin film, travels a longer distance than wave 1, which reflects from the front surface. Because of this path-length difference the two waves have different phases when they recombine after reflecting from the thin film. For example, if the film is one half wavelength thick, wave 2 travels a total of one wavelength farther than wave 1. Then the two waves should be in phase when they recombine (assuming no phase changes occur due to reflection). If the film is one-fourth of a wavelength thick, the path length difference is one-half of the wavelength. The waves are $180° = \pi$ rad out of phase when they recombine. Often those waves are said to be one half wavelength out of phase to underscore that when they add, the resulting wave has zero amplitude.

There is an additional important feature of this path-length difference. The speed $v$ of light in a medium of refractive index $n$ is $v = (c/n)$, where $c$ is the speed of light in air. If the refractive index $n$ of the thin film is greater than 1.0, then the wavelength of the light in the film is

$$\lambda_{\text{medium}} = \frac{v_{\text{medium}}}{f} = \frac{c/n_{\text{medium}}}{f} = \frac{c/f}{n_{\text{medium}}} = \frac{\lambda_{\text{air}}}{n_{\text{medium}}}$$

To determine the phase difference between the waves, we have to use the light's wavelength in the film ($\lambda_{\text{medium}}$). To determine the total phase difference, we need to account for both factors (1) and (2). Let's apply this reasoning to soap bubbles.

**Soap Bubble** The film of the soap bubble has a greater refractive index than air. Thus, light reflected from outside the air-soap bubble interface has a $180° = \pi$ rad phase change (equivalent to the one half wavelength path difference). Part of the transmitted light reflects from inside the soap bubble-air interface and has no phase change. Hence, the relative phase change from reflection between the two waves is $180° = \pi$ rad (equivalent to one half wavelength in terms of path length difference). For a *bright band* to appear on the surface of the bubble (constructive interference), the net phase change should be an integer multiple of $360° = 2\pi$ rad. Thus, the path length difference due to the light traveling back and forth through the film a distance $2t$ must also be one half wavelength, or three half wavelengths, or five half wavelengths, etc. so that when added together, those path length differences produce an integer number of whole wavelengths. Thus, $2t = (1, 3, 5, \ldots)(\lambda/n)/2$, where $\lambda/n$ is the wavelength of light inside the bubble film. We've assumed through all this that the incident light is almost perpendicular to the film surface.

For a *dark band* to appear on the surface of the bubble, destructive interference must occur between the two reflected waves. The net phase change needs to be an odd multiple of $180° = \pi$ rad or an odd number of half wavelengths in terms of path lengths. Thus, the path-length difference $2t$ must be an integer number of wavelengths: 0, 1, 2, . . . wavelengths. In short, for a dark band $2t = (0, 1, 2, \ldots)(\lambda/n)$.

Thus for a specific wavelength of light, the thickness of the bubble film determines if light will be reflected from the film. Since the film thickness varies over the bubble, some locations look shiny (constructive interference) and other locations look dark (destructive interference). We can test this reasoning experimentally (Testing Experiment **Table 23.3**).

**TESTING EXPERIMENT TABLE**

**23.3 Testing thin-film interference ideas.**

VIDEO 23.3

**Testing experiment**

Produce a soap film and shine red light on it. Predict what you will see at the top of the film just before it breaks.

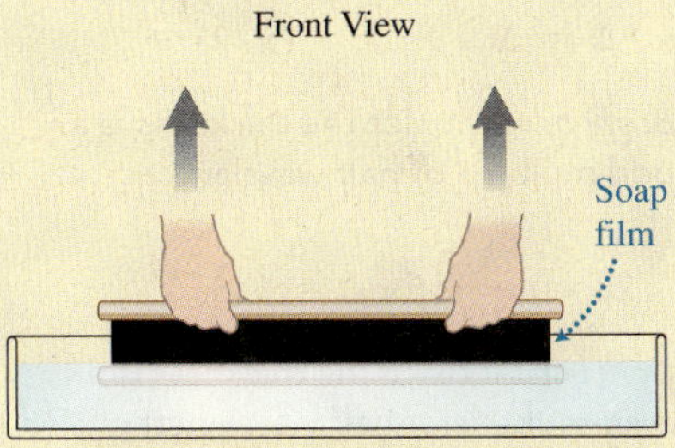


**Prediction**

The soap film thins from evaporation and dripping from the bottom. Because of gravitational effects, the film is always thinnest at the top. Just before it breaks, the thickness at the top is nearly zero ($t \approx 0$). Waves reflected off the front and back of the film are out of phase because of reflection. We predict that the top of the film appears dark just before it breaks.

**Outcome**

We observe that the top of the film is dark.

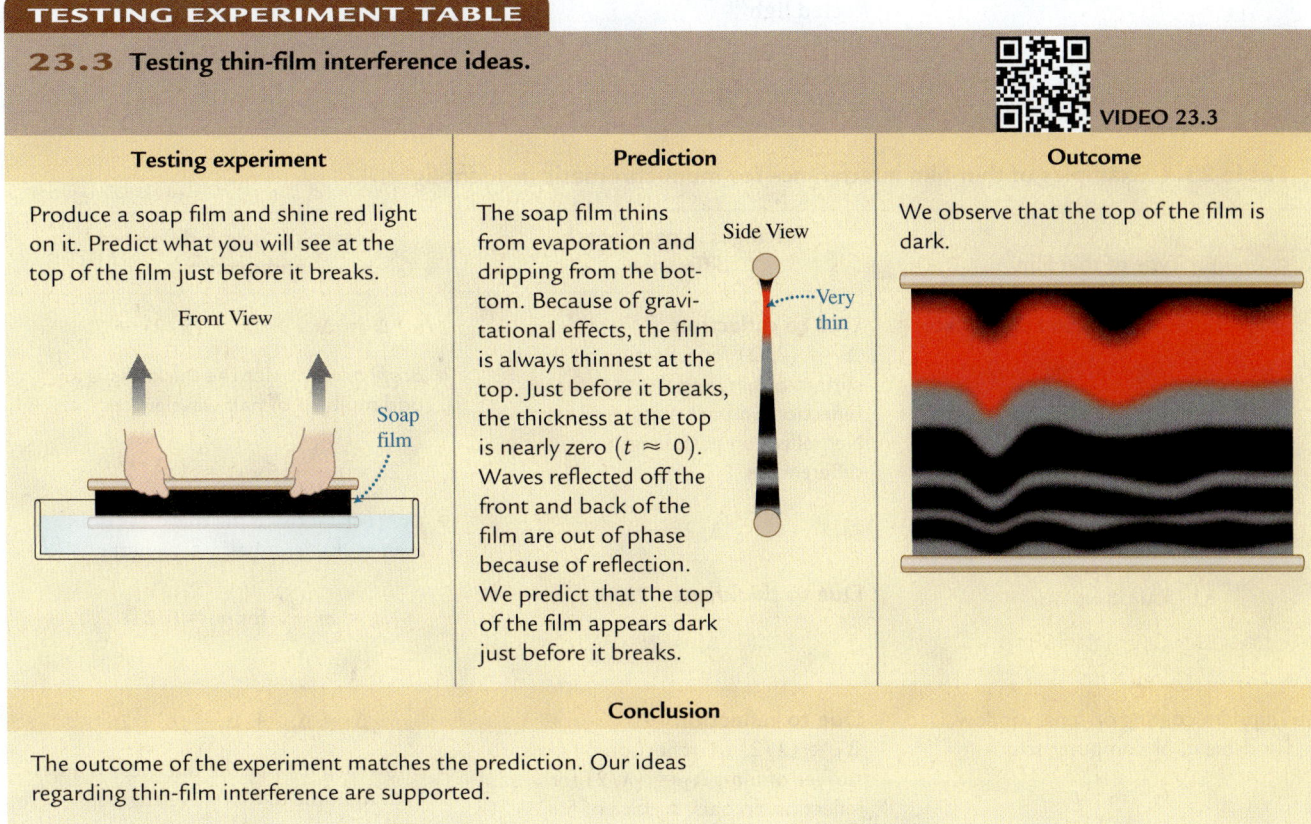

**Conclusion**

The outcome of the experiment matches the prediction. Our ideas regarding thin-film interference are supported.

Now that we are more confident in our reasoning about thin films, let's use it to develop a method for reducing the glare from glass surfaces, such as windows or camera lenses.

**Thin Film on Glass Surface** In an optical instrument with several lenses, light reflection at each air-glass interface reduces the amount of light getting to the detector, whether that be the charge-coupled device in a digital camera or your eye looking through a microscope. To reduce this effect and increase the amount of light reaching the detector, the glass surfaces are often covered with a thin film. Waves reflecting from the film interfere destructively, minimizing reflected light.

Consider light incident on a thin film with refractive index $n_{\text{film}} = 1.4$ on the front surface of a glass lens whose refractive index is $n_{\text{glass}} = 1.6$ (**Figure 23.18**). Wave 1 undergoes a $180° = \pi$ rad phase change upon reflection, since $n_{\text{film}} > n_{\text{air}}$. Wave 2 also undergoes a $180° = \pi$ rad phase change, since $n_{\text{glass}} > n_{\text{film}}$. The net effect is that the two waves remain in phase. To minimize the amount of reflected light we need to create destructive interference between these two waves. Thus, the path-length difference must equal an odd integer multiple of $180° = \pi$ rad (an odd integer number of half wavelengths of the light in terms of path-length difference). Choosing the proper thickness of the film $t$, so that $2t = (1, 3, 5, \ldots)((\lambda/n)/2)$ where $n$ is the refractive index of the film, achieves this goal. If for some reason we wanted to maximize the reflected light, then the two waves must constructively interfere. This will happen when $2t = (0, 2, 4, \ldots)((\lambda/n)/2)$.

**Figure 23.18** Light reflecting from a thin film on glass surface. Light passing through the glass is not shown.

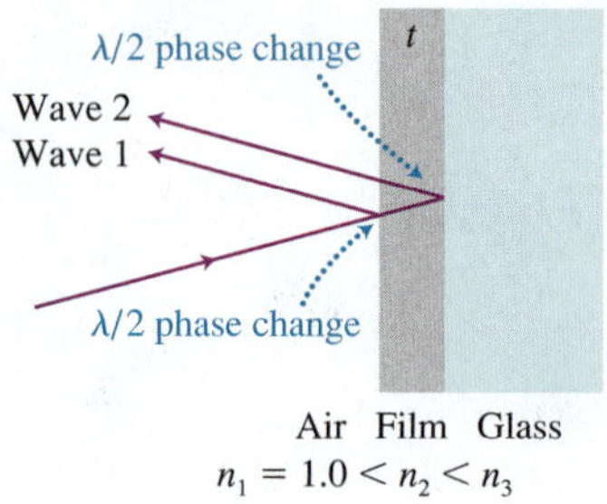


**Table 23.4** summarizes the ways in which the refractive index of a thin film and its thickness combine to produce the bright and dark bands of reflected light.

**Table 23.4 Examples of thin-film interference for monochromatic incident light.**

| Type of thin film | Changes in path length difference | Total path length difference and outcome |
|---|---|---|
| Soap bubble in air or oil film on water<br>$n_1 < n_2 > n_3$<br>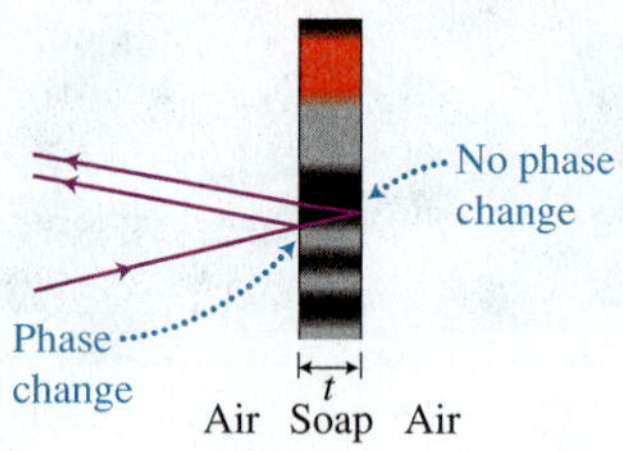 | **Due to reflection**<br>$\Delta_1 = (\lambda/2)$ for reflection on front surface of film and $\Delta_2 = 0$ for reflection on back surface of film. Net reflection path length difference is<br>$\Delta_{\text{ref}} = \dfrac{\lambda}{2}$<br>**Due to thickness**<br>$\Delta_t = 2t$ | $\Delta = \Delta_{\text{ref}} + \Delta_t = (\lambda/2) + 2t$<br>■ *Bright band* if twice the thickness is an odd multiple of half wavelengths:<br>$2t = m\dfrac{\lambda/n}{2}$ for $m = 1, 3, 5 \ldots$<br>■ *Dark band* if twice the thickness is an even multiple of half wavelengths:<br>$2t = m\dfrac{\lambda/n}{2}$ for $m = 0, 2, 4 \ldots\ldots$ |
| Thin-film coating on lens, window, windshield, or computer screen<br>$n_1 < n_2 < n_3$<br>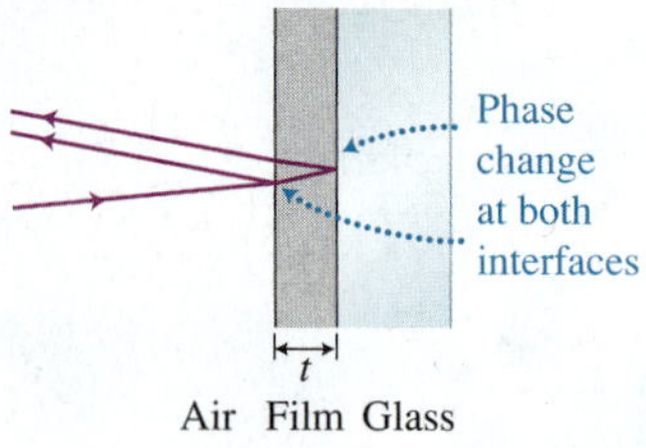 | **Due to reflection**<br>$\Delta_1 = (\lambda/2)$ for reflection on front surface of film; $\Delta_2 = (\lambda/2)$ for reflection on back surface of film. Net reflection path length difference is $\Delta_{\text{ref}} = \lambda$.<br>**Due to thickness**<br>$\Delta_t = 2t$ | $\Delta = \Delta_{\text{ref}} + \Delta_t = \lambda + 2t$<br>■ *Bright band* if twice the thickness is an even multiple of half wavelengths:<br>$2t = m\dfrac{\lambda/n}{2}$ for $m = 0, 2, 4, \ldots$<br>■ *Dark band* if twice the thickness is an odd multiple of half wavelengths:<br>$2t = m\dfrac{\lambda/n}{2}$ for $m = 1, 3, 5 \ldots$ |

## Continuous change in soap bubble band locations

If you observe a soap bubble for a while, you will see that the locations of the bright and dark bands on the surface continually change. This must mean that the phase differences between light reflected off the front and back faces of the bubble are changing with time. Since the refractive index of the film does not change with time, it must be that the thickness *t* does change—perhaps due to evaporation, dripping, or sagging at the bottom due to gravitational effects. This explanation is consistent with the observation that the light reflection patterns from thin oil films are more stable than those from soap bubbles, since oil does not evaporate as quickly as water does.

## Reflection patterns on a soap bubble in white light

**Figure 23.19** A soap bubble in white light.

Colors are the result of the absence of one or more colors from white light.

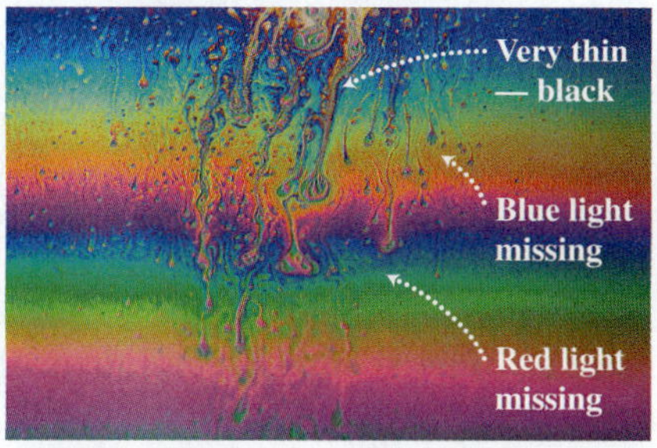


So far we have assumed that the surface of a bubble or a camera lens film is illuminated by monochromatic light (light of one wavelength). However, usually it is white light that is incident on the thin film. Consider the soap film in **Figure 23.19**. The white light incident on the film includes light wavelengths from about 400 nm to 700 nm. When the light of different wavelengths reflects and refracts, for one wavelength the net path difference might be equal to an odd number of half wavelengths. For another wavelength it could be exactly an even number of half wavelengths. For most wavelengths, however, the interference is neither completely constructive nor completely destructive. Also, when we look at a different spot on the film, the waves arriving at our eyes are incident on the film at different angles, and the film has different thicknesses at different locations. Because of all these factors, light of only a small wavelength range is destructively reduced in intensity at a particular location on the bubble when viewed in a particular direction. The color that is reduced differs from place to place. At other locations other colors are destructively reduced in intensity. As a result, each observer will see a different distribution of colors reflected from each location on the film. The white light colors that are left when light of a small wavelength range is subtracted are called **complementary colors.** Complementary colors are different from the spectrum or rainbow colors produced by a grating or a prism. These devices separate in space the primary colors that are combined spatially inside a beam of white light.

For example, if blue light is subtracted from white light, we see the remaining light as its complement, yellow. When red is subtracted from white light, we see blue-green. Long-wavelength light (red) needs a thicker bubble wall for its reflections to destructively interfere than short-wavelength light (blue). This means that as the bubble's thickness decreases, first red is canceled out, leaving blue-green; then green is canceled, leaving magenta; and finally blue is canceled, leaving yellow. Eventually, the bubble becomes so thin that all wavelengths are canceled and the bubble appears black, just as the film in the testing experiment became black at the top just before it broke.

> **TIP** Remember that it is the frequency of light that determines its color. The frequency does not change when light travels from one medium to another. The wavelength does change. Thus when you hear "the wavelength of green light is 550 nm," check whether this statement assumes that the light is propagating in a vacuum.

## Lens coatings

As we have learned, we can reduce the reflected light of a particular wavelength if we have a film of a particular thickness. The thickness of the coating on glass lenses for cameras, microscopes, and eyeglasses is usually chosen to reduce reflection of light at a wavelength of 550 nm, the center of the visible spectrum. The film reduces the reflected light from about 4% of the intensity of the incident light (without the coating) to less than 1%. The coating is less effective at reducing the reflection at extreme visible wavelengths (red at 700 nm and violet at 400 nm). A lens with a thin-film coating has a purple hue because it reflects red and violet light more than other colors; when combined, these colors appear purple.

**QUANTITATIVE EXERCISE 23.5 How thin is the film?**

In the optical industry, a thin film of magnesium fluoride ($MgF_2$) with a refractive index of 1.38 is used to coat a glass lens of refractive index 1.50. What should be the thickness of the coating to reduce the reflection of 550-nm light (the wavelength of green light in a vacuum)?

**Represent mathematically** The purpose of the coating (a thin film) is to produce destructive interference for the reflected light. The refractive index increases at the air-film interface and at the film-glass interface. Thus there is a net zero phase change due to reflection. For destructive interference, the path difference (twice the thickness $t$) should be $2t = (\lambda_{air}/2n), 3(\lambda_{air}/2n), 5(\lambda_{air}/2n)$, and so forth, where $n$ is the refractive index of the film.

**Solve and evaluate** The thinnest possible film would be when $m = 1$: $2t = (\lambda_{air}/2n)$. Thus $t = (1/4)(\lambda_{air}/n) = (1/4)(550\text{ nm}/1.38) = 100\text{ nm}$. This is very thin. Making the film slightly thicker might still achieve the purpose of canceling the reflected light and provide greater durability for the film. Possible thicknesses are 300 nm and 500 nm—any positive odd integer times the minimal thickness.

**Try it yourself:** How thick should the film in the example be to increase the amount of reflected light?

*Answer:* We could use a 200-nm thin film of the material in the above example. Alternatively, we could cover a glass surface with a thin film of greater refractive index than the glass, achieving greater light reflection with the same 100-nm film. Increasing reflection is useful when building devices that allow us to look at bright light sources, such as sun visors for astronauts.

## Limitations of the thin-film interference model

Window glass does not display color patterns. Why is that? Window glass is several millimeters thick, several thousand times the wavelength of light, like a *very* thick walled bubble. Natural sources of light such as the Sun do not emit light waves continuously. Instead, they emit light in short wave bursts over a very short time interval. Two consecutive bursts have a random phase difference between them. Imagine that light from one burst travels through the window glass and reflects back. While it was traveling inside the glass, a new burst reached the surface of the glass. As a result, the phase difference between the two waves is determined not only by reflections and path-length differences but also by the time between bursts. The bursts are therefore not coherent and do not interfere in a way that remains constant in time.

**Figure 23.20** A Morpho butterfly.

## Bird and butterfly colors

Many colors in the natural world, such as those of flower petals and leaves, are caused by organic pigments that absorb certain colors and reflect others. Chlorophyll in leaves absorbs most colors but reflects green, which makes the leaves look green. However, some natural color is the result of the thin-film interference of light. Some feathers and insect bodies consist of microscopic translucent structures that act like thin films to produce destructive and constructive interference of light. We can see the colors that result from these structures in the tail feathers of peacocks, the head feathers of hummingbirds, and the wings of Morpho butterflies (**Figure 23.20**).

**TIP** Remember that interference patterns for light appear when two coherent waves arrive at the same location at the same time. These coherent waves can be created in two ways: by using two points on the same wave front as sources (double-slit interference) or by dividing the wave front into two parts (thin-film interference).

**Review Question 23.4** If we look through a grating at a source of white light and then look at a thin film also illuminated by white light, we see different colors. What is the same and what is different about how these colors appear?

## 23.5 Diffraction of light

Earlier in the chapter we analyzed interference phenomena involving light that passed through two narrow slits. We modeled those slits as infinitely narrow, and when light shined on them we considered them to be point-like sources of light. This model explained the observed interference pattern on the screen (**Figure 23.21**). However, if you look carefully at the pattern produced by light passing through two slits on the screen, you will notice that in addition to the alternating bright and dark bands, there is also an overall periodic modulation of the brightness in the pattern. The wider the slits (not their separation, but their individual widths), the more pronounced this modulation is. In addition to the interference of light coming from the two slits, something else is going on that is related to the individual slits themselves. Consider the experiments in Observational Experiment **Table 23.5** involving light passing through a single slit.

**Figure 23.21** The double-slit interference pattern is modulated by single-slit diffraction.

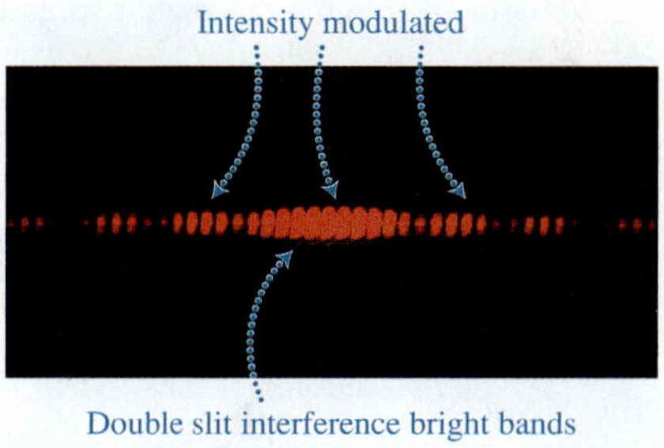


**OBSERVATIONAL EXPERIMENT TABLE**

**23.5 Behavior of light passing through a single slit.**

VIDEO 23.5

| Observational experiment | Analysis |
|---|---|
| **Experiment 1.** Using a regular lightbulb, a slit about 1 cm wide, and a screen, we shine light onto the slit. We observe on the screen an image that looks like the slit, only slightly wider and just a little fuzzy at the edges. 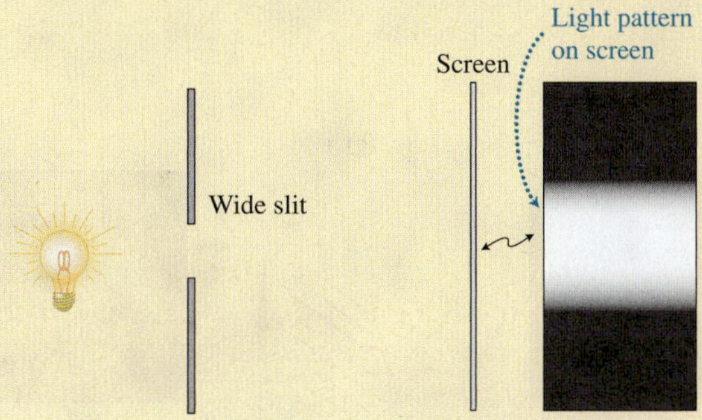 | This observation can be explained using the ray/particle model of light. No interference effects are present. |



(*continued*)

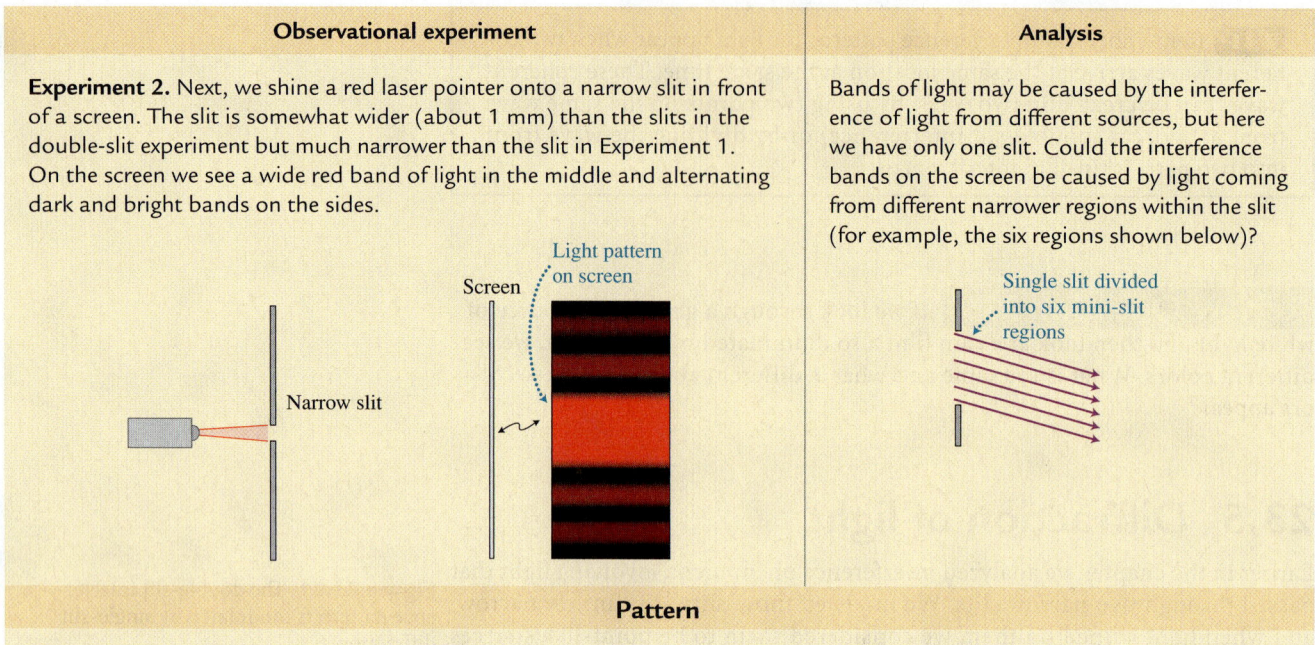

| Observational experiment | Analysis |
|---|---|
| **Experiment 2.** Next, we shine a red laser pointer onto a narrow slit in front of a screen. The slit is somewhat wider (about 1 mm) than the slits in the double-slit experiment but much narrower than the slit in Experiment 1. On the screen we see a wide red band of light in the middle and alternating dark and bright bands on the sides. | Bands of light may be caused by the interference of light from different sources, but here we have only one slit. Could the interference bands on the screen be caused by light coming from different narrower regions within the slit (for example, the six regions shown below)? |

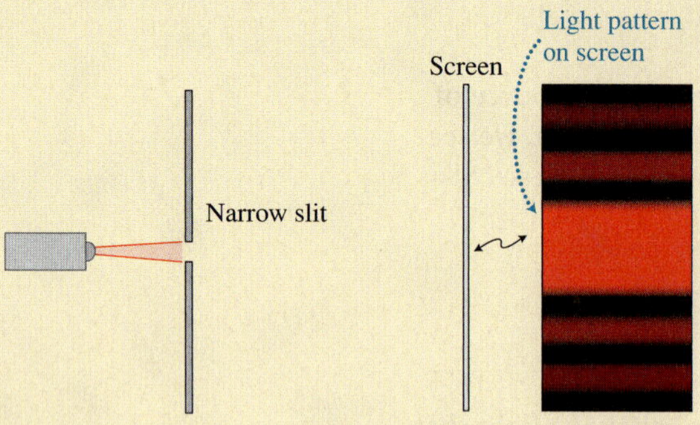


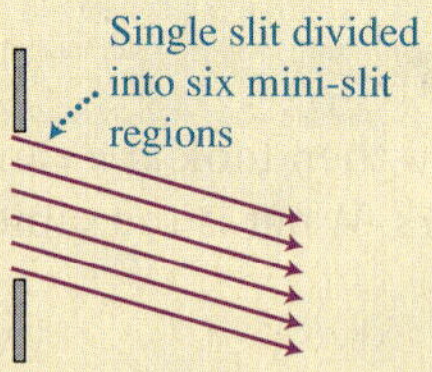


**Pattern**

We observe that laser light passing through a single narrow slit behaves differently from light passing through a wide slit. Laser light passing through a narrow slit spreads into a series of bright and dark bands beyond the slit.

In the second experiment in Table 23.5 we observed that light reaching the screen from a relatively narrow slit makes a wide interference pattern on the screen, including dark and bright bands at the sides of the central wide band. This spreading of light combined with the additional bright and dark regions is called **diffraction**. It becomes noticeable when the slit's width is roughly 1000 times the wavelength of the light or less. It is possible that this effect might be caused by the interference of wavelets produced by different mini-slit regions within the slit—an idea we will now evaluate quantitatively.

## Quantitative analysis of single slit diffraction

We have found that when light passes through a narrow slit, we can observe a series of bright and dark bands of light projected onto a screen (**Figure 23.22a**). If we use narrow slits of different widths, we find that the width of the central diffraction maximum (the central bright band on the screen) increases as the width of the slit decreases. When the slit is wider, the bright band in the center is almost equal to the actual slit width; when the slit is narrower, the band is much wider (Figure 23.22b and c).

**Figure 23.22** The single-slit diffraction pattern widens as the slit width narrows.

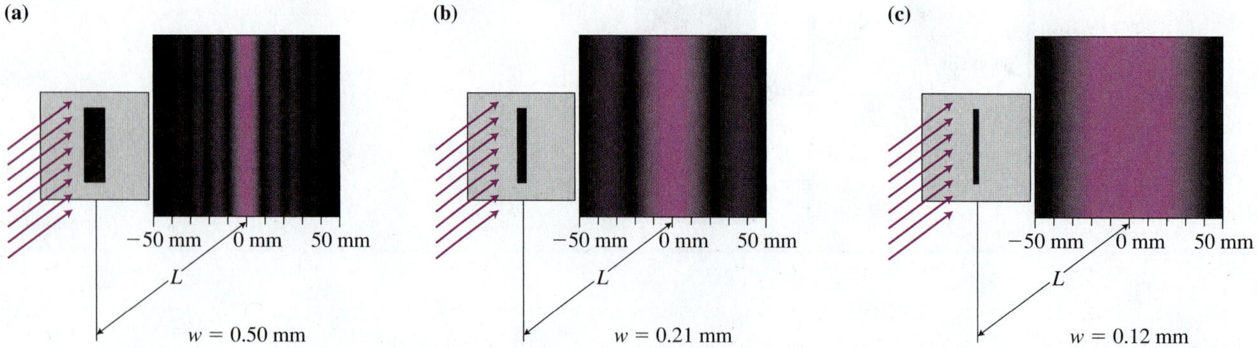

When we analyzed the interference pattern in a double-slit experiment, we modeled each slit as a point-like source of secondary wavelets on the same wave front using Huygens' principle. In the single-slit situation, the slit is not infinitely narrow. It is somewhat wider and produces an interference pattern like that shown in **Figure 23.23a**. If we see an interference pattern, then several waves might be adding to produce it. Where do those waves come from if we only have one slit? As the slit has a nonzero width, we can model it as consisting of multiple imaginary tiny mini-slit regions that become sources of the secondary waves on the same wave front when light shines on the slit. As they are on the same wave front, light waves emitted by all the mini-slits must all be in phase at the slit location. If this is the case, light emitted by those different mini-slits can interfere when reaching the screen, as they travel different distances to reach the same spot on the screen.

Imagine light shining on a single slit and then reaching the screen that is very far away. Therefore, the waves emitted by different mini-slits are traveling nearly parallel to each other when they reach a specific location on the screen. Thus, to explain the existence of a bright or a dark spot on the screen, we only need to consider the phase difference of the parallel waves traveling from the slit to the location on the screen. Consider two wavelets coming from mini-slits to the first dark band below the central bright band (position $y_1$ in Figure 23.23a). One wavelet comes from the top mini-slit of the top half of the slit (mini-slit 12 in Figure 23.23b) and the second wavelet comes from the top mini-slit of the bottom half of the slit (mini-slit 6 in Figure 23.23b). The distance between these mini-slits is $(w/2)$, where $w$ is the slit width.

If the path length difference from these two sources to the screen equals one half wavelength, then these two wavelets cancel each other, resulting in zero light at that location on the screen. If we repeat the procedure for the next pair of wavelets (11 and 5), which travel from the second mini-slit region in the top half of the slit and the second mini-slit region in the bottom half of the slit, we get the same result. The same will happen for all other pairs of waves chosen in this manner. We get total cancellation at that point on the screen—a dark band.

We see from the shaded triangle in Figure 23.23b that this first interference minimum occurs when $(w/2)\sin\theta = (\lambda/2)$ or when $w\sin\theta = \lambda$. Note that the same reasoning applies for the first dark band above the central maximum.

We designate the dark bands above the central maximum with positive signs and the dark bands below with negative signs. Thus, the first dark bands on both sides of the central bright spot occur at

$$w\sin\theta_{\pm 1} = \pm\lambda$$

We can next divide the slit into four mini-slit segments and draw lines from them to the second dark band to the side of the central maximum. We find that the angular deflection to this second dark band occurs when $(w/4)\sin\theta_2 = (\lambda/2)$ or when

$$w\sin\theta_{\pm 2} = \pm 2\lambda$$

We can repeat this process by dividing the slit into six mini-slit segments with lines drawn from them to find the third dark bands on the side of the central maximum. These dark bands occur when $w\sin\theta_{\pm 3} = \pm 3\lambda$.

We can summarize this analysis as follows. The angular positions of the dark bands relative to the central maximum are determined by

$$w\sin\theta_m = m\lambda,\text{ where } m = \pm 1, \pm 2, \pm 3\ldots \tag{23.6}$$

The first dark bands on both sides of the central maximum appear at the angles $\theta_{\pm 1}$ when $m = \pm 1$ is substituted into the above equation. All the space

**Figure 23.23** Single-slit diffraction. (a) Alternating bright and dark bands observed coming from light passing through a single slit. (b) Light from different parts of the slit cancels, causing the first dark band at $y_1$.

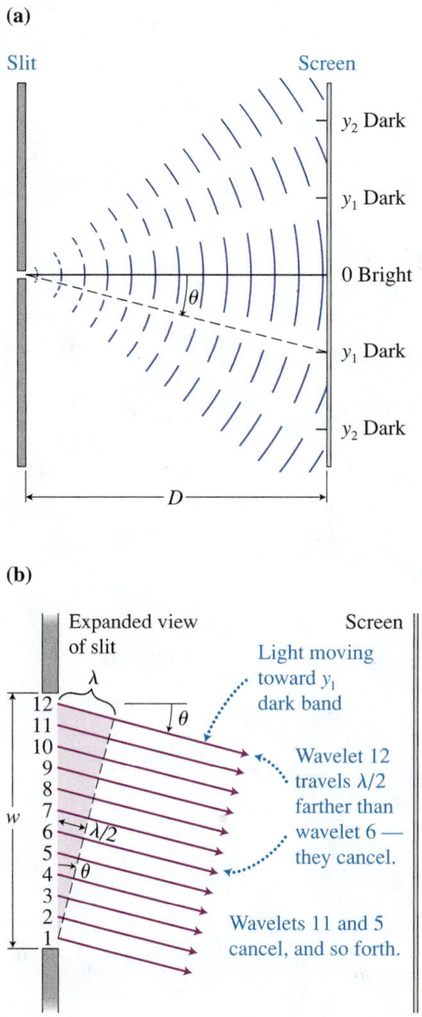

between these angles belongs to the central bright fringe. Therefore, we do not use $m = 0$ in the above equation.

Equation (23.6) $w \sin \theta_m = m\lambda$ was derived using a wave model of light and can be tested by performing an experiment using a shorter wavelength green laser with the same slit. The angular deflection to the first dark bands on the sides of the central maximum should be less, since the green laser produces light of shorter wavelength than the red laser. When you perform the experiment, you observe that the first dark bands are closer to the central bright band—the central bright band is narrower.

**TIP** Notice that $w \sin \theta_m = m\lambda$ is similar to Eq. (23.1) for double-slit interference. However, for a double slit, Eq. (23.1) describes the angles at which the maxima (bright bands) are observed, and $m = 0$ *is* allowed. For a single slit Eq. (23.6) $w \sin \theta_m = m\lambda$ describes angles at which the minima (dark bands) are observed, and $m = 0$ *is not* allowed since there is no dark band at $\theta = 0°$.

We can now describe the variation in the brightness of the double-slit interference pattern. Real slits have a finite width, which we did not account for in our original analysis. Thus, in addition to the two-slit interference pattern with the closely spaced alternating bright and dark bands, these bands are modulated in intensity by the additional single-slit diffraction due to the interference of light coming from different mini-slits within each slit. This single-slit pattern combines with the double-slit pattern, resulting in a variation in brightness of the double-slit maxima.

**Single-slit diffraction** **When monochromatic light is incident on a slit that is approximately 1000 wavelengths of light wide or less, we observe a series of bright and dark bands of light on a screen beyond the slit. The bands are caused by the interference of light from different mini-slit regions within the slit. The angle between lines drawn from the slits to the minima, the dark bands, and a line drawn from the slit to the central maximum are determined using the equation**

$$w \sin \theta_m = m\lambda \tag{23.6}$$

**where $w$ is the slit width, $\lambda$ is the wavelength of the light, and $m = \pm 1, \pm 2, \pm 3, \ldots$ (not zero). The dark bands on the screen are located at positions $y_m = L \tan \theta_m$ relative to the center of the pattern. $L$ is the slit-screen distance.**

## The Poisson spot—A historical testing of the wave model of light

The year 1818 was a decisive one for the acceptance of the wave model of light. Simeon-Denis Poisson, a talented mathematician and proponent of the particle model, suggested an experiment that would test and disprove the wave model of light. He suggested shining a narrow beam of light at a small round obstacle. Though Poisson did not himself believe in the wave model explanation, he still used it to predict the outcome. His reasoning was as follows: according to the wave model, when the obstacle is just smaller than the beam, the light should illuminate the edges of it. The edges could then be considered the sources of in-phase secondary wavelets. If so, a bright spot should also appear in the middle of the obstacle's shadow because the center of the shadow is equidistant from each secondary wavelet source. This phenomenon is similar to why a bright fringe appears at the center of a double-slit interference pattern (it is equidistant from both slits). This prediction is in direct contradiction to the particle model, which does not predict such a spot. Common sense also tells us that the bright spot should not appear at the center of the dark shadow. The French Academy of Sciences conducted the experiment and much to their (and Poisson's) surprise, they saw a bright spot right in the middle of the shadow, consistent with the wave model (**Figure 23.24**).

**Figure 23.24** Poisson's spot tested the wave theory of light.

**(a)** Side view of Poisson's experimental design (not to scale)

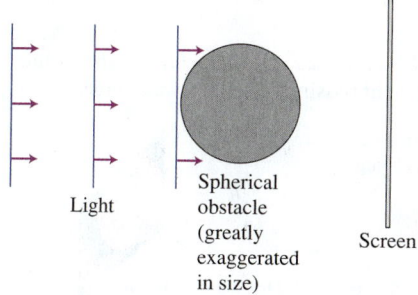

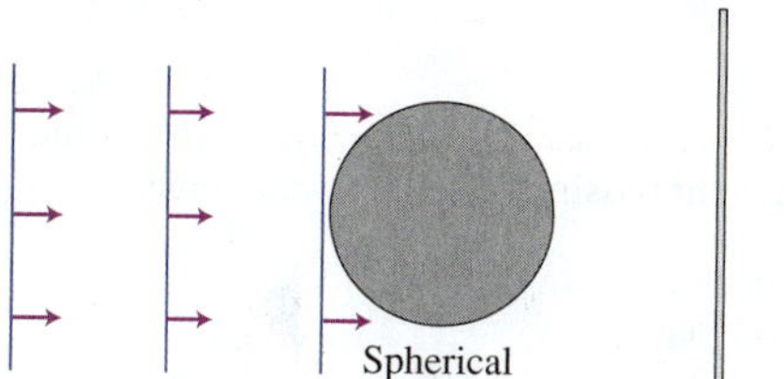


**(b)** Diffraction pattern seen on screen

A bright spot appears at the center, as predicted.

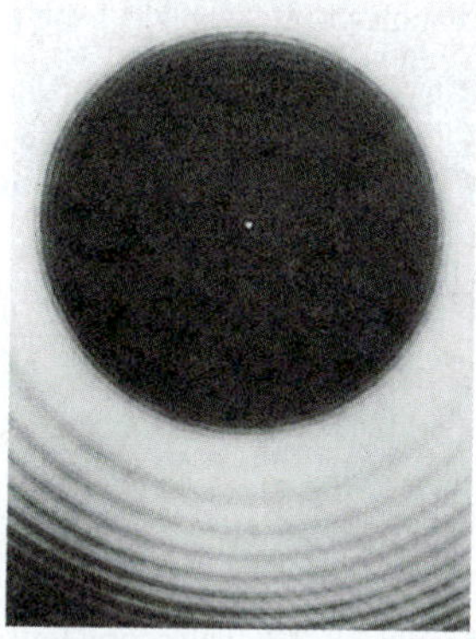

## Diffraction and everyday experience

Since the wavelengths of visible light are so small, we seldom observe light diffraction in everyday life. However, diffraction is a phenomenon that all waves exhibit, including sound waves. In Eq. (23.6) $w \sin \theta_m = m\lambda$, if $w$ is nearly equal to the wavelength of the waves, the angular position of the $m = \pm 1$ minima could reach almost 90°, in which case the waves would spread into the entire region beyond the slit with no regions of destructive interference. This full-screen diffraction seldom occurs with light, as slits are seldom that small. But the wavelengths of sound waves are much larger. The wavelength of concert A (frequency 440 Hz) is $\lambda = v_{\text{sound}}/f = (340\text{ m/s})/440\text{ Hz} = 0.77\text{ m}$, about the width of a doorway. Thus, a 90° central maximum will fill the entire region beyond the door. Diffraction explains why we can hear a person talking around a corner in another room when the door is open.

**QUANTITATIVE EXERCISE 23.6 Light and sound passing through a small slit in a door**

Sound of wavelength 0.85 m (approximately the wavelength of 400-Hz sound) and light of wavelength 400 nm (blue color) are incident on a 1.0-cm slit in a door. Will light and sound diffract significantly after passing through the slit?

**Represent mathematically** To answer the question we need to determine the angular size and the width of the central maximum for each type of wave. The angular deflection $\theta_{\pm 1}$ to the first dark bands on the side of the central maximum is determined using the equation $w \sin \theta_{\pm 1} = \pm\lambda$.

**Solve and evaluate** Rearranging the previous equation and applying it to light, we get

$$\sin \theta = \frac{\lambda}{w} = \frac{400 \times 10^{-9}\text{ m}}{1.0 \times 10^{-2}\text{ m}} = 4.0 \times 10^{-5}$$

Taking the inverse sine of both sides:

$$\theta = \sin^{-1}(4.0 \times 10^{-5}) = 0.0023°$$

The first minima appear at 0.0023° to either side of the central bright band. The diffraction is so small that it will not be noticed. You will see a sharp image of the slit opening and no interference.

For sound, $\sin \theta = \frac{\lambda}{w} = \frac{0.85\text{ m}}{1.0 \times 10^{-2}\text{ m}} = 85$. This impossible result ($\sin \theta \leq 1$) means that the central "bright" region for sound is spread over all angles on the other side of the opening and there will be no regions of destructive interference beyond the door. There will be no variation in the amplitude of sound waves behind the door due to diffraction. This phenomenon occurs whenever the wavelength is greater than the width of the opening, when $\lambda/w > 1$.

**Try it yourself:** White light is incident on a slit that is 0.10 mm wide. A screen is 3.0 m from the slit. What is the width of the red bright band adjacent to the central white band compared to the width of the corresponding blue band? The width on the screen of the first colored bright band on the side is the distance between the $m = 1$ and $m = 2$ dark bands. Consider the wavelength of blue light to be 480 nm and the wavelength of red light to be 680 nm.

*Answer:* The angular position for the first and second blue dark bands on one side of the central bright band are $\theta_1 = 0.28°$ and $\theta_2 = 0.55°$, and the distance on the screen between them is 14 mm. For red light $\theta_1 = 0.39°$ and $\theta_2 = 0.78°$, and they are separated by 20 mm.

**Review Question 23.5** Equation (23.6) $w \sin \theta_m = m\lambda$ where $m = \pm 1, \pm 2, \pm 3 \ldots$ describes the angles at which one can see dark fringes produced by light passing through a single slit. Write a similar equation that will describe the angles at which one can see dark fringes produced by light passing through two narrow slits.

**Figure 23.25** The diffraction pattern due to light passing through a small hole.

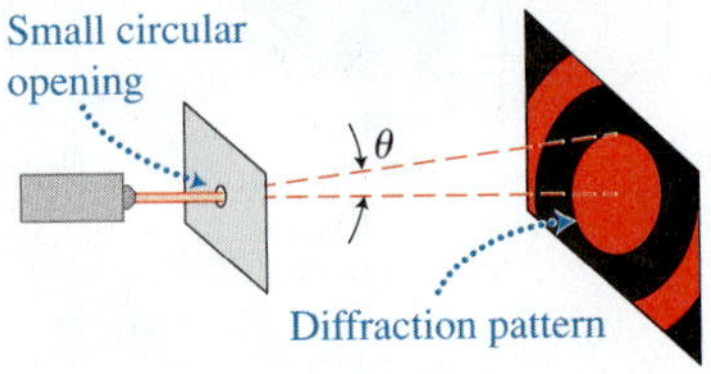


**Figure 23.26** A lens focusing light from a distant point source produces an Airy disk of radius $R$.

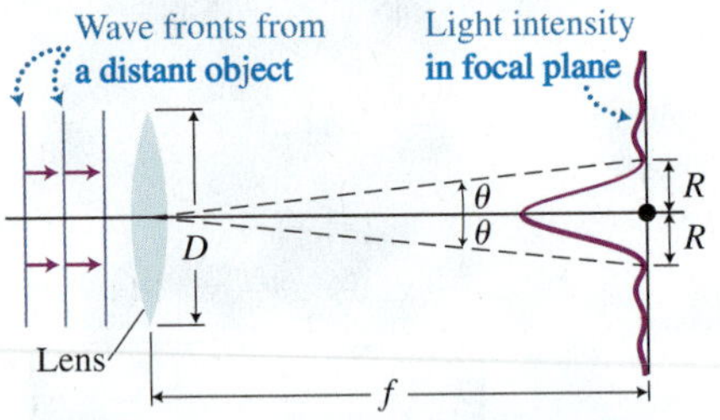


**Figure 23.27** Telescope lenses of increasing diameter resolve two stars.

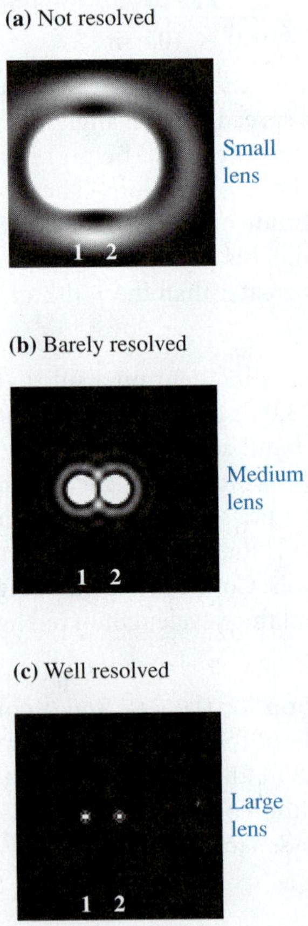


## 23.6 Resolving power: Putting it all together

The wave-like behavior of light limits our ability to see two distant closely spaced objects as separate objects or to discern the details of an individual distant object. Can we see the fine structure of a cell with a microscope? Can a reconnaissance aircraft take photographs of a missile site with sufficient detail? This section will help us answer such questions.

Consider light passing through the small circular hole shown in **Figure 23.25**. Light leaving the hole produces a diffraction pattern on a screen beyond the hole. The pattern resembles that formed by light passing through a narrow slit, except that alternating rings of bright light and darkness are formed rather than parallel bands of light and darkness. We can calculate the angle $\theta$ between a line drawn from the hole to the center of the pattern and a line drawn toward the first dark ring using the following equation (which we will not derive):

$$\sin \theta = \frac{1.22\lambda}{D} \quad (23.7)$$

where $\lambda$ is the wavelength of the wave passing through the hole and $D$ is the diameter of the hole. The shiny disk at the center of the pattern is called the Airy disk, named after George Airy (1801–1892), who first derived Eq. (23.7) in 1831.

The equation is similar to Eq. (23.6), used to calculate the angular position of the dark bands in a single-slit diffraction pattern. The factor 1.22 that appears in Eq. (23.7) and not in Eq. (23.6) is a result of the circular geometry of the opening. The intensities of the bright secondary rings around the central disk are much lower than that of the central bright spot.

### Resolving ability of a lens

How does diffraction affect the ability of a lens to produce a sharp image? Imagine that light from a very distant object enters a lens of diameter $D$ as plane waves and diffracts as it passes through the opening (**Figure 23.26**). The diffraction produces an angular spread given by Eq. (23.7): $\sin \theta = 1.22\lambda/D$. Thus, instead of passing through a focal *point* a distance $f$ from the lens, parallel light rays form a central *disk* of radius $R = f \tan \theta$ at the focal point. The radius of that central diffraction pattern depends on the diameter of the lens and the wavelength of the light. The wave-like properties of light make it *impossible* to form a perfectly sharp image.

This result is important for surveillance, astronomical studies, microscopy, and human vision. When looking at two closely spaced objects, we see central disks whose radii depend on the diameter of the aperture the light passes through (the pupil of an eye, the objective lens of a telescope, etc.). If the central disks of the diffraction patterns from the two objects overlap, we cannot perceive them as two distinct objects. Physicists say that we cannot **resolve** them. In **Figure 23.27**, we see a photograph of stars as seen through a telescope with a small-diameter objective lens. How many stars are in the field of vision? It is difficult to say by looking at Figure 23.27a. Figures 23.27b and c

**Figure 23.28** The angular separation of two sources affects the ability to resolve them.

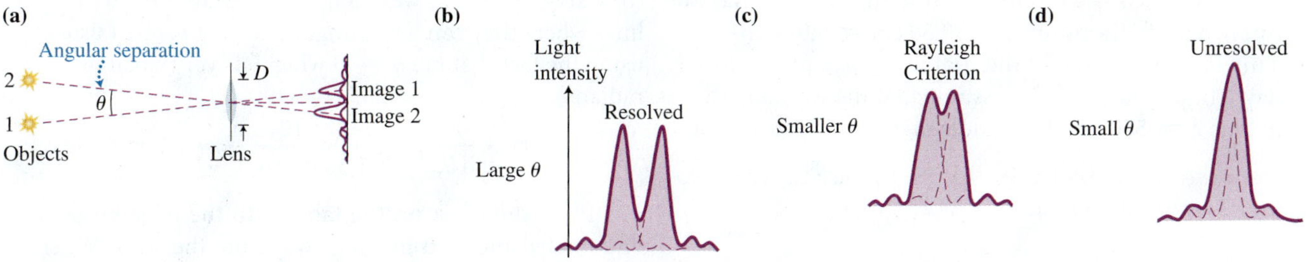


show the same stars photographed with telescopes with larger lenses that make it easier to resolve the individual stars.

What is the smallest separation of two images that can still be perceived as distinct? Consider **Figure 23.28a**, which shows the diffraction patterns of distant objects produced on a film by light of wavelength $\lambda$ passing through a lens of diameter $D$. In (b), the objects are far enough apart (large enough angular separation) that diffraction does not prevent their images from being resolved. In (c) the images can just barely be resolved. In (d), the images cannot be resolved.

According to Lord Rayleigh (1842–1919), two objects can just barely be resolved when their angular separation $\theta_{res}$ (called the **limit of resolution**) satisfies the following criterion:

**Rayleigh criterion** **The minimal angular separation of two objects that can be resolved (perceived as separate) is the limit of resolution $\theta_{res}$ of the instrument:**

$$\theta_{res} = \frac{1.22\lambda}{D} \qquad (23.8)$$

**where $D$ is the diameter of the opening through which light enters and $\lambda$ is the wavelength of the light. In this equation, $\theta_{res}$ is measured in radians.**

The equation helps us understand the role of the opening diameter $D$ of any instrument. Increasing the size of the opening of any instrument (such as a telescope) allows you to resolve more closely positioned objects or details on a single object.

**TIP** When the limit of resolution of the instrument increases, it means that its resolving power decreases. The larger $\theta_{res}$, the more difficult it is to resolve closely separated objects.

**EXAMPLE 23.7 Resolution and the human eye**

(a) What is the limit of resolution of the human eye? (b) The rectangular box shown below has vertical lines separated by 2 mm and the box to the right is solid gray. At what maximum distance (according to Rayleigh's criterion) will you be able to resolve the lines in the picture below? At greater distances both boxes would appear solid gray.

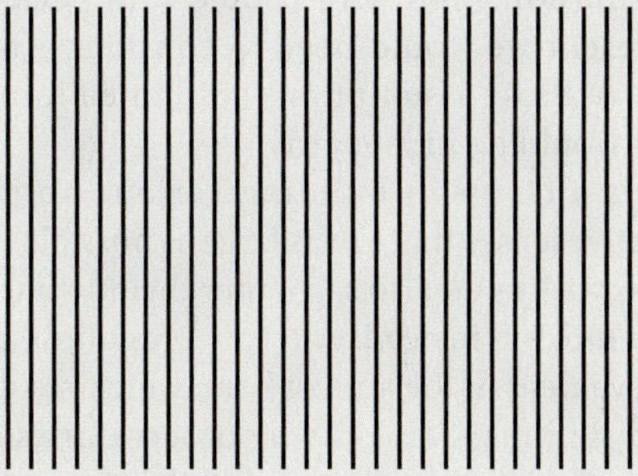

**Sketch and translate** The problem has two separate questions. We will answer the first question and then use the result to help answer the second. (a) To find the resolution limit $\theta_{res}$ of any instrument, we need to know the diameter $D$ of the aperture through which light enters and the wavelength $\lambda$ of the light. (b) After we find the resolution limit angle $\theta_{res}$, we can use geometry to find the maximum distance $s$ from which we can detect the lines separated by $\Delta y = 2\text{ mm}$.

*(continued)*

**Simplify and diagram** Assume that the eye has an opening of diameter $D = 0.50\text{ cm} = 5.0 \times 10^{-3}\text{ m}$. This is approximately the diameter of a pupil under daylight conditions. Also assume that the wavelength of light is $\lambda = 550\text{ nm}$, the middle of the visible spectrum.

**Represent mathematically** We can use Eq. (23.8) to find the resolution limit:

$$\theta_{\text{res}} = \frac{1.22\lambda}{D}$$

The distance $s$ from the lines where they can be resolved when separated by a distance $y$ can be determined using $\tan\theta_{\text{res}} = (y/s)$, or

$$s = \frac{y}{\tan\theta_{\text{res}}}$$

**Solve and evaluate**

$$\text{(a)}\ \theta_{\text{res}} = \frac{1.22\lambda}{D} = \frac{(1.22)550 \times 10^{-9}\text{ m}}{5.0 \times 10^{-3}\text{ m}} = 1.34 \times 10^{-4}\text{ rad}$$

(b) Using this value, we can find the distance from the lines where they can be distinguished as separate (also using the fact that $\tan\theta \approx \theta$ when $\theta$ is very small in radians):

$$s = \frac{y}{\tan\theta_{\text{res}}} = \frac{2.0 \times 10^{-3}\text{ m}}{1.34 \times 10^{-4}} \approx 15\text{ m}$$

Place the book on the table with the page brightly illuminated and start moving away from the book. When you get to only about 4 m away (which is much less than the 15 m that we calculated), you start seeing both rectangles as grey. This experiment shows that the resolving power is smaller than calculated above. Other factors limit our visual resolving power more than the Rayleigh criterion. For example, we assumed that light of a single wavelength was entering the eye through the pupil. The degree to which diffraction occurs depends on the wavelength of light. Different wavelengths will produce different sized central diffraction maxima on the retina, creating blurred images.

**Try it yourself:** Suppose you are 2.0 m from some vertical lines. If diffraction limited your ability to resolve the lines, approximately what would be the smallest separation of the lines that you could distinguish?

*Answer:* About 0.3 mm.

Notice that in the above example our prediction of the distance from the lines where they can be distinguished as separate was off by more than a factor of 3. Does it mean that the wave model of light is false and we need to reject it? As we have been learning throughout the book, when we make a prediction about the outcome of an experiment using a particular model, in addition to the model itself, we use many assumptions—ideas that are taken to be true. When the outcome of the experiment does not match the prediction, the first thing a physicist does is examine those assumptions. In this case, one assumption was that just one wavelength of light enters the eye. In the case of white light, this assumption is not valid and could account for the discrepancy between prediction and outcome. Air currents can cause fuzzy images. The person reporting the results may not have had perfect vision. Slight imperfections in the cornea, lens, or retina of the eye could cause slight distortions in the image. We could check this reasoning by repeating the experiment with monochromatic light, in a still room, and with people with perfect vision. If the discrepancy were still present we would need to consider other assumptions that were made. If none of those assumptions accounted for the discrepancy, the wave model of light would be in jeopardy.

The Rayleigh criterion is a factor in building various optical devices. Telescopes are built with objective lenses that are as large as possible. Large lenses are difficult to make; they break easily and also sag under their own weight. Thus, high-resolution telescopes use mirrors instead. Another limitation in producing sharp images of stars is the movement of the atmosphere, which results in images that continuously distort and shimmer. This is why observatories are built on mountains, where they are above a significant amount of the atmosphere.

Resolving the fine details of really small objects, such as parts of biological cells, is also challenging. Suppose you use a microscope to examine two organelles within an animal cell. Each produces a diffraction pattern whose central Airy disk has an angular deviation $\sin\theta = (1.22\lambda/D)$. If $\lambda \approx D$, then $\sin\theta \approx 1$ and the image for each object spreads over the entire viewing surface. You would not be able to distinguish the two objects from each other. A light-based microscope cannot resolve images of objects whose diameters are smaller than the wavelength of light ($\lambda \geq D$). To see objects that small, scientists use an electron microscope. Electron microscopes, discussed in a later chapter, resolve images using the wavelengths of electrons, which are much smaller than those of visible light. As a result, electron microscopes achieve much better resolving power than light-based microscopes.

**Review Question 23.6** Stars are so far away that they can be considered as point sources. Why then do the images of stars produced by a telescope and recorded on a film look like bright disks, not dots?

## 23.7 Skills for analyzing processes using the wave model of light

In this section we practice problem-solving skills while investigating some new phenomena. A problem-solving strategy is outlined on the left side of the table below and illustrated in Example 23.8.

### PROBLEM-SOLVING STRATEGY Analyzing processes using the wave model of light

**EXAMPLE 23.8 Purchase a grating**

You are to purchase a grating that will cause the deflection of the $m = 2$ 2nd order bright band for 650-nm wavelength red light to be 42°. How many lines per centimeter should your grating have?

**Sketch and translate**

- Visualize the situation and then sketch it.
- Identify given physical quantities and unknowns.

Known quantities:

$\lambda = 650 \times 10^{-9}$ m, $m = 2$, $\theta_2 = 42°$

Unknown quantities:

$d = ?$ lines/cm $= ?$

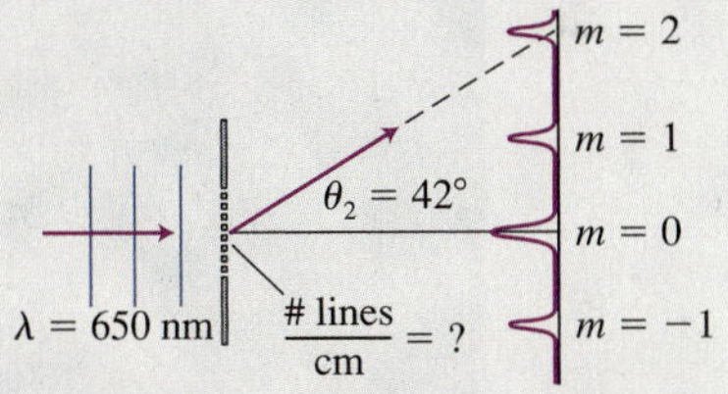


**Simplify and diagram**

- Decide if the sources in the problem produce coherent waves.
- Decide if the small-angle approximation is valid.
- Decide if the slit widths for multiple slits are wide enough that you have to consider single-slit diffraction as well as multiple-slit interference.
- If useful, represent the situation with a wave front diagram showing the overlapping crests and troughs of the light waves from different sources.

- Grating slits can be considered coherent light sources.
- With gratings the angular deflection is often 10° or more and we cannot use the small-angle approximation.
- As the slits in the grating are narrow, we can consider multiple slit interference only and ignore single slit diffraction.

(continued)

**Represent mathematically**

- Describe the situation mathematically.
- Use geometry if needed.

- Apply Eq. (23.5) to the second bright band to determine the slit separation: $\sin\theta_m = m(\lambda/d)$.
- Find the number of slits/cm: $\# = \dfrac{1\text{ cm}}{d}$

**Solve and evaluate**

- Use the mathematical description of the process to solve for the desired unknown quantity.
- Evaluate the result. Does it have the correct units? Is its magnitude reasonable? Do the limiting cases make sense?

From Eq. (23.5) with $m = 2$, we get

$$d = m\frac{\lambda}{\sin\theta_m} = 2\frac{650 \times 10^{-9}\text{ m}}{\sin 42°} = 1.9 \times 10^{-6}\text{ m} = 1.9 \times 10^{-4}\text{ cm}$$

- The number of slits in one centimeter is

$$\# = \frac{1\text{ cm}}{d} = \frac{1\text{ cm}}{1.9 \times 10^{-4}\text{ cm}} = 5100$$

The units match. This is a reasonable number of slits. A limiting case analysis is not needed, as we did not derive any new expressions.

**Try it yourself:** Suppose you looked with a spectrometer using this grating at light coming from a star and found an $m = 2$ band of light at 34°. What is the wavelength of this light?

*Answer:* $530 \times 10^{-9}\text{ m} = 530\text{ nm}$.

**EXAMPLE 23.9 Size of a red blood cell**

You shine 633-nm ($\lambda = 633 \times 10^{-9}\text{ m}$) light on a red blood cell. It produces the pattern shown below. The panel of light detectors is 0.21 m from the cell. Examine the pattern and determine the horizontal and vertical dimensions of the cell.

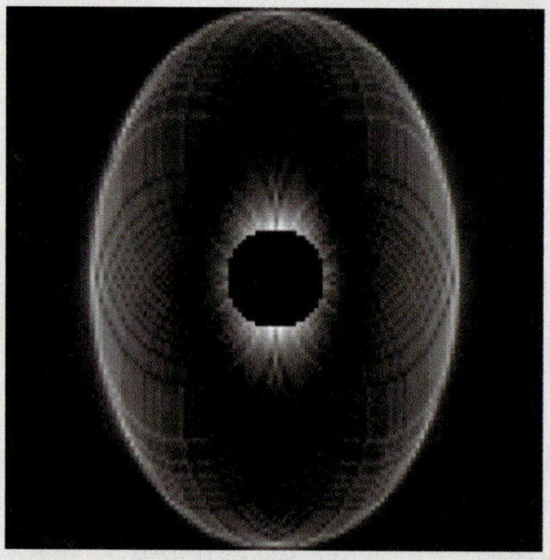

**Sketch and translate** We use Eq. (23.7) $\sin\theta = (1.22\lambda/D)$ to determine the size of the elliptical cell in each direction. The pattern should be narrower where the diameter $D$ of the cell is greater. Thus, the cell must be wider in the vertical direction and narrower in the horizontal direction. The screen is $L = 0.21$ m from the cell. We measure the distance $y_1$ of the first dark band from the center of the central maximum to be 2.3 cm horizontally and 1.3 cm vertically.

**Simplify and diagram** We draw a top view for the broader horizontal diffraction pattern and a side view for the narrower vertical diffraction pattern.

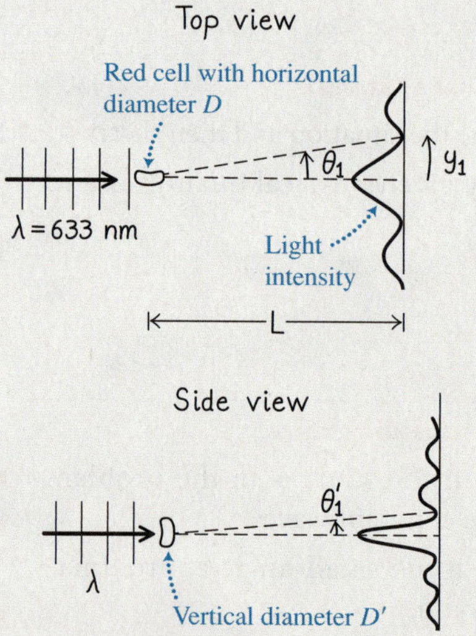


**Represent mathematically** Use Eq. (23.2) $\tan\theta_1 = (y_1/L)$ to determine the angular position $\theta_1$ for the horizontal and vertical directions. Rearrange Eq. (23.7) $\sin\theta = (1.22\lambda/D)$ to determine the dimension $D$ of the cell in each direction.

**Solve and evaluate**

*Horizontal dimension of cell*

$$\tan\theta_1 = \frac{y_1}{L} = \frac{2.3 \times 10^{-2}\,\text{m}}{0.21\,\text{m}} = 0.11 \text{ or } \theta_1 = 6.3^\circ$$

$$D = \frac{1.22\lambda}{\sin\theta_1} = \frac{1.22(633 \times 10^{-9}\,\text{m})}{\sin 6.3^\circ}$$

$$= 7.0 \times 10^{-6}\,\text{m} = 7\,\mu\text{m}$$

*Vertical dimension of cell*

$$\tan\theta_1{}' = \frac{y_1{}'}{L} = \frac{1.3 \times 10^{-2}\,\text{m}}{0.21\,\text{m}} = 0.062 \text{ or } \theta_1{}' = 3.5^\circ$$

$$D' = \frac{1.22\lambda}{\sin\theta_1{}'} = \frac{1.22(633 \times 10^{-9}\,\text{m})}{\sin 3.5^\circ}$$

$$= 12.6 \times 10^{-6}\,\text{m} = 13\,\mu\text{m}$$

Using the details of the diffraction pattern, we determined the dimensions of the cell, consistent with typical dimensions for body cells.

**Try it yourself:** Determine the diameter of the spherical bead whose diffraction pattern is shown below. The pattern was recorded with the same equipment used in this example. The distance from the center to the first dark band in the figure is approximately 1.1 cm.

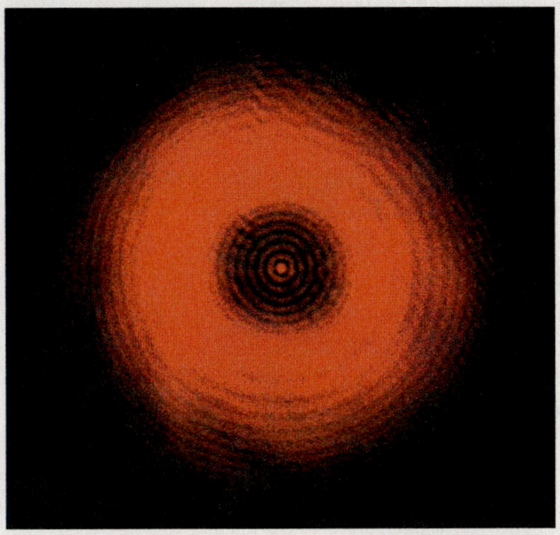

*Answer:* $D \approx 15 \times 10^{-6}\,\text{m} = 15\,\mu\text{m}$.

## A final note about light

In this chapter we applied the wave model of light to explain situations in which light passes through small openings or around small objects. The particle model of light cannot explain them. We found that the effects of light's wave-like behavior are most dramatic when the size of the holes and obstacles are comparable to the wavelength of the light. The wave model of light also explains the interaction of light with large objects (reflection and refraction). Therefore, it is tempting to accept the wave model as the "correct" model. However, we have not yet established a mechanism for this model. What is actually vibrating when light waves propagate? That question will be the topic of the next chapter.

**Review Question 23.7** Why is it especially important to keep track of units when solving problems in wave optics?

# Summary

| Words | Pictorial and physical representations | Mathematical representation |
|---|---|---|
| **Speed of light and refractive index** The speed of light is higher in a vacuum and in air than in other media. The ratio of the speed of light in air $c$ and its speed $v$ in a medium equals the index of refraction $n$ of the medium. (Section 23.2) | | $\frac{c}{v} = n$ Eq. (23.4) |
| **Monochromatic and coherent light sources** Only waves of the same constant frequency and wavelength (*monochromatic*) and of constant phase difference (*coherent*) will add to produce a constant interference pattern. White light interference involves the simultaneous interference of light of multiple frequencies, each satisfying the above conditions. (Section 23.2) | | |
| **Double-slit interference** In the wave model of light, double-slit interference is explained by the superposition of light waves passing through two closely spaced slits. The superposition of those waves creates a pattern of dark and bright bands on a screen beyond the slits (rather than the images of the slits). (Section 23.1) | 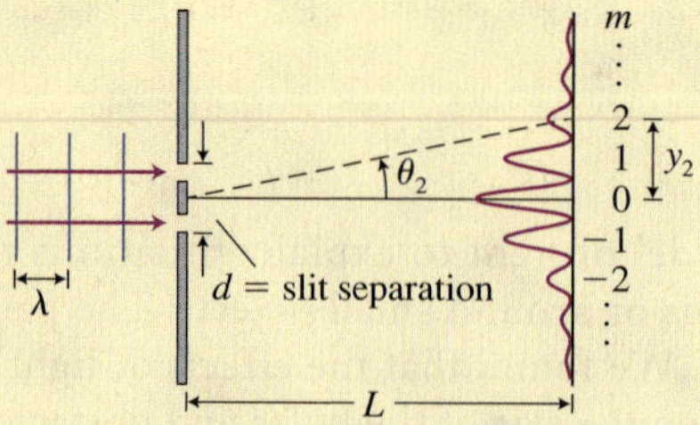 | Bright bands at<br>$d \sin \theta_m = m\lambda$ Eq. (23.1)<br>where<br>$m = 0, \pm 1, \pm 2, \ldots$<br>$L \tan \theta_m = y_m$ Eq. (23.2)<br>For small angles:<br>$\tan \theta_m \approx \sin \theta_m$ |
| **Grating** A grating has many closely spaced slits, each separated by a distance $d$ from their neighbors. Narrow bright bands (interference maxima) occur at angles and screen positions described by the same equations used for double slits. The maxima produced by gratings are much narrower and brighter than for two slits of the same width. (Section 23.3) | 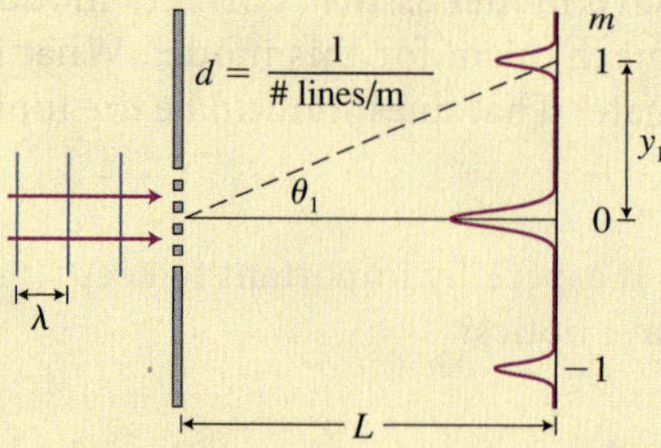 | $d \sin \theta_m = m\lambda$<br>$m = 0, \pm 1, \pm 2, \ldots$<br>$d = 1/\#\ \text{lines/m}$<br>$L \tan \theta_m = y_m$ Eq. (23.2) |
| **Thin-film interference** Thin films such as soap bubbles and lens coatings produce interference effects by reflecting light from the front and back surfaces of the film. The type of interference depends on phase differences caused by reflection and the path length difference between the waves. (Section 23.4) | 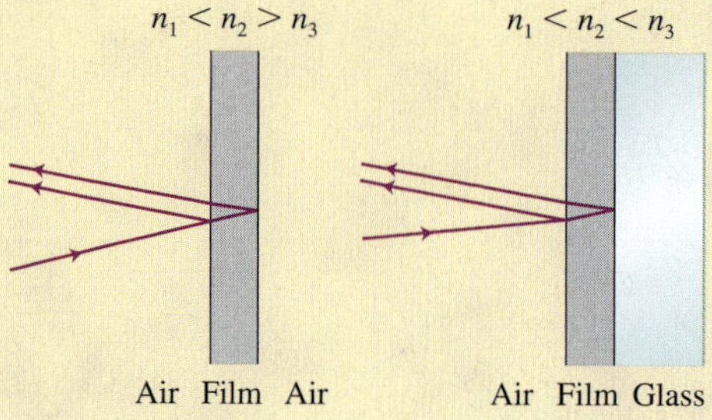 | See Table 23.4. |

**Single-slit diffraction** When monochromatic light of wavelength $\lambda$ is incident on a slit of width $w$, a series of bright and dark bands of light is formed on a screen a distance $L$ from the slit. The bands are caused by the interference of light from different mini-slit regions within the slit. (Section 23.5)

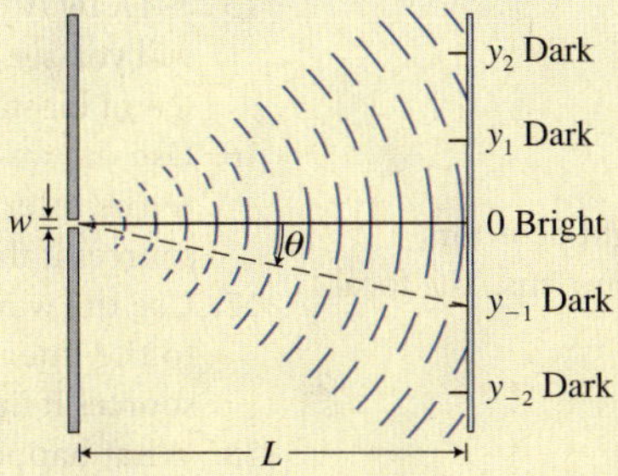


Dark bands at

$$w \sin \theta_m = m\lambda \qquad \text{Eq. (23.6)}$$

where

$$m = \pm 1, \pm 2, \pm 3 \ldots$$
$$y_m = L \tan \theta_m$$

**Diffraction by circular object** Light of wavelength $\lambda$ passing through a small hole (or around a circular obstacle) of diameter $D$ produces a diffraction pattern on a screen with the first dark band at angular position $\theta$. (Section 23.6)

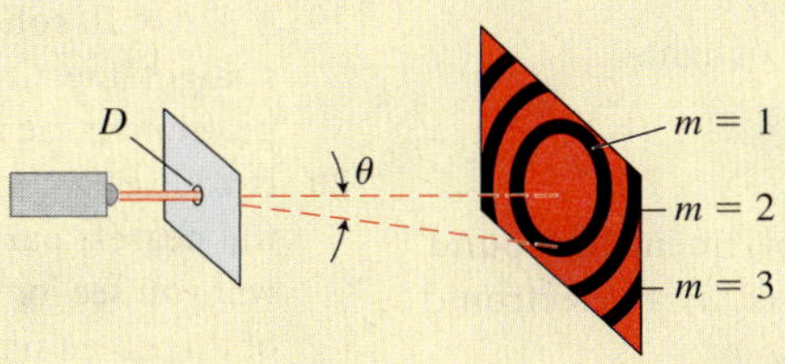


$$\sin \theta = \frac{1.22\lambda}{D} \qquad \text{Eq. (23.7)}$$

**Rayleigh criterion for resolving objects with optical instruments** The minimal angular separation for resolving two objects with an instrument of aperture diameter $D$ is $\theta_{res}$. If the objects are closer together than this, they appear as a single object (cannot be resolved). (Section 23.6)

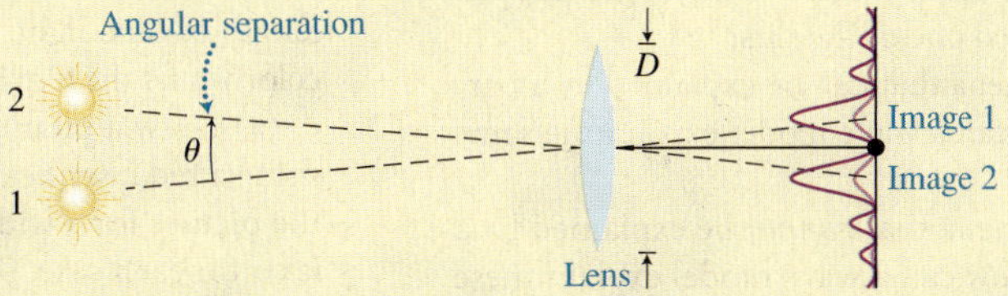

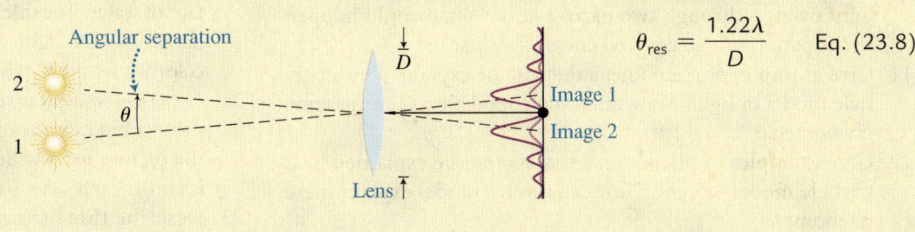

$$\theta_{res} = \frac{1.22\lambda}{D} \qquad \text{Eq. (23.8)}$$

MP For instructor-assigned homework, go to **MasteringPhysics.**

# Questions

## Multiple Choice Questions

1. You shine a flashlight on two wide slits cut in cardboard. What do you observe on a screen beyond the slits?
   (a) A sharp image of the two slits
   (b) An interference pattern consisting of bright and dark bands
   (c) A fuzzy image of the slits
   (d) Depending on the distance from the slits to the screen, the answer could be (a) or (c).
2. When you shine a very narrow beam of white light on two narrow closely spaced slits in a dark room, what do you see on a screen beyond the slits?
   (a) A sharp image of the slits on the screen
   (b) A pattern of white and dark spots of approximately equal width
   (c) A set of wide colored spectra with narrow dark spots between them
   (d) A wide white spot in the center and a set of spectra separated by narrow dark spots
3. When green light travels from air to glass, what quantities change?
   (a) Speed only
   (b) Frequency only
   (c) Wavelength only
   (d) Speed and frequency
   (e) Speed and wavelength
4. If you add a third slit to double slits and shine the same laser beam on the three slits, what will the pattern have?
   (a) More dark and bright fringes
   (b) Bright fringes farther apart
   (c) Dark fringes farther apart
   (d) The centers of the bright fringes will be the same distance from each other but their width will be narrower.
5. Why don't two flashlights about 1 m apart shining on a wall produce an interference pattern?
   (a) They are not bright enough.
   (b) The light waves are not coherent.
   (c) The fringes are too narrow to see.
6. You shine a laser beam through a grating with a known number of slits per millimeter and observe a pattern on a screen. Then you cover half of the grating with nontransparent material. What will you see?
   (a) There will be half as many bright fringes on the screen.
   (b) The bright fringes will be twice as close.
   (c) The bright fringes will be two times farther apart.
   (d) The location of the fringes will not change, but they will be a little bit wider and less bright.

7. What does the resolution limit of an optical system depend on? Choose all answers that are correct.
   (a) The wavelength of light
   (b) The diameter of the aperture
   (c) The distance to the object being viewed
   (d) The distance from the aperture to the light detectors
8. You shine light of different colors on the same lens. The focal length of the lens for blue light
   (a) is shorter than for red light.
   (b) is longer than for red light.
   (c) is the same as for red light, because focal length is a property of the lens, not the light shining on it.
   (d) depends on the location of the source of light.

## Conceptual Questions

9. Describe a double-slit interference experiment for sound waves using one or more speakers, excited by an electronic oscillator, as your sources of sound waves.
10. You are investigating a pattern produced on a screen by laser light passing through two narrow slits. What would happen to the pattern if you covered one of the slits?
11. Give examples of phenomena that can be explained by a particle model of light. How can a wave model explain the same phenomena?
12. Give examples of phenomena that *cannot* be explained by a particle model of light. How can a wave model explain these phenomena?
13. How would you explain Huygens' principle to a friend who is not taking physics but knows as much mathematics as you do?
14. Draw a point-like source of light. What is the shape of wave fronts leaving that point? Does the distance between them change as you move away from the source? What is the direction of the rays? How are the wave fronts for laser light different from those produced by the point source of light? How are rays oriented with respect to the wave fronts?
15. Draw two coherent light sources next to each other. What will you see if you place light-sensitive detectors at several different locations around the sources? How do you know?
16. Use the wave front representation to explain what happens to the interference pattern produced by two coherent light sources if their separation increases.
17. Use the wave front representation to explain what happens to the interference pattern produced by two coherent light sources if their wavelength increases.
18. What happens to the interference pattern produced by two coherent light sources if the materials are immersed in water?
19. Draw 10 coherent point-like sources of light placed along a straight line. Construct a wave front for a wave that is a superposition of the waves produced by the sources.
20. If you see green light of 520-nm wavelength when looking at the nearest part of a soap bubble of uniform thickness, why will you see longer wavelength light when looking at the sides of this green region?
21. Imagine that you have a very thin uniform oil film on the surface of water. The thickness of oil is much smaller than the wavelength of blue light. White light is shining on the film. What color would the film be if you looked at it from above? Explain.
22. (a) Draw a picture of what you will see on a screen if you shine a red laser beam through a very narrow slit. (b) Redraw the picture for a wider slit. (c) Redraw the picture for a green laser for each case. Keep the scale the same.
23. Describe three situations that you can analyze using wave fronts, rays, and disturbance-versus-position graphs for a particular time. In what situation is one of the representations more helpful than the others?
24. Why can you hear a person who is around a corner talking but not see her?
25. Astronomers often called the resolution limit described by Eq. (23.8) a "theoretical resolution limit." Why do they add the adjective "theoretical" to the description? What in real life can affect the resolution limit of a telescope?

# Problems

Below, **BIO** indicates a problem with a biological or medical focus. Problems labeled **EST** ask you to estimate the answer to a quantitative problem rather than derive a specific answer. Problems marked with a pencil icon require you to make a drawing or graph as part of your solution. Asterisks indicate the level of difficulty of the problem. Problems with no * are considered to be the least difficult. A single * marks moderately difficult problems. Two ** indicate more difficult problems.

### 23.1 and 23.2 Young's double-slit experiment and Index of refraction, light speed, and wave coherence

1. * **Sound interference** Two sources of sound waves are 2.0 m apart and vibrate in phase, producing sinusoidal sound waves of wavelength 1.0 m. (a) Use the wave front representation to explain what happens to the amplitude of sound along a line equidistant from each source and perpendicular to the line connecting the sources. (b) Use a graphical representation (pressure-versus-position graph) to explain what happens along that line. (c) Which representation is more helpful? Explain.
2. * Green light of wavelength 540 nm is incident on two slits that are separated by 0.50 mm. (a) Make a list of physical quantities you can determine using this information and determine three of them. (b) What do you need to change to double the distance between the 0th and the first bright spot on the screen?
3. * Blue light of wavelength 440 nm is incident on two slits separated by 0.30 mm. Determine (a) the angular deflection to the center of the 3rd order bright band and (b) its spatial separation from the 0th order band when the light is projected on a screen located 3.0 m from the slits. (c) Draw a sketch (not to scale) that schematically represents this situation and label all known distances and angles.
4. * Red light of wavelength 630 nm passes through two slits and then onto a screen that is 1.2 m from the slits. The center of the 3rd order bright band on the screen is separated from the central maximum by 0.80 cm. What can you determine using this information?

5. Sound from speakers Sound of frequency 680 Hz is synchronized as it leaves two speakers that are separated by 0.80 m on an open field. Draw a sketch of this arrangement and draw a line from between the speakers to a location where the sound is intense and equidistance from the two speakers (the 0th order maximum). Determine the angular deflection of a line from between the speakers to the 1st order intensity maximum to the side of this 0th order maximum. The speed of sound is 340 m/s.

6. * EST **Sketch a moving wave front** Draw a sketch of two narrow slits separated by 4.0 cm. On the sketch, show the crests of six waves of wavelength 1.0 cm that have left each slit. Draw lines from a point halfway between the slits in the directions of the center of the 0th order and 1st order intensity maxima. Draw a screen 7.0 cm from the slits and use your sketch to estimate the distance between the center of the central maximum on the screen and a 1st order bright spot to the side. Check your results using equations from the text. Were your sketch-based estimated results within 20% of the mathematical results?

7. * **Neon lamp** A neon lamp emits light that looks orange. After passing through a narrow single slit, the light strikes two very narrow slits separated by distance $d$ and located a distance $D$ from the single slit. After passing through the pair of slits, the light strikes a screen a distance $L$ away. (a) Make a sketch of the pattern that you would expect to see on the screen if light behaves like a wave. (b) Below the interference pattern, sketch the pattern that you would expect to see if light behaves like a stream of very light particles. Use correct dimensions in both cases.

8. * **Characteristics of laser light when in glass** A laser light in air has a wavelength of 670 nm. What is the frequency of the light? What is the frequency of this light when it travels in glass? In water? What is the wavelength of light in these media? (Use Table 21.7 if needed.)

9. * **Prism converts white light** Use the wave model of light to explain why white light striking a side of a triangular prism emerges as a spectrum.

## 23.3 Gratings: An application of interference

10. ** **Representing how grating works** Use the representation (wave fronts or rays) that you think is best to explain why monochromatic light forms a pattern of narrow bright and wide dark fringes on a wall when it passes through a grating as opposed to a pattern of wide bright fringes and narrow dark regions produced by a double slit.

11. Light of wavelength 520 nm passes through a grating with 4000 lines/cm and falls on a screen located 1.6 m from the grating. (a) Draw a picture (not to scale) that schematically represents the process and label all known distances and angles. (b) Determine the angular deflection of the second bright band. (c) Determine the separation of the 2nd order bright spot from the central maximum.

12. **Hydrogen light grating deflection** Light of wavelength 656 nm and 410 nm emitted from a hot gas of hydrogen atoms strikes a grating with 5300 lines per centimeter. Determine the angular deflection of both wavelengths in the 1st and 2nd orders.

13. **Purchase a grating** How many lines per centimeter should a grating have to cause a 38° deflection of the 2nd order bright band of 680-nm red light?

14. **Only half a grating** The left side of the grating from the previous problem breaks off. Half of the slits are missing. How will it affect the location of the 2nd order bright band?

15. * EST **Design** Design a quick way to estimate which one of two gratings has more lines per centimeter.

16. **Laser light on grating 1** The 630-nm light from a helium-neon laser irradiates a grating. The light then falls on a screen where the first bright spot is separated from the central maximum by 0.51 m. Light of another wavelength of light produces its first bright spot 0.43 m from its central maximum. Determine the second wavelength.

17. * **Laser light on grating 2** Light of wavelength 630 nm passes through a grating and then onto a screen located several meters from the grating. The 1st order bright band is located 0.28 m from the central maximum. Light from a second source produces a band 0.20 m from the central maximum. Determine the wavelength of the second source. Show your calculations. [Hint: If the angular deflection is small, $\tan\theta = \sin\theta$.]

## 23.4 Thin-film interference

18. * **Representing thin-film interference** (a) Draw a ray diagram for a laser beam incident from air on a thin film that has air on the other side. Make sure to take into account the processes occurring on each surface of the film. (b) Discuss in words what happens to the wave phase at each boundary. (c) Under what conditions will a person observe no reflected light?

19. * **Oil film on water** A thin film of vegetable oil ($n = 1.45$) is floating on top of water ($n = 1.33$). (a) Describe in words the processes occurring to a laser beam at the top and bottom surfaces of the oil. (b) Is it possible to have a film of water on top of the oil surface? Explain your answer.

20. * **Soap bubble 1** You look at a soap bubble film perpendicular to its surface. Describe the changes in colors of the film that you observe as the film thins and eventually breaks. Support your explanation with a ray diagram and a disturbance-versus-position graph.

21. * **Soap bubble 2** A soap bubble of refractive index 1.40 appears blue-green when viewed perpendicular to its surface (blue-green appears when red light is missing from the continuous spectrum, where $\lambda_{red}$ is about 670 nm in a vacuum). Does the light change phase when reflected from (a) the outside surface of the bubble and (b) the inside surface? (c) Determine the wavelength of red light when passing through the bubble. (d) Determine the thickness of the thinnest bubble for which the 670-nm red light reflected from the outside surface of the bubble interferes destructively with light reflected from the inside surface.

22. * **Thin-film coated lens** A lens coated with a thin layer of material having a refractive index 1.25 reflects the least amount of light at wavelength 590 nm. Determine the minimum thickness of the coating.

23. ** **Thin-film coated glass plate** A film of transparent material 120 nm thick and having refractive index 1.25 is placed on a glass sheet having refractive index 1.50. Determine (a) the longest wavelength of light that interferes destructively when reflected from the film and (b) the longest wavelength that interferes constructively.

24. * Two flat glass surfaces are separated by a 150-nm gap of air. (a) Explain why 600-nm-wavelength light illuminating the air gap is reflected brightly. (b) What wavelength of radiation is not reflected from the air gap?

## 23.5 Diffraction of light

25. * ✎ **Explain diffraction** Draw a ray diagram and show path length differences to explain how wavelets originating in different parts of a slit produce the third dark fringe on a distant screen.
26. * ✎ **How did we derive it?** Explain how we derived the equation for the first dark fringe for single-slit diffraction. Draw a ray diagram or a disturbance-versus-position graph to help in your explanation. Show the path length difference and the phase differences.
27. * **Explain a white light diffraction pattern** White light passing through a single slit produces a white bright band at the center of the pattern on a screen and colored bands at the sides. Explain.
28. Light of wavelength 630 nm is incident on a long, narrow slit. Determine the angular deflection of the first diffraction minimum if the slit width is (a) 0.020 mm, (b) 0.20 mm, and (c) 2.0 mm.
29. Light of wavelength of 120 nm is incident on a long, narrow slit of width 0.050 mm. Determine the angular deflection of the 5th order diffraction minimum.
30. * **Sound diffraction through doorway** Sound of frequency 440 Hz passes through a doorway opening that is 1.2 m wide. Determine the angular deflection to the first and second diffraction minima ($v_{sound} = 340\ m/s$).
31. Light of wavelength 624 nm passes through a single slit and then strikes a screen that is 1.2 m from the slit. The thin dark band is 0.60 cm from the central bright band. Determine the slit width.

## 23.6 Resolving power: Putting it all together

32. ** ✎ **Explain resolution** Explain the term "resolution limit." Illustrate your explanation with pictures and ray diagrams, and explain what characteristics of an optical device and the light passing through it affect the resolution limit.
33. **Resolution of telescope** A large telescope has a 3.00-m-radius mirror. What is the resolution limit of the telescope? What assumptions did you make?
34. * Laser light of wavelength 630 nm passes through a tiny hole. The angular deflection of the first dark band is 26°. What can you learn about the hole using this information?
35. * **Size of small bead** Infrared radiation of wavelength 1020 nm passes a dark, round glass bead and produces a circular diffraction pattern 0.80 m beyond it. The diameter of the first bright circular ring is 6.4 cm. What can you learn about the bead using this information?
36. * **Resolution of telescope** How will the resolution limit of a telescope change if you take pictures of stars using a blue filter as opposed to using a red filter? Explain.
37. * **Detecting visual binary stars** Struve 2725 is a double-star system with a visual separation of 5.7 arcseconds in the constellation Delphinus. Determine the minimum diameter of the objective lens of a telescope that will allow you to resolve 400-nm violet light from the two stars.
38. **Hubble Telescope resolving power** The objective mirror of the Hubble Telescope is 2.4 m in diameter. Could it resolve the binary stars *Lambda Cas* in the constellation Cassiopeia? The stars have an angular separation of 0.5 arcseconds.
39. * ✎ Draw a graphical representation of Rayleigh's criterion. Explain how this criterion relates to the concept of the resolution limit.
40. * BIO **Ability of a bat to detect small objects** Bats emit ultrasound in order to detect prey. Ultrasound has a much smaller wavelength than sound, which improves the resolution of small objects. What is the diameter of the smallest object that forms a diffraction pattern when irradiated by the $8.0 \times 10^4$-Hz ultrasound from a bat? The first diffraction minimum from the smallest object is 90°.

## 23.7 Skills for analyzing processes using the wave model of light

41. * ✎ Red light from a helium-neon gas laser has a wavelength of 630 nm and passes through two slits. (a) Draw a ray diagram to explain why you see a pattern of bright and dark bands on the screen. Show the path length difference. (b) Determine the angular deflection of the light to the first three bright bands when incident on narrow slits separated by 0.40 mm. (c) Determine the spatial separation of the centers of 0th and 2nd order bright bands when projected on a screen located 5.0 m from the slits. (d) List all of the assumptions that you made in your calculations.
42. * ✎ Red light of wavelength 630 nm is incident on a pair of slits. The interference pattern is projected on a wall 6.0 m from the slits. The fourth bright band is separated from the central maximum by 2.8 cm. (a) Draw a ray diagram to represent the situation; show the path length difference. (b) What can you learn about the slit pair using this information? (c) What can you learn about the pattern on the screen using this information?
43. * ✎ Monochromatic light passes through a pair of slits separated by 0.025 mm. On a screen 2.0 m from the slits, the 3rd order bright fringe is separated from the central maximum by 15 cm. (a) Draw a ray diagram to represent the situation. Show the path length difference. (b) What can you learn about the light source using this information?
44. * **Ratio reasoning** Two different wavelengths of light shine on the same grating. The 3rd order line of wavelength A ($\lambda_A$) has the same angular deflection as the 2nd order line of wavelength B ($\lambda_B$). Determine the ratio ($\lambda_A/\lambda_B$). Be sure to show the reasoning leading to your solution.
45. * ✎ **Design** Design an experiment to use a grating to determine the wavelengths of light of different colors. Draw a picture of the apparatus. List the quantities that you will measure. Describe the mathematical procedure that you will use to calculate the wavelengths.
46. * **Fence acts as a grating for sound** A fence consists of alternating slats and openings, the openings being separated from each other by 0.40 m. Parallel wave fronts of a single-frequency sound wave irradiate the fence from one side. A person 20 m from the fence walks parallel to it. She hears intense sound directly in front of the fence and in another region 15 m farther along a line parallel to the fence. Determine everything you can about the sound used in this problem (the speed of sound is 340 m/s).
47. * BIO **Morpho butterfly reflection grating wings** A reflection grating reflects light from adjacent lines in the grating instead of allowing the light to pass through slits, as in a transmission grating. If we assume perpendicular incidence, then we can determine the angular deflection of bright bands the same way we did for a transmission grating. White light is incident on the wing of a Morpho butterfly (whose wings act as reflection gratings). Red light of wavelength 660 nm is deflected in the 1st order at an angle of 1.2°. (a) Determine the angular deflection in the 1st order of blue light (460 nm). (b) Determine the angular deflection in the 3rd order of yellow light (560 nm).

48. * **Ratio reasoning** Laser monochromatic light is used to illuminate two different gratings. The angular deflection of the 2nd order band of light leaving grating A equals the angular deflection of the 3rd order band from grating B. Determine the ratio of the number of lines per centimeter for grating A and for grating B.

49. * **Soap bubble interference** Light of 690-nm wavelength interferes constructively when reflected from a soap bubble having refractive index 1.33. Determine two possible thicknesses of the soap bubble.

50. * **Oil film on water** A film of oil with refractive index 1.50 is spread on water whose refractive index is 1.33. Determine the smallest thickness of the film for which reflected green light of wavelength 520 nm interferes destructively.

51. **Babinet's principle** Babinet's principle states that the diffraction pattern of complementary objects is the same. For example, a rectangular slit in a screen produces the same diffraction pattern as a rectangular screen the same size as the slit; a hair should produce the same diffraction pattern as a slit of the same width. Determine the width of a hair that, when irradiated with laser light of wavelength 630 nm, produces a diffraction pattern on a screen with the first minimum 2.5 cm on the side of the central maximum. The screen is 2.0 m from the hair.

52. * **Diffraction from sound speaker** The opening of a stereo speaker is shaped like a slit. Determine the maximum width such that the first diffraction minimum of sound is at least 45° on each side of the direction in which the speaker points. Perform the calculations for sound waves of frequency (a) 200 Hz, (b) 1000 Hz, and (c) 10,000 Hz. The speed of sound is 340 m/s.

53. * The angular deflection of the 1st order bright band of light passing through double slits separated by 0.20 mm is 0.15°. Determine the angular deflection of the 3rd order diffraction minimum when the same light passes through a single slit of width 0.30 mm.

54. * EST **Diffraction of sound from the mouth** (a) Estimate the diameter of your mouth when open wide. (b) Determine the angular deflection of 200-Hz sound and of 15,000-Hz sound as it leaves your mouth. If, during your calculations, you find that $\sin\theta > 1$, explain the meaning of this result. The speed of sound is 340 m/s.

55. **Determine body cell size** Light of 630 nm wavelength from a helium-neon laser passes two different-size body cells. The angular deflection of the light as it passes the cells is (a) 0.060 radians and (b) 0.085 radians. Determine the size of each cell.

56. * EST A sound of frequency 1000 Hz passes a basketball. Estimate the angular deflection from the basketball to the first ring around the central maximum in the diffraction pattern. The speed of sound is 340 m/s.

## General Problems

57. * Monochromatic light passes through two slits and then strikes a screen. The distance separating the central maximum and the first bright fringe at the side is 2.0 cm. Determine the fringe separation when the following quantities change simultaneously: the slit separation is doubled, the wavelength of light is increased 30%, and the screen distance is halved.

58. **Sound from speakers** Two stereo speakers separated by a distance of 2.0 m play the same musical note at frequency 1000 Hz. A listener starts from position 0 (**Figure P23.58**) and walks along a line parallel to the speakers. (a) Can the listener easily hear the sound at position 0? Explain. (b) Calculate the distance from position 0 to positions 1 and 2 where intense sound is also heard. The speed of sound is 340 m/s.

**Figure P23.58**

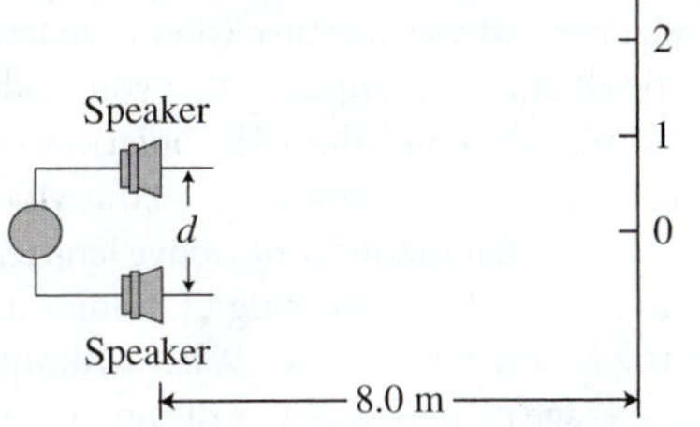


59. * **Astronomer's spectrograph** An astronomer has a grating spectrograph with 5000 lines/cm. A film 0.500 m from the grating records bands of light passing through the grating. (a) Determine the wavelength and frequency of the $H_\alpha$ line of hydrogen gas in a laboratory discharge tube that produces a 1st order band separated on the film by 17.4 cm from the central maximum. (b) Determine the wavelength and frequency of the $H_\alpha$ line coming from a galaxy in the cluster Hydra A. The 1st order band of the $H_\alpha$ line of this light is 18.4 cm from the central maximum. (c) Suggest a possible reason for the differences in the frequency of the $H_\alpha$ line from a lab source and the $H_\alpha$ line from the galaxy.

60. **Diffraction of water waves entering a harbor** The wavelength of water waves entering a harbor is 14 m. The angular deflection of the 1st order diffraction minimum of the waves in the water beyond the harbor is 38°. Determine the width of the opening into the harbor.

61. * EST **Effect of shrinking Earth** Assume that Earth, its structures, and its inhabitants are all decreased in size by the same factor. Estimate the decrease required in order that the 1st order diffraction dark band of 500 nm light entering a typical room window is at 90° (the central bright band would light most of the room). Explain all aspects of your calculations.

62. ** **Variable thickness wedge** A wedge of glass of refractive index 1.64 has a silver coating on the bottom, as shown in **Figure P23.62**. Determine the smallest distance $x$ to a position where 500-nm light reflected from the top surface of the glass interferes constructively with light reflected from the silver coating on the bottom. The light changes phase when reflected at the silver coating.

**Figure P23.62**

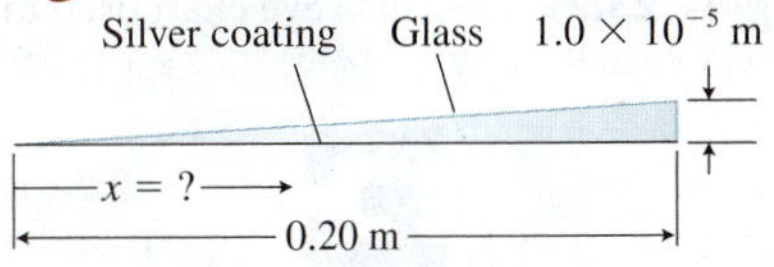


63. * EST **Resolving car headlights** Estimate the farthest away a car can be at night so that your eyes can resolve the two headlights. Indicate any assumptions you made.

64. * **Looking at Moon rocks** You have a home telescope with a 3.0-cm objective lens. Determine the closest distance between two large boulders on the Moon that you can distinguish as separate objects. Indicate any assumptions you made.

65. * BIO EST **Diffraction-limited resolving power of the eye** You look at closely spaced lines on a wall 5.0 m from your eyes. Estimate the closest the lines can be to each other and still be resolved by your eyes as separate lines. Indicate any assumptions you made in making your estimate.

66. * **Resolving sunspots** You are looking at sunspots. They usually appear in pairs on the surface of the Sun. The Sun is about $1.5 \times 10^{11}$ m from Earth. How close can two sunspots be so you can distinguish them when they are observed though an amateur telescope whose aperture (objective lens) is about 20 cm? Describe all of the assumptions that you made.

67. ** **The Moon's Mare Imbrium** The outermost ring of mountains surrounding the Mare Imbrium on the Moon has a diameter of 1300 km. What diameter objective lens telescope would allow an astronomer to see the ring of mountains as a distinct feature of the Moon's landscape? What assumptions did you make? The average center-to-center distance from the Earth to the Moon is 384,403 km, which is about 30 times the diameter of the Earth. The Moon has a diameter of 3474 km.

68. * **Can you see atoms with a light-based microscope?** Explain how you can use your knowledge of the wave model of light to explain why you cannot use an optical microscope to see atoms.

69. * **Detecting insects by diffraction of sound** A biologist builds a device to detect and measure the size of insects. The device emits sound waves. If an insect passes through the beam of sound waves, it produces a diffraction pattern on an array of sound detectors behind the insect. What is the lowest-frequency sound that can be used to detect a fly that is about 3 mm in diameter? The speed of sound is 340 m/s.

## Reading Passage Problems

**BIO What is 20/20 vision?** Vision is often measured using the Snellen eye chart, devised by Dutch ophthalmologist Hermann Snellen in 1862 (see **Figure 23.29**). With normal vision (20/20 vision) you can distinguish a letter that is 8.8 mm high from other letters of similar height at a distance of 6.1 m (20 ft) (the Snellen chart in the figure is smaller than normal size). If your vision is 20/40, the letters must be twice as high to be distinguishable. Alternatively, a person with 20/40 vision could distinguish letters from 20 ft that a person with 20/20 vision can distinguish at 40 ft. Someone with 20/60 vision could distinguish letters at 20 ft that someone with 20/20 vision could distinguish at 60 ft.

Does the Rayleigh criterion limit visual resolution? Assume that the eye's pupil is 5.0 mm in diameter for 500-nm light. The Rayleigh criterion angular deflection for such light entering the eye's pupil is

**Figure 23.29** A Snellen eye chart used to detect visual acuity.

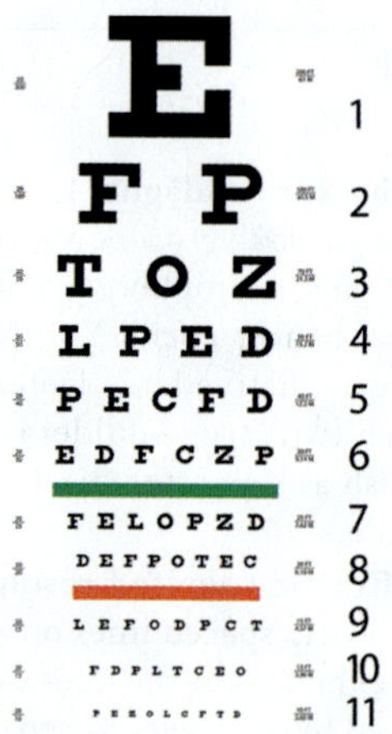


$$\theta = 1.22\frac{500 \times 10^{-9}\text{ m}}{5.0 \times 10^{-3}\text{ m}} = 1.2 \times 10^{-4}\text{ rad}$$

If the Rayleigh criterion limited visual resolution, then from a distance of 6.1 m you should be able to distinguish details in shapes of size

$$\begin{aligned} y = L\tan\theta &= (6.1\text{ m})[\tan(1.2 \times 10^{-4}\text{ rad})] \\ &= 0.7 \times 10^{-3}\text{ m} = 0.7\text{ mm} \end{aligned}$$

or about one-tenth the size of the 8.8-mm-tall letter in row 8 (20/20 vision) of the full-size Snellen eye chart. Thus, according to the Rayleigh criterion, we should easily be able to resolve different 8.8-mm-tall letters. Other factors, such as chromatic aberration, irregularities in the cornea-lens-retina shapes, the density of rods and cones in the retina, and air currents, limit visual resolution more than the Rayleigh criterion diffraction limit.

70. Which answer below is closest to the height of the letters that a person with 20/80 vision can distinguish when 20 ft from the wall chart?
(a) 2.2 mm (b) 4.4 mm (c) 8.8 mm
(d) 18 mm (e) 34 mm

71. Suppose that a person with 20/20 vision stands 30 ft from a Snellen eye chart. Which answer below is closest to the minimum height of the letters the person can distinguish?
(a) 4.4 mm (b) 6.6 mm (c) 8.8 mm
(d) 13 mm (e) 18 mm

72. What is the visual acuity of a person with 20/20 vision mainly limited by?
(a) The Rayleigh criterion
(b) Chromatic aberration
(c) Focal length of the eye lens
(d) Diameter of the eye's pupil
(e) The density of cones and rods
(f) A combination of these factors

73. A hawk's vision is said to be 20/5. If so, the hawk can distinguish 8.8-mm-tall letters from about what distance?
(a) 5 ft (b) 10 ft (c) 40 ft
(d) 80 ft (e) 120 ft

74. The hawk shown in **Figure P23.74** is only about 50 cm tall and has smaller eyes and pupils than the human eye. Estimate the Rayleigh criterion for this hawk.
(a) $1 \times 10^{-4}$ rad (b) $3 \times 10^{-4}$ rad
(c) $10 \times 10^{-4}$ rad (d) $20 \times 10^{-4}$ rad
(e) $40 \times 10^{-4}$ rad

**Figure P23.74**

75. If the vision of a hawk is actually 20/5, then the angular resolution of the hawk is closest to what angle? Assume that it can distinguish objects about 8.8 mm from a distance of 60 m. (Compare your answer to this question and the previous question to decide if the Rayleigh criterion is the limiting factor in a hawk's resolving ability.)
(a) $0.2 \times 10^{-4}$ rad (b) $0.5 \times 10^{-4}$ rad
(c) $1 \times 10^{-4}$ rad (d) $2 \times 10^{-4}$ rad

**Thin-film window coatings for energy conservation and comfort** Thin-film coatings are applied to eyeglasses, computer screens, and automobile instrument panels to reduce glare and to binocular and camera lenses to increase the amount of light transmitted. Thin-film coatings are also used to add color to architectural glass, to reduce surface friction, and to improve the energy efficiency of windows. Consider this latter application.

**Figure 23.30** shows that over 50% of the radiation reaching Earth from the Sun is long-wavelength infrared radiation, which we perceive as heat, not light. Thin-film coatings on windows serve two purposes. (1) On warm days, the coatings reflect almost all of the infrared radiation from outside while allowing almost all of the visible light through the window (the glass absorbs most of the ultraviolet). The rooms inside stay cooler and require less air conditioning. (2) On cold days, the same thin-film coating allows visible light to enter the room and reflects the thermal infrared radiation from the interior back into the room, keeping it warm. The reflectance of the thin-film window coating as a function of wavelength is shown in **Figure 23.31**.

**Figure 23.30** Over 50% of solar radiation reaching Earth's surface is infrared radiation.

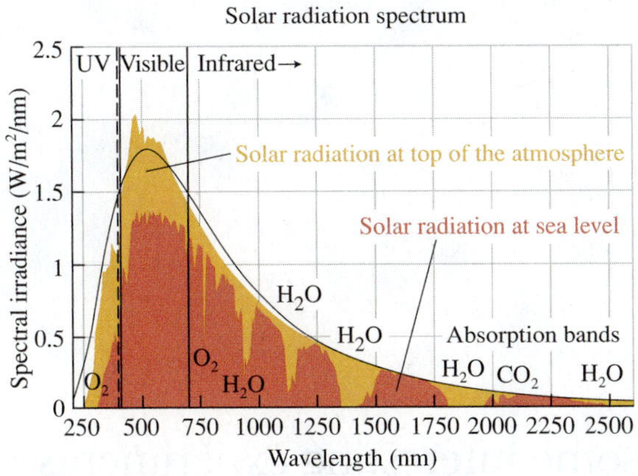


**Figure 23.31** A thin-film window coating causes reflectance of infrared radiation.

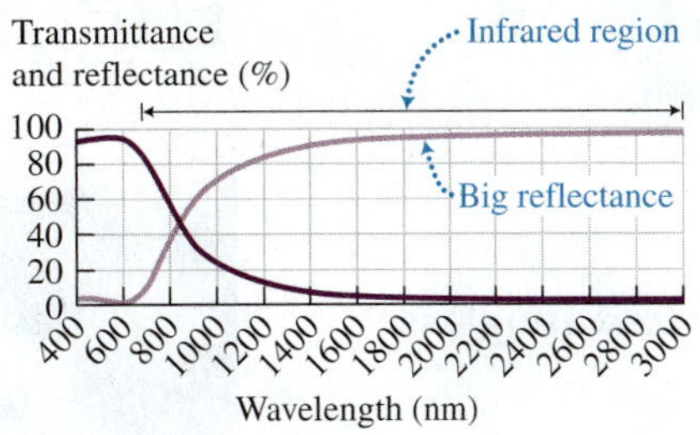


76. Which of the following are benefits of thin-film coatings on windows?
(a) They keep ultraviolet light out of the house in summer and winter.
(b) They keep infrared radiation out of the house in the summer.
(c) They keep infrared radiation in the house in the winter.
(d) b and c
(e) a, b, and c

77. A 1.25 refractive index thin film on a 1.50 refractive index glass window is made to reflect infrared radiation of wavelength 1000 nm. What is the net reflective phase change of infrared radiation reflected off the front surface of the thin film relative to the radiation reflected off the back surface of the film and returning to the front?
(a) Zero (b) 1/4 wavelength
(c) 1/2 wavelength (d) None of these

78. A 1.25 refractive index thin film on a 1.50 refractive index glass window is made to reflect infrared radiation of wavelength 1000 nm. Which wavelength below is closest to the wavelength of the infrared radiation while in the thin film?
(a) 670 nm (b) 800 nm (c) 1000 nm
(d) 1250 nm (e) 1500 nm

79. A 1.25 refractive index thin film on a 1.50 refractive index glass window is made to reflect infrared radiation of wavelength 1000 nm. Which answer is closest to the desired thickness of the thin film?
(a) 200 nm (b) 250 nm (c) 400 nm
(d) 500 nm (e) 1000 nm

80. The actual thin film built for windows reflects about what percent of the incident infrared radiation at 1000 nm?
(a) 10 (b) 30 (c) 50
(d) 70 (e) 90

# 24 Electromagnetic Waves

How do polarized sunglasses work?

How can a low-flying plane avoid radar detection?

In what way is the screen of your calculator similar to the sky?

**Be sure you know how to:**

- Explain how a capacitor and a transformer work (Sections 15.7 and 18.9).
- Explain how an electric current can be generated without a battery (Section 18.10).
- Describe wave motion using the quantities frequency, speed, wavelength, and intensity (Sections 20.2 and 20.4).

You can perform some interesting experiments with polarized sunglasses. On a sunny day, hold them at arm's length and look through them at one place in the sky (not at the Sun, though). Then slowly rotate the glasses, keeping the lenses pointing toward the same spot. The brightness of the sky changes. You can achieve the same results by looking through polarized lenses at the LCD screen of your calculator, cell phone, or laptop computer, or at sunlight reflecting off water. What do these sources of light have in common? You'll learn in this chapter.

**In the previous three chapters** we investigated light. We constructed different models of light to explain its behavior and found that a wave model explained reflection, refraction, and interference. But a mystery remains. Every other type of wave we have encountered so far (waves on a string, sound waves, etc.) involves the vibration of the medium through which the wave travels. What is vibrating in a light wave? We continue to investigate that question in this chapter. We'll resolve the question when we learn about special relativity (in Chapter 25).

## 24.1 Polarization of waves

In our previous study of mechanical waves, we neglected a phenomenon that is crucial for answering the question posed in the chapter opening. Consider the experiments in Observational Experiment **Table 24.1**.

< **Active Learning Guide**

**OBSERVATIONAL EXPERIMENT TABLE**

**24.1 Rope waves and Slinky waves.**

VIDEO 24.1

| Observational experiment | Analysis |
|---|---|
| **Experiment 1.** A rope passes through the open ends of a narrow rectangular box. Shake the rope in a vertical plane, producing a transverse wave. The long sides of the box are parallel to the shaking direction. The rope wave is unaffected by the box.<br>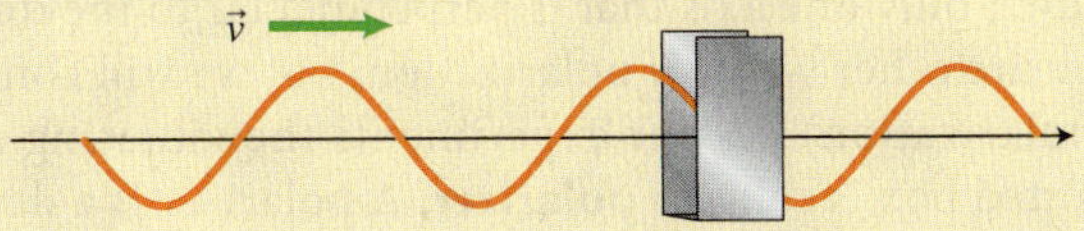<br> | Vectors represent the displacement of each section of rope at one instant. They are parallel to the long sides of the box.<br>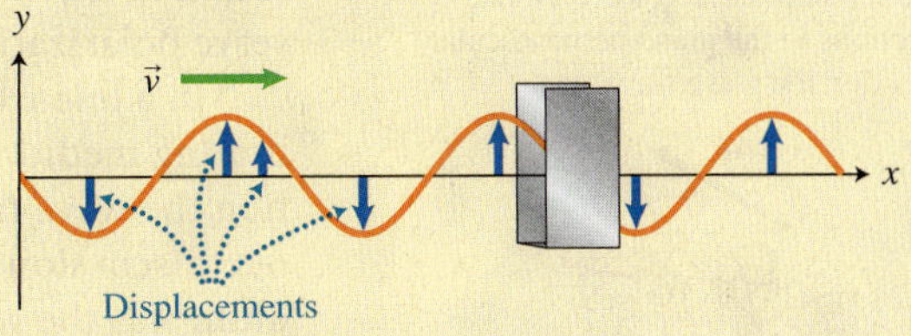<br> |
| **Experiment 2.** Rotate the box 90° with respect to the original orientation. Shake the rope the same way as in Experiment 1. The long sides of the box are perpendicular to the shaking direction. The wave does not pass through the box.<br>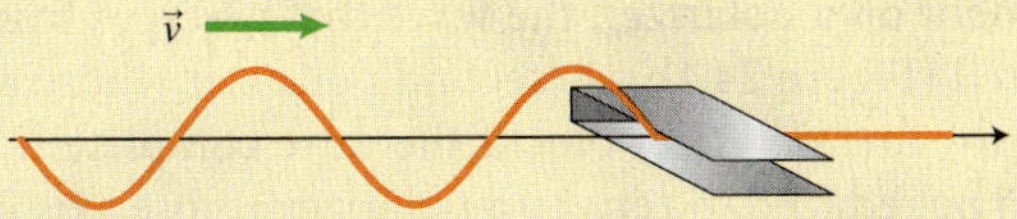<br> | The displacement vectors of the rope sections are perpendicular to the long sides of the box.<br>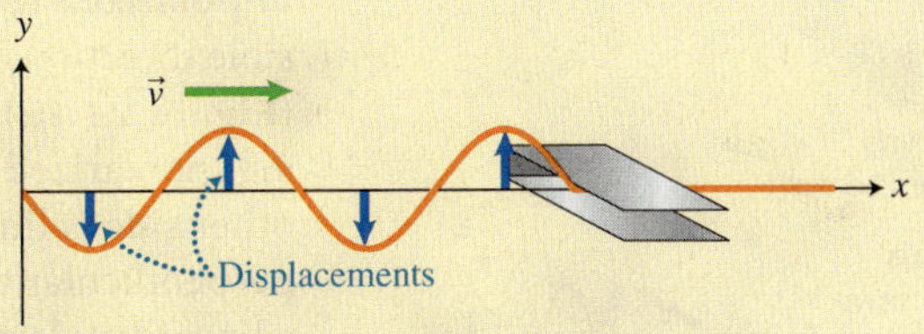<br> |
| **Experiment 3.** Now use two boxes. Rotate the first box 30° from the vertical plane. Orient box 2 perpendicular to box 1. Shake the rope as in Experiment 1. Part of the wave travels through box 1 but is eliminated by box 2.<br>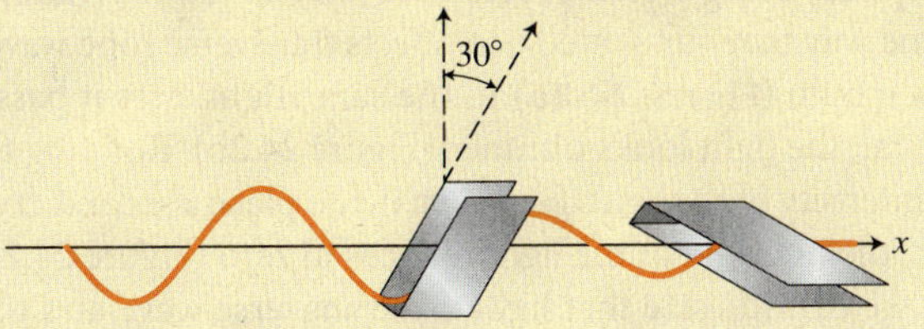<br> | The component of the rope wave parallel to the first box gets through the first box. Once the rope wave gets to the second box, its displacement vectors are perpendicular to the box.<br>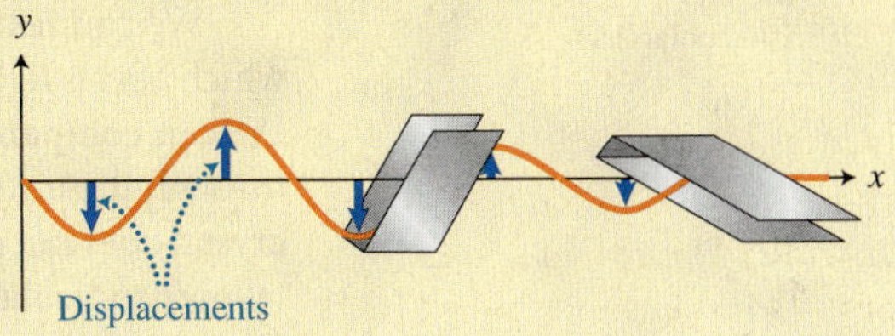<br><br>*(continued)* |

| Observational experiment | Analysis |
|---|---|
| **Experiment 4.** We achieve the same results as in Experiments 1–3 if we replace the rope with a Slinky and produce a transverse wave. However, if we compress and decompress the Slinky horizontally, producing a longitudinal wave, the wave goes through the boxes in all of the experiments. 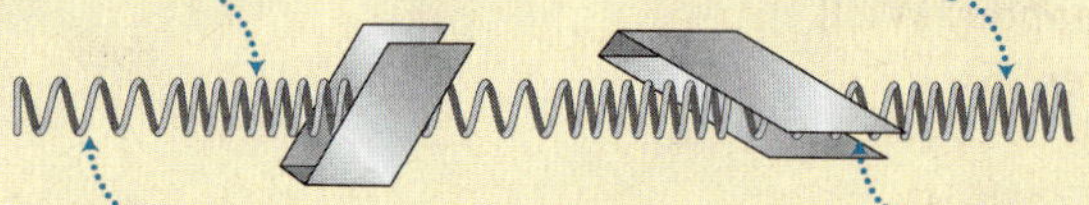  | The longitudinal displacement vectors of the Slinky coil always pass through the boxes. They point along the direction of the Slinky. 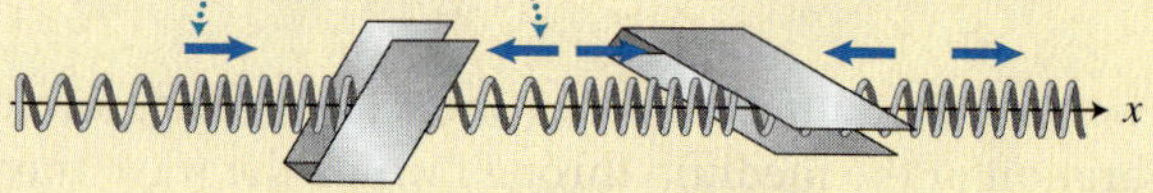  |

**Patterns**

- If the displacement vectors of a transverse wave are parallel to the slit (the box), the wave passes through undisturbed.
- If the displacement vectors are in the plane perpendicular to the slit, the wave does not pass through.
- When the displacement vectors make an angle with the slit, only the component of the vectors parallel to the slit passes through.
- The displacement vectors of a longitudinal wave always pass through the box. Longitudinal waves are not sensitive to the orientation of the box.

**Figure 24.1** The effect of polarizers on a wave that is not linearly polarized.

(a)

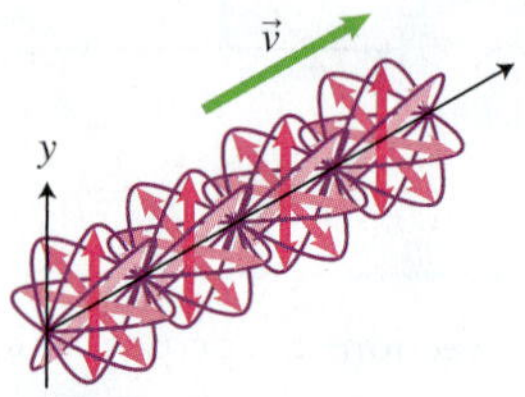


(b)

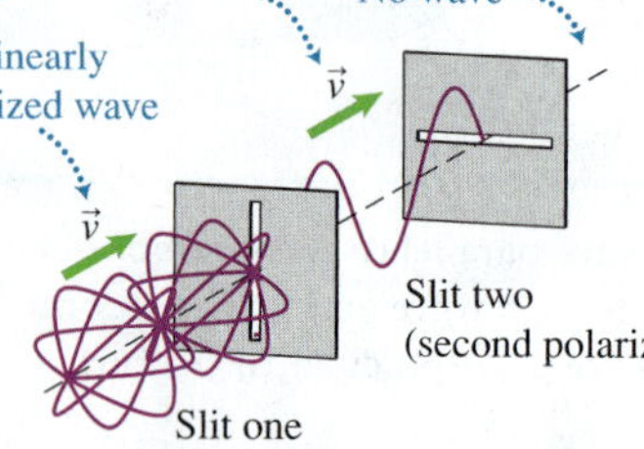


We learned in Table 24.1 that slotted boxes with the proper orientation can block transverse mechanical waves but cannot block longitudinal waves. This phenomenon is called **polarization of waves**—polarization is a property of waves that describes the orientation of the oscillations of the wave. As we found in Table 24.1, only transverse waves have this property. Thus, if we observe polarization of light waves, we know that the waves are transverse.

In a **linearly polarized** mechanical wave, the individual particles of the vibrating medium vibrate along only one axis that is perpendicular to the direction the wave travels (there are other types of polarization, but we will confine our discussion to linear polarization). In our experiments the vibrating medium was the rope. The slotted box is called a **polarizer**. A polarizer is a device that allows only a single component of transverse waves to pass through it. The component that can pass through defines the **axis** of the polarizer.

The waves in Experiments 1–3 in Table 24.1 were linearly polarized. An **unpolarized** wave is one in which the particles of the medium vibrate in all possible directions in the plane perpendicular to the direction the wave is traveling (**Figure 24.1a**), often caused by a collection of differently polarized waves. If an unpolarized wave is incident on a polarizer, the wave that emerges from the other side is linearly polarized (Figure 24.1b). A second polarizer whose axis is perpendicular to the axis of the first polarizer blocks the wave completely. Thus, a linearly polarized wave can be reduced to zero by a polarizer whose axis is perpendicular to the polarization direction of the wave. An unpolarized wave can be completely blocked by two successive perpendicularly oriented polarizers.

Are light waves transverse or longitudinal? According to our observations in Table 24.1, if we can polarize light waves, they must be transverse.

We can test this assumption using a semiprecious crystal called tourmaline, which affects light in the same way that our slotted box affects transverse rope waves. We can compare light from a bulb (**Figure 24.2a**) to the same light after it passes through the tourmaline crystal; the light looks dimmer (Figure 24.2b). Rotating the crystal does not change the intensity of the light. However, if you place a second crystal perpendicular to the first, the intensity of the light reduces to zero (Figure 24.2c). To explain this experiment we hypothesize that light is a transverse wave and that a lightbulb emits unpolarized light waves. We also hypothesize that the tourmaline

**Figure 24.2** (a) Bright light from a bulb. (b) One polarizer dims the light passing through it. (c) Two crossed polarizers completely block the light.

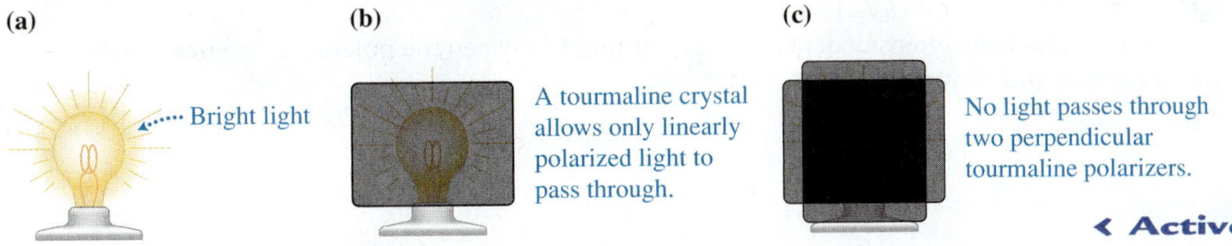


‹ Active Learning Guide

crystal allows only one component of the wave to pass through the first crystal. That component is completely blocked by the second crystal.

Polarized products consist of a sheet of material embedded with many tiny polarizing crystals. The lenses of polarized sunglasses, for instance, are covered with a film containing these crystals. Physics teaching labs often have "polarizers," sheets of plastic that are transparent but dark like sunglasses. Light seen through such a polarizer looks dimmer (**Figure 24.3**). With two crossed polarizers, no light comes through (the overlap region of the crossed polarizers in Figure 24.3). This behavior is so similar to the behavior of transverse mechanical waves passing through mechanical polarizers that it supports the hypothesis that light waves are transverse waves.

**Figure 24.3** The effect of two crossed polarizers.

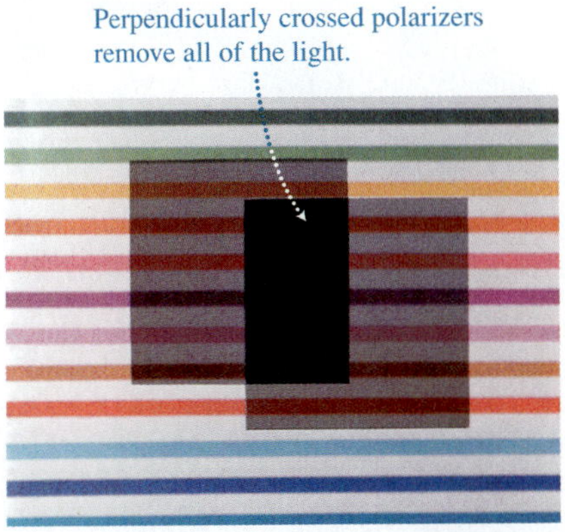

Perpendicularly crossed polarizers remove all of the light.

## Quantitative description of the effect of polarizers

Observational Experiment **Table 24.2** investigates the amount of dimming caused by one or two polarizers oriented at different angles relative to each other.

**OBSERVATIONAL EXPERIMENT TABLE**

**24.2 Effect of polarizers on light intensity.**

VIDEO 24.2

| Observational experiment | Analysis (relative light intensity) |
|---|---|
| We place a light meter a fixed distance from a lightbulb (a source of unpolarized light).  | $I_0$ |
| We place a polarizer between the bulb and the detector. 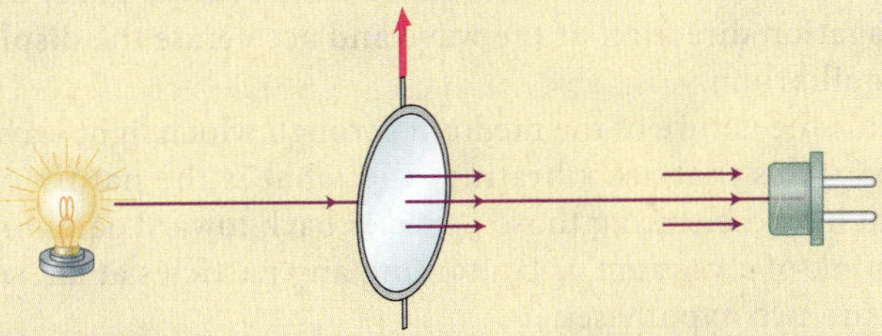 | $I_0/2$ |
| We place two polarizers between the bulb and the detector, with varying angles between their axes. 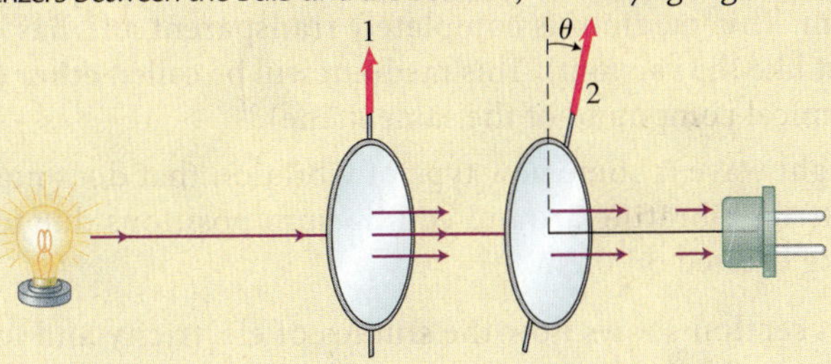  | **Angle** / **Intensity**<br>0° / $I_0/2$<br>30° / $0.75\, I_0/2$<br>45° / $0.50\, I_0/2$<br>60° / $0.25\, I_0/2$<br>90° / 0 |

*(continued)*

**Patterns**

- One polarizer reduces the light intensity by half: $I = (I_0/2)$.
- The second polarizer seems to further reduce the intensity depending on the angle between the polarizers. We need to determine the relation between this reduction and the angle.

The patterns in Table 24.2 indicate that the intensity of light passing though the second polarizer decreases as some function of the angle between the axes of the first and second polarizer. It cannot be a sine function, since $\sin 0° = 0$. Perhaps it involves the cosine function. We know that $\cos 30° = 0.86$, which is more than the 0.75 for 30° in the second row. However, $0.86^2 = 0.75$, exactly the decrease in the intensity of light caused by the second polarizer oriented at 30° relative to the first. The same relation works for all the rows. Thus the function of the angle we are looking for is $\cos^2\theta$, where $\theta$ is the angle between the axes of the two polarizers. Thus $I = I_{0/2}\cos^2\theta$, where $I_{0/2}$ is the light intensity after leaving the first filter and before the second.

You might wonder why the second polarizer reduces the light intensity by $\cos^2\theta$ and not by $\cos\theta$ or some other function of $\theta$. Recall that the intensity of sound is proportional to the amplitude squared of sound waves ($I \propto A^2$). Similarly, it appears that the polarizer reduces the amplitude of the light wave by $\cos\theta$ and thus the intensity of the light by $\cos^2\theta$.

## Hypotheses concerning light and polarization

In Observational Experiment Table 24.1 we found that polarization effects can only be observed with transverse waves and not longitudinal waves. We just found from the light polarization experiments in Table 24.2 that light seems to behave like a transverse wave with amplitude $A$ and intensity $I \propto A^2$. However, since light can travel through a vacuum (for example, when it travels from the Sun to Earth), what is actually vibrating in a light wave?

To answer this question we need to look carefully at how a transverse wave travels through a medium. Think of what is happening to a transverse wave on a Slinky when a wave travels through it. When one coil is displaced, it pulls the next coil in a direction perpendicular to the propagation direction the wave travels. That coil then does the same to the coil next to it, and so on. The individual vibrating coils (analogous to the particles of the medium) exert elastic forces on each other. These elastic forces point perpendicular to the propagation direction of the wave and accelerate the displaced coils back toward equilibrium.

What is the nature of the medium through which light waves travel? What are the particles that are vibrating, and what is the nature of the restoring forces that are accelerating those particles back toward equilibrium? We don't usually think of a vacuum as containing any particles at all, so what is going on? Here are two hypotheses.

1. Light is actually a mechanical vibration that travels through an elastic medium. This medium is completely transparent and has exactly zero mass (just like the vacuum). This medium will be called **ether** (not related to the chemical compound of the same name).
2. A light wave is some new type of vibration that does not involve physical particles vibrating around equilibrium positions due to restoring forces being exerted on them.

The next section shows how the studies of electricity and magnetism helped test the second hypothesis.

**Review Question 24.1** What is the difference between a linearly polarized mechanical wave and an unpolarized one?

## 24.2 Discovery of electromagnetic waves

Before the second half of the 19th century, the investigations of light and electromagnetic phenomena proceeded independently. However, around 1860 the work of a British physicist, James Clerk Maxwell (1831–1879), led to the unification of those phenomena and finally helped answer the question of how a transverse wave can propagate in a vacuum. We know from our study of electromagnetic induction (Chapter 18) that Michael Faraday introduced the concept of a field and the relationship between electric and magnetic fields. Recall that Faraday's law was based on the idea that *a changing magnetic field can produce an electric field*. Subsequently, in 1865 Maxwell suggested a new field relationship: *a changing electric field can produce a magnetic field*. This idea was motivated by a thought experiment devised by Maxwell in which he imagined what would happen in the space between the plates of a charging or discharging capacitor. He suggested that the changing electric field between the capacitor plates could be viewed as a special nonphysical current, but one that would still produce a magnetic field (**Figure 24.4**). This magnetic field was first detected in 1929 but not measured precisely until 1985 due to its extremely tiny magnitude. Maxwell summarized this new idea and other electric and magnetic field ideas mathematically in a set of four equations, now known as Maxwell's equations. The equations are very complicated, but we can summarize them conceptually:

1. Stationary electric charges produce a constant electric field. The $\vec{E}$ field lines representing this electric field start on positive changes and end on negative charges.
2. There are no magnetic charges.
3. A magnetic field is produced either by electric currents or by a changing electric field (Figures 24.4 and **24.5a**). The $\vec{B}$ field lines that represent the magnetic field form closed loops and have no beginnings or ends.
4. A changing magnetic field produces an electric field. The $\vec{E}$ field lines representing this electric field are closed loops (Figure 24.5b).

**Figure 24.4** A changing electric field produces a magnetic field.

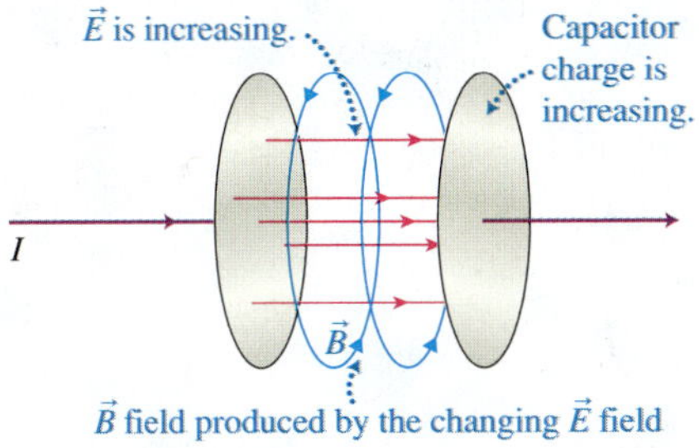

$\vec{E}$ is increasing.
Capacitor charge is increasing.
$I$
$\vec{B}$
$\vec{B}$ field produced by the changing $\vec{E}$ field

**Figure 24.5** Changing $\vec{E}$ and $\vec{B}$ fields produce $\vec{B}$ and $\vec{E}$ fields.

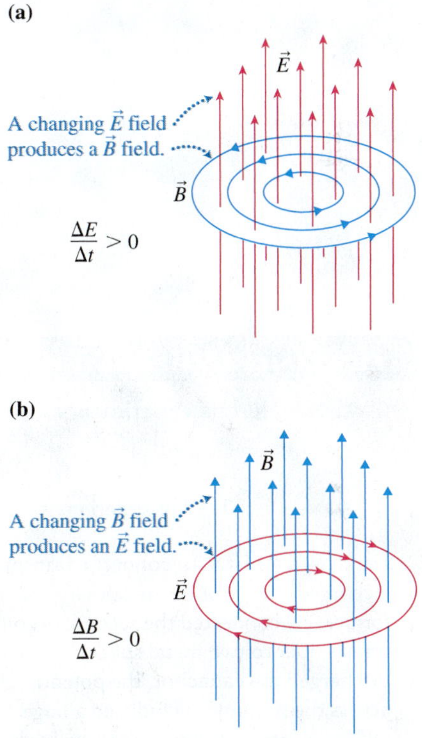

(a)
$\vec{E}$
A changing $\vec{E}$ field produces a $\vec{B}$ field.
$\vec{B}$
$\frac{\Delta E}{\Delta t} > 0$
(b)
$\vec{B}$
A changing $\vec{B}$ field produces an $\vec{E}$ field.
$\vec{E}$
$\frac{\Delta B}{\Delta t} > 0$

### Producing an electromagnetic wave

Maxwell's equations had important consequences. First, the equations led to an understanding that a changing electric field can produce a changing magnetic field, which in turn can produce a changing electric field, and on and on in a sort of feedback loop (**Figure 24.6**). This feedback loop does not require the presence of any electric charges or currents. Maxwell investigated this idea mathematically using the four equations, and to his surprise he found they led to a wave equation similar to Eq. (20.4) in which the electric and magnetic fields themselves were vibrating. The speed of propagation of these waves in a vacuum turned out to be a combination of two familiar constants: $v = (1/\sqrt{\varepsilon_0\mu_0})$, where the constants $\varepsilon_0 = 8.85 \times 10^{-12}\,\mathrm{C^2/N \cdot m^2}$ and $\mu_0 = 4\pi \times 10^{-7}\,\mathrm{N/A^2}$ relate to the electric and magnetic interactions of electrically charged particles in a vacuum. The constant $\varepsilon_0$ is the **vacuum permittivity** and is related to Coulomb's constant $k$ through the relationship

$$\varepsilon_0 = \frac{1}{4\pi k}$$

**Figure 24.6** The changing $\vec{B}$ and $\vec{E}$ fields can spread without any charges or currents.

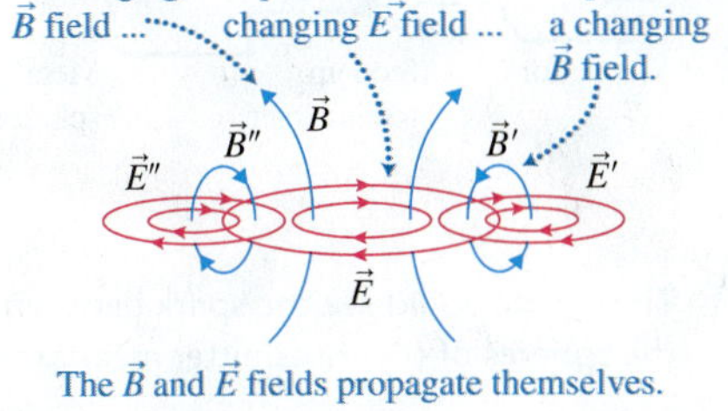

A changing $\vec{B}$ field ...
... produces a changing $\vec{E}$ field ...
... produces a changing $\vec{B}$ field.
$\vec{B}''$
$\vec{B}$
$\vec{B}'$
$\vec{E}''$
$\vec{E}'$
$\vec{E}$
The $\vec{B}$ and $\vec{E}$ fields propagate themselves.

(See Chapter 14 for more on vacuum permittivity.) The constant $\mu_0$ is the **vacuum permeability.** We discussed vacuum permeability in the chapter on magnetism (Chapter 17). When Maxwell inserted the values of the constants into the expression for the speed of electromagnetic waves, he obtained

$$v = \frac{1}{\sqrt{\varepsilon_0 \mu_0}} = \frac{1}{\sqrt{(8.85 \times 10^{-12}\,\mathrm{C^2/N \cdot m^2})(4\pi \times 10^{-7}\,\mathrm{N/A^2})}}$$

$$= \sqrt{9.00 \times 10^{16}\,\frac{\mathrm{N \cdot m^2 \cdot A^2}}{\mathrm{C^2 \cdot N}}} = \sqrt{9.00 \times 10^{16}\,\frac{\mathrm{m^2 \cdot A^2}}{\mathrm{A^2 \cdot s^2}}} = 3.00 \times 10^8\,\mathrm{m/s}$$

At the time Maxwell did this calculation, the speed of light in air had already been measured and was consistent with this value. Could it be that light is an electromagnetic wave? This was the second testable consequence of Maxwell's model. *If* the relationship between the changing electric and magnetic fields suggested by the model is correct, *then* a change in either of these fields could generate an electromagnetic wave that would propagate at the speed of light. This prediction, if consistent with the experiment, would support the hypothesis that light is an electromagnetic wave composed of vibrating electric and magnetic fields!

## Testing the hypothesis that light can be modeled as an electromagnetic wave

The German physicist Heinrich Hertz (1857–1894) was the first to test the hypothesis. In 1888 Hertz built a device called a **spark gap transmitter** and used it in his experiments. His work is summarized in Testing Experiment **Table 24.3**.

**TESTING EXPERIMENT TABLE**

**24.3 Hertz's experiments.**

| Testing experiment | Prediction | Outcome |
|---|---|---|
| Hertz used a switch to connect a transmitter with a charged capacitor to the primary coil of a transformer. He connected the secondary coil of the transformer to two metal spheres. When he discharged the capacitor, the potential difference across the primary coil induced a huge potential difference across the secondary coil, which in turn charged the spheres, causing them to spark. A receiver that consisted of two small metal spheres separated by a small gap and connected to a loop of wire was placed some distance from the metal transmitter spheres. 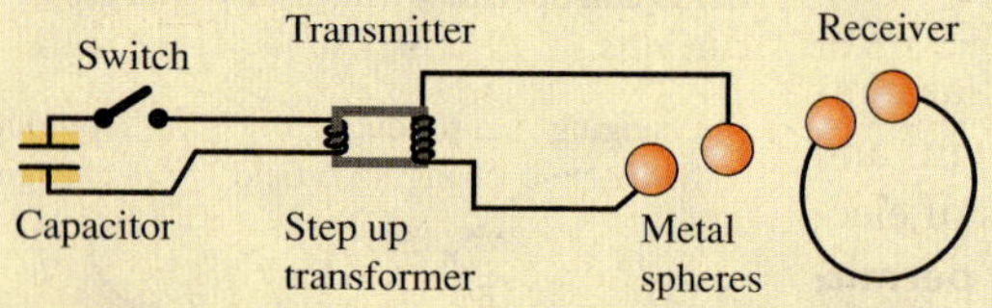 | A spark between the transmitter spheres would indicate a large changing electric field between the spheres of the transmitter. If Maxwell's hypothesis was correct, then this changing electric field would produce an electromagnetic wave. When the wave reached the receiver loop, it would induce a current in the loop and cause a weak spark between the spheres of the receiver. 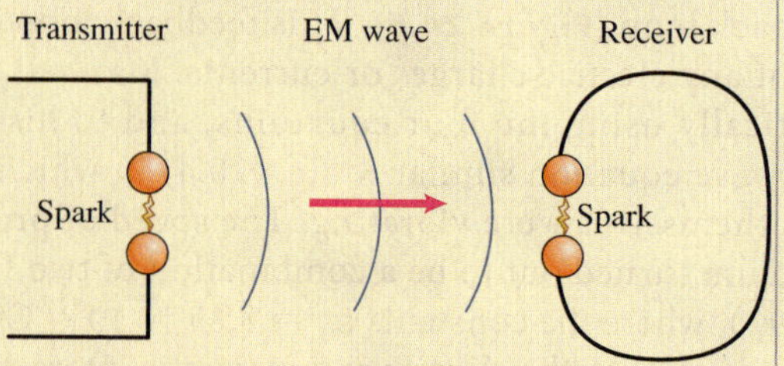 | After hours in a dark room watching for sparks, Hertz eventually saw sparks between the receiver spheres! |





**Conclusion**

That Hertz could see the spark between the spheres of the receiver in response to the spark between the spheres of the transmitter meant that something had traveled from the transmitter to the receiver.

Hertz's experiment was the first supporting evidence for the idea that electromagnetic waves existed. Hertz performed many additional experiments to determine if the electromagnetic waves generated in his experiment had the same properties as light waves. The only difference he found was in their frequency—Hertz's waves had a much smaller frequency than the light waves we are familiar with. He used metal sheets of different shapes to observe reflection. He let them pass through different media and observed refraction. He performed an analogue of a double-slit experiment and observed interference. He even observed the polarization of the waves using a metal fence. Finally, he performed experiments to measure the speed of the propagation of the waves, which turned out to be $3.00 \times 10^8 \text{ m/s}$. These experiments supported the idea that light could be modeled as a transverse wave of vibrating electric and magnetic fields.

Maxwell's model of light as an electromagnetic wave seems to resolve the problem of what is vibrating in the light wave: the $\vec{E}$ field and the $\vec{B}$ field. Each field is vibrating perpendicular to the direction of travel of the wave. Thus we have a new model of light as a transverse electromagnetic wave that travels in a vacuum and in other media. In a vacuum the speed of light is $3.00 \times 10^8 \text{ m/s}$.

The waves predicted by Maxwell and experimentally discovered by Hertz soon found a practical application. In 1892 Nikola Tesla used an improved Hertz's transmitter to conduct the first transmission of information via radio waves, one form of electromagnetic waves. By 1899 there was successful radio wave-based communication across the English Channel.

‹ Active Learning Guide

## Antennas are used to start electromagnetic waves

The most common device used to produce electromagnetic waves is an antenna. Cell phones, two-way radios, and over-the-air radio and television stations use them. A simple type of antenna (called a **half-wave electric dipole antenna**) can be made from a pair of electrical conductors, one connected to each terminal of a power supply that is producing an alternating emf (**Figure 24.7**). The alternating emf leads to the continuous charging and discharging of the two ends of the antenna (somewhat similar to the action of a capacitor). How does this antenna produce an electromagnetic wave?

**Figure 24.7** A half-wave dipole antenna.

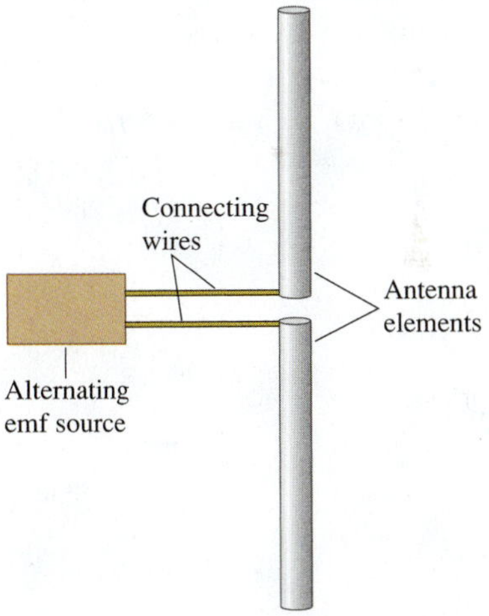


The source of alternating emf produces alternating current. Let the period of this oscillation be $T$. Assume that just after the power supply is turned on, the current is downward. This produces a $\vec{B}$ field ($t = 0$). After a very short time interval (nanoseconds for radio waves), this current causes the bottom of the antenna to become increasingly positively charged and the top increasingly negatively charged. This produces an $\vec{E}$ field ($t = T/8$). Both fields are shown in **Figure 24.8a**. Since the antenna is connected to a source of alternating emf, the current reverses direction periodically. When this happens, the current momentarily vanishes along with its corresponding $\vec{B}$ field (Figure 24.8b, $t = T/4$). When the current starts upward, the charge separation begins to decrease along with the $\vec{E}$ field produced by it (Figure 24.8c, $t = 3T/8$). Next, the charge separation vanishes along with its $\vec{E}$ field (Figure 24.8d, $t = T/2$). The process then repeats in the opposite direction and continues to repeat billions of times each second. Figure 24.8e shows what happens at one point in space during one period $T$ of oscillation. The $\vec{E}$ and $\vec{B}$ fields at that point are shown in sequence as time passes. Note that the second part in (e) corresponds to the fields shown in (a). The third part in (e) corresponds to the fields in (b), and so forth.

‹ Active Learning Guide

Near the antenna, the oscillating electric charge and current in the antenna produce the $\vec{E}$ and $\vec{B}$ fields as described above (near fields). Farther from the antenna, the fields co-create each other in the way Maxwell's equations describe (far fields). There is a noticeable difference between these two situations.

**Figure 24.8** The electric and magnetic fields produced by a dipole antenna. (a)–(d) The $\vec{E}$ and $\vec{B}$ fields at a point near the front of the dipole antenna at times $T/8$, $T/4$, $3T/8$, and $T/2$. (e) The changing $\vec{E}$ and $\vec{B}$ fields at that point are shown as time progresses. The pattern looks like that of a simple harmonic oscillator.

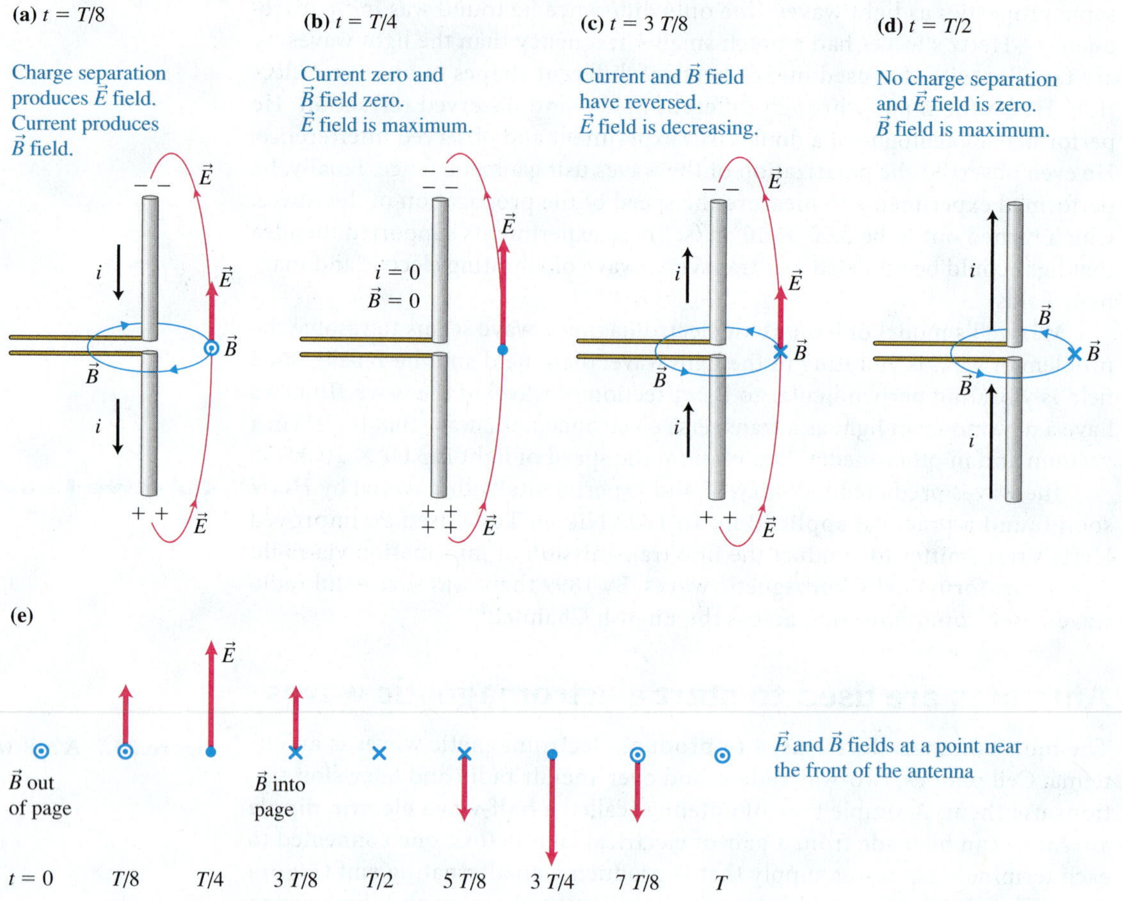


In Maxwell's electromagnetic (EM) waves, the $\vec{E}$ field and $\vec{B}$ field oscillate in phase. That is, when the $\vec{E}$ field at a point is at a maximum, so is the $\vec{B}$ field. Close to the antenna the $\vec{E}$ field and $\vec{B}$ field are out of phase, the $\vec{E}$ field being at a maximum while the $\vec{B}$ field is zero, and vice versa.

Both near and far $\vec{E}$ and $\vec{B}$ fields contribute to the net $\vec{E}$ and $\vec{B}$ fields at a particular distance from the antenna: (1) the fields produced by the oscillating charges and current in the antenna and (2) the electromagnetic waves produced by the oscillating $\vec{E}$ field and $\vec{B}$ field in the region surrounding the antenna. The amplitudes of the $\vec{E}$ field and $\vec{B}$ field produced by contribution 1 get weaker with distance $(1/r^2)$ much faster than the amplitude of the EM waves produced by contribution 2 $(1/r)$. In other words, the out-of-phase contribution becomes less and less significant farther from the antenna relative to the in-phase contribution. Far from the antenna, only Maxwell's in-phase EM waves are significant.

The size and shape of antennas are carefully selected so that they can efficiently send and receive EM waves with a particular range of wavelengths. In the case of the dipole antenna shown in Figure 24.7, the total length of the antenna is related to the desired wavelength of EM waves that it will produce. It is called a *half-wave* dipole antenna because its length is half the wavelength of the oscillating electric field within the antenna that drives the current. However, since this electric field exists in a different material (the metal antenna), its wavelength isn't the same as the wavelength of the EM wave produced by the antenna (those fields are in air), but they are related. Recall from our study of refraction (Chapter 21) that different materials have different indexes of refraction, which affects the wavelength of the EM waves passing through them. Knowing the index of refraction of the antenna material allows us to choose an antenna length that will very efficiently produce EM waves of the desired wavelengths. For example, most cell phones operate at frequencies between 850 and 1900 MHz (called a *frequency band*), corresponding to wavelengths of about 10–30 cm. Within the antenna the wavelength is about one-third of its value in air. As a result, cell phone antennas need to be roughly 3–10 cm in length, consistent with the size of modern cell phones.

### "Reflecting" on models of light

Let us review how our understanding of light has evolved over the past four chapters. First we learned about shadows and the phenomena of reflection and refraction. To explain these we used a particle model, which assumed that light consisted of tiny particles that travel in straight lines through a medium. The model did not specify the nature of these particles. Then we found new phenomena that the particle model could not explain, such as what happens when light passes through narrow openings. So we progressed to a wave model. In this chapter, observations of polarization of light led us to the idea that a light wave is a transverse wave. Maxwell's and Hertz's work led to the electromagnetic wave model of light, which proposes that in a light wave, the electric and magnetic fields themselves are vibrating. The next section gives examples of practical applications of the new model of light.

**Review Question 24.2** What needs to happen to produce an electromagnetic wave?

## 24.3 Some applications of electromagnetic waves

We use electromagnetic waves every day: they are present when we listen to the car radio or use a GPS system to find a destination, and when a police officer uses radar to catch speeding motorists.

### Radar

The EM waves produced by Hertz in his experiments are called **radio waves,** based on their frequency ($10^4$–$10^6$ Hz). Today these and even higher frequency waves are used to transmit speech, music, and video over long distances. Another practical application of radio waves is radar (an acronym of radio detection and ranging). **Radar** is a way of determining the distance to a faraway object by reflecting radio wave pulses off the object. Radar devices help locate airplanes, determine distances to planets and other objects in the solar system, find schools of fish in the ocean, and perform many other useful functions.

In radar, an antenna (called the **emitter**) emits a rapid radio wave pulse, which reflects off the surface of the object and is detected by a **receiver**. This receiver is just an antenna that is connected to a sophisticated type of ammeter that can measure the current produced when the reflected electromagnetic wave arrives. Most often the emitting antenna and the receiving antenna are actually the same antenna.

Imagine that a radar system is being used to detect aircraft. The time interval for the round trip of the pulse indicates the distance to the aircraft. The longer it takes the pulse to return, the farther away the aircraft is. By emitting a sequence of pulses and determining how the distance to and direction of the

**Figure 24.9** The radar pulse width $\tau$ and the time interval $T$ between pulses determine the near and far distances at which a radar can detect an airplane.

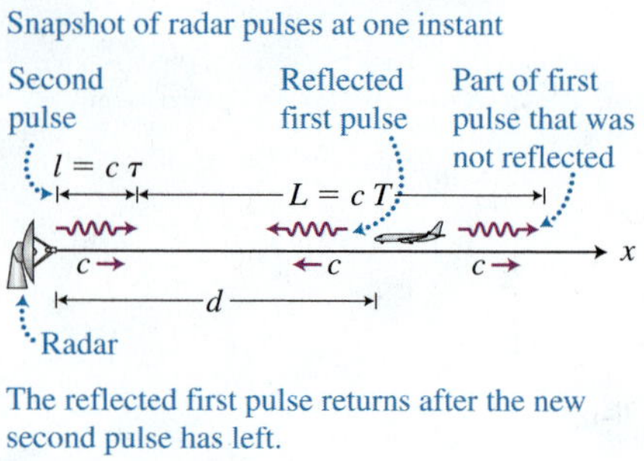


aircraft changes from one pulse to the next, the system can determine the velocity of the aircraft.

Radar has some limitations. It is less effective in tracking objects that are very close, very far away, or flying very low to the ground. The difficulty of tracking close objects is related to the time interval $\tau$ that it takes a single pulse to be emitted (called the *pulse width*) and the time interval $T$ that passes between the start of each successive pulse (called the *period* of the pulse; **Figure 24.9**). If the object is so close that the end of the pulse has not been emitted before the beginning of the pulse returns, then the distance to the object is difficult to determine accurately. This limitation has been alleviated somewhat by advanced electronic switching systems that produce very short pulse widths.

The source of the second limitation, measuring distant objects, is the time interval $T$. If the object is so far away that the reflected pulse returns after the next pulse is sent, then it becomes unclear which reflected pulse corresponds to which emitted pulse. In modern radar the period is between $T = 10^{-3}$ s and $T = 10^{-6}$ s. The overall effect of these two difficulties is that there is both a minimum and a maximum range within which an object can be accurately detected, determined by the pulse width $\tau$ and the period $T$.

The third limitation, tracking objects that are flying close to the ground, is a result of Earth's curved surface. Because of the curvature, the distance to the horizon is about 5 km as seen by a person standing on the ground. Thus, a target that is farther away than 5 km will not be visible to radar, because the ground will block the line of sight. In order to obtain a range of 20 km, the radar system must be mounted on a tall tower in order to increase the horizon distance.

## EXAMPLE 24.1 Radar

Determine the range of distances to a target that can be reliably detected by a radar system with pulse width $\tau = 10.8\ \mu s$ ($1\ \mu s = 10^{-6}$ s) and a pulse emission frequency of 7.5 kHz, that is, 7500 radio wave pulses each second.

**Sketch and translate** The pulse width of 10.8 $\mu s$ should be less than the time interval needed for the EM wave to travel to the target object and back. That way, the back of the 10.8-$\mu s$ pulse will have left the transmitter before the front of the reflected pulse returns. The pulse width determines the closest range of the target from the radar.

The pulse repetition period is $T = 1/f = 1/(7500\ \text{Hz}) = 1.33 \times 10^{-4}$ s. This period should be less than the time interval needed for the pulse to travel to the target and back. That way, the reflected pulse returns before a new pulse leaves. The period determines the maximum range of the target from the radar.

**Simplify and diagram** Assume that pulses do not reflect off other objects while traveling to or from a target. Assume also that the radio wave pulses travel at the speed of light in vacuum (they actually travel just slightly slower in air).

**Represent mathematically** If the target is a distance $d$ away, the time interval for a pulse to travel to the target and back is $\Delta t = (2d/c)$. This time interval needs to be longer than the pulse width and shorter than the pulse period in order for the radar to accurately determine the distance to the target. Mathematically,

$$\tau < \frac{2d}{c} < T$$

**Solve and evaluate** The pulse width time interval $\tau = 10.8\ \mu s = 10.8 \times 10^{-6}$ s and the period $T = 1.33 \times 10^{-4}$ s. Solving the above inequality for $d$, we get

$$\frac{c\tau}{2} < d < \frac{cT}{2}$$

The range of distances is then

$$\frac{(3.00 \times 10^8\ \text{m/s})(10.8 \times 10^{-6}\ \text{s})}{2} < d < \frac{(3.00 \times 10^8\ \text{m/s})(1.33 \times 10^{-4}\ \text{s})}{2}$$

or

$$1.62 \times 10^3\ \text{m} < d < 20.0 \times 10^3\ \text{m}$$

A target outside this 1.6-km to 20-km range will not be detectable by this radar system. However, military radars have significantly shorter pulse widths and can detect objects much closer than 1.6 km.

**Try it yourself:** What is the maximum range of a 15-kHz pulse emission rate radar system mounted on a tall tower?

*Answer:* 10 km.

## Global Positioning System

Global positioning systems enable us to determine our location using electromagnetic waves and a system of satellites. In the United States, the Global Positioning System (GPS) is operated by the U.S. Air Force and is called the Navstar Global Positioning System. Navstar has three main components:

1. A minimum of 24 satellites that orbit 20,200 km above Earth's surface and transmit signals using **microwaves**: electromagnetic waves at frequencies 1575 MHz and 1227 MHz (1 MHz $= 10^6$ Hz) (**Figure 24.10**).
2. Receivers that detect signals from multiple satellites.
3. A control system that maintains accurate information about the locations of the satellites.

**Figure 24.10** Twenty-four satellites circle Earth as part of the Global Positioning System. Three or four GPS satellites triangulate to determine your location.

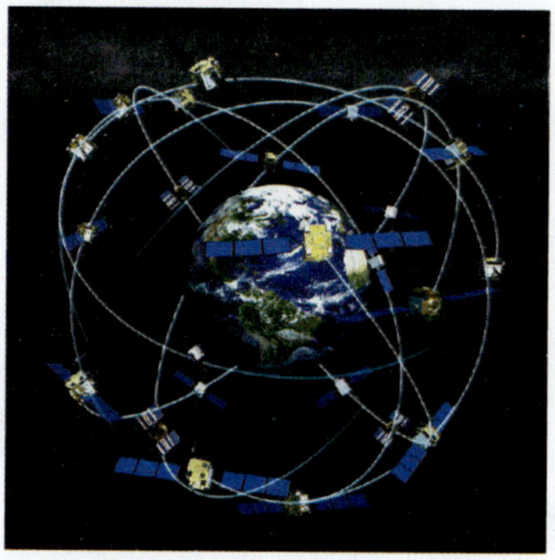

The receiver detects signals from at least three satellites in order to determine your position on the ground. With signals from four or more satellites, the receiver can also determine your altitude. The clocks on the satellites and a clock in your unit are synchronized to within a nanosecond ($10^{-9}$ s). When you turn on your GPS, it starts a measurement at the same time each satellite sends a microwave signal. Your unit measures the time of arrival of the signals from the satellites. Using the known positions of the satellites, your GPS unit is able to calculate your position by a process called **trilateration**.

Trilateration is easier to understand in a two-dimensional space than in a three-dimensional space. Suppose a visitor from overseas is traveling in the United States and is unsure of his location. He comes across a sign that says Atlanta 430 miles, Minneapolis 510 miles, and Philadelphia 580 miles. The visitor has a map of the United States and draws a circle centered on each city. The radius of each circle is the distance to that city (**Figure 24.11**). There is only one place on the map where the three circles intersect—Indianapolis.

**Figure 24.11** Surface trilateration is analogous to that used in GPS.

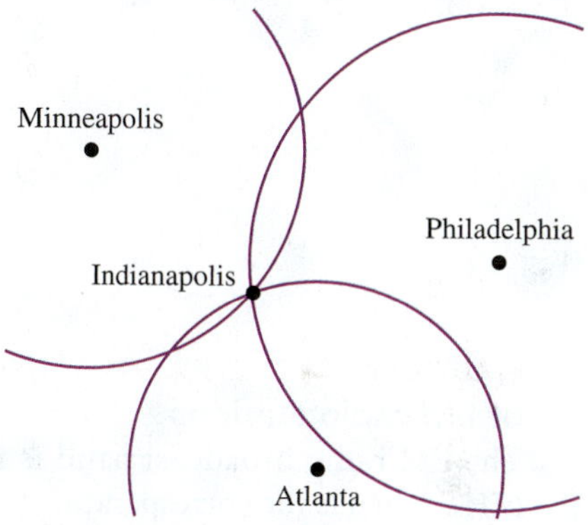


Only one place (Indianapolis) is the known distance from the other three cities.

Your GPS system locates your current position in much the same way. The unit determines the distance between you and each of at least three satellites. The unit then constructs spheres (rather than circles) around each of them with corresponding radii equal to those distances. The spheres intersect at one unique location, your location. Most handheld GPS units can determine your position within 3 to 10 m. More advanced receivers use a method called Differential GPS (DGPS) to obtain an accuracy of 1 m or better.

## Microwave cooking

Percy Spencer accidentally discovered microwave cooking in 1945 while constructing magnetron tubes at Raytheon Corporation. While standing in front of one of the tubes, Spencer suddenly felt a candy bar in his pocket start to melt. Intrigued, he placed kernels of popcorn in front of the tube, and they popped. He placed a raw egg in front of the tube and it exploded. How could this be explained?

Water, fat, and other substances in food absorb energy from microwaves in a process called **dielectric heating**. To understand what's happening, we have to think microscopically. Water is a polar molecule (**Figure 24.12**) and is a permanent electric dipole. When a microwave passes by a water molecule, the electric field of the microwave exerts oscillating electric forces on the positively and negatively charged regions of the molecule. This causes the water molecule to flip over billions of times each second (the frequency of microwaves). This molecular movement spreads to neighboring molecules through collisions (even to molecules that are not electric dipoles). The electric and magnetic energy in the microwaves is transformed into the internal energy of the food, which manifests as an increase in the food's temperature. Microwave heating is most efficient in liquid water and much less so in fats and sugars (which have smaller electric dipoles) and ice (where the water molecules are not as free to rotate).

**Figure 24.12** A water molecule is an electric dipole with more positive charge on one end and negative charge on the other.

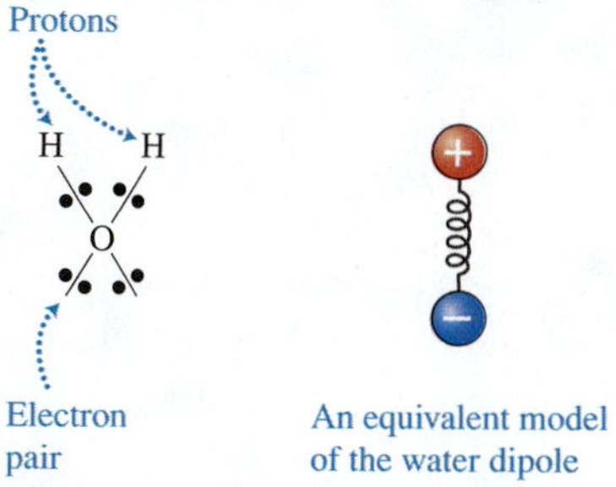

**Review Question 24.3** How are GPS and radar similar and how are they different?

## 24.4 Frequency, wavelength, speed, and the electromagnetic spectrum

Recall that waves are described by frequency $f$, speed $v$, wavelength $\lambda$, and a relationship between these three quantities:

$$\lambda = \frac{v}{f}$$

**Active Learning Guide ›**

All electromagnetic waves travel at the speed of light $c$ in a vacuum. However, in media other than a vacuum, the speed of electromagnetic waves changes to $v = c/n$, where $n$ is the refractive index of the medium. In addition, light waves of different frequencies have slightly different refractive indexes and hence travel at different speeds in nonvacuum media such as glass or water. Therefore, in the equation $v = c/n$, $n$ is not a constant for a particular medium, but has a small dependence on the frequency of the electromagnetic wave.

### Frequency and wavelength of electromagnetic waves

According to Maxwell's equations, electromagnetic waves can have essentially any frequency. AM radio stations emit relatively low frequency EM waves (about $10^6$ Hz = 1 MHz). Supernova remnants and extragalactic sources emit the highest frequency cosmic ray EM waves ever detected (up to about $10^{26}$ Hz).

**QUANTITATIVE EXERCISE 24.2 Wavelengths of FM radio stations**

The FM radio broadcast band is from 88 MHz to 108 MHz. What is the corresponding wavelength range?

**Represent mathematically** Assume that these radio waves are traveling at the speed of light in a vacuum. This is a reasonable assumption since the refractive index of air is very close to the refractive index of a vacuum (both are $\approx 1$). Apply $\lambda = c/f$ to find the range of wavelengths for FM radio waves.

**Solve and evaluate**

$$\lambda_1 = \frac{3.0 \times 10^8\,\text{m/s}}{88 \times 10^6\,\text{s}^{-1}} = 3.4\,\text{m}$$

$$\lambda_2 = \frac{3.0 \times 10^8\,\text{m/s}}{108 \times 10^6\,\text{s}^{-1}} = 2.8\,\text{m}$$

The electromagnetic waves emitted by FM radio stations have wavelengths ranging from 2.8 m to 3.4 m.

**Try it yourself:** AM radio stations broadcast in the wavelength range from 556 m to 186 m. Determine the frequency range of AM radio station broadcasting.

*Answer:* 540 kHz to 1610 kHz.

### Hearing FM radio waves

Our ears can only hear sounds with frequencies ranging from about 20 Hz to 20,000 Hz, much lower than the frequency of FM or satellite radio stations. How can we hear the music from those stations? The high-frequency EM waves used by FM radio stations are known as *carrier waves*. FM stands for "frequency modulation"; the information that gets converted into the sounds we hear is encoded as tiny variations in the frequency of the carrier wave. The receiver decodes the variations and converts them into an electric signal that a speaker can then convert into sound. A similar process occurs in AM

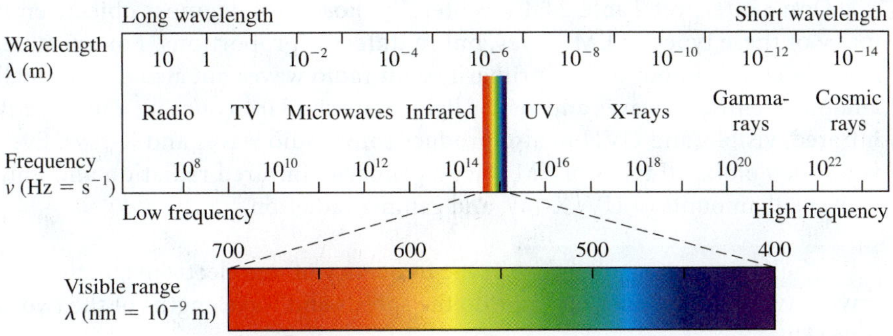


**Figure 24.13** The electromagnetic (EM) spectrum.

(amplitude modulation) radios, only in that case the amplitude of the carrier wave is varied rather than its frequency.

## The Electromagnetic spectrum

The range of frequencies and wavelengths of electromagnetic waves is called the **electromagnetic spectrum** (**Figure 24.13**). As you see, visible light occupies a very narrow range of the spectrum (wavelengths of 400 nm to 700 nm). In this chapter we've already encountered radio waves and microwaves. We encountered another type of radiation briefly (in Chapter 12) when we considered heating by radiation. Historically, scientists thought that visible light and thermal radiation (now called infrared [IR] radiation) were completely different. In other words, they thought that the light emitted by a fire in a fireplace was a different phenomenon than the warmth that the fire gave off. We now understand that they are both EM waves, just with different frequency ranges.

**Table 24.4** lists the broad categories of EM radiation and some of the ways in which it is produced and detected. We will discuss EM radiation production in more detail in later chapters.

**Table 24.4 The electromagnetic spectrum.**

| Type of EM radiation and frequency/ wavelength range | Produced by | Detected by | Used in |
|---|---|---|---|
| Radio waves and microwaves<br>$10^4$–$10^{10}$ Hz<br>$10^3$–$10^{-3}$ m | Accelerated electrically charged particles in wires with oscillating current | Metal wires (antennas) | Radio and cell phone communication<br>Microwave ovens |
| Infrared waves<br>$10^{11}$–$10^{14}$ Hz<br>$10^{-4}$–$10^{-5}$ m | Objects with a temperature between 10 K and 1000 K | Detectors whose properties (for example, electric conductivity) change with temperature | Toasters<br>Conventional ovens |
| Visible light<br>$10^{14}$–$10^{15}$ Hz<br>$10^{-6}$ m | Low-density ionized gas<br>Objects whose temperatures are between 1000 K and 10,000 K, such as typical stars | Human eyes<br>Charged coupled devices (CCDs) and complimentary metal-oxide semiconductor image sensors (CMOS) | Visual communication |
| UV light<br>$10^{16}$–$10^{17}$ Hz<br>$10^{-7}$–$10^{-8}$ m | Stars<br>Black lights | Charge coupled devices (CCDs) that can detect UV radiation | Forensics<br>Decontamination |
| X-rays<br>$10^{18}$–$10^{19}$ Hz<br>$10^{-9}$–$10^{-10}$ m | Accelerating particles in X-ray tubes<br>Very hot objects<br>Energetic stellar events such as supernovas | Photographic plates<br>Geiger counters | Medical imaging<br>Studies of material properties<br>Cosmology |
| Gamma rays<br>$10^{20}$ Hz<br>$10^{-12}$ m | Radioactive materials (such as uranium and its compounds)<br>Nuclear reactions<br>Stellar explosions | Solid-state detectors that transform gamma rays into optical or electronic signals | Detection of nuclear explosions<br>Cosmology |

Don't interpret Table 24.4 too literally. For example, most objects emit many of these types of EM waves, but in different proportions. For example, cold gas clouds in our galaxy primarily emit radio waves but also emit a small amount of infrared waves and visible light. Stars similar to our Sun emit mostly infrared, visible, and UV, but also produce some radio waves and X-rays. Even your body emits all types of EM waves: primarily infrared radiation and only very small amounts of UV, X-ray, and gamma radiation.

**Review Question 24.4** If the frequency of one electromagnetic wave is twice that of another, how do the speeds and wavelengths of the two waves differ?

## 24.5 Mathematical description of EM waves and EM wave energy

**Figure 24.14** Electromagnetic wave $\vec{E}$ and $\vec{B}$ fields (a) at one time and (b) at one location.

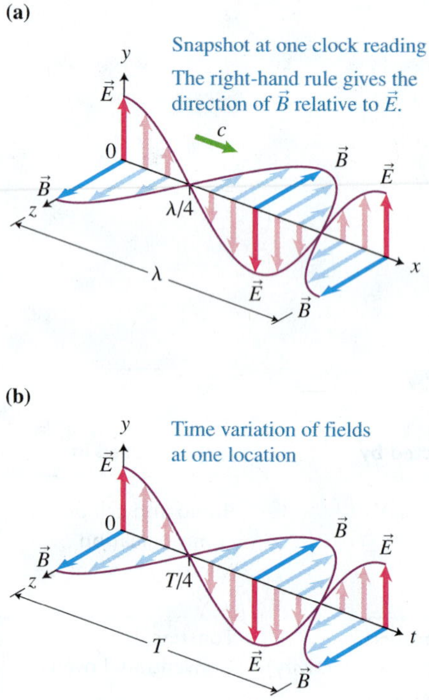


When analyzing electric dipole antennas, we found that they produce electric and magnetic fields whose $\vec{E}$ and $\vec{B}$ vectors vibrate perpendicular to each other and perpendicular to the direction the wave travels (**Figure 24.14**). Maxwell's equations also predict that the amplitudes of the changing $\vec{E}$ field and $\vec{B}$ field vectors are related:

$$E_{\text{max}} = cB_{\text{max}} \tag{24.1}$$

where $c = 3.00 \times 10^8\ \text{m/s}$ is the speed of electromagnetic waves in vacuum. In other words, not only do the electric and magnetic fields oscillate in phase but also their magnitudes are related through the speed of light.

We can check the above equation with unit analysis. The units for electric field are newtons/coulomb, or, equivalently, volts/meter. The first combination comes from the definition of the $\vec{E}$ field; the second comes from the mathematical relationship between the $\vec{E}$ field and the $V$ field (see Chapter 15). The units for magnetic field are T (tesla) = (newton · second)/(coulomb · meter). This unit can be deduced from the expression for the magnetic force exerted on a changed particle moving in a magnetic field ($F = qvB \sin \theta$ or $B = F/(qv \sin \theta)$, as discussed in Chapter 17). We can now insert the units into the right-hand side of Eq. (24.1):

$$\left(\frac{\text{m}}{\text{s}}\right)\left(\frac{\text{N} \cdot \text{s}}{\text{C} \cdot \text{m}}\right) = \frac{\text{N}}{\text{C}}$$

which are the units of the left-hand electric field side of the equation. Thus, the equation survives a unit analysis.

**TIP** The relation between $E$, $B$, and $c$ [Eq. (24.1)] is applicable only for electromagnetic waves and not for fields in general and is correct only if the three quantities are measured using SI units.

### Mathematical description of EM waves

We now have enough background to make a quantitative description of EM waves that involves their speed $c$, wavelength $\lambda$, and frequency $f$ and the relationship between the amplitudes of the vibrating $\vec{E}$ and $\vec{B}$ fields, $E_{\text{max}} = cB_{\text{max}}$. To do this, we use the mathematical descriptions of waves (developed in Chapter 20) as a guide.

Imagine the simplest EM wave—a sinusoidal wave of a single frequency. It travels along the $x$-axis in the positive direction and does not diminish in amplitude as it travels. Now, imagine that you place electric and magnetic field detectors at a point on the positive $x$-axis. As the wave passes, the detectors register the magnitude of the oscillating $\vec{E}$ field and $\vec{B}$ field as

$$E = E_{max}\cos\left(\frac{2\pi}{T}t\right) = E_{max}\cos(2\pi f t)$$

$$B = B_{max}\cos\left(\frac{2\pi}{T}t\right) = B_{max}\cos(2\pi f t)$$

‹ **Active Learning Guide**

We chose $t = 0$ at a moment when the fields have their maximum values $E_{max}$ and $B_{max}$. The frequency $f$ of the wave is the inverse of the period $T$ of the wave. A convenient right-hand rule relates the direction of the $\vec{E}$ field, the direction of the $\vec{B}$ field, and the direction of propagation of the wave. Hold your right hand flat with your thumb pointing out at a 90° angle relative to your fingers. Point your thumb in the direction of propagation of the wave. Your fingers will point in the direction of the $\vec{E}$ field, and your palm will face in the direction of the $\vec{B}$ field. For example, if the wave is traveling in the $+x$-direction at speed $c$, and at that moment the $\vec{E}$ field points in the $+y$-direction, then the $\vec{B}$ field points in the $+z$-direction (**Figure 24.15**).

The above two equations describe the wave vibrations at one specific point. However, the $\vec{E}$ field vectors and $\vec{B}$ field vectors are oscillating at every point along the wave's path, which means that the equation for the wave must be a function not only of time $t$ but also of position $x$. The wave equation that we developed in Chapter 20 [Eq. (20.4)] takes this into account. Applied to the $\vec{E}$ field and $\vec{B}$ field, the wave equation becomes

$$E_y = E_{max}\cos\left[2\pi\left(\frac{t}{T} \pm \frac{x}{\lambda}\right)\right]$$
$$B_z = B_{max}\cos\left[2\pi\left(\frac{t}{T} \pm \frac{x}{\lambda}\right)\right] \quad (24.2)$$

The minus signs are used to describe a wave traveling in the positive $x$-direction, and the plus signs are used to describe a wave traveling in the negative $x$-direction. We can represent these equations graphically (shown for a wave moving in the positive $x$-direction at $t = 0$ in (a) and at $x = 0$ in (b) in Figure 24.14).

Suppose we wish to describe light from a laser pointer that emits light of frequency $4.5 \times 10^{14}$ Hz with an oscillating $\vec{B}$ field of amplitude $6.0 \times 10^{-4}$ T. What else can you determine about this electromagnetic wave from this information?

(a) The period of the wave is

$$T = \frac{1}{f} = \frac{1}{4.5 \times 10^{14}\,\text{Hz}} = 2.2 \times 10^{-15}\,\text{s}$$

(b) The wavelength of the wave is

$$\lambda = \frac{c}{f} = \frac{3.00 \times 10^{8}\,\text{m/s}}{4.5 \times 10^{14}\,\text{Hz}} = 6.7 \times 10^{-7}\,\text{m} = 670\,\text{nm}$$

(c) A wavelength of 670 nm corresponds to red light (see Figure 24.13).

(d) The amplitude of electric field oscillation [Eq. (24.1)] is

$$E_{max} = cB_{max} = (3.00 \times 10^{8}\,\text{m/s})(6.0 \times 10^{-4}\,\text{T}) = 1.8 \times 10^{5}\,\text{N/C}$$

(e) Assuming the laser light travels in the positive $x$-direction, we can write wave equations for it:

$$E_y = (1.8 \times 10^{5}\,\text{N/C})\cos\left[2\pi\left(\frac{t}{2.2 \times 10^{-15}\,\text{s}} - \frac{x}{6.7 \times 10^{-7}\,\text{m}}\right)\right]$$

$$B_z = (6.0 \times 10^{-4}\,\text{T})\cos\left[2\pi\left(\frac{t}{2.2 \times 10^{-15}\,\text{s}} - \frac{x}{6.7 \times 10^{-7}\,\text{m}}\right)\right]$$

**Figure 24.15** (a) The $\vec{E}$, $\vec{B}$, and $\vec{v} = \vec{c}$ vectors of an electromagnetic wave. (b) The right-hand rule for the $\vec{E}$, $\vec{B}$, and $\vec{v}$ EM wave directions.

**(a)**

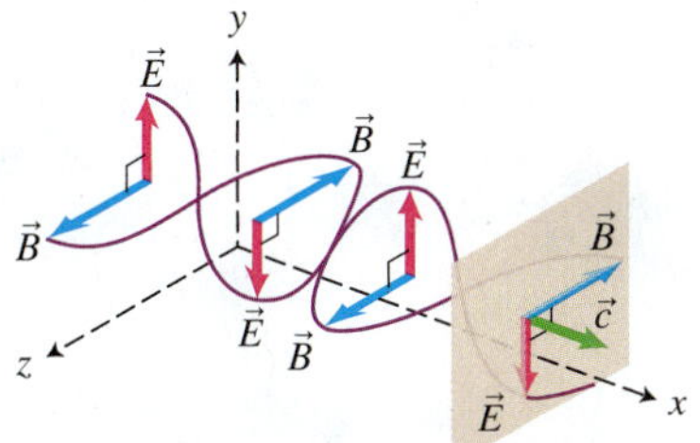


The $\vec{E}$ and $\vec{B}$ fields vibrate perpendicular to each other and in a plane perpendicular to the wave's velocity $\vec{v} = \vec{c}$.

**(b)**

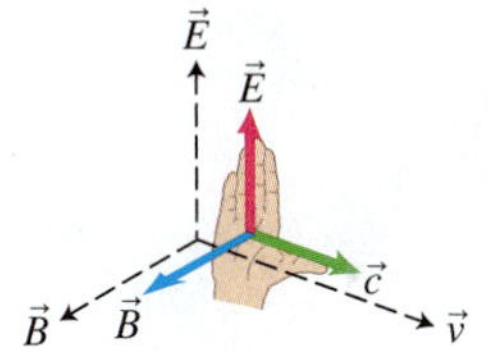

## Energy of electromagnetic waves

In previous chapters (Chapters 14 and 15) we learned that a system of electrically charged objects possesses electric potential energy, and that same idea can be expressed through the concept of electric field energy density. This energy density $u_E$ quantifies the energy stored in the electric field per cubic meter of volume:

$$u_E = \frac{\kappa}{8\pi k}E^2 \quad (15.16)$$

**Active Learning Guide ›**

Rewriting $k = \frac{1}{4\pi\varepsilon_0}$ and assuming that the region with electric field is a vacuum ($\kappa = 1$), the above becomes

$$u_E = \frac{1}{8\pi k}E^2 = \frac{1}{8\pi\left(\frac{1}{4\pi\varepsilon_0}\right)}E^2 = \frac{1}{2}\varepsilon_0 E^2$$

What about the magnetic field? Does it have energy as well? We won't develop the ideas needed to establish this, but the answer is yes. Just as with the electric field, the magnetic field also has an energy density $u_B$, the magnetic field energy per cubic meter. Thus the energy in an EM wave is carried partly by its oscillating electric field and partly by its oscillating magnetic field. It turns out that the EM wave's average electric field energy density exactly equals the average magnetic field energy density. Since for EM waves $E = cB$, the energy density of the magnetic field is

$$u_B = u_E = \frac{1}{2}\varepsilon_0 E^2 = \frac{1}{2}\varepsilon_0(cB)^2 = \frac{1}{2}\varepsilon_0 c^2 B^2$$

Remember that from Maxwell's equations, $c = \frac{1}{\sqrt{\varepsilon_0\mu_0}}$. We can substitute this expression for $c$ in the equation $u_B = \frac{1}{2}\varepsilon_0 c^2 B^2$:

$$u_B = \frac{1}{2}\varepsilon_0 c^2 B^2 = \frac{1}{2}\varepsilon_0\left(\frac{1}{\sqrt{\varepsilon_0\mu_0}}\right)^2 B^2$$

Notice that $\varepsilon_0\left(\frac{1}{\sqrt{\varepsilon_0}}\right)^2 = \varepsilon_0\frac{1}{\varepsilon_0} = 1$. Therefore,

$$u_B = \frac{1}{2\mu_0}B^2 \quad (24.3)$$

Although we arrived at Eq. (24.3) in the context of EM waves, the expression can be used to determine the magnetic field energy density in other situations as well.

The total energy density of the electric and magnetic fields in an EM wave is the sum of the energy densities of each:

$$u = u_E + u_B$$

$$\Rightarrow u = \frac{1}{2}\varepsilon_0 E^2 + \frac{1}{2\mu_0}B^2$$

However, since these two terms are equal, we can write the energy density of the EM wave as twice the first term or twice the second term:

$$u = 2\left(\frac{1}{2}\varepsilon_0 E^2\right) = \varepsilon_0 E^2 \text{ or } u = 2\left(\frac{1}{2\mu_0}B^2\right) = \frac{1}{\mu_0}B^2$$

## Average values of the energy densities

Remember that in the equations above, $E$ and $B$ are the instantaneous values of the magnitudes of the $\vec{E}$ and $\vec{B}$ fields. In an EM wave of a specific wavelength, both of those fields oscillate sinusoidally (**Figure 24.16a**). This means that the energy density at each point in the wave fluctuates as a sine-squared function since the field appears squared in the equations. During one vibration of the field, the energy density starts with a maximum value; then after one-quarter of a period the energy density is zero; and after half a period the energy is again maximal (Figure 24.16b).

Since the energy density at each point in the EM wave fluctuates with extremely high frequency, it's usually more useful to think in terms of the time-averaged energy density at each point in the EM wave. In order to do this we would need to calculate the average value of the sine-squared function shown in Figure 24.16b. This requires calculus, so we won't go through the details, but here's the result: the average value of the square of a sinusoidal function (either sine or cosine) is one-half the maximum value. Therefore, the average value (indicated by the bar above $u$) of the energy density along the direction of the EM wave is

$$\bar{u} = \frac{1}{2}u_{\text{max}} = \frac{1}{2}(\varepsilon_0 E_{\text{max}}^2) = \frac{1}{2}\varepsilon_0 E_{\text{max}}^2 \qquad (24.4)$$

**Figure 24.16** Variation of the $\vec{E}$ and $\vec{B}$ fields at one point and (b) the energy density variation with time of an EM wave at one point.

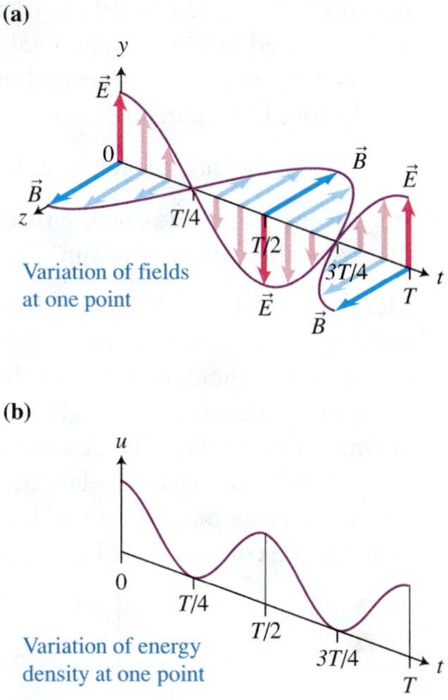


## Intensity of an EM wave

Now that we understand the average energy density of the EM wave, we can understand quantitatively what is meant by its *intensity*. When you look at a source of light, it might appear dim to you like a distant star. Or it might appear bright to you, like a car with its high beams on. The physical quantity that characterizes this brightness is called *intensity*. (We encountered this quantity in Chapter 20.) The intensity of an EM wave is its average energy density times the speed of light.

Remember that intensity is the amount of energy (not energy density) that passes through a unit area during a unit time interval [see Eq. (20.6)].

$$\text{Intensity} = \frac{\text{Energy}}{\text{time interval} \times \text{area}} = \frac{U}{\Delta t \times A} = \frac{P}{A}$$

where $P$ is the power of the wave. Technically, since the energy density of the field fluctuates, the intensity of the wave fluctuates as well, but this fluctuation is so rapid that it makes much more sense to think about the average intensity of the EM wave. The average intensity depends on the wave's average energy density $\bar{u} = (1/2)\varepsilon_0 E_{\text{max}}^2$ and the speed $c$ of the wave through the area (the faster the wave travels, the more energy passes through the area, though for EM waves this speed is always $c$).

One way to think of intensity and energy density is to imagine wind blowing into a sail. The wind speed is similar to the speed of the wave, and the kinetic energy per cubic meter of air is similar to the energy density of the wave. During a time interval $\Delta t$, the energy within the volume $V = A(c\,\Delta t)$ passes through the area $A$ (**Figure 24.17**). The intensity equals this energy divided by the time interval and by the area:

$$I = \frac{\text{Energy}}{\text{time interval} \times \text{area}} = \frac{\bar{u}V}{\Delta t A} = \frac{\bar{u}Ac\Delta t}{\Delta t A} = \bar{u}c \qquad (24.5)$$

**Figure 24.17** Intensity involves the rate of energy crossing a surface.

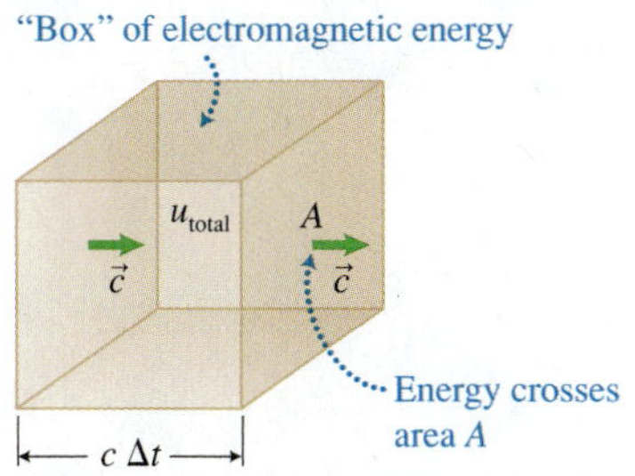

**EXAMPLE 24.3 Solar constant**

The *solar constant* is a fundamental constant in environmental science and astronomy that describes the intensity of the Sun's EM radiation when it arrives at Earth. It is expressed in terms of the total electric and magnetic field energy incident each second on a 1.0-$m^2$ surface located above the atmosphere directly facing the Sun:

$$\text{Solar constant} = 1.37\ \text{kW/m}^2$$

What is the amplitude of the $\vec{E}$ field oscillation just above the atmosphere due to the Sun's illumination of Earth?

**Sketch and translate** Imagine the Sun's radiation striking a 1.0-$m^2$ surface above Earth's atmosphere at a zero angle of incidence. Remember that we define the angle of incidence as the angle between the ray and the normal to the surface. To determine the amplitude $E_{max}$ of the $\vec{E}$ field, we need to relate the solar constant to the average energy density of the EM radiation, which in turn depends on the $\vec{E}$ field amplitude $E_{max}$.

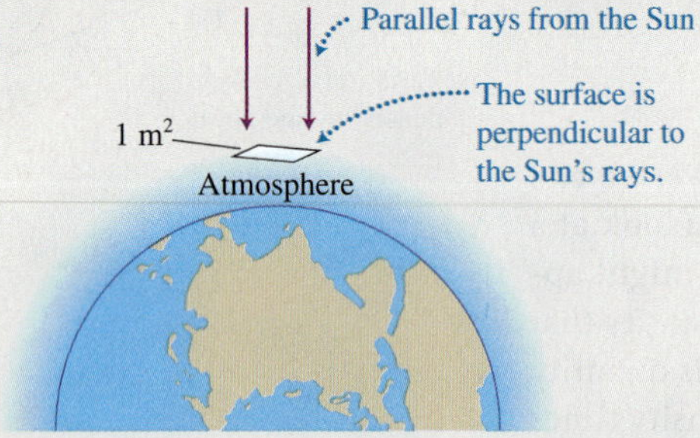


**Simplify and diagram** Assume that light from the Sun travels to Earth at speed *c*. Assume also that all of the EM radiation is of a single wavelength. This is a poor assumption since the different wavelength EM waves arriving at Earth each have different electric field amplitudes. Therefore, what we will be determining is the electric field amplitude averaged over all wavelengths.

**Represent mathematically** We know from Eq. (24.5) that $I = \bar{u}c$ or $\bar{u} = (I/c)$. Using Eq. (24.4) we have $\bar{u} = (1/2)\,\varepsilon_0 E_{max}^2$. Thus $(I/c) = (1/2)\,\varepsilon_0 E_{max}^2$.

**Solve and evaluate** We can now determine $E_{max}$ by multiplying both sides of the previous equation by 2, dividing by $\varepsilon_0$, and then taking the square root:

$$E_{max} = \sqrt{\frac{2I}{c\varepsilon_0}}$$
$$= \sqrt{\frac{2(1.37 \times 10^3\ \text{W/m}^2)}{(3.00 \times 10^8\ \text{m/s})(8.85 \times 10^{-12}\ \text{C}^2/(\text{N}\cdot\text{m}^2))}}$$
$$= 1010\ \text{N/C}$$

Is the number reasonable? We could estimate the $\vec{E}$ field amplitude a certain distance from a 60-W lightbulb in a room and compare it to this result. We expect the $\vec{E}$ field amplitude produced by the bulb to be significantly less since the Sun appears to be significantly brighter than a 60-W lightbulb. See the following Try It Yourself.

**Try it yourself:** Estimate the amplitude of the $\vec{E}$ field produced by a 60-W electric bulb if you are 2.0 m away from it. Assume that the energy emitted by the bulb each second is completely in the form of EM waves.

*Answer:* 30 N/C. As expected, this is significantly less than the $\vec{E}$ field amplitude produced by the Sun above Earth's atmosphere.

**Review Question 24.5** You are at a distance *d* from a point-like source of electromagnetic waves (similar to a lightbulb). What will be the intensity of the wave if you double your distance from the source? Explain.

## 24.6 Polarization and light reflection: Putting it all together

We now have a more powerful model of light as an electromagnetic wave. Let's use this model to understand how light is produced, how polarized sunglasses work, how to reduce glare from reflected sunlight, and the role of polarization in liquid crystal displays (LCDs) on computer and calculator screens.

### Producing unpolarized light

**Active Learning Guide ›**

Many sources of light produce unpolarized light. In this section we discuss the mechanism behind this phenomenon. To understand unpolarized light, we first review the production of polarized light by an antenna.

Electrons vibrating up and down in an antenna produce radio waves. Far from the antenna these waves take the form of plane waves with the $\vec{E}$ field vector polarized in the direction parallel to the antenna (**Figure 24.18**). A polarizer can reduce the amplitude of this radio wave to zero if the polarizer is oriented in a specific direction relative to the antenna.

We can think of light coming from an incandescent lightbulb as being produced by the motion of charged particles within the filament, each acting as a tiny "antenna." The light emitted by the bulb consists of many waves originating at random times with random polarizations (**Figure 24.19a**). If we could observe the many separate EM waves as a beam of unpolarized light moving directly toward our eyes, the oscillations of both the $\vec{E}$ and $\vec{B}$ fields would look something like a porcupine (the head-on appearance of the $\vec{E}$ field vectors is shown in Figure 24.19b). These oscillations explain why light produced by an incandescent bulb is unpolarized.

**Figure 24.18** An antenna produces $\vec{E}$ field oscillations that far from the antenna are parallel to the antenna.

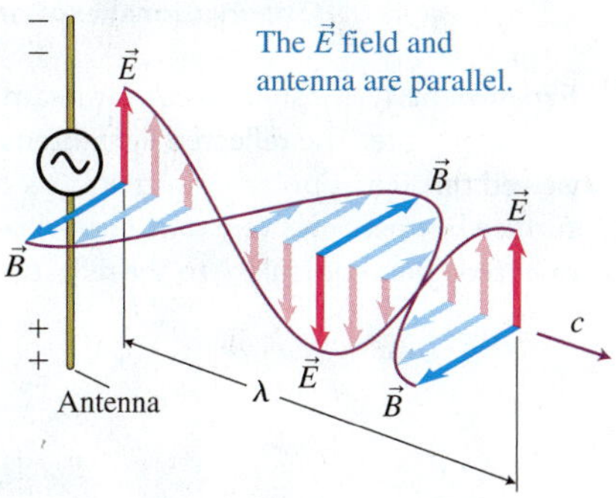
The $\vec{E}$ field and antenna are parallel.
$\vec{E}$
$\vec{B}$
$\vec{B}$
$\vec{E}$
$c$
$\lambda$
$\vec{E}$
$\vec{B}$
Antenna

**Figure 24.19** Light from a bulb is unpolarized.

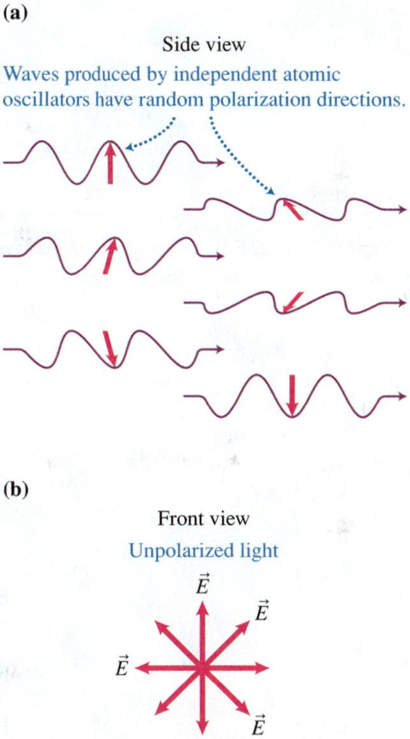
(a)
Side view
Waves produced by independent atomic oscillators have random polarization directions.
(b)
Front view
Unpolarized light
$\vec{E}$
$\vec{E}$
$\vec{E}$
$\vec{E}$

## Light polarizers

How do polarizers interact with light? A polarizer affects the electric field of the passing electromagnetic wave and not the magnetic field. The reason is that the magnetic field exerts a force only on a *moving* charged object. On the other hand, the electric field exerts a force that is independent of the motion of the object. This means that the electric field has a much more significant effect on matter since, most of the time, the protons and electrons that make up matter don't have any sort of collective motion.

A polarizer absorbs one component of the $\vec{E}$ field of the EM wave passing through it, allowing the perpendicular component to pass. To understand which component is absorbed, consider unpolarized microwaves passing through a grill of narrow metal bars (**Figure 24.20**). The component of the oscillating $\vec{E}$ field of the incident microwaves that is parallel to the grill produces an oscillating electric current in the grill. The resistive heating in the metal grill rods due to this current causes this component of the $\vec{E}$ field to diminish. The $\vec{E}$ field perpendicular to the metal grill rods does not cause a current and is not diminished. So the axis of the polarizing grill is actually perpendicular to the grill rods. The microwaves leaving the metal grill polarizer are linearly polarized in the direction perpendicular to the grill rods, and that is what defines the axis of the polarizer.

The situation is similar for light, only the dimensions are much smaller. A polarizing film such as that used in polarized sunglasses is made of long polymer molecules infused with iodine. The polymer chains are stretched in one direction to form an array of aligned linear molecules. Electrons can move along the lengths of the polymer chains but there is resistance—just like the metal wires in the microwave polarizing grill. As a result, the $\vec{E}$ field component of unpolarized light parallel to the polymer chains is diminished. The $\vec{E}$ field perpendicular to the chains is not diminished. Thus, the axis of the light polarizer is perpendicular to the direction of the stretched polymer chains.

**Figure 24.20** The $\vec{E}$ field component parallel to the metal grill wires is reduced.

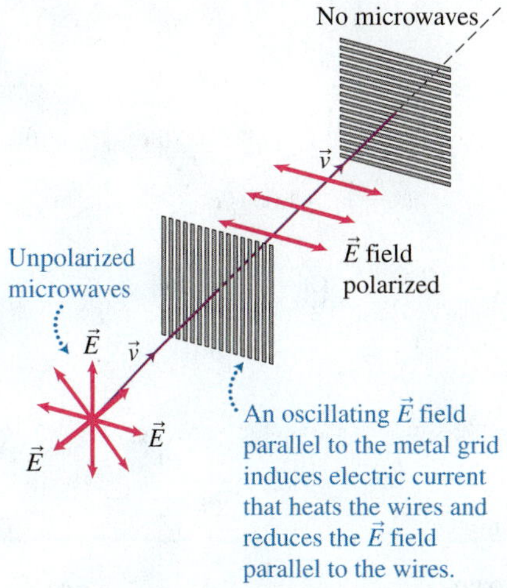
No microwaves
$\vec{v}$
$\vec{E}$ field polarized
Unpolarized microwaves
$\vec{E}$
$\vec{v}$
$\vec{E}$
$\vec{E}$
An oscillating $\vec{E}$ field parallel to the metal grid induces electric current that heats the wires and reduces the $\vec{E}$ field parallel to the wires.

## Polarization by reflection and Brewster's angle

In the chapter-opening story we considered light produced by the LCD screen of a calculator, cell phone, or laptop computer, or reflected off a body of water. If you look at this light through a polarizer or through polarized sunglasses, the intensity of reflected light varies depending on the orientation of the polarizer relative to the surfaces. This indicates that the light is partially polarized. Consider the experiments involving the polarization of reflected light in Observational Experiment **Table 24.5**.

## OBSERVATIONAL EXPERIMENT TABLE

### 24.5 Polarization of reflected light.

VIDEO 24.5

**Observational experiment**

**Experiment 1.** We shine a narrow beam of light at an angle on a glass surface. The reflected light intensity is medium bright if viewed through a polarizer with its axis oriented parallel to the surface but becomes very dim when viewed with the polarizer oriented "perpendicular" to the reflecting surface.

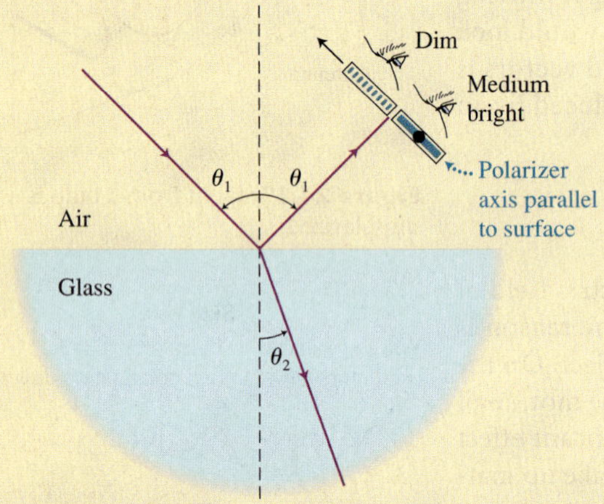


**Analysis**

The up and down "perpendicular" component of the reflected light must be reduced significantly, whereas the parallel component is not. The reflected light must be partially polarized parallel to the reflecting surface.

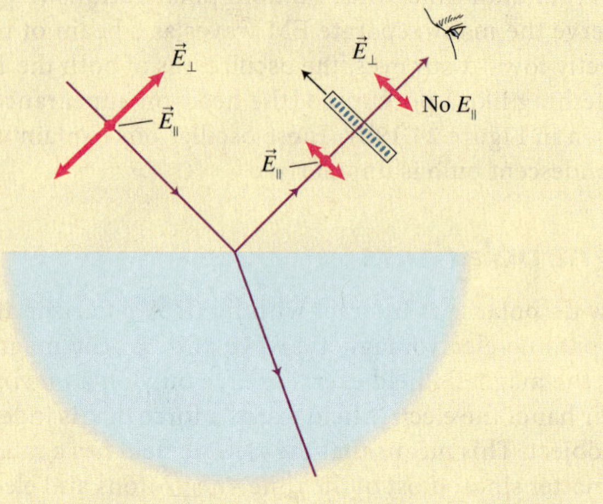


**Observational experiment**

**Experiment 2.** For one particular incident angle the reflected beam is completely blocked by the "perpendicular" polarizer.

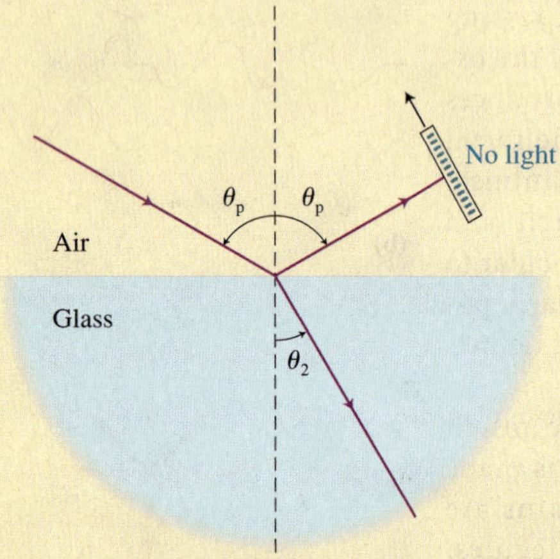


**Analysis**

The up and down "perpendicular" component of the reflected light must be completely removed by reflection. The reflected wave is completely polarized parallel to the surface when reflected at the *polarizing angle*.

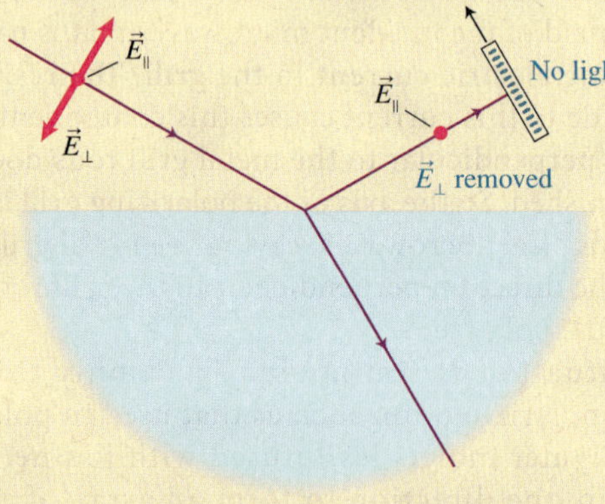


**Observational experiment**

**Experiment 3.** When light is incident at the polarizing angle, the refracted ray in the glass makes a 90° angle with the reflected ray.

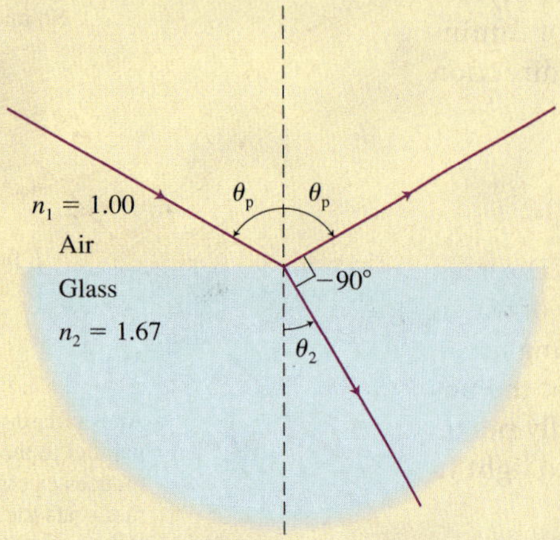


**Analysis**

The tangent of the polarizing angle $\theta_p$ is the same as the refractive index of the glass (for example $n = 1.67$).

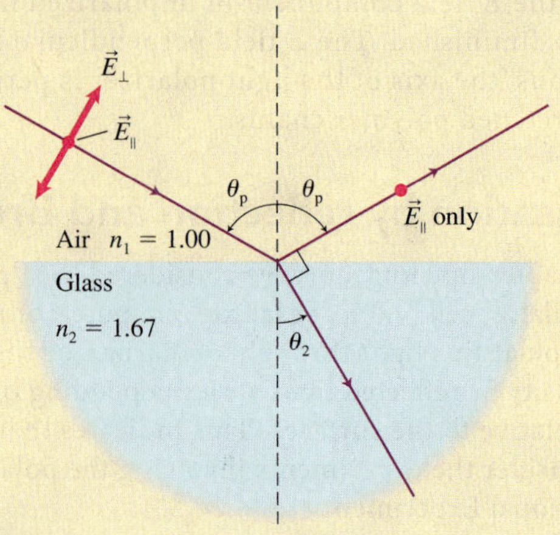

| | |
|---|---|
| **Experiment 4.** We replace glass with water and repeat the experiments. The polarizing angle is different. The refracted ray again makes a 90° angle with the reflected ray. | The tangent of the polarizing angle $\theta_p$ is $\tan\theta_p = 1.33$, the refractive index of water. |

**Patterns**

- Unpolarized light reflected off the boundary between two transparent media becomes partially linearly polarized parallel to the reflecting surface.
- When the incident light is at a particular *polarizing angle*, the reflected ray is completely polarized parallel to the surface.
- The tangent of the polarizing angle $\theta_p$ is $\tan\theta_p = n$, where $n$ is the refractive index of the reflecting surface.
- For light incident at the polarizing angle, the reflected light and the refracted light make a 90° angle relative to each other.

We found that when the angle of incidence is equal to the so-called **polarizing angle** $\theta_p$, the reflected ray is linearly polarized in the plane parallel to the reflecting surface and the reflected and the refracted rays are perpendicular to each other. Snell's law helps us understand this.

Examine the figure for Experiment 3 in Table 24.5. We see that

$$\theta_p + \theta_2 + 90° = 180°$$

$$\Rightarrow \theta_2 = 90° - \theta_p$$

We can use Snell's law to relate the angle of incidence $\theta_p$ to the angle of refraction $\theta_2$:

$$n_1 \sin\theta_p = n_2 \sin\theta_2 = n_2 \sin(90° - \theta_p)$$

From trigonometry we know that $\sin(90° - \theta_p) = \cos\theta_p$. Substituting in the above, we get

$$n_1 \sin\theta_p = n_2 \cos\theta_p$$

or

$$\frac{\sin\theta_p}{\cos\theta_p} = \tan\theta_p = \frac{n_2}{n_1} \quad (24.6)$$

This relationship is called **Brewster's law** in honor of Sir David Brewster, who found the same pattern experimentally.

**Brewster's law** **Light is traveling in medium 1 when it reflects off medium 2. The reflected light is totally polarized along an axis in the plane parallel to the surface when the tangent of the incident polarizing angle $\theta_p$ equals the ratio of the indexes of refraction of the two media:**

$$\tan\theta_p = \frac{n_2}{n_1} \quad (24.6)$$

Brewster's law allows us to explain how polarizing sunglasses help reduce glare. When you are driving your car toward the Sun, it is often difficult to see because of the glare of reflected light from the car's hood and dashboard, from the pavement, and from other cars. This reflected light is partially polarized parallel to those surfaces. Polarizing sunglasses reduce the glare by absorbing light polarized in that direction.

**EXAMPLE 24.4 Reducing glare**

You are facing the Sun and looking at the light reflected off the ocean. At what angle above the horizon should the Sun be so that you get the most benefit from your polarizing sunglasses?

**Sketch and translate** We sketch the situation below. The water surface is horizontal and perpendicular to the page. The sunlight incident on the water surface is unpolarized, represented by the dots pointing into and out of the page and the $\vec{E}$ field lines up and down perpendicular

*(continued)*

to the ray. We can answer the question by finding the polarization angle $\theta_p$ at which light reflected off the water is completely polarized parallel to the water surface—represented by the dots on the reflected ray in the sketch.

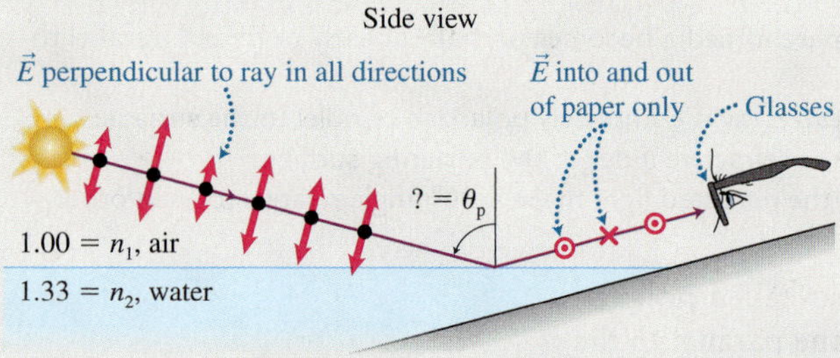


**Simplify and diagram** The polarizing axis of your glasses and the polarization of the reflected light are shown below. Assume that the water reflects rays from the Sun as if it were a mirror. The reflected light $\vec{E}$ field oscillates parallel to the water surface—in and out of the page. The reflected light will not pass though the glasses if the polarizing axis of the glasses is perpendicular to the oscillating $\vec{E}$ field.

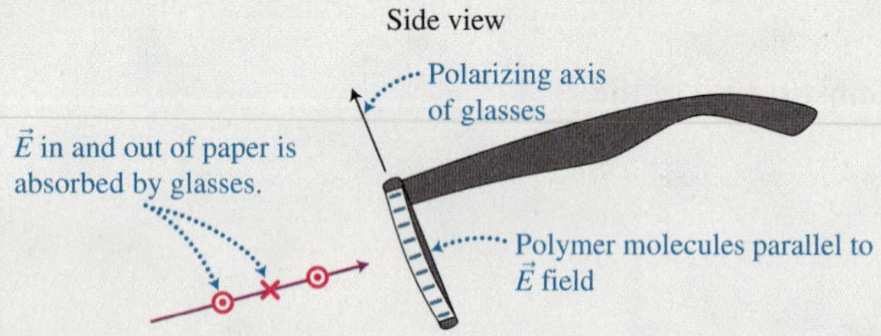


**Represent mathematically** We can use Brewster's law to find the angle at which parallel rays from the Sun that hit the water surface should produce reflected light that is completely polarized parallel to the air-water surface:

$$\tan \theta_p = \frac{n_2}{n_1}$$

**Solve and evaluate** Inserting $n_2 = 1.33$ for water and $n_1 = 1.00$ for air, we get

$$\tan \theta_p = 1.33$$

or an angle $\theta_p = 53.1°$. This is the angle between the vertical normal line and the direction of the sunlight. Thus, the direction of the sunlight above the horizontal is

$$90° - 53.1° = 36.9°$$

For all other incident angles the reflected light will be partially polarized and the glare will not be completely eliminated by the sunglasses.

**Try it yourself:** A light beam travels through glass ($n = 1.60$) into the air. At what angle of incidence is the reflected light completely polarized?

*Answer:* 32°.

**Figure 24.21** A rearview mirror with and without polarizing film.

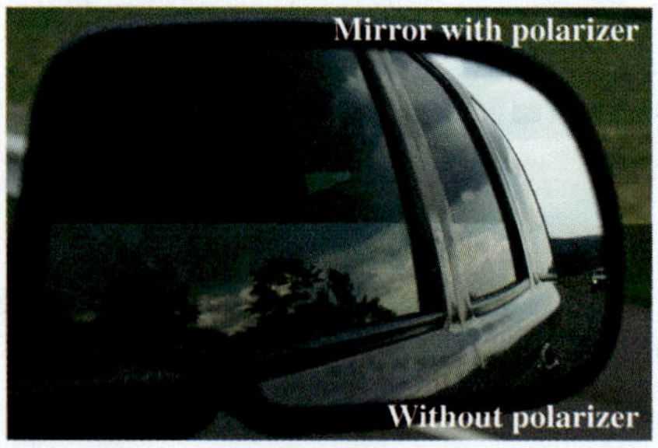


Polarized sunglasses are least effective when the Sun is very low or very high above the horizon because the angle of incidence is often far from Brewster's angle. For example, most of the time we are looking at objects that are somewhat distant, such as a car a few hundred feet ahead, or water near the horizon. When the sun is close to the horizon in these cases light often reflects toward our eyes at angles that are significantly larger than Brewster's angle which results in the light not being strongly polarized.

**Figure 24.21** illustrates polarization caused by reflection. In this photo, the top half of the rearview mirror is covered by a polarized film that is perpendicular to the direction of polarization of light reflected off the car's surface. There is little glare from light reflected from the top part of the car. The bottom half of the rearview mirror has no polarized film, and hence all light reflected from the car is seen.

## Polarization by scattering

We learned in a previous chapter (Chapter 21) that the sky appears blue from Earth's surface because the molecules in the atmosphere scatter blue light more than red into your eyes. If you look through polarized sunglasses at the clear sky in an arbitrary direction and rotate the glasses, the intensity of the light passing through the glasses changes. For example, observer A in **Figure 24.22** sees a change in the light intensity when turning polarized sunglasses while looking at the sky. This means that light scattered by the atmosphere is partially polarized. In fact, if you look in a direction that makes a 90° angle with a line toward the Sun (observer B in Figure 24.22), you can orient the glasses so the sky is almost dark. Evidently, the sunlight scattered toward you by molecules in that direction is almost completely linearly polarized.

To explain this observation, we need to assume that molecules in the atmosphere are like the tiny dipole antennas described at the beginning of this section. The Sun emits unpolarized light in which all $\vec{E}$ field orientations perpendicular to the direction the light travels are present in equal amounts. When this light strikes a molecule consisting of particles carrying opposite charges (electrons and nuclei), these particles vibrate as the EM wave passes. In **Figure 24.23a**, the electric field component $E_x$ produces molecular vibrations along the $x$-direction. Such a vibration is like that produced by a charge vibration in an antenna and produces an EM wave straight downward with an $E_x$ component (it also produces an upward wave but we are not interested in the upward wave here.). The sunlight also has an electric field component $E_y$ in the $y$-direction (See Figure 24.23b). This component causes atoms and molecules in the atmosphere to vibrate up and down in the $y$-direction. However, an antenna with vibrating electric charge (the vibrating atom or molecule) cannot emit an $\vec{E}$ field along the axis that the charge vibrates since EM waves are transverse. The result is that the downward 90° wave caused by the vibrating atoms and molecules is linearly polarized in the $x$-direction. This explains why sunlight is completely linearly polarized when you look at a point in the sky in a direction perpendicular to a line from the Sun to that point.

If you look at clouds through polarized sunglasses, the light passing through the glasses is dimmer, but varies very little in intensity as you rotate the glasses. Evidently, light scattered from clouds is unpolarized. Clouds consist of many tiny water droplets. Light entering a cloud is randomly scattered many times by water droplets before it leaves. This multiple scattering keeps the light unpolarized. Photographers take advantage of this effect to take pictures of clouds using polarized "filters." The polarizers remove more of the partially linearly polarized light from the sky and less of the unpolarized light from the clouds, making the clouds stand out (see **Figure 24.24**).

## Polarized LCDs

Nearly all computer, TV, calculator, and cell phone screens are LCDs—liquid crystal displays. Liquid crystals are made of substances that have properties of both a solid and a liquid. They can flow like a liquid, but their molecules are aligned or oriented in an orderly crystal-like manner. Polarization plays an important role in the operation of these screens.

Consider a simple calculator LCD display. When you look at the display screen, some parts are bright and other parts gray or dark. The display consists of a grid of many tiny pixels. Each pixel consists of three parallel layers: two polarizing panels with transparent electrodes and a liquid crystal between them (**Figure 24.25a**).

The axes of the polarizing panels are perpendicular to each other. If there is no liquid crystal between the polarized panels, light from behind the display cannot pass through the panels. But with the liquid crystal between the panels, light passes through even though the two polarized panels are oriented perpendicular to each other! How is this possible? The liquid crystal molecules rotate the direction of the $\vec{E}$ field of the polarized light by up to 90° so that most or all of it can pass through the second polarizer even though it is perpendicular to the first (Figure 24.25a). The screen looks bright at that pixel.

When a strong external electric field $\vec{E}_{ext}$ is applied to the liquid crystal by the transparent electrodes, it aligns all of the liquid crystal molecules, and the direction of the light's polarization (the direction of vibration of the light's

**Figure 24.22** Scattered light is partially or totally polarized.

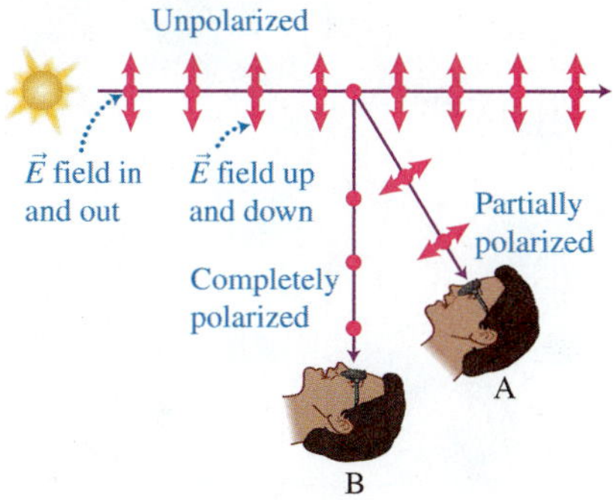


**Figure 24.23** Light scattered from a clear sky at 90° is linearly polarized.

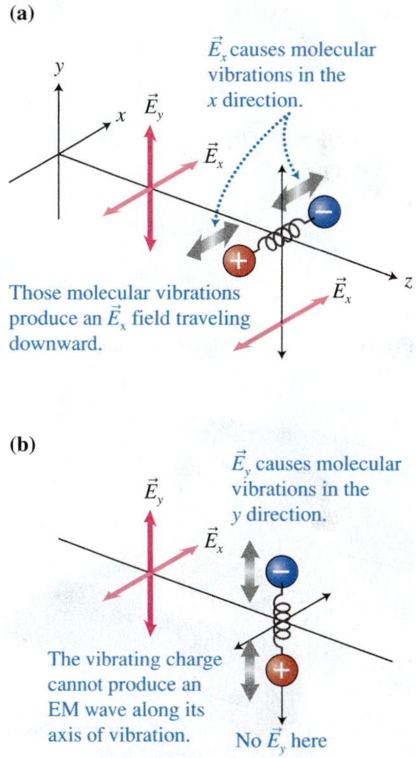


**Figure 24.24** A polarizer affects light from the clear sky but not from the clouds.

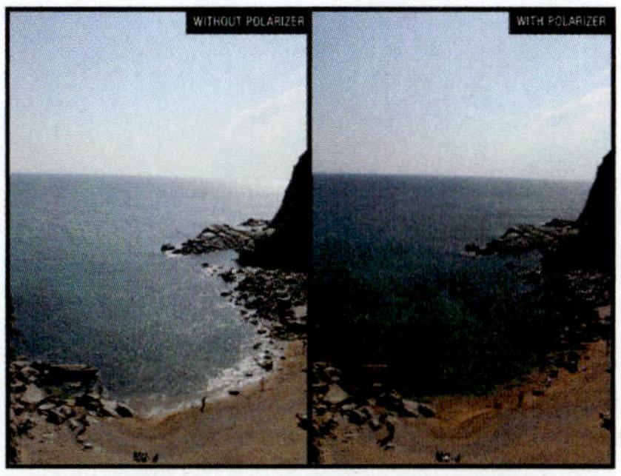

**Figure 24.25** A pixel of a liquid crystal display.

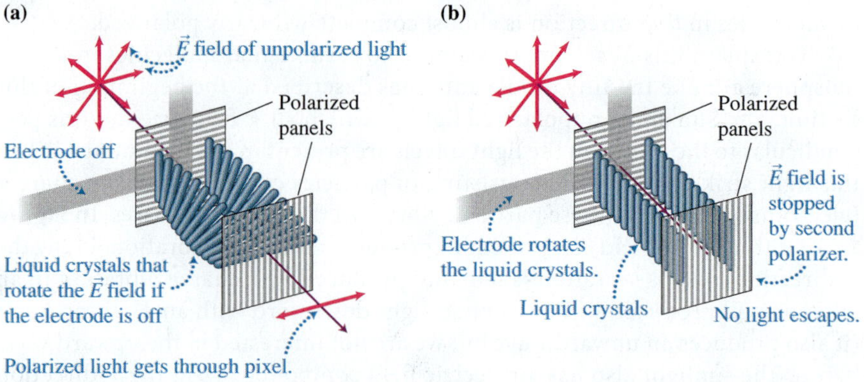


$\vec{E}$ field vector) does not change as it passes through the liquid crystal (Figure 24.25b). Thus, light does not pass through the second polarized panel, making the screen look dark at that pixel. If the external $\vec{E}_{ext}$ field does not have high enough magnitude, then some of the liquid crystal molecules align and some do not, allowing some of light to pass through. In this case, the pixel appears gray. When you press the buttons on a calculator, it turns on an electric field $\vec{E}_{ext}$ that is applied to the liquid crystals of different pixels, making some parts of the screen appear black and others bright or gray. Their combinations form the patterns of the numeric digits you see on the calculator screen.

## 3D movies

Polarization also plays a role in one way of making a viewable 3D movie. The 3D projector produces two distinct images on the screen. Each image consists of polarized light, and the two images have their polarizing axes rotated by 90° relative to each other. The two images reflect from the metallized screen to theater viewers, who wear polarized glasses. The polarization axis of the left lens of the glasses is oriented at 90° relative to the axis of the right lens (**Figure 24.26**). Thus, each eye sees a distinct image. The two images are combined in the brain to convey the 3D effect in much the same way that you see the real world without the help of 3D glasses.

**Figure 24.26** 3D glasses to view a 3D movie.

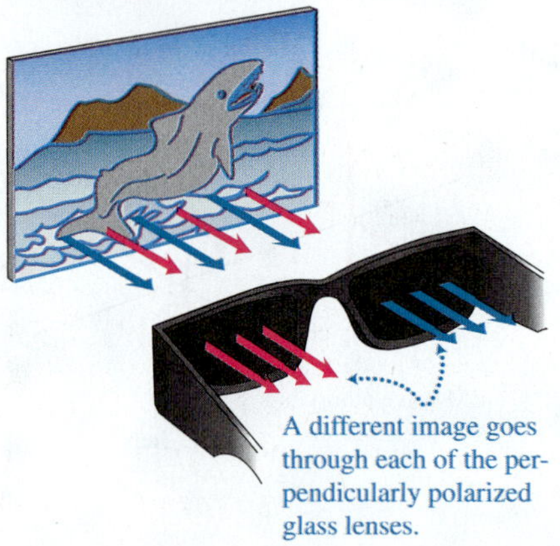

A different image goes through each of the perpendicularly polarized glass lenses.

**Review Question 24.6** Explain why polarizing glasses reduce the glare of sunlight reflected from the dashboard of your car or from the metal surface of another car.

# Summary

| Words | Pictorial and physical representations | Mathematical representation |
|---|---|---|

**Electromagnetic waves** EM radiation consists of electric and magnetic fields whose vector $\vec{E}$ field and $\vec{B}$ field vibrate perpendicular to each other and to the direction the wave travels. Accelerating charged particles, often oscillating electric dipoles, produce the waves. Once formed, the waves are self-propagating at speed $c = 3.00 \times 10^8$ m/s in a vacuum or slightly less in air. (Section 24.2, 24.5)

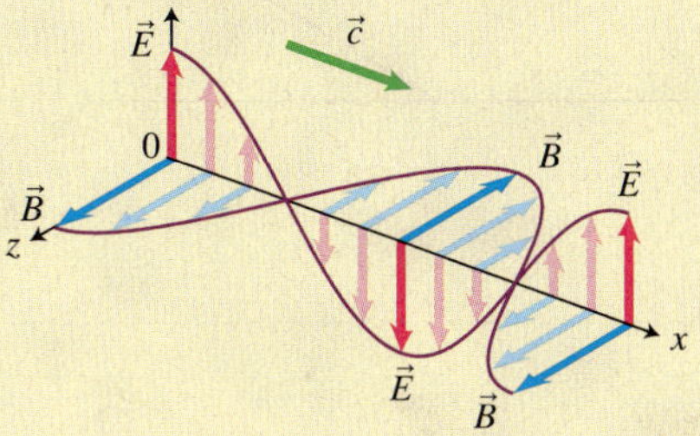


$$E_{max} = cB_{max} \quad \text{Eq. (24.1)}$$

$$E_y = E_{max} \cos\left[2\pi\left(\frac{t}{T} \pm \frac{x}{\lambda}\right)\right]$$

$$B_z = B_{max} \cos\left[2\pi\left(\frac{t}{T} \pm \frac{x}{\lambda}\right)\right] \quad \text{Eq. (24.2)}$$

**Intensity of light** The intensity $I$ of light is the electromagnetic energy that crosses a perpendicular cross-sectional area $A$ in a time interval $\Delta t$ divided by that area and time interval. (Section 24.5)

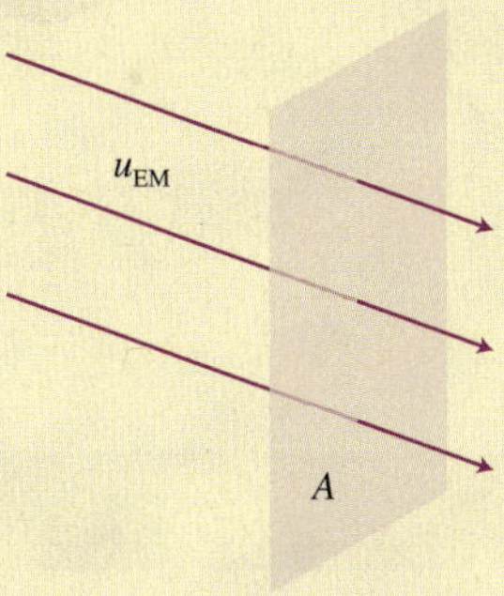


$$u_B = \frac{1}{2\mu_0}B^2 \quad \text{Eq. (24.3)}$$

$$\bar{u} = \frac{1}{2}u_{max} = \frac{1}{2}(\varepsilon_0 E_{max}^2) = \frac{1}{2}\varepsilon_0 E_{max}^2 \quad \text{Eq. (24.4)}$$

$$I = \frac{\text{Energy}}{\text{time interval} \times \text{area}} = \frac{\bar{u}V}{\Delta t A} = \frac{\bar{u}Ac\Delta t}{\Delta t A} = \bar{u}c \quad \text{Eq. (24.5)}$$

**Polarization** Randomly vibrating wave sources produce *unpolarized* light.
When light travels through a polarizer, it becomes *linearly polarized*—only the component of the $\vec{E}$ field perpendicular to the chains of molecules of the polarizer material passes through.
The intensity of linearly polarized light is reduced when it travels through a second polarizer oriented at an angle $\theta$ relative to the first. (Section 24.1 and 24.6)

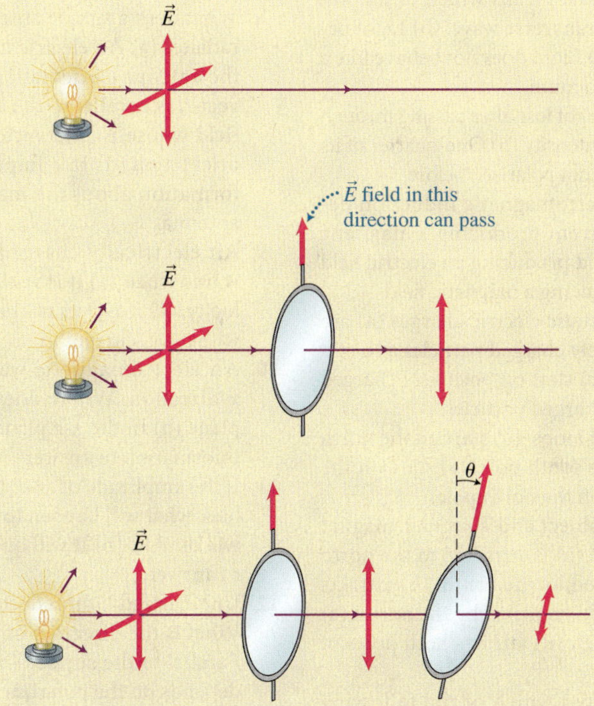


$$I = \frac{I_0}{2}$$

$$I = I_{0/2} \cos^2\theta$$

(*continued*)

| Words | Pictorial and physical representations | Mathematical representation |
|---|---|---|
| **Light polarization by reflection off a surface** Light reflected off a surface is partially polarized parallel to the surface. When light is incident at the Brewster (polarization) angle $\theta_p$, the light is totally polarized parallel to the surface. (Section 24.6) | 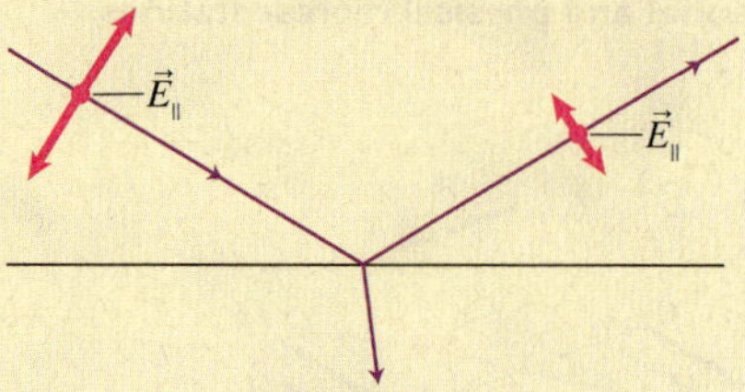  | $\frac{\sin\theta_p}{\cos\theta_p} = \tan\theta_p = \frac{n_2}{n_1}$ Eq. (24.6) where $n_1$ and $n_2$ are the refractive indexes of the incident and reflecting media. |
| **Polarization by scattering** Atmospheric dipole molecules such as water absorb and re-emit sunlight in directions perpendicular to their axes of vibration. (Section 24.6) | 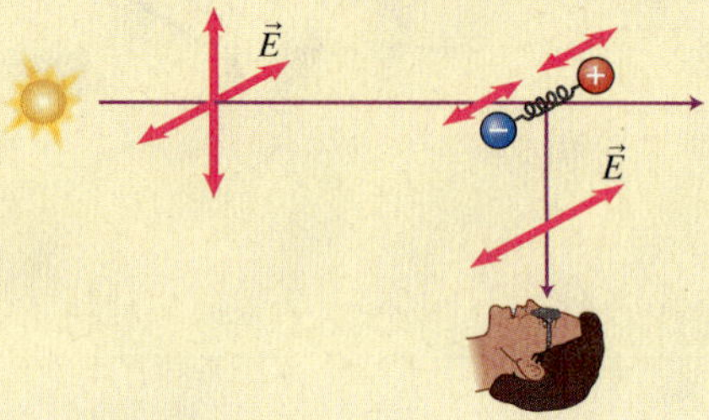  | |

MP For instructor-assigned homework, go to **MasteringPhysics.**

# Questions

## Multiple Choice Questions

1. The fact that light can be polarized means which of the following? (a) Light behaves like a transverse wave. (b) Light behaves like a longitudinal wave. (c) Light does not behave like a wave because it propagates in a vacuum.
2. What does a beam of unpolarized light lose after passing through one polarizer? (a) One-half of its intensity (b) One-quarter of its intensity (c) No intensity, as it was not polarized before
3. What does Faraday's law of electromagnetic induction describe? (a) A steady electric current producing a magnetic field (b) A changing magnetic field producing an electric field (c) A changing electric field producing a magnetic field
4. Electric field lines produced by static electric charges (a) are closed loops. (b) start on negatively charged particles and end on positively charged particles. (c) start on positively charged particles and end on negatively charged particles.
5. Magnetic field lines (a) are closed loops. (b) start on the north pole of a magnet and end on the south pole. (c) start on the south pole of a magnet and end on the north pole.
6. You have a positively charged object and a regular magnet made of iron. The charged object (a) is attracted to the north pole of the magnet and is repelled by the south. (b) attracts the south pole of the magnet and is repelled by the north. (c) does not interact with the magnet. (d) attracts both poles of the magnet.
7. Maxwell's hypothesis contributed which of the following ideas? (a) A steady electric current produces a magnetic field. (b) A changing magnetic field produces an electric field. (c) A changing electric field produces a magnetic field.
8. What does a simple antenna with a changing electric current radiate? (a) An electric field whose $\vec{E}$ field vector is parallel to the antenna's orientation (b) A magnetic field whose $\vec{B}$ field vector is parallel to the antenna's orientation (c) An electric field whose $\vec{E}$ field vector is perpendicular to the antenna's orientation (d) It is impossible to say, as we do not have information about the magnitude of the current through the antenna.
9. An electrically charged particle radiates electromagnetic waves when (a) it is stationary. (b) it moves at a constant velocity. (c) it moves at changing velocity. (d) it is inside a wire with a DC current in it.
10. An electromagnetic wave propagates in a vacuum in the $x$-direction. Where does the $\vec{E}$ field oscillate? (a) In the $x$-$y$ plane (b) In the $x$-$z$ plane (c) In the $y$-$z$ plane (d) Not enough information to answer
11. If the amplitude of an $\vec{E}$ field in a linearly polarized wave doubles, what will happen to the energy density of the wave? (a) It will double. (b) It will quadruple. (c) Not enough information to answer.
12. Unpolarized light reflected off a lake is completely polarized. What is the direction of the $\vec{E}$ field in the reflected wave? (a) Parallel to the surface (b) Perpendicular to the surface (c) It depends on the polarizer through which you observe it.

## Conceptual Questions

13. Can light phenomena be better explained by a transverse wave model or by a longitudinal wave model? Explain how you know.
14. What are two models that explain how light can propagate through air? Describe their features in detail.
15. What experimental evidence supports the transverse wave model for light?
16. Summarize Maxwell's equations conceptually and describe a way to experimentally illustrate each of them.
17. What testable predictions followed from Maxwell's equations?
18. Describe the conditions that are necessary for the emission of electromagnetic waves.
19. Describe how Hertz tested Maxwell's equations.
20. Explain how radar works to determine distances to objects.
21. What determines the nearest and farthest distances of objects that can be distinguished by pulsed radar?
22. How was the hypothesis that light can be modeled as an electromagnetic wave tested experimentally?
23. What is the difference between the following two statements: Light is a wave. Light behaves like a wave. Which one do you think is more accurate and why?
24. How do polarized glasses work?
25. You bought a pair of glasses that are marketed as having polarizing filters. How can you test this claim?
26. Why, when we use polarized glasses, is the glare from reflective surfaces reduced?
27. How does a polarizer for mechanical waves work and how does a polarizer for light waves work?
28. What is an LCD and how does it change colors?
29. Jim does not understand the wave model of light. It does not make sense to him that light can propagate in a vacuum where there is no medium to transport vibrations from one place to another. Does the wave model make sense to you? How can you help Jim reconcile his understanding of waves with the wave model of light?
30. Make a list of phenomena that can be explained by the particle model of light. Then make a list of phenomena that can be explained using the wave model of light. Now that we have a wave model of light, does it mean that we should stop using the particle model? Explain your opinion.

# Problems

Below, **BIO** indicates a problem with a biological or medical focus. Problems labeled **EST** ask you to estimate the answer to a quantitative problem rather than derive a specific answer. Problems marked with ✎ require you to make a drawing or graph as part of your solution. Asterisks indicate the level of difficulty of the problem. Problems with no * are considered to be the least difficult. A single * marks moderately difficult problems. Two ** indicate more difficult problems.

### 24.1 Polarization of waves

1. The coils of a horizontal Slinky vibrate vertically up and down; the wave propagates in the horizontal direction. The amplitude of vibrations is 20 cm. You thread the Slinky through a board with a 50-cm slot at an angle of 60° relative to the vertical. What is the amplitude of the wave that moves past the slot? In what direction do the coils now vibrate?
2. A 40-W lightbulb is 2.0 m from a screen. What is the intensity of light incident on the screen? What assumptions did you make?
3. * Assume that the bulb in Problem 2 radiates unpolarized light. You place a tourmaline crystal in front of the screen. (a) What is the intensity of light hitting the screen after passing through the crystal? (b) You place a second tourmaline crystal between the first crystal and the screen so that it is oriented at an angle of 50° relative to the axis of the first. What is the intensity of light hitting the screen? What assumptions did you make?
4. * Investigate the properties of Iceland spar and explain how it contributed to the understanding of light as a wave.
5. * Describe an experiment you can design to find out whether sound can be polarized. Discuss what outcome would help you make a conclusive judgment.

### 24.2 Discovery of electromagnetic waves

6. * ✎ Place two small electrically charged objects at a distance $d$ from each other. The charge of the first object is $+Q$; the change of the second object is $-2Q$. (a) Draw $\vec{E}$ field lines for the electric field due to these two objects. (b) Calculate the magnitude of the $\vec{E}$ field at a point that is exactly between the two objects. What assumptions did you make?
7. A current in a straight wire is 0.20 A. What is the magnitude of the $\vec{B}$ field at a point that is 30 cm from the wire? What assumptions did you make?
8. * ✎ Draw a bar magnet with marked poles. (a) Show the direction of the $\vec{B}$ field lines. Where do they start? Where do they end? (b) Imagine this magnet falling as a result of the pull of Earth with its north pole pointing down. Draw the $\vec{E}$ field lines for the induced electric field in front of the magnet and behind the magnet.
9. A uniform $\vec{B}$ field of $10^{-3}$ T decreases to zero in 30 s. What is the magnitude of the induced current in a 20-$\Omega$ closed loop of wire that is placed perpendicular to the magnetic field lines? The area of the loop is $2.00 \times 10^{-3}\ \text{m}^2$.
10. * ✎ (a) A uniform $\vec{B}$ field whose lines are oriented in the S-N direction increases steadily. Draw the $\vec{E}$ field lines of the induced electric field. (b) A uniform magnetic field whose lines are oriented E-W decreases steadily. Draw the $\vec{E}$ field lines of the induced electric field. (c) Repeat part (b) for a case when the $\vec{B}$ field changes at twice the rate.
11. * Investigate in detail how Hertz's apparatus worked and describe how it was used to produce and detect electromagnetic waves.
12. * Describe experiments that allow you to observe reflection and refraction of electromagnetic waves.

## 24.3 Some applications of electromagnetic waves

13. ** **Radar parameters** Radar is characterized by a pulse width and pulse repetition period. (a) Explain what these words mean and how you can use them to estimate the range of distances that radar can measure. (b) If you wish to be able to detect objects at a close distance, what parameter of the radar should you change and how? If you wish to be able to detect objects at a long distance, what parameter of the radar should you change and how?
14. EST **Radar distance** Estimate the range of distances at which you can detect an object using radar with a pulse width of 12 μs and a pulse repetition of 15 kHz.
15. * EST **Radar antenna height** Why do you need to raise a radar emission antenna above the ground to be able to detect objects near the surface of Earth at the farthest range of the radar? Support your answer with a sketch. Estimate how high you need to raise the radar.
16. * **More radar detection** Radar uses radio waves of a wavelength of 2.0 m. The time interval for one radiation pulse is 100 times larger than the time of one oscillation; the time between pulses is 10 times larger than the time of one pulse. What is the shortest distance to an object that this radar can detect?
17. * **Communicating with Mars** Imagine that you have a vehicle traveling on Mars. Can you use radio signals to give commands to the vehicle? The shortest distance between Earth and Mars is $56 \times 10^6$ km; the longest is $400 \times 10^6$ km. What is the delay time for the signal that you send to Mars from Earth?
18. * **TV tower transmission distance** The height of a TV tower is 500 m and the radius of Earth is 6371 km. What is the maximum distance along the ground at which signals from the tower can be received?

## 24.4 and 24.5 Frequency, wavelength, speed, and the electromagnetic spectrum and Mathematical description of EM waves and EM wave energy

19. * (a) The amplitude of the $\vec{E}$ field oscillations in an electromagnetic wave traveling in air is 20.00 N/C. What is the amplitude of the $\vec{B}$ field oscillations? (b) The amplitude of the $\vec{B}$ field oscillations in an electromagnetic wave traveling in a vacuum is $3.00 \times 10^{-3}$ T. What is the amplitude of the oscillations of the $\vec{E}$ field?
20. **Milky Way** The Sun is a star in the Milky Way galaxy. When viewed from the side, the galaxy looks like a disk that is approximately 100,000 light-years in diameter (a light-year is the distance light travels in one year) and about 1000 light-years thick (**Figure P24.20**). What is the diameter and thickness of the Milky Way in meters? In kilometers? In miles?

**Figure P24.20**

Sun and solar system's location in the Milky Way

21. We observe an increase in brightness of a star that is $5.96 \times 10^{19}$ m away. When did the actual increase in brightness take place? Is the star in the Milky Way galaxy? What assumptions did you make?
22. **Milky Way neighbors** Two neighboring galaxies to the Milky Way, the Large and Small Magellanic Clouds, are about 180,000 light-years away. How far away are they in meters? In miles?
23. * **Limits of human vision** The wavelength limits of human vision are 400 nm to 700 nm. What range of frequencies of light can we see? How do the wavelength and frequency ranges change when we are underwater? The speed of light in water is 1.33 times less than in the air.
24. **AM radio** An AM radio station has a carrier frequency of 600 kHz. What is the wavelength of the broadcast?
25. * The electric field in a sinusoidal wave changes as

$$E = (25\ \text{N/C})\cos\left[(1.2 \times 10^{11}\ \text{rad/s})t + (4.2 \times 10^{2}\ \text{rad/m})x\right]$$

(a) In what direction does the wave propagate? (b) What is the amplitude of the electric field oscillations? (c) What is the amplitude of the magnetic field oscillations? (d) What is the frequency of the wave? (e) What is the wavelength? (f) What is the speed? (g) What other information can you infer from the equation?
26. ** A sinusoidal electromagnetic wave propagates in a vacuum in the positive $x$-direction. The $\vec{B}$ field oscillates in the $z$-direction. The wavelength of the wave is 30 nm and the amplitude of the $\vec{B}$ field oscillations is $1.0 \times 10^{-2}$ T. (a) What is the amplitude of the $\vec{E}$ field oscillations? (b) Write an equation that describes the $\vec{B}$ field of the wave as a function of time and location. (c) Write an equation that describes the $\vec{E}$ field in the wave as a function of time and location.
27. ** For the previous problem determine (a) the frequency with which the electric energy in the wave oscillates; (b) the frequency at which magnetic field energy oscillates; (c) the maximum energy density; (d) the minimal energy density; (e) the average energy density; and (f) the intensity of the wave.
28. BIO **Ultraviolet A and B** UV-A rays are important for the skin's production of vitamin D; however, they tan and damage the skin. UV-B rays can cause skin cancer. The wavelength range of UV-A rays is 320 nm to 400 nm and that of UV-B rays is 280 to 320 nm. What is the range of frequencies corresponding to the two types of rays?
29. BIO **X-rays used in medicine** The wavelengths of X-rays used in medicine range from about $8.3 \times 10^{-11}$ m for mammography to shorter than $6.2 \times 10^{-14}$ m for radiation therapy. What are the frequencies of the corresponding waves? What assumption did you make in your answer?
30. ** **Power of sunlight on Earth** The Sun emits about $3.9 \times 10^{26}$ J of electromagnetic radiation each second. (a) Estimate the power that each square meter of the Sun's surface radiates. (b) Estimate the power that 1 $m^2$ of Earth's surface receives. (c) What assumptions did you make in part (b)? The distance from Earth to the Sun is about $1.5 \times 10^{11}$ m and the diameter of the Sun is about $1.4 \times 10^{9}$ m.
31. ** **Light from an incandescent bulb** Only about 10% of the electromagnetic energy from an incandescent lightbulb is visible light. The bulb radiates most of its energy in the infrared part of the electromagnetic spectrum. If you place a 100-W lightbulb 2.0 m away from you, (a) what is the intensity of the

infrared radiation at your location? (b) What is the infrared energy density? (c) What are the approximate magnitudes of the infrared electric and magnetic fields?

32. * Explain how the information about energy radiated by a lightbulb in Problem 31 can be used to compare the magnitudes of $\vec{E}$ fields oscillating at different frequencies. Pose a problem that requires the use of this information.

33. * **EST** Estimate the amplitude of oscillations of the $\vec{E}$ and $\vec{B}$ fields close to the surface of the Sun. List all of the assumptions that you made.

34. ** **BIO New laser to treat cancer** The HERCULES pulsed laser has the potential to help in the treatment of cancer, as it focuses its power on a tiny area and essentially burns individual cancer cells. The HERCULES can be focused on a surface that is $0.8 \times 10^{-6}$ m across. This pulsed laser provides 1.2 J of energy during a $27 \times 10^{-15}$-s time interval at a wavelength of approximately 800 nm. (a) Determine the power provided during the pulse. (b) Determine the magnitude of the maximum $\vec{E}$ field produced by the pulse.

35. * **Equation Jeopardy 1** Tell everything you can about the wave described by the equations below.

$$E_y = (2.4 \times 10^5\ \text{N/C})\cos\left[2\pi\left(\frac{t}{1.5 \times 10^{-15}\ \text{s}} - \frac{x}{4.5 \times 10^{-7}\ \text{m}}\right)\right]$$

$$B_z = (8.0 \times 10^{-4}\ \text{T})\cos\left[2\pi\left(\frac{t}{1.5 \times 10^{-15}\ \text{s}} - \frac{x}{4.5 \times 10^{-7}\ \text{m}}\right)\right]$$

36. * **Equation Jeopardy 2** Tell everything you can about the wave described by the equations below.

$$E_y = (9.0 \times 10^5\ \text{N/C})\cos\left[2\pi\left(\frac{t}{5.3 \times 10^{-15}\ \text{s}} - \frac{x}{1.6 \times 10^{-6}\ \text{m}}\right)\right]$$

$$B_z = B_{\text{max}}\cos\left[2\pi\left(\frac{t}{5.3 \times 10^{-15}\ \text{s}} - \frac{x}{1.6 \times 10^{-6}\ \text{m}}\right)\right]$$

37. * **Equation Jeopardy 3** Tell everything you can about the situation described below.

$$E_{\text{max}} = \sqrt{\frac{2 \cdot 640\ \text{kW/m}^2}{(3.0 \times 10^8\ \text{m/s})(8.85 \times 10^{-12}\ \text{C}^2/\text{N}\cdot\text{m}^2)}}$$

### 24.6 Polarization and light reflection: Putting it all together

38. * **Red filter** A color filter is a transparent material that permits only light of a certain color to pass through. When you place a red filter between a lightbulb and a grating, the pattern on the screen beyond the grating consists of rather wide $m = 0, 1, 2, \ldots$ red bands. Why aren't the bright bands on the screen narrow, as they are when you shine a red laser on the grating?

39. * Describe an experiment that you can perform to determine whether the light from a particular source is unpolarized or linearly polarized. If the latter, then how can you determine the polarization direction?

40. * An unpolarized beam of light passes through two polarizing sheets that are initially aligned so that the transmitted beam is maximal. By what angle should the second polarized sheet be rotated relative to the first to reduce the transmitted intensity to (a) one-half and (b) one-tenth the intensity that was transmitted through both polarizing sheets when aligned?

41. * **BIO Spider polarized light navigation** The gnaphosid spider *Drassodes cupreus* has evolved a pair of lensless eyes for detecting polarized light. Each eye is sensitive to polarized light in perpendicular directions. Near sunset, the spider leaves its nest in search of prey. Light from overhead is linearly polarized and indicates the direction the spider is moving. After the hunt, the spider uses the polarized light to return to its nest. Suppose that the spider orients its head so that one of these two eyes detects light of intensity 800 W/m$^2$ and the other eye detects zero light intensity. What intensities do the two eyes detect if the spider now rotates its head 20° from the previous orientation?

42. * Two polarizing sheets are oriented at an angle of 60° relative to each other. (a) Determine the factor by which the intensity of an unpolarized light beam is reduced after passing through both sheets. (b) Determine the factor by which the intensity of a polarized beam oriented at 30° relative to each polarizing sheet is reduced after passing through both sheets.

43. * **Light reflected from a pond** At what angle of incidence (and reflection) does light reflected from a smooth pond become completely polarized parallel to the pond's surface? How do you know? In which direction does the $\vec{E}$ field vector oscillate in this reflected light wave?

44. * **Light reflected from water in a cake pan** At what angle is reflected light from a water-glass interface at the bottom of a cake pan holding water completely polarized parallel to the surface of the glass whose refractive index is 1.65?

45. * Unpolarized light passes through three polarizers. The second makes an angle of 25° relative to the first, and the third makes an angle of 45° relative to the first. The intensity of light measured after the third polarizer is 40 W/m$^2$. Determine the intensity of the unpolarized light incident on the first polarizer.

46. ** You have two pairs of polarized glasses. Make a list of experimental questions you can answer using one of them or both. Describe the experiments you will design to answer them and discuss how you will ensure that the solutions (answers) you find make sense.

## General Problems

47. ** **EST Density of Milky Way** What is the average density of the Milky Way assuming that it contains about 200 million stars with masses comparable to the mass of the Sun? What additional information do you need to know to answer the question? What assumptions should you make?

48. ** **EST Supernova in neighboring galaxy** On February 23, 1987 astronomers noticed that a relatively faint star in the Tarantula Nebula in the Large Magellanic Cloud suddenly became so bright that it could be seen with the naked eye. Astronomers suspected that the star exploded as a supernova. Late-stage stars eject material that forms a ring before they explode as supernovas. The ring is at first invisible. However, following the explosion the light from the supernova reaches the ring. Astronomers observed the ring exactly one year after the phenomenon itself occurred (**Figure P24.48**). (a) Use geometry and the speed of light to estimate the distance to the supernova. The angular size of the ring is 0.81 arcseconds (b) How do you know whether your answer makes sense? (c) Use the width of the ring to estimate the uncertainty in the value of the distance.

**Figure P24.48**

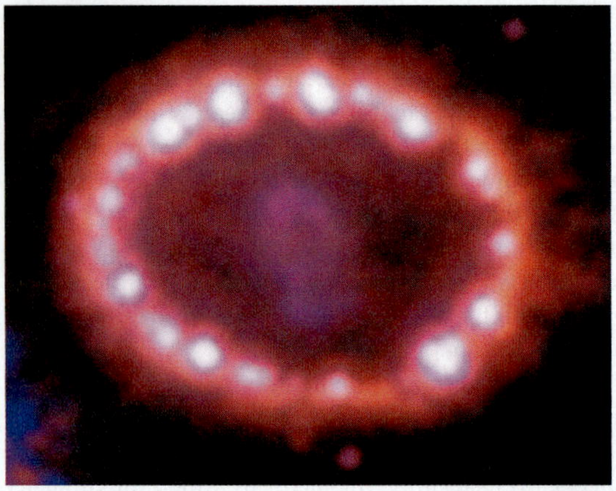

49. ** BIO EST **Human vision power sensitivity** A rod in the eye's retina can detect light of energy $4 \times 10^{-19}$ J. Estimate the power of this light that the rod can detect. Indicate any assumptions you made. You will need more information than what is provided.

50. ** **Effect of weather on radio transmission** Weather affects the transmission of AM stations but does not affect FM stations. FM stations do not broadcast in remote areas. Suggest possible reasons for these phenomena. [Hint: Find information about the wavelengths of the waves and decide how they might propagate in Earth's atmosphere.]

51. ** **Measuring speed of light** In 1849 A. Fizeau conducted an experiment to determine the speed of light in a laboratory (before that time, all methods involved astronomical distances). He used an apparatus described in **Figure P24.51**. Light from the source S went through an interrupter K and after reflecting from the mirror M returned to the rotating wheel again. If light passed between the teeth of the wheel, Fizeau could see it; if light on the way back hit the tooth of the wheel, Fizeau would not see it. He could measure the speed of light by relating the time interval between teeth crossing the beam and the distance light traveled during those time intervals. The information that Fizeau had about the system was: $L =$ the distance between the wheel and the mirror; $T =$ the period of rotation of the wheel; and $N =$ the number of teeth in the wheel (the width of one tooth was equal to the width of the gap between the teeth). Using these parameters Fizeau derived a formula for calculating the speed of light: $c = (4LN/T)$. Explain how he arrived at this equation.

**Figure P24.51**

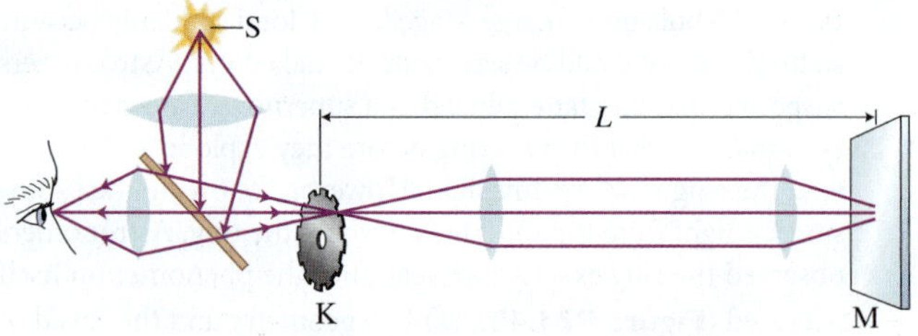


52. ** **Speed of light** In Fizeau's experiment (described in Problem 51) the distance between the wheel and the mirror was 3.733 km, the wheel had 720 teeth, and the wheel made 29 rotations per second. What was the speed of light determined by Fizeau? What were the uncertainties in this value?

53. * A sinusoidal electromagnetic wave has a 20-N/C $\vec{E}$ field amplitude. Determine everything you can about this wave.

## Reading Passage Problems

BIO **Amazing honeybees** The survival of a bee colony depends on the ability of bee scouts to locate food and to convey that information to the hive. After finding a promising food source, a honeybee scout returns to the hive and uses a waggle dance to tell its worker sisters the direction and distance to the food.

Recall that light coming directly from the Sun is unpolarized. Bees use the direction of the Sun as a reference for their travel. In the hive, the scout bee's waggle dance is in the shape of a flat figure eight (**Figure 24.27**). The upward direction in the vertical hive represents the direction toward the Sun. The middle line of the scout's waggle dance resembles a figure eight that points in the direction of the food relative to the direction of the Sun. Thus a 50° middle waggle dance to the right of the vertical in the hive would indicate a food source outside that is 50° to the right of the direction toward the Sun. The distance to the food depends on the length of time the scout takes while wiggling her tail and wings through this middle of the figure eight. The other bees know the direction to the food even if the Sun is behind a cloud—they can detect the degree of polarization in other open parts of the sky and deduce the angle to the Sun from that position, and they learned from the scout the angle from the Sun's direction to the food.

**Figure 24.27** A scout bee's waggle dance.

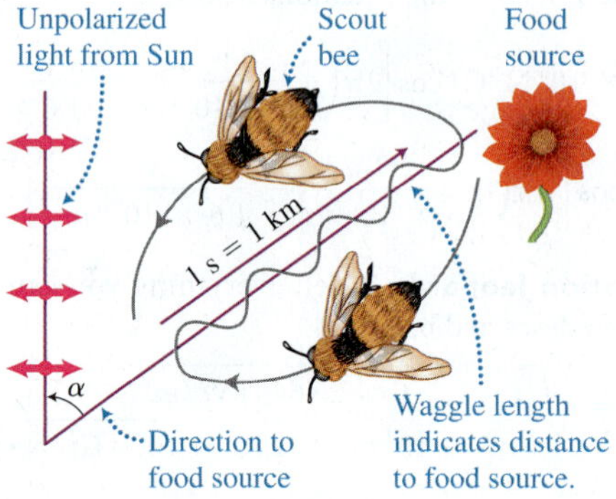


54. Bees know the direction to the Sun because
(a) direct sunlight is linearly polarized.
(b) direct sunlight is unpolarized.
(c) they detect infrared radiation from the sun.
(d) a and c.
(e) b and c.

55. Suppose that the scout bee in a waggle dance indicates that a good food source is at a 90° angle from the direction to the Sun. Bees who have never been to the food source when leaving the hive head toward a region of the sky that
(a) has a flowery odor.
(b) has completely unpolarized light coming from it.
(c) has completely linearly polarized light coming from it.
(d) has partially unpolarized light coming from it.
(e) None of these answers

56. If the direction of the middle line of the scout's waggle dance figure eight points 53° to the left side of the upward direction in the hive, in what direction to the left should bees leaving the nest for the food source head relative to light coming from the Sun?

(a) Where the light is 36% linearly polarized to the left of the Sun's direction
(b) Where the light is 60% linearly polarized to the left of the Sun's direction
(c) Where the light is 64% linearly polarized to the left of the Sun's direction
(d) Where the light is 80% linearly polarized to the left of the Sun's direction
(e) Where the light is 100% linearly polarized to the left of the Sun's direction

57. Polar molecules are caused to vibrate in all directions perpendicular to the direction of travel of sunlight. When the Sun is rising in the east and you look at the molecules directly overhead, the light from these molecules is
(a) unpolarized.
(b) linearly polarized along an axis between you and the molecules overhead.
(c) linearly polarized parallel to a plane with you, the Sun, and the molecules.
(d) linearly polarized perpendicular to a plane with you, the Sun, and the molecules.
(e) All of these are correct.

58. In what directions is the light from the sky completely unpolarized?
(a) Looking directly at the Sun
(b) Looking directly away from the Sun
(c) Looking at 90° relative to the Sun
(d) a and b
(e) b and c

**Incandescent lightbulbs—Soon to disappear** Australia, Canada, New Zealand, and the European Union started phasing out incandescent lightbulbs in 2009. The United States is scheduled to phase them out by 2014. These bulbs have provided light for the world for 90 years. What's the problem?

Incandescent lightbulbs produce light when moving electrons (electric current) move through the filament and collide with tungsten atoms and ions in the filament, causing them to vibrate violently. Accelerating charged particles emit electromagnetic radiation. This is true for vibrating and accelerating particles in the filament, in your skin, and on the Sun. The 35 °C skin of a person emits most of its radiation in the low-frequency, long-wavelength infrared—sometimes called thermal radiation. At 2500 °C, the filament of an incandescent lightbulb emits most of its radiation as higher frequency infrared radiation and just 10% as visible light (very inefficient if the goal is to produce light). At 6000 °C, the Sun emits about 45% in the infrared range and 40% in the visible light range.

Banning incandescent bulbs will reduce energy usage. According to the Department of Energy, about $8 \times 10^{17}$ J of electric energy is used each year in the United States for household lighting. Compact fluorescent lightbulbs (CFLs) and light-emitting diode (LED) bulbs use about one-fourth the energy of incandescent bulbs. Switching to CFLs and LEDs would also reduce greenhouse gas emissions by reducing the need for power produced by coal-burning electric power plants.

59. Why are the United States and other countries banning the use of incandescent lightbulbs?
(a) The bulbs get too hot.
(b) The bulbs have tungsten filaments.
(c) Ninety percent of the electric energy used is converted to thermal energy.
(d) The bulbs are only 10% efficient in converting electric energy to light.
(e) c and d

60. Which answer below is closest to the rate of visible light emission from a 100-W incandescent lightbulb?
(a) 10 W (b) 20 W (c) 50 W
(d) 100 W (e) Not enough information to make a determination.

61. What does the surface of the body at about 35 °C primarily emit?
(a) No electromagnetic radiation
(b) Long-wavelength infrared radiation
(c) Short-wavelength infrared radiation
(d) Light
(e) Invisible ultraviolet light

62. Suppose you changed all incandescent lightbulbs to more energy-efficient bulbs that used one-fourth the amount of energy to get the same light. About how many $3.3 \times 10^9$-kW·h/year electric power plants could be removed from the power grid?
(a) 3 (b) 10 (c) 20
(d) 50 (e) 100

63. How much money will you save on your electric bill each year if you replace five 100-W incandescent bulbs with five CFL or LED 25-W bulbs that produce the same amount of light? Assume that the bulbs are on 3.0 h/day and that electric energy costs \$0.12/kW·h.
(a) \$5 (b) \$10 (c) \$20
(d) \$50 (e) \$100

# 25 Special Relativity

**Does ether, as the proposed medium for light waves, exist?**

**At what speed would light pass you if you were in a spaceship moving at near light speed away from the light source?**

**How can atomic clocks on satellites help determine your location on Earth?**

**Be sure you know how to:**

- Distinguish between inertial and noninertial reference frames (Section 2.5).
- Explain why observers in noninertial reference frames cannot use Newton's laws to explain mechanics phenomena (Section 2.5).
- Calculate the change in kinetic energy of a charged particle that travels across a potential difference $\Delta V$ (Section 15.3).

In 1879 Albert Michelson published the most accurate measurement of the speed of light available at the time. He next wanted to demonstrate the existence of ether, an invisible substance that was thought to be the medium through which light traveled (like sound waves through air and water waves on the sea). In 1887 he and Edward Morley built a device to do just that. Despite their efforts, they failed to establish the existence of ether. This negative result became the experimental foundation for a new theory—the subject of this chapter.

So far in this book we have assumed that time passes the same for all observers. When a second passes for me, the same second passes for you. Similarly, we have assumed that length measurements of objects are the same for all observers. These assumptions sound so reasonable that it is tempting to think of them as "absolute truths." In this chapter we will find that we need to reconsider our ideas—not just of time and space, but also of momentum and energy.

## 25.1 Ether or no ether?

Recall (from Chapter 23) that in 1801, Thomas Young performed the double-slit experiment that strongly supported the wave model of light. All waves known at that time required a medium to travel through: ocean waves through water, sound waves through air or other materials, and so on. The purpose of the medium is to transfer the disturbance produced by a source at one location to other locations. The particles of the medium interact with each other, allowing this to occur. Light, then, it was reasoned, must also require a medium to travel. In the 19th century physicists thought that this medium was *ether*.

Ether was considered to be invisible but able to undergo shear deformations so that transverse light waves could travel through it. We know (from Chapter 24) that the work of Maxwell and Hertz led to the conclusion that light propagation could be explained by changing $\vec{E}$ and $\vec{B}$ fields that do not require any medium to travel. However, that work was done in the late 1880s. Before that physicists were searching for ether. This search, like many investigations in physics, produced an unexpected outcome that eventually changed the way we think about space and time.

### An analogy: A boat race

To understand how one of those ether-testing experiments was done, we will use a mechanics analogy. Consider a process involving two identical boats in a race on a wide river. Each boat is to travel a distance of 1.6 km relative to the shore, then 1.6 km back (**Figure 25.1**). Boat 1 travels from a starting dock to another dock 1.6 km upstream from the first, then back again to where it started. Boat 2 starts at the same dock but instead travels 1.6 km across the river to a dock exactly on the opposite side, then travels back to where it started. Each boat travels at speed 10 km/h relative to the water, and the water travels at speed 6 km/h relative to the shore. Which boat returns to the starting dock first?

You may think that the boats will arrive back at the dock at the same time. Since all the speeds are the same and are constant, wouldn't everything just balance out? As you will see shortly, the reality is something different.

**Travel time for boat 1 going upstream and back** The displacement of boat 1 with respect to the shore as it moves upstream $\vec{d}_{\text{boat 1 to shore}}$ is the result of two displacements—the displacement of the boat with respect to the water $\vec{d}_{\text{boat 1 to water}}$ and the displacement of the water with respect to the shore $\vec{d}_{\text{water to shore}}$. Mathematically:

$$\vec{d}_{\text{boat 1 to shore}} = \vec{d}_{\text{boat 1 to water}} + \vec{d}_{\text{water to shore}}$$

The time intervals during which these three displacements occur are all the same:

$$\Delta t_{\text{boat to shore}} = \Delta t_{\text{boat to water}} = \Delta t_{\text{water to shore}} = \Delta t$$

**Figure 25.1** Two identical length boat trips on a river.

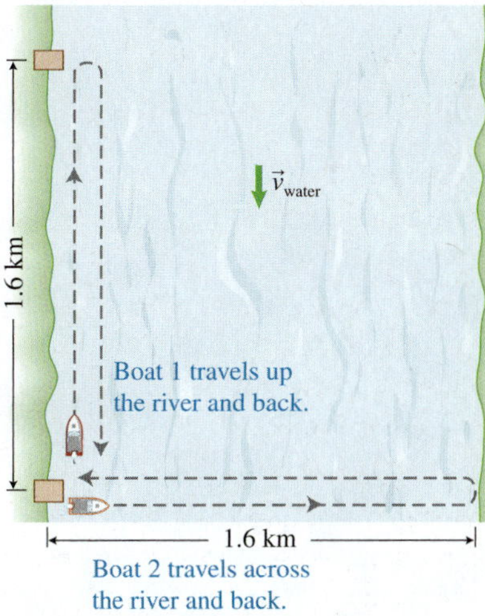

**Figure 25.2** The velocity of boat 1. Boat 1 travels against the current on its way up and with the current on its way back, which affects its speed relative to the shore in each part of the trip.

(a) Boat 1 traveling upstream

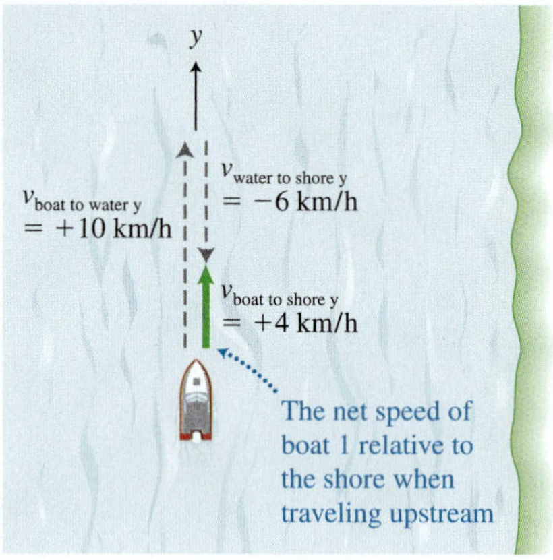


(b) Boat 1 traveling downstream

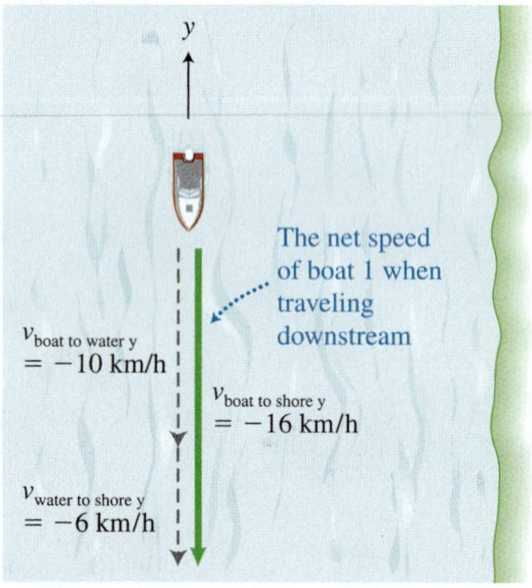


If we divide all three parts of the displacement equation by the time interval $\Delta t$ during which the displacements occurred, we obtain

$$\frac{\vec{d}_{\text{boat 1 to shore}}}{\Delta t} = \frac{\vec{d}_{\text{boat 1 to water}}}{\Delta t} + \frac{\vec{d}_{\text{water to shore}}}{\Delta t}$$

or

$$\vec{v}_{\text{boat 1 to shore}} = \vec{v}_{\text{boat 1 to water}} + \vec{v}_{\text{water to shore}}$$

Note that these are velocities, not speeds, which is why they are labeled with vector symbols. In **Figure 25.2** we summarize the situation using vector components. The upstream velocity of the boat relative to the shore is then (with the upstream direction positive)

$$v_{\text{boat 1 to shore } y} = +10\text{ km/h} + (-6\text{ km/h}) = +4\text{ km/h}$$

You can see that having the water moving against the boat's direction of travel slows the boat down relative to the shore. The time interval it takes the boat to reach the upstream dock is

$$\Delta t_{\text{upstream}} = 1.6\text{ km}/(4\text{ km/h}) = 0.4\text{ h}$$

The downstream velocity relative to the shore is

$$v_{\text{boat 1 to shore } y} = -10\text{ km/h} + (-6\text{ km/h}) = -16\text{ km/h}$$

The time interval it takes the boat to return to the downstream dock is

$$\Delta t_{\text{downstream}} = -1.6\text{ km}/(-16\text{ km/h}) = 0.1\text{ h}$$

As you would expect, the return trip takes less time because the velocity of the water relative to the shore is in the same direction as that of the boat. The total time interval for boat 1's round trip is therefore

$$\Delta t = \Delta t_{\text{upstream}} + \Delta t_{\text{downstream}} = 0.4\text{ h} + 0.1\text{ h} = 0.5\text{ h}$$

**Travel time for boat 2 going across the stream and back** The velocity of boat 2 relative to the shore is perpendicular to the velocity of the water with respect to the shore (**Figure 25.3**). In order for the boat to travel directly across the river, the 10-km/h boat velocity relative to the water has to point at an angle so that it has a 6-km/h upstream component to cancel the downstream velocity of the water relative to the shore. The magnitudes of the three velocities shown are related through the Pythagorean theorem:

$$v^2_{\text{boat 2 to water}} = v^2_{\text{boat 2 to shore}} + v^2_{\text{water to shore}}$$

or

$$\begin{aligned} v^2_{\text{boat 2 to shore}} = v^2_{\text{boat 2 to water}} - v^2_{\text{water to shore}} &= (10\text{ km/h})^2 - (6\text{ km/h})^2 \\ &= 64(\text{km/h})^2 \\ \Rightarrow v_{\text{boat 2 to shore}} &= 8\text{ km/h} \end{aligned}$$

The crossing time interval is then

$$\Delta t_{\text{crossing}} = 1.6\text{ km}/(8\text{ km/h}) = 0.2\text{ h}$$

This crossing velocity is the same in both directions. The time interval for the complete round trip, therefore, is just twice this:

$$\Delta t = 2\Delta t_{\text{crossing}} = 2(0.2\text{ h}) = 0.4\text{ h}$$

We have shown that it takes longer for boat 1 to make its round trip than it takes boat 2, even though both boats travel a distance of 3.2 km relative to the shore. The moving water affects the time intervals differently, depending on the direction of a boat's travel.

## Testing the existence of ether

An experiment analogous to the boat experiment can be used to test for the existence of ether. Imagine that ether fills the solar system and is stationary with respect to the Sun. Because Earth moves around the Sun at a speed of about $3.0 \times 10^4$ m/s, ether should be moving past Earth at this speed. Shining light waves parallel and perpendicular to the ether's motion relative to Earth is similar to sending boats parallel and perpendicular to a flowing river. Therefore, the travel time for light waves (1) up and back along the direction of the ether's motion and (2) across and back perpendicular to the ether's motion should differ if they travel the same distance. If one could measure this time interval difference experimentally, it would serve as strong support for the existence of ether as a medium for light wave propagation.

In 1887, Albert Michelson and Edward Morley set up such an experiment. They used a device called an interferometer to detect ether (see Testing Experiment **Table 25.1**). A light beam is sent to a beam splitter (a half-silvered mirror) that causes the light beam to split into two beams. One beam continues forward and the other beam reflects perpendicularly, each beam moving along a separate arm of the device. Mirrors at the end of the arms reflect the two beams back to the half-silvered mirror, where they recombine and are detected.

Because the two beams are formed by splitting a single light beam, the two beams are coherent (have a constant phase difference with respect to each other). When the beams recombine after reflection, an interference pattern results. The details of this interference pattern depend on the difference in the travel time intervals between the two beams along each arm of the interferometer. *If* the ether hypothesis is correct and we carefully rotate the interferometer (causing the direction of each beam of light to change relative to the ether), *then* the time interval it takes each beam to travel along each arm and back should change. As a result, the interference pattern produced by the recombined light should change as the device rotates.

**Figure 25.3** The velocity of boat 2. Boat 2 travels in such a way as to counter the crosscurrent that would carry it downstream; the crosscurrent is the same in both directions and the boat speed relative to the shore is less than its speed relative to the water.

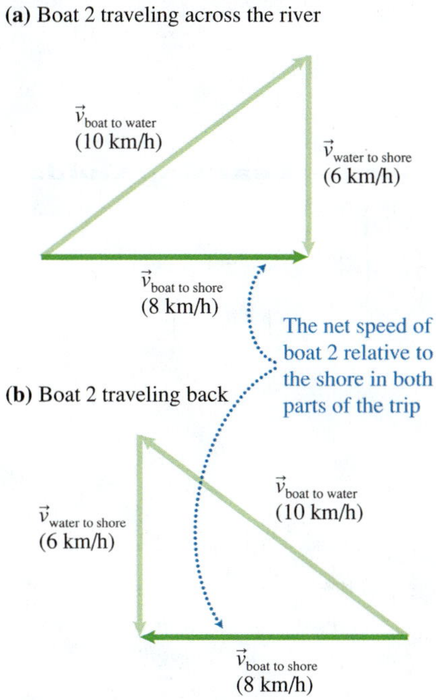


TESTING EXPERIMENT TABLE

**25.1 Testing the existence of ether.**

| Testing experiment | Prediction | Outcome |
|---|---|---|
| Assume that the ether is moving to the left past the interferometer. We shine a beam of light from the left onto the beam splitter. One beam moves parallel to the ether's velocity and the other beam moves perpendicular to its velocity. 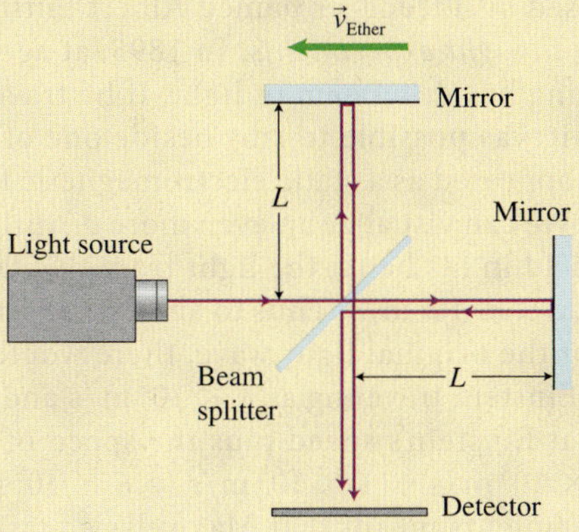 | It should take the light longer to travel against and with the ether than to travel sideways across the moving ether. Rotating the interferometer in the plane of the page should lead to a change in the interference pattern formed by the two light beams at the detector. 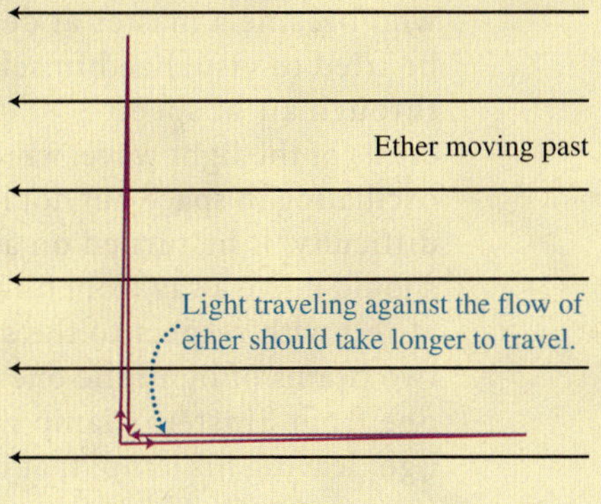 | No matter how the interferometer orientation was changed, the interference pattern did not change. |





(continued)

**Possible conclusions**

1. There is no ether through which light travels.
2. There is ether, but it is "stuck" to Earth's surface and does not move relative to the interferometer.

Michelson and Morley repeated the experiment several times, but the phase difference between the two beams never changed. It seemed that light traveled at the same speed independent of its direction. This would seem to disprove the idea of ether as a medium for light to propagate.

**Active Learning Guide ›**

Physicists, including Michelson, were reluctant to accept this no-ether result. They tried to revise their model to fit the experimental data. Perhaps the ether was dragged along through space by Earth. This would mean that the ether was at rest relative to the interferometer, explaining the null result of their experiment. However, if the ether is attached to Earth in some way, then as Earth rotates around its axis and orbits the Sun, it should cause the ether in the solar system to become twisted. This would cause light coming from stars to be slightly deflected on its way to Earth (like a water wave would be deflected by a whirlpool in the water). However, no one observed such an effect. It seemed that light did not require a medium to travel through.

**Review Question 25.1** Why did we call the Michelson-Morley experiment a testing experiment and not an observational experiment?

## 25.2 Postulates of special relativity

By the end of the 19th century, the laws of Newtonian physics had been well developed and accepted. Physicists knew that using Newton's laws to analyze a process would yield consistent results, regardless of the inertial (nonaccelerating) reference frames used. This feature is known as **invariance.**

Invariance is a result of acceleration (rather than velocity) being proportional to the net force exerted on an object. For example, if you are on an airplane moving smoothly at constant velocity, all the events occur as if the airplane is at rest on the ground. If you drop a book, it lands at your feet; if you push a suitcase, it accelerates the same way as if the airplane were parked on the runway. You would not be able to detect the airplane's motion. However, for observers in noninertial reference frames (accelerating frames), such as when an airplane is taking off, Newton's laws do not explain observed phenomena, such as a loose food cart starting to roll without being pushed.

Invariance did not apply to electromagnetism and in particular to the observation of light in different reference frames. The speed of light in different reference frames was addressed by a teenager named Albert Einstein, who became a master at designing *thought experiments.* In 1895, at age 16, he tried to visualize himself traveling beside a beam of light. If he traveled through air at speed $3 \times 10^8\ \text{m/s}$, it was possible to stay beside one of the crests of the light wave, which now appeared as a static electromagnetic field oscillating in space but not in time. We can visualize an even more disturbing difficulty. If he turned on a lamp held in his hand, the light from this lamp should move away from him at speed $3 \times 10^8\ \text{m/s}$. Thus to another observer at rest with respect to the source of the original light wave, there would be two beams of light: the one beside Einstein traveling at $3 \times 10^8\ \text{m/s}$ and the one from Einstein's lamp moving at Einstein's speed plus the speed of the light leaving his lamp, that is, at $3 \times 10^8\ \text{m/s} + 3 \times 10^8\ \text{m/s} = 6 \times 10^8\ \text{m/s}$ (**Figure 25.4**). These thought experiments meant that Maxwell's equations had to be written differently for these different observers, with a different speed of light in each case.

## Einstein's two postulates

Einstein believed that the invariance principle was fundamental to any physics idea. He also believed that Maxwell's equations were correct. To resolve this contradiction, in 1905 Einstein proposed a new theory, called **the special theory of relativity**. He could have based it on the results of the Michelson-Morley experiment that suggested that the speed of light in air or a vacuum was independent of the motion of the observer, but he did not. In fact, it is not certain whether he was aware of Michelson and Morley's results at all. Although there is some historical controversy about this, the general consensus is that Einstein based his ideas on his thought experiments.

Einstein's theory started with the following two *postulates*. A postulate is a statement that is assumed to be true (it is not derived from anything). It is usually used as the starting point for a logical argument.

1. *The laws of physics are the same in all inertial reference frames.* This was not a new idea in mechanics. Newton's second law,

$$\vec{a}_O = \frac{\sum \vec{F}_{\text{on O}}}{m_O}$$

remains the same regardless of the inertial reference frame in which you choose to apply it. But it was a new idea for Maxwell's equations of electromagnetism.

2. *The speed of light in a vacuum is measured to be the same in all inertial reference frames.* The speed of light in a vacuum measured by observers in different inertial reference frames is the same regardless of the relative motion of those observers.

**Figure 25.4** A thought experiment. Imagine a person traveling at light speed beside the crest of a light wave. His lamp is shining forward in the direction of travel. Would an observer detect one light wave traveling at $c$ and light from the moving light source traveling at $2c$?

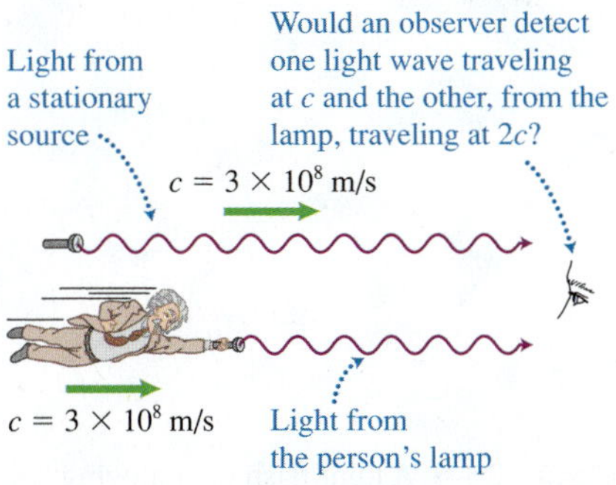

Light from a stationary source
Would an observer detect one light wave traveling at c and the other, from the lamp, traveling at 2c?
c = 3 × 10^8 m/s
c = 3 × 10^8 m/s
Light from the person's lamp

Although postulate 2 was supported by the Michelson-Morley experiment, it was a counterintuitive idea at the time. Think back to the boats in Section 25.1. The speed of the boat traveling parallel to the shore with respect to the shore depended on its velocity with respect to the water and the velocity of the water with respect to the shore. Now, suppose that an intense pulse of laser light is shot from Earth toward a spaceship moving away. A stationary observer on Earth measures the light's speed $c$ as $3.0 \times 10^8$ m/s (**Figure 25.5a**). If the spaceship is moving away from Earth at a speed $v$ that is just a little less than light speed, then our intuition tells us that the laser light will have a difficult time reaching the ship. After all, isn't the laser light's speed just $c - v$ relative to the ship (similar to the boat going upstream)? However, according to Einstein's second postulate, an observer on the spaceship should also measure the laser light speed to be $3.0 \times 10^8$ m/s (Figure 25.5b). This sounds impossible. Can we find experimental evidence supporting postulate 2?

**‹ Active Learning Guide**

**Figure 25.5** All inertial reference frame observers measure the same speed for light $c$.

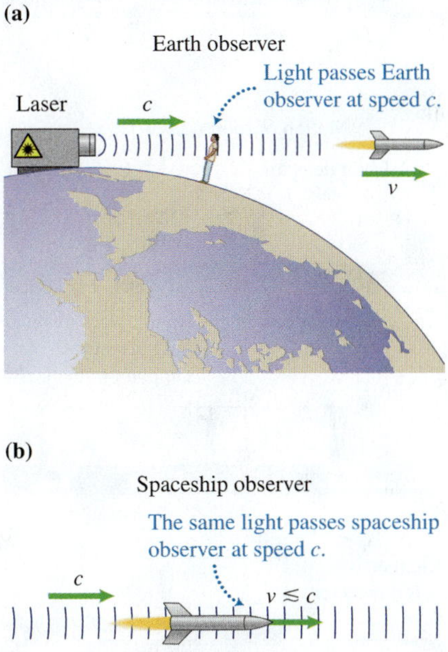

(a)
Earth observer
Laser
c
Light passes Earth observer at speed c.
v
(b)
Spaceship observer
The same light passes spaceship observer at speed c.
c
v ≲ c

## Experimental evidence for the constancy of light speed

To investigate the speed of light with respect to different observers, we need a situation where light is being produced by objects moving at very high speeds with respect to Earth. In modern experiments involving elementary particles, this happens regularly. Consider a very specific experiment involving a particle known as the neutral pion, $\pi^0$. This particle can be produced during a high-energy collision between other particles. After the collision, the newly created pion is traveling at speeds in excess of $0.999c$ (99.9% of light speed) relative to the lab where the collision occurred. The pion is very unstable and decays in less than a picosecond ($10^{-12}$ s), turning into gamma rays, a very high frequency form of EM wave. The speed of the gamma rays is measured in the lab to be precisely the speed of light, despite being produced by the decaying pion, which was already moving near light speed relative to the lab. This result supports postulate 2.

> **TIP** As strange as postulate 2 may seem, remember that our everyday ideas about how to combine velocities have come from observing objects traveling very slowly compared with light speed. We should always be ready for our ideas to require improvements when applied in new situations.

**Review Question 25.2** What is invariance, and how does it relate to light speed having a constant value?

## 25.3 Simultaneity

**Figure 25.6** A light flash on a moving train. (a) On the train, you see the light flash reach both detectors at the same time. (b) An observer on the platform sees light reach the detector on the left before it reaches the one on the right.

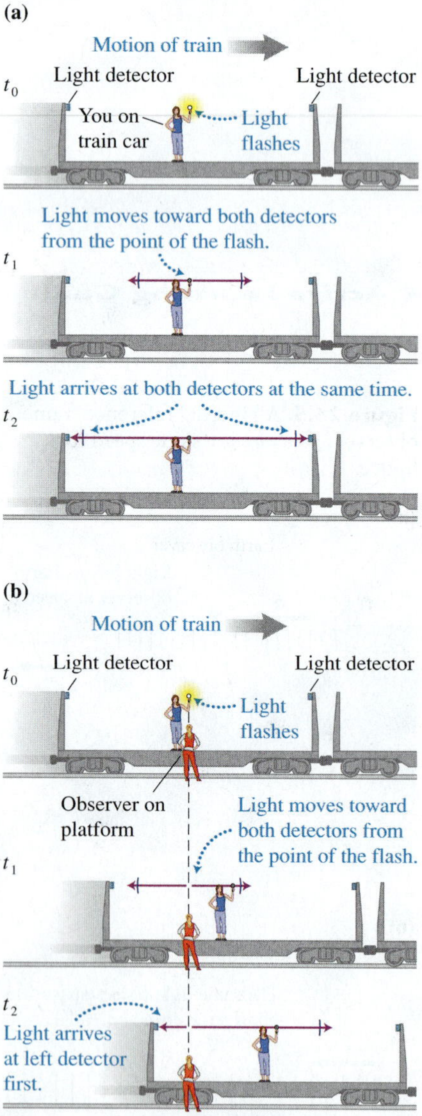


In the boats experiment, we assumed that the time interval that the boat traveled during the trip was the same independent of the reference frame (shore or water) ($\Delta t_{\text{boat to shore}} = \Delta t_{\text{boat to water}} = \Delta t$). However, this seemingly reasonable way of thinking about time turns out to contradict the second postulate proposed by Einstein (and therefore it contradicts the results of the Michelson-Morley experiment).

The second postulate of the special theory of relativity made physicists completely rethink their ideas about time. The following thought experiment suggested by Daniel Frost Comstock in 1910 and Albert Einstein in 1917 addresses one aspect of this issue.

### Simultaneity of events in two different inertial reference frames

You stand at the center of a long flatbed train car, which travels eastward. You hold a lamp that can send out a light flash in all directions. There are light detectors at each end of the car. Your friend (a second observer) stands on the platform as the train moves past. Your lamp flashes at the instant you pass your friend. You and your friend each note when the light from the lamp reaches the detectors at the ends of the car.

Since you are in the middle of the train car, and because the detectors are at rest with respect to you (that is, both you and they are in the same inertial reference frame), you note that the light reaches the front and back detectors at the same time (see **Figure 25.6a**). However, your friend on the platform sees the rear detector moving toward the lamplight while that light is moving. Your friend also sees the front detector moving away from the lamplight while the light is moving (Figure 25.6b). Because the speed of light is finite and is the same in all directions for all observers, the light headed to the back of the train travels a shorter distance before reaching the detector than the light headed to the front of the train. Thus, according to your friend on the platform, the light reaches the back detector before it reaches the front detector.

### Implications of the difference in observations

Think about what this means. We are focusing on two events: (1) the light flash reaching the front detector and (2) the light flash reaching the back detector. In your reference frame standing in the middle of the flatbed car, these two events are simultaneous. In your friend's reference frame standing on the platform beside the train track, these two events are not simultaneous.

This really occurs! Events that happen at the same time in one reference frame do not necessarily occur at the same time in another. At normal train speeds, however, the times of events won't vary by more than $10^{-15}$ to $10^{-14}$ s between observers. This difference does get larger the faster the train travels relative to the platform, but the effect only becomes significant if the train travels at a substantial fraction of light speed.

< **Active Learning Guide**

Phenomena that only become significant in high-speed circumstances like this are called **relativistic effects**. This so-called **relativity of simultaneity** is just the first of many relativistic effects that we will investigate in this chapter.

> **Relativity of simultaneity** Events that occur simultaneously for an observer in one inertial reference frame do not necessarily occur simultaneously for observers in other inertial reference frames.

**Review Question 25.3** You hear in your physics class the following statement: "Time is relative." How would you explain the meaning of this statement to your friend who is not taking physics?

## 25.4 Time dilation

In this section we examine whether the time interval between events differs in different inertial reference frames. An event is anything that happens: a person's departure for a trip, the person's return from that trip, the lamplight reaching the detectors in the train example from the previous section, and so on. To specify an event quantitatively, we need to indicate where it occurs (a position) and when it occurs (a clock reading).

Our everyday experience supports the idea that the time interval between two events is something that all observers will agree on. For example, if your physics class lasts 60 min according to your clock, won't astronauts on the International Space Station watching the webcast find that 60 min passes on their clocks as well? It shouldn't matter that the astronauts are moving at 7700 m/s (17,000 mi/h) relative to you, right? The thought experiment in the previous section, however, suggests that it might matter.

We can use another thought experiment to investigate this. You stand on a moving train. Your friend stands on the platform at the station. On the train is a laser pointing upward, a mirror located directly above the laser, and a light detector just next to the laser (**Figure 25.7a**). The train is moving to the right at speed $v$, a significant fraction of light speed relative to the platform. A flash of light is emitted from the laser (event 1), reflects off the mirror, and is then detected (event 2).

In your reference frame on the train, you see the laser light travel vertically upward, then downward (Figure 25.7b). In your friend's reference frame on the platform, the laser light must move upward at an angle in order to reflect off the mirror, then downward at an angle to hit the detector—because the train is moving (Figure 25.7c).

What is the time interval between the flash and its detection (the two events), as measured by each of the two observers? For you, the light travels at speed $c$ a distance $2L_0$, where $L_0$ is the distance from the light source to the mirror. The time interval between the flash and its detection is then

$$\Delta t_0 = \frac{2L_0}{c}$$

According to your friend on the platform, the light travels a longer distance, $2L$. Using the Pythagorean theorem, the distance $L$ is the square root

**Figure 25.7** An experiment for two observers to detect the time interval between two events. The time interval between events as seen (b) in the proper reference frame, where they occur at the same place and (c) in another reference frame, where the events occur at different places.

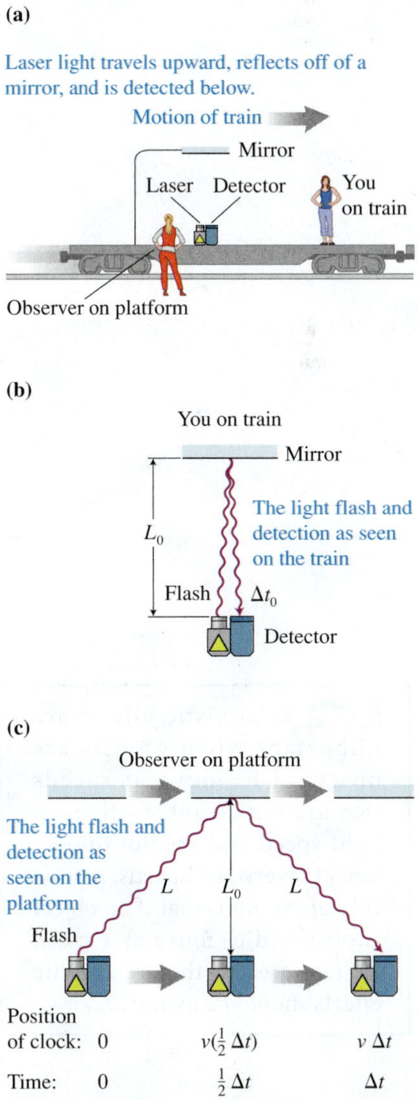

of the sum of the lengths squared of the two sides of the triangle with $L$ as its hypotenuse:

$$L = \sqrt{L_0^2 + \left(\frac{1}{2}v\Delta t\right)^2}$$

The base of the triangle $(1/2)v\Delta t$ is one half the distance the train travels at speed $v$ during the time interval $\Delta t$ that the light travels from its source up to the mirror and back down to the receiver. The time interval between the flash and its detection is

$$\Delta t = \frac{2L}{c} = \frac{2\sqrt{L_0^2 + \left(\frac{1}{2}v\Delta t\right)^2}}{c}$$

Rearrange $\Delta t_0 = \dfrac{2L_0}{c}$ to $L_0 = \dfrac{c\Delta t_0}{2}$ and substitute this into the above equation to get

$$\Delta t = \frac{2\sqrt{\frac{c^2\Delta t_0^2}{4} + \frac{v^2\Delta t^2}{4}}}{c}$$

Solving for $\Delta t$, we get

$$\Delta t = \frac{\Delta t_0}{\sqrt{1 - \frac{v^2}{c^2}}}$$

The term

$$\sqrt{1 - \frac{v^2}{c^2}}$$

is always less than or equal to 1. (It is equal to 1 only when the train is at rest with respect to the platform.) Therefore, the time interval $\Delta t$ between the flash and its detection measured by your friend on the platform is greater than the time interval $\Delta t_0$ measured by you on the train.

Notice that we used $c = 3.00 \times 10^8$ m/s for the speed of light in both reference frames (according to Einstein's second postulate). This analysis results in a striking conclusion—the time interval that the laser light was in flight differs depending on the reference frame in which it is measured. This relativistic effect is called **time dilation**.

Notice that your reference frame on the train is the one reference frame where the two events—the flash and the detection—occur at the same position. The time interval $\Delta t_0$ between the events measured in this reference frame is called the **proper time interval**, and this reference frame is known as the **proper reference frame**. The time interval $\Delta t$ between these same two events measured in any other reference frame will always be longer. That is, $\Delta t > \Delta t_0$.

We certainly don't observe this occurring in everyday life—if a particular boat trip takes 2 h according to the captain on the boat, won't it also take 2 h according to people on shore? Not if the idea of time dilation is correct. Actually, time dilation is happening all the time, though the effect only becomes substantial when the two reference frames are moving at significant fractions of light speed relative to each other. For example, for $\Delta t$ and $\Delta t_0$ to be one-millionth of a percent different, the two reference frames would have to be moving at over $4 \times 10^4$ m/s (over 90,000 mi/h) relative to each other.

**TIP** Relativistic effects are important when objects are observed to move at speeds that are significant fractions of light speed and are not important at "everyday" speeds. A good rule of thumb is that if an object is observed to move at a speed 0.1 $c$ or greater, then relativistic effects should be included.

Time dilation can be tested using what is known as a muon decay experiment, described in Testing Experiment **Table 25.2**. A muon is an elementary particle produced during reactions in high-energy particle accelerators. Once produced, muons have a **half-life** of $1.5 \times 10^{-6}$ s. This means that if 1000 muons are produced, then $1.5 \times 10^{-6}$ s later, approximately 500 muons remain.

Consider an experiment in which 1000 muons are produced and move across the lab at $0.95c$. Event 1 will be the production of the muons. Event 2 will be the decay of the 500th muon.

**TESTING EXPERIMENT TABLE**

**25.2 Testing the idea of time dilation.**

| Testing experiment | Prediction | Outcome |
|---|---|---|
| A beam of 1000 muons is produced by a source (event 1) and emitted at a speed of $0.95c$. How far from the source will the muons travel before their number is reduced to 500 (event 2)?<br>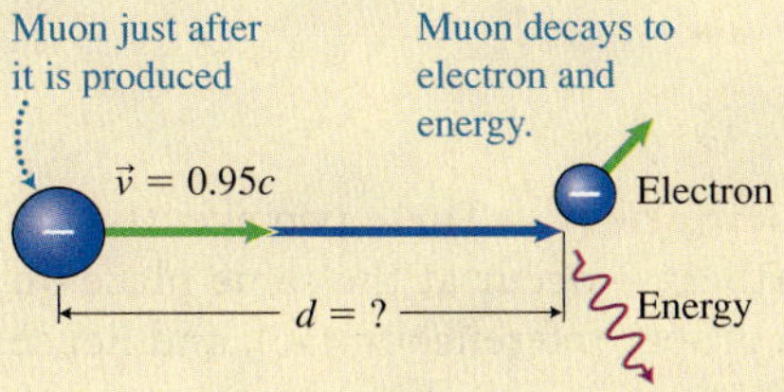<br> | *If time passes the same way for all observers (no time dilation),* then in the laboratory reference frame the muons should travel a distance equal to the product of their speed with respect to the lab reference frame times their half-life:<br>$d = (0.95c)(1.5 \times 10^{-6}\,\text{s})$<br>$= (0.95)(3.0 \times 10^{8}\,\text{m/s})(1.5 \times 10^{-6}\,\text{s}) = 430\,\text{m}$<br>*If time passes differently for observers in different reference frames and obeys the time dilation equation,* then in the laboratory reference frame one half of the muons should travel a distance equal to the product of their speed with respect to the lab reference frame times their half-life *in the lab reference frame*, which will be different from their half-life in the proper reference frame. Their half-life in the lab reference frame is<br>$\Delta t = \dfrac{1.5 \times 10^{-6}\,\text{s}}{\sqrt{1 - \dfrac{(0.95c)^2}{c^2}}} = 4.8 \times 10^{-6}\,\text{s}$<br>Because they move at speed $0.95c$ in the lab frame, one half of the muons travel<br>$d = (0.95c)(4.8 \times 10^{-6}\,\text{s})$<br>$= (0.95)(3.0 \times 10^{8}\,\text{m/s})(4.8 \times 10^{-6}\,\text{s}) = 1400\,\text{m}$<br>Thus if time dilation occurs, one half of the muons should travel a significantly greater distance through the lab than they would if there were no time dilation. | The muons actually traveled about 1400 m before their number was reduced to one-half. |

**Conclusion**

The outcome matches the prediction based on the time dilation equation. Many experiments confirm that short-lived particles travel much farther than would be possible given their proper half-lives. The idea of time dilation is not rejected. We have gained more confidence in it.

Muons are produced in nature as well. They are created when cosmic rays (high-energy protons) hit molecules in the atmosphere approximately 60,000 m above Earth's surface. If there were no time dilation, these muons would have a 1 in $10^{40}$ chance of reaching Earth from that altitude. However, because of time dilation and the fact that they are moving at nearly the speed of light, about 1 out of every 10 muons reaches Earth's surface.

‹ Active Learning Guide

**Time dilation** The reference frame in which two events occur at the same position is called the *proper reference frame* for these events. The time interval between these events measured by an observer in the proper reference frame is the *proper time interval,* $\Delta t_0$. The time interval $\Delta t$ between the same two events as measured by an observer moving at constant velocity $\vec{v}$ relative to the proper reference frame is

$$\Delta t = \frac{\Delta t_0}{\sqrt{1 - \frac{v^2}{c^2}}} \quad (25.1)$$

where $v$ is the speed of the observer relative to the proper reference frame.

The words "moving at constant velocity $\vec{v}$ relative to the proper reference frame" are especially important. Consider as an example a spaceship on which a light flashes at regular time intervals. The time interval between flashes as measured by a passenger on the ship is the proper time interval because the flashes occur at the same place relative to the passenger. For an observer on Earth, the flashes occur at different places as the spaceship moves past Earth. Therefore, Earth is not the proper reference frame. We say that Earth moves at speed $v$ relative to the proper reference frame of the spaceship.

## QUANTITATIVE EXERCISE 25.1 Heartbeat rate on spaceship

A spaceship moves past Earth at speed $2.6 \times 10^8$ m/s. The ship's captain carries a light that flashes each time his heart beats. According to the captain, a flash occurs every 1.0 s (see the first figure below). What time interval elapses between flashes according to an observer on Earth (see the second figure below)?

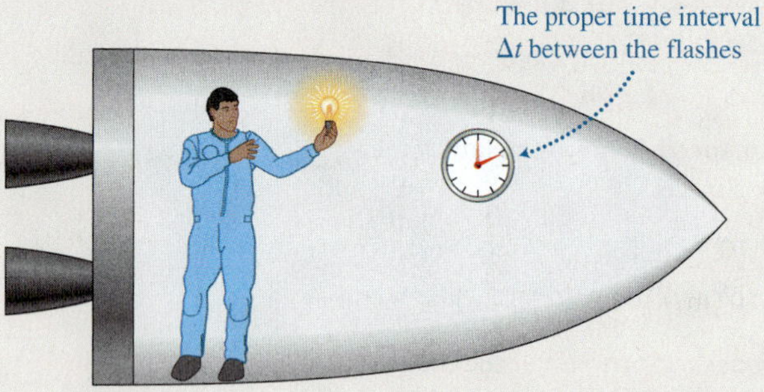


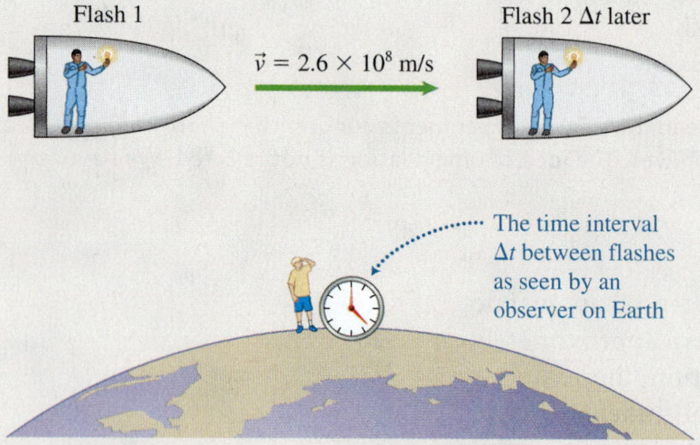


**Represent mathematically** These two events—the flash and the heartbeat— occur at the same place on the spaceship (the proper reference frame), and hence $\Delta t_0 = 1.0$ s relative to the spaceship observer. Earth moves at speed $2.6 \times 10^8$ m/s relative to the spaceship. The time interval $\Delta t$ between two flashes as measured by the Earth observer is:

$$\Delta t = \frac{\Delta t_0}{\sqrt{1 - \frac{v^2}{c^2}}}$$

**Solve and evaluate** Inserting the appropriate values:

$$\Delta t = \frac{1.0 \text{ s}}{\sqrt{1 - \left(\frac{2.6 \times 10^8 \text{ m/s}}{3.0 \times 10^8 \text{ m/s}}\right)^2}} = 2.0 \text{ s}$$

The time interval between beats of the pilot's heart is longer according to the Earth observer. This is similar to the muons in Table 25.2 whose half-life was longer in the lab reference frame.

**Try it yourself:** Suppose the captain watches the flashing light on a radio tower on Earth through a window of his ship. He observes that it flashes every 10.0 s. What is the time interval at which a person on Earth observes the tower light flashing?

*Answer:* The (proper) time between flashes as observed by an Earth observer is 5.0 s.

**Review Question 25.4** How does the muon experiment support the idea of time dilation?

## 25.5 Length contraction

The measured time interval between two events depends on the reference frame from which the measurement is made. In addition, other physical quantities that are normally thought to be independent of reference frame, such as the length of an object, in fact do depend on the frame of reference.

Consider an arrow flying across a lab that moves past a clock at rest with respect to the lab (**Figure 25.8**). As the arrow's head passes the clock (event 1) it triggers the clock to start (Figure 25.8a). As the arrow's feathers pass the clock (event 2) the clock stops (Figure 25.8b). The two events occurred at the same place in the lab reference frame, so that is the proper reference frame, and the time interval $\Delta t_0$ measured by the clock is the proper time interval between the two events. In contrast, if the arrow had a tiny clock attached to it, the time interval $\Delta t$ between the events would be

$$\Delta t = \frac{\Delta t_0}{\sqrt{1-\frac{v^2}{c^2}}}$$

where $v$ is the speed at which the lab moves past the arrow.

Could the length of the arrow depend on the reference frame from which it is measured? We'll call the arrow's **proper length,** $L_0$, its length in a reference frame in which the arrow is stationary. In this case, it's the reference frame defined by the arrow itself. Imagine that a fly has landed on the arrow and is traveling with it. Because the arrow is at rest relative to the fly, the fly would measure the length of the arrow to be the arrow's proper length $L_0$ (**Figure 25.9a**).

To an observer in the clock's (and lab's) reference frame, however, the length $L$ of the moving arrow may be different (Figure 25.9b). We can relate $L$ to $L_0$ by noting that the arrow's speed relative to the lab is the same as the lab's speed relative to the arrow (although their velocities are in opposite directions). We can calculate this speed in two different ways. In the arrow's reference frame its speed is the proper length $L_0$ divided by the time interval $\Delta t$:

$$v = \frac{L_0}{\Delta t}$$

In the lab reference frame the speed is the length $L$ divided by the proper time interval $\Delta t_0$:

$$v = \frac{L}{\Delta t_0}$$

Because these two speeds are the same, we can set them equal to each other:

$$\frac{L}{\Delta t_0} = \frac{L_0}{\Delta t}$$

Using the time dilation equation, we can substitute for $\Delta t$:

$$\frac{L}{\Delta t_0} = \frac{L_0\sqrt{1 - \frac{v^2}{c^2}}}{\Delta t_0}$$

Multiplying both sides of the above by $\Delta t_0$, we get a relationship between the proper length of the arrow, $L_0$, and its length $L$ when measured in another reference frame:

$$L = L_0\sqrt{1 - \frac{v^2}{c^2}}$$

**Figure 25.8** A moving arrow (a) starts and (b) stops a clock.

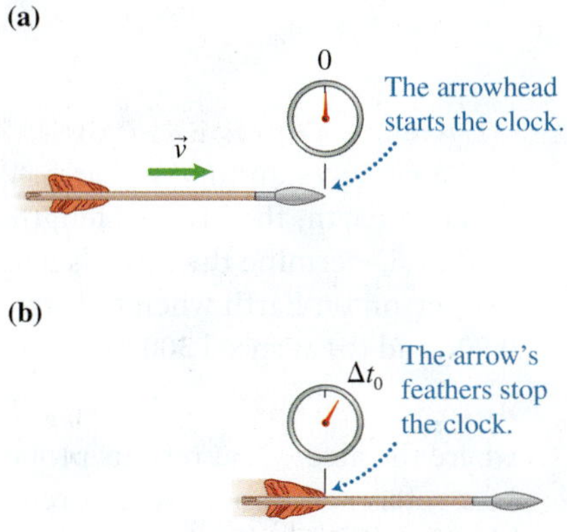
(a)
0
$\vec{v}$
The arrowhead starts the clock.
(b)
$\Delta t_0$
The arrow's feathers stop the clock.

**Figure 25.9** Measuring the arrow length (a) in the proper reference frame, where the arrow is at rest, and (b) in a reference frame, where the arrow is moving.

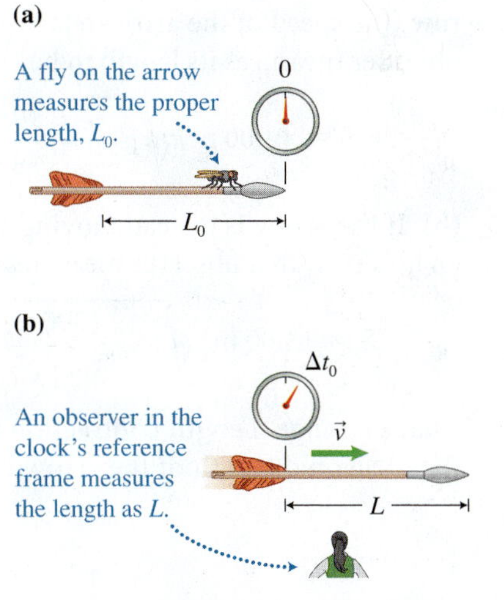
(a)
A fly on the arrow measures the proper length, $L_0$.
0
$L_0$
(b)
$\Delta t_0$
An observer in the clock's reference frame measures the length as $L$.
$\vec{v}$
$L$

The square root that appears in the above equation will always have a value less than 1. As a result, the length of the arrow measured in a reference frame where the arrow is moving will always be less than the arrow's proper length. In other words, the arrow will appear shorter along the direction that it is moving. This is known as **length contraction.**

**TIP** Notice that length contraction only occurs parallel to the direction the object is moving. In the perpendicular direction, this effect does not occur.

**Length contraction** **An object has a length $L_0$ when measured in a reference frame at rest relative to the object. This length is called the object's *proper length.* In a reference frame moving at speed $v$ relative to the proper reference frame, the object's length $L$ is**

$$L = L_0\sqrt{1 - \frac{v^2}{c^2}} \tag{25.2}$$

**QUANTITATIVE EXERCISE 25.2** An arrow flies past a person standing on Earth. When at rest with respect to Earth, the arrow's length was measured to be 1.00 m. Determine the arrow's length $L$ as measured by the person on Earth when the arrow moves (a) at speed $0.90c$ and (b) at speed 300 m/s.

**Represent mathematically** In a reference frame where the arrow is at rest, its proper length $L_0$ is 1.00 m. In a reference frame moving relative to the arrow, its length $L$ is related to its proper length $L_0$ by the length contraction equation (Eq. 25.2).

$$L = L_0\sqrt{1 - \frac{v^2}{c^2}}$$

**Solve and evaluate** (a) In a reference frame where the observer is moving at speed $0.90c$ relative to the arrow (the speed of the arrow relative to the observer), the observer measures its length to be

$$L = (1.00\text{ m})\sqrt{1 - \frac{(0.90c)^2}{c^2}} = 0.44\text{ m}$$

(b) If the arrow is instead moving at 300 m/s relative to an observer, that observer measures its length to be

$$L = (1.00\text{ m})\sqrt{1 - \frac{(300\text{ m/s})^2}{c^2}} = 1.00\text{ m}$$

That's strange. Length contraction is supposed to make the measured length of the arrow less than 1.00 m. The reason we get this answer is that most calculators do not have the precision for this calculation.

To get around this limitation we can use the binomial expansion to rewrite the length contraction equation in a form that is easier for calculation. When $x$ is very small, an application of the binomial expansion gives

$$\sqrt{1 - x} \approx 1 - \frac{1}{2}x$$

In our case, $x = v^2/c^2$, which is definitely very small in case (b). Using this approximation, then:

$$L = (1.00\text{ m})\left(1 - \frac{1}{2}\frac{(300\text{ m/s})^2}{c^2}\right)$$
$$= 1.00\text{ m} - \frac{1}{2}(10^{-12})\text{m}$$

The arrow is shortened by less than the size of an atom when seen moving at 300 m/s. This example again shows that relativistic effects are extremely tiny at the speeds of daily life.

**Try it yourself:** Suppose you live in a strange universe in which the speed of light is constant, but is only 20 m/s. Your car is 3.0 m long when parked by the curb in front of your house. How long is it measured to be when a friend drives by your car traveling at 10 m/s relative to your car?

*Answer:* 2.6 m.

**EXAMPLE 25.3 Mr. McMurphy observes a robbery**

Mr. McMurphy is on a train traveling through Wonderland where the speed of light is 10.00 m/s. The train moves at speed 9.95 m/s relative to the ground. While passing a small town, Mr. McMurphy observes a window of a fancy jewelry store breaking and simultaneously sees a man standing 2.0 m from the store window. Then, 1.0 s later, he sees a man standing in front of the broken window. Mr. McMurphy thinks this man is the robber, although he can't see him clearly. Mr. McMurphy pulls the emergency brake, stopping the train. When he gets out of the train, he goes to a police officer who is standing outside the jewelry store and tells the officer that he saw the man who robbed the store. The police officer says that it's not possible. Why would the officer say this? Remember, the speed of light in Wonderland is 10.00 m/s.

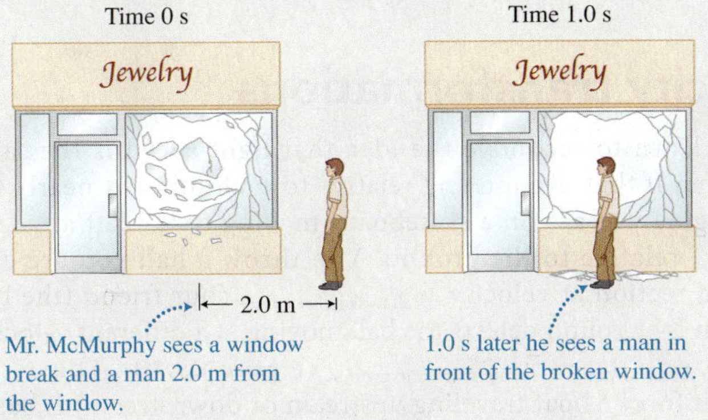

Time 0 s
Time 1.0 s
Jewelry
Jewelry
2.0 m
Mr. McMurphy sees a window break and a man 2.0 m from the window.
1.0 s later he sees a man in front of the broken window.

**Sketch and translate** There are three events to consider: (1) the store window breaking, (2) a man standing 2.0 m (measured in the train reference frame) from the store window when the window breaks, and (3) a man standing directly in front of the window 1.0 s later (again, measured in the train reference frame). Because of relativistic effects, the 2.0-m distance and the 1.0-s time interval are different when measured in the reference frame of the jewelry store. The jewelry store is the proper reference frame for events (1) and (3) because they occurred at the same place in that frame.

Mr. McMurphy is moving at speed $v = 9.95\ \text{m/s} = 0.995c$ with respect to the jewelry store. The distance Mr. McMurphy measured for the initial distance between the robber and the window, $L = 2.0$ m, is not a proper length; Mr. McMurphy is moving with respect to the store window and the robber. Also, the time interval Mr. McMurphy measured between the two events, $\Delta t = 1.0$ s, is not a proper time interval because Mr. McMurphy is not in the proper reference frame.

Let's use the equations for time dilation and length contraction to determine how this situation appears in the reference frame of the jewelry store. That should help us understand why the police officer knew that the man did not rob the jewelry store.

**Simplify and diagram** Assume that the train is moving at constant velocity relative to the jewelry store.

**Represent mathematically** The proper distance $L_0$ between the man and the window during events (1) and (2) is determine by rearranging Eq. (25.2):

$$L_0 = \frac{L}{\sqrt{1 - \dfrac{v^2}{c^2}}}$$

The proper time interval between the window breaking and the man standing in front of the window is determined by rearranging Eq. (25.1):

$$\Delta t_0 = \Delta t \sqrt{1 - \frac{v^2}{c^2}}$$

**Solve and evaluate** The proper length between the robber and the window when the window was broken was

$$L_0 = \frac{2.0\ \text{m}}{\sqrt{1 - \dfrac{(0.995c)^2}{c^2}}} = 20\ \text{m}$$

The proper time interval between the window breaking and the man standing in front of the window was

$$\Delta t_0 = (1.0\ \text{s})\sqrt{1 - \frac{(0.995c)^2}{c^2}} = 0.10\ \text{s}$$

The police officer confirms that she was standing across the street and saw the man 20 m from the window when it was broken, and therefore he could not possibly be the person in front of the window 0.10 s later. In order for that person to be the same person seen in front of the window, he would have had to move at the following speed:

$$\frac{20\ \text{m}}{0.10\ \text{s}} = 200\ \text{m/s}$$

This is not possible in Wonderland, where the speed of light is just 10 m/s. So what's going on?

The officer explains to Mr. McMurphy that the window was actually broken by a boy on a bicycle who accidentally threw a newspaper through the window while traveling in the *opposite* direction of the train. She explains further that due to the extreme length contraction of the cyclist, Mr. McMurphy probably didn't even notice him.

**Try it yourself:** Estimate the length of the bicycle of the paperboy as seen by Mr. McMurphy, assuming the bicycle is at rest with respect to the store.

*Answer:* Assume that the bicycle is 2 m long in its proper reference frame. Mr. McMurphy would measure the bicycle to be about 0.2 m or 20 cm long. Because the bicycle was moving in the opposite direction to the train, its length would be contracted even more than this estimate.

To determine the bicycle's speed in Mr. McMurphy' reference frame, we need to learn how to relativistically combine relative velocities. That is the topic of the next section.

**Review Question 25.5** You hold a 1-m-long stick. Describe the conditions under which an observer will measure the stick to be shorter than 1 m and those conditions under which an observer will measure the stick to be longer than 1 m.

## 25.6 Velocity transformations

In this section we learn to reconcile the idea that light speed is the same for all observers even if they are moving relative to each other at nearly the speed of light. Imagine you are on a skateboard moving to the left at velocity $\vec{v}_{\text{skateboarder to ground}}$ relative to the ground. You throw a ball you are carrying in the same direction at velocity $\vec{v}_{\text{ball to skateboarder}}$. Your friend (the ball catcher) standing on the ground detects the ball moving at a different velocity: $\vec{v}_{\text{ball to ground}} = \vec{v}_{\text{ball to skateboarder}} + \vec{v}_{\text{skateboarder to ground}}$ (**Figure 25.10**). This result (similar to the result for the boat traveling upstream or downstream) is based on the assumption that the time interval of the ball in flight is the same for the skateboarder and for the ball catcher on the ground (the same assumption that we made in the boat example in Section 25.1). We can generalize this equation as

$$\vec{v}_{\text{OS}} = \vec{v}_{\text{OS}'} + \vec{v}_{\text{S}'\text{S}} \qquad (25.3)$$

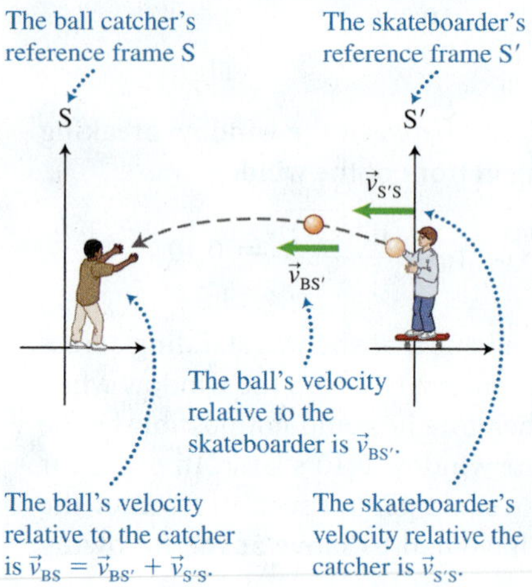
The ball catcher's reference frame S
The skateboarder's reference frame S′
S
S′
$\vec{v}_{S'S}$
$\vec{v}_{BS'}$
The ball's velocity relative to the skateboarder is $\vec{v}_{BS'}$.
The ball's velocity relative to the catcher is $\vec{v}_{BS} = \vec{v}_{BS'} + \vec{v}_{S'S}$.
The skateboarder's velocity relative the catcher is $\vec{v}_{S'S}$.

**Figure 25.10** The classical velocity transformation. Velocities are simply additive in everyday situations.

Here $\vec{v}_{\text{OS}}$ is the velocity of the object (the ball) with respect to reference frame S (the ball catcher ); $\vec{v}_{\text{OS}'}$ is the velocity of the object with respect to a second reference frame S′ (the skateboarder); and $\vec{v}_{\text{S}'\text{S}}$ is the velocity of the reference frame S′ relative to the reference frame S. Equation (25.3) is called the **classical** (or **Galilean**) **velocity transformation equation**. It is consistent with our everyday experience. The derivation of the equation is based on the assumption that the time interval during which the motion occurs is the same for all observers.

> **TIP** Because Eq. (25.3) is a vector equation, we must break it into components before using it. We have to choose a coordinate system that determines the signs of the individual vector components.

### Difficulties with the classical velocity transformation equation

**Active Learning Guide ›**

According to the ideas of special relativity and the results of the Michelson-Morley experiment, the speed of light is the same in all inertial reference frames. This idea contradicts the classical velocity transformation equation. Suppose you are in reference frame S′ moving toward the right at speed $v_{\text{S}'\text{S}} = 0.99c$ relative to another reference frame S. You shine a laser beam straight ahead. The light moves at speed $v_{\text{LS}'} = c$ forward in your S′ reference frame. According to the classical velocity transformation equation, the speed $v_{\text{LS}}$ of the light in reference frame S should be

$$v_{\text{LS}} = v_{\text{LS}'} + v_{\text{S}'\text{S}} = c + 0.99c = 1.99c$$

This result contradicts Einstein's second postulate. We need an improved velocity transformation equation that respects the postulates of relativity.

### Relativistic velocity transformation equation

Deriving this equation requires techniques beyond the scope of this book, so we'll just provide the result for one-dimensional processes.

**Relativistic velocity transformation** **Suppose that inertial reference frame S′ is moving relative to inertial reference frame S at velocity $\vec{v}_{S'S}$ (positive to the right and negative to the left). An object O moves at velocity $\vec{v}_{OS'}$ relative to reference frame S′ (also positive to the right and negative to the left). The velocity $\vec{v}_{OS}$ of O in reference frame S is then**

$$v_{OS} = \frac{v_{OS'} + v_{S'S}}{1 + \dfrac{v_{OS'}v_{S'S}}{c^2}} \qquad (25.4)$$

Notice that in Eq. (25.4) the numerator is the same as in the classical equation, but the denominator is different. Does the above equation satisfy postulate 2 of special relativity and agree with the classical equation for low speeds? Let's check some limiting cases.

## Limiting case analysis of Eq. (25.4)

*Case 1:* Suppose that an object O moves slowly in reference frame S′ ($v_{OS'} \ll c$), or that reference frame S′ moves slowly relative to reference frame S ($v_{S'S} \ll c$). In either case the denominator of Eq. (25.4) is approximately 1 and the equation becomes the classical velocity transformation Eq. (25.3).

*Case 2:* Suppose that a rocket (reference frame S′) passing Earth emits a laser pulse in the forward direction. The light travels at velocity $v_{LS'} = c$ relative to the rocket. Suppose the rocket (reference frame S′) travels at velocity $v_{S'S} = 0.99c$ relative to the Earth (reference frame S). What is the velocity $v_{LS}$ of the laser pulse relative to Earth?

$$v_{LS} = \frac{v_{LS'} + v_{S'S}}{1 + \dfrac{v_{LS'}v_{S'S}}{c^2}} = \frac{c + 0.99c}{1 + \dfrac{(c)(0.99c)}{c^2}} = \frac{(1 + 0.99)c}{(1 + 0.99)} = c$$

The laser pulse is observed to travel at speed $c$ in both inertial reference frames, consistent with the postulates of special relativity.

**Review Question 25.6** What is the meaning of the term "velocity transformation" and why is the classical velocity transformation different from the relativistic transformation?

## 25.7 Relativistic momentum

We know (from Chapter 5) that if the net impulse exerted on a system is zero, the momentum of the system is constant. This principle is very useful for analyzing collisions of various kinds. Our past use of the principle involved objects moving at nonrelativistic speeds ($v \ll c$). Can we use the classical expression for the momentum of an object

$$\vec{p} = m\vec{v} = m\frac{\Delta\vec{x}}{\Delta t}$$

and the impulse-momentum principle to analyze situations where objects are moving at relativistic velocities?

‹ Active Learning Guide

When we use the above classical definition of momentum to analyze collisions of elementary particles moving at high speed, we find that even for an isolated system, the momentum of the system is constant in some reference frames but not in others. Thus, we either have to give up momentum as a conserved quantity or find a new expression that restores consistency across all reference frames. We'll take the second approach and find a new relativistic expression for momentum.

To redefine momentum, consider the classical expression ($\vec{p} = m\vec{v} = m(\Delta\vec{x}/\Delta t)$). In that expression $m$ is the mass of the object and $\Delta\vec{x}$ is the displacement of the object during the time interval $\Delta t$, both measured in some inertial reference frame. To get an improved relativistic expression for momentum, we try replacing $\Delta t$ with the proper time interval $\Delta t_0$, where the relevant events are the object at the beginning and end of its displacement:

$$\vec{p} = m\vec{v} = m\frac{\Delta\vec{x}}{\Delta t_0}$$

If we substitute the expression for $\Delta t_0$ shown in Eq. (25.1) into the above definition, we get

$$\vec{p} = m\frac{\Delta\vec{x}}{\Delta t\sqrt{1-(v/c)^2}} = \frac{m}{\sqrt{1-(v/c)^2}}\frac{\Delta\vec{x}}{\Delta t} = \frac{m\vec{v}}{\sqrt{1-(v/c)^2}}$$

Note that the $\vec{v}$ in the numerator is the velocity of the object, while the $v$ in the denominator is its speed. This new expression reduces to the classical expression if the speed of the object is much less than the speed of light. Using this new expression for momentum it is possible to show (though we will not do it here) that when the net impulse exerted on a system is zero, the system's momentum is constant regardless of inertial reference frame.

**Relativistic momentum** **The relativistic momentum of an object of mass $m$ in a reference frame where the object is moving at velocity $\vec{v}$ is**

$$\vec{p} = \frac{m\vec{v}}{\sqrt{1-(v/c)^2}} \tag{25.5}$$

In Eq. (25.5) the denominator gets smaller and smaller as the object's speed gets closer and closer to the speed of light. This means that as speed increases toward light speed, the momentum of the object approaches infinity. This result is consistent with the idea that no object can travel faster than light speed in a vacuum. No such restriction was present in the nonrelativistic expression for the momentum of an object ($\vec{p} = m\vec{v}$).

**TIP** Equation (25.5) is an improved expression for the momentum of objects traveling at all speeds rather than just speeds small compared with light speed. However, if the speed of the object is not a significant fraction of light speed, the simpler nonrelativistic expression can be used.

Let's try to use the above ideas about relativistic momentum to analyze an everyday occurrence—although one we can't see without special equipment.

**EXAMPLE 25.4 Cosmic ray hits nitrogen**

Some cosmic rays are high-energy protons produced during the explosive collapse of stars near the ends of their lives. Suppose a cosmic ray proton traveling at $0.90c$ enters our atmosphere and collides with a nitrogen atom. After the collision, the proton recoils at speed $0.70c$ in the opposite direction. The mass of the nitrogen atom is 14 times greater than the proton's mass. How fast is the nitrogen atom moving after the collision?

**Sketch and translate** We first sketch the process (see below). The system of interest is the proton and the nitrogen atom. Chose an inertial reference frame that is at rest relative to the nitrogen atom before it is hit by the proton and choose the positive $x$-axis to be the direction the proton is initially traveling. In that case, the initial and final velocity components of the proton and nitrogen atom along that axis are $v_{\mathrm{p}ix} = +0.90c$, $v_{\mathrm{p}fx} = -0.70c$, and $v_{\mathrm{N}ix} = 0$, and $v_{\mathrm{N}fx}$ is the unknown final velocity component of the nitrogen atom. Let $m_{\mathrm{p}} = m$; then $m_{\mathrm{N}} = 14m$.

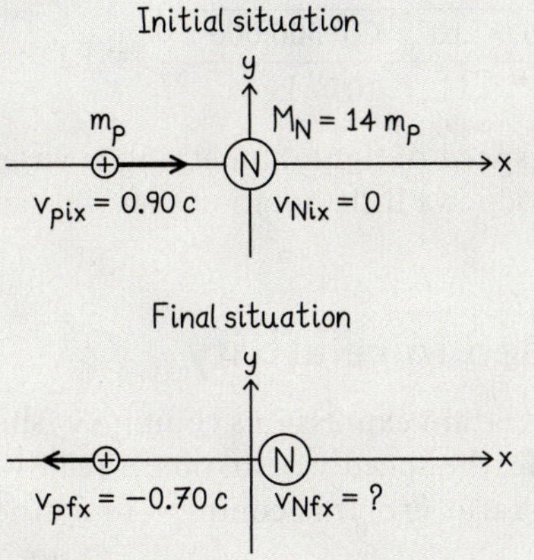


**Simplify and diagram** Assume that there are no interactions between the system and the environment so that the momentum of the system is constant. Because of the high speeds of the proton before and after the collision, we need to use the relativistic expression for the proton momentum. But we'll have to wait and see whether we need to treat the nitrogen atom relativistically; it depends on its speed relative to $c$. A momentum bar chart represents the process.

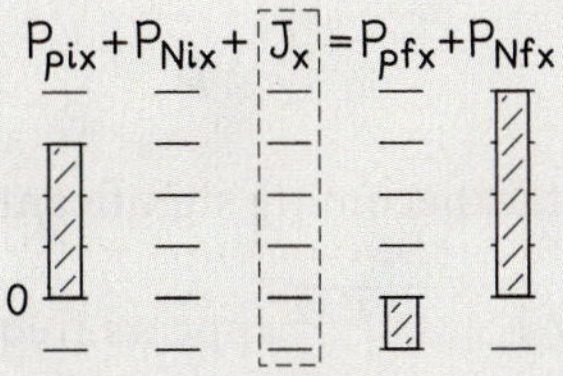


**Represent mathematically** The bar chart helps us apply the conservation of momentum principle along the $x$-axis:

$$p_{\mathrm{p}ix} + p_{\mathrm{N}ix} + J_x = p_{\mathrm{p}fx} + p_{\mathrm{N}fx}$$

or

$$\frac{m_{\mathrm{p}} v_{\mathrm{p}ix}}{\sqrt{1 - (v_{\mathrm{p}i}/c)^2}} + 0 + 0 = \frac{m_{\mathrm{p}} v_{\mathrm{p}fx}}{\sqrt{1 - (v_{\mathrm{p}f}/c)^2}} + \frac{m_{\mathrm{N}} v_{\mathrm{N}fx}}{\sqrt{1 - (v_{\mathrm{N}f}/c)^2}}$$

**Solve and evaluate** Substitute the known information and solve for $v_{\mathrm{N}f}$:

$$\frac{m(0.90c)}{\sqrt{1 - (0.90c/c)^2}} = \frac{m(-0.70c)}{\sqrt{1 - (0.70c/c)^2}} + \frac{14 m v_{\mathrm{N}f}}{\sqrt{1 - (v_{\mathrm{N}f}/c)^2}}$$

Cancel factors of $m$ and $c$ and simplify to get

$$\frac{0.90}{\sqrt{1 - (0.90)^2}} c = \frac{-0.70}{\sqrt{1 - (0.70)^2}} c + \frac{14 v_{\mathrm{N}f}}{\sqrt{1 - (v_{\mathrm{N}f}/c)^2}}$$

$$\Rightarrow \frac{14 v_{\mathrm{N}f}}{\sqrt{1 - (v_{\mathrm{N}f}/c)^2}} = \left( \frac{0.90}{\sqrt{1 - (0.90)^2}} + \frac{0.70}{\sqrt{1 - (0.70)^2}} \right) c$$

$$= (2.06 + 0.98)c$$

Finally, we get

$$\frac{v_{\mathrm{N}f}}{\sqrt{1 - (v_{\mathrm{N}f}/c)^2}} = 0.22c$$

To solve for $v_{\mathrm{N}f}$, multiply both sides by $\sqrt{1 - (v_{\mathrm{N}f}/c)^2}$ and square both sides:

$$v_{\mathrm{N}f}^2 = (0.22c)^2 (1 - (v_{\mathrm{N}f}/c)^2)$$

Now collect all terms with $v_{\mathrm{N}f}$ on the left side:

$$v_{\mathrm{N}f}^2 + (0.22 v_{\mathrm{N}f})^2 = (0.22c)^2$$

Now factor out $v_{\mathrm{N}f}^2$ and solve for it:

$$v_{\mathrm{N}f}^2 = \frac{(0.22c)^2}{1 + 0.22^2}$$

$v_{\mathrm{N}f}$ is then

$$v_{\mathrm{N}f} = \frac{0.22}{\sqrt{1 + 0.22^2}} c = 0.21c = 6.4 \times 10^7 \text{ m/s}$$

This is a significant fraction of light speed, but not at all unreasonable for a cosmic ray collision.

**Try it yourself:** Calculate the speed of the nitrogen atom after the collision using the nonrelativistic expression for the momentum of the proton and nitrogen. How does it compare to the relativistic value?

*Answer:* $3.4 \times 10^7$ m/s, compared to the relativistic value of $6.4 \times 10^7$ m/s. Relativistic effects are clearly relevant in this situation.

**Review Question 25.7** Why must the classical expression of momentum be modified in special relativity?

## 25.8 Relativistic energy

Our ideas about space, time, velocity, and now momentum have been modified to include relativistic effects. A simple thought experiment shows that ideas concerning energy must also be modified. Imagine that we accelerate an electron through a potential difference of 300,000 V (large, but still achievable). The kinetic energy of this electron will then be $(1/2)mv^2 = e\Delta V$. This means the speed of the electron once it has finished accelerating will be

$$v = \sqrt{\frac{2e\Delta V}{m}} = \sqrt{\frac{2(1.6 \times 10^{-19}\text{ C})(300{,}000\text{ V})}{9.11 \times 10^{-31}\text{ kg}}} = 3.2 \times 10^8\text{ m/s}$$

**Active Learning Guide ›**

This speed is faster than the speed of light. Something is wrong. Before we begin resolving this, let's introduce a little notation that will save time in our future analysis.

### Special notations used in relativity

You've probably noticed that certain expressions commonly show up in relativistic equations. For example, the speed $v$ of an object relative to the speed of light $c$ often occurs as the ratio $v/c$. It is common to define this ratio as beta $\beta$:

$$\beta = \frac{v}{c} \tag{25.6}$$

Relativistic effects start becoming significant when $\beta > 0.1$, that is, at speeds $v > 0.1c$.

The quantity $\sqrt{1 - (v/c)^2}$ appears frequently in relativistic expressions as well, most often in the denominator. To make relativistic equations more compact and easier to work with, physicists define the symbol gamma $\gamma$ as

$$\gamma = \frac{1}{\sqrt{1 - (v/c)^2}} = \frac{1}{\sqrt{1 - \beta^2}} \tag{25.7}$$

Using these abbreviations, the time dilation, length contraction, and momentum equations become $\Delta t = \gamma \Delta t_0$, $L = L_0/\gamma$, and $p = \gamma mv$, respectively.

### Relativistic kinetic energy

Deriving the relativistically correct expression for kinetic energy from first principles is rather complicated. Instead, we will provide the accepted expression and evaluate its low-speed limiting case to check for consistency with $(1/2)mv^2$.

The system of interest is a point-like object of mass $m$ moving at constant speed $v$ and not interacting with any other objects. The system therefore has no potential energy and no internal energy but does have kinetic energy. The correct relativistic expression for the energy of the system is

$$\gamma mc^2 = \frac{mc^2}{\sqrt{1 - \beta^2}} = \frac{mc^2}{\sqrt{1 - (v/c)^2}}$$

Note that $mc^2$ (mass times a velocity squared) has energy units. Gamma $\gamma = 1/\sqrt{1 - (v/c)^2}$ and beta $\beta = v/c$ are dimensionless numbers.

To see if this expression is consistent with $K = (1/2)mv^2$ at low speeds ($\beta \ll 1$), we use the binomial expansion to rewrite the part of the expression that involves the square root. For situations where the quantity $x$ is small,

$$\frac{1}{\sqrt{1 - x}} = (1 - x)^{-1/2} \approx 1 + \frac{1}{2}x$$

In our case $x = \beta^2$. Therefore, for small $\beta = v/c$, the expression becomes

$$\gamma mc^2 = \frac{mc^2}{\sqrt{1-\beta^2}} = mc^2(1-\beta^2)^{-1/2} \approx mc^2\left(1 + \frac{1}{2}\beta^2\right)$$
$$= mc^2\left(1 + \frac{v^2}{2c^2}\right) = mc^2 + \frac{1}{2}mv^2$$

Notice that the last term on the right is just the classical kinetic energy of an object of mass $m$ moving at speed $v$. Based on the above, it appears that

$$K = \gamma mc^2 - mc^2$$

This is the relativistically correct expression for kinetic energy $K$ that a precise derivation produces.

What is the meaning of the $mc^2$ term? Apparently, it is a kind of energy that depends only on the mass of the object. This is a new and very profound idea—an object has energy just because of its mass. The term $mc^2$ is called the **rest energy** of the object. The expression $\gamma mc^2$ represents the total energy (rest plus kinetic) of a point-like object. The kinetic energy of the object equals its total energy minus its rest energy. These three ideas are summarized below.

**Relativistic energy** A point-like object of mass $m$ has so-called rest energy because it has mass:

$$\text{Rest energy } E_0 = mc^2 \qquad (25.8)$$

The total energy of the object is

$$E = \gamma mc^2 = \frac{mc^2}{\sqrt{1-(v/c)^2}} \qquad (25.9)$$

The object's kinetic energy is the object's total energy minus its rest energy:

$$\text{Kinetic energy } K = \gamma mc^2 - mc^2 = (\gamma - 1)mc^2 \qquad (25.10)$$

You can see from Eq. (25.9) that when the speed of an object approaches the speed of light in a vacuum, its kinetic energy approaches infinity. Thus the relativistic expression for kinetic energy explains why no object can move faster than the speed of light in a vacuum—the work that needs to be done to accelerate the object to this speed is infinite.

Relativistic energy effects are most commonly relevant in the context of atomic and nuclear processes. The joule is an inconvenient energy unit to use when analyzing these processes because the energies involved are so small. Instead, we use another energy unit—the electron volt. One electron volt is the kinetic energy of an electron that accelerates from rest across a potential difference of 1 V (from lower potential at position 1 to higher potential at position 2 in **Figure 25.11**. Using the work-energy principle:

$$K_1 + U_{q1} = K_2 + U_{q2}$$
$$\Rightarrow K_2 = U_{q1} - U_{q2} = q(V_1 - V_2) = (-1.6 \times 10^{-19}\,\text{C})(0.0\,\text{V} - 1.0\,\text{V})$$
$$= 1.6 \times 10^{-19}\,\text{J}$$

**Figure 25.11** One electron volt (1 eV) is the kinetic energy gained by an electron that accelerates across a 1-V potential difference.

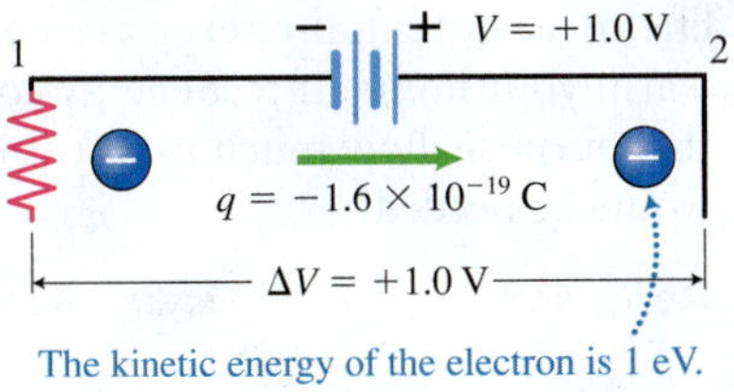


**Electron volt** An electron volt (1 eV) is the increase in kinetic energy of an electron when it moves across a 1.0-V potential difference:

$$1\,\text{eV} = 1.6 \times 10^{-19}\,\text{J}$$

In nuclear and elementary particle physics, energies much higher than this are common (but still small compared with everyday-life energies). In those cases the mega-electron volt ($1\,\text{MeV} = 1.6 \times 10^{-13}\,\text{J}$) and giga-electron volt ($1\,\text{GeV} = 1.6 \times 10^{-10}\,\text{J}$) are commonly used.

## Rest energy of particles

Any object with mass has rest energy, from elementary particles such as electrons to massive objects such as the Sun. Rest energy can be converted into other forms of energy. In fact, the rest energy of the Sun is being slowly converted via nuclear fusion reactions into internal energy, which accounts for the Sun's high temperature and brightness.

**QUANTITATIVE EXERCISE 25.5 Rest energy of an electron and of a proton**

The mass of an electron is $9.11 \times 10^{-31}$ kg. The mass of a proton is $1.67 \times 10^{-27}$ kg. Determine the electron and proton rest energies in joules and in electron volts.

**Represent mathematically** The rest energy of an object is given by Eq. (25.8):

$$E_0 = mc^2$$

**Solve and evaluate** For the electron:

$$E_{0e} = m_e c^2 = (9.11 \times 10^{-31}\,\text{kg})(3.00 \times 10^8\,\text{m/s})^2 = 8.2 \times 10^{-14}\,\text{J}$$

$$= (8.2 \times 10^{-14}\,\text{J})\left(\frac{1\,\text{eV}}{1.6 \times 10^{-19}\,\text{J}}\right)$$

$$= 510{,}000\,eV = 0.51\,\text{MeV}$$

For the proton:

$$E_{0p} = m_p c^2 = (1.67 \times 10^{-27}\,\text{kg})(3.00 \times 10^8\,\text{m/s})^2 = 1.5 \times 10^{-10}\,\text{J}$$

$$= (1.5 \times 10^{-10}\,\text{J})\left(\frac{1\,\text{eV}}{1.6 \times 10^{-19}\,\text{J}}\right)$$

$$= 939{,}000{,}000\,\text{eV} = 939\,\text{MeV}$$

This equation is also a commonly used method to express the mass of elementary particles as the rest energy equivalent of their mass.

It is difficult to evaluate whether these numbers for a single electron and single proton make sense. As you can see, the amount of energy in joules is very small compared to everyday-life energies. For example, a mole of molecular hydrogen (mass 2 g) contains $6.02 \times 10^{23}$ molecules of hydrogen, each consisting of two protons and two electrons. The total rest energy is therefore $1.8 \times 10^{14}$ J! Everyday physical processes can convert only extremely small amounts of this energy into other forms.

**Try it yourself:** Determine the rest energy of one of your 1.5-mg hairs.

*Answer:* $1.4 \times 10^{11}$ J!

**QUANTITATIVE EXERCISE 25.6 Mass equivalent of energy to warm and cool house**

On average, each year about $2 \times 10^{10}$ J of electric and chemical potential energy are converted to cool and warm your home. If rest energy could be converted for this purpose, how much mass equivalent of rest energy would be needed?

**Represent mathematically** According to Eq. (25.8):

$$m = \frac{E_0}{c^2}$$

**Solve and evaluate** Inserting the appropriate values:

$$m = \frac{2 \times 10^{10}\,\text{J}}{(3.00 \times 10^8\,\text{m/s})^2} = 2 \times 10^{-7}\,\text{kg}$$

This is approximately one-tenth the mass of one of the hairs on your head.

**Try it yourself:** Estimate the mass equivalent of the chemical potential energy that is transformed into other energy types when you drive your car for 1 year. Note that 3.8 L (1 gallon) of gasoline produces about $1.2 \times 10^8$ J.

*Answer:* Assume that you drive 13,000 km (8000 mi) in one year and get 15 km/L (35 mi/gallon). You would use 870 L (230 gallons) of gasoline or about $3 \times 10^{10}$ J; about 0.3 μg of rest energy (far less than the mass of the gasoline itself).

## Relationship between mass and energy

Earlier (in Chapter 5) we learned that mass is a conserved quantity. Now we understand that that statement is not exactly correct. Rest energy (mass) can be converted into other energy forms, so only the *total* energy of the system, $E_{\text{total}} = \Sigma E_{\text{particles}} = \Sigma \gamma mc^2$, is a conserved quantity. However, when objects are not moving at relativistic speeds, mass and energy are conserved separately almost exactly.

## Relativistic energy analysis of particle accelerator

A particle accelerator causes the speed of a particle such as an electron or proton to increase by accelerating it through a large potential difference (**Figure 25.12a**) or through a large number of smaller potential differences. In the case of an electron, if the electron is initially at rest at a point where the $V$ field is zero ($v_i = 0$; $V_i = 0$), then the initial kinetic and electric potential energies of the system are both zero. After the electron reaches a point with a large positive $V$ field value $V_f$, the system's electric potential energy has decreased and the kinetic energy has increased, but the system's total energy is still zero. This process is represented in the energy bar chart in Figure 25.12b. Assuming the electron does not reach relativistic speeds, the bar chart converts into the following equation:

$$0 = K_f + U_{qf} = \frac{1}{2}m_e v_f^2 + (-eV_f)$$

Solving for the particle's speed, we get

$$v_f = \sqrt{\frac{2eV_f}{m_e}}$$

If an electron accelerates through a potential difference of $10^4$ V or less, the preceding equation provides a fairly accurate expression for the electron's final speed. For larger potential differences, the equation makes predictions that are inconsistent with experimental results. We can resolve this discrepancy by using the relativistic kinetic energy expression.

**Figure 25.12** An electron crossing a large potential difference. (a) From an initial position on the left, the electron moves to a final position on the right, with a corresponding change in voltage ($V$). (b) A bar chart helps analyze the energy change. Notice that $0 = K_f + U_{qf}$.

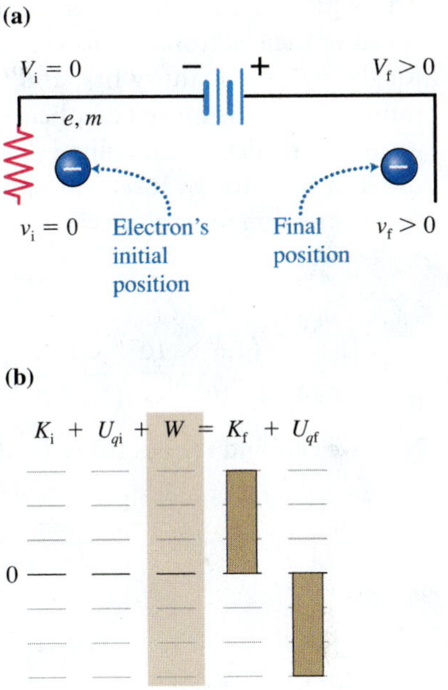


**EXAMPLE 25.7 Electron particle accelerator**

An electron in a particle accelerator accelerates through a potential difference of $10^6$ V. What is its final speed?

**Sketch and translate** The sketch in Figure 25.12a helps to visualize the motion of the electron. We know the electric charge and the mass of the electron and need to find its final speed.

**Simplify and diagram** Assume that the electron starts at rest. The system is the electron and the electric field. The initial energy of the system is the system's rest energy; the final energy is the sum of the electron's positive kinetic energy, the system's negative electric potential energy, and the system's rest energy. Because no work is done on the system by the environment, the total energy of the system is constant. The energy bar chart below represents the process (note the similarities and differences between this bar chart and the bar chart in Figure 25.12b).

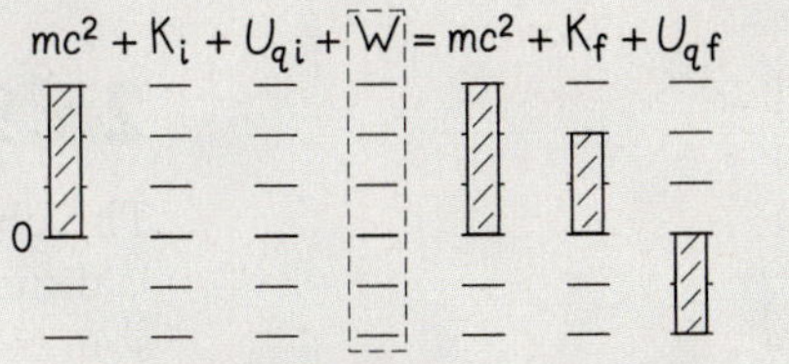

$mc^2 + K_i + U_{qi} + W = mc^2 + K_f + U_{qf}$
0


**Represent mathematically** We can now convert the energy bar chart into an equation:

$$E_{0i} + 0 + 0 + 0 = E_{0f} + K_f + U_{qf}$$

Because the rest energy of the system doesn't change during this process, we can subtract it from both sides

*(continued)*

of the equation. Inserting the appropriate expressions for each type of energy (including Eq. (25.10) for relativistic kinetic energy), we get

$$0 = (\gamma m_e c^2 - m_e c^2) + (-eV_f)$$

where $m_e$ is the mass of the electron. We can simplify this equation by dividing both sides by $m_e c^2$ and solving for $\gamma$:

$$\gamma = \frac{eV_f}{m_e c^2} + 1$$

Remember:

$$\gamma = \frac{1}{\sqrt{1 - (v_f/c)^2}}$$

**Solve and evaluate** Now we need to solve for the final speed of the electron. We usually derive the expression for the needed quantity first and then substitute in the numbers. In this more complicated case, however, it is easier to first determine a numerical value for $\gamma$ and then use it to solve for $v_f$. Insert the appropriate values in the first expression for $\gamma$ above:

$$\gamma = \frac{eV_f}{m_e c^2} + 1$$
$$= \frac{(1.6 \times 10^{-19}\,\mathrm{C})(10^6\,\mathrm{V})}{(9.11 \times 10^{-31}\,\mathrm{kg})(3.00 \times 10^8\,\mathrm{m/s})^2} + 1 = 3.0$$

Now we can find the electron's final speed:

$$\frac{1}{\sqrt{1 - (v_f/c)^2}} = 3.0$$

Invert both sides of the equation:

$$\sqrt{1 - (v_f/c)^2} = \frac{1}{3.0}$$

Solve for $v_f/c$:

$$1 - (v_f/c)^2 = \frac{1}{9.0}$$

$$\Rightarrow (v_f/c)^2 = 1 - \frac{1}{9.0}$$

$$\Rightarrow v_f/c = \sqrt{1 - \frac{1}{9.0}} = 0.94$$

or

$$v_f = 0.94c$$

Our result is reasonable—although the speed is very high, it is less than the speed of light.

**Try it yourself:** Protons are accelerated from rest to a speed of $0.993c$ in the "small" 150-m-diameter booster accelerator that is part of the Fermi National Accelerator Laboratory in Batavia, Illinois. (a) Determine the proton's total energy when it leaves the booster accelerator. (b) Determine its kinetic energy as it leaves the accelerator.

*Answer:* (a) $E_f = 1.27 \times 10^{-9}\,\mathrm{J}$; (b) $K_f = 1.12 \times 10^{-9}\,\mathrm{J}$.

**Review Question 25.8** If we did not use the relativistic expression for kinetic energy to calculate the speeds of elementary particles in accelerators, would the calculated nonrelativistic speeds be greater or less than the calculated relativistic speeds? [Hint: Estimate the values of any physical quantities that you need.]

## 25.9 Doppler effect for EM waves

The observed frequency of a mechanical wave (for example, a sound wave) is different from the frequency of the wave source if the observer and/or the source are moving relative to each other. A similar phenomenon exists for light (EM) waves; however, due to relativistic effects, the Doppler effect for light is somewhat different.

In the Doppler effect for sound [Eq. (20.18)], if the source or observer moves through the medium equal to or faster than the speed of sound, or if their motion relative to each other is equal to or faster than this speed, then the sound waves might not reach the observer at all. But the speed of EM waves, $c = 3.00 \times 10^8\,\mathrm{m/s}$, is the same for all observers no matter how they are moving relative to the source. As a result, the equation that describes the Doppler effect for EM waves differs from the equation for the Doppler effect for sound.

We won't derive the equation for the Doppler effect for EM waves. Instead, we just provide the equation and use it to analyze several processes.

**Doppler effect for EM waves**

$$f_O = f_S\sqrt{\frac{1 + v_{rel}/c}{1 - v_{rel}/c}} \qquad (25.11)$$

**where $f_O$ is the EM wave frequency detected by an observer, $f_S$ is the frequency emitted by the source, and $v_{rel}$ is the component of the relative velocity between the observer and the source along the line connecting them (positive when the observer and source are approaching each other and negative when moving apart).**

The major difference between this and the Doppler effect equation for sound is the square root. The need for this alteration comes about due to time dilation: the source and observer are in different inertial reference frames. The square root can be simplified for the Doppler effect for EM waves if $v_{rel}$ is small compared to the speed of light. Using the binomial expansion, the equation becomes the following.

**Doppler effect for EM waves (slow observer-source relative motion):**

$$f_O = f_S\left(1 + \frac{v_{rel}}{c}\right) \qquad (25.12)$$

**The sign conventions for Eq. (25.11) apply here.**

When the observer and source are approaching, the observed frequency is higher than the source frequency. When they are moving apart, the observed frequency is lower.

Suppose, for example, that the source emits light in the blue part of the spectrum and that the source and observer are moving apart. The light detected by the observer could be in the yellow or red part of the spectrum. In astronomy this is known as a **red shift,** and it was important in discovering that the universe is expanding. Conversely, if the source and observer are moving toward each other, the light detected by the observer would be shifted toward the blue end of the spectrum.

Modern police radar and sports radar measure speeds by knowledge of the source frequency and the ability to measure the frequency of a reflected wave. Let's look at such practical examples of the Doppler effect for EM waves.

## Doppler radar and speeding tickets

Police radar uses microwaves, one type of electromagnetic wave. Consider the next example.

**EXAMPLE 25.8 The speeding ticket**

Physicist Dr. R. Wood ran a red light while driving his car and was pulled over by a policeman. Dr. Wood explained that because he was driving toward the red light, he actually observed it as green due to the Doppler effect. Dr. Wood was then given a very expensive speeding ticket. Should Dr. Wood go to court to argue the speeding ticket?

**Sketch and translate** The traffic light is the source and Dr. Wood is the observer. We are interested in determining the speed of Dr. Wood's car relative to the traffic light. Dr. Wood was traveling fast enough so that the red light emitted by the traffic signal was observed by him to be green. Because the observer is moving toward the source, the relative velocity $v_{rel}$ will be positive and equal to Dr. Wood's car's speed relative to the traffic light.

*(continued)*

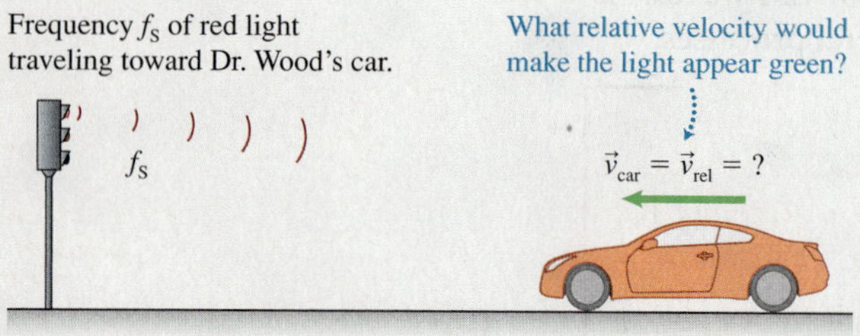
Frequency $f_S$ of red light traveling toward Dr. Wood's car.
$f_S$
What relative velocity would make the light appear green?
$\vec{v}_{car} = \vec{v}_{rel} = ?$

**Simplify and diagram** Assume that the traffic light emits red monochromatic light at a frequency of about $f_S = 4.5 \times 10^{14}$ Hz and that the frequency of green light is about $f_O = 5.4 \times 10^{14}$ Hz. Assume also that the car moves with constant velocity.

**Represent mathematically** Because we are talking about a low-speed car, we can use the low-speed Doppler effect $f_O = f_S(1 + (v_{rel}/c))$ to determine the speed of the car. Multiply both sides of the equation by $c$ and then solve for $v_{rel}$:

$$cf_O = cf_S\left(1 + \frac{v_{rel}}{c}\right) = f_S(c + v_{rel}) = f_S c + f_S v_{rel}$$

$$\Rightarrow f_S v_{rel} = cf_O - f_S c$$

$$\Rightarrow v_{rel} = \frac{c(f_O - f_S)}{f_S} = c\left(\frac{f_O}{f_S} - 1\right)$$

**Solve and evaluate** Inserting the appropriate values:

$$v_{rel} = c\left(\frac{f_O}{f_S} - 1\right) = (3 \times 10^8\ \text{m/s})\left(\frac{5.4 \times 10^{14}\ \text{Hz}}{4.5 \times 10^{14}\ \text{Hz}} - 1\right)$$

$$= 6.0 \times 10^7\ \text{m/s}$$

This speed is approaching the speed of light. Dr. Wood was clearly just trying to get out of a ticket. His explanation that the red light appeared green to him isn't reasonable.

**Try it yourself:** Can you detect the change in frequency of a red light if you are driving at a speed of 34 m/s (76 mi/h)?

*Answer:* The change is $5 \times 10^7$ Hz, which is

$$\left(\frac{5 \times 10^7\ \text{Hz}}{4.5 \times 10^{14}\ \text{Hz}}\right) \times 100\% = 1 \times 10^{-5}\%$$

which is not detectable by the human eye (although detectors do exist that can register this small change).

**Figure 25.13** How a radar detector determines ball speed. (a) The radar source emits microwaves (EM waves) with a frequency $f_S$; the moving ball "detects" the frequency as $f_D$ and reflects the waves back. (b) The radar detector (observer) detects the reflected waves at frequency $f_O$ and measures the difference from $f_S$.

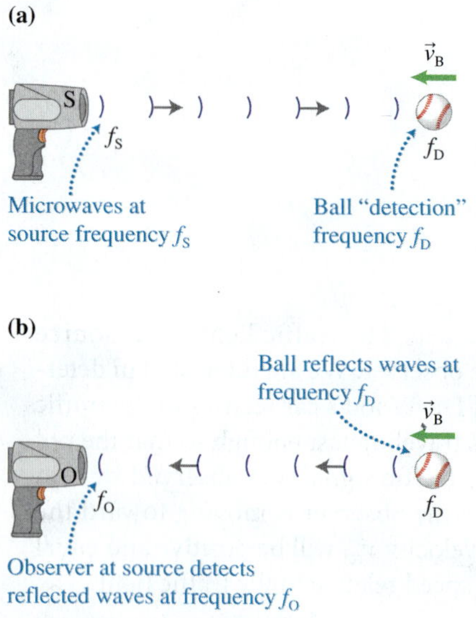
(a)
$\vec{v}_B$
S
$f_S$
$f_D$
Microwaves at source frequency $f_S$
Ball "detection" frequency $f_D$
(b)
Ball reflects waves at frequency $f_D$
$\vec{v}_B$
O
$f_O$
$f_D$
Observer at source detects reflected waves at frequency $f_O$

## Measuring car and ball speeds using the Doppler effect

A modern radar gun used by police emits microwaves (EM waves with frequencies much smaller than visible light) in the frequency range 33.4–36.0 GHz. Sports radar guns use a frequency of 10.525 GHz. The speed detection process involves three distinct steps outlined below for a baseball in flight.

1. The radar gun emits source microwaves at frequency $f_S = 10.525$ GHz. These microwaves are "detected" at frequency $f_D$ by the ball (**Figure 25.13a**).
2. The waves reflect from the ball, making the ball act as a source of microwaves at frequency $f_D$. Some of these waves are reflected back toward the radar gun, which observes or detects them at frequency $f_O$ (Figure 25.13b).
3. The observed frequency $f_O$ and source frequency $f_S$ are often combined, resulting in a beat frequency $f_{beat} = |f_O - f_S|$, which is then used to determine the speed of the baseball.

## Hubble's discovery of the expansion of the universe

At the beginning of the 20th century it was believed that the universe was static. Planets moved around the Sun, but the distances between galaxies remained constant. This static universe model seemed consistent with observations. However, because the gravitational interaction is always attractive, why doesn't the universe collapse? Even Einstein's ideas about gravity (which you will learn about in the next section) had trouble accommodating a static universe.

During this same time, astronomers Edwin Hubble and Milton Humason worked for years studying the light emitted by stars in other galaxies. They found that these stars emitted light wavelengths that were nearly the same as the light emitted from similar nearby stars—but with one major difference. With few exceptions, the light emitted by the distant galaxies was shifted toward longer wavelengths (smaller frequencies). Hubble concluded that the shift was not because the stars in other galaxies were different, but that significant relative motion existed between those stars and Earth, and that nearly all those stars and the galaxies they were in were moving away from us. The universe was expanding.

The idea of an expanding universe can be visualized as a loaf of raisin bread while it's rising. The galaxies are the raisins and the space between the galaxies is the dough (**Figure 25.14**). As the bread expands, the raisins embedded in the dough all move farther and farther apart from each other.

**Figure 25.14** The expanding universe is analogous to an expanding loaf of raisin bread.

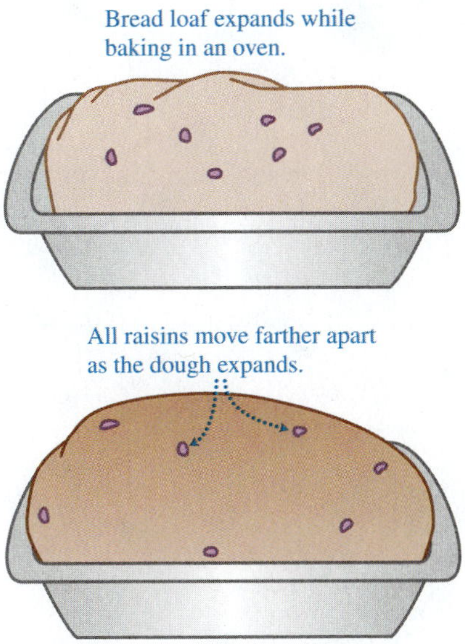

Bread loaf expands while baking in an oven.
All raisins move farther apart as the dough expands.

## Hubble's law

Hubble used the EM wave Doppler effect equations to determine the speeds of the other galaxies. They turned out to range from 0 km/s to 1500 km/s, and all but the closest galaxies were moving away from us. Additionally, Hubble found that the farther away the galaxy, the faster it was moving away. Notice in the expanding bread loaf in Figure 25.14 that the nearest two raisins on the top right changed positions from about 1 cm apart to 2 cm apart (a change in separation of 1 cm). However, the raisin farthest away from the top right raisin changed from about 2 cm apart to 4 cm apart (a change in separation of 2 cm). Hubble found a similar pattern for galaxies and described it mathematically as

$$v = Hd \tag{25.13}$$

where $v$ is the galaxy's recessional speed away from Earth, $d$ is the distance between the galaxy and Earth, and $H$ is a constant, now known as the Hubble constant. It has units of kilometers per second per unit of distance between the galaxies. In astronomy, distances to other galaxies are usually measured in megaparsecs (Mpc), with 1 parsec equal to $3.09 \times 10^{16}$ m, or about 3.3 light-years. The current accepted value for the Hubble constant coming from multiple experiments is about $70 \pm 6$ (km/s)/Mpc.

It might seem that Hubble's result implies that our galaxy is the center of the universe and all other galaxies are moving away from that center. We can see it another way, however. Going back to the analogy of the raisin bread, if Earth were one of the raisins, we would observe all other raisins moving away from us. But that's exactly what each of the other raisins would observe as well! There is no center of the universe, not our galaxy or any other.

**Hubble's Law** **The average recessional speed $v$ of two distant galaxies a distance $d$ apart is determined by**

$$v = Hd \tag{25.13}$$

**where $H = 70 \pm 6$ (km/s)/Mpc is Hubble's constant. One megaparsec (Mpc) equals $3.3 \times 10^6$ light-years (ly). Nearby galaxies do not necessarily obey this rule.**

‹ Active Learning Guide

## Age of the universe

If the galaxies are moving away from each other at measurable speeds, and we can measure the distances to them, then we can estimate how long it took the galaxies to move apart to their present positions. This would give an estimate for the age of the universe. Choose two distant galaxies a distance $d$ apart.

Assuming they have moved away from each other at a constant speed $v = Hd$, then the time interval $\Delta t$ since they were together is

$$\Delta t = \frac{d}{v} = \frac{d}{Hd} = \frac{1}{H}$$

The approximate age of the universe is the inverse of the Hubble constant. Substituting $70 \pm 6$ (km/s)/Mpc for H and converting to years from seconds, we obtain the following:

$$\Delta t = \frac{1}{H} = \frac{1}{70\frac{\text{km/s}}{\text{Mpc}}\left(\frac{1000\text{ m}}{1\text{ km}}\right)\left(\frac{1\text{ Mpc}}{10^6\text{ pc}}\right)\left(\frac{1\text{ pc}}{3.09 \times 10^{16}\text{ m}}\right)\left(\frac{3.16 \times 10^7\text{ s}}{1\text{ year}}\right)}$$
$$= 14.0\text{ billion years}$$

This result is uncertain by $(\pm 6/70) \times 100\% = \pm 8.6\%$. This result is compatible with geological studies that estimate the age of Earth at approximately $4.5 \times 10^9$ years old (4.5 billion years); that is, geological studies do not conclude that Earth is older than the universe. Rather, the findings imply that Earth formed approximately 9.5 billion years after the universe came into being.

The assumption about the universe expanding at a constant rate is questionable. In fact we have learned in the last 15 years that the universe is accelerating slightly in its expansion. The 2011 Nobel Prize in physics was awarded to Saul Perlmutter, Brian P. Schmidt, and Adam G. Riess for this stunning and very unexpected discovery.

**Review Question 25.9** Why can't we use the equations for the Doppler effect for sound to describe the Doppler effect for electromagnetic waves?

## 25.10 General relativity

**Figure 25.15** Acceleration produces the same effect as gravity.

(a)

When moving at constant velocity very far from massive objects, the person and the ball float in the spaceship.

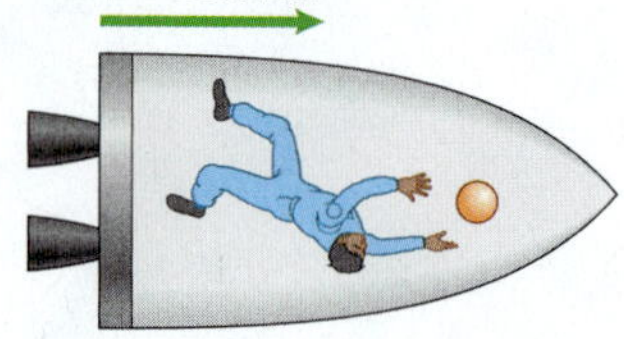


(b)

When the ship is accelerating toward the right at $\vec{a} = \vec{g}$, the left floor of the ship presses against the person, causing the same sensation as if standing on Earth's surface.

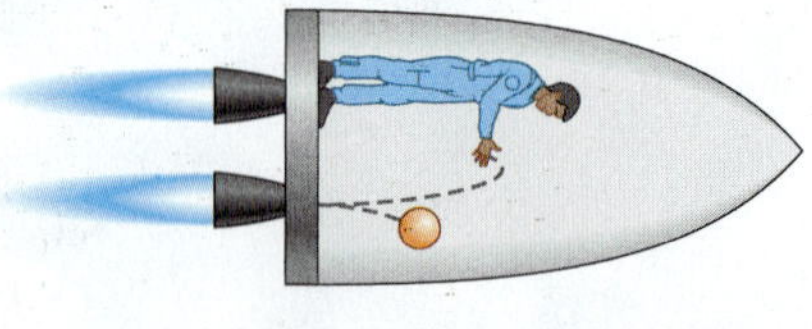


The special theory of relativity allows us to compare measurements made by observers in two reference frames that move at constant velocity relative to each other. In these inertial reference frames, the laws of physics have the same form (that is, they are said to be **invariant**). Einstein was able to generalize this invariance to all reference frames, including noninertial (accelerated) reference frames. The result is the **general theory of relativity,** which is also an improved theory of the gravitational interaction. The cornerstone of general relativity is the **principle of equivalence.**

**Principle of equivalence** **No experiment can be performed that can distinguish between an accelerating reference frame and the presence of a uniform gravitational field.**[1]

Consider a spaceship at rest or drifting at constant velocity far from any stars or planets. A passenger inside the ship would float freely (**Figure 25.15a**). Suppose now that the spaceship's rockets fired, causing the ship to accelerate at 9.8 m/s$^2$. Initially, the floating passenger "falls" to the spaceship's floor (on the left side of the ship shown in Figure 25.15b) with acceleration 9.8 m/s$^2$. From there, the passenger would be able to stand, jump "up and down," and throw a ball "upward" (away from the floor) only to have it come back "down." The passenger would have exactly the same sensations as if he or she were standing on Earth's surface.

[1]Einstein's statement of the principle of equivalence is more general than this, but this restricted version is sufficient for our discussion.

If the spaceship had no windows, and the engines provided a perfectly smooth ride, the passenger would have no way of knowing whether the ship was accelerating at $9.8\ m/s^2$ or was resting on Earth's surface in the presence of a nearly uniform gravitational field. The principle of equivalence says that there is no experiment that the passenger can perform to distinguish between these two situations.

From the principle of equivalence, Einstein reasoned that gravitation should be understood by a different mechanism entirely. He suggested that objects with mass cause the space around them to become curved and that objects moving in this curved space will not travel along straight paths at constant speed. These objects do not actually have a gravitational force exerted on them; rather, they are simply moving along natural paths through a curved space. The Moon orbits Earth not because Earth exerts a force on it, but because Earth curves space and the Moon then naturally moves in a curved path. This also very neatly explains why all objects, regardless of size, mass, or composition, move identically in a vacuum. It also explains why astronauts do not notice any forces being exerted on them as they orbit Earth: because there aren't any!

## An early testing experiment

An object of large mass such as the Sun causes space to curve more than does an object of smaller mass such as Earth (see **Figure 25.16**). In 1915 Einstein used general relativity to predict that light from distant stars that passes close to the Sun's surface would be deflected by an angle of $\theta = 4.86 \times 10^{-4}$ degrees. During a total solar eclipse, the observed position of those stars will be different because light bends as it passes close to the Sun (in **Figure 25.17**, the dashed line is the star's light path when the Sun is not present).

In 1919, astronomers under the leadership of Sir Arthur Eddington tried to measure such a deflection. They announced that the observations were in total agreement with the prediction of general relativity. Einstein and his theory of general relativity became famous almost overnight. However, despite the announcement, Eddington's results were not as conclusive as he had wanted them to be. It took many years and multiple experiments with solar eclipses until finally, by the 1960s, the measurements were convincingly in agreement with the predictions of general relativity.

**Figure 25.16** The Sun's mass causes curvature of space.

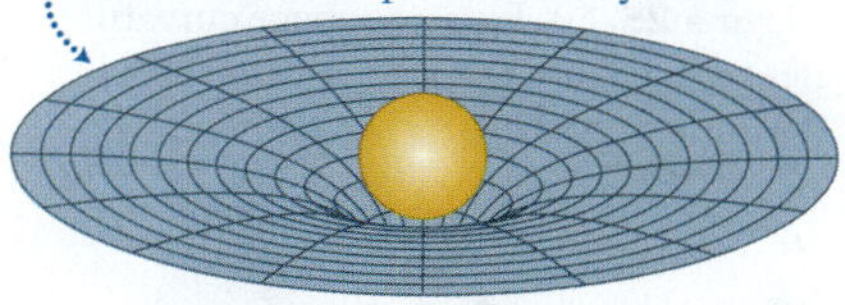


**Figure 25.17** Starlight deflection caused by Sun's space curvature.

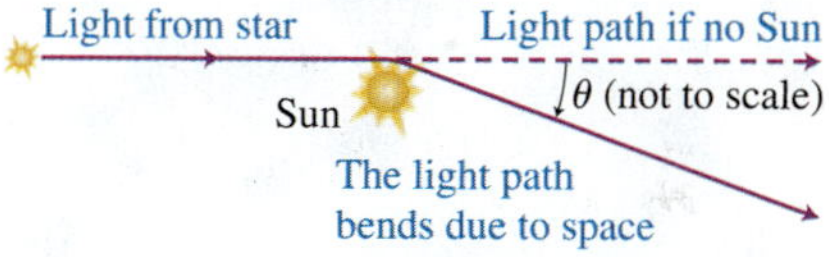


## Another testing experiment

General relativity solved another problem that had plagued astronomers since the early 1800s. The elliptical path of Mercury around the Sun was known to slowly rotate (**Figure 25.18**), a phenomenon called **precession**. This effect is so small that it takes about 3 million years for the elliptical orbit to complete one full precession. Newton's theory of gravitation predicts a precession from slight gravitational tugs due to the other planets (mostly Jupiter). However, Mercury's precession rate had been observed to be slightly larger than that predicted by Newton's theory. But if general relativity is used to predict the precession rate, the result is consistent with the observations. The additional precession results from the motion of Mercury through the curved space produced by the Sun.

**Figure 25.18** Mercury's orbit about the Sun.

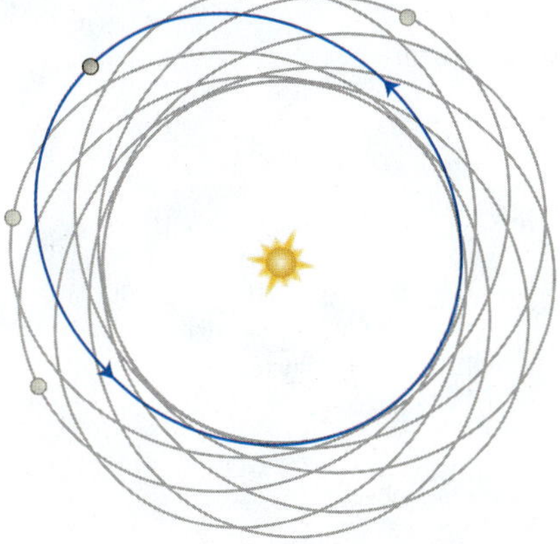


## Gravitational time dilation and red shift

Another prediction of general relativity is **gravitational time dilation.** Objects do not just curve space around them, but also alter the rates at which time passes around them as well. Time passes at different rates at different points near a massive object. The closer a point is to the massive object, the slower time passes there. For example, a person working in an office on the first floor of a tall building would age more slowly relative to an observer on the top floor. However, this effect would be only about $10^{-5}$ s over an entire human lifetime.

One consequence of gravitational time dilation is an effect known as **gravitational redshift**. If EM waves are emitted from a region closer to a massive object and observed in a region farther away, the observed frequency will be lower than the emitted frequency. In 1960, R. V. Pound and G. A. Rebka Jr. at Harvard University tested this prediction. They observed a small reduction in the vibration frequency of gamma rays emitted from radioactive nuclei on the bottom floor of a laboratory building at Harvard compared to the frequency of gamma rays emitted from the top floor—yet another testing experiment supporting general relativity.

## Gravitational waves and black holes

General relativity predicts that both space and time are curved by the presence of objects with mass. Experiments support this idea. These two ideas are unified by general relativity into the idea that space and time form a single entity known as **spacetime.** Newton imagined that space and time were a rigid background for the processes of the universe. Einstein showed that spacetime is curved by the presence of mass and changes shape when that mass moves. These changes in the curvature of spacetime can propagate as gravitational waves that ripple through the vacuum at the speed of light. General relativity predicts what the properties of these waves should be, and detecting them is a challenging area of present research.

**Figure 25.19** Extreme space curvature near a black hole causes passing objects (including light) to be sucked in.

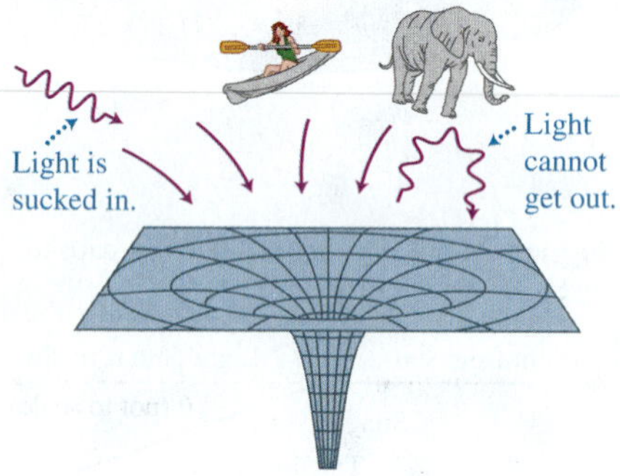
Light is sucked in.
Light cannot get out.

General relativity also predicts the existence of black holes (objects we first encountered in Chapter 6). In Newtonian mechanics, black holes are hypothetical objects whose escape speed is equal to or greater than light speed. In general relativity, black holes are regions of spacetime that are so extremely curved that all matter and even light within that region cannot escape (see **Figure 25.19**). The most common black holes form when large stars run out of nuclear fuel and collapse under their own gravity to extreme densities. Much larger black holes have been detected at the centers of most galaxies. Astronomers cannot detect black holes directly but can find them by the extreme influence they have on their immediate environment.

**Review Question 25.10** What is the general relativity explanation for why Earth orbits the Sun?

# 25.11 Global Positioning System (GPS): Putting it all together

You will likely never travel at speeds close to the speed of light, never travel to a black hole, and never worry about your biological clock becoming out of sync by a thousandth of a second over the course of your life because you live on the top floor of a building. In later chapters, however, we will explore many examples of the impact of relativistic effects on our daily lives. Meanwhile, here's an everyday example of both special and general relativity that affects your life: determining your location using the Global Positioning System (GPS).

Each GPS satellite is about 20,000 km from the surface of Earth and has an orbital speed of about 3900 m/s. An atomic clock on each satellite "ticks" every nanosecond ($1 \text{ ns} = 10^{-9} \text{ s}$), which allows it to transmit precisely timed signals. A GPS receiver in a car or on an airplane determines its position on Earth to within 5 to 10 m using these signals.

To achieve this level of precision, the clock ticks from the GPS satellites must be measured to an accuracy of 20 to 30 ns. Because the satellites move with respect to us, their clocks tick slightly slower than ours (special relativity's time dilation). Using the time dilation equation [Eq. (25.1)], we can estimate by how much the clocks on the satellites should fall behind during 1 day.

The reference frame of the satellite is the proper reference frame, and a 1-ns tick is the proper time interval. The satellite, as stated, moves at a speed of 3900 m/s. This time interval as measured from Earth will be

$$\Delta t = \frac{\Delta t_0}{\sqrt{1 - \frac{v^2}{c^2}}} = \frac{1 \text{ ns}}{\sqrt{1 - \frac{(3.90 \times 10^3 \text{ m/s})^2}{(3.00 \times 10^8 \text{ m/s})^2}}} = 1.0000000001 \text{ ns}$$

The difference between the proper time interval and the measured time interval is

$$\Delta t - \Delta t_0 = 1.0000000001 \text{ ns} - 1.0000000000 \text{ ns} = 1.0 \times 10^{-10} \text{ns}$$

This means that with each 1-ns tick of the satellite's atomic clock, that clock falls behind clocks on Earth by $1.0 \times 10^{-10}$ ns. The number $N$ of nanosecond ticks in each day is

$$N = \left(\frac{3600 \text{ s}}{\text{h}}\right)\left(\frac{24 \text{ h}}{\text{day}}\right)\left(\frac{10^9 \text{ ticks}}{\text{s}}\right) = 8.64 \times 10^{13} \text{ ticks}$$

Therefore, in 1 day time dilation will cause the satellite clock to fall behind Earth clocks a total of

$$(\Delta t - \Delta t_0)N = (1.0 \times 10^{-10} \text{ ns})(8.64 \times 10^{13} \text{ ticks}) \approx 8.6 \times 10^3 \text{ ns}$$

This result is just an estimate because the GPS satellites are not exactly inertial reference frames. But because their accelerations are low, they are very close to inertial.

In addition, the satellites are in orbits high above Earth. There, the curvature of spacetime is smaller than on Earth's surface and thus the clocks will tick faster than clocks on Earth. General relativity predicts that the space clocks get ahead by about 45 $\mu$s per day.

Together, the two effects result in the GPS satellite clocks running at about $(+45 - 8.6)$ $\mu$s per day, or about 36 to 37 $\mu$s per day faster than Earth clocks, or about 0.014 s/year. Thus, in 1 year the satellite positions (moving at about 3900 m/s) would be off by about (0.014 s) (3900 m/s) = 50 m. The position the receiver on Earth would report would be off by about the same amount.

The precise locations of the satellites are determined by tracking stations around the world. That data is collected, processed, and regularly transmitted up to the satellites, which then update their internal orbit models. Along with the orbit updates, transmissions are also sent to synchronize the atomic clocks on the satellites. This is where relativistic effects are taken into account. GPS receivers use the time and position data from multiple satellites to trilaterate their location. An understanding of both special and general relativity is essential for the GPS system to work.

**Review Question 25.11** What are two relativistic effects that must be accounted for so that the GPS system can function accurately?

# Summary

| Words | Pictorial and physical representations | Mathematical representation |
|---|---|---|
| **Postulates of special relativity**<br>1. The laws of physics are the same in all inertial reference frames.<br>2. The speed of light in a vacuum is the same in all inertial reference frames. (Section 25.2) | 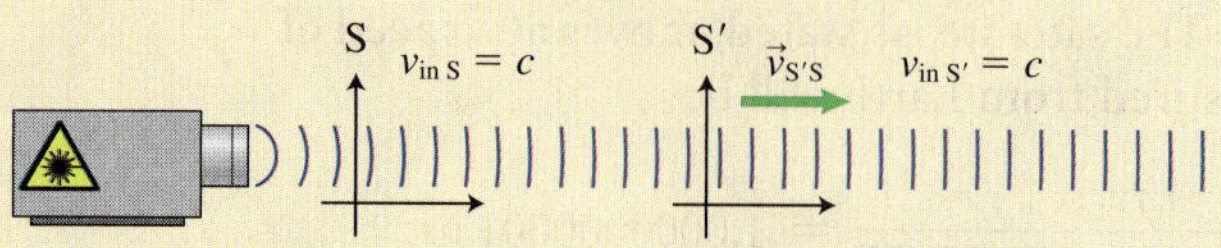 | |
| **Time dilation** In the **proper reference frame** events 1 and 2 occur at the same place. The time interval between events 1 and 2 measured by an observer in the proper reference frame is the **proper time interval** $\Delta t_0$. The time interval between events 1 and 2 as measured by an observer moving at constant speed $v$ relative to the proper reference frame is $\Delta t$. (Section 25.4) | 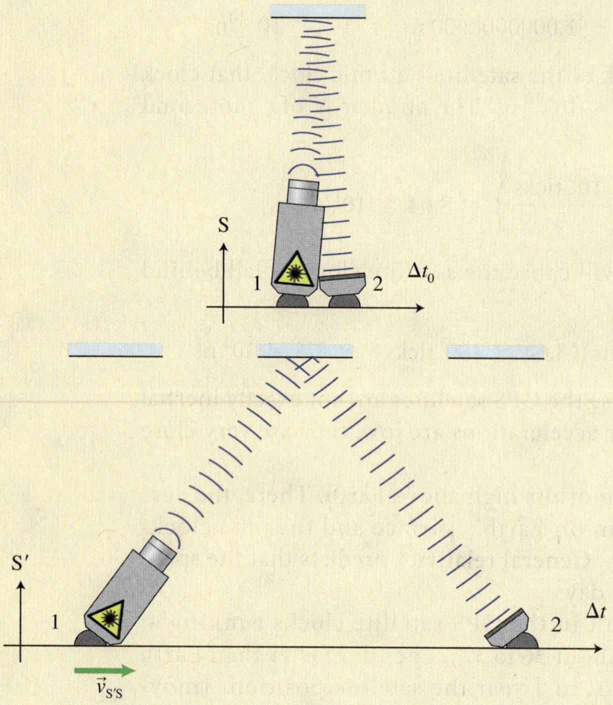 | $\Delta t = \dfrac{\Delta t_0}{\sqrt{1 - \dfrac{v^2}{c^2}}}$ Eq. (25.1) |
| **Length contraction** An object's **proper length** $L_0$ is its length measured in an inertial reference frame in which the object is at rest. The length of the object when measured in a reference frame moving at speed $v$ relative to the proper reference frame is $L$. (Section 25.5) | 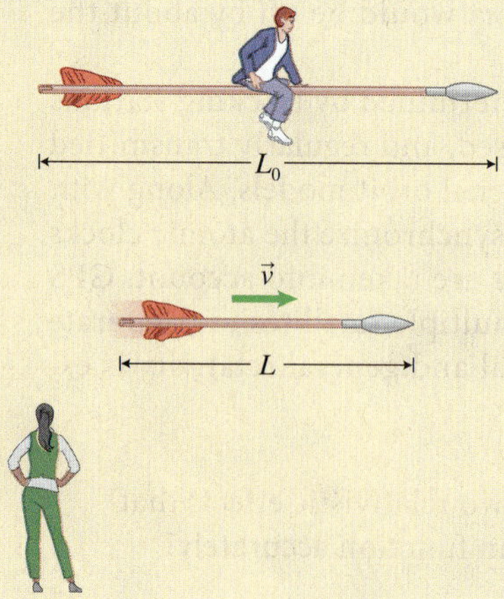 | $L = L_0\sqrt{1 - \dfrac{v^2}{c^2}}$ Eq. (25.2) |
| **Relativistic momentum** The relativistic momentum $\vec{p}$ of an object of mass $m$ must be used when an object is moving at speed $v \geq 0.1c$. (Section 25.7) | 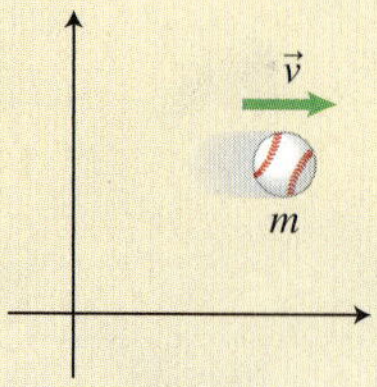 | $\vec{p} = \dfrac{m\vec{v}}{\sqrt{1 - (v/c)^2}}$ Eq. (25.5) |

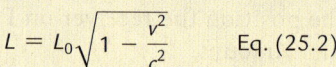

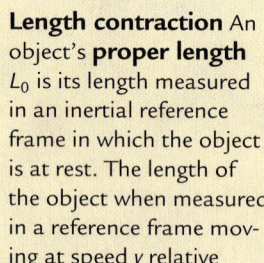

**Relativistic energy**

- **Rest energy** An object of mass $m$ has rest energy $E_0$ that depends only on the mass $m$ of the object and the speed of light $c$.
- **Kinetic energy** The relativistic kinetic energy $K$ of an object is its total energy minus its rest energy.
- **Total energy** is the total relativistic energy $E$ of an object moving at speed $v$. (Section 25.8)

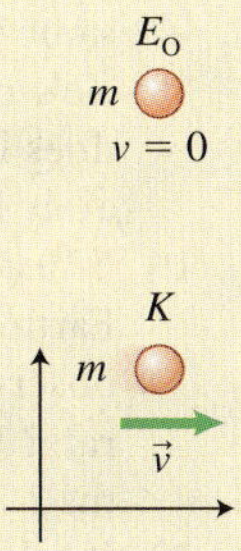


$$E_0 = mc^2 \quad \text{Eq. (25.8)}$$

$$K = \frac{mc^2}{\sqrt{1 - (v/c)^2}} - mc^2 \quad \text{Eq. (25.10)}$$

$$E = \frac{mc^2}{\sqrt{1 - (v/c)^2}} \quad \text{Eq. (25.9)}$$

**Doppler effect for electromagnetic radiation** If a source emits EM waves at frequency $f_S$ and the source and observer are moving at speed $v_{rel}$ relative to each other (+ if toward each other and − if away from each other), then the observer detects EM waves at frequency $f_O$. (Section 25.9)

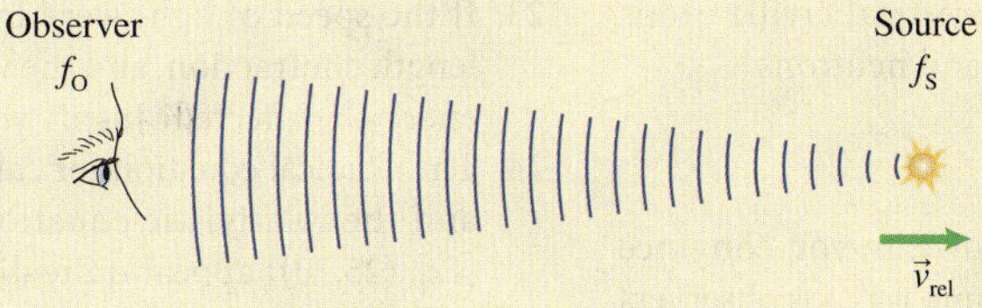


$$f_O = f_S\sqrt{\frac{1 + v_{rel}/c}{1 - v_{rel}/c}} \quad \text{for } v_{rel} \geq 0.1c \quad \text{Eq. (25.11)}$$

$$f_O = f_S\left(1 + \frac{v_{rel}}{c}\right) \quad \text{for } v_{rel} < 0.1c \quad \text{Eq. (25.12)}$$

MP® For instructor-assigned homework, go to **MasteringPhysics.**

# Questions

## Multiple Choice Questions

1. The Michelson-Morley experiment had as its goal to (a) measure the speed of light in different reference frames. (b) find out whether the Earth moves with respect to ether. (c) find whether ether exists. (d) All of the above (e) a and b only
2. On what did Michelson and Morley base their conclusions? (a) The measurement of the time intervals light takes to travel the same distance in different directions with respect to the moving ether (b) The analysis of an interference pattern for two beams of light (c) Comparison of the relative time intervals it takes light to travel in two directions
3. Physicists explained the results of the Michelson-Morley experiment by saying that (a) the motion of Earth through ether is undetectable. (b) there is no ether. (c) the length of experimental apparatus parallel to Earth's motion changes. (d) All of the above (e) None of the above
4. What is a proper time interval? (a) A correctly measured time interval (b) The shorter time interval of the two measured times (c) The time interval measured by observers with respect to whom the events occurred in the same location
5. You stand on a train platform and watch a person on a passing train through a train window. The person is eating a sandwich. The time interval it takes her to eat the sandwich as measured by you (a) is longer than measured by her. (b) is shorter than measured by her. (c) is the same. (d) depends on the orientation of the sandwich.
6. In Question 5, the length of the sandwich as measured by you (a) is longer than measured by the passenger. (b) is shorter than measured by the passenger. (c) is the same. (d) depends on the orientation of the sandwich.
7. A supernova explodes in a distant galaxy. The galaxy is moving away from us. The speed of light from the explosion with respect to Earth is (a) smaller than the speed of light with respect to the galaxy. (b) the same as light speed with respect to the galaxy. (c) larger than light speed with respect to the galaxy.

8. A supernova explodes in a distant galaxy. The galaxy is moving away from us. The speed of the elementary particles ejected by the explosion with respect to Earth is (a) smaller than the speed of light with respect to the galaxy. (b) the same as light speed with respect to the galaxy. (c) larger than light speed with respect to the galaxy.
9. We used the muon experiment to test which of the following hypotheses? (a) the existence of ether (b) time dilation (c) length contraction
10. The measurement of the coordinates of a star during a solar eclipse can be used as a testing experiment for (a) the existence of ether. (b) time dilation. (c) length contraction. (d) space-time curvature.
11. Suppose an early GPS system did not take the laws of special relativity into account. A 1-ns time interval measured by clocks on the GPS satellite would be (a) smaller than intervals measured on Earth. (b) larger than intervals measured on Earth. (c) exactly the same for both.
12. What are cosmic rays? (a) Light rays sent by cosmic objects (b) Electromagnetic radiation produced during supernova explosions (c) Signals sent by extraterrestrial civilizations (d) Subatomic particles such as protons and neutrons

## Conceptual Questions

13. What is an inertial reference frame? How can you convince someone that Newton's laws work the same way for observers in two different inertial reference frames?
14. Give an example of a phenomenon that an observer in a noninertial reference frame cannot explain using Newton's laws.
15. Explain the difference between a proper reference frame and other inertial reference frames in which the time interval between two events is measured.
16. What are the two postulates of the special theory of relativity? How would you explain them to a friend?
17. What does it mean to say that the speed of something is the same in different reference frames? Give an example.
18. You move toward a star at a speed of $0.99c$. At what speed does light from the star pass you? What if you are moving away from the star?
19. You pass Earth in a spaceship that moves at $0.99c$ relative to Earth. Do you notice a change in your heartbeat rate? Does an observer on Earth think your heart is beating at the normal rate? Do you think the heartbeat of Earth's observer is normal? Explain all three of your answers.
20. It takes light approximately $10^{10}$ years to reach Earth from the edge of the observable universe. Would it be possible for a person to travel this distance during a lifetime? Explain.
21. A person holds a meter stick in a spaceship traveling at $0.95c$ past Earth. The person rotates the meter stick so that it is first parallel and then perpendicular to the ship's velocity. Describe its changing length to (a) an observer at rest on the Earth and (b) to the person in the spaceship.
22. Name several ways in which your life would be different if the speed of light were 20 m/s rather than its actual value.
23. If the speed of light were infinite, how would time dilation, length contraction, and the ways we calculate momentum and energy be affected? Justify your answer carefully.
24. The classical equation for calculating kinetic energy [Eq. (6.5)] and the relativistic equation for calculating kinetic energy [Eq. (25.10)] appear quite different. Under what conditions is each equation appropriate? Invent a simple example to show that they produce the same result at speeds less than $0.001\ c$.
25. How did the Doppler effect for light help scientists estimate the age of the universe?
26. What is the principle of equivalence? How would you explain this principle to a friend?

# Problems

Below, **BIO** indicates a problem with a biological or medical focus. Problems labeled **EST** ask you to estimate the answer to a quantitative problem rather than develop a specific answer. Problems marked with ✎ require you to make a drawing or graph as part of your solution. Asterisks indicate the level of difficulty of the problem. Problems with no * are considered to be the least difficult. A single * marks moderately difficult problems. Two ** indicate more difficult problems.

You may need to know the following masses to work some of these problems: $m_{\text{electron}} = 9.11 \times 10^{-31}$ kg and $m_{\text{proton}} = 1.67 \times 10^{-27}$ kg.

### 25.1 Ether or no ether?

1. **Relative motion on an airport walkway** A person is walking on a moving walkway in the airport. Her speed with respect to the walkway is 2 m/s. The speed of the walkway is 1 m/s with respect to the floor. What is her speed with respect to a person walking on the floor in the opposite direction at the speed of 2 m/s?
2. **Running on a treadmill** Explain how you can run on a treadmill at 3 m/s and remain at the same location.
3. * Describe the important parts of the Michelson-Morley experimental setup and explain how this setup could help them determine whether Earth moves with respect to ether. Explain how the setup relates to the previous problem.
4. * ✎ Describe what Michelson and Morley would have observed when they rotated their spectrometer if Earth were moving through ether compared to what they would have observed if Earth were not moving through ether.

### 25.2 Postulates of special relativity

5. **Person on a bus** A person is sitting on a bus that stops suddenly, causing her head to tilt forward. (a) Explain the acceleration of her head from the point of view of an observer on the ground. (b) Explain the acceleration of her head from the point of view of another person on the bus. (c) Which observer is in an inertial reference frame?
6. * ✎ **Turning on a rotating turntable** A matchbox is placed on a rotating turntable. The turntable starts turning faster and faster. At some instant the matchbox flies off the turning table. (a) Draw a force diagram for the box when still on the rotating turntable. (b) Draw a force diagram for the box just

before it flies off. (c) Explain why the box flies off only when the turntable reaches a certain speed. In what reference frame are you when providing this explanation? (d) How would a bug sitting on the turntable explain the same situation?

7. * * Use your knowledge of electromagnetic waves to give an example illustrating that if the speed of light were different in different inertial reference frames, two inertial frame observers would see the same phenomenon differently. [Hint: Think about radio waves.]

### 25.3 and 25.4 Simultaneity and Time dilation

8. A particle called $\Sigma^+$ lives for $0.80 \times 10^{-10}$ s in its proper reference frame before transforming into two other particles. How long does the $\Sigma^+$ seem to live according to a laboratory observer when the particle moves past the observer at a speed of $2.4 \times 10^8$ m/s?
9. The $\Sigma^+$ particle discussed in the previous problem appears to a laboratory observer to live for $1.0 \times 10^{-10}$ s. How fast is it moving relative to the observer?
10. A person on Earth observes 10 flashes of the light on a passing spaceship in 22 s, whereas the same 10 flashes seem to take 12 s to an observer on the ship. What can you determine using this information?
11. A spaceship moves away from Earth at a speed of $0.990c$. The pilot looks back and measures the time interval for one rotation of Earth on its axis. What time interval does the pilot measure? What assumptions did you make?
12. * **Extending life?** A free neutron lives about 1000 s before transforming into an electron and a proton. If a neutron leaves the Sun at a speed of $0.999c$, (a) how long does it live according to an Earth observer? (b) Will such a neutron reach Pluto ($5.9 \times 10^{12}$ m from the Sun) before transforming? Explain your answers.
13. * A $\Sigma^+$ particle lives $0.80 \times 10^{-10}$ s in its proper reference frame. If it is traveling at $0.90c$ through a bubble chamber, how far will it move before it disintegrates?
14. * **Extending the life of a muon** A muon that lives $2.2 \times 10^{-6}$ s in its proper reference frame is created 10,000 m above Earth's surface. At what speed must it move to reach Earth's surface at the instant it disintegrates?
15. * **Effect of light speed on the time interval for a track race** Suppose the speed of light were 15 m/s. You run a 100-m dash in 10 s according to the timer's clock. How long did the race last according to your watch?

### 25.5 Length contraction

16. * Explain why an object moving past you would seem shorter in the direction of motion than when at rest with respect to you. Draw a sketch to illustrate your reasoning.
17. * * Explain why the length of an object that is oriented perpendicular to the direction of motion would be the same for all observers. Draw a sketch to illustrate your reasoning.
18. You sit in a spaceship moving past the Earth at 0.97 $c$. Your arm, held straight out in front of you, measures 50 cm. How long is it when measured by an observer on Earth?
19. **Length of a javelin** A javelin hurled by Wonder Woman moves past an Earth observer at $0.90c$. Its proper length is 2.7 m. What is its length according to the Earth observer?
20. At what speed must a meter stick move past an observer so that it appears to be 0.50 m long?
21. **Changing the shape of a billboard** A billboard is 10 m high and 15 m long according to a person standing in front of it. At what speed must a person in a fast car drive by parallel to the billboard's surface so that the billboard appears to be square?
22. * A classmate says that time dilation and length contraction can be remembered in a simple way if you think of a person eating a foot-long "sub" sandwich on a train (the sandwich is oriented parallel to the train's motion). The person on the train finishes the sandwich in 20 min. You, standing on the platform, observe the person eating a shorter sandwich but for a longer time interval. Do you agree with this example? Explain your answer.

### 25.6 Velocity transformations

23. * * Give examples of cases in which two observers record the motion of the same object to have different speeds, to have different directions, and to have different velocities. Provide reasonable values for the relevant velocities in your examples. Sketch each example and explain how each observer arrives at the value of the measured speed.
24. * Now repeat Problem 23, only this time instead of a moving object, use a light flash. Describe what speeds of light different observers should measure according to the second postulate of special relativity.
25. * **Life in a slow-light-speed universe** Imagine that you live in a universe where the speed of light is 50 m/s. You sit on a train moving west at speed 20 m/s relative to the track. Your friend moves on a train in the opposite direction at speed 15 m/s. What is the speed of his train with respect to yours?
26. * **More slow-light-speed universe** In the scenario described in Problem 25, you and your friend listen to music on the same radio station. What is the speed of the radio waves that your antenna is registering compared to the speed of the waves that your friend's antenna registers if the station is 100 miles to the west of you?
27. * You are on a spaceship traveling at $0.80c$ with respect to a nearby star sending a laser beam to a spaceship following you, which is moving at $0.50c$ in the same direction. (a) What is the speed of the laser beam registered by the second ship's personnel according to the classical addition of the velocities? (b) What is the speed of the laser beam registered by the second ship's personnel according to the relativistic addition of the velocities? (c) What is the speed of the second ship with respect to yours according to the classical addition of the velocities? (d) What is the speed of the second ship with respect to yours according to the relativistic addition of the velocities?
28. * * Your friend says that it is easy to travel faster than the speed of light; you just need to find the right observer. Give physics-based reasons for why your friend would have such an idea. Then explain whether you agree or disagree with him.
29. * * Your friend argues that Einstein's special theory of relativity says that nothing can move faster than the speed of light. (a) Give physics-based reasons for why your friend would have such an idea. (b) What examples of physical phenomena do you know of that contradict this statement? (c) Restate his idea so it is accurate in terms of physics.

### 25.7 Relativistic momentum

30. An electron is moving at a speed of $0.90c$. Compare its momentum as calculated using a nonrelativistic equation and using a relativistic equation.

31. * Explain why a relativistic expression is needed for fast-moving particles. Why can't we use a classical expression?
32. * If you were to bring an electron from speed zero to $0.95c$ in 10 min, what force would need to be exerted on the electron? What object could possibly exert such a force?
33. If a proton has a momentum of $3.00 \times 10^{-19}\ \text{kg} \cdot \text{m/s}$, what is its speed?

## 25.8 Relativistic energy

34. Determine the ratio of an electron's total energy to rest energy when moving at the following speeds: (a) 300 m/s, (b) $3.0 \times 10^6$ m/s, (c) $3.0 \times 10^7$ m/s, (d) $1.0 \times 10^8$ m/s, (e) $2.0 \times 10^8$ m/s, and (f) $2.9 \times 10^8$ m/s.
35. **Solar wind** To escape the gravitational pull of the Sun, a proton in the solar wind must have a speed of at least $6.2 \times 10^5$ m/s. Determine the rest energy, the kinetic energy, and the total energy of the proton.
36. * At what speed must an object move so that its total energy is 1.0% greater than its rest energy? 10% greater? Twice its rest energy?
37. * **Space travel** A 50-kg space traveler starts at rest and accelerates at 5 $g$ for 30 days. Determine the person's total energy after 30 days. What assumptions did you make?
38. * A person's total energy is twice his rest energy when he moves at a certain speed. By what factor must his speed now increase to cause another doubling of his total energy?
39. * A proton's energy after passing through the accelerator at Fermilab is 500 times its rest energy. Determine the proton's speed.
40. * A rocket of mass $m$ starts at rest and accelerates to a speed of $0.90c$. Determine the change in energy needed for this change in speed.
41. * Determine the total energy, the rest energy, and the kinetic energy of a person with 60-kg mass moving at speed $0.95c$.
42. * An electron is accelerated from rest across 50,000 V in a machine used to produce X-rays. Determine the electron's speed after crossing that potential difference.
43. * A particle originally moving at a speed $0.90c$ experiences a 5.0% increase in speed. By what percent does its kinetic energy increase?
44. * An electron is accelerated from rest across a potential difference of $9.0 \times 10^9$ V. Determine the electron's speed (a) using the nonrelativistic kinetic energy equation and (b) using the relativistic kinetic energy equation. Which is the correct answer?
45. ** A particle of mass $m$ initially moves at speed $0.40c$. (a) If the particle's speed is doubled, determine the ratio of its final kinetic energy to its initial kinetic energy. (b) If the particle's kinetic energy increases by a factor of 100, by what factor does its speed increase?
46. * Determine the mass of an object whose rest energy equals the total yearly energy consumption of the world ($5 \times 10^{20}$ J).
47. * **Mass equivalent of energy to separate a molecule** Separating a carbon monoxide molecule CO into a carbon and an oxygen atom requires $1.76 \times 10^{-18}$ J of energy. (a) Determine the mass equivalent of this energy. (b) Determine the fraction of the original mass of a CO molecule $4.67 \times 10^{-26}$ kg that was converted to energy.
48. * **Hydrogen fuel cell** A hydrogen-oxygen fuel cell combines 2 kg of hydrogen with 16 kg of oxygen to form 18 kg of water, thus releasing $2.5 \times 10^8$ J of energy. What fraction of the mass has been converted to energy?
49. ** **Mass to provide human energy needs** Determine the mass that must be converted to energy during a 70-year lifetime to continually provide electric power for a person at a rate of 1000 W. The production of the electric power from mass is only about 33% efficient.
50. * EST An electric utility company charges a customer about 6–7 cents for $10^6$ J of electrical energy. At this rate, estimate the cost of 1 g of mass if converted entirely to energy.
51. * **Mass to produce electric energy in a nuclear power plant** A nuclear power plant produces $10^9$ W of electric power and $2 \times 10^9$ W of waste heating. (a) At what rate must mass be converted to energy in the reactor? (b) What is the total mass converted to energy each year?
52. * BIO EST **Metabolic energy** Estimate the total metabolic energy you use during a day. (You can find more on metabolic rate in the reading passage in Chapter 6.) Determine the mass equivalent of this energy.
53. ** **Energy from the Sun** (a) Determine the energy radiated by the Sun each second by its conversion of $4 \times 10^9$ kg of mass to energy. (b) Determine the fraction of this energy intercepted by Earth, which is $1.50 \times 10^{11}$ m from the Sun and has a radius of $6.38 \times 10^6$ m.

## 25.9 Doppler effect for EM waves

54. **Why no color change?** Why don't the colors of buildings and tree leaves change when we look at them from a flying plane? Shouldn't the trees ahead look more bluish when you are approaching and reddish when you are receding?
55. * **Change red light to green** In a parallel universe the speed of light in a vacuum is 70.000 m/s. How fast should a driver's car move so that a red light looks green?
56. * **Effect of the Hubble constant on age and radius of the universe** How would the estimated age of the universe change if the new accepted value of the Hubble constant became (100 km/s)/mpc? How would the visible radius of the universe change?
57. * **Expanding faster** New observations suggest that our universe does not expand at a constant rate but instead is expanding at an increasing rate. How does this finding affect the estimation of the age of the universe using Hubble's law?
58. **Baseball Doppler shift** In September of 2010 Aroldis Chapman threw what may be the fastest baseball pitch ever recorded at 105 mi/h (47 m/s). What would the observed frequency of microwaves reflected from the ball be if the source frequency were 10.525 GHz? What would be the beat frequency between the source frequency and the observed frequency?
59. ** **Were you speeding?** A police officer stops you in a 29 m/s (65mi/h) speed zone and says you were speeding. The officer's radar has source frequency 33.4 GHz and observed a 3900-Hz beat frequency between the source frequency and waves reflected back to the radar from your car. Were you speeding? Explain.

# General Problems

60. * **Boat trip** A boat's speed is 10 m/s. It makes a round trip between stations A and B and then another between stations A and C. Stations A and B are on the same side of the river 0.5 km apart. Stations A and C are on the opposite sides of the river across from each other and also 0.5 km apart. The river flows at 1.5 m/s. What time interval is the round trip between stations A and B and then between A and C?

61. * **Space travel** An explorer travels at speed $2.90 \times 10^8$ m/s from Earth to a planet of Alpha Centauri, a distance of 4.3 light-years as measured by an Earth observer. (a) How long does the trip last according to an Earth observer? (b) How long does the trip last for the person on the ship?

62. ** **EST** **Extending life** Suppose that the speed of light is 8.0 m/s. You walk slowly to all of your classes during one semester while a classmate runs at a speed of 7.5 m/s during the time you are walking. Estimate your classmate's change in age, as judged by you, and your change in age according to you during that walking time. Indicate how you chose any numbers used in your estimate.

63. ** **Racecar when c is 100 m/s** Suppose that the speed of light is 100 m/s and that you are driving a racecar at speed 90 m/s. What time interval is required for you to travel 900 m along a track's straightaway (a) according to a timer on the track and (b) according to your own clock? (c) How long does the straightaway appear to you? (d) Notice that the speed at which the track moves past is your answer to part (c) divided by your answer to part (b). Does this speed agree with the speed as measured by the stationary timer?

64. ** **EST** Cherenkov radiation is electromagnetic radiation emitted when a fast-moving particle such as a proton passes through an insulator at a speed faster than the speed of light in that insulator. The Cherenkov radiation looks like a blue glow in the shape of a cone behind the particle. The radiation is named after Soviet physicist Pavel Cherenkov, who received a Nobel Prize in 1958 for describing the radiation. Estimate the smallest speed of a proton moving in oil that will produce Cherenkov radiation behind it.

65. ** A pilot and his spaceship of rest mass 1000 kg wish to travel from Earth to planet Scot ML, 30 light-years from Earth. However, the pilot wishes to be only 10 physiological years older when he reaches the planet. (a) At what constant speed must he travel? (b) What is the total energy of his spaceship and the rest energy, according to an Earth observer, while making the trip?

66. ** **Space travel** A pilot and her spaceship have a mass of 400 kg. The pilot expects to live 50 more Earth years and wishes to travel to a star that requires 100 years to reach even if she were to travel at the speed of light. (a) Determine the average speed she must travel to reach the star during the next 50 Earth years. (b) To attain this speed, a certain mass $m$ of matter is consumed and converted to the spaceship's kinetic energy. How much mass is needed? (Ignore the energy needed to accelerate the fuel that has not yet been consumed.)

67. ** (a) A container holding 4 kg of water is heated from 0°C to 60°C. Determine the increase in its energy and compare this to the rest energy when at 0°C. (b) If the water, initially at 0°C, is converted to ice at 0°C, determine the ratio of its energy change to its original rest energy.

## Reading Passage Problems

**Venus Williams's record tennis serve** Venus Williams broke the women's tennis ball serving speed record at the European Indoor Championships at Zurich, Switzerland, on October 16, 1998 with a 57 m/s (127 mi/h) serve. She has since broken her own record more than once. The Doppler radar gun used in 1998 to measure the speed had a source microwave frequency of 10.525 GHz. Answer the following questions about the radar gun used on that serve.

68. Which principle can we use to determine the frequency $f_D$ "detected" by the ball as it moved toward the source waves from the radar?
(a) The beat frequency equation
(b) The high-speed Doppler effect equation
(c) The low-speed Doppler effect equation
(d) The time dilation equation
(e) The relationship between wave speed, frequency, and wavelength

69. Which frequency is closest to the frequency $f_D$ detected by the ball as it moves toward the radar source waves?
(a) 10.525 GHz
(b) 10.525 GHz + $2.0 \times 10^{-6}$ GHz
(c) 10.525 GHz − $2.0 \times 10^{-6}$ GHz
(d) $3 \times 10^{-7}$ Hz

70. Which principle can we use to determine the frequency $f_O$ detected by the radar from waves reflected from the ball?
(a) The beat frequency equation
(b) The high-speed Doppler effect equation
(c) The low-speed Doppler effect equation
(d) The time dilation equation
(e) The relationship between wave speed, frequency. and wavelength

71. Which answer is closest to the frequency $f_O$ detected by the radar from the waves reflected from the ball?
(a) Exactly 10.525 GHz
(b) 10.525 GHz + $4.0 \times 10^{-6}$ GHz
(c) 10.525 GHz − $4.0 \times 10^{-6}$ GHz
(d) $3 \times 10^{-7}$ Hz

72. Which principle is used to determine the frequency that the radar measures of the combined source and observed waves?
(a) The beat frequency equation
(b) The high-speed Doppler effect equation
(c) The low-speed Doppler effect equation
(d) The time dilation equation
(e) The relationship between wave speed, frequency. and wavelength

73. Which answer is closest to the frequency that the radar measures of the combined source and observed waves?
(a) $2.0 \times 10^3$ Hz
(b) $4.0 \times 10^3$ Hz
(c) 10.525 GHz
(d) $3 \times 10^{-7}$ Hz
(e) $6 \times 10^{-7}$ Hz

**Quasars** In 1963, Maarten Schmidt of the California Institute of Technology found the most distant object that had ever been seen in the universe so far. Called 3C273, it emitted electromagnetic radiation with a power of $2 \times 10^{13}$ Suns, or 100 times that of the entire Milky Way galaxy! Schmidt called 3C273 a "Quasi-Stellar Radio Source," a name that was soon shortened to "quasar." Since then, astronomers have found many quasars, some substantially more powerful and distant than 3C273. Quasar 3C273 is moving away from Earth at about speed 0.16$c$. The Sun is $1.5 \times 10^{11}$ m from Earth and emits energy at a rate of $3.8 \times 10^{26}$ J/s caused by the conversion of $4.3 \times 10^9$ kg/s of mass into light and other forms of electromagnetic energy. The Hubble constant is

$$H = 70.8\,\frac{\text{km/s}}{\text{Mpc}} = 70.8\,\frac{10^3\ \text{m/s}}{3.09 \times 10^{22}\ \text{m}} = 22.9 \times 10^{-19}\ \text{s}^{-1}$$

74. What principle would you use to estimate the distance of 3C273 from Earth?
(a) The high-speed Doppler effect equation
(b) The low-speed Doppler effect equation

(c) Hubble's law
(d) The time dilation equation
(e) The relationship between wave speed, frequency, and wavelength

75. Which answer below is closest to the distance of 3C273 from the Earth in terms of the distance of the Sun from Earth?
(a) $\approx 10^3$ Sun distances (b) $\approx 10^6$ Sun distances
(c) $\approx 10^9$ Sun distances (d) $\approx 10^{12}$ Sun distances
(e) $\approx 10^{14}$ Sun distances

76. Which answer below is closest to the power of light and other forms of radiation emitted by 3C273?
(a) $\approx 10^8$ J/s (b) $\approx 10^{18}$ J/s
(c) $\approx 10^{25}$ J/s (d) $\approx 10^{32}$ J/s
(e) $\approx 10^{40}$ J/s

77. Which answer below is closest to the mass of 3C273 that is converted to light and other forms of radiation each second? By comparison, the mass of Earth is $6 \times 10^{24}$ kg.
(a) $\approx 10^{11}$ kg/s (b) $\approx 10^{15}$ kg/s
(c) $\approx 10^{19}$ kg/s (d) $\approx 10^{23}$ kg/s
(e) $\approx 10^{29}$ kg/s

78. What is the speed of light emitted by 3C273 as detected by an observer on 3C273?
(a) $1.15c$ (b) $c$
(c) $0.85c$ (d) None of these is correct.

79. If 3C273 is moving away from Earth at $0.16c$, what speed below is closest to the light speed we on Earth detect coming from 3C273?
(a) $1.15c$ (b) $c$
(c) $0.85c$ (d) None of these is correct.

# 26 Quantum Optics

What investigations started the transition from classical physics to quantum physics?

Why do some stars look red and some look blue?

How does a solar cell work?

Max Planck was a gifted young musician and composer who could play the piano, organ, and cello. However, he chose physics as a field of study when he entered the University of Munich in 1874 at age 16. Professor Philipp von Jolly discouraged him from studying physics because "almost everything is already discovered, and all that remains is to fill a few holes." Planck disregarded the advice; by 1879 he had defended his Ph.D. thesis and in 1900 he made a discovery that totally changed how physicists understood the world. Von Jolly was wrong, and around the turn of the 20th century new discoveries by physicists started pouring in "at the speed of light."

**Be sure you know how to:**

- Connect an ammeter and a voltmeter in a circuit to measure current and potential difference (Section 16.3).
- Relate the change of kinetic energy of an electrically charged particle to the potential difference across which it travels (Section 15.3).
- Relate the wavelength of a wave to its frequency and speed (Section 20.2).

We have studied the behavior of light and investigated various models of light, ultimately arriving at an electromagnetic wave model. That model explained the propagation of light in different media and its interactions with other objects—slits, gratings, small obstacles, and so on. However, the electromagnetic wave model does not explain the mechanism of light emission and absorption. That explanation required a new model of light that Planck and others helped develop. This new model initiated the revolutionary *quantum physics* that is the subject of this and subsequent chapters.

**Figure 26.1** An infrared photo taken at night with the house lights off.

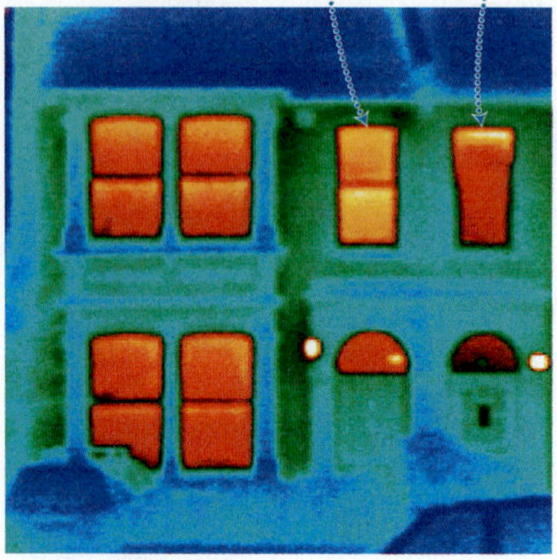

The windows look bright because of infrared emissions.

# 26.1 Black body radiation

In the last quarter of the 19th century physicists became interested in how hot objects such as stars and the coals of a fire emit thermal radiation (what we now call **infrared radiation**). Scientists studied the absorption and emission of infrared radiation and visible light from a hot object by modeling that object as a so-called **black body**. As the name indicates, a black body absorbs all incident (incoming) light and converts the light's energy into thermal energy (no light is reflected). The black body then radiates electromagnetic (EM) waves based solely on its temperature.

## What characterizes a black body?

To understand how the black body model works, imagine a small window in a house whose lights are off, or a small opening in a box, or the pupil of the human eye. Small openings like these look black to an outside observer because light entering the openings is not reflected—it is trapped inside. A small window, for example, admits sunlight. That light is absorbed inside the room, making the room warmer. As a result, more and more infrared radiation also leaves the room through the window. Eventually, the rate at which sunlight energy enters the room through the window equals the rate at which infrared radiation energy leaves the room. At this point the window becomes "a black body." In **Figure 26.1** an infrared camera has detected the infrared radiation being emitted through the windows of an unlit house.

We will model a black body as the surface of a small hole in one of the sides of a closed container (**Figure 26.2a**). Imagine that this container also has a thermometer inside. The hole looks black, and the box is at temperature $T_1 = 310\ \text{K}$. If you measure the power output per unit radiating surface area of the EM radiation coming from the hole at different wavelengths, you find that the hole emits a continuous EM spectrum (the lower curve in Figure 26.2b).

**Figure 26.2** A black body spectrum. (a) Emissions through the hole, the black body, are measured. (b) The spectrum of the emissions shows total radiated power and the peak of the power.

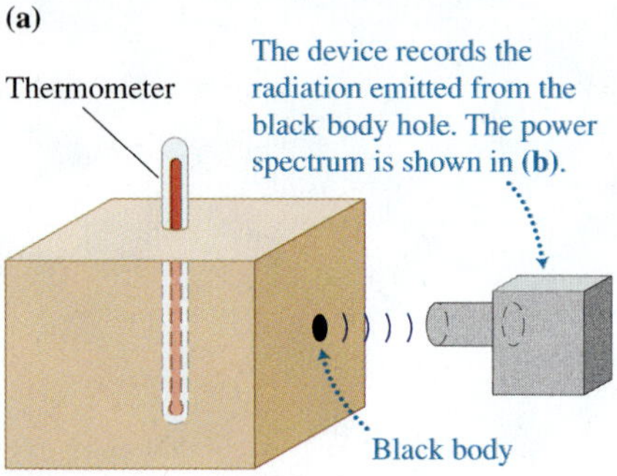


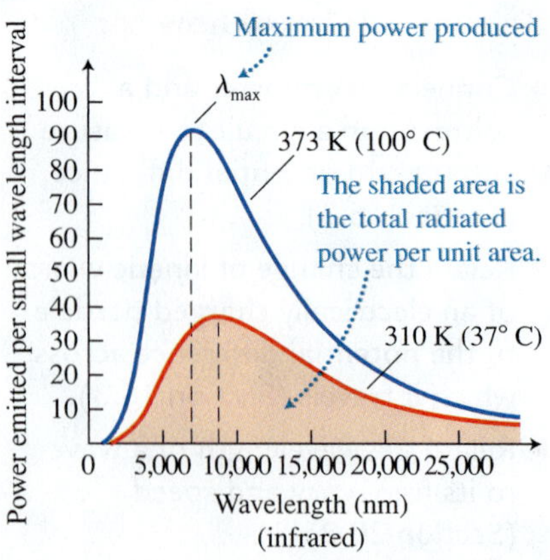


The graph in Figure 26.2b is called a **spectral curve.** The quantity on the vertical axis is the power output per unit radiating surface area per small wavelength interval of the black body as a function of wavelength. On the horizontal axis is the radiation's wavelength. The *total* power output per unit radiating surface area is the area under the black body spectral curve; this output per unit surface area is also known as the **intensity** of the EM radiation. (Notice that at a particular wavelength we see the greatest intensity of EM radiation coming from the hole.)

Now imagine that you place the box on a hot stove—the box's temperature rises and the thermometer in the box indicates a higher temperature $T_2 = 373\ \text{K}$. If we again measure the spectrum of the EM radiation being emitted from the hole, we find that

- the total power output from the hole is now greater,
- the spectral curve rises (the upper curve in Figure 26.2b) at all wavelengths, and
- the peak of the power per small wavelength interval shifts to a shorter wavelength.

As the temperature of the box increases further, these patterns persist. The higher the temperature, the shorter the wavelength at which the maximum of the black body spectral curve occurs, and the greater the power radiated at all wavelengths.

A black body is not a real object; it is a model. It turns out, however, that many real objects emit EM radiation similar to that of a black body. For example, the metal filament of an incandescent lightbulb gets very warm when electric current is passed through it. The bulb radiates some visible light and a lot of invisible infrared radiation. The spectral curve of a lightbulb is very close to the spectral curve of a black body of the same temperature. Thus, studying black body radiation might allow us to understand how objects emit and absorb light and how the radiation they emit might relate to the objects' temperatures.

## Studies of black body radiation

Several physicists in the late 1800s used the black body model to determine characteristics of EM radiation. Among their findings were the total power output of a black body (Stefan's law) and the wavelength at which the peak of the emitted radiation power occurs (Wien's law).

**Total Power Output: Stefan's Law** In 1879 a Slovenian physicist, Joseph Stefan (1835–1893), investigated the power of light emission from a black body—the total power output per unit of surface area—which is equal to the area under the black body spectral curve (see **Figure 26.3**). As the graph lines for different temperatures show, the total radiation output (power per unit area of the emitting object) increases dramatically as the temperature of the black body increases. Stefan found that the total power output per unit of surface area in Figure 26.3 is proportional to the fourth power of the temperature. For example, if you double the temperature of the object (in kelvins), the total radiation power increases by 16 times. The area under the 6000 K curve in Figure 26.3 is 16 times the area under the 3000 K curve.

**Figure 26.3** Black body radiation at different temperatures. The total power is proportional to the fourth power of the temperature in kelvins.

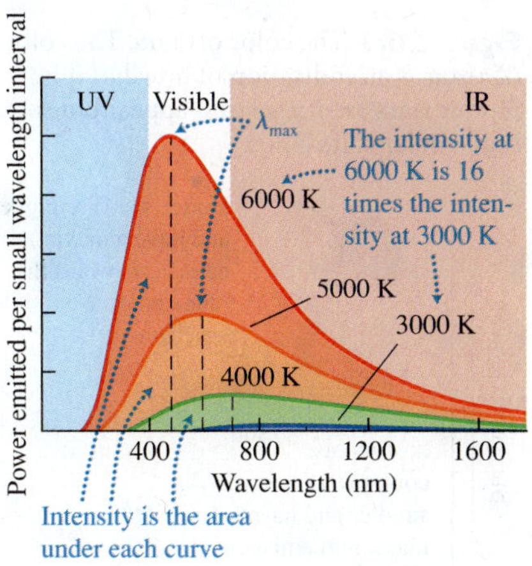


The total energy emitted by a 1-m$^2$ black body every second can be written as $\varepsilon \propto T^4$ or $\varepsilon = \sigma T^4$, where $\sigma$ (Greek letter sigma) is the coefficient of proportionality. To find the power $P$ radiated by the whole object, we multiply the intensity of its emitted radiation at its surface by the radiating surface area $A$ of the object.

Quantitatively, Stefan found the following relationship between the total emitted power $P$ in watts (W), the temperature of the black body $T$ in kelvins, and its surface area $A$ in square meters (**Stefan's law**):

$$P = \varepsilon A = \sigma T^4 A \tag{26.1}$$

where sigma ($\sigma$) stands for the coefficient of proportionality (Stefan's constant): $\sigma = 5.67 \times 10^{-8}\ \text{W/m}^2 \cdot \text{K}^4$.

> **TIP** A black body (the system) emits electromagnetic waves with power given by Stefan's law even if it is at the same temperature as the environment—but it also absorbs energy at the same rate. If the body is hotter than the environment, it radiates more energy than it absorbs. If it is cooler, it absorbs more than it radiates. Emission or absorption continues until the black body reaches thermal equilibrium with the environment.

**Wavelength at Which Maximum Intensity Occurs: Wien's Law** In 1893, the German physicist Wilhelm Wien (1864–1928) quantified a second aspect of black body radiation using data similar to that in the graphs in Figures 26.2

and 26.3. The wavelength $\lambda_{max}$ at which a black body emits radiation of maximum power per wavelength depends on the temperature of the black body:

$$\lambda_{max} = \frac{2.90 \times 10^{-3}\,\text{m} \cdot \text{K}}{T} \quad (26.2)$$

This became known as **Wien's law.** Note that the wavelength $\lambda_{max}$ at which the black body emits maximum power per wavelength becomes shorter as the black body temperature $T$ becomes greater.

**QUANTITATIVE EXERCISE 26.1 The surface temperature of the Sun**

The maximum power per wavelength of light from the Sun is at a wavelength of about 510 nm, which corresponds to yellow light. What is the surface temperature of the Sun?

**Represent mathematically** Rearrange Wien's law solving for $T$ in terms of $\lambda_{max}$:

$$T = \frac{2.90 \times 10^{-3}\,\text{m} \cdot \text{K}}{\lambda_{max}}$$

**Solve and evaluate** The temperature of a black body with its maximum power per wavelength at 510 nm = $510 \times 10^{-9}$ m is

$$T = \frac{2.90 \times 10^{-3}\,\text{m} \cdot \text{K}}{510 \times 10^{-9}\,\text{m}} = 5.69 \times 10^{3}\,\text{K}$$

This is about the temperature of the object in Figure 26.3 with the highest intensity (the uppermost curve).

At this temperature, matter is in a gaseous state. But the huge mass ($2.0 \times 10^{30}$ kg) of the Sun produces a strong gravitational field that prevents the gaseous atoms and molecules from easily leaving.

**Try it yourself:** The surface temperature of Earth is about 288 K. What is the wavelength at which Earth radiates EM waves with maximum power per wavelength?

*Answer:* Infrared radiation with a wavelength of 10,000 nm.

As we saw in the Try It Yourself question of Quantitative Exercise 26.1, the Earth radiates EM waves primarily in the infrared part of the electromagnetic spectrum, invisible to our eyes. Thus, when we see Earth and most objects on its surface we are seeing sunlight reflected from them. The same mechanism (reflection of sunlight) makes the Moon visible to us. The Sun and other stars, however, do not need an external source of EM waves to illuminate them.

A star that has a lower surface temperature than the Sun's radiates with a maximum intensity wavelength that is longer than 510 nm, and the star looks red instead of yellow. A star that is hotter than the Sun radiates with maximum intensity at shorter wavelengths, a combination of blue and other visible wavelengths that cause it to look whiter (**Figure 26.4**). Very hot stars (around 10,000 K) look blue. Stars even hotter than these are generally invisible to our eyes, but they can be detected with instruments sensitive to ultraviolet radiation.

**Figure 26.4** The color of stars. The color of a star is an indication of how hot it is. Hotter stars are bigger and appear bluer than cooler stars.

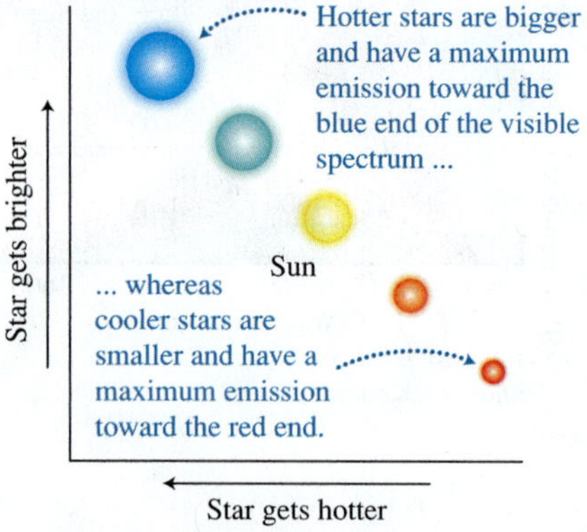
Hotter stars are bigger and have a maximum emission toward the blue end of the visible spectrum ...
Sun
... whereas cooler stars are smaller and have a maximum emission toward the red end.
Star gets brighter
Star gets hotter

**EXAMPLE 26.2 What is the power of surface radiation from the Sun?**

The radius of the Sun is about $7 \times 10^{8}$ m. Estimate the total amount of electromagnetic energy emitted by the Sun every second if its average surface temperature is about $6 \times 10^{3}$ K.

**Sketch and translate** We can visualize the Sun as a shining sphere of a given radius. If we know the total surface area of the Sun, we can use Stefan's law to calculate the total energy radiated per unit time (power).

**Simplify and diagram** To apply Stefan's law, assume that the Sun radiates as a black body. To find the surface area, assume that the Sun is a perfect sphere whose every square meter emits electromagnetic waves equally.

**Represent mathematically** According to Stefan's law the power radiated by an object with a surface area $A$ is

$$P = \sigma A T^{4}$$

The temperature is given, and the surface area of a sphere is $A = 4\pi R^2$. Thus the total power radiation by the Sun's surface is

$$P = \sigma(4\pi R^2)T^4$$

**Solve and evaluate** Now insert the appropriate values. Because we are doing an estimation, we will only calculate to one significant digit:

$$P \approx (6 \times 10^{-8}\,\mathrm{W/m^2K^4})4\pi(7 \times 10^8\,\mathrm{m})^2(6 \times 10^3\,\mathrm{K})^4$$
$$= 5 \times 10^{26}\,\mathrm{W}$$

The actual number is close to $4 \times 10^{26}$ W. Either way, this is a huge power output. Power plants on Earth typically have an output of $10^{12}$–$10^{13}$ W. The Sun has maintained (approximately) this power output for billions of years. You'll learn more about the mechanism of the Sun's power output in a later chapter (Chapter 28).

**Try it yourself:** Estimate how much solar energy reaches Earth every second. The effective surface area of Earth facing the Sun is a disk of radius $6.4 \times 10^6$ m (the radius of Earth). The average Sun-Earth distance is about $1.5 \times 10^{11}$ m.

*Answer:* $2 \times 10^{17}\,\mathrm{W} = 2 \times 10^{17}\,\mathrm{J/s}$.

## The ultraviolet catastrophe

By the end of the 19th century, physicists had devised a mechanism to explain how objects emitted electromagnetic radiation, namely, that the atoms and molecules inside objects are made of charged particles, and when these charged particles vibrate, they radiate electromagnetic waves. The higher the vibration frequency, the higher the frequency of the electromagnetic waves that are emitted. The atoms and molecules in a black body at temperature $T$ vibrate at many different frequencies centered about the frequency of the peak radiation on the black body spectrum curve.

Maxwell's electromagnetic wave theory (Chapter 24), together with some assumptions about the structure of matter, predicts that the intensity of the radiation emitted by a black body should increase with the frequency of the radiation—greater intensity for ultraviolet light than for visible light; greater intensity for X-rays than for ultraviolet light. However, actual spectral curves show that at high frequencies (short wavelengths) the emitted intensity actually drops off (**Figure 26.5**). No models built on the physics of the late 19th century made predictions that were consistent with this drop. This problem became known as the **ultraviolet catastrophe.**

**Figure 26.5** Classical black body radiation theory failed to match the experimental data at short wavelengths.

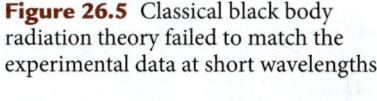

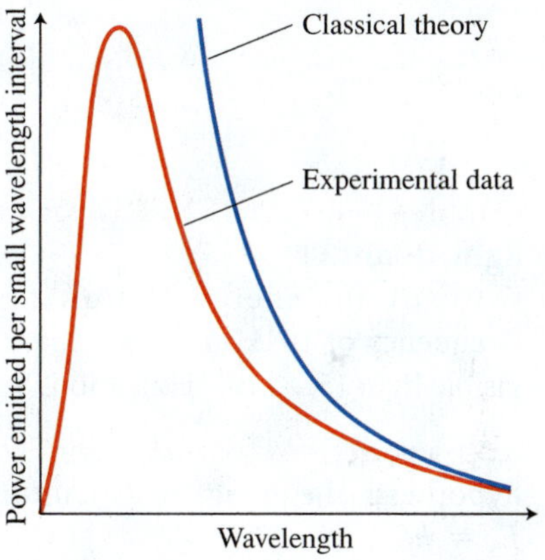


## Planck's hypothesis

After working on the problem for several years, Max Planck devised an ingenious model that correctly predicted the spectral curve for a black body. In Planck's model, the charged particles inside radiating objects were still responsible for producing electromagnetic radiation. However, they could radiate energy only in discrete portions called **quanta** (one portion is one quantum). He proposed that each microscopic oscillating charged particle had some kind of fundamental portion of energy, $E_0$, that was proportional to the frequency of its oscillation and, furthermore, that the particle could emit amounts of energy equal only to multiples of this fundamental portion: $E_0$, $2E_0$, $3E_0$, $4E_0$, etc. This idea is known as **Planck's hypothesis.**

Although the idea of this so-called **energy quantization** was completely revolutionary, the idea of quantization itself was not new to physics. Think of mechanical waves. When waves travel back and forth on a string of finite length with fixed ends, the string can vibrate only at discrete standing wave frequencies:

$$f_n = n\frac{v}{2L}$$

where $n = 1, 2, 3, \ldots$ The frequencies depend on the speed $v$ of the waves on the string and the string length $L$. The standing wave frequencies are therefore said to be **quantized**. The energy of these standing waves can take on any positive value, however, because the energy of the wave depends on the *amplitude* of the wave and not on its frequency.

Planck's hypothesis is different. He suggested that it was *the energy of the oscillator* that was quantized, rather than its frequency.

Using his energy quantization model, Planck was able to derive a mathematical function describing the black body spectral curve. The derivation is complicated, so we won't show it. When he matched the calculated values from the function to experimental results, he was able to determine the constant of proportionality between the oscillating frequency of a charged particle $f$ and its smallest fundamental energy $E_0$. The relationship is

$$E_0 = hf \tag{26.3}$$

The proportionality constant $h$ became known as **Planck's constant**. The units of $h$ in the SI system are $\text{J/Hz} = \text{J}/(1/\text{s}) = \text{J}\cdot\text{s}$. Its value was determined experimentally to be $h = 6.63 \times 10^{-34}\ \text{J}\cdot\text{s}$.

Planck viewed his energy quantization model as a mathematical trick of some sort rather than an explanation of actual physical phenomena. How could something oscillate smoothly at a particular frequency yet emit waves of that frequency only in discrete energy portions, as though it were emitting some sort of energy "particle"? He was almost embarrassed to have suggested such a model, and for much of his life he harbored doubts that real molecules or atoms radiated energy in this way. However, other scientists, including Albert Einstein, successfully applied this idea to explain other phenomena.

**QUANTITATIVE EXERCISE 26.3 Energy of light quanta**

Estimate the energy of a quantum of radio waves (frequency of $10^6$ Hz), infrared radiation ($3 \times 10^{13}$ Hz), visible light ($5 \times 10^{14}$ Hz), and UV radiation ($10^{15}$ Hz).

**Represent mathematically** According to Planck's hypothesis, the energy of a quantum of EM radiation is $E_0 = hf$.

**Solve and evaluate** The estimated energies are

$$E_{0\,\text{radio}} = hf_{\text{radio}} = (6.63 \times 10^{-34}\ \text{J}\cdot\text{s})(10^6\ \text{Hz}) = 7 \times 10^{-28}\ \text{J}$$

$$E_{0\,\text{infrared}} = hf_{\text{infrared}} = (6.63 \times 10^{-34}\ \text{J}\cdot\text{s})(3 \times 10^{13}\ \text{Hz}) = 2 \times 10^{-20}\ \text{J}$$

$$E_{0\,\text{visible}} = hf_{\text{visible}} = (6.63 \times 10^{-34}\ \text{J}\cdot\text{s})(5 \times 10^{14}\ \text{Hz}) = 3 \times 10^{-19}\ \text{J}$$

$$E_{0\,\text{ultraviolet}} = hf_{\text{ultraviolet}} = (6.63 \times 10^{-34}\ \text{J}\cdot\text{s})(1 \times 10^{15}\ \text{Hz}) = 7 \times 10^{-19}\ \text{J}$$

These are very small energies. For perspective, we can estimate how many quanta a 60-watt incandescent lightbulb emits every second. Its filament (at about $T = 2500$ K) produces very close to a black body spectrum. Using Wien's law we can estimate the wavelength of the maximum intensity radiation emitted by the filament and the corresponding frequency of that radiation:

$$\lambda_{\text{max}} = \frac{c}{f} = \frac{2.90 \times 10^{-3}\ \text{m}\cdot\text{K}}{T}$$

or

$$f = \frac{c}{\lambda_{\text{max}}} = \frac{cT}{2.90 \times 10^{-3}\ \text{m}\cdot\text{K}} = \frac{(3.0 \times 10^8\ \text{m/s})(2500\ \text{K})}{2.90 \times 10^{-3}\ \text{m}\cdot\text{K}} \approx 2.6 \times 10^{14}\ \text{Hz}$$

This frequency corresponds to infrared radiation. The energy of one quantum of infrared radiation at this frequency is approximately $2 \times 10^{-19}$ J. If we assume that all of the filament's radiated power is radiated at this frequency, we can estimate the number $n$ of quanta emitted every second as the total energy per second divided by the energy of one quantum:

$$n = \frac{60\ \text{J/s}}{2 \times 10^{-19}\ \text{J/quanta}} \approx 3 \times 10^{20}\ \text{quanta/s}$$

**Try it yourself:** Estimate how many quanta of electromagnetic radiation from the Sun reach Earth each second (use the data from the Try It Yourself question in Example 26.2) and assume that all of the Sun's energy is radiated as visible light.

*Answer:* About $10^{36}$ quanta each second—equivalent to the output of about $3 \times 10^{15}$ 60-watt bulbs.

**Table 26.1 Evolution of ideas concerning the nature of light.**

| Approximate time | Model | Consistent experimental evidence |
|---|---|---|
| 1600s | *Particle model* Light can be modeled as a stream of particles. | Reflection, shadows, and possibly refraction |
| Early 1800s | *Wave model* Particle model insufficient. Light modeled as wave. Wave mechanism unknown. | Double-slit interference, diffraction, light colors, reflection, and refraction |
| Middle 1800s | *Electromagnetic wave (ether model)* Light modeled as transverse vibrations of ether medium. | Polarization and electromagnetic wave transmission |
| Late 1800s | *Electromagnetic wave (no-ether model)* Light modeled as transverse vibrations of electric and magnetic fields. No medium required. | Michelson-Morley experiment |
| Early 1900s | *Quantum model* Electromagnetic wave model insufficient. Light modeled as stream of discrete energy quanta. | Planck's analysis of black body radiation spectral curves (and some new experiments to follow) |

Planck's quantum idea added a new twist to the study of light, and this divergence made him uncomfortable even though he knew the mathematics was correct. **Table 26.1** presents a summary of that history.

Today, Planck's quantum model is understood as the birth of quantum physics. To understand how this transformation from classical to quantum physics occurred and what model of light is currently accepted, we now turn the discussion to a new phenomenon.

**Review Question 26.1** Planck's hypothesis of light emission, like the classical explanation, stated that objects emit light through the accelerated motion of electrically charged particles. What in Planck's hypothesis brought the theory in line with observed values of light emission?

## 26.2 Photoelectric effect

Planck's quantum hypothesis contradicted the wave model of light and remained untested for several years. However, in the interim the hypothesis proved instrumental in understanding an interaction between metals and ultraviolet (UV) radiation. In order to understand this interaction we will first review what we know about the structure of metals.

### The structure of metals

In an earlier chapter (Chapter 16) you learned that metals are made of atoms whose outermost one or two electrons do not remain bound to the nucleus but are loosely bound to the whole crystal (**Figure 26.6a**). The positively charged ions (the metal atoms minus their outermost electrons) form a crystal lattice. The negatively charged electrons that have left the atoms move freely through the lattice (Figure 26.6b). These free electrons have positive kinetic energy $K = +(1/2)mv^2$. Why don't they leave the metal?

An electron and an ion have opposite electric charges ($-e$ for the electron and $+e$ for the positive ion). If we simplify the situation and consider a single electron and a single ion as the system of interest, the electric potential energy $U_q$ of this system is negative (Section 14.5):

$$U_q = k\frac{q_e q_{ion}}{r} = k\frac{(-e)(+e)}{r} = -k\frac{e^2}{r} < 0$$

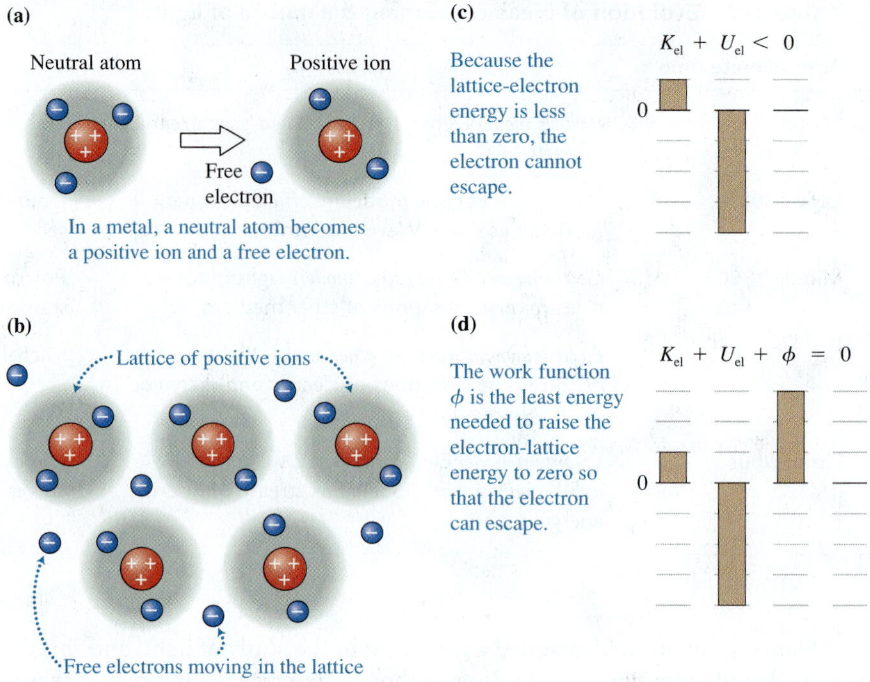
(a)
Neutral atom
Positive ion
Free electron
In a metal, a neutral atom becomes a positive ion and a free electron.
(b)
Lattice of positive ions
Free electrons moving in the lattice
(c)
Because the lattice-electron energy is less than zero, the electron cannot escape.
$K_{el} + U_{el} < 0$
0
(d)
The work function $\phi$ is the least energy needed to raise the electron-lattice energy to zero so that the electron can escape.
$K_{el} + U_{el} + \phi = 0$
0

**Figure 26.6** The structure of metals. (a) In metals, some neutral atoms become positive ions and free electrons. (b) The positive ions form a lattice, and the free electrons move around in the lattice. (c) The electric potential energy of a free electron is more negative than its kinetic energy, so it does not leave the lattice. (d) The work function ($\phi$) is the energy needed to raise the electron-lattice system energy to zero, allowing the electron to leave the lattice.

Because the energy of the system is negative, energy must be added to the system to remove the free electron from the metal (Figure 26.6c). The minimum energy needed to remove a free electron from a metal is called the **work function** $\phi$ (Greek letter phi) of the metal (Figure 26.6d).

The work function $\phi$ has a specific, known value for many metals. Although the work function has units of energy, it is not measured in joules because typical work function values are very small. Instead, we use the **electron volt** (eV), where $1\text{ eV} = 1.6 \times 10^{-19}\text{ J}$. As you may recall (from Chapter 25), one electron volt is the magnitude of the change in kinetic energy of an electron of charge $e = -1.6 \times 10^{-19}\text{ C}$ when it moves across a potential difference of 1 V.

For most metals (silver, tin, aluminum, etc), the value of $\phi$ ranges from about 4.30 eV to 4.60 eV. However, the $\phi$ values of some metals, such as cesium and lithium, are in the range of 2.10 eV to 2.90 eV. Thus it is significantly easier to remove electrons from these metals. The greater the work function of a particular metal, the more tightly the free electrons are bound to the metal as a whole, and the more energy that must be added to the electron-metal system to separate them.

**TIP** Even though $\phi$ is called the work "function," it's not a function in the mathematical sense. It is a constant that differs from material to material. And although some electrons in metals are called "free" electrons, they are not free to leave the metal; they are free to move around within the metal, however.

## Detection of the photoelectric effect

Now that we have an understanding of how metals hold on to their free electrons, let's discuss some experiments that eventually helped support Planck's hypothesis. We learned from our investigation of black body radiation that EM radiation is emitted as quanta. Is EM radiation absorbed as quanta as well? We investigate this idea in Observational Experiment **Table 26.2**. These experiments were performed originally by Wilhelm Hallwachs in Germany, in 1888. He used a gold leaf electroscope, in which two thin "leaves" of gold foil repel each other if they carry a like charge and hang straight down if they are uncharged. The first experiment is similar to one we performed earlier (in Chapter 14).

**OBSERVATIONAL EXPERIMENT TABLE**

## 26.2 Discharging an electroscope.

**Observational experiment**

**Experiment 1.** Take a plastic rod, rub it with felt, and touch it to an electroscope.

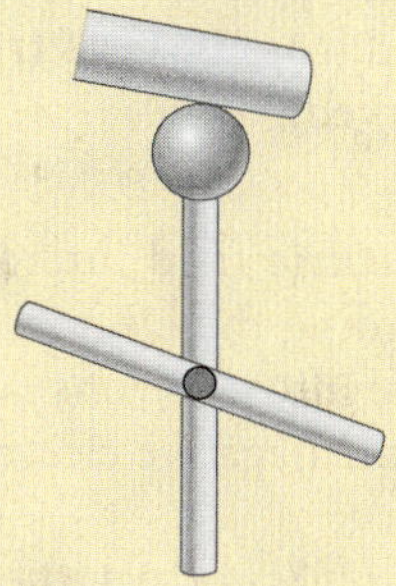

Repeat the same experiment using a glass rod rubbed with silk and an electroscope.

Observe that the leaves of both electroscopes are deflected and remain deflected.

**Analysis**

The negatively charged plastic rod rubbed with felt transfers electrons to the electroscope, making it negatively charged.

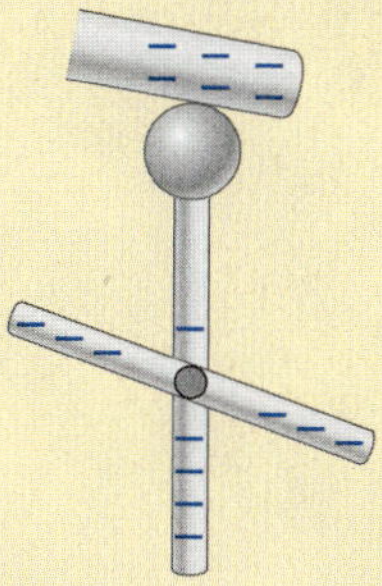

The positively charged glass rod rubbed with silk takes free electrons away from the electroscope, leaving it positively charged.

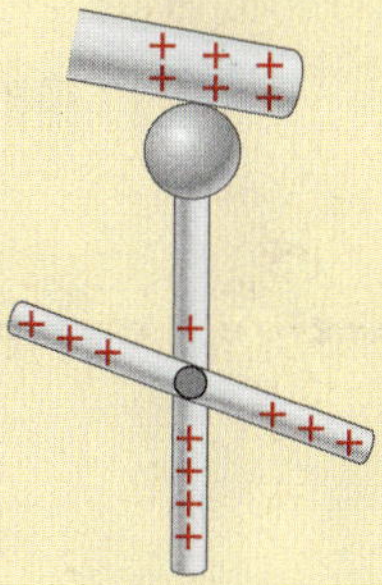

**Observational experiment**

**Experiment 2.** Shine visible light from an incandescent bulb on both the positively charged and negatively charged electroscopes. Nothing happens to them.

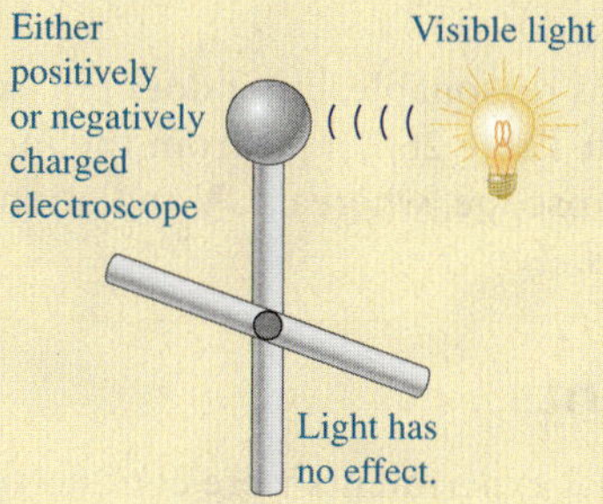


**Analysis**

Visible light does not affect the charge of the electroscope.

**Observational experiment**

**Experiment 3.** Shine UV radiation on the positively charged electroscope. The leaves of the electroscope stay deflected—the UV has no effect on the positively charged electroscope.

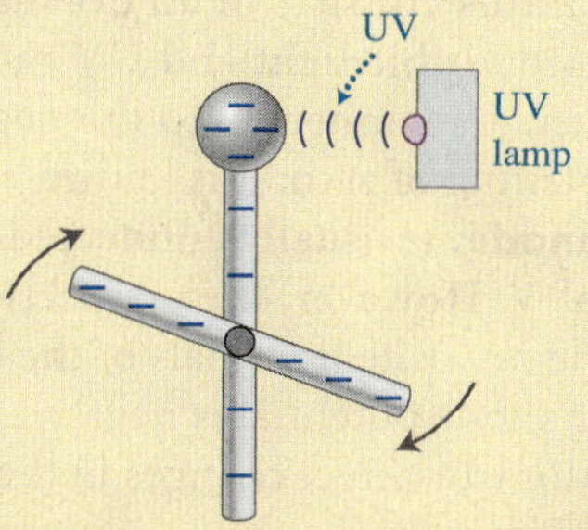


Shine UV radiation on the negatively charged electroscope. The leaves quickly return to their original uncharged position.

**Analysis**

The UV light discharges a negatively charged electroscope but does not discharge a positively charged electroscope.

**Patterns**

- UV radiation makes the negatively charged electroscope discharge but has no effect on the positively charged electroscope.
- Lower frequency visible light does not discharge either electroscope.

The top of the electroscope in Table 26.2 was made of zinc. It turns out that if it were made out of different material, cesium for example, green light would have had an effect on it similar to the effect the UV light had in the experiments above. It is the combination of the frequency of light and the material on which it shines that allows the negatively charged electroscope to discharge. A positively charged electroscope does not discharge in the presence of visible or UV light.

**Active Learning Guide ›**

The interaction described in Observational Experiment Table 26.2 is an example of the **photoelectric effect**. In the photoelectric effect, light incident on a metal causes electrons to be ejected from it. Let's try to invent an explanation for the patterns that we've observed.

*Explanation 1:* The light's vibrating electric field interacts with the electric charges in the molecules of air surrounding the electroscope. It shakes them back and forth until the molecules ionize into electrons and positively charged ions. The positive ions are attracted to the negative electroscope and attach to it, making it neutral overall.

This sounds reasonable. However, if it is correct, then both the positively and negatively charged electroscopes should discharge, because the newly freed electrons in the air should neutralize the positively charged electroscope. But this has not been observed. We can rule out this explanation for the photoelectric effect.

**Active Learning Guide ›**

*Explanation 2:* The light's vibrating electric field interacts with the electrons in the electroscope and can cause them to be ejected from it, but it cannot cause positive ions to be ejected. This hypothesis makes sense when we recall that free electrons in metals are negatively charged particles, and when we charge an object by touching, we either add or remove these electrons. The positive ions in the metal make up its crystal lattice and probably cannot be removed easily.

Explanation 2 seems reasonable, but it does not explain why the frequency of light is important: in Table 26.2 light from the incandescent lightbulb did not discharge the electroscope, whereas UV radiation did. We need more observational experiments.

## More experiments

**Figure 26.7** A device for measuring photocurrent. A device like this was used to detect the photoelectric effect and make observations about its characteristics.

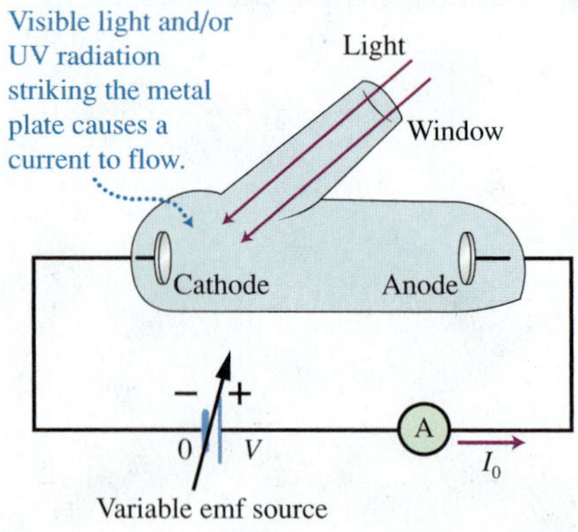


Additional observational experiments were conducted by a German physicist, Philipp von Lenard, in 1902. He used an apparatus similar to that shown in **Figure 26.7**. The apparatus consists of an evacuated glass container with a window that allows both visible light and UV radiation to shine on a metal plate. This plate is normally connected to the negative terminal of a battery and is at an electric potential of zero; it is known as a **cathode**. A corresponding plate, termed an **anode,** is usually connected to the positive terminal of the battery at potential $V$. However, the connection can be reversed by connecting the plates to the opposite terminals of the battery. When reversed, the potential of the former anode potential is negative relative to the zero cathode potential. An investigator can detect changes in the current in the circuit using the ammeter shown in Figure 26.7.

We can change the frequency of light passing through the window by inserting color filters in front of the window. We can also change the material of the cathode. When there is no light shining on the cathode, the ammeter detects no current. When the light is shining and we vary the filters, allowing light of different frequencies to illuminate the cathode, the ammeter will detect current in a circuit for some filters and not for others. For example, when the cathode is made of zinc, only UV light leads to current in the circuit (visible light does not), whereas when the cathode is made of cesium, green light leads to current (but red light does not).

‹ Active Learning Guide

Observational Experiment **Table 26.3** allows us to analyze the data that von Lenard collected.

**OBSERVATIONAL EXPERIMENT TABLE**

**26.3 Quantitative observations of the photoelectric effect.**

| Observational experiment | Analysis |
|---|---|
| **Experiment 1.** Connect the cathode on which light shines to a negative battery terminal. Find the frequency of light (using filters) for which the ammeter detects current. Keep the potential of the anode at a constant positive value relative to the zero potential at the cathode ($V_{anode} > V_{cathode} = 0$). Without changing the frequency, steadily increase the intensity of radiation shining on the cathode. The ammeter registers current. | Electric current from light exposure (the **photocurrent**) registered by the ammeter is directly proportional to the intensity of the light radiation. When no light is shining on the cathode, no current is measured by the ammeter. 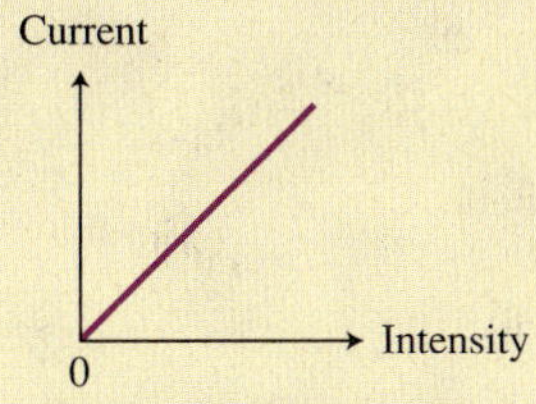  |
| **Experiment 2.** Use the frequency of EM radiation for which the photoelectric effect occurs. Using the same cathode and the same light filter, detect current in the circuit. Keeping the intensity and the frequency of the light or UV radiation constant, vary $V_{anode} - V_{cathode}$. Notice that a negative value means that the electric potential at the anode is lower than at the cathode. The ammeter detects the current. | Assume that the potential at the cathode is 0. The photocurrent increases as the positive potential at the anode increases. After the anode potential reaches a particular positive value, the current stops increasing. At a particular negative anode potential $-V_s$ (the **stopping potential**), the current stops. Note that $V_s$ is the magnitude of the stopping potential. 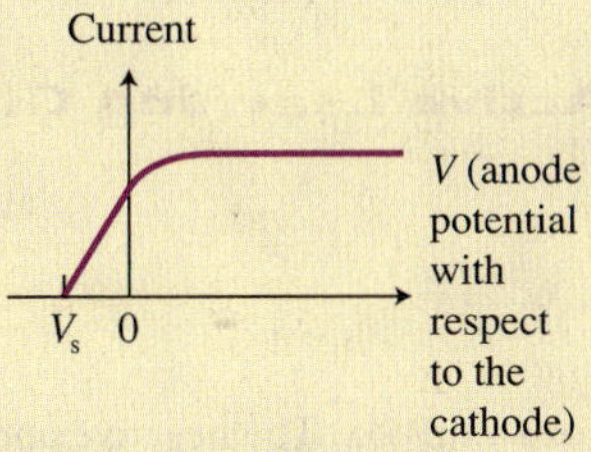  |
| **Experiment 3.** Use the frequency of EM radiation for which the photoelectric effect occurs. Repeat Experiment 2 for an increased intensity of light or UV radiation. | When the intensity of UV radiation increases, the maximum value of the photocurrent increases, but the negative stopping potential $-V_s$ does not change. 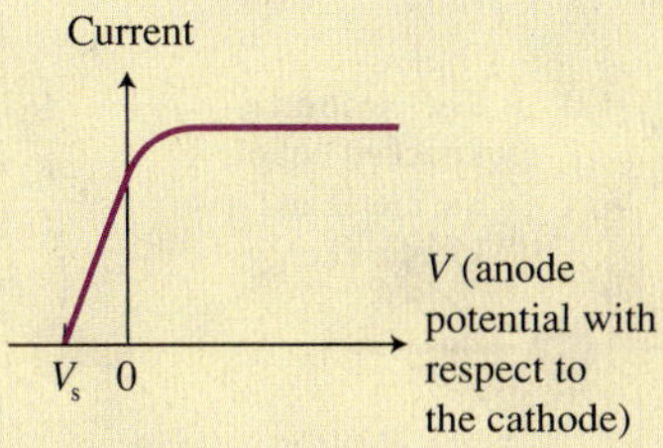  |
| **Experiment 4.** Repeat Experiment 2 with radiation in both the visible light range and UV range, using different frequencies. | The potential (now shown on the *y*-axis) is plotted against frequency (the *x*-axis). The negative stopping potential $-V_s$ depends on the frequencies of the light. When the frequency of light increases, the negative stopping potential $-V_s$ becomes more negative. For higher frequency light, it takes a greater opposing potential difference to stop the photocurrent. Also, there is no photocurrent when the frequencies are less than a value termed the **cutoff frequency**. 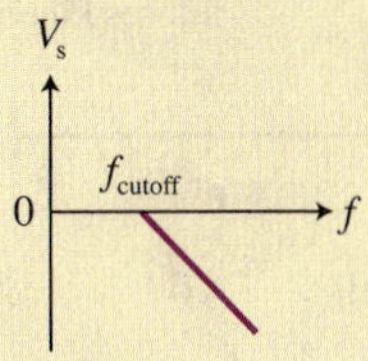  |

(*continued*)

**Patterns**

The photocurrent in the circuit depends on three factors: the potential difference across the electrodes (we assume that the potential of a cathode is zero), the intensity of the light, and the frequency of the light. If for a particular frequency of light, photocurrent is present in the circuit, then

1. For constant positive anode potential $V$, the photocurrent is directly proportional to the intensity of the radiation.
2. When the $V_{\text{anode}}$ is zero, a nonzero photocurrent $I_0$ is still present in the circuit.
3. For the same intensity and frequency of radiation, the current increases with increasing positive anode potential $V$ until it reaches some maximum value.
4. A well-defined negative stopping potential $-V_s$ stops the photocurrent.
5. Increasing the intensity of radiation increases the photocurrent $I_0$ and the maximum value of the photocurrent, but does not affect the negative anode stopping potential $-V_s$.
6. $-V_s$ is affected only by the frequency of the EM radiation incident on the cathode. The relationship is linear. When the frequency is below a certain cutoff value $f_{\text{cutoff}}$, the photocurrent is zero for any potential difference between the cathode and anode.

Philipp von Lenard noted another interesting outcome. He saw that the ammeter recorded a current in the circuit immediately after the light was turned on.

## Attempting to explain these patterns with the EM wave model

Let's try to use the EM wave model of light to explain the patterns reported for the frequencies of light for which the photoelectric effect occurred in Observational Experiment Table 26.3.

**Active Learning Guide ›**

**Pattern 1** For constant $V_{\text{anode}} - V_{\text{cathode}} > 0$, the photocurrent is directly proportional to the intensity of the radiation.

*EM wave model explanation:* Evidently, electrons in the cathode absorb enough energy from EM radiation that they are ejected from the cathode. The greater the intensity of the light, the more energy is transferred to the cathode. More electrons are freed from the cathode, causing a greater photocurrent. The wave model explains pattern 1.

**Figure 26.8** (a) The negative potential stops the electron. (b) A work-energy bar chart represents the stopping process.

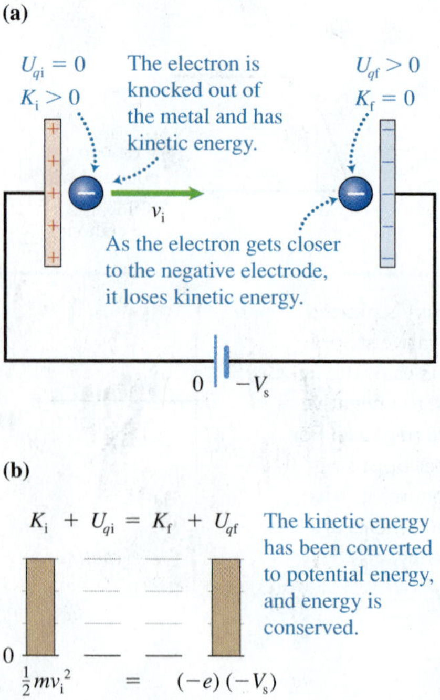
(a)
$U_{qi} = 0$
$K_i > 0$
The electron is knocked out of the metal and has kinetic energy.
$U_{qf} > 0$
$K_f = 0$
$v_i$
As the electron gets closer to the negative electrode, it loses kinetic energy.
0
$-V_s$
(b)
$K_i + U_{qi} = K_f + U_{qf}$
The kinetic energy has been converted to potential energy, and energy is conserved.
0
$\frac{1}{2}mv_i^2 = (-e)(-V_s)$

**Pattern 2** When $V_{\text{anode}} - V_{\text{cathode}} = 0$, there is still a nonzero photocurrent $I_0$ in the circuit.

*EM wave model explanation:* Evidently, the electrons in the metal gain enough energy from the EM radiation that they can escape the metal and have energy left over to reach the anode. The wave model explains pattern 2.

**Pattern 3** For the same intensity of EM radiation, the photocurrent increases with increasing positive $V_{\text{anode}} - V_{\text{cathode}}$ until the current reaches a maximum value.

*EM wave model explanation:* It would seem that with constant intensity EM radiation, the same number of electrons are ejected from the metal independent of $V_{\text{anode}} - V_{\text{cathode}}$. The electric force pulling the electrons to the anode increases as $V_{\text{anode}} - V_{\text{cathode}}$ increases. However, if the wave model is correct, it would seem that a minimum $\vec{E}$ field is needed to draw all of the electrons to the anode. Strengthening the $\vec{E}$ field beyond that would have no effect. The wave model explains pattern 3.

**Pattern 4** A well-defined negative stopping potential $-V_s$ stops the photocurrent.

*EM wave model explanation:* With a negative $-V_s$, the anode is at a lower electric potential than the cathode. The electric field now opposes the motion of electrons across the tube to the anode (**Figure 26.8a**). Suppose that $V_{\text{anode}} - V_{\text{cathode}} = (-V_s) - 0$ is the smallest negative value it can be while still preventing electrons from reaching the anode. This process can be quantified using the generalized work-energy principle—see the bar chart in Figure 26.8b. Mathematically:

$$K_i = U_{q\,f}$$

Inserting the expressions for kinetic energy and electric potential energy gives

$$\frac{1}{2} m_e v_{e\,i\,max}^2 = (-e)(-V_s)$$

where $-e$ and $m_e$ are the electron charge and mass, respectively, and $v_{e\,i\,max}$ is the maximum speed electron emitted from the metal. The magnitude of the stopping potential is then

$$V_s = \frac{m_e v_{e\,i\,max}^2}{2e}$$

The electrons have *one* particular $v_{e\,i\,max}$ and one stopping potential difference for low and high EM radiation intensity. It would seem that with high-intensity EM radiation, electrons would absorb more energy and escape at a higher speed. Thus the wave model explains pattern 4 qualitatively but does not explain why there is no intensity dependence in the expression for the stopping potential.

**Pattern 5** Increasing the intensity of EM radiation increases the photocurrent $I_0$ and the maximum value of the photocurrent but does not affect the value of $-V_s$.

*EM wave model explanation:* The greater light intensity results in more electrons knocked out of the cathode and therefore a greater photocurrent. According to the EM wave model of light, a higher intensity electromagnetic wave should transfer more energy to electrons, thus making it more difficult to stop them from reaching the anode. This is not what happens. The light intensity has no effect on the stopping potential. Thus the wave model does not explain pattern 5.

**Pattern 6** The value of $-V_s$ is only affected by the frequency $f$ of the EM radiation incident on the anode. The relationship is linear. When the frequency is below a certain cutoff value $f_{cutoff}$, the photocurrent does not appear at all.

*EM wave model explanation:* According to the wave model, the energy of an EM wave increases with the intensity of the wave; so it should be possible to use very intense but low-frequency light to knock electrons out of the metal. This does not happen. The EM wave model cannot explain pattern 6 at all and predicts something that contradicts the experiments.

In addition, for frequencies above the cutoff frequency, no delay occurs between the electromagnetic waves hitting the cathode and the release of the electrons, even for very low intensity of EM radiation (for the frequencies above the cutoff frequency for a particular cathode). This is somewhat puzzling. Shouldn't it take a while for sufficient energy to be transferred to electrons in the cathode before they start being released, especially when the EM radiation intensity is low?

To summarize, the EM wave model of light can explain some aspects of the photoelectric effect, but it can't explain the following: (a) the lack of dependence of the stopping potential $-V_s$ on the intensity of the light, (b) the presence of the cutoff frequency $f_{cutoff}$ below which there is no photocurrent (despite high intensity of light), and (c) the appearance of a photocurrent immediately after the EM radiation first strikes the cathode, even for low EM radiation intensity.

The photoelectric effect became the second phenomenon that made scientists question whether the EM wave model of light was sufficient to explain all EM wave behavior. In the next section you'll learn how to explain the photoelectric effect in its entirety. First, work the example below to become comfortable with the notion of stopping potential and an electron's kinetic energy in the photoelectric effect.

**EXAMPLE 26.4 Kinetic energy of the electrons and the stopping potential**

In one of von Lenard's experiments, the value of the stopping potential was $-0.20$ V. What was the maximum kinetic energy of the electrons leaving the cathode?

**Sketch and translate** Figure 26.7 shows a schematic of von Lenard's experiment with electrons being released from the cathode by incident UV. The situation with negative potential across the electrodes is shown in Figure 26.8a. The left electrode (the cathode) is at zero potential and the right electrode (the anode) is at $-V_s$. The electric force exerted on the electrons is toward the left, causing them to slow down and stop. The electron has zero kinetic energy when it reaches the anode.

**Simplify and diagram** Assume for simplicity that the electron travels in a straight path across the tube from the cathode to the anode. Assume also that the cathode from which an electron is ejected is at zero electric potential, and the anode that the electron flies toward is at $-0.20$ V. The work-energy bar chart in Figure 26.8b represents the process.

The electron and electric field produced by the electrodes are the system of interest. The initial state is when the electron has maximum kinetic energy just after being knocked out of the cathode and the system has zero electric potential energy; the final state is when the electron reaches the anode with zero kinetic energy. Thus, the final electric potential energy of the system is equal to the initial kinetic energy (Figure 26.8b).

**Represent mathematically** Using the generalized work-energy principle and the bar chart, we have

$$K_i = (-e)(-V_s)$$

**Solve and evaluate** The above equation can be used to determine the kinetic energy of the electron when it leaves the cathode; the energy of the electron is given in coulombs times volts ($C \cdot V$):

$$K_i = (-e)(-V_s) = (-1.6 \times 10^{-19}\,C)(-0.20\,V) = 3.2 \times 10^{-20}\,C \cdot V$$

Note that $1\,V = 1\,J/C$. Therefore, the electron's initial kinetic energy is

$$K_i = 3.2 \times 10^{-20}\,C \cdot \left(\frac{J}{C}\right) = 3.2 \times 10^{-20}\,J$$

This is a very tiny amount of energy, but is reasonable for a single electron. Converting this value to electron volts, we have

$$K_i = (3.2 \times 10^{-20}\,J)\left(\frac{1\,eV}{1.6 \times 10^{-19}\,J}\right) = 0.20\,eV$$

**Try it yourself:** Imagine that the electrodes are contained in a tube that is twice as long as the one in this example. What is the stopping potential for the electrons if they start with the same kinetic energy?

*Answer:* The stopping potential does not depend on the distance between the electrodes; it depends only on the change in the kinetic energy of the individual electrons. The potential will be the same: $-0.20$ V (with respect to the zero potential of the cathode).

**Review Question 26.2** What features of the photoelectric effect could not be explained by the electromagnetic wave model of light?

## 26.3 Quantum model explanation of the photoelectric effect

Philipp von Lenard thought that the absorption of UV radiation (and in some cases visible light) by the electron-metal system (the cathode) increased the energy of the system in accordance with the generalized work-energy principle. If the energy is large enough to raise the negative total energy of the system to zero ($-\phi + E_{light} = 0$), then an electron could just barely be ejected from the metal (see the bar chart representing the process in **Figure 26.9a**). If the system gained more energy from the EM radiation than it needed to eject the electron, then the electron would have a nonzero kinetic energy after leaving the cathode, and it might reach the anode even if the electric field produced by the electrodes pushed it in the opposite direction (Figure 26.9b). Mathematically:

$$-\phi + E_{light} = K_{ef} \tag{26.4}$$

where $K_{ef}$ is the kinetic energy of the electron immediately after leaving the metal. However, von Lenard could not explain why the stopping potential did not depend on the intensity of the incident light. In these experiments, light could mean either visible light or UV—whichever is causing the photoelectric effect. According to his understanding, increased light intensity should lead to greater energy of the light and cause the electron to leave with more kinetic energy, thus requiring a greater stopping potential to prevent it from reaching the anode.

In 1905, Albert Einstein made a breakthrough similar to Planck's hypothesis. At that time working as a clerk in a Swiss patent office, Einstein suggested that light can be viewed as a stream of quanta of energy not only when it is *emitted* by objects (as proposed by Planck), but also when it is *absorbed* by them. According to Einstein's hypothesis, an electron-lattice system (a metal) can only absorb light energy one whole quantum at a time. If the value of the light energy of a particular quantum is more than the magnitude of the metal's work function, an electron could leave the lattice. If the quantum of light energy is less than that, the electron would stay even if the number of incident quanta is very high (corresponding to high-intensity light.)

According to Planck, the energy of a quantum of EM radiation equals $E_0 = hf$. Using this idea, Einstein rewrote Eq. (26.4) as

$$-\phi + hf = K_{ef} \tag{26.5}$$

where $K_{ef}$ is the kinetic energy of a single electron just after leaving the metal. This equation explains why for most metals UV radiation causes the photoelectric effect, whereas visible light does not—each quantum of higher frequency UV radiation has sufficient energy to remove an electron from the metal, whereas a lower frequency visible light quantum does not have enough energy.

To check the feasibility of Einstein's hypothesis, let's calculate the energy of a quantum of UV radiation and compare it to typical work functions of metals. As you may remember from Section 26.2, the work function for most metals (silver, tin, aluminum, etc.) falls within the range $\phi = 4.3\text{ eV} - 4.6\text{ eV}$. The energy of one UV radiation quantum from Exercise 26.3 is about

$$E_{0\,\text{UV}} = hf_{\text{UV}} = (6.63 \times 10^{-34}\text{ J}\cdot\text{s})(1 \times 10^{15}\text{ Hz}) \approx 7 \times 10^{-19}\text{ J} \approx 4\text{ eV}$$

This is encouraging! The order of magnitude is correct. Similarly, the energy of one green light quantum is about

$$E_{0\,\text{visible}} = hf_{\text{visible}} = (6.63 \times 10^{-34}\text{ J}\cdot\text{s})(6 \times 10^{14}\text{ Hz}) = 4 \times 10^{-19}\text{ J} \approx 2.5\text{ eV}$$

which is close to the magnitude of the work function for a few metals (cesium, lithium, sodium). This supports Einstein's hypothesis. We will test it more carefully in Testing Experiment **Table 26.4**.

**Figure 26.9** The electron-lattice system gains energy from light and UV. (a) The energy gained from light offsets the negative electron-lattice energy, barely allowing the electron to escape. (b) If the energy gained from the light (or UV) is more than the amount needed to escape, then the electron has kinetic energy when it escapes.

(a)

(b)

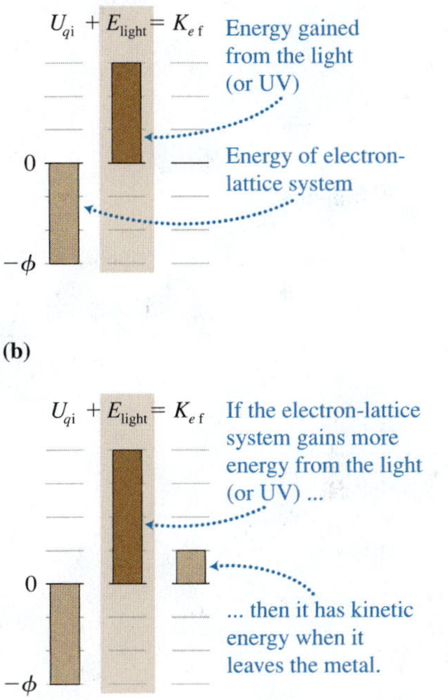
$U_{qi} + E_{\text{light}} = K_{ef}$
Energy gained from the light (or UV)
Energy of electron-lattice system
0
$-\phi$
$U_{qi} + E_{\text{light}} = K_{ef}$
If the electron-lattice system gains more energy from the light (or UV) ...
... then it has kinetic energy when it leaves the metal.
0
$-\phi$

**TESTING EXPERIMENT TABLE**

**26.4 Testing Einstein's hypothesis using the photoelectric effect.**

| Testing experiment | Prediction | Outcome |
|---|---|---|
| We repeat von Lenard's experiments, but this time use a sodium cathode with a work function 2.3 eV and use red and blue visible light. | If Einstein's hypothesis is correct, then<br>■ 1.9 eV red quanta of light (650 nm wavelength) will not have enough energy to eject electrons and thus no photocurrent will be detected in the circuit.<br>■ 2.7 eV blue photons (450 nm wavelength) should be able to eject electrons from the sodium, and thus should produce a photocurrent in the circuit. | ■ When red light shines on the cathode, the ammeter registers no photocurrent.<br>■ When blue light shines on the cathode, the ammeter registers a photocurrent. |

**Conclusion**

The outcome of the experiment matches the prediction. Thus we have not disproved Einstein's hypothesis. Instead, we have more confidence in it.

The outcome of the testing experiment supports Einstein's hypothesis. Let's now consider other features of the photoelectric effect that the wave model could not explain, as described on pages 992 and 993: (a) the lack of dependence of the stopping potential $-V_s$ on the intensity of the light, (b) the presence of the cutoff frequency $f_{cutoff}$ of light, and (c) the immediate appearance of the photocurrent once light is incident on the cathode.

**Active Learning Guide ›**

(a) *Stopping potential is independent of intensity:* The negative stopping potential ($-V_s$) of the anode with respect to the zero cathode potential depends on the kinetic energy of the electrons leaving the metal:

$$(-e)[(-V_s) - 0] = \frac{1}{2}m_e v_{ei}^2$$

According to Eq. (26.5), the kinetic energy of an electron knocked out of the metal is

$$\frac{1}{2}m_e v_{ei}^2 = -\phi + hf$$

Thus, the kinetic energy of the electron depends only on the energy $hf$ of a single quantum and not on the intensity of EM radiation incident on the cathode. Since the stopping potential depends on the initial kinetic energy of a freed electron, the stopping potential also depends only on the frequency of the radiation and not on the intensity of the radiation. Einstein's hypothesis explains this pattern.

(b) *No photocurrent below cutoff frequency:* The cutoff frequency occurs if the energy $hf$ of a light quantum equals the work function $\phi$, that is, if $hf = \phi$. We can then express the cutoff frequency in terms of the work function of the metal and Planck's constant:

$$f_{cutoff} = \frac{\phi}{h} \tag{26.6}$$

Under these circumstances the electron leaves the metal with zero kinetic energy. If the frequency of light is such that $hf < \phi$, then the energy of one quantum is less than the work function of the metal, and no photocurrent is produced. Einstein's hypothesis explains this pattern, too.

(c) *Photoelectric current produced immediately:* Absorbing one quantum of light of sufficient energy could instantly eject an electron out of the metal. Because the photoelectric effect is the result of an interaction between a single quantum of EM radiation and a single electron, energy does not need to accumulate in the metal before we see an effect. Here again, Einstein's hypothesis accounts for the experimental evidence.

Einstein's hypothesis explains the features of the photoelectric effect that the EM wave model could not. Let's see if it accounts for features that the wave model did explain.

**CONCEPTUAL EXERCISE 26.5 Apply Einstein's hypothesis**

Use Einstein's hypothesis to explain why the photocurrent in von Lenard's experiments is directly proportional to the intensity of the EM radiation incident on the cathode.

**Sketch and translate** According to Einstein's hypothesis we can visualize EM radiation as a stream of quanta, each with a specific amount of energy.

**Simplify and diagram** The quanta are represented as little "bubbles" in the figure on the top of the next page. The greater the intensity of the light, the greater the density (number per unit volume) of bubbles. Each bubble can potentially liberate one electron from the metal; thus more bubbles lead to more liberated electrons and therefore a larger photocurrent. Einstein's hypothesis does indeed predict that the photocurrent is directly proportional to the intensity of EM radiation.

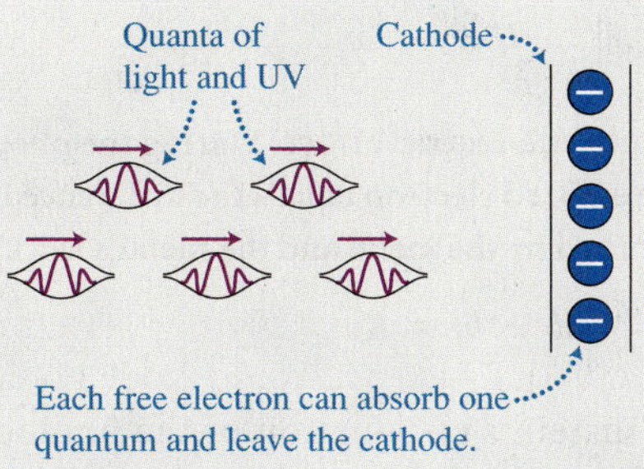
Quanta of light and UV
Cathode
Each free electron can absorb one quantum and leave the cathode.

**Try it yourself:** Explain why there is a photocurrent even when the potential difference between the electrodes in the tube is zero.

*Answer:* The kinetic energy of a liberated electron is $K_{ef} = hf - \phi$. If the energy of one quantum is more than the magnitude of the work function, then the electron will have nonzero kinetic energy after being separated from the metal. There is a chance that some of these electrons will arrive at the anode and produce the photocurrent even in the absence of a potential difference between the electrodes.

Einstein's hypothesis was successful in explaining the features of the photoelectric effect that the EM wave model for light could not explain. Einstein's hypothesis also leads to another testable prediction. Equation (26.6) $f_{\text{cutoff}} = \phi/h$ implies that for every metal, the ratio of its work function to its cutoff frequency should equal Planck's constant:

$$h = \frac{\phi}{f_{\text{cutoff}}}$$

Robert Millikan used this equation to determine the value of Plank's constant experimentally. Millikan's goal was to disprove Einstein's hypothesis and show that the absorption of light was not explained by a quantum hypothesis.

Millikan's reason for opposing Einstein's hypothesis was very similar to Planck's reason for not believing in his own quantum idea. The model of light as an electromagnetic wave could explain so many different phenomena that it was inconceivable to scientists to return to a particle-like model. As Millikan wrote himself in a paper published in 1916: "Einstein's photoelectric equation . . . cannot in my judgment be looked upon at present as resting upon any sort of a satisfactory theoretical foundation," even though "it actually represents very accurately the behavior" of photoelectricity.[1]

Later, when accepting the Nobel Prize in 1923 for this work and for his measurement of the charge of the electron, Millikan said: "this work resulted, contrary to my own expectation, in the first direct experimental proof . . . of the Einstein equation, and the first direct photoelectric determination of Planck's *h*."

The quanta of energy first invented by Planck to explain the emission of light and later applied by Einstein to explain the absorption of light later became known as **photons**.

**Photon** **A photon is a discrete portion of electromagnetic radiation that has energy**

$$E_{\text{photon}} = hf \tag{26.7}$$

**where $f$ is the frequency of the electromagnetic radiation and $h = 6.63 \times 10^{-34}\ \text{J}\cdot\text{s}$ is Planck's constant.**

[1]Millikan, R. A. (1916). A direct photoelectric determination of Planck's "*h*." *Phys. Rev.* 7, 355. http://focus.aps.org/story/v3/st23

Using the language of photons we can now summarize Einstein's photoelectric effect equation.

> **Einstein's equation for the photoelectric effect** **During the photoelectric effect, the kinetic energy $K_e$ of the emitted electron equals the difference in the energy of the photon $E_{\text{photon}} = hf$ absorbed by the metal and the metal's work function $\phi$:**
>
> $$-\phi + hf = K_e \qquad (26.5)$$

**Active Learning Guide ›** In the next example, we use Einstein's explanation of the photoelectric effect to understand some additional experimental results.

### EXAMPLE 26.6 Ejecting electrons from different metals

Light and UV shine on three different metals, each metal being a cathode in a photoelectric tube. Graph lines representing the stopping potential $-V_s$ versus the incident light or UV frequency are shown in the figure below. The work functions of the metals are: sodium (Na), 2.3 eV; iron (Fe), 4.7 eV; and platinum (Pt), 6.4 eV. Use Einstein's hypothesis of light absorption to explain the results.

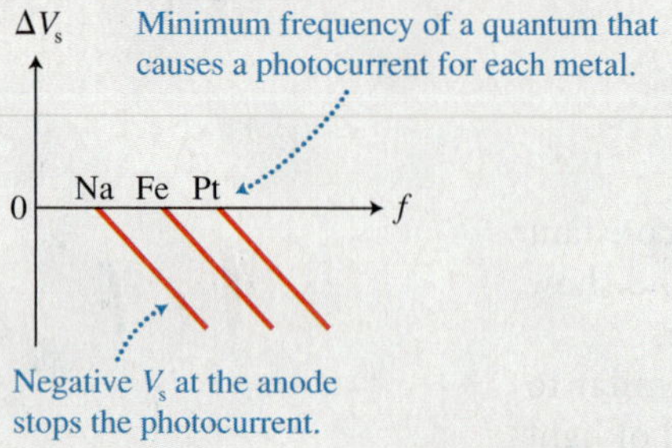


**Sketch and translate** The graph lines all have the same slope but have different intercepts on the horizontal frequency axis. Does Einstein's explanation of the photoelectric effect produce a mathematical equation that relates the observed stopping potential $-V_s$ and the frequency $f$ of the incident light and UV? The experimental setup is shown below.

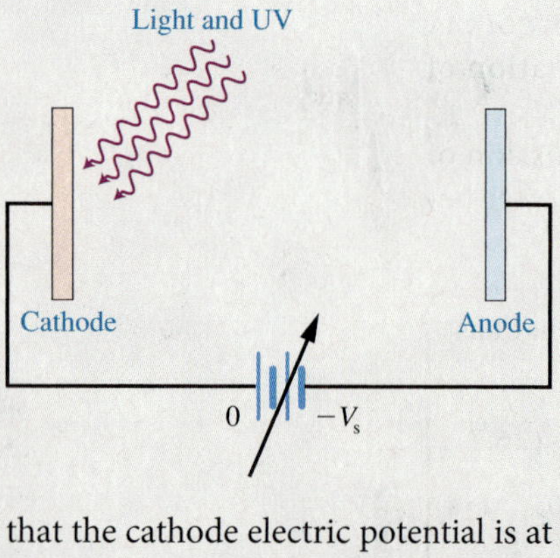


Note that the cathode electric potential is at zero and the anode potential $V$ is negative (necessary to stop the photoelectrons). The system of interest is the electron and the metal lattice—the cathode. The energy of each absorbed photon of incident light may cause an electron to be ejected from the metal. If the energy of the photon is greater than the work function of the metal, then the ejected electron has some kinetic energy after leaving the metal.

**Simplify and diagram** We represent the process with a bar chart, as shown in the first figure below. After ejection, the electron travels to the anode. We adjust the potential difference between the electrodes so that the electron stops just as it reaches the anode. The anode is at potential $-V_s$, less than the cathode (which is at zero). The travel of the ejected electron to the anode is represented by the second bar chart below.

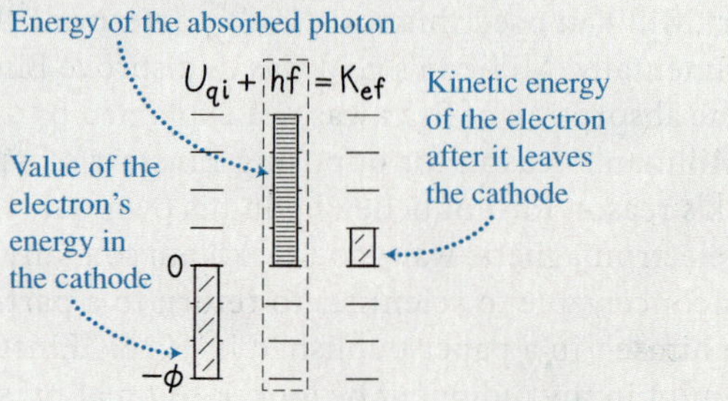


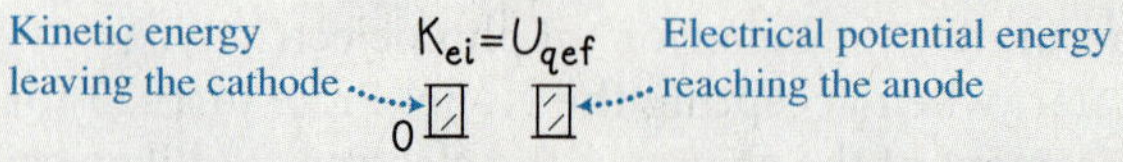


**Represent mathematically** Use the first bar chart above to represent the electron ejection process mathematically as

$$-\phi + hf = K_{ef}$$

The kinetic energy $K_{ef}$ of the electrons after ejection equals their kinetic energy $K_{ei}$ at the beginning of the next process (the movement of the electron from the cathode to the anode). Use the bar chart to represent that process mathematically:

$$K_{ei} = U_{qf} = qV_f = (-e)(-V_s)$$

where $-V_s$ is the negative anode potential. We combine the two equations to get

$$-\phi + hf = (-e)(-V_s)$$

**Solve and evaluate** Put the previous equation in a slope-intercept form so it can be compared with the graph:

$$-V_s = \left(-\frac{h}{e}\right)f + \left(\frac{\phi}{e}\right) \quad (26.8)$$

According to this equation, regardless of the metal cathode that is used, the slope will be $-h/e$. This explains why all the graph lines have the same slope, as seen in the first figure. Millikan measured these slopes to determine Planck's constant $h$. Because the photon energy $hf$ is greater in magnitude than the work function $\phi$, the stopping potential, $-V_s$, must be negative.

**Try it yourself:** If blue light of the frequency $6.7 \times 10^{14}$ Hz is incident on a sodium target, what is the value of the stopping potential?

*Answer:* $-0.5$ V.

In the above example, the $x$-intercept of each graph line is different. According to Eq. (26.8), the intercept when $V_s = 0$ depends on the work function of each metal. Put $V_s = 0$ into Eq. (26.8) to get

$$0 = \left(-\frac{h}{e}\right)f + \left(\frac{\phi}{e}\right)$$

Solving for $f$, we get

$$f_{\text{cutoff}} = \frac{\phi}{h}$$

where $f_{\text{cutoff}}$ is the cutoff frequency for a particular metal, the lowest frequency photon needed to eject an electron with no extra kinetic energy. The electron in that case is stopped with zero stopping potential $V_s = 0$. Measuring this cutoff frequency allows us to determine the work function of the metal. Or, if we know the work function, we could determine the cutoff frequency. We do this in **Table 26.5** for the three metals shown in Example 26.6.

The results in Table 26.5 make sense, because the $x$-intercept has a greater cutoff frequency for a metal with a larger work function.

**Review Question 26.3** How does Einstein's hypothesis explain the cutoff frequency observed for a particular metal cathode in a photoelectric experiment?

**Table 26.5 Cutoff frequencies for selected metals.**

| Metal | Cutoff frequency | Part of EM spectrum |
|---|---|---|
| Sodium (2.3 eV) | $f_{\text{Na}} = \frac{\phi_{\text{Na}}}{h} = \frac{(2.3\text{ eV})\left(\frac{1.6 \times 10^{-19}\text{ J}}{1\text{ eV}}\right)}{6.63 \times 10^{-34}\text{ J}\cdot\text{s}} = 5.6 \times 10^{14}\text{ Hz}$ | Visible |
| Iron (4.7 eV) | $f_{\text{Fe}} = \frac{\phi_{\text{Fe}}}{h} = \frac{(4.7\text{ eV})\left(\frac{1.6 \times 10^{-19}\text{ J}}{1\text{ eV}}\right)}{6.63 \times 10^{-34}\text{ J}\cdot\text{s}} = 1.1 \times 10^{15}\text{ Hz}$ | Ultraviolet |
| Platinum (6.4 eV) | $f_{\text{Pt}} = \frac{\phi_{\text{Pt}}}{h} = \frac{(6.4\text{ eV})\left(\frac{1.6 \times 10^{-19}\text{ J}}{1\text{ eV}}\right)}{6.63 \times 10^{-34}\text{ J}\cdot\text{s}} = 1.5 \times 10^{15}\text{ Hz}$ | Ultraviolet |

## 26.4 Photons

In the first three decades of the 20th century scientists faced a serious dilemma. On the one hand, the EM wave model of light explained a wide variety of phenomena, but not black body radiation and not the photoelectric effect. Explaining those required the construction of a new quantum model of light in which light was emitted and absorbed in discrete portions. Physicists started to think of light as composed of particle-like photons (the quanta of light). On the other hand, to explain interference and diffraction phenomena, photons had to have wave-like properties as well. This dual particle-wave property of photons is summarized in **Table 26.6**.

If we accept this model, photons must have both wave-like and particle-like behaviors. Experiments performed in 1932 tested this dual nature of photons in a dramatic way.

### Low-intensity experiments support the hypothesis of the dual nature of photons

Experiments by Soviet physicists Sergei Vavilov and Eugeny Brumberg supported the hypothesis that photons have both particle and wave properties. Vavilov and Brumberg used a double-slit setup with light of extremely low intensity. Their goal was to reduce the number of photons that pass through the slits each second to make them easier to detect as individual photons. For detection, they used the extreme sensitivity of human eyes to low intensities of light. It turns out that a person sitting in a dark room for extended periods of time develops increasing visual sensitivity to the extent that he or she can eventually see the individual photons hitting a screen that has been covered with a special material.

In Testing Experiment **Table 26.7** we describe the experimental setup, including the predictions for the outcome made by only the particle model of light, by only the wave model, and by the dual wave-particle photon model.

**Table 26.6** Wave-like and particle-like properties of photons.

| Experimental evidence | Wave-like properties $f$, $\lambda$ | Particle-like properties $E = hf$ |
|---|---|---|
| Double-slit interference | Superposition (including destructive interference) | |
| Diffraction | Single-slit interference and bending around obstacles | |
| Doppler effect | Change in frequency and wavelength | |
| Black body radiation | | Photon (quanta) emission |
| Photoelectric effect | | Photon (quanta) absorption |

**TESTING EXPERIMENT TABLE**

## 26.7 Do photons have both wave-like and particle-like properties?

### Testing experiment

Use a light source with variable intensity. Shine the light on two narrow slits. Place a fluorescent screen beyond the slits that indicates where light hits the screen.

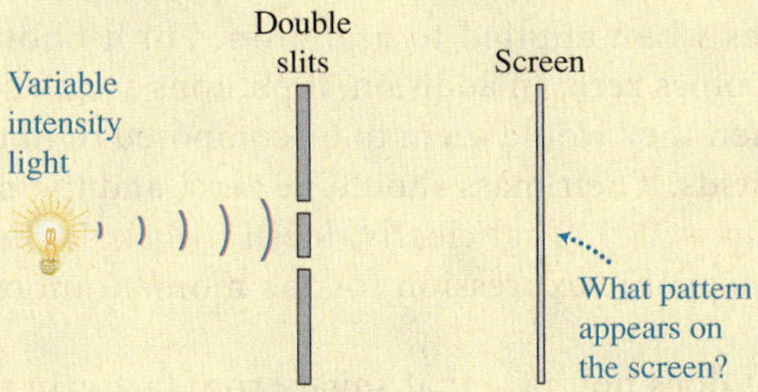


### Prediction

*Particle model:* Only two bright bands should appear—images of the slits themselves. As intensity decreases, we expect to see individual flashes caused by single photons at the slit image locations.

*Wave model:* Many alternating bright and dark bands should appear. As the intensity decreases, we expect all of the bright bands to become uniformly dimmer until they disappear.

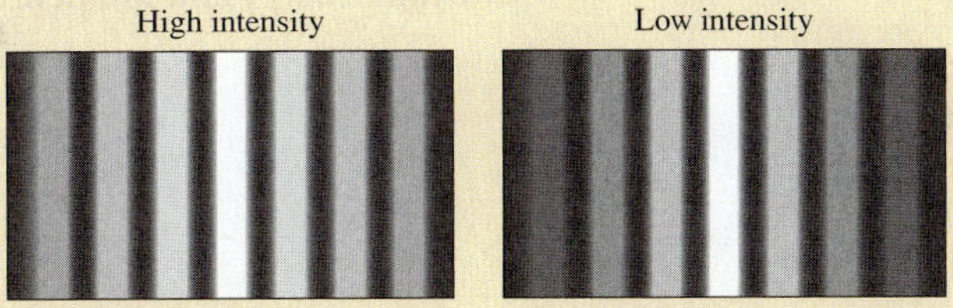


*Dual wave-particle (photon) model:* Many alternating bright and dark bands should appear. As the intensity decreases, we expect to see individual flashes due to single photons at bright band locations only. At the locations of the dark bands no flashes should be seen.

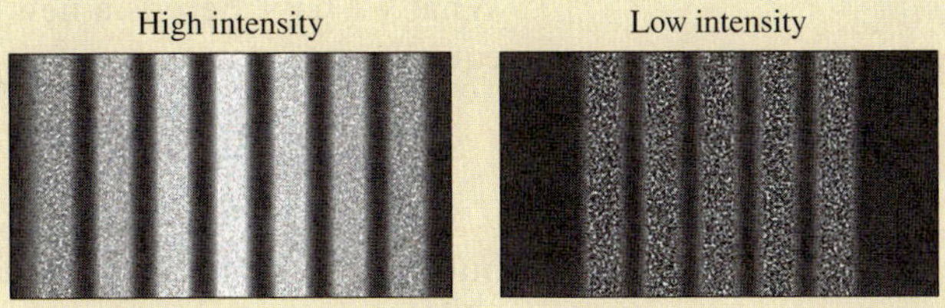


### Outcome

The experimental outcome is identical to the dual wave-particle model's prediction.

### Conclusion

- Photons hit the screen, indicated by the flashes at low intensity.
- The photons only reach the screen at places where constructive wave interference occurs.
- Photons are simultaneously exhibiting both wave-like and particle-like behaviors.

Vavilov and Brumberg's experimental results were astounding—individual photons somehow "know" that they should only reach screen locations where coherent waves from the two slits would interfere constructively. It is as if each individual photon passes through both slits, interfering with itself and thereby producing a flash only at a location on the screen where constructive interference occurs.

Earlier in this chapter we constructed the idea of a photon as a single quantum of EM radiation. If it is a single quantum, then it seems impossible that it could pass through both slits simultaneously. However, Vavilov and Brumberg's experiment supports this seemingly strange idea.

## Photon momentum

We have seen that photons participate in collisions with electrons inside metals (the photoelectric effect). This suggests that photons must have momentum. Otherwise, how could they knock an electron out of the cathode of a phototube during a collision? Because photons travel at light speed, they must be treated relativistically. The relativistic expression for the magnitude of the momentum of a particle of mass $m$ moving at speed $v$ is

**Active Learning Guide ›**

$$p = \frac{mv}{\sqrt{1 - (v/c)^2}}$$

This expression has difficulties when applied to a photon. For a photon $v = c$, and the denominator becomes zero. In addition, if photons are quanta of electromagnetic radiation, then they would seem to be composed of nothing but electric and magnetic fields. Their mass should be zero, and the numerator is also zero. So we get $p = 0/0$. That clearly doesn't work. Is there anything we can do to get a reasonable expression for the momentum of a photon?

We need an expression that does not have that square root factor in the denominator. Einstein resolved this by looking at the relativistic equations for the total energy and momentum of an object with nonzero mass:

$$E = \frac{mc^2}{\sqrt{1 - (v/c)^2}}$$

$$p = \frac{mv}{\sqrt{1 - (v/c)^2}}$$

Einstein eliminated $v$ from these equations by solving for $(v/c)^2$ in the second equation and then substituting the result into the first equation. This takes a fair amount of algebra. The final result is

$$E = \sqrt{(pc)^2 + (mc^2)^2} \tag{26.9}$$

What we have here is a new equation for the total energy of an object. It is equivalent to

$$E = \frac{mc^2}{\sqrt{1 - (v/c)^2}}$$

but has an important new feature: the mass of the object can be set to zero without the equation misbehaving:

$$E = \sqrt{(pc)^2 + (0 \cdot c^2)^2} = \sqrt{(pc)^2} = pc$$

In other words, according to Eq. (26.9), the total energy of an object with zero mass (like a photon) is $E = pc$. But we already know from Einstein's explanation of the photoelectric effect that the energy of a photon is $E = hf = (hc/\lambda)$. Therefore,

$$\frac{hc}{\lambda} = pc$$

Solving for $p$, we get an expression for the momentum of a photon:

$$p = \frac{h}{\lambda} \tag{26.10}$$

In 1922 Arthur Compton performed a testing experiment to determine if this expression for the momentum of a photon is reasonable. We will discuss Compton's experiment later in this chapter. For now, we'll consider an application of this expression.

**TIP** Notice what happens when Eq. (26.9) is used to determine the total energy of an object at rest. Then $p = 0$ and $E = mc^2$. The total energy of an object at rest just equals its rest energy, $E_0 = mc^2$.

**QUANTITATIVE EXERCISE 26.7 Revisiting $E_0 = mc^2$ with the Sun's changing mass**

Each second the Sun emits $3.8 \times 10^{26}$ J of energy, mostly in the form of visible, infrared, and UV photons. Estimate the change in mass of the Sun as a result of the energy these photons carry away each second.

**Represent mathematically** Use the idea of energy conservation for the system that includes the Sun and the emitted photons:

$$E_{0\,\mathrm{i}} = E_{0\,\mathrm{f}} + E_{\mathrm{photons}}$$

The rest energy of an object is $E_0 = mc^2$. Therefore,

$$m_{\mathrm{i}}c^2 = m_{\mathrm{f}}c^2 + E_{\mathrm{photons}}$$

**Solve and evaluate** Solving for $\Delta m = m_{\mathrm{f}} - m_{\mathrm{i}}$:

$$m_{\mathrm{i}}c^2 - m_{\mathrm{f}}c^2 = E_{\mathrm{photons}}$$
$$\Rightarrow m_{\mathrm{f}}c^2 - m_{\mathrm{i}}c^2 = -E_{\mathrm{photons}}$$
$$\Rightarrow (m_{\mathrm{f}} - m_{\mathrm{i}})c^2 = -E_{\mathrm{photons}}$$
$$\Rightarrow \Delta m = -\frac{E_{\mathrm{photons}}}{c^2}$$

Inserting the appropriate values gives

$$\Delta m = -\frac{3.8 \times 10^{26}\ \mathrm{J}}{(3.00 \times 10^8\ \mathrm{m/s})^2} = -4.2 \times 10^9\ \mathrm{kg}$$

The minus sign indicates that the Sun loses more than 4 billion kg each second! Won't the Sun "disappear" quickly while losing billions of kilograms each second?

The mass of the Sun is $m_{\mathrm{Sun}} = 2.0 \times 10^{30}$ kg. How long will it take for the Sun to "disappear"?

$$\frac{2.0 \times 10^{30}\ \mathrm{kg}}{4.2 \times 10^9\ \mathrm{kg/s}} = 4.8 \times 10^{20}\ \mathrm{s}$$

There are about $3.2 \times 10^7$ s in 1 year, which means the Sun will last

$$\frac{4.8 \times 10^{20}\ \mathrm{s}}{3.2 \times 10^7\ \mathrm{s/year}} = 1.5 \times 10^{13}\ \mathrm{years}$$

The age of the universe (as we know from Chapter 24) is approximately $13.7 \times 10^9$ years, which means the Sun will last more than 1000 times the current age of the universe if it continues to radiate energy at this rate (assuming that the mechanism behind it can continue for that long).

**Try it yourself:** Estimate the order of magnitude of the number of photons the Sun emits every second.

*Answer* Over $10^{45}$ photons/s are emitted from the Sun. Assumption: all photons have a frequency of $10^{14}$ Hz. If we make a different assumption, the estimated number will be different.

**Review Question 26.4** Explain how the outcome of the Vavilov-Brumberg experiment supports the idea that a photon has both wave-like and particle-like behaviors.

# 26.5 X-rays

The experiments and theory described in the previous section supported the idea that photons have momentum of magnitude $p = (h/\lambda)$. We will test this expression using X-rays in the next section. First, we need to learn about X-rays.

## Cathode ray tubes

The story of X-rays is another example of how persistence and attention to detail often lead to groundbreaking discoveries. At the beginning of Section 26.2 we assumed that free electrons reside inside metals, and we used this idea to help explain the photoelectric effect. In this section we'll find how the electron was discovered and how its discovery led to the accidental discovery of X-rays.

In the middle of the 19th century, physicists did not know that metals consisted of a crystal lattice and free electrons; furthermore, they did not know that electric current in metals was caused by the movement of those free electrons.

At that time, some physicists started experimenting with what are called cathode ray (Crookes) tubes—evacuated glass tubes with two electrodes embedded inside (**Figure 26.10**). The electrodes were connected to a battery that produced a potential difference across the electrodes. The tubes were similar in

**Figure 26.10** A cathode ray tube. (a) An actual cathode ray tube. (b) The electrical schematic of a cathode ray tube.

(a)

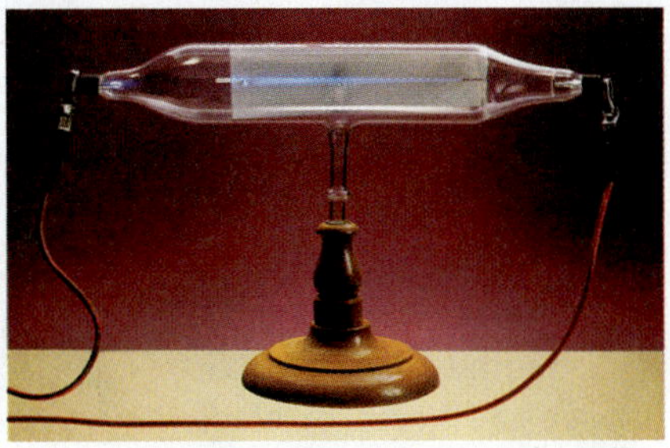

(b)

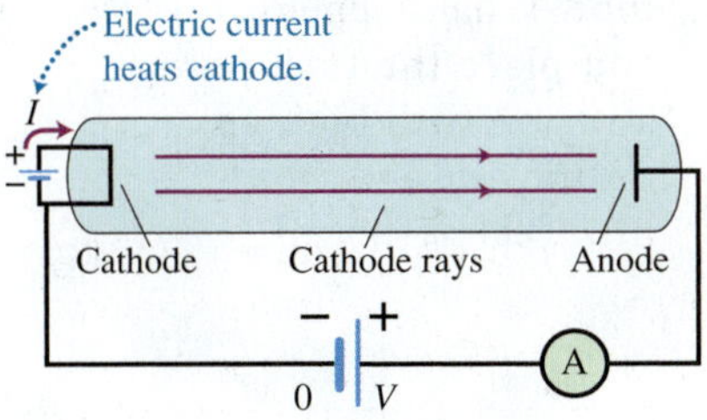

some ways to the tube that von Lenard used to study the photoelectric effect. The difference was that the cathode (connected to the negative terminal of the battery) could be heated to high temperatures, rather than being illuminated by light.

Physicists discovered that although a physical gap was present between the cathode and the anode, when the cathode was heated, a current appeared in the circuit and the tube would glow. Because the tube's interior was a vacuum, they thought that the cathode must emit some kind of rays that traveled from the cathode to the anode. These rays were called *cathode rays*.

J. J. Thomson at the Cavendish Laboratory at Cambridge University was one of the scientists studying cathode rays. He observed that when the rays hit a metal target, it became negatively charged. Through numerous experiments, Thomson found that the rays could be deflected by electric and magnetic fields as though they consisted of negatively charged particles. He determined their charge-to-mass ratio ($q/m$) and found it to be over 1000 times greater than that of a hydrogen ion, suggesting that the cathode rays either were very highly charged or had a very small mass. Later experiments by other scientists supported the latter hypothesis. The cathode rays' charge-to-mass ratio was also independent of the choice of cathode material.

This led Thomson to conclude that there was only one type of cathode ray rather than many different types. Thomson determined the charge-to-mass ratio for cathode rays, the modern value being $q/m = -1.76 \times 10^{11}\,\text{C/kg}$.

Let's model a cathode ray as a stream of charged particles and see if we can explain the behavior of the cathode ray tubes. Remember, the cathode in the tube is hot. This means that the particles inside the cathode have a large amount of kinetic energy of random motion (thermal energy) and could knock each other out of the cathode. Once outside the cathode, the electric field between the cathode and anode would cause them to accelerate toward the anode.

At about the same time, chemists were trying to find out what comprised electric currents. What was moving inside metals? They concluded that the charge carriers were tiny particles, and in 1894 George Johnstone Stoney called them "electrons." Experiments in electrochemistry helped determine the value of the smallest charge carried by these particles to be $-1.6 \times 10^{-19}\,\text{C}$.

We can now use the value of the charge of the electron found by Stoney and the charge-to-mass ratio found by Thomson to determine the mass of the electron:

$$m_e = \frac{1}{q_e/m_e} q_e = \frac{1}{-1.76 \times 10^{11}\,\text{C/kg}}(-1.6 \times 10^{-19}\,\text{C}) = 9.1 \times 10^{-31}\,\text{kg}$$

The mass of the electron is about 1/2000th the mass of a hydrogen atom.

**CONCEPTUAL EXERCISE 26.8 Deflecting cathode rays**

Imagine that you can see the beam of electrons shooting from the cathode to the anode of a cathode ray tube. What happens if you place the tube in a region with both an electric field and a magnetic field, as shown?

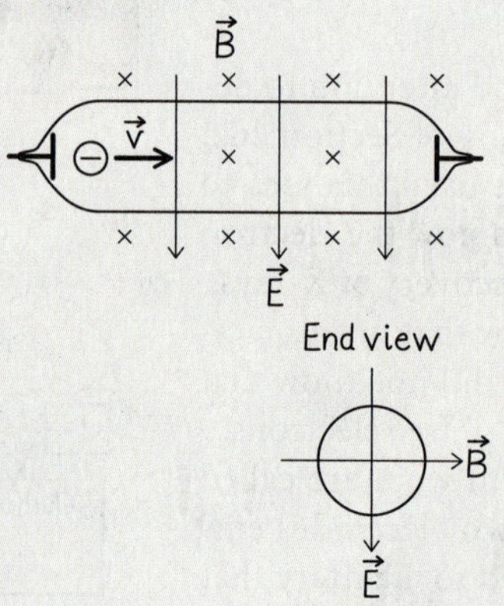


**Sketch and translate** First, determine separately the electric and magnetic forces exerted on the electrons, and then consider their combined effect.

**Simplify and diagram** Use $\vec{E}$ field and $\vec{B}$ field vectors to represent the electric and magnetic fields, and assume that they are uniform. Assume the electron is a point-like object and that the gravitational force exerted by Earth on the electron is small compared with the electric and magnetic forces exerted on it by the fields.

1. Electric force: Because electrons are negatively charged particles, the electric field shown in the figure

exerts an upward force on them. They have a vertical upward acceleration while the horizontal component of their velocity remains constant. Therefore, they move like projectiles, and their path is like an upward arcing parabola, as shown at the right.

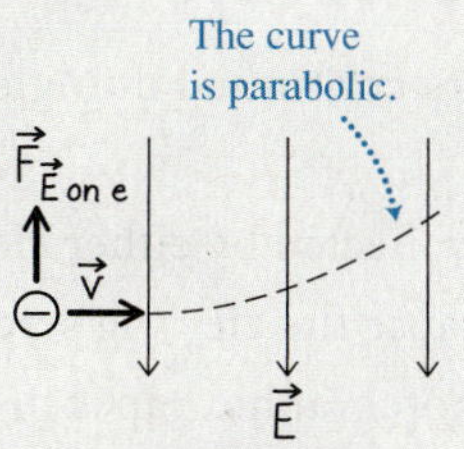


2. Magnetic force: The magnetic field exerts a force on the negatively charged electrons that is perpendicular to their velocity and to the $\vec{B}$ field. According to the right-hand rule for the magnetic force, this force points downward. Therefore, the acceleration of the electrons also points downward, as shown above, and their path will be a downward arcing circle (not a parabola). Recall that the direction of the magnetic force exerted by the field changes as the electron's velocity changes, always pointing perpendicular to that velocity. This rotation causes uniform circular motion.

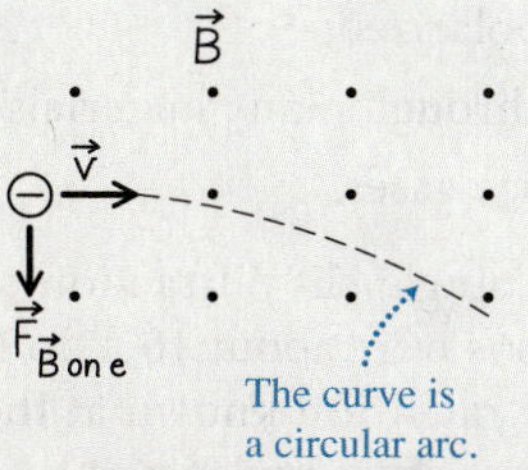


If both an electric field and a magnetic field are present as shown, then the direction of the net force exerted on the electrons depends on the magnitudes of the $\vec{E}$ field and the $\vec{B}$ field. Because the electric and magnetic forces exerted on the electrons point in opposite directions, it is possible that the net force exerted on the electrons can be zero. (This possibility is investigated in the Try It Yourself question below.) In this case, one can determine the speed of the electron and then the charge-to-mass ratio. This is exactly what Thomson did to determine $q/m$ for cathode rays. If the forces do not add to zero, the electrons follow a complicated trajectory.

**Try it yourself:** What should be the relationship between the magnitudes of the $\vec{E}$ field and the $\vec{B}$ field so that the electron travels in a straight line through this region?

*Answer:* $E = vB$.

## The accidental discovery of X-rays

Experiments with cathode ray tubes led to the accidental discovery of X-rays. In 1895 Wilhelm Roentgen, like von Lenard, was experimenting with the rays outside the vacuum tubes by letting them pass through special windows and observing them when they hit a fluorescent screen. The screen glowed when the cathode rays hit it. One night when Roentgen was ready to leave the lab, he covered the cathode ray tube with cardboard and turned off the lights. He noticed a faint glow on the fluorescent screen that was lying nearby on the lab table (**Figure 26.11a**). This was strange; the room was dark, and the cathode ray tube was covered with cardboard. Roentgen turned the lights back on and checked all of the equipment. He turned the tube off and turned off the lights again. The fluorescent screen was dark.

At this point most people would have breathed a sigh of relief and gone home. But Roentgen went back to the tube with the lights still off and turned it on. The glow on the screen appeared again even though the tube was completely covered with cardboard! When Roentgen placed his hand near the screen, he could see a strange shadow made by his hand. The shadow looked as though the flesh of his hand was completely transparent, and only his bones stopped whatever was making the screen glow.

Roentgen stayed in the lab all that night trying to figure out what rays could possibly come out of the completely covered tube and make the fluorescent screen glow. He called these mysterious rays X-rays since at the time he couldn't explain what they were. The first X-ray image of a human was made by Roentgen of his wife Anna Bertha's hand on December 22, 1895 (Figure 26.11b). Although we now understand what X-rays are, they are still called X-rays.

**Figure 26.11** Roentgen discovers X-rays. (a) Although the cathode ray tube was covered, the fluorescent screen still glowed in the dark. (b) Roentgen took an X-ray photo of his wife's hand.

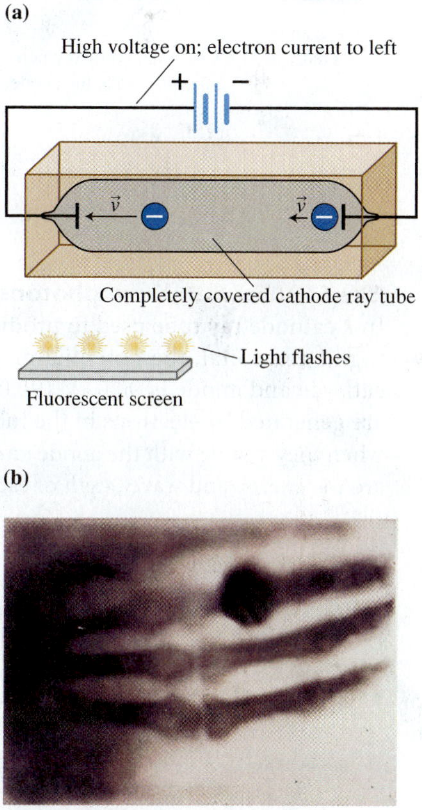

## Explanation of X-rays

Roentgen and others continued doing experiments and found the following about X-rays:

1. X-rays are not deflected by either electric or magnetic fields.
2. X-rays do not cause the viewing screen to become electrically charged.
3. X-rays cause photographic paper to darken.
4. X-rays produce a diffraction pattern of dark and bright bands on the screen after passing through a single narrow slit.
5. X-rays can be polarized.
6. X-rays can go through many materials that are opaque to visible light.
7. X-rays can ionize gases.

Analysis of the single-slit diffraction patterns of the rays showed that the wavelengths of X-rays were about $10^{-10}$ to $10^{-11}$ m, a much smaller wavelength than any other EM radiation known at the time—much smaller than visible light. Given all of these characteristics, it is reasonable to hypothesize that X-rays are streams of very high energy photons. But how are these photons produced?

## Producing X-ray photons

**Figure 26.12** Diagram of an X-ray tube. The collision of high-energy electrons with the anode produces X-ray photons.

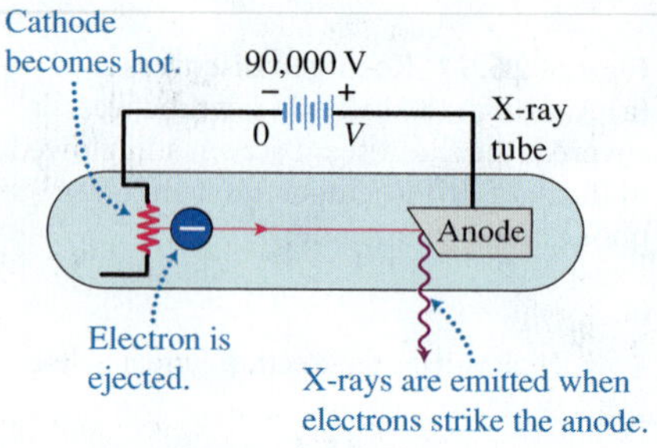


Consider **Figure 26.12**. The hot metal cathode is a source of negatively charged electrons. Due to their high random motion (the metal is very hot), the kinetic energy of some of the free electrons in the metal exceeds the metal's work function. These electrons escape the cathode and accelerate toward the anode because of the electric field within the evacuated cathode ray tube. (The electric field is produced by the negatively charged cathode and positively charged anode.) When the electrons reach the anode, they collide with it and stop abruptly. Recall that when a charged object accelerates (such as an electron stopping when it collides with the anode), it emits electromagnetic radiation (Chapter 24). The radiation emitted in this case might be in the form of X-ray photons. Let's estimate the energy of the photons produced by this mechanism and compare it to the energy of photons of visible light. We want to check whether the energy of the electrons stopping in the cathode ray tube is enough to produce an X-ray photon.

**EXAMPLE 26.9 X-ray photons**

In a cathode ray tube used to produce X-rays for imaging in a hospital, the potential difference between the cathode and anode is 90.0 kV (90,000 V). If the X-rays are generated by electrons in the tube that stop abruptly when they collide with the anode and emit photons, what are the energy and wavelength of each photon?

**Sketch and translate** The process is sketched in three steps (states) below.

**Simplify and diagram** We consider the system to consist of a single electron, the electric field produced by the electrodes, the electrodes themselves, and a single X-ray photon. We will represent each part of the process using a work-energy bar chart. The initial state shown below is just as the electron leaves the cathode. We assume that the electron has zero kinetic energy and the system has zero electric potential energy—that is, the electric potential at the cathode is zero.

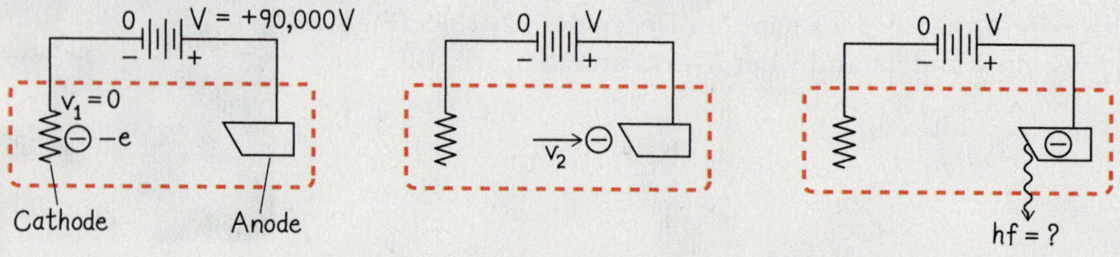

The intermediate state is just before the fast-moving electron smashes into the anode. The final state is when the electron has stopped inside the anode and just after the X-ray photon has been emitted. We assume that none of the kinetic energy of the electron is converted into the internal energy of the cathode.

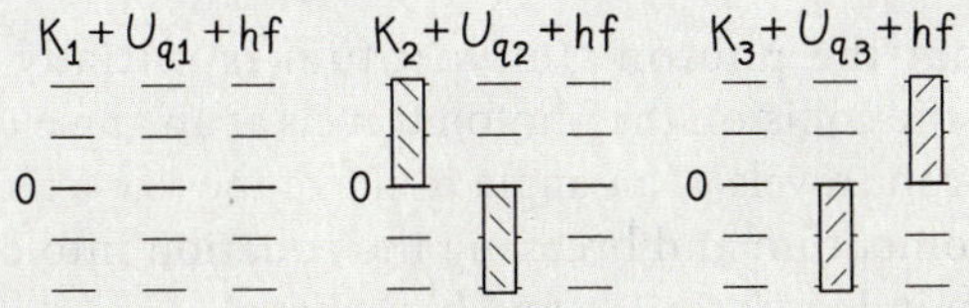


**Represent mathematically** Use the first and second state bar charts to apply the generalized work-energy principle:

$$0 + 0 + 0 = \frac{1}{2}mv_2^2 + (-e)V$$

Use the second and third state bar charts to write

$$\frac{1}{2}mv_2^2 + (-e)V = (-e)V + hf$$

where $h = 6.63 \times 10^{-34}\ \text{J} \cdot \text{s}$ is Planck's constant. Combining these two results, we get

$$0 = (-e)V + hf$$

**Solve and evaluate** The photon's energy is

$$hf = eV = (1.6 \times 10^{-19}\ \text{C})(90.0 \times 10^3\ \text{V})$$
$$= (1.44 \times 10^{-14}\ \text{J})\left(\frac{1\ \text{eV}}{1.6 \times 10^{-19}\ \text{J}}\right)$$
$$= 90 \times 10^3\ \text{V}$$

This is much greater than the energy characteristic of visible light photons. The photon's frequency is

$$f = \frac{eV}{h} = \frac{(1.6 \times 10^{-19}\ \text{C})(90.0 \times 10^3\ \text{V})}{6.63 \times 10^{-34}\ \text{J} \cdot \text{s}}$$
$$= 2.2 \times 10^{19}\ \text{Hz}$$

which is much higher than visible light frequencies. The photon's wavelength is

$$\lambda = \frac{c}{f} = \frac{3.00 \times 10^8\ \text{m/s}}{2.2 \times 10^{19}\ \text{Hz}} = 1.4 \times 10^{-11}\ \text{m}$$

**Try it yourself:** What potential difference between the electrodes in a cathode ray tube would produce $10^{-10}$ m wavelength X-rays?

*Answer:* 12,000 V.

Based on the result of our calculation we can say that in a cathode ray tube an electron can acquire enough kinetic energy to produce an X-ray photon when it stops.

**Review Question 26.5** What physical mechanism produces X-rays in a cathode ray tube?

## 26.6 The Compton effect and X-ray interference

In Section 26.4 we hypothesized that if each photon has a specific amount of electromagnetic energy, and photons possess some particle-like behavior, then each photon must also have a specific amount of momentum. We reasoned earlier that the magnitude of the momentum $p$ of a photon equals

$$p = \frac{h}{\lambda}$$

If a photon possesses momentum, then we should be able to use momentum conservation ideas to describe collision processes involving photons.

According to the particle-like aspect of the photon model, the collision of a photon with a single electron (an electron that is bound neither to an atom nor to a crystal) (**Figure 26.13a**) should be similar to a collision between two billiard balls. The electron and scattered photon move off at angles relative to the direction of motion of the incident photon (Figure 26.13b).

An X-ray photon has very large energy and momentum compared with a visible photon, and it should have a much larger effect on the electron than does a visible photon. The electron should move off at an angle at a significant speed, and the direction and magnitude of the photon's momentum (and therefore its wavelength) should change significantly.

**Figure 26.13** A photon-electron collision. (a) A photon of initial wavelength $\lambda_i$ moves in the $+x$-direction toward an electron. (b) After collision, the electron moves away at angle $\alpha$ relative to the $x$-axis, and the photon with wavelength $\lambda_f$ moves away at angle $\theta$.

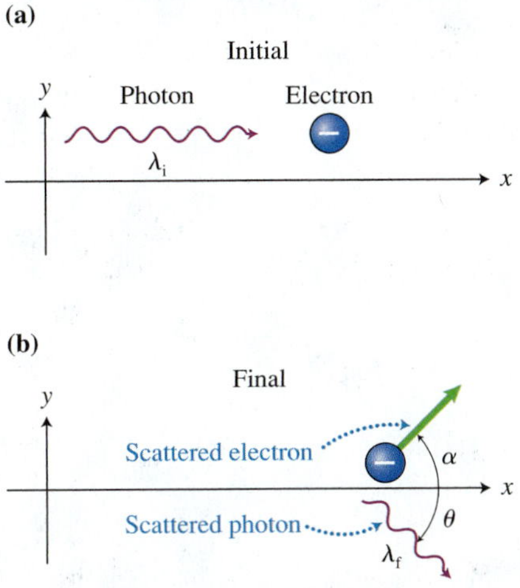

We need to describe this process mathematically. Assuming that the electron and photon interact only with each other and not with other objects, the momentum of the electron-photon system should be constant:

$$\vec{p}_{\text{p i}} + \vec{p}_{e\text{i}} = \vec{p}_{\text{p f}} + \vec{p}_{e\text{f}}$$

The initial momentum of the stationary electron is zero. Thus we have

$$\vec{p}_{\text{p i}} = \vec{p}_{\text{p f}} + \vec{p}_{e\text{f}}$$

For simplicity, assume that the photon's momentum is initially in the $+x$-direction and that after the collision the photon travels at an angle $\theta$ below the $+x$-axis, while the electron travels at an angle $\alpha$ above the $+x$-axis. Using $p = h/\lambda$ for the photon momentum and breaking the equation into components, for the process shown in Figure 26.13a and b, we have

$$x\text{-component:} \qquad \frac{h}{\lambda_{\text{i}}} = \frac{h}{\lambda_{\text{f}}}\cos\theta + p_{e\text{f}}\cos\alpha$$

$$y\text{-component:} \qquad 0 = -\frac{h}{\lambda_{\text{f}}}\sin\theta + p_{e\text{f}}\sin\alpha$$

In addition, the energy of the system should also be constant. Because the photon is neutral, the system has zero electric potential energy. Since both the photon and electron have no internal structure, the internal energy of the system is also zero. So we need to keep track of the energy of the photon and the kinetic and rest energies of the electron. Therefore,

$$E_{\text{p i}} + K_{e\text{i}} + E_{0\,e} = E_{\text{p f}} + K_{e\text{f}} + E_{0\,e}$$

As the electron is initially at rest and the rest energy of the electron doesn't change, we have

$$E_{\text{p i}} = E_{\text{p f}} + K_{e\text{f}}$$

Using $E = hc/\lambda$ for the photon energy, we get

$$\frac{hc}{\lambda_{\text{i}}} = \frac{hc}{\lambda_{\text{f}}} + K_{e\text{f}}$$

This equation can be combined with the two momentum component equations into a single relationship that describes the collision between the X-ray photon and the electron. It is a complicated derivation that must be done using relativity because of the high speeds involved. We simply provide the result:

$$\lambda_{\text{f}} - \lambda_{\text{i}} = \frac{h}{m_e c}(1 - \cos\theta) \qquad (26.11)$$

Note that if the hypothesis about X-rays behaving like particles is correct, then the wavelength of an X-ray photon after the collision should always be greater than or equal to the wavelength before the collision. The reason is that the right side of Eq. (26.11) is always non-negative (regardless of the value of $\theta$), which means that $\lambda_{\text{f}} \geq \lambda_{\text{i}}$. This also means that the magnitude of the photon's momentum decreases or remains the same. This finding is reasonable because some of the photon's momentum is transferred to the electron.

## Compton's experiment

In 1922 Arthur Compton conducted an experiment to determine whether the photon's wavelength actually changes in these collisions. He shot a beam of X-ray photons through a graphite target. Although the electrons in the target do not seem isolated, the binding energy between the electrons and the carbon atoms is about 1/1000 times the energies of the X-ray photons. As a result the electrons can be approximated as isolated and not interacting with the carbon atoms.

Compton found that the scattered photons had different wavelengths than the incident photons. The change in the photon wavelength $\lambda$ as a function of its scattering angle $\theta$ was consistent with Eq. (26.11). This phenomenon became known as the **Compton effect,** and it supported the idea that X-rays were indeed photons with both wave-like (having a wavelength) and particle-like (having momentum) properties. We can now assert this idea with greater confidence.

**Momentum of a photon** **The momentum of a photon $p_{\text{photon}}$ is Planck's constant divided by its wavelength:**

$$p_{\text{photon}} = \frac{h}{\lambda} \tag{26.10}$$

We can now summarize these ideas about photons scattering from charged particles.

**Compton effect** **If a photon of wavelength $\lambda_i$ scatters off a charged particle of mass $m$ and moves off at an angle $\theta$ relative to its initial direction of motion, the scattered photon will then have a longer wavelength $\lambda_f$ given by the following equation:**

$$\lambda_f - \lambda_i = \frac{h}{m_e c}(1 - \cos\theta) \tag{26.11}$$

Compton was awarded a Nobel Prize in 1927 for his work describing the effect that carries his name.

## Are X-rays dangerous?

Short-wavelength UV radiation, X-rays, and shorter wavelength (higher frequency) gamma ray photons are called ionizing radiation because each photon has enough energy to knock electrons out of atoms or to break bonds that hold atoms together in molecules. Ionizing radiation can damage genetic material (DNA), increasing the risk of mutations that lead to cancer. Fortunately, the body has potent DNA repair mechanisms, reducing the chance of serious harm from UV and X-ray exposure. However, people who work around ionizing radiation (dental hygienists and radiology technicians, for example) must take precautions to avoid overexposure.

### QUANTITATIVE EXERCISE 26.10 Number of photons absorbed in a medical X-ray

Determine the number of photons absorbed during a single chest X-ray if the body absorbs a total of $10^{-3}$ J of 0.025-nm wavelength X-ray photons.

**Represent mathematically** The energy of a single X-ray photon is

$$E_1 = hf = \frac{hc}{\lambda}$$

The number of photons will then be

$$N = \frac{E_{\text{total}}}{E_1}$$

**Solve and evaluate** Each photon has energy

$$E_1 = \frac{(6.63 \times 10^{-34}\ \text{J}\cdot\text{s})(3.00 \times 10^{8}\ \text{m/s})}{0.025 \times 10^{-9}\ \text{m}} = 8 \times 10^{-15}\ \text{J}$$

Thus, the number of photons absorbed in one chest X-ray is approximately

$$N = \frac{10^{-3}\ \text{J}}{8 \times 10^{-15}\ \text{J}} \approx 10^{11}\ \text{photons}$$

**Try it yourself:** Suppose the wavelengths of the X-rays were three times as long. Then approximately how many photons would pass through your body in a $10^{-3}$-J X-ray exam?

*Answer:* About $3 \times 10^{11}$ photons.

**QUANTITATIVE EXERCISE 26.11 High-energy photons absorbed from environment**

Your body is continuously exposed to natural radiation sources such as radioactive materials in the environment and in the food you eat, and from cosmic rays coming from supernovae and other objects in the universe. Each year on average you absorb about $10^{-3}$ J of this radiation per kilogram of body mass. Estimate the number of these ionizing photons absorbed each second during the year.

**Represent mathematically** Assume that your body mass $m = 70.0$ kg. The rate at which energy from this radiation is absorbed is $R = 10^{-3}$ J/kg·year. Assume for example that the radiation consists of $\lambda = 6 \times 10^{-13}$ m gamma ray photons each with energy $E_1 = 3 \times 10^{-13}$ J. The number $N$ of photons absorbed in 1 year is then

$$N = \frac{Rm}{E_1}$$

We can then determine the number absorbed each second by dividing by the number of seconds in 1 year:

$$1\text{ year}\left(\frac{365\text{ days}}{1\text{ year}}\right)\left(\frac{24\text{ h}}{1\text{ day}}\right)\left(\frac{3600\text{ s}}{1\text{ h}}\right) = 3.2 \times 10^{7}\text{ s}$$

**Solve and evaluate** Inserting the appropriate values gives

$$N = \frac{(10^{-3}\text{ J/kg}\cdot\text{year})(70.0\text{ kg})}{3 \times 10^{-13}\text{ J}}$$
$$= 2.3 \times 10^{11}\text{ photons/year}$$

The number of photons absorbed each second is then

$$\frac{2.3 \times 10^{11}\text{ photons/year}}{3.2 \times 10^{7}\text{ s/year}} \approx 7 \times 10^{3}\text{ photons/s}$$

As mentioned earlier, the body has ample repair mechanisms to take care of such radiation exposure.

**Try it yourself:** Exposure to a 1.0-s burst of 0.010-nm gamma rays depositing 1.0 J of energy per kilogram into a human body has a 10% chance of being fatal within 30 days. This level of radiation is 10 billion times greater than the ambient natural radiation our bodies are continuously exposed to. How many gamma ray photons are absorbed by the body during this exposure? (Some workers present at the 1986 Chernobyl nuclear power plant disaster experienced as much as 16 times this exposure.)

*Answer:* 3–4 × $10^{15}$ photons.

## X-ray interference

We have mostly focused on the particle-like behavior of X-ray photons. However, X-rays can exhibit wave-like behavior if they can be made to interact with objects whose size is comparable to the wavelength of the X-rays. The spacing between atoms and ions in a crystal is the same order of magnitude as the wavelength of X-ray photons. When X-rays shine on a crystal, the X-rays reflected off the crystal lattice form a pattern very similar to that of visible light passing through an interference grating. The details of the interference pattern allow scientists to determine the internal structure of the crystal. X-ray scattering can also be used to determine the structure of proteins and DNA. X-ray diffraction images produced by British biophysicist Rosalind Franklin played a key role in determining the helical structure of DNA in the early 1950s.

**Review Question 26.6** Why does the wavelength of an X-ray photon increase when it scatters off an electron?

**Figure 26.14** A photocell. Visible light and/or UV radiation knocks electrons out of the emitter. The electrons then move to the collector, thus causing a current.

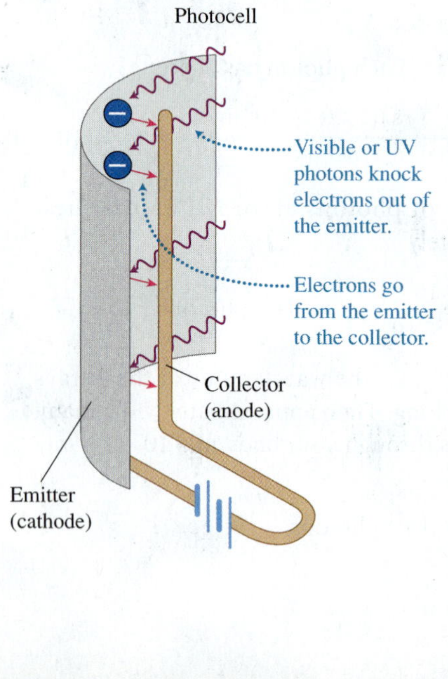
Photocell
Visible or UV photons knock electrons out of the emitter.
Electrons go from the emitter to the collector.
Collector (anode)
Emitter (cathode)

# 26.7 Photocells and solar cells: Putting it all together

The photoelectric effect has numerous applications. One of them is the **photocell**, which functions similarly to the tube that von Lenard used for his experiments. The cathode of the photocell (also called the *emitter*) is connected to the negative terminal of a battery and therefore is at a lower electric potential than the anode (called the *collector*). The cathode emitter is shaped like a cylindrical concave mirror to focus emitted electrons toward the anode, which has the shape of a thin wire that collects the electrons (**Figure 26.14**). When light of a frequency greater than the cutoff frequency shines on the emitter, electrons are ejected and absorbed by the collector, causing an electric current in the circuit. The magnitude of the current is a measure of the visible light or UV intensity being directed at the photocell. Products that use photocells include light meters (used in photography to adjust the size of a camera lens aperture), motion detectors, and photoelectric smoke detectors (**Figure 26.15**).

**Figure 26.15** A photoelectric smoke detector. Light from the source shoots across the top of the T-shaped detector. A cathode sits at the bottom of the T. If the air is clear, no light is scattered to the cathode and hence no photocurrent will be produced. However, smoke particles scatter light out of the beam, causing it to reach the cathode of the photocell and producing a photocurrent. When this photocurrent reaches a certain level, the alarm is triggered.

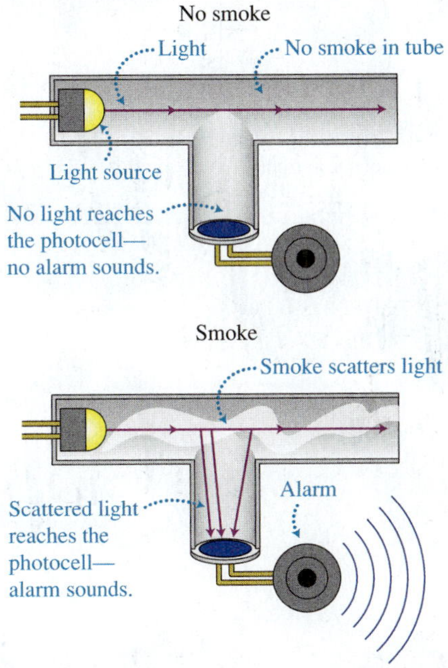
No smoke
Light
No smoke in tube
Light source
No light reaches the photocell—no alarm sounds.
Smoke
Smoke scatters light
Scattered light reaches the photocell—alarm sounds.
Alarm

## Solar cells

Solar cells, such as the rooftop panels used to generate electric current, are another application of the photoelectric effect. Solar cells are based on the semiconductor technology we discussed in an earlier chapter (Chapter 16). Semiconductors are materials that act as insulators under some conditions and conductors under others.

Silicon is a commonly used semiconductor. A silicon atom has four valence electrons that form covalent chemical bonds with neighboring atoms (each electron is shared between them). As a result, a silicon atom has no free electrons—it is an electric insulator. However, when silicon is heated or when light of sufficient frequency shines on it, some of the electrons gain enough energy to become free to roam around the crystal. The spaces vacated by these electrons lack negative charge and become positively charged "holes." These holes can be filled by other bound electrons that are not energetic enough to become free but can nevertheless move among the still-bound valence electrons. Both free electrons and holes can potentially contribute to a current if the silicon is placed in an external electric field. **Figure 26.16** shows this process—light photons deliver energy to the silicon, increasing its energy and allowing some of its valence electrons to become free. Whether electrons will be freed depends on the frequency of the light.

Adding impurities to silicon—a process called doping—is another way to turn it into a conductor. There are two types of impurities: electron donors and electron acceptors. Phosphorus, an example of an electron donor, contains five valence electrons, one more than silicon's four. The result is one "extra" electron that does not form a strong bond with a neighboring silicon atom and is essentially free to move about the bulk of the crystal (**Figure 26.17a**).

**Figure 26.16** Light shining light on silicon produces free electrons and "holes."

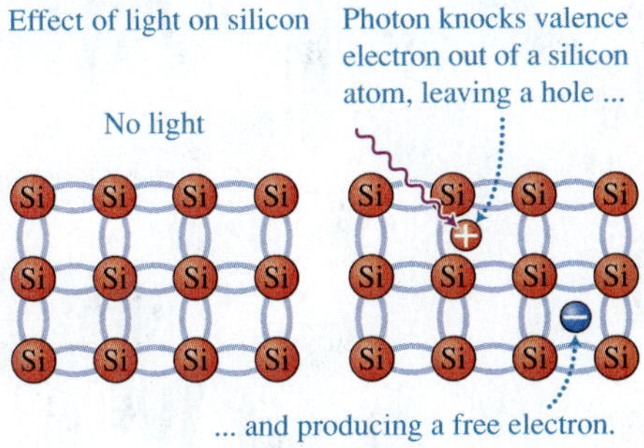
Effect of light on silicon
No light
Photon knocks valence electron out of a silicon atom, leaving a hole ...
... and producing a free electron.

**Figure 26.17** Doping of silicon. (a) Phosphorus is an n-type donor of free electrons and (b) boron is a p-type acceptor of bound electrons—a creator of "holes." (c) The contact of p- and n-type semiconductors causes charge transfer across the junction. (d) Charge transfer across the junction produces an electric field $\vec{E}$.

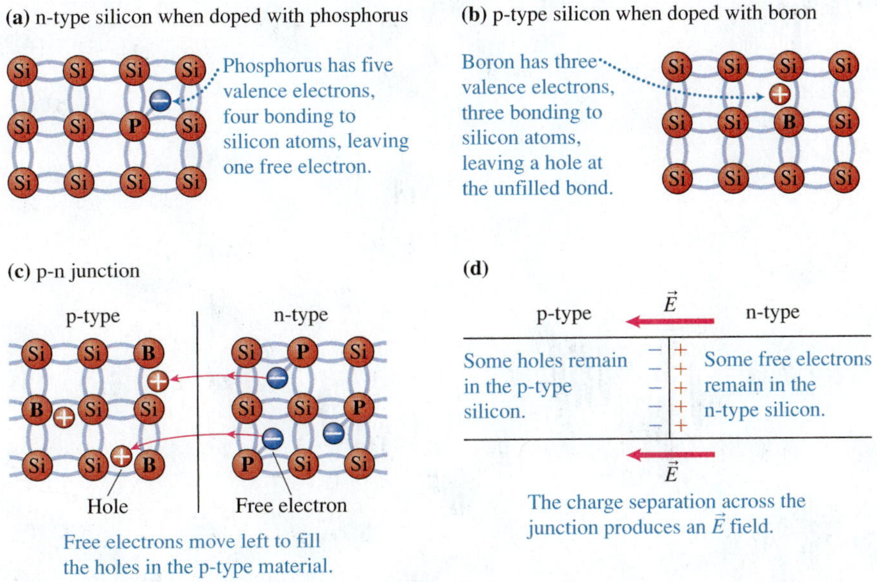
(a) n-type silicon when doped with phosphorus
Phosphorus has five valence electrons, four bonding to silicon atoms, leaving one free electron.
(b) p-type silicon when doped with boron
Boron has three valence electrons, three bonding to silicon atoms, leaving a hole at the unfilled bond.
(c) p-n junction
p-type
n-type
Hole
Free electron
Free electrons move left to fill the holes in the p-type material.
(d)
p-type
n-type
Some holes remain in the p-type silicon.
Some free electrons remain in the n-type silicon.
The charge separation across the junction produces an E field.

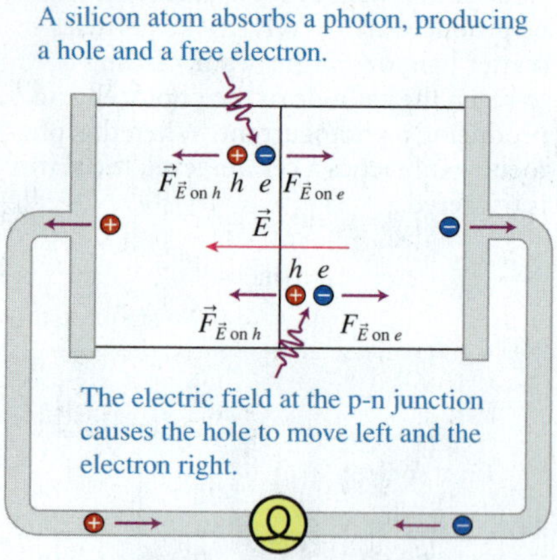
A silicon atom absorbs a photon, producing a hole and a free electron.
$\vec{F}_{\vec{E}\text{ on }h}$
$h$
$e$
$\vec{F}_{\vec{E}\text{ on }e}$
$\vec{E}$
$h$ $e$
$\vec{F}_{\vec{E}\text{ on }h}$
$\vec{F}_{\vec{E}\text{ on }e}$
The electric field at the p-n junction causes the hole to move left and the electron right.

**Figure 26.18** The free electrons and the holes produced by absorbing the Sun's photons are separated by the electric field of the p-n junction. The p-n junction in the presence of light acts like a battery.

The opposite is true for an electron acceptor. Boron, for example, has only three valence electrons. It can capture nearby electrons to form a strong fourth bond with the adjacent silicon atoms, creating extra holes (Figure 26.17b). These holes act as free positively charged particles (they lack the presence of a negatively charged electron). Silicon doped with phosphorus or any other electron donor is called an **n-type semiconductor**; the boron-doped version is called a **p-type semiconductor**. Note that these crystals are still electrically neutral, since they are still completely composed of neutral atoms.

When a p-type semiconductor and an n-type semiconductor are brought into contact, they form a **p-n junction** (Figure 26.17c). Free electrons in the n-type silicon migrate toward the p-type silicon, while the holes in the p-type silicon migrate toward the n-type silicon. In the region near the boundary, the n-type silicon becomes positively charged and the p-type silicon becomes negatively charged. This charge separation produces an electric field (Figure 26.17d).

If we shine light of sufficient frequency on the p-n junction, both halves of the junction will absorb photons whose energy is transferred to valence electrons that then become free. These electrons leave behind holes—virtual positively charged particles (**Figure 26.18**). The electric field created by the p-n junction then separates the freed electrons and holes by accelerating them in opposite directions. The p-n junction in the presence of light results in an electric current—it behaves like a battery! If we now connect a lightbulb to the two ends of the p-n junction, we will observe a current in the circuit, and the bulb will light. Thus the energy of light photons is converted into electric energy.

**Review Question 26.7** What characteristics would a material need in order to be used to make anodes in photocells?

# Summary

**Words** | **Pictorial and physical representations** | **Mathematical representation**

**Stefan's law for total radiating power** The total black body radiation power $P$ (all wavelengths) from an object with a surface of area $A$ and kelvin temperature $T$ is proportional to the fourth power of the temperature. (Section 26.1)

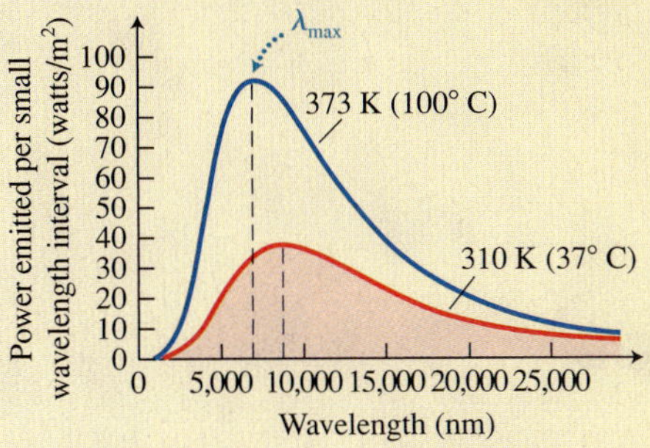


$P = \sigma A T^4$, where
$\sigma = 5.67 \times 10^{-8}\ \mathrm{W/m^2K^4}$
Eq. (26.1)

**Wien's law** The wavelength $\lambda_{max}$ at which a black body emits the maximum intensity radiation in small wavelength interval depends on the temperature $T$ in kelvins of the black body. (Section 26.1)

$$\lambda_{max} = \frac{2.90 \times 10^{-3}\ \mathrm{m \cdot K}}{T}$$
Eq. (26.2)

A **photon** is a discrete quantum of electromagnetic radiation that has both wave-like properties and particle-like properties. The energy of a photon depends on its frequency $f$ and Planck's constant $h$. (Sections 26.3 and 26.4)

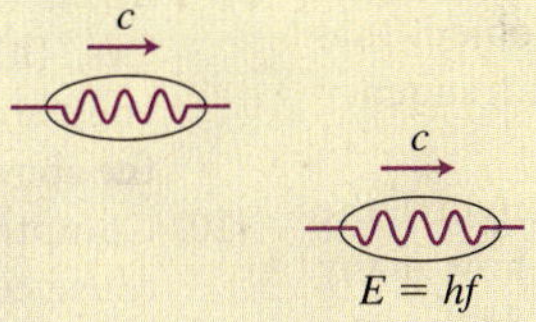


$E_{photon} = hf$
$h = 6.63 \times 10^{-34}\ \mathrm{J \cdot s}$ Eq. (26.7)

**Photoelectric effect** A photon of energy $hf$ hits a metal cathode of work function $\phi$ (the minimum energy needed to remove a free electron from the metal). The electron-lattice system absorbs the photon to free the electron from the metal. If the photon energy is greater than $\phi$, the electron has nonzero kinetic energy when it leaves the metal. (Sections 26.2 and 26.3)

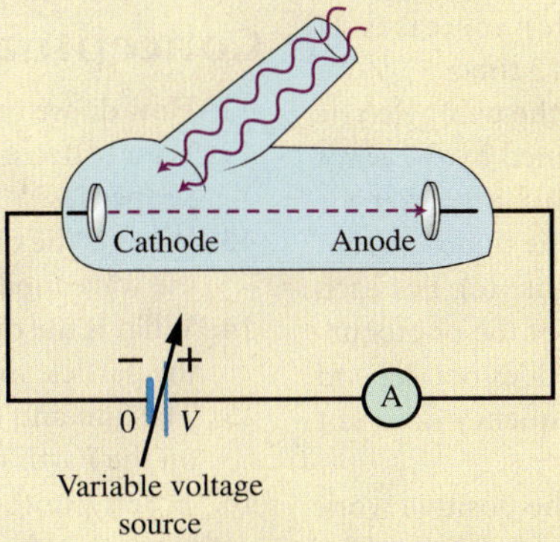


$-\phi + hf = K_e$ Eq. (26.5)

**Stopping potential** $-V_s$ is the minimum negative electric potential that stops the maximum speed electron from reaching the anode. (This assumes that the potential at the cathode is zero.) (Section 26.2)

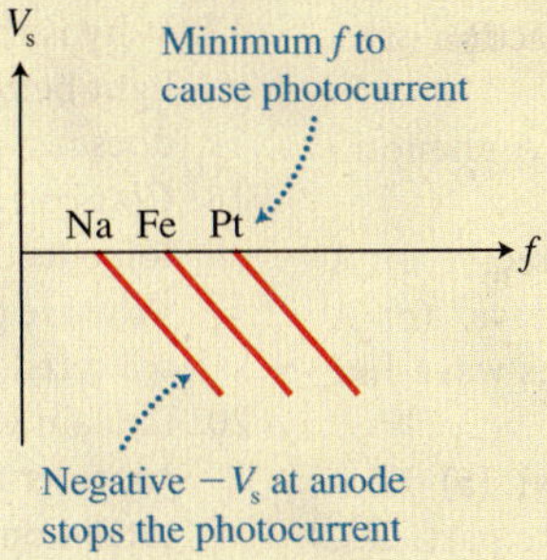


$$\frac{1}{2} m_e v_{ei\,max}^2 = (-e)(-V_s)$$

where $V_s$ is the positive magnitude of the stopping potential

**Cutoff frequency** $f_{cutoff}$ is the minimum frequency that a photon must have to remove an electron from the metal. (Sections 26.2 and 26.3)

$$f_{cutoff} = \frac{\phi}{h}$$
Eq. (26.6)

*(continued)*

| Words | Pictorial and physical representations | Mathematical representation |
|---|---|---|
| **Photon momentum** The magnitude of a photon's momentum $p_{photon}$ depends on its frequency $f$ (or wavelength $\lambda$), Planck's constant $h$, and the speed of light $c$. (Section 26.4) | 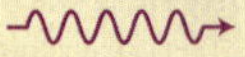 $p_{photon}$ | $p_{photon} = \dfrac{h}{\lambda} = \dfrac{hf}{c}$ Eq. (26.10) |

MP® For instructor-assigned homework, go to **MasteringPhysics.**

# Questions

## Multiple Choice Questions

1. What is a black body? (a) An object painted with black, non-reflective paint (b) An object that does not radiate any electromagnetic waves (c) A model of an object that emits electromagnetic waves with a characteristic frequency distribution
2. If you triple the temperature of a black body (in kelvins), the total amount of energy that it radiates increases how many times? (a) 3 times (b) 9 times (c) 27 times (d) 81 times
3. If you triple the temperature of a black body, the wavelength at which it emits the maximum amount of energy (a) increases 3 times. (b) increases 9 times. (c) increases 27 times. (d) increases 81 times. (e) decreases 3 times.
4. Among the items listed below, choose all of the photoelectric effect observations that could not be explained by the wave model of light. (a) Photocurrent appears instantly after the light is turned on. (b) The magnitude of the photocurrent is directly proportional to the intensity of light. (c) For each intensity of light there is a maximum value of the photocurrent. (d) The stopping potential difference does not depend on the intensity of light. (e) The cutoff frequency does not depend on the intensity of light.
5. What is the work function equal to? (a) The positive work one needs to do to remove an electron from a metal (b) The positive potential energy of interaction of an electron and a metal (c) The negative potential energy of interaction of an electron and a metal (d) a and b (e) a and c
6. What is a photon? (a) A physical quantity (b) A phenomenon (c) A model (d) A law of physics
7. Which of the following statements are not correct? (a) A photon is a particle of light. (b) A photon is a wave. (c) A photon has momentum. (d) A photon possesses wave-like properties.
8. What did the Vavilov-Brumberg experiments show? (a) Photons behave like waves. (b) Photons behave like particles. (c) Photons simultaneously behave like waves and like particles.
9. Which statement describes most accurately our present understanding of light? (a) Light is like a stream of particles. (b) Light is like an electromagnetic wave. (c) Light is a complex phenomenon that cannot be modeled by only one of the above models.
10. Compton's experiments are done with X-ray photons because compared to visible light, (a) they have short wavelength. (b) they have large momentum. (c) they have large frequency. (d) all of the above.

## Conceptual Questions

11. How do we know that photons possess wave-like properties?
12. How do we know that photons possess particle-like properties?
13. What is the difference between the photon model of light and the wave model of light?
14. What is the difference between the photon model of light and the particle model of light?
15. The Sun and other celestial objects emit X-rays. Why are we, on the Earth's surface, not concerned about them?
16. When photographic film is being developed, technicians work in a darkroom in which a special red light is used for illumination, not a regular white light. Explain why.
17. Why is the photon model of light supported by the fact that light below a certain frequency, when striking a metal plate, does not cause electrons to be freed from the metal?
18. Compare and contrast the photoelectric effect and the Compton effect.
19. Why are photons absorbed during the photoelectric effect but scattered (reflected) during the Compton effect?
20. Explain why a single electron cannot absorb a photon. [Hint: consider the kinetic energy in a reference frame in which the total momentum is zero.]

# Problems

Below, BIO indicates a problem with a biological or medical focus. Problems labeled EST ask you to estimate the answer to a quantitative problem rather than develop a specific answer. Problems marked with ✎ require you to make a drawing or graph as part of your solution. Asterisks indicate the level of difficulty of the problem. Problems with no * are considered to be the least difficult. A single * marks moderately difficult problems. Two ** indicate more difficult problems.

## 26.1 Black body radiation

1. **Wavelength of radiation from a person** If a person could be modeled as a black body, at what wavelength would his or her surface emit the maximum energy?
2. * (a) A surface at 27 °C emits radiation at a rate of 100 W. At what rate does an identical surface at 54 °C emit radiation? (b) Determine the wavelength of the maximum amount of radiation emitted by each surface.
3. **Maximum radiation wavelength from star, Sun, and Earth** Determine the wavelengths for the following black body radiation sources where they emit the most energy: (a) A blue-white star at 40,000 K; (b) the Sun at 6000 K; and (c) Earth at about 300 K.
4. * EST **Star colors and radiation frequency** The colors of the stars in the sky range from red to blue. Assuming that the color indicates the frequency at which the star radiates the maximum amount of electromagnetic energy, estimate the surface temperature of red, yellow, white, and blue stars. What assumptions do you need to make about white stars to estimate the surface temperature?
5. * EST Estimate the surface area of a 60-watt lightbulb filament. Assume that the surface temperature of the filament when it is plugged into an outlet of 120 V is about 3000 K and the power rating of the bulb is the electric energy/s it consumes (not what it radiates). Incandescent lightbulbs usually radiate in visible light about 10% of the electric energy that they consume.
6. * EST **Photon emission rate from human skin** Estimate the number of photons emitted per second from 1.0 $cm^2$ of a person's skin if a typical emitted photon has a wavelength of 10,000 nm.
7. ** **Balancing Earth radiation absorption and emission** Compare the average power that the surface of Earth facing the Sun receives from it to the energy that Earth emits over its entire surface due to it being a warm object. Assume that the average temperature of Earth's surface is about 15 °C. The distance between Earth and the Sun is about $1.5 \times 10^{11}$ m.

## 26.2 and 26.3 Photoelectric effect and Quantum model explanation of the photoelectric effect

8. (a) Explain how you convert energy in joules into energy in electron volts. (b) The kinetic energy of an electron is 2.30 eV. What is its kinetic energy in joules?
9. ** ✎ Draw a picture of a phototube and the electric circuit that you can build to study the photoelectric effect. Label all of the parts and explain the purpose of each part.
10. * (a) Describe the experimental findings for the photoelectric effect. (b) What findings could be explained by the wave model of light? (c) What experimental findings concerning the photoelectric effect could not be explained by the wave model of light?
11. The stopping potential for an ejected photoelectron is −0.50 V. What is the maximum kinetic energy of the electron ejected by the light?
12. ✎ Light shines on a cathode and ejects electrons. Draw an energy bar chart describing this process. Explain why the frequency of incident light determines whether the electrons will be ejected or not.
13. * What is the cutoff frequency of light if the cathode in a photoelectric tube is made of iron?
14. * The work function of cesium is 2.1 eV. (a) Determine the lowest frequency photon that can eject an electron from cesium. (b) Determine the maximum possible kinetic energy in electron volts of a photoelectron ejected from the metal that absorbs a 400-nm photon.
15. * Visible light shines on the metal surface of a photocell having a work function of 1.30 eV. The maximum kinetic energy of the electrons leaving the surface is 0.92 eV. Determine the light's wavelength.
16. * **Equation Jeopardy 1** Solve for the unknown quantity in the equation below and write a problem for which the equation could be a solution.

$$-3.9\text{ eV} + (6.63 \times 10^{-34}\text{ J}\cdot\text{s})f\left(\frac{1\text{ eV}}{1.6 \times 10^{-19}\text{ J}}\right) = (-e)(-1.0\text{ V})$$

17. **Camera film exposure** In an old-fashioned camera, the film becomes exposed when light striking it initiates a complex chemical reaction. A particular type of film does not become exposed if struck by light of wavelength longer than 670 nm. Determine the minimum energy in electron volts needed to initiate the chemical reaction.

## 26.4 Photons

18. **CO vibration** A vibrating carbon monoxide (CO) molecule produces infrared photons of energy 0.26 eV. Determine the frequency of CO vibration, which is the same as the frequency of the infrared radiation the molecule emits.
19. **Breaking a molecular bond** Suppose the bond in a molecule is broken by photons of energy 5.0 eV. Determine the frequency and wavelength of these photons and the region of the electromagnetic spectrum in which they are located.
20. * A 1.0-eV photon's wavelength is 1240 nm. Use a ratio technique to determine the wavelength of a 5.0-eV photon.
21. BIO **Tanning bed** In a tanning bed, exposure to photons of wavelength 300 nm or less can do considerable damage. Determine the lowest energy in electron volts of such photons.
22. * Determine the number of 650-nm photons that together have energy equal to the rest energy of an electron. [Hint: See Section 25.8.]
23. * BIO **Laser surgery** Scientists studying the use of lasers in various surgeries have found that very short $10^{-12}$-s laser pulses of power $10^{+12}$ W with 65 pulses every $200 \times 10^{-6}$ s produced much cleaner welds and ablations (removals of body tissues) than longer laser pulses. Determine the number of 10.6-$\mu$m photons in one pulse and the average power during the 65 pulses delivered in $200 \times 10^{-6}$ s.
24. * A laser beam of power $P$ in watts consists of photons of wavelength $\lambda$ in nanometers. Determine in terms of these quantities the number of photons passing a cross section along the beam's path each second.
25. * What is the mass of the photons in the previous problem?

26. * BIO **Pulsed laser replaces dental drills** A laser used for many applications of hard surface dental work emits 2780-nm wavelength pulses of variable energy (0–300 mJ) about 20 times per second. Determine the number of photons in one 100-mJ pulse and the average power of these photons during 1 s.
27. * **Light hitting Earth** The intensity of light reaching Earth is about 1400 $W/m^2$. Determine the number of photons reaching a 1.0-$m^2$ area each second. What assumptions did you make?
28. * EST **Lightbulb** Roughly 10% of the power of a 100-watt incandescent lightbulb is emitted as light, the rest being emitted as heat and longer-wavelength radiation. Estimate the number of photons of light coming from a bulb each second. What assumptions did you make? How will the answer change if the assumptions are not valid?
29. * BIO **Human vision sensitivity** To see an object with the unaided eye, the light intensity coming to the eye must be about $5 \times 10^{-12}\ J/m^2 \cdot s$ or greater. Determine the minimum number of photons that must enter the eye's pupil each second in order for an object to be seen. Assume that the pupil's radius is 0.20 cm and the wavelength of the light is 550 nm.

## 26.5 X-rays

30. * Explain how a cathode ray tube works. Draw a picture and an electric circuit. Label the important elements and explain how they work together to produce cathode rays.
31. * Explain how we know that cathode rays are low-mass light negatively charged particles. Draw pictures and field diagrams to illustrate your explanation.
32. * An X-ray tube emits photons of frequency $1.33 \times 10^{19}$ Hz or less. (a) Explain how the tube creates the X-ray photons. (b) Determine the potential difference across the X-ray tube.
33. * Electrons are accelerated across a 40,000-V potential difference. (a) Explain why X-rays are created when the electrons crash into the anode of the X-ray tube. (b) Determine the frequency and wavelength of the maximum-energy photons created.
34. ** An electron with kinetic energy $K$ moving horizontally to the right in a tube of length $L$ passes through a uniform electric field with an $\vec{E}$ field that points downward. The electron is initially moving toward the center of the screen. Develop an expression for the strength of the field so the electron hits the screen a vertical height $h$ above the center.
35. * An electron with kinetic energy $K$ moving horizontally to the right in a tube of length $L$ enters a uniform electric field that points upward. How strong and in what direction should a magnetic field be so that the electron moves straight ahead with no velocity change?
36. * A small $1.0 \times 10^{-5}$-g piece of dust falls in Earth's gravitational field. Determine the distance it must fall so that the change in gravitational potential energy of the dust-Earth system equals the energy of a 0.10-nm X-ray photon.
37. BIO **X-ray exam** While being X-rayed, a person absorbs $3.2 \times 10^{-3}$ J of energy. Determine the number of 40,000-eV X-ray photons absorbed during the exam.
38. * BIO **Body cell X-ray** (a) A body cell of $1.0 \times 10^{-5}$-m radius absorbs $4.2 \times 10^{-14}$ J of X-ray radiation. If the energy needed to produce one positively charged ion is 100 eV, how many positive ions are produced in the cell? (b) How many ions are formed in the $3.0 \times 10^{-6}$-m-radius nucleus of that cell (the place where the genetic information is stored)?

## 26.6 The Compton effect and X-ray interference

39. * **Equation Jeopardy 2** Solve for the unknown quantity in the equation below and write a problem for which the equation could be a solution.

$$\lambda_f - (100 \times 10^{-9}\ m) = \frac{(6.63 \times 10^{-34}\ J \cdot s)(1 - \cos 37°)}{(9.1 \times 10^{-31}\ kg)(3.0 \times 10^{8}\ m/s)}$$

40. * In a Compton effect scattering experiment, an incident photon's frequency is $2.0 \times 10^{19}$ Hz; the scattered photon's frequency is $1.4 \times 10^{19}$ Hz. Determine the kinetic energy increase of the electron, in units of electron volts, when the photon is scattered from it.
41. * An electron hit by an X-ray photon of energy $5.0 \times 10^{4}$ eV gains $3.0 \times 10^{3}$ eV of energy. Determine the wavelength of the scattered photon leaving the site of the collision.
42. * A laser produces a short pulse of light whose energy equals 0.20 J. The wavelength of the light is 694 nm. (a) How many photons are produced? (b) Determine the total momentum of the emitted light pulse.
43. * **Levitation with light** Light from a relatively powerful laser can lift and support glass spheres that are $20.0 \times 10^{-6}$ m in diameter (about the size of a body cell). Explain how that is possible.

## 26.7 Photocells and solar cells: Putting it all together

44. * BIO EST **Light detection by human eye** The dark-adapted eye can supposedly detect one photon of light of wavelength 500 nm. Suppose that 100 such photons enter the eye each second. Estimate the intensity of the light. Assume that the diameter of the eye's pupil is 0.50 cm.
45. * BIO EST **Fireflies** Fireflies emit light of wavelengths from 510 nm to 670 nm. They are about 90% efficient at converting chemical energy into light (compared to about 10% for an incandescent lightbulb). Most living organisms, including fireflies, use adenosine triphosphate (ATP) as an energy molecule. Estimate the number of ATP molecules a firefly would use at 0.5 eV per molecule to produce one photon of 590-nm wavelength if all the energy came from ATP.
46. * Light of wavelength 430 nm strikes a metal surface, releasing electrons with kinetic energy equal to 0.58 eV or less. Determine the metal's work function.
47. ** **Sail in laser "wind" 1** A powerful 0.50-W laser emitting 670-nm photons shines on the sail of a tiny 0.10-g cart that can coast on a horizontal frictionless track. (a) Determine the force of the light on the sail. Assume that the light is totally reflected. (b) What time interval is needed for the cart's speed to increase from zero to 2.0 m/s?
48. ** **Sail in laser "wind" 2** A powerful 0.50-W laser emitting 670-nm photons shines on a black sail of a tiny 0.10-g cart that can coast on a frictionless track. (a) Determine the force of the light on the sail. Assume that the light is totally absorbed by the sail. (b) What time interval is needed for the cart's speed to increase from zero to 2.0 m/s?
49. * **Comet tails** Comets are relatively small extraterrestrial objects that move around the Sun in highly elliptical orbits. The comet's head is made primarily of ice with a small amount of dust. When the comet is near the Sun, gases and dust evaporated from the surface of the comet form a "tail." Independent of the direction of motion of the comet, the tail always points away from the Sun. Use the photon model of light to explain why the comet's tail points away from the Sun.
50. ** **Solar cell** A 0.20-m × 0.20-m photovoltaic solar cell is irradiated with 800 $W/m^2$ sunlight of wavelength 500 nm.

(a) Determine the number of photons hitting the cell each second. (b) Determine the maximum possible electric current that could be produced. (c) Explain how a solar cell converts the energy of sunlight into electric energy.

51. * * The Sun is about 150 million km from Earth. The energy emitted by the Sun in all directions every second is about $4 \times 10^{26}$ J. Use this information to evaluate whether the value of the power per unit area provided in Problem 50 is reasonable.

## General Problems

52. * * EST **Sirius radiation power** Sirius, a star in the constellation of Canis Major, is the second brightest star of the northern sky (the brightest is the Sun). Its surface temperature is 9880 K and its radius is 1.75 times greater than the radius of the Sun. Estimate the energy that Sirius emits every second from its surface and compare this energy to the energy that the Sun emits. The radius of the Sun is about $7.0 \times 10^8$ m and the energy emitted per second is about $3.9 \times 10^{26}$ W.

53. * BIO EST **Owl night vision** Owls can detect light of intensity $5 \times 10^{-13}\ \text{W/m}^2$. Estimate the minimum number of photons an owl can detect. Indicate any assumptions you used in making the estimate.

54. * * BIO **Photosynthesis efficiency** During photosynthesis in a certain plant, eight photons of 670-nm wavelength can cause the following reaction: $6CO_2 + 6H_2O \rightarrow C_6H_{12}O_6 + 6O_2$. During respiration, when the plant metabolizes sugar, the reverse reaction releases 4.9 eV of energy per $CO_2$ molecule. Determine the ratio of the energy released (respiration) to the energy absorbed (photosynthesis), a measure of photosynthetic efficiency.

55. * * Suppose that light of intensity $1.0 \times 10^{-2}\ \text{W/m}^2$ is made of waves rather than photons and that the waves strike a sodium surface with a work function of 2.2 eV. (a) Determine the power in watts incident on the area of a single sodium atom at the metal's surface (the radius of a sodium atom is approximately $1.7 \times 10^{-10}$ m). (b) How long will it take for an electron in the sodium to accumulate enough energy to escape the surface, assuming it collects all light incident on the atom?

56. * * **Force of light on mirror** A beam of light of wavelength 560 nm is reflected perpendicularly from a mirror. Determine the force that the light exerts on the mirror when $10^{20}$ photons hit the mirror each second. [Hint: Refer to the impulse-momentum equation (Chapter 5). You may assume that the magnitude of the photons' momenta is unchanged by the collision, but their directions are reversed.]

57. * * **Force of sunlight on Earth** We wish to determine the net force on Earth caused by the absorption of light from the Sun. (a) Determine the net area of the surface of Earth exposed to sunlight (Earth's radius is $6.38 \times 10^6$ m). (b) The solar radiation intensity is $1400\ \text{J/s} \cdot \text{m}^2$. Determine the momentum of photons hitting Earth's surface each second. (c) Use the impulse-momentum equation to determine the average force of this radiation on Earth.

58. * * EST **Levitating a person** Suppose that we wish to support a 70-kg person by levitating the person on a beam of light. (a) If all of the photons striking the person's bottom surface are absorbed, what must be the power of the light beam, which is made of 500-nm-wavelength photons? (b) Estimate the person's temperature change in 1 s.

59. * * An electron that resides by itself in an open region of space is struck by a photon of light. Using nonrelativistic formulas, show that the electron cannot absorb the photon's energy and simultaneously absorb its momentum. To conserve both energy and momentum, the photon must be absorbed by the electron near another mass, which carries away some of the momentum but little of the energy.

60. * * Compton's original experiment involved scattering 0.0709-nm X-rays off a graphite target (primarily composed of carbon atoms). He observed the scattered X-rays at different angles using a spectrometer (a device that uses interference to determine the wavelength of the X-rays). What scattered the X-ray photons: the carbon nuclei or the electrons? To answer this question, determine the wavelength of the scattered photon when, after colliding with an electron or with a carbon atom, it travels at a 90° angle relative to its initial momentum. The mass of an electron is $m_e = 9.11 \times 10^{-31}$ kg and the mass of a carbon nucleus is $m_c = 19.9 \times 10^{-27}$ kg.

## Reading Passage Problems

BIO **Capturing energy from sunlight—photosynthesis** Green plants capture and store the energy of photons from the Sun to help build complex molecules such as glucose. The process starts with the photoelectric effect. Chloroplasts in plant cells contain many photosynthetic units, each of which has about 300 pigment "antenna" molecules that absorb sunlight (**Figure 26.19**).

**Figure 26.19** Photosynthetic unit inside plant chloroplasts. "Antenna" molecules absorb photons and become excited. The energy of this excitation is passed from one antenna to another until it arrives at an "acceptor" molecule. The acceptor passes the energy to an electron transport chain, where it is captured and stored in biomolecules like ATP.

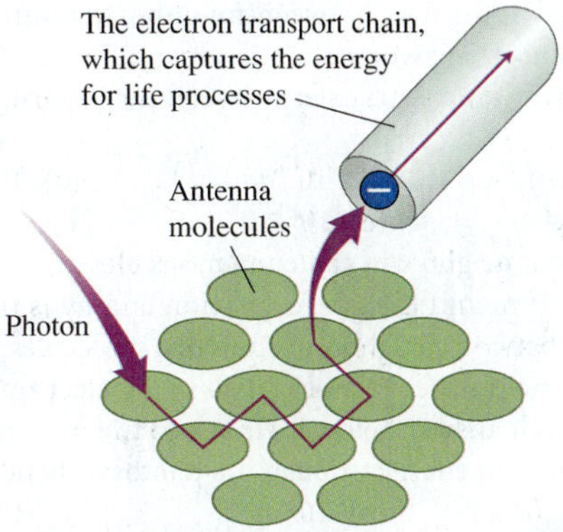


Suppose that one of the antenna molecules absorbs a photon. The electrons in the molecule are now in an excited state, which means that they are temporarily storing the extra energy gained from the photon. Typically, such a molecule reemits a photon and returns to its original energy state in about $10^{-8}$ s. If the energy of the photon is simply released, the plant has lost the Sun's energy.

In the photosynthetic units, the antenna molecules are linked closely to each other. When a photon is absorbed by one antenna molecule, its excitation energy is transferred to a neighbor antenna molecule. This excited neighbor in turn excites another of its neighbors, in a more or less random fashion. Eventually, an electron in an excited antenna molecule is transferred to a primary acceptor, carrying the extra energy of the photon along with it. This energetic electron now passes through a complex electron transport chain. At several places along this pathway, the energy is given up to help form stable chemical bonds (**Figure 26.20**).

**Figure 26.20** The electron transport chain. Electrons are passed between intermediate molecules, and energy is captured at certain steps for the ultimate production of ATP, the cellular energy molecule.

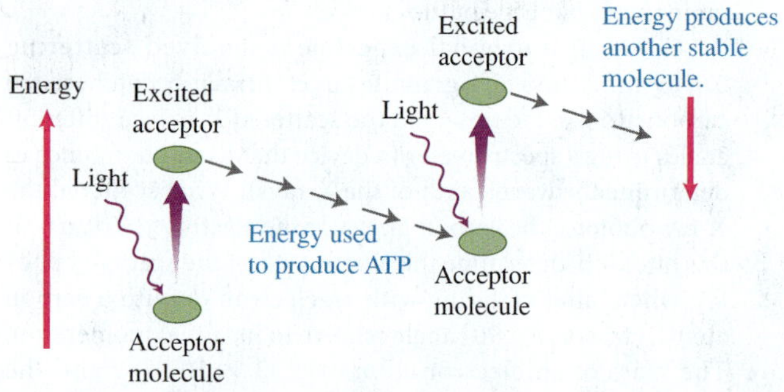


Life on Earth depends on the absorption of the photon, the "random walk" of the excitation energy from antenna molecule to antenna molecule, and the capture of the excited electron energy, all in a fraction of a second.

61. What is the number of antenna molecules that can absorb light in a photosynthetic unit?
(a) 1 (b) About 10 (c) Over 100
(d) Over 10,000 (e) About $10^8$

62. Suppose an antenna molecule absorbs a 430-nm photon and that this energy is transferred directly to the acceptor molecule. Which answer below is closest to the energy that the photoelectron brings to the electron transport chain?
(a) 1 eV (b) 2 eV (c) 3 eV
(d) 4 eV (e) 5 eV

63. Suppose that the excited energy of one antenna molecule is transferred 100 times between neighboring antenna molecules. Which answer below is closest to the maximum time interval for the transfer between neighboring antenna molecules?
(a) $10^{-3}$ s (b) $10^{-6}$ s (c) $10^{-8}$ s
(d) $10^{-10}$ s (e) $10^{-14}$ s

64. Suppose that neighboring antenna molecules are separated by about $10^{-10}$ m and that the excitation energy is transferred 100 times between neighboring antenna molecules before the photoelectric transfer of an electron to the electron transport chain. Which answer below is closest to the minimum speed of the excitation energy through the photosynthetic unit?
(a) 1 m/s (b) $10^2$ m/s (c) $10^4$ m/s
(d) $10^{-2}$ m/s (e) $10^{-4}$ m/s

65. The high-energy electron that transfers into an electron transport chain from a photosynthetic unit
(a) comes from the antenna molecule that absorbed the photon.
(b) comes from the acceptor molecule, which absorbed the photon.
(c) comes from the acceptor molecule that is excited by a nearby antenna molecule.
(d) is produced by the oxidation of a water molecule.
(e) is produced in the electron transport chain as other molecules react.

BIO **Radiation from our bodies** A person's body can be modeled as a black body that radiates electromagnetic radiation. Wien's law gives the peak wavelength of this radiation: $\lambda_{max} = (2.90 \times 10^{-3}\ \text{m} \cdot \text{K}/T)$, which is mostly infrared radiation.

The net radiated power is the difference between the power emitted and the power absorbed ($P_{net} = P_{emit} - P_{absorb}$) and is given by Stefan's law: $P_{net} = A\sigma\varepsilon(T^4 - T_0^4)$, where $\sigma = 5.67 \times 10^{-8}\ \text{W/m}^2 \cdot \text{K}^4$, $T$ is the absolute temperature of the skin, and $T_o$ is the absolute temperature of the surroundings . The surface area $A$ of a human body is about 2 $\text{m}^2$, and the emissivity $\varepsilon$ of the skin is about 1 (it is an almost perfect emitter and perfect absorber of infrared radiation). The skin temperature for a nude person is about 306 K and the temperature in a room is about 20 °C (293 K). Thus, for a nude person, $P_{net} \approx 160$ W. For a clothed person, the effective skin temperature is cooler, about 28 °C (301 K). The net radiated power is then $P_{net} \approx 100$ W.

66. The wavelength of maximum light emission from the body is closest to:
(a) 500 nm (b) 700 nm (c) 1200 nm
(d) 4500 nm (e) 9500 nm

67. During one day, the total radiative energy loss by a clothed person having a 2 $\text{m}^2$ surface area in a 20 °C room in kcal (1 kcal = 4180 J) is closest to:
(a) 0.5 kcal (b) 100 kcal (c) 400 kcal
(d) 2000 kcal (e) 3000 kcal

68. Photographs taken with a regular camera and with an infrared camera are shown in **Figure P26.68**. The man's arm is covered with a black plastic bag. Why is his arm visible in the infrared picture?
(a) Light does not pass through black plastic, but infrared radiation does.
(b) His arm is very warm under the black plastic bag and emits much more infrared radiation.
(c) The bag temperature is similar to his arm temperature.
(d) The black bag absorbs light and becomes warm and is a good thermal emitter.
(e) None of the above

**Figure P26.68**

Visible light photo Infrared photo

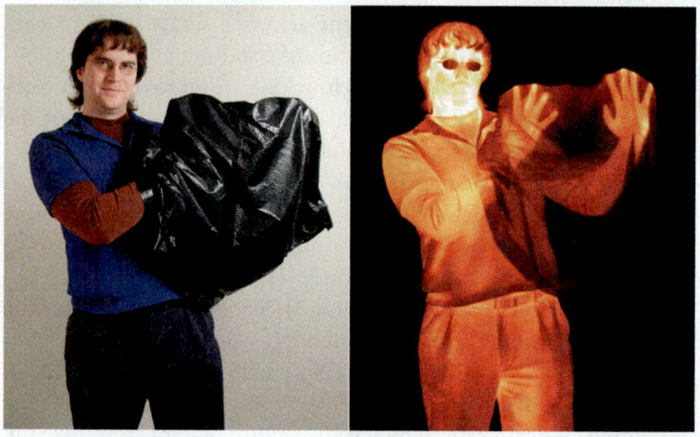

69. The man's glasses appear clear with the regular camera photo and black with the infrared camera photo in Figure P26.68. Why?
(a) Light does not pass through glass, but infrared radiation does.
(b) Light passes through glass, but infrared radiation does not.
(c) The lenses of the glasses are cool compared to the man's face, and thus they emit little infrared radiation.
(d) a and c (e) b and c

70. What is the ratio of the emitted radiative power from a 310 K surface and the same surface at 300 K closest to?
(a) 0.86 (b) 0.97 (c) 1.03
(d) 1.07 (e) 1.14

# Atomic Physics 27

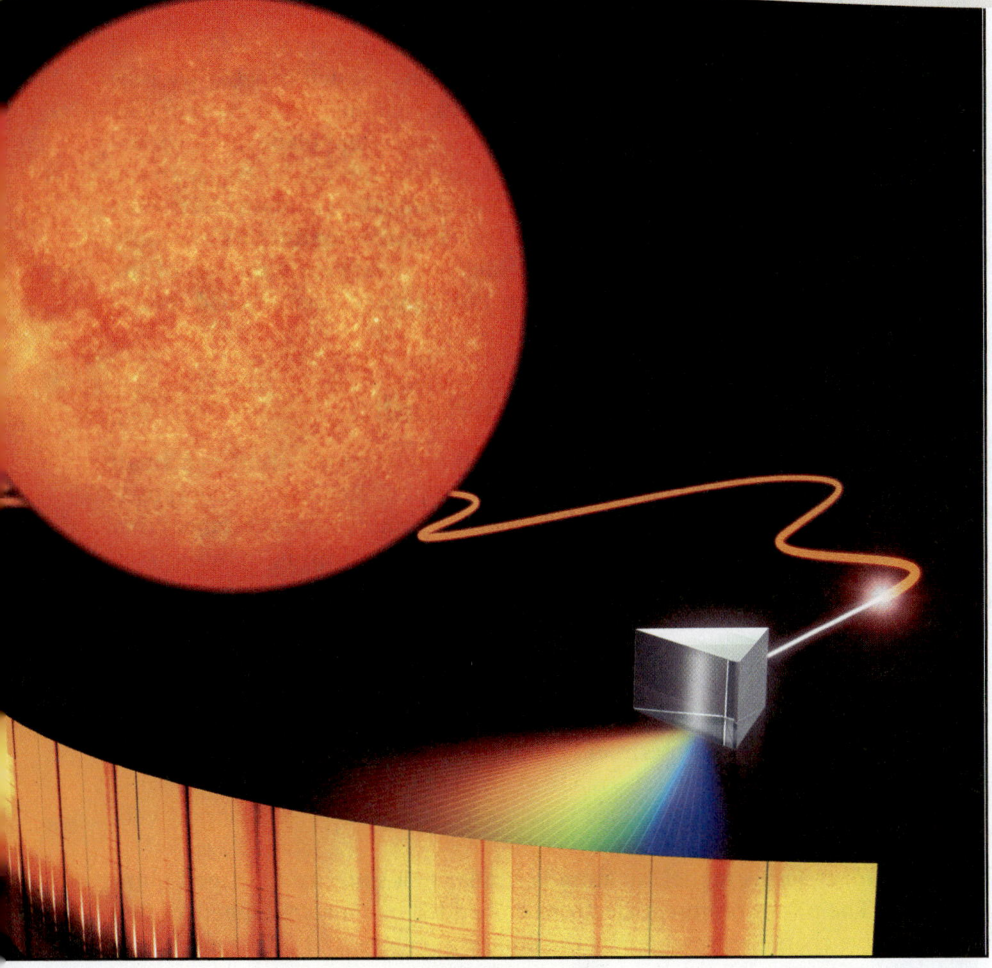

How do scientists determine the chemical composition of stars?

How do lasers work?

Why is it impossible to know exactly where an electron is?

## In the artist's rendering above, an optical fiber transmits light from the Sun to a spectroscope.

The spectroscope separates the sunlight into its many wavelengths. At first glance, we see a continuous spectrum. But on closer inspection we see narrow gaps—dark lines—in the spectrum at specific wavelengths. Little light of those wavelengths reaches Earth. These gaps are present even if we observe the Sun from a satellite, which tells us that this phenomenon is not a result of the Sun's light passing through Earth's atmosphere. What causes these gaps in the spectrum, and how do scientists use these lines to identify chemical elements present in the Sun's atmosphere and to study the Sun's magnetic field?

**Previously,** you learned about cathode ray experiments that led to the discovery of the electron, a very low mass, negatively charged particle. However, most forms of matter we encounter on Earth are electrically neutral, which implies

**Be sure you know how to:**

- Write an expression for the rotational momentum of a point-like object of mass $m$ moving at constant speed $v$ in a circular orbit of radius $r$ (Section 8.5).
- Apply Coulomb's law for the interaction between charged point-like objects (Section 14.4).
- Explain how a spectroscope works (Section 23.3).

that there must be a positively charged component as well. The relationship between these positively charged components and electrons is the subject of this chapter.

## 27.1 Early atomic models

The ancient Greek philosopher Democritus proposed that "atoms" were the smallest indivisible components of matter. Indeed, much of what we now know about matter supports some of Democritus' claims. From experiments involving evaporation we have learned that substances are made of individual point-like objects that move randomly and are surrounded by empty space. We also have learned that both dielectrics and conductors contain electrically charged particles. These particles move freely inside conductors and only a little inside dielectrics. How did these ideas motivate early models of the atomic nature of matter? By the end of the 19th century scientists had several hypotheses.

### Dalton model: Atom as billiard ball

In 1803 John Dalton proposed a billiard ball model of the atom as a small solid sphere. According to Dalton, a sample of a pure element was composed of a large number of atoms of a single kind. Compounds were substances that contained more than one kind of atom in specific integer ratios. In this model, each atom, regardless of type, contains an equal amount of positively and negatively charged subcomponents so that each atom is electrically neutral.

### Thomson model: Atom as plum pudding

In 1897 J. J. Thomson's experiments supported the model that cathode rays were electrons. Because cathodes made of different materials emitted the same electrons, Thomson hypothesized that these electrons were the negatively charged components of atoms.

According to his model, an atom is similar to a spherically shaped plum pudding. The plums represent electrons embedded in a massive sphere of diffuse positive charge represented by the pudding (**Figure 27.1**). This model explained the electrical neutrality of an atom, but it could not explain how atoms could emit light.

**Figure 27.1** The "plum pudding" model of the atom.

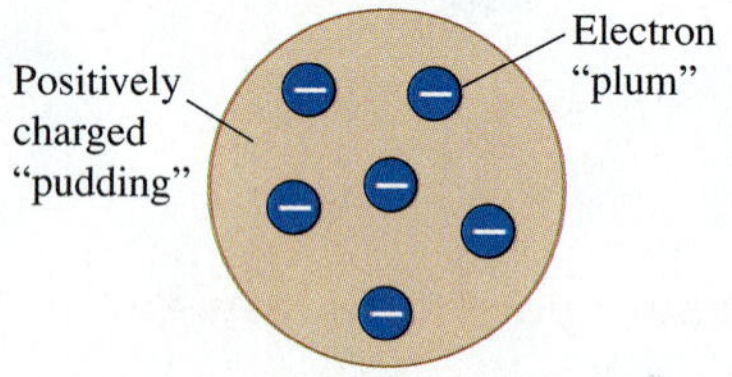


### Rutherford model: Atom as planetary system

Alpha particles are emitted by uranium and other radioactive elements. These particles have a mass four times greater than a hydrogen atom and have twice the charge of an electron, but positive, not negative. According to the Thomson model, alpha particles beamed at a thin layer of metal foil should pass nearly straight through the foil because the atoms are soft and their mass is evenly distributed. In 1909, Ernest Rutherford, a New Zealand-born British physicist, noticed that such a beam came out of the foil much wider than it had been when it entered. Thomson's model could not explain this phenomenon.

**The Atomic Nucleus** Rutherford and his postdoctoral colleague Hans Geiger hypothesized that the atom's mass was not spread uniformly throughout the atom but was instead concentrated in a small region. If this were true, then the beam should spread as observed. This new model also predicted that the alpha particles that come very close to the small dense region inside the atom should recoil backward at very large angles, possibly even in the reverse direction. However, Rutherford and Geiger did not believe that they would find this recoil. After all, how could an atom cause a very fast-moving alpha particle to almost instantaneously reverse direction? The idea seemed absurd. Geiger and Ernest Marsden set out to perform a testing experiment to disprove this hypothesis. In Testing Experiment **Table 27.1** we look at their experiment.

**TESTING EXPERIMENT TABLE**

**27.1 Structure of the atom.**

| Testing experiment | Prediction | Outcome |
| --- | --- | --- |
| The experimental apparatus is shown below. 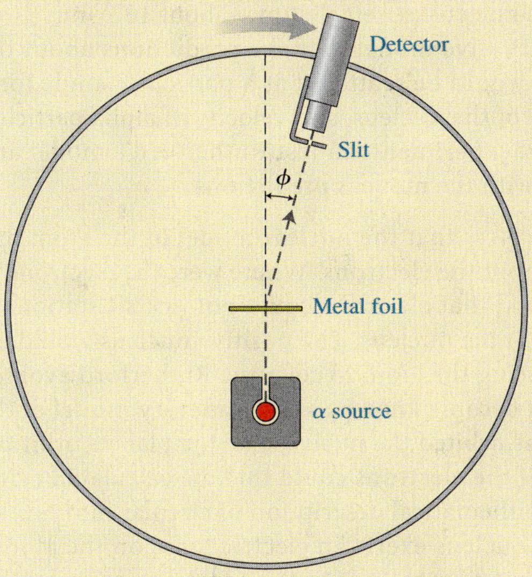  | **Thomson model:** If the atom's mass and positive charge are dispersed evenly, the alpha particles will only undergo small deflections. 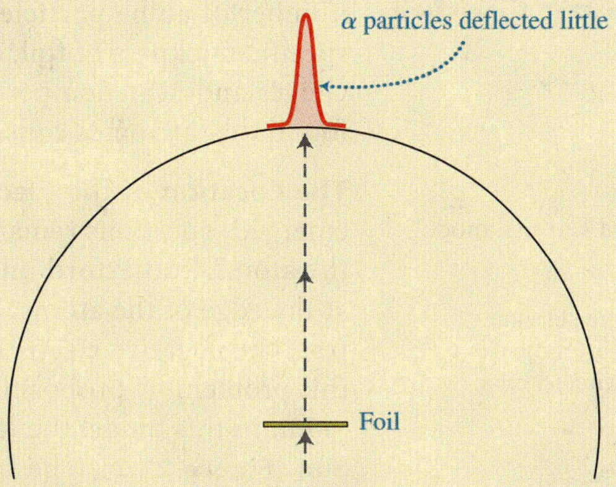  **Rutherford model:** If the atom's mass and positive charge are concentrated in a small region, then most of the alpha particles will pass straight through. But those that pass near the concentrated mass region will recoil at larger angles, resulting in a wider beam. Some will recoil at very large angles. 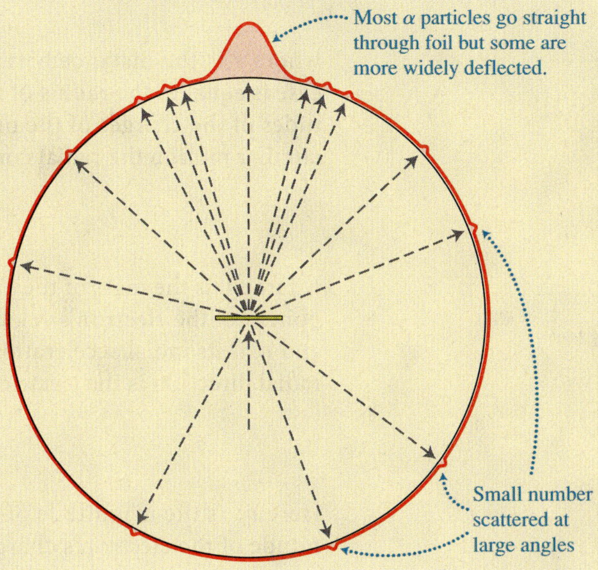  | Geiger and Marsden recorded a significant widening of the beam and a small number of alpha particles recoiling at large angles. |

*(continued)*

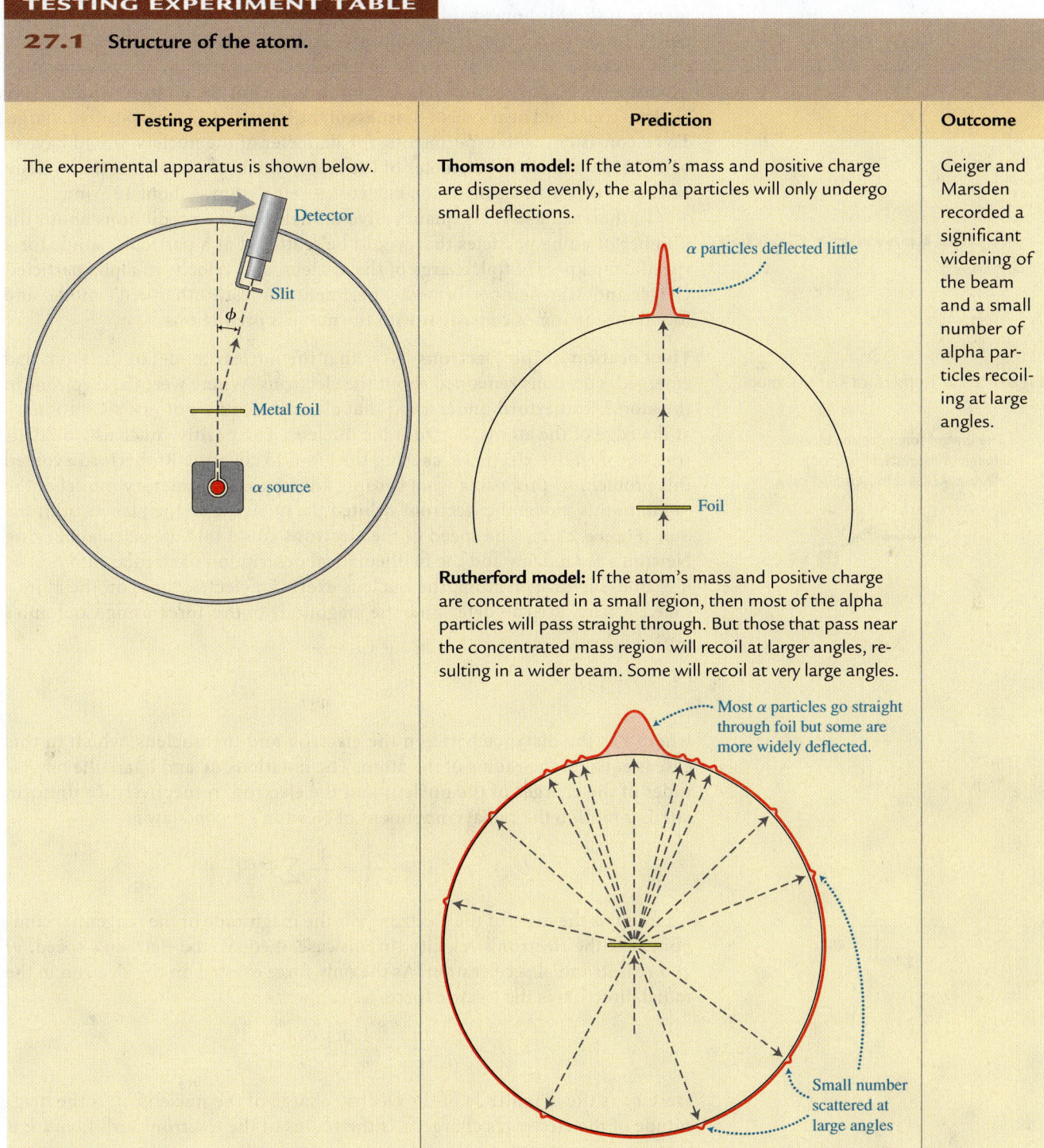

**Conclusion**

The outcome is not consistent with the prediction based on the Thomson model. The outcome is consistent with the prediction based on the Rutherford model. It seems that the mass and positive charge of the atom are concentrated in a small region, with the rest of the atom mostly empty space.

Rutherford and his colleagues were surprised to find that some alpha particles recoiled backward at such large angles. Reflecting on these experiments, Rutherford once said, "It was quite the most incredible event that happened to me in my life. It was almost as incredible as if you fired a 15-inch shell at a piece of tissue paper and it came back and hit you."[1] Following these experiments, Rutherford developed a new model of the atom in which a tiny **nucleus** contained nearly all of the mass of the atom and all of its positive charge. To be consistent with experiments, the diameter of the nucleus would have to be 1/100,000 times the diameter of the atom, which was known at that time to be about $10^{-10}$ m. This made the nucleus extremely tiny—about $10^{-15}$ m.

**Active Learning Guide›**

Rutherford used this quantitative model to make predictions about the number of alpha particles that would be scattered at a particular angle for a specific thickness of foil, charge of the nucleus, and velocity of alpha particles. Geiger and Marsden performed experiments to test Rutherford's model and found the outcomes consistent with the model's predictions.

**The Location of the Electrons** Now that the nuclear model of the atom had emerged, questions remained about the electrons. Where were the electrons in the atom? Rutherford understood that electrons could not just sit stationary at the edge of the atoms, far from the nucleus. The positive nucleus would attract the negative electrons, causing the atom to collapse. Rutherford avoided this problem by proposing what became known as the **planetary model** of the atom. In this model the electrons orbited the nucleus like the planets orbit the Sun (**Figure 27.2**). The speed of the electrons could then be calculated using Newton's second law and the mathematical description of circular motion.

**Figure 27.2** Rutherford's orbital model of the atom.

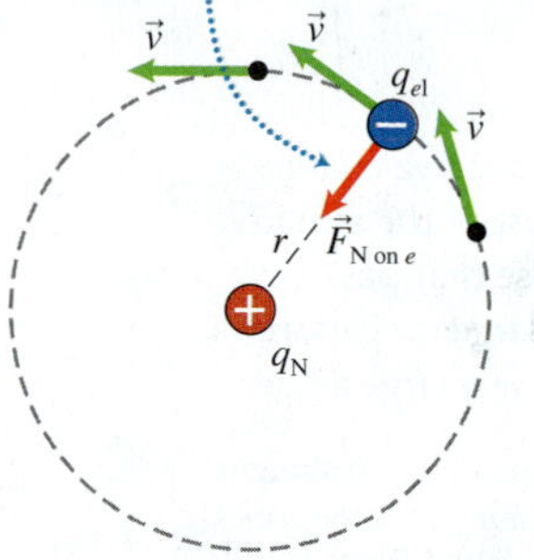

The inward force that the electric charge of the nucleus exerts on the electron keeps it in a circular orbit.
$\vec{v}$
$\vec{v}$
$q_{el}$
$\vec{v}$
$r$
$\vec{F}_{N\,on\,e}$
$q_N$

In the hydrogen atom, the nucleus exerts an electric force on the atom's lone electron. We can determine the magnitude of this force using Coulomb's law:

$$F_{N\,on\,e} = k\frac{q_N q_e}{r^2}$$

where $r$ is the distance between the electron and the nucleus, which in this case is equal to the radius of the atom. The notations $q_N$ and $q_e$ are the magnitudes of the charges of the nucleus and the electron, respectively. In uniform circular motion the radial component of Newton's second law is

$$a_r = \frac{v^2}{r} = \frac{1}{m_e}\sum F_r$$

where $m_e$ is the mass of the electron, $v$ is the magnitude of the tangential component of the electron's velocity (in this case, it equals the electron's speed, $v$) and $a_r$ is its radial acceleration. As the only force exerted on the electron in the radial direction is the electric force, we can write

$$\frac{v^2}{r} = \frac{1}{m_e}\left(k\frac{q_N q_e}{r^2}\right)$$

Here $q_N$ is the magnitude of the electric charge of the nucleus, $q_e$ is the magnitude of the electron's charge, $r$ is the radius of the electron's orbit, and $k$ is

[1] Rutherford, E. *Background to Modern Science.* New York: Macmillan, 1938.

the Coulomb's law constant. Multiplying both sides of the equation by $r$ and taking the square root, we solve for $v$:

$$v = \sqrt{\frac{k q_{\mathrm{N}} q_e}{m_e r}}$$

The above expression gives the constant speed of an electron moving in a circle at a distance $r$ from the nucleus. Thus if the electron is moving fast enough, it can avoid falling onto the nucleus.

**A Difficulty with the Planetary Model** The total energy of the nucleus-electron system, $E$, is the sum of its electric potential energy $(U_{\mathrm{q}})$ and the kinetic energy of the electron $(K_e)$:

$$E = U_{\mathrm{q}} + K_e$$

Assuming the electron is not moving relativistically:

$$E = k\frac{q_{\mathrm{N}} q_e}{r} + \frac{1}{2}m_e v^2 = -k\frac{e^2}{r} + \frac{1}{2}m_e v^2$$

For a hydrogen atom $q_{\mathrm{N}} = +e$ and $q_e = -e$. Because the atom is a bound system, its total energy should be negative (the zero level of potential energy is at infinity; if the total energy were zero or positive, the system would fly apart).

As discussed earlier (in Chapter 24), an accelerating electron emits electromagnetic (EM) radiation, and this radiation has energy. Because the electron orbiting the nucleus is continually accelerating, the electron-nucleus system should continuously lose energy. As the electron emits electromagnetic radiation, the electron should get closer to the nucleus (**Figure 27.3**). EM theory predicts that the electron would spiral into the nucleus in about $10^{-12}$ s. This definitely contradicts our experience. The Rutherford model could not explain the stability of atoms. It also could not explain another phenomenon that was well known to physicists at that time—the light emitted by low density gases.

**Figure 27.3** A difficulty with the planetary model. An accelerating electron should emit EM waves and spiral inward, causing the atom to collapse.

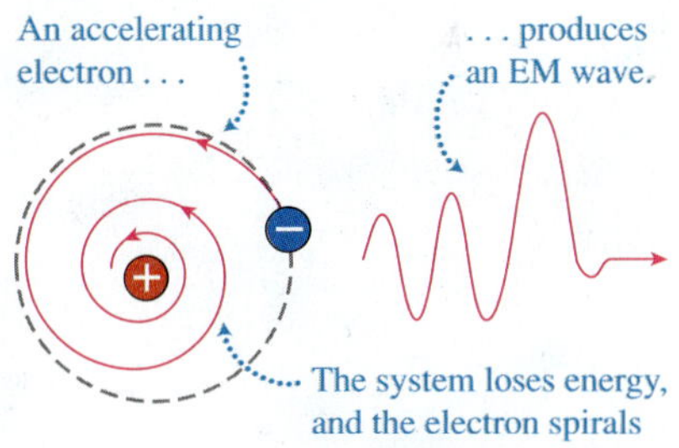


## Spectra of low-density gases and the need for a new model

By the end of the 19th century physicists knew that many objects emit a continuous spectrum of light whose properties depend on the temperature of the object—that is, the random motion of their particles. Room temperature objects primarily emit infrared photons. Objects at several thousand degrees kelvin (such as an incandescent lightbulb filament) emit a larger number of visible light photons.

If atoms are in a gaseous form and the gas's temperature is increased so that it emits visible light, we observe a **line spectrum** (**Figure 27.4**), in which only specific wavelengths of light ("lines") are present. For example, if you throw a dash of table salt into a flame and observe the emitted light with a spectroscope, you see a line spectrum.

Another way to observe a gas spectrum is to put a strong electric field across a tube filled with gas. The light produced by neon signs also has a line spectrum.

Observations show that different gases produce different sets of spectral lines. These line combinations do not depend on how the gases are made to glow—whether heated or placed in a strong electric field. When gases are mixed together, the spectral lines characteristic of each individual gas are present.

Scientists could not explain where the lines came from, but they could measure which chemical elements emitted which lines. The most carefully studied element was hydrogen because of the simplicity of its line spectra. It had just four distinct lines in the visible light region of the spectrum (**Figure 27.5**).

**Figure 27.4** A continuous spectrum and a line spectrum.

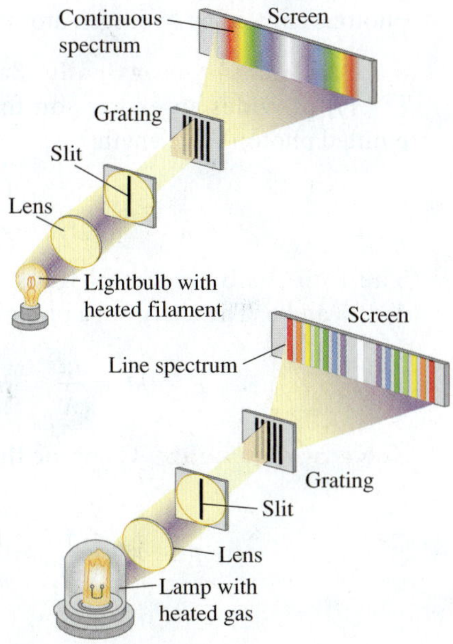

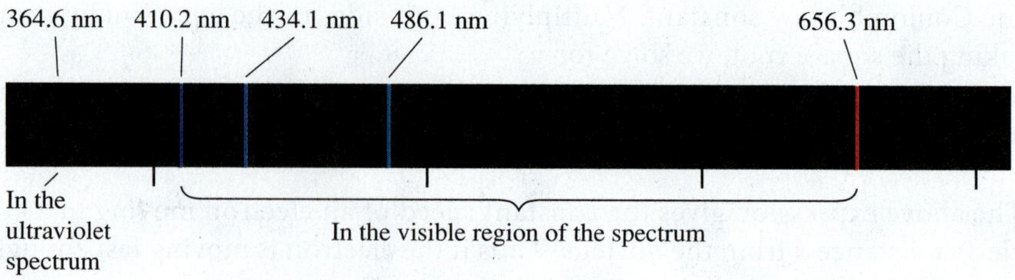


**Figure 27.5** A hydrogen atom line spectrum.

**Active Learning Guide>**

In 1885 Johann Balmer found a pattern in the wavelengths of the visible spectral lines produced by hydrogen and represented it with the following equation:

$$\frac{1}{\lambda} = R\left(\frac{1}{2^2} - \frac{1}{n^2}\right) \quad (27.1)$$

where $\lambda$ is the wavelength of the spectral line, $n$ has integer values of 3, 4, 5, and 6, and $R$ is a constant experimentally determined to equal $R = 1.097 \times 10^7\ \text{m}^{-1}$ when the wavelength is measured in meters. The constant $R$ is called the **Rydberg constant**. The Balmer equation produces the four wavelengths shown in Figure 27.5. However, Balmer could not explain why this pattern existed.

The Thomson model offered no mechanism for the emission of light by an atom because all of the components of the atom were considered stationary. The Rutherford model predicted that individual atoms should radiate a continuous spectrum until the electron spiraled into the nucleus. Physicists needed a new atomic model that was stable and that could explain the line spectra; we discuss this model in Section 27.2.

**QUANTITATIVE EXERCISE 27.1 Calculate photon energies**

Use Balmer's formula [Eq. (27.1)] to determine the energies in joules and in electron volts of the possible visible photons emitted by hydrogen atoms.

**Represent mathematically** Balmer's formula [Eq. (27.1)] provides an expression for the inverse of the emitted photon wavelength:

$$\frac{1}{\lambda} = R\left(\frac{1}{2^2} - \frac{1}{n^2}\right)$$

The four visible spectral lines correspond to $n = 3, 4, 5$, and 6. The energy of a photon is

$$E = hf = \frac{hc}{\lambda} = hc\frac{1}{\lambda}$$

**Solve and evaluate** Combine these two equations to get

$$E = hcR\left(\frac{1}{2^2} - \frac{1}{n^2}\right)$$

Inserting the appropriate values for Planck's constant $h$, the speed of light $c$, and Rydberg's constant $R$, we get for the combined constant $hcR$

$$hcR = (6.63 \times 10^{-34}\ \text{J}\cdot\text{s})(3.00 \times 10^8\ \text{m/s})(1.097 \times 10^7\ \text{m}^{-1})$$
$$= 2.18 \times 10^{-18}\ \text{J}$$

This combined constant shows up frequently when we deal with line spectra, and it is worth remembering for convenience.

Inserting the four different $n$ values and converting to electron volts ($1\ \text{eV} = 1.6 \times 10^{-19}\ \text{J}$), we get

$n = 3$:

$$E_3 = (2.18 \times 10^{-18}\ \text{J})\left(\frac{1}{2^2} - \frac{1}{3^2}\right) = 3.03 \times 10^{-19}\ \text{J}$$
$$= (3.03 \times 10^{-19}\ \text{J})\left(\frac{1\ \text{eV}}{1.6 \times 10^{-19}\ \text{J}}\right) = 1.89\ \text{eV}$$

$n = 4$:

$$E_4 = (2.18 \times 10^{-18}\ \text{J})\left(\frac{1}{2^2} - \frac{1}{4^2}\right) = 4.09 \times 10^{-19}\ \text{J}$$

$$= (4.09 \times 10^{-19}\ \text{J})\left(\frac{1\ \text{eV}}{1.6 \times 10^{-19}\ \text{J}}\right) = 2.55\ \text{eV}$$

$n = 5$:

$$E_5 = (2.18 \times 10^{-18}\ \text{J})\left(\frac{1}{2^2} - \frac{1}{5^2}\right) = 4.58 \times 10^{-19}\ \text{J}$$

$$= (4.58 \times 10^{-19}\ \text{J})\left(\frac{1\ \text{eV}}{1.6 \times 10^{-19}\ \text{J}}\right) = 2.86\ \text{eV}$$

$n = 6$:

$$E_6 = (2.18 \times 10^{-18}\ \text{J})\left(\frac{1}{2^2} - \frac{1}{6^2}\right) = 4.84 \times 10^{-19}\ \text{J}$$

$$= (4.84 \times 10^{-19}\ \text{J})\left(\frac{1\ \text{eV}}{1.6 \times 10^{-19}\ \text{J}}\right) = 3.03\ \text{eV}$$

To determine whether these results are reasonable, complete the Try It Yourself question below and see if the wavelengths correspond to the observed visible light wavelengths for hydrogen.

**Try it yourself:** Determine the wavelengths in nanometers (nm) of the above spectral lines.

***Answer:*** Using the Balmer equation, we get 656 nm for $n = 3$, 486 nm for $n = 4$, 434 nm for $n = 5$, and 410 nm for $n = 6$. All of these wavelengths fall within the visible part of the EM spectrum. Compare them to the values in Figure 27.5.

**Review Question 27.1** Rutherford made the following statement: "I was perfectly aware when I put forward the theory of the nuclear atom that . . . the electron ought to fall onto the nucleus." What were his reasons for making this statement?

## 27.2 Bohr's model of the atom: Quantized orbits

In 1913, Danish physicist Niels Bohr succeeded in creating a new model that explained the line spectra and the stability of the atom. He kept the structure of Rutherford's model in terms of the small nucleus and electrons moving around it, but imposed a restriction on the electron orbits.

‹Active Learning Guide

**TIP** Before you read the description of Bohr's ideas, review the concept of rotational (angular) momentum (Chapter 8).

Bohr's model applies only to hydrogen and to one-electron ions. Here are the fundamental ideas of Bohr's model, known as **Bohr's postulates**.

1. The atom is made up of a small nucleus and an orbiting electron. The electron can occupy only certain orbits, called **stable orbits,** which are labeled by the positive integer $n$. When in these orbits, the electron moves around the nucleus but *does not* radiate electromagnetic waves. All other orbits are prohibited. Each of these stable orbits results in a specific value of the total energy (kinetic plus electric potential) of the atom, designated $E_n$.
2. When an electron transitions from one stable orbit to another, the atom's energy changes. When the energy of the atom decreases, the atom emits a photon whose energy equals the decrease in the atom's energy ($hf = E_\text{i} - E_\text{f}$, where "i" is the energy state number for the initial state and "f" is the energy state number for the final state). For the atom's energy to increase, the atom must absorb some energy, often by absorbing a photon whose energy equals the increase in the atom's energy. Because the stable orbits are discrete, the atom can radiate or absorb only certain specific amounts of energy.

3. The stable electron orbits are the orbits where the magnitude of the electron's rotational (angular) momentum $L$ is given by

$$L = mvr = n\frac{h}{2\pi}, n = 1, 2, 3, \ldots \tag{27.2}$$

In this equation $m$ is the mass of the electron, $v$ is its speed, $r$ is the radius of its orbit, $h$ is Planck's constant, and $n$ is any positive integer. In this equation, rotational momentum $L$ is **quantized**, meaning that it can have only specific discrete values (multiples of $h/2\pi$ in this case).

While Bohr's model explained the line spectra and stability of the hydrogen atom, it does not explain why electrons do not emit EM waves when accelerating in a circular orbit. It also does not explain why the only stable orbits are those with the above rotational momentum [Eq. (27.2)]. Let's see if Bohr's model correctly predicts the size of the hydrogen atom and the specific wavelengths present in its line spectrum.

## Size of the hydrogen atom

To calculate the size of the atom using Bohr's postulates, assume that the electron and nucleus obey Newton's laws and interact only via the electrostatic (Coulomb) force, that gravitational forces exerted on the atom are negligible, and that the nucleus has charge $q_N = +e = 1.6 \times 10^{-19}$ C. A sketch of the atom and a force diagram for the electron are shown in **Figure 27.6**. The nucleus exerts an electric force on the electron, $F_{N \text{ on } e}$, of this magnitude:

**Figure 27.6** Force-acceleration analysis of the Bohr atomic model. The nucleus exerts a negligible gravitational force on the electron compared to the electrical force the nucleus exerts on the electron.

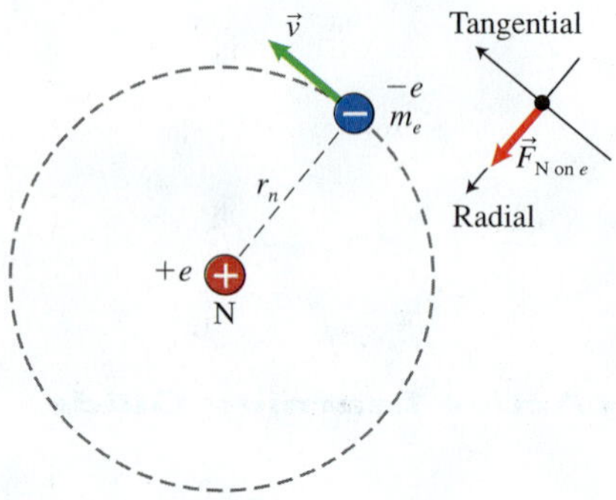
$\vec{v}$
Tangential
$-e$
$m_e$
$\vec{F}_{N \text{ on } e}$
$r_n$
Radial
$+e$
N
The electrical force that the nucleus exerts on the electron is much greater than any gravitational force that the nucleus exerts on the electron.

$$F_{N \text{ on } e} = k\frac{|q_N q_e|}{r_n^2} = k\frac{|(+e)(-e)|}{r_n^2}$$

$$\Rightarrow F_{N \text{ on } e} = k\frac{e^2}{r_n^2} \tag{27.3}$$

This force points toward the nucleus, resulting in the electron's uniform circular motion. Recall that an object's radial acceleration $a_r$ for uniform circular motion is given by the expression $a_r = (v^2/r)$. The radial $r$ component of Newton's second law is

$$a_r = \frac{1}{m_e}\sum F_{N \text{ on } e} = \frac{1}{m_e}\left(k\frac{e^2}{r_n^2}\right)$$

$$\Rightarrow \frac{v_n^2}{r_n} = \frac{ke^2}{m_e r_n^2}$$

Here $m_e$ is the mass of the electron and $a_r$ is the radial acceleration. We use the symbol $e$ for the magnitudes of the charge of both the electron and the nucleus. The radius of the electron's orbit is $r_n$, and $v_n$ is the corresponding speed; $k$ is the Coulomb's law constant. Multiplying both sides by $r_n$, we get

$$v_n^2 = \frac{ke^2}{m_e r_n} \tag{27.4}$$

Equation (27.4) relates two unknown quantities, $v_n$ and $r_n$. Bohr's third postulate involving the rotational momentum $L_n$ of the electron in the $n$th orbit provides a second relationship between these same two quantities:

$$L_n = m_e v_n r_n = n\frac{h}{2\pi}, \text{ with } n = 1, 2, 3, \ldots$$

Solving for $v_n$, we find

$$v_n = \frac{nh}{2\pi m_e r_n}, n = 1, 2, 3, \ldots$$

Insert this expression for $v_n$ into Eq. (27.4) to get

$$\left(\frac{nh}{2\pi m_e r_n}\right)^2 = \frac{ke^2}{m_e r_n}$$

Solving the above for $r_n$, we get

$$r_n = \frac{h^2}{4\pi^2 ke^2 m_e} n^2, \text{with } n = 1, 2, 3, \ldots \quad (27.5)$$

Let's check the units of this result. The units for $h$ are $\text{J}\cdot\text{s}$ and for $k$ are $(\text{N}\cdot\text{m}^2/\text{C}^2)$. Thus, the unit of the right side of the equation is

$$\frac{(\text{J}^2\,\text{s}^2)}{\left(\frac{\text{N}\cdot\text{m}^2}{\text{C}^2}\right)\text{C}^2\,\text{kg}} = \frac{(\text{N}\cdot\text{m})^2\,\text{s}^2}{\text{N}\cdot\text{m}^2\,\text{kg}} = \frac{\text{N}\cdot\text{s}^2}{\text{kg}} = \frac{\left(\frac{\text{kg}\cdot\text{m}}{\text{s}^2}\right)\cdot\text{s}^2}{\text{kg}} = \text{m}$$

which is the correct unit for a radius. Substituting the values of the known constants into Eq. (27.5), we get

$$r_n = \frac{(6.63\times10^{-34}\,\text{J}\cdot\text{s})^2}{4\pi^2\left(9.00\times10^9\,\frac{\text{N}\cdot\text{m}^2}{\text{C}^2}\right)(1.6\times10^{-19}\,\text{C})^2(9.11\times10^{-31}\,\text{kg})}n^2$$

or

$$r_n = (0.53\times10^{-10}\,\text{m})n^2, \text{for } n = 1, 2, 3, \ldots \quad (27.6)$$

The smallest possible electron orbit radius corresponds to the $n = 1$ energy state:

$$r_1 = 0.53\times10^{-10}\,\text{m}$$

This result is in agreement with the measurements of atomic sizes, about $10^{-10}$ m.

This smallest radius orbit is called the **Bohr radius** $r_1$. The other stable orbit radii can be calculated using Eq. (27.6). Notice that the radii depend on the square of $n$:

$$r_2 = 4r_1;\ r_3 = 9r_1;\ r_4 = 16r_1; \text{therefore, } r_n = n^2 r_1$$

As a result, the radius increases dramatically as $n$ increases. The term $n$ is known as a **quantum number** and must be a positive integer. Because of this, only certain radii represent stable electron orbits (for example, $4r_1$ and $9r_1$, which correspond to $n = 2$ and $n = 3$, are stable orbits, but $5r_1$ is not); a stable radius is said to be **quantized**. The radii of allowed electron orbits in the Bohr model of the hydrogen atom are represented in **Figure 27.7**. (Note that the way the nucleus is drawn in the figure dramatically exaggerates its size in comparison to the electron orbits.)

**Figure 27.7** Radii of stable Bohr electron orbits.

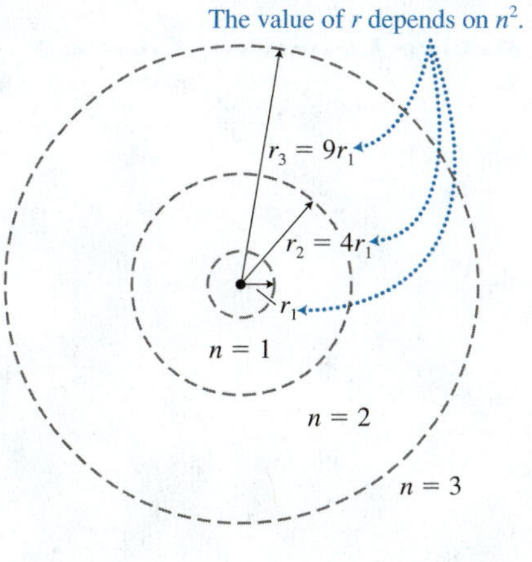


## Energy states of the Bohr model

Now that we have gained some confidence in Bohr's model, we can use it to determine the energy states of the hydrogen atom (or, put another way, the total energy of the atom for each energy state $n$). We'll make the reasonable assumption that the nucleus remains at rest as the electron orbits because of its much larger mass. We'll also assume that relativistic effects are negligible. The total energy of the electron-nucleus system is the sum of its electric potential energy plus the electron's kinetic energy (we neglect the motion of the nucleus due to the interactions with the electron). The electric potential energy $U_{q\,n}$ is

$$U_{q\,n} = k\frac{q_N q_e}{r_n} = k\frac{(+e)(-e)}{r_n} = -k\frac{e^2}{r_n}$$

The total energy of the system in the $n$th energy state is then

$$E_n = U_{q\,n} + K_{e\,n} = -k\frac{e^2}{r_n} + \frac{1}{2}m_e v_n^2 \qquad (27.7)$$

From Eq. (27.4), note that $m_e v_n^2 = k(e^2/r_n)$. Therefore, Eq. (27.7) becomes

$$E_n = -k\frac{e^2}{r_n} + \frac{1}{2}\left(k\frac{e^2}{r_n}\right) = -k\frac{e^2}{2r_n}$$

The Bohr model has predicted that the total energy of the atom is negative. This is good news because the total energy of a bound system should be negative, as explained earlier. Now use Eq. (27.5) for the radius of the atom to write an equation for its total energy:

$$E_n = -k\frac{e^2}{2r_n} = -k\frac{e^2}{2\left(\dfrac{h^2}{4\pi^2 ke^2 m_e}n^2\right)} = -\left(\frac{2\pi^2 e^4 k^2 m_e}{h^2}\right)\frac{1}{n^2} \qquad (27.8)$$

Evidently, the energy of the atom is inversely proportional to the square of the quantum number $n$. Substituting the values of the constants into the above and converting from joules to electron volts, we get

$$E_n = \frac{-13.6\text{ eV}}{n^2} \quad \text{for } n = 1, 2, 3, \ldots \qquad (27.9)$$

**Figure 27.8** Energy state diagram. The atom's energy is inversely proportional to the square of $n$.

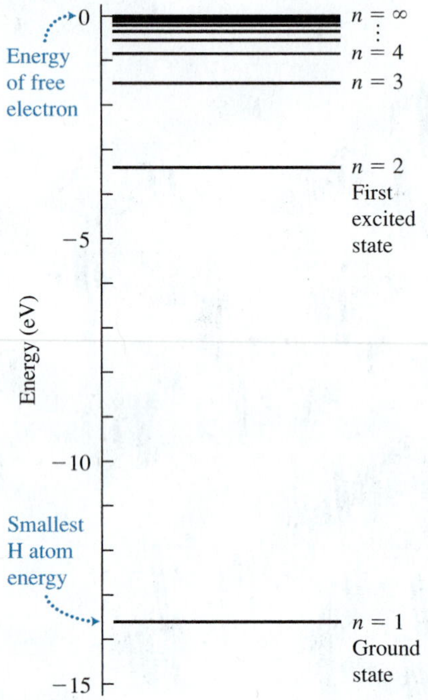


To help visualize the energy states corresponding to these possible total energy values, we use a new representation—an energy state diagram (**Figure 27.8**). On this diagram the vertical axis represents the total energy of the hydrogen atom. Each line represents a specific energy state of the atom labeled by the quantum number $n$. The −13.6 eV horizontal line at the very bottom corresponds to the $n = 1$ lowest energy state, also called the **ground state**. The next horizontal line $(-13.6\text{ eV})/2^2 = -3.4$ eV corresponds to the $n = 2$ energy state, also called the first **excited state**. This pattern repeats for the different integral values of $n$. This representation allows us to see that the higher the quantum number, the less negative the energy of the system, and also the larger the radius of the electron's orbit. The $n = \infty$ state has an infinite radius and zero energy, indicating the electron is unbound, an ionized hydrogen atom $(H^+ + e^-)$.

The energy state diagram in Figure 27.8 also allows us to represent the process of emission of light by the atom. If the atom is currently in the $n = 2$ energy state and makes a transition to the $n = 1$ energy state, the energy of the atom decreases. According to Bohr's second postulate, the energy is carried away by a newly created photon. This **emission process** is represented in **Figure 27.9** by a downward arrow pointing from the initial (higher energy) state of the atom to the final (lower energy) state. This energy, equal to the energy decrease of the atom, is indicated with the symbol $E_\gamma$. The subscript $\gamma$ (Greek letter gamma) is often used to indicate a photon.

**Active Learning Guide >**

The atom's energy can increase if it absorbs a photon. This **absorption process** is represented on the energy diagram with an upward arrow pointing from the initial state of the atom to its final state (**Figure 27.10**). An atom can also move from one energy state to another (increasing or decreasing its energy) when it collides with another atom.

> **TIP** Remember that according to Bohr's first postulate, the hydrogen atom can exist in states whose energies can be calculated by Eq. (27.8). States with energies between those values are not allowed.

We have not yet tested whether Bohr's model correctly predicts the visible light photons emitted by hydrogen; this test is the goal of Example 27.2.

**Figure 27.9** An example of photon emission. An atom transitions from the first excited state to the ground state, and a photon is emitted.

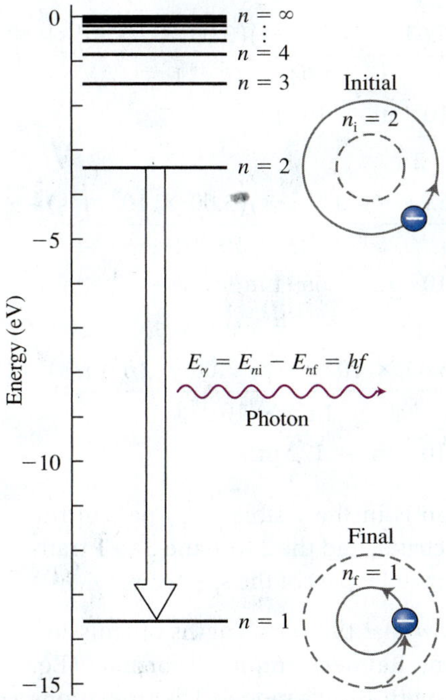


**Figure 27.10** An example of photon absorption. An atom transitions from the ground state to the first excited state by absorbing the energy of a photon.

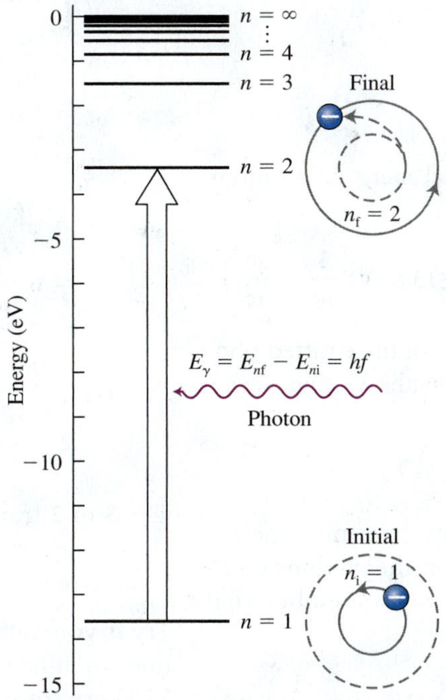


**EXAMPLE 27.2 Photons emitted by hydrogen atoms**

Use the Bohr model to predict the energies and wavelengths of photons that a group of hydrogen atoms would emit if the atoms were all initially in the $n = 3$ state. In which parts of the electromagnetic spectrum are each of these photons?

**Sketch and translate** According to Bohr's model, an atom emits photons when the energy of the atom decreases. This emission occurs when the electron in the atom transitions from an orbit with a higher quantum number ($n_i$) to an orbit with a lower number ($n_f$). Three possible transitions could occur: $3 \rightarrow 1$, $3 \rightarrow 2$, and $2 \rightarrow 1$ (after the atom has already made the $3 \rightarrow 2$ transition).

**Simplify and diagram** The three processes are represented in the energy state diagram at right. Downward vertical arrows represent the energy transitions.

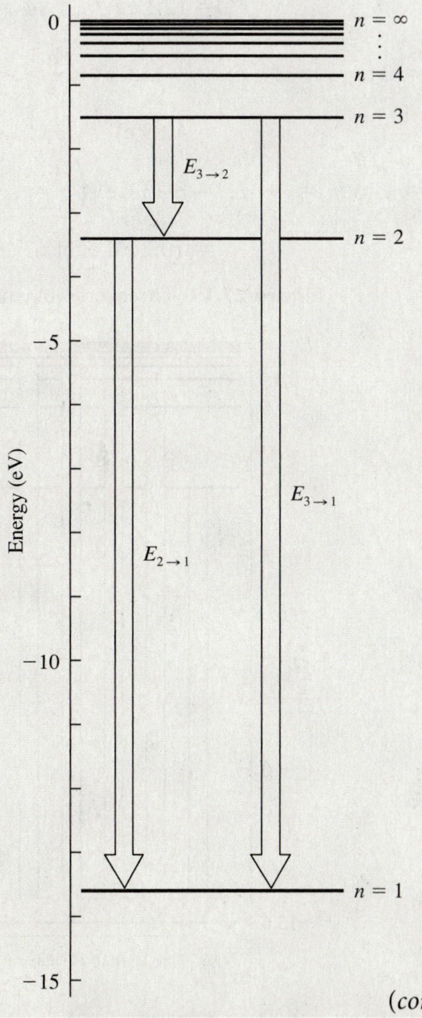


*(continued)*

**Represent mathematically** Using the principle of energy conservation, the initial energy $E_{n_i}$ of the atom must equal the final energy $E_{n_f}$ plus the photon's energy $E_\gamma$:

$$E_{n_i} = E_{n_f} + E_\gamma$$

or

$$E_\gamma = E_{n_i} - E_{n_f}$$

Inserting the expressions for the total energy of the atom, we have

$$E_\gamma = \frac{-13.6\text{ eV}}{n_i^2} - \frac{-13.6\text{ eV}}{n_f^2} = -13.6\text{ eV}\left(\frac{1}{n_i^2} - \frac{1}{n_f^2}\right)$$

Once we determine the energies $E_\gamma$ of the emitted photons, we can determine their wavelengths $\lambda$ using

$$E_\gamma = hf = \frac{hc}{\lambda} \text{ or } \lambda = \frac{hc}{E_\gamma}$$

**Solve and evaluate** We can now determine the energies in electron volts and joules of the photons emitted in each of the three processes. Remember that $1\text{ eV} = 1.6 \times 10^{-19}\text{ J}$.

$$n_i = 3 \text{ to } n_f = 1 \quad E_\gamma = -13.6\text{ eV}\left(\frac{1}{3^2} - \frac{1}{1^2}\right)$$
$$= 12.1\text{ eV} = 1.94 \times 10^{-18}\text{ J}$$

$$n_i = 3 \text{ to } n_f = 2 \quad E_\gamma = -13.6\text{ eV}\left(\frac{1}{3^2} - \frac{1}{2^2}\right)$$
$$= 1.89\text{ eV} = 3.02 \times 10^{-19}\text{ J}$$

$$n_i = 2 \text{ to } n_f = 1 \quad E_\gamma = -13.6\text{ eV}\left(\frac{1}{2^2} - \frac{1}{1^2}\right)$$
$$= 10.2\text{ eV} = 1.63 \times 10^{-18}\text{ J}$$

We can determine the photon wavelengths from their energies:

$$n_i = 3 \text{ to } n_f = 1$$
$$\lambda = \frac{hc}{E_\gamma} = \frac{(6.63 \times 10^{-34}\text{ J}\cdot\text{s})(3.00 \times 10^8\text{ m/s})}{1.94 \times 10^{-18}\text{ J}}$$
$$= 1.03 \times 10^{-7}\text{ m} = 103\text{ nm}$$

$$n_i = 3 \text{ to } n_f = 2$$
$$\lambda = \frac{hc}{E_\gamma} = \frac{(6.63 \times 10^{-34}\text{ J}\cdot\text{s})(3.00 \times 10^8\text{ m/s})}{3.02 \times 10^{-19}\text{ J}}$$
$$= 6.59 \times 10^{-7}\text{ m} = 659\text{ nm}$$

$$n_i = 2 \text{ to } n_f = 1$$
$$\lambda = \frac{hc}{E_\gamma} = \frac{(6.63 \times 10^{-34}\text{ J}\cdot\text{s})(3.00 \times 10^8\text{ m/s})}{1.63 \times 10^{-18}\text{ J}}$$
$$= 1.22 \times 10^{-7}\text{ m} = 122\text{ nm}$$

The 3 to 2 transition is in the visible (red) part of the electromagnetic spectrum, and the 2 to 1 and 3 to 1 transitions are in the ultraviolet part of the spectrum.

**Try it yourself:** Show that the wavelengths of emission lines calculated using Balmer's empirical formula (Eq. 27.1) are consistent with Bohr's model. The transitions described by Balmer's equation are indicated in **Figure 27.11**. Series of emission lines corresponding to other possible transitions are also shown.

*Answer:* Balmer's equation is

$$\frac{1}{\lambda} = R\left(\frac{1}{2^2} - \frac{1}{n^2}\right)$$

which produces the same wavelength transitions as calculated above.

**Figure 27.11** Three series of emission transitions for H atoms.

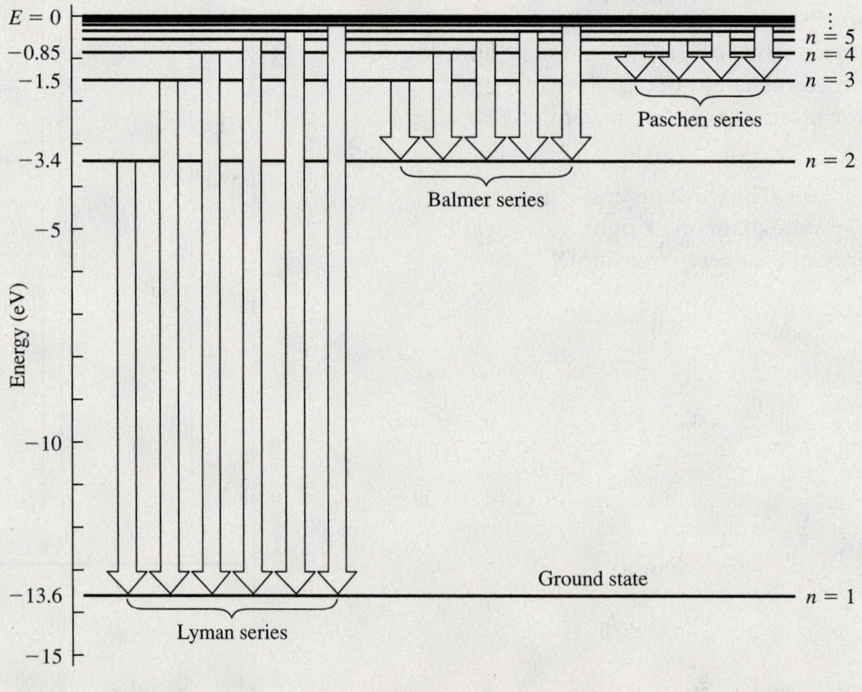

## Consistency of Balmer's equation with Bohr's model

Let's find whether Balmer's result is consistent with the Bohr model prediction for emission transitions.

$$E_\gamma = -13.6\text{ eV}\left(\frac{1}{n_\text{i}^2} - \frac{1}{n_\text{f}^2}\right)$$

If we multiply both sides of the Balmer equation by $hc$, we get

$$\frac{hc}{\lambda} = hcR\left(\frac{1}{2^2} - \frac{1}{n^2}\right)$$

The left-hand side is the energy $E_\gamma$ of the emitted photon. The constant $hcR$ must then have units of energy. This constant in electron volts is

$$hcR = (6.63 \times 10^{-34}\text{ J}\cdot\text{s})(3.00 \times 10^8\text{ m/s})(1.097 \times 10^7\text{ m}^{-1})\left(\frac{1\text{ eV}}{1.6 \times 10^{-19}\text{ J}}\right)$$

$$= 13.6\text{ eV}$$

Balmer's equation becomes

$$E_\gamma = 13.6\text{ eV}\left(\frac{1}{2^2} - \frac{1}{n^2}\right) = -13.6\text{ eV}\left(\frac{1}{n^2} - \frac{1}{2^2}\right)$$

This is exactly the result from the Bohr model with $n_\text{i} = n$ and $n_\text{f} = 2$.

Because there are infinitely many energy states ($n$ can be any positive integer), there are infinitely many possible transitions. However, only four of these transitions emit a visible wavelength photon. The 3 to 2 transition is one. The others are 4 to 2, 5 to 2, and 6 to 2. All transitions from states 7 and higher to 2 result in ultraviolet photons. Any transition to the 1 state also results in an ultraviolet photon.

Notice that the photon emitted by the 3 to 2 transition matches the 656-nm line shown in the hydrogen line spectrum in Figure 27.5. The 4 to 2, 5 to 2, and 6 to 2 transitions produce the other visible light spectral lines in Figure 27.5.

## Limitations of the Bohr model

We saw in Example 27.2 that the wavelengths of the hydrogen spectral lines predicted by Bohr's model are in good agreement with the observational evidence. Unfortunately, the model does not provide predictions that account for the spectral lines emitted by other atoms. It does, however, work well if the atoms are ionized and have only one electron (such as singly ionized helium, $He^+$, or doubly ionized lithium, $Li^{2+}$).

Also, note that Bohr's model arbitrarily imposes restrictions on the motion of the electron (its quantized rotational angular momentum), which results in a restriction on the allowed energy states of the atom. However, Bohr's model provides no explanations for these restrictions. Therefore, physicists continued searching for a new model of the atom that could provide them.

**Review Question 27.2** Why is an atom's total energy negative? Why does this make sense?

# 27.3 Spectral analysis

Bohr's model, as you have seen, helps us understand the **emission spectra** of gases (see the examples of spectra in **Figure 27.12**). These spectra allow scientists to analyze the chemical composition of different materials according

‹Active Learning Guide

**Figure 27.12** Emission spectra of gases. Each type of atom has a unique spectrum, allowing its identification.

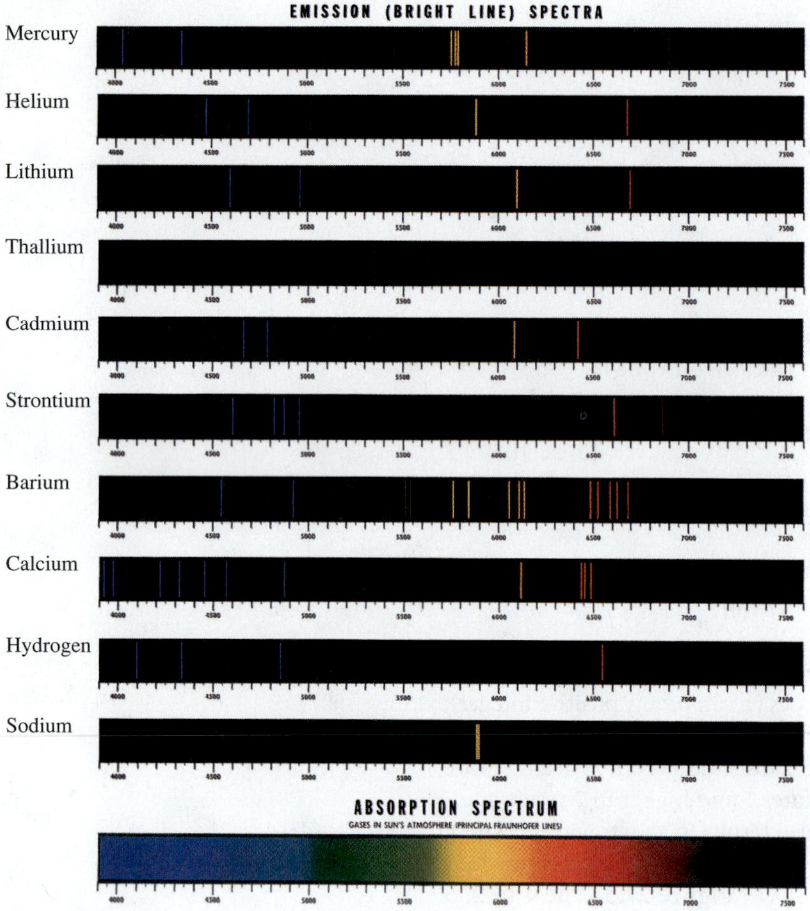


**Figure 27.13** Thermal excitation and production of a photon. (a) Random collisions at high speeds occur in a heated gas. (b) Upon collision, energy transfer may raise an atom to a higher energy state. (c) When the atom returns back to its original state, a photon is emitted.

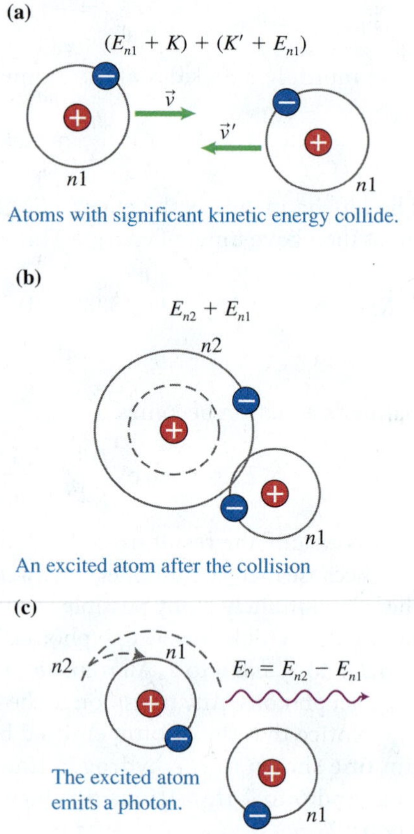


to the light they emit. So far we have discussed only the emission spectrum of hydrogen. We can understand the spectra of other atoms and molecules in a qualitatively similar way.

When the atom transitions from a higher energy state to a lower energy state, a photon of a specific wavelength is emitted. Thus, to observe an emission spectrum, some mechanism is needed to put the atom in an excited state. Many processes can cause atoms to transition to higher energy states.

**Figure 27.14** Colorful fireworks displays are the result of photon emission.

## Thermal excitation

Heating can significantly increase the random kinetic energy of atoms (**Figure 27.13a**). If the kinetic energies of the atoms are high enough, a collision can cause one or both of the atoms to transition to an excited state (Figure 27.13b). When the atom returns to a lower energy state, it emits a photon of a characteristic wavelength (Figure 27.13c). The bright yellow, red, and blue colors of fireworks come from photon emission by excited sodium, strontium, and copper compounds, respectively (**Figure 27.14**).

## Discharge tube excitation

Another process that causes atoms to transition to higher energy states involves filling a discharge tube similar to a cathode ray tube with gas. A hot cathode emits free electrons that are accelerated across an electric potential

**Figure 27.15** A discharge tube produces light. (a) Free electrons are accelerated by a potential difference from cathode to anode in a gas-filled tube. (b) Upon collisions between free electrons and atoms of gas, energy transfer may raise an atom to a higher energy state. (c) When the excited atom returns to its original state, a photon is emitted.

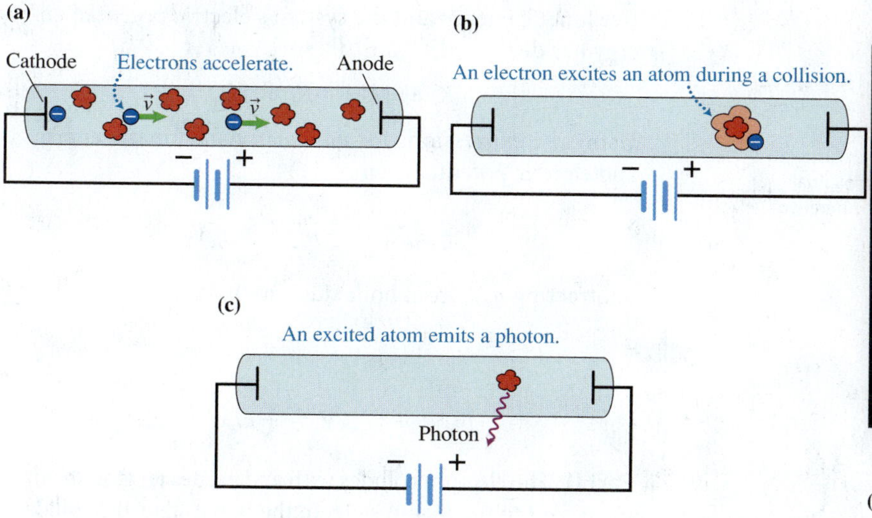
(a)
Cathode
Electrons accelerate.
Anode
(b)
An electron excites an atom during a collision.
(c)
An excited atom emits a photon.
Photon

**Figure 27.16** (a) A mercury discharge tube produces light. (b) Viewed in a spectroscope, the light shows spectral lines.

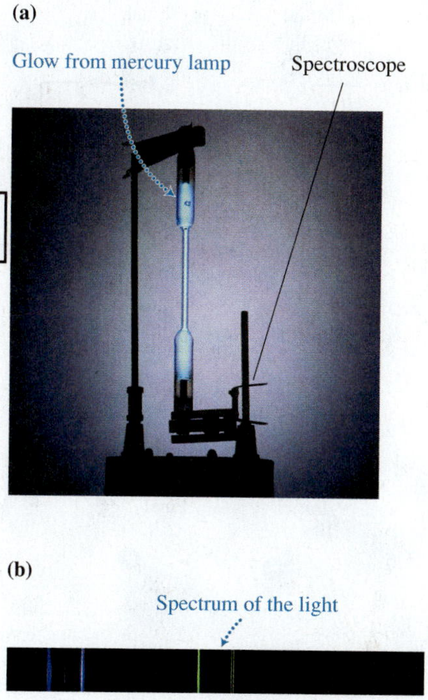
(a)
Glow from mercury lamp
Spectroscope
(b)
Spectrum of the light

difference from the cathode to the anode (**Figure 27.15a**). The electrons then collide with atoms of the gas in the tube, causing the atoms to reach excited states (Figure 27.15b). When the excited atoms return to the ground state, they emit photons whose energy equals the difference in the energy of the atom between its initial and final states (Figure 27.15c). Looking at a discharge tube, you see a glow that appears to be of a single color. But if you look at the light through a spectroscope, you see a line spectrum. Compare the mercury gas discharge tube in **Figure 27.16a** with the actual spectrum in Figure 27.16b.

### EXAMPLE 27.3 Discharge tube excitation of hydrogen

Use the results of Example 27.2 to estimate the potential difference needed between the cathode and anode in a discharge tube filled with hydrogen to produce hydrogen's visible light spectral lines.

**Sketch and translate** We sketched the gas in the discharge tube in Figure 27.15a–c, shown above. For the atom to emit a visible photon, it must transition from the $n = 3, 4, 5$, or 6 state to the $n = 2$ state. Therefore, the colliding electrons must have enough kinetic energy to induce the transition of the electrons in the hydrogen atoms from the $n = 1$ ground state to the $n = 3, 4, 5$, or 6 state. Because the transition from $n = 1$ to $n = 6$ requires the most energy, the potential difference between cathode and anode must be high enough to induce that transition. The system is the colliding electron, the electrodes and the electric field they produce, the hydrogen atoms, and the emitted photons.

The overall process has three parts.

I. The electric potential energy of the system converts into the kinetic energy of the electron, as shown in part (a), on the top of the next page.

*(continued)*

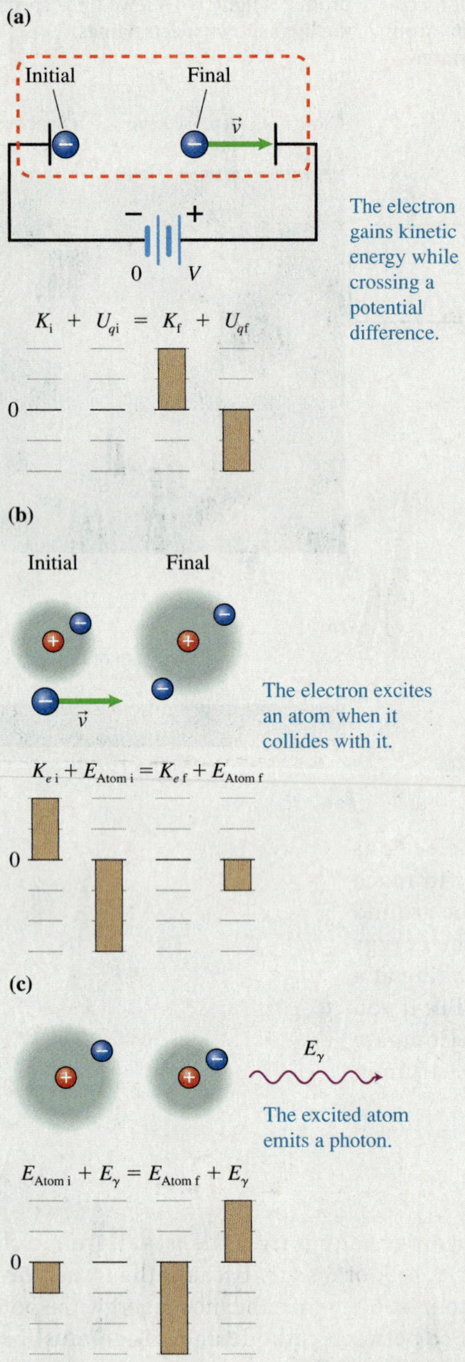

(a)
Initial
Final
$\vec{v}$
−
+
0
V
The electron gains kinetic energy while crossing a potential difference.
$K_i + U_{qi} = K_f + U_{qf}$
0
(b)
Initial
Final
$\vec{v}$
The electron excites an atom when it collides with it.
$K_{ei} + E_{\text{Atom i}} = K_{ef} + E_{\text{Atom f}}$
0
(c)
$E_\gamma$
The excited atom emits a photon.
$E_{\text{Atom i}} + E_\gamma = E_{\text{Atom f}} + E_\gamma$
0

II. The electron collides with and excites the atom (part (b)).

III. The excited atom returns to the ground state, emitting a photon (part (c)).

**Simplify and diagram** Assume first that the electron's initial kinetic energy is zero and that it increases only as a result of the decrease in electric potential energy of the system. Assume also that the electric potential at the initial position of the electron is 0 V. Before the colliding electron encounters the hydrogen atom, assume the atom is in the ground state, and that the Bohr model is a reasonable way to model the atom. Lastly, assume the electron is not moving relativistically. Energy bar charts shown earlier represent the three stages of this process.

**Represent mathematically** We can use the bar charts to help us represent these three processes mathematically.

**Part I:** In the initial state, the electron is at rest at zero electric potential. In the final state, the electron has positive kinetic energy and the system's electric potential energy has decreased.

$$K_i + U_{q\,i} = K_f + U_{q\,f}$$

Using the expressions for nonrelativistic kinetic energy and electric potential energy:

$$0 + q_e V_i = \frac{1}{2} m_e v^2 + q_e V_f$$

Subtracting $q_e V_i$ from both sides, we have

$$0 = \frac{1}{2} m_e v^2 + q_e V_f - q_e V_i$$
$$= \frac{1}{2} m_e v^2 + q_e \Delta V = \frac{1}{2} m_e v^2 - e\Delta V$$

**Part II:** The electron collides with and excites an atom from the $n = 1$ to the $n = 6$ state. In the initial state, the colliding electron has considerable kinetic energy, and the atom is in the $n = 1$ ground state. In the final state, the colliding electron stops, and the atom is in the $n = 6$ excited state.

$$K_i + E_{\text{Atom i}} = K_f + E_{\text{Atom f}}$$

Inserting the expressions for kinetic energy and the total energy of the hydrogen atom:

$$\frac{1}{2} m_e v^2 + \left(\frac{-13.6\text{ eV}}{n_i^2}\right) = 0 + \left(\frac{-13.6\text{ eV}}{n_f^2}\right)$$

**Part III:** The excited atom emits a photon and transitions to the $n = 2$ state. Initially, the atom is in the $n = 6$ state. In the final state, the atom is in the $n = 2$ state and a photon has been emitted.

$$E_{\text{Atom i}} = E_{\text{Atom f}} + E_\gamma$$

Insert the expressions for the total energy of the atom and the energy of the photon:

$$\left(\frac{-13.6\text{ eV}}{n_i^2}\right) = \left(\frac{-13.6\text{ eV}}{n_f^2}\right) + \frac{hc}{\lambda}$$

**Solve and evaluate** We are interested in finding $\Delta V$. Using our mathematical representation of Part I and solving for $\Delta V$:

$$\Delta V = \frac{1}{e}\left(\frac{1}{2} m_e v^2\right)$$

We can determine the kinetic energy from Part II. Solving for $(1/2)m_e v^2$ gives

$$\frac{1}{2} m_e v^2 = \left(\frac{-13.6\text{ eV}}{n_f^2}\right) - \left(\frac{-13.6\text{ eV}}{n_i^2}\right)$$
$$= (-13.6\text{ eV})\left(\frac{1}{n_f^2} - \frac{1}{n_i^2}\right)$$

Substituting this into the expression for $\Delta V$:

$$\Delta V = \frac{1}{e}(-13.6\ \text{eV})\left(\frac{1}{n_f^2} - \frac{1}{n_i^2}\right)$$

Inserting the appropriate values and converting to joules:

$$\Delta V = \frac{1}{1.6 \times 10^{-19}\ \text{C}}(-13.6\ \text{eV})\left(\frac{1}{6^2} - \frac{1}{1^2}\right) \times \left(1.6 \times 10^{-19}\frac{\text{J}}{\text{eV}}\right)$$

$$= 13.2\ \text{V}$$

This value is the smallest potential difference that the electrons need to cross in order to acquire enough kinetic energy to excite hydrogen atoms to the maximum $n = 6$ state so that they can emit photons in the visible spectrum. The potential difference we found is not the potential difference across the discharge tube, but rather the potential difference across the average distance that the colliding electrons travel between two consecutive collisions. The required potential difference across the cathode and anode of the tube is much greater than this.

**Try it yourself:** Estimate the minimum potential difference required to allow the hydrogen in the tube to emit photons from the $n = 4$ to $n = 2$ transition.

*Answer:* 12.8 V.

Another way to produce visible light is by the collisions of hot gas atoms with each other. In order for visible light to be produced, the collisions must cause one of the atoms to become excited to the $n = 3$ state or higher. How hot does the gas need to be in order for this process to occur?

‹Active Learning Guide

**QUANTITATIVE EXERCISE 27.4 Temperature at which an H atom emits visible photons**

Estimate the temperature of hydrogen gas at which you might observe visible line spectra.

**Represent mathematically** A collision between two hydrogen atoms must provide enough energy to excite one of the atoms to the $n = 3$ state or higher. From there it can make transitions to the $n = 2$ state and emit visible photons. Recall that the average kinetic energy (thermal energy) of a gas particle at absolute temperature $T$ is given by this expression (from Section 9.4):

$$\overline{K} = \frac{3}{2}k_B T$$

Here $k_B = 1.38 \times 10^{-23}\ \text{J/K}$, Boltzmann's constant. A hydrogen atom in the ground state ($n = 1$) with energy $E_1$ needs additional energy $\Delta E$ for it to be excited to the state with energy $E_3$ or higher. Choosing the system of interest to be one of the colliding hydrogen atoms, we can use the idea of energy conservation to represent this process mathematically:

$$E_1 + \Delta E = E_3$$

Assume that $\Delta E$ equals the average kinetic energy of a single hydrogen atom, meaning one of the colliding particles converts all of its kinetic energy to excite the other. $E_1$ and $E_3$ are the energies of the atom in the indicated states:

$$E_n = \frac{-13.6\ \text{eV}}{n^2}$$

**Solve and evaluate** Putting the above equations together, we have

$$\frac{-13.6\ \text{eV}}{1^2} + \frac{3}{2}k_B T = \frac{-13.6\ \text{eV}}{3^2}$$

Solving for $T$, inserting the appropriate values, and converting to joules we get

$$T = \frac{2(-13.6\ \text{eV})}{3(1.38 \times 10^{-23}\ \text{J/K})}\left(\frac{1.6 \times 10^{-19}\ \text{J}}{1\ \text{eV}}\right)\left(\frac{1}{3^2} - \frac{1}{1^2}\right)$$

$$= 93{,}000\ \text{K}$$

At this temperature collisions will excite almost all of the atoms into the $n = 3$ state. Even at somewhat lower temperatures, some will be excited to the $n = 3$ state. It is clearly easier to excite atoms using an electric field to accelerate free electrons in a cathode ray tube (which then collide with the atoms) than it is to raise the temperature of all of the atoms.

**Try it yourself:** Estimate the temperature at which an average hydrogen atom is ionized.

*Answer:* At a temperature of $10^5$ K or less, the collisions between atoms are violent enough to easily transfer the 13.6 eV of energy needed to ionize an atom.

The Try It Yourself question of the last exercise indicates that the temperature at which hydrogen ionizes as a result of collisions with other atoms is almost the same as the temperature needed for hydrogen atoms to become excited to higher energy states so they can emit visible light. Ionized hydrogen gas no longer emits spectral lines, instead emitting a continuous spectrum. This is because the energy states of free electrons can have any positive value of energy, so when they combine with protons to form neutral hydrogen again the emitted photons do not have discrete wavelengths. A gas with most of its atoms ionized due to collisions is called a *high-temperature plasma*. The cores of stars are at millions of kelvins and consist mostly of high-temperature plasma. Gas can also be ionized if placed in a strong electric field, producing a *low-temperature plasma* such as exists in neon lights or discharge tubes.

## Continuous spectra from stars

The Sun consists almost entirely of hydrogen and helium atoms, so we would expect its light to consist primarily of hydrogen and helium emission lines. Instead, light from the Sun and many other high-temperature objects is close to a continuous black body spectrum that depends only on the object's surface temperature (**Figure 27.17**).

**Figure 27.17** As the temperature increases, the black body radiation increases in intensity and the peak moves toward shorter wavelength.

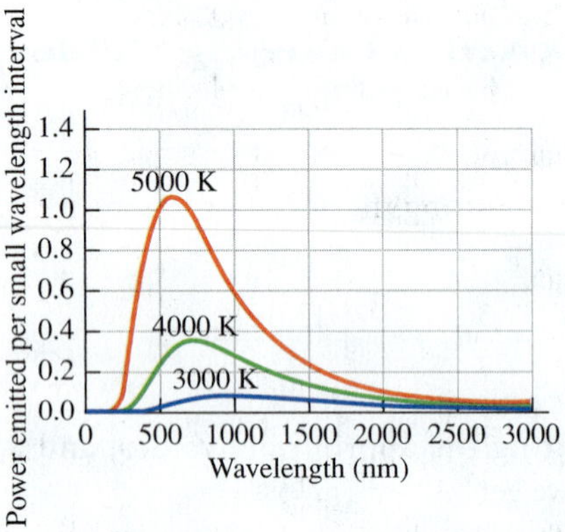


The continuous spectrum from these objects must depend on the distribution of the random kinetic energy (thermal energy) and temperature of the surface particles and not on their atomic compositions. On the hot Sun (5700 K on the surface), the material is dense and the atoms are continually colliding with each other. Evidently, these collisions distort the shapes of atoms during the collisions and excite them in unpredictable ways so that the light emitted when they de-excite is characteristic of the temperature of the gas and not of the types of atoms in the gas. These continuous (black body) spectra are the same for all objects at a particular temperature, but they differ significantly for objects at different temperatures (Figure 27.17).

Other stars emit similar black body radiation with the peak intensity at a wavelength that depends on their temperatures. However, if you look closely at radiation from the Sun and other stars, you observe some gaps in the continuous spectrum—narrow regions where very little light is emitted. Observational Experiment **Table 27.2** helps us analyze these findings.

**OBSERVATIONAL EXPERIMENT TABLE**

**27.2 Stellar spectra.**

| Observational experiment | Analysis | |
|---|---|---|
| We use a spectrograph to photograph and graphically represent the spectrum of the Sun (intensity versus wavelength). | In the visible region, we see a continuous spectrum but with several dark lines. The figure shows a small part of the spectrum. | 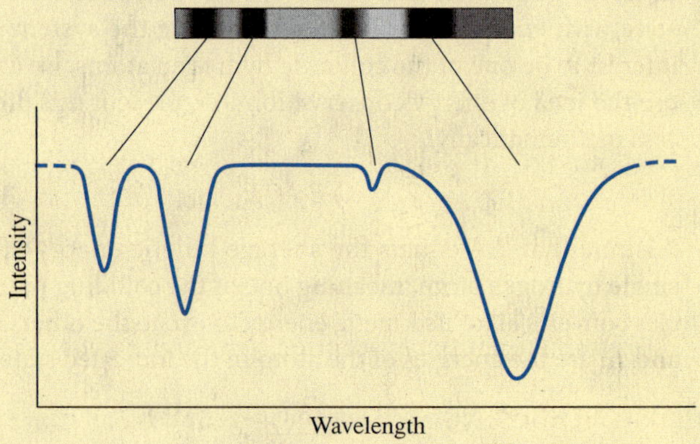 |

If we look at a narrow part of the visible spectrum of the Sun, we see dips in the intensity at particular wavelengths of the intensity-versus-wavelength graph. The wavelengths of the dips in the solar spectrum match the wavelengths of the line spectra emitted by a sodium vapor lamp in a lab.

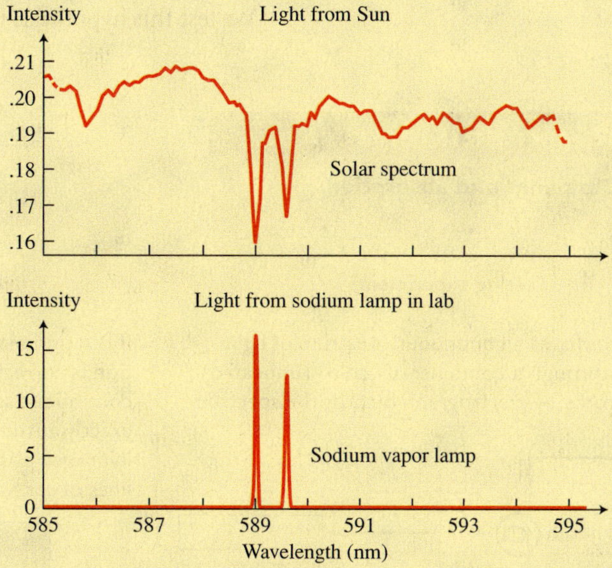


**Pattern**

If we observe a narrow wavelength region (516.6–517.4 nm) of the solar spectrum, we see dips that match the wavelengths of the line spectra of specific elements (iron, magnesium, etc.). The spectra of other stars are similar—continuous emission spectra with specific dark lines.

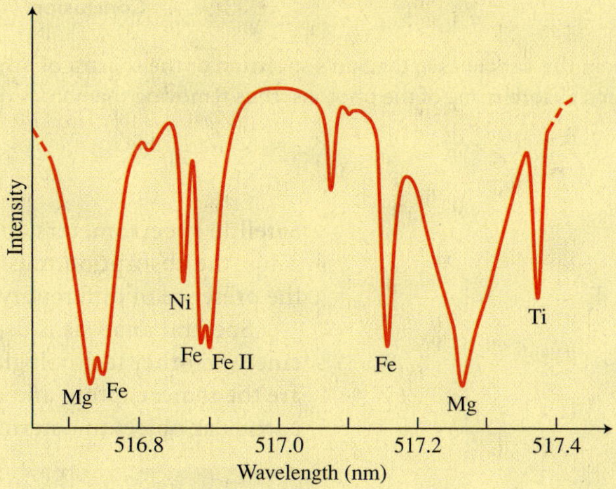


Why would the photons be missing at particular wavelengths? The first explanation that might come to mind is that sodium is not present on the surface of the Sun. However, this explanation does not work very well, because we find dark lines matching the emission lines for almost all elements known on Earth.

Another possible explanation is that the dark lines are not the result of missing elements, but of the absorption of photons traveling from the Sun to Earth. Imagine that some "cold" sodium atoms exist in the region between the Sun and the observer on Earth. Sodium atoms can absorb those photons coming from the Sun's surface whose energies match possible energy transitions in the atoms (see Figure 27.10). Soon after absorption, the atoms reemit these photons but in random directions. Thus the original photons that were absorbed would be "missing" from the continuous spectrum arriving at Earth. If we take a picture of the arriving spectrum, it should have dark lines at the locations of the missing photons. This is known as an **absorption spectrum**—a continuous spectrum with missing photons of specific wavelengths.

We test this hypothesis for the dark lines in Testing Experiment **Table 27.3**.

**TESTING EXPERIMENT TABLE**

**27.3 Dark lines and absorption.**

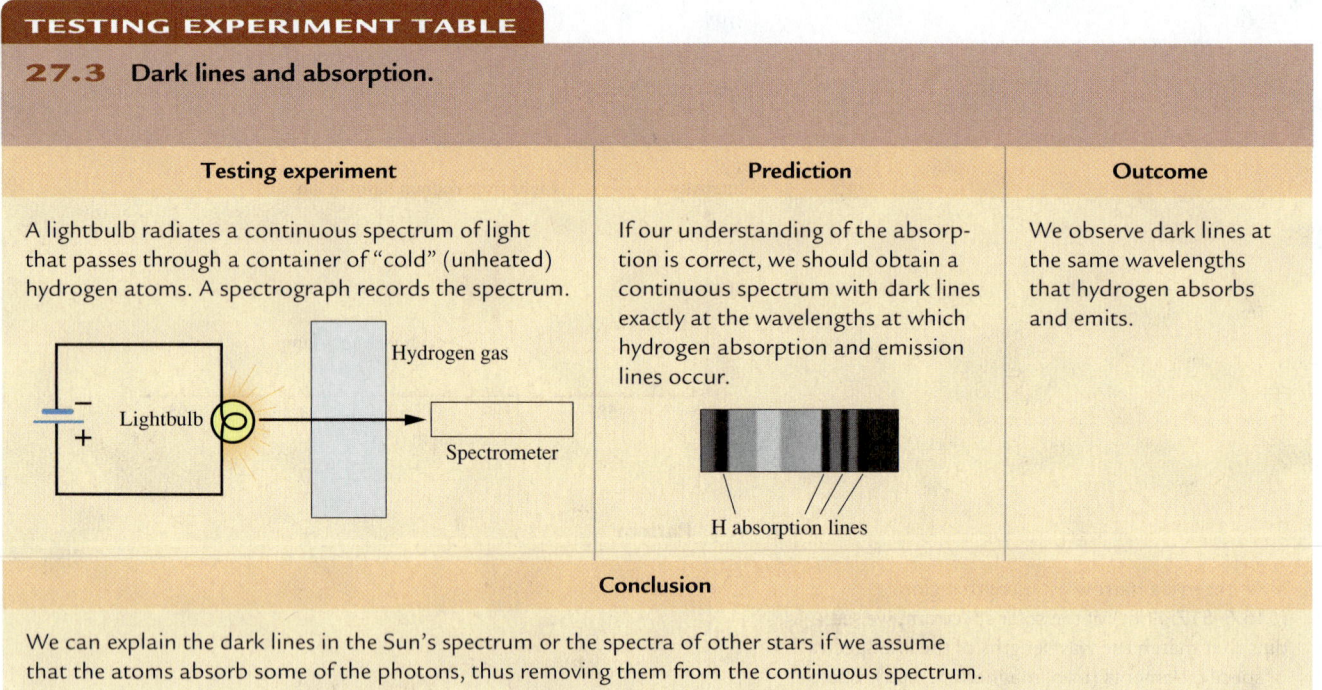

| Testing experiment | Prediction | Outcome |
|---|---|---|
| A lightbulb radiates a continuous spectrum of light that passes through a container of "cold" (unheated) hydrogen atoms. A spectrograph records the spectrum.<br>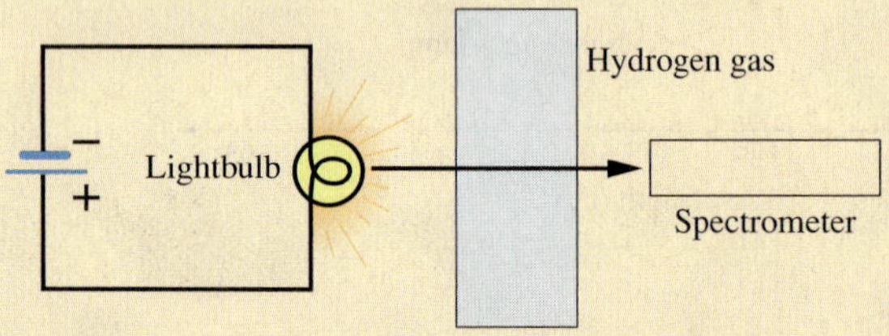<br> | If our understanding of the absorption is correct, we should obtain a continuous spectrum with dark lines exactly at the wavelengths at which hydrogen absorption and emission lines occur.<br>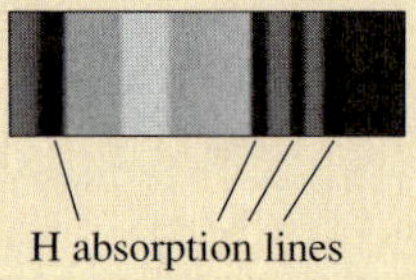<br> | We observe dark lines at the same wavelengths that hydrogen absorbs and emits. |

**Conclusion**

We can explain the dark lines in the Sun's spectrum or the spectra of other stars if we assume that the atoms absorb some of the photons, thus removing them from the continuous spectrum.

Satellite spectrometers above Earth's atmosphere show the same dark lines. Thus, the absorption must occur just outside the Sun's hot surface, indicating the presence of different types of atoms in the gas surrounding the Sun.

Spectral analysis is used in chemistry, engineering, transportation, medicine, and other technological fields. The main principles for these applications are the same: capture and analyze the wavelengths of the photons emitted by a particular object to determine its composition.

**Review Question 27.3** What materials and measuring instruments do you need in order to observe the absorption spectrum of a gas?

## 27.4 Lasers

When excited gas atoms emit light, they produce photons at different times and in different directions; thus the corresponding waves are incoherent (have different phases) and spread out in many directions. This process is called **spontaneous emission**. Lasers, on the contrary, emit very narrow, almost coherent beams of light. How is this accomplished?

In a laser, many excited atoms return from the same excited state to the same ground state simultaneously. The result is emission of in-phase photons traveling in the same direction. This process, called **stimulated emission**, was first suggested by Einstein. The word *laser* is an acronym for "light amplification by stimulated emission of radiation."

**Figure 27.18** Stimulated emission and production of laser light. (a) In the initial state, a photon of exactly the energy difference of the excited and ground states passes through, stimulating the excited atom. (b) In the final state, the excited atom's electron drops to the lower energy state, emitting a photon. Now two synchronized photons have the same energy.

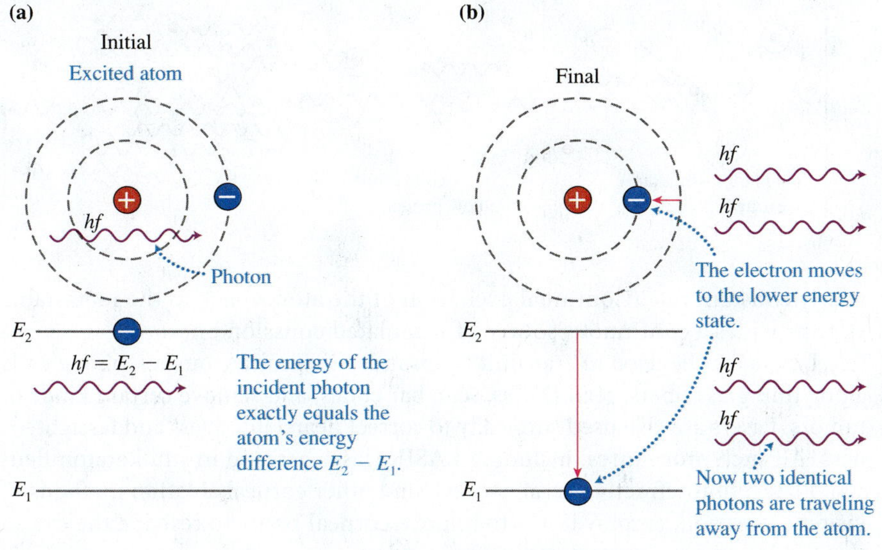
(a)
Initial
Excited atom
hf
Photon
E2
hf = E2 − E1
The energy of the incident photon exactly equals the atom's energy difference E2 − E1.
E1
(b)
Final
hf
hf
The electron moves to the lower energy state.
E2
hf
hf
Now two identical photons are traveling away from the atom.
E1

**Figure 27.19** Steps leading to the production of a laser beam. (a) Atoms are excited from energy state 1 to energy state 3. (b) Most atoms drop to metastable energy state 2. (c) Transitions from state 2 to state 1 are suppressed, so atoms stay in state 2 (population inversion). (d) Eventually, an atom transitions from 2 to 1, emitting a photon. (e) Stimulated emission begins a "chain reaction" with other atoms in state 2.

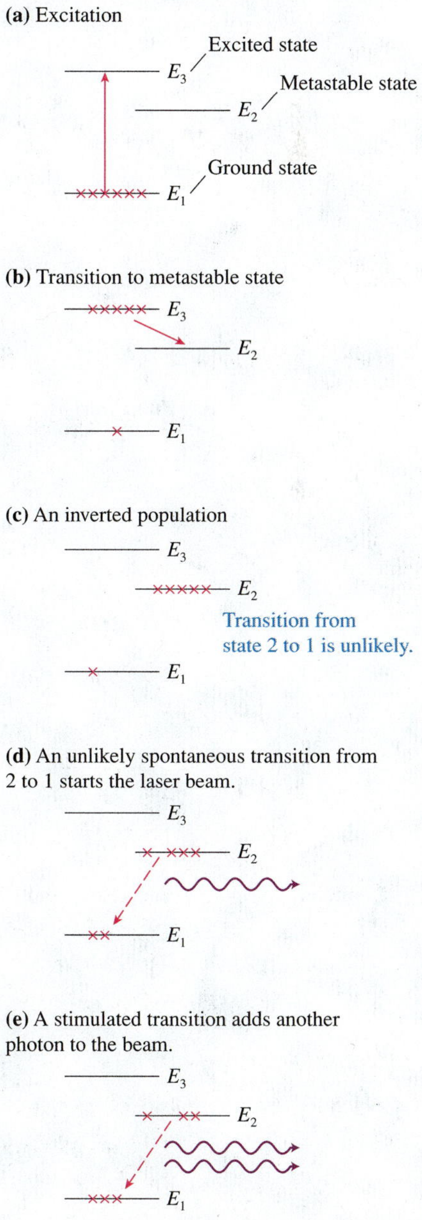
(a) Excitation
Excited state
E3
Metastable state
E2
Ground state
E1
(b) Transition to metastable state
E3
E2
E1
(c) An inverted population
E3
E2
Transition from state 2 to 1 is unlikely.
E1
(d) An unlikely spontaneous transition from 2 to 1 starts the laser beam.
E3
E2
E1
(e) A stimulated transition adds another photon to the beam.
E3
E2
E1

## How lasers work

Imagine that you have an atom in an excited state. You also have a photon whose energy equals exactly the energy difference between the atom's excited state and the ground state (**Figure 27.18a**). This photon can cause the excited atom to "de-excite" and emit a photon of exactly the same frequency (Figure 27.18b) —the stimulated emission mentioned above. The result is two photons of the same frequency traveling in the same direction.

To make stimulated emission effective, many more atoms must be in the same excited state than in the ground state, a situation known as **population inversion.** It is generally difficult to create population inversion because the average time that an atom spends in an excited state is very short (about $10^{-8}$ s). However, some atoms enter excited states called **metastable states**, in which they can remain for a significantly longer time (about $10^{-3}$ s). These atoms are good candidates for population inversion.

Suppose that the atoms in a material make the transition from ground state 1 to excited state 3 (**Figure 27.19a**) by absorbing photons. Instead of quickly returning to state 1, they make a more probable spontaneous transition to a metastable state 2 (Figure 27.19b). These atoms will remain in that state for a relatively long time since transitions from state 2 to state 1 are comparatively unlikely to occur. Thus, more atoms will be in state 2 than in the ground state 1 (Figure 27.19c). This is an example of population inversion.

One of these excited atoms eventually transitions to ground state 1 and emits a photon of energy $E_2 - E_1$ (Figure 27.19d). That photon passes through the material, many of whose atoms are in excited metastable energy state 2. This photon causes another atom in the metastable state 2 to de-excite via stimulated emission, resulting in two synchronized photons moving through the material (Figure 27.19e). Each of these photons can then stimulate another transition, resulting in a total of four photons. This process continues and creates an avalanche of new photons by stimulated emission—an intense laser beam (**Figure 27.20**).

The *lasing* material is placed between two mirrors that reflect the emitted photons back and forth through the material. In continuous lasers, one mirror is only partially reflecting and allows photons to leave the lasing region at a

**Figure 27.20** Stimulated emission produces an increasing number of in-phase photons in the laser cavity.

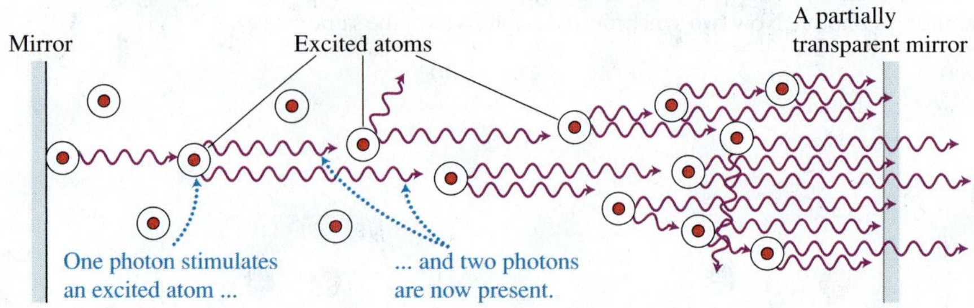


nearly constant rate. Continual excitation of the atoms back to the metastable state provides a continuous source of stimulated emission photons.

Lasers can be used to transmit thousands of simultaneous telephone calls along thin glass fibers, read DVDs, scan bar codes, and remove certain kinds of tumors. Lasers are also used surgically to correct nearsightedness and farsightedness. All such procedures, including LASIK (laser-assisted in situ keratomileusis), PRK (photorefractive keratectomy), and other corneal ablation methods of vision correction, employ lasers to remove corneal tissue to reshape the eye so that it can focus an image directly on the retina.

**Review Question 27.4** What is the main difference between spontaneous and stimulated emission of light?

# 27.5 Quantum numbers and Pauli's exclusion principle

Bohr's model was successful in explaining certain features of hydrogen's spectrum and structure, but it could not explain the structure of multi-electron atoms or some of the fine details of the hydrogen spectrum. We describe one of these details next.

## The Zeeman effect

In 1896 Dutch physicist Pieter Zeeman performed an experiment in which light from a known source passed through a region with sodium atoms in a strong magnetic field and then through a grating. He observed that when the magnetic field was on, a particular sodium spectral line was split into three closely spaced lines. This phenomenon, which became known as the Zeeman effect, could only be explained after Bohr's model was extended.

## Sommerfeld's extension of the Bohr model and additional quantum numbers

In 1915 Arnold Sommerfeld, a German theoretical physicist, modified Bohr's model to account for the Zeeman lines. Later, his modifications helped explain the spectral lines of more complex atoms. One of his innovations was the suggestion that electrons could have elliptical orbits. He proposed this through the introduction of two new quantum numbers related to the motion of the electron in the atom. (Prior to this, only one quantum number, $n$, was specified, and it was related to the energy and the size of the atom.) The new quantum numbers and the associated motion could explain the Zeeman effect.

In the Sommerfeld model, the $n$ quantum number remains an indicator of the size and energy of the atom in a particular state, but not of the rotational

**Figure 27.21** The allowed electron $n = 1$ to 4 orbits in the Sommerfeld model.

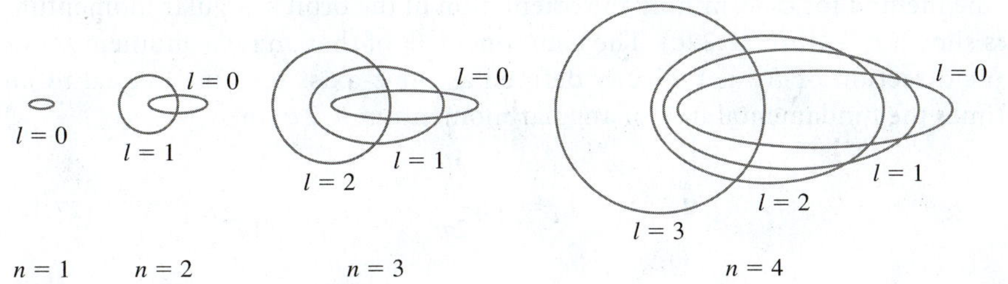


(angular) momentum of the state. A new quantum number $l$, called the **orbital angular momentum quantum number,** identifies the angular momentum of a state, whose value is $L = \sqrt{l(l+1)}(h/2\pi)$, where $l = 0, 1, 2, \ldots, n-1$ and $h/2\pi$ acts as a fundamental unit of angular momentum (see **Figure 27.21**).

For the $n = 1$ energy state, $l$ can only equal zero. Thus the angular momentum of that state could only be $L = \sqrt{0(0+1)}(h/2\pi) = 0$. For the $n = 2$ state, $l$ can be zero or 1, resulting in two possible states $L = 0$ or $L = \sqrt{2}(h/2\pi)$. These two states correspond to electron orbits that look very different (see the shapes of the different $n$ and $l$ orbits in Figure 27.21). The orbit of the $l = 0$ state is a flat ellipse. Recall that the angular momentum of an object moving in an orbit is $L = m r v \sin\theta$, where $\theta$ is the angle between the $r$ and $v$ vectors. For the flat $L = 0$ orbit, $r$ and $v$ are nearly parallel or anti-parallel and $\theta$ is zero. Thus, $L = 0$. For the more circular higher $L$ orbit, $r$ and $v$ are perpendicular and $\sin\theta$ is approximately 1.

According to the Sommerfeld model, in an $n = 2$ state, an electron zipping around the $l = 1$ orbit is the equivalent of an electric current loop, which therefore produces a magnetic dipole field. The flat $l = 0$ orbit should not produce any magnetic field. This differs from Bohr's model, in which an electron in the $n = 1$ ground state has angular momentum and moves in a circular orbit, thus producing a circular current and a magnetic field. Sommerfeld's prediction is consistent with experimental results, which show that the ground state of hydrogen does not have this magnetic field. But how does the Sommerfeld model help explain the Zeeman effect?

Consider hydrogen. The electrons in most hydrogen atoms in a low-temperature gas are in the $n = 1$ ground state, the orbit nearest the nucleus. Suppose this cool gas is irradiated by photons of just the right frequency to excite an atom from the $n = 1$ state to one of the two possible $n = 2$ states of the atom. If the gas is placed in a strong magnetic field, the atoms in states with nonzero angular momentum would behave like tiny magnets (with their own dipole moments) and interact with the magnetic field. This would result in the slightly different energies of the two $n = 2$ states. The stronger the magnetic field, the greater was the difference in energies of the states.

Zeeman's experiments on hydrogen gas using sensitive equipment and strong magnetic fields ($\vec{B}_{ex}$) produced three distinct spectral lines for transitions from the $n = 2$, $l = 1$ state to the $n = 1$, $l = 0$ state. This result could be explained if the $\vec{B}$ field due to the electron's motion in the $n = 2$, $l = 1$ state could be oriented in three discrete directions (see **Figure 27.22a**). This caused the $n = 2$, $l = 1$ state to split into three separate states (Figure 27.22b) due to different orientations of the electron orbit relative to the external magnetic field. The stronger the external magnetic field, the greater was the splitting.

This reasoning led to the introduction of a third quantum number, the **magnetic quantum number** $m_l$. If this quantum number is allowed to have values $m_l = 0, \pm 1, \pm 2, \ldots, \pm l$, then the Zeeman effect for the $n = 2$, $l = 1$ to $n = 1$, $l = 0$ transition is explained: Three slightly different energy photons would be emitted, as shown in Figure 27.22b. Apparently, $m_l$ is related

**Figure 27.22** Explanation for the Zeeman effect. (a) Possible orbit orientations in a $\vec{B}_{ex}$ field. (b) In an external magnetic field, the $n = 2$, $l = 1$ state splits into three energy states, leading to three spectral lines during the transition to the $n = 1$, $l = 0$ state. (c) The direction of the angular momentum vector due to an electron orbit.

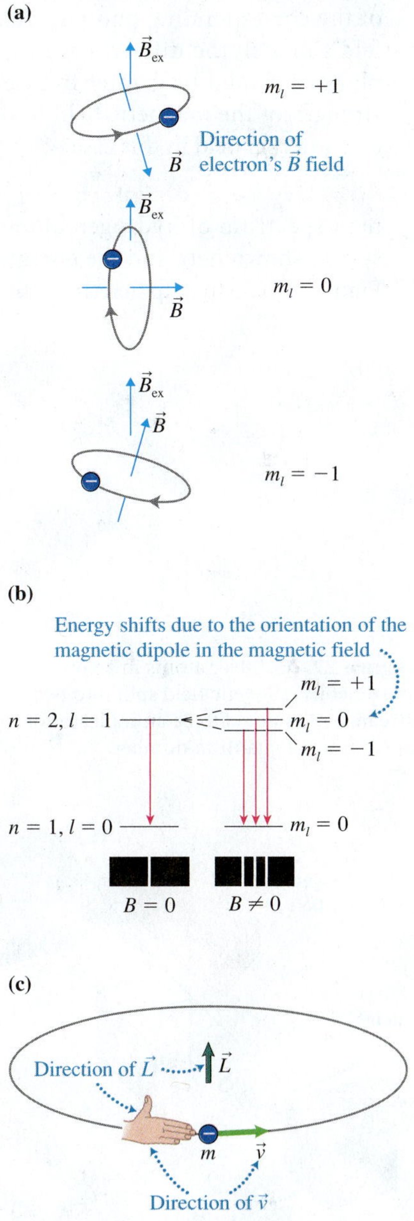

to the orientation of the angular momentum vector of the electron in its orbit (the method for determining the orientation of the orbit's angular momentum is shown in Figure 27.22c). The component $L_z$ of that angular momentum in the direction of $\vec{B}_{ex}$ is typically defined as the $z$-axis. $L_z$ is then equal to $m_l$ times the fundamental unit of angular momentum $h/2\pi$, or

$$L_z = m_l \frac{h}{2\pi} \tag{27.10}$$

**CONCEPTUAL EXERCISE 27.5 Magnetic field in the Sun's atmosphere**

How can we use the Zeeman spectral lines for a particular atom to determine the magnitude of the $\vec{B}$ field in the atmosphere of the Sun or of a distant star?

**Sketch and translate** If we can measure the wavelengths of these three lines, we can determine the energy of the corresponding photons. According to Sommerfeld's model, the differences in energy between these photons should be greater in direct proportion to the strength of the magnetic field in the Sun's atmosphere (the external field in this case).

**Simplify and diagram** We could use the Zeeman spectrum of hydrogen atoms in the Sun's atmosphere shown here and the energy state diagrams from Figure 27.22b to help determine the magnetic field.

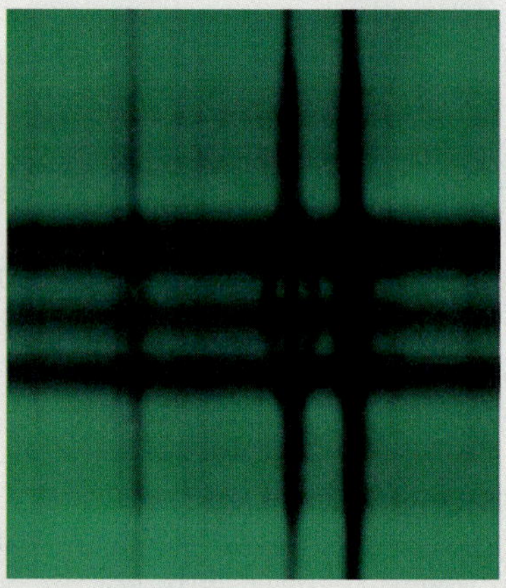

Three hydrogen lines from the Sun are split by its strong magnetic field.

**Try it yourself:** How would the spectrum in the figure in this exercise differ if there were no $m_l$ magnetic quantum number?

*Answer:* We would observe just one spectral line instead of three—like the zero magnetic field transition on the left side of Figure 27.22b.

In the exercise above we outlined a method for determining the $\vec{B}$ field at the surface of a star. When we apply this method to our Sun, we find that its surface magnetic field is about 4 T (tesla), roughly 100,000 times the magnitude of the $\vec{B}$ field near Earth's surface.

**Figure 27.23** Silver atoms in a nonhomogeneous magnetic field split into two streams, considered to be an indicator of an additional quantum number.

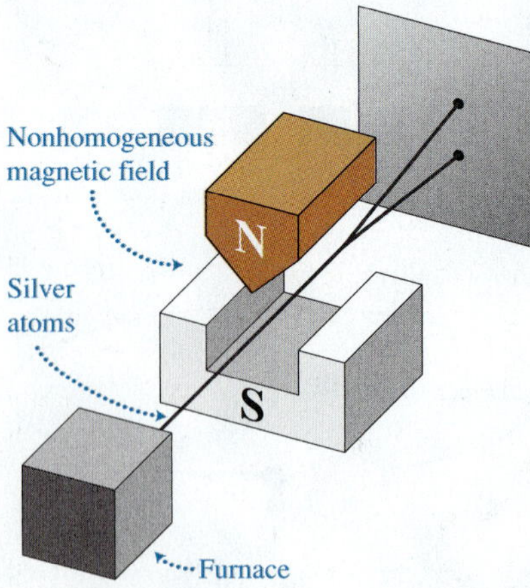


## Another difficulty with observed spectral lines

At about the same time that Zeeman made his observations, Thomas Preston performed Zeeman effect experiments with an even better magnet. He observed that some of the spectral lines for many elements did not split into three lines when a magnetic field was applied, but instead split into four, six, or more lines. This became known as the "anomalous" Zeeman effect. Sommerfeld's model could not explain it.

Around 1920 Sommerfeld and Alfred Landé suggested that the nucleus was affecting the angular momentum of the electron, thus producing this anomalous Zeeman effect. They incorporated this by assigning a quantum number to the nucleus, an ad hoc mathematical proposal with little physical model in mind.

Otto Stern and Walther Gerlach decided to test this idea by passing silver atoms through a nonhomogeneous magnetic field (**Figure 27.23**). If the Sommerfeld and Landé theories were correct, then the silver atoms should split into two streams, corresponding to the two values for the quantum number of the nucleus, rather than scatter randomly (the classical physics prediction). The experimental outcome matched the prediction based on the Sommerfeld and Landé theories.

## A new fourth quantum number

In 1923–1924, however, Wolfgang Pauli disproved the Sommerfeld and Landé theories using the theory of relativity. With relativistic corrections, the Landé and Sommerfeld theories consequently made predictions that were contradicted by the Stern-Gerlach experiments.

In trying to explain the anomalous Zeeman effect, Pauli uncovered many unexplained patterns in the periodic table of the elements. Why did elements in the same column (group) have similar properties when they had very different numbers of electrons but very different properties from adjacent columns? Why didn't the properties of atoms change gradually with additional electrons? Why were atoms with an even number of electrons less chemically active than those with odd numbers? Why were elements in the periodic table with 2, 8, or 18 electrons almost chemically inert? Finally, and perhaps most importantly, why at low temperatures weren't nearly all electrons in multi-electron atoms in the $n = 1$ ground state?

Pauli realized that these patterns in the periodic table and the anomalous Zeeman effect could be explained by a single, simple rule, namely, that each available state in the atom could hold at most *one* electron, an idea called the **Pauli exclusion principle**. This principle required the introduction of a fourth quantum number, but a different one from that of the Sommerfeld and Lande theories. The fourth quantum number was about the electron itself and was called $m_s$, the **spin quantum number** because it was related to the electron's intrinsic angular momentum (called **spin**). Pauli reasoned that in a magnetic field, an electron's spin could orient with the external field $\vec{B}_{ex}$ or against the external field $\vec{B}_{ex}$ but not in any other directions—like a compass needle that could only point precisely north or south. These two electron spin orientations correspond to values of $\pm 1/2$ for the quantum number $m_s$.

By suggesting this additional electron property, Pauli could account for the anomalous Zeeman effect. Together with the exclusion principle, which states that the allowed states of an atom could be completely characterized by the four quantum numbers $n$, $l$, $m_l$, and $m_s$, and that only one electron could exist in each state, Pauli's ideas could account for many of the patterns observed in the periodic table (we will discuss the patterns in the periodic table in Section 27.7).

## An attempt at a classical explanation for the spin quantum number

In 1925 Samuel Goudsmit and George Uhlenbeck attempted to explain the physical meaning of the $m_s$ quantum number. For the electron to have angular momentum, it would have to be like a tiny spinning sphere, similar to Earth spinning about its axis once every 24 hours (**Figure 27.24**). However, small charged parts $\Delta q$ at the edge of this small spinning spherical shell of charged material would have to move faster than light speed to have the necessary angular momentum. This assumption contradicts special relativity. The spin of the electron seemed to be some sort of new property that could not be understood using classical physics.

**Figure 27.24** An incorrect classical model of electron spin. The angular momentum needed would require charge at the edge of the spinning electron to move faster than the speed of light, which violates special relativity.

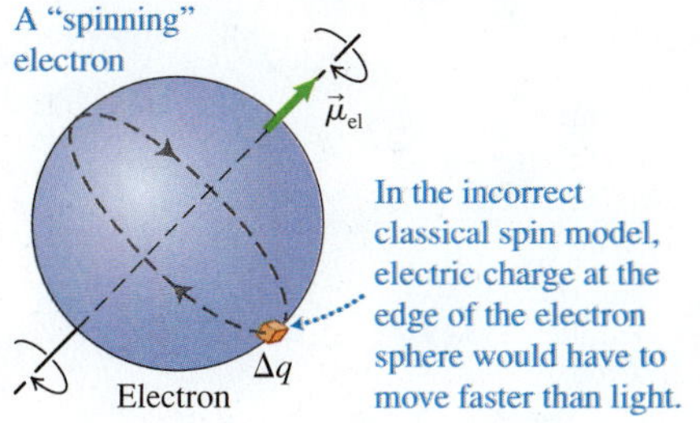
A "spinning" electron
$\vec{\mu}_{el}$
In the incorrect classical spin model, electric charge at the edge of the electron sphere would have to move faster than light.
$\Delta q$
Electron

## Summary of quantum numbers

By 1925 scientists had four quantum numbers that could describe the state of an electron in an atom. Also, Pauli's exclusion principle accounted for the distribution of electrons in multiple-electron atoms. Let's summarize the four quantum numbers:

1. *Principal quantum number n* can be any positive integer:

$$n = 1, 2, 3, \ldots \tag{27.11}$$

   and it indicates the energy of the electron in the atom. The larger the value of $n$, the larger the energy of the atom.

2. *Orbital angular momentum quantum number l* characterizes the orbital angular momentum of the electron. For a particular value of $n$, the allowed $l$ values are

$$l = 0, 1, 2, \ldots, n - 1 \tag{27.12}$$

   The magnitude of the angular momentum of the electron is $L = \sqrt{l(l+1)}\dfrac{h}{2\pi}$

3. *Magnetic quantum number $m_l$* describes the component of the electron's orbital angular momentum along an external magnetic field (Figure 27.22a). The energy of each $l$ state splits according to the state's $m_l$ value, which results in the additional lines in the spectra observed by Zeeman (Figure 27.22b). The magnetic quantum number $m_l$ can have values that depend on the quantum number $l$:

$$m_l = 0, \pm 1, \pm 2, \ldots, \pm l \tag{27.13}$$

   The component of the angular momentum in the $z$-direction is $L_z = m_l\,(h/2\pi)$.

4. *Spin magnetic quantum number $m_s$* Each electron has an intrinsic spin and corresponding magnetic dipole. The spin magnetic quantum number $m_s$ indicates the two directions that this spin dipole can be oriented relative to an external magnetic field:

$$m_s = \pm\frac{1}{2} \tag{27.14}$$

   The spin quantum number doubles the number of available states available for an electron in an atom.

Pauli's exclusion principle states a relationship between these quantum numbers and the electrons in an atom, regardless of the type of atom.

**Pauli exclusion principle** **Each electron in an atom must have a unique set of quantum numbers $n$, $l$, $m_l$, and $m_s$.**

Consider the case of an $n = 2$ state for electrons in an atom. The possible values of $l$ are 0 and 1. For $l = 0$ the only possible value of $m_l$ is 0. For $l = 1$ the possible values of $m_l$ are –1, 0, and 1. For each of these possibilities $m_s$ can be $\pm 1/2$. Thus, there are eight possible $n = 2$ states, as listed in **Table 27.4**.

**Table 27.4 The possible $n = 2$ states of an atom.**

| | |
|---|---|
| $l = 0,\ m_l = 0,\ m_s = +1/2$ | $l = 0,\ m_l = 0,\ m_s = -1/2$ |
| $l = 1,\ m_l = -1,\ m_s = +1/2$ | $l = 1,\ m_l = -1,\ m_s = -1/2$ |
| $l = 1,\ m_l = 0,\ m_s = +1/2$ | $l = 1,\ m_l = 0,\ m_s = -1/2$ |
| $l = 1,\ m_l = +1,\ m_s = +1/2$ | $l = 1,\ m_l = +1,\ m_s = -1/2$ |

An electron with $n = 2$ is said to be in the $n = 2$ shell of the atom. If a particular type of atom has eight $n = 2$ electrons, the atom is said to have a filled $n = 2$ shell—the shell can hold no more electrons.

You can use similar reasoning for other electron shells of atoms. You should find that the $n = 3$, $l = 2$ subshell of an atom can hold 10 electrons. The $n = 3$, $l = 1$ subshell can hold 6 electrons and the $n = 3$, $l = 0$ subshell can hold 2 electrons. Thus, the $n = 3$ shell can hold a maximum of 18 electrons.

Pauli's ideas of atomic structure explain the possible states of electrons in atoms, and also the reason that not all electrons in multi-electron atoms are in the lowest energy state (the $n = 1$ state). However, these ideas do not explain how to calculate the values of the quantities that are based on these quantum numbers (energy, orbital angular momentum, etc.). Physicists needed to construct an entirely new theory for describing this subatomic world, the subject of the next section.

**Review Question 27.5** In what energy states are the 10 electrons in the atom neon? Label each by its set of four quantum numbers.

## 27.6 Particles are not just particles

While the work of Bohr, Sommerfeld, and Pauli explained many observed properties of atoms, they did not explain why quantum numbers existed. Why were the energies of atoms and the orbital and spin angular momenta of the electrons quantized in the first place? Physicists began arriving at answers in the mid-1920s.

### A proposal for the wave nature of particles

The theoretical foundation for this effort began in 1924 with a young French physicist, Louis de Broglie. Knowing that light had both wave and particle behaviors, he thought it seemed reasonable that the elementary constituents of matter might also have these behaviors. Given that the wavelength $\lambda$ of a photon is $\lambda = h/p_{\text{photon}}$, de Broglie suggested that elementary particles also had a wavelength that could be determined in a similar way:

$$\lambda = \frac{h}{p} \tag{27.15}$$

where $h$ is Planck's constant and $p$ is the particle's momentum (which could potentially be relativistic).

If matter has wave-like properties, then why don't macroscopic objects such as human beings have them as well? Using Eq. (27.15), the wavelength of a 50-kg person walking at speed 1 m/s would be $1.33 \times 10^{-35}$ m. This is a hundred billion billion times smaller than the nucleus of an atom. Thus, there is no real-world situation that would cause a macroscopic object to exhibit wave-like behavior. However, the faster the object moves and the smaller its mass, the larger its wavelength. For example, for an electron (mass $m = 9.11 \times 10^{-31}$ kg) moving at a speed of 1000 m/s (typical speeds for electrons in cathode ray tubes), its wavelength would be

$$\lambda = \frac{6.63 \times 10^{-34}\ \text{J}\cdot\text{s}}{(9.11 \times 10^{-31}\ \text{kg})(1000\ \text{m/s})} = 7.3 \times 10^{-7}\ \text{m}$$

This is just a little longer than the wavelength of red light. Thus, the wave-like properties of atomic-sized particles might be very important for understanding their behavior.

**‹Active Learning Guide**

**Figure 27.25** (a) Constructive and (b) destructive electron waves.

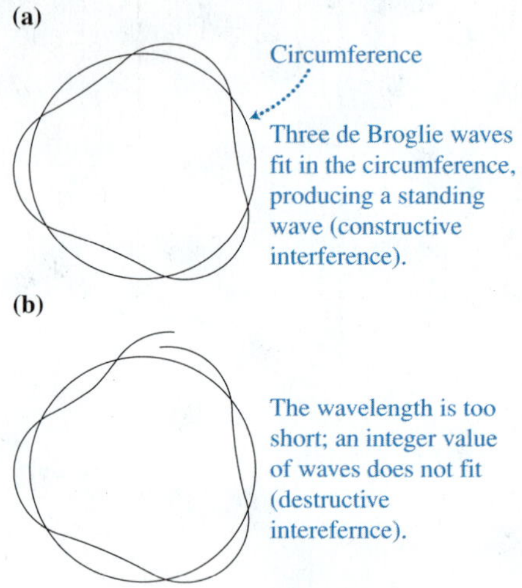

(a)
Circumference
Three de Broglie waves fit in the circumference, producing a standing wave (constructive interference).
(b)
The wavelength is too short; an integer value of waves does not fit (destructive intereference).


**Figure 27.26** Electron standing waves.

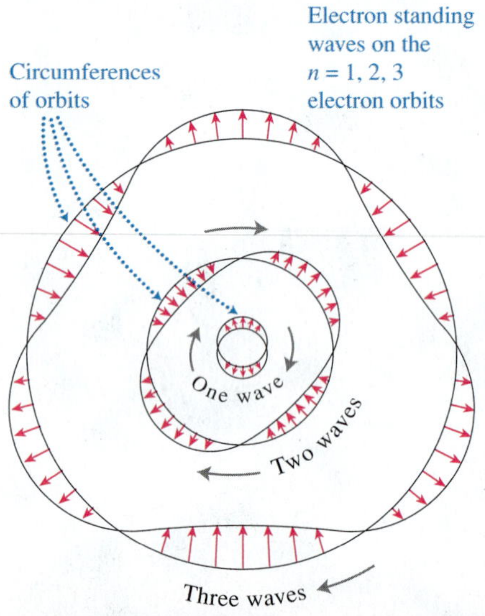


## De Broglie waves and Bohr's third postulate

De Broglie's radical idea contributed to building a new model of subatomic physics. Consider an electron moving around the nucleus of an atom in a stable Bohr orbit. Instead of modeling the electron as a point-like particle, de Broglie's ideas allow us to model it as a matter wave vibrating around the nucleus.

As the wave returns to the starting point in its orbit around the nucleus, it can interfere constructively with itself (**Figure 27.25a**); it can interfere destructively with itself; or some intermediate interference could occur (Figure 27.25b). If anything other than perfect constructive interference occurs, the wave will ultimately dampen and vanish as it circles many times around the nucleus. Thus, for the electron wave to be stable, an exact integer number $n$ of electron wavelengths must be wrapped around the nucleus (**Figure 27.26**), with a circumference of $2\pi r$. Mathematically (and assuming the electron is not moving relativistically):

$$2\pi r = n\lambda = n\frac{h}{p} = n\frac{h}{mv}, n = 1, 2, 3, \ldots$$

Because $n$ must be a positive integer, this equation represents a quantization condition for the hydrogen atom. If we take this equation, multiply both sides by $mv$, and divide both sides by $2\pi$, we get

$$mvr = n\frac{h}{2\pi}$$

This is precisely Bohr's third postulate [Eq. (27.2)]! Therefore, de Broglie's idea of the electron behaving like a standing wave explains Bohr's third postulate, which originally seemed rather arbitrary except that it led to the correct states of the hydrogen atom.

## Testing de Broglie's matter wave idea

Recall our earlier experiment with a low-intensity photon beam irradiating a double slit (Section 26.4). The individual flashes of single photons on the screen gradually built up an interference pattern, suggesting that even individual photons had wave-like behavior. In Testing Experiment **Table 27.5**, we replicate that experiment using electrons instead of photons.

**TESTING EXPERIMENT TABLE**

**27.5 Do electrons behave like waves?**

| Testing experiment | Prediction | Outcome |
|---|---|---|
| We shoot electrons one by one toward a screen with two closely spaced narrow slits (similar to Young's double-slit experiment). The electrons that pass through the slits reach a screen that lights up (flashes) when hit. The electron speed is about 1000 m/s. 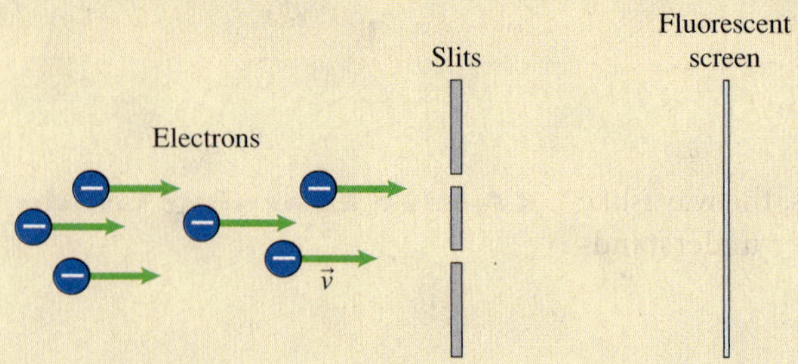 | If the electrons behave only like particles, there should be two images of the slits on the screen 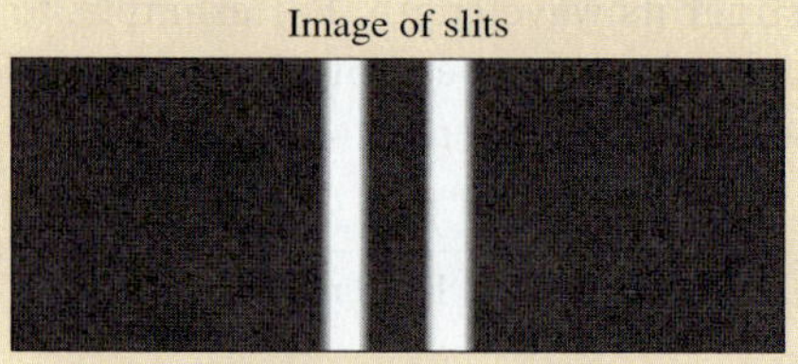 | An alternating bright and dark interference pattern gradually appears. |

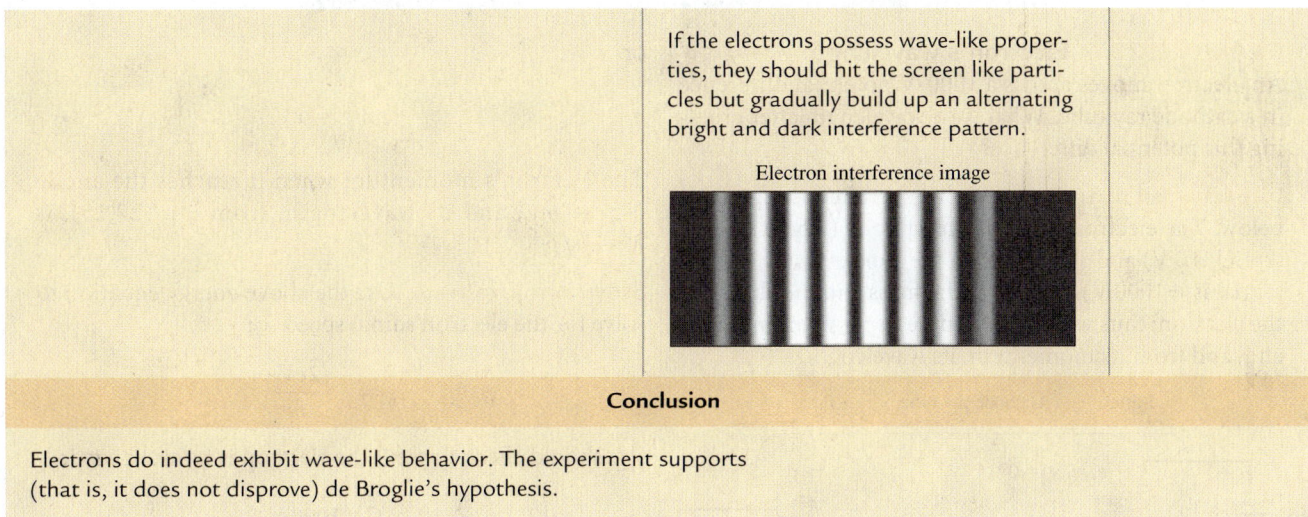

If the electrons possess wave-like properties, they should hit the screen like particles but gradually build up an alternating bright and dark interference pattern.

**Conclusion**

Electrons do indeed exhibit wave-like behavior. The experiment supports (that is, it does not disprove) de Broglie's hypothesis.

## Another test of de Broglie waves

De Broglie's hypothesis had another test a few years later. Shortly after de Broglie proposed his matter wave hypothesis, two scientists in New York, Clinton Davisson and his young colleague Lester Germer, were probing the structure of the atom by firing low-speed electrons at a nickel target made of small crystals and observing the resulting electron scattering (**Figure 27.27a-b**). The experiments did not give them any interesting results until they had a small accident in the laboratory. In 1925 their equipment malfunctioned, and the nickel target previously consisting of small crystals suddenly melted and resolidified into one large crystal.

When Davisson and Germer resumed the experiments with the large crystal, they saw a new pattern (compare Figure 27.27c with b). As they changed the position of the detector, they found alternating directions at which electrons were scattered and not scattered. The results made no sense to them until they heard about de Broglie's hypothesis. The large crystal was behaving like a three-dimensional interference grating, and they had observed the wave-like behavior of the electrons. This story is a perfect example of how data sometimes make sense only when scientists have an underlying model that can explain it. Later, similar experimen.ts were performed with hydrogen nuclei and alpha particles, supporting the idea that all subatomic particles exhibit wave-like properties.

**Figure 27.27** Davisson and Germer's experiment and accidental finding. (a) An electron beam excites a sample, and scattering is detected. (b) The scattering from a sample made up of small crystals. (c) The scattering from a sample accidentally turned into a large crystal.

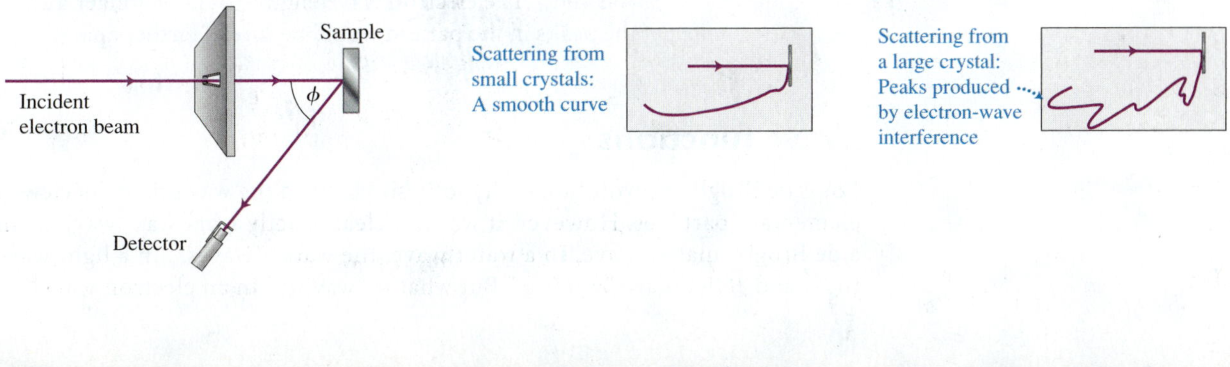

**EXAMPLE 27.6 Electron's wavelength**

An electron moves across a 1000-V potential difference in a cathode ray tube. What is its wavelength after crossing this potential difference?

**Sketch and translate** We've sketched the process below. The electron starts at the cathode (where the potential is 0 V) and accelerates to the anode (where the potential is +1000 V). We know the mass and the charge of the electron; thus we can determine its momentum at the end, and from its momentum, its wavelength.

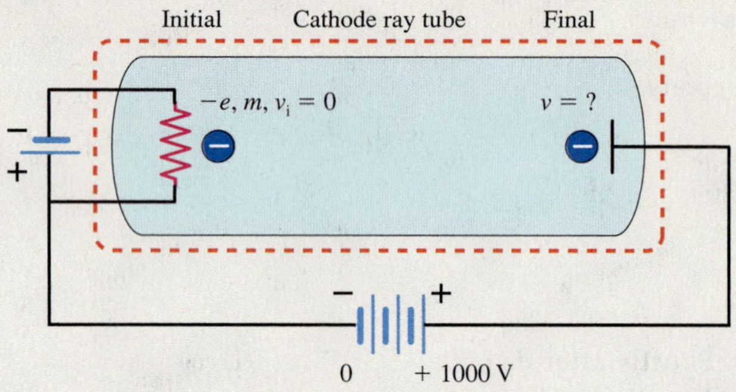


**Simplify and diagram** We assume that the kinetic energy of the electron is zero at the cathode where it begins and that it moves nonrelativistically. We can represent the process with an energy bar chart. The system of interest is the electron and the electric field produced by the cathode ray tube. In the initial state the system has zero kinetic and electric potential energy (since $V = 0$ at the cathode). In the final state the system has positive kinetic energy and negative electric potential energy (since the electron's electric charge is negative).

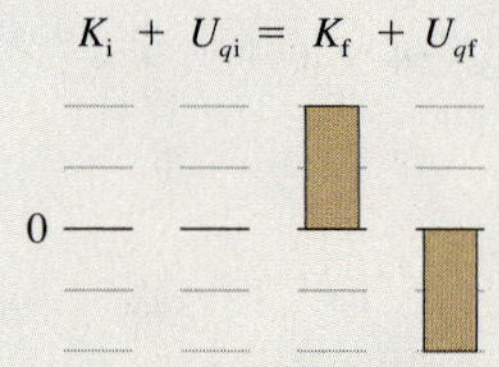


**Represent mathematically** We can use the bar chart to apply energy constancy to the system ($-e$ represents the charge of the electron):

$$K_i + U_{q\,i} = K_f + U_{q\,f}$$

Inserting the appropriate energy expressions:

$$0 + 0 = \frac{1}{2}mv_f^2 + (-e)V_f$$

or

$$\frac{1}{2}mv_f^2 = eV_f$$

The electron's momentum when it reaches the anode is $p = mv_f$, and its wavelength, from Eq. (27.15), is $\lambda = h/p$.

**Solve and evaluate** Use the above energy equation to solve for the electron's final speed, $v_f$:

$$v_f = \sqrt{\frac{2eV_f}{m}}$$

We should check to see if the electron is nonrelativistic:

$$v_f = \sqrt{\frac{2(1.6 \times 10^{-19}\,\text{C})(1000\,\text{V})}{9.11 \times 10^{-31}\,\text{kg}}} = 1.9 \times 10^7\,\text{m/s}$$

This speed is extremely fast, but only about 6% of light speed, so using the low-speed expression for momentum was justified. Combining this with the expression for the momentum and wavelength of the electron, we get

$$\lambda = \frac{h}{p} = \frac{h}{mv_f} = \frac{h}{m\sqrt{\dfrac{2eV_f}{m}}} = \frac{h}{\sqrt{2meV_f}}$$

Inserting the appropriate values:

$$\lambda = \frac{6.63 \times 10^{-34}\,\text{J}\cdot\text{s}}{\sqrt{2(9.11 \times 10^{-31}\,\text{kg})(1.6 \times 10^{-19}\,\text{C})(1000\,\text{V})}}$$

$$= 3.9 \times 10^{-11}\,\text{m}$$

This is a tiny wavelength, the same order of magnitude as the wavelength of X-rays and the size of an atom! This knowledge of the value of the electron's wavelength led to the development of microscopes that use electrons to image individual atoms, a task that would be impossible with visible light.

**Try it yourself:** Imagine that the electrons in the example above were used to investigate the structure of some crystal. When the electrons reflect off the crystal, they form an interference pattern. What would happen to the distances between the peaks in the pattern if the electrons were accelerated by a potential difference that is half what was used in the example?

*Answer:* The electron wavelength would be longer and the peaks in the pattern would be spread farther apart.

## Wave functions

Louis de Broglie's revolutionary hypothesis changed the way scientists viewed elementary particles. However, it was not clear exactly what was "waving" in a de Broglie matter wave. In a water wave, the water "waves." In a light wave the $\vec{E}$ and $\vec{B}$ fields are "waving." But what is "waving" in an electron wave?

De Broglie himself thought of the wave as some sort of vibrating shroud surrounding the true electron (which he still thought of as a point-like particle). This idea is known as de Broglie's **pilot wave model**; the wave guides the actual particle within it. Others thought the electron wasn't point-like and was "smeared" over a certain volume; it was this smeared electron that was vibrating as it moved through space. If this were true, then the electron would occupy an extended region (such as being wrapped around the nucleus of an atom). In this case, thinking of the electron as being in a specific place wouldn't be reasonable because it was actually spread out.

**TIP** In the previous paragraphs (and throughout the rest of this section) we use the word "particle" to refer to a fundamental constituent of matter (an electron, for example), even though we have learned that these constituents have both wave and particle behaviors.

In 1926 German scientist Max Born suggested that a particle's matter wave is a mathematical distribution related to the likelihood of measuring the particle to be at various locations at a specific time. If the value (technically, the square of the value) of this matter wave at a particular location is near zero, then there is almost zero chance of detecting the particle at that location at that time. If the value is large, then there is a large chance of finding the electron there.

Erwin Schrödinger coined the term **wave function** for these matter waves and developed an equation for determining them. The field of **quantum mechanics** is based on understanding the particles of the micro-world in terms of wave functions and the Schrödinger equation (the mathematics of which is beyond the scope of this textbook).

**Figure 27.28** Probability distributions (wave functions squared) for $n = 2$ states.

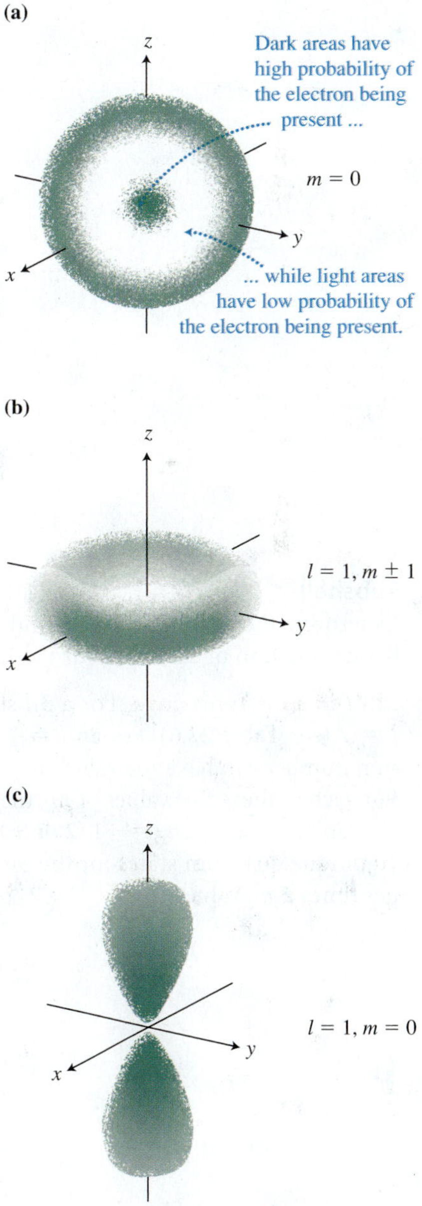
(a)
z
Dark areas have high probability of the electron being present ...
m = 0
y
x
... while light areas have low probability of the electron being present.
(b)
z
l = 1, m ± 1
y
x
(c)
z
l = 1, m = 0
y
x

## Atomic wave functions

Understanding matter in terms of wave functions lets us construct a better model of the hydrogen atom. When the Schrödinger equation is used to determine the electron wave functions, three of the quantum numbers ( $n$, $l$, and $m_l$) appear naturally, without needing to be put in by hand. The fourth number, the spin magnetic quantum number $m_s$, appears naturally when quantum mechanics is extended so that it incorporates Einstein's theory of special relativity. This model resulted in a detailed understanding of the periodic table of the elements based on fundamental physics ideas. We will learn later in the book (Chapter 29) that this theory of relativistic quantum mechanics predicts the existence of antimatter and ultimately evolves into the standard model of particle physics, a theory that describes all the known elementary particles and their interactions.

**Review Question 27.6** How did the Davisson and Germer experiment serve as a testing experiment for the idea that subatomic particles possess wave-like properties?

# 27.7 Multi-electron atoms and the periodic table

Together, the four quantum numbers $n$, $l$, $m_l$, and $m_s$ specify the wave functions of possible electron states in atoms. Examples of the square of the $n = 2$ wave functions (that is, the values that generate the probability waves) are shown in **Figure 27.28**. The more darkly shaded regions indicate where the electron's probability wave has larger values and therefore a greater probability of finding the electron at that location. The $l = 0$ probability wave actually has a greater probability than the $l = 1$ and 2 waves of the electron's being at the nucleus—the dark region at the center of its probability wave. This depiction provides an explanation for the lower energy in multi-electron atoms of the $l = 0$ electron states.

**Table 27.6 Letter designations of different $l$ states and maximum number of electrons in that subshell.**

| $l$ value | 0 | 1 | 2 | 3 | 4 | 5 | 6 |
|---|---|---|---|---|---|---|---|
| Letter | $s$ | $p$ | $d$ | $f$ | $g$ | $h$ | $i$ |
| Maximum number of electrons | 2 | 6 | 10 | 14 | 18 | 22 | 26 |

## Subshells and number of electrons in them

Early in the study of the emission and absorption of light by atoms, scientists used letters to designate the different states of orbital angular momentum quantum number $l$. This notation is still widely used today. **Table 27.6** summarizes the letter designations. A state with $n = 1$ and $l = 0$ is called a $1s$ state; an $n = 4$ and $l = 2$ state is called a $4d$ state.

Specifying the $n$ and $l$ quantum numbers specifies a group of states called a **subshell**. Specifying just the $n$ quantum number specifies a group of subshells that is called a **shell**. The $l$ quantum number also determines the maximum number of electrons that can populate that subshell.

How is the maximum number of electrons in the third row of Table 27.6 determined? Let's look at the $n = 1$, $l = 0$ $s$ subshell. In this subshell, $m_l$ can only be zero and $m_s$ can be $+1/2$ or $-1/2$. Thus, there are two different states with $n = 1$ and $l = 0$. According to Pauli's exclusion principle, each state can hold just one electron; so the $1s$ shell can hold at most two electrons. This reasoning holds for any $s$ state.

A subshell is said to be filled if each state in the subshell is occupied by an electron. The number of states in a subshell depends on the number of different allowed combinations of $m_l$ and $m_s$ for the given value of $l$. In the following exercise we consider the number of states in an $l = 2$ (or $d$) subshell.

**CONCEPTUAL EXERCISE 27.7 States in $3d$ subshell**

Determine the number of states and the quantum number designation of each state for the $3d$ subshell.

**Sketch and translate** For a $3d$ subshell, $n = 3$ and $l = 2$ (see Table 27.6). For an $l = 2$ state, the $m_l$ quantum number can have the values $m_l = -2, -1, 0, 1,$ or $2$. For each of these five values of $m_l$, the $m_s$ quantum number can have values $m_s = -1/2$ or $+1/2$. Thus, there are 10 unique quantum states for the $3d$ subshell (5 $m_l$ values times 2 $m_s$ values).

**Simplify and diagram** The 10 states and their corresponding quantum numbers are summarized in **Table 27.7**. If one electron resides in each of these states, such as occurs in a copper atom, the atom is said to have a filled $3d$ subshell.

**Try it yourself:** How many electrons can be in a $4f$ subshell?

*Answer:* 14 because $l = 3$, $m_l = -3, -2, -1, 0, 1, 2,$ or $3$, and $m_s = -1/2$ or $+1/2$.

## Electron configuration for the ground state of an atom

The electrons in an atom normally occupy the states with lowest energy. In this case the atom is said to be in its ground state. An electron configuration for the ground state of an atom indicates the subshells occupied by its electrons and the number of electrons in each subshell. For example, a neutral

**Table 27.7** **The number of states and the quantum number designation of each state for the 3*d* subshell.**

| $n$ | $l$ | $m_l$ | $m_s$ |
|---|---|---|---|
| 3 | 2 | −2 | +1/2 |
| 3 | 2 | −1 | +1/2 |
| 3 | 2 | 0 | +1/2 |
| 3 | 2 | 1 | +1/2 |
| 3 | 2 | 2 | +1/2 |
| 3 | 2 | −2 | −1/2 |
| 3 | 2 | −1 | −1/2 |
| 3 | 2 | 0 | −1/2 |
| 3 | 2 | 1 | −1/2 |
| 3 | 2 | 2 | −1/2 |

**Table 27.8** **Atomic subshells from lowest to highest energy (approximate).**

| Quantum Numbers | | | Number of Quantum States | |
|---|---|---|---|---|
| $n$ | $l$ | $m_l$ | In the Subshell | Total |
| 1 | 0 (*s*) | 0 | 2 | 2 |
| 2 | 0 (*s*) | 0 | 2 | |
| | 1 (*p*) | −1, 0, +1 | 6 | 8 |
| 3 | 0 (*s*) | 0 | 2 | |
| | 1 (*p*) | −1, 0, +1 | 6 | 18 |
| | 2 (*d*) | −2, −1, 0, +1, +2 | 10 | |
| 4 | 0 (*s*) | 0 | 2 | |
| | 1 (*p*) | −1, 0, +1 | 6 | 32 |
| | 2 (*d*) | −2, −1, 0, +1, +2 | 10 | |
| | 3 (*f*) | −3, −2, −1, 0, +1, +2, +3 | 14 | |

sodium atom has 11 electrons distributed as follows: $1s^2 2s^2 2p^6 3s$. The superscript represents the number of electrons in each subshell: $1s^2$ indicates that two electrons occupy the 1*s* subshell, $2p^6$ indicates that six electrons occupy the 2*p* subshell, and so forth. The subshell 3*s* has only a single electron. The maximum number of electrons that can be in a subshell and their approximate order of filling are listed in **Table 27.8**. We see that for sodium, 10 electrons are needed to fill the 1*s*, 2*s*, and 2*p* subshells. The 11th electron in sodium occupies a state with the next available lowest energy subshell, the 3*s* subshell. When sodium is ionized to form $Na^+$, it loses the electron from the 3s subshell since that electron is most weakly bound to the nucleus.

## The periodic table

Many of the properties of an atom, such as its tendency to form positive or negative ions and its tendency to combine with certain other atoms to form molecules, depend on the number of electrons in its outermost subshell and on the type of subshell it is. In addition, groups of atoms with the same number of outer electrons in subshells with the same letter designation (*s, p, d,* etc.) usually exhibit similar properties. This leads naturally to the classification of the atoms as presented in the **periodic table of the elements** (**Table 27.9**). In the table, elements in the same column have similar electron configurations for their outer electrons and therefore similar chemical properties.

**Review Question 27.7** What is the difference between a shell and a subshell?

# 27.8 The uncertainty principle

Quantum mechanics provided a theoretical basis for observations of spectral lines. It also explained how different chemical elements could have similar chemical properties. Another profound prediction of quantum mechanics concerned the determinability of certain physical quantities.

Think of a classical particle. The ability to determine its velocity and its position would appear to be limited only by the precision of the measuring

**Table 27.9 Periodic table of the elements***

| I | II | | | | | | | | | | | III | IV | V | VI | VII | 0 |
|---|---|---|---|---|---|---|---|---|---|---|---|---|---|---|---|---|---|
| 1 H 1.0080 | | | | | | | | | | | | | | | | | 2 He 4.0026 |
| 3 Li 6.941 | 4 Be 9.0122 | | | | | | | | | | | 5 B 10.81 | 6 C 12.011 | 7 N 14.0067 | 8 O 15.9994 | 9 F 18.9984 | 10 Ne 20.179 |
| 11 Na 22.9898 | 12 Mg 24.305 | | | | Transition elements | | | | | | | 13 Al 26.9815 | 14 Si 28.086 | 15 P 30.9738 | 16 S 32.06 | 17 Cl 35.453 | 18 Ar 39.948 |
| 19 K 39.102 | 20 Ca 40.08 | 21 Sc 44.956 | 22 Ti 47.90 | 23 V 50.941 | 24 Cr 51.996 | 25 Mn 54.9380 | 26 Fe 55.847 | 27 Co 58.9332 | 28 Ni 58.71 | 29 Cu 63.54 | 30 Zn 65.37 | 31 Ga 69.72 | 32 Ge 72.59 | 33 As 74.9216 | 34 Se 78.96 | 35 Br 79.909 | 36 Kr 83.80 |
| 37 Rb 85.467 | 38 Sr 87.62 | 39 Y 88.906 | 40 Zr 91.22 | 41 Nb 92.906 | 42 Mo 95.94 | 43 Tc (99) | 44 Ru 101.07 | 45 Rh 102.906 | 46 Pd 106.4 | 47 Ag 107.870 | 48 Cd 112.40 | 49 In 114.82 | 50 Sn 118.69 | 51 Sb 121.75 | 52 Te 127.60 | 53 I 126.9045 | 54 Xe 131.30 |
| 55 Cs 132.906 | 56 Ba 137.34 | 57 La 138.906 | 72 Hf 178.49 | 73 Ta 180.948 | 74 W 183.85 | 75 Re 186.2 | 76 Os 190.2 | 77 Ir 192.2 | 78 Pt 195.09 | 79 Au 196.967 | 80 Hg 200.59 | 81 Tl 204.37 | 82 Pb 207.2 | 83 Bi 208.981 | 84 Po (210) | 85 At (210) | 86 Rn (222) |
| 87 Fr (223) | 88 Ra 226.03 | 89 Ac 227.028 | 104 Rf (261) | 105 Db (262) | 106 Sg (266) | 107 Bh (264) | 108 Hs (269) | 109 Mt (268) | 110 Ds (271) | 111 Rg (272) | 112 Cn (285) | | 114 Fl (289) | | 116 Lv (293) | | |

| Lanthanide series | | | | | | | | | | | | | | |
|---|---|---|---|---|---|---|---|---|---|---|---|---|---|---|
| | 58 Ce 140.12 | 59 Pr 140.908 | 60 Nd 144.24 | 61 Pm (147) | 62 Sm 150.4 | 63 Eu 151.96 | 64 Gd 157.25 | 65 Tb 158.925 | 66 Dy 162.50 | 67 Ho 164.930 | 68 Er 167.26 | 69 Tm 168.934 | 70 Yb 173.04 | 71 Lu 174.97 |

| Actinide series | | | | | | | | | | | | | | |
|---|---|---|---|---|---|---|---|---|---|---|---|---|---|---|
| | 90 Th 232.038 | 91 Pa 231.036 | 92 U 238.029 | 93 Np 237.048 | 94 Pu (242) | 95 Am (243) | 96 Cm (248) | 97 Bk (249) | 98 Cf (249) | 99 Es (254) | 100 Fm (257) | 101 Md (258) | 102 No (259) | 103 Lr (260) |

*Atomic weights of stable elements are those adopted in 1969 by the International Union of Pure and Applied Chemistry. For those elements having no stable isotope, the mass number of the "most stable" isotope is given in parentheses.

instruments. However, quantum mechanics predicts something quite different for subatomic particles.

Recall that if we send a laser beam through a single slit, it forms a pattern of bright and dark fringes on the screen behind the slit. What if you send a beam of electrons though the same horizontally oriented single slit and use a scintillating screen to register the electrons after they pass through the slit (**Figure 27.29**)? Electrons hit the screen at different locations. Instead of a sharp horizontal image of the slit as shown in Figure 27.29a, we see a set of "bright" and "dark" bands in the $y$-direction, perpendicular to the slit, as in Figure 27.29b. The pattern looks similar to the diffraction pattern that light produced when passing through a slit. Electron waves are being diffracted by the slit. The angular position $\theta_1$ of the first dark region (where no electrons are detected) above or below the central maximum is determined by the light diffraction equation [Eq. (23.6) from Section 23.5]:

$$\sin\theta_1 = \frac{\lambda}{w}$$

Here $w$ is the slit width and $\lambda$ is the electron's wavelength (**Figure 27.30**).

Now, let's consider a single electron passing through the slit. Because the slit has width $w$, the electron's $y$-coordinate position can be determined only to within $w$. However, in order to hit the screen in any location other than directly across from the slit, the electron must have some $y$-component of momentum once it has passed through the slit. (We're considering the particle's momentum rather than velocity because the momentum determines its wavelength.)

> **TIP** Remember that a significant diffraction pattern is being produced because of the wave-like properties of the electron combined with the narrowness of the slit.

A connection exists between the narrowness of the slit and the apparent $y$-component of the electron's momentum once it passes through. The narrower the slit, the larger the $y$-component of the electron's momentum might be. To put it another way, the more precisely we attempt to determine the $y$-coordinate of the electron's position $y$ as it passes through the slit (by using a narrower slit), the less precisely we can determine the $y$-component of the electron's momentum $p_y$. It seems there is a limit on how to simultaneously determine the position $y$ and momentum component $p_y$ of the electron.

Let's see if we can make this idea more quantitative. The angular location of the first dark fringe is

$$\sin\theta_1 = \frac{\lambda}{w}$$

Inserting the expression for the wavelength of the electron, we have

$$\sin\theta_1 = \frac{(h/p)}{w}$$

Use Figure 27.30 to write $\sin\theta_1$ in terms of momentum:

$$\sin\theta_1 = \frac{p_y}{p} = \frac{(h/p)}{w} = \frac{h}{wp}$$

Combining and rearranging the second and fourth terms above, we get

$$wp_y = h$$

Note that $w$ is the width of the slit and quantifies how determinable the $y$-position of the electron is (all we know is that the electron went through the slit). Assuming that nearly all of the electrons are detected between the first dark fringes on either side of the central maximum, $p_y$ quantifies how determinable the $y$-component of the momentum of an electron is once it has passed through the slit (all we know is that it reaches the screen between the two dark fringes). We indicate this by saying $w = \Delta y$ and $p_y = \Delta p_y$. The relationship then becomes

$$\Delta y \Delta p_y \approx h$$

Since the right-hand side of this equation is not zero, neither $\Delta y$ nor $\Delta p_y$ is exactly determinable. These limitations did not come from using a particular measuring device. They result solely from the electron having wave-like

**Figure 27.29** Electron diffraction patterns. (a) The expected pattern for electrons as particles. (b) The observed pattern, which supports electrons as waves.

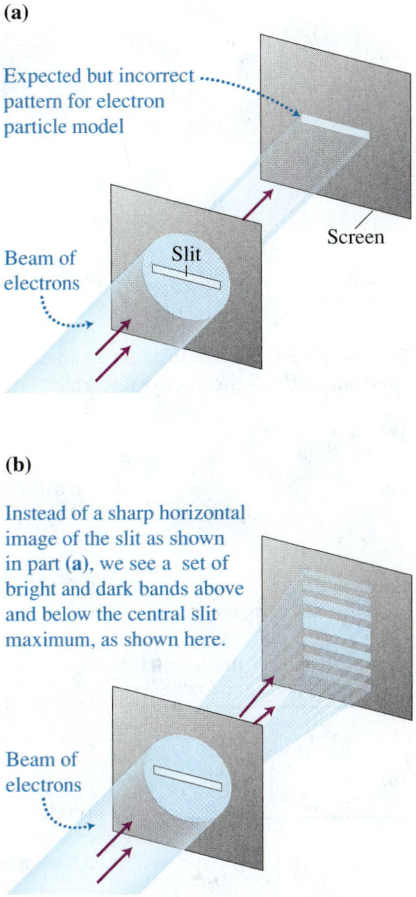
(a)
Expected but incorrect pattern for electron particle model
Screen
Slit
Beam of electrons
(b)
Instead of a sharp horizontal image of the slit as shown in part (a), we see a set of bright and dark bands above and below the central slit maximum, as shown here.
Beam of electrons

**Figure 27.30** Single-slit diffraction with electrons. The angle between the maximum and the first dark region can be determined just as for light diffraction.

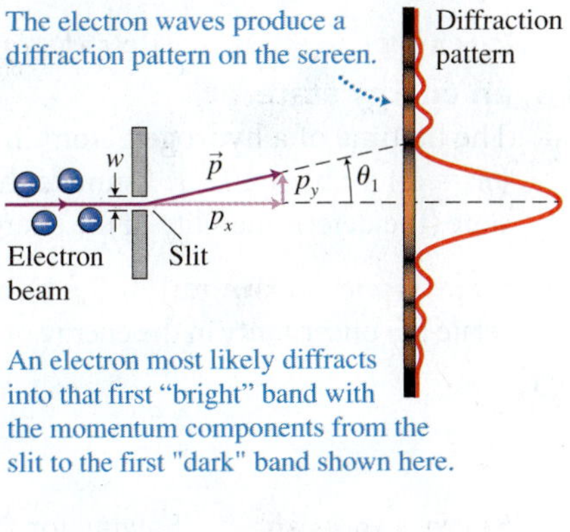
The electron waves produce a diffraction pattern on the screen.
Diffraction pattern
$w$
$\vec{p}$
$p_y$
$\theta_1$
$p_x$
Electron beam
Slit
An electron most likely diffracts into that first "bright" band with the momentum components from the slit to the first "dark" band shown here.

properties. Quantum mechanics sets a fundamental limit on how well the position and momentum of a particle in a particular direction can be determined simultaneously.

A more detailed analysis of the situation leads to an improved relationship (we switched to the $x$-direction, as it is the typical way to write this expression):

$$\Delta x \Delta p_x \geq \frac{h}{4\pi} \tag{27.16}$$

The factor $4\pi$ is a result of a more careful statistical analysis. Equation (27.16) is one example of what is known as the **uncertainty principle** of quantum mechanics, formulated by German physicist Werner Heisenberg in 1925.

## Uncertainty principle for energy

**Figure 27.31** The lifetime $\Delta t$ in an excited state affects the energy uncertainty.

(a)

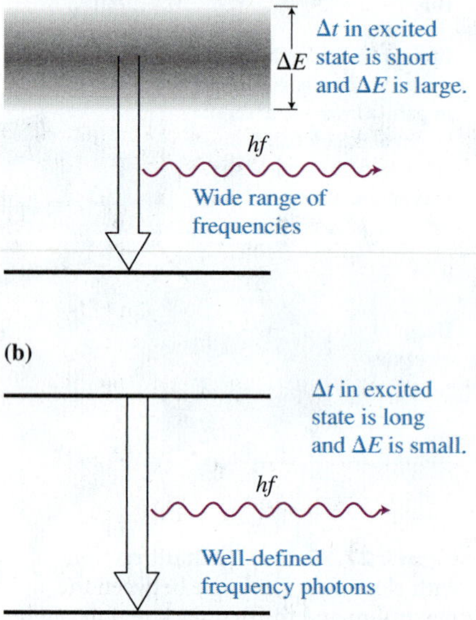


Imagine that we have an atom that can potentially emit a photon of energy $E$. Remember that $E = pc$ for a photon. But because a fundamental limit exists on the determinability of the momentum of a particle, a fundamental limit should also exist on the determinability of the energy of the photon:

$$\Delta E = (\Delta p)c$$

Combining this with the uncertainty principle, we have

$$\Delta x\left(\frac{\Delta E}{c}\right) \geq \frac{h}{4\pi}$$

Photons travel at the speed of light, so they travel a distance $\Delta x$ in a time interval $\Delta t = (\Delta x/c)$. Therefore, we can express $\Delta x$ equivalently as $c\Delta t$. The equation then becomes

$$\Delta E \Delta t \geq \frac{h}{4\pi} \tag{27.17}$$

In words, if a system (such as an atom in an excited state) exists for a certain time interval $\Delta t$ before it emits a photon, then the energy of the state is only determinable within a range $\Delta E$. Thus, the energy states of an atom should be represented as "bands" whose thickness is determined by how long the atom can be in this state—the shorter-lived the state, the greater the energy uncertainty and the wider its energy spread (**Figure 27.31**).

**QUANTITATIVE EXERCISE 27.8 Width of an energy state**

The lifetime of a hydrogen atom in the first excited state ($n = 2$) is about $10^{-9}$ s. Estimate the *width* of this energy state (the determinability of the energy of the state).

**Represent mathematically** Using Eq. (27.17) we can relate the uncertainty in the energy of the state to its lifetime:

$$\Delta E \Delta t \geq \frac{h}{4\pi}$$

**Solve and evaluate** Solving for $\Delta E$, we have

$$\Delta E = \frac{h}{4\pi \Delta t}$$

We have changed the inequality to an equal sign, so we are calculating the minimum determinability of the energy of the state. Inserting the appropriate values gives

$$\Delta E = \frac{6.63 \times 10^{-34}\ \text{J} \cdot \text{s}}{4\pi(10^{-9}\ \text{s})} = 5.28 \times 10^{-26}\ \text{J} \approx 5 \times 10^{-26}\ \text{J}$$

Comparing this uncertainty to the value of the energy when in that energy state, $5.4 \times 10^{-19}$ J, we see that the determinability of the energy is less than one-millionth of the value of the energy. This means that the energy of the $n = 2$ state is sharply defined, though not exactly determinable.

**Try it yourself:** Using the uncertainty principle, determine the minimum determinability of the position of a bowling ball whose momentum is $50 \pm 2\ \text{kg} \cdot \text{m/s}$ in the direction of motion.

*Answer:* $\approx 3 \times 10^{-35}$ m. This is 100 billion billion times smaller than the size of an atomic nucleus. From this we can conclude that the uncertainty principle does not place any practical limits on the determinability of the physical quantities of a macroscopic system.

When we perform any measurements in physics we always encounter experimental uncertainty, although classical physics places no fundamental limits on how precise a measurement can be. Moreover, the uncertainty in one measurement generally is not related to the uncertainty in some other measurement. In the subatomic world the situation is different. The wavelike properties of particles make many pairs of physical quantities not simultaneously exactly determinable: the components of position and momentum of a particle in a particular direction, or the energy and lifetime of a quantum state.

## Tunneling

The energy-time interval uncertainty relationship, $\Delta E \Delta t \geq (h/4\pi)$, predicts that for a short time interval $\Delta t$ the energy of a quantum system is only determinable to within $\Delta E \approx (h/4\pi\Delta t)$. In other words, energy conservation can be violated by an amount $\Delta E$ for a short time $\Delta t < (h/4\pi\Delta E)$. This means that the energy even of an isolated system is not constant in time but fluctuates. This result leads to an important phenomenon called **quantum tunneling**, which allows particles to have nonzero chances of passing through locations where classical physics forbids their travel.

Consider a one-way mirror used during a police interrogation. The mirror allows an observer to see into the interrogation room but does not let anyone in the room see out. The interrogation room is bright, while the observation room is dark (**Figure 27.32**). The mirror is made of glass and a thin sheet of polished metal. If the metal surface is very thin, some of the light can pass through to the observation room even though metal reflects light. This effect of light passing through a forbidden region is an example of quantum tunneling.

**Figure 27.32** A one-way mirror and quantum tunneling. Some photons can tunnel through a very thin metal surface on glass.

The same effect happens to electrons and other subatomic particles. In classical mechanics, the energy of a system is exactly determinable for all practical purposes because it is a macroscopic system. (Imagine a skier not having enough speed to go over a hill—he stops on the side of the hill before getting over the top.) However, a subatomic system's energy is not exactly determinable, so a particle that seems to be totally trapped actually isn't.

Consider a particle interacting with other particles so that the potential energy $U$ of the particle-surroundings system as a function of position $x$ looks as shown in **Figure 27.33** (the rectangular hump is the energy barrier). The horizontal line represents the system's energy $E$. The particle's kinetic energy at each position is then $K = E - U$. A classical particle on the left side of the barrier bounces back and forth between positions 0 and $x_1$. It cannot escape that region because its kinetic energy would be negative between positions $x_1$ and $x_2$ (the region where $U > E$), and that's impossible. In quantum mechanics, however, the square of a wave function determines the probability of finding the particle at different locations. Solving the Schrödinger equation for this situation reveals that the amplitude of the wave function decreases exponentially into the barrier region, rather than immediately becoming zero. This result means there is a nonzero probability that the particle will be detected on the right of the barrier. It will have "tunneled through" the barrier.

**Figure 27.33** A particle can pass through an energy barrier. For a particle to move from $x_1$ to $x_2$ it must cross a negative kinetic energy barrier; in quantum physics, a nonzero chance exists that the particle can pass through that barrier.

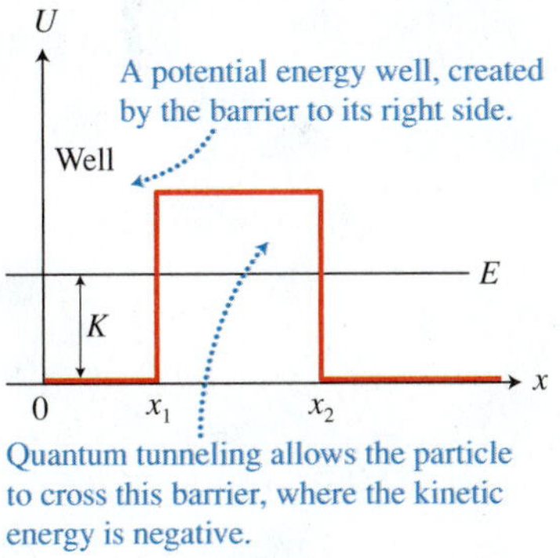

U
A potential energy well, created by the barrier to its right side.
Well
E
K
x
0
x1
x2
Quantum tunneling allows the particle to cross this barrier, where the kinetic energy is negative.

## Hydrogen bonds and proton tunneling in DNA

A DNA molecule, which stores the genetic code, is made up of two complementary strands that coil about each other in a helical form. The two strands pair with each other through hydrogen bonding between adenine (A) and thymine (T) molecules and between cytosine (C) and guanine (G) molecules. A hydrogen bond forms when a hydrogen atom bonds with a more electronegative atom (such as oxygen or fluorine).

In each strand of the DNA molecule an electron pair extends out from either a nitrogen or an oxygen atom (consider the top bond in the A-T pair of molecules in **Figure 27.34**). The proton nucleus of a hydrogen atom, H in the figure, resides in a potential well created by the electron pair, represented by the two dots (:) on the A side of the figure. On the T side is another electron pair where the electric potential well is not as deep. The proton usually resides on the A side, but it can tunnel through the barrier between the wells to briefly reside on the T side (see **Figure 27.35**). In the next bond down in Figure 27.34, the well is deeper on the T side, and the proton spends more time there than on the A side.

A new DNA molecule is formed when these H bonds are unzipped and new A, T, C, or G molecules bond with the complementary molecule on each side of the unzipped DNA to form two identical DNA molecules. However, there is a slight chance that a proton will have tunneled to the wrong side during the unzipping process. In that case, a different code is produced in the new DNA—a mutation occurs.

Let's consider the implications of this situation. Because the proton is a tiny particle that must be described using the principles of quantum mechanics, it is *impossible* for the genetic code to be 100% stable. A finite probability exists that because of proton tunneling, the proton will be on the wrong side of its bond when the DNA is replicated. This approximately 1-nm displacement of protons through classically forbidden barriers plays a small part in the evolution of living organisms.

**Review Question 27.8** What experimental evidence supports the uncertainty principle?

**Figure 27.34** Hydrogen bonding between the two strands of a DNA molecule. An H nucleus (a proton) associated with one strand may sometimes cross to the other strand due to tunneling.

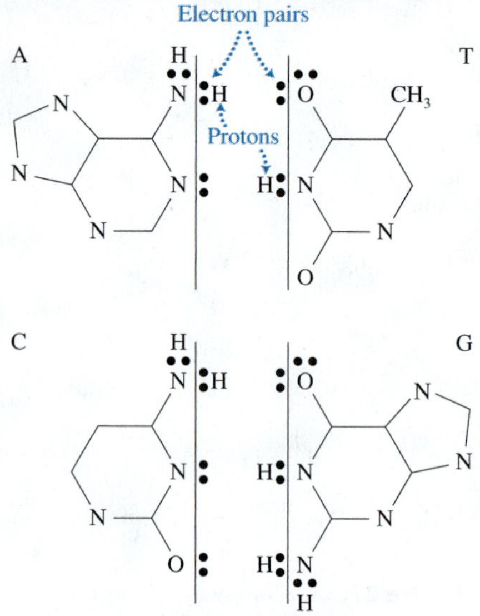


**Figure 27.35** Energy wells for a hydrogen nucleus associated with DNA strands. In a pairing between A and T, some small probability exists that a hydrogen nucleus can cross the kinetic energy barrier between the wells.

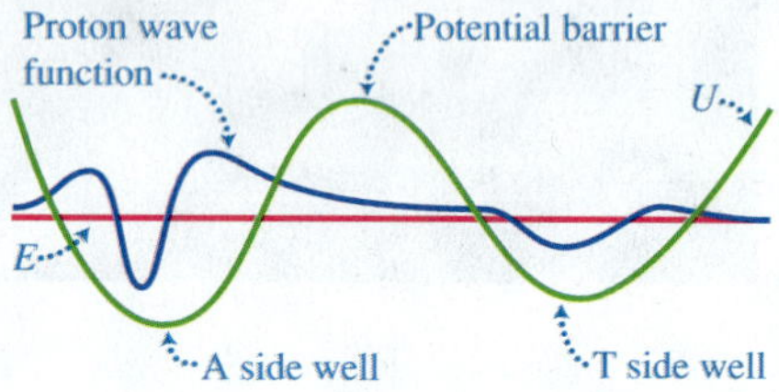

# Summary

| Words | Pictorial and physical representations | Mathematical representation |
|---|---|---|

**Nuclear model of atom** The scattering of alpha particles at large angles from a thin foil indicated that the mass of an atom must be localized in a tiny nucleus (at most $10^{-14}$ m in radius) at the center of the atom. Low-mass electrons circle the nucleus. (Section 27.1)

**Bohr's model of one-electron atom**

- A negatively charged electron moves in a stable circular orbit about a positively charged nucleus.
- In the stable orbits the angular momentum of the electron is quantized and equal to a positive integer $n$ times $h/2\pi$.
- When an electron transitions from one stable orbit to another, the atom's energy changes and a photon is emitted or absorbed. (Section 27.2)

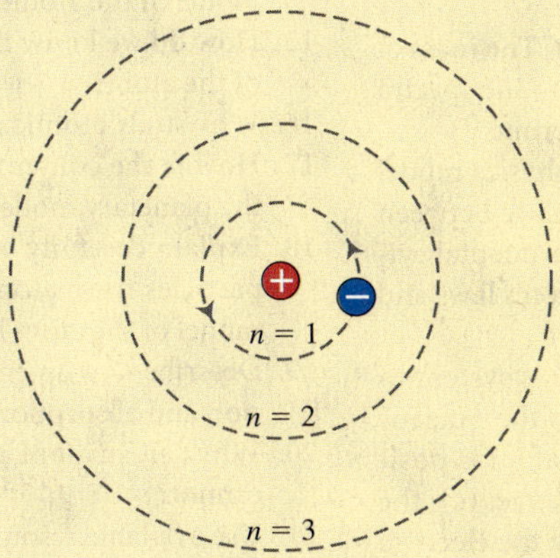


$$mvr = n\frac{h}{2\pi}, n = 1, 2, 3 \ldots \quad \text{Eq. (27.2)}$$

$$hf = E_i - E_f$$

**Wave-particle duality** EM radiation has a dual wave-particle (photon) behavior. Subatomic particles also have dual wave-particle behaviors. The wavelength $\lambda_{particle}$ depends on the momentum $p$ of the particle. (Section 27.6)

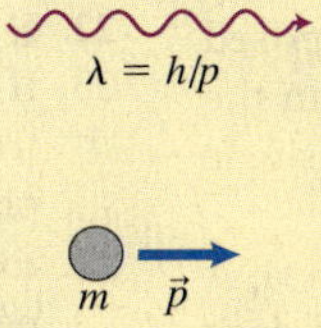


$$\lambda_{particle} = h/p \quad \text{Eq. (27.15)}$$

**Quantum numbers** Electron states in atoms are described by four quantum numbers:

*Principal quantum number n* indicates the energy of the state.

*Orbital angular momentum quantum number l* indicates the orbital angular momentum $L$ of the electron wave and its shape.

*Magnetic quantum number $m_l$* indicates the projection $L_z$ of the electron's angular momentum on the direction of an external $\vec{B}_{ex}$ field.

*Spin magnetic quantum number $m_s$* indicates the two directions that the electron's spin can be oriented relative to an external $\vec{B}_{ex}$ field. (Section 27.5)

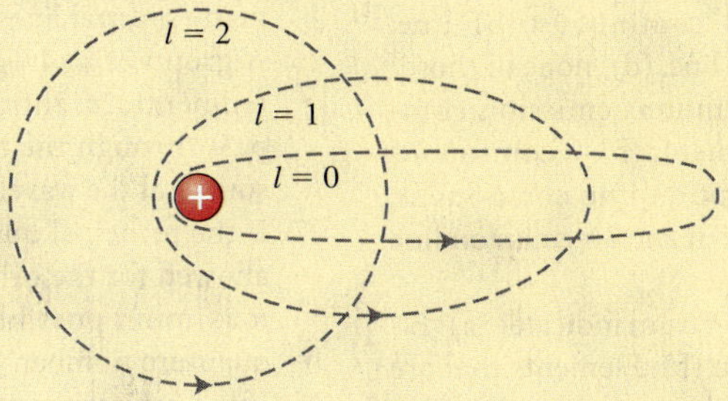


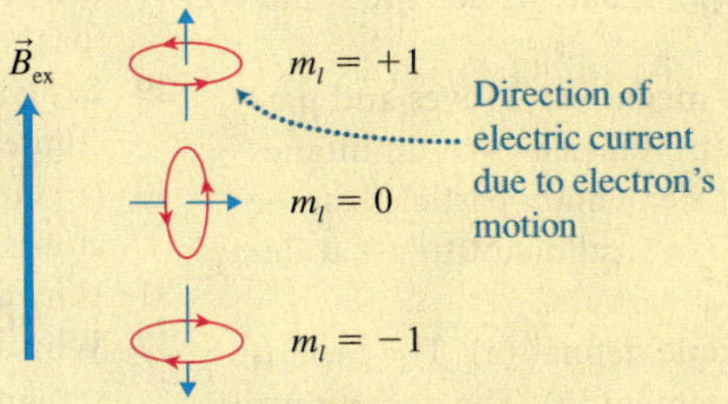


$$n = 1, 2, 3, \ldots \quad \text{Eq. (27.11)}$$

$$l = 0, 1, 2, \ldots, n - 1 \quad \text{Eq. (27.12)}$$

$$m_l = 0, \pm1, \pm2, \ldots \pm l \quad \text{Eq. (27.13)}$$

$$m_s = \pm 1/2 \quad \text{Eq. (27.14)}$$

**Pauli exclusion principle** Each electron in an atom must have a unique set of quantum numbers $n$, $l$, $m_l$, and $m_s$. (Section 27.5)

**Uncertainty principle** The products of the uncertainties of position and momentum and of energy and lifetime of a state must be greater than $h/4\pi$. (Section 27.8)

$$\Delta x \Delta p_x \geq \frac{h}{4\pi} \quad \text{Eq. (27.16)}$$

$$\Delta E \Delta t \geq \frac{h}{4\pi} \quad \text{Eq. (27.17)}$$

# Questions

## Multiple Choice Questions

1. What could the plum pudding model of the atom account for? (a) The mass of the atom (b) The electric charge of the atom (c) Line spectra (d) The stability of the atom (e) Scattering experiments
2. What could the planetary model account for? (a) The mass of the atom (b) The electric charge of the atom (c) Line spectra (d) The stability of the atom (e) Scattering experiments
3. What was Bohr's model? (a) A revolutionary physics model that established modern physics (b) A compromise between classical and modern physics to account for experimental evidence (c) A classical model, as it relied on Newton's laws and Coulomb's law
4. If you consider the energy of an electron-nucleus system to be zero when the electron is infinitely far from the nucleus and at rest, then (a) the energy of a hydrogen atom is positive. (b) the energy of a hydrogen atom is negative. (c) the sign of the energy depends on where in the atom the electron is located.
5. If you choose the electron to be the system (as opposed to an electron-nucleus system), its potential energy in the atom (a) is positive. (b) is negative. (c) is zero. (d) a single particle does not have potential energy.
6. What does an emission spectrum of gases consist of ? (a) A set of bright lines (b) A set of dark lines (c) A set of dark lines on the continuous bright spectrum
7. The Sun's emission spectrum is (a) continuous. (b) line. (c) a combination of continuous and line. (d) none of those.
8. The Sun's spectrum has (a) a continuous emission component and an emission line component. (b) a continuous emission component and an absorption line component. (c) a continuous absorption component and an emission line component.
9. What do the dark lines in the Sun's spectrum indicate? (a) Elements that are missing from the Sun (b) Elements that are present in the Sun's atmosphere (c) Elements that are somewhere between Earth and the Sun
10. What does wave-particle duality mean? (a) Waves and particles have the same properties. (b) Particles are simultaneously waves. (c) The behavior of elementary particles can be described simultaneously using physical quantities that describe the behavior of waves and of particles.
11. What does the uncertainty principle define? (a) The experimental uncertainty of our measurement (b) The uncertainty of our knowledge of some physical quantities that we try to determine simultaneously (c) The limits of our knowledge about nature.
12. Why does the term "electron orbit" have no meaning in quantum mechanics? (a) The electron is considered to be at rest. (b) The electron does not follow a particular orbit. (c) The electron is a cloud smeared in a three-dimensional space.

## Conceptual Questions

13. How do we know that the plum pudding model is not an accurate model of an atom?
14. How do we know that the planetary model is not an accurate model of the atom?
15. How do we know that Bohr's model is not an accurate model of the atom?
16. Why study atomic models that we do not think are accurate?
17. How is the contemporary model of the atom different from the planetary model? From Bohr's model?
18. Explain carefully why the scattering at large angles of alpha particles from atoms is inconsistent with the plum pudding model of the atom but is consistent with the nuclear model.
19. Describe how an energy state diagram helps explain the emission and absorption of photons by an atom.
20. Why can an atom absorb a photon but an individual electron cannot?
21. Use available resources to learn what the Lyman and Paschen series are. Draw an energy diagram describing the series.
22. How can one determine if the Sun contains iron?
23. If you do not see helium absorption lines in the Sun's spectrum, does it mean that helium is missing from the Sun? Explain your answer.
24. Explain carefully why gas at room temperature emits little or no light whereas the Sun's surface emits considerable light.
25. Explain how an atom with only one electron can produce so many spectral lines.
26. A group of hydrogen atoms is contained in a tube at room temperature. Infrared radiation, light, and ultraviolet rays pass through the tube, but only the ultraviolet rays are absorbed at the wavelengths of the Lyman series. Explain.
27. If the principal quantum number $n$ is 5, what are the values allowed for the orbital quantum number $l$? Explain how the maximum possible $l$ quantum number is related to the $n$ quantum number.
28. If the orbital quantum number is 3, what are possible values for the magnetic quantum number?
29. Explain why an electron cannot move in an orbit whose circumference is 1.5 times the electron's de Broglie wavelength.
30. If Planck's constant were approximately 50% bigger, would atoms be larger or smaller? Explain your answer.
31. Why do fluorine and chlorine exhibit similar properties?
32. Why do atoms in the first column of the periodic table exhibit similar properties?
33. How do astronomers know that atoms of particular elements are present in a star?
34. Explain how some elements can go undetected when they are actually present in a star.
35. What are the differences between spontaneous emission of light and stimulated emission?
36. What property of laser light makes it useful for (a) cutting metal and (b) carrying information?

# Problems

Below, **BIO** indicates a problem with a biological or medical focus. Problems labeled **EST** ask you to estimate the answer to a quantitative problem rather than develop a specific answer. Problems marked with ✎ require you to make a drawing or graph as part of your solution. Asterisks indicate the level of difficulty of the problem. Problems with no * are considered to be the least difficult. A single * marks moderately difficult problems. Two ** indicate more difficult problems.

## 27.1 and 27.2 Early atomic models and Bohr's model of the atom: Quantized orbits

1. * The electron in a hydrogen atom spends most of its time $0.53 \times 10^{-10}$ m from the nucleus, whose radius is about $0.88 \times 10^{-15}$ m. If each dimension of this atom was increased by the same factor and the radius of the nucleus was increased to the size of a tennis ball, how far from the nucleus would the electron be?
2. * **EST** A single layer of gold atoms lies on a table. The radius of each gold atom is about $1.5 \times 10^{-10}$ m, and the radius of each gold nucleus is about $7 \times 10^{-15}$ m. A particle much smaller than the nucleus is shot at the layer of gold atoms. Roughly, what is its chance of hitting a nucleus and being scattered? (The electrons around the atom have no effect.)
3. ** (a) Determine the mass of a gold foil that is 0.010 cm thick and whose area is 1 cm × 1 cm. The density of gold is 19,300 kg/m$^3$. (b) Determine the number of gold atoms in the foil if the mass of each atom is $3.27 \times 10^{-25}$ kg. (c) The radius of a gold nucleus is $7 \times 10^{-15}$ m. Determine the area of a circle with this radius. (d) Determine the chance that an alpha particle passing through the gold foil will hit a gold nucleus. Ignore the alpha particle's size and assume that all gold nuclei are exposed to it; that is, no gold nuclei are hidden behind other nuclei.
4. * An object of mass $M$ moving at speed $v_0$ has a direct elastic collision with a second object of mass $m$ that is at rest. Using the energy and momentum conservation principles (Chapters 5 and 6), show that the final velocity of the object of mass $M$ is $v = (M - m)v_0/(M + m)$. Using this result, determine the final velocity of an alpha particle following a head-on collision with (a) an electron at rest and (b) a gold nucleus at rest. The alpha particle's velocity before the collision is $0.010c$; $m_{\text{alpha}} = 6.6 \times 10^{-27}$ kg; $m_{\text{electron}} = 9.11 \times 10^{-31}$ kg; and $m_{\text{gold nucleus}} = 3.3 \times 10^{-25}$ kg. (c) Based on your answers, could an alpha particle be deflected backward by hitting an electron in a gold atom?
5. * ✎ Describe what happens to the energy of the atom and represent your reasoning with an energy bar chart when (a) a hydrogen atom emits a photon and (b) a hydrogen atom absorbs a photon.
6. * How do we know that the energy of the hydrogen atom in the ground state is −13.6 eV?
7. * Determine the wavelength, frequency, and photon energies of the line with $n = 5$ in the Balmer series.
8. * Determine the wavelengths, frequencies, and photon energies (in electron volts) of the first two lines in the Balmer series. In what part of the electromagnetic spectrum do the lines appear?
9. * **Invent an equation** An imaginary atom is observed to emit electromagnetic radiation at the following wavelengths: 250 nm, 2250 nm, 6250 nm, 12,250 nm, . . . . Invent an empirical equation for calculating these wavelengths; that is, determine $\lambda = f(n)$, where $f$ is an unknown function of an integer $n$, which can have the values 1, 2, 3, 4, . . . .
10. * Write three basic equations to derive the expressions for the allowed radii and the allowed energy of electron states in the Bohr model of the atom.
11. * If we know the value of $n$ for the orbit of an electron in a hydrogen atom, we can determine the values of three other quantities related to either the electron's motion or the atom as a whole. Briefly describe these quantities.
12. * **EST** Is it possible for a hydrogen atom to emit an X-ray? If so, describe the process and estimate the $n$ value for the energy state. If not, indicate why not.
13. * A gas of hydrogen atoms in a tube is excited by collisions with free electrons. If the maximum excitation energy gained by an atom is 12.5 eV, determine all of the wavelengths of light emitted from the tube as atoms return to the ground state.
14. * ✎ Some of the energy states of a hypothetical atom, in units of electron volts, are $E_1 = -31.50$ $E_2 = -12.10$, $E_3 = -5.20$, and $E_4 = -3.60$. (a) Draw an energy diagram for this atom. (b) Determine the energy and wavelength of the least energetic photon that can be absorbed by these atoms when initially in their ground state.
15. * An atom in an excited state usually remains in that state only about $10^{-8}$ s before transitioning to a lower energy state. How many times will an electron in the $n = 3$ state of hydrogen move around the $n = 3$ orbit before the atom transitions to the $n = 2$ or $n = 1$ state?
16. * Determine the speed and frequency of an electron moving around the first Bohr orbit in hydrogen. According to classical physics, the atom should emit electromagnetic radiation at this frequency. In what portion of the electromagnetic spectrum is this frequency?
17. ** Show that the frequency of revolution of an electron around the nucleus of a hydrogen atom is $f = (4\pi^2 k^2 e^4 m/h^3)(1/n^3)$.
18. * Are we justified in using nonrelativistic energy equations in the Bohr theory for hydrogen? (That is, is the electron's speed smaller than $0.1c$?)
19. * Determine the ratio of the electric force between the nucleus and an electron in the ground state of the hydrogen atom and the gravitational force between the two particles. Based on your answer, are we justified in ignoring the gravitational force in the Bohr theory?

## 27.3 Spectral analysis

20. * A group of hydrogen atoms in a discharge tube emit violet light of wavelength 410 nm. Determine the quantum numbers of the atom's initial and final states when undergoing this transition.
21. * ✎ Draw an energy state diagram for a hydrogen atom and explain (a) how an emission spectrum is formed and (b) how an absorption spectrum is formed.
22. * Explain what spectral lines could be emitted by hydrogen gas in a gas discharge tube with an 11.5-V potential difference across it. What assumptions did you make?
23. * The fractional population of an excited state of energy $E_n$ compared to the population of the ground of energy $E_o$ is $N_n/N_o = e^{-(E_n - E_o)/kT}$. At approximately what temperature $T$ are 20 percent of hydrogen atoms in the first excited state?
24. * ✎ Draw an energy bar chart that describes the ionization process for a hydrogen atom (a) due to collisions with other atoms and (b) when placed in an external electric field.

### 27.4 Lasers

25. (a) A laser pulse emits 2.0 J of energy during $1.0 \times 10^{-9}$ s. Determine the average power emitted during that short time interval. (b) Determine the average light intensity (power per unit area) in the laser beam if its cross-sectional area is $8.0 \times 10^{-9}\ \text{m}^2$.
26. A pulsed laser used for welding produces 100 W of power during 10 ms. Determine the energy delivered to the weld.
27. **BIO Welding the retina** A pulsed argon laser of 476.5-nm wavelength emits $3.0 \times 10^{-3}$ J of energy to produce a tiny weld to repair a detached retina. How many photons are in the laser pulse?
28. ** **BIO More welding the retina** A laser used to weld the damaged retina of an eye emits 20 mW of power for 100 ms. The light is focused on a spot 0.10 mm in diameter. Assume that the laser's energy is deposited in a small sheet of water of 0.10-mm diameter and 0.30-mm thickness. (a) Determine the energy deposited. (b) Determine the mass of this water. (c) Determine the increase in temperature of the water (assume that it does not boil and that its heat capacity is $4180\ \text{J/kg} \cdot \text{C}°$).

### 27.5 and 27.6 Quantum numbers and Pauli's exclusion principle and Particles are not just particles

29. * (a) An electron moves counterclockwise around a nucleus in a horizontal plane. Assuming that it behaves as a classical particle, what is the direction of the magnetic field produced by the electron? (b) The same one-electron atom is placed in an external magnetic field whose $\vec{B}$ field points horizontally from right to left. Draw a picture showing what happens to the atom.
30. * **EST** Estimate the wavelength of a tennis ball after a good serve.
31. ** What is the wavelength of the electron in a hydrogen atom at a distance of one Bohr radius from the nucleus?
32. * **EST** Estimate the average wavelength of hydrogen atoms at room temperature.
33. * Describe how you will determine the wavelength of an electron in a cathode ray tube if you know the potential difference between the electrodes.
34. * An electron first has an infinite wavelength and then after it travels through a potential difference has a de Broglie wavelength of $1.0 \times 10^{-10}$ m. What is the potential difference that it traversed? Draw a picture of the situation and an energy bar chart, and explain why the electron's wavelength decreased.
35. * Describe two experiments whose outcomes can be explained using the concept of the de Broglie wavelength.
36. How does the wavelength of an electron relate to the radius of its orbit in a hydrogen atom? Why?
37. * Discuss the similarities and differences in the way a hydrogen atom is pictured in Bohr's model and in quantum mechanics.
38. A high-energy particle scattered from the nucleus of an atom helps determine the size and shape of the nucleus. For best results, the de Broglie wavelength of the particle should be the same size as the nucleus (approximately $10^{-14}$ m) or smaller. If the mass of the particle is $6.6 \times 10^{-27}$ kg, at what speed must it travel to produce a wavelength of $10^{-14}$ m?
39. * (a) Use de Broglie's hypothesis to determine the speed of the electron in a hydrogen atom when in the $n = 1$ orbit. The radius of the orbit is $0.53 \times 10^{-10}$ m. (b) Determine the electron's de Broglie wavelength. (c) Confirm that the circumference of the orbit equals one de Broglie wavelength.
40. * Repeat Problem 39 for the $n = 3$ orbit, whose radius is $4.77 \times 10^{-10}$ m. Three de Broglie wavelengths should fit around the $n = 3$ orbit.
41. * A beam of electrons accelerated in an electric field is passing through two slits separated by a very small distance $d$ and then hits a screen that glows when an electron hits it. What do you need to know about the electrons to be able to predict where on the screen, which is $L$ meters from the slits, you will see the brightest and the darkest spots?
42. (a) An electron is in an $n = 4$ state. List the possible values of its $l$ quantum number. (b) If the electron happens to be in an $l = 3$ state, list the possible values of its $m_l$ quantum number. (c) List the possible values of its $m_s$ quantum number.
43. (a) An electron in an atom is in a state with $m_l = 3$ and $m_s = +1/2$. What are the smallest possible values of $l$ and $n$ for that state? (b) Repeat for an $m_l = 2$ and $m_s = -1/2$ state.
44. (a) An atom is in the $n = 7$ state. List the possible values of the electron's $l$ quantum number. (b) Of these different states, the electron occupies the $l = 4$ state. List the possible values of the $m_l$ quantum number.
45. ** Draw schematic orbits and arrows representing the different $m_l$ quantum states for an $l = 2$ atomic electron in a magnetic field.

### 27.7 Multi-electron atoms and the periodic table

46. List the $n$, $l$, $m_l$, and $m_s$ states available for an electron in a $4p$ subshell.
47. List the $n$, $l$, $m_l$, and $m_s$ states available for an electron in a $4f$ subshell.
48. Identify the values of $n$ and $l$ for each of the following subshell designations: $3s$, $2p$, $4d$, $5f$, and $6s$.
49. (a) Determine the electron configuration of sulfur (its atomic number is 16). (b) Why are sulfur and oxygen (atomic number 8) in the same group on the periodic table?
50. (a) Determine the electron configuration of silicon (atomic number 14). (b) Why are carbon and silicon in the same group of the periodic table?
51. Determine the electron configuration for iron (atomic number 26).
52. Manganese (atomic number 25) has two $4s$ electrons. How many $3d$ electrons does it have? Explain your answer.
53. * Determine the electron configurations of four elements of group I in the periodic table. Explain why these elements are likely to have similar properties. Note that a higher $s$ shell fills before the next lower $d$ shell—the electron in the $s$ shell spends more time on average closer to the nucleus.
54. * Determine the electron configurations of three elements in group VI of the periodic table. Explain why these elements are likely to have similar properties.

### 27.8 The uncertainty principle

55. * Describe the experimental evidence that supports the concept of the uncertainty principle.
56. ** If you assume that the uncertainty in our knowledge of Bohr's radius of the atom is 1% of the value of the radius, then what is the minimum uncertainty in our knowledge of the electron's momentum? What component of momentum is it—radial or tangential? What does this uncertainty mean for our interpretation of the atomic orbits?
57. * The lifetime of the hydrogen atom in the $n = 3$ second excited state is $10^{-9}$ s. What is the uncertainty of that energy state of the atom? Compare this uncertainty with the magnitude of the $-1.51$ eV energy of the atom in that state.
58. * Use the uncertainty principle to discuss whether lasers can emit 100% monochromatic light.

## General Problems

59. * (a) Determine the radii and energies of the $n = 1, 2$, and 3 states in the $He^+$ ion (it has two protons in its nucleus and

one electron). (b) Construct an energy state diagram for this ion. Indicate any assumptions that you made.

60. * (a) Determine the radii and energies of the $n = 1, 2$, and 3 states of a sodium ion in which 10 of its electrons have been removed. (b) Construct an energy state diagram for the ion. Indicate any assumptions that you made.

61. EST* A uranium atom with $Z = 92$ has 92 protons in its nucleus. It has two electrons in an $n = 1$ orbit. Estimate the radius of this orbit. Indicate any assumptions that you made.

62. EST* Estimate the energy needed to remove an electron from (a) the $n = 1$ state of iron ($Z = 26$ is the number of protons in the nucleus) and (b) the $n = 1$ state of hydrogen ($Z = 1$).

63. An electron in a hydrogen atom changes from the $n = 4$ to the $n = 3$ state. Determine the wavelength of the emitted photon.

64. An electron in a $He^+$ ion changes its energy from the $n = 3$ to the $n = 1$ state. Determine the wavelength, frequency, and energy of the emitted photon.

65. Determine the energy, frequency, and wavelength of a photon whose absorption changes a $He^+$ ion from the $n = 1$ to the $n = 6$ state.

66. * A helium ion $He^+$ emits an ultraviolet photon of wavelength 164 nm. Determine the quantum numbers of the ion's initial and final states.

67. The average thermal energy due to the random translational motion of a hydrogen atom at room temperature is $(3/2)kT$. Here $k$ is the Boltzmann constant. Would a typical collision between two hydrogen atoms be likely to transfer enough energy to one of the atoms to raise its energy from the $n = 1$ to the $n = 2$ energy state? Explain your answer. [Note: Earth's free hydrogen is in the molecular form $H_2$. However, the above reasoning is similar for atomic and molecular hydrogen.]

## Reading Passage Problems

BIO **Electron microscope** The *scanning electron microscope (SEM)* uses high-energy electrons to form three-dimensional images of the surfaces of biological, geological, and integrated circuit samples. The electrons in the SEM have much shorter wavelengths than the light used in visible wavelength microscopes and consequently can produce high-resolution images of a surface, revealing details of about 1 to 5 nm in size with magnification ranging from 25 to 250,000.

A color-enhanced image of the inner surface of a lung is shown in **Figure 27.36**. The cavities are alveoli where inhaled oxygen is exchanged into the blood for carbon dioxide, a waste product removed in the lungs during exhalation.

**Figure 27.36** A scanning electron microscope photograph of lung tissue showing alveoli.

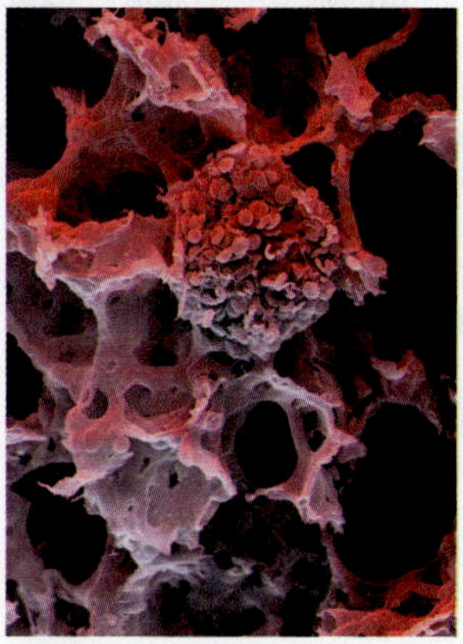

A schematic diagram shows how a SEM works (see **Figure 27.37**). Electrons are emitted from a hot tungsten cathode (the electron gun) and accelerated across a potential difference of a few hundred volts to 40 kV. Two or more condenser lenses, each consisting of a coil of wire through which current flows, produce a magnetic field that bends and focuses the electron beam so that it irradiates a small 0.4-nm to 5-nm spot on the sample. The focused beam is then deflected so that it scans over a rectangular area of the sample surface.

**Figure 27.37** A schematic of a scanning electron microscope. Secondary electrons from the sample are recorded by the detector, producing a three-dimensional surface image.

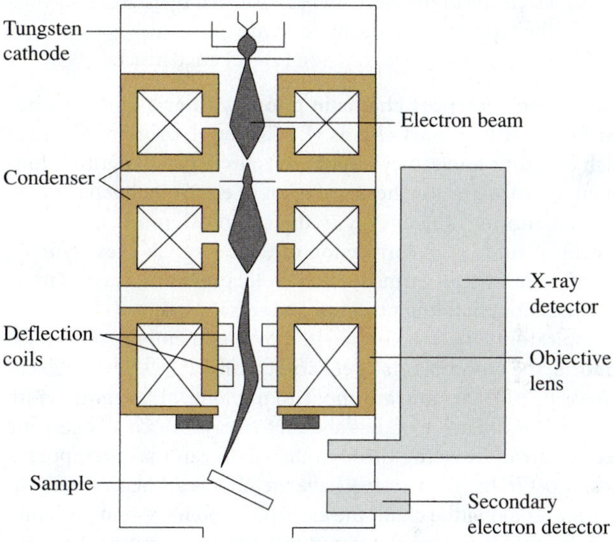


The focused electron beam knocks electrons out of surface atoms. These so-called secondary electrons have low energy ($< 50$ eV) and are caught by a secondary-electron detector. The number of secondary electrons produced depends on the orientation and nature of the exposed surface, thus forming a vivid three-dimensional image. The electron beam can also knock inner electrons out of the sample atoms. When outer electrons move into these vacant inner electron orbits, X-ray photons are emitted. The wavelengths of these photons are characteristic of the particular type of atom and thus provide information about the types of atoms in the sample.

68. Which answer below is closest to the speed of an electron accelerated from rest across a 100-V potential difference?
(a) 600 m/s (b) $6 \times 10^4$ m/s
(c) $6 \times 10^6$ m/s (d) $6 \times 10^7$ m/s

69. Which answer below is closest to the wavelength of an electron accelerated from rest across a 100-V potential difference?
(a) $1 \times 10^{-11}$ m (b) $1 \times 10^{-10}$ m
(c) $1 \times 10^{-9}$ m (d) $1 \times 10^{-8}$ m

70. Compared to the electron in Problem 69, the wavelength of an electron accelerated from rest across a 10,000-V potential difference would be
(a) the same length (b) 10 times longer
(c) 1000 times longer (d) $1/10$ times as long
(e) 1/1000 times as long

71. An electron in the SEM electron beam is moving parallel to the paper and downward toward the bottom of the page. In which direction should a magnetic field point to deflect the electron toward the right as seen while looking at the page?
(a) Right (b) Left
(c) Toward the top of the page

(d) Toward the bottom of the page
(e) Into the paper (f) Out of the paper

72. The SEM can also detect the types of atoms in the sample by measuring which of the following?
(a) Energy of the secondary electrons
(b) The wavelengths of X-rays emitted from the sample
(c) Both (a) and (b)

73. The secondary electrons detected by the SEM detector are
(a) outer electrons knocked out of atoms on or near the surface of the sample.
(b) X-rays produced when outer electrons fall into vacant inner electron orbits.
(c) electrons in the electron beam slowed by passing through the top layer of the sample.
(d) none of the above. (e) (a)–(c) are correct.

**BIO Electron transport chains in photosynthesis and metabolism** Electron transport chains (ETCs), large proteins through which electrons move, play important roles in two of nature's fundamental processes: (1) the conversion of electromagnetic energy from the Sun into the energy in the chemical bonds in glucose molecules and (2) the conversion of the energy in these glucose molecules into useful forms for metabolic processes, such as muscle contraction, building proteins, and respiration.

Respiratory ETCs are located in the inner membranes of mitochondria, the power plants of eukaryotic cells. ETCs have molecular mass of 800,000 amu, are about 20 nm long, and vary in width from 3 to 6 nm. Each ETC has about 20 sites at which an energetic "free" electron traveling through the chain can make temporary stops. Most of the resting sites have metal ions at their centers. At three places along the chain the electron-protein system goes into a significantly lower energy state ($\approx 0.4$ eV). The released energy catalyzes a reaction that converts two other molecules into ATP, which carries the energy gained to other parts of the cell for different processes that require chemical energy.

How does a lone energetic electron traverse such a long distance along the bumpy electrical potential hills and valleys of a large protein and not have its energy transformed into thermal energy or light? The answer is still uncertain but several possibilities have been proposed: (1) the ground-state electron wave functions at resting sites are spread out (delocalized) over neighboring bonded atoms, which in turn are spread over and overlap with the next resting site; (2) an electron-atom system is excited into a higher energy, very delocalized state, after which the electron comes back down from this conduction state into a new location; or (3) the electron tunnels through the potential energy barrier between the resting sites. We'll examine briefly possibilities (2) and (3).

Explanation (2): Consider the visible spectra of metal atoms at the resting sites. If the excited state is delocalized over many atoms, as occurs in a semiconductor, this excited so-called conduction band is much broader than the excited state of a single atom. Thus, the visible spectra of a metal ion with a spread out excited conduction band should be somewhat broader than the excited state of a localized resting site with no conduction band. However, the observed visible spectra of the metal ion resting sites in ETCs look similar to metal ion sites found in other proteins not involved in electron conduction, with no apparent broader energy conduction band. Another problem with this explanation is the difficulty of exciting an electron at one site into the higher energy conduction band. The random kinetic energy at room temperature is about 0.025 eV, whereas visible spectra indicate that higher energy bands are 2–3 eV above the ground state. At room temperature, it is almost impossible to "promote" the electron to the conduction band.

Explanation (3): Tunneling can occur between the ground state of the atom at one site and an energy state at a nearby site—a state that is empty and slightly above the ground state. After the electron transfer, the electron-local atom system at the new site moves down to the slightly lower energy ground state, which prevents the electron from tunneling back. The system can also transition into a somewhat lower energy state and catalyze a reaction that converts the site's extra energy into the stable bond of some other molecule. Tunneling times seem consistent with measured electron transfer rates. But the actual transfer mechanism is still an open question.

74. If all of the electron resting sites were in a line along the length of an ETC, what would be the approximate distance from one site to the next?
(a) 400 nm (b) 20 nm
(c) 1 nm (d) (1/20) nm

75. Electron tunneling involves electrons
(a) jumping to a conduction band and then falling down at a different place.
(b) diffusing with a small protein from one place to another.
(c) converting to a photon and moving at light speed to another site.
(d) passing through a potential barrier to a different location.
(e) having an uncertain energy for a short time interval.
(f) (d) and (e)

76. An electron with mass $9.1 \times 10^{-31}$ kg and kinetic energy 20 eV is in a potential well that is 0.15 nm wide. Which answer is closest to the electron's speed?
(a) $3 \times 10^3$ m/s (b) $3 \times 10^4$ m/s
(c) $3 \times 10^5$ m/s (d) $3 \times 10^6$ m/s
(e) $3 \times 10^7$ m/s

77. The electron from Problem 76 is in a potential well that is 0.15 nm wide. Which answer is closest to the frequency with which the electron hits one side of the well?
(a) $9 \times 10^{16}\ \text{s}^{-1}$ (b) $9 \times 10^{15}\ \text{s}^{-1}$
(c) $9 \times 10^{14}\ \text{s}^{-1}$ (d) $9 \times 10^{13}\ \text{s}^{-1}$
(e) $9 \times 10^{12}\ \text{s}^{-1}$

78. Suppose the barrier height above the electron energy ($U_{\text{barrier}} - E_{\text{electron}}$) is 1 eV and that the barrier is 2 nm wide. According to the uncertainty principle, what is the minimum time interval that the electron's energy is in this classically forbidden region?
(a) $3 \times 10^{-16}$ s (b) $3 \times 10^{-15}$ s
(c) $3 \times 10^{-14}$ s (d) $3 \times 10^{-13}$ s
(e) $3 \times 10^{-12}$ s

79. Electron transport chains are fundamental parts of:
(a) The conversion of glucose into energetic molecules used in the body
(b) Molecules used for fuel in trains and other vehicles
(c) The photosynthetic conversion of light into stable chemical compounds
(d) The passing of electric current in nerve cells
(e) (a) and (c) (f) (a), (c), and (d)

80. The conduction band model of electron transport in electron transport chains may not work because:
(a) The electron transport chain is not entirely metal ions.
(b) The conduction band is too localized.
(c) The electron has little chance of being "promoted" into the conduction band.
(d) The visible spectrum of the resting site is not a broad band as is expected.
(e) (c) and (d) (f) (a), (b), (c), and (d)

# 28 Nuclear Physics

How are new elements created?

What are the natural sources of ionizing radiation?

How does carbon dating work?

Pierre and Marie Curie made great strides in the discovery and characterization of atomic radiation in the late 1800s and early 1900s, primarily using salts of the element uranium. In the course of their studies, Marie Curie made an interesting, and accidental, observation. While extracting uranium from uranium ore, she noted that the residual material was producing more radiation than even the pure uranium.

Marie Curie hypothesized that in addition to uranium, the ore contained new chemical elements that also produced radiation. However, testing this hypothesis required processing several tons of the uranium ore. After 4 years of laborious, painstaking work, she isolated two new elements. She called them polonium (named after her native Poland)

**Be sure you know how to:**

- Use the right-hand rule for magnetic force to determine the direction of the force exerted by a magnetic field on a moving charged particle (Section 17.4).
- Relate mass to energy using the special theory of relativity (Section 25.8).

and radium. For their work on radioactivity, the Curies shared the 1903 Nobel Prize in physics with Henri Becquerel. In 1911, Marie Curie was awarded a second Nobel Prize in chemistry for her discoveries of radium and polonium.

The mechanism by which radiation was produced was not understood for some time. In this chapter we consider the structure of the atomic nucleus and what is responsible for radioactivity.

By 1911, the atomic nucleus was known to scientists from experiments in which a small fraction (one in 8000) of positively charged alpha particles passing through a thin gold foil would "bounce" back after colliding with the gold nuclei. These scattering experiments indicated that the nucleus occupied about $10^{-12}\%$ of the atom's volume, and yet contained 99.97% of the atom's mass. In this chapter we will investigate several questions about the nucleus: What is the structure of the nucleus, and what processes do nuclei undergo? Do these processes occur only in stars or in huge particle accelerators, or do they happen every day and perhaps even in our bodies?

## 28.1 Radioactivity and an early nuclear model

Wilhelm Roentgen's discovery of what became known as X-rays left open the question of what X-rays are. In this section we review the findings that followed this discovery.

### Becquerel and the emissions from uranyl crystals

**Active Learning Guide ›**

Henri Becquerel, the third generation of a family of French scientists, was very interested in Roentgen's X-rays. In 1896, Becquerel worked with potassium uranyl sulfate crystals (crystals containing the element uranium). These crystals would glow in the dark after exposure to sunlight. In addition, Becquerel found that the crystals would produce images on photographic plates. At that time, scientists thought that photographic plates could only be exposed by light, ultraviolet rays, and X-rays. Consequently, Becquerel hypothesized that the uranium crystals absorbed the Sun's energy, which then caused them to emit X-rays. However, Becquerel also discovered that crystals that had not been exposed to sunlight still formed images on photographic plates. This meant that the uranium emitted radiation without an external source of energy.

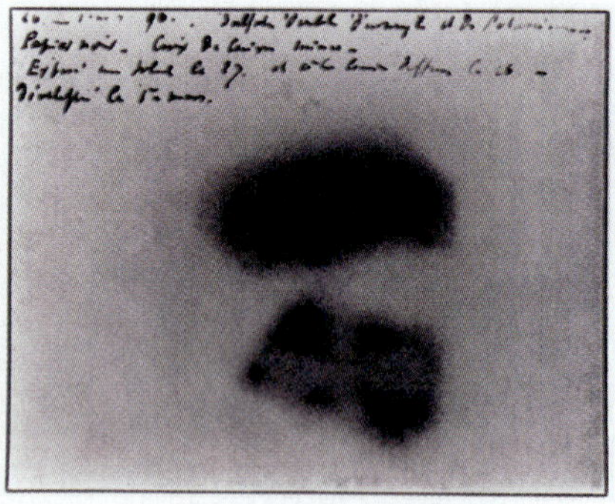

This image of the uranium-laden cross that Becquerel used for his investigations appeared on a photographic plate.

Did the crystals emit electrically neutral high-energy X-rays—electromagnetic waves? If they did, then an external magnetic field would not deflect them. However, when Becquerel passed the rays through a region with a magnetic field, he found that the field deflected the rays. Therefore, they were electrically charged particles, not X-rays. Strangely, the magnetic field deflected the rays in two opposite directions, as if both negatively and positively charged particles were being emitted. He also found that the rays could discharge an electroscope independently of the sign of its initial charge, also consistent with the idea that the rays contained both positively and negatively charged particles.

### Pierre and Marie Curie and the particles responsible for Becquerel's rays

In early 1896, Marie and Pierre Curie continued the work of Becquerel. They found a way to measure the rate of emission of "Becquerel rays" by using an electrometer, a device similar to an electroscope and invented by Pierre (see Observational Experiment **Table 28.1**). A charged electrometer maintains its charge in a dry room for a relatively long time (minutes) because dry air normally contains mostly electrically neutral particles. However, an electrometer discharges much more quickly when placed near a sample containing uranium salts because uranium rays ionize air molecules. By recording the time it takes the electrometer to discharge, one can infer the ion concentration in the air.

**‹ Active Learning Guide**

**OBSERVATIONAL EXPERIMENT TABLE**

**28.1 Marie Curie's studies of uranium salts with an electrometer.**

| Observational experiment | Analysis |
|---|---|
| **Experiment 1.** Marie Curie used uranium salts of different masses and measured the discharge time of the electrometer. Doubling the mass of the same salt led to discharge in half the time. | The ion concentration measured by the electrometer was proportional to the mass of uranium present. |
| **Experiment 2.** She used various uranium salts in which the concentration of uranium was known and measured the time of discharge of the electrometer. As long as the total uranium mass in the salt stayed the same, the discharge time did not change. | The ion concentration depended only on the presence of the uranium and not on the particular uranium salt that was used. |
| **Experiment 3.** She varied the amount of light to which the same sample of uranium salt was exposed and measured the discharge time of the electrometer. The discharge time did not change. | The ion concentration did not change. |
| **Experiment 4.** She varied the temperature of the same sample of the uranium salt and measured the discharge time of the electrometer. The discharge time did not change. | The ion concentration did not change. |
| **Experiment 5.** She varied the wetness of the same sample of uranium and measured the discharge time of the electrometer. The discharge time did not change. | The ion concentration did not change. |

**Pattern**

Assuming that the ion concentration measured by the electrometer was due to the radiation from uranium salts, then the amount of radiation depended only on the amount of the uranium used and not on the chemical composition of the salt, its temperature, the presence of water, or the amount of light shining on it.

From the experiments in Table 28.1 we see that chemical changes or changes in the amount of light shining on a sample did not lead to changes in the amount of radiation produced by uranium salts. These findings suggested that the electrons in the atoms were not responsible for the rays. Marie and Pierre Curie concluded that the Becquerel rays must come from the nuclei of atoms.

**Figure 28.1** A schematic of Rutherford's uranium radiation detection device.

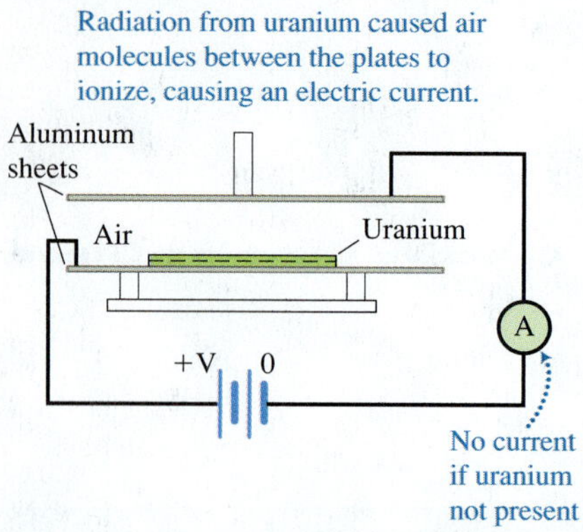


## Rutherford and experiments investigating the charge of emitted particles

In 1899 Ernest Rutherford and his colleagues in England investigated the ability of uranium salts to ionize air. He set up two parallel metal plates with a potential difference between them. Because the air between the plates was a dielectric, there was no current in the circuit. However, when a uranium sample was placed between the plates, a considerable current was detected. The current was due to the ions and free electrons created by the uranium's radiation, which were then pulled to the plates by the electric field (**Figure 28.1**).

Rutherford then covered the uranium sample with thin aluminum sheets to investigate how metal layers affected the amount of radiation. He observed that the amount of current decreased as he added more sheets, but only up to a point. After this point, no further decrease in radioactivity was observed, even with the addition of more aluminum sheets. He proposed that the radiation consisted of at least two components, one of which was not absorbed by the aluminum sheets. He reasoned that the component of radiation that was absorbed by the aluminum sheets consisted of charged particles. If true, then these particles should be deflected by a magnetic field. In contrast, if there were no charged particles in the rays, the magnetic field should not affect the rays.

**Active Learning Guide ›**

In 1903, Rutherford tested his reasoning by passing rays from radium through a magnetic field. In Testing Experiment **Table 28.2**, we describe the experiment.

**TESTING EXPERIMENT TABLE**

**28.2 Deflection of radiation by magnetic fields.**

| Testing experiment | Prediction | Outcome |
|---|---|---|
| Radiation emanating from the radioactive sample passes through a $\vec{B}$ field pointing into the page. A scintillating screen (which glows when hit with radiation) is in the plane perpendicular to the beam. | If the radiation contains positively charged particles, then according to the right-hand rule, the magnetic field should deflect them upward with respect to the original direction.   If the radiation contains negatively charged particles, the magnetic field should deflect them downward with respect to the original direction.   | The screen glowed in three places: one straight ahead, one deflected up, and one down. The location of the downward-deflected radiation was much farther from the original beam than for the upward-deflected radiation. |

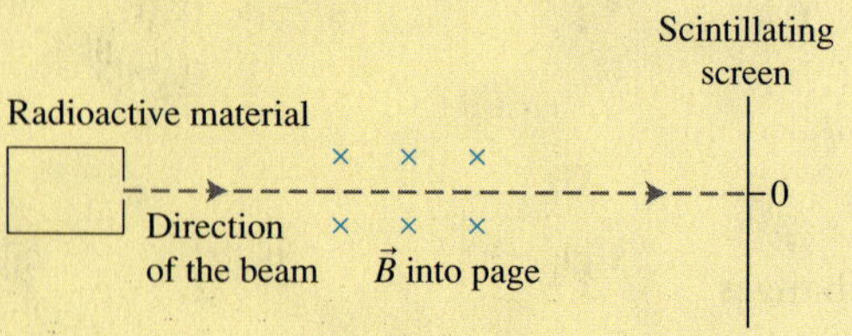


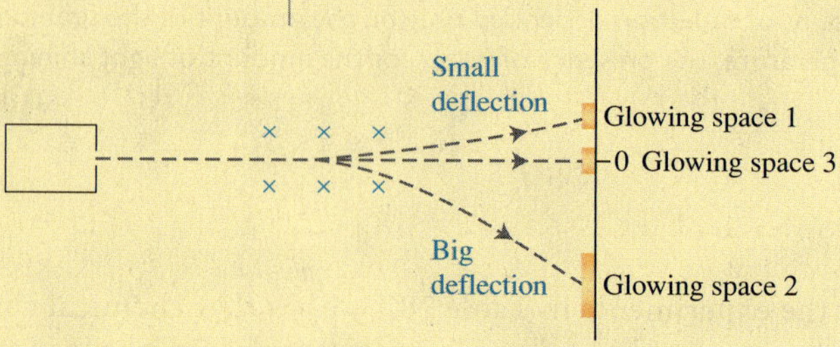


**Conclusion**

The radioactive beam consisted of positively charged particles, negatively charged particles, and particles with no electric charge.

Rutherford reasoned that the greater downward deflection than upward deflection could be the result of the downward-deflected negatively charged particles having a smaller mass-to-charge ratio than the upward-deflected positively charged particles. In the same year, using more powerful magnets, Rutherford found that the upward-bending particles were positively charged, with a mass-to-charge ratio twice that of a hydrogen ion. These positively charged massive particles were called **alpha rays** or **alpha particles.** These are the same alpha particles that Rutherford and his colleagues used later to probe the structure of the atom (Chapter 27). The downward-bending negatively charged particles had the same mass-to-charge ratio as that of the electron and were called **beta rays**. The radiation that was not deflected by the magnetic field was thought to be high-energy electromagnetic waves, called **gamma rays**.

Subsequent studies revealed that many elements with high atomic numbers were radioactive; all of them emitted alpha, beta, and/or gamma rays. Some elements with low atomic numbers were also radioactive, but generally these emitted only beta rays and gamma rays.

### The early model of the nucleus

Based on these experiments, scientists developed a provisional explanation for radioactivity: the nucleus of an atom is made of positively charged alpha particles and negatively charged electrons. Their electrostatic attraction holds them together. When a nucleus contains a large number of alpha particles, they start repelling each other more strongly than the electrons can attract them, and the alpha particles leave the nucleus, a process called **alpha decay**. Alpha decay leaves behind a large number of electrons that also repel each other; thus beta rays are emitted (**beta decay**). After each transformation, the nucleus is left in an excited state and emits a high-energy photon, a gamma ray (**gamma decay**). This model provided a start for nuclear physics and was later significantly modified when new findings emerged.

A hydrogen atom, however, is lighter than an alpha particle. What, then, is the composition of its nucleus? In 1918, Rutherford noticed that when alpha particles were shot into nitrogen gas, particles that were knocked out of the nitrogen nuclei moved in curved paths that indicated they were positively charged. Further testing indicated that the particles had the same magnitude charge as an electron (only positive) and a mass that equaled that of a hydrogen nucleus. The particle was called a **proton**. Protons must be the nuclei of hydrogen atoms. The proton became an important part of a new model of the nucleus that began to emerge.

**Review Question 28.1** How do we know that radioactive emission consists of three components: positively charged particles, negatively charged particles, and radiation with no electric charge?

## 28.2 A new particle and a new nuclear model

Werner Heisenberg's uncertainty principle (discussed in Chapter 27) had a profound effect on the nuclear model, in which alpha particles and electrons were thought to be the primary nuclear constituents.

### Size of the nucleus: Too small for an electron

Consider an electron confined to a carbon nucleus whose radius is approximately $2.7 \times 10^{-15}$ m. The position determinability $\Delta x$ of such an electron can be no greater than the size of the nucleus (**Figure 28.2**). According to the uncertainty principle, this electron must have a momentum determinability of

$$\Delta p \geq \frac{h}{4\pi \Delta x} = \frac{6.63 \times 10^{-34}\ \text{J}\cdot\text{s}}{4\pi(2.7 \times 10^{-15}\ \text{m})} = 1.95 \times 10^{-20}\ \text{kg}\cdot\text{m/s}$$

**Figure 28.2** According to the uncertainty principle, if an electron is confined within a nucleus, its speed would be greater than light speed. Thus, it cannot be within the nucleus.

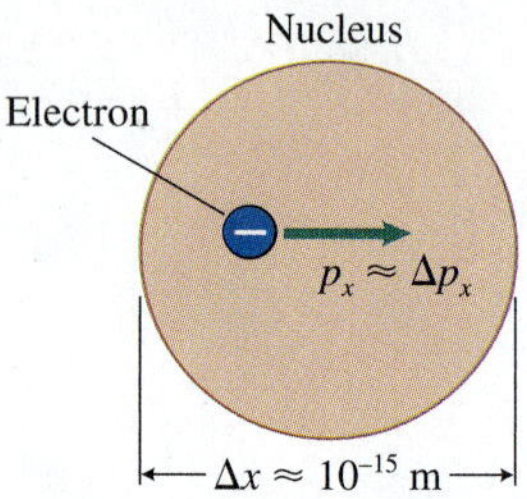


If an electron is in a nucleus, its position uncertainty is the size of the nucleus and its momentum can be estimated by the uncertainty principle.

We can use this value as an estimate for the momentum of the electron confined to the nucleus. Using the equation $p = mv$, we would find that the speed of the electron is hundreds of times the speed of light, so we need to treat the electron relativistically. Using energy ideas, we find that the kinetic energy of the electron is the total energy of the electron minus its rest energy:

$$K = E_{\text{total}} - E_{\text{rest}}$$

Inserting the appropriate expressions:

$$K = \sqrt{(pc)^2 + (mc^2)^2} - mc^2$$

We can evaluate this expression using our estimate for the momentum of the electron, the mass of the electron, and the speed of light:

$$K = \sqrt{\begin{aligned}&[(1.95 \times 10^{-20}\ \text{kg}\cdot\text{m/s})(3.00 \times 10^{8}\ \text{m/s})]^2 \\ &+ [(9.11 \times 10^{-31}\ \text{kg})(3.00 \times 10^{8}\ \text{m/s})^2]^2\end{aligned}}$$
$$- (9.11 \times 10^{-31}\ \text{kg})(3.00 \times 10^{8}\ \text{m/s})^2$$
$$= 5.77 \times 10^{-12}\ \text{J}$$

How does this result compare to the electric potential energy between the electron and the nucleus? When studying the Bohr model of the hydrogen atom, we found that the total energy of a system (potential plus kinetic) must be negative for the system to be bound (Chapter 27). Let's think, for example, about a carbon nucleus. Carbon is assigned atomic number 6, so its nucleus has a charge of $+6e$. If there are both protons and electrons in the nucleus, there must be 6 more protons than electrons. We'll make a rough estimate of the electric potential energy between an electron in a carbon nucleus and the rest of the nucleus by assuming that the electron is an average distance from the protons equal to half the diameter of the nucleus, or about $2.7 \times 10^{-15}$ m. The electric potential energy between the electron and rest of the nucleus (total charge +7) is approximately

**Active Learning Guide ›**

$$U_q = k\frac{q_{\text{nucleus}}q_e}{r} = \left(9 \times 10^9\ \frac{\text{N}\cdot\text{m}^2}{\text{C}^2}\right)\frac{(7\cdot 1.6 \times 10^{-19}\ \text{C})(-1.6 \times 10^{-19}\ \text{C})}{2.7 \times 10^{-15}\ \text{m}}$$
$$\approx -6.0 \times 10^{-13}\ \text{J}$$

This value is less than one-tenth the kinetic energy of the electron. In other words, the total energy of the system (kinetic plus potential) is positive! This result means that an electron in the carbon nucleus would very rapidly escape the nucleus. This form of carbon is stable, however, and does not emit electrons. Similar reasoning applies to other nuclei.

Thus we have determined that, because of the uncertainty principle, electrons cannot be components of nuclei.

## The search for a neutral particle

**Active Learning Guide ›**

Yet another mystery existed: an alpha particle had the charge of two protons ($+2e$) but had four times the mass of a proton. Thus, the alpha could not be made simply of two protons. In 1920, Rutherford suggested that a neutral particle with the approximate mass of a proton should exist, produced by the capture of an electron by a proton. This hypothesis stimulated a search for the particle. However, its electrical neutrality complicated the search because almost all experimental techniques of the time could detect only charged particles.

In 1928, a German physicist, Walter Bothe, and his student Herbert Becker took the initial step in the search. They bombarded beryllium atoms with alpha particles and found that a neutral radiation left the beryllium atoms. This radiation was initially thought to be high-energy gamma ray photons.

## Detecting neutral radiation with paraffin

In 1932, Irène Joliot-Curie, one of Marie Curie's daughters, and her husband, Frédéric Joliot-Curie, used a stronger source of alpha particles to repeat and extend Bothe's experiment. They bombarded beryllium atoms with alpha particles, as Bothe had done. However, in their experiment they placed a block of paraffin (a wax-like substance) beyond the beryllium and a particle detector beyond the paraffin. They found that the neutral radiation leaving the beryllium ejected protons from the paraffin. The Joliot-Curies knew that ultraviolet and high-frequency visible photons could knock electrons from a metal surface (the photoelectric effect discussed in Chapter 26) and made the reasonable assumption that high-frequency gamma ray photons could knock protons out of paraffin.

## The neutral radiation is not a gamma ray photon

In England, James Chadwick repeated the Joliot-Curie experiments. He placed a variety of materials such as helium and nitrogen (not just paraffin) beyond the beryllium (**Figure 28.3**). By comparing the energies and momenta of the particles knocked out of these different atoms by the neutral rays from the beryllium atoms, he found that the rays were not gamma ray photons but instead uncharged particles with a mass approximately equal to that of the proton. He called the new particle the **neutron**. In 1935, Chadwick received the Nobel Prize in physics for this work.

**Figure 28.3** Chadwick's apparatus for detecting neutrons.

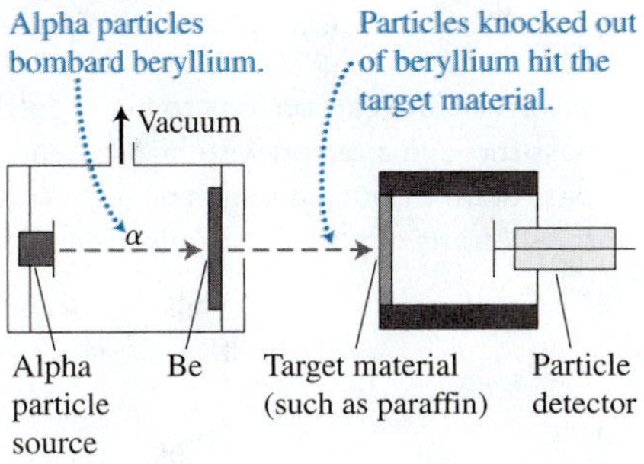


## Revising ideas of the structure of the nucleus

The discovery of the neutron was a major factor in revising ideas of nuclear structure, as was the realization that electrons were definitely not constituents of the nucleus.

Alpha particles could not account for the electric charge of many nuclei. For example, the hydrogen nucleus has an electric charge $+e$, whereas the alpha particle has charge $+2e$. The lithium nucleus has a charge $+3e$, while the boron nucleus has a charge of $+5e$. There was no way that a boron nucleus or any other odd-charged nuclei could be made from one or more alpha particles (**Figure 28.4a**).

A new model of nuclear constituents evolved that involved protons and neutrons. The protons, each of charge $+e$, accounted for the electric charge of the nucleus. The uncharged neutrons accounted for the extra mass. For example, two protons and two neutrons made a helium nucleus; five protons and five neutrons made a boron nucleus (Figure 28.4b). The most common form of uranium was composed of 92 protons and 146 neutrons, with electric charge $+92e$, and 238 protons and neutrons.

**Figure 28.4** (a) Nuclei with an odd number of charges (3, 5, 7, ...) cannot be constructed from alpha particles. (b) They can be constructed from protons and neutrons.

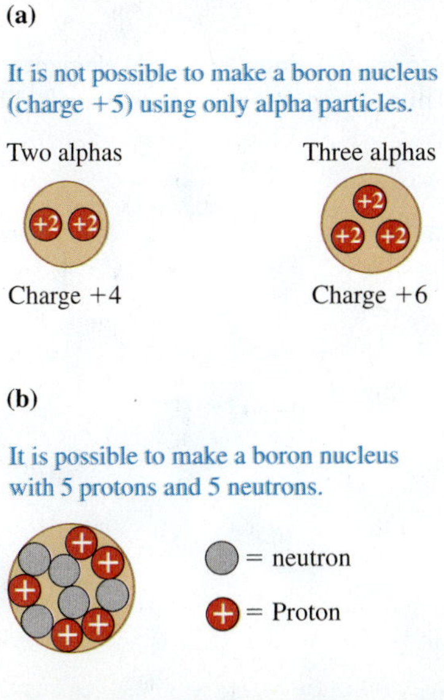


## Describing atomic nuclei

Atoms are now described with the notation ${}^{A}_{Z}X$, where X is the letter abbreviation for the type of atom (e.g., Fe for iron, Na for sodium), and $Z$ is the number of protons in the nucleus and is called the **atomic number**. $N$ (not shown in the notation) is the number of neutrons in the nucleus. $A = Z + N$ is the **mass number,** the total number of **nucleons** (protons plus neutrons). Thus, helium with two protons and two neutrons is shown as ${}^{4}_{2}He$, and the most common form of uranium with 92 protons and 146 neutrons (a total number of protons and neutrons of $A = 92 + 146 = 238$) is ${}^{238}_{92}U$.

The masses of atoms and nuclei are very small if given in kilograms. Thus, another useful mass unit, called the **atomic mass unit (u),** was developed.

> **One atomic mass unit (u)** equals one-twelfth of the mass of a carbon atom with six protons and six neutrons in the nucleus ($^{12}_{6}C$), including the six electrons in the atom. In terms of kilograms,
>
> $$1\ u = 1.660539 \times 10^{-27}\ kg$$

## Isotopes

In an earlier chapter we described the mass spectrometer—a device that allows scientists to measure the mass of an elementary particle by observing its motion in a magnetic field (Chapter 17). Scientists use mass spectrometers to measure the mass of ionized atoms. Their observations are described in Observational Experiment **Table 28.3**.

**OBSERVATIONAL EXPERIMENT TABLE**

**28.3 Measurement of atomic mass**

| Observational experiment | Analysis |
|---|---|
| We accelerate carbon ions to high speeds by letting them pass through a region with high potential difference and then pass them through a magnetic field, observing the paths they take. Atoms of a single chemical element take different paths.<br>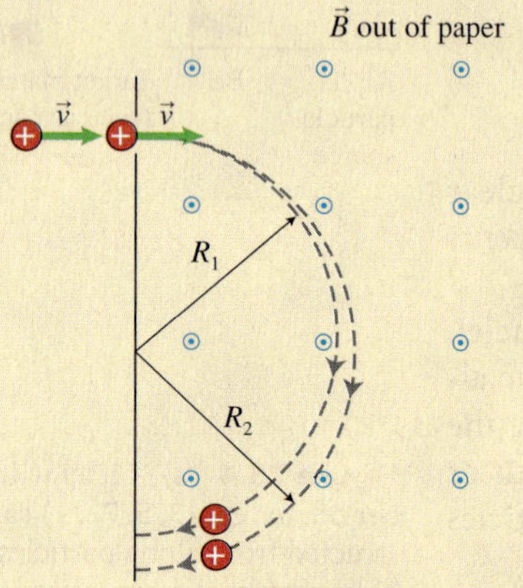 | The force exerted by the magnetic field on the positive ions is perpendicular to their velocity. Thus, ions move in a semicircle. We assume that the force exerted on the ions by the magnetic field is the only force exerted on them. Newton's second law in radial component form becomes $(mv^2/R) = qvB$ or $R = (mv/qB)$. Ions entering the region with magnetic field have the same kinetic energy: $K = (mv^2/2) = q\Delta V$. Thus their speeds are<br>$v = \sqrt{\frac{2q\Delta V}{m}}$<br>Therefore the radius of curvature for the ions of different mass is determined by<br>$R = \frac{mv}{qB} = \frac{m}{qB}\sqrt{\frac{2q\Delta V}{m}} = \sqrt{\frac{m^2 2q\Delta V}{q^2B^2m}} = \sqrt{\frac{2m\Delta V}{qB^2}}$ |

**Pattern**

Atoms of a single chemical element with the same atomic number have different masses. Thus, different atoms must have a different number of neutrons in the nucleus.

Because atoms of a particular element have the same number of protons, the differences in mass among them can be explained only if we assume that the number of neutrons is different. Atoms with a different number of neutrons are called **isotopes** of that particular element. For example, carbon comes in three naturally occurring isotopic forms: $^{12}_{6}C$, $^{13}_{6}C$, and $^{14}_{6}C$, with six, seven, and eight neutrons, respectively. Each isotope has six protons.

The electronic structure of an element's isotopes is the same, which means their chemical behaviors are almost identical. However, the nuclei behave quite differently. Generally, only a small number (usually one or two) of the possible isotopes of an element are stable. The rest undergo radioactive decay, which we will consider shortly.

**CONCEPTUAL EXERCISE 28.1 Reading symbolic descriptions of atoms**

Determine the number of protons and neutrons in each of the following nuclei: ${}^{6}_{3}\mathrm{Li}$, ${}^{16}_{8}\mathrm{O}$, ${}^{27}_{13}\mathrm{Al}$, ${}^{56}_{26}\mathrm{Fe}$, ${}^{64}_{30}\mathrm{Zn}$, and ${}^{107}_{47}\mathrm{Ag}$.

**Sketch and translate** For lithium $Z = 3$, so it has three protons in the nucleus. Lithium has $A = 6$ nucleons; thus there are three neutrons ($N = 3$).

**Simplify and diagram** We use the same strategy to determine the number of protons ($Z$) and the number of neutrons ($N$) in the other elements and summarize the results in the table below:

| ${}^{6}_{3}\mathrm{Li}$ | | ${}^{16}_{8}\mathrm{O}$ | | ${}^{27}_{13}\mathrm{Al}$ | | ${}^{56}_{26}\mathrm{Fe}$ | | ${}^{64}_{30}\mathrm{Zn}$ | | ${}^{107}_{47}\mathrm{Ag}$ | |
|---|---|---|---|---|---|---|---|---|---|---|---|
| $Z$ | $N$ | $Z$ | $N$ | $Z$ | $N$ | $Z$ | $N$ | $Z$ | $N$ | $Z$ | $N$ |
| 3 | 3 | 8 | 8 | 13 | 14 | 26 | 30 | 30 | 34 | 47 | 60 |

**Try it yourself:** Determine $Z$ and $N$ for potassium-39, ${}^{39}_{19}\mathrm{K}$.

*Answer:* $Z = 19$ and $N = 20$.

**CONCEPTUAL EXERCISE 28.2 Reading and understanding the symbols in the periodic table**

Determine the chemical elements represented by the symbol X for each of the following atoms: ${}^{20}_{10}\mathrm{X}$, ${}^{52}_{24}\mathrm{X}$, ${}^{59}_{27}\mathrm{X}$, and ${}^{93}_{41}\mathrm{X}$.

**Sketch and translate** To determine the element, we use the periodic table (Table 27.9) and the lower left number ($Z$) in the symbol to determine the element.

**Simplify and diagram** The results for the four symbols are shown in the table below.

| ${}^{20}_{10}\mathrm{X}$ | ${}^{52}_{24}\mathrm{X}$ | ${}^{59}_{27}\mathrm{X}$ | ${}^{93}_{41}\mathrm{X}$ |
|---|---|---|---|
| Ne (neon) | Cr (chromium) | Co (cobalt) | Nb (niobium) |

**Try it yourself:** Determine the chemical element with the following $Z$ and $A$: ${}^{63}_{29}\mathrm{X}$.

*Answer:* An isotope of copper, that is, copper-63.

**TIP** The $A$ number in the periodic table might not match the number for the same element in the examples above. The reason is that in the periodic table, the $A$ number is constructed to represent an average mass number over all isotopes, accounting for their relative abundances as well. As a result, it is generally not an exact integer. You can look up the individual isotopes in the appendix.

**Review Question 28.2** Explain why the nucleus cannot contain any electrons.

## 28.3 Nuclear force and binding energy

How can positively charged protons stay so close together in the nucleus? Let's compare the magnitudes of the repulsive force that two protons exert on each other when inside a nucleus with the force that a proton exerts on an electron in the atom. The size of the nucleus is about $10^{-15}$ m and the distance between the nucleus and an atomic electron is about $10^{-10}$ m (the radius of a typical atom.) Using Coulomb's law (Chapter 14):

$$\frac{F_{\text{proton on proton}}}{F_{\text{proton on electron}}} = \frac{k\dfrac{|q_p q_p|}{r^2_{\text{p to p}}}}{k\dfrac{|q_p q_e|}{r^2_{\text{p to }e}}} = \frac{k\dfrac{e^2}{r^2_{\text{p to p}}}}{k\dfrac{e^2}{r^2_{\text{p to }e}}} = \frac{\dfrac{1}{r^2_{\text{p to p}}}}{\dfrac{1}{r^2_{\text{p to }e}}} = \frac{r^2_{\text{p to }e}}{r^2_{\text{p to p}}} = \frac{(10^{-10}\text{ m})^2}{(10^{-15}\text{ m})^2} = 10^{10}$$

The electric force exerted by two protons on each other (**Figure 28.5a**) is about 10 billion times greater than the force that the nucleus exerts on one

**Figure 28.5** The nuclear force. (a) Protons exert an electric repulsion on each other. (b) This repulsion is balanced by the nuclear force. (c) Nearby nucleons exert a strong force and more distant nucleons do not.

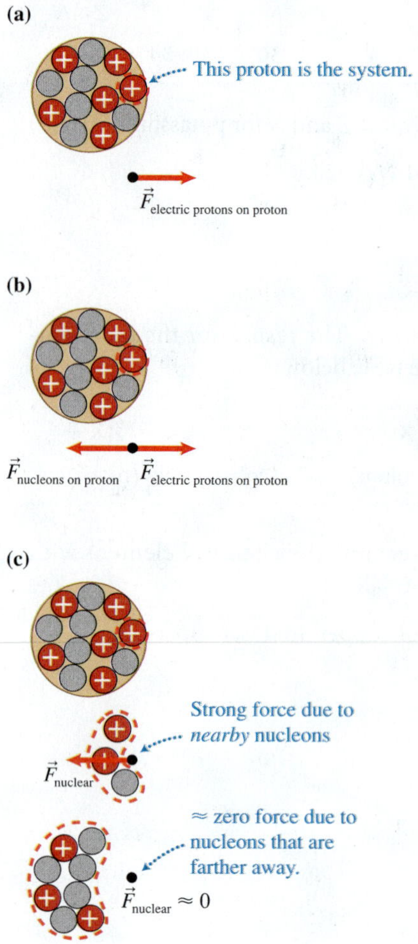


of the atom's electrons. How can the protons stay bound together when they repel each other so strongly?

## Nuclear force

Some attractive force must balance this electrical repulsive force and must attract neutrons as well—it has to be an attractive force for both protons and neutrons. (Figure 28.5b). We call this attractive force a **nuclear force.**

The nuclear force must also weaken to nearly zero extremely rapidly with increasing distance between nucleons. If it didn't, then nuclei of nearby atoms would be attracted to each other, clumping together into ever-larger nuclei. Since this does not occur, the nuclear force must be very short range—no greater than about $2 \times 10^{-15}$ m. When nucleons are farther apart, the nuclear force they exert on each other is essentially zero (Figure 28.5c).

This $2 \times 10^{-15}$ m distance limit is important since $2 \times 10^{-15}$ m is actually smaller than the size of many of the higher-$Z$ nuclei. Therefore, not every nucleon is attracted to every other nucleon, because some of them are farther away than this maximum range. If they were attracted equally strongly, then a massive uranium nucleus with 238 protons and neutrons should be exceedingly more stable than a helium nucleus with only 4 protons and neutrons simply because of the greater number. However, experiments show that these two nuclei are approximately equally stable. Thus, a proton in a small nucleus must be bound by the nuclear force to about the same number of protons and neutrons as a proton in a large nucleus. Evidently, the nuclear force involves only nearest neighbor nucleons (Figure 28.5b and c).

However, in a nucleus with many protons, the repulsive force exerted on a particular proton by all the other protons must be strong; this repulsion might exceed the strong nuclear attraction between a proton and its neighbors. If this model of the nuclear force is correct, then we predict that such nuclei might be unstable. Could it be that we have found a mechanism for radioactive decay? Before we explore this possibility, let's consider another prediction based on the hypothesis of this attractive nuclear force.

## Mass defect

Recall from our study of the Bohr model of the hydrogen atom (Chapter 27) that we must add energy to a system in order to separate the electron from the proton, ionizing the atom. When the sum of added energy, electrical potential energy, and kinetic energy reaches zero, the electron becomes unbound from the nucleus. The magnitude of the energy that must be added to ionize the atom is known as the **binding energy** of the hydrogen atom.

We could make similar statements about the nucleus itself, composed of protons and neutrons. The nucleus is a bound system, and therefore its **nuclear potential energy** plus electric potential energy (positive since the protons repel each other) plus kinetic energy of the protons and neutrons must be negative. The binding energy of the nucleus is the energy that must be added to the nucleus to separate it into its component protons and neutrons.

Recall that all objects with mass have a corresponding rest energy given by $E_0 = mc^2$ (Chapter 25). The rest energy of the nucleus has multiple components. First, there is the rest energy of the protons and neutrons that compose it. But there is also the nuclear potential energy of the nucleus, the electric potential energy of the nucleus, and the kinetic energy of the protons and neutrons, which when added together produce a negative number. The rest energy of the nucleus is the sum of these four contributions.

According to Einstein's equation, the mass $m$ of an object equals $E_0/c^2$. Because the other three forms of energy of the nucleus add to a negative value, the rest energy of the nucleus should be less than the rest energy of the separated protons and neutrons. Consequently, the mass of the nucleus should be less than the total mass of its constituents. To check whether this prediction matches experimental evidence, we need to collect data on the masses of the nuclei and their constituents. We do this in Testing Experiment **Table 28.4** using the helium nucleus (composed of two protons and two neutrons) as an example.

**TESTING EXPERIMENT TABLE**

**28.4 Comparison of the mass of a nucleus to the masses of its protons and neutrons.**

| Testing experiment | Prediction | Outcome |
|---|---|---|
| We use a mass spectrometer to measure the masses of a helium atom, ${}^4_2\text{He}$, and a hydrogen atom, ${}^1_1\text{H}$. We use the neutron ${}^1_0\text{n}$ mass determined from earlier neutron scattering experiments. We compare the mass of the helium atom to the masses of its constituents (two hydrogen atoms, ${}^1_1\text{H}$, and two neutrons ${}^1_0\text{n}$). | If the hypothesis about the energy content of the nucleus is correct, we should see a difference in the total mass of the atom and the masses of its separate constituents. We assume that the electrical binding energy of the nuclei and electrons in individual atoms is negligible. | $m_{\text{helium}} = 4.002602\text{ u}$<br>$m_{\text{hydrogen}} = 1.007825\text{ u}$<br>$m_{\text{neutron}} = 1.008665\text{ u}$<br>*Mass of helium constituents:*<br>$2m_{\text{hydrogen}} + 2m_{\text{neutron}}$<br>$= 2(1.007825\text{ u}) + 2(1.008665\text{ u})$<br>$= 4.032980\text{ u}$<br>*Missing mass:*<br>$(2m_{\text{hydrogen}} + 2m_{\text{neutron}}) - m_{\text{helium}}$<br>$= 4.032980\text{ u} - 4.002602\text{ u}$<br>$= 0.030378\text{ u}$<br>The missing mass is on the order of 1% of the mass of the nucleus! |

**Conclusion**

Our prediction matched the outcome of the experiment. The mass of the helium nucleus is less than the total mass of the protons and neutrons that compose it.

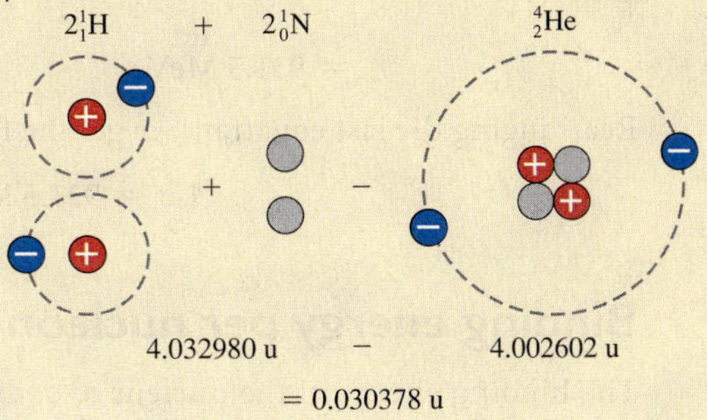


Helium is not unique in having a smaller mass than its constituents. For example, the nucleus of lithium ${}^7_3\text{Li}$ is made of three protons and four neutrons. The atom also has three electrons. Thus we can compare it to three hydrogen atoms ${}^1_1\text{H}$ and four neutrons. The total mass of the three hydrogen atoms and four neutrons is 7.058135 u, whereas the mass of a lithium atom is 7.016003 u, or 0.042132 u less than the mass of its constituents. On the basis of such findings, we can now define a new physical quantity called **mass defect**.

‹ **Active Learning Guide**

**Mass defect** **The mass defect of a certain type of nucleus is the difference in mass between the constituents of the atom (as hydrogen atoms and neutrons) and the mass of the atom itself. The mass defect $\Delta m$ is**

$$\Delta m = [Zm_{\text{hydrogen atom}} + (A - Z)m_{\text{neutron}}] - m_{\text{atom}} \qquad (28.1)$$

The reason the mass of the hydrogen atom is used rather than the mass of the proton is to account for the mass of the electrons. When the mass of the atom is subtracted, the mass of the electrons cancels. In practice, it is also much easier to measure the masses of atoms than the masses of just their nuclei, which is why it's more useful to define mass defect in terms of the masses of atoms.

## Binding energy *BE* of the nucleus

In Testing Experiment Table 28.4 we determined the mass defect of a helium nucleus. This defect is related to the binding energy of the atomic nucleus.

**Binding energy** **The binding energy of a nucleus is the rest energy equivalent of its mass defect:**

$$BE = \Delta m \cdot c^2 \tag{28.2}$$

**The binding energy represents the total energy needed to separate the nucleus into its component nucleons.**

The binding energy of a nucleus is most easily expressed in units of million electron volts (MeV), where

$$1\,\text{MeV} = 10^6\,\text{eV} = 1.602 \times 10^{-13}\,\text{J}$$

Recall that mass defect is easily expressed in atomic mass units (u). It is convenient to be able to quickly convert the mass defect of a nucleus into the corresponding binding energy. Because 1 u has units of mass, $1\,\text{u} \cdot c^2$ has units of energy. The goal is to convert $1\,\text{u} \cdot c^2$ into MeV:

$$\begin{aligned} 1\,\text{u} \cdot c^2 &= 1(1.660539 \times 10^{-27}\,\text{kg})(2.9979 \times 10^8\,\text{m/s})^2 \\ &= 1.4924 \times 10^{-10}\,\text{J} \\ &= (1.4924 \times 10^{-10}\,\text{J})\left(\frac{1\,\text{MeV}}{1.6022 \times 10^{-13}\,\text{J}}\right) \\ &= 931.5\,\text{MeV} \end{aligned}$$

Rearranging the last equation, we get the following conversion factor:

$$1\,\text{u} = 931.5\,\text{MeV}/c^2 \tag{28.3}$$

## Binding energy per nucleon

The binding energy of the nucleus of sodium-23 ($^{23}_{11}\text{Na}$) is 187 MeV; for lithium-7 ($^{7}_{3}\text{Li}$), the binding energy is 39.2 MeV; and earlier we found that the binding energy of helium-4 was 28.3 MeV. These differences means that more energy is needed to separate a sodium-23 atom into hydrogen atoms and neutrons than to separate a lithium-7 atom into hydrogen atoms and neutrons, and both require more energy than is needed to separate a helium-4 atom into hydrogen atoms and neutrons.

Does this mean that sodium is more stable than lithium, and that lithium is more stable than helium? Not necessarily. Sodium has an "unfair" advantage in total binding energy because it has more nucleons. Instead, a better measure of stability is the binding energy of a nucleus divided by its number of nucleons, *A*. For sodium-23, its binding energy per nucleon is $(187\,\text{MeV})/(23\,\text{nucleons}) = 8.1\,\text{MeV/nucleon}$. For lithium-7, it is $(39.2\,\text{MeV})/(7\,\text{nucleons}) = 5.6\,\text{MeV/nucleon}$, and for helium-4, it is $(28.3\,\text{MeV})/(4\,\text{nucleons}) = 7.1\,\text{MeV/nucleon}$. With this comparison, we can see that sodium is a more stable nucleus than helium, and that helium is more stable than lithium. Put another way, more energy is needed per nucleon to separate the sodium nucleus into its constituents than is needed for the helium nucleus, and the least stable is the lithium nucleus. Binding energy per nucleon is a much better indicator of nuclear stability than total binding energy.

**Binding energy per nucleon** **Binding energy per nucleon is the binding energy *BE* of the atom divided by the number of nucleons *A* in the atom's nucleus:**

$$\text{Binding energy per nucleon} = BE/A \tag{28.4}$$

**Binding energy per nucleon is a good indicator of the stability of a nucleus; the larger the binding energy per nucleon the more stable is the nucleus.**

**Review Questions 28.3** What is the difference between nuclear binding energy and the energy needed to ionize an atom when in the ground state (atomic binding energy)?

## 28.4 Nuclear reactions

In 1919 Ernest Rutherford became the first person to transmute one element into another when he shot helium nuclei (alpha particles) at nitrogen and produced some oxygen-17 atoms (an isotope of oxygen that has nine neutrons and eight protons). This process, called a **nuclear reaction,** can be represented as follows:

$$^{4}_{2}\text{He} + ^{14}_{7}\text{N} \rightarrow ^{17}_{8}\text{O} + ^{1}_{1}\text{H}$$

Rutherford's work was followed by many remarkable advances in the study of nuclear physics.

< **Active Learning Guide**

### Representing nuclear reactions

Nuclear reactions, like chemical reactions, involve the transformation of reactants into different products. In a chemical reaction, molecules are transformed into other molecules, but the number of each type of atom remains the same. In a nuclear reaction, nuclei are transformed into different nuclei, resulting in different elements. In these reactions two nuclei may interact to form one or more new nuclei (e.g., $^{1}_{1}\text{H} + ^{7}_{3}\text{Li} \rightarrow ^{4}_{2}\text{He} + ^{4}_{2}\text{He}$). Or a single nucleus may divide into two or more new nuclei, or perhaps emit a small particle, thus leaving behind a different nucleus (e.g., the radioactive decay experiments observed by Curie and others).

The advantage of writing nuclear reactions as shown above is that atomic masses found in atomic mass tables can be used to analyze the energy transformations that occur during the reactions. Even though atomic masses are used, the energy transformations are associated almost entirely with the rest energies of the reactant and product nuclei.

### Rules for nuclear reactions

In balancing chemical reactions the number of atoms of each type must be the same both before the reaction and after. Similar rules apply to nuclear reactions. Consider Observational Experiment **Table 28.5**, which describes reactions that have been observed to take place and reactions that have never been observed to occur.

**OBSERVATIONAL EXPERIMENT TABLE**

## 28.5 Deducing rules for allowed nuclear reactions.

| Observational experiment | Analysis |
|---|---|
| *Nuclear reactions that have been observed:*<br>(a) ${}^{1}_{1}\text{H} + {}^{7}_{3}\text{Li} \rightarrow {}^{4}_{2}\text{He} + {}^{4}_{2}\text{He}$<br>(b) ${}^{226}_{88}\text{Ra} \rightarrow {}^{222}_{86}\text{Rn} + {}^{4}_{2}\text{He}$<br>(c) ${}^{14}_{6}\text{C} \rightarrow {}^{14}_{7}\text{N} + {}^{0}_{-1}\text{e}$<br>In the above reactions, the products are observed to have significantly more kinetic energy than the reactants. | *Nuclear reactions that have been observed:*<br>(a) $A$: 1 + 7 = 4 + 4   $Z$: 1 + 3 = 2 + 2<br>(b) $A$: 226 = 222 + 4   $Z$: 88 = 86 + 2<br>(c) $A$: 14 = 14 + 0   $Z$: 6 = 7 − 1<br>Both $A$ and $Z$ numbers are equal on both sides of the equations. |
| *Nuclear reactions that have never been observed:*<br>(d) ${}^{4}_{2}\text{He} + {}^{27}_{13}\text{Al} \rightarrow {}^{32}_{15}\text{P} + {}^{1}_{0}\text{n}$<br>(e) ${}^{2}_{1}\text{H} + {}^{3}_{1}\text{H} \rightarrow {}^{4}_{2}\text{He} + {}^{1}_{1}\text{H}$ | *Nuclear reactions that have never been observed:*<br>(d) $A$: 4 + 27 ≠ 32 + 1   $Z$: 2 + 13 = 15 + 0<br>(e) $A$: 2 + 3 = 4 + 1   $Z$: 1 + 1 ≠ 2 + 1<br>Either $A$ or $Z$ numbers are not equal on both sides of the equations. |

**Patterns**

Two patterns appear in the observed nuclear reactions:

- The total numbers of nucleons $A$ of the reactants and of the products are equal.
- The total $Z$ numbers of the reactants and of the products are equal.
- $Z$ is the electric charge of the nuclei involved in the reaction.

**Active Learning Guide ›**

The second pattern, in which $Z$ of the reactants is equal to $Z$ of the products, is the result of electric charge conservation: in an isolated system, the total electric charge of the reactants equals that of the products.

**TIP** The symbol for an electron is ${}^{0}_{-1}e$. It has charge $Z = -1$ and has zero nucleons, $A = 0$. It can be included in the reaction rules using this notation—see Reaction (c) in Table 28.5.

Why is the number of nucleons (protons plus neutrons) on both sides of the reactions equal? If a neutron vanished, it wouldn't violate charge conservation, but neutrons don't seem to do that. Evidently, they can transform into protons (see Reaction (c) in Table 28.5), but they cannot vanish. At the moment, we don't have a reason for why this pattern exists (but we will return to it in Chapter 29). For now, the rule is given a name: **nucleon number conservation**. This conservation rule explains the stability of matter in the universe—nucleons, and therefore nuclei, cannot simply disappear. Here are the two conservation rules summarized.

**Rule 1: Electric charge conservation** **The total electric charge of the reacting nuclei and particles equals the total charge of the nuclei and particles produced by the reaction. This condition is satisfied if the total atomic number $Z$ of the reactants equals the total atomic number of the products, including $Z = -1$ for free electrons.**

**Rule 2: Nucleon number conservation** **The total number of nucleons $A$ (protons plus neutrons) of the reactants always equals the total number for the products.**

**QUANTITATIVE EXERCISE 28.3 Determine the missing product in the following reactions.**

(a) ${}^{4}_{2}\text{He} + {}^{12}_{6}\text{C} \rightarrow {}^{15}_{7}\text{N} + ?$

(b) ${}^{2}_{1}\text{H} + {}^{3}_{1}\text{H} \rightarrow {}^{4}_{2}\text{He} + ?$

(c) ${}^{1}_{0}\text{n} + {}^{235}_{92}\text{U} \rightarrow {}^{140}_{54}\text{Xe} + ? + 2{}^{1}_{0}\text{n}$

(d) ${}^{137}_{55}\text{Cs} \rightarrow ? + {}^{0}_{-1}e$

**Represent mathematically** To determine the products we need to use the rules that the $A$ and $Z$ numbers remain constant throughout the reaction.

(a) $4 + 12 = 15 + A$
$2 + 6 = 7 + Z$

(b) $2 + 3 = 4 + A$
$1 + 1 = 2 + Z$

(c) $1 + 235 = 140 + A + 2 \cdot 1$
$0 + 92 = 54 + Z + 2 \cdot 0$

(d) $137 = A + 0$
$55 = Z + (-1)$

**Solve and evaluate** Solving for $A$ and $Z$ in each case, we find the following:

(a) $A = 1$, $Z = 1$. The unknown product must be a hydrogen nucleus ${}^{1}_{1}\text{H}$.

(b) $A = 1$, $Z = 0$. The unknown product must be a neutron ${}^{1}_{0}\text{n}$.

(c) $A = 94$, $Z = 38$. The unknown product must be a strontium nucleus ${}^{94}_{38}\text{Sr}$.

(d) $A = 137$, $Z = 56$. The unknown product must be a barium nucleus ${}^{137}_{56}\text{Ba}$.

**Try it yourself:** Determine the missing product in this reaction: ${}^{14}_{6}\text{C} \rightarrow ? + {}^{0}_{-1}e$.

*Answer:* ${}^{14}_{7}\text{N}$.

## Energy conversions in nuclear reactions

In the three observed reactions in Table 28.5, the products had significantly more kinetic energy than the reactants. How did the products get this kinetic energy? One hypothesis is that some of the rest mass energy of the reactants was converted to kinetic energy. If this is correct, then the mass of the reactants should be greater than the mass of the products. Let's test this hypothesis using the reaction described in Testing Experiment **Table 28.6**.

**TESTING EXPERIMENT TABLE**

**28.6 Accounting for the extra kinetic energy.**

| Testing experiment | Prediction | Outcome |
|---|---|---|
| In the following reaction, the products have more kinetic energy than the reactants:<br>${}^{1}_{1}\text{H} + {}^{7}_{3}\text{Li} \rightarrow {}^{4}_{2}\text{He} + {}^{4}_{2}\text{He} + \text{kinetic energy}$<br>Is this consistent with the idea that some rest mass energy of the reactants was transformed into kinetic energy of the products? The atomic masses of the atoms involved are<br>$m({}^{1}_{1}\text{H}) = 1.007825\ \text{u}$<br>$m({}^{7}_{3}\text{Li}) = 7.016003\ \text{u}$<br>$m({}^{4}_{2}\text{He}) = 4.002602\ \text{u}$ | Mass of reactants:<br>$m({}^{1}_{1}\text{H}) + m({}^{7}_{3}\text{Li}) = 1.007825\ \text{u} + 7.016003\ \text{u} = 8.023828\ \text{u}$<br>Mass of products:<br>$m({}^{4}_{2}\text{He}) + m({}^{4}_{2}\text{He}) = 4.002602\ \text{u} + 4.002602\ \text{u} = 8.005204\ \text{u}$<br>The energy equivalent of this mass difference is<br>$\Delta mc^2 = (8.023828\ \text{u} - 8.005204\ \text{u})c^2\left(\frac{931.5\ \text{MeV}}{\text{u} \cdot c^2}\right) = 17.3\ \text{MeV}$<br>We predict that the products will have 17.3 MeV more kinetic energy than the reactants. | Through a complex process, the products are found to have 17.3 MeV more kinetic energy than the reactants, as predicted. |

**Conclusion**

This experiment supports the idea that some of the rest energy of the reactants is converted into kinetic energy of the products.

These kinds of reactions, where rest energy is converted into kinetic energy, are called **exothermic.** The opposite can occur as well. In certain nuclear reactions the kinetic energy of the reactants is greater than the kinetic energy of the products. In this case, the rest energy of the products is greater than that of the reactants. In these **endothermic** reactions, some of the kinetic energy of the reactants is converted into rest energy of the products. Both types of reactions can be understood in terms of energy conservation.

**Rule 3: Energy conservation** **The rest energy of the reactants equals the rest energy of the products plus any change in kinetic energy of the system:**

$$\sum_{\text{reactants}} mc^2 = \sum_{\text{products}} mc^2 + \Delta K$$

**$\Delta K$ can be either positive (if the products' rest energy is less that of the reactants) or negative (if the products' rest energy is greater). $\Delta K$ is also known as the reaction energy $Q$. Rearranging, we find an expression for $Q$:**

$$Q = \left( \sum_{\text{reactants}} m - \sum_{\text{products}} m \right) c^2 \qquad (28.5)$$

We will have several opportunities to use this idea in the following sections.

**Review Question 28.4** Is it possible for the nuclear reaction $^4_2\text{He} + {}^6_3\text{Li} \rightarrow {}^{11}_6\text{C} + 2\,{}^{\;0}_{-1}e$ to occur? Explain your answer.

## 28.5 Nuclear sources of energy

Ernest Rutherford once said, "Anyone who expects a source of power from the transformation of the atom is talking moonshine."

Rutherford had an enormous impact on early atomic and nuclear physics, but he missed it with this statement. Let's consider more carefully why nuclear reactions can be a source of power.

### Binding energy and energy release

**Figure 28.6** is a graph of the binding energy per nucleon $(BE/A)$ versus mass number $A$ for atomic nuclei. Notice that nuclei with $A \approx 60$ have the greatest binding energy per nucleon, while very small and very large nuclei have less binding energy per nucleon. The higher the binding energy per nucleon, the more energy is needed to split the nucleus into its constituent protons and neutrons; therefore, the nuclei with the highest binding energy per nucleon—nuclei around $A \approx 60$—should be the most stable.

**Figure 28.6** A graph of binding energy versus number of nucleons $(BE/A)$.

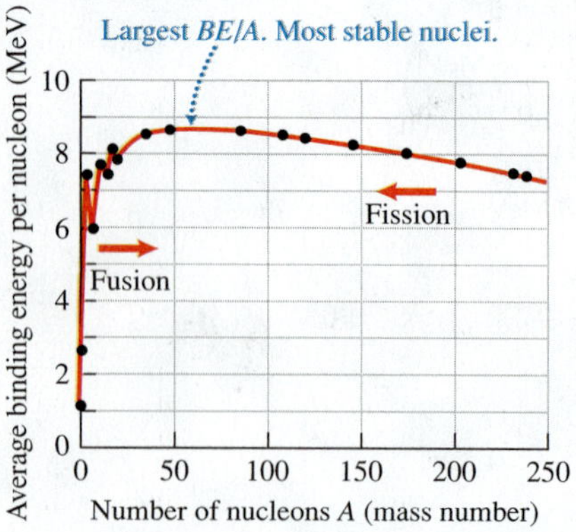


The graph in Figure 28.6 allows us to predict two new phenomena. First, note that the binding energy per nucleon for nuclei with small atomic numbers is less than that for nuclei with larger atomic numbers (up to about 60). Therefore, if two smaller nuclei were to combine to make a heavier one, the difference in their total binding energy should be released. Second, the graph shows that nuclei with very large atomic numbers have less binding energy per nucleon than those with atomic numbers near 60. Thus if there was a way to break one of these high-$Z$ nuclei into two or more smaller-$Z$ nuclei, the difference in binding energy could be released.

Now we have two predictions that are based on the patterns in the graph:

- When two small nuclei combine, energy should be released.
- When a large nucleus breaks apart, energy should be released.

It turns out that both of these predicted processes do occur in nature.

## Fusion and chemical elements

Small nuclei ($A < 60$) do not spontaneously join to form heavier ones because the nuclei repel each other due to their positive electric charges. For the attractive nuclear force to be exerted, the nuclei must be very close to each other—less than $10^{-14}$ m apart. In solids and liquids the nuclei of neighboring atoms are about $10^{-10}$ m apart, far too distant from one another to have a chance to fuse. However, if a material is so hot that it ionizes completely (that is, it becomes a plasma), and the pressure is high enough, a small percentage of nuclei will come close enough so that the nuclear force pulls them together and a huge amount of energy is released (**Figure 28.7**). This process, called **fusion**, occurs naturally in stars, which begin their lives with mostly hydrogen and a little helium. Fusion could potentially be a source of clean energy on Earth; however, the temperatures and pressures needed test the limits of current technology.

The combining of light nuclei into heavier ones happens in the cores of stars where the temperature and pressure is very high. Stars with masses comparable to that of the Sun produce elements from helium to carbon through fusion. More massive stars produce elements up to iron. These stars finish their lives by cooling down, and the heavy elements produced in their cores remain there. In contrast, stars that are much more massive than the Sun explode at the end of their lives. These explosions, known as supernovas, produce elements heavier than iron.

Supernovas explosions contribute to the chemical composition of the universe in two ways. First, the elements lighter than iron that are produced in stars' cores before the explosion are ejected into space. (Lighter stars that do not explode as supernovas produce those elements, too, but those elements remain in their cores.) Second, elements heavier than iron that are produced during the explosion are also ejected into space. The Sun and Earth have a high abundance of elements heavier than helium; thus the Sun and Earth are made of elements produced long ago inside stars that exploded as supernovas. This means that many of the atoms in our bodies came from these supernovas.

**Figure 28.7** A greatly simplified example of a fusion reaction. In plasma, ions are moving so fast that they can get close enough to each other during a collision that the nuclear force binds them together.

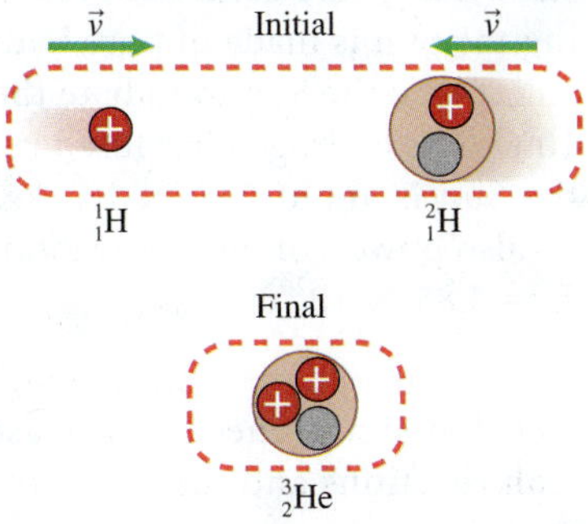


**‹ Active Learning Guide**

**QUANTITATIVE EXERCISE 28.4 Fusion energy in the Sun**

The energy released by the Sun comes from several sources, including the so-called proton-proton chain of fusion reactions. This chain of reactions can be summarized as follows:

$$2\,^1_1\text{H} + 2\,^1_0\text{n} \rightarrow {}^4_2\text{He} + \text{energy}$$

Determine the energy released in this chain of reactions in MeV. Use the masses $^1_1$H (1.007825 u), $^1_0$n (1.008665 u), and $^4_2$He (4.002602 u) to determine the rest energy converted to other forms.

**Represent mathematically** The energy released, the reaction energy $Q$, is determined using Eq. (28.5):

$$Q = \left(\sum_{\text{reactants}} m - \sum_{\text{products}} m\right)c^2$$

**Solve and evaluate** Inserting the appropriate values and converting to MeV gives

$$Q = [2(1.007825\text{ u}) + 2(1.008665\text{ u}) - 4.002602\text{ u}]c^2\left(\frac{931.5\text{ MeV}}{\text{u}\cdot c^2}\right)$$
$$= 28\text{ MeV}$$

This value is approximately a million times more energy than is released in a typical chemical reaction.

**Try it yourself:** What fraction of the total reactant rest energy in the proton-proton chain reactions is converted to other forms of energy during the above fusion reaction?

*Answer:* About $(0.03\text{ u})/(4.0\text{ u}) = 0.008$, or 0.8%.

Just as in any reaction, once the reactants have been used up, the reaction stops. This means that eventually the Sun will run out of hydrogen and start burning helium into carbon. After helium is exhausted, fusion will cease (as the Sun is not massive enough to be able to fuse carbon), and our star will slowly go dark. How long will the Sun be able to shine?

**EXAMPLE 28.5 How long will the Sun shine?**

According to the present models of stellar structure, the central parts of stars, which contain about 10% of the star's mass, have conditions favorable for fusion. Assuming the Sun is made of pure hydrogen, how much time will it take the Sun to radiate the energy released when 10% of its hydrogen has fused into helium? The mass of the Sun is about $2.00 \times 10^{30}$ kg. The total luminosity $L$ (also power, or energy/second) emitted by the Sun is $L = 3.85 \times 10^{26}$ W.

**Sketch and translate** Imagine the Sun as an object that simultaneously releases energy through fusion reactions and then radiates it as electromagnetic radiation.

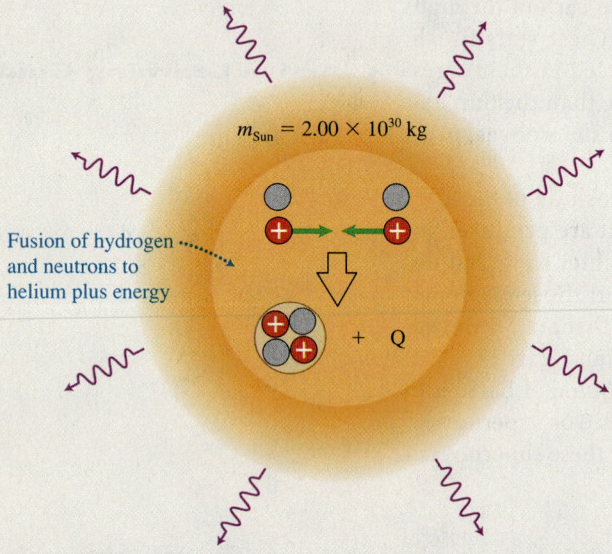


**Simplify and diagram** Assume that during a fusion reaction in the Sun's interior, two protons and two neutrons join together to form a helium nucleus. The amount of energy released is equal to approximately 0.008 of the rest energy of the participating particles (see Exercise 28.4). Further, assume that there are no other sources of energy, so that the rate at which energy is produced inside the Sun by fusion reactions is the same as the rate at which the energy is radiated.

**Represent mathematically** The total energy released by fusion reactions throughout the lifetime of the Sun is equal to 0.8% of the rest energy of the 10% of the Sun's mass energy that is available for fusion:

$$E_{\text{released}} = (0.008)(0.10)m_{\text{Sun}}c^2$$

This equals the luminosity $L$ of the Sun times the time interval $\Delta t$ during which the Sun will shine, or

$$E_{\text{released}} = L\Delta t$$

**Solve and evaluate** Setting the above two expressions equal to each other, solving for $\Delta t$, and inserting the appropriate values, we estimate the expected lifetime of the Sun:

$$\begin{aligned}\Delta t &= \frac{E_{\text{released}}}{L}\\ &= \frac{(0.008)(0.10)(2.00 \times 10^{30}\text{ kg})(3.00 \times 10^{8}\text{ m/s})^2}{3.85 \times 10^{26}\text{ W}}\\ &= 3.7 \times 10^{17}\text{ s}\\ &= 3.7 \times 10^{17}\text{ s}\left(\frac{1\text{ year}}{3.16 \times 10^{7}\text{ s}}\right) = 1.2 \times 10^{10}\text{ years}\end{aligned}$$

This result is about 2.5 times the age of Earth. Nuclear fusion is a possible mechanism for powering the Sun for billions of years.

**Try it yourself:** Studying the energy emitted by stars of different masses, astronomers found that the energy that a typical star emits every second is proportional to its mass raised to the power of 3.5, that is, $L \propto m^{3.5}$. Will a 10 solar mass star take more or less time than the Sun to exhaust its nuclear fuel, assuming that both the star and the Sun have 10% of its mass in hydrogen available for fusion?

*Answer:* The star will take significantly less time—about 0.003 times the time for the Sun:

$$\begin{aligned}\frac{\Delta t_{\text{star}}}{\Delta t_{\text{Sun}}} &= \frac{\left(\dfrac{E_{\text{released}}}{L}\right)_{\text{star}}}{\left(\dfrac{E_{\text{released}}}{L}\right)_{\text{Sun}}} = \frac{\dfrac{(0.008)(0.10)(10m_{\text{Sun}})c^2}{L_{\text{star}}}}{\dfrac{(0.008)(0.10)m_{\text{Sun}}c^2}{L_{\text{Sun}}}}\\ &= \frac{10\,L_{\text{Sun}}}{1\,L_{\text{Star}}} = \frac{10\,m_{\text{Sun}}^{3.5}}{1\,m_{\text{Star}}^{3.5}} = \frac{10}{1}\left(\frac{m_{\text{Sun}}}{10m_{\text{Sun}}}\right)^{3.5}\\ &= 0.0032\end{aligned}$$

## Fission and nuclear energy

Active Learning Guide >

We saw from the graph in Figure 28.6 that energy is released when very heavy nuclei split into nuclei of atoms that reside near the middle of the periodic table. However, the heavy nuclei do not just spontaneously split. One way the

**Figure 28.8** Nuclear fission. (a) A neutron approaches a U-235 nucleus. (b) The collision forms an excited U-236*. (c) The U-236* nucleus is unstable. (d) The excited nucleus splits into two smaller nuclei and two neutrons. (e) The newly released neutrons can go on to collide with other U-235 nuclei.

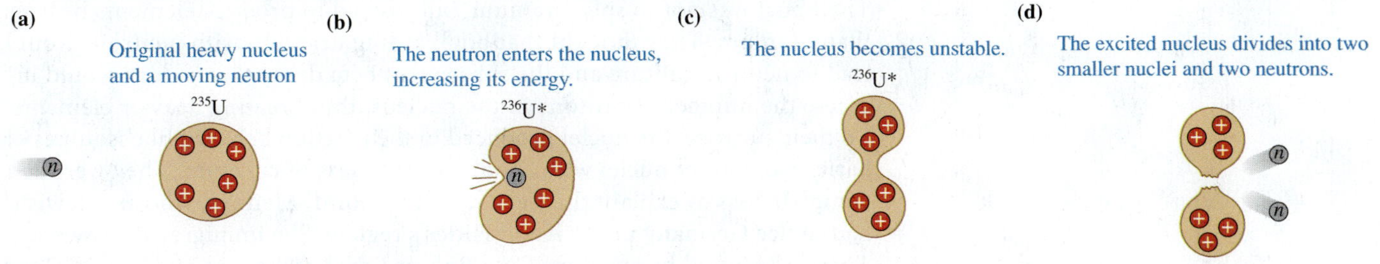
(a)
Original heavy nucleus and a moving neutron
235U
n
(b)
The neutron enters the nucleus, increasing its energy.
236U*
(c)
The nucleus becomes unstable.
236U*
(d)
The excited nucleus divides into two smaller nuclei and two neutrons.

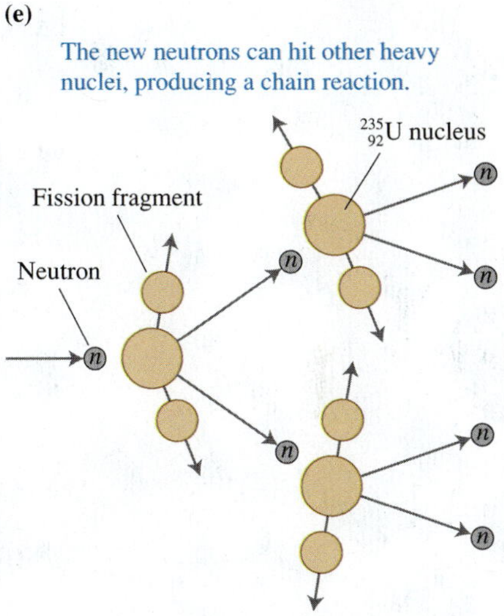
(e)
The new neutrons can hit other heavy nuclei, producing a chain reaction.
235 92U nucleus
Fission fragment
Neutron
n

process can be initiated is the collision of a moving neutron with a heavy nucleus, which disturbs the equilibrium of the nucleus and causes it to fall apart into two smaller nuclei. This process is called **fission** (**Figure 28.8a**–d). Heavy nuclei have a high neutron-to-proton ratio; however, nuclei near the middle of the periodic table (which are the typical product nuclei of these kinds of reactions) usually have a somewhat lower ratio. Thus splitting a heavy nucleus usually releases several neutrons, which can then catalyze the splitting of other nuclei, causing a chain reaction (Figure 28.8e). This reaction results in a huge release of energy.

A common fission reaction is

$$^{1}_{0}\text{n} + ^{235}_{92}\text{U} \rightarrow ^{141}_{56}\text{Ba} + ^{92}_{36}\text{Kr} + 3^{1}_{0}\text{n} + \text{energy}$$

Notice that one neutron absorbed by one uranium nucleus produces two smaller nuclei plus three new neutrons that can cause three new fission reactions. Let's calculate the energy released by the above fission reaction.

### QUANTITATIVE EXERCISE 28.6 Energy from a fission reaction

Determine the energy released in the above fission reaction. Use the following masses: $^{1}_{0}\text{n}$ (1.008665 u), $^{235}_{92}\text{U}$ (235.043924 u), $^{141}_{56}\text{Ba}$ (140.914411 u), $^{92}_{36}\text{Kr}$ (91.926156 u). One neutron collides with and joins momentarily with the U-235 atomic nucleus to form an excited U-236, which quickly splits into Ba-141, Kr-92, and three neutrons.

**Represent mathematically** The energy $Q$ released in the reaction is the energy equivalent to the mass difference between the reactants and products.

$$\begin{aligned} Q &= [m(^{1}_{0}\text{n}) + m(^{235}_{92}\text{U})]c^2 \\ &\quad - [m(^{141}_{56}\text{Ba}) + m(^{92}_{36}\text{Kr}) + 3m(^{1}_{0}\text{n})]c^2 \\ &= [m(^{235}_{92}\text{U}) - m(^{141}_{56}\text{Ba}) - m(^{92}_{36}\text{Kr}) - 2m(^{1}_{0}\text{n})]c^2 \end{aligned}$$

**Solve and evaluate** Inserting the appropriate values and converting to MeV, we get

$$\begin{aligned} Q &= [235.043924\text{ u} - 140.914411\text{ u} - 91.926156\text{ u} \\ &\quad - 2(1.008665\text{ u})]c^2\left(\frac{931.5\text{ MeV}}{\text{u}\cdot c^2}\right) \\ &= 173\text{ MeV} \end{aligned}$$

This is a typical energy release during fission reactions. This reaction produces over 300 million times more energy than burning one octane molecule.

**Try it yourself:** Your friend says that nuclear power plants could release energy by the fission of iron-56 (very abundant) instead of the much less abundant uranium-235. What do you say?

*Answer:* Iron-56 has about the highest binding energy per nucleon of any nucleus. This means energy must be added to the nucleus to cause it to break into smaller nuclei with lower $BE/A$—exactly the opposite of what is desired in a nuclear power plant.

## Bohr's liquid drop model of the nucleus

Quantitative Exercise 28.6 represents a historical experiment performed by German scientists Lise Meitner, Otto Hahn, and Fritz Strassmann in the 1930s. Their goal was not to split uranium, but instead to produce elements heavier than uranium. They thought that bombarding uranium with neutrons would lead to neutron capture and the subsequent beta decay that in turn would increase the number of protons in the nucleus, thus creating heavier elements. To their surprise, the nuclei produced in the reaction behaved like isotopes of barium and other nuclei with about half the mass of uranium. They were at a complete loss to explain this result. At this point, Meitner, who was Jewish, had to flee Germany under Adolf Hitler's regime. She immigrated to Sweden. There, she asked her nephew Otto Robert Frisch (who was working in Denmark with Niels Bohr) to help with the explanation of these results.

Frisch and Meitner decided that Bohr's **liquid drop model** of a nucleus could explain the strange results of the experiments (see Figure 28.8). In this model the nucleus was modeled as a drop of water in which surface tension holds the water drop together, though in this case it is the nuclear forces that hold the nucleons together. However, in heavy nuclei, such as uranium, there are too many electrically charged protons present. The protons would repel each other and overwhelm the effect of the "surface tension," especially if the nucleus was not spherical. The model suggested that the nucleus could stretch itself and divide into two smaller pieces. This meant that the uranium nucleus would be very unstable and ready to split with the slightest provocation, such as being hit by a neutron. Meitner and Frisch had come up with a model for fission.

**Review Question 28.5** Explain how fusion can lead to a release of energy. Does this release mean that energy is not conserved in nuclear fusion reactions?

# 28.6 Mechanisms of radioactive decay

Now that we understand the stability of nuclei in terms of their binding energy per nucleon and the energy released in nuclear reactions (the $Q$ of the reactions), we can explain why some nuclei undergo radioactive decay while others do not.

## Alpha decay

Of the small nuclei, helium ($^4_2$He) is one of the most stable. Because of this, in a larger nucleus, groups of two protons and two neutrons tend to form helium nuclei within the larger nucleus (like a small group of friends joined together in a large crowd). In addition to the attractive forces that the two protons and two neutrons in a helium nucleus exert on each other, a repulsive electric force is exerted on the helium nuclei by the other protons in the larger nucleus. When the nucleus becomes too large, the electric repulsion becomes significant compared to the nuclear attraction. Then there is a chance that one of these helium nuclei will leave the larger nucleus. This is the emission of an alpha particle.

When one of the alpha particles leaves, this emission reduces the number of protons in the original nucleus and also reduces the electric repulsion between the protons that remain. Because a helium nucleus also contains two neutrons, the total number of nucleons $A$ in the original nucleus decreases by four. **Figure 28.9** illustrates the alpha decay of radium-226. When radium undergoes this decay, it becomes radon-222, and an alpha particle is released.

**Figure 28.9** Alpha decay of radium-226. The result is a new element plus release of an alpha particle.

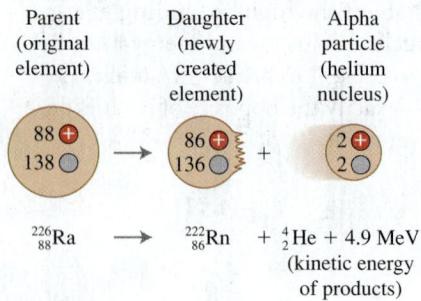

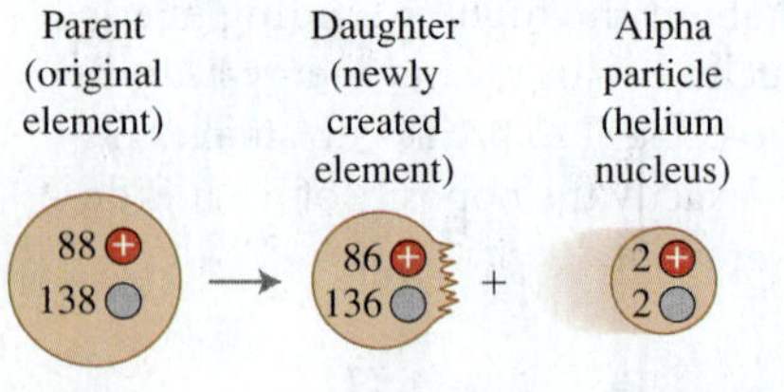


$^{226}_{88}$Ra ⟶ $^{222}_{86}$Rn + $^4_2$He + 4.9 MeV (kinetic energy of products)

The mass of the new radon nucleus (called the **daughter nucleus**) plus the alpha particle is less than the original radium nucleus (called the **parent nucleus**). Some of the rest energy of the radium nucleus was converted into the kinetic energy of the products.

$$^{226}_{88}\text{Ra} \rightarrow {}^{222}_{86}\text{Rn} + {}^{4}_{2}\text{He} + 4.9\ \text{MeV}$$

**QUANTITATIVE EXERCISE 28.7 Kinetic energy produced during alpha decay**

Determine the kinetic energy of the product nuclei when polonium-212 undergoes alpha decay. The masses of the nuclei involved in the decay are $m\,(^{212}_{84}\text{Po}) = 211.9889$ u; $m\,(^{208}_{82}\text{Pb}) = 207.9766$ u, and $m(^{4}_{2}\text{He}) = 4.0026$ u.

**Represent mathematically** The alpha decay of polonium-212 is

$$^{212}_{84}\text{Po} \rightarrow {}^{208}_{82}\text{Pb} + {}^{4}_{2}\text{He} + \text{energy}$$

Use Eq. (28.5) to find the energy $Q$ released during this reaction:

$$Q = [m(^{212}_{84}\text{Po}) - m(^{208}_{82}\text{Pb}) - m(^{4}_{2}\text{He})]c^2$$

**Solve and evaluate** Insert the appropriate values and convert to MeV:

$$Q = (211.9889\ \text{u} - 207.9766\ \text{u} - 4.0026\ \text{u})c^2$$
$$= (0.0097\ \text{u})c^2\left(\frac{931.5\ \text{MeV}}{\text{u}\cdot c^2}\right)$$
$$= 9.0\ \text{MeV}$$

Of this 9.0 MeV of released energy, 8.8 MeV is converted to the kinetic energy of the alpha particle ($^{4}_{2}$He). Most of the remaining 0.2 MeV is converted to the kinetic energy of the recoiling lead-208 nucleus. In addition, a small fraction of the energy may be released as an additional product in the reaction: a gamma ray (a high-energy photon).

**Try it yourself:** Determine the product nucleus when thorium-232 ($^{232}_{90}$Th) undergoes alpha decay.

*Answer:* Radium-228 ($^{228}_{88}$Ra).

## Beta decay

The particle emitted from a nucleus during beta decay is an electron. But we know from the uncertainty principle that electrons cannot reside in the nucleus. How can we explain this apparent contradiction? One explanation is that a neutron inside a nucleus can spontaneously decay into a proton and an electron. The proton stays inside the nucleus while the electron created during the decay leaves. This neutron decay reaction can be written as follows:

$$^{1}_{0}\text{n} \rightarrow {}^{1}_{1}\text{p} + {}^{\;\,0}_{-1}e$$

where p represents a proton. Notice that both the total electric charge and the total nucleon number on both sides of the equation are equal.

If this explanation is correct, then during beta decay, a particular element should transform to an element with a $Z$ number that is larger by one. This is exactly what Ernest Rutherford and his colleague Frederick Soddy predicted while studying radioactive decay. Another prediction that follows from this explanation is that the heavier isotopes of a particular element should produce more beta radiation than the lighter isotopes of that element. This is exactly what is observed. For example, carbon-14 is a heavy isotope of carbon that has two more neutrons than protons, and it is unstable, undergoing beta decay (**Figure 28.10**), whereas carbon-12 and carbon-13 are not. However, when scientists studied beta decay in greater detail, they encountered two difficulties.

**Figure 28.10** Beta-minus decay. In this example, an isotope of carbon decays to nitrogen, releasing an electron (beta-minus particle).

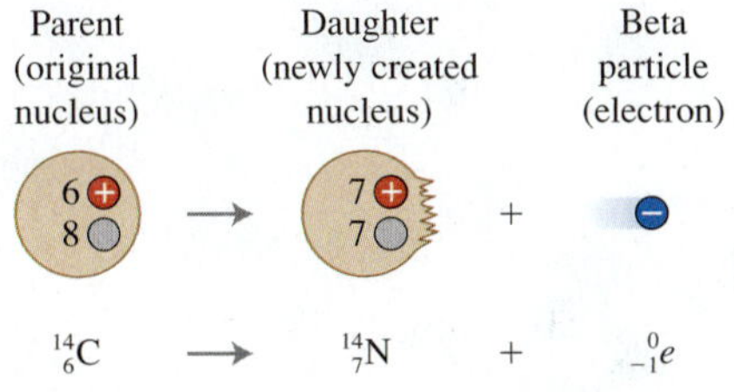


**The problem with spin conservation** You learned in Chapter 27 that electrons have an intrinsic spin quantum number. This spin quantum number was observed in many other experiments to be a conserved quantity. However, in beta decay, it seemed not to be. For example, during the beta decay of a neutron, all three participating particles have spin values equal to $\pm 1/2$. But because there is only one particle on one side of the reaction but two particles on the other side, it is not possible that the total spin on both sides could be equal. Either spin is not a conserved quantity, or our understanding of beta decay is flawed.

**The problem with energy conservation** The other problem involved energy conservation. It, too, seemed to not be conserved in beta decay. The total energy of the products was always observed to be less than the total energy of the reactants. This pattern was so persistent that originally Niels Bohr suggested that maybe energy was not always conserved in the subatomic world. However, this idea was too radical for most physicists to take seriously.

In 1930, Wolfgang Pauli proposed an explanation for beta decay that did not require abandoning energy conservation or spin number conservation. He hypothesized that some unknown particle carried away the missing energy and accounted for the discrepancy in spin number. This particle had zero electric charge, zero mass, and a spin number of either $+1/2$ or $-1/2$. Enrico Fermi called the particle a **neutrino**, meaning "little neutral one." Because the neutrino was thought to have zero mass, it was expected to travel at the speed of light.

Although this sounds like the description of a photon, there is a subtle difference between a photon and a neutrino. An atom can absorb a photon, and as a result the atom changes its energy state. These allowed energy transitions occur only between states that differ by 1 in the angular momentum quantum number $l$. The photon's intrinsic spin must then be equal to the change in angular momentum of the atom, in this case 1. A neutrino's intrinsic spin, however, must be one-half.

The existence of neutrinos was confirmed 25 years after Pauli proposed their existence. Large numbers of neutrinos produced by the Sun and other astronomical objects continually pass through our bodies. Yet they cause no damage and leave almost no trail because of their extremely small likelihood of interacting with the atoms in their path.

The examples of beta decay reactions shown in **Figure 28.11** and represented mathematically below always result in the production of a neutrino $\nu$ or an antineutrino $\bar{\nu}$ (we will discuss neutrinos and antimatter in more detail in Chapter 29). The first reaction is called **beta-minus decay,** as it produces a negative electron:

$$^{14}_{6}\text{C} \rightarrow {}^{14}_{7}\text{N} + {}^{\;\,0}_{-1}e + \bar{\nu}$$

$$^{22}_{11}\text{Na} \rightarrow {}^{22}_{10}\text{Ne} + {}^{0}_{1}e + \nu$$

The second beta decay is an example of a somewhat less common decay known as **beta-plus decay,** which produces a positron ($^{0}_{1}e$), which is otherwise identical to an electron but has positive charge $e$ instead of the negative charge $-e$

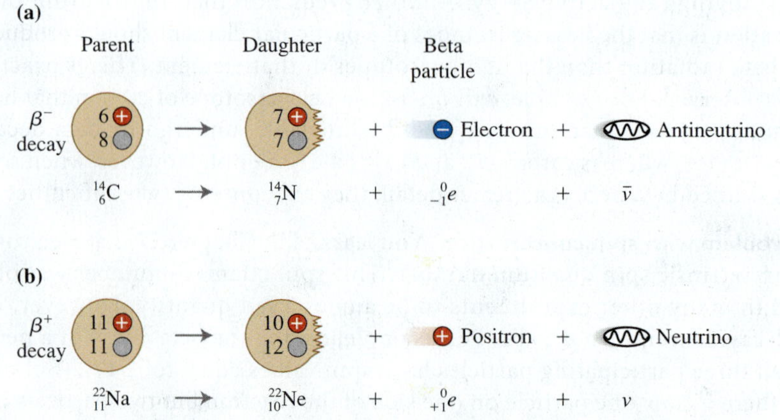


**Figure 28.11** Two types of beta decay. (a) Beta-minus decay, releasing an electron and an antineutrino. (b) Beta-plus decay, releasing a positron and a neutrino.

of the electron. The positron is the antiparticle to the electron (we will discuss positrons more in Chapter 29).

What is the mechanism behind beta decay? Because neutrinos interact with atoms only extremely rarely, it's unlikely that the mechanism is electromagnetic in nature. For this and other technical reasons, Enrico Fermi proposed a new interaction in 1934 called the **weak nuclear interaction,** which explained beta decay. Like the nuclear interaction, the weak interaction is very short range and is much weaker (hence its name).

A free neutron is not a stable particle. It spontaneously decays rather quickly (in about 15 min on average), producing a proton, an electron, and an antineutrino. The weak interaction is responsible for this decay. When neutrons are bound inside nuclei they are very stable.

**Figure 28.12** Gamma decay. (a) After beta decay, the daughter nucleus is left in an excited state and drops to the ground state, emitting a gamma ray photon. (b) An energy level diagram for the gamma ray emission process.

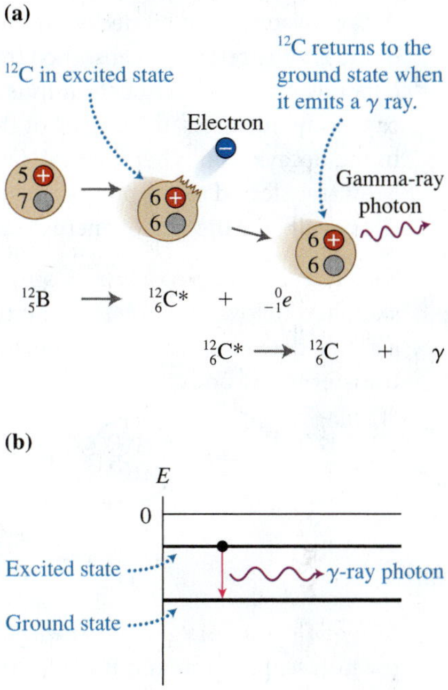


## Gamma decay

We already know that alpha and beta decays are usually accompanied by rays that are not deflected by a magnetic field but that are easily detected by their interactions with atoms. These rays are **gamma ($\gamma$) rays**—photons that have higher energy and therefore shorter wavelength than X-rays.

One explanation for why gamma rays accompany alpha and beta decay is that the energy of the nucleus is quantized in a similar way to the energy in atoms. If this is true, then after alpha or beta decay the nucleus could be left in an excited state from which it then emits one or more photons to return to its ground state. Because the binding energy of a nucleus is much higher than that of the atom's electrons, the energy differences between states are also likely to be much larger. As a result, the energy of the emitted photons is much larger. Gamma decay happens, for example, when boron-12 undergoes beta decay to form carbon-12. The carbon-12 that is produced is in an excited state ($^{12}_{6}C^*$) (the asterisk indicates that the nucleus is in an excited state). When it returns to its lowest energy state, a gamma ray photon is emitted (**Figure 28.12a** and b):

$$^{12}_{5}B \rightarrow ^{12}_{6}C^* + ^{\ 0}_{-1}e$$

$$^{12}_{6}C^* \rightarrow ^{12}_{6}C + \gamma$$

**QUANTITATIVE EXERCISE 28.8 Alpha and beta decay**

The following nuclei undergo different types of radioactive decay. Determine the daughter nucleus for each and write an equation representing each decay reaction. (a) $^{239}_{94}Pu$ alpha decay, (b) $^{144}_{58}Ce$ beta-minus decay, (c) and $^{65}_{30}Zn$ beta-plus decay. The latter produces a positron.

**Represent mathematically**

(a) We have to subtract $A = 4$ and $Z = 2$ (the $^{4}_{2}He$ alpha particle) from the $^{239}_{94}Pu$. Thus, the daughter nucleus has $A = 239 - 4 = 235$ and $Z = 94 - 2 = 92$, which is uranium-235.

(b) We have to subtract $A = 0$ and $Z = -1$ (the $^{\ 0}_{-1}e$ beta minus particle, an electron) from the $^{144}_{58}Ce$. Thus, the daughter nucleus has $A = 144 - 0 = 144$ and $Z = 58 - (-1) = 59$, which is praseodymium-144.

(c) We have to subtract $A = 0$ and $Z = +1$ (the $^{0}_{1}e$ beta plus particle, a positron) from the $^{65}_{30}Zn$. Thus, the daughter nucleus has $A = 65 - 0 = 65$ and $Z = 30 - 1 = 29$, which is copper-65.

**Solve and evaluate** The reactions are then:

(a) $^{239}_{94}Pu \rightarrow ^{235}_{92}U + ^{4}_{2}He + \text{energy}$

(b) $^{144}_{58}Ce \rightarrow ^{144}_{59}Pr + ^{\ 0}_{-1}e + \bar{\nu} + \text{energy}$

(c) $^{65}_{30}Zn \rightarrow ^{65}_{29}Cu + ^{0}_{1}e + \nu + \text{energy}$

"Energy" here means the kinetic energy of the products and possibly one or more gamma rays.

**Try it yourself:** Identify the daughter nucleus and write a reaction equation for the beta-minus decay of $^{131}_{53}I$.

*Answer:* $^{131}_{53}I \rightarrow ^{131}_{54}Xe + ^{\ 0}_{-1}e + \bar{\nu} + \text{energy}$.

**EXAMPLE 28.9 Beta decay in our bodies**

The body normally contains about 7 mg of radioactive potassium-40 $^{40}_{19}\text{K}$. It is present in some of the foods we eat (e.g., bananas). Each second, about $2 \times 10^3$ of these potassium nuclei undergo either beta-plus or beta-minus decay. Assume that all decays are beta-plus and that 40% of the energy released is absorbed by the body (the percent depends on the ratio of beta-minus decay to beta-plus decay, body thickness, the energy of the neutrinos produced in the decays, and other factors). Determine the energy in MeV transferred to the body each second by $^{40}_{19}\text{K}$ beta-plus decay. What is the rate of energy transfer in watts?

**Sketch and translate** A single beta-plus decay is sketched below. If we determine the energy released by a single decay, we can then determine the rate of energy transfer to the body.

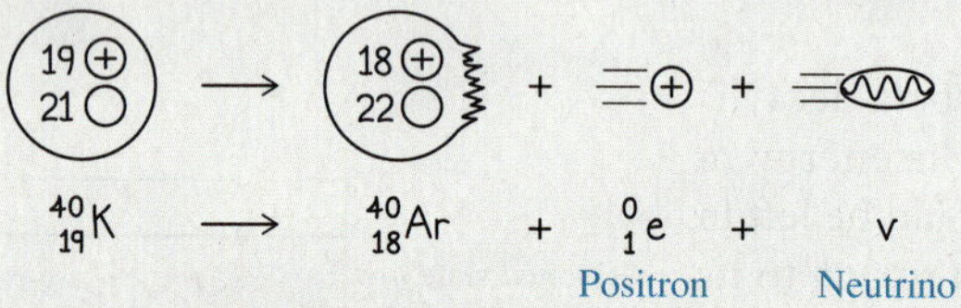


**Simplify and diagram** The system of interest will be the potassium atom and its decay products.

**Represent mathematically** The $Q$ of a single beta plus decay is

$$Q = [m(^{40}_{19}\text{K}) - m(^{40}_{18}\text{Ar}) - m(^{0}_{1}\text{e})]c^2$$

The rate of energy transferred to the body $(\Delta U/\Delta t)$ will be 40% of $Q$ times the number $(\Delta N/\Delta t)$ of decays per second, or

$$\frac{\Delta U}{\Delta t} = \frac{\Delta N}{\Delta t}(0.4\,Q)$$

**Solve and evaluate** From the appendix of selected isotopes we find the masses of the atoms in atomic mass units (u):

$$Q = (39.964000\text{ u} - 39.962384\text{ u} - 5.4858 \times 10^{-4}\text{ u})c^2$$
$$= (0.001068\text{ u})c^2\left(\frac{931.5\text{ MeV}}{\text{u}\cdot c^2}\right)$$
$$= 0.995\text{ MeV}$$

The energy transferred to the body per second is then

$$\frac{\Delta U}{\Delta t} = (2000\text{ decays/s})(0.995\text{ MeV})(0.4) = 800\text{ MeV/s}$$

In watts this is

$$\frac{\Delta U}{\Delta t} = (800\text{ MeV/s})\left(\frac{1.6 \times 10^{-13}\text{ J}}{\text{MeV}}\right)$$
$$= 1.3 \times 10^{-10}\text{ W}$$
$$\approx 1 \times 10^{-10}\text{ W}$$

This is extremely small compared to the average human metabolic rate of 100 W. We'll see in a later section whether these beta particles cause any negative health effects.

**Try it yourself:** Is more or less energy released by potassium-40 beta-minus decay than beta-plus decay?

*Answer:* The product of the beta-minus decay is $^{40}_{20}\text{Ca}$, which has mass 39.962591 u. This is slightly greater than the 39.962383-u $^{40}_{18}\text{Ar}$ product mass of the beta-plus decay. The beta-plus decay releases more energy.

**Review Question 28.6** What was the observational evidence related to beta decay that made scientists think that a new particle was involved?

## 28.7 Half-life, decay rate, and exponential decay

**Active Learning Guide ›**

Radioactive materials are used for medical diagnosis, and they are also part of the world in which we live. Radioactive isotopes can also help determine the age of bones and other archeological artifacts through radioactive dating. For isotopes to serve these purposes, we need a quantitative description of how the number of radioactive nuclei in a sample changes and the rate at which radioactive decay occurs. These quantitative measures include half-life, the decay rate of nuclei, and the application of the exponential function to decay.

## Half-life

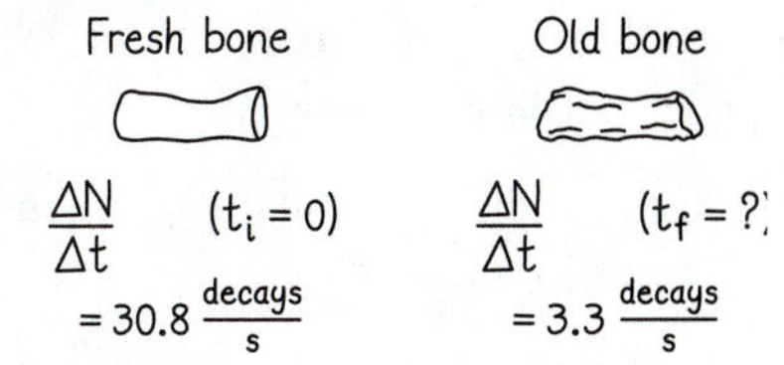


Suppose a radioactive sample has a number $N_0$ of radioactive nuclei at time $t_0 = 0$. We can determine this number by dividing the mass of the sample by the mass of one atom. Using a particle detector such as a Geiger counter, we can measure the number of nuclei that decay in a short time interval. By continually subtracting the number of decays/time interval from the initial number, we can determine the number $N$ of radioactive nuclei that remain in the sample as a function of time. In Observational Experiment **Table 28.7** we describe an experiment to look for a pattern in the variation of $N$ with time $t$.

**OBSERVATIONAL EXPERIMENT TABLE**

**28.7 The number $N$ of radioactive nuclei versus time $t$.**

**Observational experiment**

We measure the decay rate in a radioactive sample and determine the number of radioactive nuclei in the sample as a function of time:

| Number $N$ of radioactive nuclei | Clock reading $t$ (min) |
|---|---|
| $N_0$ | 0 |
| 0.50 $N_0$ | 1 |
| 0.25 $N_0$ | 2 |
| 0.13 $N_0$ | 3 |
| 0.06 $N_0$ | 4 |
| 0.03 $N_0$ | 5 |
| undetectable | 10 |

**Analysis**

- The number decreased from $N_0$ to 0.50 $N_0$ from time 0 to 1 min.
- The number decreased from 0.50$N_0$ to 0.25$N_0$ from time 1 min to 2 min.
- The number decreased from 0.25$N_0$ to 0.13$N_0$ from time 2 min to 3 min.
- The number decreased from 0.13$N_0$ to 0.06$N_0$ from time 3 min to 4 min.
- The number decreased from 0.06$N_0$ to 0.03$N_0$ from time 4 min to 5 min.

**Pattern**

For this sample, the number $N$ of radioactive nuclei in the sample decreased by half during each 1-min time interval.

The data in Table 28.7 are plotted in **Figure 28.13**. As noted in the table, the number $N$ of radioactive nuclei present in this particular radioactive sample decreased by half during each 1-min time interval. The reduction by one-half in a particular time interval, which varies depending on the type of nuclei, is a general property of radioactive samples called the **half-life** of the sample.

**Figure 28.13** A graph of radioactive decay. The number of radioactive nuclei decreases by one-half during each half-life time interval.

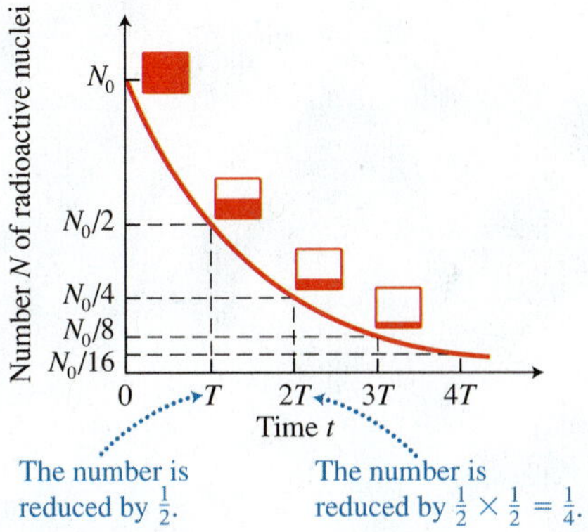


**Half-life** **The half-life $T$ of a particular type of radioactive nucleus is the time interval during which the number of nuclei in a given sample is reduced by one-half. After $n$ half-lives, the fraction of radioactive nuclei that remain is**

$$\frac{N}{N_0} = \frac{1}{2^n} \text{ at } t = nT \tag{28.6}$$

Note that after one half-life, the ratio $N/N_0 = 1/2$. During the next half-life, there is another 50% reduction in the number of nuclei. Thus, after two half-lives $N/N_0 = 1/2^2 = 1/4$, and so forth. For the radioactive sample in Table 28.7, the half-life is 1 min. According to Eq. (28.6), the fraction of radioactive nuclei that remains at $t = 4$ min $= 4T$ is $N/N_0 = 1/2^4 = 1/16 = 0.06$, in agreement with the data in the table.

This method can be used for fractional half-lives. For example, after 4.5 min (4.5 half-lives), the fraction of nuclei that remain is $N/N_0 = 1/2^{4.5} = 1/22.6 = 0.044$. The half-lives for a variety of nuclei are listed in **Table 28.8**.

**Table 28.8 Half-lives and decay constants of some common nuclei.**

| Isotope | Half-life | Decay constant ($s^{-1}$) |
|---|---|---|
| $^{87}_{37}Rb$ | $4.75 \times 10^{10}$ years | $4.62 \times 10^{-19}$ |
| $^{238}_{92}U$ | $4.47 \times 10^{9}$ years | $4.91 \times 10^{-18}$ |
| $^{40}_{19}K$ | $1.28 \times 10^{9}$ years | $1.72 \times 10^{-17}$ |
| $^{239}_{94}Pu$ | $2.41 \times 10^{4}$ years | $9.11 \times 10^{-13}$ |
| $^{14}_{6}C$ | 5730 years | $3.83 \times 10^{-12}$ |
| $^{226}_{88}Ra$ | 1600 years | $1.37 \times 10^{-11}$ |
| $^{137}_{55}Cs$ | 30.0 years | $7.32 \times 10^{-10}$ |
| $^{90}_{38}Sr$ | 28.9 years | $7.60 \times 10^{-10}$ |
| $^{3}_{1}H$ | 12.3 years | $1.79 \times 10^{-9}$ |
| $^{60}_{27}Co$ | 5.27 years | $4.17 \times 10^{-9}$ |
| $^{131}_{53}I$ | 8.03 day | $9.99 \times 10^{-7}$ |
| $^{11}_{6}C$ | 20.4 min | $5.66 \times 10^{-4}$ |

## Determining the source of carbon in plants

The complex photosynthesis process in plant growth can be summarized by the following chemical reaction:

$$6CO_2 + 6H_2O + \text{sunlight} \rightarrow C_6H_{12}O_6 + 6O_2$$

What is the source of the $CO_2$ in this process—is it part of the minerals found in the ground in which the plants grow, such as calcium carbonate ($CaCO_3$), or is it the gaseous $CO_2$ in the air? A radioactive decay experiment was originally used to answer this question.

The carbon isotope carbon-11 ($^{11}_{6}C$) is radioactive with a half-life of 20 min. Carbon-11 was incorporated into the $CO_2$ in the air surrounding growing barley plants in a controlled environment. Investigators found that radioactive carbon-11 became part of the carbohydrate molecules produced by photosynthesis in the barley. This result is evidence that carbon in plants comes from $CO_2$ in the atmosphere. More evidence could be obtained by repeating the experiment with carbon-11 incorporated into soil minerals.

## Decay rate

In working with radioactive samples, we seldom know or measure directly the number of radioactive nuclei in a sample. Instead, we measure the number of nuclei that decay, $\Delta N$, during a certain time interval $\Delta t$. From this measurement we can calculate the number of radioactive nuclei that still remain in the sample, because the number $\Delta N$ of nuclei that decay during time interval $\Delta t$ is proportional to the starting number $N$ of radioactive nuclei in the sample, and is proportional to $\Delta t$:

$$\Delta N = -\lambda N \Delta t \qquad (28.7)$$

This relationship makes sense. The number that decay is greater if the number $N$ in the sample is greater. The number that decay during a certain time interval should also be proportional to that time interval $\Delta t$ as long as the time interval is short compared to the half-life of the radioactive sample. $\lambda$ is a proportionality constant, called the **decay constant,** that depends on the type of radioactive nucleus (see Table 28.8). The greater the value of $\lambda$, the greater the rate at which the radioactive nuclei decay. The ratio of $\Delta N$ and $\Delta t$ is called the **decay rate,** or the **activity,** of a sample of radioactive material.

**Decay rate (activity) $A$** **The ratio of the number of nuclei that decay $\Delta N$ during a certain time interval $\Delta t$ and that time interval is called the decay rate or activity of a sample of that type of radioactive nucleus:**

$$\text{Decay rate (activity) } A = \frac{\Delta N}{\Delta t} = -\lambda N \quad (28.8)$$

**where $\lambda$ is the decay constant for that particular type of nucleus. The minus sign indicates that the number of radioactive nuclei in the sample has decreased by $\lambda N$ during that time interval.**

The unit of decay rate is the becquerel (Bq) and equals 1 decay/s. An older unit of decay rate is the curie (Ci), where $1\text{ Ci} = 3.70 \times 10^{10}\text{ Bq}$. This unusual number for the curie was chosen because it represents roughly the activity of 1 g of pure radium, the radioactive element that Marie Curie isolated from tons of uranium ore. Radioactive tracers used in medicine usually have activities of tens of kilobecquerels or microcuries $\mu$Ci ($1\ \mu\text{Ci} = 10^{-6}\text{ Ci}$). Radiation detection devices such as Geiger counters are used to measure the activity of radioactive samples.

## Exponential function and decay

The $N$-versus-$t$ data in Table 28.7 and in Figure 28.13 are characteristic of a variety of phenomena that can be represented mathematically by what is called the exponential function. We can use Eq. (28.8) to derive an expression for $N$ as a function of time $t$. To start, we turn the $\Delta t$ in Eq. (28.8) into a very short time interval $\Delta t'$ during which the number $N$ changes by $\Delta N'$. We then rearrange the equation to

$$\frac{\Delta N'}{N} = -\lambda \Delta t'$$

These short time changes can be summed up for the time interval starting from time 0 when $N_0$ nuclei are present to some later time $t$ when $N$ nuclei remain. The result of this summation is the exponential function

$$N = N_0 e^{-\lambda t}$$

Because the activity $A$ of a sample is proportional to $N$, the activity is also described by an exponential function:

$$A = \frac{\Delta N}{\Delta t} = -\lambda N = -\lambda N_0 e^{-\lambda t}$$

The number of nuclei present in the sample decreases exponentially with time, as does the activity of the sample. The minus sign simply indicates that the number of nuclei is decreasing with time.

**Exponential decay** **The number $N$ of radioactive nuclei that remain at time $t$ if the number at time zero was $N_0$ is given by an exponential function:**

$$N = N_0 e^{-\lambda t} \quad (28.9)$$

**where $\lambda$ is the decay constant of the nucleus, and $e = 2.718\ldots$ is the base of the natural logarithms. The activity $A$ of the sample also decreases exponentially:**

$$A = \frac{\Delta N}{\Delta t} = -\lambda N_0 e^{-\lambda t}$$

$N$ is an exponentially decreasing function of time and is plotted in the graph shown in Figure 28.13. We can now determine the number $N$ of radioactive nuclei that remain in a sample at time $t$ compared to the number $N_0$ at time zero by using either Eq. (28.6) or Eq. (28.9). The former equation, $(N/N_0) = (1/2^n)$, is often easier to use. However, rearrangement of the exponential function $(N = N_0 e^{-\lambda t})$ leads to an important equation for calculating the unknown age of a sample, which we use when we discuss radioactive dating in the next section.

Equations for radioactive decay have numerous applications. For example, they can be used in medicine to measure the volume of blood in a patient. The process for this measurement is illustrated in the following example.

**EXAMPLE 28.10 Estimating blood volume**

When a small amount of radioactive material is placed in 1.0 $cm^3$ of water, it has an activity of 75,000 decays/min. Imagine that an identical amount of radioactive material is injected into an individual's bloodstream. Later, after the material has spread throughout the blood, a 1.0-$cm^3$ sample of blood taken from the individual has an activity of 16 decays/min. Determine the individual's total blood volume.

**Sketch and translate** The total blood volume can be thought of as a large container of unknown volume and the 1.0-$cm^3$ sample as a small portion of that. We know that a direct relationship exists between the decays/min in the sample and the decays/min in the total volume of blood. Therefore, we can use the value of one to determine the value of the other.

**Simplify and diagram** Assume that during this process the person does not drink any water (so that their blood volume doesn't change), that all the radioactive material is absorbed by the blood, and that none is filtered out by the kidneys. Assume also that the material disperses evenly throughout the blood. Lastly, assume that the half-life of that type of material is much longer than the time interval needed to complete the experiment so that the activity of the radioactive material can be considered constant.

**Represent mathematically** We use a ratio method to determine the blood volume by comparing the activity $A$ and volume $V$ of the two amounts of blood (the whole body and the sample):

$$\frac{V_{\text{blood}}}{V_{\text{sample}}} = \frac{A_{\text{blood}}}{A_{\text{sample}}}$$

**Solve and evaluate** Solve for $V_{\text{blood}}$ and substitute the known quantities into the above to determine the blood volume:

$$V_{\text{blood}} = \frac{A_{\text{blood}}}{A_{\text{sample}}} V_{\text{sample}} = \left(\frac{75{,}000\ \text{decays/min}}{16\ \text{decays/min}}\right)(1.0\ \text{cm}^3)$$
$$= 4.7 \times 10^3\ \text{cm}^3$$

where $(4.7 \times 10^3\ \text{cm}^3)(1\ \text{L}/10^3\text{cm}^3) = 4.7\ \text{L}$. This value is within the 4- to 6-L range that is considered a normal human blood volume.

**Try it yourself:** Suppose the radioactive sample had a half-life of 1.0 h and you waited 2 h for the injected sample to distribute uniformly before measuring the activity of 1 $cm^3$ of blood. What would the activity of that blood sample be?

*Answer:* In two half-lives, the activity would be reduced to $(1/2^2)$ (16 decays/min) = 4 decays/min.

## Decay rate and half-life

The values of the half-life and decay constant for several different radioactive nuclei were reported in Table 28.8. Because these two quantities are both related to how quickly a sample of radioactive material decays, we expect that a relationship exists between them. What is that relationship?

At time $t = T$ (one half-life), the number $N$ of radioactive nuclei remaining is one-half the number $N_0$ at time zero. Using the equation $N = N_0 e^{-\lambda t}$, we can write

$$\frac{1}{2} N_0 = N_0 e^{-\lambda T}$$

or

$$\frac{1}{2} = e^{-\lambda T}$$

Take the natural logarithm of both sides of this equation and solve for the half-life $T$:

$$\ln\left(\frac{1}{2}\right) = -\lambda T \Rightarrow T = -\frac{1}{\lambda}\ln\left(\frac{1}{2}\right)$$

or

$$T = \frac{0.693}{\lambda} \quad (28.10)$$

This relationship between the half-life of the material and its decay constant makes sense. If the decay constant $\lambda$ is large, then the material decays rapidly and consequently has a short half-life $T$.

**Review Question 28.7** What does it mean that the half-life of a particular isotope is 200 years?

## 28.8 Radioactive dating

In the previous section, you saw how to calculate the fraction $N/N_0$ of radioactive nuclei that remain in a sample after a known time interval. Archeologists and geologists, in contrast, are interested in the reverse calculation—determining the age (that is, time interval that has passed) of a radioactive sample from the known fraction $N/N_0$ of radioactive nuclei that remain in a sample. Their method is based on a rearrangement of Eq. (28.9), $N = N_0 e^{-\lambda t}$.

First, divide both sides of the equation by $N_0$ and then take the natural logarithm of each side to get

$$\ln\left(\frac{N}{N_0}\right) = -\lambda t$$

Now substitute $\lambda = 0.693/T$ [Eq.( 28.10)] and rearrange to get an expression for the age of a radioactive sample:

$$t = -\frac{\ln(N/N_0)}{0.693}T \quad (28.11)$$

In this equation $T$ is the half-life of the radioactive material and $t$ is the sample's age when $N$ radioactive nuclei remain. The sample started at time zero with $N_0$ radioactive nuclei.

**TIP** Notice that in Eq. (28.11) you can use whatever units you want for $t$ and $T$ (seconds, years, thousands of years) provided you are consistent.

### Carbon dating

An important type of radioactive dating, called carbon dating, is used to determine the age of objects that are less than 40,000 years old. Most of the carbon in our environment consists of the isotope carbon-12, but about 1 in $10^{12}$ carbon atoms has a radioactive carbon-14 nucleus with a half-life of about 5700 years. Carbon-14 is continuously produced in our atmosphere by collisions of neutrons from the solar wind with nitrogen nuclei in the atmosphere, resulting in the reaction

$$^{14}_{7}\text{N} + ^{1}_{0}\text{n} \rightarrow ^{14}_{6}\text{C} + ^{1}_{1}\text{H}$$

Atmospheric carbon-14 radioactively decays at the same rate as it is created by the above reaction: a stable equilibrium exists.

Any plant or animal that metabolizes carbon incorporates about one carbon-14 atom into its structure for every $10^{12}$ carbon-12 atoms it metabolizes. Because carbon is no longer absorbed and metabolized by the organism after death, the carbon-14 in the bones starts to transform by negative beta decay into nitrogen-14. After 5700 years the carbon-14 concentration decreases by one-half. Two half-lives, or 11,400 years after death, the concentration has dwindled to one-fourth what it was when the organism died. A measurement of the current carbon-14 concentration indicates the age of the remains.

**EXAMPLE 28.11 Age of bone**

A bone found by an archeologist contains a small amount of radioactive carbon-14. The radioactive emissions from the bone produce a measured decay rate of 3.3 decays/s. The same mass of fresh cow bone produces 30.8 decays/s. Estimate the age of the sample.

**Sketch and translate** The situation is sketched below.

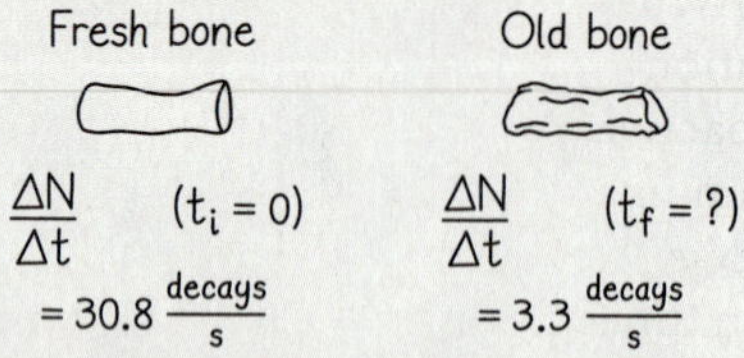


**Simplify and diagram** We assume that the concentration of radioactive carbon-14 in the atmosphere is the same today as it was at the time of death of the animal being studied.

**Represent mathematically** The decay rate and number of radioactive nuclei are proportional to each other ($\Delta N/\Delta t = -\lambda N$). Therefore, we can determine the ratio of carbon-14 in the old bone and what it was when the animal died.

$$\frac{(\Delta N/\Delta t)_{\text{now}}}{(\Delta N/\Delta t)_{\text{at death}}} = \frac{-\lambda N_{\text{now}}}{-\lambda N_{\text{at death}}} = \frac{N_{\text{now}}}{N_{\text{at death}}} = \frac{N}{N_0}$$

From here we can use Eq. (28.11) to determine the age of the bone.

$$t = -\frac{\ln(N/N_0)}{0.693}T \qquad (28.11)$$

**Solve and evaluate** The fraction of carbon-14 atoms remaining in the old bone is

$$\frac{N}{N_0} = \frac{3.3}{30.8} = 0.107$$

Substituting this value into Eq. (28.11), we find the age of the bone:

$$t = -\frac{\ln(0.107)}{0.693}(3 \text{ years}) = 18{,}500 \text{ years}$$

The unit is correct and the order of magnitude is reasonable.

**Try it yourself:** The ratio of carbon-14 in an old bone compared to the number in a fresh bone of the same mass is 0.050. Determine the age of the bone.

*Answer:* 25,000 years.

## Radioactive decay series

Many of the nuclei produced in radioactive decays are themselves radioactive. In fact, the decay of the original parent nucleus may start a chain of successive decays called a **decay series**. An example of a decay series is illustrated in **Figure 28.14**. First, the alpha decay of $^{238}_{92}\text{U}$ (top right) leads to the formation of $^{234}_{90}\text{Th}$. Thorium-234 then undergoes beta decay to form $^{234}_{91}\text{Pa}$. The series continues until $^{218}_{84}\text{Po}$ is formed. The series branches at this point, as polonium-218 can decay by either alpha or beta emission. Branching occurs at several other points as well. Eventually, the stable lead isotope $^{206}_{82}\text{Pb}$ is formed (lower left) and the series ends.

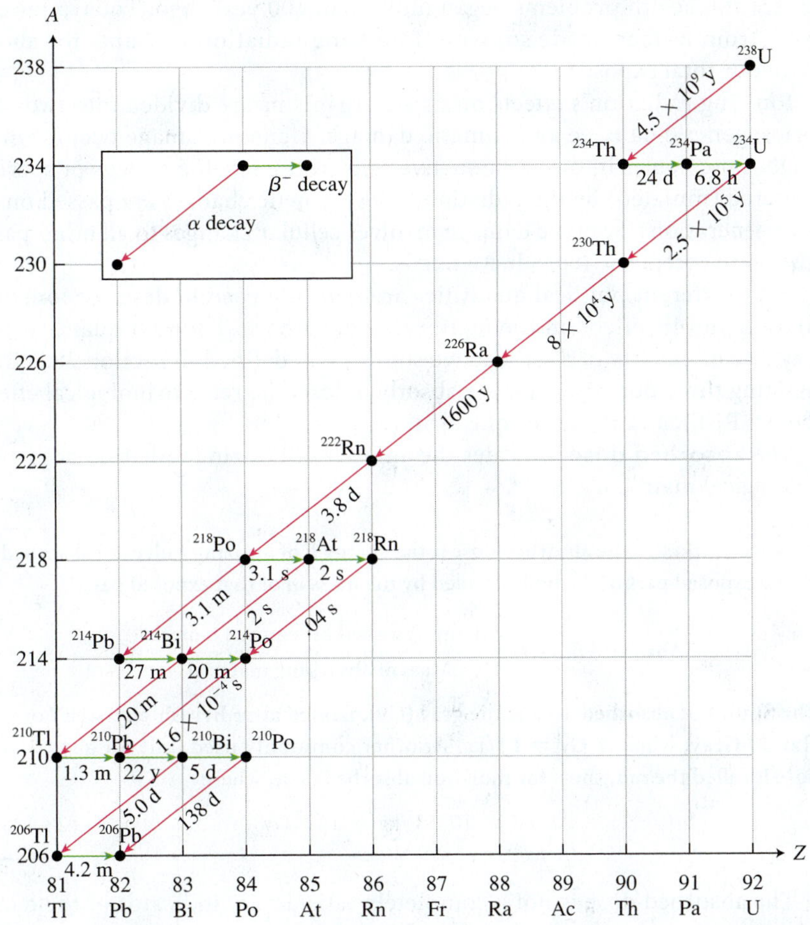


**Figure 28.14** A radioactive decay series. The sequence begins in the upper right with uranium-238.

Radioactive series replenish our environment with nuclei that would normally have disappeared long ago. For example, radium ($^{226}_{88}$Ra) has a half-life of 1600 years. During the approximately $5 \times 10^9$ years of our solar system's existence, the original abundance of radium-226 would have long since been depleted by radioactive decay. However, the supply is continually replenished as uranium-238, with an extremely long $4.5 \times 10^9$-year half-life, decays and leads to the production of radium-226 via a series of reactions.

**Review Question 28.8** How does the presence of radioactive carbon-14 help us determine the age of the bone of an animal that died thousands of years ago?

## 28.9 Ionizing radiation and its measurement

Ionizing radiation consists of photons and moving particles whose energy exceeds the 1 to 15 eV needed to ionize atoms and molecules. Ionizing radiation exists in many forms, such as ultraviolet, X-ray, and gamma ray photons, alpha and beta particles emitted during the radioactive decay of nuclei, and cosmic rays (high-energy particles) that reach Earth from space.

Life has evolved on Earth in the presence of a steady background of radiation, most of it coming from emissions of radioactive nuclei in Earth's crust and from cosmic rays and their collision products passing down through Earth's atmosphere. Using ionizing radiation for the purpose of diagnosing and treating health problems began more than 100 years ago. Today, our exposure from human-made sources of ionizing radiation accounts for about 40% of our total exposure.

Ionizing radiation's effects on living organisms are divided into two categories: genetic damage and somatic damage. Genetic damage occurs when the DNA molecules in the reproductive cells (those leading to eggs or sperm) are altered (mutated) by the radiation. These genetic changes are passed on to future generations. Somatic damage involves cellular changes to all other parts of the body except the reproductive cells.

Four different physical quantities are typically used to describe ionizing radiation and its effect on the matter that absorbs it. The first quantity, the decay rate or activity of a radioactive source, was defined in Section 28.7. The remaining three quantities are the absorbed dose, the relative biological effectiveness (RBE), and the dose equivalent.

The **absorbed dose** indicates the amount of ionizing radiation absorbed by biological material.

**Absorbed dose** **The absorbed dose is the energy $E$ of ionizing radiation absorbed by the exposed part of the body divided by the mass $m$ of that exposed part:**

$$\text{Absorbed dose} = \frac{\text{Energy absorbed by exposed material}}{\text{Mass of absorbing material}} \quad (28.12)$$

**The SI unit of absorbed dose is the gray (Gy), named after British physicist Louis Harold Gray, where 1 Gy = 1 J/kg. Another commonly used unit for absorbed dose is called the rad, short for radiation absorbed dose, where**

$$1 \text{ rad} = 10^{-2}\,\text{J/kg} = 10^{-2}\,\text{Gy} \quad (28.13)$$

The absorbed dose is not a completely satisfactory indicator of the damage we might expect when a living organism absorbs ionizing radiation. The reason is that when two different forms of ionizing radiation deposit the same amount of energy into organic material, the damage they cause is generally different. For example, a sample that absorbs 1 rad of alpha particles experiences about 20 times more damage than the same sample does when it absorbs 1 rad of X-rays. Alpha particles move through matter more slowly and slow down more gradually than other forms of radiation, and as a result, they interact with a larger number of atoms. X-rays and gamma rays, by contrast, deposit a large fraction of their energy within a small number of atoms.

Because of the variation in damage caused by different types of radiation, it is useful to define a quantity called **relative biological effectiveness (RBE)**. This measure indicates approximately the relative damage caused by different types of ionizing radiation. The RBE of several different types of radiation is listed in **Table 28.9**.

**Table 28.9 The approximate relative biological effectiveness (RBE) of different types of radiation.**

| Type | RBE |
|---|---|
| X-rays and gamma rays | 1 |
| Electrons | 1–2 |
| Neutrons | 4–20 |
| Protons | 10 |
| Alpha particles | 20 |
| Heavy ions | 2–5 |

**Relative biological effectiveness (RBE)** **Relative biological effectiveness is a dimensionless number that indicates the relative damage caused by a particular type of radiation compared with $2 \times 10^5$ eV X-rays. The approximate RBE for a particular type of radiation can be found in tables.**

The last quantity is the **dose equivalent,** or simply **dose**—an indicator of the net biological effect of ionizing radiation.

**Dose or dose equivalent** **This quantity is a measure of the net biological effect (damage) of a particular exposure to ionizing radiation. The dose is the product of the absorbed dose and the relative biological effectiveness:**

$$\text{Dose or dose equivalent} = (\text{Absorbed dose})(\text{RBE}) = \left(\frac{\text{Energy absorbed}}{\text{Mass of absorbing material}}\right)(\text{RBE}) \quad (28.14)$$

**The SI unit of dose is the sievert (Sv), named after Swedish medical physicist Rolf Sievert, where 1 Sv = (1 Gy)(RBE). Another commonly used unit is the rem (roentgen equivalent in man or in mammal) where 1 rem = (1 rad)(RBE).**

**QUANTITATIVE EXERCISE 28.12 Ions produced by a chest X-ray**

In a typical chest X-ray about 10 mrem ($10 \times 10^{-3}$ rem) of radiation is absorbed by about 5 kg of body tissue. The X-ray photons used in such exams each have energy of about 50,000 eV. Determine about how many ions are produced by this X-ray exam.

**Represent mathematically** We can estimate first the number of X-ray photons absorbed and then try to use that number to estimate the number of ions formed. We can accomplish this using Eq. (28.14):

$$\text{Dose or dose equivalent} = (\text{Absorbed dose})(\text{RBE}) = \left(\frac{\text{Energy absorbed}}{\text{Mass of absorbing material}}\right)(\text{RBE})$$

**Solve and evaluate** The RBE of X-rays is 1. Thus, the dose of X-rays in rem equals the absorbed dose in rad. For an absorbed dose of 10 mrad, we can insert the known values and solve the equation to determine the energy absorbed per kilogram of exposed tissue:

$$10\ \text{mrad} = (10 \times 10^{-3}\ \text{rad})\left(\frac{10^{-2}\ \text{J/kg}}{1\ \text{rad}}\right) = 10^{-4}\ \text{J/kg}$$

Because 5 kg of tissue receives this dose, the total absorbed energy is

$$(5\ \text{kg})(10^{-4}\ \text{J/kg}) = 5 \times 10^{-4}\ \text{J}$$

We can determine the number of X-ray photons by dividing this result by the energy of a single photon:

$$\frac{5 \times 10^{-4}\ \text{J}}{\left(50{,}000\ \dfrac{\text{eV}}{\text{photon}}\right)\left(\dfrac{1.6 \times 10^{-19}\ \text{J}}{\text{eV}}\right)} = 6 \times 10^{10}\ \text{photons}$$

or 60 billion photons. Because the energy of each photon is so large compared to the typical ionization energy of atoms and molecules, it's reasonable to suggest that each photon will produce approximately 100 ions. Therefore, the number of ions is approximately

$$6 \times 10^{10}\ \text{photons} \times 100\ \text{ions/photon} = 6 \times 10^{12}\ \text{ions}$$

**Try it yourself:** What number of ions may be produced by a bilateral mammogram, assuming that each breast has a mass of 0.5 kg, the absorbed dose for both breasts is 0.2 rad, and 50,000-eV X-ray photons are used?

*Answer:* $\approx 3 \times 10^{11}$ photons or $3 \times 10^{13}$ if each photon causes 100 ions.

## Sources of ionizing radiation

The U.S. Environmental Protection Agency estimated that the average dose of ionizing radiation received by a person in the United States or in Canada is about 300–400 mrem/year. The sources of this radiation can be divided into two broad categories: natural and human-made.

**Natural sources** The three major natural sources of radiation are:

1. radioactive elements in the Earth's crust, such as uranium-238, potassium-40, and radon-226. Small amounts of radioactive radon, an inert gaseous atom, diffuse out of the soil into buildings, exposing the occupants;

2. foods that contain radioactive isotopes; and
3. cosmic rays. A cosmic ray is an elementary particle moving at almost the speed of light. The original source of cosmic rays is primarily supernova explosions of stars in our galaxy.

**Human-made sources** Human-made radiation comes predominantly from medical applications such as X-rays used in diagnostic procedures, the medical use of radioactive nuclei as tracers, and gamma rays used in cancer treatment. Together, these sources account for a dose of about 60 mrem/year (see **Table 28.10**). Smoking and exposure to radon significantly increase the dose/year.

**Review Question 28.9** Why are X-rays, gamma rays, and fast-moving electrically charged particles potentially harmful to our bodies?

**Table 28.10 Human-made ionizing radiation sources and doses.**

| Exposure mechanism | Typical dose |
|---|---|
| Having a smoke detector | 0.008 mrem/year[a] |
| Living near nuclear plant | 0.009 mrem/year[b] |
| Living near coal-fired plant | 0.03 mrem/year[a] |
| 5-hour airplane flight | 2.5 mrem[a] |
| Natural gas stoves and heaters | 6–9 mrem/year[c] |
| Dental X-ray | 0.5 mrem/X-ray[a] |
| Chest X-ray | 10 mrem/X-ray[a] |
| Smoking (one half-pack per day) | 18 mrem/year[a] |
| Radon | 200 mrem/year[b] |

Sources:
[a]American Nuclear Society. [b]PBS Frontline. [c]Lawrence Berkeley Lab.

# Summary

| Words | Pictorial and physical representations | Mathematical representation |
|---|---|---|
| **Terminology** Atoms are described with the notation ${}^A_Z\text{X}$, where X is the letter abbreviation for that element, the atomic number $Z$ is the number of protons in the nucleus, $N$ is the number of neutrons in the nucleus, and the mass number $A = Z + N$ is the number of nucleons (protons plus neutrons) in the nucleus. (Section 28.2) | 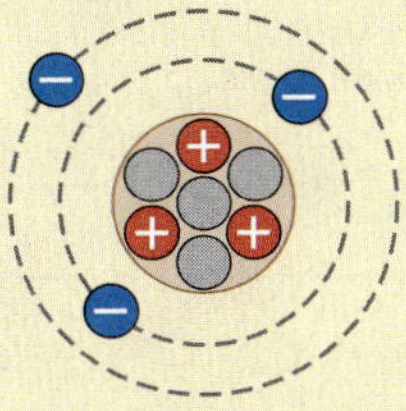 ${}^7_3\text{Li}$ | |
| **Isotopes** Atoms with a common number of protons but differing numbers of neutrons are called isotopes. (Section 28.2) | | |
| **Mass defect** is the difference between the total mass of the proton and neutron components of a nucleus and its mass when the components are bound together. | $\Delta m_{\text{Li}} = [3m_{\text{H}} + 4m_{\text{n}}] - m_{\text{Li}}$ | mass defect $= \Delta m = [Z \cdot m_{\text{hydrogen atom}} + (A - Z)m_{\text{neutron}}] - m_{\text{atom}}$ Eq. (28.1) |
| **Binding energy** of a nucleus is the rest energy equivalent of its mass defect. The binding energy represents the total energy needed to separate an atom and its nucleus into hydrogen atoms and neutrons. | 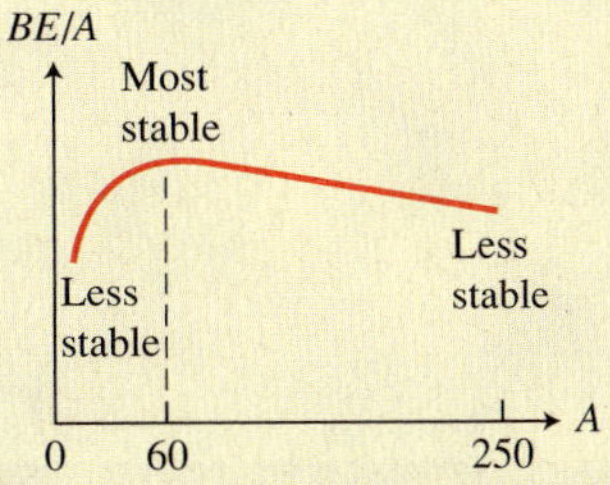 | $BE = \Delta m \cdot c^2$ Eq. (28.2) |
| **Binding energy per nucleon** is the binding energy of the nucleus divided by its number of nucleons. It is a reasonable measure of the stability of a nucleus. (Section 28.3) | | Binding energy per nucleon $= BE/A$ Eq. (28.4) |
| **Types of radioactive decay** **Alpha decay** A heavy nucleus emits an alpha particle ($\alpha$), the nucleus of a helium atom ${}^4_2\text{He}$. | 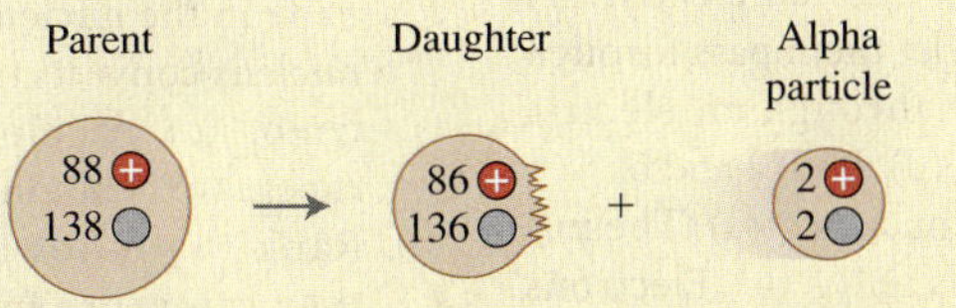 ${}^{226}_{88}\text{Ra} \rightarrow {}^{222}_{86}\text{Rn} + {}^4_2\text{He} + 4.9\text{ MeV}$ (kinetic energy of products) | $\alpha$ decay: ${}^A_Z\text{X} \rightarrow {}^{A-4}_{Z-2}\text{X} + {}^4_2\text{He} + \text{energy}$ |
| **Beta decay** The nucleus ejects a beta particle ($\beta^-$ electron or $\beta^+$ positron) and an antineutrino $\bar{\nu}$ or neutrino $\nu$. |  ${}^{14}_6\text{C} \rightarrow {}^{14}_7\text{N} + {}^0_{-1}e + \bar{\nu}$ 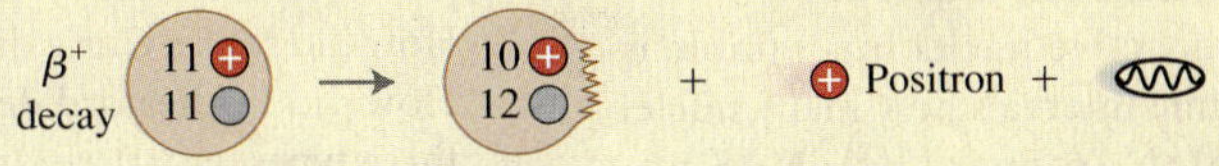 ${}^{22}_{11}\text{Na} \rightarrow {}^{22}_{10}\text{Ne} + {}^0_{+1}e + \nu$ | $\beta^-$ decay: ${}^A_Z\text{X} \rightarrow {}_{Z+1}^{A}\text{X} + {}^0_{-1}e + \bar{\nu} + \text{energy}$ $\beta^+$ decay: ${}^A_Z\text{X} \rightarrow {}_{Z-1}^{A}\text{X} + {}^0_1e + \nu + \text{energy}$ |

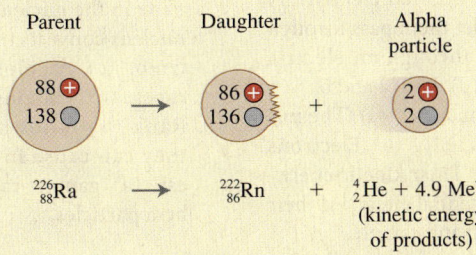

**Gamma decay** A nucleus in an excited state ($^{A}_{Z}X^{*}$) emits a high-energy photon called a gamma ray. (Sections 28.1 and 28.6)

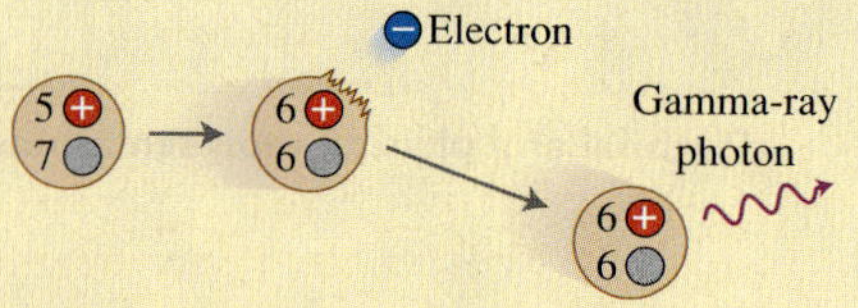


$^{12}_{5}B \rightarrow {}^{12}_{6}C^{*} + {}^{0}_{-1}e$

$^{12}_{6}C^{*} \rightarrow {}^{12}_{6}C + \gamma$

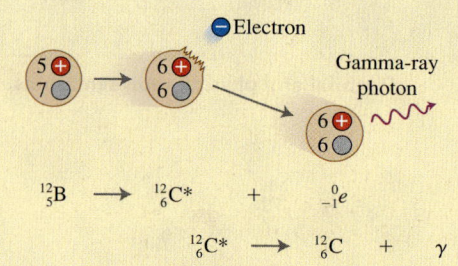

$\gamma$ decay : $^{A}_{Z}X^{*} \rightarrow {}^{A}_{Z}X + \gamma$

**Half-life** The half-life $T$ of a particular type of radioactive nucleus is the time interval during which the number of these nuclei in a given sample is reduced by one-half. (Section 28.7)

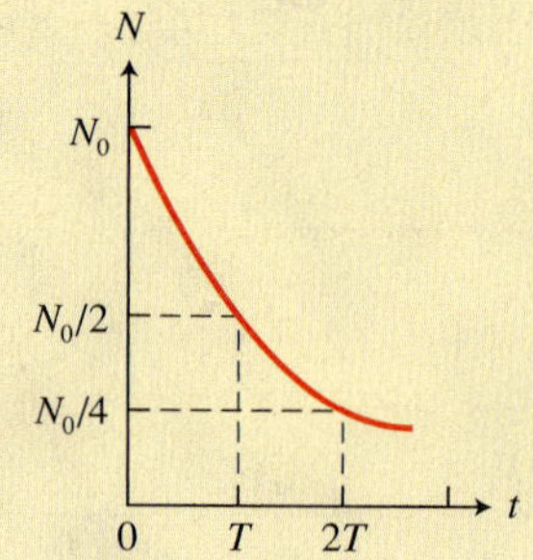


$N/N_0 = 1/2^n$ at $t = nT$, where $N_0$ is the number at time $t = 0$ and $N$ is the number at time $t$.

For instructor-assigned homework, go to **MasteringPhysics.**

# Questions

## Multiple Choice Questions

1. What were Becquerel rays? (a) Positively charged particles (b) Negatively charged particles (c) High-energy photons (d) All of the above
2. To decide whether Becquerel rays contained electrically charged particles, scientists could (a) let them pass through a magnetic field. (b) let them pass through an electric field. (c) study the existing literature. (d) both a and b
3. Why are there are no electrons in the nucleus? (a) The nucleus consists of protons and neutrons only. (b) Electrons do not exert a strong nuclear force. (c) Their kinetic energy is greater than the magnitude of the potential energy of their interaction with the nucleus. (d) Both b and c apply.
4. What is radioactive decay? (a) A physical quantity (b) An observable physical phenomenon (c) A law of physics (d) A unit
5. Elements with higher decay constants decay (a) slower. (b) faster. (c) The decay constant does not relate to how fast the substance decays.
6. One can speed up the rate of natural radioactive decay of a substance by (a) increasing its temperature. (b) illuminating it with high energy photons. (c) turning it into a powder. (d) No physical or chemical changes affect the rate of decay.
7. If the original number of radioactive nuclei in a sample is $N_0$, then after three half-life time intervals how many nuclei are left? (a) $N_0/3$ nuclei (b) $N_0/6$ nuclei (c) $N_0/8$ nuclei (d) $N_0/9$ nuclei
8. During alpha decay, (a) a neutron decays into an electron, proton, and antineutrino. (b) a photon is emitted by the nucleus changing its energy state. (c) a helium nucleus is emitted from the nucleus.
9. What happens during beta decay? (a) Electrons originally in the nucleus escape. (b) A neutron originally in the nucleus converts into a proton, an electron, and an antineutrino. (c) An alpha particle originally in the nucleus escapes. (d) Photons are emitted by the excited nucleus.
10. Rank the following types of radiation for the damage they can cause in the human body from highest to lowest: (a) gamma rays; (b) alpha particles; (c) electrons or beta particles.

## Conceptual Questions

11. How did scientists figure out that radioactive materials emit radiation from the nucleus rather than from the electronic structure of atoms and molecules?
12. Describe the experiments that helped scientists understand that uranium salts produce fast-moving charged particles or high-energy photons.
13. Why did Becquerel have burns on his chest near the vest pocket in which he carried a test tube containing radium?
14. How did Marie Curie discover radium?
15. How did Rutherford determine that radioactivity consists of three types of particles/photons?
16. How did scientists devise the concept of half-life?

17. Estimate the density of an atomic nucleus.
18. Estimate the radius of a sphere that would hold all of your mass if its density equaled the density of a nucleus.
19. Determine the chemical elements represented by the symbol X in each of the following: (a) ${}^{23}_{11}X$; (b) ${}^{56}_{26}X$; (c) ${}^{64}_{30}X$; and (d) ${}^{107}_{47}X$.
20. Explain why heavier isotopes of light nuclei are unstable.
21. The half-life of radium-226 is 1600 years. Earth is about 4 billion years old, so essentially all of the radium-226 present originally should have decayed. How can you explain the fact that radium-226 still remains in our environment in moderate abundance?
22. Why is a high temperature usually needed for fusion to occur?
23. Design an experiment in which radioactive nuclei are used to test the ability of different detergents to remove dirt from clothes.

# Problems

Below, **BIO** indicates a problem with a biological or medical focus. Problems labeled **EST** ask you to estimate the answer to a quantitative problem rather than develop a specific answer. Asterisks indicate the level of difficulty of the problem. Problems with no * are considered to be the least difficult. A single * marks moderately difficult problems. Two ** indicate more difficult problems.

### 28.1 and 28.2 Radioactivity and an early nuclear model and A new particle and a new nuclear model

1. **EST** Estimate the magnitude of the repulsive electrical force between two protons in a helium nucleus. How are they held together in the nucleus?
2. * **EST** Estimate the magnitude of the repulsive electrical force that the rest of a gold nucleus (${}^{197}_{79}Au$) exerts on a proton near the edge of the nucleus. State any assumptions used in your estimation. This is an order-of-magnitude estimate, so do not become bogged down in details.
3. Determine the number of protons, neutrons, and nucleons in the following nuclei; ${}^{9}_{4}Be$, ${}^{16}_{8}O$, ${}^{27}_{13}Al$, ${}^{56}_{26}Fe$, ${}^{64}Zn$, and ${}^{107}Ag$.
4. * **EST** Suppose the radius of a copper nucleus is $10^{-14}$ m and that the radius of the copper atom is about 0.1 nm. Estimate the atom's size if the nucleus was the size of a tennis ball.
5. ** **BIO EST Electrons and nucleons in body** Estimate the total number of (a) nucleons and (b) electrons in your body. (c) Indicate roughly the volume in cubic centimeters occupied by these nucleons.

### 28.3 Nuclear force and binding energy

6. * Measurements with a mass spectrometer indicate the following particle masses: ${}^{7}_{3}Li$ (7.016003 u), ${}^{1}_{1}H$ (1.007825 u), and ${}^{1}_{0}n$ (1.008665 u). Compare the mass of the lithium atom to the mass of the particles of which it is made. What do you conclude? Note: $1\text{ u} = 1.660539 \times 10^{-27}$ kg.
7. * A 5.5-g marble initially at rest is dropped in the Earth's gravitational field. How far must it fall before its decrease in gravitational potential energy is 938 MeV, the same as the rest mass energy of a proton?
8. Determine the rest mass energies of an electron, a proton, and a neutron in units of mega-electron volts.
9. **EST** Use Figure 28.6 to estimate the total binding energy of ${}^{197}_{79}Au$.
10. * Determine the total binding energy and the binding energy per nucleon of ${}^{12}_{6}C$.
11. * Determine the binding energies per nucleon for ${}^{238}_{92}U$ and ${}^{120}_{50}Sn$. Based on these numbers, which nucleus is more stable? Explain.
12. * Determine the energy that is needed to remove a neutron from ${}^{7}_{3}Li$ to produce ${}^{6}_{3}Li$ plus a free neutron. [Hint: Compare the mass of ${}^{7}_{3}Li$ and that of ${}^{6}_{3}Li + {}^{1}_{0}n$].

### 28.4 and 28.5 Nuclear reactions and Nuclear sources of energy

13. Insert the missing symbol in the following reactions.
    (a) ${}^{4}_{2}He + {}^{12}_{6}C \rightarrow {}^{15}_{7}N + ?$
    (b) ${}^{2}_{1}H + {}^{3}_{1}H \rightarrow {}^{4}_{2}He + ?$
    (c) ${}^{1}_{0}n + {}^{235}_{92}U \rightarrow {}^{140}_{54}Xe + ? + 2{}^{1}_{0}n$
    (d) ${}^{3}_{1}H \rightarrow ? + {}_{-1}^{0}e$
14. Explain why the following reactions violate one or more of the rules for nuclear reactions.
    (a) ${}^{4}_{2}He + {}^{27}_{13}Al \rightarrow {}^{32}_{15}P + {}^{1}_{0}n$
    (b) ${}^{2}_{1}H + {}^{3}_{1}H \rightarrow {}^{4}_{2}He + {}^{1}_{1}H$
    (c) ${}^{1}_{0}n + {}^{238}_{94}Pu \rightarrow {}^{140}_{54}Xe + {}^{96}_{40}Zr + 2{}^{1}_{0}n$
    (d) ${}^{14}_{6}C \rightarrow {}^{14}_{7}N + {}^{0}_{1}e$
15. * Explain why the following reaction does not occur spontaneously: ${}^{4}_{2}He \rightarrow {}^{3}_{1}H + {}^{1}_{1}H$.
16. * The following reaction occurs in the Sun: ${}^{3}_{2}He + {}^{4}_{2}He \rightarrow {}^{7}_{4}Be$. How much energy is released?
17. * **Nuclear reaction in the Sun** Oxygen is produced in stars by the following reaction: ${}^{12}_{6}C + {}^{4}_{2}He \rightarrow {}^{16}_{8}O$. How much energy is absorbed or released by the reaction in units of mega-electron volts?
18. * **Another reaction in the Sun** One part of the carbon-nitrogen cycle that provides energy for the Sun is the reaction ${}^{12}_{6}C + {}^{1}_{1}H \rightarrow {}^{13}_{7}N + 1.943$ MeV. Using the known masses of ${}^{12}C$ and ${}^{1}H$ and the results of this reaction, determine the mass of ${}^{13}N$.
19. * Determine the missing nucleus in the following reaction and calculate how much energy is released: ${}^{232}_{92}U \rightarrow ? + {}^{4}_{2}He + \text{energy}$.
20. * Determine the missing nucleus in the following reaction and calculate its mass: $? \rightarrow {}^{211}_{83}Bi + {}^{4}_{2}He + 8.20$ MeV.
21. * Determine (a) the number of protons and neutrons in the missing fragment of the reaction shown below and (b) the mass of that fragment:
    $${}^{1}_{0}n + {}^{235}_{92}U \rightarrow ? + {}^{136}_{54}Xe + 12{}^{1}_{0}n + 126.5\text{ MeV}.$$
22. * **More energy for the Sun** A series of reactions in the Sun leads to the fusion of three helium nuclei (${}^{4}_{2}He$) to form one carbon nucleus (${}^{12}_{6}C$). (a) Determine the net energy released by the reactions. (b) What fraction of the total mass of the three helium nuclei is converted to energy?

23. * **Another Sun process** A series of reactions that provides energy for the Sun and stars is summarized by the following equation: $6\ {}^{2}_{1}H \rightarrow 2\ {}^{1}_{1}H + 2\ {}^{1}_{0}n + 2\ {}^{4}_{2}He$. (a) Determine the net energy released by the reaction. (b) Convert this answer to units of joules per kilogram of deuterium (${}^{2}_{1}H$).

24. * **Equation Jeopardy** Equations for determining the mass defect for two nuclear reactions are shown below. Represent each reaction in symbolic form (as in the previous problems). Decide whether each reaction results in energy release or energy absorption.
(a) $235.0439\ u + 1.0087\ u \rightarrow 95.9343\ u + 137.9110\ u + 2(1.0087\ u) + \text{energy}$
(b) $3(4.002602\ u) \rightarrow 12.000000\ u + \text{energy}$

## 28.6 Mechanisms of radioactive decay

25. * In 1913, Frederick Soddy collected the following data related to the radioactive transformation of uranium (**Figure P28.25**). The first product that appears in the sample is thorium, then protactinium, then another isotope of uranium, and so on. Examine the series of the transformation found by Soddy and explain using your knowledge of alpha and beta decays. Discuss what quantities are conserved in each process.

**Figure P28.25**

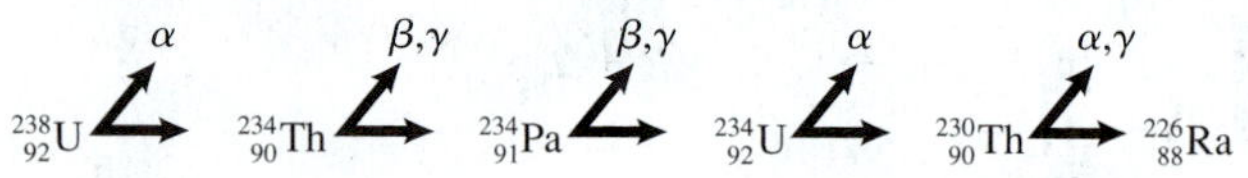


26. * In the 1930s, Meitner, Hahn, and Strassmann did experiments irradiating uranium with neutrons. They predicted three possible outcomes:
- Production of a new element (if the neutron undergoes beta decay in the nucleus).
- Production of a heavier isotope of uranium (if the neutron stays in the nucleus).
- Production of a slightly lighter nucleus (if the neutron captured by the nucleus causes one or two alpha particles to leave).

Instead, the nuclei produced in the reaction included isotopes of barium and other nuclei with about half the mass of uranium. How could they explain their findings?

27. * The following nuclei produced in a nuclear reactor each undergo radioactive decay. Determine the daughter nucleus formed by each decay reaction: (a) ${}^{239}_{94}Pu$ alpha decay; (b) ${}^{144}_{58}Ce$ beta-minus decay; (c) ${}^{129}_{53}I$ beta-minus decay; and (d) ${}^{60}_{30}Zn$ beta-plus decay.

28. Potassium-40 (${}^{40}_{19}K$) undergoes both beta-plus and beta-minus decay. Determine the daughter nucleus in each case.

29. * Radon-222 (${}^{222}_{86}Rn$) is released into the air during uranium mining and undergoes alpha decay to form ${}^{218}_{84}Po$ of mass 218.0090 u. Determine the energy released by the decay reaction. Most of this energy is in the form of alpha particle kinetic energy.

30. * Carbon-11 (${}^{11}_{6}C$) undergoes beta-plus decay. Determine the product of the decay and the energy released.

31. * (a) Determine the total binding energy of radium-226 (${}^{226}_{88}Ra$). (b) Determine and add together the binding energies of a radon-222 (${}^{222}_{86}Rn$) and an alpha particle. (c) Determine the difference of these numbers, which equals the energy released during alpha decay of ${}^{226}Ra$.

32. * BIO **Potassium decay in body** The body contains about 7 mg of radioactive ${}^{40}_{19}K$ that is absorbed with the foods we eat. Each second, about $2.0 \times 10^{3}$ of these potassium nuclei undergo beta decay (either beta-minus or beta-plus). About how many potassium nuclei are in the average body and what fraction decay each second?

33. * A radioactive ${}^{60}Co$ nucleus emits a gamma ray of wavelength $0.93 \times 10^{-12}$ m. If the cobalt was initially at rest, use the conservation of momentum equation to determine its speed following the gamma ray emission.

## 28.7 Half-life, decay rate, and exponential decay

34. * BIO **$O_2$ emitted by plants** Design an experiment to determine whether $O_2$ emitted from plants comes from $H_2O$ or from $CO_2$, the basic input molecules that lead to plant growth.

35. A radioactive sample initially undergoes $4.8 \times 10^{4}$ decays/s. Twenty-four hours later, its activity is $1.2 \times 10^{4}$ decays/s. Determine the half-life of the radioactive species in the sample.

36. * Cesium-137, a waste product of nuclear reactors, has a half-life of 30 years. Use two different methods to determine the fraction of ${}^{137}Cs$ remaining in a reactor fuel rod: (a) 120 years after it is removed from the reactor, (b) 240 years after, and (c) 1000 years after.

37. * A sample of radioactive technetium-99 of half-life 6 h is to be used in a clinical examination. The sample is delayed 15 h before arriving at the lab for use. Use two methods to determine the fraction of radioactive technetium that remains.

38. ** BIO **Radiation therapy** If 120 mg of radioactive gold-198 with half-life 2.7 days is administered to a patient for radiation therapy, what is the gold-198 activity 3 weeks later if none is eliminated from the body by biological means?

39. **Fuel rod decay** How many years are required for the amount of krypton-85 (${}^{85}_{36}Kr$) in a spent nuclear reactor fuel rod to be reduced by a factor of 1/8? 1/32? 1/128? The half-life of ${}^{85}Kr$ is 10.8 years.

40. * BIO **Wisconsin glacier tree sample** A tree sample was uprooted and buried 60,000 years ago during part of the Wisconsin glaciation. How many years after it was buried was the radioactive carbon-14 (${}^{14}C$) in the root reduced by a factor of (a) 1/2, (b) 1/4, and (c) 1/8? (d) What fraction remained after 60,000 years? The carbon-14 was not replenished after the tree stopped growing.

41. * How many years are required for the amount of strontium-90 (${}^{90}Sr$) released from a nuclear explosion in the atmosphere to be reduced by a factor of (a) 1/16, (b) 1/64, and (c) 0.010?

42. ** BIO **Student swallows radioactive iodine** A student accidentally swallows 0.10 $\mu g$ of iodine-131 while pipetting the radioactive material. (a) Determine the number of ${}^{131}I$ atoms swallowed (the atomic mass of ${}^{131}I$ is approximately 131 u). (b) Determine the activity of this material. The half-life of ${}^{131}I$ is 8.02 days. (c) What is the mass of radioactive iodine-131 that remains 21 days later if none leaves the body?

43. * An unlabeled container of radioactive material has an activity of 90 decays/min. Four days later the activity is 72 decays/min. Determine the half-life of the material. When will its activity reach 9 decays/min?

44. * BIO EST **Estimate number of ants in nest** To estimate the number of ants in a nest, 100 ants are removed and fed sugar made from radioactive carbon of a long half-life. The ants are returned to the nest. Several days later, of 200 ants taken from the nest, only 5 are radioactive. Roughly how many ants are in the nest? Explain your calculation technique.

45. * One gram of pure, radioactive radium produces 130 J of energy per hour due to radium decay only and has an activity of 1.0 Ci. Determine the average energy in electron volts released by each radioactive decay.

### 28.8 Radioactive dating

46. **Age of mallet** A mallet found at an archeological excavation site has 1/16 the normal carbon-14 decay rate. Determine the mallet's age.
47. The $^{235}U$ in a rock decays with a half-life of $7.04 \times 10^8$ years. A geologist finds that for each $^{235}U$ now remaining in the rock, 2.6 $^{235}U$ have already decayed to form daughter nuclei. Determine the age of the rock.
48. A sample of water from a deep, isolated well contains only 30% as much tritium as fresh rainwater. How old is the water in the well?
49. * The decay rate of $^{14}C$ from a bone uncovered at a burial site is 12.6 decays/min, whereas the decay rate from a fresh bone of the same mass is 1610 decays/min. Approximately how old is the uncovered bone?
50. ** BIO The tree sample described in Problem 40 contains 50 g of carbon when it is discovered. (a) If 1 in $10^{12}$ carbon atoms in a fresh tree sample are carbon-14, how many carbon-14 atoms. would be in 50 g of carbon from a fresh tree? (b) Calculate the carbon-14 activity of the sample. (c) Determine the age of the buried tree if its 50 g of carbon has an activity of $-2.2\ s^{-1}$.
51. * A radioactive series different from those shown in Figures 28.14 and P28.25 begins with $^{232}_{90}Th$ and undergoes the following series of decays: $\alpha\beta^-\beta^-\alpha\alpha\alpha\alpha\beta^-\alpha\beta^-$. Determine each nucleus in the series.

### 28.9 Ionizing radiation and its measurement

52. BIO **Radiation dose** A 70-kg person receives a 250-mrad whole-body absorbed dose of radiation. (a) How much energy does the person absorb? (b) Is it better to absorb 250 mrad of X-rays or 250 mrad of beta rays? Explain. (c) What is the dose in each case?
53. BIO **Dose due to different types of radiation** Determine the dose equivalent of a 70-mrad absorbed dose of the following types of radiation: (a) X-rays, (b) beta rays, (c) protons, (d) alpha particles, and (e) heavy ions.
54. * BIO **Potassium decay in body** The yearly whole-body dose caused by radioactive $^{40}K$ absorbed in our tissues is 17 mrem. (a) Assuming that $^{40}K$ undergoes beta decay with an RBE of 1.4, determine the absorbed dose in rads. (b) How much beta ray energy does an 80-kg person absorb in one year? [Note: $^{40}K$ also emits gamma rays, many of which leave the body before being absorbed. Because fatty tissue has low potassium concentration and muscle has a higher concentration, gamma ray emissions indicate indirectly a person's fat content.]
55. ** BIO **More potassium** Determine the number of $^{40}K$ nuclei in the body of an 80-kg person using the information provided in the previous problem and the fact that $^{40}K$ has a radioactive half-life of $1.28 \times 10^9$ years. Assume that each $^{40}K$ beta decay results in 1.4 MeV of energy that is deposited in the person's tissue.
56. * BIO **X-ray exam** During an X-ray examination a person receives a dose of 80 mrem in 4.0 kg of tissue. The RBE of X-rays is 1.0. (a) Determine the total energy absorbed by the 4.0 kg of body tissue. (b) Determine the energy in joules of each 40,000-eV X-ray photon. (c) Determine the number of photons absorbed by that tissue.

## General Problems

57. * EST Estimate the temperature at which two protons can come close enough together to form an isotope of a helium nucleus.
58. ** EST The mass of a helium nucleus is less than the mass of the nucleons inside it. (a) Explain how this observation led scientists to the idea that it is possible to convert hydrogen into helium to produce thermal energy. (b) Will this process mean that energy is not conserved? Explain. (c) Why do you think scientists need very high temperatures and high pressures for this reaction? (d) Estimate the temperature at which two protons will join together due to their nuclear attraction. Remember that nuclear forces are effective at distances less than or equal to $10^{-15}$ m. [Hint: Use an energy approach, not a force approach.] (e) Suggest possible ways of containing ionized hydrogen to make the reaction possible (remember that all solid containers will melt at this temperature).
59. ** **World energy use** World energy consumption in 2005 was about $4 \times 10^{20}$ J. (a) Determine the number of deuterium nuclei ($^2_1H$) that would be needed to produce this energy. The fusion of two deuterium nuclei releases about 4 MeV of energy. (b) Determine the volume of water of density 1000 kg/m$^3$ needed to supply the energy. One mole of water has a mass of 18 g, and about one in every 6700 $^1_1H$ atoms in water is a deuterium $^2_1H$.
60. * Convert the 200 MeV per nucleus energy that is released by $^{235}U$ fission to units of joules per kilogram. By comparison, the energy release by burning coal is $3.3 \times 10^7$ J/kg.
61. ** (a) Determine the energy used by a 1200-W hair dryer in 10 min. (b) Approximately how many $^{235}U$ fissions must occur in a nuclear power plant to provide this energy if 35% of the energy released by fission produces electrical energy? The energy released per fission is about 200 MeV.
62. * EST Estimate the number of $^{235}U$ nuclei that must undergo fission to provide the energy to lift your body 1 m. Indicate any assumptions made.
63. ** **Comparing nuclear and coal power plants** Suppose that a nuclear fission power plant and a coal-fired power plant both operate at 40% efficiency. Determine the ratio of the mass of coal that must be burned in 1 day of operating a 1000-MW plant compared to the mass of $^{235}U$ that must undergo fission in the same plant. The energy released by burning coal is $3.3 \times 10^7$ J/kg. The energy released per $^{235}U$ fission is 200 MeV.
64. ** **Uranium energy potential** The world's uranium supply is approximately $10^9$ kg ($10^6$ tons), 0.7% of which is $^{235}U$. (a) How much energy is available from the fission of this $^{235}U$? (b) The world's energy consumption rate for production of electricity is about $10^{20}$ J/year. At this rate, how many years would the uranium last if it were used to provide all our electrical energy?
65. * A radioactive sample contains two different types of radioactive nuclei: A, with half-life 5.0 days, and B, with half-life 30.0 days. Initially, the decay rate of the A-type nucleus is 64 times that of the B-type nucleus. When will their decay rates be equal?
66. ** After a series of alpha and beta-minus decays, plutonium-239 ($^{239}_{94}Pu$) becomes lead-207 ($^{207}_{82}Pb$). Determine the number of alpha and beta-minus particles emitted in the complete decay process. Explain your method for determining these numbers.

67. ** BIO **Ion production due to ionizing radiation** A body cell of $1.0 \times 10^{-5}$ m radius and density 1000 kg/m$^3$ receives a radiation absorbed dose of 1 rad. (a) Determine the mass of the cell (assume it is shaped like a sphere). (b) Determine the energy absorbed by the cell. (c) If the average energy needed to produce one positively charged ion is 100 eV, determine the number of positive ions produced in the cell. (d) Repeat the procedure for the $3.0 \times 10^{-6}$-m-radius nucleus of the cell, where its genetic information is stored.

## Reading Passage Problems

BIO **Nuclear accident and cancer risk** If 10,000 people are exposed to an average radiation dose of 1 rem, it is estimated that six cancer deaths will result. In March, 2011, a 9.0-magnitude earthquake and subsequent tsunami rocked Japan, causing hydrogen explosions in four reactors at the Fukushima Daiichi Nuclear Power Station. As a result, radioactive materials such as cesium-137 and iodine-131 were released into the atmosphere. About 2 million people living within 80 km of the Fukushima reactors were each exposed to an average dose of approximately 1 rem.

68. Which answer below is closest to the Fukushima total person-rem exposure?
(a) $2 \times 10^4$ person · rem (b) $2 \times 10^6$ person · rem
(c) $2 \times 10^8$ person · rem (d) $2 \times 10^{10}$ person · rem

69. Which answer below is closest to the number of statistically likely cancer deaths to be caused by this event?
(a) 10 (b) 100 (c) 1000
(d) 10,000 (e) 100,000

70. About 4000 of the 2 million residents die each year from cancer caused by other factors, such as smoking. If those Fukushima accident cancer deaths occurred during the 5 years following the accident, which number below is closest to the ratio of Fukushima-caused cancer deaths to normal cancer deaths for the exposed population?
(a) 1/2000 (b) 1/20 (c) 1 (d) 20/1 (e) 200/1

71. Which answer below is closest to the number of cancer deaths in the United States due to an estimated 20 mrad/year average radon exposure to each person?
(a) 50 (b) 500 (c) 5000 (d) 50,000 (e) 500,000

72. Which is the most important reason why it is difficult statistically to identify cancer deaths due to radiation exposure from the Fukushima accident?
(a) An unknown number of people stayed in their homes following the accident.
(b) The radiation detection devices were inaccurate.
(c) The fraction of people who left the area was unknown.
(d) The normal cancer rate significantly exceeded the cancer caused by the accident.
(e) Authorities often distort information they provide about radiation exposure.

**Was the sculpture of David from Notre Dame?** Neutron activation analysis (NAA) is a sensitive analytical technique used in environmental studies, semiconductor quality control, forensic science, and archeological studies to identify trace and rare elements and determine the effect of rare elements on chronic diseases. In NAA

**Figure 28.15** NAA comparison showing that the relative amount of different elements in stone from Cathedral of Notre Dame, Paris, and Amiens Cathedral are distinctively different.

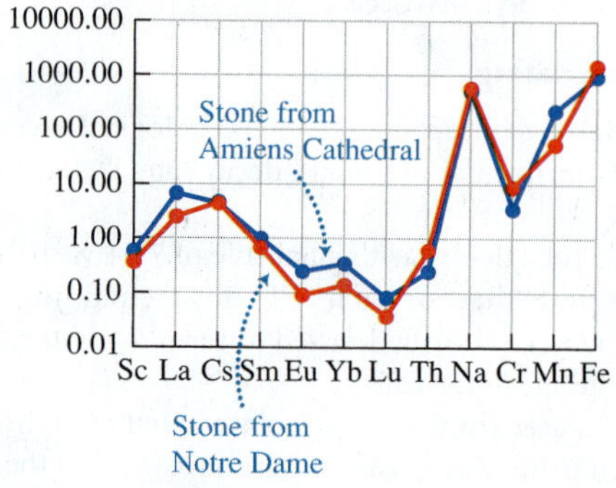


a tiny specimen under investigation is irradiated with neutrons. The neutrons are absorbed by nuclei in the sample, thus creating artificial radioactive nuclei. These nuclei decay via the emission of alpha particles, beta particles, and gamma rays. The energy and wavelength of the gamma rays are characteristic of the element from which they were emitted.

NAA has been used to help identify the sources of art and sculpture by an elemental analysis of the paint or stones used. For example, the Metropolitan Museum of Art recently determined that stone used in the mid-13th-century *Head of King David* sculpture that they acquired in 1938 matched stone from the cathedral of Notre Dame in Paris. Previously, it had been assumed that the sculpture came from Amiens Cathedral north of Paris. **Figure 28.15** shows the elemental content of stone from Notre Dame (the red line) and from Amiens Cathedral (the blue line). Although the patterns on this logarithmic scale look similar, they are significantly different for some of the 16 elements tested.

73. In neutron activation analysis (NAA) a particular elemental concentration in a sample is determined by measuring which of the following?
(a) Alpha particles (b) Beta particles
(c) Gamma rays (d) Neutrons
(e) Two of these choices

74. Which answer below is closest to the ratio of the relative amount of lutetium (Lu) in stone from Notre Dame compared to stone from Amiens Cathedral?
(a) 0.5 (b) 1 (c) 2 (d) 10 (e) 1000

75. What does neutron absorption by sodium ($^{23}_{11}Na$) produce?
(a) $^{22}_{11}Na$ (b) $^{24}_{11}Na$ (c) $^{24}_{12}Mg$ (d) $^{23}_{12}Mg$ (e) $^{22}_{10}Ne$

76. Neutron absorption by potassium produces an excited $^{40}_{19}K$ nucleus, which emits a beta-minus and a gamma ray. Which nucleus below is a product of this decay?
(a) $^{40}_{19}K$ (b) $^{40}_{18}Ar$ (c) $^{40}_{20}Ca$ (d) $^{41}_{19}K$ (e) $^{39}_{19}K$

77. Rank the elements Cr, Mn, and Na in stone samples from Notre Dame compared to stone from Amiens Cathedral as the best indicators for distinguishing the two types of stone (rated from best to worst).
(a) Cr, Mn, Na (b) Mn, Cr, Na
(c) Na, Mn, Cr (d) Mn, Na, Cr

# Particle Physics 29

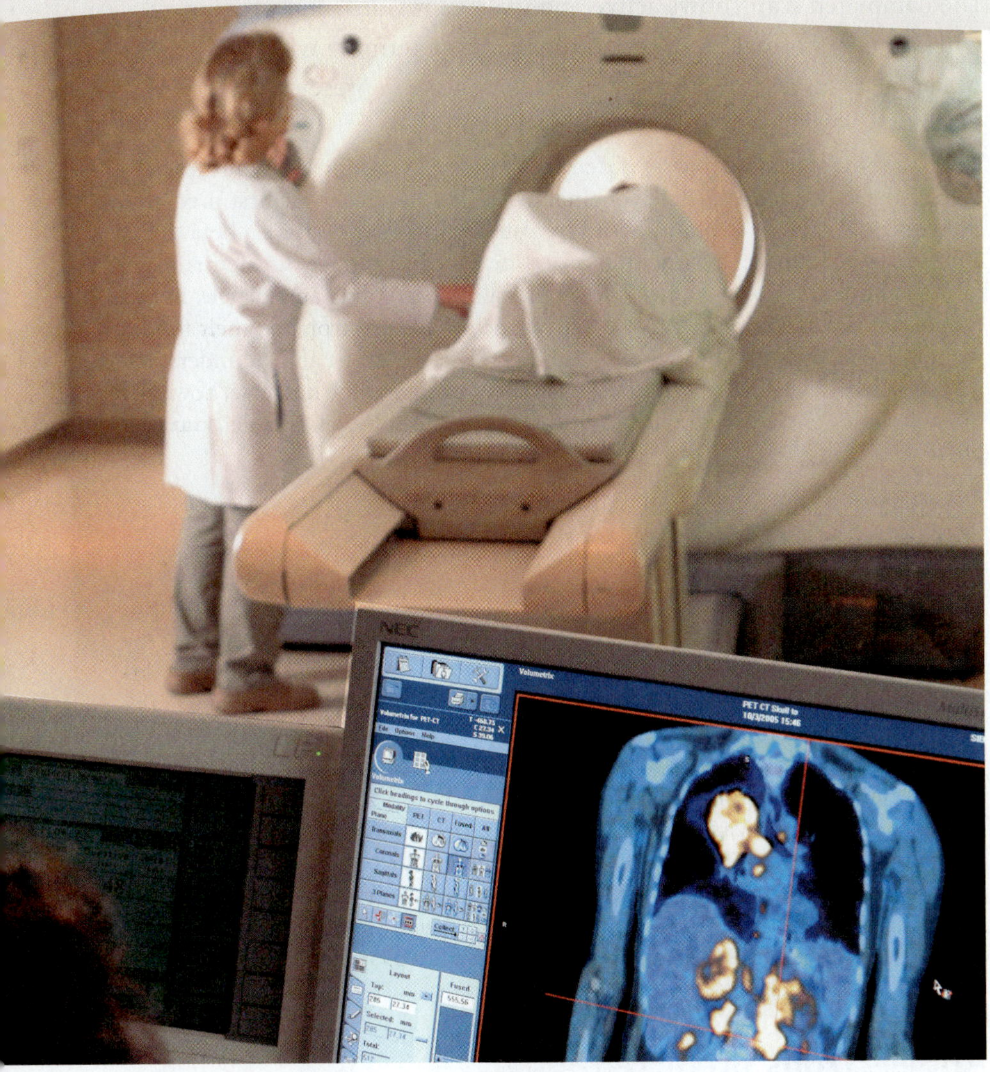

What is antimatter?

What are the fundamental particles and interactions in nature?

What was the Big Bang, and how has the universe evolved since?

James, 82 years old, started forgetting things and getting lost in familiar places. His doctor suspected that James was developing Alzheimer's disease. To test this hypothesis, the doctor prescribed PET (positron emission tomography), which allows doctors to obtain a three-dimensional image of the brain. Before the PET scan, James received an injection of a glucose-like molecule containing a radioactive isotope of fluorine. An hour after injection, the substance had been absorbed into James's brain. Radioactive decay of the fluorine nuclei produced positrons.

**Be sure you know how to:**

- Use the right-hand rule to determine the direction of the magnetic force exerted by a magnetic field on a moving charged particle (Section 17.4).
- Explain beta decay (Section 28.6).
- Write an expression for the rest energy of a particle (Section 25.8).

Immediately after positron emission, each positron interacts with a nearby electron, producing a pair of photons traveling in opposite directions. Detectors inside the PET machine determine where in the brain these photons originated, indicating an active part of the brain. The completed scan shows active and inactive regions, and comparison with scans of normal brains helps physicians determine whether Alzheimer's disease or other forms of dementia are present.

Understanding radioactive beta decay (Chapter 28) required hypothesizing the existence of a new particle—the neutrino. The neutrino was first predicted by Wolfgang Pauli to explain the mysterious loss of energy during beta decay and was later found experimentally. Beta decay can produce antineutrinos, a form of antimatter. Every known particle has a corresponding antiparticle. In this chapter we investigate elementary particles such as the positron and their fundamental interactions. By the end of this chapter, you should be able to understand the physics behind PET as well as the basic components of the universe, how they interact, and how they were discovered. This area of physics is called **particle physics.**

## 29.1 Antiparticles

As you may recall, the word "atom" means indivisible. Physicists thought at one time that atoms were the smallest constituents of matter. However, in the late 19th century physicists learned that atoms had an internal structure consisting of negatively charged electrons and a positively charged nucleus. The discovery of radioactivity and its subsequent explanation indicated that the nucleus itself has a complex structure. In addition, the investigation of black body radiation, the photoelectric effect, and Compton scattering led scientists to conclude that light can be modeled as a photon—an object that has particle-like properties. By 1930, physicists had identified four particles—the electron, the proton, the neutron, and the photon. At that time, these were the only known truly **elementary particles**, a description used to indicate the simplest and most basic particles. This view changed with the proposal and discovery of so-called antiparticles.

**Active Learning Guide ›**

### Antielectrons predicted

At the end of the 1920s physicists solving the Schrödinger equation for the possible electron wave functions in the hydrogen atom produced solutions that were labeled by three quantum numbers: $n$, $l$, and $m_l$. However, the Schrödinger equation did not explain the existence of the fourth quantum number of the electron in the atom, its spin $m_s$.

In 1928 Paul Dirac extended quantum mechanics to incorporate Einstein's theory of special relativity. Dirac's model of relativistic quantum mechanics successfully predicted the existence of the spin quantum number for the electron. However, along with this prediction came an unexpected complication. Dirac's model also predicted that free electrons (electrons that are not interacting with other objects or fields) had an infinite number of possible quantum states with negative total energy. As a result, a free electron in a positive energy state should be able to transition to one of these negative energy states by emitting a photon whose energy equals the difference between those two states. All free electrons in the universe would transition to increasingly negative energy states, continually releasing electromagnetic energy. This certainly is inconsistent with the observed behavior of electrons. How could a free electron have negative total energy? (Remember that only certain types of potential energies, such as electric potential energy, can be negative.)

Dirac suggested that these negative energy states were occupied by an infinite number of so-called virtual electrons. Because only one electron can occupy an electron state according to the Pauli exclusion principle (Chapter 27), free electrons with positive energy could not then transition into these states. Over the next three years Dirac modified his model and eventually proposed that one of these virtual electrons in a negative energy state could be lifted out of its negative energy state to become a positive energy free electron. The empty state left behind would behave like a positively charged electron—like a particle with an electric charge of $+1.6 \times 10^{-19}$ C. Many quantum physicists disagreed with many aspects of his model. But Dirac argued that such particles should exist. He called them *antielectrons*, a new type of particle that had not yet been observed.

**Figure 29.1** (a) The cloud chamber photograph taken by Anderson. (b) The direction of the force that the magnetic field would exert on an electron if it were moving from the top to the bottom. (c) The direction of the force that the magnetic field would exert on a positron if it were moving from the bottom to the top.

(a)

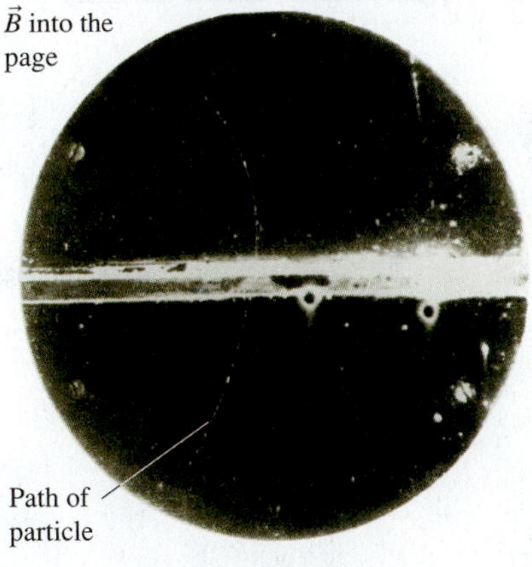


(b) The magnetic force if the path was caused by an electron moving from the top to the bottom

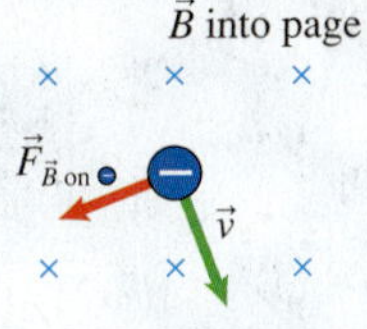


(c) The magnetic force if the path was caused by a positron moving from the bottom to the top

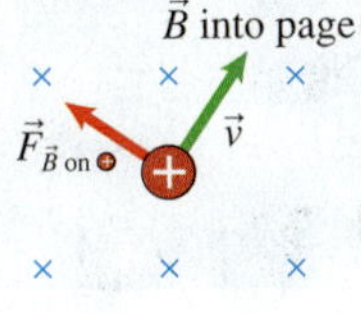


## Antielectrons detected

To check Dirac's prediction, physicists had to search for a particle that behaved exactly like an electron but had a positive electric charge. To determine the sign of elementary particles at that time, scientists used cloud chambers. Inside the chamber was a gas that was supersaturated with vaporized water or alcohol. A charged particle like an electron or alpha particle passing through the vapor caused the vapor to condense into visible droplets, leaving a trail marking the particle's trajectory.

In 1932, American physicist Carl Anderson was studying cosmic rays, protons, and other particles coming to Earth from the Sun and from supernova explosions. He used a cloud chamber placed in a magnetic field. One of his photographs (**Figure 29.1a**) included the curved path of a particle passing through the chamber. The magnetic field pointed into the plane of the page.

The direction of travel of the particle and its charge were not immediately apparent from its path. If it entered the chamber from the top, then using the right-hand rule for magnetic force, it must be a negatively charged particle in order to curve as shown (Figure 29.1b). However, if it entered the chamber from the bottom, then it must be a positively charged particle (Figure 29.1c). How could Anderson distinguish between these two possibilities?

Notice that the lower part of the path is less curved than the upper part. Recall that a uniform magnetic field causes a charged particle to travel in a circular path (Chapter 17). The radius of its path can be expressed as

$$r = \frac{mv}{|q|B}$$

This means that the radius of the curved path is largest when the speed is largest—on the bottom half of the path. The particle passed through a thin lead plate in the middle of the chamber. This caused the particle's speed to decrease. Thus, the particle must have been traveling from the bottom of the chamber to the top, meaning it was a positively charged particle.

Anderson was able to determine the charge-to-mass ratio of the particle and found that it was the same as for an electron. But it was positively charged! Anderson realized that the cloud chamber trace must have been produced by one of Dirac's antielectrons. Anderson found many more such tracks in other cloud chamber photographs. He called the new particle a **positron**.

## Pair production

Electrons are abundant in our world, but there are few positrons. Under most circumstances positrons exist for very short time periods. How are they produced? One way is during the radioactive decay of certain nuclear isotopes—for example, ${}^{11}_{6}\text{C}$, ${}^{13}_{7}\text{N}$, ${}^{15}_{8}\text{O}$, ${}^{22}_{11}\text{Na}$, and ${}^{40}_{19}\text{K}$ (we discussed this topic in Chapter 28).

**Active Learning Guide ›**

Positrons can also be produced during the interaction of a high-energy gamma ray photon $\gamma$ with matter. The photon has zero rest mass—you can't measure a photon's mass on a scale because the photon cannot exist at rest. However, under the right conditions this photon can simultaneously produce an electron ${}^{\ 0}_{-1}e$ and a positron ${}^{\ 0}_{+1}e$ (particles with nonzero rest mass). The process is called **pair production** and is represented by the reaction below:

$$\gamma \rightarrow {}^{\ 0}_{-1}e + {}^{\ 0}_{+1}e$$

**Figure 29.2** (a) A gamma ray photon converts to an electron and a positron. (b) If in a magnetic field, the magnetic force causes the particles to move apart in spiral circles. (c) A cloud chamber photograph of this pair production.

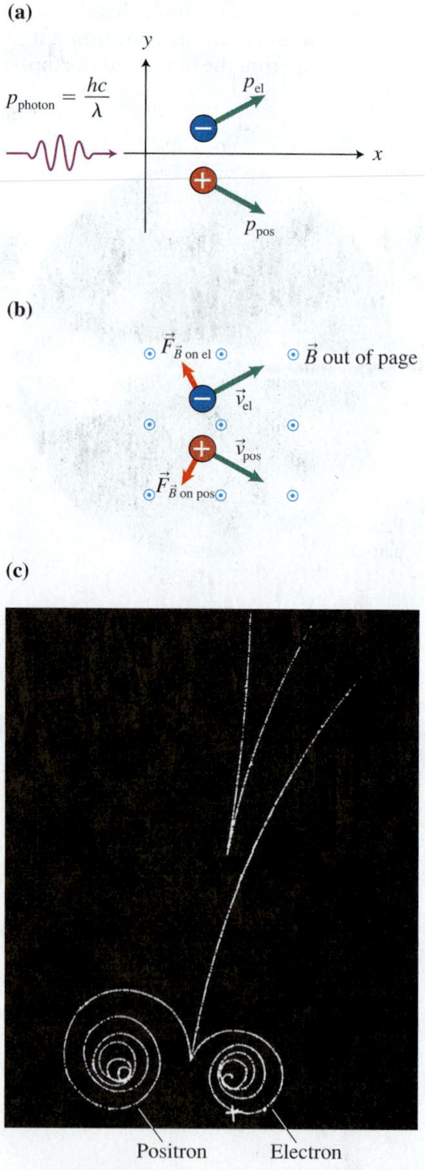


For the photon to simultaneously produce both an electron and a positron, its energy must at least equal the combined rest energies of the electron and the positron. For the minimal energy calculation, we assume the kinetic energies of the electron and positron after their production are zero. The rest energy of the electron (and of the positron) is $E_0 = mc^2$. Thus, the photon must have double this energy:

$$E_{\text{photon}} = 2mc^2 = 2(9.11 \times 10^{-31}\ \text{kg})(3.00 \times 10^8\ \text{m/s})^2 = 1.64 \times 10^{-13}\ \text{J}$$

In electron volts, this is

$$E_{\text{photon}} = (1.64 \times 10^{-13}\ \text{J})\left(\frac{1\ \text{eV}}{1.6 \times 10^{-19}\ \text{J}}\right) = 1.02 \times 10^6\ \text{eV}$$

The frequency of such a photon is

$$f = \frac{E_{\text{photon}}}{h} = \frac{1.64 \times 10^{-13}\ \text{J}}{6.63 \times 10^{-34}\ \text{J}\cdot\text{s}} = 2.47 \times 10^{20}\ \text{Hz}$$

and its wavelength is

$$\lambda = \frac{c}{f} = \frac{3.00 \times 10^8\ \text{m/s}}{2.47 \times 10^{20}\ \text{Hz}} = 1.21 \times 10^{-12}\ \text{m}$$

This wavelength is about 1/100 times the wavelength of an X-ray photon, making it a gamma ray photon.

If the positron and electron produced are to possess some kinetic energy, then an even higher energy gamma ray is needed. If this pair production occurs in a magnetic field, the two particles should spiral in opposite directions.

Consider the photon as the system—it has momentum in the $x$-direction before producing the pair (**Figure 29.2a**). Immediately after pair production the photon no longer exists, so the momentum of the two particles should be the same as the momentum of the original photon. The electron and positron should fly apart at an angle relative to each other, so that the $y$ components of momentum perpendicular to the original momentum of the photon cancel each other.

If we place the system in a magnetic field, the field exerts a force on the electron and on the positron in opposite directions because they have opposite electric charge (Figure 29.2b). The force exerted on them by the field will be perpendicular to their velocities so they should spiral in opposite directions. In addition, the radius of their circular paths should decrease as their speeds decrease, resulting in an inward spiraling motion. The photographs of pair production by gamma ray photons obtained by Anderson are consistent with this reasoning—see Figure 29.2c.

## Pair annihilation

We have just learned that high-energy gamma ray photons can be used to create electron-positron pairs. Is the reverse process possible? If an electron and positron meet, will one or more photons be produced? The answer is yes, and the process is called **pair annihilation.**

**CONCEPTUAL EXERCISE 29.1 Pair annihilation**

Imagine that an electron and a positron meet and annihilate each other. Assume that they are moving directly toward each other at constant speed. Will one or two photons be produced? Write a reaction equation for this process. In what directions do these photons move relative to each other after the process?

**Sketch and translate** If one photon is produced, the reaction for this process is

$$^{0}_{-1}e + ^{0}_{+1}e \rightarrow \gamma$$

If two photons are produced, the reaction is

$$^{0}_{-1}e + ^{0}_{+1}e \rightarrow \gamma + \gamma$$

**Simplify and diagram** Assume that the process occurs in isolation, which means the total electric charge and momentum of the system must remain constant. Electric charge is constant in both versions of the process. Because the two particles were moving directly toward each other at constant speed, the total initial momentum of the system is zero. In the first version of the process, however, momentum is not conserved. If only one photon (with momentum $p = E/c$) is produced, the system's final momentum after the annihilation cannot be zero. Therefore, for the momentum of the system to be constant in this process, two photons must be produced, and they must travel in opposite directions, as shown in the figure.

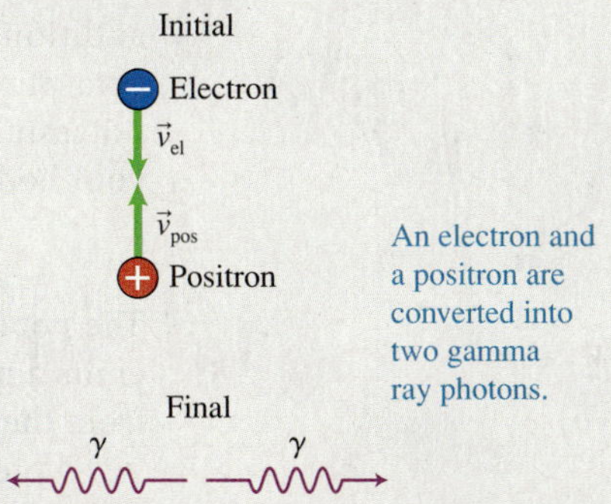


**Try it yourself:** In this pair annihilation two gamma rays were produced. But in pair production an electron-positron pair was produced by only one gamma ray photon. How is this possible?

*Answer:* The momentum of the single gamma ray photon can be converted to the combined momentum of the two particles (see Figure 29.2).

The previous exercise involved a positron. We have already discussed the creation of a positron and an electron by a high-energy gamma ray photon as well as the production of positrons by radioactive decay. Let's consider another possibility. Recall the explanation of the beta decay of a neutron in nuclear physics (Chapter 28). A free neutron is an unstable particle with half-life of about 10 min. A free neutron decays into a proton, an electron, and an antineutrino (the antiparticle of a neutrino):

$$^{1}_{0}\mathrm{n} \rightarrow ^{1}_{1}\mathrm{p} + ^{0}_{-1}e + ^{0}_{0}\overline{\nu}$$

The electric charge of the system during this process is constant $(0 = +1 - 1 + 0)$, as is its nucleon number $(+1 = +1 + 0 + 0)$. The rest energy of the neutron is greater than the combined rest energies of the proton and electron; therefore, the system's energy can possibly be constant as well, provided the proton, electron, and neutrino have sufficient kinetic energy. Finally, an appropriate combination of the momenta of the three particles in the final state can equal the initial momentum of the neutron—the momentum of the system can be constant. This process could happen without interaction with the environment. Is there a similar process by which a positron is produced instead of an electron?

## Beta-plus decay: Transforming a proton into a neutron

Since the rest energy of the neutron is greater than the rest energy of the proton, it seems that transforming a proton into a neutron cannot occur. However, if a proton captures a gamma ray photon, the energy of the excited state proton ${}^{1}_{1}\mathrm{p}^{*}$ may be great enough to produce a neutron and the other particles according to the reaction below:

$$\gamma + {}^{1}_{1}\mathrm{p} \rightarrow {}^{1}_{1}\mathrm{p}^{*} \rightarrow {}^{1}_{0}\mathrm{n} + {}^{0}_{+1}e + {}^{0}_{0}\nu$$

The proton absorbs the photon and then decays into a neutron, a positron, and a neutrino. This process is called **beta-plus decay** to indicate that a positron (not an electron) is produced as the result of the process.

Beta-plus decay can occur without the proton absorbing a photon. It can happen inside a nucleus if some of the energy of the nucleus converts into the additional rest energy of the products. Examples of nuclei that can undergo beta-plus decay include carbon-11, nitrogen-13, oxygen-15, fluorine-18, and potassium-40. In fact, about 20,000 positrons are produced each minute in your body from the beta-plus decay of potassium-40:

$${}^{40}_{19}\mathrm{K} \rightarrow {}^{40}_{18}\mathrm{Ar} + {}^{0}_{+1}e + {}^{0}_{0}\nu$$

The potassium nucleus transforms into an argon nucleus and in the process emits a positron and a neutrino. The potassium-40 nuclei in your body come from the potassium in foods you eat, such as bananas.

Note that the positively charged positrons produced in your body by this radioactive decay travel infinitesimal distances. They are attracted to negatively charged electrons and abruptly undergo pair annihilation, producing high-energy gamma rays, most of which leave the body. This process is what makes positron emission tomography possible.

## Positron emission tomography (PET)

**Figure 29.3** Positron emission tomography (PET).

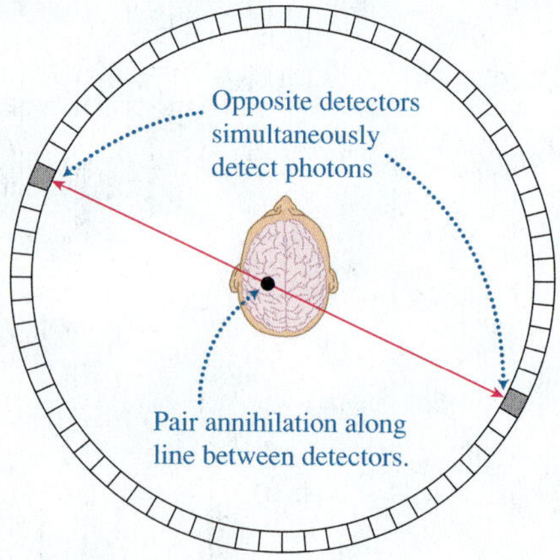
Opposite detectors
simultaneously
detect photons
Pair annihilation along
line between detectors.

In the chapter opening we described positron emission tomography (PET), a process for imaging the brain using radioactive fluorine-18 isotopes injected into a person's body. The isotopes undergo beta-plus decay continually, producing positrons. The positrons immediately meet electrons and annihilate, producing a pair of gamma ray photons that move in opposite directions. The PET chamber surrounding the patient's head detects the pairs of simultaneously appearing photons (**Figure 29.3**) and determines where in the brain they were emitted. Combining many pairs of gamma rays produces a three-dimensional image of the active parts of the brain—the places where more photons are produced.

This technique not only allows doctors to find the active and inactive regions of the brain in sick patients, but also helps brain researchers identify parts of the brain that are active when a person performs a particular task, such as listening to a poem being read or thinking about a mathematical problem.

## Other antiparticles

Other particles have antiparticles as well. The positively charged proton that is part of all nuclei has a negatively charged antiproton of the same mass but opposite electric charge. Even though the neutron has zero electric charge, it too has an antimatter counterpart; other properties besides charge differentiate the neutron from the antineutron (see Section 29.3 on quarks). Occasionally, a particle is its own antiparticle; the photon is an example. Later in this chapter we consider why our universe is composed almost entirely of ordinary particles and almost no antiparticles.

**Review Question 29.1** Why does proton decay occur only inside nuclei and not when a proton is free?

## 29.2 Fundamental interactions

In our studies of physics we have learned about many types of interactions (gravitational, electromagnetic, elastic, etc.). Some of these interactions are *fundamental* and others are *nonfundamental.* Fundamental interactions are the most basic interactions known, such as the electromagnetic interaction between charged particles. Nonfundamental interactions, such as friction, can be understood in terms of fundamental interactions. Friction is a macroscopic manifestation of the electromagnetic interaction between the electrons of the two surfaces that are in contact. We use the term "interaction" rather than "force" when speaking in general terms because interactions can also be represented using energy ideas. The four fundamental interactions are the gravitational interaction, the electromagnetic interaction, the strong interaction, and the weak interaction.

‹ Active Learning Guide

**Gravitational interaction** All objects in the universe participate in **gravitational interactions** due to their mass (and according to general relativity, also due to their energy content through $m = E/c^2$). The gravitational force that two objects exert on each other is proportional to the product of their masses and the inverse square of their separation $1/r^2$. This interaction is important for mega-objects (planets, stars, galaxies, etc.) but much less important for objects in our daily lives (e.g., the gravitational force that one human exerts on another) and is extremely insignificant for microscopic objects (atoms, electrons, etc.).

**Electromagnetic interaction** Electrically charged objects participate in **electromagnetic interactions.** The interaction is electric if the objects are at rest or in motion with respect to each other. The interaction is magnetic only if the objects are moving with respect to each other. Similar to the gravitational interaction, the electromagnetic interaction between two charged particles also decreases as the inverse square of the distance between them. The electromagnetic interactions between nuclei and electrons are important in understanding the structure of atoms.

The electromagnetic interaction is tremendously stronger than the gravitational interaction. For example, the electrostatic force that an electron and a proton exert on each other in an atom is about $10^{39}$ times greater than the gravitational force that they exert on each other (**Figure 29.4**).

**Figure 29.4** The electric force that atomic size objects exert on each other is much greater than the gravitational force they exert on each other.

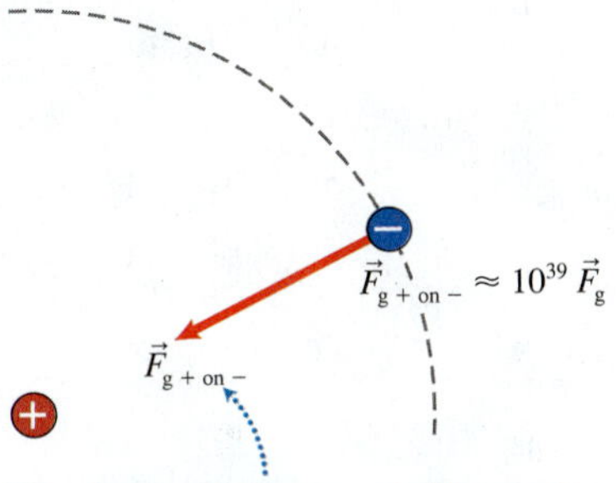
$\vec{F}_{g\,+\,on\,-} \approx 10^{39}\,\vec{F}_g$
$\vec{F}_{g\,+\,on\,-}$
The electrostatic force exerted by a proton on an electron in a hydrogen atom is about $10^{39}$ times greater than the gravitational force.

Atoms are electrically neutral, but because they are composed of distributions of charged particles (protons and electrons), they do participate in electromagnetic interactions with each other. These interactions contribute to the formation of molecules and to holding liquids and solids together. Because these interactions occur between overall electrically neutral objects, they are called **residual interactions.** Interactions such as friction, tension, and buoyancy are nonfundamental interactions and also are macroscopic manifestations of residual electromagnetic interactions.

**Strong interaction** Atomic nuclei are made up of protons that repel each other via electromagnetic interactions and of neutrons that do not significantly interact electromagnetically. The interaction that binds protons and neutrons together into a nucleus is a residual interaction of the **strong interaction.** The strong interaction is a very short range interaction, exerted by protons and neutrons only on their nearest neighbors within the nucleus (a range of $\sim 3 \times 10^{-15}$ m).

**Weak interaction** The weak interaction is responsible for beta decay. Protons, neutrons, electrons, and neutrinos all participate in it. The **weak interaction** is (as the name implies) significantly weaker than the strong interaction and has a significantly shorter range, about $10^{-18}$ m or less.

## Mechanisms of fundamental interactions

How do elementary particles interact with each other? Consider an analogy. Humans interact using speech. A vibration in a person's larynx produces a sound wave that travels through the air to another person. That sound wave vibrates the eardrums of the other person. This vibration is then converted into an electrical signal that travels to the brain, where it is interpreted. How can we use this idea to understand the microscopic mechanism behind the four fundamental interactions?

Earlier in this text we developed a field model for the interactions between electrically charged objects (Chapter 15). The idea was that electrically charged particles produce a disturbance in the region around them called an electric field. When a charged particle moves, the "signal" produced by that movement ripples through the electric field at a finite speed (light speed), and only when that signal reaches other charged particles does the force exerted on them by the field change.

We later used the field model to explain the mechanism behind the magnetic interaction, ultimately predicting the existence of electromagnetic waves of which visible light is just one example. We also discussed how electromagnetic waves could be modeled as streams of photons and showed that this model was necessary to explain microscopic phenomena such as the photoelectric effect.

During the first half of the 20th century physicists realized that the photon actually played a more central role in electromagnetic phenomena. Researchers learned that the mechanism behind all microscopic electromagnetic phenomena was the exchange of photons between electrically charged particles.

This particle exchange mechanism has since been successful in describing the weak and strong interactions, and it has had some success in describing the gravitational interaction (although there are still many unresolved difficulties). In each of the four fundamental interactions, when two particles interact, one of the two particles emits a particle that is then absorbed by the other. This emitted and absorbed particle is called a **mediator**. For the electromagnetic interaction, the mediator is the photon. Two electrons repel each other because one emits a photon, which travels at light speed to the other electron, where it is absorbed. The emitting electron recoils when the photon is emitted; the absorbing electron recoils when the photon is absorbed.

But the particle exchange mechanism of particle interactions appears to have a serious problem. Where does the energy required to create the mediating particle come from? To answer this question we need to think about the uncertainty principle (Chapter 27):

$$\Delta E \Delta t \geq \frac{h}{4\pi}$$

The energy of an atomic-scale system is only determinable to within a range $\Delta E$. When the system is involved in a process, the smaller the time interval $\Delta t$ in which the process occurs, the larger the energy uncertainty of the system during the process:

$$\Delta E \approx \frac{h}{4\pi \Delta t}$$

This helps us understand how the mediator particle can exist. The energy of an atomic-scale system is not determinable exactly; it fluctuates from instant to instant. In the case of two electrons interacting, these fluctuations manifest as what are known as **virtual photons** being exchanged between the electrons. These mediator photons are called "virtual" because they exist only as extremely small energy fluctuations of the system. These virtual photons do not have independent energy of their own that could initiate a chemical change in your retina, or in an electronic detector. As a result, virtual photons (and other virtual particles in general) cannot be experimentally detected in any direct way.

We see from the uncertainty principle that for interactions that are mediated over long distances and therefore require the mediator particles to exist for long time intervals $\Delta t$ to travel between the particles, the energy uncertainty $\Delta E$ of the system must be small (large $\Delta t$, small $\Delta E$). This means that photons mediating long-range electromagnetic interactions should have low total energy. This is possible because photons have zero rest energy and can have very low frequencies (recall $E_{\text{photon}} = hf$).

## Interaction mediators

We now have four fundamental interactions that we can now think of as exchange processes of four different mediators. The mediators of the electromagnetic, strong, and weak interactions have all been discovered. The hypothetical mediator for the gravitational interaction, the so-called **graviton**, has not. Below we summarize the four types of interaction mediators.

**Photons (Electromagnetic Interaction)** The electromagnetic interaction can be modeled as an exchange of photons. Photons are massless particles and always travel at the speed of light. They have zero electric charge and are their own antiparticles (the photon and antiphoton are the same particle).

**Gluons (Strong Interaction)** Gluons are massless and have zero electric charge but interact strongly with each other. As a result, the strong interaction has a very short range. There are eight different kinds of gluons. The exchange of gluons mediates the interaction of quarks and allows them to be bound together into protons, neutrons, and more exotic particles. (More on quarks below.)

**$W^{\pm}$ and $Z^0$ (Weak Interaction)** Three particles mediate the weak interaction. These particles are known as the positively charged $W^+$, its antiparticle, the negatively charged $W^-$, and the neutral $Z^0$ (which is its own antiparticle). The existence of these particles was predicted in the 1960s by a theory that unified the electromagnetic and weak interactions. This electroweak theory of Sheldon Glashow, Abdus Salam, and Steven Weinberg remained largely theoretical until 1983, when the $W^{\pm}$ and $Z^0$ particles were discovered at the European Organization for Nuclear Research (CERN) in Geneva, Switzerland.

Electroweak theory predicted (correctly) that these mediators have a very large rest energy. Producing them, therefore, required the colliding particles to have extremely high total energy. When the $W^{\pm}$ and $Z^0$ particles were first predicted, the accelerator at CERN could accelerate protons and antiprotons only to energies of tens of giga-electron volts (1 GeV is one billion electron volts), which was well below the rest energy of the weak interaction mediators. This explains why it took so many years to confirm the mediators' existence—technology had to catch up.

**Gravitons (Gravitational Interaction)** Based on some fairly widely accepted assumptions about what a quantum theory of gravity should be, the graviton is predicted to be electrically neutral and have zero mass and therefore to travel exclusively at the speed of light.

Our current understanding of the four fundamental interactions is summarized in **Table 29.1**. For reference, recall that the mass of a proton is $1.67 \times 10^{-27}$ kg.

**TIP** Although the properties of the graviton seem identical to the photon, they are very different particles. In this text we won't discuss the properties that distinguish them.

**Table 29.1 Fundamental interactions.**

| Interaction (Force) | Range | Mediating particle | | Relative strength |
|---|---|---|---|---|
| Electromagnetic | Infinity $\left(F \sim \frac{1}{r^2}\right)$ | Photon | | $10^{-2}$ |
| | | Mass 0 kg | Electric charge 0 | |
| Strong | Short (about $10^{-15}$ m) | Gluons | | 1 |
| | | Mass 0 kg | Electric charge 0 | |
| Weak | Short (about $10^{-18}$ m) | $W^{\pm}$ | | $10^{-13}$ |
| | | Mass $1.43 \times 10^{-25}$ kg | Electric charge $+e$ for $W^+$ $-e$ for $W^-$ | |
| | | $Z^0$ | | |
| | | Mass $1.62 \times 10^{-25}$ kg | Electric charge 0 | |
| Gravitational | Infinity $\left(F \sim \frac{1}{r^2}\right)$ | Graviton | | $10^{-38}$ |
| | | Mass 0 kg | Electric charge 0 | |

As you can see, the mediator particles are massless except for the $W^{\pm}$ and $Z^0$, which have masses of about 100 times that of a proton. This huge mass explains why the weak interaction has such a short range. When produced as virtual particles, the $W^{\pm}$ and $Z^0$ can only last very briefly, so briefly that even if they are traveling at near light speed they can travel only an average of $10^{-18}$ m.

In Table 29.1 we listed the masses of the interaction mediators in kilograms. This is not the common practice in particle physics, since the values are so small in those units. In fact, mass is not generally used at all. Instead, the "mass" of a particle is given in terms of its rest energy $E_0 = mc^2$ measured in electron volts (eV). Typical particle "masses" are in the mega-electron volt ($1\ \text{MeV} = 10^6\ \text{eV}$) or giga-electron volt ($1\ \text{GeV} = 10^9\ \text{eV}$) range.

**QUANTITATIVE EXERCISE 29.2 Units in particle physics**

Convert the masses of the $W^{\pm}$ and $Z^0$ particles into electron volts.

**Represent mathematically** First we need to determine the rest energies of the particles in joules, then convert them into electron volts, where $1\ \text{eV} = 1.6 \times 10^{-19}\ \text{J}$.

**Solve and evaluate** The rest energy of the $W^{\pm}$ particles is

$$E_0 = mc^2 = (1.43 \times 10^{-25}\ \text{kg})(3.00 \times 10^8\ \text{m/s})^2 = 1.29 \times 10^{-8}\ \text{J}$$

Converting to eV, we get

$$(1.29 \times 10^{-8}\ \text{J})\left(\frac{1\ \text{eV}}{1.6 \times 10^{-19}\ \text{J}}\right) = 80.6 \times 10^9\ \text{eV} = 80.6\ \text{GeV}$$

Using the same method, the rest energy of the $Z^0$ particle is 91.3 GeV.

**Try it yourself:** Determine the mass in mega-electron volts of the $1.67 \times 10^{-27}$ kg proton and of the $9.11 \times 10^{-31}$ kg electron.

*Answer:* 938 MeV and 0.51 MeV.

**Review Question 29.2** Why do two protons repel each other—for example, in an ionized gas of hydrogen—but attract each other when they are inside a nucleus?

# 29.3 Elementary particles and the Standard Model

Since the early 20th century, physicists have been discovering new elementary particles using particle accelerators such as the electron-positron collider at the Stanford Linear Accelerator Center (SLAC), the proton-antiproton collider at Fermilab in Illinois, and the Large Hadron Collider (LHC) at CERN in Europe. Because the total energy of these accelerated particles is significantly greater than the electron and proton rest energies, it is possible for additional particles to be produced in the collisions. The properties of these additional particles can be determined using elaborate detectors. Most of the particles produced are not stable. They quickly decay into other particles until eventually only stable particles remain.

< **Active Learning Guide**

The interaction mediators we discussed in the previous section are the mechanism behind the four fundamental interactions. These form one category of elementary particles. The building blocks of matter form two more categories, **leptons** and **hadrons**, which are the topic of this section.

## Leptons

Leptons interact only through the weak, electromagnetic (if electrically charged), and gravitational interactions (though very weakly), but not through the strong interaction. The electron is an example of a lepton, as is the electron neutrino $\nu_e$, which is produced in beta-plus decay (Chapter 28). The electron neutrino is electrically neutral. These two particles form what is called a generation (or family) of leptons.

In 1936 the first member of a second lepton generation was discovered. The **muon** $\mu^-$ is similar to the electron in that it is negatively charged, but it is approximately 200 times more massive. It also is unstable, with a half-life of $1.5 \times 10^{-6}$ s. The corresponding **muon neutrino** $\nu_\mu$ was discovered in 1962. A third generation of leptons was later discovered, namely, the **tau** $\tau^-$ and the **tau neutrino** $\nu_\tau$ in 1974 and 2000, respectively. The tau is 17 times more massive than the muon and has an even shorter half-life of $2.0 \times 10^{-13}$ s. All six of these particles (the electron, muon, tau, and their corresponding neutrinos) have their own distinct antiparticles. The electron and the three neutrinos are stable particles and do not decay. Some of the properties of leptons are shown in **Table 29.2**.

**Table 29.2 Lepton generations (families).**

| | Electron generation | | Muon generation | | Tau generation | |
|---|---|---|---|---|---|---|
| | Mass (GeV) | Charge $1.6 \times 10^{-19}$C | Mass (GeV) | Charge $1.6 \times 10^{-19}$C | Mass (GeV) | Charge $1.6 \times 10^{-19}$C |
| Particle | Electron $e$ | | Muon $\mu$ | | Tau $\tau$ | |
| | $5.1 \times 10^{-4}$ | −1 | 0.106 | −1 | 1.777 | −1 |
| Antiparticle | Positron $e^+$ | | Antimuon $\overline{\mu}$ | | Antitau $\overline{\tau}$ | |
| | $5.1 \times 10^{-4}$ | +1 | 0.106 | +1 | 1.777 | +1 |
| Paired neutrino | Electron neutrino $\nu_e$ | | Muon neutrino $\nu_\mu$ | | Tau neutrino $\nu_\tau$ | |
| | $<2 \times 10^{-9}$ | 0 | $<1.9 \times 10^{-4}$ | 0 | $<1.8 \times 10^{-2}$ | 0 |
| Paired antineutrino | Electron antineutrino $\overline{\nu}_e$ | | Muon antineutrino $\overline{\nu}_\mu$ | | Tau antineutrino $\overline{\nu}_\tau$ | |
| | $<2 \times 10^{-9}$ | 0 | $<1.9 \times 10^{-4}$ | 0 | $<1.8 \times 10^{-2}$ | 0 |

*Source:* Young, H. D (2012). *College Physics*, Ninth edition, p. 1032. Pearson Education, Inc.

Until 1998 physicists had no evidence to suggest that the masses of the neutrinos were anything other than zero (though the possibility was considered as early as the 1950s). In 1998 the Super-Kamiokande neutrino detector in Japan established that the weak interaction allows for processes that can convert one type of neutrino into another, a process called **neutrino oscillation**. This process is only possible if one or more of the neutrinos have nonzero mass. Since then, additional evidence from both particle physics and astrophysics experiments suggests that all three neutrinos have a rest energy of about 1 eV. Experiments are underway to measure this more precisely. Because neutrinos do not interact via the electromagnetic or strong interaction, they are extremely difficult to detect.

## Hadrons

Two different types of hadrons can be distinguished—**baryons** and **mesons**. You already know two examples of baryons, the proton and the neutron, but there are many more. In 1949–1952 new higher mass baryons were discovered in interactions between cosmic rays and the atmosphere: the lambda particle $\Lambda^0$ and a set of four similar baryons known as the delta particles $\Delta^-$, $\Delta^0$, $\Delta^+$, and $\Delta^{++}$. Additional baryons were discovered in later years.

The first example of a meson was Hideki Yukawa's suggestion in 1935 of the existence of new particles that mediated the strong interaction. He called these particles mesons. The motivation for Yukawa's ideas came from combining relativity theory with quantum theory. The masses of Yukawa's mesons were predicted to be about 200 times the electron mass. In 1947 physicists discovered a meson in cosmic rays that participated in strong interactions and had the correct properties to be Yukawa's meson (they called it a pi-meson or pion for short). Later, additional mesons were discovered, such as the $K^+$ and $K^0$.

The difference between baryons and mesons is in their internal structure and is discussed in the next subsection. With the exception of the proton, all hadrons are unstable. Their half-lives range from 610 s for the neutron to $10^{-24}$ s for the shortest lived. A small sample of some hadrons and their properties is provided in **Table 29.3**. Hundreds of hadrons have been discovered since 1950.

**Table 29.3 Properties of hadrons.**

| | Particle and antiparticle | Mass (GeV) | Electric charge (in $e$) | Half-life (s) |
|---|---|---|---|---|
| *Baryons* | | | | |
| Proton | $p, \bar{p}$ | 0.938 | 1, −1 | Infinity |
| Neutron | $n, \bar{n}$ | 0.940 | 0, 0 | 610 |
| Lambda | $\Lambda^0, \bar{\Lambda}^0$ | 1.116 | 0, 0 | $1.8 \times 10^{-10}$ |
| Sigma | $\Sigma^+, \bar{\Sigma}^-$ | 1.189 | 1, −1 | $5.6 \times 10^{-11}$ |
| | $\Sigma^0, \bar{\Sigma}^0$ | 1.193 | 0, 0 | $5.1 \times 10^{-20}$ |
| | $\Sigma^-, \bar{\Sigma}^+$ | 1.197 | −1, 1 | $1.0 \times 10^{-10}$ |
| Omega | $\Omega^-, \Omega^+$ | 1.672 | −1, 1 | $5.7 \times 10^{-11}$ |
| *Mesons* | | | | |
| Pion | $\pi^+, \pi^-$ | 0.140 | 1, −1 | $1.8 \times 10^{-8}$ |
| | $\pi^0$, self | 0.135 | 0, 0 | $5.9 \times 10^{-17}$ |
| Eta | $\eta^0$, self | 0.548 | 0, 0 | $<10^{-18}$ |
| Kaon | $K^+, K^-$ | 0.494 | 1, −1 | $8.6 \times 10^{-9}$ |

## Quarks

The large number of hadrons and the differences in their properties led to the belief that they were not truly elementary particles. In 1964 Murray Gell-Mann and his collaborators in the United States and Kazuhiko Nishijima in Japan independently proposed a new model of hadrons. In this model the hadrons are complex objects made of a small number of more fundamental particles that combine in different combinations to make all of the existing hadrons. The idea is similar to combining protons, neutrons, and electrons in different combinations to make the elements of the periodic table.

The concept of hadrons having internal structure helped explain the results of experimental investigations in which electron beams with energies of about 25 GeV (25 billion eV) were shot into a sample of liquid hydrogen. Electrons were scattered at various angles, similar to the alpha particles shot by Rutherford's colleagues at gold foil atoms. Electrons that were scattered at large angles had energy changes as if they had collided with tiny particles within the protons (**Figure 29.5**) rather than with the protons themselves.

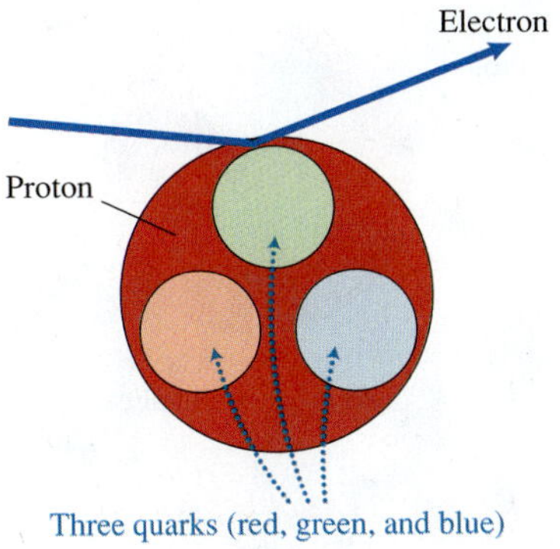


**Figure 29.5** Electron scattering from protons indicates that they are made of three quarks.

These particles had to be electrically charged for the proton to be charged. Surprisingly, the experiments showed that the particles had *fractional* electric charge. These particles became known as **quarks**. When the quark model was first invented, only three quarks were needed to build all the hadrons that were known. These quarks were given the whimsical names **up**, **down**, and **strange**. Since then, three more quarks have been proposed (called **charm**, **top**, and **bottom**) to explain the structure of the ever more massive hadrons being produced in increasingly powerful accelerators. This brings the total number of quarks to six, all of which have since been discovered experimentally.

These different quark types are known in the physics community as **flavors** (having nothing to do with the sense of taste). Quarks interact through the strong (via gluons), weak (via the $W^{\pm}$ and $Z^0$), and electromagnetic (via photons) interactions. Because the quarks have nonzero mass, they also interact gravitationally, but only extremely weakly. The difference between baryons and mesons comes from their quark content: baryons are bound states of three quarks, while mesons are bound states of one quark and one antiquark.

Recall that leptons come in three generations, each with a negatively charged member (electron, muon, or tau) and a neutrino for a total of six leptons. The quarks also fall into three generations (up/down, charm/strange, top/bottom) for a total of six quarks. This is no coincidence—physicists have learned there is a deep connection between the leptons and the quarks.

Particular properties of particles determine whether they participate in particular interactions. For example, objects with nonzero electric charge participate in the electromagnetic interaction. The property that determines if a particle participates in the strong interaction is known as **color charge**. It is not related to the colors we observe with our eyes, but rather is a technical term that is used as a name for this property. Because quarks participate in the strong interaction, they have color charge. A particle with color charge can interact with another color-charged particle by exchanging gluons.

Unlike electric charge, color charge comes in three varieties, which are known as **red**, **green**, and **blue** (again, these do not correspond to the colors we see with our eyes). Just as neutral atoms have a net electric charge of zero, composite particles such as the proton and neutron are made of three quarks with complementary colors, which results in an overall color-neutral object. In other words, a proton has one red quark, one blue quark, and one green quark. This neutrality is analogous to the effect of shining complimentary beams of red, blue, and green light on a surface. That surface glows white, which corresponds to color neutral (sometimes called colorless).

**Figure 29.6** A proton is made of three charged quarks, with colors red, green, and blue.

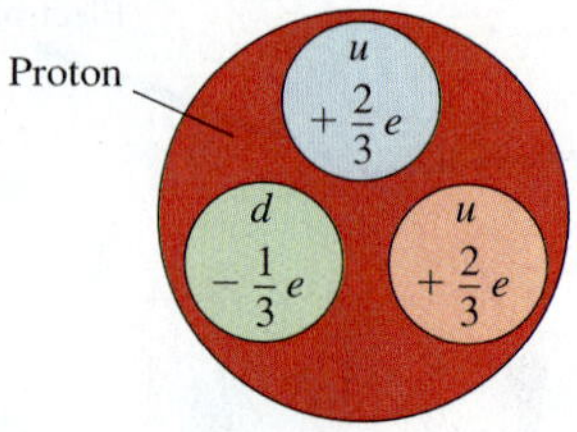


The net charge is $+e$ and is color neutral.

We mentioned earlier that quarks have fractional electric charge. The up quark has charge $(2/3)e$, where $e$ is the magnitude of the charge of the electron. The down quark has charge $(-1/3)e$. It is reasonable to suggest, then, that the proton is composed of two up quarks and one down quark, or *uud* for short (**Figure 29.6**). This results in a total electric charge of $(2/3)e + (2/3)e + (-1/3)e = e$, consistent with the charge of the proton. The three quarks also have color charge, one red, one green, and one blue, so that the proton is color neutral. In the next example we determine the quark content of a neutron.

**CONCEPTUAL EXERCISE 29.3 Making a neutron**

What combination of quarks will combine to have the correct properties to be a neutron?

**Sketch and translate** Because the electric charge of the neutron is zero, the charges of its component quarks must add to zero. The neutron is a baryon, so it is composed of three quarks (and no antiquarks).

**Simplify and diagram** This is accomplished if we include one up quark and two down quarks.

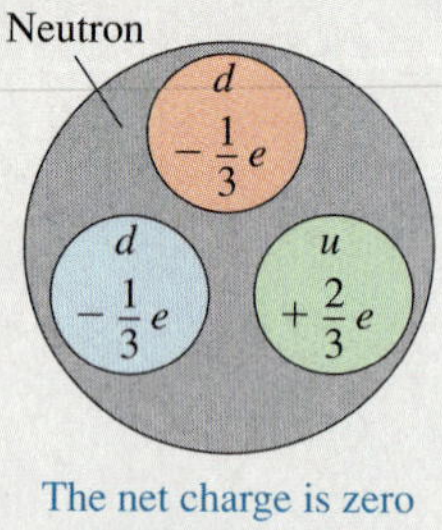


The net charge is zero and is color neutral.

$$2(-1/3) + (2/3) = 0$$

Two *d* quarks — One *u* quark

**Try it yourself:** How does the proton composition differ from the neutron composition? What would have to happen for a neutron to be converted into a proton?

*Answer:* The proton has two up quarks and one down quark. A neutron could be converted into a proton if one down quark were converted to an up quark. This sounds very similar to what occurs in beta decay.

We stated earlier that an antineutron has properties besides charge that differentiate it from a neutron. An antineutron is composed of three antiquarks, namely, one up antiquark and two down antiquarks.

**Figure 29.7** summarizes our understanding of the structure of matter and the role of quarks in that structure. All of the matter that makes up our local world is composed of electrons and two types of quarks (up and down).

**Figure 29.7** A summary of particles (matter) and their interactions.

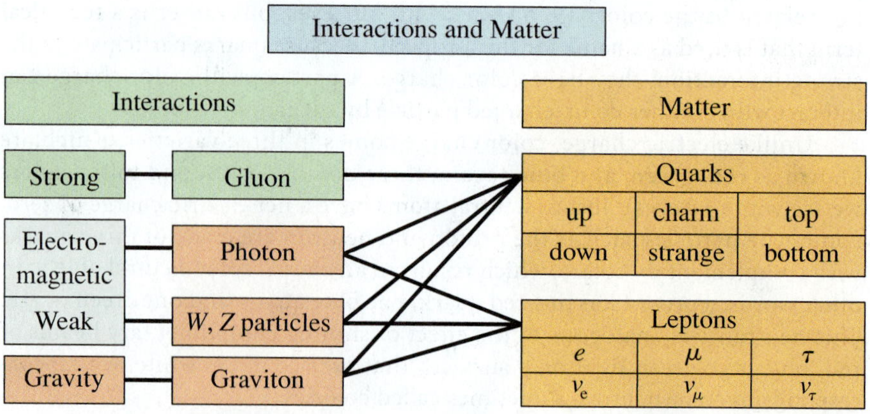

## Confinement

No experiment has ever produced a quark in isolation. Every quark and antiquark ever produced has always been part of a hadron (either a three-quark baryon or a quark-antiquark meson). Trying to split a proton into quarks by shooting a high-energy particle into it produces additional quark-antiquark pairs rather than separate quarks. These pairs then combine with the original proton's quarks to form new baryons and mesons. This phenomenon is called **confinement**, and it is an indication of a feature of the strong interaction that is very different from the other three interactions. The strong interaction between quarks is weakest when they are close together and gets stronger the farther apart the quarks are. The reasons why only the strong interaction exhibits confinement are beyond the scope of this book, but the feature is crucial to explaining the structure and stability of the protons and neutrons in every atomic nucleus in your body.

## Development of the Standard Model

With this basic understanding of elementary particles (interaction mediators, leptons, and hadrons) we now can discuss the **Standard Model**, the name given to the combined theory of the building blocks of matter and their interactions. Its first versions were constructed during the first half of the 20th century, and it has continued to evolve ever since.

In the late 1940s physicists Richard Feynman, Julian Schwinger, and Sin-Itiro Tomonaga independently combined the ideas of special relativity and quantum mechanics into a single model of the electromagnetic interaction that explains all electromagnetic phenomena. (The three of them were jointly awarded the Nobel Prize in physics in 1965 for this work.) Their model is known as **quantum electrodynamics** (QED) and is a cornerstone of the Standard Model. In 1957–1959 Julian Schwinger, Sidney Bludman, and Sheldon Glashow, working separately, proposed a model in which the weak interaction was mediated by two massive charged particles, which were later called the $W^+$ and $W^-$ particles. Chen-Ning Yang and Robert Mills developed the mathematical framework needed to describe this model.

In 1967 Sheldon Glashow, Abdus Salam, and Steven Weinberg independently put forth a model that unified the electromagnetic and weak interactions into a single interaction, which they called the **electroweak model**. This model made several striking predictions. First, it predicted that in the very distant past when the universe was much smaller and very much hotter, all particles were massless. Second, it predicted the existence of a particle, which became known as the **Higgs particle** after physicist Peter Higgs. Third, it predicted that as the universe cooled, the Higgs particle began interacting significantly with other elementary particles, reconfiguring them into the familiar forms they have today (most having nonzero mass). This is known as the **Higgs mechanism**.

In 1973, using Yang's and Mills' mathematical framework, Harald Fritzsch and Murray Gell-Mann formulated **quantum chromodynamics** (QCD), a mathematical model of the strong interaction in terms of the exchange of gluons between quarks. By the end of 1974, results from the first experiments testing QCD began arriving. Among them was the discovery of the so-called $J/\psi$ (J-psi) particle, a meson composed of one charm and one anticharm quark. The existence of the J-psi particle was a successful prediction of QCD. Between 1976 and 1979 scientists discovered the tau lepton and bottom quark and found direct evidence for gluons.

The 1980s brought the discoveries of the predicted weak interaction mediators $W^{\pm}$ and $Z^0$. The 1990s gave us the top quark, and in 2000 the tau neutrino was discovered. In July 2012 CERN announced the discovery of a particle that may be the long-sought-after Higgs particle.

Although the predictions of the Standard Model of particle physics are very consistent with the outcomes of particle physics experiments, several unanswered questions remain and are the focus of active research:

1. Can the strong interaction be unified with the electroweak interaction, just as the weak and electromagnetic interactions were unified?
2. Why are there only three families of quarks/leptons?
3. Are the Standard Model particles truly fundamental, or do they have internal structure?
4. Are there additional particles beyond those predicted by the Standard Model? Although the newly discovered particle at CERN appears to be a Higgs particle, more work is needed to determine if it is the Higgs particle predicted by the Standard Model or the lightest of a series of Higgs particles predicted by theories that go beyond the Standard Model.

### Summary of the Standard Model

The Standard Model is the currently accepted model of fundamental particles and their interactions (see Figure 29.7). It includes the following:

1. Quarks and leptons make up the matter of the universe. The quarks participate in all four fundamental interactions. The leptons participate in all but the strong interaction.
2. The theory of strong interactions (QCD) mediated by gluons.
3. The theory of electromagnetic (QED) and weak interactions mediated by photons and the $W^{\pm}$ and $Z^0$ particles.
4. The Higgs particle, which explains, through the Higgs mechanism, why some of the fundamental particles have nonzero mass.

The Standard Model does not include the gravitational interaction. How to combine this interaction with the Standard Model is a very challenging unsolved problem in physics. One framework for potentially doing this is *string theory*, and while much progress has been made throughout the past 40 years, no one knows yet whether string theory can achieve this goal.

**Review Question 29.3** Using what you have learned about particle physics, describe as many differences as you can between a proton and an electron.

## 29.4 Cosmology

**Active Learning Guide ›**

As we have seen, almost all elementary particles have distinct antiparticles. Furthermore, when particles are produced they are always produced in matter-antimatter pairs. Why, then, is our universe not filled with equal numbers of particles and antiparticles? Why is there an imbalance? These questions are answered in part by particle physics and by **cosmology**—a branch of physics that studies the composition and evolution of the universe as a whole.

Recall that independently of which direction we look, distant galaxies are moving away from us—the universe is expanding (Chapter 25). If we could run this expansion in reverse, the universe would get smaller, denser, and hotter. About 13.7 billion years ago, the entire universe would have been in an almost unimaginably hot and dense state of nearly zero size, which then rapidly expanded. This initial expansion is called the **Big Bang** (see the timeline from the Big Bang to the present in **Figure 29.8**). The Big Bang model of cosmology

**Figure 29.8** The development of the universe since the Big Bang 13.7 billion years ago.

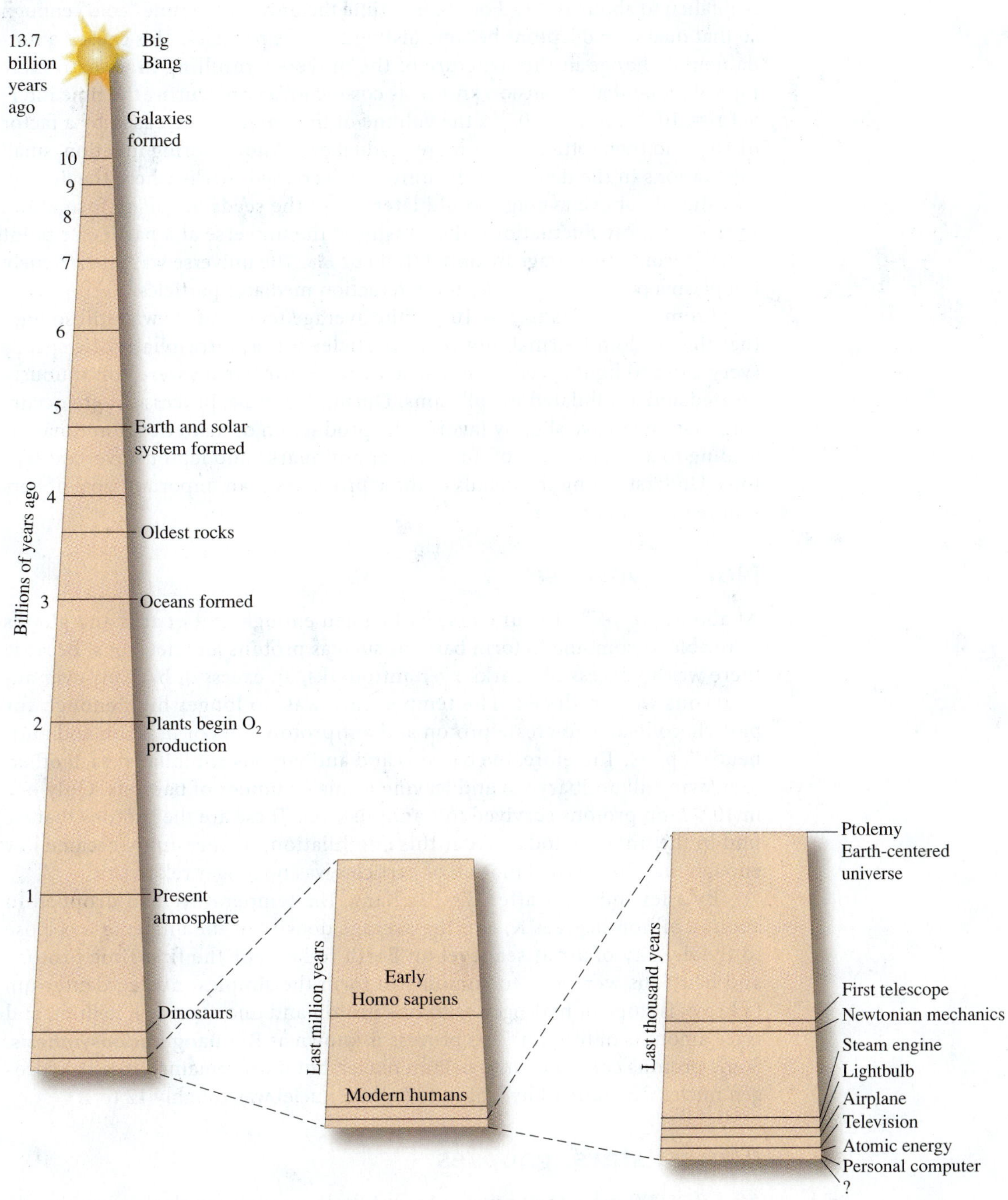


was first proposed in 1927 by a Belgian Catholic priest, Georges Lemaître, and more fully developed by George Gamow, Ralph Alpher, and Robert Herman in the late 1940s. The model explained the observed red shifts of spiral nebulae, now called galaxies. It also accounted for the fact that the red shifts were greater for more distant nebulae than for closer nebulae. Below is a brief summary of what we know about the history of the universe.

## Inflation

From $t = 0$ s to $t = 10^{-43}$ s, the average temperature of the universe was around $10^{32}$ K. At that time it was so hot that quarks and leptons could easily be converted into each other. At about $t = 10^{-36}$ s the temperature of the universe had fallen to about $10^{28}$ K. For the first time the universe became "cold" enough so that quarks and leptons became distinguishable particles. This caused a fundamental change in the structure of the universe, resulting in an extremely rapid exponential expansion known as cosmic **inflation**. During the time interval $t = 10^{-36}$ s to $t = 10^{-32}$ s the volume of the universe increased by a factor of $10^{26}$ and then settled into a more gradual expansion. During inflation, small fluctuations in the density of the universe decreased. Areas where the density was slightly above average would later act as the seeds of galaxy formation. Even with these fluctuations, the density of the universe at a particular point differed from the average by only 1/1000 of 1%. The universe was an extremely hot plasma of quarks, leptons, and interaction mediator particles.

From $t = 10^{-32}$ s to $t = 10^{-12}$ s the average temperature was still so high that the random thermal motion of particles was at ultra-relativistic speeds (very close to light speed), and particles and antiparticles were continuously created and annihilated in collisions. During this time, processes were occurring that must have slightly favored the production of matter over antimatter, leading to a small excess of quarks over antiquarks and leptons over antileptons. Understanding the details of these processes is an important goal of current research in physics.

## Nucleosynthesis

At about $t = 10^{-6}$ s the universe had cooled enough that quarks and gluons were able to combine to form baryons such as protons and neutrons. Because there was an excess of quarks over antiquarks, an excess of baryons over antibaryons was produced. The temperature was no longer high enough for particle collisions to create proton and antiproton pairs or neutron and antineutron pairs. Therefore, the baryons and antibaryons annihilated each other, destroying all antibaryons and leaving a small number of baryons. Only one in 10 billion protons survived this annihilation. These are the protons that we find in the universe today. After this annihilation, temperatures became low enough that the thermal motion of particles was no longer relativistic.

By a few minutes after the Big Bang, the temperature had dropped to about a billion degrees K, and the average density of the universe was close to the density of air at sea level on Earth today. For the first time protons and neutrons were able to combine to form the simplest nuclei: deuterium (a heavy isotope of hydrogen with one proton and one neutron), helium, and trace amounts of lithium. This process is known as **Big Bang nucleosynthesis**. Some protons combined into helium nuclei, but most remained free as hydrogen nuclei. The ratio of hydrogen to helium nuclei was roughly 12 to 1.

## Atoms, stars, galaxies

After 370,000 years the universe became cold enough for electrons to finally combine with nuclei to form the first neutral atoms. When this happened, the universe became transparent to electromagnetic radiation. Prior to this, the universe was a plasma, an ionized gas of nuclei and electrons. Plasmas do not allow photons to travel freely. Now that the universe was transparent, the photons that were produced when the neutral atoms formed were able to travel freely. These photons are still present in the universe today in the form of the **cosmic microwave background** (CMB), the afterglow of a process that took place more than 13 billion years ago. Due to the expansion of the universe, these photons have been red-shifted so that they have an effective temperature today of about 2.7 K. This can be thought of as the ambient temperature of the present universe.

The discovery of the CMB by Arno Penzias and Robert Wilson in 1965 was one of the most significant pieces of supporting evidence for the Big Bang model. The existence of this radiation was a prediction of the model. The model also made predictions about the relative abundances of hydrogen, helium, and lithium that resulted from nucleosynthesis. These predictions are consistent with experiments as well.

When the universe had cooled enough, the gravitational interaction became the dominant driver of its further evolution. Density fluctuations caused some regions to begin contracting. This process led to the formation of the first galaxies and stars just 500,000 years after the Big Bang. These early stars went through their life cycles, some ending in a violent collapse and explosion known as a **supernova**. Through nuclear fusion processes, heavier elements such as carbon, oxygen, iron, and gold were produced in supernovae and released into space to become parts of planets such as our own.

The present matter on Earth consists of quarks that were part of the early universe. These quarks combined to form protons and neutrons in the early universe, which then formed complex nuclei and atoms during the life cycles of stars. Thus, our bodies really are composed of matter that was created near the dawn of time and processed in stellar explosions before becoming part of us.

In the latter part of the 20th century two serious problems with this picture of the universe arose that have yet to be resolved. First, galaxies aren't nearly massive enough to explain the way stars move within them. Second, the universe is speeding up in its expansion rather than slowing down, as the gravitational interaction would predict. What possible explanations have physicists devised to explain these unexpected experimental results?

**Review Question 29.4** Explain how the Big Bang model predicts the existence of the cosmic microwave background (CMB).

## 29.5 Dark matter and dark energy

Earlier (in Chapter 4), we learned that Earth's speed around the Sun is determined by the radius of Earth's orbit and the mass of the Sun. In a similar way, the solar system's speed around the galaxy is determined by the radius of the solar system's orbit and the total mass of everything that lies between that orbit and the center of the galaxy.

When astronomers measure the mass of all the stars and gas that they can see, however, they find that the total mass is only about one-tenth of the mass needed to account for the speed of the solar system around the center of the galaxy. Similarly, when we look at the motion of galaxies relative to each other, we find the same thing; there is far too little visible mass to account for the galaxies' motion. Apparently, the universe is "missing" about 90% of the mass needed to account for the observed motion of stars and galaxies. How can this contradiction be resolved?

‹ Active Learning Guide

### Dark matter

The first evidence of the problem came in 1933 when astrophysicist Fritz Zwicky looked at the galaxies in the Coma cluster. He found that the galaxies at the edge of the cluster were traveling far too fast to remain part of the cluster. He speculated that there must be some unseen **dark matter** present in the Coma cluster, but for about 40 years his observation was the only evidence for its existence.

In the 1970s astronomer Vera Rubin presented further evidence. She looked at individual galaxies and found that stars near the edge of a galaxy were also traveling too fast to remain part of the galaxy. It was at this point that the dark matter explanation started to become more widely accepted.

Scientists suggested a new form of unseen matter and devised experiments to detect it as directly as possible. But as of 2012, direct detection of dark matter has not been accomplished. Astronomers are actually more certain about what this unseen dark matter is not than about what it is. First, it does not emit photons or otherwise participate in the electromagnetic interaction (this is why it is called "dark"). As a result, dark matter cannot be some sort of dark cloud of protons or gaseous atoms, because these could be detected by the scattering of radiation passing through them. Several hypotheses have been proposed, however, for what this dark matter might be. Two of these go by the names MACHOs (massive compact halo objects) and WIMPs (weakly interacting massive particles).

**MACHOs (Massive Compact Halo Objects)** These objects could be black holes, neutron stars, or brown dwarfs (objects that were not massive enough to achieve nuclear fusion and become stars). Astronomers have been detecting MACHOs using their gravitational effects on the light from distant objects (a phenomenon called gravitational micro-lensing). When one of these MACHOs passes in front of a distant object, such as a star or another galaxy, the light bends around the MACHO and is "focused" for a short time interval, making the light appear brighter. The MACHO Project has observed about 15 such lensing events over a span of 6 years of observations. This small number of events translates into MACHOs accounting for at most 20% of the dark matter in our galaxy. There must be another (or an additional) explanation.

**WIMPs (Weakly Interacting Massive Particles)** These objects are more exotic in nature. WIMPs are elementary particles, but not the quarks and leptons that make up ordinary matter. They are "weakly interacting" because they can pass through ordinary matter with almost no interaction, and light is not absorbed or emitted by them. They are "massive" in that their mass is not zero. Prime candidates for WIMPs include neutrinos, axions, and neutralinos, described next (the latter two are not Standard Model particles and hence require the Standard Model to be extended to accommodate their existence).

Cosmologists have a good idea of how many neutrinos there are in the universe. The Standard Model gives a zero mass to the neutrinos, but the model can be extended to include nonzero masses. Recent experiments have strongly suggested that neutrinos have a rest energy in the range of 0.1 to 1.0 eV, a million times smaller than the electron. For many years astronomers hoped that if neutrinos had a sufficient mass they could be the dark matter. However, it appears neutrinos are too light for this. Some **grand unified theories** (theories that combine the strong, weak, and electromagnetic interactions into a single interaction) predict the existence of a so-called "sterile neutrino" that could be even less interacting than standard-model neutrinos and also far more massive. Physicists do not know how to detect such a particle, but if it exists in sufficient abundance it could account for the dark matter.

Physicists originally proposed the existence of **axions** to explain a mystery about the properties of the neutron. These proposed axions have very small mass (less than that of the electron), no electric charge, and very little interaction with Standard Model particles. However, they would have been produced abundantly in the Big Bang. Current searches for axions include Earth-based laboratory experiments and astronomical searches in the halo of our galaxy and in the Sun. Axions have never been experimentally observed.

The **neutralino** is an example of another type of hypothetical particle known as a **superpartner**. Several outstanding problems in theoretical physics are understood better by suggesting the existence of what is known as **supersymmetry**. Supersymmetry is an extension of the Standard Model that effectively doubles the number of elementary particles and gives insight into the cosmological constant problem (to be described shortly), allows for a more precise understanding of the unification of interactions in grand unified theories, and gives a potential candidate for dark matter. None of the superpartners have ever been detected, but their detection is one of the primary design goals of the Large Hadron Collider.

Supersymmetry predicts that the lightest superpartner is stable and very weakly interacting. There are many versions of supersymmetry, but in some of them the lightest superpartner is known as the neutralino and is predicted to have a mass of 100 to 1000 times the mass of a proton. Astronomers and physicists hope to detect neutralinos by either using underground detectors, searching the universe for signs of their interactions, or producing them in particle accelerators.

Because none of the particles suggested as a solution to the dark matter problem have been detected, the mystery of the missing matter of the universe is still largely unsolved.

Although the dark matter problem has been around since the 1930s, one thing that seemed irrefutable through the 20th century was that the gravitational interaction between all the massive objects in the universe would gradually slow the expansion rate of the universe. Then in 1998, two independent experiments using the Hubble Space Telescope produced observations of very distant supernovae that showed that the universe is currently expanding more rapidly than it was 5–10 billion years ago. The expansion of the universe has not been slowing; it has been speeding up! No one expected this, and no one at the time knew how to explain it. The 2011 Nobel Prize in physics was awarded to Saul Perlmutter, Brian P. Schmidt, and Adam G. Riess for the discovery of the accelerating universe. This led to extensions of the Big Bang model that could potentially explain this accelerated expansion. The most widely accepted of these is known as **dark energy**.

## Dark energy

Physicists came up with many ideas to explain the accelerating expansion of the universe. Here are a few:

- Invoke a discarded feature of Einstein's general theory of relativity (our current best model of the gravitational interaction) known as the **cosmological constant.**
- Suggest the existence of a strange kind of energy-fluid that fills space and has a repulsive gravitational effect.
- Propose a modified version of general relativity that includes a new kind of field that creates this cosmic acceleration.

Let's look at each of these ideas.

**The Cosmological Constant Model** The cosmological constant is a term that Einstein included in general relativity to allow for a static universe. At the time he constructed general relativity there was no experimental evidence for the Big Bang or the expansion of the universe. The dominant model of the universe was known as the **steady state model,** which asserted that the universe essentially didn't change in any major way as time passed. But general relativity predicted that a static universe was unstable. Einstein introduced the cosmological constant into general relativity in an attempt to allow the theory to accommodate a steady state universe.

Einstein could have predicted the expansion of the universe 10 years before it was discovered, but even he couldn't accept such a radical idea. It is said that he considered this his greatest blunder. However, because the cosmological constant introduces a repulsive effect into the equation to balance the attractive effect of gravity, it can also be used to describe the accelerated expansion. An idea introduced and discarded by Einstein 80 years earlier has been resurrected to explain recent observations.

**The Dark Energy Model** The second idea is actually a suggestion about what the cosmological constant term means physically. It seems to represent a type of **dark energy** that is present at every point in space with equal density. Even as the universe expands, this density does not decrease because it is a property of space itself. What's even stranger is that this energy has a negative pressure. In general relativity this produces a gravitationally repulsive effect on space.

As the universe expands, the matter density decreases, but the dark energy density remains constant. This means that as time passes, the attractive gravitational effect of the matter decreases while the repulsive gravitational effect of the dark energy remains the same. In other words, as time goes on, the repulsion gets stronger relative to the attraction. This is precisely what astronomers have observed—the expansion is more rapid today than it was in the past.

**Modified General Relativity** The third idea is in some sense the least radical. Just as general relativity is an improvement on Newton's view and is able to make more accurate predictions, the idea is that there is some better theory of the gravitational interaction that would make even better predictions than general relativity. The challenge has been to construct the new theory in such a way that it doesn't make predictions that contradict experiments that have already been done. Thus far physicists have been unsuccessful in doing this.

**Mysteries** Of these three ideas, dark energy has become the favored one. Various versions have been set forth, and the simplest of them do represent the dark energy mathematically in general relativity as a cosmological constant. Dark energy models have a problem, though, and it comes from a basic feature of the combination of relativity and quantum mechanics.

This combined model predicts that each elementary particle in the Standard Model is actually an excitation of an associated quantum field, similar to how the photon is one quantum of excitation of the electromagnetic field. Each of these quantum fields has a certain minimum energy, called its **zero point energy**. In many dark energy models, the dark energy is the sum of the zero point energies of all the quantum fields in the universe. But when physicists make estimates of what value these models predict for the cosmological constant, they get a result that is $10^{120}$ times the observed value. It's been said that this is the largest disagreement between prediction and experiment in all of science. This so-called **cosmological constant problem** is a major unsolved problem in physics.

**Figure 29.9** The proportion of matter, dark matter, and dark energy in the universe.

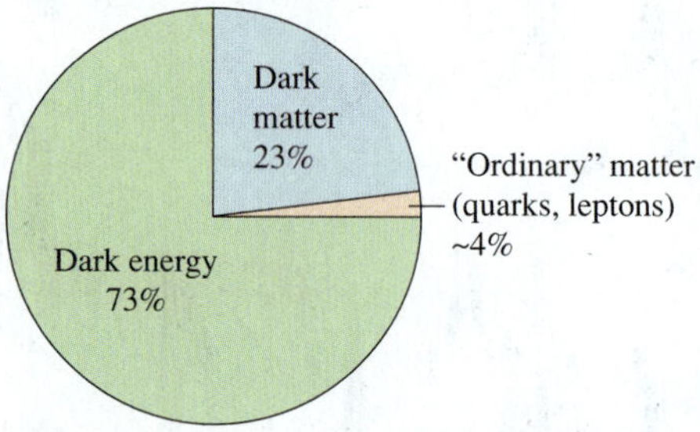


Supersymmetry, which predicts a doubling of all the types of elementary particles, suggests a possible solution. Each of these superpartner particles also has an associated quantum field. But the zero point energies of each of these additional quantum fields contribute oppositely to the dark energy compared to their non-super counterparts. This means the total dark energy density would be zero. That's not consistent with observations, because we do observe an accelerated expansion, but the dark energy density is measured to be very small, just not quite zero. That's a much better prediction than $10^{120}$ times the observed value! None of the superpartner particles have been observed, however, which means supersymmetry is not present in its full form in this universe. This idea allows for a small but nonzero dark energy density, possibly consistent with observation. One of the goals of the Large Hadron Collider is to produce some of these superpartner particles and see whether supersymmetry can resolve the cosmological constant problem.

Explaining the accelerated expansion of the universe is an even greater puzzle than explaining dark matter. We know how much dark energy there is because we know how it affects the universe's expansion. Other than that, it is a complete mystery. It turns out that roughly 73% of the total energy of the universe is dark energy. The dark matter's rest energy makes up about 23%. The rest—everything on Earth, everything ever observed with all of our instruments, all normal matter—adds up to ~4% of the energy in the universe (see **Figure 29.9**). We understand only a few percent of what comprises the universe we live in. That is a humbling but highly motivating realization.

**TIP** Dark matter and dark energy are ideas invented by physicists to explain patterns observed in nature. Because neither of these has yet been detected, it is possible that one or both of these ideas could be disproven.

**Review Question 29.5** How do we know that much more mass exists in the Milky Way galaxy than is present in the galaxy's stars and interstellar gas?

## 29.6 Is our pursuit of knowledge worthwhile?

In this textbook, we have been building models that describe the behavior of only 4% of the contents of our universe. The nature of the remaining 96% of our universe currently remains an unsolved problem. Is it worthwhile trying to learn about these other mysterious components of our universe?

In the late 1800s, the prime minister of England asked Michael Faraday what use there was for his idea of electromagnetic induction. Faraday could not say. But today we have electric power generators, microphones, credit card readers, electric guitar pickups, and electromagnetic braking systems for hybrid vehicles. All are based on electromagnetic induction.

In 1897, J. J. Thomson discovered and characterized the electron—he could not have imagined how this understanding would lead to the computing, communication, and entertainment devices that pervade our everyday lives.

In the 1930s, physicists at MIT were studying microwaves. What use would they have? Many historians believe that microwave radar saved England in World War II. Now, microwaves, a form of nonionizing radiation, are involved in satellite communications and GPS, and we use microwaves to cook our food.

Will our eventual knowledge of the other 96% of the universe someday make people's lives better? It's impossible to say for sure, but history suggests that it very likely will. Perhaps it will lead to new sustainable energy sources, the ability to easily travel to other planets, or ways to protect us from cosmic events that potentially threaten life on Earth. The most amazing future applications of understanding our universe are the ones no one has yet devised.

# Summary

**Words and pictorial and physical representations**

**Antiparticles** Each particle has an antiparticle with identical mass but opposite electric charge, color, etc. Some particles (such as the photon) are their own antiparticles. (Section 29.1)

**Fundamental interactions** All interactions between objects are gravitational, electromagnetic, weak, or strong. All other interactions are nonfundamental and can be understood in terms of these. (Section 29.2)

**Elementary particles** Every known particle is a quark (up, down, etc.), a lepton (electron, neutrino, etc.), or an interaction mediator (photon, gluon, etc.)

**Standard Model** The elementary particles together with the four fundamental interactions and the Higgs particle comprise the Standard Model of particle physics. (Section 29.3)

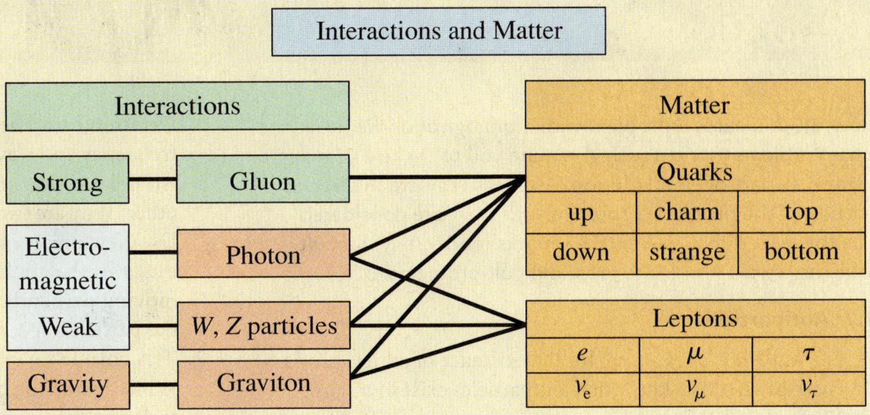


(continued)

**Cosmology** The universe rapidly expanded from a hot, dense state 13.7 billion years ago, an event called the Big Bang. As it expanded and cooled, nuclei formed, followed by atoms, stars, and galaxies. (Section 29.4)

**Dark matter and dark energy** Most of the energy in the universe is thought to be in two mysterious forms. Dark matter interacts only gravitationally and clumps around galaxies. Most of the mass of galaxies is proposed to be in their dark matter. Dark energy is thought to be spread uniformly throughout the universe and is responsible for its accelerated expansion. (Section 29.5)

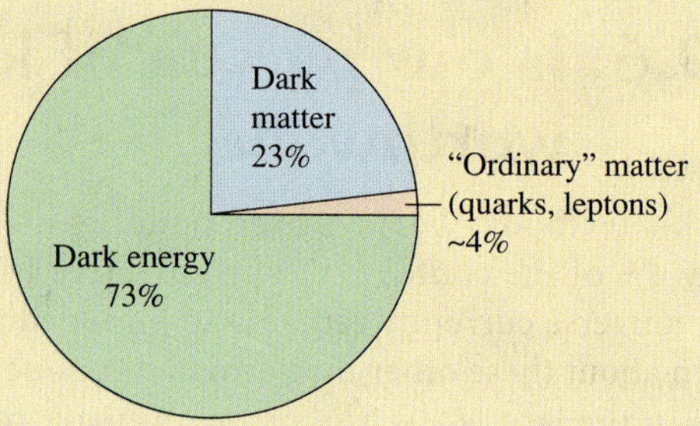


MP For instructor-assigned homework, go to MasteringPhysics.

# Questions

## Multiple Choice Questions

1. What is the composition of all baryons? (a) Three different color quarks (b) Three quarks of any color (c) Two quarks of any color (d) Three gluons (e) Two quarks of complementary color
2. What is the composition of all mesons? (a) Three different color quarks (b) Three quarks of any color (c) Two quarks of any color (d) Three gluons (e) Two quarks of complementary color
3. What is the weak interaction mediated by? (a) Gluons (b) The $W^+$, $W^-$, or $Z^0$ particles (c) Photons (d) Gravitons (e) Mesons
4. What is the strong interaction mediated by? (a) Gluons (b) The $W^+$, $W^-$, or $Z^0$ particles (c) Photons (d) Gravitons (e) Mesons

## Conceptual Questions

5. Give three examples of particles that are elementary (as far as is currently known) and three examples of those that are not.
6. What differs between a particle and its antiparticle?
7. Billions of neutrinos continually pass through each square centimeter of your body each second. Why don't you notice them?
8. How do we know that protons are very stable particles?
9. Free neutrons (those that are not part of a nucleus) decay into protons plus other particles with a half-life of about 10 min. Free protons, however, do not decay into neutrons plus other particles. Explain why.
10. What particle(s) could potentially not have an antiparticle or could be said to be its own antiparticle? Explain your reasoning.
11. Give an example of a long-range interaction and a short-range interaction. What is the difference between the mechanisms for each?
12. What are the components of the Standard Model?
13. Why do scientists need particle accelerators to discover new particles and to study the properties of known particles?
14. In 1930 the proton and neutron were thought to be elementary particles. We now know this is not true. How could physicists have been mistaken?

# Problems

Below, **BIO** indicates a problem with a biological or medical focus. Problems marked with ✎ require you to make a drawing or graph as part of your solution. Asterisks indicate the level of difficulty of the problem. Problems with no * are considered to be the least difficult. A single * marks moderately difficult problems. Two ** indicate more difficult problems.

### 29.1 Antiparticles

1. Explain Dirac's reasoning for the existence of antielectrons.
2. How do scientists know that antiparticles exist in nature?
3. **BIO** Explain how positron emission tomography (PET) works.
4. A proton and an antiproton, both with negligible kinetic energy, annihilate each other to produce two photons. Determine the energy in electron volts of each photon and the frequency and wavelength of each photon.
5. An isolated slow-moving electron and positron annihilate each other. Why are two photons produced instead of just one?
6. Use Newtonian circular motion concepts to show that the radius $r$ of the circle in which a charged particle spirals while moving perpendicular to a magnetic field is proportional to the particle's speed $v$.
7. ✎ A cosmic ray photon enters a cloud chamber from above. While inside the chamber the photon converts into an electron-positron pair. Draw a sketch of this situation. Label the photon, electron, and positron tracks and indicate the direction of the magnetic field within the cloud chamber.

8. A particle enters a cloud chamber from above traveling at $0.50c$ and leaves a curving track. The magnetic field within the cloud chamber has magnitude 0.120 T and points out of the page. The track curves to the right and has a radius of 0.020 m. Determine everything you can about this particle.

### 29.2 Fundamental interactions

9. Compare and contrast the fundamental interactions.
10. Even though the electromagnetic interaction is so much stronger than the gravitational interaction, the gravitational interaction clearly is much more relevant in everyday life. Do you agree with this statement? Explain either way.
11. Explain the mechanism behind elementary particle interactions.
12. * Explain how the energy to produce interaction mediators is provided.
13. What is the difference between a real particle and a virtual particle?
14. * Make an analogy between the interactions of elementary particles and some real-life interactions. Indicate which object in one system corresponds to each object in the other. Describe the limitations of your analogy.

### 29.3 Elementary particles and the Standard Model

15. Why are neutrinos difficult to detect?
16. Describe the evidence for neutrinos having nonzero mass.
17. What criteria do scientists use to classify elementary particles?
18. Compare and contrast leptons and hadrons.
19. How do we know that baryons are made of quarks? What is the experimental evidence that supports this idea?
20. Compare and contrast mesons and baryons.
21. How do quarks interact with each other inside a baryon?
22. * In what way is the interaction of electrically neutral atoms similar to the interactions of color-neutral protons and neutrons?
23. Describe the phenomenon of confinement.
24. What were four important steps in the building of the Standard Model?
25. ** What is the significance of the Higgs particle and the Higgs mechanism?
26. * What major piece of physics is not part of the Standard Model?
27. * Describe several open questions of the Standard Model.

### 29.4 Cosmology

28. * What is inflation, and what eventually happened as a result of the density fluctuations in the universe?
29. * What is the origin of the elementary particles?
30. * How would the universe be different if the number of particles in it had equaled the number of antiparticles?
31. * What process produced the cosmic microwave background?
32. * Our bodies contain significant amounts of carbon, oxygen, nitrogen, and other "heavy" elements. If only "light" elements such as hydrogen and helium were produced during Big Bang nucleosynthesis, how were the heavier elements produced?

### 29.5 Dark matter and dark energy

33. * What is the evidence that a large proportion of the mass of the universe is in the form of dark matter? Explain carefully.
34. * Describe as many hypotheses as you can about the nature of dark matter.
35. * What is the experimental evidence for dark energy? Explain carefully.
36. ** Describe as many hypotheses as you can that explain the accelerated expansion of the universe.
37. ** What is supersymmetry, and why is it a useful idea in physics?
38. ** What is the cosmological constant problem? Describe a possible resolution for it.

### 29.6 Is our pursuit of knowledge worthwhile?

39. What are the potential benefits of continued research in physics?

## General Problems

40. * An electron and a positron are traveling directly toward each other at a speed of $0.90c$ (90% of light speed) with respect to the lab in which the experiment is being performed. The electron and positron collide, annihilate, and produce a pair of photons. Determine the wavelength of each photon. If you think they will have the same wavelength, explain why.
41. * In order to discover the $W^+$ and $W^-$ particles, a large particle accelerator at CERN collided protons and antiprotons together at extremely high speeds. What is the minimum speed the proton and antiproton would need to produce a $W^\pm$ pair?

## Reading Passage Problems

**The solar neutrino problem** Our Sun continually converts hydrogen into helium through nuclear fusion. The chain of reactions that occurs can be combined into the single reaction shown here:

$$4{}^1_1\text{H} \rightarrow {}^4_2\text{He} + 2e^+ + 2\nu_e$$

Four hydrogen nuclei fuse into one helium nucleus. In the process two positrons and two electron neutrinos are produced. These positrons interact immediately with electrons present in the Sun and annihilate with them, producing gamma ray photons. Because the Sun is a plasma, these photons interact strongly with it and push outward, supporting the Sun against gravitational collapse. The neutrinos, however, have only an extremely small chance of interacting with matter and therefore stream out of the Sun's core unhindered, making their way to Earth and beyond. So many neutrinos are produced in the fusion reactions that $\sim 100$ billion of them pass through each square centimeter of your skin each second.

Solar neutrinos act as a "window" into the fusion reactions occurring in the Sun that normally would be obscured from our view. For example, if we can measure the rate at which solar neutrinos reach Earth we can estimate the nuclear reaction rate in the Sun. We also have an independent way to measure the nuclear reaction rate—by measuring the intensity of the Sun's radiation. The higher the intensity of the Sun's radiation, the greater the nuclear reaction rate at its core. These two methods should produce consistent results. But they don't. The amount of electron neutrinos detected is less than half the amount expected. This is the **solar neutrino problem**.

The resolution of the problem comes from suggesting that some of the electron neutrinos convert into other types of neutrinos (muon or tau) on their way to Earth. This phenomenon is called **neutrino oscillation**. For this to be possible, the neutrinos cannot have zero mass. The experiment that first detected solar neutrinos could not detect muon or tau neutrinos. In 2001 a new detector was used that could detect all three types of neutrinos. About two-thirds of the electron neutrinos had in fact oscillated into muon and tau neutrinos. This resolved the solar neutrino problem.

42. Why don't you notice the solar neutrinos passing through your body?
    (a) The Sun is very far away so the neutrinos have very little energy by the time they get to Earth.
    (b) Nearly all the neutrinos oscillate into other types that can pass easily through your body.
    (c) Solar neutrinos only participate in nuclear fusion reactions and since nuclear fusion doesn't happen in your body you don't notice the neutrinos.
    (d) Neutrinos only interact very weakly with atoms.
    (e) Only a very small number of them pass through your body each second.
43. In 1987 a supernova was detected in the Large Magellanic Cloud. Neutrinos coming from the supernova were detected, but there were fewer than expected. What might be the reason for that?
    (a) The supernova actually occurred farther from Earth than astronomers thought.
    (b) Some of the neutrinos were converted into other types of neutrinos as they traveled to Earth.
    (c) The model used to predict the number of neutrinos produced in the supernova makes some unreasonable assumptions.
    (d) The detector used was not as efficient as the designers thought.
    (e) All of these are possible explanations.
44. Originally a different way to resolve the solar neutrino problem was suggested—the nuclear reaction rate wasn't as high as astrophysicists thought. What would have to be true for this suggestion to be reasonable?
    (a) The mass of the Sun would have to be significantly different than thought.
    (b) The brightness of the Sun would have to be significantly different than thought.
    (c) The size of the Sun would have to be significantly different than thought.
    (d) The orbits of the planets would have to be significantly different than thought.
    (e) Both (b) and (c) are correct.

# APPENDIX A
# Mathematics Review

A study of physics at the level of this textbook requires some basic math skills. The relevant math topics are summarized in this appendix. We strongly recommend that you review this material and become comfortable with it as quickly as possible so that, during your physics course, you can focus on the physics concepts and procedures that are being introduced, without being distracted by unfamiliarity with the math that is being used.

## A.1 Exponents

Exponents are used frequently in physics. When we write $3^4$, the superscript 4 is called an **exponent** and the **base number** 3 is said to be raised to the fourth power. The quantity $3^4$ is equal to $3 \times 3 \times 3 \times 3 = 81$. Algebraic symbols can also be raised to a power—for example, $x^4$. There are special names for the operation when the exponent is 2 or 3. When the exponent is 2, we say that the quantity is **squared**; thus, $x^2$ is $x$ squared. When the exponent is 3, the quantity is **cubed**; hence, $x^3$ is $x$ cubed.

Note that $x^1 = x$, and the exponent is typically not written. Any quantity raised to the zero power is defined to be unity (that is, 1). Negative exponents are used for reciprocals: $x^{-4} = 1/x^4$. The exponent can also be a fraction, as in $x^{1/4}$. The exponent $\frac{1}{2}$ is called a **square root**, and the exponent $\frac{1}{3}$ is called a **cube root**.

For example, $\sqrt{6}$ can also be written as $6^{1/2}$. Most calculators have special keys for calculating numbers raised to a power—for example, a key labeled $y^x$ or one labeled $x^2$.

Exponents obey several simple rules, which follow directly from the meaning of raising a quantity to a power:

1. When the product of two powers of the same quantity are multiplied, the exponents are added:
$$(x^n)(x^m) = x^{n+m}.$$
For example, $(3^2)(3^3) = 3^5 = 243$. To verify this result, note that $3^2 = 9$, $3^3 = 27$, and $(9)(27) = 243$.
A special case of this rule is $(x^n)(x^{-n}) = x^{n+(-n)} = x^0 = 1$.
2. The product of two different base numbers raised to the same power is the product of the base numbers, raised to that power:
$$(x^n)(y^n) = (xy)^n.$$
For example, $(2^4)(3^4) = 6^4 = 1296$. To verify this result, note that $2^4 = 16$, $3^4 = 81$, and $(16)(81) = 1296$.
3. When a power is raised to another power, the exponents are multiplied:
$$(x^n)^m = x^{nm}.$$
For example, $(2^2)^3 = 2^6 = 64$. To verify this result, note that $2^2 = 4$, so $(2^2)^3 = (4)^3 = 64$.

If the base number is negative, it is helpful to know that $(-x)^n = (-1)^n x^n$, and $(-1)^n$ is $+1$ if $n$ is even and $-1$ if $n$ is odd.

**EXAMPLE A.1 Simplifying an exponential expression**

Simplify the expression $\dfrac{x^3y^{-3}xy^{4/3}}{x^{-4}y^{1/3}(x^2)^3}$, and calculate its numerical value when $x = 6$ and $y = 3$.

**Represent mathematically, solve, and evaluate**

We simplify the expression as follows:

$$\frac{x^3x}{x^{-4}(x^2)^3} = x^3x^1x^4x^{-6} = x^{3+1+4-6} = x^2;$$

$$\frac{y^{-3}y^{4/3}}{y^{1/3}} = y^{-3+\frac{4}{3}-\frac{1}{3}} = y^{-2}.$$

$$\frac{x^3y^{-3}xy^{4/3}}{x^{-4}y^{1/3}(x^2)^3} = x^2y^{-2} = x^2\left(\frac{1}{y}\right)^2 = \left(\frac{x}{y}\right)^2.$$

For $x = 6$ and $y = 3$, $\left(\dfrac{x}{y}\right)^2 = \left(\dfrac{6}{3}\right)^2 = 4$.

If we evaluate the original expression directly, we obtain

$$\frac{x^3y^{-3}xy^{4/3}}{x^{-4}y^{1/3}(x^2)^3} = \frac{(6^3)(3^{-3})(6)(3^{4/3})}{(6^{-4})(3^{1/3})([6^2]^3)}$$

$$= \frac{(216)(1/27)(6)(4.33)}{(1/1296)(1.44)(46{,}656)} = 4.00,$$

which checks.

This example demonstrates the usefulness of the rules for manipulating exponents.

**EXAMPLE A.2 Solving an exponential expression for the base number**

If $x^4 = 81$, what is $x$?

**Represent mathematically, solve, and evaluate**

We raise each side of the equation to the $\frac{1}{4}$ power: $(x^4)^{1/4} = (81)^{1/4}$. $(x^4)^{1/4} = x^1 = x$, so $x = (81)^{1/4}$ and $x = +3$ or $x = -3$. Either of these values of $x$ gives $x^4 = 81$.

Notice that we raised *both sides* of the equation to the $\frac{1}{4}$ power. As explained in Section A.3, an operation performed on both sides of an equation does not affect the equation's validity.

## A.2 Scientific Notation and Powers of 10

In physics, we frequently encounter very large and very small numbers, and it is important to use the proper number of significant figures when expressing a quantity. We can deal with both these issues by using **scientific notation**, in which a quantity is expressed as a decimal number with one digit to the left of the decimal point, multiplied by the appropriate power of 10. If the power of 10 is positive, it is the number of places the decimal point is moved to the right to obtain the fully written-out number. For example, $6.3 \times 10^4 = 63{,}000$. If the power of 10 is negative, it is the number of places the decimal point is moved to the left to obtain the fully written-out number. For example, $6.56 \times 10^{-3} = 0.00656$. In going from 6.56 to 0.00656, the decimal point is moved three places to the left, so $10^{-3}$ is the correct power of 10 to use when the number is written in scientific notation. Most calculators have keys for expressing a number in either decimal (floating-point) or scientific notation.

When two numbers written in scientific notation are multiplied (or divided), multiply (or divide) the decimal parts to get the decimal part of the result, and multiply (or divide) the powers of 10 to get the power-of-10 portion of the result. You may have to adjust the location of the decimal point in the answer to express it in scientific notation. For example,

$$(8.43 \times 10^8)(2.21 \times 10^{-5}) = (8.43 \times 2.21)(10^8 \times 10^{-5})$$
$$= (18.6) \times (10^{8-5}) = 18.6 \times 10^3$$
$$= 1.86 \times 10^4.$$

Similarly,

$$\frac{5.6 \times 10^{-3}}{2.8 \times 10^{-6}} = \left(\frac{5.6}{2.8}\right) \times \left(\frac{10^{-3}}{10^{-6}}\right) = 2.0 \times 10^{-3-(-6)} = 2.0 \times 10^3.$$

Your calculator can handle these operations for you automatically, but it is important for you to develop good "number sense" for scientific notation manipulations.

## A.3 Algebra

### Solving Equations

Equations written in terms of symbols that represent quantities are frequently used in physics. An **equation** consists of an equals sign and quantities to its left and to its right. Every equation tells us that the combination of quantities on the left of the equals sign has the same value as (is equal to) the combination on the right of the equals sign. For example, the equation $y + 4 = x^2 + 8$ tells us that $y + 4$ has the same value as $x^2 + 8$. If $x = 3$, then the equation $y + 4 = x^2 + 8$ says that $y = 13$.

Often, one of the symbols in an equation is considered to be the *unknown,* and we wish to solve for the unknown in terms of the other symbols or quantities. For example, we might wish to solve the equation $2x^2 + 4 = 22$ for the value of $x$. Or we might wish to solve the equation $x = v_0t + \frac{1}{2}at^2$ for the unknown $a$ in terms of $x$, $t$, and $v_0$.

An equation can be solved by using the following rule:

**An equation remains true if any operation performed on one side of the equation is also performed on the other side.** The operations could be (a) adding or subtracting a number or symbol, (b) multiplying or dividing by a number or symbol, or (c) raising each side of the equation to the same power.

**EXAMPLE A.3 Solving a numerical equation**

Solve the equation $2x^2 + 4 = 22$ for $x$.

**Represent mathematically, solve, and evaluate** First we subtract 4 from both sides. This gives $2x^2 = 18$. Then we divide both sides by 2 to get $x^2 = 9$. Finally, we raise both sides of the equation to the $\frac{1}{2}$ power. (In other words, we take the square root of both sides of the equation.) This gives $x = \pm\sqrt{9} = \pm 3$. That is, $x = +3$ or $x = -3$. We can verify our answers by substituting our result back into the original equation: $2x^2 + 4 = 2(\pm 3)^2 + 4 = 2(9) + 4 = 18 + 4 = 22$, so $x = \pm 3$ does satisfy the equation.

Notice that a square root always has *two* possible values, one positive and one negative. For instance, $\sqrt{4} = \pm 2$, because $(2)(2) = 4$ and $(-2)(-2) = 4$. Your calculator will give you only a positive root; it's up to you to remember that there are actually two. Both roots are correct mathematically, but in a physics problem only one may represent the answer. For instance, if you can get dressed in $\sqrt{4}$ minutes, the only physically meaningful root is 2 minutes!

**EXAMPLE A.4 Solving a symbolic equation**

Solve the equation $x = v_0t + \frac{1}{2}at^2$ for $a$.

**Represent mathematically, solve, and evaluate** We subtract $v_0t$ from both sides. This gives $x - v_0t = \frac{1}{2}at^2$. Now we multiply both sides by 2 and divide both sides by $t^2$, giving $a = \dfrac{2(x - v_0t)}{t^2}$.

As we've indicated, it makes no difference whether the quantities in an equation are represented by variables (such as $x$, $v$, and $t$) or by numerical values.

### The Quadratic Formula

Using the methods of the previous subsection, we can easily solve the equation $ax^2 + c = 0$ for $x$:

$$x = \pm\sqrt{\frac{-c}{a}}.$$

For example, if $a = 2$ and $c = -8$, the equation is $2x^2 - 8 = 0$ and the solution is

$$x = \pm\sqrt{\frac{-(-8)}{2}} = \pm\sqrt{4} = \pm 2.$$

The equation $ax^2 + bx = 0$ is also easily solved by factoring out an $x$ on the left side of the equation, giving $x(ax + b) = 0$. (To *factor out* a quantity means to isolate it so that the rest of the expression is either multiplied or divided by that quantity.) The equation $x(ax + b) = 0$ is true (that is, the left side equals zero) if either $x = 0$ or $x = -\frac{b}{a}$. Those are the equation's two solutions. For example, if $a = 2$ and $b = 8$, the equation is $2x^2 + 8x = 0$ and the solutions are $x = 0$ and $x = -\frac{8}{2} = -4$.

But if the equation is in the form $ax^2 + bx + c = 0$, with $a$, $b$, and $c$ all nonzero, we cannot use our standard methods to solve for $x$. Such an equation is called a **quadratic equation**, and its solutions are expressed by the **quadratic formula**:

**Quadratic formula** **For a quadratic equation in the form $ax^2 + bx + c = 0$, where $a$, $b$, and $c$ are real numbers and $a \neq 0$, the solutions are given by the quadratic formula:**

$$x = \frac{-b \pm \sqrt{b^2 - 4ac}}{2a}$$

In general, a quadratic equation has two roots (solutions). But if $b^2 - 4ac = 0$, then the two roots are equal. By contrast, if $b^2 < 4ac$, then $b^2 - 4ac$ is negative, and both roots are complex numbers and cannot represent physical quantities. In such a case, the original quadratic equation has mathematical solutions, but no physical solutions.

**EXAMPLE A.5 Solving a quadratic equation**

Find the values of $x$ that satisfy the equation $2x^2 - 2x = 24$.

**Represent mathematically, solve, and evaluate** First we write the equation in the standard form $ax^2 + bx + c = 0$: $2x^2 - 2x - 24 = 0$. Then $a = 2$, $b = -2$, and $c = -24$. Next, the quadratic formula gives the two roots as

$$x = \frac{-(-2) \pm \sqrt{(-2)^2 - 4(2)(-24)}}{(2)(2)}$$

$$= \frac{+2 \pm \sqrt{4 + 192}}{4} = \frac{2 \pm 14}{4},$$

so $x = 4$ or $x = -3$. If $x$ represents a physical quantity that takes only nonnegative values, then the negative root $x = -3$ is nonphysical and is discarded.

As we've mentioned, when an equation has more than one mathematical solution or root, it's up to *you* to decide whether one or the other or both represent the true physical answer. (If neither solution seems physically plausible, you should review your work.)

## Simultaneous Equations

If a problem has two unknowns—for example, $x$ and $y$—then it takes two independent equations in $x$ and $y$ (that is, two equations for $x$ and $y$, where one equation is not simply a multiple of the other) to determine their values uniquely. Such equations are called **simultaneous equations** (because you solve them together). A typical procedure is to solve one equation for $x$ in terms of $y$ and then substitute the result into the second equation to obtain an equation in which $y$ is the only unknown. You then solve this equation for $y$ and use the value of $y$ in either of the original equations in order to solve for $x$. In general, to solve for $n$ unknowns, we must have $n$ independent equations.

**EXAMPLE A.6 Solving two equations in two unknowns**

Solve the following pair of equations for $x$ and $y$:

$$x + 4y = 14$$
$$3x - 5y = -9$$

**Represent mathematically, solve, and evaluate**

The first equation gives $x = 14 - 4y$. Substituting this for $x$ in the second equation yields, successively, $3(14 - 4y) - 5y = -9$, $42 - 12y - 5y = -9$, and $-17y = -51$. Thus, $y = \frac{-51}{-17} = 3$. Then $x = 14 - 4y = 14 - 12 = 2$. We can verify that $x = 2$, $y = 3$ satisfies both equations.

An alternative approach is to multiply the first equation by $-3$, yielding $-3x - 12y = -42$. Adding this to the second equation gives, successively, $3x - 5y + (-3x) + (-12y) = -9 + (-42)$, $-17y = -51$, and $y = 3$, which agrees with our previous result.

As shown by the alternative approach, simultaneous equations can be solved in more than one way. The basic method we describe is easy to keep straight; other methods may be quicker, but may require more insight or forethought. Use the method you're comfortable with.

A pair of equations in which all quantities are symbols can be combined to eliminate one of the common unknowns.

**EXAMPLE A.7 Solving two symbolic equations in two unknowns**

Use the equations $v = v_0 + at$ and $x = v_0t + \frac{1}{2}at^2$ to obtain an equation for $x$ that does not contain $a$.

**Represent mathematically, solve, and evaluate**

We solve the first equation for $a$:

$$a = \frac{v - v_0}{t}.$$

We substitute this expression into the second equation:

$$x = v_0t + \frac{1}{2}\left(\frac{v - v_0}{t}\right)t^2 = v_0t + \frac{1}{2}vt - \frac{1}{2}v_0t$$
$$= \frac{1}{2}v_0t + \frac{1}{2}vt = \left(\frac{v_0 + v}{2}\right)t.$$

When you solve a physics problem, it's often best to work with symbols for all but the final step of the problem. Once you've arrived at the final equation, you can plug in numerical values and solve for an answer.

## A.4 Logarithmic and Exponential Functions

The base-10 logarithm, or **common logarithm** (log), of a number $y$ is the power to which 10 must be raised to obtain $y$: $y = 10^{\log y}$. For example, $1000 = 10^3$, so $\log(1000) = 3$; you must raise 10 to the power 3 to obtain 1000. Most calculators have a key for calculating the log of a number.

Sometimes we are given the log of a number and are asked to find the number. That is, if $\log y = x$ and $x$ is given, what is $y$? To solve for $y$, write an equation in which 10 is raised to the power equal to either side of the original equation: $10^{\log y} = 10^x$. But $10^{\log y} = y$, so $y = 10^x$. In this case, $y$ is called the **antilog** of $x$. For example, if $\log y = -2.0$, then $y = 10^{-2.0} = 1.0 \times 10^{-2.0} = 0.010$.

The log of a number is positive if the number is greater than 1. The log of a number is negative if the number is less than 1, but greater than zero. The log of zero or of a negative number is not defined, and $\log 1 = 0$.

Another base that occurs frequently in physics is the quantity $e = 2.718\ldots$ The **natural logarithm** (ln) of a number $y$ is the power to which $e$ must be raised to obtain $y$: $y = e^{\ln y}$. If $x = \ln y$, then $y = e^x$. Most calculators have keys for $\ln x$ and for $e^x$. For example, $\ln 10.0 = 2.30$, and if $\ln x = 3.00$, then $x = 10^{3.00} = 20.1$. Note that $\ln 1 = 0$.

Logarithms with any choice of base, including base 10 or base $e$, obey several simple and useful rules:

1. $\log(ab) = \log a + \log b$.
2. $\log\left(\frac{a}{b}\right) = \log a - \log b$.
3. $\log(a^n) = n\log a$.

A particular example of the second rule is

$$\log\left(\frac{1}{a}\right) = \log 1 - \log a = -\log a,$$

since $\log 1 = 0$.

**EXAMPLE A.8 Solving a logarithmic equation**

If $\frac{1}{2} = e^{-\alpha T}$, solve for $T$ in terms of $\alpha$.

**Represent mathematically, solve, and evaluate**

We take the natural logarithm of both sides of the equation: $\ln\left(\frac{1}{2}\right) = -\ln 2$ and $\ln\left(e^{-\alpha T}\right) = -\alpha T$. The equation thus becomes $-\alpha T = -\ln 2$, and it follows that $T = \frac{\ln 2}{\alpha}$.

The equation $y = e^{\alpha x}$ expresses $y$ in terms of the exponential function $e^{\alpha x}$. The general rules for exponents in Appendix A.1 apply when the base is $e$, so $e^x e^y = e^{x+y}$, $e^x e^{-x} = e^{x+(-x)} = e^0 = 1$, and $(e^x)^2 = e^{2x}$.

**Figure A.1**

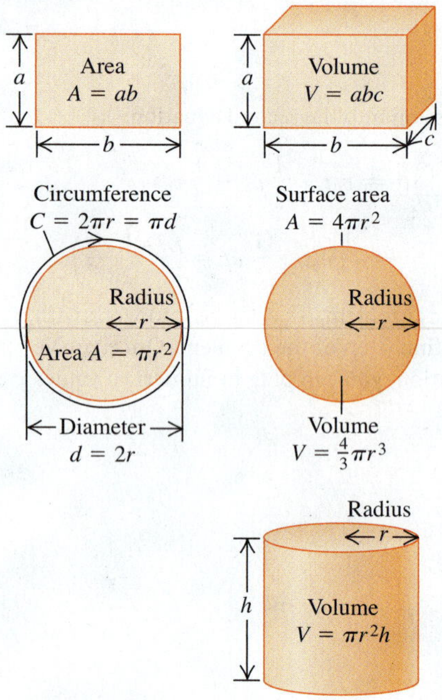


## A.5 Areas and Volumes

**Figure A.1** illustrates the formulas for the areas and volumes of common geometric shapes:

- A rectangle with length $a$ and width $b$ has area $A = ab$.
- A rectangular solid (a box) with length $a$, width $b$, and height $c$ has volume $V = abc$.
- A circle with radius $r$ has diameter $d = 2r$, circumference $C = 2\pi r = \pi d$, and area $A = \pi r^2 = \pi d^2/4$.
- A sphere with radius $r$ has surface area $A = 4\pi r^2$ and volume $V = \frac{4}{3}\pi r^3$.
- A cylinder with radius $r$ and height $h$ has volume $V = \pi r^2 h$.

## A.6 Plane Geometry and Trigonometry

Following are some useful results about angles:

1. Interior angles formed when two straight lines intersect are equal. For example, in **Figure A.2**, the two angles $\theta$ and $\phi$ are equal.
2. When two parallel lines are intersected by a diagonal straight line, the alternate interior angles are equal. For example, in **Figure A.3**, the two angles $\theta$ and $\phi$ are equal.
3. When the sides of one angle are each perpendicular to the corresponding sides of a second angle, then the two angles are equal. For example, in **Figure A.4**, the two angles $\theta$ and $\phi$ are equal.
4. The sum of the angles on one side of a straight line is 180°. In **Figure A.5**, $\theta + \phi = 180°$.
5. The sum of the angles in any triangle is 180°.

**Figure A.2**

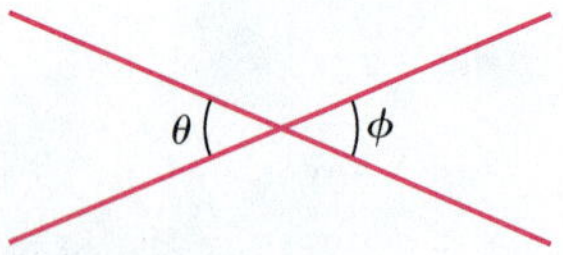


Interior angles formed when two straight lines intersect are equal: $\theta = \phi$

**Figure A.3**

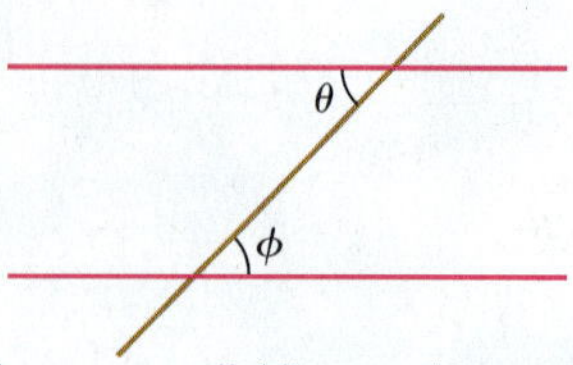


When two parallel lines are intersected by a diagonal straight line, the alternate interior angles are equal: $\theta = \phi$

### Similar Triangles

Triangles are **similar** if they have the same shape, but different sizes or orientations. Similar triangles have equal angles and equal ratios of corresponding sides. If the two triangles in **Figure A.6** are similar, then $\theta_1 = \theta_2$, $\phi_1 = \phi_2$, $\gamma_1 = \gamma_2$, and $\frac{a_1}{a_2} = \frac{b_1}{b_2} = \frac{c_1}{c_2}$.

If two similar triangles have the same size, they are said to be **congruent**. If triangles are congruent, one can be rotated to where it can be placed precisely on top of the other.

**Figure A.4**

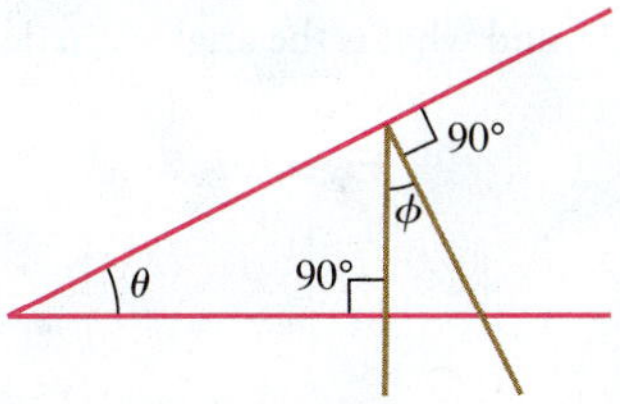


When the sides of one angle are each perpendicular to the corresponding sides of a second angle, then the two angles are equal: $\theta = \phi$

## Right Triangles and Trig Functions

In a **right triangle**, one angle is 90°. Therefore, the other two acute angles (*acute* means less than 90°) have a sum of 90°. In **Figure A.9**, $\theta + \phi = 90°$. The side opposite the right angle is called the **hypotenuse** (side $c$ in the figure). In a right triangle, the square of the length of the hypotenuse equals the sum of the squares of the lengths of the other two sides. For the triangle in Figure A.9, $c^2 = a^2 + b^2$. This formula is called the **Pythagorean theorem**.

If two right triangles have the same value for one acute angle, then the two triangles are similar and have the same ratio of corresponding sides. This true statement allows us to define the functions **sine**, **cosine**, and **tangent** that are ratios of a pair of sides. These functions, called **trigonometric** functions or **trig functions**, depend only on one of the angles in the right triangle. For an angle $\theta$, these functions are written $\sin\theta$, $\cos\theta$, and $\tan\theta$.

In terms of the triangle in Figure A.9, the sine, cosine, and tangent of the angle $\theta$ are as follows:

$$\sin\theta = \frac{\text{opposite side}}{\text{hypotenuse}} = \frac{a}{c}.$$

$$\cos\theta = \frac{\text{adjacent side}}{\text{hypotenuse}} = \frac{b}{c}.$$

$$\tan\theta = \frac{\text{opposite side}}{\text{adjacent side}} = \frac{a}{b}.$$

Note that $\tan\theta = \dfrac{\sin\theta}{\cos\theta}$. For angle $\phi$, $\sin\phi = \dfrac{b}{c}$, $\cos\phi = \dfrac{a}{c}$, and $\tan\phi = \dfrac{b}{a}$.

In physics, angles are expressed in either degrees or radians, where $\pi$ radians $= 180°$. Most calculators have a key for switching between degrees and radians. Always be sure that your calculator is set to the appropriate angular measure.

Inverse trig functions, denoted, for example, by $\sin^{-1} x$ (or $\arcsin x$), have a value equal to the angle that has the value $x$ for the trig function. For example, $\sin 30° = 0.500$, so $\sin^{-1}(0.500) = \arcsin(0.500) = 30°$. Note that $\sin^{-1} x$ does *not* mean $\dfrac{1}{\sin x}$.

**Figure A.5**

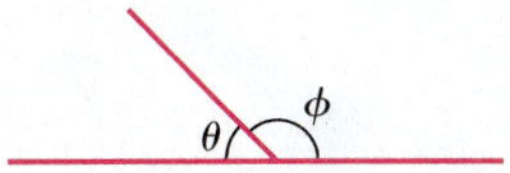


The sum of the angles on one side of a straight line is 180°: $\theta + \phi = 180°$

**Figure A.6**

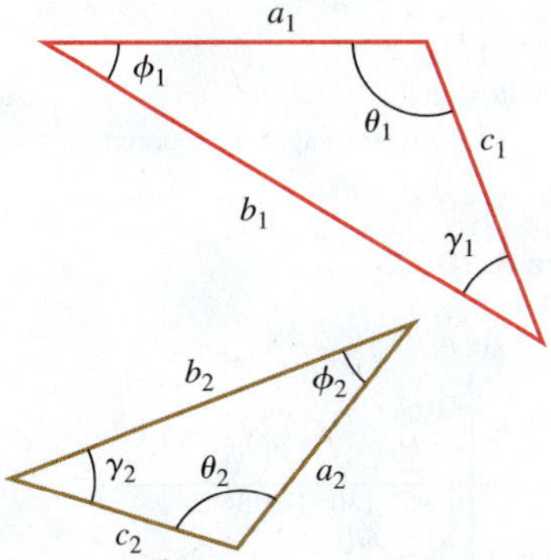


Two similar triangles: Same shape but not necessarily the same size

### EXAMPLE A.9 Using trigonometry I

A right triangle has one angle of 30° and one side with length 8.0 cm, as shown in **Figure A.7**. What is the angle $\phi$, and what are the lengths $x$ and $y$ of the other two sides of the triangle?

**Figure A.7**

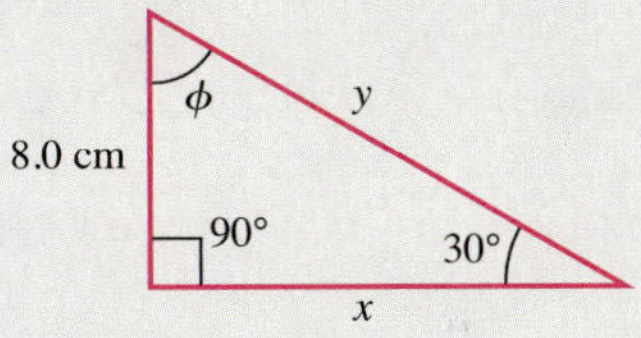


**Represent mathematically, solve, and evaluate** $\phi + 30° = 90°$, so $\phi = 60°$.

$$\tan 30° = \frac{8.0\text{ cm}}{x}, \text{ so } x = \frac{8.0\text{ cm}}{\tan 30°} = 13.9\text{ cm}.$$

To find $y$, we use the Pythagorean theorem: $y^2 = (8.0\text{ cm})^2 + (13.9\text{ cm})^2$, so $y = 16.0$ cm.

Or we can say $\sin 30° = 8.0\text{ cm}/y$, so $y = 8.0\text{ cm}/\sin 30° = 16.0$ cm, which agrees with the previous result.

Notice how we used the Pythagorean theorem in combination with a trig function. You will use these tools constantly in physics, so make sure that you can employ them with confidence.

**EXAMPLE A.10 Using trigonometry II**

A right triangle has two sides with lengths as specified in **Figure A.8**. What is the length $x$ of the third side of the triangle, and what is the angle $\theta$, in degrees?

**Figure A.8**

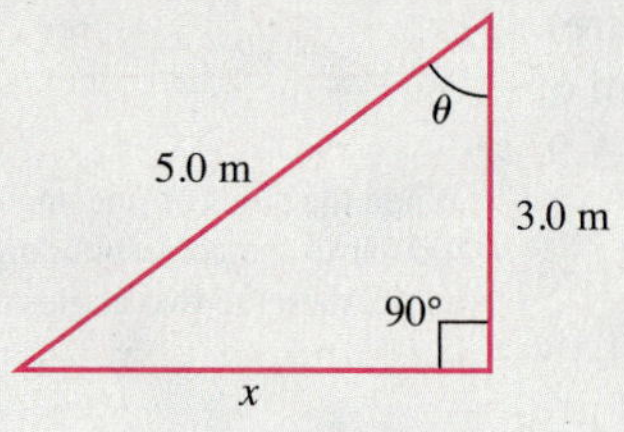


**Represent mathematically, solve, and evaluate** The Pythagorean theorem applied to this right triangle gives $(3.0\text{ m})^2 + x^2 = (5.0\text{ m})^2$, so $x = \sqrt{(5.0\text{ m})^2 - (3.0\text{ m})^2} = 4.0\text{ m}$. (Since $x$ is a length, we take the positive root of the equation.) We also have

$$\cos\theta = \frac{3.0\text{ m}}{5.0\text{ m}} = 0.600,\ so\ \theta = \cos^{-1}(0.600) = 53.1°.$$

In this case, we knew the lengths of two sides, but none of the acute angles, so we used the Pythagorean theorem first and then an appropriate trig function.

**Figure A.9**

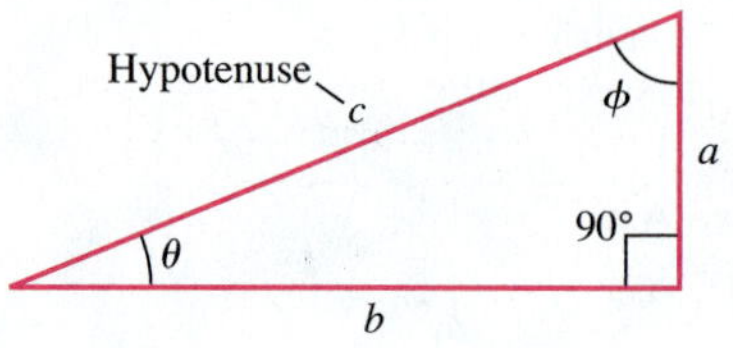


For a right triangle:
$\theta + \phi = 90°$
$c^2 = a^2 + b^2$ (Pythagorean theorem)

**Figure A.10**

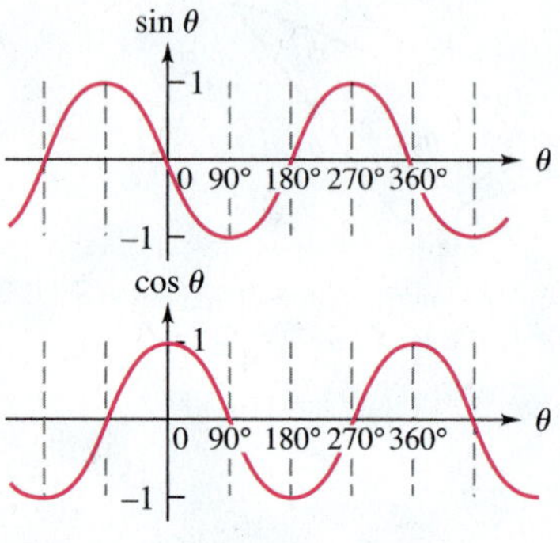


**Figure A.11**

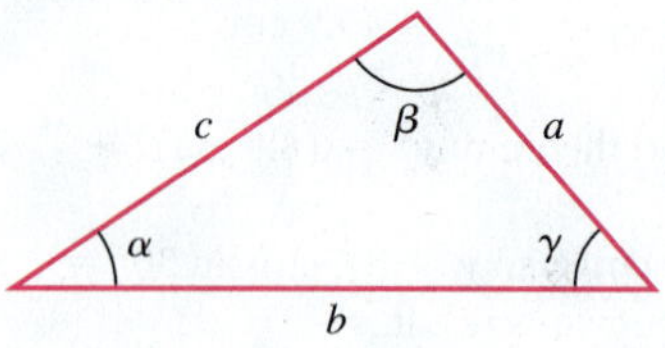


In a right triangle, all angles are in the range from 0° to 90°, and the sine, cosine, and tangent of the angles are all positive. This must be the case, since the trig functions are ratios of lengths. But for other applications, such as finding the components of vectors, calculating the oscillatory motion of a mass on a spring, or describing wave motion, it is useful to define the sine, cosine, and tangent for angles outside that range. Graphs of $\sin\theta$ and $\cos\theta$ are given in **Figure A.10**. The values of $\sin\theta$ and $\cos\theta$ vary between $+1$ and $-1$. Each function is periodic, with a period of 360°. Note the range of angles between 0° and 360° for which each function is positive and negative. The two functions $\sin\theta$ and $\cos\theta$ are 90° out of phase (that is, out of step). When one is zero, the other has its maximum magnitude.

For any triangle (see **Figure A.11**)—in other words, not necessarily a right triangle—the following two relations apply:

1. $\dfrac{\sin\alpha}{a} = \dfrac{\sin\beta}{b} = \dfrac{\sin\gamma}{c}$ (law of sines).

2. $c^2 = a^2 + b^2 - 2ab\cos\gamma$ (law of cosines).

Some of the relations among trig functions are called trig identities. The following table lists only a few, those most useful in introductory physics:

**Useful trigonometric identities**

$\sin(-\theta) = -\sin(\theta)$ ($\sin\theta$ is an odd function)
$\cos(-\theta) = \cos(\theta)$ ($\cos\theta$ is an even function)
$\sin 2\theta = 2\sin\theta\cos\theta$
$\cos 2\theta = \cos^2\theta - \sin^2\theta = 2\cos^2\theta - 1 = 1 - 2\sin^2\theta$
$\sin(\theta \pm \phi) = \sin\theta\cos\phi \pm \cos\theta\sin\phi$
$\cos(\theta \pm \phi) = \cos\theta\cos\phi \mp \sin\theta\sin\phi$
$\sin(180° - \theta) = \sin\theta$
$\cos(180° - \theta) = -\cos\theta$
$\sin(90° - \theta) = \cos\theta$
$\cos(90° - \theta) = \sin\theta$

## APPENDIX B

# Working with Vectors

## Graphical Representation of Vectors

We use arrows to graphically represent vector quantities. The arrow's direction indicates the direction of the vector, and the arrow's length indicates the vector's magnitude. We need to use an appropriate scale when drawing the length of the arrow. Imagine that you need to walk from point M to point N (**Figure B.1a**). The distance between the points is 200 m and the direction is 25° North of East. We represent your walk with an arrow or a vector called a displacement vector $\vec{D}$. If 1.0 cm represents a displacement of 50 m, then the length of the vector will be 4.0 cm and its direction is 25° North of East (Figure B.1b). This is the first very important observation about vectors—it is not important where the vector starts or ends, as long as it reflects the same direction and magnitude as the quantity it represents. Notice the difference between the vectors $\vec{A}$ and $\vec{B}$ and $\vec{A}$ and $\vec{C}$ in **Figures B.2a** and b. $\vec{A}$ and $\vec{B}$ in Figure B.2a have the same magnitude but different directions and $\vec{A}$ and $\vec{C}$ in Figure B.2b have the same direction but different magnitudes.

**Figure B.1** Representing a displacement with a vector.

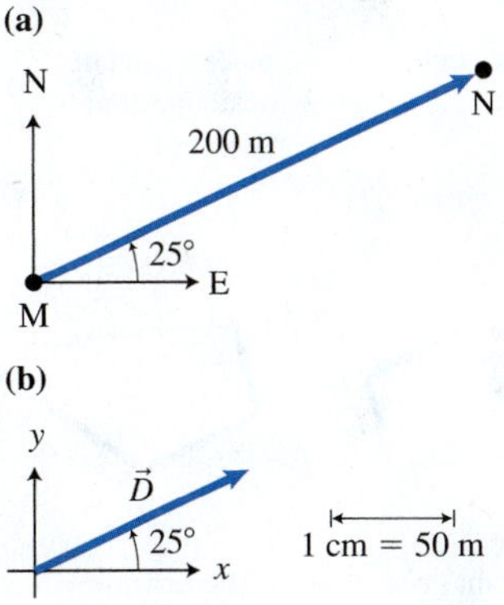


## Graphical Addition and Subtraction of Vectors

*Addition* Because vector quantities have both magnitude and direction, we cannot use the normal rules of algebraic addition and subtraction to add or subtract them. Suppose, for example, that you take a two-day trip. The first day you travel 500 km toward the east along a straight road. The second day involves another 500-km displacement, but not necessarily in the same direction. In what direction and how far from your starting position is your final location? The answer, of course, depends on the direction of the second day's trip. You could be 1000 km east of your starting position if you continued traveling east during the second day (**Figure B.3a**). However, if you traveled west on the second day, your net displacement would be zero as you would be back where you started (Figure B.3b). If you traveled north during the second day, your net displacement would be a little over 700 km to the northeast of your starting position (Figure B.3c). When adding displacement vectors or vectors of any type, we are concerned only with their net result.

To add two vectors we can use the following graphical technique. Suppose we want to add the two vectors $\vec{A}$ and $\vec{C}$ shown in **Figure B.4a**. To add

**Figure B.2** (a) Equal magnitude but different direction vectors. (b) Equal direction but different magnitude vectors.

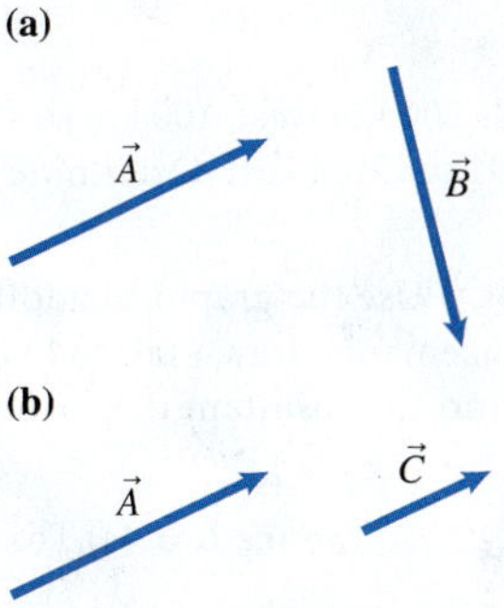


**Figure B.3** The net displacement for two 500-km successive trips depends on the relative direction of each day's trip.

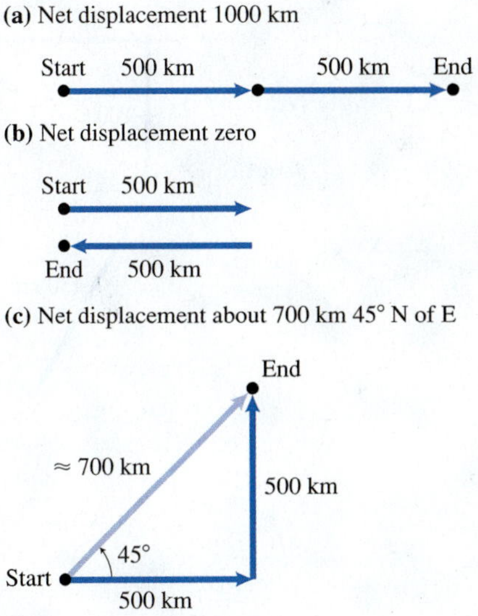


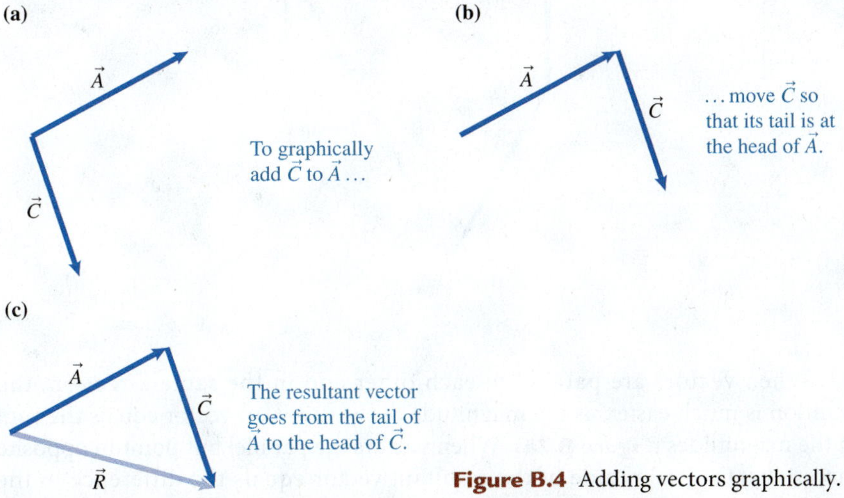


**Figure B.4** Adding vectors graphically.

**Figure B.5** Graphical vector addition.

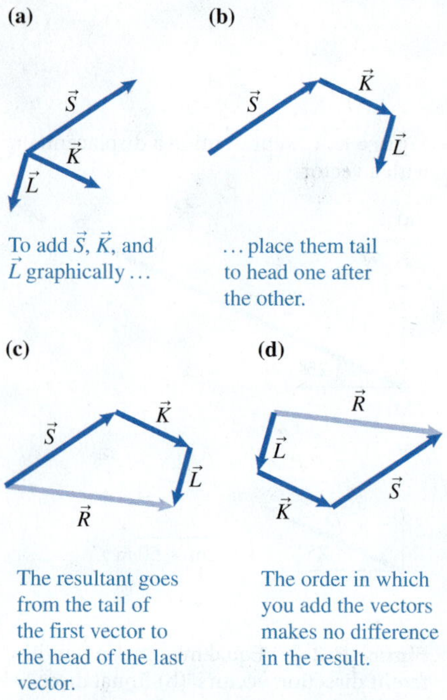


them graphically, we place the tail of $\vec{C}$ at the head of $\vec{A}$, as in Figure B.4b. We can move a vector from one location to another as long as we do not change its magnitude or direction; therefore moving vector $\vec{C}$ as shown did not change it. Having moved vector $\vec{C}$, we draw another vector $\vec{R}$ from the tail of $\vec{A}$ to the head of $\vec{C}$, as in Figure B.4c. This vector $\vec{R}$ represents the result of the addition of two vectors $\vec{A}$ and $\vec{C}$. We can write the resultant vector as a mathematical equation: $\vec{R} = \vec{A} + \vec{C}$. As you see in Figure B.4c, the magnitude of the resultant vector is not equal to the sum of the magnitudes of $\vec{A}$ and $\vec{C}$.

(a)
(b)
$\vec{S}$
$\vec{K}$
$\vec{L}$
To add $\vec{S}$, $\vec{K}$, and $\vec{L}$ graphically . . .
. . . place them tail to head one after the other.
(c)
(d)
$\vec{R}$
The resultant goes from the tail of the first vector to the head of the last vector.
The order in which you add the vectors makes no difference in the result.

**Graphical addition of vectors:** To graphically add two or more vectors (**Figure B.5a**), place the vectors tail to head one at a time, as illustrated in Figure B.5b. The magnitudes and directions of the vectors must not be changed as they are moved about. The resultant vector $\vec{R}$ goes from the tail of the first vector to the head of the last vector, as illustrated in Figure B.5c, and equals the sum of the vectors $\vec{R} = \vec{S} + \vec{K} + \vec{L}$. The order in which you add the vectors makes no difference (for example, $\vec{S} + \vec{K} + \vec{L} = \vec{L} + \vec{K} + \vec{S}$, as seen by comparing Figures B.5c and d).

Usually, the graphical technique for adding vectors is used as a rough check on another vector addition technique introduced later. However, if done with care using a ruler and protractor, the graphical vector addition technique is a fairly accurate method for determining the resultant of several vectors.

**EXAMPLE B.1**

A car travels 200 km west, 100 km south, and finally 150 km at an angle 60° south of east. Determine the net displacement of the car.

**Reasoning** Use the graphical addition technique with the three displacements drawn tail to head, as shown in **Figure B.6a**. To find the resultant displacement $\vec{R}$, draw an arrow from the tail of the first displacement to the head of the last (Figure B.6b). We measure the magnitude of the resultant with a ruler and find that its length is 5.2 cm. Since each centimeter represents 50 km, the magnitude of the resultant displacement is $(5.2\text{ cm})(50\text{ km/cm}) = 260\text{ km}$. Using a protractor, we confirm that the direction of the resultant is 60° south of west.

**Figure B.6** (a) Three displacement vectors placed tail to head. (b) The resultant displacement.

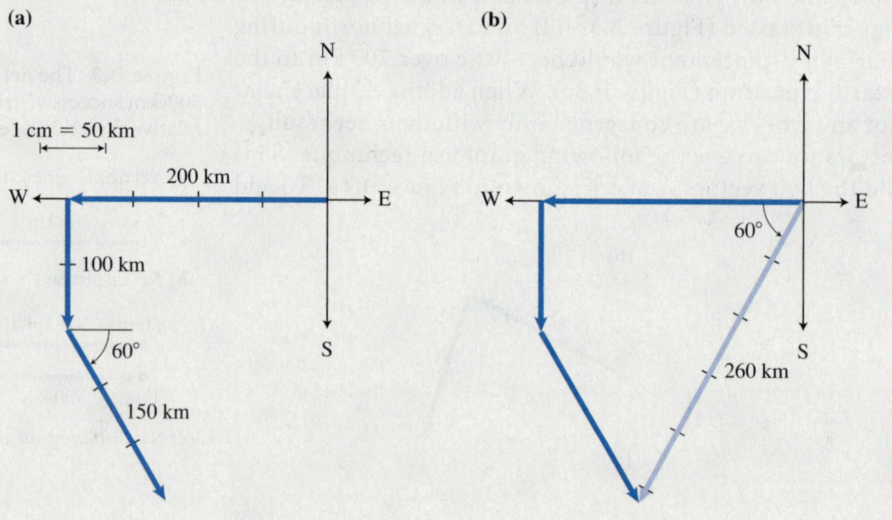


When vectors are parallel to each other and in the same direction, the addition is much easier, as the magnitude of the resultant vector equals the sum of the magnitudes (**Figure B.7a**). When vectors are parallel but point in opposite directions, the magnitude of the resultant vector equals the difference in the magnitudes of the vectors being added (Figure B.7b).

*Subtraction* Sometimes we need to subtract two vectors ($\vec{A} - \vec{B}$). We can view the procedure of subtraction as the familiar procedure of addition where instead of adding a vector $\vec{B}$ to the vector $\vec{A}$, we add $(-\vec{B})$ to $\vec{A}$: $\vec{R} = \vec{A} - \vec{B} = \vec{A} + (-\vec{B})$. To draw $(-\vec{B})$, we simply reverse the direction of vector $\vec{B}$. Vector subtraction is illustrated in **Figure B.8**, where vector $\vec{B}$ is subtracted from $\vec{A}$.

**Graphical subtraction of vectors:** To graphically subtract vector $\vec{B}$ from vector $\vec{A}$ (see Figure B.8a), first reverse the direction of $\vec{B}$, thus producing $-\vec{B}$ (Figure B.8b). Then add $\vec{A}$ and $-\vec{B}$ (Figure B.8c). The resultant vector goes from the tail of $\vec{A}$ to the head of $-\vec{B}$ (Figure B.8d) and is the difference of $\vec{A}$ and $\vec{B}$:

$$\vec{R} = \vec{A} + (-\vec{B}) = \vec{A} - \vec{B}$$

**Figure B.7** (a) The magnitudes of parallel vectors can be added. (b) The magnitudes of anti-parallel vectors can be subtracted.

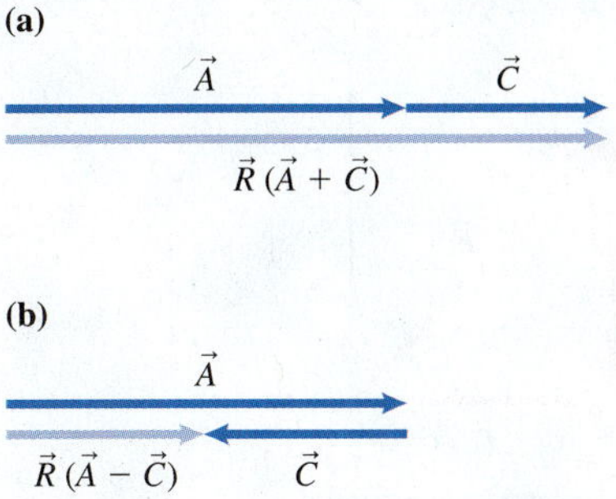


## Components of a Vector

The graphical method of adding and subtracting vectors takes considerable time if done accurately. There is a different vector addition technique, which we describe in this and the next sub-sections. It is more productive for calculating the length and direction of the resultant vector and is also faster and more convenient for solving a variety of interesting problems involving vectors. The component method, as it is called, uses a principle that any vector can be represented as the sum of two vectors, called *vector components*, which are perpendicular to each other (**Figure B.9**).

A graphical method for finding these two vectors whose sum equals vector $\vec{A}$ is illustrated in **Figure B.10a**. Draw a coordinate system so its $x$-axis points east and $y$-axis points north. The $x$ component of $\vec{A}$ is a vector $\vec{A}_x$ that points along the $x$-axis and whose length equals the projection of $\vec{A}$ on that axis, as shown in Figure B.10b. The $y$ component of the vector $\vec{A}_y$ is the projection of $\vec{A}$ on the $y$-axis in Figure B.10b. It is important to notice that the vector sum of $\vec{A}_x$ and $\vec{A}_y$ equals $\vec{A}$, as shown in Figure B.10c.

We often work with vectors whose tails are located at the origin of a coordinate system—see **Figure B.11**. This is especially true for force vectors representing the forces that other objects exert on an object of interest. The vector components of these vectors can be converted to scalar components, which represent vectors using numbers for the magnitude and the signs for direction. For example, in Figure B.11 the head of the $x$ component $\vec{F}_x$ is located 5 units in the negative $x$ direction. The scalar $x$ component $F_x = -5$ units. The head of the $y$ component $\vec{F}_y$ is located 5 units in the positive $y$ direction. The scalar $y$ component $F_y = +5$ units. Note that if a vector component points in the positive $x$ or $y$ direction, its scalar component is positive; if the vector component points in the negative $x$ or $y$ direction, the scalar component is negative. We can calculate the scalar

**Figure B.8** Graphical subtraction of vectors.

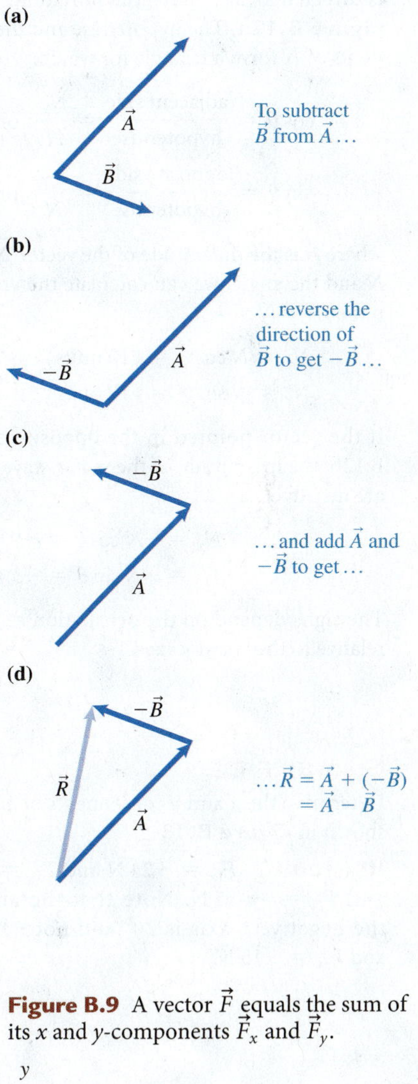


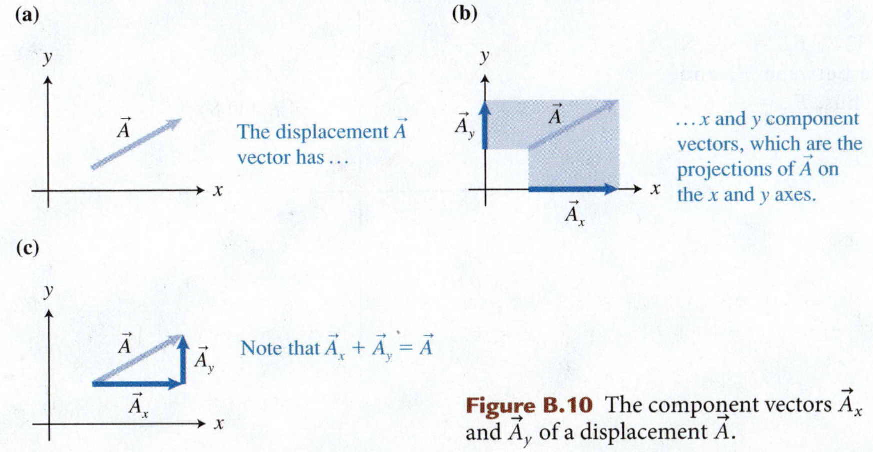


**Figure B.10** The component vectors $\vec{A}_x$ and $\vec{A}_y$ of a displacement $\vec{A}$.

**Figure B.9** A vector $\vec{F}$ equals the sum of its $x$ and $y$-components $\vec{F}_x$ and $\vec{F}_y$.

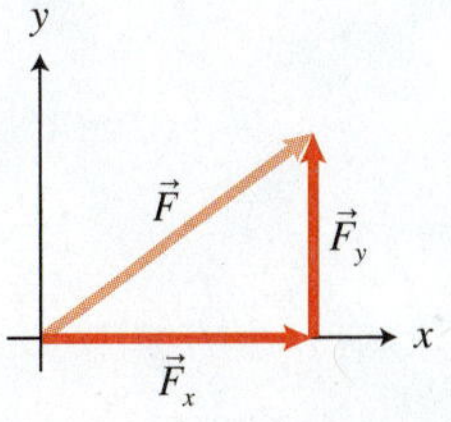

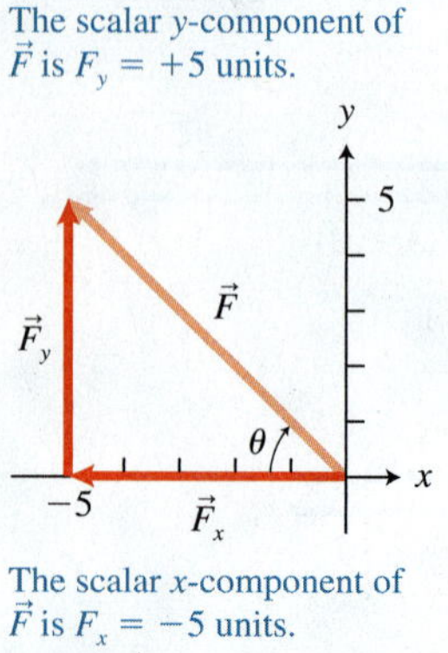


**Figure B.11** The scalar components of a vector.

components using the triangle shown in **Figure B.11** and the trigonometric sin and cos functions.

***x* and *y* scalar components of a vector:** The $x$ and $y$ scalar components of a vector are:

$$F_x = \pm F \cos\theta$$
$$F_y = \pm F \sin\theta$$

where $F$ is the magnitude of the vector and $\theta$ is the angle (90° or less) that $F$ makes with the $x$ axis. $F_x$ is positive if $\vec{F}_x$ points in the positive $x$ direction and negative if $\vec{F}_x$ points in the negative $x$ direction. $F_y$ is either positive or negative, depending on the direction of $\vec{F}_y$ relative to the $y$ axis.

**EXAMPLE B.2**

Suppose we have a vector $\vec{N}$ of the magnitude of 10 units that is directed at 20° above the horizontal direction as shown in **Figure B.12a**. The hypotenuse and the $x$ and $y$ scalar components of $\vec{N}$ form a triangle for which:

$$\cos\theta = \frac{\text{adjacent side}}{\text{hypotenuse}} = \frac{N_x}{N} \quad \text{or} \quad N_x = N\cos\theta$$

$$\sin\theta = \frac{\text{opposite side}}{\text{hypotenuse}} = \frac{N_y}{N} \quad \text{or} \quad N_y = N\sin\theta$$

where $N$ is the magnitude of the vector $\vec{N}$. Using the magnitude $N$ and the angle, we can calculate the values of the scalar components of $\vec{N}$:

$$N_x = N\cos\theta = (10 \text{ units})\cos 20° = 9.4 \text{ units}$$
$$N_y = N\sin\theta = (10 \text{ units})\sin 20° = 3.4 \text{ units}$$

If the vector pointed in the opposite direction, as in Figure B.12b, the magnitude of the scalar stays the same but the signs are negative:

$$N_x = -N\cos\theta = -9.4 \text{ units}$$
$$N_y = -N\sin\theta = -3.4 \text{ units}$$

The signs depend on the orientation of the scalar components relative to the $x$ and $y$ axes.

**Figure B.12** $N_x$ and $N_y$ can be calculated from known $N$ and $\theta$.

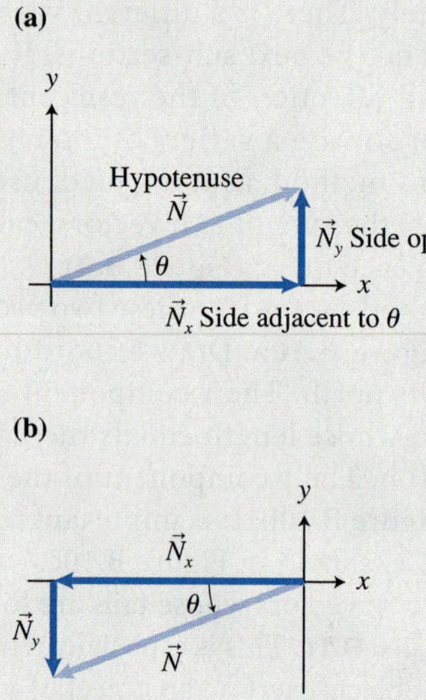


**EXAMPLE B.3**

Determine the $x$ and $y$ components of each of the force vectors shown in **Figure B.13**.

**Reasoning** $F_{1x} = +24$ N and $F_{1y} = +32$ N; $F_{2x} = -35$ N and $F_{2y} = +35$ N. Note that the angle between $\vec{F}_3$ and the negative $x$ axis is 30° and not 60°. Thus, $F_{3x} = -26$ N and $F_{3y} = -15$ N.

**Figure B.13**

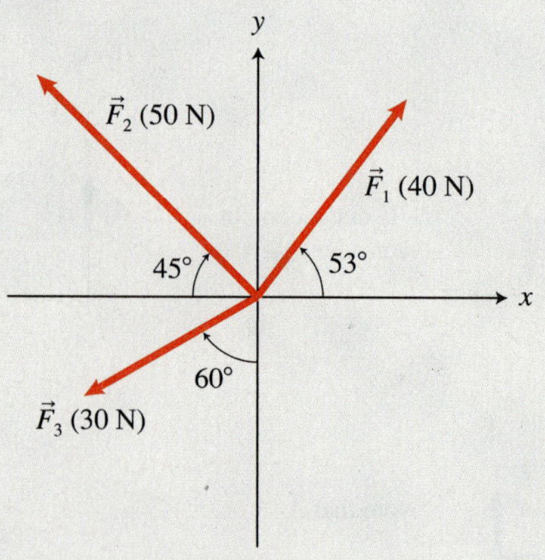

## Vector addition by components

Let us see how the vector and scalar components allow us to add vectors. In **Figure B.14a** we add two vectors $\vec{K}$ and $\vec{L}$ to form the resultant vector $\vec{R}$. We first determine the scalar components of the vectors being added (the scalar components have been given in Figure B.14b—normally we would have to calculate the scalar components from the known directions and magnitudes of $\vec{K}$ and $\vec{L}$). Suppose that for vector $\vec{K}, K_x = +3$ units and $K_y = (-1)$ unit. For vector $\vec{L}, L_x = +2$ units and $L_y = (-3)$ units.

Next, we add the scalar components of $\vec{K}$ and $\vec{L}$ to find the scalar components of vector $\vec{R}$ (Figure B.14c):

$$R_x = K_x + L_x = 3 + 2 = 5 \text{ units}$$
$$R_y = K_y + L_y = (-1) + (-3) = (-4) \text{ units.}$$

The negative sign for the $y$ component of the resultant vector means it points in the negative direction relative to the $y$-axis, as seen in Figure B.14d.

To find the length of the resultant vector we can use our knowledge of geometry. Examine the right triangle (Figure B.14d) that is formed by the scalar components of $\vec{R}$: $R_x$ and $R_y$. In this triangle, $R_x$ and $R_y$ are the sides and the magnitude $R$ of $\vec{R}$ is the hypotenuse. Using the Pythagorean theorem we find the length of the hypotenuse if we know the length of the sides: $R^2 = R_x^2 + R_y^2$ or $R = \sqrt{R_x^2 + R_y^2}$. Apply this reasoning to our situation:

$$R = \sqrt{R_x^2 + R_y^2} = \sqrt{5^2 + (-4)^2} = \sqrt{25 + 16} = 6.4 \text{ units.}$$

Notice that all numbers here have two significant digits.

**Figure B.14** Scalar component vector addition.

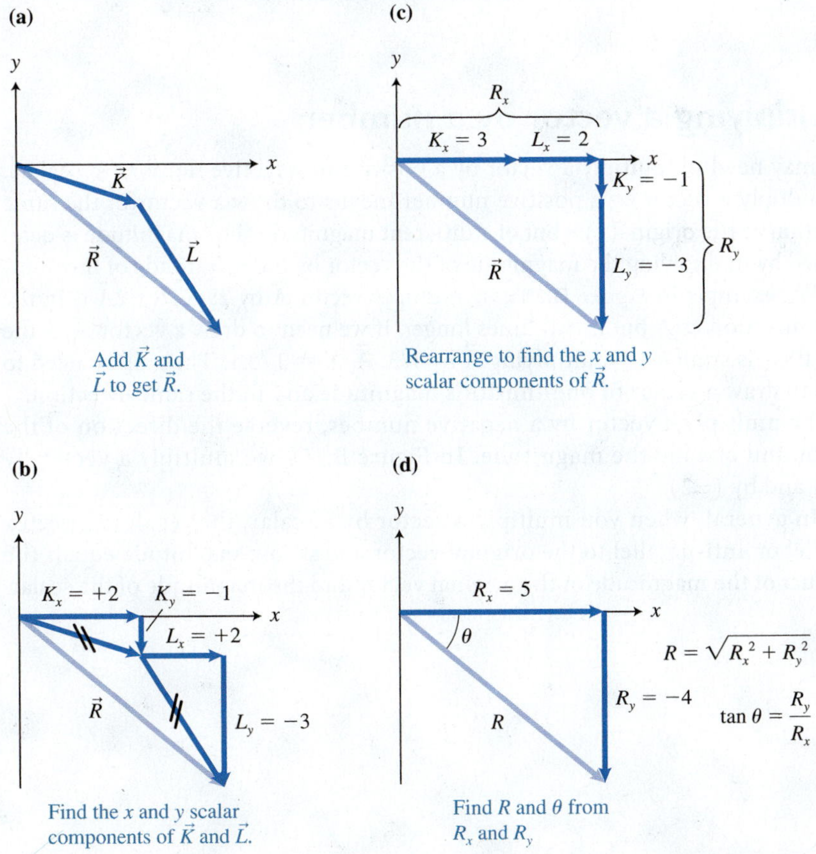

The tan function allows us to determine the angle between vector $\vec{R}$ and the positive or negative $x$-axis: $\tan\theta = R_y/R_x = (-4)/5 = (-0.80)$. Using the arctan function on the calculator, we find that the angle whose tan is $(-0.80)$ equals $-39°$. Therefore the vector $\vec{R}$ points 39° *below* the direction of the $+x$ axis. Recall that in the right triangle (Figure B.13d) formed by vector $\vec{R}$ and its components $\vec{R}_x$ and $\vec{R}_y$, $R_x$ is the adjacent side and $R_y$ is the opposite side.

**To determine a vector if its scalar components are known:** If the scalar components $B_x$ and $B_y$ are known, the magnitude of the vector $\vec{B}$ is

$$B = \sqrt{B_x^2 + B_y^2}$$

and the angle $\theta$ that $\vec{B}$ makes with the horizontal direction is

$$\theta = \arctan\frac{B_y}{B_x}.$$

We can now summarize this operation.

**EXAMPLE B.4**

Determine the sum of the three forces shown in Figure B.13. You can use the results of the component calculation in Example B.3.

**Reasoning** The $x$ and $y$ components of the resultant force are:

$$R_x = F_{1x} + F_{2x} + F_{3x} = +24\text{ N} + (-35\text{ N}) + (-26\text{ N})$$
$$= -37\text{ N}$$

$$R_y = F_{1y} + F_{2y} + F_{3y} = +32\text{ N} + 35\text{ N} + (-15\text{ N})$$
$$= +52\text{ N}$$

The magnitude of the resultant force is:

$$R = \sqrt{(-37\text{ N})^2 + (52\text{ N})^2} = 64\text{ N}.$$

The resultant vector is in the second quadrant (it has a negative $x$ component and a positive $y$ component) and makes the following angle above the negative $x$ axis:

$$\theta = \arctan\frac{52\text{ N}}{37\text{ N}} = \arctan 1.41 = 75°.$$

Thus, the sum of the three forces has a magnitude of 64 N and points 75° above the negative $x$-axis.

**Figure B.15** Multiplying a vector by a scalar (a) changes its length (a positive number) or (b) changes its length and reverses its direction (a negative number).

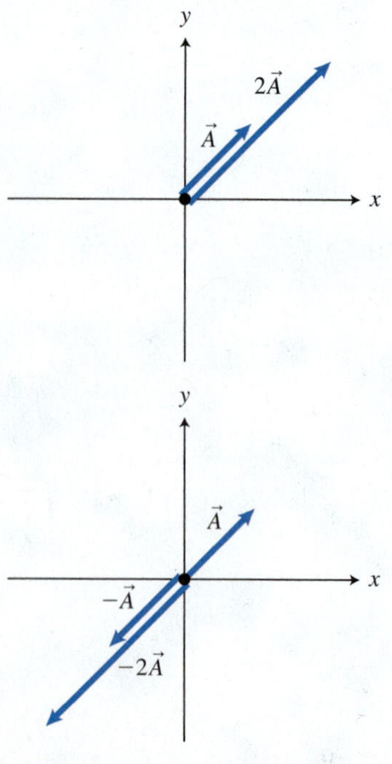


## Multiplying a vector by a number

You may need to multiply a vector by a positive or negative number (a scalar). To multiply a vector by a positive number means to draw a vector in the same direction as the original one but of a different magnitude. The magnitude is determined by multiplying the magnitude of the vector by the magnitude of the number. For example in **Figure B.15a** we multiply vector $\vec{A}$ by 2: vector $2\vec{A}$ is in the same direction as $\vec{A}$ but is two times longer. If we need to draw a vector $\vec{A}/3$ the operation is similar to multiplying $\vec{A}$ by 1/3: $\vec{A}/3 = 1/3\vec{A}$. Thus all we need to do is to draw a vector of one-third $\vec{A}$'s magnitude and in the same direction.

To multiply a vector by a negative number, reverse the direction of the vector and change the magnitude. In Figure B.15b we multiply a vector by $(-1)$ and by $(-2)$.

In general, when you multiply a vector by a scalar, the result is a vector parallel or anti-parallel to the original vector and whose magnitude equals the product of the magnitude of the original vector and the magnitude of the scalar.

# APPENDIX C

# Base Units of SI System

The SI system of units is built on the seven base units listed below. All other derived units are combinations of two or more of these base units.

**Table C.1 Base Units of the SI System**

| Base Quantity | Name | Abbreviation |
|---|---|---|
| Length | meter | m |
| Mass | kilogram | kg |
| Time | second | s |
| Electric current | ampere | A |
| Temperature | kelvin | K |
| Amount of substance | mole | mol |
| Luminous intensity | candela | cd |

An example of a derived unit is the unit for kinetic energy, the joule (J), which is expressed in terms of the kg, m, and s:

$$1\ \text{J} = 1\ \text{kg}\left(\frac{1\ \text{m}}{1\ \text{s}}\right)^2$$

## Definitions of the base units

**meter (m)** The distance that light travels in a vacuum in a time interval of $\left(\frac{1}{299{,}792{,}458}\right)$s.

**kilogram (kg)** The mass of an international standard cylindrical platinum-iridium alloy stored in a vault in Sèvres, France by the International Bureau of Weights and Measures.

**second (s)** The duration of 9,192,631,770 vibrations of radiation emitted by a particular transition of a cesium-133 atom.

**ampere (A)** The constant electric current that, when flowing in two very long parallel straight wires placed 1 m apart in a vacuum, causes them to exert a force on each other of $2 \times 10^{-7}$ newton per meter length of wire.

**kelvin (K)** 1/273.16 of the thermodynamic temperature of the triple point of water.

**mole (mol)** The amount of a substance that contains as many elementary entities as there are carbon atoms in 0.012 kg of the carbon isotope carbon 12. The elementary entities could be atoms, molecules, ions, electrons, other particles, or specified groups of such particles.

**candela (cd)** The luminous intensity in a given direction of a light source that emits monochromatic radiation of frequency $540 \times 10^{12}$ Hz and that has a radiant intensity in that direction of 1/683 W per steradian.

## APPENDIX D

# Atomic and Nuclear Data

| Atomic number ($Z$) | Element | Symbol | Mass number ($A$) | Atomic mass (u) | Percent abundance | Decay mode | Half-life $t_{1/2}$ |
|---|---|---|---|---|---|---|---|
| 0 | (Neutron) | n | 1 | 1.008 665 | | $\beta^-$ | 10.4 min |
| 1 | Hydrogen | H | 1 | 1.007 825 | 99.985 | stable | |
| | Deuterium | D | 2 | 2.014 102 | 0.015 | stable | |
| | Tritium | T | 3 | 3.016 049 | | $\beta^-$ | 12.33 yr |
| 2 | Helium | He | 3 | 3.016 029 | 0.000 1 | stable | |
| | | | 4 | 4.002 602 | 99.999 9 | stable | |
| | | | 6 | 6.018 886 | | $\beta^-$ | 0.81 s |
| 3 | Lithium | Li | 6 | 6.015 121 | 7.50 | stable | |
| | | | 7 | 7.016 003 | 92.50 | stable | |
| | | | 8 | 8.022 486 | | $\beta^-$ | 0.84 s |
| 4 | Beryllium | Be | 7 | 7.016 928 | | EC | 53.3 days |
| | | | 9 | 9.012 174 | 100 | stable | |
| | | | 10 | 10.013 534 | | $\beta^-$ | $1.5 \times 10^6$ yr |
| 5 | Boron | B | 10 | 10.012 936 | 19.90 | stable | |
| | | | 11 | 11.009 305 | 80.10 | stable | |
| | | | 12 | 12.014 352 | | $\beta^-$ | 0.020 2 s |
| 6 | Carbon | C | 10 | 10.016 854 | | $\beta^+$ | 19.3 s |
| | | | 11 | 11.011 433 | | $\beta^+$ | 20.4 min |
| | | | 12 | 12.000 000 | 98.90 | stable | |
| | | | 13 | 13.003 355 | 1.10 | stable | |
| | | | 14 | 14.003 242 | | $\beta^-$ | 5 730 yr |
| | | | 15 | 15.010 599 | | $\beta^-$ | 2.45 s |
| 7 | Nitrogen | N | 12 | 12.018 613 | | $\beta^+$ | 0.011 0 s |
| | | | 13 | 13.005 738 | | $\beta^+$ | 9.96 min |
| | | | 14 | 14.003 074 | 99.63 | stable | |
| | | | 15 | 15.000 108 | 0.37 | stable | |
| | | | 16 | 16.006 100 | | $\beta^-$ | 7.13 s |
| | | | 17 | 17.008 450 | | $\beta^-$ | 4.17 s |
| 8 | Oxygen | O | 14 | 14.008 595 | | EC | 70.6 s |
| | | | 15 | 15.003 065 | | $\beta^+$ | 122 s |
| | | | 16 | 15.994 915 | 99.76 | stable | |
| | | | 17 | 16.999 132 | 0.04 | stable | |
| | | | 18 | 17.999 160 | 0.20 | stable | |
| | | | 19 | 19.003 577 | | $\beta^-$ | 26.9 s |
| 9 | Fluorine | F | 17 | 17.002 094 | | EC | 64.5 s |
| | | | 18 | 18.000 937 | | $\beta^+$ | 109.8 min |
| | | | 19 | 18.998 404 | 100 | stable | |
| | | | 20 | 19.999 982 | | $\beta^-$ | 11.0 s |
| 10 | Neon | Ne | 19 | 19.001 880 | | $\beta^+$ | 17.2 s |
| | | | 20 | 19.992 435 | 90.48 | stable | |
| | | | 21 | 20.993 841 | 0.27 | stable | |
| | | | 22 | 21.991 383 | 9.25 | stable | |

| Atomic number ($Z$) | Element | Symbol | Mass number ($A$) | Atomic mass (u) | Percent abundance | Decay mode | Half-life $t_{1/2}$ |
|---|---|---|---|---|---|---|---|
| 11 | Sodium | Na | 22 | 21.994 434 | | $\beta^+$ | 2.61 yr |
| | | | 23 | 22.989 770 | 100 | stable | |
| | | | 24 | 23.990 961 | | $\beta^-$ | 14.96 hr |
| 12 | Magnesium | Mg | 24 | 23.985 042 | 78.99 | stable | |
| | | | 25 | 24.985 838 | 10.00 | stable | |
| | | | 26 | 25.982 594 | 11.01 | stable | |
| 13 | Aluminum | Al | 27 | 26.981 538 | 100 | stable | |
| | | | 28 | 27.981 910 | | $\beta^-$ | 2.24 min |
| 14 | Silicon | Si | 28 | 27.976 927 | 92.23 | stable | |
| | | | 29 | 28.976 495 | 4.67 | stable | |
| | | | 30 | 29.973 770 | 3.10 | stable | |
| | | | 31 | 30.975 362 | | $\beta^-$ | 2.62 hr |
| 15 | Phosphorus | P | 30 | 29.978 307 | | $\beta^+$ | 2.50 min |
| | | | 31 | 30.973 762 | 100 | stable | |
| | | | 32 | 31.973 908 | | $\beta^-$ | 14.26 days |
| 16 | Sulfur | S | 32 | 31.972 071 | 95.02 | stable | |
| | | | 33 | 32.971 459 | 0.75 | stable | |
| | | | 34 | 33.967 867 | 4.21 | stable | |
| | | | 35 | 34.969 033 | | $\beta^-$ | 87.5 days |
| | | | 36 | 35.967 081 | 0.02 | stable | |
| 17 | Chlorine | Cl | 35 | 34.968 853 | 75.77 | stable | |
| | | | 36 | 35.968 307 | | $\beta^-$ | $3.0 \times 10^5$ yr |
| | | | 37 | 36.965 903 | 24.23 | stable | |
| 18 | Argon | Ar | 36 | 35.967 547 | 0.34 | stable | |
| | | | 38 | 37.962 732 | 0.06 | stable | |
| | | | 39 | 38.964 314 | | $\beta^-$ | 269 yr |
| | | | 40 | 39.962 384 | 99.60 | stable | |
| | | | 42 | 41.963 049 | | $\beta^-$ | 33 yr |
| 19 | Potassium | K | 39 | 38.963 708 | 93.26 | stable | |
| | | | 40 | 39.964 000 | 0.01 | $\beta^+$ | $1.28 \times 10^9$ yr |
| | | | 41 | 40.961 827 | 6.73 | stable | |
| 20 | Calcium | Ca | 40 | 39.962 591 | 96.94 | stable | |
| | | | 42 | 41.958 618 | 0.64 | stable | |
| | | | 43 | 42.958 767 | 0.13 | stable | |
| | | | 44 | 43.955 481 | 2.08 | stable | |
| | | | 47 | 46.954 547 | | $\beta^-$ | 4.5 days |
| | | | 48 | 47.952 534 | 0.18 | stable | |
| 24 | Chromium | Cr | 50 | 49.946 047 | 4.34 | stable | |
| | | | 52 | 51.940 511 | 83.79 | stable | |
| | | | 53 | 52.940 652 | 9.50 | stable | |
| | | | 54 | 53.938 883 | 2.36 | stable | |
| 26 | Iron | Fe | 54 | 53.939 613 | 5.9 | stable | |
| | | | 55 | 54.938 297 | | EC | 2.7 yr |
| | | | 56 | 55.934 940 | 91.72 | stable | |
| | | | 57 | 56.935 396 | 2.1 | stable | |
| | | | 58 | 57.933 278 | 0.28 | stable | |

| Atomic number ($Z$) | Element | Symbol | Mass number ($A$) | Atomic mass (u) | Percent abundance | Decay mode | Half-life $t_{1/2}$ |
|---|---|---|---|---|---|---|---|
| 27 | Cobalt | Co | 59 | 58.933 198 | 100 | stable | |
| | | | 60 | 59.933 820 | | $\beta^-$ | 5.27 yr |
| 28 | Nickel | Ni | 58 | 57.935 346 | 68.08 | stable | |
| | | | 60 | 59.930 789 | 26.22 | stable | |
| | | | 61 | 60.931 058 | 1.14 | stable | |
| | | | 62 | 61.928 346 | 3.63 | stable | |
| | | | 64 | 63.927 967 | 0.92 | stable | |
| 29 | Copper | Cu | 63 | 62.929 599 | 69.17 | stable | |
| | | | 65 | 64.927 791 | 30.83 | stable | |
| 37 | Rubidium | Rb | 96 | 95.93427 | | | 33 min |
| 38 | Strontium | Sr | 90 | 89.907 320 | | $\alpha$ | 28.9 y |
| 47 | Silver | Ag | 107 | 106.905 091 | 51.84 | stable | |
| | | | 109 | 108.904 754 | 48.16 | stable | |
| 48 | Cadmium | Cd | 106 | 105.906 457 | 1.25 | stable | |
| | | | 109 | 108.904 984 | | EC | 462 days |
| | | | 110 | 109.903 004 | 12.49 | stable | |
| | | | 111 | 110.904 182 | 12.80 | stable | |
| | | | 112 | 111.902 760 | 24.13 | stable | |
| | | | 113 | 112.904 401 | 12.22 | stable | |
| | | | 114 | 113.903 359 | 28.73 | stable | |
| | | | 116 | 115.904 755 | 7.49 | stable | |
| 50 | Tin | Sn | 120 | 119.902 197 | 32.4 | | |
| 53 | Iodine | I | 127 | 126.904 474 | 100 | stable | |
| | | | 129 | 128.904 984 | | $\beta^-$ | $1.6 \times 10^7$ yr |
| | | | 131 | 130.906 124 | | $\beta^-$ | 8.03 days |
| 54 | Xenon | Xe | 128 | 127.903 531 | 1.9 | stable | |
| | | | 129 | 128.904 779 | 26.4 | stable | |
| | | | 130 | 129.903 509 | 4.1 | stable | |
| | | | 131 | 130.905 069 | 21.2 | stable | |
| | | | 132 | 131.904 141 | 26.9 | stable | |
| | | | 133 | 132.905 906 | | $\beta^-$ | 5.4 days |
| | | | 134 | 133.905 394 | 10.4 | stable | |
| | | | 136 | 135.907 215 | 8.9 | stable | |
| 55 | Cesium | Cs | 133 | 132.905 436 | 100 | stable | |
| | | | 137 | 136.907 078 | | $\beta^-$ | 30 yr |
| | | | 138 | 137.911 017 | | $\beta^-$ | 32.2 min |
| 56 | Barium | Ba | 131 | 130.906 931 | | EC | 12 days |
| | | | 133 | 132.905 990 | | EC | 10.5 yr |
| | | | 134 | 133.904 492 | 2.42 | stable | |
| | | | 135 | 134.905 671 | 6.59 | stable | |
| | | | 136 | 135.904 559 | 7.85 | stable | |
| | | | 137 | 136.905 816 | 11.23 | stable | |
| | | | 138 | 137.905 236 | 71.70 | stable | |
| 79 | Gold | Au | 197 | 196.966 543 | 100 | stable | |
| | | | 198 | 197.968 242 | | $\beta^-$ | 2.7 d |
| 81 | Thallium | Tl | 203 | 202.972 320 | 29.524 | stable | |
| | | | 205 | 204.974 400 | 70.476 | stable | |
| | | | 207 | 206.977 403 | | $\beta^-$ | 4.77 min |

| Atomic number ($Z$) | Element | Symbol | Mass number ($A$) | Atomic mass (u) | Percent abundance | Decay mode | Half-life $t_{1/2}$ |
|---|---|---|---|---|---|---|---|
| 82 | Lead | Pb | 204 | 203.973 020 | 1.4 | stable | |
| | | | 205 | 204.974 457 | | EC | $1.5 \times 10^7$ yr |
| | | | 206 | 205.974 440 | 24.1 | stable | |
| | | | 207 | 206.975 871 | 22.1 | stable | |
| | | | 208 | 207.976 627 | 52.4 | stable | |
| | | | 210 | 209.984 163 | | $\alpha, \beta^-$ | 22.3 yr |
| | | | 211 | 210.988 734 | | $\beta^-$ | 36.1 min |
| 83 | Bismuth | Bi | 208 | 207.979 717 | | EC | $3.7 \times 10^5$ yr |
| | | | 209 | 208.980 374 | 100 | stable | |
| | | | 211 | 210.987 254 | | $\alpha$ | 2.14 min |
| | | | 215 | 215.001 836 | | $\beta^-$ | 7.4 min |
| 84 | Polonium | Po | 209 | 208.982 405 | | $\alpha$ | 102 yr |
| | | | 210 | 209.982 848 | | $\alpha$ | 138.38 days |
| | | | 215 | 214.999 418 | | $\alpha$ | 0.001 8 s |
| | | | 218 | 218.008 965 | | $\alpha, \beta^-$ | 3.10 min |
| 85 | Astatine | At | 218 | 218.008 685 | | $\alpha, \beta^-$ | 1.6 s |
| | | | 219 | 219.011 294 | | $\alpha, \beta^-$ | 0.9 min |
| 86 | Radon | Rn | 219 | 219.009 477 | | $\alpha$ | 3.96 s |
| | | | 220 | 220.011 369 | | $\alpha$ | 55.6 s |
| | | | 222 | 222.017 571 | | $\alpha, \beta^-$ | 3.823 days |
| 87 | Francium | Fr | 223 | 223.019 733 | | $\alpha, \beta^-$ | 22 min |
| 88 | Radium | Ra | 223 | 223.018 499 | | $\alpha$ | 11.43 days |
| | | | 224 | 224.020 187 | | $\alpha$ | 3.66 days |
| | | | 226 | 226.025 402 | | $\alpha$ | 1 600 yr |
| | | | 228 | 228.031 064 | | $\beta^-$ | 5.75 yr |
| 89 | Actinium | Ac | 227 | 227.027 749 | | $\alpha, \beta^-$ | 21.77 yr |
| | | | 228 | 228.031 015 | | $\beta^-$ | 6.15 hr |
| 90 | Thorium | Th | 227 | 227.027 701 | | $\alpha$ | 18.72 days |
| | | | 228 | 228.028 716 | | $\alpha$ | 1.913 yr |
| | | | 229 | 229.031 757 | | $\alpha$ | 7 300 yr |
| | | | 230 | 230.033 127 | | $\alpha$ | 75.000 yr |
| | | | 231 | 231.036 299 | | $\alpha, \beta^-$ | 25.52 hr |
| | | | 232 | 232.038 051 | 100 | $\alpha$ | $1.40 \times 10^{10}$ yr |
| | | | 234 | 234.043 593 | | $\beta^-$ | 24.1 days |
| 91 | Protactinium | Pa | 231 | 231.035 880 | | $\alpha$ | 32.760 yr |
| | | | 234 | 234.043 300 | | $\beta^-$ | 6.7 hr |
| 92 | Uranium | U | 232 | 232.03713 | | $\alpha$ | 72 y |
| | | | 233 | 233.039 630 | | $\alpha$ | $1.59 \times 10^5$ yr |
| | | | 234 | 234.040 946 | | $\alpha$ | $2.45 \times 10^5$ yr |
| | | | 235 | 235.043 924 | 0.72 | $\alpha$ | $7.04 \times 10^8$ yr |
| | | | 236 | 236.045 562 | | $\alpha$ | $2.34 \times 10^7$ yr |
| | | | 238 | 238.050 784 | 99.28 | $\alpha$ | $4.47 \times 10^9$ yr |
| 93 | Neptunium | Np | 236 | 236.046 560 | | EC | $1.15 \times 10^5$ yr |
| | | | 237 | 237.048 168 | | $\alpha$ | $2.14 \times 10^6$ yr |
| 94 | Plutonium | Pu | 238 | 238.049 555 | | $\alpha$ | 87.7 yr |
| | | | 239 | 239.052 157 | | $\alpha$ | $2.412 \times 10^4$ yr |
| | | | 240 | 240.053 808 | | $\alpha$ | 6 560 yr |
| | | | 242 | 242.058 737 | | $\alpha$ | $3.73 \times 10^6$ yr |

APPENDIX E

# Answers to Review Questions

## Chapter 1

**1.1** It is true because an observer A can see an object moving and observer B can see the same object not moving. When you are sitting on a train, you are not moving with respect to the train but are moving with respect to the trees on the ground.
**1.2** We can decide if the object is moving at a constant rate, moving faster and faster, or slowing down. We can also decide if the direction of the motion is changing.
**1.3** Ten kilometers is the distance, and 16 km is the path length.
**1.4** To find the position we need to find the value of $x$ corresponding to the value of $t$. We find that the object's position is $x = 60$ m at $t = 2.0$ s and $x = 20$ m at $t = 5.0$ s.
**1.5** Because the area of a rectangle is equal to the product of length and width. In this case it is $v_x \cdot \Delta t$, which is equal to the displacement.
**1.6** (a) A person (the object of interest) standing in a parking lot as seen by a person in a car that is leaving the lot at increasing speed (the forward direction is positive). (b) While standing in a parking lot, you observe a car moving past you in the negative direction and slowing down. The direction of the car's acceleration is opposite its velocity in the positive direction.
**1.7** Mike is right about the initial position, $x_0 = -48$ m, but is not right about the acceleration; it is $a_x = -4\text{ m/s}^2$.
**1.8** You can choose the positive direction for your coordinate axis as either up or as down at your convenience. If the axis points down, the component of acceleration is positive ($g_y = +9.8\text{ m/s}^2$); if the axis points up, the component of acceleration is negative ($g_y = -9.8\text{ m/s}^2$).
**1.9** The first car starts to slow down at the same time that the tailgating car sees the first car's brake lights. Only after the reaction time does the tailgating car's speed start to decrease. The front car is now moving slower and continues to move slower than the tailgating car until they collide.

## Chapter 2

**2.1** Earth, the floor, and the air surrounding you, although the interaction with the air as you are sliding might be very weak.
**2.2** About $64 \pm 1$ units.
**2.3** We conducted experiments with the bowling ball and the person on rollerblades and analyzed them using force diagrams and motion diagrams. Our analysis showed that when the sum of the forces exerted on the object was zero, the object moved at constant velocity.
**2.4** The magnitude of the upward force exerted by the cable on the elevator $F_{\text{C on El}}$ is exactly equal to the magnitude of the downward force exerted by Earth on the elevator $F_{\text{E on El}}$. The cable and Earth are the only two objects interacting with the elevator. If the forces they exert on the elevator have the same magnitudes and opposite directions, the sum of the force is zero, and the elevator should move at constant velocity.
**2.5** Observers in inertial reference frames find that if an object does not interact with any other objects or all interactions add to zero, the object's velocity does not change. However, an observer in a noninertial reference frame can find an object whose velocity changes even when there are no other objects interacting with it or when all interactions are balanced (add to zero). For example, a passenger on a bus observes her purse slide off her lap (accelerating away from her) without any extra objects pushing the purse. A person standing beside the bus sees the purse moving forward at constant speed.
**2.6** $\vec{a}$ is a measure of how fast velocity changes and is a consequence of the sum of all forces exerted on the object. The product of mass and acceleration is *not* an additional force and thus should not be on the force diagram.
**2.7** The force that Earth exerts on any object is proportional to the object's mass. The acceleration is inversely proportional to the object's mass. Thus the effect of the mass cancels.
**2.8** Mike is correct; the scale reads the force that it exerts on the person. We know this because the reading of the scale changes as the person rides the elevator while the gravitational force that Earth exerts on the person (the weight of the person) does not change. It does not read the sum of the forces. If it did, then the reading should be zero when the elevator is at rest.
**2.9** The rollerblader pushes the floor, and the floor pushes the rollerblader ($\vec{F}_{\text{R on F}} = -\vec{F}_{\text{F on R}}$). You push the refrigerator and the refrigerator pushes you ($\vec{F}_{\text{Y on R}} = -\vec{F}_{\text{R on Y}}$). The truck pulls the car and the car pulls the truck ($\vec{F}_{\text{T on C}} = -\vec{F}_{\text{C on T}}$).
**2.10** By extending the stopping distance and consequently reducing the acceleration during the stopping process.

## Chapter 3

**3.1** A scalar component is positive if the vector projects in the positive direction on the axis and negative if the vector projects in the negative direction on the axis.
**3.2** P—person, S—surface, E—Earth

$$x\text{: } ma_x = +F_{\text{P on C}}\cos 30^\circ + N_{\text{S on C}}\cos 90^\circ + F_{\text{E on C}}\cos 90^\circ$$
$$\text{or } ma_x = +F_{\text{P on C}}\,0.87 + 0 + 0$$
$$y\text{: } ma_y = -F_{\text{P on C}}(\sin 30^\circ) + N_{\text{S on C}}\sin 90^\circ + (-F_{\text{E on C}}\sin 90^\circ)$$
$$\text{or } m(0) = (-F_{\text{P on C}}\,0.50) + N_{\text{S on C}} - F_{\text{E on C}}$$

**3.3** For objects moving along inclined surfaces, the acceleration is parallel to the inclined surface. Thus, if we choose the $x$-axis parallel to the surface, the acceleration is entirely along that axis, and is zero in the direction of the $y$-axis, which is perpendicular to the surface. This makes problem solving much easier than if we choose horizontal and vertical axes, in which case there is a nonzero component of the acceleration along each axis.
**3.4** The friction force is zero if no one is pushing or pulling the refrigerator. However if you do need to move it, you will have to push rather hard since the force needed to get it moving will have to be greater than $f_{s\,\text{max}} = \mu_s N = \mu_s(mg) = 350$ N (or about 80 lb).
**3.5** We analyze the constant velocity horizontal motion and the constant acceleration vertical motion independently. We need to use the appropriate initial velocities (the velocity components) for each motion.
**3.6** The engine is a part of the system; thus it cannot exert an external force that will accelerate the car system. The ground is

the main external object interacting with the car that exerts a force on it.

## Chapter 4

**4.1** The velocity vector is tangent to the circle. Using the velocity change technique, we find that the velocity change vector $\Delta\vec{v}$ points toward the center. The acceleration vector equals the velocity change vector divided by the time interval during which the change occurred: $\vec{a} = \Delta\vec{v}/\Delta t$. So the acceleration vector points in the same direction as the velocity change vector, toward the center of the circle.

**4.2** At all times during the ball's motion, the downward force exerted by Earth on the ball and the upward normal force exerted by the surface on the ball balance so that there is no net force in the vertical direction. But in the horizontal plane of the table while the ball is in contact with the semicircular ring, the ring exerts an inward force on the ball toward the center of the circle. This force causes the ball to accelerate toward the center of the circle while it contacts the semicircle. Before and after the semicircle, the ball is not accelerating, and the net force that other objects exert on the ball is zero. After the ball exits the barrier it will move in direction B.

**4.3** For radial acceleration in terms of speed and radius, we have $a = v^2/r$. The dimensions on the right are $(\mathrm{L/T})^2/\mathrm{L} = \mathrm{L/T}^2$, which are the correct dimensions for acceleration: length $L$ over time squared $T^2$. Centripetal acceleration in terms of radius and period is $a = 4\pi^2 r/T^2$. The dimensions are $\mathrm{L/T}^2$, which again are correct for acceleration.

**4.4** The woman's velocity is tangent to the drum. She would normally fly forward and fall (like a projectile). But the drum's surface intercepts her forward path and pushes in on her, causing her to move in a circular path. The friction force exerted by the drum prevents her from falling.

**4.5** The friend can imagine that we take two point-like objects each with a mass of 1 kg and place them apart at a distance of 1 m. They will attract each other, exerting a force of magnitude $6.67 \times 10^{-11}$ N—a very tiny force. It looks like such a force should not affect the interactions of objects of masses comparable to masses of people, houses, trees, etc. However, when the mass of one of the objects is on the order of $10^{26}$ kg, as the mass of Earth is, the force becomes significant even though the separation between the center of Earth and any object is huge—more than 6000 km.

**4.6** The Moon actually does fall toward Earth all the time, if the word "fall" implies the motion of an object when the only force exerted on it is the force exerted by Earth. But it also flies forward all the time. Thus, the Moon combines two motions, flying forward and at the same time falling toward Earth. The net result is that it continually "lands" on its circular path around Earth.

## Chapter 5

**5.1** We can define the system as the log and all of the air that participated in burning.

**5.2** Momentum is a vector quantity. Before the collision, the carts of the same mass were moving in the opposite directions at the same speeds. The vector sum of their momenta was zero. After the collision, it was zero again because both carts had zero velocities.

**5.3** The momentum of the apple increases because of the impulse of the force exerted on it by Earth. If you consider the apple and Earth as a system, then the momentum is constant: the apple gains a downward momentum and Earth gains an upward momentum. We do not observe Earth moving upward because its mass is huge.

**5.4** *Sad ball:* $m_b v_i + 0 = m_b \cdot 0 + m_{\text{board}} v_{\text{board fx}}$

**5.4** *Happy ball:* $m_b v_i + 0 = m_b(-v_i) + m_{\text{board}}(v_{\text{board fx}})$

**5.5** We considered the block and the bullet as a system. The force exerted on the bullet by the block is an internal force.

**5.6** A 2.0-kg skateboard was rolling in the negative direction at a speed of 8 m/s when Marsha (58 kg) jumped on it. What is the new speed of the skateboard and Marsha together?

**5.7** When we consider the momentum of a system, mass is as important as velocity. The mass of Earth is huge compared to the mass of the meteorite. Consider the meteorite and Earth to be the system. When the meteorite hits Earth, they both continue moving in the direction of motion of the meteorite, so that the initial momentum of the meteorite-Earth system is constant before and after the collision; however, the velocity of the system is tiny due to the huge mass of Earth.

## Chapter 6

**6.1** (1) You push a heavy crate sitting on the floor and it does not move—the work is zero because the displacement is zero. (2) Earth does zero work on the orbiting Moon because the force that Earth exerts on the Moon is perpendicular to the Moon's displacement.

**6.2** Work involves *a process* that occurs when a force is exerted by some external object on the object in the system as the system object moves. Thus the system does not possess work. On the other hand, the state of a system can be characterized by the amount of each type of energy in the system.

**6.3** We can do it in one of two ways. (1) We include Earth in the system and consider the change in gravitational potential energy of the system that changes as the elevation of the object relative to Earth changes. (2) We exclude Earth from the system and include the work done on an object by the external force that Earth exerts on the object as it changes elevation. The first method is usually easier to use in problem solving.

**6.4** The force that you would exert on the spring while stretching it is not constant in magnitude. It increases linearly from 0 to $kx$, and the average force is $(1/2)kx$.

**6.5** We could exclude a stationary surface from the system and determine the negative work done by friction—a common practice. However, this method has been shown not to account for the thermal energy change of the touching surface of an object in the system. It is easier, we think, to include the surfaces in the system and account for the energy change as an internal energy change—a change that we can see and feel.

**6.6** The system would have gravitational potential energy but Earth would do work on the system—the work-energy bar chart would have changed and instead of $U_{gi} = U_{sf}$ we would write $W_{\text{Earth on system}} = U_{sf}$. However, as the work done by Earth is equal to $mgy_i$, the final answer will not change.

**6.7** Calculate the total kinetic energy of the system (the two colliding objects) before the collision and after the collision. If the numbers are the same, then no kinetic energy went into internal energy—the collision was elastic. If the kinetic energy of the system changed, then the collision was partially or totally inelastic.

**6.8** If we neglect rolling friction and air friction (which are pretty small in this case), there is no work done on Jim by any forces and there are no energy changes (no matter what system we choose). So the power is zero.

**6.9** The potential energy of two objects can be negative if the objects exert attractive forces on each other, and we choose their zero energy reference separation to be very far apart—for example, zero potential energy when infinitely far apart. An example is the gravitational potential energy of two objects that attract each other. Positive work is required to pull them far apart where their energy is approximately zero. Thus, they must have started with negative energy.

## Chapter 7

**7.1** The nail should be along a vertical line that passes through the center of mass of the painting when it is in the correct orientation. The force that Earth exerts on the painting will then not cause the painting to rotate.

**7.2** (a) The torque of the force that Earth exerts on an any object about its center of mass is zero, but it is not zero about any other point. Thus when we support an object at the center of mass, it is in equilibrium, but if we put the support at any other location, it tips. (b) Suspend a meter stick and pull it along the direction of the meter stick. (c) Two people sitting on a seesaw so that their torques with respect to the fulcrum are the same in magnitude but opposite in signs.

**7.3** We can choose any point as a possible axis of rotation to write the torque condition of equilibrium (because when the object is in equilibrium, it does not rotate about any axis at all, so we can choose the one we want to use). However, it is useful to choose a location for which the largest number of forces has zero torque.

**7.4** One way to do it is to put the person face down on a big exercise ball and ask him to keep his muscles tense so that his body remains straight. Then slowly roll the ball under him until it is in a position where the person can balance horizontally on the ball. The ball will be under his center of mass.

**7.5** The torque exerted by the muscle needs to balance the torque exerted by the backpack. The straps of the backpack are longer than the muscle; thus the distance from the force exerted by the backpack to the axis of rotation is larger than the distance between the force exerted by the muscle and the axis of rotation. In addition, the trapezius exerts a force at an angle, thus diminishing its rotational effect. Both factors lead to the increase of the magnitude of the force exerted by the muscle to balance the effect of the force exerted by the backpack.

**7.6** To answer this question, consider the forces exerted on the ball and on the pencil when in equilibrium (Figures 7.43a and c) and when the equilibrium is disturbed (Figures 7.43b and d). Earth exerts a gravitational force on the string-ball system, which returns the ball toward its stable equilibrium position. The gravitational force exerts a torque on the pencil, which moves the pencil farther away from its unstable equilibrium position.

**7.7** The distance between the muscle force exerted on the bone and the axis of rotation is somewhat less than the distance between the force exerted by the load being lifted or supported and the axis of rotation.

## Chapter 8

**8.1** When an ice skater is rotating faster and faster in a clockwise direction, her rotational velocity $\omega$ is negative and rotational acceleration $\alpha$ is negative. When she starts slowing down, the rotational velocity is still negative, but the rotational acceleration becomes positive.

**8.2** We tested both ideas in the experiments with the rotating cylinder when we exerted the same magnitude force at different locations for the axis of rotation. If the force determines the magnitude of the rotational acceleration, the outcomes of both experiments would be the same, but this is not what we found. When the same force was exerted farther from the axis of rotation, the acceleration was larger. From these experiments we concluded that it was the torque that affects angular acceleration.

**8.3** The wooden ball has more mass farther from the axis of rotation and thus a higher rotational inertia. Thus the same torque will produce a smaller change in motion in the wooden ball.

**8.4** Both laws represent a cause-effect relationship between an interaction of an object with external objects, the object's inertial properties, and the change in its motion due to this external interaction. The change in motion (rotational or translational acceleration) is directly proportional to the measure of the interaction (external torque or force) and inversely proportional to the inertial properties of the object (rotational inertia or mass).

**8.5** A person jumping on the merry-go-round right before landing had zero rotational velocity. Her landing increases the rotational inertia of the system; thus the rotational speed decreases. A person jumping off is originally moving with the carousel and has rotational momentum. The total rotational momentum of the merry-go-round and the person was the sum of both. When the person steps off, she takes her rotational momentum with her, and the rotational momentum of the carousel alone remains the same.

**8.6** The chicken soup can is similar to the bottle filled with water, in which the mass of the water does not rotate as the bottle rolls down the ramp. The clam chowder can is similar to a bottle filled with ice, which rotates as it rolls down. The chicken soup rolls faster!

**8.7** Choose the solid Earth and the ocean water as the system but not the tidal bulges that rise due to the Moon's greater gravitational force on water nearer the Moon. The tidal bulges rubbing against water below cause an external friction force that slows Earth's rotation.

## Chapter 9

**9.1** Moist objects dry when randomly moving liquid particles leave the object's surface.

**9.2** The distances by themselves do not tell us much. We need to compare them to the sizes of the particles. As the size of the particles is about $10^{-8}$ cm, the average $3 \times 10^{-7}$-cm distance between particles is about 30 times larger than their sizes.

**9.3** We can think of the $m_p v^2$ as the product of two terms—as $m_p v$ times $v$. If the momentum $m_p v$ is greater, then the particles exert a greater force when hitting the wall. A greater force per collision causes greater pressure. Finally, if the particles move back and forth faster (larger $v$), they have more frequent collisions, leading to greater pressure. So the equation seems reasonable.

**9.4** Assuming the same particle temperature (the same average kinetic energy $1/2m_p v^2$ of the molecules), the less massive molecules (smaller $m_p$) will have higher average speed (greater $v^2$).

**9.5** The right side of the first equation involves microscopic quantities that are not measured directly, while the right side of the second equation involves macroscopic quantities that can be measured directly.

**9.6** Stern's experiment tested the quantitative relations of the theory. Stern had the clear goal of designing an experiment to see if the particle speeds matched the predictions based on the ideal gas model and Newton's laws.

**9.7** When the mass of the gas is constant and one of the following parameters ($P$, $V$, or $T$) stays constant, then one can use gas laws to describe the behavior of the gas.

**9.8** We calculated that the time during which the Sun has emitted thermal energy is about 40 million years. The age of Earth is about 4.5 billion years. Thus, the Sun must have some other source of energy that would allow it to have emitted light for at least 4.5 billion years.

## Chapter 10

**10.1** Determine the mass of the object in kilograms using a scale. Then submerge the object in water in a graduated cylinder and note the change in the water level. Use the volume of the displaced water to determine the volume of the object in cubic meters. Divide mass by volume to find density.

**10.2** When you squeeze the closed end of the tube, you exert an additional pressure that is transferred uniformly in all directions.

**10.3** The pressure differs at different heights because liquid layers at lower elevations support the liquid above it. An increased pressure in one part of the liquid (for example, by a plunger) causes a uniform increase throughout the liquid. At the same height, the pressure is the same in all directions—up, down, right, and left.

**10.4** It means that a column of mercury of density $\rho = 13{,}600\ \text{kg/m}^3$ and height $h = 760\ \text{mm}$ exerts the same pressure $P = \rho g h$ as Earth's atmosphere (about $1.0 \times 10^5\ \text{N/m}^2$).

**10.5** The pressure that the fluid exerts on the walls of a container increases as the depth of the fluid increases. When an object is submerged in a fluid, the upward pressure of the fluid on the bottom surface of the object is greater than the downward fluid pressure on the top of the object.

**10.6** The forces are the same and are independent of the mass of the object. The forces are determined by the volume of the displaced fluid.

**10.7** You can measure how much water the ship displaces if it submerges to the waterline. The mass of this water will be exactly equal to the mass of cargo that you can put on the ship for it to submerge to the waterline level.

## Chapter 11

**11.1** The lightbulbs came together because the pressure of the air in between them is lower than the pressure outside. Thus the outside pressure pushes them closer together.

**11.2** The cross-sectional area of the river before the outlet is much greater than the area of the cross section at the outlet. Water flows at higher speed through the outlet with the smaller cross-sectional area.

**11.3** In most cases we would prefer streamline flow. There is less friction-like resistance to this flow and less pressure needed to cause the flow—the heart does not have to work as hard.

**11.4** They are both based on the work-energy equation. However, the work-energy charts apply mainly to processes involving solid objects and the Bernoulli charts to fluid processes. The latter involve energy densities in the fluids and pressures that cause the fluid energy density to change. They also apply to the pressure and energy densities at particular positions in the fluid, whereas the work-energy bar charts apply to initial and final situations in a process.

**11.5** The water is open to the air pressure at both levels. We neglect the change in the atmospheric pressure with height because the bottle is very small compared to the height of the atmosphere.

**11.6** The blood flows more easily; thus your blood pressure can be lower as your heart needs to do less work in pumping blood. You may also have fewer heartbeats per minute.

**11.7** The air pressure above is reduced because the air's kinetic energy density is greater and the pressure is lower. Thus, the normal pressure from inside the house pushes the roof up and off the house.

**11.8** For an object to move at constant velocity with respect to an inertial reference frame (in this case, Earth), the magnitudes of the forces exerted on it in opposite directions should be equal. In the case of skydiving, the net force is zero and the diver moves down at constant terminal speed. Note that we are neglecting the buoyant force exerted by the air on the diver. If we take the buoyant force into account, the sum of the upward drag and buoyant forces should exactly equal the magnitude of the gravitational force that Earth exerts downward.

## Chapter 12

**12.1** $W = -P_{\text{atm}}\Delta V$

**12.2** About 1674 Joules.

**12.3** The work-energy equation allows us to find the final energy of a system from knowing its initial energy and the work done on the system. The first law of thermodynamics also tells us how to include the energy transferred to the system through heating.

**12.4** When a hot aluminum or iron block is added to cool water of the same mass, the water and block reach the same final temperature, which is much closer to the initial temperature of the water. Because of its much greater specific heat, the water temperature changes much less.

**12.5** Imagine that the same amount of energy is provided by heating a gas at constant volume and then at constant pressure. At constant volume, the gas reaches a higher temperature than at constant pressure, because the environment does negative work on the gas while the gas expands. Because of this negative work, the change in the internal energy of the gas in the latter case is less than in the former; hence the change in the temperature is less.

**12.6** Specific heat characterizes how much energy should be supplied to a unit mass of a substance to change its temperature by 1 °C while heat of fusion or vaporization characterizes energy supplied at constant temperatures to a unit mass of a substance to change its state. The unit of specific heat is J/kg · °C. The unit of heat of fusion or vaporization is J/kg.

**12.7** Conduction is efficient in transferring energy from atom to atom or molecule to molecule in solids with close contact between neighboring particles. Convection is very efficient in transferring energy through liquids and gases if the warm parts of the liquid or gas move toward cooler parts. Radiation is the best way to transfer energy through a vacuum, in which there are no particles to transfer energy by interactions or by moving from one place to another.

**12.8** Carbon dioxide absorbs long-wavelength infrared radiation but does not absorb much short-wavelength infrared radiation and visible light. This means that carbon dioxide does not reduce energy Earth receives from the Sun, but it reduces energy emitted by Earth. This reduces the cooling rate of Earth and its atmosphere and thus contributes to the increase of Earth's temperature.

## Chapter 13

**13.1** The gasoline molecules are relatively large and contain considerable chemical potential energy in their bonds. When they burn with oxygen, chemical products are formed that have somewhat less chemical energy. Part of the energy released is converted by the car's engine into kinetic energy, and about 80% is converted into thermal energy. Thus, the final energy is much less organized, even though a relatively small part is converted to organized kinetic energy.

**13.2** The lightbulb converts electrical energy into some light and considerable thermal energy (energy is conserved). The final state of the energy is more disordered with greater entropy. Energy was conserved, but entropy increased.

**13.3** Entropy is a state function; the equation $Q/T$ characterizes the change in entropy, not the entropy itself.

**13.4** The second law is formulated for isolated systems. In a refrigerator or air conditioner, there is an external object (a pump) that does work on the system.

**13.5** The engine efficiency is limited by a second law of thermodynamics expression that depends on the difference in temperature between the hot and cold heating reservoirs. In addition, there are other factors that limit the efficiency, such as friction in moving parts (generators and compressors) and burning of the fuel used for the hot reservoir.

## Chapter 14

**14.1** Oppositely charged objects do attract each other, but uncharged objects also attract other charged objects. Repulsion occurs only between like charged objects.

**14.2** Electrically neutral objects have equal numbers of positively and negatively charged particles so that their net electric charge is zero. When some of the electrons leave an electrically neutral object, it has more positively charged particles than negatively charged particles and becomes positively charged. The process can be expressed mathematically as $0 - (-2) = +2$.

**14.3** The human body is an electric conductor that allows electrons to transfer to/from the object. This keeps the metal object grounded and therefore electrically neutral.

**14.4** The ratio is 1. The forces are the same in magnitude because of Newton's third law.

**14.5** If the charges are the same—both positive or both negative—the energy can only be reduced by increasing the distance between them. If the charges are opposite, the energy can only be reduced by moving them closer together.

**14.6** If we chose only the escaping $\alpha$ particle as the system, then this system would have no electric potential energy. However, the radon nucleus would do positive work on the $\alpha$ particle, causing its kinetic energy to increase. The final result would not change.

**14.7** Some of the electric potential energy of the dome-metal ball system is converted into light energy.

## Chapter 15

**15.1** Draw a sketch showing the $\vec{E}$ field vectors due to each source charge at the point of interest. Add them as vectors using the head-to-tail method and find the direction of $\vec{E}_{net}$.

**15.2** Both charges contribute positive $x$-components to the field at the point of interest. The $-2Q$ charge contributes a negative $y$-component while the $+Q$ charge contributes a positive $y$-component. Since the point of interest is equidistant from both charges and the $-2Q$ charge has greater magnitude, the net $y$-component of the field at the point of interest is in the negative $y$-direction. Hence, the $\vec{E}$ field will point toward the right and slightly downward.

**15.3** The work is the same and equals the change in electric potential from A to B and from C to D.

**15.4** Assume that the potential difference $\Delta V_x$ between the doorknob and your finger is independent of the distance $\Delta x$ between them. As you approach the knob, $\Delta x$ becomes smaller and the magnitude of $|E_x| = |\Delta V|/|\Delta x|$ increases and can reach the breakdown value.

**15.5** If there is a nonzero electric field inside, the field exerts forces on free electrons, which rearrange themselves until the net field is zero. The free electrons create their own $\vec{E}_{cond}$ field, which cancels the external field $\vec{E}_{ext}$. The net $\vec{E}_{net}$ field inside is $\vec{E}_{ext} - \vec{E}_{cond} = 0$.

**15.6** Inside a conducting material placed in an external electric field, the net electric field is zero; inside a dielectric material, the field is weaker but not entirely zero. The difference can be explained microscopically. In the conductors there are freely moving electrons that can redistribute to cancel the external field. Inside a dielectric, the molecules change their shape and orientation but do not move translationally; thus the field is diminished but not cancelled.

**15.7** According to the superposition principle, the net $\vec{E}_{net}$ field is the sum of the $\vec{E}$ fields created by the plates. The $\vec{E}$ field vectors due to the plates are in the same direction inside the capacitor and in the opposite directions outside.

**15.8** You must make sure that the rod is grounded. This can be done by running a wire on the outside of the house connecting the rod to the ground.

## Chapter 16

**16.1** (1) The two locations must continuously have different electric potentials ($V$ field values). (2) There have to be electrically charged particles that can travel from one location to the other. (3) There has to be a continuous path for the particles to travel.

**16.2** The potential in the circuit changes from high to low outside the battery and from low to high inside the battery. The negatively charged electrons move from lower potential to higher potential outside the battery and from higher potential to lower potential inside the battery.

**16.3** When a circuit is complete you can trace a path of an imaginary positive charge moving from the positive terminal of the battery to the negative terminal; the path passes along conducting material at every location.

**16.4** As the physical quantity of current measures how much electric charge is transferred through a cross-sectional area of a conductor per unit time, it makes sense that the charge passing through all circuit elements is the same in this experiment. Electric charge is a conserved quantity. There is no place for the charge to leave the path and no extra paths for the charge to come from.

**16.5** When you turn off one of the appliances, the rest do not go off. When you turn on extra appliances, the bulbs that are on do not get dimmer.

**16.6** The 60-W bulb has a higher resistance than the 100-W bulb. When they are connected in parallel with the same potential difference across the bulbs (as connected in your home), the 60-W bulb is dimmer than the 100-W bulb ($P = \Delta v^2/R$ is less for the higher resistance bulb). In series, the 60-W bulb should be brighter: the same current goes through the two bulbs and the one with the greater resistance will convert more electric energy into thermal energy ($P = I^2R$).

**16.7** The potential is higher on the side where the current enters the resistor than on the side where it leaves (again, we consider positive charges flowing in the direction of the current). There is almost no potential difference across a conducting wire, as its resistance is negligible.

**16.8** Brightest to least bright: $4 > 1 > 2 = 3$

**16.9** It means that the direction of the current is opposite to the direction that we chose as positive.

**16.10** As the current passes through the lightbulb, the filament becomes warmer and its resistivity increases, leading to the increase in resistance—a lightbulb is *not* an ohmic device. One of the models of resistivity explains its increase with temperature as the result of interactions of the free electrons with the more violently vibrating ions of the crystal lattice.

**16.11** To get the same electric current through the system, the potential difference across the battery would have to increase (a higher emf). In terms of the circulatory system, the blood pressure produced by the heart in pumping blood would have to increase.

## Chapter 17

**17.1** The north pole of a compass needle is attracted to both positively and negatively charged objects, as is the other pole. Thus, poles must not be positively or negatively charged.

**17.2** The $\vec{B}$ at this location is zero.

**17.3** $F_{\vec{B}\text{ on W max}}$ is the maximum force exerted on the wire of length $L$ with current $I$ in it when it is placed in a region of uniform magnetic field $\vec{B}$ whose origin is unknown. The current and wire length are independent variables. The force exerted on the wire depends on both of them. The magnitude of the magnetic field is independent of current, wire length, and force exerted on the wire, because it is determined by the sources that created that magnetic field.

**17.4** The magnetic force is always perpendicular to the direction of the magnetic field. Thus, the work $W = F\,\Delta s\cos\theta$ is zero for each displacement $\Delta s$ along the circular path, as $\cos 90° = 0$.

**17.5** Consider current-carrying wire 1 to be the source of the magnetic field and wire 2 to be the test object. We can determine the magnitude of the $\vec{B}$ field created by 1 in the vicinity of 2 (at a distance of 1.0 m) and then determine the magnetic force exerted on 2 by this magnetic field—see the following equation.

$$F_{\vec{B}_1\text{ on W}_2} = I_2L\left(\frac{\mu_0}{2\pi}\frac{I_1}{r}\right)\sin\theta = \frac{\mu_0}{2\pi}\frac{I_1I_2}{r}L\sin\theta$$

$$= \frac{(4\pi\times 10^{-7}\,\text{T}\cdot\text{m/A})(1.0\,\text{A})(1.0\,\text{A})}{2\pi(1.0\,\text{m})}(1.0\,\text{m})(\sin 90°)$$

$$= 1.0\times 10^{-7}\,\text{N}$$

**17.6** The right-hand rule for magnetic force indicates the direction of the force that a magnetic field exerts on a moving particle or on a current-carrying wire. The right-hand rule for the $\vec{B}$ field indicates the direction of the magnetic field caused by the current-carrying wire.

**17.7** In an MHD generator, free electrons and positively charged ions traveling at high speeds (for example, in a horizontal direction from right to left) enter a magnetic field region in which the $\vec{B}$ field points vertically down (Figure 17.32). The magnetic field exerts oppositely directed forces on positive ions and on negative ions, which will accumulate on two vertical plates producing an electric field that exerts a force opposite the magnetic force. This leads to a condition causing a constant electric field and a constant potential difference across the plates.

**17.8** An external magnetic field causes atoms in diamagnetic materials to become slightly magnetized opposite the magnetic field, thus slightly reducing the net field, whereas paramagnetic atoms become oriented in the same direction as the field and add slightly to the field. This difference is due to the differences in the atomic structure of diamagnetic and paramagnetic materials.

## Chapter 18

**18.1** When a bar magnet moves with respect to a coil, the magnetic field through the coil changes—thus a current should be induced. However, one can change the magnetic field without motion. Place one coil on top of another, with one of them connected to a battery through a switch and the other one connected to a galvanometer. When you close the switch in the first coil, the galvanometer detects current in the second coil, with no relative motion of the coils.

**18.2** Place the ring with its surface parallel to the magnetic field lines leaving the N pole. The $\vec{B}$ field passes by the ring's surface, but not *through* that surface.

**18.3** Suppose you have a coil and a bar magnet, and you move the north pole of the magnet slowly toward the opening of the coil. The magnetic field through the coil would increase and induce an electric current that would cause an induced magnetic field in the same direction as the external bar magnet magnetic field. This would cause an even larger increasing magnetic field, and an even larger induced electric current, which would cause an even larger induced magnetic field, and so forth. The result would be a runaway induced current that would melt the coil.

**18.4** The area and the orientation are both factors in the rate of flux. In the case of Observational Experiment Table 18.3 we wanted to focus on the change in the magnetic field that contributes to the change in flux rather than other physical quantities that contribute to the flux change.

**18.5** The expression for the flux that we used to calculate the emf is only valid when the field is uniform throughout the area surrounded by the coil.

**18.6** The tissue in the brain is a reasonable electric conductor. A changing current in the coil placed on the scalp causes a changing magnetic field through the scalp and into the brain. This changing magnetic field in the brain produces circular electric currents in the brain—similar to eddy currents.

**18.7** It has to move at a speed that is 3.75 times larger than the speed in the example. Although the emf induced is the same, the current in the circuit will be larger than in the example as the length of the wire decreased, and consequently the resistance decreased too.

**18.8** As the coil rotates, the angle between the normal line to the coil and the direction of the external magnetic field changes. Thus the flux through the coil changes. When the flux through the coil changes, the emf is induced.

**18.9** The difference in the number of turns in the two transformer coils is the main reason. An alternating emf across the primary coil produces changing flux passing through the secondary coil (the core confines the magnetic field produced by electric current in the primary coil so that it passes through the second coil), creating a larger or smaller alternating emf across the secondary coil, depending on the relative number of loops in each coil.

**18.10** (a) The current is induced only if there is a component of the velocity of the loop perpendicular to the $\vec{B}$ field lines. A magnetic force is exerted on the electric charges in the moving wire. This force causes a charge separation—similar to that produced by a battery. These unbalanced charges produce an electric field that exerts forces on other charges in the wire creating the current. (b) A changing magnetic field is accompanied by an induced electric field that circulates around the loop and ends on itself. This electric field causes free electric charge in the loop to move around the loop in a coordinated fashion even though there is no battery or other power source in the loop.

## Chapter 19

**19.1** The book passes through the same locations in opposite directions during one cycle of motion—this might remind Brian of vibrational motion. However there is no specific location along this path that is the equilibrium—as long as he is pushing the book, the book is never in the equilibrium position. There is no restoring force and there is no change in the forms of energy.

**19.2** The quantities are inversely related by definition, not by cause-effect relationship. Period and frequency are just two different ways to characterize the vibration. The quantities depend on other features of the vibrating system not yet discussed.

**19.3** The amplitude of the velocity is 0.84 m/s; the period of vibrations is about 1.5 seconds. The object starts vibrating at the maximum positive displacement. The ratio of the mass of the cart to the spring constant is 0.057 kg·m/N. The amplitude of the vibration is 0.2 m.

**19.4** The period will decrease by a factor of 1.4.

**19.5** The cart's 2.0-s period means that every 2.0 s the cart passes the same position moving in the same direction. Suppose that at time $t = 0$, the cart has its maximum positive displacement ($x = +A$). Its potential energy at this position is maximum. At time $t = 0.50$ s the cart passes the equilibrium position, and its potential energy is zero. After another 0.50 s it is again stretched the maximum amount, only in the negative direction ($x = -A$). Its potential energy is again maximum $[0.5k(-A)^2]$. Then 1.0 s later the cart is back at $x = +A$ and the potential energy is again maximum. Thus, the elastic potential energy is maximum twice during each vibration period—every 1.0 s for this cart. If we neglect friction, the total mechanical energy of the system is constant—it does not change. Thus, the period of the total energy change is infinity.

**19.6** You need to move the bob closer to the clock, making the effective pendulum length smaller.

**19.7** The springs that obey Hooke's law exert a force on an object stretching or compressing them. This force varies linearly with the stretch or compression of the spring and points in the direction opposite to the stretch or the compression of the spring. These properties of the restoring force are necessary for the system to undergo SHM.

**19.8** If a system undergoes damped vibrational motion, the amplitude of vibration decreases and the period of vibration increases slightly.

**19.9** *Description:* Resonance is a physical phenomenon in which the amplitude of vibration of a system increases rapidly in response to an external periodic stimulus. *Explanation:* Resonance occurs because the external interaction with the system causes a force that always does positive work, thus increasing the total mechanical energy of the vibrating system. The external source should vibrate at the same frequency as the stimulated system.

**19.10** The $CO_2$ absorbs some of the infrared radiation leaving Earth (resonant absorption) and remits it back to the Earth, thus reducing the energy leaving Earth. Earth warms.

## Chapter 20

**20.1** For a longitudinal wave, vibrate the end of the Slinky back and forth parallel to the Slinky. For the transverse wave, vibrate the end of the Slinky up and down perpendicular to the Slinky.

**20.2** The speed of a vibrating object is the instantaneous vibrational speed of a particle in the medium where the wave propagates. This speed depends on the frequency and amplitude of the wave but does not depend on the speed of wave propagation. The speed of the wave is the speed that the disturbance travels through the medium. This speed is independent of the frequency or amplitude of the propagating wave.

**20.3** One difference is due to the mass of the string; the G string is thicker than the A string. Additionally, the A string is probably pulled tighter (greater force) than the G string.

**20.4** The intensity of the wave decreases proportional to the distance squared.

**20.5** As any Slinky is finite in length, you will always have a reflected wave propagating back.

**20.6** You could create a transverse pulse on one end of a Slinky and let your friend create an oppositely oriented pulse from the other end. The pulses would cancel as they pass in the middle.

**20.7** Loudness is related to the amplitude of sound waves and consequently their intensity. It also depends on the sensitivity of a human ear to the sounds of different frequencies. So the loudness of sound depends on both the amplitude and frequency.

**20.8** The resultant sound is a complex periodic wave made of more than one sinusoidal wave. We can look at the sound using a spectrum analyzer that registers the sound pressure amplitude at different frequencies.

**20.9** You will see that at certain frequencies, you get large-amplitude standing wave vibrations and at other frequencies, there is small erratic motion.

**20.10** Not all frequencies of vibration produce standing waves. But for those frequencies that do, the wavelengths of the standing waves do depend on the vibration frequency and the speed of the disturbance in the air inside the pipe.

**20.11** You detect a higher frequency sound when the ambulance is moving toward you. The waves get compressed in the decreasing distance between you and the ambulance, causing a shorter wavelength and thus a higher frequency. You detect a lower frequency sound when the ambulance is moving away from you. The waves get stretched in an increasingly longer space with a longer wavelength and thus a lower frequency.

## Chapter 21

**21.1** Each point of a light source emits light rays in all directions. But only one ray from each point passes though the small opening. The rays emitted by the bottom of the object have to travel up to pass through the opening; the rays emitted by the top of the object have to travel down to pass through the opening (see figure in Exercise 21.2).

**21.2** We can arrange two or more mirrors and measure the angles between them. Then we can predict in which direction a laser beam should be reflected after bouncing off these mirrors using the law of reflection and perform the actual experiment to see if the prediction is close to the outcome.

**21.3** When a ray of light meets a border with a different medium, it changes its path in two different ways: some of it reflects back into the first medium according to the law of reflection, and some of it bends and travels in a different direction in the second medium.

**21.4** The phenomenon of total internal reflection occurs due to the refraction of light going from an optically denser medium to optically less dense medium. Total internal reflection is an application of refraction.

**21.5** Water is less optically dense than glass; thus there is no total internal reflection for light going from water to glass.

**21.6** The sky is blue because the chemical composition of the atmosphere causes it to reflect blue light in all directions. The clouds are white because their tiny water droplets reflect all colors in the same way.

**21.7** The particle model uses an analogy with a stream of bullets and the wave model uses the analogy with waves. The former predicts that the speed of light in water is higher than in the air and the wave model predicts a lower speed.

## Chapter 22

**22.1** You will see the same amount.

**22.2** Hold the mirror near a wall facing the window opposite to it. Take a small cardboard piece and hold it in front of the mirror (parallel to it) so light from the window hits the mirror. Move the cardboard slowly away from the mirror until you see a sharp image of the window. The distance between the cardboard and the mirror will be approximately the focal length.

**22.3** Concave mirrors produce different types of images depending on the object distance relative to the focal length. If the object is between the mirror and the focal point, the image is virtual, upright, and enlarged; if the object is behind the focal point, the image is real and inverted. Thus knowing where to place the screen for a real image to appear on it is very important. Use the strategies described in Review Question 22.2.

**22.4** An object emits light in all directions, represented by an infinite number of rays. To draw an image, we choose the rays whose paths we can easily predict.

**22.5** You need to place it at a distance equal to double the focal length. The image will be the same size as the object.

**22.6** The ray diagram helps predict where the image will be located and helps estimate its magnification. You also need it to evaluate your calculation.

**22.7** In the camera, the distance from the lens to the image can change; in the eye, the focal length of the system lens-cornea (eye optical system) changes.

**22.8** A person who is farsighted will get the greater magnification. Because his near point is farther, he can look at the image farther away—the image will be bigger.

**22.9** A telescope magnifies the angular size of an object but does not magnify the object itself.

## Chapter 23

**23.1** According to the wave model of light, light leaving each narrow slit moves outward in all directions; each slit is a source of circular wavelets. Beyond the slits, the wavelets from each slit can interfere constructively at many different places. Thus we can see "bright light" at many places.

**23.2** A refraction index of glass of 1.6 means that the speed of light in glass is 1/1.6 times the speed in vacuum (air). The frequency of light stays the same when it travels through media where its speed is different. The wavelength of light of the same frequency changes when light travels to a different medium: in glass it is less than in the vacuum (air).

**23.3** The locations of the centers of the maxima do not change; the width of the bright peaks decreases.

**23.4** The colors produced by both a grating and a film are due to the interference of light. However, the colors produced by the grating are primary colors—thin bands of light of a particular color due to constructive interference—while the colors of thin films are complementary colors—bands of white light with one color missing due to destructive interference.

**23.5** The similar equation for two slits will be $\sin\theta = (2 + 1)\frac{\lambda}{2}$, where $m = \pm 0, 1, 2, 3, \ldots$.

**23.6** What we see on the film is not the image of a star produced by a lens. It is instead a diffraction pattern produced by light passing through the telescope's circular opening.

**24.7** Because in wave optics there are several commonly used units of length: nanometers (for the wavelength of light), millimeters (for slit width), and meters (for the distances between the slits/gratings and the screen). It is important to use the same units for the length-related quantities in the same equation.

## Chapter 24

**24.1** In a linearly polarized wave the particles vibrate along *one* direction in a plane perpendicular to the direction of the wave propagation. In an unpolarized wave the particles vibrate along *all* axes in that plane perpendicular to the direction of wave propagation.

**24.2** A changing $\vec{E}$ or $\vec{B}$ field will produce the other field ($\vec{B}$ or $\vec{E}$, respectively). To produce an electromagnetic wave, one or both of the fields at a particular location must have a nonconstant rate of change. This happens, for example, in an antenna where charged particles vibrate back and forth in a coordinated way.

**24.3** They both measure the time interval for an electromagnetic wave to travel to/from a target object. However, radar sends a wave to one target and determines the distance by the time interval for the round trip. In GPS, EM waves travel from three or more satellites to the target. The travel times of those signals are converted to distances, and trilateration is used to estimate the location of the target object.

**24.4** The speeds of the waves in a vacuum are the same. The wavelength of the first wave will be half the wavelength of the second wave.

**24.5** At a distance $2d$ from the source, the intensity of the electromagnetic wave will be one-fourth of the intensity at a distance of $d$. This happens because the same energy is spread over the surface area of a sphere that is four times bigger. Recall that the surface area of a sphere is directly proportional to the square of its radius.

**24.6** The reflected light is partially polarized—most of its electric field vectors are in the plane parallel to the horizontal plane of the dashboard. The glasses have a polarized film that lets only the vertical component of the $\vec{E}$ field pass through. Thus the intensity of the reflected light is reduced.

## Chapter 25

**25.1** We called it a testing experiment because these two physicists set out expecting a particular result based on their understanding of ether.

**25.2** Invariance is a principle of Newtonian physics that states that results of the laws of physics are identical in all inertial reference frames. For Maxwell equations to be the same in all inertial frames, the speed of light has to have a single value regardless of the inertial reference frame in which it is measured.

**25.3** One explanation can be that the events seen as happening simultaneously by one observer might not seem to be happening simultaneously for some other observer.

**25.4** In the muon experiment, the particles traveled much farther than they should have during their short lifetimes. Their travel distance could only be explained if one assumed that for Earth-based observers, the muon lifetime increased according to the time dilation equation.

**25.5** An observer who measures the stick to be shorter than 1 m must be moving with respect to you. No observers will see the stick as being longer than 1 m; relativistic effects only shorten lengths, not elongate them.

**25.6** "Velocity transformation" is a rule that prescribes how to find the velocity of an object with respect to a "stationary" reference frame, if one knows its velocity with respect to a moving reference frame and the velocity of the moving reference frame with respect to the stationary frame. The relativistic velocity transformation incorporates the limitations of the second postulate of the theory of special relativity, which the classical transformation does not.

**25.7** We need to modify the classical expression of momentum in special relativity theory because if we use the classical expression, the momentum of an isolated system is not constant.

**25.8** For an electron that moves at about $0.93c$, the value of relativistic kinetic energy is considerably larger ($1.41 \times 10^{-13}$ J) than the kinetic energy calculated using the classical equation ($1.35 \times 10^{-14}$ J).

**25.9** Light can travel without a medium, and the speed of light is independent of the observer. The Doppler equations for sound waves involve speeds relative to a medium, and the speed of sound varies for observers in different inertial reference frames.

**25.10** In classical terms, the Sun exerts a force on the Earth. In general relativity terms, by contrast, the Sun curves space, and Earth then naturally moves along a curved path.

**25.11** Two effects are the motion of the satellites with respect to the Earth and their position high above Earth's surface.

## Chapter 26

**26.1** Planck suggested that electromagnetic energy is radiated by charged particles in discrete portions—quanta—thus making it impossible to radiate an arbitrary amount of energy or an amount of energy that is smaller than the energy of one quantum. According to classical physics a charged particle could emit any amount of electromagnetic energy.

**26.2** The independence of the stopping potential on the intensity of light, the independence of the cutoff frequency on the intensity of light, and the immediate detection of the photocurrent.

**26.3** The work function of the metal is the minimal energy that the free electron-lattice system needs to absorb in order to make the energy of the system zero—to free the electron. The photon needs to provide this minimal energy $E_{\text{mimimum}} = hf_{\text{cutoff}}$ or more in order to free an electron.

**26.4** At low light intensity, Vavilov and Brumberg saw individual flashes of light on the screen (particle-like photons hitting the screen). But the individual flashes of light on the screen only occurred in locations where light waves produced interference maxima and did not occur where light waves interfered to produce minima.

**26.5** The electrons traveling across the cathode ray tube stop abruptly at the anode (large acceleration). Since electrons are electrically charged particles, they radiate electromagnetic waves when they accelerate. Since the acceleration was so large they radiate high-energy X-ray photons.

**26.6** The momentum of a photon is inversely proportional to its wavelength $p_{\text{photon}} = h/\lambda$. When the photon scatters off an initially stationary electron, the electron gains momentum, and consequently the photon must lose momentum. Thus, its wavelength must increase.

**26.7** The materials should have a small work function so that the cutoff frequency is that of red light. From Eq. (26.6), $f_{\text{cutoff}} = \phi/h$. Therefore, appropriate materials should have a work function $\phi < hf_{\text{red}}$.

## Chapter 27

**27.1** The electron is an electrically charged particle. When it moves in a circle, it emits electromagnetic energy. Due to this continuous emission, the energy of the atom would decrease and the size of the atom decreases accordingly. As the atom loses energy, the electron's orbit would become smaller.

**27.2** The zero point reference level of the electron-nucleus electric potential energy is when they are infinitely far apart. When closer together, the oppositely signed charges (electron and nucleus) have negative energy, which is greater in magnitude than the positive kinetic energy of the electron. The total net negative energy is therefore reasonable for two particles bound together.

**27.3** You need a light source that produces a continuous spectrum, such as an incandescent lightbulb; a spectroscope or a spectrograph—any instrument that allows you to see light of different colors in different locations; and a container holding the gas between the bulb and the spectroscope.

**27.4** Photons emitted through spontaneous emission travel in different directions and are not coherent. That is, their wave forms do not have a constant phase difference. Photons radiated though stimulated emission travel in the same direction and are coherent.

**27.5** The electron configuration is $1s^2 2s^2 2p^6$ with electron quantum numbers $n = 1, l = 0, m_l = 0, m_s = \pm 1/2$ (the two 1$s$ electrons); $n = 2, l = 0, m_l = 0, m_s = \pm 1/2$ (the two 2$s$ electrons); and $n = 2, l = 1, m_l = \pm 1, 0, -1, m_s = \pm 1/2$ (the six 2$p$ electrons).

**27.6** The peaks and troughs in the numbers of electrons at different angles, which Davisson and Germer observed accidentally, could have been predicted using de Broglie's idea of the wave nature of particles. If the particles behave like waves, then when passing through small openings (like the spaces between the atoms in a lattice) interference should lead to peaks and troughs.

**27.7** A shell includes all the electron states for a particular principle quantum number $n$. A subshell includes all the electron states for a particular angular momentum quantum number $l$.

**27.8** Diffraction of electrons passing through a single slit, interference of electrons passing through two slits, and the nonzero width of the spectral lines.

## Chapter 28

**28.1** We know it by passing the radiation through a magnetic field that is perpendicular to the velocity of the particles/photons. The particles/photons with no electric charge do not deflect; the charged particles deflect in a direction consistent with the direction of the force exerted by the magnetic field on a moving charged particle.

**28.2** If the electron is contained in the tiny nucleus, the magnitude of the potential energy of nucleus-electron system is only 1/10 the magnitude of the kinetic energy of the electron. Thus total energy of the system (kinetic plus potential) is positive. This means an electron in any nucleus would very rapidly escape the nucleus. Electrons cannot be components of nuclei.

**28.3** The ionization energy is the energy needed to remove an electron from the atom (13.6 eV for hydrogen). The binding energy is primarily the energy needed to separate the nucleus into its constituents. The latter is much larger.

**28.4** No. The mass number $A$ of the reactants and the products is not conserved $(4 + 6 \neq 11 + 0)$. Also, the atomic number $Z$ of the reactants and products is not conserved $(2 + 3 \neq 6 + 2(-1))$.

**28.5** The mass of the product nuclei is less than the mass of the reactants. Thus, a small fraction of the reactant mass is converted to other forms of energy. Energy is conserved if we include the rest mass energy of the particles involved in the process.

**28.6** Spin angular momentum and energy were not constant during the reaction.

**28.7** It means that after 200 years, the number of remaining radioactive nuclei is half the number of radioactive nuclei that existed 200 years earlier.

**28.8** There is a particular ratio of regular carbon to radioactive carbon in the body that stays constant as we inhale carbon. When the body dies, the radioactive carbon decays but the new atoms do not enter the body, thus the ratio changes. The smaller the ratio is, the older the specimen.

**28.9** They ionize the atoms and molecules in our bodies, possibly causing genetic problems or perhaps cancer.

## Chapter 29

**29.1** The mass of a free proton is less than the mass of the neutron, so a free proton has too little energy to decay into a neutron and the other smaller products. When within certain nuclei, the proton can acquire enough additional energy from other nuclear constituents to allow it to decay into a neutron, positron, and neutrino.

**29.2** Although the attractive strong interaction between protons is greater than the repulsive electric interaction between then, the strong interaction has a very short range. In ionized hydrogen gas, the protons are far enough apart that the electric repulsion dominates. Within a nucleus the strong attraction dominates.

**29.3** The major difference is that the electron is a truly elementary particle (as far as we know), whereas a proton has internal structure—it is made of quarks. The electron is a lepton, while the proton is a baryon. The electron is negatively charged, and the proton is positively charged. The mass of the electron is ~1700 times smaller than the mass of the proton.

**29.4** As the universe expanded, the average temperature eventually became cool enough (~370,000 years after the Big Bang) for neutral atoms to form (mostly hydrogen). When an electron combines with a proton to form a hydrogen atom, one or more photons are produced. It is the photons that were produced at this time in the history of the universe that make up the cosmic microwave background.

**29.5** By applying Newton's laws of motion and gravitation to stars orbiting near the edge of the galaxy, we can estimate the mass of everything within the interior of that orbit. That mass is about 10 times greater than the mass of the detectable objects in the galaxy.

# APPENDIX F

# Answers to Select Odd-Numbered Problems

## Chapter 1

### Multiple-Choice Questions

1. (c) 3. (b) 5. (b) 7. (d) 9. (b) 11. (c) 13. (b) 15. (c)

### Problems

1.
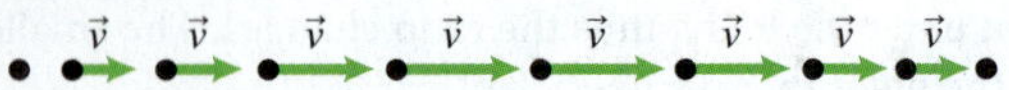


11. (a) 36 mph = 57.9 km/h = 16.1 m/s; (b) 349 km/h = 217 mph = 96.9 m/s; (c) 980 m/s = 3528 km/h = 2192 mph

13. (a) 60 mph

17. $(1.4 \pm 0.1)$ km

19. $3.99 \times 10^{16}$ m; uncertainty: $0.009 \times 10^{16}$ m

21. (a) Gabriele: 160 m, Xena: 120 m; (b) 30 s; (c) $60\text{ m} - (2.0\text{ m/s})t$

25. (a) Gabriele: $3.0 \times 10^3\text{ m} - (8.0\text{ m/s})t$; Xena: $1.0 \times 10^3\text{ m} + (6.0\text{ m/s})t$; (b) 1.9 km; (c) $-2.0 \times 10^3\text{ m} + (14\text{ m/s})t$

27. 3.6 mph

29. impossible to reach this average speed

31. (a) $x_1(t) = 30\text{ m} + (-8.33\text{ m/s})t$, $x_2(t) = -10\text{ m}$, $x_3(t) = -10\text{ m} + (5.0\text{ m/s})t$, $x_4(t) = -10\text{ m} + (-3.33\text{ m/s})t$

37. (a)
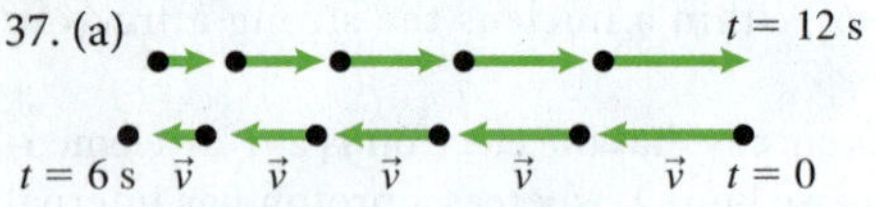


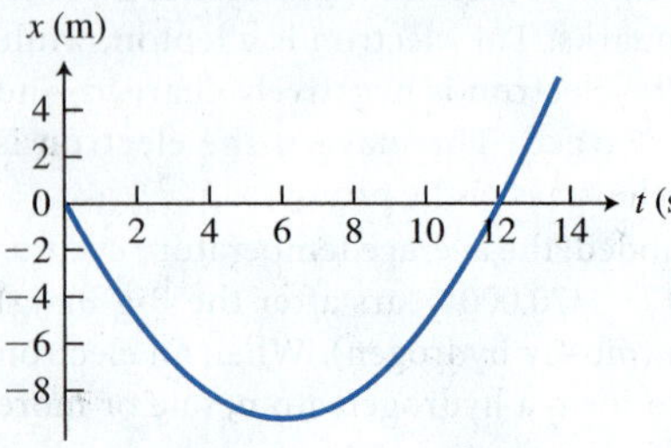


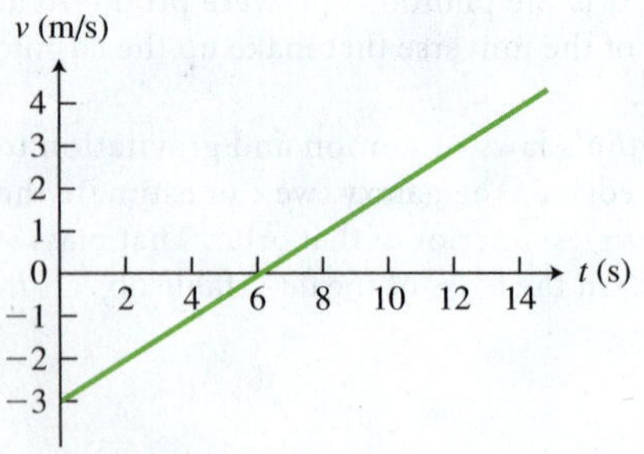


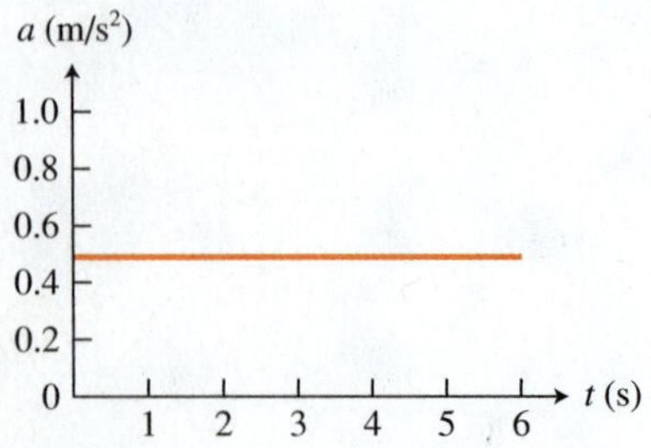


39. $-154\text{ m/s}^2$

41. $57.2\text{ m/s}^2; -9.2\text{ m/s}^2$

43. The distance between the runners will increase.
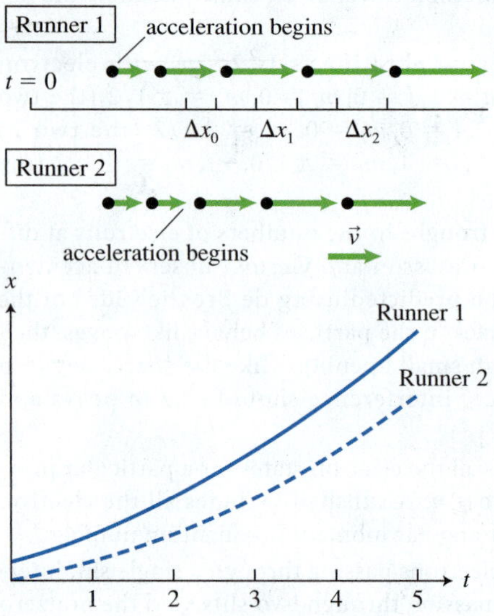


45. $4.0 \times 10^3\text{ m/s}^2$, 2.0 mm

49. (a) $5.6\text{ m/s}^2$; (b) 11.2 m; (d) 7.93 s; (e) 9.9 s, the uncertainty is 0.1 s

51. 42 m

53. yes, $a = 2.0\text{ m/s}^2$

55. $x_A(t) = 200\text{ m} - (20\text{ m/s})t$, $x_B(t) = -200\text{ m} + (10\text{ m/s})t$

57. (a) constant acceleration; (b) $x_0 = -100$ m, $v_0 = +30$ m/s, $a = 6.0\text{ m/s}^2$; (d)
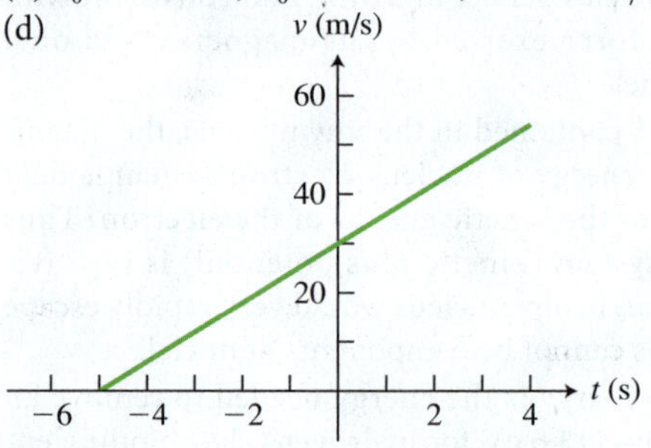


(e) $v(t) = 30\text{ m/s} + (6.0\text{ m/s}^2)t$, $v = 0$ at $t = -5.0$ s

59. 2.2 s, 1.2 m
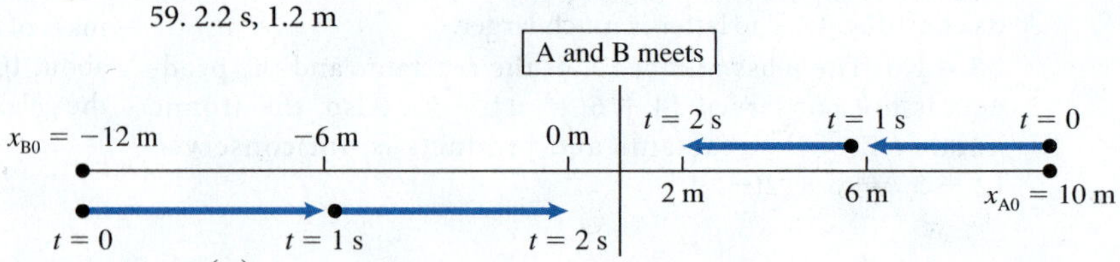


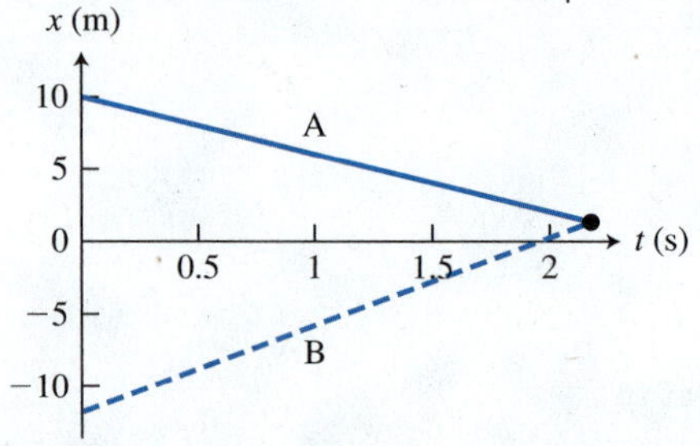

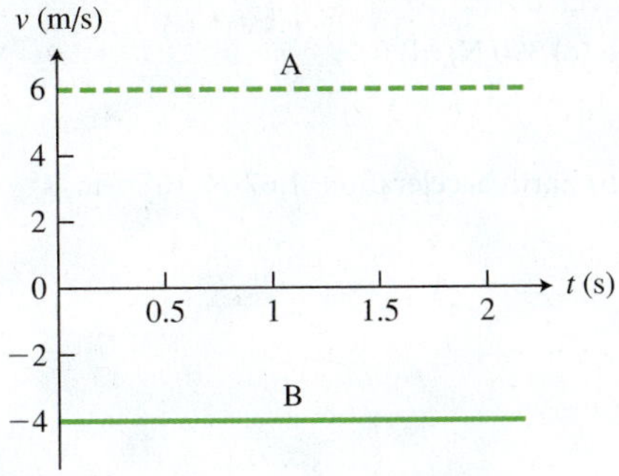


61. (b) 83.3 m; (c) 14 m/s
63. (a) 3.5 s; (b) 1.0 s
65. (a) 1.75 s; (b) 17.1 m/s
69. 9.9 m/s, 2.02 s
71. 0.156 s
73. 5.33 m/s
75. 22.2 m/s
79. 5.0 m (two floors) above you
81. 24.7 m/s$^2$
85. (c)
87. (a)
89. (d)
91. (a)
93. (c)

## Chapter 2

### Multiple-Choice Questions

1. (c) 3. (a) 5. (c) 7. (d) 9. (a) 11. (a) 13. (b) 15. (c) 17. (d)

### Problems

1. (1) (d); (2) (a); (3) (d); (4) (b); (5) (a)

3. (a)

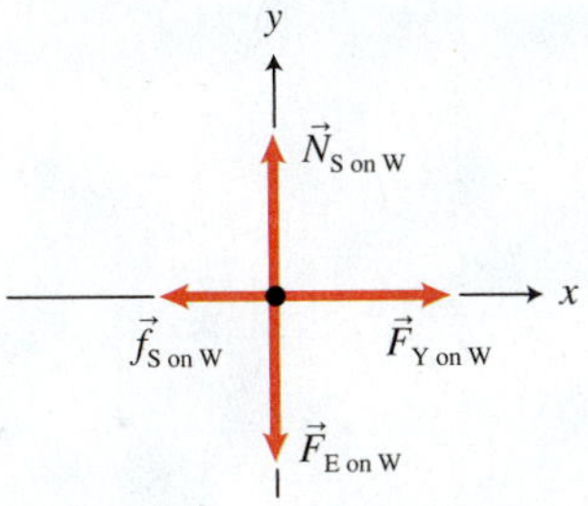


(b)

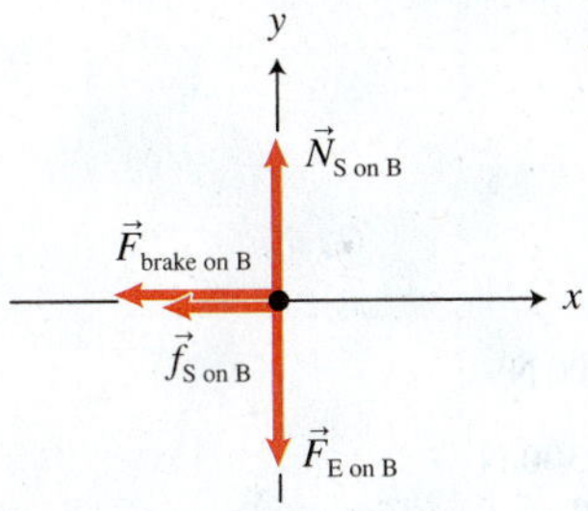


(c)

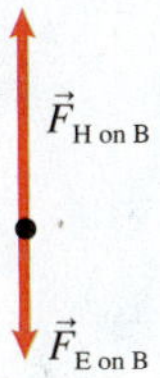


5. (a)

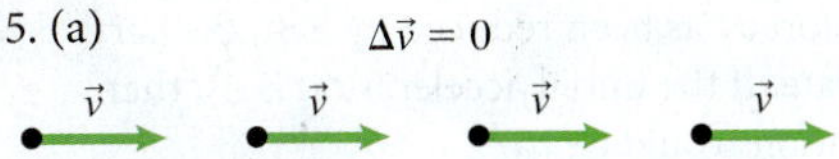


(b)

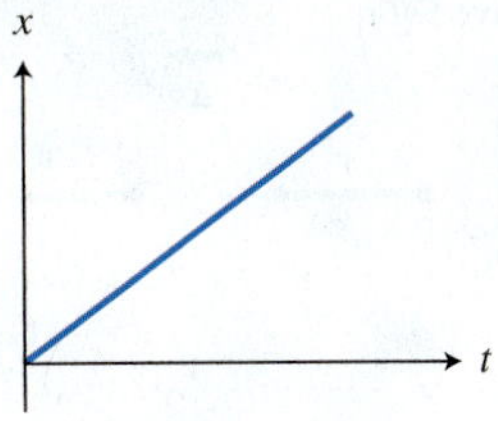


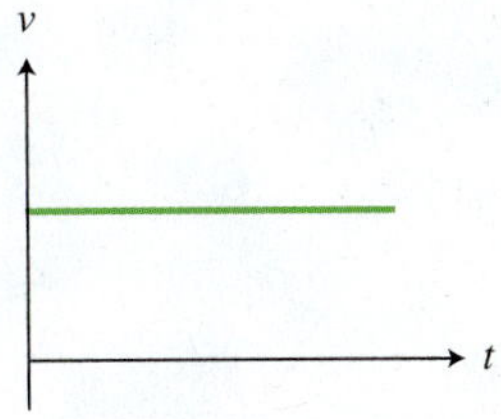


(c)

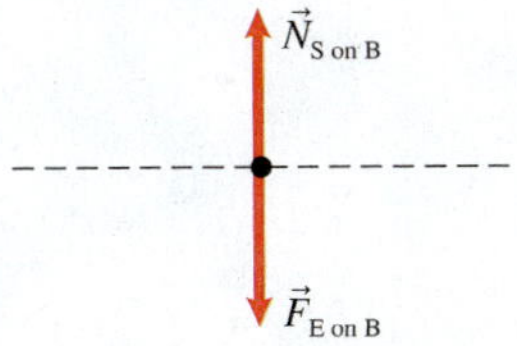


7. (a) After the pulling stops, the cart continues to move with constant velocity, in accordance with Newton's first law.

(b)

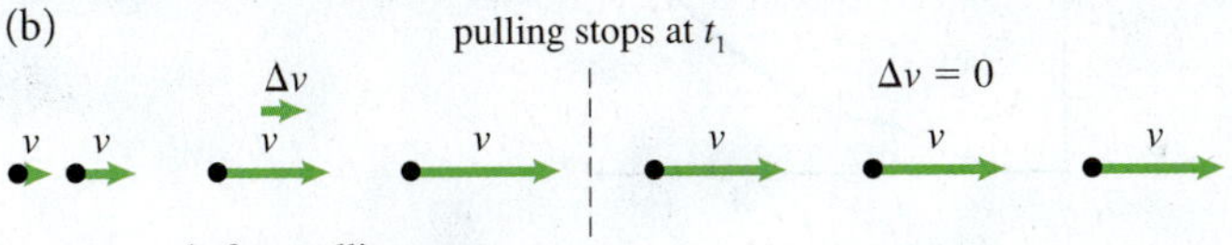


before pulling stops

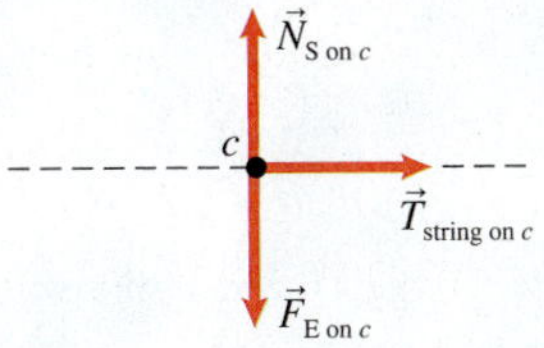


after pulling stops

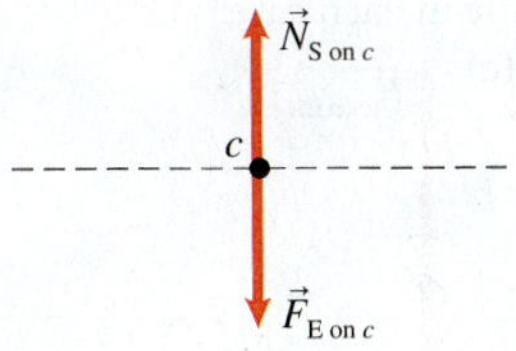


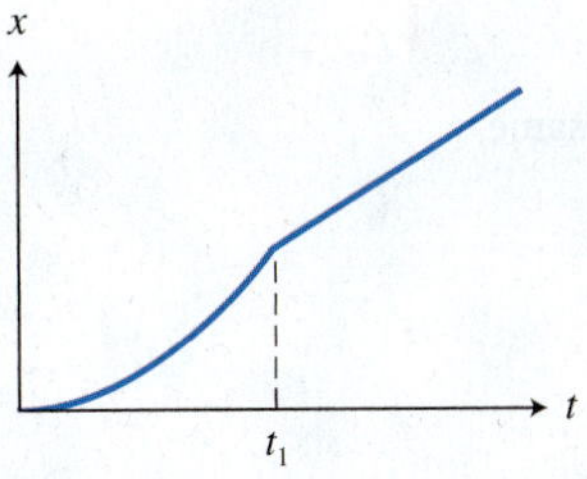


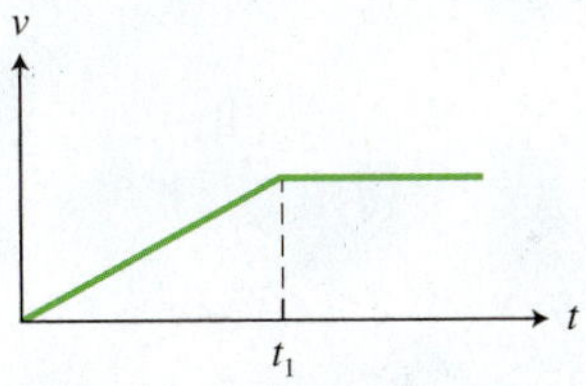

9. (a) After the applied force has been reduced by half, the cart will continue to accelerate. If the initial acceleration is $a_0$, then the subsequent acceleration would be $a_0/2$.

(b)

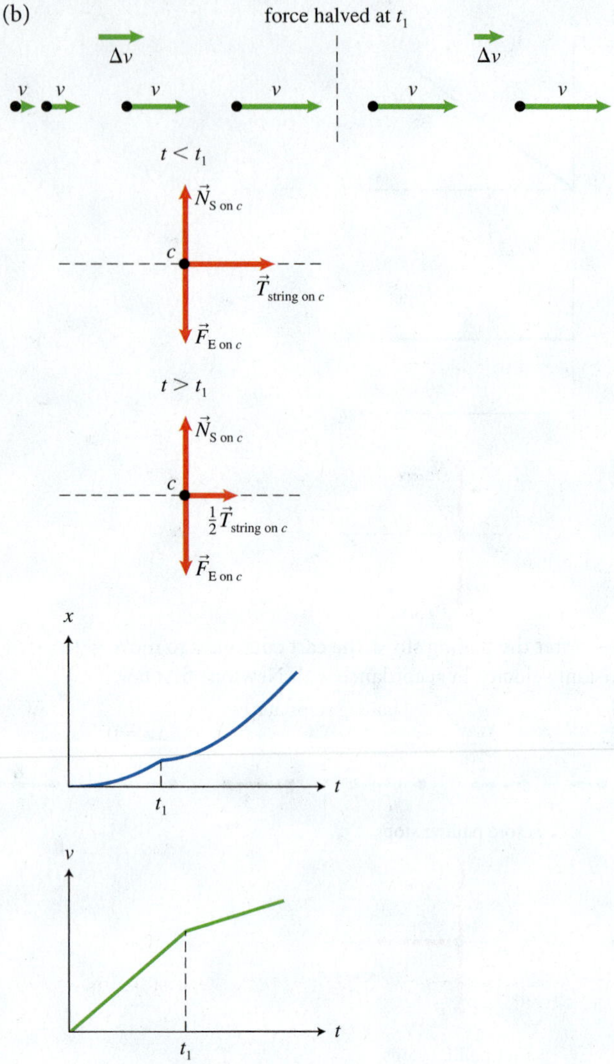


11. Both force diagrams are the same in each case.

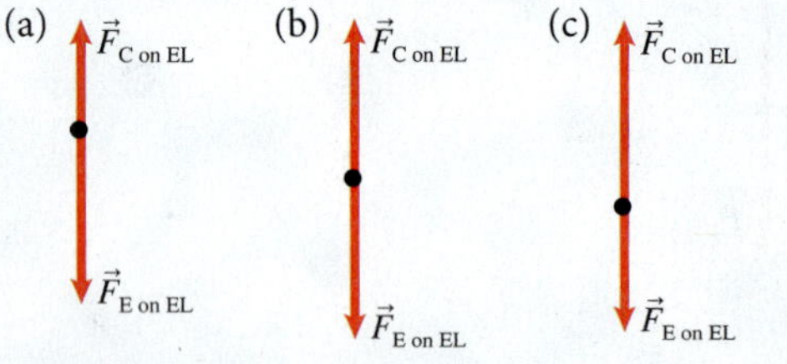


15. All force diagrams are the same.

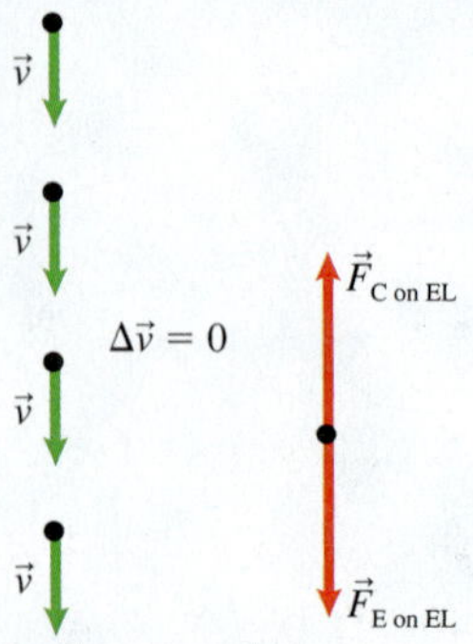


21. 10 N

29. (a) 490 N; (b) 590 N; (c) 390 N; (d) 0

31. 168 N

37. 33 N

39. (a) 1.0 N, upward; (b) Earth acceleration: $1.67 \times 10^{-25}\ \text{m/s}^2$

45. (b) 0.67 m

47. (a) 0.9 N; (b) 30%

49. $-1.55 \times 10^3$ N

55. (d)

57. (b)

59. (c)

61. (b)

63. (a)

65. (d)

## Chapter 3

### Multiple-Choice Questions

1. (c) 3. (b) 5. (b) 7. (a) 9. (a) 11. (b)

### Problems

1. $N_{\text{S on C}\,x} = 0$, $N_{\text{S on C}\,y} = 250\ \text{N}$, $F_{\text{E on C}\,x} = 0$, $F_{\text{E on C}\,y} = -150\ \text{N}$, $F_{\text{P on C}\,x} = 173\ \text{N}$, $F_{\text{P on C}\,y} = -100\ \text{N}$

3. $A_x = -7.07\ \text{m}$, $A_y = 7.07\ \text{m}$, $B_x = 5.0\ \text{m}$, $B_y = 0$

5. (a) 223.6 N, 153°

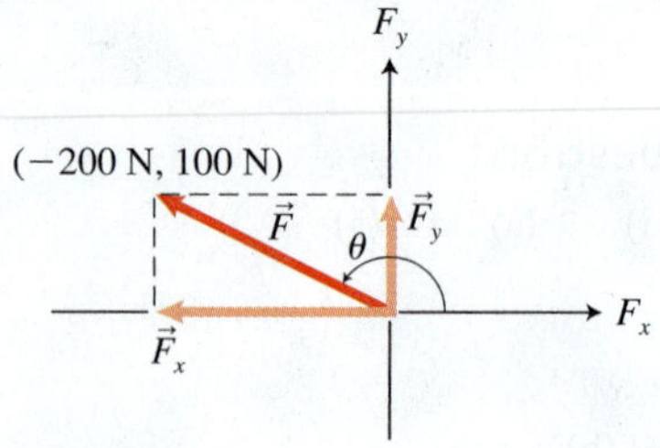


(b) 500 N, 53°

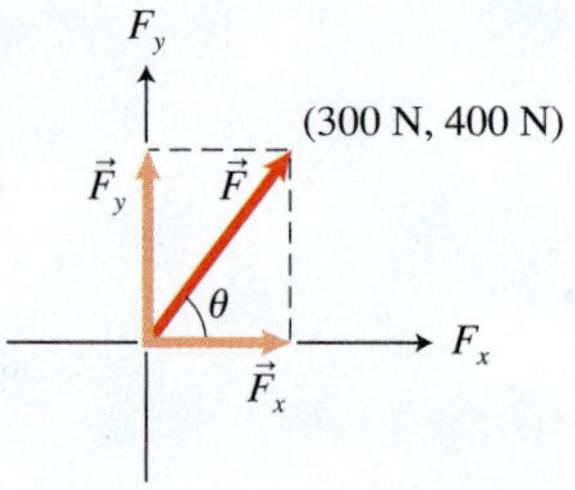


(c) 500 N, 37° below $+x$ axis

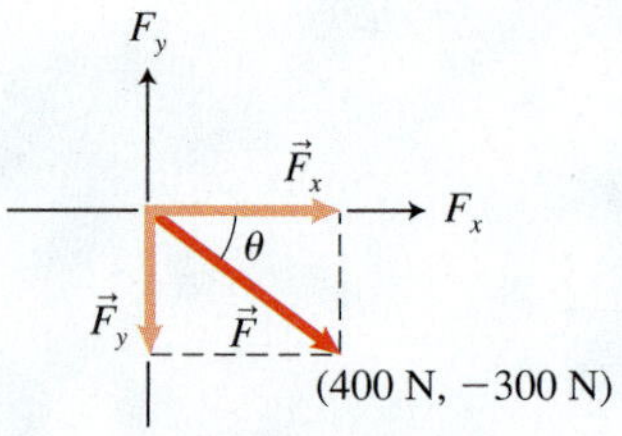


7. $T_{2\,\text{on K}} = 500\ \text{N}$, $T_{3\,\text{on K}} = 300\ \text{N}$

9. (a) $\sum F_{\text{on W}\,x} = T_{\text{R on W}}\cos\theta - f_{\text{S on W}} = m_{\text{W}}a_x$,
$\sum F_{\text{on W}\,y} = T_{\text{R on W}}\sin\theta + N_{\text{S on W}} - F_{\text{E on W}} = 0$;
(b) $\sum F_{\text{on B}\,x} = -f_{\text{S on B}} = m_{\text{B}}a_x$,
$\sum F_{\text{on B}\,y} = N_{\text{S on B}} - F_{\text{E on B}} = m_{\text{B}}a_y = 0$;
(c) $\sum F_{\text{on Y}\,x} = F_{\text{E on Y}}\sin\theta - f_{\text{S on Y}} = ma_x$,
$\sum F_{\text{on Y}\,y} = N_{\text{S on Y}} - F_{\text{E on Y}}\cos\theta = m_{\text{W}}a_y = 0$;

(d) $\sum F_{\text{on B } y} = F_{\text{Y on B}} - F_{\text{E on B}} = m_{\text{B}} a_y = 0$;
(e) $m_1$: $\sum F_{\text{on m1 } x} = T_{\text{R on m1}} - T_{\text{m2 on m1}} - f_{\text{S on m1}} = m_1 a_x$,
$\sum F_{\text{on m1 } y} = N_{\text{S on m1}} - F_{\text{E on m1}} = m_1 a_y = 0$,
$m_2$: $\sum F_{\text{on m2 } x} = T_{\text{m1 on m2}} - f_{\text{S on m2}} = m_2 a_x$,
$\sum F_{\text{on m2 } y} = N_{\text{S on m2}} - F_{\text{E on m2}} = m_2 a_y = 0$
11. (a) $\sum F_{\text{on B } y} = F_{\text{Y on B}} - F_{\text{E on B}} = 0$;
(b) $\sum F_{\text{on sled } x} = T_{\text{R on sled}} \cos\theta - f_{\text{S on sled}} = m_{\text{sled}} a_x$,
$\sum F_{\text{on sled } y} = T_{\text{R on W}} \sin\theta + N_{\text{S on sled}} - F_{\text{E on sled}} = m_{\text{sled}} a_y = 0$;
(c) $\sum F_{\text{on sled } x} = F_{\text{R on sled}} - F_{\text{E on sled}} \sin\theta - f_{\text{S on sled}} = m_{\text{sled}} a_x$,
$\sum F_{\text{on sled } y} = N_{\text{S on sled}} - F_{\text{E on sled}} \cos\theta = m_{\text{sled}} a_y = 0$;
(d) $\sum F_{\text{on SD } y} = F_{\text{air on SD}} - F_{\text{E on SD}} = m_{\text{SD}} a_y = 0$
13. $\sum F_{\text{on O } x} = N_{\text{R2 on O } x} + F_{\text{E on O } x} + F_{\text{R1 on O } x} =$
$0 + 0 + (30\text{ N})\cos 245° = -12.7\text{ N}$,
$\sum F_{\text{on O } y} = N_{\text{R2 on O } y} + F_{\text{E on O } y} + F_{\text{R1 on O } y}$
$= 40\text{ N} + (-10\text{ N}) + (30\text{ N})\sin 245° = 2.8\text{ N}$
19. 5.94 m/s
23. 101.5 N
25. 1350 N, 1690 N
27. 150.3 N, 60 N
31. 1.73 $\text{m/s}^2$; 242 N; 3.4 s
33. 0.67 $\text{m/s}^2$; 104.7 N
39. 0.41
41. 0.76, static
45. $g/\mu_s$
47. 98.8 N, $\mu_k = 0.68$
49. 0.84
51. 2.43 $\text{m/s}^2$
55. $m_2 = \mu_k m_1$
57. On table

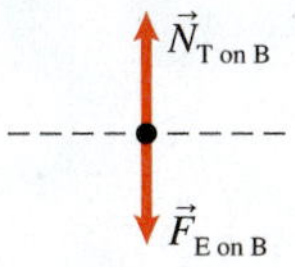


In air

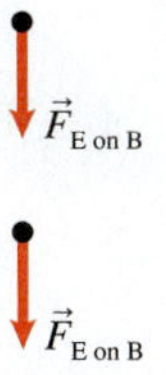


59.

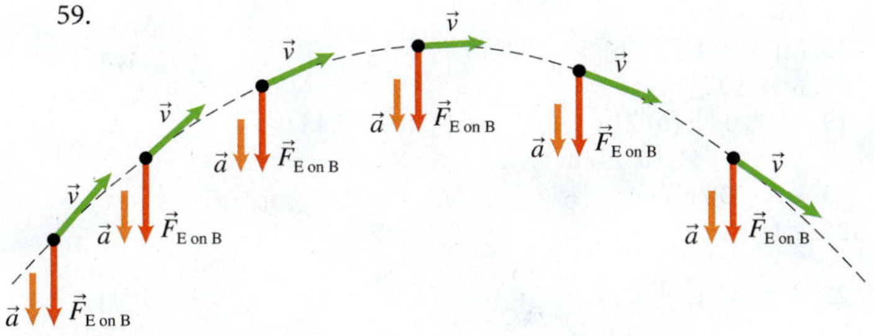


61. 44.6 m/s
63. 57.1°, 23.0 m/s
65. (a) from a horizontal distance of 723 m; (b) later
67. Projectile fired at $\theta_2 = 60°$ has longer flight time, and hence, greater air resistance.
69. 4.9 m
71. (a) Frictional force by road surface on tires propels minivan to move forward; (b) 3420 N
75. 6.13 $\text{m/s}^2$
77. $a = 0$, no motion
81. 5.35 s
89. (c)
91. (a)
93. (b)
95. (e)
97. (a)

## Chapter 4

### Multiple-Choice Questions

1. (d) 3. (b) 5. (a) 7. (a) 9. (a) 11. (b) 13. (a)

### Problems

1. bottom: $mg + mv^2/r$; top: $mg - mv^2/r$
5. $5.9 \times 10^{-3}\text{ m/s}^2$
7. 92.2 m/s
9. 2.20 $\text{m/s}^2$, or 0.22 g
13. person at outermost radius
15.

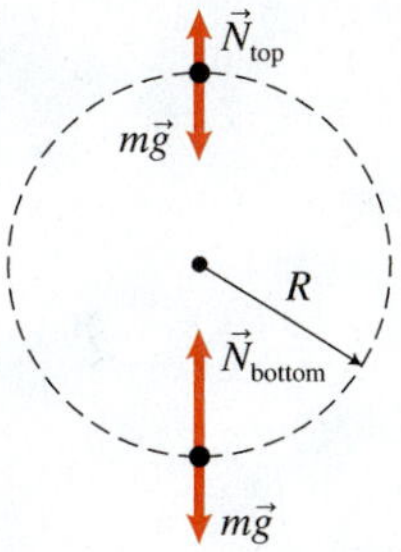


19. 138 N
21. 2.70 m/s
25. 214 N
27. 9.3 m/s
29. 90 N, radially inward
31. 9000 N
35. (a) $v = \sqrt{rg\tan\theta}$; (b) Speed is independent of the mass of the object.
39. (a) $4.3 \times 10^{20}$ N; (b) $2.0 \times 10^{20}$ N; (c) $2.0 \times 10^{20}$ N
41. 200 N
43. 27.4 days
45. 37.6% of what it is on Earth
47. $1.8 \times 10^{27}$ kg
49. 1.67 h
51. $2.0 \times 10^{7}$ m above Earth's surface
53. 45 s
55. 1764 N
65. 1.4 h
67. 1.7 km/s
69. (c)
71. (b)
73. (c)
75. (a)
77. (a)
79. (b)

## Chapter 5

### Multiple-Choice Questions

1. (b) 3. (e) 5. (a) 7. (c) 9. (d) 11. (b) 13. (d)

## Problems

1. (a) 1.71 kg·m/s; (b) 5.34 m/s; (c) 298.3 kg·m/s
3. the ball that rebounds
7. $v_2 = -\frac{10v}{7}$
11. 0.313 N·s, 0.13 N·s
15. (a) $4.0 \times 10^5$ N; (b) 100 s
19. 160 N
21. 0.22 N·s, 22 N
23. (a) 25%; (b) −10.6%
31. $p_{Bix} + J_{W\text{ on }Bx} = p_{Bfx}$
Earth is the object of reference.

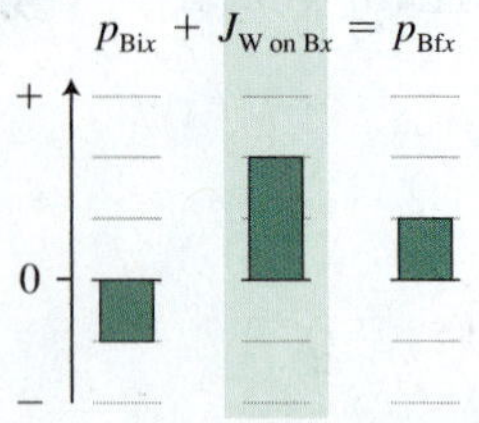


33.

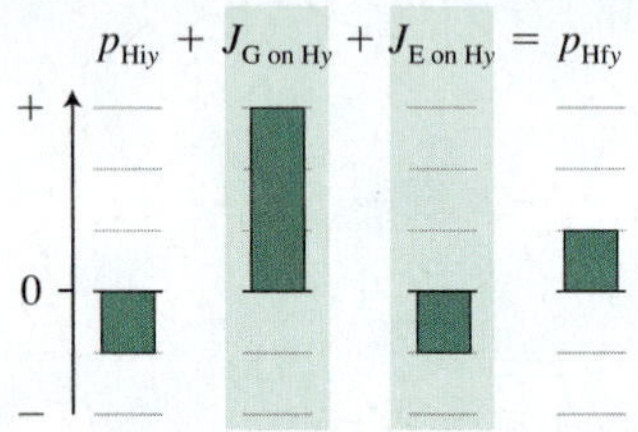


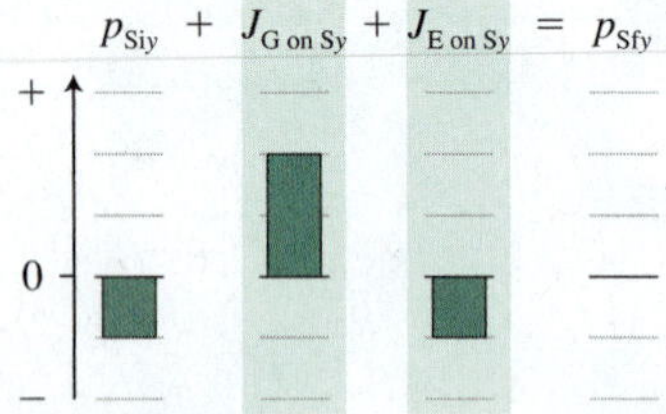


37. (a) 6.12 kg·m/s upward; (b) 56.9 N
39. 43.8 m/s
41. −0.19 m/s
43. 1.33 m/s
47. 165 km/h
49. $2 \times 10^4$ N
53. (a) 7500 N; (b) yes; (c) 0.167 s
57. $-1.4 \times 10^7$ m/s
59. 8.27 m/s, 14.7° north of east
61. 56.3° north of east, 6.18 m/s
63. (a) 0.75 m/s; (b) 13.3 s
65. (a)

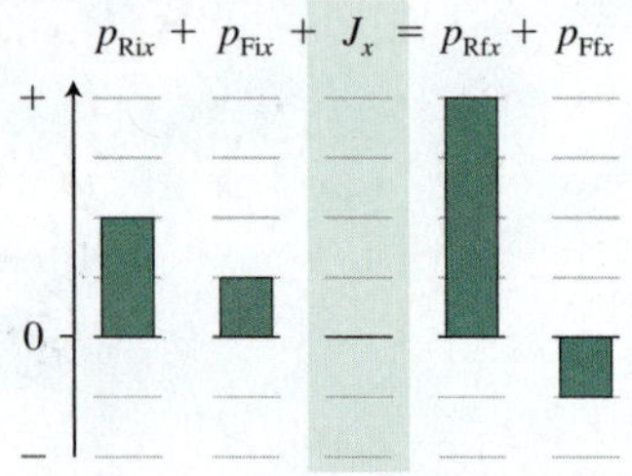


(b)

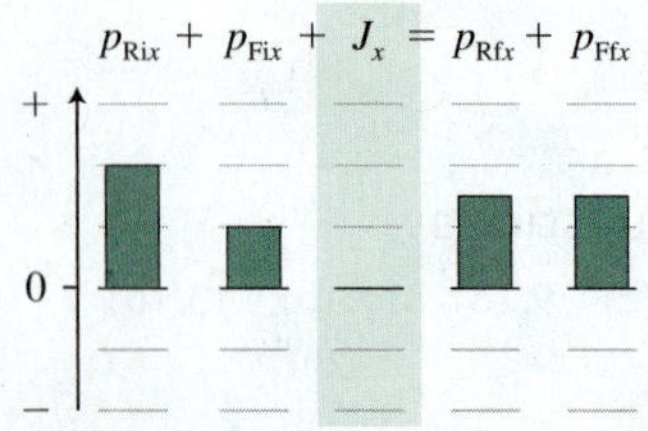


(c)

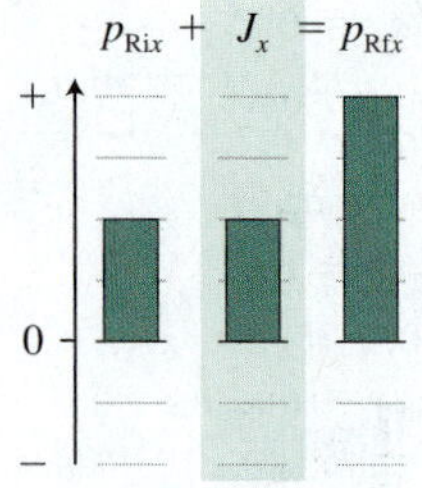


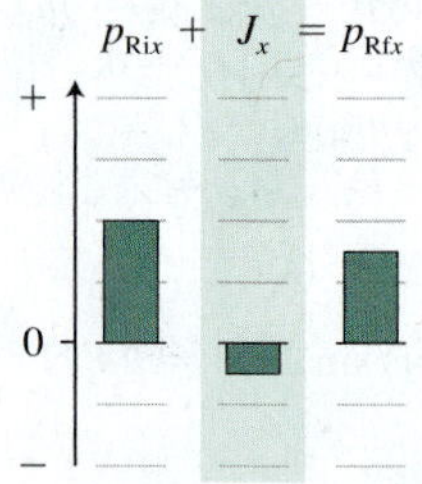


67. $1 \times 10^{-14}$ m/s
77. $\tan\theta = \frac{m_1 v_1}{m_2 v_2}$, $d = \frac{1}{2\mu_k g}\frac{m_1^2 v_1^2 + m_2^2 v_2^2}{(m_1 + m_2)^2}$
81. (d)
83. (e)
85. (b)
87. (c)
89. (e)
91. (c)

# Chapter 6

## Multiple-Choice Questions

1. (b) 3. (c) 5. (b) 7. (b) 9. (e) 11. (c)

## Problems

1. 540 J
3. (a) 7500 J; (b) −7500 J; c) 7500 J
5. lifting 196 J, carrying 0, setting down −196 J, total 0
7. 6.2 m/s
9. (a) $\frac{K_P}{K_C} = \frac{m_C}{m_P} = \frac{1100\text{ kg}}{2268\text{ kg}} = 0.485$; (b) $K_P/K_C = 1$
11. yes
15. (a) $3 \times 10^9$ J; (b) $5 \times 10^6$ m
17. $6 \times 10^{-4}$ m
19. (a) 720 N; (b) 21.6 J; (c) −16.2 J; (d) −5.4 J
21. $k = 8.9 \times 10^5$ N/m
23. $k = 330$ N/m
25. 4.4 m/s
27. $v_f = \sqrt{v_i^2 + 2gh - \frac{2fl}{m}}$
29. 6.4 m/s, 46 m, increase if the friction force remains the same
33. 8.4 m/s
43. $v_f = 22$ m/s
45. (b) decrease of 33.3%
47. (a) 0.635 m/s; (b) 0.28 m
49. 222 m/s
51. 36 kg, −55 J
53. (a) 90 m/s; (b) 540 J; (c) Falcons have strong feet that can deliver a greater force (impulse) to strike their prey.

55. $v_{1f} = \dfrac{m_1 - m_2}{m_1 + m_2}v$, $v_{2f} = \dfrac{2m_1}{m_1 + m_2}v$
57. (a) 9.8 W; (b) 196 W; (c) 196 W; (d) 196 W
59. (a) $1.15 \times 10^7$ J; (b) 47 s
61. (a) $4.34 \times 10^5$ J; (b) 60.3 W
65. 1 hp
67.

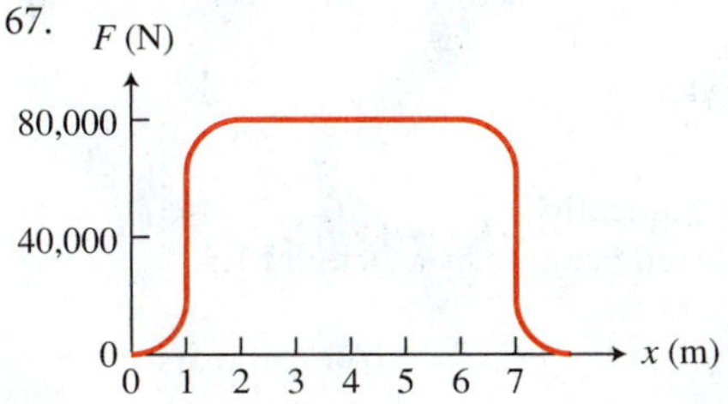


69. (a) 1; (b) 4; (c) more hydrogen molecules can attain speeds higher than $v_{esc}$
71. $4.21 \times 10^4$ m/s
77. (a) 11 m/s; (b) 1.8 $mg$
79. 690 N
81. (a) $3.3 \times 10^{-6}$ m/s; (b) $1.1 \times 10^{20}$ N; (c) $1.1 \times 10^{23}$ J $= 1.7 \times 10^9$ atomic bombs
85. (c) and (d)
87. (c)
89. (b)
91. (a)
93. (d)
95. (b)

# Chapter 7

## Multiple-Choice Questions

1. (b) 3. (a) 5. (a) 7. (d) 9. (c)

## Problems

1. $\tau_1 = 0$, $\tau_2 = -183.85$ N·m, $\tau_3 = +240$ N·m, $\tau_4 = -240$ N·m
3. $\pm 10.2$ N·m
7. 218 N, 113°
9. $T_1 = 11.8$ N, $T_2 = 19.5$ N, $T_3 = 15.6$ N
11. $m = 7.82$ kg, $T_1 = 77$ N, $T_3 = 64$ N
13. 3250 N, 15° above horizontal
15. (a)

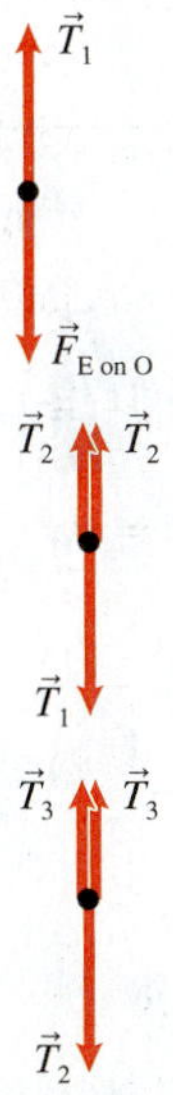

(b) $T_1 = 9800$ N, $T_2 = 4900$ N, $T_3 = 2450$ N, $T_4 = 4900$ N
17. 10.2°
19. (a) 823 N·m; (b) 1.56 m
21. 0.56 m from the center, towards Tahreen
23. 250 N
27. left 617 N, right 274 N
29. 0.77 m
31. (a) 2.1 m; (b) It did not move.
33. $-\dfrac{r^2 a}{R^2 - r^2}$
37. $x_{cm} = \dfrac{m_b b/2}{m_a + m_b}$, $y_{cm} = \dfrac{m_a a/2}{m_a + m_b}$
39. 550 N
41. 1250 N, 1050 N
43. 220 lb, 200 lb
45. 171 N
47. $T = 37.3$ N, $F_{H\,on\,B\,x} = 37.3$ N, $F_{H\,on\,B\,y} = 39.2$ N, $F_{H\,on\,B} = 54.1$ N, $\phi = 46.4°$
49. 980 N
53. 35 cm from the clay
57. (a) $T = 1570$ N, $F = 1620$ N
59. 2240 N
61. 1650 N, 1630 N
67. 5.0 m up the ladder
71. (a)
73. (c)
75. (a)
77. (b) or (c)
79. (e)

# Chapter 8

## Multiple-Choice Questions

1. (a) 3. (b) 5. (d) 7. (b) 9. (d)

## Problems

1. (a) 0.105 rad/s; (b) 0.021 m/s; (c) zero
3. (a) $-0.058$ rad/s²; (b) 0.231 rad/s²; (c) 0.625 rad/s²; (d) 0.094 m/s²
5. 8.38 s
7. 300 rad/s
9. (a) 200 m/s²; (b) 133 rad/s²
13. 8300
15. (a) $\alpha = \dfrac{\omega^2}{4\pi}$; (b) $\Delta t_1 = \dfrac{4\pi}{\omega}$; (c) $\Delta t_2 - \Delta t_1 = \dfrac{4\pi}{\omega}(\sqrt{2} - 1)$; (d) $\Delta s = 4\pi l$
17. $-1.2 \times 10^{-14}$ rad/s²
19. 0.040 N
21. 205 N·m
23. (a) $\dfrac{T_2}{T_1} = \dfrac{4}{3}$; (b) $T_2 > \dfrac{4}{3}T_1$; (c) $T_2 < \dfrac{4}{3}T_1$
25. (a) $\alpha = 4a$; (b) $\omega = 16a$
29. $8m$
31. $5.9 \times 10^4$ N·m

33. 1.9 s
35. 1.2 kg · m$^2$
37. −780 N
39. (a)

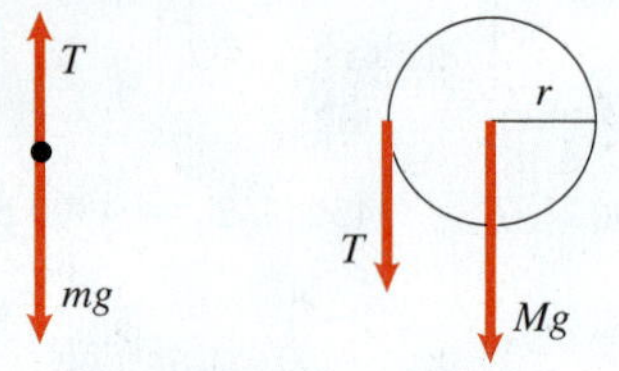


(b) $mg - T = ma$; (c) $rT = I\alpha$;
(d) $a = 3.27\ \text{m/s}^2$, $T = 196$ N
41. (a)

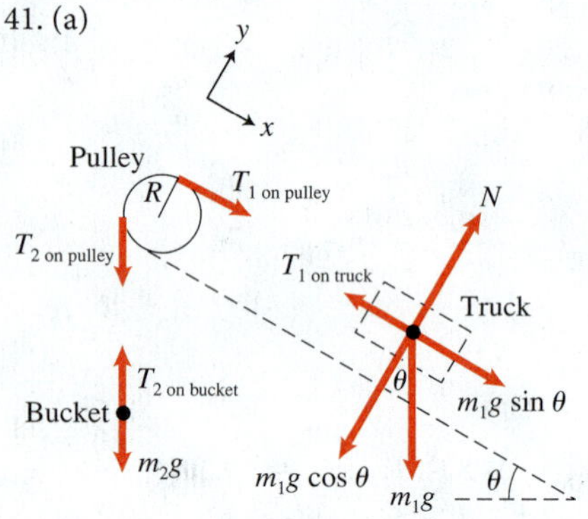


(b) $T_{1\text{ on truck}} = m_1g\sin\theta$,
$T_{2\text{ on bucket}} = m_2g$; (c) the same
43. 3.2 kg · m$^2$
47. $7.8 \times 10^6$ s ≈ 90 days
49. $\omega = \sqrt{\dfrac{2K}{I}}$
51. 96%
53. (a) 12.3 rad/s; (b) $-1.74 \times 10^{-3}$ J; (c) Negative work is done as the beetle moves from the center to the edge.
55. 16.8 J
59. (a) 1.78 rad/s; (b) −35.6 J; (c) −67.2 J; (d) +31.6 J;
(e) $\Delta K = \Delta K_b + \Delta K_m$
61. (a) 1.67 rad/s; (b) −80 J; (c) −97.8 J; (d) +17.8 J;
(e) $\Delta K = \Delta K_b + \Delta K_m$
63. (a) 1.28 J; (b) 9.23 rad/s; (c) 1.38 m/s; (d) −0.43 J; (e) +0.19 J;
(f) −0.24 J
65. 142 N
67. $7.7 \times 10^3$N
71. (a) it does not move; (b) to the left; (c) to the right
73. 0.035 rad/s
75. (a)
77. (c)
79. (e)
81. (c)
83. (b)

# Chapter 9

## Multiple-Choice Questions

1. (c) 3. (c) 5. (d) 7. (c) 9. (c) 11. (b) 13. (b) 15. (d)

## Problems

3. $8.4 \times 10^{56}$
5. (a) $2.68 \times 10^{25}$ molecules/m$^3$; (b) $D/d \approx 30$; (c) yes
7. $3.0 \times 10^{-26}$ kg, $4.8 \times 10^{-26}$ kg
9. $6.2 \times 10^{-3}$ kg
11. (a) 4.8 N; (b) 1.6 N/m$^2$
15. 440 m/s
17. $P_1 = 2P_2$
21. 198 °F, atmospheric pressure is lower
23. 254 °C
25. $5.65 \times 10^{-21}$ J
27. 0.041 mol
29. $v_{1f} = v_{2i}$, $v_{2f} = v_{1i}$
31. 7.2 cm$^3$
33. $1.4 \times 10^{-16}$ N/m$^2$
43. 1.1 m
45. about 3 times more frequently
49. decrease the absolute temperature by a factor of 1.5
51. $6.6 \times 10^6$ N/m$^2$
57. (a) $2.1 \times 10^{-18}$ J; (b) $6.2 \times 10^{12}$ s $= 2.0 \times 10^5$ years
59. $U = \dfrac{3mkT}{2M}$, $P_2 = \frac{1}{2}P_1$, temperature and thermal energy are unchanged
67. $2.6 \times 10^{21}$ $O_2$ molecules per breath
71. 68.5 °C
73. (d)
75. (a)
77. (e)
79. (b)
81. (b)

# Chapter 10

## Multiple-Choice Questions

1. (c) 3. (c) 5. (d) 7. (b) 9. (a) 11. (c) 13. (a)
15. (b) 17. (c) 19. (b)

## Problems

1. 0.047 m$^3$
3. 7850 m
5. 6.9% decrease
7. Density of oil is 900 kg/m$^3$.

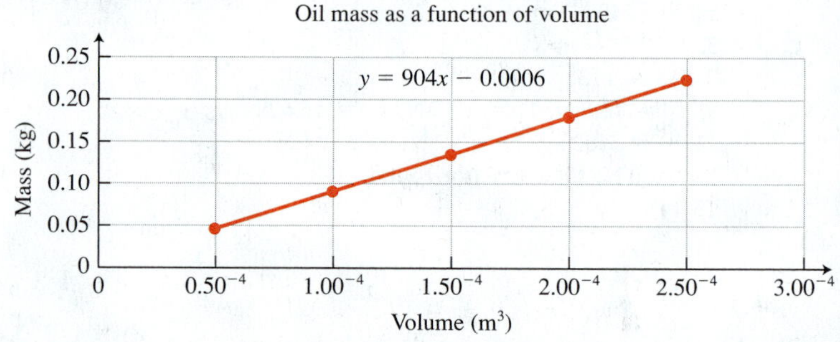


9. (a) $\rho_A = \rho_B = \rho_C$ (b) $V_A = a^3$, $V_B = (2a)^3 = 8a^3$, and $V_C = (3a)^3 = 27a^3$. Thus, $V_A < V_B < V_C$.
(c) $S_A = 6a^2$, $S_B = 6(2a)^2 = 24a^2$, and $S_C = 6(3a)^2 = 54a^2$. Thus, $S_A < S_B < S_C$.
(d) $A_A = a^2$, $A_B = (2a)^2 = 4a^2$, and $A_C = (3a)^2 = 9a^2$. Thus, $A_A < A_B < A_C$.
(e) $m_A = \rho V_A = \rho a^3$, $m_B = \rho V_B = 8\rho a^3$, and $m_C = \rho V_C = 27\rho a^3$. Thus, $m_A < m_B < m_C$.
11. 200 kg/m$^3$
13. 2300 kg/m$^3$
17. both are wrong; $6 \times 10^3$ N
19. $8.0 \times 10^{-3}$ m$^2$
21. $\dfrac{A_1}{A_2} = 0.034$; 8.82 m
25. $2.32 \times 10^3$ N

29. $P_A = P_B = P_C$; $F_A > F_C > F_B$
31. $5.1 \times 10^5\ \mathrm{N/m^2}$
35. $-2.2 \times 10^4\ \mathrm{N/m^2}$
39. 0.336 m
43. 14.4 cm
47. 10.74 m
51. Liquid B has higher density than liquid A.
57. The upward buoyant force exerted by the water on the brick makes it feel lighter when you hold it.

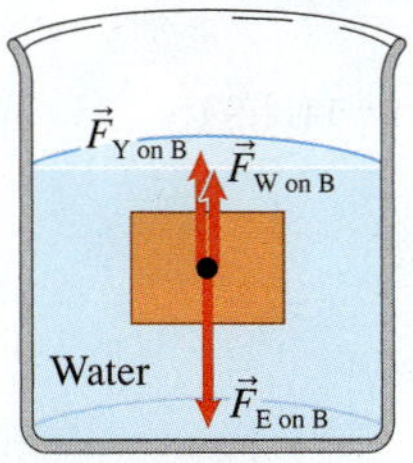


61. 0.774 N
63. The kerosene side will tilt down, and the water side will tilt up.
65. $400\ \mathrm{kg/m^3}$
69. 582.6 N; The person will sink.
71. The pressure will not change.
75. 200 kg
77. $\frac{1}{4}(\rho_{\text{water}} - \rho_{\text{log}})\pi d^2 L$
79. 92%
85. $8.6 \times 10^5$ N
91. $1.05 \times 10^4\mathrm{m^2}$
93. $V_2 = \frac{5}{8}V_1$
95. (d)
97. (c)
99. (d)
101. (a)
103. (c)
105. (d)

# Chapter 11

## Multiple-Choice Questions

1. (b) 3. (a) 5. (c) 7. (b) 9. (c) 11. (a)

## Problems

1. (a) $1.45 \times 10^{-4}\ \mathrm{m^3/s}$; (b) 0.46 m/s
5. $6.6 \times 10^{-3}\mathrm{cm/s}$
7. $K_1 + P_1 = P_2 + K_2 + U_{g2}$

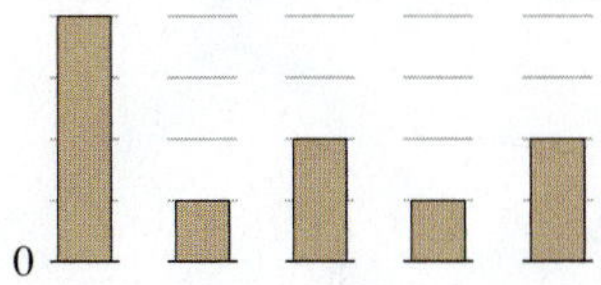


$P_1 + K_1 + U_{g1} = P_2 + K_2 + U_{g2}$ or

$P_1 + \frac{1}{2}\rho v_1^2 + 0 = P_2 + \frac{1}{2}\rho v_2^2 + \rho g y_2$

17.

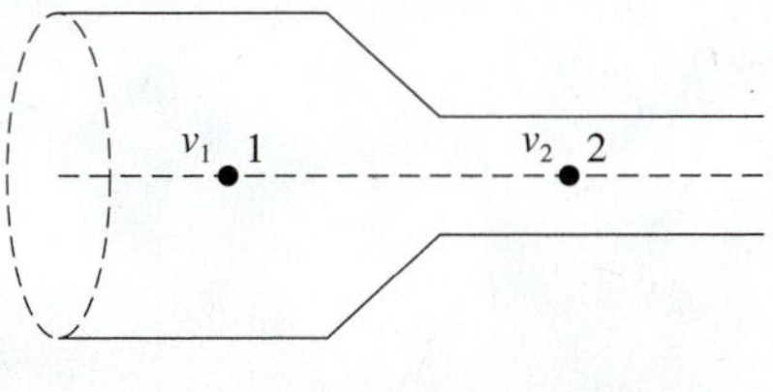


$K_1 + U_{g1} + (P_1 - P_2) = K_2 + U_{g2}$

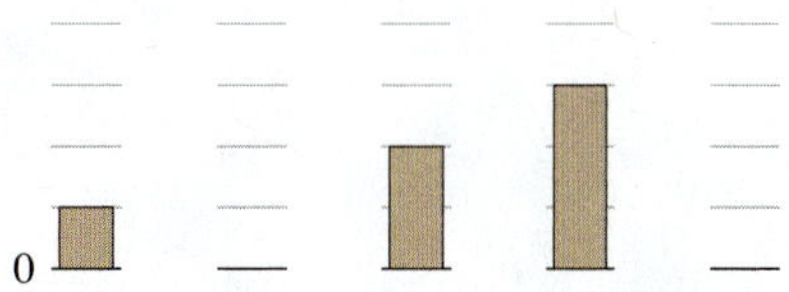


$P_1 + \frac{1}{2}\rho v_1^2 = P_2 + \frac{1}{2}\rho v_2^2$

19. (a) 5.24 m/s; (b) 1.41 cm; (c) The speed remains unchanged.
21. (a) 4.77 m/s; (b) 5.3 m/s; (c) $4.95 \times 10^5\mathrm{N/m^2}$
25. 0.0041 m
27. $9.22 \times 10^3\ \mathrm{N/m^2}$
31. 0.50 cm
33. 46 m/s
35. $\frac{1}{32} = 0.03125$
39. (a) 630 N; (b) $1.0 \times 10^3$ N
41. $1.9 \times 10^{-12}$N
43. $2.0 \times 10^{-5}$ m
45. $v_{\text{terminal}} = \sqrt{\frac{mg}{0.03}} = 18.1\sqrt{m}$
53. $33\ \mathrm{m^2}$
59. (c)
61. (d)
63. (c)
65. (c)
67. (c)
69. (a)

# Chapter 12

## Multiple-Choice Questions

1. (b), (d) 3. (e) 5. (b) 7. (c) 9. (c)

## Problems

3. (a) zero; (b) The temperature remains the same, the density and pressure each decrease by half, the average kinetic energy and the thermal energy remain the same.
5. (a) −1300 J; (b) 1.23 mol; (c) 208 K
9. (a) $2.1 \times 10^4$ J; (b) $1.2 \times 10^4$ J; (c) $2.3 \times 10^3$ J
11. (b) $\frac{v_i^2}{4(130\ \mathrm{J/kg \cdot {}^\circ C})}$
13. 6.1 °C
15. $2.1 \times 10^5$ J
17. 75.2 °C
19. The material is likely to be iron.
29. 13 °C
31. no; 0.33 kg
33. 0.011 kg
35. $1.57 \times 10^4$s = 4.36 h
37. (a) $2.26 \times 10^5$ J; (b) $8.54 \times 10^4$ J
39. steam at $1 \times 10^5$ °C
41. (a) 443 kg/s; (b) $3.94 \times 10^7$ kg/day
43. (a) −410 Btu/h = −120 W; (b) −252 Btu/h = −73.8 W; (c) −384 Btu/h = −113 W
45. − (negative); You will perspire until your body temperature becomes normal.
47. 16 J/s
49. 2.0 kg
51. −11 °C
53. 84 W
59. $1.33 \times 10^{-4}$ kg/s

61. (a) $1.1 \times 10^{23}$ J; (b) $9.8 \times 10^{16}$ kg; (c) $1.1 \times 10^{14}$ m$^3$; (d) up to 0.27 m
63. 5.8 kW
65. 0.55 kW
73. (b)
75. (b)
77. (d)
79. (b)
81. (c)
83. (d)

# Chapter 13

## Multiple-Choice Questions

1. (a) 3. (c) 5. (b) 7. (c) 9. (a) 11. (d) 13. (c)

## Problems

5. 4.0 °C; irreversible
7. (a) (6, 0), (5, 1), (4, 2), (3, 3), (2, 4), (1, 5), (0, 6)
(b) and (c)

| Macrostate $(n_L, n_R)$ | $W_{n_L} = W(n_L, n_R) = \frac{n!}{n_L! n_R!}$ | $S_i = k \ln W_i$ |
|---|---|---|
| (0, 6) | $W_0 = W(0, 6) = \frac{6!}{0!6!} = 1$ | 0 |
| (1, 5) | $W_1 = W(1, 5) = \frac{6!}{1!5!} = 6$ | $k \ln 6 = 1.79k$ |
| (2, 4) | $W_2 = W(2, 4) = \frac{6!}{2!4!} = 15$ | $k \ln 15 = 2.71k$ |
| (3, 3) | $W_3 = W(3, 3) = \frac{6!}{3!3!} = 20$ | $k \ln 20 = 3.00k$ |
| (4, 2) | $W_4 = W(4, 2) = \frac{6!}{4!2!} = 15$ | $k \ln 15 = 2.71k$ |
| (5, 1) | $W_5 = W(5, 1) = \frac{6!}{5!1!} = 6$ | $k \ln 6 = 1.79k$ |
| (6, 0) | $W_6 = W(6, 0) = \frac{6!}{6!0!} = 1$ | 0 |

9. (a) $\frac{1}{8} = 0.125$; (b) $\frac{1}{64} = 0.016$
11. (a) 972.4; (b) 25.2; (c) The ratio will be very large.
13. (a)

| $n$ | 2 | 3 | 4 | 5 | 6 | 7 | 8 | 9 | 10 | 11 | 12 |
|---|---|---|---|---|---|---|---|---|---|---|---|
| $W_n$ | 1 | 2 | 3 | 4 | 5 | 6 | 5 | 4 | 3 | 2 | 1 |

(b) $n = 7$; (c) $n = 2$ and $n = 12$
17. (a) 52.5 °C; (b) +7.7 J/K
19. $+1.94 \times 10^4$ J/K
21. +0.853 J/K
23. (a) 0.50; (b) 0.19; (c) 0.047
25. (a) 0.648; (b) $1.54 \times 10^9$ W
27. (a) $+2.7 \times 10^5$ J; (b) $-9.0 \times 10^4$ J; (c) $+9.0 \times 10^4$ J; (d) $T_1 = 1.4 \times 10^4$ K, $T_2 = 5.8 \times 10^4$ K, $T_3 = 7.2 \times 10^3$ K; (e) $U_{thermal\,1} = 1.8 \times 10^5$ J, $U_{thermal\,2} = 7.2 \times 10^5$ J, $U_{thermal\,3} = 9.0 \times 10^4$ J, $\Delta U_A = 5.4 \times 10^5$ J, $\Delta U_B = -6.3 \times 10^5$ J, $\Delta U_C = 9.0 \times 10^4$ J; (f) $Q_A = +9.0 \times 10^5$ J, $Q_B = -9.0 \times 10^5$ J, $Q_C = +9.0 \times 10^4$ J; (g) 0.091
29. (a) $6.71 \times 10^4$ J; (b) $4.9 \times 10^3$ J; (c) $7.2 \times 10^4$ J
31. 0.36, 1800 J
33. 0.42, 700 MW
37. (b)
39. (a)
41. (d)

# Chapter 14

## Multiple-Choice Questions

1. (d) 3. (f) 5. (a) (c) (d) (f) 7. (b) 9. (c) 11. (c) (f)

## Problems

1. $1.25 \times 10^{18}$
3.

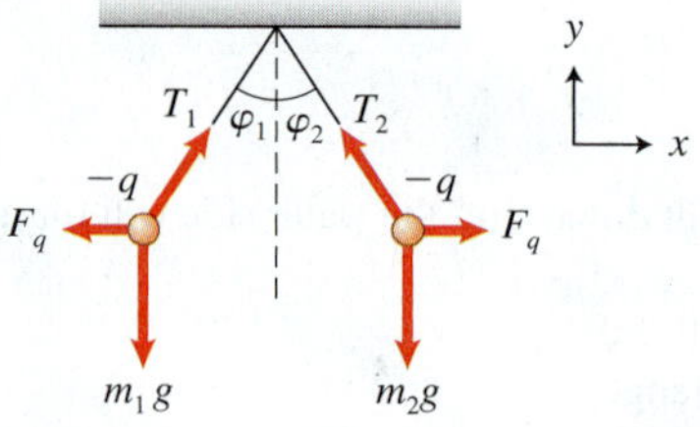


$$m_1 \tan \varphi_1 = m_2 \tan \varphi_2$$

5. (a) $9.6 \times 10^{-18}$ C; (b) $1.1 \times 10^8$ electrons, $1.8 \times 10^{-11}$ C
7. 58 N
11. $3.6 \times 10^{-10}$N
13. They will be attracted with force $3.7 \times 10^{21}$ N.
15. (a) zero; (b) zero; (c) $1.8 \times 10^6$ N to the right
17. $2.2 \times 10^{-9}$ N to the left
25. (a) positive; (b) Increase the separation;
(c)

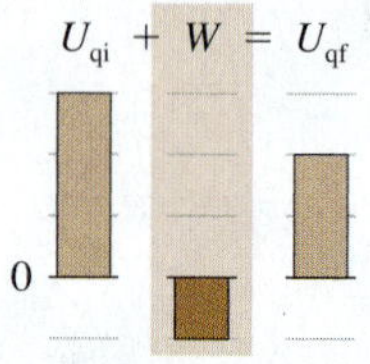


27. (a) negative; (b) Decrease the separation;
(c)

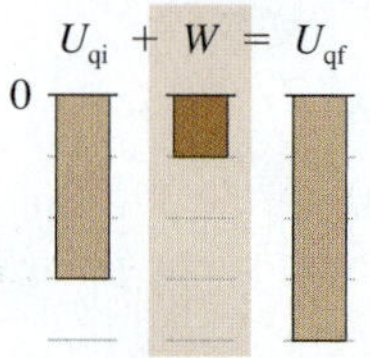


29. away from the sphere
31. (a) $U_q = W = 405$ J; (b) $U_q = W = -1350$ J
41. $F_{on\,1} = 0.839 \frac{kq^2}{a^2}$
45. (a) $1.49 \times 10^{-6}$ C
47. (a) 6600 m
51. $2.2 \times 10^{-5}$ C
53. (a) $1.86 \times 10^6$ m/s
55. $2.35 \times 10^7$m/s
57. 3.2 C
59. 8.3 m/s
63. (a) $6.25 \times 10^9$; (b) $4 \times 10^{-19}$
65. (c)
67. (e)
69. (a)
71. (d)
73. (e)
75. (e)

## Chapter 15

### Multiple-Choice Questions

1. (c) 3. (c)

### Problems

1.

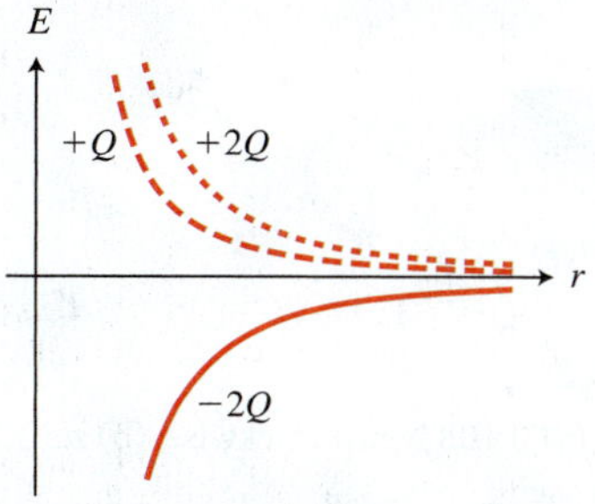


7. 0.27 m from the object with the larger charge
9. $E = 2.43 \times 10^4$ N/C, $\theta = 242°$
11. (a) $E_1 = E_0 + \dfrac{kq}{d^2}$; (b) $E_2 = E_0 - \dfrac{kq}{d^2}$
13. (a) 0.553 s; (b) 0.866 m
15. It will slow down with constant acceleration until it comes to a stop and then pick up speed in the reverse direction,
$v = v_0 - \dfrac{|e|E}{m_e}t$
17. upward with a magnitude $9.8 \times 10^5$ N/C
21. $-1.2 \times 10^9$ J
23. $V = \dfrac{2kq}{d}$
25. $E = 0,\ V = -\dfrac{4\sqrt{2}kq}{d}$
27. (a) $-5.4 \times 10^5$ V; (b) $-2.03 \times 10^5$ V
29. $-3.33 \times 10^4$ V
35. none
39. The spheres become polarized. The negative charges move to the side closer to sphere A, while the positive charges move to the side further away from sphere A.

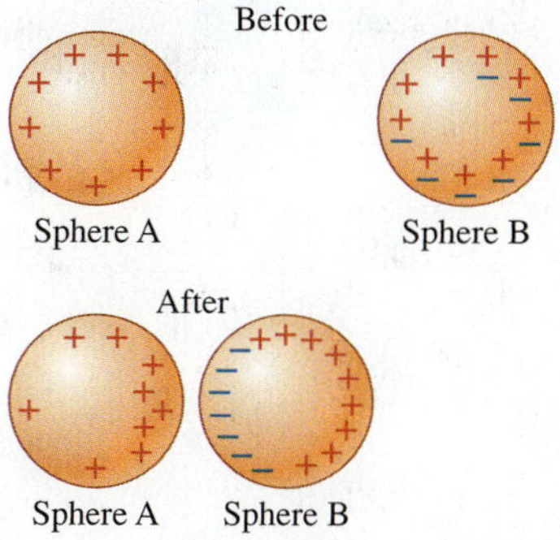


43. $E = 645$ N/C, $\theta = 110°$, $F = 1.03 \times 10^{-16}$ N, $\theta = -70°$
47. (a) $2.4 \times 10^4$ V/m; (b) $4.0 \times 10^{-5}$ m
49. (a) $q = C|\Delta V|,\ U_q = \dfrac{1}{2}C|\Delta V|^2$;
(b) $q' = q,\ U'_q = \dfrac{1}{2}\kappa C\left|\dfrac{\Delta V}{\kappa}\right|^2 = \dfrac{U_q}{\kappa}$;
(c) You do negative work to insert the dielectric into the region between the plates.
51. (a) $1.67 \times 10^{-7}$ F; (b) $1.17 \times 10^{-8}$ C; (c) $4.1 \times 10^{-10}$ J
55. $9.1 \times 10^3$ V
61. $2.7 \times 10^{-6}$ N/C, yes
63. (a) $7.52 \times 10^6$ J; (b) $6.27 \times 10^3$ s = 1.74 h
65. (a) $3.3 \times 10^{-8}$ F; (b) $2.65 \times 10^{-9}$ C; (c) $1.06 \times 10^{-10}$ J; (d) $1.0 \times 10^7$ V/m; (e) $2.65 \times 10^3$ J/m$^3$; (f) $1.06 \times 10^{-10}$ J, They are the same.
67. (c)
69. (b)
71. (a)
73. (e)
75. (d)
77. (c)

## Chapter 16

### Multiple-Choice Questions

1. (c) 3. (c) 5. (c) 7. (d) 9. (c) 11. (c) 13. (b)

### Problems

3. $3.4 \times 10^{22}$
5.

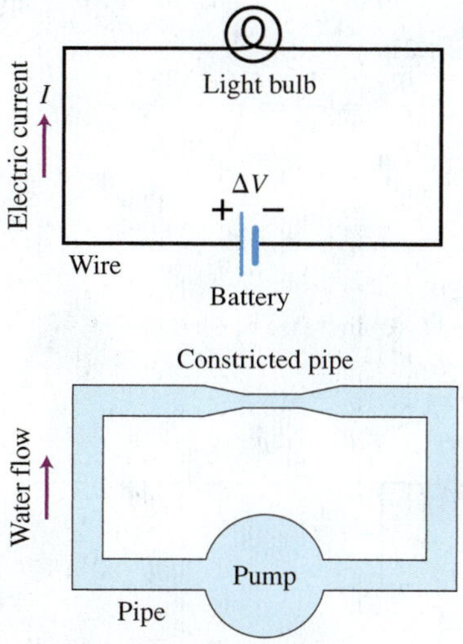


7. 2, in series or in parallel; another constricted pipe
11. $I_{dry} = 1.2$ mA, $I_{wet} = 24$ mA, dry no, wet yes
15. (a) $I_1 = I_2 = I_3$; (b) $\Delta V_3 > \Delta V_2 > \Delta V_1$
17. (a) 1.56 W; (b) more
19. 123 W
23. (a) $I_2 + I_3 = I_1, \varepsilon - I_1R_1 - I_2R_2 = 0, -I_2R_2 + I_3R_3 = 0$; (b) $I_1 = 10$ A, $I_2 = 3.33$ A, $I_3 = 6.67$ A
25. $I_2 = 1.6$ A, $I_3 = 0.40$ A, $\varepsilon = 24$ V
27. (a) $\varepsilon_1 - IR_3 - \varepsilon_2 - IR_2 - IR_1 = 0$; (b) 0.20 A; (c) $\Delta V_{R_1} = -6.0$ V, $\Delta V_{\varepsilon_1} = +20$ V, $\Delta V_{R_3} = -2.0$ V, $\Delta V_{\varepsilon_2} = -8.0$ V, $\Delta V_{R_2} = -4.0$ V; (d) zero, in agreement with Kirchhoff's loop rule
29. (a) 27 V; (b) −25 V;

(c)

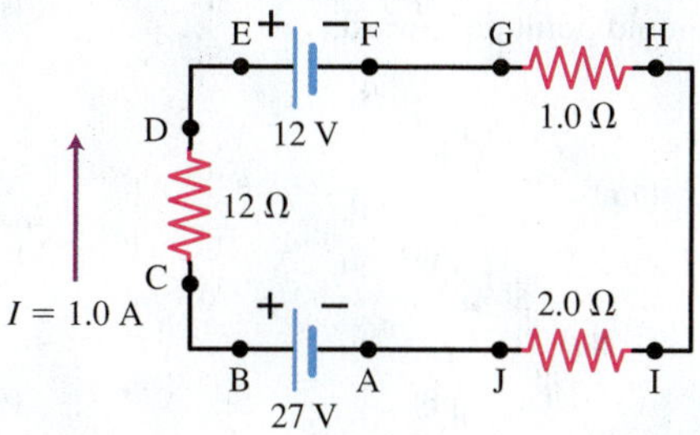

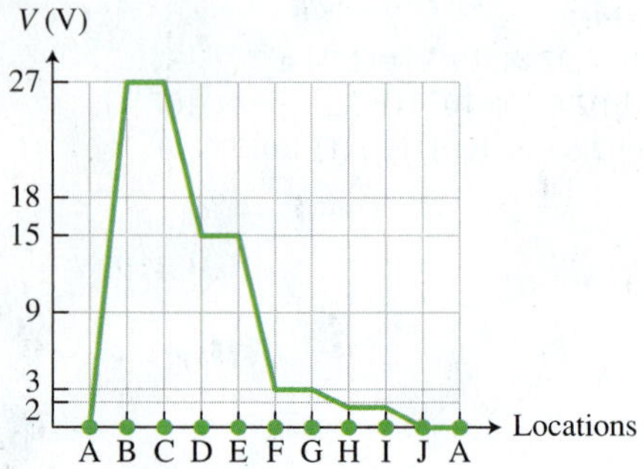


31. 33.75 Ω
33. (a) 10 Ω; (b) 2.0 A; (c) $I_2 = 0.67$ A, $I_3 = 1.33$ A
35. 56 Ω
39. (a) $I_2 + I_3 = I_1$, $\varepsilon_1 - I_1R_1 - I_3R_3 = 0$, $-I_2R_2 + \varepsilon_2 + I_3R_3 = 0$; (b) $I_1 = 0.15$ A, $I_2 = 0.023$ A, $I_3 = 0.127$ A; (c) 2.54 V
41. 15.25 A
43. (a) 3000 Ω; (b) 0.0044 A
45. 135 W
49. 1.96 m
51. increases by a factor of 1.21
53. (a) $\frac{R_A}{R_B} = \frac{L_A}{L_B} = 2$; (b) $\frac{R_A}{R_B} = \frac{A_B}{A_A} = \left(\frac{r_B}{r_A}\right)^2 = \frac{1}{4}$; (c) $\frac{R_A}{R_B} = \frac{\rho_{Cu}}{\rho_{Al}} = 0.607$
55. (a) 30 V, 15 Ω
57. no
59. (a) 0.010 Ω; (b) 0.015 Ω; (c) increase by 50%
61. $9 \times 10^8$
63. (a) $3.13 \times 10^{19}$; (b) $9.34 \times 10^{-10}$ m$^3$
69.

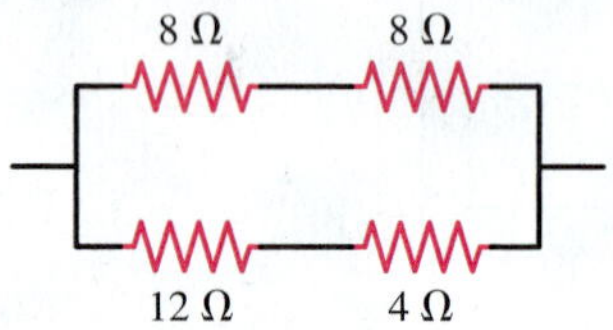


75. (e)
77. (e)
79. (d)
81. (d)
83. (e) or (d)
85. (b)

## Chapter 17

### Multiple-Choice Questions

1. (c) 3. (b) 5. (a), (b), (c), (e) and (f) 7. (a) 9. (a)

### Problems

3. No, the needle should point downward.
7. $B \geq 2.9$ T
9. (a)

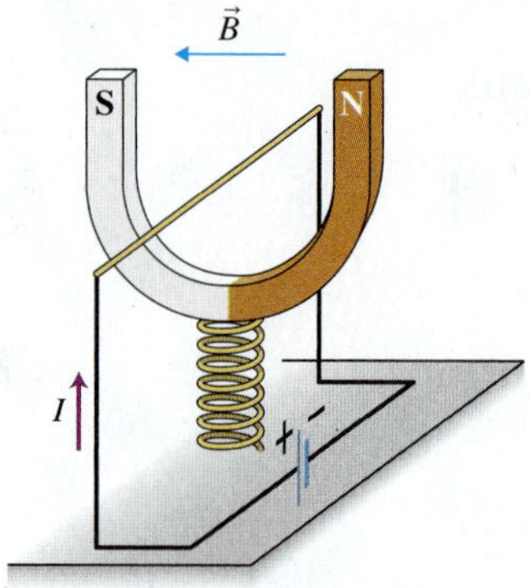


(b) Reverse the battery.
13. perpendicular to the coil (a) 0.108 N · m clockwise; (b) zero; (c) 0.086 N · m counterclockwise
15. (a) toward the west; (b) 70 N · m
17. (a) $3.0 \times 10^{-3}$ N · m; (b) 0.1875 rad = 10.7°
19. $F_{\text{B on a}} = 100$ N, $F_{\text{B on b}} = 0$, $F_{\text{B on c}} = 100$ N, $F_{\text{B on d}} = 70.7$ N

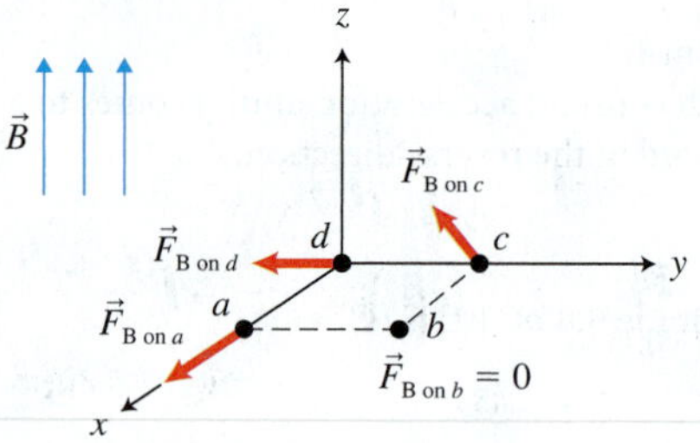


21. $1.86 \times 10^{-18}$ T
23. (a) $-2.88 \times 10^{-14}$ N; (b) $2.88 \times 10^{-21}$ N; (c) no
27. $1.0 \times 10^{-5}$ T to the north
29. $F_{1\text{ on }2} = \frac{\mu_0 I_1 I_2 L}{2\pi d}$, attractive if the currents are in the same direction and repulsive if they are in opposite directions. The force will be doubled.
37. (a) $1.2 \times 10^{-5}$ T; (b) $4.0 \times 10^{-6}$ T
41. (a) $7.68 \times 10^{-17}$ N upward; (b) 4000 m/s to the right
45. $m = \frac{qB^2r^2}{2(\Delta V)}$
49. 0.624 V
51. (d)
53. (e)
55. (d)
57. (b)
59. (c)
61. (a)
63. (b)

## Chapter 18

### Multiple-Choice Questions

1. (d) 3. (b) 5. (a) 7. (b) 9. (b) 11. (d) 13. (b)

## Problems

3. (a) no; (b) yes
9. (a) 0.096 $T \cdot m^2$; (b) 0.096 $T \cdot m^2$; (c) zero
11. $3 \times 10^{-2}\,T \cdot m^2$
13. (a) from a to b; (b) no current; (c) from b to a
17. (a)

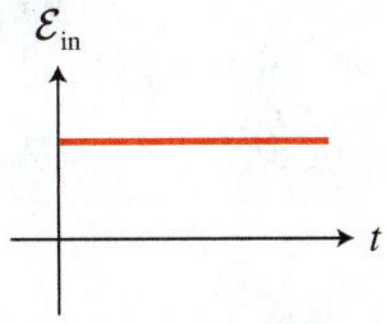


(b)

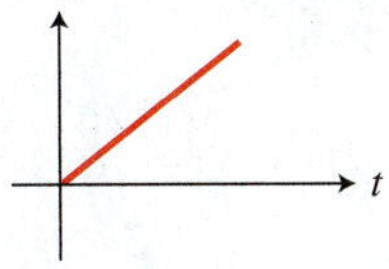


(c)

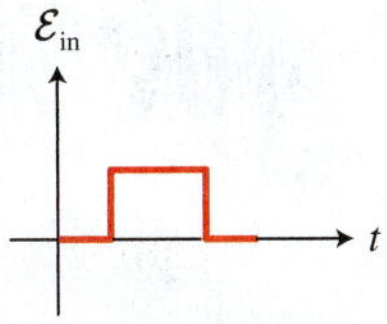


19. (a) $\varepsilon_1 = 2\varepsilon_2$; (b) Rotate the first coil so that the normal of the coil makes a 60° angle with respect to the external field.
23. $1.9 \times 10^{-5}$ V
25. (a) $1.18 \times 10^{-3}$ V; (b) b to a in the galvanometer (clockwise as viewed from above)
29. 0.089 V
37. (a) 0.057 V; (b) 0.141 V
39. 27 V
43. (a) 25.2 $cm^2$; (b) $\varepsilon_{in} = (38\text{ V})\sin[(2\pi \times 80/\text{s})t]$; (c) 26 V
45. 100:1
47. Use a step-up transformer with a turn ratio 13.33:1. Flip the switch on and off.

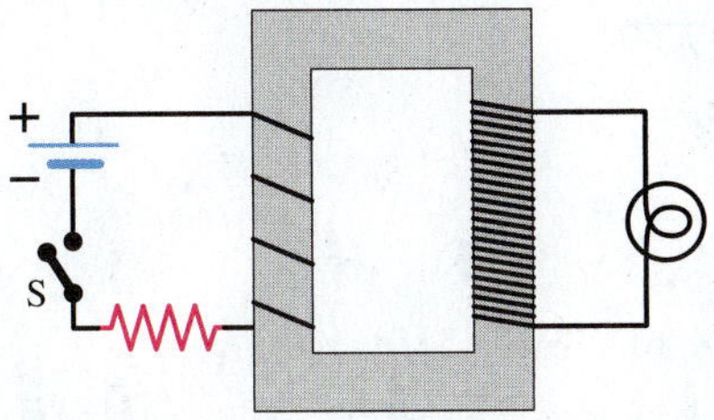


49. (a) 0.377 J; (b) 0.377 V; (c) 3.77 Ω
51. $4.1 \times 10^{-5}$ V
59. (a) An ordinary 120 V bulb will not glow;

(b)

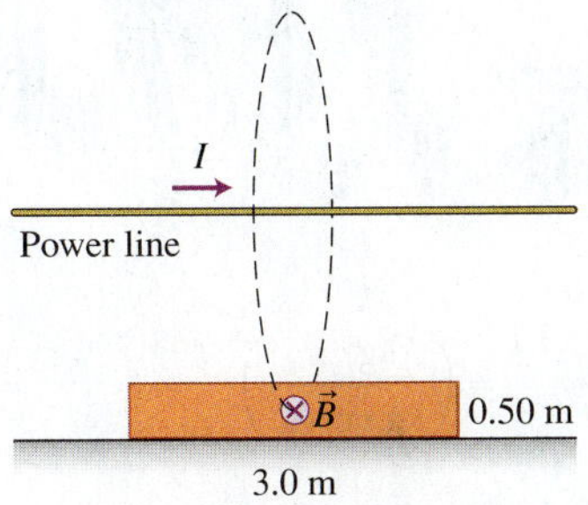


(c) 0.12 V
61. (a) 0.36 V; (b) 0.072 A; (c) clockwise, as viewed from above
65. (d)
67. (a)
69. (e)
71. (d)
73. (e)

# Chapter 19

## Multiple-Choice Questions

1. (c) (d) 3. (c) 5. (b) 7. (a) 9. (d)

## Problems

3. (a) 250 N/m
5. (a) yes;

(b)

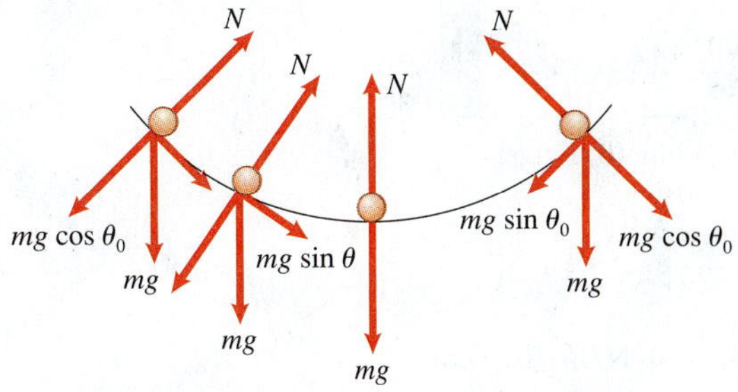


(c)

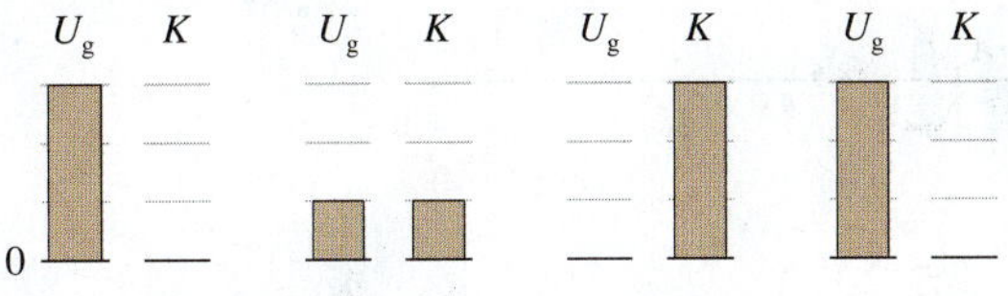


7.

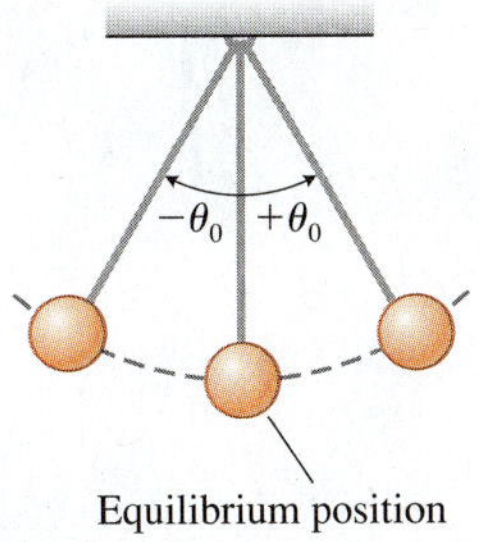


9. (a) 20 Hz; (b) $5.0 \times 10^{-5}$ s
11.

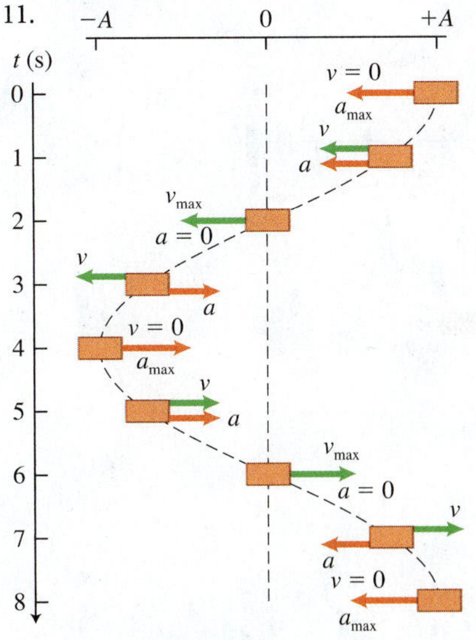


19. $k = 400$ N/m, $\Delta x_1 = 0.25$ m
21. 2.25 Hz
23. $3m$
27. (a) 250 N/m; (b) 0.45 J; (c) $x(t) = (0.060\text{ m})\cos(2\pi f t)$
29. 2
31. 1.05

33. 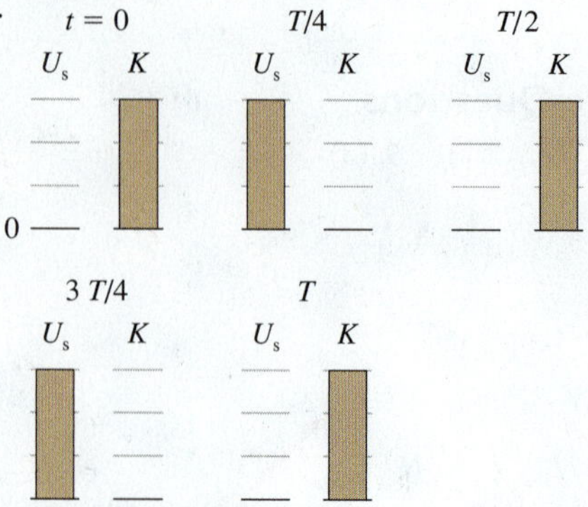


35. (a) 0.248 m; (b) longer
37. 0.063 m/s
39. 4.48 m/s
41. $g' = 9.751\ \mathrm{m/s^2}$
43. 5.62 s
47. (a) $8 \times 10^7$ N/m; (b) 5 cm
49. 27.6 m/s
55. (a)

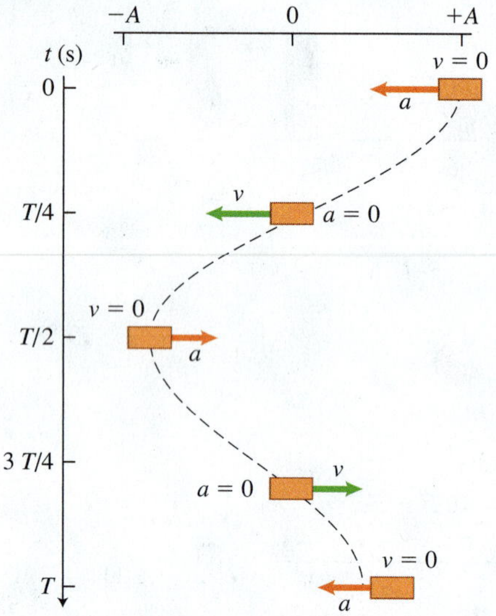


(b)

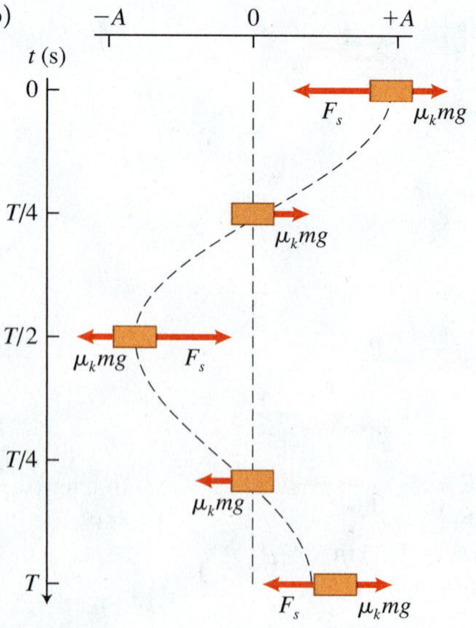


(c) 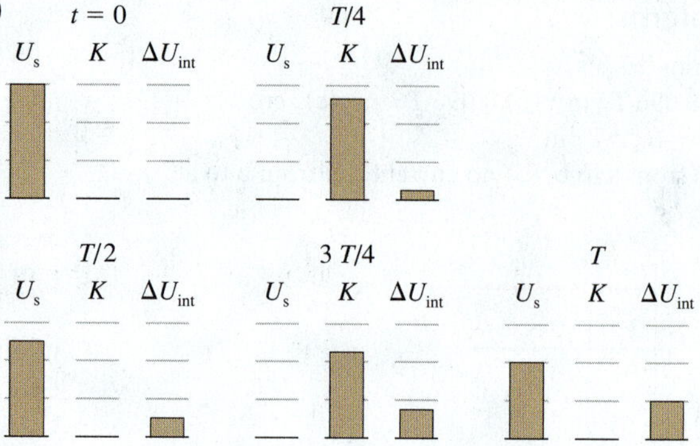


(d) and (e)

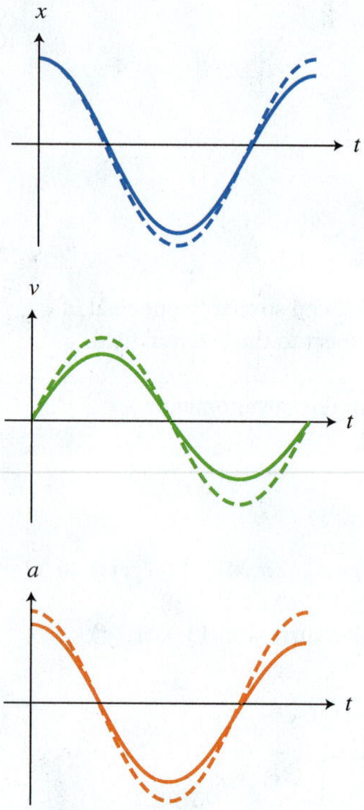


57. (a) 2.17 m/s
59. 1.06 m/s
61. (a) 659 N/m; (b) $3.54 \times 10^{-12}$ m
63. $T = 2\pi\sqrt{m/k}$

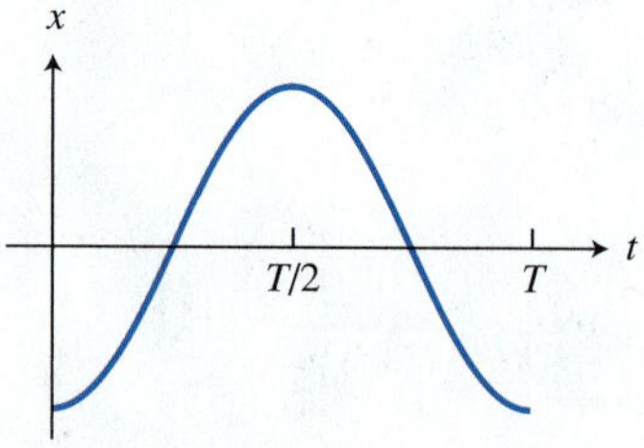


65. yes, 1.4 h, about one-thirtieth
67. $[T] = \left[\frac{m}{k}\right]^{1/2} = \left(\frac{\mathrm{kg}}{\mathrm{N/m}}\right)^{1/2} = \left(\frac{\mathrm{kg \cdot m}}{\mathrm{kg \cdot m/s^2}}\right)^{1/2} = (\mathrm{s}^2)^{1/2} = \mathrm{s}$
69. 120 m/s
73. $T = 2\pi\sqrt{\frac{m}{\rho g A}}$

75. (d)
77. (a)
79. (a)
81. (c)
83. (c)

## Chapter 20

### Multiple-Choice Questions

3. (c) 5. (c) 7. (b) 9. (c) 11. (b) 13. (b)

### Problems

1.

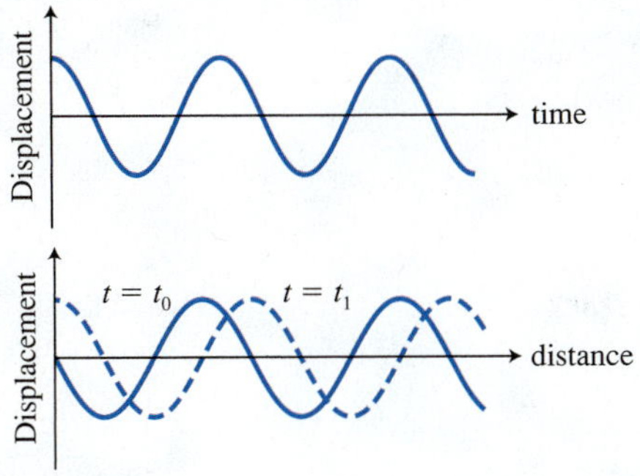


7. 0.75 m
9. 9.6 m
11. 17 m, 17 mm
13. 2.8 km
15. 1.7
17. tightened
23. $\sim 2 \times 10^{11}$ MJ
25. $\sim 2 \times 10^{12}$ m
29. 7.1 m, out of phase
31.

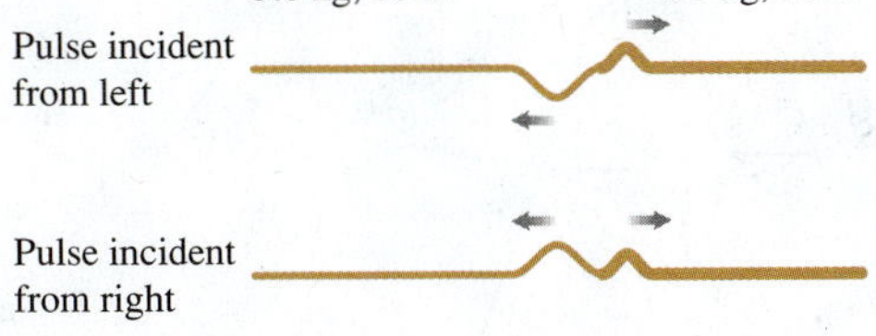


33.

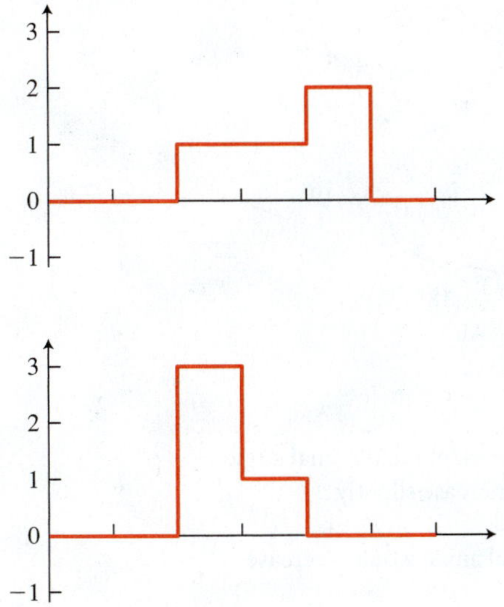


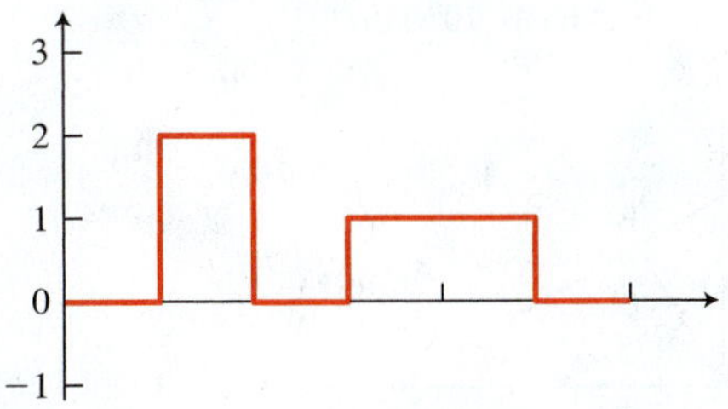


35.

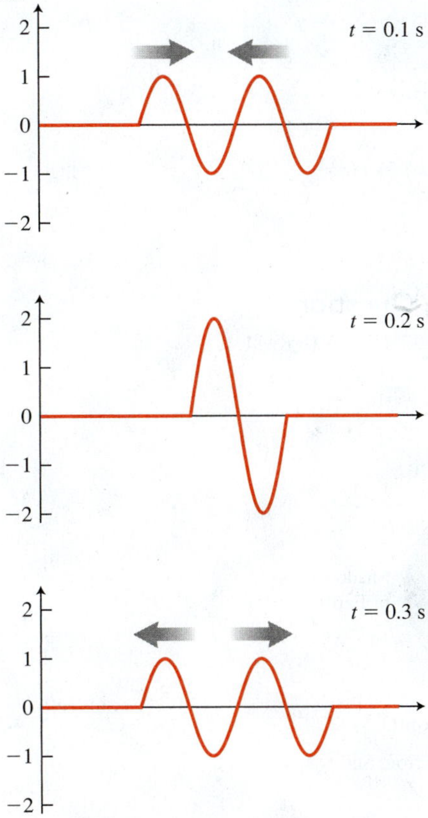


37.

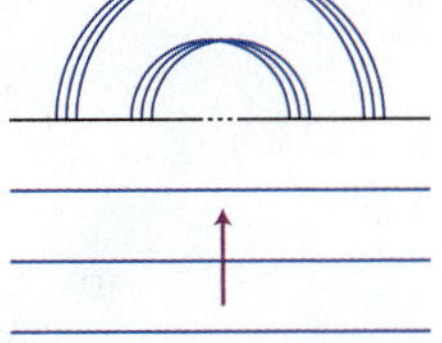

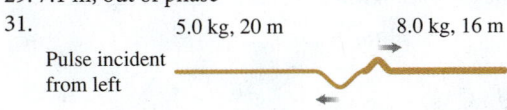

39. no vibrations at 3.5 m from Source A along the line joining Source A and Source B, maximum vibration at 3.0 m from Source A along the line joining Source A and Source B
43. $\sim$3.5% faster
45. 700 m
47. 40 dB, 100 dB
49. 9, 19, 29
51. 0.07 m
53. linear mass density, wave speed, wavelength, and frequency, $\mu = 0.080$ kg/m, $v = 39$ m/s, $\lambda = 3.3$ m, $f = 12$ Hz
57. 4.9%
59. (a) 0.061 Hz; (b) $n \geq 327$
61. (a) closed, $L = 0.71$ m; (b) 240 Hz, 480 Hz
63. 350 m/s
65. 100,040 Hz

67. (a) 337 Hz; (b) 266 Hz
69. (a) $4.28 \times 10^5$ Hz; (b) $4.65 \times 10^5$ Hz
71. 44 m/s
73.

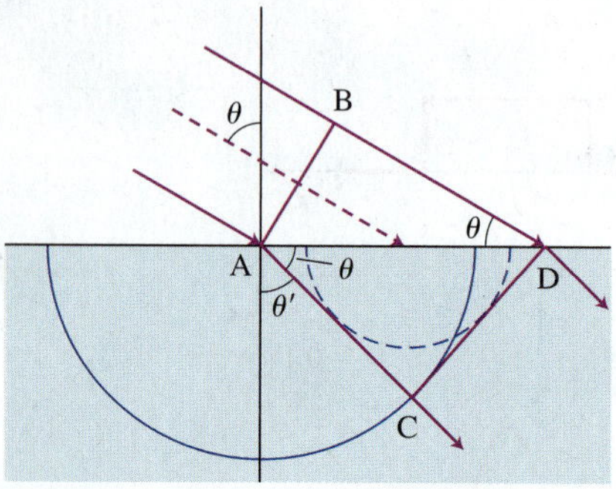


75. ~0.2 s
77. (a) 100,053 Hz; (b) 53 Hz

# Chapter 21

## Multiple-Choice Questions

1. (d) 3. (a) 5. (b) 7. (c) 9. (b) 11. (c)

## Problems

1. 17 m
3.

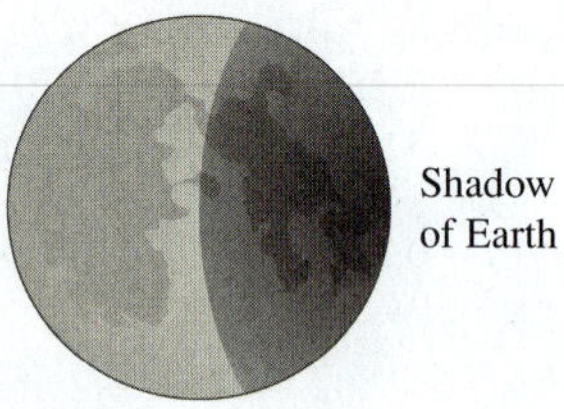


9. 65° below the horizontal

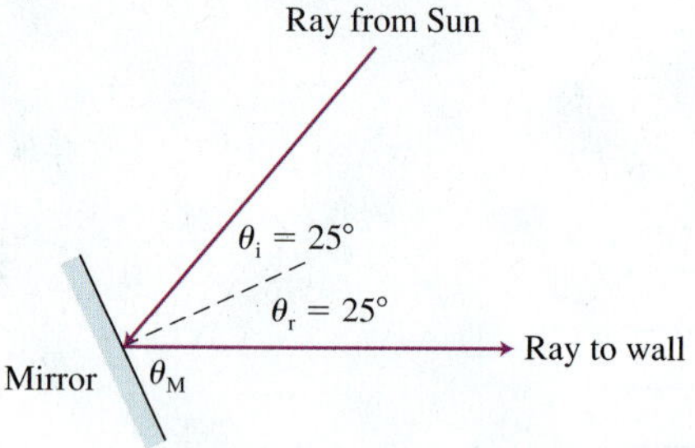


13.

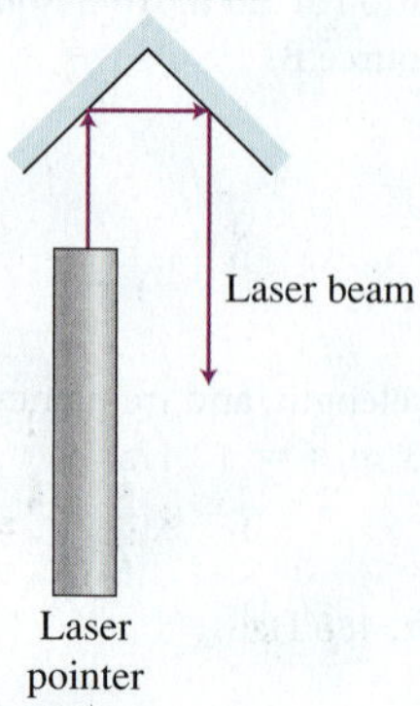


15.

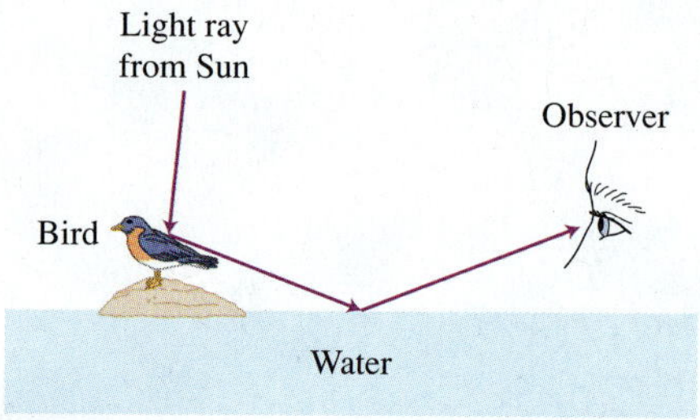


19. (a) 25°; (b) 16°
(c)

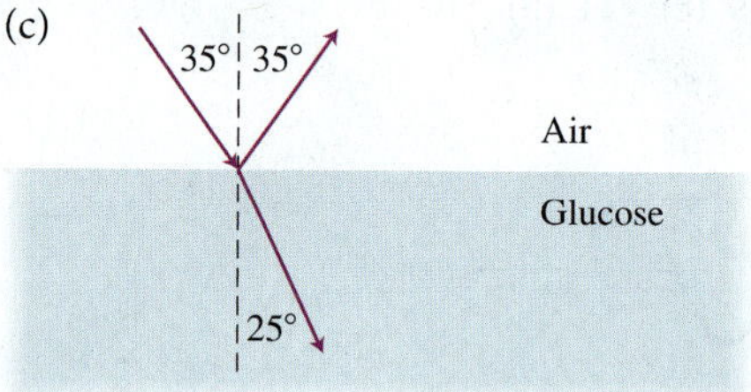


(d)

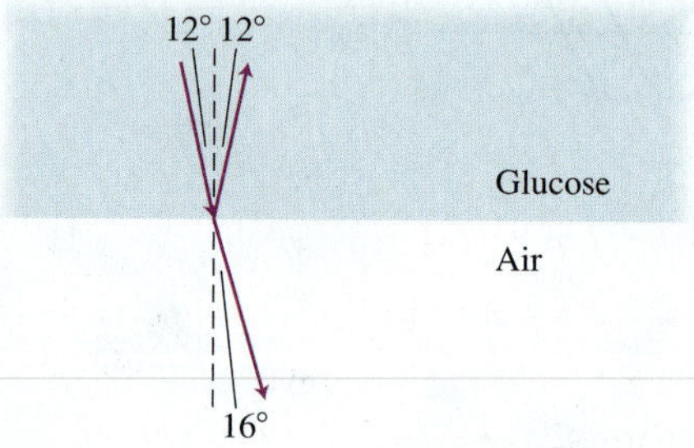


21. 1.4
23. (a) The laser beam will emerge horizontal, but displaced in height. (b) The laser beam will be deflected upward, as shown.

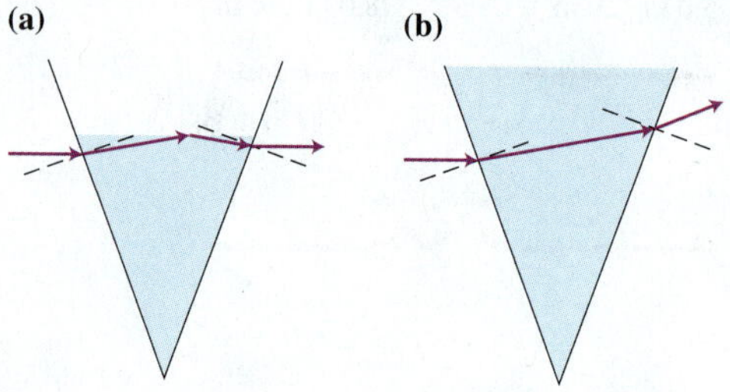


25. 62.9°
27.

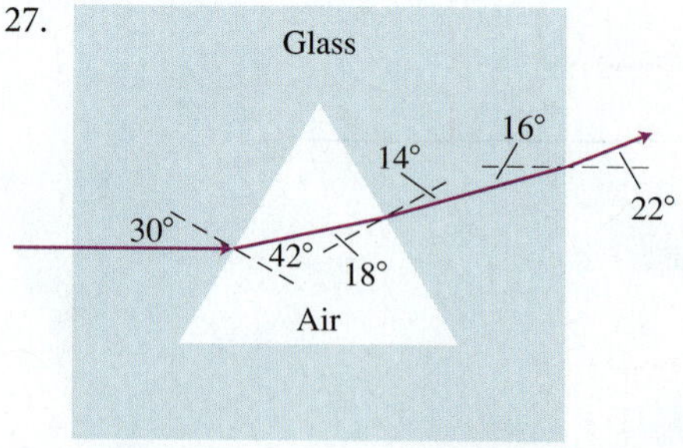


29. 32° above the horizontal, the final angle of refraction would increase slightly
31. 48°
33. 1.48, the critical angle would decrease
35. 50°
37. 1.651
39. 2.3 m
41. (a) 39.3°; (b) no
45. $\theta_i = \tan\left(\frac{n_1}{n_2}\right)$

47. Normal to mirror is 2.6° above the horizontal.

51. $\theta_1 = 60°, \theta_2 = 33°, \theta_3 = 27°, \theta_4 = 47°$

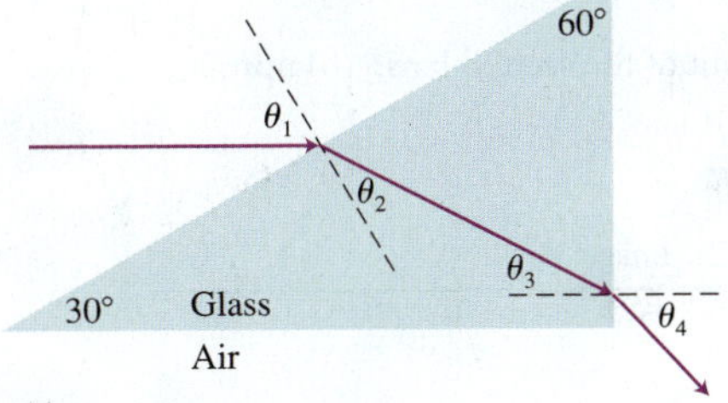


53. 0.19 cm

57. 1.9 m

59. 1.1 m

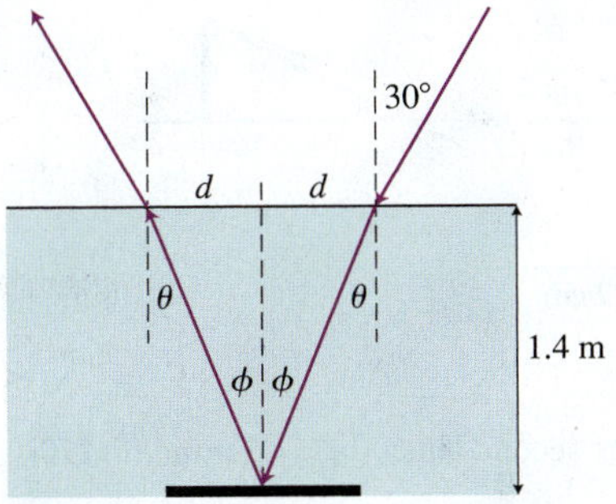


# Chapter 22

## Multiple-Choice Questions

1. (c) 3. (b) 5. (a) 7. (c) 9. (b) 11. (e) 13. (c) 15. (d)

## Problems

5. 95 cm

15. (a) $s' = -18.2$ cm, virtual, upright

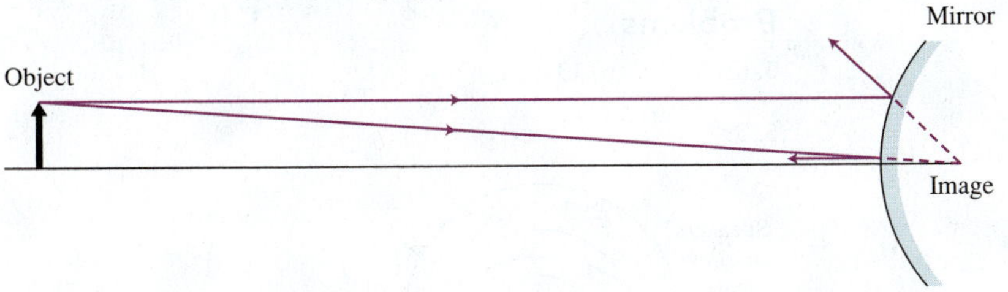


(b) $s' = -13.3$ cm, virtual, upright

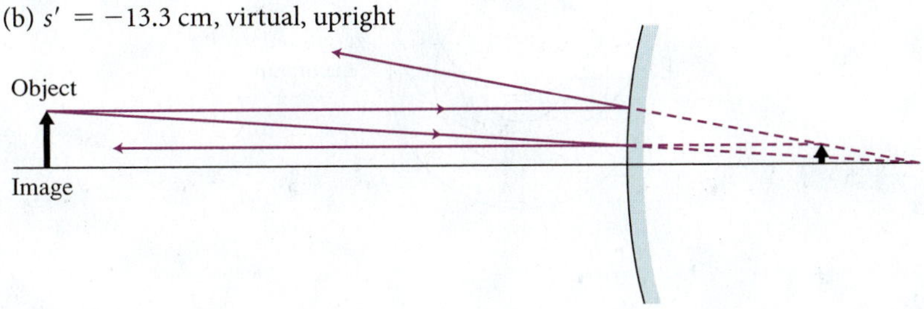


(c) $s' = -6.67$ cm, virtual, upright

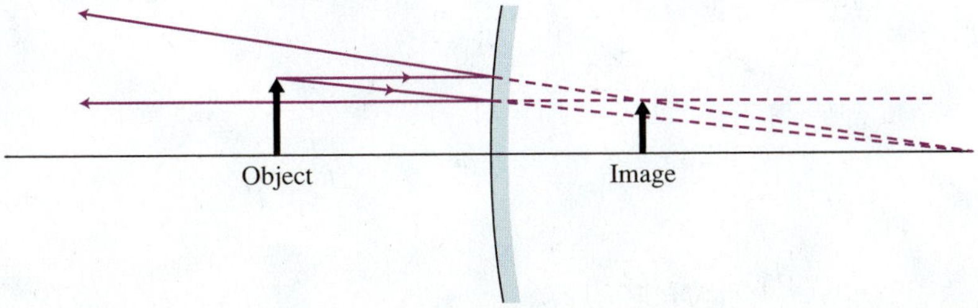


17. $f \sim 150$ m, $R \sim 300$ m, $d \sim 10$ cm

19. (a) 0.86 m behind the mirror; (b) 0.49 m

21. 6.7 cm from the mirror on the optic axis

23. $s = 1.0$ cm, $s' = -8.0$ cm, $f = 1.1$ cm, $R = 2.2$ cm, virtual upright image, $m = 8.0$, concave mirror

27. (a) $s' \approx 15$ cm, real, inverted, reduced

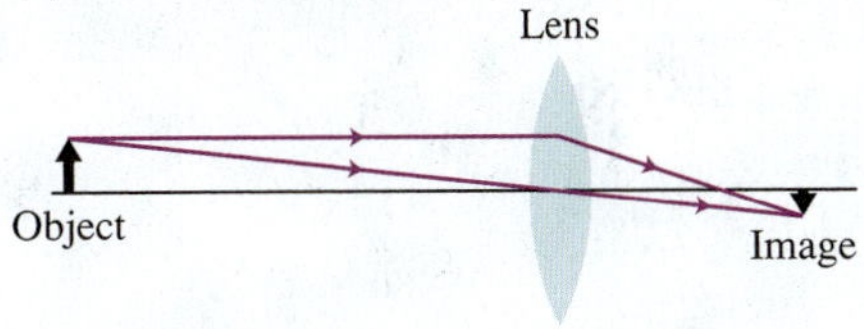


(b) $s' \approx 38$ cm, is real, inverted, enlarged

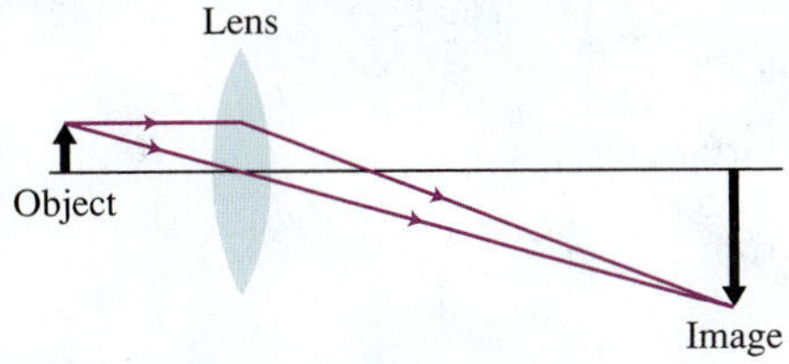


(c) $s' \approx -10$ cm, virtual, upright, enlarged

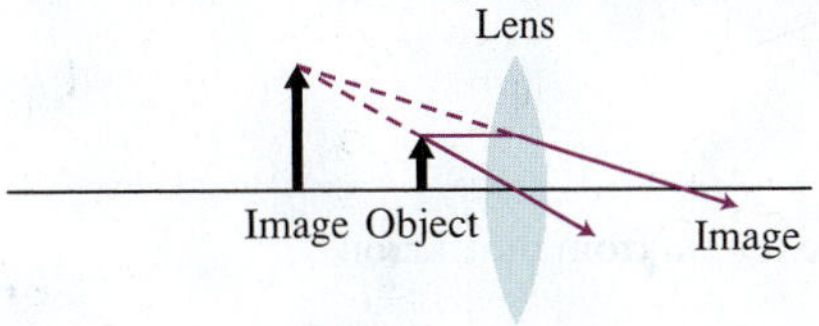


29. (a) $s' \approx -23$ cm, virtual, upright, enlarged

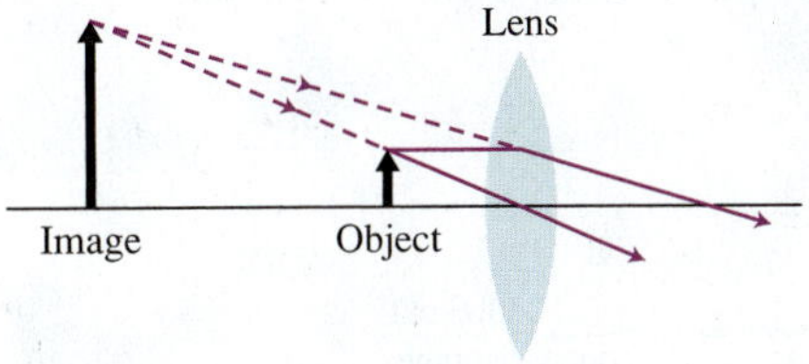

(b) $s' \approx -4$ cm, virtual, upright

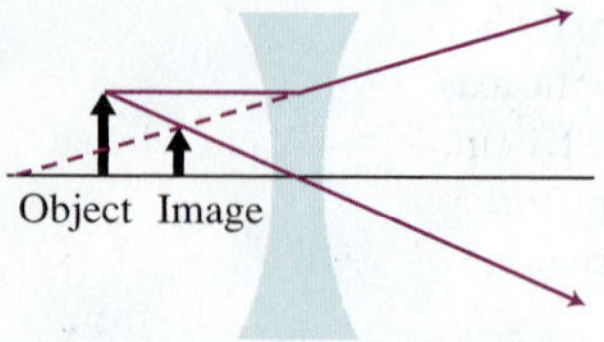


33. (a) $s' \approx 12$ cm, real, inverted

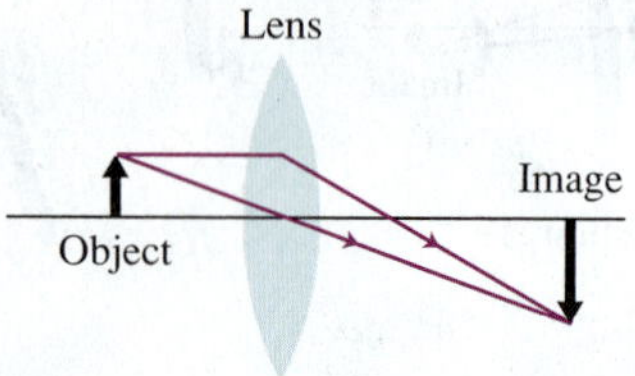


(b) $s' \approx -2.7$ cm, virtual, upright

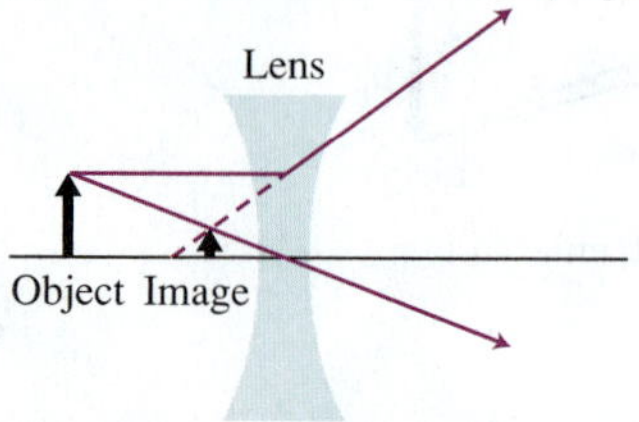


(c) 12 cm, −2.7 cm
37. Position your face 6.7 cm from the mirror.
41. 6.1 cm
43. (a) 2.52 m; (b) 0.40 m
45. 25 cm
47.

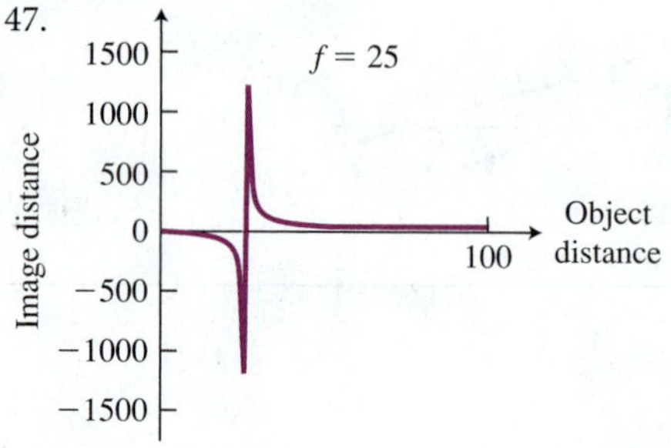


49.

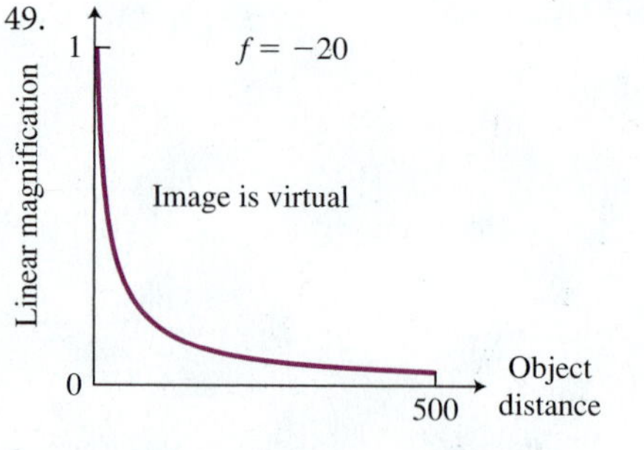


51. (a) 2.10 cm; (b) 2.11 cm; (c) 2.29 cm

53. (a) −2.4 m; (b) 03.6 m
55. (a) hyperopia, 0.50 m, infinity; (b) myopia, 3.5 m, 25 cm
57. 5.8, increase, decrease
59. (a) 4.8 cm; (b) 31
61. (a) 20 cm to the right of the second lens; (b) upright; (c) real
63. (a)

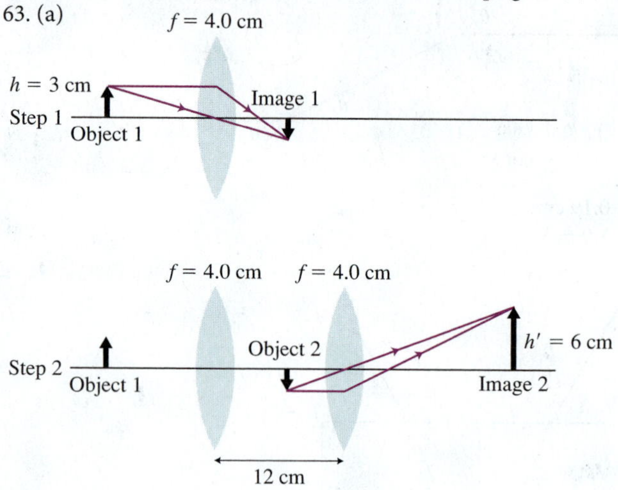


(b) $m = 2$
65. −17 cm
67. (a) 1.1 m to the left of the second lens; (b) −1.5 mm; (c) 270
69. 12.4 m, 4.4
71. (a) 8.1 cm; (b) −20
73. $d = 46$ cm, $s = 1.0$ cm
79. 24 cm to 18 m

# Chapter 23

## Multiple-Choice Questions

1. (d) 3. (e) 5. (b) 7. (a), (b)

## Problems

3. (a) 0.25°; (b) 13 mm

5. 39°

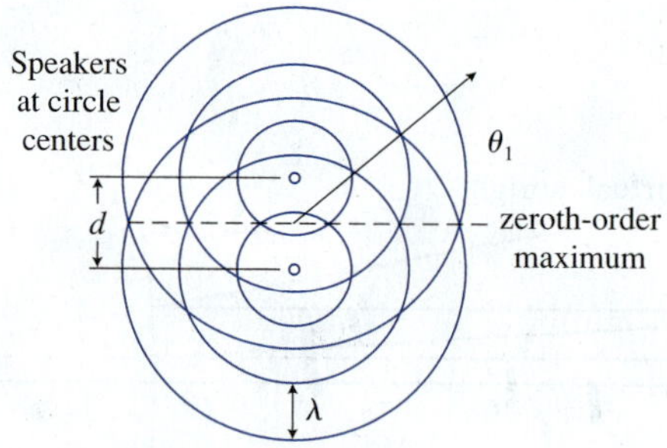

7.

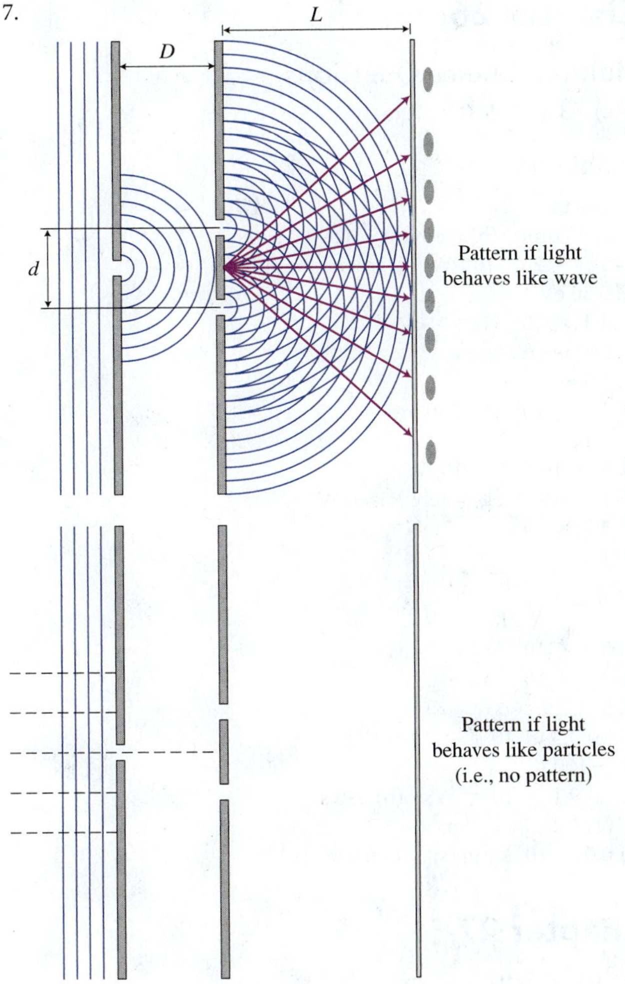


(b) 25°; (c) 73 cm

13. 4500 $cm^{-1}$

17. 450 nm

21. (a) yes; (b) no; (c) 480 nm; (d) 240 nm

23. (a) 600 nm; (b) 300 nm

29. 0.69°

31. 0.125 mm

33. $1.7 \times 10^{-7}$ rad

35. Bead diameter is 31 $\mu$m.

37. 1.8 cm

41. (a)

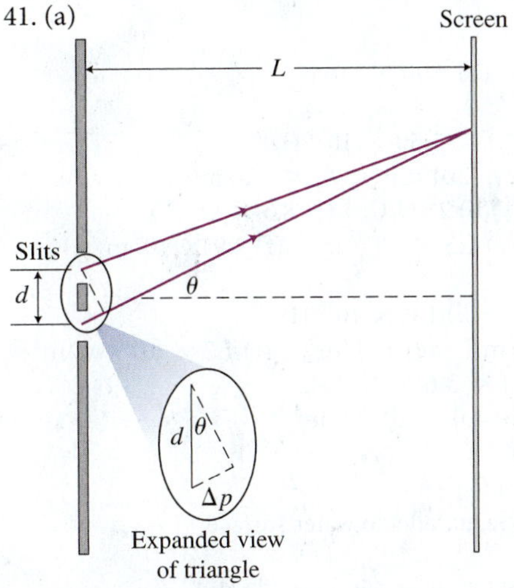


(b) 0.090°, 0.18°, 0.27°; (c) 1.6 cm

(d) coherent source and $L \gg d \gg \lambda$

43. (a)

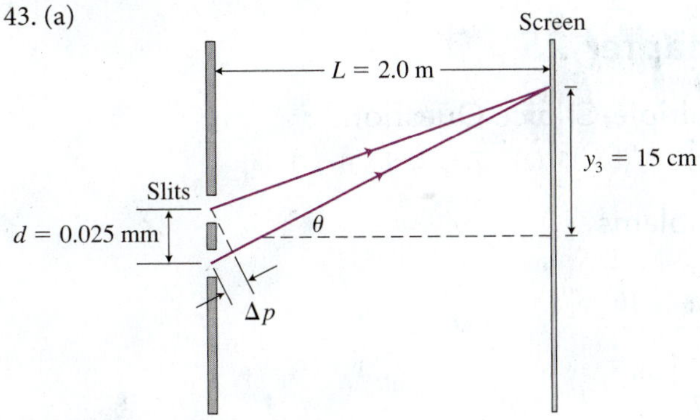


(b) coherent light source, $\theta_3 = 4.29°$, $\lambda = 620$ nm

47. (a) 0.84°; (b) 3.1°

49. 130 nm, 389 nm

53. 0.30°

57. 0.65 cm

59. (a) 657 nm, $f = 4.57 \times 10^{14}$ Hz;
(b) 993 nm, $f = 3.02 \times 10^{14}$ Hz; (c) The galaxy Hydra is receding from Earth.

63. ~10 km

65. ~0.61 mm

67. 0.18 mm

69. $f_{min} = 1 \times 10^5$ Hz

11. (a)

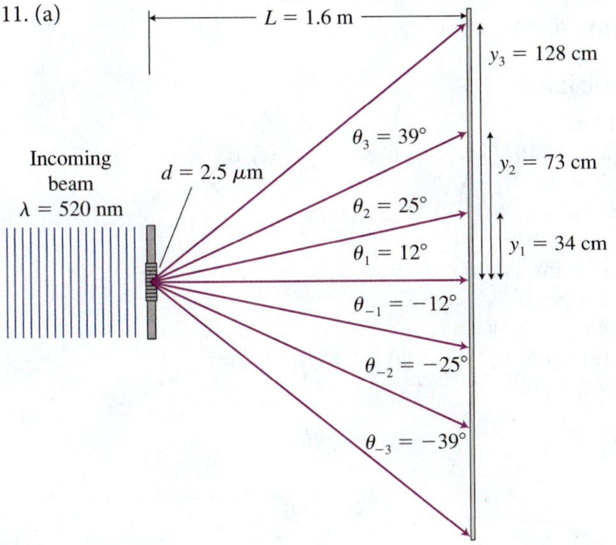

# Chapter 24

## Multiple-Choice Questions

1. (a) 3. (b) 5. (a) 7. (c) 9. (c) 11. (b)

## Problems

1. 10 cm, vertical
3. (a) $0.40\ \text{W/m}^2$; (b) $0.17\ \text{W/m}^2$
7. $1.3 \times 10^{-6}\ \text{T}$
9. 3.3 nA
17. no, 3 minutes
19. (a) $6.67 \times 10^{-8}$; (b) $9.00 \times 10^5\ \text{N/C}$
21. 6300 y, yes
23. $4.29 \times 10^{14}\ \text{Hz} < f < 7.50 \times 10^{14}\ \text{Hz}$
no change in frequency; $301\ \text{nm} < \lambda' < 526\ \text{nm}$
25. (a) $-x$ direction; (b) 25 N/C; (c) $8.3 \times 10^{-8}\ \text{T}$
(d) $1.9 \times 10^{10}\ \text{Hz}$; (e) $1.5 \times 10^{-2}\ \text{m}$; (f) $2.9 \times 10^8\ \text{m/s}$;
(g) linearly polarized
27. (a) $1.0 \times 10^{16}\ \text{Hz}$; (b) $1.0 \times 10^{16}\ \text{Hz}$;
(c) $80\ \text{J/m}^3$; (d) $0\ \text{J/m}^3$; (e) $40\ \text{J/m}^3$; (f) $1.2 \times 10^{10}\ \text{W/m}^2$
29. $4.8 \times 10^{21}\ \text{Hz} < f < 3.6 \times 10^{18}\ \text{Hz}$
31. (a) $1.8\ \text{W/m}^2$; (b) $6.0 \times 10^{-9}\ \text{J/m}^3$;
(c) 37 N/C, $1.2 \times 10^{-7}\ \text{T}$
41. $710\ \text{W/m}^2$, $94\ \text{W/m}^2$
43. 37°, Brewster's angle, parallel to water surface
45. $110\ \text{W/m}^2$
53. $B_{\text{max}} = 6.7 \times 10^{-8}\ \text{T}$, $\bar{u} = 1.8 \times 10^{-9}\ \text{J/m}^3$,
$u_{\text{max}} = 3.6 \times 10^{-9}\ \text{J/m}^3$, $I = 0.53\ \text{W/m}^2$

# Chapter 25

## Multiple-Choice Questions

1. (c) 3. (b) 5. (a) 7. (b) 9. (b) 11. (a)

## Problems

1. 5 m/s
9. $1.8 \times 10^8\ \text{m/s}$
11. 7.1 d
13. 5.0 cm
15. 7.5 s
21. $2.2 \times 10^8\ \text{m/s}$
25. 31 m/s
27. (a) $0.7c$; (b) $c$; (c) $0.30c$; (d) $0.50c$
35. $1.5 \times 10^{-10}\ \text{J}$, $3.2 \times 10^{-16}\ \text{J}$, $1.5 \times 10^{-10}\ \text{J}$
37. $4.9 \times 10^{18}\ \text{J}$
39. 299,791,858 m/s
41. $1.7 \times 10^{19}\ \text{J}$, $5.4 \times 10^{18}\ \text{J}$, $1.2 \times 10^{19}\ \text{J}$
43. 59%
45. (a) 7.3; (b) 2.5
47. (a) $1.96 \times 10^{-35}\ \text{kg}$; (b) $4.2 \times 10^{-10}$
49. $7.4 \times 10^{-5}\ \text{kg}$
51. (a) $3.3 \times 10^{-8}\ \text{kg/s}$; (b) 1.1 kg
53. (a) $3.6 \times 10^{26}\ \text{J}$; (b) $4.5 \times 10^{-10}$
55. 1.20 m/s
57. increase in estimated life of the universe
59. no
61. (a) 4.45 y; (b) 1.14 y
63. (a) 10 s; (b) 4.4 s; (c) 390 m; (d) yes
65. (a) $2.8 \times 10^8\ \text{m/s}$; (b) $2.9 \times 10^{20}\ \text{J}$, $9.0 \times 10^{19}\ \text{J}$

# Chapter 26

## Multiple-Choice Questions

1. (c) 3. (e) 5. (e) 7. (a), (b) 9. (c)

## Problems

1. 9.4 μm
3. (a) 73 nm; (b) 0.48 μm; (c) 9.7 μm
7. $P_{\text{Sun}} = 2 \times 10^{17}\ \text{W}$, $P_{\text{Earth}} = 2 \times 10^{17}\ \text{W}$
11. 0.50 eV
13. $1.1 \times 10^{15}\ \text{Hz}$
15. 560 nm
17. 1.9 eV
19. $1.2 \times 10^{15}\ \text{Hz}$, 250 nm
21. 4.1 eV
23. $5 \times 10^{19}$, $3 \times 10^5\ \text{W}$
25. $P \times 10^{-17}\ \text{kg}$, where $P$ is in W
27. $3.7 \times 10^{21}$
29. 170
35. $B = E\sqrt{\dfrac{m_e}{2K}}$
37. $5.0 \times 10^{11}$
41. $2.6 \times 10^{-11}\ \text{m}$
45. 5
47. (a) $3.3 \times 10^{-9}\ \text{N}$; (b) 17 h
53. $\sim 100\ \text{s}^{-1}$
55. (a) $9.1 \times 10^{-22}\ \text{W}$; (b) 390 s
57. (a) $1.28 \times 10^{14}\ \text{m}^2$;
(b) $6.0 \times 10^8\ \text{kg·m/s}$; (c) $6.0 \times 10^8\ \text{N}$

# Chapter 27

## Multiple-Choice Questions

1. (b) 3. (b) 5. (c) 7. (a) 9. (b) 11. (b)

## Problems

1. 2 km
3. (a) $2 \times 10^{-4}\ \text{kg}$; (b) $6 \times 10^{20}$; (c) $1.1 \times 10^{-28}\ \text{m}^2$;
(d) $7 \times 10^{-4}$
7. 435 nm, $6.90 \times 10^{14}\ \text{Hz}$, 2.86 eV
9. $\lambda = 250(2n - 1)^2\ \text{nm}$
13. 658 nm, 122 nm, 103 nm
15. $2.4 \times 10^6$
23. $4.6 \times 10^{23}\ \text{K}$
25. (a) $2.0 \times 10^9\ \text{W}$; (b) $2.5 \times 10^{17}\ \text{W/m}^2$
27. $7.2 \times 10^{15}$
29.

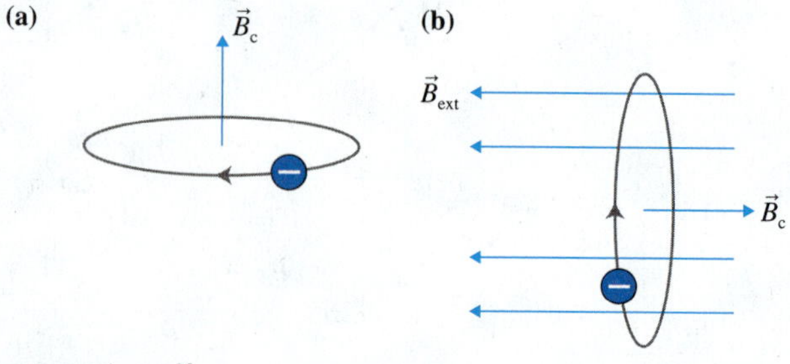


31. $3.3 \times 10^{-10}\ \text{m}$
39. (a) $2.2 \times 10^6\ \text{m/s}$; (b) $3.3 \times 10^{-10}\ \text{m}$; (c) $3.3 \times 10^{-10}\ \text{m}$

41. You need to know the potential difference through which electrons were accelerated.
43. (a) $l = 3, n = 4$; (b) $l = 2, n = 3$
45.

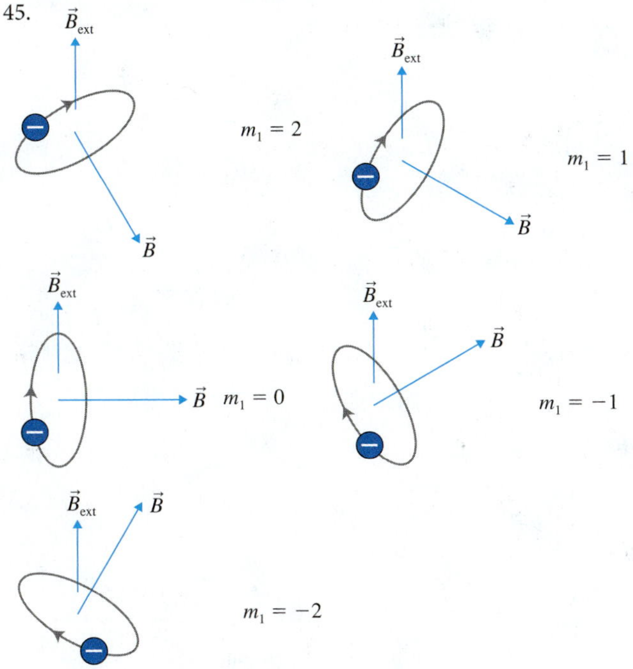


47. $(n,l,m_l,m_s) = \left(4,3,\pm 3, \pm\frac{1}{2}\right), \left(4,3,\pm 2, \pm\frac{1}{2}\right), \left(4,3,\pm 1, \pm\frac{1}{2}\right), \left(4,3,0, \pm\frac{1}{2}\right)$
49. (a) $1s^2, 2s^2, 2p^6, 3s^2, 3p^4$; (b) Sulfur and oxygen have the outermost subshell ($p$) and the same number of electrons (4) in this shell.
51. $1s^2, 2s^2, 2p^6, 3s^2, 3p^6, 4s^2, 3d^6$
53. H: $1s^1$, Li: $1s^2, 2s^1$, Na: $1s^2, 2s^2, 2p^6, 3s^1$, K: $1s^2, 2s^2, 2p^6, 3s^2, 3p^6, 4s^1$
59. (a) −54.4 eV, 0.027 nm; −13.6 eV, 0.106 nm; −6.04 eV, 0.239 nm
(b)

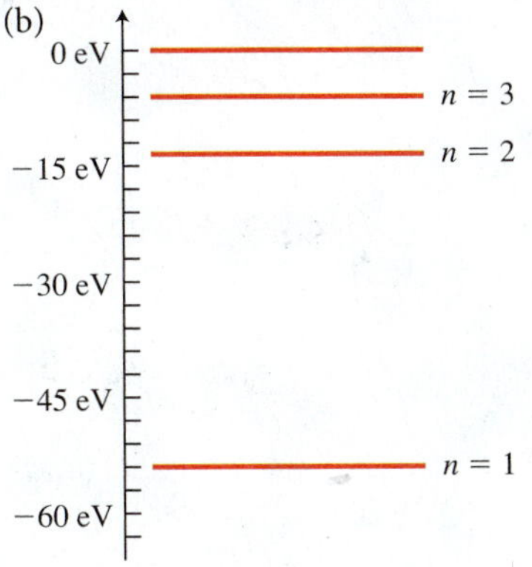


61. $\sim 5.8 \times 10^{-13}$ m
63. 1.88 $\mu$m
65. 52.9 eV, 23.5 nm, $1.28 \times 10^{16}$ Hz

# Chapter 28

## Multiple-Choice Questions

1. (d) 3. (c) 5. (b) 7. (c) 9. (b)

## Problems

5. (a) $\sim 4 \times 10^{28}$; (b) $\sim 2 \times 10^{28}$; (c) $\sim 2 \times 10^{-10}$ cm$^3$
7. 2.8 nm
9. 1537 MeV
13. (a) ${}^{1}_{1}$H; (b) ${}^{1}_{0}n$; (c) ${}^{94}_{38}$Sr; (d) ${}^{3}_{2}$He
17. 7.16 MeV
19. ${}^{228}_{90}$Th, 5.41 MeV
21. (a) 38, 50; (b) 87.905586 u
23. (a) 43.2 MeV; (b) $3.45 \times 10^{14}$ J/kg
27. (a) ${}^{235}_{92}$U; (b) ${}^{144}_{59}$Pr; (c) ${}^{129}_{54}$Xe; (d) ${}^{60}_{29}$Cu
29. 5.65 MeV
33. $7.2 \times 10^{-3}$ m/s
35. 12 h
37. 0.177
39. 32.4 y, 54 y, 75.6 y
41. (a) 115 y; (b) 173 y; (c) 191 y
43. 12 d, 41 d
45. 6.09 MeV
47. $1.3 \times 10^9$ y
49. 40,000 y
53. (a) $7.0 \times 10^{-4}$ Sv; (b) $1.2 \times 10^{-3}$ Sv; (c) $7.0 \times 10^{-3}$ Sv; (d) $7.0 \times 10^{-3}$ Sv; (e) $1.4 \times 10^{-2}$ Sv
59. (a) $1 \times 10^{33}$; (b) $1.3 \times 10^8$ m$^3$
61. (a) $7.2 \times 10^5$ J; (b) $6.4 \times 10^{16}$
63. $\frac{m_{coal}}{m_{^{235}U}} = 2.5 \times 10^6$
65. 36 d
67. (a) $4.2 \times 10^{-12}$ kg; (b) $4.2 \times 10^{-14}$ J; (c) $2.6 \times 10^3$; (d) $1.1 \times 10^{-13}$ kg, $1.1 \times 10^{-15}$ J, 71

# Chapter 29

## Multiple-Choice Questions

1. (a) 3. (b)

## Problems

7.

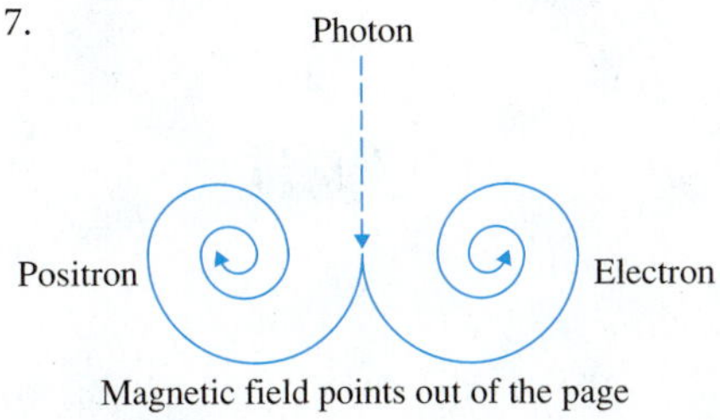


17. Elementary particles are classified based on their fundamental interactions.
25. Higgs gives mass to the other particles.
31. The cosmic background radiation was produced during the period when neutral atoms first formed, about 370,000 years after the Big Bang.
41. $2.9977 \times 10^8$ m/s

# Credits

## Introducing Physics

Opener: Reuters/Sergei Ilnitsky/Pool; p. xxxiv: John Reader/Photo Researchers, Inc.; p. xxxv: Gary Yim/Shutterstock; p. xxxviii (top); Goddard Space Flight Center/NASA; p. xxxviii (middle): Reuters/Frank Polich; p. xxxviii (bottom): Reuters/Toby Melvile; p. xlii: Sadequl Hussain/Shutterstock; p. xliii: Fotolia.

## Chapter 1

Opener: Transtock Inc./Alamy.

## Chapter 2

Opener: Reuters/ADAC; Fig. 2.6: Reuters/Anwar Mirza; p. 71: Cheryl Power/Photo Researchers, Inc.; Fig. 2.8: fStop/Alamy; Fig. P2.33: Seaman Justin E. Yarborough/U.S. Navy; Fig. P2.52: Berc/iStockphoto.

## Chapter 3

Opener: AP Photo/The News Tribune, Janet Jensen; p. 93: PhotoStock-Israel/Alamy; Fig. 3.17: HP Canada/Alamy; Fig. 3.18: Reuters/Stefan Wermuth.

## Chapter 4

Opener: vesilvio/Shutterstock.

## Chapter 5

Opener: NASA; Fig. 5.5: Ted Kinsman/Photo Researchers, Inc.; Fig. 5.8: NASA; Fig. 5.9: Walter G Arce/Shutterstock.

## Chapter 6

Opener: J-L Charmet/Photo Researchers, Inc.; p. 194: Vladimir Wrangel/Shutterstock; Fig 6.10: Ted Foxx/Alamy; Fig. P6.23: EPA/Horacio Villalobos/Newscom; Fig. P6.78: Exactostock/SuperStock; Fog. P6.79: BrandonR; Fig. P6.82: billdayone/Alamy.

## Chapter 7

Opener: Leo Mason sports photos/Alamy; Fig 7.1: Ayakovlev/Shutterstock; Fig. 7.13: Reuters/Dylan Martinez; Fig 7.15: GIPhotoStock/Photo Researchers, Inc.; Leo Mason sports photos/Alamy; Fig. 7.21: Hogar/Shutterstock; Fig. 7.23: Reuters/Stringer; p. 257: Pearson Science/Eric Schrader; Fig. 7.24: 4 × 6/iStockphoto; p. 265: Hogar/Shutterstock.

## Chapter 8

Opener: Matt Tilghman/Shutterstock; p. 280: NASA; Fig. 8.8: Ivonne Wierink; Fig. 8.19: Associated Press/Aman Sharma; Fig 8.21: JP5/ZOB/WENN/Newscom.

## Chapter 9

Opener: F1online digitale Bildagentur GmbH/Alamy; Fig. 9.9: Richard Megna/Fundamental Photographs; Fig 9.19: imagebroker/Alamy.

## Chapter 10

Opener: Steve Bower/Shutterstock; Fig. 10.15: SuperStock/SuperStock; Fig. 10.16: Richard Megna/Fundamental Photographs; Fig. P10.44: John Kershner/Shutterstock.

## Chapter 11

Opener: Biophoto Associates/Photo Researchers, Inc.; Fig. 11.1: Pearson Science/Eric Schrader; Fig. 11.7: Thomas Otto/Fotolia; Fig. 11.8: Dmitry Naumov/Fotolia; Fig. 11.9: Ustyujanin/Shutterstock.

## Chapter 12

Opener: Bruce Mitchell/Getty; Fig. 12.1 (Top): USGS; Fig. 12.1 (Bottom): USGS; p. 445: Dorling Kindersley Media Library; p. 446: Marc Mueller/dpa/picture-alliance/Newscom; Fig. 12.14: Ted Kinsman/Photo Researchers, Inc.

## Chapter 13

Opener: Richard Megna/Fundamental Photographs; Fig. 13.1: iPics/Fotolia; Fig. 13.5: misu/Fotolia.

## Chapter 14

Opener: Image Source/Alamy; Fig. 14.19: sciencephotos/Alamy; p. 518 (Bottom): Ted Kinsman/Photo Researchers, Inc.; Fig. 14.21: Clive Streeter/Dorling Kindersley; Fig. P14.19: blickwinkel/Alamy.

## Chapter 15

Opener: SuperStock/SuperStock; Fig. 15.18: AP Photo/Westinghouse; Fig. 15.30: Gary Dublanko/Alamy.

## Chapter 16

Opener: moodboard/Alamy; Fig. 16.35 (Left): D. Hurst/Alamy; Fig. 16.35 (Right): Robert Asento/Shutterstock.

## Chapter 17

Opener: Pi-Lens/Shutterstock; Fig. 17.8: Stephen Oliver/Dorling Kindersley; Fig. 17.27: Pi-Lens/Shutterstock; Fig. 17.37: Thomas Dobner 2007/Alamy.

## Chapter 18

Opener: Phanie/SuperStock; Fig. 18.10: Dreamworld Tower of Terror II, Gold Coast, Australia; Fig. P18.51: Erik Charlton, Used under a Creative Commons license: http://creativecommons.org/licenses/by/2.0/deed.en.

## Chapter 19

Opener: DBURKE/Alamy; Fig. 19.3: Jim Lopes/Shutterstock; Fig. 19.9: Pal Teravagimov/Shutterstock; p. 713: David Hancock/Alamy; Fig. 19.13: NASA; Fig. 19.15: imagebroker/Alamy; Fig. 19.16: Cathy Keifer/Shutterstock; Fig. 19.19a–b: AP images; Fig. 19.20: Eric Schrader/Fundamental Photographs.

## Chapter 20

Opener: Merlin D. Tuttle/Photo Researchers, Inc.

## Chapter 21

Opener: Richard Megna/Fundamental Photographs; Fig. 21.1: malekas/Fotolia; Fig. 21.4: Detail Heritage/Alamy; Fig. 21.8: Artem Mazunov/Shutterstock; Fig. 21.9: Eugenia Etkina; Fig. 21.12: Alexander Tsiaras/Photo Researchers, Inc.; Fig. 21.15: Loskutnikov/Shutterstock; Fig. P21.15: StockPhotoAstur/Shutterstock; Fig. P21.37: dcwcreations/Shutterstock; Fig. 21.23: Pedro Salaverría/Shutterstock.

## Chapter 22

Opener: Sergej Razvodovskij/Shutterstock; Fig. 22.3: Major Pix/Alamy; Fig. 22.20: Vladislav Lebedinski/Fotolia; Fig. 22.22: David Paul Morris/Bloomberg via Getty Images; Fig. P22.40: Kevin Norris/Shutterstock; Fig. 22.36: Anion/Fotolia.

## Chapter 23

Opener: Richard Megna/Fundamental Photographs; Fig. 23.4: Dieter Zawischa/www.itp.uni-hannover.de/~zawischa/ITP/multibeam.html; Fig. 23.7: sciencephotos/Alamy; Fig. 23.9: Berenice Abbott/Photo Researchers, Inc.; Fig. 23.12: GIPhotoStock/Photo Researchers, Inc.; Fig. 23.13: langdu/Shutterstock; Fig. 23.15: Kelley/Fotolia; Fig. 23.16a: Andrew Lambert Photography/Photo Researchers, Inc.; Fig. 23.16b: Eric Schrader/Fundamental Photographs; Fig. 23.19: Adam Filipowicz/Shutterstock; Fig. 23.20: Winelover/Fotolia; Fig. 23.21: Richard Megna/Fundamental Photographs; Fig. 23.24b: P.M. Rinard, American Journal of Physics, Vol. 44, #1, 1976, p. 70. Copyright 1976 American Association of Physics Teachers; Fig. 23.27a–c: Nick Strobel; p. 881: Andrew Lambert Photography/Photo Researchers, Inc.; Fig. 23.29: cphoto/Fotolia; Fig. P23.74: avs_lt/fotolia.

## Chapter 24

Opener: Christian Lambert; Fig. 24.3: George Resch/Fundamental Photographs; Fig. 24.10: NOAA; Fig. 24.21: Stefano Bittante; Fig. 24.24: Thomas Heylan; Fig. P24.20: NASA/JPL-Caltech/R. Hurt; Fig. P24.48: P. Challis, R. Kirshner (CfA), and B. Sugerman (STScI), NASA.

## Chapter 25

Opener: Pixellover RM 10/Alamy.

## Chapter 26

Opener: Hubble Heritage Team (AURA/STScI/NASA/ESA); Fig. 26.1: Ted Kinsman/Photo Researchers, Inc.; Fig. 26.4: ESO; Fig. 26.10: Richard Megna/Fundamental Photographs; Fig. 26.11b: Wilhelm Roentgen; Fig. P26.68: NASA/JPL-Caltech/R. Hurt (SSC).

## Chapter 27

Opener: European Southern Observatory; Fig. 27.12: Wabash Instrument Company/Fundamental Photographs, Inc.; Fig. 27.14: Patryk Kosmider/Fotolia; Fig. 27.16a: Richard Megna/Fundamental Photographs; Fig. 27.16b: Jan Homann; p. 1020: NSO/AURA/NSF; Fig. 27.32: Pathe Films/courtesy Everett Collection; Fig. 27.36: Stefan Diller/Photo Researchers, Inc.

## Chapter 28

Opener: Bettmann/Corbis; p. 1042: SPL/Photo Researchers, Inc.

## Chapter 29

Opener: Michael Ventura/Alamy; Fig. 29.1a: Lawrence Berkeley National Laboratory; Fig. 29.2c: Lawrence Berkeley National Laboratory.

# Index

## G

## N

## Q



## R

## S

## T